SECOND EDITION

Illustrated Dictionary of Mycology

Miguel Ulloa
Departamento de Botánica
Instituto de Biología
Universidad Nacional Autónoma de México

and

Richard T. Hanlin
Mycological Herbarium
Museum of Natural History
University of Georgia

with the technical assistance of

Samuel Aguilar and **Elvira Aguirre Acosta**
Departamento de Botánica
Instituto de Biología
Universidad Nacional Autónoma de México

APS PRESS
The American Phytopathological Society
St. Paul, Minnesota

Drawings by Miguel Ulloa.
Samuel Aguilar provided technical assistance for formatting and editing this book.

This book has been reproduced directly from electronic files submitted in final form to APS PRESS by the authors. No editing has been done by APS PRESS.

Library of Congress Catalog Card Number: 2011945640
International Standard Book Number: 978-0-89054-400-6

Printed in The United States of America on acid-free paper.

The American Phytopathological Society
3340 Pilot Knob Road
St. Paul, Minnesota 55121, U.S.A.

Table of Contents

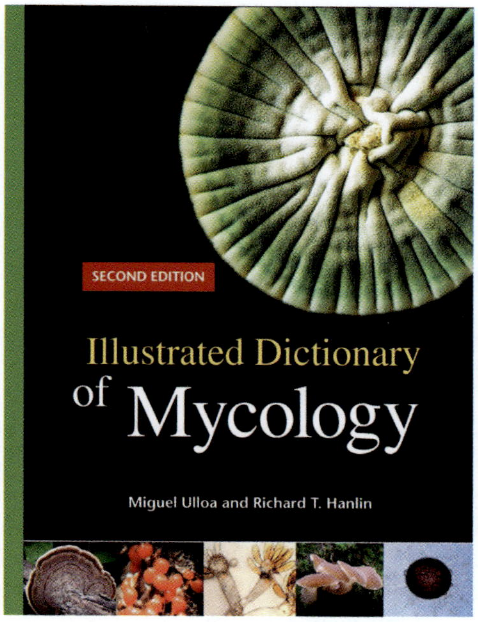

Preface to the *Illustrated Dictionary of Mycology, Second Edition* (2012)

Following publication of the *Illustrated Dictionary of Mycology* by APS PRESS in July, 2000, work was begun on a revised, expanded version to be published in Spanish as an aid to the study of mycology by students and professionals in Spanish-speaking countries. This effort culminated in the publication of the *Nuevo Diccionario Ilustrado de Micología* by APS PRESS in April, 2006. The evolution of these works is outlined in the Preface to the *Nuevo Diccionario*, which appears in translation on the following pages.

Because of the large number of new terms and illustrations included in the *Nuevo Diccionario Ilustrado de Micología*, compared to the English edition, it was felt desirable to revise the latter to include these additions. The present volume represents the result of this effort. As in the previous editions, each entry occurs in boldface, along with the plural and orthographic variations, where appropriate, the etymological derivation of the term, followed by the definition(s) and one or more examples of fungi to which the term is applied. Nearly 5,000 terms are defined in this new English edition, including many of the Greek and Latin prefixes and suffixes used in the formation of scientific terms, as well as in composing generic names and specific epithets. Complementing these terms are 2,708 illustrations, representing the same number of fungal scientific names, mostly determined to the level of species, plus several to genus only, and a few names to other taxonomic categories, which provide a visual component to the definitions. Most of the photographs were taken by the authors over the years; others have been kindly provided by colleagues. Included after each of the fungal scientific names is the name of the term (in bold) being illustrated. Photo credits are included in the caption of each photograph, and a list of contributors and their affiliation is provided at the end of the volume. The color drawings (672) were all made by Miguel Ulloa. A unique feature of this volume is the printing of the photographs and drawings in color, which hopefully will enhance their usefulness. These additions have expanded the volume to 782 pages which reveal the immense diversity of the kingdom *Fungi* and of the fungus-like microorganisms (kingdoms *Chromista* and *Protozoa*).

To produce this book, the text of the *Nuevo Diccionario Ilustrado de Micología* was translated back into English. The separate files in Spanish for each letter of the alphabet were opened in Adobe PageMaker 7.0, and each entry was then translated directly into English, retaining the original formatting. At the end of this project (December, 2010), the authors undertook the task of reformatting the text using InDesign, a program that provides better spacing of the words and an overall improvement in the appearance of the text. All binomials were checked in Index Fungorum to make the names as current as possible. While this was being done, the color transparencies were digitized with a

Polaroid Sprint Scan 35 Plus scanner at a resolution of 1,350 pixels, then optimized for contrast, size and definition in Photoshop 5.0. After completion of the translation, the formatting of the English text was undertaken. Because of differences in the spelling of words in the two languages (e.g., *espora* versus *spore*), it was necessary to move many terms, and the accompanying illustrations, to other sections of the book, a time-consuming process. As this was done, recently published new terms were inserted into the appropriate places in the text. The completed manuscript was then provided to APS PRESS as electronic files.

The final product of this project is a comprehensive, up-to-date compendium of mycological terms, with their origins and definitions, illustrated with beautiful color photographs and drawings, that should be useful to anyone interested in the exciting and fascinating world of the fungi. Poster exhibitions and conferences have been given in scientific congresses to explain the evolution of this mycological dictionary, e.g., the VI Congreso Latinoamericano de Micología (Mar del Plata, Argentina, Ulloa *et al.*, 2008) and the X Congreso Nacional de Micología (Guadalajara, Mexico, Ulloa, 2009); this has been done to receive feedback and improve the work of this project.

This book represents the culmination of more than four decades of teaching and research by the authors. As with any such work, we are indebted to our students, colleagues and others for helpful comments, suggestions and criticisms that have contributed to improving the volume. The final product, however, is our responsibility.

Special recognition must be made here of Ricardo Valenzuela, past president of the Mycological Society of Mexico (2007-2010), and currently the Curator of the Mycological Herbarium, National Polytechnic Institute of Mexico, who gave us copies of his large collection of fungal digital images, taken in their habitats in widespread regions of Mexico; from this collection (he gave us more than 3,500 digital images) we selected 297 color photographs of fungi for illustrating several concepts of this dictionary. We owe a debt of gratitude to other colleagues for fungal photographs, and to two mycologists from the University of Tlaxcala, Mexico, Arturo Estrada (the current president of the Mycological Society of Mexico, 2010-2013) and Mercedes Rodríguez Palma, who provided many specimens of Myxomycetes collected on the Volcano La Malinche, State of Tlaxcala, Mexico, already identified to species or genus, of which Miguel Ulloa and Calixto Benavides took several photographs in color included in this volume. The authors wish to express their deep gratitude to two biology colleagues, Samuel Aguilar and Elvira Aguirre Acosta, of the Institute of Biology, Autonomous National University of Mexico. Both collaborated closely and diligently during the six years required to complete the project. Samuel contributed mainly in helping to prepare the images for insertion into the text and in the final formatting of the book. Elvira participated, with painstaking care, in the many revisions of the text and in locating all of the genera and orders of fungi that were employed as examples to complement the different concepts, so that this information could be incorporated into the system of classification and the Appendix that were included in the dictionary.

This dictionary is not primarily taxonomic, so to assist users in placing the fungi mentioned, it follows the fungal classification scheme of *Introductory Mycology*, 4th edition, by Alexopoulos *et al.*, 1996, a textbook frequently employed in the teaching of mycology; this scheme was modified by Ulloa and Hanlin in their book *Nuevo Diccionario Ilustrado de Micología*, 2006, using the classification schemes of Hawksworth *et al.*, 1999 and Kirk *et al.*, 2001. When the *Illustrated Dictionary of Mycology, Second Edition*, was being compiled, two new mycology books were produced: 1) *Fungal Families of the World* (Cannon and Kirk, 2007), which is an informative companion to the *Dictionary of Fungi*; it covers 536 currently recognized families and depicts nearly 900 example species, describing their morphology, ecology, phylogeny, distribution and economic importance, mostly illustrated in color. 2) One year later, a new edition of *Ainsworth and Bisby's Dictionary of the Fungi* (10th edition, by Kirk *et al.*, 2008) was published, with major changes in the systematics of the fungi. These books considered that some parts of the previous classifications of fungi, including the scheme of Alexopoulos *et al.*, had species of fungi forming unnatural groupings, resulting in an artificial classification. Based upon new research, comprising new structural and ultrastructural studies, in addition to molecular characters, some taxa of fungi were determined to be heterogeneous, e.g., Blastomycetes, Hyphomycetes, Coelomycetes (deuteromycetes), Plectomycetes, Pyrenomycetes, Discomycetes, Loculoascomycetes (Ascomycota), Hymenomycetes and Gasteromycetes (Basidiomycota). The names of these fungal taxa are also treated in the present book, to give their etymology, definition and usages in mycology. In the Appendix there is a table that compares the classification of fungi by Alexopoulos *et al.*, 1996, modified by Ulloa and Hanlin, 2006, to that of Kirk *et al.*, 2008, so the users of this dictionary will have the information to make the transition from the old scheme to the new one, and be able to see the changes of some orders, classes, phyla, and even to the level of kingdom, of some groups of fungi and fungus-like microorganisms; the goal of these taxonomic rearrangements is to achieve a phylogenetic classification. The organisms classified in the three kingdoms (Fungi, Stramenopila and Protista), of the modified outline of Alexopoulos *et al.*, have been transferred to three kingdoms

(Fungi, Chromista and Protozoa), in the system of Kirk *et al.*, with the concurrent changes of the names of taxa. In Protozoa the taxon names have the endings used in zoological taxonomy, e.g., Acrasida (Acrasiales), Protostelida (Protosteliales), and Physarida (Physarales).

We are also grateful to Dr. María de los Ángeles Herrera-Campos, Faculty of the Institute of Biology, who provided photographs and data about lichens, and to Julio César Montero and Samuel Aguilar, also from the Institute of Biology, who produced an initial version of the cover of the book. We acknowledge the editorial advice of Antonio Bolívar, Managing Editor of Redacta, S.A., of Mexico City.

Finally, we would like to thank the staff of APS PRESS for their interest in publishing the present work. We recognize the efforts of Mary Palm, Senior Editor, and the two senior mycologists selected by the APS PRESS in this project, who spent several months reviewing the new edition. Special thanks go to Margery Daughtrey, Editor in Chief of APS PRESS, whose editorial skills and personal interest also contributed to making this a better dictionary.

Miguel Ulloa
Mexico City, Mexico

Richard T. Hanlin
Athens, Georgia

October, 2011

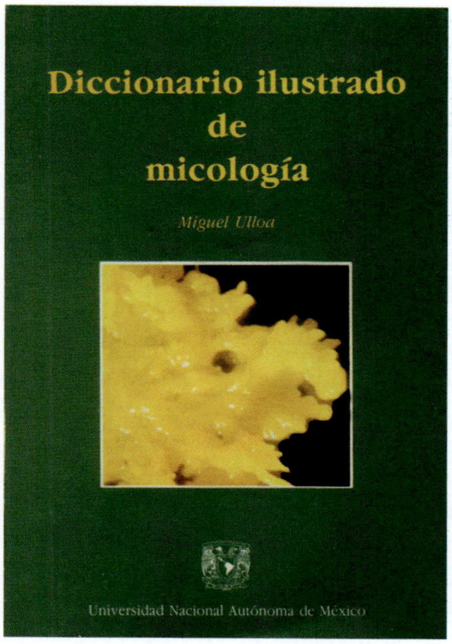

Preface to the *Diccionario ilustrado de micología* (1991)
(translated from Spanish into English)

The idea of doing this *Diccionario ilustrado de micología* occurred to me in July, 1984, while I was working on the book *El reino de los hongos, micología básica y aplicada*, and was due principally to the uneasiness I frequently felt on finding in the mycological bibliography written in Spanish many technical terms that were poorly translated or inappropriately applied. With this in mind, and taking into account the convenience and necessity of assembling a mycological lexicon of broad coverage, and complementing it with a considerable number of high quality illustrations, I dedicated myself to undertaking the first work of this type written in Spanish, with the object of contributing to filling a vacuum that had existed in this aspect of mycology, in Mexico as well as in other Spanish-speaking countries.

The dictionary brings together around 2,000 terms, the majority of them mycological, which I considered to be of major usage in the instruction and investigation of fungi. Many terms are not strictly mycological, since they are employed in the nomenclature of various biological sciences, but I have included them because they frequently are employed in various specialities of mycology.

For each term is given the equivalent in English, the Greek (transliterated into Spanish) or Latin etymology (sometimes both), the definition, one or more examples of fungi (generally a genus or species) to which the term is applied, the taxonomic order to which the examples belong, one or more figures, when possible, and when appropriate, the opposite or related term, which can be consulted under the corresponding entries.

Since the major part of the mycological bibliography that is consulted in our field is in English, it is understandable that students of the fungi are more or less familiar with their lexicology in this language, and that at times they may encounter certain difficulties in trying to use them translated into Spanish; for them, a list of the terms in English is included at the end of the dictionary, followed by their equivalents in Spanish. When the ending of the words in English changes in its plural form, this is given in parentheses, e.g., *perithecium* (*perithecia*).

In case the user of this book needs to know the taxonomic position of the fungi used to illustrate various concepts, he/she can consult the alphabetical list of taxonomic orders and classes to which the genera belong, as well as the taxonomic scheme, included at the end of the work. The correct placement of the taxa will permit the reader to refer to other works to obtain additional information about the terms consulted.

With the objective of making the work more useful, didactic, and attractive, I have placed special interest on including a considerable quantity of figures (766); whenever it was convenient, or possible, I illustrated the fungal structures and the biological processes or phenomena in which the fungi are involved, and which I considered most important.

Of the total of 611 photographs, more than half (326) are mine; all of the drawings (158) are drawings in black and white made by me (many are based on or adapted from drawings or photographs already published, and which are found in the works cited in the bibliography). The initials of the authors of the photographs are indicated in the captions, and the names corresponding to these initials, as well as the institutions in which these authors work, are presented in alphabetical order in the credits of the photographs at the end of the work. The scale at which the figures are printed also is given in the caption of each figure.

Although the review I made of the mycological bibliography did not pretend to be exhaustive, I am sure that it was sufficiently extensive to attain an important coverage of the subject. The orthography of the terms in Spanish and their Greek or Latin etymologies is based principally on the *Diccionario de botánica* by P. Font Quer (1963), although the selection and editing of the definitions was derived from the analysis of various possible interpretations, principally of those that are considered in the specialized dictionaries of mycology written in English (which are noted below). Inasmuch as certain information was obtained from Font Quer and from two other works done in Mexico, which are the *Glosario para Spermatophyta, Flora mesoamericana*, by M. Sousa and S. Zárate (1983), and the *Glosario botánico ilustrado*, by N. P. Moreno (1984), the last two designed to be utilized in the study of higher plants, some data were useful in the preparation of the present mycological dictionary. The orthography of the terms in English and the major part of the information that was utilized in the editing of the definitions was derived from a complete and careful review of the following works: *A Glossary of Mycology*, by W. H. Snell and E. A. Dick (1957), *Collegiate Dictionary of Botany*, by D. Swartz (1971), and *Ainsworth and Bisby's Dictionary of the Fungi*, by D. L. Hawksworth, B. C. Sutton and G. C. Ainsworth (1983), as well as from the other works cited in the bibliography. In the majority of cases, the definitions given by the different authors agree in essence and they served as the basis for editing them according to my own criteria, trying always to include representative examples. In the available botanical and mycological dictionaries, the inclusion of examples is not common. It is appropriate to make clear that the present book is basically a compilation of terms relative to the macro- and micromorphology, reproduction, physiology and ecology of fungi, since terms referring to metabolic products (such as organic acids, antibiotics, pigments and others) and of properties related to the color, taste or odor of these products, are deliberately excluded, terms that, e.g., are described in Snell and Dick. Except for the names of classes of fungi (such as Chytridiomycetes, Zygomycetes, etc.), which for practical reasons are explained, the names of genera and other suprageneric taxonomic categories are also excluded, entries which are considered in Hawksworth *et al.* In the appendix are given some terms common in taxonomy.

The principal purposes of this *Diccionario ilustrado de micología* are to collect the major part of the basic words utilized in the exercise of mycology and related sciences, to help in the translation and interpretation of the nomenclature that is utilized in English, and to standardize the usage of the terminology in Spanish and Engish, trying to present in a uniform manner the modern use of a lexicology as extensive and specific as that employed in the study of the fungi, and enriching the information with the aid of a good number of pertinent illustrations. Additionally, the book shows the graphic focus that I have always followed in the exercise of my profession.

I am grateful for the collaboration of Ricardo Valenzuela, who besides providing various photographs, also supplied me with herbarium specimens for use in taking various slides; of Patricia Lappe, who helped me prepare the list of terms in English and to assemble the measurements of the fungal structures illustrated; of Calixto Benavides, who with great care collaborated with me in the reviewing and arranging of the figures and their legends, as well as their corresponding scales, besides the meticulous task of reviewing the text proofs of the dictionary; of Laura Estrada Cuéllar and Magdalena Reyes Téllez, for painstakingly typing the text; and of Evangelina Pérez Silva and Teófilo Herrera, for their suggestions of species of fungi used to represent various terms. I also express my gratitude, in a special way, to Elvira Aguirre Acosta, for her constant interest in the progress of the dictionary, and her valuable assistance involved in various aspects of the work, particularly in the search for herbarium specimens to be photographed, and in the preparation of the list of taxonomic orders and classes to which the genera of fungi utilized as examples of the included terms belong. Finally, I recognize the important assistance of Antonio Lot Helgueras, Director of the Institute of Biology of the Autonomous National University of Mexico (UNAM), as well as of the personnel of Redacta, S.A. for their painstaking and professional participation in composing this work.

Miguel Ulloa
Mexico City, Mexico

May, 1991

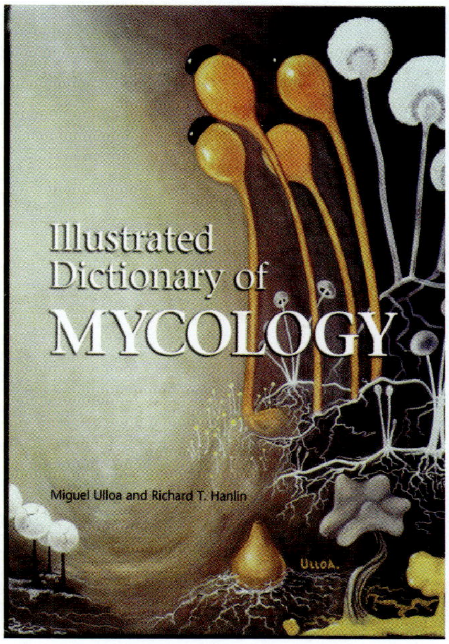

Preface to the *Illustrated Dictionary of Mycology* (2000)

When the *Diccionario Ilustrado de Micología* first appeared in 1991, it provided, for the first time, a reference which Spanish-speaking students of mycology could consult in their own language. The *Diccionario* contained some 2,000 technical terms, many of which were illustrated by drawings and/or photographs that were integrated into the text. These illustrations served to assist the student in understanding and interpreting the various terms they depicted. In addition, the etymology was given for each entry term, thus enabling the student to learn the derivation of these terms. That the *Diccionario* filled a need is demonstrated by the fact that it was soon out of print.

After reviewing the *Diccionario* soon after its publication, it was apparent that English-speaking/reading students also could benefit from its unique features, but few possess the language skills necessary to do so. Because of our earlier collaboration on mycological works, it was decided to undertake the laborious task of translating it into English. As the translation slowly progressed, however, new terms appeared in the literature, and it was felt desirable to include them to make the dictionary as current as possible. This also led us to include some older terms that had been omitted from the Spanish edition. The result is that while this work is nominally based on the Spanish version, it is much expanded, with nearly twice as many terms and some 80% more illustrations.

The text of the dictionary was initially entered into the computer in Word Perfect 6.0, but as the text increased greatly in size, it was migrated to Adobe Page Maker 6.5 for ease of handling. The color transparencies were digitized with a Polaroid Sprintscan slide scanner at 1,350 dpi resolution and the images obtained in black and white were optimized with Photoshop 5.0. The same procedure was followed for the black and white transparencies of the drawings. Once the text was completed, the illustrations were then inserted near the terms they represent. The completed manuscript was furnished as film-ready copy on diskette for printing.

Although all of the drawings and the majority of the photographs were made by the authors, we are grateful to various individuals for permission to include additional illustrations belonging to them; they are acknowledged at the end of the book. Special mention must be made of two colleagues, Samuel Aguilar and Elvira Aguirre, of the Instituto de Biología, Universidad Nacional Autónoma de México. Both spent many months assisting with various phases of the project. The former contributed greatly in editing the book. Without their help, the project would have taken much longer to complete. We also thank María C. González, who helped calculate the scales of many of the illustrations, and María Ángeles Herrera-Campos, who provided some data on lichens, both of the Instituto de Biología. Finally, we acknowledge the technical advice of Antonio Bolívar, the Managing Editor of Redacta, S. A., of Mexico City.

We are indebted to Héctor Hernández, Director of the Instituto de Biología, for his continuous support, and to APS PRESS for their willingness to publish the present work. Their encouragement and support as we learned the details of computer publishing eased the task of preparing the final manuscript.

Miguel Ulloa
Mexico City, Mexico

Richard T. Hanlin
Athens, Georgia

December, 1999

Preface to the *Nuevo Diccionario Ilustrado de Micología* (2006)
(translated from Spanish into English)

In May of 1991, the Institute of Biology of the Autonomous National University of Mexico published the *Diccionario Ilustrado de Micología* (by Miguel Ulloa), consisting of 310 pages, with etymologies and definitions of some 2,000 technical terms relating to the study of the fungi, and complemented by 769 illustrations (611 photographs and 158 drawings), to better understand and interpret the written information. This was the first mycological dictionary published in Spanish, and it was also the first to include photographs and drawings, incorporating the illustrations as close as possible to the corresponding terms. When the first and only edition of this dictionary was exhausted, it was decided that, instead of making a second edition, it would be more convenient to undertake a version in English, one more extensive and up-to-date, with the object of filling a void in this area and of complementing other existing mycological dictionaries. In July of 2000, APS PRESS published the *Illustrated Dictionary of Mycology* (by Miguel Ulloa and Richard T. Hanlin), whose characteristics surpassed those of its Spanish predecessor, since it achieved a broader coverage of the discipline, in a larger and more easily legible format, hard bound, and, especially, with better quality printing, therefore of greater utility for mycological teaching and research. Basically following the style and focus of the mycological dictionary in Spanish, the *Illustrated Dictionary of Mycology* also provided the etymological derivation of each term, its definition, one or more examples of species of fungi to which the term in question applied, and referring in many cases to terms opposite or related to the main entry, as well as a large number of high quality illustrations that permitted a greater understanding of the concepts. This English mycological dictionary contains 448 pages, and includes more than 3,800 terms and 1,322 illustrations (774 photographs and 548 drawings).

Taking into consideration the fact that the mycological dictionary in English greatly surpassed the original edition in Spanish that was its predecessor, and with the aim of continuing to increase the level of teaching of mycology in Spanish-speaking countries, it was necessary to rely on a work of this type, in Spanish, of equal or better quality than that published in English. Since the appearance of the latter in July, 2000, work was initiated on a new version in Spanish that would serve as a reference book for students and professionals of mycology in the Spanish-speaking countries, and also for readers of Spanish that live in the United States and other places. It must be pointed out that this *Nuevo Diccionario Ilustrado de Micología* is not a mere translation of its predecessor (the *Illustrated Dictionary of Mycology*), but it constitutes an up-to-date work of greater coverage, and with a greater number of terms and illustrations, since it includes nearly 5,000 terms and 1,766 illustrations (1,129 photographs and 637 drawings), compared to the 3,800 terms and 1,322 illustrations of the English version, published in 2000. The distance between the new mycological dictionary in Spanish and the original version of 1991 is even greater, since the latter contains only 2,000 terms and 769 illustrations. For this reason, the title *Nuevo Diccionario Ilustrado de Micología* is fully justified, as it has 50% more pages and around 30% more terms and 30% more illustrations than the English version published in 2000.

The entire text of the *Illustrated Dictionary of Mycology* was translated into Spanish and then relevant, up-to-date information was added, as well as a considerable number of new illustrations. In order to form the new mycological dictionary, the text was entered into the computer in Times New Roman font, with the program Page Maker 6.5. With the objective of including the largest possible quantity of terms, a detailed review of the *Illustrated Dictionary of Mycology* was made, collating all of the terms included in it with all of those contained in the 9th ed. of the *Dictionary of the Fungi* (edited by Kirk *et al.*, 2001), as well as those in the *Diccionario de Botánica* (Font Quer, 1963). The review and comparison of these three works, which up to now are the most complete and relevant on the topic, along with a review of scientific mycological journals, made possible the incorporation in the new mycological dictionary of many terms that had been omitted from both the Spanish and English editions, respectively published in 1991 and 2000, as well as other terms that have appeared in the recent mycological literature: all of this with the purpose of making the new mycological dictionary the most complete and accurate work possible. Many of the additional entries (*ca.* 1,200) represent the prefixes and suffixes that are employed in the formation of compound or derived terms. Also, special care was taken to interrelate some terms with others, making the necessary cross references among them, in order to achieve, in an encyclopedic manner, a better integration of the information. In general, the spelling of the terms agrees with the *Diccionario de Botánica*, by Font Quer, although the accentuation of many terms in Mexico differs from that in Spain; thus, for example, in the present book *ascospora, basidiospora, zoospora*, etc., are written and pronounced as grave words, and not *ascóspora, basidióspora, zoóspora*, etc., words accented on the antepenultimate syllable, which, according to Font Quer, also are etymologically correct. Other terms, such as *mitocondria* and *vacúola*, which actually correspond to the plurals, and which in the singular must be written as *mitocondrion* and *vacúolo* (etymologically more correct), are now, by custom, in such general use in Mexico and other Spanish-speaking countries, that it will be difficult to change. On the other hand, some terms that are utilized in English were left intact in Spanish, as their literal translation would have resulted in obviously forced or inadequate words, as in the words *flush* and *gossamer*.

After the publication of the *Illustrated Dictionary of Mycology* there appeared 13 published reviews of it, the majority of them positive, indicating that the book was well accepted, although some of these reviews made various criticisms and gave certain recommendations to improve future editions, such as including a list of the species of fungi illustrated, due to its didactic utility. Heeding this suggestion, which was considered pertinent, the present book includes at the end a list of the more than 1,100 species of fungi illustrated by means of photographs or drawings. In this list, arranged alphabetically, also are given the author credits for the illustrations; the initials of the authors of the photographs appear after the scientific names of the species of fungi; names not followed by any initial represent drawings made by Miguel Ulloa.

Another important aspect is the incorporation of an up-to-date English-Spanish vocabulary, which is based on the one included in the original version of the dictionary, published in 1991; however, in this *Nuevo Diccionario Ilustrado de Micología* the vocabulary is much more extensive, as it comprises nearly 5,000 terms for each language. Hopefully, the list of species of fungi illustrated, as well as the English-Spanish vocabulary, will result in better use of the dictionary. The list indicates under which term(s) one can see a structure, a particular biological phenomenon, or other characteristic, which represents one species in particular. The vocabulary can be useful for students, professors and researchers involved in the field of mycology, because, even though their native language is Spanish, it is common practice and necessary to consult literature written in English, so that the English-Spanish vocabulary can help to standardize the mycological terminology that is used in both languages. Also, for the same reason, and following the focus of the original version of 1991, all of the entries of the new dictionary contain the equivalent terms in English (written with a different font - Helvetica - for better distinction), including their plurals (when the plural ending is different from the singular, as, e.g., ascus, pl. asci). Therefore, in these last two aspects, the present book has a bilingual function. Once the formation of the text was completed, the illustrations, with their respective captions, were inserted near the terms that they represent. For the final printing, the completed manuscript was delivered to APS PRESS in the form of electronic files (film-ready files), on compact disks.

All of the drawings in the present dictionary were made by Miguel Ulloa, and the majority of the photographs were taken by Miguel Ulloa and Richard T. Hanlin, although also included are additional photographs provided by various persons, to whom credits are given in the figure captions, as well as in the list of species of fungi illustrated, where only the initials of the names of the persons that contributed photographs are provided; however, at the end of the book is included a list of credits in which is indicated the complete name of the contributors, the institutions where they work, as well as the number of photographs provided by each of them.

Special recognition is made here to two biology collegues, Samuel Aguilar and Elvira Aguirre Acosta, of the Institute of Biology, Autonomous National University of Mexico. Both collaborated closely and efficiently during the more than five years required to complete the project. Samuel contributed mainly in the formation of the book, utilizing the program Page Maker 6.5, and in handling the images; the color transparencies were digitized with a Polaroid Sprint Scan 35 Plus scanner at a resolution of 1,350 pixels, obtaining black and white images that were optimized for contrast, definition and other characteristics, with the program Photoshop 5.0 before printing them at 1,440 pixels; the same procedure was followed for the black and white transparencies of the drawings. Elvira participated, with painstaking care, in the many revisions of the text and in locating all of the genera of fungi that were employed as examples to complement the different concepts, with the aim of incorporating these additional genera in the system of classification that was followed in the dictionary. To assist the reader in the placement of the genera of fungi mentioned in the text, a list of taxonomic orders and classes is included, which can serve as the first step before consulting the system of classification. Elvira also helped in the updating and preparation of the system of classification, including the two appendices. We are also grateful for the assistance of other members of the staff of the Institute of Biology, Drs. María C. González and María de los Ángeles Herrera-Campos; the first collaborated in the early stages of the search for new terms, and the second provided photographs and some data about lichens; the designer Julio César Montero participated, with Miguel Ulloa and Samuel Aguilar, in the design of the cover of the book. Finally, the interest of APS PRESS in publishing the present work is acknowledged.

Miguel Ulloa
Mexico City, Mexico

Richard T. Hanlin
Athens, Georgia

April, 2005

Abbreviations and Symbols

abbr.	-	abbreviation
adj.	-	adjective
AF.	-	Anglo-French
Ar.	-	Arabic
Azt.	-	Aztec (from Nahuatl)
ca.	-	about, approximately (< L. *circa*)
cf.	-	compare (< L. *confer*)
Chin.	-	Chinese
cm	-	centimeter
des.	-	desinence (ending)
diam.	-	diameter
dim.	-	diminutive
Du.	-	Dutch
E.	-	English
ed.	-	edition
eds.	-	editors
e.g.	-	for example (< L. *exempli gratia*)
et al.	-	and others (< L. *et alii*, *aliorum*)
etc.	-	and so forth (< L. *et cetera*)
fam.	-	family
fem.	-	feminine
Fr.	-	French
f. sp.	-	*forma specialis*
G.	-	German
Ga.	-	Gaelic
gen.	-	genus (singular) and genera (plural)
genit.	-	genitive
Gr.	-	Greek
Hind.	-	Hindi
IE.	-	Indo-European
i.e.	-	that is (< L. *id est*)
inc. sed.	-	*incertae sedis* (uncertain position)
Indo.	-	Indonesian
ISV.	-	International Scientific Vocabulary
It.	-	Italian
Jap.	-	Japanese
L.	-	Latin
LL.	-	Late Latin
LME.	-	Late Middle English
m	-	meter (< L. *metrum*, measure)
Ma.	-	Malay
MD.	-	Middle Dutch

ME.	-	Middle English
Med.	-	medical
MF.	-	Middle French
MHG.	-	Middle High German
ML.	-	Middle Latin
mm	-	millimeter
µm	-	micrometer
mycol.	-	mycology
NL.	-	Neo Latin
nm	-	nanometer
OE.	-	Old English
OF.	-	Old French
OHG.	-	Old High German
OIr.	-	Old Irish
OIt.	-	Old Italian
ON.	-	Old Norse (Norwegian)
ONF.	-	Old North French
OPr.	-	Old Provençal
Pers.	-	Persian
pg.	-	pages
pl.	-	plural
Por.	-	Portuguese
poss.	-	possibly
pp.	-	past participle
pref.	-	prefix
prob.	-	probably
prp.	-	present participle
pt.	-	past tense
Rus.	-	Russian
Scand.	-	Scandinavian
sing.	-	singular
Skt.	-	Sanskrit
Sp.	-	Spanish
sp.	-	species
spp.	-	species (plural)
subsp.	-	subspecies
suf.	-	suffix
syn.	-	synonym
Turk.	-	Turkish
var.	-	variation, variety
VL.	-	Vulgar Latin
<	-	derived from, less than
>	-	giving rise to, greater than
=	-	equals, equivalent to
°C	-	degrees Celsius

Pictorial Guide to the

Illustrated Dictionary of Mycology

Second Edition

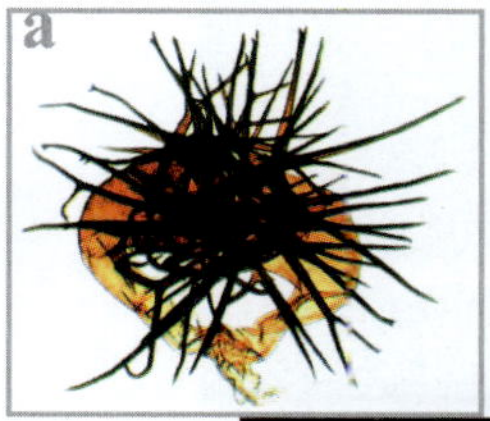

1

53

79

145

173

215

241

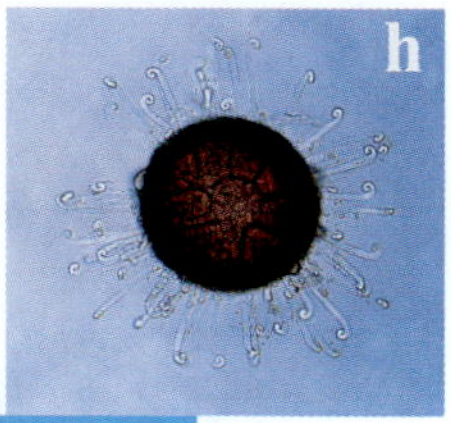

265

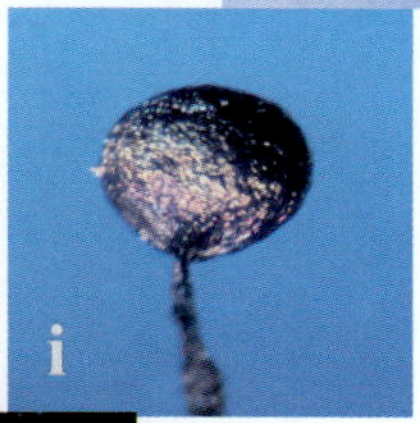

299

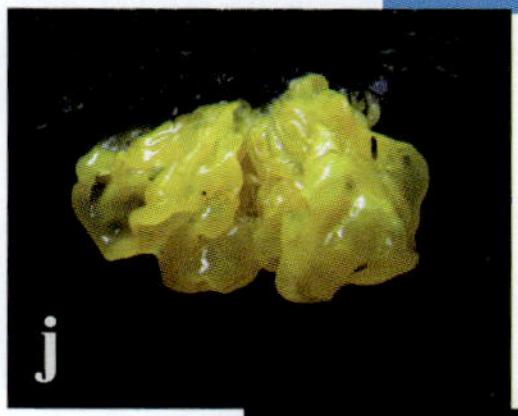

317

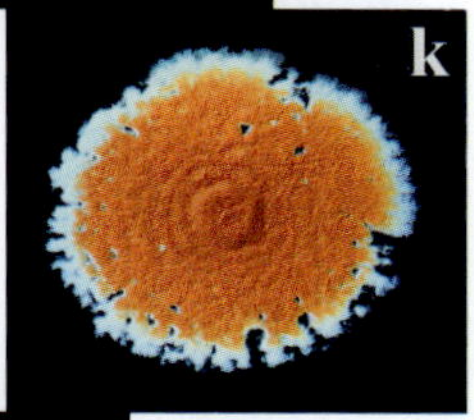

321

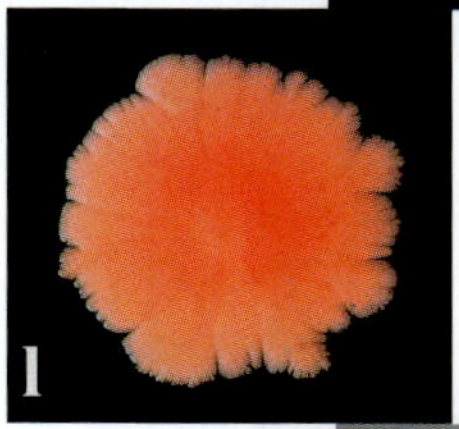

325

355

415

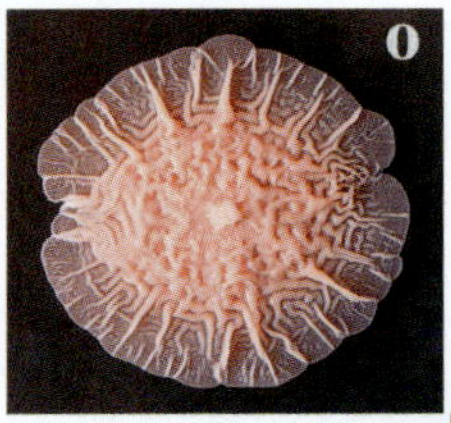

425

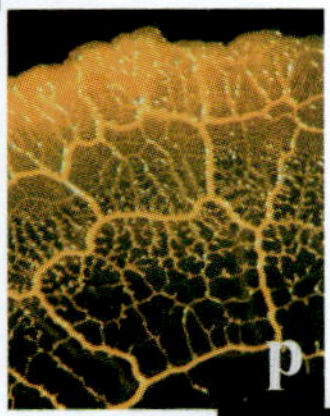

441

529

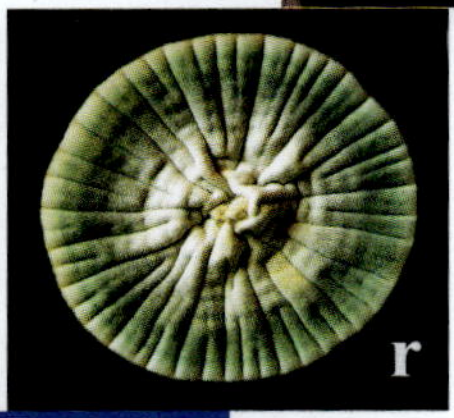

531

561

637

667

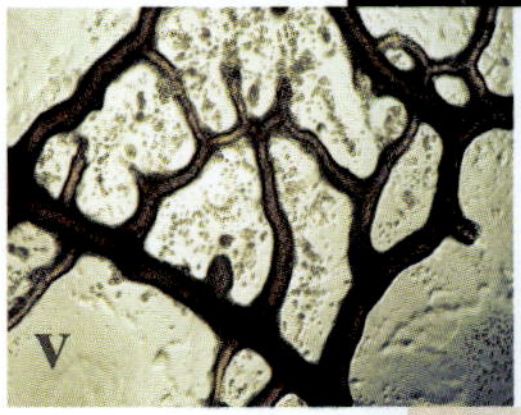

679

695

699

703

709

a

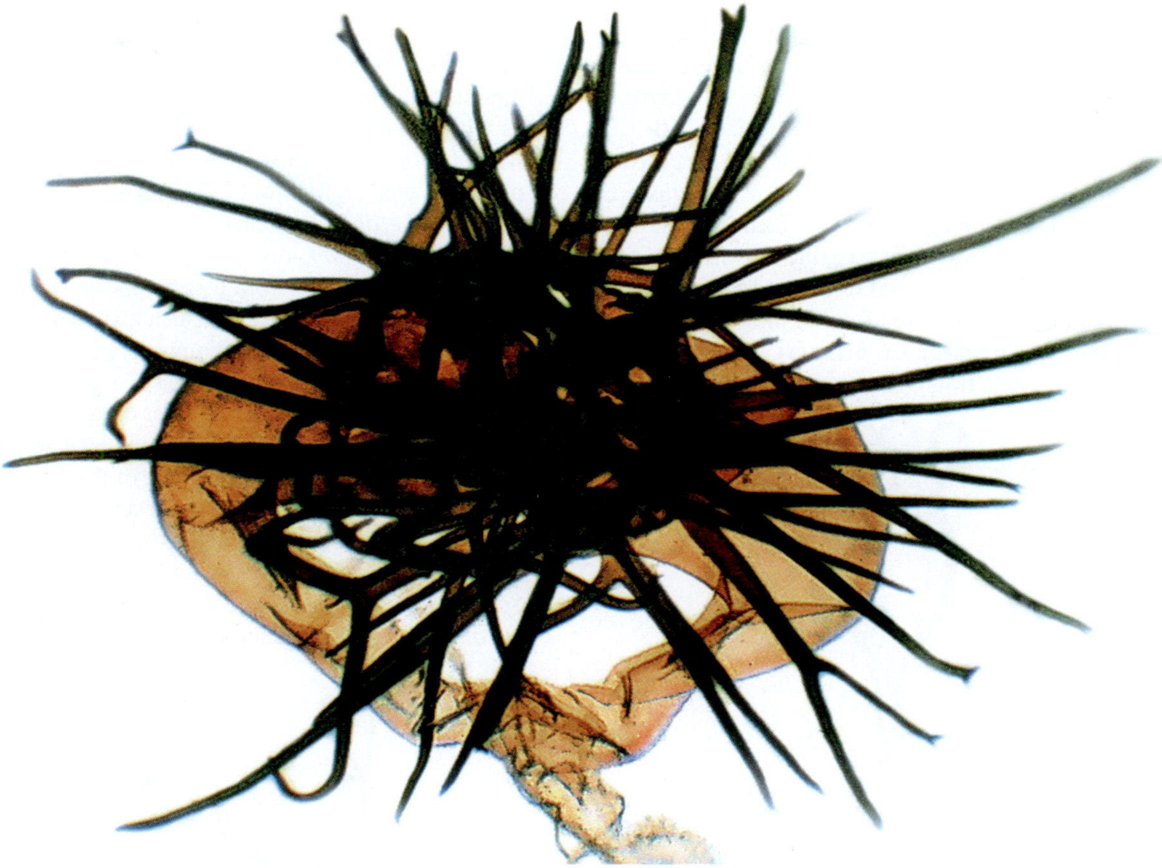

Appendages on the suspensors of *Phycomyces nitens* (*MU*).

a

a- (Gr. *a-*, exclusive): an inseparable particle that expresses deprivation or negation of something, without, as in *acellular,* not having cells.

ab-, **a-**, **abs-** (L. prep. that corresponds to Gr. *ápo-*, opposite, away from, far from): a pref. employed in numerous compounds, such as *abaxial*. Cf. **ad-**.

abaxial (NL. *abaxialis* < L. *ab-*, opposite, away from, far from + *axialis*, axial < *axis*, axis + suf. *-alis* > E. *-al*, relating to or belonging to): facing away from the main or central axis; e.g., the abaxial side of a basidiospore is the ventral or convex side that faces away from the longitudinal axis of the basidium, e.g., *Amanita* (Agaricales). Cf. **adaxial**.

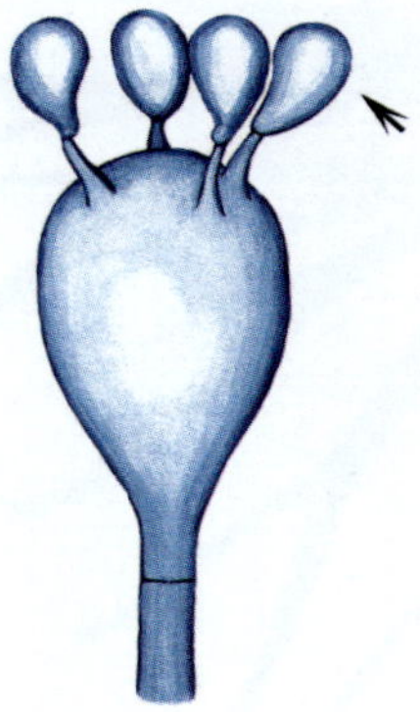

Abaxial side of the basidiospores of *Amanita* sp., x 1,100 (*MU*).

aberrant (L. *aberrans*, genit. *aberrantis*, wandering < *aberrare*, to deviate from + E. suf. *-ant*, one that is, one who performs): differing in major or minor respects from the common fundamental character of the taxon in which it is included; e.g., a certain isolate of a mold can differ in some structural or cultural characteristics of a particular species or genus, making it difficult to assign to a species or genus.

abhymenial (L. pref. *ab-*, opposite, away from, far from + *hymenium* < Gr. *hyménion*, dim. of *hymén*, membrane + L. suf. *-alis* > E. *-al*, relating to or belonging to): opposite the hymenial or spore-producing surface.

abiotic (Gr. *a-*, without + *biotikós*, vital < *bíos*, life + suf. *-tikós* > L. *-ticus* > E. *-tic*, relation, aptitude, inclination or capacity): contrary to life, either because the conditions to support life are lacking, as in high regions of the atmosphere, or because life is destroyed, as with certain radiations. It is also applied to non-living factors of the environment, such as temperature, pH, light, etc. Cf. **biotic**.

abjection (L. *abjectus*, cast away, discharged, set free + E. suf. *-tion*, result of an action, state of): separation of a spore from a sporophore, sterigma, conidiophore, etc., by means of an active mechanism of the fungus; e.g., in the basidiomycetes the discharge of a basidiospore from the sterigma of a basidium is a process of abjection, which involves the formation of a microscopic drop of water that is produced between the hilum of the spore and the apex of the sterigma, as is observed in *Calocera cornea* (Dacrymycetales). Also, the ballistoconidia or ballistospores of various species of sporobolomycetaceous yeasts (such as *Sporobolomyces roseus* and *Itersonilia perplexans*) are discharged by abjection.

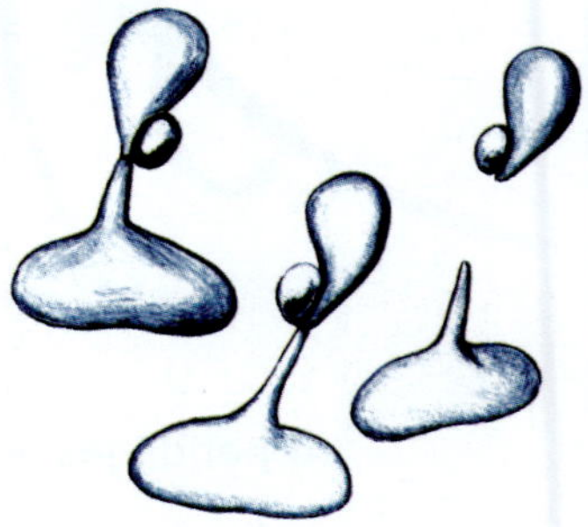

Abjection of the ballistospores of *Itersonilia perplexans*, x 3,160 (*MU*).

abjunction (L. *abjungere*, to separate, to disarticulate < pref. *ab-*, opposite, away from, far from + *junctus*, articulated, united + E. suf. *-tion*, result of an action, state of): the delimitation and separation of spores from certain portions of hyphae through the formation of septa, such as occurs in the production of arthrospores (arthroconidia) in *Geotrichum* and of blastospores in *Monilia* (moniliaceous asexual fungi). Considered by some as a syn. of **abstriction**.

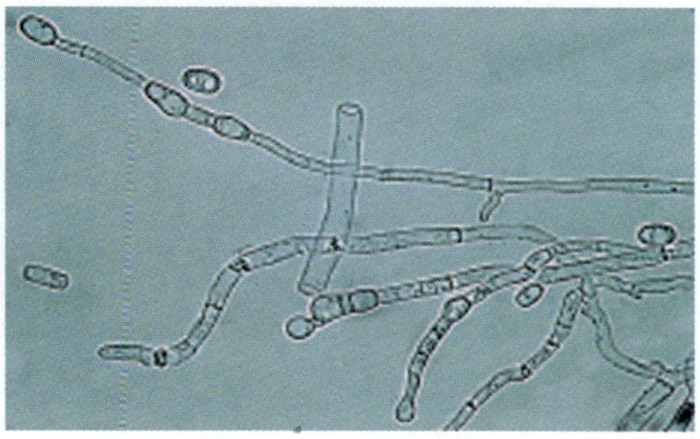

Abjunction and **abscission** (successive stages of **abstriction**) of the arthrospores of *Geotrichum candidum*, x 450 (*MU*).

ablastic, ablastous (Gr. *ablastóo*, to not germinate or germinate poorly + suf. *-íkos* > L. *-icus* > E. *-ic*, belonging to or relating to): incapable of germinating, as happens with some spores or other type of fungal propagules, which under certain environmental conditions do not produce any germ tube or bud initiating development.

aboospore (L. pref. *ab-*, opposite, away from, far from + Gr. *oón*, egg or ovule + *sporá*, spore): an oospore that results from the parthenogenetic development of an oosphere, as occurs in some species of aquatic zoosporic microorganims of the order Saprolegniales.

abortive (L. *abortatus*, *abortivus*, born prematurely < *abortare*, to miscarry < pref. *ab-*, far from, without + *ortus*, birth + suf. *-ivus* > E. *-ive*, quality or tendency, fitness): incompletely developed; an organ or structure whose development remains in a rudimentary state, like the primordia of the basidiocarps of *Entoloma* (=*Rhodophyllus*) *abortivum* (Agaricales) which do not reach maturity and, therefore, do not form a fertile hymenophore. See **atrophy**.

Abortive basidiocarps of *Entoloma* (=*Rhodophyllus*) *abortivum*, x 0.5 (*TH*).

abraded (L. *abradere*, to scrape off, to rub or wear away, erode): *Lichens*. Having the thallus surface worn, eroded.

abrupt (L. *abruptus*, abrupt, precipitous < *abrumpere*, to separate, to break off < pref. *ab-*, away from, far from + *rumpere*, to break, to burst): terminating abruptly, not attenuated or gradual, appearing as if it had been cut transversally, like the stipe of some basidiomycetes (e.g., of the gen. *Hohenbuehelia*, of the Agaricales). Syn. of **truncate**.

abscission (L. *abscission* < *abscissio*, genit. *abscissionis*, the act of cutting off, breaking off < *abscindere*, to tear off, separate + suf. *-io*, genit. *-ionis* > E. *-ion*, state, result of): separation, as when the partition or the layer that unites two cells or two organic structures breaks up, e.g., a spore and its sporophore.

absorption (L. *absorption* < *absorptio*, genit. *absorptionis*, the act of absorbing or swallowing < *absorptus* < *absorbere*, to swallow, engulf + suf. *-io*, genit. *-ionis* > E. *-ion*, result of an action, state of): the action and effect of absorbing, which is the passage of solid, liquid or gaseous materials (generally nutritive) dissolved in water into or through living cells, by means of the process of osmosis. The fungi typically acquire their nutrients by absorption through the cell wall and plasma membrane. Cf. **ingestion**.

abstriction (NL. *abstrictio*, genit. *abstrictionis*, the act of binding or pressing together < pref. *ab-*, opposite, away from, far from + *strictus*, tightly united + suf. *-io*, genit. *-ionis* > E. *-ion*, result of an action, state of): *Sporogenesis*. The production of spores in a sporogenous filament through the successive formation of transverse walls, so that the spores remain grouped in short chains, separated by more or less distinct constrictions, as occurs, e.g., in *Geotrichum* and *Monilia* (moniliaceous asexual fungi). In abstriction *abjunction* occurs first, and then *abscission*, especially by constriction.

abyssal (Gr. *ábyssos*, abyss, the depths of the sea, an immeasurable space + L. suf. *-alis* > E. *-al*, relating to or belonging to): pertaining to an abyss, in particular referring to those organisms that inhabit marine waters at great depths; e.g., *Periconia abyssa* (dematiaceous asexual fungi), which is found on wood submerged at depths of 3,975-5,315 m, where the hydrostatic pressure is 50 atmospheres or greater.

acanthocyte (Gr. *ákantha*, spine + *kýtos*, cavity, here cell): a spiny cell produced on a short branch on the mycelium of *Stropharia* spp. (Agaricales).

acanthophysis, pl. **acanthophyses** (Gr. *ákantha*, spine + *phýsis*, action of springing forth or giving birth, a growth): *Basidiomycetes*. A botryose, clavate or coralloid cystidium or paraphysis, covered with small needles or spines, that is found on the basidiocarp of the Thelephoraceae (Aphyllophorales), such as *Aleurodiscus mirabilis*, *Pseudofistulina* (=*Fistulina*) *radicata* and *Vararia* spp. In some Agaricales acanthophysoid cystidia are present, such as the cheilocystidia and pileocystidia of *Mycena brownii* and some other species of *Mycena*.

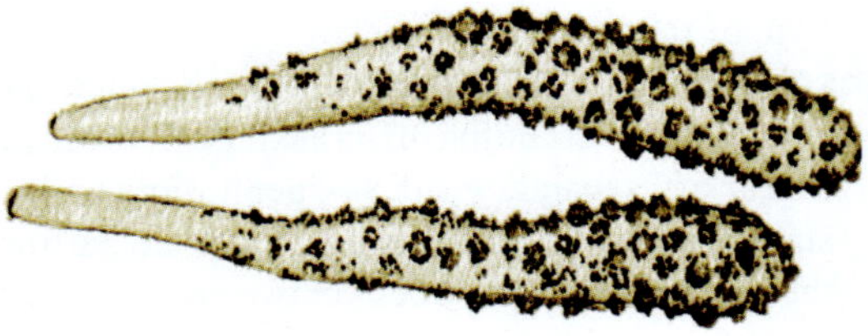

Acanthophyses from the hymenophore of *Pseudofistulina* (= *Fistulina*) *radicata*, x 350 (*MU*).

acaryallagic (Gr. *a-*, not, without + *káryon*, nucleus + *allagé*, permutation, commutation + suf. *-íkos* > L. *-icus* > E. *-ic*, belonging to or relating to): see **akaryallagic**.

acaudate (L. *acaudatus*, without a tail < *a-*, not, without + *cauda*, tail + suf. *-atus* > E. *-ate*, provided with or likeness): without a tail. Cf. **caudate**.

acaulescent, **acauline**, **acaulose** (L. *acaulis*, without a stem < *a-*, not, without + *caulis*, stem, pedicel + suf. *-escens*, genit. *-escentis* > E. *-escent*, that which turns, beginning to, slightly; or + E. suf. *-ose*, like): without a stem or pedicel, sessile, like the sporiferous apparatus of a group of Myxomycetes, which for lacking a pedicel or peduncle are referred to as *acauline* (probably referring to those species that have aethalia, such as those of *Lycogala*, of the Liceales).

Acaulescent sporangium of *Lycogala epidendrum*, x 8 (*MU*).

accumbent (L. *accumbens*, genit. *accumbentis*, reclining, laying down < *accumbere*, to recline, lay down + suf. *-entem* > E. *-ent*, being, creature; existence): reclining or laying against something, like the basidiocarps of *Poria*, *Corticium* and other gen. of Aphyllophorales.

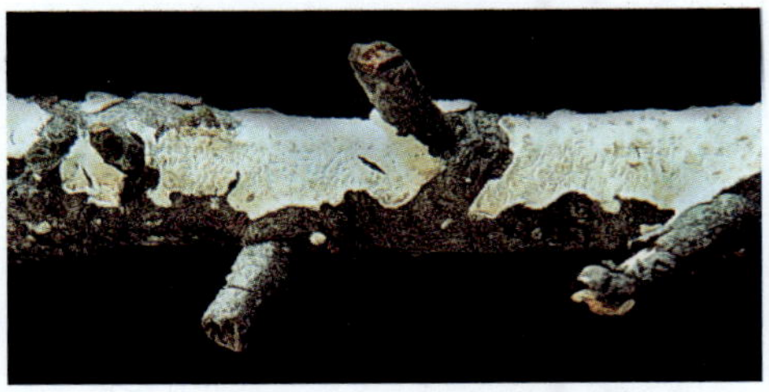

Accumbent basidiocarp of *Corticium* sp., x 0.3 (*MU*).

-aceae (L. *-aceae*, fem. pl. adjectival suf.): ending of taxonomic families, e.g., Mucoraceae, Saccharomycetaceae, etc.

acellular (Gr. *a-*, not, without + NL. *cellula*, cell, living cell, dim. of *cella*, small room, compartment, cell of a honeycomb + L. suf. *-aris* > E. *-ar*, like, relating to): not being divided into cells, like the plasmodium of the Myxomycetes; not constituted of cells, like the peridium of the sporangia of the Myxomycetes. Cf. **cellular**.

-aceous (L. suf. *-aceus* > E. *-aceous*, of or pertaining to, with the nature of): a suf. used to indicate a relation to something, or having the nature of; e.g., micaceous and crustaceous respectively indicate that they are similar to mica and to a crust. Other examples are pucciniaceous, ustilaginaceous and agaricaceous, to refer to fungi having the fundamental characteristics of the families Pucciniaceae, Ustilaginaceae and Agaricaceae, respectively.

acephalous (Gr. *aképhalos*, without a head < *a-*, not, without + *kephalé*, head): without a head; applied to structures that do not have a head or enlargement on the upper end. Cf. **capitate**.

acerose, **acerous** (L. *acerosus*, needle-shaped < *acer*, sharp, rigid, stinging + E. suf. *-ose*, like; or L. suf. *-osus* > E. *-ous*, quality of, full of, augmented): referring to a structure with the shape of a needle, of hard consistency and ending in a point, like pine leaves. E.g., the phialides of *Penicillium verruculosum*, *P. purpurogenum* and *Isaria fumosorosea* (=*Paecilomyces fumosoroseus*), moniliaceous asexual fungi, are acerose.

acervate, **acervatus** (L. *acervatus*, heaped up, piled < *acervare*, to heap up, pile + suf. *-atus* > E. *-ate*, provided with, likeness): growing in groups, cushion-like piles or masses. Cf. **solitary.**

acervulus, pl. **acervuli** (NL. *acervulus* < dim. of L. *acervus*, pile, prominence): a pseudoparenchymatous aggregation of hyphae in the shape of a small erumpent cushion on which are formed short, crowded conidiophores on the tissues (subcuticle, subepidermis or parenchyma) of the plants parasitized by the fungi that form these structures. The acervulus is the asexual sporophore characteristic of the melanconiaceous asexual fungi, e.g., *Colletotrichum* and *Spilocaea*.

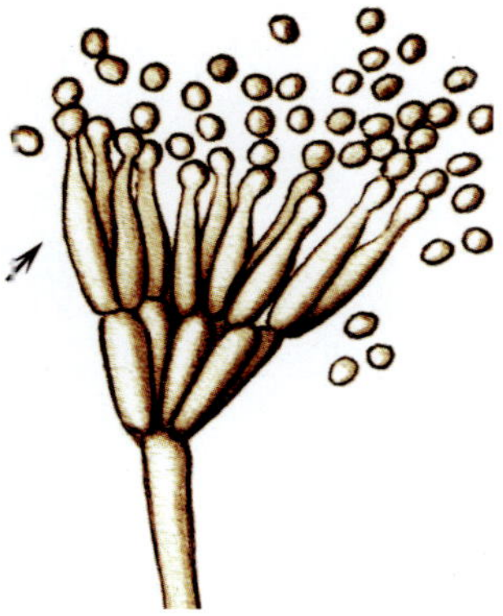

Acerose phialides of *Penicillium purpurogenum*, x 700 (*MU*).

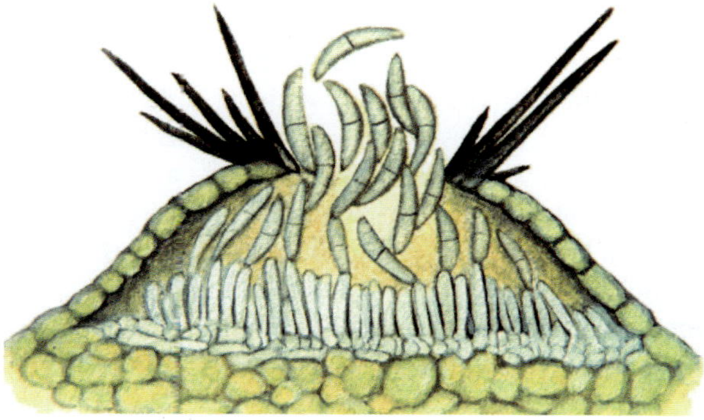

Acervulus of *Colletotrichum falcatum*, x 1,100 (*MU*).

Acervulus of *Colletotrichum circinans*, x 350 (*MU*).

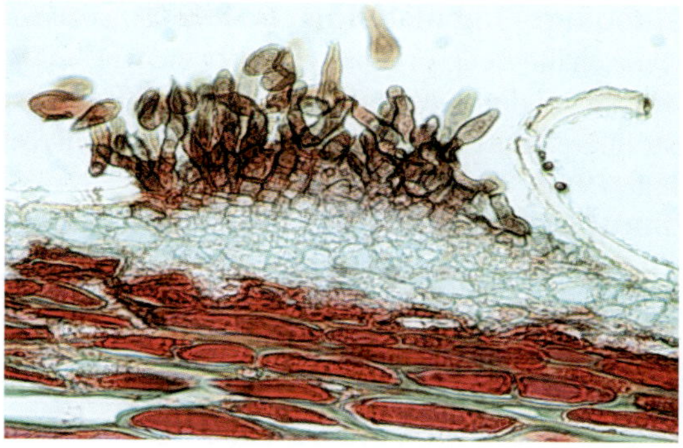

Acervulus of *Spilocaea pomi*, x 300 (*MU*).

acetabulate (L. *acetabulum*, receptacle; wide, shallow goblet, originally used for vinegar < *acetum*, vinegar + dim. suf. *-ulum* + suf. *-atus* > E. *-ate*, provided with, likeness): syn. of **acetabuliform**.

acetabuliform (NL. *acetabuliformis* < L. *acetabulum*, receptacle; wide, shallow cup, originally used for vinegar + dim. suf. *-ulum* + suf. *-formis* < *forma*, shape): with the shape of a shallow, wide cup, like the apothecia of various lichens and Pezizales, e.g., *Cookeina sulcipes* (Pezizales).

Acetabuliform apothecium of *Cookeina sulcipes*, x 1 (*EPS*).

Acetabuliform apothecia of *Cookeina sulcipes*, x 1 (*TH*).

acetabulum, pl. **acetabula** (L. *acetabulum*, receptacle; wide, shallow cup, originally used for vinegar < *acetum*, vinegar + dim. suf. *-ulum*): a shallow, cup-shaped structure, like the apothecia of certain Discomycetes, e.g., *Peziza* and *Helvella* (Pezizales).

achlorophyllous (Gr. *a-*, not, without + *chlorós*, green, the color of grass + *phýllon*, leaf): lacking chlorophyll, the photosynthetic pigment of green plants.

achroic, achrous (Gr. *a-*, not, without + *chrõs* or *chroía*, color or superficial appearance of a thing + suf. *-íkos* > L. *-icus* > E. *-ic*, belonging or relative to; or + L. suf. *-osus* > E. *-ous*, full of, with the qualities of): without color or pigment.

acicular (L. *acicularis*, narrow, stiff and sharp-pointed < *acicula*, dim. of *acus*, needle, point + suf. *-aris* > E. *-ar*, like, pertaining to): a long, very slender, sharp-pointed structure, like certain types of cystidia, as in *Pouzarella nodospora* (Agaricales). In *Mycena acicula* (=*M. longiseta*), Agaricales, the tips of the appendages of the cheilocystidia are acicular.

aciculate (NL. *aciculatus* < L. *acicula*, dim. of *acus*, needle, sharp end + suf. *-atus* > E. *-ate*, provided with, similar): **1**. With small needles, or structures similar to needles, like the ascospores of *Lophodermium juniperinum* (Rhytismatales). **2**. Structures that have fine grooves on the surface, as though made by a small needle.

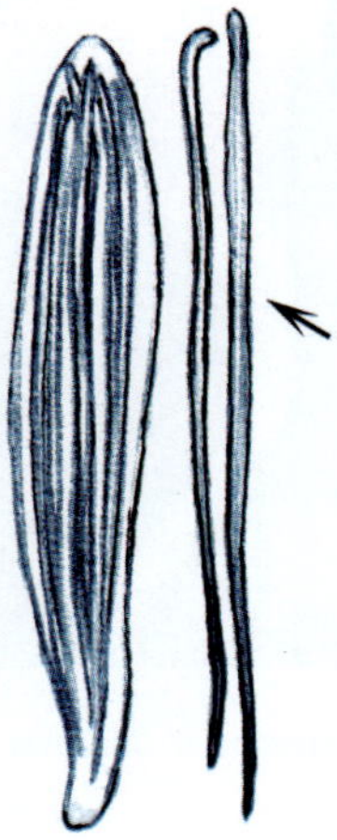

Aciculate ascospores of *Lophodermium juniperinum*, x 590 (*MU*).

aciculiform (L. *acicula*, dim. of *acus*, needle, point + suf. *-formis* < *forma*, shape): having the shape of a needle. Syn. of **acicular**.

acidophilous (L. *acidus*, acid, sour + Gr. *phílos*, have an affinity for): **1**. An organism that grows in acid substrates; in particular it refers to certain types of lichens that grow in peaty soils. **2**. A cellular organelle that stains in acidic dyes. Cf. **basophil**.

acinaciform (L. *acinaciformis* < *acinaces*, sabre, scimitar, curved sword < Gr. *akinákes*, short sword + L. suf. *-formis* < *forma*, shape): in the shape of a sword, like the appendages that cover the zygosporangium of *Absidia spinosa* (Mucorales).

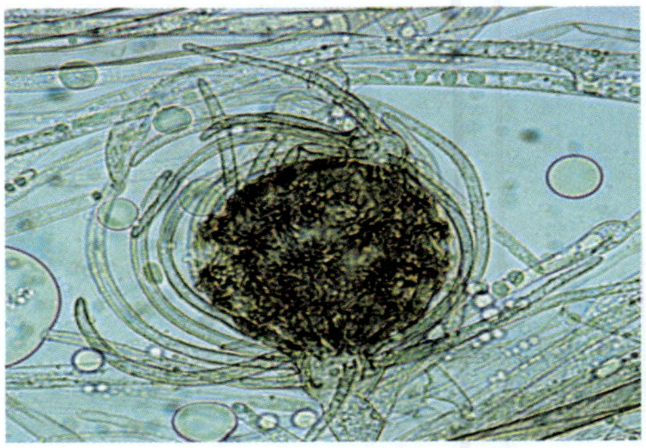

Acinaciform appendages of *Absidia spinosa*, x 180 (*EAA*).

acinose, acinosus, acinous (L. *acinosus*, like grapes < *acinus*, grape seed + E. suf. *-ose*, like; or L. *-osus* > E. *-ous,* full of, with the qualities of): in the shape of a grape seed or with a granular structure similar to a cluster of grapes. See **botryose**.

acolumellate (Gr. *a-*, not, without + L. *columna*, column + dim. suf. *-ella* + suf. *-atus* > E. *-ate*, provided with, likeness): lacking a columella. *Gasteromycetes*. Refers to a fruiting body in which the gleba lacks a basal columella, as occurs in *Astraeus* (Sclerodermatales).

acrasids (Gr. *akrásia*, bad mixture, or *akrátos*, immiscible): acrasids, or the cellular slime fungi, whose myxamoebae (cellular vegetative phase) unite to constitute a *pseudoplasmodium*, in which these myxamoebae retain their individuality. These organisms are classified in the phylum Acrasiomycota of the kingdom Protista. See **pseudoplasmodium**.

acrasin (Gr. *akrásia*, bad mixture, or *akrátos*, immiscible + NL. *-ina*, suf. used in chemistry to denote an activator or a compound): a chemotactic compound produced by *Dictyostelium* and other gen. of Dictyosteliomycota, which controls the aggregation of the myxamoebae to form the pseudoplasmodium.

Acrasiomycetes (Gr. *akrásia*, bad mixture, or *akrátos*, immiscible + L. *-mycetes*, ending of a taxonomic class < Gr. *mýkes*, genit. *mýketos*, fungus): a group of viscous or mucilaginous fungi (Acrasiomycetes) with myxamoebae that aggregate in pseudoplasmodia, which are transformed into fruiting bodies called sorocarps. In the classification outline of Alexopoulos *et al*., 1996, modified by Ulloa and Hanlin, 2006, this class was placed in the phylum Acrasiomycota of the kingdom Protista. See the Appendix to compare the transfer of this group of fungus-like organisms of the order Acrasiales, kingdom Protista, to the order Acrasida, class Heterolobosea, phylum Percolozoa, kingdom Protozoa (Kirk *et al.*, 2008). These taxonomic rearrangements, with the concurrent changes of the names of taxa, represent a constant process of study and evaluation, in order to achieve a phylogenetic classification of the fungi. See **Protozoa**.

acro-, **acr-** (Gr. *ákros*, on the end, apical, terminal): a pref. often employed to introduce the idea that something is on the end or on the upper part, or that it tends toward it, as in *acroauxic*, that which grows on the apex.

acroauxic (Gr. *ákros*, on the upper part, the end of something, apex + *aúxo*, to grow, develop, increase + suf. *-íkos* > L. *-icus* > E. *-ic*, belonging or relative to): having growth in length restricted to the apical region, as occurs in many conidiophores. Cf. **basauxic**.

acrochroic (Gr. *ákros*, on the upper part, the end of something, apex + *chrõs*, color + suf. *-íkos* > L. *-icus* > E. *-ic*, belonging or relative to): having pigmentation concentrated principally in the apex, as is observed in the hyphae of the Clavariaceae (basidiomycetes).

acrogenous (Gr. *ákros*, on the upper part, the end of something, apex + *génos*, origin, birth): borne or formed on the apex, like the acrogenous spores of certain fungi, e.g., *Monilia* and *Cladosporium* (moniliaceous and dematiaceous asexual fungi, respectively).

acrokont, **acrocont**, **acrokontous**, **acrocontous** (Gr. *ákros*, on the upper part, the end of something, apex + *kontós*, oar, flagellum): a motile cell which has the flagellum or swimming organ on the anterior or front part, as occurs in the zoospores of the aquatic fungi of the phylum Hyphochytriomycota, of the kingdom Stramenopila, e.g., *Rhizidiomyces* (Hyphochytriales). Cf. **opisthokont** and **pleurokont**.

acronema, pl. **acronemata** (Gr. *ákros*, on the upper part, the end of something, apex + *nêma*, genit. *nêmatos*, filament): the tip of the flagellum consisting only of the extension of the two central microtubules, i.e., without the nine peripheral triplets of microtubules.

acronym (Gr. *ákros*, on the upper part, the end of something, apex; beginning, edge + *ónyma*, *ónoma*, name): a word formed from the initial letter or letters of each of the successive or principal parts of a compound term. The reference collections (herbaria, cultures, genetic resources, etc.) have their corresponding acronyms, e.g., *IMI*, for the *International Mycological Institute*.

acropetal (Gr. *ákros*, apex + L. *petere*, to direct toward, grow toward): refers to a structure that develops from the base toward the apex; e.g., in a chain of conidia with acropetal succession, the conidium at the tip is the youngest and of most recent formation. *Monilia* and *Cladosporium* (moniliaceous and dematiaceous asexual fungi, respectively) are examples of gen. of fungi that produce conidia in this manner. Cf. **basipetal**.

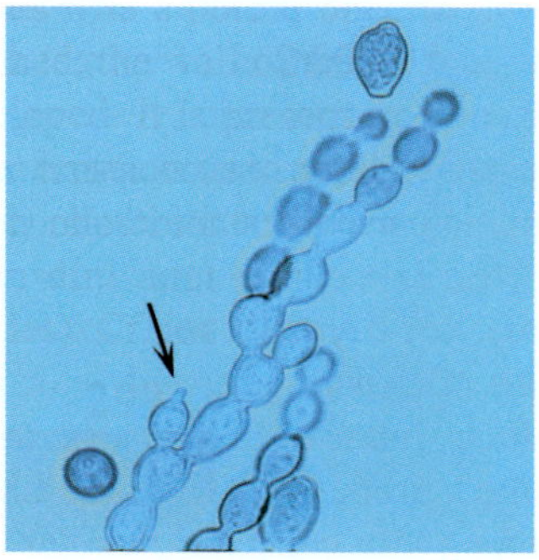

Acropetal succession of the conidia of *Monilia fructicola*, x 320 (*MU*).

acropleurogenous (Gr. *ákros*, on the upper part, the end of something, apex + *pleurá*, side, flank + *génos*, origin, birth): that which originates at the apex and later turns toward one side due to indeterminate growth of the apex; e.g., the conidia of *Trichothecium* (moniliaceous asexual fungi) are acropleurogenous. Also called **pleuroacrogenous**.

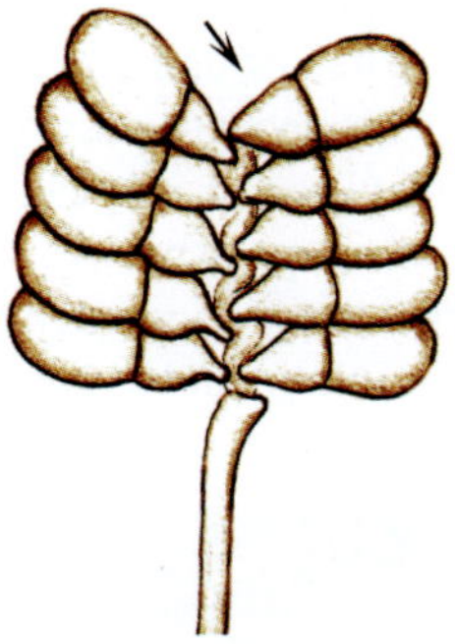

Acropleurogenous conidia of *Trichothecium roseum*, x 1,260 (*MU*).

acrospore (Gr. *ákros*, on the upper part, the end of something, apex + *sporá*, spore): an apical spore.

acrosporogenous (Gr. *ákros*, on the upper part, the end of something, apex + *sporá*, spore + *génos*, origin, birth): a type of maturation in which a spore (conidium, spermatium, etc.) forms on the end of a mother cell or sporophore; i.e., the cells continue delimiting and maturing in sequence from the base toward the apex, with the youngest spore on the apex.

acrosporous (Gr. *ákros*, on the upper part, the end of something, apex + *sporá*, spore + L. suf. *-osus* > E. *-ous*, full of, with the qualities of): see **acrosporogenous**.

acroton (Gr. *akrotónos*, strained to the utmost, completely extended < *ákros*, on the upper part, the end of something, apex + *tónos*, tension): *Lichens*. A needle-like structure with lateral branches.

actin (Gr. *aktîs*, genit. *aktînos*, radius, ray + L. *-in*, suf. used to denote a chemical compound): *actin,* together with *myosin,* are the proteins associated with the contraction and relaxation of animal muscle. They are also found in other eukaryotic organisms; e.g., they are responsible for the reversible and rhythmic protoplasmatic currents of the plasmodium of some Myxomycetes, as occurs in *Physarum polycephalum* (Physarales). See **myosin**.

actinogyrose (Gr. *aktîs*, genit. *aktînos*, radius, ray + *gýros*, circular, round + L. suf. *-osus* > E. *-ose*, full of, augmented, prone to): a lichen apothecium of the gyrodisc type (with concentric circles on the upper surface) but with radial furrows or folds and without a proper margin, like that of *Umbilicaria cylindrica* (Lecanorales).

actinolichen (Gr. *aktîs*, genit. *aktînos*, radius, ray + *leichén*, lichen): lichenized association between an actinobacterium (*Streptomyces*) and an alga (*Chlorella xanthella*).

actinomycete (Gr. *aktîs*, genit. *aktînos*, radius, ray + L. *-mycetes*, ending of a taxonomic class < Gr. *mýkes*, genit. *mýketos*, fungus): a filamentous, gram positive bacterium. The actinomycetes, which include saprobic species and parasites of animals and humans, have sometimes been erroneously classified as filamentous conidial fungi; they are correctly called actinobacteria. These prokaryotic organisms are the producers of various important antibiotics, such as amphotericin, cycloheximide, nystatin and streptomycin.

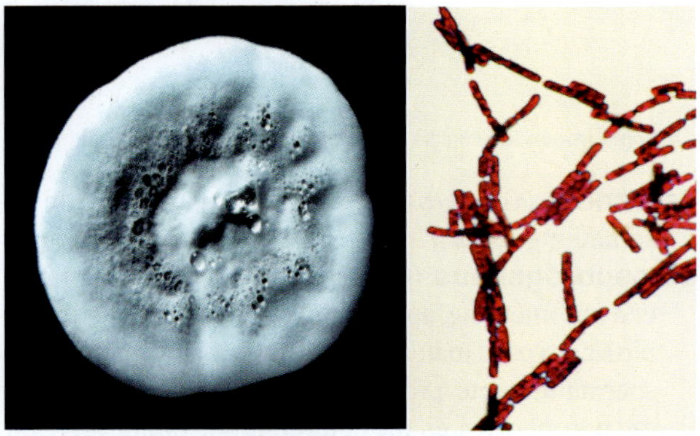

Colony (fungus-like) of an undetermined **actinomycete**, on agar, x 2 (*MU*). Catenate cells of the **actinomycete**, x 1,300 (*MU*).

aculeate (L. *aculeatus*, provided with prickles or spines < *aculeus*, spur, spine < *acus*, needle + suf. *-atus* > E. *-ate*, provided with, likeness): **1**. Having spurs or spines, like the columella of the sporangium of *Mucor spinosus* (Mucorales) and the basidiospores of *Laccaria laccata* (Agaricales). **2**. Shaped like a needle or spine, like the synnemata of *Cephalotrichum* (*=Doratomyces*) *stemonitis*, of the stilbellaceous asexual fungi , and the hymenophore of hydnaceous basidiomycetes (*Hydnum*, *Sarcodon*, *Hymenochaete* and other gen. of Aphyllophorales). Sometimes the term *triquiform* is used for the aculeate caulocystidia of *Laccaria laccata* and other Agaricales. See this last term.

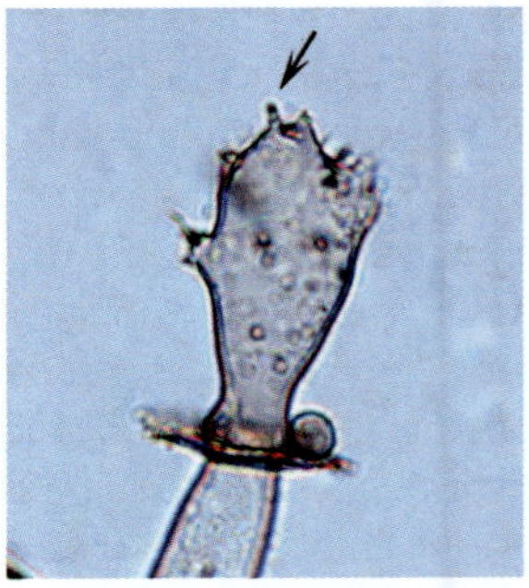

Aculeate columella of the sporangium of *Mucor spinosus*, x 800 (*MU*).

Aculeate synnemata of *Cephalotrichum* (*=Doratomyces*) *stemonitis*, x 27 (*MU*).

aculeolate (L. *aculeus*, spur, spine < *acus*, needle + dim. suf. *-ulus* + suf. *-atus* > E. *-ate*, provided with or likeness): having spines or *aculei*.

aculeus, pl. **aculei** (L. *aculeus*, spur, spine < *acus*, needle + dim. suf. *-ulus*): each of the sharp spines or teeth of the hymenophore of *Hydnum* and *Sarcodon*, among other Hydnaceae (Aphyllophorales).

acuminate (L. *acuminatus*, ending in a point, the point of anything < *acuminare*, to sharpen + suf. *-atus* > E. *-ate*, provided with, likeness): ending in a point, regardless of its consistency. E.g., the conidia of

Alatospora acuminata (moniliaceous asexual fungi), the apex of the perithecial stroma of *Cordyceps stylophora* (Hypocreales), the cystidia of *Psathyrella squamosa*, *Mycena alcalina* and *Macrocystidia cucumis* (Agaricales), and the apex of the pileus in *Weraroa* (Hymenogastrales) are all acuminate.

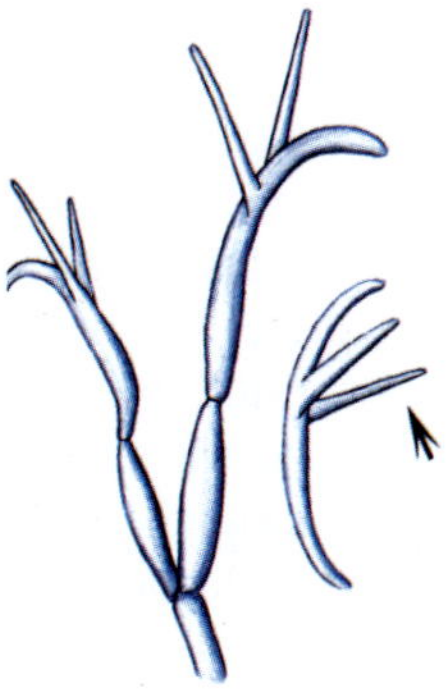

Acuminate conidia of *Alatospora acuminata*, x 1,250 (*MU*).

Acuminate apex of the perithecial stroma of *Cordyceps stylophora*, x 4 (*MU*).

acute (L. *acutus*, sharp < *acuere*, to sharpen < *acus*, needle): with the edges or margins forming an acute angle, less than 90°, at the apex; the sides of this angle are essentially straight or slightly convex. Also used for a solid structure ending in a point. E.g., the distal ends of the aciculate, aculeate, or acuminate cystidia of various Agaricales are acute.

acyanophilic (Gr. *a*-, not, without + *kýanos*, dark blue + *phílos*, have an affinity for): not staining blue; refers to the negative reaction of the walls of cells or spores when placed in 1% cotton blue in lactic acid. Cf. **cyanophilic**.

ad- (L. prep. used as a pref. to denote proximity, toward, near): a pref. used in compounds such as *adaxial*. Cf. **ab-**.

adaxial (NL. *adaxialis* < L. pref. *ad*-, toward + *axialis*, axial < *axis*, axis, axle + suf. *-alis* > E. *-al*, relative to or belonging to): the structure or side that is closest to an axis; e.g., the adaxial side of a basidiospore is the dorsal or concave side that is nearest the longitudinal axis of the basidium, e.g., *Amanita* (Agaricales). Cf. **abaxial**.

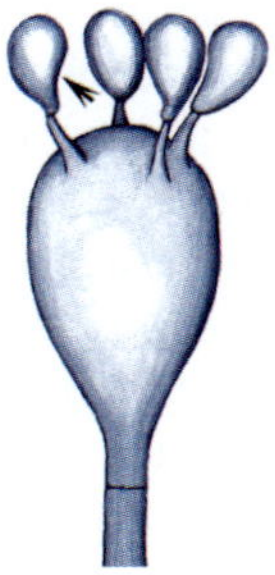

Adaxial side of the basidiospores of *Amanita* sp., x 1,170 (*MU*).

adelomycete (Gr. *adélos*, hidden, unknown + *mýkes*, genit. *mýketos*, fungus): a fungus whose sexual state is not known.

adelphogamy (Gr. *adelphós*, brother, close relative + *gámos*, sexual union + E. suf. *-y*): a particular type of pseudomictic sexual reproduction, which consists of the copulation of a mother cell and an independent daughter cell, as happens, e.g., in yeasts of the gen. *Zygosaccharomyces* (Endomycetales). Cf. **pedogamy**.

adelphotaxis, pl. **adelphotaxes** (Gr. *adelphós*, brother, close relative + *táxis*, movement): a tactic phenomenon of mutual attraction exhibited by the zoospores of *Achlya* (Saprolegniales), which may result in their auto-aggregation. This chemotactic process is initiated by certain exogenous attractants and is amplified by signals emitted by the aggregated zoospores.

adeno-, **aden-** (Gr. *adén*, genit. *adénos*, gland): a term used in the formation of generic names, nouns, and adjectives, to denote that something has glands or has the shape of a gland.

adenose (Gr. *adén*, genit. *adénos*, gland + L. suf. *-osus* > E. *-ose*, full of, augmented, prone to): having glands; similar to glands.

adherence (ML. *adherencia*, action and effect of adhering < *adhaere*, to be united + *-entia* > E. *-ence*, suf. that signifies state, quality or action): the quality of certain fungicides, and other protectants that are applied to crops, to indicate their ability to adhere to the leaf surface. Cf. **retention**.

adherent (L. *adhaerent* < *adhaerens* < pref. *ad*-, near, toward + *haerere*, to stick, adhere + suf. *-entem* > E. *-ent*, being): syn. of **adnate**.

adhesorium, pl. **adhesoria** (L. *adhesus*, adhesion + suf. *-orium* > E. *-orium*, the place of a thing): a tubular, bulbous extension derived from the germination of the cyst of *Plasmodiophora brassicae* (Plasmodiophorales), from which the *Stachel* (stylet) is discharged through the cell wall into the cytoplasm of a root hair of the host plant (cabbage); in a few seconds the cytoplasm of the parasitic fungus enters through the hole made by the *Stachel*, where it is caught up by the cytoplasmic currents of the host plant; subsequently, the cytoplasm of the fungus develops into a plasmodium that continues the infection process in the host plant. In *Polymyxa betae*, also in the Plasmodiophorales, a similar infection mechanism occurs, with an adhesorium that is derived from the evagination of the sheath (the *Rohr*), which contains the *Stachel*; the Stachel is discharged from the *Rohr*, it perforates the host cell wall, and in a period of one second an infective amoeboid unit is injected . The total time that elapses from formation of the adhesorium to penetration of the host is ca. one minute.

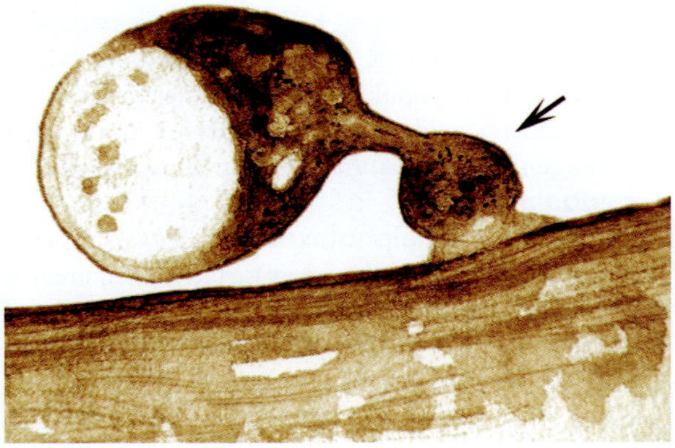

Adhesorium of *Plasmodiophora brassicae.* Drawing based on a transmission electron micrograph published by Beckett *et al.* in their *Atlas of Fungal Ultrastructure*, 1974, x 13,470 (*MU*).

adiaspore (Gr. *adiáspore* < *a-*, not, without + *diá*, through, separating, dividing + *sporá*, spore): a spore that increases notably in size but which does not divide. *Med. mycol.* This term is applied to the large, spherical chlamydospore that is derived from a conidium of *Sporotrichum* (=*Chrysosporium*) *pruinosum*, of the moniliaceous asexual fungi, which increases greatly in size on being inhaled and introduced into the bronchial tubes and lungs of rodents and other small, warm-blooded wild animals, in which the adiaspores cause the disease known as *adiasporomycosis*. Adiaspores also can form from conidia incubated *in vitro* at temperatures near 37ºC.

adiasporomycosis, pl. **adiasporomycoses** (Gr. *adiáspore* < *a-*, not, without + *diá*, through, separating, dividing + *sporá*, spore + *mýkes* > L. *myces*, fungus + Gr. suf. *-osis*, condition or state of a thing): a mycotic infection of warm-blooded animals caused by adiaspores of *Sporotrichum* (=*Chrysosporium*) *pruinosum*, of the moniliaceous asexual fungi, that are introduced into the lungs and bronchi of these animals.

adnate (L. *adnatus*, broadly attached < pref. *ad-*, in, near + *natus*, born): *Agaricology*. The small gills of the agaricaceous fungi are said to be adnate or *adnexed* when they extend to the foot or stipe and adhere to it throughout their width, as in the gen. *Laccaria* (Agaricales). Also called **adherent** or **concrescent**, which also is used to designate a film or certain scales that are strongly adhered to the surface of a pileus. Cf. **decurrent**, **free** and **sinuate**.

Adnate gills of *Laccaria proxima*, x 3 (*MU*).

adpressed, **appressed** (L. *adpressus*, pressed closely against something < *apprimere*, to press): refers to scales, fibers, hairs, etc., that are closely applied or flattened against the surface of the organ or part on which they are inserted. E.g., the surface of the pileus of *Agaricus placomyces* and *Macrolepiota procera* (Agaricales) is adpressed (appressed) squamulose, in that the fibrils of the points of the hyphae agglutinate, forming applanate scales.

Adpressed fibrous scales of the pileus of *Agaricus placomyces*, x 1 (*MRO*).

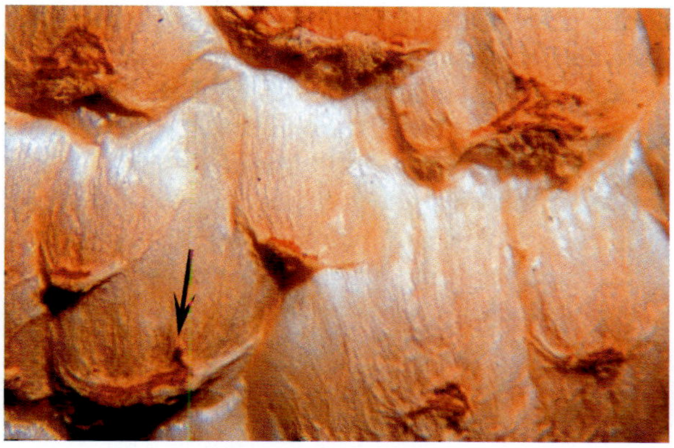

Adpressed scales of the pileus of *Macrolepiota procera*, x 5 (*MU*).

adspersed (L. *ad-*, towards + *sparsus*, spread out < *spargere*, to scatter): scattered; of wide distribution.

adunc, **aduncate**, **aduncous** (L. *aduncus*, hooked, bent < pref. *ad-*, toward + *uncus*, hook + suf. *-atus* > E. *-ate*, provided with, likeness): doubled or curved like a hook. See **unciform**.

adventitious (L. *adventicius*, added from another source and not inherent or innate, supplemental; arising or occurring sporadically or in other than the usual location < ML. *adventus*, coming from outside, with the E. suf. *-ous*, having, possessing the qualities of < OF. *-ous*, *-eus* < L. *-osus*): appearing in an abnormal or unusual position; e.g., an *adventitious septum* is one that forms independently or in the absence of nuclear division, particularly in relation to the movement of cytoplasm from one part to another of the hypha of the fungus. In contrast is the *primary septum*, which forms in direct association with nuclear division. See **adventitious ramification** and **adventitious septum**.

adventitious ramification (L. *adventicius*, accidental, unexpected, supplementary < ML. *adventus*, coming from outside; L. *ramificatio*, production of branches from an axis or lower branch < *ramus*, branch, pl. *rami* + *-ationem*, action, state or condition, or result > E. suf. *-ation*): branching that occurs outside the normal pattern in the branches of the lichen *Cladonia* (Lecanorales), as a regeneration following damage.

adventitious septum, pl. **septa** (L. *adventicius*, accidental, unexpected, supplementary < ML. *adventus*, coming from outside; L. *septum*, barrier, partition): a septum that does not form in association with nuclear division, as when a hypha is broken, either naturally or artificially, and the formation of this septum avoids the loss of protoplasm. See **septum**.

aecidiole, **aecidiolum**, pl. **aecidiola** (NL. *aecidiolum*, dim. of *aecidium* < Gr. *aikídion* < *aikía*, lesion, injury, wound, alluding to the symptoms on infected host plants + dim. suf. *-ídion* > L. *-idium*; the L. dim. suf. *-olum* > E. *-ole*): *Uredinales*. Syn. of **spermogonium**.

aecidium, pl. **aecidia** (L. *aecidium* < Gr. *aikídion* < *aikía*, lesion, injury, wound, alluding to the symptoms on the leaves and stems of plants infected by these fungi + dim. suf. *-ídion* > L. *-idium*): an alternative name given to the **aecium**, which is the preferred term.

aeciospore (L. *aecium* < Gr. *aikía*, wound, damage + *sporá*, spore): a binucleate spore produced in an **aecium**. Some examples are *Gymnosporangium clavipes* and *Cronartium conigenum* (Uredinales).

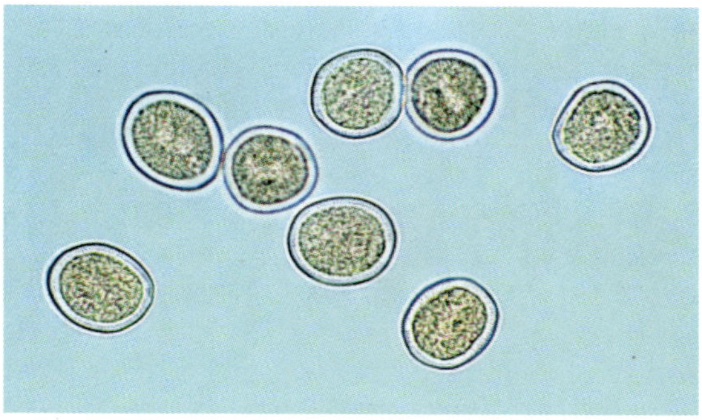

Aeciospores of *Gymnosporangium clavipes*, x 740 (*CB*).

Mass of **aeciospores** of *Cronartium conigenum* on a pine cone, x 0.5 (*MU*).

aecium, pl. **aecia** (L. *aecium* < Gr. *aikía*, wound, lesion + dim. suf. *-íon* > L. *-ium*): also called **aecidium**; *Uredinales* (the plant pathogenic fungi commonly known as rusts). A goblet- or cup-shaped sorus (stage I of the demicyclic or macrocyclic rusts) in which originate the binucleate aeciospores (aecidiospores) capable of infecting a plant of a different species; e. g., in *Puccinia graminis tritici*, the aecia form on the lower surface of the leaves of barberry (*Berberis vulgaris*), and in *Aecidium* sp. they are produced on the leaves of *Diospyros digyna*.

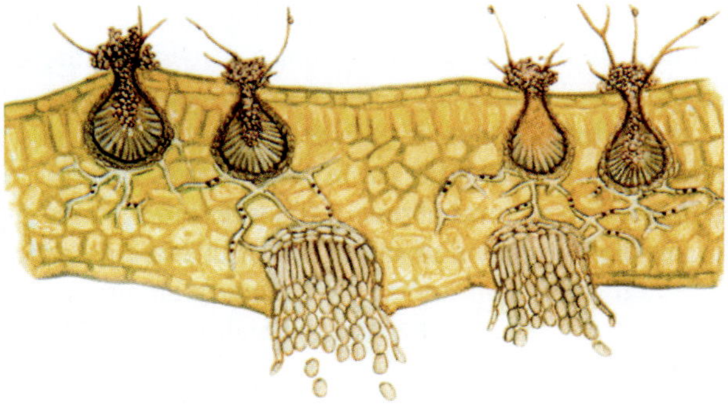

Spermagonia (upper surface) **aecia** (lower surface); the latter with aeciospores of *Puccinia graminis tritici*. Drawing made from a longitudinal section of a barberry leaf, x 80 (*MU*).

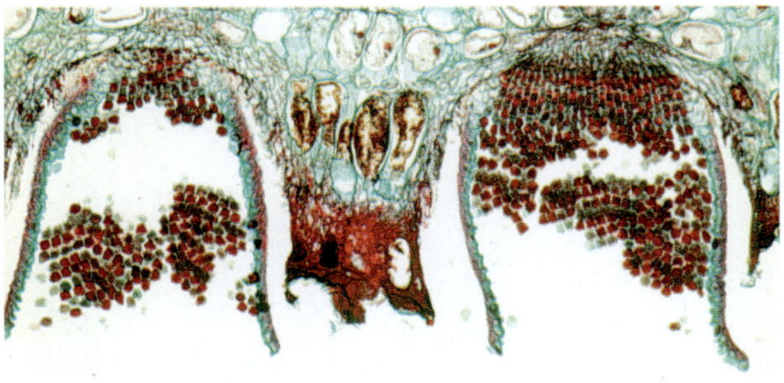

Aecium and aeciospores of *Puccinia graminis tritici*, seen in longitudinal section of a barberry leaf, x 100 (*MU*).

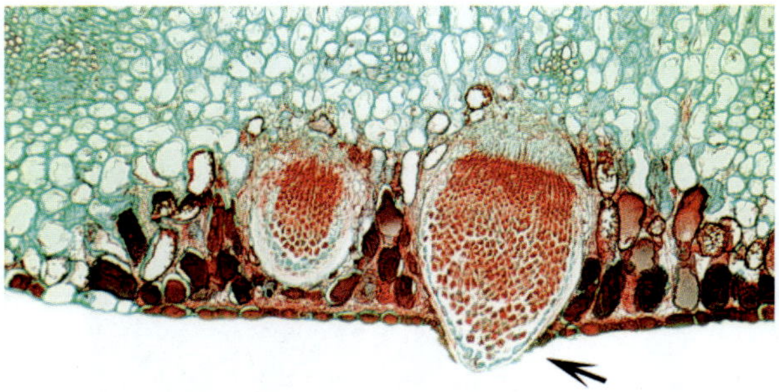

Aecium and aeciospores of *Puccinia graminis tritici*, seen in longitudinal section of a barberry leaf, x 125 (*MU*).

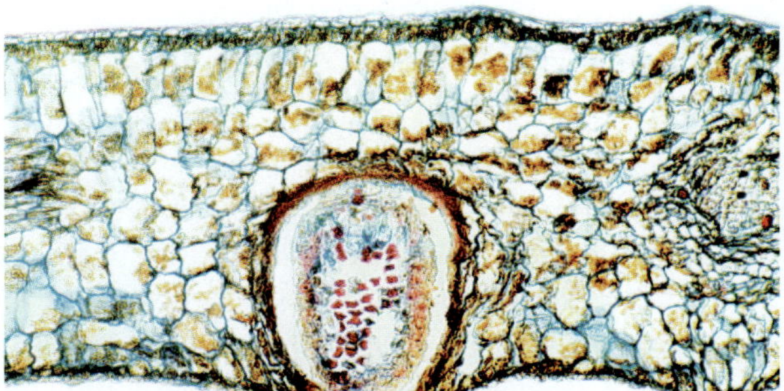

Aecium and aeciospores of *Aecidium* sp., seen in longitudinal section of a host leaf (*Diospyros digyna*), x 160 (*MU*).

aequi- (L. *aequs*, equal): pref. meaning equal. Equivalent to the Gr. pref. **iso-**. Cf. **inaequi-**.

aequihymenial (L. *aequus*, equal + *hymenium*, membrane + L. suf. *-alis* > E. *-al*, relating to or belonging to): see **aequihymeniiferous**.

aequihymeniiferous (L. *aequus*, equal + *hymenium*, membrane + *-ferous*, bearer < *ferre*, to bear, carry + L. *-osus* > OF. *-ous*, *-eus* > E. *-ous*, having, possessing the qualities of): bearer of a hymenium that develops in a uniform manner, in which the basidia mature and liberate their basidiospores almost simultaneously over all the surface of the gills. This type of hymenial development also is called **aequihymenial**, and it is characteristic of all of the gen. of Agaricales, except *Coprinus*. Cf. **inaequihymeniiferous**.

aero-, **aer-** (L. *aer* < Gr. *aér*, genit. *aéros*, air): a pref. employed to form compound words to introduce the idea that something is in the air, or is full of air, etc.

aeroallergen (L. *aer* < Gr. *aér*, genit. *aéros*, air + *allergen* < Gr. *állos*, another + *érgon*, work + *génos*, origin < *gennáo* to engender, produce): a particle, viable or not, present in the atmosphere, that produces an allergy in humans and higher animals, generally respiratory, such as asthma and rhinitis. The spores of many fungi, e.g., the conidia of *Penicillium* and *Aspergillus* (moniliaceous asexual fungi) and the ascospores, basidiospores and other types of fungal propagules, together with pollen grains, are among the most important aeroallergens.

Colonies of *Penicillium aurantiogriseum* growing on an agar plate; their radial arrangement corresponds to the position of the holes in one of the metallic covers used in a six-stage Andersen air sampler. The mold colonies developed from the spores trapped from the air and which, after passing through the holes, were impacted on the agar surface. These types of spores are among the most common **aeroallergens**, x 0.8 (*MU*).

aeroaquatic (L. *aer* < Gr. *aér*, genit. *aéros*, air + L. *aqua*, water + Gr. suf. *-tikós* > L. *-ticus* > E. *-tic*, relation, fitness, inclination or ability): aquatic hyphomycetes found in still and stagnant, non-marine, or occasionally marine, waters. Usually aeroaquatic fungi survive in water and mud with low oxygen content, but they apparently need higher levels of oxygen to colonize new substrates. The conidia of these fungi often have three-dimensional shapes, such as coils, spirals and many-celled cage-like structures, that are formed at the air-water interface, the attribute for which they are named. Some examples of this type of fungus are *Helicoon* and *Varicosporina* (moniliaceous asexual fungi).

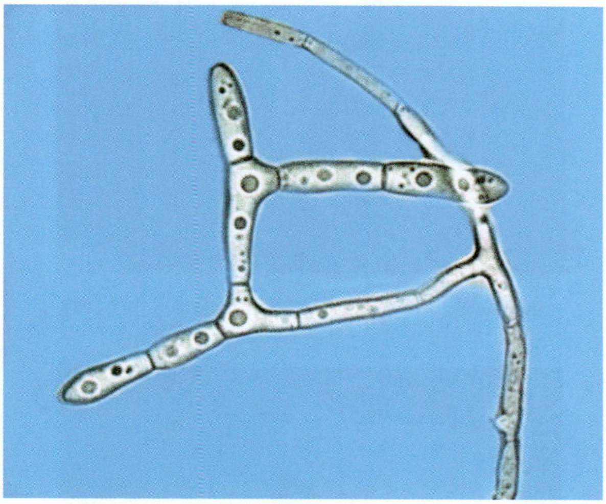

Aeroaquatic tridimensional conidium of *Varicosporina ramulosa*, x 980 (*MCG*).

aerobe (L. *aer* < Gr. *aér*, genit. *aéros*, air + Gr. *bíos*, life): an organism that requires free oxygen to live. The majority of fungi are aerobes and obtain their energy by means of respiration maintained by the oxygen in the air. Cf. **anaerobe**.

aerogenic (L. *aer* < Gr. *aér*, genit. *aéros*, air + Gr. *génos*, origin < *gennáo*, to engender, produce): an organism that produces detectable quantities of gas (CO_2) during the breakdown of carbohydrates, as yeasts and molds can do.

aerohygrophilous (L. *aer* < Gr. *aér*, genit. *aéros*, air + Gr. *hygrós*, humid, wet + *phílos*, have an affinity for): applied to species of lichens whose thallus has a hydrophobic surface and which inhabit ecological niches where the water is only available as vapor or in liquid form in the substrate, such as *Ramalina maciformis* (Lecanorales), a species that lives in hot deserts and which can absorb a sufficient quantity of water from the air to remain physiologically active.

aeromycology (L. *aer* < Gr. *aér*, genit., *aéros*, air + *mýkes*, fungus + E. *-logy*, combining form used in naming sciences < ME. *-logie* < L. *-logía*, discourse, treatise < Gr. *lógos*, word < *légo*, I say): the branch of aerobiology that treats of the study of fungi in the air, in the various aspects related to the biology of the dispersal of fungal propagules by the air and its implications, basic as well as applied, in agriculture, medicine and industry.

A six-stage Andersen air sampler, one of the spore trappers most commonly used in **aeromycology**. Each stage includes a metallic plate with 200 holes of uniform diameter (which decreases at each lower stage to simulate the human respiratory tract), through which air is drawn at 28 liters/min, to impact on petri dishes containing agar, which are placed immediately below, x 0.1 (*CC*).

A Burkard sampler, a device used to trap spores and other fungal propagules, as well as pollen grains and other viable and nonviable particles, from the air. Sampling periods may be for 24 hours or several days, and the particles are trapped on a moving tape covered with glycerine/hexane. This type of sampler is used to monitor the airspora in agricultural fields and in other environments that are studied in **aeromycology**, and in aerobiology in general, x 0.04 (*CC*).

Burkard sampler (**aeromycology**); the tape used to trap spores and other particles is attached to a rotating drum, x 0.1 (*CC*).

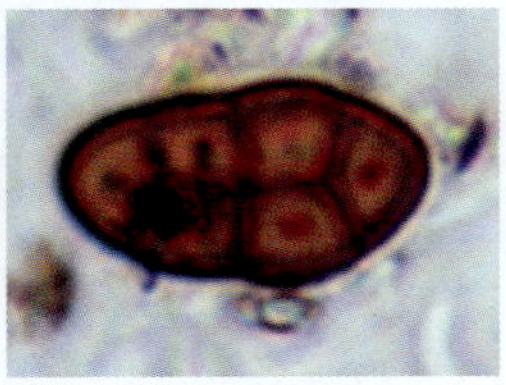

Ascospore of *Pleospora* (Pleosporales) trapped by a Burkard sampler (**aeromycology**), along with nonviable particles, x 430 (*CC*).

aerotaxis, pl. **aerotaxes** (L. *aer* < Gr. *aér*, genit. *aéros*, air + Gr. *táxis*, disposition, movement): a chemotactic phenomenon in which the air, oxygen or other gas acts as an excitant. An example of this is shown in the zoospores of *Rhizophydium pollinis-pini* (Chytridiales), which are extremely sensitive to the presence of free oxygen; when mounted in water under a cover glass, the zoospores immediately move toward the edge of the drop or toward air bubbles in the preparation.

aerotropism (L. *aer* < Gr. *aér*, genit. *aéros*, air + Gr. *trópos*, a turn, change in orientation < *tropé*, a gyration, turn < *trépo*, to gyrate, revolve toward + suf. *-ismós* > L. *-ismus* > E. *-ism*, state, phase, tendency, action): movement stimulated by air, oxygen or another gas. In the presence of small fragments of iron, the sporangiophores of *Phycomyces nitens* (Mucorales) turn toward them, apparently in response to small quantities of ozone liberated by the oxidation of the iron. Aluminum foil or fragments of glass also induce the curvature of the sporangiophores of this fungus, due to the presence of ozone adsorbed to these materials. See **tropism**.

aethalium, pl. **aethalia** (Gr. *aíthalos*, soot + dim. suf. *-íon* > L. *-ium*): a semiglobose, pulvinate or irregular fructification, relatively large, that results from the transformation of all or most of a plasmodial mass, in which various sporangia remain partially or totally enveloped by a common peridium; typical of some Myxomycetes, such as *Lycogala*, *Reticularia* (Liceales) and *Fuligo* (Physarales).

Aethalia of *Lycogala epidendrum* on dead wood, x 6 (*MU*).

Aethalia of *Lycogala epidendrum* on dead wood, x 9 (*MU*).

Aethalium of *Fuligo septica*, x 0.5 (*EPS*).

Aethalium of *Lycogala flavofuscum* on dead wood, x 5 (*RV*).

aetiology (ML. *aetiologia*, statement of causes < Gr. *aitíologia* < *aitía*, cause + *lógos*, word, discourse, treatise + *-y*, E. suf. of concrete nouns): see **etiology**.

agamic, **agamous** (Gr. *ágamous*, unwed < *a-*, not, without + *gámos*, sexual union + suf. *-íkos* > L. *-icus* > E. *-ic*, belonging to or relating to; or L. *-osus* > OF. suf. *-ous* > E. *-ous*, having, possessing the qualities of): asexual, without sexual union, lacking sexual organs; not reproducing sexually.

agamogenesis, pl. **agamogeneses** (Gr. *ágamous*, unwed < *a-*, not, without + *gámos*, sexual union + *génesis*, engendering, origin): agamic or asexual generation, i.e., vegetative, which occurs without nuclear fusion.

agar (Ma. *agaragar*, marine alga from which a gelatin is obtained, agar): a mucilaginous substance (galactan) obtained from *Gracilaria lichenoides*, various species of the gen. *Gelidium*, and other marine algae, principally from Asiatic seas. It is frequently employed as a gel for various solid and semisolid culture media for fungi, bacteria and other organisms.

agaric (L. *agaricum*, a fungus < Gr. *agarikón*, fungus, mushroom, excrescence + ending *-ic*, which denotes relation): a member of the Agaricales, one of the largest groups of basidiomycetes (ca. 6,000 spp.), with terrestrial, lignicolous, saprobic, mycorrhizal, edible, hallucinogenic and poisonous species, rarely pathogenic. Cf. **bolete**.

agaricicolous (L. *agaricum*, a fungus < Gr. *agarikón*, fungus, mushroom, excrescence + ending *-ico*, which denotes relation + L. suf. *-cola*, inhabitant < *colere*, to inhabit + L. suf. *-osus* > E. *-ous*, having, possessing the qualities of) living on agarics, like the carpophores of *Asterophora* (=*Nyctalis*) *parasitica*, which are mycoparasitic on the basidiocarps of *Russula* and *Lactarius* (Agaricales).

agaricoid (< gen. *Agaricus* < Gr. *agarikón*, fungus, mushroom, excrescence + L. suf. *-oide*, like < Gr. *-oeídes*, similar to): *Agaricology*. Like the fruiting body of the gen. *Agaricus* (Agaricales). A fruiting body in the shape of a parasol, with the hymenophore arranged in lamellae (gills), a clear evidence of a particular affinity to some family of the Agaricales. See **lepiotoid**, which is the type of fruiting body that includes *Agaricus*, in modern agaric taxonomy. Cf. **gasteroid**.

agarolytic (Ma. *agaragar*, a marine alga from which a gelatin is obtained, agar + Gr. *lytikós*, able to disolve < *lýtos*, dissolvable, broken + suf. *-íkos* > L. *-icus* > E. *-ic*, belonging to or relating to): capable of causing lysis and liquefaction of agar, the solidifying agent of many culture media. The production of agarolytic enzymes, although a relatively rare phenomenon among microorganisms, has been reported from bacteria (*Pseudomonas*, *Alginomonas*, *Agarbacterium*, *Nocardia* and others), mostly from marine habitats. Recently, an agarolytic activity has been reported in the plasmodium of *Fuligo septica* (Physarales), apparently the first eukaryotic microorganism in which this property is known.

agglutinate (L. *agglutinatus* < *agglutinare*, to stick, to glue < *ag-* < *ad-*, a pref. used to denote proximity, toward, near + *gluten*, adhesive + suf. *-atus* > E. *-ate*, provided with, likeness): to cause to adhere. A clump of agglutinated material; applied to something that is united but not fused. E.g., the spores of *Acremonium*, *Phialophora* (=*Rhinocladiella*=*Fonsecaea*) and *Fusarium* (moniliaceous, dematiaceous and tuberculariaceous asexual fungi, respectively), are agglutinated by a mucilage, but they can be separated without altering their integrity.

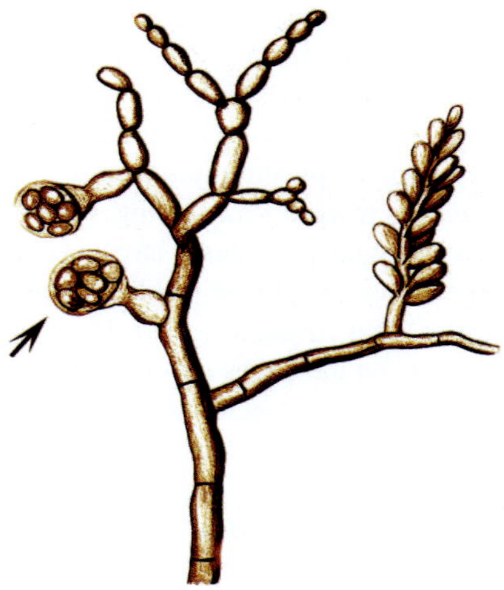

Agglutinated conidia of *Phialophora* (=*Rhinocladiella*=*Fonsecaea*) *pedrosoi*, x 500 (*MU*).

agnotobiotic (Gr. *a-*, not, without + *gnótos*, well known + *bioté*, life, mode of living + suf. *-íkos* > L. *-icus* > E. *-ic*, belonging to or relative to): refers to an impure culture or mixture whose specific identity is unknown. Cf. **axenic**, **gnotobiotic** and **monoxenic**.

agonomycetaceous (Gr. *a-*, without + *gónos*, offspring, seed + *mýkes*, genit. *mýketos*, fungus + L. suf. *-aceus* > E. *-aceous*, of or pertaining to, with the nature of): term applied to the fungi without asexual or sexual reproduction by spores, that reproduce by non-dehiscent propagules (bromatia, bulbils, chlamydospores, etc.). The agonomycetaceous fungi (Agonomycetales) may be states of basidiomycetes, ascomycetes or other anamorphic fungi. They are an unnatural group in the gen. *Cenococcum*, *Papulaspora*,

Rhizoctonia and *Sclerotium*, that includes important species of plant pathogens. See **Hyphomycetes**.

akaryallagic (Gr. *a*-, not, without + *káryon*, nucleus + *allagé*, permutation, commutation + suf. *-íkos* > L. *-icus* > E. *-ic*, belonging to or relating to): a type of asexual reproduction, i.e., vegetative multiplication, in which the nucleus remains unchanged. Cf. **karyallagic**.

akaryote (Gr. *a*-, not, without + *káryon*, nucleus; lacking a nucleus). *Plasmodiophoromycetes*. A stage (prophase) in nuclear division, prior to meiosis, during which the nucleus has no or very little chromatin, for which reason it does not stain and consequently is not evident.

akinete (Gr. *akínetos*, motionless): a thick-walled, nonmotile spore of cyanobacteria present in some lichen thalli.

akinetospore (Gr. *akínetos*, motionless + *sporá*, spore): a motionless spore, lacking flagella, that goes through a rest period before germinating.

-al (L. *-alis*, belonging to or relating to): a suf. that is added to nouns to form many terms, e.g., *hyphal*, *mycelial*, *plasmodial*, etc., to indicate, respectively, that something belongs to or is related to, the hypha, the mycelium, the plasmodium, etc.

alate (L. *alatus*, winged < *ala*, wing + suf. *-atus* > E. *-ate*, provided with or likeness): having wings or wing-like projections. E.g., the stipe of *Boletellus russellii* (Agaricales) has broken and velvety extensions, called wings, on the edges or reticulations. Also, the basidiospores of this species are winged, as are those of *Lactarius pterosporus* (Agaricales), because their edges are so large that they give the spores a winged appearance. Also called **pterate**.

Alate basidiospores of *Boletellus russellii*, x 450 (*MU*).

alepidote (Gr. *a*-, not, without + *lepidotós*, covered with scales): without scales, entirely smooth. Cf. **squamulose** and **lepidote**.

-ales (L. *-alis*, pl. *-ales*, fem. suf.): the ending of taxonomic orders, e.g., Blastocladiales, Entomophthorales, etc.

aleuriospore (Gr. *áleuron*, wheat flour + *sporá*, spore): an asexually reproduced spore that originates by swelling of a terminal or lateral cell of a hypha; characteristic of some conidial fungi, such as the gen. *Humicola*, *Nigrospora* and *Thielaviopsis basicola* (=*Chalara elegans*), dematiaceous asexual fungi.

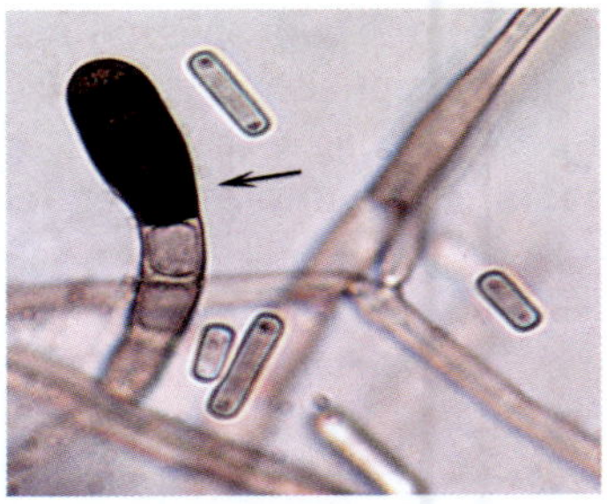

Aleuriospore and conidia
of *Thielaviopsis basicola* (=*Chalara elegans*) , x 600 (*MU*).

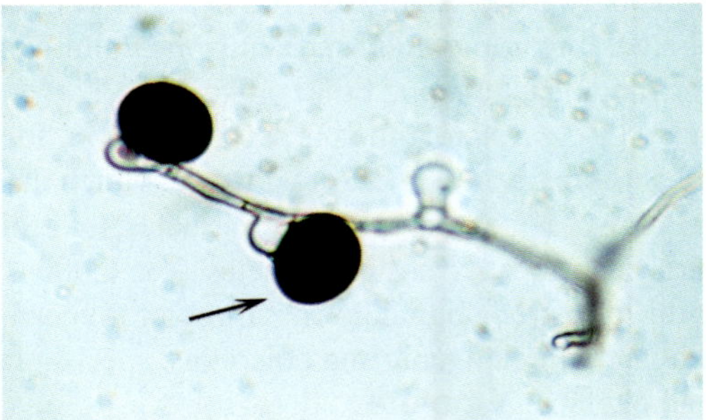

Aleuriospores of *Nigrospora* sp., x 800 (*MU*).

algicolous (L. *alga*, alga, in general, any pigmented, photosynthetic thallophyte + suf. *-cola*, inhabitant + suf. *-osus* > E. *-ous*, having, possessing the qualities of): living or growing on algae, whether as a parasite or symbiont, such as *Phlyctidium anatropum* (Chytridiales), which parasitizes the thallus of the alga *Stigeoclonium* sp.; *Spathulospora phycophila* (Spathulosporales), a laboulbeniomycete that lives as a parasite on the thallus of *Ballia callitricha* and *B. scoparia* (marine rhodophyceous algae), and *Mycophycias* (=*Mycosphaerella*) *ascophylli* (Dothideales), a loculoascomycete that has established a symbiotic mode of life that qualifies it as mycophycobiosis, i.e., an obligate symbiotic association between a systemic marine fungus and a macroalga (*Ascophyllum nodosum*, of the Phaeophyceae).

aliform (L. *aliformis*, shaped like a wing < *ala*, wing + suf. *-formis* < *forma*, shape): similar to a wing, like the projections of the ascospore wall of *Emericella variecolor* (Eurotiales), giving it a stellate shape.

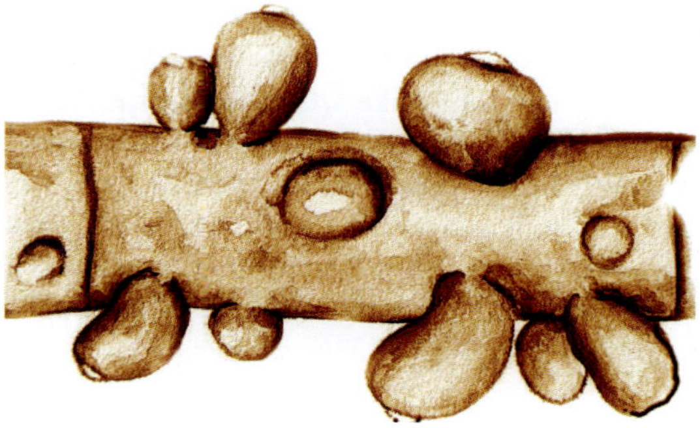

Algicolous thalli of *Phlyctidium anatropum* parasitizing the thallus of the alga *Stigeoclonium* sp., x 1,000 (*MU*).

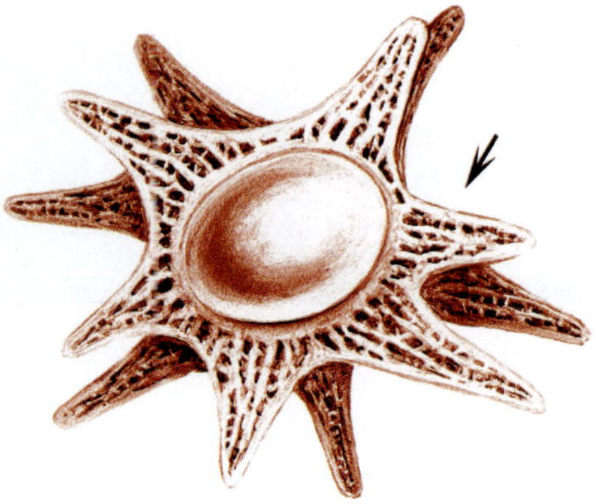

Aliform projections of the ascospore wall of *Emericella variecolor*, x 4,250 (*MU*).

allantoid (L. *allantoideus*, sausage-shaped < Gr. *allantoeidés*, similar to a sausage < *allás*, genit. *allantós*, sausage + suf. *-oeídes*, similar to): with the shape of a sausage, like some of the Hülle cells of *Aspergillus puniceus* (moniliaceous asexual fungi) and the ascospores of the Diatrypaceae (Xylariales). Equal to **botuliform**.

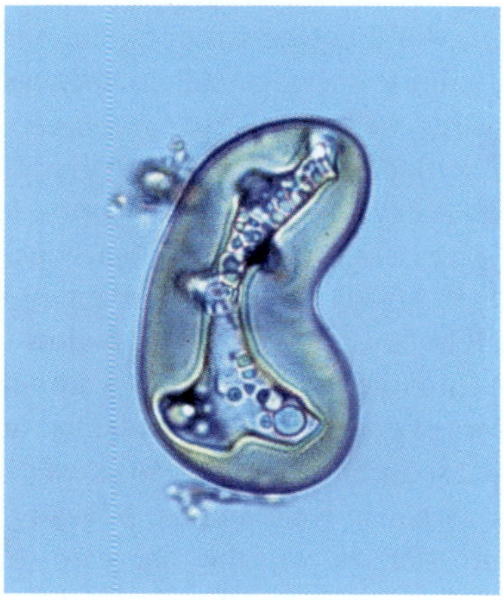

Allantoid Hülle cell of *Aspergillus puniceus*, x 700 (*MU*).

allantospore, **allantoid spore** (Gr. *allantoeidés*, similar to a sausage < *allás*, genit. *allantós*, sausage + *sporá*, spore + suf. *-oeídes*, similar to): a spore that is sausage-shaped, slightly arcuate, with rounded ends; a Saccardoan term applied originally to conidial fungi to designate sausage-shaped spores, such as the conidia of *Dendrodochium* (moniliaceous asexual fungi), the asexual state of *Nectria magnusiana* (Hypocreales). The basidiospores of *Pleurotus* spp. and of *Panellus stypticus* (Agaricales) also are allantoid.

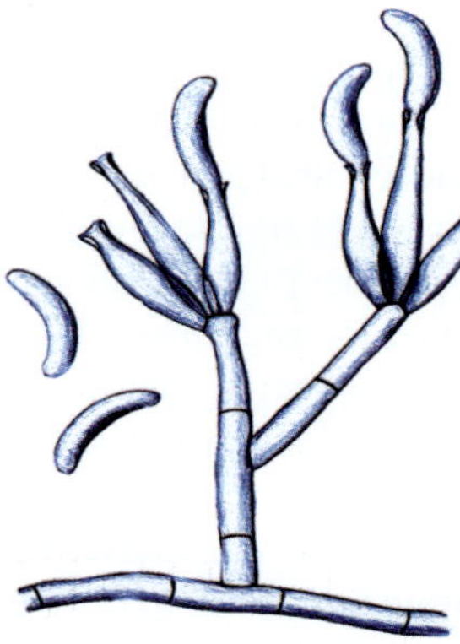

Allantospores of *Dendrodochium* sp., x 600 (*MU*).

allelo- (Gr. *allélo* < *allélon*, of/to one another, reciprocally; pref. employed in the formation of various biological terms to give the idea that something occurs in two or more forms; mutually): *Genetics*. Also called **allelomorph**; a factor or gene located on a site or locus on a chromosome, and which together with the other equal allele, located on the same locus of the other homologous chromosome, forms a pair of alleles that control the same type of characters. In fungi with bipolar sexual compatability, the latter is controlled by a pair of allelic genes or allelomorphs; in those that have tetrapolar compatability, two pairs of alleles determine sexual union.

allelomorph (Gr. *allélo* < *allélon*, of/one to another, reciprocally + *morphé*, form): see **allelo-**.

allergen (Gr. *állos*, another + *érgon*, work + *génos*, origin < *gennáo*, to engender, produce): a substance that induces allergy. See **aeroallergen**.

allochronic (Gr. *állos*, other + *chronikós*, of or concerning time < *chrónos*, time + suf. *-ikos* > L. *-icus* > E. *-ic*, belonging to or relative to): occurring at different time periods, e.g., contemporary and fossil species of fungi.

allochrous, **allochroous** (Gr. *allóchroos*, that which acquires another color < *állos*, another + *chrõs*, genit. *chroíakos*, colored): changing from one color to another, like the basidiocarps of *Psilocybe*

(Agaricales), which on being rubbed or damaged change from brown to bluish and greenish tones; in this particular case they are said to be *caerulescent*.

allochthonous (Gr. *állos*, other + *thón*, genit. *thonós*, the earth, country): refers to an organism that is not native to the country in which it is growing; also called **exotic** and **exochthonous**. E.g., *Amanita phalloides* (Agaricales) is considered allochthonous because it was introduced into the United States from Europe. Cf. **autochthonous**.

allocyst (Gr. *állos*, other + *kýstis*, bladder; here, vessel, cell): *Agaricales*. A cystidium formed in the mycelium; an obovoid, claviform or pyriform cell with a thin or slightly thickened wall, that is formed individually or in a chain in the terminal part of a haploid mycelium, or the submerged, dikaryotic mycelium, of *Pholiota gummosa* (Agaricales). Germination of allocysts has not been observed, for which reason they are considered a form of food reserve; they appear to correspond to what have been called *chlamydospores*.

allopatric (Gr. *állos*, other + L. *patris*, a father, one's country + Gr. suf. *-íkos* > L. *-icus* > E. *-ic*, belonging to or relating to): occurring in different geographic regions Cf. **sympatric**.

alpha-conidium, **α-conidium**, pl. **conidia** (*alpha*, α, first letter of the Gr. alphabet; NL. *conidium* < Gr. *konídion* < *kónis*, dust + dim. suf. *-idion* > L. *-idium*): a term applied to species that produce two morphological types of spores, in particular, the fusoid to oblong, biguttulate spore of *Phomopsis* (sphaeropsidaceous asexual fungi) and other conidial states of the Diaporthales. These fungi also generally produce **beta-conidia**. Alpha-conidia often form first, are usually more common, and are germinable.

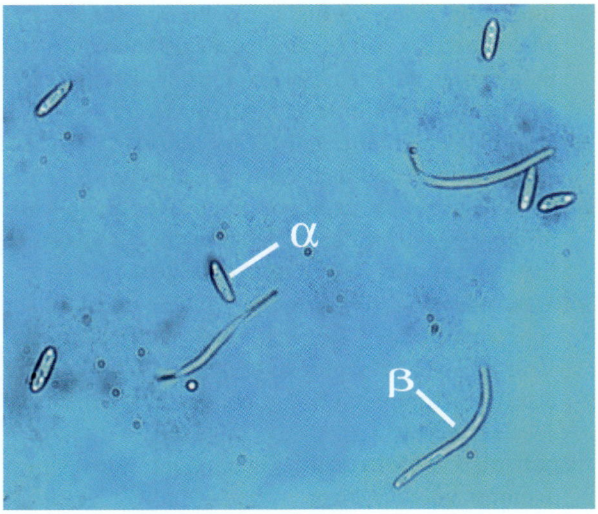

Alpha- (α) and beta- (β) **conidia** from a pycnidium of *Phomopsis* sp., x 600 (*RTH*).

alternarioid (< gen. *Alternaria* < L. *alternare*, to alternate, do by turns + suf. *-ia*, which denotes quality + suf. *-oide*, like < Gr. *-oeídes*, similar to): a dictyosporous chlamydospore, i.e., with transverse and longitudinal septa, resembling the dictyospores of *Alternaria* (dematiaceous asexual fungi); also present in some species of *Phoma* (sphaeropsidaceous asexual fungi), such as *Ph. glomerata* and *Ph. jolyana*.

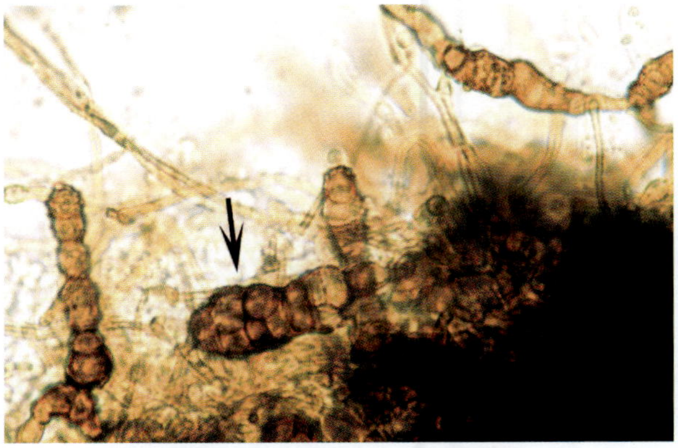

Alternarioid chlamydospore *of Phoma* sp., x 600 (*MU*).

alternate host (L. *alternatus*, alternate < *alternus* < *alter*, occurring or succeeding by turns + suf. *-atus* > E. *-ate*, with the property of; *hospes*, a living organism affording subsistence or lodgment to a parasite): a host plant in which the parasite (generally a micromycete of the Uredinales) carries out the complementary phases of its life cycle. For example, barberry is the alternate host of *Puccinia graminis tritici* (Uredinales), the causal agent of wheat rust. Cf. **intermediate host**, **primary host** and **bridging host**.

alutaceous (LL. *alutaceus*, *alutacius*, soft leather < *aluta*, leather + suf. *-aceus* > E. *-aceous*, of or pertaining to, with the nature of): **1**. *Color*. Having the color of leather, light brown or pinkish-cinnamon. **2**. *Texture*. Of tough consistency, but with a certain flexiblity, like leather; also called **coriaceous**. Some basidiocarps, such as those of *Lentinus badius* (of the order Aphyllophorales), and the thallus of the lichen *Masophalea richardsonia* (=*Cetraria richardsonii*, of the order Lecanorales), are alutaceous.

alveolate (L. *alveolatus*, hollowed, pitted < *alveolus*, pit, small hollow, dim. of *alveus*, hollow + suf. *-atus* > E. *-ate*, provided with, likeness): with small cavities or hollows of any concavity; similar to a honeycomb. E.g., the ascospores of *Gelasinospora* (Sordariales) have an alveolate wall, and the pileus of *Morchella esculenta* (Pezizales), the ascocarp of *Cyttaria hariotii* (Cyttariales), and the stipe of *Boletellus frostii* (Agaricales) are alveolate. Also called **faveolate**.

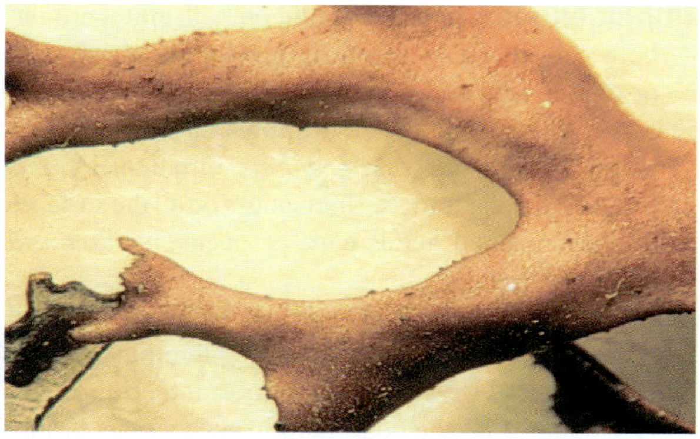

Alutaceous thallus of *Masophalea richardsonia* (=*Cetraria richardsonii*), x 2 (*MU*).

Alveolate ascocarp of *Cyttaria hariotii*, x 7.5 (*MU*).

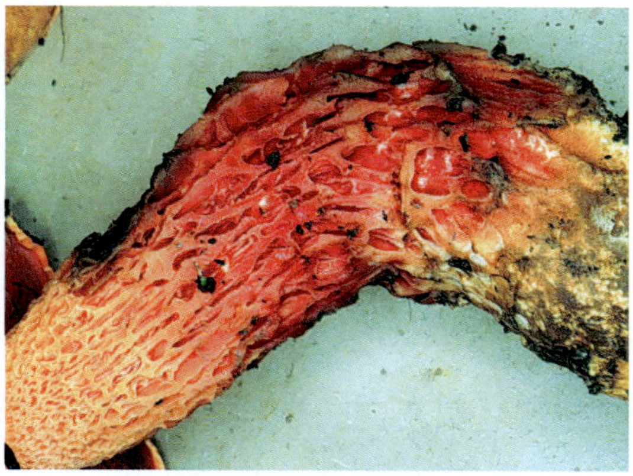

Alveolate stipe of the basidiocarp of *Boletellus frostii*, x 1(*JC*).

alveolus, pl. **alveoli** (L. *alveolus*, small hollow, pit, dim. of *alveus*, hollow): a pit or small cavity in the surface of spores, stipes and other structures, which for having alveoli are said to be *alveolate*, and whose appearance is that of a honeycomb. Formerly the term alveolus was applied to each pore of the hymenophore of the polypores.

amanitoid (< gen. *Amanita* < Gr. *amanós*, the name of a mountain in Asia, or < Gr. *amánores*, pustules + L. suf. *-oide*, like < Gr. *-oeídes*, similar to): *Agaricology*. A type of fruiting body (13 main types are recognized) with the gills free or finely attached, and with a ring and volva, represented by *Amanita* (Agaricales).

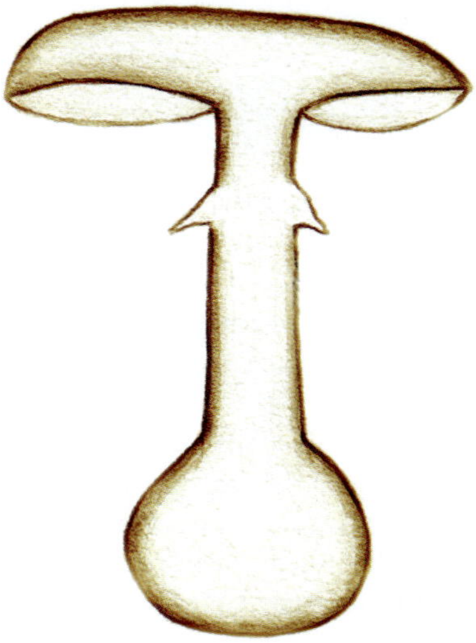

Amanitoid basidiocarp of *Amanita* sp. Diagram of a longitudinal section, x 0.5 (*MU*).

ambrosia (Gr. *ambrosía*, food of the Gods < *ámbrotos*, immortal): refers to certain fungi that some scolytic coleopterans (*Xyleborus*, *Corthylus*) cultivate in the galleries that they open in the wood, and on which they feed.

amend (ME. *amenden* < OF. *amender* < L. *emendare* < *e*, *ex*, out + *menda*, fault): to change or alter, not necessarily to correct a fault or error. Cf. **emend**.

amensalism (Gr. *a-*, not, without + L. *mensa*, table + Gr. suf. *-ismós* > L. *-ismus* > E. *-ism*, state, phase, tendency, action): the condition in which a species is favored by its ability to secrete metabolites that inhibit other fungi.

amerospore (Gr. *a-*, not, without + *méros*, part, portion + *sporá*, spore): a unicellular spore; a spore not divided into parts by septa; a Saccardoan term applied originally to the conidial fungi. E. g., the conidia of *Oedocephalum* (moniliaceous asexual fungi), among many other gen., are amerospores.

amitosis, pl. **amitoses** (Gr. *a-*, not, without + *mítos*, filament + suf. *-osis*, condition): simple, direct nuclear division by fragmentation; the nucleus constricts until it separates into two parts, often unequal. In this type of division the characteristic figures of mitosis, which is the formation of chromosomes, are not observed; it occurs in old cells and in lower organisms.

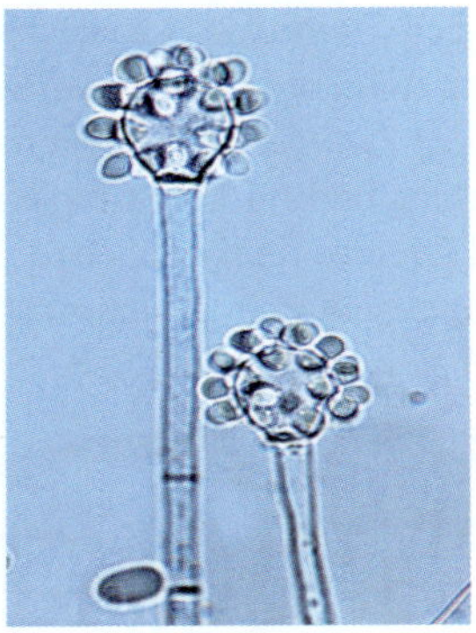

Amerospores of *Oedocephalum* sp., x 720 (*MU*).

amixis, pl. **amixes** (Gr. *a-*, not, without + *míxis*, mixture, intimate union): lack of mixis. A primitive cellular state in which there is no sexual reproduction or amphimixis. Also called **apomixis**. Cf. **amphimixis**.

ammophilous (Gr. *ámmos*, sand + *phílos*, have an affinity for): sand loving, growing preferentially in sandy habitats, such as beaches, e.g., like *Peziza ammophila* (Pezizales). See **arenicolous**.

amoeboid (NL. *amoeba* < Gr. *amoibé*, change, transformation + L. suf. *-oide*, like < Gr. *-oeídes*, similar to): similar to an amoeba. E.g., the myxamoebae of the Dictyosteliales and the plasmodia and myxamoebae of the Myxomycetes are amoeboid, as are the zoospores of certain aquatic fungi that lack a cell wall and move by creeping, extending pseudopodia or prolongations into which the protoplasm flows, continually changing shape with the successive formation of the pseudopodia.

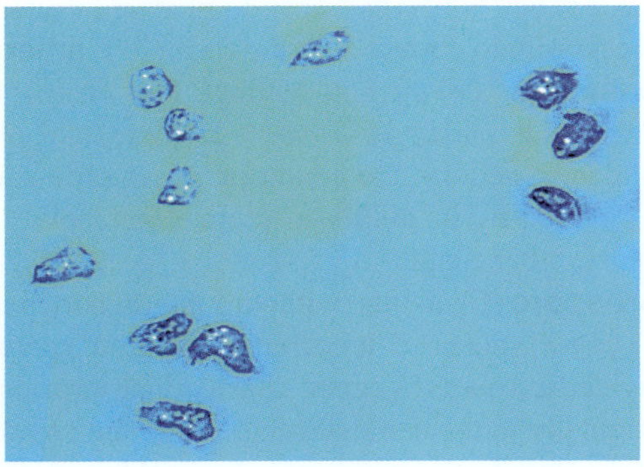

Amoeboid myxamoebae of *Dictyostelium discoideum*, x 840 (*RTH*).

amorphous (Gr. *amorphós* < *a-*, not, without + *morphé*, shape): without a defined shape or structure.

amphi-, amph- (Gr. *amphí*, both, on both sides, around both sides): a pref. used in many terms that implies the idea of both, as in *amphigenous*, forming on both sides.

amphigenous (Gr. *amphí*, both, on both sides, around both sides + *génos*, origin): developing all around, or on one side and the other, on both parts or faces; e.g., a parasitic fungus that grows on both surfaces of a leaf (like *Puccinia graminis*, Uredinales), or a basidiocarp that forms an hymenium on all the surfaces of the branched aerial part, as occurs in *Ramaria* (Aphyllophorales).

amphigynous (Gr. *amphí*, both, on both sides, around both sides + *gyné*, female; here, feminine sexual organ): *Peronosporales.* Refers to an antheridium that is penetrated by the oogonium in such a manner that the antheridium finally envelops the pedicel and the base of the oogonium. Characteristic of some species of *Phytophthora*, such as *Ph. infestans* and *Ph. capsici* (Peronosporales). Cf. **paragynous.**

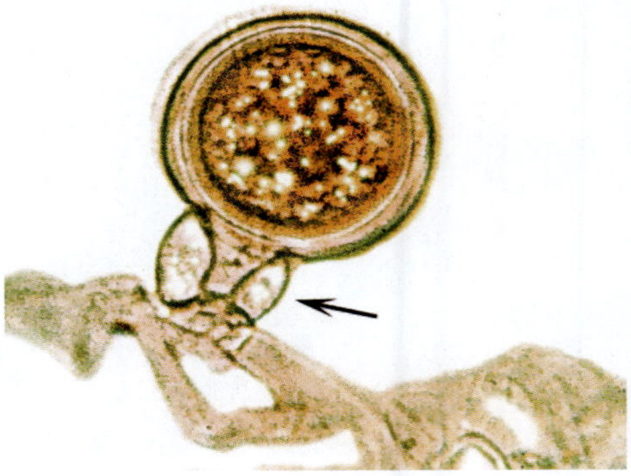

Amphigynous antheridium of *Phytophthora capsici*, x 200 (*JG*).

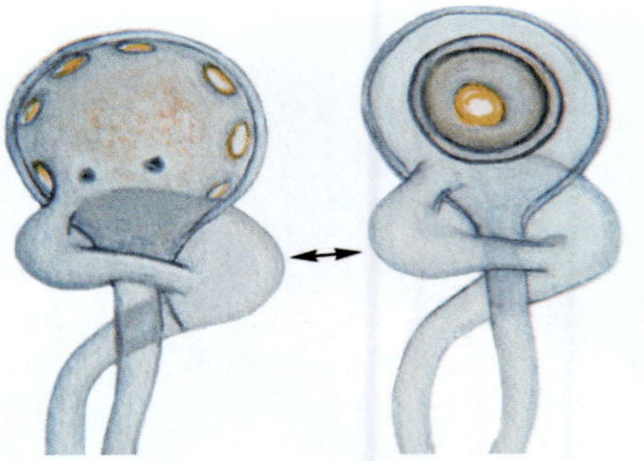

Amphigynous antheridium of *Phytophthora infestans*, x 300 (*MU*).

amphimixis, pl. **amphimixes** (Gr. *amphí*, both, on both sides, around both sides or on all sides + *míxis*, mixture, intimate union): the union of two germ plasms at the moment of fertilization, during which the two gametes unite into a single cell, which is the zygote. In the simplest case, amphimixis is reduced to the union of two nuclei. Cf. **amixis** and **apomixis**. See **apogamy**, **automixis** (**parthenogamy** and **autogamy**), **parthenogenesis**, **pseudomixis** or **pseudogamy** (**adelphogamy** and **pedogamy**).

amphispore (Gr. *amphí*, both, on both sides, around all sides + *sporá*, spore): *Uredinales*. A special type of urediniospore, with a thickened wall, sometimes darkly pigmented, capable of passing through a more or less long period of unfavorable ecological conditions in a dormant state; i.e., a urediniospore with the dormant or resistant functions of a teliospore or resting spore. Also called **mesospore**.

amphithallism (Gr. *amphí*, both, on both sides, around or by both sides + *thallós*, thallus + suf. *-ismós* > L. *-ismus* > E. *-ism*, state, phase, tendency, action): the phenomenon in which spores derived from the same sporophore give rise to both homothallic and heterothallic mycelia, as occurs in some species of *Coprinus* (Agaricales).

amphithecium, pl. **amphithecia** (NL. *amphithecium* < Gr. *amphí*, both, on both sides, around or by both sides + *thekíon* < *thêke*, case, box; here, of the asci): *Lichens*. **1**. In the discolichens, it refers to the zone of the thallus that surrounds the apothecium; it usually bears gonidia and is limited on the outside by the thalline cortex and the thalline margin. **2**. In the pyrenocarpous lichens, it refers to the layer of hyaline or light-colored tissue that surrounds the hymenium of the perithecium embedded in the thallus, e.g., as in *Dermatocarpon* (of the order Verrucariales).

amplectant, amplective (L. *amplectens* < *amplectere*, to embrace, envelop, encircle + suf. *-ant* > E. *-ant*, being, to carry out; or + L. *-ivus* > E. *-ive*, quality or tendency, fitness): that which embraces or envelops, such as the volva of the basidiocarp of certain species of *Amanita*, e.g., *A. hemibapha* (Agaricales), and the fruiting body of *Phylloporia fruticum* (Aphyllophorales), which envelopes the host twig.

Amplectant fruiting body of *Phylloporia fruticum*, x 0.5 (*RV*).

amplicon (E. *amplify* < L. *amplificare* < *amplus*, extensive + *facere*, to make + *con*, abbrev. of *codon*, a sequence of three nucleotide bases that code for the synthesis of an amino acid): an amplified portion of DNA obtained by the polymerase chain reaction (PCR). See **molecular biology**.

ampulla, pl. **ampullae** (NL. *ampulla*, a small Roman glass vessel, oval in shape < Gr. *ámphora* + L. dim. suf. *-la*): a swollen conidiogenous cell formed at the apex of a conidiophore, on which the conidia are produced simultaneously, as occurs in *Ostracoderma* (=*Chromelosporium*), *Oedocephalum*, and other gen. of moniliaceous asexual fungi. See **botryoblastospore**.

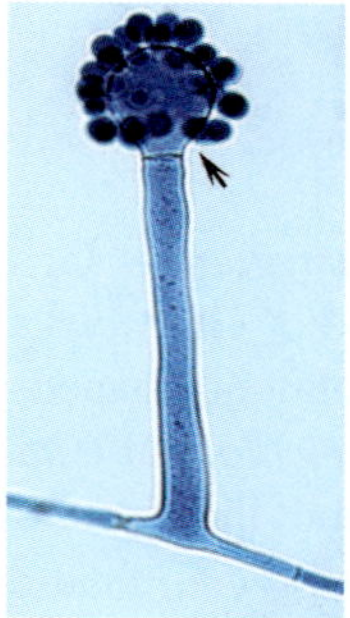

Ampulla of the conidiophore of *Oedocephalum* sp., x 480 (*MU*).

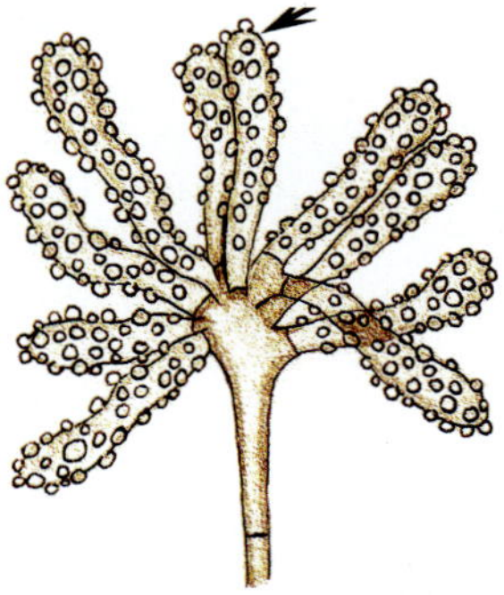

Ampullae of the conidiophore of *Ostracoderma* (=*Chromelosporium*) *ollare*, x 870 (*MU*).

ampullate (L. *ampulla*, a small Roman glass vessel, oval in shape < Gr. *ámphora* + L. *-atus* > E. *-ate* provided with or likeness): with *ampullae*.

ampulliform (L. *ampulliformis* < *ampulla*, a small Roman glass vessel, oval in shape < Gr. *ámphora* + L. dim. suf. *-la* + suf. *-formis* < *forma*, shape): with the shape of a bottle, narrow above and swollen in the basal part. It is applied especially to the conidiogenous cell called a *phialide*, like that of *Penicillium citrinum* and *Trichoderma viride* (moniliaceous asexual fungi). Also see **raduliform** and **rachiform**. In the basidiomycetes, the cheilocystidia of *Galerina ampullaceocystis* and *Agrocybe pediades*, among other Agaricales, are also ampulliform.

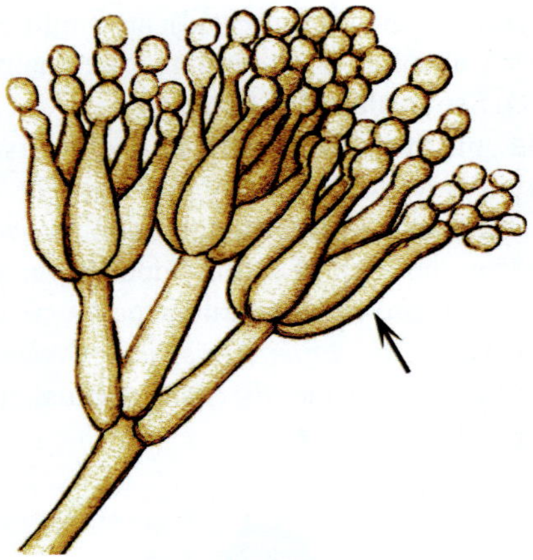

Ampulliform phialides of *Penicillium citrinum*, x 1,000 (*MU*).

amygdaloid (L. *amygdala* < Gr. *amygdále*, almond + L. suf. *-oide*, like < Gr. *-oeídes*, similar to): similar to an almond, almond-shaped, like the ascospores of *Thielavia terricola* and *Chaetomium incomptum* (Sordariales), and the basidiospores of *Coprinus comatus* and *Panaeolina foenisecii* (Agaricales).

Amygdaloid ascospores of *Thielavia terricola*, x 650 (*MU*).

Amygdaloid basidiospores of *Coprinus comatus*, x 1,500 (*MU*).

amyloid (L. *amyloideus*, resembling starch < Gr. *ámylon*, starch + L. suf. *-oide*, like < Gr. *-oeídes*, similar to): turning blue on reacting with Melzer's reagent (potassium iodide, iodine, chloral hydrate and water), presumably due to containing amylaceous substances; e.g., the basidiospores of *Lactarius* and *Russula* (Agaricales) are amyloid. Cf. **anamyloid**, **dextrinoid** and **pseudoamyloid**.

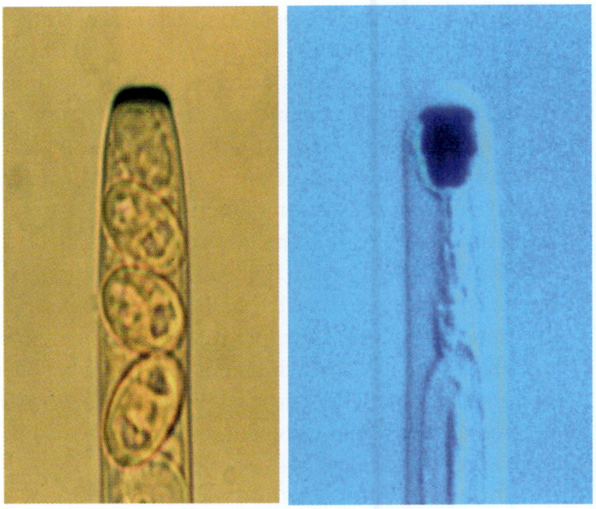

Amyloid ascus apices of *Peziza ammophila* (Pezizales), x 500 (*RTH*), and *Rosellinia herpotrichoides* (Xylariales), x 450 (*RTH*).

an-, **a-** (Gr. *an-*, *a-*, negation): a pref. employed in many compound biological terms to denote that something is lacking, as in *anaerobe*, which lives without air.

ana-, **an-** (Gr. *aná*, above, ascendent; anew, again, to come back again): a pref. employed in the formation of many compound biological terms of Greek origin, such as *anabiosis*, that which returns to life.

anabiosis, pl. **anabioses** (Gr. *aná*, to come back again, + *bíosis* < *bíos*, life + suf. *-osis*, condition or state): to revive; recovery of active metabolism and growth from a condition of suspended life, similar to death; revival from a dry or frozen state, like cultures that have been lyophilized or maintained in silica gel or in liquid nitrogen.

anaerobe (Gr. *an-*, not, without + *aér*, genit. *aéros*, air + *bíos*, life): an organism capable of living in the absence of free oxygen; anaerobic organisms obtain their energy through the decomposition of various organic substances. An anaerobe is considered obligate when even the smallest amount of oxygen makes life impossible; on the other hand, if it can resist it (as in some yeasts) it is considered a facultative or discretional anaerobe. Numerous fungi can ferment sugars anaerobically, but since this capacity is not essential for growth, they cannot be called anaerobes. There exist, however, some species of Chytridiomycetes that are strictly anaerobic, such as *Piromyces spiralis*, *P. minutus*, *P. mae* and *P. dumbonicus* (Neocallismasticales), which live in the intestine of goats, deer, horses and elephants, respectively. These zoosporic fungi (known as rumen fungi) are the only group that lacks mitochondria and grow only in the absence of oxygen (obligate anaerobes), although they can tolerate oxygen during the passage from one host to another. They utilize diverse substrates and produce formate, acetate, lactate, ethanol, succinate, carbon dioxide and hydrogen. Cf. **aerobe**.

anaholomorph (Gr. pref. *aná-*, anew, again + *hólos*, entire + *morphé*, shape): a fungus that reproduces asexually, which appears to lack a teleomorph. Anaholomorphs also are called **asexual fungi**, **conidial fungi**, **deuteromycetes**, **Fungi Imperfecti** or **mitosporic fungi**.

anamorph (Gr. *anamorphóo*, to transform): the asexual, conidial, or imperfect state of a fungus, which produces its spores by mitosis, in contrast to the **teleomorph**, which is the sexual or perfect state (ascogenous or basidiogenous) and whose spores are produced after meiosis. The fungus in all its forms is called the **holomorph**. See **synanamorph**.

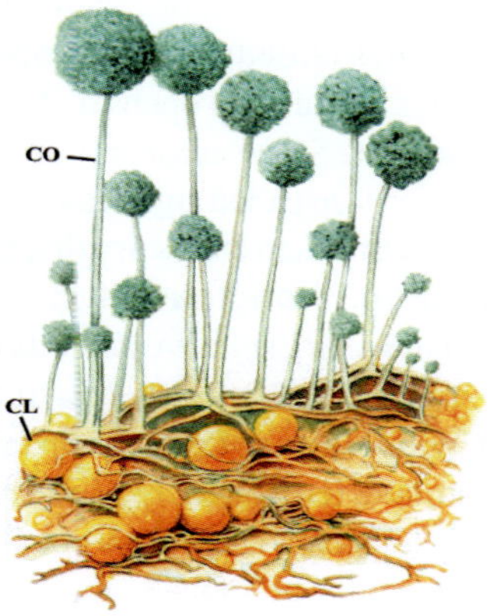

Anamorph (*Aspergillus rubrobrunneus*) of *Eurotium rubrum* (teleomorph); shown are the corresponding conidiophores (CO) and cleistothecia (CL), x 66 (*MU*).

anamorph-genus, pl. **genera** (Gr. *anamorphóo*, to transform; L. *genus*, genit. *generis*, genus, race, kind): a term used to replace **form-genus**, which was applied to a gen. of fungi that reproduces asexually (deuteromycete).

anamorph-species (Gr. *anamorphóo*, to transform; L. *species*, species, kind): a term used to replace **form-species**, which was applied to a fungal species that reproduces asexually (deuteromycete).

anamyloid (Gr. pref. *an-*, not, without + *ámylon*, starch + L. suf *-oide*, like < Gr. *-oeídes*, similar to): not amyloid. Applied to the color reaction of spores or cellular elements of fungi when treated with Melzer's reagent (potassium iodide, iodine, chloral hydrate and water), which take on a yellow color or are almost colorless. Cf. **amyloid**, **dextrinoid** and **pseudoamyloid**.

anaphysis, pl. **anaphyses** (Gr. *anáphysis*, action of being reborn, new growth): a sterigmatoid filament that arises from the apothecium of certain lichens, as in the gen. *Ephebe* (Lichinales).

anascosporic (Gr. pref. *an-*, not, without + *askós*, wineskin, sac, ascus + *sporá*, spore + suf. *-íkos* > L. *-icus* > E. *-ic*, belonging to or relating to): commonly used as a syn. of **anascosporogenous**. Cf. **ascosporic**.

anascosporogenous (Gr. pref. *an-*, not, without + *askós*, wineskin, sac, ascus + *sporá*, spore + *génos*, origin < *gennáo*, to engender, produce): incapable of producing ascospores. Cf. **ascosporogenous**.

anastomosis, pl. **anastomoses** (Gr. *anastómosis*, opening, fusion, connection): fusion of two cells or hyphae in contact that, on reabsorbing their walls, fuse into one; this has great functional importance in the fungi, in which are recognized vegetative, sexual and parasitic anastomoses. An example of the first is observed in fusion of the hyphae of *Oedocephalum* (moniliaceous asexual fungi).

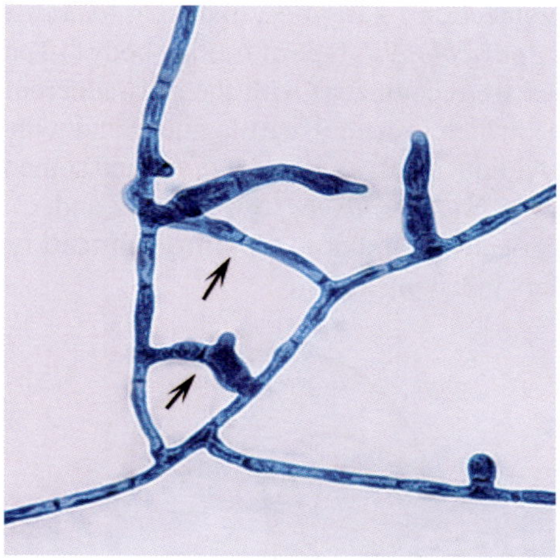

Anastomosis of somatic hyphae of *Oedocephalum* sp., x 200 (*MU*).

anastral (Gr. pref. *an-*, not, without + *astér*, star + L. suf. *-alis* > E. *-al*, relating to or belonging to): without astral rays; applied to nuclear division in the Myxomycetes and some fungi that lack these rays.

andro-, **andr-** (Gr. *andrós*, male): a pref. employed in biology in the sense of masculine or relative to the male sex, as in *androphore*, bearing male sexual organs.

androgynous (L. *androgynus* < Gr. *andrós*, male + *gyné*, female, taken here as the masculine and feminine sexual organs): applied to zoosporic microorganisms with an antheridium and an oogonium on the same hypha, as in *Pythium* (Peronosporales), *Achlya* and *Saprolegnia* (Saprolegniales), among other Oomycota.

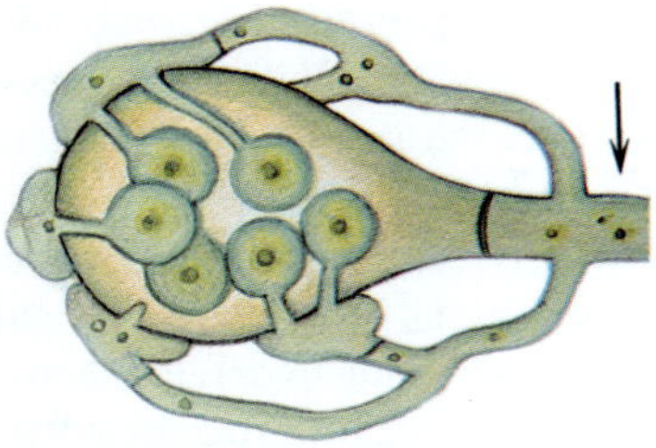

Androgynous hypha of *Saprolegnia parasitica*, with an oogonium and several antheridia, x 400 (*MU*).

andromorph (Gr. *andrós*, male, taken here as the male sexual organ + *morphé*, shape): applied to the morphological structure corresponding to the spermatial state of a fungus.

androphore (Gr. *andrós*, male, taken here as the male sexual organ + suf. *-phóros*, bearer): forming and bearing male sexual organs, like the hyphal branch that produces antheridia in certain ascomycetes (e.g., *Pyronema*, of the Pezizales).

anellarioid (< gen. *Anellaria* < L. *anellus*, *annellus*, dim. of *anus*, ring + suf. *-aria*, which indicates possession or connection + suf. *-oide*, like < Gr. *-oeídes*, similar to): *Agaricology*. A type of fruiting body (13 principal types are recognized) with the gills adherent to the stipe, which is central, cartilaginous and with a ring, but without a volva, as in *Anellaria* and some species of *Galerina*, *Psilocybe*, *Psathyrella* and *Coprinus* (Agaricales). Similar to the **omphalinoid** type, but differs in having a ring.

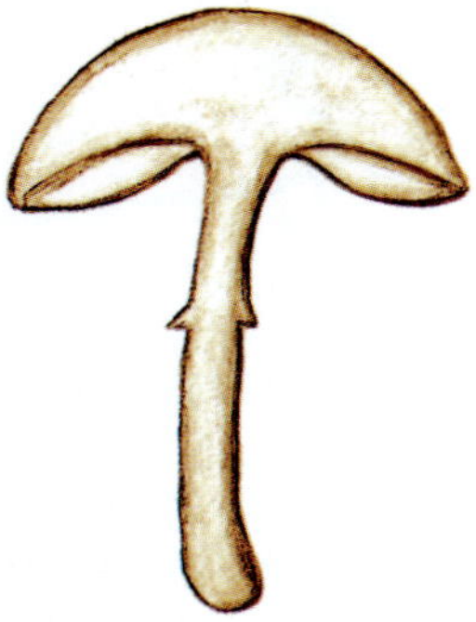

Anellarioid basidiocarp of *Anellaria* sp. Diagram of a longitudinal section, x 0.5 (*MU*).

anemo-, anem- (Gr. *ánemos*, wind): a pref. used to denote the idea of wind or air, as in *anemochore*, dispersed by the wind.

anemochore, anemochoric (Gr. *ánemos*, wind + *choréo*, change of place, to move away): disseminated by the wind, depending on the wind for dispersal, like the majority of fungi with dry spores (e.g., *Alternaria* and *Cladosporium*, dematiaceous asexual fungi, and *Aspergillus* and *Penicillium*, moniliaceous asexual fungi). Also called **anemophilous**. Cf. **hydrochoric**.

anemophilous (Gr. *ánemos*, wind + *phílos*, have an affinity for): syn. of **anemochore**.

anemophily (Gr. *ánemos*, wind + *phílos*, have an affinity for): dispersal of spores or other propagules by wind.

aneuploid (Gr. *an-*, a pref. used to indicate that something is lacking + *euploide*, an international scientific term < Gr. *eû*, true): refers to the number of chromosomes of any species when this number deviates from the normal number, whether haploid, diploid or polyploid. E.g., if a nucleus has 2n +1, 2n+2, 2n+3, etc., chromosomes, it is aneuploid.Cf. **euploid**.

angiocarpic, angiocarpous (L. *angio-*, vessel, closed receptacle < Gr. *angeîon*, small vessel < *ángos*, vessel + dim. suf. *-eíon* + Gr. *karpós*, fruit + suf. *-íkos* > L. *-icus* > E. *-ic*, belonging to or relating to): a fruiting body that remains closed during the major part of its development; at maturity it opens in different ways to liberate the spores. Some ascocarps and basidiocarps are angiocarpic. Also called angiocarpic development. Cf. **gymnocarpic** and **hemiangiocarpic**.

angiole, angiolum, pl. **angiola** (NL. *angiolum*, dim. of *angium* < Gr. *angeîon*, small vessel, closed receptacle + L. dim. suf. *-olum* > E. *-ole*): syn. of **peridiole**.

ang-kah (Chin. *ang-kah*, red rice): the name of Chinese red rice, a product fermented by *Monascus purpureus* (Eurotiales). Ang-kah, also called anka, ang-kak, red rice, Chinese red rice, ankak, angquac, beni-koji, and aka-koji, is utilized to add color and flavor to various foods, including fish, rice wine, soy cheese, pickled vegetables, and salted meats. The red pigment produced by the fungus is consistent and stable, soluble in water, and lacks toxicity or carcinogenicity. According to Chinese herbalists, ang-kah is an effective cure for infant urinary incontinence and asthma. See **fermentation**.

Grains of **ang-kah**, or red rice, produced by fermentation with *Monascus purpureus*, x 20 (*MU*).

Ang-kah grain sown on agar, from which the mycelium of *Monascus purpureus* is developing, x 20 (*MU*).

anguilliform (L. *anguilliformis* < *anguilla*, eel + dim. suf. *-illa* + suf. *-formis* < *forma*, shape): with the shape of an eel; long, slender and curved, like the beta conidia of *Phomopsis* (sphaeropsidaceous asexual fungi), the conidia of *Anguillosporella* (=*Anguillospora*) *vermiformis* (moniliaceous asexual fungi), or the spores of certain lichens. Also called **vermiform**.

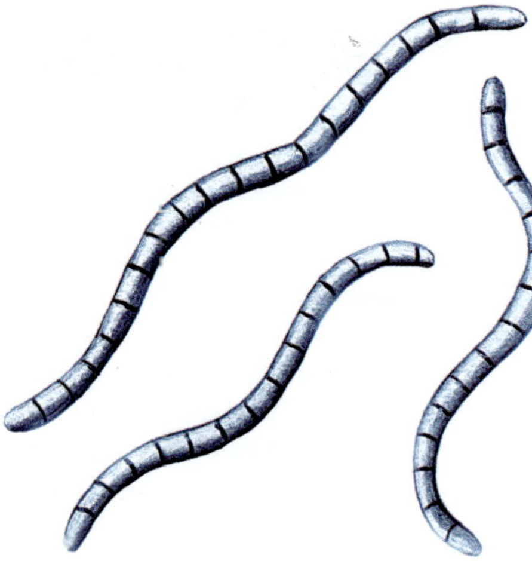

Anguilliform conidia of *Anguillosporella* (= *Anguillospora*) *vermiformis*, x 530 (*MU*).

anguilluliform (L. *anguilluliformis* < *anguillula*, small eel < *anguilla* + dim. suf. *-ula* + suf. *-formis* < *forma*, shape): dim. of **anguilliform**.

angular (L. *angularis*, having angles < *angulus*, angle, corner + suf. *-aris* > E. *-ar*, pertaining to): refers to a structure whose contour is not rounded but in the form of angles. These can be triangular, like the ascospores of *Microascus trigonosporus* (Microascales); quadrangular, like the ascospores of *Chaetomium quadrangulatum* (Sordariales); rhomboid, like the conidia of *Beltrania rhombica* (dematiaceous asexual fungi); hexagonal, like the basidiospores of *Psilocybe aerugineomaculans* (Agaricales), or polygonal, like the peridium of the sporangia of *Diderma radiatum* (Physarales) and the sporangiospores of *Umbelopsis vinacea* (=*Mortierella ramanniana* var. *angulispora*, Mucorales).

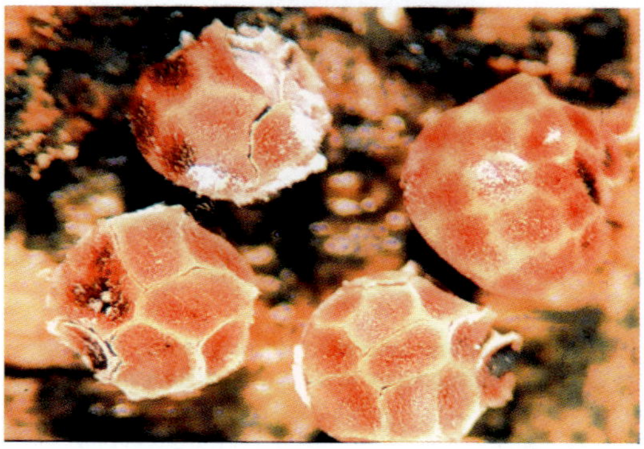

Angular peridium (polygonal) of the sporangia of *Diderma radiatum* growing on dead wood, x 20 (*MU*).

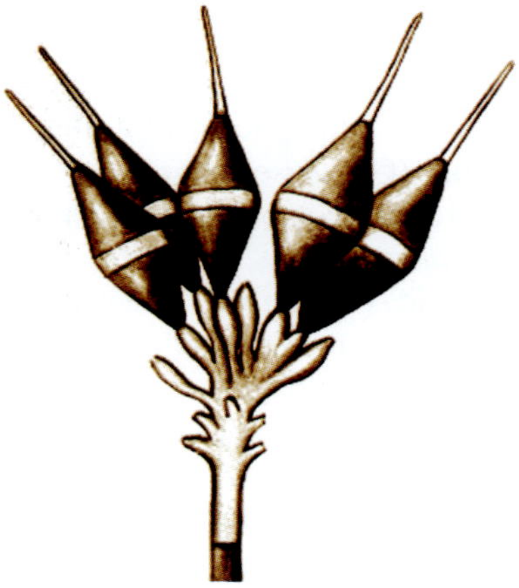

Angular conidia (rhombic) of *Beltrania rhombica*, x 700 (*MU*).

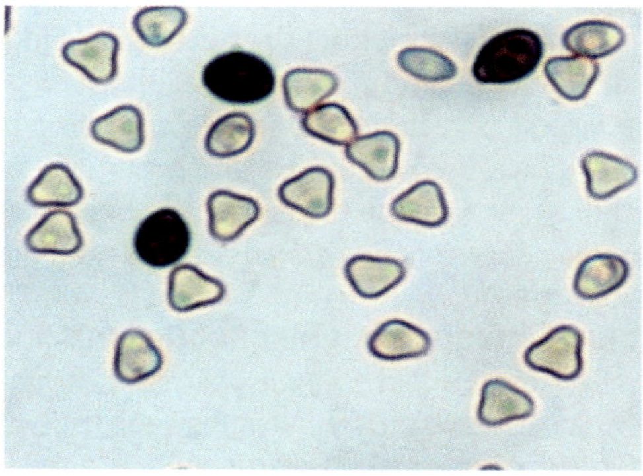

Angular ascospores (triangular) of *Microascus trigonosporus*, x 530 (*CB*).

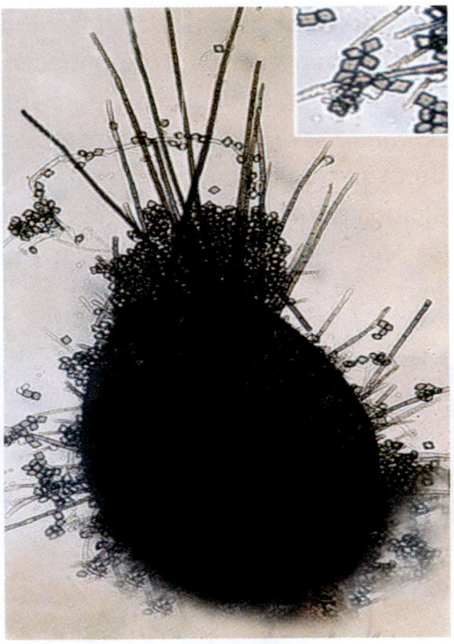

Angular ascospores (quadrangular) of *Chaetomium quadrangulatum*, x 530; insert x 2,120 (*EPS*).

angustate (L. *angustus*, narrow, reduced < *anguste*, within narrow limits): narrow.

angusti- (L. *angustus*, narrow): a pref. that alludes to something narrow, as in *angustifolium,* a narrow leaf. Cf. **lati-**.

anheliophilous (Gr. pref. *an-*, without + *hélios*, sun + *philos*, have an affinity for): see **heliophobic**.

aniso-, **anis-** (Gr. *ánisos*, unequal): a pref. often employed to form compounds that denote inequality, as in *anisodiametric*.

anisodiametric (Gr. *ánisos*, unequal + *diámetros*, diameter < *diá*, through + *métron*, measure + *-íkos* > L. *-icus* > E. *-ic*, belonging to or relating to): refers to a structure that is longer than wide, such as the hyphae that comprise the prosenchyma of various fungal organs. Cf. **isodiametric**.

anisogamete (Gr. *ánisos*, unequal + *gamétes*, husband, spouse): one of two types of gametes that differ phenotypically. See **anisogamy**.

anisogamy (Gr. *ánisos*, unequal + *gámos*, sexual union > E. suf. *-gamy*): fusion of two gametes similar in shape and structure, but differing in size (*anisogametes*); the largest is regarded as female and the other as male; generally, both have flagella and are motile, and they are found in the aquatic fungi of the order Blastocladiales, such as *Allomyces*. Cf. **heterogamy** and **isogamy**.

anisokont, **anisocont**, **anisokontic**, **anisocontic** (Gr. *ánisos*, unequal + *kontós*, oar, here, flagellum): having flagella of different lengths or of distinct type. Cf. **heterokont** and **isokont**.

anisosporic, **anisosporous** (Gr. *ánisos*, unequal + *sporá*, spore + suf. *-íkos* > L. *-icus* > E. *-ic*, belonging to or relating to; or + L. suf. *-osus* > E. *-ous*, having, possessing the qualities of): having spores of more than one type (*anisospory*). See **heterosporic**.

anisotomic, **anisotomous** (Gr. *ánisos*, unequal + *tómos*, cut + suf. *-íkos* > L. *-icus* > E. *-ic*, belonging to or relating to; or + L. suf. *-osus* > E. *-ous*, having, possessing the qualities of): a type of branching in which one of the branches (when branching is dichotomous) is more robust and forms a stem or main axis, so that the other branch of the dichotomy appears to be lateral, as happens in the lichen thallus of *Alectoria ochroleuca* (Lecanorales). Cf. **isotomic**.

annellate, **annulate** (L. *annulatus*, ringed < *annulus*, ring + suf. *-atus* > E. *-ate*, provided with or likeness): having a ring. **1**. *Conidiogenesis*. The phialides of some asexual fungi, e.g., *Exophiala jeanselmei* (dematiaceous asexual fungi), are annellate and the spores they produce are called annellospores. **2**. *Agaricales*. The stipe of the basidiocarp of various gen. (*Agaricus*, *Amanita*, *Lepiota*, etc.) is annellate. Cf. **exannellate**.

Annellate phialides of *Exophiala jeanselmei*, a black yeast, x 2,500 (*MU*).

annellation, **annulus**, pl. **annuli** (L. *annulus*, ring): **1**. *Conidiogenesis*. A scar, similar to a ring or collar, that remains on the apex of an annellate phialide or *annellide* with the secession of each conidium formed. **2**. *Agaricales*. The **annulus** refers to the fragments of the partial veil (which covers the hymenophore during development) that remain partially adhered to the upper portion of the foot or stipe of the basidiocarp, as happens in species of *Agaricus*, *Amanita*, *Chlorophyllum*, *Coprinus*, *Lepiota*, *Macrolepiota* and other gen.

annellide (L. *anellus*, dim. of *anus*, ring): a type of conidiogenous cell (annellate phialide or annellophore) that produces blastic conidia in basipetal succession (*annellospores*); with the production of each conidium, the conidiogenous apex of the phialide, which is not fixed, extends distally, and as each mature conidium detaches a scar is formed, similar to a ring or collar in the outer wall of the mouth of the phialide; the repetition of this process gives rise to the annellate phialide, as occurs, e.g., in *Scopulariopsis brevicaulis* and *Spilocaea pomi* (moniliaceous and dematiaceous asexual fungi, respectively).

Annulus of the basidiocarp of *Chlorophyllum* sp., x 0.5 (*RV*).

Annulus of the basidiocarp of *Chlorophyllum molybdites*, x 0.5 (*EPS*).

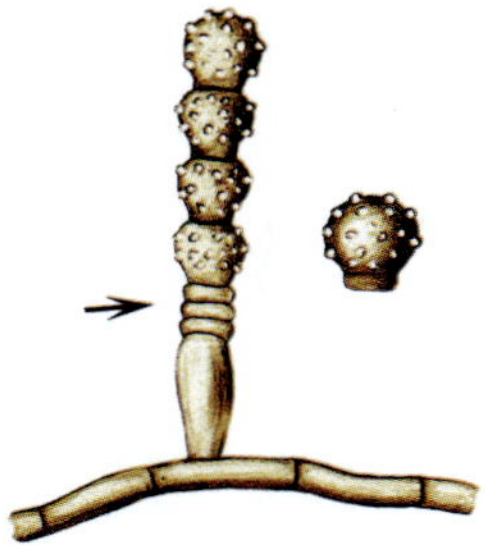

Annellide or annellate phialide of *Scopulariopsis brevicaulis*, x 550 (*MU*).

annellidic (L. *anellus*, dim. of *anus*, ring + suf. *-ídikos* > L. *-idicus* > E. *-idic*, ones own, belonging to itself): enteroblastic conidiogenesis in which the conidiogenous cell (*annellide* or *annellophore*) produces consecutive conidia (*annellospores*) by means of basipetal percurrent proliferation, in which the conidiogenous cell forms scars or transverse bands (*annellides*).

annellophore (L. *anellus*, dim. of *anus*, ring + Gr. *phóros*, bearer < *phéro*, to carry, bear): equal to **annellide**.

annellospore (L. *anellus*, dim. of *anus*, ring + Gr. *sporá*, spore): an asexual spore (phialospore) that originates from an annellate conidiogenous cell (annellate phialide or *annellide*), as in *Saccharomycodes ludwigii* (Saccharomycetales). *Scopulariopsis brevicaulis* and *Spilocaea pomi* (moniliaceous and dematiaceous asexual fungi, respectively) are examples of asexual fungi with this type of conidium.

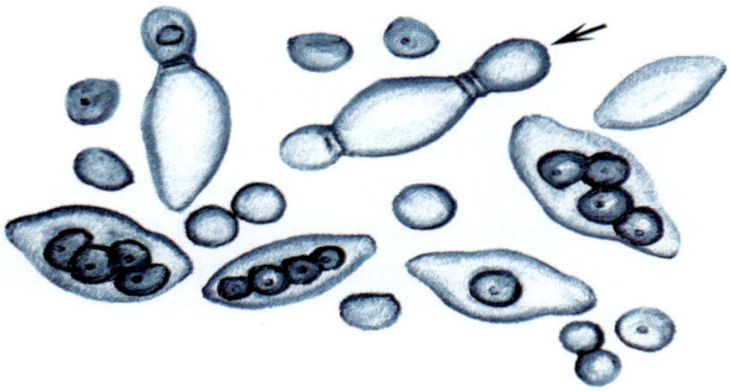

Annellospores of *Saccharomycodes ludwigii*, x 1,000 (*MU*).

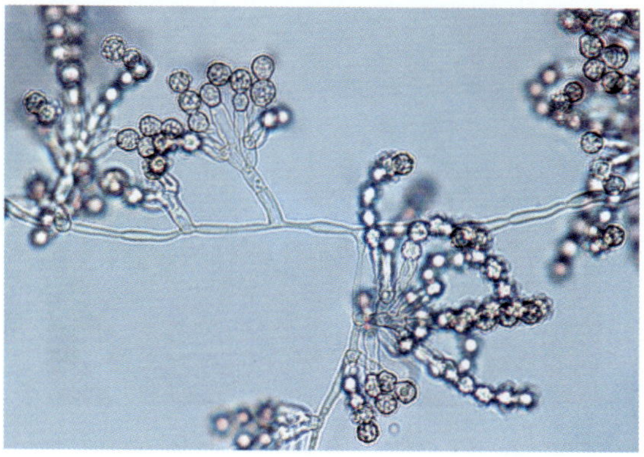

Annellospores of *Scopulariopsis brevicaulis*, x 1,200 (*MU*).

annexed (L. *annexus* < *annectere*, to unite or attach one thing to another that depends on it): united or attached to another thing dependent on it. See **adnate**.

annual (LL. *annualis* < *annuus*, yearly < *annus*, year + suf. *-alis* > E. *-al*, relating to or belonging to): refers to an organism that completes its life cycle in a single period that does not exceed one year. In the fungi it refers to the durability of the fruiting body, not to the assimilative mycelium immersed in the substrate, which in the majority of species is annual. See **biennial** and **perennial**.

annular (L. *anularis* < *anellus*, ring, arc + suf. *-aris* > E. *-ar*, belonging to or relating to): concerning a ring, or ring-like figure.

antagonism (Gr. *antagónisma*, dispute, opposition < pref. *ant-*, against, opposed + *agón*, fight, dispute + suf. *-ismós* > L. *-ismus* > E. *-ism*, state, phase, tendency, action): *Ecology*. A relationship between two species of organisms in which one of the two affects the life of the other, either inhibiting partially or totally its growth, including killing it; e.g., in dual agar cultures, in which *Agrobacterium azotophilum* inhibits the growth *of Rhizopus stolonifer* (=*Rh. nigricans*, Mucorales), or *Dendrostilbella* (stilbellaceous asexual fungi), which inhibits the growth of *Phymatotrichopsis omnivora* (a moniliaceous asexual fungus that is capable of attacking more than 2,000 different host species). The parasitism that occurs in multiple relations between a parasite and its host are considered as an antagonistic symbiosis or antagonism. The opposing organisms are referred to as antagonists. See **commensalism**, **helotism**, **inhibition**, **metabiosis**, **mutualism**, **parabiosis**, **parasymbiosis**, **symbiosis** and **synergy**.

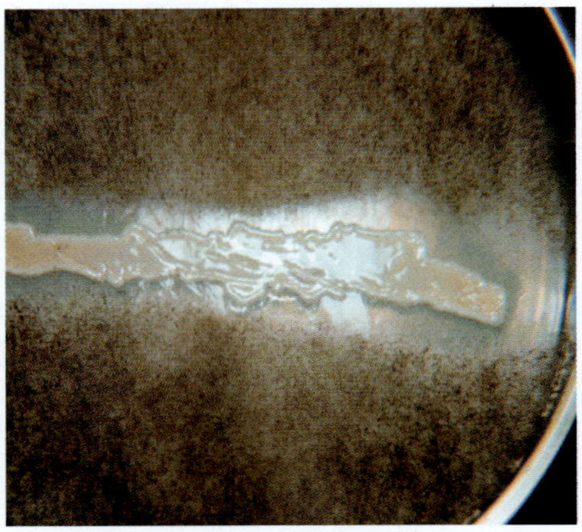

Antagonism *in vitro* of *Agrobacterium azotophilum* towards *Rhizopus stolonifer* (=*Rh. nigricans*), x 1 (*TH*).

Antagonism *in vitro* of *Dendrostilbella* sp. (the smaller colony) towards *Phymatotrichopsis omnivora*, x 0.6 (*MU*).

antagonistic symbiosis, pl. **symbioses**, or **parasitism** (Gr. *antagonistés* < *antí*, against + *agonistés*, fighter + suf. *-íkos* > L. *-icus* > E. *-ic*, belonging to, relating to; *symbíosis*, life in common; *parásitos*, parasite + *-ismós* > L. *-ismus* > E. *-ism*, state, phase, tendency, action): see **symbiosis**.

anterior (L. *anterior*, preceding in place or time; oriented in that direction): referring to the hymenial gills of the agaricaceous fungi, it is the part that is found next to the border of the pileus. Cf. **posterior**.

antheridiol (L. *antheridium* < NL. *anthera*, anther, here, alluding to the male sexual organ < Gr. *antherós*, flowery + dim. suf. *-ídion* > L. *-idium*, with the suf. *-ol* < alcohol, a chemical compound that contains hydroxyl): a sexual hormone (sterol) of *Achlya bisexualis* (Saprolegniales) that induces the formation of antheridia in male thalli. Cf. **oogoniols**.

antheridium, pl. **antheridia** (L. *antheridium* < NL. *anthera*, anther, here, alluding to the male sexual organ < Gr. *antherós*, flowery + dim. suf. *-ídion* > L. *-idium*): the male gametangium, i.e., the organ in which are generated the antherozoids, as in *Monoblepharis polymorpha* (Monoblepharidales), or the gametic nuclei, as in the Saprolegniales and Peronosporales, as well as in the ascomycetes.

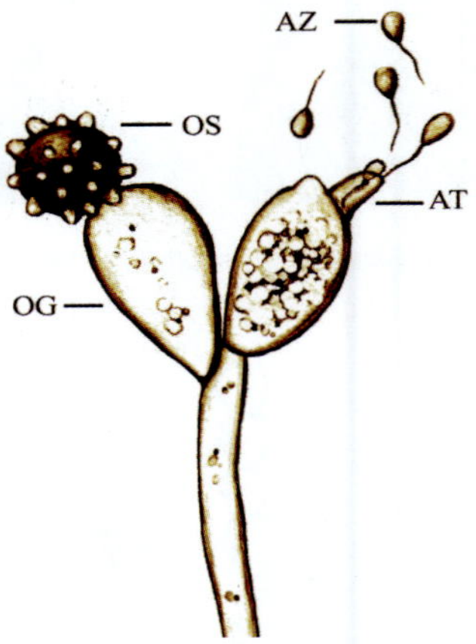

Antheridium (AT) and antherozoids (AZ) of *Monoblepharis polymorpha*. The oogonium (OG) and the oospore (OS) are also shown, x 350 (*MU*).

antherozoid (NL. *anthera*, anther, here, alluding to the male sexual organ < Gr. *antherós*, flowery + Gr. *zõon*, animal + L. suf. *-oide*, like < Gr. *-oeídes*, similar to): a motile, uniflagellate gamete of the Monoblepharidales. Syn. of **spermatozoid**.

anthracnose (Gr. *ánthrax*, coal, carbuncle + *nósos*, illness + L. suf. *-osis*, condition or state of a thing, a process, action): a term initially utilized to designate a disease of plants caused by some species of melanconiaceous asexual fungi, such as

Colletotrichum. Now the term is applied to necrotic diseases caused by gen. such as *Elsinoë* (Myriangiales), *Glomerella* (Phyllachorales), and *Sphaceloma* and *Marssonina* (=*Gloeosporium*), melanconiaceous asexual fungi, among others.

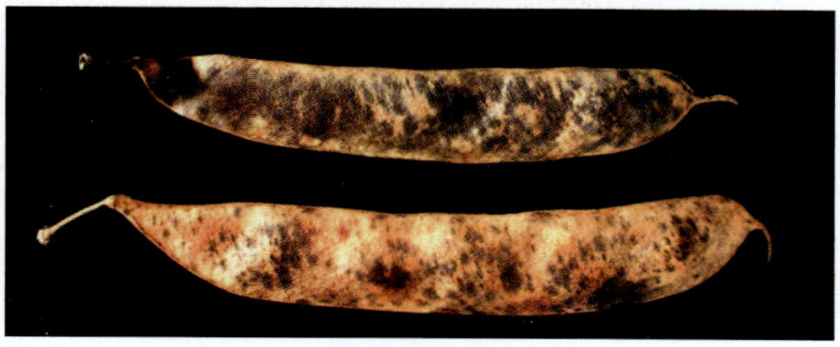

Anthracnose of green bean pods caused by *Colletotrichum lindemuthianum,* x 0.5 (*SA*).

anthracobiontic (Gr. *ánthrax*, coal, carbuncle + *bíos*, life + *óntos*, genit. of *ón*, a being + suf. *-ikos* > L. *-icus* > E. *-ic*, belonging to or relating to): refers to a fungus that obligately inhabits burned soil or wood and whose fructification is found on these types of substrates. See **carbonicole**, **phoenicoid fungus** and **pyrophilous**.

anthropophilic (Gr. *ánthropos*, man + *phílos*, have an affinity for + suf. *-íkos* > L. *-icus* > E. *-ic*, belonging to or relating to): *Med. mycol.* A fungus that preferentially infects humans instead of other higher animals, as happens with the dermatophyte *Epidermophyton floccosum* (moniliaceous asexual fungi), the causal agent of "athletes foot". Cf. **geophilic** and **zoophilic**.

anti-, **ant-** (Gr. *antí*, against): a pref. employed in the formation of biological terms, to give the idea of opposition, as in *antibiosis*, the inhibitory effect of one microorganism on another.

antibiogram (Gr. *antí*, against + *bíos*, life + *grámma*, describe): a laboratory method used to determine the effect of different antibiotics on the growth of bacteria, usually pathogenic, seeded on solid agar media. A translucent halo around a certain antibiotic means that the growth of the seeded bacterium has been inhibited by that antibiotic; the absence of such a halo indicates that the seeded bacterium is not sensitive to the antibiotic. See **antibiosis** and **antibiotic**.

antibiosis, pl. **antibioses** (Gr. *antí*, against + *bíos*, life + L. suf. *-osis*, condition, state): the inhibitory effect of one organism on the physiological processes of another, e.g., the inhibitory effect of an extract of *Oidiodendron tennuissimum* (dematiaceous asexual fungi) on the growth of *Trichophyton tonsurans* (moniliaceous asexual fungi). See **antibiogram** and **antibiotic**.

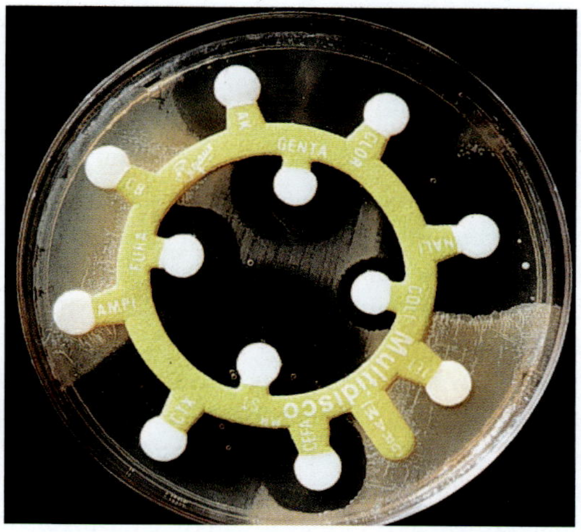

Antibiogram of *Escherichia coli.* A multidisc ring with differnt antibiotics was placed on an agar plate, which had been seeded with the bacterium. The effect of different antibiotics on the growth of *E. coli* is evident, e.g., it was inhibited by chloramphenicol, but not by ampicillin, x 0.6 (*MU*).

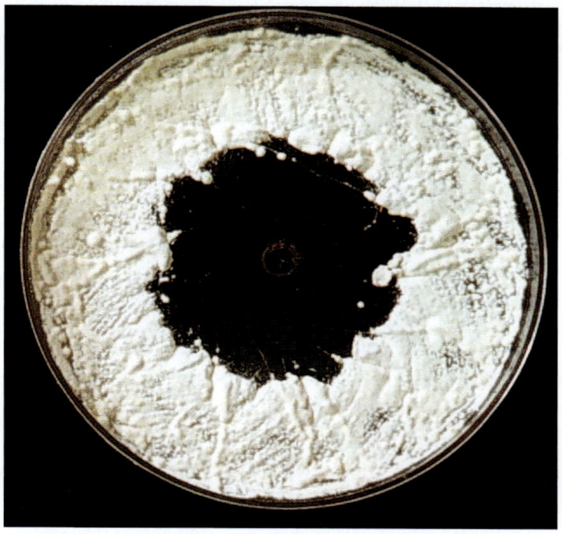

Antibiosis of an extract of *Oidiodendron tennuissimum* on the growth of *Trichophyton tonsurans* seeded on an agar plate. A halo around the central well where the antibiotic extract was placed is indicative of antibiosis, x 0.6 (*MU*).

antibiotic (Gr. *antí*, against + *biotikós*, of or relating to life < *bíos*, life + suf. *-tikós* > L. *-ticus* > E. *-tic*, relation, fitness, inclination or ability): a substance produced by a living organism that damages or kills another living organism, especially microorganisms; it refers, above all, to the substances produced by various species of molds (mainly species of *Penicillium* and *Aspergillus*, moniliaceous asexual fungi) and actinobacteria (*Actinomyces* and *Streptomyces*) that are harmful to other microorganisms. Some examples of antibiotics are: *penatin*, also called *corylophyline*, *notatin* and *penicillin B* (produced by *Penicillium chrysogenum*

(=*P. notatum*), with antibacterial properties; *penicillic acid* (a tetronic acid, antibacterial, produced by *P. aurantiogriseum*); *penicillin* (one of a group of antibiotic substances produced by *P. chrysogenum*), active against Gram + bacteria and of low toxicity for humans; *phomine* (a cytostatic antibiotic produced by *Phoma*, of the sphaeropsidaceous asexual fungi); *proliferine* (an antituberculosis antibiotic produced by *Aspergillus proliferans*). See **antibiogram** and **antibiosis**.

anticlinal (Gr. *antiklín*, to lean against each other < *antí*, against, opposed + *klínein*, to incline, revolve + L. suf. *-alis* > E. *-al*, relating to or belonging to): a membrane or cell wall of a structure that is perpendicular to the main growth axis of that structure. Cf. **periclinal**.

antigen (Gr. *antí*, against, opposed + *génos*, origin < *gennáo*, to engender, produce): a foreign substance that on being introduced into the body of a vertebrate animal, stimulates the production of an antibody.

antiseptic (Gr. *antí*, against, opposite + *septikós*, causing decay or damage < *sépsis*, putrefaction + suf. *-tikós* > L. *-ticus* > E. *-tic*, relation, fitness, inclination or ability): an antiputrefactant; a chemical substance that is utilized to destroy or restrict the growth of pathogenic or infectious microorganisms.

antitumoral (Gr. *antí*, against, opposite + L. *tumor*, genit. *tumoris*, tumor, abnormal mass of tissue + suf. *-alis* > E. *-al*, relating to or belonging to): a substance that inhibits the development of certain tumors, such as *poricin*, an antitumoral antibiotic produced by *Oxyporus* (=*Poria*) *corticola* (Aphyllophorales).

antrorse (NL. *antrorsus*, directed upward < L. *ante*, in front + *versum*, toward): directed forward or upward. Cf. **retrorse**.

ap. (L. abbrev. of *ápud*, in, in the writings of): an abbrev. that is used to indicate that what is cited is according to a particular author, e.g., *ap.*, or *ápud*, Lister, i.e., according to, in the works of Lister.

apandrous (Gr. pref. *ap-* < *apó*, estrangement + *andrós*, male, here, male sexual organ): an organism in which the phenomenon of *apandry* occurs.

apandry (Gr. pref. *ap-* < *apó*, estrangement + *andrós*, male, here, male sexual organ + suf. *-ia* > E. *-y*, which denotes condition or quality): apogamy with abortion of the male sexual elements, as occurs in some fungi that form oospores without the formation of antheridia.

aphanoplasmodium, pl. **aphanoplasmodia** (Gr. *aphanés*, invisible + L. *plasmodium* < Gr. *plásma*, formation, bland material with which a living being is formed + L. suf. *-odium*, with resemblance or similarity to < Gr. *ode*, like < *-oeídes*, similar to): *Myxomycetes*. A plasmodium that in its initial stages is constituted of a network of very fine, transparent protoplasmatic filaments that are not obviously differentiated into ectoplasm and endoplasm, as occurs in the plasmodia of other Myxomycetes, in which case they are called phaneroplasmodia. In the aphanoplasmodia the protoplasm is not very granulose, although the protoplasmic currents are rapid and reversibly rythmic. Characteristic of *Stemonitis* and related gen. (Stemonitales). Cf. **phaneroplasmodium** and **protoplasmodium**.

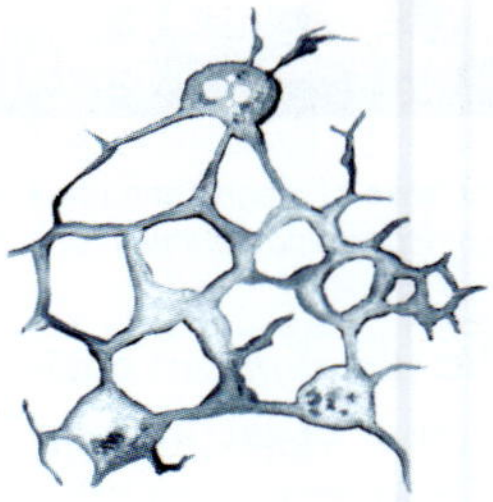

Aphanoplasmodium of *Stemonitis axifera*, x 100 (*MU*).

aphysoclastic (Gr. *a-*, not, without + *physáo*, *physéo*, *physõ*, to throw + *klástos*, broken < *kláo*, to break + suf. *-íkos* > L. *-icus* > E. *-ic*, belonging to or relating to): applied to an ascus that liberates its spores without a previous rupture of the external wall layer (which is less elastic) and also without undergoing an extension of the inner wall layer. It is a term utilized in marine mycology; this type of ascus is present, e.g., in *Ceriosporopsis* and *Carbosphaerella* (Halosphaeriales), fungi saprobic on wood and other floating vegetable remains in the intertidal zone. Cf. **physoclastic**.

apical (L. *apicalis*, at the apex < *apex*, genit. *apicis*, apex, tip > *apic-* + suf. *-alis* > E. *-al*, relating to or belonging to): located at the top or apex, such as the apical paraphyses of the ascocarp of *Haematonectria* (=*Nectria*) *haematococca* (Hypocreales). Cf. **basal**.

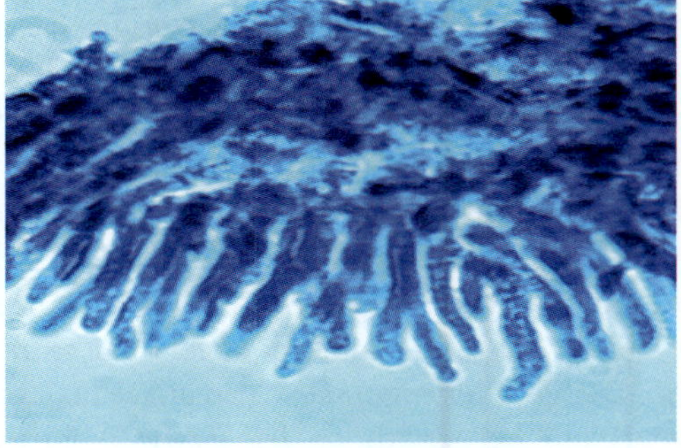

Apical paraphyses of the ascocarp of *Haematonectria* (=*Nectria*) *haematococca*, x 1,500 (*RTH*).

apiculate (NL. *apiculatus*, very sharp-pointed < *apiculus*, dim. of L. pref. *apic-*, *apex*, genit. *apicis*, apex, tip + suf. *-atus* > E. *-ate*, provided with, likeness): ending in a small, distinct point. E.g., the ascospores of *Hypomyces* (Hypocreales) and *Chaetomium* (Sordariales) are apiculate. See **apiculus** and **biapiculate**

Apiculate ascospores of *Chaetomium nigricolor*, x 1,500 (*MU*).

apiculus, pl. **apiculi** (NL. *apiculus* < L. pref. *apic-*, *apex*, genit. *apicis*, apex, tip + dim. suf. *-ulus* > E. *-ulus*): a short, sharp tip, but not rigid; generally refers to the short, sharp diverticulum found in the basal portion of basidiospores, by means of which they adhere to the sterigma of the basidium, and from which exudes a droplet at the moment of discharge of the basidiospore. See **hilar appendix**, which is a more correct term than apiculus to designate the diverticulum described, which is characteristic of all Agaricales.

apileate (Gr. pref. *a-*, not, without + L. *pileatus*, provided with a pileus < *pileus*, pileus or hat + suf. *-atus* > E. *-ate*, provided with or likeness): applied to fungi whose basidiocarps lack a pileus, e.g., the resupinate species, such as *Poria*, *Stereum* and other thelephoraceous members of the order Aphyllophorales. Cf. **pileate**.

apiospore (Gr. *ápion*, pear + *sporá*, spora): a bicellular, pear-shaped spore, with one cell much smaller than the other, like the pyriform ascospores of *Apiosporina collinsii* (Pleosporales).

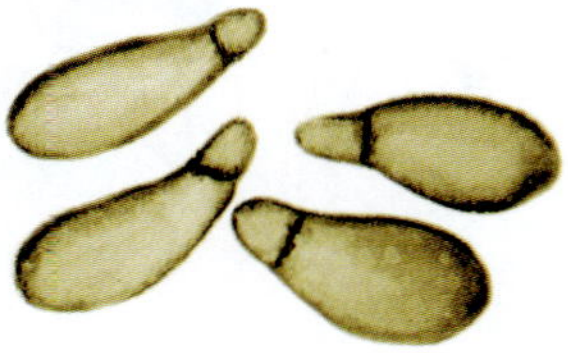

Apiospores of *Apiosporina collinsii*, x 1,000 (*MU*).

aplanetic (Gr. pref. *a-*, not, without + *planétes*, wanderer, + suf. *-íkos* > L. *-icus* > E. *-ic*, belonging to or relating to): immotile; refers to spores that are not motile. The condition of having non-motile spores instead of zoospores, as occurs in *Geolegnia* (Saprolegniales), is termed *aplanetism*. See **aplanospore**.

aplanogamete (Gr. pref. *a-*, not, without + *plános*, wandering + *gamétes*, spouse): an immotile gamete, i.e., lacking flagella. Cf. **planogamete**.

aplanospore (Gr. *aplanés*, immotile, fixed + *sporá*, spore): an asexual, immotile spore produced individually or in variable numbers inside a sporangium, as happens in molds of the order Mucorales. See **aplanetic**.

aplerotic (Gr. pref. *a-*, not, without + *plerotikós* < *pleróo*, to fill + suf. *-tikós* > L. *-ticus* > E. *-tic*, relation, fitness, inclination or ability): *Peronosporales*. Not filled up; e.g., in some species of Pythiaceae, such as *Pythium echinulatum*, the oospore is aplerotic because it does not fill the oogonium. Cf. **plerotic**.

Aplerotic oospores of *Pythium echinulatum*, x 430 (*MU*).

apo-, **ap-** (Gr. *apo-* < the prep. *apó*, away from): a pref. employed in the formation of various terms that imply the idea of something lacking, a defect or liberation, separation, distance, the end or loss of something, transformation, negation, as in *apobasidium*, a basidium different from the typical basidium. Corresponds to L. *a-*, *ab-*, *abs-*.

apobasidium, pl. **apobasidia** (Gr. *apó*, away from + L. *basidium*, small pedestal < Gr. *basídion* < *básis*, base + dim. suf. *-ídion* > L. *-idium*): *Gasteromycetes*. A type of basidium that is characterized by having its spores terminal and symmetrical or orthotropic on the axial prolongations of the sterigmata, like those of *Sclerogaster* (Hymenogastrales). The fungi with apobasidia are called apobasidiomycetes. Cf. **autobasidium**.

apocyte (Gr. *apó*, away from, here it expresses the idea of something separated or hidden + *kýtos*, cavity, here, cell): a multinucleate cell in which each nucleus is surrounded by a protoplasmic mass, thus constituting true cellular units, but without separation by means of walls. The apocyte, which is an accidental, transitory or secondary cellular condition in ascomycetes and basidiomycetes, is morphologically similar to a syncytium, but differs in that the former comes from an original single cell in which the nuclei have multiplied without establishing cellular separation of the protoplasts, whereas the syncytium originates by the fusion of two or more distinct cells.

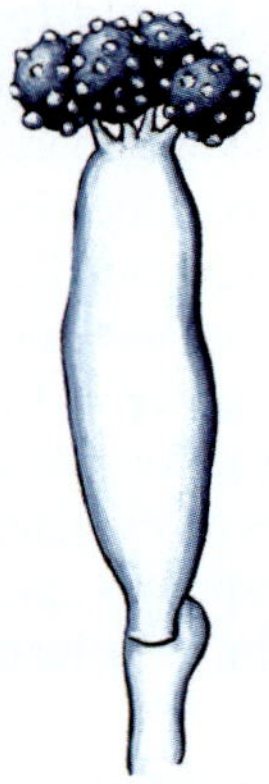

Apobasidium of the hymenium of *Sclerogaster* sp., x 1,270 (*MU*).

apodal, **apodial**, **apodous** (Gr. *ápous*, genit. *ápodos* < pref. *a-*, not, without + *poús*, genit. *podós*, foot + L. suf. *-alis* > E. *-al*, relating to or belonging to): without a foot; without a pedicel or stipe; sessile. Cf. **stipitate**, **pedicellate** and **sessile**.

apogamy (Gr. *apó*, defect, distance, something lacking, end or loss of something + *gámos*, sexual union > E. suf. *-gamy*): away from sexuality; apomixis or apomictic development of diploid cells. See **apomixis**. Cf. **amphimixis**.

apomictic (Gr. *apó*, away from, negation, lack of something + *miktós*, mixture): relating to the phenomenon of *apomixis*.

apomixis, pl. **apomixes** (Gr. *apó*, away from, negation, lack of something + *míxis*, mixture, intimate union): the development of sexual cells in spores, thalli, etc., without previous fertilization. Two types of apomixis are known: **parthenogenesis** and **apogamy**. See **amixis**. Cf. **amphimixis**.

apophysate (NL. *apophysis* < Gr. *apóphysis*, offspring < *apó*, transformation + *phýsis*, to bring forth, bear < *phýein*, to produce, bring forth + L. suf. *-atus* > E. *-ate*, provided with, likeness): having an *apophysis*.

apophysis, pl. **apophyses** (NL. *apophysis* < Gr. *apóphysis* < *apó*, offspring + *phýsis*, to bring forth, bear < *phýein*, to produce, bring forth): **1**. The swollen portion that is found beneath the sporocarp of some Protosteliales, such as *Nematostelium*, *Schizoplasmodium* and *Protostelium*. **2**. The subsporangial swelling present in the sporangium of certain aquatic fungi, such as *Phlyctochytrium* (Chytridiales). **3**. The widening apical part of the sporangiophore of some Mucorales, such as *Chlamydoabsidia*, *Gongronella* and *Rhizopus*. **4**. The ventral swelling of the endoperidium of *Geastrum* (Lycoperdales).

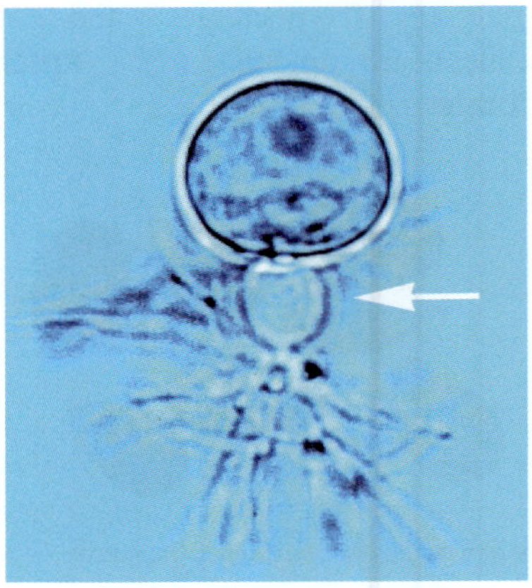

Apophysis of the sporangium of *Phlyctochytrium* sp., x 850 (*RTH*).

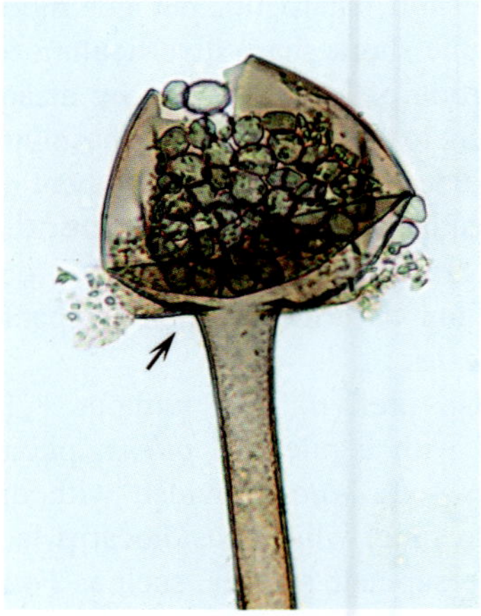

Apophysis of the sporangium of *Rhizopus oryzae* (=*Rh. arrhizus*), x 400 (*PL*).

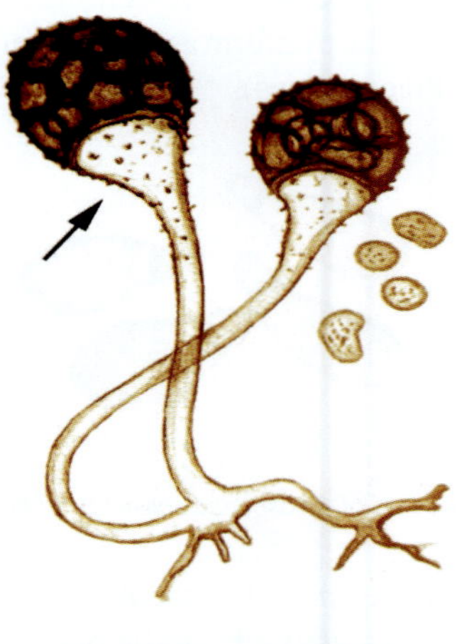

Apophysis of the sporangia of *Rhizopus microsporus* (=*Rh. oligosporus*), x 300 (*MU*).

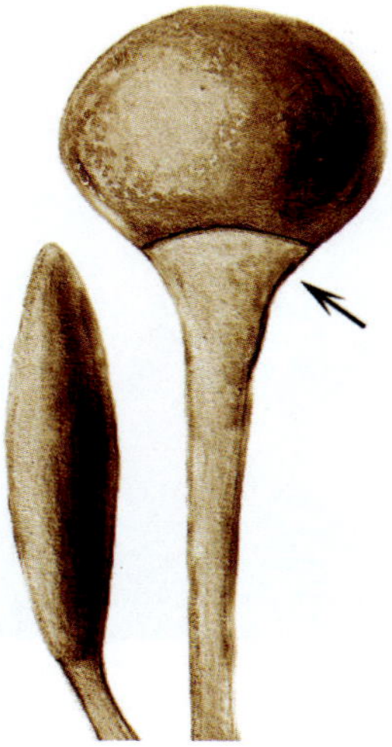

Apophysis of the sporangia of *Chlamydoabsidia padenii*, x 1,100 (*MU*).

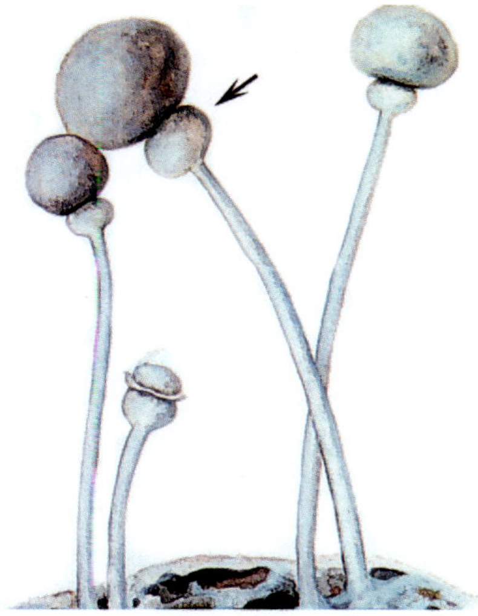

Apophysis of the sporangia of *Gongronella butleri*, x 460 (*MU*).

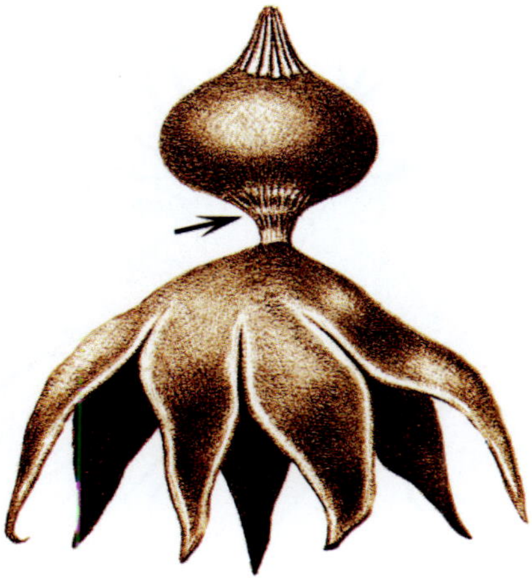

Apophysis of the fruiting body of *Geastrum pectinatum*, x 0.7 (*MU*).

apoplasmodial (Gr. *apó*, away from + *plasmodial* < L. *plasmodium* < Gr. *plásma*, formation, bland material of which living beings are formed + L. suf. *-odium*, to resemble, look like < Gr. *ode*, similarity, likeness < *-oeídes*, similar to + L. suf. *-alis* > E. *-al*, relating to or belonging to): refers to an organism that does not form true plasmodia, due to lack of fusion of myxamoebae, as occurs in the pseudoplasmodium of the Acrasiomycota.

apoplastic (Gr. *apó*, away from + *plastós*, formed, an organized particle or body + suf. *-íkos* > L. *-icus* > E. *-ic*, belonging to or relating to): the movement of substances via the cell walls, without entering the cytoplasm of the cell. Cf. **symplastic**.

apospory (Gr. *apó*, away from + *sporá*, spore): the incorporation of a diploid oogonial or antheridial nucleus and cytoplasm directly into a spore without the occurrence of meiosis.

apothecioid (NL. *apothecium* < Gr. *apothekíon*, dim. of *apothêke*, granery < *apó*, something that separates, end or loss of something + *thêke*, box; here, of the asci + L. suf. *-oide*, like < Gr. *-oeídes*, similar to): refers to the ascostroma of certain loculoascomycetes, such as *Schizothyrioma* (=*Schizothyrium*) *ptarmicae* (Dothideales) and *Myriangium* (Myriangiales), which on maturing acquire a shape similar to an apothecium.

Apothecioid multilocular ascostroma of *Myriangium* sp., seen in longitudinal section, x 50 (*MU*).

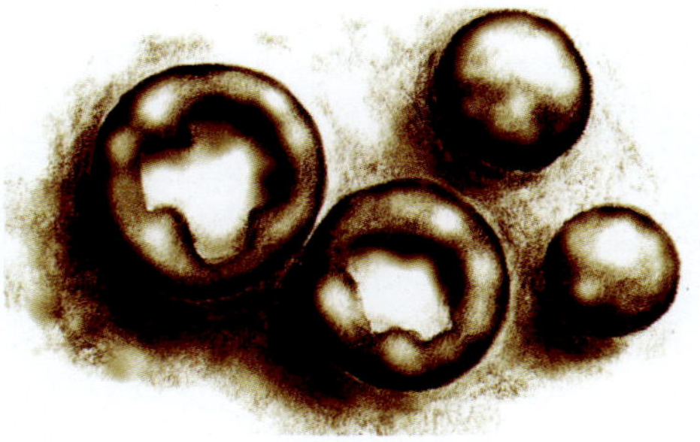

Apothecioid pseudothecia of *Schizothyrioma* (=*Schizothyrium*) *ptarmicae*, two with the scutellum open, on a leaf, x 80 (*MU*).

apothecium, pl. **apothecia** (NL. *apothecium* < Gr. *apothekíon* < *apothêke*, granary < *apó*, separated or away from + *thêke*, box; here, of the asci + dim. suf. *-íon* > L. *-ium*): a type of ascocarp (fruiting body) which at maturity opens widely to expose the hymenium of asci; it commonly assumes the shape of a cup, plate or disk, although it can have other shapes. It is the characteristic ascocarp of the discomycetes, such as *Aleuria*, *Cookeina*, *Helvella*, *Peziza* (Pezizales) and *Chlorociboria*, *Monilinia* and *Sclerotinia* (Helotiales). When it has the shape of a disk or plate it is also called a **discocarp**.

Apothecia of *Chlorociboria aeruginascens*, x 5 (*RTH*).

Apothecia of *Cookeina tricholoma*, x 1 (*RV*).

Apothecium of *Aleuria aurantia* in side view, x 3 (*CB*).

Apothecium of *Peziza hemisphaerica*, x 1 (*RV*).

appendage, appendix, pl. **appendixes, appendices** (L. *appendix*, appendage, supplement, or *appendage* < *appendere*, to add, annex, to adorn): a part projecting from another structure, generally accessory. Distinctive structures present in very diverse fungal organs are called appendages. E.g., the appendages that form on the suspensors and which more or less cover the zygosporangia of many species of Mucorales, such as *Absidia glauca*, *Phycomyces nitens*, etc. Other examples of appendages are some of the hymenial elements of various Agaricales (such as the digitate, echinulate or diverticulate cystidia, the coprinoid appendages, etc.), the piliform appendages of the sporangiospores of *Choanephora cucurbitarum* (Mucorales), of the trichospores of *Smittium* (Harpellales) and of the conidia of *Chaetospermum camillae* and *Entomosporium maculatum* (melanconiaceous asexual fungi), as well as the membranaceous ones of the conidia of *Myrothecium verrucaria* (tuberculariaceous asexual fungi).

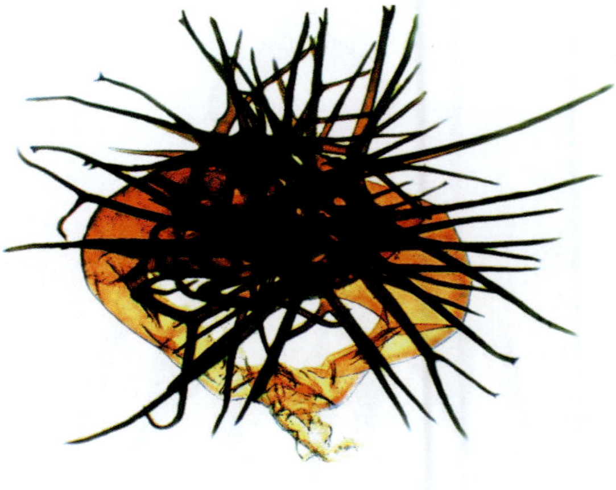

Appendages on the suspensors of *Phycomyces nitens*, x 100 (*MU*).

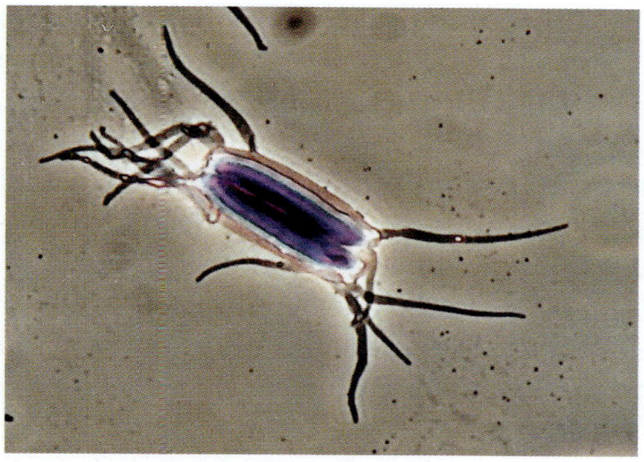

Appendages of the conidia of *Chaetospermum camillae*, x 1,000 (*ERG* & *MCG*).

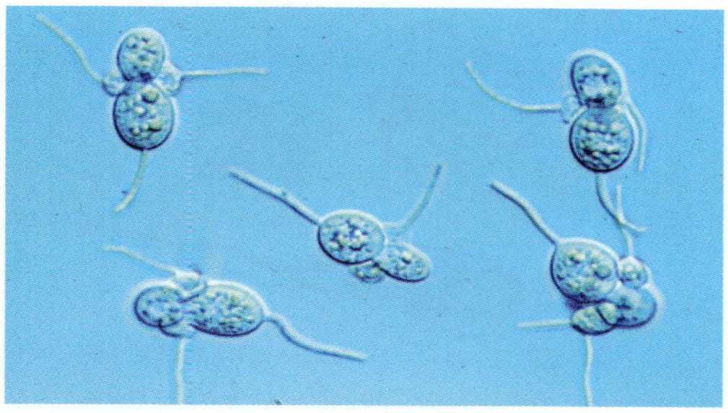

Appendages of the conidia of *Entomosporium maculatum*, x 1,000 (*RTH*).

appendiculate (L. *appendiculatus*, furnished with appendages < *appendiculus*, small appendage < *appendix*, appendage + dim. suf. *-culus* + suf. *-atus* > E. *-ate*, provided with, likeness): having appendages or protruding parts, generally accessory or remnant. In fungi of the gen. *Hydnum* and *Sarcodon* (Aphyllophorales), each of the teeth of the sporophore, covered by hymenium, is called an *appendiculus*. Also called an appendiculus is the portion of the partial veil that remains hanging from the edge of the pileus after its rupture and separation, as happens in *Psathyrella*, *Cystoderma* (Agaricales) and *Polyporus tricholoma* (Aphyllophorales), whose pileus is appendiculate.

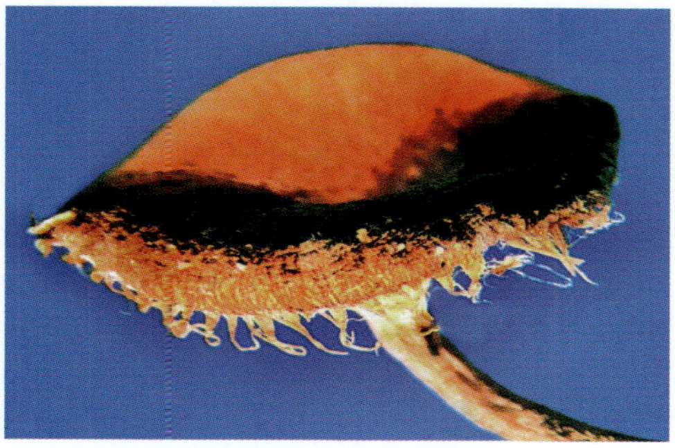

Appendiculate pileus of *Polyporus tricholoma*, x 2 (*MU*).

Appendiculate pileus of *Cystoderma* sp., x 0.5 (*TRO*).

appendiculus, pl. **appendiculi** (L. *appendiculus*, small appendage < *appendix*, genit. *appendicis*, appendage + dim. suf. *-culus* > E. *-culus*): a small appendage. See **appendiculate**.

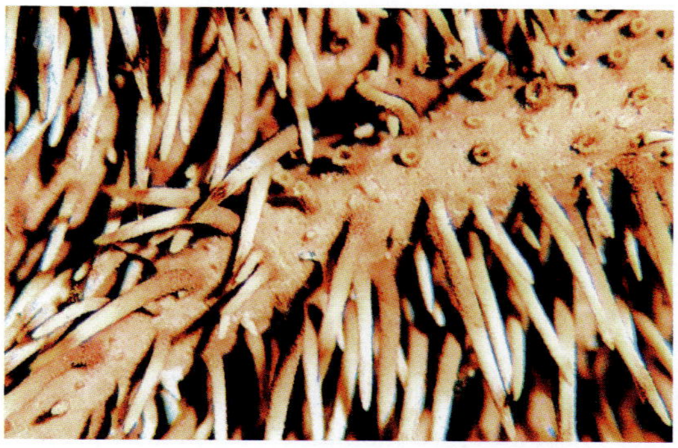

Appendiculi of the hymenophore of *Sarcodon* sp., x 30 (*MU*).

apposed (MF. *aposer* < OF. < L. *appositus*, united, placed near, applied): placed in juxtaposition, side by side. See **parallel**.

apposition (LL. *apposition*, genit. *appositionis*, the action of placing or drawing near < L. *appositus*, united, placed near, applied < pref. *ap-*, in + *positus*, placed + *-io*, genit. *-ionis* > E. suf. *-ion*, result of an action, state of): the growth in thickness of a cell wall by successive deposits of new molecules on the internal or external surface. Endogenous apposition is apposition in the strict sense. Exogenous apposition involves growth in thickness through centrifugal deposition on the exterior wall; this occurs in the spores of many fungi, which develop various ornamentations in this manner. Cf. **intususception**.

appressorium, pl. **appressoria** (NL. *appressorium* < *appress*, kept down, compressed + L. suf. *-orium*, the place of a thing): a flat swelling that forms on the germ tube of a spore or on a vegetative hypha, and which adheres to the surface of a host before penetrating it with an infection hypha that originates from the swelling; characteristic of plant pathogenic zoosporic microorganisms, such as *Phytophthora* (Peronosporales), *Cercosporidium* (dematiaceous asexual fungi), and *Colletotrichum* (melanconiaceous asexual fungi), and of parasites of lower animals, such as *Entomophthora* (Entomophthorales) and *Metarhizium* (moniliaceous asexual fungi).

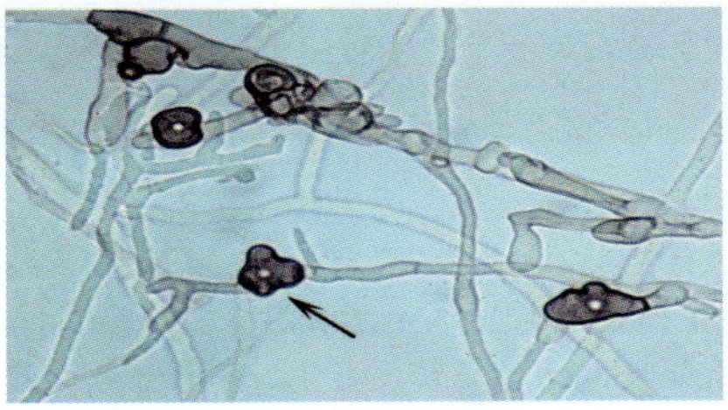

Appressoria of *Colletotrichum* sp., x 420 (*MU*).

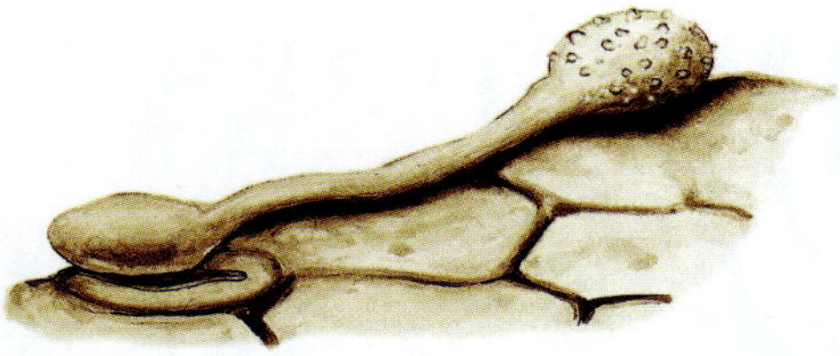

Appressorium of *Cercosporidium personatum*, formed on the end of the germ tube that grew over the stoma of a leaf, x 1,600 (*MU*).

apud (L. *ápud*, in, in the writings of): a term, abbrev. **ap.**, that is used before the name of a person to indicate that what is cited in a text is according to the criteria or the works of that person.

aquatic molds, **hydrofungi**, sing. **hydrofungus**, **water molds** (L. *aqua*, water + suf. *-ticus* < Gr. *-tikos* > E. *-tic*, relation, fitness, inclination or capacity; ME. *mowlde* < *mowled* < *moulen*, *malwen*, to grow moldy; L. *hydro* < Gr. *hýdro-* < *hýdor*, water; L. *fungi*, pl. of *fungus*, fungus): fungi which live in water, especially fresh water, in contrast to marine fungi. The main zoosporic microorganisms that live in fresh water belong to the Chytridiomycota and Oomycota, especially Chytridiales and Saprolegniales, respectively. There are also many hyphomycetes (asexual or mitosporic fungi), ascomycetes and saxicolous lichens (some of the latter can remain submerged), which live in fresh water, as well as a variety of fungi that are found in contaminated waters and sewages. Cf. **primary marine fungus** and **secondary marine fungus**.

aqueous (L. *aquosus*, having a large proportion of water < *aqua*, water + suf. *-osus* > E. *-ous*, abundant in, full of): relative to or similar to water; made of, with or by water.

aqui- (L. *acui-* < *aqua*, water): a pref. employed to denote that something is from an aquatic habitat or that it has to do with water, as in *aquicolous*, living in water.

arachnoid (NL. *arachnoides* < Gr. *arachnoeidés*, similar to a cobweb < *aráne* < *aráchne*, spider, cobweb + L. suf. *-oide*, like < Gr. *-oeídes*, similar to): similar to a cobweb, like the medullary layer of lichens when it is formed of loosely intertwined hyphae. The conidia of *Arachnophora fagicola* (moniliaceous asexual fungi) are arachnoid, since their lateral protuberances have appendages that are incurved, spinose or claw-shaped, which confers on them a peculiar aspect, as if they were small arachnids, a characteristic implicit in the generic name. Also called **araneose**.

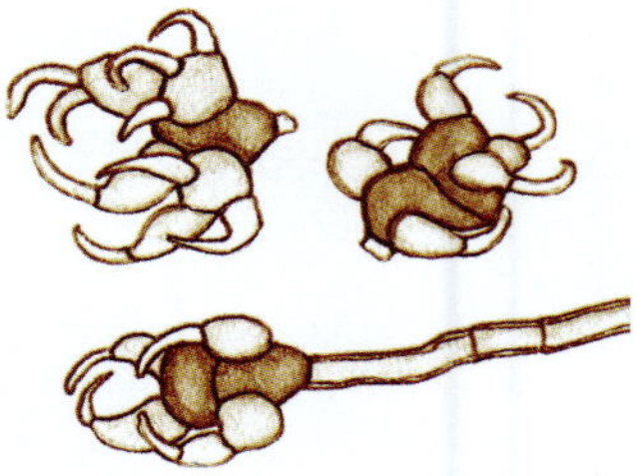

Arachnoid conidia of *Arachnophora fagicola*, x 1,000 (*MU*).

Arachnoid stipe of *Cortinarius* sp. (Agaricales), x 0.5 (*RV*).

araneose, **araneous** (L. *araneosus*, pertaining to a cobweb < Gr. *aráne* < *aráchne*, spider, cobweb + L. suf. *-osus*, full of, augmented, prone to > E. *-ose*, *-ous*, having the qualities of): similar to a cobweb. See **arachnoid**.

arboricolous (L. *arbor*, genit. *arboris*, tree + suf. *-cola*, inhabitant < *colere*, to inhabit): living on trees, like many lichens.

arbuscle, **arbuscule** (L. *arbusculum* < *arbor*, tree + dim. suf. *-culum* > E. *-cule*): a group of hyphae with a shrubby or coralline aspect which mycorrhizal fungi form inside the root cells of the associated plant. Arbuscles are one of the forms (vesicles constitute the other) produced by endotrophic mycorrhizae (endomycorrhizae). An example of an arbuscle is

Pythium ultimum (Peronosporales) in cortical cells of onion root. Many Glomales, such as *Glomus* and *Gigaspora*, form vesicular-arbuscular mycorrhizae in many species of higher plants.

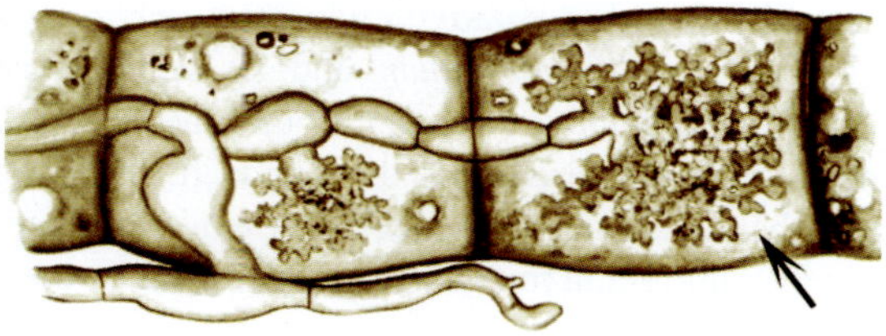

Arbuscles of *Glomus diaphanum* in root cells of red clover, x 370 (*MU*).

archaeomycete (Gr. *arché*, ancient + *mýkes*, genit. *mýketos*, fungus): a fungus, especially a plant pathogen, introduced into a new geographic region in prehistoric or early historic times, becoming established before such events were observed and recorded.

arche-, **archi-** (Gr. *arché*, primitive, origin, point of departure): a pref. employed in biology to give the idea of primitiveness, from a phylogenetic (as in *Archiascomycetes*) as well as an ontogenetic (e.g., in *archicarp*) point of view.

Archiascomycetes (Gr. *arché*, primitive, origin, point of departure + *askós*, sac, ascus + L. *-mycetes*, ending of a class < Gr. *mýkes*, genit. *mýketos*, fungus): a class of fungi, formerly referred to in part as Hemiascomycetes, in which the asci are borne free (i.e., not contained in any type of fruiting body or ascocarp), and without ascogenous hyphae; the thallus is simple, and when mycelium is present it is poorly developed. The Archiascomycetes, together with the classes Saccharomycetes, Plectomycetes, Pyrenomycetes, Discomycetes and Loculoascomycetes (along with some other ascomycetes, such as the orders Saccharomycetales, Erysiphales, Laboulbeniales and Spathulosporales) constitute the phylum Ascomycota of the kingdom Fungi. See the Appendix to compare the classification changes of this class in the outline of Alexopoulos *et al.*, 1996, modified by Ulloa and Hanlin, 2006, to that of Kirk *et al.*, 2008. In the latter classification scheme, the orders of this class (Taphrinales and Schizosaccharomycetales) were moved to the subphylum Taphrinomycotina. These fungi consist of heterogeneous species, forming unnatural groupings, and therefore are considered an artificial classification. These taxonomic rearrangements, with the concurrent changes of the names of taxa, represent a constant process of study and evaluation, in order to achieve a phylogenetic classification of the fungi.

archicarp, **archecarpium**, pl. **archecarpia** (Gr. *arché*, primitive, origin, point of departure + *karpós*, fruit): *Ascomycetes*. The ascogonium or female sexual organ, which forms the first initial of the carpophore or ascocarp. The term is used indistinctly to designate the whole female organ, the branch, the primary cell, the group of initial cells, or the coiled hypha that develops into the fruiting body or ascocarp, as is seen in *Podosphaera* (Erysiphales).

Archimycetes (Gr. *arché*, primitive, origin, point of departure + L. *-mycetes*, ending of a class < Gr. *mýkes*, genit. *mýketos*, fungus): a class recognized by Gäumann and Dodge (1928) and Gäumann (1952) for certain unicellular, holocarpic chytrids which lack a cell wall in the early stages of development; this characteristic was used to distinguish this class from the rest of the chytrids, which were retained in the class Chytridiomycetes.

archontosome (Gr. *árchon*, genit. *árchontos*, a ruler + *sõma*, body): an electron-dense body that occurs near nuclei during the stages of ascosporogenesis in *Xylaria polymorpha* (Xylariales), from the formation of the croziers to the development of the young ascospores.

arcuate (L. *arcuatus*, bent like a bow, curved < *arcus*, bow + suf. *-atus* > E. *-ate*, provided with or likeness): curved in an arc, like the pileus of the basidiocarp of *Clitocybe infundibuliformis* (Agaricales), which first develops upward and later bends downward, forming an arc in the lower part. The sporangia of *Stemonitopsis typhina* (=*Comatricha typhoides*, Stemonitales) also are arcuate.

Arcuate sporangia of *Stemonitopsis typhina* (=*Comatricha typhoides*), x 55 (*MU*).

ardella, pl. **ardellae** (L. *ardella*, little spot < Gr. *árda*, spot + L. dim. suf. *-ella*): a small apothecium in the shape of a round spot, characteristic of lichens of the family Arthoniaceae (Arthoniales), such as *Arthothelium spilomatoides*. If the apothecia are elongate, they are called **lirellae**.

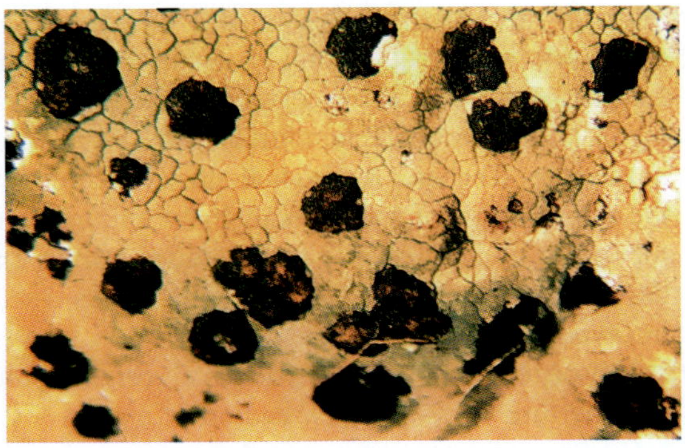

Ardellae (apothecia) of the thallus of *Arthothelium spilomatoides*, x 7 (*MU*).

arenarious (L. *arenarius*, pertaining to sand < *arena*, sand + suf. *-arius* > E. *-arious*, belonging to): living in sand. Syn. of **arenicolous** and **psammophilous**.

arenicolous (L. *arena*, sand + suf. *-cola*, inhabitant): an organism that lives among or on the grains of sand, whether in a desert or on a beach. Also called **psammophilous** and **sabulicolous**. In mycology it is applied most commonly to fungi that inhabit sand grains or their interstices on a marine beach, such as *Corollospora intermedia* and *Lulworthia lignoarenaria* (Halosphaeriales), among other species. See **ammophilous**, **arenarious** and **endopsammon**.

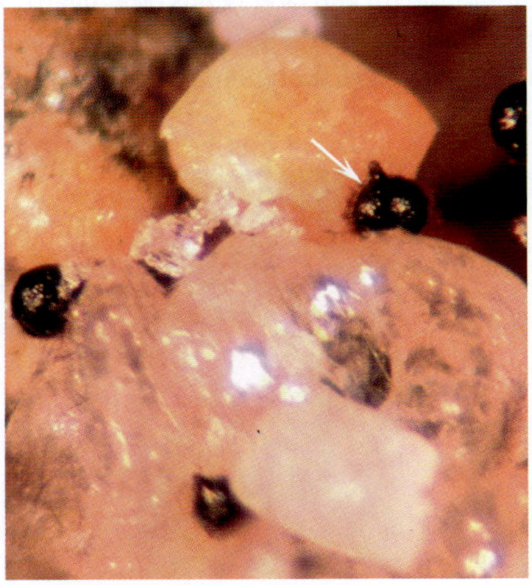

Arenicolous perithecia of *Corollospora intermedia* (arrow) adhered to grains of sand, x 23 (*MU*).

areola, pl. **areolae** (L. *areola*, dim. of *area*, halo, open space): a small space or area delimited on a surface, distinct in color, texture or other characteristic from the rest of the surface; e.g., an areola can be separated from others by fissures or cracks. See **areolate**.

areolate, **areolar** (L. *areolatus*, with areolae < *areola*, dim. of *area*, halo, open space + suf. *-atus* > E. *-ate*, provided with or likeness; or + suf. *-aris* > E. *-ar*, like, belonging to): pertaining to or relative to an *areola*; having a halo or ring of a different color. E.g., in *Annulohypoxylon thouarsianum* and *A. michelianum* (=*Hypoxylon thouarsianum* and *H. michelianum*, Xylariales) halos are present around the ostioles of the perithecia embedded in the stroma. In the lichens, the thallus is said to be areolate when its surface is divided into small rounded or angular areas by means of cracks, as occurs in *Rhizocarpon geographicum* (Lecanorales) and *Thelomma santessonii* (Caliciales).

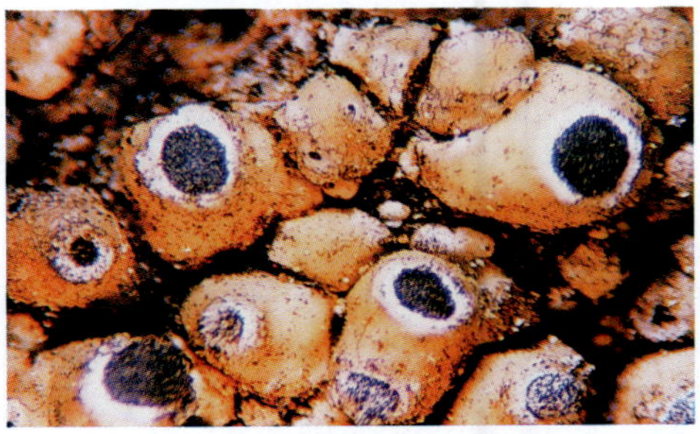

Areolate apothecia of the thallus of *Thelomma santessonii*, x 10 (*CB*).

arescent (L. *arescens*, genit. *arescentis* < *arescere*, to become dry, not deliquescent < *areere*, to be dry + suf. *-escens*, genit. *-escentis* > E. *-escent*, that which turns, beginning to, slightly): dry, not moistened, like the gills of many agaricaceous fungi. Cf. **deliquescent**.

argentate, **argenteous** (L. *argenteus*, silvery < *argentum*, silver): refers to organs or structures whose surface has a certain luster like silver, due to being covered by abundant hairs or soft, appressed scales, like the basidiocarps of *Geastrum argenteus* (Lycoperdales), *Amanita argentea* and *A. spreta* (Agaricales), and the peridiola of the fruiting body of *Cyathus intermedius* (Nidulariales), among others.

Argentate peridiola of the fruiting body of *Cyathus intermedius*, x 7 (*CB* & *MU*).

argentophilous (L. *argentum*, silver + Gr. *phílos*, have an affinity for): living on substrates containing silver or other heavy metals, which not only can resist them but capture the silver ions in their cells, such as *Cryptococcus albidus* var. *albidus* (cryptococcaceous asexual yeasts), an encapsulated yeast that has been isolated from the mud discharged from silver mines; these yeasts capture the silver and accumulate it in their capsules.

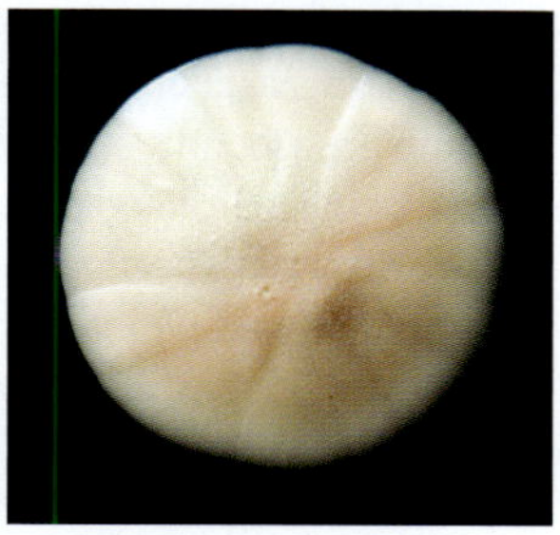

Argentophilous giant colony of *Cryptococcus albidus* var. *albidus*, on agar, x 1 (*MU*).

arithmotype (Gr. *arithmós*, number + *typós*, type, mark, model): a specimen recorded as an isotype of a taxon, with the same collector's name, but belonging to a taxon different from the first, to which the holotype belongs.

armilla, pl. **armillae** (L. *armilla*, bracelet, ring): a ring situated on the upper part of the foot of certain agaricaceous fungi, as in *Armillaria mellea*. See **ring**.

armillarioid (< gen. *Armillaria* < L. *armilla*, ring + suf. *-aria*, which indicates possession or connection + L. suf. *-oide*, like < Gr. *-oeídes*, similar to): *Agaricology*. A fruiting body characterized by having gills adhering to the stipe, which is fibrous-fleshy or chalky in consistency, and centrally attached to the pileus, with a ring, but lacking a volva. Typified by *Armillaria* and certain species of *Pholiota* (=*Flammula*) and *Gomphidius*, among other gen. of Agaricales.

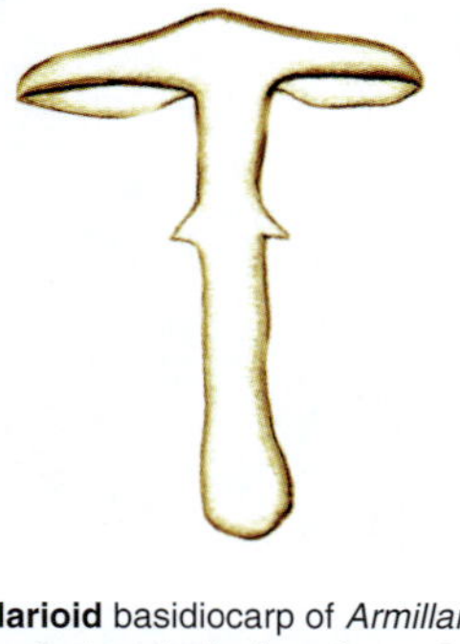

Armillarioid basidiocarp of *Armillaria* sp. Diagram of a longitudinal section, x 0.5 (*MU*).

armillate (L. *armilla*, bracelet, ring + suf. *-atus* > E. *-ate*, provided with or likeness): provided with an *armilla* or ring.

arrect (L. *arrectus*, straight, vertical, set upright < *arrigere*, to erect): erect, directed upward, very stiff, like the hairs of the basidiocarp of *Trametes hirsuta* (=*Polyporus hirsutus*, Aphyllophorales).

arthonioid (< gen. *Arthonia*, of the order Arthoniales + L. suf. *-oide*, like, similar to < Gr. *-oeídes*, similar to): *Lichens*. A type of ascoma, usually apothecioid but at times with a poroid opening, often elongated and branched; the wall of the ascoma can be little or well developed. It is present in species with varied thalli, from crustaceous to poorly developed or absent, with green photobionts; lichenicolous or saprobic; with pycnidial anamorphs, as *Arthonia cyanea* (Arthoniales). Cf. **lecanorine** and **lecideine**.

Arthonioid ascomata of a thallus of *Arthonia cyanea*, x 20 (*MAH*).

arthric (Gr. *árthron*, joint, articulation + suf. *-ikos* > L. *-icus* > E. *-ic*, belonging to or relating to): a conidium or spore that results from the fragmentation and disarticulation of fertile hyphae with determinate growth. In this type of thallic conidial development two types are distinguished: **holoarthric** and **enteroarthric**.

arthro-, **arthr-** (Gr. *árthron*, articulation): a pref. employed to form compounds that indicate articulation, the juncture of one member with another, as in *arthrospore*, each one of the spores that result from the disarticulation of a small chain of cells.

arthrocatenate (Gr. *árthron*, articulation + L. *catenatus*, in a chain < *catena*, chain + suf. *-atus* > E. *-ate*, provided with or likeness): applied to thalloconidia formed in chains by the simultaneous or random fragmentation of a hypha.

arthroconidium, pl. **arthroconidia** (Gr. *árthron*, articulation + *kónis*, dust, powder + dim. suf. *-ídion* > L. *-idium* > E. *-idium*): a type of holoarthric conidium that is characterized by the conversion of a preexisting hyphal element, of determinate growth, that detaches by disarticulation, as occurs in *Geotrichum* (moniliaceous asexual fungi). Syn. of **arthrospore**.

arthrospore (Gr. *árthron*, articulation + *sporá*, spore): *Conidiogenesis*. A spore that results from the fragmentation of a hypha; in current terminology of conidial ontogeny, arthrospores are called holoarthric conidia and they are present, e.g., in some moniliaceous asexual fungi, such as *Trichosporon beigelii* and *Geotrichum candidum* (the teleomorph of the last species is *Galactomyces* (=*Endomyces*) *geotrichum*, of the Saccharomycetales). Syn. of **arthroconidium**.

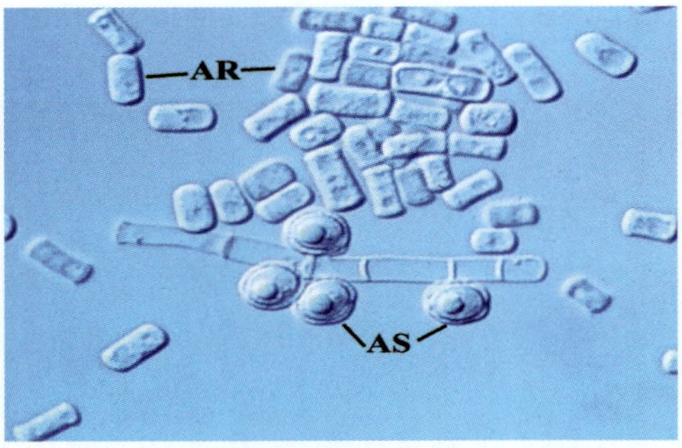

Arthrospores (AR) and ascospores (AS) of *Galactomyces* (=*Endomyces*) *geotrichum*, x 550 (*RTH*).

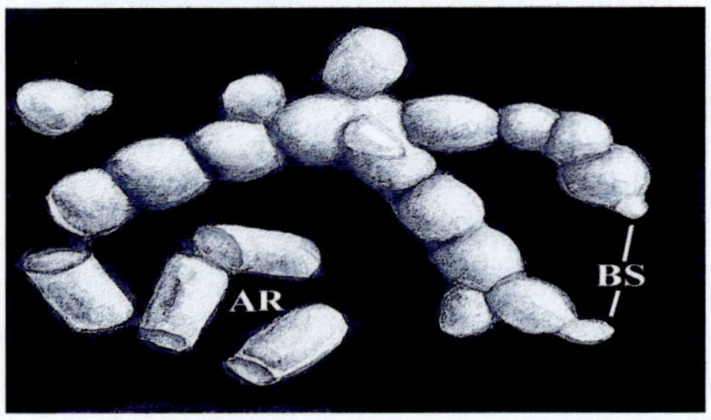

Arthrospores (AR) and blastospores (BS) of *Trichosporon beigelii*, x 1,450 (*MU*).

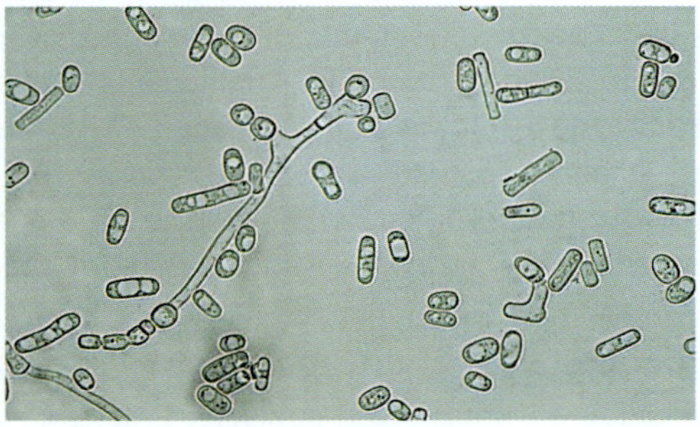

Arthrospores of *Geotrichum candidum*, x 450 (*MU*).

arthrosterigma, pl. **arthrosterigmata** (NL. *arthrosterigma* < Gr. *árthron*, articulation + *stérigma*, support): the septate sterigma or the multicellular filaments found in the interior of the pycnidia of some lichens that produce spermatia or pycnoconidia.

articulate (L. *articulatus*, divided into joints or nodes and internodes < *articulus*, dim. of *artus*, articulation, node + suf. *-atus* > E. *-ate*, provided with or likeness): with articulations; having segments, like the hyphae that have nodes and internodes.

ascendent, **ascendant** (ML. *ascendent,* that which goes up < *ascendens*, ascending, climbing up < *ascendere*, to climb up, ascend + suf. *-entem* > E. *-ent*, one that performs, being): directed or curved upward. E. g., gills that are arranged vertically on the pileus, rather than horizontally, and which remain separated from the stipe, as in *Agaricus* and other Agaricales. Conidiophores, pedicels, etc., are said to be ascendent when they first take a horizontal direction or nearly so, and then curve upward to an approximately vertical position. The sorocarp of *Polysphondylium* (Dictyosteliomycota) is ascendent, as during its development it raises up on its horizontal base (the pseudoplasmodium).

asciferous (L. *ascifer* < *asci*, pl. of *ascus*, sac, wineskin, ascus + suf. *-ferous*, bearer < *ferre*, to carry, bear): *Ascomycetes.* Bearer of asci; applied to the form of the fungus that produces asci (in contrast to the conidiophorous form, which produces conidia), or to the hyphal layer of the fruiting body that gives rise to the asci; the asciferous layer bears the asci, as in *Haematonectria* (=*Nectria*) *haematococca* (Hypocreales). Equal to **ascigerous**.

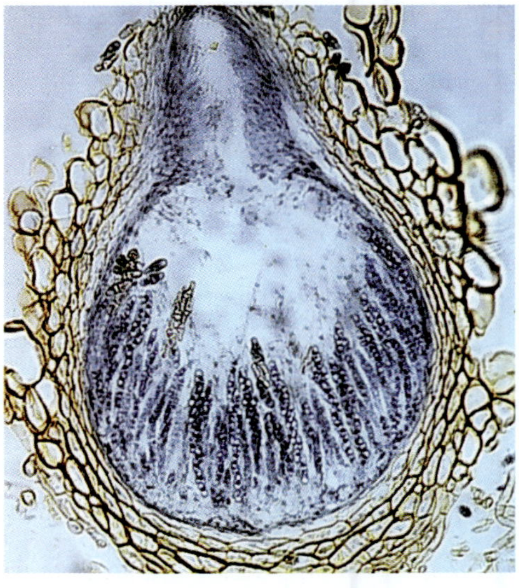

Asciferous hymenium of the perithecium of *Haematonectria* (=*Nectria*) *haematococca*, seen in longitudinal section, x 140 (*RTH*).

ascigerous (L. *asci*, pl. of *ascus*, sac, ascus + suf. *-gero*, to bear, carry): *Ascomycetes*. The structure or phase that produces and bears asci; the ascigerous state is the part of the life cycle in which the asci are produced, as in *Apiosporina morbosa* (=*Dibotryon morbosum*, Pleosporales). Equal to **asciferous**.

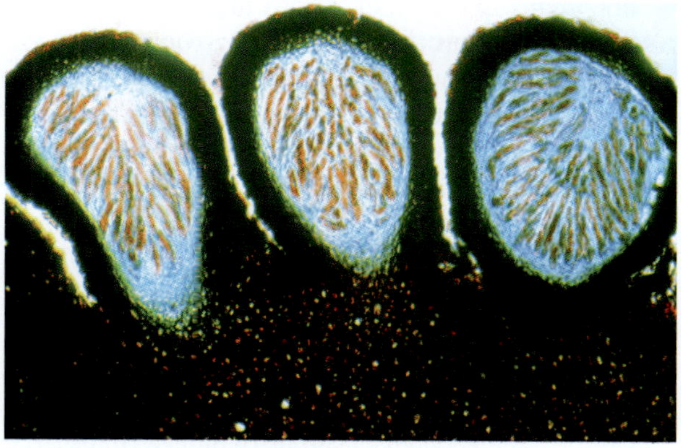

Ascigerous locules of *Apiosporina morbosa* (=*Dibotryon morbosum*), seen in longitudinal section, x 160 (*RTH*).

asco- (Gr. *askós*, sac, ascus): a pref. that refers to the ascus, as in *ascocarp*, a fruiting body with asci.

ascocarp (Gr. *askós*, sac, ascus + *karpós*, fruit): the sporiferous apparatus with asci and ascospores of the ascomycetes, whatever its form, e.g., the apothecia of *Hyaloscypha hyalina* (Helotiales). Also called **ascoma**.

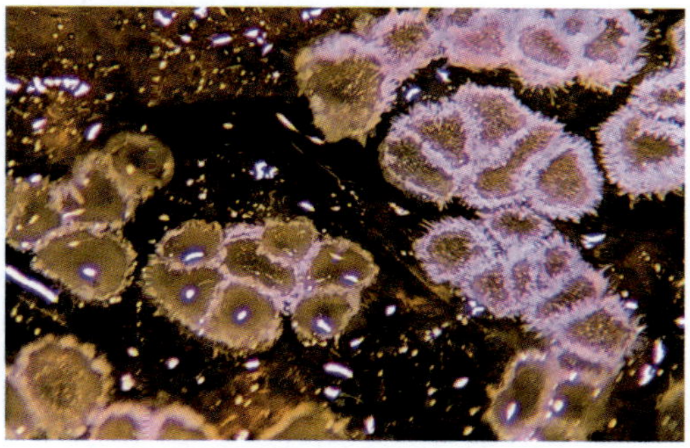

Ascocarps of *Hyaloscypha hyalina*, x 12 (*CB*).

ascoconidiophore (Gr. *askós*, sac, ascus + NL. *conidium* < Gr. *konídion* < *kónis*, dust, powder + dim. suf. *-ídion* > L. *-idium*): applied to the phialide of *Ascoconidium* (the anamorph of *Sageria*, of the Helotiales), bearer of an ascoconidium.

ascoconidium, pl. **ascoconidia** (Gr. *askós*, sac, ascus + NL. *conidium* < Gr. *konídion* < *kónis*, dust, powder + dim. suf. *-ídion* > L. *-idium*): a blastic, round or ovoid conidium produced by budding of an ascospore, as occurs, e.g., in *Taphrina deformans* (Taphrinales), whose ascospores may bud before being released from the asci and continue budding after they are released onto the substrate. In other ascomycetes, such as *Claviceps purpurea* (Hypocreales), ascospore germination occurs not by formation of a germ tube, but by the production of conidia directly, without an intervening mycelial state. This process has been termed iterative germination or **microcyclic conidiation**.

ascogenous (Gr. *askós*, sac, ascus + *génos*, origin, birth): *Ascomycetes*. Applied to the specialized structures that produce asci, like those, e.g., of *Protomyces* (Taphrinales). Ascogenous hyphae generally terminate in the shape of a hook (the *uncinulum* or *crozier*), as is seen, e.g., in *Neurospora*, *Thielavia* (Sordariales) and *Nectria* (Hypocreales). Ascogenous hyphae are homologous with the basidiogenous hyphae of the basidiomycetes.

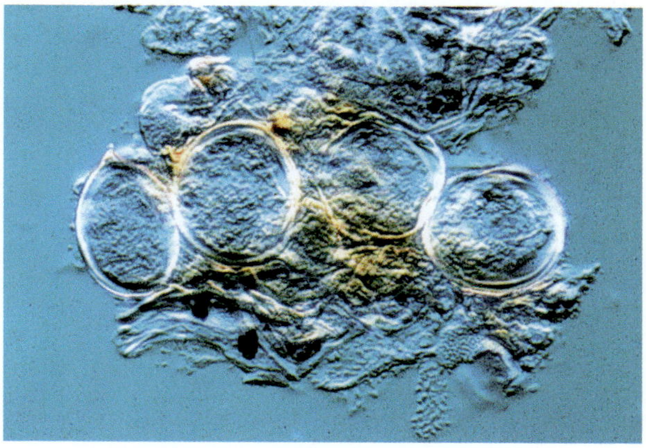

Ascogenous cells of *Protomyces* sp., x 1,000 (*RTH*).

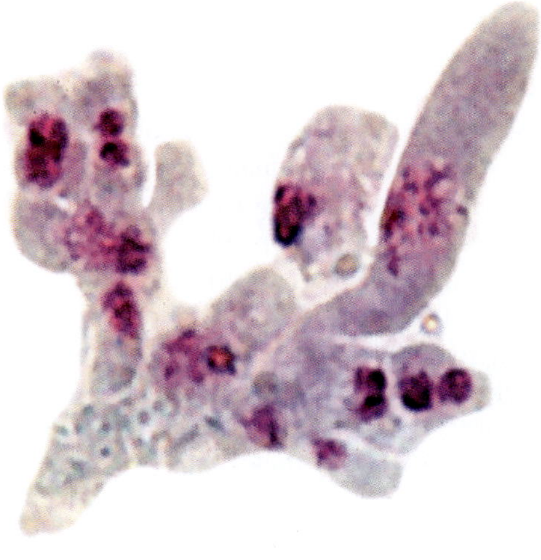

Ascogenous hyphae of *Haematonectria* (=*Nectria*) *haematococca*, x 1,200 (*RTH*).

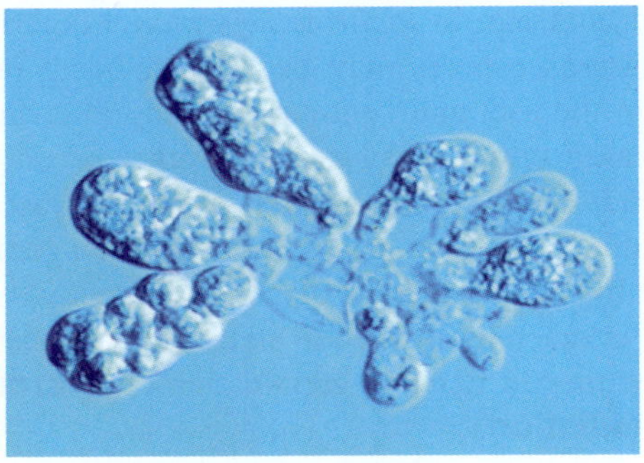

Ascogenous hyphae of *Thielavia terricola*, x 1,200 (*RTH*).

ascogonium, pl. **ascogonia**, **ascogone** (Gr. *askós*, sac, ascus + *gónos*, *góne*, the engendered, progeny, descendents + L. dim. suf. *-ium* > E. *-ium*): *Ascomycetes*. **1**. The cell or cells in an ascocarp from which the asci are ultimately derived. **2**. The female gametangium, in those ascomycetes that form female and male (antheridium) gametangia. Ascogonia may be of various shapes, depending upon the species, but they are frequently coiled, as in *Eurotium* (Eurotiales), or clavate, as in *Pyronema* (Pezizales).

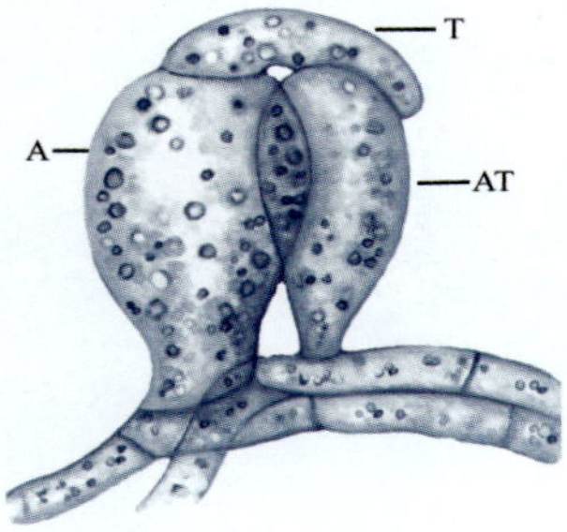

Ascogonium (A), antheridium (AT), and trichogyne (T) of *Pyronema omphalodes*, x 800 (*MU*).

ascohymenial (Gr. *askós*, sac, ascus + L. *hymenium* < Gr. *hyménion*, dim. of *hymén*, membrane + L. suf. *-alis* > E. *-al*, relating to or belonging to): *Ascocarp ontogeny.* A type of ascocarp development in which ascocarp formation originates from special, usually coiled, hyphal branches (the ascogonial initial), which are usually enveloped by the hyphae that will form the wall of the ascocarp. An antheridium may or may not be present. This type of development is characteristic of the Plectomycetes, Pyrenomycetes and Discomycetes. Cf. **ascolocular**.

ascolocular, **ascoloculate** (Gr. *askós*, sac, ascus + L. *loculus*, dim. of *locus*, place, cavity, locule + L. suf. *-aris* > E. *-ar*, like, pertaining to): *Ascocarp ontogeny*. A type of ascocarp development in which ascocarp formation is initiated as a small stroma, in which the ascogenous cells are later differentiated. The asci develop in cavities or locules whose wall consists of stromal tissue. This type of development is characteristic of the Loculoascomycetes. Cf. **ascohymenial**.

ascoma, pl. **ascomata** (Gr. *askóma*, leather cushion < *askós*, sac, ascus + suf. *-oma*, which implies the idea of entirety): *Ascomycetes*. Any sporocarp or fruiting body with asci. Syn. of **ascocarp**.

ascomycete, **ascomycote** (L. *ascomycete* < Gr. *askós*, sac, ascus + *mýkes*, genit. *mýketos*, fungus): a member of the Ascomycota, the largest group of fungi, considered as a phylum of the kingdom Fungi in various books, e.g., *Five Kingdoms*, 2nd. ed., by Margulis and Schwartz, 1988, and *Introductory Mycology*, 4th. ed., by Alexopoulos *et al.*, 1996; in this dictionary the system of classification of the latter is followed.

ascomycetes (L. *ascomycetes* < Gr. *askós*, sac, ascus + L. *-mycetes*, ending of taxonomic class < Gr. *mýkes*, genit. *mýketos*, fungus): a group of fungi characterized by the formation of asci. Once considered as a single taxonomic class of higher fungi (Ascomycetes), in the classification system followed in this dictionary, it is equivalent to the phylum Ascomycota (of the kingdom Fungi), which includes the classes Archiascomycetes, Plectomycetes, Pyrenomycetes, Discomycetes and Loculoascomycetes, as well as the ascosporogenous yeasts (Saccharomycetales) and other filamentous ascomycetes, such as the Erysiphales, Laboulbeniales and Spathulosporales. *Ascomycetes* is used here as a general name and not as a taxonomic class, for which reason it is written in lower case; the same applies to the word *basidiomycetes*. See the Appendix to compare the classification changes of this phylum in Alexopoulos *et al.*, 1996, modified by Ulloa and Hanlin, 2006, to that of Kirk *et al.*, 2008. In the latter classification scheme, the orders of ascomycetes have been transferred to different subphyla, classes and subclasses. These fungi are made up of heterogeneous species, forming unnatural groupings, and therefore are considered an artificial classification. These taxonomic rearrangements, with the concurrent changes of the names of taxa, are in a constant process of study and evaluation, in order to achieve a phylogenetic classification of the fungi.

ascophore (Gr. *askós*, sac, ascus + suf. *-phóros*, bearer): *Ascomycetes*. A structure that produces or supports asci; in particular, it refers to the erect, specialized hypha of *Cephaloascus* and *Ascoidea* (Saccharomycetales), which form asci at their apex.

ascoplasma (Gr. *askós*, sac, ascus + *plásma*, formation, bland material with which a living being is formed): syn. of **ectoplasm**.

ascospore (Gr. *askós*, sac, ascus + *sporá*, spore): *Ascomycetes*. A sexual haploid spore that is formed in the interior of an ascus, usually as the result of meiosis; characteristic of the ascomycetous fungi. See the figures that represent some examples of asci with ascospores.

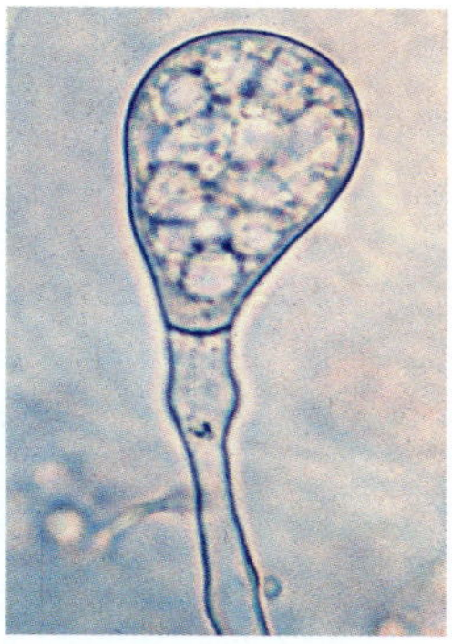

Ascophore of *Ascoidea* sp.; it represents the hypha that supports a young, terminal, multispored ascus, x 570 (*RTH*).

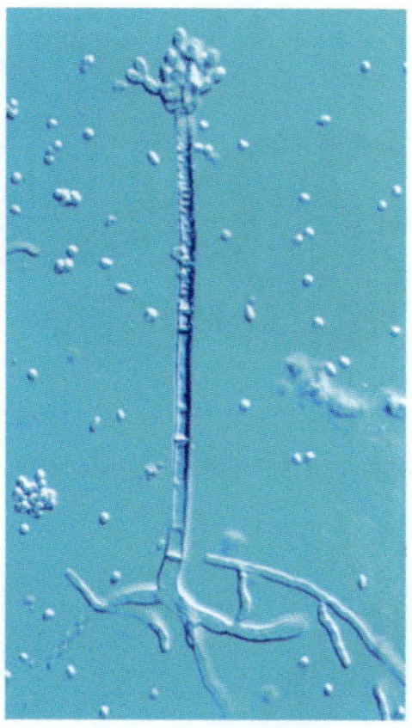

Ascophore with asci of *Cephaloascus fragrans*, x 570 (*RTH*).

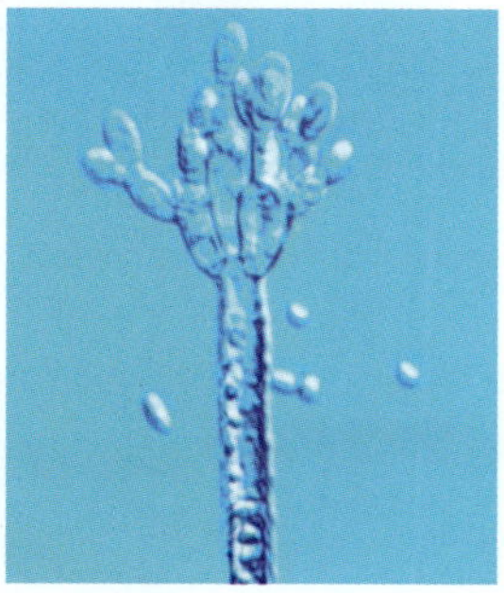

Ascophore with asci of *Cephaloascus fragrans*, x 1,000 (*RTH*).

ascosporic (Gr. *askós*, sac, ascus + *sporá*, spore + suf. *-íkos* > L. *-icus* > E. *-ic*, belonging to or relating to): syn. of **ascosporogenous**. Cf. **anascosporic**.

ascosporogenous (Gr. *askós*, sac, ascus + *sporá*, spore + *génos*, origin < *gennáo*, to engender, produce): capable of producing ascospores. Cf. **anascosporogenous**.

ascostome (Gr. *askós*, sac, ascus + *stóma*, mouth): the pore in the apex of an ascus.

ascostroma, pl. **ascostromata** (Gr. *askós*, sac, ascus + *strõma*, cushion): *Ascomycetes*. **1**. A stromatic ascocarp with ascolocular development, with the asci produced in locules or cavities, whose wall consists only of stromatic tissue, i.e., without forming a separable wall around the locules. If the ascostroma has a single cavity, it is **unilocular**; if it has several cavities, it is **multilocular**. The ascostroma is characteristic of the class Loculoascomycetes, such as *Preussia* (=*Sporormiella*) (Pleosporales) and *Myriangium* (Myriangiales). **2**. Any stroma containing asci. For species with ascohymenial ontogeny it is preferable to use the term *perithecial stroma*.

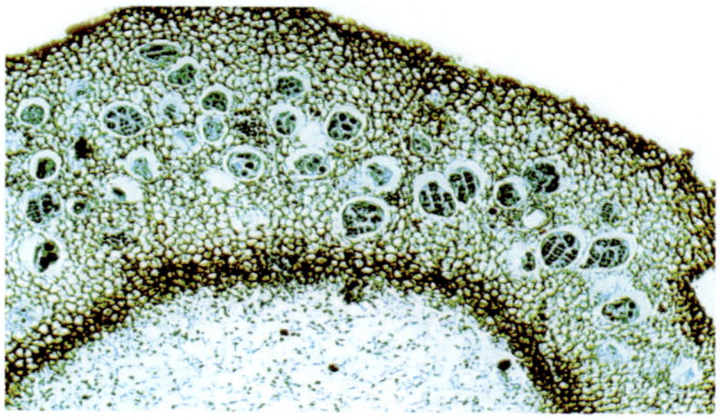

Ascostroma (multilocular) of *Myriangium duriaei*, seen in vertical section, x 100 (*RTH*).

ascus, pl. **asci** (Gr. *askós*, sack, ascus): *Ascomycetes*. A specialized cell in which the ascospores are formed internally by a process called free-cell formation, usually following karyogamy and meiosis. The ascus is the definitive character of the ascomycetes; it generally contains a definite number of ascospores, typically eight. Some representative examples of the fungi with asci are *Saccharomyces*, *Dipodascopsis* (Saccharomycetales), *Octosporomyces* (=*Schizosaccharomyces,* Schizosaccharomycetales), *Emericella* (Eurotiales), *Podospora*, *Sordaria* (Sordariales), *Glomerella* (Diaporthales) and *Morchella* (Pezizales). Asci are homologous to the basidia of the basidiomycetes, but they differ in that they form their spores endogenously, whereas basidia form their spores exogenously. Asci vary greatly in structure. Initially they were separated into three broad groups: *prototunicate asci*, with deliquescent walls; *unitunicate asci*, with a single functional wall layer; and *bitunicate asci*, with two functional wall layers. Further studies, however, showed this to be an oversimplification. Sherwood, e.g., recognized nine different types of asci that were distinguishable with light microscopy, and Eriksson distinguished seven types of dehiscence among the bitunicate asci. Even within these types there is further variation; Bellemere recognized three predehiscence and 11 dehiscence categories among the asci. These variations involve the behavior of the ascus at the time of ascospore discharge. E.g., among unitunicate asci the ascospores may be discharged through a preformed apical pore, or through the opening left when an apical cap (operculum) dehisces, or the ascus apex may rupture in various ways to release the spores. In bitunicate asci the outer, rigid wall layer (ectotunica) may rupture, allowing the inner, elastic wall layer (endotunica) to extend through the opening, carrying the ascospores along to be forcibly shot off. In other bitunicate forms, however, the mode of ascospore discharge may be quite different. Additional differences among asci involve the nature of the ascus apex, whether it is differentiated or not, and if so, the nature of the apical structures (such as pores, rings, swellings, etc.) that are present. Numerous terms have been developed to describe these structures. The reaction of the ascus apex to different stains, especially iodine, is an important taxonomic character in identifying ascomycetes, particularly among lichenized forms. It should be noted that most information on ascospore discharge has been obtained under the artificial conditions of observing them in water mounts with light microscopy. Due to technical difficulties, ascospore discharge under natural conditions has been observed in relatively few species.

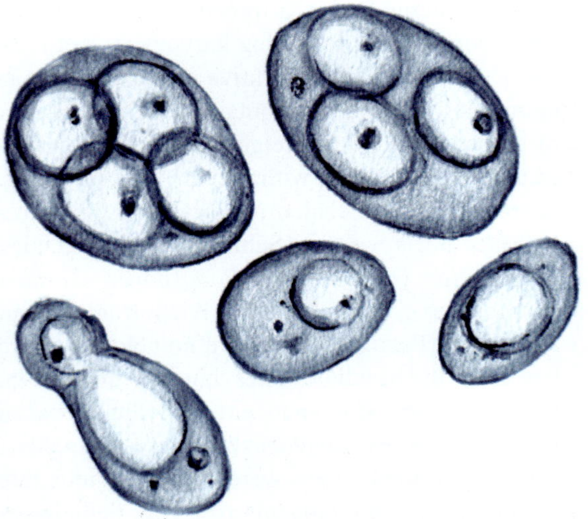

Asci (tri- and tetrasporate) of *Saccharomyces cerevisiae*, x 2,000 (*MU*).

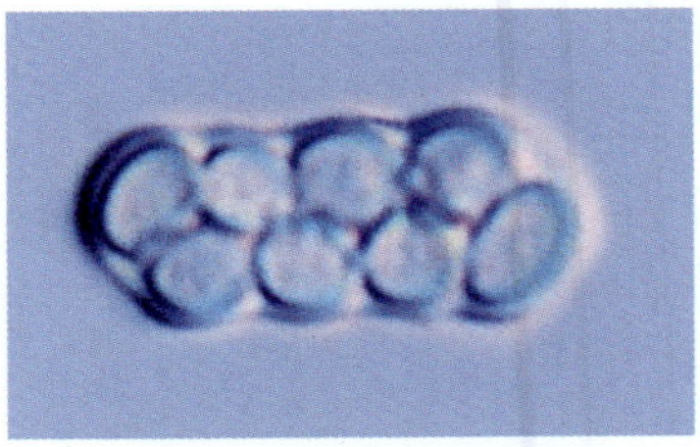

Ascus (octosporate) of *Octosporomyces* (=*Schizosaccharomyces*) *octosporus*, x 1,250 (*MU*).

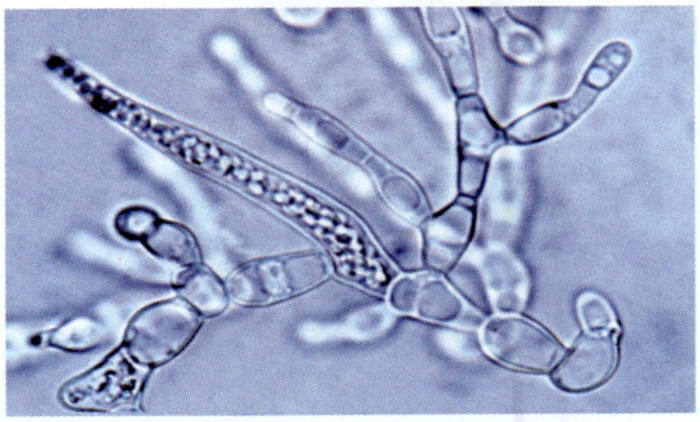

Ascus (multisporate) of *Dipodascopsis uninucleata*, x 460 (*MU*).

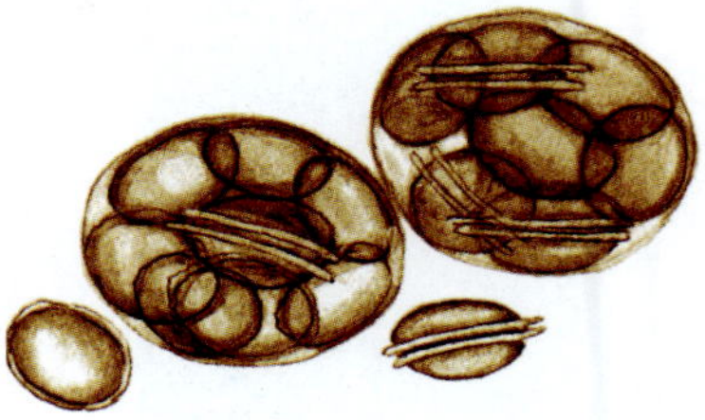

Asci (octosporate) of *Emericella nidulans*, x 3,500 (*MU*).

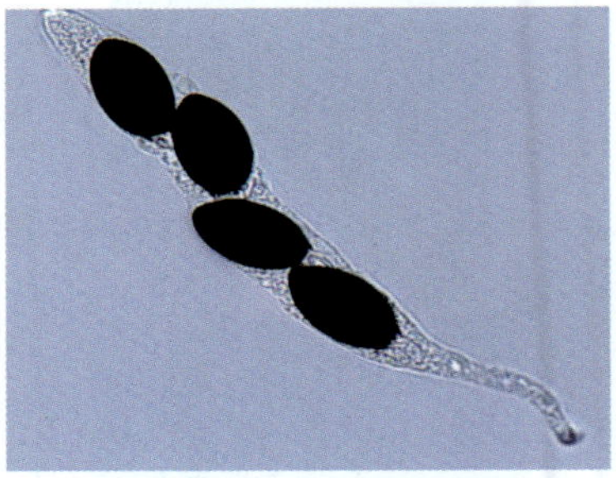

Ascus (tetrasporate) of *Podospora comata*, x 230 (*EAA*).

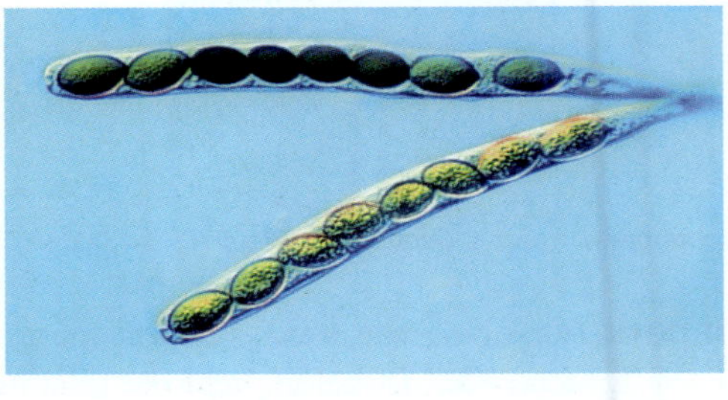

Asci (octosporate) of *Sordaria fimicola*, x 340 (*RTH*).

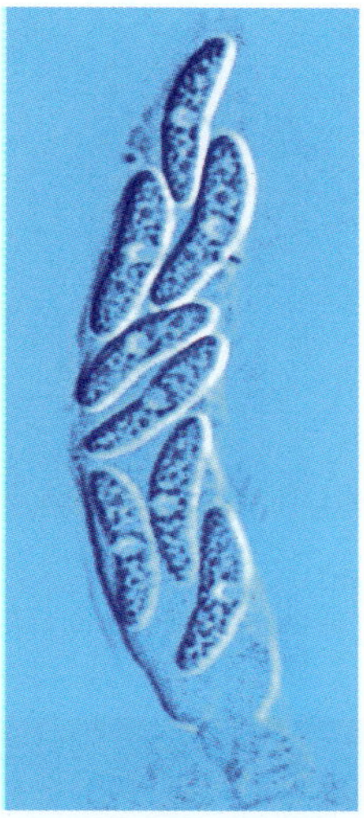

Ascus (octosporate) of *Glomerella* sp., x 1,120 (*RTH*).

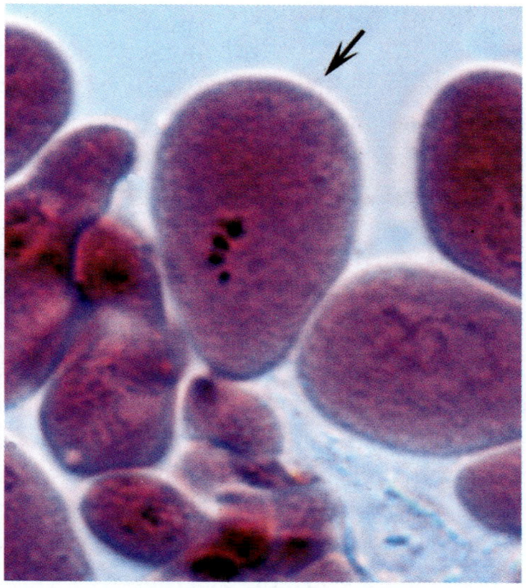

Ascus mother cells of *Thielavia terricola*, x 1,700 (*RTH*).

ascus mother cell (Gr. *askós*, sac; ME. *mother*, *moder* < OE. *modor* < L. *mater*, genit. *matris* < Gr. *máter*, mother; L. *cella*, storeroom > ML. *cella* > OE. *cell*, cell, compartment): generally applied to the cells of yeasts that by budding give rise to one or several buds that constitute daughter cells. In the ascomycetes the diploid cell of the hook is called the ascus mother cell, which by meiosis forms the haploid nuclei of the future ascospores, as one sees, e.g., in *Haematonectria* (=*Nectria*) *haematococca* (Hypocreales) and *Thielavia terricola* (Melanosporales).

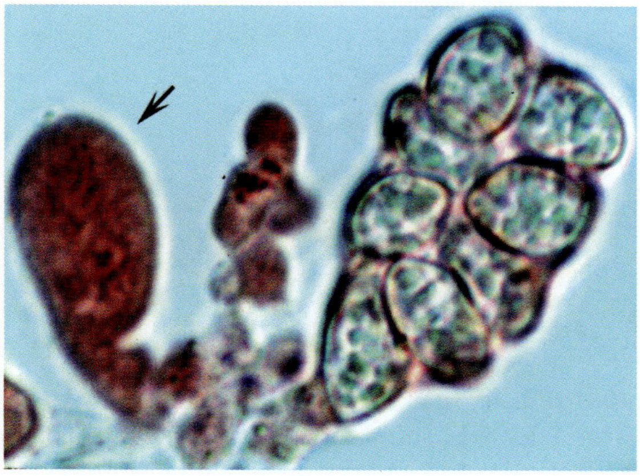

Ascus mother cells of *Thielavia terricola*, x 1,700 (*RTH*).

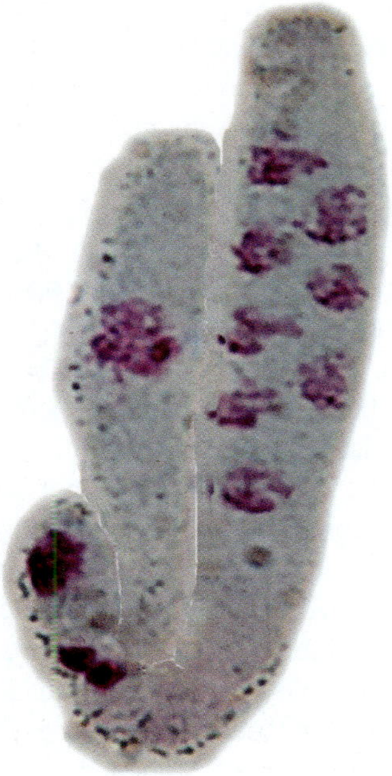

Ascus mother cell, with a single diploid nucleus, and young ascus, with eight haploid nuclei, of *Haematonectria* (=*Nectria*) *haematococca*, x 1,840 (*RTH*).

aseptate (L. *aseptatus* < Gr. pref. *a-*, not, without + L. *septatus*, septate, divided < *septum*, septum, wall, partition + suf. *-atus* > E. *-ate*, provided with or likeness): without partitions (transverse walls or septa). See **cenocytic**. Cf. **septate**.

aseptic, **asepticous** (Gr. pref. *a-*, not, without + *septikós*, causing putrefaction < *sépsis*, decomposition + suf. *-tikós* > L. *-ticus* > E. *-ticous*, relation, fitness, inclination or capability): not septic; without contamination. Cf. **septic**.

asexual (Gr. pref. *a-*, not, without + L. *sexus*, sex + suf. *-alis* > E. *-al*, relating to or pertaining to): a type of reproduction that does not involve karyogamy and meiosis; vegetative multiplication. Apomixis (if it lacks fertilization or amphimixis, although sexual organs or cells are present) and amixis (the primitive type, without organs or sexual cells, and without amphimixis) also are asexual. Cf. **sexual**.

asexual fungus, pl. **fungi** (Gr. *a-*, without + L. *sexus*, sex + suf. *-alis* > E. *-al*, relating to or belonging to; L. *fungus*, fungus < Gr. *spóngos*, *sphóngos*, sponge): a fungus that lacks a means of sexual reproduction. Asexual fungi are also called **anaholomorphs**, **conidial fungi**, **deuteromycetes**, **Fungi Imperfecti** and **mitosporic fungi**.

asperate, **asperous**, **asperulous**, **asperulate** (L. *asperum*, *asperulus*, dim. of *asper*, rough or uneven + suf. *-atus* > E. *-ate*, provided with or likeness; or + suf. *-osus* > E. *-ous*, having, possessing the qualities of): with a lightly roughened or uneven surface due to the presence of small spines, warts or other irregularities; generally applied to spores, conidiophore walls, etc., as seen under the microscope. E.g., the spores of *Hemitrichia calyculata* (Trichiales), the basidiospores of *Russula* spp. and *Lepiota asperula* (Agaricales), and of *Clavaria* (Aphyllophorales), and the surface of the spore sac of *Lycoperdon pulcherrimum* (Lycoperdales) are all asperulate.

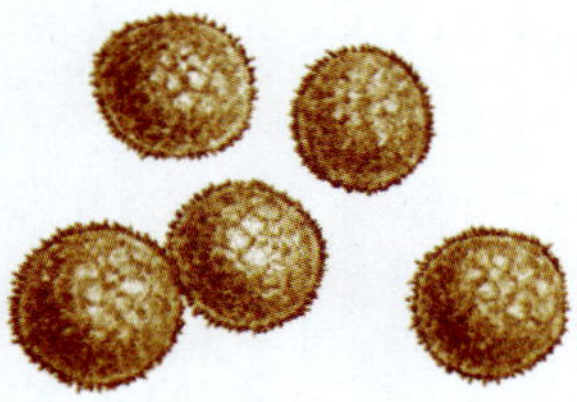

Asperate spores of *Hemitrichia calyculata*, x 1,800 (*MU*).

aspergilloid (< gen. *Aspergillus* < L. *aspergillo*, hyssop, sprinkler + L. suf. *-oide*, like < Gr. *-oeídes*, similar to): applied to a sporiferous structure that resembles the conidiophore of *Aspergillus* (moniliaceous asexual fungi), as is seen in the conidiophore of *Penicillium aspergilloides* (also of the moniliaceous asexual fungi).

aspergillosis, pl. **aspergilloses** (< gen. *Aspergillus* < L. *aspergillo*, hyssop, sprinkler + Gr. suf. *-osis*, state or condition): a wide range of animal and human diseases caused by several cosmopolitan species of *Aspergillus* (moniliaceous asexual fungi), particularly *A. fumigatus*, *A. flavus*, and *A. niger*, which, as opportunistic invaders, provoke secondary infections in immunosuppressed patients (e.g., those weakened by cancer and its chemotherapy). The symptoms of pulmonary aspergillosis can be acute (bronchopneumonia) or chronic (tuberculoid); other types are the so-called *fungus ball* (localized in preexisting lung cavities), allergic aspergillosis, onychomycosis, otomycosis and the systemic infections that affect various organs of the body (heart, brain, nasal sinuses, etc.). See **mycosis**.

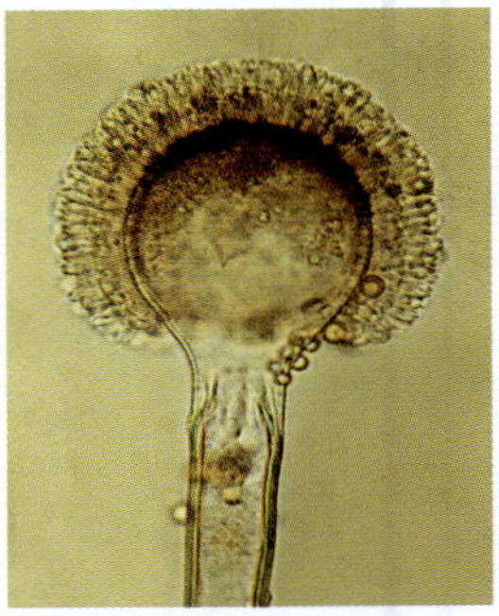

Conidial head of *Aspergillus flavus*, a causal agent of **aspergillosis**, x 350 (*MU*).

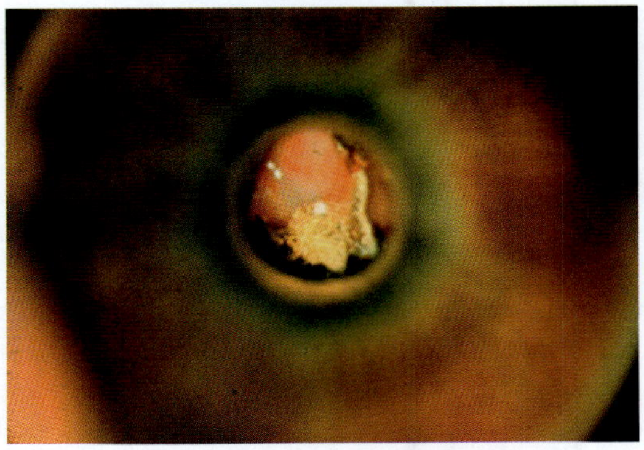

Aspergillosis of the ear. With the otoscope hyphae of *Aspergillus flavus* are observed adhered to the eardrum and to the walls of the external auditory canal, x 10 (*RLM*).

aspicilioid (< gen. *Aspicilia* < Gr. *aspís*, genit. *aspídos*, shield + L. suf. *-oide*, like < Gr. *oeídes*, similar to): *Lichens* (fam. Hymeneliaceae). Refers to lecanorine apothecia that are more or less immersed in the thallus, at least when young. Cf. **lecanorine** and **lecideine**.

asporogenous (L. *asporogenous* < Gr. pref. *a-*, not, without + *sporá*, spore + *génos*, origin, birth): not producing spores, i.e., sterile. Cf. **sporogenous**.

assimilative (ML. *assimilativus* < L. *assimilatus*, likened to, made like + suf. *-ivus* > E. *-ive*, quality or tendency, fitness): to take in and incorporate, absorb and assimilate. See **vegetative**.

astatocoenocytic (Gr. *astatós*, not fixed, unstable + *koinós*, common + *kýtos*, cell, hollow + suf. *-íkos* > L. *-icus* > E. *-ic*, belonging to or relating to): *Basidiomycetes*. Refers to the nuclear behavior of the mycelium as it passes through different phases; the cells of the haploid mycelium are uninucleate and coenocytic, those of the diploid mycelium are binucleate, coenocytic and lack clamp connections when aeration is insufficient, and those of the mycelium of the basidioma are binucleate. Cf. **holocoenocytic**.

asterinoid (< gen. *Asterina* < L. *aster*, star + suf. *-ina*, which indicates possession or appearance + L. suf. *-oide*, like < Gr. *-oeídes*, similar to): a type of ectoparasitism in which the fungus develops a black, superficial mycelium in a radial arrangement on the parts of the living host plant from which it obtains its nutrients by means of haustoria and later produces a uniloculate ascostroma with a radial structure. Asterinoid forms are present in *Asterina veronicae* (Asterinaceae), the Microthyriaceae and related families of Loculoascomycetes that occur in humid climates.

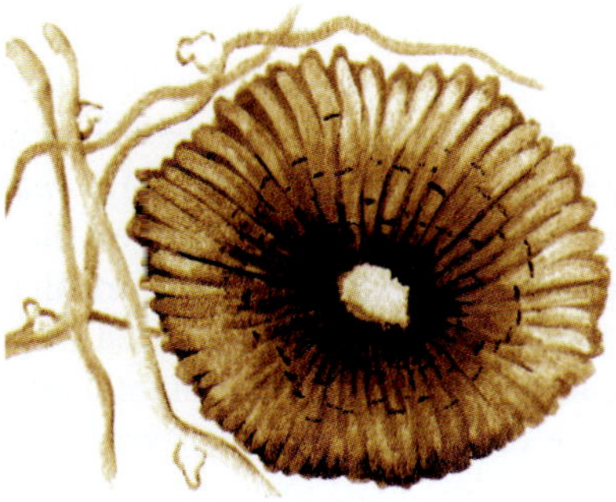

Asterinoid ascostroma of *Asterina veronicae*. Surface view of an ascostroma parasitic on a leaf of *Veronica*, x 500 (*MU*).

asterophysis, pl. **asterophyses** (Gr. *astér*, star + *phýsis*, a growth): a stellate, thick-walled, pigmented, cystidioid hypha with radiate branches that is characteristic of the trama and hymenium of some gen. of Aphyllophorales, such as *Asterodon* and *Asterostroma* (Asterostromataceae). Also called **asteroseta**.

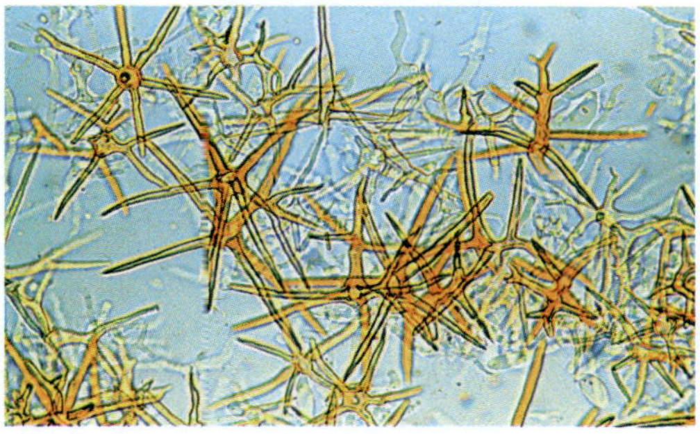

Asterophyses of the hymenophore of *Asterostroma* sp., x 130 (*MU*).

asteroseta, pl. **asterosetae** (Gr. *astér*, star + L. *seta*, bristle, stiff hair): a cellular, star-shaped element with a thick, pigmented wall that is present in the cuticle and hymenophore of certain Aphyllophorales, as in *Asterostroma cervicolor*. Also called **asterophysis**.

astomous, **astomate** (Gr. *ástomos*, without a mouth or opening, or + L. suf. *-atus* > E. *-ate*, provided with or likeness): applied to structures lacking an ostiole. Formerly referred to as closed ascomata (cleistothecia) of the Erysiphales, the term *chasmothecium* has been proposed for this type of ascoma, as on maturing it opens by a vertical slit that permits ascospore release.

astrogastraceous (Gr. *ástron*, heavenly body, star, generally employed alluding to the star-shape of an organ or a structure + *gastér*, stomach, used when one needs to introduce the idea of something that is contained in an organ or closed apparatus + L. suf. *-aceus* > E. *-aceous*, of or pertaining to, with the nature of): refers to the gasteroid members of some Russulaceae.

asymmetric, **asymmetrical** (Gr. *asymmetría*, lack of symmetry < pref. *a-*, not, without + *symmetría*, symmetry + suf. *-íkos* > L. *-icus* > E. *-ic*, pertaining to or relating to; or + L. suf. *-alis* > E. *-al*, relating to or belonging to): an organ or structure that is not identical on both sides of a common axis, so that neither side can form an exact mirror image of the other. E.g., the ascospores of *Xylaria*, *Hypoxylon* and *Annulohypoxylon* (Xylariales) are asymmetrical, since one side is convex and the other flat (see **inaequilateral**). Also, some bicellular ascospores with guttulae in only one cell are asymmetric. Cf. **symmetrical**.

Asymmetric ascospores of *Nemania* (=*Hypoxylon*) *serpens*, x 1,100 (*MU*).

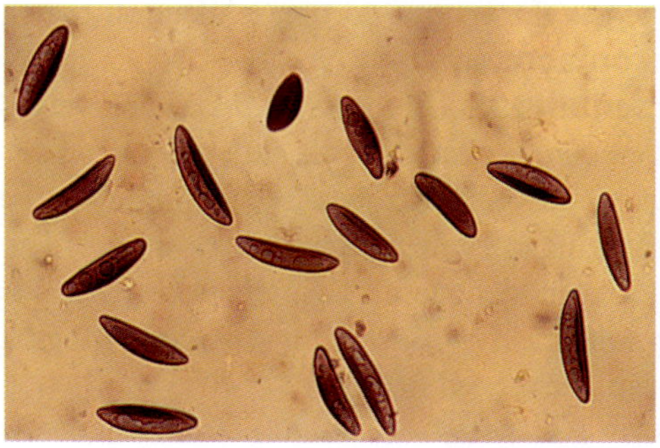

Asymmetric ascospores of *Annulohypoxylon* (=*Hypoxylon*) *thouarsianum*, x 1,000 (*MU*).

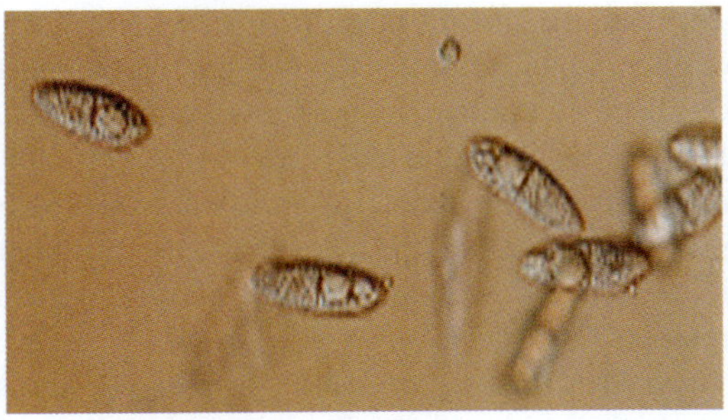

Asymmetric bicellular ascospores of *Halosarpheia heteroguttulata* (Microascales), x 1,000 (*ERG* & *MCG*).

-ate (L. *-atus*, provided with or likeness): a suf. that is added to nouns to form adjectives that signify provided with or with the aspect of, e.g., *pileate* (with a pileus or having the shape of a pileus), *turbinate* (with the shape of a top), etc.

atomate (L. *atomatus*, sprinkled with minute particles < Gr. *átomos*, not divisible + L. suf. *-atus* > E. *-ate*, provided with or likeness): refers to the surface of a pileus or stipe of a basidiocarp that is covered with tiny, sharp, brilliant particles, as occurs in *Psathyrella atomata* (Agaricales).

atopic (Gr. *atópos*, out of place, uncommon < pref. *a-*, not, without + *tópos*, place + suf. *-ikos* > L. *-icus* > E. *-ic*, belonging to or relating to): predisposed to certain conditions or materials, as in some allergic reactions; an atopic organism has an allergy, probably hereditary, characterized by symptoms (such as asthma, hay fever and hives) produced by exposure to the exciting antigen, such as fungal spores, without having had an inoculation. See **aeromycology**.

atractosome (Gr. *átraktos*, spindle, shaft + *sõma*, body): a globose structure present in the organelle-free zone of the septal pore apparatus in some members of Atractiellales (e.g., *Atractiella*). Atractosomes form from cisternae of the endoplasmic reticulum when the edge of a cisterna curves around to form a small globule that becomes free. The periphery of the globule is covered by electron-opaque material, whereas the central portion is electron translucent. Cf. **colacosome**.

atramentarious (L. *atramentarius*, inkwell < NL. *atramentum*, black beard, dark-colored ink < *atra*, black + *mentum*, chin, beard + suf. *-arius* > E. *-arious*, belonging to): with the appearance of a beard or black ink, like the deliquescent hymenophore of *Coprinopsis atramentaria* (=*Coprinus atramentarius*, Agaricales) and other related fungi.

atrophy (Gr. *atrophía*, lack of nutriment): decrease in size or attrition of a tissue or organ; suspension of the development, decline or loss of a part or organ due to the lack of nutrition, diminution in activity due to old age, or as a result of a parasitosis. When a cell, a tissue or an organ suffer atrophy they are said to be atrophied. Atrophied is often confused with **aborted**. Cf. **distrophy** and **hypertrophy**.

attenuate (L. *attenuatus*, tapered < *attenuare*, to taper < pref. *ad-*, toward + *tenuis*, thin + suf. *-atus* > E. *-ate*, provided with or likeness): **1**. Tapering gradually toward the apex; e.g., certain perithecial necks, phialides and many other structures. **2**. Applied to pathogens that have lost their virulence, or that exhibit reduced or diminshed pathogenicity.

aulaeate development (Gr. *aulaós*, fissure, furrow, to extend by narrowing + suf. *-atus* > E. *-ate*, provided with or likeness; MF. *développer* < OF. *desveloper* < *dés-* < L. *de-*, from, down, away + MF. *enveloper*, to enclose + L. *mentum* > OF. *-ment* > ME. *-ment*, action, process): *Gasteromycetes*. A type of fruiting body development characterized by a hymenium formed on the surface of furrows comprised of primordial tissue that grows from the peridium downward and toward the center of the glebal chamber, as in the Hymenogastrales and Lycoperdales. Cf. **forate development**, **homogeneous development**, **lacunar development** and **pileate development**. See **aulaeothecium**.

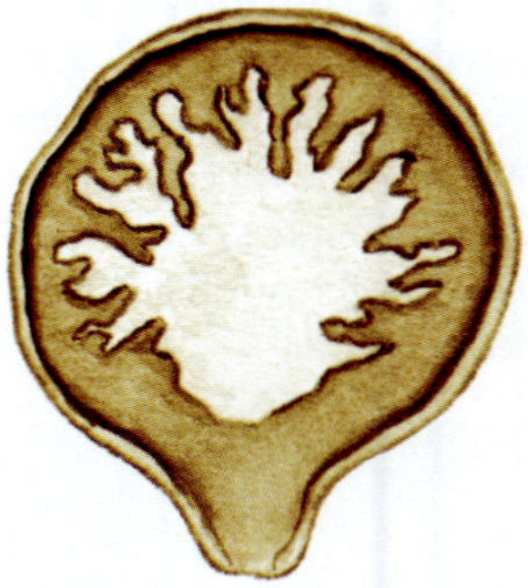

Aulaeate type of **development** of the hymenium in the fruiting body of Hymenogastrales (*Gastrosporium*) and Lycoperdales. Diagram of a longitudinal section, x 1 (*MU*).

aulaeothecium, pl. **aulaeothecia** (NL. *aulaeothecium* < Gr. *aulaós*, fissure, furrow + *thekíon*, dim. of *thêke*, box; here, of the basidia + dim. suf. *-íon* > L. *-ium* > E. *-ium*): a type of growth in which the primordial tissue grows downward and toward the center of the glebal cavity during *auleate development*.

auriculiform (L. *auricula*, dim. of *auris*, ear + suf. *-formis* < *forma*, shape): having the shape of an ear, like the ascocarps of *Flavoscypha cantharella*, *Otidea onotica* and *O. leporina* (Pezizales) and the basidiocarp of *Auricularia* (Auriculariales).

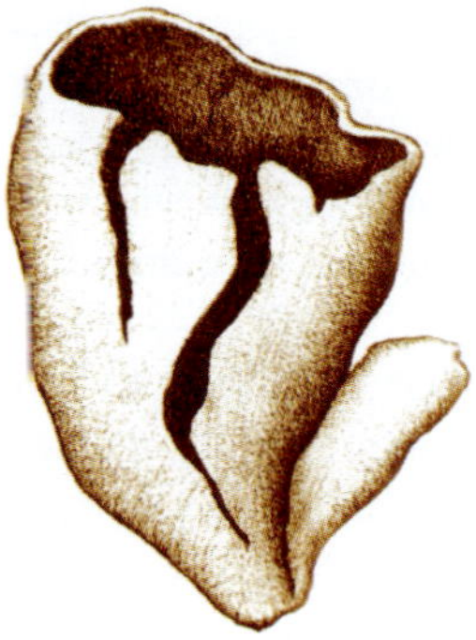

Auriculiform ascocarp of *Otidea onotica*, x 1.5 (*MU*).

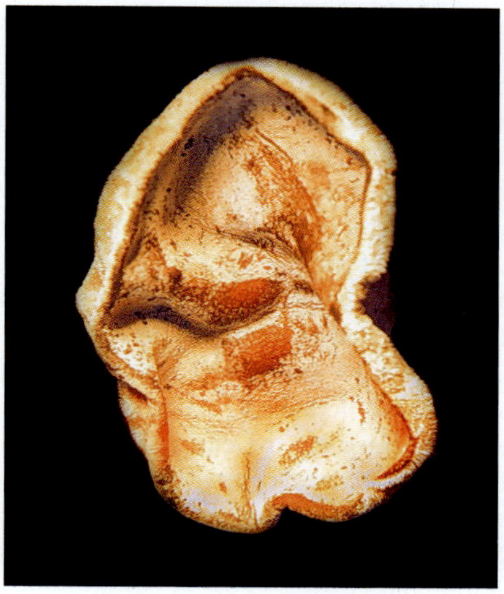

Auriculiform basidiocarp of *Auricularia polytricha*, x 1 (*MU*).

Auriculiform ascocarps of *Otidea onotica*, x 1 (*EPS*).

Auriculiform ascocarps of *Otidea leporina*, x 1 (*MRO*).

Auriculiform basidiocarp of *Auricularia delicata*, x 1 (*RV*).

autapomorphy (Gr. *autós*, same, by itself, own + *apó*, away from + *morphé*, shape + E. suf. *-y*, which denotes quality): a term used in cladistics to refer to a derived, unique character state that is restricted to a single taxon. E.g., the Amoebidiales (Zygomycota) are unique among the organisms classified as Fungi in their autapomorphic production of amoeboid cells.

auteuform (Gr. *autós*, same, by itself, own + *eû*, true + L. *forma*, shape): *Uredinales*. Applied to the *euform* of a rust when it is autoecious, i.e., whose life cycle includes the four classes of spores (basidiospores, aeciospores, urediniospores and teliospores) produced on the same host plant.

auto-, aut- (Gr. *autós*, same, by itself, own): a pref. that is employed with the meaning of "self" in various biological words, as in *autodeliquescent*, dissolving by itself.

autobasidium, pl. **autobasidia** (Gr. *autós*, same, by itself, own + *basídion*, small base): a basidium in which the spores are borne asymetrically (heterotropic) on the sterigmata and which are forcefully liberated, as occurs in the Agaricales. Cf. **apobasidium**.

autochthonous (Gr. *autóchthon* < *autós*, same, by itself, own + *thón*, genit. *thonós*, land, country): indigenous; from the country, not introduced or naturalized, like the "horse's tooth of maize" (*Claviceps gigantea*, of the Hypocreales), which is native to Mexico. Cf. **allochthonous** and **exotic**.

autodeliquescent (Gr. *autós*, same, by itself, own, + L. *deliquescens*, genit. *deliquescentis*, liquify, to become fluid on maturing < *deliquescere*, to become liquid + suf. *-escens*, genit. *-escentis* > E. *-escent*, that which turns, beginning to, slightly): becoming liquid on maturing by a process of autodigestion or *autolysis*, like the gills and pileus of *Coprinus* (Agaricales). Equal to **deliquescent**.

autoecious (Gr. *autós*, same, by itself, own + *oîkos*, house or dwelling): *Uredinales*. Rust fungi that develop their complete life cycle on hosts belonging to the same genus, or the same species, like, e.g., *Phragmidium mucronatum* (rose rust). The capacity of a fungus to complete its life cycle on single host plants is called *autoecism*. Cf. **heteroecious**.

autoecism (Gr. *autós*, same, by itself, own + *oîkos*, house, dwelling + suf. *-ismós* > L. *-ismus* > E. *-ism* > state, phase, tendency, action): *Uredinales*. The capacity to complete the entire life cycle on single host plants. Species of rusts exhibiting autoecism are called *autoecious*. Cf. **heteroecism**.

autogamous (Gr. *autós*, same + *gámos*, sexual union): exhibiting *autogamy*.

autogamy (Gr. *autós*, same, by itself, own + *gámos*, sexual union + E. suf. *-y*, which denotes quality): sexual union of two closely related nuclei in a cell of the female organ, without the occurrence of fusion with cells of the opposite sex; the union of two nuclei, of presumed different sexual value, derived from the division of a primitive nucleus in the interior of a cell. See **automixis** and **parthenogamy**.

autolysis, pl. **autolyses** (Gr. *autós*, same, by itself, own + *lýsis*, dissolution, disintegration): autodigestion, like that present in the gills of the basidiocarp of *Coprinus* (Agaricales), which causes the elimination of those parts of the gills that have already liberated their spores, permitting the pileus to continue opening and creating the spaces necessary for the spores to continue being detached from the basidia. This autolysis or autodeliquescence produces a black fluid mass that contributes to the dispersion of the spores. Also called *deliquescence.*

automixis, pl. **automixes** (Gr. *autós*, same, by itself, own + *míxis*, mixture, union): reproduction in which there is self-fertilization as a consequence of the fusion of two sexual cells of close parentage, or simply of two sexual nuclei. The first case is called **parthenogamy**, and the second **autogamy**. Automixis also is referred to as *automictic sexual reproduction* (Gr. *autós* + *míxis* + suf. *-tíkos* > L. *-ticus* > E. *-tic*, relation, fitness, inclination or ability).

autotroph (Gr. *autós*, same, by itself, own + *trophós*, that which nourishes, which serves as food): an organism which, endowed with chlorophyll or an analogous pigment, is capable of synthesizing carbohydrates from carbon anhydride, so that it does not need to acquire them already constituted, but which nourishes itself, without the direct assistance of others. Cf. **heterotroph** and **mixotrophic**.

autotropism (Gr. *autós*, same, by itself, own + *trópos*, a turn, change in direction < *tropé*, a turning < *trépo*, to revolve, turn towards + suf. *-ismós* > L. *-ismus* > E. *-ism*, state or condition): an evasive response between neighboring hyphae at the margins of fungal colonies. Cf. **aversion**.

auxanogram (Gr. *auxáno*, to grow, develop + *grámma*, describe): literally, a diagram of growth; it is applied to the culture of yeasts, bacteria and other microorganisms, principally on plates of solid agar media, when there is deposited a determined quantity of certain substances with the object of seeing its effect on the growth of the microorganism seeded. If the latter is stimulated around the substance added, the auxanogram is positive; if, on the contrary, development is inhibited or not stimulated, the auxanogram is negative. Auxanograms are utilized to determine the types of carbon or nitrogen sources that various species of yeasts can assimilate. Also called the auxanographic method. Cf. **zymogram**.

auxotroph (Gr. *aúxo*, to grow, develop + *trophós*, that which nourishes, that serves as food): refers to an organism which, in order to grow, requires an external source of certain amino acids or vitamins, i.e., it is not capable of synthesizing these compounds, as occurs with specific isolates which because of mutation differ from the wild type; e.g., there are isolates of *Schizophyllum commune* (Aphyllophorales) that are auxotrophic for tryptophane. See **prototroph**.

aversion (L. *aversus*, genit. *aversionis*, withdrawal, turned away + suf *-io*, *-ionis* > E. *-ion*, result of an action, state of): inhibition of the growth in the marginal zones of adjacent colonies of fungi or bacteria, as occurs, e.g., in *Aspergillus flavofurcatus* (moniliaceous asexual fungi). Cf. **autotropism**.

Aversion between colonies of *Aspergillus flavofurcatus* on agar, x 0.8 (*MU*).

axenic (Gr. pref. *a-*, not, without + *xénos*, stranger + suf. *-íkos* > L. *-icus* > E. *-ic*, belonging to or relating to): **1**. A culture free of contamination, i.e., which contains only one species and is free of any other living organism. **2**. An organism that is not capable of serving as a host to a parasite, i.e., that exhibits inhospitality or axenia, which is considered as passive resistance of the host toward the invading organism. Cf. **agnotobiotic**, **gnotobiotic** and **monoxenic**.

axoneme, axonema, pl. **axonemata** (Gr. *áxon*, genit. *áxonos*, axis + *nêma*, thread, filament): the central axial filament of a flagellum, composed of microtubules. In the mastigonemate flagella of the zoospores of the Hyphochytriomycota and Oomycota (such as *Rhizidiomyces*, of the Hyphochytriales, and *Achlya*, of the Saprolegniales), numerous secondary fibrils (the *mastigonemes*) are implanted on the axoneme.

azonate (L. *a-*, not, without + *zona* < Gr. *zóne*, band, stripe + L. suf. *-atus* > E. *-ate*, provided with or likeness): without distinguishable zones. Cf. **zonate**.

azotodesmic (Gr. *azóton*, nitrogen + *desmós*, fixation, union + suf. *-íkos* > L. *-icus* > E. *-ic*, belonging to or relating to): an organism with the ability to fix elemental atmospheric nitrogen. Some cyanobacteriophilic lichens (whose photobionts are cyanobacteria) are azotodesmic, as, e.g., *Collema* (Lecanorales)

azygosporangium, pl. **azygosporangia** (Gr. *a-*, not, without + *zygón*, yoke, pair + *sporá*, spore + *angeîon*, vessel, closed receptacle): a structure equivalent to a zygosporangium, but which contains an *azygospore*, which is a zygospore developed parthenogenetically, as occurs, e.g., in *Mucor bainieri* (Mucorales).

Azygosporangium of *Mucor bainieri*, x 200 (*MU*).

azygospore (Gr. *a-*, not, without + *zygón*, yoke, pair + *sporá*, spore): *Zygomycetes*. A spore morphologically equivalent to a zygospore, but which develops parthenogenetically, as happens in many species of Mucorales and in some species of Entomophthorales, including *Entomophthora muscae*. The structure that contains an azygospore is called an *azygosporangium*.

b

Bulbous stipe of the basidiocarps of *Leucocoprinus cepistipes* (*SA*).

b

baccate (L. *baccatus* < *bacca*, *baca*, berry + suf. *-atus* > E. *-ate*, provided with or likeness): similar to a berry, whether in shape or consistency; soft and fleshy like a berry. Also refers to something that has a pearly or nacreous appearance, like the basidiocarp of *Amanita baccata* (Agaricales).

bacillar (L. *bacillus*, small stick, genit. *bacillaris*, dim. of *baculus*, staff, support + suf. *-aris* > E. *-ar*, like, belonging to): shaped like a little stick or rod. Applied to cylindrical structures whose length: width ratio is about 3:1. E.g., the conidia of *Myrothecium verrucaria* (tuberculariaceous asexual fungi) and *Polyscytalum fecundissimum* (moniliaceous asexual fungi) are bacillar. Also called **bacilliform**, **baculate** and **baculiform**.

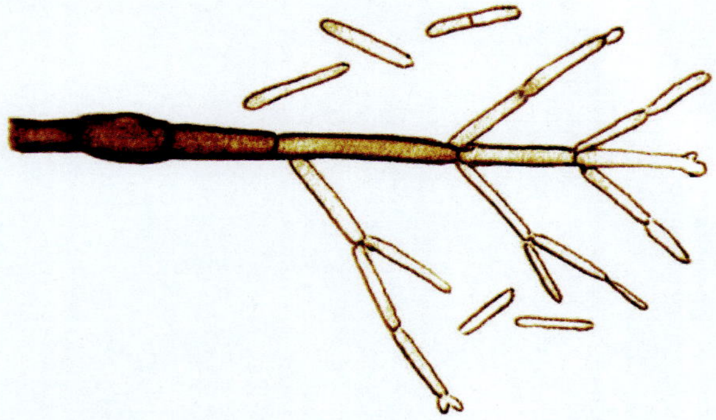

Bacillar conidia of *Polyscytalum fecundissimum*, x 500 (*MU*).

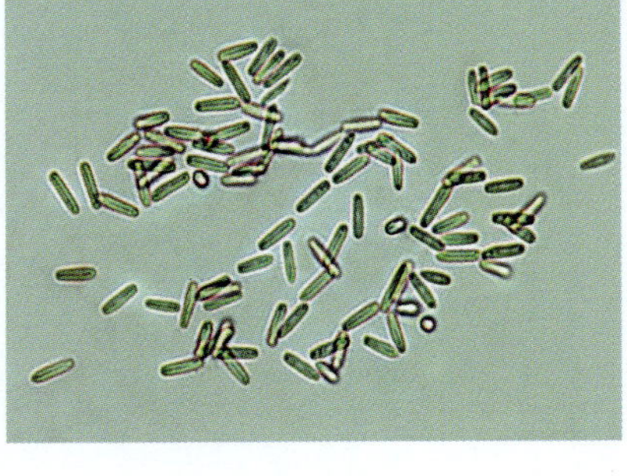

Bacillar conidia of *Polyscytalum fecundissimum*, x 500 (*MU*).

bacilliform (L. *bacillus*, small stick, dim. of *baculus*, staff, support + suf. *-formis* < *forma*, shape): shaped like a small stick or rod. Also called **bacillar**, **baculate** and **baculiform**.

bactericide (L. *bacterium* < Gr. *baktérion*, dim. of *báktron*, staff, rod + E. des. *-cide* < L. *caedere*, to cut, kill): capable of causing the death of bacteria. Some antibiotics of fungal origin are bactericides.

bacteriform, bactriform (L. *bacterium* < Gr. *baktérion*, dim. of *báktron*, staff, rod + L. suf. *-formis* < *forma*, shape): with the shape of a bacterium. See **bacillar**.

bacteriostatic (L. *bacterium* < Gr. *baktérion*, dim. of *báktron*, staff, rod + Gr. *statós*, that which stops, holds back + Gr. suf. *-ikos* > L. *-icus* > E. *-ic*, belonging to or relating to): a substance that inhibits the growth of bacteria, or a bactericide that at a lower concentration only arrests their growth, without killing them.

bacterium, pl. **bacteria** (L. *bacterium* < Gr. *baktérion*, dim. of *báktron*, staff, rod): the bacteria constitute an extensive and heterogeneous group of prokaryotic organisms, usually unicellular. Some species are pathogens of plants, others of animals and man, and a few are pathogens of fungi. Among the last are, e.g., *Bacillus polymyxa* (the cause of lesions, in the form of small perforations, on cultivated mushrooms); *Pseudomonas agarici* (the causal agent of bacterial spot of the gills of cultivated mushrooms), and *Pantoea agglomerans* pv. *uredovora* (which parasitizes rust uredinia). Other species of bacteria, in particular actinobacteria, *Bacillus* and *Pseudomonas*, produce a variety of biologically active substances (antibiotics, enzymes and others) that have been utilized as biological control agents against phytopathogenic fungi.

bacterivorous (L. *bacterium* < Gr. *baktérion*, dim. of *báktron*, staff, rod + L. *voro* < *vorare*, to eat, devour): an organism that feeds on bacteria, such as the plasmodia and myxamoebae of the Myxomycota. Within the kingdom Fungi, the phenomenon of *bacterivory* is only known in certain Thraustochytriales (phylum Labyrinthulomycota), epibiotic on marine algae, such as *Thraustochytrium*.

bacterized medium, pl. **media** (L. *bacterium* < Gr. *baktérion*, dim. of *báktron*, stick, rod + suf. *-ized*; L. *medium*, medium, substrate on which an organism lives): a fluid or solid (agar) medium that contains living or dead bacteria as a source of food. Myxamoebae of the dictyostelids (Dictyosteliomycota) utilize a medium of this nature as a source of nutrients.

bactobiont (L. *bacterium* < Gr. *baktérion*, dim. of *báktron*, staff, rod + Gr. *bíos*, life + *óntos*, genit. of *ón*, a being): the bacterial component (cyanobacteria) of a lichen symbiosis. See **photobiont**.

baculate, **baculiform** (L. *baculatus*, with the shape of a stick < *baculus*, staff, rod + suf. *-atus* > E. *-ate*, provided with or likeness; *baculus* + suf. *-formis* < *forma*, shape): having the shape of a staff or rod, like the ascospores of certain lichens. Also called **bacillar** and **bacilliform**.

bait (ME. < ON. *beit, beita*, fish bait, to put to pasture): food used to lure animals to a trap or hook. In mycology, a bait is a substrate placed in or on water or soil to selectively recover fungi from these media. After the bait is colonized, the fungi can be studied and isolated. The kind of fungi attracted to a particular bait depends on its chemical composition, i.e., whether cellulosic, keratinous, chitinous or ligneous. Common baits used for fungi include dead insects, defatted hair, fingernails, hemp seeds, pollen grains, snake skin, cellophane, small pome fruits and wood. Baits are especially important in isolating fungi from aquatic habitats, e.g., Chytridiales and Saprolegniales.

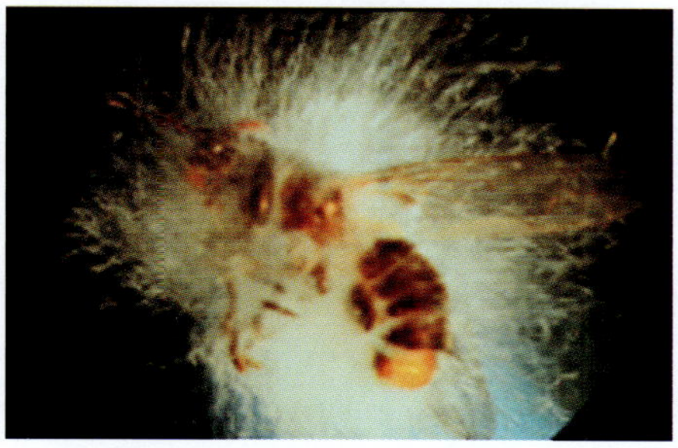

A dead insect used to **bait** *Saprolegnia parasitica* (Saprolegniales); the developing thallus shows the mycelium and zoosporangia, x 5 (*MU*).

balanoid (Gr. *bálanos*, acorn + L. suf. *-oide*, like < Gr. *-oeídes*, similar to): acorn-shaped.

ballistospore (L. *ballista*, catapult < Gr. *ballistás* < *ballein*, to throw + *sporá*, spore): a spore that is forcefully discharged from the sterigma of the basidium that supports it; this generally involves the explosion of droplets of liquid that are formed on the spores. Yeasts of the gen. *Sporobolomyces* and *Itersonilia* (sporobolomycetaceous asexual yeasts), all of the Hymenomycetes (Agaricales and Aphyllophorales), and the species *Schizoplasmodium cavostelioides* (a primitive member of the mucilaginous molds of the order Protosteliales) have ballistospores. Cf. **statismospore**.

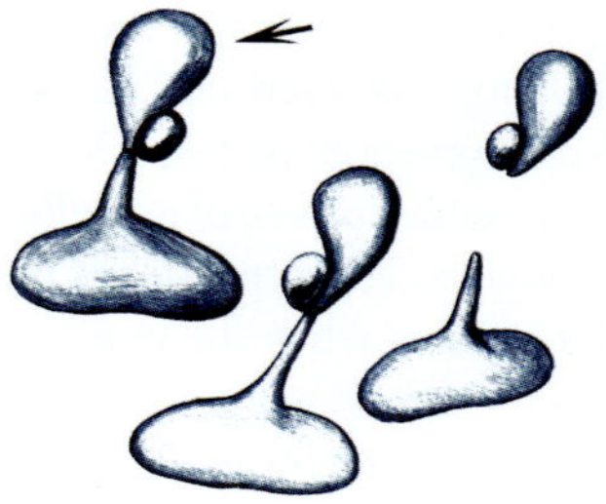

Ballistospores of *Itersonilia perplexans*, x 2,000 (*MU*).

ballistosporic (L. *ballista*, catapult < Gr. *ballistás* < *ballein*, to throw + *sporá*, spore + Gr. suf. *-íkos* > L. *-icus* > E. *-ic*, belonging to or relating to): forming *ballistospores*.

barbate (L. *barbatus* < *barba*, beard + suf. *-atus* > E. *-ate*, provided with or likeness): provided with hairs; bearded, like the thallus of the lichen *Usnea barbata* (Lecanorales).

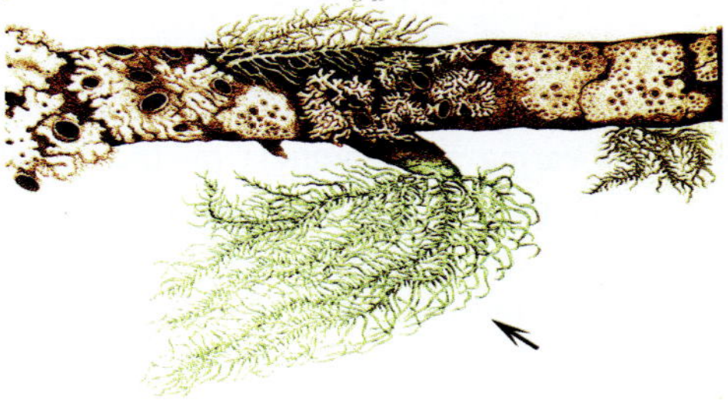

Barbate thallus of *Usnea* sp., x 0.1 (*MU*).

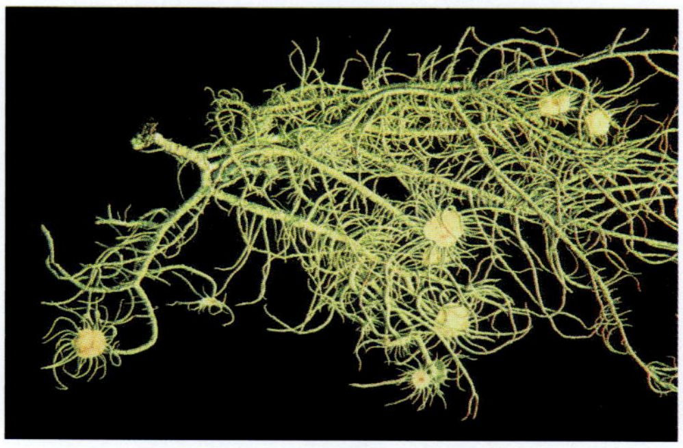

Barbate thallus of *Usnea* sp., with apothecia, x 0.5 (*MU*).

barm, yeast (ME. *berme* < OE. *beorma*, yeast; ME. *yest* < OE. *gist*, yeast): the foam formed on the surface of fermenting malt; beer yeast. Used to ferment bread dough and alcoholic beverages.

barophilic (Gr. *barós*, weight; pressure + *phílos*, have an affinity for): capable of living at great depths, like the marine *Bathyascus* (Halosphaeriales) and *Oceanitis* (an ascomycete of uncertain affinities) that grow on submerged wood, enduring high hydrostatic pressures at depths of 1,720 and 3,975 m, respectively.

barrage (Fr. *barrage*, barrier): the phenomenon of mutual repulsion between certain haploid isolates (haplonts) of Hymenomycetes, which results in the formation of a zone deprived of mycelium between the two paired isolates. This repulsion is manifested only in tetrapolar species when the two haplonts possess the *a* factor in common, while one haplont possesses the *b* factor and the other the *B* factor.

basal (L. *basalis*, basal < *basis*, base + suf. *-alis* > E. *-al*, relating to or belonging to): near the base or relative to it; what is located on the lower end or near the point of attachment or adhesion. Cf. **apical**.

basal body (L. *basalis*, basal < *basis*, base + suf. *-alis* > E. *-al*, relating to or belonging to; ME. < OE. *bodig*, body): see **blepharoplast** and **kinetosome**.

basal cell (L. *basalis*, basal < *basis*, base + suf. *-alis* > E. *-al*, relating to or belonging to; L. *cellula*, cell, living cell, dim. of *cella*, small room, compartment, cell of a honeycomb): the cell that supports the thallus in the aquatic fungi of the order Blastocladiales (such as *Blastocladia* and *Blastocladiella*); in the latter gen., the basal cell adopts the general shape of a trunk, with branches bearing zoosporangia on the apical part and rhizoids or attachment organs on the base. Also applied to the cell that supports the conidiophore of *Aspergillus* (moniliaceous asexual fungi).

basal piece (L. *basal*, basal < *basis*, base + suf. *-alis* > E. *-al*, relating to or belonging to): see **sheath**.

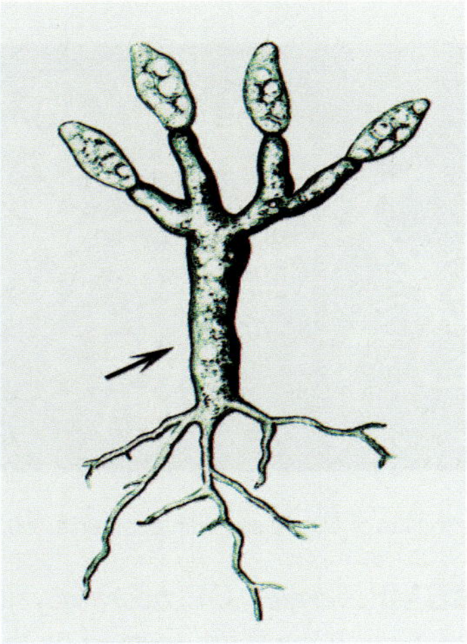

Basal cell of a thallus with zoosporangia of *Blastocladia ramosa*, x 50 (*MU*).

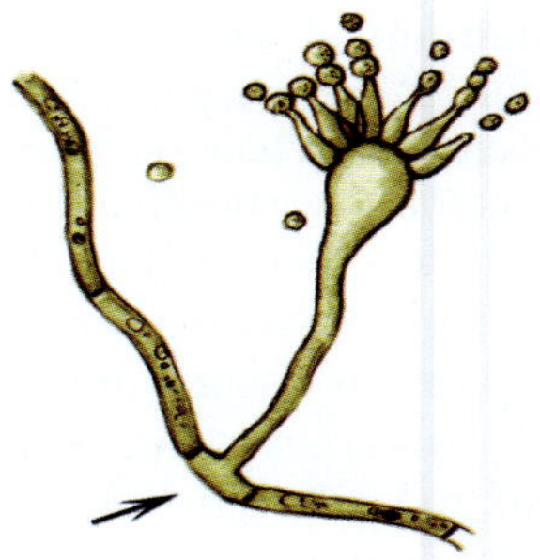

Basal cell of the conidiophore of *Aspergillus japonicus*, x 200 (*MU*).

basauxic (L. *basis*, base + Gr. *aúxo*, to grow, develop, increase + Gr. suf. *-íkos* > L. *-icus* > E. *-ic*, belonging to or relating to): *Conidiogenesis*. A type of development of a conidiogenous cell in which there is elongation of a basal growth point of a fertile cell, after a blastic conidium has formed at its apex. In the fungi that produce a basipetal succession of conidia, the conidiogenous cell can continue elongating from the basal point of growth after formation of the primary conidium, as occurs in *Spegazzinia tessarthra* (moniliaceous asexual fungi). Cf. **acroauxic**.

Basauxic conidiogenous cell of *Spegazzinia tessarthra*, x 1,300 (*MU*).

basidial (L. *basidium* < Gr. *basídion*, dim. of *básis*, base + L. suf. *-alis* > E. *-al*, relating to or belonging to): pertaining to or relative to basidia; possessing basidia.

basidiocarp (Gr. *basídion* < *básis*, base + dim. suf. *-ídion* + *karpós*, fruit): *Basidiomycetes*. A sporophore or fruiting body on which or in which are formed basidia and basidiospores. See the illustrations under the terms **adnate**, **adpressed**, **appendage**, **auriculiform**, **bulbous**, **calyptra**, **campanulate**, **clathrate**, **claviform**, **comate**, **conchate**, **connate**, **cortina**, **costate**, **crenate** and **crispate**, for examples of basidiocarps showing various characteristics. Some of the best known and attractive basidiocarps are those of *Volvariella bombycina* and *Amanita flavivolvata* (Agaricales). Also see **basidiome**.

Basidiocarps of *Volvariella bombycina*, x 0.3 (*RV*).

Basidiocarps of *Amanita flavivolvata*, x 0.5 (*EPS*).

basidiogenous (Gr. *basídion* < *básis*, base + dim. suf. *-ídion* + *génos*, origin): a structure that produces *basidia*. Cf. **ascogenous**.

basidiole, basidiolum, pl. **basidiola** (NL. *basidiolum*, dim. of *basidium* < Gr. *basídion* < *básis*, base + dim. suf. *-ídion* > *basis*, base + dim. suf. *-idium*; L. dim. suf. *-olum* > E. *-ole*): *Basidiomycetes*. A young basidium, still binucleate, like those in the hymenium of the Agaricales, e.g., in *Amanita*, *Boletus* and *Leptonia*. Basidiola are considered either as stages of development of basidia or as aborted basidia. Generally they are clavate or widely clavate and similar in shape to basidia; the basidiola, like the basidia, typically originate in the hymenium.

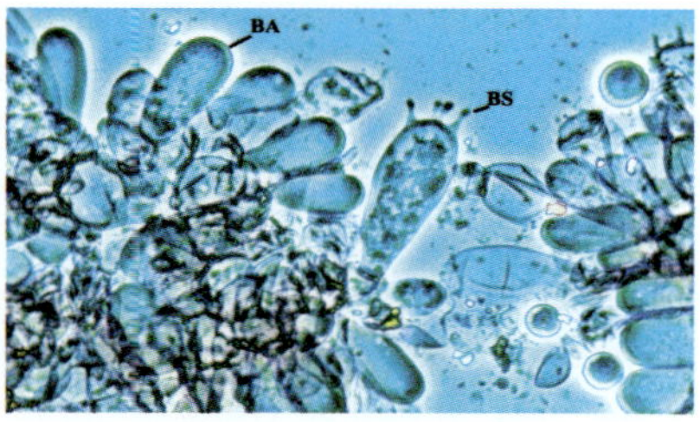

Basidiola (BA) and **basidia** (BS) of the hymenium of *Amanita* sp., x 300 (*RMA*).

basidiome, basidioma, pl. **basidiomata** (Gr. *basídion*, dim. of *básis*, base + *-oma*, suf. that implies the idea of entirety): *Basidiomycetes*. In general, any structure that produces basidia, whether it be a simple layer of basidia or a complex sporophore. E.g., in the Exobasidiales the basidia are formed superficially on the leaves of parasitized plants, without formation of a true fruiting body, whereas in the Agaricales the basidia are formed on a fruiting body of varying complexity. Consequently, the term basidioma is more precise and is preferable to **basidiocarp**, which cannot be applied to all types of structures that bear basidia.

basidiomycete, basidiomycote (L. *basidiomycete* < *basidium* + *mycete* < Gr. *basídion*, dim. of *básis*, base + *mýkes*, genit. *mýketos*, fungus): a representative of the Basidiomycota, a phylum of the kingdom Fungi in the system of classification adopted in some books, e.g., *Five Kingdoms*, 2nd. ed., by Margulis and Schwartz, 1988, and *Introductory Mycology*, 4th. ed., by Alexopoulos *et al.*, 1996; the latter is the one followed in this dictionary. The commonest fruiting body is setiform, with a stipe and a pileus, the latter with the hymenium, basidia and basidiospores as the meiotic or sexual stage.

basidiomycetes (NL. *basidiomycetes* < L. *basidium* < Gr. *basídion*, dim. of *básis*, base + L. *-mycetes*, ending of a taxonomic class < Gr. *mýkes*, genit. *mýketos*, fungus): a group of fungi characterized by the formation of basidia. Formerly considered as a single taxonomic class of higher fungi (Basidiomycetes), in the system of classification adopted in this dictionary it is equivalent to the phylum Basidiomycota, which includes the classes Hymenomycetes, Gasteromycetes, Ustilaginomycetes and Urediniomycetes. See the Appendix to compare the classification changes of this fungal group in Alexopoulos *et al.*, 1996, modified by Ulloa and Hanlin, 2006, to that of Kirk *et al.*, 2008. In the latter classification scheme, the orders of basidiomycetes have been transferred to different subphyla and classes. Please note that some important orders, such as Aphyllophorales and Nidulariales, have disappeared, and most of the orders were placed in the subphylum Agaricomycotina. These fungi are made up of heterogenous species, forming unnatural groupings, and therefore are considered an artificial classification. These taxonomic rearrangements, with the concurrent changes of the names of taxa, are under constant study and evaluation, in order to achieve a phylogenetic classification of the fungi.

basidiospore (Gr. *basídion*, dim. of *básis*, base + *sporá*, spore): a sexual spore typical of the basidiomycetes, formed on the exterior of a basidium, 1-2 generally 4 or more in number, depending on the species. Basidiospores are haploid and are formed individually and acrogenously on the sterigmata of the basidia, as the result of karyogamy and meiosis; on germinating they produce primary mycelium, which forms the secondary and tertiary mycelia, and the latter gives rise to the fruiting bodies that produce the basidia. E.g., in *Lycoperdon* (Lycoperdales), *Ramaricium polysporoideum*, *Ramariopsis corniculata* and *R. rubripermanens* (Aphyllophorales).

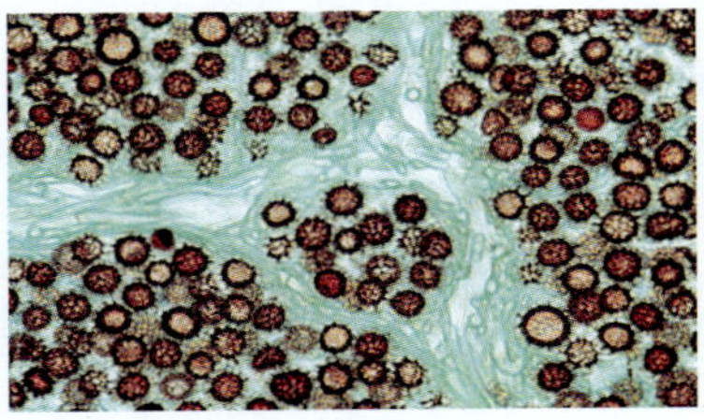

Basidiospores from a basidiocarp of *Lycoperdon* sp., x 300 (*MU*).

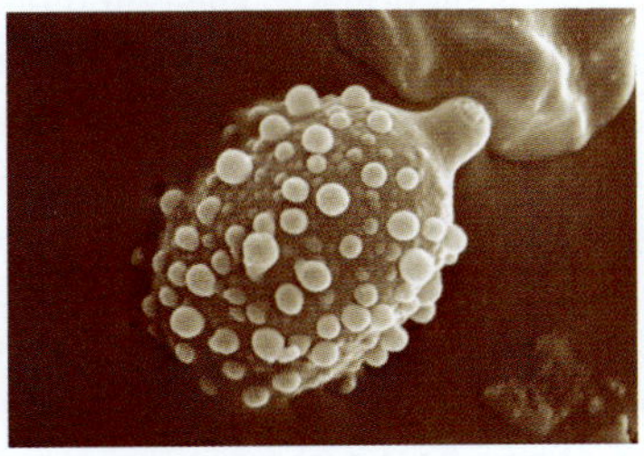

Basidiospore of *Ramaricium polyporoideum,* seen with the scanning electron microscope, x 5,000 (*MV*).

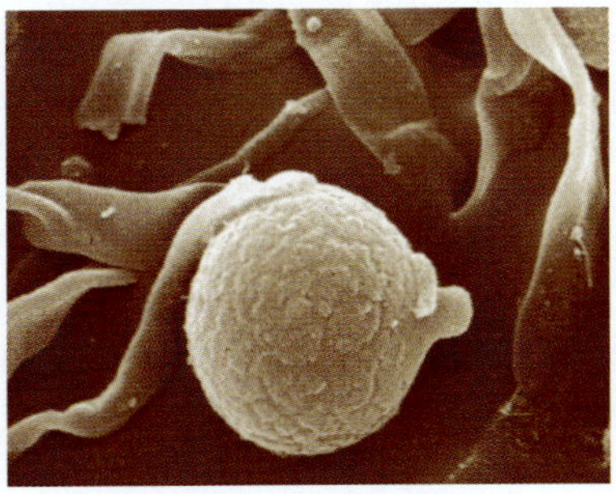

Basidiospore of *Ramariopsis corniculata,* seen with the scanning electron microscope, x 4,000 (*MV*).

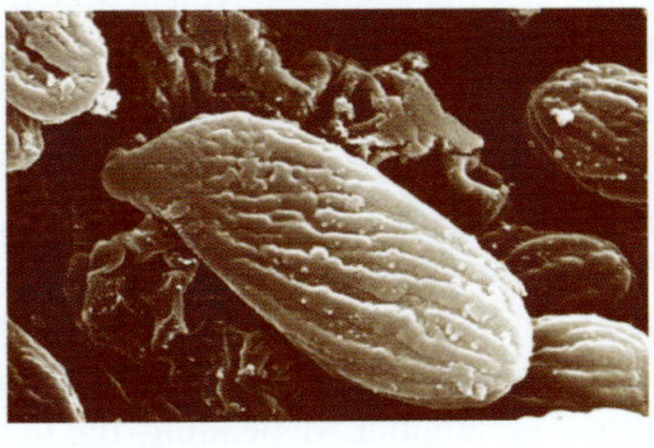

Basidiospore of *Ramariopsis rubripermanens,* seen with the scanning electron microscope, x 7,000 (*MV*).

basidium, pl. **basidia** (L. *basidium*, small base; here, of the spores < *basis*, base + dim. suf. *-idium* < Gr. *basídion* < *básis*, base + dim. suf. *-idion* > L. *-idium*): *Basidiomycetes*. A specialized cell, characteristic of the basidiomycetes, on which are formed the basidiospores supported by the sterigmata, following karyogamy and meiosis. The basidium is homologous to the ascus with regard to the mother cell and to the cytological processes (karyogamy and meiosis) that occur in it, although they differ essentially in that the basidium forms exogenous spores, whereas the ascus forms them endogenously. See the figures for **abaxial** and **apobasidium**, where examples of basidia with basidiospores are illustrated, and the basidia of *Gymnopilus* (Agaricales).

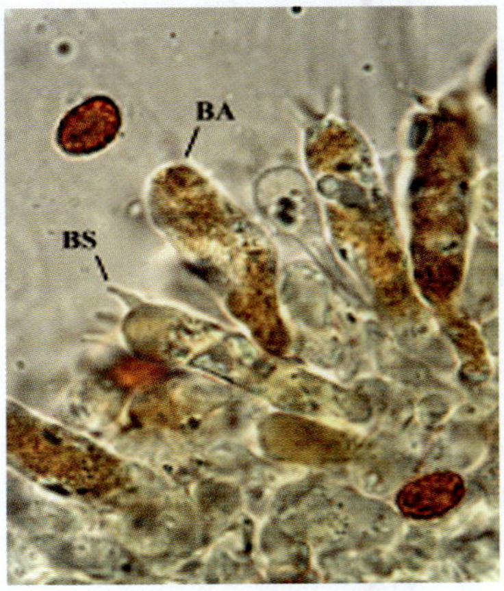

Basidium (BS) and **basidiola** (BA) of the hymenium of *Gymnopilus* sp., x 300 (*LG*).

basifugal (L. *basis*, base + *fugus* < *fugere*, to flee, move away from): applied to any structure that develops from the base toward the apex. Syn. of **acropetal**. Cf. **basipetal**.

basionym, basonym (Gr. *básis*, base + *ónyma*, *ónoma*, name): in botanical and mycological nomenclature, it is the name that bears the specific or varietal epithet, etc., of a new combination to designate a species; i.e., it is the original name that must be maintained by reason of priority when proposing a new combination of a species in a genus or a variety in a species. The epithet can be conserved unaltered or with ending modified; e.g., *Entomophthora coronata* became a synonym of *Conidiobolus coronatus* (Entomophthorales), *E. coronata* being the basionym.

basipetal (L. *basis*, base + *petere*, to go towards, grow toward): *Conidiogenesis*. Developing or maturing from the apex toward the base, i.e., the degree of growth is greater with increased distance from the base; e.g., in a chain of conidia with basipetal succession, the conidium at the base is the youngest and of most recent formation. *Aspergillus*, *Basipetospora*, *Penicillium* and *Scopulariopsis* (moniliaceous asexual fungi) are examples of gen. of fungi that produce their conidia in this manner. *Albugo* (Peronosporales) also forms sporangia with basipetal succession. Cf. **acropetal**.

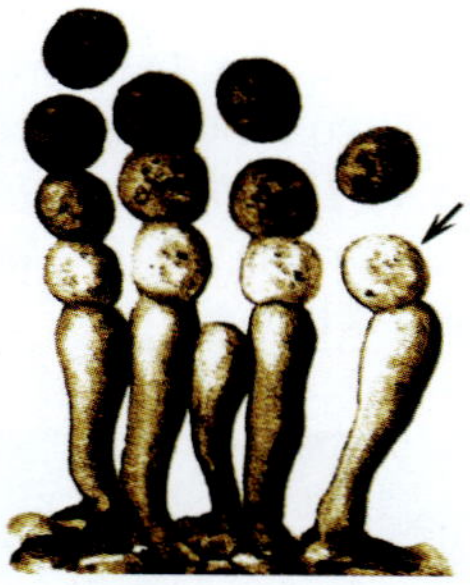

Basipetal succession of the sporangia of *Albugo candida*, x 160 (*MU*).

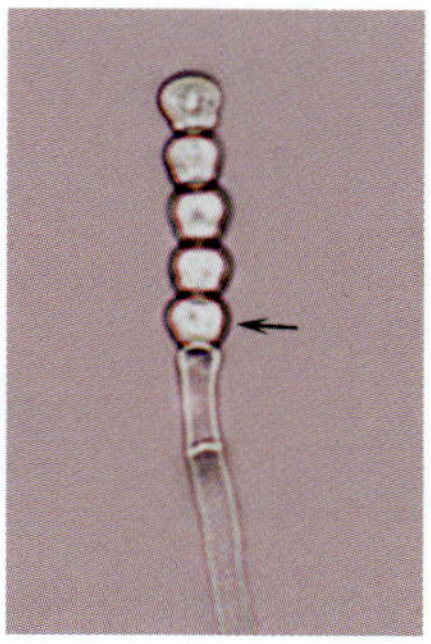

Basipetal succession of the conidia of *Basipetospora rubra*, x 460 (*CB*).

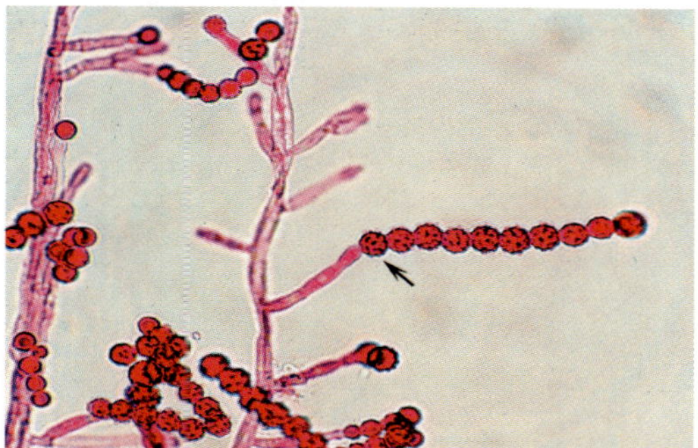

Basipetal succession of the conidia of *Scopulariopsis brevicaulis*, x 460 (*CB*).

basocatenate (Gr. *básis*, base + L. *catenatus*, linked together, arranged in a chain < *catena*, chain + suf. *-atus* > E. *-ate*, provided with or likeness): applied to conidia formed in chains, with the youngest conidium at the basal end of the chain.

basophil, **basophile**, **basophilic** (Gr. *básis*, base + *phílos*, have an affinity for): **1.** Having a preference for basic media or substrates. **2.** Also used for cellular organelles or bodies that stain with basic dyes (carmine, hematoxylin, thionine). Cf. **acidophilous**.

beard moss (ME. *berd* < OE. *beard*, beard; ME. *mos*, moss): species of *Alectoria*, *Bryoria*, *Ramalina* and *Usnea* (Lecanorales), characterized by long, pendant thalli.

beef-steak fungus, pl. **fungi** (E. *beef* < ME. < OF. *buef*, ox, beef < L. *bov-*, *bos*, head of cattle; E. *steak* < ME. *steke* < ON. *steik*, to roast on a stake or spit; L. *fungus*, fungus < Gr. *spóngos*, *sphóngos*, sponge): the common name applied to the edible basidioma of *Fistulina hepatica* (Aphyllophorales), the causal agent of brown rot of oak. Also called *liver fungus*.

benthic (Gr. *bénthos*, depth, generally marine + suf. *-íkos* > L. *-icus* > E. *-ic*, pertaining to or relative to): forming part of the benthos, relative to the benthos, i.e., adapted to live on a solid substrate, whether in the littoral or abyssal benthos, in contrast to the planktonic organisms, which live in suspension in an aquatic medium; an example of an abyssal benthic fungus is *Periconia abyssa* (dematiaceous asexual fungi), which lives on submerged wood in the sea at depths of 3,975-5,315 m.

beta-conidium, **ß-conidium**, pl. **conidia** (*beta*, *ß*, second letter of the Gr. alphabet; NL. *conidium* < Gr. *konídion* < *kónis*, dust, powder + dim. suf. *-ídion* > L. *-idium*): a term applied to species that produce two morphological types of spores, in particular to the filiform spore of *Phomopsis* (sphaeropsidaceous asexual fungi) and other conidial states of the Diaporthales. Generally these fungi also produce *alpha conidia*, which are fertile and oblong or fusoid in shape. Beta conidia often form later than alpha conidia, they are less common, and they do not germinate.

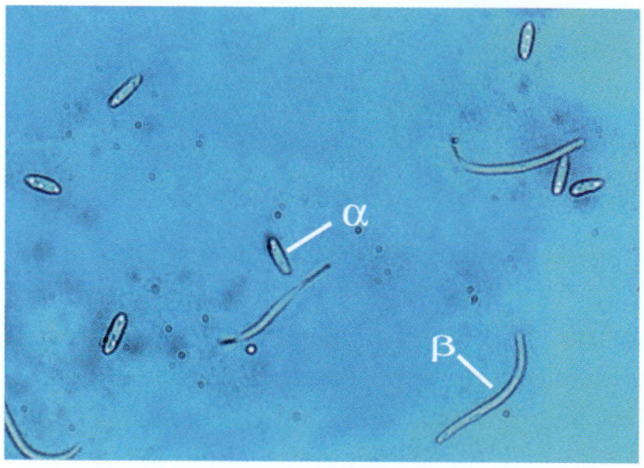

Beta-(ß) and alpha-(α) **conidia** of a pycnidium of *Phomopsis* sp., x 800 (*RTH*).

bi- (L. *bis*, two, twice): a pref. often employed in biology to form terms in Latin, or derived from Latin, when one wishes to indicate the doubleness of something, as in *biallelic*, with two alleles per locus, along with the following terms.

biallelic (L. *bis*, two, twice + Gr. *allélo* < *allélon*, one to the other, reciprocally + Gr. suf. *-íkos* > L. *-icus* > E. *-ic*, pertaining to or relating to): with two alleles per locus; applied in cases of sexual incompatibility. See **allele**.

biapiculate (L. *bis*, two, twice + *apiculatus*, provided with an apiculus < *apiculum*, dim. of *apex*, apex, end + suf. *-atus* > E. *-ate*, provided with or likeness): refers to a globose or elliptical structure with an apiculus at each end, i.e., more or less limoniform, like the ascospores of *Chaetomium cochlioides* (Sordariales), *Melanospora* (Melanosporales) and *Hypomyces lactifluorum* (Hypocreales). See **apiculate**.

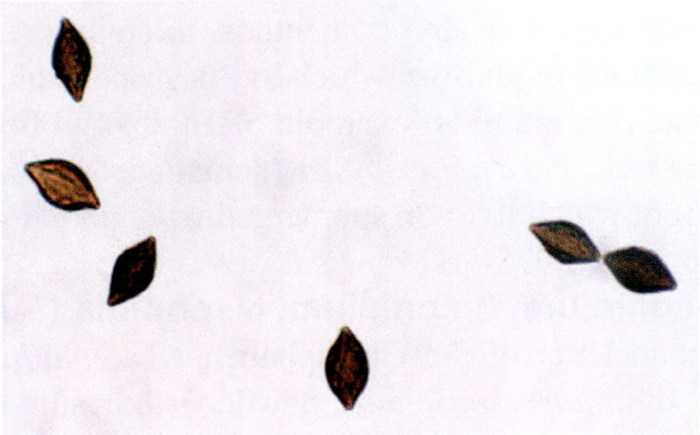

Biapiculate ascospores of *Melanospora* sp., x 1,200 (*RTH*).

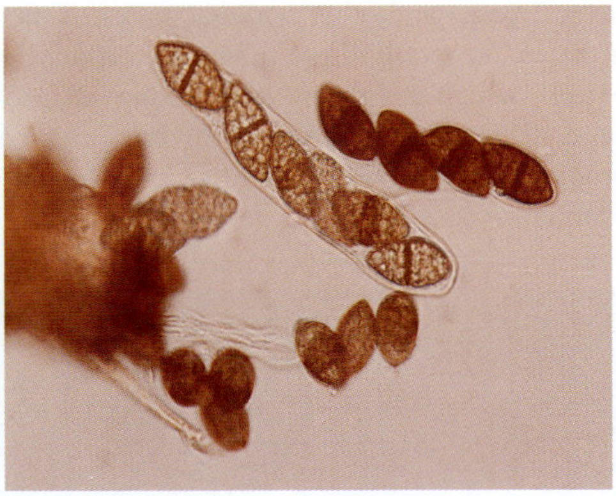

Biapiculate ascospores of *Chaetomium* sp., x 800 (*CB*).

biatorine (NL. *biatora*, name of unknown meaning to designate a gen. of mosses, *Biatora*, + L. suf. *-inus* > E. *-ine*, belonging to, like): in the lichens it is applied to apothecia of the lecideine type, light in color and soft in consistency, like those of *Biatorella* (Lecanorales).

bibulous (L. *bibulus,* fond of drinking, absorbent + suf. *-osus* >OF. *-ous* > E. *-ous*, possessing the qualities of): capable of absorbing moisture, as in the surface of the pileus of some boletes (*Boletus*, *Suillus*) and agarics (*Hygrophorus*) of the Agaricales.

bicampanulate (L. *bis*, two, twice + *campanulatus*, bell-shaped < *campanula*, small bell + suf. *-atus* > E. *-ate*, provided with or likeness): like two bells arranged mouth to mouth.

biconic, biconical (L. *bis*, two, twice + Gr. *konikós*, conical): having two cones opposed at the base, like the teliospores of *Puccinia graminis* (Uredinales). The perithecial stroma of *Phylacia poculiformis* (Xylariales) also is biconic.

Biconic perithecial stroma of *Phylacia poculiformis*, x 8 (*MU*).

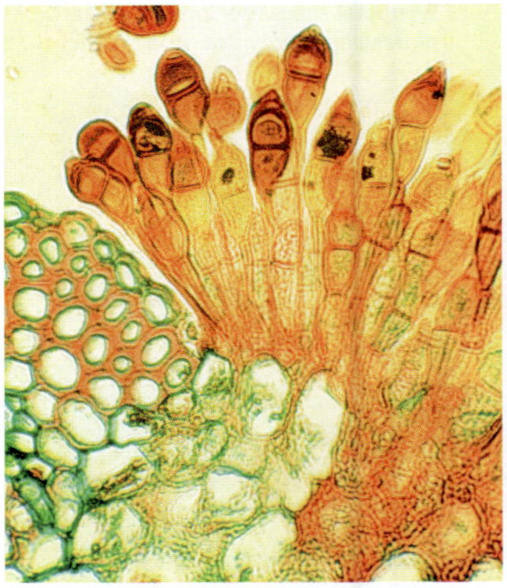

Biconic teliospores of *Puccinia graminis tritici*, seen in longitudinal section of a wheat leaf, x 250 (*GGA*).

biennial (L. *biennalis* < *bis*, two, twice + *annus*, year + suf. *-alis* > E. *-al*, relating to or pertaining to; occurring every two years, continuing or lasting for two years): refers to an organism that lives more than one year without passing two; in the fungi it refers to the duration of the fruiting body, not to the vegetative mycelium immersed in the substrate. Cf. **annual** and **perennial**.

bifarious (L. *bifarius*, in two lines or series): arranged in two rows or lines. Equivalent to **distichous**.

bifid (L. *bifidus*, divided into two parts): with a fissure or division near the middle part. Equal to **bifurcate**. Cf. **multifid**.

biforate (L. *biforatus*, *biforus*, having two doors or openings < *bis*, two, twice + *forare*, to perforate + suf. *-atus* > E. *-ate*, provided with or likeness): with two pores or openings, like the urediniospores of *Uromyces appendiculatus*, the causal agent of bean rust (Uredinales), which have two pores or equatorial orifices.

bifurcate (ML. *bifurcatus*, *bifurcus*, with two points < *bifurcare*, to fork < *bis*, two, twice + *furca*, pitchfork + suf. *-atus* > E. *-ate*, provided with or likeness): forked or divided into two; dichotomous, like the basidium with two sterigmata of *Dacrymyces deliquescens* (Dacrymycetales), or the sporangiophore branches of *Piptocephalis lepidula* (Zoopagales). See **bipartite**.

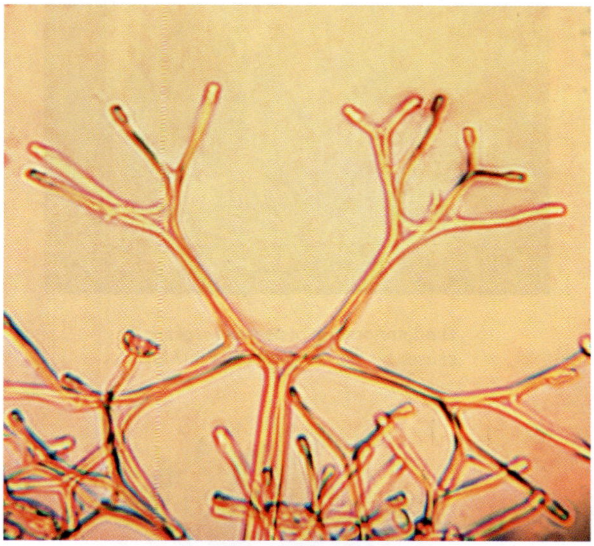

Bifurcate branches of the sporangiophore of *Piptocephalis lepidula*, x 900 (*EAA*).

biguttulate (L. *bis*, two, twice + *guttulatus*, having droplets or droplet-like masses < *guttula*, dim. of *gutta*, drop + suf. *-atus* > E. *-ate*, provided with or likeness): provided with two droplets or small globules of oil. The conidia of *Phoma glomerata* (sphaeropsidaceous asexual fungi) and the ascospores of some species of *Xylaria* (Xylariales) are *biguttulate*. See **guttulate**.

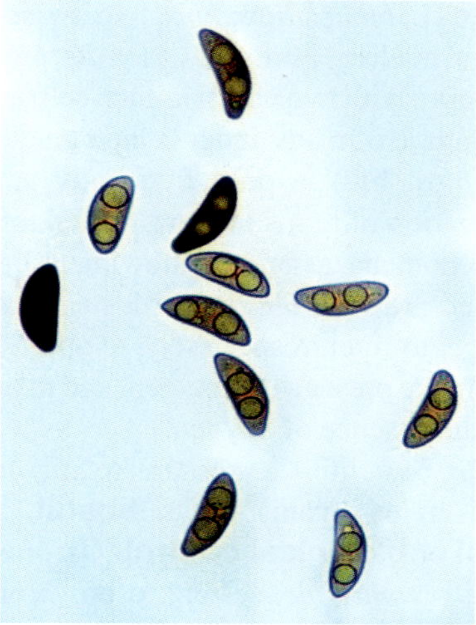

Biguttulate ascospores of *Xylaria* sp., x 1,170 (*RTH*).

bilabiate (L. *bis*, two, twice + *labiatus*, labiate < *labium*, lip + suf. *-atus* > E. *-ate*, provided with or likeness): having two lips, like some hysterothecia, e.g., those of *Hysterium* (Dothideales), and a type of ascus present in some lichens, such as *Pertusaria* (Pertusariales), in which the ectotunica splits apically, forming a fissure bordered by a lip on each side, exposing the endotunica.

bilateral (L. *bilateralis* < *bis*, two, twice + *later-*, *latus*, side + suf. *-alis* > E. *-al*, relating to or pertaining to): having two sides or arranged in two rows on both sides of an organ, as in the so-called *bilateral symmetry* of many spores (with two equal sides), or as in the bilateral gill trama of certain Agaricales (e.g., *Amanita muscaria*), which is characterized by having a central zone of columnar parallel cells (hyphae) that project downward from the pileus (the middle layer), and from which diverge obliquely two zones of cells, one on each side, called the lateral strata; this type of hymeniiferous trama is known as *bilaterally divergent* and is present in Boletaceae (*Boletus*, *Paxillus*) and Gomphidiaceae (*Gomphidius*). See **divergent**.

binate (L. *binatus* < *bini*, a pair + suf. *-atus* > E. *-ate*, provided with or likeness): composed of a pair; in two parts.

binding hypha, pl. **hyphae** (ME. *binden* < OE. *bindan*, to fasten, secure, bind + E. suf. *-ing*, action or process; NL. *hypha*, hypha < Gr. *hyphé*, spider web; hypha): *Basidiomycetes*. Highly branched, thin- or thick-walled cylindrical and tortuous hyphae that intertwine with other hyphae of the same type to form a firm but flexible tissue in the sporophores of some Aphyllophorales, such as *Polyporus squamosus* and *Lentinus badius*. See **generative hypha** and **skeletal hypha**.

binucleate (L. *binucleatus* < *bis*, two, twice + *nucleus*, small nut, nucleus + suf. *-atus* > E. *-ate*, provided with or likeness): with two nuclei in each cell or spore. The dikaryophase of many fungi is the binucleate phase.

bio- (Gr. *bíos*, life): a pref. frequently employed in the formation of terms that are related to life and to vital phenomena, as in the following terms. It also is employed as a suf., *-obe*, in words such as *aerobe* and *anaerobe*, to refer, respectively, to an organism that thrives in the presence of oxygen, and to one that can live in the absence of oxygen.

biocide (Gr. *bíos*, life + L. *caedere*, to kill): a substance that kills living organisms. Cf. **biostat**.

biocontrol or **biological control** (Gr. *bíos*, life + Fr. *controle* < *contrerole*, search; to prove or verify by means of evidence, or of experiments, to exercise a restrictive or directing influence over something; or *bíos* + *lógos*, word or discourse + Gr. suf. *-ikos* > L. *-icus* > E. *-ical*, pertaining to or relating to): the utilization of one or more organisms to reduce the number of, or eliminate, other organisms that are considered undesirable (such as a pest), by means of interference in their ecology. At present various species of pathogenic fungi, parasites and antagonists are utilized to control a variety of agricultural pests, including arthropods (principally insects and mites), nematodes, weeds and other organisms that attack crops, causing diseases. See **entomogenous**, **mycoparasite**, **mycopesticide** and **nematophagous**.

bioconversion (Gr. *bíos*, life + L. *conversion*, genit. *conversionis*, the act and effect of converting or being converted < *con-*, with + *versus*, turned, changed + *-io*, genit. *-ionis* > suf. E. *-ion*, result of an action, state of): the conversion of organic materials, usually wastes, into one or more products of increased value (e.g., the production of ethanol from lignocellulosic residues with the use of living organisms). Cf. **biodegradation**, **biodeterioration**, **bioremediation** and **biotransformation**.

biodegradation (Gr. *bíos*, life + L. *degradatio*, genit. *degradationis*, the act and effect of degrading or being degraded < *degradare* < *de-*, from, downward, far from + *gradus*, step, grade + suf. *-ationem* > E. *-ation*, action, state or condition, result): the utilization of living organisms for the beneficial degradation of some materials, with the goal of obtaining more useful, innocuous or acceptable products from residues or wastes. An example of biodegradation is the use of fungal enzymes in fermentations in a solid or liquid state to improve the digestibility of lignocelluloses in ruminants fed with residues. Cf. **bioconversion**, **biodeterioration**, **bioremediation** and **biotransformation**. See **taninolytic**.

biodeterioration (Gr. *bíos*, life + L. *deterioratio*, genit. *deteriorationis*, act and effect of deteriorating or being deteriorated < LL. *deterioratus* < *deteriorare*, to spoil, to impair, to put something in a worse condition + suf. *-ationem* > E. *-ation*, action, state or condition, result): the undesirable change in the properties of a material caused by the activities of living organisms. Fungi are important in biodeterioration and their activities often damage grains and seeds, meat, all kinds of foods, construction materials, wood, stone, combustibles, electrical and optical equipment, leather, paper, textiles and many other natural and man-made materials. Examples of fungi involved in biodeterioration are *Monilia fructicola* (mold of peach, moniliaceous asexual fungi), *Myxotrichum* (mold of paper, Myxotrichaceae, ascomycete *inc. sed.*) and *Chaetomium* (mold of paper, Sordariales). Cf. **bioconversion**, **biodegradation**, **bioremediation** and **biotransformation**.

Biodeterioration of peach, a molding caused by *Monilia fructicola*, x 1 (*MU*).

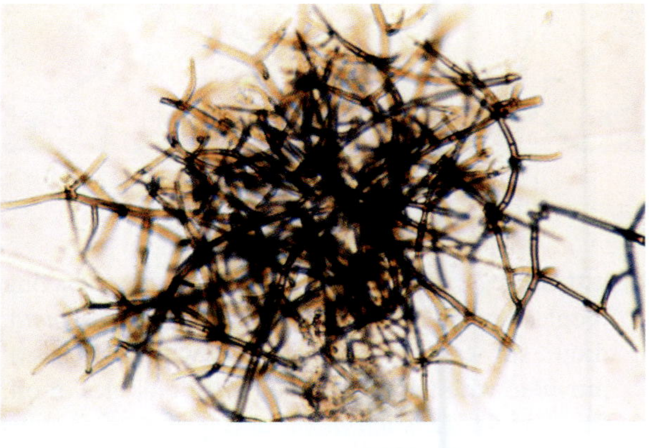

Cleistothecium of *Myxotrichum* sp., one of the molds of paper (**biodeterioration**), x 450 (*MU*).

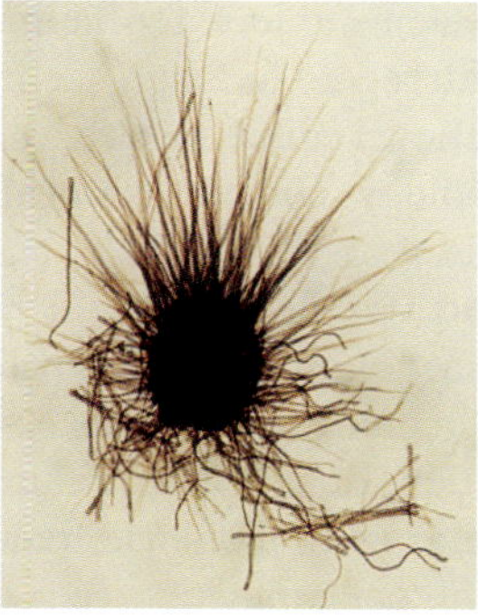

Perithecium of *Chaetomium funicolum*, one of the molds of paper (**biodeterioration**), x 450 (*MU*).

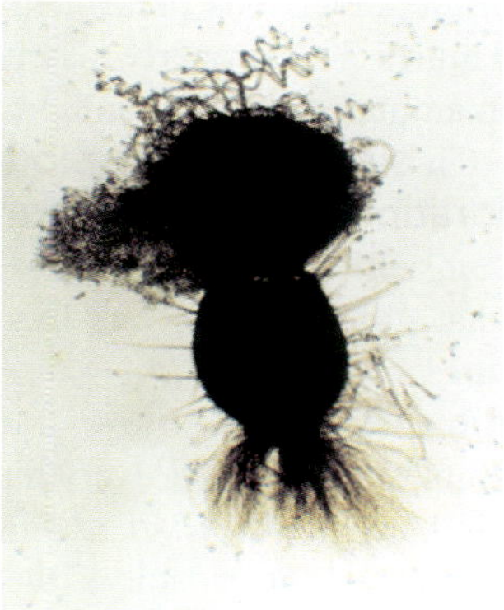

Perithecium of *Chaetomium* sp., one of the molds of paper (**biodeterioration**), x 500 (*MU*).

biodiversity (Gr. *bíos*, life + L. *diversitas*, genit. *diversitatis*, variety, dissimilar, difference; abundance, confluence of various distinct things): the variety and value of life on Earth, from the genetic to the organismal and ecological levels; the natural biotic resources of the planet.

biogenous (Gr. *bíos*, life + *génos*, origin < *gennáo*, to engender, produce): produced by living organisms; living on another living organism; a **parasite**.

biohazard (Gr. *bíos*, life + ME. *hasard* < Ar. *z-zahr*, a source of danger): a biological agent or condition (such as an infectious organism or insecure and risky laboratory conditions) that constitutes a hazard to humans or their environment.

bioindication (Gr. *bíos*, life + L. *indicatio*, genit. *indicationis*, action and effect of indicating < *indicare*, to point out or indicate, signify something with marks or signs + suf. *-ationem* > E. *-ation*, action, state or condition; result): the use of one or more organisms (bioindicators) that express particular symptoms or responses to indicate and monitor changes in some environmental influence. E.g., certain lichens, such as *Usnea ceratina* (Lecanorales), are utilized as bioindicators of atmospheric contamination in urban and suburban environments, especially by sulfur dioxide; some lichens and other fungi are sensitive to acid rain, and still others are bioindicators of radioactivity in a given area, since they can absorb radioactive sustances.

bioinsecticide (Gr. *bíos*, life + L. *insectum* < neuter of *insectus* < *insecare*, to cut, segment < *in*, in + *secare*, to cut, section; insect + E. des. *-cide* < L. *caedere*, to kill): a biological insecticide that does not damage the environment, as the chemical insecticides do. Examples of fungi involved in bioinsecticides are *Verticillium* and *Metarhizium* (moniliaceous asexual fungi). See **mycoinsecticide**.

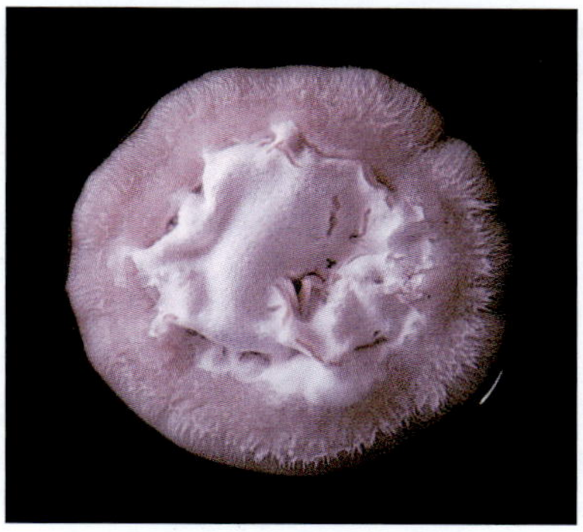

Colony of *Lecanicillium* (=*Verticillium*) *lecanii* on agar, a species used as a **bioinsecticide**, x 1 (*MU*).

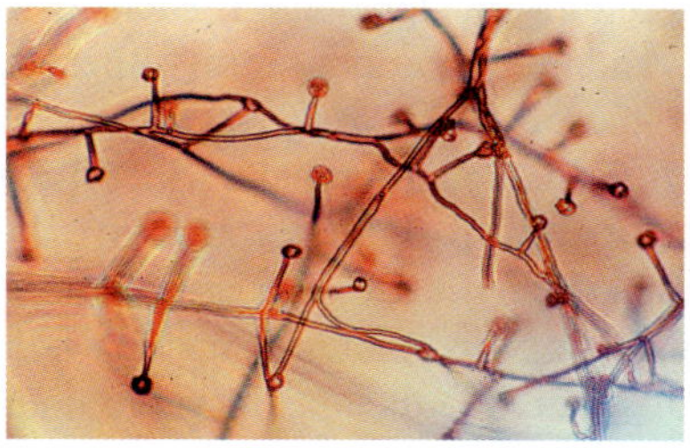

Verticillate conidiophores of *Lecanicillium* (=*Verticillium*) *lecanii* on agar, a species used as a **bioinsecticide**, x 700 (*MU*).

Colony of *Metarhizium anisopliae* on agar, a species used as a **bioinsecticide**, x 5 (*MU*).

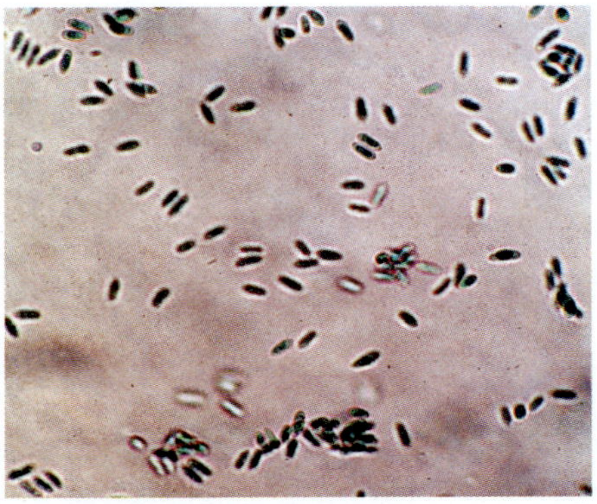

Spores of *Metarhizium anisopliae* on agar, a species used as a **bioinsecticide**, x 500 (*MU*).

biological form (Gr. *bíos*, life + *lógos*, word < *légo*, I say + suf. *-íkos* > L. *-icus* > E. *-ical*, belonging to or relating to; L. *forma*, form): see **physiological form**.

bioluminescent (Gr. *bíos*, life + NL. *luminescens*, genit. *luminescentis* < *lumen*, light + *-escens*, genit. *-escentis* > E. *-escent*, that which turns, that begins to, slightly): a living organism that produces light of low intensity, without being incandescent, i.e., it exhibits the phenomenon of *luminescence*, due to slow oxidations, mediated by special enzymes that produce light but do not elevate the temperature (for this reason also called *chemoluminescence* and, when treating of living organisms, *bioluminescence*). The mycelium of some fungi, such as *Armillaria mellea*, and the sporophores of other Agaricales, such as that of *Omphalotus olearius* or of *Mycena lux-coeli*, are bioluminescent. See **fluorescent**, **phosphorescent** and **luminescent**.

biomass (Gr. *bíos*, life + L. *massa*, volume, ensemble, reunion; quantity or amount of material): the amount of living material; the quantity (volume, weight, number of cells) of living organisms or material that there is in a particular environment, e.g., of yeasts or of filamentous fungi in the soil, in a fermented food or in any other substrate.

biont (Gr. *bíos*, life + *óntos*, genit. of *ón*, a being): an organism; commonly used as a suf. of a word that indicates the nature or position of the biont, e.g., see **diplobiont**, **haplobiont** and **symbiont**.

biophagous (Gr. *bíos*, life + *phágos*, that which eats < *phágesis* < *phágia*, to eat): that which feeds on living organisms; a **parasite**.

biophilous (Gr. *bíos*, life + *phílos*, have an affinity for): living on another living organism; a **parasite**.

bioprospecting (Gr. *bíos*, life + ME. *prospecting* < *prospect* < L. *prospectus,* view, prospect, to look forward, explore): the exploration or examination of natural ecosystems to look for biologically valuable products. With regard to fungi, such products might include, e.g., novel edible species, valuable enzymes for biotechnology companies, or new biological control agents.

bioremediation (Gr. *bíos*, life + L. *reparatio*, genit. *reparationis*, action and effect of repairing < *reparare*, to compose, remedy, reestablish + suf. *-ationem* > E. *-ation*, action, state or condition; result): the use of microorganisms to eliminate, reduce or improve environmental contamination, e.g., micromycetes that absorb heavy metals from the soil. *Cryptococcus albidus* var. *albidus* (cryptococcaceous asexual yeasts) is capable of capturing, and keeping in its capsule, the silver ions present in the soil near abandoned mines. This property may be exploited to remedy contaminated soils that inhibit the growth of vegetation. Cf. **bioconversion**, **biodegradation**, **biodeterioration** and **biotransformation**.

biostat (Gr. *bíos*, life + *statós*, stopped, detained): a substance that stops the growth of living organisms. Cf. **biocide**.

biosystematics (Gr. *bíos*, life + *systematikós* < *sýstema*, complete, aggregated, everything, a methodic compendium + suf. *-tíkos* > L. *-ticus* > E. *-tic*, relation, adaption, inclination or capacity): the study of the relationships and the classification of the living world or biota in an ordered system or conceptual hierarchy. It is based on the species as the fundamental unit; the species, however, can be constituted of lower hierarchies, such as subspecies, races, varieties and forms. The species, in turn, are grouped into higher entities or systematic categories, which are genera, families, orders, classes, etc. In this way systems or methods of classification of the organisms are constituted, which in modern systematics are intended to arrange them according to their true affinities, ideally taking into account their morphological characteristics, their anatomy, their genetic characters, their ecology, geographical distribution, fossils, molecular characteristics and other traits, in order to form an appropriate judgment about the true affinities or degree of relationship that exists among the diverse groups of organisms.

biota (Gr. *bioté*, nature, the condition of life < *bíos*, life): all of the living things of a locality, a country, or of the Earth, consisting of the plants (flora), animals (fauna), microorganisms (microbiota) and the fungi (mycobiota).

biotechnology (Gr. *bíos*, life + *technología*, systematic treatment of an art; applied science < *techné*, art + *logy*, discourse, treatise < *lógos*, word < *légo*, I say): applied biological science by which products are obtained from raw materials with the aid of living organisms (such as bioengineering and recombinant DNA technology, as well as all areas of industrial mycology). See **bioconversion**, **biodegradation**, **bioprospecting**, **bioremediation** and **biotransformation**.

biotic (Gr. *bíos*, life + suf. *-tíkos* > L. *-ticus* > E. *-tic*, relation, fitness, inclination or capacity): pertaining to or relating to life or biosis. Referring to the organic entities that comprise the biosphere. Caused by or produced by living beings. Cf. **abiotic**.

biotransformation (Gr. *bíos*, life + L. *transformatio*, genit. *transformationis*, action and effect of transforming or being transformed < *transformare*, to undergo a change of shape + suf. *-ationem* > E. *-ation*, action, state or condition; result): also known as *biological transformation* or *microbial transformation*, and more generally as **bioconversion**; it involves the use of microorganisms to modify, by means of enzymatic activities, organic compounds to produce products of industrial, medical or environmental importance. Among the most important reactions where a substrate is biotransformed are oxidations, reductions, hydrolysis and condensation, among others, some of which are utilized industrially (e.g., the transformation of antibiotics, steroids and sterols). *Rhizopus stolonifer* (Mucorales) has been used to produce 11a-hydroxyprogesterone from progesterone, and *Penicillium chrysogenum* (=*P. notatum*, moniliaceous asexual fungi) has been utilized for the production of penicillin. Cf. **biodegradation**, **biodeterioration** and **bioremediation**.

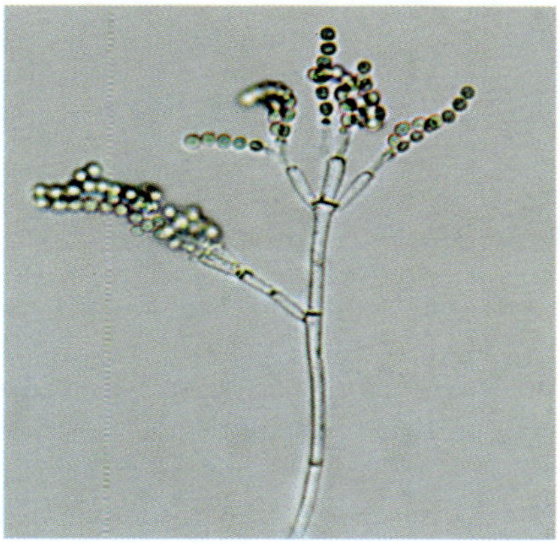

Conidiophore of *Penicillium chrysogenum* (=*P. notatum*), a mold used to produce penicillin (**biotransformation** or **bioconversion**), x 450 (*MU*).

biotroph, **biotrophic** (Gr. *bíos*, life + *trophós*, that which nourishes, serves as food; or + Gr. suf. *-íkos* > L. *-icus* > E. *-ic*, pertaining to or related to): a parasite that can grow and sporulate only on living cells of a host. E.g., the piptocephalidaceous fungi that parasitize other Mucorales (such as *Piptocephalis,* which lives on *Mucor*), the phytoparasitic Peronosporales known as downy mildews and white rusts (such as *Peronospora parasitica* and *Albugo candida*, respectively), the powdery mildews, such as *Blumeria* (=*Erysiphe*) *graminis* and *Phyllactinia* of the Erysiphales (as well as their conidial states *Oidium* and *Ovulariopsis*, of the moniliaceous asexual fungi), the Meliolales (such as *Meliola floridensis*), and the rusts (such as *Puccinia graminis*, of the Uredinales) are all biotrophs. Syn. of **obligate parasite**. Cf. **necrotroph** and **perthotroph**.

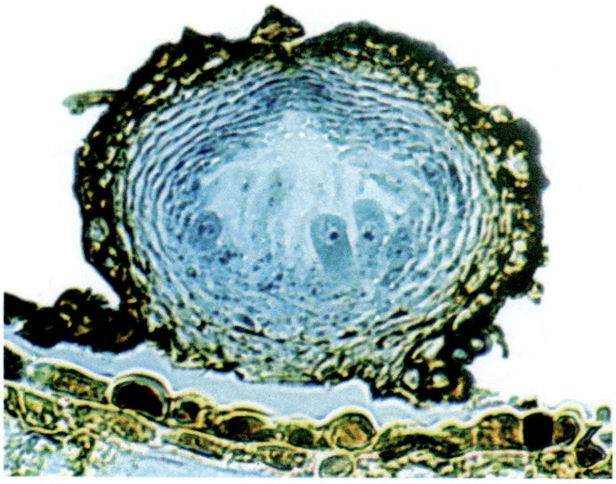

Ascocarp of the **biotroph** *Meliola floridensis* on a host leaf, x 200 (*RTH*).

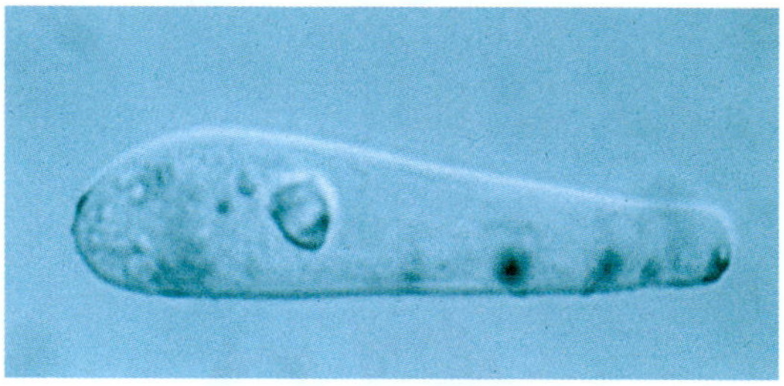

Conidium of the **biotroph** *Ovulariopsis* sp.; *Platymiscium* is the host plant, x 600 (*RTH*).

biotype (Gr. *bíos*, life + L. *typus*, type): a population of individuals of the same genotype that comprise a *forma specialis* (see **form**). The term *race* is given to a group of biotypes that have a similar pattern of virulence-avirulence on a specific group of plants. See **race**.

bipartite (L. *bis*, two + *partitus*, divided < *partire*, to divide): divided into two, like the heterobasidium of *Dacrymyces* (Dacrymycetales). See **bifurcate**.

bipartition (L. *bipartitio*, divide into two < *bis*, two + *partitus*, divided + *-io*, genit. *-ionis* > E. suf. *-ion*, result of an action, state of): division into two equal parts, such as multiplication by cellular bipartition. See **fission** and **schizogenesis**.

bipolar (L. *bis*, two + *polaris*, polar < *polus*, pole + suf. *-aris* > E. *-ar*, like, pertaining to): **1**. That which is found on the two poles or ends of a cell, like the flagella of certain bacterial cells. **2**. A type of sexuality in which the compatibility factors are of only two classes (AB, ab), as occurs in the ascomycetes. Cf. **tetrapolar** and **octopolar**. **3**. A type of spore germination, or budding of a cell, in which a germ tube or bud is formed on each end or pole of the spore or cell. E.g., bipolar germination occurs in the conidia of *Bipolaris* and *Drechslera* (dematiaceous asexual fungi), and bipolar budding in yeasts of the gen. *Kloeckera* (cryptococcaceous asexual yeasts) and *Nadsonia* and *Saccharomycodes* (Saccharomycetales), which form a bud on each end of the mother cell. Cf. **monopolar**, **multipolar** and **unipolar**.

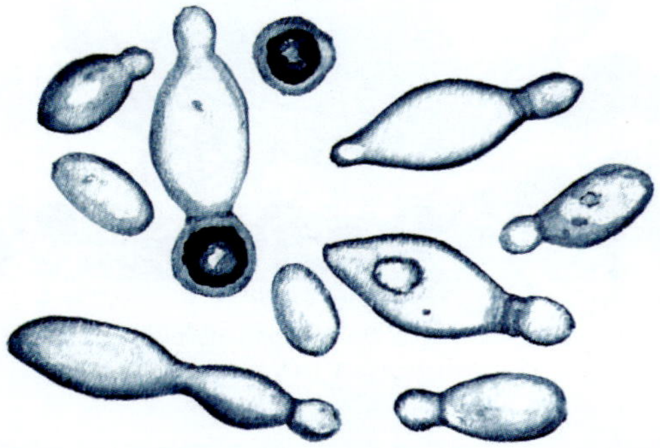

Bipolar budding of the vegetative cells of *Nadsonia fulvescens*, x 1,100 (*MU*).

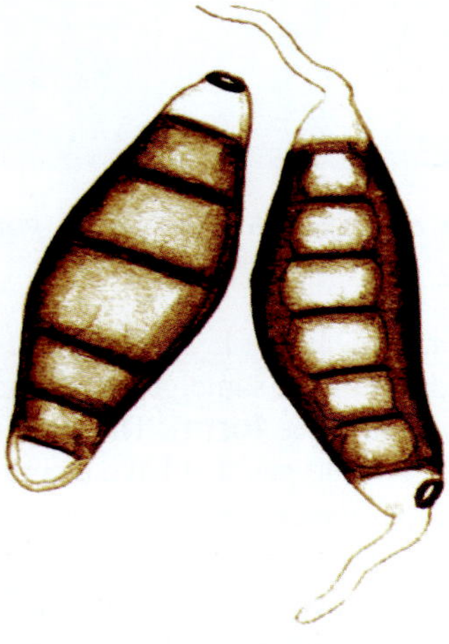

Bipolar germination of the conidia of *Drechslera iridis*, x 980 (*MU*).

bird´s nest fungus, pl. **fungi** (ME. < OE. *bridd*, bird; ME. < OE. *nest*, nest; L. *fungus*, fungus): a member of the Nidulariales (such as *Cyathus*, *Crucibulum* and *Nidularia*), with small, cupulate sporophores containing peridioles, thus resembling a bird´s nest with eggs.

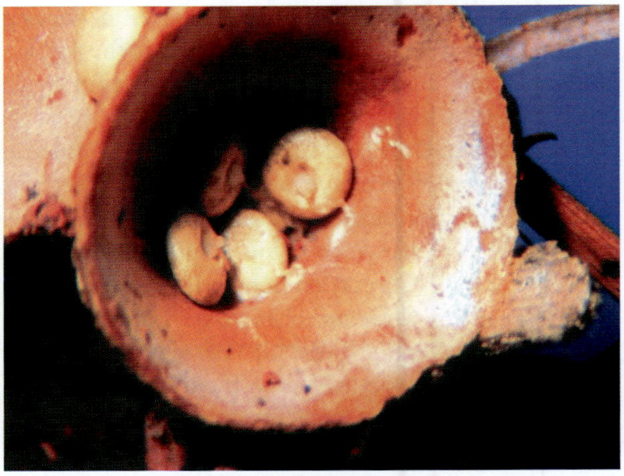

Bird´s nest fungus, basidiocarp of *Crucibulum laeve*, x 10 (*CB*).

Bird´s nest fungi, basidiocarps of *Cyathus striatus*, x 1 (*TH*).

biscoctiform (ML. *biscoctus*, seaman's bread or biscuit < L. *bis*, two, twice + *coctus* < *coquere*, to cook + suf. *-formis* < *forma*, shape): with the shape of a biscuit, like the skin lesions produced by some dermatophytic fungi, and the colonies of the mold *Oidiodendron tenuissimum* (dematiaceous asexual fungi) when grown on certain culture media.

biserial, **biseriate** (NL. *biseriatus* < *bis*, two, twice + *series*, series, row + suf. *-atus* > E. *-ate*, provided with or likeness; or *bis* + *series* + suf. *-alis* > E. *-al*, relative to or pertaining to): arranged in two series or rows; e.g., in some species of *Aspergillus* (*A. candidus*, *A. flavus*, *A. niger*, *A. versicolor*, moniliaceous asexual fungi) the conidial head is biseriate because it has a row of supporting cells on the vesicle of the conidiophore, on which is a series of phialides with conidia. Syn. of **distichous**. Cf. **uniserial** and **monostichous**.

Biscoctiform colonies of *Oidiodendron tenuissimum* on agar, x 20 (*MU*).

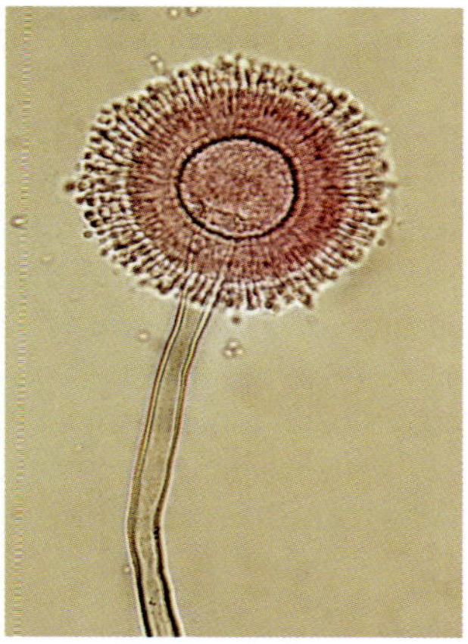

Biserial conidial head of *Aspergillus candidus*, x 300 (*CB*).

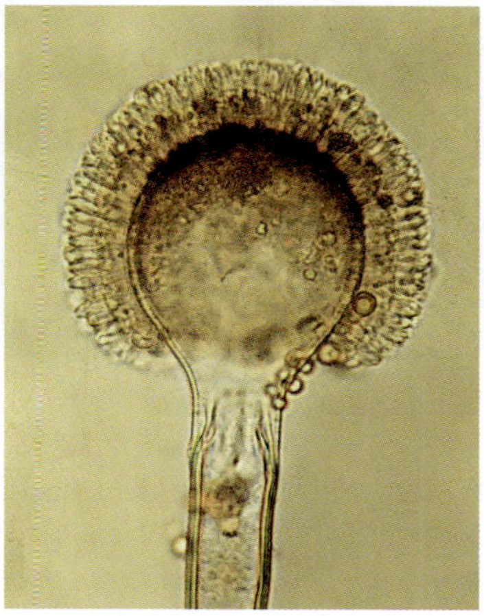

Biserial conidial head of *Aspergillus flavus*, x 400 (*MU*).

Biserial conidial head of *Aspergillus niger*, x 400 (*PL*).

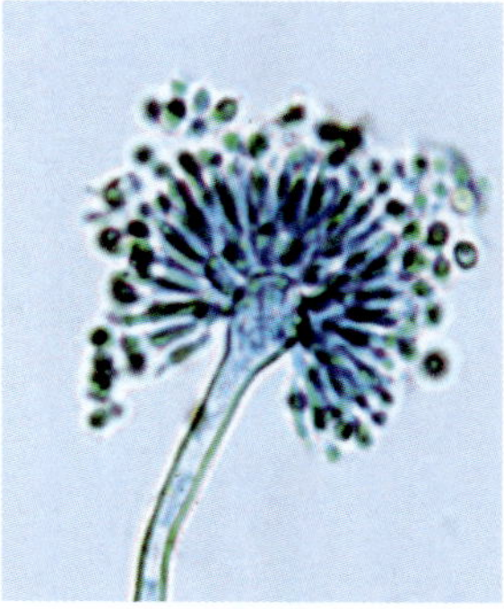

Biserial conidial head of *Aspergillus versicolor*, x 400 (*PL*).

bisporic, **bisporous** (L. *bisporus* < *bis*, two, twice + Gr. *sporá*, spore + suf. *-íkos* > L. *-icus* > E. *-ic*, belonging to or relating to; or L. *-osus* > OF. *-ous* > E. *-ous*, having or possessing the qualities of): with two spores, like the basidia of *Agaricus bisporus* (Agaricales). Another example is the bisporic basidium of *Brachybasidium pinangae* (Exobasidiales). See **unispored** and **multisporate**.

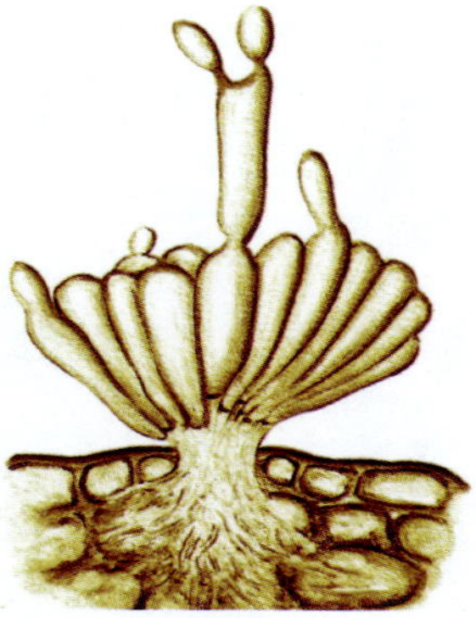

Bisporic basidia of *Brachybasidium pinangae*, and young basidia, formed from probasidia in an hymenium erumpent from a stroma on leaves of the pinang palm, x 400 (*MU*).

bitunicate (L. *bitunicatus* < *bis*, two, twice + *tunica*, cloak, cover + suf. *-atus* > E. *-ate*, provided with or likeness): *Ascomycetes*. A type of ascus whose wall has two functional layers, an internal wall layer (endotunica), that is elastic and which extends beyond the external wall layer (ectotunica) at the time of spore liberation; bitunicate asci are characteristic of ascomycetes with ascolocular ascomal ontogeny, i.e., the Loculoascomycetes, such as *Sporormiella* (=*Preussia*) *australis* (Pleosporales) and *Leptosphaerulina crassiasca* (Dothideales). Cf. **prototunicate** and **unitunicate**.

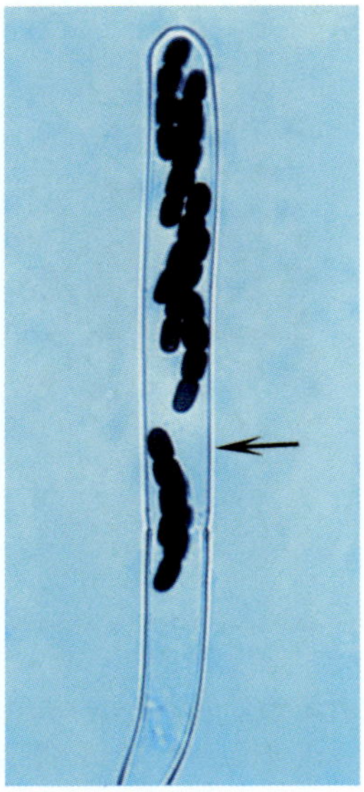

Bitunicate ascus of *Sporormiella* (=*Preussia*) *australis*, with the endotunica extended, 300 (*RTH*).

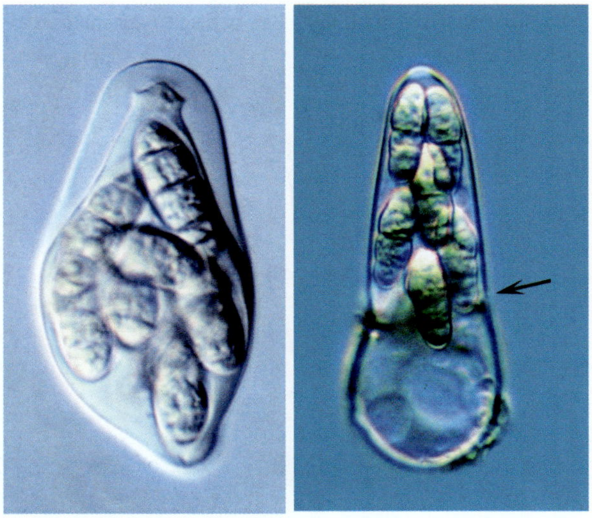

Bitunicate asci of *Leptosphaerulina crassiasca*, unextended (left) and extended; the arrow marks this phenomenon, x 400 (*RTH*).

bivalve (L. *bivalvis* < *bis*, two, twice + *valva*, valve, door): having two valves or convex faces, like the ascospores of *Eurotium chevalieri* and *E. rubrum* (Eurotiales).

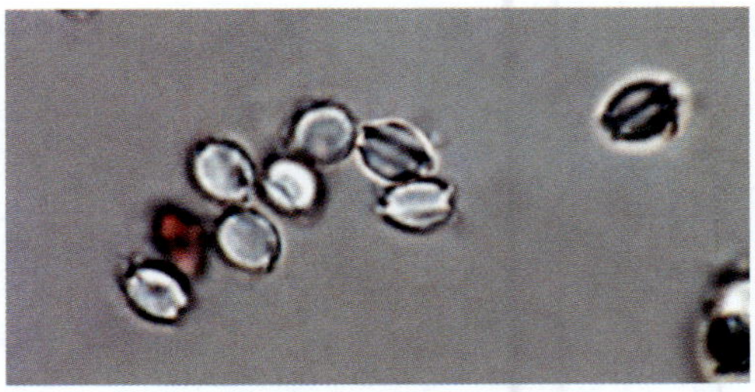

Bivalve ascospores of *Eurotium chevalieri*, x 2,000 (*PL*).

Bivalve ascospores of *Eurotium rubrum*, x 4,000 (*MU*).

biverticillate (L. *biverticillatus* < *bis*, two, twice + *verticillus*, verticil + suf. *-atus* > E. *-ate*, provided with or likeness): branched at two levels, i.e., with supporting cells and phialides, like the conidiophores of some species of *Penicillium* (*P. claviforme*, *P. islandicum*, *P. chrysogenum* (=*P. notatum*, of the moniliaceous asexual fungi) which Pitt places in the subgenus *Biverticillium*. See **monoverticillate**, **triverticillate**, **quadriverticillate** and **polyverticillate.**

Biverticillate conidiophore of *Penicillium* sp., x 1,000 (*CM*).

black knot (ME. *black* < OE. *blaec*, black; ME. < OE. *cnotta*, knot, to press): the black ascostromata of *Apiosporina morbosa* (=*Dibotryon morbosum*), of the Pleosporales, formed on branches of plum and cherry trees.

Black knot, ascostromata of *Apiosporina morbosa* (=*Dibotryon morbosum*) on branch of a plum tree, x 2 (*RTH*).

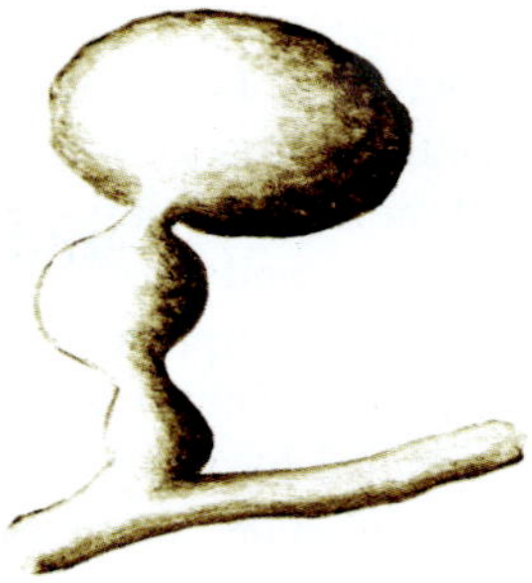

Blastic (**holoblastic**) conidium of *Acremoniella verrucosa*, produced from an inflated conidiogenous cell, x 2,350 (*MU*).

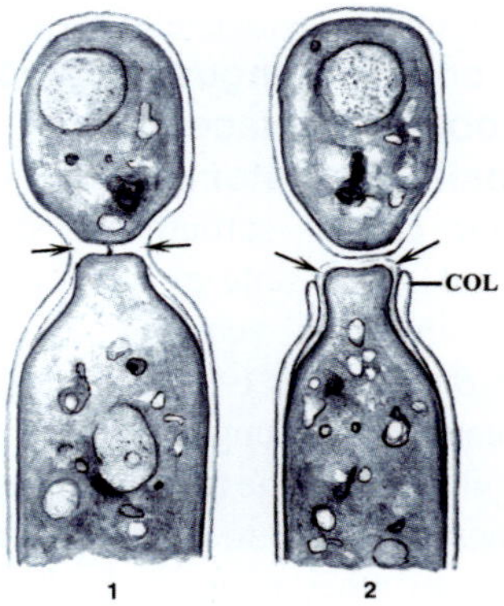

Blastic (**enteroblastic**) conidia of *Metarhizium anisopliae*. Longitudinal sections through the basal septa of differentiated conidia and apices of the phialides, before initiation of the new conidium (1, x 8,850) and at an early stage of secondary conidium development (2, x 7,450). The arrows in 1 indicate the continuity of the inner phialide wall and outer conidial wall. The ring of torn wall is termed the collarette (COL). The arrows in 2 indicate the newly differentiated inner wall layer which will surround the following conidium being formed. Drawing based on transmission electron micrographs published by Cole and Samson in their book *Patterns of Development in Conidial Fungi*, 1979 (*MU*).

black yeasts (ME. *black* < OE. *blaec*, *black*; ME. *yest* < OE. *gist*, yeast): yeast-like states of *Aureobasidium*, *Cladosporium* and *Moniliella*, among other dematiaceous asexual fungi, especially anamorphs of Herpotrichiellaceae, including *Exophiala*, *Ramochloridium* and *Rhinocladiella*, which form black colonies in culture.

blasteniospore (< gen. *Blastenia*, syn. of *Caloplaca*, of the Teloschistales + Gr. *sporá*, spore): a polarilocular lichen spore.

blastesis, pl. **blasteses** (Gr. *blástesis*, germination, development): germination of yeast cells by production of a germ tube instead of a bud, which results in the formation of a membranous phase. Also applied to thallus development in the lichens.

blastic (Gr. *blastós*, sprout, bud, shoot + suf. *-íkos* > L. *-icus* > E. *-ic*, belonging to or relating to): *Conidiogenesis*. A type of conidium development from a portion of a conidiogenous cell in which the conidial primordium undergoes an obvious enlargement before it is delimited by a septum. When all of the layers of the wall of the conidiogenous cell are involved in the synthesis of the wall of the conidium, the latter is called *holoblastic*, as occurs, e.g., in *Acremoniella verrucosa* (dematiaceous asexual fungi); if the outer layer of the wall of the conidiogenous cell is perforated during formation of the conidium (by evagination) and does not contribute to the synthesis of the conidium wall, it is said to be *enteroblastic*, as occurs, e.g., in *Metarhizium anisopliae* (moniliaceous asexual fungi). Blastic conidia include **blastospores**, **botryoblastospores**, **sympodulospores**, **aleuriospores**, **phialospores**, **porospores** and **meristem blastospores**. Blastic conidium ontogeny or conidiogenesis is one of the two principal modes of conidium formation among the conidial fungi, the other being the **thallic** mode.

blastidium, pl. **blastidia** (NL. *blastidium* < Gr. *blastídion* < *blastós*, sprout, germ, shoot + dim. suf. *-ídion* > L. *-idium*): *Lichens*. A vegetative propagule, composed of a mycobiont and a photobiont, which segment in a yeast-like manner, in the thallus of *Physcia opuntiella* (Lecanorales).

blastoconidium, pl. **blastoconidia** (Gr. *blastós*, sprout, bud, germ + NL. *conidium* < Gr. *konídion* < *kónis*, dust, powder + dim. suf. *-ídion* > L. *-idium*): a blastic conidium. See **blastic**.

Blastomycetes (Gr. *blastós*, sprout, bud, germ + L. *-mycetes*, ending of taxonomic class < Gr. *mýkes*, genit. *mýketos*, fungus): a class of conidial fungi (Blastomycetes) that reproduces mainly by budding. It includes the imperfect or anascosporogenous yeasts of the orders Sporobolomycetales and Cryptococcales. Together with the Hyphomycetes and Coelomycetes, the Blastomycetes were formerly considered, in some classification systems, as the

Subdivision Deuteromycotina, with the general name of deuteromycetes. In the classification system followed in this dictionary, the asexual yeasts are no longer assigned the rank of class and are classified as cryptoccocaceous and sporobolomycetaceous asexual yeasts. See the Appendix to compare the classification changes of this class in Alexopoulos *et al*., 1996, modified by Ulloa and Hanlin, 2006, to that of Kirk *et al*., 2008. These asexual fungi are made up of heterogeneous species, forming unnatural groupings, and therefore are considered an artificial classification. These taxonomic rearrangements, with the concurrent changes of the names of taxa, are in order to achieve a phylogenetic classification of the fungi. See **conidial fungus**, **cryptoccocaceous** and **sporobolomycetaceous**.

blastomycosis, pl. **blastomycoses** (< gen. *Blastomyces* < Gr. *blastós*, sprout, bud, shoot + *mýkes*, fungus + suf. *-osis*, state or condition): a primary pulmonary infection acquired by inhalation of conidia of *Blastomyces* (*=Ajellomyces*) *dermatitidis* (moniliaceous asexual fungi), a dimorphic species that inhabits soil. Initially, the fungus causes a pulmonary blastomycosis similar to tuberculosis, but it can disseminate to other parts of the body, causing a cutaneous or systemic infection. See **mycosis**.

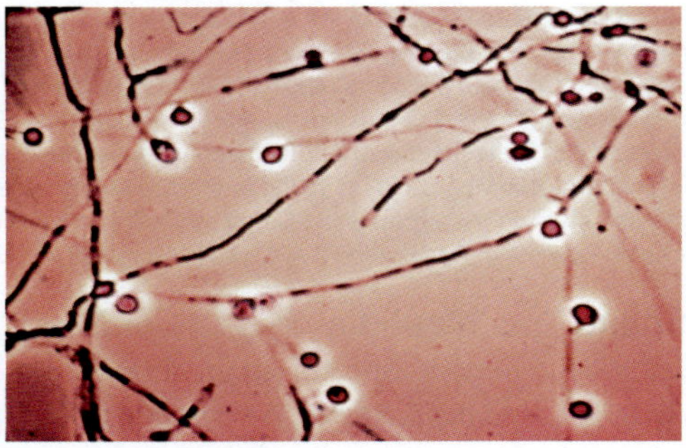

Cells (capable of budding) of the parasitic phase of *Blastomyces* (*=Ajellomyces*) *dermatitidis*, the causal agent of **blastomycosis**, x 750 (*RLM*).

blastospore (Gr. *blastós*, sprout, bud, germ + *sporá*, spore): *Conidiogenesis*. An asexual reproductive spore that originates by budding, as in the yeasts and some filamentous fungi. *Torulopsis*, *Kloeckera* and *Candida* (cryptococcaceous asexual yeasts), and *Geotrichum*, *Monilia* and *Cladosporium* (moniliaceous and dematiaceous asexual fungi, respectively), are examples of gen. of fungi that reproduce asexually by means of blastospores. See **budding**.

blematogen, **blematogenous** (Gr. *blêma*, genit. *blêmatos*, cover + *génos*, origin, birth < *gennáo*, to engender, produce): see **protoblem**.

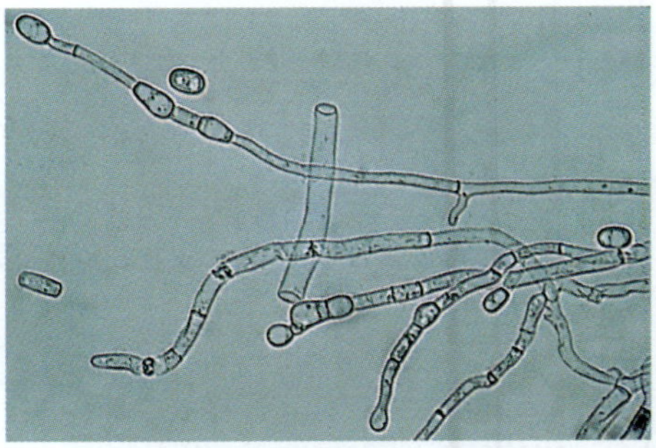

Arthrospores and **blastospores** of *Geotrichum candidum*, x 450 (*PL*).

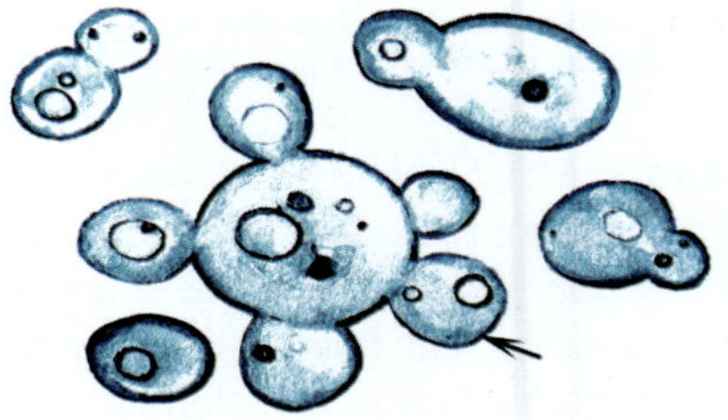

Blastospores of the somatic cells of *Torulopsis taboadae*, x 1,100 (*MU*).

blepharoplast (NL. *blepharoplastus* < Gr. *blepharís*, eyelash + *plastós*, modeled, formed; an organized body or particle): see **basal body** and **kinetosome**.

blight (etymological origin unknown): a necrotizing disease or injury of potato plants resulting in withering, cessation of growth, and death of parts, caused by *Phytophthora infestans* (*late blight*) and by *Alternaria solani* (*early blight*), of the Peronosporales and the dematiaceous asexual fungi, respectively. A common term for other plant diseases, as chestnut blight and pole blight.

blue (**bleu**) **cheese** (ME. < OF. *blou*, blue; ME. *chese* < OE. *cese* < L. *caseus*, cheese): a type of cheese that is ripened and flavored by *Penicillium roqueforti* (moniliaceous asexual fungi), e.g., Roquefort, Stilton, Gorgonzola, Danish Blue and others.

Blue cheese produced by *Penicillium roqueforti*, x 0.5 (*MU*).

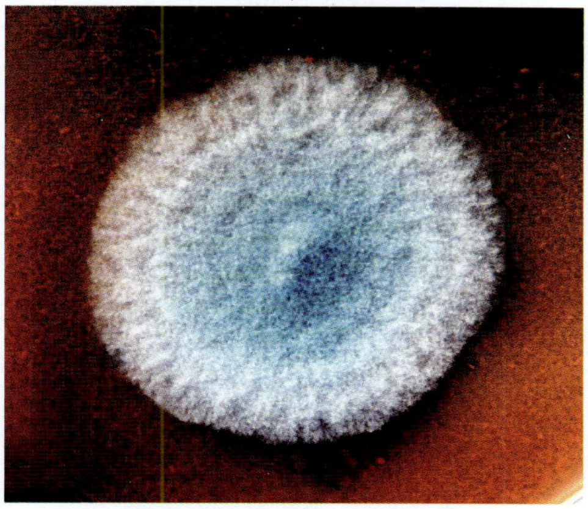

Colony of *Penicillium roqueforti* on agar, a species used to produce **blue cheese**, on agar, x 1 (*MU*).

blue stain (ME. < OE. *blou*, blue; ME. *steyne* < MF. *desteindre*, to discolor, and ON. *steina*, to paint): a blue-gray coloration of wood caused by the growth of fungal hyphae, e.g., *Ceratocystis* (Microascales) and *Ophiostoma* (Ophiostomatales), in the tissues of the wood, but this coloration is not a wood decay.

Blue stain of pine wood caused by *Ceratocystis* infection; cross section of a trunk, x 0.2 (*RTH*).

bolete (< gen. *Boletus* < Gr. *bolítes*, an old name for certain edible mushrooms and roots < *bõlos*, clod, gleba, mass, lump): a fleshy, agaricoid fungus with a tubular hymenophore, in the family Boletaceae. The boletes (*Boletus* spp.) include terrestrial fungi, usually saprobic, occasionally fungicolous or mycorrhizal, mostly edible, although a few are poisonous, of wide geographic distribution. Cf. **agaric**.

boletinoid (< gen. *Boletinus* < gen. *Boletus* < Gr. *bolítes*, an old name for certain edible mushrooms and roots < *bõlos*, clod, gleba, mass, lump + L. suf. *-oide*, like < Gr. *-oeídes*, similar to): applied to an hymenophore with a structure intermediate between pores and gills, as occurs in *Boletinus* and *Gyrodon*, of the Boletaceae (Agaricales).

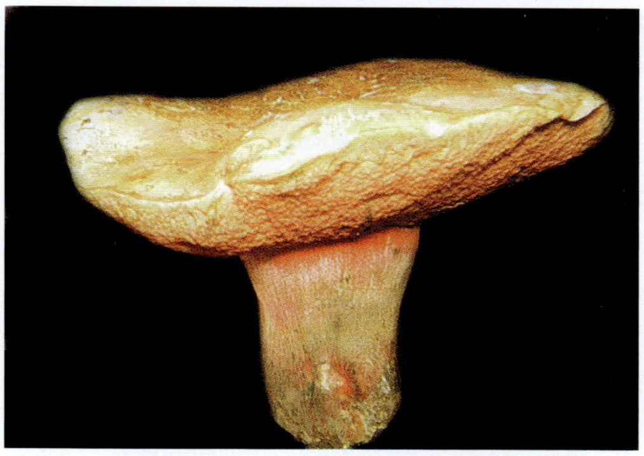

Bolete fruiting body of *Boletus satanas*, x 0.5 (*TH*).

boletoid (< gen. *Boletus* < Gr. *bolítes*, an old name for certain edible mushrooms and roots < *bõlos*, clod, gleba, mass, lump + L. suf. *-oide*, like < Gr. *-oeídes*, similar to): a mushroom with a capitate-stipitate fruiting body, but with a poroid hymenophore, as in *Boletus*, *Boletellus*, *Suillus* and *Leccinum*, among other gen. of putrescent Agaricales.

Boletoid fruiting bodies of *Leccinum aurantiacum*, x 0.5 (*RV*).

bombycine (L. *bombycinus*, of silk < *bombyx*, the silkworm, and silk cloth): refers to a structure whose surface has very fine hairs or threads, like silk.

bothrosome (Gr. *bóthros*, a hole, trench + *sõma*, body): an invaginated organelle on the cell surface that connects the plasma membrane to the network of membranes in the Labyrinthulomycota. Equal to **sagenogen**.

botryoaleuriospore (Gr. *bótrys*, raceme of grapes + *áleuron*, wheat flour + *sporá*, spore): each one of the aleuriospores that, in the shape of a raceme, are formed basipetally on the apex of a conidiogenous cell.

botryoblastospore (Gr. *bótrys*, raceme of grapes + *blastós*, germ, shoot, bud + *sporá*, spore): *Conidiogenesis*. A spore that is formed simultaneously with others by budding on an *ampulla* or vesicle of a conidiophore. E.g., *Botryosporium*, *Botrytis*, *Ostracoderma* (=*Chromelosporium*) and *Oedocephalum* (moniliaceous asexual fungi) are all gen. that form botryoblastospores.

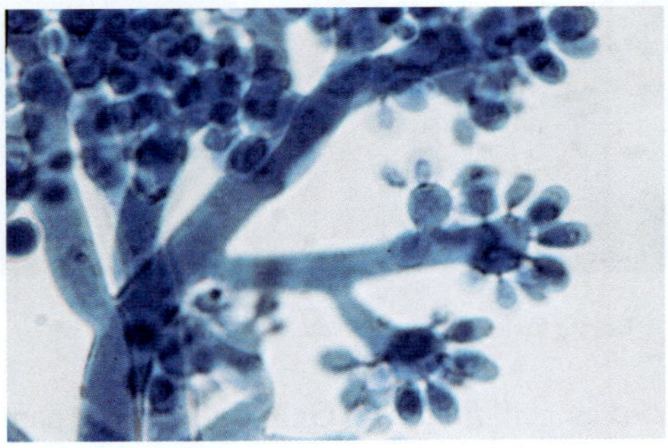

Botryoblastospores on conidiogenous cells (ampullae) of *Botrytis cinerea*, x 580 (*MU*).

botryose, **botryomorphic**, **botryomorphous** (Gr. *bótrys*, raceme of grapes + L. suf. *-osus* > E. *-ose*, full of, increased, prone to; or + *bótrys* + *morphé*, shape + suf. *-íkos* > L. *-icus* > E. *-ic*, pertaining to or related to; or *bótrys* + *morphé* + L. suf. *-osus* > E. *-ous*, full of, increased, prone to): arranged like a cluster of grapes; racemose, racemiform, like the arrangement of the conidia of *Botrytis cinerea*, *Botryosporium* and *Oedocephalum* (moniliaceous asexual fungi) and the sporangia of the Myxomycete *Trichia botrytis* (Trichiales). See **acinose**.

botuliform (NL. *botuliformis* < *botulus*, sausage + suf. *-formis* < *forma*, shape): sausage-shaped, i.e., cylindrical, plump, slightly curved, with both ends rounded. Equal to **allantoid**.

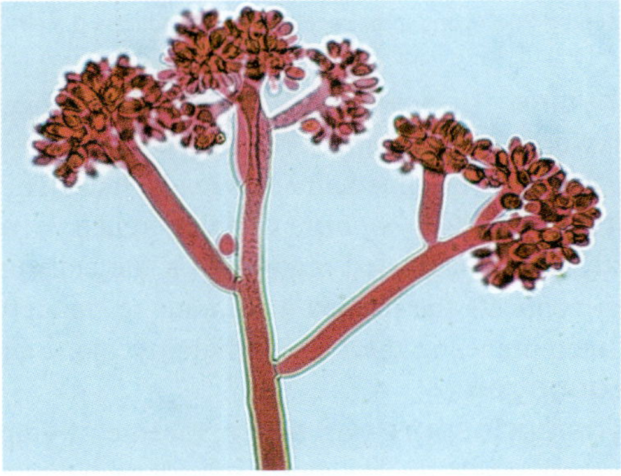

Botryose or **botryomorphic** arrangement of conidia of *Botrytis cinerea*, x 240 (*CB*).

Botryose conidia on an ampulla of *Oedocephalum* sp., x 300 (*MU*).

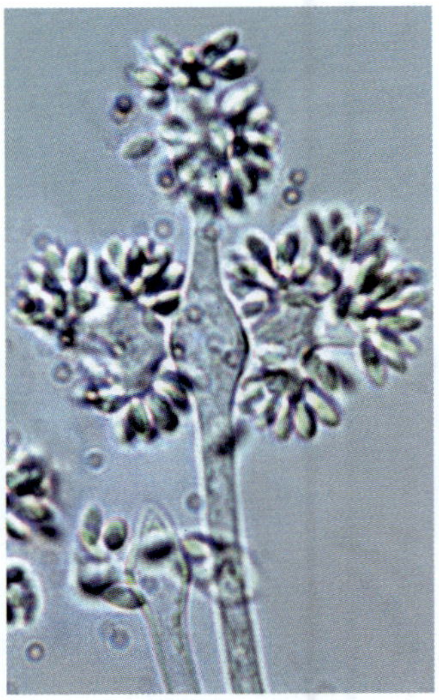

Botryose conidia on conidiogenous cells (ampullae) of *Botryosporium* sp., x 600 (*MU*).

Botryose sporangia of *Trichia botrytis*, x 30 (*MU*).

bourrelet (Fr. *bourrelet*, rim, flange): ascus border. See **ascus**.

bovistelloid (< gen. *Bovistella* < gen. *Bovista* < G. *Bovista*, common puffball + L. suf. *-oide*, like < Gr. *-oeídes*, similar to): *Basidiomycetes*. A type of fruiting body development (*forate*) in which a cavity is formed with a sterile base and a capillitium composed of separate filaments, which taper toward the apices, as seen in *Bovistella* (Lycoperdales). See **forate development**.

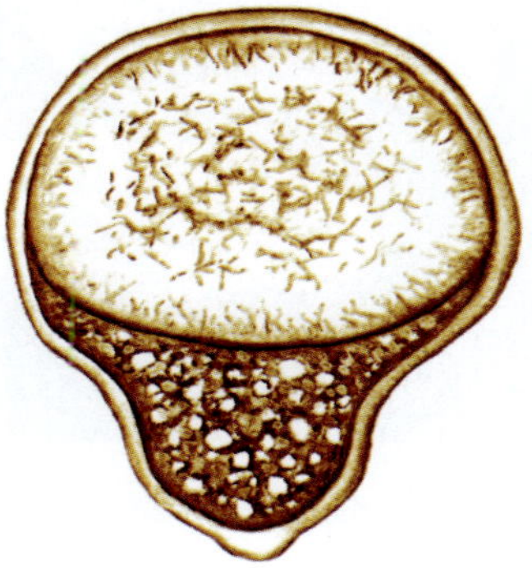

Bovistelloid fruiting body development of *Bovistella* sp., seen in longitudinal section, x 1 (*MU*).

bovistoid (< gen. *Bovista* < G. *Bovista*, common puffball + L. suf. *-oide*, like < Gr. *-oeídes*, similar to): *Basidiomycetes*. A type of fruiting body development with a usually ovoid shape and lacking a sterile base, but with a capillitium composed of separate branched filaments with tapered apices, as in *Bovista* (Lycoperdales).

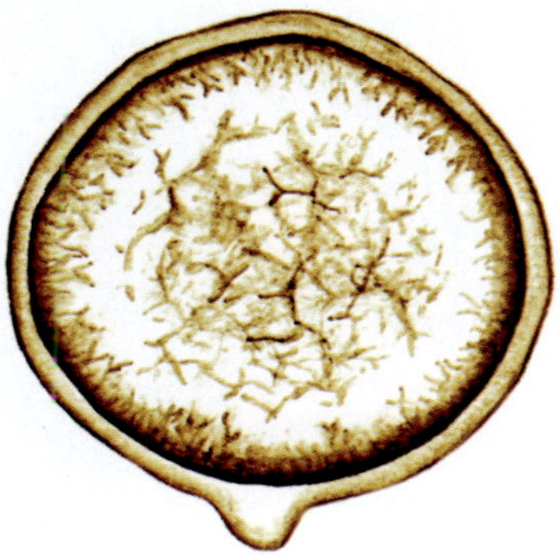

Bovistoid development of the fruiting body of *Bovista* sp., seen in longitudinal section, x 1 (*MU*).

brachial (L. *brachialis*, relating to arms < *brachium*, arm + suf. *-alis* > E. *-al*, pertaining to or relating to): in general, belonging to or relating to the arm.

brachiate (L. *brachiatus*, with arms < *brachium*, arm + suf. *-atus* > E. *-ate*, provided with or likeness): having arms. E.g., the fertile hyphae, conidiophores, of *Euryancale sacciospora* (Zoopagales), are shaped like arms and develop outside the nematode parasitized by this fungus.

brachy- (Gr. *brachýs*, short): a pref. often employed in biological terminology to form technical words taken from the Greek to give the idea of shortness, as in *brachybasidiole*. Cf. **longi-**.

Brachiate conidiophore branches and conidia of *Euryancale sacciospora*, arising from a parasitized nematode (of the gen. *Bunonema*), x 800 (*MU*).

brachybasidiole, brachybasidiolum, pl. **brachybasidiola** (Gr. *brachýs*, short + NL. *basidiolum* < L. *basidium* + dim. suf. *-olum* > E. *-ole*): *Basidiomycetes*. An hymenial element, especially in the gen. *Coprinus* (Agaricales), similar to a basidiolum in structure, but slightly more differentiated; it differs from a basidiolum mainly in being larger and more inflated. The brachybasidiola dominate to form a layer through which the basidia project; apparently the brachybasidiola act as spacing elements for the basidia. This type of hymenial structure is known as coprinoid. It also is called **brachybasidium** and **brachycystidium**. Besides species of *Coprinus* (*C. comatus*), *Coprinopsis* (*Co. atramentarius*) and *Coprinellus* (*Cpr. micaceus*), brachybasidiola also are present in *Conocybe lactea* and *Bolbitius vitellinus*, among other Agaricales.

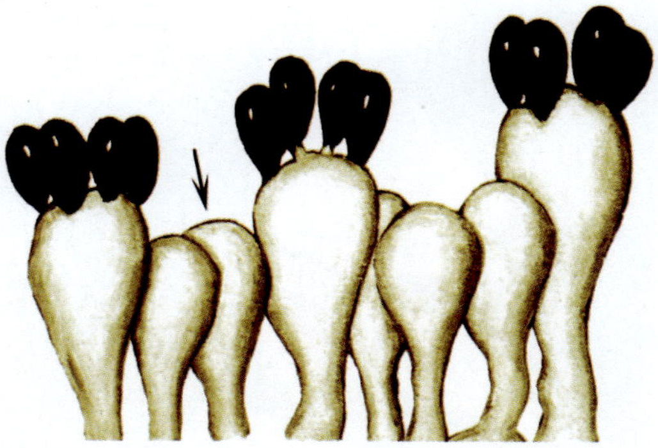

Brachybasidiola of the hymenium of *Coprinus comatus*, x 400 (*MU*).

brachybasidium, pl. **brachybasidia** (Gr. *brachýs*, short + L. *basidium* < Gr. *basídion* < *básis*, base + dim. suf. *-ídion* > L. *-idium*): see **brachybasidiole**.

brachycystidium, pl. **brachycystidia** (Gr. *brachýs*, short + *kystídion* < *kýstis*, bladder, vesicle, cell + dim. suf. *-ídion* > L. *-idium*): see **brachybasidiole**.

brachyrostrate (Gr. *brachýs*, short + *rostratus*, rostrate < *rostrum*, beak, snout + suf. *-atus* > E. *-ate*, provided with or likeness): see **brevirostrate**.

bracket fungus, pl. **fungi** (MF. *braguette*, codpiece < L. *braca*, breeches; an overhanging member that projects from a structure (as a wall) and is usually designed to support a vertical load or to strengthen an angle; L. *fungus*, fungus): a common name for the basidiocarps of Aphyllophorales, which are in the form of a codpiece from the trunks or logs where they are growing, e. g., in *Pyrofomes demidoffii*, *Fomes*, *Hexagonia*, *Inonotus*, *Trametes* and other gen.

Bracket fungus, a basidiocarp of *Pyrofomes demidoffii*, x 0.5 (*RV*).

Bracket fungi, the basidiocarps of *Trametes maxima*, x 0.2 (*RV*).

Bracket fungi, the basidiocarps of *Hexagonia tenuis*, x 0.3 (*RV*).

brevi- (L. *brevis*, brief, short): a pref. used to give the idea of short, as *brevirostrate*. Cf. **longi-**.

brevirostrate (L. *brevis*, brief, short + *rostratus*, rostrate < *rostrum*, beak, snout + suf. *-atus* > E. *-ate*, provided with or likeness): with a short beak or rostrum; applied to certain pycnidia or perithecia whose ostiolar neck is short, as *Lophotrichus brevirostratum* (Sordariales). Also called **brachyrostrate**.

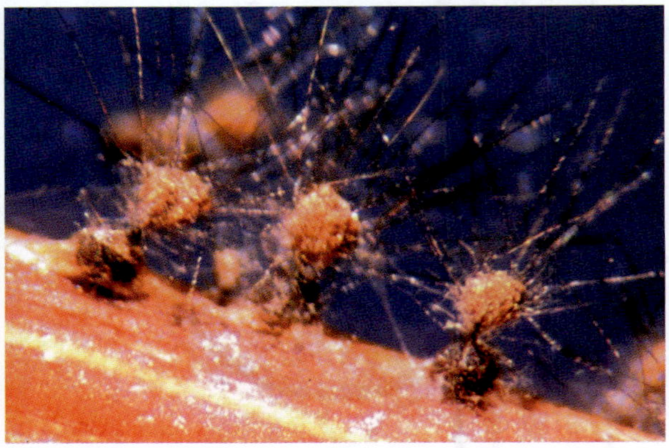

Perithecia of *Lophotrichus brevirostratum*, with a short rostrum (**brevirostrate**) and hairs, x 80 (*MU*).

bridging host or **gephyrophyte** (ME. *brigge* < OE. *brycg*, pier, connecting structure + E. suf. *-ing*, denotes the action of the verb; L. *hospes*, genit. *hospitis*, a living organism that provides food and lodging to a parasite; one who receives a stranger as guest or visitor; Gr. *géphyra*, bridge + *phytón*, plant): a host plant that permits a parasitic fungus to infect a plant it could not otherwise infect, as occurs in some Erysiphaceae. E.g., a fungus that can infect plant *a* but not plant *c* becomes able to infect *c* after passing through plant *b*, which serves as the bridging host. Cf. **alternate host**, **intermediate host** and **primary host**.

bromatium, pl. **bromatia** (NL. *bromatium* < Gr. *bromátion* < *bróma*, genit. *brómatos*, food + dim. suf. *-íon* > L. *-ium*): any of the swollen tips of hyphae that serve as food for ants that cultivate fungus gardens. *Lepiota* (Agaricales) is an example of a genus of fungi that forms bromatia. Also called bromatia are the masses of yeasts with the appearance of cheese that ants of the species *Cyphomyrmex rimsus* cultivate. Bromatia are also called **gongyli** and **staphyla**.

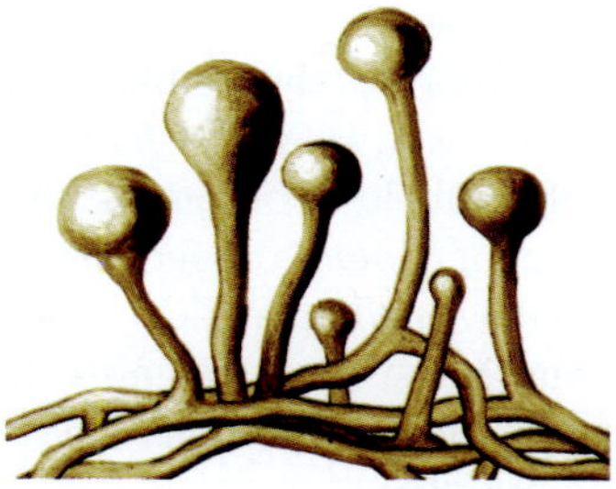

Bromatia of *Lepiota* sp., x 300 (*MU*).

broom cell, **brush cell** (OE. *brom* > ME. *brome*, broom or OF. *broce* > MF. *broisse* > ME. *brusshe*, brush; NL. *cellula*, cell, living cell, dim. of *cella*, small room, compartment, cell of a honeycomb): *Basidiomycetes*. Each one of the cystidia that have on the apex various protuberances or simple or branched appendages that are sharp-pointed and dark in color, clavate or sphaeropedunculate, and hirsute, echinulate, or arborescent, like those in *Marasmius plicatilis* and *Mycena rulantiformis* (Agaricales).

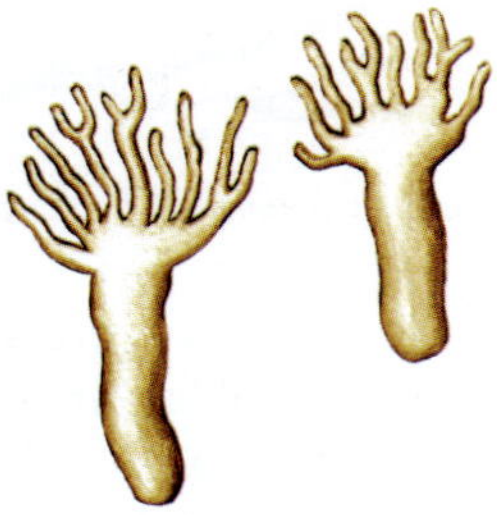

Broom cells of the hymenium of *Marasmius plicatilis*, x 1,120 (*MU*).

bryophilous (Gr. *brýon*, moss + *phílos*, have an affinity for): an organism that grows in or on mosses. Syn. of **muscicolous**. Various species of Myxomycetes, e.g., are bryophilous, like *Craterium minutum* (Physarales), among others, although this species also grows on wood.

budding (E. *budding*, to set or put forth buds < OE. *budda*, beetle, a small lateral or terminal protuberance; something not yet mature or at full development): see **gemmation**.

bulbil (L. *bulbilus*, dim. of *bulbus*, bulb): a small sclerotium, almost microscopic, formed of a few cells, like those of *Papulaspora* (agonomycetaceous asexual fungi) and *Minimedusa polyspora,* an asporogenous fungus with sterile mycelium with an affinity to the basidiomycetes, considered as the anamorph of *Sistotrema* (Aphyllophorales). Also called **microsclerotium** and **papulospore**.

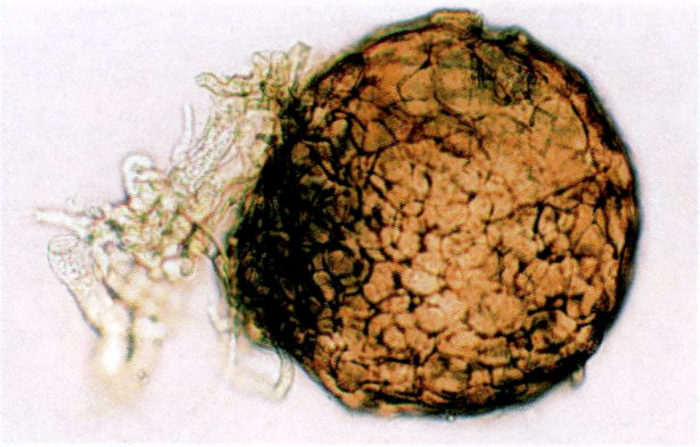

Bulbil of *Minimedusa polyspora*, x 1,120 (*CB*).

bulbillate (L. *bulbilus*, dim. of *bulbus*, bulb + suf. *-atus* > E. *-ate*, provided with or likeness): having *bulbils*. Also applied to a stipe with a small bulb (*bulbulo* < L. *bulbulus*, dim. of *bulbus*), or a slightly delimited bulb at the base.

bulbillosis, pl. **bulbilloses** (L. *bulbilus*, dim. of *bulbus*, bulb + Gr. suf. *-osis*, condition): *Basidiomycetes.* A phenomenon observed in the gills of the agaricaceous fungi and which consists of their disintegration into rounded fragments, in the mode of bulbils, in whose cells meiosis occurs; i.e., bulbillosis is the formation of bulbils instead of basidiospores.

bulbous (L. *bulbosus*, full of bulbs, having bulbs, bulb-shaped < *bulbus*, bulb + suf. *-osus* > E. *-ous*, full of, increased, prone to): enlarged basally in the manner of a bulb, like the stipe of *Leucocoprinus cepistipes*, *Inocybe calospora* and *Lepiota* sp. (Agaricales).

Bulbous stipe of the basidiocarp of *Leucocoprinus cepistipes*, x 0.3 (*SA*).

bulgar grain, **kefir grain** (ML. *bulgarus* < Gr. *boulgaros* < Turk. *bulghar*, of mixed origin < *bulgamaq*, to mix; Rus. *kefir*, milk curdled in an animal skin): a *microbioglea,* generally known as *kefir grain*, that is composed of lactic bacteria and yeasts that is utilized in many parts of the world as a starter culture to ferment milk and obtain the so-called *kefir*. See **microbioglea**.

Bulgars (microbiogleae) cultivated in milk, x 0.5 (*MU*).

Bulgars (microbiogleae) cultivated in milk, x 1 (*MU*).

Bulgar (microbioglea) cultivated in milk, x 5 (*MU*).

bullate (L. *bullatus*, blistered, vesiculate < *bulla*, bubble, blister + suf. *-atus* > E. *-ate*, provided with or likeness): having the surface covered with slight, irregular elevations that give it a blistered appearance. See **bulliform**.

Buller phenomenon (named after discoverer Canadian mycologist A. H. R. Buller; LL. *phaenomenon* < Gr. *phainómenon*, appearance < *phaíno*, to show + *ménon*, disposition, force): the diploidization of a unisexual mycelium by a dikaryotic mycelium in ascomycetes and basidiomycetes.

bulliform (L. *bulliformis*, shaped like a bubble < *bulla*, bubble, blister + suf. *-formis* < *forma*, shape): with the shape of a bubble or blister, like the cystidia of certain species of *Leptonia* (Agaricales), *Peniophora* and *Corticium* (Aphyllophorales), whose apices are bulliform and evanescent, leaving a kind of collar that consists of the basal portion of the original wall. Also called **bullate.**

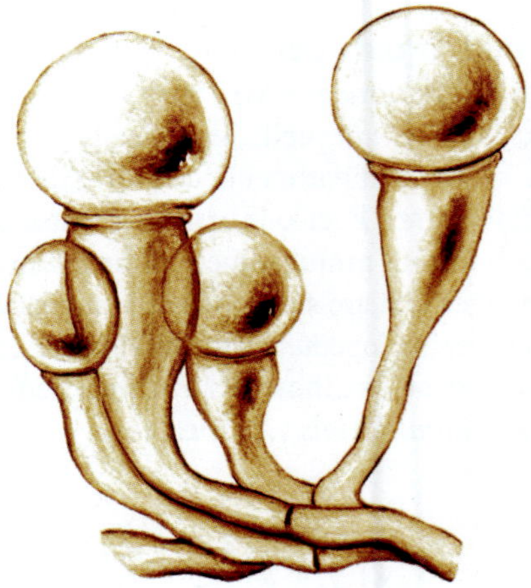

Bulliform cystidia of the hymenium of *Leptonia* sp., x 720 (*MU*).

bursiculate (L. *bursiculatus*, provided with small bags, or similar to a small bag or pouch < *bursicula*, VL. dim. *bursa*, purse, bag + suf. *-atus* > E. *-ate*, provided with or likeness): having small bags or small bladders, or with a shape similar to a small bag. Equal to **bursiform**.

bursiform (L. *bursiformis*, shaped like a bag or pouch VL. *bursa*, purse, pouch + suf. *-formis* < *forma*, shape): equal to **bursiculate**.

button (ME. *boton* < MF. *boton* < OF. *boter*, to thrust, throw, project): *Basidiomycetes*. The bud, shoot or primordium of an organ; in general, it is applied to the early stage of a developing fruiting body, such as the mushrooms (*Agaricus bisporus* and *Pleurotus ostreatus*), or any other cultivated or wild agaric, when it is still unopened and the gills are not exposed.

Primordia of several basidiocarps (**buttons**) of *Pleurotus ostreatus*, x 3 (*MU*).

butyraceous (L. *butyraceus*, buttery < *butyrum*, butter + suf. *-aceus* > E. *-aceous*, of or pertaining to, with the nature of): having the appearance or consistency of butter, like many cultures of yeasts on solid media, which appear moist and opaque or semibrilliant, like those of *Candida lipolytica* and *Rhodotorula rubra* (cryptococcaceous asexual yeasts) and *Saccharomyces cerevisiae* (Saccharomycetales).

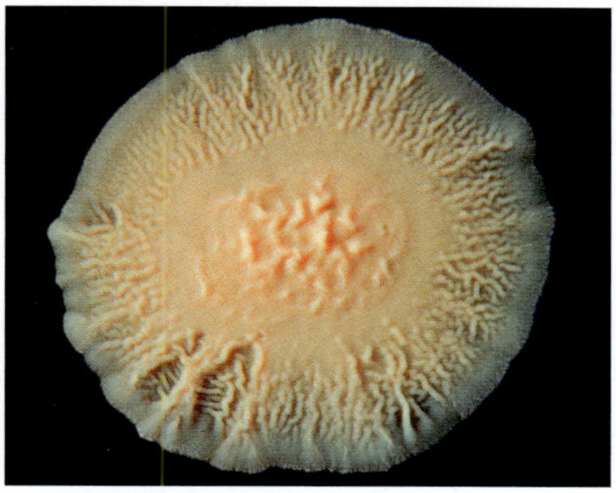

Butyraceous giant colony of *Candida lipolytica*, on agar, x 1 (*MU*).

Butyraceous giant colony of *Rhodotorula rubra*, on agar, x 1 (*MU*).

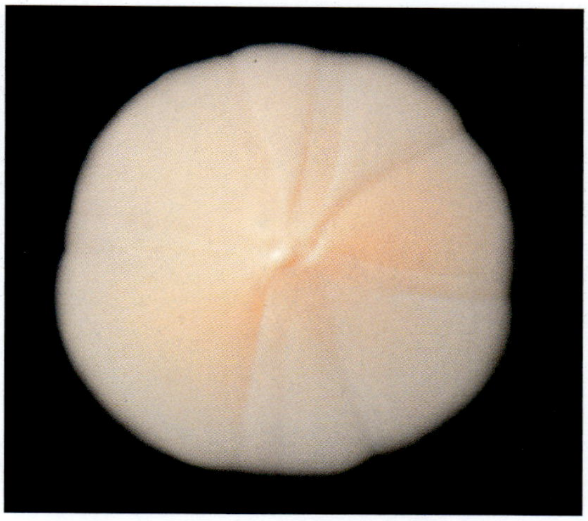

Butyraceous giant colony of *Saccharomyces cerevisiae*, on agar, x 1 (*MU*).

byssaceous (L. *byssus* < Gr. *býssos*, linen cloth of very fine threads + L. suf. *-aceus* > E. *-aceous*, of or pertaining to, with the nature of): equal to **byssoid**.

byssoid (L. *byssus* < Gr. *býssos*, linen cloth of very fine threads + L. suf. *-oide*, like < Gr. *-oeídes*, similar to): composed of very fine, delicate filaments, cottony or floccose, like the mycelium of many fungi. E.g. the asci of *Byssochlamys nivea* (Eurotiales) are found scattered among the intertwined, cottony vegetative hyphae of the mycelium. The delicate pattern of radial hyphae (=fibrils) characteristic of the gleba of *Sclerogaster* (Hymenogastrales), and the apothecial ascomata of the lichen *Byssoloma subdiscordans* (Lecanorales), with brilliant colors and poorly developed margin composed of loosely intertwined hyphae, which at times gives them a lanose appearance, also are byssoid.

Byssoid mycelium, with asci, of *Byssochlamys nivea*, x 1,500 (*MU*).

byssoid

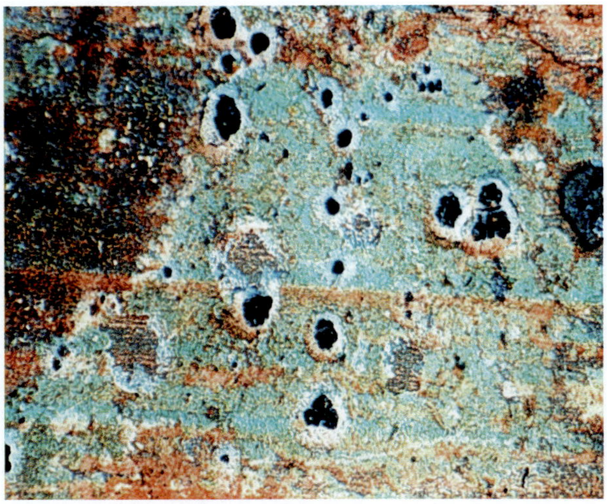

Byssoid apothecial ascomata of the lichen *Byssoloma subdiscordans*, x 20 (*MAH*).

C

Cancellate sporangia of *Cribraria cancellata* (*MU*).

C

cactophilic (L. *cactus*, cactus < Gr. *káktos*, cardoon + *phílos*, have an affinity for + suf. *-íkos* > L. *-icus* > E. *-ic*, belonging to or relating to): developing preferentially on cactaceous plants, like the myxomycete *Badhamia gracilis* (Physarales), which forms its sporangia in the tissue of the cladodes of the prickly pear (*Opuntia ficus indica*), when this tissue begins to decompose and forms a kind of humid chamber where the plasmodia of this organism can grow. *Badhamia gracilis* and other species of Myxomycetes are considered cactophilic because they are often found associated with Cactaceae, although sporadically they can develop on other substrates.

caducous (L. *caducus*, perishable, deciduous, falling early < *cadere*, to fall + *-osus* > OF. *-ous, -eus* > E. *-ous*, possessing the qualities of): detaching, sloughing off or falling off early, like the peridium of the sporangium of *Physarum polycephalum* (Physarales), and the exoperidium of *Lycoperdon* (Lycoperdales) at maturity.

Caducous peridium of the sporangium of *Physarum polycephalum*, x 20 (*MU*).

caeoma, pl. **caeomata** (NL. *caeoma* < Gr. *kaío*, to burn, be consumed by fire + suf. *-óma*, a diseased state): *Uredinales*. A type of aecium that is surrounded by paraphyses instead of a peridium, or which lacks both structures, as in the gen. *Caeoma* of the Uredinales.

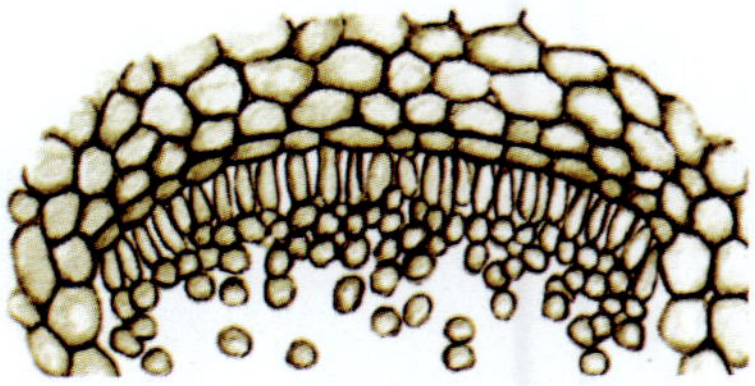

Caeoma of *Caeoma* sp. Drawing of a longitudinal section of an aecium of this fungus, formed on a leaf of a host plant, x 45 (*MU*).

caerulescent, **coerulescent** (L. *caeruleous*, dark blue, bluish + suf. *-escens*, genit. *-escentis* > E. *-escent*, that which turns, that begins to, slightly): turning bluish or bluish-green, like the basidiocarps of the hallucinogenic fungi of the gen. *Psilocybe* (Agaricales), which stain these colors when they are rubbed or damaged, which appears to be an oxidation reaction.

Caerulescent fruiting bodies of *Psilocybe caerulescens*, x 1 (*TH*).

Caerulescent fruiting bodies of *Psilocybe* sp., x 1 (*TH*).

caespitose, caespitous, cespitose (NL. *caespitosus*, tufted like sod-grass < *caespes*, genit. *caespitis*, turf, grass): aggregated or clustered, but not fused, like the sporangia of *Stemonitis smithii* (Stemonitales) and the basidiocarps of *Coprinellus* (=*Coprinus*) *micaceus* and *Laccaria* sp. (Agaricales). Cf. **solitary**.

Caespitose sporangia of *Stemonitis smithii*, x 10 (*MU*).

Caespitose basidiocarps of *Coprinellus* (=*Coprinus*) *micaceus*, x 0.3 (*TH*).

Caespitose basidiocarps of *Laccaria amethystina*, x 1 (*RV*).

calcareous (L. *calcarius*, composed of calcium carbonate < *calx*, genit. *calcis*, lime, chalk, of calcium carbonate + suf. *-arius* > E. *-areous*, belonging to): made up of or containing calcium carbonate; of a grayish-white color. The peridium, columella or capillitium of the sporangia of various species of Myxomycetes is calcareous in nature (*Badhamia*, *Didymium*, *Mucilago* and *Physarum*, among other Physarales).

Calcareous capillitium of the sporangia of *Badhamia dearnessii*, x 35 (*MU*).

Calcareous peridium of the sporangium of *Didymium* sp., x 30 (*MU*).

Calcareous peridium of the sporangium of *Didymium minus*, x 30 (*MU*).

Calcareous peridium of the aethalium of *Mucilago crustacea*, x 15 (*MU*).

Calcareous peridium of the sporangia of *Physarum* sp., x 15 (*MU*).

Calcareous peridium of the sporangium of *Physarum* sp., x 30 (*MU*).

calcariferous (L. *calcar*, genit. *calcaris*, spur, spine + suf. *-ferous*, bearer < *ferre*, to bear, carry): having or bearing spines.

calcariform (L. *calcariformis*, similar to a spine < *calcar*, genit. *calcaris*, spur, spine + suf. *-formis* < *forma*, shape): with the shape of a spine, like the appendages of the conidia of *Harpagomyces lomnickii* (moniliaceous asexual fungi).

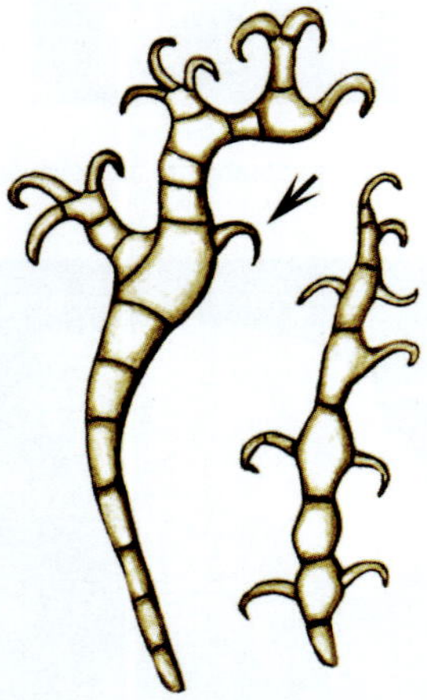

Calcariform appendages of the conidia of *Harpagomyces lomnickii*, x 2,160 (*MU*).

calceiform (L. *calceiformis*, similar to a shoe < *calceus*, shoe + suf. *-formis* < *forma*, shape): with the shape of a shoe, like the conidia of *Calceispora hachijoensis* (moniliaceous asexual fungi).

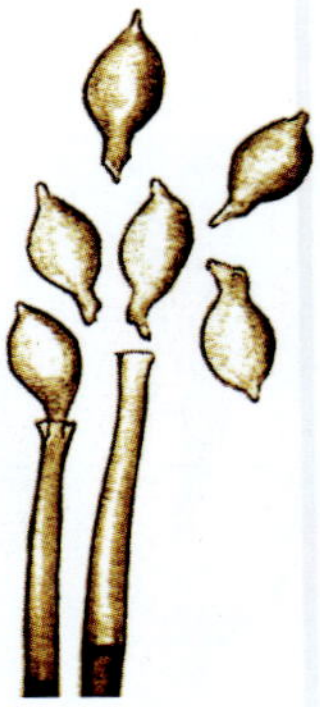

Calceiform conidia of *Calceispora hachijoensis*, x 350 (*MU*).

calcicolous (L. *calx*, genit. *calcis*, lime, chalk, calcium carbonate + suf. *-cola*, inhabitant < *colere*, to inhabit): an organism that lives on substrates rich in calcium carbonate, especially calcareous or gypseous rocks and soils.

calciferous (L. *calx*, genit. *calcis*, lime, chalk, calcium carbonate + suf. *-ferous*, bearer < *ferre*, to bear, carry): having calcium carbonate, like the capillitium of the sporangia of the Physarales (Myxomycetes).

calcivorous (L. *calx*, genit. *calcis*, lime, chalk, calcium carbonate + suf. *-voro*, that which eats < *vorare*, to eat, devour): consuming the calcium carbonate from the calcareous rocks on which it lives, like the lichen *Lecidea calcivora* (Lecanorales), which gradually dissolves this type of substrate.

caligate (L. *caliga*, militay boot + suf. *-atus* > E. *-ate*, provided with or likeness): having a boot; wrapped in a veil. Syn. of **ocreate** and **peronate**.

callosity, pl. **callosities** (L. *callositas*, genit. *callositatis*, hardness, swelling): applied to the thickenings in hyphal cell walls in response to penetration by parasitic fungi.

callus, pl. **calli** (L. *callus*, hardened skin): **1**. *Agaricales*. The apical region of a basidiospore, which is thinner-walled, more or less convex and protuberant, in which the germ pore is located, as is seen in many species of *Galerina* and in *Psathyrella* (=*Lacrymaria*) *rigidipes*, among other Agaricales. **2**. Tissue that forms around wounds in woody plant stems and branches, and which serves to protect the inner tissues. Callus cells are large, thin-walled and undifferentiated.

calo- (Gr. *kalós*, beautiful): a pref. used in many generic names, such as *Calomyxa* (Trichiales), *Calonectria* (Hypocreales) and *Calocera* (Dacrymycetales).

calvescent (L. *calvescens*, genit. *calvescentis*, becoming bald < *calvescere*, to become bald < *calvus*, without hair, smooth + suf. *-escens*, genit. *-escentis* > E. *-escent*, that which turns, that begins to, slightly): becoming hairless.

calvous (L. *calvus*, bald): without hair or pubescence.

calyciform (L. *calyciformis*, like a calyx < *calyx*, genit. *calycis*, cup, goblet < Gr. *kályx*, wrapper, cover + L. suf. *-formis* < *forma*, shape): in the shape of a cup or goblet; e.g., the apothecia of many Discomycetes (*Cookeina*, *Peziza*, *Helvella* and other Pezizales), the basidiocarp of *Cotylidia aurantiaca* (Aphyllophorales), or the sporangia of some Myxomycetes (*Craterium* and other Physarales) are calyciform.

Calyciform apothecium of *Cookeina sulcipes*, x 1 (*MU*).

Calyciform apothecium of *Cookeina tricholoma*, x 2 (*MU*).

calycular (L. *calycularis* < *calyculus*, dim. of *calyx*, calyx, cup + suf. *-aris* > E. *-ar*, belonging to or relating to): pertaining or relating to the *calyculus*. Goblet-shaped.

calyculus, pl. **calyculi** (L. *calyculus* < *calyx*, genit. *calycis*, calyx, goblet + dim. suf. *-culus*): a structure shaped like a goblet, e.g., the calyciform base of the sporangia of various common species of Myxomycetes, such as *Arcyria denudata*, *Hemitrichia calyculata* (Trichiales) and related forms; the calyculus remains as a continuation of the pedicel after the upper portion of the sporangial peridium disappears.

Calyculus of the sporangia of *Hemitrichia calyculata*, x 13 (*CB*).

calyptra, pl. **calyptrae** (NL. *calyptra* < Gr. *kalýptra*, cover): a structure or tissue that covers something; e.g., the perisporial cover that partially surrounds the basidiospore in certain Agaricales, such as *Galerina hypnorum* (= *G. calyptrospora*); also called calyptra are the remnants of the universal veil that remains like a partial cover on the pileus of some Agaricales, such as *Amanita calyptratoides* and *A. fulva*, as well as the tissue that covers the apex of the fruiting body of *Itajahya galericulata* (Phallales), and to which the gleba is attached. Cf. **utriculate**.

Calyptra of the basidiocarp of *Amanita fulva*, x 0.3 (*EPS*).

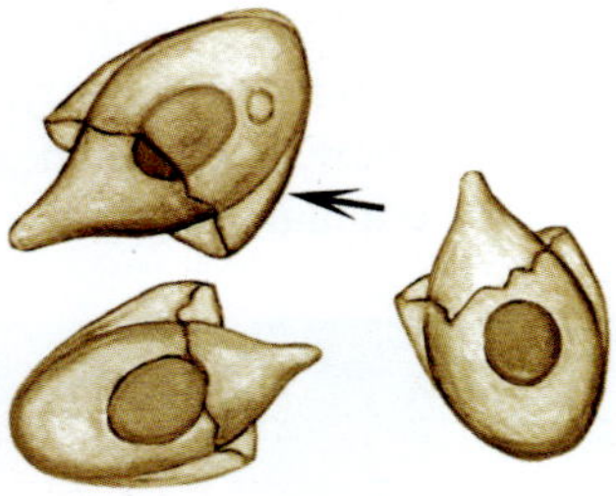

Calyptra of the basidiospores of *Galerina hypnorum* (=*G. calyptrospora*), x 1,570 (*MU*).

calyptrate (NL. *calyptratus*, with a calyptra < *calyptra* < Gr. *kalýptra*, cover + suf. *-atus* > E. *-ate*, provided with or likeness): provided with a *calyptra*.

campanulate (NL. *campanulatus*, like a bell < L. *campana*, bell + suf. *-atus* > E. *-ate*, provided with or likeness): shaped like a bell, like the pileus of certain basidiomycetes, e.g., *Panaeolus campanulatus*, *P. semiovatus* and *P. sphinctrinus* (Agaricales).

Campanulate pileus of the basidiocarps of *Panaeolus campanulatus*, x 1 (*RV*).

Campanulate pileus of the basidiocarps of *Panaeolus campanulatus*, x 1 (*MU*).

campestroid (L. *campestris*, rural, peasant + suf. *-oide*, like < Gr. *-oeídes*, similar to): refers to an agaricoid fruiting body with a diametrical proportion between the pileus and the stipe of 1:1 or > 1:1 Cf. **placomycetoid**.

campylidium, pl. **campylidia** (L. *campylidium* < Gr. *kampylídion* < *kampýlos*, curved, bent + dim. suf. *-ídion* > L. *-idium*): a term utilized initially to designate a type of asexual fructification that occurs in certain lichens, such as *Cyphella* (=*Aleurodiscus*) *aeruginascens* (Aphyllophorales), but recognized later as a fungal parasite of this supposed lichen, since now the latter is regarded as a corticiaceous fungus. Campylidia have been described in the anamorphic phases of the thalli of *Badimia dimidiata* and *Sporopodium xantholeucum* (Lecanorales), tropical foliicolous lichens.

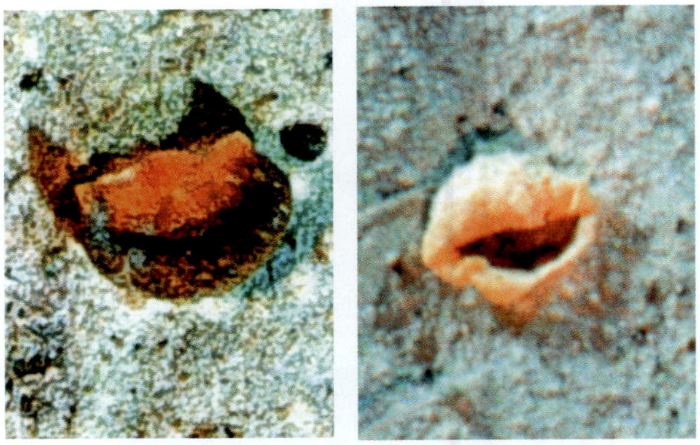

Campylidia of *Badimia dimidiata* and *Sporopodium xantholeucum*, crustaceous foliicolous lichens, x 10 (*MAH*).

canal (L. *canalis* < *canna*, reed, cane + suf. *-alis* > E. *-al*, belonging to or relating to; conduit or bore of a gun, blow pipe): sometimes applied to the pore that extends between the two cells of a polarilocular ascospore.

canaliculate (L. *canaliculatus*, channelled, grooved < *canaliculus*, dim. of *canalis*, canal, groove, furrow + suf. *-atus* > E. *-ate*, provided with or likeness): provided with one or more channels.

canaliculus, pl. **canaliculi** (L. *canaliculus* < *canalis*, canal, groove, conduit + dim. suf. *-culus*): a narrow passage, similar to a small canal, like those formed in the cell wall of various structures and through which pass the *plasmodesmata* that connect one cell with another.

cancellate (L. *cancellatus*, made in the style of an iron-work screen or window lattice < *cancellus*, grating + suf. *-atus* > E. *-ate*, provided with or likeness): similar to a window lattice, like the peridium of the sporangia of *Cribraria cancellata* (=*Dictydium* *cancellatum,* Liceales), which partially decomposes and leaves behind an ensemble of longitudinal ribs that are interconnected by delicate transverse bands, so that the mature sporangia have the appearance of a mesh or lattice.

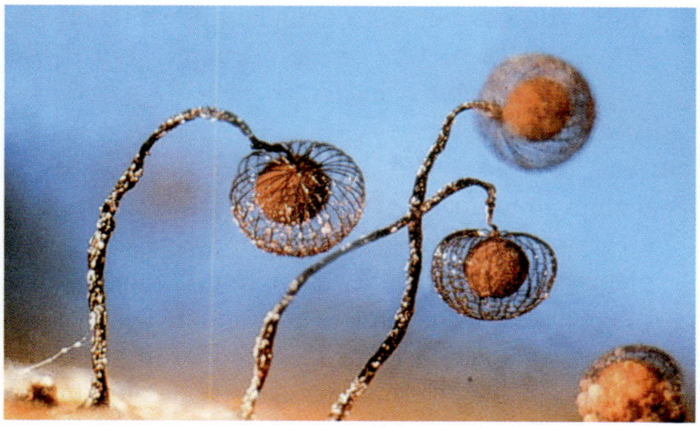

Cancellate sporangia of *Cribraria cancellata* (=*Dictydium cancellatum*), x 20 (*MU*).

Cancellate young sporangium of *Cribraria cancellata*, x 32 (*MU*).

Cancellate sporangium of *Cribraria cancellata*, x 32 (*MU*).

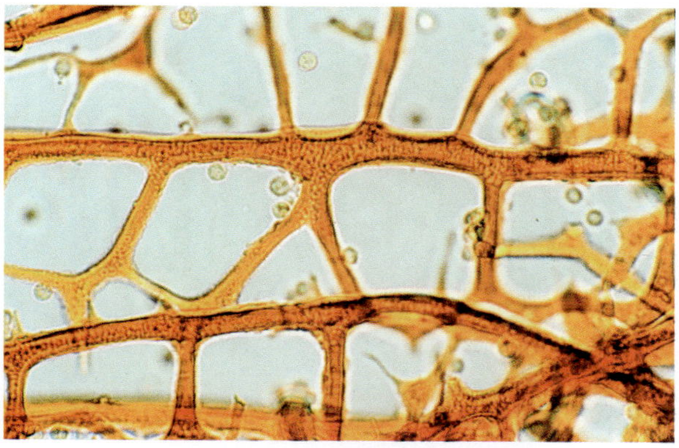

Cancellate peridial net of the sporangium of *Cribraria cancellata*, x 200 (*CB*).

candelabrum, pl. **candelabra** (L. *candelabrum*, candelabra < *candela*, candle, candelstick + *labrum*, lip): a structure that is branched like a candelabrum, like the conidiophore of *Candelabrella* and the dichotomous hyphae, called favic candelabra, characteristic of *Trichophyton schoenleinii* (moniliaceous asexual fungi), the causal agent of *tinea favosa* in man. See **tinea**.

candidiasis (< gen. *Candida* < L. *candidus*, white + Gr. *-iasis*, suf. of names of diseases, although it can also denote an action or process): a superficial infection of the skin, especially of the intertriginous areas (axillary, inframammary, perianal, interdigital and others) or of the face, nails and mucous membranes of the mouth and vagina. In weakened patients the yeasts that cause this mycosis (*Candida albicans*, *C. tropicalis* and *C. parapsilosis*, among other species of cryptococcaceous asexual yeasts) can become systemic. See **thrush.**

Candidiasis of the tongue of a child, caused by *Candida albicans*, x 0.3 (*RLM*).

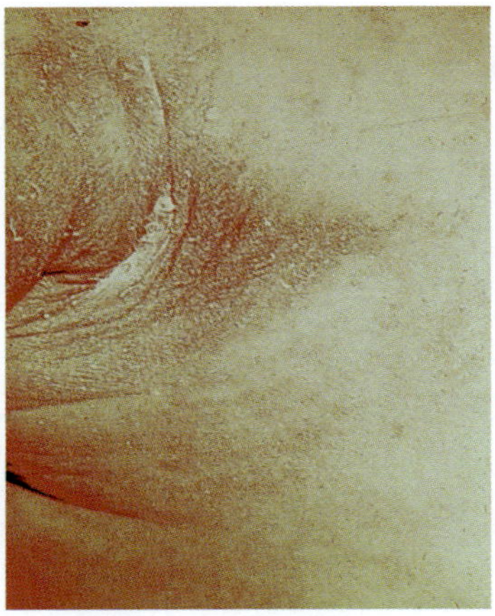

Candidiasis of the armpit of a woman, caused by *Candida albicans*, x 0.3 (*RLM*).

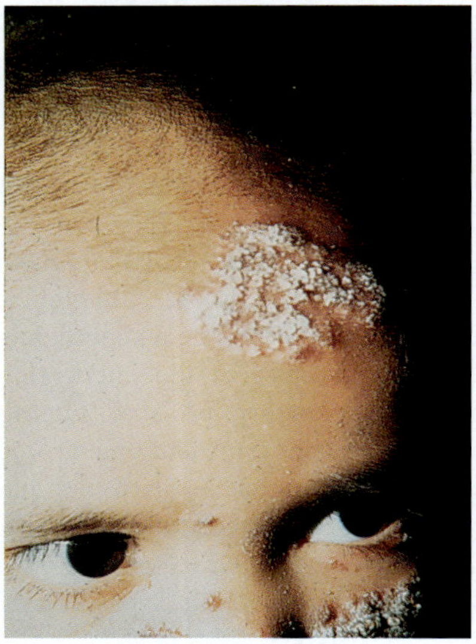

Candidiasis of the forehead of a child, caused by *Candida albicans*, x 0.3 (*RLM*).

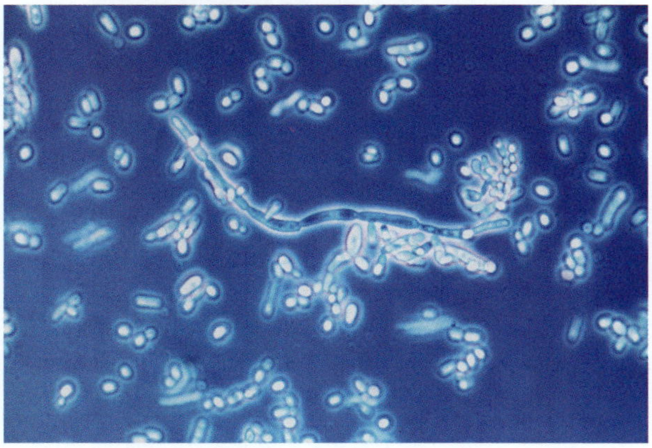

Pseudomycelium and blastospores of *Candida albicans*, the causal agent of **candidiasis**, x 500 (*RLM*).

canescent, **hoary** (L. *canescens*, genit. *canescentis* < *canescere*, to become gray < *canus*, white, hoary, gray or white hair + suf. *-escens*, genit. *-escentis* > E. *-escent*, that which turns, beginning to, slightly; ME. *hor* < OE. *har*, hoarfrost, gray with age + *-y*, characterized by, inclined to > E. *hoary*, gray or white with age): covered with abundant down, short and silky, like the surface of the pileus of *Coprinopsis* (=*Coprinus*) *lagopus* and *Hygrocybe* (=*Hygrophorus*) *canescens* (Agaricales).

canker (ME. < ONF. *cancre* < L. *cancer*, cancer, malignant tumor): a well defined, localized necrotic area on the stems, branches or twigs of woody plants that is surrounded by living tissue. Cankers often appear sunken due to continued growth of the living tissue around them. In addition, callus tissue may form around the canker, making it more sunken. The causal agents of cankers can be bacteria (such as *Agrobacterium tumefaciens*) or fungi, such as *Apiosporina morbosa*, of the Pleosporales, and *Neonectria* (=*Nectria*) *galligena*, of the Hypocreales, and by *Calosphaeria* sp. (Calosphaeriales). See **tumor**.

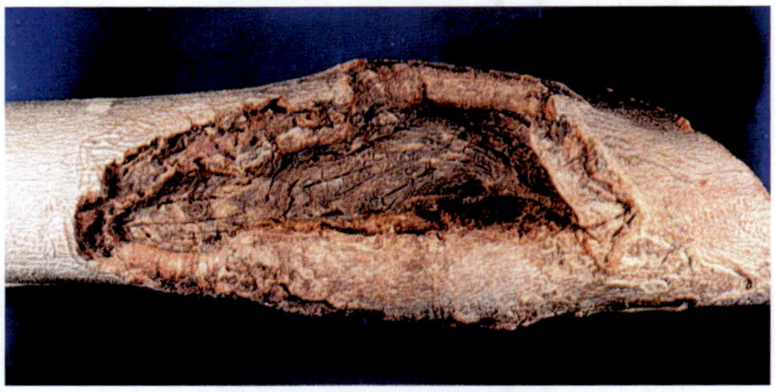

Canker on a branch of maple due to infection by *Nectria* sp., x 0.5 (*RTH*).

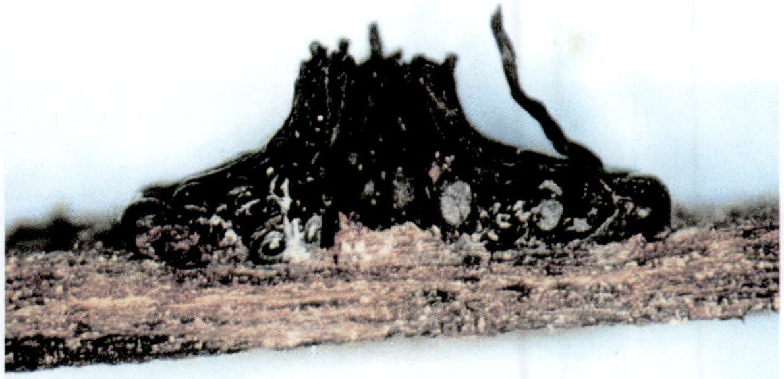

Canker on a peach stem caused by *Calosphaeria* sp., x 10 (*RTH*)

cantharelloid (< gen. *Cantharellus* < L. *cantharus* < Gr. *kántharos*, cup, pitcher + L. dim. suf. *-ellus* + L. suf. *-oide*, like < Gr. *-oeídes*, similar to):

similar to the fruiting body of *Cantharellus cibarius* (Aphyllophorales), i.e., more or less turbinate or goblet-shaped, with decurrent gills.

Cantharelloid basidiocarp of *Cantharellus cibarius*, x 0.5 (*MRO*).

capillaceous (L. *capillaceus*, similar to a hair < *capillus*, hair + suf. *-aceus* > E. *-aceous*, of or pertaining to, with the nature of): see **capilliform**.

capillary (L. *capillaris*, similar to a hair < *capillus*, hair + suf. *-aris* > E. *-ary*, similar to or pertaining to): fine like a hair.

capillary spore (L. *capillaris*, similar to a hair < *capillus*, hair + suf. *-aris* > E. *-ar*, similar to or belonging to; Gr. *sporá*, spore): refers to the only known reproductive structure of *Stylopage* (Zoopagales), a common gen. on dung, where it is predaceous on amoebae. The capillary spore is almost identical to the *capilliconidium* of some entomophthoraceous fungi. Capillary spores possess a hapteron that develops at its tip when moisture is abundant; by means of this structure it attaches to mites which in nature function in dispersal of these fungi. The mites, in turn, are dispersed by dung beetles. This mode of transport is known as *phoresy*. Cf. **capilliconidium**.

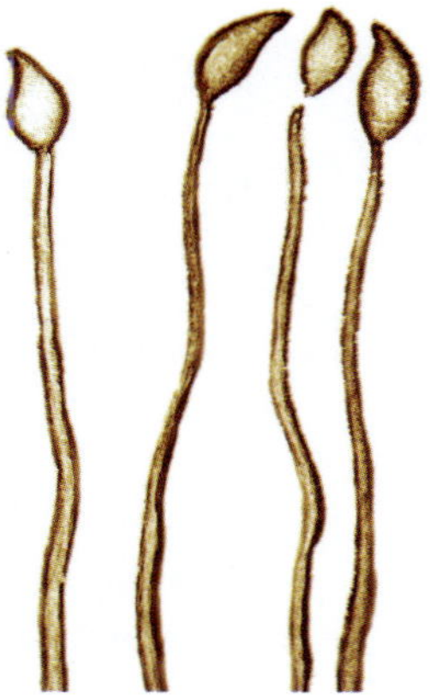

Capillary spores of *Stylopage rhynchospora*, x 300 (*MU*).

capilliconidium, pl. **capilliconidia** (L. *capillus*, hair + NL. *conidium* < Gr. *konídion* < *kónis*, dust, powder + dim. suf. *-ídion* > L. *-idium*): *Entomophthorales*. A secondary conidium formed on the end of a capilliform filament, which results from the germination of a primary conidium, as happens in species of *Conidiobolus* and *Basidiobolus*. Cf. **capillary spore**.

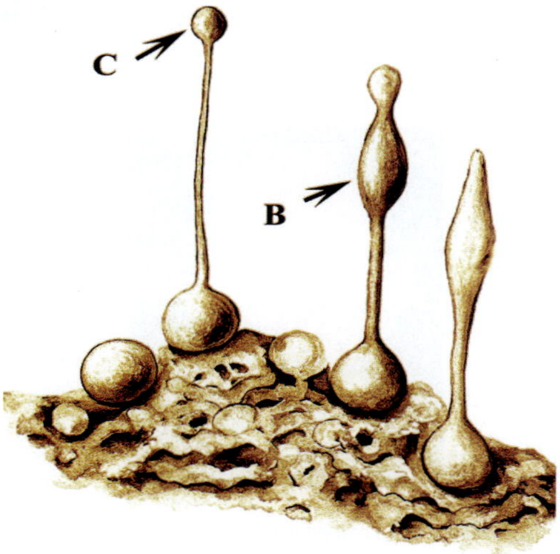

Capilliconidia of *Conidiobolus* sp. (C) and *Basidiobolus* sp. (B), x 1,200 (*MU*).

capilliform (L. *capilliformis*, hair-shaped < *capillus*, hair + suf. *-formis* < *forma*, shape): fine like a hair; hair-like, long and slender.

capillitium, pl. **capillitia** (L. *capillitium*, long hair, in the collective sense < *capillus*, hair + *-itium*, adjectival ending that signifies provided with) **1**. *Myxomycetes*. Sterile filaments, free or anastomosed, that are found in the interior of many sporangia. In many Myxomycetes, such as *Hemitrichia calyculata*, *Arcyria cinerea*, *Metatrichia* and *Trichia* (Trichiales), *Diachea* and *Elaeomyxa* (Stemonitales), the capillitium constitues a cottony or spongy, elastic mass that appears to have an important function in spore dispersal. In other species the capillitium is calcareous, as in many Physarales, e.g., *Badhamia gracilis*. Cf. **pseudocapillitium**. **2**. *Gasteromycetes*. In the gleba or fertile part of the basidiocarps, such as those of *Lycoperdon candidum* and *L. perlatum* (Lycoperdales) and *Scleroderma* spp. (Sclerodermatales), there also exist some filaments that constitute a capillitium, but in this case they are hyphal filaments and not inert structures as in the myxomycete capillitium. Such filaments are coarse, thick-walled cells that are generally branched and which mature late, and whose secondary wall can have pores or slits. Cf. **paracapillitium**.

capillitium

Calcareous **capillitium** of the sporangia of *Badhamia gracilis*, x 30 (*MU*).

Capillitium of the sporangium of *Hemitrichia calyculata* (=*H. stipitata*), x 20 (*MU*).

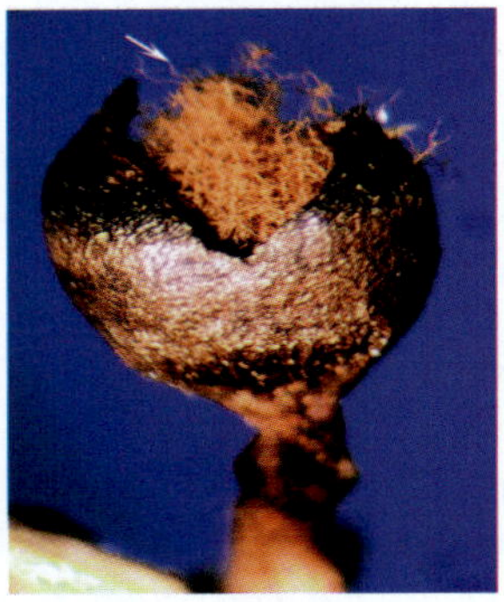

Capillitium of the sporangium of *Metatrichia* (=*Trichia*) *floriformis*, x 27 (*MU*).

Capillitium of the sporangium of *Diachea bulbillosa*, x 20 (*MU*).

Capillitium of the sporangium of *Elaeomyxa cerifera*, x 27 (*MU*).

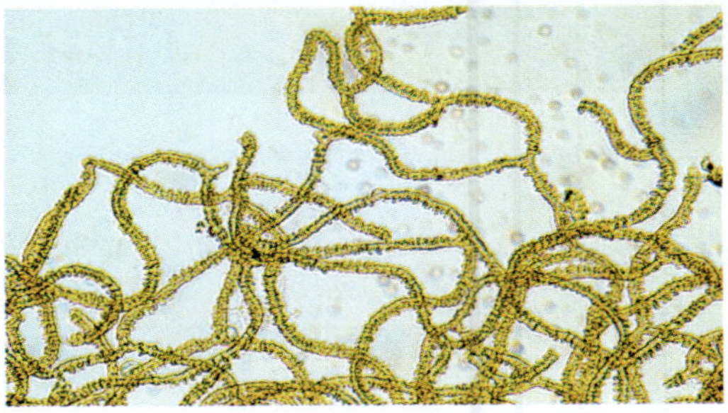

Capillitium of the sporangium of *Arcyria* sp., x 400 (*MU*).

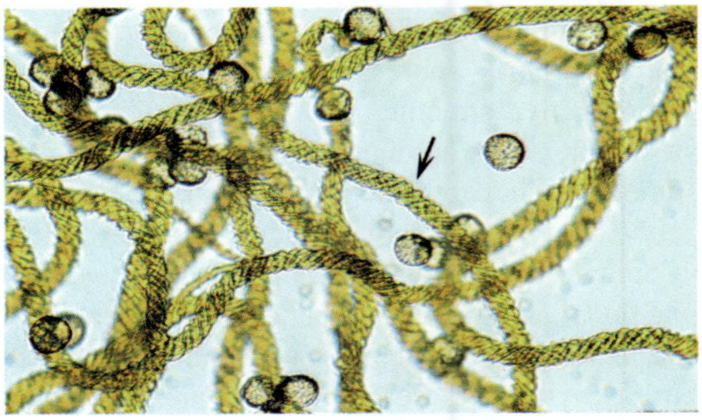

Capillitium and spores of the sporangium of *Hemitrichia calyculata* (=*H. stipitata*), x 600 (*MU*).

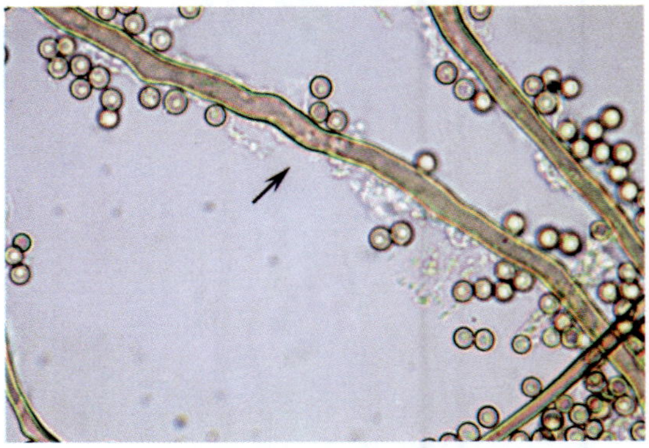

Capillitium and basidiospores of the basidiocarp of *Lycoperdon candidum*, x 600 (*ACV*).

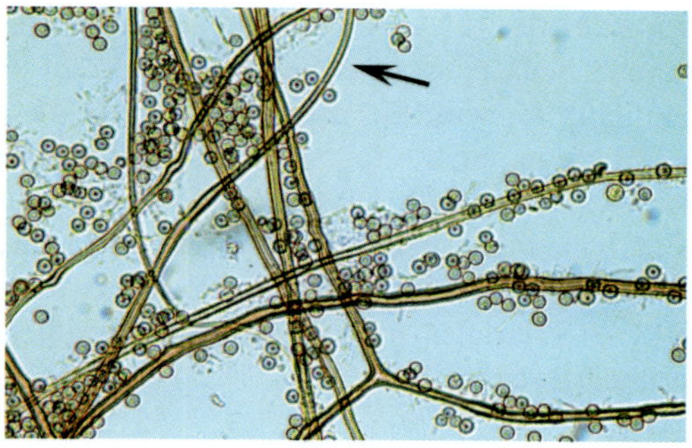

Capillitium and basidiospores of the basidiocarp of *Lycoperdon perlatum*, x 375 (*ACV*).

capitate (L. *capitatus*, having a head < *caput*, genit. *capitis*, of the head + suf. *-atus* > E. *-ate*, provided with or likeness): having an enlarged apex, like a small head; e.g., like some synnemata, stromata, cystidia and spores. The ascocarps of *Leotia* (Helotiales) and the synnemata of *Antromycopsis smithii* (stilbellaceous asexual fungi) and the perithecial stromata of *Cordyceps melolonthae* var. *rickii* (Hypocreales) are capitate. Cf. **acephalous**.

Capitate apothecia of *Leotia atrovirens*, x 1 (*RV*), and **capitate** synnema of *Antromycopsis smithii* on agar, x 10 (*CB* & *MU*).

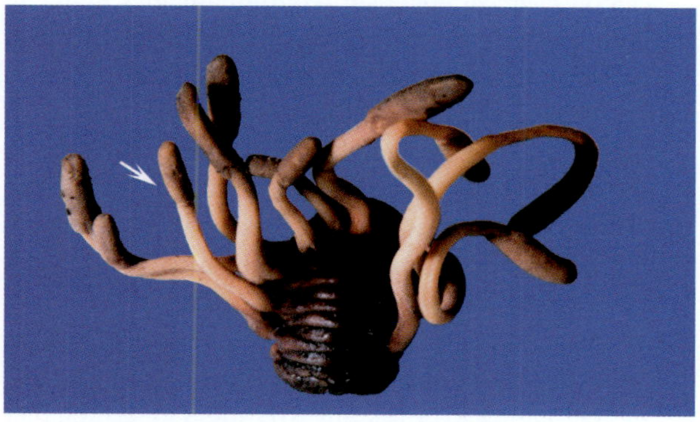

Capitate perithecial stromata of *Cordyceps melolonthae* var. *rickii* on a parasitized coleopteran larva, x 1 (*TH*).

capitate-fastigiate (L. *capitatus*, capitate; provided with a head, or with the shape of a head < *capita*, head + suf. *-atus* > E. *-ate*, provided with or similar to; L. *fastigiatus*, ending in a point < *fastigium*, summit, peak + suf. *-atus*): the term applied to the cortex of a thallus that has erect, parallel hyphae with swollen, pigmented apical cells that occur in certain macrolichens.

capitellum, pl. **capitella** (L. *caput*, genit. *capitis*, head + dim. suf. *-ellum*): a small head.

capitulum, pl. **capitula** (L. *capitulum*, small head < *caput*, genit. *capitis*, head + dim. suf. *-ulum*): the apical, globose, pedunculate apothecium of lichens of the order Caliciales (*Calicium*). In some lichens, such as *Pertusaria* (Pertusariales), the soredia do not form on the surface of the thallus, but are arranged in sorediferous capitula that occupy all of the excipulum of the apothecia, which abort and do not produce asci. Also called **sphaeridium**.

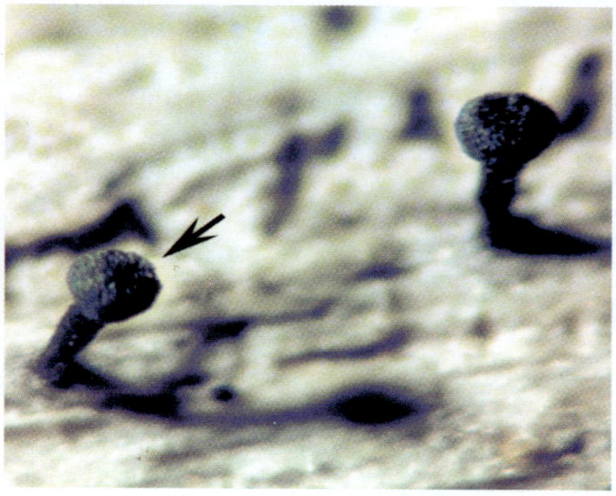

Capitula (apothecia) of the thallus of *Calicium farietinum*, x 13 (*MU*).

capnodic, **capnoid** (Gr. *kapnós*, smoke, or *kapnódes*, smoky + suf. *-ikos* > L. *-icus* > E. *-ic*, belonging to or relating to; or + L. suf. *-oide*, like < Gr. *-oeídes*, similar to): smoke-color, blackish. Applied to the dark mycelium of various fungi, known as *sooty molds*, that cover branches and leaves, and which develop on the thick, sweet liquid called honeydew that can be of plant origin or an exudate of certain insects, such as plant lice and mealybugs. These fungi represent species of *Capnodium* (Capnodiales), which grow mainly on citrus, willows and poplars.

capsule (L. *capsula* < *capsa*, box + dim. suf. *-ule*): a hyaline, gelatinous covering that surrounds the exterior of the cell wall of certain yeasts (*Cryptococcus albidus*, *C. neoformans*, cryptococcaceous asexual yeasts), or of the ascospores of some ascomycetes (*Sordaria fimicola*, Sordariales; *Leptosphaerulina crassiasca*, Pleosporales), and *Aliquandostipite* (Jahnulales).

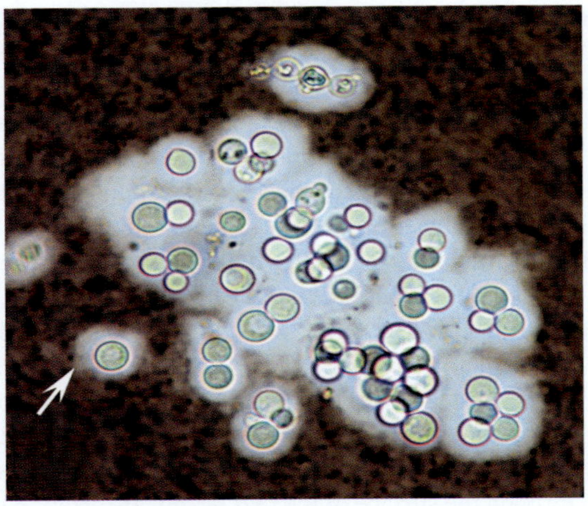

Capsule of the vegetative cells of *Cryptococcus albidus*, x 1,000 (*MU*).

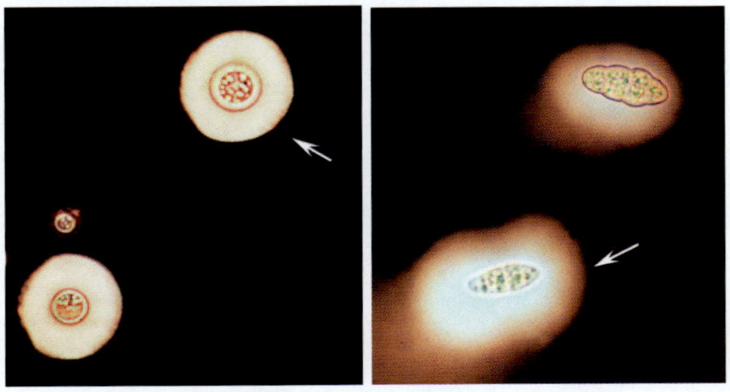

Capsule of the cells of *Cryptococcus neoformans*, x 630 (*RLM*), and of the ascospores of *Leptosphaerulina crassiasca*, x 200 (*RTH*).

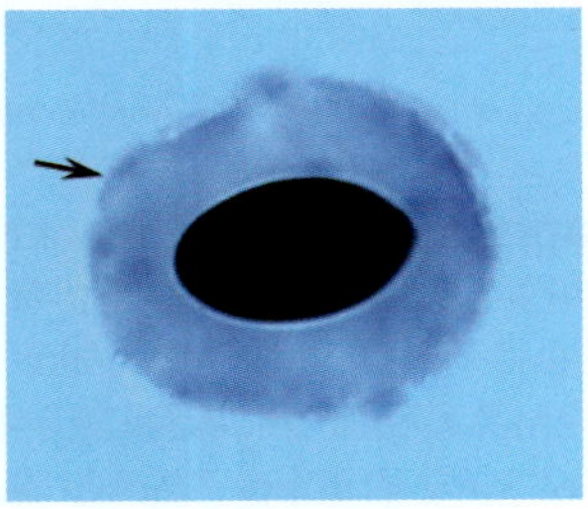

Capsule of an ascospore of *Sordaria fimicola*, x 550 (*RTH*).

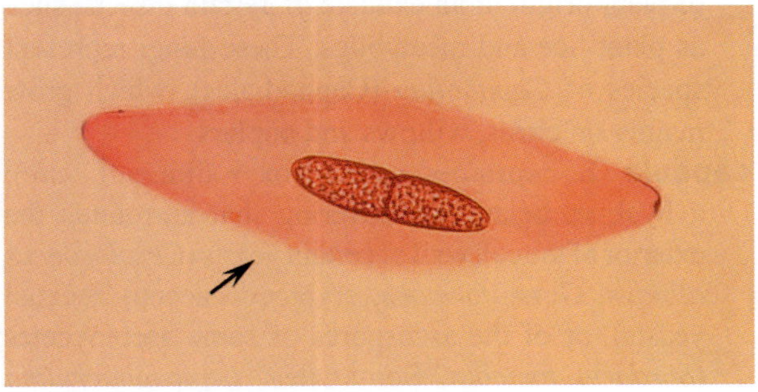

Capsule of the ascospore of *Aliquandostipite* sp., x 1,000 (*ERG* & *MCG*).

carbonaceous (L. *carbo*, genit. *carbonis*, carbon, charcoal, coal + suf. *-aceus* > E. *-aceous*, of or belonging to, with the nature of): similar to coal, i.e., dark in color, almost black, and fragile or brittle; also used for substances composed mainly of carbon. Many pyrenomycetes in genera such as *Camillea*, *Hypoxylon*, *Phylacia*, *Poroleprieuria* and *Xylaria* (Xylariales) have carbonaceous perithecia or perithecial stromata.

Carbonaceous perithecial stroma of *Poroleprieuria rogersii*, x 10 (*CB*).

Carbonaceous perithecia of *Phylacia sagraeana*, x 14 (*CB*).

carbonicolous, carbonicole (L. *carbo*, genit. *carbonis*, carbon, charcoal, coal + suf. *-cola*, inhabitant): living on burned wood or soil, or on coal. See **anthracobiontic**, **phoenicoid fungus** and **pyrophilous**.

carinate (L. *carinatus*, keel-shaped < *carina*, keel + suf. *-atus* > E. *-ate*, provided with or likeness): equal to **cymbiform**, **scaphoid** and **navicular**.

cariose (L. *cariosus*, deteriorated, rotted < *caries*, putrefaction + suf. *-osus* > E. *-ose*, full of, increased, inclined to): very decomposed or rotted.

carminophilic (Fr. *carmine*, carmine + Gr. *phílos*, have an affinity for + Gr. suf. *-íkos* > L. *-icus* > E. *-ic*, belonging to or relating to): refers to basidia that become densely granular after being treated with acetocarmine stain. Cf. **siderophilous**.

carneous (L. *carneus*, having flesh, or the appearance of flesh < *caro*, genit. *carnis*, flesh): with the color of flesh. Cf. **carnose**.

carnose, **carnous** (L. *carnosus*, fleshy, pulpy < *caro*, genit. *carnis*, of flesh + suf. *-osus* > E. *-ose*, full of, increased, inclined to): fleshy, with the consistency of flesh, like the ascomata of species of *Helvella* and *Discina* (Pezizales), and the majority of the Agaricales, whose basidiocarps have a fleshy pileus. Cf. **carneous**.

Carnose ascoma of *Helvella* (=*Paxina*) *acetabulum*, x 1 (*RV*).

Carnose ascoma of *Discina perlata*, x 1 (*RV*).

carp (Gr. *karpós*, fruit): *carp-*, is employed as a pref., e.g., in *carpogenous* and in the terms below. As a suf., *-carp*, is often used in the same way in words like *ascocarp*, *basidiocarp*, *sorocarp*, etc., as well as in many botanical terms (*endocarp*, *pericarp*, etc.).

carpogenic, **carpogenous** (Gr. *karpogenés* < *karpós*, fruit + *génos*, origin, birth < *gennáo*, to engender, produce): **1**. The basal cell of the *carpogonium* in the thalli of Laboulbeniaceae that eventually develops into the ascogonium. **2**. Applied to fungi that grow on fruits, such as *Monilinia fructicola* (Helotiales). See **fructicole**.

carpogonium, pl. **carpogonia** (Gr. *karpós*, fruit + *gónos*, procreation, progeny, seed + L. dim. suf. *-ium*): the female sex organ of the thallus of the Laboulbeniaceae, which is composed of three cells: the *carpogenic*, the *trichophore*, and the *trichogyne* (the latter is terminal and also is called *gynotrichous*, a more appopiate term for this structure). The carpogonium results from cell divisions of the developing ascocarp. The basal cell of the carpogonium (the carpogenic cell) eventually produces the ascogonium.

carpoid (Gr. *karpós*, fruit + suf. *-oeídes* > L. *-oide*, similar to): refers to the galls formed by species of *Biatoropsis* (a gen. of basidiomycetes of the order Platygloeales) on the thallus of the lichen *Usnea* (Lecanorales).

carpomycetes, **carpomyceteae** (Gr. *karpós*, fruit + L. *-mycetes*, ending of a taxonomic class < Gr. *mýkes*, genit. *mýketos*, fungus): a broad term formerly applied to the ascomycetes and basidiomycetes, which have more or less complex fruiting bodies (*carpophores*), in contrast to the fungi considered simpler and less advanced (phycomycetes).

carpophore (Gr. *karpós*, fruit + suf. *-phóros*, bearer < *phéro*, to bear, sustain): a term used by French mycologists to refer to the fruiting body of the higher fungi, in which the conidiophores, ascophores and basidiophores tend to assemble to form more complex fructifications that are called, respectively, *conidiocarps*, *ascocarps* and *basidiocarps*. For that reason, these fungi are called *carpomycetes*.

carpophoroid (Gr. *karpós*, fruit + suf. *-phóros*, bearer < *phéro*, to bear, sustain + L. suf. *-oide*, like < Gr. *-oeídes*, similar to): a false carpophore, by reason of being sterile, that is found in some Agaricales, as occasionally happens in the basidiocarps of *Asterophora* (=*Nyctalis*) *parasitica*, that develop as mycoparasites on basidiocarps of *Russula* and *Lactarius* (Agaricales).

Carpophoroids of *Asterophora* (=*Nyctalis*) *parasitica* on a basidiocarp of *Russula* sp., x 0.5 (*MU*).

cartilagineus, cartilaginous (L. *cartilagineus*, cartilaginous < *cartilago*, genit. *cartilaginis*, cartilage + suf. *-osus* > E. *-ous*, having or possessing the qualities of): refers to any structure with a consistency similar to that of animal cartilage, such as the peridium of the sporangia of *Leocarpus fragilis* (Physarales), the stipe of the basidiocarps of *Auricularia delicata* (Auriculariales), *Clitocybe cartilaginea* (Agaricales), the basidioma of *Sparassis* (Aphyllophorales), and the peridium of *Calostoma cinnabarinum* (Tulostomatales).

Cartilagineus peridium of the sporangia of *Leocarpus fragilis*, x 40 (*CB*).

Cartilagineus basidiocarp of *Auricularia delicata*, x 1 (*RV*).

Cartilagineus fruiting bodies of *Sparassis* sp., x 0.5 (*RV*), and of *Calostoma cinnabarinum*, x 1 (*HB*).

cassidiform (NL. *cassidiformis*, helmet-shaped < *cassidium*, dim. of *cassis*, helmet + suf. *-formis* < *forma*, shape): equal to **galeriform**.

cata- (Gr. *katá*, downward, inferior): a pref. used in the formation of various terms when it is desired to introduce the idea of descent or of inferior, external, distant, opposed position, as in *cata-species*.

catahymenium, pl. **catahymenia** (NL. *catahymenium* < Gr. *katá*, downward, inverted + *hyménion*, dim. of *hymén*, membrane): a type of hymenium in which the hyphidia are the first elements formed; the basidia are embedded at various levels and on elongating reach the surface, but without forming a palisade. Cf. **euhymenium**.

cataphysis, pl. **cataphyses** (Gr. *katá*, downward, inverted + *phýsis*, action of bringing forth, to be born, a growth): equal to **pseudoparaphysis**.

cata-species (Gr. *katá*, inferior, distant, external; L. *species*, a particular form or class, a type): a term that at times is applied to the species of rusts (Uredinales) that in their life cycle have all of the spore stages except the spermogonium or pycnium.

catathecium, pl. **catathecia** (NL. *catathecium* < Gr. *katá*, downward, inverted + *thekíon*, dim. of *thêke*, box; here, of the asci + Gr. dim. suf. *-íon* > L. *-ium*): syn. of **thyriothecium**.

catenate (L. *catenatus*, chained, arranged in chains < *catena*, chain + suf. *-atus* > E. *-ate*, provided with or likeness): arranged in chains, like the latent sporangia of *Catenaria allomycis*, which develop endoparasitically in the hyphae of *Allomyces*, both aquatic fungi of the order Blastocladiales. See **catenulate**.

Catenate latent sporangia of *Catenaria allomycis* in a hypha of *Allomyces* sp., x 900 (*MU*).

catenophysis, pl. **catenophyses** (L. *catena*, chain + Gr. *phýsis*, a being, growth): a persistent chain of thin-walled, utriculate cells formed by the vertical separation of the pseudoparenchyma in the centrum of some marine ascomycetes of the family

Halosphaeriaceae, such as species of *Aniptodera* and *Lignincola* (Halosphaeriales). As the number of asci in the centrum of the maturing perithecium increases, the catenophyses are pushed against the perithecial wall, so that at the time of ascospore discharge, they are no longer evident.

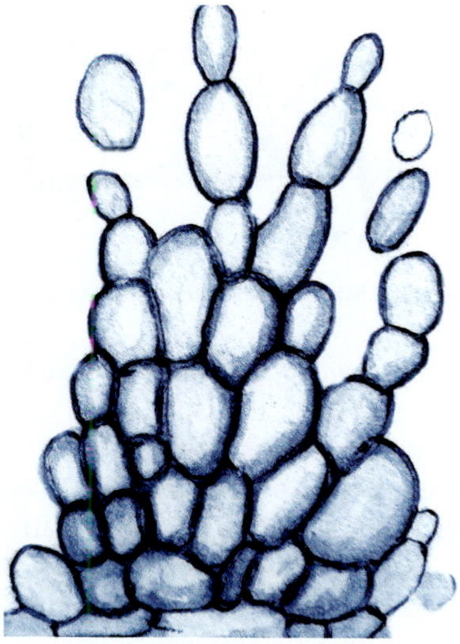

Catenophyses of the perithecium of *Aniptodera chesapeakensis*, x 325 (*MU*).

catenulate (L. *catenulatus*, arranged in chains < *catenula*, dim. of *catena*, chain + suf. *-atus* > E. *-ate*, provided with or likeness): forming a small chain or similar to one, like the resting sporangia of the aquatic fungus of the gen. *Catenaria* (Blastocladiales), or the conidia of some moniliaceous asexual fungi, e.g., *Aspergillus*, *Paecilomyces*, *Penicillium* and *Scopulariopsis*, and dematiaceous fungi, such as *Alternaria alternata*. See **catenate**.

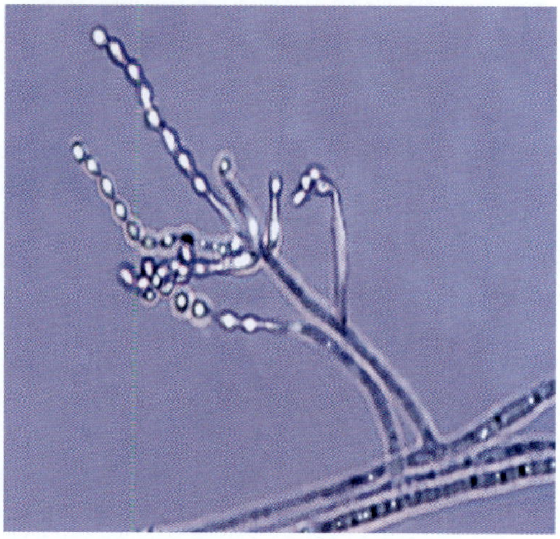

Catenulate conidia of *Paecilomyces* sp., x 770 (*MU*).

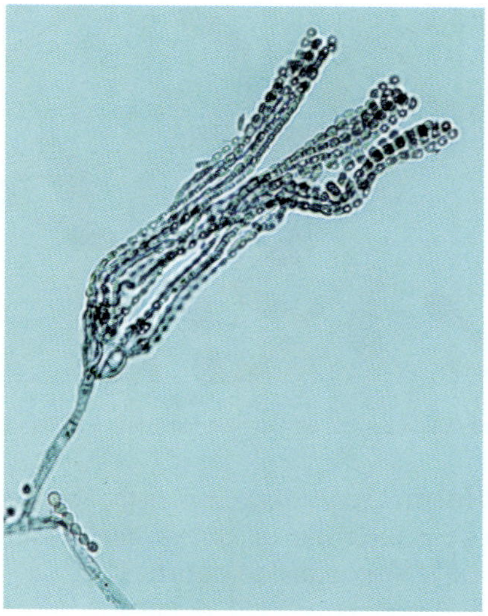

Catenulate conidia of *Penicillium* sp., x 650 (*MU*).

Catenulate conidia of *Aspergillus flavofurcatus* (=*A. flavo-furcatis*), x 770 (*MU*).

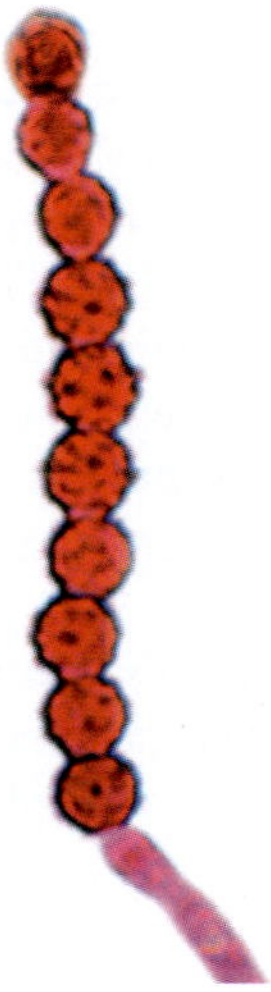

Catenulate conidia of *Scopulariopsis brevicaulis*, x 1,000 (*CB*).

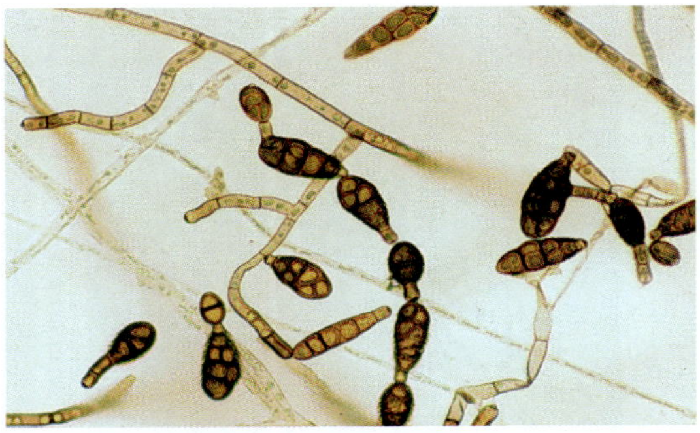

Catenulate conidia of *Alternaria alternata*, x 500 (*MU*).

catenuliform (L. *catenuliformis*, in the shape of a small chain < *catenula*, dim. of *catena*, chain + suf. *-formis* < *forma*, shape): equal to **catenulate**.

cauda (L. *cauda*, tail): a tail, or tail-like appendage.

caudate (L. *caudatus* < *cauda*, tail + suf. *-atus* > E. *-ate*, provided with or likeness): with a tail, like the pseudoplasmodium of *Dictyostelium* (Dictyosteliales), which on moving across a substrate leaves a tail or trail of mucilage behind it. Cf. **acaudate**.

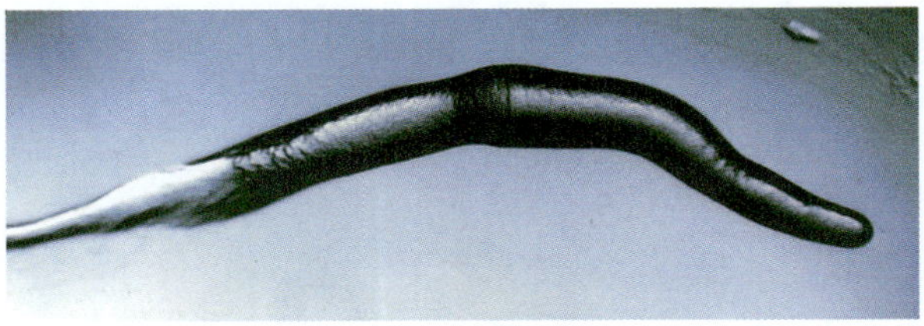

Caudate pseudoplasmodium of *Dictyostelium discoideum*, x 30 (*RTH*).

caulescent (NL. *caulescens*, genit. *caulescentis* < *caulescere*, to sprout < *caulis*, stem + suf. *-escens*, genit. *-escentis* > E. *-escent*, that which turns, begins to, slightly): capable of forming a stem or pedicel.

cauli- (L. *caulis*, stem): see **caulo-**.

caulicole, **caulicolous** (L. *caulis*, stem + suf. *-cola*, inhabitant < *colere*, to inhabit): living or developing on the stem or branches of a host plant that serves as support. Many parasitic and saprobic fungi, as well as various lichens, are caulicolous.

cauligenous, **caulogenous** (L. *caulis* < Gr. *kaulós*, stem + Gr. *génos*, origin): arising from or forming on the stems of plants.

caulo-, **caul-** (Gr. *kaulós*, stem, trunk): a pref. that is applied to whatever has a relationship to a stem or support, or that has or implies the same nature or origin as the stem.

caulocystidium, pl. **caulocystidia** (Gr. *kaulós*, stem, trunk; here, stipe + *kystídion* < *kýstis*, bladder, vesicle, cell + dim. suf. *-ídion* > L. *-idium*): a dermatocystidium that is found on the surface of the stipe of certain agarics, such as *Entoloma versatile* (=*Pouzarella versatilis*) and *Panaeolus sphinctrinus* (Agaricales), and Clavariaceae, such as *Pterula* spp. (Aphyllophorales). Cf. **pileocystidium**.

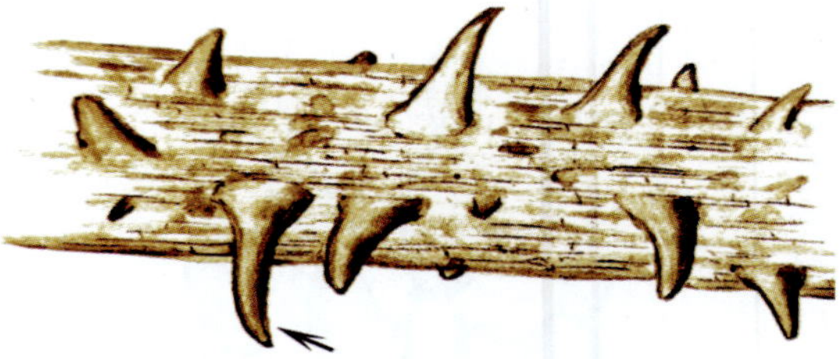

Caulocystidia of the filiform basidiocarp of *Pterula* sp., x 500 (*MU*).

cauloplane (L. *caulis* < Gr. *kaulós*, stem + L. *planus*, plane, flat surface): the surface of a stem. Cf. **rhizoplane** and **rhizosphere**.

cavernose, **cavernous** (L. *cavernosus*, having hollows or cavities < *caverna*, cavity, hollow + suf. *-osus* > E. *-ose*, having or possessing the qualities of): having holes or cavities, like the fruiting bodies of hypogeous fungi called truffles (Pezizales), and those of *Cyttaria hariotii* (Cyttariales). Cf. **farctate**, **fistular** and **solid**.

Cavernose ascocarp of *Cyttaria hariotii*, growing on *Nothofagus dombeyi*, x 5 (*MU*).

cavernula (L. *cavernula*, dim. of *caverna*, cave, cavity): a small cavity; in particular it refers to the cavities in the lower cortex of the lichen *Cavernularia* (Lecanorales).

cecidium, pl. **cecidia** (NL. *cecidium* < Gr. *kekís*, genit. *kekídos*, gall): the formation or hypertrophy of plant tissues due to the action of another plant (*phytocecidium*), an animal (*zoocecidium*) or a fungus (*mycocecidium*).

cell (NL. *cellula*, cell, living cell, dim. of L. *cella*, small room, compartment, cell of a honeycomb): a unit of cytoplasm delimited by a membrane (the cellular or plasma membrane) that contains various organelles and one or more nuclei, and which, in the fungi, is usually enclosed by a wall (the cell wall), whose structure and composition vary according to the group of fungi being considered. In the Oomycota the cell wall is composed mainly of cellulose (although recently, in *Saprolegnia monoica*, of the Saprolegniales, chitin and chitin synthetase have been found) and (1-3)-ß/(1-6)-ß-glucan. In the Chytridiomycota the polymers of the wall are chitin and glucan; in the Zygomycota the polymers are chitosan and polyglucuronic acid, whereas in the Ascomycota and Basidiomycota the polymers are chitin and ((1-3)-ß/(1-6)-ß-glucan. In the kingdom Fungi the cell is *eukaryotic*.

cell wall, **wall** (NL. *cellula*, dim. of *cella*, small room, cell of a honeycomb; ME. < OE. *weall*, rampart, wall): a dynamic structure that surrounds each fungal cell or hypha, which is subject to change and modification during the different stages of the life cycle of a fungus. The cell wall is responsible for the majority of the unique distinctive characters of the fungi. The capacity of the wall to maintain the turgidity of the fungal cells in a secure manner appears to be a primordial reason for the survival and evolution of the fungi. Besides maintaining turgidity, the cell wall carries out other important functions in the life of a fungus. E.g., the wall determines the shape of the hyphae, acts as a filter that up to a certain point controls what enters the fungal protoplasm, protecting it from environmental dangers, and it functions in the recognition of events associated not only with sexual reproduction but also with various interactions of the fungi with potential plant and animal symbionts. The chemical composition of the fungal cell wall varies according to the group of species. Basically it is composed of microfibrillar or skeletal elements located on the inner surface of the wall and usually embedded in an amorphous matrix that extends to the outer surface of the wall. In the Myxomycota only the spores have a cell wall and it contains cellulose. In the Chytridiomycetes the zoospores lack a cell wall, but the cell wall of the thallus contains chitin and glucan; the cell wall of the hyphae and spores contains chitin and chitosan in the Zygomycota, cellulose and chitin in the Oomycota, and chitin, protein and glucan in the Ascomycota and Basidiomycota. In the majority of the fungi the skeletal component consists of chitin, which can be in combination with various polysaccharides (e.g., hemicellulose) and small quantities of lipids. In some groups of fungi, chitin can be absent and the wall is composed mainly of cellulose (e.g., in the Oomycota), chitosan, ß-glucan, mannan, glycogen, galactan and polymers of galactosamine. Fungal cell walls are stratified, usually composed of two or more layers of microfibrils embedded in an amorphous matrix, with the inner layer in contact with the plasma membrane.

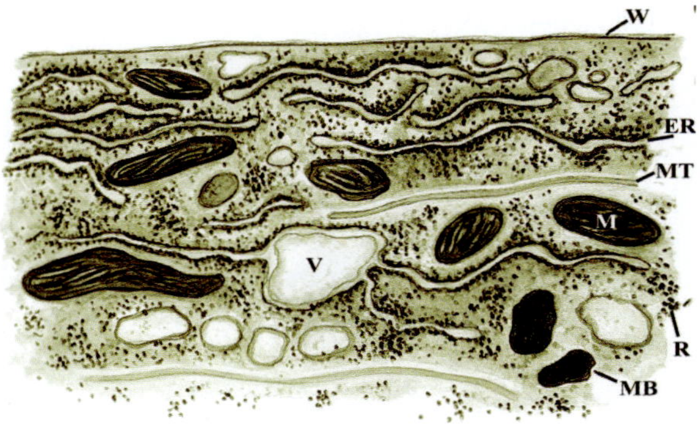

Cell wall (W) of a young hypha of *Armillaria mellea* (Agaricales), seen in longitudinal section; also shown are the endoplasmic reticulum (ER), microbodies (MB), mitochondria (M), microtubules (MT), ribosomes (R), and vesicles (V). Drawing composed from several transmission electron micrographs published by Beckett *et al.* in their *Atlas of Fungal Ultrastructure*, 1974, x 27,000 (*MU*).

cellular (NL. *cellula*, cell, living cell, dim. of L. *cella*, small room, compartment, cell of a honeycomb + L. suf. *-aris* > E. *-ar*, like, concerning, of the nature of): pertaining to or relative to cells or formed of cells. Cf. **acellular**.

cellular dermis (NL. *cellularis* < *cellula*, cell, living cell, dim. of *cella*, small room, compartment, cell of a honeycomb + suf. *-aris* > E. *-ar*, like, belonging to; Gr. *dérma*, skin, leather): see **cystoderm**.

cellular slime mold (NL. *cellularis* < *cellula*, cell, living cell, dim. of *cella*, small room, compartment, cell of a honeycomb + suf. *-aris* > E. *-ar*, like, pertaining to; ME. *slyme* < OE. *slim*, slime; ME. *mowlde*, mold): the common name of a member of the phyla Acrasiomycota or Dictyosteliomycota, with pseudoplasmodia formed by aggregation of myxamoebae, which retain their individuality, i.e., they do not fuse to form a true plasmodium.

cellulin (NL. *cellula*, cell, living cell, dim. of L.*cella*, small room, compartment, cell of a honeycomb + NL. *-ina*, suf. employed in chemistry to denote an activator or compound): a polysaccharide in the form of granules or refringent disks that transitorily plug the hyphae of the Oomycota of the order Leptomitales (such as *Leptomitus* and *Apodachlya*), mainly at the level of the constrictions. These plugs of cellulin constitute what are called *pseudosepta*.

cellulolytic (NL. *cellulose* < *cellula*, cell, living cell, dim. of L. *cella*, small room, compartment, cell of a honeycomb + suf. *-ose*, ending for carbohydrates + Gr. *lytikós*, able to loosen, release, separate < *lýtos*, dissolvable, broken, loose + suf. *-íkos* > L. *-icus* > E. *-ic*, belonging to or relating to): having the capacity to degrade cellulose in substrates where it occurs; many fungi are cellulolytic, like the species of *Eidamella* (Onygenales), *Chaetomium, Chaetomidium* (Sordariales) and *Trichoderma* (moniliaceous asexual fungi), among many others.

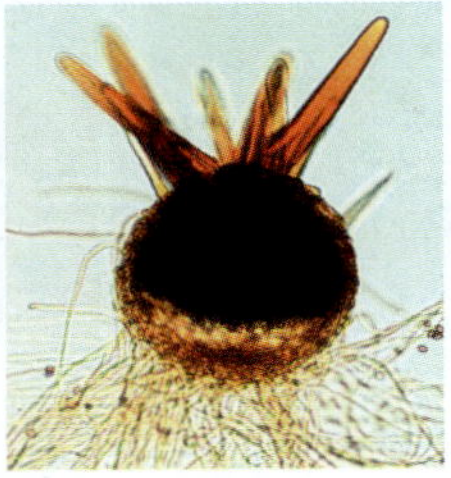

Perithecium of *Chaetomidium* sp., a **cellulolytic** fungus, x 150 (*MU*).

ceno- (Gr. *kainós*, new, recent): used in the formation of words such as Cenozoic.

central body (L. *centralis*, central, situated in the center < *centrum*, center + suf. *-alis* > E. *-al*, belonging to or relating to; ME. < OE. *bodig*, body): *Ascomycetes*. The cellular structure (central apparatus) from which the astral rays emanate that initiate the excision of the cytoplasm during cell division.

centri-, centr- (L. *centrum*, center): L. pref. equivalent to Gr. *kéntron*.

centric (L. *centrum* < Gr. *kéntron*, center + Gr. suf. *-ikos* > L. *-icus* > E. *-ic*, belonging to or relating to): **1**. *Aquatic fungi*. A type of oospore in which the ooplast is located in the center and the reserve substances of the periplasm (oil droplets) are distributed regularly around the ooplast, as is observed in *Saprolegnia hypogyna* (Saprolegniales). **2**. *Basidiomycetes*. A stipe that is united with the central part of the pileus of a basidiocarp. Cf. **eccentric** and **subcentric**.

Centric oospores in the oogonium of *Saprolegnia hypogyna*, x 125 (*MU*).

centrifugal (NL. *centrifugus*, fleeing from the center < L. *centrum*, center + *fugus* < *fugere*, to flee, to move away from): moving outward from the center toward the periphery; maturing or developing from the center outward. In lichenology it refers to apothecia that preferentially occupy the borders of the thallus, as happens in *Psora crenata* (Lecanorales). Cf. **centripetal**.

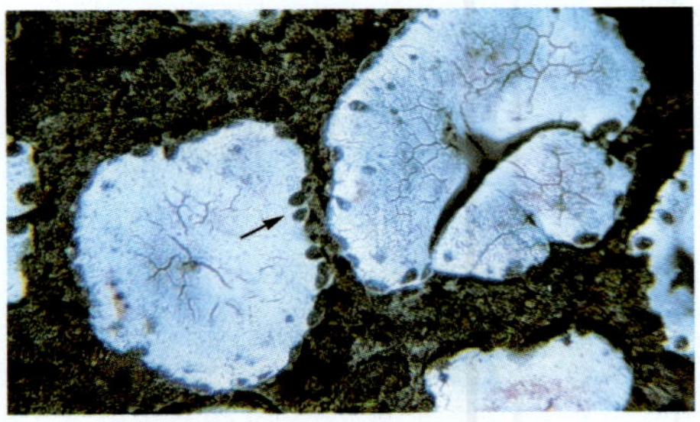

Centrifugal apothecia on a thallus of *Psora crenata*, x 10 (*MU*).

centriole, centriolum, pl. **centriola** (L. *centriolum* < *centrum*, center + dim. suf. *-olum* > E. *-ole*): a cellular organelle that during mitosis forms the polar bodies of the achromatic spindle. In the zoosporic fungi it gives rise to the kinetosome or blepharoplast, in the base of the flagellum. Cf. **spindle pole body**.

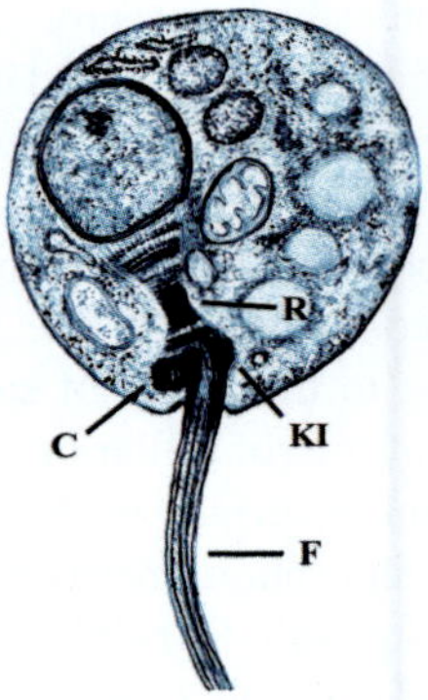

Centriole (C) of the zoospore of *Rhizophlyctis rosea* (Spizellomycetales). Also shown are the kinetosome (KI), the rhizoplast (R), and the flagellum (F), x 14,000 (*MU*).

centripetal (NL. *centripetus*, directed toward the center < L. *centrum*, center + *petus*, directed toward < *petere*, to direct toward, to go for): moving inward from the periphery toward the center; maturing or developing from the outside inward. In lichenology it refers to apothecia that are mainly grouped in the central part of the thallus. The aggregation currents of the myxamoebae that form the pseudoplasmodium of the Dictyosteliales (*Dictyostelium*, *Polysphondylium*, etc.) also are centripetal. Cf. **centrifugal**.

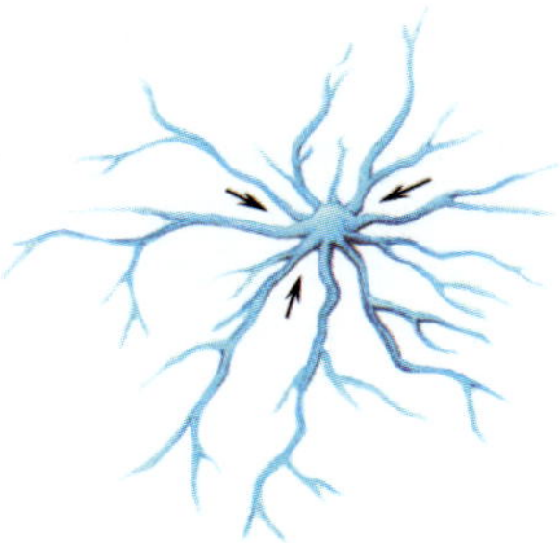

Centripetal aggregation currents of the myxamoebae of *Dictyostelium discoideum* forming a center that later gives rise to a pseudoplasmodium, x 1,000 (*MU*).

centrum, pl. **centra** (L. *centrum* < Gr. *kéntron*, center): *Ascomycetes*. The complex of fertile and sterile structures that occupy the interior of an ascocarp. The type of centrum present in different ascomycetes has been a fundamental criterion in modern classification of these fungi. The developing centrum in a young perithecium can be seen in *Sordaria fimicola* (Sordariales) and *Hypocrea schweinitzii* (Hypocreales).

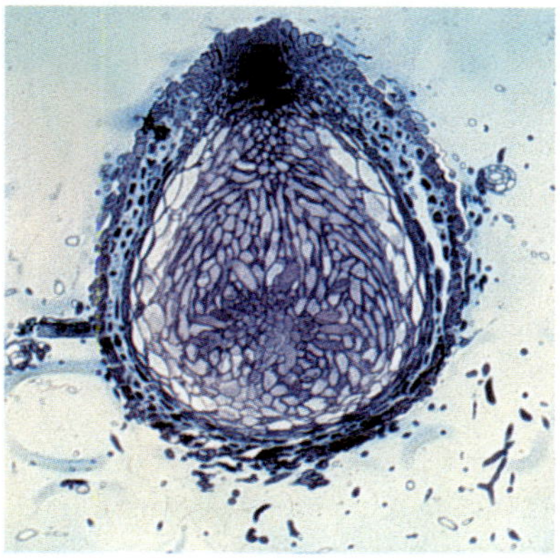

Centrum of a developing perithecium in *Sordaria fimicola*, x 130 (*RTH*).

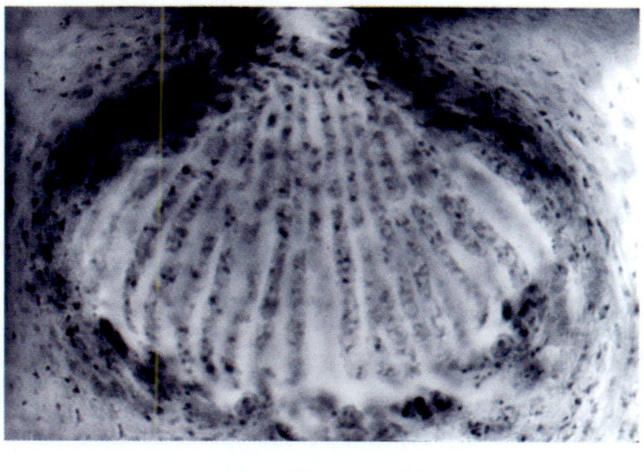

Centrum of a developing perithecium in *Hypocrea schweinitzii*, x 150 (*RTH*).

cephalo-, cephal- (Gr. *kephalé*, head): a pref. employed in biological terms alluding to a head or capitulum, or to a globulose or capituloid organ, such as *cephalodium*, and the gen. *Cephaloascus* (Saccharomycetales).

cephalodium, pl. **cephalodia** (Gr. *kephalé*, head + L. suf. *-odium*, to resemble < Gr. *ode*, like < *-oeídes*, similar to): *Lichens*. A small globose body, irregular or verrucose, that contains symbionts different from those present in the major part of the thallus of a lichen; i.e., the symbionts of the cephalodia are cyanobacteria (they contribute to the nutrition of the lichen by fixing atmospheric nitrogen) and the algae in the rest of the thallus are Chlorophyceae; at times, the cephalodia function as propagules of vegetative multiplication. E.g., *Stereocaulon alpinum*, *S. saxatile* and *S. ramulosum* (Lecanorales) have cephalodia. Cf. **paracephalodium**.

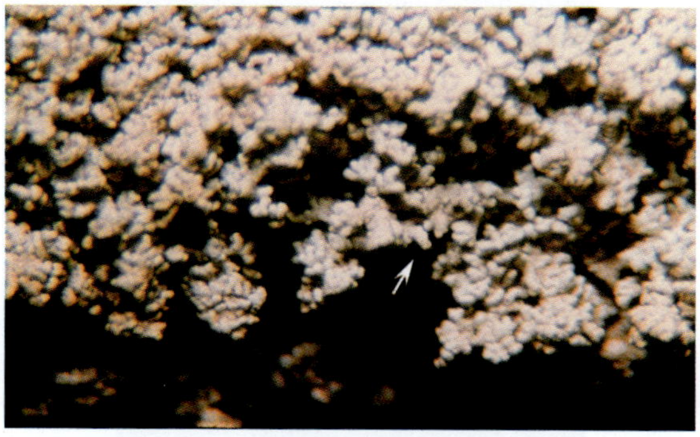

Cephalodia of the thallus of *Stereocaulon saxatile*, x 12 (*MU*).

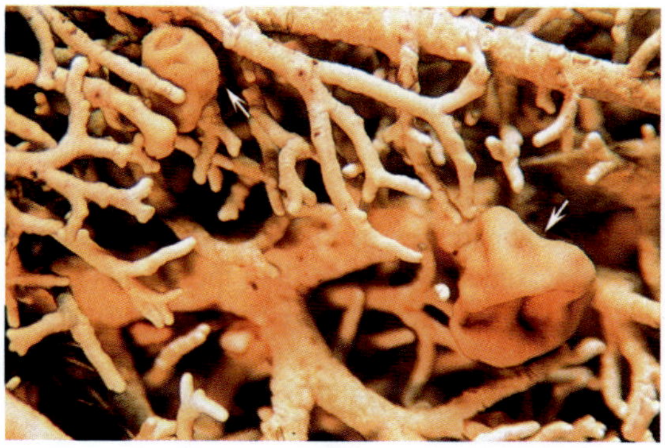

Cephalodia of the thallus of *Stereocaulon ramulosum* f. *elegans*, x 20 (*MU*).

cephalothecoid (< gen. *Cephalotheca* < Gr. *kephalé*, head + *thêke*, box; here, of the asci + L. suf. *-oide*, like < Gr. *-oeídes*, similar to): having a peridium in which the cells split along preformed sutures, as in the cleistothecial ascocarps of *Cephalotheca* (probably related to *Pseudeurotium*) of the Eurotiales, and a few unrelated gen.; the peridium is composed of radial plates of cells that originate from various meristematic zones, as in some species of *Zopfiella* (Lasiosphaeriaceae of the Sordariales), a marine saprobe.

ceraceous (L. *ceraceus*, waxy < *cera*, wax + suf. *-aceus* > E. *-aceous*, of or belonging to, with the nature of): having the appearance or consistency of wax, like the ascocarps of some species of *Aleuria* (Pezizales) and the basidiocarps of some species of *Laccaria* (Agaricales), which are delicate and soft.

Ceraceous apothecium of *Aleuria aurantia*, x 1 (*RTH*).

Ceraceous lamellae of the basidiocarp of *Laccaria amethystina*, x 1 (*RV*).

cerebriform, **cerebroid**, **cerebrose** (L. *cerebriformis* < *cerebrum*, brain + suf. *-formis* < *forma*, shape, or L. *cerebrum* + suf. *-oide*, like < Gr. *-oeídes*, similar to): with folds or creases, bends, and undulations that give the appearance of a brain, like the thallus of *Psora cerebriformis* (Lecanorales). Syn. of **convolute**.

Cerebriform thalli of *Psora cerebriformis*, x 10 (*CB*).

cernuous (L. *cernuus*, to bend, incline, fall): inclined toward the earth, drooping from the head.

chaeto-, **chaet-** (Gr. *chaíte*, crest, mane): a pref. that alludes to a group of hairs, as in the term *chaetomiaceous*, or to a fungus (*Chaetomium*, of the Sordariales) whose perithecia are covered with characteristic hairs.

chalaroplectenchyma, pl. **chalaroplectenchymata** (Gr. *chaláros*, loose, lax + *plektós*, intertwined, interwoven + *énchyma*, stuffed): *Lichens*. A tissue composed of loosely interwoven hyphae, with distinctive holes, present in the medulla of the thallus of some species. The crystals of various lichen products often are found incrusted on the walls of these medullary hyphae. Cf. **paraplectenchyma** and **prosoplectenchyma**.

character (ME. *caracter* < MF. *caractere* < L. *character* < Gr. *charactér*, sign, mark, notable or distinctive quality): a distinctive and variable feature or characteristic of a particular taxon.

chartaceous (L. *chartaceus*, parchment-like < *charta*, paper + suf. *-aceus* > E. *-aceous*, of or belonging to, with the nature of): with the consistency of paper or parchment, like the basidiocarp of *Trichaptum biforme* (=*Polyporus pergamenus*), of the Aphyllophorales.

chasmathecium, **chasmothecium**, pl. **chasmathecia**, **chasmothecia** (Gr. *chásma*, genit. *chásmatos*, opening, open mouth + *thêke*, box; here, of the asci): a fruiting body that is initially closed, without a predefined opening, but with the asci, arranged in a single basal fascicle, liberated at maturity through a broad rupture of the ascoma wall, as occurs, e.g., in the ascoma of *Erysiphe* (Erysiphales).

chasmo-, **chasm-** (Gr. *chásma*, opening): a pref. employed to refer to any opening, or to the manner in which a structure opens, as in *chasmothecium*.

cheilo -, **cheil-** (Gr. *cheílos*, lip, border or margin): a pref. employed for something that has the shape of lips, or that is found on the borders, as in *cheilocystidium*.

cheilocystidium, pl. **cheilocystidia** (Gr. *cheílos*, lip, border, margin + *kystídion* < *kýstis*, bladder, vesicle; here, cell + dim. suf. *-ídion* > L. *-idium*): a marginal cystidium, of various shapes, depending upon the species, that arises from the edge of the gills, as in *Entoloma* (=*Nolanea*) *hirtipes* and *Marasmius* (Agaricales), in which case they are constricted or tortuous.

cheiroid, **chiroid** (Gr. *cheír*, genit. *cheirós*, hand + L. suf. *-oide*, like < Gr. *-oeídes*, similar to): shaped like a hand, like the spores of *Cheiromyces* (dematiaceous asexual fungi) or the haustoria of *Blumeria* (=*Erysiphe*) *graminis* (Erysiphales). Cf. **palmate**.

cheiropterocoprophilous, chiropterocoprophilous (Gr. *cheír*, genit. *cheirós*, hand + *pterón*, wing + *kópros*, dung + *phílos*, have an affinity for): applied to organisms that live on bat dung in caves, such as species of *Histoplasma* (moniliaceous asexual fungi), *Guanomyces* (Sordariales), and *Coprinus* (Agaricales), among many others.

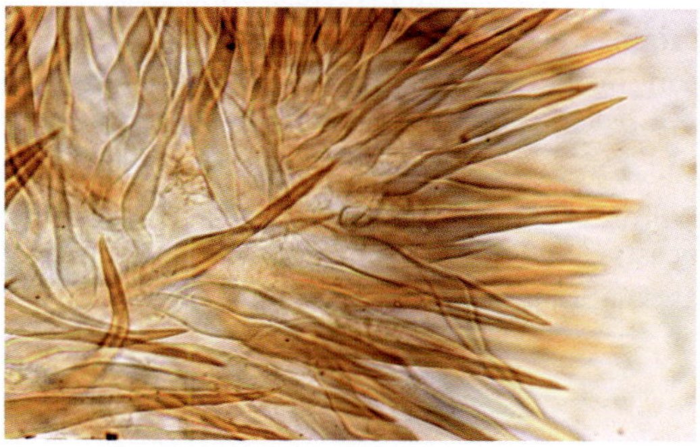

Cheilocystidia of the hymenophore of *Marasmius* sp., x 230 (*EPS*).

cheirospore, cheiroid or **chiroid spore, chirospore** (Gr. *cheír*, genit. *cheirós*, hand + *sporá*, spore): a multicellular conidium whose appearance resembles a hand with the fingers more or less together, as occurs in *Cheiromyces stellatus* and *Dictyosporium elegans* (dematiaceous asexual fungi).

Cheirospores of *Cheiromyces stellatus*, x 1,000 (*MU*).

Cheirospores of *Dictyosporium elegans*, x 325 (*CC*).

chemotaxis, pl. **chemotaxes** (Gr. *chymós*, substance + *táxis*, orientation): a tactic phenomenon determined by the action of a chemical agent, such as is observed in the zoospores of *Saprolegnia* (Saprolegniales), which swim toward aqueous extracts of dead insects and toward aqueous solutions of the decomposition products of these insects, such as proteins, phosphates, urea and organic acids; this chemotaxis is present when these products are adequately diluted, since a substance can be an attractant if its concentration is weak, but if the concentration increases then it becomes repellent. The chemotactic reactions to this type of external stimulus can be related to the distribution and nutrition of the zoospores. Similar responses occur in the approach and fusion of the sexual organs, e.g., of antheridia and oogonia in the Oomycota, of progametangia in Zygomycota, and of gametangia in ascomycetes, responses that evidently are of biological importance. See **chemotropism**.

chemotropism (Gr. *chymós*, substance + *trópos*, gyration, change in position or manner < *tropé*, turn < *trépo*, to gyrate, revolve, turn towards + suf. *-ismós* > L. *-ismus* > E. *-ism*, state or condition): movement or orientation of an organism, or part of an organism, due to the influence of a chemical agent. In essence it is equal to chemotaxis, but this latter implies the free movement of the organisms (e.g., zoospores) to position themselves tactically. Chemotropism, like tropisms in general, is negative or positive, depending on how the organism responds in its orientation, whether it turns toward the point from which the stimulus emanates, or moves away from it. An example of negative chemotropism is the response of hyphae and germ tubes that grow away from areas of high concentrations of the products of their own metabolism, such as ammoniac and potassium bicarbonate. It is known that, in general, depending on the concentration, germ tubes and hyphae grow toward sources of salts, sugars and other nutritive substances (positive chemotropism), and away from acids, alkalis and alcohols. See **chemotaxis** and **tropism**.

chiasto-, chiast- (Gr. *chiastós*, crossed, formed in the shape of a cross): a pref. used in terms alluding to the shape of a cross, as in *chiastobasidium*.

chiastobasidium, pl. **chiastobasidia** (Gr. *chiastós*, crossed, in the form of a cross + L. *basidium* < Gr. *basídion* < *básis*, base + dim. suf. *-ídion* > L. *-idium*): a type of basidium without septa (holobasidium), apically widened, in which the achromatic spindle is arranged perpendicular to the main axis during the second meiotic division of the nucleus. This type of basidium is present in all of the Agaricaceae and Boletaceae. Cf. **stichobasidium**.

chimeroid (L. *chimaera* < Gr. *chímaira*, goat, chimera, monster of Greek mythology + L. suf. *-oide*, like < Gr. *-oeídes*, similar to): refers to lichen thalli, due to their being composed of apparently abnormal parts.

chiono-, **chion-** (Gr. *chíon*, genit. *chíonos*, snow): a pref. used in terms such as *chionophilous*, having a preference for, or requiring, snow, in order to grow.

chionophilous (Gr. *chíon*, genit. *chiónos*, snow + *phílos*, have an affinity for): living in snow; also called **nivicolous**.

chitinoclastic (Gr. *chitón*, tunica, alluding to the exoskeleton of insects, chitin + *klástos*, broken + suf. *-íkos* > L. *-icus* > E. *-ic*, pertaining to or relating to): having the capacity to degrade the chitinous substrates on which it grows, as do certain aquatic fungi of the order Chytridiales. Also called **chitinolytic**. *Abyssomyces hydrozoicus* (Halosphaeriales) is a deep-sea marine fungus, of uncertain affinities, that lives as a parasite (or perhaps as a saprobe) on the chitinous exoskeleton of hydrozoans. The hyphae of this fungus form canals on the surface of the hydrozoan colonies and penetrates them, suggesting that it is an important degrader of chitin in the deep sea.

chitinolytic (Gr. *chitón*, tunica, alluding to the exoskeleton of insects; chitin + *lytikós*, capable of loosening, to set free, separate < *lýtos*, dissoluble, broken, loose + suf. *-íkos* > L. *-icus* > E. *-ic*, pertaining to or relating to): see **chitinoclastic**.

Sporangiophores of *Saprolegnia parasitica* (Saprolegniales) on an insect submerged in water. This zoosporic fungus is **chitinolytic**, x 20 (*MU*).

chitosome (Gr. *chitón*, tunica, alluding to the exoskeleton of insects; chitin + *sõma*, body): a small spheroidal structure (40-70 nm in diam.) present in many fungi, which contains the zymogen chitin synthetase, which is involved in cell wall synthesis.

chlamydocyst (Gr. *chlamýs*, genit. *chlamýdos*, cape, cloak + NL. *cystis* < Gr. *kýstis*, bag, bladder, vesicle): *Blastocladiales*. A long-lasting or latent zoosporangium with two enveloping walls that is found within a hypha.

chlamydospore (Gr. *chlamýs*, genit. *chlamýdos*, cape, cloak + *sporá*, spore): an asexual spore with a thick, durable cell wall that functions as a resistant or latent spore; with respect to ontogeny, chlamydospores develop like holothallic conidia, but their function is survival rather than propagative. In the past this term was applied incorrectly to the teliospores of the heterobasidiomycetes, which are of sexual origin. Many fungi form chlamydospores of some sort, including, e.g., Mucorales (*Mucor racemosus*), *Fusarium oxysporum* (tuberculariaceous asexual fungi) and *Harposporium anguillulae* (moniliaceous asexual fungi), *Candida albicans* (cryptococcaceous asexual yeasts), and *Phoma eupyrena* (sphaeropsidaceous asexual fungi).

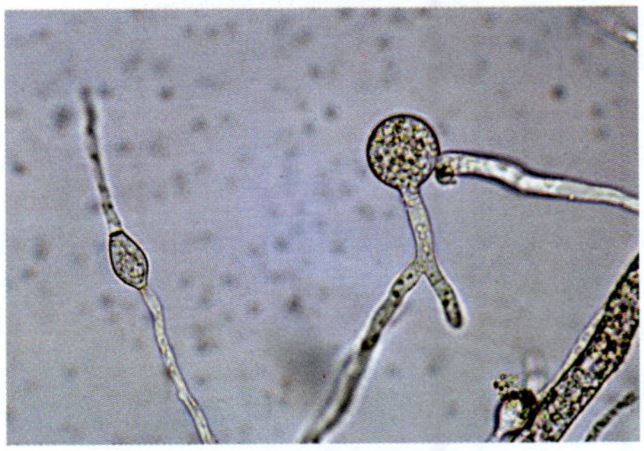

Chlamydospores of *Mucor racemosus*, x 680 (*MU*).

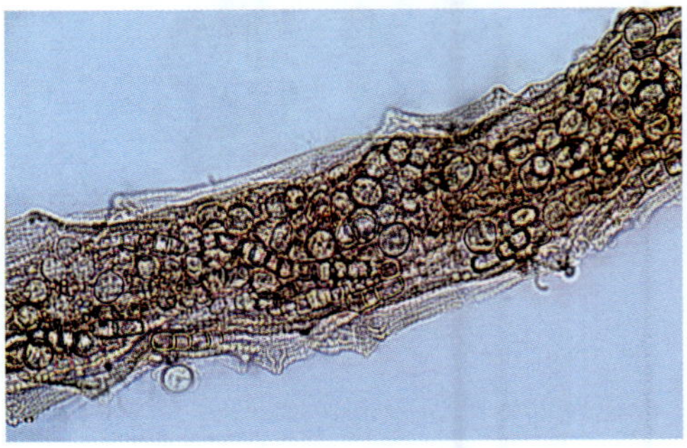

Chlamydospores of *Harposporium anguillulae* inside a parasitized nematode, x 200 (*PL*).

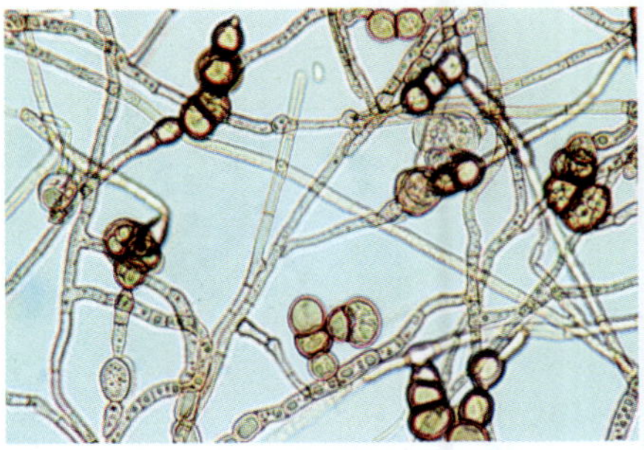

Chlamydospores of *Fusarium oxysporum*, x 250 (*MU*).

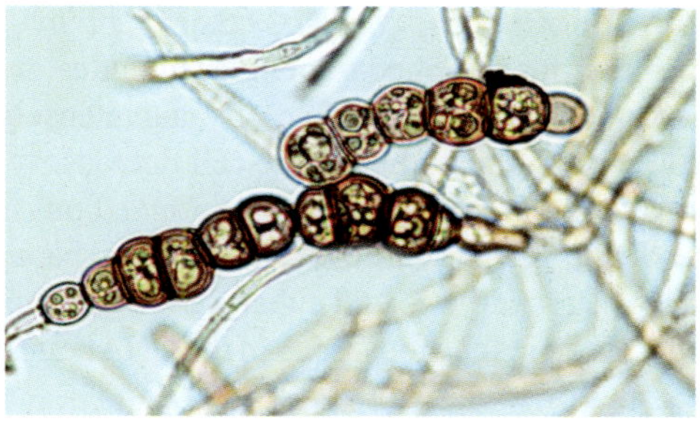

Chlamydospores of *Fusarium* sp., x 250 (*MU*).

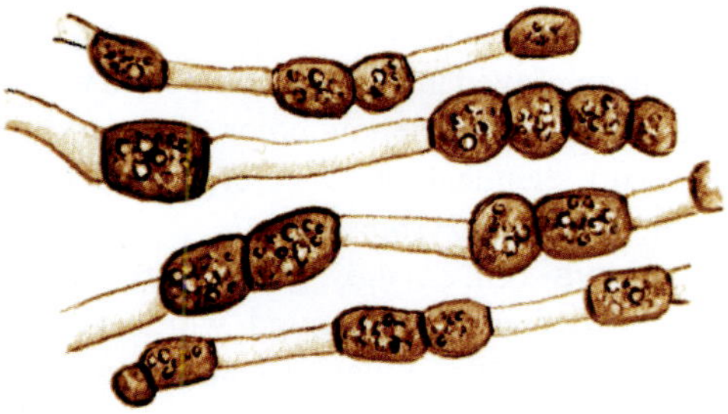

Chlamydospores of *Phoma eupyrena*, x 670 (*MU*).

chloro-, **chlor-** (Gr. *chlorós*, green or greenish-yellow): a pref. used in the formation of terms that allude to the color green or to that of chlorophyll, as in *chlorophycophilous*.

chlorophycophilous (Gr. *chlorós*, green + *phýkos*, alga + *phílos*, have an affinity for): *Lichens*. A thallus in which the photobiont belongs to the Chlorophyceae, unicellular algae that live in symbiosis with ascomycetous or basidiomycetous fungi. Cf. **cyanophycophilous**. See **diphycophilous** and **phycophilous**.

chlorosis, pl. **chloroses** (Gr. *chlorós*, greenish-yellow + suf. *-osis*, condition or state): a pathologic state of plants that is characterized by a yellowish color of normally green plant tissues. The cause derives from a lack of activity of the chloroplasts. The etiological agents can be viruses, bacteria, fungi, insects, etc., but the chlorotic spots also can be physiological in origin, such as the lack of iron in the substrate. See **leaf spot**.

chondroid (Gr. *kóndros*, cartilage + L. suf. *-oide*, like < Gr. *-oeídes*, similar to): *Lichens*. Refers to the medullary layer of the thallus when the hyphae that constitute it are intimately united and have a cartilaginous consistency, as in *Usnea* (Lecanorales).

chondroid axis, pl. **axes** (Gr. *kóndros*, cartilage + L. suf. *-oide*, like < Gr. *-oeides*, similar to; L. *axis*, axis): *Lichens*. The cartilaginous axis occupying the central portion of the medulla in the thallus of *Usnea* (Lecanorales).

-chore (Gr. *choréo*, to change place, move away from): a suf. employed in terms that refer to the dispersal of spores and other propagules, and to the dispersal media, as in *anemochore*, *hydrochore*, *zoochore*, etc., i.e., dispersed by air, water, animals, etc.

chromato-, **chromat-** (Gr. *chrõma*, genit. *chrómatos*, color): a pref. employed in forming terms to introduce the idea of color, as in *chromatography* (a process used to separate the components of a chemical mixture, carried out in liquid or in gas, that depends on the differential distribution of solutes as they flow through a stationary liquid or a solid phase) and *chromatoplasm* (the cortical plasma in cells of the cyanobacteria provided with assimilating pigments).

Chromista (Gr. *chrõma*, color + L. suf. *-ista*, which denotes what or who carries it out or has the capacity to): this group of organisms was previously included in the kingdom Protoctista (or Protista), but now is considered as a kingdom separate from the Protozoa. It includes 10 phyla (according to Corliss, 1994) or 5 phyla and 7 subphyla (according to Cavalier-Smith, 1998). It includes a wide range of brown algae, diatoms, chrysophytes and cryptomonads (but no chlorophytes or rhodophytes, which are classified in the kingdom Plantae). It is believed that the three phyla of organisms considered as *fungoid*, which are classified in Chromista (in the subkingdom Heterokonta), resulted from the loss of chloroplasts. The three *fungoid* phyla belonging to the Chromista, which are studied by mycologists and are accepted in the 9th ed. of the *Dictionary of the Fungi* (Kirk *et al*., 2001), are the Hyphochytriomycota, Labyrinthulomycota and Oomycota, whose flagellate forms have at least one mastigonemate flagellum, i.e., a rigid, tubular flagellum provided with tripartite flagellar hairs or mastigonemes. In another classification scheme proposed (Dick, 2001), the three phyla are placed in the kingdom *Stramenopila*. These organisms are characterized by being predominantly unicellular, filamentous, or colonial and primarily phototrophic; with cell walls lacking chitin and beta glucan (they are often cellulosic); the chloroplasts (when present) are localized in the lumen of a rough endoplasmatic reticulum; they lack starch and have a double enveloping membrane, within a periplastidic membrane, with chlorophylls *a* and *c* (when present); the mitochondria generally have tubular cristae; Golgi bodies and peroxisomes always present; when flagella are present, at least one has flagellar hairs or tripartite mastigonemes (except in haptophytes); the majority are free-living organisms, many microscopic in size (with exceptions, such as the brown algae). In the case of the phyla of fungi, it is thought that they have lost their chloroplasts in a secondary manner; the synthesis of lysine is by the diaminopimelic acid metabolic pathway. See **Fungi**, **fungal** and **fungoid**. In the Appendix, the user of this dictionary can see the comparison between the outline of Alexopoulos *et al.*, 1996, modified by Ulloa and Hanlin, 2006, to that of Kirk *et al.*, 2008. In the latter classification scheme, the phyla of kingdom Stramenopila are now in the kingdom Chromista. These taxonomic rearrangements are part of a constant process of study and evaluation, in order to achieve a phylogenetic classification of the fungi. See **Stramenopila**.

chromo- (Gr. *chrõma*, color): equal to **chromato-**, which is preferable.

chromogenic, **chromogenous** (Gr. *chrõma*, color + *génos*, origin): an organism, cell, etc., capable of producing a colored material or substance. This phenomenon is called *chromogenesis*, pl. *chromogeneses* (Gr. *chrõma* + *génesis*, origin, production).

chromomycosis, pl. **chromomycoses** (Gr. *chrõma*, color + *mýkes*, fungus + suf. *-osis*, condition or state): a chronic, subcutaneous, granulomatous mycosis of man and higher animals caused by the traumatic implantation in the skin of any of several dematiaceous fungi from exogenous sources (wood, soil and plant remains). In general, it is a localized infection of the skin and subepidermal tissues, especially of the lower limbs (although it may occur elsewhere in the body), that is characterized by the formation of verrucoid nodules and tumor-like masses. Among the principal causal agents of chromomycosis are *Phialophora verrucosa*, *Phialophora* (=*Rhinocladiella*=*Fonsecaea*) *pedrosoi* and *Cladophialophora* (=*Cladosporium*) *carrionii* (dematiaceous asexual fungi). See **mycosis**.

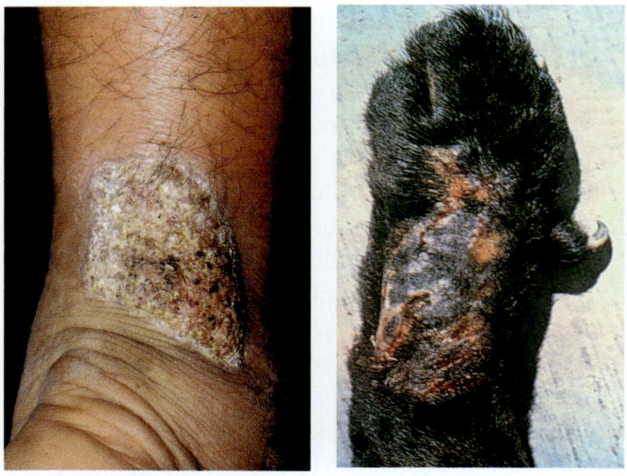

Chromomycosis on the ankle of a human patient, x 0.5 (*AB*), and the same mycosis on a foot of a dog, caused by *Phialophora* (=*Rhinocladiella*=*Fonsecaea*) *pedrosoi*, x 0.5 (*ECN*).

chromophilous (Gr. *chrõma*, color + *phílos*, have an affinity for): applied to cellular granules or other diverse substances that stain easily and intensely, like the chromophilic bodies of the cell nucleus.

chromosome (Gr. *chrõma*, color + *sõma*, body): each one of the linear or sometimes circular basophilic bodies of prokaryotic and eukaryotic organisms that contain proteins and most or all of the DNA comprising the genes of an individual.

chrono-, **chron-** (Gr. *chrónos*, time, season of the year): a pref. used to refer to time, as in *chronology*, the science that treats of the measurement of time.

chryseous (Gr. *chrysós* > L. pref. *chryso-*, gold, generally employed in forming compound words that allude to the golden-yellow color of some organ or organic part, chemical substance, etc.): golden-yellow in color, like the *chrysocystidia*.

chryso-, **chrys-** (Gr. *chrysós*, gold): a pref. employed in the formation of compound words that allude to the golden-yellow color of some organ, organic part, substance, etc., as in *chrysocystidium* and *chrysosporus*.

chrysocystidium, pl. **chrysocystidia** (Gr. *chrysós*, golden, golden-yellow + L. *cystidium* < Gr. *kystídion* < *kýstis*, bladder; here, vesicle + dim. suf. *-ídion* > L. *-idium*): *Agaricales*. A clavate or fusiform-clavate pseudocystidium with a yellow inclusion that accentuates on being treated with ammonia; it is characteristic of species of *Hypholoma* (=*Naematoloma*), *Psilocybe* and *Stropharia*, among other gen. Cf. **lamprocystidium**, **leptocystidium** and **macrocystidium.**

Chrysocystidia of the hymenium of *Stropharia* sp., x 500 (*MU*). Basidiocarp of *Phellinus linteus* (Aphyllophorales) with yellow spores (**chrysosporus**), x 0.2 (*RV*).

chylo-, **chyl-** (Gr. *chylós*, juice): a pref. used in terms such as *chylophyte*, *chylophyllous*, etc., a succulent plant, plant with succulent leaves, etc.

chymo- (Gr. *chymós*, juice): a pref. that alludes to plant juices or to their *chemical products*, as in *chymophore*, that makes juice, that is juicy (chymophore vessel), and *chymokinesis*, pl. *chymokineses*, kinesis due to the action of chemical substances, like that which phosphoric acid and phosphates exert on the zoospores of *Saprolegnia* (Saprolegniales).

chytrid (Gr. *chytrídion*, dim. of *chytrís*, pot): a member of the Chytridiomycetes (class), or of the Chytridiomyota (phylum) of the kingdom Fungi in some systems of classification (e.g., Margulis and Schwartz, 1988, and Alexopoulos *et al.*, 1996).

Zoosporangium of *Chytriomyces hyalinus* (Chytridiales), a saprobic **chytrid** on plant and animal detrita, and a parasite of diatoms, x 700 (*MU*).

Chytridiomycetes (Gr. *chytrídion*, dim. of *chytrís*, kettle + L. *-mycetes*, ending of taxonomic class < Gr. *mýkes*, genit. *mýketos*, fungus): Alexopoulos *et al.*, 1996, consider this group of aquatic organisms, freshwater and marine, in the phylum Chytridiomycota of the kingdom Fungi, with uniflagellate zoospores on the posterior end, lacking mastigonemes, a unicellular or filamentous thallus, and sexual reproduction by planogametic copulation. The class Chytridiomycetes includes the orders Spizellomycetales, Olpidiopsidales, Neocallimastigales, Chytridiales, Blastocladiales and Monoblepharidales. See the Appendix to compare the classification of this class in Alexopoulos *et al.*, 1996, modified by Ulloa and Hanlin, 2006, to that of Kirk *et al.*, 2008. In the latter classification scheme, the order Monoblepharidales was raised to the class Monoblepharidomycetes, phylum Chytridiomycota. The order Blastocladiales was raised to the class Blastocladiomycetes, phylum Blastocladiomycota, and Neocallimastigales has been moved to a different class and phylum, Neocallimastigomycetes, phylum Neocallimastigomycota. The order Olpidiopsidales (previously in the class Chytridiomycetes) has been moved to the class Oomycetes, phylum Oomycota. These taxonomic rearrangements are part of a constant process of study and evaluation, in order to achieve a phylogenetic classification of the fungi.

cicatricose (L. *cicatricosus* < *cicatrix*, *cicatricis*, scar + suf. *-osus* > E. *-ose*, full of, increased, prone to): with scars or longitudinal borders.

ciliate (L. *ciliatus*, provided with cilia < *cilium*, the edge of the eyelid where the eyelashes are implanted; generally applied to the eyelash proper or to a very fine hair + suf. *-atus* > E. *-ate*, provided with or likeness): provided with *cilia*; with very fine threads in the border.

cincinnatus (L. *cincinnatus*, rolled up, curled < *cincinnus*, curl, ringlet of hair + suf. *-atus* > E. *-atus*, provided with or likeness): provided with curls or ringlets.

cinereous (L. *cinereus*, ash-colored): with the color of ashes, like the sporangia of *Arcyria cinerea* (Trichiales) and *Physarum cinereum* (Physarales).

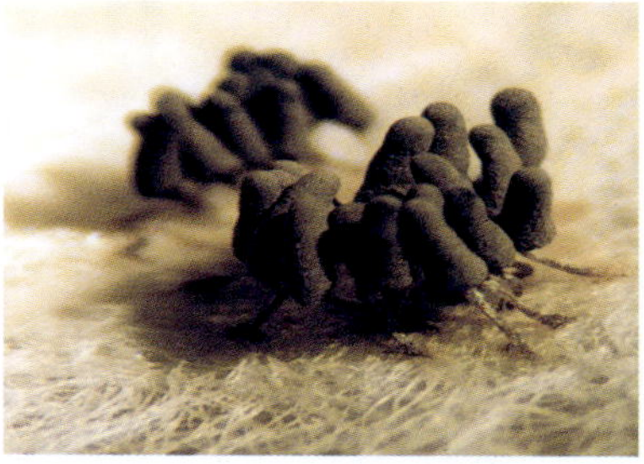

Cinereous sporangia of *Arcyria cinerea*, x 10 (*MU* & *CB*).

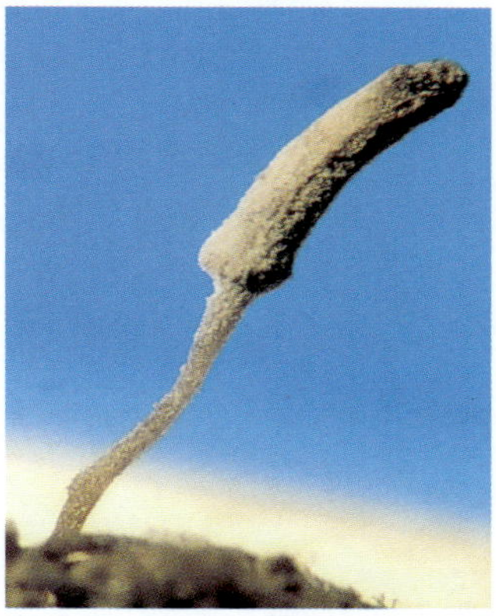

Cinereous sporangium of *Arcyria cinerea*, x 20 (*MU* & *CB*).

Cinereous sporangia of *Physarum cinereum*, x 20 (*MU*).

cingulate (L. *cingulatus*, belted < *cingulum*, strap, belt + suf. *-atus* > E. *-ate*, provided with or likeness): provided with a girdle, i.e., with a border or a belted zone, like a sash all around, like the ascospores of *Glomerella cingulata* (Phyllachorales), *Corollospora intermedia* and *C. maritima* (Halosphaeriales), which have a slight central constriction.

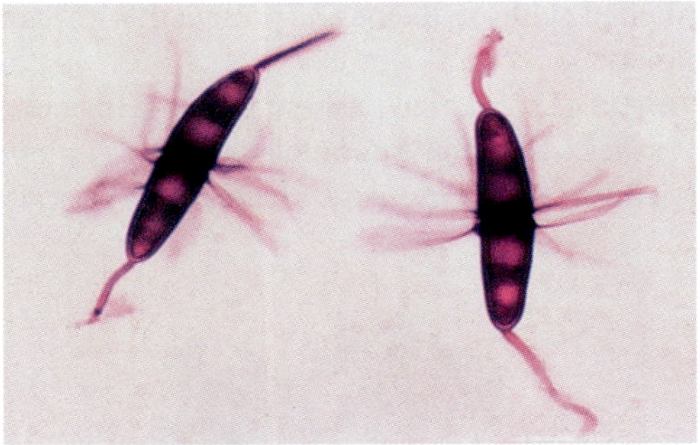

Cingulate ascospores of *Corollospora intermedia*, x 750 (*MCG*).

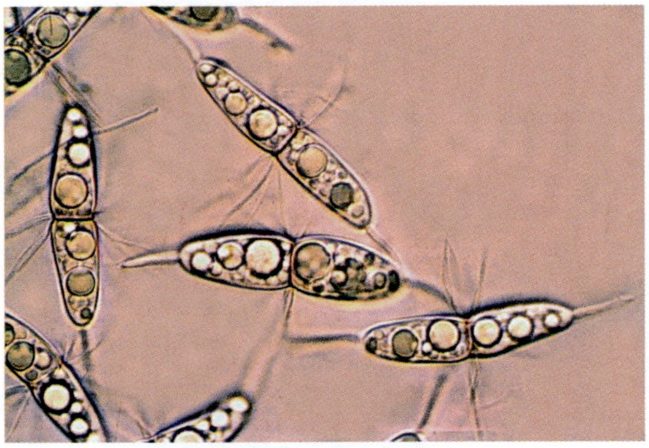

Cingulate ascospores of *Corollospora maritima*, x 750 (*MCG*).

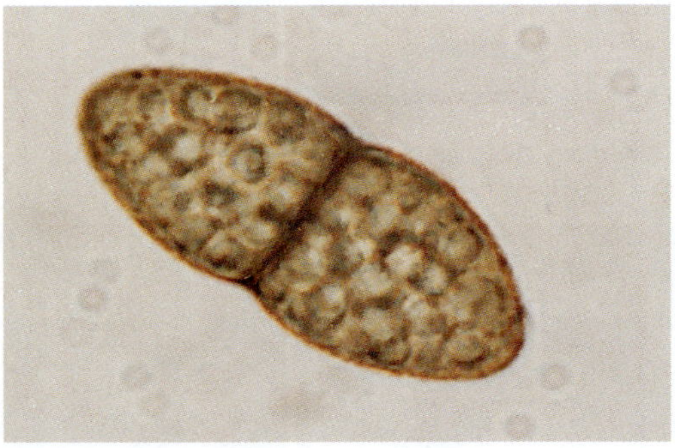

Cingulate ascospore of an undetermined ascomycete, x 2,000 (*MCG*).

circadian rhythm (L. *circa*, near or around + *dies*, day + suf. *-anus* > E. *-ian*, pertaining to; Gr. *rhytmós*, rhythm, rhythmic movement): periodic and cyclic modification, such as cell division, sporulation, the maximum rate of photosynthesis, bioluminiscence, or the production of enzymes, among other phenomena present in living cells, whose periodicity is ca. 24 hours. See **diel**.

circinate (L. *circinatus*, rounded off, rolled up < *circinare*, to make round, to wind around + suf. *-atus* > E. *-ate*, provided with or likeness): coiled inward from the apex, like the sporangiophores of *Circinella umbellata* and *C. muscae* (Mucorales), the ascocarp appendages of *Uncinula* (Erysiphales) and the perithecial hairs of *Chaetomium* spp. (Sordariales).

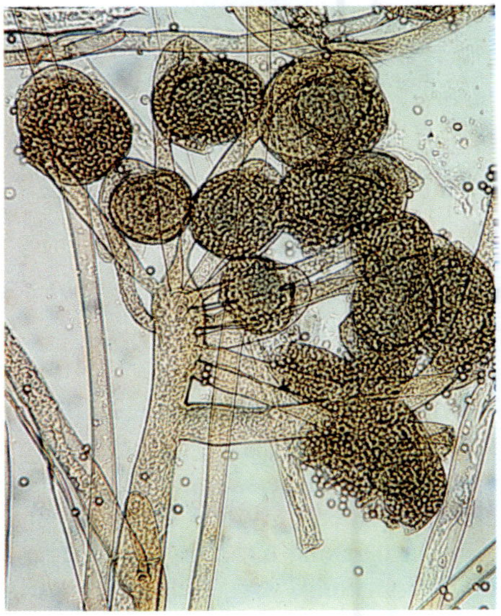

Circinate sporangiophores of *Circinella umbellata*, x 100 (*EAA*).

Circinate sporangiophores of *Circinella muscae*, x 100 (*MCG*).

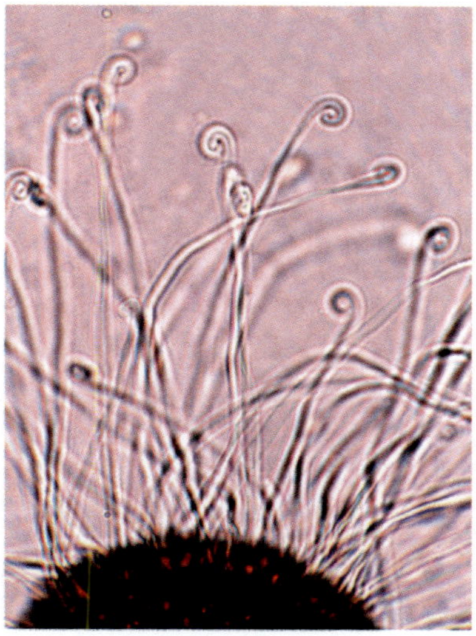

Circinate ascocarp appendages of *Uncinula* sp., x 300 (*MU*).

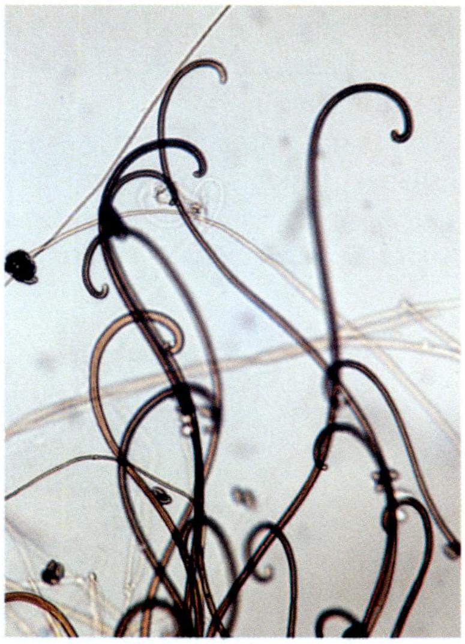

Circinate perithecial hairs of *Chaetomium* sp., x 200 (*MU*).

circum- (L. *circum*, around): a pref. that signifies around, as in *circumpolar*, around the pole, and *circumscription*, limitation, restriction; periphery.

circumcinct (L. *circumcinctus*, surrounded by a band in the middle < *circum*, around + *cinctum*, belt, sash): with a band or belt around the middle part, as in some spores.

circumscissile, circumscissile dehiscence (NL. *circumcisilis* < *circumcisus*, cut all around < *circum*, around + *scissile*, capable of breaking open or splitting < *scindere*, to cut + suf. *-ilis* > E. *-ile*, capable, with the character of; L. *dehiscentia*, opening < *dehiscere*, to open itself or to part by itself + suf. *-entia* > E. *-ence*, state, quality of or action): opening or separating along a circular, equatorial or longitudinal line; e.g., the sporangia of the Myxomycete *Craterium minutum* (Physarales) and of the Zygomycetes *Choanephora cucurbitarum* and *Gilbertella persicaria* (Mucorales), as well as the basidiocarp of *Cyathus* (Nidulariales), have circumscissile dehiscence.

Circumscissile dehiscence of the fruiting body of *Cyathus* sp., x 6.5 (*MU*).

cirrate, cirrhate (L. *cirratus*, curled < *cirrus*, ringlet of hair + suf. *-atus* > E. *-ate*, provided with or likeness): with curls or shaped like a curl. Also called **cirriferous**.

cirrhus, cirrus, pl. **cirrhi, cirri** (L. *cirrus*, ringlet of hair): a slender column of spores, often curled and held together by mucilage, that emerges from the ostiole of certain fungal fruiting bodies, as is seen in the pycnidia of *Phoma* and *Phomopsis* (sphaeropsidaceous asexual fungi), and the perithecia of *Microascus* (Microascales) and *Chaetomium* (Sordariales).

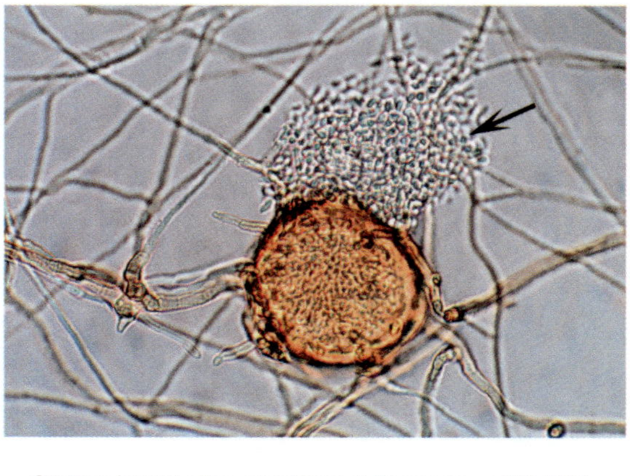

Cirrhus (arrow) of the pycnidium of *Phoma* sp., x 150 (*MU*).

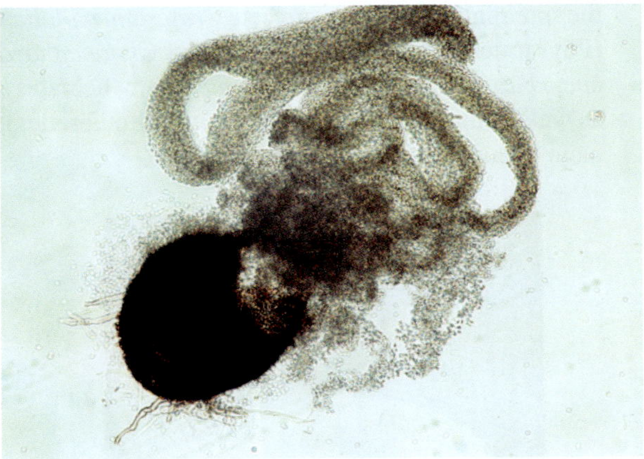

Cirrhus of the pycnidium of *Phoma sorghina*, x 150 (*PL*).

Cirrhus of the perithecium of *Microascus trigonosporus*, x 90 (*CB*).

Cirrhus of the pycnidium of *Phoma* sp., x 220 (*RTH*).

Cirrhus of the perithecia of *Microascus trigonosporus*, x 70 (*CB*).

cirriferous (L. *cirrus*, ringlet of hair + suf. *-ferous*, bearer < *fero*, to bear, carry): provided with a cirrhus. Syn. of **cirrate**.

citriform (L. *citrus*, citron, lemon + suf. *-formis* < *forma*, shape): lemon-shaped, like the sporangia of *Phytophthora infestans* and *Pseudoperonospora* (=*Peronoplasmopara*) *canabina* (Peronosporales).

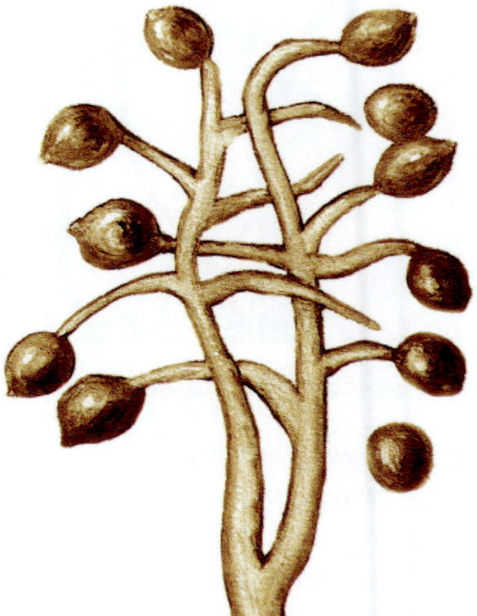

Citriform sporangia of *Pseudoperonospora* (=*Peronoplasmopara*) *canabina*, x 80 (*MU*).

cladistics (Gr. *kládos*, branch + NL. suf. *-istic* > E. *-istic*, concerning it): a method of systematics based on phylogenetic relationships, whose objective is to reconstruct the genealogical pathway of organisms by means of objective and repeatable analysis, and from this to propose an hypothesis of natural or phylogenetic classification. The product of cladistic analysis is a branching diagram (*cladogram*) that shows the pattern of relationships among the organisms based on the characters utilized. The recognition and hierarchal arrangement of the taxa in clades constitutes the practice of cladistics. See **classification**, **phylogeny**, **nomenclature**, **parsimonious**, **systematics** and **taxonomy**.

clado-, **clad-** (Gr. *kládos*, branch): a pref., as well as a suf., *-cladium*, often employed in biology to give the idea of something that involves one or more branches, as in *sporocladium*.

cladogram (Gr. *kládos*, branch + *grámma*, to describe): see **cladistics**.

clamp connection (ME. *clamp* < MD. *clampe*, clamp; L. *connexion*, connection, union < *connexus*, united, connected + suf. *-io*, genit. *-ionis* > E. *-ion*, result of an action, state of): *Basidiomycetes*. A slender hyphal branch that connects two adjacent cells that are separated by a septum. The passage created by this hyphal bridge permits one of the nuclei resulting from mitosis in one of the cells to move into the adjacent cell; e.g., *Sclerotium coffeicola* (agonomycetaceous asexual fungi). Also called **fibula**.

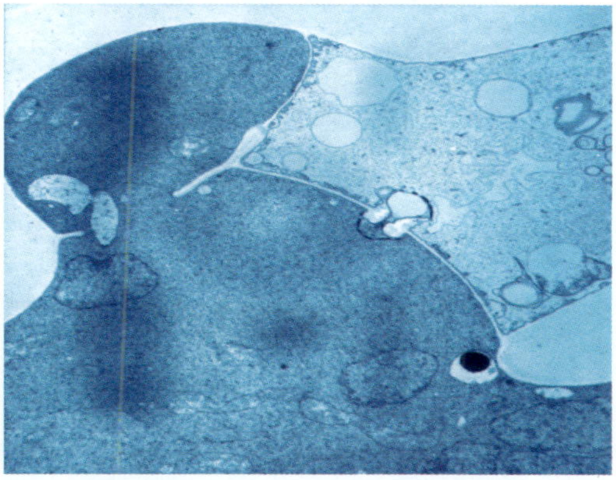

Clamp connection of *Sclerotium coffeicola*, x 1,000 (*RTH*).

class (L. *classis*): a taxon or taxonomic category intermediate between division (or subdivision) and order. In mycology the ending for class is *-mycetes* (Gr. *mýkes*, genit. *mýketos*, fungus), as in Zygomycetes.

classification (L. *classificare*, to classify; *classificatio*, genit. *classificationis*, the act of classifying or grouping < *classis*, group + suf. *-ationem* > E. *-ation*, action, state or condition, result): the systematic arrangement into groups or categories according to established criteria; *taxonomy*. The application of scientific names to the categories into which fungi may be placed; the relative order of these categories is governed by the International Code of Botanical Nomenclature (ICBN). See **nomenclature**, **systematics** and **taxonomy.**

clathrate, **clathroid** (L. *clathratus*, sieve-like, resembling a lattice, trellis < *clathrare*, to surround with lattice-work < *clathra*, bar, small rod < Gr. *klãthra*, pl. of *kléithron*, bar + L. suf. *-atus* > E. *-ate*, provided with or likeness): *Gasteromycetes*. A latticed or interconnected receptaculum, like a cage, that supports the gleba, like the fruiting bodies of species of *Clathrus* and *Laternea* (Phallales).

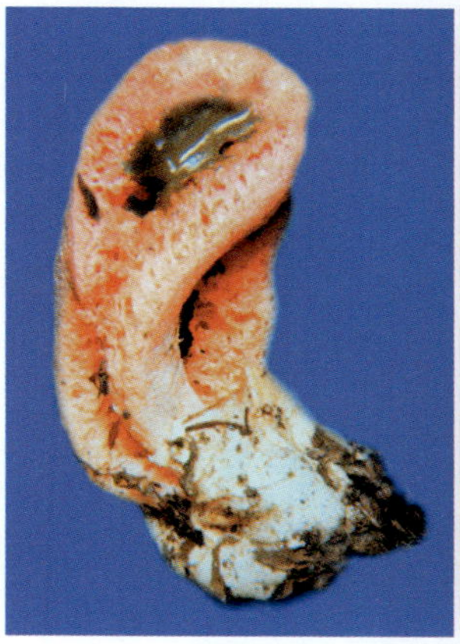

Clathrate fruiting bodies of *Clathrus ruber*, x 1 (*TH*) and x 1 (*SBH*).

Clathrate fruiting body of *Clathrus ruber*, x 0.5 (*JC*).

clavate, **claviform** (NL. *clavatus*, club-shaped < L. *clava*, club, mallet + suf. *-atus* > E. *-ate*, provided with or likeness; L. *claviformis* < *clava* + suf. *-formis* < *forma*, shape): shaped like a club, like the conidial head of *Aspergillus giganteus*, *A. clavatus* and *Penicillium vulpinum* (=*P. claviforme*), moniliaceous asexual fungi, the synnemata of *Hirsutella saussurei* (stilbellaceous asexual fungi), and the fruiting body of *Clavariadelphus truncatus* (Aphyllophorales). Cf. **obclavate**.

Clavate conidial heads of *Aspergillus giganteus*, x 40 (*MU*).

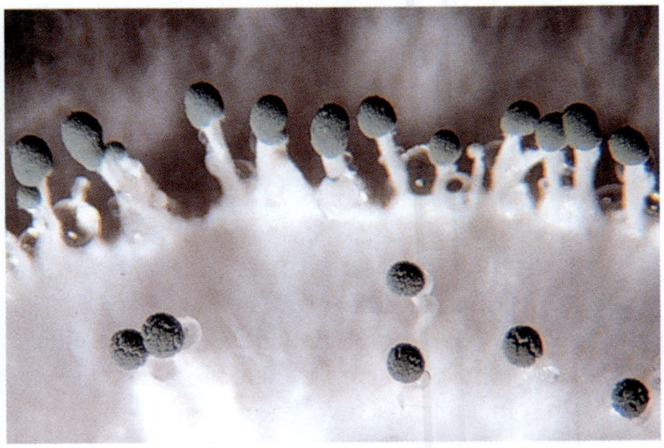

Clavate conidiophores of *Penicillium vulpinum* (=*P. claviforme*), x 2 (*MU*).

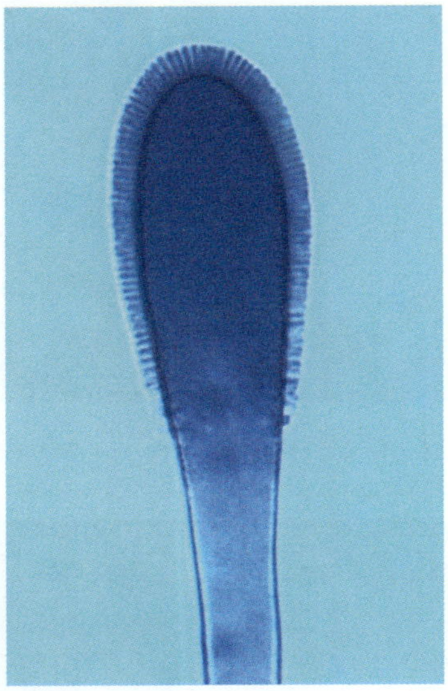

Clavate conidial head of *Aspergillus clavatus*, x 400 (*RTH*).

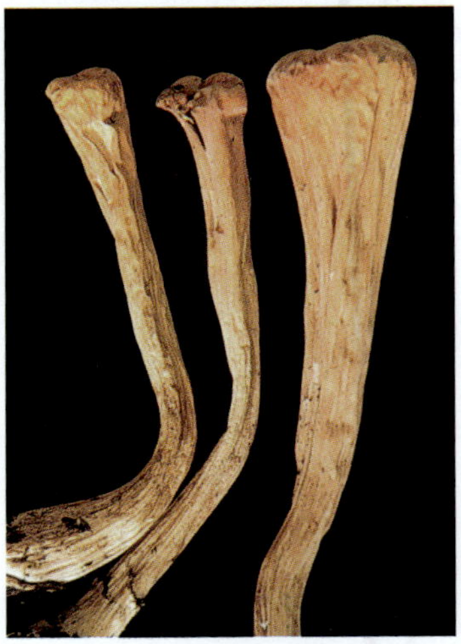

Clavate basidiocarps of *Clavariadelphus truncatus*, x 1 (*MU*).

Clavate synnema of *Hirsutella saussurei*, x 220 (*MU*).

Clavate basidiocarps of *Clavariadelphus truncatus*, x 0.3 (*MRO*).

clavule (L. *clavula*, small club < *clava*, club, mallet + dim. suf. *-ula* > E. *-ule*): a claviform or clavulate fruiting body of certain fungi, like that of *Calocera cornea* (Dacrymycetales), *Clavaria pistillaris* and *Clavulina amoena* (Aphyllophorales).

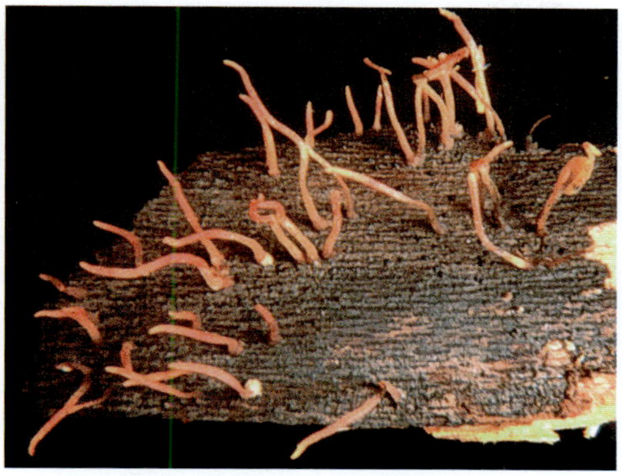

Clavules (fruiting bodies) of *Calocera cornea*, x 1 (*BL*).

Clavules (fruiting bodies) of *Clavulina amoena*, x 0.5 (*TH*).

cleisto-, **cleist-** (Gr. *kleistós*, closed): a pref. employed in words that allude to something that does not open, as in *cleistothecium*.

cleistocarp (Gr. *kleistós*, closed + *karpós*, fruit): an enclosed ascocarp, without a pore, whose ascospores are only liberated after rupture or disintegration of the wall or peridium, as occurs in the truffles (*Tuber* and other gen. of Tuberaceae, Pezizales).

cleistohymenial (Gr. *kleistós*, closed + L. *hymenium* < Gr. *hyménion*, dim. of *hymén*, membrane + L. suf. *-alis* > E. *-al*, relating to or belonging to): one of the three types of apothecial development in the Discomycetes, characterized by remaining closed during most of the process. Sterile hyphae growing out of the stalk cell of the ascogonium or from adjacent mycelial hyphae completely surround the ascogonium in a loose weft, forming a ball with the ascogenous hyphae in the center. In some species the ascocarp remains permanently closed, as in the hypogeous forms; in others the cortical layer over the developing hymenium is ruptured prior to maturity. This type of development has been studied in *Inermisia* (=*Byssonectria*), of the Pezizales. Cf. **eugymnohymenial** and **paragymnohymenial**.

cleistothecioid (Gr. *kleistós*, closed + *thêke*, box + L. suf. *-oide*, like < Gr. *-oeídes*, similar to): **1**. Any enclosed (nonostiolate) ascoma. **2**. An enclosed ascoma, generally spherical, similar to a cleistothecium, but differing in having the asci arranged in an hymenium or fascicle, as in *Brasiliomyces malachrae*, species of *Podosphaera* and *Uncinula* (Erysiphales). It represents a perithecium or apothecium that lacks the typical opening.

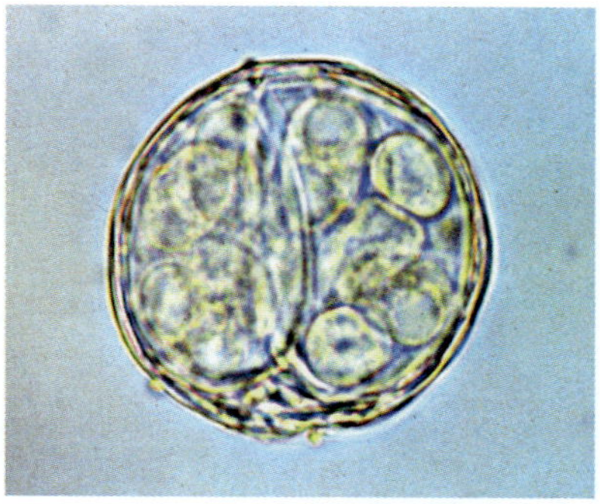

Cleistothecioid perithecium of *Brasiliomyces malachrae*, x 400 (*RTH*).

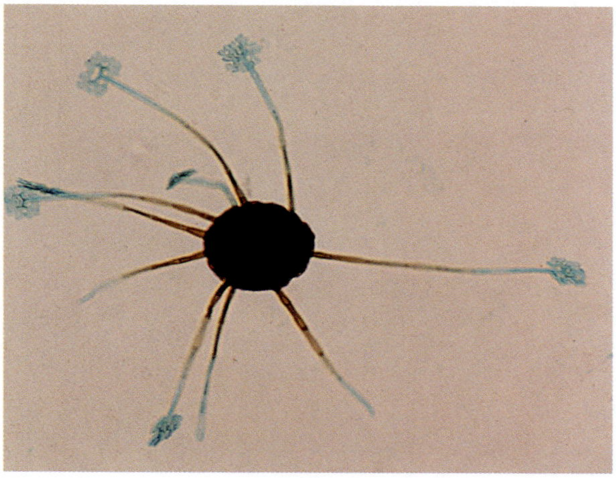

Cleistothecioid perithecium of *Podosphaera* sp., x 300 (*RTH*).

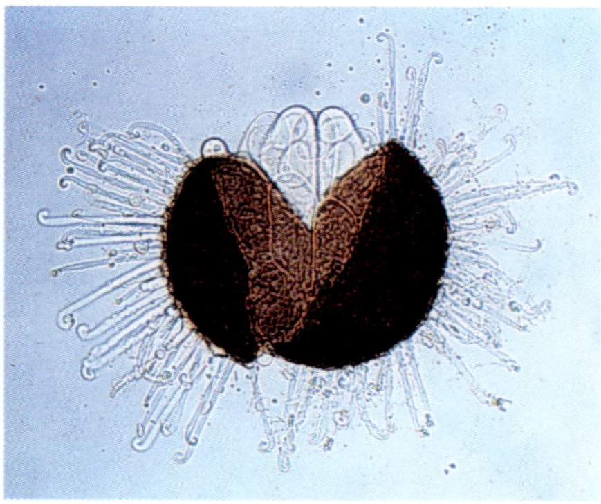

Cleistothecioid perithecium of *Uncinula macrospora*, x 400 (*RTH*).

cleistothecium, pl. **cleistothecia** (NL. *cleistothecium* < Gr. *kleistós*, closed + *thekíon* < *thêke*, box; here, of the asci + dim. suf. *-íon* > L. *-ium*): **1**. An enclosed ascocarp (fruiting body), i.e., lacking a pore, and delimited by a wall that ruptures irregularly or disintegrates on maturing to liberate the ascospores. **2**. *Plectomycetes*. A type of ascocarp with ascohymenial ontogeny and small, spherical asci irregularly distributed throughout the ascocarp, which is surrounded by a peridium or wall lacking a pore, as, e.g., in *Emericella variecolor*, *Eurotium chevalieri* and *Monascus purpureus* (Eurotiales).

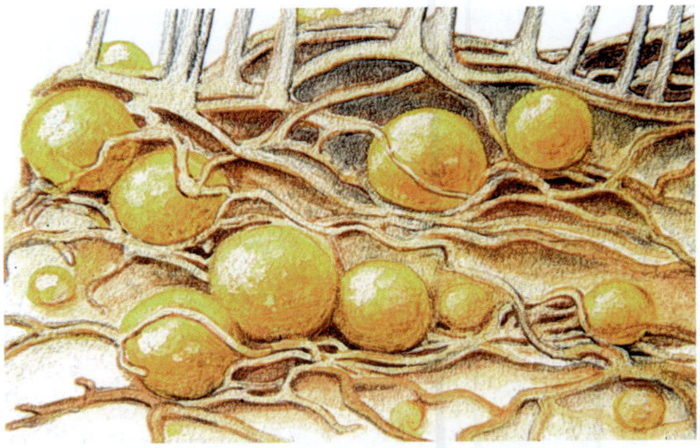

Cleistothecia of *Eurotium rubrum*, x 50 (*MU*).

Cleistothecia of *Emericella variecolor* on agar, x 35 (*CB*).

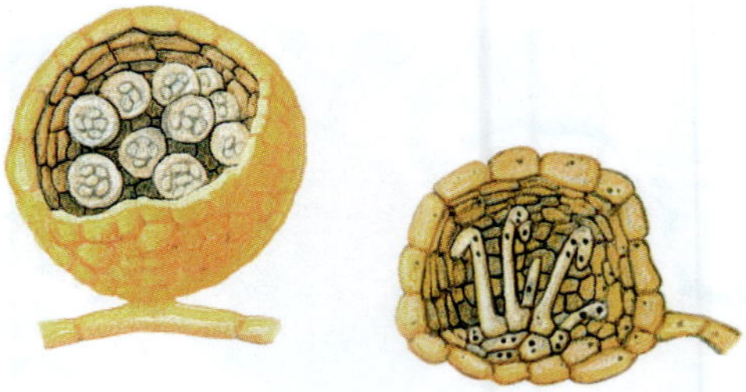

Cleistothecia of *Eurotium rubrum*. The young cleistothecium on the right is shown in longitudinal section to show the croziers inside. The one on the left shows the globose asci in the interior, x 80 (*MU*).

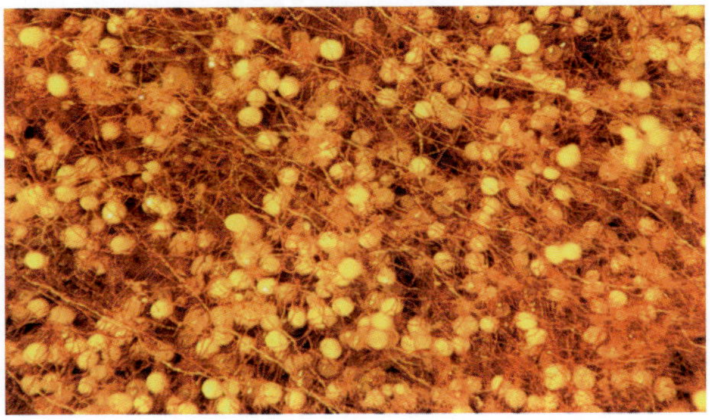

Cleistothecia of *Eurotium rubrum*, growing on agar, x 8 (*MU*).

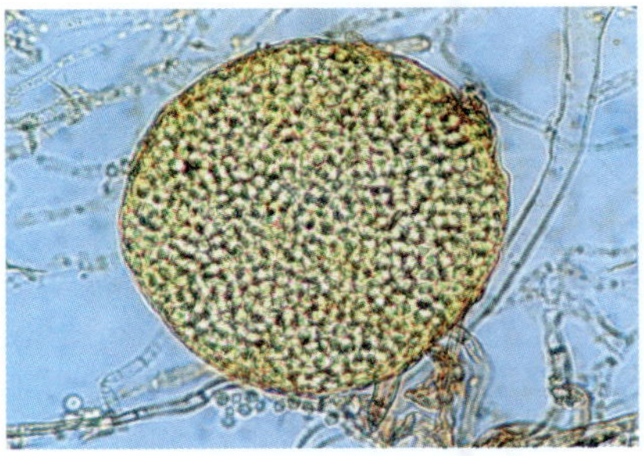

Cleistothecium of *Eurotium chevalieri*, x 120 (*PL*).

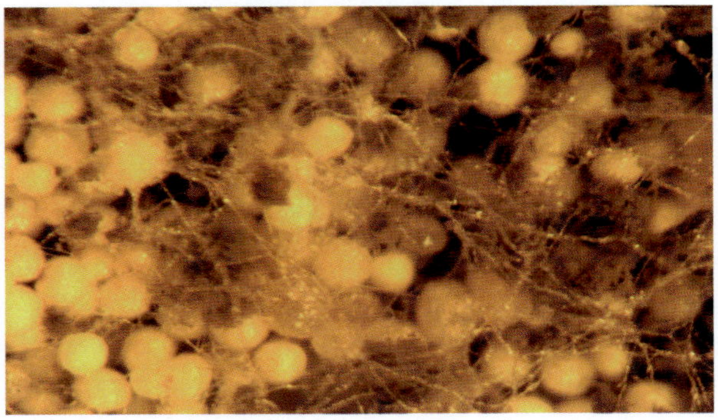

Cleistothecia of *Eurotium rubrum*, growing on agar, x 20 (*MU*).

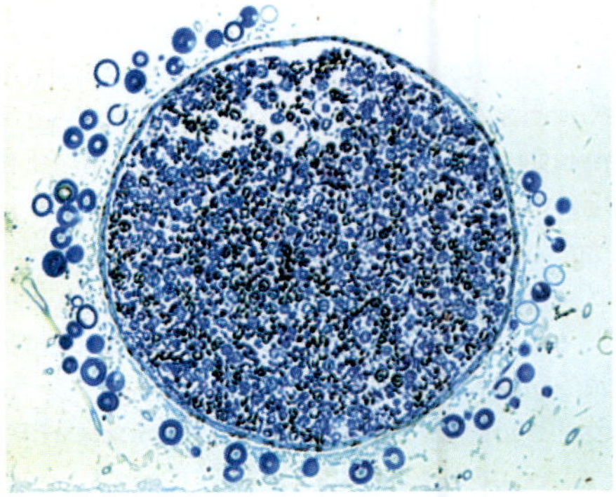

Cleistothecium of *Emericella nidulans* shown in cross section to show the asci in the interior and the Hülle cells on the outside of the wall, x 130 (*RTH*).

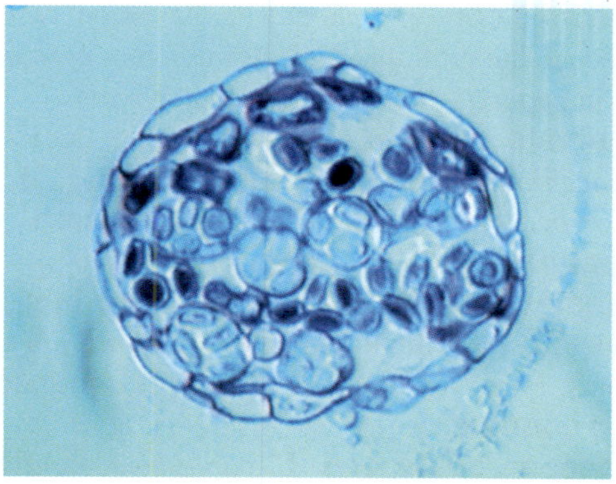

Cleistothecium of *Eurotium* sp., seen in cross section to reveal the asci in the interior, x 95 (*RTH*).

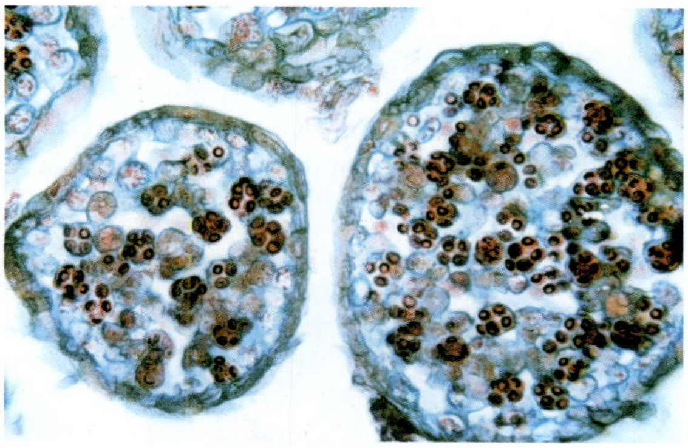

Cleistothecia of *Eurotium rubrum*, seen in cross section. The octospored asci are scattered throughout the interior, x 150 (*RTH*).

cleptotype (Gr. *kleptés*, thief + *týpos*, type, mark, model): a fragment taken from a holotype specimen.

clino-, **clin-**, **-clino** (Gr. *klíno*, to incline): a pref. that indicates inclination. When derived from the Gr. *klíne*, bed, it is utilized as a suf. in terms such *monoclinous* and *diclinous*.

clitocyboid (< gen. *Clitocybe* < Gr. *klitós*, inclined + *kýbe*, head + L. suf. *-oide*, like < Gr. *-oeídes*, similar to): *Agaricales*. A type of fruiting body (of 13 principal types recognized) with adherent, decurrent or subdecurrent gills on the stipe, which is fibrous, fleshy or chalky in consistency, centrally attached to the pileus and lacking a ring and volva, like that of species of *Clitocybe* and some species of *Russula*, *Lactarius* and *Inocybe*, among other gen.

clonal culture (Gr. *klón* > E. *clone*, progeny, offspring + L. suf. *-alis* > E. *-al*, relating to or belonging to; L. *cultus*, culture, action and effect of cultivating): the cultivation of genetically identical individuals, all derived from an original individual by asexual multiplication (e.g., fission, budding, cell division, parthenogenesis), as happens with monosporic cultures.

clone (Gr. *klón*, progeny, descendents): the offspring or descendents produced from a single parental cell by means of asexual reproduction.

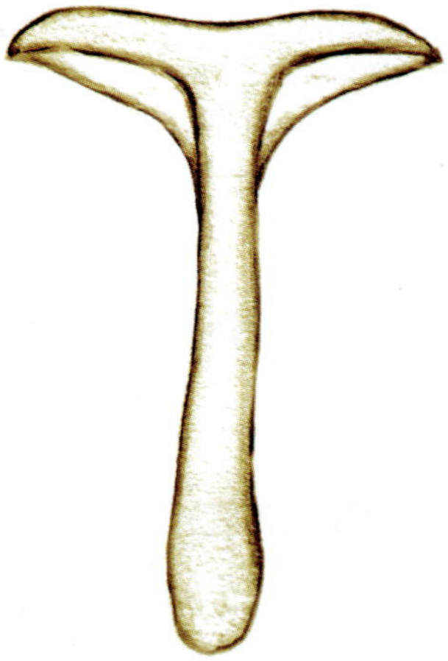

Clitocyboid basidiocarp of *Clitocybe* sp.
Diagram of a longitudinal section, x 0.6 (*MU*).

closterospore (Gr. *klostér*, spindle + *sporá*, spore): a multinucleate, fusiform, phragmospore (macroconidium), like that of *Trichophyton rubrum*, *Microsporum canis* and *M. cookei* (moniliaceous asexual fungi).

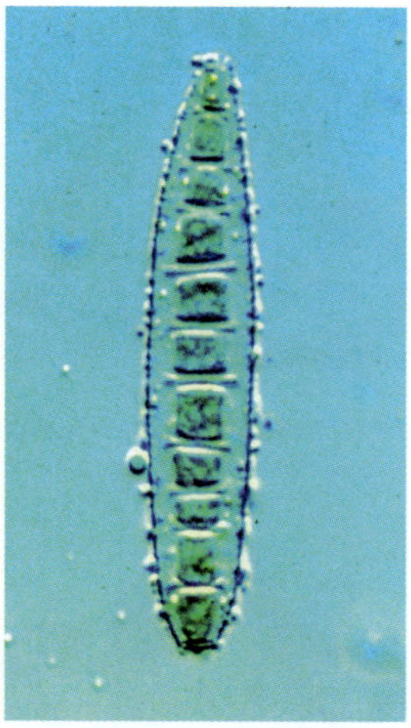

Closterospore (macroconidium) of *Microsporum cookei*, x 150 (*RTH*).

club root (ME. *clubbe* < ON. *klubbe*, club; ME. < ON. *rot*, root): a disease of cabbage and other crucifers caused by *Plasmodiophora brassicae* (Plasmodiophorales), so named because the tumors on the hypertrophied roots are often club-shaped.

clypeate (L. *clypeatus*, armed with a shield < *clypeus*, shield + suf. *-atus* > E. *-ate*, provided with or likeness): having a *clypeus*. See **scutiform** and **peltate**.

clypeus, pl. **clypei** (L. *clypeus*, round shield): pseudostromatic tissue, often in the shape of a shield, generally dark brown or black in color, that surrounds the ostiole of conidiomata and perithecia of certain ascomycetes (e.g., *Phyllachora cibotii* and *Ophiodothella vaccinii*, Phyllachorales), as well as of some pycnidia (sphaeropsidaceous asexual fungi). In some species the clypeus also forms beneath the perithecium.

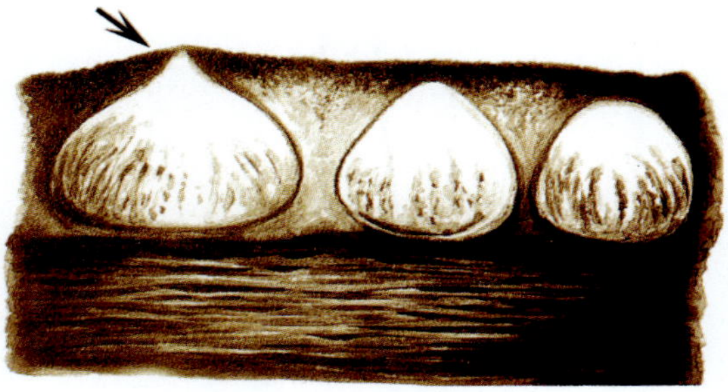

Clypeus above the perithecia of *Phyllachora cibotii*. Diagram of a longitudinal section of a perithecial stroma formed on a host leaf, x 70 (*MU*).

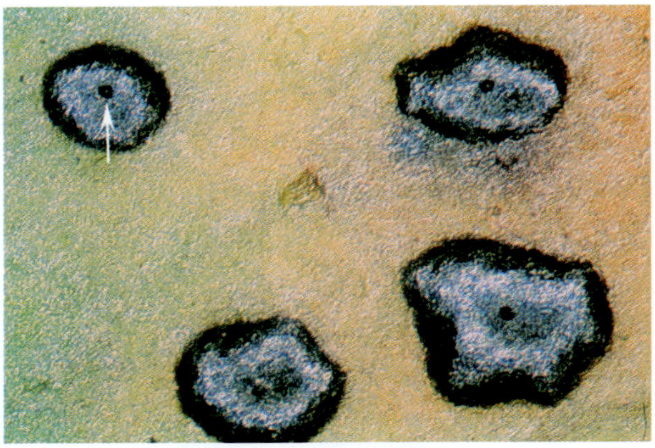

Clypeus around the ostiole of the conidiomata of *Ophiodothella vaccinii* on a parasitized leaf, x 40 (*RTH*).

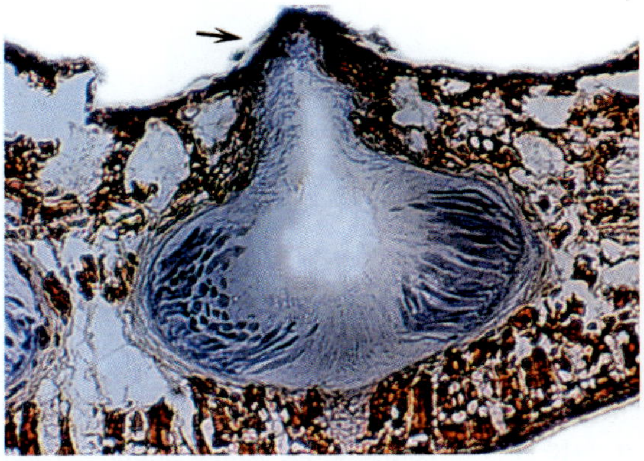

Clypeus of the perithecium of *Ophiodothella vaccinii*, seen in longitudinal section of a parasitized leaf, x 140 (*RTH*).

co-, **com-**, **con-** (L. *cum*, with, in union with): an inseparable pref. that indicates union or companion, as in *coadnate*, *compaginate* and *connate*.

coacervate (L. *coacervatus*, piled up, amassed < *coacervare*, to join, to pile up + suf. *-atus* > E. *-ate*, provided with or likeness): united, coherent. Equal to **coadnate** and **connate**.

coadnate (L. *coadnatus*, united < *co-*, union or companion + *adnatus*, broadly united < pref. *ad-*, in, near + *natus*, born): united, joined. Equal to **coacervate** and **connate**.

coalescent, **coalesced** (L. *coalescere*, to grow together, to unite + suf. *-escens*, genit. *-escentis* > E. *-escent*, that which turns, that begins, slightly): syn. of **concrescent**, **confluent** and **connate**. To grow together; e.g., the sporangia of the Myxomycete *Metatrichia vesparium* (Trichiales), which constitute a pseudoaethalium, have their pedicels united at the base. The gloeoid heads of the synnemata of *Antromycopsis smithii* (stilbellaceous asexual fungi) also are coalescent.

Coalescent sporangia of *Metatrichia vesparium*, x 25 (*MU*).

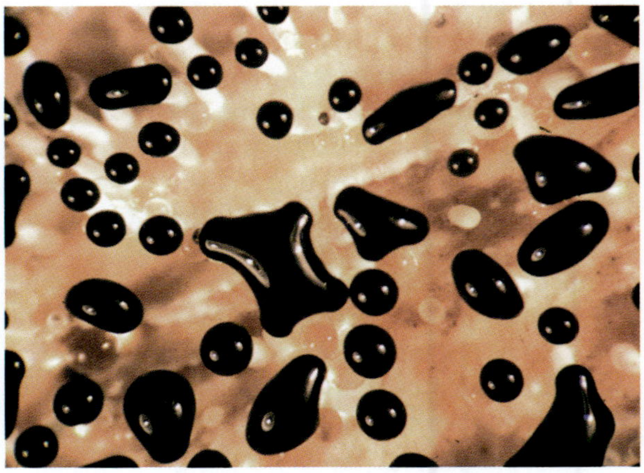

Coalescent gloeoid heads of the synnemata of *Antromycopsis smithii* on agar, x 5 (*CB* & *MU*).

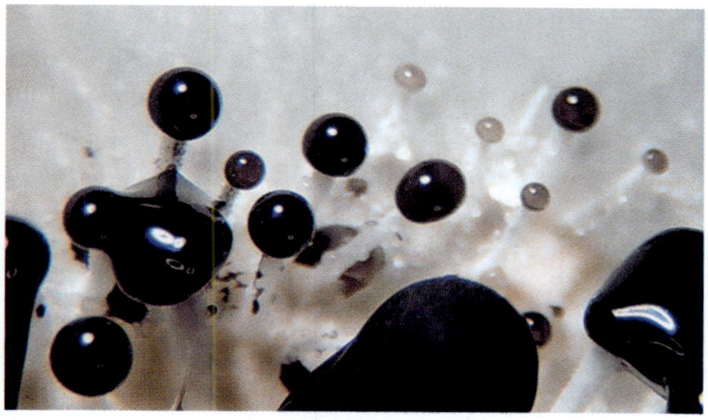

Coalescent gloeoid heads of the synnemata of *Antromycopsis smithii* on agar, x 5 (*CB* & *MU*).

Coalescent gloeoid heads of the synnemata of *Antromycopsis smithii* on agar, x 10 (*CB*).

coarctate (L. *coarctatus*, confined, constrained): piled up, crowded; deformed or broken by compression.

coccidioidomycosis, pl. **coccidioidomycoses** (<gen. *Coccidioides* < L. *coccidium* < Gr. *kokkíon*, little ball + L. suf. *-oides*, like < Gr. *-oeídes*, similar to, i.e., similar to coccidia, since the spherules with endospores that this dimorphic fungus develops in its parasitic phase were initially interpreted as sporiferous structures of coccidia, sporozoans that live as endocellular parasites of various animals + L. *mycosis* < Gr. *mýkes*, fungus + *-osis*, state or condition): a mycosis caused by the dimorphic fungus *Coccidioides immitis* (of uncertain taxonomic position, since it has been classified in the Synchytriaceae, of the order Chytridiales, and also among the moniliaceous asexual fungi), a saprobic inhabitant of soil in arid and semiarid zones of northern Mexico and the southwestern USA. It is a pulmonary infection that is acquired by inhalation of the arthrospores of the fungus, which is considered as one of the most virulent pathogenic fungi that exists. The disease can be acute and self-limiting, but when the fungus is disseminated through the blood stream, the infection becomes chronic and progressive, affecting the skin, bones, joints, lymphatic ganglia, suprarenal glands, central nervous system and other organs, conditions that can be fatal. See **mycosis**.

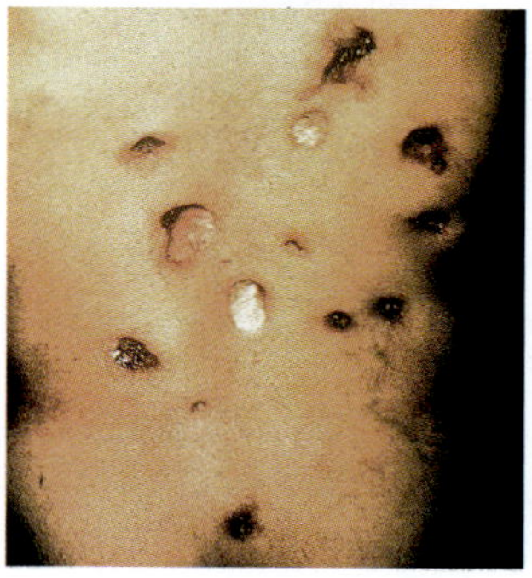

Coccidioidomycosis disseminated in the back of a patient, caused by *Coccidioides immitis*, x 0.1 (*RLM*).

coccus, pl. **cocci** (L. *coccus* < Gr. *kókkos*, grain, coccus): a spherical bacterium.

cochleariform (L. *cochleariformis* < *cochleare*, spoon + suf. *-formis* < *forma*, shape): having the shape of a spoon, like the basidiocarp of *Dacryopinax elegans* and *D. spathularia* (Dacrymycetales), which resemble a spoon or miniature spatula.

Cochleariform basidiocarps of *Dacryopinax elegans*, x 1 (*TH*).

Cochleariform basidiocarp of *Dacryopinax elegans*, x 6 (*MU*).

cochleate (L. *cochleatus*, spiral < *cochlea*, snail, snail shell < Gr. *kochlías*, snail + L. suf. *-atus* > E. *-ate*, provided with or likeness): spirally coiled, like the shell of a snail; the perithecial hairs of *Chaetomium cochliodes* (Sordariales) are cochleate.

Cochleate perithecial hairs of *Chaetomium cochliodes*, x 50 (*MU*).

coelo- (Gr. *koîlos*, hollow, cave, hole > L. *coelo*, hollow): a pref. employed in the formation of terms that denote something hollow, not solid, as in *coelonema*, a hollow filament of the capillitium of some Myxomycetes. Cf. **stereo-**.

Coelomycetes (Gr. *koîlos*, hollow, cave, hole > L. *coelo*, hollow + L. *-mycetes*, ending of taxonomic class < Gr. *mýkes*, genit. *mýketos*, fungus): a class of asexual fungi that produce their conidia on conidiophores contained in a fruiting body, either an acervulus, a pycnidium or a pycnothyrium. In some classifications the Coelomycetes (which includes the orders Melanconiales and Sphaeropsidales), together with the Blastomycetes and Hyphomycetes, constitute the Subdivision Deuteromycotina, or deuteromycetes in a general sense. In the classification adopted in this dictionary, this group is considered as sphaeropsidaceous and melanconiaceous asexual fungi, without class rank. See the Appendix to compare the classification changes of this class in Alexopoulos *et al.*, 1996, modified by Ulloa and Hanlin, 2006, to that of Kirk *et al.*, 2008. These asexual fungi are made up of heterogeneous species, forming unnatural groupings, and therefore are considered an artificial classification. These taxonomic rearrangements, with the concurrent changes of the names of taxa, are in order to achieve a phylogenetic classification of the fungi. See **conidial fungus**, **sphaeropsidaceous**, **pycnothyriaceous** and **melanconiaceous**.

coelonema, pl. **coelonemata** (Gr. *koîlos*, hollow, cave, hole + *nêma*, genit. *nêmatos*, filament): *Myxomycetes*. A hollow or tubular filament that constitutes a type of capillitium in the interior of the sporophores of many species, like those of the gen. *Hemitrichia* (Trichiales); in some very evolved forms the coelonematoid capillitium has various reliefs or ornamentations (*elaters*) that contribute to spore dispersal. Cf. **stereonema**.

coeno-, **coen-** (Gr. *koinós*, common): a pref. employed in the formation of compound words to give the idea of community, as in *coenogamete* and *coenozygote*, or Gr. *kenós*, empty, as in *coenocarp*.

coenocentrum, pl. **coenocentra** (Gr. *koinós*, common + *kéntron*, center): the multinucleate portion of the cytoplasm that surrounds the oosphere during oogonial development in species of certain leptomitaceous zoosporic fungi, such as *Apodachlya* and *Apodachlyella* (Leptomitales).

coenocytic (Gr. *koinós*, common + *kýtos*, cell, hollow + suf. *-íkos* > L. *-icus* > E. *-ic*, belonging to or relating to): aseptate; a thallus whose nuclei are contained in a common, continuous cytoplasm, without being separated by cross-walls or transverse septa. The majority of aquatic fungi are coenocytic, such as species of *Saprolegnia* (Saprolegniales), as well as those of some lower terrestrial forms, such as *Rhizopus* (Mucorales), and *Hyphomyces destruens* (*nomen invalidum*). See **aseptate**. Cf. **septate**.

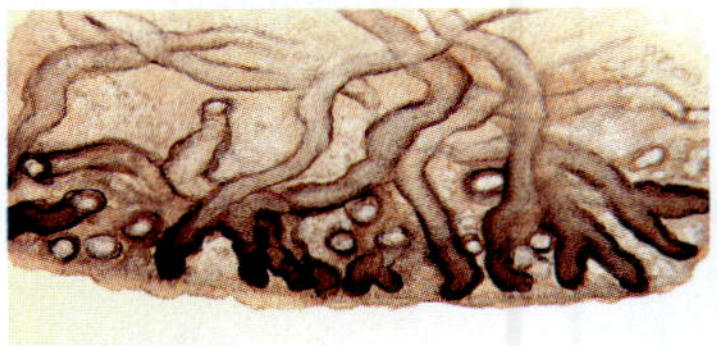

Coenocytic hypha of *Hyphomyces destruens*, in an abscess of a horse, x 300 (*MU*).

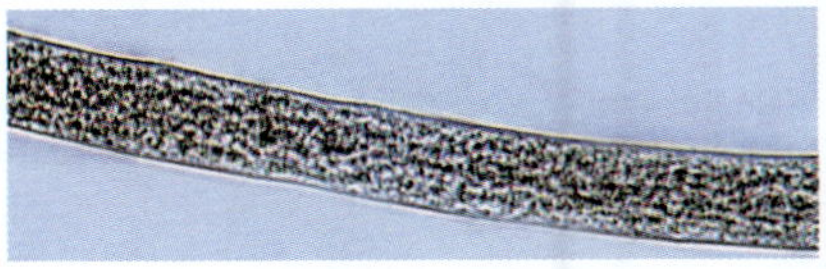

Coenocytic hypha of *Rhizopus stolonifer* (=*Rh. nigricans*), x 330 (*MU*).

coenogamete (Gr. *koinós*, common + *gamétes*, spouse, gamete): a gamete with more than one nucleus. In the gametangium of *Mucor* (Mucorales) there are from 20 to 40 nuclei, which are undifferentiated; all of them together constitute a coenogamete. The fusion of two coenogametes gives rise to a *coenozygote*.

coenozygote (Gr. *koinós*, common + *zygotós*, united by a yoke < *zygo-* < *zygóun*, to unite, to join < *zýgon*, yoke, pair): a zygote resulting from the copulation of two *coenogametes*.

coetaneous (L. *co-*, together, at one time + *aetas*, age): of the same age; existing for the same length of time.

colacosome (Gr. *kólax*, genit. *kólakos*, sycophant, parasite + *sõma*, body): a vesicular structure formed by cells of mycoparasitic hyphae growing internally in a host, when they come into contact with a host cell. A colacosome forms when the plasmalemma in a cell of the parasite pushes into the cytoplasm and recurves, delimiting a globose to subglobose segment of cytosplasm, 0.3-0.4 μm in diam. The central portion of the globule is electron opaque, bounded by a membrane, and surrounded by an electron transparent zone. A thin secondary wall is formed and the plasmalemma covers the colacosome. Two types of colacosomes are known. In one, the electron opaque central region intrudes through the cell wall of the parasite and through the cell wall of the host, where the host plasmalemma is invaginated. In the second type, the central region is less opaque and forms a pore 7-14 nm in diam. through both the parasite and host cell walls. Colacosomes thus far have been found only in certain heterobasidiomycetes. Cf. **atractosome**.

-cole (L. *-cola*, inhabitant < *colere*, to inhabit): employed as an indeclinable suf. in the formation of compounds that imply the idea of living in some place or on a particular substrate: *fimicole*, *lignicole*, *rupicole*, etc.

coleo- (Gr. *koleós*, sheath): a pref. used in words that allude to a sheath or cover, as in the generic name *Coleosporium* (Uredinales), which means sheathed spore.

collabent (L. *collabens*, genit. *collabentis*, to collapse, to fall to pieces, to contract, to shrink): collapsing at maturity, as occurs in many different types of sporiferous structures.

collar (L. *collare*, collar, neckband): **1**. *Gasteromycetes*. The fringe or edge of tissue that surrounds the pedicel of the fruiting body of some species of *Geastrum* (Lycoperdales), or the small, cup-like swelling (the "little cup" referred to by Coker) on the base of the basidiospores of *Rhizopogon* (Hymenogastrales) and *Hysterangium* (Phallales), formed when the basidiospores separate from the pedicels on the basidia. **2**. *Uredinales*. The more or less circular deposit on the host cell wall that is formed around the neck of the haustorium in *Melampsora*.

Collar of the orthotropic basidiospores of *Hysterangium* sp., x 1,400 (*MU*).

collarette (L. *collarette* < *collare*, collar, neckband + dim. suf. *-etta*, *-ette*): **1**. *Conidial fungi*. A cup or goblet-shaped structure present on the apex or subterminal portion of a phialide; characteristic of many gen., such as *Phialophora*, *Cyphellophora* and *Phialogeniculata* (dematiaceous asexual fungi). In the case of *Penicillium* (moniliaceous asexual fungi), the length of the collarette (more correctly called *collulum* or *neck* in L. and E.) is considered an important taxonomic character in species distinction. **2**. *Mucorales*. The remains of the sporangial peridium that persist on the base of the columella, as is seen in *Mucor*.

Collarette of the phialides of *Phialophora parasitica*, x 1,200 (*MU*).

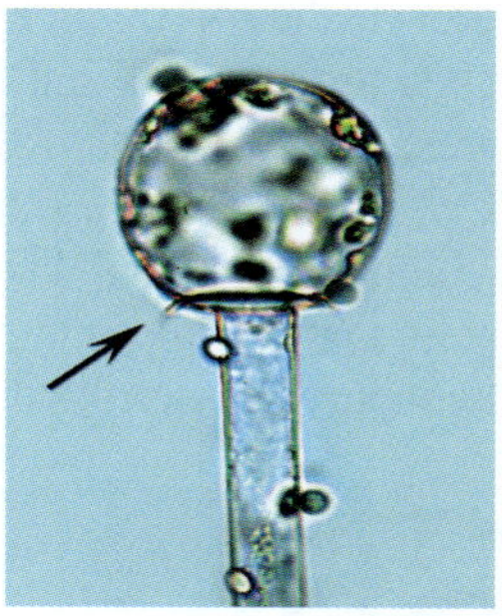

Collarette of the sporangium of *Mucor* sp., x 320 (*MU*).

collarium, pl. **collaria** (L. *collare*, collar, neckband + suf. *-arium* > E. *-arium*, a suf. added to names to form adjectives that denote the place of something): a ring of tissue that adheres to the proximal edges of the gills on the lower side of a basidiocarp, as in *Marasmius* (Agaricales).

colliculose, colliculous (L. *colliculus*, little hill < *collis*, hill, mound + dim. suf. *-culus* + suf. *-osus* > E. *-ose*, full of, augmented, prone to, having, with the qualities of): covered with small, rounded elevations, like the surface of the giant colony of *Candida colliculosa* (cryptococcaceous asexual yeasts).

collulum, pl. **collula** (L. *collulum* < *collum*, neck + dim. suf. *-ulum*): the part of the phialide of *Penicillium* (moniliaceous asexual fungi) located between the mouth and the base, i.e., the subapical region where the phialide tapers; the length of the phialide neck is considered an important taxonomic character in distinguishing species of this gen. of fungi. See **collarette**.

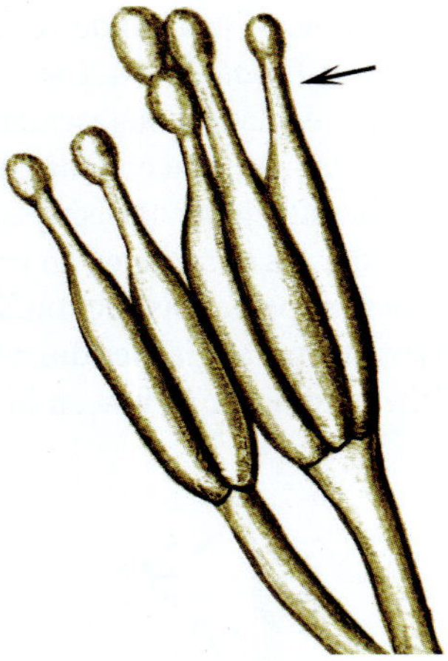

Collulum of the phialides of *Penicillium solitum* var. *crustosum* (=*P. crustosum*), x 1,500 (*MU*).

collybioid (< gen. *Collybia* < Gr. *kóllybos*, a small coin + L. suf. *-oide*, like < Gr. *-oeídes*, similar to): *Agaricales*. A type of fruiting body (of 13 main types recognized) with gills adherent to the stipe, which is cartilaginous, centrally attached to the pileus and lacks a ring and volva; it is represented by *Collybia* and some species of *Marasmius*, *Psilocybe* and *Coprinus*, among other gen. The collybioid type is similar to the mycenoid type but differs in that the pileus is convex, with the margin initially incurved or inrolled.

Collybioid basidiocarps of *Collybia* sp.
Diagram of a longitudinal section, x 0.7 (*MU*).

colony (L. *colonia* < *colonus*, colonist < *colere*, to inhabit, cultivate): a group of individuals of the same species that live close together. In the fungi the term generally refers to the mass of hyphae (mycelium) or individual cells that proliferate from a central point (centrifugally) to form an entity with the characteristics typical of the species.

columella, pl. **columellae** (L. *columella*, small column < *columna*, column + dim. suf. *-ella*): a sterile, often columnar, supporting structure or tissue present in the interior of some types of sporangia or other fructifications; frequently it is an extension of the stalk that is surrounded by sporiferous tissues. The sporangia of various Myxomycetes, such as species of *Diachea* (Physarales) and *Stemonitis* (Stemonitales), and Mucorales, such as *Mucor*, *Rhizopus* and others, have a columella, as does the sporiferous apparatus of certain Gasteromycetes (*Lycoperdon*, of the Lycoperdales; *Scleroderma*, of the Sclerodermatales). See **dendritic columella** or **percurrent columella**, and **stipe-columella**.

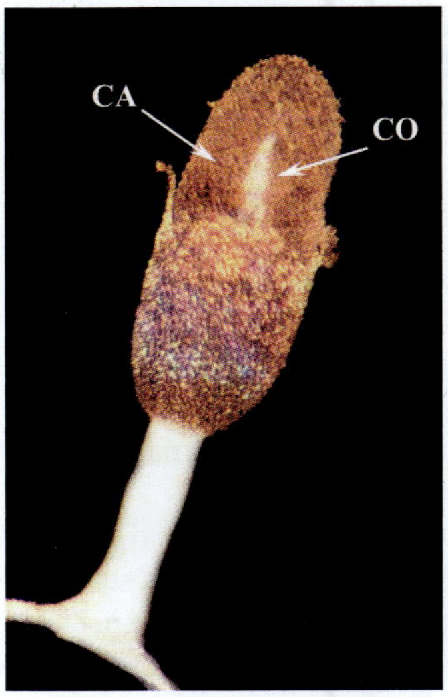

Columella (CO) and capillitium (CA) of the sporangium of *Diachea leucopoda*, x 45 (*MU*).

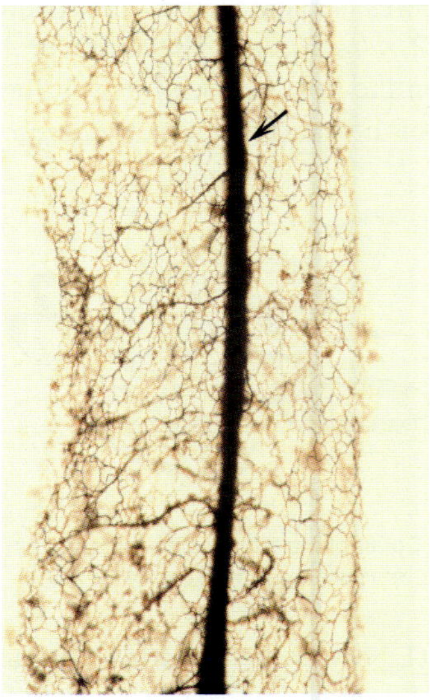

Columella of the sporangium of *Stemonitis* sp., x 35 (*MU*).

Columella of the germ sporangia
of *Rhizopus oryzae* (=*Rh. arrhizus*), x 200 (*MU*).

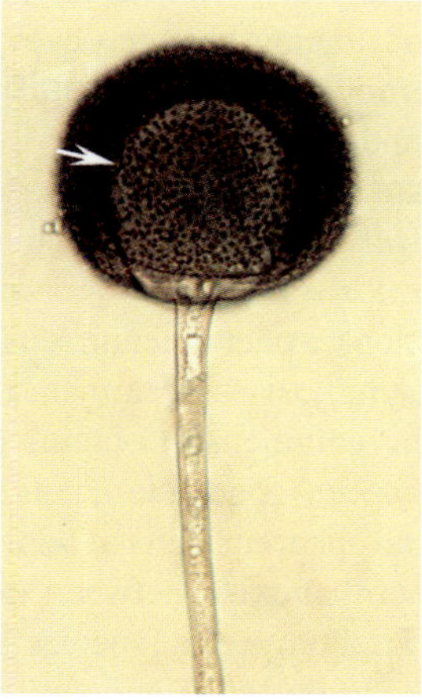

Columella of the sporangium
of *Rhizopus stolonifer* (=*Rh. nigricans*), x 170 (*MU*).

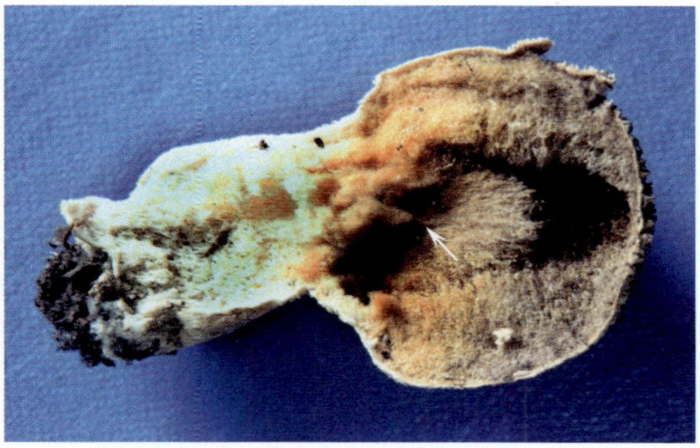

Columella of the basidiocarp of *Lycoperdon stellare*,
seen in longitudinal section, x 0.6 (*MU*).

column (L. *columna*, column): *Gasteromycetes*. Each of the various pillars that support the sporiferous sac of the fruiting body of *Myriostoma* (Lycoperdales), separating it from the exoperidium. The conidiophores of *Metarhizium anisopliae* (moniliaceous asexual fungi) also form characteristic columns.

Columns of the basidiocarp of *Myriostoma coliforme*, x 0.3 (*MU*).

Columns of the basidiocarp of *Myriostoma coliforme*, x 8 (*CB*).

Columns of conidiophores and conidia
in *Metarhizium anisopliae*, x 5 (*MU* & *CB*).

Columns of conidiophores and conidia
in *Metarhizium anisopliae*, x 8 (*MU*).

comate (L. *comatus*, tufted, crested < *coma*, tuft of hair, plume, tassel < Gr. *kóme* + L. suf. *-atus* > E. *-ate*, provided with or likeness): provided with hairs grouped in woolly tufts, like the pileus of *Coprinus comatus*, *Coprinopsis echinospora* (=*Coprinus echinosporus*) and *Coprinopsis* (=*Coprinus*) *lagopus*, of the Agaricales.

Comate basidiocarps of *Coprinopsis echinospora* (=*Coprinus echinosporus*), x 2 (*TH*).

Comate basidiocarps of *Coprinopsis* (=*Coprinus*) *lagopus*, x 2 (*MU*).

commensal (L. *commensalis* < *co-*, with + *mensa*, table + suf. *-alis* > E. *-al*, relating to or belonging to; one who shares a table with another): *Mycology*. A fungus that lives in, on or with an animal without injury to either organism. The fungus may be attached to the outer surface of the host animal (*ectocommensal*), as occurs with *Amoebidium parasiticum* (Amoebidiales), a Trichomycete inhabitant of the antennae of *Daphnia*, a crustacean of the order Cladocera. A fungus found in the interior of the host animal is called *endocommensal*, as occurs in some species of Trichomycetes of the order Eccrinales, e.g., *Parataeniella armadillidii*, which lives inside scale insects, and *Enterobryus* sp., which inhabits the posterior intestinal tract of myriapods, as well as the order Asellariales, e.g., *Ramacrinella raibautii,* whose thalli live attached to the proctodeum of crustaceous marine amphipods. Commensalism (a phenomenon relative to the living together of commensals) is most clearly seen in the association that the Trichomycetes establish with various types of arthropods (insects, myriapods, crustaceans, etc.), an association in which the fungal commensal obtains refuge and transport in the host arthropod, feeding on the nutrients that pass through the anterior digestive tract of the animal, or on the nutrients that remain without being absorbed in the posterior tract, but without causing apparent injury to the host. On the other hand, it is not known with certainty if the host receives any benefit from the commensal, with respect to food, mating, change, development, maturation and reproduction. See **parasitism**, **parasite**, **symbiont** and **symbiosis**.

commensalism (L. *commensalis* < *co-*, with + *mensa*, table + suf. *-alis* > E. *-al*, relating to or belonging to + Gr. suf. *-ismós* > L. *-ismus* > E. *-ism*, state, phase, tendency, action): a phenomenon relating to the living together or symbiosis of commensals, in which an organism (the commensal) obtains certain benefits (shelter, transport, food, etc.), and the other (the host) shows no apparent signs of benefit or harm. The living together that exists between various members of the class Trichomycetes (species of *Amoebidium*, *Harpella*, *Smittium*, etc.) and multiple species of mandibulate arthropods (especially insects, myriapods and crustaceans) qualifies it as commensalism. See **antagonism**, **helotism**, **metabiosis**, **mutualism**, **parabiosis**, **parasymbiosis**, **symbiosis** and **synergy**.

commissure (L. *commissura*, joining, union of two things bound together): the surface or line along which two or more organic organs are joined, whether separable or not.

community (L. *communitas*, congregation, community): in biology, a more or less complex group of organisms that occupy a particular area, independently of the character or of the complexity and structure of their association.

comose (L. *comosus*, adorned with plumes < *coma*, head of hair, crest): having groups or tufts of hair. Syn. of **comate**.

compaginate (L. *compaginatus*, strongly united < *com-*, together, with + *pagina*, sheet, page + suf. *-atus* > E. *-ate*, provided with or likeness): applied to flat structures whose surfaces are stacked one above the other.

companion cell (LL. *companion*, mess-mate > OF. *compaignon* > ME. *compainoun* > E. *companion*, one who is in the company of another or of others; NL. *cellula*, cell, living cell, dim. of L. *cella*, small room, compartment, cell of a honeycomb): refers to the adjacent donor gametangium that is present in the zoosporic fungi of the order Olpidiopsidales. These organisms are heterothallic; during sexual reproduction the larger thalli function as oogonia and the smaller ones as antheridia (companion cells).

compatible (L. *compatibilis*, genit. of *compati*, to go well with; having the ability or opportunity to unite or concur on the same place or subject): applied to mating strains or types that are capable of intercrossing; that are fertile on crossing.

complanate (L. *complanatus*, level, flattened < *complanare*, to make level + suf. *-atus* > E. *-ate*, provided with or likeness): depressed, flattened and smooth. The majority of basidiospores (e.g., in *Coprinus comatus*, of the Agaricales) are slightly flattened in lateral view. See **compressed**.

Complanate basidiospores of *Coprinus comatus*, x 1,500 (*MU*).

complex (L. *complexus*, joining or union of two or more things < *complecti*, to bind): sometimes used to designate a group of closely related species.

complicate (L. *complicatus*, folded or rolled on itself < *com-*, together, with + *plica*, fold, crease + suf. *-atus* > E. *-ate*, provided with or likeness): doubled or folded on itself.

compost (E. *compost* < L. *compositus*, composite): a mixture of decaying organic materials, generally prepared by fermentation of straw and horse manure, in which the mycelium of the mushroom *Agaricus bisporus* (=*A. brunnescens*), Agaricales, is grown in industrial cultivation.

compound (L. *compositus*, compound < *componere*, to compound): made up of various parts; an aggregate of various things that comprise a total.

compressed (L. *comprimere*, to squeeze together): flattened laterally. See **complanate**.

concatenate (L. *concatenatus*, linked together < *con-*, together, with + *catena*, chain +suf. *-atus* > E. *-ate*, provided with or likeness): united in a chain. Equal to **catenate**.

concave (L. *concavus*, concave < *con-*, with + *cavus*, hollow, cavity): having the surface depressed in the center. Cf. **convex**.

concentric (L. *con-*, with, together + *centrum* < Gr. *kéntron*, center, central point of a circle + Gr. suf. *-íkos* > L. *-icus* > E. *-ic*, belonging to or relating to): applied to structures that have rings or zones arranged in parallel circles around a common center, as is seen, e.g., in the zonate colonies formed by some molds, such as *Aspergillus* and *Penicillium* (moniliaceous asexual fungi), or in the perithecial stromata of *Daldinia concentrica* (Xylariales), which in vertical section show a series of concentric zones, with the perithecia embedded in the peripheral zone.

Concentric stromatic tissues of *Daldinia concentrica*, seen in longitudinal section, x 2 (*MU*).

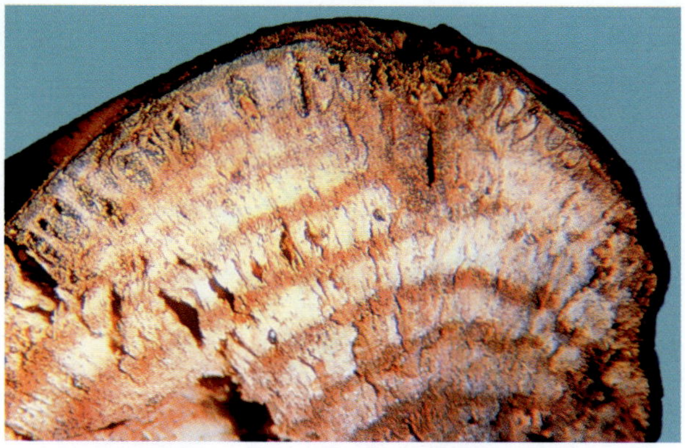

Concentric stromatic tissues of *Daldinia concentrica*, seen in longitudinal section, x 5 (*MU*).

concentric body (L. *con-*, with, together + *centric* < *centrum*, center + Gr. suf. *-íkos* > L. *-icus* > E. *-ic*, belonging to or relating to; ME. < OE. *bodig*, body): a spherical concentric structure present in the hyphae of many species of lichenized and some nonlichenized fungi. Concentric bodies are found in an area of the cytoplasm that is devoid of other organelles, e.g., in species of *Cercospora* (dematiaceous asexual fungi). In species of *Allomyces* (Blastocladiales) the concentric bodies are arranged in the form of granules.

conceptacle (L. *conceptaculum*, receptacle or vessel, a place where something is conceived or engendered < *conceptus*, conceived + dim. suf. *-culum* > E. *-cle*): any cavity or locule that opens to the exterior and in which are produced and lodged propagative cells; applied to pycnidia, ascigerous locules, structures that contain propagules in the lichens and other types of structures. The term is little used.

conchate (L. *conchatus*, with the shape of a sea-shell or provided with a shell < *concha*, shell of bivalve mollusks < Gr. *kónche*, mussel, mytilus + L. suf. *-atus* > E. *-ate*, provided with or likeness): syn. of **conchiform** and **conchoid**; having the shape of a bivalve mollusk shell, like the basidiocarp of *Panus conchatus* (Aphyllophorales), *Pleurotus ostreatus* and *P. cornucopioides* (Agaricales), and the thallus of the lichen *Dictyonema pavonium* (=*Cora pavonia*, Aphyllophorales). See **oyster mushroom**.

Conchate basidiocarps of *Pleurotus cornucopioides*, x 0.5 (*EPS*).

Conchate thalli of *Dictyonema pavonium* (=*Cora pavonia*), x 1.5 (*MU*).

conchiform (L. *concha*, the shell of bivalve mollusks < Gr. *kónche*, mussel, mytilus + L. suf. *-formis* < *forma*, shape): with the shape of a mollusk shell. See **conchate**.

conchoid (L. *concha*, the shell of bivalve mollusks < Gr. *kónche*, mussel, mytilus + L. suf. *-oide* < Gr. *-oeídes*, similar to): similar to the shell of a mollusk. See **conchate**.

concinnous (L. *concinnus*, neatly arranged; elegant, agreeable): elegant, harmonious, natural. Generally applied to language, as in a discourse.

concolor, concolorous (L. *con-*, with + *color*, color): refers to structures that have the same color throughout, in all parts; e.g., gills concolorous with the pileus. Cf. **discolorous**.

concrescent (L. *concrescere*, to conglutinate, coagulate, unite < *con-*, with + *crescere*, to increase, grow + suf. *-escens*, genit. *-escentis* > E. *-escent*, that which turns, that begins to, slightly): syn. of **confluent** and **connate**. See **coalescent**.

Concrescent basidiocarps of *Boletellus russellii* (Agaricales), x 0.5 (*RV*).

concrete (L. *concretus*, concrete, having concretion, i.e., the accumulation of various particles to form a mass): united by growth.

condidium, pl. **condidia** (L. *condere*, to hide + dim. suf. *-idium* > E. *-idium*): an old term for any fructification that bears conidia.

conferted (L. *confertus*, stuffed, packed): densely packed or piled up.

confervoid (< gen. of algae *Conferva* + L. suf. *-oide*, like < Gr. *-oeídes*, similar to): composed of septate filaments with short cells, similar to those of the thallus of *Conferva*, a heterokontic, filamentous alga.

confluent (L. *confluere*, to unite, join + suf. *-entem* > E. *-ent*, being): see **coalescent**.

congeneric (L. *con-*, with + *genus*, genit. *generis*, race, lineage + Gr. suf. *-íkos* > L. *-icus* > E. *-ic*, belonging to or relating to): belonging to the same genus, a systematic entity in the taxonomic classification, composed of species.

conglobate (L. *conglobatus*, gathered into a ball < *con*, with + *globus*, globe, ball + suf. *-atus* > E. *-ate*, provided with or likeness): united into more or less globose, compact masses or groups, like the bases of the stipes of some polyporaceous basidiomycetes, e.g., *Grifola frondosa* (=*Polyporus frondosus*), of the Aphyllophorales, and the ascospores of various ascomycetes, e.g., *Emericellopsis terricola* (Hypocreales).

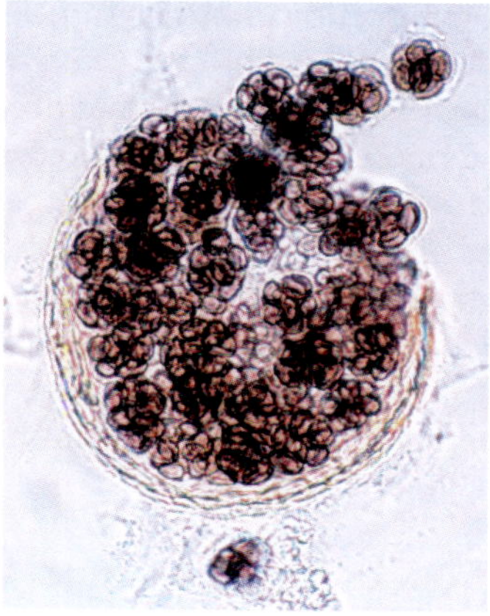

Conglobate asci of *Emericellopsis terricola*, x 550 (*MU* & *CB*).

conglomerate (L. *conglomeratus*, united, joined < *conglomerare*, to unite, to be grouped into a compact mass): equal to *agglomerate*.

conglutinate (L. *conglutinatus*, sticky < *con-*, together, with + *gluten*, genit. *glutinis*, adhesive + suf. *-atus* > E. *-ate*, provided with or likeness): glued together. Applied to the paraphyses of the lichenized ascomata of the order Lecanorales.

congregate (L. *congregatus* < *con-*, with + *gregare*, to unite + suf. *-atus* > E. *-ate*, provided with or likeness): *Gasteromycetes*. Refers to a type of fruiting body, like that of *Broomeia congregata* (Lycoperdales), that is composed of numerous sporiferous sacs (5-900) arranged in alveoli on the upper surface of a thick, fleshy stroma. At maturity the sporiferous sacs, which correspond to the endoperidium, are surrounded by the lower half of the exoperidium.

Congregate pycnidia of *Sphaeronema* sp. (sphaeropsidaceous asexual fungi), on agar, x 20 (*MU*).

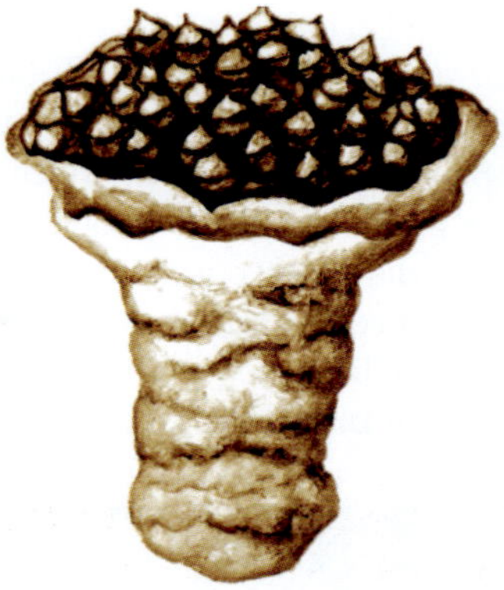

Congregate sporiferous sacs of the fruiting body of *Broomeia congregata*, x 1 (*MU*).

conic, conical (L. *conicus*, conical < Gr. *konikós*, cone-shaped < *kónos*, cone + Gr. suf. *-ikos* > L. *-icus* > E. *-ic*, belonging to or relating to): cone-shaped, i.e., a structure whose height is greater than its width, with a sharp-pointed apex, like the pileus of the basidiocarp of *Hygrocybe conica* and *Macrolepiota* sp. (Agaricales).

Conic pileus of *Macrolepiota* sp., x 0.2 (*ACV*).

conidange, conidangium, pl. **conidangia** (NL. *conidangium* < *conidium* < Gr. *kónis*, dust, powder + dim. suf. *-ídion* > L. *-idium* + Gr. *angeîon*, vessel, receptaculum): an obsolete term that was applied to the thin-walled pycnidium found in some lichen thalli.

conidial fungi, sing. **fungus** (NL. *conidium* < Gr. *kónis*, powder, dust + dim. suf. *-ídion* > L. *-idium* + L. suf. *-alis* > E. *-al*, relating to or belonging to; L. *fungus*, fungus < Gr. *spóngos*, *sphóngos*, sponge): fungi that reproduce by means of conidia. The conidial fungi also are called **anaholomorphs**, **asexual fungi**, **deuteromycetes**, **Fungi Imperfecti** and **mitosporic fungi**.

conidial head (NL. *conidium* < Gr. *kónis*, dust, powder + dim. suf. *-ídion* + L. suf. *-alis* > E. *-al*, relating to or belonging to; ME. *hed*, head): the upper part of a conidiophore where the conidiogenous cells are produced; e.g., in *Aspergillus* (moniliaceous asexual fungi) the conidial head consists of a vesicle on whose surface are formed phialides (conidiogenous cells) or metulae and phialides, depending on the species. See **uniseriate** and **biseriate**.

Conidial head of the conidiophore of *Aspergillus nidulans*, x 550 (*CM*).

conidic (NL. *conidium* < Gr. *kónis*, dust, powder + dim. suf. *-ídion* > L. *-idium* + Gr. suf. *-íkos* > L. *-icus* > E. *-ic*, belonging to or relating to): relating to or pertaining to conidia.

conidiogenesis, pl. **conidiogeneses** (NL. *conidium* < Gr. *kónis*, dust, powder + dim. suf. *-ídion* > L. *-idium* + L. *genesis* < Gr. *génesis*, engenderment, origin, production): the process by which conidia are formed by a conidiogenous cell or an undifferentiated hypha. See **blastic** and **thallic**.

conidiogenous locus, pl. **loci** (NL. *conidium* < Gr. *kónis*, powder, dust + dim. suf. *-ídion* > L. *-idium* + Gr. *génos*, origin; L. *locus*, place): the particular site or point on a hypha or cell where conidia are generated; e.g., the point or conidiogenous locus of the phialides of *Chloridium chlamydosporis* (moniliaceous asexual fungi).

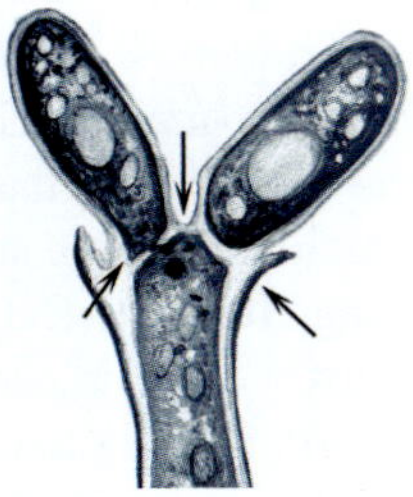

Conidiogenous loci of the conidiogenous cell of *Chloridium chlamydosporis*. Note the formation of the transverse septum inside the collarette. Drawing based on a transmission electron micrograph published by Beckett *et al.* in their *Atlas of Fungal Ultrastructure*, 1974, x 10,760 (*MU*).

conidiole, **conidiolum**, pl. **conidiola** (NL. *conidiolum* < *conidium* < Gr. *kónis*, dust, powder + dim. suf. *-ídion* > L. *-idium*, + L. dim. suf. *-olum* > E. *-ole*): a secondary conidium; a small conidium formed from another conidium, as in *Entomophthora* (Entomophthorales).

conidioma, pl. **conidiomata** (NL. *conidium* < Gr. *kónis*, dust, powder + dim. suf. *-ídion* > L. *-idium* + Gr. suf. *-óma*, which implies the idea of entirety): any structure formed by hyphae that produces conidia, whether a synnema, sporodochium, acervulus or pycnidium. This term was proposed initially for species whose fructification is variable, like that of *Pestalotiopsis* (melanconiaceous asexual fungi). The pycnidium of *Zythia mougeotii* (sphaeropsidaceous asexual fungi) is an example of a pycnidial conidioma.

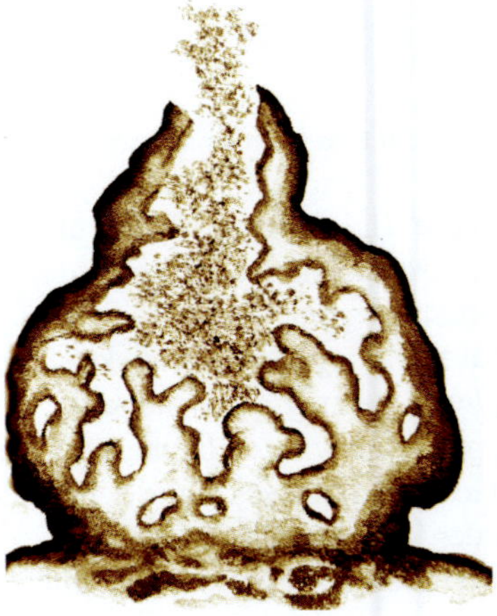

Pycnidial **conidioma** (pycnidium) of *Zythia mougeotii*, x 90 (*MU*).

conidiophore (NL. *conidium* < Gr. *kónis*, dust, powder + dim. suf. *-ídion* > L. *-idium* + Gr. suf. *-phóros*, bearer): a hypha, simple or branched, that is morphologically and/or physiologically differentiated from a somatic hypha to produce and bear conidia; these are generally formed on specialized *conidiogenous cells*, which can be arranged in many ways. In some species the terms conidiophore and conidiogenous cell are used interchangeably. Illustrations of conidiophores can be found in species of *Oedocephalum* (moniliaceous asexual fungi), *Periconia* and *Cladosporium* (dematiaceous asexual fungi), under the figures of the different types of conidia (e.g., **meristem arthrospore**, **blastospore**, **meristem blastospore** and **catenulate**), and of certain types of conidiophore branching (e.g., **biverticillate**, **monoverticillate**, **triverticillate** and **polyverticillate**), or of other structures related to conidiophores (e.g., **foot cell**).

Conidiophores of *Oedocephalum* sp., x 300 (*MU*).

conidiosporangium, pl. **conidiosporangia** (NL. *conidium* < Gr. *kónis*, dust, powder + dim. suf. *-ídion* > L. *-idium* + NL. *sporangium* < Gr. *sporá*, spore + *angeîon*, vessel, receptacle): *Peronosporales*. A deciduous sporangium, oval or limoniform, which on maturing can either germinate directly by means of a germ tube to form a mycelium (as do the conidia of higher fungi), or can produce biflagellate, reniform zoospores, thus behaving like a zoosporangium. This phenomenon is observed, e.g., in *Phytophthora infestans*, *Basidiophora entospora* and *Albugo candida*. The mode of germination is controlled by environmental conditions, temperature in particular; low temperatures (optimum of 12°C) favor the production of zoospores, and higher temperatures (optimum of 24°C) promote the formation of germ tubes.

Conidiophores of *Periconia byssoides*, x 450 (*CB*).

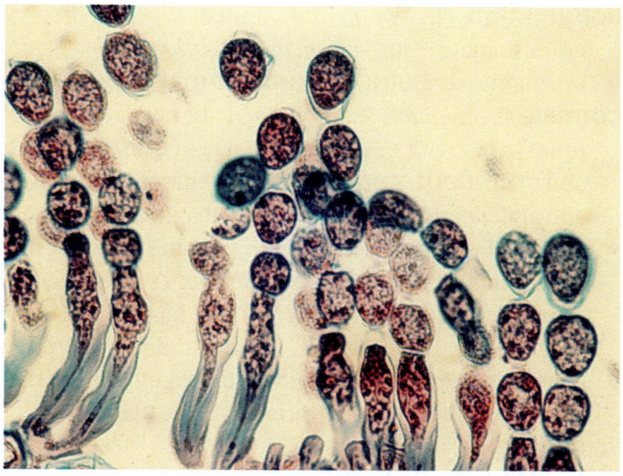

Conidiosporangia of *Alburo candida*, x 500 (*MU*).

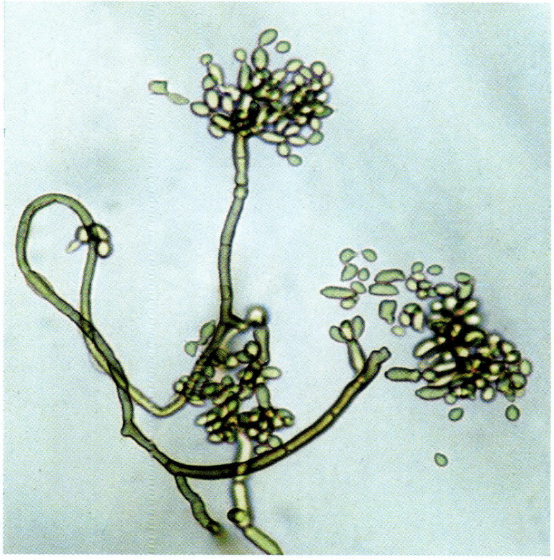

Conidiophores of *Cladosporium* sp., x 400 (*MU*).

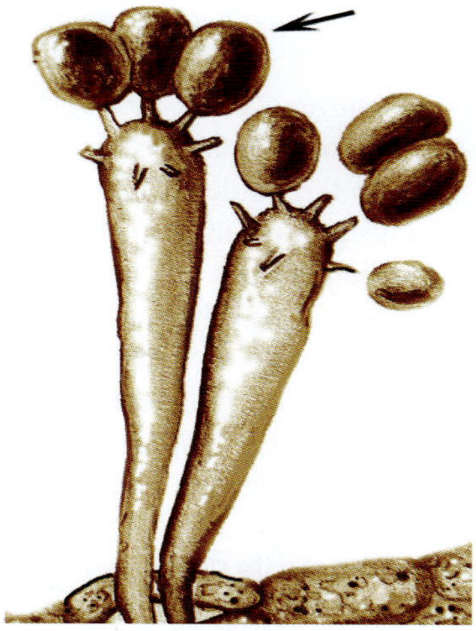

Conidiosporangia (zoosporangia similar to conidia) of *Basidiophora entospora*, produced on two zoosporangiophores that project from a stoma of a host leaf, x 110 (*MU*).

conidium, pl. **conidia** (NL. *conidium* < Gr. *kónis*, dust, powder + dim. suf. *-ídion* > L. *-idium*): syn. of **conidiospore**: a nucleate, asexual, immotile spore that is generally formed at the apex or side of a specialized sporogenous (conidiogenous) cell. Conidia are of two basic types, *blastic* and *thallic*. Blastic conidia (the majority) are generated *de novo*, whereas thallic conidia form from preexisting hyphal cells. Conidia can be one, two or multicellular, dry or mucose, but they are never produced inside a structure with a rigid cell wall, as occurs with sporangiospores. Conidia are the asexual spores of ascomycetes and basidiomycetes; typically they have been grouped in the deuteromycetes, now called *mitosporic* or *conidial fungi*.

conigenous (Gr. *kõnos*, pine cone + *génos*, origin): arising from or living on pine cones, like the basidiocarps of *Auriscalpium vulgare* (Aphyllophorales).

conjugation (L. *conjugatus*, united < *conjugare*, to unite, to mate + suf. *-ationem* > E. *-ation*, action, state or condition, result): see **gametangial copulation**.

connate (L. *connatus* < *connascor*, to be born at the same time < *con-*, with + *natus*, born): syn. of **concrescent** and **confluent**. Applied to organs that form together and appear more or less united with one another; i.e., that have congenital adherence, like the stipes of the synnemata of *Antromycopsis smithii* (stilbellaceous asexual fungi), the stipes of the basidiocarps of *Lyophyllum connatum*, *L. decastes* and *Pleurotus ostreatus* (Agaricales), which grow together from the substrate. Connate is a variant of **caespitose**. Cf. **solitary**.

connective (L. *connectivum*, that which can unite or join one thing with another < *connectere*, to connect, bind): that which establishes a connection or link, like the connective hyphae of the context of the basidiocarps.

connivent (L. *connivens*, genit. *conniventis*, to conceal < *connivere*, approach, touch + suf. *-entem* > E. *-ent*, being): refers to the border of the pileus in certain types of basidiocarps (such as *Coprinus*, of the order Agaricales) which through convergence touches the stipe but without fusing with it.The minute fibrillae on the exoperidium of *Lycoperdon* and *Calvatia* (Lycoperdales) also are connivent, since their apices unite in a point, leaving a space below, giving the appearence of a cupule or vault to each group of fibrillae.

consortium, pl. **consortia** (L. *consortium*, consortium, association): an obsolete term that was applied to the entity of morphological and physiological types constituted by two distinct organisms that live in symbiosis, like the lichen thallus.

Connate basidiocarps of *Pleurotus ostreatus* var. *florida*, x 0.5 (*MU*).

Connate synnemata of *Antromycopsis smithii* grown in pure culture, x 9 (*MU*).

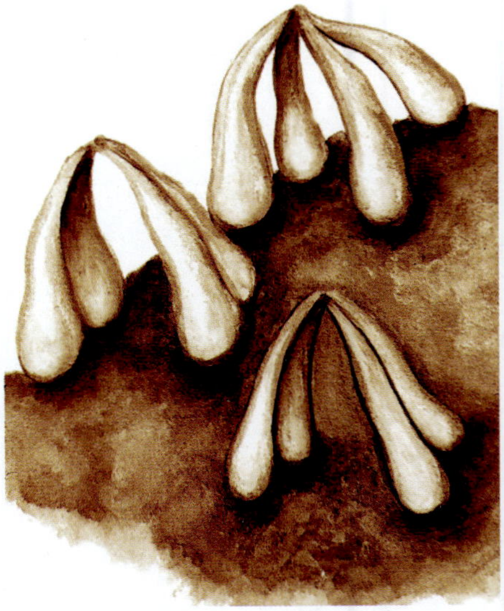

Connivent fibrillae of the pyramidal scales formed on the exoperidium of the basidiocarp of *Lycoperdon* sp., x 10 (*MU*).

conspecific (L. *con-*, with, together + LL. *specificus*, specific < L. *species*, species, class; a category in the biological classification hierarchy immediately below the genus and subgenus that includes related organisms or populations, potentially capable of interbreeding, and which is designated with a binomial that consists of the generic name, which is capitalized, followed by a specific epithet comprised of a Latin word or latinized adjective that is written in lower case and agrees grammatically with the generic name): the same species; two or more taxa that are considered as belonging to the same species.

constipate, constipated (ML. *constipatus* < L. *constipare*, to crowd, to press, to pile up + suf. *-atus* > E. *-ate*, provided with or likeness): piled up, to tighten, replete.

constricted (L. *constrictus* < *constringere*, to make narrow or draw together; to tie up): narrowed down, with a constriction, like the hyphae near the septa in *Rhizoctonia* (agonomycetaceous asexual fungi) and the volva of the basidiocarp of *Amanita constricta* (Agaricales).

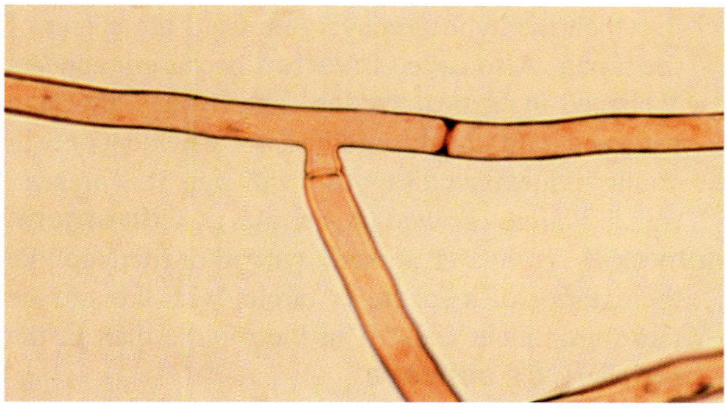

Constricted hypha in *Rhizoctonia* sp., x 450 (*CB*).

constrictive (L. *constrictivus*, having the capacity to constrain < *constringere*, to tighten and close, oppress): closing and oppressing, like the nematode trapping rings formed by some species of fungi, e.g., *Arthrobotrys dactyloides* (moniliaceous asexual fungi). See **vermivorous**.

contaminate (L. *contaminare*, to corrupt, infect): to make impure or inadequate by the introduction of undesirable elements, e.g., an impure culture, seeds or soil with fungal spores, etc. Cf. **infect** and **infest.**

context (L. *contextus*, interwoven, united, connected < *contexere*, to intertwine, interweave): *Basidiomycetes*. The fibrous or fleshy tissue comprising the body of the pileus and stipe of a basidiocarp, not including the cortex and hymenium. The context of the pileus corresponds to the hymeniiferous trama, which bears the hymenium. See **trama**.

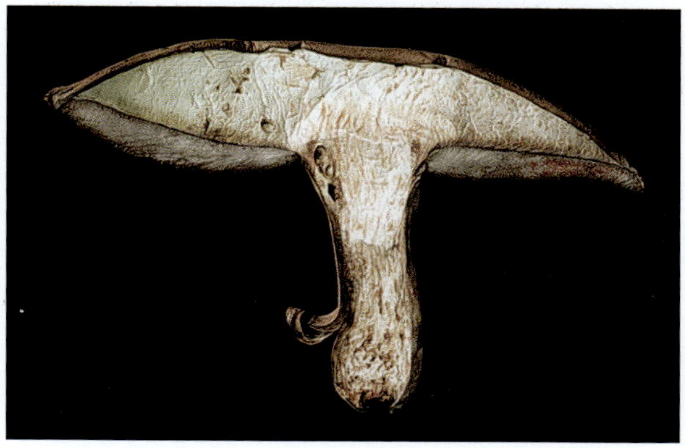

Context of the basidiocarp of *Boletus* sp. (Agaricales), seen in longitudinal section, x 0.6 (*JC*).

contiguous (L. *contiguus*, near together, neighbor, adjoining < *contingere*, have contact with): being in contact with; union, joining.

contingent (L. *contingent* < *contingere*, have contact with + suf. *-entem* > E. *-ent*, being): in contact.

continuous (L. *continuus*, continuous, continued, uninterrupted): **1**. A unicellular spore, hypha, or capillitial filament lacking septa. **2**. A stipe that merges imperceptibly with the pileus, peridium, etc., and which is composed of the same tissue. Cf. **septate**.

continuous culture (L. *continuus*, continuous, continued, uninterrupted; L. *cultus*, culture, action and effect of cultivating): the cultivation of organisms or cells in a vessel in which the culture medium or substrate is constantly maintained through the continuous addition of new media and the separation of the spent medium from the medium that contains the organisms or cells.

contorted (L. *contortus*, twisted, crooked): twisted or crooked, like the sporangiolar branches of *Cokeromyces recurvatus* (Mucorales) and the perithecial hairs of *Chaetomium crispatum* (=*C. tortile*), Sordariales, with spirals in opposite directions.

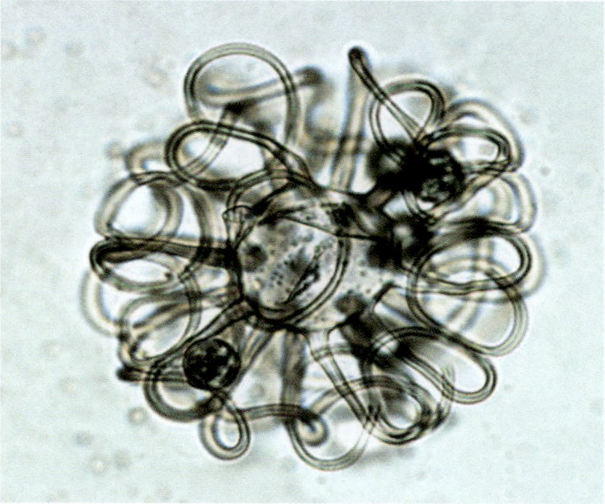

Contorted sporangiophores of *Cokeromyces recurvatus*, x 460 (*MU*).

Contorted perithecial hairs
of *Chaetomium crispatum* (=*C. tortile*), x 1,500 (*MU*).

Contorted perithecial hairs
of *Chaetomium crispatum* (=*C. tortile*), x 2,000 (*MU*).

contortuplicate (L. *contortuplicatus* < *contortus*, contorted + *plicatus*, folded): contorted and folded at the same time.

control (ME. *controllen*, to regulate, have power over, prevent or retard): to restrain, hold in check through inspection and examination; e.g., to monitor crop plants in the field for diseases and infestations and to devise treatments to combat them.

convergent (L. *convergere*, to move toward a point, proceeding from various points + suf. *-entem* > E. *-ent*, being): *Agaricology*. A type of hymeniiferous trama in which the hyphae converge toward the center of the trama. Also called **inverted** because it appears V-shaped in vertical section. The hyphae are quite wide and long, and the middle layer is evident only in young tissue, then disappears with age. It is present, e.g., in *Pluteus cervinus* (Agaricales). Cf. **divergent**.

convex (L. *convexus*, arched): curved or rounded like the exterior of a sphere or circle; with the surface more prominent (raised) in the middle than in the periphery. Cf. **concave**.

convolute, **convoluted** (L. *convolutus*, rolled up, in-rolled): syn. of **cerebriform**; with folds, bends and undulations that give a brain-like appearance, like the colony of some strains of *Sporothrix schenckii* (moniliaceous asexual fungi) cultivated on agar, the pileus of *Gyromitra esculenta* and *Helvella infula* (Pezizales), and the basidiocarp of *Tremella* (Tremellales).

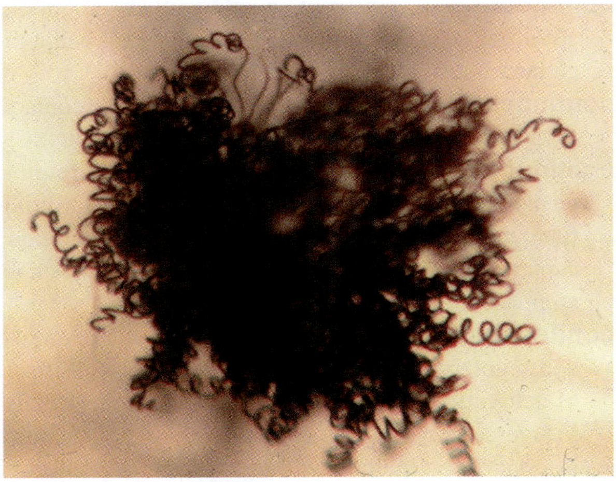

Contorted perithecial hairs
of *Chaetomium crispatum* (=*C. tortile*), x 1,000 (*MU*).

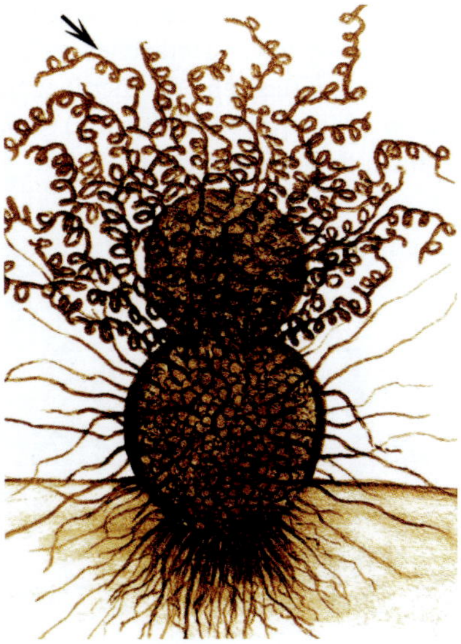

Contorted perithecial hairs
of *Chaetomium crispatum* (=*C. tortile*), x 1,000 (*MU*).

Convolute colony of *Sporothrix schenckii*, on agar, x 10 (*MU*).

Convolute pileus of the ascocarp of *Gyromitra esculenta*, x 0.5 (*TRO*).

Convolute basidiocarp of *Tremella* sp., x 1.6 (*TH*).

coprinoid (< gen. *Coprinus* < Gr. *kópros*, dung + L. suf. *-inus*, which indicates relation + suf. *-oide*, like < Gr. *-oeídes*, similar to): a type of basidium present in *Coprinus* and other Agaricales. See **brachy-basidiole**.

copro-, **copr-** (Gr. *kópros*, dung): a pref. employed in terms that allude to dung, as in *coprophilous*, *coprogen*, etc.

coprogen (Gr. *kópros*, dung + *génos*, born, origin): a growth factor found in dung and which is required by *Pilobolus* spp. (Mucorales); it has been identified as an organic iron compound, similar to the heme of the blood of vertebrates.

coprophilous (Gr. *kópros*, dung + *phílos*, have an affinity for): a fungus that grows preferentially on dung or on manured soils, as occurs with species of *Pilaira* and *Pilobolus* (Mucorales), *Coprinus*, *Paenaeolus*, *Psilocybe coprophila* (Agaricales), and *Guanomyces* (Sordariales), among many other fungi.

Coprophilous sporangiophores of *Pilobolus crystallinus* var. *kleinii* (=*P. kleinii*), growing on cow dung in a damp chamber, x 1.5 (*MU*).

Coprophilous sporangiophores of *Pilobolus crystallinus* var. *kleinii* (=*P. kleinii*), growing on cow dung in a damp chamber, x 20 (*MU*).

Coprophilous basidiocarps of *Panaeolus semiovatus*, x 0.5 (*RV*).

Coprophilous basidiocarps of *Psilocybe coprophila*, x 0.5 (*MU*).

Perithecium of *Guanomyces polythrix* on agar (**coprophilous**), x 50 (*MU*).

Coprophilous basidiocarps of *Psilocybe coprophila*, x 0.5 (*MU*).

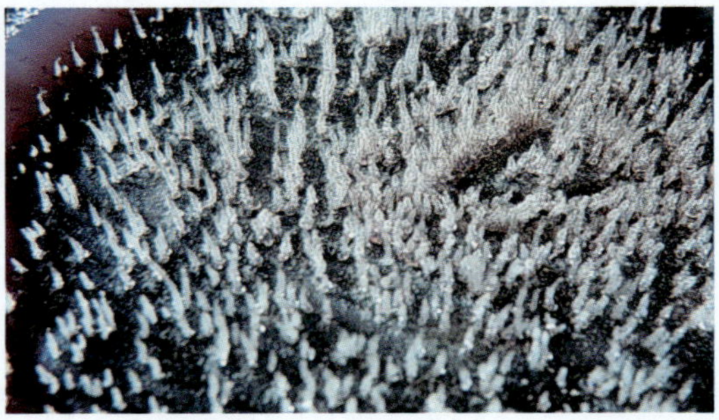

Perithecia of *Guanomyces polythrix* on agar (**coprophilous**), x 3 (*MU*).

Perithecia of *Guanomyces polythrix* on agar (**coprophilous**), x 30 (*MU*).

coralloid (L. *corallium* < Gr. *korállion*, coral, the skeletal calcareous or corneous deposit produced by anthozoans and hydrozoans + suf. *-oide*, like < Gr. *-oeídes*, similar to): similar to coral in the manner of branching, like the basidiocarps of *Calocera viscosa*, *Clavicorona pyxidata*, *Clavaria vermicularis*, *Ramaria stricta* and other clavariaceous species (Aphyllophorales), or like the coralloid subhymenium of *Amanita sublutea* (Agaricales). In some Gasteromycetes, it also refers to a type of fruiting body development, known as coralloid or **forate development**.

corbicula, pl. **corbiculae** (L. *corbicula*, little basket): *Uredinales*. A peripheral, paraphysis-like structure present in the telia of some rusts; it consists of cells with a thick, brown wall that adhere to the internal wall of a host epidermal cell, and which fuse laterally with others to form a basket-like structure that protects during sporogenesis and aids in dissemination of the teliospores.

Coralloid basidiocarps of *Clavaria* sp., x 0.5 (*MRO*).

Coralloid basidiocarps of *Ramaria stricta*, x 0.5 (*TH*).

Coralloid basidiocarps of *Ramaria flava*, x 0.5 (*RV*).

Coralloid basidiocarp of *Ramaria* sp., x 0.5 (*TRO*).

cordate, cordiform (L. *cordatus* < *cordis*, heart + suf. *-atus* > E. *-ate*, provided with or likeness; L. *cordiformis* < *cordis* + *-formis* < *forma*, shape): heart-shaped, like the vesicle of the sporangiophore of *Syncephalis cordata* (Zoopagales) and the basidiospores de *Psathyrella hirsutosquamulosa* (Agaricales).

Cordate basidiospores of *Psathyrella hirsutosquamulosa*, x 1,500 (*MU*).

coremiform (NL. *coremiformis* < *coremium* < Gr. *kórema*, broom + L. suf. *-formis* < *forma*, shape): having the shape of a coremium (synnema), like the conidiophores of *Penicillium vulpinum* (=*P. claviforme*), moniliaceous asexual fungi. Also applied to the texture or appearence of the mycelium of coremioid fungi.

Coremiform texture of colonies of *Penicillium vulpinum* (=*P. claviforme*) on agar, x 2 (*TH*).

Coremiform mycelium of a colony of *Penicillium vulpinum* (=*P. claviforme*) on agar, x 5 (*MU*).

coremium, pl. **coremia** (NL. *coremium* < Gr. *kórema*, broom + L. dim. suf. *-ium*): an erect bundle of loosely united conidiophores, with the apex similar to a broom, like that of *Penicillium vulpinum* (=*P. claviforme*), moniliaceous asexual fungi. A coremium is a type of *synnema*.

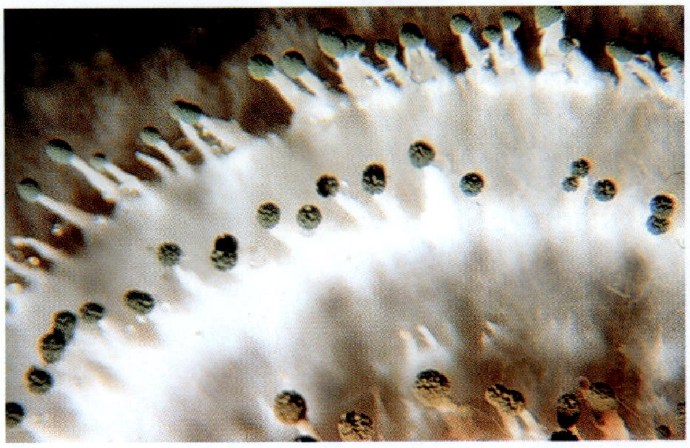

Coremia of *Penicillium vulpinum* (=*P. claviforme*) on agar, x 5 (*MU*).

Coremia of *Penicillium vulpinum* (=*P. claviforme*) on agar, x 10 (*MU* & *CB*).

coriaceous (L. *coriaceus*, leather-like < *corium*, leather, skin + suf. *-aceus* > E. *-aceous*, of or belonging to, with the nature of): with the consistency of leather, tough but with a certain flexibility, like that of the fruiting bodies of some species of Helotiales, such as *Bulgaria*, and of the Aphyllophorales, especially of the families Corticiaceae and Stereaceae, as well as of certain lichen thalli, e.g., *Psora crenata* (Lecanorales).

corky (L. *cortex*, genit. *corticis*, cortex, cork + E. suf. *-y*, having the quality of): syn. of **suberose**.

corneous (L. *corneus*, similar to horn < *cornu*, horn): of a hard consistency, like horn.

corniculate (L. *corniculatus*, with horns < *corniculum*, dim. of *cornu*, horn + suf. *-atus* > E. *-ate*, provided with or likeness): having small, horn-like structures or ending in a small, horn-shaped protuberance, like the cystidia of *Pluteus cervinus* (Agaricales), which have apical appendages in the shape of little horns.

Coriaceous ascocarps of *Bulgaria* sp., x 0.3 (*TH*).

Coriaceous thalli of the lichen *Psora crenata*, x 1 (*MU*).

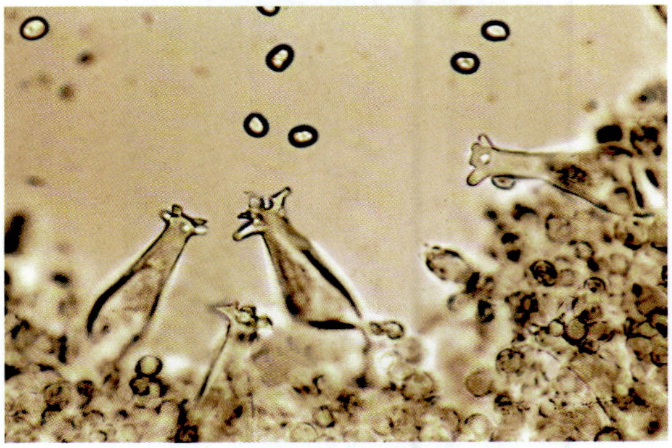

Corniculate cystidia of the hymenium of *Pluteus cervinus*, x 300 (*MU*).

corniculiform (L. *corniculiformis* < *corniculum*, little horn + suf. *-formis* < *forma*, shape): having the shape of a little horn, like the sclerotia of *Claviceps purpurea* (Hypocreales).

corniform (L. *corniformis* < *cornu*, a horn + suf. *-formis* < *forma,* shape): shaped like a horn, like the appendages of the suspensors of *Phycomyces nitens* (Mucorales).

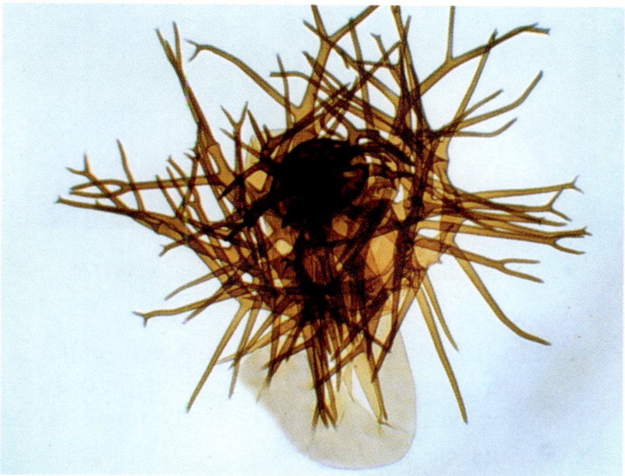

Corniform appendages of the suspensors of *Phycomyces nitens*, x 90 (*MU*).

cornute (L. *cornutus*, having horns or the shape of horns): *Uredinales*. A type of aecium, horn-like and sharp-pointed, with the peridium composed of characteristically imbricated cells that break by longitudinal fissures, like that of *Roestelia*.

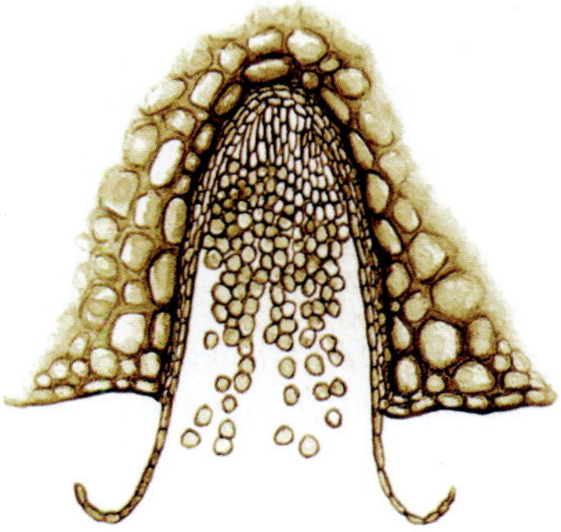

Cornute aecium of *Roestelia* sp. on a host leaf, seen in longitudinal section, x 110 (*MU*).

coronate (L. *coronatus*, crowned < *coronare*, to crown < *corona*, crown − suf. *-atus* > E. *-ate*, provided with or likeness): **1**. Provided with a crown, like the conidia of *Conidiobolus coronatus* (Entomophthorales), the teliospores of *Puccinia coronata* (Uredinales), or the ascocarp of *Phyllactinia guttata* (=*Ph. corylea*), Erysiphales; the latter has a crown composed of short, branched, gleoid, apical appendages. **2**. Splitting to form a crown-like structure, like the exoperidium of *Sphaerobolus* (Nidulariales).

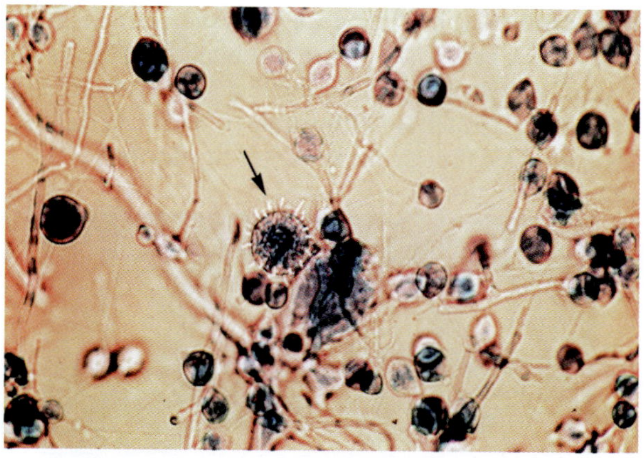

Coronate conidium (provided with denticles on which are formed secondary conidia) of *Conidiobolus coronatus*, x 190 (*RLM*).

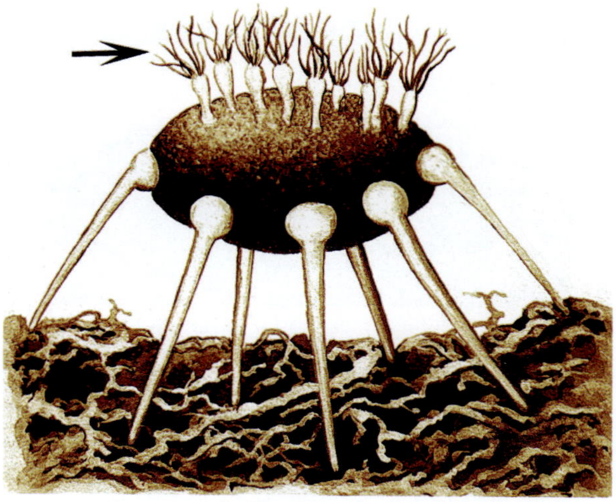

Coronate ascocarp of *Phyllactinia guttata* (=*Ph. corylea*) on the surface of a host leaf, x 190 (*MU*).

corrugate (L. *corrugatus* < *corrugare* < *co-*, with + *ruga*, wrinkle + suf. *-atus* > E. *-ate*, provided with or likeness): wrinkled or folded, with alternating ridges and grooves.

cortex, pl. **cortexes**, **cortices** (L. *cortex*, cortex, bark): the outer layer or layers of a sporocarp or of a somatic structure, including the peridium; e.g., in the Nidulariales, it corresponds to the main wall of the peridiole. In the heteromerous lichen thalli with horizontal stratification, there is an external cortex and an internal one; in heteromerous thalli with radial stratification, there is only an external cortex.

cortical (L. *corticalis* < *cortex*, cortex, genit. *corticis*, bark + suf. *-alis* > E. *-al*, relating to or belonging to): relating to the bark, belonging to it.

corticate (L. *corticatus* < *cortex*, cortex, genit. *corticis*, bark + suf. *-atus* > E. *-ate*, provided with or likeness): provided with bark.

corticicolous, **corticolous** (L. *cortex*, cortex, genit. *corticis*, bark + suf. *-cola*, inhabitant): living on the bark of trees and shrubs, as with many Myxomycetes, such as *Craterium* (Physarales), and lichens, e.g., *Caloplaca* (Teloschistales). See **lignicole**.

Corticicolous sporangia of *Craterium minutum*, x 10 (*MU*).

Corticicolous thallus of *Caloplaca* sp., x 7 (*MU*).

cortina pl. **cortinae** (LL. *cortina*, veil, curtain): *Agaricology*. A thin, cloth-like tissue that hangs from the edge of the pileus of the basidiocarp in some agarics (such as in the gen. *Cortinarius*). The cortina originates from the partial veil that initially unites the edge of the pileus with the stipe of the basidiocarp, but which because of its delicate consistency breaks and disappears, leaving at most a filamentous peripheral remnant on the pileus and another on the stipe. Cf. **pellicular veil**.

Cortina of the basidiocarp of *Cortinarius* sp., x 0.5 (*RV*).

Cortina of the basidiocarp of *Cortinarius* sp., x 0.8 (*TH*).

cortinate (LL. *cortina*, veil, curtain + suf. *-atus* > E. *-ate*, provided with or likeness): having a cobweb-like, fibrillose partial veil; typically found in the Cortinariaceae (Agaricales).

corymbiferous, **corymbose** (L. *corymbus* < Gr. *kórymbos*, cluster of flowers + suf. *-ferous*, bearer < *ferre*, to bear, carry; or + L. suf. *-osus* > E. *-ose*, having, augmented, prone to): borne in a corymb, i.e., with the sporogenous branches arising at different points along the vertical axis but all terminating at the same level at the apex; e.g., like the sporangiophore of *Mycocladus corymbiferus* (=*Absidia corymbifera*), Mucorales.

coscinocystidium, pl. **coscinocystidia** (Gr. *kóskinion*, sieve + *kystídion* < *kýstis*, bladder, vesicle, cell + dim. suf. *-ídion* > L. *-idium*): a special type of gloeocystidium with a cribose or porose wall and a sponge-like interior. The coscinocystidia are the terminal cells of a *coscinoid*. They are present, e.g., in species of *Gloeocantharellus* (Aphyllophorales), but are absent in the Agaricaceae.

coscinoid (Gr. *kóskinion*, sieve + L. suf. *-oide*, like < Gr. *-oeídes*, similar to): a specialized conductive hypha, brown in color, with a cribose surface and spongy contents; characteristic of *Gloeocantharellus* (=*Linderomyces*) *lateritius* (Aphyllophorales). See **coscinocystidium**.

cosmopolitan, **cosmopolite** (Gr. *kósmos*, universal + *polités*, citizen, inhabitant): found naturally in many parts of the world, under various ecological conditions. Cf. **endemic**.

costate (L. *costatus*, ribbed < *costa*, rib, flank + suf. *-atus* > E. *-ate*, provided with or likeness): having ribs or rib-like borders, like the basidiocarp of *Cantharellus cibarius* (Aphyllophorales).

cothurnate (L. *cothurnatus*, blunt < *cothurnus*, boot, laced shoe, laced boot + suf. *-atus* > E. *-ate*, provided with or likeness): *Agaricales*. Having a tissue that covers the base of the stipe, generally inrolled (like a sock) at the top, as in *Amanita pantherina* var. *velutipes*.

Costate basidiocarp of *Cantharellus cibarius*, x 1 (*MU*).

cottony (E. *cotton* < ME. *coton* < OF. < OIt. *cottone* < Ar. *qutun*, *qutn*, cotton + E. suf. *-y*, having the quality of): with long, white, soft hairs, like cotton. The mycelium of many fungi is cottony; e.g., such as that of *Penicillium camemberti* and *Aspergillus tamarii* (moniliaceous asexual fungi).

Cottony colonies of *Penicillium camemberti* on agar, x 1 (*MU*).

Cottony colony of *Aspergillus tamarii* on agar, x 1 (*MU*).

cotyliform (Gr. *kotýle*, concavity, a cup-shaped cavity > L. *cotyla*, *cotula*, bowl + L. suf. *-formis* < *forma*, shape): syn. of **cotylimorphous**. Shaped like a bowl, concave, like the apothecia of *Peziza* (Pezizales), which have the aspect of a plate or cup, with the edge erect or turning upward.

cotylimorphous (Gr. *kotýle*, concavity, a cup-shaped concavity + *morphé*, shape): equal to **cotyliform**.

cotype (L. *co-*, with + Gr. *týpos*, type, mark, model): *Taxonomy*. When a systematic taxon is described and the description is based on the characteristics of two or more specimens, the cotype refers to any of these specimens, independently of the one that represents the holotype. Etymologically, a specimen that accompanies the holotype. This term is somewhat ambiguous, as it has been used as equivalent to syntype, and sometimes poorly employed for isotype or paratype.

crateriform (L. *crateriformis* < *crater*, vessel or cup + *-formis* < *forma*, shape): shaped like a glass or large bowl, hemispherical and concave, like the sporangia of *Craterium minutum* (Physarales).

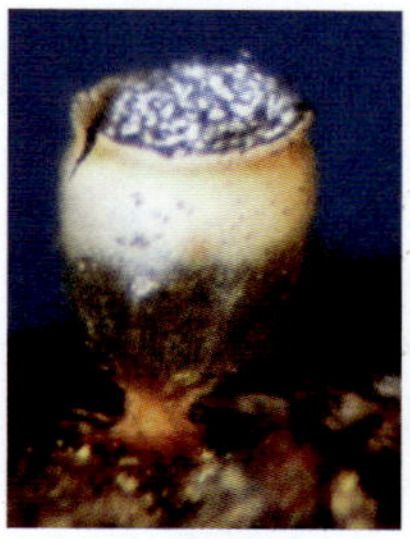

Crateriform sporangium of *Craterium minutum*, x 13 (*MU*).

crenate (L. *crenatus*, with notches, festooned < *crena*, notch + suf. *-atus* > E. *-ate*, provided with or likeness): with the margin edged with notches, like the giant colony of some strains of yeasts when grown on solid media (e.g., *Saccharomyces cerevisiae* of the Saccharomycetales), or like the margin of the gills of certain Agaricales (e.g., *Cortinarius alboviolaceus*). Also called **festooned**.

Crenate border of the giant colony of *Saccharomyces cerevisiae*, on agar, x 2 (*MU*).

crenulate (L. *crenulatus*, minutely notched < *crenula*, dim. of *crena*, notch or low cut + suf. *-atus* > E. *-ate*, provided with or likeness): like crenate, but with smaller notches, like the margin of the pileus of *Mycena pura* (Agaricales).

crescentiform, **crescentic** (L. *crescere*, increase, grow + suf. *-escens*, genit. *-escentis* > E. *-escent*, beginning to, slightly + L. suf. *-formis* < *forma*, shape): shaped like the crescent half-moon, quarter or half moon. Syn.of **lunulate**.

cretaceous (L. *cretaceus*, chalky < *creta*, chalk, gypsum + suf. *-aceus* > E. *-aceous*, of or belonging to, with the nature of): with the color or consistency of chalk or gypsum, like the medulla of many crustose lichens, which is white, compact and pimply, with few hyphae.

cribrate, **cribrose** (L. *cribrum*, sieve, sifter + L. suf. *-atus* > E. *-ate*, provided with or likeness; or + L. suf. *-osus* > E. *-ose*, full of, augmented, prone to): perforated; sometimes used to describe the perforated and reticulate appearance of some boletes, a condition that results from the decurrence of the walls of the tubes in the hymenophore of the pileus. Also called cribate is the shot hole disease of stone fruits (peach, almond, cherry, plum and apricot) caused by the fungus *Clasterosphaeria* sp. (Diaporthales), in its conidial form *Clasterosporium* sp. (dematiaceous asexual fungi). The disease takes its name from the perforations left in the leaves when the infected lesions fall out, leaving holes.

cribri- (L. *cribrum*, sieve, sifter): a pref. that implies a pierced sieve or plate, as in *cribroform*, with the shape of a sieve. Words with postcomponent Greek become *cribro-*, *cribr-*, as in *cribroparenchyma*, parenchyma that accompanies the cribrose tubes or vessels.

crinate, **crinite** (L. *crinatus*, hairy, with long hair < *crinus*, hair + suf. *-atus* > E. *-ate*, provided with or likeness): having long hair or similar structures, like the fruiting body of *Lentinus* (=*Panus*) *crinitus* (Aphyllophorales).

crispate, **crisped** (L. *crispatus*, curled < *crispus*, curly, wavy + suf. *-atus* > E. *-ate*, provided with or likeness): finely curled or wavy, like the gills of the fungi of the gen. *Trogia* (Agaricales).

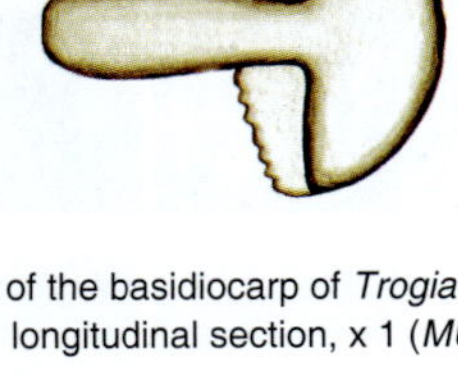

Crispate gills of the basidiocarp of *Trogia* sp. Diagram of a longitudinal section, x 1 (*MU*).

crista, pl. **cristae** (NL. *crista*, crest, relief): each of the inwardly projecting folds of the inner membrane of a mitochondrion; cristae, which can be tubular, pouch-like, or shelf-like, depending on the group of fungi, are the site of ATP production during aerobic metabolism. See **mitochondrion**.

cristate (L. *cristatus* < *crista*, crest, relief + suf. *-atus* > E. *-ate*, provided with or likeness): having one or more crests.

crozier (variant of G. *crosier* < ME. *croser*, crosier bearer < MF. *crossier* < *crosse*, crosier): a structure with a hooked end. See **uncinulum**. The term is commonly applied to the initial stage in ascus formation of most ascomycetes.

cruciate, **cruciform** (L. *cruciformis* < *crux*, genit. *crucis*, cross + suf. *-formis* < *forma*, shape): a figure in the shape of a cross; e.g., like the conidia of *Valdensia heterodoxa* (moniliaceous asexual fungi), or like the spindle in the nuclear division (cruciform division) of *Plasmodiophora* and other Plasmodiophorales, such as *Sorosphaera*, which adopt this figure because the chromosomes form a ring that surrounds perpendicularly the elongated nucleolus, which appears as a cross when viewed from one side. This type of nuclear division also is called **promitosis**.

Cruciform conidium of *Valdensia heterodoxa*, x 650 (*MU*).

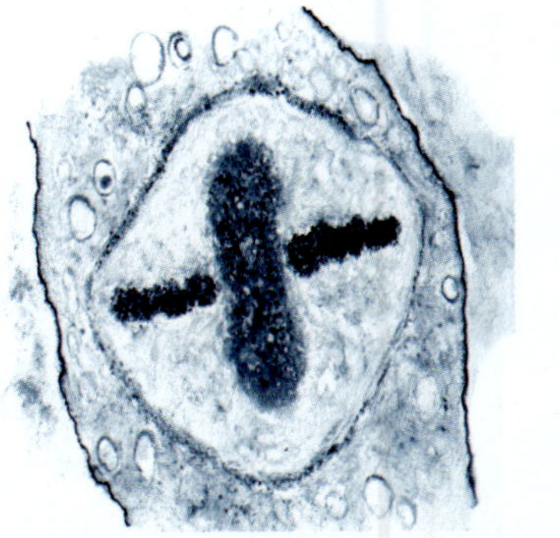

Cruciform nuclear division in the endobiotic plasmodium of *Sorosphaera veronicae*, x 12,500 (*MU*).

crust (L. *crusta*, crust, hard superficial layer): a general term to designate a hard superficial layer, like the lichen thallus. Some phytopathogenic fungi cause diseases of fruits that produce crustose spots; e.g., apples infected by the fly-spot fungus, *Schizothyrium pomi* (Microthyriales), form crustose ascomata on the surface. The black tar spots of maple leaves, caused by *Rhytisma acerinum* (Rhytismatales), and the red spots of plum leaves caused by *Polystigma rubrum* (Phyllachorales) also are crustose. See **crustaceous.**

crustaceous, crustose (L. *crustaceus*, having a shell, rind, or crust < *crusta*, crust, cortex + suf. *-aceus* > E. *-aceous*, of or pertaining to, with the nature of): in the lichens it refers to a thallus with the appearance of a powdery, warted, areolate or granulose crust, closely attached to the substrate and lacking a lower cortex and rhizines, as, e.g., in *Ochrolechia parella* (Pertusariales), *Caloplaca*, *Xanthoria* (=*Caloplaca*) *elegans* (Teloschistales), *Haematomma subpuniceum* and *Lecanora frustulosa* (Lecanorales). The aethalium of *Mucilago crustacea* (Physarales) is another example of the application of this term. See **crust**.

Crustaceous thallus of *Caloplaca saxicola* on rock, x 5 (*CB*).

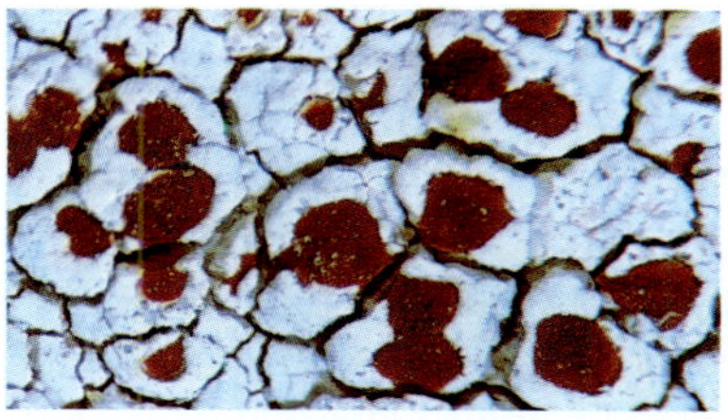

Crustaceous thallus of *Haematomma subpuniceum* on rock; the dark bodies are apothecia, x 5 (*MU*).

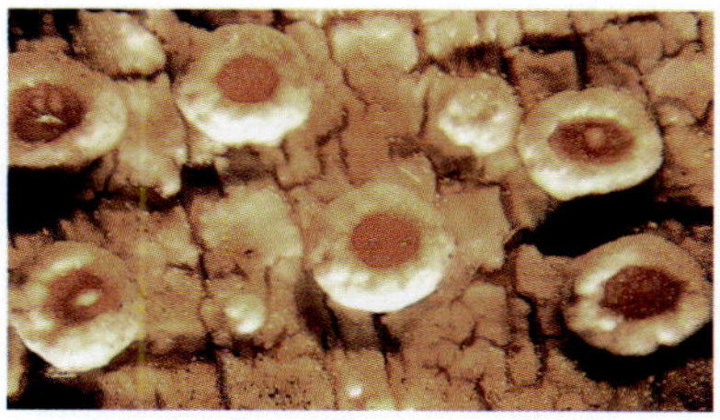

Crustaceous thallus of *Ochrolechia parella* on tree bark; the apothecia are evident, x 7 (*MU*).

Crustaceous aethalium of *Mucilago crustacea* on a grass stem, x 7 (*MU*).

cryo- (Gr. *krýos*, cold): a pref. employed in words alluding to cold, as in *cryophilous*, an organism that prefers cold temperatures.

cryopreservation (Gr. *krýos*, cold, frozen, ice + E. *preservation*, action and effect of preserving < L. *preservare* < *prae*, before + *servare*, to keep): one of the most efficient and preferred methods of preserving pure cultures of fungi, bacteria and other microorganisms for a prolonged time (several years), utilizing ultralow temperatures (-140 to -150°C) and storage in ultralow freezers. Cryopreservation and other methods of preserving strains of microbial cultures, such as lyophilization and storage on silica gel crystals, among others, are designed to minimize changes that can cause instability of the cultures, such as spontaneous mutations, loss of plasmids, loss of viability and genetic recombination (due to the occurrence of heterokaryons, the parasexual cycle or normal sexual events). The preservation method selected can depend on the type and quantity of microbial cultures to be conserved, on the installations and equipment available, on the costs of material and labor, and on the level of stability and longevity required. See **strain** and **lyophilization**.

crypta, pl. **cryptae** (L. *crypta* < Gr. *kryptós*, hidden, covered): a sleeve-like formation around the roots of certain trees developed by some Agaricales, and which serve as shelter for various epibiotic, symbiotic or parasitic insects, or with a relationship to the tree that involves a combination of these modes of life.

cryptic (L. *crypta* < Gr. *kryptós*, hidden, covered + suf. *-tikós* > L. *-ticus* > E. *-tic*, relation, aptitude, inclination or ability): inconspicuous, not very evident, hidden, like the cells of *Cryptococcus* (cryptococcaceous asexual yeasts), that are covered by a mucilaginous capsule. The term also refers to species that are "hidden", i.e., that are not recognizable on the basis of morphological characters. Such species can now be separated out by utilizing molecular data to distinguish them.

crypto-, **crypt-** (Gr. *kryptós*, hidden, covered): a pref. used to allude to something that is not very evident, less ostensible than ordinary, as in *cryptobiosis*. Cf. **phanero-**.

cryptobiosis, pl. **cryptobioses** (Gr. *kryptós*, hidden, covered + *bíosis*, act of living < *bios*, life + suf. *-osis*, condition or state): the condition of suspended life, similar to death, that generally occurs through starvation, desiccation or freezing, a condition that can revert through *anabiosis*.

cryptococcaceous (G. *kryptós*, hidden, covered + *kokkíon*, small ball + L. suf. *-aceus* > E. *-aceous*, of or pertaining to, with the nature of): term applied to yeasts that reproduce by budding, not by sexual ascospores or basidiospores. The cryptococcaceous asexual yeasts is an unnatural group in the gen. *Candida*, *Cryptococcus*, *Kloeckera*, *Pityrosporum* (=*Malassezia*) and *Trichosporon*, some examples of these kinds of fungi that reproduce by the formation of blastospores (the old family Cryptococcaceae, order Cryptococcales, which includes the asporogenous or imperfect yeasts, with ascomycetous or basidiomycetous teleomorphs). See **Blastomycetes**.

cryptococcosis, pl. **cryptococcoses** (< gen. *Cryptococcus* < Gr. *kryptós*, hidden, covered + *kókkos*, grain < *kokkíon*, small ball + suf. *-osis*, state or condition): a subacute or chronic infection of the central nervous system of man, with meningitis and cerebral tumors, although it can affect other organs (skin, bones, lungs). The infection is acquired by the inhalation of dust containing cells of the causal agent, the yeast *Cryptococcus neoformans* (cryptococcaceous asexual yeasts), an ubiquitous inhabitant of soil, especially that enriched with bird droppings (chickens, doves and pigeons). See **mycosis**.

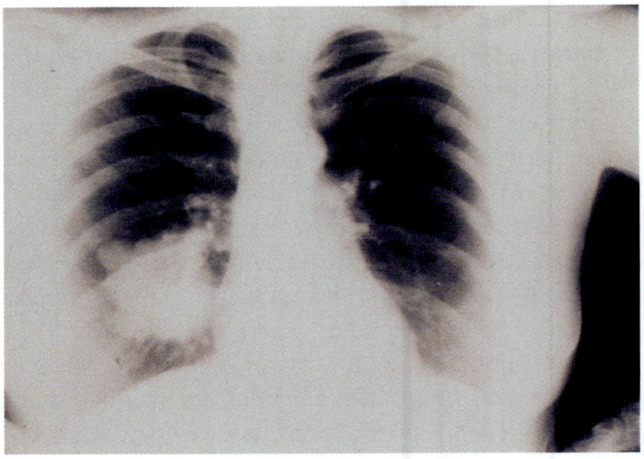

Pulmonary **cryptococcosis** in lungs of a man, caused by *Cryptococcus neoformans*, x 0.2 (*RLM*).

Vegetative cells of *Cryptococcus neoformans*, the causal agent of **cryptococcosis** of a man, x 630 (*RLM*).

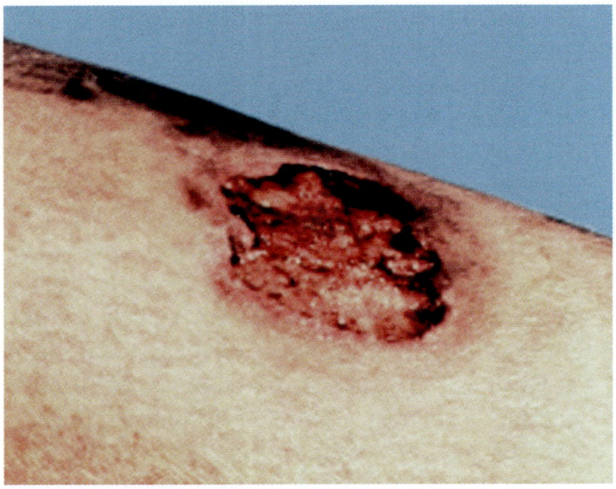

Cutaneous **cryptococcosis** on the forearm of a man, caused by *Cryptococcus neoformans*, x 0.5 (*RA* & *RLM*).

cryptoendolithic (Gr. *kryptós*, hidden, covered + *éndon*, inside + *líthos*, stone + suf. *-íkos* > L. *-icus* > E. *-ic*, belonging to or relating to): an organism, especially a lichen, that survives at low temperatures through modification of the thallus, so that it can exist inside a rock between the rock crystals, as with, e.g., *Lecidea phillipsiana* (Lecanorales). Cf. **endolithic** and **epilithic**.

cryptogamy (Gr. *kryptós*, hidden, covered + *-gamy* < *gámos*, sexual union): Class XXIV of the sexual system of Linnaeus, which included plants "whose fructification is scarcely visible or not clearly distinguished", namely, the ferns, mosses, algae and fungi, together called "*cryptogams*". As a systematic group, cryptogamy is unsupportable due to lack of homogeneity of its contents, and it has been rejected. However, in current language, the use of this word persists to refer to any plant not belonging to the *anthophytes* or plants with flowers (L. *anthophyta* < Gr. *ánthos*, flower + *phytón*, plant) or *phanerogams* (Gr. *phanerós*, visible, apparent + *gámos*), i.e., plants with visible reproductive organs. Cf. **phanerogamy**.

cteinomycete (Gr. *kteíno*, to kill + *mýkes*, genit. *mýketos*, fungus): a parasitic fungus that causes the immediate death of the cells that it penetrates to obtain food.

ctenoid (Gr. *ktenoeídes* < *kten-*, *kteis*, comb + suf. *-oeídes*, similar to): having a dentate border, similar to the teeth of a comb. Syn. of **pectinate**.

cuboid (L. *cubus* < Gr. *kýbos*, cube + L. suf. *-oide*, like < Gr. *-oeídes*, similar to): similar to a cube; applied to the type of brown rot of wood caused by Aphyllophorales, which results in the formation of cuboid wood fragments.

cucullar, **cucullate** (L. *cucullatus*, hooded < *cuculla*, cowl, hooded cloak + suf. *-atus* > E. *-ate*, provided with or likeness): with the shape of a cowl or cloak, hooded, like the pileus of *Phaeocollybia christinae* (Agaricales). Also called **cuculliform**.

Cucullate basidiocarps of *Phaeocollybia christinae*, x 1 (*GL*).

cuculliform (NL. *cuculliformis* < *cuculla*, cowl, hooded cloak + suf. *-formis* < *forma*, shape): see **cucullar**.

culmicole, **culmicolous** (L. *culmus*, grass stem + suf. *-cola*, inhabitant < *colere*, to inhabit): living and developing on the stems of cereals and other grasses, like many of the fungi commonly called rusts (Uredinales). Also called **culmigenous**.

culmigenous (L. *culmus*, grass stem + *-gena*, the idea of producing or engendering < *genos*, origin): see **culmicole**.

cultrate, **cultriform** (L. *cultratus*, like a knife < *culter*, knife + suf. *-atus* > E. *-ate*, provided with or likeness; L. *cultriformis* < *culter* + suf. *-formis* < *forma*, shape): knife-shaped.

culture (L. *cultus*, culture, action and effect of cultivating): **1**. A population of organisms, especially micro-organisms, maintained in the laboratory, where they are grown on suitable substrates (culture media). **2**. The act and effect of growing organisms on appropriate culture media in the laboratory, where they are transferred to other media by inoculation or seeding. See **agnotobiotic**, **isolate**, **isolation**, **axenic**, **strain** and **monoxenic**.

cumulate (L. *cumulatus*, accumulated, piled up < *cumulare* < *cummulus*, heap, mass or pile + suf. *-atus* > E. *-ate*, provided with or likeness): accumulated, piled up; combined into a single mass.

cuneiform, **cuneate** (L. *cuneiformis*, wedge-shaped < *cuneus*, wedge + suf. *-formis* < *forma*, shape; or + L. suf. *-atus* > E. *-ate*, provided with or likeness): wedge-shaped or similar to a wedge in longitudinal section, when used for laminar structures. E.g., the basidiospores of *Lepiota cuneatospora* (Agaricales) are cuneiform.

Cuneiform basidiospores of *Lepiota cuneatospora*, x 2,000 (*MU*).

cuniculate (L. *cuniculatus*, tube-shaped < *cuniculus*, cavern, den, burrow + suf. *-atus* > E. *-ate*, provided with or likeness): traversed on one end by a long passage.

cup-shaped fungus, pl. **fungi** (ME. *cuppe* < LL. *cuppa*, alteration of L. *cupa*, tub; an open bowl; ME. *shapen*, to shape; form, create; L. *fungus*, fungus): having the shape of a tub or vat, like the fruiting bodies of *Cookeina tricholoma* and *Peziza hemisphaerica* (Pezizales). See **cupulate**.

cupulate, cupular (L. *cupulatus*, tub-shaped, trough or vat < *cupula*, dim. of *cupa*, vat + suf. *-atus* > E. *-ate*, provided with or likeness): having the shape of a tub or vat, like the aecia of *Aecidium* and *Puccinia graminis* (Uredinales), with a peridium in the shape of a cup or goblet, cylindric, with a recurved edge. The ascocarps of *Cookeina aurantiaca*, *C. sulcipes* and *C. tricholoma* (Pezizales) and the podetia of the lichens *Cladonia chlorophaea* and *C. pyxidata* (Lecanorales) also are cupulate. See **cup-shaped fungus**.

Cup-shaped fungi, ascocarps of *Cookeina tricholoma*, x 0.5 (*RV*).

Cupulate ascocarps of *Cookeina sulcipes*, x 0.5 (*EPS*).

Cup-shaped fungi, ascocarps of *Peziza hemisphaerica*, x 0.5 (*RV*).

Cupulate podetia of the thallus of *Cladonia chlorophaea*, x 9 (*MU*).

Cup-shaped fungi, ascocarps of *Peziza hemisphaerica*, x 0.5 (*RV*).

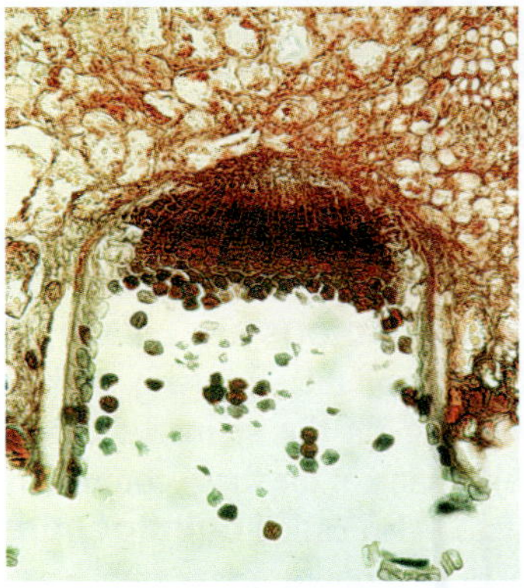

Cupulate aecium of *Puccinia graminis tritici*, seen in longitudinal section of a host leaf, x 175 (*MU*).

curved (L. *curvare*, to have or take a turn, change, or deviation from a straight line or plane surface without sharp breaks or angularity): bent or formed into a curve, like the vegetative cells of *Cryptococcus curvatus* (cryptococcaceous asexual yeasts).

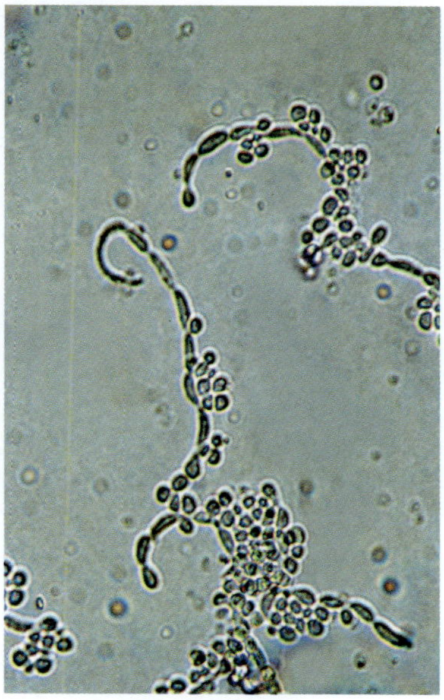

Curved vegetative cells of *Cryptococcus curvatus*, x 450 (*GA*).

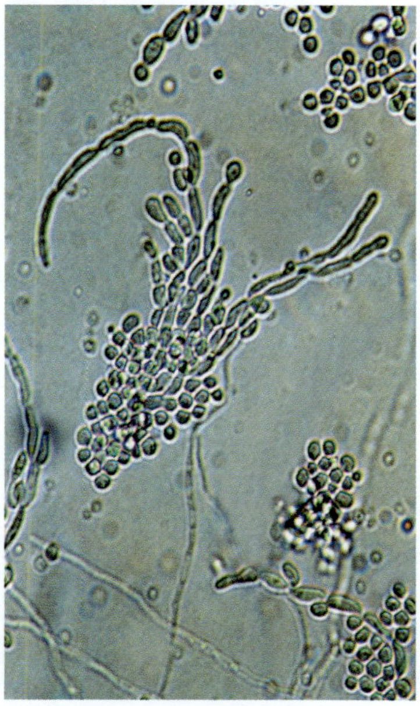

Curved vegetative cells of *Cryptococcus curvatus*, x 450 (*GA*).

cuspidate (NL. *cuspidatus*, ending in a point < L. *cuspis*, genit. *cuspidis*, point + suf. *-atus* > E. *-ate*, provided with or likeness): ending in a sharp point, like the ascospores of *Metschnikowia bicuspidata* (Saccharomycetales), the pileus of the basidiocarp of *Phaeocollybia*, or the cystidia of *Panaeolus cyanescens* (Agaricales).

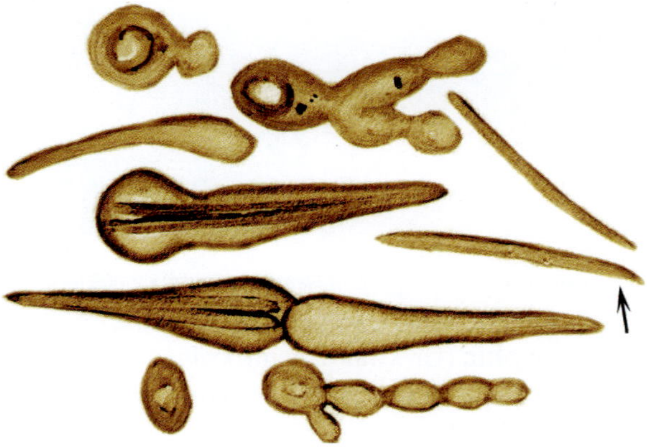

Cuspidate ascospores of *Metschnikowia bicuspidata*, x 1,000 (*MU*).

cyanescent (Gr. *kýanos*, genit. *kyáneos*, a substance dark bluish-green in color + L. suf. *-escens*, genit. *-escentis* > E. *-escent*, that which turns, beginning to, slightly): turning a bluish color.

cyano-, cyan- (Gr. *kýanos*, genit. *kyáneos*, a substance dark bluish-green in color): a pref. often employed to introduce the idea of blue, as in the following terms.

cyanobiont (Gr. *kýanos*, genit. *kyáneos*, a substance dark bluish-green in color + *bíos*, life + *óntos*, genit. of *ón*, a being): the photosynthetic cyanobacterial symbionts of a lichen. Among the most common gen. are *Nostoc*, *Scytonema*, *Gleocapsa* and *Stigonema*. See **photobiont**.

cyanophycophilous (Gr. *kýanos*, genit. *kyáneos*, a substance dark bluish-green in color + *phýkos*, alga + *phílos*, have an affinity for < *philein*, to love, prefer): *Lichens*. Refers to thalli that have cyanobacteria as photobionts, as occurs in the gen. *Collema* (Leca-norales). Cf. **chlorophycophilous**. See **diphycophilous** and **phycophilous**.

cyanophilic, cyanophilous (Gr. *kýanos*, genit. *kyáneos*, a substance dark bluish-green in color + *phílos*, have an affinity for < *philein*, to love, prefer): refers to fungal structures, especially cell or spore walls, that readily stain dark blue in 1% cotton blue in lactic acid (Amman's solution, commonly known as cotton blue or aniline blue) or with gentian violet. E.g., the hyphae of *Crinipellis piceae*, the basidia of *Lyophyllum multiceps* (Agaricales) and the basidiospores of *Gomphus* and *Ramaria grandis* (Aphyllophorales) are all cyanophilic. In some *Gomphus* spores the ornamentation of the wall stains distinctly and contrasts with the interior which stains lightly.

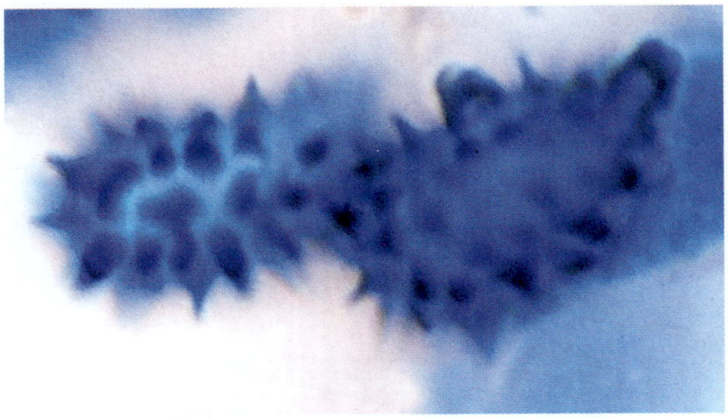

Cyanophilic basidiospores of *Ramaria grandis*, x 2,000 (*MV*).

cyanotrophic (Gr. *kýanos*, genit. *kyáneos*, a substance dark bluish-green in color + *trophós*, what nourishes, serves as food + suf. *-íkos* > L. *-icus* > E. *-ic*, belonging to or relating to): refers to fungi, especially those that form lichens, that obtain nutrients (particularly atmospheric nitrogen) by means of their association with free-living cyanobacteria, or lichenized cyanobacteria that form part of the cephalodia of the lichenized thallus. See **cephalodium**.

cyathiform (L. *cyathus*, cup < Gr. *kyáthos*, ladle + L. suf. *-formis* < *forma*, shape): cup-shaped, like the receptacles of the Gasteromycetes of the gen. *Cyathus* (Nidulariales) and the podetia of some lichens of the gen. *Cladonia*, such as *C. chlorophaea* (Lecanorales).

Cyathiform podetium of *Cladonia chlorophaea*, x 25 (*MU*).

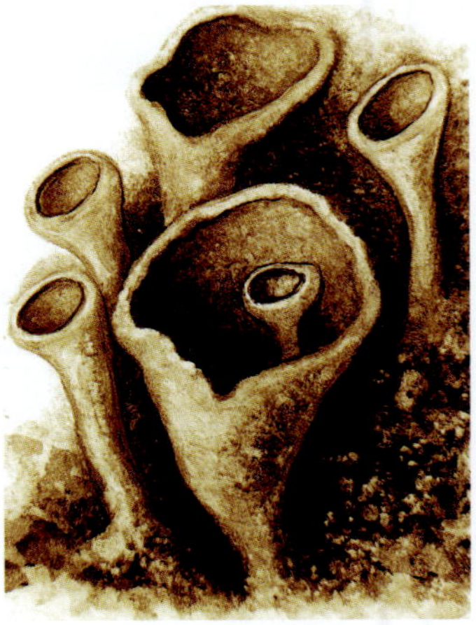

Cyathiform podetia of *Cladonia chlorophaea*, x 5 (*MU*).

Cyathiform podetium of the thallus of *Cladonia chlorophaea*, x 9 (*MU*).

Cyathiform basidiocarps of *Cyathus striatus*, x 1 (*MU*).

Cyathiform basidiocarp of *Cyathus* sp., x 5 (*CB*).

cyclo-, cycl- (Gr. *kýklos*, circle): a pref. employed in the formation of generic names (*Cyclographa*, a lichen of the order Arthoniales) and of technical terms that involve a circle or a rounded shape (*cyclomoid*), or a circular movement (*cyclosis*), etc.

cyclomoid (< gen. *Cyclomyces* < Gr. *kýklos*, circle + L. suf. *-oide*, like < Gr. *-oeídes*, similar to): *Basidiomycetes*. With the hymenium formed on structures similar to gills or plates arranged concentrically, and which represent transformed pores, not true gills, as in the gen. *Cyclomyces* and *Coltricia*, of the Aphyllophorales.

cyclosis, pl. **cycloses** (Gr. *kýklosis*, action of enclosing, surrounding): the cytoplasmatic circulatory current characteristic of eukaryotic cells; this internal motility depends on the complex of actinomyosin fibers, which are distinct from those present in animal muscle.

cygneous (L. *cycnus* < Gr. *kýknos*, swan): curved like the neck of a swan. The generic name of the ascosporogenous yeast *Schwanniomyces* (G. *Schwan*, swan) owes its name to the shape of the conjugating mother and daughter cells during formation of the ascus, which takes the shape of a swan neck. Presently, the species of *Schwanniomyces* have been placed in other gen. of Saccharomycetales, such as *Debaryomyces* and *Torulaspora*.

cylindric, cylindrical, cylindroidal (NL. *cylindricus* < Gr. *kylindrikós* < *kýlindros* > L. *cylindrus*, roller, cylinder < *kylíndein*, to roll, to roll up + suf. *-íkos* > L. *-icus* > E. *-ic*, belonging to or relating to): having the shape of a cylinder, with a length:width ratio of between 2:1 and 3:1. E.g., the sporangiola of *Syncephalis depressa* (Zoopagales) and the conidia of *Cylindrocarpon olidum*, *Geosmithia* (moniliaceous asexual fungi) and *Blumeriella haddenii* (Helotiales) are cylindrical.

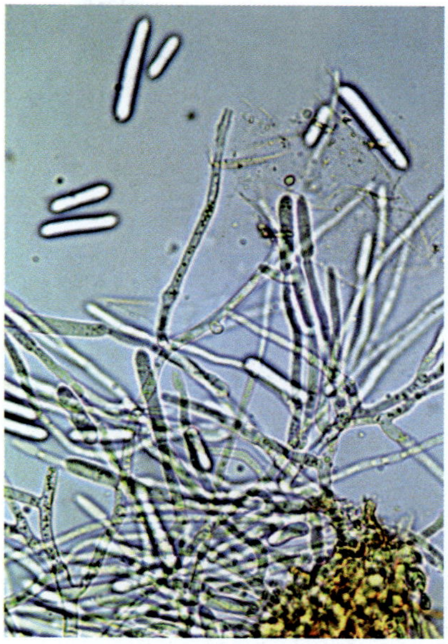

Cylindrical conidia of *Cylindrocarpon olidum*, x 250 (*JAS*).

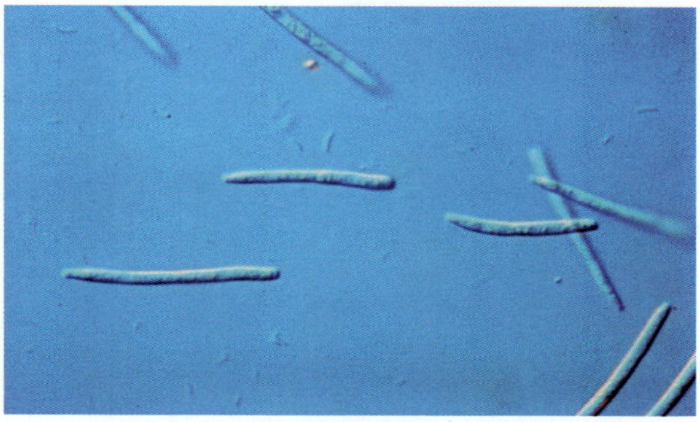

Cylindrical conidia of *Blumeriella haddenii*, x 1,300 (*RTH*).

cymbiform (NL. *cymbiformis*, boat-shaped < *cymba*, ship, boat + suf. *-formis* < *forma*, shape): with the shape of a boat; equal to **carinate**, **navicular** and **scaphoid**.

cyme (L. *cyma*, green cabbage < Gr. *kýma*, swelling, bud): raceme; a cluster of sporangial hyphae in which each main axis ends in a sporangium; each of the secondary and tertiary axes also can end in a sporangium, as is seen in some Oomycota sporangiophores. In some dematiaceous asexual fungi, such as *Dicyma* (anamorph of *Ascotricha*, of the Sordariales), the conidiophores also are arranged in cymes.

cyphella, pl. **cyphellae** (L. *cyphella*, the concavity of the ear): *Lichens*. The small cavities or rounded depressions, delimited by a cortical layer, that are only found on the lower surface of the majority of the lichens of the family Lobariaceae (*Sticta*), of the Peltigerales. See **cyphelloid** and **pseudocyphella**.

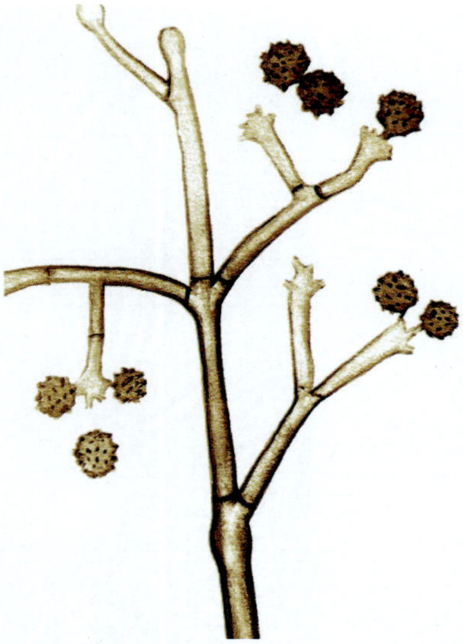

Cymes of the conidiophore of *Dicyma chartarum*, x 750 (*MU*).

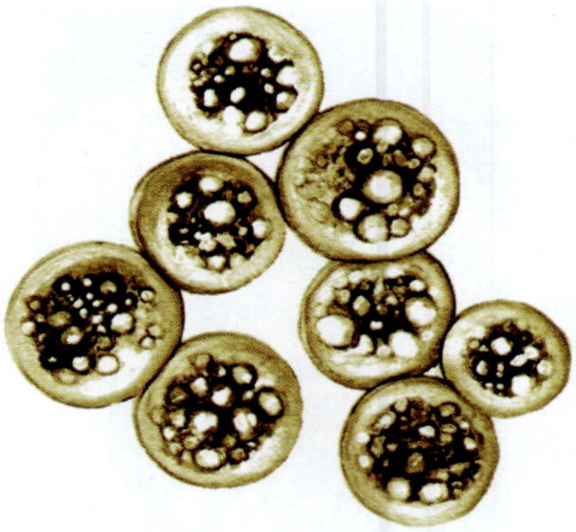

Cysts of *Amoebidium parasiticum*, x 730 (*MU*).

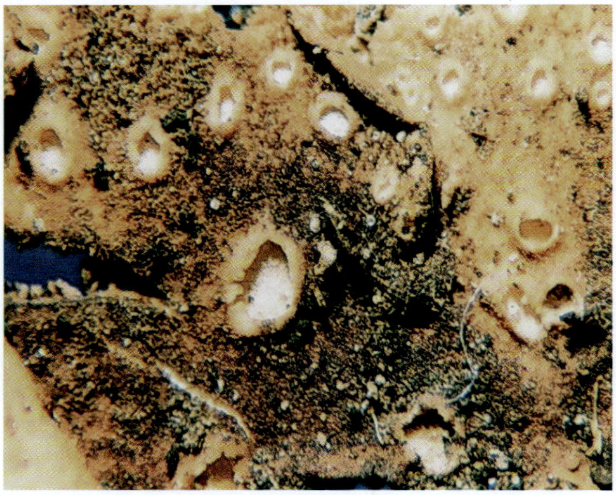

Cyphellae on the lower surface of the thallus of *Sticta weigelii*, x 5 (*MU*).

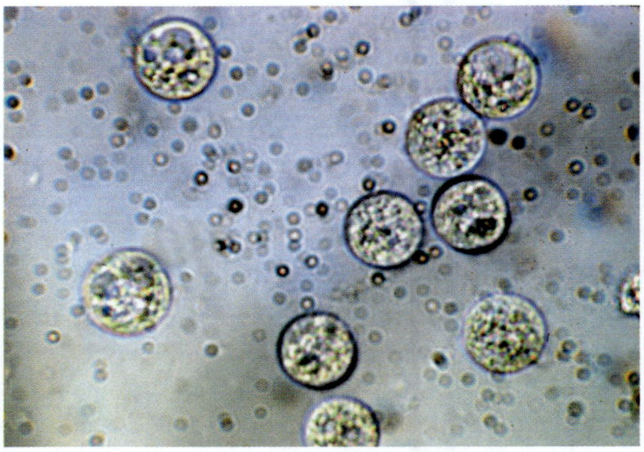

Cysts of *Saprolegnia parasitica*, x 600 (*MU*).

cyphelloid (L. *cyphella*, the concavity of the ear + suf. *-oide*, like < Gr. *-oeídes*, similar to): *Lichens*. A concavity or pit in the upper surface of the thallus of some lichens. Cf. **cyphella** and **pseudocyphella**.

cyst (NL. *cystis* < Gr. *kýstis*, bag, bladder; here, cell, vesicle): a spore protected by a resistent cell wall; it is especially applied to the structure that forms when a zoospore encysts, as happens, e.g., in the aquatic fungi of the orders Chytridiales and Saprolegniales (such as *Phlyctochytrium* and *Saprolegnia*, respectively), or to the structure resulting from the encystment of an amoeba, as in *Amoebidium parasiticum* (Amoebidiales). Also utilized as a suf., e.g., in *heterocyst*, *macrocyst*, *microcyst*, etc.

cystesium, pl. **cystesia** (Gr. *kýstis*, genit. *kystéos*, bag, bladder, vesicle, cell + dim. suf. *-ión* > L. *-ium*): a hymenial cell that is differentiated to adhere to a cystidium that originates in the opposite hymenium.

cystidiole (NL. *cystidiolum* < *cystidium* + dim. suf. *-olum* < Gr. *kystídion*, dim. of *kýstis*, bag, bladder, vesicle, cell + dim. suf. *-ídion*): a primordial cystidium, of superficial origin, that originates on the same level as the basidia in the hymenium of some Agaricales (such as *Conocybe tenera*); cystidioles differ slightly in size and shape from the basidia and brachycystidia.

cystidium, pl. **cystidia** (NL. *cystidium* < Gr. *kystídion* < *kýstis*, bag, bladder, vesicle, cell + dim. suf. *-ídion* > L. *-idium*): *Agaricales*. A large, inflated sterile cell that grows among the basidia and paraphyses; cystidia are usually larger in size, so that they project from the hymenial layer. Their function is not well understood, although in some cases they appear to serve as spacers between the basidia. See **brachycystidium**, **caulocystidium**, **cheilocystidium**, **chrysocystidium**, **coscinocystidium**, **gleocystidium**, **lamprocystidium**, **leptocystidium** and **pleurocystidium**.

cysto-, **cyst-** (Gr. *kýstis*, bladder): a pref. employed in biological terminology to give the idea of something vesiculose, as in *cystosorus*. Also used as a suf. *-cyst,* as in *macrocyst*.

cystochroic (NL. *cystis* < Gr. *kýstis*, bag, bladder, vesicle, cell − *chrós* or *chroía*, color or the superficial appearance of something + suf. *-íkos* > L. *-cus* > E. *-ic*, belonging to or relating to): with the pigment disseminated in the cell vacuoles. See **achroic** and **cytochroic**.

cystoderm (NL. *cystis* < Gr. *kýstis*, bag, bladder, vesicle, cell + *derma* < Gr. *dérma*, skin): *Agaricales*. A type of pellicle (*pellis*, in particular of the *suprapellis*), composed of a single layer of vesiculose and subglobose or globose cells (*sphaerocysts*) that cover the pileus of certain species, such as *Cystoderma amianthinum*, *Coprinellus micaceus* and *Coprinus stercorarius*. The term **cellular dermis** is also applied to this cuticle or pilear cortex (*pileipellis*).

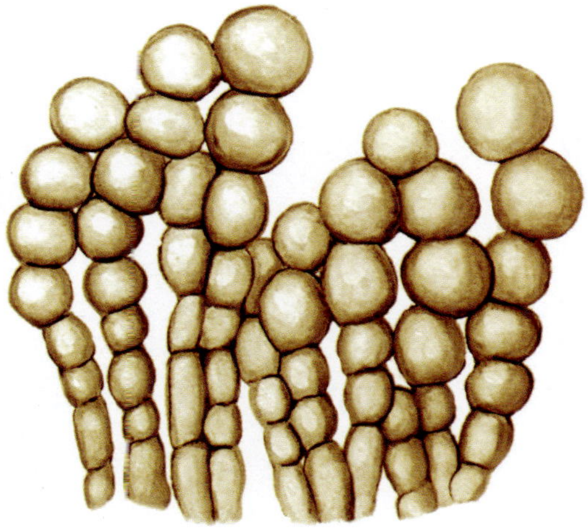

Cystoderm of the pileus of *Cystoderma amianthinum*, x 600 (*MU*).

cystogenous (NL. *cystis* < Gr. *kýstis*, bag, bladder, vesicle, cell + *génos*, origin < *gennáo*, to engender, produce): *Plasmodiophorales*. Refers to the plasmodium of the sexual phase that is transformed into immotile spores or cysts, which can remain separate or grouped into sori; some species form a definite number of cysts, e.g., four in *Tetramyxa* and eight in *Octomyxa*, whereas other gen. have numerous cysts (*Polymyxa* and *Spongospora*). Cf. **sporangiogenous**.

cystosorus, pl. **cytosori** (NL. *cystis* < Gr. *kýstis*, bag, bladder, vesicle, cell + NL. *sorus* < Gr. *sorós*, heap, pile): a more or less compact aggregate of latent spores or sporangiocysts, which are derived from a single protoplast, as occurs in *Woronina* and *Spongospora* (Plasmodiophorales).

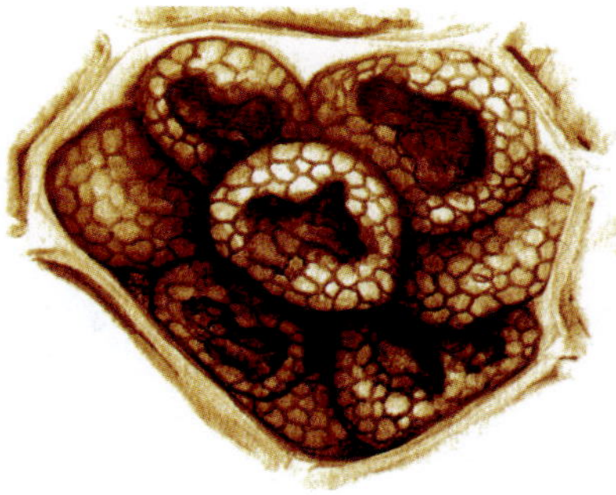

Cystosorus of *Spongospora subterranea* inside a cell of a potato tuber, x 600 (*MU*).

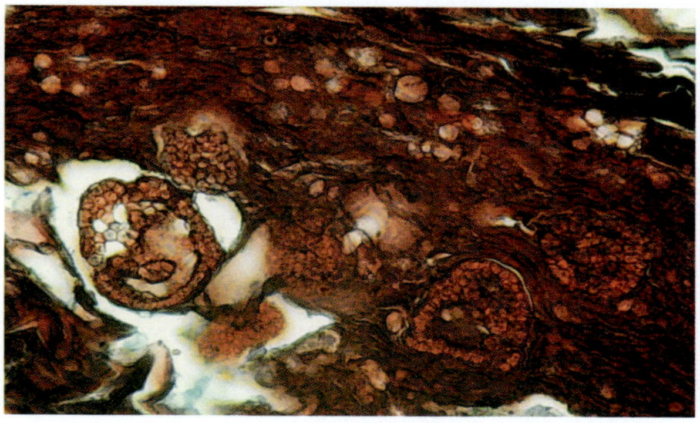

Cystosorus of *Spongospora subterranea* inside a cell of a potato tuber, x 600 (*MU*).

cystospore (NL. *cystis* < Gr. *kýstis*, bag, bladder, vesicle, cell + *sporá*, spore): *Trichomycetes*. The spore that results from the encystment of an amoeba, as happens, e.g., in *Amoebidium* (Amoebidiales). Also used for the encysted zoospore of certain Chytridiales.

cyto- (Gr. *kýtos*, cavity, cell): a pref. used in words concerning the cell, as in *cytology*, the science involving the study of the cell. Also used as a suf., *-cyte*, in *sporocyte*, *gametocyte*, etc.

cytochroic (Gr. *kýtos*, hollow receptacle, now often utilized to refer to a cell + *chrós* or *chroía*, color or the superficial appearance of something + suf. *-íkos* > L. *-icus* > E. *-ic*, belonging to or relating to): with the pigment disseminated in the cytoplasm of a cell. See **achroic**, **cystochroic**.

cytolysis, pl. **cytolyses** (Gr. *kýtos*, hollow receptacle, now often utilized to refer to a cell + *lýsis*, dissolution, disintegration): disintegration of cells or tissues by dissolution of the cell wall or of the middle lamella, which occurs, e.g., in various white rots.

cytoskeleton (Gr. *kýtos*, hollow receptacle, now often utilized to refer to a cell + *skeletós*, skeleton < *skéllo*, to dry, desiccate): an intracellular network of proteinaceous filaments insoluble in non-ionic detergents.

d

Dimension of the microcosmos of fungi (*MU*).

d

dacryo-, **dacry-** (Gr. *dácryon*, tear): a pref. used in the formation of terms of Gr. origin that allude to the shape or appearance of tears, as in *dacryoid*.

dacryoid (Gr. *dácryon*, tear + L. suf. *-oide*, like < Gr. *-oeídes*, similar to): tear-shaped, more or less rounded on one end and prolonged and sharp-pointed on the other, like the conidia of *Spiniger meineckellus* (moniliaceous asexual fungi) and the basidiospores of *Clitocybe gibba* (Agaricales). Also called **lacrimoid**.

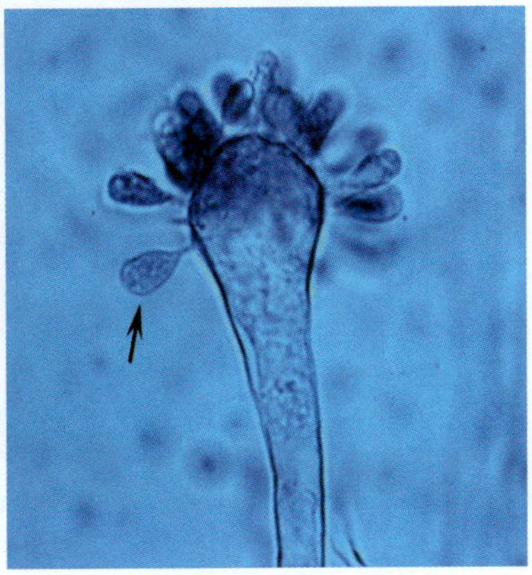

Dacryoid conidia on a conidiogenous cell (ampulla) of *Spiniger meineckellus*, x 500 (*RTH*).

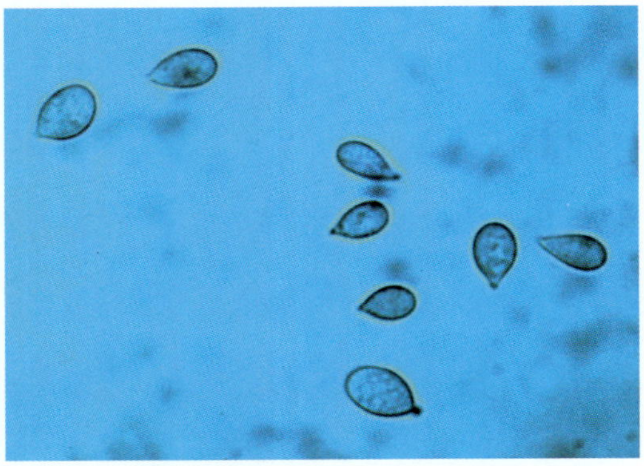

Dacryoid conidia of *Spiniger meineckellus*, x 590 (*RTH*).

dactylo-, **dactyl-** (Gr. *dáktylos*, finger): a pref. used in the formation of terms and generic names that allude to the shape or appearance of fingers, as in *dactyloid* and *Dactylospora* (Lecanorales).

dactyloid (Gr. *dáktylos*, finger + L. suf. *-oide*, like < Gr. *-oeídes*, similar to): divided into finger-like structures, like the haustoria of *Blumeria* (=*Erysiphe*) *graminis* (Erysiphales). Also called **digitate**.

daedaleoid, **daedaloid** (< gen. *Daedalea* < Gr. *daidáleos*, labyrinth + L. suf. *-oide*, like < Gr. *-oeídes*, similar to): with the tube mouths enlarged and sinuous, labyrinthiform, like the hymenophore of *Daedalea* (Aphyllophorales).

damping-off (E. *damp*, moist < MD. < MHG. *damph*, vapor; OE. *off*, removal): a disease condition of seedlings or cuttings caused by fungi (e.g., *Rhizoctonia*, of the agonomycetaceous asexual fungi), and characterized by collapse and death of the host plants.

dasy- (Gr. *dasýs*, dense, leafy, or with a full beard, hairy): a pref. employed to form words of Gr. origin that allude to a hairy structure, as in *Dasypyrena* (Microthyriales) and *Dasyspora* (Uredinales).

data, sing. **datum** (L. *datum*, something given): items of information, facts, statistics, etc., that are necessary to arrive at an exact understanding of something or to provide the basis for making decisions; information derived from scientific experiments. Frequently used as an adjective, e.g., data base, data bank, data matrix, etc.

de-, **des-** (L. *de-*, from, beneath, far away): a pref. that denotes negation, opposition, removal or elimination of something, deviation, separation, out of, excess; e.g., *decortication*. See **di-**, **dis-**.

de novo (L. *de novo*, new, newly): an expression that implies new, recent, renewed. *Conidiogenesis*. Cells that are formed from newly synthesized protoplasm (*blastic*), and not from existing cells, like the pseudomycelium of *Saccharomyces* (Saccharomycetales). See **arthric.**

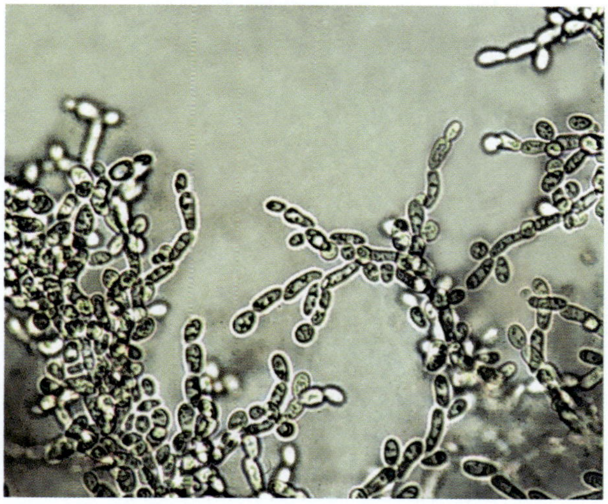

De novo, blastospores of the pseudomycelium of *Saccharomyces cerevisiae*, x 300 (*MU*).

deciduous (L. *deciduus*, ready to fall, deciduous < *decidere*, to fall): being easily detached from the structure that produces and supports it, like the sporangia of the higher Peronosporales (*Peronospora*, *Sclerospora*). Cf. **persistent**.

deciduous wart (L. *deciduus*, caducous, deciduous < *decidere*, to fall; ME. *werte* < OE. *woerte*, wart, prominence): a wart that detaches easily from the exoperidium and which generally leaves a visible scar, as happens in the fruiting body of *Lycoperdon* (Lycoperdales).

declinate (L. *declinatus*, bent downward, turned toward one side < *declinare*, to incline, bend + suf. *-atus* > E. *-ate*, possessing the qualities of or likeness): bent or tilted downward. See **decurvate**.

decortication (L. *decorticatio*, genit. *decorticationis*, action of removing the bark): the natural or deliberate removal of the bark of a tree, branch, etc. Also termed *debarking*, the removal of bark from a log.

decumbent (L. *decumbens*, genit. *decumbentis*, reclining, inclined, lying down < *decumbere*, to recline + suf. *-entem* > E. *-ent*, a being): not upright, but lying down or with a tendency to stretch out on the substrate, with the margin tending to turn upward, like the fruiting body of the corticiaceous and stereaceous fungi of the order Aphyllophorales.

decurrent (L. *decurrens*, genit. *decurrentis*, running down or extended downward < *decurrere*, to run downward + suf. *-entem* > E. *-ent*, a being): applied to gills that adhere to the stipe and which extend toward its base, as in the basidiocarp of *Clitocybe dealbata*, *Omphalotus olearius*, *O. mexicanus* (Agaricales) and *Lentinus badius* (Aphyllophorales). Cf. **adnate**, **free** and **sinuate**.

Decurrent gills of the basidiocarps of *Omphalotus olearius*, x 0.5 (*RV*).

Decurrent gills of the basidiocarps of *Omphalotus mexicanus*, x 0.5 (*EPS*).

Decurrent gills of the basidiocarps of *Lentinus badius*, x 1.7 (*MU*), and *Hygrophorus* sp. (Agaricales), x 0.5 (*RV*).

decurvate, decurved (L. *decurvatus*, curved or bent downward < *de-*, pref. that denotes from, down, far from + *curvus*, curve + suf. *-atus* > E. *-ate*, provided with or likeness): bent or curved downward, like the margin of the pileus of *Boletus* (=*Xerocomus*) *chrysenteron*, *Boletus edulis* and other species of the gen. *Boletus* (Agaricales), since the edge is curved toward the bottom, more or less parallel to the stipe or pointed slightly toward it.

Decurvate pileus of the basidiocarp
of *Boletus* (=*Xerocomus*) *chrysenteron*, x 0.5 (*MU*).

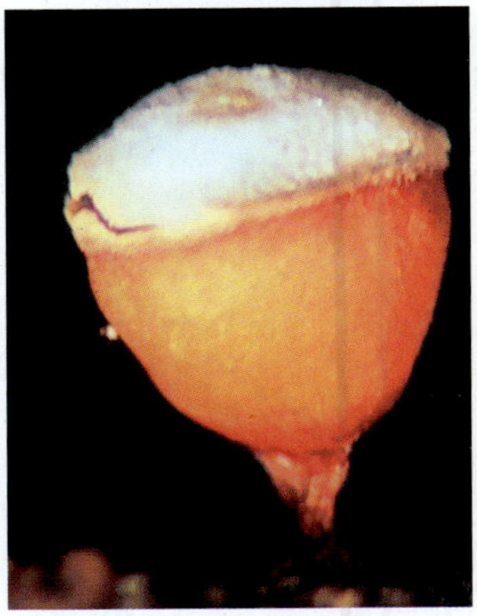

Circumscissile **dehiscence** of the sporangium
of *Craterium minutum*, x 20 (*SA*).

dediploidization (L. *de-*, pref. that denotes below, from, far from, out of; also signifies loss, to undo + Gr. *diplóos*, double + L. suf. *-oide*, like < Gr. *-oeídes*, similar to + Gr. suf. *-ationem* > E. *-ation*, action, state or condition, result): in the ascomycetes and basidiomycetes it refers to the formation of haploid cells or hyphae from a diploid dikaryotic cell or mycelium.

dehiscence (NL. *dehiscentia* < L. *dehiscere*, to open itself, to part + suf. *-entia* > F. *-ence* > E. *-ence*, state, quality or action): the process by which an organ opens spontaneously on reaching maturity to liberate its contents. In the fungi, many types of sporangia, gametangia, asci, and various sporiferous structures have some type of characteristic dehiscence that permits the liberation of the spores, gametes, etc. Among the modes of dehiscence are the *circumscissile*, the *stellate,* and the *longitudinal*, as, e.g., in *Craterium minutum*, *Diderma floriforme* (Physarales), *Utharomyces epallocaulus* (Mucorales), and *Lophodermium* sp. (Rhytismatales), respectively. In some cases spore liberation is by *irregular fragmentation* of the wall, as occurs with the sporangia of *Physarum polycephalum* (Physarales) and the asci of *Hanseniaspora valbyensis* and *Pichia membranifaciens* (Saccharomycetales). The peridium of the mature fruiting bodies of many gasteromycetes (e.g., *Scleroderma*, of the Sclerodermatales) also break irregularly to expose the gleba and facilitate the liberation of the spores. In other cases, the peridium of many lycoperdaceous species undergoes dehiscence by a pore (e.g., *Bovista*, *Lycoperdon*) or by several pores (as in *Myriostoma*), or it opens by circumscissile dehiscence (as in *Cyathus*, of the Nidulariales). Cf. **indehiscence**.

Stellate **dehiscence** of the sporangium
of *Diderma floriforme*, x 20 (*MU*).

Stellate **dehiscence** of the sporangia
of *Utharomyces epallocaulus*, x 250 (*MU*).

dehiscent, dehiscing (L. *dehiscens*, genit. *dehiscentis*, parting, dividing, opening < *dehiscere*, to divide, to split, open by itself + suf. *-entem* > E. *-ent*, a being): opening at maturity, most commonly by a pore or a slit, although there exist various types of *dehiscence*. Cf. **indehiscent**.

delimited peristome (Fr. *délimiter* < L. *delimitare*, to limit; Gr. *perí*, around, in turn + *stóma*, mouth): *Gasteromycetes*. A zone around the peristome, generally small and somewhat raised, distinct from the exoperidium and frequently of different color.

delineate (L. *delineatus*, demarcated, marked by lines < *delineare*, to delineate < *de-*, pref. that denotes from, beneath, far from + *linea*, line + suf. *-atus* > E. *-ate*, provided with or likeness): outlined; a figure that is delimited by lines.

deliquescent (L. *deliquescens*, genit. *deliquescentis*, the process of becoming fluid at maturity < *deliquescere*, to become liquid + suf. *-escens*, genit. *-escentis* > E. *-escent*, that which turns, that begins to, slightly): an organ or structure that undergoes a gradual dissolution and liquefaction at maturity through the attraction and absorption of water, becoming a liquid mass through autodigestion or autolysis, e.g., like the basidiocarps of the gen. *Coprinus* (Agaricales), which dissolve into an inky fluid. The phenomenon is called *deliquescence*. Cf. **arescent**. See **autolysis**.

delitescent (L. *delitescens*, genit. *delitescentis*, hidden < *delitescere*, to hide, conceal + suf. *-escens*, genit. *-escentis* > E. *-escent*, that which turns, that begins to, slightly): concealed, hidden.

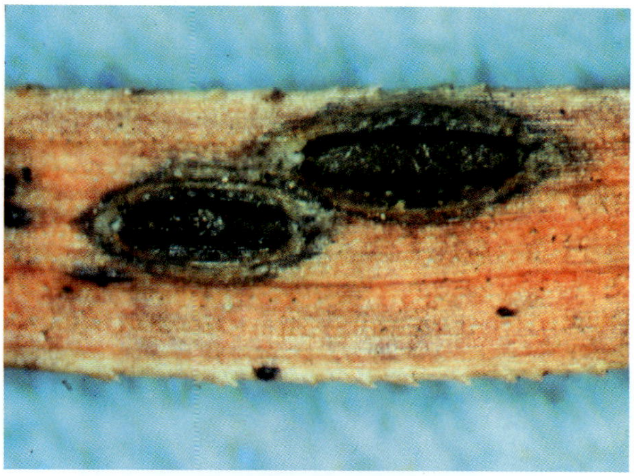

Longitudinal **dehiscence** of the elongate apothecia of *Lophodermium* sp., growing on a pine leaf, x 20 (*RTH*).

Irregular **dehiscence** of the sporangia of *Physarum* sp., x 100 (*MU*).

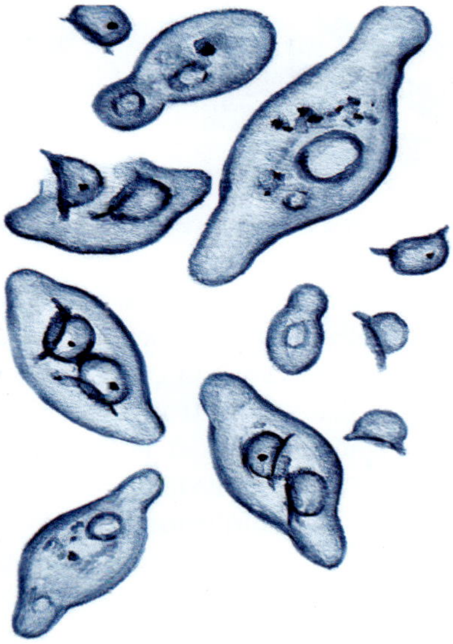

Dehiscence by irregular fragmentation of the asci of *Hanseniaspora valbyensis*, x 2,000 (*MU*).

Deliquescent basidiocarps of *Coprinus* sp., x 0.5 (*RV*), and *Coprinus comatus*, x 0.5 (*MU*).

Deliquescent basidiocarp of *Coprinus comatus*, x 2 (*TH*).

deltoid (L. *delta* < Gr. *delta*, Δ, fourth letter of the Gr. alphabet + L. suf. *-oide*, like < Gr. *-oeídes*, similar to): triangular, similar to the shape of the letter delta, like the conidia of *Anthopsis deltoidea* (moniliaceous asexual fungi) and the ascocarp of *Coccomyces delta* (Rhytismatales).

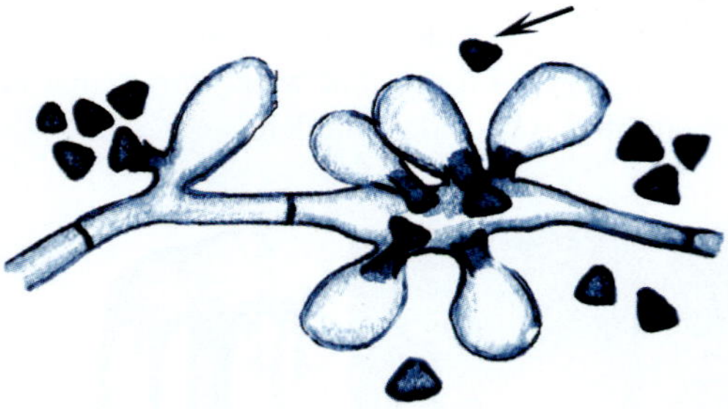

Deltoid conidia of *Anthopsis deltoidea*, x 1,500 (*MU*).

Deltoid ascocarp of *Coccomyces delta*, x 10 (*RTH*).

dematiaceous (L. *dematiaceae* < gen. *Dematium* < Gr. dim. *demátion*, band, bundle + L. suf. *-aceus* > E. *-aceous*, of or belonging to): a fungus of the old family Dematiaceae, filamentous, mitosporic, characterized by having dark-colored hyphae and/or conidia, and with the conidiophores not united in fascicles. Cf. **moniliaceous**. The gen. *Alternaria*, *Cladosporium*, *Polyschema*, among many others, have these characteristics. See **Hyphomycetes**.

Dematiaceous colony of *Cladosporium* sp. on agar, x 1 (*MU*).

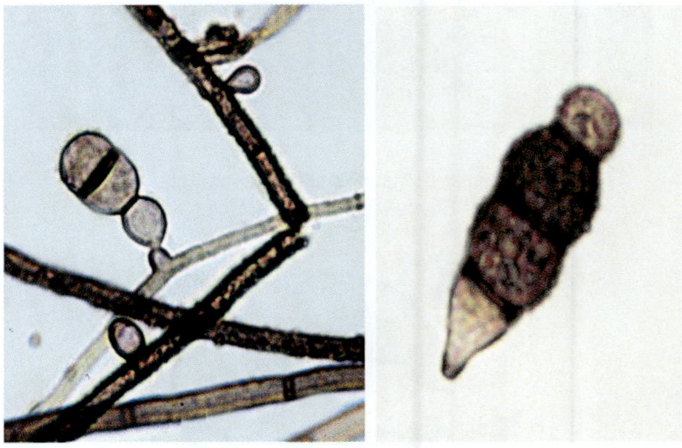

Dematiaceous conidia of *Polyschema indica*, x 1,000 and x1,500 (*MU*).

dematioid (< gen. *Dematium* < Gr. *demátion*, small bundle or sheaf + L. suf. *-oide*, like < Gr. *-oeídes*, similar to): similar to *Dematium* (a genus of dematiaceous asexual fungi that represents a *nomen confusum*), i.e., dark in color, almost black, filamentous, with a layer or felt of dark hyphae, like those of the sooty molds (Dothideales). See **fuliginous**.

demicyclic (L. *demi*, half + Gr. *kyklikós*, circular, cyclic < *kýklos*, circle + suf. *-íkos* > L. *-icus* > E. *-ic*, belonging to or relating to): *Uredinales*. A rust in whose life cycle there is no uredinial state (stage II), but with the spermogonial, aecial, telial and basidiospore states (stages 0, I, III and IV, respectively), as occurs in *Gymnosporangium juniperi-virginianae*, the cedar-apple rust; also called **hemicyclic**. Cf. **macrocyclic** and **microcyclic**.

dendritic (L. *dendriticus*, branched like a tree < Gr. *dendrítes*, fem. *dendrítis*, of or relating to a tree, similar to a tree + Gr. suf. *-íkos* > L. *-icus* > E. *-ic*, belonging to or relating to): see **dendroid**.

dendritic columella, pl. **columellae** (L. *dendriticus*, branched like a tree < Gr. *dendrítes*, fem. *dendrítis*, of or relating to a tree, similar to a tree + Gr. suf. *-íkos* > L. *-icus* > E. *-ic*, belonging to or relating to; L. *columella*, dim. of *columna*, column): *Gasteromycetes*. Sterile tissue with numerous branches inside the gleba that generally form from a solid basal tissue. Also called **percurrent columella** and **stipe-columella**.

dendrohyphidium, pl. **dendrohyphidia** (Gr. *déndron*, tree + NL. *hyphidium*, small hypha < *hypha*, tissue, spiderweb; hypha + dim. suf. *-idium* > E. *-idium*): syn. of **dendrophysis**.

dendroid (Gr. *déndron*, tree + L. suf. *-oide*, like < Gr. *-oeídes*, similar to): with tree-like branches, like the sporangiophores of *Peronospora destructor* (Peronosporales), the plasmodium of *Physarum polycephalum* (Physarales), and the conidiophores of *Botrytis cinerea*, *Metarhizium anisopliae* and *Monilia fructicola* (moniliaceous asexual fungi). Also called **dendritic** and **dendromorphous.**

Dendroid phaneroplasmodium of *Physarum polycephalum* on agar; the red pigment comes from the medium, x 1 (*MU*).

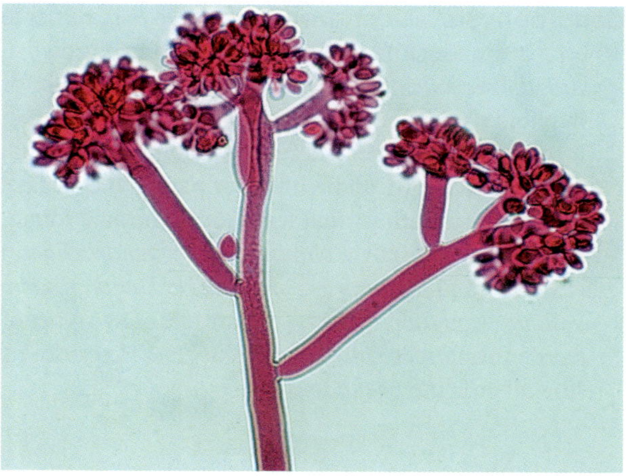

Dendroid conidiophore of *Botrytis cinerea*, x 50 (*MU*).

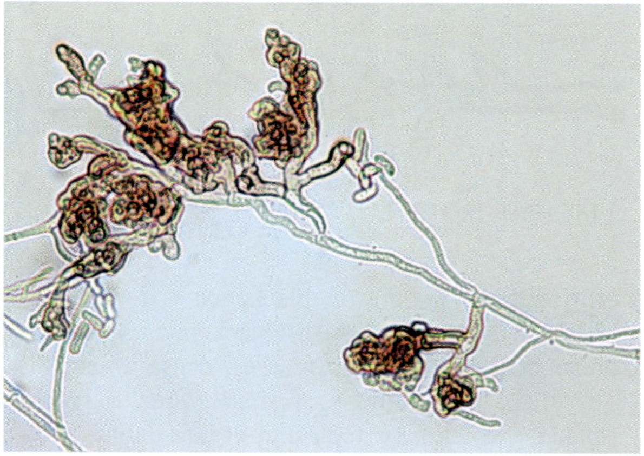

Dendroid conidiophore of *Metarhizium anisopliae*, x 270 (*MU*).

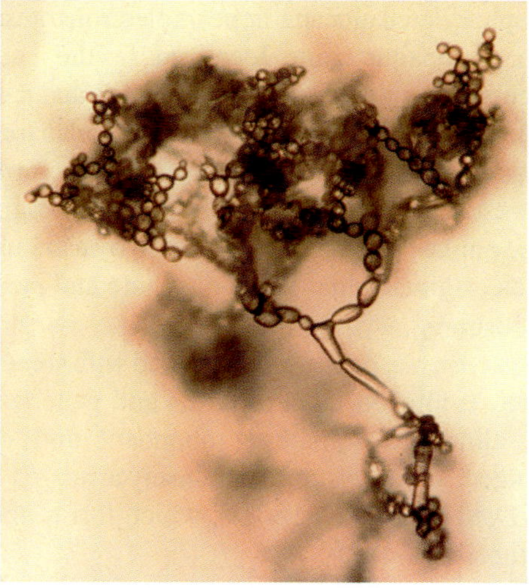

Dendroid conidiophore of *Monilia fructicola*, x 250 (*MU*).

dendromorphous (Gr. *déndron*, tree + *morphé*, shape): see **dendroid**.

dendrophysis, pl. **dendrophyses** (Gr. *déndron*, tree + *phýsis*, action of budding or being born; a growth): *Aphyllophorales*. A type of paraphysis, cystidium or hyphoid present in the hymenium; dendrophyses can be simple or branched, more or less compact, of equal or unequal length, and with all or part of the surface provided with spiny ornamentations. They are found, e.g., in *Aleurodiscus*, *Echinochaete* (=*Dendrochaete*) *russiceps* and *Laetiporus persicinus* (=*Dendrochaete vallata*), Aphyllophorales. They are one of two types of sterile formations belonging to the gen. *Aleurodiscus* (the other is the *pseudophysis*).

Dendrophyses of the hymenium of *Echinochaete* (=*Dendrochaete*) *russiceps*, x 650 (*MU*).

denigrate (L. *denigratus*, blackened < *denigrare*, to blacken, to spot, to tarnish): blackened, tarnished.

dense body vesicle (L. *densus*, dense, thick; ME. < OE. *bodig*, body; L. *vesicula*, dim. of *vesica*, bladder vesicle): a cytoplasmic vesicle delimited by a membrane that is present in some cells of Oomycota (Saprolegniales) and other heterokont organisms. This organelle, which is related to the metabolism of phosphorylated glucans, appears electron transparent, with an opaque zone localized in the center or eccentrically, at times with compact and alternate laminar formations, with electron transparent and opaque zones. At the moment of oospore formation, the vesicles of the dense body coalesce to give rise to the ooplast (inclusion surrounded by a membrane), which originates first in the oosphere and persists in the oospore.

densi- (L. *densus*, dense, thick, packed with something): a pref. employed to form terms and generic names that allude to something abundant and densely arranged, as in *densifolium*, densely arranged leaves, and *Densocarpa* (Pezizales), which forms abundant apothecia. Cf. **laxi-**.

dentate (L. *dentatus*, dentate, with teeth < *dens*, tooth + suf. *-atus* > E. *-ate*, provided with or likeness): **1**. Applied to solid structures with hymenophores represented by tooth-like protuberances, as in basidiocarps of *Auriscalpium vulgare*, *Hericium erinaceum*, *Sarcodon* (Aphyllophorales) and *Pseudohydnum gelatinosum* (Tremellales), or to the lamellate structures that have them on the border, resembling the teeth of a saw, but less pointed, as is seen in the gills of the basidiocarps of *Entoloma serrulatum* (=*Leptonia serrulata*), Agaricales. **2**. Used for the margin of certain colonies of fungi grown on solid media.

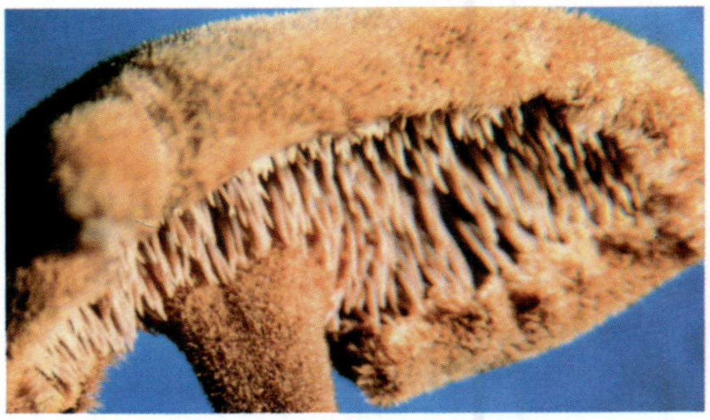

Dentate hymenophore of *Auriscalpium vulgare*, x 4 (*MU*).

Dentate hymenophore of *Hericium erinaceum*, x 1 (*RV*).

Dentate hymenophore of *Sarcodon* sp., x 6 (*MU*).

Dentate hymenophore of *Pseudohydnum gelatinosum*, x 1 (*RV*).

denticle (L. *denticulus*, small tooth < *dens*, genit. *dentis*, tooth + dim. suf. *-culus* > E. *-cle*): *Conidiogenesis*. A small projection similar to a tooth, on which is borne a spore Denticles are characteristic, e.g., of the conidiogenous cells known as *ampullae*, which form botryoblastospores, like those of *Botrytis cinerea*, *Oedocephalum glomerulosum*, *Spiniger meineckellus* and other moniliaceous asexual fungi, although they also are present in other species. See **rachiform** and **raduliform.**

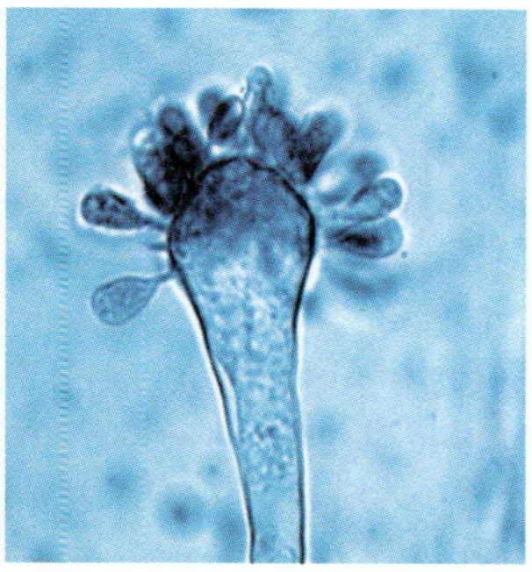

Denticles on a conidiogenous cell (ampulla) supporting conidia of *Spiniger meineckellus*, x 650 (*RTH*).

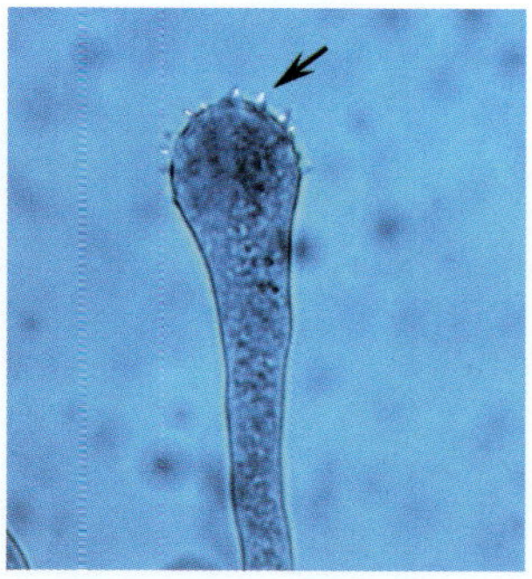

Denticles on a conidiogenous cell (ampulla) of *Spiniger meineckellus*, after releasing the conidia, x 650 (*RTH*).

denticulate (L. *denticulatus*, provided with small teeth < *dens*, genit. *dentis*, tooth + dim. suf. *-culus* > E. *-ate*, provided with or likeness): having small teeth.

denudate (L. *denudatus*, bare, naked < *des*, intensive + *nudus*, bare + suf. *-atus* > E. *-ate*, provided with or likeness): applied to structures that lack hairs, scales, spines, etc.; e.g., a denudate pileus, or a mature sporangium that has shed the peridium, like that of *Arcyria denudata* (Trichiales), in which the capillitium is exposed. The asci of the Taphrinales also are naked, since on maturing they are freely exposed on the tissues of the host plant, i.e., they are not enclosed in an ascocarp. See **naked**.

Denudate mature sporangia (now without the peridium) of *Arcyria denudata*, on dead wood, x 20 (*MU*).

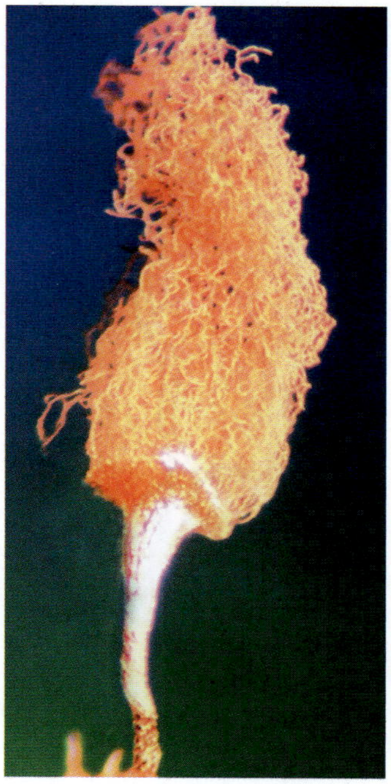

Denudate mature sporangium (now without the peridium) of *Arcyria denudata*, on dead wood, x 30 (*MU*).

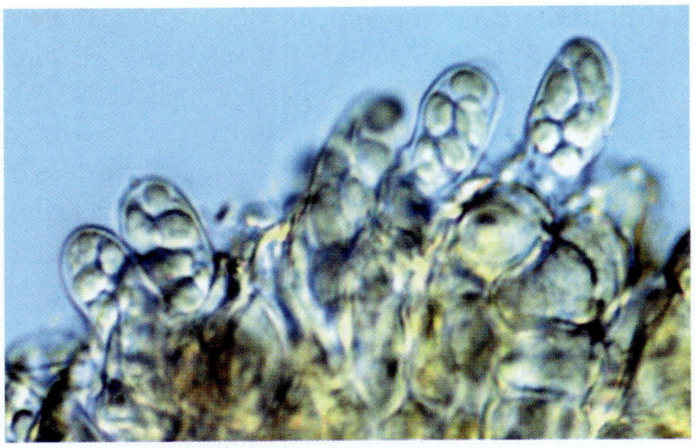

Denudate mature asci of *Taphrina robinsoniana*, on tissue of *Alnus serrulata*, x 20 (*RTH*).

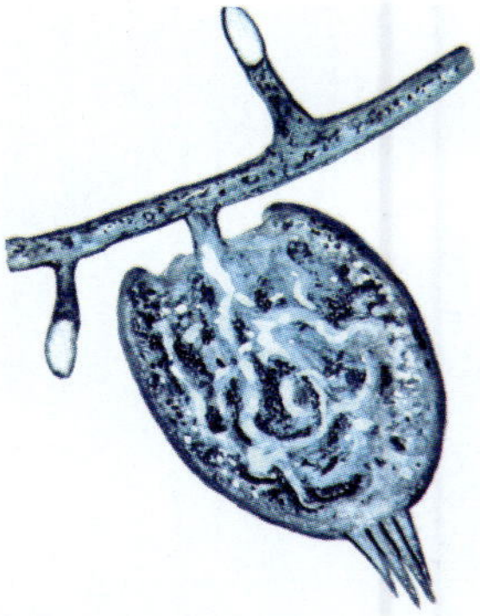

Depredation fungus, *Zoophagus insidians*, capturing a rotifer, x 180 (*MU*).

depauperate (ML. *depauperatus*, impoverished < L. *depauperare*, to become impoverished + suf. *-atus* > E. *-ate*, provided with or likeness): impoverished, in a metaphoric sense; poorly developed.

dependent (L. *dependens*, genit. *dependentis*, that which hangs): hanging.

deplanate (L. *deplanatus*, flattened, spread out): flattened, leveled.

depredation fungus, pl. **fungi** (LL. *depredatus*, pp. of *depraedari* < L. *de*, from, down, away + *praedari*, to plunder + L. *-ation* > E. *-ation*, action or process; L. *fungus*, fungus < Gr. *spóngos*, *sphóngos*, sponge): to engage in plunder, ravage; e.g., fungi that trap, kill, and feed on soil amoebae and rotifers, like *Sommerstorffia* (an oomycote), and *Zoophagus* and *Amoebophilus* (Zoopagales).

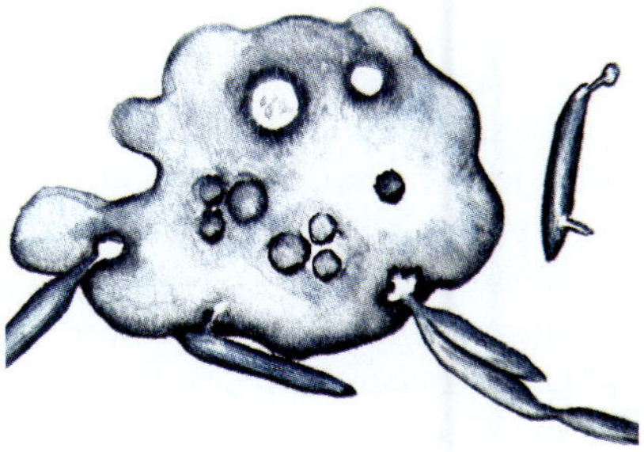

Depredation fungus, *Amoebophilus sicyosporus*, with the conidia adhered to an amoeba, x 920 (*MU*).

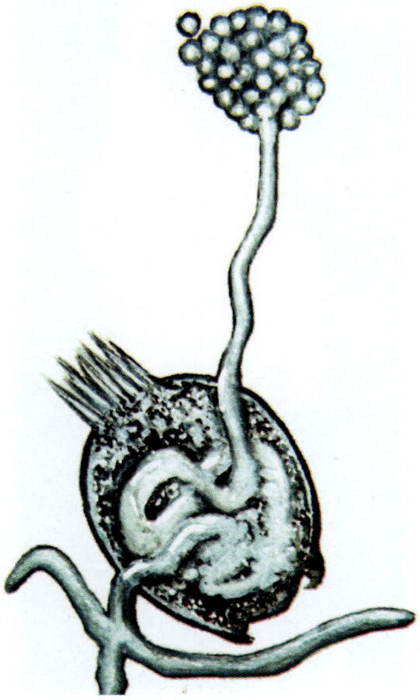

Depredation fungus, *Sommerstorffia spinosa* (an oomycote), feeding on a rotifer, x 170 (*MU*).

depressed (L. *depressus*, depressed, compressed < *deprimere*, to diminish the volume of a body by means of pressure): compressed, but in such a manner as if by pressure exerted in the direction of the axis of the organ; flattened, wider than high. *Mycology*. A pileus with a central depression or concavity, but without becoming infundibuliform or omphalloid; it can be greatly or slightly depressed, as is seen in basidiocarps of *Laccaria laccata* and *Russula rosea* (=*R. lepida*), Agaricales, respectively.

Depressed pileus of a basidiocarp of *Russula* sp., x 0.5 (*EPS*).

derived, derivate (L. *derivatus*, derived, coming from): applied to species that appear to be derived from an earlier species; also applied to a character that has changed from the form in which it is found in an ancestor. Cf. **primitive**.

dermato-, dermat- (Gr. *dérma*, genit. *dérmatos*, skin, leather): a pref. used to form compounds or structures that allude to the skin or cuticle, as in *dermatobasidium* and *dermatocystidium*.

dermatobasidium, pl. **dermatobasidia** (Gr. *dérma*, genit. *dérmatos*, skin, leather + *basídion* < *básis*, base + dim. suf. *-ídion* > L. *-idium*): *Basidiomycetes*. A cell with the shape of a basidium that is present in the cuticular layer of the basidiocarp, not in the hymenium. In the pileus of various species of *Russula* and in the stipe of other species of Agaricales (such as *Leccinum scabrum*) dermatobasidia are frequent.

dermatocystidium, pl. **dermatocystidia**, **dermatocyst** (Gr. *dérma*, genit. *dérmatos*, skin, leather + *kystídion* < *kýstis*, bladder; here, vessel or cell + dim. suf. *-ídion* > L. *-idium*): *Basidiomycetes*. A cystidium that is formed on the cuticular layer of a basidiocarp, not in the hymenium, such as occurs in *Pluteus*, *Panaeolus*, *Coprinus* and other gen. of Agaricales. Two types of dermatocystidia are known, *pileocystidia* (on the surface of the pileus) and *caulocystidia* (on the surface of the stipe).

dermatomycete (Gr. *dérma*, genit. *dérmatos*, skin, leather + *mýkes*, genit. *mýketos*, fungus): see **dermatophyte**.

dermatomycosis, pl. **dermatomycoses** (Gr. *dérma*, genit. *dérmatos*, skin, leather + *mýkes*, fungus + suf. *-osis*, condition or state of something; here, the state of the skin subjected to the action of the parasitic fungi): a broad gamut of cutaneous infections of man and higher animals caused by very diverse species of molds and yeasts from the soil; at times they are present as a result of a secondary invasion from a systemic mycosis. See **mycosis**.

dermatophyte (Gr. *dérma*, genit. *dérmatos*, skin, leather + *phytón*, plant): a member of the group of moniliaceous fungi that grow on the skin, hair and nails of man and other higher animals; these keratinolytic fungi are classified in three gen.: *Microsporum*, *Trichophyton* and *Epidermophyton*, and they cause a series of diseases known as *dermatophytoses*. The term dermatomycete is more accurate, considering that the fungi do not belong to the plant kingdom, although the word dermatophyte is of very general use. See **mycosis** and **tinea**.

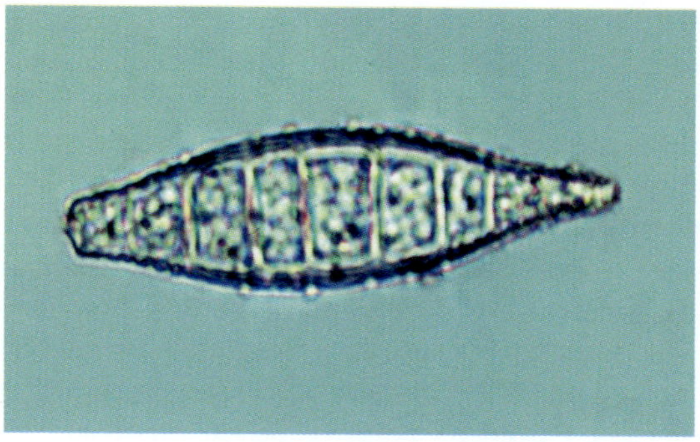

A macroconidium of *Microsporum canis*, a **dermatophyte** causing tinea capitis in man, x 920 (*MU*).

dermatophytid (*dermatophyte* < Gr. *dérma*, genit. *dérmatos*, skin, leather + *phytón*, plant + Gr. suf. *-id*, a patronymic ending that means son of, often utilized to indicate descent or relation): an allergic reaction expressed as pruritic, pustular or vesicular lesions of the skin (*id-reaction*) at a zone distant from the primary infection caused by a dermatophyte; it is interpreted as a reaction due to the presence of antigens produced by the pathogen in the circulatory system. See **mycosis**.

dermatophytosis, pl. **dermatophytoses**, **ringworm** (Gr. *dérma*, genit. *dérmatos*, skin, leather + *phytón*, plant + suf. *-osis*, state or condition; OE. > ME. *hring* > E. *ring*, a circular area + OE. > ME. *wyrm*, serpent, worm): a series of infectious diseases of the skin, hair and nails of man and other higher animals, caused by *dermatophytes* (gen. *Epidermophyton*, *Microsporum* and *Trichophyton*, of the moniliaceous asexual fungi), which produce lesions on the skin characterized by the formation of circular, colored (usually reddish) areas covered with vesicles and scales. See **dermatomycosis** and **tinea**.

dermo-, derm- (Gr. *dérma*, skin, leather): see **dermato-**.

descendant, descending (ME. *descenden* < OF. *descendre* < L. *descendere*, to go down, move from a higher to a lower place, to fall, to flow, to drain): refers to the ring of a basidiocarp that on maturing detaches from the outer edge of the pileus as well as from the stipe, so that it moves freely and falls to the base of the stipe, as in *Macrolepiota procera* (Agaricales). See **ascendent**.

desmato-, desmat- (Gr. *désma*, genit. *désmatos*, cord, rope, cable and anything that serves to bind): a prefixed form equivalent to **desmo-**, with the same meaning, employed to suggest the idea of something that is united or bound, as in *Desmaturus* (=*Lysurus*, of the Phallales).

desmo-, desm- (Gr. *désma*, cord, rope, cable and anything that serves to bind): see **desmato-**.

desquamate (L. *desquamatus* < *desquamare* < *de-, des-,* a pref. that denotes privation or elimination of something + *squama*, scale + suf. *-atus* > E. *-ate*, likeness): lacking scales. Equal to **alepidote**. Cf. **squamulose** and **lepidote**.

determinate (L. *determinatus*, limited, that ends < *determinare*, to end, terminate, finalize + suf. *-atus* > E. *-ate*, with the property of): used for a sporogenous structure that produces a propagule but which does not continue developing after the propagule has detached from its point of origin; there are, e.g., determinate sporangiophores and conidiophores that produce sporangia and conidia only once, such as the sporangiophores of *Plasmopara viticola*, *Peronospora destructor* and other species of the family Peronosporaceae (Peronosporales). Cf. **indeterminate**.

detersile (L. *detersilis*, genit. of *detersus*, clean < *detergere*, to clean + suf. *-ilis* > E. *-ile*, capable of, with the character of): refers to something that detaches, such as the pubescence of the pileus of certain agarics, which when mature are almost glabrous or bare due to the loss of their pubescence.

deutero-, **deuter-** (Gr. *déuteros*, second, secondary, inferior): a pref. employed in the formation of technical terms that allude to something secondary, supposedly of lesser importance, as in *deuteroconidium*.

deuteroconidium, pl. **deuteroconidia** (Gr. *déuteros*, second, secondary, inferior + *kónis*, dust, powder + dim. suf. *-ídion*): a sporoid cell that results from the division of a protoconidium in the dermatophytes.

deuterogamy (Gr. *déuteros*, second, secondary, inferior + *-gamia* < *gámos*, sexual union, reproduction): a general name for the ensemble of secondary processes of sexual reproduction, which in the fungi are exceptionally varied.

deuteromycetes (Gr. *déuteros*, second, secondary, inferior + L. *-mycetes*, ending of a taxonomic class < Gr. *mýkes*, genit. *mýketos*, fungus): conidial fungi that lack a means of sexual reproduction (although they manifest the phenomenon of *parasexuality*); they represent the asexual states of ascomycetes and basidiomycetes. Formerly considered as a single taxonomic class (Deuteromycetes), and as *Fungi Imperfecti*; in some classifications they were frequently separated into three classes, Blastomycetes, Hyphomycetes and Coelomycetes, in the subdivision Deuteromycotina. In this dictionary the deuteromycetes are considered mainly as asexual ascomycetes, together with other asexual fungi, some with basidiomycetous teleomorphs, but not as a valid and recognized taxonomic group of class rank. Some recent authors, however, no longer recognize a separate, formal taxonomic status for these fungi, since they are in fact the anamorphs of teleomorphic fungi, preferring instead to classify them in appropriate teleomorphic orders. Taking into consideration the fact that the deuteromycetes are a large, widespread group of important fungi, which for a long time have been classified as different families, orders and classes, mainly following Saccardoan criteria of practical value, in this dictionary the orders Sporobolomycetales and Cryptococcales of the Blastomycetes are referred to, respectively, as **sporobolomycetaceous** and **cryptococcaceous** asexual yeasts. The families of the order Moniliales of the Hyphomycetes, Agonomycetaceae, Moniliaceae, Dematiaceae, Stilbellaceae and Tuberculariaceae, are herein considered as **agonomycetaceous**, **moniliaceous**, **dematiaceous**, **stilbellaceous**, and **tuber-culariaceous** asexual fungi, respectively. The orders Melanconiales and Sphaeropsidales of past classifications are referred to as **melanconiaceous**, **sphaeropsidaceous** and **pycnothyriaceous** asexual fungi, respectively. See the Appendix to compare the classification changes of this class in Alexopoulos *et al.*, 1996, modified by Ulloa and Hanlin, 2006, to that of Kirk *et al.*, 2008. These asexual fungi are made up of heterogeneous species, forming unnatural groupings, and therefore are considered an artificial classification. The goal of these taxonomic rearrangements is to achieve a phylogenetic classification of the fungi. The deuteromycetes also are called **anaholomorphs**, **asexual fungi**, **conidial fungi** and **mitosporic fungi**.

deuto-, **deut-** (a pref. that is a contraction of *déuteros*): it has the same meaning as **deutero-**.

dextrinoid (L. *dexter*, right, referring to the rotation to the right of the plane of polarized light through crystals + suf. *-oide*, like < Gr. *-oeides*, similar to): applied to the color reaction of spores or cell walls that stain yellowish-brown or reddish-brown in Melzer's reagent (iodine, potassium iodide, chloral hydrate and water). This color reaction also is called *hemiamyloid*, or *pseudoamyloid*. For example, the basidiospores of *Macrolepiota procera* and the hyphae of *Mycena* (Agaricales) are dextrinoid or pseudoamyloid. Cf. **amyloid** and **anamyloid**.

dextro- (L. *dexter*, right hand): a pref. employed in terms such as *dextrorotatory*, which rotates the plane of polarized light toward the right, as does dextrose or d-glucose. Cf. **levo-**.

di- (L. *dis* < Gr. *dís*, two times, double): a pref. used in numerous terms of Gr. origin to give the idea of doubleness, as in *dikaryon* and *dispira*. Equivalent to the L. pref. *bi-*, employed when the word is derived from Latin.

dia-, di- (Gr. *diá*, through): a pref. used in the formation of compound terms of Gr. origin to introduce the idea of "through something" or of separating or dividing, as in *diageotropism*.

diageotropism (Gr. *diá*, through + *geo* < *gê*, earth, soil + *trópos*, a turn, change in position < *tropé*, gyration < *trépo*, to revolve, to gyrate, change in direction + suf. *-ismós* > L. *-ismus* > E. *-ism*, state, phase, tendency, action): a transverse geotropism in which the organ affected by the geotropic stimulus remains in a perpendicular position with respect to the stimulus, i.e., horizontal. The pilei of many basidiomycetes, including Agaricales and Aphyllophorales, are diageotropic, since on developing they expand in a direction parallel to the surface of the soil or substrate, whereas the stipe of the fruiting body exhibits negative geotropism. This phenomenon is also known as **plagiotropism**; the organ or organism affected is called *plagiotropic*. Cf. **orthogeotropism**. See **tropism**.

diagnosis, pl. **diagnoses** (Gr. *diágnosis*, action of discerning, of judging): an abbreviated description of a species, gen., etc., in which are indicated, in a succinct manner, the characteristics that distinguish it from its relatives, gen. from other related gen. According to Article 38 of the International Code of Botanical Nomenclature, the names of new taxa of plants (including fungi, but excluding the bacteria) are not considered validly published unless accompanied by a Latin diagnosis. In phytopathology and in medical and veterinary mycology, the diagnosis includes the symptoms of the disease caused by the pathogenic fungus and its specific identity. The characteristics of a particular taxon that are taken into account to reach an identification are known as *diagnostic*.

diaphanous (ML. *diaphanous* < Gr. *diaphanés*, transparent < *diá*, through + *phaíno*, to make appear, to show < *phanós*, light, lantern): transparent or almost so, like many of the hyaline structures of the fungi, e.g., the perithecia of species of *Melanospora* (Sordariales).

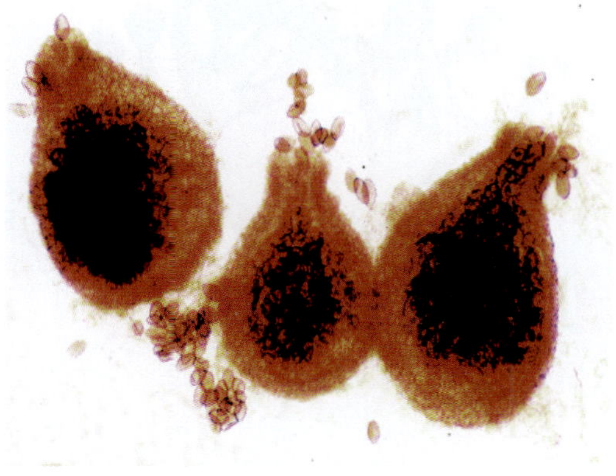

Diaphanous perithecia of *Melanospora* sp., x 350 (*MU*).

diaphragm (LL. *diaphragma* < Gr. *diáphragma*, separation, partition): *Gasteromycetes*. A membrane or homogeneous wall, composed of hyphae, that separates the gleba from the sterile base or subgleba in the fruiting body of species of *Calvatia*, *Lycoperdon*, *Vascellum* and other Lycoperdales.

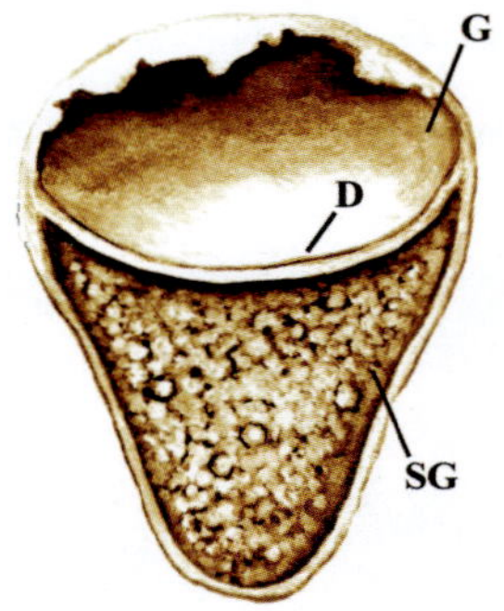

Diaphragm (D) of the fruiting body of *Vascellum pratense*, separating the gleba (G) from the subgleba (SG), seen in longitudinal section, x 1 (*MU*).

diaspore (Gr. *diasporá*, dissemination, dispersal): a propagule such as an isidium or a soredium, produced by a lichen thallus. In general, it is used for any propagule or other portion of a fungus that constitutes a dispersal phase, with the capacity to produce a new organism.

diastatic (Gr. *diastatikós*, relating to or having the properties of diastase < *diástasis*, separation + suf. *-íkos* > L. *-icus* > E. *-ic*, belonging to or relating to; an enzyme or ferment that converts starch into sugar): applied to any natural, biochemical, protein activator of a living cell (enzyme) that causes fermentation. See **zymogenous**.

diastole (LL. *diastole* < Gr. *diastolé*, expansion, dilatation): the dilatation movement of a contractile or pulsating vacuole, such as occurs in certain plasmodia and zoospores. Cf. **systole**.

diatrypoid (< gen. *Diatrype* < Gr. *diatrypáo*, to perforate + L. suf. *-oide*, like < Gr. *-oeídes*, similar to): *Ascomycetes*. A type of stroma composed of both fungal and host plant tissues (*pseudostroma*), with a poorly developed ectostroma that often bears the conidial state, and a well developed endostroma in which the perithecia are embedded. The perithecial necks extend straight up and project through the erumpent upper part of the stroma (the disc), as occurs in Pyrenomycetes of the gen. *Diatrype* (Xylariales), common on dry branches of many trees. Cf. **eutypoid** and **valsoid**.

Diatrypoid, erumpent perithecial stromata of *Diatrype disciformis*, x 25 (*CB*).

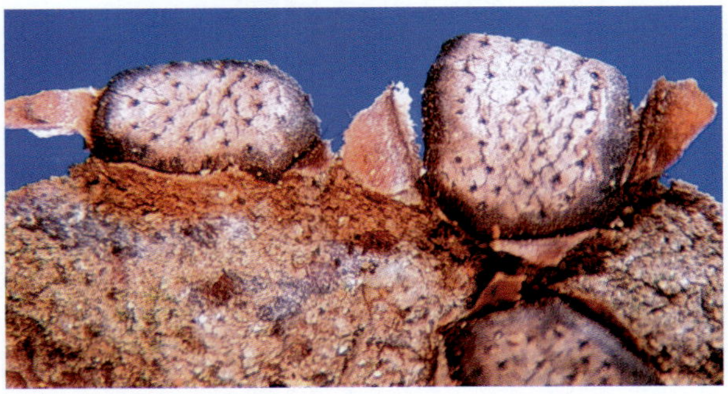

Diatrypoid, erumpent perithecial stromata of *Diatrype disciformis*, x 25 (*CB*).

Diatrypoid, erumpent perithecial stromata of *Diatrype disciformis*, x 25 (*CB*).

dicho- (Gr. *dícho* < *dícha*, in two, two times, distinct or differently): a pref. employed in various biological terms of Gr. origin to introduce the idea of something that has two distinct modes or manners, as in *dichophysis*.

dichohyphidium, pl. **dichohyphidia** (Gr. *dícho* < *dícha*, in two, two times, distinct or differently + NL. *hyphidium* < *hypha*, tissue, spiderweb, hypha + dim. suf. *-idium* > E. *-idium*): syn. of **dichophysis**.

dichophysis, pl. **dichophyses** (Gr. *dícho* < *dícha*, in two, two times, distinct or differently + *phýsis*, a growth): *Basidiomycetes*. A sterile cellular element or hyphoid, with dichotomous branching, thick, pigmented wall and narrow lumen that is present in the hymenium of the thelephoraceous Aphyllophorales, in species of Lachnocladidaceae, such as in *Vararia* sp. and *Vararia* (=*Asterostromella*) *investiens*; frequently the dichophyses have the appearance of antlers.

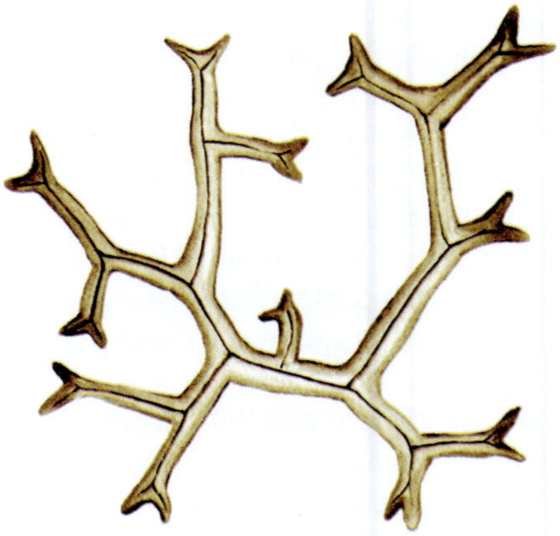

Dichophysis of the hymenium of *Vararia* sp., x 600 (*MU*).

dichotomous (LL. *dichotomos* < Gr. *dichótomos*, cut into two < *dícha*, in two parts + *tómos*, a cut): a type of branching in which an axis is divided into two equal parts, so that it produces a fork of approximately equal branches, at least at first, as is seen in the sporangiophore of *Dicranophora fulva* (Mucorales), *Piptocephalis cylindrosporus* (=*P. virginiana*, Zoopagales) and in the perithecial appendages of *Erysiphe* (section *Microsphaera*, Erysiphales) and *Chaetomium* (Sordariales). Cf. **trichotomous**, **tetrachotomous** and **polychotomous**. See **polytomous**.

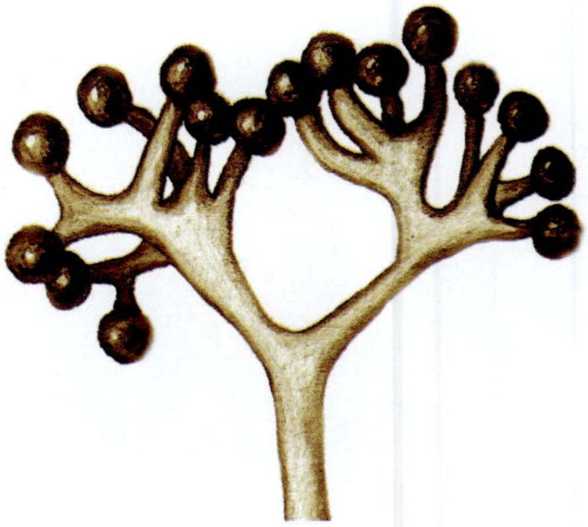

Dichotomous branching of the sporangiophore of *Dicranophora fulva*, x 165 (*MU*).

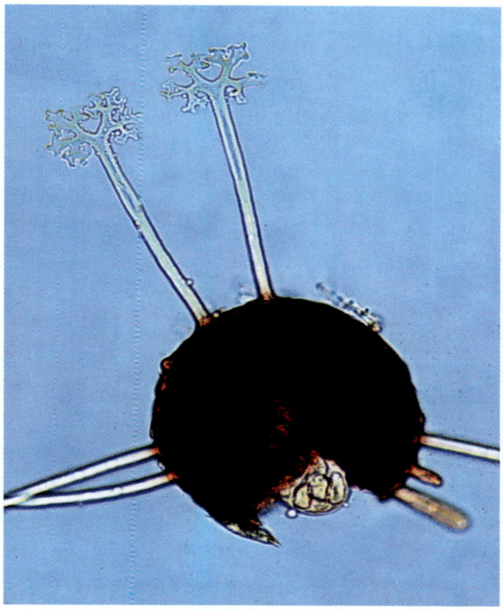

Dichotomous perithecial appendages of *Erysiphe* (section *Microsphaera*), x 100 (*MU*).

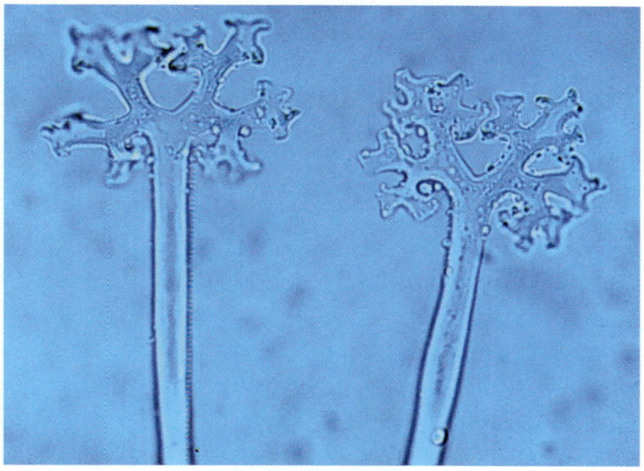

Dichotomous perithecial appendages of *Erysiphe* (section *Microsphaera*), x 260 (*MU*).

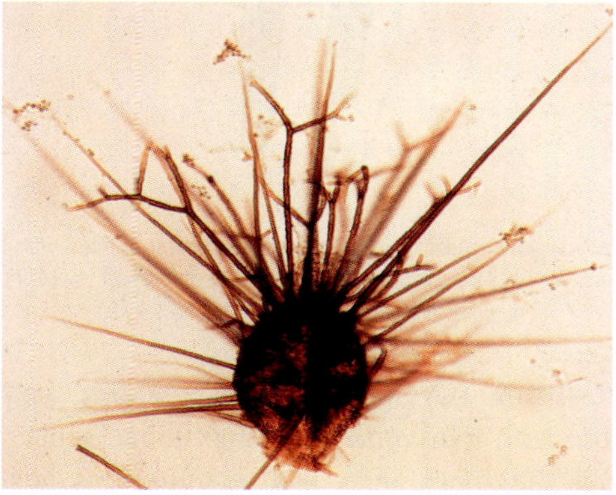

Dichotomous perithecial appendages of *Chaetomium funicolum*, x 300 (*MU*).

diclinous (Gr. *dís*, two + *klíne*, bed, bridal chamber; here, thallus): refers to fungi that have an antheridium on one hypha or branch and the oogonium on another, as happens in various aquatic species, e.g., *Achlya bisexualis* (Saprolegniales). Cf. **monoclinous**.

dictionary (L. *dictio*, genit. *dictionis*, diction, word; the manner of speaking or writing, considered as good or bad only through the correct or incorrect use of the words and constructions; manner of pronouncing + *-arius* > E. *-ary*, a suf. added to nouns to denote belonging to or relating to): a book that contains and explains all of the words of one or more languages, commonly in alphabetical order, or of a class of words, such as of a particular science or other subject. See **glossary**, **lexicon** and **vocabulary**.

dictydine (< gen. *Dictydium* < Gr. *díktyon*, net + L. dim. suf. *-idium* + L. suf. *-inus* > E. *-ine*, of or belonging to): microscopic structures, generally dark in color, separate, aggregated or arranged in rows, that comprise the so-called dictydine granules or plasmodial granules that are found on the surface of the peridium or peridial net, and also frequently on the spores of the cribrariaceous Myxomycetes, such as *Cribraria purpurea* (=*Dictydium*, of the Liceales).

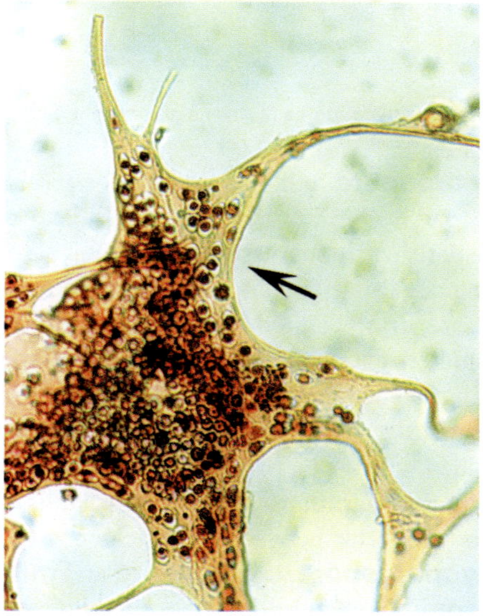

Dictydine (or plasmodial) granules on the peridial net of the sporangium of *Cribraria purpurea*, x 300 (*MU*).

dictyo- (Gr. *díktyon*, net): a pref. used to give the idea of something that forms a reticulum, as in *dictyochlamydospore*.

dictyochlamydospore (Gr. *díktyon*, net + *chlamýs*, genit. *chlamýdos*, cloak, mantle + *sporá*, spore): a dictyosporous chlamydospore, i.e., with transverse and longitudinal septa, resembling a net, like the chlamydospores formed by *Pochonia chlamydosporia* (=*Verticillium chlamydosporum*), moniliaceous asexual fungi, and *Phoma glomerata* (sphaeropsidaceous asexual fungi). Also called alternarioid chlamydospore due to its resemblance to the conidia of species of *Alternaria* (dematiaceous asexual fungi). A distinctive characteristic of this multicellular chlamydospore is that the outer wall layer is separable from the other layers of the wall.

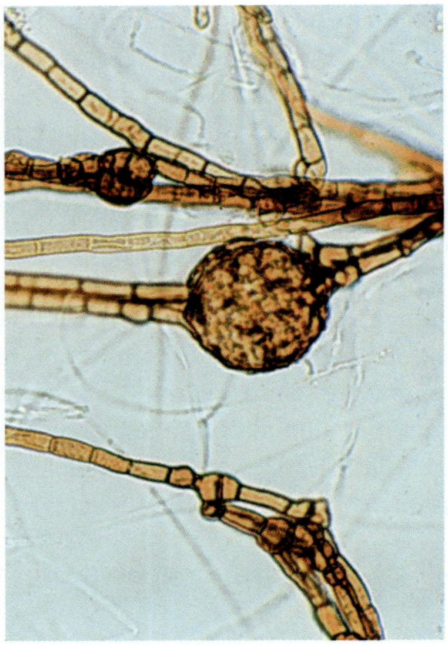

Dictyochlamydospores of *Phoma glomerata*, x 140 (*CB*).

dictyoporospore (Gr. *díktyon*, net + *póros*, pore, way, road + *sporá*, spore): a multicellular, deciduous porospore, with transverse and longitudinal septa (and sometimes also with oblique septa), whose component cells are firmly united and are not surrounded by a common external wall, as is seen in species of *Alternaria* (dematiaceous asexual fungi).

dictyoseptate (Gr. *díktyon*, net + L. *septatus*, provided with septa, partioned): with transverse and longitudinal partitions (septa or cross walls), like the layers of mortar between the bricks of a wall. See **dictyoporospore**, **dictyospore** and **muriform**.

dictyosome (Gr. *díktyon*, net + *sõma*, body): each of the bodies that form part of the *Golgi apparatus*, more or less globose and comprised of a series of cisternae or cavities. Dictyosomes are complexes of lipids and proteins and are found in young cells involved in the elaboration of cytoplasm and secretions. During the apical growth of a hypha the precursor materials for the synthesis of the cell wall are transferred in small vesicles from the endoplasmic reticulum to the dictyosomes, where the vesicles coalesce to form a cisterna. From the cisternae of the dictyosomes the materials and membranes migrate as secretory vesicles toward the hyphal apex, where they fuse with the plasma membrane and liberate their contents to contribute to the formation of the cell wall.

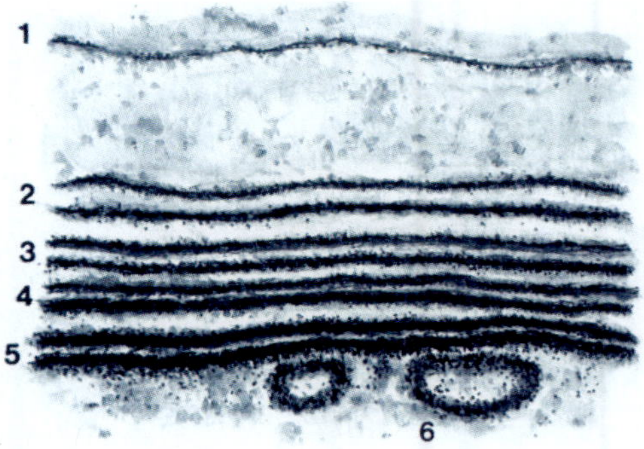

Dictyosomes of the Golgi apparatus of a hyphal cell of *Pythium ultimum* (Peronosporales). Transverse section of a Golgi body showing the way in which the structure of the membrane changes from that of the nuclear envelope (1), which is comparable to that of the endoplasmic reticulum, passing through the cisternae of the Golgi body (2-5), and finally to that of the Golgi vesicle (6). Drawing based on a transmission electron micrograph published by Beckett *et al.* in their *Atlas of Fungal Ultrastructure*, 1974, x 150,000 (*MU*).

dictyosporangium, pl. **dictyosporangia** (Gr. *díktyon*, net + *sporá*, spore + *angeîon*, vessel, receptacle): *Saprolegniales*. A term applied to the zoosporangium of *Dictyuchus*, so named because the cell walls of the encysted zoospores remain inside the zoosporangium in the form of a polyhedric net after the zoospores have been liberated.

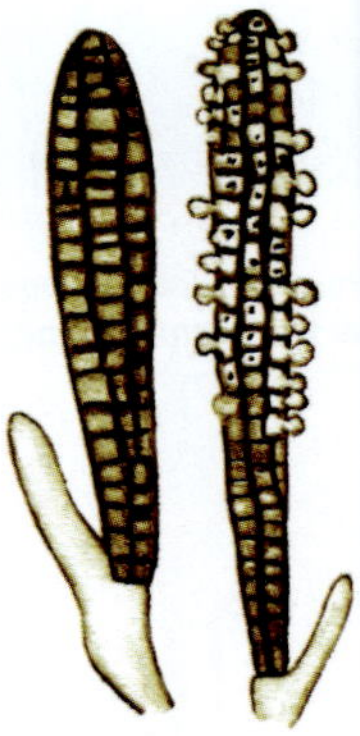

Dictyosporangia of *Dictyuchus monosporus*, the left with the zoospores not liberated, the right in the processs of liberating the zoospores, x 200 (*MU*).

dictyospore (Gr. *díktyon*, net + *sporá*, spore): a spore with transverse and longitudinal septa, which gives the appearance of a reticulum, as is seen in *Alternaria alternata* (dematiaceous asexual fungi). See **dictyoporospore** and **muriform**.

Dictyospores of *Alternaria alternata*, x 390 (*MU*).

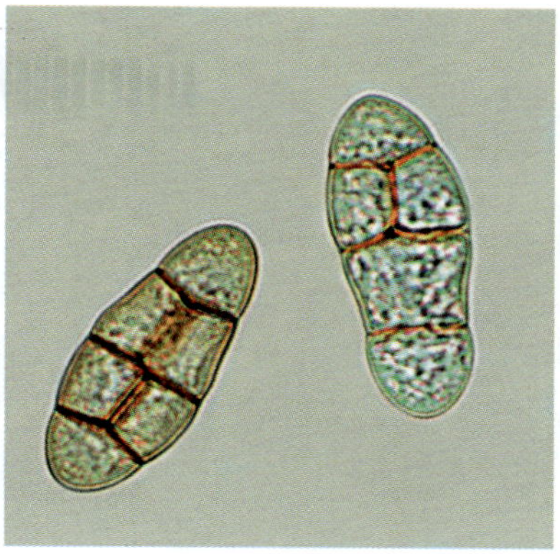

Dictyosporic ascospores of *Pyrenophora* sp., x 410 (*CB*).

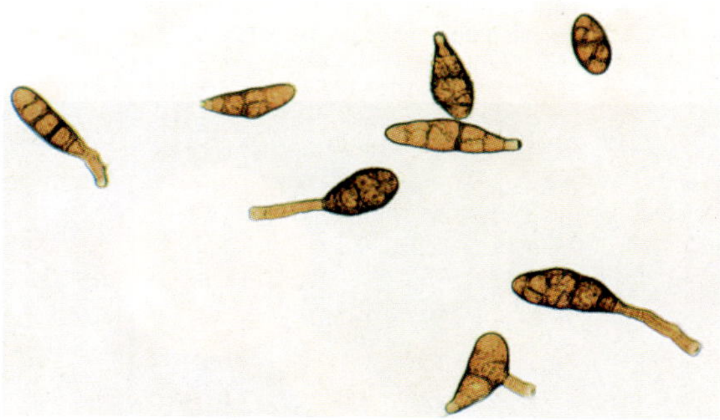

Dictyospores of *Alternaria alternata*, x 390 (*MU*).

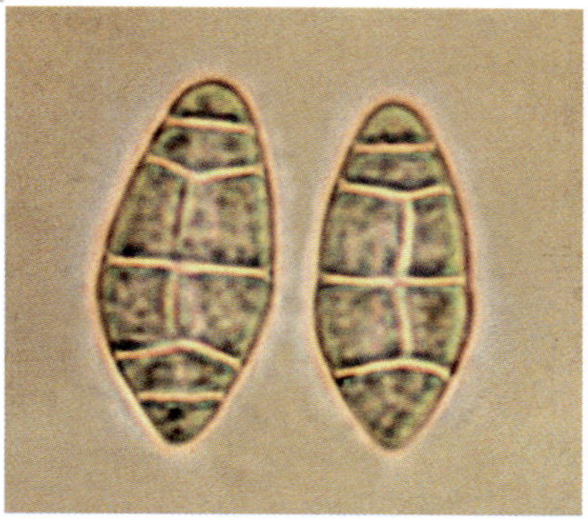

Dictyosporic ascospores
of *Leptosphaerulina crassiasca*, x 660 (*RTH*).

dictyosporic (Gr. *díktyon*, net + *sporá*, spore + suf. *-íkos* > L. *-icus* > E. *-ic*, belonging to or relating to): applied to fungi that have spores with transverse and longitudinal septa, i.e., *dictyospores*, like the conidia of *Alternaria*, *Stemphylium* and other gen. of dematiaceous asexual fungi, and the ascospores of *Hysterographium* (Hysteriales), *Pyrenophora* (Pleosporales) and *Leptosphaerulina crassiasca* (Dothideales).

didymospore (NL. *didymospora* < Gr. *dídymos*, twin + *sporá*, spore): a bicellular spore (divided by a septum); a Saccardoan term that is applied mainly to conidial fungi, as, e.g., in the conidia of *Arthrobotrys oligospora*, *Trichothecium roseum* (moniliaceous asexual fungi), and *Lasiodiplodia theobromae* (sphaeropsidaceous asexual fungi), although it is also applied to a bicellular ascospore, like that of *Venturia liriodendri* (Dothideales).

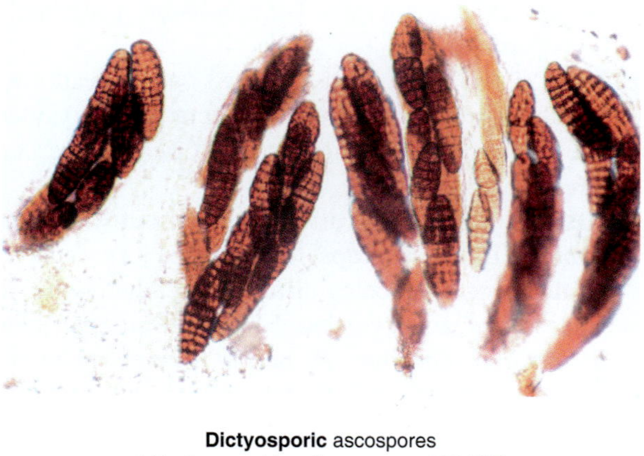

Dictyosporic ascospores
of *Hysterographium flexuosum*, x 660 (*CB*).

Didymospores (pycnidiospores) of *Lasiodiplodia theobromae*, x 320 (*CB*).

Didymospores (pycnidiospores) of *Lasiodiplodia theobromae*, x 650 (*MCG*).

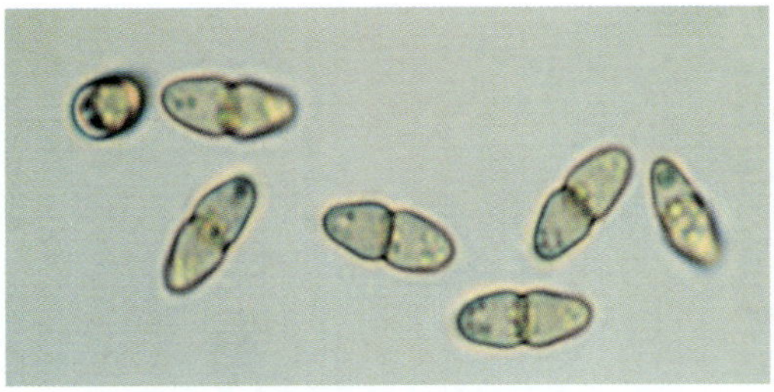

Didymospores (ascospores) of *Venturia liriodendri*, produced on *Liriodendron*, x 320 (*RTH*).

didymous (Gr. *dídymos*, twin, double): applied to structures that are formed in pairs; double.

diel (L. *dialis*, one day): refering or relating to one day; diurnal, with a periodicity of 24 hours. See **circadian rhythm**.

diffluent (L. *diffluens*, genit. *diffluentis*, dissoluble, soluble < *diffluere*, to dissolve, to scatter, to extend + suf. *-entem* > E. *-ent*, a being): something that dissolves or degrades easily in the presence of water, like the peridium of the sporangia of *Mucor hiemalis*, *Rhizomucor pusillus* (=*M. pusillus*) and other species of *Mucor*, and other Mucorales, such as *Phycomyces blakesleanus*.

Diffluent sporangia of *Phycomyces blakesleanus*, isolated from mouse dung, x 320 (*MU*).

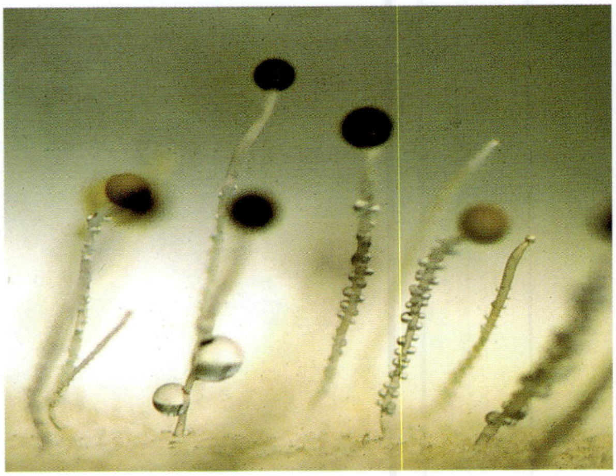

Diffluent sporangia of *Phycomyces blakesleanus*, isolated from mouse dung, x 320 (*MU*).

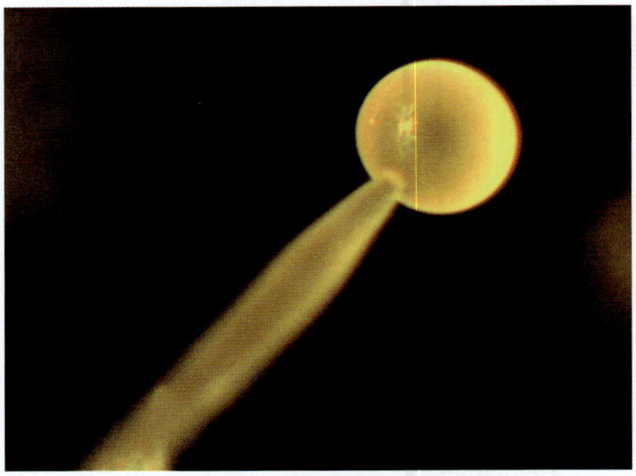

Diffluent sporangium of *Phycomyces blakesleanus*, isolated from mouse dung, x 750 (*MU*).

diffracted (L. *diffractus*, broken up, separated < *diffringere*, to break, divide into parts, break into small pieces): refers to the surface of a fruiting body that is fragmented into areolas, like the exoperidium of *Handkea utriformis* (=*Lycoperdon caelatum*), Lycoperdales.

diffuse (L. *diffusus*, diffuse, spread out): spread out, scattered, without a definite border or boundary, like the thallus of crustose lichens, e.g., *Caloplaca cerina* (Teloschistales).

diffusible (L. *diffusus*, diffuse, spread out + *-ible* > *-able* < L. *-abilis*, capable or worthy of): capable of diffusing or of being diffused, e.g., the pigments and organic acids produced by fungi on agar, like *Penicillium citrinum*, *P. purpurogenum* (moniliaceous asexual fungi), *Brettanomyces* spp. (cryptococcaceous asexual yeasts), and *Saccharomyces exiguus* (Saccharomycetales).

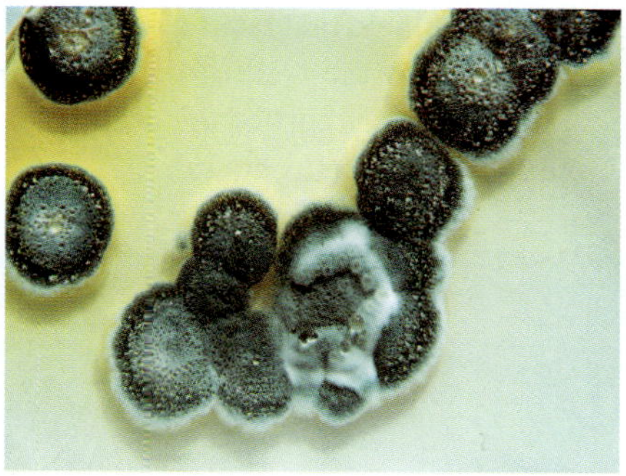

Diffusible yellow pigment of *Penicillium citrinum* on agar, x 1 (*MU*).

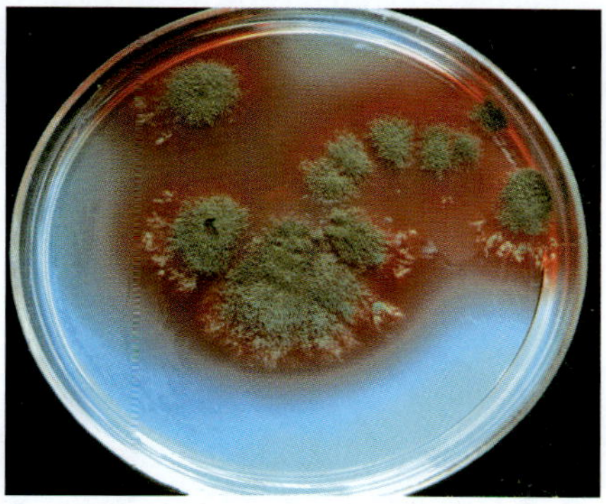

Diffusible red pigment of *Penicillium purpurogenum* on agar, x 1 (*MU*).

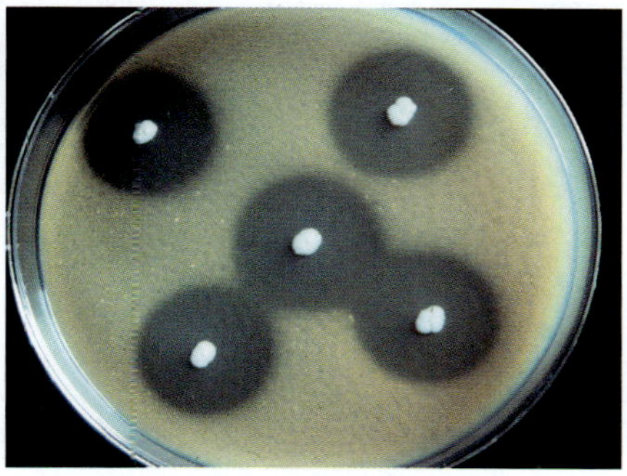

Diffusible acetic acid of *Brettanomyces* sp. on agar, x 1 (*MU*).

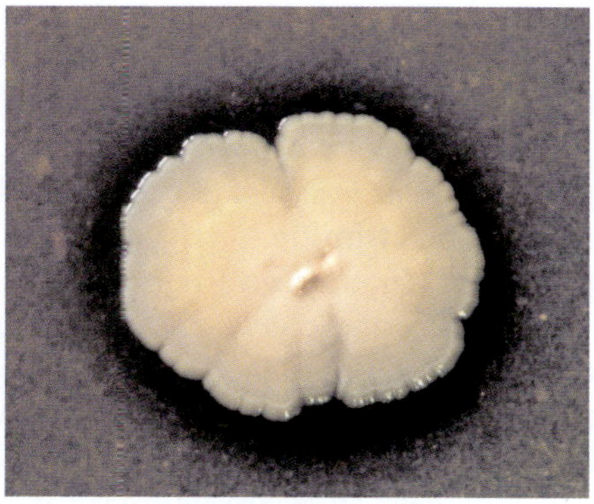

Diffusible organic acid of *Saccharomyces exiguus* on agar, x 1 (*MU*).

digitate, **digitiform** (L. *digitatus*, having fingers < L. *digit*, finger + suf. *-atus* > E. *-ate*, provided with or likeness; or + L. suf. *-formis* < *forma*, shape): provided with apical, finger-like protuberances, like the macroconidia of *Histoplasma capsulatum* (moniliaceous asexual fungi), the haustoria of *Blumeria* (=*Erysiphe*) *graminis* (Erysiphales) and the cystidia of *Mycena epipterygia* and *Agrocybe arvalis* (Agaricales). Also called **dactyloid**.

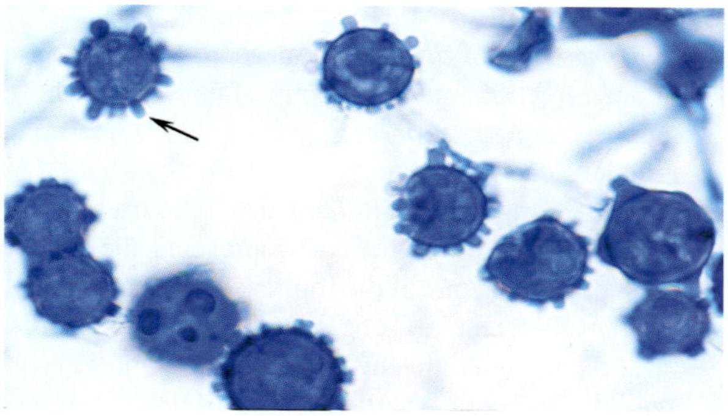

Digitate macroconidia of *Histoplasma capsulatum*, with projections on the wall, x 2,000 (*CB*).

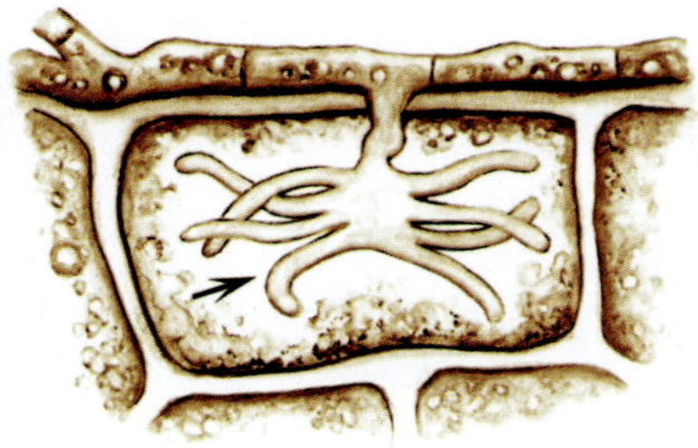

Digitate haustorium of *Blumeria* (=*Erysiphe*) *graminis*, x 2,000 (*MU*).

digiti- (L. *digiti-*): a pref. that implies the idea of something that resembles the fingers of an open hand, as in *digitate* and *digitiform*.

dikaryon, **dicaryon** (Gr. *dís*, two + *káryon*, nucleus): see **dikaryotic**. Cf. **monokaryon**.

dikaryotic, **dicaryotic**, **dikaryontic**, **dicaryontic** (Gr. *dís*, two + *káryon*, nucleus + suf. *-íkos* > L. *-icus* > E. *-ic*, belonging to or relating to): containing a *dikaryon* (n + n), i.e., two nuclei capable of complementing each other genetically; some examples of dikaryotic structures are the ascogenous hyphae of the Plectomycetes (such as *Eurotium rubrum*, of the Eurotiales) and Pyrenomycetes (such as *Neurospora crassa*, of the Sordariales), the young teliospores and the infective mycelium of *Ustilago maydis* (Ustilaginales), and the secondary mycelium and young basidia of *Coprinus comatus* (Agaricales). Cf. **monokaryotic**.

dikaryotization (Gr. *dís*, two + *káryon*, nucleus + suf. *-ationem* > E. *-ation*, action, state or condition, result): the conversion of two homokaryons into a dikaryon by means of the fusion or plasmogamy of two compatible homokaryons, without karyogamy. See **Buller phenomenon**.

dilacerate (L. *dilaceratus*, *laceratus*, torn, cut to pieces < *dilacerare*, *lacerare*, to tear, to cut up + suf. *-atus* > E. *-ate*, provided with or likeness): cut up, divided into pieces; divided unequally and deeply.

dimension (ME. < MF. < L. *dimension*, *dimensio* < *dimensus*, pp. of *dimetri*, to measure out): measure in one direction; one of a group of properties whose number is necessary and sufficient to determine uniquely each element of a system; the quality of spatial extension; magnitude, size. The drawing of micromycetes (hypothetical) represents the bodily form or proportions of the immense microcosmos of fungi. Robert Hooke (1665) was the first microscopist to see and illustrate the microscopic sporangia of *Mucor* (Mucorales) and the telia of *Phragmidium mucronatum* (Uredinales).

An hypothetical drawing of micromycetes, a **dimension** of the microcosmos of fungi, x 0.1 (*MU*). This image is the cover of the *Illustrated Dictionary of Mycology* (2000).

dimerous (Gr. *dís*, two + *méros*, part): composed of two parts; generally applied to basidia that have a constriction between the probasidium (lower part) and the metabasidium (upper part), as in species of *Brachybasidium* (Exobasidiales).

dimidiate (L. *dimidiatus*, divided in half < *dimidiare*, to divide + suf. *-atus* > E. *-ate*, provided with or likeness): **1**. *Basidiomycetes*. A semicircular basidiocarp, with one side more developed than the other, as is seen in *Ganoderma applanatum*, *Heterobasidion annosum*, *Gloeophyllum saepiarium* (=*Lenzites saepiaria*) and other Aphyllophorales. **2**. *Ascomycetes*. An ascocarp in the shape of a shield that lacks a basal wall; in some lichens, it refers to the wall of the ascostroma that covers only the upper part of the ascoma, as occurs in *Parmularia hankei* (Lecanorales).

Dimidiate basidiocarp of *Gloeophyllum saepiarium* (=*Lenzites saepiaria*), x 0.5 (*MU*).

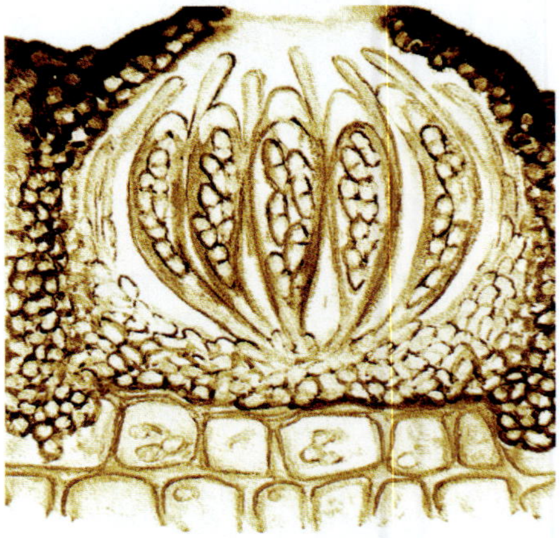

Dimidiate ascostroma of *Parmularia hankei* in longitudinal section, erumpent from a hypostroma, in the tissues of a leaf of *Styrax* sp.; visible are the clavate, bitunicate asci with bicellular ascospores, and the paraphyses, x 350 (*MU*).

dimitic (Gr. *dís*, two + *mítos*, filament; here, hypha + suf. *-íkos* > L. *-icus* > E. *-ic*, belonging to or relating to): *Basidiomycetes*. A fructification with two types of hyphae, generative or fertile (producing basidia) and either skeletal or binding hyphae, as in *Lentinus tigrinus* and *Ramaria stricta* (Aphyllophorales). Cf. **monomitic** and **trimitic**.

dimitic system (Gr. *dís*, two + *mítos*, filament; here, hypha + suf. *-íkos* > L. *-icus* > E. *-ic*, pertaining to or relating to; LL. *systemat-*, *systema* < Gr. *sýstema*, composed of various parts): *Gasteromycetes*. Used to describe the development of true capillitium and of paracapillitium inside the same gleba of a fruiting body.

dimorphic (Gr. *dís*, two + *morphé*, shape + suf. *-íkos* > L. *-icus* > E. *-ic*, belonging to or relating to): **1**. An aquatic oomycote that has two morphological types of zoospores (primary or pyriform, and secondary or reniform) in the life cycle, as occurs e.g., in species of *Saprolegnia* (Saprolegniales). Cf. **monomorphic**. **2**. A fungus that is capable of growing in mycelial or yeast-like form, depending upon the conditions it encounters. The term is usually applied to pathogenic fungi that form mycelium in the host and a yeast-like colony in culture, as is seen in the plant pathogens *Taphrina* (Taphrinales) and *Ustilago* (Ustilaginales). In *Histoplasma capsulatum* (moniliaceous asexual fungi), a pathogen of man and animals (the causal agent of histoplasmosis), the yeast state is formed in the host, and the mycelium grows in laboratory culture as well as in the substrate where it normally grows as a saprobe.

Colony on agar of the mycelial phase of *Histoplasma capsulatum*, a **dimorphic** fungal pathogen of man, x 1 (*MU*).

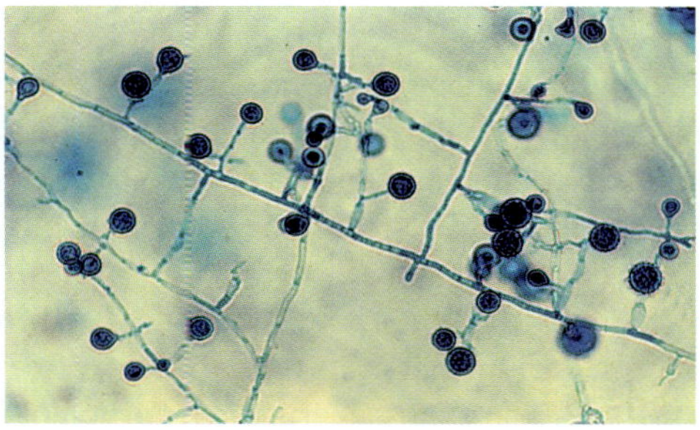

Conidia of the mycelial phase of *Histoplasma capsulatum*, a **dimorphic** fungus cultured on agar, x 1,000 (*MU*).

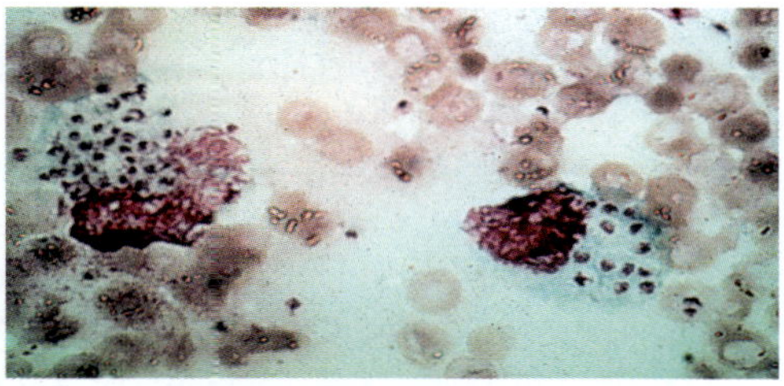

The parasitic yeast phase of *Histoplasma capsulatum*, a **dimorphic** fungus, inside leucocytes of a human lymphatic ganglion, x 750 (*RLM*).

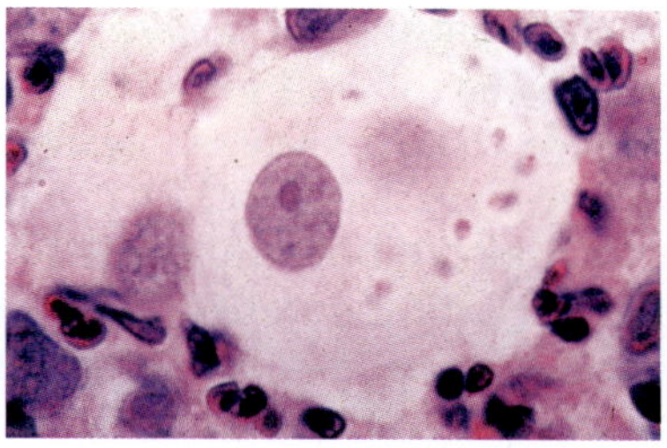

The parasitic yeast phase of *Histoplasma capsulatum*, a **dimorphic** fungus, inside leucocytes of a human lymphatic ganglion, x 1,000 (*RLM*).

dioecious (Gr. *dís*, two + *oikía*, house or inn; here, thallus): an organism in which a thallus produces either male or female sexual organs, but not both; i.e., with sexes segregated on different individuals. Also called **unisexual**. Cf. **monoecious** and **hermaphrodite**.

diorchidioid (Gr. *dís*, two + *órchis*, testicle + L. suf. *-oide*, like < Gr. *-oeídes*, similar to): a bicellular spore with the septum arranged vertically or, in large part, diagonally, like the teliospores of *Diorchidium spinulosum* (Uredinales).

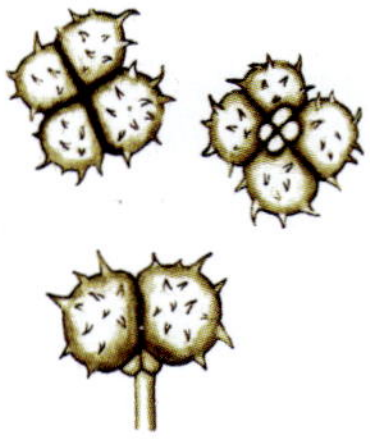

Diorchidioid teliospores of *Diorchidium spinulosum*, x 250 (*MU*).

diphycophilous (Gr. *dís*, two + *phýkos*, alga + *phílos*, have an an affinity for): *Lichens*. A lichen thallus with two types of photobionts, one chlorophyceous and the other cyanophyceous (now considered as cyanobacteria). In *Peltigera aphthosa* (Peltigerales), e.g., the associated green alga (*Coccomyxa*) is found throughout the thallus, whereas the cyanophyceous *Nostoc* is confined to the cephalodia. Cf. **chlorophycophilous** and **cyanophycophilous**.

diplanetic (Gr. *dís*, two + *planétes*, roving, wandering + suf. *-íkos* > L. *-icus* > E. *-ic*, belonging to or relating to): *Oomycota*. A species in whose life cycle there are two morphological types of zoospores (dimorphic), primary and secondary, and two periods of swimming (diplanetic), one for each type of zoospore, as occurs in some species of *Saprolegnia* (Saprolegniales). Some monomorphic species, i.e., with only one type of zoospore, also can be diplanetic. Cf. **monoplanetic** and **polyplanetic**.

diplobiont, **diplobiontic** (Gr. *diplóos*, double + *bíos*, life + *óntos*, genit. of *ón*, a being): an organism in whose life cycle both the haploid and diploid phases occur as free-living individuals. Cf. **haplobiont**. Distinguish from **diplont** and **haplont**.

diploconidium, pl. **diploconidia** (Gr. *diplóos*, double + *kónis*, dust, powder + dim. suf. *-ídion* > L. *-idium*): *Basidiomycetes*. A binucleate conidium that is formed in the fruiting bodies of some jelly fungi (Tremellales). Cf. **haploconidium**.

diplodiosis, pl. **diplodioses** (< gen. *Diplodia*, of the sphaeropsidaceous asexual fungi, + Gr. suf. *-osis*, state or condition): a paretic neuromuscular syndrome of bovine and ovine cattle caused by ingestion of maize infected by *Diplodia maydis*, syn. of *Stenocarpella maydis* (sphaeropsidaceous asexual fungi).

diploid (Gr. *diplóos*, double + L. suf. *-oide*, like < Gr. *-oeídes*, similar to): having a double set of chromosomes (2n). Cf. **haploid**.

diploidization (Gr. *diplóos*, double + L. suf. *-oide*, like < Gr. *-oeídes*, similar to + Gr. suf. *-ationem* > E. *-ation*, action, state or condition, result): the action and effect of diploidizing or being diploidized. In mycology it refers to the process by which a haploid cell or mycelium is converted to diploid (dikaryotic), with conjugated nuclei.

diplokaryon (Gr. *diplóos*, double + *káryon*, nucleus): a cell nucleus with 2n chromosomes. Cf. **haplokaryon**.

diplomitotic (Gr. *diplóos*, double + *mitosis* < Gr. *mítos*, filament + suf. *-osis*, condition or state + suf. *-tíkos* > L. *-ticus* > E. *-tic*, relation, adaption, inclination or capacity): refers to a nuclear cycle that has two mitotic phases of different ploidy: karyogamy - mitosis - meiosis - mitosis - karyogamy.

diplont (Gr. *diplóos*, double + *óntos*, genit. *ón*, of a being): an organism whose existence is spent all, or mostly, in the diploid phase (diplophase), as happens in the Oomycota and in some yeasts, e.g., *Saccharomycodes* (Saccharomycetales). Cf. **haplont**. Distinguish from **diplobiont** and **haplobiont**.

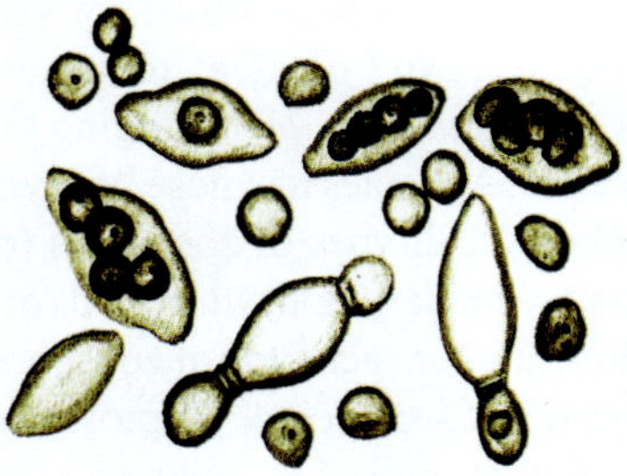

Diplont cells of the yeast *Saccharomycodes ludwigii*, x 1,000 (*MU*).

diplophase (Gr. *diplóos*, double + *phásis*, phase, appearance, aspect < *phaíno*, to shine): the part of the life cycle during which the cells are diploid. Cf. **haplophase**.

diplospory (Gr. *diplóos*, double + *sporá*, spore + des. *-ia* > E. *-ry*, which denotes quality or state of being): *Oomycota*. A term used to define the condition in which a diploid nucleus is incorporated into the cytoplasm of a coenocytic gametangium, which is influenced by adjacent meioses, and which subsequently gives rise to the oospores.

diplostichous (Gr. *diplóos*, double + *stichós*, line, thread): syn. of **distichous**.

diplostromatic (Gr. *diplóos*, double + *strõma*, cushion, small pillow + suf. *-íkos* > L. *-icus* > E. *-ic*, belonging to or relating to): *Ascomycetes*. Having both ectostroma and endostroma, as in certain Xylariales and Diatrypales. Cf. **haplostromatic**. See **placodium**.

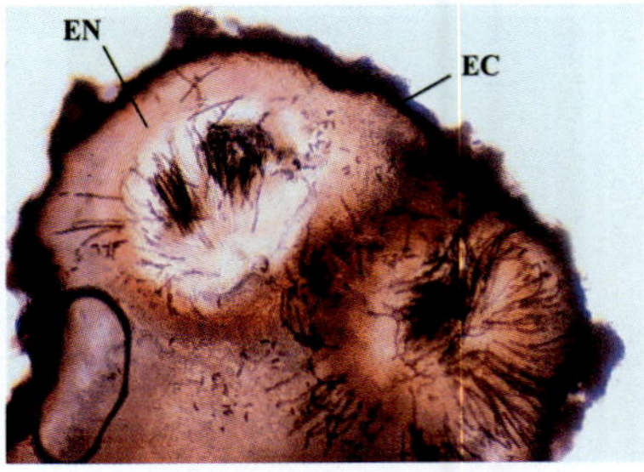

Xylaria sp., a **diplostromatic** pyrenomycete, with ectostroma (EC) and endostroma (EN), x 250 (*MU*).

dis-, **di-** (L. *dis-* and *di-*): a pref. that denotes negation, difficulty, or opposition, as in *dislocate*, something outside its normal place, and *displasia*, *displasy*, alteration in the creative or formative activity of an organism, which determines the production of poorly constituted organs or parts.

disc (Gr. *dískos*, disc): **1**. The curved or flattened portion of an apothecium that bears the hymenium. See **disk fungus**. **2**. The exposed, erumpent portion of a diatrypoid stroma. See also **delimited peristome**.

disciform (L. *disciformis*, disc-shaped < pref. *disci-* < *discus*, disc + suf. *-formis* < *forma*, shape): syn. of **discoid**.

discocarp (Gr. *dískos*, disc + *karpós*, fruit): see **apothecium**.

discoid (Gr. *dískos*, disc + L. suf. *-oide*, like < Gr. *-oeídes*, similar to): similar to a disc, flat and circular, like the sporangia of the Myxomycete *Diderma hemisphaericum* (Physarales), the cleistothecial ascomata of *Amazonia* (Meliolales), and the apothecia of many discomycetes, e.g., species of *Poronia* (Xylariales), *Phillipsia*, *Sarcoscypha* (Pezizales) and *Lecanora* (Lecanorales). See **lecanorine**.

Discoid head of the sporangia of *Diderma haemisphaericum*, x 11 (*MU*).

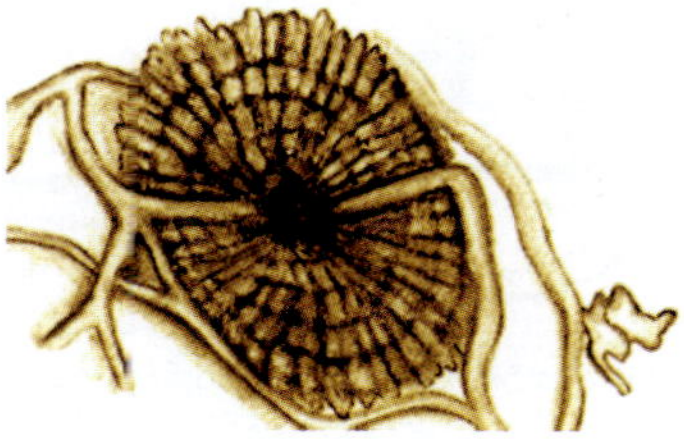

Discoid ascoma of *Amazonia* sp., x 500 (*MU*).

Discoid ascomata of *Poronia punctata*, x 8 (*MU*).

Discoid ascomata of *Phillipsia* sp., x 1 (*TH*) and *Sarcoscypha coccinea*, x 1 (*RV*).

discolichens (L. *discus* < Gr. *dískos*, disc + L. *lichen* < Gr. *leichén*, lichen): a subgroup of ascolichens that is characterized by apothecial fruiting bodies, with a structure analogous to that of the Discomycetes. The great majority of lichen species are discolichens, with discoid apothecia.

discolorous (L. *discolorus*, of different colors < *dis*-, a pref. that denotes negation or opposition + *color*, color): of two or several colors, mottled, variegated, like the two surfaces of a foliose lichen thallus. Cf. **concolorous**.

Discomycetes (Gr. *dískos*, disc, tray + L. *-mycetes*, ending of taxonomic class < Gr. *mýkes*, genit. *mýketos*, fungus): a class of ascomycetes that is characterized by an apothecial ascocarp in which the asci are formed in an hymenium that is exposed at maturity. In some classifications they are considered as a subclass (Discomycetidae) of the Euascomycetes. In this dictionary the Discomycetes includes the orders Medeolariales, Rhytismatales, Ostropales, Cyttariales, Helotiales, Neolectales, Gyalectales, Lecanorales, Lichinales, Peltigerales, Pertusariales, Teloschistales, Caliciales and Pezizales. Together with the ascosporogenous yeasts, the plectomycetous (Plectomycetes), pyrenomycetous (Pyrenomycetes) and loculoascomycetous (Loculoascomycetes) forms, they constitute the phylum Ascomycota of the kingdom Fungi. In the Appendix, the user of this dictionary can see the comparison between the outline of Alexopoulos *et al.*, 1996, modified by Ulloa and Hanlin, 2006, to that of Kirk *et al.*, 2008. The class Discomycetes has been transferred to the subphylum Pezizomycotina, phylum Ascomycota. The remaining orders of ascomycetes have been transferred to the different subphyla, classes and subclasses. These fungi are made up of heterogeneous species, forming unnatural groupings, and therefore are considered an artificial classification. These taxonomic rearrangements, with the concurrent change of the names of taxa, are part of a constant process of study and evaluation, in order to achieve a phylogenetic classification of the fungi.

discothecium, pl. **discothecia** (NL. *discothecium* < Gr. *dískos*, disc + *thekíon*, dim. of *thêke*, box; here, of the asci): a type of ascostroma (an apothecial pseudothecium), similar to an apothecium but bearing bitunicate asci; it differs from a hysterothecium in that its covering tissue disintegrates at maturity, as is seen in Loculoascomycetes of the family Patellariaceae (of the Patellariales), e.g., *Rhytidhysterium rufulum*.

Discothecia of *Rhytidhysterium rufulum* on *Acacia*, x 10 (*RTH*).

discrete (L. *discretus*, separated, discontinuous): a conidiogenous cell that is clearly distinct from the conidiophore. Cf. **integrated**.

disinfectant (Fr. *désinfectan* < L. pref. *de-*, *des-*, which denotes privation or elimination of something + *infectare*, to corrupt, to infect): something that disinfects or is capable of disinfecting. A chemical or physical agent that destroys infectious or pathogenic microorganisms.

disjunctor (L. *disjunctor*, separator < *disjungere*, to separate, disunite): each of the elements (connective cells or cell wall material) that is found between the spores arranged in moniliform chains, and which serve as a link, often ephemeral. The conidial chains of *Amblyosporium botrytis* (moniliaceous asexual fungi) and the species of *Oidiodendron* (dematiaceous asexual fungi), have disjunctor cells between adjacent conidia. The aeciospores of *Puccinia sorghi* (Uredinales) have disjunctors every two spores.

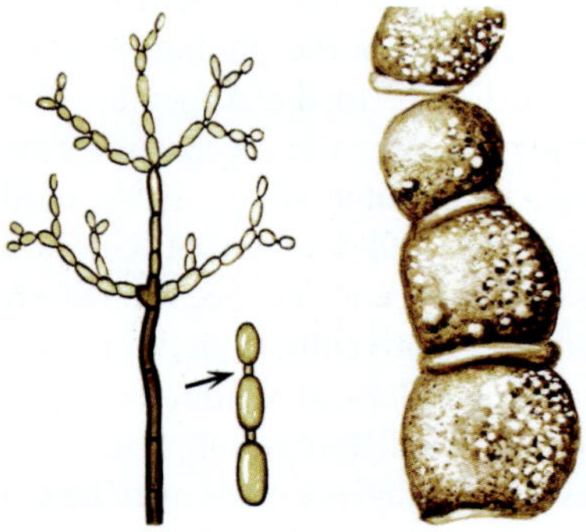

Disjunctors between the conidia on a conidiophore of *Oidiodendron* sp., x 1,600 (*MU*); **disjunctors** of the aeciospores of *Puccinia sorghi*, x 800 (*MU*).

disk fungus, pl. **fungi** (L. *discus* < Gr. *dískos*, disk, a thin circular object; L. *fungus*, fungus): a common name for fungi which have disk-shaped apothecia, e.g., like those of *Scutellinia scutellata* (Pezizales).

Apothecia of *Scutellinia scutellata*, so-called **disk fungi**, x 2 (*RV*).

dispirous (Gr. *dís*, two + *speîra*, spiral): with a double spiral, like the sporangiolar branches of *Dispira* (Dimargaritales), which are markedly twisted or recurved.

Dispirous branches of the sporangiophore of *Dispira simplex*, x 850 (*MU*).

dissepiment (L. *dissepimentum*, partition, division; what separates two contiguous things < *dissaepiere*, to divide + suf. *-mentum* > E. *-ment*, result, object, agent of an action): *Hymenomycetes*. Tissue (plectenchyma) that is found between the pores of the poroid Hymenomycetes, such as *Boletus* (Agaricales) or *Polyporus* (Aphyllophorales).

distal (L. *distare*, away from the point taken as the base or origin + suf. *-alis* > E. *-al*, relating to or belonging to): situated far from the center of a body or from the point of origin; terminal. Cf. **proximal**.

distichous (Gr. *dístichos*, arranged in two rows or lines < *dís*, two + *stichós*, line, thread): arranged in two rows, like the ascospores inside the ascus of *Hymenoscyphus* (=*Helotium*) *herbarum* (Helotiales), *Ascobolus michaudii* and *Saccobolus truncatus* (Pezizales) and *Verruculina* (Pleosporales), which are arranged partially or totally in two rows of four spores each. Also called **biseriate**. Cf. **monstichous**, **tristichous** and **uniseriate**.

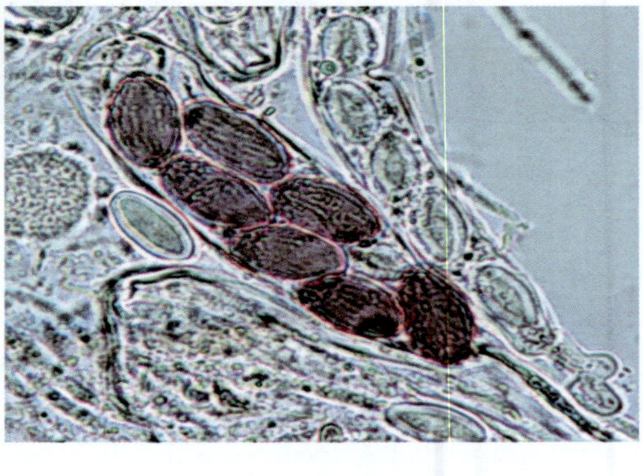

Distichous ascus of *Ascobolus michaudii*, x 500 (*EAA*).

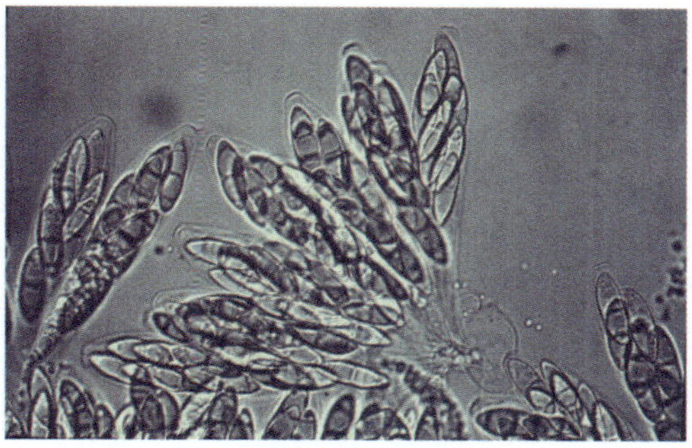

Distichous asci of *Verruculina* sp., x 500 (*MCG*).

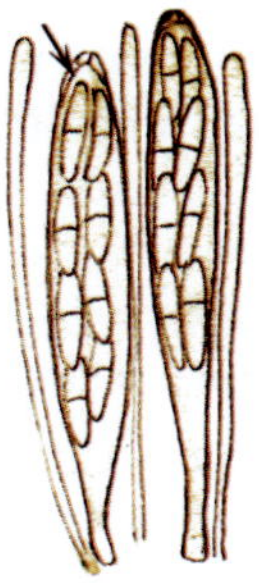

Distichous ascospores inside the asci of *Hymenoscyphus* (=*Helotium*) *herbarum*, x 650 (*MU*).

distorted (Gr. *dýs*-, a pref. that expresses the idea of difficulty or inferiority + L. *torsio*, genit. *torsionis*, action and effect of twisting or being twisted + suf. *-atus* > E. *-ate*, provided with or likeness): twisted or rolled up, so that the normal shape is lost; deformed.

distoseptate (L. *disto*, to be distant, to be away from, separated + *septatus*, provided with septa, partitioned < *septum*, septum, wall + suf. *-atus* > E. *-ate*, provided with or likeness): *Conidial fungi.* A type of multicellular conidium in which each cell is surrounded by a thick secondary wall (*distoseptum*), which is easily separable from the external wall of the conidium, so that when the outer wall is broken, the individual thick-walled cells inside separate; this occurs, e.g., in *Drechslera avenae* and *Setosphaeria rostrata* (=*Helminthosporium rostratum*, dematiaceous asexual fungi), and *Pyrenophora* (Pleosporales). Cf. **euseptate**.

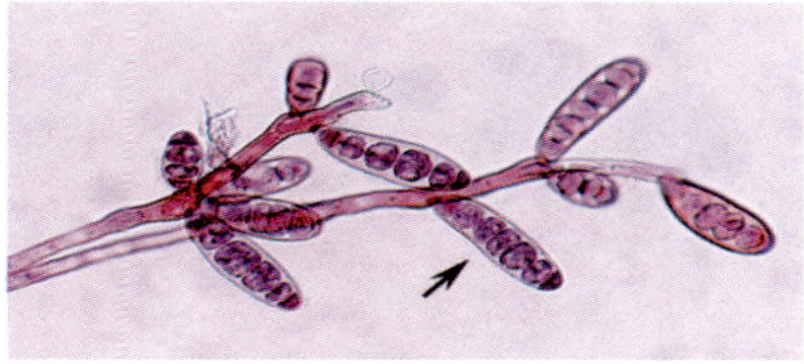

Distoseptate conidia on a conidiophore of *Drechslera avenae*, x 540 (*CB*).

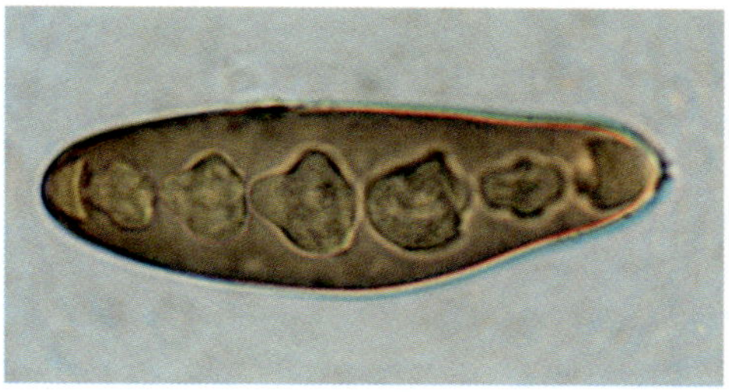

Distoseptate conidium of *Drechslera avenae*, x 1,340 (*CB*).

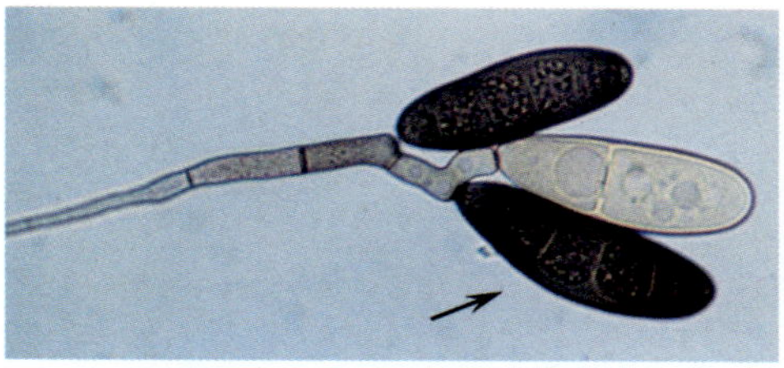

Distoseptate conidia on a conidiophore of *Helminthosporium* sp., x 630 (*CB*).

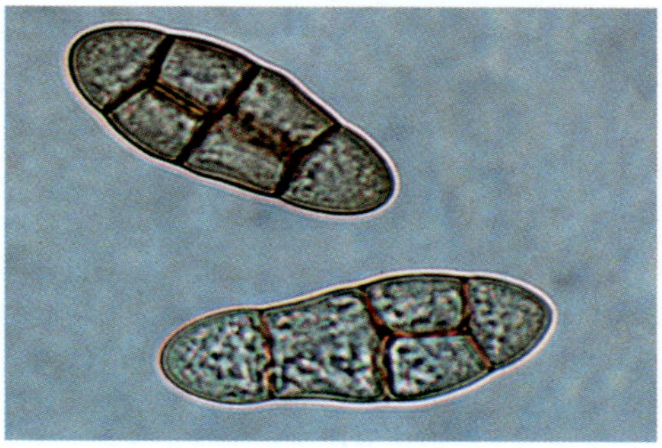

Distoseptate ascospores of *Pyrenophora* sp., x 410 (*CB*).

distoseptum, pl. **distosepta** (L. *disto*, to be distant, to be far from, separated + *septum*, septum, wall, partition): equal to **pseudoseptum**. See **distoseptate**.

distrophic (Gr. *distrophía*, malnutrition < *dýs*-, pref. that expresses the idea of difficulty or inferiority + *trophós*, what nourishes or serves as food): applied to aquatic media that are very poor in dissolved substances available for algal nutrition, but which are rich in humic acids, giving the water a rather high acidity. Peat-bogs are distrophal or distrophic. Cf. **eutrophic** and **oligotrophic**.

distrophy (Gr. *distrophía*, malnutrition < *dýs*-, pref. that expresses the idea of difficulty or inferiority + *trophós*, what nourishes or serves as food): an abnormality or structural modification of an organism that results from poor or inappropriate nutrition. Cf. **hypertrophy**. See **atrophy**.

diurnal (L. *diurnus*, pertaining to a day): refers to a period of one day. Cf. **circadian rhythm**.

divaricate (L. *divaricatus*, spread apart < *divaricare*, to spread apart, to move away + suf. *-atus* > E. *-ate*, provided with or likeness): extremely divergent; branching at a wide angle, like the conidiophore of *Penicillium oxalicum*, *P. citrinum*, *P. melinii*, *P. waksmanii* and other species of the subgenus *Furcatum* (moniliaceous asexual fungi), whose branches form very wide angles with the main axis.

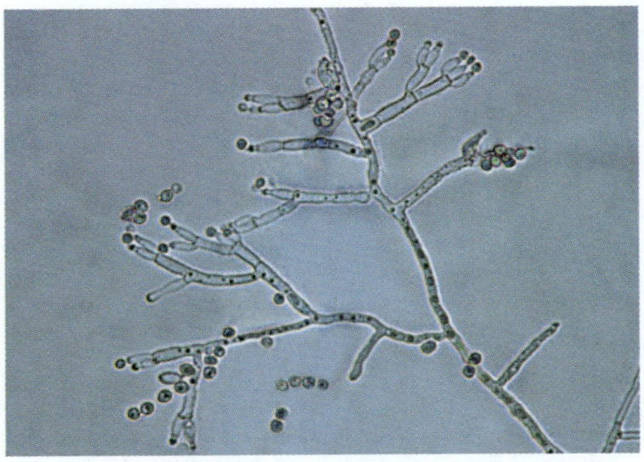

Divaricate conidiophore of *Penicillium waksmanii*, x 250 (*PL*).

divergent (L. *divergens*, genit. *divergentis*, widespread, bending and separating < *di-*, two, twice + *vergere*, to bend + suf. *-entem* > E.*-ente*, a being): *Agaricology*. A type of hymeniiferous trama in which there is a central zone of hyphae (the middle layer) from which diverge obliquely two rows of cells (the lateral layers). It is present, e.g., in *Amanita muscaria* and *Hygrophorus bakerensis* (Agaricales). Cf. **bilateral** and **convergent**.

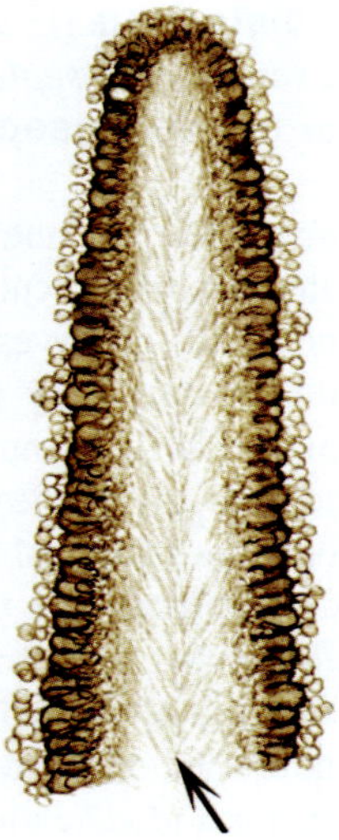

Divergent bilateral hymeniiferous trama of *Amanita muscaria*, seen in longitudinal section of a gill, x 160 (*MU*).

diverticulum, pl. **diverticula** (L. *diverticulum*, byway, turning from the main route, tributary < pref. *de-*, from, away from, away + *vertare*, to turn + dim. suf. *-culum* > E. *-culum*): a structure in the shape of a sac or blind appendage that originates from a larger or more important cavity, as occurs in the mycelium of *Pythium* (Peronosporales), or in the four sterigmata of a typical basidium, which are formed from four apical diverticula.

division (L. *divisio*, genit. *divisionis* < *di-*, two, double + *visus*, evident, view + suf. *-io*, genit. *-ionis* > E. *-ion*, result of an action, state of): in the classification of taxonomic categories, it is the systematic unit between the kingdom and the class. It is equivalent to *phylum*, which is the preferred term. In mycology the phylum (division) ending is *-mycota*, as in Myxomycota and Eumycota.

dolabriform (L. *dolabriformis* < *dolabra*, keg, barrel + suf. *-formis* < *forma*, shape): having the shape of a barrel.

dolicho-, dolich- (Gr. *dolichós*, long): a pref. employed in the formation of compound nouns to add the qualification of long, as in *dolichospore*.

dolichospore (Gr. *dolichós*, long + *sporá*, spore): a long spore.

doliiform (L. *doliiformis* < *dolium*, keg, barrel + suf. *-formis* < *forma*, shape): barrel-shaped, like the arthrospores of *Geotrichum candidum* (moniliaceous asexual fungi), the dictyospores of *Pithomyces chartarum* (dematiaceous asexual fungi) and the catenulate conidia of *Ceratocystis paradoxa* (Microascales).

Doliiform conidia of *Pithomyces chartarum*, x 500 (*CB*).

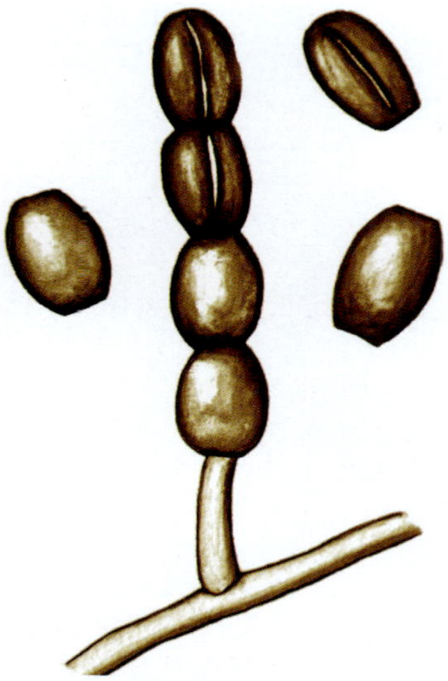

Doliiform conidia of *Ceratocystis paradoxa*, x 600 (*MU*).

dolipore septum, pl. **septa** (L. *dolium*, cask, barrel + *porus* < Gr. *póros*, pore, passage; L. *septum*, septum, wall, partition): a septum or partition with a central pore surrounded by a swelling of the septal wall in the shape of a barrel, and covered on both sides by perforated membranes called *septal pore caps* or *parenthesomes*; common in many basidiomycetes, e.g., *Heterobasidion annosum* (Aphyllophorales).

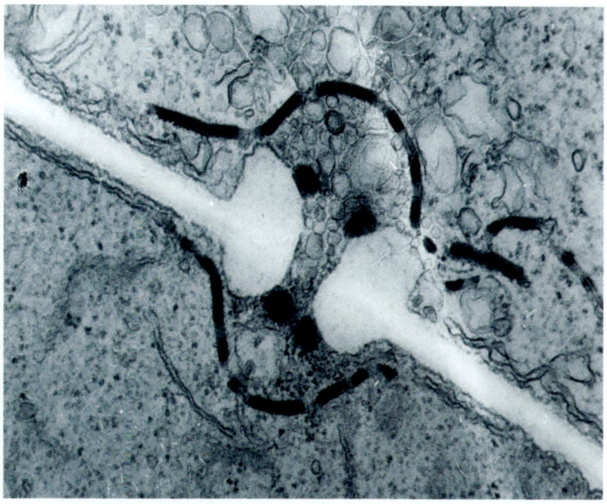

Dolipore septum of a hypha of *Heterobasidion annosum*, seen in longitudinal section with the transmission electron microscope, x 64,500 (*RTH*).

domain (L. *dominium*, domain, sphere of influence or coverage): see **superkingdom**.

dome (L. *domus*, dome, vault, cupola): the apical part of a bitunicate (fissitunicate) ascus, where the wall is thickened and forms a vault in which there is a digitiform projection (the *nasse*) which extends from the endotunica. The dome can be seen clearly, e.g., in the ascus of *Leptosphaerulina* (Dothideales).

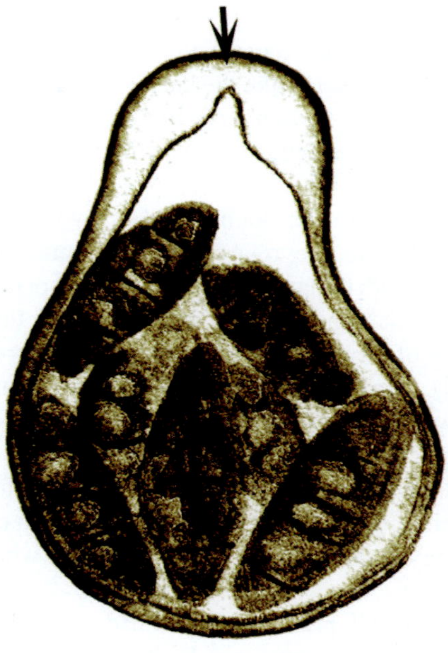

Dome of the bitunicate ascus of *Leptosphaerulina crassiasca*, x 600 (*MU*).

dormancy (L. *dormire*, to sleep < *dormiens*, genit. *dormientis*, sleeping, resting): the act of being **dormant**.

dormant (L. *dormire*, to sleep < *dormiens*, genit. *dormientis*, sleeping, resting): a biological state characterized by lack of growth and minimal metabolic activity; it can be the result of adverse environmenal conditions or be part of the normal life cycle of an organism.

dorsal (L. *dorsalis*, corresponding to *dorso* < L. *dorsum*, back, loin + suf. *-alis* > E. *-al*, belonging to or relating to): in dorsiventral organs, it refers to the back; what is at the back or reverse of something; the back or upper surface, which is opposite to the axis. At times it is used to refer to the upper surface of foliose lichen thalli. Cf. **ventral**.

downy mildew (ME. *downe* < ON. *dunn*, soft plumage + E. suf. *-y*, characterized by or inclined to; OE. *mildeaw* < *mil-*, honey + *deaw*, dew): the common name for plant diseases caused by members of the Peronosporaceae and characterized by the growth of downy growth of conidiophores on the underside of leaves, arising from the stomates. Also called false mildews. It was to control the downy mildew of grape in France, caused by *Plasmopara viticola* (Peronosporales), that led to the development of the first fungicide (Bordeaux mixture).

druse (G. *Druse*, mineral crystals): a stellate or rounded group of incomplete crystals (generally composed of calcium oxalate) that are formed in the thallus of some lichens.

dryophilous (Gr. *drýs*, genit. *drýos*, tree, especially the oak + *phílos*, have an affinity for): living on, beneath or near oak trees, like the basidiocarps of *Gymnopus dryophilus* (=*Collybia dryophila*), Agaricales.

dulcious (L. *dulcis*, sweet): with a sweet flavor or taste.

duplex (L. *duplex*, double < *duo*, two + *plex*, fold, twofold): double; refers to a context made up of two layers, with the layer adjacent to the gills or tubes being harder than the one over it, as is seen in the fruiting body of *Phylloporia fruticum* (Aphyllophorales).

duri- (L. *durus*, hard): a pref. that denotes hardness of any structure, as in the gen. *Durispora* (Diaporthales), which means hard spore.

duvet (Fr. *duvet*, down, plush): a thick, soft layer of hyphae, quilted, similar to combed cotton cloth, that some colonies of dermatophytes form when grown on agar.

dysgonic (Gr. pref. *dýs*-, apart, bad, difficult + *gónos*, generation, progeny + suf. *-íkos* > L. *-icus* > E. *-ic*, belonging to or relating to): applied to strains of dermatophytes that grow more slowly in culture, often with less aerial mycelium than normal strains (*eugonic*).

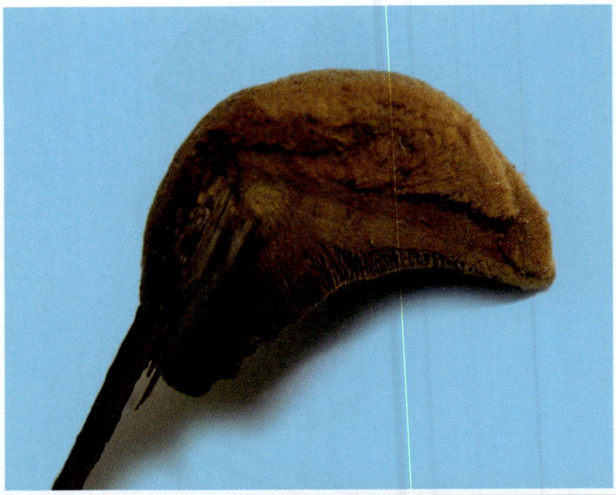

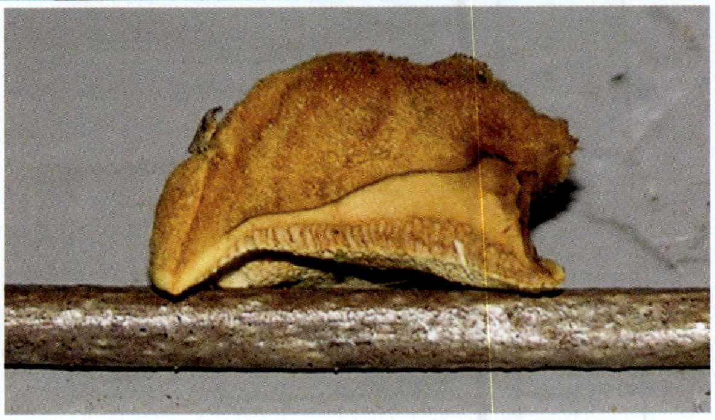

Duplex fruiting bodies of *Phylloporia fruticum* and *Phylloporia* sp. (a longitudinal sections), x 1 (*RV*).

Entomogenous perithecial stromata of *Cordyceps melolonthae* var. *rickii* on a parasitized coleopteran larva (*TH*).

e

e- a Latin particle employed as a pref. that implies the idea of the loss of something, as in *ecaudate*, i.e., without a tail. See **ec-**.

earth ball (OE. *eorthe* > ME. *erthe* > E. *earth*, earth; ME. *ball*, *balle* < OF. < G. *ballax*, spherical structure): common name for members of the Sclerodermatales (*Scleroderma*) and Hymenogastrales (*Sclerogaster*), which form globose fructifications in the soil. See **puffball**.

Earth ball (fruiting body) of *Scleroderma polyrhizum*, x 1 (*RTH*).

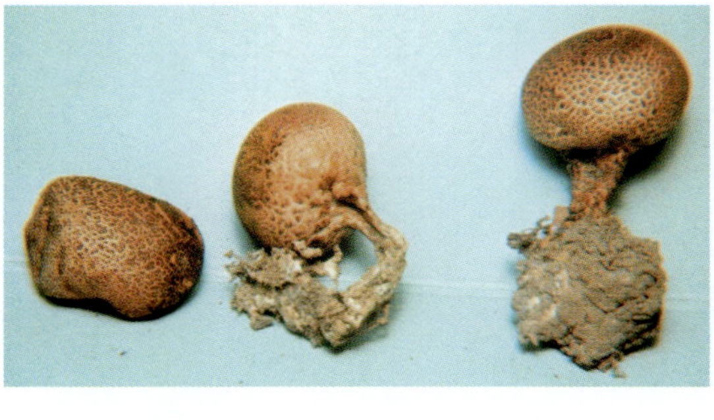

Earth balls (fruiting bodies) of *Scleroderma lycoperdoides*, x 0.3 (*TH*).

earth star (OE. *eorthe* > ME. *erthe* > E. *earth*, earth; ME. *sterre* > OE. *steorra*, star): a basidioma of *Geastrum* (Lycoperdales), in which the exoperidium splits into several more or less triangular segments that fold back in a star-like manner.

Earth stars, fruiting bodies of *Geastrum* sp., x 0.5 (*RV*).

earth tongue (OE. *eorthe* > ME. *erthe* > E. *earth*, earth; ME. *tunge*, tongue): the common name of the ascocarp of species of *Geoglossum* and *Trichoglossum* (Helotiales).

Earth tongues of *Geoglossum nigritum* and *Geoglossum* sp., x 1 (*RV*).

Earth tongues (ascocarps) of *Geoglossum* sp., x 0.5 (*RV*).

ec-, ex- (Gr. *ek*): a variable pref., *ec-*, before a consonant, or *ex-*, before a vowel, is a pref. that alludes to something that moves away, changes, or is lacking, as in the generic name *Eccilia* (=*Entoloma*, of the Agaricales, which signifies an uncommon veil).

eccentric, **excentric** (Gr. *ékkentros*, not in the center + suf. *-íkos* > L. *-icus* > E. *-ic*, belonging to or relating to): **1**. *Oomycota*. A type of oospore in which the ooplast is displaced from the center, in contact with the internal wall of the oogonium, for which reason it is not completely surrounded by the oil droplets that constitute the reserve substances, as is observed, e.g., in *Scoliolegnia* (=*Saprolegnia*) *asterophora* (Saprolegniales). **2**. *Basidiomycetes*. A stipe that is not attached to the center of the pileus of the basidiocarp, as occurs in some agaricaceous and polyporaceous fungi, e.g., *Amauroderma* sp. and *Ganoderma curtisii*, of the Aphyllophorales. Cf. **centric** and **subcentric**.

Eccentric stipe of the basidiocarp of *Amauroderma* sp., x 0.5 (*MU*).

eccrinid (< gen. *Eccrinidus* < Gr. *ekkríno*, to separate, expel + L. suf. *-idus*, state or action): a member of the order Eccrinales, e.g., *Eccrinidus flexilis*, Trichomycetes that live as endocommensals in the digestive tract of arthropods.

Eccrinidus flexilis, an endocommensal **eccrinid** of myriapods and diplopods, x 20 (*MU*).

echinate (L. *echinatus*, having spines or structures similar to spines < *echinus* < Gr. *echínos*, porcupine, sea urchin + L. suf. *-atus* > E. *-ate*, provided with or likeness): covered with spines or bristles that recall the porcupine. E.g., the conidia of *Aspergillus carbonarius* (moniliaceous asexual fungi), the ascospores of *Tuber melanosporum* (Pezizales), and the basidiocarps of *Lycoperdon echinatum* (Lycoperdales) are echinate. See **echinulate**.

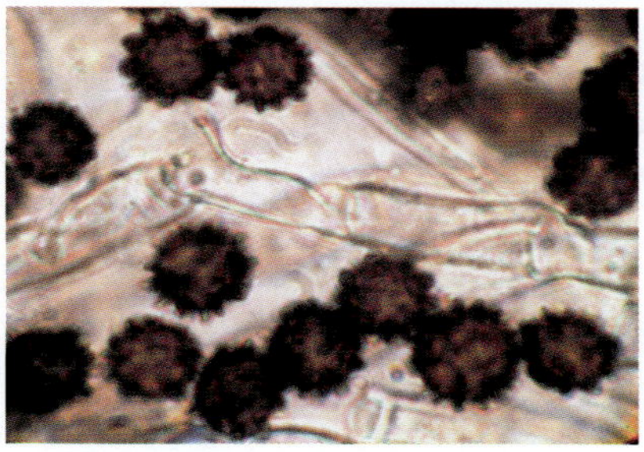

Echinate conidia of *Aspergillus carbonarius*, x 900 (*MU*).

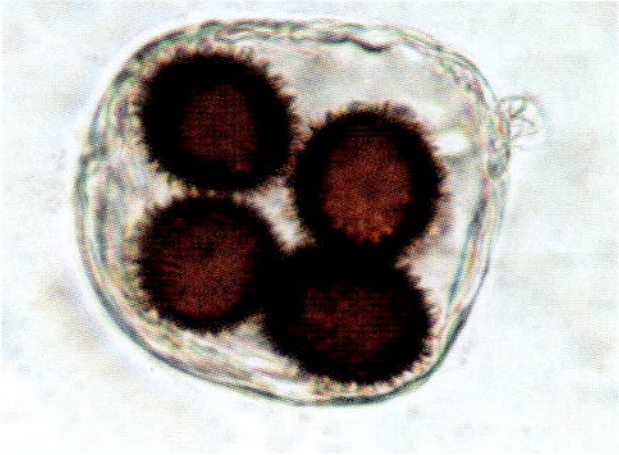

Echinate ascospores inside an ascus of *Tuber melanosporum*, x 1,200 (*EAA*).

Echinate basidiocarps of *Lycoperdon echinatum*, x 0.5 (*EPS*).

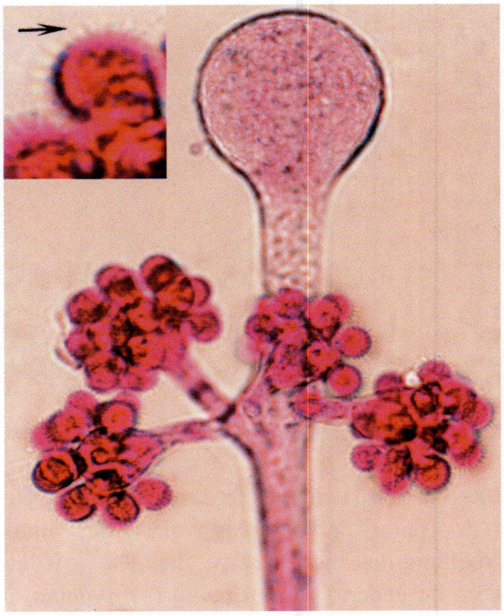

Echinulate sporangiola of *Cunninghamella echinulata*, x 300, insert, x 1,200 (*MU*).

echinid, echinidium, pl. **echinidia** (L. *echinus* < Gr. *echínos*, porcupine, sea urchin + L. dim. suf. *-idium*): a type of marginal hair characteristic of the covering of the pileus of species of *Mycena* and *Marasmius* (Agaricales), not of the hymenium, although at times it is present in portions next to the hymenium due to an invasion; it varies in shape from clavate to sphaeropedunculate, fusiform or lageniform, and has spines, diverticula or bristles that give it a peculiar appearance.

echinulate (NL. *echinulatus*, provided with small spines or with the appearance of small spines < *echinus* < Gr. *echínos*, porcupine, sea urchin + L. suf. *-atus* > E. *-ate*, provided with or likeness): like *echinate*, but with the spines or bristles smaller and less rigid. Applied mainly when it refers to microscopic structures or to microorganisms. E.g., the sporangial wall of *Actinomucor elegans* and the sporangiola of *Cunninghamella echinulata* (Mucorales), the conidia of *Penicillium echinulatum* (moniliaceous asexual fungi), and the urediniospores of *Puccinia graminis tritici* (Uredinales) are echinulate.

Echinulate sporangiola of *Cunninghamella echinulata*, x 900 (*MU*).

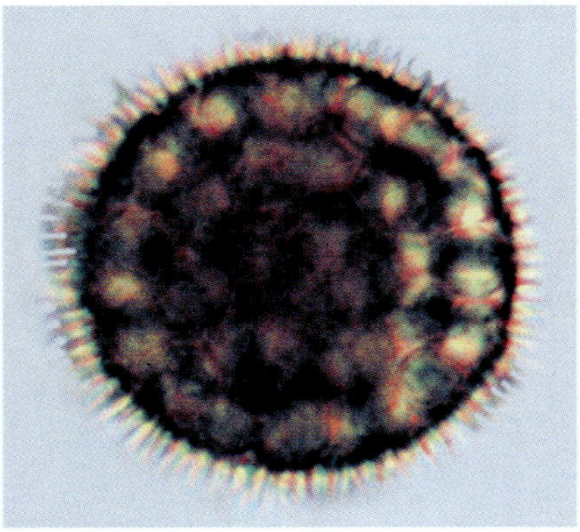

Echinulate sporangium of *Actinomucor elegans*, x 850 (*MU*).

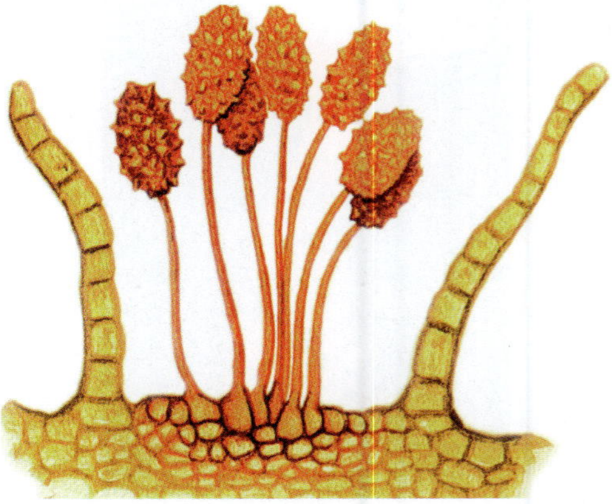

Echinulate urediniospores of *Puccinia graminis tritici*, x 400 (*MU*).

eclosion (Fr. *éclosion* < *éclore*, to hatch, prob. < VL. *exclaudere*, alteration of L. *excludere*, to hatch, mature, to expel, to throw out): an explosive series of movements that result in the release of a germinating inner spore from a rigid exosporium, as occurs in the ascospores of *Hypoxylon fragiforme* (Xylariales).

eco- (Gr. *oîkos* > L. *oeco-*, house or inn): a pref. employed in the formation of compound terms that refer to the environment or the natural habitat to which an organism is related, as in *ecology* (the branch of science that treats of the interrelations of organisms and their environment) and *ecospecies* (subdivision of a *coenospecies*, more or less equivalent to a taxonomic species, capable of undergoing a free interchange of genes among its members without impairing fertility, but less capable of achieving fertile crosses with members of other subdivisions), etc.

ecorticate (*e*, L. particle employed as a pref., *e-*, implies the idea of the loss of something + L. *corticatus*, having bark < *corticea*, des. fem. of *corticeus*, bark, + suf. *-atus* > E. *-ate*, provided with or likeness): without bark.

ecotype (Gr. *oîkos*, house or inn + *týpos*, type): a part of an ecospecies whose morphological, physiological and other characteristics appear to be genetically determined but influenced by environmental selection and isolation. Some consider an ecotype as a taxonomic subspecies, whereas others do not think that it has important taxonomic significance.

ect-, **ecto-** (Gr. *ektós*, outside, external): a pref. employed in various compound terms of Gr. origin to allude to something external or that is related to the exterior of another thing, as in *ectangial* and *ectoascus*. Equal to **exo-**. Cf. **endo-**.

ectal (Gr. *ektós*, outside, external + L. suf. *-alis* > E. *-al*, relating to or belonging to): external, the outermost, superficial. Cf. **ental**.

ectangial, **ectoangial** (Gr. *ektós*, outside, external + *angeîon*, vessel, receptacle + L. suf. *-alis* > E. *-al*, relating to or belonging to): formed or developed outside of a conceptacle, like the zoospores in the vesicle of *Pythium* (Peronosporales). Cf. **entangial**.

ectoascus, pl. **ectoasci** (L. *ecto* < Gr. *ektós*, outside, external + L. *ascus* < Gr. *askós*, sack, pouch): the outer layer of the ascus wall, which on breaking permits the extension of the *endoascus*. It is present in fungi of the Loculoascomycetes, such as *Sporormiella* (*=Preussia*) *australis* (Pleosporales). See **bitunicate**.

ectochroic (Gr. *ektós*, outside, to the exterior + *chrõs*, color + suf. *-ikos* > L. *-icus* > E. *-ic*, belonging to or relating to): refers to a hypha that has pigmentation on the exterior. Cf. **endochroic**.

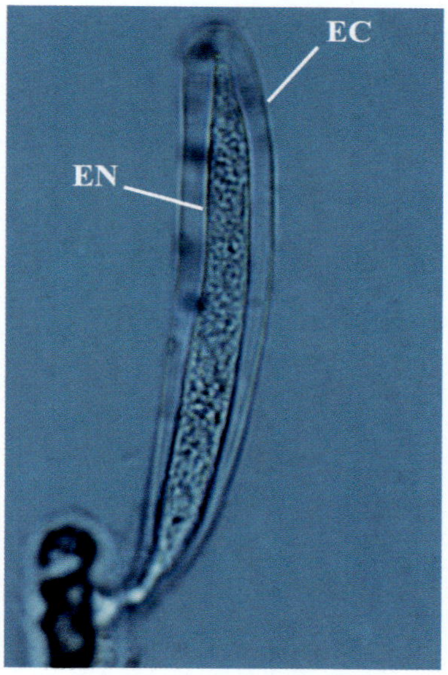

Ectoascus (EC) and **endoascus** (EN) of a young bitunicate ascus of *Sporormiella* (=*Preussia*) *australis*, x 600 (*RTH*).

ectocommensal (Gr. *ektós*, outside, external + L. *commensalis*, one who shares a table < *cum*, with + *mensa*, table + suf. *-alis* > E. *-al*, relating to or belonging to): an exterior commensal. See **commensal**.

ectomycorrhiza, pl. **ectomycorrhizae** (Gr. *ektós*, outside, external + *mýkes*, fungus + *rhíza*, root): a mycorrhiza in which the hyphae of the associated fungus only grow intercellularly, never within the cells of the associated plant; e.g., the symbiosis between *Pisolithus tinctorius* (Sclerodermatales) and the roots of pine seedlings. Cf. **endomycorrhiza**.

Ectomycorrhizae of pine roots formed by *Pisolithus tinctorius*, x 0.6 (*MU*).

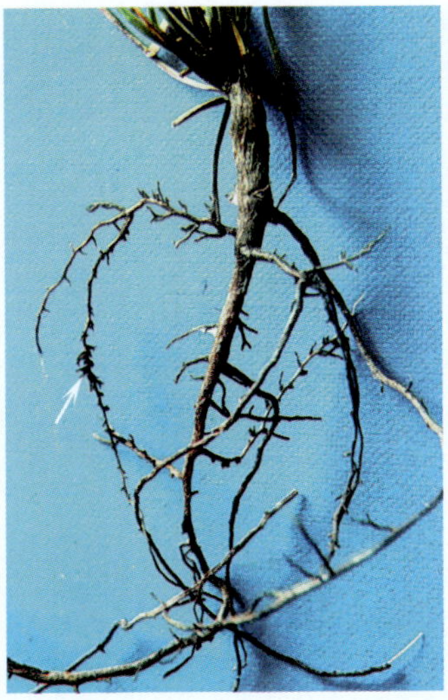

Ectomycorrhizae of pine roots formed by *Pisolithus tinctorius*, x 1 (*MU*).

Ectomycorrhizae of pine roots formed by *Pisolithus tinctorius*, x 3 (*DM*).

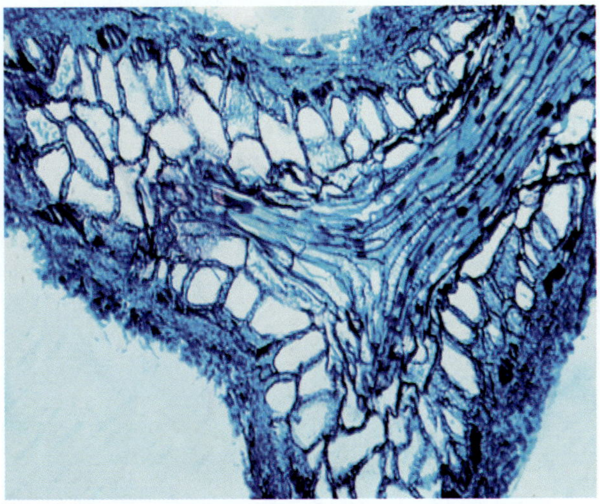

Ectomycorrhizae of pine roots formed by *Pisolithus tinctorius*, seen in longitudinal section, x 300 (*DM*).

ectoparasite, **ectoparasitic** (Gr. *ektós*, outside + *parásitos* < *pará*, at the side of, together + *sîtos*, flour, bread, food in general; or + suf. *-íkos* > L. *-icus* > E. *-ic*, belonging to, relating to): a parasitic organism which for the most part develops on the exterior of another organism (the host), from which it is nourished by means of haustoria that penetrate the cells or tissues of the host. For example, the Laboulbeniales (such as *Rickia*, e.g., *R. apiculifera*, common on coleopterous passalids, and *Laboulbenia*, such as *L. formicarum*, which lives on ants, and *L. elongata*), are ectoparasites of arthropods; the powdery mildews (Erysiphales), which attack various vascular plants, also are ectoparasites. Cf. **endoparasite**. See **epibiotic**, **endobiotic** and **interbiotic.**

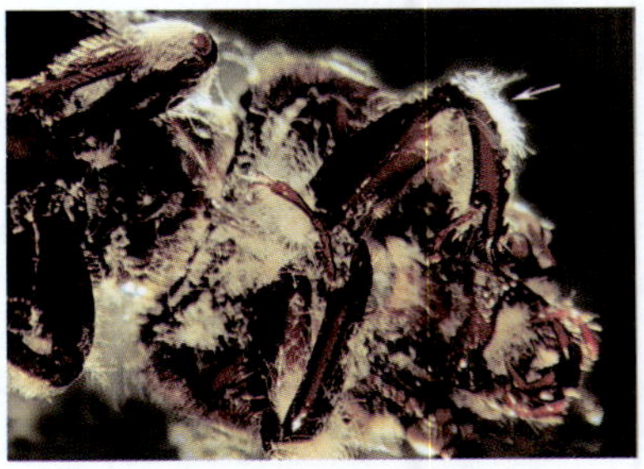

Ectoparasitic thalli (arrows) of *Rickia apiculifera* on the body of a coleopterous passalid (*Chondrocephalus debilis*), x 10 (*MU*).

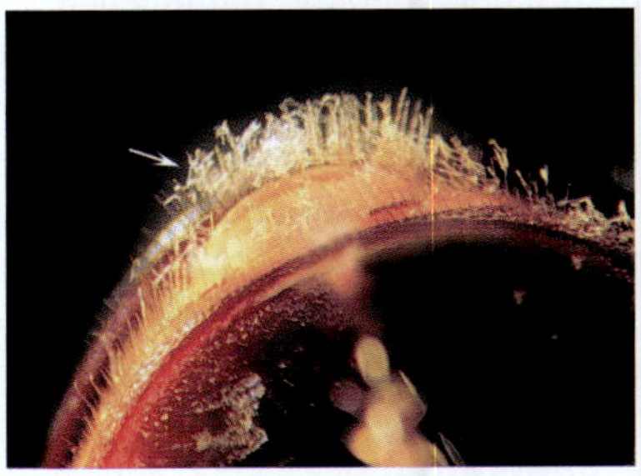

Ectoparasitic thalli of *Rickia apiculifera* on the abdomen of a coleopterous passalid (*Chondrocephalus debilis*), x 20 (*MU*).

Ectoparasitic thalli of *Rickia apiculifera* on the body of a coleopterous passalid (*Chondrocephalus debilis*), x 30 (*MU*).

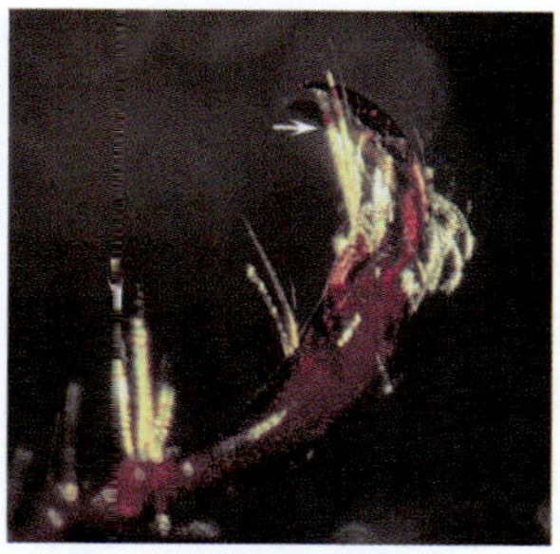

Ectoparasitic thalli of *Rickia apiculifera* on the leg of a coleopterous passalid (*Chondrocephalus debilis*), x 40 (*MU*).

Ectoparasitic thalli of *Rickia apiculifera* on the leg of a coleopterous passalid (*Chondrocephalus debilis*), x 40 (*MU*).

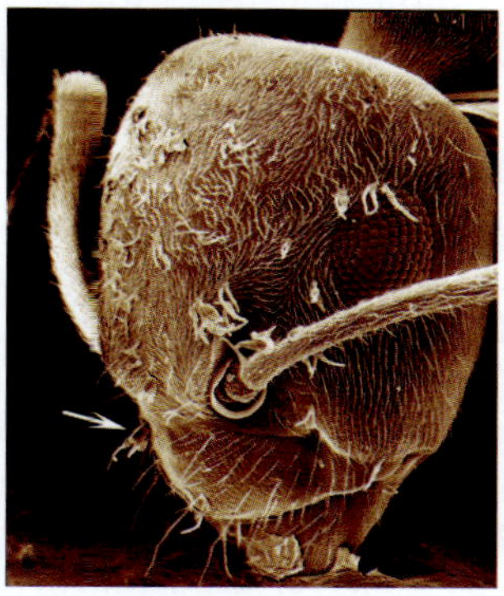

Ectoparasitic thalli of *Laboulbenia formicarum* on the head of an ant, x 65 (*GVD*).

Ectoparasitic thalli of *Laboulbenia formicarum* on the body of an ant, x 130 (*GVD*).

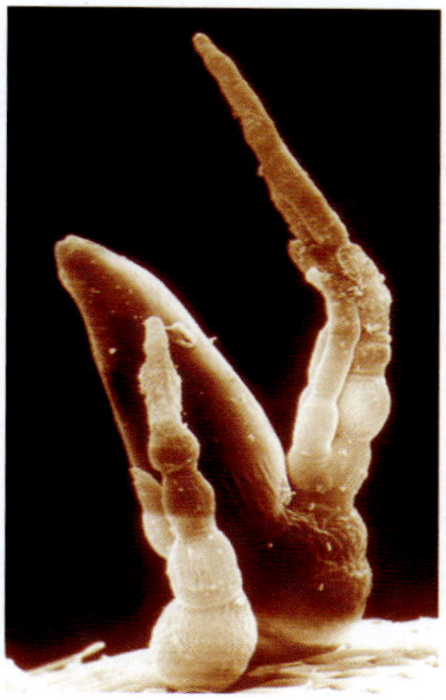

Ectoparasitic thalli of *Laboulbenia formicarum* on the body of an ant, x 250 (*GVD*).

Ectoparasitic thalli of *Laboulbenia elongata*, x 250 (*RTH*).

ectophloeodic, **ectophloeodal**, **ectophloeic** (Gr. *ektós*, outside, external + *phloiós* > L. *phloeo*, bark): refers to the thallus of crustaceous lichens that are found on the surface of tree bark. Syn. of **epiphloeodic**. Cf. **endophloeodic**.

ectoplacodium, pl. **ectoplacodia** (Gr. *ektós*, outside + NL. *placodium* < Gr. *plakós*, plaque + L. suf. *-odium*, resembling < Gr. *ode*, like < *-oeídes*, similar to): see **placodium**.

ectoplasm (Gr. *ektós*, outside, to the exterior + *plásma*, formation, bland material of which a living being is formed): an external layer of the cytoplasm, relatively rigid and free of granules, that can revert from a gel to a sol state, as occurs in the plasmodial veins of the Myxomycetes. Cf. **endoplasm**.

ectospore (Gr. *ektós*, outside, to the exterior + *sporá*, spore): an exogenous spore. A term formerly used to designate a basidiospore. Syn. of **exospore**.

ectostroma, pl. **ectostromata** (Gr. *ektós*, outside + *strõma*, bed, small cushion): in the stromatic Pyrenomycetes, such as species of *Eutypa* (Xylariales) and *Cryphonectria* (Diaporthales), it represents the stroma situated in the periderm of the host plant, above the *endostroma* (internal stroma in which the perithecia form); generally, the ectostroma produces conidia. In other stromatic Pyrenomycetes, such as the species of *Xylaria* (Xylariales), the ectostroma (which is usually dark in color) is penetrated by the ostiolar necks of the perithecia and is clearly separated from the endostroma, which is much less pigmented. Cf. **endostroma**.

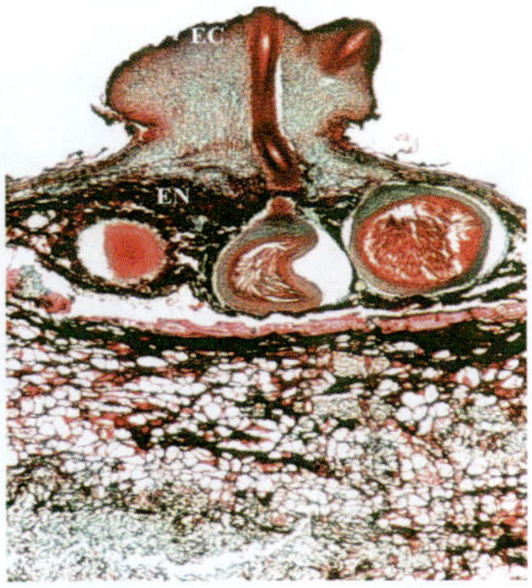

Ectostroma (EC) and **endostroma** (EN) of *Cryphonectria* (=*Endothia*) *parasitica*; the perithecia are embedded in the endostroma and their ostiolar necks protrude from the ectostroma, x 25 (*MU*).

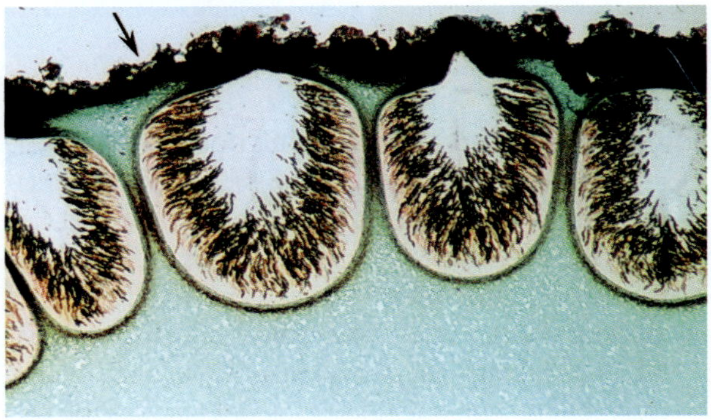

Ectostroma (arrow) and endostroma of the perithecial stroma of *Xylaria* sp., seen in transverse section, x 80 (*MU*).

ectothecal (Gr. *ektós*, outside + L. *thecicus*, belonging to or relating to the theca < Gr. *thêke*, box; here, of the asci + L. pref. *-alis* > E. *-al*, belonging to or relating to): *Ascomycetes*. Having the hymenium exposed.

ectothrix (Gr. *ektós*, outside + *thríx*, hair): one of the two modes of invasion by fungi of the hair of man and animals, in which the parasite forms a layer of arthrospores in a mosaic outside the axis of the hairs, as happens in the ectothritic species *Microsporum audouinii* and *M. canis* (moniliaceous asexual fungi). Cf. **endothrix**.

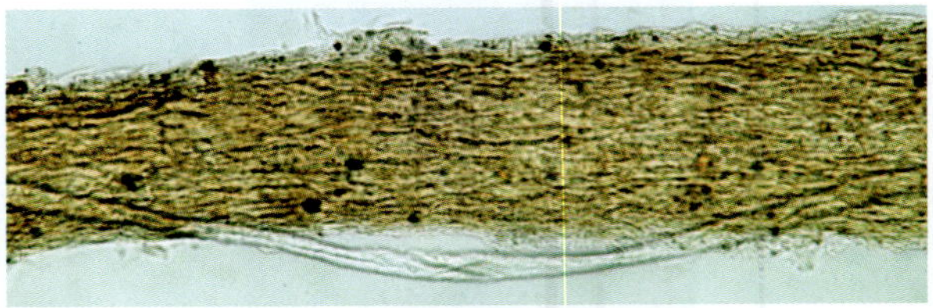

Ectothrix invasion of a human hair by *Microsporum canis*, x 260 (*MU*).

ectotroph (Gr. *ektós*, outside + *trépho*, to feed, nourish): **1**. *Mycorrhizae*. A mycorrhizogenous fungus that does not penetrate the cells of the host plant, but remains outside the root, or at most, insinuates itself between the cells; also applied to a mycorrhiza formed by a fungus with this behavior. The name ectotroph is due to the fungus obtaining from the external medium the nutrients that it provides for the host. **2.** *Lichens*. A cephalodium that is formed on the thallus. Cf. **endotroph**.

ectotropic (Gr. *ektós*, outside, on the exterior + *trépo*, to revolve, to turn): something that curves away from its point of origin. Cf. **endotropic**.

ectotunica (Gr. *ektós*, outside, on the exterior + L. *tunica*, covering, wrapper): the external layer of a bitunicate ascus. Syn. of *exotunica*. Cf. **endotunica**.

ectozoic (Gr. *ektós*, outside + *zõon*, animal + suf. *-íkos* > L. *-icus* > E. *-ic*, belonging to or relating to): fungi that develop on the exterior of living animals as commensals, symbionts or parasites, as happens, e.g., in the Laboulbeniales. Cf. **endozoic**.

edaphic (Gr. *édaphos*, soil + suf. *-íkos* > L. *-icus* > E. *-ic*, belonging to or relating to): relating to the soil.

edaphosphere (Gr. *édaphos*, soil + *sphaîra*, sphere, scope): the area of the soil that is free of any effect produced by the roots of plants, beyond the rhizosphere. The microbial composition of the edaphosphere differs qualitatively and quantitatively from that of the rhizosphere; the latter is generally richer in fungal species, including mycorrhizae. See **rhizosphere**.

edible, **esculent** (LL. *edibilis* < L. *edere*, to eat; L. *esculentus*, edible, good to eat < *esca*, food < *edere*, to eat + suf. *-ulentus*, that abounds in): edible; not causing damage after being ingested. There are many species of edible fungi, mainly among the macroscopic forms, wild as well as cultivated. *Morchella esculenta* (Pezizales), e.g., is an edible species much appreciated for its culinary properties; in fact, the specific epithet reflects this property (L. *esculenta*, edible). Other edible species are in the gen. *Amanita*, *Volvariella*, *Lactarius* (Agaricales) and *Hypomyces lactifluorum* (Hypocreales), the latter parasitizes *Russula delica*. Cf. **poisonous fungus**.

Edible ascocarps of *Morchella esculenta*, x 0.2 (*SA*).

Edible basidiocarps of *Amanita caesarea*, x 0.3 (*RV*).

Edible basidiocarps of *Volvariella bombycina*, x 0.3 (*RV*).

Edible basidiocarp of *Lactarius indigo*, x 0.5 (*RV*).

Edible basidiocarps of *Russula delica* (Agaricales) parasitized by *Hypomyces lactifluorum*, x 1 (*RV*).

edible fungi, sing. **fungus** (LL. *edibilis* < *edere*, to eat + L. suf. *-ibilis* > E. *-ible*; L. *fungus*, fungus < Gr. *spóngos*, *sphóngos*, sponge): a wide variety of macromycetes are utilized as an important source of food in many regions of the world, especially fleshy basidiomycetes of the order Agaricales, e.g., *Agaricus*, *Boletus*, *Coprinus*, *Lactarius*, *Lepista*, *Macrolepiota*, *Pleurotus*, *Termitomyces* and *Tricholoma*, as well as of other orders, such as *Cantharellus* (Aphyllophorales) and *Lycoperdon* (Lycoperdales). Accurate species identification is important, but the genera of edible species (e.g., *Amanita*, *Boletus*, *Coprinus*) contain others that are nonedible or poisonous. There also are various edible ascomycetes, mainly in the gen. *Morchella* (the morels) and *Helvella* (the saddle fungus), as well as the different varieties of truffles (*Tuber*, *Terfezia* and *Tirmania*, of the Pezizales), some of which are cultivated. In the basidiomycetes, the edible part is generally the basidioma, but in the Aphyllophorales *Laccocephalum* (=*Polyporus*) *mylittae* (of Australia), *Wolfiporia extensa* (=*Poria cocos*), of North America, and some tropical species of *Lentinus*, the sclerotium is consumed as food, since it can be quite large (in the latter case it can weigh 10 kilos or more). In the Ustilaginales, *Yenia* (=*Ustilago*) *esculenta*, which grows on *Zizania aquatica* (Canadian rice), is consumed in Taiwan, and in Mexico cuitlacoche, *Ustilago maydis*, which parasitizes ears of corn (maize), is a much appreciated food. Among the Agaricales cultivated industrially are the common agaric *Agaricus bisporus* (=*A. brunnescens*), the oyster mushroom (*Pleurotus ostreatus*), the Japanese shiitake (*Lentinula edodes*), and the straw mushroom (*Volvariella volvacea*). Although many species of basidiomycetes have the prospect of being cultivated in the future, only a few more besides those mentioned, are cultivated on an industrial scale, among them the Agaricales: *Flammulina velutipes* (the inokitake of Japan), *Pholiota nameko* (the Japanese nameko or nametake, which means viscous mushroom), and *Tricholoma matsutake* (the Japanese pine mushroom); as well as *Auricularia polytricha*, of the Auriculariales (consumed in China, Japan and Taiwan), and *Tremella fuciformis*, of the Tremellales (the white jelly fungus, much esteemed in China for its medicinal qualities and edibility). The yeasts *Saccharomyces cerevisiae* (Saccharomycetales) and *Candida utilis* (cryptococcaceous asexual yeasts), as well as the mycelium (mycoprotein) of a non-pathogenic strain of *Fusarium graminearum* (tuberculariaceous asexual fungi), are utilized to prepare human foods. Wild mushrooms also are a source of food for animals (see **mycophagy**). Some lichens, such as *Bryoria* and *Cladonia* (Lecanorales) are very important in the diet of reindeer and caribou, but in general the majority of lichens are not edible for humans. Cf. **poisonous fungus**.

effete (L. pref. *effetus*, spent, decrepit < *ex-*, out of, without + *fetus*, progeny, fruit): no longer fertile, overmature, exhausted, as when a fruiting body becomes empty.

efflorescent (L. *efflorescens*, genit. *efflorenscentis*, flourishing < *efflorescere*, to flower + suf. *-escens*, genit. *-escentis* > E. *-escent*, that which turns, that begins, slightly): emerging or budding suddenly and vigorously.

effuse (L. *effusus*, spread out, extended < *effundere*, emptied, poured out): extended and flat, spread out or scattered, especially toward one side, like the thallus of some lichens, e.g., *Cryptothecia rubrocincta* (Arthoniales) and various species of *Lecanora* (Lecanorales), which do not have a well-defined outline. The term *effused-reflexed* is commonly applied to the pileus of certain basidiomycetes in which the margin is extended and bent backwards, as happens in some species of *Corticium* and *Schizopora* (Aphyllophorales).

Effuse basidiocarp of *Schizopora flavipora*, x 1 (*TRO*).

Effuse basidiocarp of *Auricularia mesenterica* (Auriculariales), x 1 (*RV*).

egg (ME. *egge* < ON. *egg* < L. *ovum* < Gr. *oón*, egg): **1**. The female gamete of the Oomycota, also called the **oosphere**. **2**. A young basidioma (primordium) of the phallaceous fungi, and of *Amanita* and other Agaricales, before the volva breaks.

Egg (primordium) of a fruiting body of *Amanita tuza*, x 0.5 (*MU*).

eguttulate (L. pref. *e-*, absence + *guttulatus*, provided with small droplets or with that appearance < *guttula*, dim. of *gutta*, drop + suf. *-atus* > E. *-ate*, provided with or likeness): lacking oil or lipid droplets, like the conidia of *Sphaeropsis malorum* (sphaeropsidaceous asexual fungi) and *Marssonina juglandis* (melanconiaceous asexual fungi), among many such fungi. Many species of fungi do possess guttulate conidia. Cf. **guttulate**.

elasmo-, elasm- (Gr. *elasmós*, lamina, thin plate): a pref. employed in names such as *Elasmomyces* (syn. of *Macowanites*, of the Agaricales), referring to the gills of the fruiting body.

elastic (NL. *elaster*, spring, elasticity < Gr. *elastikós* < *elatés*, *elatér*, a driver, dispersant + suf. *-tikós* > L. *-ticus* > E. *-tic*, relation, fitness, inclination or ability): the ability of a body to recover more or less completely its shape and size once the action that altered it ceases, as occurs, e.g., with the ascoma of *Helvella elastica* (Pezizales) and the stipe of the basidiocarp of *Marasmius ferruginosus* (Agaricales).

Elastic ascomata of *Helvella elastica*, x 1 (*TH*).

Elastic stipe of the basidiocarps of *Marasmius ferruginosus*, x 1 (*TH*) and *Marasmius* sp. x 1 (*RV*).

elater (Gr. *elatér*, a driver, dispersant): each of the filaments or fibriform tubes, free, sharp-pointed and ornamented with engravings, spiral thickenings or various reliefs that facilitate spore dispersal by means of hygroscopic movements, as occurs, e.g., with the capillitium of the sporiferous apparatus of the Myxomycetes of the order Trichiales (such as *Hemitrichia serpula* and *Arcyria denudata*), or the gleba of *Battarrea* (Tulostomatales).

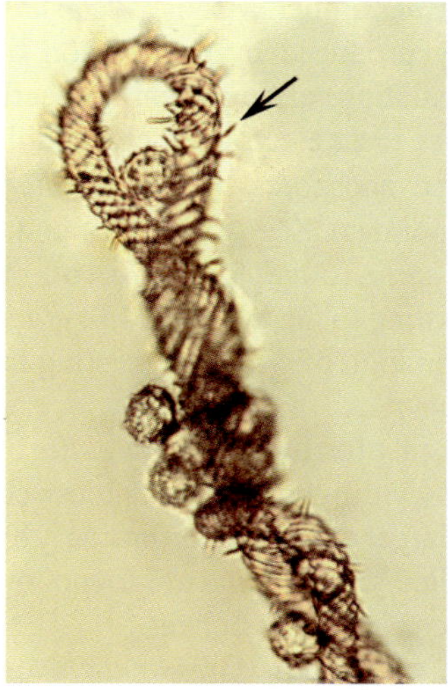

Elaters of the capillitium of *Hemitrichia serpula*, x 610 (*MU*).

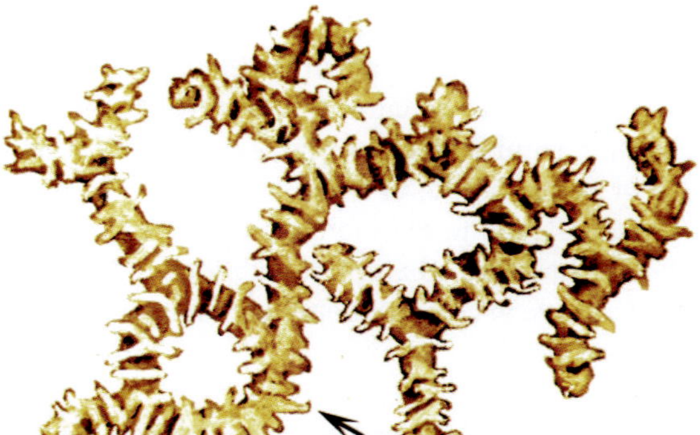

Elaters of the capillitium of *Arcyria denudata*, x 650 (*MU*).

electrophoresis, pl. **electrophoreses** (Gr. *élektron* > L. *electrum*, amber > NL. *electricus*, electrical, electricity produced by the friction of amber + Gr. *phorésis*, act of carrying, transmission): a useful method of chromatography in which particles with an electric charge (especially DNA, RNA and proteins) are separated by an electric field, due to their differential migration in an inert matrix. For DNA and RNA an agarose matrix is utilized, which is suitable because it has large pores, and for proteins a matrix of polyacrilamide and starch is used, as it has smaller pores. This method is used to analyze total protein patterns and for specific enzymes (inter and extracellular), in particular the variations of an enzyme among different isolates of fungi, especially plant pathogenic species; the variants are isozymes (isoenzymes), e.g., esterases, dehydrogenases, pectinases and phosphatases.

electrotaxis, pl. **electrotaxes** (Gr. *élektron* > L. *electrum*, amber > NL. *electricus*, electrical, electricity produced by the friction of amber; Gr. *táxis*, regulating, arrangement): a tactic movement in response to an electrically charged field; e.g., like that of the zoospores of *Phytophthora palmivora* (Peronosporales). See **gravitaxis** and **taxis**.

eleo- (Gr. *elaía*, olive, or *alaío*, oil, fat): a pref., corresponding to Gr. *élaio-* and L. *elaeo-*, employed in the formation of words suggesting the idea of oil, as in *eleohypha*.

eleohypha, pl. **eleohyphae** (Gr. *elaía*, olive, or *alaío*, oil, fat + *hyphé*, hypha, tissue, spiderweb): *Lichens*. A hypha of larger diameter than ordinary hyphae, whose cells is filled with oil, that are found in the thallus of endolithic calcicolous lichens.

eleuthero-, **eleuther-** (Gr. *eleútheros*, free): a pref. employed in the formation of compound terms of Gr. origin that allude to something that is free, i.e., not adhered or attached to another organ or organic part, as in the generic name *Eleutherascus* (Pezizales).

ellipsoid, **ellipsoidal**, **elliptic** (L. *ellipsis*, ellipse < Gr. *éllepsis*, closed curve + L. suf. *-oide*, like < Gr. *-oeídes*, similar to; or + L. suf. *-alis* > E. *-al*, relating to or belonging to; or + Gr. suf. *-íkos* > L. *-icus* > E. *-ic*, relating to or belonging to): refers to a solid body that forms an ellipse in the longitudinal plane and a circle in cross section. Many fungal spores are ellipsoidal.

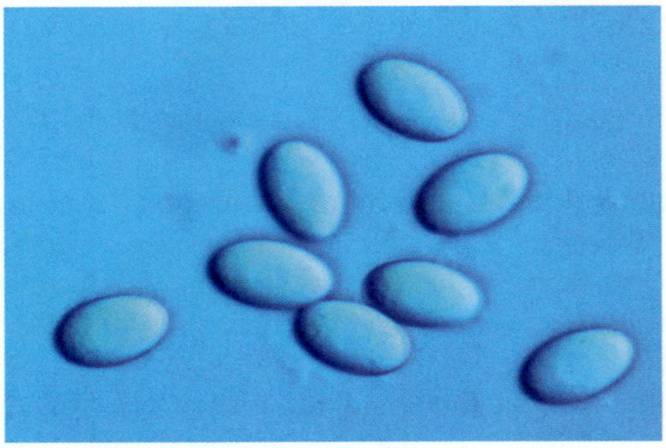

Ellipsoidal ascospores of *Pyronema domesticum* (Pezizales), x 600 (*RTH*).

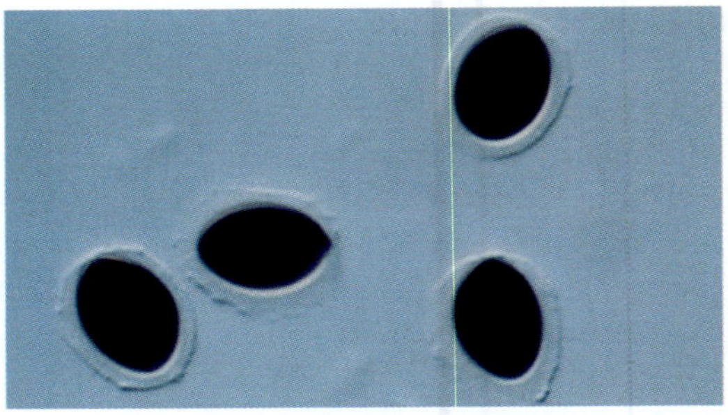

Ellipsoidal ascospores of *Sordaria fimicola* (Sordariales), x 570 (*RTH*).

-em (Gr. *-ema*, that made, created): a suf. employed to form compounds that denote something created, as in *meristem*, which signifies that which divides or multiplies (Gr. *meríso*, to divide), and *xylem*, to be transformed into wood, to become lignified (Gr. *xylón*, log, wood).

emarginate (L. *emarginatus*, without a margin < pref. *e-*, without + *marginatus*, margined < *margo*, margin + suf. *-atus* > E. *-ate*, provided with or likeness): lacking a margin; with a shallow notch or groove. Generally used for gills that have a notch near their attachment to the stipe, as occurs in *Tricholoma equestre* (=*T. flavovirens*), Agaricales. Cf. **marginate**.

Emarginate gills of the basidiocarp of *Tricholoma* sp., x 0.6 (*JC*).

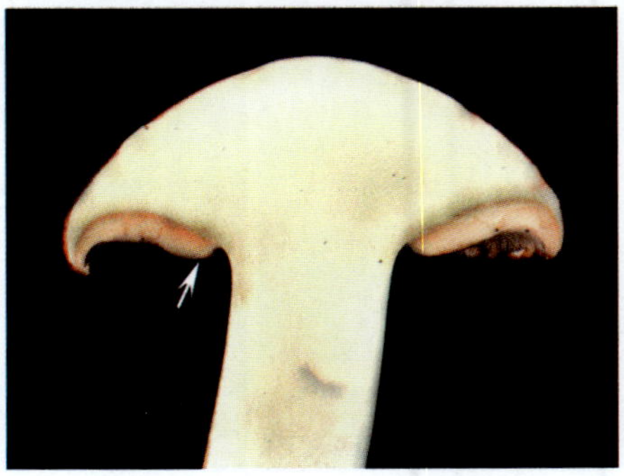

Emarginate gills of a basidiocarp of *Tricholoma* sp., seen in longitudinal section, x 0.6 (*JC*).

embedded (*em-*, var. of *en-* < ME. < OF. < L. *in*, in + *bedded*, pp. of *bed* < ME. < OE. *bedd*, bed): immersed, completely enclosed or surrounded; e.g., the perithecia in the stroma of *Claviceps purpurea* and *Epichloë typhina* (of the Hypocreales), as well as the pycnidia of the sphaeropsidaceous asexual fungi, can be partially or totally embedded in the tissues of the host plant, in the substrate on which they grow, or in a fungal stroma.

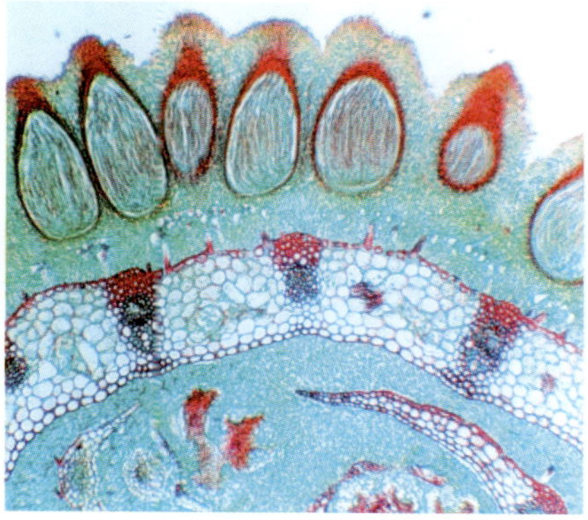

Embedded perithecia in a stroma of *Epichloë typhina*, seen in longitudinal section, x 60 (*RTH*).

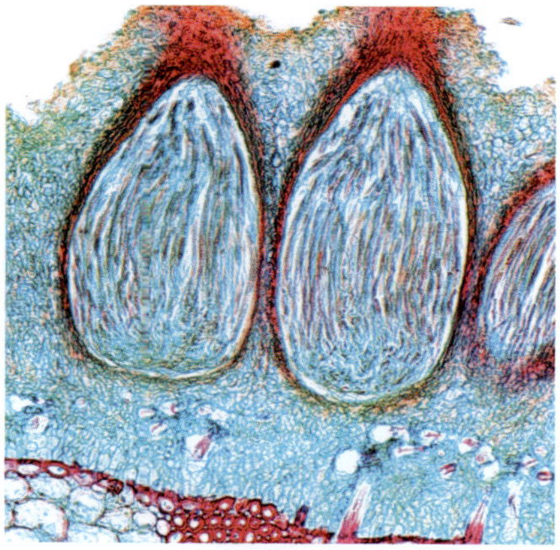

Embedded perithecia in a stroma of *Epichloë typhina*, seen in longitudinal section, x 125 (*RTH*).

Sclerotia of *Claviceps gigantea*, a fungal parasite of corn, **endemic** to Mexico, x 0.5 (*TH*).

Sclerotium of *Claviceps gigantea*, a fungal parasite of corn, **endemic** to Mexico, x 1 (*MU*).

emend (L. *emendare*, to correct, remove defects, amend < pref. *e-*, *ex-*, away, without + *menda*, fault, defect): to correct an error, fault or defect; to change or amend.

endemic (Gr. *endemía*, endemism < *en*, in + *démos*, people + suf. *-íkos* > L. *-icus* > E. *-ic*, belonging to or relating to): native to a particular region; found exclusively in one place. The pyrenomycete *Claviceps gigantea* (Hypocreales), e.g., is endemic to the region near the volcanos Popocatépetl and Iztaccíhuatl, in Mexico. Cf. **cosmopolitan**.

endo-, **end-** (Gr. *éndon*, inside, internal): a pref. that signifies inside, internal, as in *endoascus* and the terms that follow. Equal to **ento-**. Cf. **ecto-** and **exo-**.

endoascus, pl. **endoasci** (L. *endo* < Gr. *éndon*, inside + L. *ascus* < Gr. *askós*, sack, ascus): the internal wall layer of a bitunicate ascus, present in Loculoascomycetes, e.g., *Sporormiella* (=*Preussia*) *australis* (Pleosporales); the endoascus expands before liberating the ascospores when the *ectoascus* breaks. See **bitunicate**.

endobasidium, pl. **endobasidia** (Gr. *éndon*, inside, internal + *basídion*, dim. of *básis*, base): a basidium formed inside a basidioma, like that of the gen. *Lycoperdon* (Lycoperdales). See **exobasidial**.

endobiotic (Gr. *éndon*, inside + *biotikós*, vital < *bíos*, life + suf. *-tikós* > L. *-ticus* > E. *-tic*, relation, fitness, inclination or ability): a parasitic or saprobic organism that develops its reproductive as well as its somatic structures in the interior of the host or substrate. E.g., *Plasmodiophora brassicae* and *Polymyxa graminis* (Plasmodiophorales) form the spores inside the cells of cabbage and wheat roots, respectively, and *Synchytrium endobioticum* and *Endochytrium ramosum* (Chytridiales) do the same in the cells of potato tubers and *Cladophora*; *Rozella allomycis* (Spizellomycetales) lives in the cells of the aquatic fungus *Allomyces javanicus* (Blastocladiales), and *Nucleophaga amoebae* (Spizellomycetales) forms its sporangia inside the nucleus of *Amoeba verrucosa*. Equivalent to **intramatrical**. Cf. **epibiotic** and **interbiotic**.

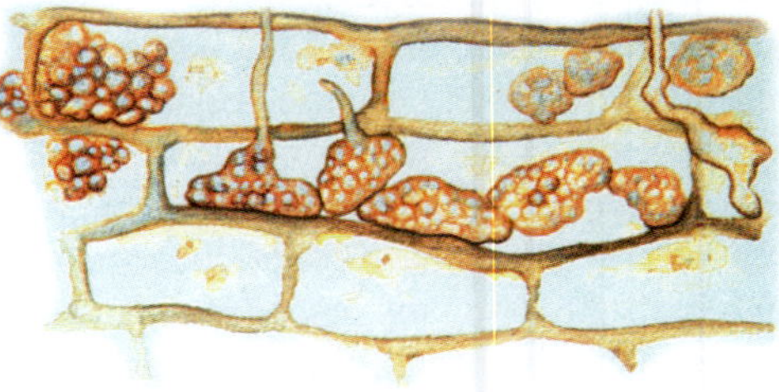

Endobiotic zoosporangia of *Polymyxa graminis* in cells of wheat root (*Triticum sativum*), x 400 (*MU*).

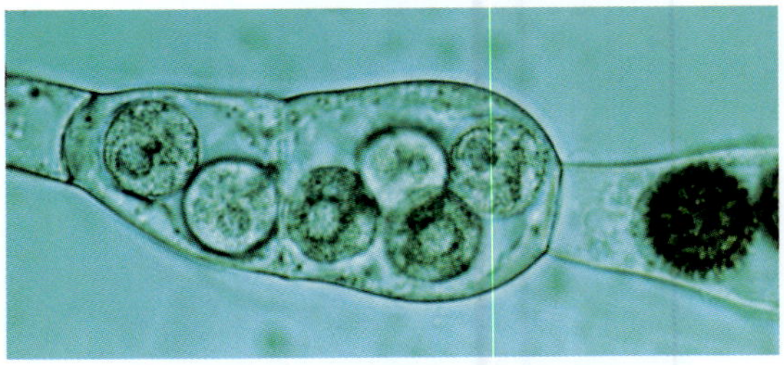

Endobiotic young thalli of *Rozella allomycis* in a hyphal cell of *Allomyces javanicus*, x 410 (*RTH*).

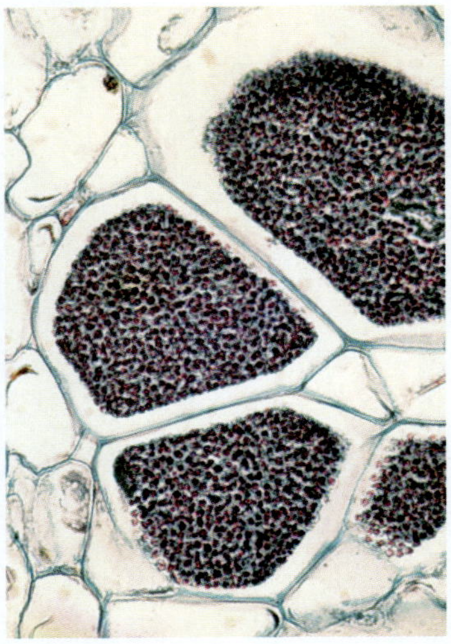

Endobiotic spores of *Plasmodiophora brassicae*, in root cells of cabbage, seen in transverse section, x 1,000 (*MU*).

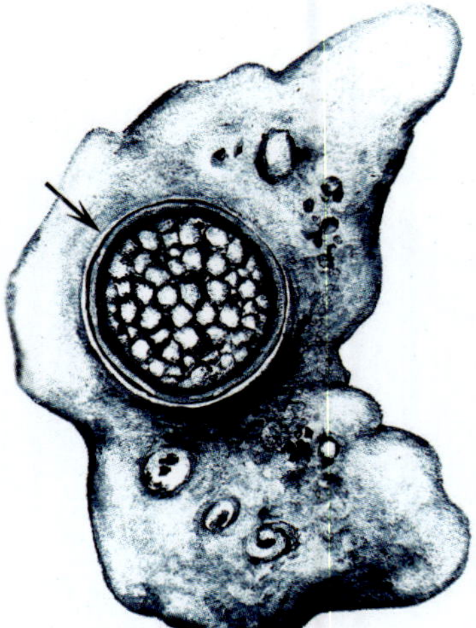

Endobiotic sporangium of *Nucleophaga amoebae* inside the nucleus of the host (*Amoeba verrucosa*), x 500 (*MU*).

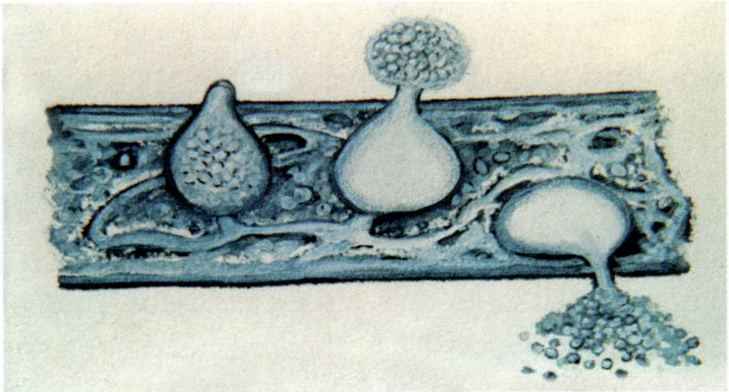

Endobiotic zoosporangia of *Endochytrium ramosum* inside the thallus of *Cladophora* sp., x 300 (*MU*).

endocarpic (Gr. *éndon*, inside + *karpós*, fruit + suf. *-íkos* > L. *-icus* > E. *-ic*, belonging to or relating to): *Lichens*. Applied to species with perithecia inserted in the thallus, as is seen in *Endocarpon* and *Dermatocarpon* (Verrucariales). See **endocarpinoid**.

endocarpinoid (< gen. *Endocarpon*, of the order Verrucariales < Gr. *éndon*, inside, interior + L. suf. *-carpon* < Gr. *karpós*, fruit): refers to the lichenized perithecia that are embedded in the thallus tissues, as in *Endocarpon*. Also called **endocarpic**.

endochroic (Gr. *éndon*, inside, interior + *chrōs*, color + suf. *-ikos* > L. *-icus* > E. *-ic*, belonging to or relating to): having pigmentation inside the cells. Cf. **ectochroic**.

endocommensal (Gr. *éndon*, inside + L. *commensalis*, one who shares a table with another < *cum*, with + *mensa*, table + suf. *-alis* > E. *-al*, relating to or belonging to): a fungus that lives in the interior of an animal host, as happens with some species of Trichomycetes of the order Eccrinales, e.g., *Parataeniella armadillidii*, which lives inside scale insects, and *Enterobryus* sp., which inhabits the posterior intestinal tract of myriapods, as well as the order Asellariales, e.g., *Ramacrinella raibautii*, whose thalli live attached to the proctodeum of marine crustacean amphipods. See **commensal.**

Enterobryus sp., an **endocommensal** eccrinid, with thalli attached to the chitinous wall of the posterior intestinal tract of a myriapod, x 20 (*MU*).

Ramacrinella raibautii, an **endocommensal** eccrinid of crustaceans, x 20 (*MU*).

endoconidium, pl. **endoconidia** (Gr. *éndon*, inside + *kónis*, dust, powder + dim. suf. *-ídion* > L. *-idium* > E. *-idium*): a conidium produced in the interior of a conidiogenous cell, as in *Thielaviopsis basicola* (=*Chalara elegans*, dematiaceous asexual fungi), whose phialides contain chains of endoconidia that emerge through the mouth of the phialides on maturing.

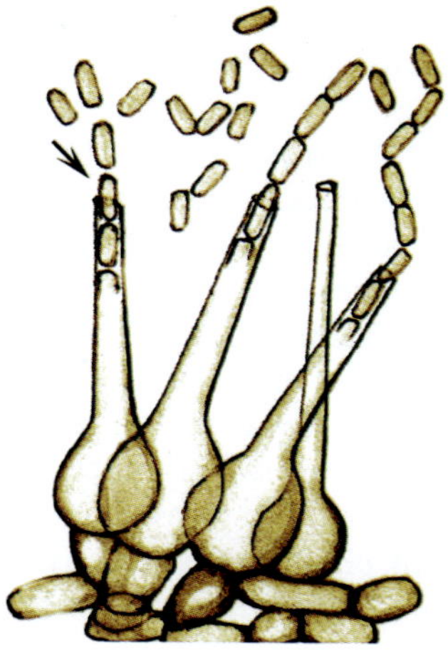

Endoconidia inside the phialides of *Thielaviopsis basicola* (=*Chalara elegans*), x 190 (*MU*).

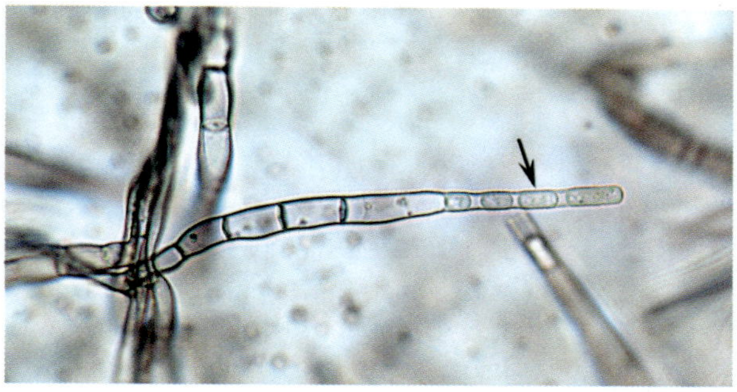

Endoconidia inside a phialide of *Thielaviopsis basicola* (=*Chalara elegans*), x 275 (*MU*).

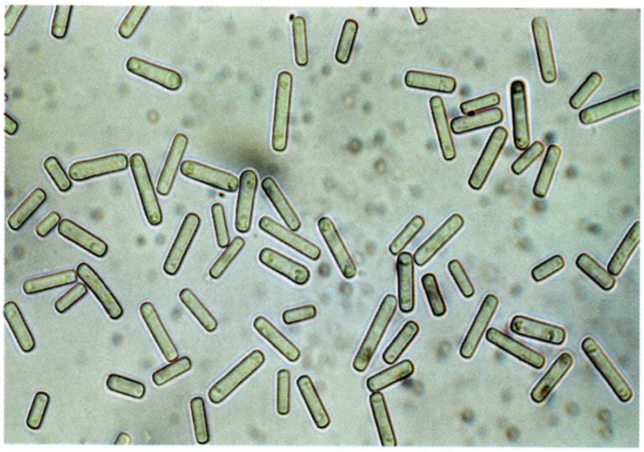

Endoconidia liberated from the phialides of *Thielaviopsis basicola* (=*Chalara elegans*), x 275 (*MU*).

endocyanosis, pl. **endocyanoses** (Gr. *éndon*, inside + *kýanos*, a substance dark blue in color + suf. *-osis*, state or condition): the condition in which the cells of an organism contain cyanobacteria, as happens in the gen. *Geosiphon* (a member of the Zygomycota, of uncertain affinity), the only known example of endocyanosis in the fungi. This organism is sometimes considered lichenized, as it has cyanobacteria inside vesicles.

endocyclic rust fungus, pl. **fungi** (Gr. *éndon*, inside + *kyklikós*, circular, cyclic < *kýklos*, circle + suf. *-íkos* > L. *-icus* > E. *-ic*, pertaining to or relating to): *Uredinales*. A species of microcyclic rust that lacks states I and II (spermogonia and aecia) and a primary host, and whose teliospores are morphologically identical to aeciospores; notwithstanding, on germinating, these teliospores give rise to basidia and basidiospores instead of forming germ tubes.

endocystidium, pl. **endocystidia** (Gr. *éndon*, inside + *kystídion* < *kýstis*, bladder, vesicle, cell + dim. suf. *-ídion* > L. *-idium* > E. *-idium*): *Agaricales*. A chrysocystidium that is embedded in the trama of the pileus, stipe or hymenophore of many Agaricales. E.g., in some species of *Pholiota*, *Stropharia* and *Hypholoma*, endocystidia are present in the trama of the hymenophore.

endoectothrix (Gr. *éndon*, inside, interior + *ektós*, outside + *thríx*, genit. *trichós*, hair): developing in the interior and on the exterior of a hair, like some dermatophytic fungi.

endoexogenous (Gr. *éndon*, inside + *éxo*, outside + *génos*, origin < *gennáo*, to engender, produce): a mode of growth and development of aquatic fungi of the order Chytridiales, which is characterized first by a flow of cytoplasm and food and a direction of growth from the zoospore toward an extramatrical rhizoidal system, but later this flow of protoplasm and food reverses and is directed from the rhizoids toward the developing zoospore.

endogenous (Gr. *éndon*, inside + *génos*, origin < *gennáo*, to engender, produce): any structure produced or developed in the interior of another structure; e.g., spores produced in sporangia, pycnidia, asci, etc., are endogenous. Cf. **exogenous**.

endohypha, pl. **endohyphae** (Gr. *éndon*, inside + *hyphé*, spiderweb, hypha): the intraconidiophorous hypha of *Acrogenospora sphaerocephala* (=*Brachysporiella*=*Monotosporella sphaerocephala*, moniliaceous asexual fungi), that originates from the internal wall of the conidiophores (which is electron transparent), and which on growing gives the conidiophores the superficial appearance of an annellide or annellate phialide.

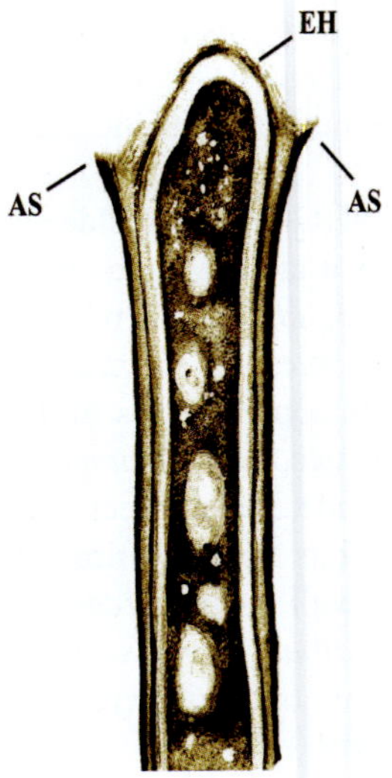

Endohypha (EH) in a conidiophore of *Acrogenospora sphaerocephala* (=*Brachysporiella*=*Monotosporella sphaerocephala*). Longitudinal section of a conidiophore that shows the proliferation as the result of the penetration by an endohypha through the remnant of the septum, which delimited the previously formed conidium. Note the annular scar (AS). Drawing based on a transmission electron micrograph published by Beckett *et al.* in their *Atlas of Fungal Ultrastructure*, 1974, x 1,900 (*MU*).

endokapylic (Gr. *éndon*, inside + *kapyleío*, to falsify, adulterate + suf. *-íkos* > L. *-icus* > E. *-ic*, belonging to or relating to): the thallus of a lichenized fungus that does not form a morphologically distinctive lichenized structure. Cf. **epikapylic**.

endolichen fungi, sing. **fungus** (Gr. *éndon*, inside + Gr. *leichén*, lichen + *-tikós* > L. *-ticus* > E. *-ic*, belonging to or relating to; L. *fungi*, pl. of *fungus*, fungus): fungi that grow asymptomatically in lichen thalli.

endolithic (Gr. *éndon*, inside + *líthos*, stone + suf. *-íkos* > L. *-icus* > E. *-ic*, belonging to or relating to): *Lichens*. A crustaceous lichen that develops in the interior of rocks, such as *Verrucaria calciseda* (Verrucariales) and other related species. Cf. **epilithic**. See also **cryptoendolithic**.

endomycobiont (Gr. *éndon*, inside, interior + *mýkes*, fungus + *bíos*, life + *óntos*, genit. of *ón*, a being): a fungal biont in a symbiosis, completely immersed in the tissues of the host (e.g., fungal component of a mycophycobiosis); also used for the mycobiont of some mycorrhizae.

endomycorrhiza, pl. **endomycorrhizae** (Gr. *éndon*, inside + *mýkes*, fungus + *rhíza*, root): a mycorrhiza in which the hyphae of the associated fungus penetrate the cells of the associated plant; also called *arbuscular mycorrhiza*, like the symbiosis between certain species of *Endogone*, *Acaulospora* (Endogonales) and the roots of various plants. Cf. **ectomycorrhiza**.

Endomycorrhiza, intracellular vesicle of *Endogone* sp. in onion root, x 245 (*SP*).

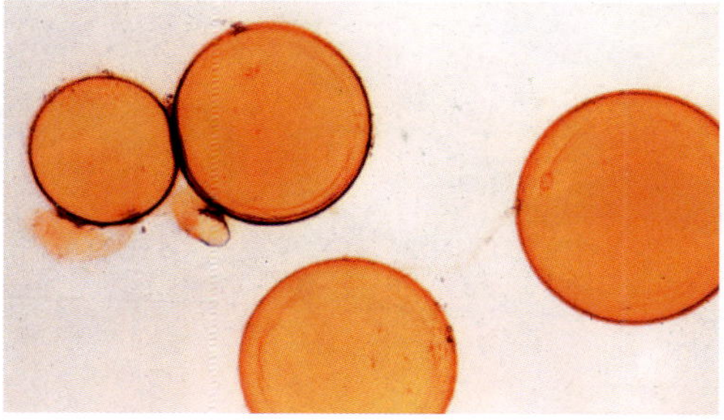

Azygospores of *Acaulospora laevis*; this fungus forms **endomycorrhizae** with various plants (gramineous, legumes and palms), x 760 (*LVF* & *AE*).

endoparasite, **endoparasitic** (Gr. *éndon*, inside + *parásitos* < *pará*, at the side of, together + *sîtos*, flour, bread, food in general; or + suf. *-ikos* > L. *-icus* > E. *-ic*, belonging to, relating to): a parasitic organism that develops in the interior of a cell (intracellular parasite) or of the tissues (intercellular parasite) of another organism (the host), on which it feeds. Some primitive aquatic fungi (Chytridiales) are endoparasites of algae, of plants and of other fungi. E.g., *Synchytrium endobioticum* is endoparasitic in potato tubers. *Harposporium anguillulae* (moniliaceous asexual fungi) is an endoparasite of nematodes, which are infected when the animals ingest the conidia of the fungus. Among the Saccharomycetales of the family Metschnikowiaceae, *Coccidiascus legeri* represents a rare and interesting endoparasite that lives in the intestinal cells of the fly *Drosophila funebris*. Cf. **ectoparasite**.

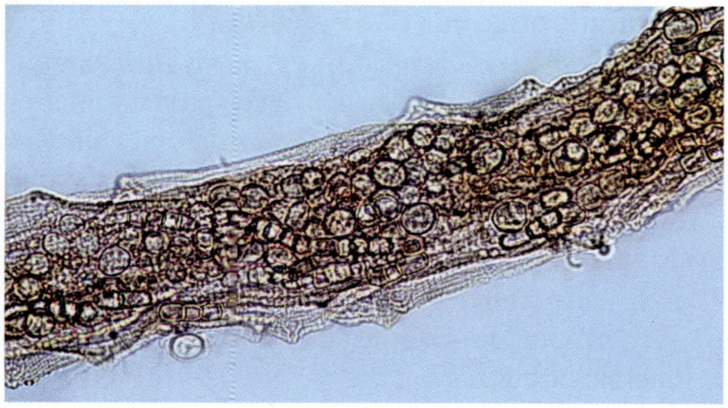

Nematode with chlamydospores of *Harposporium anguillulae*, an **endoparasitic** fungus, x 400 (*PL*).

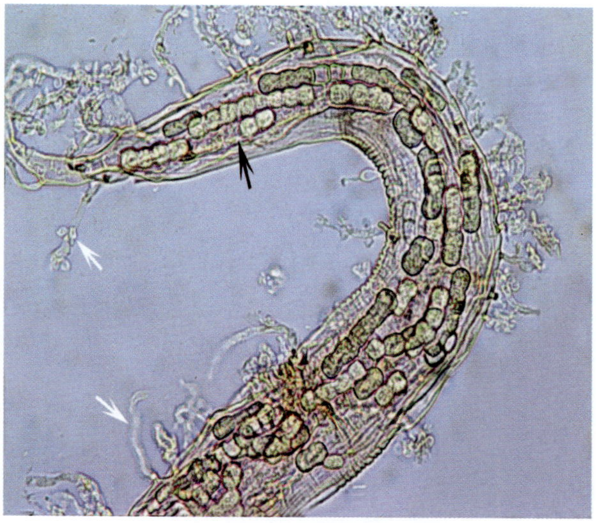

Nematode with chlamydospores (black arrow) and conidiophores (white arrow) with conidia of *Harposporium anguillulae*, an **endoparasitic** fungus, x 400 (*PL*).

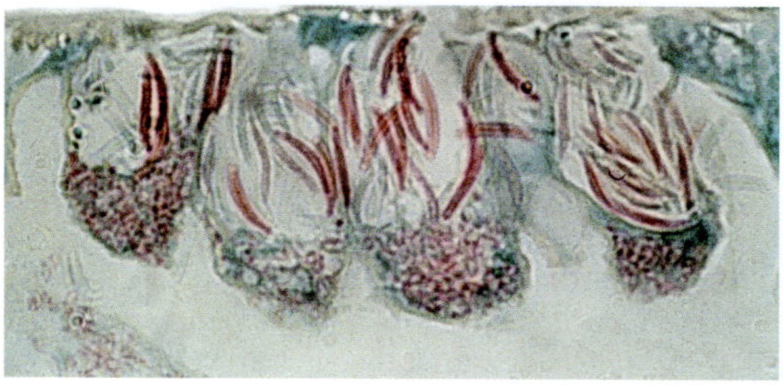

Asci of *Coccidiascus legeri*, **endoparasitic** in the intestinal cells of the fly *Drosophila funebris*, x 400 (*RTH*).

endoperculation (Gr. *éndon*, inside, toward the interior + L. *operculum*, operculum, lid, cover + Gr. suf. *-ationem* > E. *-ation*, action, state or condition, result): refers to a type of sporangial dehiscence in the Chytridiales, in which the operculum is pushed outward by the emerging sporogenous contents. Cf. **exoperculation**.

endoperidium, pl. **endoperidia** (Gr. *éndon*, inside + *péridion*, small leather purse): the inner wall layer of a multilayered peridial wall, as in the Gasteromycetes of the gen. *Geastrum* (Lycoperdales), e.g., *G. sac-catum* and *G. triplex*; the outer wall layer is the *exoperidium*.

Endoperidium (EN) and exoperidium (EX) of the basidiocarp of *Geastrum saccatum*, x 0.5 (*TH*).

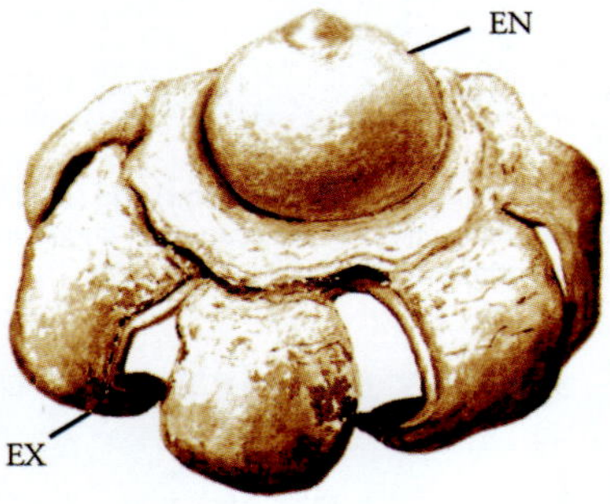

Endoperidium (EN) and exoperidium (EX) of the basidiocarp of *Geastrum triplex*, x 1 (*MU*).

endophloeodic, endophloeodal, endophloeic (Gr. *éndon*, inside + *phloiós* > L. *phloeo*, bark): refers to the thallus of crustaceous lichens that are almost completely immersed in the bark of trees. Syn. of **hypophloeodic**. Cf. **ectophloeodic**.

endophyllous (Gr. *éndon*, inside + *phýllon*, leaf): living inside a leaf; beneath the cuticle.

endophyte (Gr. *éndon*, inside + *phytón*, plant): **1**. A fungus that lives in plant tissues without showing external signs of its presence, like certain fungal parasites of pasture grasses that are toxic to animals that feed on them. **2**. The internal, systemic mycelium of clavicipitaceous fungi, such as *Epichloë* (Hypocreales).

endoplacodium, pl. **endoplacodia** (Gr. *éndon*, inside + NL. *placodium* < Gr. *plakós*, plaque + L. *-odium*, resembling < Gr. *ode*, like < *-oeídes*, similar to): see **placodium**.

endoplasm (Gr. *éndon*, inside + *plásma*, formation, soft material with which a living being is formed): the inner layer of the cytoplasm, relatively fluid, like the plasmodial veins of the Myxomycetes. Cf. **ectoplasm**.

endoplasmic reticulum, pl. **reticula** (Gr. *éndon*, inside + *plásma*, genit. *plásmatos*, formation, soft material with which a living being is formed + suf. *-íkos* > L. *-icus* > E. *-ic*, pertaining to or relating to; L. *reticulum*, small net < *rete*, net + dim. suf. *-culum*): a network of double membranes, connected to the nuclear envelope, that is arranged in flattened, perforate, laminar or tubular chambers called cisternae, that are spread out in the cytoplasm. The membranes of the endoplasmic reticulum can have ribosomes on the cytoplasmic side (rough endoplasmic reticulum) or lack them (smooth endoplasmic reticulum). Generally abbreviated as ER.

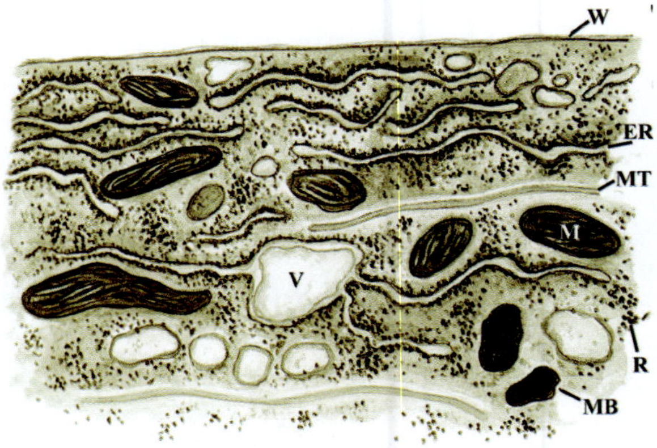

Endoplasmic reticulum (ER) of a young hypha of *Armillaria mellea* (Agaricales). Also observed are ribosomes (R) of the cisternae of the ER, microbodies (MB), mitochondria (M), microtubles (MT), vesicles (V), and the cell wall (W). Drawing based on several transmission electron micrographs published by Beckett *et al.* in their *Atlas of Fungal Ultrastructure*, 1974, x 22,200 (*MU*).

endopropagule (Gr. *éndon*, inside + NL. *propagulum*, propagule < *propagare*, to propagate, disseminate): *Med. mycol*. A propagule produced inside the body of the host.

endopsammon (Gr. *éndon*, inside + *psámmos*, sand): the microhabitat among the grains of sand in marine beaches, where arenicolous and psammophilous organisms, including some fungi, such as *Corollospora*, *Carbosphaerella* (Halosphaeriales) and *Nia* (Melanogastrales) are found. See **arenicolous** and **psammophilous**.

endosaprobe (Gr. *éndon*, inside + *saprós*, rotten + *bíos*, life): a saprobic organism that develops in the interior of a dead host; an example of *endosaprobiosis* (+ Gr. suf. *-osis*, state or condition) is the destruction of the alga by the fungus of a lichen thallus.

endosclerotium, pl. **endosclerotia** (Gr. *éndon*, inside + NL. *sclerotium*, sclerotium < Gr. *sklerotés*, hardness < *sklerós*, hard + L. dim. suf. *-ium*): a sclerotium of internal origin.

endospore, **endosporium**, pl. **endosporia** (Gr. *éndon*, inside + *sporá*, spore + L. dim. suf. *-ium*): **1**. The innermost layer, adjacent to the cytoplasmic membrane, of the five layers that form the wall of the most complex basidiospore. The endospore can vary from very thick (in which case it can be subdivided into internal and external portions) up to apparently or truly absent; e.g., the spores of *Tubaria*, Agaricales, are thin-walled and lack an endospore. Over the endospore are four other layers: *mesospore*, *epispore*, *exospore* and *perispore*. **2**. Also applied to the asexual spores formed by internal budding of the hyphae of some species of yeasts with true mycelium, such as *Trichosporon capitatum* and *Geotrichum fermentans* (=*Trichosporon fermentans*), cryptococcaceous asexual yeasts. In these yeasts 1-15 and 2-3 endospores, respectively, are formed in intercalary or terminal hyphal segments, either from the septa or in the tip of tubular intrahyphal growths. In *Aureobasidium pullulans* (dematiaceous asexual fungi) one can also observe endospores or endoconidia in the hyphae.

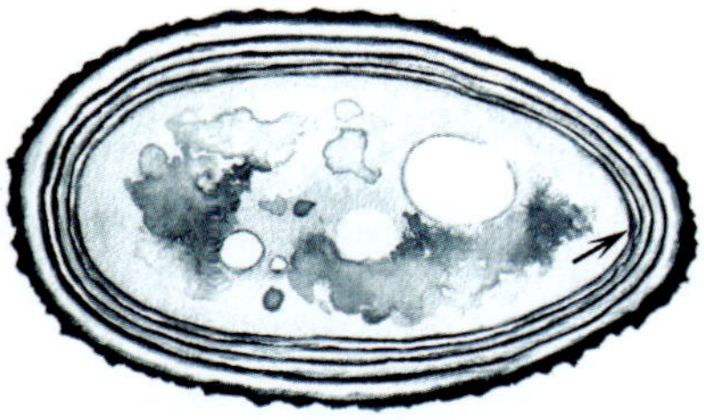

Endospore, mesospore, epispore, exospore and perispore of the cell wall of a basidiospore, seen in this diagram of a longitudinal section, x 5,500 (*MU*).

endostroma, pl. **endostromata** (Gr. *éndon*, inside + *strõma*, small pillow, cushion): *Pyrenomycetes*. The portion of the perithecial stroma situated beneath the *ectostroma*; generally it is sunken in the tissues of the host plant and surrounds the perithecia. Stromatic species, such as *Diatrype*, *Eutypa* and *Xylaria*, of the Xylariales, have an endostroma. Cf. **ectostroma**.

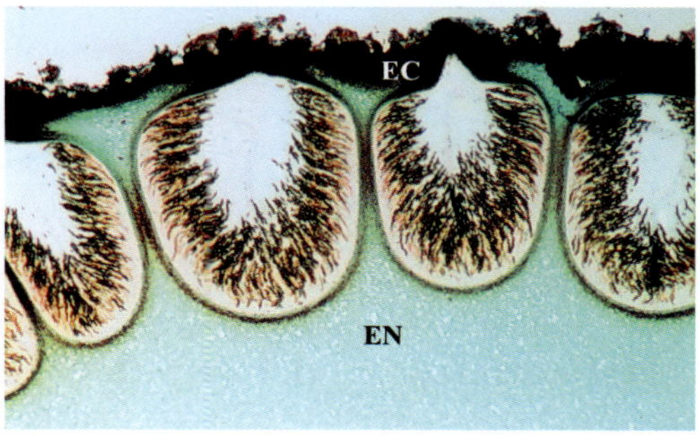

Endostroma (EN) and ectostroma (EC) of the perithecial stroma of *Xylaria* sp., seen in transversal section; the perithecia are seen in longitudinal section, x 80 (*MU*).

endosymbiont (Gr. *éndon*, inside + *symbióntos* < *symbióo*, life in common + *óntos*, genit. of *ón*, a being): an organism that lives in mutualistic symbiosis inside the cells of another organism.

endothrix (Gr. *éndon*, inside + *thríx*, hair): one of the two modes of invasion by fungi of the hair of man and animals, in which the parasite forms parallel chains of arthrospores inside the axis of the hairs, as is seen in the endothritic species *Trichophyton tonsurans*, *T. mentagrophytes* and *T. violaceum* (moniliaceous asexual fungi). Cf. **ectothrix**.

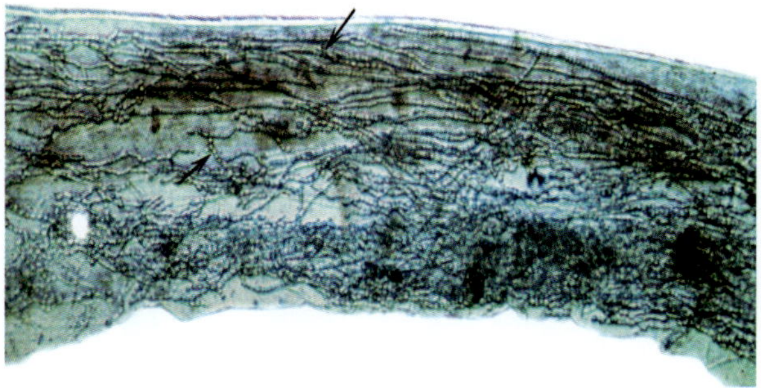

Endothrix invasion of a human hair by *Trichophyton tonsurans*, x 400 (*RLM*).

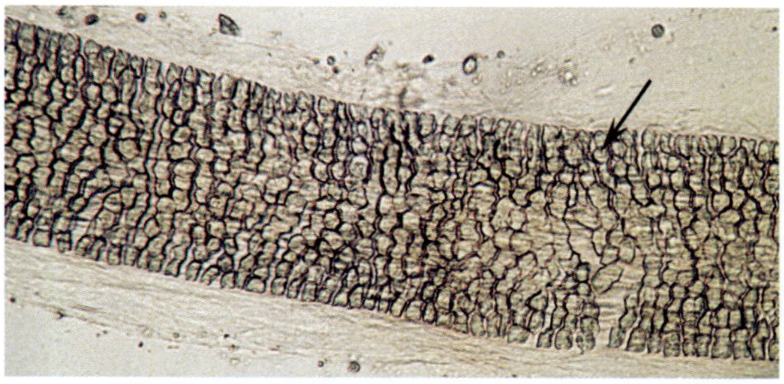

A type of **endothrix** infection caused by *Trichophyton mentagrophytes* in a hair of white rat; the arthrospores of the fungus are seen inside the shaft of the hair, x 115 (*MU*).

endotroph (Gr. *éndon*, inside + *trépho*, to nourish): **1**. *Mycorrhizae*. A mycorrhizal fungus that penetrates the cells of the host plant, and of the mycorrhiza that results from the association of both (endotrophic mycorrhiza). The exchange of nutrients between the two organisms occurs inside the cells of the host. Cf. **ectotroph**. **2**. *Lichens*. A cephalodium that originates in the interior of the thallus.

endotropic (Gr. *éndon*, inside + *trópos*, to turn, change of position or manner): curving inward from the point of origin. Cf. **ectotropic**.

endotunica (Gr. *éndon*, inside + L. *tunica*, covering, wrapper): the inner wall layer of a bitunicate ascus. Equal to **endoascus**. Cf. **ectotunica**.

endozoic (Gr. *éndon*, inside + *zõon*, animal + suf. *-íkos* > L. *-icus* > E. *-ic*, belonging to or relating to): a fungus that grows as a commensal, symbiont or parasite in the interior of living animals, like the majority of the species of Trichomycetes, e.g., *Smittium* (Harpellales). Other examples of endozoic species are found in some Oomycota of the order Peronosporales, such as *Myzocytium zoophthorum*, which forms its zoosporangia and gametangia inside the rotifers that it parasitizes, as well as in many Zoopagales, such as *Bdellospora helicoides* and *Endocochlus asteroides*, which invade terricolous amoebae. Cf. **ectozoic**.

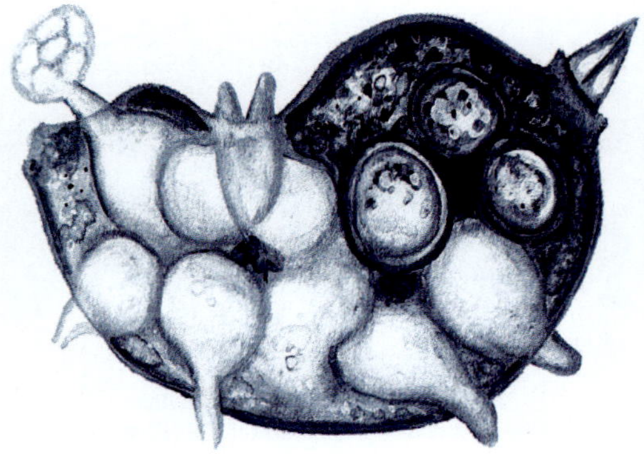

Rotifer with zoosporangia and gametangia of *Myzocytium zoophthorum*, an **endozoic** fungus, x 180 (*MU*).

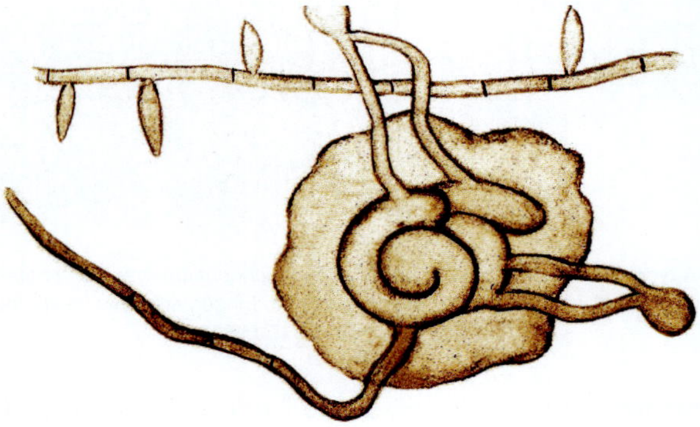

An amoeba, *Amoeba terricola*, with helicoid thalli of *Endocochlus asteroides*, an **endozoic** species, from which are emerging two zygosporangia and a hypha, x 350 (*MU*).

energid (Gr. *energós*, active, operative + L. des. *-id*, which denotes quality of, one that has or does something): a vital entity constituted of a nucleus and a certain enveloping protoplasmic mass controlled by it. In all true cells there is an energid. See **polyenergid**.

enphytia, **enphytotia** (Gr. pref. *en-*, in + *phytón*, plant + *-ia*, des. that denotes quality or state of being): an endemic disease whose effects are constant from one year to another. Cf. **epiphytotic**.

enphytotic (Gr. pref. *en-*, in + *phytón*, plant + suf. *-tíkos* > L. *-ticus* > E. *-tic*, relation, fitness, inclination or ability): belonging to or relating to **enphytia** or **enphytotia**.

ensiform, **ensate** (L. *ensis*, sword + suf. + *-formis* < *forma*, shape): having the shape of a sword, i.e., long and flat with parallel, sharp edges, and ending in a point, like some hymenial cystidia of certain Aphyllophorales. Also called **gladiate** and **xiphioid**.

-ent (NL. *-ant*, *-ans*, one who does or promotes something): a suf. in adjectives and nouns; in adjectives it signifies being, that which is or what is, as in *pendent*, that which dangles or hangs, and in nouns it signifies that which does something, as in *detergent*, a cleaning agent. At times equivalent to *-ant* (with the same etymology), a suf. employed in terms such as *sporulant*, etc.

ental (Gr. *entós*, inside + L. suf. *-alis* > E. *-al*, relating to or belonging to): of, or in, the interior. Cf. **ectal**.

entangial, **entoangial** (Gr. *entós*, inside + *angeîon*, vessel, receptacle + L. suf. *-alis* > E. *-al*, relating to or belonging to): formed or developed inside a conceptacle, like the sporangiospores of the Chytridiomycota (e.g., *Phlyctochytrium*, of the Chytridiales) and Zygomycota (e.g., *Rhizopus*, of the Mucorales). Cf. **ectangial**.

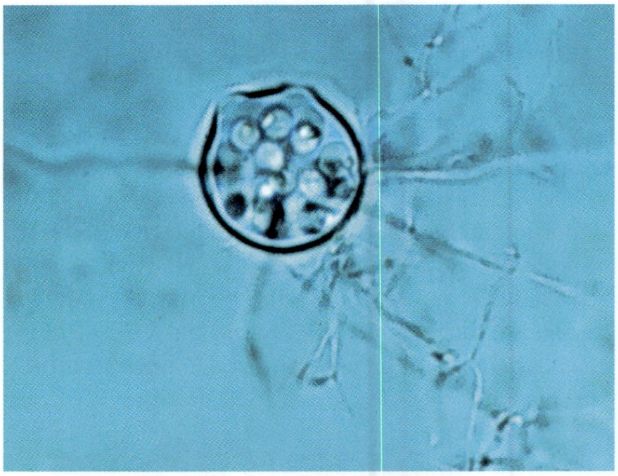

Entangial development of the sporangiospores of *Phlyctochytrium* sp., x 500 (*RTH*).

enteroarthric (Gr. *énteron*, intestine + *árthron*, articulation + suf. *-ikos* > L. *-icus* > E. *-ic*, belonging to or relating to): *Conidiogenesis*. A type of conidial development in which the outer wall of the conidiogenous cell generally separates from the wall of the conidium that forms in its interior, as occurs in *Sporendonema* (moniliaceous asexual fungi). Cf. **holoarthric**.

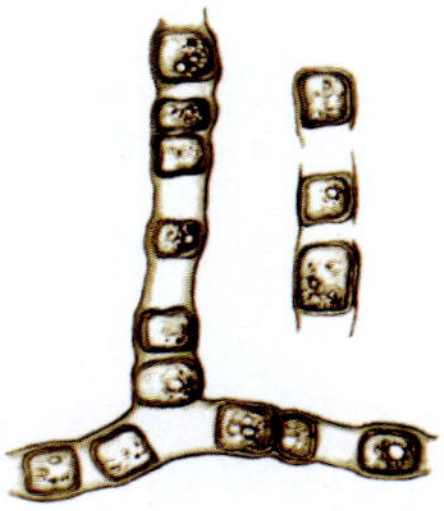

Enteroarthric conidia in hyphae of *Sporendonema purpurascens*, x 1,100 (*MU*).

enteroblastic (Gr. *énteron*, intestine + *blastós*, sprout, bud, germ + suf. *-íkos* > L. *-icus* > E. *-ic*, belonging to or relating to): see **blastic**.

enterophilous (Gr. *énteron*, intestine + *phílos*, have an affinity for): an organism that lives in the intestine of animals, like the Trichomycetes of the gen. *Enterobryus* (Eccrinales), which live as endocommensals, attached to the chitinous lining of the intestine of coleopterans, diplopods and decapods.

enterothallic (Gr. *énteron*, intestine + *thallós*, shoot, thallus + suf. *-íkos* > L. *-icus* > E. *-ic*, belonging to or relating to) see **thallic**.

entheogen, entheogenic (Gr. pref. *en-*, in + *theós*, God + *génos*, engenderment; or + suf. *-íkos* > L. *-icus* > E. *-ic*, belonging to, relating to): literally, "God generated within". A term used for hallucinogens (plants, fungi), with a strong religious connotation, as part of a theory maintaining that, very early in the evolution of human beings, the strange and frightening effects of hallucinogenic substances (such as the psilocybin of *Psilocybe*, Agaricales, and lysergic acid of *Claviceps purpurea*, Clavicipitales) induced man to believe in the supernatural, giving rise to the prototypical ideas of the religious experience. See **ethnomycology**, **hallucinogenic fungi**, **magic mushrooms** and **psychotomimetic**.

Fruiting bodies of *Psilocybe mexicana*, an **entheogenic** fungus, x 1 (*TH*).

Sclerotia of *Claviceps purpurea* on a rye spike, an **entheogenic** fungus, x 1 (*MU*).

entire (ME. *entere* < MF. *entier* < L. *integrum* < *integer*, continuous, complete, whole): having absolute marginal integrity, i.e., without teeth, notches, or indentations that make it uneven; applied to the margin of a colony on culture media, to the edge of gills on an agaric pileus, etc. E.g., the colony margin of *Cryptococcus albidus* var. *albidus* and *Candida catenulata* (cryptococcaceous asexual yeasts), on agar is entire.

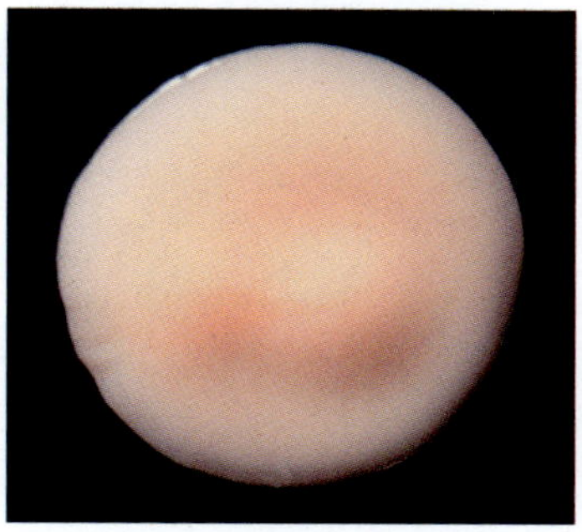

Entire margin of a colony of *Cryptococcus albidus* var. *albidus*, on agar, x 1 (*MU*).

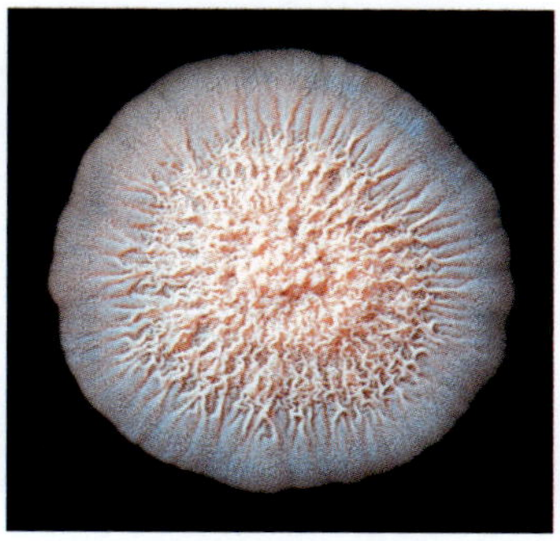

Entire margin of a colony of *Candida catenulata*, on agar, x 1 (*MU*).

ento-, **ent-** (Gr. *entós*, inside, in the interior): equal to **endo-**.

entomo-, **entom-** (Gr. *éntomos*, cut, divided into segments, insect): a pref. employed in the formation of terms of Gr. origin to introduce the idea of insect, as in *entomogenous*.

entomogenous (Gr. *éntomos*, cut, divided into segments, insect + *génos*, origin < *gennáo*, to engender): growing on insects or obtaining nutrients from them (*entomophagous* < Gr. *phágos*, which eats). The entomogenous fungi include a wide range, from commensals or mutualists to ectoparasites that do not seriously affect their arthropod hosts, to pathogens that are lethal. E.g., species of the gen. *Septobasidium* (Septobasidiales) form symbiotic associations with scale insects. The Trichomycetes and Laboulbeniales appear to have little effect on their hosts, but species of *Coelomomyces* (Blastocladiales) and *Lagenidium giganteum* (Lagenidiales), both mainly on mosquito larvae, as well as *Entomophthora*, *Erynia*, *Massospora* (Entomophthorales), *Beauveria*, *Metarhizium*, *Paecilomyces*, *Verticillium*, *Lecanicillium* (moniliaceous asexual fungi), *Hirsutella* (stilbellaceous asexual fungi), and *Cordyceps* (Hypocreales), are important pathogens of various insects. Cf. **entomophilous** and **entomopathogenic**.

Entomogenous synnemata of *Hirsutella saussurei*, arising from a parasitized wasp, x 2 (*MU* & *CB*).

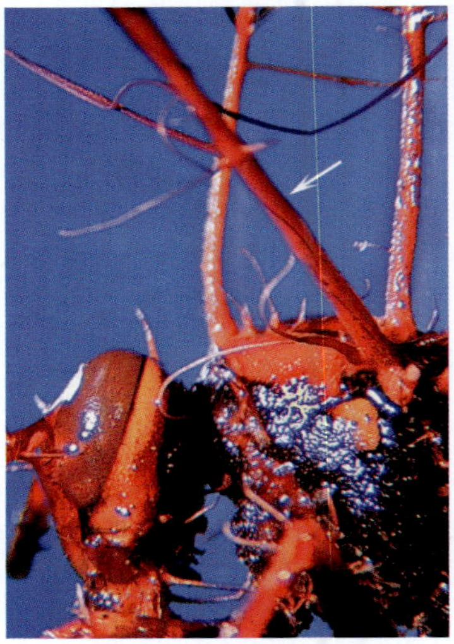

Entomogenous synnemata of *Hirsutella saussurei*, arising from a parasitized wasp, x 5 (*MU* & *CB*).

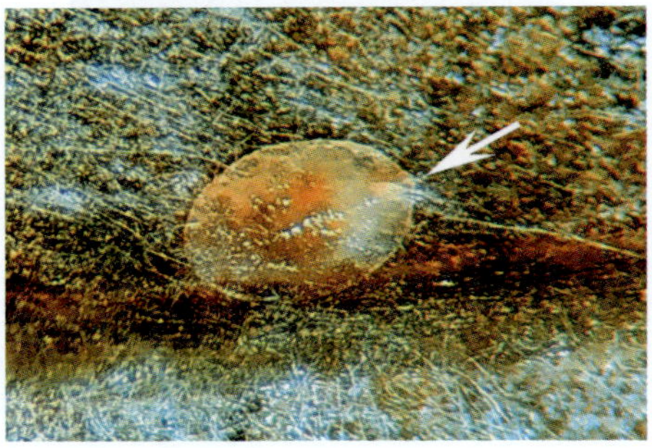

A whitefly nymph parasitized by *Lecanicillium* (=*Verticillium*) *lecanii*, on the lower surface of a bean leaf. The fungus is **entomogenous** and is utilized for the biological control of this insect, x 20 (*MU*).

Conidial columns of *Metarhizium anisopliae* var. *anisopliae*, on an agar culture. The fungus is **entomogenous** and is utilized for the biological control of various insects, x 5 (*MU*).

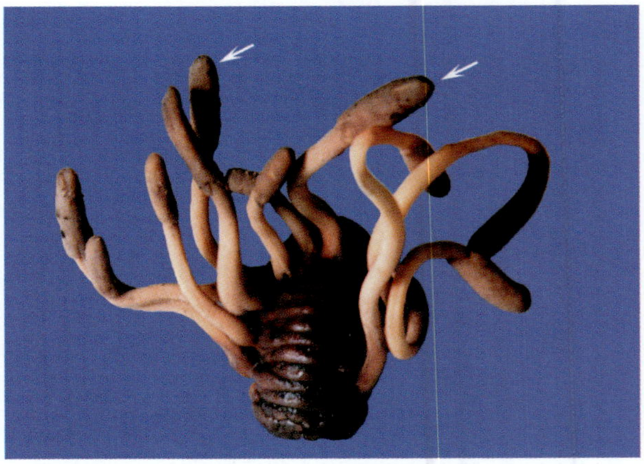

Entomogenous perithecial stromata of *Cordyceps melolonthae* var. *rickii*, on a parasitized coleopteran larva, x 2 (*TH*).

Entomogenous perithecial stroma of *Cordyceps sphecocephala*, on a parasitized wasp, x 2 (*MU*).

Entomogenous perithecial stromata of *Cordyceps stylophora*, arising from a parasitized wasp, x 1.5 (*MU*).

entomopathogenic (Gr. *éntomos*, cut, divided into segments, insect + *pathós*, disease + *génos*, origin < *gennáo*, to engender): applied to a fungus that causes disease in insects and other arthropods. This term is interrelated with the terms **entomophilous** (because the spores of the pathogenic fungus are disseminated by the insect attacked) and **entomogenous** (because the somatic and sporiferous structures of the fungus develop from the body of the infected insect). Examples of these species are *Hirsutella* (stilbellaceous asexual fungi), *Lecanicillium*, *Metarhizium*, *Isaria*, *Paecilomyces* (moniliaceous asexual fungi) and *Rickia* (Laboulbeniales).

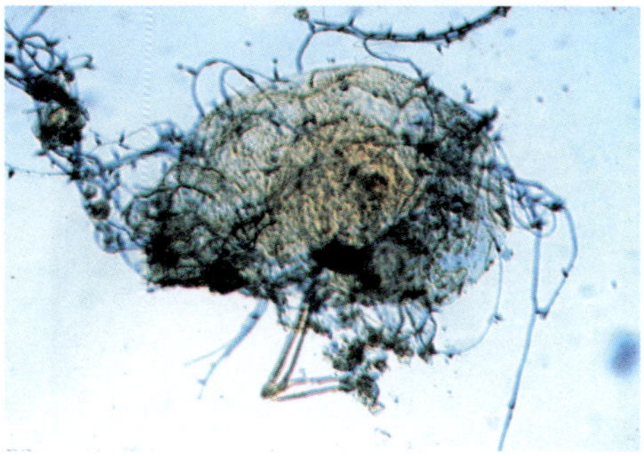

A citrus mite (*Tetranychus* sp.) parasitized by *Hirsutella thompsonii*. The fungus is **entomopathogenic** and is utilized for the biological control of this mite, x 100 (*LS*).

A whitefly nymph parasitized by *Lecanicillium* (=*Verticillium*) *lecanii*. The fungus is **entomopathogenic** and is used to control this insect, x 5 (*CT*).

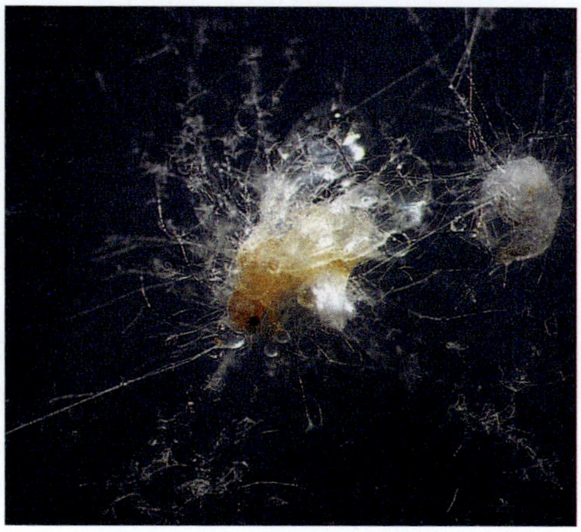

An adult whitefly parasitized by *Lecanicillium* (=*Verticillium*) *lecanii*. The fungus is **entomopathogenic** and is used to control this insect, x 5 (*CT*).

A colony of *Lecanicillium* (=*Verticillium*) *lecanii*, an **entomopathogenic** fungus, on agar, x 1 (*MU*).

A locust parasitized by *Metarhizium anisopliae* var. *acridum*. The fungus is **entomopathogenic** and is used to control this insect, x 1 (*CT*).

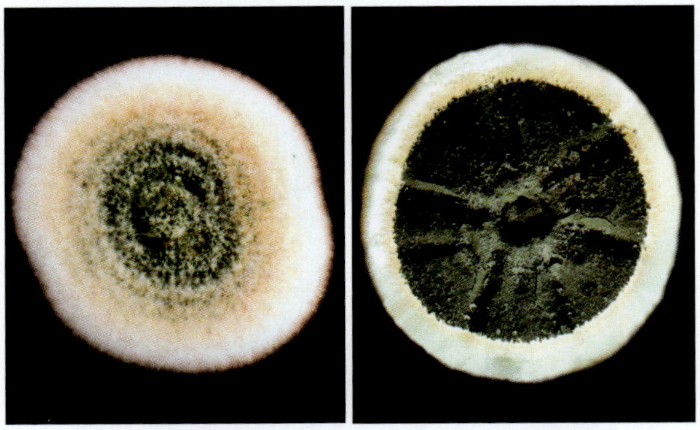

Early and late colonies of *Metarhizium anisopliae* var. *anisopliae* on agar. The fungus is **entomopathogenic**, x 1 (*MU*) & x 0.5 (*CT*).

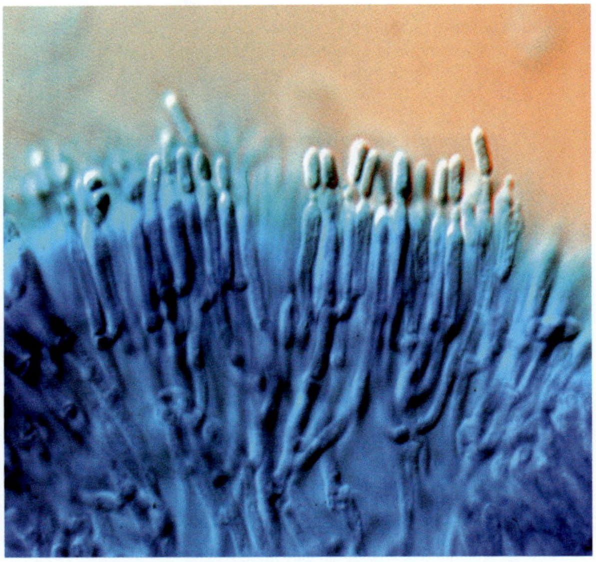

Palisade of conidiophores of *Metarhizium anisopliae* var. *anisopliae* from an agar culture. The fungus is **entomopathogenic**, x 1,200 (*CT*).

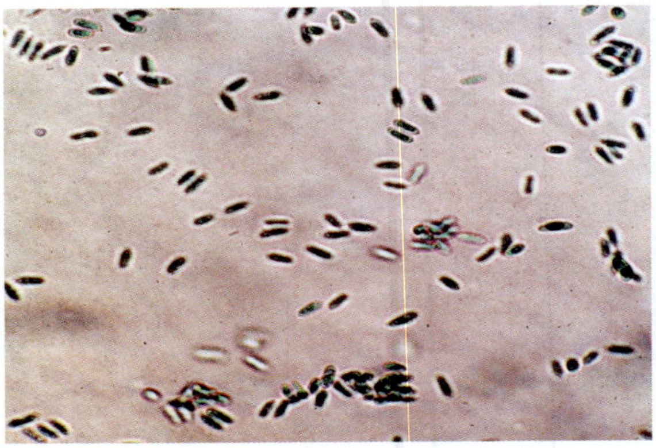

Conidia of *Metarhizium anisopliae* var. *anisopliae*, an **entomopathogenic** fungus, x 1,000 (*MU*).

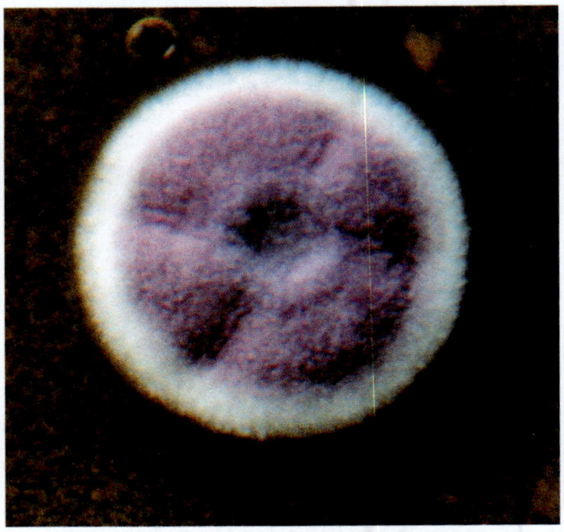

Colony of *Isaria fumosorosea* (=*Paecilomyces fumosoroseus*), an **entomopathogenic** fungus, on agar, x 1 (*MU*).

Entomopathogenic thalli of *Rickia apiculifera* on a body of a coleopterous passalid (*Chondrocephalus debilis*), x 5 (*MU*).

entomophagous (Gr. *éntomos*, cut, divided into segments, insect + *phágos*, which eats): feeding on living or dead insects.

entomophilous (Gr. *éntomos*, cut, divided into segments, insect + *phílos*, have an affinity for): refers to fungi whose spores are disseminated by insects, such as *Entomophthora muscae* and *Conidiobolus coronatus* (Entomophthorales), pathogenic species of the domestic fly and of the spittlebug, respectively. The second species also is the causal agent of entomophthoramycosis in man, a mycosis so-named because *C. coronatus* was formerly considered as *Entomophthora coronata*. As these types of fungi are pathogens of insects they also can be called **entomopathogenic**, and **entomogenous**, due to developing in the body of insects; besides, their spores are disseminated by these animals. Another example of the application of this term is *Cordyceps stylophora* (Hypocreales), a fungal pathogen of wasps.

Conidia of *Entomophthora muscae*, an **entomophilous** fungus (entomogenous or entomopathogenic) of the domestic fly; the conidia are visible among the hairs of a fly foot, x 30 (*MU*).

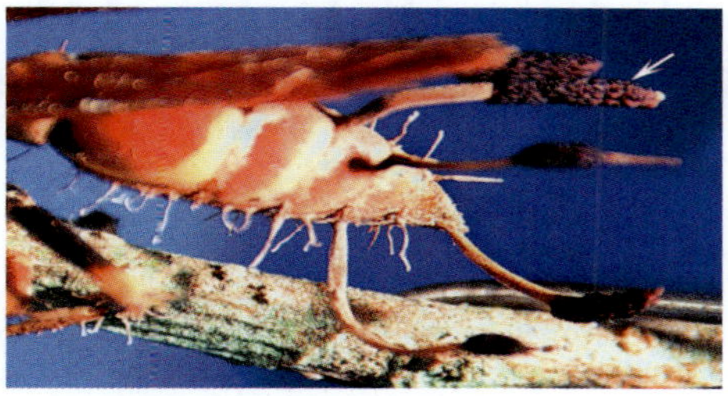

Entomophilous perithecial stromata (entomogenous or entomopathogenic) of *Cordyceps stylophora*, on a parasitized wasp, x 5 (*MU*).

eo- (Gr. *eós*, day break): a pref. that alludes to precosity, to something that springs forth in the early days of any organization, as in the generic name *Eosphaeria* (Sordariales).

epapillate (L. pref. *e-* without, lacking + *papilla*, papilla, nipple + suf. *-atus* > E. *-ate*, provided with or likeness): without a papilla; e.g., in the gen. *Didymella* (Dothideales) there are marine species with an epapillate ascocarp (such as *D. fucicola*, a perthotroph that grows in the middle veins and in the air vesicles of the thallus of *Fucus*), and species with an epapillate ascocarp or with a short papilla (as in *D. gloiopeltidis*, a parasite of the red alga *Gloiopeltis furcata*). Cf. **papillate**.

ephemeromycete (Gr. *ephémeros*, short-lived + *mýkes*, genit. *mýketos*, fungus): a fungus that is introduced into a new geographic region but is unable to establish itself and soon disappears.

epi-, **ep-** (Gr. *epí*, on, above): a pref. frequently employed in biological lexicology to form compounds of Gr. origin to introduce the idea of something that is above something else, as in the following terms. Cf. **hyper-** and **hypo-**.

epibasidium, pl. **epibasidia** (Gr. *epí*, on, above + *basídion* < *básis*, base + dim. suf. *-ídion* > L. *-idium*): see **heterobasidium**.

epibiotic (Gr. *epí*, on + *biotikós*, vital < *bíos*, life + suf. *-tikós* > L. *-ticus* > E. *-tic*, relation, fitness, inclination or ability): an organism that develops its reproductive organs on the surface of the host or substrate, although part or all of the thallus, or of the somatic structures, are found in the interior; many aquatic fungi of the class Chytridiomycetes (e.g., *Phlyctochytrium* and *Chytriomyces*, of the Chytridiales) and of the order Hyphochytriales, are epibiotic. Cf. **endobiotic** and **interbiotic**.

epiblastesis, pl. **epiblasteses** (Gr. *epí*, on + *blástesis*, germination, development): the growth of lichens from gonidia that are developed on the mother lichen.

epichroic (Gr. *epí*, on, after + *chrõs*, or *chroía*, color): refers to the pigmentation of hyphae or fruiting bodies that appears after injury or damage, as happens on cutting the basidioma of some boletes, whose tissue changes color from white or yellow to a bluish-green tone. Cf. **euchroic**.

epicortex (Gr. *epí*, on + L. *cortex*, cortex, bark): *Lichens*. A thin layer of polysaccharides that is found on the surface of the upper cellular cortex in the thallus of some parmeliaceous lichens, visible with the scanning electron microscope, in which it can have regular pores that function in gas exchange. Cf. **eucortex**.

epicutis, pl. **epicutes** (Gr. *epí*, on + L. *cutis*, skin): the layer external to the cuticle of the pileus, when the cuticle or cuticular layer is composed of two or more layers.

epidemic (LL. *epidemia* < Gr. *epidemía*, among the people < *epí*, on + *démos*, people, public): a disease that suddenly and extensively affects many people. When an infectious disease causes true epidemics or plagues in cultivated or wild plants it is called an **epiphytotic**; if the epidemic affects animals it is called an **epizootic**.

epidemiology (LL. *epidemia* < Gr. *epidemía*, among the people < *epí*, on + *démos*, people, public + *-logía* > E. *-logy*, discourse, treatise < *lógos*, word < *légo*, I say): the study of the incidence, distribution and control of diseases.

epigeal, **epigean**, **epigeic** (L. *epigeus* < Gr. *epí*, on + *geo*, earth, soil): growing or developing on the surface of the soil. Cf. **hypogeal**.

epigynous (Gr. *epí*, on + *gyné*, female; here, the female sexual organ): *Aquatic fungi*. The formation of the antheridium on the oogonium, as occurs in *Monoblepharis polymorpha* (Monoblepharidales), or the male gametangium on the female gametangium, as in *Allomyces javanicus* (Blastocladiales). Cf. **hypogynous**. See **exogynous**.

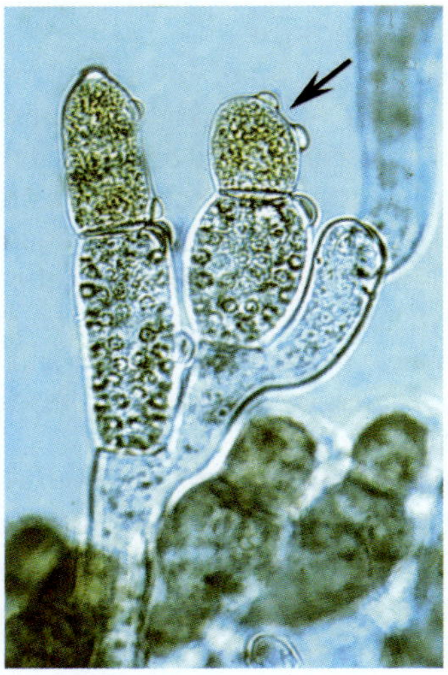

Epigynous male gametangia on a thallus of *Allomyces javanicus*, x 575 (*RTH*).

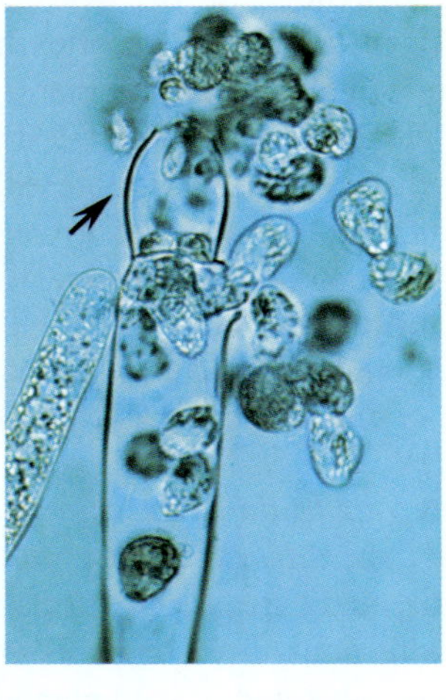

Epigynous male gametangium on a thallus of *Allomyces javanicus*. The male and female gametes are being released through a pore in the cell wall of the gametangia, x 575 (*RTH*).

epihypothallic (Gr. *epí*, on + L. *hypothallus* < Gr. *hypó*, beneath + *thallós*, offspring, thallus + suf. *-ikos* > L. *-icus* > E. *-ic*, belonging to or relating to): refers to a type of sporocarp development present in Myxomycetes of the order Stemonitales (also known as *stemonitoid development*); e.g., the species of *Stemonitis* and *Comatricha* are characterized by the formation of a hypothallus on the lower surface of the plasmodium, which separates it from the substrate. Certain equidistant points on the hypothallus are converted into centers of deposition or accumulation of materials for the formation of pedicels, each of which elongates upward, carrying a mass of prespore protoplasm that continues secreting more pedicelar material at the apex. This differs from the subhypothallic development of the orders Liceales, Trichiales, Echinosteliales and Physarales, in which there is no continuity of the hypothallus with the pedicel and the peridium of the sporocarp; the hypothallus forms beneath the plasmodium, and any connection between the pedicel and the peridium of the mature sporangium is secondary. It is called epihypothallic development because the sporangium is formed on the hypothallus. Cf. **subhypothallic**.

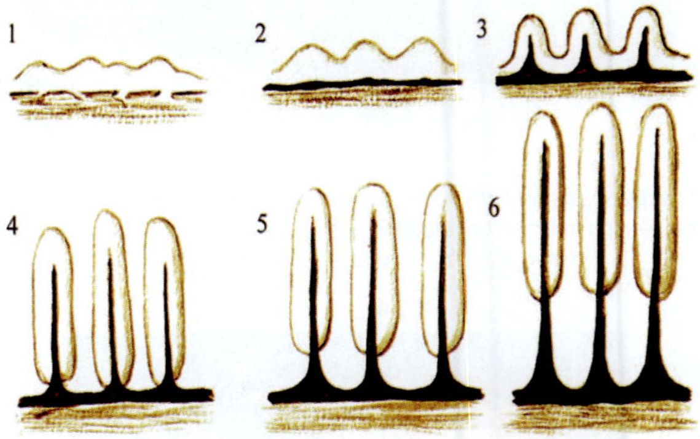

Epihypothallic sporangial development in the species of Stemonitales (Myxomycetes). 1. Emerging plasmodium. 2. Formation of the subplasmodial hypothallus. 3. Differentiation into primordia; the primordium of the stalk is secreted internally as the protoplasm forms mounds. 4. Segregation of the protoplasm into separate primordia, each secreting a stalk. 5. Continuous deposition of material at the apex of the stalk causes stalk elongation and the elevation of the prespore mass of protoplasm. 6. Full-size sporangia prior to capillitial formation; the interior part of the stalk is continuous with the subplasmodial hypothallus. The peridium, if present, is secondary. Diagram based on figures and information published by I. K. Ross in *Mycologia* *65*:477-485, 1973, x 10 (*MU*).

epikapylic (Gr. *epí*, on + *kapyleío*, to falsify, adulterate + suf. *-ikos* > L. *-icus* > E. *-ic*, belonging to or relating to): the thallus of a lichenicolous fungus that develops a morphologically distinctive lichenized structure. Cf. **endokapylic**.

epilithic (Gr. *epí*, on + *líthos*, stone + suf. *-íkos* > L. *-icus* > E. *-ic*, belonging to or relating to): growing on rocks. Also called **petricolous**, **rupicola** and **saxicolous**. Among the lichens there exist many species that inhabit rocks, e.g., *Lecanora frustulosa* (Lecanorales). Cf. **endolithic**.

Epilithic thallus of *Lecanora frustulosa*, x 1 (*MU*).

epinecral (Gr. *epí*, on + *nekrós*, dead + L. suf. *-alis* > E. *-al*, relating to or belonging to): refers to the layer of dead hyphae, with indistinct lumina, that are found near the cortex of the lichen thallus, above the algal layer. Cf. **hyponecral**.

epiphloeodic, **epiphloedal** (Gr. *epí*, on + *phloiós*, cortex, bark > L. *phloeo*, cortex, bark + Gr. suf. *-oeídes*, similar to, like + suf. *-íkos* > L. *-icus* > E. *-ic*, belonging to or relating to): *Lichens*. A crustaceous thallus that grows on the surface of tree bark, such as *Opegrapha*, of the Arthoniales. Syn. of **ectophloeodic**. Cf. **hypophloeodic**.

epiphragm (Gr. *epíphragma*, stopper, plug): a round and membranous apical lid that covers the upper part of the peridium of the receptacle or carpophore of the nidulariaceous Gasteromycetes, and which tears at maturity to permit liberation of the peridioles, which contain the basidiospores, as happens, e.g., in *Cyathus olla* and *Crucibulum laeve* (=*C. vulgare*), Nidulariales.

Epiphragm of the fruiting body of *Cyathus* sp., x 10 (*MU*).

Epiphragm of the fruiting bodies of *Cyathus* sp., x 10 (*MU*).

Epiphragm of the fruiting body of *Cyathus* sp., x 12 (*MU*).

epiphyllous (L. *epiphyllus* < Gr. *epí*, on + *phýllon*, leaf): what lives, is borne, or develops on the leaves of plants, such as the lichens of the gen. *Strigula* (Chaetothyriales), which live on tropical plants, and *Porina epiphylla* (Trichotheliales) which is found on fern leaves. The apothecia of certain lichens of the gen. *Cladonia* (Lecanorales) also are epiphyllous, as they are produced on the leaflets of the primary thallus, instead of on the podetia. Cf. **hypophyllous**. See **foliicole**.

Epiphyllous thallus of *Porina epiphylla* on a fern leaf; the pseudothecia of the lichen are evident, x 25 (*MU*).

epiphyte (Gr. *epí*, on + *phytón*, plant): an organism that lives on a plant without obtaining nutrients from it, since the plant only provides support; many lichens are epiphytes. See **epiphyllous**.

epiphytotic (Gr. *epí*, on + *phytón*, plant): an infectious disease that suddenly and destructively affects the wild or cultivated plants in a particular locality. Cf. **epidemic** and **epizootic**.

epiplasm (Gr. *epí*, on + *plásma*, formation, soft material with which a living being is formed): the portion of the cytoplasm in an ascus that is not incorporated into the ascospores during their delimitation; it may contribute to their nutrition during their development, or condense to form ornamentations of the epispore. Epiplasm can be seen, e.g., in the asci of many Sordariales, such as *Neurospora crassa* and *Sordaria fimicola*, and Rhytismatales, such as *Lophodermella sulcigena*. Cf. **sporoplasm**.

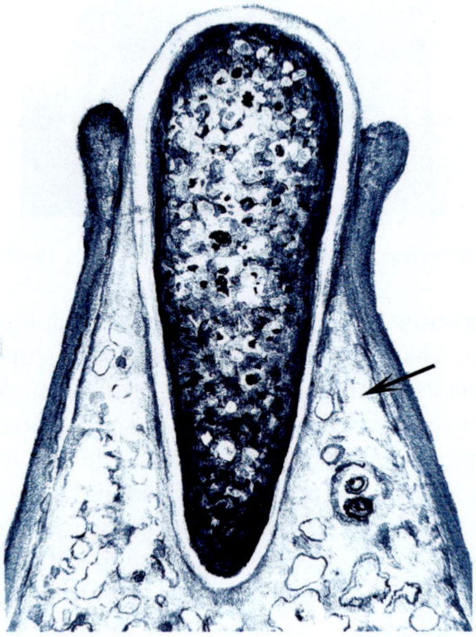

Epiplasm of the ascus apex of *Lophodermella sulcigena*, evident at the time of ascospore release. Drawing based on a transmission electron micrograph published by Beckett *et al.* in their *Atlas of Fungal Ultrastructure*, 1974, x 9,500 (*MU*).

epispore, **episporium**, pl. **episporia** (Gr. *epí*, on + *sporá*, spore): the wall layer of the basidiospore just beneath the *perispore* and the *exospore*. It is the thickest layer of the spore wall, and gives the spore its shape and rigidity, as well as its color and ornamentation. In the basidiomycetes the epispore, which is a continuation of the outer wall of the sterigma and basidium, is outside the *mesospore* and the *endospore*, the two innermost wall layers of the most structurally complex basidiospores.

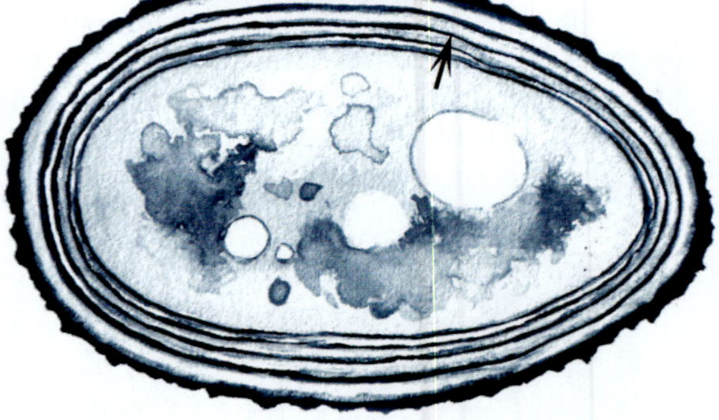

Epispore of the cell wall of the basidiospore; the endospore, mesospore, exospore and perispore also are shown in this diagram of a longitudinal section, x 5,500 (*MU*).

epistroma, pl. **epistromata** (Gr. *epí*, on + *strõma*, bed, small pillow, cushion): see **ectostroma**.

epithalline (Gr. *epí*, on + *thallós*, thallus): belonging to or relating to the *epithallus*, pl. *epithalli*, which is the outermost cortical layer of a lichen thallus.

epithecium, pl. **epithecia** (NL. *epithecium* < Gr. *epí*, on + *thekíon*, dim. of *thêke*, box; here, of the asci): *Discomycetes*. A layer of tissue above the surface of the hymenium of an apothecium, formed by the union of the tips of the paraphyses that project from among the asci, as is seen, e.g., in *Peziza* (Pezizales) and in the apothecia of many lichens. Cf. **hypothecium**.

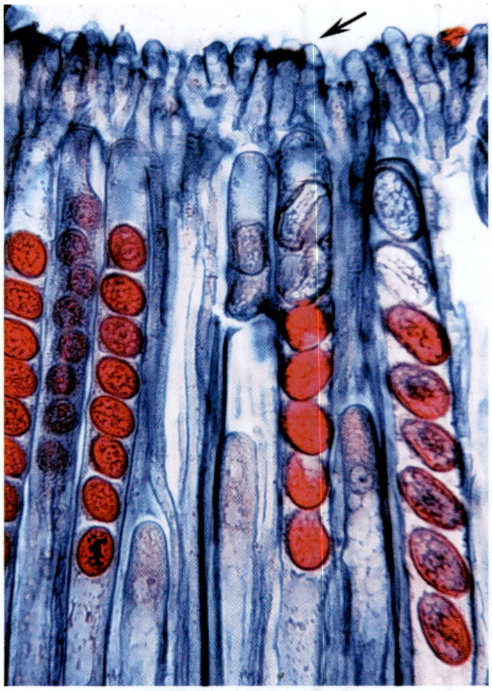

Epithecium of the apothecium of *Peziza* sp., seen in longitudinal section, x 200 (*RTH*).

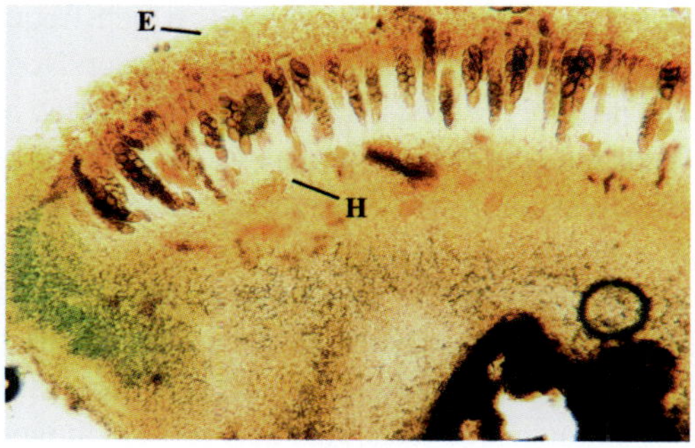

Epithecium (E) and hypothecium (H) of a lecanorine lichen apothecium, seen in longitudinal section, x 50 (*RV*).

epithelium, pl. **epithelia** (NL. *epithelium* < Gr. *epí*, on + *thelé*, nipple + L. dim. suf. *-ium*): *Agaricales*. A trichoderm composed of small globose or subglobose elements arranged in one or several layers.

epithet (L. *epitheton* < Gr. *epítheton*, something added): an adj. or participle that is designed to characterize or describe a particular quality of a name. In scientific terminology it represents the second (the specific) word in a Latin or Latinized binomial, and also, when appropriate, the third or fourth word (that of a variety).

epitunica, pl. **epitunicae** (Gr. *epí*, on + L. *tunica*, tunica, covering or wrapper): see **exospore**.

epixylous (Gr. *epí*, on + *xýlon*, wood): syn. of **lignicolous**; living or growing on wood; both terms refer only to the presence of a fungus on wood, and not to its nutrition. In the case of fungi that degrade wood, i.e., take nourishment from it, it is more accurate to use the term **xylophagous**. Many Aphyllophorales are epixylous.

epizoic (Gr. *epí*, on + *zõon*, animal + suf. *-íkos* > L. *-icus* > E. *-ic*, belonging to or relating to): a fungus that grows as a commensal, saprobe or parasite on living animals, as is seen, e.g., in *Amoebidium parasiticum* (Trichomycetes, Amoebidiales), which lives on the antennae of crustaceans of the gen. *Daphnia*. Cf. **endozoic**.

epizootic (Gr. *epí*, on + *zõon*, animal): a disease that suddenly and extensively affects many animals. Cf. **epidemic** and **epiphytotic**.

epruinose (L. pref. *e-*, abbrev. of *ex-*, without + *pruinosus*, frosted, covered with frost < *pruina*, frost + suf. *-osus* > E. *-ose*, full of, augmented, inclined to): lacking a frosty covering or bloom.

equi- (L. *aequus*, equal): a pref. that signifies equal. Equivalent to the Gr. pref. **iso-**. Cf. **inaequi-**.

equilateral (LL. *aequilateralis*, with equal sides < *aequus*, equal + pref. *later-* < *latus*, side + suf. *-alis* > E. *-al*, relating to or belonging to): *Of spores*. A spore that is bilaterally symmetrical, i.e., that has two identical sides. Examples of spores with equilateral shapes are globose, dacryoid, ovate, pyriform, amygdaliform, lenticular, cylindrical and bacilliform, among others. Cf. **inequilateral**.

erasure phenomenon (L. *erasus*, pp. of *eradicare* < *e-*, *ex-* out + *radere*, to scratch, scrape, delete; LL. *phaenomenon* < Gr. *phainómenon*, appearance < *phaíno*, to show): refers to the myxamoebae of *Dictyostelium* (Dictyosteliomycota), which lose their capacity to recapitulate and aggregate to form the pseudoplasmodium when their developing cultures are disaggregated and placed on a new culture medium.

erect (L. *erectus*, upright): in a vertical position; straight, not curved.

eremo-, **erem-** (Gr. *éremos*, deserted, also solitary, isolated): a pref. employed in the formation of some names, as in the gen. *Eremothecium* (Saccharomycetales).

ergastic (Gr. *ergastikós*, able to work < *ergásomai*, to work < *érgon*, work): constituting the nonliving by-products of protosplasmic activity, e. g., oxalate crystals, starch granules or protein crystals.

ergot (Fr. *ergot*, ergot, spur, little horn): a term applied to the disease of rye and of other cultivated grains, such as wheat, barley, millet and rice, caused by various species of *Claviceps*, mainly *C. purpurea* (Hypocreales); the term ergot also is applied to the fungus in general, and to the sclerotia of the fungus, which have the shape of a little horn. Among the wild grasses attacked by this fungus are various species of pasture grasses in the gen. *Agrostis*, *Bromus*, *Agropirus*, *Poa*, *Festuca* and *Lolium*, among others.

Ergot, sclerotia of *Claviceps paspali*, x 4 (*MU* & *CB*).

Ergot, sclerotium of *Claviceps purpurea* on a barley spike, x 2 (*MU*).

Ergot, sclerotium of *Claviceps purpurea* on a barley spike, x 5 (*MU*).

ergotism (Fr. *ergot*, ergot, spur, little horn + Gr. suf. *-ismós* > E. *-ismo*, state or condition): an intoxication of humans and herbivorous animals caused by the ingestion of cereals contaminated with the sclerotia of *Claviceps purpurea* or other species of *Claviceps* (Hypocreales). Ergotism was one of the plagues of the Middle Ages, and due to its being attributed to a supernatural origin, the disease was called Sacred Fire, Infernal Fire, Saint Anthony's Fire and Saint Marcial's Fire. The sclerotia of the fungus, which frequently grow in rye seeds, were accidentally ground together with healthy rye seeds to make flour for baking bread. There are two clinical forms of ergotism: the acute, convulsive or cerebral (psychic disruptions, among them hallucinations), and the chronic or gangrenous (which can result in the total or partial loss of body extremities without bleeding due to vasoconstriction). The toxins of these fungi, which are responsible for the pathological effects, vary according to the species, but generally they are alkaloids that frequently contain *lysergic acid* in their molecules. The alkaloids of *C. purpurea* have been intensively studied; many of them have applications in therapy and others produce psychotropic effects or serve as the basis for the preparation of hallucinogenic drugs (such as LSD, lysergic acid diethylamide). Of the alkaloids that have therapeutic application in medicine, *ergotamine* and its derivatives (*dihydroergotamine* and *ergonovine* or *ergometrine*) are especially important, as well as *ergotoxine*, which acts as a vasoconstrictor and causes contraction of the smooth muscle fibers (uterus, blood vessels); it is applied to cure migraine and certain hives, as an abortive and to facilitate childbirth and impede hemorrhaging, due to its oxytocic and antihemorrhagic properties. On the other hand, other alkaloids of ergot, such as *ergocristine*, *ergocornine* and *ergocryptine*, have vasodilatatory action and serve to correct some peripheric vascular disruptions.

ericoid mycorrhiza, pl. **mycorrhizae**: see **monotropoid mycorrhiza**.

erinaceous (Gr. *érion*, wool; *erineous*, woolly; *erin,* hedgehog, woolen; L. *er*, genit. *eris*, porcupine; *erinaceus*, spiny < *ericius*, porcupine, hedgehog + suf. *-aceous* > E. *-aceous*, of or belonging to): spiny like a porcupine.

erio-, **eri-** (Gr. *érion*, wool): a pref. used to form compounds that allude to something woolly, as in the gen. *Erionema* (Physarales), a myxomycete with a woolly capillitium.

erogen (Gr. *éros*, love + *génos*, origin, birth): a substance that controls the induction and differentiation of sexual organs, such as *erotactin*.

erose, **eroded** (L. *erosus*, eroded < *erodere*, eaten away < pref. *ex-*, contracted to *e-*, out, beyond + *rodere*, to gnaw): having unequal borders, with small teeth or sinuosities, as if nibbled or gnawed, like the border of the giant colonies of the yeasts *Kluyveromyces marxianus* (Saccharomycetales) and *Candida lambica* (cryptococcaceous asexual yeasts), when grown on certain solid media.

Erose border of a giant colony of *Candida lambica* on agar, x 1 (*MU*).

erotactin (Gr. *éros*, love + *taktós*, handling, touch + NL. *-in* > E. *-in*, suf. used in chemistry to denote an activator or a compound): a substance attractant to male gametes. See **sirenin**.

erratic, **vagrant**, **wandering** (L. *erraticus*, unsteady, hesitant < *erratus* < *errare*, to fail, err, vacilate + Gr. suf. *-tikós* > L. *-ticus* > E. *-tic*, relation, fitness, inclination or ability; LME. *vagaraunt* < ME. *vagren* < L. *vagari*, to wander; ME. *wanderen* < OE. *wanderian*, to wander + suf. *-ing*, expresses the action of the verb): without a fixed course; refers to lichen thalli that are not attached to the substrate and are blown around by the wind, such as *Xanthoparmelia* (=*Chondropsis*) *semiviridis* and *Aspicilia* (=*Sphaerothallia*) *esculenta* (manna), of the Lecanorales.

erubescent, **rubescent** (L. *erubescens*, genit. *erubescentis*, becoming red < *erubescere*, reddening, blush, or L. *rubescens*, genit. *rubescentis*, turning red + suf. *-escens*, genit. *-escentis* > E. *-escent*, that which turns, that begins to, slightly): turning red. Erubescence or rubescence is present in some fungi when they are cut or bruised, such as *Amanita rubescens* (Agaricales); in plant pathology it is equivalent to *eritrosis*.

erumpent (L. *erumpens*, genit. *erumpentis*, breaking out, bursting forth < *erumpere*, to break, bring forth + suf. *-entem* > E. *-ent*, a being): an internal sporiferous structure which on maturing opens a passage through the substrate or, when the fungus is a parasite, through the overlying host tissues, to liberate the spores; as happens, e.g., with the acervuli of the melanconiaceous asexual fungi (such as *Colletotrichum*), with the perithecial stromata of *Cryphonectria* (=*Endothia*) *parasitica* (Diaporthales) and *Diatrype disciformis* (Xylariales), the ascostroma of *Botryosphaeria ribis* (Dothideales), and the aecia and uredinia of the Uredinales. Equal to **perrumpent**.

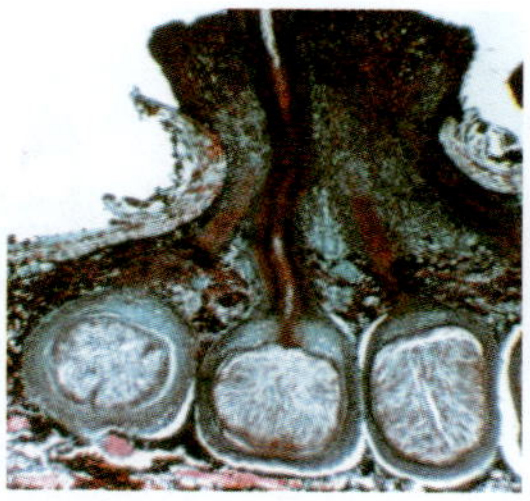

Erumpent perithecial stromata of *Cryphonectria* (=*Endothia*) *parasitica*, x 75 (*RTH*).

Erumpent perithecial stromata of *Diatrype disciformis*, x 15 (*CB*).

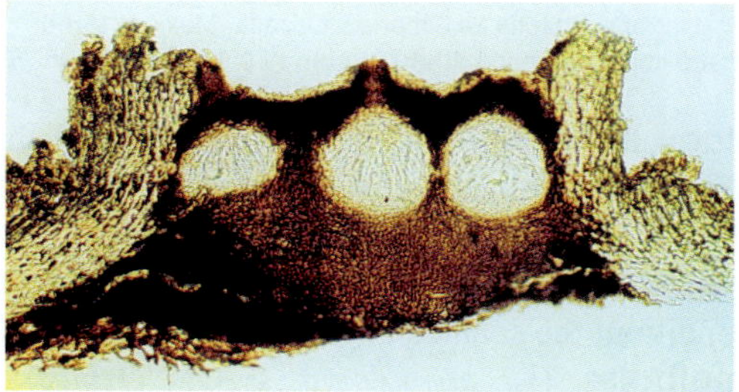

Erumpent ascostromata of *Botryosphaeria ribis*, x 40 (*RTH*).

erythro-, **erythr-** (Gr. *erythrós*, red): a pref. that forms a part of compounds that allude to the color red, as in the generic name *Erythrodecton* (Arthoniales).

-escent (L. *-escens*, genit. *-escentis*, that which turns, that begins to change, slightly): a suf. employed to denote that something has changed in appearance, color, etc., e.g., *caerulescent*, turning bluish-green in color, like the fruiting body of *Psilocybe* (Agaricales) when it is rubbed or damaged.

ethnomycology (Gr. *éthnos*, people, race + *mýkes*, fungus + des. *-logía* > E. *-logy*, discourse, treatise < *lógos*, word < *légo*, I say): a special branch of science that links mycology with ethnology, and which studies the various aspects related to ethnic groups and the uses they make of fungi, as food, in medicine, and ceremonies, etc. See **entheogen**, **hallucinogenic fungi**, **magic mushrooms** and **psychotomimetic.**

María Sabina †, a Mazatecan folk healer (*curandera*), who cured with the intervention of hallucinogenic fungi. Behind is Dr. Teófilo Herrera. Huautla de Jiménez, Oaxaca, Mexico (1969). Both are important in the **ethnomycology** of Mexico, x 0.05 (*EPS*).

etiology, **aetiology** (ML. *aetiologia*, the establishment of causes < Gr. *aitíologia* < *aitía*, cause + des. *-logía* > E. *-logy*, discourse, treatise < *lógos*, word < *légo*, I say): the cause or origin of a disease or abnormal condition; the science of the causes of diseases.

etymologist (L. *etymologia* < Gr. *étymos*, true, authentic + des. *-logía* > E. *-logy*, discourse, treatise < *lógos*, word < *légo*, I say + L. suf. *-ista* > E. *-ist*, one who practices some discipline or activity; who has the ability or capacity to carry out an activity or profession): a specialist in etymology. Cf. **glossarist**.

etymology (L. *etymologia* < Gr. *étymos*, true, authentic + des. *-logía* > E. *-logy*, discourse, treatise < *lógos*, word < *légo*, I say): the study, treatise or science of the true or authentic meaning of words through knowledge of their origin, structure and transformations or changes. The origin of words and their significance, the reason for their existence, and their form. The etymological art is an essential part of linguistics, and this knowledge is indispensable to speak and write correctly, with the accuracy, clarity, precision and elegance that distinguishes academic and scientific language. The technical and scientific vocabulary is derived from ancient words, mainly Greek and Latin, from which have been derived innumerable words or neologisms. Etymology is the most natural and constant orthographic norm; it expresses the essential basis of the definition; it is a powerful aid to memory to more easily remember the meaning of a word, and indispensable for the serious study of the grammar of a particular language. See **Greek**, **Latin** and **neologism**.

eu- (Gr. *eû*, true): a pref. employed to express the idea of legitimate, authenticity, as in *Euascomycetes* and the following terms. Cf. **pseudo-**.

Euascomycetes (Gr. *eû*, true + *askós*, sack, ascus + L. *-mycetes*, ending of taxonomic class < Gr. *mýkes*, genit. *mýketos*, fungus): in some classification systems, the Euascomycetes were considered as a class of fungi with well developed, septate mycelium and unitunicate asci produced in some type of ascocarp (cleistothecium, perithecium, or apothecium) formed by the ascohymenial pattern of ascocarp development. It included the subclasses Plectomycetidae, Pyrenomycetidae and Discomycetidae, which together with the Hemiascomycetes, Laboulbeniomycetes and Loculoascomycetes, comprised the subdivision Ascomycotina. In the classification followed in this dictionary, these subclasses are recognized as classes (Plectomycetes, Pyrenomycetes and Discomycetes); the Hemiascomycetes now correspond to the Archiascomycetes and Saccharomycetes; the Laboulbeniomycetes are regarded only as the order Laboulbeniales, and the Loculoascomycetes remain as such. In the Appendix, the user of this dictionary can see the comparison between the outline of Alexopoulos *et al.*, 1996, modified by Ulloa and Hanlin, 2006, to that of Kirk *et al.*, 2008. In the latter classification scheme, this class has disappeared; the class Archiascomycetes has been moved to subphylum Taphrinomycotina, and the class Saccharomycetes to the subphylum Saccharomycotina. The classes Plectomycetes, Pyrenomycetes, Discomycetes and Loculoascomycetes have been transferred to the subphylum Pezizomycotina. The fungi in the euascomycetes are made up of heterogeneous species, forming unnatural groupings, and therefore are considered an artificial classification. These taxonomic rearrangements, with the concurrent change of the names of taxa, are part of a constant process of study and evaluation, in order to achieve a phylogenetic classification of the fungi.

eucarpic (Gr. *eû*, true + *karpós*, fruit + suf. *-ikos* > L. *-icus* > E. *-ic*, belonging to or relating to): an organism in which only one part of the thallus is transformed into reproductive organs, and the thallus itself continues carrying out its somatic functions; the great majority of the fungi are eucarpic, and are said to reproduce by *eucarpy*. In general they are considered as more evolved differentiated forms than the *holocarpic* species, which reproduce by *holocarpy*. Cf. **holocarpic**.

Some **eucarpic** fungi: mycelium (somatic phase) and some basidiocarps (reproductive phase) of *Inocybe geophylla*, Agaricales, x 1 (*RV*), and *Cotylidia aurantiaca*, Aphyllophorales, x 1 (*EPS*).

eucarpy (Gr. *eû*, true + *karpós*, fruit > E. suf. *-carpy*): the condition of being *eucarpic*.

euchroic (Gr. *eû*, true + *chrõs*, color + suf. *íkos* > L. *-icus* > E. *-ic*, belonging to or relating to): refers to fruiting bodies and hyphae that have the same pigmentation, including after having been damaged. Cf. **epichroic**.

eucortex (Gr. *eû*, true + L. *cortex*, cortex, covering): refers to the cortex of a lichen thallus composed of well differentiated hyphae. Cf. **epicortex**.

euform (Gr. *eû*, true + L. *forma*, shape): see **auteuform**.

eugonic (Gr. *eû*, true + *gónos*, generation, offspring + suf. *-íkos* > L. *-icus* > E. *-ic*, belonging to or relating to): a dermatophyte strain that grows faster in culture, often with more aerial mycelium than an abnormal (*disgonic*) strain.

eugymnohymenial (Gr. *eû*, true + *gymnós*, naked + L. *hymenium* < Gr. *hyménion*, dim. of *hymén*, membrane + L. suf. *-alis* > E. *-al*, relating to or belonging to): a type of development in some Discomycetes, in which the ascocarp is completely open from the beginning; the ascogonium is never covered above by hyphae, and the investing hyphae grow upward as a palisade that develops into a layer of paraphyses. Some species form an excipulum, but many others completely lack an excipulum. Cf. **cleistohymenial** and **paragymnohymenial**.

euhymenium, pl. **euhymenia** (Gr. *eû*, true + L. *hymenium*, hymenium < Gr. *hyménion* < *hýmen*, membrane + dim. suf. *-ión* > L. *-ium*): an hymenium in which the basidia and their sterile homologs (hyphidia) are the first elements formed. Cf. **catahymenium**.

eukaryote, eukaryont, eucaryote, eucaryont (Gr. *eû*, true + *káryon*, nucleus + *-ote*, NL. suf., which indicates possesion): an organism composed of one or more cells with clearly evident nuclei. Cf. **prokaryotic**.

eukaryotic, eukaryontic, eucaryotic, eucaryontic (Gr. *eû*, true + *káryon*, nut, nucleus + suf. *-tikós* > L. *-ticus* > E. *-tic*, relation, fitness, inclination or ability): an organism whose cells possess true nuclei (surrounded by a perforate double membrane), mitochondria, endoplasmic reticulum, dictyosomes, vacuoles and ribosomes, among other characteristics; i.e., with a level of cellular organization more complex than that of a *prokaryotic* organism. Only the organisms of the kingdom Monera are prokaryotic; those of the kingdoms Protoctista, Fungi, Plantae and Animalia are eukaryotic. Cf. **prokaryotic**.

Eukaryota (Gr. *eû*, true + *káryon*, nut, nucleus + NL. suf. *-ota*, which denotes possession; also used as the ending of a name of a group of organisms): a taxonomic category above the kingdom, in this case the *superkingdom* (although at present the term *domain* is utilized more) to which belong all of the eukaryotic organisms, which are distinguished by having one or more true nuclei in their cells that are delimited by a perforated nuclear membrane, with pairs of chromosomes that contain DNA, and with a series of complex cellular organelles (e.g., mitochondria and Golgi bodies). Cf. **Prokaryota**.

eumorphic (Gr. *eû*, true + *morphé*, shape + suf. *-íkos* > L. *-icus* > E. *-ic*, belonging to or relating to): well conformed, normal.

eumycetoma, pl. **eumycetomata** (Gr. *eû*, true + *mýkes*, genit. *mýketos*, fungus + *-óma*, a suf. that refers to a pathological state, generally a tumor): a clinical entity of mycetoma, a mycotic infection in man, especially in tropical areas, that affects the foot and other parts of the body, causing swellings that contain characteristic granules of different color, depending on the etiologic agent involved. Eumycetoma is caused by different species of fungi (a difference from actinomycetoma, which is caused by actinobacteria). See **mycetoma** and **mycosis**.

Eumycota (Gr. *eû*, true + L. *-mycota*, ending of division or phylum < Gr. *mýkes*, genit. *mýketos*, fungus): the phylum (division) of the kingdom Fungi that contains the true fungi. In some classifications the division Eumycota, together with the division Myxomycota, constitute the kingdom Fungi. In the classification adopted in this dictionary, the Eumycota has been replaced by the various phyla that constitute the kingdom Fungi (Chytridiomycota, Zygomycota, Ascomycota and Basidiomycota). The other phyla, which are now considered separate from the kingdom Fungi, belong to the kingdom Stramenopila (Oomycota, Hyphochytriomycota and Labyrinthulomycota) and to the kingdom Protista (Plasmodiophoromycota, Dictyosteliomycota, Acrasiomycota and Myxomycota). In the Appendix, the user of this dictionary can see the comparison between the outline of Alexopoulos *et al*., 1996, modified by Ulloa and Hanlin, 2006, to that of Kirk *et al*., 2008. In the latter classification scheme, the phyla Chytridiomycota, Zygomycota, Ascomycota and Basidiomycota have been retained in the kingdom Fungi, with the addition of the phyla Blastocladiomycota, Neocallimastigomycota and Glomeromycota. These taxonomic rearrangements, with the concurrent change of the names of taxa, are under constant study; evaluation of these groups of organisms, is underway, in order to achieve a phylogenetic classification of the fungi.

euploid (an international scientific term < Gr. *eû*, true + *ploide* < *ploidía*, degree of repetition of the basic chromosome number): having a chromosome number that is an exact multiple of the basic or monoploid number. Cf. **aneuploid**.

eury- (Gr. *eurýs*, ample, wide, broad): a pref. employed in the formation of compound terms that allude to the extent of certain structures, or to the wide fluctuation between the limits of certain factors that affect the development of some organisms, such as the concentration of salts, temperature, etc.; e.g., a *euryhaline* organism is one that tolerates wide limits in the concentration of haloid salts in the medium. Cf. **steno-**.

eurychoric (Gr. *eurýs*, broad, wide + *chóra*, country): a cosmopolitan species, i.e., one that is widely distributed throughout the world. Cf. **stenochora**.

euryoecious (Gr. *eurýs*, broad, wide + *oîkos*, house or inn): applied to an organism found in the most diverse ecological conditions, like the common molds of the gen. *Aspergillus*, *Penicillium* and *Cladosporium*, among other moniliaceous and dematiaceous asexual fungi. Cf. **stenoecious**.

euryxenous (Gr. *eurýs*, broad, wide + *xénos*, stranger, host): an organism capable of parasitizing a more or less large number of hosts; among the fungi, e.g., the moniliaceous asexual fungus *Phymatotrichopsis omnivora* has the potential of parasitizing more than 2,000 species of dicotyledoneous plants. Cf. **stenoxenous**.

euseptate (Gr. *eû*, true + L. *septatus*, having partitions or septa, partioned < *septum*, septum, wall + suf. *-atus* > E. *-ate*, provided with or likeness): a type of multicellular conidium in which the cells that compose it are separated by true septa, which consist of diaphragms that are united peripherally with the lateral wall of the conidium, as occurs, e.g., in *Alternaria alternata* and many other dematiaceous conidial fungi. The ascospores of *Pyrenophora* (Pleosporales) also are euseptate. Cf. **distoseptate**.

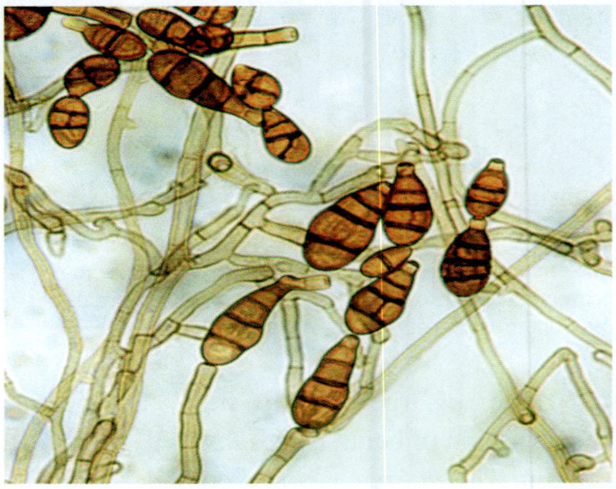

Euseptate conidia of *Alternaria alternata*, x 450 (*MU*).

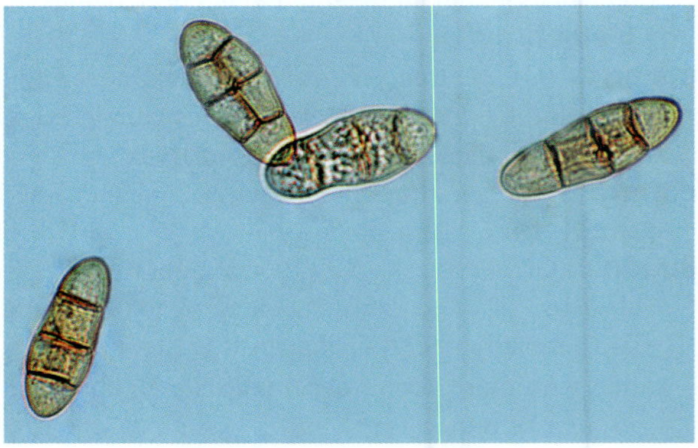

Euseptate ascospores of *Pyrenophora* sp., x 270 (*CB*).

eustroma, pl. **eustromata** (Gr. *eû*, true + *strõma*, small pillow, bed): a stroma that is composed only of fungal tissue, as occurs in the majority of species of stromatic fungi. Cf. **pseudostroma**.

euthecium, pl. **euthecia** (NL. *euthecium* < Gr. *eû*, true + *thekíon*, dim. of *thêke*, box; here, of the asci): an ascoma (*cleistothecium*, *perithecium*, *apothecium*) of an euascomycete. Cf. **pseudothecium**.

euthyplectenchyma, pl. **euthyplectenchymata** (Gr. *euthýs*, right, straight, direct + *plektós*, braided, interwoven + *énchyma*, stuffed): hyphal tissue in which there is no evident cellular structure.

eutrophic (Gr. *eû*, true + *trophé*, action of feeding + suf. *-íkos* > L. *-icus* > *-ic*, belonging to or relating to): applied to organisms that live in media rich in calcium, nitrogen and phosphorus, neutral or slightly alkaline (e.g., a lake), which are most favorable for good vegetative development. Also refers to a biotic medium that can increase nutritive materials through the contributions of nutrient elements that it receives from outside, a phenomenon known as *eutrophication*. This phenomenon of nutrient enrichment in a medium, especially aquatic, is generally caused by direct or indirect human influence. Cf. **distrophic** and **oligotrophic**.

eutypoid (< gen. *Eutypa* < Gr. *eû*, true + *typé*, genit. *týpos*, mark, scar + L. suf. *-oide*, like < Gr. *-oeídes*, similar to): a fungus or sporiferous apparatus in which the perithecia are embedded in an effuse endostroma; the perithecial necks project from the ectostroma without converging toward a central point, as occurs in Pyrenomycetes of the gen. *Eutypa* (Xylariales). This type of stroma generally is formed on decorticated branches of many trees. Cf. **diatrypoid** and **valsoid**.

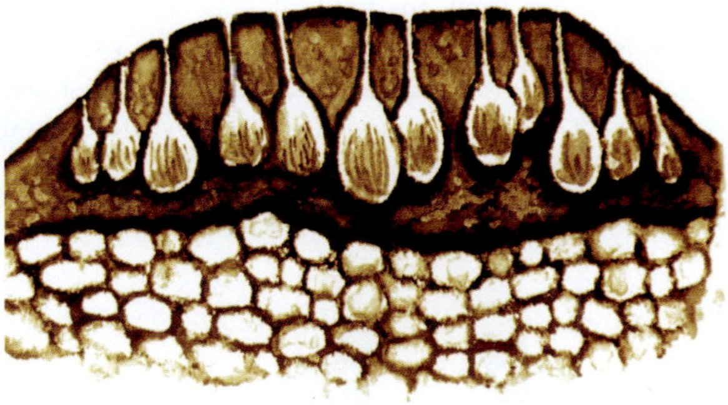

Eutypoid perithecial stroma of *Eutypa armeniacae* on host tissue, drawn as seen in longitudinal section, x 5 (*MU*).

evanescent (L. *evanescens*, genit. *evanescentis*, what vanishes, loses strength, that withers < *evanescere*, to wither, fade + suf. *-escens*, genit. *-escentis* > E. *-escent*, that which turns, that begins to, slightly): ephemeral or of short duration, like the asci of the fungi of the gen. *Chaetomium* (Sordariales), or the thin, fragile exoperidium of species of *Gautieria* (Gautieriales).

eversion (L. *evertion* < *evertere*, with that inside turning outward, reverse < *ex-*, out + *vertere*, to turn + suf. *-io*, genit. *-icnis* > E. *-ion*, state, result of): the act of turning inside out, like the membrane found in the base of the sporiferous sack of *Sphaerobolus* (Nidulariales), which is initially concave but suddenly becomes convex at the moment of discharging the sack into the air.

Eversion of the fruiting body of *Sphaerobolus* sp., x 8 (*CL*).

everted (L. *evertus*, turn inside out < *evertere*, with that inside thrown outward < *ex-*, out + *vertere*, to turn): with the interior thrown outward.

ex (Gr. *ex* > L. *ex*, out of, from): in bibliographic citations it signifies from, e.g., *Clathrus* Micheli *ex* Persoon; which was validly published by the second author, but based on specimens or materials of the first author.

ex situ (L. *ex situ* < *ex*, out of, from; *situ*, site, place): an organism taken from its natural habitat, like the living cultures isolated from nature and maintained in culture collections. The term also is applied to non-living materials kept in reference collections. Cf. **in situ**.

exannulate (L. pref. *e-*, *ex-*, without + *annulus*, ring + suf. *-atus* > E. *-ate*, provided with or likeness): lacking a ring, like the stipe of *Russula brunneoviolacea* (Agaricales). Cf. **annellate**.

excavate (L. *excavatus*, hollow < *excavare*, to excavate, to take from something solid part of its mass or bulk, making a hole or cavity in it): with a visible hole or cavity, as if it had been uncovered by excavation.

excipulum, pl. **excipula** (L. *excipulum*, receptacle, receptaculum): generally, it is a layer of hyphae on which rests the asci or basidia of the hymenium (hymenophore); in the apothecium of the Discomycetes the excipulum stands out because of its dense structure or dark color and comprises most of the tissue of the apothecium, as in *Ascobolus* and *Peziza* (Pezizales). In the lichens the excipulum is sometimes referred to as the *hypothecium*, and other times as the *parathecium* or the *amphithecium*, but always to the parts that envelop the *thecium* (hymenium or hymenial layer). The excipulum of *Ascobolus* (Pezizales) and the thalline excipulum of the discolichens (such as *Physcia*, of the Lecanorales) is the basal zone of the apothecium on which the hypothecium rests. The so-called marginal excipulum or proper excipulum corresponds to the parathecium. The term excipulum also is applied to the hyphal tissue of the pycnidial wall of the sphaeropsidaceous asexual fungi.

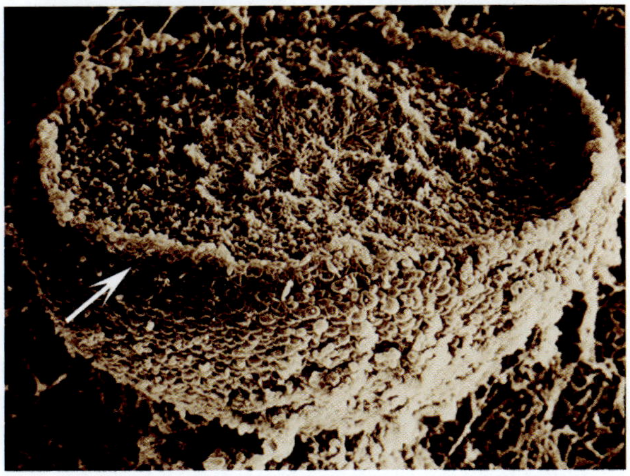

Excipulum of the apothecium of *Ascobolus furfuraceus*, x 25 (*KO*).

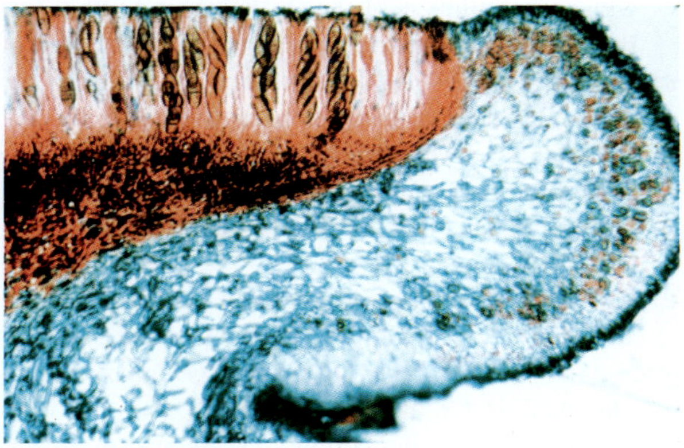

Excipulum of the apothecium of *Physica* sp., seen in longitudinal section, x 130 (*RTH*).

exohabitant (Gr. *éxo*, outside, on the exterior + E. *habitant*, one who inhabits > L. *habitare*, to live, live in a place or house): a term applied to an integral organism of a lichen symbiosis that forms the external tissues of the thallus, i.e., mycobiont or fungal symbiont. Cf. **inhabitant**.

exigynous (NL. *exigynus* < L. *ex*, out of, from + Gr. *gyné*, woman; here, female sexual organ): *Aquatic fungi*. With the antheridial stalk originating directly from the oogonial cell, above the basal septum, as occurs in *Monoblepharis polymorpha* and *Monoblepharella taylorii* (Monoblepharidales). See **epigynous** and **hypogynous**. Cf. **exogynous**.

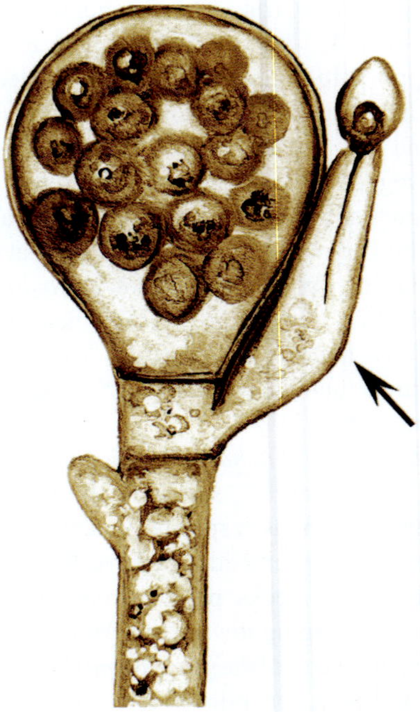

Exigynous antheridium of *Monoblepharella taylorii*, x 350 (*MU*).

exit tube (L. *exitus*, departure, way out < L. *exire*, to go out; L. *tubus*, tube, conduit with a narrow lumen): a tubular extension of the sporangium, through which the zoospores are liberated outside the host or substrate, as occurs, e.g., in *Olpidium* (Spizellomycetales) and *Pontisma* (Lagenidiales).

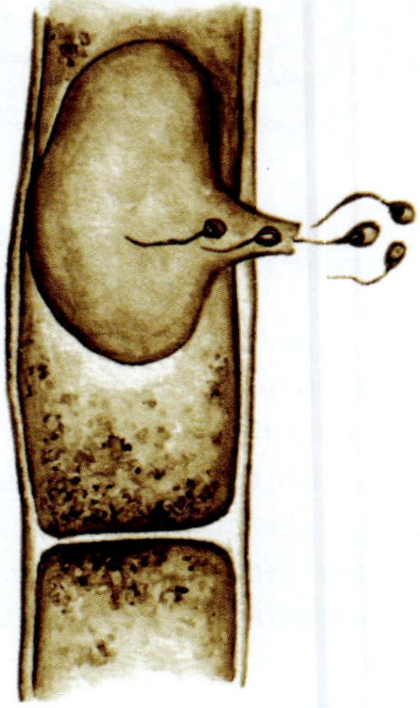

Exit tube of the zoosporangia of *Olpidium endogenum*, endoparasitic in a cell of the alga *Zygnema*, x 400 (*MU*).

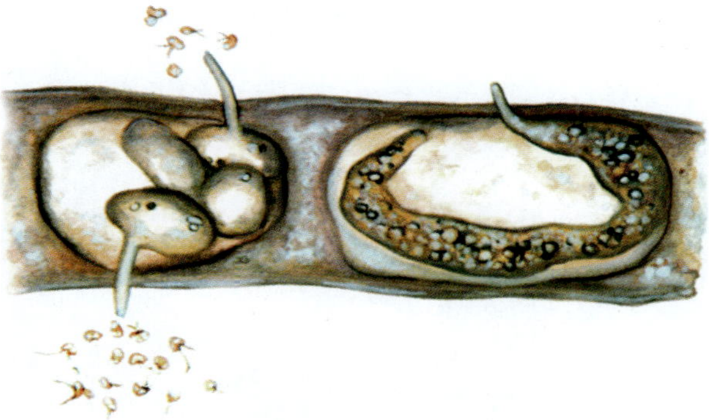

Exit tube of the zoosporangia of *Pontisma lagenidioides*, endoparasitic in the cells of the marine alga *Ceramium*, x 1,100 (*MU*).

exo- (Gr. *éxo*, outside, on the exterior): a pref. that forms terms that allude to something outside or on the exterior, as in the following terms. Equal to **ecto-**. Cf. **endo-**.

exobasidial (Gr. *éxo*, outside, on the exterior + *basídion*, dim. of *básis*, base + L. suf. *-alis* > E. *-al*, relating to or concerning): pertaining or relating to the exobasidium; with uncovered, external basidia, like the fungi of the order Exobasidiales, which are plant parasites, forming galls, e.g., *Exobasidium* sp., which grows on azaleas. Also applied to something that is separated from a basidium by a wall or partition. And formerly, it also referred to lichen pycnidia in which the pycnoconidia are formed only on the apex of simple sterigmata or on the apices of branched arthrosterigmata.

exochthonous (Gr. *éxo*, outside, on the exterior + *thón*, genit. *thonós*, earth, country): syn of **allochthonous**.

exogenization (Gr. *éxo*, outside, on the exterior + *génos*, origin < *gennáo*, to generate, produce + suf. *-ationem* > E. *-ation*, action, state or condition, result): a hypothetical process by means of which spores that are formed endogenously acquire an exogenous mode of formation; a mechanism proposed to support the evolution of the basidiomycetes from ancestral ascomycetes.

exogenous (Gr. *éxo*, outside, on the exterior + *génos*, origin < *gennáo*, to generate, produce): any structure produced or developed on the exterior of another structure; e.g., the spores of the Myxomycetes of the order Ceratiomyxales are exogenous, as are the conidia of the majority of asexual fungi. In *Ceratiomyxa fruticulosa*, Ceratiomyxales, this kind of spore is an *external spore* (L. *externus*, external, superficial + Gr. *sporá*, spore), because they develop on the outside of the many white, columnar structures, arising from a common hypothallus that comprises the sporophore. The exogenous, also called exosporous, species of Ceratiomyxales differ from the five orders of endosporous Myxomycetes (Liceales, Echinosteliales, Trichiales, Physarales and Stemonitales). Each spore has a cell wall of its own, but previous interpretations have emphasized the absence of a peridium enveloping all the spores as occurs in endosporous species. However, it has been concluded that the spores of *Ceratiomyxa* are unispored sporangia and are homologous to the sporangia of the endosporous species. In support of this conclusion, it has been demonstrated that the spores of *Ceratiomyxa* actually are enveloped in an electron-dense layer that may be equivalent to the peridium of the endosporous Myxomycetes. This finding casts doubt on the use of the term exosporous to describe this type of development. Cf. **endogenous**.

White columns of *Ceratiomyxa fruticulosa*, which produces **exogenous** spores, x 10 (*RTH*).

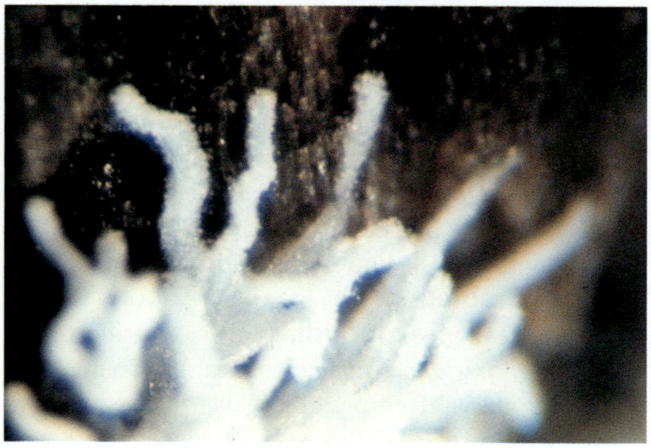

White columns of *Ceratiomyxa fruticulosa*, wich produces **exogenous** spores, x 30 (*RTH*).

exogynous (NL. *exogynus* < Gr. *éxo*, outside, on the exterior + *gyné*, woman; here, female sexual organ): *Aquatic fungi*. The zygote of *Monoblepharis* (Monoblepharidales) which migrates from the interior of the oogonium and becomes situated on top of it, remaining at rest until it matures and is transformed into an oospore. Cf. **exigynous**.

Exogynous oospore of *Monoblepharis polymorpha*, x 500 (*MU*).

exolete (L. *exoletus* < *exolescere*, to stop growing, degenerate, fall into disuse): applied to pycnidia, perithecia and other sporophores that remain empty and unused after maturing and liberating their spores.

exomycology (Gr. *éxo*, outside, on the exterior + *mýkes*, fungus + des. *-logía* > E. *-logy*, discourse, treatise < *lógos*, word < *légo*, I say): mycology of outer space.

exon (E. *exon*, expressed sequence): a term used in molecular biology to refer to a sequence of nucleotides in DNA or RNA that is expressed as all or as part of the polypeptide chain of a protein. Cf. **intron**.

exoperculation (Gr. *éxo*, outside, on the exterior + L. *operculum*, operculum, cap, lid, cover + Gr. suf. *-ationem* > E. *-ation*, action, state or condition, result): refers to a type of sporangial dehiscence in the Chytridiales, in which the operculum is attached by a hinge on the edge of the pore of the sporangium; also known as true operculation. Cf. **endoperculation**.

exoperidium, pl. **exoperidia** (Gr. *éxo*, outside, on the exterior + *péridion*, small leather bag): the outer layer of the multilayered peridial wall of a basidiocarp, as in *Geastrum*, *Disciseda* (Lycoperdales) and *Astraeus* (Sclerodermatales); the internal layer is the **endoperidium**.

Exoperidium (EX) of *Geastrum saccatum*, after stellate dehiscence; the endoperidium (EN) contains the basidiospores, x 0.5 (*TH*).

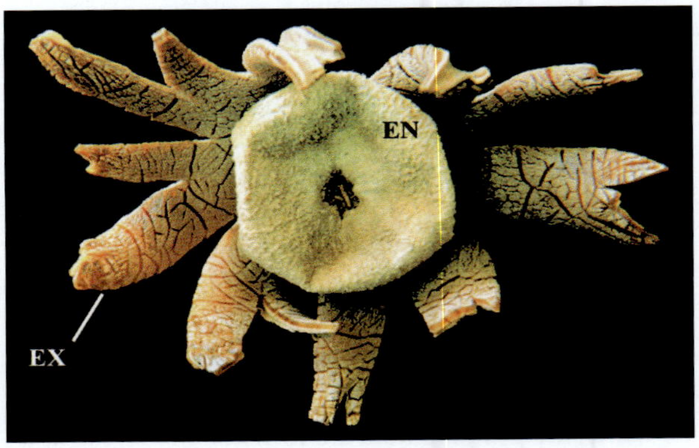

Exoperidium (EX) of *Astraeus hygrometricus*, after stellate dehiscence; the endoperidium (EN) contains the basidiospores, x 1 (*MU*).

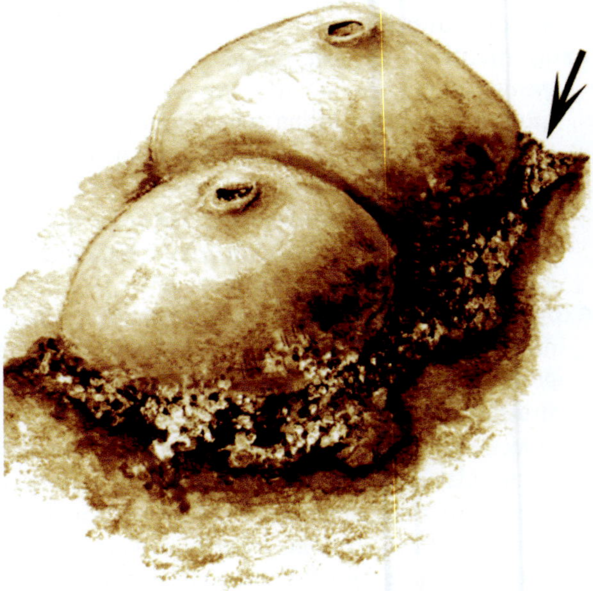

Exoperidium of the fruiting body of *Disciseda candida*, x 1 (*MU*).

exospore, **exosporium**, pl. **exosporia** (Gr. *éxo*, outside, on the exterior + *sporá*, spore): the second layer, beneath the *perispore*, in the cell wall of the most complex basidiospore, which is composed of five layers. When the perispore disappears, the exospore becomes the outer layer and generally forms the spore ornamentation. The other three layers, the *epispore*, *mesospore* and *endospore* are beneath the exospore, which usually lacks pigments but frequently can be distinguished chemically from the other adjacent layers.

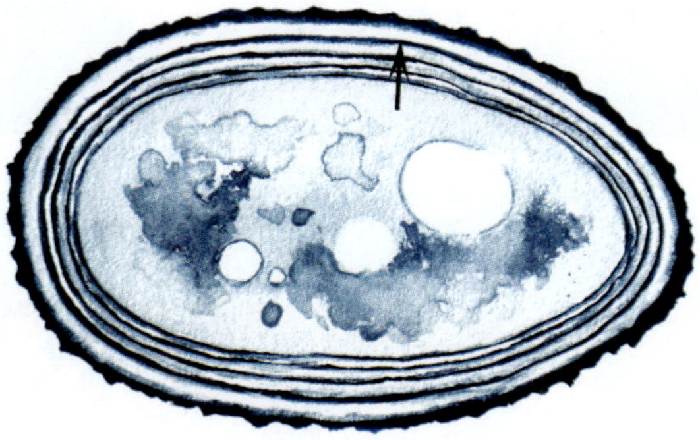

Exospore of the cell wall of the basidiospore. The other four layers: endospore, mesospore, epispore and perispore also are shown in this diagram of a longitudinal section, x 5,500 (*MU*).

exotic, **exotical** (Gr. *exotikós*, from the outside < *éxo*, outside, on the exterior + suf. *-tikós* > L. *-ticus* > E. *-tic*, relation, fitness, inclination or ability): see **allochthonous**. Cf. **autochthonous**.

expallant (L. pref. *ex-*, out of, from + ME. *pallen* < L. *pallidus* < L. *pallere*, to be pale + suf. *-ant* < E. *-ant*, one that performs): becoming pale on drying, like the pileus of the basidiocarp of species of *Laccaria* (Agaricales).

expersate (L. pref. *ex-*, out of, from + *sparsus*, scattered, extended, separated < *spargere*, to scatter + suf. *-atus* > E. *-ate*, provided with or likeness): refers to the oospores of Saprolegniaceae that have a refringent oil globule surrounded by a homogeneous cytoplasm.

explosive (L. *explosion*, to drive off by clapping < *explosio*, genit. *explosionis*, action of exploding + *-ivus* > E. *-ive*, quality, tendency, fitness): a type of ascus dehiscence that allows the rapid, usually violent discharge of the ascospores, as in *Sordaria* (Sordariales).

exserted (L. *exsertus*, thrown out, apparent, uncovered): refers to asci that stick out of the epithecium of the ascocarp, since their tips project to discharge the ascospores, as happens in *Ascobolus michaudii*, *A. stercorarius* (=*A. furfuraceus*) and *Saccobolus truncatus* (Pezizales).

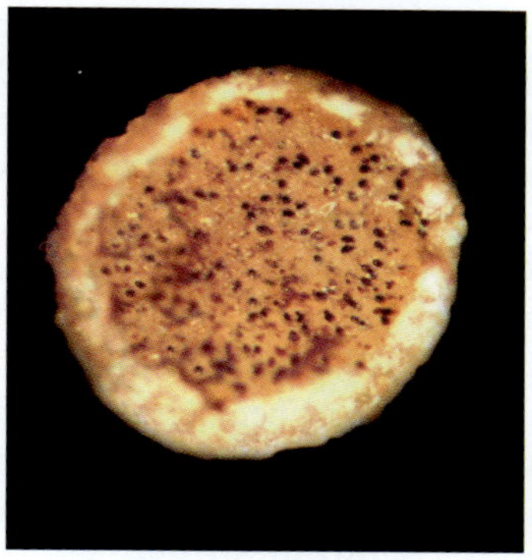

Exserted asci of an apothecium of *Ascobolus michaudii*, x 20 (*EAA*).

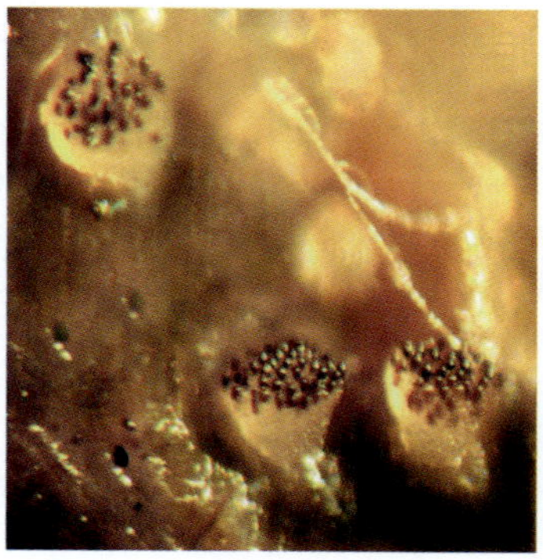

Exserted asci on three apothecia of *Saccobolus truncatus*, x 37 (*EAA*).

Exserted asci of *Ascobolus* sp., projecting from the surface of the apothecium, x 100 (*MU*).

exsiccatus, pl. **exsiccata**, **exsiccatae**, **exsiccati** (L. *exsiccare*, to desiccate + suf. *-atus*, likeness): applied to properly dried and labelled herbarium specimens. *Exsiccata* refers to a collection of plant specimens, and *exsiccati* is utilized for a collection of fungi. Note that the terms employed are in Latin.

extenditunicate (L. *extendere*, to spread out, increase its surface, occupy more space than before + L. *tunicatus*, covered by a tunica < *tunica*, covering + suf. *-atus* > E. *-ate*, provided with or likeness): a term applied to a type of bitunicate ascus whose walls extend without splitting at the moment of ascospore discharge, as happens with the fissitunicate ascus. Cf. **fissitunicate**.

extra- (L. *extra*, out of, beyond, outside): an inseparable prep. employed to form terms that allude to something on the exterior, as in *extramatrical*; in other terms, such as *extramature*, it signifies that it is more than mature, excessively so.

extramatrical (L. *extra*, out of, beyond, outside + *matrix*, matrix, uterus; often used in the sense of place where something is generated + L. suf. *-alis* > E. *-al*, relating to or belonging to): outside of the matrix or substrate, like the extramatrical filaments of an epibiotic fungus, which grow on the surface of the substrate. Cf. **intramatrical**.

extrorse (L. *extrorsus* < *extra*, out of, beyond, outside + *versus*, toward): arranged toward the outside, toward the exterior or edge, like the margin of the pileus of *Coprinus* (Agaricales). Cf. **introrse**.

extrusome (L. *extrudere*, to expell, to throw out + Gr. *sõma*, body): a structure derived from the system of Golgi vesicles, connected to the cell membrane by means of proteinaceous particles, whose contents are expelled in response to certain stimuli.

exudate (L. *exudare* < pref. *ex-*, out of, from + *sudare*, to sweat + suf. *-atus* > E. *-ate*, provided with or likeness): any liquid or fluid that, through exudation, oozes out of the hyphae of many molds, accummulating in droplets of various shades according to the species and their growth conditions. The characteristics of the exudate are taken into account in species determination of isolates of *Penicillium*, *Aspergillus* and other gen. of moniliaceous asexual fungi, as well as for other types of fungi, as *Pestalotia* (melanconiaceous asexual fungi). E.g., on the tips of the hairs of the perithecial necks of *Guanomyces polythrix* (Sordariales) droplets of a hyaline exudate are produced, which gives it a characteristic ornamentation. The colonies of various higher fungi also produce distinctive exudates under certain culture conditions, as is the case with species of *Suillus* and *Boletus* (Agaricales).

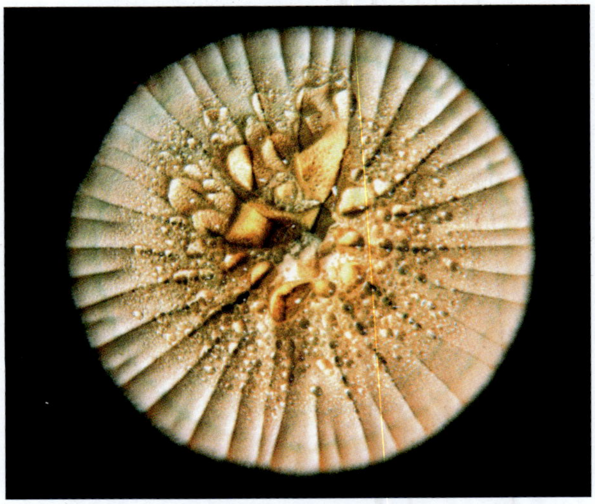

Exudate of a colony of *Penicillium aurantiogriseum* growing on agar, x 2 (*MU*).

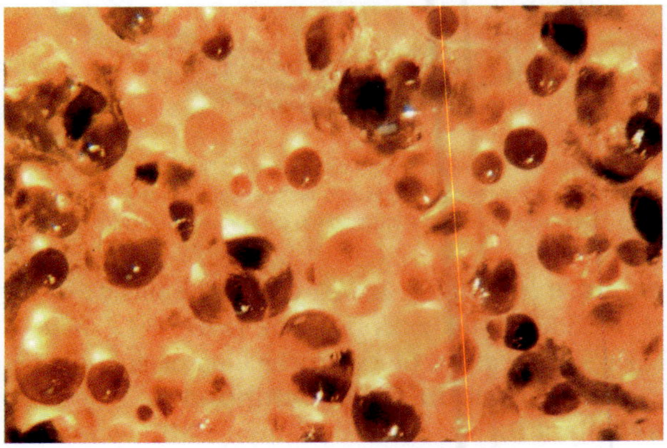

Exudate of a colony of *Aspergillus candidus* growing on agar, x 10 (*MU*).

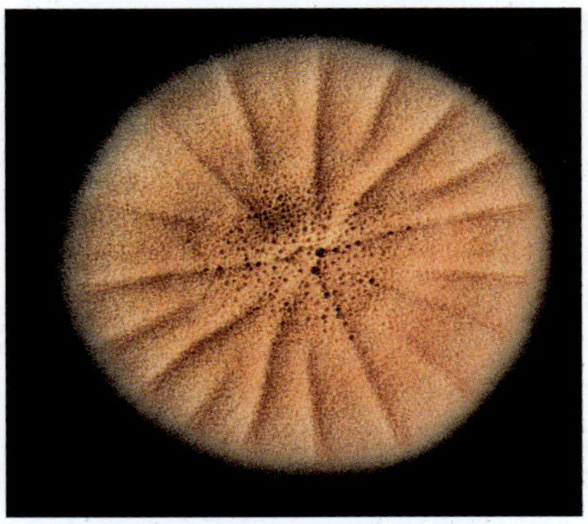

Exudate of a colony of *Aspergillus terreus* growing on agar, x 1 (*MU*).

Exudate of the colonies of *Aspergillus versicolor* growing on agar, x 1 (*TH*).

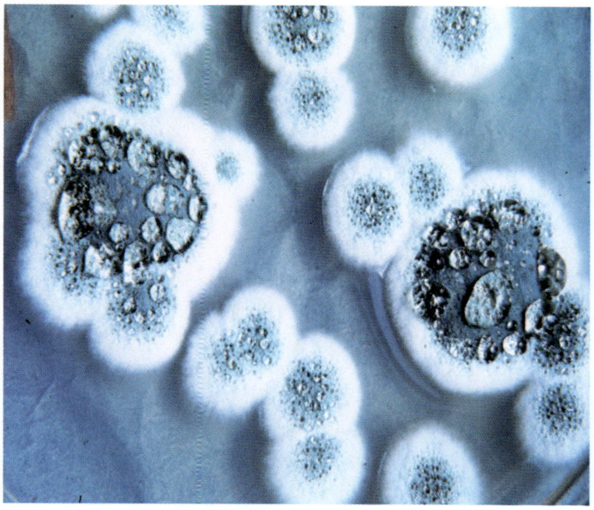

Exudate of the colonies of *Penicillium aurantiogriseum* growing on agar, x 10 (*TH*).

Exudate of a colony of *Suillus* sp. on agar, x 7.5 (*MU*).

Exudate of a colony of *Pestalotia* sp. growing on agar, x 1 (*RTH*).

Exudate of a basidiocarp of *Phanerochaete* sp. (Aphyllophorales) on bark, x 2 (*TRO*).

Exudate of the perithecial hairs, apically bulbose, of *Guanomyces polythrix* on agar, x 25, x 50 (*RTH*).

Exudate of a basidiocarp of *Boletus frostii*, x 0.5 (*RV*).

Flabellate fruiting bodies of *Schizophyllum commune* (*RV*).

f

f. (L. initial for *filius*, son): an initial used after an authors name, meaning son of, e.g., *L. f.*, son of Linnaeus. In genetics, *filial* signifies generation. Also employed as an abbrev. of *form*, as in *f. sp.*, *forma specialis*.

fabiform (L. *faba*, broad bean + suf. *-formis* < *forma*, shape): with the shape of a broad bean, like the ascospores of *Petriella* sp. (Microascales). See **reniform**.

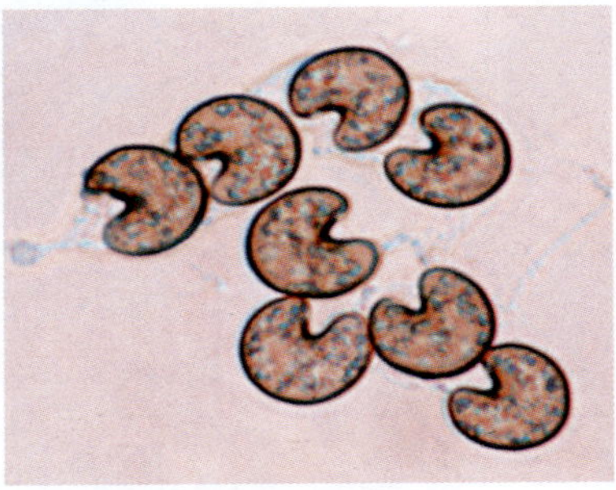

Fabiform ascospores of *Petriella* sp., x 1,450 (*CB*).

facultative (NL. *facultativus* < L. *facultas*, capacity, ability + suf. *-ivus* > E. *-ive*, quality or tendency, fitness): having the capacity to live under diverse environmental conditions; e.g., fungi that can live saprobically or as parasites (*facultative parasites*), or microorganisms that can live in a medium without air (oxygen), commonly being aerobic (*facultative aerobes*), or vice versa, if they can tolerate the presence of air, when they normally live without it (*facultative anaerobes*).

facultative parasite (NL. *facultativus* < L. *facultas*, capacity, ability + suf. *-ivus* > E. *-ive*, quality or tendency, fitness; Gr. *parásitos* < *pará*, together, at the side of + *sîtos*, flour, bread, food in general): it refers to an organism that is capable of invading and living on another living organism, or of living on dead organic material, depending on the circumstances; in the fungi, this capacity is restricted to a smaller number of species than the opposite condition (*facultative saprobe*). Cf. **obligate parasite**.

facultative saprobe (NL. *facultativus* < L. *facultas*, capacity, ability + suf. *-ivus* > E. *-ive*, quality or tendency, fitness; Gr. *saprós*, rotten + *bíos*, life): an organism that has the capacity to live on organic material or to infect another living organism, depending on the circumstances; the majority of the fungi have this capacity. See **facultative parasite** and **obligate parasite**.

fairy ring (ME. *fairie*, fairyland, fairy people < OF. *faerie* < *feie, fee,* fairy, a fantastic being represented in the form of a woman, to whom magic powers were attributed; ME. < OE. *hring*, ring): a ring or circle formed by fungus fruiting bodies, generally of basidiomycetes, which are produced on the periphery of a mycelial mass that has developed centrifugally from an initial point of growth, in particular of the species *Marasmius oreades* (Agaricales). Some 60 species have been recorded as frequent formers of these rings in lawns and meadows; they are less frequently found in forests.

Fairy ring of mushrooms on a lawn, x 0.02 (*BL*).

falcate (L. *falcatus* < *falx*, genit. *falcis*, sickle + suf. *-atus* > E. *-ate*, provided with or likeness): with a flat, curved shape, like a sickle; like the conidia of

Harposporium species (moniliaceous asexual fungi) and the macroconidia of *Fusarium oxysporum* (tuberculariaceous asexual fungi). Also called **falciform** and **lunate**.

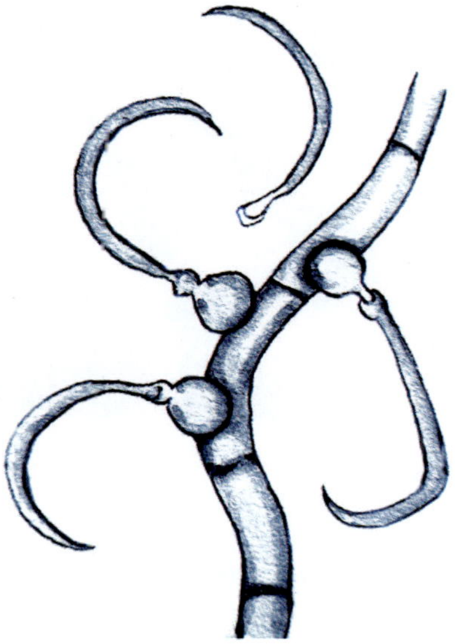

Falcate conidia of *Harposporium* sp., produced on ampulliform conidiogenous cells, x 1,170 (*MU*).

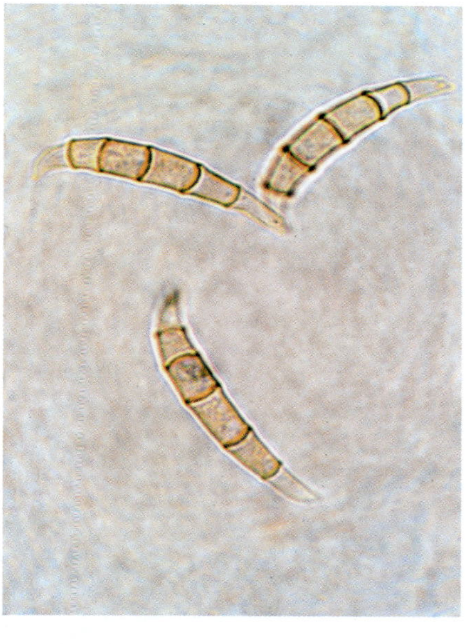

Falcate conidia of *Fusarium* sp., x 840 (*CB*).

falciform (L. *falciformis*, sickle-shaped < *falx*, genit. *falcis*, sickle + suf. *-formis* < *forma*, shape): see **falcate** and **lunate**.

falciphore (L. *falx*, genit. *falcis*, sickle + suf. *-phóros*, bearer): a hypha bearing a sickle-shaped conidiogenous cell, characteristic of the gen. *Zygosporium* (moniliaceous asexual fungi).

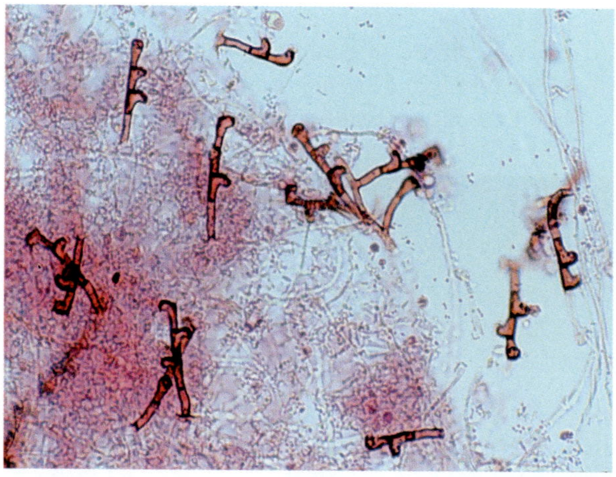

Falciphores of the conidiophores of *Zygosporium masonii*, x 200 (*MU*).

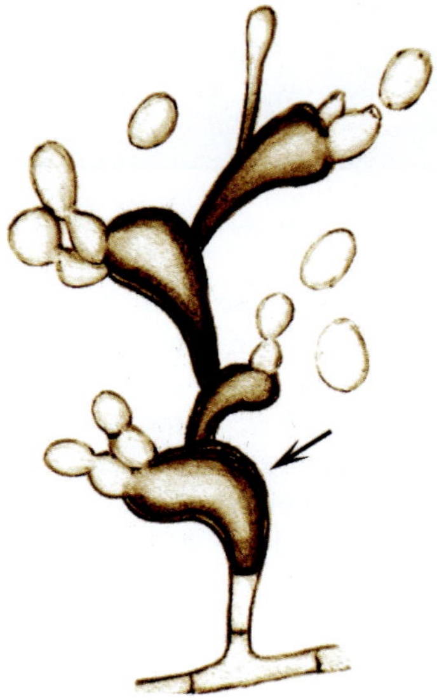

Falciphores of the conidiophore of *Zygosporium masonii*, x 790 (*MU*).

family (L. *familia,* a household): a systematic unit of classifications by taxonomic categories that includes a group of genera, all of which have in common various important characters. This entity occupies a position intermediate between order and genus, and its name ends in *-aceae*, as in Agaricaceae and Polyporaceae.

farctate (L. *farctus*, stuffed): turgent, solid; used for the stipe of a basidiocarp that is not hollow, but filled with more or less compact hyphae. Also called **solid**. Cf. **cavernose** and **fistular**.

farinaceous, **farinose** (L. *farinaceus*, similar to flour in appearance or color; mealy < *farina*, flour, meal + suf. *-aceus* > E. *-aceous*, of or pertaining to, with the nature of): covered with a mealy powder. The surface of the colonies of some fungi on agar have a farinaceous texture, e.g., those of *Paecilomyces farinosus*, those of certain dermatophytes (such as *Microsporum* and *Trichophyton*) and species of other moniliaceous asexual fungi (such as *Geotrichum*).

Farinaceous colony of *Microsporum nanum*, on agar, x 1 (*MU*).

fasciate, **fasciated** (L. *fasciatus*, bundled; a group of things united by a common axis < *fascia*, raceme, band, ribbon, sash + suf. *-atus* > E. *-ate*, provided with or likeness): **1**. Concrescent, with the tissues intimately continuous, as is seen in some fungi whose stipes or pilei grow united together. **2**. With wide ribbon or band-like structures, like the thallus of certain fruticose lichens whose branches are flat (e.g., *Ramalina homalea* and *Cetraria islandica*, of the Lecanorales).

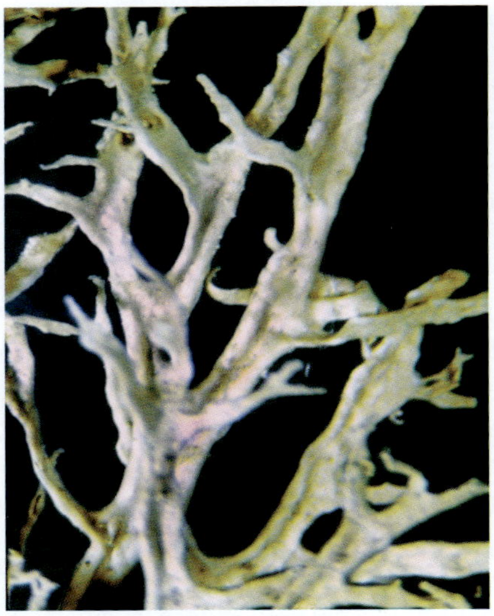

Fasciate branches of the thallus of *Ramalina homalea*, x 2 (*MU*).

Fasciate branches of another thallus of *Ramalina homalea*, x 2 (*MU*).

fascicle (L. *fasciculus*, dim. of *fascis*, raceme, bundle, faggot): a bundle or cluster of hyphae, conidiophores, asci, etc. E.g., *Oidiodendron tenuissimum* (dematiaceous asexual fungi) forms fascicles of hyphae with conidiophores; also called *funiculose*.

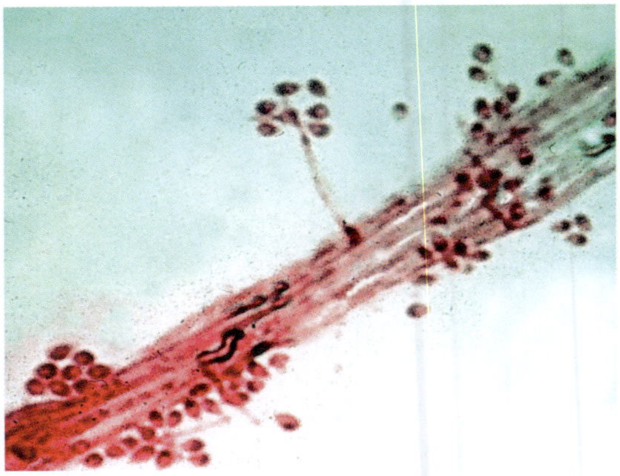

Fascicle of hyphae with conidiophores of *Sporothrix schenckii* (moniliaceous asexual fungi), x 550 (*RLM*).

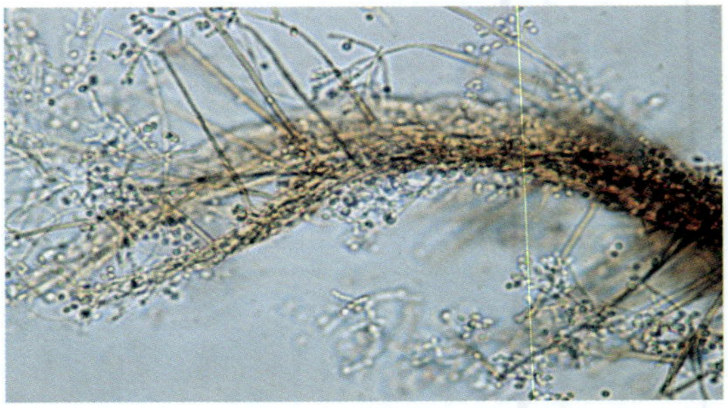

Fascicle of hyphae with conidiophores of *Oidiodendron tenuissimum*, x 350 (*MU*).

fasciculate (NL. *fasciculatus*, arranged in small bundles or racemes < *fasciculus*, dim. of *fascis*, raceme, bundle, faggot + suf. *-atus* > E. *-ate*, provided with or likeness): grouped into bundles or clusters, like the sporangia of *Stemonitis axifera* (Stemonitales), the hyphae of *Oedocephalum*, *Acremonium*, some species of *Penicillium* (moniliaceous asexual fungi), and *Stachybotrys* (dematiaceous asexual fungi); the conidiophores of *Cercospora coffeicola* (also dematiaceous), the asci of *Emmonsiella capsulata* (Onygenales) and the basidiocarps of *Hypholoma* (=*Naematoloma*) *fasciculare*, *Oudemansiella canarii* (Agaricales) and *Lentinus tigrinus* (Aphyllophorales), are other examples of fasciculate structures.

Fasciculate sporangia of *Stemonitis axifera*, x 4.5 (*MU*).

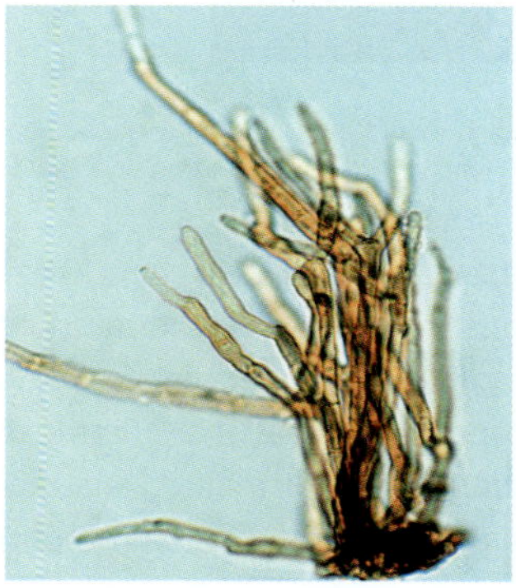

Fasciculate conidiophores of *Cercospora coffeicola*, x 750 (*CB*).

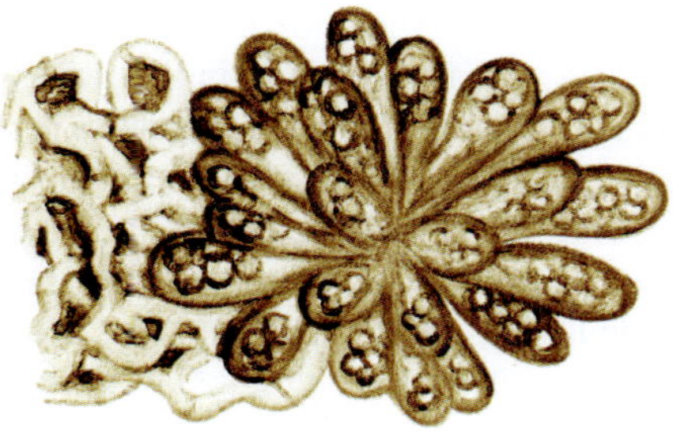

Fasciculate asci of *Emmonsiella capsulata*, x 750 (*MU*).

Fasciculate basidiocarps of *Hypholoma* (=*Naematoloma*) *fasciculare*, x 0.5 (*TRO*).

Fasciculate basidiocarps of *Lentinus tigrinus*, x 0.2 (*RV*).

Fasciculate basidiocarps of *Oudemansiella canarii*, x 0.5 (*TH*).

fastigiate, **fastigiated** (NL. *fastigiatus*, ending in a point < L. *fastigium*, narrowing toward the tip, with erect branches generally arranged in clusters + suf. *-atus* > E. *-ate*, provided with or likeness): **1**. Applied to adherent structures that taper to a point; e.g., hyphae, conidiophores, etc., whose branches, stalks, etc., end in a point. E.g., in *Inocybe rimosa* (=*I. fastigiata*), Agaricales, the fibers on the outer surface of the pileus are fastigiate. **2**. *Lichens*. Refers to a cortical zone whose hyphae are perpendicular to the main axis of the thallus. Cf. **fibrous**.

Fastigiate fibers of the pileus of *Inocybe rimosa* (=*I. fastigiata*), x 0.5 (*MU*).

faveolate (L. *faveolatus*, honey-combed < *faveolus*, dim. of *favus*, honeycomb + suf. *-atus* > E. *-ate*, provided with or likeness): syn. of **alveolate**; with honeycomb-like cavities or cells, like the ascocarp of *Morchella esculenta* (Pezizales) and the hymenophore of the basidiomycetes *Hexagonia tenuis* and *H. variegata* (=*H. papyracea*), Aphyllophorales. See **foveolate**.

Faveolate ascocarps of *Morchella* spp. x 0.5 (*MRO*).

Faveolate hymenophore of *Hexagonia tenuis*, x 0.7 (*CB*).

Faveolate hymenophore of *Hexagonia variegata* (=*H. papyracea*), x 10 (*MU*).

favus, pl. **favi** (L. *favus*, honeycomb): syn. of *tinea favosa*. *Med. mycol.* A clinical entity characterized by the formation of a dense mass of mycelium and arthrospores that originate in the hair follicles and which acquire the shape of a rhomb or inverted cone; this type of tinea in man and animals is caused by *Trichophyton shoenleinii*, *T. violaceum* and *Microsporum gypseum* (moniliaceous asexual fungi). See **mycosis** and **tinea**.

fenestrate (L. *fenestratus*, having windows < *fenestra*, window + suf. *-atus* > E. *-ate*, provided with or likeness): provided with symmetrically arranged perforations or gaps, like the indusium of *Dictyophora duplicata* (Phallales).

Fenestrate indusium of the fruiting body of *Dictyophora duplicata*, x 0.3 (*TH*), and drawing of the same species, x 0.3 (*MU*).

fermentation (L. *fermentatio*, genit. *fermentationis* < *fermentare*, to undergo fermentation < *fermentum*, yeast, enzyme + suf. *-ationem* > E. *-ation*, action, state or condition, result): a chemical change with effervescence (if the fermented sustrate is liquid); an enzymatically controlled anaerobic breakdown of an energy-rich compound (such as a carbohydrate to carbon dioxide and alcohol, or to an organic acid). See **zymosis**. In a broad sense, fermentation is the enzyme-mediated transformation of an organic compound, such as that carried out in the elaboration of beer, bread and other fermented products made with the activities of yeasts, as well as in the production of antibiotics, acids, steroids, and a broad range of chemical substances obtained through the use of molds and other microorganisms that are utilized traditionally (such as the production of indigenous fermented foods and beverages, i.e., traditional fermentations), or on a larger scale and with a certain control of the process, that is undertaken in the pharmaceutical and food-processing industries with the use of starter cultures or inoculum, composed of a single microorganism or of a mixture of microorganisms, depending on the requirements of the process. Bread, beer and wine, obtained through fermentation by the yeast *Saccharomyces cerevisiae* (Saccharomycetales), are the best known fermented industrial products in the world, and which initially originated as a traditional fermentation process. Throughout the world, occidental as well as oriental, there exists a wide variety of traditional fermentations, industrialized or not, from which are derived many products that different ethnic groups utilize as food, stimulants, medicine, or in rituals. As an example, among the oriental fermented foods, whose production is undertaken industrially and is therefore more or less controlled, are found soybean products, such as *miso* paste, *miso* soup and *natto* in Japan, and rice products, such as *ang-kah*, also called *anka*, *ang-kak* or Chinese red rice. The essential microorganisms for the production of miso are a mold, *Aspergillus flavus* var. *oryzae* (=*A. oryzae*), moniliaceous asexual fungi, and a yeast, *Zygosaccharomyces* (=*Saccharomyces*) *rouxii* (Saccharomycetales). The bacteria *Pediococcus halophilus* and *Streptococcus faecalis* sometimes are also involved in the same fermentation. The fermenting microorganism of natto is the bacterium *Bacillus natto*, closely related to *B. subtilis*. *Ang-kah* is a product of the fermentation of rice carried out by *Monascus purpureus* (Eurotiales). The list of products derived from traditional fermentations is so extensive that the extant information greatly exceeds the extent and coverage of the present dictionary. However, below are included some names of traditional fermented products about which additional information can be obtained under the corresponding entries. See **ang-kah**, **bulgar grain**, **kefir grain**, **microbioglea**, **mother of vinegar**, **pozol**, **pulque**, **shoyu**, **tapé ketella**, **tea fungus**, **tempeh**, **tepache**, **tesgüino**, **tibi grain**, and **tuba**.

Colonies of *Aspergillus flavus* var. *oryzae* (=*A. oryzae*) on agar; it is used in the **fermentation** to produce the *miso* paste, x 1 (*MU*).

Some industrialized oriental foods derived from traditional **fermentations**. Shown are commercial packages of the products: *miso* soup, *miso* paste, *natto* and *ang-kah*, x 0.5 (*MU*).

-fer, **-ferous** (L. *fero* < *ferre*, to carry, bear, bring, produce): a suf. that denotes something that bears or produces, as in *aquiferous* and *carboniferous*, i.e., that bears or produces water and carbon, respectively; *asciferous*, bearing asci.

fertilization (L. *fertilis*, fertile < *ferre*, to carry, bear + suf. *-ationem* > E. *-ation*, action, state or condition, result): the action and effect of fertilizing; the fusion of gametes.

fertilization tube (L. *fertilis*, fertile + suf. *-ationem* > E. *-ation*, action, state or condition, result; L. *tubus*, tube): a tube that originates from the male gametangium (antheridium) and penetrates the female (oogonium), after which the male gametic nucleus unites with the female nucleus to produce the zygote nucleus, as occurs, e.g., in *Saprolegnia parasitica* (Saprolegniales).

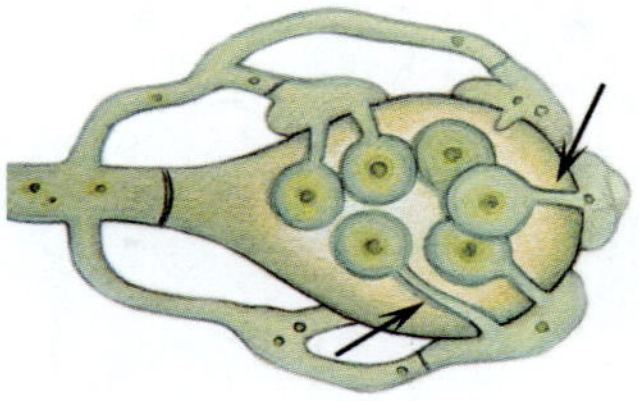

Fertilization tubes of the antheridia of *Saprolegnia parasitica*, x 260 (*MU*).

festooned (L. *festa*, party, fiesta, wavy adornment at the edge of something): syn. of **crenate**; i.e., edged with scallops or indentations, like the gill margin of *Cortinarius alboviolaceus*, or the margin of the pileus of *Laccaria laccata* (Agaricales).

fibercle (L. *fibra*, fiber, filament, thread + *tuberculum*, tubercle, dim. of *tuber*, mass, lump, prominence): each of the protuberant scars that are formed on the branches of some lichen thalli after the fibrils have broken, as is seen in *Usnea nashii* (Lecanorales).

fibril (NL. *fibrilla*, dim. of *fibra*, fiber, filament, thread): a very thin fiber. **1**. *Lichens*. A small, short, simple branch, containing both the mycobiont and photobiont, that arises perpendicularly from the main branches of a fruticose thallus, as in *Usnea* (Lecanorales); it serves as a vegetative propagule. **2**. *Gasteromycetes*. Exoperidial ornamentation in the shape of a fine, hair-like structure; typical of *Lycoperdon* and *Calvatia* (Lycoperdales).

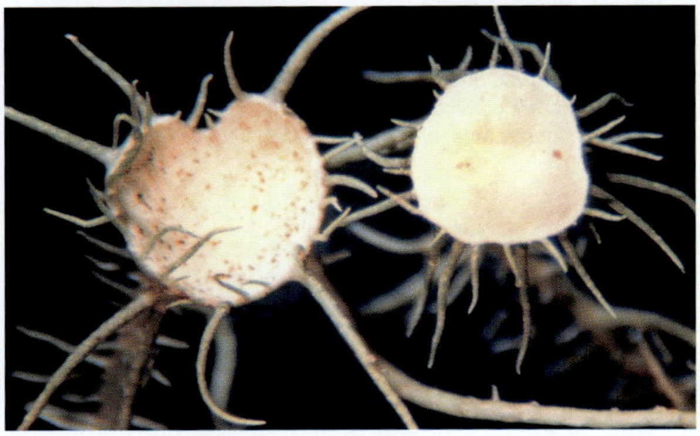

Fibrils of the branches and apothecia of *Usnea* sp., x 3 (*CB* & *MU*).

fibrillose (L. *fibrillosus*, having fibers < *fibrilla*, dim. of *fibra*, fiber, filament + suf. *-osus* > E. *-ose*, full of, augmented, prone to): having thin fibers, with the appearance of very fine silk threads, like the surface of the pileus of *Amanita phalloides* and *Volvariella lepiotospora* (Agaricales).

Fibrillose pileus of the basidiocarp of *Volvariella lepiotospora*, x 0.5 (*SBH*).

fibrillose-truncate wart (L. *fibrillosus*, having fibrils < *fibrilla*, dim. of *fibra*, fiber, filament + suf. *-osus* > E. *-ose*, full of, augmented, prone to; *truncatus*, truncate, cut off > *truncare*, to lop off + suf. *-atus* > E. *-ate*, provided with or likeness; ME. < OE. *wearte*, wart): a wart with the upper part flat and adorned with fibrils.

fibrosin bodies (L. *fibra*, fiber, filament, thread + *-in*, des. of many chemical terms; ME. < OE. *bodig*, body): structures whose exact nature is unknown that are found in the cytoplasm of the conidia of some powdery mildews (Erysiphales).

fibrous (L. *fibrosus*, having fibers or slender like a fiber < *fibra*, fiber, filament, thread + suf. *-osus* > E. *-ose*, full of, augmented, prone to): provided with fibers or having the appearance of a fiber. In the lichens, it refers to the cortical zone that is composed of hyphae parallel to the thalline axis; in this latter case it is the opposite of **fastigiate**.

fibula, pl. **fibulae** (L. *fibula*, buckle, clasp, clamp): also called *clamp connection* or hook; it is a bridge-like hyphal connection between two adjacent cells through which nuclei migrate to maintain the dikaryon; it is characteristic of the secondary or dikaryotic mycelium of many basidiomycetes (such as *Coprinus comatus*, of the Agaricales).

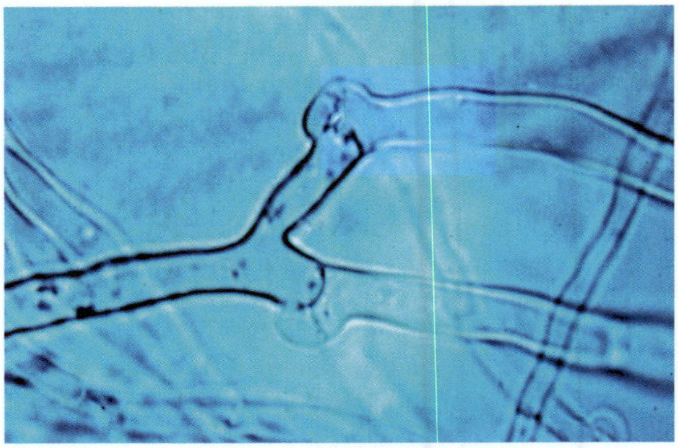

Fibulae of the secondary mycelium of a basidiomycete, x 500 (*DM*).

filament (ML. *filamentum*, filament < *filare*, to spin < *filum*, linen thread + suf. *-amentum* > E. *-ament*, strap, band): a fine thread; a term loosely used to refer to a hypha or any other fungal element that is thread-like in appearance.

filamentous (ML. *filamentosus*, composed of filaments < *filare*, to spin < *filum*, linen thread + suf. *-amentum* > E. *-ament*, strap, band + L. suf. *-osus* > E. *-ous*, provided with, with the qualities of): adj. which means composed of filaments, like the mycelium of the majority of the true fungi; with the appearance of a filament, like the fungal hyphae.

filiform (L. *filiformis* < *filum*, linen thread + suf. *-formis* < *forma*, shape): with the shape of a thread, fine and slender, like a linen fiber, like the ascospores of *Epichloë typhina* (Hypocreales) and *Ophiodothella* (Phyllachorales), the appendages of the ascospores of *Cercophora palmicola* (Sordariales), the ß-conidia of *Phomopsis* (sphaeropsidaceous asexual fungi), and the apical and basal appendages of the conidia of *Monochaetia carissae* (melanconiaceous asexual fungi).

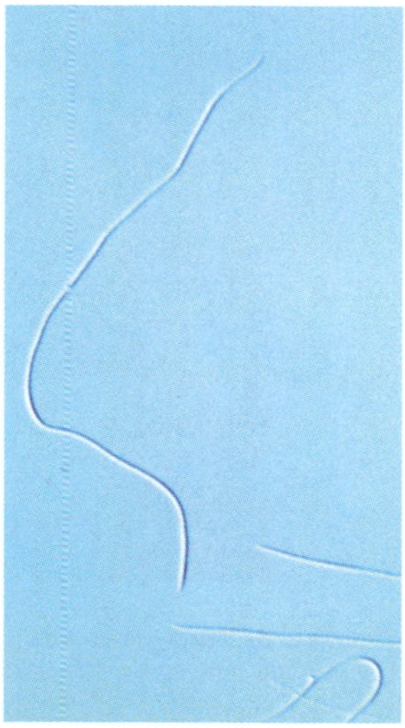

Filiform ascospore of *Epichloë typhina*, x 550 (*RTH*).

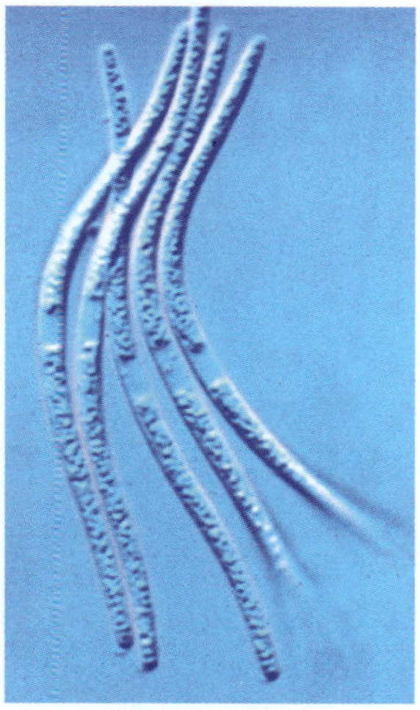

Filiform ascospores of *Ophiodothella vaccinii*, x 650 (*RTH*).

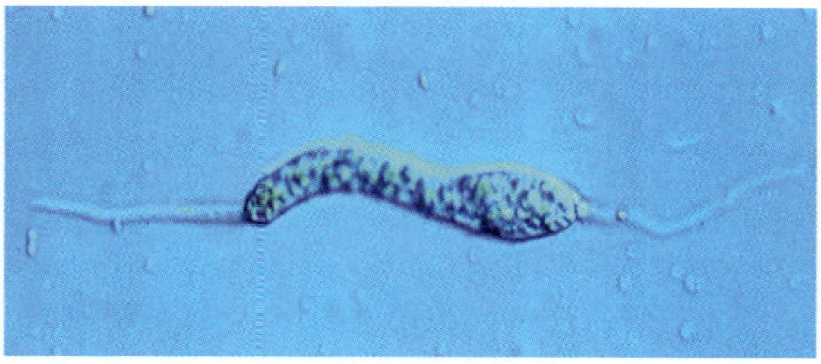

Filiform polar appendages of a young ascospore of *Cercophora palmicola*, x 770 (*RTH*).

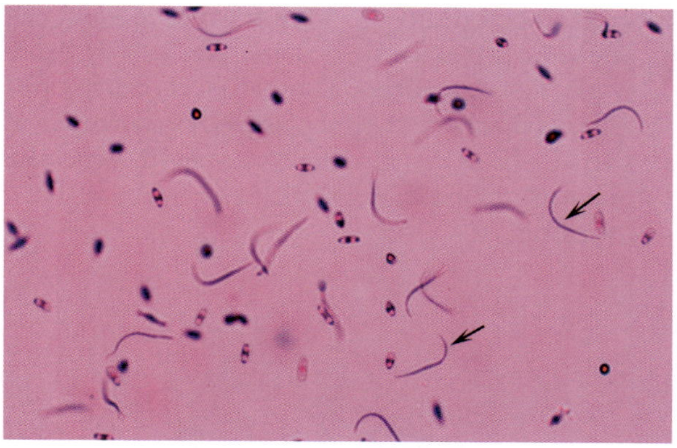

Filiform ß-conidia from a pycnidium of *Phomopsis* sp., x 450 (*CB*).

Filiform polar appendages of the conidia of *Monochaetia carissae*, x 1,000 (*MU*).

filoplasmodium, pl. **filoplasmodia** (L. *filum*, linen thread + Gr. *plásma*, formation, soft material with which a living being is formed + L. suf. *-odium*, resembling < Gr. *ode*, like < *-oeídes*, similar to): a reticular, filamentous pseudoplasmodium composed of a colony of individual fusiform cells that move and grow within the filaments or tunnels of mucilage that they produce. Characteristic of the Labyrinthulales, such as *Labyrinthula marina*. Also known as a *mucilaginous pseudoplasmodium*.

Filoplasmodium of *Labyrinthula marina*, x 370 (*MU*).

filopod, **filopodium**, pl. **filopodia** (L. *filum*, linen thread + NL. *podium* < Gr. *pódion*, dim. of *poús*, genit. *podós*, foot): a pseudopodium or filiform extension of a plasmodium or of a protoplast; the myxamoebae of *Dictyostelium* (Dictyosteliomycota) and the reticulate plasmodia of the Labyrinthulales form filopodia. Cf. **lobopod**.

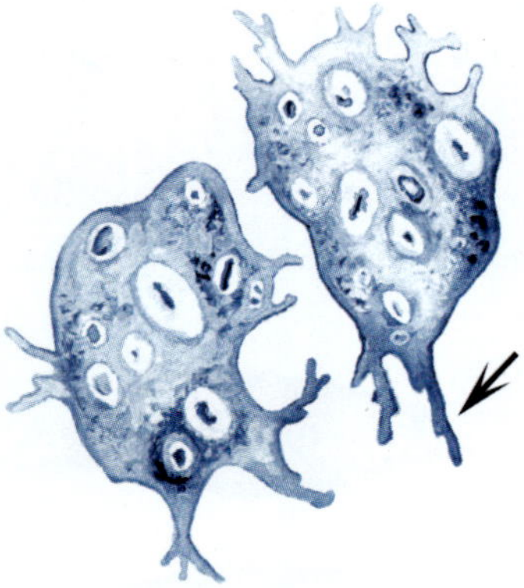

Filopodia of the myxamoebae of *Dictyostelium discoideum*, x 500 (*MU*).

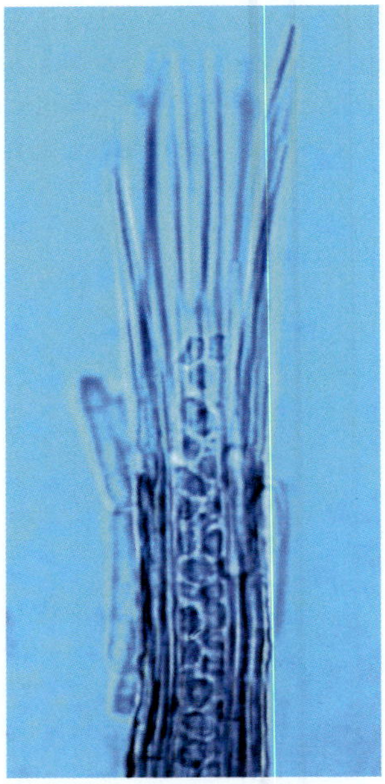

Fimbriate ostiole of the ascocarp of *Ceratocystis fimbriata*, x 300 (*RTH*).

fimbriate (L. *fimbriatus*, fibrous, fringed, with hairs or fibers on the margin < *fimbria*, fiber, fringe, end, extremity + suf. *-atus* > E. *-ate*, provided with or likeness): **1**. Provided with a fringe, like the border of the colony of some yeasts (such as *Candida tropicalis* and *C. krusei*, of the cryptococcaceous asexual yeasts), which have slender lobes or very fine segments with a pointed tip. The ostioles of the perithecia of *Ceratocystis fimbriata* (Microascales) and *Melanospora zamiae* (Melanosporales) also are fimbriate. **2**. Finely toothed; often applied in reference to a stoma. Also called **laciniate**.

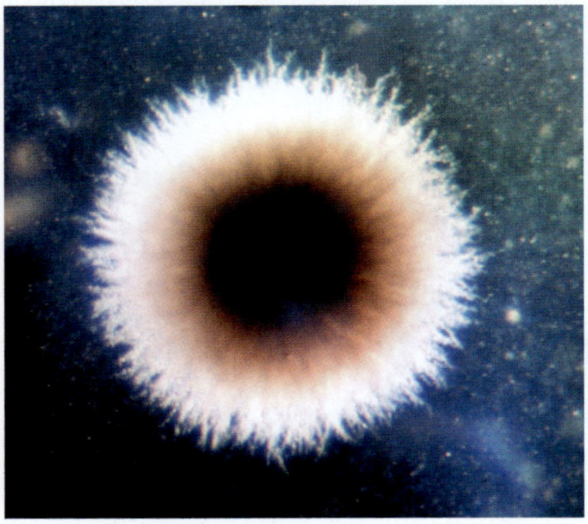

Fimbriate border of a colony of *Candida krusei* on agar, x 25 (*TH*).

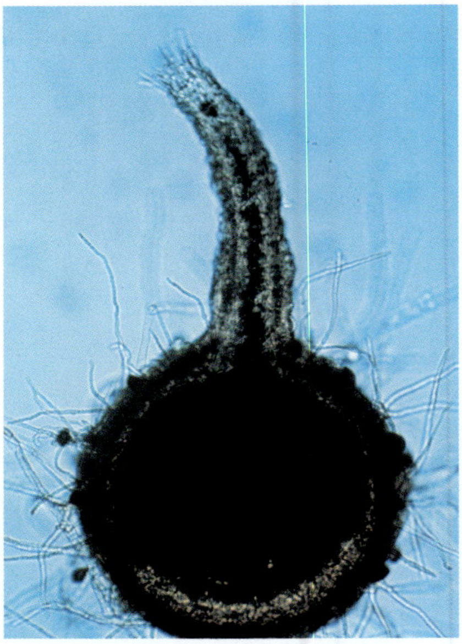

Fimbriate ostiole of the perithecium of *Melanospora zamiae*, x 100 (*RTH*).

fimbrillate (NL. *fimbrillatus*, bordered with minute fibers or hairs < *fimbrilla*, dim. of *fimbria*, fiber, fringe, end, extremity + suf. *-atus* > E. *-ate*, provided with or likeness): minutely *fimbriate*.

fimicole, **fimicolous** (L. *fimus*, *fimo*, dung + suf. *-cola*, inhabitant < *colere*, to inhabit): living or growing on dung, like many Mucorales (*Mucor*, *Pilobolus*, etc.) and other fungi that belong to various groups. See **coprophilous.**

Fimicole apothecium of *Coniochaeta scatigena* (=*Ascobolus scatigenus*, Pezizales), x 3 (*EAA*).

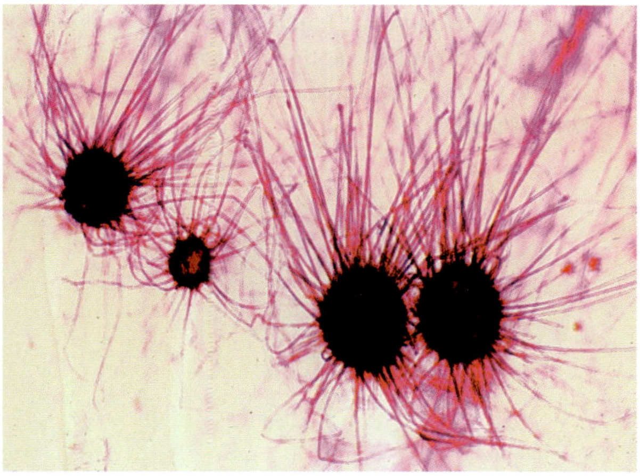

Fimicole perithecia of *Chaetomidium fimeti* (Sordariales), x 400 (*MU*).

fission (L. *fissio*, genit. *fissionis*, partition, splitting + suf. *-io*, genit. *-ionis* > E. *-ion*, result of an action, state of): see **bipartition** and **schizogenesis**.

fissitunicate (L. *fissio*, genit. *fissionis*, partition, splitting + *tunicatus*, covered by a tunic < *tunica*, cover, cloak + suf. *-atus* > E. *-ate*, provided with or likeness): a type of *bitunicate ascus* (a term from Luttrell) in which the elastic internal layer, or layers, of the wall (endotunica) is separated from the rigid outer layer or layers (exotunica, ectotunica); the endotunica extends through the ectotunica before liberating the ascospores, as occurs, e.g., in *Leptosphaerulina crassiasca* (Dothideales). The fissitunicate ascus corresponds to the *physoclastic ascus* of the literature on marine ascomycetes. The *non-fissitunicate ascus* (*aphysoclastic ascus* of marine mycology) lacks the separation and extension of the ascus wall layers before liberation of the ascospores. Inasmuch as electron microscopy has shown that all ascus walls are multilaminate, it is preferable to use the terms *physoclastic* and *fissitunicate* instead of *bitunicate*, and *non-fissitunicate* and *aphysoclastic* instead of *unitunicate*, to refer to these types of ascus dehiscence. Cf. **semifissitunicate** and **rostrate dehiscence**.

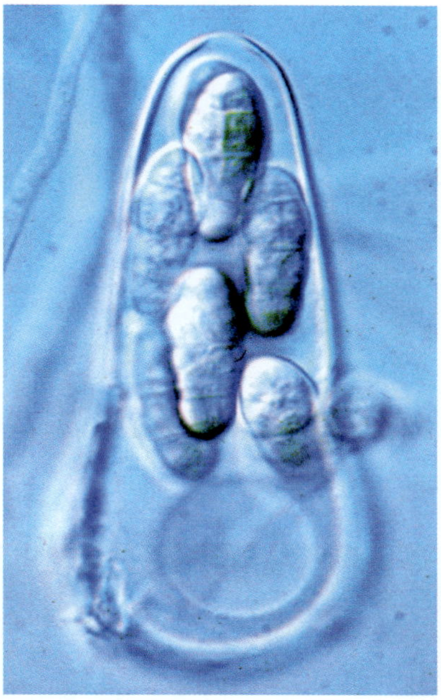

Fissitunicate ascus of *Leptosphaerulina crassiasca*, x 600 (*RTH*).

fistular, **fistulose** (L. *fistularis*, like a pipe < *fistula*, fistula, a hollow stem, pipe, tube + suf. *-aris* > E. *-ar*, like, belonging to; L. *fistulosus*, full of pipes < *fistula* + suf. *-osus* > E. *-ose*, full of, augmented, prone to): having one or several tubes, i.e., with a tubular structure, hollow, like the stipe of *Amanita caesarea* (Agaricales). Cf. **cavernose**, **farctate** and **solid**.

Fistular stipes of the basidiocarps of *Amanita caesarea*, x 0.4 (*RMA*).

flabellate, **flabelliform** (L. *flabellatus*, having a fan-like structure < *flabellum*, fan + suf. *-atus* > E. *-ate*, provided with or likeness; *flabellum* + suf. *-formis* < *forma*, shape): with the shape of a fan, like the fruiting bodies of *Schizophyllum commune*, *Panellus* sp., and the segments of the fruiting body of *Sparassis crispa* (Aphyllophorales).

Flabellate fruiting bodies of *Schizophyllum commune*, x 0.1 (*RV*).

Flabellate fruiting bodies of *Panellus* sp., x 0.5 (*RV*).

Flabellate segments of the fruiting body of *Sparassis crispa*, x 0.5 (*RV*).

flaccid (L. *flaccidus*, *flaccus*, loose, limp, without consistency): weak, limp, lacking strength or turgor, incapable of remaining erect, hanging due to its own weight.

flagellate (L. *flagellatus* < *flagellum*, flagellum, whip + suf. *-atus* > E. *-ate*, provided with or likeness): provided with one or more *flagella*.

flagellum, pl. **flagella** (L. *flagellum*, flagellum, whip): a motile, flexible protoplasmic filament in the form of a little whip, with or without mastigonemes, that serves to propel a motile cell, such as zoospores and planogametes. Some authors apply the term flagellum to the propelling filaments of prokaryotic cells (without a 9 + 2 composition of microtubules, composed of a single protein called flagelline), whereas they use *undulipodia* for the propelling filaments of the eukaryotic cells (with the 9 + 2 composition of microtubules, composed of more than 100 proteins). The amoeboid cells of some species of Myxomycota (such as the myxozoospores of *Physarum*, of the Physarales) and the zoospores of the Chytridiomycetes (such as *Rhizophlyctis*, of the Spizellomycetales, and *Allomyces*, of the Blastocladiales), Hyphochytriomycota (such as *Hyphochytrium*, of the Hyphochytriales) and Oomycota (such as *Saprolegnia*, of the Saprolegniales) also have flagella. Cf. **undulipode**.

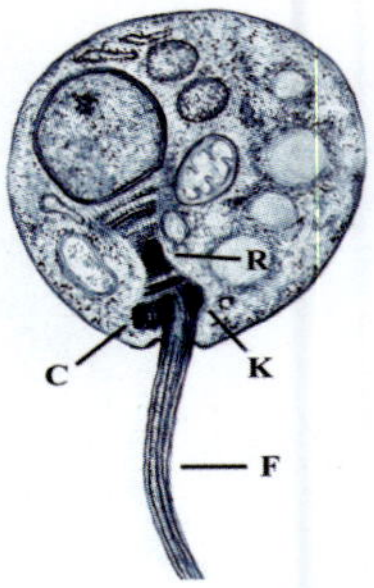

Flagellum (F) of the zoospore of *Rhizophlyctis rosea*. The centriole (C), kinetosome (K), and the rhizoplast (R) also are shown, x 16,000 (*MU*).

flexible (L. *flexibilis*, flexible): not rigid or firm, bending easily, like the basidiocarp of *Collybia* (Agaricales).

flexuose, **flexuous** (L. *flexuosus*, twisted, tortuous < *flexus*, bent, turned, curved + suf. *-osus* > E. *-ose*, full of, augmented, prone to): twisted or bent alternately in opposite directions, in a zigzag, like the stalk of the sporangium of *Physarum polycephalum* (Physarales).

flimmer hairs, **flimmers** (G. *flimmer* < *flimmern*, to shimmer; ME. *heer, hare, heir* < OE. *haer*, hair, cylindrical filament): equivalent to **mastigonemes**.

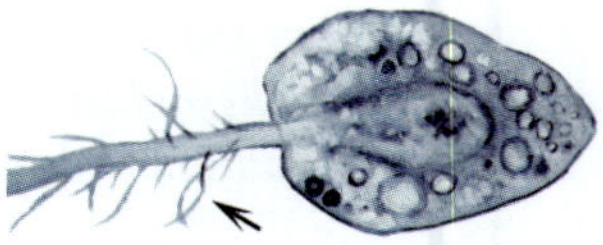

Flimmer hairs of the flagellum of a zoospore of *Rhizidiomyces apophysatus* (Hyphochytriales), x 4,800 (*MU*).

flimmergeissel (G. *flimmer*, to vibrate, tremble + *Geissel,* flagellum): a flagellum bearing two rows of vibratory, tripartite, tubular hairs or filaments. Cf. **straminopilous**. See **tripartite tubular hair**.

floccose (L. *floccosus*, full of flocks of wool < *floccus*, flock or tuft of wool + suf. *-osus* > E. *-ose*, full of, augmented, prone to): tomentose, loosely cottony or woolly, or more densely agglomerated into small bundles like flannel. The colonies of many fungi, formed under natural conditions as well as in culture, have a floccose appearance, as occurs in *Aspergillus flavofurcatus* (=*A. flavo-furcatis*), *Chrysonilia* (=*Monilia*) *sitophila* and *Microsporum nanum* (moniliaceous asexual fungi), among others. The basidiocarps of *Coprinopsis nivea* (=*Coprinus niveus*) and *Cystolepiota*, Agaricales, also are noticeably floccose.

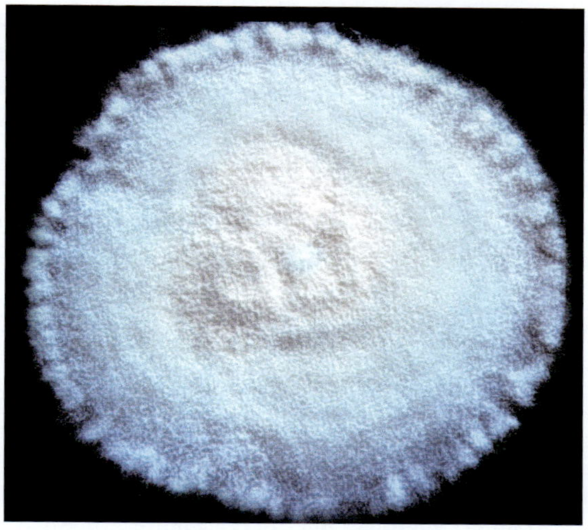

Floccose texture of a colony of *Microsporum nanum*, on agar, x 2 (*MU*).

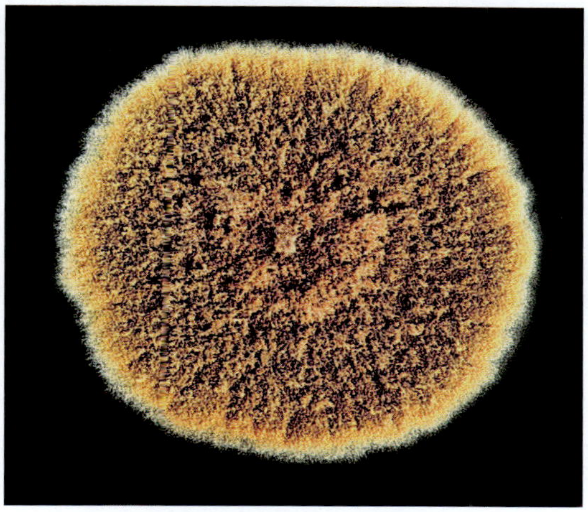

Floccose texture of a colony of *Aspergillus flavofurcatus* (=*A. flavo-furcatis*), on agar, x 1 (*MU*).

Floccose texture of a basidiocarp of *Cystolepiota* sp., growing on bark, x 0.5 (*RV*).

Floccose texture of a colony of *Microsporum nanum*, on agar, x 1 (*MU*).

flocculose, **flocculent** (LL. *flocculosus*, full of diminutive flocks or tufts of wool < *flocculus*, dim. of *floccus*, tuft, tassel, or small bundle of wool + suf. *-osus* > E. *-ose*, full of, augmented, prone to; L. *flocculentus*, slightly floccose): minutely or slightly *floccose*. In cultures of yeasts or bacteria grown in liquid media, it refers to the small clusters or agglomerations of microorganisms distributed throughout the medium or accumulated in the bottom.

flock (ME. < OF. *floc* < L. *floccus*, tuft of wool): a group of hyphae that resemble a small tuft of cotton or wool and which together form the floccose mycelium of many fungi.

-flora, **-florous** (L. *flor*, *flos*, *florus*, flower): a suf. employed in words such as *triflorous*, with three flowers, or, when it alludes to *Flora*, the Roman Goddess of flowers, in reference to the ensemble of plants in an area. *Mycoflora* is a term in disuse, having been replaced by the more correct **mycobiota**.

flori- (L. *flor*, *flos*, *florus*, flower): a pref. that alludes to flowers, as in *floricole*, an organism that inhabits flowers.

floricole, **floricolous** (L. *flor*, *flos*, *florus*, flower + suf. *-cola*, inhabitant < *colere*, to inhabit): an organism that lives on flowers or in inflorescences that are ecologically compatible with it, like certain yeasts that live in the nectaries of flowers.

flower of tan (L. *flor*, *flos*, *florus*, flower; ML. *tannare* < *tanum*, *tannum*, oak bark, tannin): applied to the aethalium of the Myxomycete *Fuligo septica* (Physarales), which is sometimes found growing on the tannin-containing tubs used to cure hides.

fluorescent (NL. *fluor*, one of the minerals utilized as fluxes, including fluorite < L. *fluere*, to flow + suf. *-escens*, genit. *-escentis* > E. *-escent*, that which turns, that begins to, slightly): having *fluorescence*, i.e., the property of emitting light in the presence of certain radiations, because of which the structure or compound appears different in color than in reflected or transmitted light. E.g., the aflatoxins of *Aspergillus flavus* (moniliaceous asexual fungi) and the mycotoxins of other fungi have a characteristic fluorescence in a chromatogram under ultraviolet light. Cf. **bioluminescent** and **phosphorescent**.

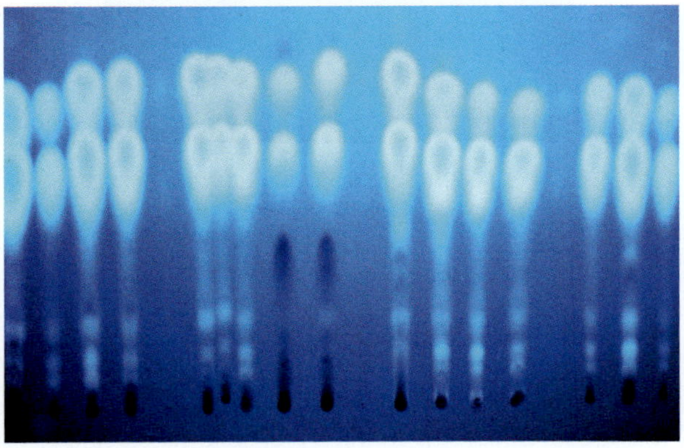

Fluorescent spots (greenish-blue) of aflatoxins of *Aspergillus flavus*, in a chromatogram viewed under ultraviolet light, x 1 (*MU*).

flush (MF. *flus*, *fluz* < L. *fluxus*, flux, flow): the sudden and usually abundant development of mycelium or the periodic emergence of fruiting bodies, especially of edible agaricaceous fungi in culture, such as *Agaricus bisporus* and *Pleurotus ostreatus*.

A **flush** of basidiocarps in a mushroom farm; *Pleurotus ostreatus* growing on wheat straw, x 0.05 (*CB*).

fly agaric, **fly mushroom** (ME. *flie* < OE. *fleoge*, to fly; a winged insect; L. *agaricum*, a fungus < E. *agaric*, one of the Agaricales; ME. *musseroum* < MF. *mousseron* < LL. *mussirion*, mushroom): the basidioma of *Amanita muscaria* (Agaricales), which has insecticidal properties.

fly fungus, pl. **fungi** (ME. *flie* < OE. *fleoge*, to fly; a fly, a winged insect; L. *fungus*, fungus < Gr. *spóngos*, *sphóngos*, sponge): refers to *Entomophthora muscae* (Entomophthorales), the entomopathogenic fungus that attacks the domestic fly, generally in places where these insects rest.

foetid (L. *foetidus*, bad smelling < *foetere*, to smell bad): stinking; with a nauseating odor, like some gen. of Phallaceae, e. g., *Mutinus* and *Phallus*.

foliicole, **foliicolous** (L. *folium*, leaf + suf. *-cola*, inhabitant < *colere*, to inhabit): an organism that lives or develops on leaves; there are many kinds of foliicolous fungi, e.g., species of the gen. *Erysiphe* and other Erysiphales, as well as various lichens, e.g., *Strigula orbicularis* (=*S. complanatum*), Chaetothyriales, and *Porina epiphylla*, Trichotheliales. See **epiphyllous** and **hypophyllous**.

Foetid basidiocarps of *Mutinus caninus* and *Phallus hadrianii*, x 0.5 (*RV*).

Foliicolous lichen thalli of *Strigula orbicularis* (=*S. complanatum*) on a leaf, x 1 (*TH*).

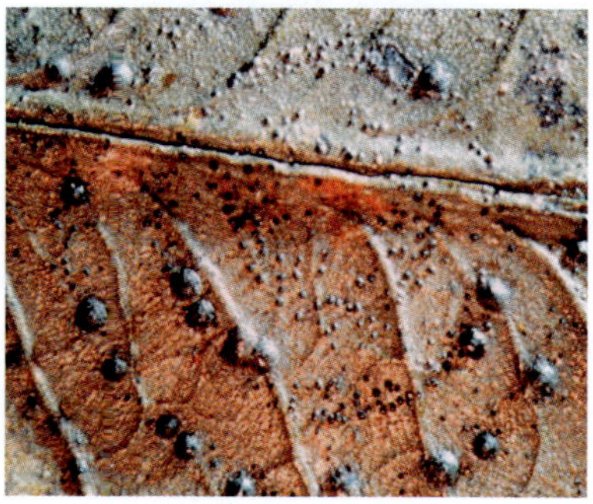

Foliicolous lichen thalli of *Strigula subtilis* on a leaf, x 1 (*MAH*).

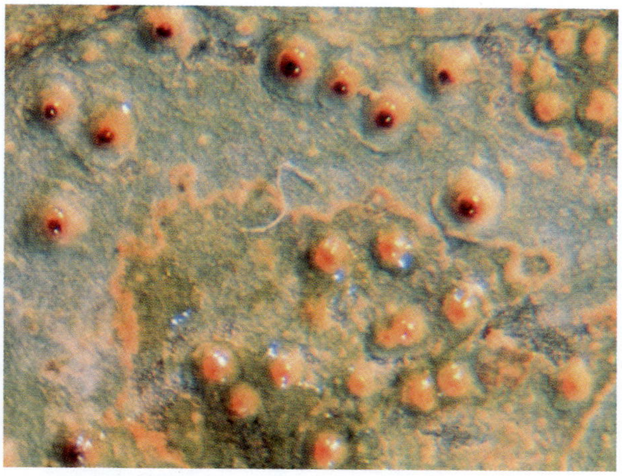

Foliicolous lichen thalli of *Porina epiphylla*, on a fern leaf, x 1.2 (*MU*).

folio-, foli- (L. *folium*, leaf): a pref. that means leaf, as in *foliose*, with abundant leaves, leafy.

foliole, foliolum, pl. **foliola** (L. *foliolum* < *folium*, leaf + dim. suf. *-olum* > E. *-ole*): each one of the small outgrowths similar to leaves that form on the surface of a foliose lichen, such as *Hypotrachyna* (=*Parmelia*) *imbricatula* (Lecanorales).

Folioles of the thallus of *Parmelia* sp., x 1.5 (*MU*).

Folioles of the thallus of *Hypotrachyna* (=*Parmelia*) *imbricatula*, x 1 (*MU*).

foliose (L. *foliosus*, leafy < *folium*, leaf + suf. *-osus* > E. *-ose*, full of, augmented, prone to): *Lichens*. A type of thallus composed of numerous small leaflike lobes that extend in a more or less circular pattern from a center of growth; *Hypotrachyna* (=*Parmelia*) *imbricatula*, *Anaptychia palmatula,* Lecanorales, and many other lichens have this type of thallus.

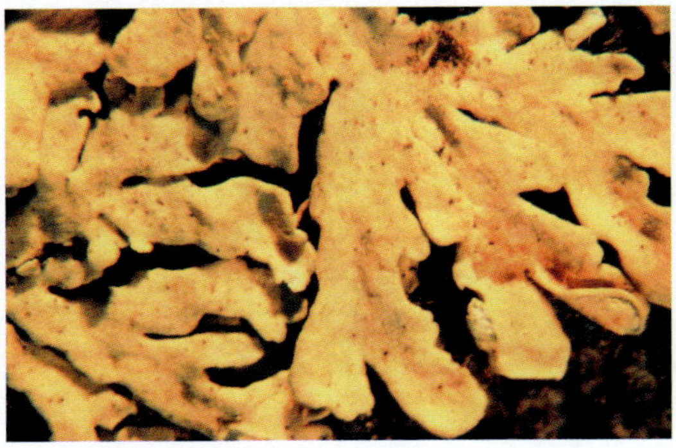

Foliose thallus of *Anaptychia palmatula* on bark, x 9 (*MU*).

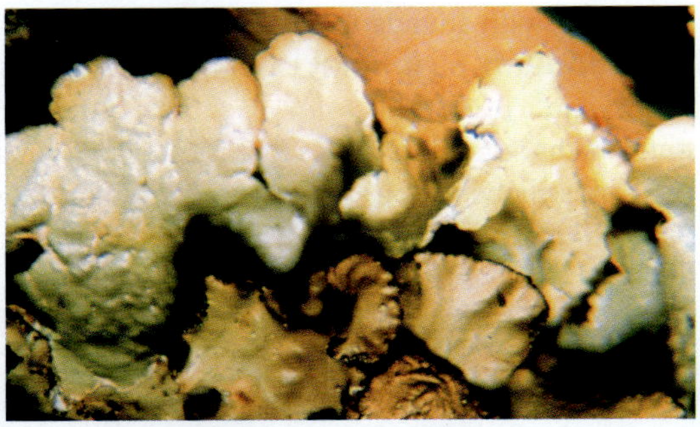

Foliose thallus of *Punctelia* (=*Parmelia*) *subrudecta*, x 2 (*MU*).

follicular (L. *folliculus*, dim. of *follis*, leather purse + suf. *-aris* > E. *-ar*, like, belonging to): relating to a *follicle*; having follicles or similar to a follicle or small bag.

fomite (L. *fomitis,* genit. of *fomes*, tinder): **1**. Combustible tinder of a woody consistency, like the basidiocarps of *Fomes fomentarius* (Aphyllophorales), which are utilized to produce fire by rubbing them with other hard objects; in fact, *Fomes* (L. *fomes*) means tinder. **2**. An inanimate object that is used to transmit the infectious states of parasites or pathogens.

foot cell, basal cell (ME. < OE. *fot*, foot; L. *basalis*, basal < *basis*, base + suf. *-alis* > E. *-al*, belonging to or relating to; L. *cellula*, cell, living cell, dim. of *cella*, small room, compartment, cell of a honeycomb): a hyphal cell that supports a sporogenous cell or a thallus; e.g., a foot or basal cell supports the conidiophore in some species of *Aspergillus* (moniliaceous asexual fungi), the macroconidium of *Fusarium solani* (tuberculariaceous asexual fungi), and the thallus of the aquatic fungi of the Blastocladiales (such as *Blastocladia* and *Blastocladiella*), which in the latter gen. adopts the general form of a trunk, with branches bearing the zoosporangia on the apical part, and rhizoids or attachment organs on the base.

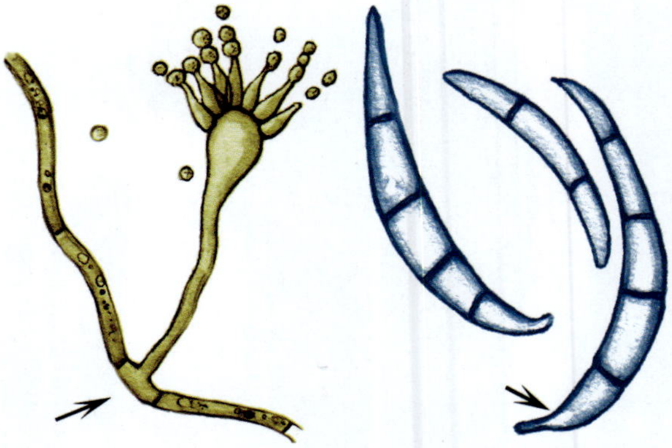

Foot cell of the conidiophore of *Aspergillus japonicus*, x 200, and **foot cell** of the macroconidia of *Fusarium solani*, x 1,800 (*MU*).

foramen (L. *foramen*, orifice, hole < *forare*, to perforate): a small orifice or opening.

forate development (L. *forare*, to perforate + suf. *-atus* > E. *-ate*, provided with or likeness; OF. *desveloper* > MF. *développer* + E. *-ment* < L. *-mentum*, denoting action or resulting state): *Gasteromycetes*. A type of fruiting body development in which the differentiation of the plectobasidia occurs among the primordial tissues, inside minute locules that are formed near the peridium and which mature centripetally. The development occurs from the periphery of the gleba, near the peridium or in a center of the basal sterile region, and continues toward the interior of the fructification. It is characteristic of the orders Gautieriales and Lycoperdales (such as *Lycoperdon*). Also called *coralloid development*. Cf. **auleate development, homogeneous development, lacunar development** and **pileate development**.

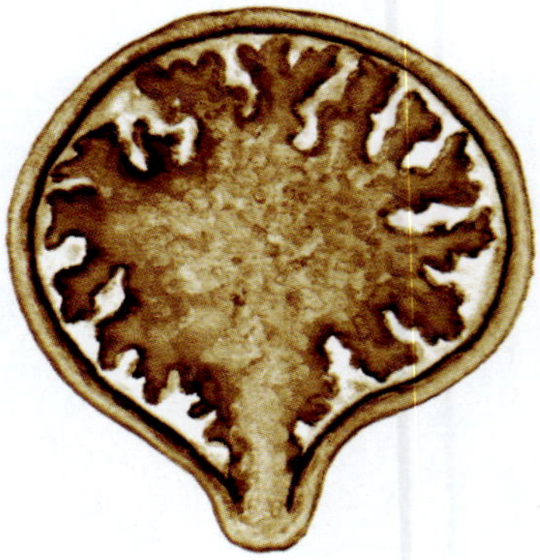

Forate development of the hymenium of the fruiting body of *Lycoperdon* (Lycoperdales), seen in longitudinal section, x 1.2 (*MU*).

forcipate, forcipiform (L. *forceps*, forceps, tong + suf. *-atus* > E. *-ate*, provided with or likeness; L. *forceps* + suf. *-formis* < *forma*, shape): tong-shaped, like the gametangia of *Phycomyces nitens* (Mucorales).

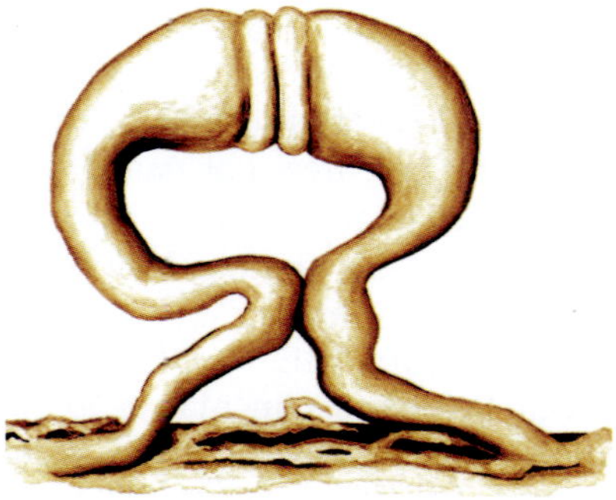

Forcipate gametangia of *Phycomyces nitens*, x 45 (*MU*).

form (L. *forma*, form, shape): *Botany*. A somewhat imprecise systematic category, generally considered below variety and subvariety. The term *forma specialis* (abbrev. *f. sp.*) is utilized when referring to parasitic fungi that cannot be distinguished morphologically, but which are adapted physiologically to particular hosts. It is recommended that the special form bear the Latin name of the respective host in the genitive; e.g., *Puccinia graminis* f. sp. *tritici* (i.e., of *Triticum aestivum*), the wheat rust, or *Fusarium oxysporum* f. sp. *lycopersici* (i.e., of *Lycopersicon esculentum*), a causal agent of tomato wilt. The f. sp. of the fungi corresponds to the *pathovar* and *pathotype* that are used for pathogenic bacteria, when it is desirable to designate a subdivision of a species that is distinguished by its common pathogenic characteristics.

-form (L. *-formis* < *forma*, form, shape): a suf. often employed in forming compound terms that allude to the shape of something, as in *claviform*, club-shaped, and many other terms.

form-genus, pl. **genera** (L. *forma*, form; L. *genus*, genit. *generis*, genus, race, stirp, class): see **anamorph genus**.

form-species (ME. *forme* < OF. < L. *forma*, form; L. *species*, species, class, stirp): see **anamorph-species**.

fornicate, forniculate (NL. *fornicatus*, arched-over, arched < *fornix*, vault, arch; NL. *forniculatus* < *forniculus*, dim. of *fornix* + suf. *-atus* > E. *-ate*, provided with or likeness): arched-over, arched; refers to the fruiting body of some geastraceous fungi, such as *Geastrum fornicatum* (Lycoperdales), which on maturing has fibrous and fleshy layers of the peridium arched over the cup-shaped mycelial layer, forming a kind of vault.

Fornicate fruiting body of *Geastrum fornicatum*, x 0.8 (*MU*).

fornicate ray (NL. *fornicatus*, arched, vaulted < *fornix*, genit. *fornicis*, vault, arch + suf. *atus* > E. *-ate*, provided with or likeness; L. *radius*, ray): *Gasteromycetes*. An arched ray of a fruiting body that is formed when the mesoperidium separates from the exoperidium, and is attached only at the margin; the arched rays support the elevated endoperidium, as is seen in *Geastrum fornicatum* (Lycoperdales).

fossil fungus, pl. **fungi** (L. *fossilis*, excavated, genit. of *fossus* < *fodere*, to excavate; L. *fungus*, fungus < Gr. *spóngos*, *sphóngos*, sponge): a remnant, impression or trace of a fungus from past geological eras that has remained preserved in the earth's crust. The fossil record of fungi dates from the early Phanerozoic to the Proterozoic age. The generic names of fossil fungi are generally based on the names of extant taxa, modified with a suf., usually *-ites*, e.g., *Boletellites*, *Papulosporonites*, etc.

Fossil fungus (fruiting body of *Geastrum* sp.). Shown are portions of the stellate peridium (arrows), x 0.2 (*SM*).

foveate (L. *foveatus*, pitted, perforated < *fovea*, hole, concavity, pit + suf. *-atus* > E. *-ate*, provided with or likeness): with pits or perforations.

foveolate (L. *foveolatus*, with minute perforations < *foveola*, dim. of *fovea*, pit, hole, concavity + suf. *-atus* > E. *-ate*, provided with or likeness):

refers to surfaces that have engravings in the shape of small pits, as is seen in the ascospore wall of *Gelasinospora* (Sordariales), and the ascocarps of *Morchella* (Pezizales) and *Cyttaria* (Cyttariales). See **faveolate**.

Foveolate ascospore of *Gelasinospora* sp., x 960 (*MU*).

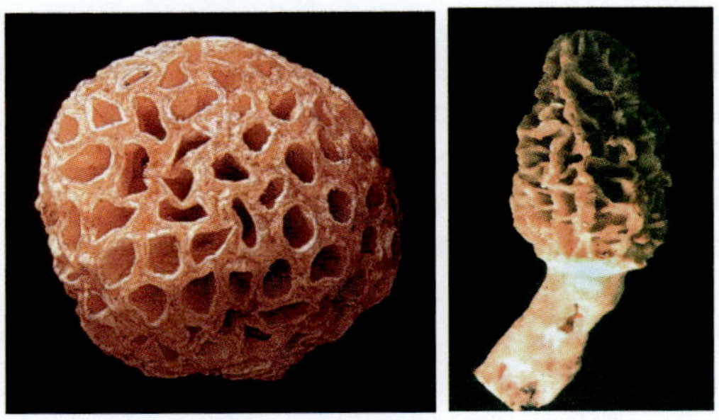

Foveolate ascocarps of *Cyttaria* sp., x 1 (*MU*), and *Morchella esculenta*, x 0.5 (*MU*).

Foveolate basidiocarp of *Favolus tenuiculus* (Aphyllophorales), x 0.3 (*RV*).

-fraga,-frago (L. *frangere*, to break, destroy): a suf. employed to express the idea of something that ends up broken or destroyed, as in *septifraga* (L. *septum*, septum, fence or partition) dehiscence, by which is meant that the dissepiments of the fruiting body of a fungus are split with this manner of opening. Also used as a pref., *frago-*, *frag-*, in words such as *Fragosphaeria* (Sordariales), for the type of dehiscence of the perithecium.

fragmentation (L. *fragmentatio*, genit. *fragmentationis*, to reduce to fragments < *fragmentum*, piece, a broken and detached part + suf. *-ationem* > E. *-ation*, action, state or condition, result): in the fungi it refers to the separation of a thallus into fragments, each of which has the capacity to grow and form a new individual; it is a type of asexual reproduction or vegetative propagation, which occurs under natural as well as artificial conditions; mycelial fragmentation is commonly utilized in the laboratory to transfer the cultures of fungi to various media.

free (ME. < OE. *freo*, free, not united or attached): *Agarics*. Applied to gills or tubes that are not united with the stipe of the fruiting body. Cf. **adnate**, **decurrent** and **sinuate**.

free cell formation (ME. < OE. *freo*, free; NL. *cellula*, cell, living cell, dim. of L. *cella*, small room, compartment, cell of a honeycomb; L. *forma*, form + suf. *-ationem* > E. *-ation*, action, state or condition, result): *Ascomycetes*. The process by which ascospores are formed in the asci, as occurs in *Ascodesmis* and *Ascophanus*, of the Pezizales. Following meiosis the eight nuclei, each with some surrounding cytoplasm, are delimited by portions of the ascus vesicle that encloses them; this is followed by the deposition of cell wall material to form the separate ascospores in the mature ascus.

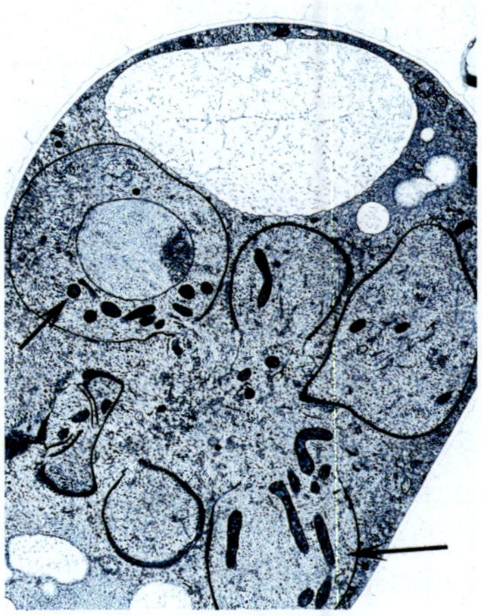

Free cell formation of ascospores in the ascus of *Ascodesmis* sp., seen in longitudinal section, x 2,350 (*CM*).

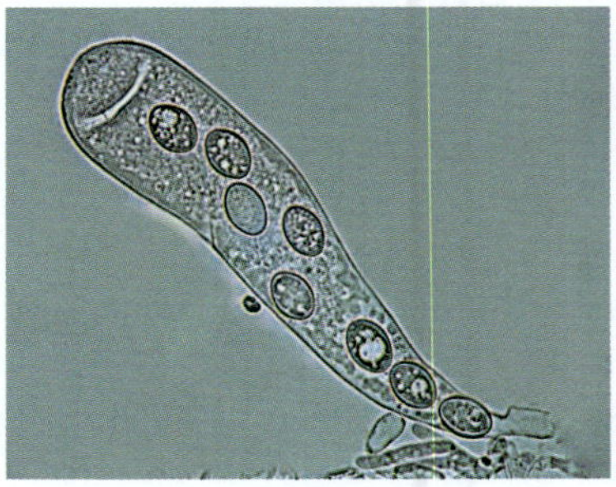

Free cell formation of ascospores in the ascus of *Ascophanus carneus*, x 350 (*EAA*).

friable (L. *friabilis*, easily broken into small pieces < *friare*, to break up, crumble + suf. *-abilis* > E. *-able*, with a tendency to, capable of being): easily broken up into small pieces, like the volva of certain species of *Amanita* (Agaricales), which is fugacious, and therefore rarely leaves any trace on the mature stipe.

fructicole, **fructicolous** (L. *fructus*, fruit + suf. *-cola*, inhabitant < *colere*, to inhabit): an organism that lives or develops in or on fruits, such as *Ciboria carunculoides* and *Monilinia fructicola* (Helotiales), which parasitize the fruits of mulberry and peach, respectively. See **carpogenic**.

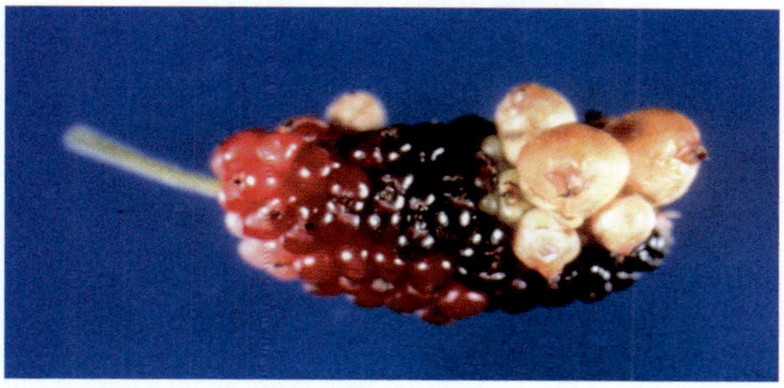

Stromatized ovaries on a mulberry fruit infected by *Ciboria carunculoides*, a **fructicole** fungus, x 2 (*RTH*).

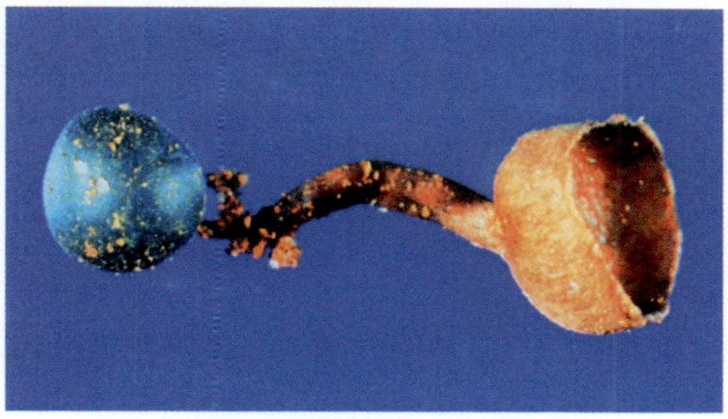

Stromatized ovary of a mulberry fruit with an apothecium of *Ciboria carunculoides*, a **fructicole** fungus, x 2.5 (*RTH*).

Pseudosclerotium of *Monilinia fructicola*, a **fructicole** fungus, with apothecia, x 2.5 (*MU*).

fructification (LL. *fructificatio*, genit. *fructificationis*, related to the bearing of fruit < L. *fructificare*, to bear fruit < *fructus*, fruit + suf. *-ationem* > E. *-ation*, action, state or condition, result): **1**. The action and effect of forming or producing fruit. **2**. The formation of sporiferous apparatus or fruiting bodies of fungi. The fructification is any fungal structure that produces or bears spores, whether sexual or asexual. Syn. of **fruiting body**.

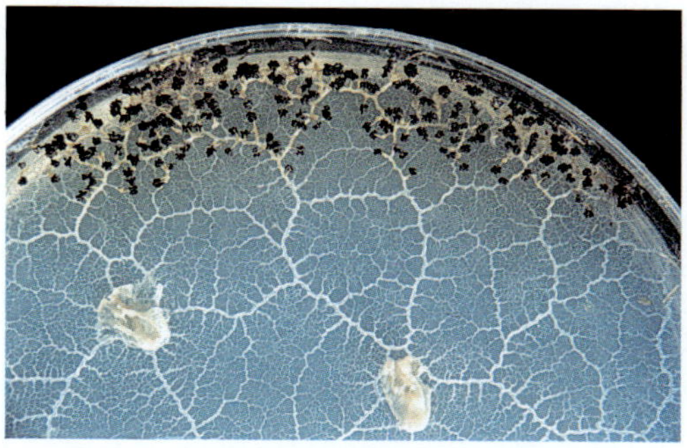

Fructification, the formation of sporangia by *Physarum polycephalum* (Physarales), from a mature plasmodium, x 2 (*MU*).

Fructification, the formation of sporangia by *Physarum polycephalum* (Physarales), from a mature plasmodium, x 10 (*MU*).

Fructification, the formation of basidiocarps by *Pleurotus ostreatus* (Aphyllophorales), from the mycelium on agar, x 2 (*MU*).

fruiting body (L. *fructus*, fruit + ME. < OE. suf. *-ing*, one of a specified kind; product of an action or process; ME. < OE. *bodig*, body): a general term applied to spore-bearing organs in both macro- and micromycetes. The more precise terms **apothecium**, **perithecium**, **ascostroma**, **conidioma**, **basidioma**, etc., are used for specific types of fruiting bodies.

frustulose (L. *frustulosus*, composed of small pieces < *frustum*, piece or small fragment + suf. *-osus* > E. *-ose*, full of, augmented, prone to): with the surface cracked or fissured in a more or less regular pattern, like the polygonal pyramids on the surface of a pileus, spore, etc. E.g., the fruiting bodies of *Boletus frustulosus* (Agaricales) and *Xylobolus frustulatus* (=*Stereum frustulatum*), Aphyllophorales, as well as the thallus of *Arthothelium spilomatoides* (Arthoniales), are frustulose.

Frustulose basidiocarp of *Boletus frustulosus*, x 0.5 (*RV*).

Frustulose thallus of *Arthothelium spilomatoides*, x 8.5 (*MU*).

fruticole, **fruticolous** (L. *frutex*, shrub + suf. *-cola*, inhabitant < *colere*, to inhabit): living or developing on trees or shrubs.

fruticose (L. *fruticosus*, similar to a shrub, shrubby < *frutex*, shrub + suf. *-osus* > E. *-ose*, full of, augmented, prone to): *Lichens*. A thallus that is attached to the substrate by a single point and which develops branches and becomes shrub-like; many Lecanorales have this type of thallus, which can be upright, as in *Cladonia rangiferina*, *Alectoria* and *Pseudevernia*, pendulous, as in *Usnea florida*, or intermediate, like *Teloschistes* (Teloschistales).

fruticulose, **fruticulous** (L. *fruticulosus*, with the aspect of a small shrub < *fruticulus*, dim. of *frutex*, shrub + suf. *-osus* > E. *-ose*, full of, augmented, prone to): *Lichens*. Equal to *fruticose*, but with a smaller thallus.

Fruticose thallus of *Pseudevernia* sp., x 1.3 (*MU*).

Fruticose thallus of *Cladonia rangiferina*, x 6 (*MU*).

Fruticose thallus of *Teloschistes exilis*, with branches that end in apothecia, x 6 (*MU*).

Fruticose thallus of *Alectoria ochroleuca*, x 4 (*MU*).

Fruticose thallus of *Pseudevernia* sp., x 1.3 (*MU*).

Fruticose thallus of *Cladonia stellaris*, x 5 (*MU*).

Fruticose thallus of *Cladonia stellaris*, x 7 (*MU*).

fugaceous (L. *fugax*, genit. *fugacis*, fleeting, unstable + suf. *-aceus* > E. *-aceous*, of or belonging to, with the nature of): a structure of very short duration, like the ring on the basidiocarp of *Coprinus comatus* (Agaricales), which disappears very soon after becoming evident during development (deciduous, ephemeral).

-fuge (L. *fugere*, to flee, move away): a suf. that means to move away, as in *centrifuge* (L. *centrum*, center), moving away from the center.

fulcrum, pl. **fulcra** (L. *fulcrum*, prop or support < *fulcire*, support): **1**. A term used for the sporophore of lichens. **2**. A supporting structure, like the perithecial appendages of *Erysiphe*, *Blumeria*, *Phyllactinia* and other Erysiphales.

fuliginose, **fuligineous**, **fuliginous** (L. *fuliginosus*, full of soot < *fuligo*, genit. *fuliginis*, soot + suf. *-osus* > E. *-ose*, full of, augmented, prone to): blackish, like soot smudge. In phytopathology it is applied to plants with a dunnish or black covering on the leaves, branches and fruits, due to the superficial growth of certain Loculoascomycetes of the gen. *Capnodium* (Capnodiales) and *Dothidea* (Dothideales), among others, commonly known as smut or sooty mold, this last from the gen. *Fumago* (a *nomen confusum*, considered as the asexual or conidial phase of *Capnodium*).

Fuliginose mycelial growth of *Capnodium walteri*, parasitizing the upper surface of a leaf of *Arbutus menziesii*, x 1 (*MU*).

fumaceous, **fumacious**, **fumaginous** (L. *fumus*, smoke + suf. *-aceus* > E. *-aceous*, of or belonging to, with the nature of): smoke-colored, dark gray.

fumagoid (L. *fumago*, similar to smoke < *fumus*, smoke + NL. suf. *-ago*, resembling + L. suf. *-oide*, like < Gr. *-oeides*, similar to): *Med. mycol.* Refers to a type of rounded cells, isolated or in groups, with thick walls pigmented gray or dark brown, that are developed by *Phialophora* and other gen. of dematiaceous asexual fungi (causing chromomycosis in man and higher animals) in infected tissues.

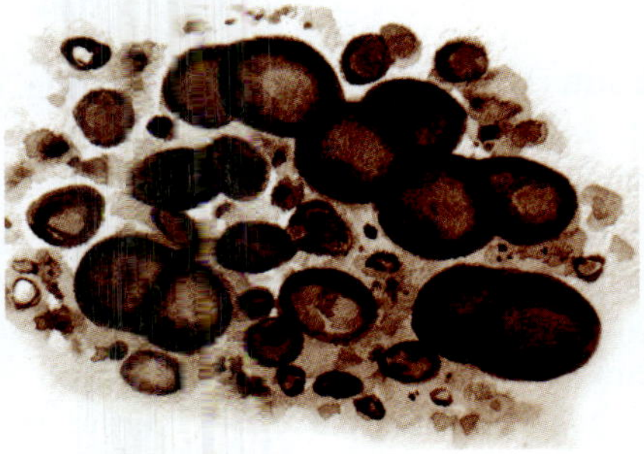

Fumagoid budding cells of *Phialophora pedrosoi*, infecting human tissues, x 1,500 (*MU*).

fungaemia, **fungemia** (L. *fungus*, fungus + NL. suf. *-emia* < Gr. *haíma*, genit. *haímatos*, blood): the invasion of the blood stream by pathogenic fungi. Equal to **mycohemia**.

fungal (L. *fungus*, fungus + suf. *-alis* > E. *-al*, belonging to or relating to): belonging to the fungi or relating to them; e.g., fungal species. Cf. **fungoid**.

Fungi (L. *fungi*, pl. of *fungus*, fungus): in former classifications the taxonomic category *Fungi* was considered as one of the five kingdoms into which all living beings were classified (Whittaker, 1969; Margulis and Schwartz, 1988): *Monera*, *Protista* (or *Protoctista*), *Fungi*, *Plantae* and *Animalia*, with the *Monera* being the only kingdom of prokaryotic organisms and the others with eukaryotic organisms. However, with advances in studies of the ultrastructure, biochemistry, and especially of molecular biology of the fungi, to consider the fungi as one of the five kingdoms of life has become unsustainable. In some recent classifications (such as that followed in the 9th ed. of Ainsworth & Bisby's *Dictionary of the Fungi*, 2001), the organisms studied by mycologists are now considered as polyphyletic, i.e., with different phylogenies, and belonging to three different kingdoms: the *Protozoa*, *Chromista* and *Fungi*. Only the kingdom *Fungi* consists exclusively of true fungi, whose mitochondrial cristae are flat and lysine synthesis is by the aminoadipic acid metabolic pathway; the other two kingdoms contain phyla of organisms called *fungoid*, because they are not considered as true fungi, and whose mitochondrial cristae are tubular, and lysine synthesis is by the diaminopimelic acid pathway. Some authors unite the *Chromista* and *Protozoa* in a single polyphyletic kingdom, the *Protoctista* (or *Protista*), but this conclusion is not supported by molecular, biochemical and other indole evidence. Others segregate the kingdom *Stramenopila* (more recently named *Straminipila*) to form the *Chromista*. In the system of classification followed in this dictionary, the phyla of fungi and fungus-like organisms are included in the kingdoms *Fungi* (Chytridiomycota, Zygomycota, Ascomycota and Basidiomycota, together with the asexual ascomycetes and other asexual fungi), *Stramenopila* (Oomycota, Hyphochytriomycota and Labyrinthulomycota), and *Protista* (Plasmodiophoromycota, Dictyosteliomycota, Acrasiomycota and Myxomycota). See **Chromista**, **Protista**, **Protoctista**, **Protozoa** and **Stramenopila**, and **fungal** and **fungoid**. In the Appendix, the user of this dictionary can see the comparison between the outline of Alexopoulos *et al.*, 1996, modified by Ulloa and Hanlin, 2006, to that of Kirk *et al.*, 2008. In the latter classification scheme, the phyla Chytridiomycota, Zygomycota, Ascomycota and Basidiomycota have been retained in the kingdom Fungi, with the addition of three new phyla, Blastocladiomycota, Neocallimastigomycota and Glomeromycota. The phyla of the kingdom Stramenopila (Oomycota, Hyphochytriomycota and Labyrinthulomycota) are now in the kingdom Chromista, and the phyla of kingdom Protista (Acrasiomycota, Dictyosteliomycota, Myxomycota and Plasmodiophoromycota) have been moved to the kingdom Protozoa (phyla Percolozoa, Amoebozoa and Cercozoa, respectively), with the concurrent change of taxa with zoological endings. These taxonomic variations are part of a constant process of study and evaluation of these groups of organisms, in order to achieve a phylogenetic classification of the fungi. See **fungus**.

Fungi Imperfecti, **imperfect fungi**, sing. **fungus** (L. *fungi*, pl. of *fungus*, fungus; L. *imperfectus*, pl. *imperfecti* < pref. *im-*, *in-*, not + *perfectus*, perfect): imperfect (asexual) or conidial fungi. See **anaholomorph**, **deuteromycetes**, **asexual fungi**, **conidial fungi** and **mitosporic fungi**.

fungicide (L. *fungus*, fungus + L. suf. *-cida*, killer < *caedere*, to kill): any natural or synthetic substance capable of destroying or killing fungi, such as the chemicals captan, maneb, thiabendazole and others. Cf. **fungistatic**.

fungicole, **fungicolous** (L. *fungus*, fungus + suf. *-cola*, inhabitant < *colere*, to inhabit): living or developing on fungi; there exist many fungicolous organisms (bacteria, nematodes, insects), including other fungi, like those of the gen. *Hypomyces* (Hypocreales), whose perithecial stromata are formed on the basidiocarps of *Russula* and *Lactarius* (Agaricales), or like the carpophoroids of *Asterophora* (=*Nyctalis*) *parasitica*, which grow on the basidiocarps of *Russula* (both gen. in the Agaricales).

Fungicolous carpophoroids of *Asterophora* (=*Nyctalis*) *parasitica* on a basidiocarp of *Russula* sp., x 1 (*MU*).

fungiferous (L. *fungiferous*, bearer of fungi < *fungus*, fungus + suf. *-ferous*, bearer < *ferre*, to carry, bear): any organism or structure that bears fungi, like the fungiferous layer of mycorrhizae.

fungiform (L. *fungiformis*, like a fungus < *fungus*, fungus + suf. *-formis* < *forma*, shape): with the shape of an agaricaceous fungus, i.e., with a stipe and pileus.

funginous (L. *funginus*, relating to or pertaining to the fungi): belonging to the fungi; syn. of **fungal** and **fungous**.

fungistatic (L. *fungus*, fungus + Gr. *statós*, that which stops + Gr. suf. *-ikos* > L. *-icus* > E. *-ic*, pertaining to or relating to): any natural or synthetic substance that inhibits the sporulation or growth of a fungus, but does not kill it. Syn. of **mycostatic**. Cf. **fungicide**.

fungistic (L. *fungus*, fungus + NL. suf. *-istic* > E. *-istic*, pertaining to or relating to): relating to or pertaining to the fungi. A term utilized to refer to the taxonomic study of the fungi in a particular geographic region, which is made with the objective of inventorying the systematic entities, giving the area of each one and information relative to their habitat, abundance or scarcity, time of fructification and other data. Earlier the term *mycofloristic* (the floristic study of fungi) was utilized, but considering that the fungi are not plants, and even more, that they lack flowers, the use of the term fungistic is more appropiate.

fungivorous , **fungivore** (L. *fungus*, fungus + *voro*, that which devours < *vorare*, to devour, eat): that which eats fungi. Syn. of **mycetophagist**.

fungoid (L. *fungus*, fungus + suf. *-oide*, like < Gr. *-oeídes*, similar to): similar to a fungus in texture, morphology, physiology or other characteristic. Also called **fungous**. Cf. **fungal** and **funginous.**

fungology (L. *fungus*, fungus + Gr. *lógos*, word > E. *-logy*, study, treatise): the study of fungi. An old E. term for mycology.

fungoma, pl. **fungomata** (L. *fungus*, fungus + Gr. suf. *-óma*, which implies the presence of a pathological state, especially a tumor): see **aspergillosis**, **fungus ball** and **mycosis**.

fungosity (L. *fungositas*, *fungositatis* < *fungus*, fungus + suf. *-itas* > E. *-ity*, quality, state or grade): spongy flesh, with the consistency of a fungus. In phytopathology it is applied to the mycelium of the agaricaceous fungus *Armillaria mellea,* which parasitizes the roots of various plants, although it can also refer to other parasitic or saprobic fungi (such as *Rosellinia necatrix*, of the Xylariales).

fungous (L. *fungus*, fungus + L. suf. *-osus* > E. *-ous*, possessing the qualities of): pertaining to or relating to the fungi; it generally refers to the nature of the flesh of the sporiferous apparatus of many fungi; spongy, soft. In this case, it is also called **fungoid**. Syn. of **fungal.**

fungus, pl. **fungi** (L. *fungus*, fungus < Gr. *spóngos*, *sphóngos*, sponge): **1**. An old term used by Pliny to designate all the fungi that were developed outside a covering called the volva. **2**. In some modern systems of classification of living beings (such as the five kingdoms of Whittaker, 1969, and of Margulis, 1983), *Fungi* is the name of the kingdom that includes the higher fungi, excluding the Myxomycota and all of the zoosporic fungi. These classifications, among others, broke the tradition of a three-kingdom system of classification for all living organisms: *prokaryotes*, *animals* and *plants* (including the *fungi*). The addition of the kingdoms Fungi and Protista attempted to place organisms into kingdoms that more accurately reflected their presumed evolutionary relationships. This was an important beginning toward establishing monophyletic groups (groups that contain an ancestor and all of its descendents), and to develop a hierarchical classification that reflects the relationships of these groups, i.e., a phylogenetic classification. In an attempt to develop monophyletic groups, the organisms previously classified as fungi now are considered in three different groups: two monophyletic phyla, *Fungi* and *Stramenopila*, and four phyla of *Protista*. This is the system of classification adopted in this dictionary (Alexopoulos *et al*., 1996), which recognizes the fact that the organisms that have been called *fungi* are not all closely related. Although these organisms do not share a common evolutionary history, they do constitute a compact group on the basis of their morphology, modes of nutrition and ecology. With advances in ultrastructural, biochemical, and especially molecular biology studies, the organisms studied by mycologists are now regarded as polyphyletic (i.e., with different phylogenies) and must be placed in at least three distinct kingdoms: the kingdom *Fungi*, which includes the phyla Chytridiomycota, Zygomycota, Ascomycota and Basidiomycota; the kingdom *Stramenopila* with the phyla Oomycota, Hyphochytriomycota and Labyrinthulomycota, and the kingdom *Protista*, which includes the phyla Dictyosteliomycota, Acrasiomycota, Myxomycota and Plasmodiophoromycota. **3**. The organisms included in the category of fungi are so diverse that it is difficult to give a concise differential diagnosis. The following definition applies to all of the phyla mentioned in **2**: all are heterotrophs (never photosynthetic) and absorbent with respect to nutrition. The thallus varies from a myxamoeba or plasmodium lacking a cell wall, to a unicellular or filamentous thallus (*hypha*) delimited by a rigid cell wall. In species that have a filamentous thallus (*mycelium*), the latter can be septate or not, but those with septa are functionally coenocytic because the septa are perforate. Thalli can be inside or on a host or substrate. The thallus wall is typically chitinous, but occasionally it is cellulosic (in Oomycota); only rarely do chitin and cellulose occur together. Depending upon the group, they can have other polysaccharides (mannan, glucan, etc.) in the wall. Fungi are generally immotile, although they have cytoplasmic flow within the mycelium; in some groups motile structures (zoospores) are produced. The nucleus is eukaryotic. Thalli can be uni- or multinucleate, with cells homo- or heterokaryotic, haploid, dikaryotic, or diploid, the latter generally of short duration. The life cycle can be simple or complex, and reproduction asexual, sexual and/or parasexual. Fungi can live in diverse types of habitats, as saprobes, parasites or symbionts. See **Fungi**.

fungus ball (L. *fungus*, fungus; ME. *bal* > E. *ball*, a round object or mass): a clinical type of aspergillosis characterized by a rounded mass of hyphae of *Aspergillus* (especially *A. fumigatus*, of the moniliaceous asexual fungi) and cellular debris formed in a preexisting cavity in the lung or in the bronchial tube. See **aspergillosis** and **mycosis**.

funicular, **funiculose** (L. *funiculus*, little cord, string, dim. of *funis*, cord + L. *funicularis*, pertaining to or relating to the funiculus + L. suf. *-aris* > E. *-ar*, like, belonging to; or + L. suf. *-osus* > E. *-ose*, full of, augmented, prone to): being aggregated into little cords or bundles, like those that give rise to conidiophores, as on the mycelium of *Acremonium polychromum* and of some species of *Penicillium* (e.g., *P. funiculosum*, of the moniliaceous asexual fungi); also applied to the structures provided with a funiculus, such as the peridioles of the nidulariaceous basidiomycetes (such as *Cyathus striatus* and *Crucibulum laeve*), which have a pedicel that unites them to the internal peridium of the receptacle.

Funiculose mycelium of *Acremonium polychromum*, x 1 (*JAS*).

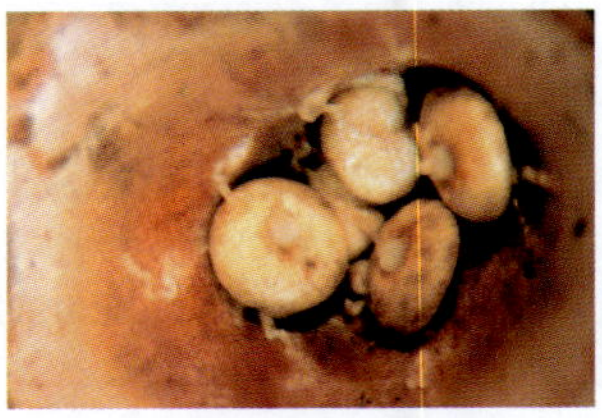

Funiculose peridioles inside the fruiting body of *Crucibulum laeve*, x 8 (*CB*).

funiculus, pl. **funiculi**, **funicle** (L. *funiculus*, little cord, string < *funis*, string + dim. suf. *-culus*): *Gasteromycetes* (order Nidulariales). In fungi such as *Cyathus*, it refers to the slender mycelial connection (composed of the sheath, the middle piece and the purse), that attaches the peridioles to the inside of the peridium (endoperidium) of the receptacle or basidiocarp. In certain lichens, such as those in the gen. *Umbilicaria* (Lecanorales), it corresponds to the *umbilical cord*.

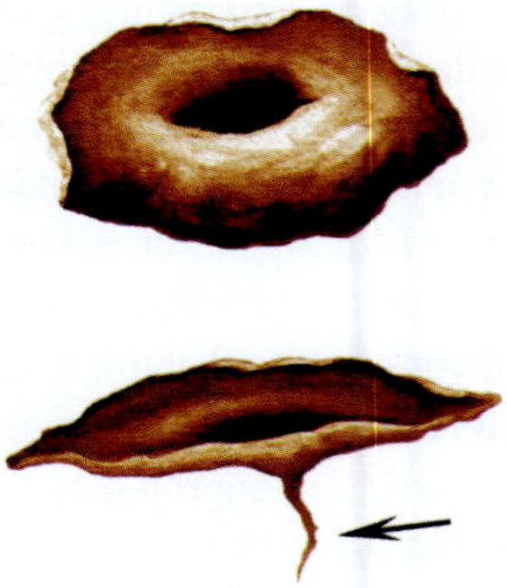

Funiculus of the thallus of *Umbilicaria* sp., x 0.5 (*MU*).

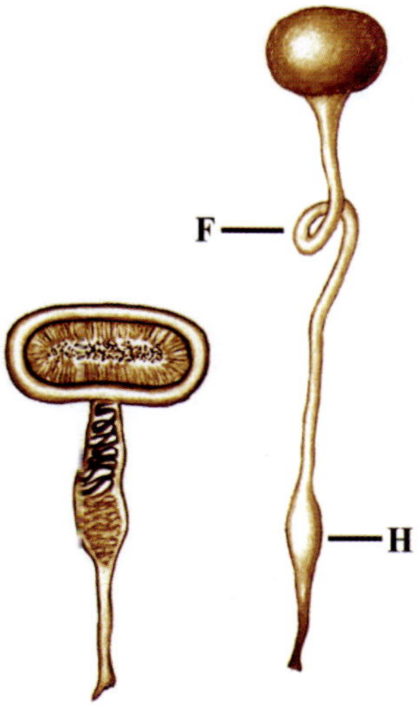

A peridiole and **funiculus** of *Cyathus striatus* (on the left) is seen in longitudinal section. The peridiole on the right is shown with the funicular cord (F) and hapteron (H) expanded, x 8 (*MU*).

funoid (L. *funis*, string, cord + suf. *-oide*, like < Gr. *-oeídes*, resembling, similar to): composed of filaments or fibers resembling cords. Syn. of **funiculate**.

furcate (L. *furcatus*, forked < *furca*, pitchfork, fork + suf. *-atus* > E. *-ate*, provided with or likeness): forked or bifurcate, like the cystidia of *Paxillus involutus* (Agaricales), or like the basidia of fungi in the gen. *Dacrymyces* (Dacrymycetales). Syn. of **lituate**.

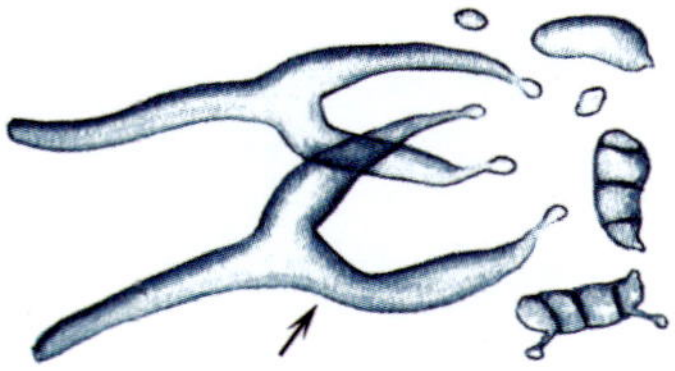

Furcate basidia of *Dacrymyces capitatus* (=*D. ellisii*), x 450 (*MU*).

furfuraceous (LL. *furfuraceus*, scaly < *furfur*, bran, dandruff + suf. *-aceus* > E. *-aceous*, of or pertaining to, with the nature of): covered with small flakes similar to bran or dandruff, like the basidiocarps of *Tubaria furfuracea* and *Lepiota clypeolaria* (Agaricales).

fuscous- (L. *fuscus*, fuscous, dark, blackish): a pref. that alludes to a dark brown color, almost black, like the ascomata of *Fuscoscypha* (Helotiales).

fusiform (L. *fusiformis*, spindle-shaped < *fusus*, spindle + suf. *-formis* < *forma*, shape): like a spindle, tapered at the ends, like the conidia of *Paecilomyces fumosoroseus*, *Trichophyton ajelloi* (moniliaceous asexual fungi) and *Fusarium* (tuberculariaceous asexual fungi), the ascospores of *Melanospora zamiae* (Melanosporales) and *Leptosphaeria* species (Pleosporales), or like the basidiospores of *Pleurotus ostreatus* (Agaricales). Spores that are somewhat fusiform are termed *fusoid.*

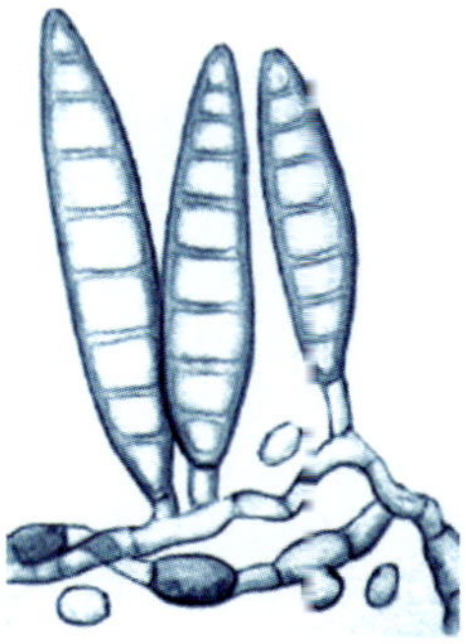

Fusiform macroconidia of *Trichophyton ajelloi*, x 1,000 (*MU*).

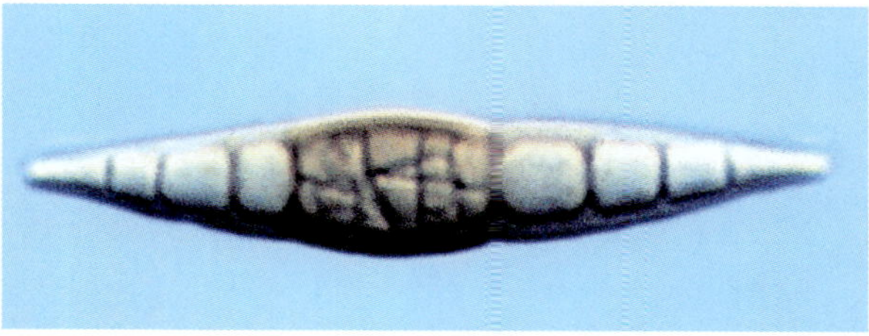

Fusiform ascospore of *Leptosphaeria* sp., x 1,700 (*RTH*).

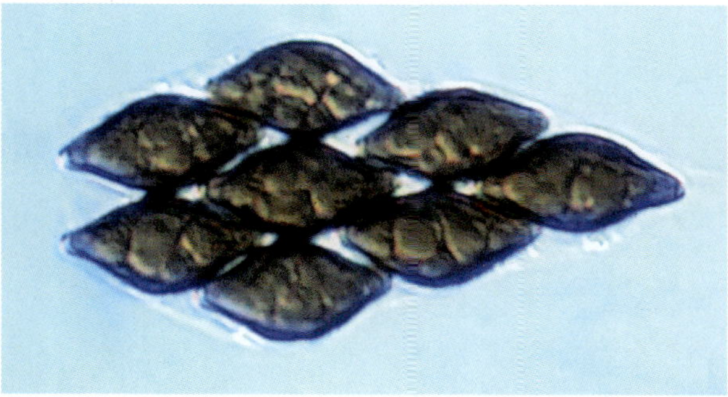

Fusiform ascospores of *Melanospora zamiae*, x 1,250 (*RTH*).

fusion biotrophism (L. *fusion-*, *fusio* < *fusus*, union, fusion, as if it were by melting; incorporation of elements into a unified whole + suf. *-io*, genit. *-ionis* > E. *-ion*, result of an action, state of; Gr. *bíos*, life + *trophós*, that which nourishes, serves as food + suf. *-ismós* > L. *-ismus* > E. *-ism*, state, phase, tendency, action): a type of interaction exhibited by some species of *Melanospora* (Melanosporales), whose hyphae can cause the erosion of the cell walls of a host plant, followed by fusion with the protoplasts of the plant; in this interaction the fungus absorbs nutrients from the host plant, causing diminished growth in it.

fusoid (L. *fusus*, spindle + suf. *-oide*, like < Gr. *-oeídes*, similar to): see **fusiform**.

g

Gelatinous basidiocarps of *Psedohydnum gelatinosum* (*RV*).

g

galacto-, galact- (Gr. *gála*, genit. *gálactos*, milk): a pref. that alludes to milk, as in *galactose*, a dextrorotatory hexose, less soluble and less sweet than glucose, and in *Galactopus* (=*Mycena*, of the Agaricales), a basidiomycete with a white stipe.

galeate (L. *galeatus*, covered with a helmet < *galea*, helmet, hat + suf. *-atus* > E. *-ate*, provided with or likeness): provided with a helmet-shaped structure. Cf. **galeiform**.

galeiform (L. *galea*, helmet, hat + suf. *-formis* < *forma*, shape): with the shape of a helmet or hat, like the ascospores of *Endomyces fibuliger* (=*Saccharomycopsis fibuliger*), of the Saccharomycetales, and *Ceratocystis fimbriata* (Microascales). Cf. **galeate**.

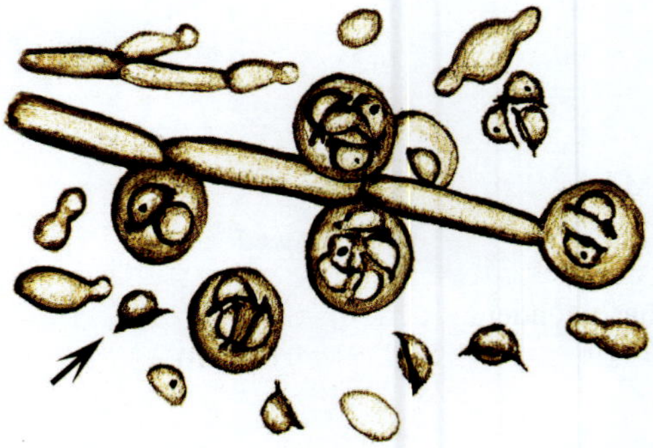

Galeiform ascospores
of *Endomyces fibuliger* (=*Saccharomycopsis fibuliger*), x 900 (*MU*).

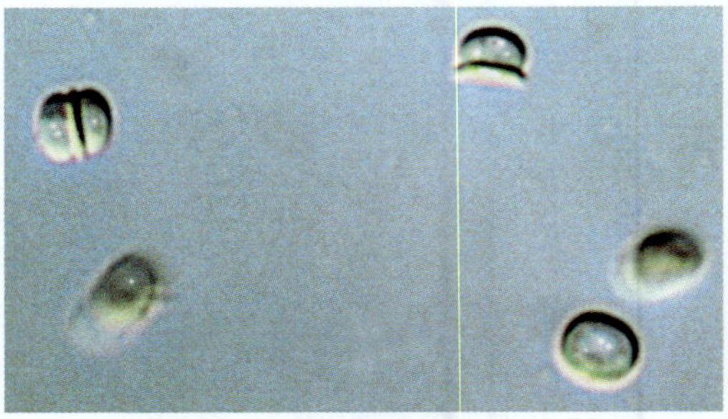

Galeiform ascospores
of *Ceratocystis fimbriata*, x 1,000 (*RTH*).

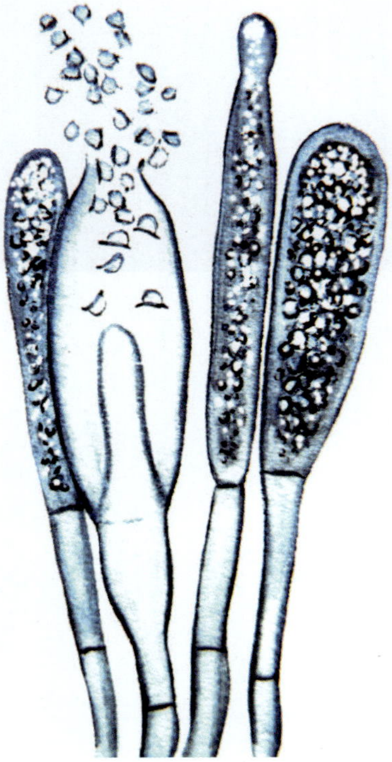

Galeiform ascospores
of *Ascoidea rubescens* (Endomycetales), x 450 (*MU*).

galeriform (L. *galerus* or *galerum*, cap, helmet-like covering for the head < *galea*, helmet, hat + suf. *-formis* < *forma*, shape): having the shape of a cap or helmet, like the pileus of the basidiocarp of *Galerina* (Agaricales).

gall (ME. *galle* < L. *galla*, gallnut): a globose tumor (cecidium) that forms in leaves, stems, roots and other plant tissues in response to the presence of a parasitic organism. The cecidogenous parasites or gall-producers can be animals (especially insects, mites and nematodes), fungi or bacteria, because of which three categories of galls are distinguished: *zoocecidia*, *mycocecidia*, and *bacteriocecidia*. The galls caused by insects also are called *entomocecidia*, and those produced by mites (acari) *acarocecidia*. The fungi that produce galls directly on plants include Exobasidiales and Uredinales. The water flies that live in galls (Cecidomyiidae) have mycophagous larvae implicated in the formation of the galls; some larvae are solely mycophagous and do not cause gall formation. Some galls of lichen thalli are produced by lichenicolous fungi, by nematodes, or by mites, or may be of unknown origin. See **cecidium** and **mycocecidium**.

galvano-, galvan- (It. *galvanismo* < Luigi Galvani): a pref. that alludes to electric current, as in *galvanotropism*.

galvanotropism (E. *galvanic* < Fr. *galvanisme* < It. *galvanismo* < Luigi Galvani; galvanic current, a direct current of electricity, especially when produced by chemical action + Gr. *trópos*, a turning, change in position or manner < *tropé*, a turning < *trépo*, to gyrate, revolve, turn towards + suf. *-ismós* > L. *-ismus* > E. *-ism*, state, phase, tendency, action): a tropic reaction produced by an electrical field. See **tropism**.

gamma particle (*gamma*, γ, third letter of the Greek alphabet; L. *particula* < *part*, part, piece + dim. suf. *-cle*): each of the small globose structures, delimited by membranes, that contain a dense body approximately 0.5 nm in diam., and which appear to be involved in the formation of the wall of a cyst or encysted zoospore; characteristic of the zoospores of *Blastocladiella emersonii* (Blastocladiales).

gametangial contact (NL. *gametangium* < Gr. *gamétes*, spouse, gamete + *angeîon*, vessel, receptacle + L. suf. *-alis* > E. *-al*, relating to or belonging to; L. *contactus* < *contingere*, have contact with): a type of sexual reproduction in which two gametangia come in contact but do not fuse, at least not totally (both gametangia more or less retain their individuality); the male nucleus migrates through a pore or a fertilization tube and penetrates the female gametangium, as happens, e.g., in aquatic Oomycota of the order Saprolegniales (*Achlya americana*, *Saprolegnia parasitica*, *Dictyuchus monosporus*). Cf. **gametangial copulation**.

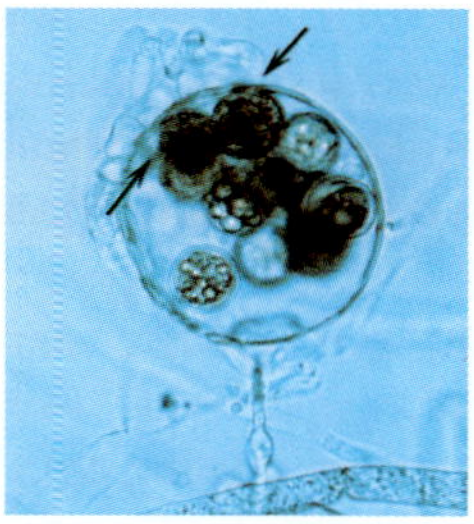

Gametangial contact in *Achlya americana*, x 490 (*RTH*).

gametangial copulation (NL. *gametangium*, gametangium < Gr. *gaméte*, spouse, gamete + *angeîon*, vessel, receptacle + L. suf. *-alis* > E. *-al*, relating to or belonging to; L. *copulatio*, genit. *copulationis*, union, link < *copula*, band, link + suf. *-ationem* > E. *-ation*, action, state or condition, result): a type of sexual reproduction in which the two gametangia or their protoplasts fuse (*plasmogamy*) and give rise to a zygote (*karyogamy*) that develops into a latent spore (the zygospore), as occurs, e.g., in molds such as *Phycomyces nitens*, *Rhizopus stolonifer* (=*Rh. nigricans*) and *Rhizomucor* (=*Mucor*) *miehei* (Mucorales), and in yeasts such as *Zygosaccharomyces bailii* (Saccharomycetales) and *Octosporomyces* (=*Schizosaccharomyces*) *octosporus* (Schizosaccharomycetales). This type of sexual union also is known as **conjugation**. Cf. **gametangial contact**.

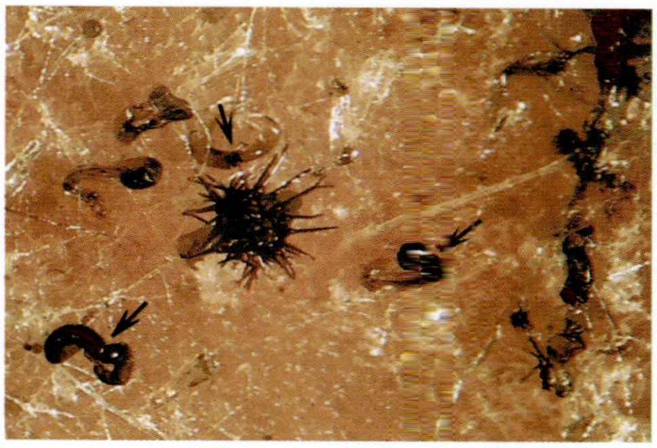

Gametangial copulation in *Phycomyces nitens*, x 40 (*MU*).

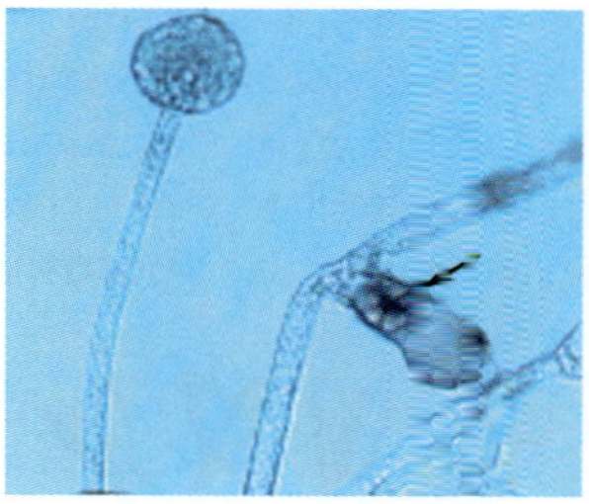

Gametangial copulation in *Rhizomucor* (=*Mucor*) *miehei*, x 840 (*MU*).

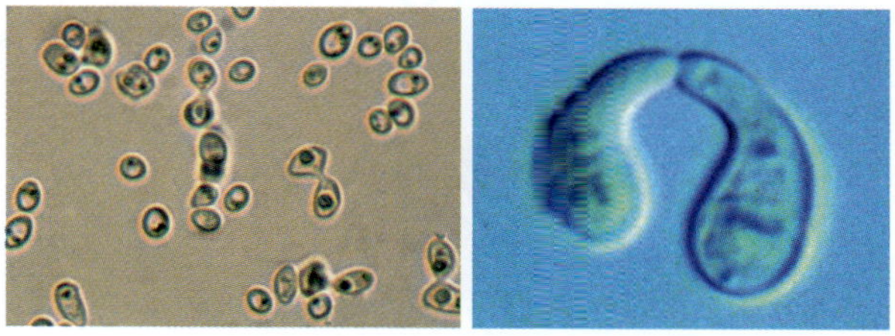

Gametangial copulation in the yeast *Zygosaccharomyces bailii*, x 1,000 (*MU*), and of two cells of the yeast *Octosporomyces* (=*Schizosaccharomyces*) *octosporus*, x 2,000 (*RTH*).

gametangium, pl. **gametangia** (NL. *gametangium*, gametangium < Gr. *gamétes*, spouse, gamete + *angeîon*, vessel, receptacle): a sexual organ that contains gametes, which can be motile cells (planogametes) or gametic nuclei, depending on the species.

gametangy (NL. *gametangium*, gametangium < Gr. *gamétes*, spouse, gamete + *angeîon*, vessel, receptacle): a sexual process in which two compatible gametangia fuse; these can be equal or not in size and shape. Two types of gametangy are the *gametangial contact*, which occurs in the Oomycota (e.g., Saprolegniales and Peronosporales), and *gametangial copulation*, which occurs in the Zygomycota (e.g., the Mucorales). The process also is known as *gametangiogamy*. See **-gamy**.

gamete (Gr. *gamétes*, spouse, gamete): a sexual cell or nucleus that fuses with another complementary cell or nucleus in sexual reproduction to give rise to the zygote. See **aplanogamete** and **planogamete**.

gameto-, **gamet-** (Gr. *gamétes*, spouse, gamete): a pref. that means gamete, a sexual cell or nucleus, as in the following terms.

gametogenesis, pl. **gametogeneses** (Gr. *gamétes*, spouse, gamete + *génesis*, engenderment): the production of gametes.

gametophyte (Gr. *gamétes*, spouse, gamete + *phytón*, plant): in plants with alternation of generations, it refers to the generation with haploid cells (haplophase), which produces sexual reproductive cells (gametes). Cf. **sporophyte**.

gametothallus, pl. **gametothalli** (Gr. *gamétes*, spouse, gamete + *thallós*, sprout, thallus): a thallus that produces gametes; it is present in aquatic fungi with an alternation of generations in the life cycle, as in *Allomyces javanicus* (Blastocladiales), which has gametothalli that alternate with sporothalli, which produce asexual spores. See **sporothallus**.

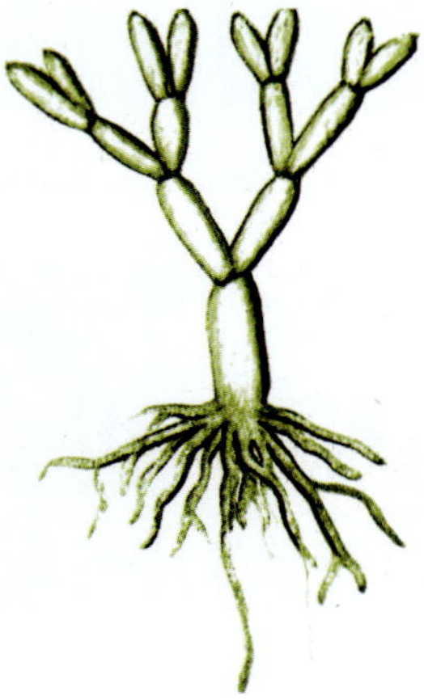

Gametothallus of *Allomyces javanicus*, x 480 (*MU*).

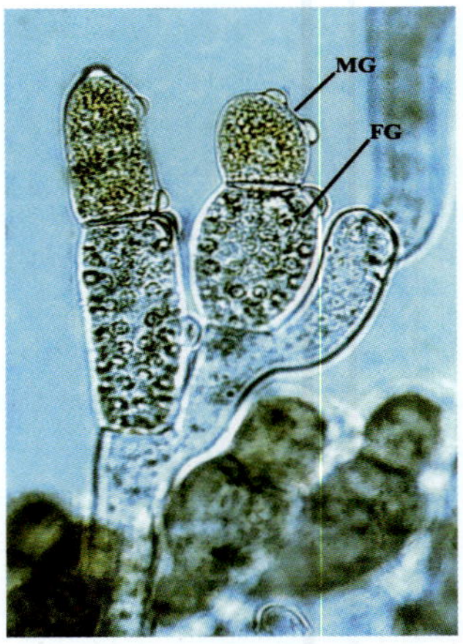

Gametothallus of *Allomyces javanicus*, with male gametangia (MG) and female gametangia (FG), x 840 (*RTH*).

gamo-, **gam-** (Gr. *gámos*, sexual union): a pref. employed in the formation of terms that refer to sexual union, as in *gamont*.

gamont (Gr. *gámos*, sexual union + *óntos*, genit. of *ón*, a being): an individual that produces gametes, especially in a holocarpic form, or a separate part of one that produces gametes.

-gamy (Gr. *gamía* < *gámos*, sexual union): a suf. that denotes sexual union, as in *karyogamy*, the union of two nuclei to form a zygote nucleus.

gangliform (Gr. *gánglion*, swelling, nodule, tumor + suf. L. *-formis* < *forma*, shape): having swellings or knots; knotted, like the hyphae of some fungi.

gangliospore (Gr. *gánglion*, swelling, nodule, tumor + *sporá*, spore): a term used to designate an aleuriospore, which by its origin is a holoblastic conidium. See **aleuriospore**.

gastero-, **gaster-** (Gr. *gastér*, stomach, abdomen, belly): a pref. employed to give the idea of something that is contained in an enclosed organ or apparatus, as in the following terms. Equal to **gastro-**.

gasteroconidium, pl. **gasteroconidia** (Gr. *gastér*, stomach, abdomen, belly + NL. *conidium* < Gr. *kónis*, powder, dust + dim. suf. *-ídion* > L. *-idium*): syn. of **gasterospore**.

gasteroid (Gr. *gastér*, stomach, abdomen, belly + L. suf. *-oide*, like < Gr. *-oeídes*, similar to): with the hymenium in cavities, i.e., enclosed, not arranged on gills, or without an hymenium, as occurs in the Gasteromycetes. Equal to **gastroid**. Cf. **agaricoid**.

gasteromorphic (Gr. *gastér*, stomach, abdomen, belly + *morphé*, shape + suf. *-íkos* > L. *-icus* > E. *-ic*, belonging to or relating to): having partially or completely the habit, ontogeny, shape and characteristics of gasteromycete spores.

Gasteromycetes (Gr. *gastér*, stomach, abdomen, belly + L. *-mycetes*, ending of taxonomic class < Gr. *mýkes*, genit. *mýketos*, fungus): a group of basidiomycetes with the spores contained in the interior of the sporiferous apparatus, enclosed during the major part of its development or indehiscent, like the Lycoperdaceae. Corresponds to the class Gasteromycetes, of the phylum Basidiomycota, and includes the orders Lycoperdales, Tulostomatales, Sclerodermatales, Phallales, Nidulariales and Hymenogastrales, as well as the orders of Hymenomycetes. Together with the Urediniomycetes and Ustilaginomycetes, and other orders, they constitute the phylum Basidiomycota of the kingdom Fungi. In the Appendix, the user of this dictionary can see the comparison between the outline of Alexopoulos *et al.*, 1996, modified by Ulloa and Hanlin, 2006, to that of Kirk *et al.*, 2008. In the latter classification scheme, the class Gasteromycetes has disappeared, and its orders have been moved to the subphylum Agaricomycotina, phylum Basidiomycota. These taxonomic rearrangements, with the concurrent change of the names of taxa, are part of a constant process of study and evaluation, in order to achieve a phylogenetic classification of the fungi.

gasterospore (Gr. *gastér*, stomach, abdomen, belly + *sporá*, spore): a globose spore, with a thick wall (chlamydospore), probably formed by apomixis in the interior of the tissues or tubes of a fruiting body, not on the exterior, as happens in *Ganoderma* (Aphyllophorales).

gastro-, gastr- (Gr. *gastrós*, stomach, abdomen, belly): equal to **gastero-**.

gastroid (Gr. *gastrós*, stomach, abdomen, belly + L. suf. *-oide*, like < Gr. *-oeídes*, similar to): a general adj. applied to all basidiomycetes that have angiocarpic development. Equal to **gasteroid**. Cf. **agaricoid**.

gel tissue (ML. *gelatina*, gelatine, gel < L. *gelatus*, ice, frozen < *gelare*, to freeze; L. *texere*, tissue): **1**. *Leotiales* and *Tremellales*. A mixture of gel and hyphae present in some fruiting bodies that can originate by direct secretion or by disintegration of hyphae. Cf. **gliatope**. **2**. *Phallales*. A translucid, pectinose matrix, with or without embedded hyphae, like that of the gen. *Kobayasia*.

gelatinous (ML. *gelatina*, gelatin, gel < L. *gelatus*, frozen < *gelare*, to freeze + suf. *-osus* > E. *-ous*, having, possessing the qualities of): similar to gelatin; having the appearance and consistency of gelatin, a bland and spoilable substance, glutinous, which is obtained by prolonged boiling of animal tissues. In the fungi it is applied to tissues whose hyphae become glutinous and partially dissolve under conditions of high relative humidity; on being mounted in water and viewed under the microscope, they appear more transparent, wider and looser than non-hydrated tissues. Fungi of the order Tremellales, such as *Tremella fuciformis* and *T. mesenterica*, are commonly called *gelatinous fungi* or *jelly fungi*. The fruiting bodies of some exidiaceous fungi also have a gelatinous consistency, such as that of *Pseudohydnum gelatinosum*. The gleba of the Gasteromycetes of the order Phallales also is gelatinous, as well as the hygroscopic hymenium of some lichens, e.g., *Collema* and *Leptogium* (Lecanorales).

Gelatinous basidiocarps of *Tremella fuciformis*, x 1 (*TH*).

Gelatinous basidiocarp of *Tremella mesenterica*, x 2 (*RV*).

Gelatinous basidiocarps of *Pseudohydnum gelatinosum*, x 0.8 (*RV*).

Gelatinous basidiocarp of *Pseudohydnum gelatinosum*, x 1 (*RV*).

Gelatinous thallus with apothecia of *Collema coilocarpum*, x 4 (*MU*).

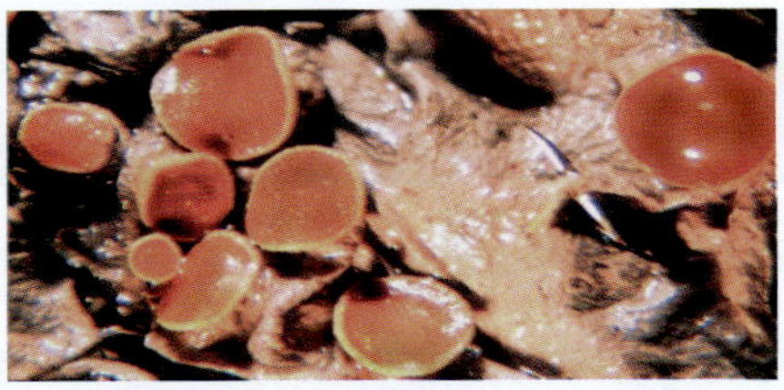

Gelatinous thallus with apothecia of *Leptogium corticola*, x 10 (*MU*).

gelatinous peridium, pl. **peridia** (ML. *gelatina*, gelatin, gel < L. *gelatus*, frozen < *gelare*, to freeze + L. suf. *-osus* > E. *-ous*, having, possessing the qualities of; NL. *peridium* < Gr. *péridion*, small leather purse < *péra*, purse + dim. suf. *-ídion* > L. *-idium* > E. *-idium*): a peridium composed of a hyaline, pectinous matrix, with embedded hyaline hyphae or hyphal fragments, as happens in *Calostoma cinnabarinum* (Tulostomatales).

Gelatinous peridium of the fruiting bodies of *Calostoma cinnabarinum*, x 0.5 (*EPS*).

gemma, pl. **gemmae**, **germling** (L. *gemma*, bud < Gr. *yémo*, to be full, because it is full of life): **1**. A bud or young propagule that is produced by budding in the early stages of development, as in the yeasts, capable of growing into an adult organism. **2**. In the aquatic saprolegniaceous Oomycota (such as *Saprolegnia*), the structures called gemmae are formed like the chlamydospores of true fungi, from intercalary or terminal cells in the mycelium; in this case they also function as thick-walled propagules of asexual reproduction.

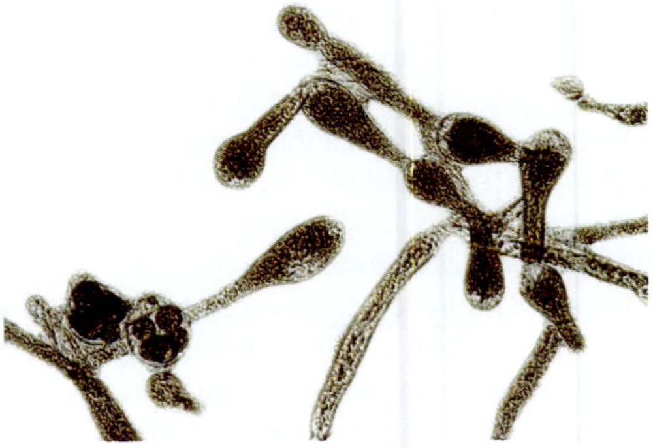

Gemmae of the mycelium of *Saprolegnia parasitica*, x 200 (*MU*).

Gemmae of the mycelium of *Saprolegnia parasitica*, x 280 (*MU*).

gemmation, **budding** (L. *gemmatio*, genit. *gemmationis*, action of budding < *gemmare*, to bud, sprout < *gemma*, bud, sprout < Gr. *yémo*, to be full, because it is full of life + L. suf. *-ationem* > E. *-ation*, action, state or condition, result): a type of cellular asexual reproduction characterized by the formation of small superficial protuberances or evaginations in the generative cells; the protuberances enlarge, become individual, and on separating from the mother cell, they constitute independent individuals. Budding is typical of the yeasts, such as *Pichia* (=*Hansenula*) *anomala*, *Nadsonia fulvescens* and *Saccharomyces cerevisiae* (of the Saccharomycetales), but also is present in many molds. The spores or cells produced by budding are known as *blastospores*. See **bipolar**, **monopolar**, **multipolar** and **unipolar**, for budding with two, one or several buds. Cf. **germination**.

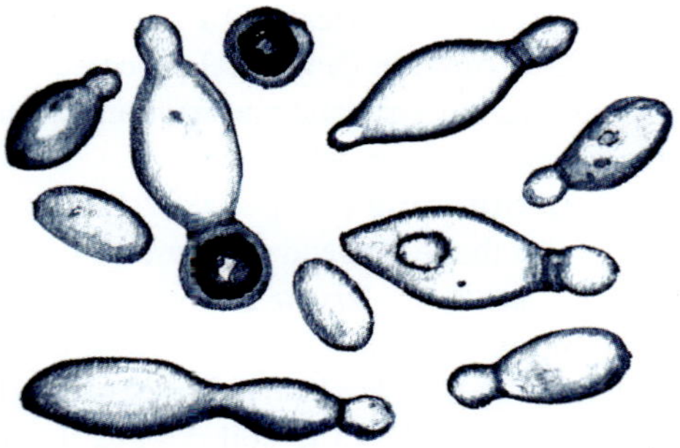

Bipolar **gemmation** of the vegetative cells and asci of *Nadsonia fulvescens*, x 1,000 (*MU*).

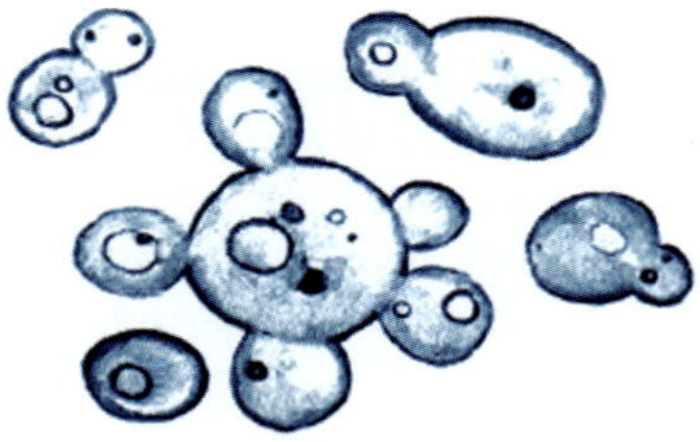

Multipolar **gemmation** of the somatic cells of *Saccharomyces cerevisiae*, x 2,000 (*MU*).

gemmifer (L. *gemmifer* < *gemma*, bud, sprout + *-fero*, *-fer* < *ferre*, to carry, bear): bearing a bud; occasionally applied to the small and incompletely developed pileus of a basidiocarp of *Omphalina luteovitellina* (=*O. flavida*), Agaricales, which functions as a bud.

gemmule (L. *gemmula*, dim. of *gemma*, bud, sprout): a minute cellular protuberance formed in multiplication by budding.

gene (Gr. *gen*, short for *pangen* < *pan*, all, completely + suf. *-gen*, *-genes*, born, or *génos*, origin, birth, stirp, class, descendent): an element of the germ plasm with a specific function in inheritance that is determined by a specific sequence of bases (purines and pyrimidines) in DNA, or at times in RNA, and which serves to control the transmission of an hereditary character by specifying the specific structure of a particular protein, such as an enzyme, or by controlling the function of other genetic material.

generative hypha, pl. **hyphae** (L. *generatum* < *generare*, to generate, engender + L. suf. *-ivus* > E. *-ive*, quality or tendency, fitness; NL. *hypha*, hypha < Gr. *hyphé*, tissue, spiderweb; hypha): *Basidiomycetes*. A slender (1.5-10 µm diam.), branched, cylindrical, undifferentiated hypha, with a thin wall, regularly septate, and capable of generating fertile cells, i.e., the basidia of the hymenium and related tissues, like those of *Trametes* (=*Polyporus*) *versicolor* (Aphyllophorales). They stain intensely in cotton blue. See **binding hypha** and **skeletal hypha**.

generitype (L. *genus*, genit. *generis*, genus, race, stirp, class + *typus* < Gr. *týpos*, mark, model): the type species of a genus.

genestasis, pl. **genestases** (Gr. *gennáo*, to engender + *stásis*, immotility): see **genistat**.

-genic (Gr. *génos*, race, origin < *gennáo*, to engender, produce + suf. *íkos* > L. *-icus* > E. *-ic*, pertaining to or relating to): a suf. equivalent to **-genous**.

geniculate (L. *geniculatus*, provided with a knot or protuberance similar to a knee or elbow; knotted < *geniculum*, dim. of *genu*, knee + suf. *-atus* > E. *-ate*, provided with or likeness): having knee-like joints; applied to the part of a hypha or conidiophore that forms bends due to changes in direction resulting from sympodial growth, as is seen in the *Drechslera* asexual state of *Cochliobolus bicolor* (Pleosporales), and in species of *Helminthosporium*, *Alternaria* and *Cercospora*, among many other dematiaceous asexual fungi.

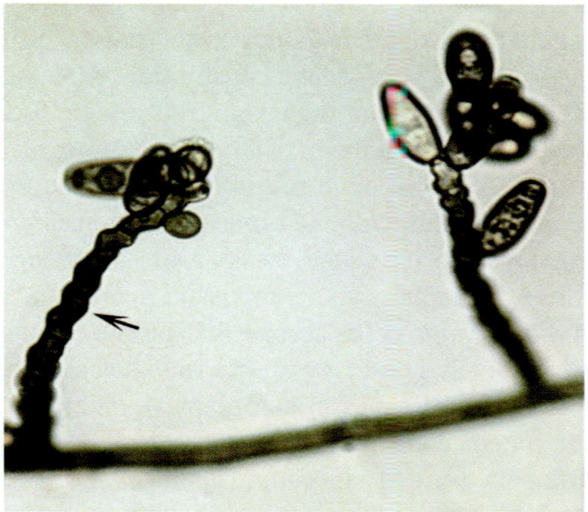

Geniculate conidiophores of *Drechslera* sp., x 380 (*JAS*).

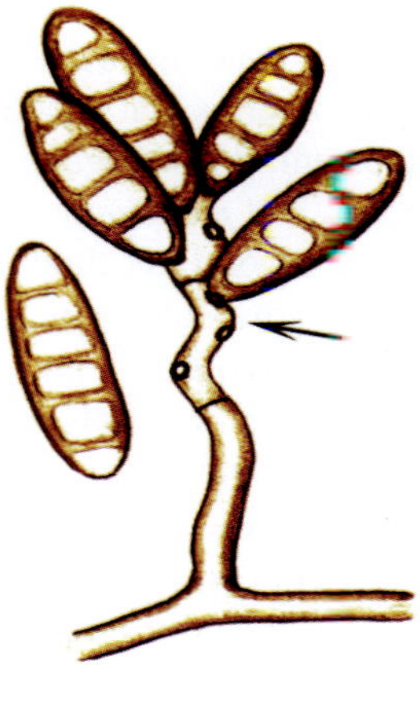

Geniculate conidiophore of the *Drechslera* asexual state of *Cochliobolus bicolor*, x 630 (*MU*).

genistat (Gr. *gennáo*, to engender + *statikós*, what detains): a substance that prevents or reduces the sporulation of fungi, but without affecting the growth of the somatic or assimilative hyphae; i.e., which produces *genestasis*. This phenomenon can be observed among some species of moniliaceous asexual fungi; e.g., *Phymatotrichopsis omnivora* inhibits the sporulation of *Trichoderma viride*, and the formation of microsclerotia by species of *Papulaspora* (of the agonomycetaceous asexual fungi), when they are grown in dual cultures on agar.

genome (G. *Genom* < *gen*, gene + des. *-om* < *chromosom*, chromosome; or < Gr. *gen*, *génos*, origin, birth, stirp, class, descendent + suf. *-óma* > E. *-ome*, which implies entirety): the totality of the genetic material of an organism that can be inherited; the haploid complement of chromosomes with the genes that these contain.

genotype (Gr. *génos*, race, origin < *gennáo*, to engender, produce + *týpos* > L. *typus*, type, mark, model): the complex of hereditary factors of an organism which regulates all of its reactions in the presence of the environment. The genetic constitution, expressed as well as latent, of an organism. The genetic formula that an individual inherits. Cf. **phenotype**.

-genous (Gr. *génos*, race, origin < *gennáo*, to engender, produce): a suf. with the same application as *-genic*, employed to introduce the idea of something that forms or produces an organ, tissue, spore, etc., as in *conidiogenous cell*, the cell that produces conidia. Also used as a pref., *geno-*, in terms such as *genome*, the totality of genes in an organism, and *genotype*, the genetic constitution of an organism.

genus, pl. **genera** (L. *genus*, genit. *generis*, genus, race, stirp, class): the systematic category below the family in taxonomic classifications that is composed of similar species. The generic name generally is a noun and does not require a particular ending.

-geny (Gr. *géneia*, engenderment, the act of being born): a suf. employed in terms that denote engenderment, origin, development, as in *ontogeny*, the development of a living being, and *phylogeny*, the genesis of the phyla of the organic world, i.e., the probable derivation of some organisms from others, in order to construct a genealogical tree of life.

geo-, **ge-** (Gr. *geo* < *gê*, the Earth, as a planet, and as the soil or land): a pref. used in the formation of compound terms that allude to the earth or to the force of gravity, as in *geophilic*, of an organism that prefers the soil, and *geotropism*, the movement of an organism in response to gravity.

geomycology (Gr. *geo* < *gê*, the Earth, as a planet, and as the soil or land + NL. *mycologia* < Gr. *mýkes*, fungus + *-lógos*, study, treatise + E. suf. *-y*): the study of the role of fungi in geological processes, such as element recycling, rock and mineral transformations, and bioweathering, among others.

geophilic, **geophilous** (Gr. *geo* < *gê*, the Earth, as a planet, and as the soil or land + *phílos*, have an affinity for + suf. *-íkos* > L. *-icus* > E. *-ic*, pertaining to or relating to; or + L. suf. *-osus* > E. *-ous*, possessing the qualities of): **1**. *Med. mycol.* Refers to a fungus that naturally inhabits the soil, where it probably grows as a saprobe on keratinous residues; e.g., *Microsporum cookei* (moniliaceous asexual fungi) which, although it is related to species of keratinophilic dermatophytes, is not pathogenic to man and higher animals, or is very rarely and accidentally pathogenic. Cf. **anthropophilic** and **zoophilic**. **2**. Also used for fungi that produce their fruiting bodies underground from assimilative mycelium that also develops underground, such as is observed in the truffles (Pezizales).

geotrichosis, pl. **geotrichoses** (gen. *Geotrichum* < Gr. *geo* < *gê*, soil, land + *thríx*, genit. *trichós*, hair + L. suf. *-um* + suf. *-osis*, condition or state): an infrequent infection caused by *Geotrichum candidum* (moniliaceous asexual fungi), a species of wide distribution that can cause infection at the broncopulmonary, bronchial, buccal or cutaneous level, and on rare occasions, not well documented, at the digestive level. See **mycosis**.

geotropism (Gr. *geo* < *gê*, soil, land + *trópos*, a turn, change in position or manner < *tropé*, a turn < *trépo*, to gyrate, revolve, turn towards + suf. *-ismós* > L. *-ismus* > E. *-ism*, state, phase, tendency, action): the movement or orientation of an organism, or part of an organism, in which gravity is the stimulating factor; this can be positive or negative, depending on whether the organism moves toward or away from the earth. The sporangiophores of *Phycomyces* and of *Mucor* (Mucorales) have negative geotropism, although the mycelium appears to be indifferent to the influence of gravity. The fructifications of many ascomycetes and basidiomycetes have a negatively geotropic stalk, but the tramal layers of the fertile region are positively geotropic, e.g., the teeth of the hydnaceous fungi and the tubes of the polyporaceous species (Aphyllophorales). In some species, such as *Abortiporus biennis* (=*Polyporus rufescens*), it is notable that on occasion abnormal fruiting bodies develop, larger and more or less spherical, due to the loss of the capacity to respond to the morphogenic stimulus of gravity; these abnormal fructifications lack a stipe and differentiated pileus and the hymenial tubes are arranged irregularly over the whole surface. Normal fruiting bodies have stipes with negative geotropism, pilei with diageotropism, and hymenial tubes with positive geotropism. Also called **gravitropism**. See **tropism**.

gephyro-, **gephyr-** (Gr. *géphyra*, bridge): a pref. that means bridge, in terms such as *gephyrophyte*, which is applied as follows: when an erysiphaceous micromycete is parasitic on plant *a* and is also capable of infecting plant *b*, but incapable of infecting plant *c*, it sometimes happens that after passing through host *b*, it acquires the capacity to attack *c*. In such a case, plant *b* is a *gephyrophyte*, i.e., a bridge plant or bridge host that permits the fungus to pass from *a* to *c*.

germ pore (L. *germinalis*, pertaining to or relating to the germ < *germen*, bud, shoot, sprout, rudimentary beginning of a new organic being; L. *porus* < Gr. *póros*, pore, passage): an orifice or opening in the wall of a spore (conidium, ascospore, basidiospore, etc.) through which it germinates, producing a germ tube that proceeds to develop a hypha and eventually the mycelium if conditions are favorable. Generally the germ pore of the basidiospores is apical and oblique, as in *Coprinus comatus* and other species of the gen., or apical and central, as in *Panaeolina foenisecii* (Agaricales). In the ascospores of *Sordaria fimicola* and *Podospora* sp. (Sordariales) the germ pore is apical. In some spores there is a germ slit instead of a pore, as happens in the ascospores of species of *Hypoxylon* (Xylariales). Cf. **germ slit**.

Germ pore of the ascospores of *Podospora* sp., x 360 (*MU*).

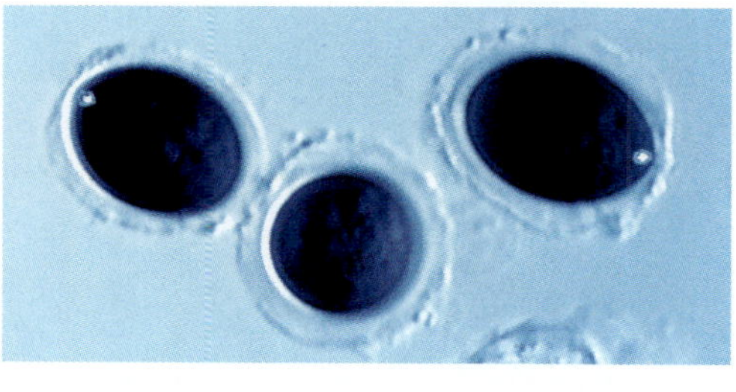

Germ pore of the ascospores of *Sordaria fimicola*, x 570 (*RTH*).

germ slit (L. *germinalis*, pertaining to or relating to the germ < *germen*, bud, shoot, offspring, rudimentary beginning of a new organic being + suf. *-alis* > E. *-al*, belonging to or relating to; ME. *slitte* < *slitten*, to split along a line): a longitudinal fissure in a spore wall through which the germ tube emerges upon germination of the spore, as occurs in the conidia of *Wardomyces humicola* (dematiaceous asexual fungi) and in the ascospores of *Nemania* (=*Hypoxylon*) *serpens* and *Rosellinia necatrix* (Xylariales). Cf. **germ pore**.

Germ slit of the conidia of *Wardomyces humicola*, x 660 (*MU*).

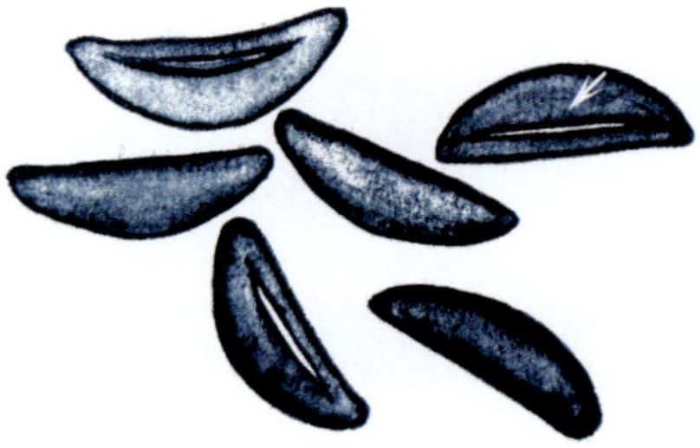

Germ slit of the ascospores of *Nemania* (=*Hypoxylon*) *serpens*, x 570 (*MU*).

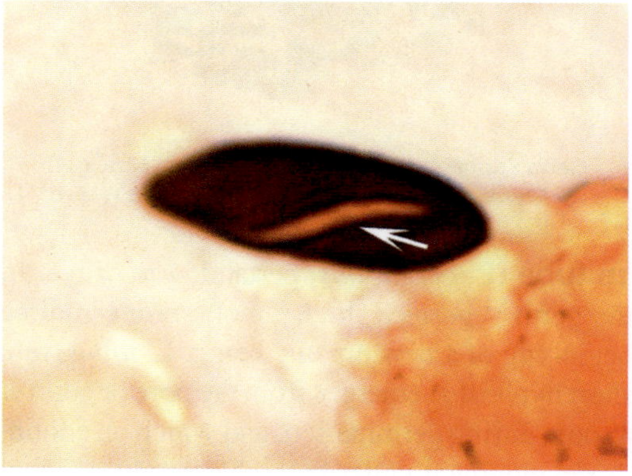

Germ slit of the ascospore of *Hypoxylon* sp., x 1,000 (*CB*).

germ sporangium, pl. **sporangia** (L. *germinalis*, pertaining to or relating to the germ < *germen*, germ, bud, offshoot, rudimentary beginning of a new organic being; L. *sporangium* < Gr. *sporá*, spore + *angeîon*, vessel, receptacle): a sporangium formed on the end of a germ tube or hypha that originates from the germination of an oospore (as in *Phytophthora infestans*, Peronosporales) or of a zygosporangium, as in *Rhizopus stolonifer* (=*Rh. nigricans*), Mucorales.

Germ sporangium of a germinating oospore of *Phytophthora infestans*, x 450 (*MU*).

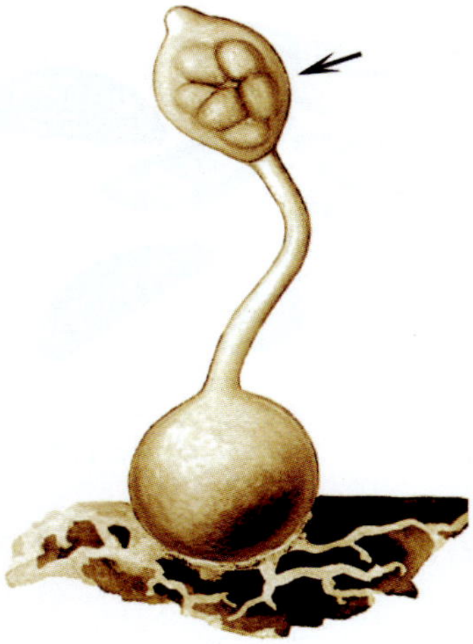

Germ sporangium of a germinating zygosporangium of *Rhizopus stolonifer* (=*Rh. nigricans*), x 180 (*MU*).

germ tube (L. *germinalis*, pertaining to or relating to the germ < *germen*, bud, shoot, sprout, rudimental beginning of a new organic being; L. *tubus*, tube): a short hypha that emerges from the pore or germ slit of a spore during germination, and which on continuing its development under favorable conditions forms a hypha of larger size, and then a mycelium. In many plant pathogenic fungi, e.g., the germ tube forms an *appresorium*, from which the infective hypha that penetrates the host tissues originates. The germ tube represents the beginning of a new assimilative phase of the fungus; it can be seen clearly, e.g., in the germinating conidia of *Conidiobolus thromboides* (Entomophthorales), the germinating ascospores of *Leptosphaerulina crassiasca* (Dothideales), and in the germinating conidia of *Stachybotrys chartarum* (=*S. atra*, dematiaceous asexual fungi).

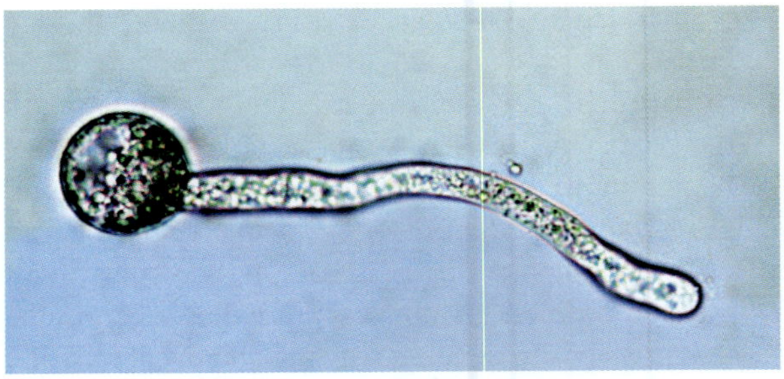

Germ tube of a germinating conidium of *Conidiobolus thromboides*, x 480 (*MU*).

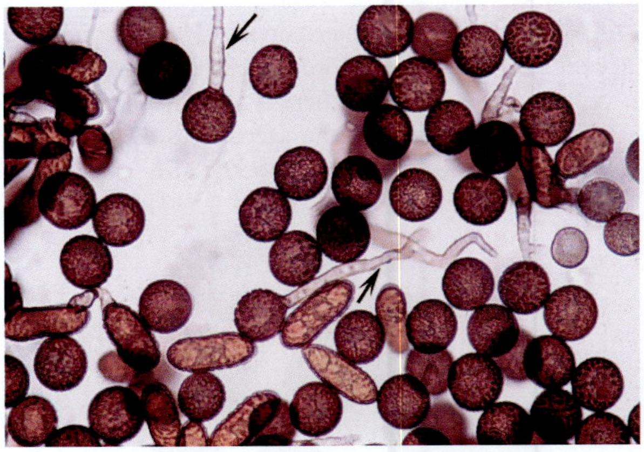

Germ tube of the germinating conidia of *Stachybotrys chartarum* (=*S. atra*); the cylindric conidia of this species are immature, x 800 (*MU*).

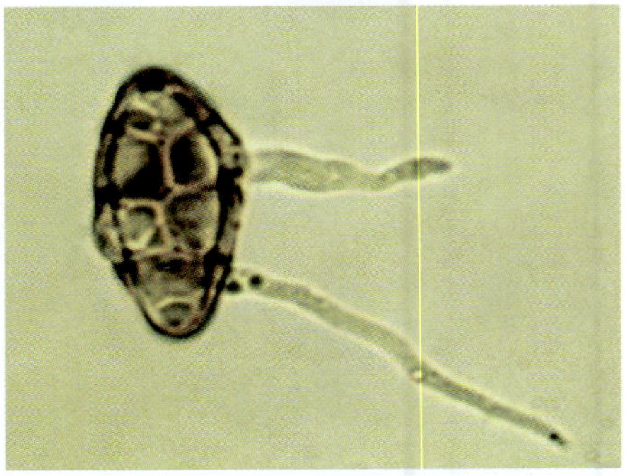

Germ tubes of a germinating ascospore of *Leptosphaerulina crassiasca*, x 560 (*RTH*).

germicide (L. *germen*, bud, sprout, shoot, rudimentary beginning of a new organic being, germ + suf. *-cide* > E. des. *-cide*, what kills < *caedere*, to kill): a substance that kills microorganisms, although the term is utilized mainly for bacteria. See **fungicide**.

germination (L. *germinatio*, genit. *germinationis*, budding, action of germinating): the complex of phenomena relating to organisms that germinate. Germination can be *unipolar* or *multipolar*, whether one or several germ tubes are formed from a single spore, whether sexual (ascospore) or asexual (conidium, chlamydospore); e.g., germination is multipolar in the conidia of *Massospora cicadina* (Entomophthorales) and of *Alternaria* (dematiaceous asexual fungi). See **gemmation**.

germling (L *gemma* < Gr. *gémo*, to be full, because it is full of life): a bud or young propagule in the early stages of development, capable of growing into an adult organism.

-gerous (L. *-ger* < *gerere*, to produce, engender): a suf. employed in the formation of terms such as *ascigerous*, bearing or producing asci.

gibbose, gibbous (L. *gibbosus*, hunched < *gibbus*, bent, hunched + suf. *-osus* > E. *-ose*, full of, augmented, prone to): with one or several asymmetrical convexities, like the sporangium of *Physarum polycephalum* (Physarales). Also used for a structure that has a convex upper part and a flat lower part, or that has an umbo or swelling, like the pileus of the basidiocarp of *Psilocybe mazatecorum* (Agaricales).

Multipolar **germination** of the conidia of *Massospora cicadina*, x 800 (*MU*).

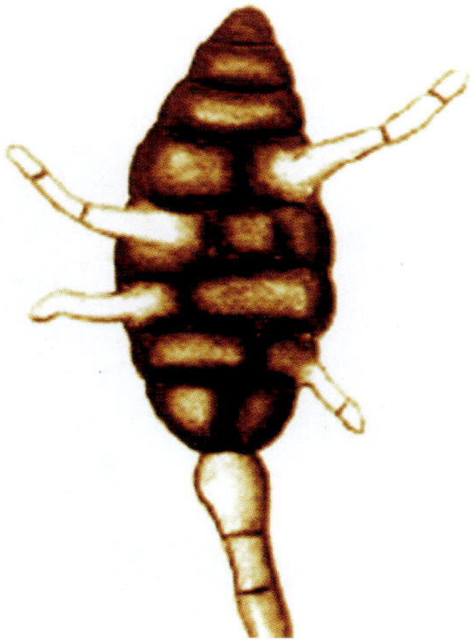

Multipolar **germination** of a conidium of *Alternaria* sp., x 800 (*MU*).

Gibbose, young sporangia of *Physarum polycephalum*, x 17 (*MU*).

Gibbose, mature sporangia of *Physarum polycephalum*, x 17 (*MU*).

Gibbose, mature sporangium of *Physarum polycephalum*, x 20 (*MU*).

glabrescent (L. *glabrescens*, genit. *glabrescentis* < *glabrescere*, to stay hairless < *glabrare*, to become smooth < *glaber*, glabrous, without hair, smooth + suf. *-escens*, genit. *-escentis* > E. *-escent*, that which turns, that begins to, slightly): becoming glabrous, like certain areas of the pilear surface of the basidiocarp of *Trametes hirsuta* (=*Polyporus hirsutus*), which initially is hirsute (Aphyllophorales).

glabrous (L. *glaber*, glabrous, without hair, smooth < *glabrare*, to become smooth, to divest of hair and bristles): lacking hair, bald, beardless. Cf. **pubescent**.

gladiate (L. *gladiatus*, provided with a sword or similar to a sword < *gladius*, sword + suf. *-atus* > E. *-ate*, provided with or likeness): syn. of **ensiform** and **xiphioid**.

glaireous (E. *glair* or *glaire* < ME. *gleyre*, egg white < MF. *glaire*, modification of VL. *claria* < L. *clarus*, clear): viscous or mucilaginous, suggestive of an egg white, like some myxomycete plasmodia.

glaucous (L. *glaucus* < Gr. *glaukós*, sea-green in color): with a bluish-gray tone, like the colonies of *Aspergillus glaucus* (moniliaceous asexual fungi).

gleba, pl. **glebae** (L. *gleba*, lump, mass): the central, internal portion of the fruiting body of the Gasteromycetes, composed of a fertile hymeniiferous part that produces the basidiospores, and a sterile part consisting of a trama of pseudotissue. As examples of various types of gleba are the glebae of the basidiocarp of *Lycoperdon*, *Vascellum* (Lycoperdales) and *Scleroderma* (Sclerodermatales), which correspond to the central zone. In the Nidulariales (such as *Cyathus*) the gleba is arranged into glebal chambers or *peridioles*, and in the Phallales it is present as a viscous mass in the receptaculum, located in the apex of the stipe of the basidiocarp (as in *Phallus*), or in the *glebifer* or *lantern*, which is found in the interior of the arch of the receptaculum, which in this case is a clathrate basidiocarp (as in *Clathrus* and *Laternea*).

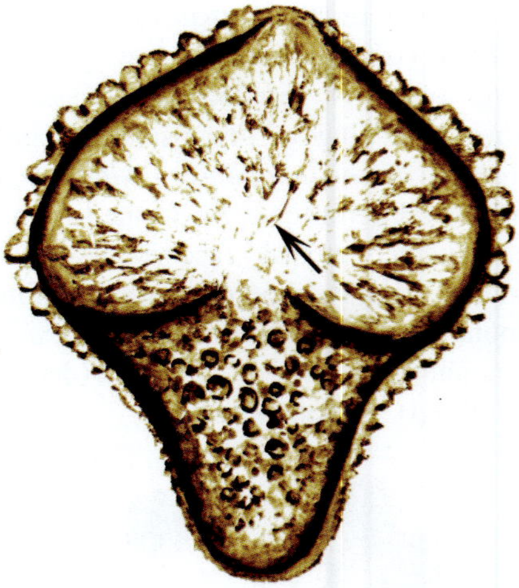

Gleba of the fruiting body of *Lycoperdon perlatum*, seen in longitudinal section, x 1 (*MU*).

Gleba of the fruiting bodies of *Scleroderma verrucosum*, x 1 (*RV*).

Gleba of the fruiting body of *Scleroderma verrucosum*, x 1 (*TH*).

Gleba of the fruiting body of *Scleroderma verrucosum*, x 1 (*RV*).

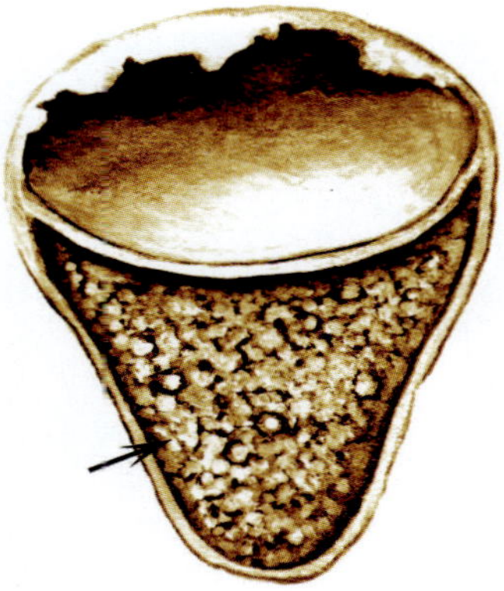

Gleba of the fruiting body of *Vascellum pratense*, seen in longitudinal section, x 1 (*MU*).

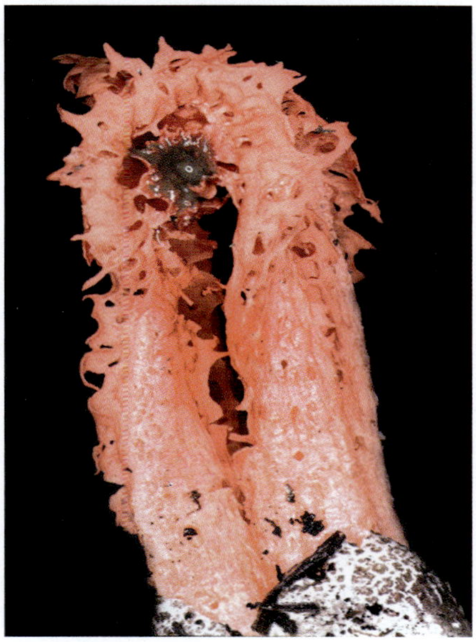

Gleba (lantern or glebifer) of the fruiting body of *Laternea pusilla*, seen in the arch of a receptaculum, x 1 (*JC*).

glebal chamber (L. *gleba*, lump, mass + suf. *-alis* > E. *-al*, relating to or belonging to; ME. *chambre* < OF.< L. *camara*, room, vault): a cavity within the gleba, whether empty or full of gel, that is usually lined with hymenial elements, as occurs, e.g., in *Hymenogaster* (Hymenogastrales) and *Melanogaster* (Melanogastrales). In other gasteromycetes the glebal chambers correspond to the *peridioles* (as in *Cyathus* and *Crucibulum*, of the Nidulariales).

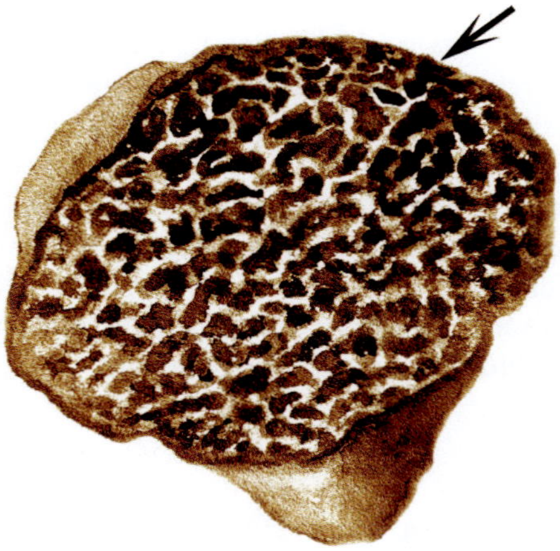

Glebal chambers of a basidiocarp of *Hymenogaster sulcatus*, seen in longitudinal section, x 4.5 (*MU*).

glebifer (L. *gleba*, lump, mass + suf. *-ferous*, bearer < *-fer* < *ferre*, to bear, carry): *Gasteromycetes*. A special hymenial structure that is formed in the arch of the receptaculum; also called **lantern**. The glebifer is held in place by either a piece of tissue or by *trabeculae* that anchor it to the arch, as in *Laternea pusilla* (Phallales).

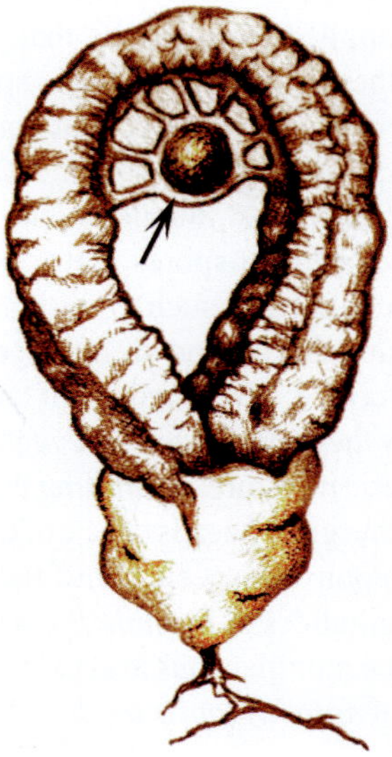

Glebifer or lantern of the fruiting body of *Laternea pusilla*, x 1 (*MU*).

glebose, **glebous** (L. *glebosus*, full of lumps < *gleba*, lump, mass + suf. *-osus* > E. *-ose*, full of, augmented, prone to): full of lumps; lumpy.

gleo-, **gloeo-** (Gr. *gloiós*, gum and other glutinous substances): a pref. employed in terms that allude to something sticky, viscous or mucilaginous, as in the following terms.

gleocystidium, pl. **gleocystidia**, **gloeocystidium**, pl. **gloeocystidia** (Gr. *gloiós*, gum and other glutinous substances + *kystídion* < *kýstis*, bladder + dim. suf. *-ídion* > L. *-idium* > E. *-idium*): a cystidium with refractive contents present in some species of *Russula*, such as *R. polyphylla* (Agaricales) and other fungi. With cresil blue the gloeocystidia stain intensely blue, except for the wall, which acquires a pale violet tone. The gloeocystidia are versiform cystidia, of which four types are known: **pseudocystidium** or **macrocystidium**, **chrysocystidium**, **phaeocystidium** and **coscinocystidium**.

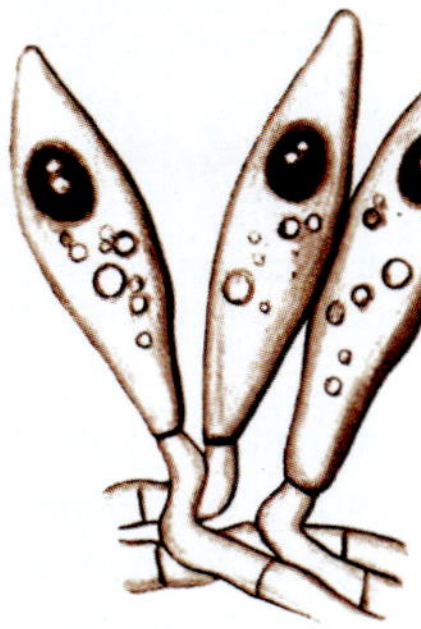

Gleocystidia of the hymenium of *Russula polyphylla*, x 740 (*MU*).

gleoid (Gr. *gloiós*, gum, glue and other sticky substances + suf. *-oeides* > E. *-oid*, like, similar to): any glutinous substance; applied to structures that are gelatinous, viscid or adhesive, such as hyphae, spores, etc.

gleoid head (Gr. *gloiós*, gum, glue and other sticky substances + suf. *-oeídes* > E. *-oid*, similar to; ME. *hed*, head): the mucilaginous globose mass containing embedded spores that forms at the apex of conidiogenous cells, such as certain phialides. The conidia accumulate around the mouth of the phialides as they are formed. The mucilage of the gleoid heads, which keeps the spores together, eventually dissolves in atmospheric moisture, permitting dissemination of the spores. The gleoid heads can be of different colors, depending upon the species, and they are present, e.g., on the phialides of *Lecanicillium* (=*Verticillium*) *lecanii*, of the moniliaceous asexual fungi, and on the synnemata of *Antromycopsis smithii*, *Didymostilbe* sp., *Stilbella flavidum* and *Graphium* sp. (stilbellaceous asexual fungi). See **slime spore**.

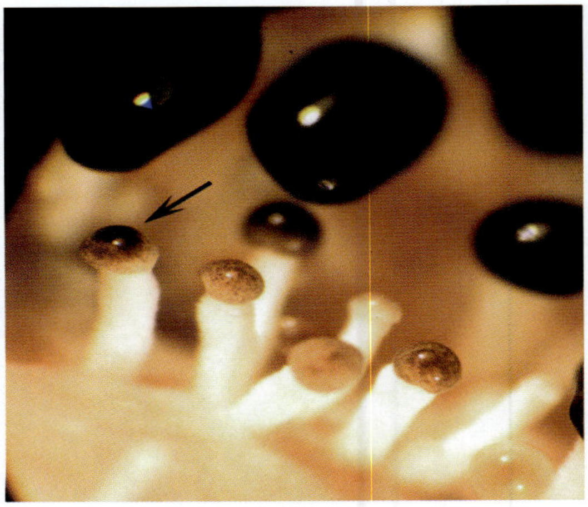

Gleoid head of the synnemata of *Antromycopsis smithii* on agar, x 7 (*RV*).

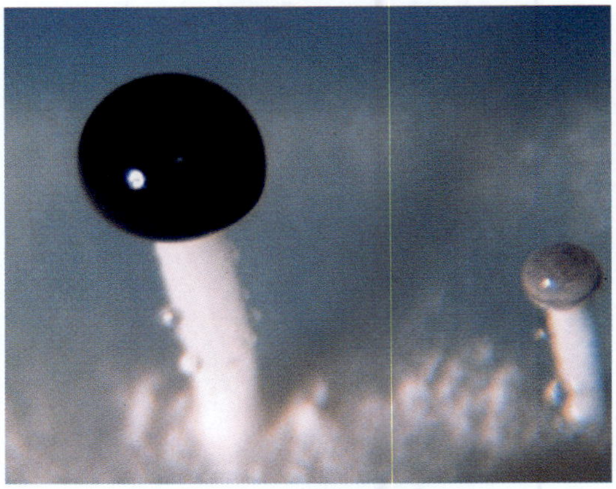

Gleoid head of the synnemata of *Antromycopsis smithii* on agar, x 7 (*CB* & *MU*).

Gleoid head of a synnema of *Stilbella flavidum*, with embedded conidia, x 48 (*RTH*).

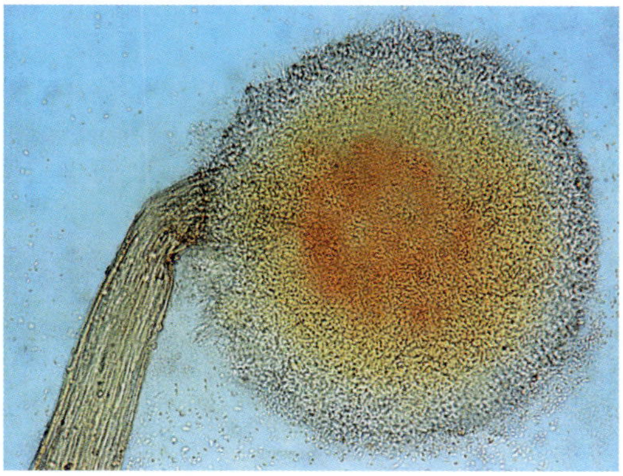

Gleoid head of a synnema of *Stilbella flavidum*, with embedded conidia, x 105 (*RTH*).

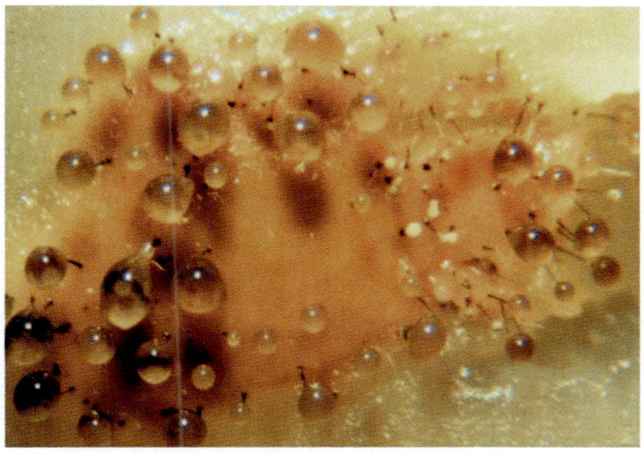

Gleoid head of the synnemata of *Didymostilbe* sp. on agar, x 1.5 (*CB*).

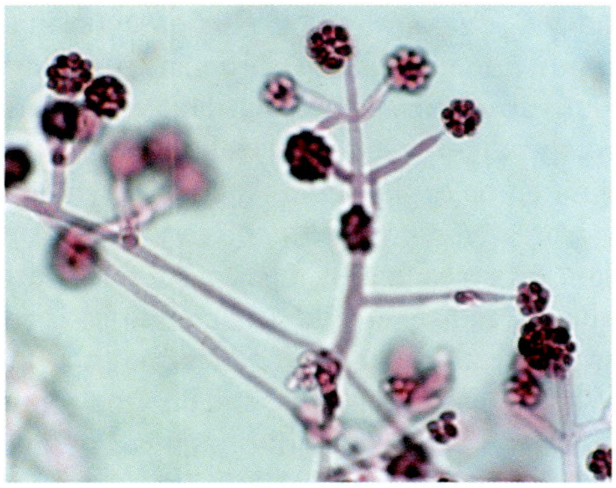

Gleoid heads on a conidiophore of *Lecanicillium* (=*Verticillium) lecanii*, x 600 (*CB*).

Gleoid head on the synnemata of *Graphium* sp., x 600 (*CB*).

gleolichen, **gloeolichen** (Gr. *gloiós*, gum, mucilage and other glutinous substances + *leichén* > L. *lichen*, lichen): a homoiomerous lichen whose gonidia or algal cells belong to *Chroococcus*, *Gleocapsa* and other Chroococcales, which have a mucilaginous capsule.

gleoplerous, **gloeoplerous hypha**, pl. **hyphae** (Gr. *gloiós*, gum and other glutinous substances + *plerotikós* < *plerés*, to fill; NL. *hypha*, hypha < Gr. *hyphé*, tissue, spiderweb, hypha): refers to hyphae with long cells that contain numerous oil droplets in the cytoplasm.

gleospore, **gloeospore** (Gr. *gloiós*, gum and other glutinous substances + *sporá*, spore): a viscous or mucilaginous, moist spore, like that of *Clonostachys rosea* (=*Gliocladium roseum*), of the moniliaceous asexual fungi, and *Marssonina* (=*Gloeosporium*), of the melanconiaceous asexual fungi, adapted for dissemination by water droplets, simple contact, insects and other similar vectors. Cf. **xerospore**.

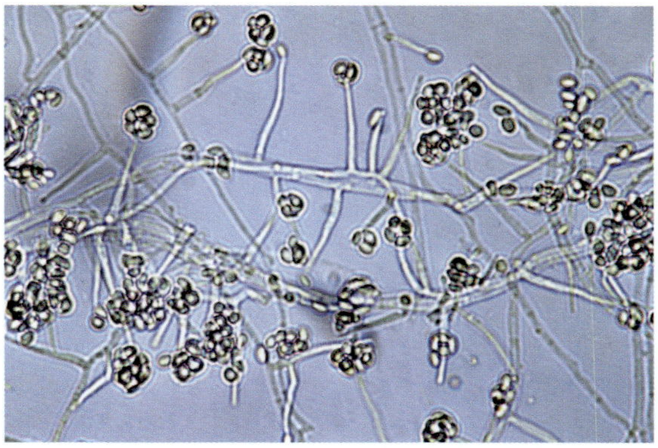

Gleospores on phialides of *Cephalosporium* sp. (moniliaceous asexual fungi), x 620 (*JAS*).

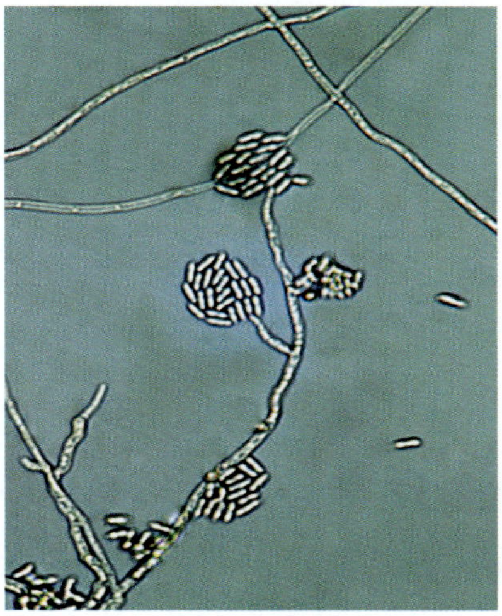

Gleospores on phialides of *Fusarium* sp. (tuberculariaceous asexual fungi), x 620 (*MU*).

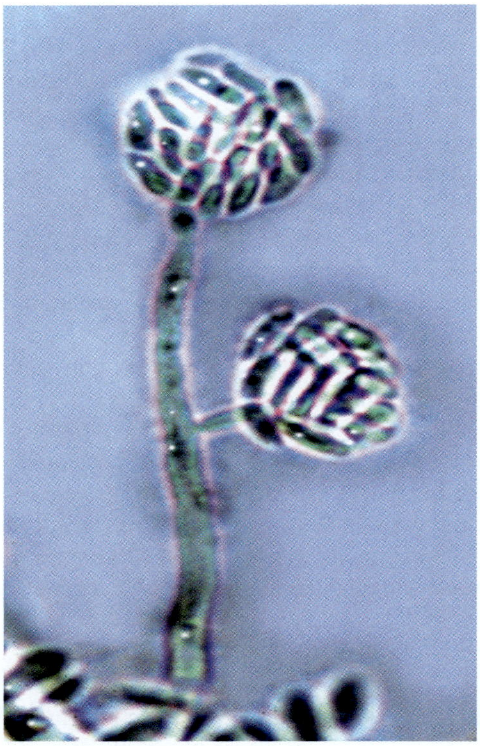

Gleospores on phialides of *Clonostachys rosea* (=*Gliocladium roseum*), x 620 (*JAS*).

gliatope (NL. *glia* < Gr. *glía*, adhesive; a gelatinous substance that absorbs water to form a viscous solution + *tópos*, place, site): the site of abundant gel production in gelatinous fungi. Cf. **gel tissue**.

glio-, glia- (Gr. *glía*, glue, glutinous material): a pref. that forms a part of terms related to something sticky, as in *Gliocladium* (moniliaceous asexual fungi), a gen. of fungi with conidiophores that produce viscous spores, and *gliatope*, a site of abundant gel production.

globose, globoid, globular (L. *globosus*, round like a ball < *globus*, ball, globe + suf. *-osus* > E. *-ose*, full of, augmented, prone to; or *globus* + L. suf. *-oide*, like < Gr. *-oeídes*, similar to; or L. *globulus*, dim. of *globus* + L. suf. *-aris* > E. *-ar*, like, relative to): having the shape of a globe or globule; spherical or almost spherical. Applied to structures (spores, sporocarps, etc.) whose length:width ratio is between 1:1.0 and 1:1.05.

glochid, glochidium, pl. **glochidia** (NL. *glochidium* < Gr. *glochídion* < *glochís*, arrowhead or trident + dim. suf. *-ídion* > L. *-idium*): a hair or bristle with small barbs.

Glomeromycota (NL. *glomerulus*, dim. of *glomer*, *glomus*, ball of thread or yarn + Gr. *mýkes*, fungus + NL. *-ota*, a suf. that signifies to have, utilized as the ending for the taxonomic category of phylum): a phylum recently segregated from the Zygomycota, as a clade that consists of vesicular-arbuscular mycorrhizal fungi and related forms. It has been considered as a basal group in the same clade as the Ascomycota and Basidiomycota. The glomeromycetes are capable of forming vesicular-arbuscular mycorrhizae or of living in mutualistic associations with cyanobacteria. Their diversification appears to have occurred at the time it is believed that plants began to colonize the terrestrial environment. See Appendix.

glomerospore (NL. *glomerulus*, dim. of *glomer*, *glomus*, ball of thread or yarn + Gr. *sporá*, spore): the unique asexual reproductive structures formed by species of vesicular-arbuscular mycorrhizal fungi (Glomeromycota), the formation of which can vary depending upon the species. In *Gigaspora* the spore is terminal and forms on the end of a subtending hypha. In *Acaulospora* the spore forms laterally at the end of a short hypha that arises from a sporiferous saccule, and in *Entrophospora* the spore is intercalary in the hypha subtending the sporiferous saccule.

glomerulus, pl. **glomeruli** (NL. *glomerulus*, dim. of *glomer*, *glomus*, ball of thread or yarn): a dense mass formed by an aggregate of clustered cells or spores, as is observed on the surface of certain lichens. Also used for the haustoria of some species of *Septobasidium* (Septobasidiales), and for the mass of hyphae formed by mycorrhizogenous fungi in the interior of the cells of the associated plant, in which case it is also known as a **peloton**.

glossary (L. *glossarium* < *glossa*, obscure language < Gr. *glõssa*, tongue, as a buccal organ and as language + L. suf. *-arium* > E. *-ary*, which denotes belonging to): a compendium or list of specialized terms, with their definition or explanation. Although etymologically the words *glossary* and *dictionary* have a different origin, their meaning is similar, and in practice they are employed indistinctly. In general, glossaries contain less information than dictionaries, they are at the end of a book, like an appendix, and they may or may not include etymologies and illustrations. For the title of this treatise on mycological lexicology it was preferred to employ the word *dictionary*, taking into account the coverage, focus, format, etymologies, images, and the other distinctive characteristics of the work. See **dictionary**, **lexicon** and **vocabulary**.

glossarist (L. *glossarium* < *glossa*, obscure language < Gr. *glõssa*, tongue, as a buccal organ and as language + L. suf. *-arium*, which denotes belonging to + L. suf. *-ista* > E. *-ist*, one who practices some discipline or activity; one who has the ability or capacity to carry out an activity or profession): the compiler of a glossary. Cf. **etymologist**.

glosso-, **gloss-** (Gr. *glõssa*, tongue, as a buccal organ and as language): a pref. employed in terms such as *glossoid*, with a shape similar to a tongue. See **glossology**.

glossoid cell (Gr. *glõssa*, tongue + L. suf. *-oide*, like < Gr. *-oeídes*, similar to; NL. *cellula*, cell, living cell, dim. of L. *cella*, small room, compartment, cell of a honeycomb): an elongate, tongue-shaped cell in *Haptoglossa* (Myzocytiopsidales), a member of the Oomycota. See **trichocyst**.

glossology (Gr. *glõssa*, tongue, as a buccal organ and as language + *lógos*, word < *légo*, saying > E. des. *-logy*, that said, narrated, study, treatise): a treatise on language; *mycological glossology*, the part of glossology dedicated to the study of the terms most adequate to represent all of the mycological concepts. See **lexicology** and **orismology**.

gluco- (Gr. *glukús*, sweet): a pref. that alludes to sugar, as in *glucogen*, which gives or produces sugar, the reserve substance in fungi.

gluten (L. *gluten*, genit. *glutinis*, glue, adhesive): a viscous substance, very sticky (glutinous), that is derived from the dissolution of the gelatinous hyphae that form part of certain fungal tissues, such as those of the pileus cuticle and stipe, or like those of the universal veil, present in fungi of the gen. *Suillus* (Agaricales).

glutinous, **glutinose** (L. *glutinosus*, sticky < *glutin-*, *gluten*, genit. *glutinis*, glue + suf. *-osus* > E. *-ose*, full of, augmented, prone to): with an adhesive surface; with the consistency of egg white or of liquid glue, like the colony of *Rhodotorula glutinis* (cryptococcaceous asexual yeasts) and the basidiocarp of *Gomphidius glutinosus* (Agaricales). Equivalent to **viscid**.

Glutinous colony of *Rhodotorula glutinis*, on agar, x 1 (*MU*).

glyco- (Gr. *glukús*, sweet): see **gluco-**.

glypholecideous (Gr. *glyphé*, sculpture, engraving, carving < *glyphó*, to carve, engrave + *lekidós*, genit. of *lekís*, dim. of *lekós*, plate, pot): refers to the thallus of the lichen *Glypholecia* (Lecanorales), which has markedly labyrinthiform lirellae or lirelline apothecia.

-gnomy (Gr. *gnómon*, that which distinguishes): a suf. employed in Gr. terms, such as *physiognomy* (< *phýsis*, nature + *gnómon* > L. *physiognomia* > E. *physiognomy*), the particular appearance of a person's face, which results from the combination of its features.

gnotobiotic (Gr. *gnótos*, well known + *biotikós*, relating to life < *bíos*, life + suf. *-tikós* > L. *-ticus* > E. *-tic*, relation, fitness, inclination or ability): a term applied to an axenic culture when it is certain that it contains a single species and is free of contamination; pure. Cf. **agnotobiotic**, **axenic** and **monoxenic**.

Golgi apparatus, **Golgi body** (named in honor of discoverer Camilo Golgi; L. *apparatus*, apparatus, equipment < *apparare*, to prepare; ME. < OE. *bodig* > E. *body*, body): a group of flat cisternae delimited by double membranes, often with the surface dilated or vesiculose, from which are generated secretory vesicules that are involved in the synthesis of the plasma membrane and of the cell wall in the subapical region of a growing hypha. These stacked cisternae also are called **dictyosomes**.

gongylidius, pl. **gongylidia** (dim. form of Gr. *gongýlos*, round): a bulbous or globular structure that develops in the mycelium of fungi cultivated by termites.

gongylus, pl. **gongyli** (Gr. *gongýlos*, round, alluding to the spherical shape of the majority of the spores): **1**. A globular propagative body that is formed in the thallus of certain lichens; an example is the *soredium*. **2**. A swollen cell that is formed on the apices of hyphae and which serves as food for the ants that cultivate fungal gardens. See **bromatium** and **staphylum**.

gonidium, pl. **gonidia** (NL. *gonidium* < Gr. *gonídion* < *gónos*, progeny, offspring, seed + dim. suf. *-ídion* > L. *-idium*): *Lichens*. A lichen photobiont (preferably a Chlorophyceae, although in some lichens it is a cyanobacterium) to which is attributed a role in the sexual reproduction of the lichen thalli, a process that only the mycobiont undergoes; notwithstanding, there are hymenial gonidia that are found in the hymenium of the reproductive bodies of the fungus and which are disseminated at the same time as the fungal spores, thus assuring the formation of a new lichen, since these spores germinate together with the appropriate algal or bacterial cells to establish symbiosis with them. See **gonimium**.

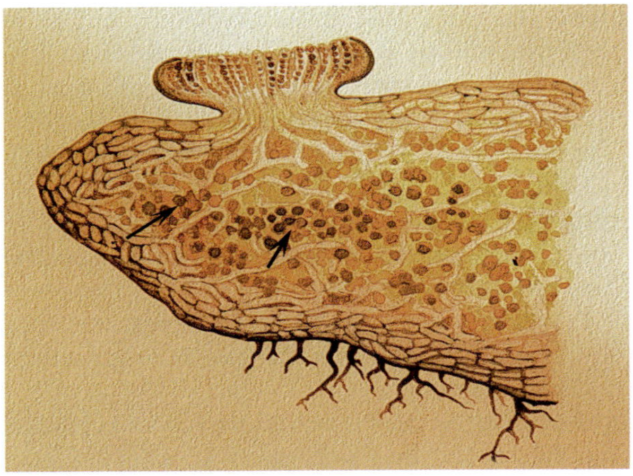

Gonidia of the thallus of *Hypotrachyna* (=*Parmelia*) *imbricatula*, seen in longitudinal section, x 10 (*MU*).

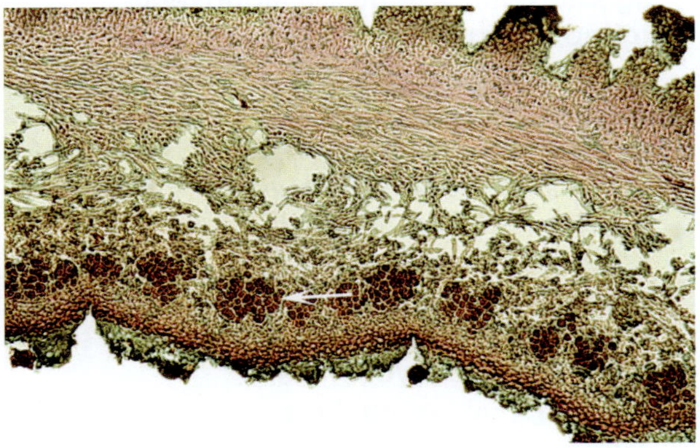

Gonidia of the thallus of *Hypotrachyna* (=*Parmelia*) *imbricatula*, seen in longitudinal section, x 10 (*MU*).

gonimium, pl. **gonimia** (Gr. *gónimos*, fertile, fecund, capable of producing + L. dim. suf. *-ium*): *Lichens*. A bacterial cell (gonidium) with a thin cell wall and blue color, belonging to the cyanobacteria, that is found in certain lichen thalli. See **gonidium**.

goniocyst (NL. *goniocyst* < *gonium* < Gr. *gónos*, progeny + *kýstis*, bladder, vesicle, cell): *Lichens*. A vegetative propagule of the thallus of some tropical foliicolous lichens, composed of a cell of the photobiont and its progeny enveloped by the hyphae of the mycobiont; it differs from a soredium in that it is formed in a special organ called the *goniocystangium*.

goniocystangium, pl. **goniocystangia** (NL. *goniocystangium* < *goniocyst* < Gr. *gónos*, progeny + *kýstis*, bladder, vesicle, cell + *angeîon*, vessel, receptacle): *Lichens*. A special organ of some foliicolous lichens in which *goniocysts* are produced.

goniospore (Gr. *gonía*, angle + *sporá*, spore): an angular spore, like the basidiospores of *Inocybe* (Agaricales).

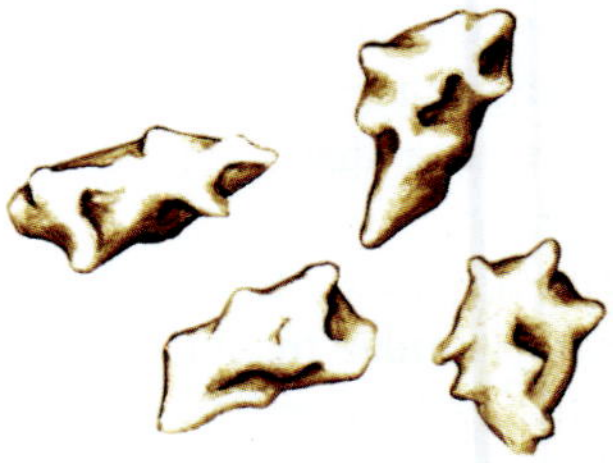

Goniospores of *Inocybe* sp., x 1,800 (*MU*).

-gonium, pl. **-gonia**, **gonio-** (L. *gonium* < Gr. *gónos*, that engendered, the child): a suf. employed when one wishes to allude to the reproductive organs, as in *oogonium*, the female sexual organ of the Oomycetes. As a pref., *gonio-*, is used in terms such as *goniospore* and *goniocyst*.

gono-, **gon-** (Gr. *gónos*, progeny, procreation). a pref. that implies the idea of sexuality or of engenderment, as in the following words.

gonoplasm (Gr. *gónos*, progeny, procreation + *plásma*, bland material with which an organ or living being is formed): *Peronosporales*. The central mass of protoplasm of the antheridium that passes through the fertilization tube into the oogonium, where it fuses with the oosphere.

gonotaxis, pl. **gonotaxes**, **gonotropism** (Gr. *gónos*, progeny, procreation; as a pref., *gono-*, it implies the idea of sexuality or of engenderment + *táxis*, arrangement, disposition; or *gónos* + *trópos*, a turn, change in position or manner < *tropé*, a turning < *trépo*, to gyrate, revolve, turn towards + suf. *-ismós* > L. *-ismus* > E. *-ism*, state, phase, tendency, action): the movement of antherozoids toward the female sexual organ, which contains the oosphere, supposedly due to the unilateral effect of a stimulating factor, as occurs in *Monoblepharis polymorpha* (Monoblepharidales); because the movement is toward the center from which the excitation emanates, it is considered as a positive tropism. See **taxis** and **tropism**.

gonotocont, **gonotokont** (Gr. *gónos*, progeny, procreation, which here alludes to the *gonia*, the offspring + *tókos*, childbirth): a diploid cell or organ in which, by reduction division or meiosis, the progeny are formed.

-gony (Gr. *-gonía* < *goneío*, to engender): a suf. employed in terms such as *theogony*, the generation or engenderment of gods (see **entheogen**).

gossamer (ME. *gossomer* < *gos*, goose + *somer*, summer; a spiderweb or enticement that floats in mild, calm air; something light, delicate, tenuous): the flimsy or minute networks of mycelium (floating films) produced by some fungi in liquid culture media lacking a source of carbon.

grain (L. *granum*, grain, seed, small particle): each of the minute granules formed by various species of actinobacteria (e.g., *Nocardia brasiliensis* and *Streptomyces madurae*) and pathogenic fungi (such as *Madurella mycetomi* and *M. grisea*, of the dematiaceous asexual fungi) that cause mycetoma in man and higher animals. These granules are contained in the pus of the lesions; they vary in size from microscopic up to more than 2 mm in diameter. The size, color, shape and texture of the granules, as well as the dimensions of the fungal hyphae within the granule, vary considerably, depending on the species of the fungus; the characteristics of the granules are utilized in the differential diagnosis of the causal agents. E.g., in *M. mycetomi* the granules have a uniformly brown pigmentation of the matrix.

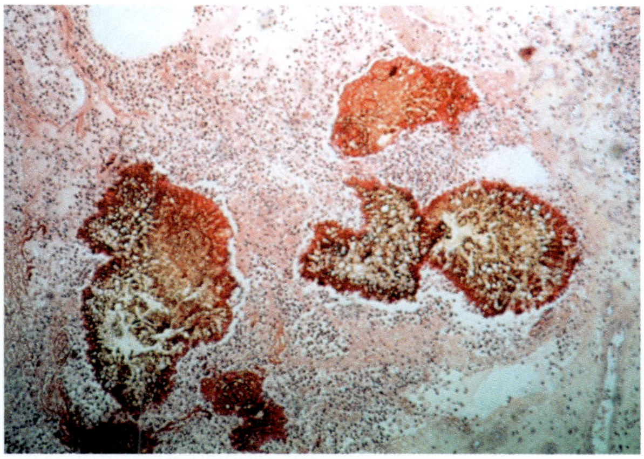

Grains of *Madurella mycetomi*, seen in transverse section of infected host tissue, x 30 (*RLM*).

Gram stain (< *Gram*, a Danish bacteriologist; ME. *steynen*, partly < MF. *desteindre*, to discolor, and < Scand., akin to ON. *steina*, to paint; to color by processes affecting chemically or otherwise the material itself): a staining method for bacteria developed by Gram, who found that certain bacteria stained with crystal violet, followed by Gram's iodine solution, retained their color after being washed with absolute alcohol. The organisms that retain the violet color are gram positive, and those that do not are gram negative, usually abbreviated as gram + and gram –.

graminicole, graminicolous (L. *gramen*, pasture grass + suf. *-cola*, inhabitant): living and developing on grasses, like the perithecial stromata of *Balansia claviceps*, *B. henningsiana* and *Claviceps purpurea* (Hypocreales), and the zoosporangia of *Polymyxa graminis* (Plasmodiophorales) in cells of wheat root, and the mycelia of *Rhizoctonia solani* (moniliaceous asexual fungi) in grains of wheat.

Graminicolous sclerotia of *Claviceps purpurea* (ergot), growing on a spike of rye, x 1 (*MU*).

Graminicolous perithecial stromata of *Balansia claviceps*, in a leaf of grass, x 5 (*RTH*).

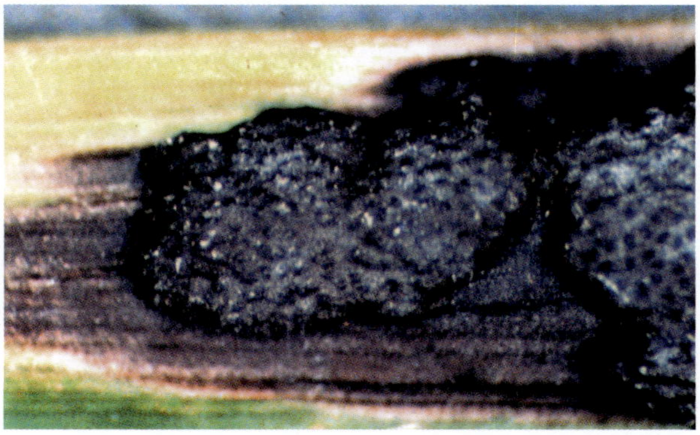

Graminicolous perithecial stromata of *Balansia henningsiana*, in a leaf of grass, x 47 (*RTH*).

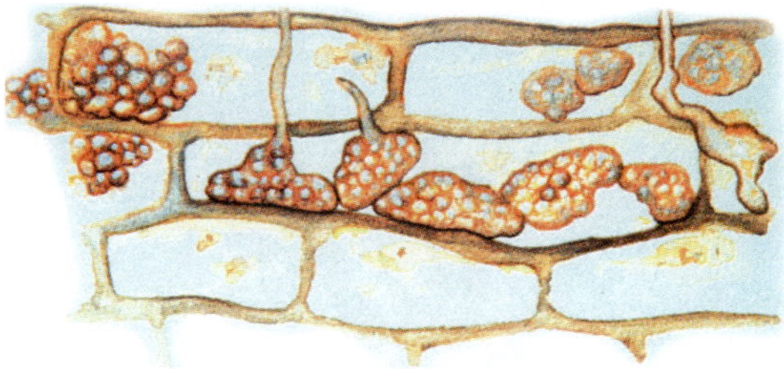

Graminicolous zoosporangia of *Polymyxa graminis*, in cells of wheat root, x 350 (*MU*).

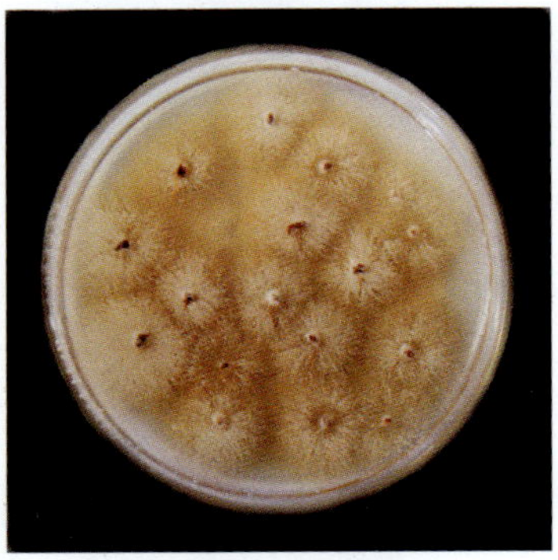

Graminicolous colonies of *Rhizoctonia solani*, growing from pieces of wheat, x 0.5 (*JAS*).

granular, **granulate**, **granulose** (L. *granulus*, dim. of *granum*, grain, seed of wheat, barley and other cereals + L. suf. *-aris* > E. *-ar*, like, relative to; or + L. suf. *-atus* > E. *-ate*, provided with or likeness; or + L. suf. *-osus* > E. *-ose*, full of, augmented, prone to): having grains or composed of grains.

granular core (L. *granulus*, dim. of *granum*, grain, seed, small kernel, small particle + L. suf. *-aris* > E. *-ar*, like, pertaining to; ME. *core*, a central and often foundational part usually distinct from the enveloping part by a difference in nature): refers to the central part of the fruiting body of *Inonotus dryophilus* (Aphyllophorales), which has a granular consistency, different from the surrounding tissues.

Granular core, in the central part of the fruiting body of *Inonotus dryophilus*, seen in longitudinal section, x 0.5 (*RV*).

-graphy (Gr. *-graphía* < *grápho*, to write, describe). a suf. employed in terms such as *geography*, the science that treats of the description of the Earth. Also used as a pref. *graphy-*, *graph-*, in terms such as *graphydaceous*, a family of crustaceous ascolichens, and *Graphis*, a gen. of these ascolichens of the order Ostropales, whose ascomata form figures similar to symbols or hieroglyphs.

gravitaxis, pl. **gravitaxes** (L. *gravis*, heavy + Gr. *táxis*, arrangement, disposition): a tactic movement in response to the gravitational attraction of the mass of the Earth, like that of the zoospores of *Phytophthora palmivora* (Peronosporales). See **electrotaxis** and **taxis**.

gravitropism (L. *gravis*, heavy + Gr. *trópos*, a turn, change in position or manner < *tropé*, a turning < *trépo*, to gyrate, revolve, turn towards + suf. *-ismós* > L. *-ismus* > E. *-ism*, state, phase, tendency, action): a reaction to the influence of gravity, like that seen in the sporangiophores of Mucorales (negative), and in the stipes, gills and tubes of basidiocarps (positive). Syn. of **geotropism**. See **tropism**.

Greek (Gr. *graikós* > L. *graecus*, Greek; a native of Greece; the language of Greece): the Greek language, whose alphabet consists of 24 letters, is considered among the richest and most perfect known. In ancient or classical Greece several dialects were spoken: the *Achaean* dialects (which include the *Aeolic,* the *southern* dialects, like that of *Arcadia*, and of *Cyprus*), the *Doric* dialect, the *Ionic* dialect (with several variations, including the *Attic*, the language of Plato and Aristotle), the *Common* dialect (used throughout the empire conquered or influenced by the Hellenists), *Biblical* Greek, and *Byzantine* Greek. *Modern* Greek is the simple form derived from Common or Hellenistic Greek and is a more analytical than synthetic language, a difference from Latin, which is more synthetic. Innumerable words in Spanish, English, French and other modern languages are derived from Greek, generally through Latin, and usually are formed from the roots of Greek words and Greek prefixes, suffixes or endings, or a combination of the three elements. Words of Greek and Latin origin are used abundantly in modern languages, by virtue of the Greeks, and to a lesser extent the Romans, having been the forefathers of culture, science and modern technology. Greek roots are used in sciences such as biology, medicine, physics, chemistry, mathematics, astronomy, etc., whose terminology or specialized vocabulary mainly comes directly from the Greek. The Greeks were the most advanced scientists of classic antiquity; the great authors of treatises in medicine and biology, such as Hippocrates, Aristotle and Galen, were Greek. See **etymology** and **Latin**.

gregarious (L. *gregarius*, grouped < *grex*, genit. *gregis*, flock, group + suf. *-arius* > E. *-arious*, belonging to): an organism that lives in close proximity to others in a small area, not very scattered. E.g., the sporangia of the majority of the Myxomycetes, such as *Perichaena* sp. (Trichiales), *Diachea leucopoda*, *Leocarpus fragilis* and *Physarum cinereum* (Physarales) are gregarious, as are many Agaricales and some Gasteromycetes, such as *Hypholoma* (=*Naematoloma*) *fasciculare* and *Mycena*, and *Lycoperdon pyriforme* (Lycoperdales). The myxamoebae of the Dictyosteliomycota also are gregarious, giving rise to the common name of *grex* for the pseudoplasmodium formed by the aggregation of the myxamoebae. Cf. **solitary**.

Gregarious sporangia of *Physarum cinereum* on a leaf, x 9 (*MU*).

Gregarious sporangia of *Perichaena* sp., x 25 (*MU*).

Gregarious basidiocarps of *Hypholoma* (=*Naematoloma*) *fasciculare*, x 0.5 (*TH*).

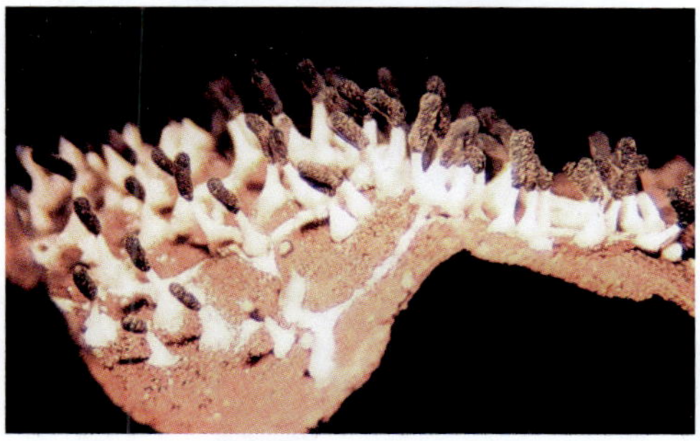

Gregarious sporangia of *Diachea leucopoda*, x 10 (*MU*).

Gregarious basidiocarps of *Hypholoma marginatum*, x 1 (*TRO*).

Gregarious sporangia of *Leocarpus fragilis*, x 6 (*MU* & *CB*).

Gregarious basidiocarps of *Lycoperdon pyriforme*, x 0.4 (*RV*).

grumose, **grumous** (L. *grumus*, bunch, lump, clot, pile, mass + suf. *-osus* > E. *-ose*, full of, augmented, prone to): piled up; composed of accumulated grains or granules; with bunches or masses of hyphae or cells, as is seen in some cultures of filamentous fungi or yeasts in liquid media.

gummose (L. *gummosus*, gummy < *gummi*, gum + suf. *-osus* > E. *-ose*, full of, augmented, prone to): gummy, mucilaginous, like the exudate of various sources produced by some plants infected by fungi, bacteria, viruses, etc. *Gummosis*, pl. *gummoses* (< *gummi* + suf. *-osis*, state or condition), a disease of cucumber caused by *Cladosporium cucumerinum* (dematiaceous asexual fungi), has as a characteristic symptom a gummy secretion.

gusset (ME. < OF. *gousset* < *gousse*, pod, husk): each of the triangular segments resulting from the splitting of the fruiting body of *Montagnea* (Hymenogastrales). The gussets are black, hard, recurved plates on which the spores are borne; they represent portions of the tramal plate that have hardened at the point of contact with the peridium or with another *tramal plate*.

Gussets of the fruiting body of *Montagnea arenaria*, x 1 (*MU*).

guttulate (L. *guttulatus*, having droplets or droplet-like masses < *guttula*, dim. of *gutta*, drop + suf. *-atus* > E. *-ate*, provided with or likeness): provided with droplets or small globules of oil. There are *uni-*, *bi-*, *tri-* or *multiguttulate* structures. E.g., the ascospores of *Halosphaeria salina* (Halosphaeriales) and of *Podospora comata* (Sordariales) are *uniguttulate*; the conidia of *Phoma glomerata* (sphaeropsidaceous asexual fungi) and the ascospores of some species of *Xylaria* (Xylariales) are *biguttulate*, since they contain a droplet of oil in each end. Among the basidiomycetes, the basidiospores of *Bolbitius vitellinus* (Agaricales) are uniguttulate, and, among the conidial fungi, the conidia of *Pestalotia* (melanconiaceous asexual fungi) occasionally have a droplet of oil in each one of the two central pigmented cells. Cf. **eguttulate**.

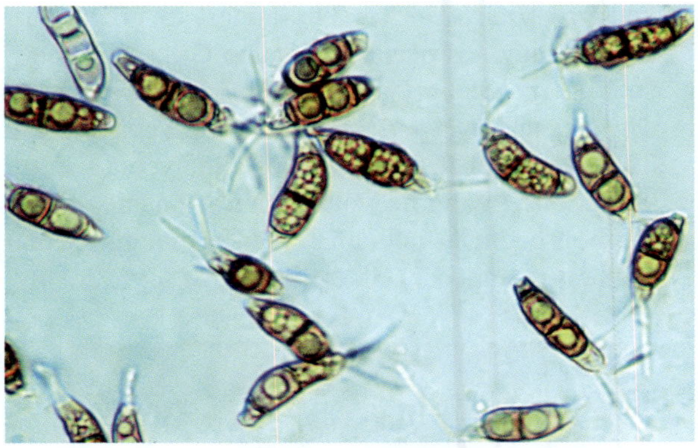

Guttulate (each cell is uniguttulate) conidia of *Pestalotia* sp., x 830 (*MU*).

Guttulate (uniguttulate) ascospores of *Podospora comata*, x 530 (*CB*).

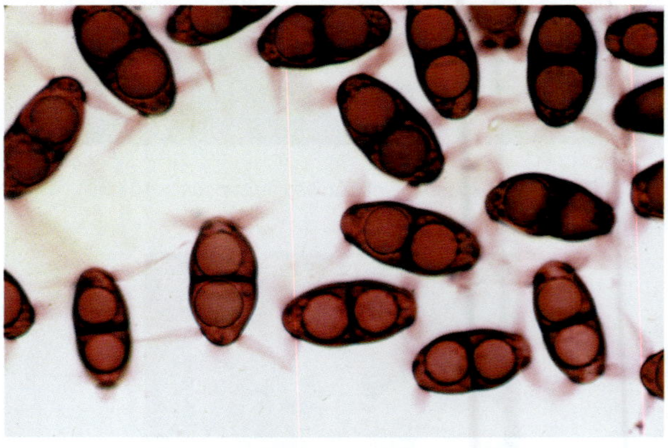

Guttulate (each cell is uniguttulate) ascospores of *Halosphaeria salina*, x 500 (*MCG*).

gyga- (Gr. *gýgas*, genit. *gýgantos*, gigantic, of great magnitude): a pref. that implies large size, as in the generic name *Gigaspora* (Glomales), which refers to the large size of the azygospores. See **macro-**.

gymno-, **gymn-** (Gr. *gymnós*, nude, to lay bare, without protection): a pref. employed to give the idea of something that lacks protection, as in *gymnoascaceous*, a fungus in the family Gymnoascaceae, whose asci are surrounded by a peridium formed of loosely interwoven hyphae.

gymnocarpic, **gymnocarpous** (Gr. *gymnós*, nude + *karpós*, fruit + suf. *-íkos* > L. *-icus* > E. *-ic*, pertaining to or relating to): a fruiting body that is open from the early stages of its development, exposing the fertile layer that produces the spores; some ascomata (apothecia), such as those of *Leotia* and *Geoglossum* (Helotiales), and basidiocarps, like those of *Ramaria* and *Clavaria* (Aphyllophorales), are gymnocarpic. Cf. **angiocarpic** and **hemiangiocarpic**.

gymnothecium, pl. **gymnothecia** (NL. *gymnothecium* < Gr. *gymnós*, nude + *thekíon*, dim. of *thêke*, box; here, of the asci): an ascoma composed of small clusters of asci surrounded by a loose network of hyphae, with or without appendages of various types, depending on the gen.; gymnothecia are characteristic of the families Arthrodermataceae, Myxotrichaceae and Gymnoascaceae (Onygenales), which include, among others, the gen. *Arthroderma*, *Myxotrichum*, *Eidamella*, *Gymnoascus* and *Pseudogymnoascus*.

Gymnothecium of *Gymnoascus reessii*, x 430 (*MU*).

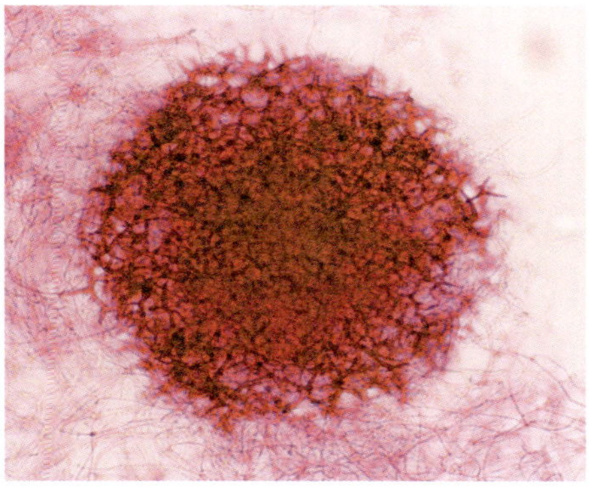

Gymnothecium of *Pseudogymnoascus roseus*, x 260 (*CB*).

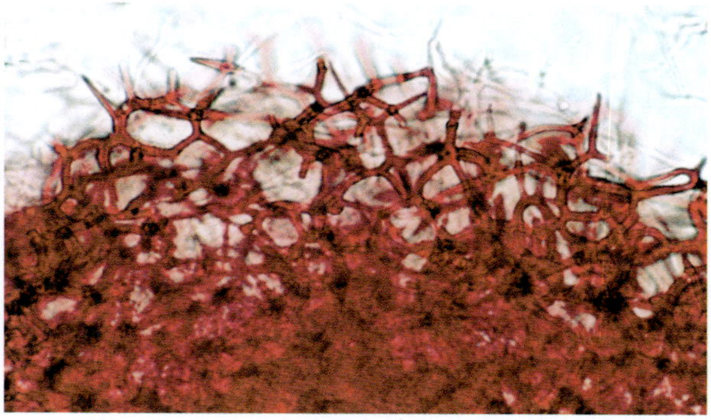

Gymnothecium of *Pseudogymnoascus roseus*; close-up of the peripheral area, x 760 (*CB*).

gynophore (Gr. *gyné*, female, here, female sexual organ + suf. *-phóros*, bearer): bearing or borne by the female sexual organ; accomplished in a feminine way. Applied to the fructification of *Pyronema* (Pezizales), originating from the multinucleate female structure, which develops numerous helicoid ascogonia.

gynotrichous (Gr. *gyné*, female, female sexual organ + *thríx*, genit. *trichós*, hair): see **trichogyne** and **carpogonium**.

-gynous (Gr. *gyné*, female, here, the female sexual organ): a suf. employed in terms that allude to the female sexual organ, as in *androgynous*, having an antheridium and oogonium on the same hypha, and *epigynous*, with the antheridium above the oogonium on a hypha, terms that correspond respectively, as adjectives, to the nouns *androgyny* and *epigyny*. Also used as a pref., *gyno-*, in words such as *gynophore*, which bears or is borne by the female sexual organ.

-gyny (Gr. *gyné*, female, here, the female sexual organ): a suf. employed in nouns such as *androgyny*, a phenomenon relating to the coexistence of male and female sexual organs; this term is sometimes employed as a syn. of *hermaphroditism*. See **-gynous**.

gyrodisc (L. *gyrus* < Gr. *gýros*, round, circular; rotation + *dískos*, disc): *Lichens*. An apothecium that has concentric circles on the upper surface, as is seen in *Umbilicaria cylindrica* (Lecanorales).

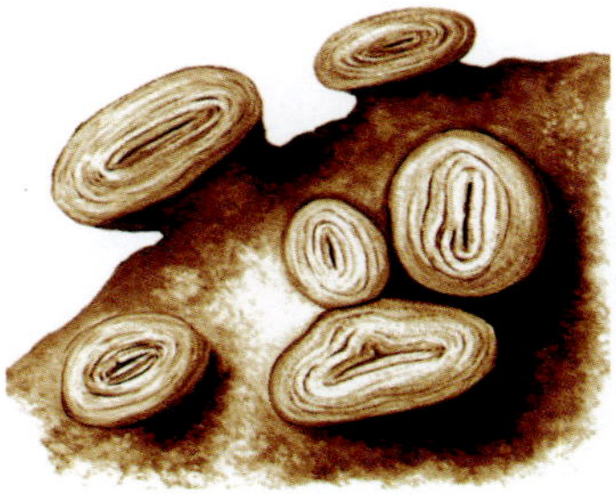

Gyrodiscs on the thallus of *Umbilicaria cylindrica*, x 7.5 (*MU*).

gyroma, pl. **gyromata** (Gr. *gyrós*, round, circular; rotation + *-óma*, a suf. that implies entirety): an apothecium with concentric circular ridges, as in the gen. *Gyrophora* (syn. of *Umbilicaria*, a lichen of the order Lecanorales). Appears to be equivalent to **gyrodisc**.

gyrose, **gyrate** (Gr. *gyrós*, round, circular; rotation + L. suf. *-osus* > E. *-ose*, full of, augmented, prone to): curved backward and then forward; folded and undulate, cerebriform or convoluted, like the sporangium of *Physarum polycephalum* (Physarales), the pileus of *Gyromitra* (Pezizales) and the stipe of the fruiting body of *Tulostoma* (Tulostomatales).

Gyrose pileus of the ascoma of *Gyromitra esculenta*, x 0.5 (*MU*).

Gyrose sporangial head of *Physarum polycephalum*, x 17 (*MU*).

Gyrose stipe of the fruiting body of *Tulostoma* sp., x 6.5 (*MU* & *CB*).

h

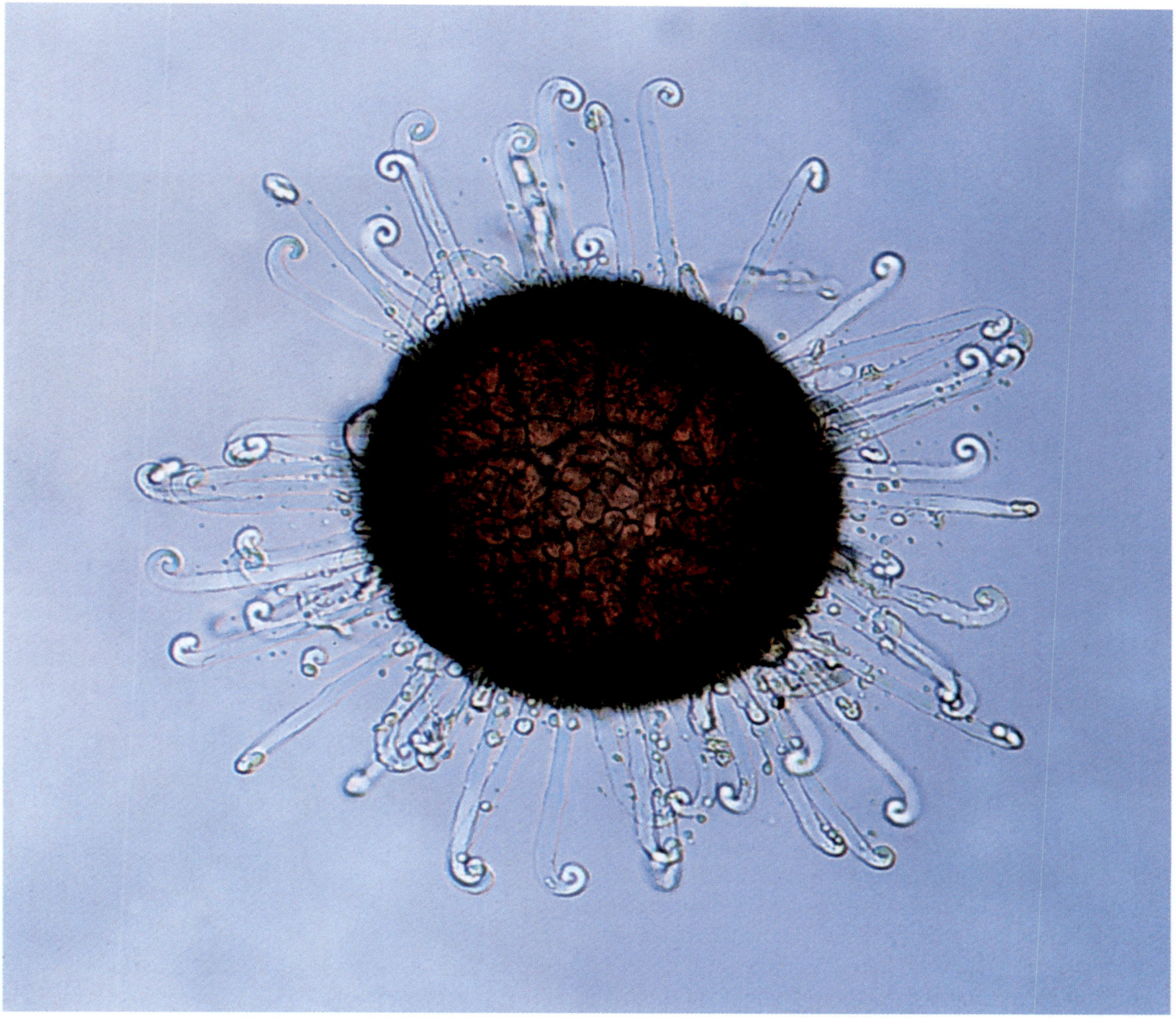

Hamate ascomal appendages of *Erysiphe* (=*Uncinula*) *macrospora* (*RTH*).

h

habit, **habitus** (L. *habitus*, outward appearance, condition, character, aspect, nature): the general external appearance, characteristics, or manner of growth of a fungus.

habitat (L. *habitat* < *habitare*, to inhabit, to dwell, reside): the natural place where an organism lives and grows.

hadro-, **hadr-** (Gr. *hadrós*, thick): a pref. used in terms that allude to the thickness of something, as in *hadromycosis*, pl. *hadromycoses*, a disease produced by fungi that selectively attack or invade woody tissues whose cell walls are thick.

haerangium, pl. **haerangia** (L. *haerangium* < *haerere*, to be adhered + Gr. *angeîon*, vessel, receptacle): a sporulating organ in certain ascomycetes, such as *Ceratostomella* (Xylariales), in which the eight ascospores that are developed from the *octophore* remain covered by a membrane and are surrounded by a circle of hairs, the *tenacle*, around the ostiole of the perithecium.

haline (Gr. *halís*, genit. *halós*, salt, alluding to the sea + L. suf. *-inus* > E. *-ine*, of or concerning): belonging to the marine environment; growing in saline media.

hallucinogenic fungi, sing. **fungus** (L. *hallucinatus* < *hallucinari*, *allucinari*, to hallucinate, chat, dream < Gr. *alýein*, to be anguished or distressed, delirious + Gr. *génos*, engenderment; L. *fungus*, fungus < Gr. *spóngos*, *sphóngos*, sponge): applied to the basidiomata of *Psilocybe* (Agaricales), mainly *P. mexicana*, *P. caerulescens*, *P. aztecorum*, *P. cubensis,* and *P. mazatecorum*, among others, which are ingested by various ethnic groups in Mexico during religious ceremonies to induce fantastic visions and imaginary perceptions that are interpreted as magic. See **entheogen**, **ethnomycology** and **mushroom stones**. The main active compounds of these fungi, *psilocybin* and *psilocin*, are derived from the same basic compound (tryptamine) as serotonin, which is a cerebral neurotransmitter. This amazing relationship can help explain the psychotropic potency of these hallucinogens. Also called *sacred mushrooms* and **magic mushrooms**, among other names.

Sculptures of mushroom stones, from the Mayan Culture (Guatemala), have been related to the cult of **hallucinogenic fungi.** From left to right: a sculpture with an animal effigy (Early and Middle Preclassic); an anthropomorphic stone mushroom (Middle Preclassic); a figure without an effigy, and a tripod base (Advanced Preclassic); a sculpture with a feline effigy (Early and Middle Preclassic), and a figure without an effigy, and a tripod base (Early and Middle Preclassic), x 0.2 (*MU*).

Basidiocarps of *Psilocybe caerulescens* (**hallucinogenic fungi**), x 0.05 (*TH*).

Psilocybe cubensis (**hallucinogenic fungi**) cultivated on horse manure in the laboratory; note the developing basidiocarps, x 0.5 (*TH*).

Basidiocarps of *Psilocybe cubensis* (**hallucinogenic fungi**) growing from cow´s manure, x 0.5 (*TH*).

Basidiocarps of *Psilocybe cubensis* (**hallucinogenic fungi**) growing from manured soil in a flowerpot, x 0.5 (*EPS*).

Basidiocarps of *Psilocybe cubensis* (**hallucinogenic fungi**) growing from cow´s manure, x 0.5 (*EPS*).

halo, pl. **haloes** (L. *halos* < Gr. *halós*, disk, halo; a differentiated zone surrounding a central object): a clear area surrounding a colony, a spore, etc. See **diffusible**.

halo- (Gr. *halís*, genit. *halós*, salt, alluding to saline soils or water): a pref. used in terms such as *halophilic*, an organism that only grows in saline media (soil or water).

halonate, **haloniferous** (L. *halos*, halo, aureola + suf. *-atus* > E. *-ate*, provided with or likeness; or + L. suf. *-ferous*, bearer < *ferre*, to bear, carry): applied to certain lichen spores that appear to be surrounded by a halo, due to the presence of a mucilaginous, hyaline epispore.

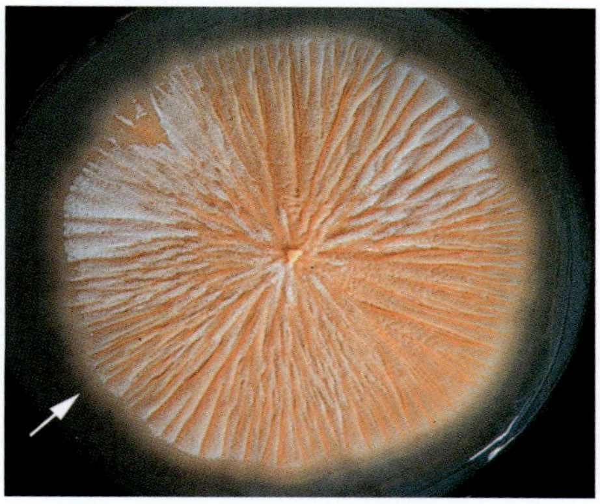

A colony of a yeast (*Trichosporon moniliiforme*, cryptococcaceous asexual fungi) in agar; note the **halo** around the colony, x 1 (*MU*).

halophilic, **halophilous** (Gr. *halós*, salt, especially when alluding to saline soils or water + *phílos*, have an affinity for): preferring saline media (soil or water). The fungi of marine and brackish water are halophilic, e.g., *Catenochytridium carolinianum*, a zoosporic fungus (of the Chytridiales), parasitic on the marine alga *Cladophora japonica*, as well as some species of yeasts, e.g., *Candida marina* (cryptococcaceous asexual yeasts). Cf. **halophobic**.

halophobic (Gr. *halós*, salt + *phóbos*, fearing or shunning < *phobéo*, to fear): not tolerating the presence of soluble salts, even in small quantities; the majority of lichens are halophobic. Cf. **halophilic**.

hamate, **hamose**, **hamous** (L. *hamatus*, with hooks or the shape of a hook or fishhook < *hamus*, fishhook + suf. *-atus* > E. *-ate*, provided with or likeness; or + L. suf. *-osus* > E. *-ose*, full of, augmented, prone to): having hooks, being hook-shaped, like a fishhook; uncinate, hooked, like the ascomal appendages of *Erysiphe* (=*Uncinula*) *macrospora* (Erysiphales). Also called **hamulate** and **hamulose**.

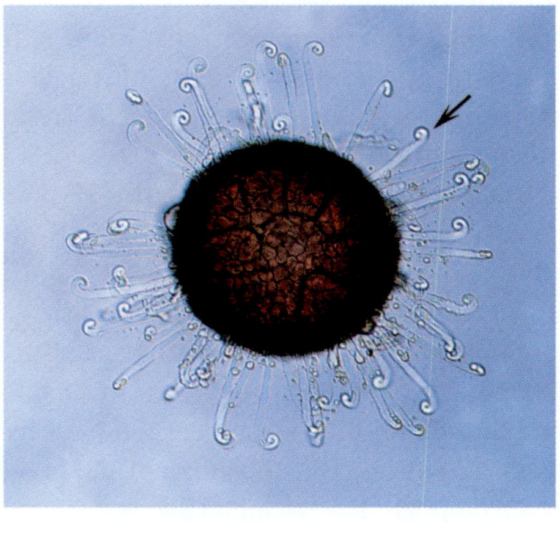

Hamate ascomal appendages of *Erysiphe* (=*Uncinula*) *macrospora*, x 200 (*RTH*).

hamathecium, pl. **hamathecia** (NL. *hamathecium* < Gr. *háma*, all together, complete + *thekíon*, dim. of *thêke*, box; here, of the asci): a general term that is applied to the tissues that separate the asci in the ascomycetes, and whose nature depends on the ontogeny of the ascoma. The different types of hamathecium, which are present in distinct taxonomic groups of ascomycetes, are: *interascal pseudoparenchyma*, *paraphyses*, *paraphysoids*, *pseudoparaphyses*, *periphysoids* and *periphyses*; or the hamathecium may be absent. See these terms.

hamulate, hamulose (L. *hamulatus*, with fishhooks or small hooks < *hamulus*, dim. of *hamus*, fishhook, hook + suf. *-atus* > E. *-ate*, provided with or likeness; or < L. *hamulosus* < *hamulus* + suf. *-osus* > E. *-ose*, full of, augmented, prone to): see **hamate**.

hanging collar (ME. < OE. *hangande*, hanging < ME. *hangen* < OE. *hangian*, to hang; L. *collare*, collar): *Gasteromycetes*. A border of tissue that surrounds the pedicel of the fruiting body and is attached to the base of the endoperidium, typical of some species of *Geastrum* (Lycoperdales).

haplo-, **hapl-** (Gr. *haplóos*, simple, single): a pref. employed to refer to something that is single, when it also could be double or in multiples, as in the following terms.

haplobiont, haplobiontic (Gr. *haplóos*, simple, single + *bíos*, life + *óntos*, genit. of *ón*, a being; or + Gr. suf. *-tikós* > L. *-ticus* > E. *-tic*, relation, fitness, inclination or capacity): an organism whose life cycle has a single type of cell or thallus, haploid in nature, like the species of conidial fungi (deuteromycetes) that reproduce only by means of asexual mechanisms, or at most by *parasexual* reproduction (never truly sexual, i.e., without meiosis). Cf. **diplobiont**. Do not confuse with **haplont**.

haploconidium, pl. **haploconidia** (Gr. *haplóos*, simple, single + *kónis*, powder, dust + dim. suf. *-ídion* > L. *-idium*): a conidium with a haploid nucleus that is formed on the mycelium of species of the Tremellales. Cf. **diploconidium**.

haplodiplobiont, **haplodiplobiontic** (Gr. *haplóos*, simple, single + *diplóos*, double + *bíos*, life + *óntos*, genit. of *ón*, a being; or + Gr. suf. *-tikós* > L. *-ticus* > E. *-tic*, relation, fitness, inclination or capacity): an organism that has alternation of generations and develops its life cycle in two phases, one haploid and the other diploid; this occurs, e.g., in the Chytridiomycete *Allomyces* (Blastocladiales) and in the yeast *Saccharomyces cerevisiae* (Saccharomycetales). Since it has two types of cells (*n* and *2n*), it is a *diplobiontic* organism (literally, two modes of life).

haploid (Gr. *haplóos*, simple, single, alluding to the number of chromosomes + L. suf. *-oide*, like < Gr. *-oeídes*, similar to): an organism, or phase in the life cycle (such as the gametes), in which the cells have only one set (n) of chromosomes; normally this is one-half the diploid number (2n). Cf. **diploid**.

haplokaryon (Gr. *haplóos*, simple, single + *káryon*, nucleus): a nucleus or a cell with *n* chromosomes. Cf. **diplokaryon**.

haplomitotic (Gr. *haplóos*, simple, single + *mítos*, filament + suf. *-tikós* > L. *-ticus* > E. *-tic*, relation, fitness, inclination or capacity): the occurrence of a haploid mitotic phase in the nuclear cycle (type A), i.e., with the sequence: karyogamy - meiosis - mitosis - karyogamy; or of a diploid mitotic phase (type B), with the sequence: karyogamy > mitosis > meiosis > karyogamy.

haplont (Gr. *haplóos*, simple, single + *óntos*, genit. of *ón*, a being): the thallus of the haplophase; the gametothallus of the aquatic fungi that have alternation of generations in the life cycle, such as *Allomyces* (Blastocladiales). Also applied to an organism whose existence occurs entirely, or in major part, in the haploid phase (haplophase), as happens, e.g., in yeasts of the gen. *Schizosaccharomyces* (Schizosaccharomycetales). Cf. **diplont**. Do not confuse with **diplobiont**.

haplophase (Gr. *haplóos*, simple, single + *phásis*, phase, appearance, aspect < *phaíno*, to shine): the phase in the life cycle of an organism during which the cells are haploid. Cf. **diplophase**.

haplostromatic (Gr. *haplóos*, simple, single + *strõma*, mattress, small pillow + suf. *-tikós* > L. *-ticus* > E. *-tic*, relation, fitness, inclination or capacity): composed of only one type of stroma; e.g., in certain Xylariales the placodium is formed only of the ectostroma. Cf. **diplostromatic**. See **placodium**.

hapteron (Gr. *hápto*, to fix, fasten, grasp, touch, adhere): **1**. *Nidulariales*. A globoid mass of adhesive hyphae that forms an attachment organ, located at the base of the funicular cord of the peridioles (as in *Cyathus striatus*). **2**. *Entomophthorales*. An adhesive appendage on the apex of the conidium, as in *Basidiobolus* and *Neozygites*. **3**. *Lichens*. Filaments that serve to fasten the thallus to the substrate in some fruticose species, such as *Cladonia*, *Usnea* and *Ramalina* (Lecanorales). They differ from rhizines and from lichen hairs (which also are attachment structures) by their place of origin. See **funiculus**.

hapto- (Gr. *hápto*, to fix, fasten, grasp, touch, adhere): a pref. employed to form terms that imply the idea of grasping or touching something, as in *haptomorphosis*. Equal to **thigmo-**.

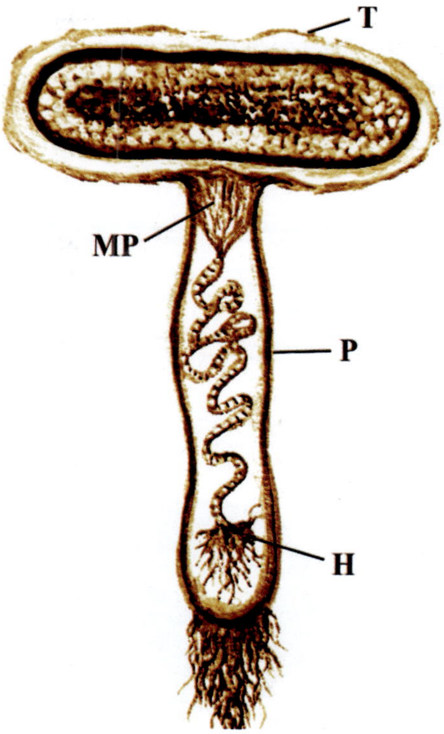

Hapteron (H) of a peridiole of *Cyathus striatus*. In the diagram of a longitudinal section also are shown the middle piece (MP), the purse (P) and tunica (T), x 20 (*MU*).

Hastate synnemata of *Cephalotrichum* (=*Doratomyces*) *stemonitis* on agar, x 8 (*MU*).

Hastate synnemata of *Cephalotrichum* (=*Doratomyces*) *stemonitis* on agar, x 33 (*CB*).

haptomorphosis, pl. **haptomorphoses** (Gr. *hápto*, to fix, fasten, grasp, touch, adhere + *mórphosis*, action of taking shape, of giving form to): a structure whose shape is determined by a contact stimulus, as in the formation of the appressorium and of various sexual organs.

haptonema, pl. **haptonemata** (Gr. *hápto*, to fix, fasten, grasp, touch, adhere + *nêma*, filament): in the flagellated forms, it refers to the filamentous appendage, usually coiled, that consists of the plasma membrane, a sheath of endoplasmic reticulum and a core of microtubules anchored near the kinetosome.

hastate (L. *hastatus*, armed with a spear < *hasta*, spear or pike + suf. *-atus* > E. *-ate*, provided with or likeness): a flat, sharp-pointed structure shaped like a lance. The *Stachel* found in the encysted zoospore of *Plasmodiophora brassicae* (Plasmodiophorales), has a hastate apex. The synnemata of *Cephalotrichum* (=*Doratomyces*) *stemonitis* (stilbellaceous asexual fungi) also are hastate.

haustorial cap (NL. *haustorium* < L. *haustor*, a drawer of water + dim. suf. *-ium* + suf. *-alis* > E. *-al*, relating to or belonging to; ME. *cappe* < LL. *cappa*, hooded cloak, cap): an electron-dense, cap-like mass found at the end of a lobe of the haustorium of *Exobasidium camelliae* (Exobasidiales).

haustorial mother cell (NL. *haustorium* < L. *haustor*, a drawer of water + dim. suf. *-ium* + suf. *-alis* > E. *-al*, relating to or belonging to; ME. *moder,* mother; NL. *cellula*, dim. of L. *cella*, cell of a honeycomb, chamber, compartment, living cell): the cell from which the haustorium of plant pathogenic fungi originates, as in *Melampsora* (Uredinales). This cell comes in contact with the host cell wall and a layer of adhesive material is deposited between them. The infective hypha originates from the inner layer of the haustorial mother cell and effects the invagination of the parasitized host cell. During the development of the haustorium a *haustorial neck* is formed, in which three distinctive zones are differentiated: **1**. A zone that originates from the wall of the haustorial mother cell, which it resembles; **2**. A ring or *collar*, which is dark in color, and **3**. The lower haustorial neck, which is surrounded by the *sheath*, which is the structure that separates the cytoplasm of the haustorium from the cytoplasm of the host cell, and which is the zone of the host-parasite interface.

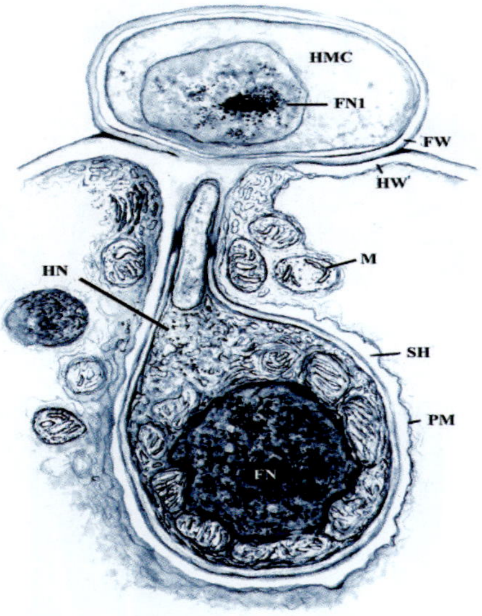

Haustorium of *Melampsora lini*. Longitudinal non-median section through the **haustorial mother cell** (HMC), at the point where it penetrates the cell wall of the host (flax, *Linum usitatissimum*). The fungus wall (FW) is in contact with the host cell wall (HW). The haustorium and the haustorial neck (HN) are surrounded by the invaginated host plasma membrane (PM). One fungal nucleus (FN) is in the haustorium, and the other fungal nucleus (FN1) appears to be partly in the haustorial neck and partly in the haustorial mother cell. The mitochondria of the haustorial mother cell (M) and the sheath (SH) also are shown. Drawing based on a transmission electron micrograph published by Beckett *et al.* in their *Atlas of Fungal Ultrastructure*, 1974, x 9,800 (*MU*).

haustorium, pl. **haustoria** (L. *haustor*, a drawer of water + dim. suf. *-ium* > E. *-ium*): an absorbent organ that originates from the hypha of a parasitic fungus and which penetrates a host cell; the host protoplast is not perforated by the haustorium, but is only invaginated. Haustoria are formed mainly by obligately parasitic fungi, such as *Melampsora lini* (Uredinales), but also by facultative parasites and mycorrhizae. The shape of the haustorium varies depending upon the species, from bulbous to pyriform or digitate; e.g., in *Blumeria* (=*Erysiphe*) *graminis*, of the Erysiphales, they are digitate or dactyloid. Some Zygomycetes parasitic on terricolous amoebae, such as *Bdellospora helicoides* (Zoopagales), also form haustoria with which they obtain their nutrition from the host cells; the conidia of this predacious fungus adhere to the membrane of the amoeba, which they perforate and develop an haustorium within; the amoeba is then transformed into a swollen body from which emerge conidiophores and zygophores.

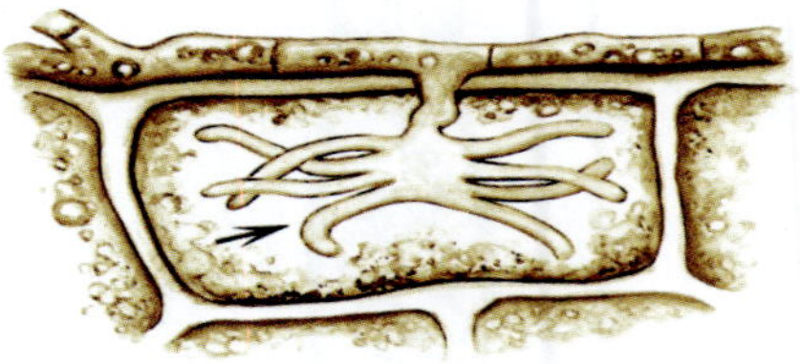

Digitate **haustorium** of *Blumeria* (=*Erysiphe*) *graminis* inside a host cell, x 1,250 (*MU*).

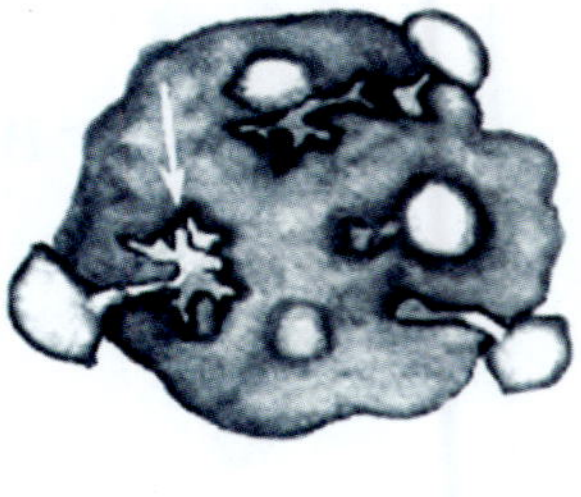

Haustoria of *Bdellospora helicoides* inside an amoeba host cell, x 700 (*MU*).

hebe- (Gr. *Hébe*, Goddess of youth > *hébe*, adolescence, and, by extension, signs of puberty, such as pubescence): a pref. employed to give the idea of something downy, as in the generic name *Hebeloma* (Agaricales), whose basidiocarp has a downy border (Gr. *lõma*, genit. *lómatos*, fringe, edge).

helicospore (Gr. *hélix*, genit. *hélikos*, spiral + *sporá*, spore): a cylindrical spore coiled into a spiral or helix; helicospores are commonly multicellular, like those of *Helicosporium linderi*, *Ceratophorum helicosporum* (moniliaceous asexual fungi), *Helicoma* sp. and *Zalerion varia* (dematiaceous asexual fungi).

helio-, **heli-** (Gr. *hélios*, sun): a pref. employed in terms that allude to the sun, as a source of light or as a heavenly body, e.g., in *heliophilous*, an organism that prefers sunshine.

Helicospore of *Helicoma* sp., x 1,300 (*CC*).

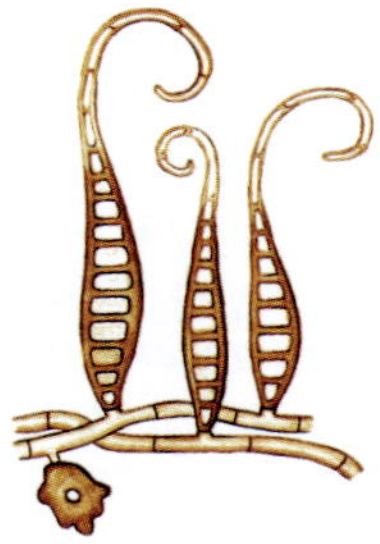

Helicospores of *Ceratophorum helicosporum*, x 260 (*MU*).

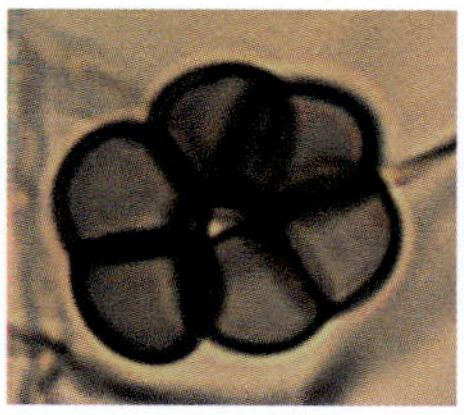

Helicospore of *Zalerion varia*, x 1,300 (*MCG*).

heliophilous (Gr. *hélios*, sun + *phílos*, have an affinity for): preferring sunny habitats, or requiring them to grow, such as the species of *Panaeolus* (Agaricales). Cf. **heliophobic** and **anheliophilous**.

heliophobic, **heliophobous** (Gr. *hélios*, sun + *phóbos*, fearing or avoiding < *phobéo*, to fear): avoiding light and requiring shade, like the species of *Hypholoma marginatum* (Agaricales). Also called **anheliophilous**. Cf. **heliophilous**.

Basidiocarps of *Hypholoma marginatum*, an **heliophobic** fungus, x 1 (*TRO*).

heliotropic (Gr. *hélios*, sun + *trópos*, gyration, change in position or manner < *tropé*, a turn < *trépo*, to gyrate, be directed toward): an organism that exhibits *heliotropism*.

heliotropism (Gr. *hélios*, sun + *trópos*, gyration, change in position or manner < *tropé*, turn < *trépo*, to gyrate, be directed toward + suf. *-ismós* > L. *-ismus* > E. *-ism*, state, phase, tendency, action): a tropism exhibited by heliotropic organs and organisms, which change position or move on being stimulated by sunlight. Positive heliotropism occurs, e.g., in the curvature of the stipes of the majority of the coprinaceous fruiting bodies (Agaricales), the sporangiophores of *Mucor*, *Phycomyces* and *Pilobolus* (Mucorales), the asci of the Ascobolaceae (Pezizales) and the perithecial necks of the Sordariaceae (Sordariales). See **tropism**.

helmintho-, **helminth-** (*hélmis*, genit. *hélminthos*, worm): a pref. employed in names such as *Helminthosporium* (dematiaceous asexual fungi), that alludes to its worm-shaped spores.

helo- (Gr. *hélos*, marsh): a pref. employed in terms that allude to marshes or flooded soils, as in *heloplankton*, plankton composed mainly of Chlorophyceae and cyanobacteria, that live in these environments.

helotism (Gr. *Helos*, a city in Peloponesia < *heilótis*, *hilota* or *ilota*, a Greek slave + suf. *-ismós* > L. *-ismus* > E. *-ism*, state, phase, tendency, action): a type of association between living beings in which one of the members obtains the major benefit; the association that exists between a fungus and an alga to form a lichen has at times been considered as helotism, and other times as mutualistic symbiosis. Cf. **antagonism**, **commensalism**, **metabiosis**, **mutualism**, **parabiosis**, **parasymbiosis**, **symbiosis** and **synergy**.

hemat-, **hemato-**, **haemato-**, **haemat-** (Gr. *haíma*, genit. *haímatos*, blood): a pref. that alludes to blood or to the color of blood, as in the generic name *Haematomma* (Lecanorales), a lichen with blood-red apothecia.

hemi- (Gr. *hémi-*, half, in the middle): a pref. that alludes to something that is in the middle, or in certain cases expresses partial or imperfectly, as in *Hemiascomycetes*, a class of fungi with free asci, i.e., not contained in any type of fruiting body or ascocarp. Corresponds to L. **semi-**.

hemiamyloid (Gr. *hémi-*, half, in the middle + *ámylon*, starch + L. suf. *-oide*, like < Gr. *-oeídes*, resembling, similar to): equal to **dextrinoid** and **pseudoamyloid**.

hemiangiocarpic (Gr. *hémi-*, half, in the middle + *angeîon*, vessel, receptacle + *karpós*, fruit + suf. *-íkos* > L. *-icus* > E. *-ic*, belonging to or relating to): a type of ontogeny in ascocarps (Pezizales) and basidiocarps (Agaricales) whose development is initially angiocarpic, but later becomes gymnocarpic. Cf. **angiocarpic** and **gymnocarpic**.

Hemiascomycetes (Gr. *hémi-*, half, in the middle + *askós*, wine-skin, sac, ascus + L. *-mycetes*, ending of taxonomic class < Gr. *mýkes*, genit. *mýketos*, fungus): in some classifications, a class of fungi (Hemiascomycetes) with asci that are borne free (i.e., not contained in any type of fruiting body or ascocarp) and without ascogenous hyphae; the thallus is simple and when mycelium is present it is poorly developed. In the system of classification adopted in this dictionary, the Hemiascomycetes correspond in part to the class Archiascomycetes (which includes the orders Taphrinales and Schizosaccharomycetales); together with the classes Plectomycetes, Discomycetes and Loculoascomycetes, plus some other filamentous ascomycetes (such as the Eurotiales, Laboulbeniales and Spathulosporales) they constitute the phylum Ascomycota of the kingdom Fungi. In the Appendix, the user of this dictionary can see the comparison between the outline of Alexopoulos *et al.*, 1996, modified by Ulloa and Hanlin, 2006, to that of Kirk *et al.*, 2008. In the latter classification scheme, the orders of the class Hemiascomycetes have been moved to the subphyla Taphrinomycotina and Saccharomycotina, phylum Ascomycota. These taxonomic rearrangements are part of a constant process of study and evaluation, in order to achieve a phylogenetic classification of the fungi.

hemibiotroph, **hemibiotrophic** (Gr. *hémi-*, in the middle, half + *bíos*, life + *trophós*, what nurtures, serves as food; or + suf. *-íkos* > L. *-icus* > E. *-ic*, belonging to or relating to): applied to the entomogenous fungi, such as the Clavicipitaceae (Hypocreales), with a parasitic phase composed of yeast-like cells that inhabit the hemocoel of the insect host, and a well delimited saprobic phase corresponding to the mycelial colonization of the entire insect body after it has died due to the infection. The parasitized insect is transformed into a hard, sclerotiform structure in which the fungus can survive for long periods and produce synnemata and perithecial stromata when conditions are favorable and while food reserves last. This situation is observed, e.g., in the species of *Cordyceps* that infect larvae of coleopteran Scarabaeidae.

Cordyceps sobolifera, an **hemibiotrophic** fungus that is developing perithecial stromata from a hypogeal cicada larva, x 1 (*RV*).

hemicyclic (Gr. *hémi-*, half, in the middle + *kyklikós*, circular, cyclic < *kiklós*, circle + suf. *-íkos* > L. *-icus* > E. *-ic*, belonging to or relating to): *Uredinales*. Species of rust fungi in whose life cycle one of the dikaryotic spore stages is lacking; e.g., *Gymnosporangium juniperi-virginianae*, the cedar-apple rust, lacks urediniospores. These rusts also are called **demicyclic**. Cf. **macrocyclic** and **microcyclic**.

hemiparasite (Gr. *hémi-*, half, in the middle + *parásitos* < *pará*, together, at the side of + *sítos*, bread, food): equal to **facultative parasite**.

hemispore (Gr. *hémi-*, half + *sporá*, spore): in the dermatophytes it refers to a cell that is formed on the apex of a hypha and then divides to become a deuteroconidium. Also called **protoconidium**. In some ascomycetes it is applied to one of the two cells produced by formation of a transverse septum in an ascospore.

hepaticolous (L. *hepaticae*, hepatic plant < *hepaticus*, hepatic, for the resemblance of some to the liver + suf. *-cola*, inhabitant): an organism (it can be a fungus) that grows on liverworts, a group of bryophytic plants. See **bryophilous**.

hepta-, **hept-** (Gr. *hépta*, seven): a pref. that forms part of biological terms derived from Gr. that indicate that something is repeated seven times, as in *heptandrous flower*, a flower with seven stamens. Some generic names of fungi also allude to the number seven, as in *Heptameria* (Dothideales) and *Heptaster* (dematiaceous asexual fungi).

herbarium, pl. **herbaria** (L. *herbarium*, a collection of dried plants < L. *herba*, herb, grass, plant + suf. *-arium* > E. *-arium*, which denotes the place of something): an ordered collection of dried plants, and the place where they are stored, destined for study and instruction in botany. The dried plants are maintained under proper conditions that best preserve the shape and position of their organs as in the living state. Frequently the term also refers to a reference collection of dried fungi, especially when these are curated together with specimens of plants.

herbicolous (L. *herba*, herb, grass, plant + suf. *-cola*, inhabitant): living on herbs.

hermaphrodite, **hermaphroditic** (Gr. *hermaphróditos*, son of Hermes and Aphrodite, with the attributes of both sexes): a species or thallus having both male and female sexual organs, which may or may not be compatible; also called **monoecious**. E.g., the gametothallus of *Allo-myces javanicus* (Blastocladiales) is hermaphroditic or monoecious, since it has female gametangia and on these, male gametangia; *Achlya americana* and *Saprolegnia parasitica*

(Saprolegniales), among others, also have antheridia and oogonia on the same thallus. In other groups of fungi there also are hermaphroditic thalli, as in many Laboulbeniales, e.g., *Ceratomyces contortus*, a species parasitic on aquatic beetles. Cf. **dioecious**.

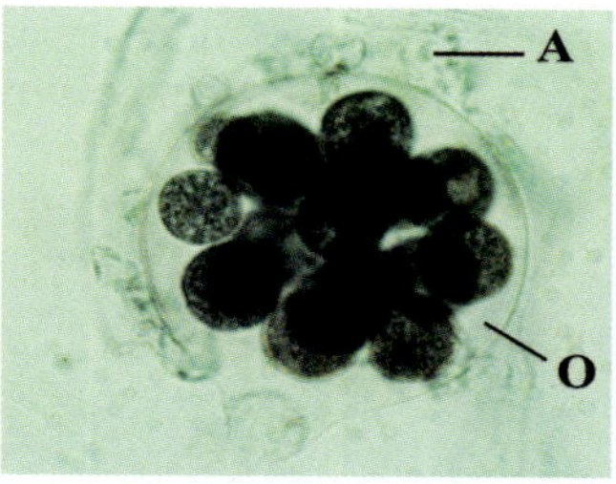

Hermaphroditic thallus of *Saprolegnia parasitica*, with oogonium (O) and antheridium (A), x 660 (*MU*).

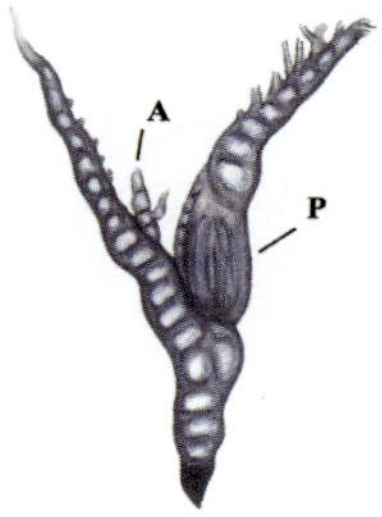

Hermaphroditic thallus of *Ceratomyces contortus*, with a perithecium (P) and various appendages (A), x 200 (*MU*).

herpo-, **herp-** (Gr. *hérpo*, to crawl or go forward lightly on the floor, to meander): a pref. employed in names such as *Herpomyces* (Laboulbeniales), a fungal ectoparasite that lives attached to the exoskeleton of cockroaches.

hetero-, **heter-** (Gr. *héteros*, another, distinct): a pref. often employed to form terms that refer to heterogeneity, which implies something different from another thing, as in the following terms.

Heterobasidiomycetes (Gr. *héteros*, another, distinct + *basídion*, dim. of *básis*, base + L. *-mycetes*, ending of taxonomic class < Gr. *mýkes*, genit. *mýketos*, fungus): in some classifications, a class of fungi that includes those basidiomycetes with heterobasidia and basidiospores that typically germinate by forming secondary spores instead of mycelium. In the classification adopted in this dictionary, the heterobasidiomycetes are included in the classes Ustilaginomycetes (with the order Ustilaginales) and Urediniomycetes (with the order Uredinales) of the phylum Basidiomycota. Other orders that include forms with heterobasidia, but which are not considered within a recognized taxonomic class, are the Tremellales, Auriculariales, Dacrymycetales, Ceratobasidiales, Tulasnellales, Sporidiales, Septobasidiales and Exobasidiales. In the Appendix, the user of this dictionary can see the comparison between the outline of Alexopoulos *et al.*, 1996, modified by Ulloa and Hanlin, 2006, to that of Kirk *et al.*, 2008. In the latter classification scheme, the orders of the class Heterobasidiomycetes have been moved to the subphyla Agaricomycotina, Pucciniomycotina and Ustilaginomycotina, phylum Basidiomycota. These taxonomic rearrangements, with the concurrent change of the names of taxa, are part of a constant process of study and evaluation, in order to achieve a phylogenetic classification of the fungi.

heterobasidium, pl. **heterobasidia** (Gr. *héteros*, different + *basídion* < *básis*, base + dim. suf. *-ídion* > L. *-idium*): a type of basidium with two different parts, a lower portion separated by a septum, or *hypobasidium*, and an upper portion, or *epibasidium*, on which the spores are borne. Because this basidium is divided or separated by septa, it is called a *phragmobasidium*. This type of basidium is present, e.g., in *Puccinia* (Uredinales), *Ceratobasidium* (Ceratobasidiales) and *Ustilago* (Ustilaginales). Some authors also consider the bifurcate basidium of the Dacrymycetales as a type of heterobasidium. Cf. **holobasidium** and **homobasidium**.

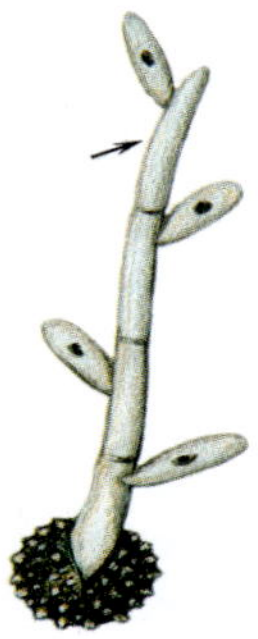

Heterobasidium of a germinating teliospore of *Ustilago maydis*, x 1,360 (*MU*).

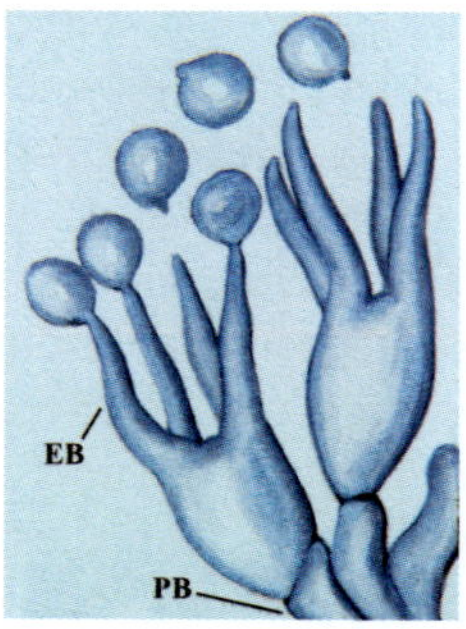

Heterobasidia of *Scotomyces subviolaceum* (=*Ceratobasidium plumbeum*); each epibasidium (EB) is derived from a probasidium (PB) and produces basidiospores on the apex, x 1,540 (*MU*).

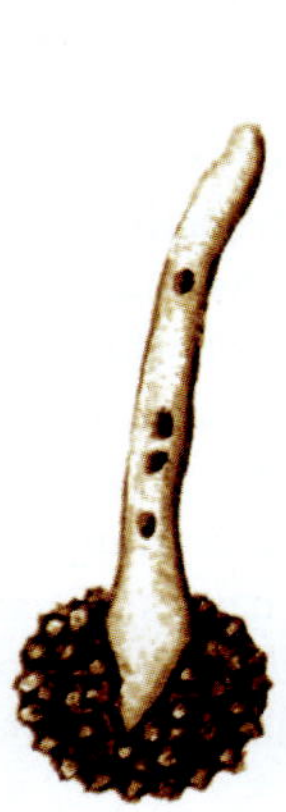

Heterobasidia (germinating teliospores) of *Ustilago maydis*, x 1,250 (*MU*).

heterocyst (Gr. *héteros*, another, different + *kýstis*, bladder, vesicle; here, cell): *Lichens*. A special cell that originates from a vegetative cell in some families of cyanobacteria, by a thickening of the wall and the disappearance of the assimilative pigment; heterocysts, like hormogonia, are involved in the multiplication of the cyanobacteria that constitute the photobiont of some lichens, but in addition it is believed that they function in the fixation of atmospheric nitrogen by the lichen thallus.

heterocytic (Gr. *héteros*, another, different + *kýtos*, concavity; here, cell + suf. *-íkos* > L. *-icus* > E. *-ic*, belonging to or relating to): refers to nuclear behavior in basidiomycetes, when the haplont is regularly coenocytic. See **astatocoenocytic** and **holocoenocytic**.

heteroecious (Gr. *héteros*, another, different + *oîkos*, house or inn): a parasitic fungus that develops one part of its life cycle on one host and the other part on a different host species; e.g., in the wheat rust, *Puccinia graminis* (Uredinales), the urediniospores and teliospores develop on cereals, and the pycniospores (or spermatia) and aeciospores develop on barberry; the phenomenon of requiring two host species to complete the life cycle of the parasite is called *heteroecism*. **Heteroxenous** is a syn. of heteroecious. Cf. **autoecious**.

heteroecism (Gr. *héteros*, another, different + *oîkos*, house or inn): a phenomenon relating to the parasitic fungi called *heteroecious*, which require two different species of host plants in order to complete their life cycle. Many rust fungi (Uredinales) are heteroecious. Cf. **autoecious**.

heterogametangium, pl. **heterogametangia** (Gr. *héteros*, another, different + *gamétes*, spouse + *angeîon*, vessel, receptacle): applied to male and female gametangia that differ morphologically, whether in size or shape; e.g., the gametangia of *Zygorhynchus moelleri* (=*Z. vuilleminii*), Mucorales, differ in size, whereas the gametangia that are differentiated into antheridia and oogonia, like those of the Oomycota, and the antheridia and ascogonia of the ascocarpic ascomycetes, often are distinct in both size and shape. Cf. **isogametangium**.

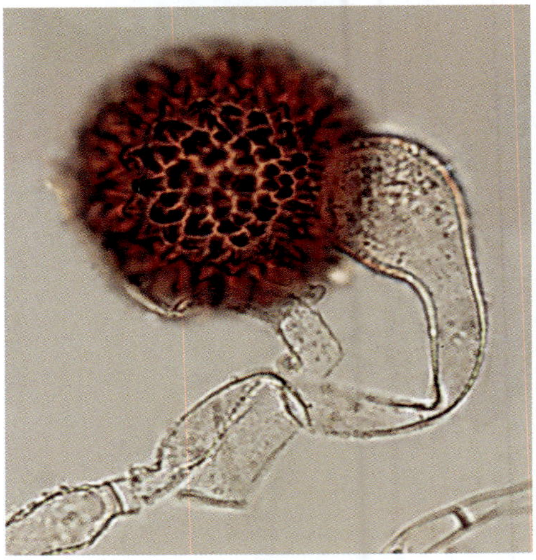

Heterogametangia of *Zygorhynchus moelleri* (=*Z. vuilleminii*), x 360 (*MU*).

heterogamete (Gr. *héteros*, another, different + *gamétes*, spouse): refers to male and female gametes that are different morphologically, as occurs in the aquatic fungi of the order Monoblepharidales (such as *Monoblepharis polymorpha*), which have motile male gametes (antherozoids produced in antheridia) and immotile female gametes (oospheres contained in oogonia). Cf. **isogamete**. This type of sexual reproduction is called *heterogamy*.

heterogamy (Gr. *héteros*, another, different + *gámos* > E. des. *-gamy*, sexual union): the union of two dissimilar gametes. Cf. **anisogamy** and **isogamy**. See **heterogamete**.

heterokaryosis, pl. **heterokaryoses**, **heterocaryosis**, pl. **heterocaryoses** (Gr. *héteros*, another, different + *káryon*, nucleus + suf. *-osis*, state or condition): the condition in which genetically different nuclei are associated in the same protoplast, or in the same mycelium. Such organisms are said to be *heterokaryotic*. Cf. **homokaryosis**.

heterokaryotic, **heterokaryontic**, **heterocaryotic**, **heterocaryontic** (Gr. *héteros*, another, different + *káryon*, nucleus + suf. *-tikós* > L. *-ticus* > E. *-tic*, relation, fitness, inclination or capacity): an individual that exhibits *heterokaryosis*. Cf. **homokaryotic**.

heterokont, **heterocont**, **heterokontic**, **heterocontic** (Gr. *héteros*, another, different + *kontós*, oar; here, flagellum): with flagella of different length or of a distinct type. In many algae and some fungi zoospores are formed with one flagellum longer than the other; e.g., the Plasmodiophoromycetes (*Plasmodiophora*, *Spongospora*, *Woronina*, etc.) have zoospores with two unequal flagella, both of the whiplash type (amastigonemate), located on the anterior end of the zoospores. Also considered heterokont are the zoospores of the Oomycota (such as *Achlya*, *Saprolegnia*, *Plasmopara*, etc.), which have one whiplash flagellum and one mastigonemate flagellum. Equal to **anisokont**. Cf. **isokont**.

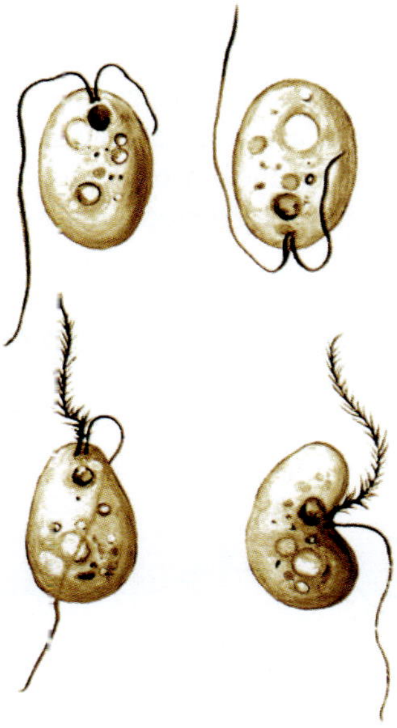

Heterokont zoospores of *Plasmodiophora brassicae* (top) and of *Saprolegnia parasitica* (bottom); for the latter species a primary zoospore is shown on the left, and a secondary zoospore, on the right, x 4,600 (*MU*).

heteromerous (Gr. *héteros*, another, different + *méros*, part): **1**. *Agaricales* (*Russulaceae*). A type of context or trama in the gills of the basidiocarp, which is composed of clusters of spherical cells (*sphaerocysts*) intermixed with hyphae (as in the gen. *Russula*), or of tissue composed of normal hyphae and *laticiferous* hyphae (latex conductors), as in the gen. *Lactarius*. **2**. *Lichens*. A thallus in which the *mycobiont* and the *photobiont* (*phycobiont*) are arranged in definite layers or strata, whether with horizontal stratification or with radial stratification. The majority of crustaceous, squamulose and foliose lichens are heteromerous, with the layers of hyphae and algal cells well delimited, as in *Hypotrachyna* (Lecanorales). Cf. **homoiomerous**.

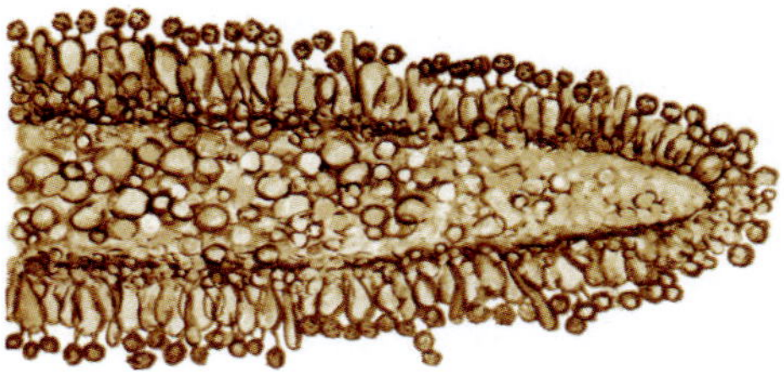

Heteromerous hymeniiferous trama of a gill of *Russula* sp., seen in longitudinal section, x 180 (*MU*).

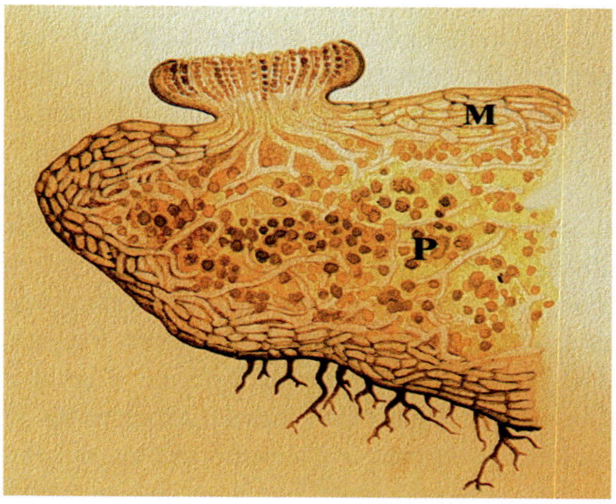

Heteromerous thallus of the foliose lichen *Hypotrachyna* (=*Parmelia*) *imbricatula*, seen in longitudinal section. Mycobiont (M) and photobiont (P) are shown, x 12.5 (*MU*).

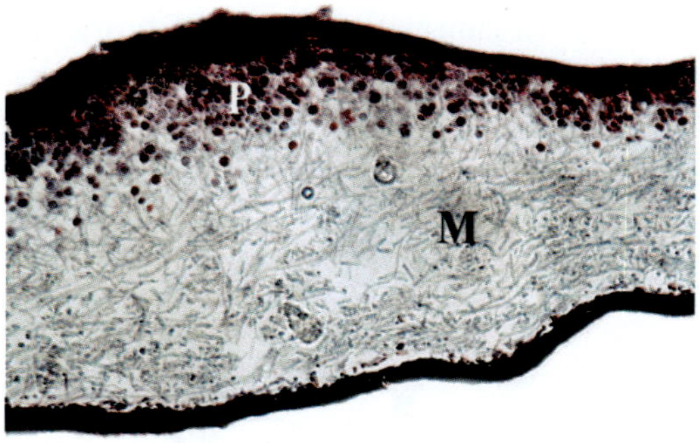

Heteromerous thallus of a foliose lichen, seen in longitudinal section. Mycobiont (M) and photobiont (P) are shown,x 12.5 (*MU*).

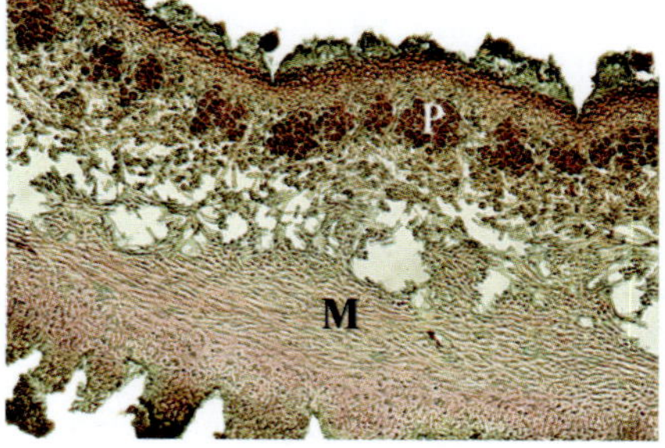

Heteromerous thallus of a foliose lichen, seen in longitudinal section. Mycobiont (M) and photobiont (P) are shown, x 12.5 (*MU*).

heteromorphic, heteromorphous (Gr. *héteros*, another, different + *morphé*, shape + suf. *-íkos* > L. *-icus* > E. *-ic*, belonging to or relating to): something that varies from the structure considered normal (of different size, with different elements, etc.). In general, it is applied to organs or structures different in shape or size, such as thalli, gametangia, etc. In agaricology it refers to the margins of the gills or the borders of the tubes that are sterile or predominantly sterile, due to the presence of a type of cystidium, or of a type of marginal hair that does not occur on the sides of the gills or in the interior of the tubes. Cf. **homomorphic**.

heterosporic, heterosporous (Gr. *héteros*, another, distinct + *sporá*, spore + suf. *-íkos* > L. *-icus* > E. *-ic*, belonging to or relating to): producing asexual spores of more than one type (*heterospory*); e.g., *Phialophora pedrosoi* (=*Rhinocladiella*=*Fonsecaea*), of the dematiaceous asexual fungi, a human pathogen, produces three types of conidia (phialospores, catenulate blastospores and blastospores on a rachis), which are characteristic of the gen. *Phialophora*, *Cladosporium* and *Rhinocladiella*, respectively. These conidia are formed in the same colony, or even on the same conidiophore. The term also is applied to heterothallic fungi (e.g., Mucorales) that have spores of different mating types (+ and -), as well as to some higher fungi, such as *Inocybe* (Agaricales), that have a polymorphism of the basidiospores associated with extreme environmental conditions. Cf. **homosporic** and **isosporic**.

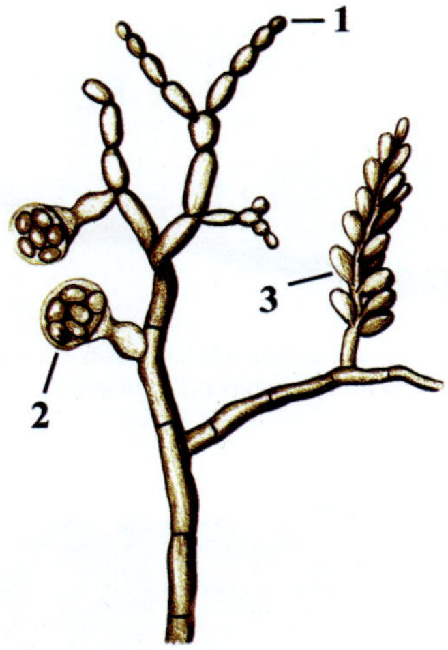

Heterosporic conidiophore of *Phialophora* (=*Rhinocladiella*=*Fonsecaea*) *pedrosoi*, with three types of conidia: 1) *Cladosporium*, 2) *Phialophora* and 3) *Rhinocladiella*, x 600 (*MU*).

heterothallic (Gr. *héteros*, another, distinct + *thallós*, thallus + suf. *-íkos* > L. *-icus* > E. *-ic*, belonging to or relating to): this term has been defined in two different ways: **1**. An organism with the male and female sexual organs segregated on different thalli, so that two thalli are required for sexual reproduction to occur , i.e., it is *dioecious*. **2**. An organism in which the male and female sexual organs are borne on the same thallus (i.e., *hermaphrodite* or *monoecious*), but they are self-sterile (self-incompatible), because of which it is necessary to have two compatible thalli (cross fertilization) to achieve sexual reproduction. Cf. **homothallic**.

heterothallism (Gr. *héteros*, another, distinct + *thallós*, thallus + suf. *-ismós* > L. *-ismus* > E. *-ism*, state, phase, tendency, action): the condition of being *heterothallic*. This phenomenon can be observed clearly in the laboratory when two sexually compatible strains, e.g., + and - strains of *Mucor hiemalis* (a heterothallic species, of the Mucorales) are cultivated together on agar, since in the zone on the agar plate where the two mycelia meet a dark-colored band is formed, which represents the production of zygospores, resulting from successful sexual reproduction between the two heterothallic strains. Cf. **homothallism**.

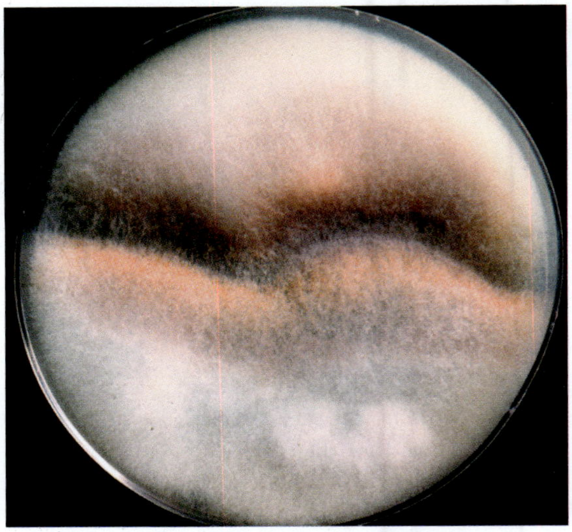

Heterothallism, as demonstrated in an agar culture of two compatible strains (+ and -) of the heterothallic species *Mucor hiemalis*. Note the dark band in the center where the two strains meet, which is where the zygospores are formed, x 1 (*MU*).

heterotroph, heterotrophic (Gr. *héteros*, another, different + *trophós*, what nurtures, serves as food; or + suf. *-íkos* > L. *-icus* > E. *-ic*, belonging to or relating to): an organism that must feed on organic material elaborated by other organisms, because it is incapable of synthesizing carbohydrates from inorganic elements. All fungi are heterotrophs, because of which they live as saprobes or as parasites. Cf. **autotroph** and **mixotroph.**

heterotropic (Gr. *héteros*, another, distinct + *trópos*, rotation, gyration, turn + suf. *-íkos* > L. *-icus* > E. *-ic*, belonging to or relating to): *Basidiomycetes*. A type of basidiospore attachment on the basidium. Heterotropic basidiospores, like those of the Agaricales (e.g., *Amanita*), from which spore prints can be obtained, are arranged obliquely on the sterigmata of the basidium, so that the longitudinal axis of each one differs from that of the sterigma, this facilitating forcible discharge. Cf. **orthotropic**.

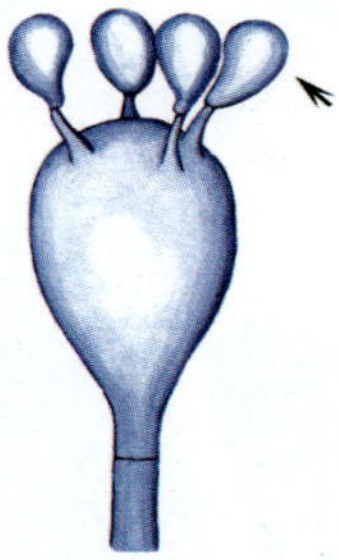

Heterotropic basidiospores on a basidium of *Amanita* sp., x 1,100 (*MU*).

heterotypic (Gr. *héteros*, another, different + *týpos*, type, mark, model + suf. *-íkos* > L. *-icus* > E. *-ic*, belonging to or relating to): a heterogeneous type. In biological nomenclature, it is a synonym whose name is based on a type different from the one that served as the basis for the original description of the taxon; also called *taxonomic synonym* or *"facultative" synonym*. Cf. **homotypic**.

heteroxenous (Gr. *héteros*, another, different + *xénos*, host): with another host. Syn. of **heteroecious**.

heterozygotic, **heterozygous** (Gr. *héteros*, another, different + *zygotós*, united by marriage, yoked; the cell resulting from the copulation of two gametes + suf. *-íkos* > L. *-icus* > E. *-ic*, belonging to or relating to): pertaining to or relating to a *heterozygote*, i.e., a zygote resulting from the union of genetically distinct gametes. A heterozygotic organism contains two different factors or alleles of a pair of allelomorphs or homologous chromosomes; it has heterokaryosis as the result of the fusion of gametes. *Heterozygotism* (with the same roots + Gr. suf. *-ismós*, state or condition) is the state of being heterozygotic. Cf. **homozygotic**.

hexa-, **hex-** (Gr. *héxa*, six): a pref. used to form compounds that denote something repeated six times.

hexagonal (Gr. *héxa*, six + *gónia*, angle + L. suf. *-alis* > E. *-al*, relating or belonging to): with six angles, as in the name *Hexagonia* (Aphyllophorales), for the hexagonal shape of the pores of the hymenophore.

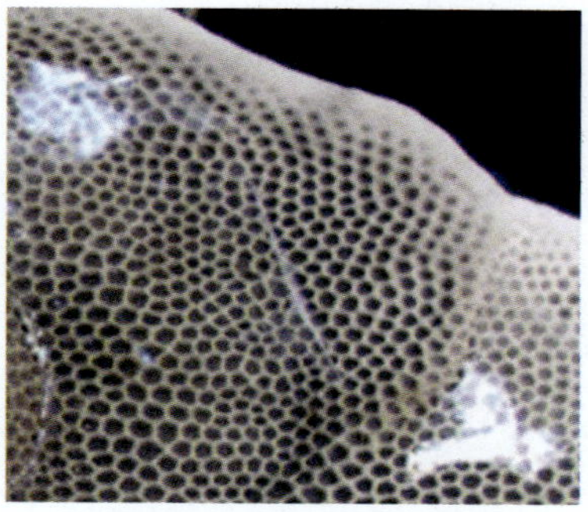

Hexagonal shape of the pores of a hymenophore of *Hexagonia tenuis*, x 0.3 (*RV*).

hiascent (L. *hiascens*, opening + suf. *-escens*, genit. *-escentis* > E. *-escent*, that which turns, that begins to, slightly): opening widely.

higher fungi, sing. **fungus** (ME. *heigh*, var. of *hegh* < OE. *heah, heh*, elevated + E. suf. *-er*, which denotes degree; L. *fungus*, fungus < Gr. *spóngos*, *sphóngos*, sponge): the general name for the fungi classified in the phyla Ascomycota and Basidiomycota. Cf. **lower fungi**.

hilar appendix, pl. **appendixes** or **appendices** (L. *hilum*, a mask or navel; the scar on a bean seed; here, the scar on the wall of a spore that indicates its point of attachment to the sporophore, sterigma, etc. + suf. *-aris* > E. *-ar*, like, pertaining to; L. *appendix*, appendix, a part projecting from another structure, generally accessory): *Basidiomycetes*. A minute diverticulum or appendage that protrudes from the basidiospore, adjacent to the hilum of the spore. This term is preferable to **apiculus**, as the hilar appendix is basal and not apical. It can be observed clearly in the basidiospores of *Amanita abrupta* (Agaricales). The area above the hilar appendix, in which the ornamentation of the wall is reduced or absent, is known as the **suprahilar depression**, **suprahilar disc** or **suprahilar spot**; it is found especially in the basidiospores of *Lactarius*, *Russula*, *Phaeocollybia* and *Galerina* (Agaricales).

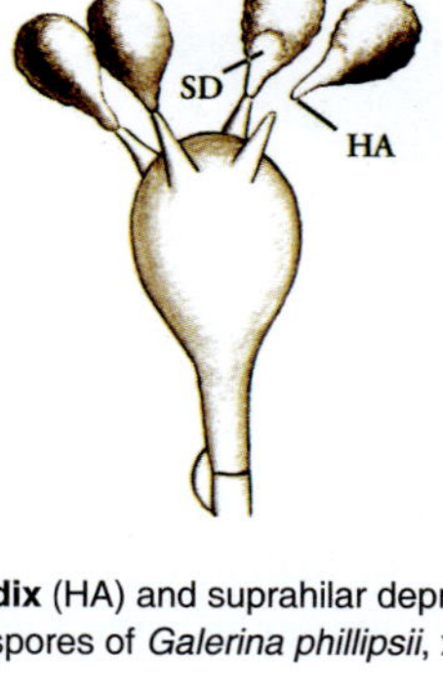

Hilar appendix (HA) and suprahilar depression (SD) of the basidiospores of *Galerina phillipsii*, x 1,530 (*MU*).

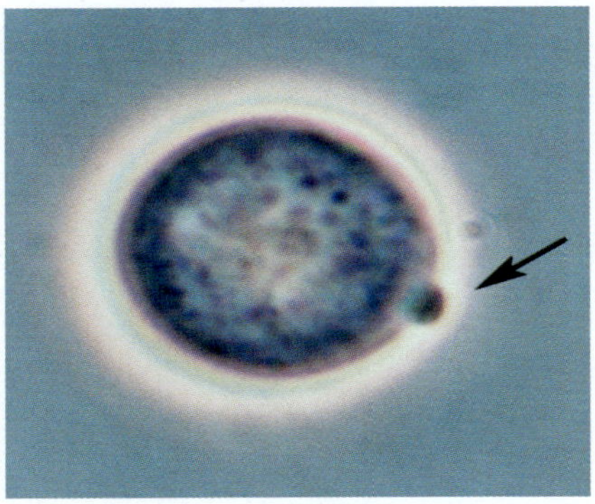

Hilar appendix of the basidiospore of *Amanita abrupta*, x 4,600 (*RMA*).

hilum, pl. **hila** (L. *hilum*, navel or scar of a bean seed; here, the scar on the spore wall that indicates its point of attachment to the sporophore, sterigma, etc.): a scar, point or mark on spores (conidia, basidiospores, etc.) that indicates the point of attachment to the sporophore, sterigma, etc. The hilum is a distinctive structure, like a dark-colored ring, e.g., on the conidia of species of *Alternaria*, *Bipolaris*, *Exserohilum* and *Helminthosporium* (dematiaceous asexual fungi).

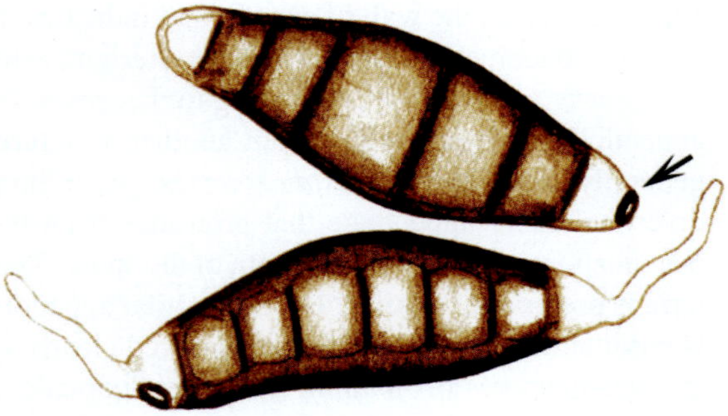

Hilum of the conidia of *Bipolaris* (=*Drechslera*) *iridis*, x 1,100 (*MU*).

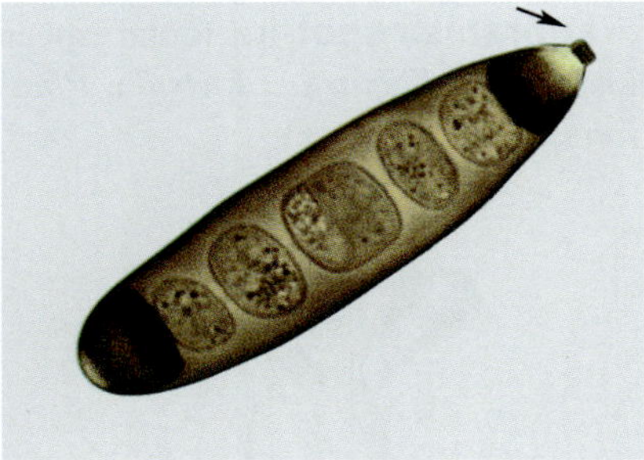

Hilum of a conidium of *Setosphaeria rostrata* (=*Exserohilum rostratum*), x 1,100 (*MCG*).

himantoid (< gen. *Himantia* < Gr. *himás*, genit. *himántos*, belt or leather strap + L. suf. *-oide*, like < Gr. *-oeídes*, similar to): with spreading, fan-like cords of mycelium, like those of *Cylindrobasidium* (=*Himantia*), of the family Hyphodermataceae (Aphyllophorales).

hirsute (L. *hirsutus*, bristly, hairy): covered with long, rigid hairs that are rough to the touch. E.g., the base of the stipe of some species of *Peziza* (Pezizales) and of *Marasmius alliaceus* (Agaricales), as well as the pileus of *Lentinus velutinus* (Aphyllophorales), are hirsute. Also called **hispid** and **strigose**.

Hirsute pileus of the basidiocarp of *Lentinus velutinus*, x 0.8 (*MU*).

Hirsute pileus (seen in close-up) of the basidiocarp of *Lentinus velutinus*, x 12 (*MU*).

Hirsute basidiocarps of *Phellinus sarcites* (Aphyllophorales), x 0.1 (*RV*).

hirtous (L. *hirtus*, rough, hairy > *hirsutus*, hairy, downy, bristly): applied to a surface that has straight and rigid hairs; employed in the terms *hirtous*, *hirsute* and *hispid* to express a gradual increase in the length and rigidity of the hairs, in the order in which they are written.

hispid (L. *hispidus*, coarsely hairy): with long, very stiff hairs, almost sharp, very rough to the touch, like the pileus of *Pluteus hispidulus* (Agaricales), *Hexagonia hydnoides* and *Lentinus strigosus* (Aphyllophorales). Syn. of **hirsute** and **strigose**.

Hispid pileus of the basidiocarp of *Hexagonia hydnoides*, x 0.3 (*RV*).

Hispid pili of the basidiocarps of *Lentinus strigosus*, x 1 (*RV*).

hispidulous (L. *hispidulus*, dim. of *hispidus*, covered with very stiff hairs that are rough to the touch, almost sharp): with small, short, rigid hairs.

histo-, hist- (Gr. *hístos*, tissue): a pref. employed in a metaphoric sense, referring to organic tissues, such as in the following terms.

histogenous (Gr. *hístos*, tissue + *génos*, origin < *gennáo*, to engender, produce; literally, what forms tissue): **1**. The production of basidiospores by conversion of the masses of tissue of *Ustilago maydis* (Ustilaginales) that form in the host plant. **2**. The formation of openings and cavities as the result of tissue development.

histolysis, pl. **histolyses** (Gr. *hístos*, tissue + *lýsis*, dissolution): the process of tissue destruction by dissociation of its elements, like that of the necrobiosis of plant tissues caused by pathogenic fungi.

histoplasmosis, pl. **histoplasmoses** (< gen. *Histoplasma* < Gr. *hístos*, tissue + *plásma*, primordial material of living beings + suf. *-osis*, state or condition): a primarily pulmonary infection that is acquired by inhalation of conidia and hyphal fragments of the dimorphic fungus *Histoplasma capsulatum* (moniliaceous asexual fungi), abundant in some soils, especially those in caves contaminated with bird and bat dung. Histoplasmosis can be benign or severe, even mortal, and can infect various organs (lymphatic ganglia, spleen, liver, kidneys, central nervous system and oral and intestinal mucous membranes). See **mycosis**.

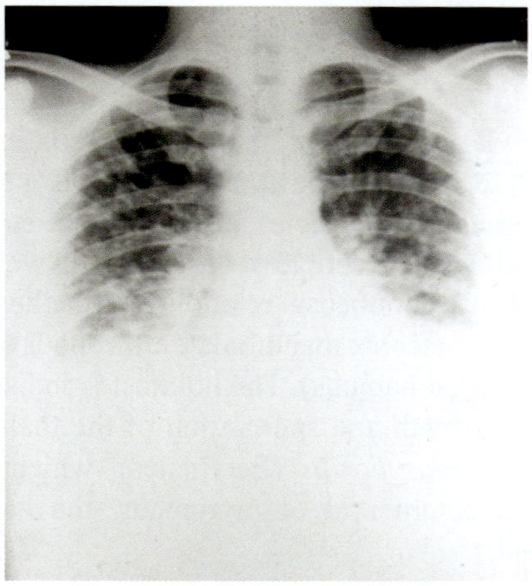

Histoplasmosis of the lungs seen in an x-ray of the thorax of a patient, x 0.1 (*RLM*).

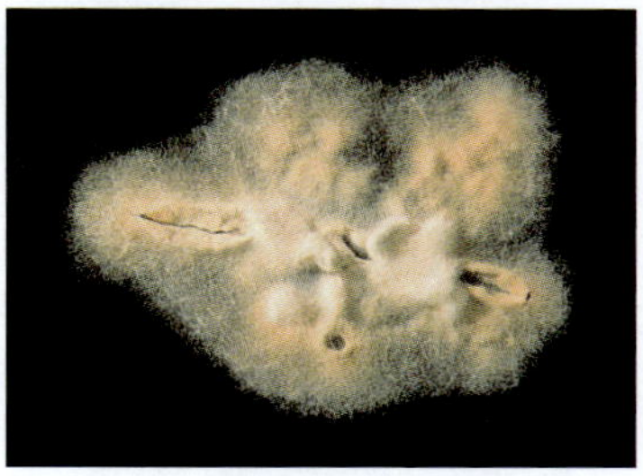

Mycelial colony of *Histoplasma capsulatum* on agar, the fungus that causes **histoplasmosis**, x 1 (*MU*).

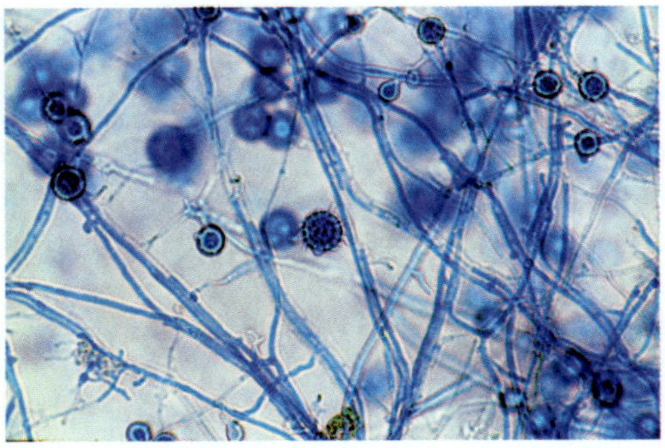

Macroconidia of the mycelial phase of *Histoplasma capsulatum* (the species that causes **histoplasmosis**), on agar, x 1,000 (*MU*).

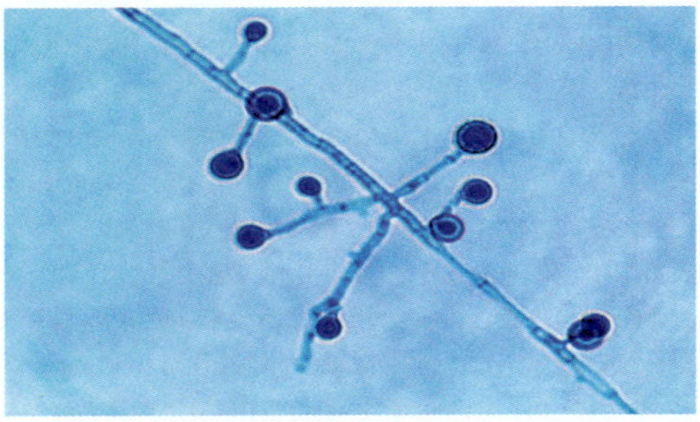

Macroconidia of the mycelial phase of *Histoplasma capsulatum* (the species that causes **histoplasmosis**), on agar, x 1,000 (*MU*).

holdfast (E. *holdfast* < ME. *holden* < OE. *halden*, to hold + ME. < OE. *faest*, fixed): *Trichomycetes*. The basal structure of the thallus, mucilaginous and adhesive, which can be discoid, branched or pectinate, depending on the species, and which serves to attach the thallus to the cuticle of the host, with which it has an obligate commensal relationship (the hosts of the Trichomycetes are mandibulate arthropods of aquatic or terrestrial habitats). The holdfast is indispensable for the growth and sporulation of the thallus. E.g., in *Eccrinidus flexilis* (Eccrinales), which lives as an endocommensal of myriapods, the holdfast is discoid.

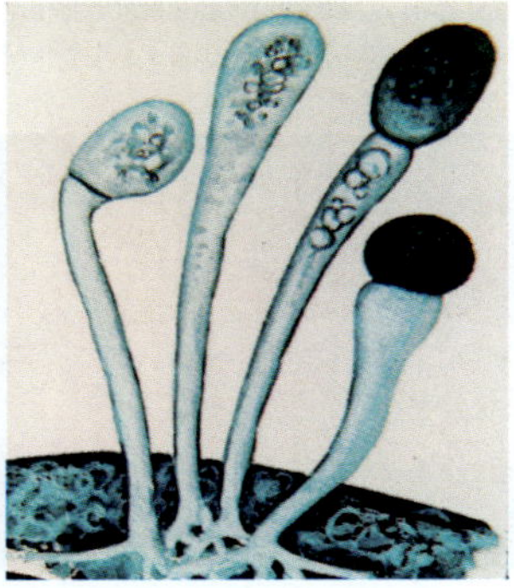

Holdfast of the thalli of *Blastocladiella simplex* (Blastocladiales), x 20 (*MU*).

Discoid **holdfast** of the thalli of *Eccrinidus flexilis*, x 20 (*MU*).

Holdfast of the thalli of *Rickia passalina* (Laboulbeniales), x 20 (*MU*).

holo-, **hol-** (Gr. *hólos*, entire, entirely, completely): a pref. employed to introduce the idea of totality, as in the following terms.

holoarthric (Gr. *hólos*, entire + *árthron*, articulation, joint + suf. *-íkos* > L. *-icus* > E. *-ic*, belonging to or relating to): *Conidiogenesis*. A type of conidium development in which all of the wall layers of the conidiogenous cell participate in the formation of the conidium wall, as occurs, e.g., with the arthrospores of *Geotrichum* (moniliaceous asexual fungi). Cf. **enteroarthric**.

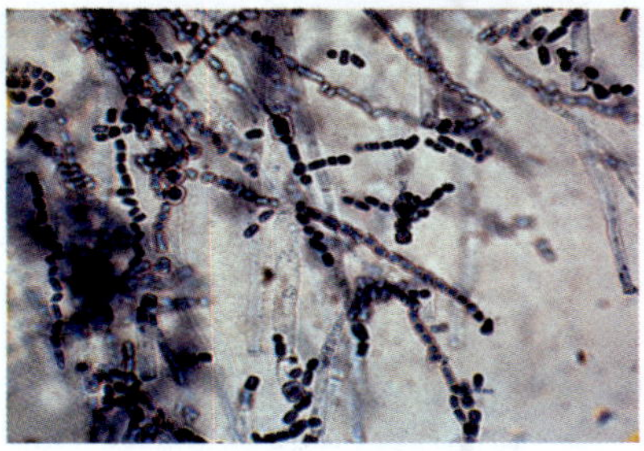

Holoarthric conidia (arthrospores) of *Geotrichum candidum*, x 300 (*MU*).

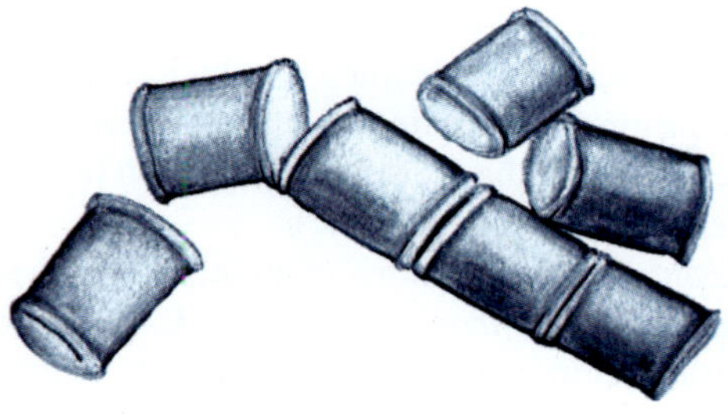

Holoarthric conidia (arthrospores) of *Geotrichum candidum*, x 1,800 (*MU*).

Holobasidiomycetes (Gr. *hólos*, entire, complete + *basídion*, dim. of *básis*, base + L. *-mycetes*, ending of taxonomic class < Gr. *mýkes*, genit. *mýketos*, fungus): also called **Homobasidiomycetes**. In some classifications, a class of fungi with holobasidia (homobasidia) and basidiospores that germinate by forming a mycelium instead of secondary spores. In the classification followed in this dictionary, these fungi are not recognized as a taxonomic class, but they correspond to the basidiomycetes of the orders Agaricales and Aphyllophorales of the class Hymenomycetes, as well as the orders of Gasteromycetes (Lycoperdales, Tulostomatales, Sclerodermatales, Phallales, Nidulariales and Hymenogastrales). Together with the classes Ustilaginomycetes and Urediniomycetes, along with the orders Tremellales, Auriculariales, Dacrymycetales, Ceratobasidiales and Tulasnellales (which in this system are included in the class Hymenomycetes), they constitute the phylum Basidiomycota of the kingdom Fungi. In the Appendix, the user of this dictionary can see the comparison between the outline of Alexopoulos *et al.*, 1996, modified by Ulloa and Hanlin, 2006, to that of Kirk *et al.*, 2008. In the latter classification scheme, the class Holobasidiomycetes has disappeared, and its orders have been moved to the subphylum Agaricomycotina, phylum Basidiomycota. These taxonomic rearrangements, with the concurrent change of the names of taxa, are part of a constant process of study and evaluation, in order to achieve a phylogenetic classification of the fungi.

holobasidium, pl. **holobasidia** (Gr. *hólos*, entire, complete + *basídion* < *básis*, base + suf. *-ídion* > L. *-idium*): also called **homobasidium**. A unicellular basidium, typically club-shaped; characteristic of the holobasidiomycetes. Cf. **heterobasidium**.

holoblastic (Gr. *hólos*, entire, complete + *blastós*, sprout, bud, germ + suf. *-íkos* > L. *-icus* > E. *-ic*, belonging to or relating to): see **blastic**.

holocarpic (Gr. *hólos*, entire, complete + *karpós*, fruit + suf. *-íkos* > L. *-icus* > E. *-ic*, belonging to or relating to): an organism whose thallus is completely converted into one or more reproductive organs, so that the somatic and reproductive phases do not coexist; this condition is called *holocarpy*. The aquatic, zoosporic fungi of the families Olpidiaceae (Spizellomycetales) and Synchytriaceae (Chytridiales), e.g., are holocarpic. Cf. **eucarpic**.

Holobasidium of the hymenium of *Lactarius fuliginosus* (Agaricales), x 1,400 (*MU*).

holocarpy (Gr. *hólos*, entire, complete + *karpós* > E. des. *-carpy*, fruit): the condition of being *holocarpic*.

holocoenocytic (Gr. *hólos*, entire, complete + *koinós*, common + *kýtos*, cell, hollow + suf. *-ikos* > L. *-icus* > E. *-ic*, belonging to or relating to): refers to nuclear behavior in the basidiomycetes, whose haplonts and diplonts are coenocytic, and only the basidia are binucleate. Cf. **astatocoenocytic**.

hologamy (Gr. *hólos*, entire, complete + *gámos* > E. des. *-gamy*, sexual union): a type of fertilization consisting of the union of two entire individuals, i.e., the entire thallus is converted into a gametangium, as is observed in the fusion of two mature individuals of *Polyphagus euglenae* (Chytridiales).

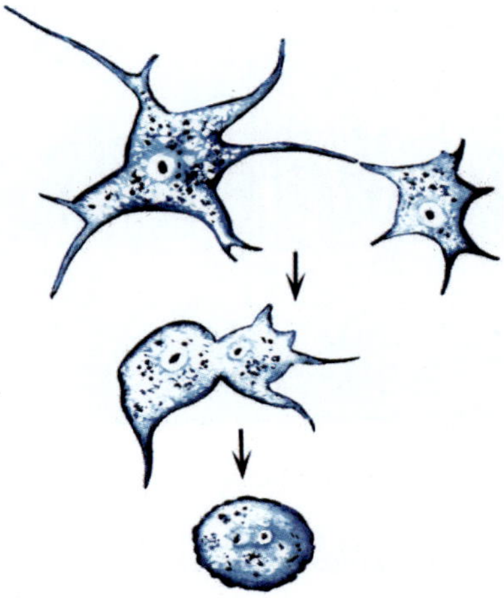

Hologamy in *Polyphagus euglenae*, x 620 (*MU*).

holomorph (Gr. *hólos*, entire, complete + *morphé*, shape, form): the fungus as a total organism, i.e., in all its forms and phases. The holomorph includes the *teleomorph* (the sexual form, characterized by the presence of an ascoma, or of a basidioma, depending on the species being considered) and the *anamorph* (the asexual phase, represented by some type of conidioma), with their corresponding binomials. E.g., *Neosartorya fischeri* (Eurotiales) is the teleomorph, and *Aspergillus fischeri* (moniliaceous asexual fungi) is the anamorph.

Holomorph, *Neosartorya fischeri*, with its two reproductive phases, the sexual (teleomorph), represented by asci and ascospores, and the asexual (anamorph), by conidiophores and conidia, x 620 (*MU*).

holosporic, holosporous (Gr. *hólos*, entire, complete + *sporá*, spore + suf. *-íkos* > L. *-icus* > E. *-ic*, belonging to or relating to; or + L. suf. *-osus* > E. *-ous*, having, possessing the qualities of): a type of conidial maturation in which the conidium attains its final size and shape before the cells that compose it are delimited and it matures as a whole. The conidia of *Microsporum* and *Trichophyton* (moniliaceous asexual fungi) are examples of holosporic conidia.

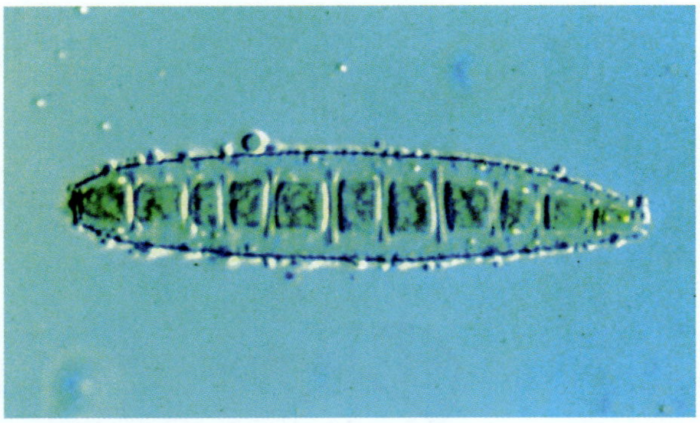

Holosporic conidium of *Microsporum cookei*, x 170 (*RTH*).

holothallic (Gr. *hólos*, entire, complete + *thallós*, shoot, thallus + suf. *-íkos* > L. *-icus* > E. *-ic*, belonging to or relating to): see **thallic**.

holotype (Gr. *hólos*, entire, complete + *týpos*, mark, model, type): the specimen on which an author bases the original description of a new taxon and which then serves in the correct interpretation of that taxon (genus, species, variety, etc).

holozoic (Gr. *hólos*, entire, complete + *zoikós* < *zõon*, animal + suf. *-íkos* > L. *-icus* > E. *-ic*, belonging to or relating to): an organism that ingests solid food particles by means of phagocytosis, as do the amoebae and other protozoans; e.g., the myxamoebae of *Cavostelium*, *Protostelium*, *Ceratiomyxella* and many other Protosteliales are holozoic. Also called **phagotroph**.

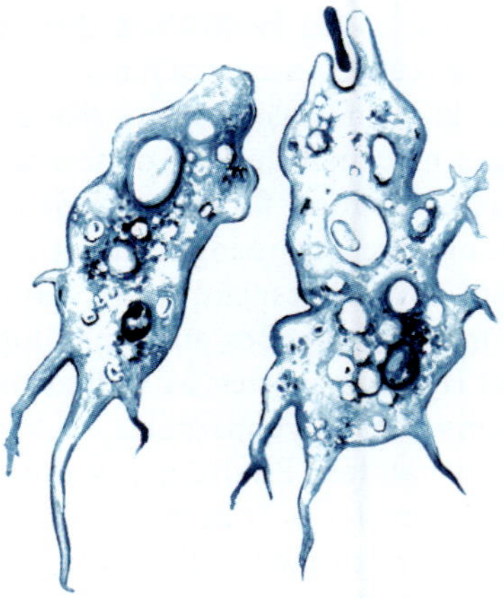

Holozoic myxamoebae of *Cavostelium bisporum*, x 1,280 (*MU*).

homalo-, homal- (Gr. *homalós*, plane, with a surface without prominences, equal, uniform): a pref. employed in names such as *Homalopeltis* (an anamorphic ascomycete), which means small flat shield.

homeostasis, pl. **homeostases** (Gr. *homós*, similar, same + *stásis*, fixed, permanent, stable): a relatively stable state of equilibrium, or a tendency to attain this state, of the chemical and physical conditions within a living organism or a group of organisms.

homo- (Gr. *homós*, analogous, similar, same): a pref. used in many biological terms, such as the following.

Homobasidiomycetes (Gr. *homós*, similar, same + *basídion*, dim. of *básis*, base + L. *-mycetes*, ending of taxonomic class < Gr. *mýkes*, genit. *mýketos*, fungus): see **Holobasidiomycetes**.

homobasidium, pl. **homobasidia** (Gr. *homós*, similar, same + *basídion* < *básis*, base + dim. suf. *-ídion* > L. *-idium*): see **holobasidium**.

homobium, pl. **homobia** (NL. *homobium* < Gr. *homós*, similar, same + *bios*, life): an autotrophic symbiotic association of an alga and a fungus, as in the lichens.

homogeneous development (ML. *homogeneus* < Gr. *homogenés*, of the same class; OF. *desveloper* > MF. *développer* + E. *-ment* < L. *-mentum*, denoting action or resulting state): *Gasteromycetes*. The differentiation of basidia from primordial tissue in a uniform manner throughout the gleba; it is typically found in species of *Bovista* (Lycoperdales). Cf. **auleate development**, **forate development**, **lacunar development** and **pileate development**.

homokaryosis, homocaryosis, pl. **homokaryoses, homocaryoses** (Gr. *homós*, similar, same + *káryon*, nut, nucleus + suf. *-osis*, condition): the condition in which genetically similar nuclei occur in the same protoplast or in the same mycelium. Cf. **heterokaryosis**.

homokaryotic, homokaryontic, homocaryotic, homocaryontic (Gr. *homós*, similar, same + *káryon*, nut, nucleus + suf. *-tikós* > L. *-ticus* > E. *-tic*, relation, fitness, inclination or capacity): an individual or cell that has genetically similar nuclei. Cf. **heterokaryotic**.

homologous (Gr. *homólogos*, congruent < *homós*, similar, same + *lógos*, reason): **1**. Refers to organs or organic parts that have *homology* (< *homós* + E. des. *-logy*, discourse < Gr. *lógos*), i.e., that have the same origin; they can be similar in shape or structure, although not necessarily in function, which is considered as evidence of an evolutionary relationship. **2**. Homologous chromosomes are those that contain homologous genes that are paired during meiosis.

homoiomerous (Gr. *homós*, similar, same + *méros*, part): **1**. *Agaricales*. A type of context or trama in the gills of a basidiocarp that is composed of only one type of hypha, as, e.g., in *Coprinus*. **2**. *Lichens*. A type of thallus in which the cells of the cyanobacterial photobiont *Nostoc* are distributed uniformly throughout the thallus, intermixed with hyphae of the mycobiont, without forming a definite layer; the homoiomerous thallus is characteristic of the cyanobacterial photobiont gelatinous lichens, such as *Leptogium* and *Collema* (Lecanorales), although not exclusively so. Cf. **heteromerous**.

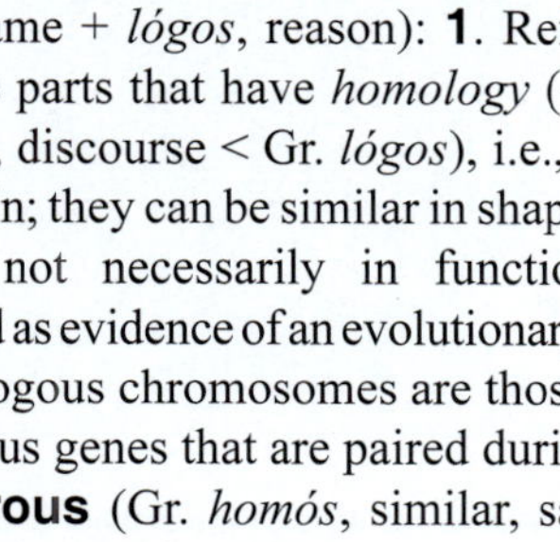

Homoiomerous thallus of *Collema tenax*, seen in longitudinal section, x 470 (*MU*).

homomorphic, homomorphous (Gr. *homós*, similar, same + *morphé*, shape + suf. *-íkos* > L. *-icus* > E. *-ic*, belonging to or relating to; of similar shape): *Agaricales*. Applied to the margins of gills or tube edges that have the same tissue composition as the hymenium. Equal to **isomorphic**. Cf. **heteromorphic**.

homonym (L. *homonymus* < Gr. *homónymos* < *homós*, similar, same + *ónoma*, *ónyma*, name): a name that (in accordance with the International Code of Botanical Nomenclature) cannot be validly used because it was previously used in a different sense; i.e., it was written exactly like another validly published name (with the same taxonomic rank), but based on a different nomenclatural type.

homosporic, homosporous (Gr. *homós*, similar, same + *sporá*, spore + suf. *-íkos* > L. *-icus* > E. *-ic*, belonging to or relating to): having only one kind of spore. Also called **isosporic** or **isosporous**; the phenomenon relating to the homosporic organisms is *homospory*. Cf. **heterosporic**.

homothallic (Gr. *homós*, similar, same + *thallós*, shoot, thallus + suf. *-íkos* > L. *-icus* > E. *-ic*, belonging to or relating to): a fungus in which sexual reproduction occurs in a single thallus, which is therefore self compatible; this phenomenon is known as *homothallism*. Cf. **heterothallic.**

homothallism (Gr. *homós*, similar, same + *thallós*, shoot, thallus + suf. *-ismós* > L. *-ismus* > E. *-ism*, state, phase, tendency, action): the condition of being *homothallic*. Cf. **heterothallism**.

homotypic (Gr. *homós*, similar, same + *týpos*, type, mark, model + suf. *-íkos* > L. *-icus* > E. *-ic*, relating to or concerning): a homogeneous type. In biological nomenclature, it refers to a synonym that is based on the same type that served as the basis of the original description of the taxon. Also called a *nomenclatural synonym* or "*obligate*" *synonym*. Cf. **heterotypic**.

homozygotic, homozygous (Gr. *homós*, similar, same + *zygotós*, united by a yoke, yoked; a cell resulting from the copulation of two gametes + suf. *-íkos* > L. *-icus* > E. *-ic*, belonging to or relating to): pertaining to or relating to the *homozygote*, i.e., a zygote resulting from the union of genetically identical gametes. A homozygotic organism. *Homozygotism* (of the same roots + Gr. suf. *-ismós*, state or condition) is the state of homozygotic organisms. Cf. **heterozygotic**.

hormesis, pl. **hormeses** (Gr. *hormáo*, to excite, stimulate + suf. *-esis*, tendency): the exitation or stimulus produced in an organism by means of a toxic substance utilized in innocuous concentrations.

hormo-, **horm-** (Gr. *hórmos*, collar): a pref. that alludes to something in the shape of a collar, as in the following terms.

hormocystangium, pl. **hormocystangia** (Gr. *hórmos*, collar + *kýstis*, bladder, vesicle, cell + *angeîon*, vessel, receptacle): see **hormocyst**.

hormocyst (Gr. *hórmos*, collar + *kýstis*, bladder, vesicle, cell): *Lichens*. A propagule or *diaspore* composed of hyphae of the fungus with a few algal cells that is produced in a special organ called the *hormocystangium*; it is found in certain species of gelatinous lichens of the family Collemataceae, e.g., *Lempholemma cladodes* and *L. vesiculiferum* (Lichinales).

hormogonium, pl. **hormogonia** (NL. *hormogonium* < Gr. *hórmos*, collar + NL. *gonium* < Gr. *gónos*, progeny + L. dim. suf. *-ium*): *Lichens*. A fragment, generally motile, of the trichome of a cyanobacterium (previously considered as Cyanophyceae), composed of a variable number of cells, which can fluctuate between two and several hundred and which separates naturally from the thallus by virtue of a process of budding, or proceeds directly from the germination of a hypnocyst. The hormogonia are destined for the propagation of the species, as they are capable of growing and forming a new filament or thallus; the cyanobacteria that constitute the bacteriobionts (photobionts) of some lichens can multiply by means of hormogonia.

host (L. *hospes*, genit. *hospitis*, a living organism that provides food and lodging to a parasite; one who receives a stranger as a guest or visitor): *Biology*; *Mycology*. A plant, animal or fungus from which a parasite obtains its nourishment, whether internally or externally. The hosts or victims of the parasitic fungi are, in order of frequency: angiosperms, gymnosperms, pteridophytes, insects and other arthropods, mollusks, other fungi, algae, bryophytes, nematodes, protozoans, man, and higher animals. An *alternate host* is a plant in which the parasite (generally a uredinial micromycete) carries out the complementary phases of its life cycle; an *intermediate host* is one that serves as a link for the infection by a fungus of two different species of plants; a *primary host* is the plant on which the parasitic fungus develops its sexual state, and which almost always represents the plant of major economic importance.

Hülle cell (G. *Hülle*, covering, envelopment; NL. *cellula*, cell, living cell, dim. of L. *cella*, small room, compartment, cell of a honeycomb): a cell with a thick wall and a narrow lumen, terminal or intercalary, large numbers of which surround the peridium of the cleistothecia of certain plectomycetous ascomycetes, such as those of *Emericella nidulans* (Eurotiales), and in the mycelium of *Aspergillus puniceus* (moniliaceous asexual fungi).

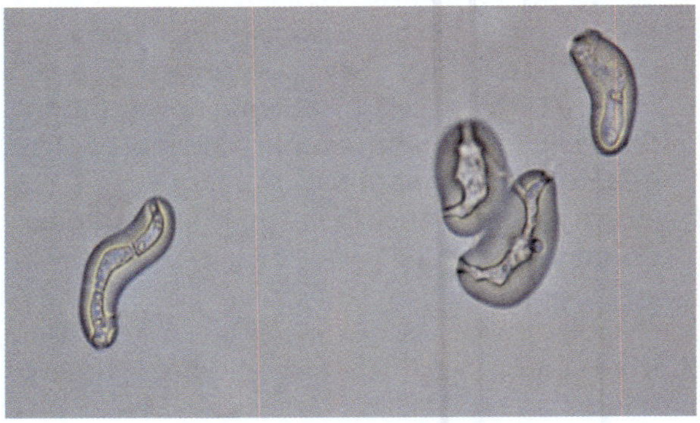

Hülle cells of *Aspergillus puniceus*, x 400 (*MU*).

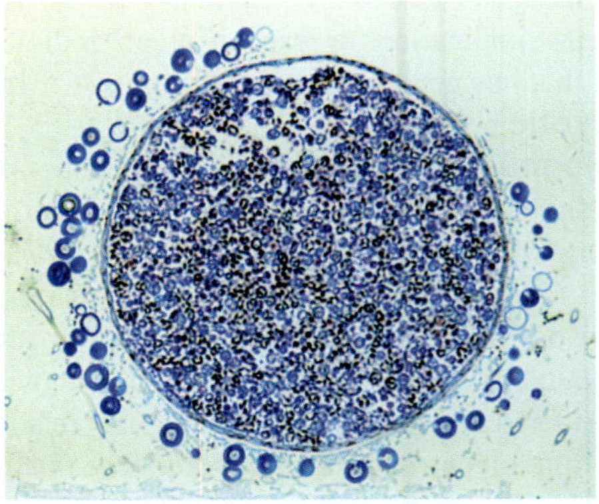

Hülle cells of *Emericella nidulans* on the exterior of the cleistothecial wall, seen in transverse section, x 250 (*RTH*).

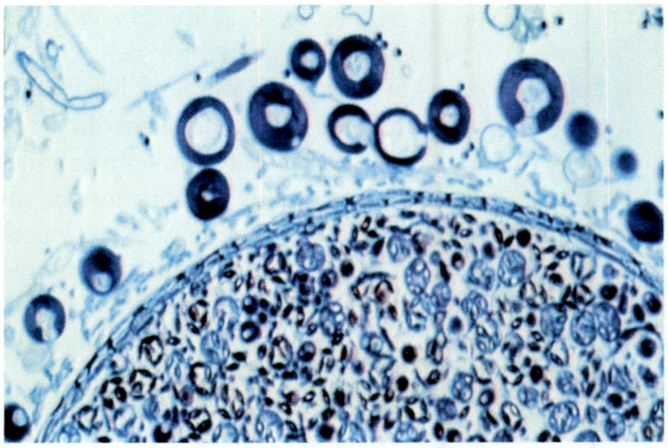

Hülle cells of *Emericella nidulans* on the exterior of the cleistothecial wall, seen in transverse section, x 530 (*RTH*).

humicole, **humicolous** (L. *humus*, earth, soil + suf. *-cola*, inhabitant < *colere*, to inhabit): living or growing in humus or in soil, as, e.g., the species *Humicola fuscoatra* (dematiaceous asexual fungi).

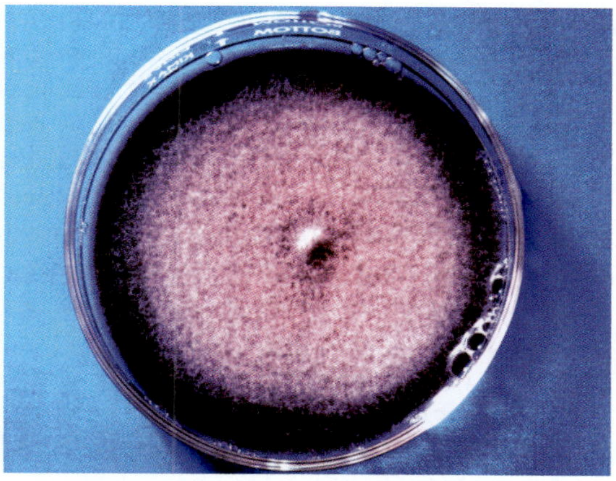

Colony of *Humicola fuscoatra* on agar, an **humicole** fungus, x 0.5 (*MU*).

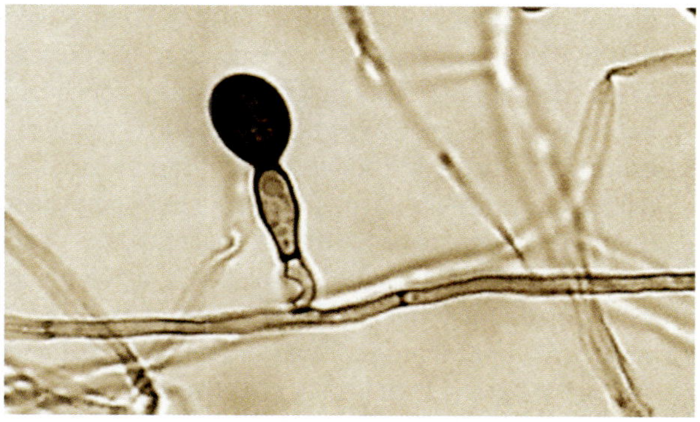

Conidium of *Humicola fuscoatra*, an **humicole** fungus, x 530 (*MU*).

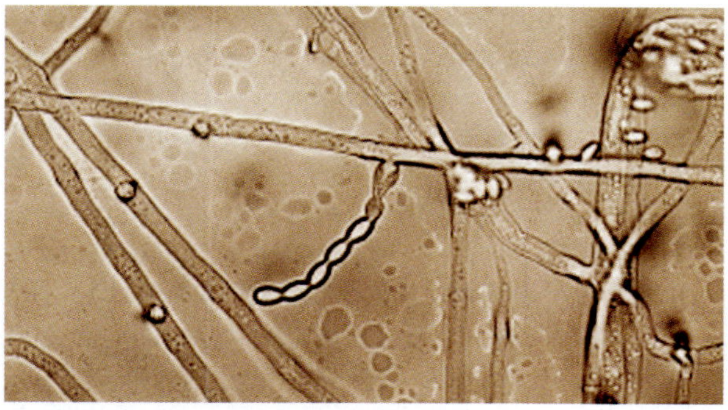

Conidia of *Humicola alopallonella*, an **humicole** fungus, x 530 (*MCG*).

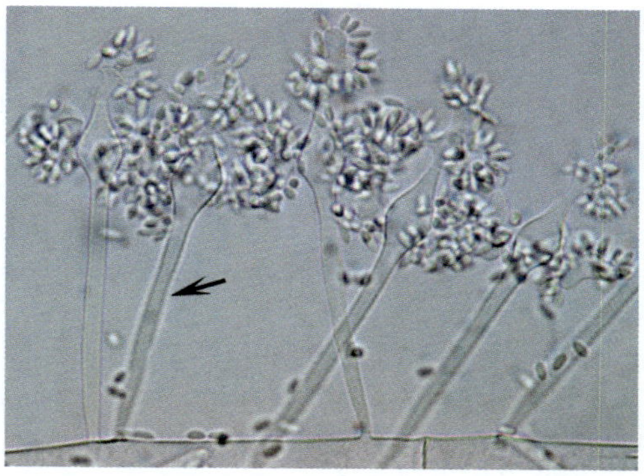

Hyaline conidiophores of *Botryosporium* sp., x 300 (*MU*).

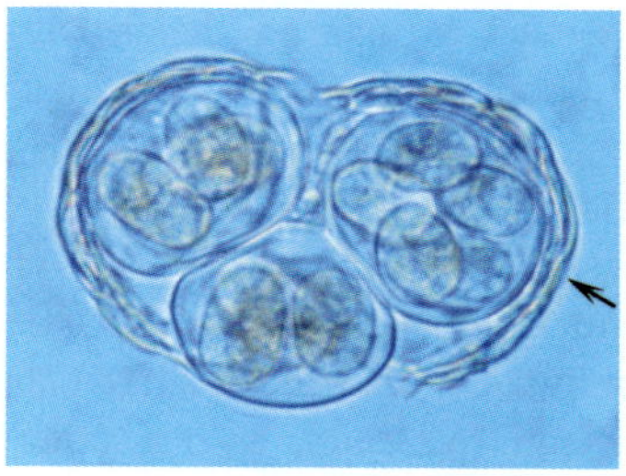

Hyaline ascocarp of *Brasiliomyces malachrae*, x 505 (*RTH*).

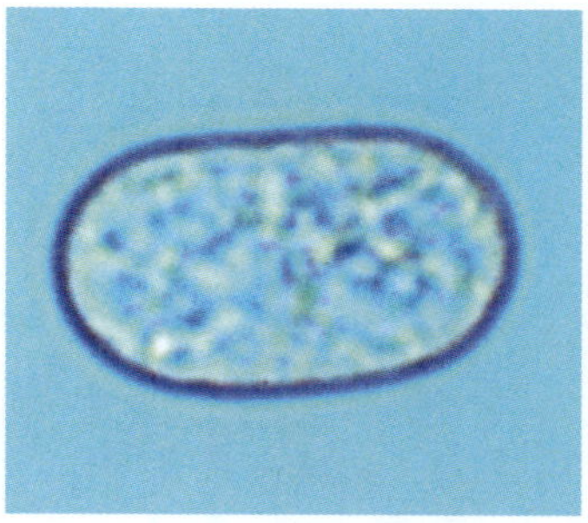

Hyaline ascospore of *Brasiliomyces malachrae*, x 1,345 (*RTH*).

hyaline (L. *hyalinus* < Gr. *hyálinos*, hyaline, transparent < *hýalos*, crystal + L. suf. *-inus* > E. *-ine*, of or pertaining to): transparent and colorless, as if it were glass, or at least diaphanous. Initially, hyphae, spores and other types of structures are hyaline in the majority of the fungi, and in many they remain so at maturity, as happens with the conidiophores and conidia of species of *Botryosporium* (moniliaceous asexual fungi) and the ascocarp of *Brasiliomyces malachrae* (Erysiphales).

hyalo-, **hyal-** (Gr. *hýalos*, crystal): a pref. employed to form compounds that allude to something transparent or translucent, as in *hyaline*, transparent as if it were glass.

hyalohyphomycosis, pl. **hyalohyphomycoses** (L. *hyalinus* < Gr. *hyálinos*, hyaline, genit. of *hýalos*, crystal + *mýkes*, fungus + suf. *-osis*, condition or state): mycosis of man and animals caused by a non-dematiaceous fungus. Cf. **phaeohyphomycosis**.

hyalospore (L. *hyalinus* < Gr. *hyálinos*, hyaline, genit. of *hýalos*, crystal + *sporá*, spore): a hyaline asexual reproductive spore; a Saccardoan term that is applied mainly to conidial fungi. The conidia of *Botrytis*, *Beauveria* and many other gen. of moniliaceous asexual fungi, e.g., are hyaline at maturity. Cf. **phaeospore**.

hybrid (L. *hybrida*, a cross between two different species < *hýbris*, progeny of different races, of people from different countries): the result of a cross between organisms belonging to different taxa that results in a viable progeny. Hybridization can occur from the level of genus downward; e.g., *Endomyces fibuliger* (=*Saccharomycopsis fibuliger*) x *Yarrowia lipolytica* (Saccharomycetales), *Cladonia grayi* x *Cladonia merochlorophaea* (Lecanorales). The International Code of Botanical Nomenclature has rules that define the different types of hybrids and their naming.

hydathode (Gr. *hydatódes*, aqueous, full of water < *hýdatid*, genit. *hýdatis*, aqueous vesicle < *hýdatos*, *hýdor*, water): a secretory organ that secretes very dilute aqueous solutions, a little less than pure water, like the trichomes of certain Agaricales.

hydnoid (< gen. *Hydnum* < Gr. *hýdnon*, an old name applied to a certain type of edible fungi + L. suf. *-oide*, like < Gr. *-oeídes*, similar to): a type of fruiting body with a dentate hymenophore, i.e., arranged on teeth or prongs, stipitate or not (sessile), like that of *Hydnum*, *Sarcodon* and *Hericium*, among other Hydnaceae (Aphyllophorales).

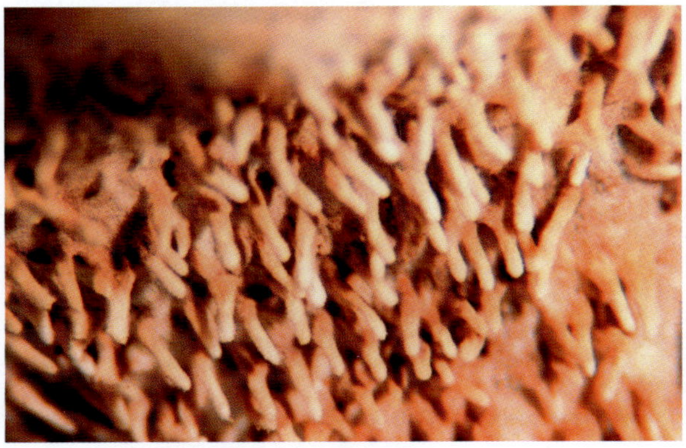

Hydnoid hymenophore of *Sarcodon imbricatus*, x 10 (*MU*).

hydro-, **hydr-** (L. *hydro* < Gr. *hýdro-* < *hýdor*, water): a pref. often employed to form compound terms that introduce the idea of water, as in the following terms.

hydrocarbonoclastic (L. *hydro* < Gr. *hýdro-* < *hýdor*, water + L. *carbo*, genit. *carbonis*, carbon, coal; any substance composed exclusively of carbon and hydrogen + Gr. *klástos*, broken + suf. *-ikos* > L. *-icus* > E. *-ic*, belonging to or relating to): capable of decomposing carbohydrates, like the yeasts *Trichosporon* (cryptococcaceous asexual yeasts) and *Pichia ohmeri* (Saccharomycetales), isolated from oil and petroleum spills and contaminated marine environments.

hydrochoric (Gr. *hýdro-* < *hýdor*, water + suf. *-coro* < *choréo*, to change place, move away from + suf. *-íkos* > L. *-icus* > E. *-ic*, belonging to or relating to): applied to organisms that are disseminated by water. Many fungi have appendaged spores that facilitate their flotation in water (such as the marine species *Orbimyces spectabilis*, of the dematiaceous asexual fungi); other species depend on the kinetic energy of raindrops for dispersal of their spores (e.g., species of *Cyathus* and *Nidularia*, of the Nidulariales). For the zoospores of the aquatic fungi, water is an essential factor in their dispersal. Cf. **anemochore.**

hydrofungi, sing. **hydrofungus**, **aquatic molds**, **water molds** (Gr. *hýdor* < E. *hydro-*, *hydr-*, water; L. *fungus*, fungus < Gr. *spóngos*, *sphóngos*, sponge; L. *aqua*, water + suf. *-ticus* < Gr. *-tíkos* > E. *-tic*, relation, fitness, inclination or capacity): fungi and oomycetes that live in water, especially fresh water, in contrast to the marine fungi. The main fresh water zoosporic microorganisms are Chytridiomycota and Oomycota, especially the Chytridiales and Saprolegniales, respectively. There also are many hyphomycetes (asexual or mitosporic fungi), ascomycetes and saxicolous lichens (some of the latter can remain submerged), that live in fresh water, as well as a variety of fungi that are found in contaminated and sewage waters. Cf. **primary marine fungus** and **secondary marine fungus**.

hydrophilous (L. *hydrophilus*, with a preference for water < Gr. *hýdro* < *hýdor*, water + *phílos*, have an affinity for): from an aquatic habitat; being disseminated by water, although the term *hydrochoric* is more appropriate to define this latter concept. The zoosporic fungi are hydrophilous. Cf. **hydrophobic**.

hydrophobic (L. *hydrophobus*, with an aversion to water < Gr. *hýdro* < *hýdor*, water + *phóbos*, fear, aversion): avoiding or repelling water, like the dry conidia of many fungi (e.g., *Penicillium* and *Aspergillus*, of the moniliaceous asexual fungi), which are not easily wettable, and the thallus of numerous crustaceous and foliose lichens (such as *Ramalina maciformis*, of the Lecanorales) which live in ecological niches (such as hot deserts) where the water is only available in the form of water vapor or as liquid water in the substrate, from which they absorb it to become physiologically active; because of this property they are called *aerohygrophilic* lichens. Cf. **hydrophilous**.

hydrotropism (Gr. *hýdro* < *hýdor*, water + *trópos*, gyration, change in manner or direction < *tropé*, a turn < *trépo*, to gyrate, be directed toward + suf. *-ismós* > L. *-ismus* > E. *-ism*, state, phase, tendency, action): a chemotrophic phenomenon in which water is the stimulating agent. The sporangiophores of the Mucorales, e.g., adopt a position perpendicular to the substrate due to negative hydrotropism. In air saturated with moisture, the sporangiophores of *Phycomyces* are oriented irregularly and at times are slightly curved, moving away from moist paper when this is placed vertically near them. See **chemotropism**, **galvanotropism**, **gravitropism**, **heliotropism** and **phototropism**.

hygro-, **hygr-** (Gr. *hygrós*, wet, humid): a pref. that alludes to moisture or humidity, as in *hygrophanous*.

hygrophanous (Gr. *hygrós*, wet, humid + *phaíno*, to show < *phanós*, light, lantern): becoming translucent when embedded in water; in certain Agaricales (such as *Psathyrella*, *Oudemansiella*, *Hygrocybe* and *Hygrophorus*) the pileus and other fungal organs acquire a translucent appearance due to imbibition. See **pellucid**.

Hygrophanous fruiting body of *Oudemansiella canarii*, x 0.5 (*RV*).

Hygrophanous pileus of the basidiocarps of *Hygrocybe psittacina* (=*Hygrophorus psittacinus*), x 1.5 (*MU*).

hygrophilous (L. *hygrophilus* < Gr. *hygrós*, wet, humid + *phílos*, have an affinity for): an organism that prefers very humid media. Cf. **xerophile**.

hygroscopic, **hygroscopical** (Gr. *hygrós*, wet, humid + *skópein* < *skopiáo*, to observe, see + suf. *-íkos* > L. *-icus* > E. *-ic*, belonging to or relating to): absorbing water readily (from the atmosphere), thus becoming soft and expanded so as to facilitate spore dissemination, as occurs, e.g., in the capillitium of some Myxomycetes (*Trichia*, *Arcyria* and other Trichiales), or like the basidiocarp of *Astraeus hygrometricus* (Sclerodermatales), in which the exoperidium recurves to enclose the endoperidium when dry, but which opens when moistened, to expose and elevate the inner sporiferous sac.

Hygroscopic apothecia of *Lophodermium* sp. (Rhytismatales), on a pine leaf, x 20 (*RTH*).

Hygroscopic fruiting bodies of *Astraeus hygrometricus*, x 1 (*MU*).

hymenial veil (L. *hymenium*, hymenium < Gr. *hyménion*, dim. of *hymén*, membrane + L. suf. *-alis* > E. *-al*, relating to or pertaining to; L. *velum*, veil, curtain, cloth): a curtain or cloth that covers something. A hymenial veil, also called *partial veil*, is a hyphenchymatous covering of the gills (in the hemiangiocarpic Agaricales), and is clearly seen in *Agaricus silvicola* and *Amanita strobiliformis*. See **veil**.

Hymenial veil of the basidiocarp of *Amanita strobiliformis*, x 1 (*RV*).

Hymenial veil of the basidiocarp of *Amanita strobiliformis*, x 3 (*RV*).

Hymenial veil of the basidiocarp of *Amanita strobiliformis*, x 1 (*RV*).

hymeniform (L. *hymenium*, hymenium < Gr. *hyménion*, dim. of *hymén*, membrane + suf. *-formis* < *forma*, shape): refers to a layer of narrowly clavate cells that appear to be similar to basidiola, and which form part of the hymenium.

hymeniiferous (L. *hymeniifer*, that which bears the hymenium < *hymenium*, hymenium + suf. *-ferous*, bearer < *ferre*, to bear, carry): see **hymenophore**.

hymeniiferous trama, pl. **tramae** (L. *hymeniifer*, what bears the hymenium < *hymenium*, hymenium + suf. *-ferous*, bearer < *ferre*, to bear, carry; L. *trama*, trama, tissue, warp): see **context** and **trama**.

Hymeniiferous trama of a gill of *Amanita muscaria* (Agaricales), seen in longitudinal section, x 160 (*MU*).

hymenium, pl. **hymenia** (L. *hymenium*, hymenium < Gr. *hyménion* < *hymén*, membrane + dim. suf. *-íon* > L. *-ium*): a layer or stratum of highly diverse organization but always constituted of hyphae specialized for the production of spores (ascogenous, or producers of asci, in the ascomycetes, or basidiogenous, for the production of basidiospores, in the basidiomycetes), arranged in the shape of a palisade and frequently intermixed with sterile elements (paraphyses in the ascomycetes, and cystidia in the basidiomycetes); the hymenium covers particular parts of the sporophores, called *hymenophores*, that can be gills (in Agaricales, such as *Coprinus*), tubes (in Agaricales, such as *Boletus*), or teeth or spines (in Aphyllophorales, such as *Hydnum*).

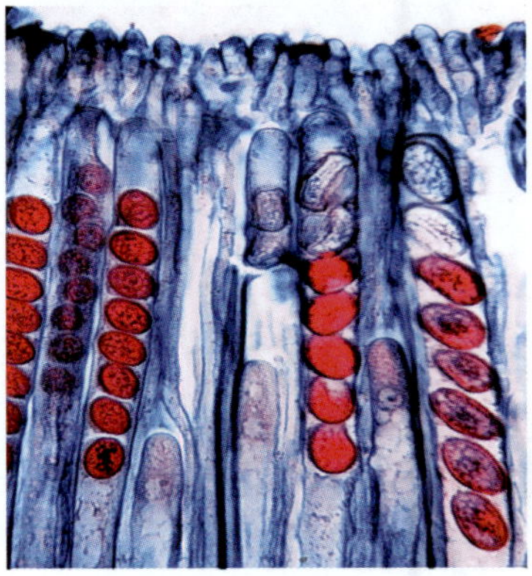

Hymenium of an apothecium of *Peziza* sp., seen in longitudinal section, x 200 (*RTH*).

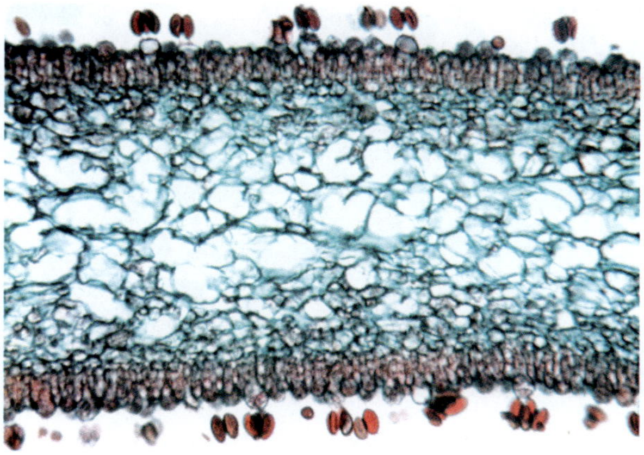

Hymenium of a gill of *Coprinus* sp., seen in longitudinal section, x 200 (*MU*).

Hymenolichen, thallus of *Dictyonema pavonium* (=*Cora pavonia*), x 1 (*TRO*).

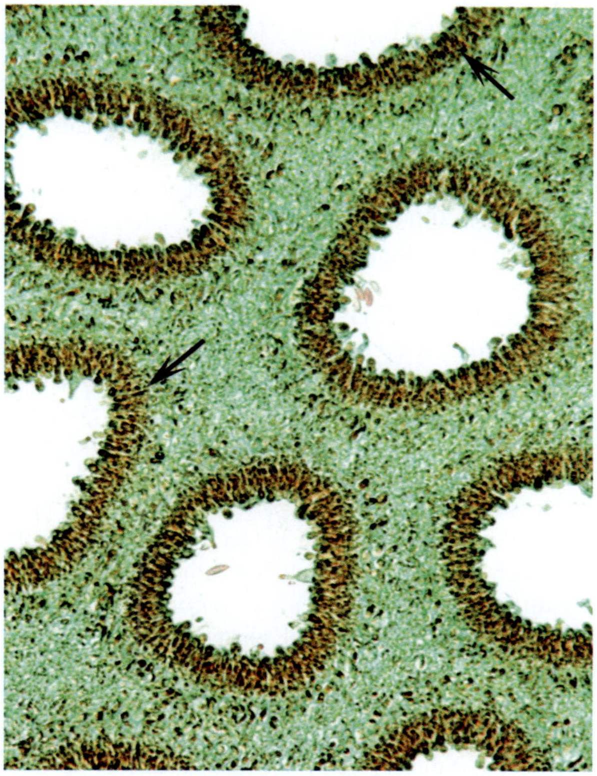

Hymenium of the tubes of *Boletus* sp., seen in transverse section, x 265 (*MU*).

Hymenolichen, thallus of *Dictyonema pavonium* (=*Cora pavonia*), x 2 (*TRO*).

hymenolichen (L. *hymenolichen* < Gr. *hyménion*, dim. of *hymén*, membrane + L. *lichen* < Gr. *leichén*, lichen): a member of a subgroup of basidiolichens in which a hymenomycete is associated with an alga. E.g., the species *Dictyonema pavonium* (=*Cora pavonia*, association of a meruliaceous fungus with algal cells of *Croococcus*); in other species of *Dictyonema* the meruliaceous fungus is associated with algal cells of *Scytonema*; both are hymenolichens.

Hymenomycetes (L. *hymenium*, hymenium < Gr. *hyménion*, dim. of *hymén*, membrane + L. *-mycetes*, ending of taxonomic class < Gr. *mýkes*, genit. *mýketos*, fungus): in the system of classification followed in this dictionary, the Hymenomycetes comprise a class of basidiomycetes with the basidia formed on an hymenium that remains exposed at maturity. The hymenium can be formed freely on the surface of a host plant or on some type of fruiting body or basidiocarp. The Hymenomycetes includes the orders Agaricales and Aphyllophorales, as well as the orders of Gasteromycetes. Together with the Urediniomycetes and Ustilaginomycetes, and other orders, they constitute the phylum Basidiomycota of the kingdom Fungi. In the Appendix, the user of this dictionary can see the comparison between the outline of Alexopoulos *et al.*, 1996, modified by Ulloa and Hanlin, 2006, to that of Kirk *et al.*, 2008. In the latter classification scheme, this class has disappeared, and its orders have been moved to the subphylum Agaricomycotina, phylum Basidiomycota; some important orders, such as Aphyllophorales and Nidulariales, have also disappeared. These taxonomic rearrangements are part of a constant process of study and evaluation, in order to achieve a phylogenetic classification of the fungi.

hymenophore (L. *hymenium*, hymenium < Gr. *hyménion*, dim. of *hymén*, membrane + Gr. suf. *-phóros*, bearer <*phéro*, to bear, support): part of the sporiferous apparatus, composed of sterile tissue, on which the hymenium is borne, such as the trama of the gills of the Agaricales (e.g., *Chlorophyllum*), the teeth of the Auriscalpiaceae (such as *Auriscalpium*), and the pores of Strobilomycetaceae (such as *Strobilomyces*). Also called **hymeniiferous**. The term hymenophore is sometimes applied incorrectly to refer to the whole basidiocarp.

Hymenophore (gills) of the pileus of *Chlorophyllum molybdites*, x 0.5 (*EPS*).

Hymenophore (teeth) of the pileus of *Auriscalpium vulgare*, x 6 (*MU*).

Hymenophore (tubes) of the pileus of *Strobilomyces strobilaceus* (=*S. floccopus*), x 9 (*MU*).

hymenopode, hymenopodium, pl. **hymenopodia** (L. *hymenium* < Gr. *hyménion*, dim. of *hymén*, membrane + NL. *podium* < Gr. *pódion* <*poús*, genit. *podós*, foot, support + dim. suf. *-íon* > L. *-ium*): the layer of tissue found between the trama and the subhymenium in the pezizaceous ascomycetes and the agaricaceous basidiomycetes; it is composed of thin, filiform, connective elements densely intertwined and little differentiated, as in the lateral layer in forms with a bilateral trama. In some species, the hymenopodium can be gelatinized, as in *Pholiota decorata* (Agaricales). Syn. of **hypothecium** in the ascomycetes.

hyper- (Gr. *hypér*, above, on): a pref. employed to give the idea of something that is found on or beyond what is normal, in a strict as well as metaphoric sense, as in the following terms. Cf. **hypo-**.

hyperepiphyllous (Gr. *hypér*, above, on + *epí*, on + *phýllon*, leaf): living on lichen or bryophyte thalli, which in turn are epiphyllic or foliicolous.

hyperparasite (Gr. *hypér*, above, on, or, in a metaphoric sense, beyond the normal, exaggerated + *parásitos*, parasite <*pará*, next to + *sîtos*, bread, food): a parasite that lives at the expense of another parasite, on or inside it. E.g., the fungi of the gen. *Septobasidium* (Septobasidiales) are hyperparasites, as they develop on coccids parasitic on vascular plants. *Lecanicillium* (=*Verticillium*) *lecanii* and *Zygosporium gibbum* (moniliaceous asexual fungi) grow, respectively, on *Hemileia vastatrix*, the coffee rust, and *Puccinia oxalidis*, the trefoil rust (Uredinales). *Phycomelaina laminariae* (Phyllachorales) is a fungal parasite of the marine species *Laminaria agardhii*, *L. digitata* and others (Phaeophyceae) and its ascocarps are at the same time hyperparasitized by thalli of Oomycota of the gen. *Petersenia* (Myzocytiopsidales) and *Atkinsiella* (Salilagenidiales), whose presence impedes the formation of asci in *Ph. laminariae*.

Zygosporium gibbum (black arrow) growing as a **hyperparasite** on *Puccinia oxalidis* (white arrow), infecting a trefoil leaf, x 1 (*GC*).

Zygosporium gibbum (black arrows) growing as a **hyperparasite** on *Puccinia oxalidis* (white arrow), infecting a trefoil leaf, x 1 (*GC*).

Zygosporium gibbum (black arrow) growing as a **hyperparasite** on *Puccinia oxalidis* (white arrow), infecting a trefoil leaf, x 10 (*GC*).

hyperplasia, pl. **hyperplasiae** (Gr. *hypér*, on, excessive + *plásis*, action of shaping or forming): an abnormal increase in cell multiplication, which can cause an increase in the size of the organs affected due to the large number of cells produced. This condition can be induced by certain plant parasitic fungi, such as *Plasmodiophora brassicae* (Plasmodiophorales), the causal agent of club root of cabbage, *Synchytrium endobioticum* (Chytridiales), the causal agent of black wart of potato, and *Taphrina deformans* (Taphrinales), the species that causes peach leaf curl disease. Hyperplasia is usually accompanied by *hypertrophy*, or abnormal increase in cell size. Cf. **hypoplasia**.

Hyperplasia and **hypertrophy** of the leaf tissues of peach, infected by *Taphrina deformans*, x 1 (*RTH*).

Hyperplasia and **hypertrophy** of the leaf tissues of peach, infected by *Taphrina deformans*, x 1 (*RTH*).

hypersaprobe (Gr. *hypér*, above, on, or, in a metaphoric sense, beyond the normal, exaggerated + *saprós*, rotted + *bíos*, life): a saprobe that develops only on substrates that have been invaded by other saprobes, as is observed, e.g., in species of *Lasiosphaeria* (Sordariales) and *Nectria* (Hypocreales). Other examples of hypersaprobic fungi are the marine saprobes that grow on other saprobic fungi, such as *Massarina* (Dothideales), whose ascocarps are sometimes associated with species of the sphaeropsidaceous asexual fungi *Coniothyrium*, *Ceratophoma* and *Microsphaeropsis* (anamorphs of Leptosphaeriaceae), among other saprobic fungi, and *Halosphaeria hamata* (Halosphaeriales), which develops its ascocarps as a hypersaprobe inside the old or dead, decomposing fruiting bodies of *Leptosphaeria oreamaris* (Pleosporales), a marine saprobe.

hypersensitivity (Gr. *hypér*, above, on, or, in a metaphoric sense, beyond the normal, exaggerated + L. *sensibilitas*, sensibility, faculty of feeling, property of living beings) : a resistant reaction in plants to attack by biotrophic parasites that involves the rapid death of a limited number of host cells, which becomes visible as flecklike lesions. The hypersensitivity of the host can effectively prevent the establishment of a biotrophic parasite, such as a rust fungus, on a particular plant.

hypertonic (Gr. *hypér*, above the normal, excessive + *tonikós* < *tónos*, tension, tone + suf. *-íkos* > L. *-icus* > E. *-ic*, belonging to or relating to): applied to a culture medium or substrate that has an osmotic pressure higher than that of the organism being cultivated. Cf. **hypotonic**.

hypertrophy (Gr. *hypér*, above the normal, excessive + *trophía*, to enlarge, fatten): an abnormal increase in the size of cells, resulting in the swelling of the affected tissue or organ and the formation of a tumor or canker; hypertrophy usually is accompanied by *hyperplasia* (abnormal cell multiplication). Certain plant parasitic fungi (Plasmodiophoromycetes, Chytridiomycetes and Taphrinales) are capable of causing this pathological process. Cf. **distrophy**.

hypha, pl. **hyphae** (NL. *hypha*, hypha < Gr. *hyphé*, tissue, spiderweb; hypha): a tubular filament that represents the structural entity (thallus) of the majority of the fungi. Hyphae can be somatic or fertile and from their differentiation are derived a great diversity of structures related to their assimilative and reproductive functions, including, e.g., rhizoids, stolons, haustoria, appressoria and a wide range of asexual and sexual sporophores, such as sporangiophores, conidiophores, ascocarps and basidiocarps, all of them manifesting very diverse forms depending upon the species. The two general forms of hyphae present in the fungi are the *coenocytic* (in many of the so-called lower fungi) and the *septate* (in the Zygomycetes and Trichomycetes of the Zygomycota, in the asexual fungi, Ascomycota and Basidiomycota). See **binding hypha**, **generative hypha**, **gleoplerous hypha**, **inflated hypha**, **lactiferous hypha**, **oleiferous hypha**, **primary hypha**, **racket hypha**, **receptive hypha**, **secondary hypha**, **skeletal hypha** and **spiral hypha**.

hyphal (NL. *hypha*, hypha < Gr. *hyphé*, tissue, spiderweb; hypha + L. suf. *-alis* > E. *-al*, relating to or belonging to): belonging to a hypha or relating to it.

hyphal body (NL. *hypha* < Gr. *hyphé*, tissue, spiderweb, hypha + L. suf. *-alis* > E. *-al*, relating to or belonging to; ME. < OE. *bodig*, body): *Zygomycetes*. Each of the cells resulting from the fragmentation of the mycelium of the Entomophthorales (such as *Entomophthora muscae* and *Conidiobolus thromboides*). Hyphal bodies reproduce asexually by means of conidia, or they may fuse in pairs to form zygosporangia.

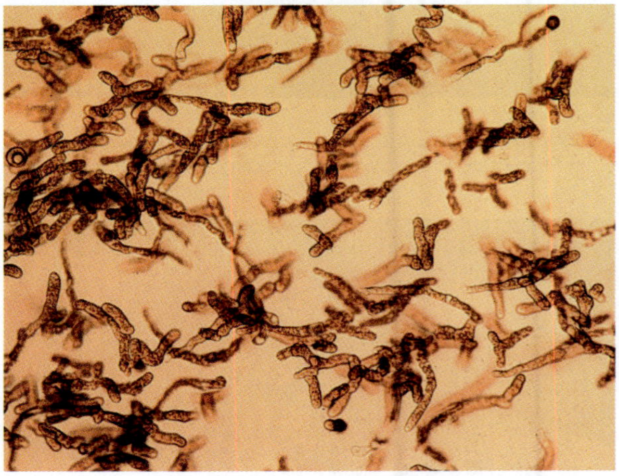

Hyphal bodies of *Conidiobolus thromboides*, x 40 (*MU*).

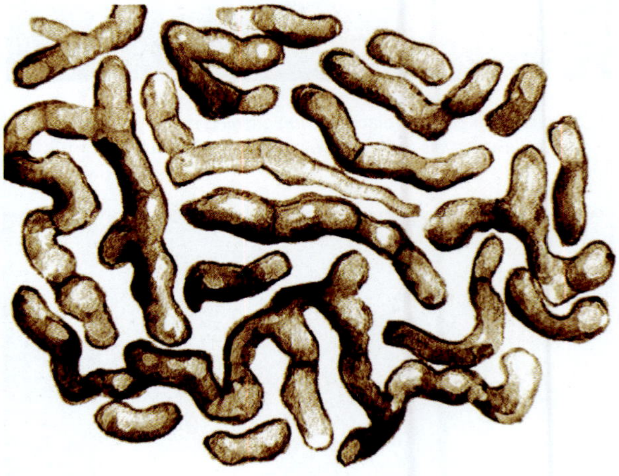

Hyphal bodies of *Conidiobolus thromboides*, x 69 (*MU*).

hyphal net (NL. *hypha* < Gr. *hyphé*, tissue, spiderweb; hypha + L. suf. *-alis* > E. *-al*, relating to or pertaining to; ME. < OE. *net, nett*, net): an adhesive organ of some squamulose (placodioid) lichens, such as *Psora decipiens* (Lecanorales), consisting of a delicate network of hyphae that penetrate the substrate. Also known as *rhizoidal string*.

hyphal peg (NL. *hypha* < Gr. *hyphé*, tissue, spiderweb; hypha; ME. *pegge*, peg): **1**. A fascicle of hyphae that projects from the hymenium, consisting of two or more parallel or intertwined hyphae that are incrusted or gelatinized; present in some Aphyllophorales, such as *Lentinus*, *Megasporoporia* and *Dichomitus mexicanus* (=*Megasporoporia mexicana*). **2**. A hyphal projection or extension that fuses or anastomoses with another hypha.

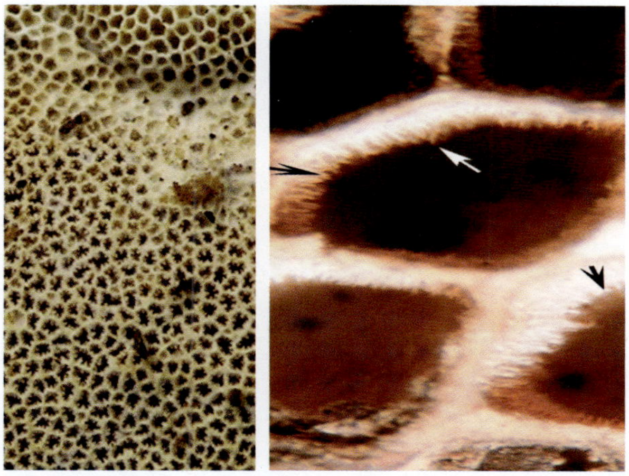

Hyphal pegs of the porose hymenophores of *Megasporoporia setulosa*, x 2 (*RV*) and *Dichomitus mexicanus* (=*Megasporoporia mexicana*), x 12 (*CB*).

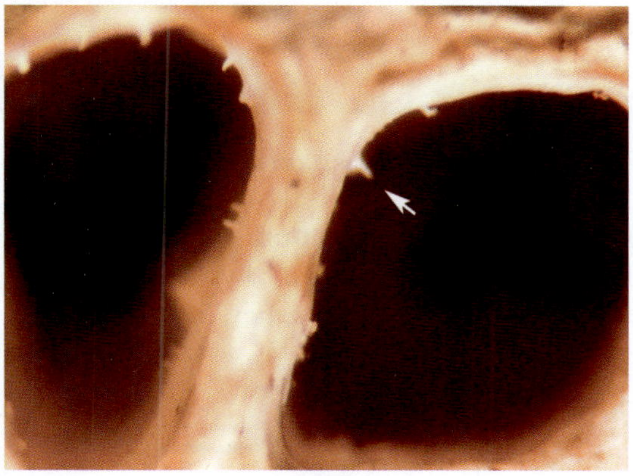

Hyphal pegs of the porose hymenophore of *Dichomitus mexicanus* (=*Megasporoporia mexicana*), x 21 (*CB*).

hyphal string (NL. *hypha* < Gr. *hyphé*, tissue, spiderweb, hypha + L. suf. *-alis* > E. *-al*, relating to or belonging to; ME. *string* < OE. *streng*, string): a loose cord of soft hyphae that extends into the central internal part of the cavity of the stipe, from the apex toward the base, in the fruiting body of *Coprinus comatus* and *C. sterquilinus* (Agaricales). As the stipe continues growing, the hyphal cord stretches, becomes thinner, and breaks.

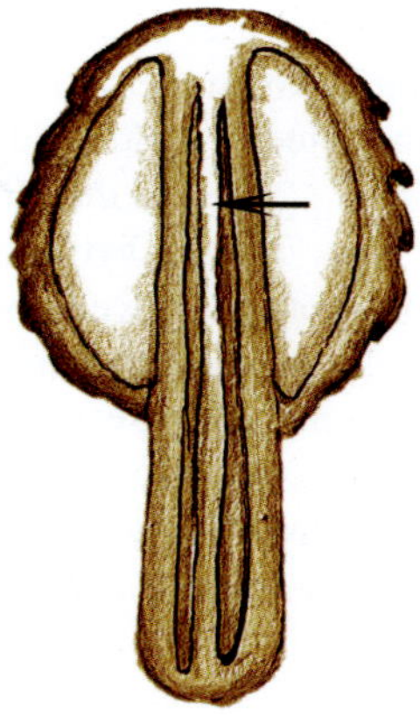

Hyphal string (cord) of the basidiocarp of *Coprinus comatus*, seen in longitudinal section, x 1.5 (*MU*).

hyphidium, pl. **hyphidia** (NL. *hyphidium*, small hypha < *hypha*, hypha < Gr. *hyphé*, tissue, spiderweb; hypha + dim. suf. *-idium*): syn. of **hyphoid**. **1**. *Aphyllophorales*. Each one of the characteristic hyphae found in the *catahymenium* of some basidiomycetes (such as *Aleurodiscus*), with a thick or thin wall and generally lacking cytoplasmic contents, frequently branched in a distinct and intricate manner, producing complex forms. The hyphidia, which include a variety of cystidia, are developed in early stages and envelop the basidia in distinct levels, because of which the basidia have to open a passage in order to reach the surface of the catahymenium. Several types of hyphidia are recognized: *acanthophyses*, *asterophyses* and *dendrophyses*. **2**. Another term for *spermatium*.

hyphochytrid (NL. *hypha*, hypha < Gr. *hyphé*, tissue, spiderweb; hypha + *chytrídion*, dim. of *chytrís*, pot): a member of the Hyphochytriomycota, a phylum of the kingdom Stramenopila. This phylum, with the single order Hyphochytriales, is included in the kingdom Protoctista by Margulis and Schwartz (1988) and in the kingdom Stramenopila by Alexopoulos *et al*. (1996).

Hyphochytriomycetes (NL. *hypha*, hypha < Gr. *hyphé*, tissue, spider web; hypha + *chytrídion*, dim. of *chytrís*, pot + L. *-mycetes*, ending of taxonomic class < Gr. *mýkes*, genit. *mýketos*, fungus): in some classifications, a class of fungi (Hyphochytriomycetes, also called Hyphochytridiomycetes) with anteriorly uniflagellate zoospores (mastigonemate flagellum); the general structure of the thallus is similar to that of the Chytridiomycetes. There is a single order, Hyphochytriales. In the classification adopted in this dictionary (Alexopoulos *et al*., 1996), this group is regarded as the phylum Hyphochytridiomycota, which together with the phyla Labyrinthulomycota and Oomycota, form part of the kingdom Stramenopila.

In the Appendix, the user of this dictionary can see the comparison between the outline of Alexopoulos *et al*., 1996, modified by Ulloa and Hanlin, 2006, to that of Kirk *et al*., 2008. In the latter classification scheme, the class Hyphochytriomycetes (phylum Hyphochytriomycota), as well as the other two phyla of the kingdom Stramenopila (Oomycota and Labyrinthulomycota) are now in the kingdom Chromista. These taxonomic rearrangements are part of a constant process of study and evaluation of these groups of fungi, in order to achieve a phylogenetic classification.

hyphoid (NL. *hypha*, hypha < Gr. *hyphé*, tissue, spiderweb; hypha + L. suf. *-oide*, like < Gr. *-oeídes*, similar to): **1**. Similar to a hypha. **2**. *Uredinales*. A type of aecium, characterized by having colorless, branched hyphae that form a mold-like colony on the epidermis of the host plant, with each hypha producing a colorless aeciospore on the apex, as is observed, e.g., in *Dasyspora*.

Hyphomycetes (NL. *hypha*, hypha <Gr. *hyphé*, tissue, spiderweb; hypha + L. *-mycetes*, ending of taxonomic class < Gr. *mýkes*, genit. *mýketos*, fungus): in some classifications, it corresponds to a class of conidial fungi characterized by the formation of asexual spores (conidia) on conidiophores that are not contained in a fruiting body; i.e., without acervuli or pycnidia, although some form sporodochia and synnemata. The Hyphomycetes (with the single order Moniliales), together with the Blastomycetes and the Coelomycetes, comprise the subdivision Deuteromycotina, or deuteromycetes in a general sense. However, in the classification adopted in this dictionary all of these forms are considered as asexual ascomycetes and other asexual fungi, without class rank; the Moniliales have been changed to agonomycetaceous, moniliaceous, dematiaceous, tuberculariaceous and stilbellaceous asexual fungi, that respectively correspond to the families Agonomycetaceae, Moniliaceae, Dematiaceae, Tuberculariaceae and Stilbellaceae of the order Moniliales. The same has been done with the Blastomycetes and Coelomycetes, which herein are not recognized as taxonomic classes, but are treated as cryptococcaceous and sporobolomycetaceous asexual yeasts, and as sphaeropsidaceous and melanconiaceous asexual fungi, respectively.

hyphophore (NL. *hypha*, hypha < Gr. *hyphé*, tissue, spiderweb; hypha + Gr. suf. *-phóros*, bearer): an erect, pedicellate, centrally peltate, asexual sporophore, characteristic of tropical foliicolous lichens, such as *Gyalectidium catenulatum* and *G. eskuchii* (family Gomphillaceae).

Hyphophores of *Gyalectidium catenulatum*, x 40 (*MAH*).

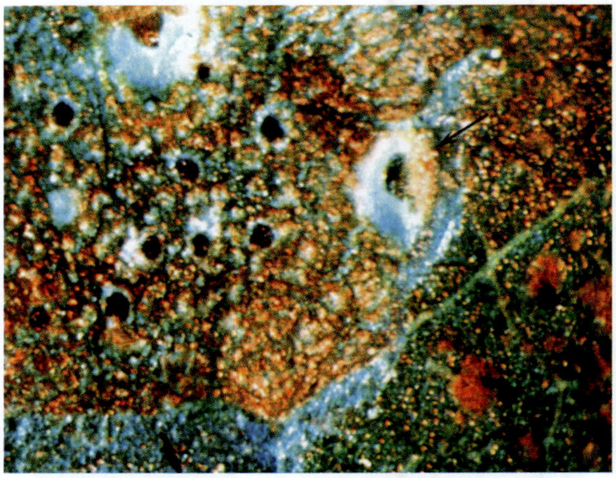

Hyphophores of *Gyalectidium eskuchii*, x 40 (*MAH*).

hyphopodium, pl. **hyphopodia** (NL. *hypha*, hypha < Gr. *hyphé*, tissue, spiderweb; hypha + NL. *podium* < Gr. *pódion* < *poús*, genit. *podós*, foot + dim. suf. *-íon*

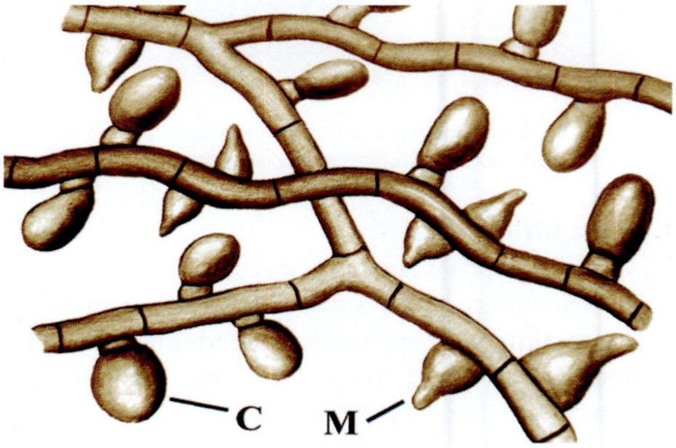

Hyphopodia of the capitate (C) and mucronate (M) types formed by *Meliola caesariae*, x 350 (*MU*).

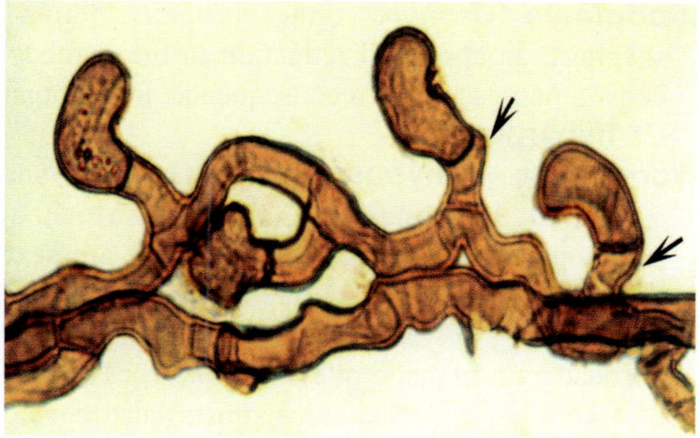

Hyphopodia of *Meliola* sp., x 480 (*RTH*).

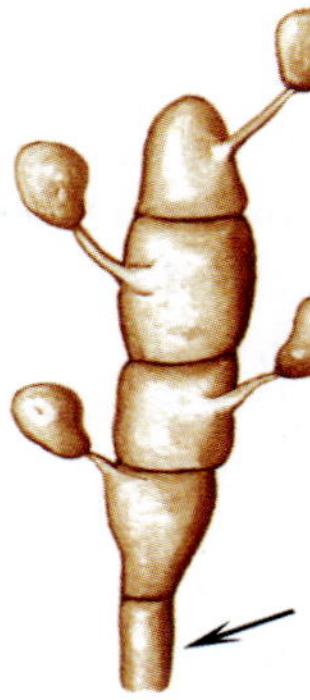

Hypobasidium, the lower cell, of a heterobasidium of *Auricularia* sp., x 1,770 (*MU*).

> L. -*ium*): a short, lateral hyphal branch consisting of one or two cells, characteristic of the obligately parasitic fungi belonging to the order Meliolales (such as *Meliola caesariae* and *Asteridiella homalii-angustifolii*). Two types of hyphopodia are formed: the *capitate hyphopodium* is flattened, sometimes lobed, with a rounded end cell that adheres to the host surface; it gives rise to a tube that penetrates the cuticle to reach an epidermal cell, in which it develops an haustorium that absorbs nutrients, thus functioning as an appressorium. The *mucronate hyphopodium* is an erect, tapered cell that is conidiogenous. See **stigmatocyst** and **stigmatopod**.

hypno-, **hypn-** (Gr. *hýpnos*, sleep): a pref. employed to allude to something in a latent state of life, as in the following terms.

hypnocyst (Gr. *hýpnos*, sleep + *kýstis*, vesicle, cell): a group of cells similar to spores of *Alternaria* (dematiaceous asexual fungi), but which are latent.

hypnospore (Gr. *hýpnos*, sleep + *sporá*, spore): a thick-walled resistant spore (such as chlamydospores, resting spores and teliospores), that does not germinate immediately, but which must pass through a resting period before developing. Also called **memnospore**. Cf. **xenospore**.

hypo-, **hyp-** (Gr. *hypó*, under): a pref. employed to introduce the idea of something that is beneath another thing, subordinate to it or nearly so, as in the following terms. Cf. **epi-** and **hyper-**.

hypobasidium, pl. **hypobasidia** (Gr. *hypó*, under + *basídion*, basidium < *básis*, base + dim. suf. - *ídion* > L. -*idium*): the lower cell of a **heterobasidium**, e.g., in *Auricularia* (Auriculariales).

hypochnoid (< gen. *Hypochnus* < Gr. *hypó*, under + *chnóos*, wool + L. suf. -*oide*, like < Gr. -*oeídes*, similar to): similar to *Hypochnus*, a syn. of *Tomentella* (Aphyllophorales), which has an effuse, resupinate and dry basidioma, composed of more or less loosely intertwined hyphae, like a tomentum.

hypocreaceous (< gen. *Hypocrea* < Gr. *hypó*, under + *kréas*, flesh + L. suf. -*aceus* > E. -*aceous*, of or belonging to, with the nature of): fleshy and brightly colored (orange, red), like the perithecial stromata and perithecia of species of *Hypocrea* (Hypocreales).

hypoderm (Gr. *hypó*, under + *dérma*, skin): see **subcutis**.

hypogeal, **hypogean**, **hypogeic**, **hypogeous** (Gr. *hypó*, under + *geo*, soil, earth): growing or developing subterraneally, as occurs in the truffles (e.g., species of *Elaphomyces,* of the Elaphomycetales, and *Tuber*, of the Pezizales), the false truffles (such as *Rhizopogon* and *Hymenogaster*, of the Hymenogastrales, among other gasteromycetes), and a few Zygomycetes (such as *Endogone*, of the Endogonales), the majority of them mycorrhizal. Cf. **epigeal**.

hypogynous (Gr. *hypó*, under + *gyné*, woman; here, female sexual organ): an arrangement in which the male gametangium is located beneath the female gametangium, as happens, e.g., in *Allomyces arbuscula* (Blastocladiales); this condition is known as *hypogyny*. Cf. **epigynous**.

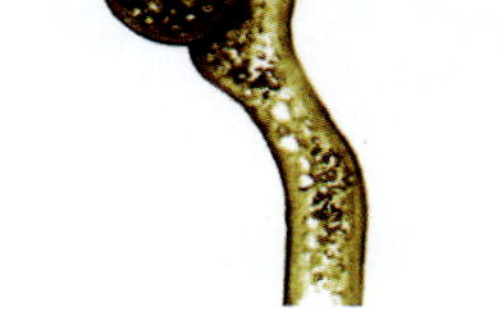

Hypogynous male gametangia of *Allomyces arbuscula*, x 500 (*MU*).

hyponecral (Gr. *hypó*, under + *nekrós*, dead + L. suf. *-alis* > E. *-al*, relating to or pertaining to): refers to the layer of dead hyphae, with indistinct lumens, that is found near the cortex of the lichen thallus, beneath the algal layer. Cf. **epinecral**.

hypoparasite (Gr. *hypó*, under + *pará*, together + *sîtos*, bread, food): a hidden parasite; a pathogen dispersed with another pathogen, like the mycovirus in *Olpidium viciae* (the chytrid vector, of the Spizellomycetales), which affects plants of *Vicia unijuga*.

hypophloeodic, **hypophloeodal** (Gr. *hypó*, under + *phloiós* > L. *phloeo*, cortex, bark + Gr. *-oeídes*, like, similar to + suf. *-íkos* > L. *-icus* > E. *-ic*, belonging to or relating to): *Lichens*. Refers to crustaceous thalli that grow almost immersed in the bark of trees, like various species of *Thelenella* (=*Microglaena*) and other gen. of Thelenellaceae, a family of ascomycetes of uncertain affinities, previously classified in the Pyrenulales. Cf. **epiphloeodic**.

hypophoretic (Gr. *hypó*, under + suf. *-phóros*, bearer < *phéro*, to bear, carry + suf. *-tikós* > L. *-ticus* > E. *-tic*, relation, fitness, inclination or capacity): applied to fungi that are carried by mites, and which in turn, are transported by insects, such as *Pyxidiophora* (Laboulbeniales). See **phoresy**.

hypophyllous (Gr. *hypó*, under + *phýllon*, leaf): found beneath a leaf or on the lower surface, like the parasitic fungi whose fructifications appear on the underside of the leaves, e.g., the apothecia of *Phacidium lacerum*, of the Helotiales, which are formed subepidermally on fallen pine needles, and the aecia of *Puccinia graminis tritici*, of the Uredinales, which develop on the underside of barberry leaves. There are also lichens whose foliicolous thalli are typically found on the lower surface of leaves, such as *Coenogonium hypophyllum* (Gyalectales). Cf. **epiphyllous**.

hypoplasia (Gr. *hypó*, under + *plásis*, action of shaping): an abnormal reduction in the number of cells in tissues and, as a consequence, in the organs. Cf. **hyperplasia**.

hypostroma, pl. **hypostromata** (Gr. *hypó*, under + *strõma*, bed, mattress, small cushion): the pseudoparenchymatous base on which are seated the ascocarps of some marine Dothideales, such as *Manglicola guatemalensis*, which inhabit the bark of the dead roots of mangrove (*Rhizophora*). The thallus of *Spathulospora* (Spathulosporales) also develops a hypostroma that is introduced into the cells of the algal host, like a foot formed by infective cells that unite to make the space that the assimilative cells require.

hypothallus, pl. **hypothalli** (Gr. *hypó*, under + *thallós*, shoot, thallus): **1**. *Myxomycetes*. A thin, transparent pellicle or membrane, commonly with calcareous deposits, formed by the plasmodium on the substrate, at the base of the fructifications. In some species the hypothallus is not very conspicuous, but in others it can be red, as in *Metatrichia vesparium* (Trichiales), yellow, as in *Leocarpus fragilis* and *Physarum polycephalum* (Physarales), or greenish-yellow, as in *Stemonitis nigrescens* (Stemonitales). **2**. *Lichens*. In the crustaceous species, it corresponds to the lower layer of the thallus, blackish in color, that produces the rhizines, as occurs in *Hypotrachyna* (=*Parmelia*) *imbricatula* (Lecanorales). **3**. *Clavicipitaceae* (Hypocreales). The thin layer of fungal tissue formed on the host surface and upon which the perithecial stroma develops, as in *Balansia*.

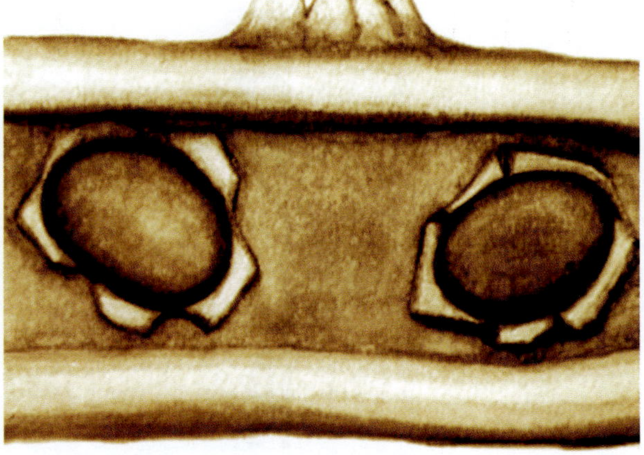

Hypophyllous apothecia of *Phacidium lacerum*, with radial dehiscence, on a fallen leaf of *Pinus sylvestris*, x 5 (*MU*).

Hypothallus of the sporangium of *Leocarpus fragilis*, x 20 (*CB*).

Hypothallus of the sporangium of *Physarum polycephalum*, on agar, x 26 (*MU*).

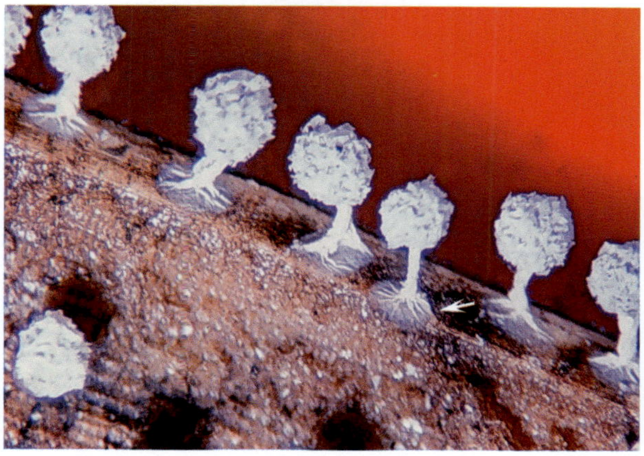

Hypothallus of the sporangia of *Didymium* sp., on the surface of a grass, x 12 (*MU*).

Hypothallus of the sporangium of *Didymium* sp., x 15 (*MU*).

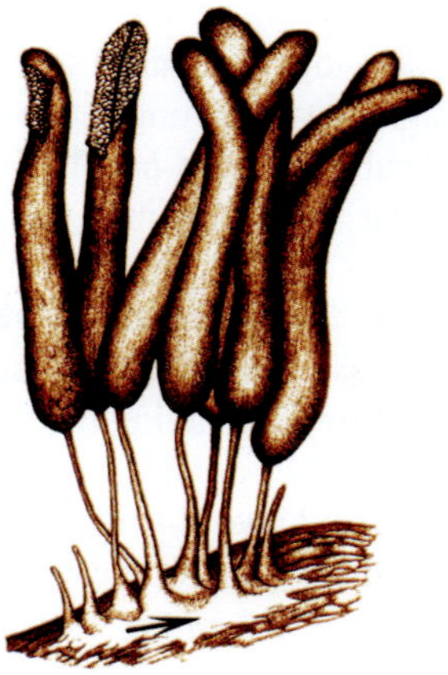

Hypothallus of the sporangia of *Stemonitis nigrescens*, x 15 (*MU*).

hypothecium, pl. **hypothecia** (NL. *hypothecium*, hypothecium < Gr. *hypó*, under + *thekíon*, dim. of *thêke*, box; here, of the asci): **1**. *Discomycetes*. The thin upper layer of the apothecial tissues on which the asci rest, as in the Pezizales and related fungi. **2**. *Discolichens*. The subhymenial layer that extends laterally to form the wall of the apothecium (*parathecium*). Cf. **epithecium**.

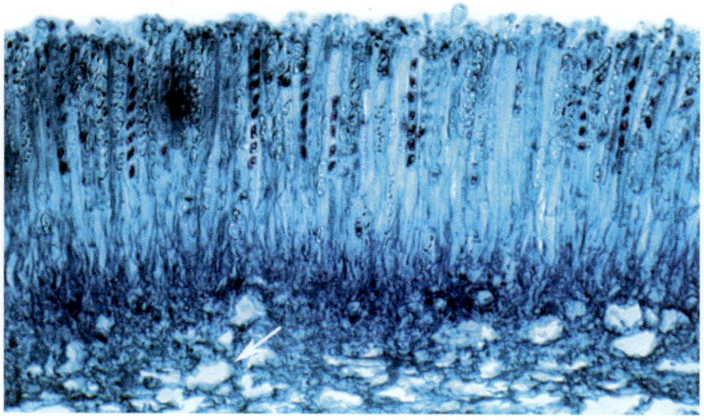

Hypothecium of the apothecium of *Peziza* sp., seen in longitudinal section, x 80 (*RTH*).

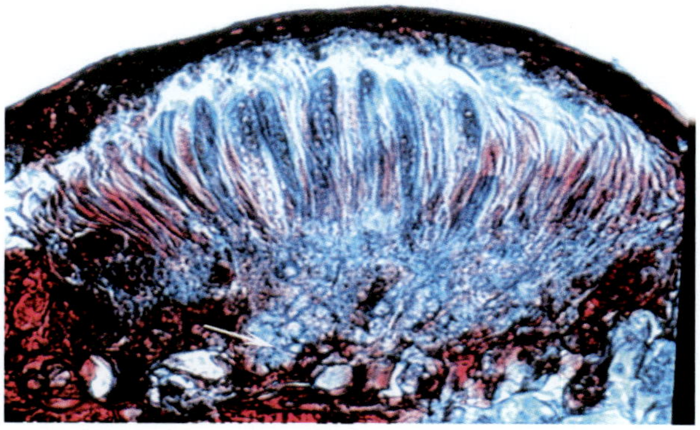

Hypothecium of the apothecium of *Pseudopeziza medicaginis* (Helotiales), seen in longitudinal section, x 80 (*RTH*).

hypotonic (Gr. *hypó*, under, below the normal + *tonikós* < *tónos*, tension, tone, pressure + suf. *-íkos* > L. *-icus* > E. *-ic*, belonging to or relating to): applied to a culture medium or substrate that has an osmotic pressure lower than that of the organism being cultured. Cf. **hypertonic**.

hysteriaceous, **hysterioid**, **hysteriiform** (Gr. *hýstera*, uterus,womb, cavity + L. suf. *-aceous* > E. *-aceous*, belonging to, having the nature of; or + L. suf. *-oide*, like < Gr. *-oeídes*, similar to; or + L. suf. *-formis* < *forma*, shape): having an elongated shape with a longitudinal fissure, like a *hysterothecium*.

hystero-, **hyster-** (Gr. *hýstera*, uterus, womb, cavity; or Gr. *hýsteros*, posterior, last, what comes after): a pref. employed in terms such as *hysterothecium*, referring to the typical shape of the ascocarp of the Hysteriales, and *hysterogenous*, when it refers to the tannin receptacle of some plants, which originates by secondary divisions of the parenchyma cells.

hysterothecium, pl. **hysterothecia** (NL. *hysterothecium*, hysterothecium < Gr. *hýstera*, uterus, womb, cavity + *thekíon*, dim. of *thêke*, box; here, of the asci): an elongated ascocarp, boat-shaped, that is closed at first, then opens on maturing by means of a longitudinal fissure along a line of dehiscence, exposing the hymenium of bitunicate asci. The hysterothecium is characteristic of the Loculoascomycetes of the order Dothideales, such as *Actidium*, *Aulographum*, *Glonium*, *Hysterium* and *Hysterographium*, and of certain loculoascolichens (Arthoniales and Pyrenulales).

Stellate **hysterothecia** of *Actidium hysterioides*, with a longitudinal slit on each arm, on conifer bark, x 3 (*MU*).

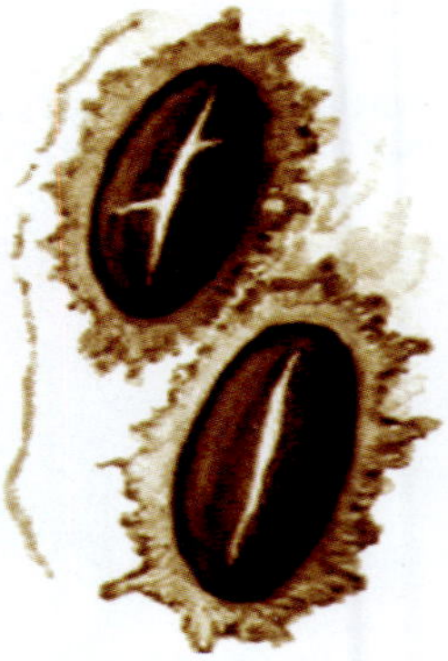

Ellipsoidal **hysterothecia** of *Aulographum hederae* (= *A. vagum*), with a longitudinal slit and a margin of irregularly branched, brown hyphae, on a fallen leaf of English ivy (*Hedera helix*), x 3 (*MU*).

i

Sporangium of *Lamproderma columbinum*, with an **iridiscent** peridium (*MU*).

i

ianthinosporous (Gr. *iánthinos*, purplish-blue + *sporá*, spore + L. suf. *-osus* > OF. *-ous, -eus* > E. *-ous*, having, possessing the qualities of): applied to fungi with spores that are purplish-brown in color when seen in mass in a spore print, like those of the basidiocarps of some species of Agaricaceae, Strophariaceae and Coprinaceae (Agaricales).

icon, pl. **icones** (Gr. *eikón*, image): printed figures, plates or illustrations, generally in color. An example of the use of this term in mycological works is the *Icones fungorum hucusque cognitorum* (*Illustrations of fungi known until now*) by A. C. J. Corda, published in six parts, from 1837 to 1854.

iconography (ML. *iconographia* < Gr. *eikonographía*, sketch, description < *eikón*, image + *graphía*, writing or representation < *grápho*, to describe): a descriptive treatise based mainly on images; a pictorial or illustrative work.

iconotype (Gr. *eikón*, image + *typós*, mark, model, type): refers to a specimen that is not from the type locality, but whose characters agree with those of the accepted type of the species, variety, etc.

id (L. *-id*, patronymic fem. suf.): a secondary effect that is manifested as a rash or eruption of the skin, as the result of an allergic reaction to the presence of spores or toxins produced by an infectious agent. If this agent is a dermatophyte the *id* reaction is called a *dermatophytid*; it can be *epidermophytid*, a *trichophytid* or a *microsporid* (respectively caused by species of *Epidermophyton*, *Trichophyton* and *Microsporum*, of the moniliaceous asexual fungi).

-ide, **-ite** (Gr. *ités*): a suf. that denotes one of a group, as in *glucoside*. The ending *-ite* is used to denote something that is tightly united to, or is part of, another thing, as in *exopodite*, and also to refer to a natural fossil, as in *trilobite*.

idio- (Gr. *ídios*, one's own, special, particular, peculiar): a pref. employed in terms that express a peculiarity, as in *idiomorph* and *idiotaxonomy*.

-idium, pl. **-idia** (Gr. *ídion* > L. *-idium* > E. *-idium*): dim. des. of many mycological terms, as in *antheridium*, the male gametangium or sexual organ of the lower fungi; *conidium*, a specialized asexual spore formed by the asexual conidial fungi, etc.

idiomorph (Gr. *ídios*, one's own, special, particular, peculiar + *morphé*, form, shape): in the ascomycetes and basidiomycetes that have been studied, the genes at the mating type loci are completely dissimilar and cannot be regarded as homologous alleles. The term idiomorph was coined to describe such alternative forms of a locus that lack significant sequence homology but which occupy the same site in the genome.

idiotaxonomy (Gr. *ídios*, one's own, special, peculiar + *táxon*, to order, constitute < *táxis*, to put in order + *nómos* > *nomía*, legitimate according to the law, normal): the taxonomy of organisms. Cf. **syntaxonomy**.

igniarious (L. *igniarius*, pertaining to or relating to fire < *igneus*, fire + suf. *-arius* > E. *-arious*, pertaining to or relating to): tinder; very dry, easily flammable material, such as thistle, pieces of wood or dried fungi; of the latter, *Phellinus* (=*Fomes*) *igniarius* and other woody Aphyllophorales, wood decomposers, as well as some Lycoperdaceae, are examples of this type of tinder or punk.

Basidiocarp of *Phellinus rimosus*, used as **igniarious**, x 0.7 (*RV*).

illegitimate (L. *illegitimus* < pref. *in-*, exclusive > *-il* before *l*, + *legitimus*, legitimate, certain, genuine, true, conforming to the law): refers to any name or epithet that does not conform to the rules of the International Code of Botanical Nomenclature.

illustrate, illustrated (L. *illustrare*, to give light to understanding, clarify a point or material): to include illustrations or images (drawings, prints, photographs, etc.) in a work to complement and enrich the written information.

imbricate, imbricated (L. *imbricatus*, with roof tiles or the shape of one < *imbricare*, to cover with roof tiles + suf. *-atus* > E. *-ate*, provided with or likeness): superimposed, like the tiles in a tiled roof or the scales of fish that overlap one another. E.g., the colonies of many molds, such as *Phoma destructiva* (sphaeropsidaceous asexual fungi) and the leaflets that constitute the foliose thallus of lichens such as *Parmelia* (Lecanorales), as well as the surface of the basidiocarps of *Sarcodon imbricatus* (=*Hydnum imbricatum*), *Trametes* (=*Polyporus*) *versicolor* and *Trametes cervina* (Aphyllophorales) are imbricate.

Imbricate colony of *Phoma destructiva* on agar, x 0.8 (*MU*).

Imbricate basidiocarps of *Phellinus sarcites*, x 0.5 (*RV*).

Imbricate basidiocarps of *Trametes cervina*, x 0.7 (*RV*).

Imbricate basidiocarps of *Trametes* (=*Polyporus*) *versicolor*, x 0.7 (*MU*).

immaculate (L. *immaculatus*, without spots < pref. *in-*, exclusive + *maculatus* < *macula*, spot + suf. *-atus* > E. *-ate*, provided with): without spots. Cf. **maculate**.

immersed (L. *immersus*, immersed, embedded < *immergere* < *in-*, in + *mergere*, to unite, fuse; to blend, absorb, mix): found beneath the surface; embedded, sunken or submerged in the parts that surround it. E.g., the perithecia of *Xylaria*, *Daldinia* and other Xylariales are immersed in the substrate or stromatic tissue, as well as the perithecial stromata of *Diaporthe eres* (Diaporthales), which are immersed in elm branches (and which on maturing erupt through the bark to liberate the spores), and the ascostromata of *Dothiora* (=*Dothidella*) *ribesia* (Dothideales), which grow immersed in red currant stems.

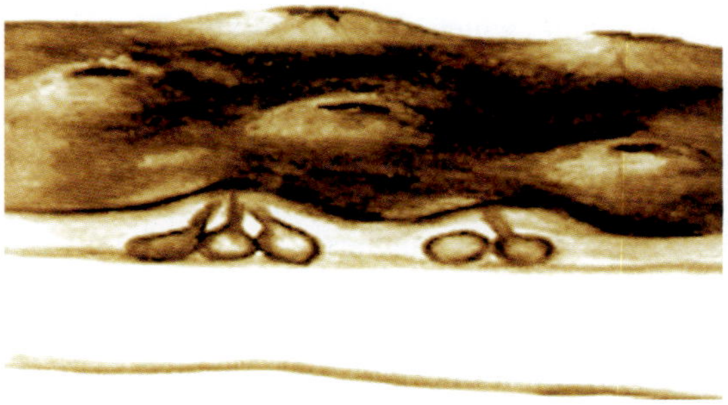

Perithecial stromata of *Diaporthe eres*, **immersed** in an elm branch, x 10 (*MU*).

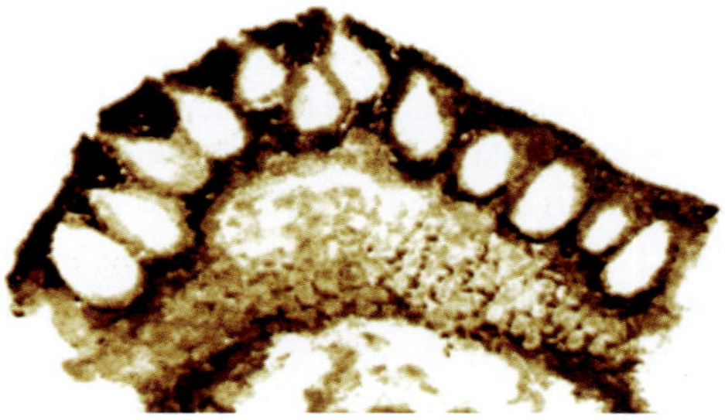

Ascostroma of *Dothiora* (=*Dothidella*) *ribesia*, with ascigerous locules, **immersed** in a stem of red currant (*Ribes* sp.), x 75 (*MU*).

immune (L. *immunis*, free, exempt, privileged): not susceptible to becoming ill when infected by a particular pathogenic germ. Having immunity.

immunize (L. *immunis*, free, exempt, privileged + LL. suf. *-izare* > E. *-ize*, to cause, conform, to turn over, transform): to make immune; the process of increasing the resistance of, or giving resistance to, a living organism (but not in plants) to a particular disease; i.e., to promote the development of this quality during the lifetime of the organism in question, generally as the result of having suffered a disease either naturally or experimentally. This is known as acquired immunity, as contrasted to natural immunity, which is innate in an organism.

immunosuppressant (L. *immunis*, free, exempt, privileged + ME. *suppressen* < L. *supprimere*, to press down): a substance that partially or totally suppresses the immune system of an organism; in medicine it is applied to prevent rejection of implanted organs. Cyclosporine, a metabolite of *Tolypocladium inflatum* (an asexual anamorph of Hypocreales) is used for this purpose.

in L. prep. generally employed with the meaning of in, inside, within (e.g., in the bibliographic citation: Alexopoulos, C.J. 1973. Morphogenesis in the Myxomycetes. *In* G.C. Ainsworth and A.S. Sussman (eds.), *The Fungi* Vol. II. Academic Press, New York, pg. 39-60).

in- (L. *in-*, which denotes negation or inversion of meaning): a pref. that has an exclusive meaning, as in *indefinite*, which means not limited, without end. Before *b* or *p* it is converted to *im-* (*imperfect*, not perfect), to *ir-*, before *r* (*irregular*, not according to the rules, anomalous), and to *i-* when before *l* (*illegitimate*, false).

-in, -ine (NL. *-ina*, likeness, derived from): a suf. employed in chemical terms, as in the alkaloids (*strychnine*, *quinine*), polysaccharides (*chitin*), proteins (*gelatin*), glucosides (*digitalin*), etc.

in situ (L. *in situ* < *in*, in, within + *situ*, site, place): in the natural or original place, like an organism that is living in its natural habitat. Cf. **ex situ**.

in vitro (NL. *in vitro*, literally, in or within glass): outside a living body, or outside its natural habitat and in an artificial environment, like the microbial cultures that are grown in glass vessels in the laboratory. Cf. **in vivo**. See **in situ** and **ex situ**.

in vivo (NL. *in vivo*, literally, in living condition): found in the living body of a plant or animal, like the parasitic microorganisms that are maintained in contact with the living cells of their hosts for purposes of research. Cf. **in vitro**. See **in situ** and **ex situ**.

inaequi- (L. *inaequi-*, unequal < *in-*, not + *aequus*, equal): a pref. used to form terms that reflect inequality. Cf. **equi-**.

inaequihymeniiferous (L. *inaequi*, unequal < pref. *in-*, not + *aequus*, equal + *hymenium*, hymenium, membrane + suf. *-ferous*, bearer < *ferre*, to bear, carry): *Agaricales*. Bearing an hymenium that develops zones of different age, i.e., inaequihymenial, as happens in the gen. *Coprinus* (Agaricales). Cf. **aequihymeniiferous** and **aequihymenial**.

inaequilateral (L. *inaequi*, unequal < pref. *in-*, not + *aequus*, equal + *later*, *latus*, side + suf *-alis* > E. *-al*, relating to or pertaining to): *Of spores*. A spore that is bilaterally asymmetrical, i.e., with one side different from the other, as, e.g., the allantoid and reniform spores, such as the conidia of *Dendrodochium* (moniliaceous asexual fungi) and the ascospores of *Petriella* (Microascales). The ascospores of *Xylaria* (Xylariales), which have one side convex and the other flat (see **asymmetric**) also are inaequilateral. Cf. **equilateral**.

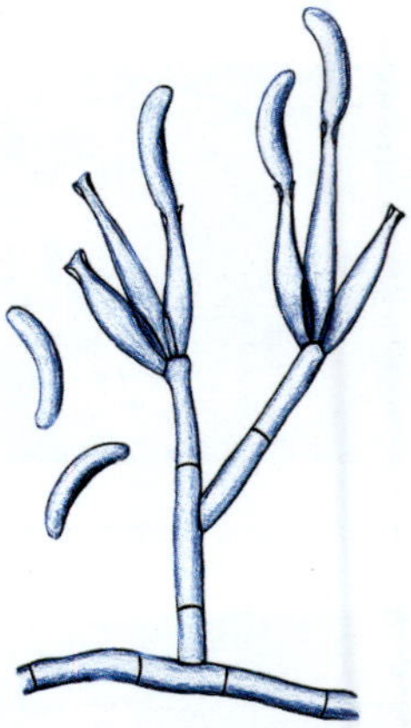

Inaequilateral (or allantospores) conidia of *Dendrodochium* sp., x 600 (*MU*).

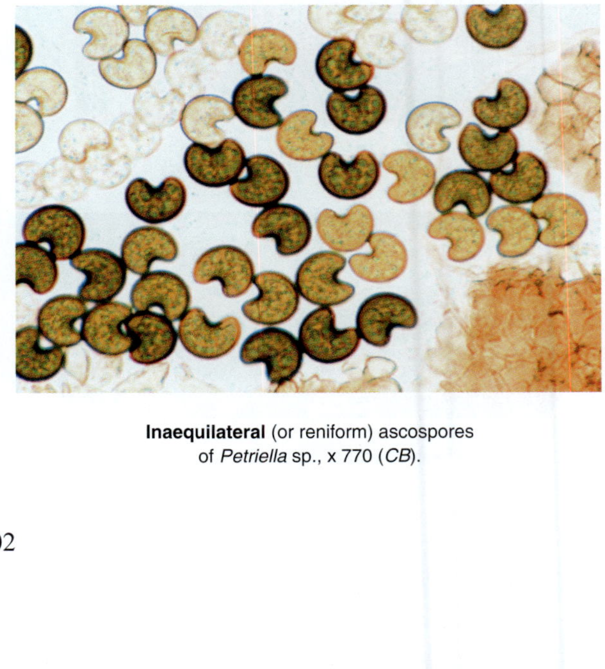

Inaequilateral (or reniform) ascospores of *Petriella* sp., x 770 (*CB*).

incertae sedis (L. phrase, genit. of *incerta sedes*, uncertain place (with regard to taxonomy). *Species incertae sedis*, i.e., a species of doubtful placement in the system of classification, when the author or another taxonomist is uncertain where to place it.

incised (L. *incisus*, cut): refers to the margin of a pileus that on maturing is divided into deep, irregular segments, as in *Oudemansiella canarii*, *Crinipellis* sp. and *Inocybe rimosa* (=*I. fastigiata*), Agaricales, or also to the lobules of the thallus of a foliose lichen, such as *Parmelia* (Lecanorales).

Incised margin of the pileus of *Oudemansiella canarii*, x 0.9 (*MU*).

Incised margin of the pileus of *Crinipellis* sp., x 1 (*RV*).

incompatible (L. *in-*, pref. that denotes negation + E. *compatible* < L. *compatibilis* < *compati*, to go well with): not compatible. With regard to sex in the fungi, it refers to an incompatible thallus or strain, one that cannot cross with another due to fertility barriers, i.e., which is not of the appropriate mating type. Vegetative mycelium also can have incompatibility, and therefore, cannot form a stable heterokaryon due to genetic differences at one or more of the loci that regulate this process.

incrassate (L. *in*, in, toward + *crassus*, thick + suf. *-atus* > E. *-ate*, provided with or likeness): thickened, enlarged; becoming thick.

incrusted (L. *incrustare*, to incrust, cover with a crust): hardened by the interposition of mineral materials among the organic particles. E.g., the wall or peridium of the sporangium of some species of *Mucor*, such as *M. mucedo* and *M. racemosus*, as well as *Actinomucor elegans* (Mucorales), is incrusted with aciculiform crystals of calcium oxalate. Also, the sporangial peridium of certain Myxomycetes, such as *Badhamia gracilis* and *Physarum compressum* (Physarales), is incrusted with calcium carbonate, forming a calcareous crust on the surface. In the Agaricales there exist various types of hyphae whose wall is incrusted with different materials that give rise to distinctive ornamentations (spirals, warts, spines, ribs, needles, etc.).

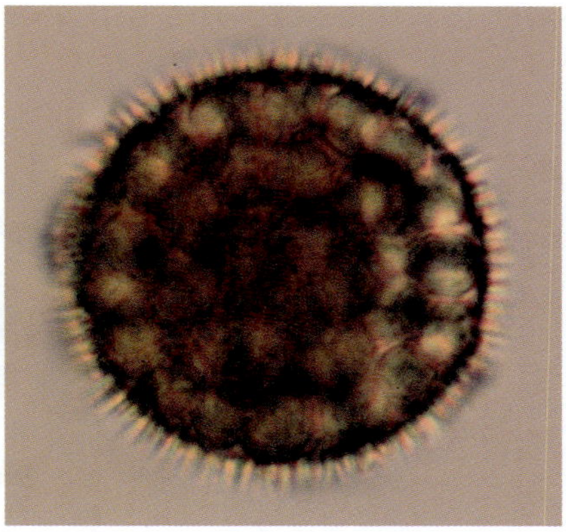

Incrusted peridium of the sporangium of *Actinomucor elegans*, x 850 (*MU*).

Incrusted peridium of the sporangium of *Badhamia gracilis*, x 50 (*MU*).

incurvate, **incurved** (L. *incurvatus*, incurved < *incurvare*, to bend in or curve inward < *in*, in, within + *curvare*, to curve < *curvus*, curved + suf. *-atus* > E. *-ate*, provided with or likeness): syn. of **inflexed**; curved inward, i.e., with the concavity on the inside, like the border of the pileus of *Entoloma sinuatum* (=*E. lividum*), *Lactarius indigo*, *L. vellereus*, *Russula delica* and *Chlorophyllum* (Agaricales). Cf. **recurved** and **reflexed**.

Incurved pileus of the basidiocarp of *Entoloma sinuatum* (=*E. lividum*), x 0.8 (*MU*).

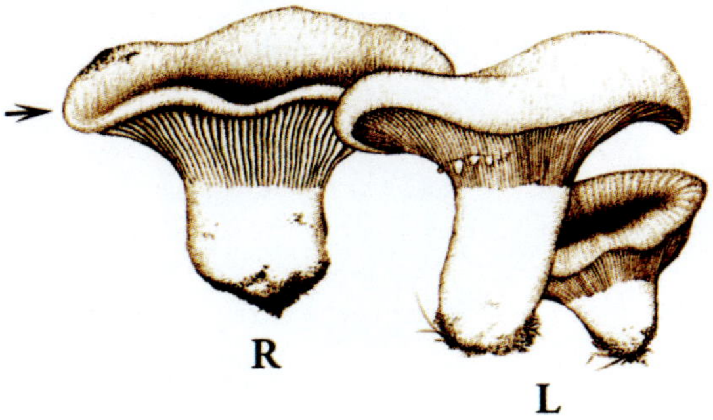

Incurved pileus of the basidiocarps of *Russula brevipes* (R) and *Lactarius piperatus* (L), x 0.3 (*MU*).

Incurved pileus of the basidiocarp of *Chlorophyllum* sp., x 0.5 (*RV*).

indefinite (L. *indefinitus*, without a specific limit or end): refers to structures too numerous to be easily counted, or which do not have a specific limit.

indehiscence (NL. *in-*, pref. that denotes negation + *dehiscentia*, dehiscence, bursting open < L. *dehiscere*, to open, be exposed + suf. *-entia* > E. *-ence*, state, quality or action): the condition of not rupturing or splitting at maturity; often refers to the peridium. Cf. **dehiscence**.

indehiscent (L. *indehiscens*, genit. *indehiscentis* < pref. *in-*, not + *dehiscere*, to open, be exposed + suf. *-entem* > E. *-ent*, to be): not opening naturally. Cf. **dehiscent**.

indeterminate (LL. *indeterminatus*, not determined, indefinite < *in-*, not + *determinatus*, delimited, defined < *determinare*, to determine, delimit + suf. *-atus* > E. *-ate*, provided with or likeness): **1**. The continuous development of sporangia or conidia due to the elongation of the producing hypha; i.e., no terminal sporangia or conidia are formed, thus permitting indefinite growth. E.g., in the Peronosporales of the family Pythiaceae (such as *Pythium* and *Phytophthora*) the sporangia are formed on somatic hyphae or on sporangiophores of indeterminate growth. **2**. *Lichens*. Refers to certain crustaceous thalli whose margin is difficult to determine due to its being immersed in the substrate where it grows, as in some species of *Lecanora*, of the Lecanorales. Cf. **determinate**.

indigenous (L. *indigena*, native, indigenous, not foreign): belonging to the place or country where it is found; not having come from outside. Syn. of **autochtonous**.

indumentum, pl. **indumenta** (L. *indumentum*, indument, a garment < *induere*, to dress): the ensemble of hairs, scales, etc., that cover the surface of specific organs or structures, as occurs in the basidiocarp of *Trametes* (=*Polyporus*) *versicolor* and other Aphyllophorales.

indurate, indurated trama, pl. **tramae** (L. *induratus*, hardened < *indurare* < *in*, in, toward + *durare*, to harden < *durus*, hard + suf. *-atus* > E. *-ate*, provided with or likeness; L. *trama*, trama, tissue): a trama that hardens with age, like the tramal plates (gussets) in the mature sporocarps of *Montagnea arenaria* (Hymenogastrales).

indusium, pl. **indusia** (L. *indusium*, tunic, shirt): a white, net-like veil that unfolds like a petticoat hanging from the receptacle, beneath the gleba, of the expanded fruiting body of *Phallus indusiatus* (=*Dictyophora indusiata*), a gasteromycete of the order Phallales.

inerm, **inermous** (L. *inermis*, unarmed, without arms): without spines or thorns.

Indurate tramal plates (gussets) of the fruiting body of *Montagnea arenaria*, x 1 (*MU*).

Indusium of the fruiting body of *Phallus indusiatus* (=*Dictyophora indusiata*), x 0.8 (*TH*).

infect (L. *infectare*, *inficionare*, to corrupt, infect): *Pathology*. The process by which an organism enters, invades and establishes a pathogenic relationship with a host organism; to communicate a pathogen or a disease; to carry out an attack on an organism. *Infected*: the result of an attack by a pathogen on an organism. *Infection*: the action and effect of infecting; the process during which a pathogen extracts nutrients from its host. *Infectious*: capable of causing infection; communicable by infection; able to be passed on by contact (contagious) or by inoculation; capable of spreading rapidly from one host to another. *Infective*: a pathogen capable of attacking a living organism; vector, medium, etc., capable of carrying out the transmission of a pathogen. Cf. **contaminate** and **infest**.

infest (L. *infestare*, to infest, cause damage < *infestus*, hostile): to disseminate and invade in a harmful manner, as with insects, rodents and other animals that attack agricultural crops. Sometimes used in the sense of *contaminate*, for fungi in soil and other substrates. Cf. **infect**.

inflated (L. *inflatus*, inflated < *inflare*, to inflate, swell up < *in*, inward + *flare*, to blow + suf. *-atus* > E. *-ate*, provided with or likeness): swollen like a vesicle, like some basidiomycete cystidia, and the base of the appendages of the ascocarp of *Phyllactinia guttata* (=*Ph. corylea*, Erysiphales), which are inflated or bulbous.

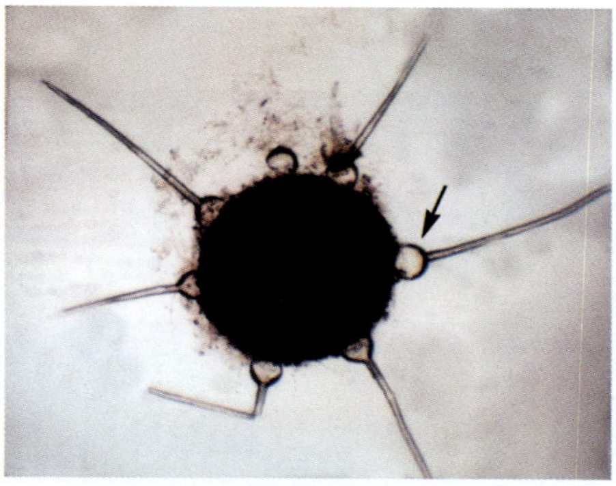

Inflate, or bulbous, base of the appendages of the ascocarp of *Phyllactinia guttata* (=*Ph. corylea*), x 190 (*MU*).

inflated hypha, pl. **hyphae** (L. *inflatus*, inflated < *inflare*, to inflate, swell < *in*, inward + *flare*, to blow + suf. *-atus* > E. *-ate*, provided with or likeness; NL. *hypha*, hypha < Gr. *hyphé*, tissue, spiderweb; hypha): *Basidiomycetes*. A hypha with subapical cells that enlarge and undergo the rapid growth that characterizes the majority of the basidiocarps of the Agaricales and Gasteromycetes.

inflexed (L. *inflexus*, curved inward): syn. of **incurved**; curved inward with respect to the pedicle, branch, etc., on which it is borne; e.g., the margin of the pileus of the basidiocarp of some species of *Clitopilus*, *Clitocybe* and *Coprinus* (Agaricales) is inflexed. Cf. **reflexed**.

Inflexed pileus of the basidiocarps of *Clitopilus prunulus*, x 0.8 (*RV*).

infra-, infr- (L. *infra*, beneath, inferior): a pref. employed in the formation of terms referring to something that is beneath something else, as in *infract*, to violate a commitment; broken, weakened; and as in *infrageneric* and *infraspecific*, to refer to taxonomic categories that are below genus and species, respectively.

infundibuliform (L. *infundibulum*, funnel + suf. *-formis* < *forma*, shape): with the shape of a funnel, like the fruiting body of *Clitocybe infundibuliformis* (Agaricales), *Cantharellus cibarius*, *Lentinus badius*, *Polyporus tuberaster* (Aphyllophorales) and *Guepinia helvelloides*, of the Tremellales, among others.

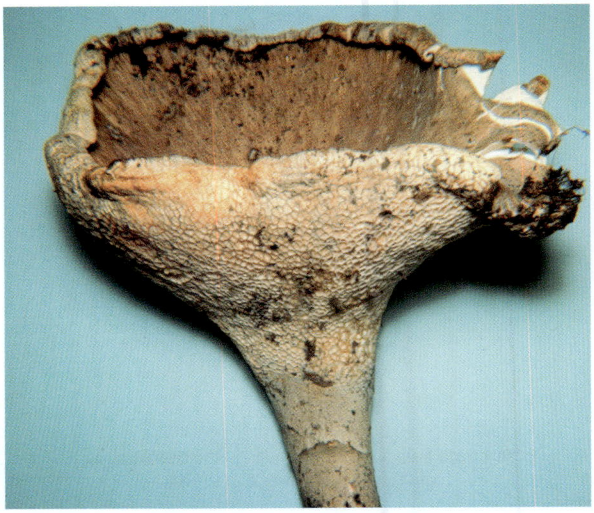

Infundibuliform basidiocarp of *Polyporus tuberaster*, x 1 (*TH*).

Infundibuliform basidiocarps of *Clitocybe infundibuliformis*, x 0.5 (*EPS*).

Infundibuliform basidiocarps of *Clitocybe gibba*, x 0.5 (*RV*).

Infundibuliform basidiocarp of *Lentinus badius*, x 2 (*CB*).

Infundibuliform basidiocarps of *Guepinia helvelloides*, x 1 (*MRO*).

Infundibuliform basidiocarps of *Cotylidia aurantiaca* (Aphyllophorales), x 1 (*RV*).

ingestion (L. *ingestio*, genit. *ingestionis*, the action of introducing food or medication into the mouth): in mycology, it refers to obtaining food by means of *phagocytosis*. Cf. **absorption**.

ingroup (ME. < OE. *in* < L. *in* < Gr. *en* + Fr. *groupe* < It. *gruppo*, group, an assemblage of beings or things that form a complex that shares some unifying character): a term used in systematics to refer to a group of taxa that generally are considered as monophyletic; usually it is the group being studied. Cf. **outgroup**.

inhabitant (L. *inhabitare*, to live in a place or house): a term applied to an integral organism of a lichen symbiosis that is included in the internal tissues of the thallus, i.e., the photobiont or algal symbiont. Cf. **exohabitant**.

inhibition (L. *inhibitus*, to prohibit, restrict or impede the functioning of something): the act of inhibiting; the state of being inhibited. Applied to the effect caused by a physical, chemical or biochemical agent on the growth or other function of an organ or organism, as happens, e.g., with antibiotics and other substances produced by the metabolism of an organism, which have an inhibitory effect on other organisms.

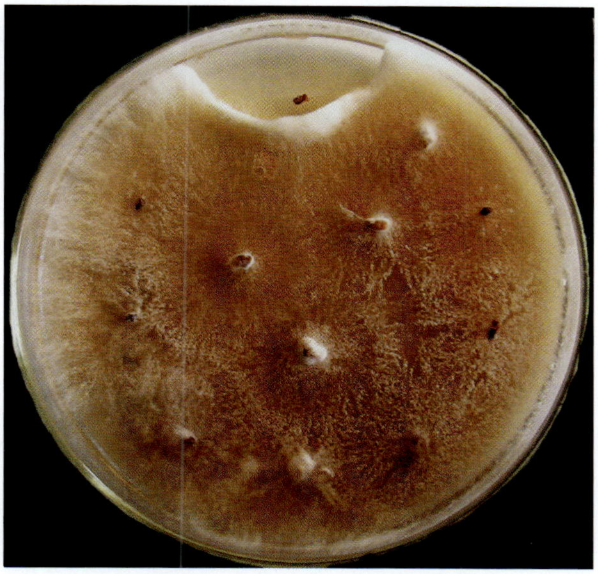

Inhibition of the growth of *Rhizoctonia solani* (agonomycetaceous asexual fungi, growing from portions of wheat grains) caused by an antibiotic released from a colony of a bacterium, x 1 (*JAS*).

ink-cap fungus, pl. **fungi** (ME. *enke*, ink < OF. < LL. *encaustum* < L. *encaustus*, burned in; ME. *cappe* < OE. *caeppe* < LL. *cappa*, head covering, cap; L. *fungus*, fungus < Gr. *spóngos*, *sphóngos*, sponge): the deliquescent or autodigesting basidiocarp of *Coprinus* spp. (Agaricales). See **autolysis** and **deliquescence**.

Ink-cap fungi, the basidiocarps of *Coprinopsis atramentaria* (=*Coprinus atramentarius*), x 1 (*RV*).

innate (L. *innatus*, natural, innate, inborn < *innasci*, born in): applied to scales, fibers, etc., that form part of a tissue; i.e., that are immersed in the tissue, not on its surface.

inoculation (L. *inoculatio*, genit. *inoculationis* < *inoculare*, to implant, inject, to sow; to comunicate a contagious disease + suf. *-ation* > E. *-ation*, action, state, condition or result): the act or process of inoculating; especially the introduction of a pathogen into a living organism. See **inoculum**.

inoculate (L. *inoculare*, to implant, graft, sow; to communicate a contagious disease): to introduce or put a microorganism (usually a pathogen), or a substance that contains it, into an organism (the host) or on a substrate.

inoculum, pl. **inocula** (E. *inoculum*, a small portion of a germ, generally a pathogen, that is transferred to an organism or substrate to inoculate it, where it can grow and develop < L. *inoculare*, to implant, inject, seed; to communicate a contagious disease): fungal material (mycelium, spores, etc.) that is transferred to an organism, substrate or culture medium for the purpose of propagating it. In phytopathology and medical mycology it refers to the portion of the pathogen that is transferred to the host, whether artificially or naturally, with the object of causing the infection of the host.

inoperculate (L. *in-*, pref. that denotes negation + *operculatus*, operculate < *operculum*, lid, operculum + suf. *-atus* > E. *-ate*, provided with or likeness): lacking an operculum, like the asci and sporangia whose dehiscence is not by means of an operculum, but irregular or of another type. In the Discomycetes, the asci of the gen. classified in the orders Rhytismatales (such as *Rhytisma*), Leotiales (such as *Vibrissea*) and Helotiales (such as *Monilinia*) are inoperculate. Cf. **operculate**, **pseudoperculate** and **suboperculate**.

inserted (L. *insertus* < *inserere*, to insert, join): refers to a structure that originates directly from the substrate, without being attached by rhizoids or fibrils of any type, like the stipe of the basidiocarp of *Hygrophorus chrysodon* (Agaricales).

integrated (L. *integratus* < *integrare*, to integrate, form, coordinate or complete a funcional whole + suf. *-atus* > E. *-ate*, provided with or likeness): *Conidiogenesis*. A conidiogenous cell that is incorporated into the main axis or branches of the conidiophore; it generally resembles the conidiophore. Cf. **discrete**.

inter- (L. *inter*, prep. which means among, in the middle of, in between, intermediate, between the two, mixed with quality, situation or intermediate state): a pref. that denotes the situation or state between two or more things or actions, as in *intercross* (to cross two or more things with each other, to hybridize, intertwine) and *intercellular* (that found between the cells). Cf. **intra-**.

interbiotic (L. *interbioticus* < *inter*, among + *bioticus* < Gr. *biotikós*, relating to or belonging to life or biosis < *bíos*, life + suf. *-tikós* > L. *-ticus* > E. *-tic*, relation, fitness, inclination or capacity): a parasitic organism whose situation with respect to the host is intermediate between endobiotic and epibiotic; if the organism is saprobic it refers to its position in the substrate. Thus there are some aquatic fungi, such as *Chytridium olla*, *Chytriomyces hyalinus* and *Rhizophydium sphaerotheca* (Chytridiales), that are interbiotic. The first is a parasite of algae; it forms sporangia on the exterior and rhizoids on the interior of the algal thallus. The second is a chitinolytic saprobe; it forms a rhizoidal system inside insect exuviae or fragments of chitin, and its sporangia on the exterior of these substrates. The third is parasitic on pollen grains; it forms rhizoids inside and sporangia outside the grains. The Laboulbeniales also are interbiotic, since their thalli or receptacles are found on the surface of the exoskeleton of the host insects, but the haustoria are located in the interior of the host tissues. Cf. **endobiotic** and **epibiotic**.

intercalary (L. *intercalaris*, of or for insertion < *intercalare*, to insert): interposed between the apex and the base, i.e., neither terminal nor basal. It refers to resistant or reproductive structures, such as chlamydospores, sporangia, asci, etc., when they are bordered on both ends by vegetative cells. Also applied to growth that is not apical, but located in the base or, rather, in the middle part of the organ or structure being considered. E.g., the asci of *Nematospora* (=*Ashbya*) *gossypii* (Endomycetales) and, occasionally, those of *Pichia membranifaciens* (Saccharomycetales), are intercalary.

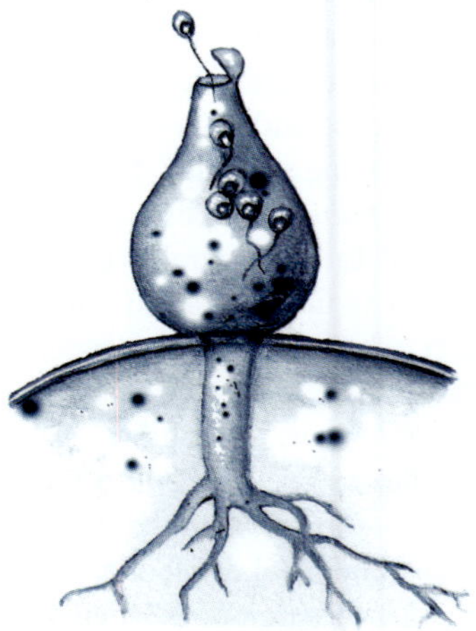

Interbiotic thallus of *Chytridium olla* in an algal host; the sporangium represents the epibiotic part of the thallus, and the rhizoids the interbiotic part, x 475 (*MU*).

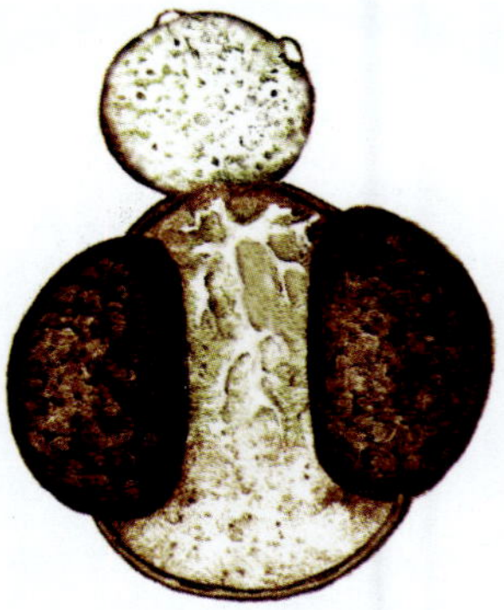

Interbiotic thallus of *Rhizophydium sphaerotheca*, growing on a pollen grain, x 300 (*MU*).

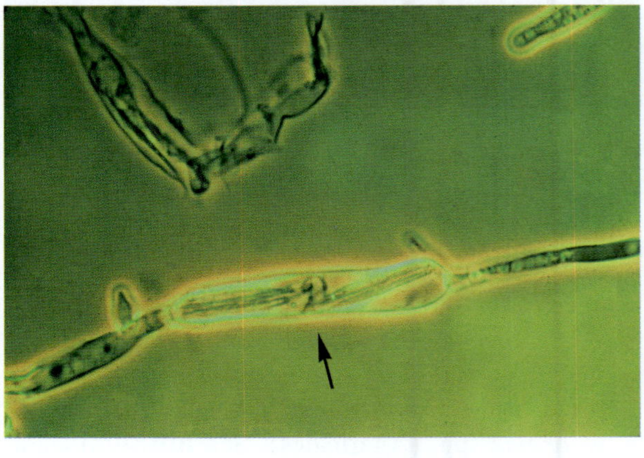

Intercalary asci of *Nematospora* (=*Ashbya*) *gossypii*, x 920 (*RTH*).

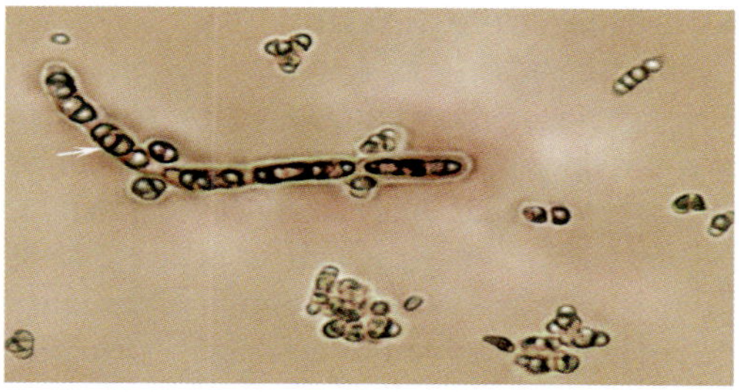

Intercalary asci of *Pichia membranifaciens*, x 920 (*MU*).

intercellular (L. *intercellularis* < *inter*, among or in the middle of two or more things + *cellularis*, pertaining to or relating to the cell < NL. *cellula*, cell, living cell, dim. of *cella*, small room, compartment, cell of a honeycomb): something that is found between the cells, like the hyphae of many phytopathogenic fungi, which penetrate through the stoma or directly through the cuticle and epidermis of the host plant, later invading the intercellular spaces of the tissues. Cf. **intracellular**.

intermediate host (L. *intermedius* < pref. *inter-*, between + *medius*, middle, in the middle; L. *hospes*, genit. *hospitis*, a living organism that provides food and lodging to a parasite; one who receives a stranger as guest or visitor): a host plant that enables a fungus to infect a species of plant that it otherwise could not infect. This type of host allows certain rusts (Uredinales) to adapt to species of host plants distinct from the one usually infected; also called **bridging host**, or *gephyrophyte*. Cf. **alternate host** and **primary host**.

internode (L. *internodium*, internode): the portion encompassed between two nodes or consecutive joints, as occurs between the radulae of the percurrent conidiophore of *Arthrobotrys oligospora* (moniliaceous asexual fungi) and *Gonatobotryum apiculatum* (dematiaceous asexual fungi).

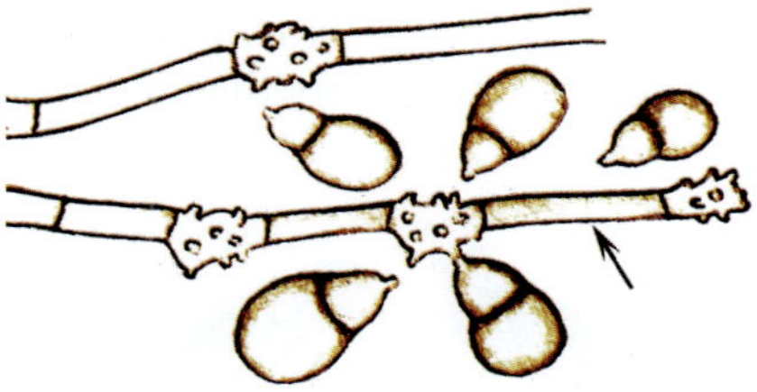

Internodes of the conidiophore of *Arthrobotrys oligospora*, x 1,000 (*MU*).

interspace (L. *interspatum* < *inter*, between, among + *spatium*, space): the space between two gills of a sporophore of the agaricaceous fungi; *lamellar interspace* is the more precise term for this.

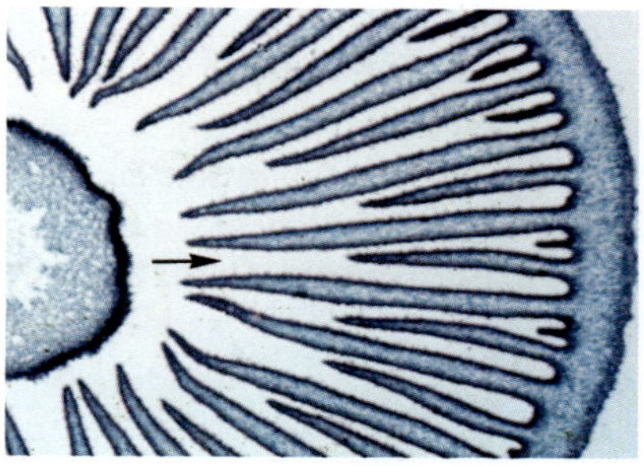

Lamellar **interspaces** of the basidiocarp of *Coprinus* sp. (Agaricales), seen in transverse section, x 25 (*MU*).

interstice (LL. *interstitium* < L. *interstitus* < *intersistere*, found in the middle): a small space in the middle between two bodies or between two parts of the same body.

interthecial (L. *inter*, between + *thecicus* < *theca* < Gr. *thêke*, box; here, of the asci + L. suf. *-alis* > E. *-al*, relating to or pertaining to): found between the asci, or between the ascigerous locules, such as the plectenchyma of the ascostroma of the Myriangiales.

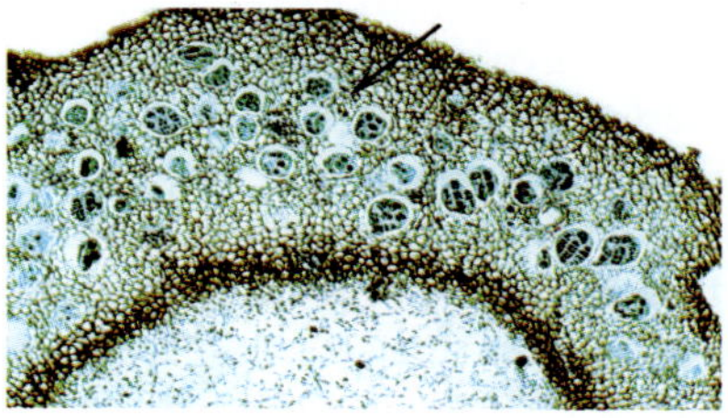

Interthecial plectenchyma of an ascostroma of *Myriangium duriaei*, x 80 (*RTH*).

intervenose (L. *intervenosus* < *inter*, between + *venosus*, venous, full of veins < *vena* + suf. *-osus* > E. *-ose*, full of, augmented, prone to): applied to agaric fruiting bodies that have veins in the interspaces.

intra- (L. *intra-*, in, inside): a pref. that means within, in and among, as in the following terms.

intracellular (L. *intra-*, a pref. that denotes in, within + *cellularis*, pertaining to or relating to the cell < NL. *cellula*, cell, living cell, dim. of *cella*, small room, compartment, cell of a honeycomb + suf. *-aris* > E. *-ar*, pertaining to): found within the cell, like the haustoria of many phytopathogenic fungi. Cf. **intercellular**.

intrahyphal (L. *intra-*, a pref. that denotes in, inside + NL. *hypha* < Gr. *hyphé*, tissue, spiderweb; hypha + L. suf. *-alis* > E. *-al*, relating to or pertaining to): present inside a hypha; e.g., the development of one hypha inside another, as occurs in the somatic hyphae of *Myrioconium* (moniliaceous asexual fungi).

intramatrical (L. *intra-*, a pref. that denotes inside, in, among + *matrix*, matrix, a place where something is generated, or the host plant + L. suf. *-alis* > E. *-al*, relating to or pertaining to): an organism that lives in the interior of the host, within the substrate. Equivalent to **endobiotic**. Cf. **extramatrical**.

intraparietal (L. *intra-*, a pref. that denotes within, in, among + *parietalis* < *paries*, genit. *parietis*, wall + suf. *-alis* > E. *-al*, relating to or pertaining to): occurring in the wall or membrane; e.g., like the segments that occur in the interior of the hyphal wall in the shape of spirals, rings or irregular heaps, resulting in a certain type of internal ornamentation, as in *Panaeolus sphinctrinus* (Agaricales).

intrasporangial (L. *intra-*, inside, in, within + NL. *sporangium*, pertaining to or relating to the sporangium < Gr. *sporá*, spore + *angeîon*, receptacle + L. suf. *-alis* > E. *-al*, relating to or pertaining to): developing inside a sporangium. See **proliferation**.

intricate (L. *intricatus*, intrincate, entangled, complicated < *intricare*, to entangle + suf. *-atus* > E. *-ate*, provided with or likeness): entangled, but not fused; having many complexly interrelated parts or elements. In lichenology it is applied to a type of thalline cortex composed of hyphae that are twisted and entangled with one another.

intron (E. *inter*vening, to come between + E. des. *-on*, in contact with or supported by): a term employed in molecular biology to refer to a sequence of polynucleotides in a nucleic acid that does not code information for protein synthesis and is removed before translation of the messenger RNA. Cf. **exon**.

introrse (L. *introrsus*, toward the inside): next to the central axis, toward the inside. Cf. **extrorse**.

intumescence (L. *intumescens*, intumescent, swollen < *intumescere*, to swell + suf. *-escens*, genit. *-escentis* > E. *-escence*, that which turns, that begins to, slightly): a swelling due to the flow of watery liquids into a tissue.

intususception (L. *intus*, within, toward the inside + *susceptio*, genit. *susceptionis*, to receive, accept < *susceptus*, undertaken < *suscipiere*, to take, perceive + suf. *–io*, *-ionis* > E. *-ion*, result of an action, state of): applied to growth in thickness of the cell wall by intercalation among the molecules or compounds that constitute it, by other elements of the same nature. The growth occurs, therefore, without superficial internal or external deposits of molecules or layers; the new wall materials are laid down within the original thickness of the wall. The subapical growth of fungal hyphae occurs by intususception. Cf. **apposition**.

invaginated, **invagination** (ML. *invaginatus*, sheathed, enclosed < L. *invaginare*, to cover, envelop < *in*, in + *vagina*, covering, sheath + suf. *-atus* > E. *-ate*, provided with or likeness): **1**. Contained in a covering or sheath, like the stylet (originally called *Stachel*, in German), which is inside a tube (the *Rohr*, in German), which in turn is in the interior of the encysted zoospore of *Plasmodiophora brassicae* (Plasmodiophorales). **2**. *Gasteromycetes*. An inward growth from the peridium, as occurs in forate development, forming multiple cavities in which the basidia are found. See **forate**.

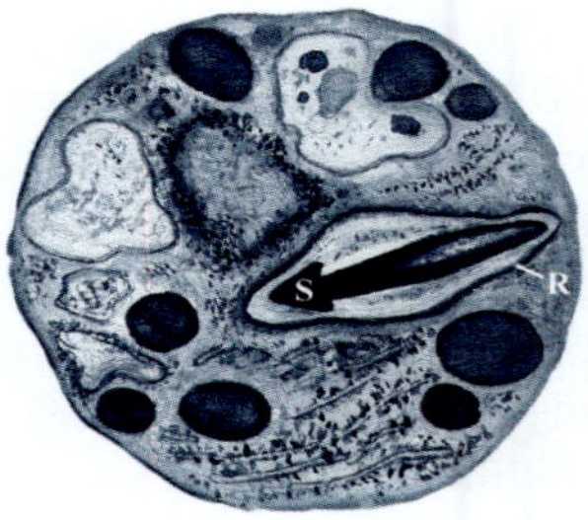

Invaginated stylet (Stachel, S) in the tube (Rohr, R) of an encysted zoospore of *Plasmodiophora brassicae*. Drawing based on a transmission electron micrograph published by Beckett *et al.* in their *Atlas of Fungal Ultrastructure*, 1974, x 20,500 (*MU*).

inverse (L. *inversus*, opposite in order, nature or effect): placed so that the lower side or face is on top, or vice versa. See **convergent**.

involucellum, pl. **involucella**, **involucrellum**, pl. **involucrella** (L. *involucellum* < *involucrum*, wrapper, covering + dim. suf. *-ellum*): the upper region of the ascocarp of some marine ascomycetes, such as *Pharcidia* (=*Stigmidium*) *laminariicola*, of the Dothideales, whose peridium is especially thick around the ostiole, and which is composed of thick-walled, subglobose or somewhat flattened cells in the central part of the pseudothecium. Some loculoascolichens, such as *Verrucaria* and *Staurothele* (Verrucariales), have an involucellum that extends outward from the ostiole, forming a carbonaceous, scutiform layer.

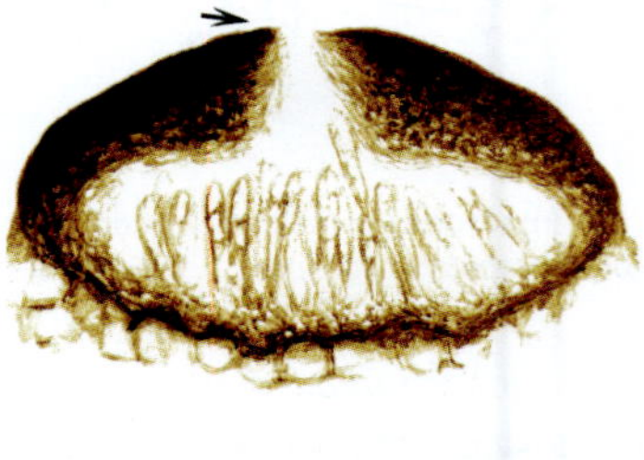

Involucellum of the peridium of a pseudothecium of *Pharcidia* (=*Stigmidium*) *laminariicola* parasitizing the alga *Laminaria digitata*. Drawing of a longitudinal section, x 520 (*MU*).

involute (L. *involutus*, turned inward, coiled in a spiral; entangled < *involvere*, coiled, rolled up): *Agaricales*. A pileus in which the borders are more or less rolled toward the upper face of the pileus; involute is like *decurvate*, but rolled up, and it can be exemplified by the pileus of the basidiocarp of some species of Agaricales of the gen. *Marasmius*, *Coprinus*, and *Coprinopsis* (=*Coprinus*) *lagopus*. Cf. **revolute**.

Involute margin of the pileus of *Coprinopsis* (=*Coprinus*) *lagopus*, x 1 (*EPS*).

Involute margin of the pileus of *Coprinopsis* (=*Coprinus*) *lagopus*, x 1 (*MU*).

involute ray (L. *involutus*, turned inward, coiled in a spiral; entangled < *involvere*, coiled, rolled up; L. *radius*, radius, ray): *Gasteromycetes*. A ray curved or coiled upward and inward at the border, as is seen in the fruiting body of *Astraeus hygrometricus* (Sclerodermatales), on dehydrating.

Involute rays of the exoperidium of the fruiting body of *Astraeus hygrometricus*, x 1.5 (*MU*).

ionomidotic (Gr. *ión*, genit. *iontós*, violet in color, like the gen. *Ionomidotis*, of the Helotiales + Gr. suf. *-ikos* > L. *-icus* > E. *-ic*, pertaining to or relating to): refers to a pigment dark violet in color that is liberated when a fragment of the fruiting body of *Ionomidotis* sp. (syn. of *Ameghiniella*) is mounted in an aqueous solution of potassium hydroxide. This reaction is useful in the taxonomic determination of some apothecia of dark-colored Helotiales, e.g., *Claussenomyces*.

iridescent (L. *iris*, genit. *iridis*, iris, rainbow + suf. *-escens*, genit. *-escentis* > E. *-escent*, that which turns, that begins to, slightly): showing a range of colors similar to those of a rainbow, like the peridium of the sporangia of Myxomycetes of the gen. *Prototrichia* (Trichiales), *Diachea* (Physarales) and *Lamproderma* (Stemonitales), which have the property of reflecting various colors of light.

Sporangia of *Prototrichia metallica*, with an **iridescent** peridium, x 20 (*MU*).

Sporangia of *Lamproderma columbinum*, with an **iridescent** peridium, x 20 (*MU*).

Sporangium of *Lamproderma columbinum*, with an **iridescent** peridium, x 20 (*MU*).

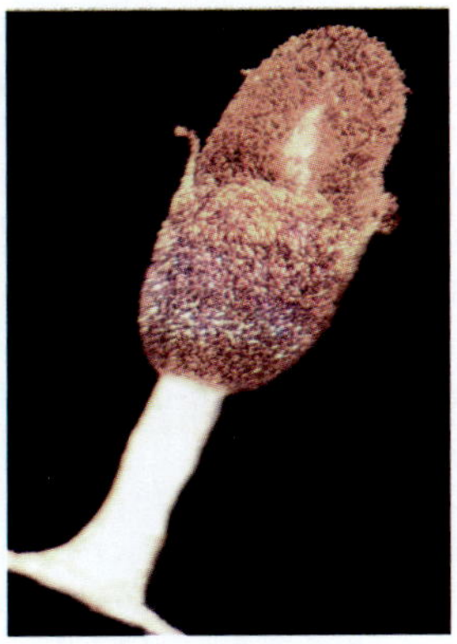

Sporangium of *Diachea leucopoda*, with an **iridescent** peridium, x 20 (*MU*).

Sporangia of *Diachea splendens*, with an **iridescent** peridium, x 20 (*RS*).

Sporangia of *Elaeomyxa cerifera*, with an **iridescent** peridium, x 20 (*MU*).

irpiciform, **irpiciniform** (< gen. *Irpex* < L. *irpex*, genit. *irpicinis*, rake or harrow with iron teeth, dentate + suf. *-formis* < *forma*, shape): with flattened teeth, similar to those of the hymenophore of *Irpex*; e.g., *Paxillus*, *Tapinella* and *Hydnochaete* (Aphyllophorales).

Irpiciform hymenophore of *Tapinella* (=*Paxillus*) *panuoides*, x 0.5 (*RV*).

Irpiciform hymenophore of *Tapinella* (=*Paxillus*) *panuoides*, x 2 (*RV*).

Irpiciform hymenophore of *Hydnochaete olivacea*, x 1 (*RV*).

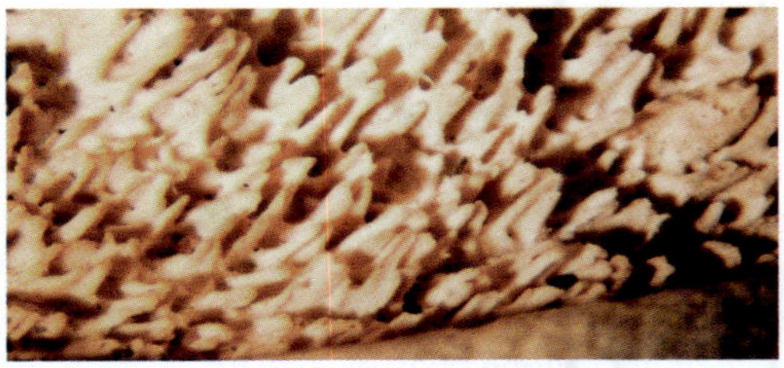

Irpiciform hymenophore of *Irpex lacteus*, x 20 (*MU*).

isarioid (< gen. *Isaria* < Gr. *ísis*, coral + L. suf. *-aria*, similar, equal + L. suf. *-oide*, like < Gr. *-oeídes*, similar to): similar to the synnemata of *Isaria* (stilbellaceous asexual fungi), in the shape of a cylinder or bundle of hyphae.

Isarioid synnemata of *Isaria* sp., x 5 (*TH*).

isidiate (NL. *isidium* < Gr. *ísis*, coral + L. dim. suf. *-idium* + suf. *-atus* > E. *-ate*, provided with or likeness): having *isidia*.

isidiiferous (NL. *isidium* < Gr. *ísis*, coral + L. dim. suf. *-idium* + L. suf. *-ferous*, bearer < *ferre*, to bear, carry): a lichen thallus bearing isidia.

isidium, pl. **isidia** (NL. *isidium* < Gr. *ísis*, coral + L. dim. suf. *-idium*): *Lichens*. A microscopic outgrowth or papilla formed on the surface of the thalli of some lichens that serves to vegetatively propagate the species. Isidia differ from soredia by being covered by the cortical layer of the thallus, as well as in shape, which can be cylindrical, globose, as in *Hypotrachyna* (=*Parmelia*) *imbricatula* (of the Lecanorales), claviform (as in *Pseudevernia consocians*, of the Lecanorales), squamiform, disciform, cupuliform (as in *Phyllophiale alba*, of the Trichotheliales) or coralloid. Cf. **soredium**.

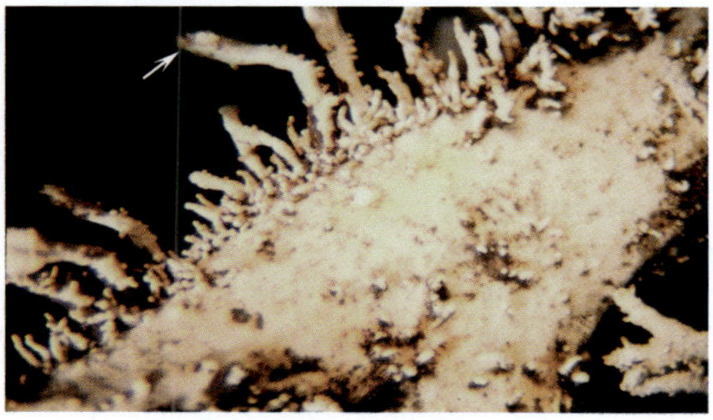

Claviform **isidia** of the thallus of *Pseudevernia* sp., x 60 (*MU*).

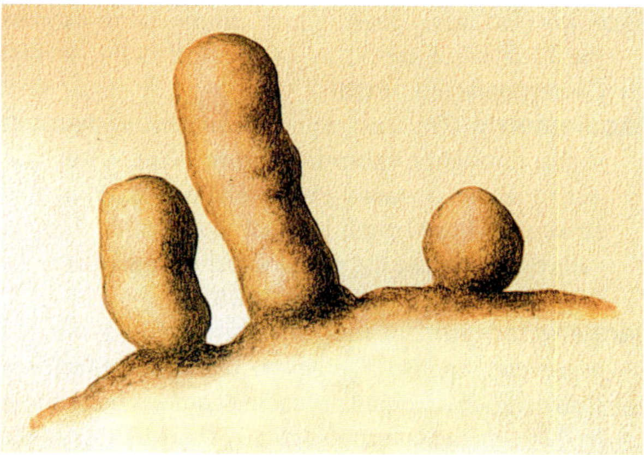

Cylindrical **isidia** of the thallus of *Hypotrachyna* (=*Parmelia*) *imbricatula*, x 175 (*MU*).

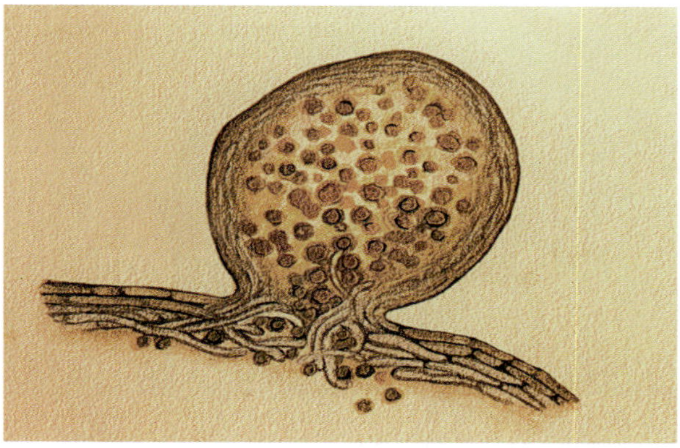

Globose **isidium** of *Hypotrachyna* (=*Parmelia*) *imbricatula*, seen in longitudinal section, showing the gonidia inside, x 170 (*MU*).

Cupuliform, disciform, **isidia** of *Phyllophiale alba*, x 17 (*MAH*).

-ism (Gr. *-ismós* > L. *-ismus*, state, phase, habit, tendency, action, doctrine, system): a suf. employed in terms such as *antagonism*, which is the phenomenon relating to antagonists, understanding as such the parasite and the host in an antagonistic symbiosis.

iso-, **is-** (Gr. *ísos*, equal): a pref. employed to express the idea of equality, as in the following terms. Corresponds to L. **equi-**.

isodiametric (Gr. *ísos*, equal + *diámetron*, diameter < *diá*, through + *métron*, measure + suf. *-íkos* > L. *-icus* > E. *-ic*, pertaining to or relating to): refers to a structure whose length is equal to its width, like the cells that compose the pseudoparenchyma of various fungal organs. Cf. **anisodiametric**.

isogametangium, pl. **isogametangia** (Gr. *ísos*, equal + *gamétes*, spouse + *angeîon*, vessel, receptacle): a gametangium or organ producing gametes that cannot be distinguished morphologically from another of the opposite sex, as is observed in *Rhizopus* and other Mucorales, such as *Rhizomucor* (=*Mucor*) *miehei*. Cf. **heterogametangium**.

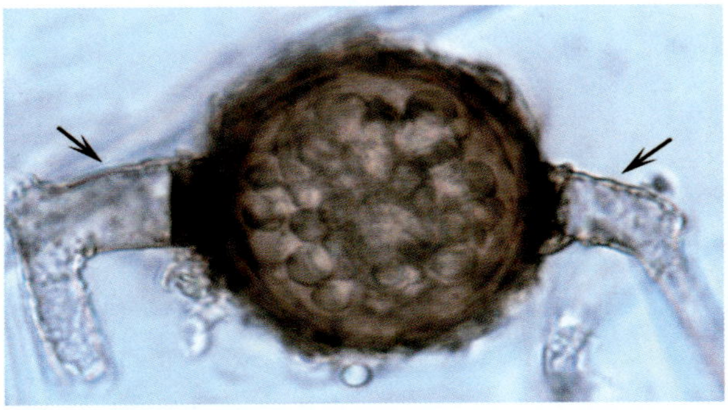

Isogametangia of *Rhizomucor* (=*Mucor*) *miehei*, x 350 (*MU*).

isogamete (Gr. *ísos*, equal + *gamétes*, spouse): a gamete (cell or sexual nucleus) that cannot be distinguished morphologically from another of the opposite sex; generally isogametes are flagellate and motile (isoplanogametes) and they are present in aquatic fungi, such as *Olpidium* (Spizellomycetales). Cf. **anisogamete** and **heterogamete**.

isogamy (Gr. *ísos*, equal + *gámos* > E. *-gamy*, wedding, sexual union): the fertilization of two gametes that are similar in shape, size and structure (*isogametes*). The gametes are indistinguishable from one another, for which reason they are called + and - isogametes; they are commonly flagellate and motile, and are formed in the aquatic fungi, such as *Olpidium* (Spizellomycetales) and *Synchytrium* (Chytridiales). Cf. **anisogamy** and **heterogamy**.

isohaplont (Gr. *ísos*, equal + *haplonte* < Gr. *haplóos*, simple, single + *óntos*, genit. of *ón*, a being): a haplont that has genetically equal nuclei. Cf. **mictohaplont**.

isokont, **isocont**, **isokontous**, **isocontous** (Gr. *ísos*, equal + *kontós*, oar; here, flagellum): with flagella of equal length, like the zoospores of *Ectrogella bacillariacearum* (Myzocytiopsidales) and of some other species of aquatic zoosporic fungi. It should be noted that although in this case the two flagella are more or less equal in length, only one of them is mastigonemate. Cf. **anisokont** and **heterokont**.

isolate (E. *isolated*, separated, set apart, alone < Fr. *isolé* < It. *isolato* < *isola* < L. *insula*, island, something separated from others): a strain or population of an organism that is maintained in pure culture, either active or preserved, under laboratory conditions, for research, teaching, or technological development for industrial production. Cf. **strain**.

isolation (E. *isolated*, separated, set apart, alone < Fr. *isolé* < It. *isolato* < *isola* < L. *insula*, island, something separated from others + E. suf. *-ion*, result of an action, state of): the process of obtaining a fungus or other organism in pure culture.

isomorphic (Gr. *ísos*, equal + *morphé*, shape + suf. *-íkos* > L. *-icus* > E. *-ic*, pertaining to or relating to): applied to organs or structures that are similar in shape, such as thalli, gametangia, etc. Equal to **homomorphic**. Cf. **heteromorphic**.

isoneotype (Gr. *ísos*, equal + *néos*, new + *týpos*, mark, model, type): a duplicate of a neotype.

isonym (Gr. *ísos*, equal + *-ónyma*, *ónoma*, name): one of two or more names (new names or new combinations) based on the same nomenclatural type.

isoparatype (Gr. *ísos*, equal + *pará*, at the side, near, parallel, belonging + *týpos*, mark, model, type): a duplicate of a paratype.

isoplanogamete (Gr. *ísos*, equal + *gamétes*, spouse): a motile, flagellate isogamete.

isosporic, **isosporous** (Gr. *ísos*, equal + *sporá*, spore + suf. *-íkos* > L. *-icus* > E. *-ic*, pertaining to or relating to; or + L. *-osus* > E. *-ous*, having, possessing the quality of): having equal spores. Equal to **homosporic**. Cf. **heterosporic**.

isosyntype (Gr. *ísos*, equal + *sýn*, with + *týpos*, mark, model, type): a duplicate of a syntype.

isotomic (Gr. *ísos*, equal + *tómikos*, of or for cutting < *tómos*, to cut, slice, divide + suf. *-íkos* > L. *-icus* > E. *-ic*, pertaining to or relating to): refers to the length of the branches of a dichotomy, when it is almost the same, even in the oldest parts of the thallus, as occurs in the type of isotomic-dichotomous branching of the lichen *Cladonia evansii* (Lecanorales). Cf. **anisotomic**.

isotype (Gr. *ísos*, equal + *týpos*, mark, model, type): a duplicate or part of a type, other than the holotype.

isthmospore (L. *isthmus* < Gr. *isthmós*, isthmus; narrowed in the middle part + *sporá*, spore): an asexual spore constituted of four cells with a thick wall and separated by thin-walled cells, as in *Isthmospora* *spinosa* (dematiaceous asexual fungi), the asexual state of *Trichothyrium asterophorum* (Dothideales). The ascospores of *Diatractium cordianum* (*incertae sedis*, Diaporthales) also are isthmospores; their end cells are connected by a narrow cell.

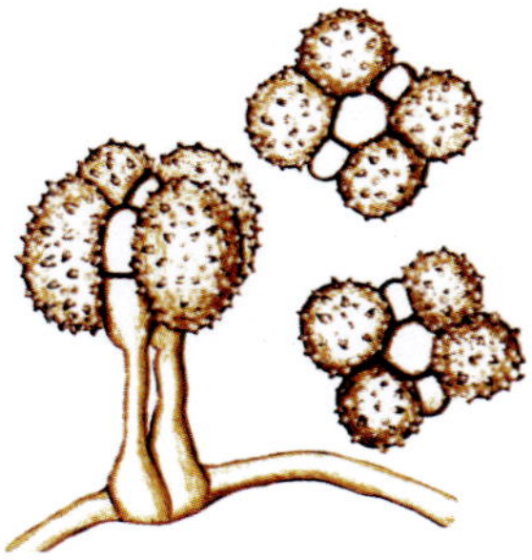

Isthmospores of *Isthmospora spinosa*, x 740 (*MU*).

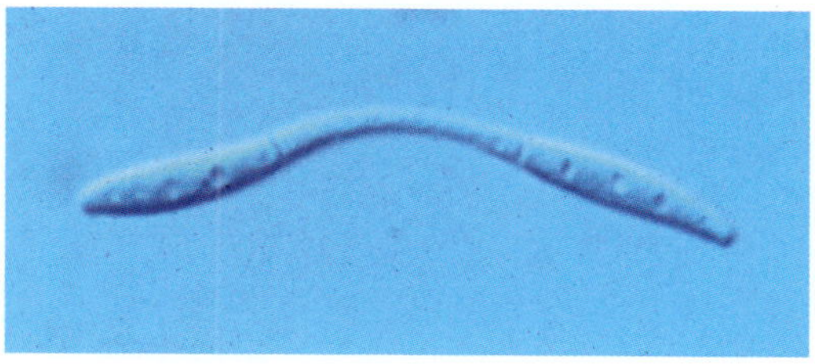

Isthmospore of *Diatractium cordianum*, x 1,000 (*RTH*).

isthmus, pl. **isthmi** (L. *isthmus* < Gr. *isthmós*, isthmus; narrowed at the middle): *Lichens*. With the middle portion narrowed and the ends swollen, as is seen in the perforate septum of the polocellate ascospore of *Caloplaca citrina* and *Teloschistes chrysophthalmus*, of the order Teloschistales.

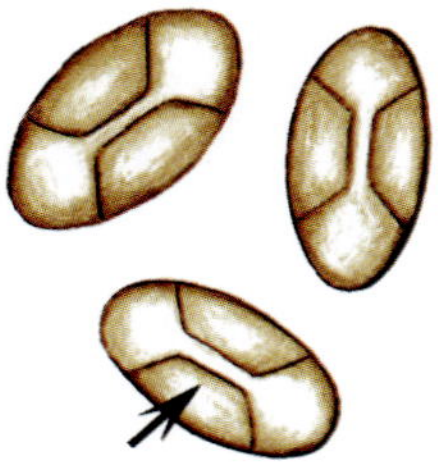

Isthmus of the ascospores of *Teloschistes chrysophthalmus*, x 900 (*MU*).

ixo- (Gr. *íxos*, viscosity): a pref. used in terms that allude to something viscous, mucilaginous or sticky, as in the following terms.

ixocutis (Gr. *íxos*, viscosity + L. *cutis*, skin): see **ixotrichoderm**.

ixotrichoderm (Gr. *íxos*, viscosity + *thríx*, genit. *trichós*, hair + *dérma*, skin): a pileus covering composed of the apices of more or less erect hyphae, undulate and intertwined, that become gelatinized to a certain degree and make the surface of the pileus turn viscous, sticky or mucilaginous, as happens, e.g., in *Suillus* and *Hygrophorus* (Agaricales).

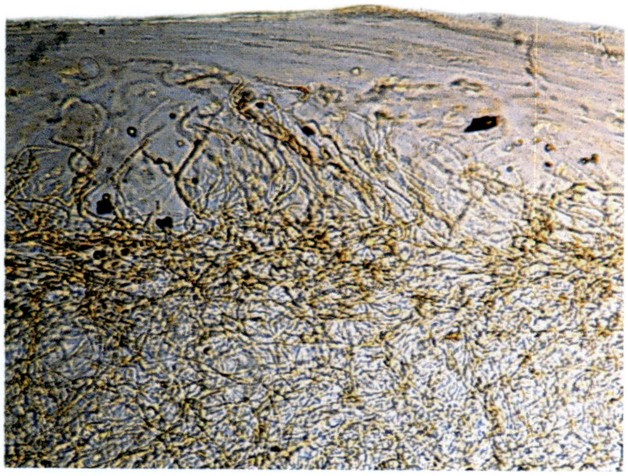

Ixotrichoderm of the pileus of *Hygrophorus olivaceoalbus*, x 30 (*RV*).

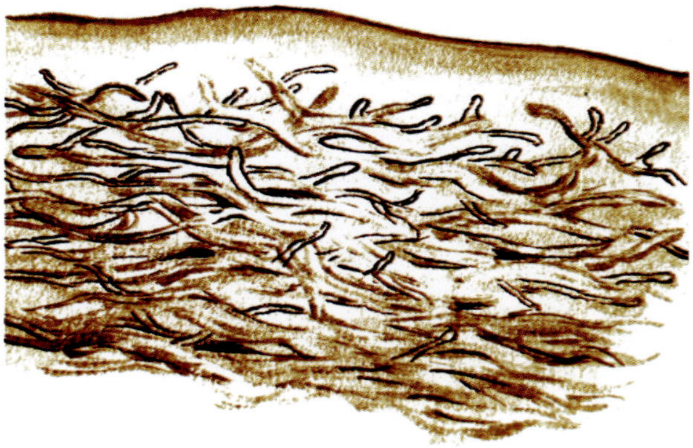

Ixotrichoderm of the pileus of *Hygrophorus* sp., x 55 (*MU*).

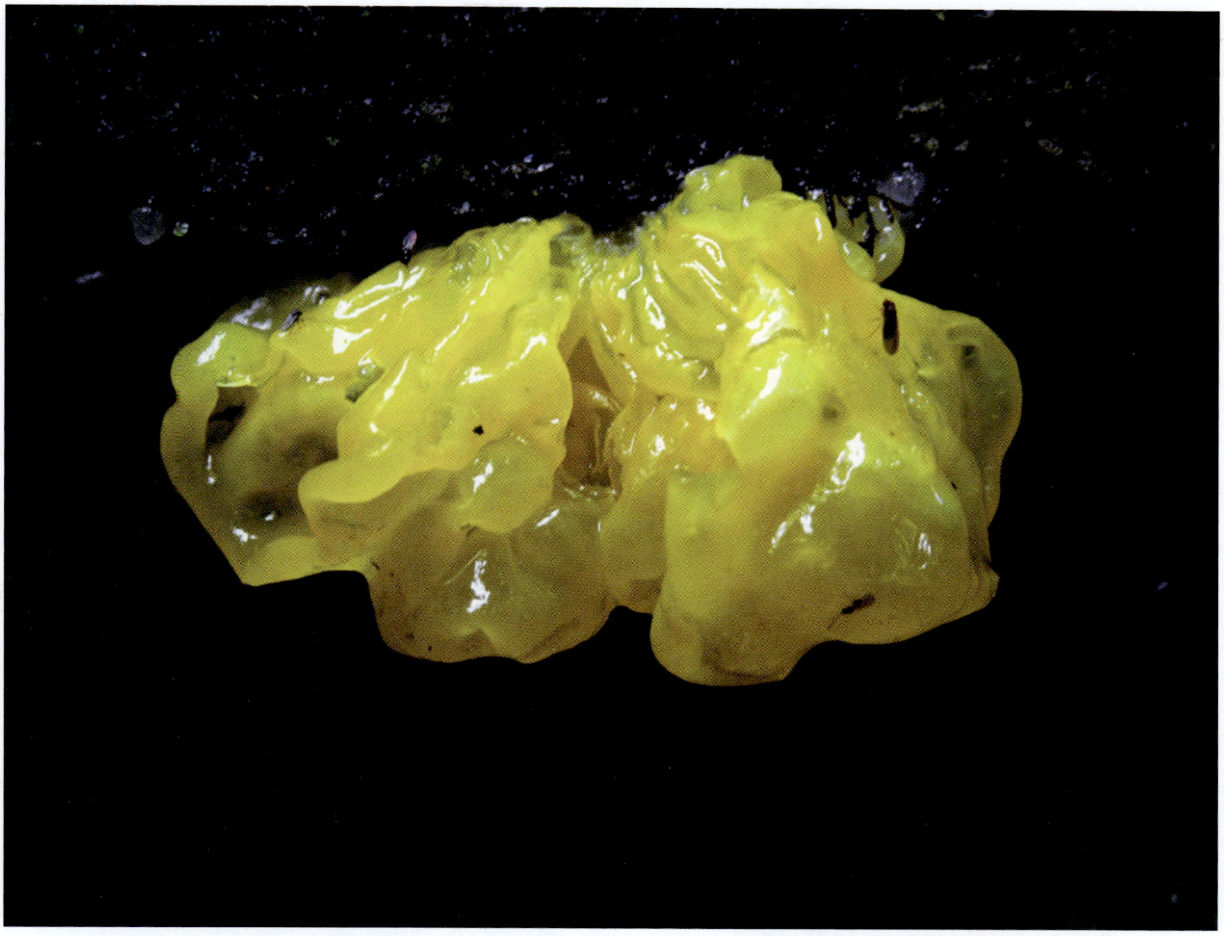

Convolute basidiocarp of *Tremella mesenterica,* a luteous **jelly fungus** (*RV*).

j

jelly fungus, pl. **fungi** (ME. *gelly* < MF. *gelee* < fem. of *gelé*, pp. of *geler*, to freeze, congeal < L. *gelare*; L. *fungus*, fungus): a common name for the Tremellales, also called the trembling fungi, due to their soft, elastic consistency, similar to gelatin or jelly, e.g., the basidiocarps of *Dacrymyces deliquescens*, *D. palmatus*, *Dacryopinax elegans*, *Tremella fuciformis*, *T. mesenterica*, *Exidia recisa* and *Guepinia helvelloides* (Dacrymycetales), which grow on rotten wood.

Orbicular, yellow basidiocaps of *Dacrymyces deliquescens*, the so-called **jelly fungi**, x 1 (*TRO*).

Young basidiocarp of *Dacrymyces palmatus*, a bright yellow, palmate **jelly fungus**, x 1 (*RV*).

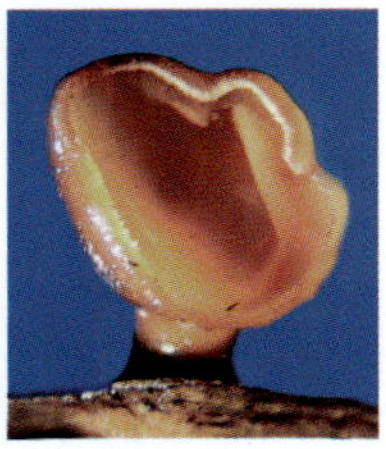

Cochleariform basidiocarp of *Dacryopinax elegans*, a **jelly fungus** with a peculiar shape, x 2 (*MU*).

Small basidiocarps of *Exidia recisa*, a reddish-brown **jelly fungus**, x 1 (*BL*).

Infundibuliform basidiocarps of *Guepinia helvelloides*, a pink-yellow **jelly fungus**, x 1 (*MRO*).

Convolute, white basidiocarps, of *Tremella fuciformis*, the common **jelly fungi**, x 2 (*TH*).

Convolute basidiocarp of *Tremella fuciformis*, a white **jelly fungus**, x 1 (*RV*).

Convolute basidiocarps of *Tremella mesenterica*, a common, yellow **jelly fungus**, x 2 (*TH*).

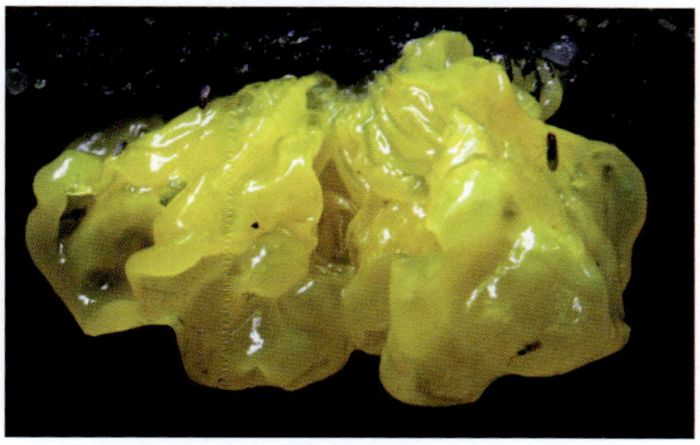

Convolute basidiocarp of *Tremella mesenterica*, a luteous **jelly fungus**, x 1 (*RV*).

Jod (G. *Jod,* iodine < Gr. *iódes*, violetlike < *íon*, a violet + L. suf. *-oide* < Gr. *-oeídes*, similar to): often used in European descriptions of discomycetes to indicate a positive (J+) or negative (J-) reaction of the asci to treatment with an iodine solution; in a positive (amyloid) reaction, a portion of the ascus apex takes a blue stain. In some recent works the English I+ or I- is used instead.

juvenescence (L. *juvenilis*, juvenile, young, immature + suf. *-escens*, genit. *-escentis* > E. *-escence*, what turns, begins to, slightly): the process of maturation of a structure at a stage when it normally is immature.

k

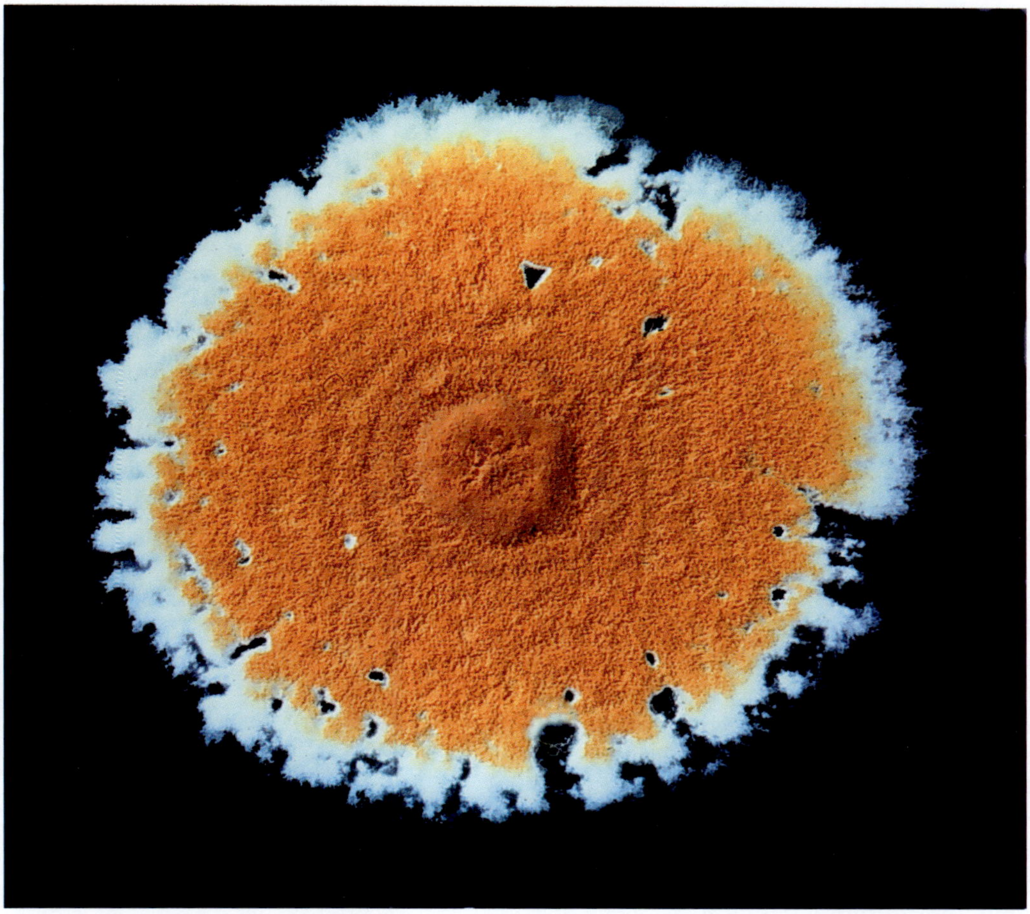

Colony of *Malbranchea aurantiaca* on agar, a **keratolytic** fungus (*MU*).

k

K-selection (L. *selectio*, genit. *selectionis*, selection of one organism or things from among others): the adaptation of certain species of organisms to the long-term colonization of habitats already occupied by other species; in the fungi, reduced quantities of long-lived, sexually produced propagules generally are involved, e.g., thick-walled ascospores. Cf. **r-selection**.

karyallagic (L. *caryallagicus* < Gr. *káryon*, nucleus + *allagé*, permutation, commutation + suf. *-ikos* > L. *-icus* > E. *-ic*, belonging to, relating to): refers to reproduction that occurs by means of exchange of nuclear material, i.e., true or sexual reproduction. Cf. **akaryallagic**.

karyo-, **kary-** (Gr. *káryon*, nut, and everything similar to a nut, like the heart of some object; nucleus): a pref. employed in the formation of various generic names (*Caryospora*, of the Pleosporales, *Caryosporella*, of the Dothideales, etc.) with the meaning of nut or hard endocarp, and in some technical terms that allude to the cell nucleus (*karyochorisis*, *karyogamy*, etc.).

karyochorisis, pl. **karyochorises** (Gr. *káryon*, nucleus + *chóris*, to separate, divide + suf. *-sis*, act of): somatic nuclear division that results from the constriction of the nuclear membrane.

karyogamy, **caryogamy** (Gr. *káryon*, nucleus + *gámos* > E. des. *-gamy*, sexual union): the fusion of two nuclei to form the zygote nucleus; it represents the second phase of sexual reproduction, following *plasmogamy* and prior to *meiosis*.

kb (kilobases < *kilo-* < Gr. *chilioi*, thousand): abbrev. for 1,000 base pairs of DNA or 1,000 bases of RNA.

kefir grains (Rus. *kefir*, beverage made from fermented cow's milk; L. *granum*, grain): granular masses (microbiogleae) composed of lactic bacteria and yeasts in symbiosis, embedded in *kefiran* (a polysaccharide of bacterial origin, soluble in water), that is utilized as a starter culture or inoculum for fermenting milk in order to obtain *kefir*, a slightly alcoholic, effervescent, lactic acid beverage that is consumed mainly in Russia and other European countries. Cf. **bulgar grain**, **tea fungus** and **tibi grains**. See **fermentation** and **microbioglea**.

keratinophilic, **keratinophilous** (Gr. *kéras*, genit. *kératos*, horn, with the nature of horn + *phílos*, have an affinity for + suf. *-ikos* > L. *-icus* > E. *-ic*, pertaining to or relating to): see **keratolytic**.

keratitis, pl. **keratitides** (Gr. *kéras*, genit. *kératos*, horn, with the nature of horn + suf. L. *-itis*, disease or inflammation): inflammation of the cornea of the eye caused by *Fusarium* sp. (tuberculariaceous asexual fungi), an opportunistic fungus that provokes disease on the human host, when his defenses are diminished. See **mycosis**.

Kefir grains (microbiogleae) cultivated on cow's milk; they are a symbiotic association of lactic bacteria and yeasts, x 1 (*MU*).

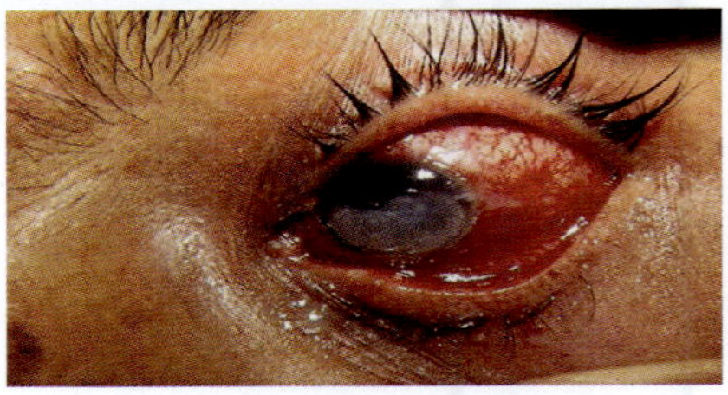

Keratitis, or inflammation of the cornea of the eye of a man, caused by *Fusarium* sp., x 1 (*AB*).

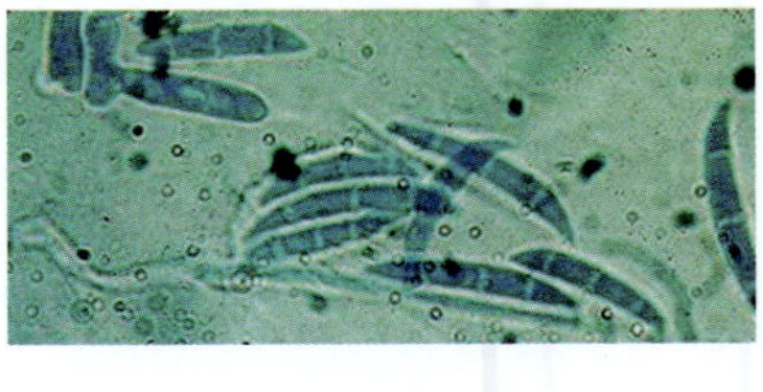

Conidia of *Fusarium* sp. isolated from the diseased cornea of a man (**keratitis**), x 550 (*AB*).

kerato-, kerat- (Gr. *kéras*, genit. *kératos*, horn, with the nature of horn): a pref. employed in terms that allude to keratin, a proteinaceous compound with a corneous nature that is found in the skin, hair and nails of man and higher animals, as in *keratolytic*.

keratolytic (Gr. *kéras*, genit. *kératos*, horn, with the nature of horn + *lytikós*, capable of dissolving or dissociating < *lýtos*, dissolved, degraded + suf. *-íkos* > L. *-icus* > E. *-ic*, pertaining to or relating to): a fungus that has the capacity to break down keratin and assimilate it for growth, as do members of the gen. *Epidermophyton*, *Microsporum*, *Malbranchea* and *Trichophyton* (moniliaceous asexual fungi), and *Malassezia* (=*Pityrosporum*, cryptococcaceous asexual yeasts), which infect man and higher animals, causing various types of tinea on hair, skin and nails.

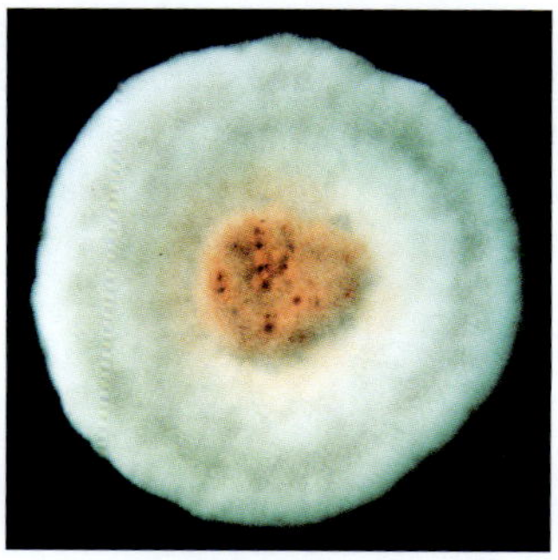

Twelve day old colony of *Malbranchea aurantiaca* on malt extract agar; the fungus is **keratolytic**, x 2 (*MU*).

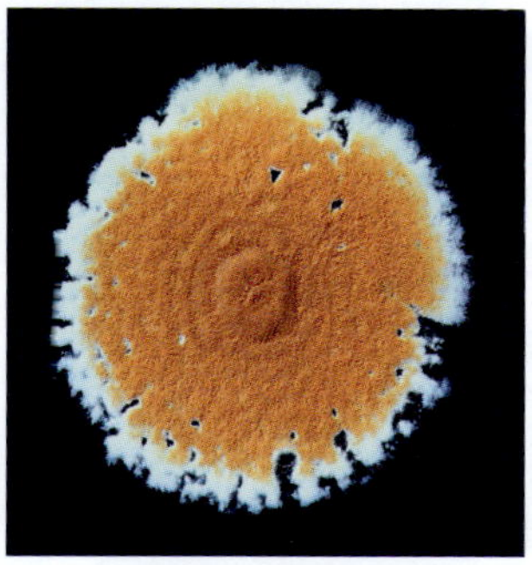

Two week old colony of *Malbranchea aurantiaca* on corn-meal agar; the fungus is **keratolytic**, x 2 (*MU*).

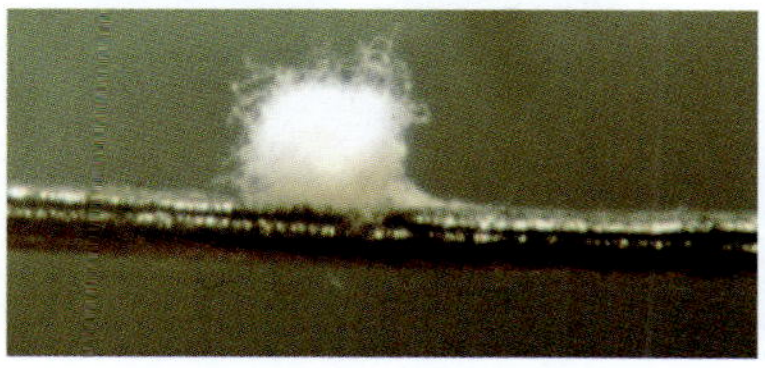

Mycelium of *Malbranchea aurantiaca* on human hair. Culture *in vitro* to determine its **keratolytic** property, x 20 (*MU*).

Mycelium of *Malbranchea aurantiaca* on human hair. Culture *in vitro* to determine its **keratolytic** property, x 20 (*MU*).

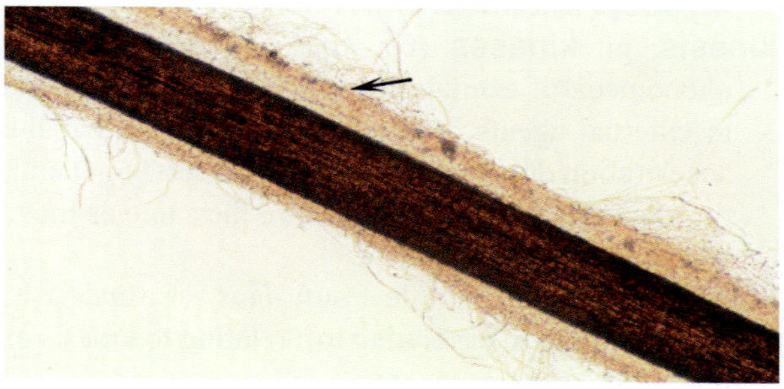

Arthrospores of *Malbranchea aurantiaca* on human hair. Ten day old culture *in vitro* to determine its **keratolytic** property, x 100 (*MU*).

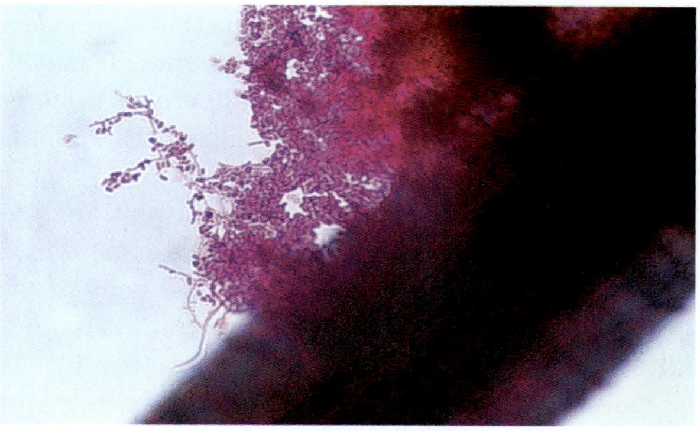

Arthrospores of *Trichophyton mentagrophytes* on rat hair. This dermatophyte is **keratolytic**, x 60 (*MU*).

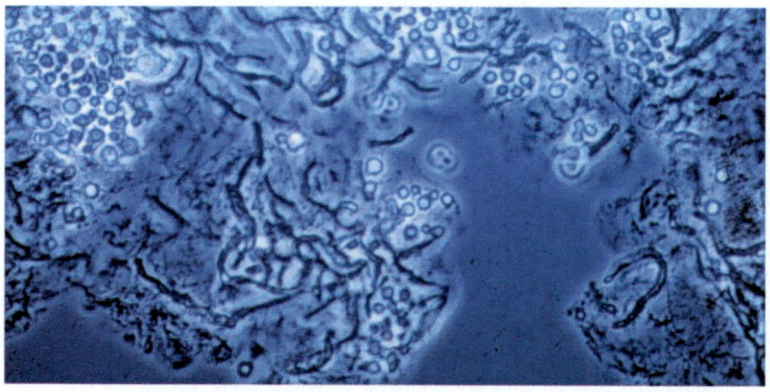

Mycelium and blastospores of *Malassezia* (=*Pityrosporum*) *furfur* on human skin; this yeast is **keratolytic**, x 400 (*RLM*).

keratomycosis, pl. **keratomycoses** (Gr. *kéras*, genit. *kératos*, horn, with the nature of horn + *mýkes*, fungus + suf. *-osis*, condition or state): an infection of the cornea of the eye by opportunistic fungi, mainly of the gen. *Fusarium*, such as *F. solani* (tuberculariaceous asexual fungi), the gen. *Aspergillus*, such as *A. fumigatus* (moniliaceous asexual fungi) and other gen., e.g., *Candida* (cryptococcaceous asexual yeasts), *Curvularia* (dematiaceous asexual fungi) and *Cephalosporium* (=*Acremonium*, of the moniliaceous asexual fungi). Also known as *mycotic keratitis*, pl. *keratitides*. See **mycosis**.

kerion (Gr. *kérion*, honeycomb): an inflammatory form of tinea of hairy skin; *tinea kerion*. See **favus**, **mycosis** and **tinea**.

kinesis, pl. **kineses** (Gr. *kínesis*, movement): the phenomena of excitability of motile organisms due to external agents, which can determine either the acceleration or slowing of the movement. In general, the excitants of kinetic agents are the same as those found in tactic phenomena.

kinetic (Gr. *kinetós*, motile + suf. *-íkos* > L. *-icus* > E. *-ic*, belonging to or relating to): relating to kinesis or pertaining to it.

kinetid (Gr. *kinetós*, motile + L. suf. *-id*, with the quality of, that which has or does): basal apparatus; flagellar (undulipodial) apparatus. Kinetids always consist of at least one kinetosome, but can have a pair of kinetosomes or occasionally more than two. The structural details of the kinetids are essential for taxonomic and evolutionary studies of the motile protoctists.

kineto-, **kinet-** (Gr. *kinetós*, motile): a pref. used to give the idea of movement, as in *kinetosome*, which is equivalent to blepharoplast or basal body of the flagellum.

kinetosome (Gr. *kinetós*, motile + *sõma*, body): the base of the zoospore flagellum, which consists of a cylinder with nine triplets of microtubules; also called **blepharoplast** or **basal body**. Kinetosomes generally originate from the centrioles and are connected to the nuclear membrane by means of a *rhizoplast*. They are present, e.g., in *Rhizophlyctis* of the Spizellomycetales, and in *Blastocladia*, of the Blastocladiales.

kingdom (ME. < OE. *cyningdom* < *cyning*, king + suf. *-dom*, domain): in natural history, each of the large groups into which all of the living beings are placed on the basis of their common characters. In the system of classification adopted in this dictionary (Alexopoulos *et al.*, 1996) the fungi are placed in the kingdoms *Fungi* (phyla Chytridiomycota, Zygomycota, Ascomycota, Basidiomycota, together with the asexual ascomycetes and other asexual fungi), *Stramenopila* (phyla Oomycota, Hyphochytriomycota and Labyrinthulomycota) and *Protista* (phyla Plasmodiophoromycota, Dictyosteliomycota, Acrasiomycota and Myxomycota). In the Appendix, the user of this dictionary can see the comparison between the outline of Alexopoulos *et al.*, 1996, modified by Ulloa and Hanlin, 2006, to that of Kirk *et al.*, 2008. In the latter classification scheme, the kingdom Fungi has seven phyla (Chytridiomycota, Blastocladiomycota, Neocallimastigomycota, Zygomycota, Glomeromycota, Ascomycota and Basidiomycota), with the consequent changes of the names of taxa. The kingdom Stramenopila is now called the kingdom Chromista, and the phyla of kingdom Protista (Acrasiomycota, Dictyosteliomycota, Myxomycota and Plasmodiophoromycota) have been moved to the kingdom Protozoa. These taxonomic rearrangements, with the concurrent change of the names of taxa, are part of a constant process of study and evaluation of these groups of fungi, or fungus-like organisms, in order to achieve a phylogenetic classification. See **Protozoa**.

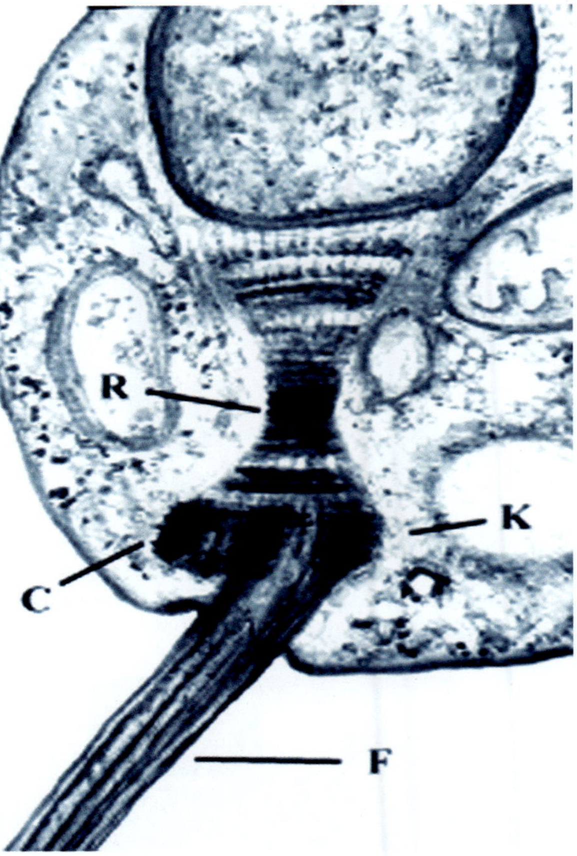

Kinetosome (K) of the zoospore of *Rhizophlyctis rosea*. Also shown are the centriole (C), the flagellum (F), and the rhizoplast (R), x 30,000 (*MU*).

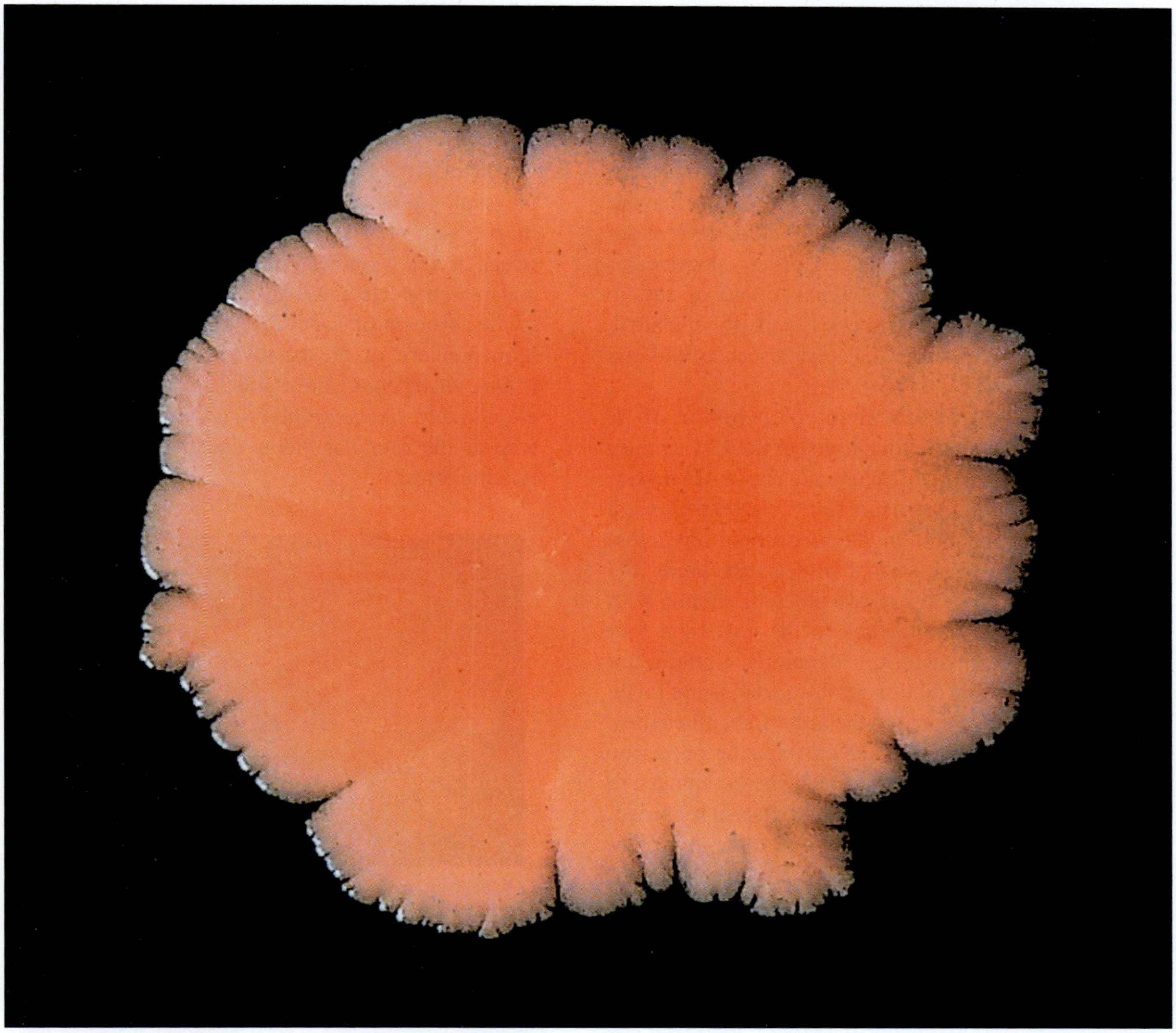

Lobulate giant colony of the yeast *Rhodotorula rubra*, on agar (*MU*).

l

labiate (NL. *labiatus*, with lips < L. *labium*, lip + suf. *-atus* > E. *-ate*, provided with or likeness): having lips, like the hysterothecia of the Loculoascomycetes of the family Hysteriaceae, order Dothideales (such as *Hysterium* and *Glonium*), and of certain loculoascolichens (Arthoniales and Pyrenulales), which have a longitudinal line of dehiscence bordered by a lip on each side.

Laboulbeniomycetes (named in honor of *A. Laboulbène*, French entomologist of the 19th century + L. *-mycetes*, ending of taxonomic class < Gr. *mýkes*, genit. *mýketos*, fungus): in some classifications, a class of perithecial fungi (Laboulbeniomycetes) that live as ectoparasites of insects, arachnids and myriapods. The thallus or receptacle is simple, with few cells, and it is attached to the host by a foot from which an haustorium develops. Conidia are not present. In this dictionary this group is not given class rank, but is treated as the order Laboulbeniales. The Spathulosporales, which are parasites of marine Rhodophyceae, and which were formerly included in the class Laboulbeniomycetes, are herein regarded, along with the Laboulbeniales, as separate orders within the phylum Ascomycota of the kingdom Fungi. In the Appendix, the user of this dictionary can see the comparison between the outline of Alexopoulos *et al.*, 1996, modified by Ulloa and Hanlin, 2006, to that of Kirk *et al.*, 2008. In the latter classification scheme, the order Laboulbeniales is in the class Laboulbeniomycetes, and the order Spathulosporales, now named Lulworthiales, belongs to the class Sordariomycetes; both classes are in the subphylum Pezizomycotina, phylum Ascomycota. These taxonomic rearrangements, with the concurrent change of the names of taxa, are part of a constant process of study and evaluation, in order to achieve a phylogenetic classification of the fungi.

labriform (L. *labriformis* < *labrum*, lip + suf. *-formis* < *forma*, shape): with the shape of a lip; generally applied to the terminal soralia of some lichens.

labyrinthiform (L. *labyrinthiformis*, with the shape of a labyrinth < *labyrinthos*, labyrinthine, tortuous passage, maze + suf. *-formis* < *forma*, shape): refers to long, irregular and tortuous cavities, repeatedly branched and interconnected, as, e.g., those of the hymenophore of the ascocarp of *Tuber texense* (Pezizales) and the hymenophore of the basidiocarp of *Daedaleopsis* (=*Daedalea*) *confragosa* and *Lenzites* (=*Daedalea*) *elegans* and *Trametes incerta* (Aphyllophorales), as well as the irregular glebal chambers of *Gautieria* (Gautieriales).

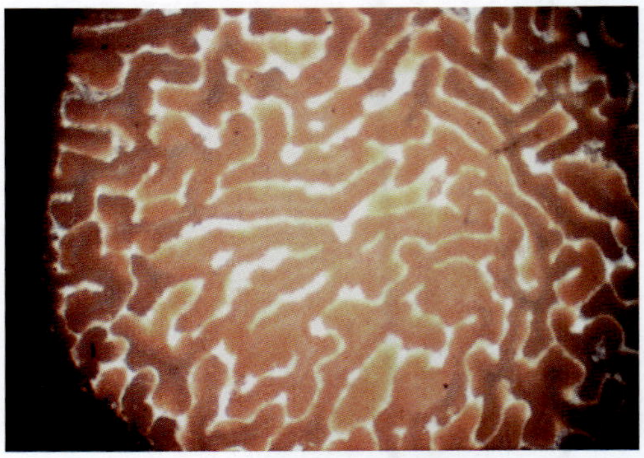

Labyrinthiform hymenophore of the ascocarp of *Tuber texense*, seen in transverse section, x 2 (*RTH*).

Labyrinthiform hymenophore of the basidiocarp of *Daedalea* sp., x 1 (*RV*).

Labyrinthiform hymenophore of the basidiocarp of *Daedaleopsis confragosa*, x 0.5 (*RV*).

laccate (L. *laccatus*, varnished < Hind. *lakh*, reddish resin < Skt. *laksha*, the lac insect + suf. *-atus* > E. *-ate*, provided with or likeness): polished, burnished, as if varnished, like the upper surface of the basidiocarp of *Laccaria laccata* (Agaricales) and *Ganoderma lucidum* (Aphyllophorales). See **lucid**.

Laccate surface of the basidiocarp of *Ganoderma curtisii*, x 0.5 (*RV*).

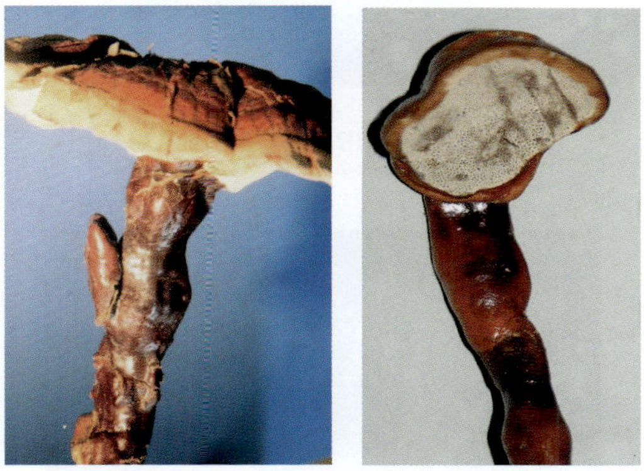

Laccate surface of the basidiocarp of *Ganoderma lucidum*, x 0.5 (*MU*), and *Ganoderma* sp., x 0.5 (*RV*).

lacerate, **lacerated** (L. *laceratus*, lacerate, torn to pieces < *lacerus*, broken up, torn + suf. *-atus* > E. *-ate*, provided with or likeness): refers to a ring, pileus, scales, etc., that appear as if they had been torn up, like the scales of the pileus of *Lepiota clypeolaria* (Agaricales) and the pore walls of the hymenophore of *Irpex* (Aphyllophorales).

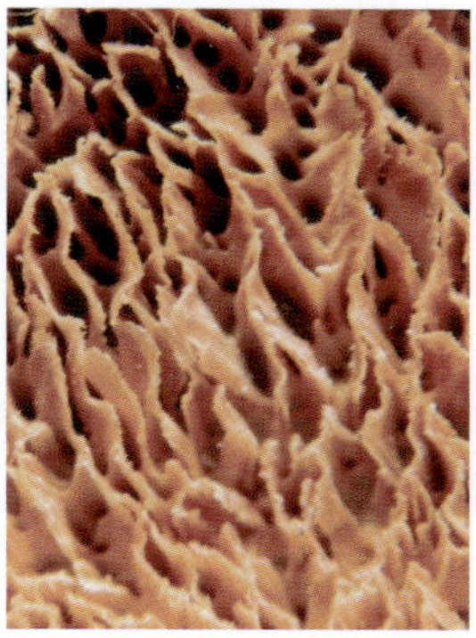

Lacerate walls of the tubes of the hymenophore of *Irpex* sp., x 5 (*MU*).

lacerate peristome (L. *laceratus*, lacerate, torn into pieces < *lacerus*, broken, torn + suf. *-atus* > E. *-ate*, provided with or likeness; Gr. *perí*, around, in turn + *stóma*, mouth): a peristome with an irregularly torn border around the mouth (*stoma*), like that of *Geastrum* (Lycoperdales).

lacinia, pl. **laciniae** (L. *lacinia*, fringe, ragged edge, strip): a delicate branch of a foliose lichen thallus, with the typical anatomy of a foliose lichen.

laciniate, **laciniated** (L. *laciniatus*, having a fringe < *lacinia*, fringe, ragged edge, strip + suf. *-atus* > E. *-ate*, provided with or likeness): with the border segmented or split into narrow lobules, generally deep and sharp-pointed. Applied to foliar or laminar structures, the border of some fruiting bodies (e.g., the apothecia of *Hyaloscypha hyalina*, of the Helotiales), and the colony margin of some molds and yeasts when grown on solid media (e.g., *Malbranchea aurantiaca*, of the moniliaceous asexual fungi). Also called **fimbriate**.

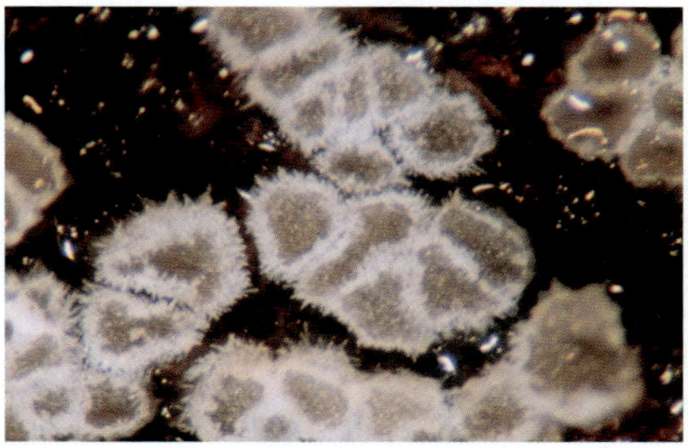

Laciniate border of the apothecia of *Hyaloscypha hyalina*, x 40 (*CB*).

Laciniate border of a colony of *Malbranchea aurantiaca*, on agar, x 0.5 (*MU*).

lacrimiferous (L. *lacrima*, tear + suf. *-ferous*, bearer < *ferre*, to bear, carry): having small droplets similar to tears, like the gills of the basidiocarps of *Lacrymaria* (=*Psathyrella*) *lacrymabunda* (Agaricales).

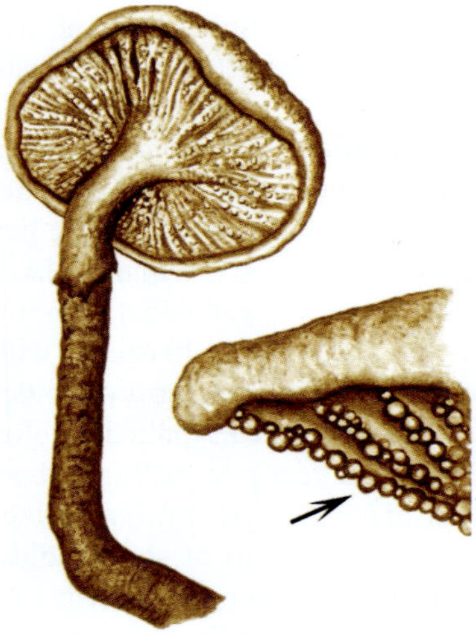

Lacrimiferous gills of the basidiocarp of *Lacrymaria* (=*Psathyrella*) *lacrymabunda*, x 1; the close-up, x 3 (*MU*).

lacrimoid (L. *lacrima*, tear drop + suf. *-oide*, like < Gr. *-oeídes*, similar to): shaped like a tear drop. Also called **dacryoid**.

lactarioid (< gen. *Lactarius* < L. *lactarius*, lacteal, chylous < *lacteus*, of milk + L. suf. *-oide*, like < Gr. *-oeídes*, similar to): with characteristics similar to those of *Lactarius* (Agaricales), i.e., with fragile gills, sphaerocysts in the trama, lactiferous ducts and amyloid spores, among others.

lacteous (L. *lacteus*, pertaining to milk or similar to it): similar to milk, like the latex of some species of *Lactarius* (Agaricales).

lactescent (L. *lactescens*, genit. *lactescentis* < *lactescere*, producing milk < *lact-*, *lac*, milk + suf. *-escens*, genit. *-escentis* > E. *-escent*, that which turns, that begins to, slightly): milky; a structure that secretes latex on being cut, like the basidiocarps of *Lactarius* (Agaricales). See **latex**.

lactiferous hypha, pl. **hyphae** (L. *lacti-*, *lac*, milk + suf. *-ferous*, bearer < *ferre*, to bear, carry; NL. *hypha*, hypha < Gr. *hyphé*, tissue, spiderweb; hypha): *Basidiomycetes*. A hypha that contains latex or a milky liquid, as in *Lactarius* (Agaricales).

lacunar, **lacunose**, **lacunous** (L. *lacuna*, pit, hole, gap + L. suf. *-aris* > E. *-ar*, like, concerning; L. *lacunosus*, with abundant pits or gaps < *lacuna* + suf. *-osus* > E. *-ose*, full of, augmented, prone to; or + *-osus* > E. *-ous*, having, with the qualities of): having gaps or holes, like the stipe of the apothecium of *Gyromitra esculenta*, *Helvella lacunosa* and *H. crispa* (Pezizales), or the basidiospores of *Strobilomyces strobilaceus* (=*S. floccopus*), Agaricales; in the latter the projecting striations in the wall are quite large and constitute interconnected borders, forming a network with large holes or pits between the borders. The basidiospores of *Calostoma* (Tulostomatales) also are lacunose, with pitted walls.

Lacunar stipe of the ascocarp of *Gyromitra esculenta*, x 0.5 (*TRO*).

Lacunar stipe of the ascocarps of *Helvella crispa*, x 0.5 (*MRO*).

Lacunar stipe of the ascocarp of *Helvella crispa*, x 3 (*CB*).

lacunar development (L. *lacunosus*, with abundant gaps < *lacuna*, gap, hole, hollow + suf. *-osus* > E. *-ar*, provided with or likeness; OF. *desveloper* > MF. *développer*, development + L. *-mentum* > E. *-ment*, denoting action or resulting state): *Gasteromycetes*. A type of fruiting body development in which the gleba is differentiated in a series of chambers more or less regular in shape and lined by a well developed hymenium; it is characteristic of the Gautieriales (such as *Gautieria*), the Melanogastrales (such as *Melanogaster*), the Nidulariales (such as *Nidularia*), and the Sclerodermatales (such as *Pisolithus*). Cf. **auleate development**, **forate development**, **homogeneous development** and **pileate development**

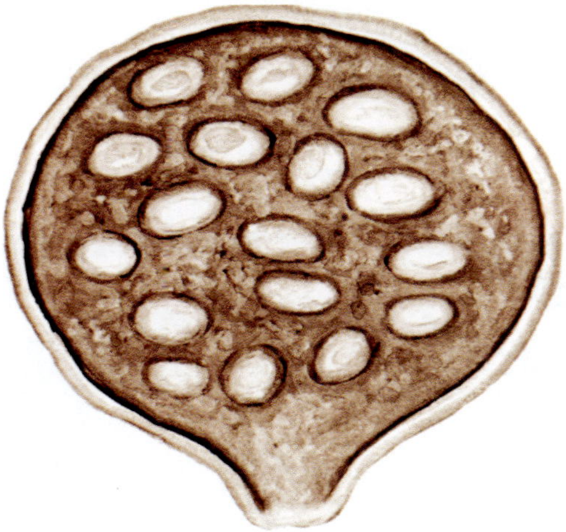

Lacunar development of the hymenium of the fruiting body of Melanogastrales and Nidulariales. Diagram of a longitudinal section, x 1 (*MU*).

laevigate, levigate (L. *levigatus*, polished, burnished < *levigare*, to polish, make smooth + suf. *-atus* > E. *-ate*, provided with or likeness): having a smooth, polished surface, like the top of the pileus of *Hydnum laevigatum* (Aphyllophorales).

lageniform (L. *lageniformis* < *lagena*, bottle, jar + suf. *-formis* < *forma*, shape): applied to an organ or structure enlarged at the base and with the upper part narrower, like the young sporangia of *Arcyria* (Trichiales), the zoosporangia of *Lagenidium* (Lagenidiales), and the ascocarps of *Microascus* (Microascales) and *Lagenulopsis* (Coryneliales). Also called **sicyoid**, **utriculose** and **ventricose-rostrate**. See **lecythiform**.

Lageniform ascocarps of *Lagenulopsis bispora*, borne on a stroma formed on a leaf of the host plant (*Podocarpus milanjianus*), x 28 (*MU*).

Young **lageniform** sporangia of *Arcyria* sp., x 6 (*CB* & *MU*).

Young **lageniform** sporangia of *Arcyria* sp., x 9 (*CB* & *MU*).

lamella, pl. **lamellae**, **gill** (L. *lamina*, plate, sheet; ME. *gile*, *gille*, of Scand. origin; akin to Gr. *chelyne*, lip, jawbone): **1**. *Agaricales*. Each one of the thin, radial plate or knife-like structures found on the lower side of the pileus of mushroom basidiocarps, on which are borne the hymenium of basidia and basidiospores. Gill characters are important in delimiting taxonomic categories, as, e.g., in *Chlorophyllum molybdites*, *Macrolepiota procera*, *Crepidotus mollis* and *Mycena pura*. **2**. *Lichens*. The main part of the thallus of the foliose lichens; the epithecium together with the hymenium and subhymenium in an apothecium.

Lamellae of the basidiocarp of *Mycena pura*, x 4 (*MU*).

Lamellae of the basidiocarp of *Chlorophyllum molybdites*, x 0.3 (*EPS*).

lamellate (L. *lamellatus*, provided with gills < *lamella*, small gill or plate + suf. *-atus* > E. *-ate*, provided with or likeness): composed of or provided with small plates or sheets, i.e., little gills, like the pileus of the Agaricales.

lamellula, pl. **lamellulae** (L. *lamellula*, small gill or plate): a small gill that does not extend all the way from the edge of the pileus to the stipe, as occurs, e.g., in *Amanita caesarea*, *Oudemansiella canarii*, *Pleurotus ostreatus* and *Xeromphalina tenuipes* (among other Agaricales).

Lamellae of the basidiocarp of *Macrolepiota procera*, x 1 (*EPS*).

Lamellulae of the basidiocarps of *Pleurotus ostreatus* var. *florida*, x 2 (*MU*).

Lamellae of the basidiocarps of *Crepidotus mollis*, x 1 (*RV*).

Lamellulae of the basidiocarp of *Amanita strobiliformis*, x 5 (*RV*).

Lamellulae of the basidiocarp of *Amanita* sp., x 5 (*RMA*).

Lamellulae of the basidiocarp of *Oudemansiella canarii*, x 2 (*RV*).

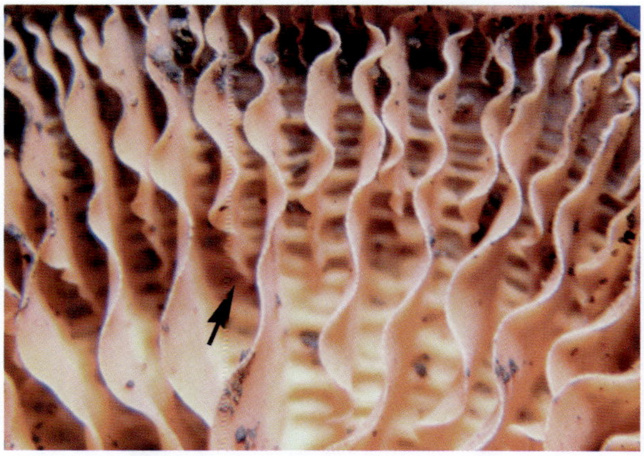

Lamellulae of the basidiocarp of *Xeromphalina tenuipes*, x 4 (*MU*).

laminal, laminar (L. *lamina* + suf. *-alis* > E. *-al*, relating to or pertaining to; or + L. suf. *-aris* > E. *-ar*, like, pertaining to): on plates, arranged in plates, consisting of plates, or resembling plates.

laminal hymenophore (L. *lamina* + suf. *-alis* > E. *-al*, relating to or pertaining to; L. *hymenium*, hymenium < Gr. *hyménion*, dim. of *hymén*, membrane + Gr. suf. *-phóros*, bearer < *phéro*, to bear, support): hymenophore arranged on plates or gills, which are the producers of the basidia and basidiospores, like, e.g., in *Panellus stipticus*, *Hygrocybe coccinea* and *Hygrophorus russula* (Agaricales).

Laminal hymenophore of the basidiocarps of *Panellus stipticus*, x 2 (*RV*).

Laminal hymenophore of the basidiocarps of *Hygrocybe coccinea*, x 1 (*TRO*).

Laminal hymenophore of the basidiocarp of *Hygrophorus russula*, x 0.5 (*RV*).

lamprocystidium, pl. **lamprocystidia** (Gr. *lamprós*, brilliant + *kystídion*, cystidium < *kýstis*, bladder; here, cell + dim. suf. *-ídion*): a type of cystidium, close to the leptocystidium, with a thick, refringent, frequently spiny wall, with crystals on the upper portion in some species, and with a hamate, corniform or proliferating apex; it is characteristic of the gen. *Inocybe*, but also is found in other gen. of Agaricales, such as *Galerina*, *Hohenbuehelia*, *Psathyrella* (=*Drosophila*) and *Pluteus*, and of Aphyllophorales, like *Auriporia* sp. Cf. **chrysocystidium**, **leptocystidium** and **macrocystidium**.

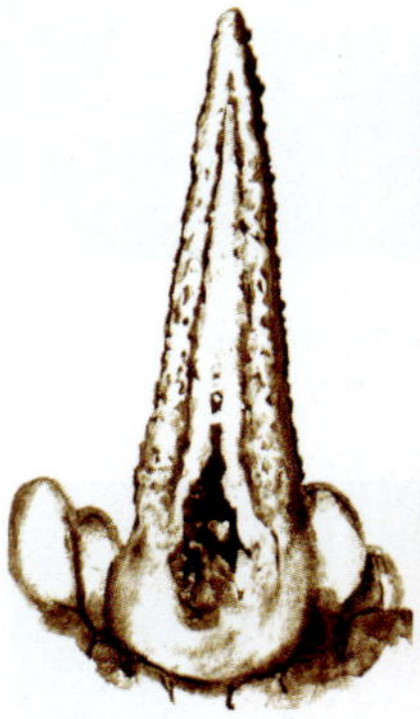

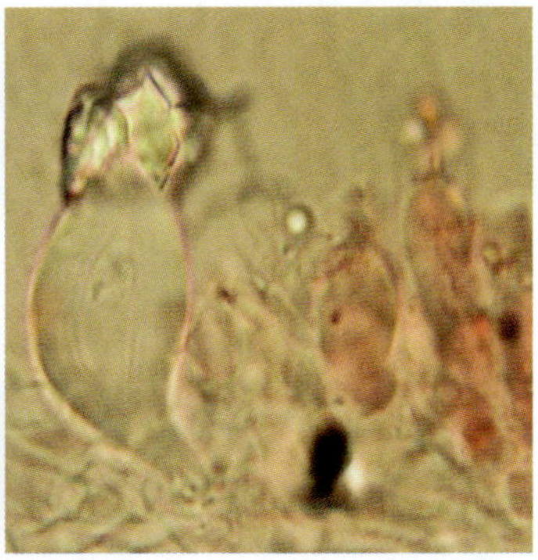

Lamprocystidia of the hymenia of *Hohenbuehelia niger*, x 1,520 (*MU*), and of *Auriporia* sp., x 1,200 (*TRO*).

lanate (L. *lanatus*, woolly < *lana*, wool + suf. *-atus* > E. *-ate*, provided with or likeness): see **lanose** and **lanuginose**.

lanceolate (L. *lanceolatus*, with the shape of a lance or provided with a lance < *lanceola*, dim. of *lancea*, lance + suf. *-atus* > E. *-ate*, provided with or likeness): applied to laminar structures that are shaped like a lance, narrowly elliptical and pointed at both ends, like certain spores, phialides, cystidia, etc., e.g., the cystidia of *Mycena corticaticeps* and *Marasmius* (Agaricales).

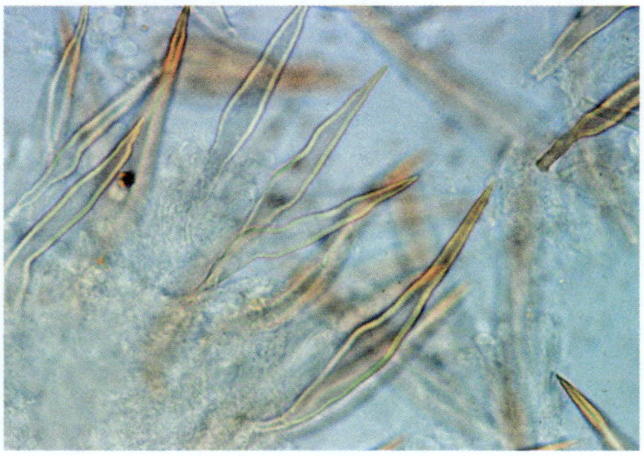

Lanceolate pleurocystidia of the hymenium of *Marasmius* sp., x 250 (*EPS*).

languid (L. *languidus*, feeble, weak, fatigued): weak, withered.

lanose (L. *lanosus*, woolly, with soft down or nap < *lana*, wool + suf. *-osus* > E. *-ose*, full of, augmented, prone to): covered with fibers similar to wool, like the colonies of *Linderina pennispora* (Kickxellales), *Botryosporium*, *Malbranchea* (moniliaceous asexual fungi), *Fusarium* (tuberculariaceous asexual fungi) and many other fungi.

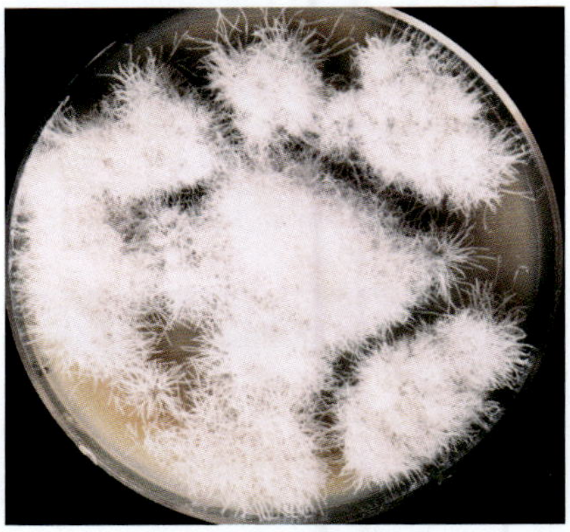

Lanose mycelium of a colony of *Botryosporium* sp., on agar, x 0.5 (*MU*).

Lanose mycelium of a colony of *Linderina pennispora*, on agar, x 0.8 (*MU*).

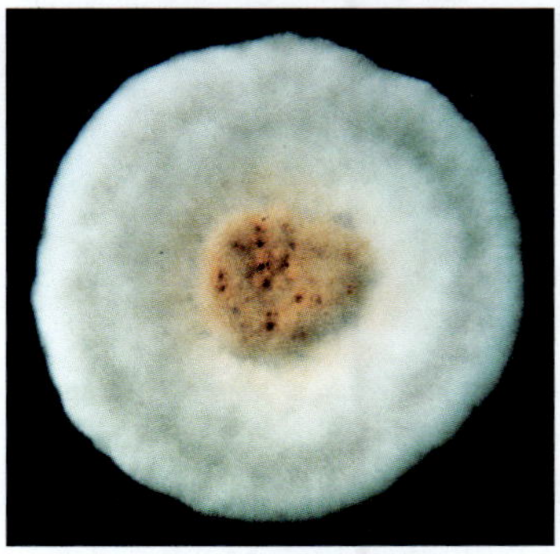

Lanose mycelium of a colony of *Malbranchea aurantiaca*, on agar, x 1 (*MU*).

lantern (L. *lanterna* < Gr. *lamptér*, lantern, lamp): *Gasteromycetes*. An ovoid, deep-orange colored *glebifer* suspended from the columnar receptacle and held in place by *trabeculae*, often numerous, which on maturing are covered by a greenish-black basidial layer, as is seen in *Laternea pusilla* (Phallales). See **glebifer**.

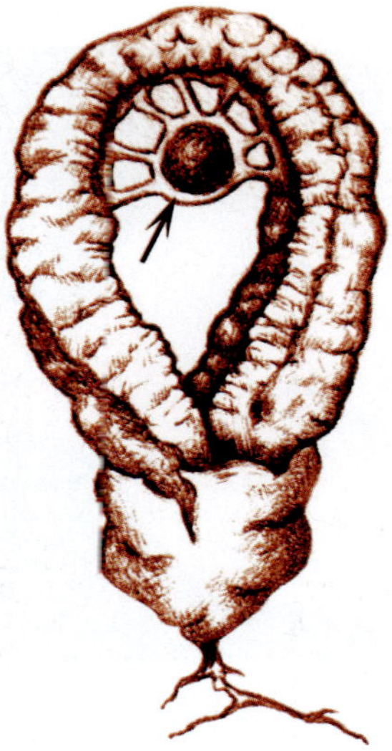

Lantern of the fruiting body of *Laternea pusilla*, x 1 (*MU*).

lanuginose, **lanuginous** (L. *lanuginosus*, downy, woolly < *lanugo*, woolly substance, down + suf. *-osus* > E. *-ose*, full of, augmented, prone to; or + *-osus* > E. *-ous*, having, with the qualities of): covered with soft, downy hair, like the basidiocarp of *Inocybe lanuginosa* (Agaricales).

larviform (L. *larva*, larva, spectre, shadow + suf. *-formis* < *forma*, shape): with the appearance of an insect larva, like the conidia of *Polyschema larviformis* (dematiaceous asexual fungi).

Larviform conidia of *Polyschema larviformis*, x 480 (*MU*).

lasio-, **lasi-** (Gr. *lásios*, hairy, downy): a pref. employed in the formation of botanical terms, such as lasiocarp, a downy fruit, and of generic names, such as *Lasiosphaeria* (Sordariales), which alludes to its hairy perithecium.

latent spore or **latent sporangium**, **resting spore** or **resting sporangium**, pl. **sporangia** (L. *latens*, genit. *latentis*, latent, hidden < *latere*, to hide; NL. *spora* < Gr. *sporá*, spore; Gr. *sporá* + *angeîon*, vessel, closed recepacle; E. *resting*, dormant < ME. *resten* < OE. *restan*, to rest): *Aquatic fungi*. A spore or sporangium, derived from the transformation of a zygote, that remains inactive; i.e., that is latent or resting for a time before germinating, as happens, e.g., in *Octomyxa achlyae*, *Polymyxa graminis*, *Woronina polycystis* (Plasmodiophorales) and *Rozella* (Spizellomycetales), *Allomyces* (Blastocladiales) and *Chytriomyces* (Chytridiales). In some cases the resting sporangium also is called a *resistant sporangium* (e.g., in *Rozella*), and the supposed zygospore of the Trichomycetes appears to correspond to a resting spore.

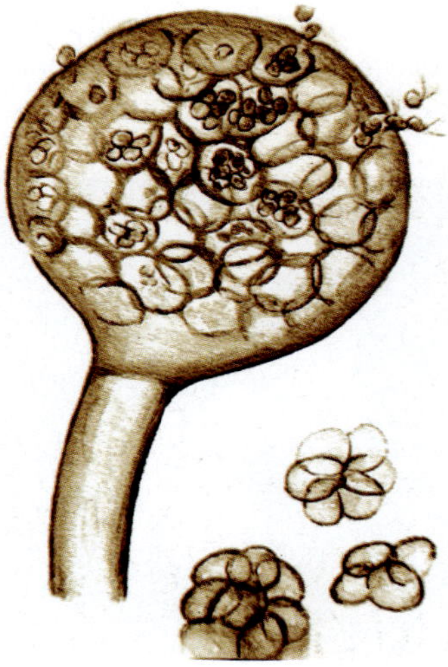

Latent or **resting spores** of *Octomyxa achlyae* inside a hypha of *Achlya* sp., x 400 (*MU*).

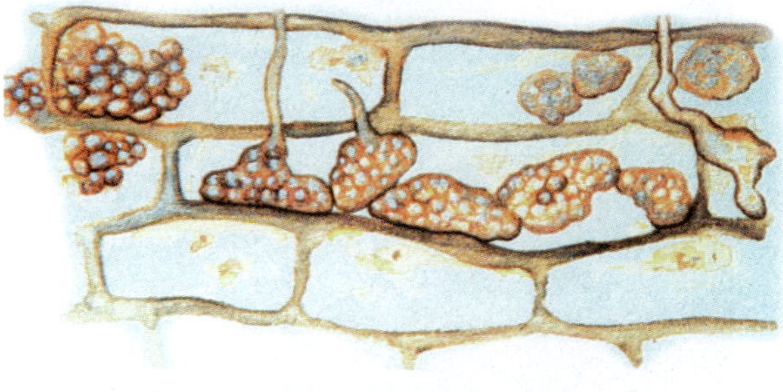

Latent or **resting spores** of *Polymyxa graminis* inside cells of wheat root, x 400 (*MU*).

Latent or **resting spores** of *Woronina polycystis* inside a hypha of *Achlya* sp., x 400 (*MU*).

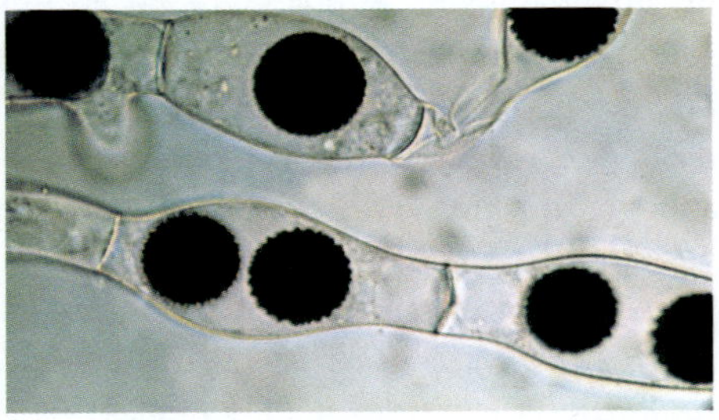

Latent or **resting spores** of *Rozella allomycis* inside a hypha of *Allomyces javanicus*, x 400 (*RTH*).

lateral (L. *lateralis*, of the side < *later*, genit. *latus*, side + suf. *-alis* > E. *-al*, relating to or pertaining to): pertaining to the side or situated on it; the flank of something. In the fungi it is applied, e.g., to the arrangement of the stipe on the fructification, which is lateral in *Ganoderma* (Aphyllophorales), *Pleurotus* (Agaricales) and other basidiomycetes.

Lateral stipe of the basidiocarp of *Ganoderma curtisii*, x 0.5 (*RV*).

Lateral stipe of the basidiocarp of *Pleurotus ostreatus*, x 0.5 (*MU*).

latex, pl. **latices** or **latexes** (NL. *latic-*, milk < L. *latex*, fluid, juice): a fluid, generally milky and white in color, although at times it can be yellow or other colors, that flows or is exuded from various parts of certain fungi when these are cut or damaged, as happens in the gen. *Lactarius* (Agaricales), whose species have latex of distinctive colors and properties. See **lactescent**.

Latex exuded from the cut gills of *Lactarius volemus*, x 0.5 (*VBM*).

lati- (L. *latus*, wide): a pref. used in terms that allude to something wide, as *latifolium*, pl. *latifolia*, a wide leaf. Cf. **angusti-**.

Latin (< *latinus* > Latin, a native of Latium, a province in central Italy): Latin was originally the dialect spoken in Latium, whose capital was Rome. It was established as the official language of the Roman Empire, extending it throughout the vast area dominated by the Romans and giving rise to the Neo-Latin or romance languages, to which Spanish belongs. Its alphabet consists of 25 letters. Literary and scientific Latin was the language of the educated Romans, whereas Vulgar Latin was the language of the middle class; there exist differences in the lexicology, phonetics, morphology and syntax between the two. The majority of the common Spanish words are derived from Vulgar Latin. Fewer words have their origin in literary or classical Latin and these are called educated or learned words. Latin, because it is a synthetic language, synthesizes in a single word ideas that in other languages need various terms, giving it an expressive force suitable for conciseness and precision; in addition, it has been the principal vehicle in the transmission of a multitude of Greek roots that form the numerous technical and scientific terms that are used in various modern languages. For this reason, since before the time of Christ, all of the sciences (and other non-scientific disciplines), such as biology, anatomy, zoology, botany, medicine, astronomy, philosophy, religion, and law, were systematically written in Latin during 19 centuries, and were taught in Latin up to the 18th century, as the universal language of culture. Bacon, Copernicus, Descartes, Newton, Linnaeus, Gauss, Matthioli, Clusius, Bauhin, Tournefort, Micheli, Corda, Tulasne and Saccardo, among many others, wrote their scientific works in Latin. A basic knowledge of Latin is useful for any biologist, particularly for specialists in taxonomy, due to the fact that the scientific names of all taxa are Latinized. At present, the International Code of Botanical Nomenclature, which is applied in mycology, requires Latin descriptions or diagnoses to validate the names of fungi in all formal categories. See **etymology** and **Greek**.

laxi- (L. *laxus*, loose, slack, lax): a pref. used in the formation of compounds that alludes to something that is not very tight, as in the generic name *Laxitextum* (Aphyllophorales), with loose hyphal tissues. Cf. **densi-**. **Lax** is used to refer to the colonies of *Mucor* (Mucorales) and *Aspergillus tamarii* (moniliaceous asexual fungi) on agar.

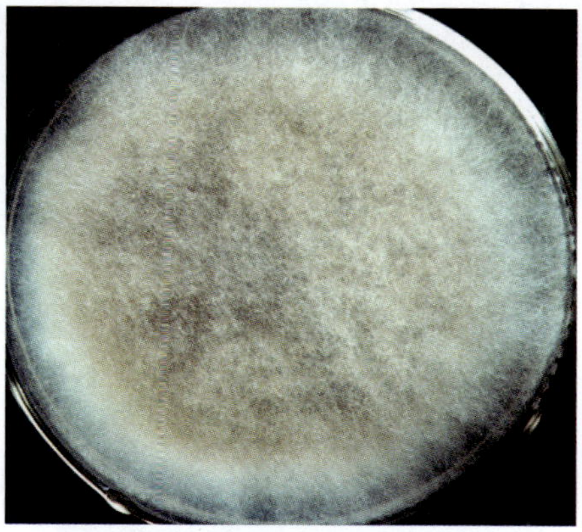

Lax colony of *Mucor* sp., cultured on agar, x 0.8 (*MU*).

Lax colony of *Aspergillus tamarii*, cultured on agar, x 1 (*MU*).

leaf curl (ME. *leef* < OE. *leaf,* leaf < L. *liber*, bast, book; ME. *curlen* < *crul*, curly, prob. < MD., to form into coils or ringlets): a disease symptom of peach and almond leaves infected by *Taphrina deformans*, and of cherry (by *T. minor*), of the Taphrinales. Also called **verrucosis**.

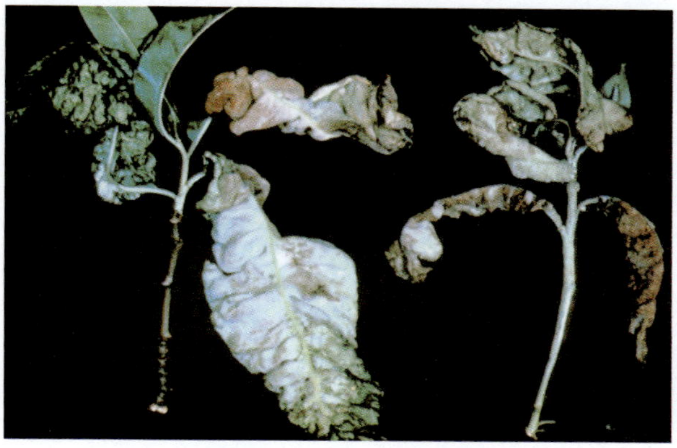

Leaf curl of peach leaves caused by *Taphrina deformans*, x 0.5 (*RTH*).

leaf spot (ME. *leef* < OE. *leaf,* leaf; ME. *spotte* < *spotten*, to stain, mark): one of the extensive variety of chlorotic or necrotic spots caused by various pathogenic fungi on the leaves of many host plants, e.g., the yellow leaf spot of carnation, produced by *Septoria dianthi* (sphaeropsidaceous asexual fungi); the spot blotch of leaves of cereal grains, caused by *Bipolaris sorokiniana* (=*Helminthosporium sativum*, of the dematiaceous asexual fungi); the chocolate spot of leaves of faba bean, caused by *Botrytis fabae* and *B. cinerea* (moniliaceous asexual fungi), and the small dark spots caused by *Blumeriella jaapii* (Helotiales) on the leaves of *Prunus serotina*, among many more phytopathogenic species.

Leaf spots on leaves of *Prunus serotina*, caused by *Blumeriella jaapii*; conidia and ascomata form in these dark spots, x 1 (*RTH*).

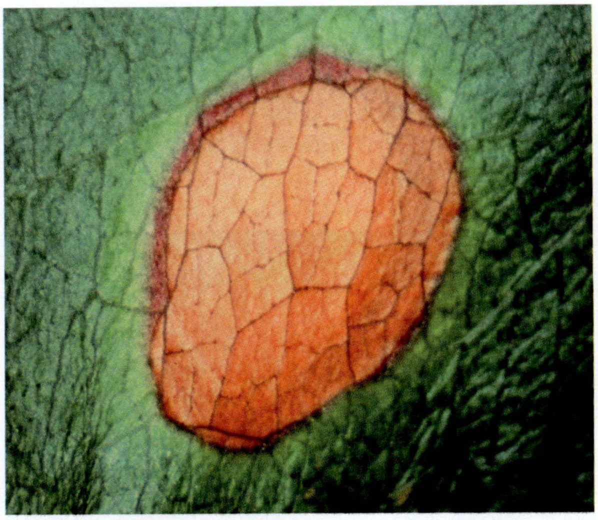

Leaf spot caused by *Blumeriella jaapii* on a leaf of *Prunus serotina*, x 15 (*RTH*). In old lesions the central area falls out, leaving a hole.

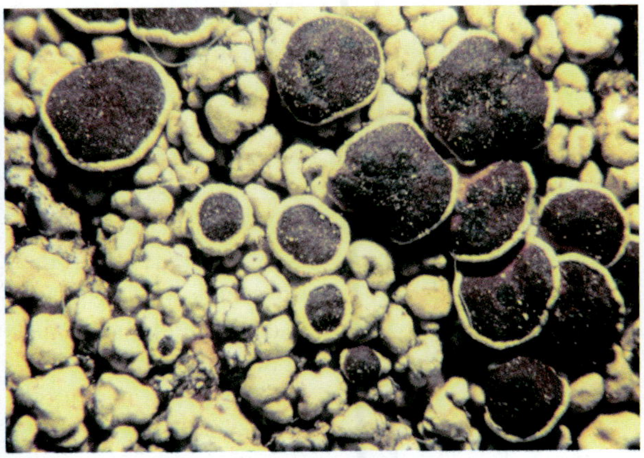

Lecanorine apothecia of the thallus of *Lecanora subfusca*, x 12 (*MU*).

Lecanorine apothecia of the thallus of *Candelariella* sp., x 12 (*TRO*).

lecanorine (< gen. *Lecanora* < Gr. *lekáne*, dim. of *lekánion*, kettle, plate, bowl + L. suf. *-ora* < Gr. *hóra*, genit. *horaîos*, beautiful, a crustaceous lichen of the order Lecanorales + L. suf. *-inus* > E. *-ine*, of or belonging to): *Lichens*. A rounded apothecium bordered by a protruding thalline margin, like that in *Lecanora frustulosa*, *L. subfusca* and *Candelariella*. The type of ascus in these lichens is termed *lecanoralean*; a lecanoralean ascus is essentially bitunicate in structure, generally thick-walled, with the apex obviously thickened, and with dehiscence by a rostrate eversion of the endoascus. The lecanorine apothecium also is known as a *zeorine* apothecium, from the gen. *Zeora* (=*Lecanora*). Cf. **lecideine**. See **discoid**.

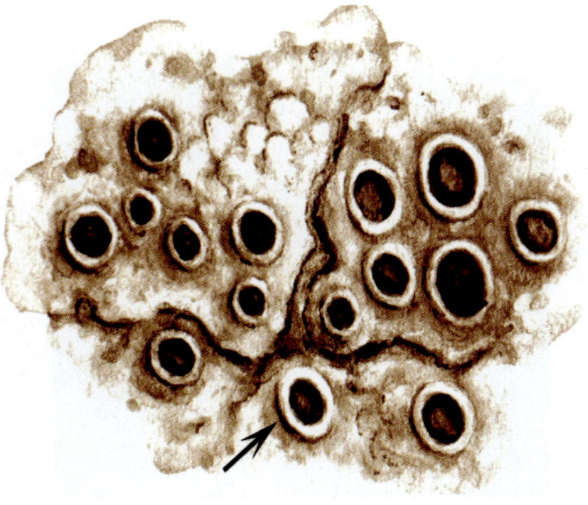

Lecanorine apothecia of the thallus of *Lecanora frustulosa*, x 10 (*MU*).

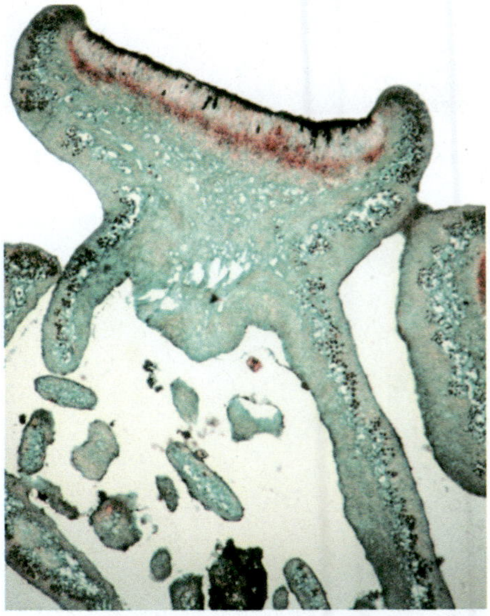

Lecanorine apothecium of the thallus of a discolichen, seen in longitudinal section, x 57 (*MU*).

lecideine (< gen. *Lecidea* < Gr. *lékos*, plate, shield + suf. *-ídes*, similarity, a lichen of the order Lecanorales + L. suf. *-inus* > E. *-ine*, of or pertaining to): *Lichens*. An orbicular apothecium that lacks a thalline margin; commonly the margin is composed of the excipulum itself, as is observed in *Lecidea albocaerulescens* and *L. atrobrunnea* and *Psora decipiens*. Cf. **lecanorine**.

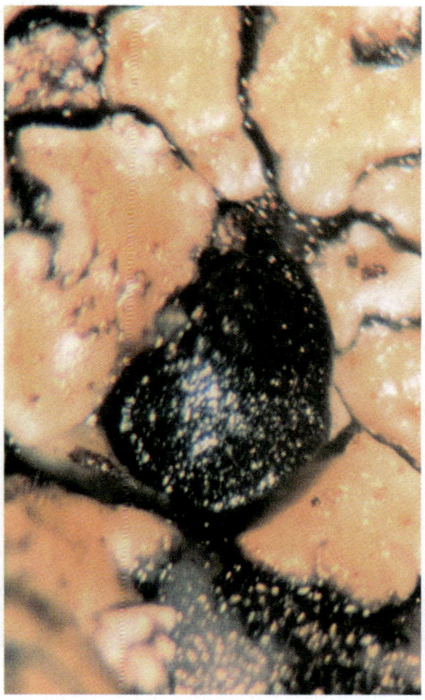

Lecideine apothecia of the thallus of *Lecidea atrobrunnea*, x 35 (*MU*).

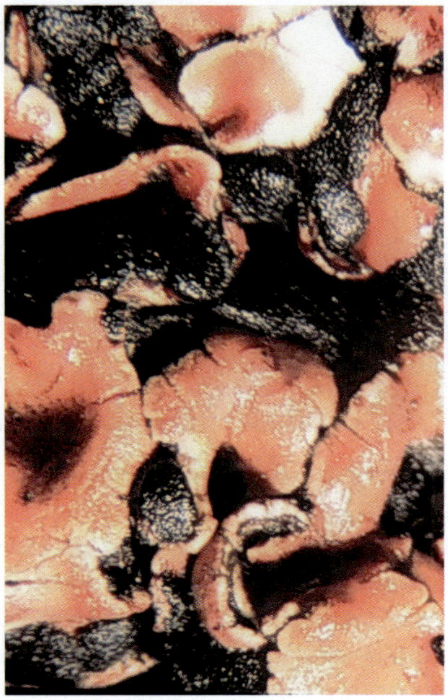

Lecideine apothecia of the thallus of *Psora decipiens*, x 30 (*MU*).

lectotype (L. *lecto*, selected < Gr. *léktos*, chosen < *légo*, to elect, choose + *týpos*, type, mark, model): a type specimen selected (by a subsequent author) from among the original elements (a specimen, figure or description) on which the name of the taxon was based, to serve as the nomenclatural type when the original holotype no longer exists; also applied to the element (syntype) selected in a later work as the holotype when this was not designated originally.

lecythiform (Gr. *lékythos*, oil jar + L. suf. *-formis* < *forma*, shape): with the shape of a bottle having a narrow base, an ellipsoid body, a narrow neck and a mouth bordered by an enlargement in the shape of a stopper; applied to cystidia (lamprocystidia) of certain basidiomycetes (such as those of the gen. *Agrocybe* and *Conocybe*, Agaricales), which are markedly ventricose in the basal part and attenuated in the upper part, where they form a neck that ends in a rounded or enlarged button shaped like a small pear. The sporangium of *Saksenaea vasiformis* (Mucorales) also is lecythiform. See **lageniform**.

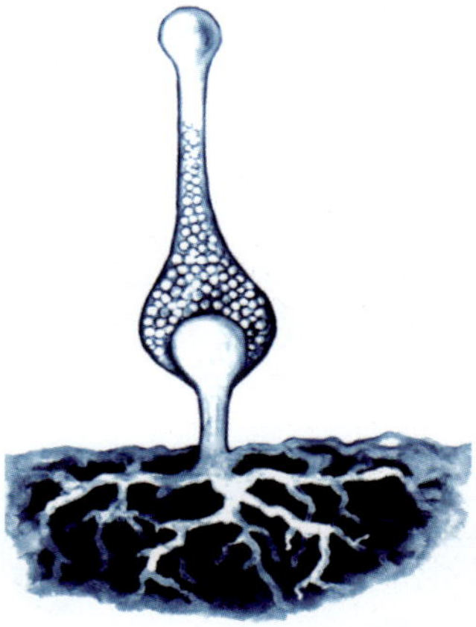

Lecythiform sporangium of *Saksenaea vasiformis*, x 470 (*MU*).

legitimate (L. *legitimus*, conforms to the laws, certain, genuine, true): to consider legitimate, to make legitimate. Applied to indicate that a name or epithet is considered in conformity with the rules of the International Code of Botanical Nomenclature.

leio-, lei- (Gr. *leîos*, smooth): a pref. employed in terms that allude to something smooth, as in the following.

leiodisc (Gr. *leîos*, smooth + *dískos*, disc): *Lichens*. An apothecium with the appearance of a smooth, shiny disc, as is observed in *Umbilicaria coriacea* (=*U. rigida*), Lecanorales.

leiospore (Gr. *leîos*, smooth + *sporá*, spore): a spore that is characterized by having a smooth wall. See **leiosporic**.

leiosporic, **leiosporous** (Gr. *leîos*, smooth + *sporá*, spore + suf. *-íkos* > L. *-icus* > E. *-ic*, pertaining to or relating to; or + L. suf. *-osus* > E. *-ous*, having, with the qualities of): a fungus whose spores have a smooth wall.

lentic (L. *lentus*, sluggish + suf. *-íkos* > L. *-icus* > E. *-ic*, pertaining to or relating to): of, relating to, or living in still waters (such as lakes, ponds or swamps). Cf. **lotic**.

lenticular (L. *lenticularis*, of or pertaining to a lentil < *lenticula*, lentil, dim. of *lens*, genit. *lentis*, disc, lens + suf. *-aris* > E. *-ar*, like, concerning): having the shape of a disc or watch glass, like a double convex lens, wider than ellipsoid but not as long. Also called **lentiform**. E.g., the conidia of *Myrothecium verrucaria* (tuberculariaceous asexual fungi) and the basidiospores of *Psilocybe atrorufa* and *Coprinus comatus* (Agaricales) are lenticular or lentiform.

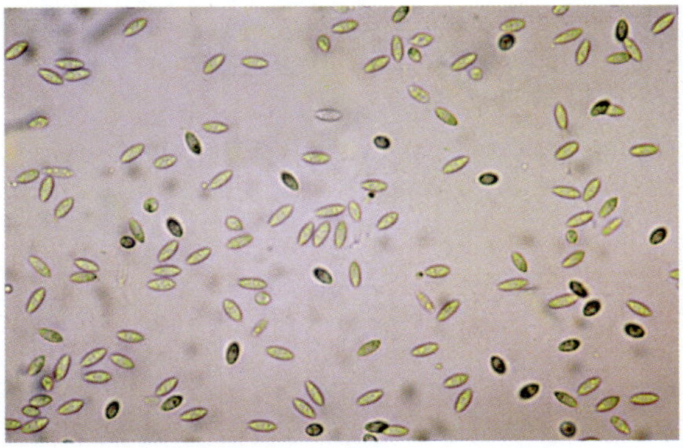

Lenticular conidia of *Myrothecium verrucaria*, x 650 (*MU*).

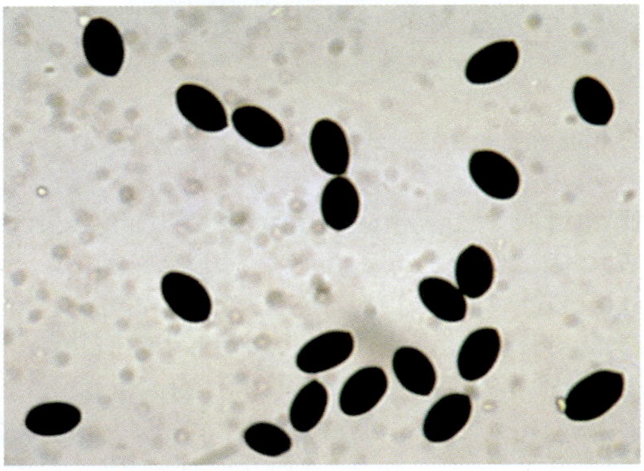

Lenticular basidiospores of *Coprinus comatus*, x 650 (*MU*).

lentiform (L. *lens*, genit. *lentis*, disc, lentil, lens + suf. *-formis* < *forma*, shape): see **lenticular**.

lentiginose, **lentiginous** (L. *lentiginosus*, freckled < *lentigo*, mole, freckle + suf. *-osus* >E. *-ose*, full of, augmented, prone to): having small spots, similar to freckles, like the thallus of the lichen *Melaspilea lentiginosula* (of the Melaspileaceae, a family of ascomycetes of uncertain affinity, previously included in the Hysteriales), which has minute apothecia on its surface.

lepido-, **lepid-** (Gr. *lepís*, genit. *lepídos*, scale): a pref. used to form compounds related to something scaly, such as *lepidote*.

lepidote (NL. *lepidotus* < Gr. *lepidotós*, covered with scales): scaly or covered with scale-like hairs, like the pileus of the basidiocarp of *Lepiota* and *Macrolepiota* (Agaricales). Equivalent to **squamulose**. Cf. **alepidote**.

Lepidote pileus of the basidiocarps of *Lepiota* sp., x 0.6 (*RV*).

Lepidote pileus of the basidiocarp of *Macrolepiota mastoidea*, x 1 (*RV*).

lepiotoid (< gen. *Lepiota* < Gr. *lepís*, scale + L. suf. *-otus*, *-ota*, which indicates possession or likeness + L. suf. *-oide*, like < Gr. *-oeídes*, similar to): *Agaricales*. A type of fruiting body (13 principal types are recognized) with free or finely attached gills, and with a ring but no volva, represented by *Lepiota* and some species of *Agaricus* and *Coprinus*.

Lepiotoid fruiting body of *Lepiota* sp., x 0.7 (*MU*).

leprose (L. *leprosus*, scurfy, scaly < *lepras* < Gr. *lépra*, leprosy < *leprós*, rough + L. suf. *-osus* > E. *-ose*, full of, augmented, prone to): *Lichens*. Applied to a type of thallus whose surface has a large number of soredia, or when all of the thallus disintegrates and gives rise to soredia, as occurs in *Leproloma membranaceum* (=*Lepraria membranacea*, Lichenes Imperfecti, Deuterolichenes or imperfect lichens).

Leprose thallus of *Leproloma membranaceum* (=*Lepraria membranacea*), x 10 (*MU*).

-leptic (Gr. *leptós* < *lambáno*, to grab, catch; literally, what grabs, catches or "senses" something + suf. *-íkos* > L. *-icus* > E. *-ic*, belonging to or relating to): a suf. employed in terms such as *baroleptic*, *organoleptic* and *photoleptic*, to refer, respectively, to something that senses gravity, the sensory organs, and light. Note: here *leptós* is written in Gr. with *eta*; cf. **lepto-**.

lepto-, **lept-** (Gr. *leptós*, in this case written with *épsilon*, minute, fine, delicate, graceful, thin): a pref. employed in terms such as *leptocystidium*. Cf. **-leptic**.

leptocystidium, pl. **leptocystidia** (Gr. *leptós*, minute, fine, delicate + *kystídion* < *kýstis*, bladder; here, vesicle, cell + dim. suf. *-ídion* > L. *-idium*): a cellular element, similar to a *cystidiole*, that originates from the lower part of the subhymenium or from the trama itself, at a level lower than the basidia. When it has a thickened, pigmented wall it is called a *seta*, *setula*, or *setuloid cystidium*. Leptocystidia are found in various species of *Amanita* (such as *A. herrerae*), *Psathyrella* (=*Drosophila*) and *Leptonia* (Agaricales), and *Auriporia* (Aphyllophorales). Cf. **chrysocystidium**, **lamprocystidium** and **macrocystidium**.

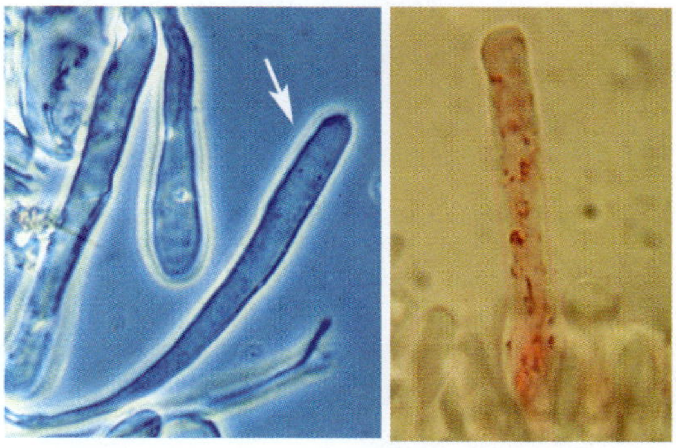

Leptocystidia of the hymenium of *Amanita herrerae*, x 360 (*RMA*), and of *Auriporia* sp., x 350 (*TRO*).

leptoderm, **leptodermatous**, **leptodermous** (Gr. *leptós*, minute, delicate, fine + *dérma*, skin, cortex + L. suf. *-osus* > E. *-ous*, having, with the qualities of): a hypha with the outer wall thinner than the lumen. Cf. **mesoderm** and **pachyderm**.

leptospore (Gr. *leptós*, minute, delicate, fine + *sporá*, spore): a spore adapted to germinate quickly, without having to pass through a latent period, as occurs in some Uredinales.

leptotichous (Gr. *leptós*, minute, delicate, fine + *teichós*, wall): refers to tissues whose cells have thin walls.

lesion (L. *laesion*, *laesio*, lesion, damage, injury, genit. *laesionis* < *laesus* < *laedere*, to damage, injure + suf. *-io*, genit. *-ionis* > E. *-ion*, result of an action, state of): detriment, injury or abnormal change in the structure of an organ or part due to an injury or disease, in particular an injury whose area is well delimited or circumscribed. E.g., the lesions caused by *Blumeriella jaapii* (Helotiales) on the leaves of *Prunus serotina* are known by the common name of "shot-hole", because the center of old lesions falls out, leaving a hole. *Schizothyrium pomi* (Dothideales) produces tiny lesions on fruits of apple that are called "fly speck". Other examples of lesions are *Ophiodothella vaccinii* and *Phyllachora fusicarpa* (Phyllachorales) on leaves of *Vaccinium arboreum* and *Duranta repens*, respectively.

Lesions on leaves of *Vaccinium arboreum* caused by *Ophiodothella vaccinii*; the dark spots are conidiomata of the fungus, x 1 (*RTH*).

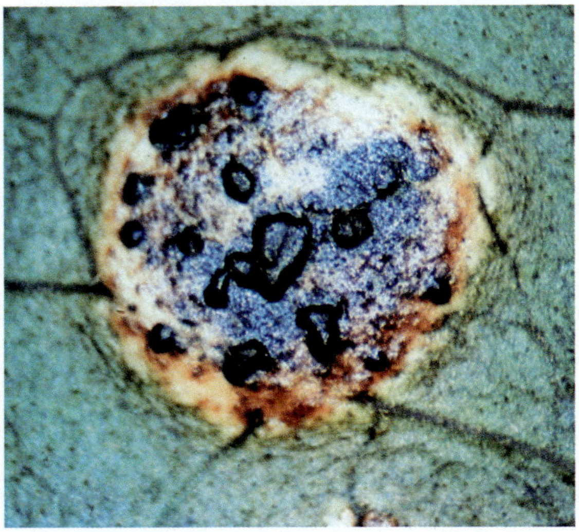

Lesion caused by *Phyllachora fusicarpa* on the lower surface of a leaf of *Duranta repens*, x 15 (*RTH*).

leuco-, **leuc-** (Gr. *leukós*, brilliant, clear, limpid or transparent, and also white): a pref. employed in the formation of terms that allude to transparency or lack of color, or to the whiteness of something, as in *leucosporous*.

leucosporous (Gr. *leukós*, white + *sporá*, spore + L. suf. *-osus* > E. *-ous*, having, with the qualities of): having white spores, like the basidiocarps of *Amanita* (Agaricales).

levo- (L. *laevus*, left): a pref. employed in terms that allude to the left side, as in *levorotatory*, rotating the plane of polarized light toward the left, as occurs with levulose or fructose. Cf. **dextro-**.

lexicographer (Gr. *lexikográphos* < *lexikón*, neutral of *lexikós*, of words < *léxis*, word, language, discourse + *grápho*, to write, describe): a collector of all of the words that will form a lexicon or dictionary; the author or editor of a dictionary.

lexicography (Gr. *lexikón*, neutral of *lexikós*, of words < *léxis*, word, language, discourse + *graphía* > E. *-graphy*, writing or representation < *grápho*, to write, describe): the art of compiling lexica or dictionaries, or rather, of collecting all of the words of a language, to investigate, discover and establish the sense and use of each of them.

lexicology (Gr. *lexikón*, neutral of *lexikós*, of words < *léxis*, word, language, discourse + *lógos*, word < *légo*, I say > E. des. *-logy*, discourse, treatise): a treatise or special study related to the true meaning (etymology) of words, especially those that will be included in a lexicon or dictionary. The branch of linguistics concerning the meaning and application of words. A *lexicologist* is one versed in lexicology. See **glossology** and **orismology**.

lexicon, pl. **lexica** or **lexicons** (Gr. *lexikón*, neutral of *lexikós*, of words < *léxis*, word, language, discourse + suf. *-íkos* > E. *-ic*, pertaining to or relating to): a dictionary of the Greek language; by extension, a dictionary of any other language. A book that contains the words of a language and their definitions, arranged alphabetically. The vocabulary of a language, an orator or group of orators, or of a material or science. The total inventory of morphemes of a language; an inventory of words. See **dictionary**, **lexicography** and **lexicology**.

lichen (L. *lichen* < Gr. *leichén*, lichen): **1**. A dual organism, constituted, in the majority of species, of the symbiotic association of fungi and algal and/or bacterial photobionts; this union sometimes results in a type of parasitism of the fungus on the alga. The majority of lichens belong to the ascomycetes (ascolichens); a minority are basidiomycetes (basidiolichens), and a few are deuteromycetes (deuterolichens or imperfect lichens). The photobionts that participate in the lichen association belong, according to the species of lichen, to the Chlorophyceae and the cyanobacteria. As examples of lichens are the fruticulose thallus of *Cladonia* and the foliose thalli of *Lasallia* and *Peltigera* (Lecanorales). **2**. A cutaneous disease caused by dermatophytes (formerly called *ringworm*), which due to its scaly appearance resembles the thallus of certain lichens.

Fruticulose thalli of *Cladonia* sp., a beautiful **lichen** ("British soldiers"), x 3 (*TRO*).

Foliose thalli of the **lichen** *Lasallia* sp., x 3 (*TRO*).

Foliose thalli of the **lichen** *Peltigera* sp., x 3 (*TRO*).

lichenicolous (L. *lichen* < Gr. *leichén*, lichen + L. suf. *-cola*, inhabitant < *colere*, to inhabit + L. suf. *-osus* > E. *-ous*, having, with the qualities of): living on lichens, as a parasite, parasymbiont, or saprobe. More than 1,000 species of lichenicolous fungi are known, belonging to very diverse groups; there are even lichenicolous lichens.

licheniferous (L. *lichen* < Gr. *leichén*, lichen + suf. *-ferous*, bearer < *ferre*, to bear, carry): bearing lichens, like certain beetles in New Guinea, including *Gymnopholus lichenifera*, on whose back grow foliose and crustose lichen thalli, which also have been found on the shell of the giant tortoise (*Geochelone elephantopus*) in the Galapagos Islands.

lichenin (L. *lichen* < Gr. *leichén*, lichen + NL. *-in,* suf. that denotes belonging to): a linear glucose polymer, with 1-3 and 1-4 bonds, that is found in the hyphal wall of many fungi that form lichens. It takes on a red coloration in iodine.

lichenoid (L. *lichen* < Gr. *leichén*, lichen + L. suf. *-oide*, like < Gr. *-oeídes*, similar to): a type of symbiotic association present in some primitive marine lichens, in which the algal components are capable of living free, i.e., independently of the fungal component. The algae involved in the lichenoid association are usually microscopic and the fungi belong to gen. such as *Chadefaudia* (Halosphaeriales) and *Pharcidia* (=*Stigmidium*), of the Dothideales. *Pharcidia* (=*Stigmidium*) *laminariicola* is associated with epiphytic phaeophyceous algae of the species *Ectocarpus fasciculatus*, on pedicels of *Laminaria digitata*. *Chadefaudia corallinarum* is associated with epiphytic crustaceous rhodophyceous algae, of the gen. *Dermatolithon* and *Epilithon*. The fact that this fungus can live with different photobionts indicates that the symbiosis is not very strict. This association is only found in submerged conditions and the fungus grows on macroalgae (Chlorophyceae, Phaeophyceae and Rhodophyceae) and leaves of marine grasses, whose tissues are never penetrated by hyphae of the fungus, although ascocarps frequently are found partially covered by the calcareous thallus of the photobiont.

lichenometry (L. *lichen* < Gr. *leichén*, lichen + L. *metria* < Gr. *métrein*, to measure): a method that utilizes lichens to measure the minimum elapsed time of the exposure of a substrate since the lichens first developed on the substrate in question. Lichenometry is important in dating relatively recent geomor-phological or archeological events, such as the deposition of glacial moraines or the erection of the stone heads on Easter Island. This method is based on accurate data of growth rates of long-lived species of lichens under specific conditions.

life history (OE. *lif* < ME. *life* > E. *lifer*, the quality that distinguishes a living, functional being from a dead body; L. *historia* < Gr. *istoría*, investigation, history, information < *hístor*, *ístor*, to instruct, knowledgeable, learned, erudite): in the fungi it refers to the series of stages (often characterized by different spore states) through which a fungus passes from the formation of one spore type to the development of the same type of spore again. Also called *life cycle* (Gr. *kýklos*, circle, cycle).

ligneous (L. *ligneus*, woody < *lignum*, wood + L. suf. *-osus* > E. *-ous*, abundant, full of, prone to): having the consistency or the nature of wood, like the fruiting bodies of *Fomes*, *Ganoderma*, *Gloeophyllum*, *Lenzites*, *Phellinus* and other Aphyllophorales.

Ligneous basidiocarps of *Phellinus nilgheriensis*, x 0.2 (*RV*).

Ligneous basidiocarp of *Phellinus linteus*, x 0.2 (*RV*).

Ligneous basidiocarp of *Gloeophyllum* (=*Lenzites*) *sepiarium*, x 1 (*MU*).

lignicole, lignicolous (L. *lignicola* < *lignum*, wood + suf. *-cola*, inhabitant < *colere*, to inhabit + L. suf. *-osus* > E. *-ous*, having, with the qualities of): living on wood, but not necessarily deriving nourishment from it, as would a *xylophagous* organism. Many Myxomycetes, mitosporic fungi, ascomycetes, basidiomycetes and lichens are lignicolous, such as *Trichia* (Trichiales), *Taeniolella* (dematiaceous asexual fungi), *Auricularia* (Auriculariales), *Crepidotus*, *Leucocoprinus*, *Volvariella* (Agaricales), *Schizophyllum*, *Hexagonia*, *Poria*, *Pycnoporus* (Aphyllophorales), *Evernia* and certain species of *Lecanora* (Lecanorales). Equal to **epixylous**. Cf. **xylosaprobic**.

Lignicolous sporangia of *Trichia decipiens*, x 15 (*MU*).

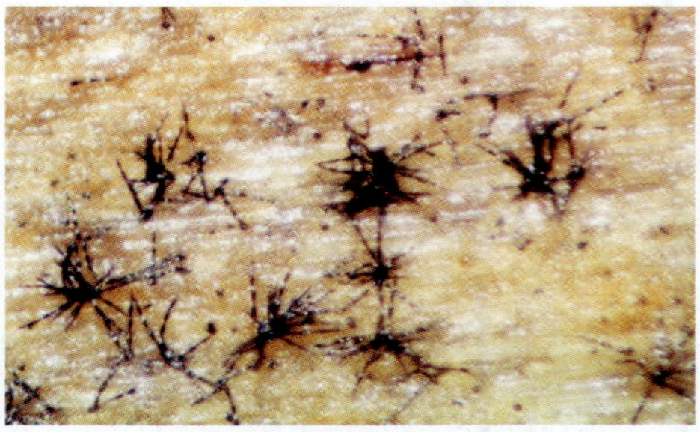

Lignicolous conidiophores of *Taeniolella* sp. growing on *Pinus* wood, x 15 (*MCG* & *AC*).

Lignicolous conidiophores of *Taeniolella* sp. growing on *Pinus* wood, x 15 (*MCG* & *AC*).

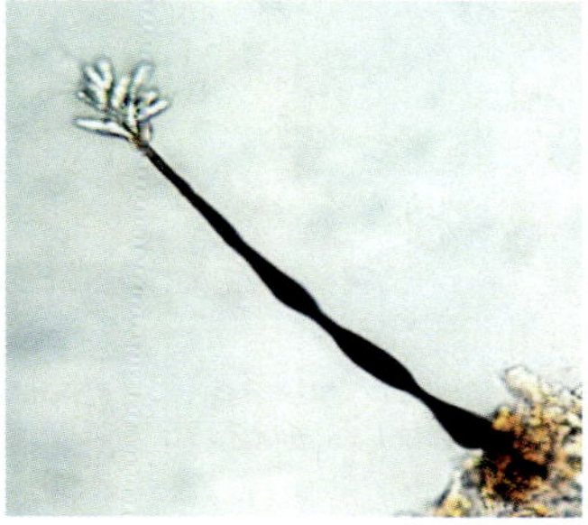

Lignicolous conidiophore of *Taeniolella* sp. growing on *Pinus* wood, x 50 (*MCG* & *AC*).

Lignicolous basidiocarps of *Auricularia auricula*, x 1 (*TH*).

Lignicolous basidiocarps of *Crepidotus mollis*, x 1 (*RV*).

Lignicolous basidiocarps of *Leucocoprinus birnbaumii*, x 0.5 (*HT*).

Lignicolous basidiocarps of *Volvariella bombycina*, x 0.5 (*RV*).

Lignicolous basidiocarps of *Schizophyllum commune*, x 0.2 (*RV*).

Lignicolous basidiocarp of *Schizophyllum commune*, x 5 (*MU*).

Lignicolous basidiocarps of *Hexagonia tenuis*, x 0.2 (*RV*).

Lignicolous basidiocarp of *Pycnoporus sanguineus*, x 0.5 (*MU*).

ligulate, liguliform, lingulate (L. *ligula*, small tongue < *lingua*, tongue + suf. *-atus* > E. *-ate*, provided with or likeness; or *lingua* + suf. *-formis* < *forma*, shape): with the shape of a tongue, flat like a belt, like the long and narrow thalline branchlets (the so-called *lorulae*) of some lichens (e.g., certain species of *Evernia*, *Letharia* and *Ramalina*, of the Lecanorales), and the apical projection of the perithecia of *Rickia passalina* (Laboulbeniales). Also called **loriform**.

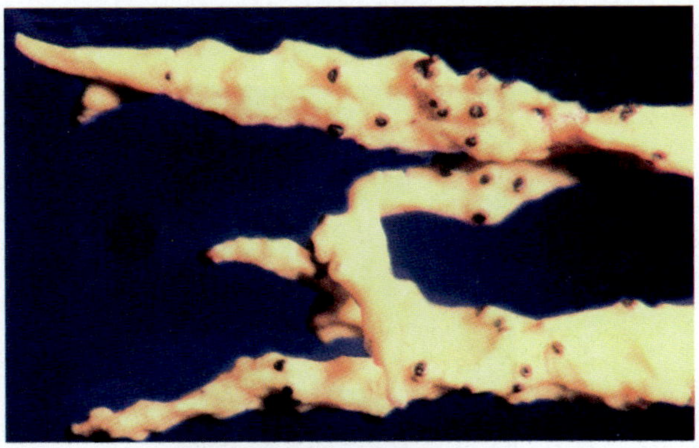

Ligulate branches of the thallus of *Letharia vulpina* subsp. *californica*, x 16 (*MU*).

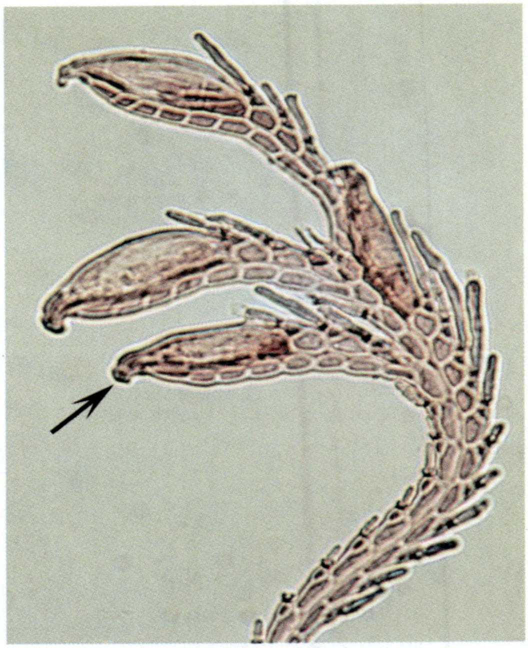

Liguliform projection on the apex of the perithecia of *Rickia passalina*, x 700 (*MU*).

limb (L. *limbus*, hem, sash, fringe, border): a border or margin that is distinguished in a structure.

limbate (LL. *limbatus*, bordered < L. *limbus*, hem, sash, fringe, border + suf. *-atus* > E. *-ate*, provided with or likeness): with the border edged with another color, like the pileus of the basidiocarp of *Psilocybe coprophila* and *P. pelliculosa*, the volva of *Amanita plumbea* (Agaricales), the basidiocarps of *Gloeoporus dichrous* and *Flaviporus hydrophilus* (Aphyllophorales).

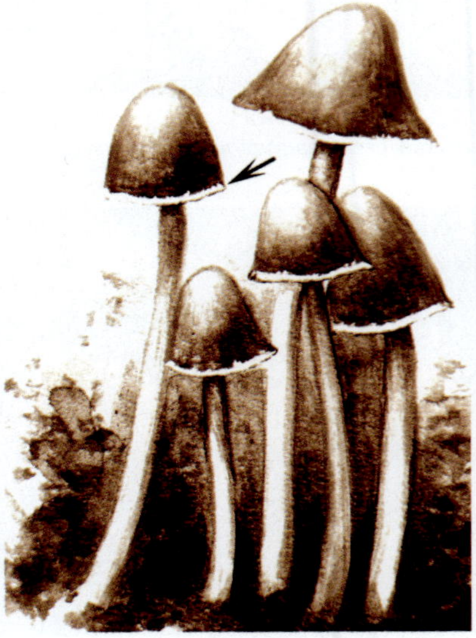

Limbate margin of the pileus of *Psilocybe pelliculosa*, x 1 (*MU*).

Limbate margin of the basidiocarps of *Gloeoporus dichrous*, x 1 (*RV*).

Limbate margin of the basidiocarp of *Flaviporus hydrophilus*, x 0.3 (*RV*).

limno-, **limn-** (Gr. *límne*, lake, swamp): a pref. used in terms that allude to fresh water, as in *limnology*, the science that studies lakes, and, by extension, all kinds of fresh water, including the organisms that grow in them, mainly from an ecological point of view.

limoniform (LL. *limoniformis* < ML. *limon*, lemon + suf. *formis* < *forma*, shape): lemon-shaped, i.e., broadly ellipsoidal with apiculate ends, like the fruit of the lemon tree (*Citrus*), because of which it is also called **citriform**. E.g., the sporangia of *Phytophthora infestans* (Peronosporales), the ascospores of *Melanospora* (Melanosporales) and the basidiospores of *Phaeocollybia* sp. (Agaricales) are limoniform or citriform.

Limoniform sporangia of *Phytophthora infestans*, x 114 (*MU*).

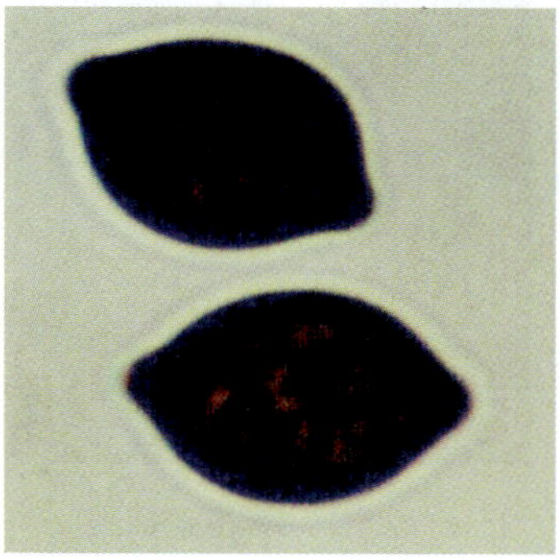

Limoniform ascospores of *Melanospora* sp., x 800 (*RTH*).

lineal (L. *linealis* < *linea*, line + suf. *-alis* > E. *-al*, pertaining to): pertaining to one or more lines. See **linear**, which is the preferable term.

linear (L. *linearis* < *linea*, line + suf. *-aris* > E. *-ar*, pertaining to): pertaining to one or more lines. Elongated and narrow, with parallel or almost parallel borders.

lineate (L. *lineatus*, lined < *lineare*, to mark with lines < *linea*, line + suf. *-atus* > E. *-ate*, provided with or likeness): having lines, generally lengthwise and parallel, distinguishable by their color or texture; striped, like the conidia of *Stachybotrys cylindrospora* (dematiaceous asexual fungi) and the pileus of the basidiocarps of *Xeromphalina* and *Marasmius* (Agaricales).

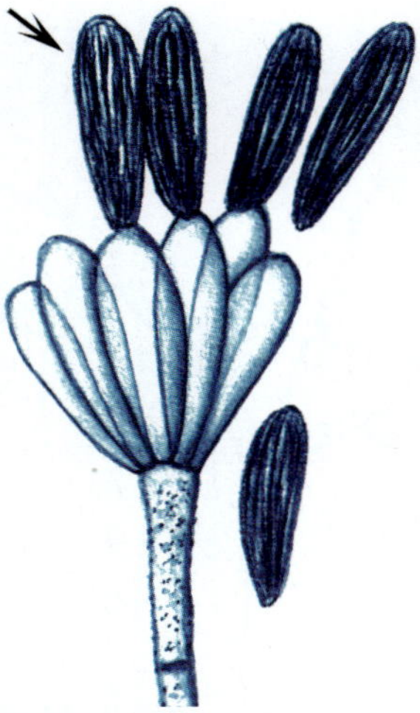

Lineate wall of the conidia of *Stachybotrys cylindrospora*, x 800 (*MU*).

Lineate pileus of the basidiocarps of *Xeromphalina campanella*, x 1 (*TRO*).

Lineate pileus of the basidiocarps of *Marasmius* sp., x 1 (*RV*).

lineolate (L. *lineolatus* < *lineola*, dim. of *linea*, line + suf. *-atus* > E. *-ate*, provided with or likeness): being marked with small lines.

lipo- (Gr. *leípo*, to leave, abandon, lack): a pref. employed to form terms that indicate something lacking, defective or deficient, as in *lipochloroplasy*, the morbid reduction or suppression of chlorophyll.

lipo-, **lip-** (Gr. *lípos*, fat): a pref. that alludes to something fatty, as in *lipochroic*.

lipochroic (Gr. *lípos*, fat + *chrõs*, color + suf. *-íkos* > L. *-icus* > E. *-ic*, pertaining to or relating to): having pigmentation in the lipid globules of the cellular cytoplasm.

lipolytic (Gr. *lípos*, fat + *lytikós*, capable of loosening, to release, separate < *lýtos*, dissoluble, broken, loose + suf. *-íkos* > L. *-icus* > E. *-ic*, pertaining to or relating to): having the capacity to degrade lipids in the substrates where it occurs; some molds are lipolytic, such as *Trichophyton mentagrophytes* (moniliaceous asexual fungi) and some cryptococcaceous asexual yeasts, such as *Malassezia* (=*Pityrosporum*), among others.

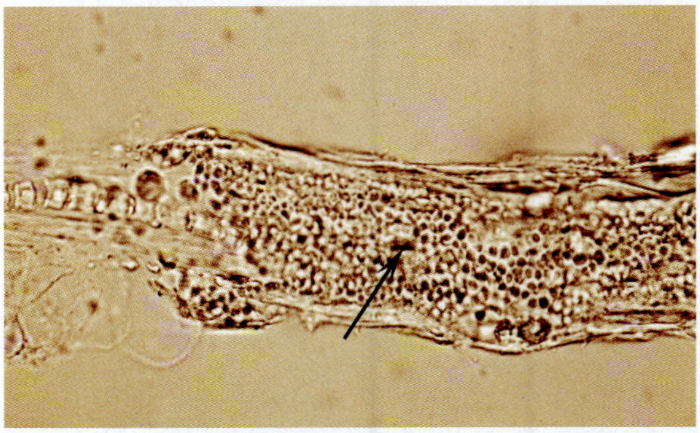

Arthrospores of *Trichophyton mentagrophytes*, a **lipolytic** species, on the hair of a white rat, x 800 (*MU*).

lipsanenchyma, pl. **lipsanenchymata** (Gr. *leipsánon*, remnant, relic + *énchyma*, tissue): the primordial tissue of a basidiocarp that is found between the stipe and the pileus covering the hymenium, but which does not correspond to the universal veil.

lirella, pl. **lirellae** (NL. *lirella*, dim. of L. *lira*, groove): the lirellae, also called *hysterothecia* or *lirelline apothecia*, are linear ascocarps, straight or curved and flexuous, simple or branched and sunken or somewhat erumpent, with a longitudinal slit and a well-defined proper excipulum, unitunicate asci and simple paraphyses, or bitunicate asci and pseudoparaphyses. The lirellae are characteristic of crustaceous lichens of the orders Ostropales (such as *Graphis scripta*) and Arthoniales (such as *Chiodecton*).

Lirellae of the thallus of *Graphis scripta*, x 14 (*MU*).

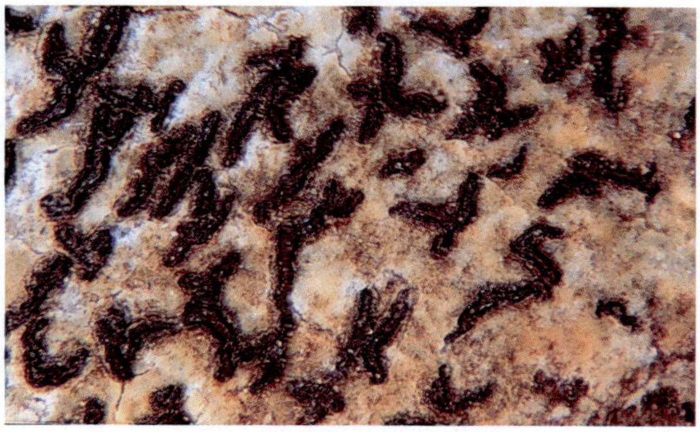

Lirellae of the thallus of *Graphis* sp., x 15 (*MU*).

litho-, **lith-** (Gr. *líthos*, stone, rock): a pref. employed in the formation of compounds that allude to stone, as in *lithotroph*.

lithotroph (Gr. *líthos*, stone, rock + *trophós*, what nourishes): capable of utilizing rocks as a source of nourishment, as do some lichens, e.g., *Caloplaca saxicola* (Teloschistales) and *Candelina* (Lecanorales). See **epilithic**, **petricolous**, **rupicola** and **saxicolous**.

Lithotroph thalli of *Candelina* sp., x 1 (*TRO*).

litmus (ME. *litmose* < Scand. origin, akin ON. *litmosi*, dyer's herb < *litr*, color, dye + *mosi*, moss): an amphoteric coloring matter obtained from lichens (such as *Roccella montagnei*, of the Arthoniales), generally incorporated into small strips of paper, which turns red in acidic solutions and blue in alkaline solutions, because of which it is used as an indicator of the level of acidity or alkalinity.

lituate (NL. *lituatus*, forked < L. *lituus*, augural baton, battle trumpet, bugle, hunting horn + suf. *-atus* > E. *-ate*, provided with or likeness): forked, with the points slightly arching outward, like the basidium of *Dacrymyces* (Tremellales). Also called **furcate**.

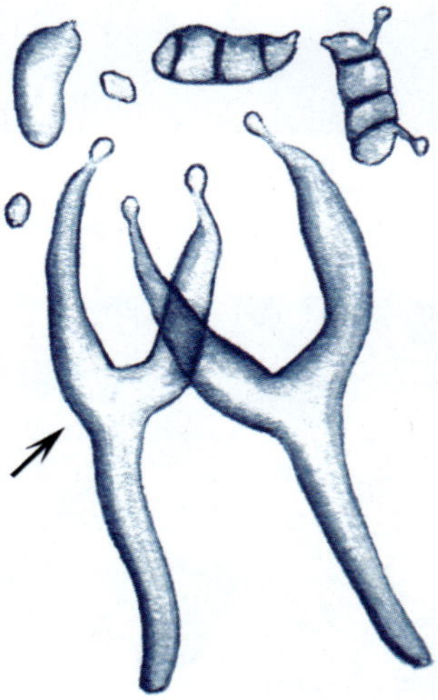

Lituate basidia of the hymenium of *Dacrymyces capitatus* (=*D. ellisii*), x 490 (*MU*).

liver fungus, pl. **fungi** (ME. < OE. *lifer*, liver; L. *fungus*, fungus < Gr. *spóngos*, *sphóngos*, sponge): a common name of the fruiting body of *Fistulina hepatica* (Aphyllophorales). Also called **beef-steak fungus**.

lobate (NL. *lobatus*, provided with lobes or with the appearance of a lobe < LL. *lobus* < Gr. *lobós*, lobe + L. suf. *-atus* > E. *-ate*, provided with or likeness): divided into lobes, i.e., in portions not very deep and more or less rounded; applied to laminar organs as well as to larger structures. When the lobes are smaller (*lobules*) they are said to be *lobulate*. E.g., the basidiospores of *Inocybe grammata* (Agaricales) are lobate and those of *I. oblectabilis* are lobulate. Also applied to the border of the macrocolonies of yeasts in culture, like those of certain strains of *Saccharomyces cerevisiae* (Saccharomycetales), *Rhodotorula* (cryptococcaceous asexual yeasts), and to the thallus of some lichens, such as *Lobaria scrobiculata* (Peltigerales), among other structures. See **lobulate**.

Lobate border of a giant colony of *Saccharomyces cerevisiae* on agar, x 2 (*MU*).

Lobate thallus of the lichen *Lobaria scrobiculata*, x 0.5 (*MU*).

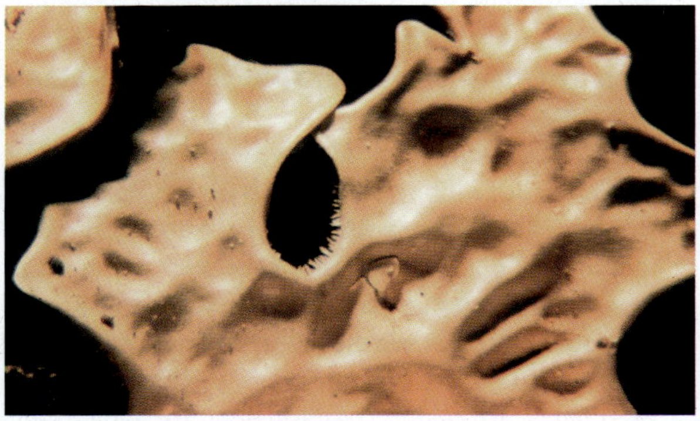

Lobate thallus of the lichen *Lobaria scrobiculata*, x 0.5 (*MU*).

lobomycosis, pl. **lobomycoses** (< gen. *Loboa*, dedicated to the Brazilian physician Jorge Lobo, + NL. *mycosis* < Gr. *mýkes*, fungus + suf. *-osis*, state or condition): a chronic subcutaneous, localized infection of man characterized by nodular and verrucose, tumor-like lesions in the arms, legs, face and ears, which contain the causal agent, *Loboa loboi*, a species of uncertain taxonomic position. Besides man, lobomycosis has been reported only from dolphins. See **mycosis**.

lobopod, lobopodium, pl. **lobopodia** (Gr. *lobós*, lobe + NL. *podium* < Gr. *pódion* < *poús*, genit. *podós*, foot): a pseudopodium or thick and rounded protoplasmatic prolongation of a protoplast or of a plasmodium; e.g., the myxamoebae of *Acrasis rosea* (Acrasiomycota) move and feed (phagocytosis) by means of lobopods. Cf. **filopod**.

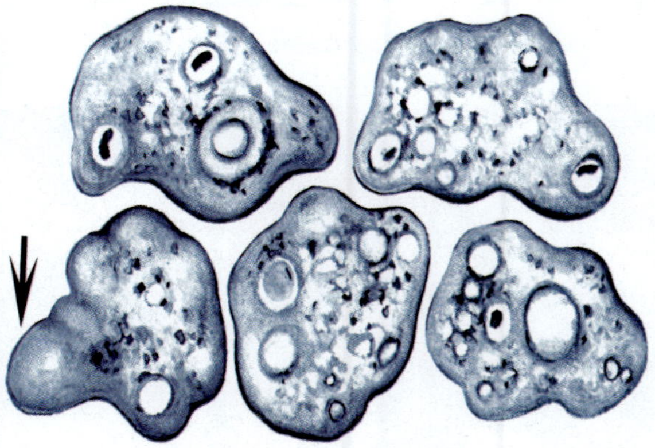

Lobopodia of the myxamoebae of *Acrasis rosea*, x 1,000 (*MU*).

lobulate (NL. *lobulatus*, divided into lobules < *lobulus*, dim. of LL. *lobus*, lobe + suf. *-atus* > E. *-ate*, provided with or likeness): provided with lobules, like the margin of the colonies of *Rhodotorula rubra* (cryptococcaceous asexual yeasts) and *Kluyveromyces marxianus* var. *marxianus* (Saccharomycetales) on agar. See **lobate**.

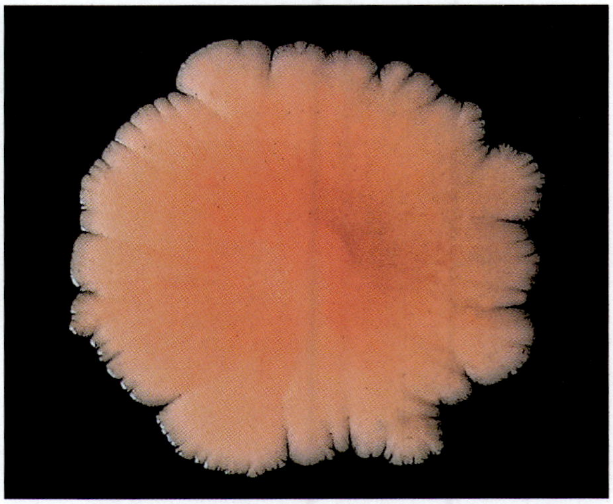

Lobulate giant colony of the yeast *Rhodotorula rubra* on agar, x 1 (*MU*).

Lobulate giant colony of the yeast *Kluyveromyces marxianus* var. *marxianus* on agar, x 1 (*MU*).

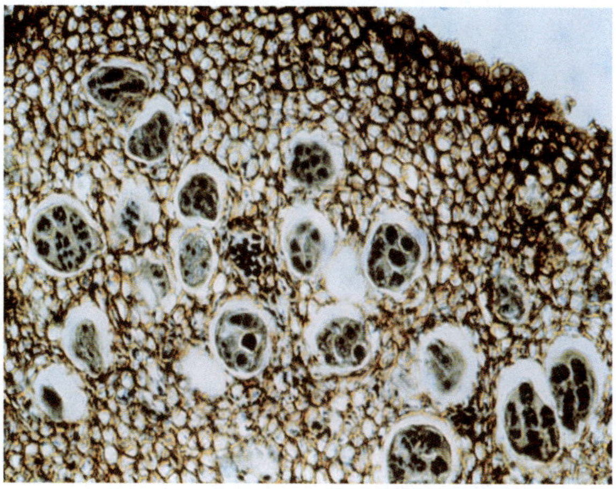

Locules of an apothecioid ascostroma of *Myriangium duriaei*, seen in longitudinal section; each locule contains one ascus, x 140 (*RTH*).

loc. cit. (L. abbrev. of *loco*, place, site; pl. *loci*; L. abbrev. of *citato*, cited): literally, in the place or passage cited. An abbrev. used, especially in older literature, to refer to a reference that was cited in full earlier in the text.

locomotion (L. *locus*, place, site + *motio*, genit. *motionis*, movement): in biology, the action of moving an organism from one place to another. E.g., in the fungi with zoospores the locomotion occurs by means of flagella, and in the Myxomycetes, besides the locomotion of the flagellate cells (myxozoospores), the myxamoebae move by sliding, through the extension of pseudopodia.

locotype (L. *locus*, place, site + *typus* < Gr. *týpos*, mark, model, type): a specimen of a taxon that comes from the same locality as the nomenclatural type of the taxon. Syn. of **topotype**.

loculate (L. *loculus*, dim. of *locus*, point, place, cavity + suf. *-atus* > E. *-ate*, provided with or likeness): provided with *locules* or chambers.

locule, loculus, pl. **loculi** (L. *loculus* < *locus*, point, place, cavity + dim. suf. *-ulus* > E. *-ule*): a cavity or chamber without a wall inside a stroma where asci are formed (*ascigerous locule*); depending on the species, the stroma can be *uniloculate* or *multiloculate*, and the locules can be uniascal or multiascal. Locules are characteristic of the Loculoascomycetes. E.g., in *Myriangium* (a gen. of the Myriangiales that parasitizes scale insects), the locules are uniascal, whereas in *Dothidea* (Dothideales) and *Apiosporina* (=*Dibotryon*, of the Pleosporales) the locules are multiascal. The term locule also is applied to the cavity of the pycnidium of the sphaeropsidaceous asexual fungi, in which the conidiophores and conidia are produced.

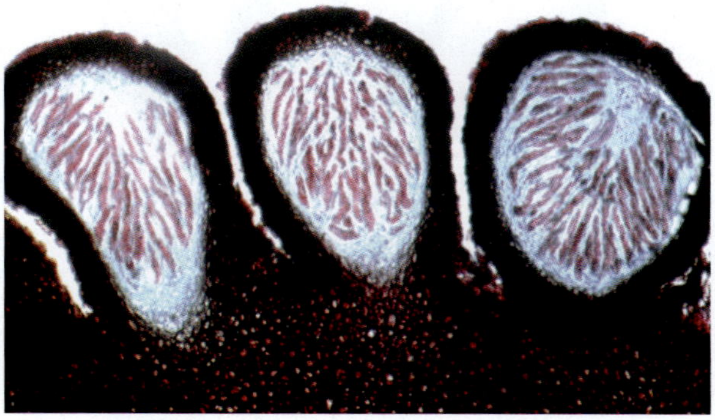

Locules of an ascostroma of *Apiosporina morbosa* (=*Dibotryon morbosum*), seen in longitudinal section; the locules are multiascal, x 140 (*RTH*).

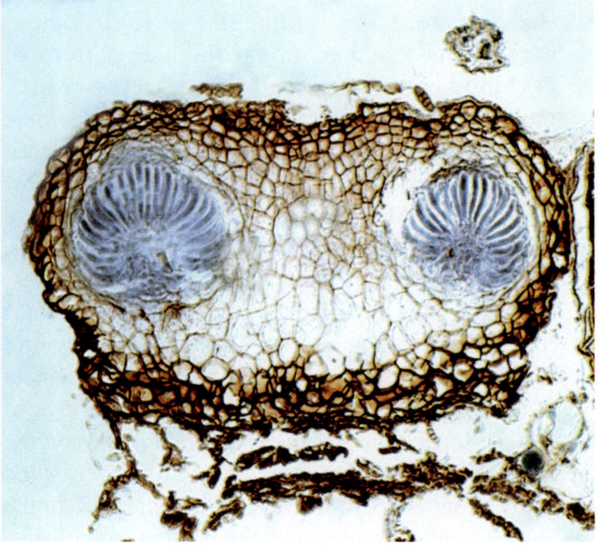

Locules of an ascostroma of *Dothidea* sp., seen in longitudinal section; the locules are multiascal, x 140 (*RTH*).

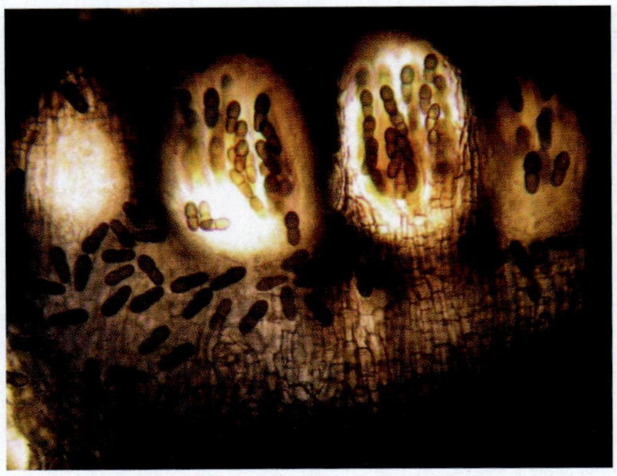

Locules of an ascostroma of an unidentified loculoascomycete, seen in longitudinal section; the locules are multiascal, x 140 (*RV*).

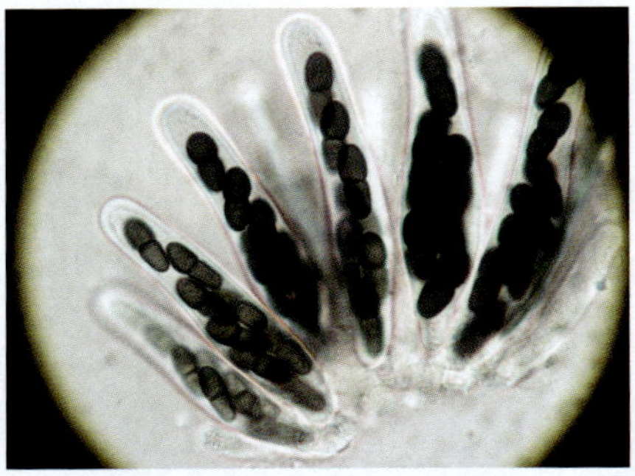

Locule of an ascostroma of an unidentified loculoascomycete, seen in longitudinal section; the locule is multiascal, x 350 (*RV*).

Loculoascomycetes (L. *loculus*, dim. of *locus*, point, place, cavity + Gr. *askós*, sac + L. *-mycetes*, ending of taxonomic class < Gr. *mýkes*, genit. *mýketos*, fungus): a class of fungi with bitunicate asci produced in a unilocular or multilocular ascostroma that has an ascolocular pattern of development. The Loculoascomycetes (which includes the orders Coryneliales, Dothideales, Myriangiales, Arthoniales, Pyrenulales, Asterinales, Capnodiales, Chaetothyriales, Patellariales, Pleosporales, Melanommatales, Trichotheliales, and Verrucariales) together with the Archiascomycetes, Saccharomycetes, Plectomycetes, Pyrenomycetes, Discomycetes, and other filamentous ascomycetes (such as the Erysiphales, Laboulbeniales and Spathulosporales), constitute the phylum Ascomycota of the kingdom Fungi. In the Appendix, the user of this dictionary can see the comparison between the outline of Alexopoulos *et al.*, 1996, modified by Ulloa and Hanlin, 2006, to that of Kirk *et al.*, 2008. In the latter classification scheme, the orders Coryneliales, Pyrenulales, Chaetothyriales and Verrucariales, have been moved to the class Eurotiomycetes; the order Trichotheliales moved to the class Lecanoromycetes, and the order Arthoniales has been moved to the class Pezizomycetes. The three classes are in the subphylum Pezizomycotina, phylum Ascomycota. These taxonomic rearrangements, with the concurrent change of the names of taxa, are part of a constant process of study and evaluation, in order to achieve a phylogenetic classification of the fungi.

loculomycetes (L. *loculus* < *locus*, point, place, cavity + dim. suf. *-ulus* > E. *-ulo* + Gr. *mýkes*, genit. *mýketos*, fungus): the name given by the mycologist Gorovij to a class of fungi whose sexual sporulation was considered of a type different from that previously known, in which he included the gen. *Zerovaemyces*, an agaric similar to *Coprinus*, in which the gills of the fruiting body are replaced by spherical, multiloculate structures that contain spores; each chamber or locule contains a *loculospore*.

loculospore (L. *loculus* < *locus*, point, place, cavity + dim. suf. *-ulus* > E. *-ulo* + *sporá*, spore): see **loculomycetes**.

locus, pl. **loci** (L. *locus*, point, place): **1**. *Genetics*. A point or site on a chromosome in which is situated a gene. **2**. *Conidiogenesis*. The particular site on a hypha or a cell, i.e., conidiogenous locus, that represents the fertile point where the conidia are generated. E.g., in a phialide the conidiogenous locus is located in the subapical part, almost in the mouth through which the conidia emerge.

-logy (Gr. *-logía* > E. *-logy*, treatise, discourse < *lógos*, word, language < *légo*, I say, in the sense of narrating, describing): a suf. employed in forming terms that allude to something spoken or narrated, to discourse, to a doctrine. In *biology*, e.g., *bio-* < Gr. *bíos*, life + *-logía*, the term *biology* is the narration of everything that refers to life, to living beings.

lomasome (Gr. *lõma*, border, edge, fringe + *sõma*, body): a polymorphic structure, most commonly composed of membranous vesicles or tubules that are found between the plasma membrane and the cell wall, although at times the vesicles can be embedded in the cell wall. It has been considered that lomasomes are implicated in the synthesis of the cell wall, although it also has been suggested that they represent excess plasma membrane in relation to that necessary to line the cell wall, in which case their function would be that of normal cell membrane. Cf. **plasmalemasome**.

lomato-, **lomat-** (Gr. *lõma*, genit. *lómatos*, fringe, edge, border): a pref. infrequently employed in names such as *Lomatina* (syn. of *Cytidia*, of the family Corticiaceae, order Aphyllophorales). The pref. *loma-* is more common, being used in terms such as *lomasome*.

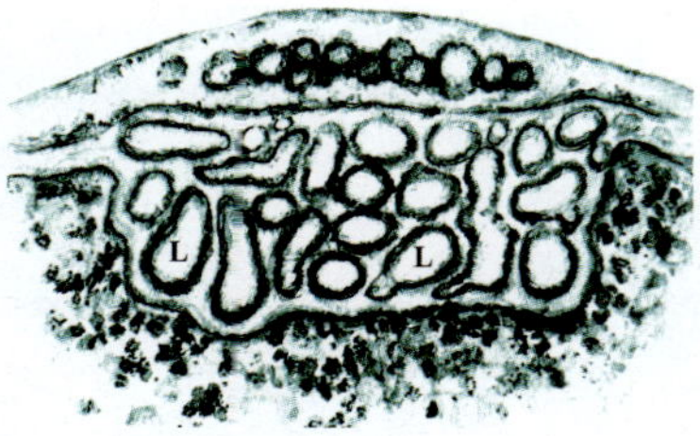

Lomasome (L) of a hyphal cell of *Saprolegnia ferax* (Saprolegniales) seen in longitudinal section. Drawing based on a transmission electron micrograph published by Beckett *et al.* in their *Atlas of Fungal Ultrastructure*, 1974, x 61,000 (*MU*).

loncho-, lonch (Gr. *lónche*, lance, spear): a pref. used to allude to something that is lance-shaped, as in *Lonchospermella* (a dubious name of an anamorphic fungus), which alludes to the shape of the spores.

longevity (L. *longaevitas*, genit. *longaevitatis*, long life): the period of life of an organism.

longi- (L. *longus*, long): a pref. used to emphasize the length of a structure, e.g., *longicaule*, *longicollous*, *longiseptum*, pl. *longisepta*, etc., for the length of a stem, neck, septum, etc. Cf. **brachy-**.

longitudinal (L. *longitudinalis*, pertaining to or relating to length < *longitudo*, *longitudin-*, length, longitude + suf. *-alis* > E. *-al*, relating to or pertaining to): formed or positioned parallel to the longitudinal axis of a structure, like the longitudinal dehiscence of the ascocarp of *Lophodermium* (Rhytismatales) and the longitudinal septa of many types of spores. When these septa are united with transverse septa they form dictyospores (e.g., in *Alternaria*, dematiaceous asexual fungi). A longitudinal section is one made parallel to the major dimension of the cell, organ, etc. Cf. **transverse**.

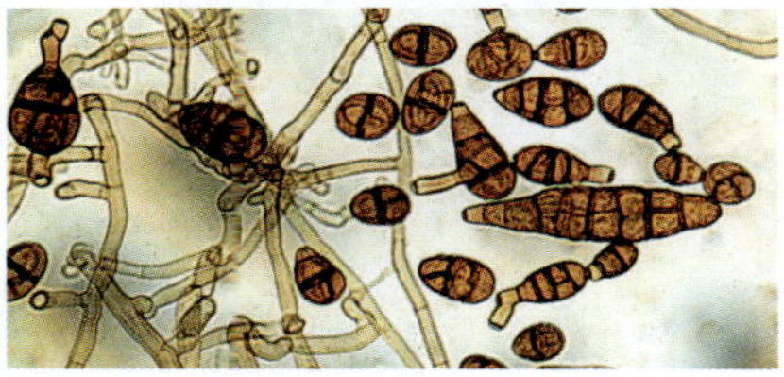

Longitudinal and transverse septa of the dictyospores of *Alternaria alternata*, x 400 (*MU*).

Longitudinal dehiscence of an ascocarp of *Lophodermium* sp. growing on *Myrica*, x 25 (*RTH*).

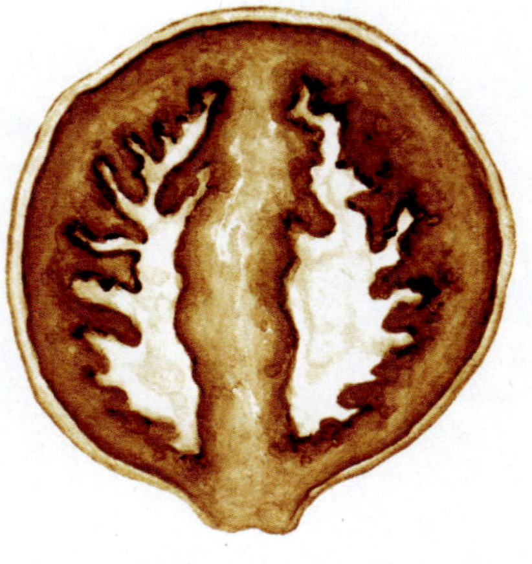

Longitudinal section of the fruiting body of *Podaxis pistillaris* (Hymenogastrales), x 2 (*MU*).

lopado-, lopad- (Gr. *lópas*, genit. *lópados*, flat plate): a pref. used in names such as *Lopadostroma* (flattened stroma), of the Xylariales, and *Lopadiopsis* (syn. of *Gyalectidium*), of the Ostropales.

lopho-, loph- (Gr. *lóphos*, tuft, plume, crest): a pref. employed in terms such as *lophotrichous*, with a polar tuft of flagella; applied mainly to certain bacterial cells.

lorate (L. *loratus*, held together by a leather strap, now interpreted as having the shape of a strap or bridle): see **loriform**.

loriform (L. *loriformis* < *lorum*, bridle, strap + suf. *-formis* < *forma*, shape): long and narrow like a strap. See **ligulate**.

lotic (L. *lotus*, washed + suf. *-íkos* > L. *-icus* > E. *-ic*, pertaining to or relating to): of, relative to, or living in running water. Cf. **lentic**.

lower fungi, sing. **fungus** (ME. *lowe, lohe* < ON. *lagr*, low + suf. *-er*, denotes comparative degree; L. *fungus*, fungus < Gr. *spóngos*, *sphóngos*, sponge): the general name for the fungi classified in the phyla Chytridiomycota, Hyphochytriomycota, Myxomycota, Oomycota, Plasmodiophoromycota and Zygomycota. Cf. **higher fungi**.

loxo-, lox- (Gr. *loxós*, oblique): a pref. employed in names such as *Loxospora* and Loxosporaceae (lichens of the order Lecanorales), that allude to the oblique arrangement of the ascospores in the asci.

lubricous (L. *lubricus*, smooth and slippery < *lubricare*, to make slippery + L. *-osus* > OF. *-ous*, *-eus* > E. *-ous*, quality, full of, augmented): a surface, usually shiny, that feels waxy or oily to the touch, slippery, like the ascocarp of *Coryne atrovirens* and *Leotia lubrica* (Helotiales), the basidiocarp of *Mycena galericulata*, *Boletus pinophilus* and *Hygrocybe conica* (Agaricales). See **oleaginous**.

Lubricous apothecia of *Coryne atrovirens*, x 2 (*RV*).

Lubricous apothecia of *Leotia lubrica*, x 1 (*RV*).

Lubricous basidiocarp of *Boletus pinophilus*, x 1 (*RV*).

Lubricous basidiocarp of *Hygrocybe conica*, x 1 (*TRO*).

lucid (L. *lucidus*, clear, brilliant, luminous): a term often employed in the sense of dazzling, more than bright, brilliant, as applied in the name *Ganoderma lucidum* (Aphyllophorales), referring to the fruiting body, in which the surface of the pileus is lustrous, as if it were varnished with lacquer. See **laccate**.

lumen, pl. **lumina** or **lumens** (L. *lumen*, light, window, extent of an orifice or conduit): a cavity enclosed by the walls of a cell or a hypha; also applied to the interior space of the tubes of a polypore, as is seen in *Dichomitus mexicanus* (=*Megasporoporia mexicana*), Aphyllophorales.

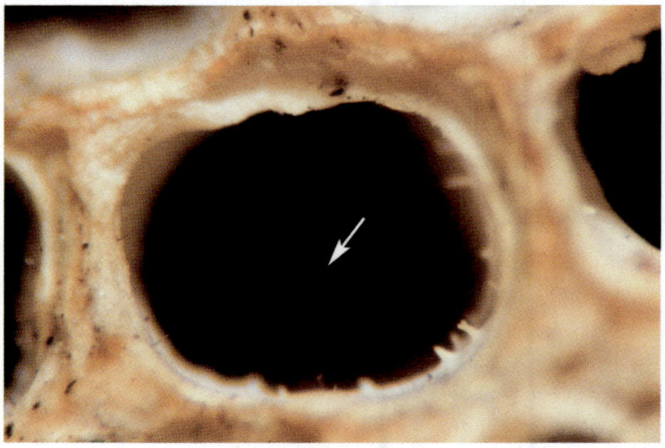

Lumen of a tube in the porose hymenophore of *Dichomitus mexicanus* (=*Megasporoporia mexicana*), x 23 (*MU*).

luminescent (L. *luminescens*, genit. *luminescentis* < *lumen*, light + suf. *-escens*, genit. *-escentis* > E. *-escent*, that which turns, that begins to, slightly): an organism that exhibits the phenomenon of *luminescence*; when describing living beings it is also called *bioluminescence*, like the basidiocarps of *Mycena lux-coeli* (Agaricales). See **bioluminescent**, **fluorescent** and **phosphorescent**.

Luminescent basidiocarps of *Mycena lux-coeli*, x 3 (*MU*).

lunate (L. *luna*, moon + suf. *-atus* > E. *-ate*, provided with or likeness): having the shape of a half moon, or lunar crescent, like the macroconidia of *Fusarium* (tuberculariaceous asexual fungi) in side view. Also called **crescentiform**, **lunulate**, **falcate** and **falciform**.

lunulate (L. *lunula*, dim. of *luna*, moon + suf. *-atus* > E. *-ate*, provided with or likeness): minutely lunate.

lurid (L. *luridus*, pale yellow): pale yellow in color, like the basidiocarp of *Boletus luridus* (Agaricales).

luteous (L. *luteus*, golden-yellow): yellow, like the basidiocarps of *Omphalotus olearius* (Agaricales).

Luteous basidiocarps of *Omphalotus olearius*, causing a white rot of oak roots, x 1 (*RV*).

lycoperdoid (< gen. *Lycoperdon* < Gr. *lýkos*, lobe + *pérdo*, wind + L. suf. *-oide*, like < Gr. *-oeídes*, similar to): refers to a globose or subglobose fruiting body that has a powdery gleba at maturity, usually supported by a sterile tissue; characteristic of *Lycoperdon*, *Bovista*, *Calvatia* and other related gen. of Lycoperdales.

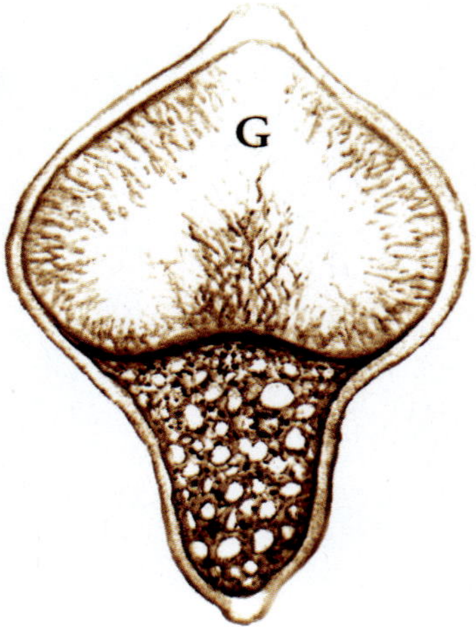

Lycoperdoid type of development. Diagram of a longitudinal section of a fruiting body of *Lycoperdon* sp., showing the gleba (G) or fertile part and a supporting stalk, x 1.5 (*MU*).

lyma-, lym- (Gr. *lýma*, filth, dirtiness, impurity): a pref. employed in the formation of compounds that allude to sewage, as in *lymabiont*.

lymabiont (Gr. *lýma*, filth, dirtiness, impurity + *bíos*, life + *óntos*, genit. of *ón*, a being): an organism that is only found in sewage.

lymaphile (Gr. *lýma*, filth, dirtiness, impurity + *phílos*, have an affinity for): an organism commonly found in sewage, like *Leptomitus lacteus* (Leptomitales).

lymaphobe (Gr. *lýma*, filth, dirtiness, impurity + *phóbos*, fearing or avoiding < *phobéo*, to fear): an organism that is never found in sewage.

lymaxene (Gr. *lýma*, filth, dirtiness, impurity + *xénos*, stranger): an organism that is rarely found in sewage.

lyocystidium, pl. **lyocystidia** (Gr. *lyó*, to loosen, release + NL. *cystidium* < Gr. *kystídion* < *kýstis*, purse, bladder, vesicle, cell + dim. suf. *-ídion* > L. *-ium*): a hyaline, cylindrical to conical cystidium with a very thick wall, with the apex abruptly tapered, like that in species of the polyporaceous gen. *Tubulicrinis* (Aphyllophorales).

lyophilization (Gr. *lyó*, to dissolve, disperse + *phílos*, have an affinity for + L. suf. *-ationem* > E. *-ation*, action, state or condition, result): a technique that is utilized to preserve fruiting bodies and cultures of fungi and other microorganisms in a state of suspended animation (*lyophilized*); it involves a process of freezing the fruiting body, or a liquid medium containing a suspension of the organism, usually as spores, followed by drying the suspension under vacuum (sublimation). Also known as *freeze-drying*.

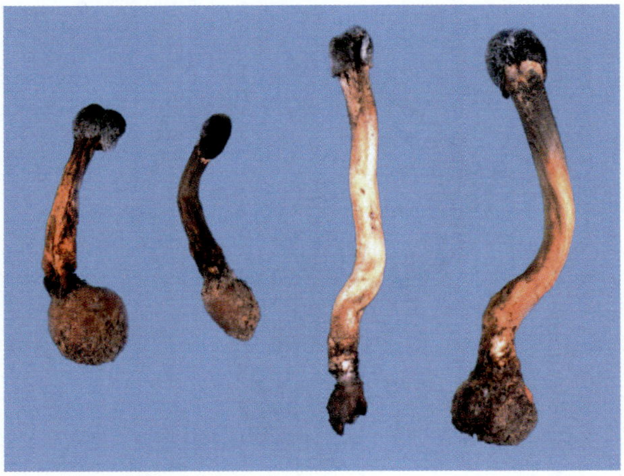

Lyophilized perithecial stromata of *Cordyceps militaris* (Hypocreales). The two stromata on the left were air-dried, x 1 (*RTH*).

Culture of *Sordaria fimicola* (Sordariales) on agar; the perithecia on agar were **lyophilized** for their preservation, x 0.5 (*RTH*).

Lyophilized basidiocarps of *Exidia recisa*, x 1 (*RTH*).

Lyophilized pileus of a basidiocarp of *Russula emetica* (Agaricales), x 1 (*RTH*).

lyrate (L. *lyratus* < *lyra*, lyre or lute + suf. *-atus* > E. *-ate*, provided with or likeness): lyre-shaped, i.e., with the narrowest part at the apex, similar to the musical instrument.

lysigenous (Gr. *lýsis*, dissolution, disintegration + *génos*, origin): forming by means of the dissolution of cells, as happens during the formation of the ostiolar canal of many ascomycetes, e.g., *Haloguignardia* (Phyllachorales), and of the cavities (locules) that originate during the morphogenesis of the ascocarps of the loculoascomycetes. Cf. **schizolytic**.

-lysis, pl. **-lyses** (Gr. *lýsis*, dissolution, disintegration): the destruction of cells or tissues by dissolution or dissociation of their elements. A suf. used in words such as *autolysis*, *bacteriolysis*, etc., to refer to self disintegration, the decomposition of bacteria, etc. See **autolysis**.

A sculpture of an anthropomorphic **mushroom stone**, from the Mayan culture (*MU*).

m

macro-, **macr-** (Gr. *makrós*, long, wide, deep, large): a pref. used in numerous compound terms, such as *macroconidium*, *macrocyclic*, etc., for large conidium, long life cycle, etc., and for names of genera, such as *Macrocybe* (large head), Agaricales. Cf. **micro-**.

Basidiocarp of *Macrocybe titans*; the pref. **macro-** in the genus was used for obvious reasons, x 0.05 (*AA* & *RA*).

macrocephalic (Gr. *makrós*, large + *kephalé*, head + suf. *-íkos* > L. *-icus* > E. *-ic*, pertaining to or relating to): literally, and figuratively, with a large head. It is applied to dictyosporous ascospores in which septation proceeds from the central primary septum toward the poles, so that immature spores have larger end cells. Cf. **microcephalic**.

macroconidium, pl. **macroconidia** (Gr. *makrós*, large + *kónis*, powder, dust + dim. suf. *-ídion*): a conidium or asexually reproduced spore that is distinguished from a microconidium by its larger size as well as by being multicellular; macroconidia are present in some asexual fungi, e.g., in various species of *Fusarium* (such as *F. moniliforme* and *F. subglutinans*, tuberculariaceous species), in dermatophytes such as *Epidermophyton floccosum* and *Trichophyton mentagrophytes* (moniliaceous asexual fungi), and in *Neurospora* (Sordariales). Cf. **microconidium**.

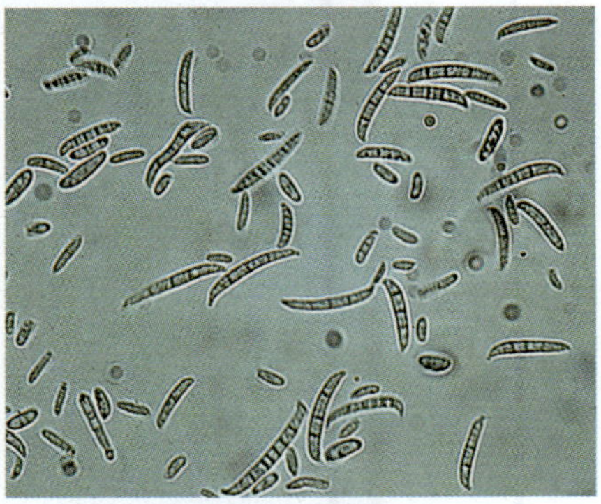

Macroconidia and microconidia of *Fusarium subglutinans* (=*F. moniliforme* var. *subglutinans*), x 350 (*PL*).

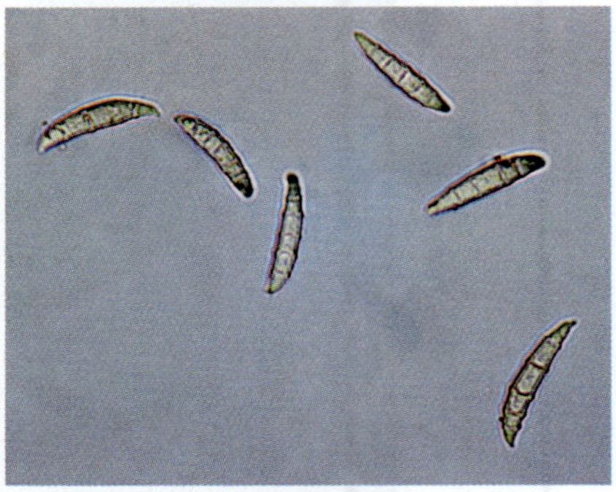

Macroconidia of *Fusarium* sp., x 450 (*MU*).

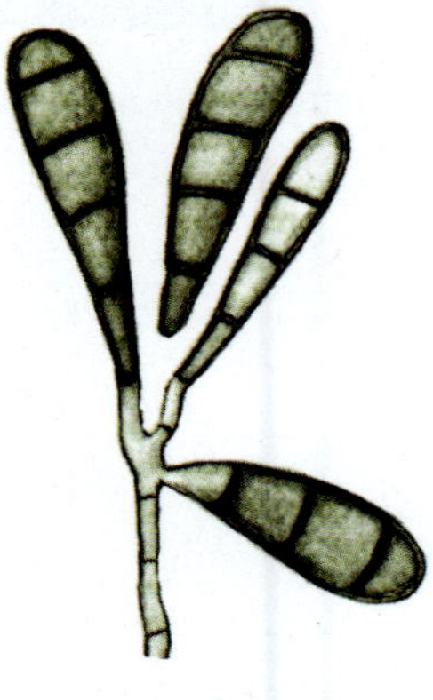

Macroconidia of *Epidermophyton floccosum*, x 1,000 (*MU*).

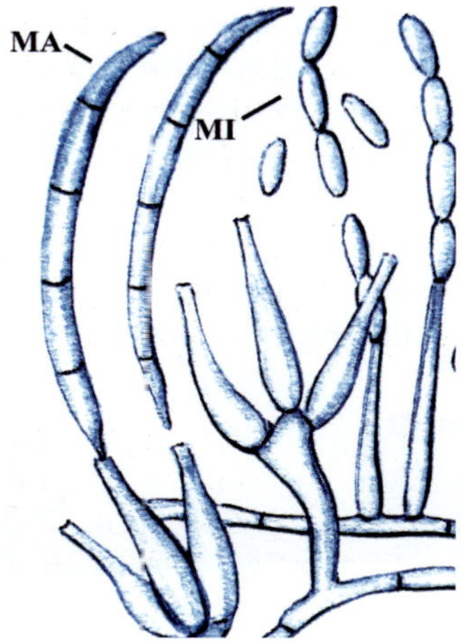

Macroconidia (MA) and microconidia (MI) produced on phialides of *Fusarium moniliforme*, x 790 (*MU*).

macrocyclic (Gr. *makrós*, large + *kyklikós*, circular, cyclic < *kiklós*, circle + suf. *-íkos* > L. *-icus* > E. *-ic*, pertaining to or relating to): *Uredinales*. Applied to species of rusts with a long life cycle, which encompasses five phases or stages of development, each with a characteristic type of spore (spermatia, aeciospores, urediniospores, teliospores and basidiospores), as, e.g., in *Puccinia graminis tritici*, the wheat rust. The macrocyclic rusts can be autoecious or heteroecious, depending upon whether they complete their life cycle on a single host or on two different hosts, respectively. Cf. **microcyclic** and **hemicyclic**.

macrocyst (Gr. *makrós*, large + *kýstis*, bladder; here, vesicle, cell): **1**. *Dictyosteliomycota*. A cell with a thick cellulosic wall, that originates from the fusion of a pair of haploid myxamoebae, and in which karyogamy and meiosis occur to form again the haploid myxamoebae of the vegetative phase. Thus the macrocysts represent the sexual state of these organisms, such as *Dictyostelium discoideum* (Dictyosteliales). **2**. An old term for the ascogonium of *Pyronema* (Pezizales). Cf. **microcyst** and **paracyst**.

macrocystidium, pl. **macrocystidia** (Gr. *makrós*, large + *kystídion*, cystidium < *kýstis*, bladder; here, vesicle, cell + dim. suf. *-ídion*): *Agaricales*. A type of cystidium that originates from the deep zone of the trama; it is very long, fusiform or claviform, often with an apical appendage of variable shape, and with a peduncle that generally is connected to a lactiferous hypha in the trama; it is characteristic of *Macrocystidia*, *Lactarius* and *Russula* (Agaricales). The macrocystidium is considered a very evolved form of pseudocystidium, which contains refringent structures such as needles or masses, and at times is muricated with crystals that are soluble in ammoniac. Cf. **chrysocystidium**, **lamprocystidium** and **leptocystidium**.

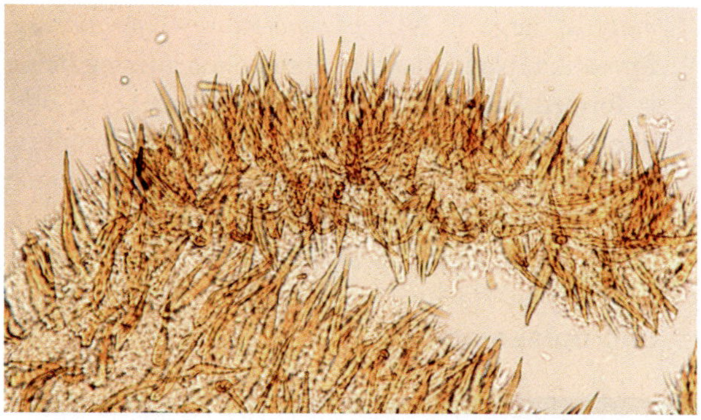

Macrocystidia of the hymenium of *Macrocystidia* sp., x 96 (*MU*).

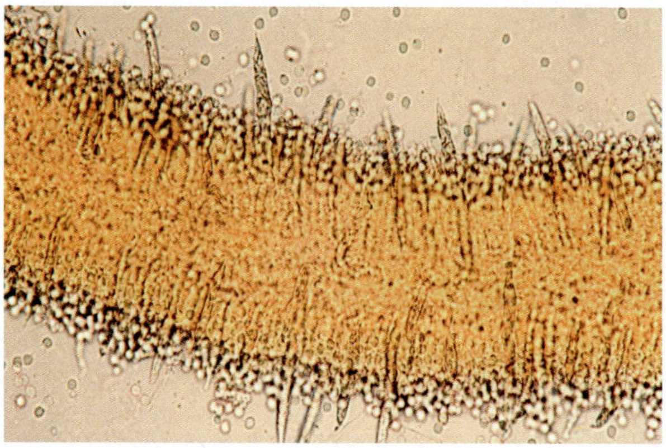

Macrocystidia of the hymenium of *Russula* sp., x 110 (*MU*).

macrolichen (Gr. *makrós*, large + L. *lichen* < Gr. *leichén*, lichen): a lichen with a large thallus, including those with a squamulose, foliose or fruticulose habit, e.g., *Cladonia*, *Parmelia* and *Usnea* (Lecanorales), respectively. Cf. **microlichen**.

macromycete (L. *macromycete* < Gr. *makrós*, large + *mýkes*, genit. *mýketos*, fungus): a fungus with a macroscopic reproductive body or sporiferous apparatus; the macromycetes represent the higher fungi, including certain ascomycetes (mainly Xylariales, Tuberales and Pezizales) and the major part of the basidiomycetes (Tremellales, Agaricales, Aphyllophorales and the orders of Gasteromycetes). Cf. **micromycete**.

macronematous (Gr. *makrós*, large + *nêma*, genit. *nêmatos*, filament + L. suf. *-osus* > E. *-ous*, having, abundant in, with the qualities of): refers to a specialized conidiophore that is distinguished morphologically from the assimilative hyphae in the rest of the mycelium. Cf. **micronematous**. These terms, as well as *mononematous* and *synnematous*, are utilized frequently to describe species of asexual fungi, especially dematiaceous species.

macroplasmodium, pl. **macroplasmodia** (Gr. *makrós*, large + NL. *plasmodium* < Gr. *plásma*, formation, bland material with which a living being is formed + L. suf. *-odium*, resembling < Gr. *ode*, like < *-oeídes*, similar to): applied to the plasmodium (*phaneroplasmodium*) of *Physarum polycephalum* and other species (Physarales), which attains great size, at times up to a meter or more; another example is *Hemitrichia serpula* (Trichiales). Cf. **microplasmodium**.

Macroplasmodium of *Physarum polycephalum* growing on the bark of a tree, x 1 (*RV*).

Plasmodiocarp, a sporiferous apparatus, which retains the shape of a **macroplasmodium** of *Hemitrichia serpula*, x 1 (*RV*).

macroscopic (Gr. *makrós*, large + *skópein*, to see + suf. *-íkos* > L. *-icus* > E. *-ic*, pertaining to or relating to): something sufficiently large to be seen without magnification, without the aid of a microscope. Cf. **microscopic**.

macrospore, **macrosporous** (Gr. *makrós*, large + *sporá*, spore): with large spores. A large spore. Cf. **microspore**.

maculate (L. *maculatus*, spotted < *macula*, spot + suf. *-atus* > E. *-ate*, provided with or likeness): with spots, spotted. In lichenology it refers to species in which the apothecia on the upper surface of the thallus appear as small spots, as in *Arthonia radiata* (Arthoniales) and *Melaspilea maculosa* (of the Melaspileaceae, a family of ascomycetes of uncertain affinities, previously included in the Hysteriales). Cf. **immaculate**.

Maculate thallus of *Arthonia radiata*, x 1.7 (*MU*).

maculicole, **maculicolous** (L. *macula*, spot + suf. *-cola*, inhabitant < *colere*, to inhabit): occurring on spots; many fungi sporulate on leaf spots caused by other plant pathogens on a host plant. E.g., *Fusarium stoveri* (tuberculariaceous asexual fungi) has been found on leaf spots of banana caused by the sigatoka disease, whose etiologic agent is *Cercospora musicola* (dematiaceous asexual fungi).

magic mushrooms (L. *magice*, *magicus* < Gr. *magiké*, fem. of *magikós*, magic < *mágos*, magician, sorcerer + suf. *-íkos* > L. *-icus* > E. *-ic*, belonging to or relating to a magician, pertaining to magic, marvelous; ME. *musseroum* < MF. *mousseron* < LL. *mussirion*, mushroom): fungi that contain some hallucinogenic or psychotomimetic substance, like the species of the gen. *Psilocybe* (Agaricales), e.g., *P. mexicana*, *P. caerulescens*, *P. zapotecorum* and *P. aztecorum*, among others. See **hallucinogenic fungi**.

Fruiting bodies of *Psilocybe caerulescens* and *P. mexicana*, the so-called **magic mushrooms**, x 0.5 (*TH*).

malaco-, **malac-** (Gr. *malakós*, bland, soft): a pref. employed in terms that allude to something bland or mucilaginous, such as *malacoid*.

malacoid (Gr. *malakós*, bland, soft + L. suf. *-oide*, like < Gr. *-oeídes*, similar to): with a bland or mucilaginous consistency, like the microbiogleae called *tibicos* in Mexico (*tibi grains* or *sugary kefir grains*, in other areas), which are macrocolonies of yeasts and bacteria in symbiosis, embedded in dextrans, that are utilized popularly to produce fermented beverages known as tepache and vinegar of tibicos. The disease called *leucoencephalomalacia*, which is caused by some toxins of *Fusarium moniliforme* (tuberculariaceous asexual fungi), receives its name from the white and mucilaginous appearance of the encephalic lesions of horses that have suffered this type of mycotoxicosis due to the ingestion of contaminated feed.

malt (OE. *mealt*, malt, beer < *meltan*, to melt, dissolve, soften): cereal grain (mainly barley) which, after being soaked in water, is allowed to germinate; this process is called malting, and consists of the transformation of maltose into glucose through activity of the enzyme maltase. Afterwards, the malt is dried, then boiled, and the must (malt liquor) undergoes a process of fermentation carried out by yeasts, mainly *Saccharomyces cerevisiae* (Saccharomycetales). Malt is used especially in the manufacture of beer or, through distillation, some liquors, such as whiskey.

mammiform (L. *mammiformis*, breast-shaped < *mamma*, breast + suf. *-formis* < *forma*, shape): shaped like a breast, i.e., conical or rounded, with a protruding apex, like the ostioles of the perithecia of *Entoleuca* (=*Hypoxylon*) *mammatum* (Xylariales), the apothecia of the thallus of *Thelomma santessonii* (Caliciales), and the pileus of the basidiocarp of *Psilocybe caerulescens* and *P. sanctorum* (Agaricales).

Mammiform pileus of the mature basidiocarps of *Psilocybe caerulescens*, x 0.9 (*MU*).

Mammiform pileus of the basidiocarp of *Psilocybe barrerae*, x 1 (*RV*).

Mammiform ostiole of the perithecium of *Entoleuca* (=*Hypoxylon*) *mammatum*, x 12 (*EPS*).

Mammiform apothecia of the thallus of *Thelomma santessonii*, x 10 (*CB*).

mange, scab (ME. *manjewe* < MF. *mangeue*, an eating, itching < *mangier*, to eat < L. *manducare*, to chew, eat; ME. < Scand. origin < L. *scabies*, mange < *scabere*, to scratch): any of the contagious diseases of the skin, caused by mites and distinguished by an eczematose inflammation and loss of hair, that affects domestic animals and occasionally man. At times, the lesions of mange are infected by bacteria and dermatophytic fungi. In plant pathology the term **scab** is employed to refer to analogous lesions in plants.

manglicolous (Sp. *mangle,* mangrove + L. suf. *-cola*, inhabitant + L. suf. *-osus* >E.*-ous*, having, abundant in, with the qualities of): living on mangrove trees (species of the gen. *Rhizophora*), such as *Manglicola guatemalensis* (Dothideales), a saprobic marine fungus that forms its ascocarps on the bark of dead roots of red mangrove (*Rhizophora mangle*). *Mycosphaerella pneumatophorae* (Dothideales) and *Lopadostroma* sp. (Xylariales) also are manglicolous, since the first species is parasitic (or perhaps saprobic) on the bark of living pneumatophores of the mangroves *Avicennia africana* and *A. germinans*, whereas the second develops its perithecial stromata on leaves of red mangrove.

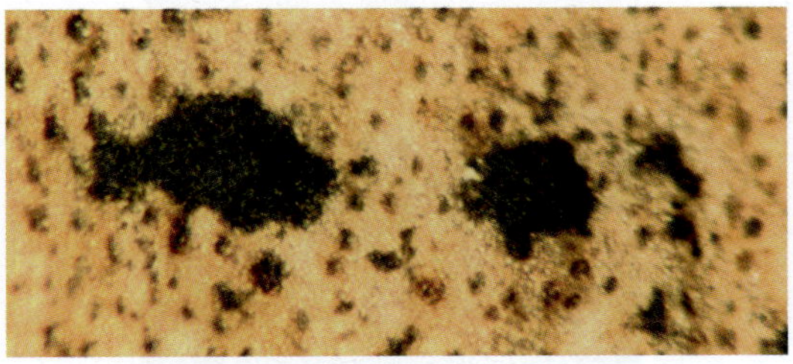

Perithecial stromata of *Lopadostroma* sp., a **manglicolous** species, on a leaf of red mangrove, x 10 (*MU*).

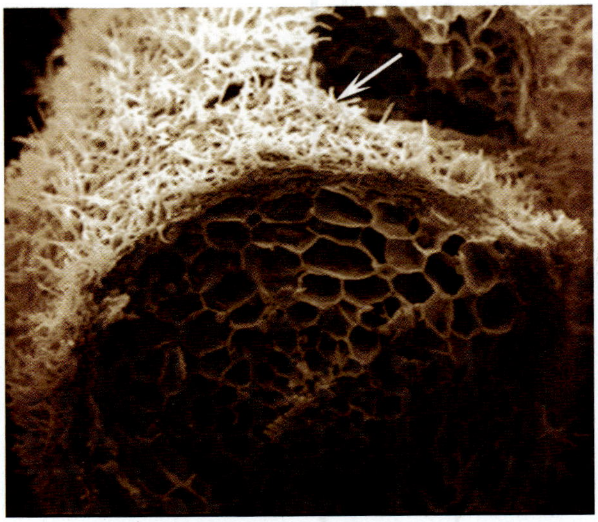

Scanning electron photograph of the root of *Monotropa uniflora*, in cross section, which shows the **mantle** (arrow) encapsulating the entire root, with cystidia-like hyphae, x 2,000 (*NH*).

Mantle (arrow) of the mycorrhiza of *Lactarius* sp. (Agaricales) with *Pinus sylvestris*, x 5 (*LV*).

manna (L. *manna*, manna, miraculous food; a miraculous food sent by God from heaven in the form of frost to feed the people of Israel in the desert): some types of manna described in the Bible appear to be manna lichens, corresponding to the thalli of *Sphaerothallia* (=*Aspicilia*) *esculenta* (Lecanorales), with which one can make bread.

manocyst (Gr. *manós*, thin + *kýstis*, bladder, cell): *Peronosporales*. An oogonial papilla that extends toward the interior of the antheridium during gametangial contact in species of the gen. *Phytophthora*.

mantle (L. *mantellum*, cape, cloak, cover): a compact layer of hyphae that envelops the short feeder roots of ectomycorrhizal plants; it is connected to the Hartig net on the inside, and to the extramatrical hyphae on the exterior, acting as a nutrient reservoir.

marcescent (L. *marcescens*, genit. *marcescentis*, becoming dry or withered < *marcens*, withered, dehydrated + suf. *-escens*, genit. *-escentis* > E. *-escent*, that which turns, that begins to, slightly): refers to structures that become dry and remain attached to the organs that gave rise to them, or to those that after being dehydrated can revive and recover their normal appearance when moistened, as occurs with the basidiocarps of *Marasmius siccus* and *M. haematocephalus* (Agaricales).

marginal cilium, pl. **cilia** (ML. *marginalis*, situated on the edge or margin < L. *margo*, genit. *marginis*, edge, margin, border + suf. *-alis* > E. *-al*, relating to or belonging to; L. *cilium*, the edge of the eyelid where the eyelashes are implanted; generally applied to the eyelash proper or to a very fine hair): *Lichens*. Each of the fine prolongations, composed of hyphae, present on the edges of the laminar or foliose borders of the thalli of certain species, such as *Parmotrema chinense* (=*Parmelia perlata*), Lecanorales.

Marcescent basidiocarps of *Marasmius haematocephalus*, x 2 (*TH*).

Marcescent basidiocarp of *Marasmius siccus*, x 3 (*MU* & *CB*).

Marginal cilia of the thallus of a foliose lichen, x 3 (*MU*).

marginal hair (ML. *marginalis*, situated on the edge or margin < L. *margo*, genit. *marginis*, edge, margin, border + suf. *-alis* > E. *-al*, relating to or pertaining to; OE. *haer*, hair, a slender threadlike outgrowth of the epidermis of an animal): hairs that occur on the margin of the apothecia of *Cookeina tricholoma* (Pezizales) and of some basidiocarps, as ocurr in *Polyporus tricholoma* (Aphyllophorales).

Marginal hairs on the edge of the apothecia of *Cookeina tricholoma*, x 1 (*RV*).

Marginal hairs on the margin of the basidiocarps of *Polyporus tricholoma*, x 1 (*RV*).

marginal veil (ML. *marginalis*, situated on the edge or margin < L. *margo*, genit. *marginis*, edge, margin, border + suf. *-alis* > E. *-al*, relating to or pertaining to; L. *velum*, veil, curtain, cloth): a secondary proliferation of the hyphae of the margin of the pileus that is destined to protect the developing hymenium, as occurs in agarics and boletes. See **partial veil**.

marginate (L. *marginatus*, bordered, marginal < *marginare*, to provide with a border < *margin*, *margo*, edge, margin, border + suf. *-atus* > E. *-ate*, provided with or likeness): having a well marked edge or border, like the reddish border of the gills of

Mycena rubromarginata (Agaricales), the basal bulb of the stipe of *Leucocoprinus bulbiger* (Agaricales), which has a border similar to a dropper or canal, the volva of *Amanita abrupta* (Agaricales), which separates abruptly from the base of the stipe, and the margin of the basidiocarp of *Fomitopsis pinicola* (=*Fomes marginatum*) and *Hexagonia variegata* (=*H. papyracea*, Aphyllophorales). Cf. **emarginate**.

Marginate volva of the basidiocarp of *Amanita abrupta*, x 0.8 (*RMA*).

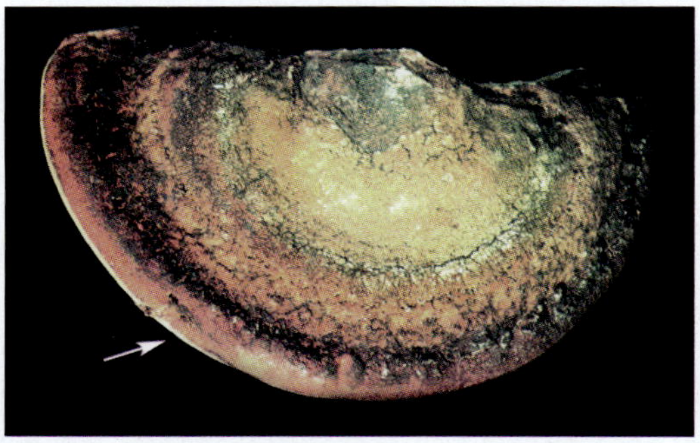

Marginate basidiocarp of *Fomitopsis pinicola* (=*Fomes marginatum*, Aphyllophorales), x 0.5 (*MU*).

Marginate basidiocarp of *Hexagonia variegata* (=*H. papyracea*, Aphyllophorales), x 0.2 (*EAA*).

marine fungi, sing. **fungus** (L. *marinus*, pertaining to the sea; L. *fungus*, fungus < Gr. *spóngos*, *sphóngos*, sponge): a large group of fungi (800-1000 species) widely distributed in this habitat. They parasitize algae or marine animals or are saprobic on residues of wood, algae, marine grasses, cysts of protozoans, corals, sea foam, and other substrates. The spores of many of these fungi have special appendages to attach them to different substrates. Other fungi are found in brackish, salty or marshy waters, mangrove swamps, and sand; of the latter, a good example is *Corollospora pulchella* (Halosphaeri-ales). And among the lichens there are species that form distinctive bands on silicaceous rocks of marine coasts. See **primary marine fungus**, **secondary marine fungus** and **thalassiomycetes**.

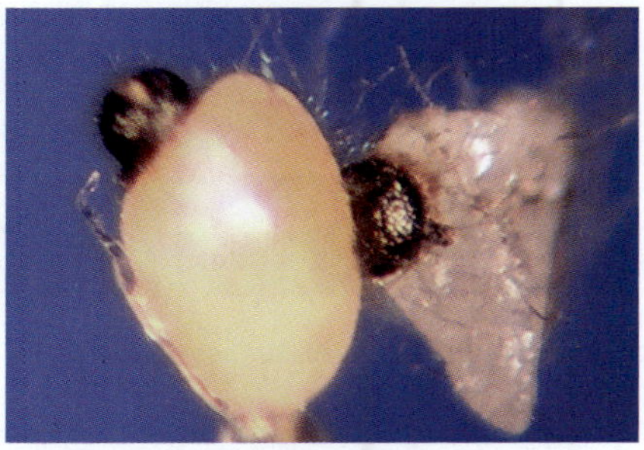

Dark fruiting bodies of *Corollospora pulchella*, a species of **marine fungus**, arenicolous, growing on sand grains, x 40 (*MU*).

mastigo-, **mastig-** (Gr. *mástix*, genit. *mástigos*, whip): a pref. used in terms that allude to flagella, as in *mastigoneme*.

mastigonemate (Gr. *mástix*, genit. *mástigos*, whip + *nêma*, filament + suf. *-atus* > E. *-ate*, provided with or likeness): provided with *mastigonemes*.

mastigoneme (Gr. *mástix*, genit. *mástigos*, whip + *nêma*, filament): each one of the small and numerous secondary, piliform fibrils of a tinsel flagellum that are implanted on the *axoneme* (main axis); the *mastigonemate* flagella, which also are called *pantonemate*, are present in the zoospores of the Hyphochytriomycota, such as *Hyphochytrium* and *Rhizidiomyces*, and the Oomycota, such as *Saprolegnia*. Mastigonemes also are known as **flimmer hairs**. See **axoneme** and **pantonema**.

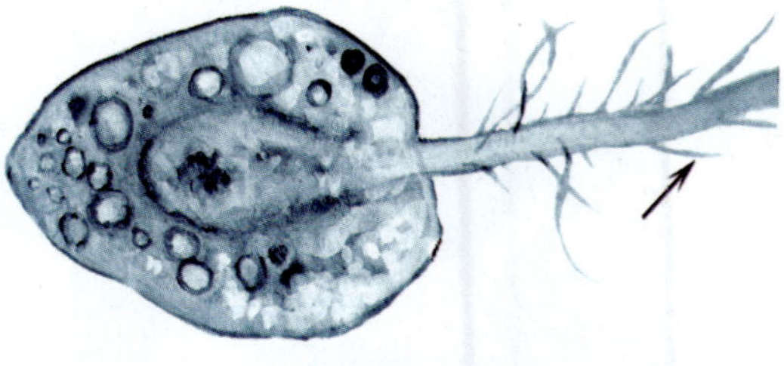

Mastigonemes of the flagellum of a zoospore of *Rhizidiomyces apophysatus*, x 4,800 (*MU*).

mating types (ME. *maten* < MF. *mater* < OF. *mat*, checkmate, associate, companion; either of two matched objects + ME. < OE. *-ing*, one of a specified kind; product or result of an action or process; L. *typus*, type, mark, model): the equivalent of the sexes of a fungus species. The ascomycetes have a sexuality in which the compatibility factors are of only two kinds (*bipolar* and *biallelic*). The basidiomycetes have sex factors in four groups (*tetrapolar* and *multiallelic*), some having as many as 28,000 different mating types. See **bipolar** and **tetrapolar**.

matrix, pl. **matrices** or **matrixes** (L. *matrix*, matrix, womb; a place where something is generated < *matr*, *mater*, mother): **1**. The substrate in which an organism is living. **2**. Mucilaginous material in which spores (ascospores, conidia) are embedded, and which influences their dissemination, survival, germination, etc. See **extramatrical** and **intramatrical**.

mazaedium, pl. **mazaedia** (L. *mazaedium* < Gr. *máza*, kneaded flour or any other dough + L. *-odium*, resembling < Gr. *ode*, like < *-oeídes*, similar to): refers to the mucilaginous and powdery mass that appears in the ascocarp of certain ascomycetes, such as *Onygena corvina* (Onygenales), composed of the tips of the paraphyses and their secretions, and that normally contains large quantities of ascospores. The mazaedium is also a type of fructification of the lichens of the order Caliciales.

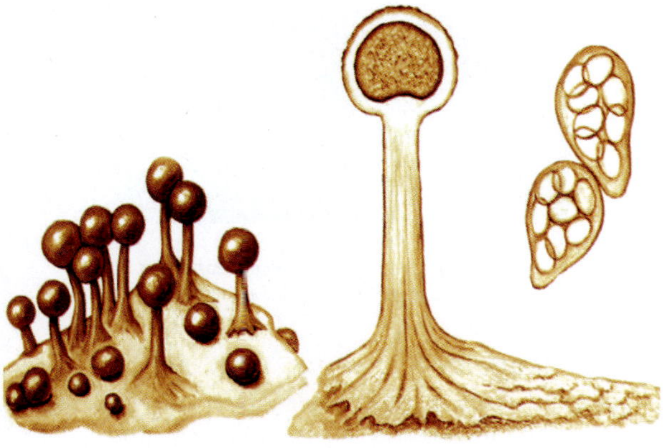

Mazaedium of *Onygena corvina*, seen in the center in longitudinal section, x 10. Also shown is a group of mazaedia, x 3, and two octosporate asci, x 120 (*MU*).

medi- (L. *medius*, middle, intermediate): a pref. that indicates something that is found in the middle, that corresponds to the center of a thing, or that is of medium size, neither very large nor very small. Equivalent to Gr. *mésos*.

medical mycology (L. *medicus*, pertaining to or relating to medicine; NL. *mycologia* < Gr. *mýkes*, fungus + *lógos*, word > E. des. *-logy*, study, treatise): the science that studies diseases of man caused by fungi. See **phytopathology**.

medicinal fungi, sing. **fungus** (L. *medicinalis*, pertaining to medicine < *medicina*, the science and art of preventing and curing diseases of the human body; medicament; the things that have salutary virtue and are adverse to a disease or ailment; L. *fungus*, fungus < Gr. *spóngos*, *sphóngos*, sponge): species of fungi with certain medicinal properties, mainly in traditional Chinese medicine, although in many other parts of the world they also are utilized for various purposes. E.g., the sclerotia of *Claviceps purpurea* (Clavicipitales), or ergot, contain alkaloids (the most important of which are derived from lysergic acid), as well as ergometrine, ergotamine and ergotoxin, including the active principles that give ergot not only its poisonous properties, but also its therapeutic applications, e.g., for the treatment of migraine, in obstetrics and as hallucinogens. The fruiting body of *Fomes officinalis* (Aphyllophorales) is utilized as a purgative; the spores and capillitium of *Lycoperdon* and *Calvatia* (Lycoperdales) are used to stop hemorrhages, and the perithecial stromata of *Cordyceps sinensis* (Clavicipitales), together with the insect larvae that they parasitize, are consumed in China to increase athletic and sexual vigor. In naturalistic medicine and the traditional indigenous medicine of Mexico and other countries, there are many other examples of the use of medicinal fungi, among them *Ganoderma applanatum*, *G. lucidum* and *Hericium erinaceus* (Aphyllophorales), as well as *Tricholoma matsutake* and *Lentinula edodes* (Agaricales), as antiviral and antitumor agents for the prevention and cure of cancer. Among the lichens there are also examples of medicinal uses, in accordance with the so-called medieval *Doctrine of Signatures*, a doctrine based on the similarity that exists in the appearance of the organism to be used as medicine with the shape of the organ to be treated; e.g., some lichens, such as *Lobaria lobata* (Peltigerales), whose thallus is lobate and alveolate, and supposedly similar in shape to the lungs, have been utilized to treat infections of these organs.

medipellis (L. *medius*, medium, intermediate + *pellis*, skin): see **pellicle**.

mediostratum, pl. **mediostrata** (NL. *mediostratum* < L. *medius*, medium, intermediate + *stratum*, stratum, layer): the central portion of the hymenophorous trama of the gills (in agarics) or of the wall of the tubes (in boletes and polypores), which is distinct from the lateral layers. In *Boletus*, e.g., the mediostratum is found in the center of a bilateral trama.

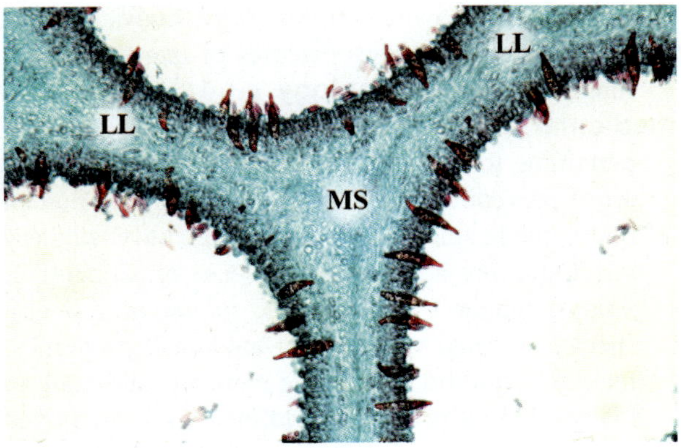

Mediostratum (MS) and lateral layers (LL) in the tube walls of *Boletus* sp., seen in transverse section, x 85 (*MU*).

medium, pl. **media** (L. *medium*, neuter of *medius*, middle, intermediate; in this case, an enveloping or surrounding substance; a culture medium): a nutrient system or substrate used for the artificial cultivation of fungi and other microorganisms. In mycology a wide variety of liquid and solid media are used to stimulate the growth, sporulation and preservation of fungi. See **minimal medium**.

Seven day old colonies of *Penicillium chrysogenum* (=*P. notatum*, moniliaceous asexual fungi) grown on culture **media**; the colonies on the left are on Czapek's **medium**, and the colonies on the right are on malt extract agar **medium**, x 0.5 (*CB*).

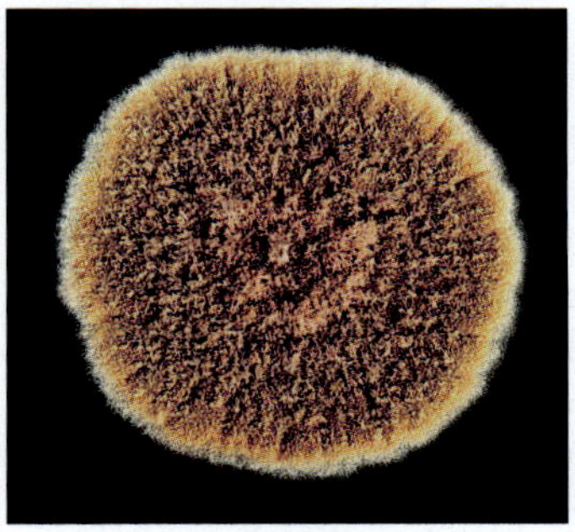

Ten day old colony of *Aspergillus flavofurcatus* (=*A. flavo-furcatis*, moniliaceous asexual fungi) on a culture **medium**, malt extract agar, x 0.5 (*MU*).

Ten day old colony of *Aspergillus variecolor* (moniliaceous asexual fungi) on Czapek's agar, the culture **medium** used to grow this fungus in the laboratory, x 1 (*MU*).

medulla, pl. **medullae** (L. *medulla*, medulla, marrow, heart, center): **1**. *Agaricales*. The internal tissues of a thallus or of an organ, soft, spongy or of another consistency in nature, but different from the outermost or peripheral zone; in *Amanita caesarea* (Agaricales), e.g., the medulla of the stipe of the basidiocarp has hyphae with inflated, claviform cells. **2**. *Lichens*. In species with a heteromerous thallus, the medulla is the central layer, beneath the layer of photobionts, and it is usually composed of looser hyphae than in the outer layers, i.e., the upper and lower cortices.

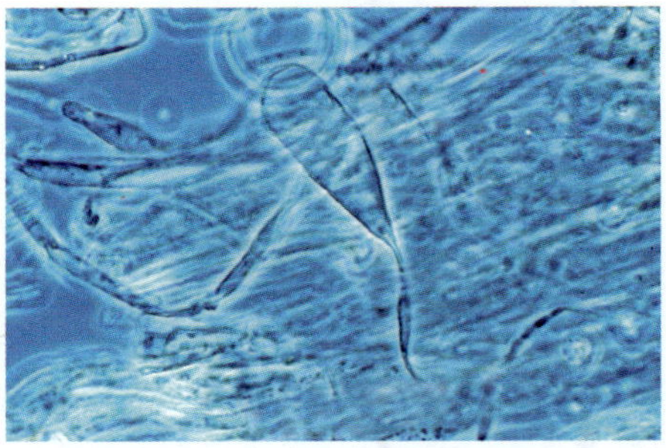

Medulla of the stipe of *Amanita caesarea*, with inflated, claviform cells, x 540 (*RMA*).

mega-, **megalo-** (Gr. *mégas*, genit. *mégalon*, large, of great size): a pref. employed in many biological terms to indicate that a particular structure is large, although in mycology **macro-** is used more. Cf. **micro-**. It also forms part of generic names, such as *Megachytrium* (Chytridiales) and *Megalonectria* (Hypocreales), which, respectively, allude to a large sporangium and to large perithecia.

meiocyte (Gr. *meíosis*, diminution + *kýstis*, bladder, vesicle, cell): a cell in which meiosis occurs. See **gonotocont**.

meiophase (Gr. *meíosis*, diminution + *phásis*, phase, appearance, aspect < *phaíno*, to shine): the phase in the life cycle of an organism during which the diploid nuclei of the cells undergo reduction in the number of chromosomes (meiosis) and become haploid. Cf. **diplophase**.

meiosis, pl. **meioses** (Gr. *meíosis*, diminution, alluding to the number of chromosomes): a series of two successive nuclear divisions in which the number of chromosomes is reduced by half, from the diploid to the haploid number. This represents the last phase of sexual reproduction, after *plasmogamy* and *karyogamy*. Cf. **mitosis**.

meiosporangium, pl. **meiosporangia** (Gr. *meíosis*, diminution + *sporá*, spore + *angeîon*, vessel, receptacle): a sporangium in which meiosis occurs. It refers to thick-walled, latent sporangia that are formed on the sporothallus of the aquatic fungus *Allomyces* (Blastocladiales), in which the diploid nuclei undergo meiosis and produce haploid zoospores (meiospores); these initiate a new haploid generation consisting of a gametothallus with gametangia, since in this fungus there is alternation of generations. The sporothallus also forms a second type of sporangium (the *mitosporangium*), with a thin wall, in which diploid zoospores are produced.

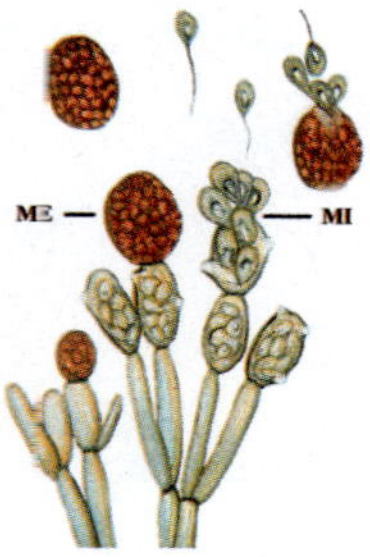

Meiosporangia (ME) of a sporothallus of *Allomyces macrogynus*. Mitosporangia (MI) also are shown, x 600 (*MU*).

meiospore (Gr. *meíosis*, diminution + *sporá*, spore): a spore from a *meiosporangium*. Also applied to an ascospore or basidiospore that is the product of meiosis. Cf. **mitospore**.

melanconiaceous (gen. *Melanconium* < Gr. *mélas*, genit. *mélanos*, black < *melanía*, blackness + *kónis*, dust + suf. L. *-ium,* added to noun and verb-stems denoting offices and groups + suf. *-aceus* > E. *-aceous*, of or pertaining to, with the nature of): term applied to asexual or anamorphic fungi that form their black or dark-colored spores in acervular conidiomata. The gen. *Colletotrichum*, *Gloeosporium*, *Entomosporium*, *Monochaetia* and *Pestalotia*, are examples of the old family Melanconiaceae, order Melanconiales, class Coelomycetes, subdivision Deuteromycotina (or deuteromycetes in a general sense). See **Coelomycetes**.

melanin (Gr. *mélas*, genit. *mélanos*, black, or *melaíno*, to blacken, darken, or *melanía*, blackness + NL. *-in* > E. *-in*, suf. used in chemistry to denote an activator or a derivative): a dark brown or black pigment derived from tyrosine, synthetized by fungi and animals. The dematiaceous asexual fungi are characterized by having melanized hyphae and spores.

melano-, **melan-** (Gr. *mélas*, genit. *mélanos*, black): a pref. employed in terms that allude to a black or dark brown color, as in *melanosporous*, with black spores or spore print, as well as the generic name *Melanospora* (Melanosporales).

melanosporous (Gr. *mélas*, genit. *mélanos*, black + *sporá*, spore + L. suf. *-osus* > E. *-ous*, having, with the qualities of): with black or dark brown spores; e.g, the ascospores of *Tuber melanosporum* (Pezizales), basidiocarps of *Coprinus comatus* and *Coprinopsis atramentaria* (=*Coprinus atramentarius*, Agaricales) form black or dark gray spores and spore prints.

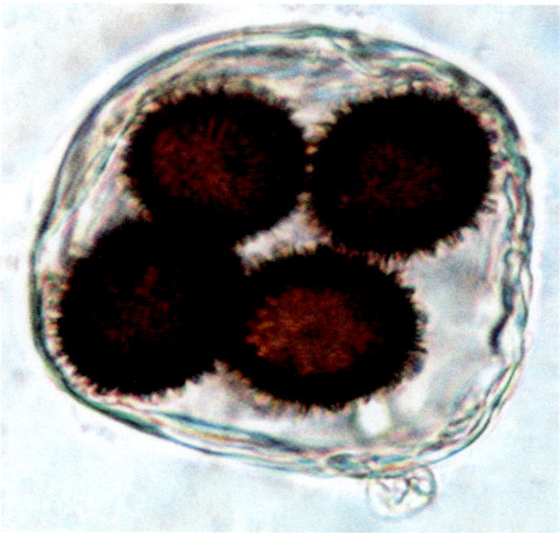

Echinate ascospores of *Tuber melanosporum* (**melanosporous**), x 1,200 (*EAA*).

Basidiocarps of *Coprinopsis atramentaria* (=*Coprinus atramentarius*) with black spores (**melanosporous**), x 2 (*RV*).

membranaceous (L. *membranaceus*, similar to a membrane < *membrana*, membrane, skin, parchment + suf. *-aceus* > E. *-aceous*, of or concerning, with the nature of): similar to a membrane, i.e., a thin, flexible sheet or film, like the hypothallus of *Physarum polycephalum* and the peridium of *Badhamia capsulifera* (Physarales), as well as the polar appendage of the conidia of *Myrothecium verrucaria* (tuberculariaceous asexual fungi). Cf. **membranous**.

Membranaceous peridium of the sporangia of *Badhamia capsulifera*, x 25 (*MU*).

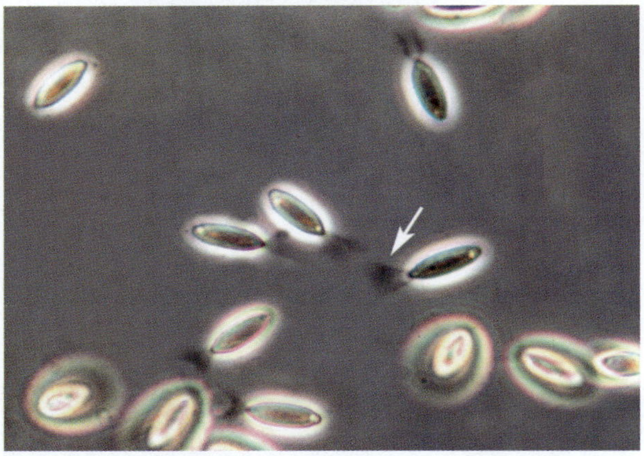

Membranaceous appendage of the conidia of *Myrothecium verrucaria*, x 830 (*MU*).

membranous (L. *membranosus*, provided with a membrane < *membrana*, membrane, skin, parchment + suf. *-osus* > E. *-ous*, full of, abundant in, with the qualities of): having one or more membranes, like the fruiting body of *Vascellum* (Lycoperdales), which has a membrane between the gleba and the subgleba. Cf. **membranaceous**.

memnonious (L. *memnonius*, black, brownish-black, dull gray): dark brown in color, almost black, like the conidial chains of *Memnoniella* (dematiaceous asexual fungi), a gen. segregated from *Stachybotrys*.

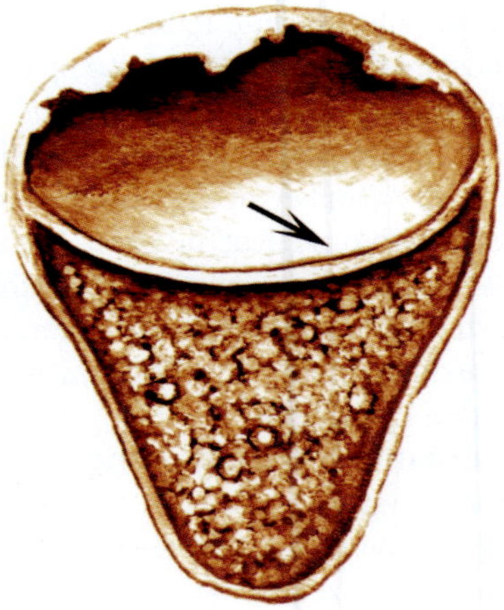

Membranous diaphragm of the fruiting body of *Vascellum pratense*, x 1 (*MU*).

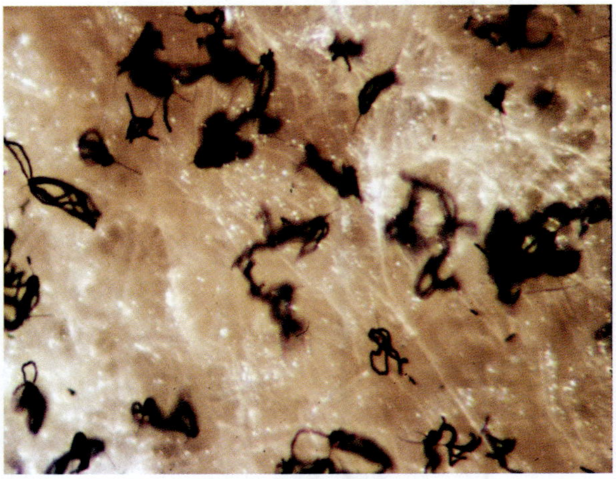

Memnonious chains of conidia of *Memnoniella* sp. on agar, x 40 (*MU*).

memnospore (Gr. *mémnon* < *méno*, to remain, persist + *sporá*, spore): a term sometimes applied to a latent spore that is surrounded by a thick, resistent wall, like the chlamydospores (asexual) and teliospores (sexual), that can remain at rest for a long time (often more than one year); these spores do not participate in the dispersal of the species as do other types of spores (conidia, sporangiospores, etc.), generally thin-walled, that disseminate the fungus during the same season. Also called **hypnospore**. Cf. **xenospore**.

mer-, **meri-** (Gr. *merís*, part): a pref. used to allude to something that is part of a whole or that is incomplete, compared with what is considered normal, as in *merispore*.

merenchyma, pl. **merenchymata** (Gr. *méros*, part or partial + *énchyma*, stuffed): a type of tissue (parenchyma) derived from the division of cells in several planes.

merido-, **merid-** (Gr. *merís*, genit. *merídos*, part): a pref. with the same meaning as *meri-* and *mero-*, but which is rarely employed in botany.

merismatoid, **merismoid** (Gr. *mérisma*, genit. *merísmatos*, part < *merís*, portion, and *merismós*, division + L. suf. *-oide*, resembling < Gr. *-oeídes*, similar to): applied to a pileate fructification that is formed from many small pilei, or which is divided into slender lobes. Sometimes the first term is used for polypores and the second for agarics.

merispore (Gr. *merís*, portion + *sporá*, spore): each one of the spores that comprise a compound spore or spore ball. See **sporodesm**.

meristem arthrospore (Gr. *merísto*, to divide + L. suf. *-ema*, that made or created; Gr. *árthron*, articulation, joint + *sporá*, spore): *Conidiogenesis*. An asexual spore that originates by meristematic growth of the apical region of the conidiophore, and its concurrent basipetal conversion into spores, as occurs, e.g., in *Oidium*, the moniliaceous asexual state of *Blumeria* (=*Erysiphe*) *graminis* (Erysiphales).

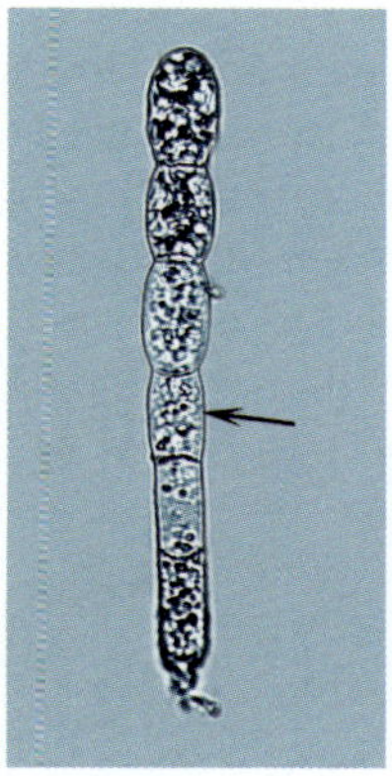

Meristem arthrospores of the conidiophore of *Oidium* sp., the asexual state of *Blumeria* (=*Erysiphe*) sp., x 670 (*MU*).

Meristem arthrospores of the conidiophores of *Oidium monilioides*, the asexual state of *Blumeria* (=*Erysiphe*) *graminis*, x 590 (*WKW*).

meristem blastospore (Gr. *merísto*, to divide + L. suf. *-ema*, that made or created; Gr. *blastós*, germ, bud, shoot + *sporá*, spore): *Conidiogenesis*. An asexual reproductive spore that originates by budding of a conidiophore that extends from a basal meristematic zone (*basauxic conidiophore*), as in *Arthrinium phaeospermum* (dematiaceous asexual fungi).

Meristem blastospores of the conidiophores of *Arthrinium phaeospermum*, x 1,100 (*MU*).

meristem spore (Gr. *merísto*, to divide + L. suf. *-ema* > E. *-em*, that made or created; Gr. *sporá*, spore): *Conidiogenesis*. A spore that detaches after having matured in basipetal succession from the tip of a conidiophore, or from a phialide or a hypha, in such a manner that the tip is considered as an open growing point.

meristematic (Gr. *merísto*, to divide + L. suf. *-ema*, that made, that created + Gr. suf. *-tikós* > L. *-ticus* > E. *-tic*, relation, fitness, inclination or capacity): belonging to the meristem, relating to it. See **meristem arthrospore**.

meristo-, **merist-** (Gr. *merísto*, to divide): a pref. used in terms that allude to the division of a cell, an organ, etc., as in *meristogenous*.

meristogenous (Gr. *merísto*, to divide + *génos*, origin < *gennáo*, to engender, reproduce): a type of development that results from repeated divisions of hyphae (meristogenesis); generally applied to the development of a pycnidium. In simple meristogenous development the pycnidium is formed by the growth and division of a single hypha, sometimes including branches of the same hypha, to form a stroma in which a cavity is produced, as occurs, e.g., in *Phoma exigua* (=*Ph. herbarum*), sphaeropsidaceous asexual fungi. In compound meristogenous development, which is present in *Phoma pyrina*, several hyphae aggregate to form a sphere of tissue in which a cavity is subsequently differentiated. Cf. **symphogenous**.

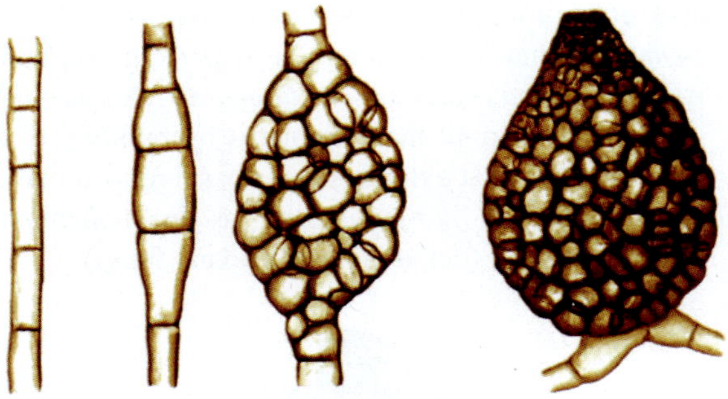

Meristogenous (simple) development of the pycnidium of *Phoma exigua* (=*Ph. herbarum*), x 145 (*MU*).

mero-, mer- (Gr. *méros*, part, portion): a pref. employed in the formation of terms that refer to the parts or elements of a complex, as in the following terms.

merogamy (Gr. *méros*, part, portion + *gámos* > E. des. *-gamy*, sexual union): a particular case in amphimictic multiplication when the sexual cells are gametes formed in gametangia, as in *Synchytrium* (Chytridiales).

meront (Gr. *méros*, part, portion + suf. *ón*, genit. *óntos*, a being): a part of something, but it refers especially to a daughter myxamoeba that is separated from the periphery of the mother myxamoeba, a phenomenon that occurs repeatedly in the myxamoebae and plasmodia of the Myxomycetes.

merosporangiophore (Gr. *méros*, part, portion + *sporá*, spore + *angeîon*, vessel, receptacle + suf. *-phóros*, bearer): a specialized hypha that produces and supports one or more merosporangia; it is characteristic of *Syncephalastrum racemosum* (Mucorales) and is equivalent to the sporangiophore of other Mucorales.

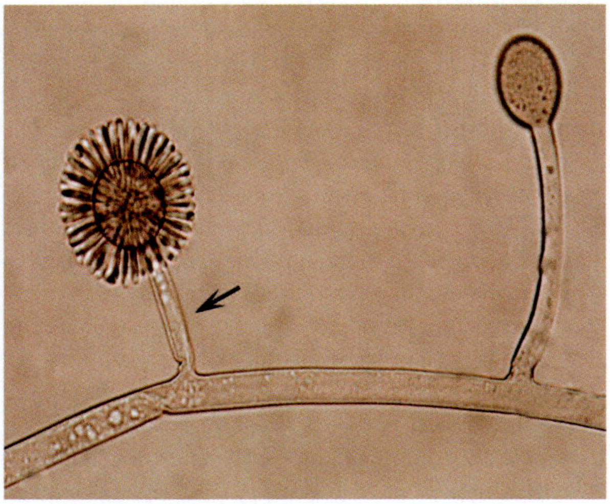

Merosporangiophores of *Syncephalastrum racemosum*, x 140 (*MU*).

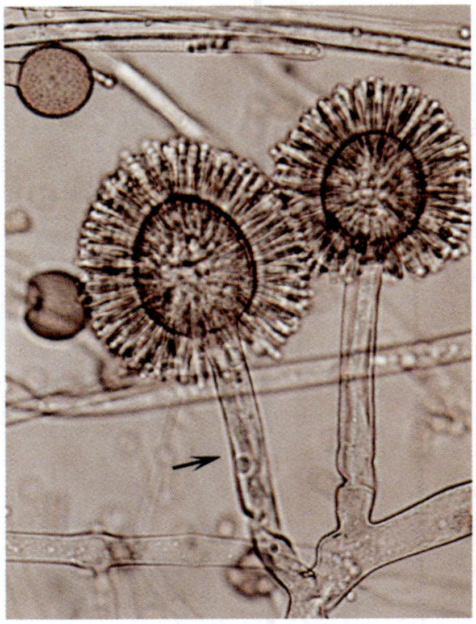

Merosporangiophores of *Syncephalastrum racemosum*, x 200 (*MU*).

merosporangium, pl. **merosporangia** (Gr. *méros*, part, portion + *sporá*, spore + *angeîon*, vessel, receptacle): a cylindrical sporangiolum, with the spores in a row, that is produced on the swollen apex of a sporangiophore (*merosporangiophore*), as is seen in *Syncephalastrum racemosum* (Mucorales).

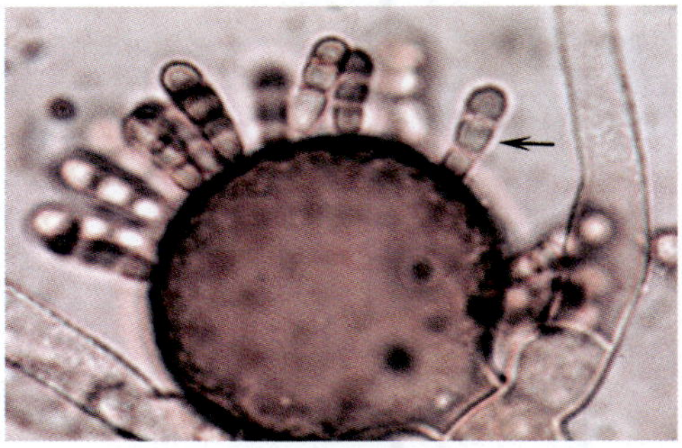

Merosporangia of *Syncephalastrum racemosum*, x 680 (*MU*).

mesendogenous (Gr. *mésos*, middle, intermediate + *éndon*, within + *génos*, origin < *gennáo*, to engender, produce): *Conidiogenesis*. A mode of conidium formation in which the secondary cell wall of the conidium is formed inside the primary cell wall of the phialide or conidiogenous cell, as occurs in *Penicillium* and *Aspergillus* (moniliaceous asexual fungi).

meso-, mes- (Gr. *mésos*, middle, intermediate, median): a pref. employed when forming terms that allude to something that is found in the middle, between two things, or which is of median size, as in the following terms.

mesochroic (Gr. *mésos*, middle, intermediate + *chrõs*, color + suf. *-íkos* > L. *-icus* > E. *-ic*, which denotes relationship): having pigmentation in the hyphal walls.

mesoderm, mesodermous, mesodermatous (Gr. *mésos*, middle, intermediate + *dérma*, skin, bark + L. suf. *-osus* > E. *-ous*, having, with the qualities of): a hypha in which the outer wall and the lumen have almost the same thickness. Cf. **leptoderm** and **pachyderm**.

mesoperidium, pl. **mesoperidia** (Gr. *mésos*, middle, intermediate + *péridion*, small leather purse): the intermediate layer of a peridium composed of three layers: *exoperidium*, *mesoperidium* and *endoperidium*, as occurs in the fruiting bodies of many Gasteromycetes, e.g., *Disciseda candida*, *Geastrum saccatum* and *G. triplex* (Lycoperdales).

Mesoperidium of the fruiting bodies of *Disciseda candida*, x 1.5 (*MU*).

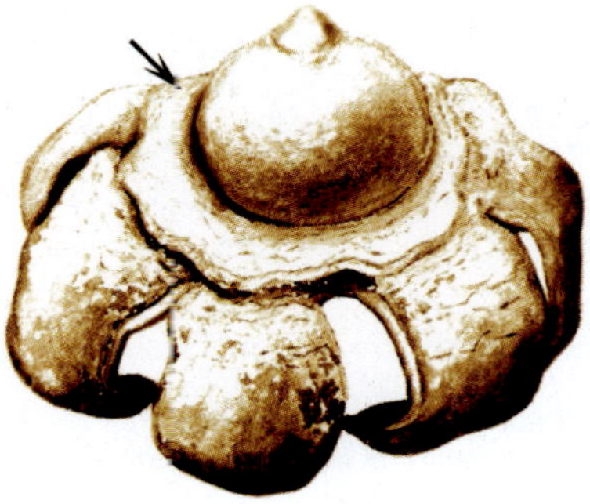

Mesoperidium of the fruiting body of *Geastrum saccatum*, x 1 (*MU*).

mesophile, mesophilic, mesophilous (Gr. *mésos*, middle, intermediate + *phílos*, have an affinity for): an organism that grows well at medium temperatures; the majority of the fungi are mesophiles, since they grow at temperatures between 10 and 40ºC, with an optimum around 25 to 35ºC. Cf. **psychrophile** and **thermophile**.

mesospore (Gr. *mésos*, middle, intermediate + *sporá*, spore): **1**. *Uredinales*. A unicellular teliospore that is found among normal bicellular teliospores, as sometimes occurs in *Puccinia*. The term also is applied to the intermediate cell of tricellular spores, and to urediniospores of the rusts that germinate only after a latent period. **2**. The latter type of mesospore also is called an **amphispore**, for having a double function, that of germinating quickly and that of germinating after a latent period. **3**. In the wall of the most complex basidiospores, which is composed of five layers, the *mesospore*, or *mesosporium*, is a delicate and scarcely distinguishable structure; it is found beneath the three outer layers (*perispore*, *exospore* and *epispore*) and above the *endospore*.

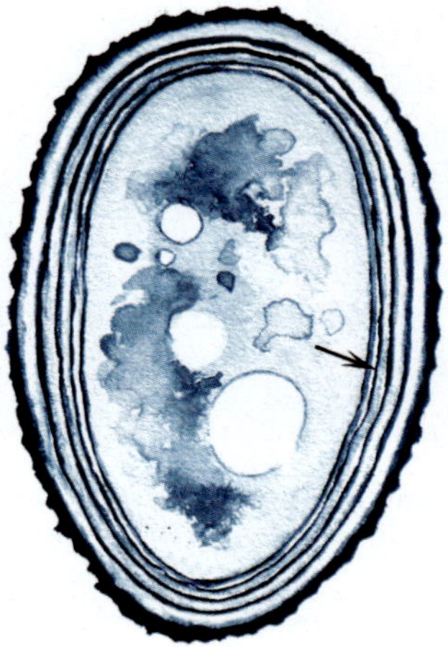

Mesospore of the most complex basidiospore wall. In this diagram of a transverse section, the other four layers: endospore, epispore, exospore and perispore, also are shown, x 8,000 (*MU*).

meta-, met- (Gr. *metá*, between, with, afterwards, beyond): a pref. used to form compound terms that express a change of place, alteration of form or of condition (later position, subsequent transformation, a more advanced evolutionary condition or degenerate), e.g., *metabasidium*, a basidium that is modified or degenerate.

metabasidium, pl. **metabasidia** (Gr. *metá*, afterwards, beyond + *basídion*, basidium < *básis*, base + dim. suf. *-ídion*): a basidium in the phase of development in which the diploid zygote nucleus divides by meiosis and gives rise to the four haploid nuclei that migrate through the sterigmata to the basidiospores, as occurs, e.g., in *Amanita* and *Coprinus* (Agaricales).

metabiosis, pl. **metabioses** (Gr. *metá*, afterwards, beyond + *bíosis*, condition of being alive < *bíos*, life + suf. *-osis*, state, condition): the interdependence or association of two organisms that act one after the other, i.e., one of them prepares a substrate which later can accomodate or support the second; this can occur with the former as a symbiont or it can develop only when the first has disappeared due to incompatibility with the now altered medium. An example of metabiosis is seen in the preparation of *shoyu sauce*; the transformation for starch into sugar by *Aspergillus oryzae* (moniliaceous asexual fungi), making it available as a medium to the yeasts that grow and produce the alcoholic fermentation of the sugar. Cf. **antagonism**, **commensalism**, **helotism**, **mutualism**, **parabiosis**, **parasymbiosis**, **symbiosis** and **synergy**.

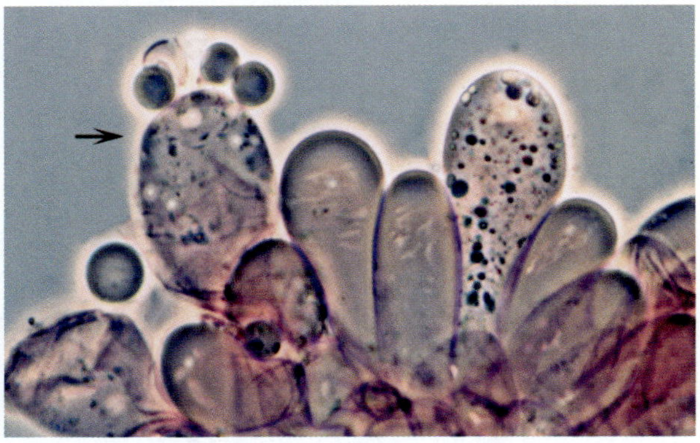

Metabasidium and basidiola of the hymenium of *Amanita* sp., x 700 (*RMA*).

metachroic (Gr. *metá*, afterwards, beyond + *chrõs*, color + suf. *-íkos* > L. *-icus* > E. *-ic*, which denotes relationship): changing color due to the appearance of a new pigment in more mature tissue.

metachromatic (Gr. *metá*, afterwards, beyond + *chromatikós*, concerning the color < *chrõma*, color + suf. *-tikós* > L. *-ticus* > E. *-tic*, relation, fitness, inclination or capacity): applied to a structure, such as a cell wall or the inner spore wall, that stains violet or reddish in cresyl blue.

metachromic (Gr. *metá*, afterwards, beyond + *chrõma*, color + suf. *-íkos* > L. *-icus* > E. *-ic*, which denotes relationship): the red color present in various fungal structures on reacting with cresyl blue; used in the taxonomy of the Agaricales.

metalophilic (ME. < L. *metallum* < Gr. *métallon*, quarry, metal + *phílos*, have an affinity for): growing on metal surfaces, such as railroad tracks, like some lichens.

metamorphosis, pl. **metamorphoses** (Gr. *metamórphosis* < *metá*, aftewards, beyond + *morphé*, shape + suf. *-osis*, state, condition): all of the morphological and structural modifications that an organism or an organ undergoes during the course of its development. E.g., the myxozoospores of the Myxomycetes can originate either by direct germination of the spores or by metamorphosis of the myxamoebae, which can be transformed into myxozoospores, depending on environmental conditions.

metastasis, pl. **metastases** (Gr. *metástasis*, removal, moving, emigration): deviation or change of an organ to an abnormal position, or the transport of a pathogen from a primary point of infection to other parts of the host.

metathallus, pl. **metathalli** (Gr. *metá*, afterwards, beyond + *thallós*, thallus, shoot): the assimilative part of a lichen thallus (which contains the photobiont), especially when there is also a *prothallus*.

metoecious (Gr. *metá*, afterwards, beyond + *oîkos*, house or inn): a term used by de Bary, equivalent to **heteroecious**, i.e., having different parasitic phases on two distinct hosts, like the Uredinales.

metula, pl. **metulae** (L. *metula*, dim. of *meta*, pyramid, tip of something): a small branch of a conidiophore that gives rise to phialides or conidiogenous cells, as occurs in the gen. *Penicillium* and *Gliocladium* (moniliaceous asexual fungi), among others. The metula also has been called a **prophialide**, since it is found beneath the phialide. The term "primary sterigma" was formerly used for the metula, but this is no longer considered appropriate. Cf. **sterigma**.

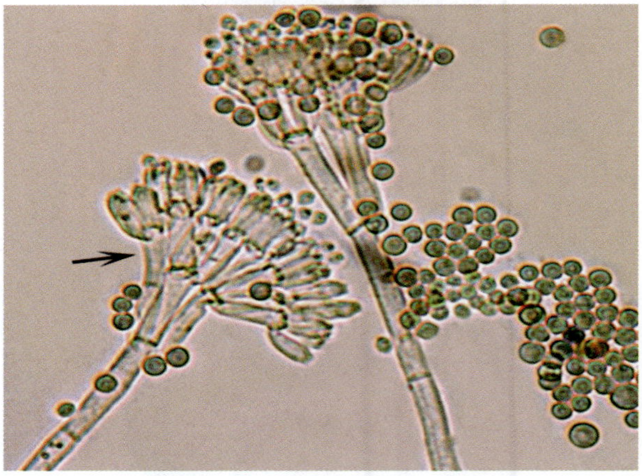

Metulae of the conidiophores of *Penicillium aurantiogriseum* (=*P. lanosocoeruleum*), x 500 (*PL*).

metuloid (L. *metula*, dim. of *meta*, pyramid, tip of something + suf. *-oide*, like < Gr. *-oeides*, similar to): *Basidiomycetes*. A cystidium incrusted with calcium carbonate, which is deeply inserted in the trama; it has a thickened wall and is generally hyaline or lightly straw-colored. It is present, e.g., in *Peniophora*, *Lentinus strigosus* (=*L. rudis*) and some species of *Hydnum* (Aphyllophorales), as well as in certain Agaricales, such as *Panaeolus cyanescens*.

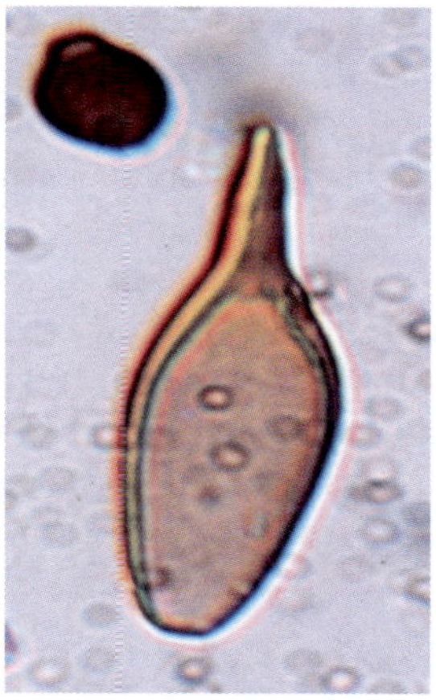

Metuloid cystidium of the hymenium of *Panaeolus cyanescens*, x 890 (*MU*).

micaceous (L. *micaceus*, similar to mica < *mica*, glittering sand powder + suf. *-aceus* > E. *-aceous*, of or pertaining to, with the nature of): covered with shiny particles, like the surface of the pileus of *Coprinellus* (=*Coprinus*) *micaceus* (Agaricales).

micro- (Gr. *mikrós*, small, minute): a pref. employed in numerous words that denote small size, like *microbe*, a minute living being. Cf. **macro-** and **mega-**.

microaerophilic (Gr. *mikrós*, small, minute + *aér*, *aéros*, air + *phílos*, have an affinity for + suf. *-íkos* > L. *-icus* > E. *-ic*, pertaining to or relating to): living preferentially in environments with low concentrations of oxygen, such as *Aqualinderella fermentans* (Leptomitales), which grows on submerged fruits in stagnant waters with high concentrations of carbon dioxide and fermentable organic matter.

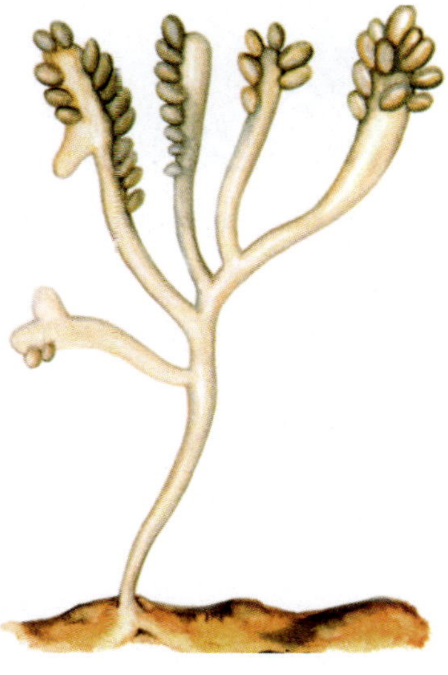

Thallus of *Aqualinderella fermentans*, a **microaerophilic** species, x 25 (*MU*).

microbe (Gr. *mikrós*, small, minute + *bíos*, life): a **microorganism**.

microbial (Gr. *mikrós*, small, minute + *bíos*, life): relating to or pertaining to microbes.

microbiogloea, pl. **microbiogloeae** (Gr. *mikrós*, small, minute + *bíos*, life + *gloiós*, glutinous, viscous): a complex of bacteria and yeasts that comprise a yellowish-white macrocolony of gelatinous consistency, like the one that forms the so-called *tibi grains*, which are utilized in Mexico as inoculum or starter culture to make tepache and vinegar of tibi, which involves a lactic acid-alcohol-acetic acid fermentation of sugary liquids. In the United States the tibi grains are known as *American bees*, *California bees*, *ginger beer plant* and *sugary kefir grains*. Another example of microbiogloeae are the so-called *kefir grains* in the former USSR, and *bulgaro grains* in Mexico, which are utilized to ferment milk in the manufacture of *kefir*. These grains are composed of lactic acid bacteria (*Lactobacillus* spp., *Streptococcus* spp.) and yeasts (*Saccharomyces cerevisiae* and *Kluyveromyces marxianus*, of the Saccharomycetales, *Candida kefyr*, of the cryptococcaceous asexual yeasts, and others), embedded in a matrix of polysaccharides produced by the lactic acid bacteria; they are bright white in color and their consistency, although somewhat gelatinous, is quite elastic. See **bulgaro grain**, **fermentation**, **kefir grains**, **tea fungus** and **tibi grains**.

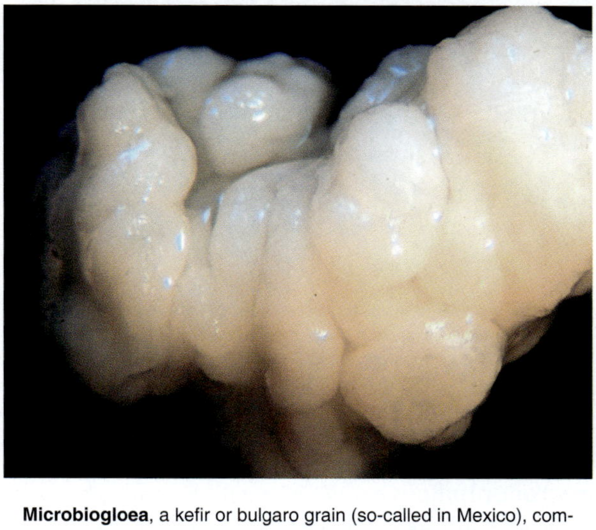

Microbiogloea, a kefir or bulgaro grain (so-called in Mexico), composed of the symbiotic association of lactic bacteria and yeasts that is utilized to ferment milk, x 5 (*MU*).

microbiologist (Gr. *mikrós*, small, minute + *bíos*, life + *lógos*, word > E. des. *-logy*, discourse, treatise): a biologist dedicated to the study of microorganisms or microscopic forms of life.

microbiology (Gr. *mikrós*, small, minute + *bíos*, life + *lógos*, word > E. des. *-logy*, discourse, treatise): the branch of the biological sciences that studies microorganisms. The biology of microbes.

microbiota (Gr. *mikrós*, small, minute +*bioté*, biota, nature or condition of living beings or of life): the totality of the microorganisms present in an area or specific habitat, including algae, bacteria, protozoans and fungi. See **mycobiota**.

microbiotic (Gr. *mikrós*, small, minute + *biotikós*, vital < *bíos*, life + suf. *-tikós* > L. *-ticus* > E. *-tic*, relation, fitness, inclination or ability): applied to the crusts of free-living lichens, bryophytes and cyanobacteria found on the soil of certain ecosystems, such as deserts, the heaths of *Stereocaulon* (Lecanorales) of the tundra, and the forests of old growth Douglas fir, where the lichens and cyanobacterial photobionts are important in fixing atmospheric nitrogen, contributing significant quantities of this element to the ecosystem, exceeding the input of nitrogen from other sources.

microbody (Gr. *mikrós*, small, minute + ME. < OE. *bodig*, body): each one of the subcellular organelles, also called *peroxysomes*, of eukaryotic organisms that have essential functions in the metabolism of *n*-alkanes, fatty acids, methanol and various nitrogen containing compounds.

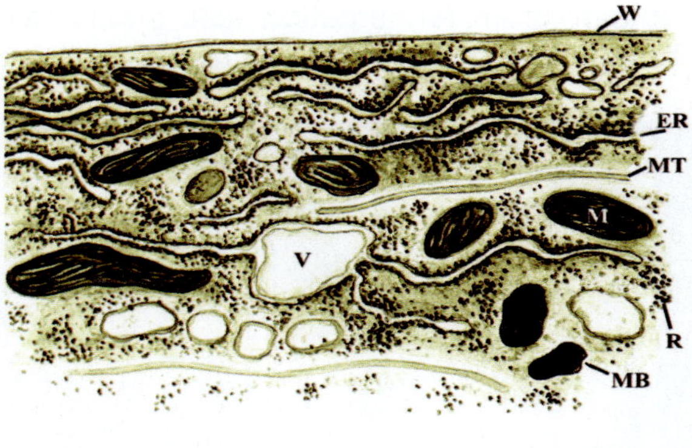

Microbodies (MB) of a young hypha of *Armillaria mellea* (Agaricales) seen in longitudinal section. Also observed are ribosomes (R) of the cisternae of the endoplasmic reticulum (ER), mitochondria (M), microtubules (MT), vesicles (V), and cell wall (W). Drawing based on several transmission electron micrographs published by Beckett *et al.* in their *Atlas of Fungal Ultrastructure*, 1974, x 22,200 (*MU*).

microcephalic (Gr. *mikrós*, small, minute + *kephalé*, head + suf. *-íkos* > L. *-icus* > E. *-ic*, pertaining to or relating to): literally, and figuratively, with a small head. Applied to dictyosporous ascospores whose transverse septa divide each segment into two parts that are more or less equal in size, so that the end cells of the ascospore are not larger than the other cells. Cf. **macrocephalic**.

microconidium, pl. **microconidia** (Gr. *mikrós*, small, minute + *kónis*, dust, powder + dim. suf. *-ídion*): a small conidium, generally unicellular. In the conidial fungi microconidia function as asexual spores, e.g., in *Fusarium moniliforme* (tuberculariaceous asexual fungi) and in *Clonostachys rosea* (=*Gliocladium roseum*), and in the dermatophytes, such as *Trichophyton* (moniliaceous asexual fungi). In certain ascomycetes, such as *Neurospora* (Sordariales), the microconidia can also function as spermatia in sexual reproduction. Cf. **macroconidium.**

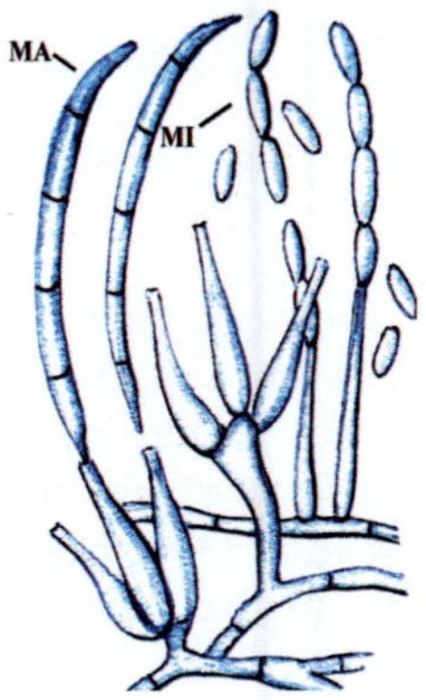

Microconidia (MI) and macroconidia (MA) formed on phialides of *Fusarium moniliforme*, x 940 (*MU*).

Microconidia on phialides of *Clonostachys rosea* (=*Gliocladium roseum*), x 620 (*JAS*).

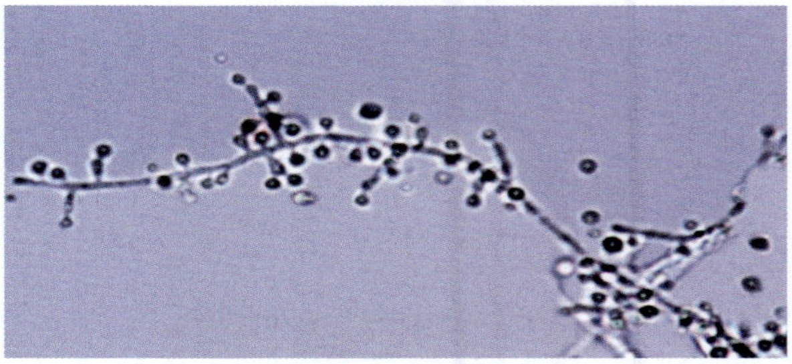

Microconidia on hyphae of *Trichophyton mentagrophytes*, cultured on agar, x 350 (*MU*).

microcrystal (Gr. *mikrós*, small, minute + *krýstallos*, crystal): each one of the microscopic crystals, diagnostic in shape, that are produced by recrystallization of phenolic compounds of lichens dissolved in various solvents. This method of microcrystal formation on slides is extremely sensitive and under microscopic examination some compounds can be detected at concentrations near the resolution obtained with thin-layer chromatography.

microculture (Gr. *mikrós*, small, minute + L. *cultus*, the action and effect of cultivating): the culture of an organism, especially a microorganism, for observation with a microscope. Microcultures of fungi are commonly employed, e.g., for the study of conidiogenesis in mitosporic fungi, since one can follow the development of the sporiferous structures on the selected culture medium without altering them too much, i.e., without breaking them, as happens when making a preparation for microscopic examination.

microcyclic conidiation (Gr. *mikrós*, small, diminutive + *kyklikós*, circular, cyclic < *kýklos*, circle, cycle + suf. *-íkos* > L. *-icus* > E. *-ic*, belonging to or relating to; NL. *conidium* < Gr. *kónis*, powder, dust + dim. suf. *-ídion* + Gr. suf. *-ationem* > E. *-ation*, action, state or condition, result): the production of conidia directly from a spore without the intervention of hyphal growth.

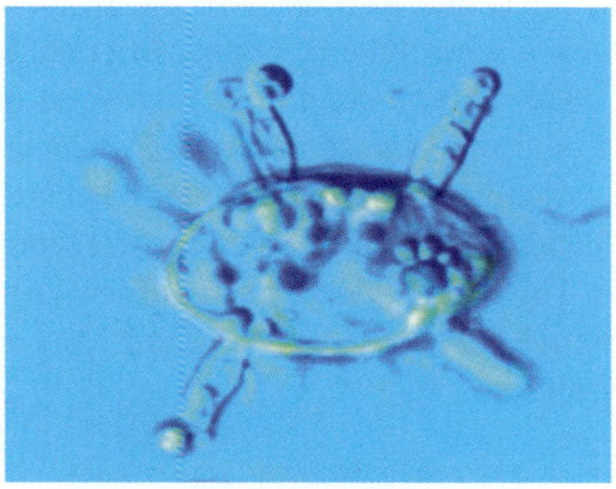

Microcyclic conidiation of an ascospore of *Cookeina sulcipes* (Pezizales), x 1,000 (*RTH*).

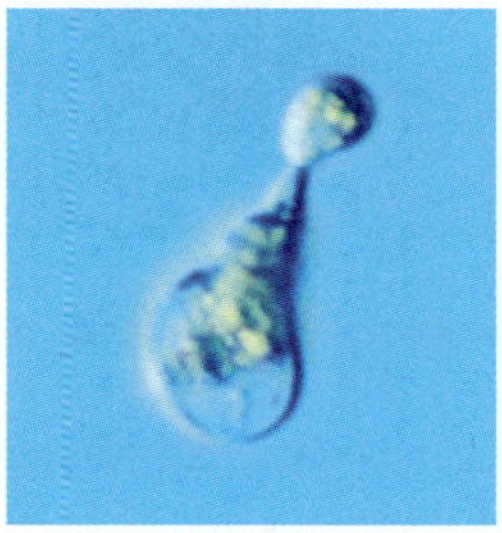

Microcyclic conidiation of an aeciospore of *Gymnosporangium juniperi-virginianae* (Uredinales), x 2,000 (*RTH*).

microcyclic (Gr. *mikrós*, small, minute + *kyklikós*, circular, cyclic < *kýklos*, circle, cycle + suf. *-íkos* > L. *-icus* > E. *-ic*, pertaining to or relating to): refers to species of rusts (Uredinales) in whose life cycles the aeciospores and urediniospores (and sometimes the spermogonia) are not present; the teliospores are the only dikaryotic spores that are developed, as occurs in *Puccinia delicatula*, the salvia rust. The microcyclic rusts are *autoecious*, i.e., they develop their entire life cycle on a single host. The plant pathogenic smut fungi (Ustilaginales) also are microcyclic. Cf. **hemicyclic** and **macrocyclic**.

microcyst (Gr. *mikrós*, small, minute + *kýstis*, bladder; here, cell): an encysted protoplast, protected by a wall resistant to conditions of unfavorable moisture; the term is generally applied to the encysted myxamoebae of the Acrasiomycota, such as *Copromyxa arborescens* (Acrasiales), and to the Myxomycota, although at times also to the encysted zoospore of the aquatic fungi (Chytridiomycetes). Cf. **macrocyst** and **paracyst**.

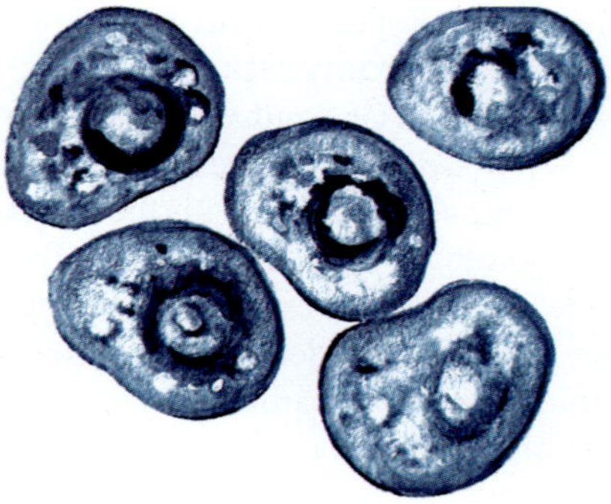

Microcysts (encysted myxamoebae) of *Copromyxa arborescens*, x 1,300 (*MU*).

microendospore (Gr. *mikrós*, small, minute + *éndon*, inside + *sporá*, spore): a minute cytoplasmic particle in a hypha or in a spore that is capable of developing into a structure typical of that from which it originated; it has been described in *Graphium ulmi* (stilbellaceous asexual fungi).

microflora (Gr. *mikrós*, small, minute + L. *flora*, alluding to *Flora*, the godess of flowers, the totality of plants in a region): a term that is sometimes used, inappropriately, to refer to all of the microorganisms present in a site or specific habitat; it is better to use the term **microbiota** for this, and **mycobiota** for fungi, in particular.

microform (Gr. *mikrós*, small, minute + L. *forma*, form): a name given by Schröter to the life cycle of the rust *Puccinia heterospora* (Uredinales), which has stages 0, III and IV (in which are produced spermatia, teliospores and basidiospores, respectively). Referring to the life cycle of this same rust, other authors use the terms macrocyclic, demicyclic and microcyclic. The microform cycle also is called *micro-Puccinia*.

microfungi, sing. **microfungus** (Gr. *mikrós*, small, minute + L. *fungus*, fungus): syn. of **micromycetes**.

microlichen (Gr. *mikrós*, small, minute + L. *lichen* < Gr. *leichén*, lichen): one of the lichens whose totality of morphological characteristics can be observed only with magnification of 10 x or more, e.g., many of the crustose, foliicolous, tropical thalli, such as the gomphilaceous *Gyalectidium* (Ostropales). Cf. **macrolichen**.

micrometer (Gr. *mikrós*, small, minute + *métron*, measure, meter): abbrev. µm; a unit of measure equal to 1/1,000 part of a millimeter or 1/1,000,000 of a meter. Equivalent to the *micron*, abbrev. µ, which was used previously. Cf. **nanometer**.

micromycete (L. *micromycete* < Gr. *mikrós*, small, minute + *mýkes*, genit. *mýketos*, fungus): a fungus of very small size, usually with microscopic sporiferous structures; micromycetes occur in all the main groups of fungi. Cf. **macromycete**.

micronematous (Gr. *mikrós*, small, minute + *nêma*, genit. *nêmatos*, filament + L. suf. *-osus* > E. *-ous*, having, abundant in, with the qualities of): a conidiophore that is not distinguished morphologically from the hyphae in the rest of the mycelium; e.g., in the gen. *Scytalidium* (dematiaceous asexual fungi), the conidiophores with arthrospores are very similar to the hyphae that do not generate spores. Cf. **macronematous**.

microorganism (Gr. *mikrós*, small, minute + *órganon*, organ, instrument, tool, similar to *érgon*, work + suf. *-ismós* > L. *-ismus* > E. *-ism*, state, phase, tendency, action): a microbe; an organism microscopic or ultramicroscopic in size, which cannot be seen with the unaided eye; it requires examination with a microscope and growth in pure culture to be studied and identified. Microorganisms comprise all of the prokaryotes and unicellular eukaryotes, as well as some multicellular eukaryotes, i.e., algae, bacteria, fungi (including the yeasts), protozoans and viruses. The term microorganism is sometimes used only for prokaryotes (bacteria and viruses).

microphyllous (Gr. *mikrós*, small, minute + *phýllon*, leaf): with small leaves. In mycology it is applied to foliose lichen thalli that have small lobes or scales.

microplasmodium, pl. **microplasmodia** (Gr. *mikrós*, small, minute + NL. *plasmodium*, plasmodium < Gr. *plásma*, formation, bland material with which a living being is formed + L. suf. *-odium*, resembling < Gr. *ode*, like < *-oeídes*, similar to): a minute or microscopic plasmodium; e.g., the protoplasmodium of *Echinostelium minutum* (Echinosteliales). See **protoplasmodium**. Cf. **macroplasmodium**.

micropycnidium, pl. **micropycnidia** (Gr. *mikrós*, small, minute + NL. *pycnidium*, pycnidium < Gr. *pyknós*, dense, compact, concentrated + dim. suf. *-ídion*): a small pycnidium that develops from a dictyochlamydospore or alternarioid chlamydospore, as occurs in *Phoma destructiva* and *Ph. glomerata* (sphaeropsidaceous asexual fungi).

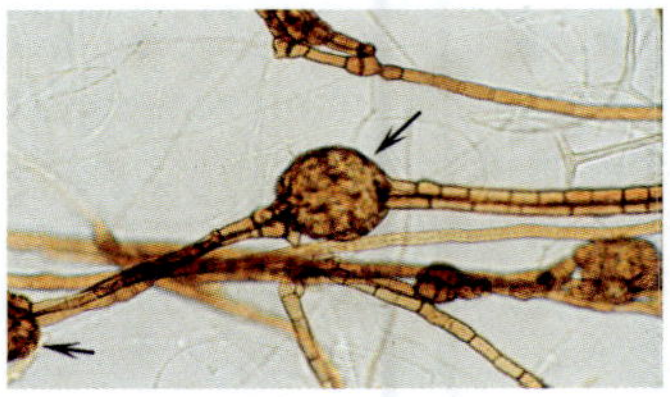

Micropycnidia and alternarioid chlamydospores of *Phoma destructiva*, x 200 (*CB*).

microsclerid (Gr. *mikrós*, small, minute + *sklerós*, hard + suf. *-eídos*, form, appearance): equivalent to a versiform endolamprocystidium.

microsclerotium, pl. **microsclerotia** (Gr. *mikrós*, small, minute + L. *sclerotium*, sclerotium < Gr. *sklerotés*, hardness < *sklerós*, hard + dim. suf. *-ium*): a small group of dark-colored cells, with a more or less thickened wall, each of which is viable and can form a germ tube or hypha. Microsclerotia can form under natural conditions as well as on culture media, and they are characteristic of many asexual fungi, such as *Verticillium* (moniliaceous asexual fungi) and *Rhizoctonia* (agonomycetaceous asexual fungi). The *papulospores* or *bulbils* of *Papulaspora* (agonomycetaceous asexual fungi) also are called microsclerotia.

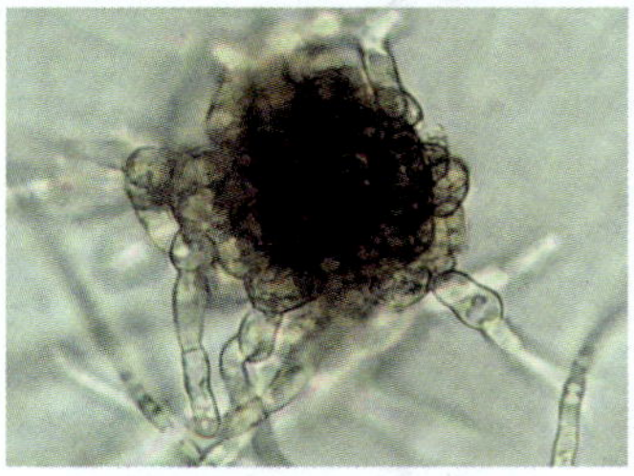

Microsclerotium of *Papulaspora* sp., x 320 (*MU*).

microscopic (Gr. *mikrós*, small, minute + *skopéo*, to see, examine + suf. *-ikos* > L. *-icus* > E. *-ic*, pertaining to or relating to): of sufficiently small size that a microscope is needed to see it clearly. Cf. **macroscopic**.

microscopy (Gr. *mikrós*, small, minute + *skopéo*, to see, examine): the use of, or investigation with, the microscope. There exist two main categories of microscopes, *light microscopes* and *electron microscopes*. The first includes the *compound microscope*, with two sets of lenses (objectives and one or two oculars) that magnify the object observed up to a maximum near 1,000 x, and the *stereomicroscope* (also called a *dissecting microscope*), with lower magnification capacity (more or less up to 250 x), but with a longer focal distance to work with the specimen. To observe a specimen with the compound microscope it is necessary to mount a small portion of it on a slide beneath a cover glass in a mounting medium that is clear or with the addition of a stain. The most common stains in mycology are cotton blue or fuchsin in lactic acid, or lactic acid alone if it is not necessary to stain the specimen. Air bubbles in the preparation can be eliminated by gently heating the slide. In order to maintain the preparations for a longer time, the cover glass is sealed with nail polish or a similar sealer. The usual method of observing specimens with the compound microscope is *bright field*, in which the light passes from below, through the condenser and the specimen, illuminating the translucid structures. With *Nomarski differential interference contrast* optics one can obtain images that are almost tridimensional, making it very useful for observing hyaline structures; this type of microscopy is based on an arrangement of special polarizing prisms, configured according to the design of Georges Nomarski, in addition to a special condenser and high quality objectives. The *phase contrast* microscope also is used to observe hyaline structures, without having to stain them, and although the halo produced around the object observed helps to clarify certain minute details, it makes precise measurement of structures difficult. The *dark field* microscope prevents the light from passing through the opening of the objective; the image is formed by the scattering of light from the features of the specimen, which appears bright in contrast to the dark field. With *fluorescence* microscopy the image is formed by fluorescence emitted from the specimen when it is illuminated with ultraviolet radiation. The sample can autofluoresce naturally, but generally fluorochromes are utilized, such as mithramycin and calcafluor, to make specific structures stand out. Electron microscopes are of two types, the *scanning electron microscope* (*SEM*) and the *transmission electron microscope* (*TEM*). With the SEM one can obtain images of the superficial morphology of a specimen magnified up to 80,000 x, with great depth of field. With the TEM images similar to those achieved with the light microscope are obtained, but at much higher magnifications (more than 100,000 x), which permits the observation of ultrastructural details. The specimens are prepared by means of fixation, dehydration, staining, embedding in resin, and cutting ultrafine sections on an ultramicrotome.

microspore (Gr. *mikrós*, small, minute + *sporá*, spore): a small spore, especially when spores of two sizes are present. Cf. **macrospore**.

Microsporidia (Gr. *mikrós*, small, minute + *sporídia*, pl. of *sporídion*, dim. of *sporá*, spore): eukaryotic, spore-forming, lacking flagella, obligately parasitic organisms of all groups of animals (protozoa, insects, crustaceans, fish, humans, and other organisms), with two or three different types of spores. They have been extensively studied as biocontrol agents, and one species (*Nosema locustae*) is commercially available for the control of locusts and grasshoppers. The microsporidians are a phylum related to the kingdoms Protozoa and Fungi. The recent classifications treat the suprageneric ranking differently and no classification can accommodate all established taxa. See the **Appendix**.

microtubule (Gr. *mikrós*, small, minute + L. *tubulus*, tubule < *tubus*, tube + dim. suf. *-ulus* > E. *-ule*): a proteinaceous tubule, apparently rigid, from 20-25 nm diam., found in the cytoplasm, in the achromatic spindles and in the axonemes of the flagella. In the hyphae, the microtubules are oriented parallel to the major axis; they are commonly associated with the nuclei and mitochondria, which suggests that there is an interaction between these organelles and the microtubules, related to the mobility of the organelles, which is independent of the cytoplasmic currents. The microtubules also can constitute a cytoskeleton in various fungal cells, e.g., in zoospores and basidia.

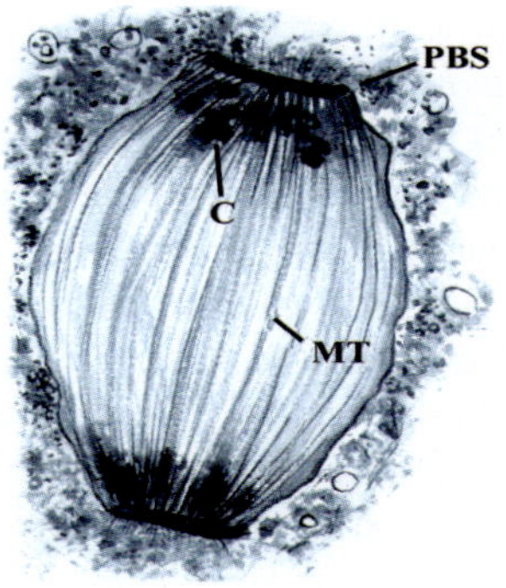

Microtubules (MT) of the achromatic spindle (of nuclear division) of *Ascobolus stercorarius* (Pezizales). A longitudinal section through a nucleus in anaphase; note the relationship between the microtubules of the spindle, the polar bodies of the spindle (PBS), and the chromosomes (C). Drawing based on an electron micrograph published by Beckett *et al.*, in their *Atlas of Fungal Ultrastructure*, 1974, x 7,620 (*MU*).

middle piece (L. *medius*, middle, central; L. *petia*, *petius*, part, piece): **1**. The central region that connects the two ends of a bipolar achromatic spindle; it is a band of electron opaque material, similar in shape and position to the structure known as the *plaque bridge*, which interconnects the two globular daughter halves that result from the replication of the spindle pole body during mitosis in yeast cells, such as *Saccharomyces* (Saccharomycetales). The middle piece is typical of the basidiomycetes. **2.** A stout, short tissue that anchors the purse to the sheath in peridioles of some species of Nidulariales, such as *Cyathus*.

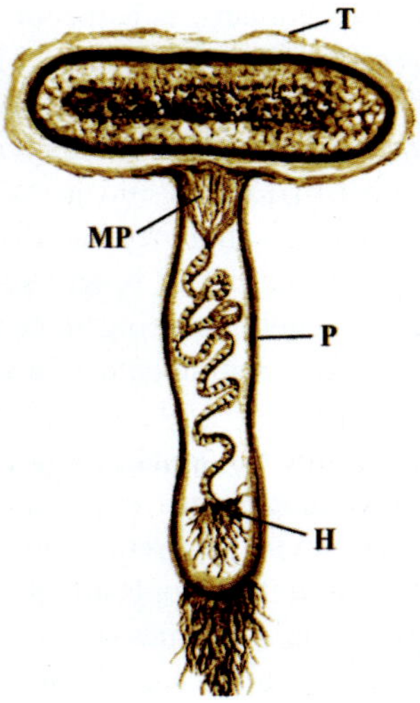

Middle piece (MP) of a peridiole of *Cyathus striatus*. In this diagram of a longitudinal section, the hapteron (H), purse (P) and tunica (T) also are shown, x 20 (*MU*).

miktohaplont (Gr. *miktón*, neuter form of *miktós*, mixture, blend + *haplóos*, simple, single + *ón*, genit. *óntos*, a being, one who is, that which is): a haplont whose cells have genotypically heterogeneous nuclei. Cf. **isohaplont**.

mildew (ME. *mildew* < OE. *meledeaw* > Fr. *mildiou* > Por. *míldio*, honeydew): superficial mycelial growth, usually whitish, produced by fungi on organic material or living plants. A *powdery mildew* (L. *pulverulentus*, powdery, full of or covered with powder < *pulver*, genit. *pulveris*, powder, dust + suf. *-entus*, abundant in, full of) is caused by a species of Erysiphaceae, e.g., *Erysiphe* (=*Uncinula*) *necator* on grape, *Erysiphe cichoracearum* on cucurbits and *Blumeria* (=*Erysiphe*) *graminis* on wheat; a *downy mildew* (ME. *downe*, soft pubesence) by a member of the Peronosporaceae, e.g., *Peronospora hyoscyani* f. sp. *tabacina* (=*P. tabaci*) on tabacco; a *dark* or *black mildew* (ME. *blac*, black) by one of the Meliolales (such as *Meliola palmicola*, on palms).

Powdery **mildew**, *Oidium monilioides*, conidial state of *Blumeria* (=*Erysiphe*) *graminis*, on the surface of a wheat leaf; note the erect conidiophores with an acropetal chain of conidia, x 900 (*WKW*).

minimal medium, pl. **media** (L. *minimus*, the minor, minimum; L. *medium*, neuter of *medius*, medium, intermediate; in this case, an enveloping or surrounding substance; culture medium): refers to the simplest culture medium, with regard to chemical composition, on which a species of organism called a *prototroph* (wild type or strain) can grow; this minimal medium must be supplemented by one or more specific substances (such as a certain vitamin or a particular amino acid) in order for mutant strains derived from the wild type (which are *auxotrophs*) to be able to grow on it. See **medium**.

mischoblastiomorph (Gr. *míschos*, pedicel, peduncle + *blastós*, germ, shoot, bud + *morphé*, shape): applied to a type of ascospore present in some lichens, such as *Rinodina sophodes* (Lecanorales), which is characterized by having a thick wall and two Y-shaped cellular compartiments or locules, opposed at the base, resembling a gymnasium weight. This type of ascospore is similar to the *polarilocular*, but without a septum, or with only an incomplete septum.

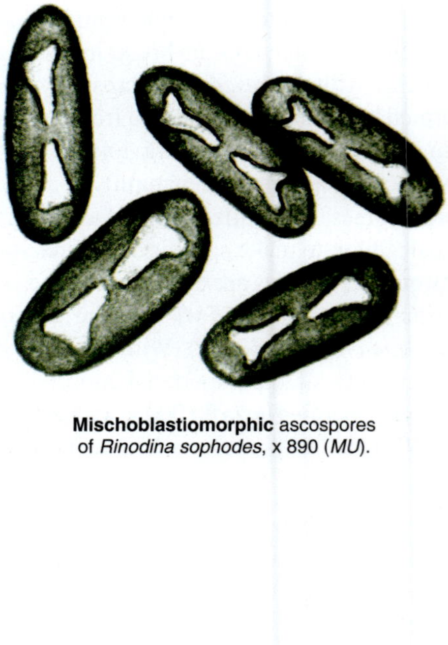

Mischoblastiomorphic ascospores of *Rinodina sophodes*, x 890 (*MU*).

mitochondrion, pl. **mitochondria** (NL. *mitochondrium*, mitochondrion < Gr. *mítos*, filament + *kondríon*, dim. of *kóndros*, grain): an intracellular organelle, generally granular, in which part of the enzymatic system necessary for respiration and the production of energy is located. It is delimited by a double membrane, the outer one smooth and the inner one convoluted, forming digitate or laminar invaginations called *cristae*.

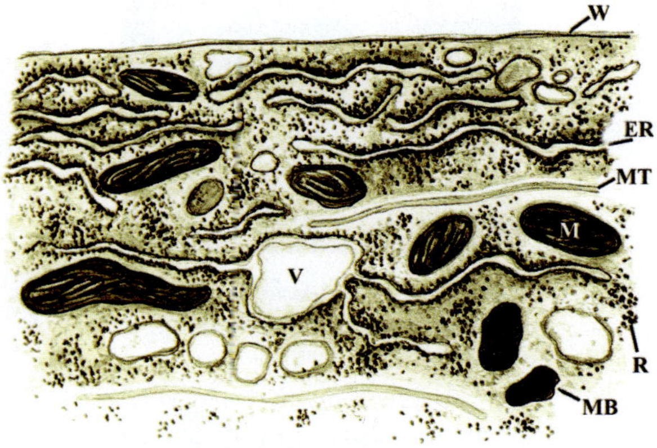

Mitochondria (M) in a young hypha of *Armillaria mellea* (Agaricales). Also shown are the endoplasmic reticulum (ER), ribosomes (R) of the ER cisternae, microbodies (MB), microtubules (MT), vesicles (V), and the cell wall (W). Drawing based on several transmission electron micrographs published by Beckett *et al.*, in their *Atlas of Fungal Ultrastructure*, 1974, x 15,800 (*MU*).

mitosis, pl. **mitoses** (Gr. *mítos*, filament + suf. *-osis*, condition, state): normal division of the nucleus in which the chromosomes are differentiated and these divide and split, resulting in two daughter nuclei with the same number of chromosomes as the mother nucleus. Cf. **meiosis**.

mitosporangium, pl. **mitosporangia** (Gr. *mítos*, filament + *sporá*, spore + *angeîon*, vessel, receptacle): a sporangium whose nuclei divide by mitosis. In the aquatic fungus *Allomyces* (Blastocladiales) the mitosporangia, which are thin-walled (zoosporangia), are formed on the sporothallus, because of which the nuclei are diploid. Mitosis results in additional diploid nuclei that form the zoospores (mitospores), which repeat the diploid generation after encysting and germinating to give rise to another sporothallus. A second type of sporangium, the *meiosporangium*, also is formed on the sporothallus. Cf. **meiosporangium**.

mitospore (Gr. *mítos*, filament + *sporá*, spore): a spore produced in a *mitosporangium*. Also applied to any propagule of ascomycetes and basidiomycetes that is not an ascospore or a basidiospore, since these are the product of meiosis. Cf. **meiospore**.

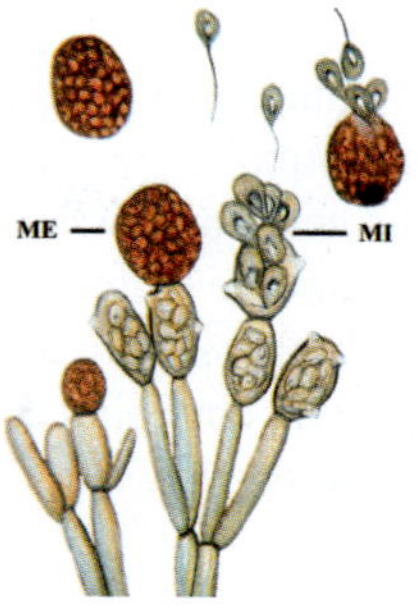

Mitosporangia (MI) and meiosporangia (ME) of the sporothallus of *Allomyces macrogynus*, x 630 (*MU*).

mitosporic fungi, sing. **fungus** (Gr. *mítos*, filament + *sporá*, spore + suf. *-íkos* > L. *-icus* > E. *-ic*, belonging to or relating to; L. *fungus*, fungus < Gr. *spóngos*, *sphóngos*, sponge): an artificial grouping of fungi that constitutes the second largest group of these organisms (ca. 2,600 gen. + 1,500 syn., 15,000 spp.); also known as deuteromycetes, Fungi Imperfecti, Deuteromycotina, imperfect fungi, asexual fungi, and conidial fungi. They are called mitosporic fungi because they reproduce asexually and/or parasexually by mitosis or presumed mitosis (i.e., they lack a meiotic state). Many, however, have been connected with teleomorphs in the ascomycetes and basidiomycetes, so that the mitosporic fungi can be regarded as the anamorphs or anamorphic states of these groups. See **anaholomorph**, **deuteromycetes**, **Fungi Imperfecti**, **asexual fungi** and **conidial fungi**.

mitriform (L. *mitra*, mitre, head-dress + suf. *-formis* < *forma*, shape): shaped like a mitre, like the ascocarps of the gen. *Mitrula* (Helotiales).

Mitriform pileus of the apothecium of *Mitrula* sp., x 1 (*MU*).

mixotrophic (Gr. *mixo-* < *míxis*, mixture, combination + *trépho*, to feed + suf. *-íkos* > L. *-icus* > E. *-ic*, pertaining to or relating to): an organism that exhibits the phenomenon of *mixotrophy*, i.e., that has a type of heterotrophic nutrition that is not absolute, but combined, as occurs in the majority of the fungi, which are capable of taking in inorganic matter as part of their nutrition. See **autotroph** and **heterotroph**.

molar (L. *mola*, millstone): a structure composed of mycelium intertwined with particles of soil (like a pseudosclerotium) that *Termitomyces* (Agaricales) forms in termite nests and which give rise to the basidiocarps that emerge from the nest. Molars vary in shape and size, but generally they are more or less globose, with a rough, convoluted surface, similar to the millstones used to grind cereal grains.

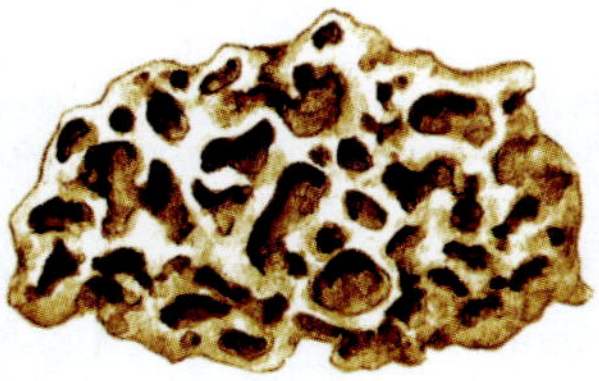

Molar of *Termitomyces microcarpus*, x 1 (*MU*).

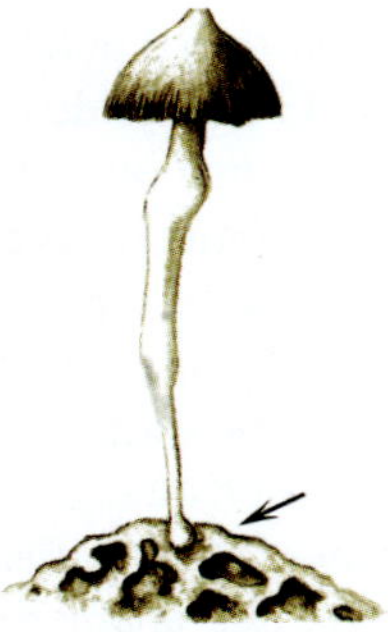

Molar of *Termitomyces striatus* giving rise to a basidiocarp, which has a basal insertion disk in the pseudorhiza, x 1 (*MU*).

mold, **mould** (ME. *mowlde* < *mowled*, pp. of *moulen, malwen,* to grow moldy): applied to any micromy-cete, principally of the Mucoraceae (Zygomycetes), mainly the gen. *Mucor* and *Rhizopus*, and to the conidial asexual fungi (agonomycetaceous, moniliaceous, dematiaceous, tuberculariaceous and stilbellaceous hyphomycetes), especially the moniliaceous gen. *Penicillium*, *Aspergillus*, *Monilia* and *Botrytis*, which grow on decomposing organic matter, covering it with a more or less dense mycelium, commonly with a downy, velvety or powdery appearance, due to the mass of hyphae and spores. The following are examples of names of common molds: *blue mold of citrus fruits*: *Penicillium italicum*; *apple blue mold*: *P. expansum*; *white mold of cheese*: *P. camemberti* (=*P. candidum* =*P. caseicola*); *common mold*: *Mucor mucedo*, which, together with other Mucorales and certain moniliaceous asexual fungi, cause the decay of many kinds of fruit; *snow mold*: *Sclerotinia borealis* (Helotiales), sexual state of *Monilia*; *gray mold*: *Botrytis cinerea* on fruits of grape and flowers of many plants (*Rosa*, *Dahlia*, etc.); *black mold*: *Aspergillus niger*; *black bread mold*: *Rhizopus stolonifer* (=*Rh. nigricans*); *green mold*: *Penicillium digitatum* on citrus fruits, and *P. glaucum* on cheese; *pink bread mold*: *Monilia sitophila*; *spiderweb mold*: *Corticium koleroga* (Aphyllophorales) of coffee, also known as spider-catcher of coffee. See **moldy**.

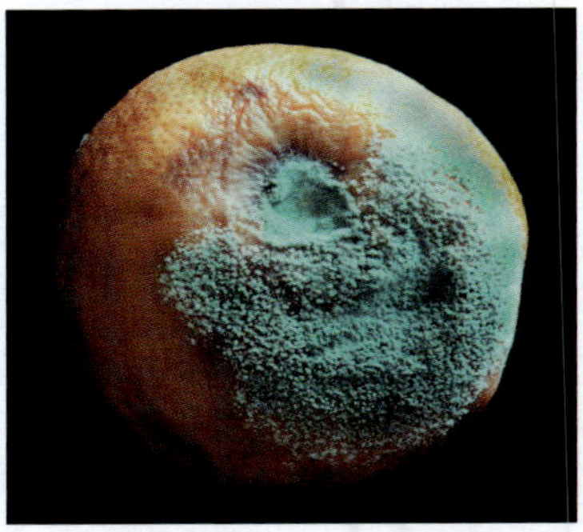

Blue **mold** of citrus (*Penicillium italicum*) growing on an orange, x 0.7 (*MU*).

Conidiophore of the gray **mold** of grapes (*Botrytis cinerea*), x 250 (*MU*).

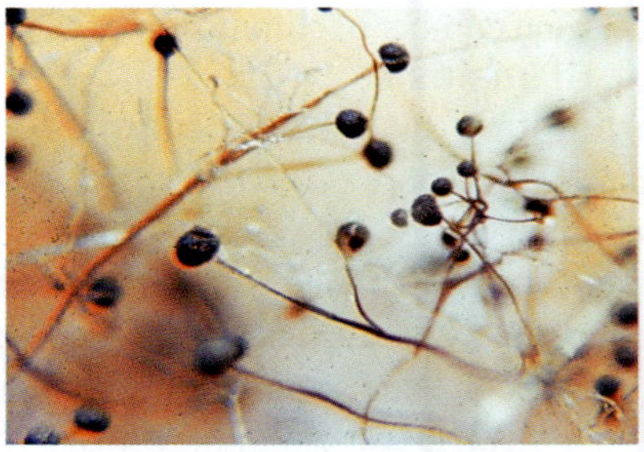

Sporangia of *Rhizophus stolonifer* (=*Rh. nigricans*), known as black bread **mold**, x 150 (*MU*).

moldy (ME. *mowlde* < *mowled*, pp. of *moulen, malwen*, to grow moldy + E. suf. *-y*, characterized by or inclined to): covered with mold; degraded or decomposed by molds. The molding of plants and dead animals is common when environmental conditions, especially temperature and humidity, favor the growth of ubiquitous filamentous fungi (such as the species of *Penicillium* and *Aspergillus*, of the moniliaceous asexual fungi, and *Mucor* spp., of the Mucorales) on organic remains, e.g., insects, fruits, wood, etc. See **mold**.

Moldy dead fly covered with *Penicillium* sp., x 10 (*MU*).

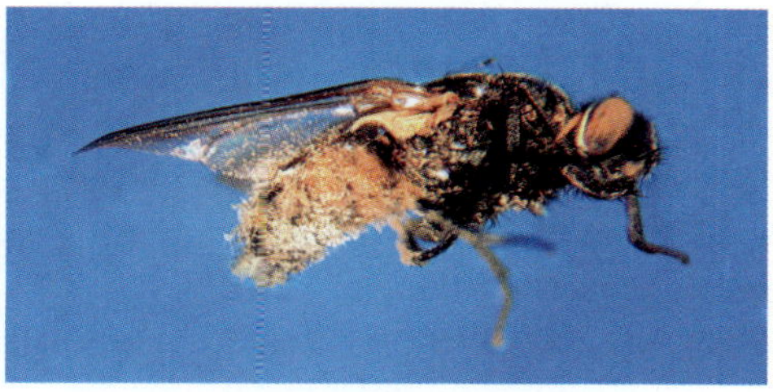

Moldy dead fly invaded by *Penicillium* sp., x 10 (*MU*).

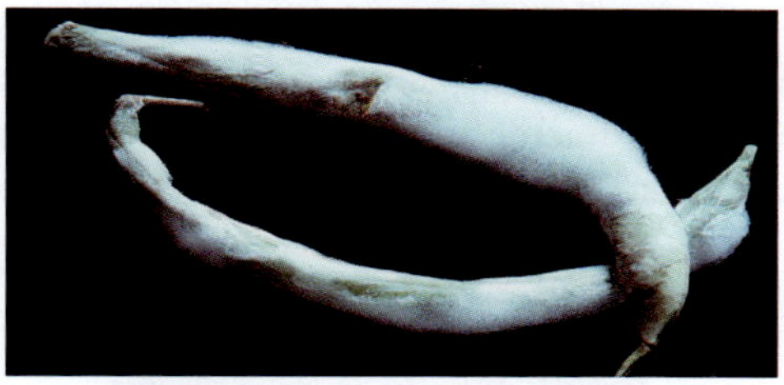

Moldy string bean pods covered with *Mucor* sp., x 0.5 (*MU*).

Moldy ravioli covered with *Aspergillus flavus* and *Penicillium* sp., x 0.5 (*MU*).

molecular biology (*molecular*, of, relative to or produced by molecules < Fr. *molécule* < NL. *molecula*, dim. of L. *moles*, mass + L. suf. *-aris* > E. *-ar*, pertaining to or relating to; Gr. *bíos*, life + *logía*, discourse, treatise < *lógos*, word < *légo*, I say; the science of life): the branch of biology that studies the most fundamental physicochemical organization of living material, especially the organization of the molecular basis of inheritance and of the synthesis of proteins. The techniques of molecular biology are varied and at present they are making important contributions to our understanding of fungal biology and to the phylogenetic relationships among the diverse groups of fungi. The majority of the techniques are based on the analysis of DNA, which first must be extracted from the cells. The technique used will depend on the specific purpose of the study. The principal study methods for DNA are: *DNA-DNA hybridization*, often utilized earlier for yeasts and molds; it involves the separation of the double-stranded DNA from the two fungi to be compared into separate strands. These strands are then allowed to reassociate or pair up; the degree of successful pairing is expressed as the % of relationship of the DNA of the two fungi involved. The technique of *DNA fingerprinting* is based on repeated sequences, dispersed throughout the genome, which have a high variability; it is useful in the manufacture of specific probes for the genotype of species or races pathogenic to plants or man. *DNA probes* are marked fragments of DNA that are used to identify particular regions of DNA in other organisms, including fungi. A *DNA probe* is used to generate *restriction fragment length polymorphisms* (*RFLPs*) from digested DNA. In order to identify specific regions in particular species, special DNA probes are elaborated. The *mol % G+C* (guanine + cytosine) content in DNA is determined by means of profiles, expressed as percentage, obtained by thermal denaturation; this technique was frequently utilized in yeast taxonomy, but it has been replaced by more sensitive techniques. The *polymerase chain reaction* (*PCR*) consists of a series of steps involving heating and cooling that permit the amplification of DNA. The DNA is denatured (the two strands of the double helix are separated) and the primers bind to the separated strands; afterwards complementary strands (dimers) are formed from the primers, with the addition of triphosphate deoxynucleotides (dTPNs) and thermostable DNA polymerase. With successive heatings these new dimers are denatured and the process is repeated through many cycles. RFLPs are utilized for nuclear as well as mitochondrial DNA, through the use of restriction enzymes that cut the molecules into fragments of a certain size; the size of the fragments varies and these are separated by electrophoresis of the total cellular DNA, producing a flow of fragments of different sizes; the polymorphisms associated with individual genes or groups of genes require hybridization with a marked DNA primer. The analysis of the *random amplified polymorphic DNA* (*RAPD*) utilizes PCR with short primers to produce sufficient quantities of particular portions of a blank sequence of DNA or RNA, which can then be compared on agarose gels; this technique is especially useful in studies of closely related fungi, in which can be found particular patterns of the species being analyzed, as well as for the study of small samples, such as a single spore or an old type specimen. *DNA sequencing* involves the comparison of the sequences of bases in particular portions of the molecules of DNA (or of RNA); groups of ribosomal genes have been used by mycologists for phylogenetic studies, since the sequences of bases can be analysed cladistically.

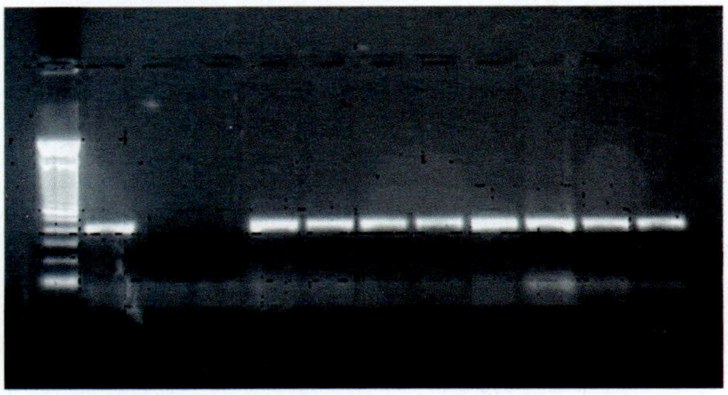

Detection of airborne conidia of *Metarhizium anisopliae* var. *acridum* (moniliaceous asexual fungi), with DNA probes (PCR: a technique of **molecular biology**), in Coatepec, Puebla, Mexico, x 0.5 (*CGG* & *CC*).

mollicute (L. *mollis*, smooth, bland + *cutis*, skin): see **mycoplasma**.

monad (L. *monas*, genit. *monadis*, unity < Gr. *monás*, genit. *monádos* < *mónos*, alone, only): an amoeboflagellate microorganism belonging to the Protozoa; also called Proteomyxa and Hydromyxales. See **myxomonad**.

monandrous (Gr. *mónos*, alone, only + *andrós*, male; here, male sexual organ): *Oomycota*. Applied to oospores that develop when they are fertilized by a single functional antheridium, as occurs in some gen., e.g., *Ectrogella* (Myzocytiopsidales) and *Pythiella* (Lagenaceae, *familia incertae sedis*). Cf. **polyandrous**.

monaxial (Gr. *mónos*, alone, only + L. *axialis*, axial < *axis*, axis + suf. *-alis* > E. *-al*, relating to or pertaining to): see **uniaxial**.

Monera (Gr. *monéres*, single): a special group within the kingdom Protista proposed by Haeckel in 1866, that includes the bacteria. In 1956, in his system of classification of living organisms into four kingdoms, Copeland placed the bacteria and Cyanophyceae or cyanobacteria (prokaryotes) in the kingdom Monera, which is still sustained in the five kingdom systems of classification of Whittaker (1959) and Margulis and Schwartz (1988).

moniliaceous (L. *moniliaceus* < gen. *Monilia* < *monile*, genit. *monilis*, chain, collar + suf. *-aceus* > E. *-aceous*, of or pertaining to, with the nature of): applied to hyphae, spores, etc., that are hyaline or brightly colored. Cf. **dematiaceous**. Moniliaceous asexual fungi, of the old family Moniliaceae, order Moniliales, class Hyphomycetes, subdivision Deuteromycotina, or deuteromycetes in a general sense, are an unnatural group of anamorphic fungi in the gen. *Monilia*, *Aspergillus*, *Penicillium*, among many others. See **mucedinaceous**, **Hyphomycetes**.

moniliasis, pl. **moniliases** (< gen. *Monilia* < L. *monile*, genit. *monilis*, chain, collar + Gr. suf. *-iasis*, ending of names of diseases): another of the names for candidiasis, also called *candidosis* and *candidamycosis*; it has been called moniliasis because previously the generic name *Monilia* (which correctly refers to a mold) was used to refer to yeasts of the present gen. *Candida* (cryptococcaceous asexual yeasts). See **candidiasis**.

moniliform (L. *moniliformis* < gen. *Monilia* < L. *monile*, genit. *monilis*, chain, collar + suf. *-formis* < *forma*, shape): in a chain, composed of a series of segments more or less rounded and linked together, like the conidiophore of *Monilia* (moniliaceous asexual fungi), composed of branched chains of blastic conidia, or like the microconidia of *Fusarium moniliforme* (tuberculariaceous asexual fungi).

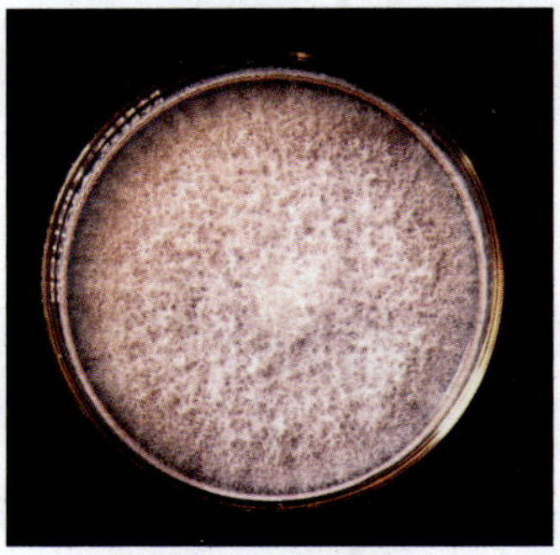

Colony of *Fusarium moniliforme* on agar; the specific name is based on the **moniliform** microconidial chains, x 0.5 (*MU*).

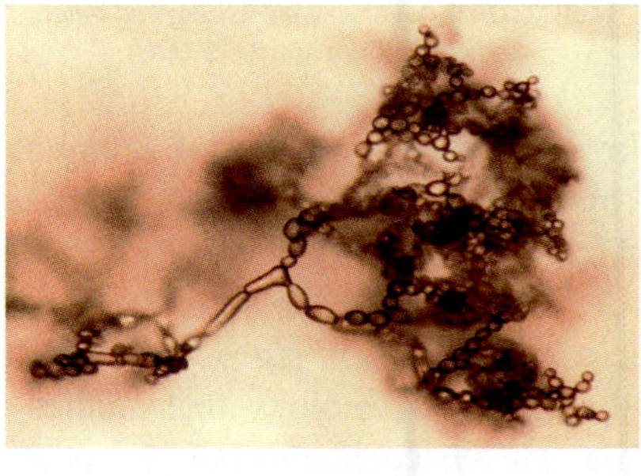

Moniliform conidial chains of *Monilia sitophila*, x 600 (*MU*).

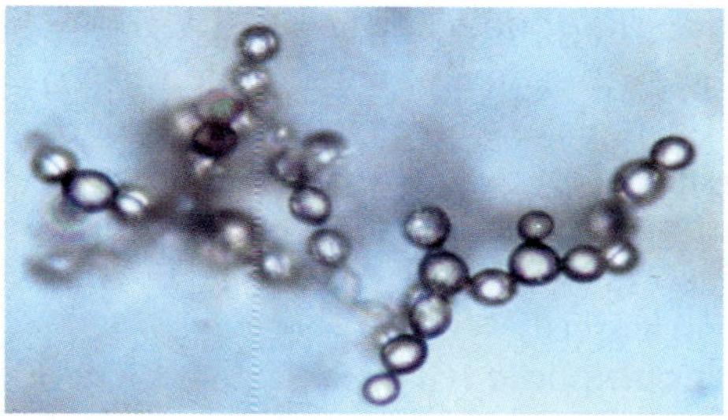

Moniliform conidial chains of *Monilia sitophila*, x 1,200 (*MU*).

mono- (Gr. *mónos*, one, only): a pref. that is used to form a multitude of generic names and biological terms when one wishes to give the idea that something is unique or single, as in the following words. Corresponds to L. **uni-**.

monoblastic (Gr. *mónos*, only, one + *blastós*, sprout, bud, germ + suf. *-íkos* > L. *-icus* > E. *-ic*, pertaining to or relating to): a type of holoblastic conidiogenous cell that produces a conidium at only one point or site, as occurs in *Acrogenospora*, *Humicola* and other gen. of dematiaceous asexual fungi. See **monopodial**. Cf. **polyblastic**.

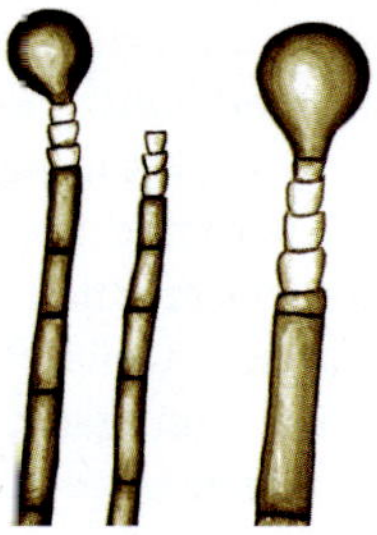

Monoblastic conidiogenous cells of *Acrogenospora sphaerocephala* (=*Monotospora sphaerocephala*), x 320 (*MU*).

monocarpic (Gr. *mónos*, single, one + *karpós*, fruit + suf. *-íkos* > L. *-icus* > E. *-ic*, pertaining to or relating to): refers to infections caused by *Exobasidium* (Exobasidiales), that are circumscribed and annual. Cf. **polycarpic**.

Monocarpic basidiocarp of *Exobasidium* sp. growing on a branch of an oak tree, x 0.5 (*RV*).

monocentric (Gr. *mónos*, single, one + *kéntron*, center + suf. *-íkos* > L. *-icus* > E. *-ic*, pertaining to or relating to): a thallus that has a single center of growth, in which is formed a single reproductive organ (sporangium or resting spore), as is observed in *Chytriomyces* (Chytridiales) and other aquatic fungi. Equal to **monogenocentric**. See **reproductocentric**. Cf. **polycentric**.

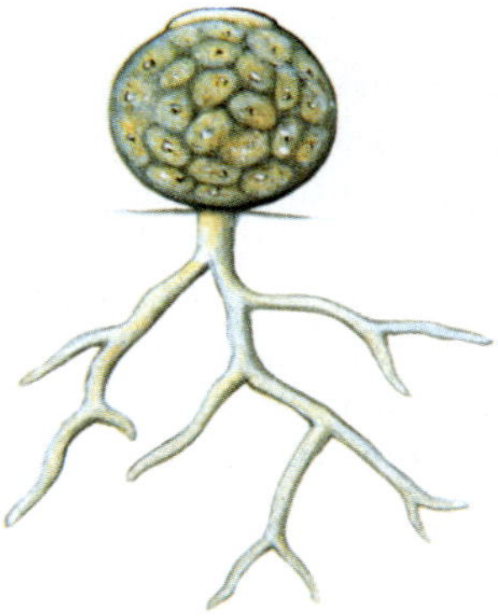

Monocentric thallus of *Chytriomyces hyalinus*, x 525 (*MU*).

monocephalous (Gr. *mónos*, single, one + *kephalé*, head): with a single head or capitulum.

monoclinous (Gr. *mónos*, single, one + *klíne*, bed; here, chamber, thallus > L. *clinare*, to lean, incline): a fungus that has the antheridium on the same branch or pedicel that bears the oogonium, as occurs in many aquatic species, e.g., *Monoblepharis polymorpha* (Monoblepharidales). In a strict sense, monoclinous is syn. of *hermaphrodite*. Cf. **diclinous**.

Monoclinous thallus of *Monoblepharis polymorpha*; the antheridium (A) is on the same branch as the oogonium (O), x 440 (*MU*).

monoecious (Gr. *mónos*, single, one + *oikía*, house): literally, with a single house, alluding to fungi that form male and female organs on the same mycelium or thallus, for which reason they are called bisexual or hermaphrodites. Cf. **dioecious**. See **hermaphrodite**.

monogenocentric (Gr. *mónos*, single, one + *génos*, origin < *gennáo*, to engender, produce + *kéntron*, center + suf. *-íkos* > L. *-icus* > E. *-ic*, pertaining to or relating to): developing a single reproductive structure in the center of gravity of the thallus, as in *Chytridium* (Chytridiales). Equal to **monocentric**. Cf. **polycentric**. See **reproductocentric**.

monograph (Gr. *mónos*, single, one + *grápho*, to draw, sketch, represent by means of lines, write > *graphía*, to write, describe): the description of a specific topic in any science. In systematics it refers to a detailed study of a group of organisms, of a family or of a genus, or even of a single species.

monokaryon, **monocaryon** (Gr. *mónos*, one, only + *káryon*, nucleus): a haploid cell nucleus. See **monokaryotic**. Cf. **dikaryon**.

monokaryotic, **monokaryontic**, **monocaryotic**, **monocaryontic** (Gr. *mónos*, only, one + *káryon*, nut, nucleus + suf. *-tikós* > L. *-ticus* > E. *-tic*, relation, fitness, inclination or ability): referring to an organism whose cells contain a single nucleus, like the primary mycelium of the Agaricales. Cf. **dikaryotic**.

monomitic (Gr. *mónos*, single, one + *mítos*, filament; here, hypha + suf. *-íkos* > L. *-icus* > E. *-ic*, pertaining to or relating to): referring to a fructification with a single type of hypha, the generative, which has a thin wall, septa, frequent clamp connections, and is branched and fertile, since it has the capacity to produce basidia. Examples of monomitic fructifications are *Clavariadelphus* (Aphyllophorales) and all of the fleshy agarics. Cf. **dimitic** and **trimitic**.

monomorphic, **monomorphous** (Gr. *mónos*, single, one + *morphé*, shape + suf. *-íkos* > L. *-icus* > E. *-ic*, pertaining to or relating to): *Oomycota*. Said of species of zoosporic microorganisms in whose life cycle there is only one type of zoospore (called primary), as occurs in the gen. *Pythiopsis* (Saprolegniales). If a species produces only secondary zoospores, it also is considered monomorphic. Cf. **dimorphic**.

monomycelial, **monomycelic** (Gr. *mónos*, single, one + NL. *mycelium*, mycelium < Gr. *mýkes*, fungus + L. suf. *-alis* > E. *-al*, relating to or concerning; or + Gr. suf. *-íkos* > L. *-icus* > E. *-ic*, pertaining to or relating to): applied to an isolate that comes from one spore or a single hyphal tip.

mononematous (Gr. *mónos*, single, one + *nêma*, genit. *nêmatos*, thread, filament + L. suf. *-osus* > E. *-ous*, what has, abundant in, possessing the qualities of): said of a conidiophore composed of a single hypha or filament, as in the majority of the gen. of hyphomycetous asexual fungi. Cf. **synnematous**.

monophagous (Gr. *mónos*, single, one + *phágos*, what eats): in mycology, it refers to a species of Chytridiales whose thallus is developed inside a single host cell. Cf. **polyphagous**.

monophagy (Gr. *mónos*, single, only + *phageín*, to eat): the phenomenon relating to *monophagous* organisms.

monophialidic (Gr. *mónos*, single, one + *phialís*, genit. *phialídes*, dim. of *phiále*, glass, jar + suf. *-íkos* > L. *-icus* > E. *-ic*, relating to): applied to a phialide (conidiogenous cell) that has a single site or locus through which conidia are produced. Cf. **polyphialidic**.

monophyletic (Gr. *mónos*, single, one + *phyletikós* < *phyletés*, of the same tribe, race, stock < *phylé*, tribe + suf. *-tikós* > L. *-ticus* > E. *-tic*, relation, fitness, inclination or ability): derived from a single common ancestral form; from a single line of descent. See **phylum**. Cf. **paraphyletic** and **polyphyletic**.

monophyllous (Gr. *mónos*, single, one + *phýllon*, leaf): a foliose lichen thallus with a single leaf or foliolum. Cf. **polyphyllous**.

monopileate (Gr. *mónos*, single, one + L. *pileatus*, with a hat or with the appearance of a hat < *pileus*, pileus, hat, cap + suf. *-atus* > E. *-ate*, provided with or likeness): a type of fruiting body development in certain Gasteromycetes, such as *Podaxis* (Hymenogastrales), in which a single stipe bearing the gleba is formed, which is arranged as in the embryonic button of an agaric, with hymeniiferous plates. The term monopileate also is applied to the fruiting body of *Dictyophora*, *Phallus* and *Mutinus* (among other Phallales), whose gleba forms a cylinder of hymenial tissue that is differentiated around a central sterile stipe. Cf. **multipileate**.

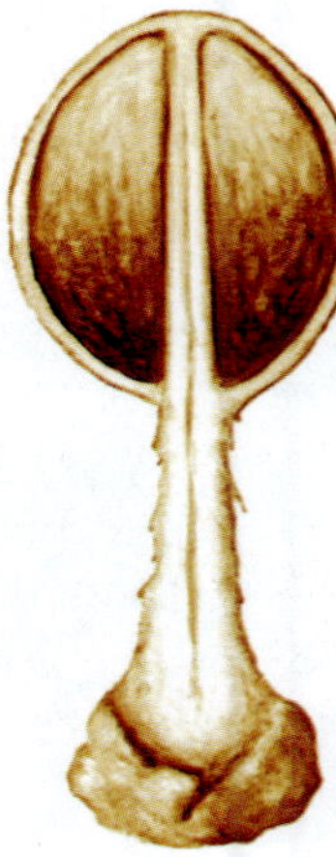

Monopileate basidiocarp of *Podaxis pistillaris*, seen in longitudinal section, x 1.5 (*MU*).

Monopileate basidiocarp of *Phallus hadrianii*, x 0.5 (*RV*).

Monopileate basidiocarp of *Mutinus caninus*, x 0.5 (*RV*).

monoplanetic (Gr. *mónos*, single, one + *planétes*, wandering, vagabond + suf. *-ikos* > L. *-icus* > E. *-ic*, pertaining to or relating to): *Oomycota*. A species in whose life cycle there is a single type of zoospore (called primary) with a single period of swimming; as happens, e.g., in the gen. *Pythiopsis* (Saprolegniales). However, this is not necessarily so; *Aphanomyces* (Saprolegniales) has monomorphic and monoplanetic secondary zoospores. Cf. **diplanetic** and **polyplanetic**.

monoplanetism (Gr. *mónos*, single, only + *planétes*, wandering, vagabond + suf. *-ismós* > L. *-ismus* > E. *-ism*, state or condition): the phenomena relating to *monoplanetic* organisms.

monopodial (Gr. *mónos*, single, only + NL. *podium* < Gr. *pódion*, dim. of *poús*, genit. *podós*, foot; here, support + L. suf. *-alis* > E. *-al*, relating to or pertaining to): adj. of *monopodium*.

monopodium, pl. **monopodia** (NL. *monopodium* < Gr. *mónos*, single, only + NL. *podium* < Gr. *pódion*, dim. of *poús*, genit. *podós*, foot; here, support): a type of branching in which a hypha continues growing at the apex without changing direction, while lateral branches of the same type are produced below in acropetal succession. Among the fungi, there are many kinds of sporangiophores and conidiophores that have monopodial branching, e.g., the conidiophore of *Botryosporium* (moniliaceous asexual fungi) and *Acrogenospora sphaerocephala* (=*Monotospora sphaerocephala*), a dematiaceous asexual fungus. Cf. **sympodium**.

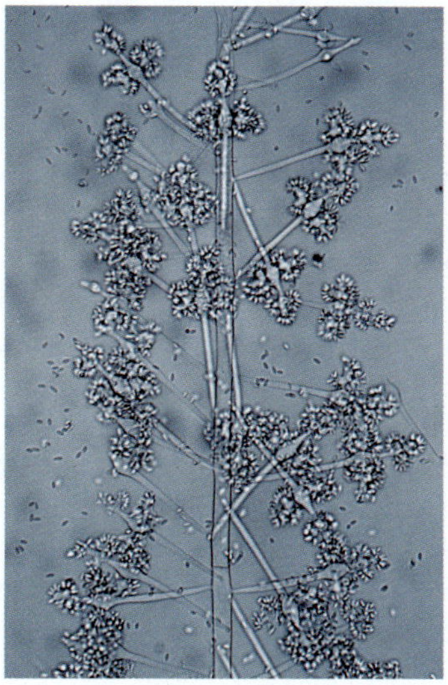

Monopodium, the conidiophore of *Botryosporium* sp., x 300 (*MU*).

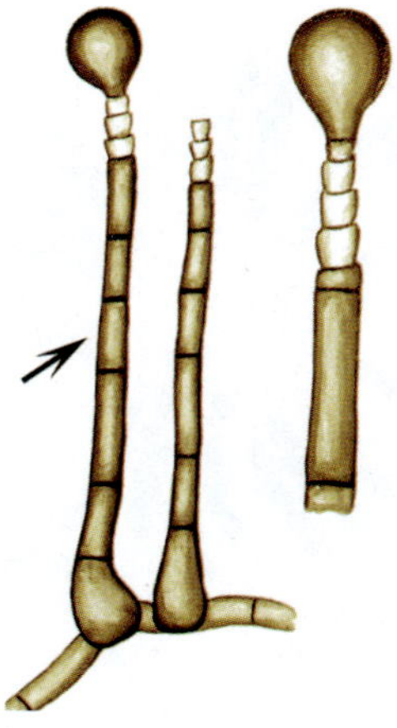

Monopodium, the conidiophore of *Acrogenospora sphaerocephala* (=*Monotospora sphaerocephala*), x 300 (*MU*).

monopolar (Gr. *mónos*, single, one + *polaris*, polar < *polus*, pole + suf. *-aris* > E. *-ar*, like, belonging to): **1**. A type of germination in which a spore forms a single germ tube; characteristic of the majority of fungi. **2**. A type of budding in yeasts in which a single bud forms at one end of the mother cell, as in *Malassezia* (=*Pityrosporum*), of the cryptococcaceous asexual yeasts. Cf. **bipolar**, **multipolar** and **unipolar**.

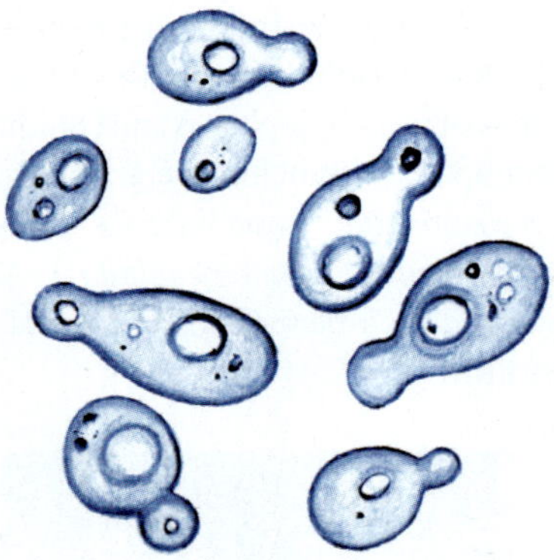

Monopolar budding of the somatic cells of *Malassezia* (=*Pityrosporum*) sp., x 1,900 (*MU*).

monospore (Gr. *mónos*, single, only + *sporá*, spore): the only spore that matures on a bispored basidium.

monosporous, monosporic (Gr. *mónos*, single, only + *sporá*, spore + suf. *-íkos* > L. *-icus* > E. *-ic*, pertaining to or relating to): having a single spore.

monostichous, **monostic**, **monostichate** (Gr. *mónos*, single, only + *stíchos*, line, row, file + L. suf. *-osus* > E. *-ous*, having, with the qualities of; or + suf. *-íkos* > L. *-icus* > E. *-ic*, pertaining to or relating to; or + L. suf. *-atus* > E. *-ate*, provided with or likeness): arranged in a single row, like the ascospores within the asci of *Podospora* and *Sordaria* (Sordariales). Syn. of **uniseriate**. Cf. **distichous** and **tristichous**.

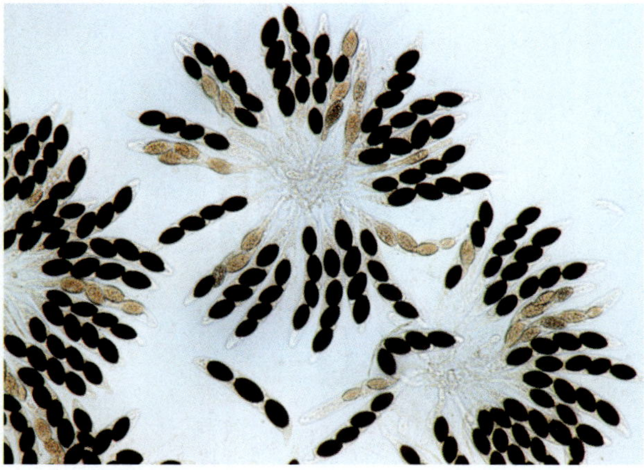

Monostichous asci of *Podospora comata*, x 150 (*EAA*).

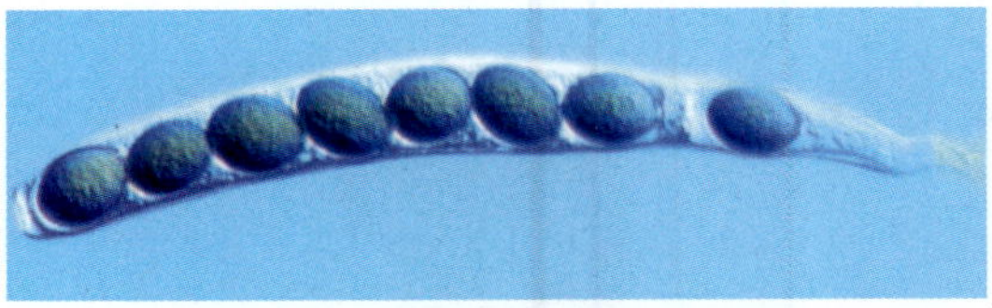

Monostichous ascus of *Sordaria fimicola*, x 270 (*RTH*).

monotretic (Gr. *mónos*, single, only + *tretós*, perforate + suf. *-íkos* > L. *-icus* > E. *-ic*, pertaining to or relating to): *Conidiogenesis*. A type of enteroblastic conidiogenous cell that produces conidia through a single canal or perforation in the outer wall; the inner wall of the conidiogenous cell protrudes through the perforation and becomes the outer wall of the young conidium. These conidia are generally solitary, but sometimes form acropetal chains. E.g., the conidiogenous cells of *Corynespora cassiicola* (dematiaceous asexual fungi) are monotretic. Cf. **polytretic**.

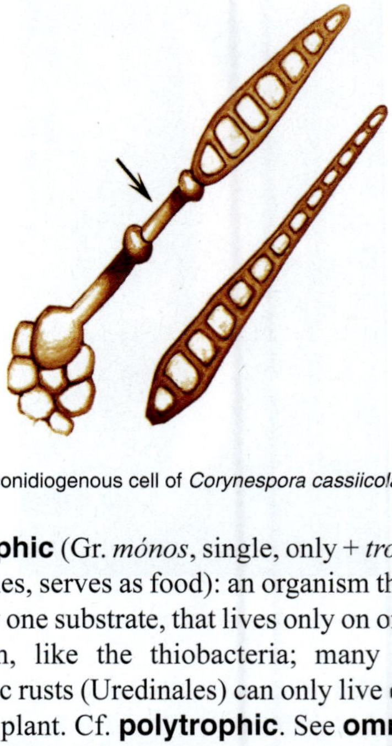

Monotretic conidiogenous cell of *Corynespora cassiicola*, x 310 (*MU*).

monotrophic (Gr. *mónos*, single, only + *trophós*, what nourishes, serves as food): an organism that can feed on only one substrate, that lives only on one nutritive medium, like the thiobacteria; many autoecious parasitic rusts (Uredinales) can only live on one type of host plant. Cf. **polytrophic**. See **omnivorous**.

monotropoid mycorrhiza, pl. **mycorrhizae** (< *Monotropa,* gen. of plants + L. suf. *-oide*, like < Gr. *oeídes*, similar to; Gr. *mýkes*, fungus + *rhíza*, root): the term by which the so-called *ericoid mycorrhiza* also is known, named for the associated plants, including *Vaccinium* (blueberry) and various parasitic angiosperms, such as *Monotropa*, all belonging to the order Ericales. Actually, the parasitic angiosperms are connected to their host plants by means of a bridge of ericoid ectendomycorrhizal fungi that includes inoperculate species of *Thelebolus* (=*Pezizella*) and hypogeous *Elaphomyces* (Pezizales).

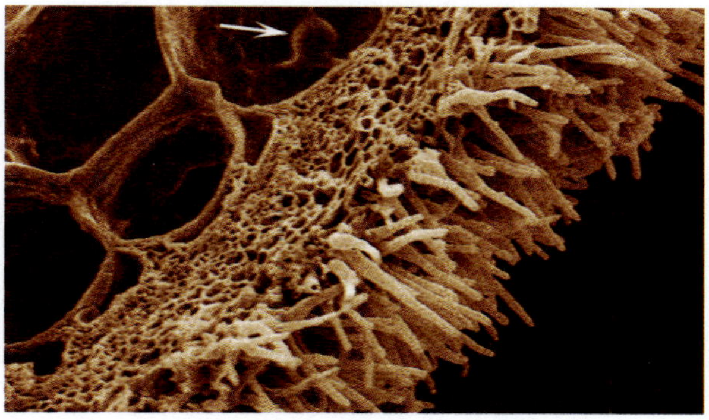

Scanning electron photograph of the roots of *Monotropa uniflora*, in cross section, which shows the key character of the **monotropoid mycorrhizae**: the fungal peg (arrow), x 4,000 (*NH*).

monotype (Gr. *mónos*, single, only + *týpos*, type, mark, model): a single element on which a taxon is based, a case in which the author did not explicitly indicate that this element was to be considered as the nomenclatural type of the taxon. Also refers to a genus that is monotypic when it is established with a single species; a monotypic family is one established with a single genus.

monotypic (Gr. *mónos*, single, only + L. *typicus* < Gr. *typikós*, conforming to a type + suf. *-ikos* > L. *-icus* > E. *-ic*, pertaining to or relating to): having a single representative; a monotypic gen. has a single species; a monotypic family has a single gen., etc.

monoverticillate (Gr. *mónos*, single, only + NL. *verticillatus*, verticillate < *verticillus*, verticil + suf. *-atus* > E. *-ate*, provided with or likeness): generally applied to a conidiophore that strictly or predominantly has a single point of branching between the pedicel and the chain of conidia; i.e., that bears a single verticil of phialides, which are formed directly on the pedicel; very rarely the phialides are formed on metulae in some species. Some examples of monoverticillate conidiophores are those of *Penicillium citreonigrum*, *P. glabrum* and *P. restrictum* (moniliaceous asexual fungi), which certain authors consider within the subgenus *Aspergilloides*. Cf. **biverticillate**, **triverticillate**, **quadriverticillate** and **polyverticillate**.

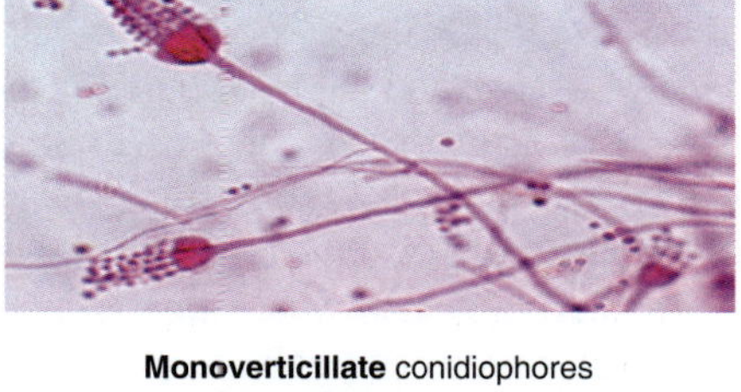

Monoverticillate conidiophores of *Penicillium citreonigrum*, x 585 (*CB*).

monoxenic, **monoxenous** (Gr. *mónos*, single, only + *xénos*, stranger + suf. *-íkos* > L. *-icus* > E. *-ic*, pertaining to or relating to): **1**. The growth of two species of organisms together in the laboratory; it can involve, e.g., a ciliate with another organism, such as a bacterium, an alga, a yeast, or another ciliate. The second organism can be an undesirable contaminant, or it can serve as food for the first one. Cf. **agnotobiotic**, **axenic** and **gnotobiotic**. **2**. Having a single host, completing the life cycle on a single host, like some plant pathogenic fungi (e.g., the autoecious rusts of the order Uredinales). In this sense it is a syn. of *autoecious*. Cf. **heteroxenous** or **heteroecious**.

morbose (L. *morbosus*, sickly < *morbus*, sickness + suf. *-osus* > E. *-ose*, full of, augmented, prone to): illness; causing or related to illness.

Morbose peach tree, killed by *Phymatotrichopsis omnivora* (moniliaceous asexual fungi), x 0.01 (*JAS*).

morel (Fr. *morille* < E. *morel*, morel): the common name for the edible ascoma of *Morchella*, especially *M. esculenta* (Pezizales).

Morels, the edible ascomata of *Morchella esculenta*, x 0.3 (*SA*).

Morels, the edible ascomata of *Morchella* sp., x 1 (*RV*).

moriform (L. *moriformis* < *morus*, mulberry + suf. *-formis* < *forma*, shape): shaped like a mulberry, like the clusters of spores of some fungi, e.g., *Oedocephalum* (moniliaceous asexual fungi), and the groups of warted perithecia of *Bertia moriformis* (Coronophorales).

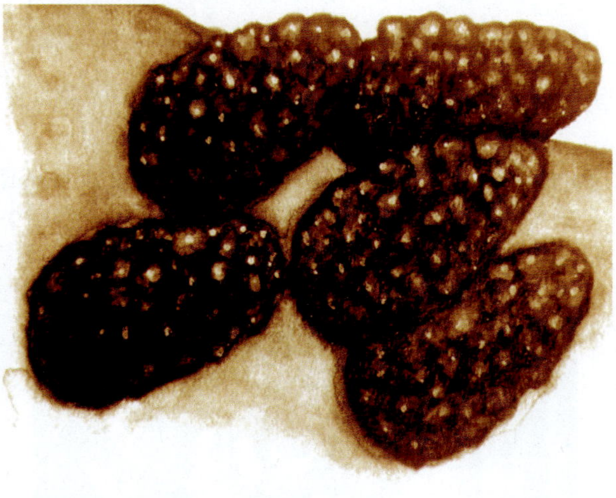

Gregarious (**moriform**) perithecia of *Bertia moriformis*, on decorticated beech wood, x 20 (*MU*).

morpho-, **morph-** (Gr. *morphé*, shape): a pref. employed to allude to the shape of something, as in *morphogenesis*. Also employed as a suf., *-morph*, to form adjectives, as in *actinomorph*, *polymorph*, *zygomorph*, etc., which signify, respectively, shaped like a ray, with many shapes, having bilateral symmetry, etc.

morphogenesis, pl. **morphogeneses** (Gr. *morphé*, shape + *génesis*, origin): the ensemble of phenomena relating to the creation of the shape of an organism, as well as to the evolution of morphological characters. The study of morphogenesis of the ascocarp, e.g., has served to better define the affinities or the differences among the groups of ascomycetes.

morphogenetic (Gr. *morphé*, shape + *génesis*, engenderment + suf. *-tikós* > L. *-ticus* > E. *-tic*, relation, fitness, inclination or ability): relating to *morphogenesis* or belonging to it.

morphogenous (Gr. *morphé*, shape + *génos*, race, origin): capable of modifying shape. E.g., a morphogenous stimulus is what determines the development of qualitative morphological processes, and consequently, is capable of altering the morphology of the organs or structures.

morphological (Gr. *morphé*, shape + *lógos*, word > E. des. *-logy*, study or treatise + suf. *-alis* > E. *-al*, pertaining to or relating to): concerning morphology, as opposed to physiology and biology. Systems of classification based on morphological characters complement those from phylogenetics, and with molecular characters, reflect the constitution of the genome and contribute to an understanding of phylogeny.

morphology (Gr. *morphé*, shape + *lógos*, word > E. des. *-logy*, study or treatise): the science of shape.

morphotype (Gr. *morphé*, shape + *týpos*, type, mark, model): in polymorphic species of unknown taxonomic significance, the morphotype is any morphological state with definite differentiated characters.

mother (ME. *mother, moder* < OE. *modor* < L. *mater*, genit. *matris* < Gr. *máter*, mother): in mycology, and in biology in general, it is often employed with the meaning of cause, root, origin or source of something, e.g., the *ascus mother cell*, the *mother of vinegar*, etc.

mother cell (ME. *mother, moder* < OE. *modor* < L. *mater*, genit. *matris* < Gr. *máter*, mother; NL. *cellula*, cell, living cell, dim. of L. *cella*, small room, compartment, cell of a honeycomb): generally applied to the cells of yeasts that by budding produce one or several buds that constitue daughter cells. In the ascomycetes the diploid cell of the uncinulum or crozier is known as the ascus mother cell, which after meiosis forms the haploid nuclei of the future ascospores, as is observed, e.g., in *Haematonectria* (=*Nectria*) *haematococca* (Hypocreales).

mother of vinegar (L. *mater*, genit. *matris* < Gr. *máter*, mother; L. *vinum*, wine + *acre*, sour): a thick film or scum, fibrous and gelatinous, that grows on the surface of wines and ciders during its conversion to vinegar. The film consists of a cellulosic matrix synthesized by the bacterium *Acetobacter aceti* subsp. *xylinum*, which oxidizes the ethanol to acetic acid. In the process of fermentation, this bacterium grows symbiotically, embedded in the cellulosic matrix, with various species of yeasts, especially *Saccharomyces* sp., *Pichia membranifaciens* (Saccharomycetales), *Candida guilliermondii* and *Kloeckera apiculata* (cryptococcaceous asexual yeasts), although the microbial composition will vary according to the type of substrate, the conditions of the fermentation process and the origin. The mother of vinegar, which is traditionally produced in many parts of the world, is closely related to various types of microbial films or scums that also are cultivated in order to produce traditional fermented beverages, such as the *tea fungus* (of Japan and Indonesia), which is not developed in wine but in an infusion of tea, also known as *kombucha*, *teeschwamm*, *wunderpilz*, *hongo*, *fungus Japonicus*, *Manchurian tea*, and other names that are used to refer to this slightly sweet and carbonated beverage, with an acetic acid flavor, that is widely consumed in China, Russia, Japan, Poland, Bulgaria, Germany, Manchuria and Indonesia. Another cellulosic film, which also is produced by the same microorganism involved in the formation of mother of vinegar and the tea fungus (*A. aceti* subsp. *xylinum*), is the so-called *Philippine scum*, which is cultivated on coconut milk, to prepare a sweet beverage very popular in the Philippines. Cf. **tea fungus** and **tibi grains**. See **fermentation**.

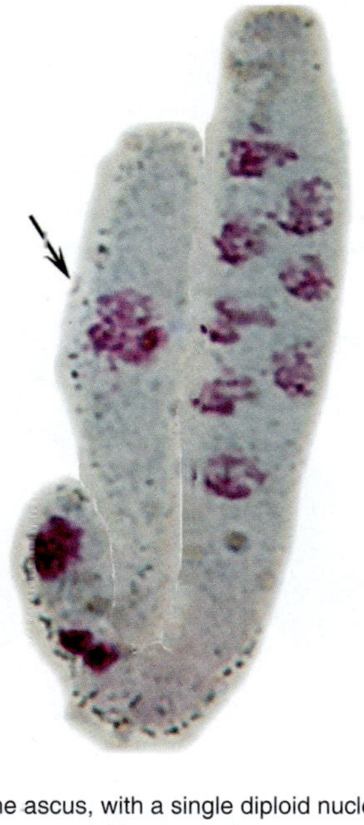

Mother cell of the ascus, with a single diploid nucleus, and a young ascus, with eight haploid nuclei, in *Haematonectria* (=*Nectria*) *haematococca*, x 1,840 (*RTH*).

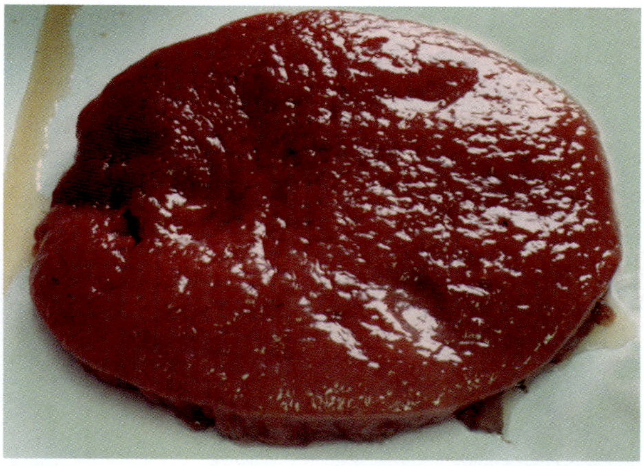

Mother of vinegar developed on red wine, x 0.2 (*MU*).

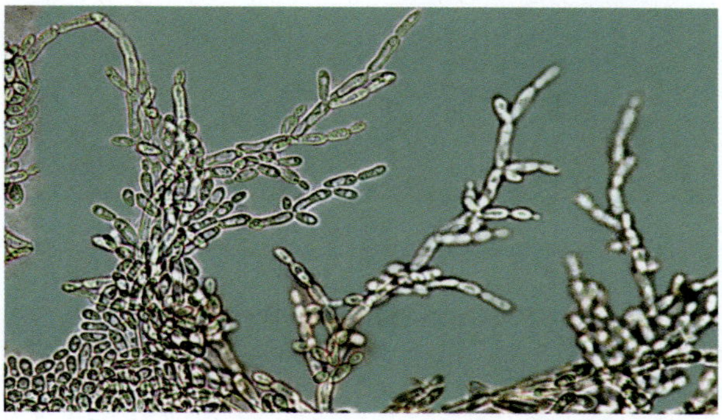

Pseudomycelium of *Pichia membranifaciens* on agar, one yeast of the **mother of vinegar**, x 680 (*MU*).

mucedinaceous (L. *mucedinaceae* < old gen. *Mucedo*, of the Mucorales + suf. *-aceus* > E. *-aceous*, of or pertaining to): equal to **mucedinoid**.

mucedinoid, **mucedinous** (L. *mucedus*, *mucidus*, moldy < *muceere*, to become moldy < *mucus*, mucus + *-oide*, like < Gr. *-oeides*, similar to): applied to fungi that have a white, cottony, loose, filamentous mycelium, as well as to hyaline or more or less intensely colored hyphae, conidia and conidiophores characteristic, e.g., of *Mucor mucedo* (Mucorales), *Botrytis* and *Verticillium* (moniliaceous asexual fungi). See **moniliaceous**.

mucid (L. *mucedus*, *mucidus*, moldy, old, rancid < *mucus*, mucus): moldy; mucilaginous or mucous when moistened.

mucilaginous, **slimy** (LL. *mucilaginosus* < *mucilagin*, mucilage + L. suf. *-osus* > E. *-ous*, what has, abundant in, with the properties of; ME. *slyme* < OE. *slim*, slime): containing mucilage or mucus, or some of its properties; viscous, sticky or mucous when wet, due to its chemical composition (mucic acid, pentoses and hexoses); in water the mucilage becomes viscous or swells to form a gelatinous pseudosolution. These materials are present in many fungi, e.g., in members of Acrasiomycota, Dictyosteliomycota and Myxomycota, which are known as the mucilaginous fungi or slime molds (fungus-like microorganisms). The colonies of *Cryptococcus albidus* (cryptococcaceous asexual yeasts) and *Aureobasidium pullulans* (dematiaceous asexual fungi) also are mucilaginous or viscous.

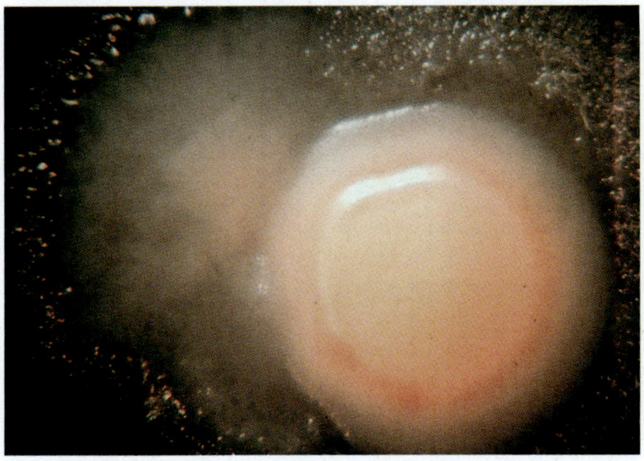

Mucilaginous colony of *Aureobasidium pullulans* on agar, x 3 (*MU*).

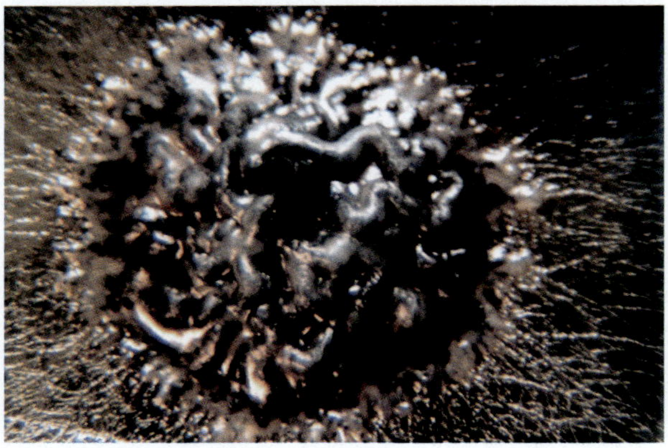

Mucilaginous colony of *Aureobasidium pullulans* on agar, x 2.5 (*MU*).

mucoid, **mucous** (L. *mucus*, mucus + suf. *-oide*, like < Gr. *-oeídes*, similar to; or + L. suf. *-osus* > OF. *-ous*, *-eus* > E. *-ous*, what has, abundant in, with the properties of): with the appearance and consistency of mucilage.

mucormycosis, pl. **mucormycoses** (< gen. *Mucor* < L. *mucor*, mold < *muceere*, to be moldy + Gr. *mýkes*, fungus + suf. *-osis*, state or condition): an opportunistic infection of the rhinocerebral tissues, from which it may disseminate to the brain and meninges, therefore causing the death of the patient. Other types of mucormycosis vary from the subcutaneous (localized) to the gastrointestinal (systemic). The most common causal agents are the widely distributed *Mucor circinelloides*, *Rhizomucor pusillus* (=*M. pusillus*), *M. racemosus*, *Rhizopus oryzae* (=*Rh. arrhizus*) and *Mycocladus* (=*Absidia*) *corymbifera* (Mucorales). These types of infections are more common in immunosupressed patients (due to chemotherapy with corticosteroids and antimicrobial substances). See **mycosis**.

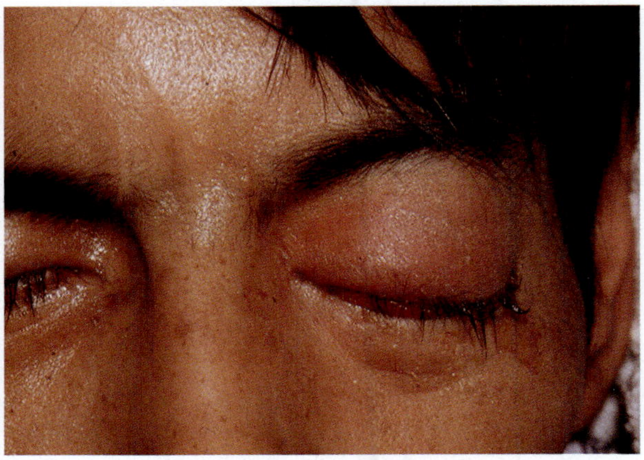

Mucormycosis of the eyes of a young man, caused by *Rhizopus oryzae* (=*Rh. arrhizus*), x 0.5 (*AB*).

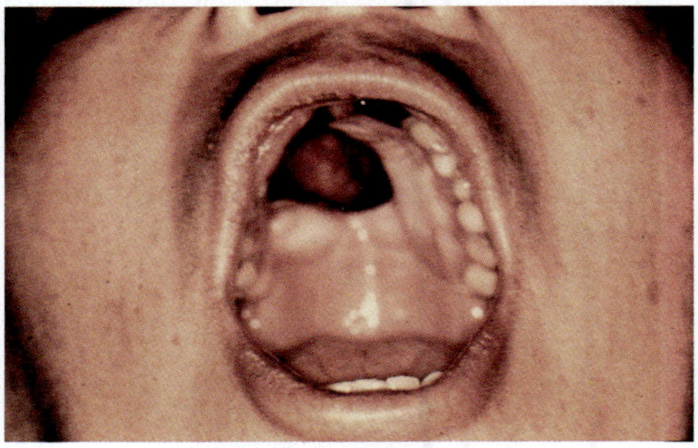

Mucormycosis (palatal perforation) in an immunosuppressed patient, caused by *Rhizopus oryzae* (=*Rh. arrhizus*), x 0.5 (*RLM* & *EM*).

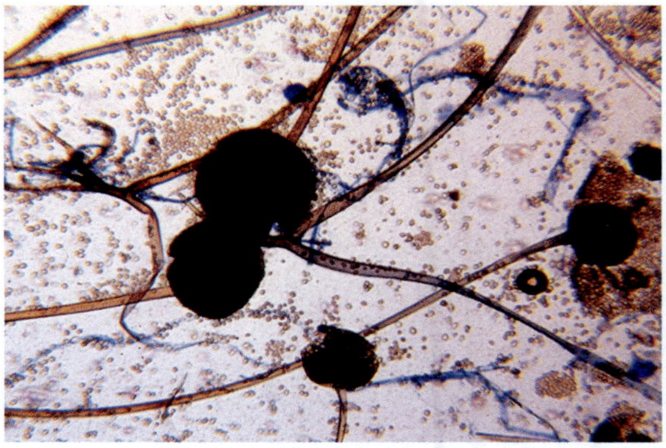

Sporangia of *Rhizopus oryzae* (=*Rh. arrhizus*), an etiological agent of **mucormycosis**, x 150 (*AB*).

mucronate (L. *mucronatus* < *mucro*, genit. *mucronis*, point + suf. *-atus* > E. *-ate*, provided with or likeness): ending abruptly in a point, often sharp, like the hyphopodia of *Meliola caseariae* (Meliolales), the pileus of the basidiocarps of *Phaeocollybia*, *Weraroa* and *Galeropsis*, or the cheilocystidia of *Mycena haematopus* and *Panaeolus cyanescens* (Agaricales).

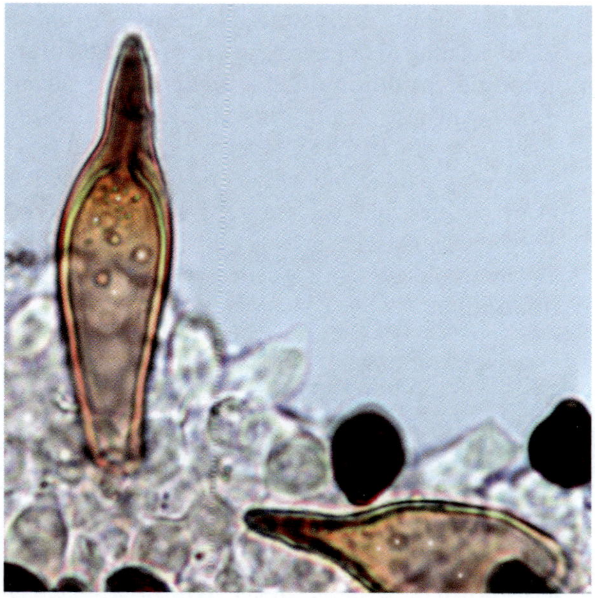

Mucronate cystidia, and basidiospores, of the hymenium of *Panaeolus cyanescens*, x 700 (*EAA*).

multi- (L. *multus*, many): a pref. employed in terms such as *multiallelic*, with more than two alleles per chromosomic locus; *multiaxial*, with many axes; *multifarious*, arranged in several rows, etc., and in the following words.

multiallelic (L. *multus*, many + Gr. *allélon*, one and another, one and the other, reciprocally + Gr. suf. *-íkos* > L. *-icus* > E. *-ic*, pertaining to or relating to): also called *multiple allelomorph*, it refers to a system of sexual incompatibility in which there are more than two alleles per locus on a chromosome. See **allelo-**.

multiaxial (L. *multus*, many + *axialis*, relating to the axis < *axis*, axis + suf. *-alis* > E. *-al*, pertaining to or relating to): refers to a type of structure that has several axes, more or less radially branched and close together, so that they meet in a general compact mass. The fruiting bodies of some ascomycetes, such as *Cordyceps sobolifera* (Hypocreales), and basidiomycetes, such as *Clavicorona pyxidata*, *Ramaria flava* and *R. stricta* (Aphyllophorales), are multiaxial, since their stipes are divided into many branches or axes. Cf. **uniaxial**.

Multiaxial perithecial stromata of *Cordyceps sobolifera*, x 0.5 (*RV*).

Multiaxial basidiocarp of *Clavulinopsis* sp., x 1 (*RV*).

Multiaxial basidiocarp of *Ramaria* sp., x 1 (*TRO*).

Multiaxial basidiocarp of *Ramaria* sp., x 1 (*RV*).

Multiaxial basidiocarp of *Ramaria* sp., x 1 (*TH*).

Multiaxial basidiocarp of *Ramaria* sp., x 1 (*RV*).

multifarious (L. *multifarius*, occuring in several rows < *multi* < *multus*, many + suf. *-farius* < *farior*, to speak < Gr. *phánai*, to speak; used in the sense of being arranged into several rows): equal to **polystichous**.

multifid (L. *multifidus*, multifissured < *multi* < *multus*, many + *findo* < *fidi*, to split, to part): divided into several lobes or lobules. Cf. **bifid**.

multiguttulate (L. *multus*, many + *guttulatus*, with many droplets < *guttula*, dim. of *gutta*, drop, spot similar to a drop + suf. *-atus* > E. *-ate*, provided with or likeness): containing many droplets or globules of oil. See **guttulate**.

multilateral (L. *multilateralis*, with many sides < *multi* < *multus*, many + *later*, *latus*, side, flank + suf. *-alis* > E. *-al*, relating to or pertaining to): see **multipolar**.

multilocular, **multiloculate** (L. *multilocularis* < *multi* < *multus*, many + *loculus*, dim. of *locus*, cavity, locule + suf. *-aris* > E. *-ar*, like, concerning; or + suf. *-atus* > E. *-ate*, provided with or likeness): with several or many cavities, like the ascostroma of *Elsinoë* and *Myriangium* (Myriangiales), which have numerous ascigerous locules. Also called **plurilocular**. Cf. **unilocular**.

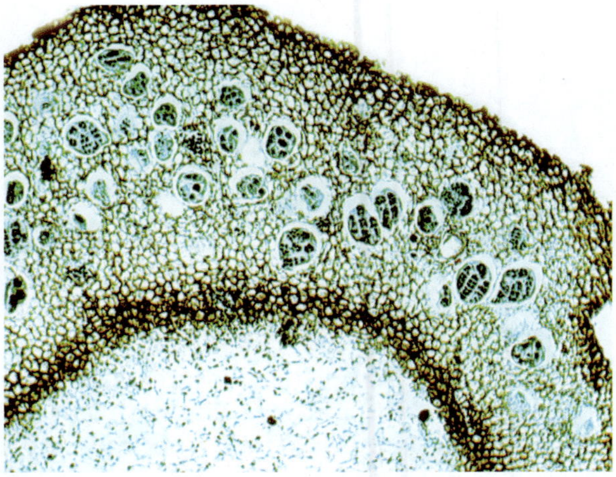

Multilocular ascostroma of *Myriangium duriaei*, seen in longitudinal section, x 80 (*RTH*).

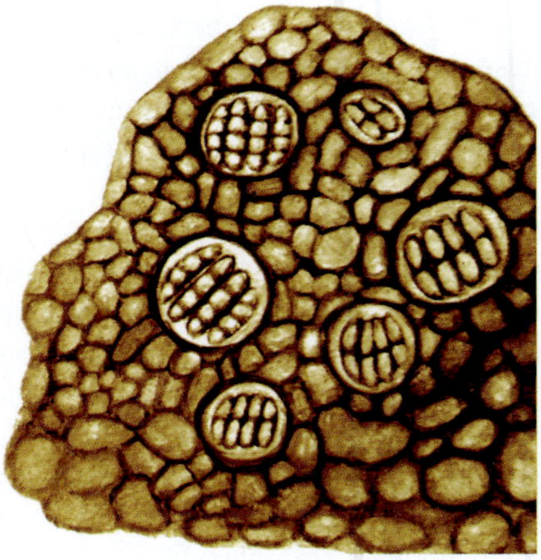

Multilocular ascostroma of *Elsinoë ampelina*, x 400 (*MU*).

multiperforate (L. *multus*, many + *perforatus*, perforate): applied to a septum with many pores that interconnect two adjacent cells, as occurs in some moniliaceous conidial fungi, e.g., *Geotrichum*.

multipileate (L. *multus*, many + *pileatus*, provided with a pileus or hat < *pileus*, pileus, cap, hat, helmet + suf. *-atus* > E. *-ate*, provided with or likeness): a type of fruiting body development characterized by the formation of three or more secondary growths composed of sterile tissue, which originate from a central stipe. Each of the secondary stipes or pedicels generates its own hymenial tissues, whether a thin layer or a concentrated glebal mass. Characteristic of *Clathrus* and *Aseroë*, among other Phallales of the family Clathraceae. Cf. **monopileate**.

Multipileate type of fruiting body development, characteristic of *Clathrus* sp. Diagram of a longitudinal section, x 1(*MU*).

multipileate development (L. *multus*, many, multiple + *pileatus*, with a hat < *pileus*, pileus, cap, hat, helmet + suf. *-atus* > E. *-ate*, provided with or likeness; OF. *desveloper* > MF. *développer* + L. *-mentum* > E. *-ment*, denoting action or resulting state): *Agaricales*. Development in which many individual stalks are formed from a single stalk; each of the individual stalks has a small pileus, with its own hymenial tissue.

multiplication (L. *multiplicatio*, genit. *multiplicationis*, action and effect of multiplying): equivalent to asexual reproduction or vegetative multiplication, because the organisms reproduce without recourse to sex. See **conidiogenesis**, **schizogenesis** or **bipartition**, and **budding**.

multipolar (L. *multus*, many, multiple + *polaris*, polar < *polus*, pole – suf. *-aris* > E. *-ar*, like, concerning): syn. of **multilateral**. **1**. The formation of multiple buds on the surface of the mother cell in certain yeasts, like those of the gen. *Candida*, *Torulopsis* (cryptococcaceous asexual yeasts) and *Hansenula* (Saccharomycetales). **2**. The formation of several germ tubes during ascospore germination, as occurs in *Leptosphaerulina* (Dothideales), or of conidia, as in *Alternaria* (dematiaceous asexual fungi). Cf. **bipolar** and **monopolar**.

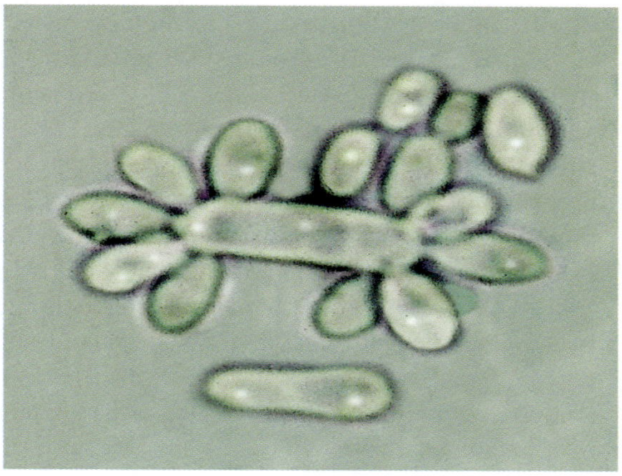

Multipolar budding of the somatic cells of *Torulopsis taboadae*, x 1,500 (*MU*).

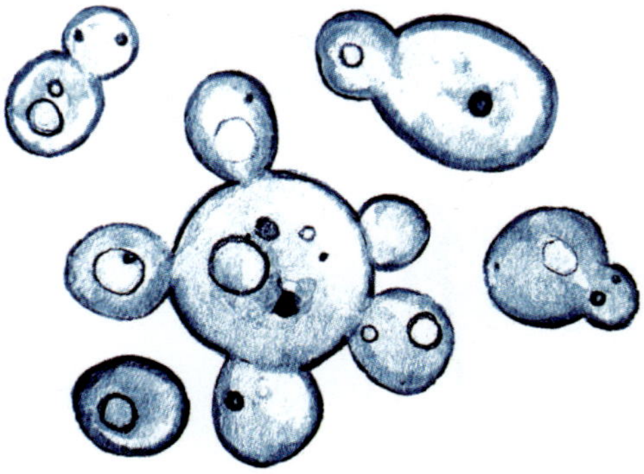

Multipolar budding of the somatic cells of *Torulopsis taboadae*, x 1,500 (*MU*).

Multipolar germination of a conidium of *Alternaria* sp., x 1,200 (*MU*).

multiseptate (L. *multus*, many + *septatus*, septate, partitioned < *septum*, septum, partition + suf. *-atus* > E. *-ate*, provided with or likeness): with many septa or dissepiments; applied to hyphae, spores and other structures.

multisporate, **multisporous** (L. *multus*, many + Gr. *spóros*, spore + L. suf. *-atus* > E. *-ate*, provided with or likeness; + L. suf. *-osus* > E. *-ous*, having, with the qualities of): having many spores. Equal to **myriosporous**. Cf. **bisporate** and **unispored**.

multivesicular bodies (L. *multus*, many + *vesiculatus*, provided with vesicles < *vesicula*, dim. of *vesica*, bladder + suf. *-atus* > E. *-ar*, provided with or likeness; ME. < OE. *bodig*, body): small vesicles delimited by a membrane that originate from the endoplasmic reticulum in *Monilinia* (=*Sclerotinia*) *fructigena* (Helotiales); they are possibly related to the secretion of extracellular enzymes.

muricate, **muricated** (L. *muricatus*, sharp-pointed, pierced < *murex*, genit. *muricis*, sharp-pointed stone, prickle + suf. *-atus* > E. *-ate*, provided with or likeness): covered with short, hard, sharp points, like the cystidia of *Inocybe* and *Melanoleuca melaleuca* (Agaricales), which are incrusted with crystals of oxalate.

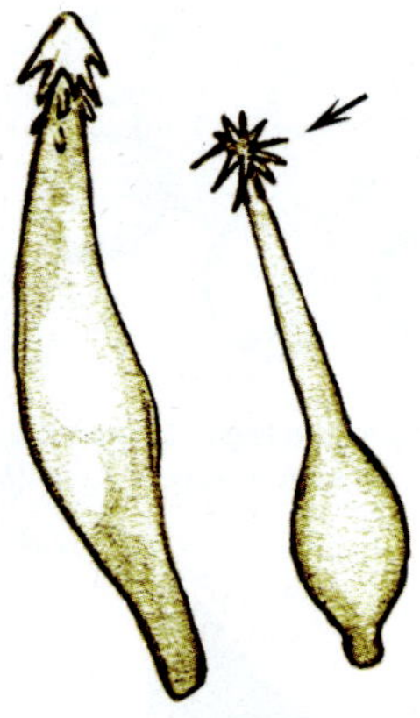

Muricate cystidia of the hymenium of *Melanoleuca* sp., x 1,360 (*MU*).

muriculate (L. *muriculatus*, dim. of *muricatus*, sharp-pointed, pierced < *murex*, genit. *muricis*, sharp-pointed stone + suf. *-atus* > E. *-ate*, provided with or likeness): delicately muricate; with minute prongs.

muriform (L. *muriformis*, in the form of a wall < *murus*, wall + suf. *-formis* < *forma*, form, shape): with transverse and longitudinal septa, arranged like the partitions of a brick wall, like the conidia of *Pithomyces chartarum* (dematiaceous asexual fungi) and *Epicoccum nigrum* (=*E. purpurascens*), tuberculariaceous asexual fungi. See **dictyospore**.

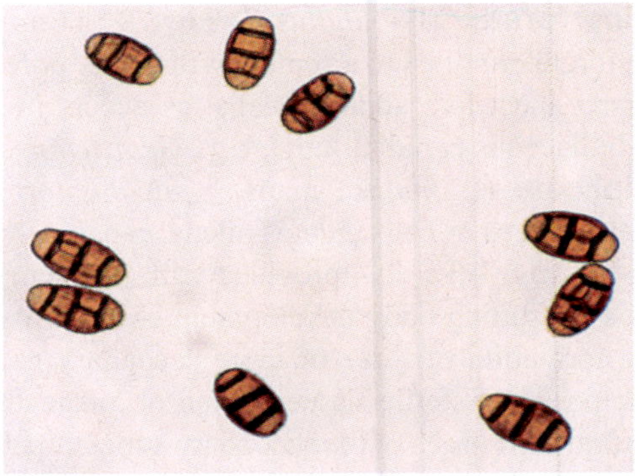

Muriform conidia of *Pithomyces chartarum*, x 800 (*CB*).

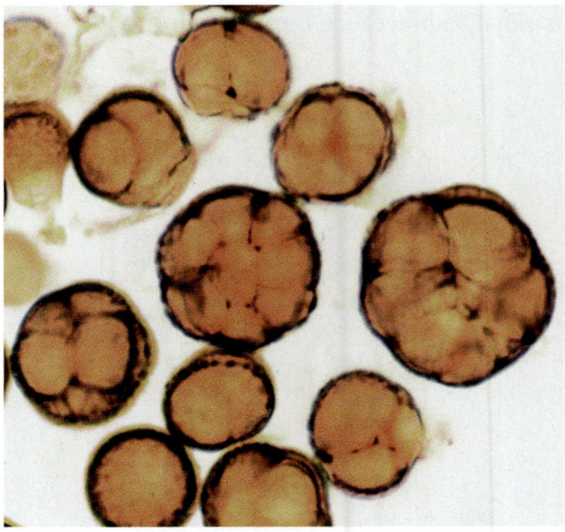

Muriform conidia of *Epicoccum nigrum* (=*E. purpurascens*), x 800 (*CB*).

muriform cell (L. *muriformis* < *murus*, wall + suf. *formis* < *forma*, shape; NL. *cellula*, cell, living cell, dim. of L. *cella*, small room, compartment, cell of a honeycomb): a dark, thick-walled, dictyosporous cell, frequently referred to as a *sclerotic cell*, that is found in tissues affected by chromoblastomycosis.

murogenous (L. *murus*, wall + Gr. *génos*, origin): *Conidiogenesis*. A type of blastic conidium development in which the tip of a hypha enlarges and is then delimited by a septum. The width of the point of attachment of the conidium is equal to that of the hypha.

muscardine fungus, pl. **fungi** (Fr. *muscardin* < L. *muscus*, musk < Gr. *móskos*, akin to Skt. *mushka*, testicle, originally, a small mouse; L. *fungus*, fungus < Gr. *spóngos*, *sphóngos*, sponge; musk is a substance contained in an egg-shaped sac situated beneath the skin of the abdomen of the male musk deer. Dry musk has a penetrating odor and is used as a fixative in perfumery. However, the use of musk in this sense may derive from plant musk; the fruit of the muscadine grape, *Vitus rotundifolia*, e.g., has a musky aroma): the muscardinic fungi are those that parasitize silkworms, i.e., they are fungal pathogens of silkworms; the green muscardine fungus is *Metarhizium anisopliae*, the yellow muscardine fungus is *Paecilomyces farinosus*, and the white muscardine fungus is *Beauveria bassiana*, all of the moniliaceous asexual fungi.

muscicole, **muscicolous** (L. *muscus*, moss + suf. *-cola*, inhabitant): living on or among mosses. See **bryophilous**.

mushroom (ME. *musseroum* < MF. *mousseron* < LL. *mussirion*, mushroom): *Basidiomycetes*. **1**. The name generally applied to the fruiting body of the higher fungi, especially edible species (an agaric or a bolete), although it is also used for the fruiting bodies of poisonous and hallucinogenic species that typically are parasol-shaped, with a pedicel or stipe bearing a pileus or cap. **2**. Any agaric; a macroscopic fungus with a distinctive fruiting body (whether epigeous or hypogeous), sufficiently large to be seen with the unaided eye and collected by hand. Cf. **mold**.

mushroom stones (ME. *musseroum* < MF. *mousseron* < LL. *mussirion*, mushroom; ME. < OE. *stan*, stone): a wide variety of stone sculptures in the shape of a mushroom (more than 300 are known), in particular of agarics, with different effigies sculpted on the pedicel or support (including anthropomorphic or animal figures). These sculptures have been related to the cult of hallucinogenic fungi, which the Maya and other ethnic groups of Mesoamerica have worshiped since before the time of Cortez, especially in Guatemala and the southeast of Mexico. See **mycolatry**.

Mushroom stones from the Mayan Culture. From left to right: a sculpture with a feline effigy (Early Preclassic, Kaminaljuyú, Guatemala); an anthropomorphic stone mushroom (Middle Preclassic, Guatemala), and a sculpture with an animal effigy (Early and Middle Preclassic, Kaminaljuyú, Guatemala), x 0.2 (*MU*).

musiform (< gen. *Musa*, banana + L. suf. *-formis* < *forma*, shape): shaped like a banana, like the basidiospores of *Exobasidium* (Exobasidiales).

must (L. *mustum*, juice): juice expressed from fruit, especially grapes, before fermenting it to make wine. Also applied to the fermentable liquid obtained from agave leaves that are crushed and cooked to make tequila and mescal.

mutagen (L. *mutation*, genit. *mutationis*, change, mutation + *génos*, origin < *gennáo*, to engender, produce): a physical, chemical or biological agent capable of producing a mutation: alteration of the genes, chromosomes or hereditary constituents of the plasma. Among the most important physical mutagenic agents are the penetrating radiations (e.g., X-rays and gamma rays). Also mutagenic are sharp changes in cold or heat. Among the chemical agents there is a whole series of very disparate compounds: formaldehyde, urethane, sulphonamides, mustard gas, and some carcinogenic phenanthrene substances, among others. Also called *mutationigerous* (L. *mutation*, genit. *mutationis* + suf. *-ger*, *-gero* < *gerere*, to produce, engender).

mutant (L. *mutare*, to change): an individual (or strain) with a genotypic character different from that of its progenitors, caused by an induced or natural mutation.

muticate, **muticous** (L. *muticus*, cut, mutilated, without hair or beard): refers to a structure that lacks a point; blunt, like the conidia of *Drechslera*, among other dematiaceous asexual fungi.

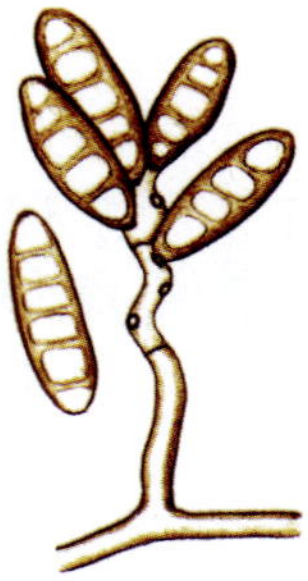

Muticate conidia on a conidiophore of the *Drechslera*, asexual state of *Cochliobolus bicolor*, x 700 (*MU*).

Muticate conidia of *Drechslera* sp., x 1,360 (*CB*).

mutualism (L. *mutuus*, mutual, reciprocal + suf. *-alis* > E. *-al*, pertaining to or relating to + Gr. suf. *-ismós* > L. *-ismus* > E. *-ism*, state, phase, tendency, action): a type of association between two living beings that provides mutual benefits; it represents a type of mutualistic symbiosis, such as exists in the lichens. Cf. **antagonism**, **commensalism**, **helotism**, **metabiosis**, **parabiosis**, **parasymbiosis**, **symbiosis** and **synergy**.

mutualistic symbiosis, pl. **symbioses** (L. *mutuus*, mutual, reciprocal + suf. *-alis* > E. *-al*, pertaining to or relating to + NL. suf. *-istic* > E. *-istic*, concerning; Gr. *symbíosis*, life in common): see **symbiosis**.

myc-, **mycet-**, **-mycete**, **myceto-** (L. pref. *myc-*, suf. *-mycete*, < Gr. *mýkes*, genit. *mýketos*, fungus): a fungus or relating to or pertaining to a fungus; prefixes and sufixes that are employed in forming terms such as *mycangium*, *mycetism*, *ascomycete*, *basidiomycete*, *mycetophagus*, etc.

mycangium, pl. **mycangia** (Gr. *mýkes*, fungus + *angeîon*, vessel, receptacle): each one of the small pouches, located in various parts of the exoskeleton of certain insects (ambrosia beetles and wood wasps), that contain the fungal spores (or yeast-like cells) with which the particular insect has established a mutualistic relationship. In the wood boring beetles (*Xyleborus*), the symbiotic fungi contained in the mycangia (and which, depending upon the species of beetle, can be located in the base of each elytron or in the oral cavity) include *Graphium* spp. (stilbellaceous asexual fungi) and *Acremonium* (=*Cephalosporium* spp., moniliaceous asexual fungi). Species of *Stereum* (Aphyllophorales) are disseminated by wood wasps.

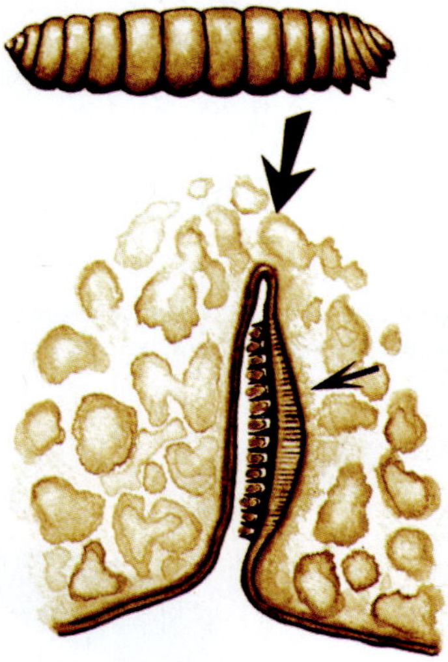

Mycangium of *Stereum* sp. between the first two abdominal segments of a wood wasp larva, x 28 (*MU*).

mycelial (NL. *mycelium*, mycelium < Gr. *mýkes*, fungus + L. suf. *-alis* > E. *-al*, relating to or concerning): concerning or relating to the mycelium.

mycelial crosspiece (NL. *mycelium*, mycelium < Gr. *mýkes*, fungus + L. suf. *-alis* > E. *-al*, relating to or belonging to; ME. < OE. *cros* < ON. *kross* < OIr. *cros* < L. *crux*, cross + ME. *pece* < OF. < VL. *pettia*, piece, portion of something): a portion of a hypha, in the shape of a cross, which *Phymatotrichopsis omnivora* (moniliaceous asexual fungi) forms in culture.

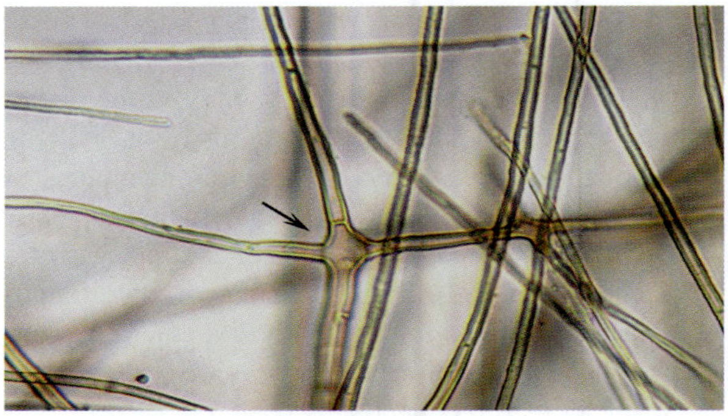

Mycelial crosspiece of *Phymatotrichopsis omnivora* in culture, x 920 (*JAS*).

mycelial strand (NL. *mycelium*, mycelium < Gr. *mýkes*, fungus + L. suf. *-alis* > E. *-al*, relating to or belonging to; ME. *strond*, fibers or filaments twisted, plaited, or laid parallel to form a unit for further twisting or plaiting into yarn, thread, rope, or cordage): a compact group of parallel hyphae that form a bundle or packet, with apical growth analogous to that of vascular plant roots; mycelial strands are present in the majority of basidiomycetes, as well as in other groups. When these mycelial strands have a more complex structure, with a hard cortex, in which the hyphae lose their individuality, they are called *rhizomorphs*.

Mycelial strands of *Phymatotrichopsis omnivora* (moniliaceous asexual fungi) on walnut wood, x 1 (*JAS*).

Mycelial strands of *Marasmius* sp. (Agaricales) between the basidiocarps, growing on bark, x 1 (*RV*).

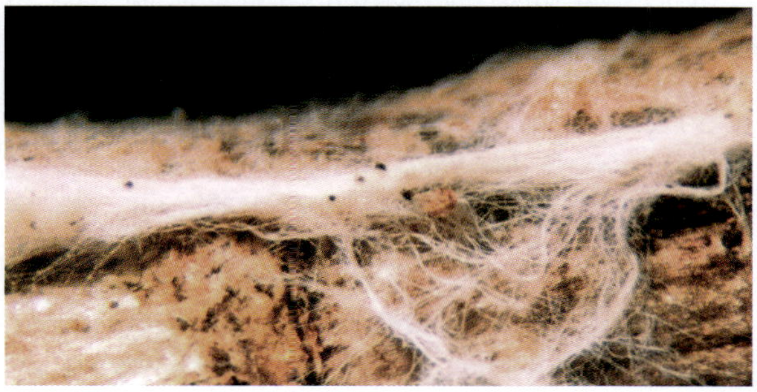

Mycelial strand of a basidiomycete, on bark, x 2 (*SA*).

mycelioid (NL. *mycelium*, mycelium < Gr. *mýkes*, fungus + L. suf. *-elis*, relating to or concerning + L. suf. *-oide*, like < Gr. *-oeídes*, similar to): with the characteristics of a mycelium, similar to a mycelium, like the perithecial appendages of *Blumeria* (=*Erysiphe*) *graminis* (Erysiphales).

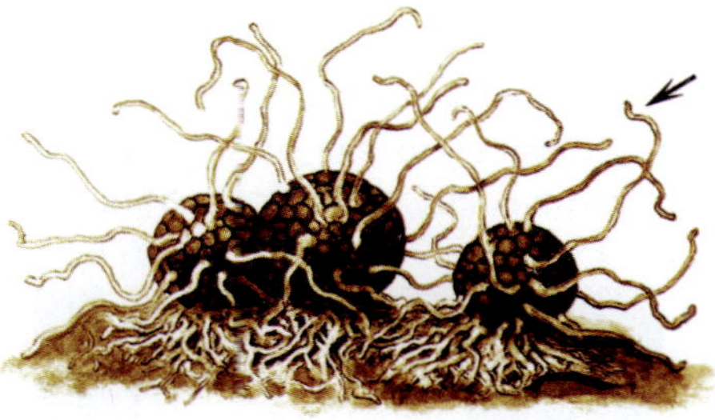

Mycelioid perithecial appendages of *Blumeria* (=*Erysiphe*) *graminis*, x 190 (*MU*).

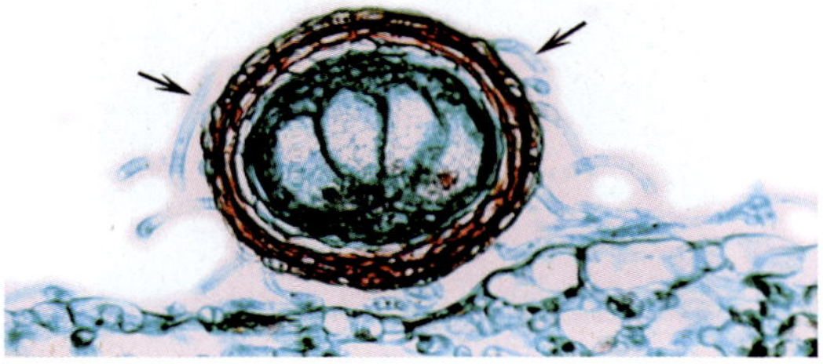

Mycelioid appendages of *Blumeria* (=*Erysiphe*) *graminis*, x 230 (*RTH*).

mycelium, pl. **mycelia** (NL. *mycelium* < Gr. *mýkes*, fungus + L. suf. *-elis*, concerning + L. dim. suf. *-ium*): the entire mass of hyphae that constitutes the vegetative body or thallus of a fungus.

Mycelium of a colony of *Aureobasidium pullulans* on agar, x 3 (*MU*).

Mycelium at the base of the basidiocarps of *Cotylidia aurantiaca*, growing on a twig, x 1 (*EPS*).

myceloconidium, pl. **myceloconidia** (NL. *mycelium* < Gr. *mýkes*, fungus + L. suf. *-elis*, concerning + L. dim. suf. *-ium* + Gr. *kónis*, powder, dust + *-ídion*, dim. suf.): see **stylospore**.

mycenoid (< gen. *Mycena* < Gr. *mýkes*, fungus + *kenós*, hollow + L. suf. *-oide*, like < Gr. *-oeídes*, similar to): *Agaricales*. A type of fruiting body (13 principal types are recognized) with gills adhered to the stipe, which is cartilaginous, central and lacks a ring and volva, like that of *Mycena* and some species of *Marasmius*, *Nolanea*, *Psilocybe* and *Panaeolus*, among other gen. The mycenoid type is similar to the collybioid type but differs in its conical to campanulate pileus with an initially decurved margin.

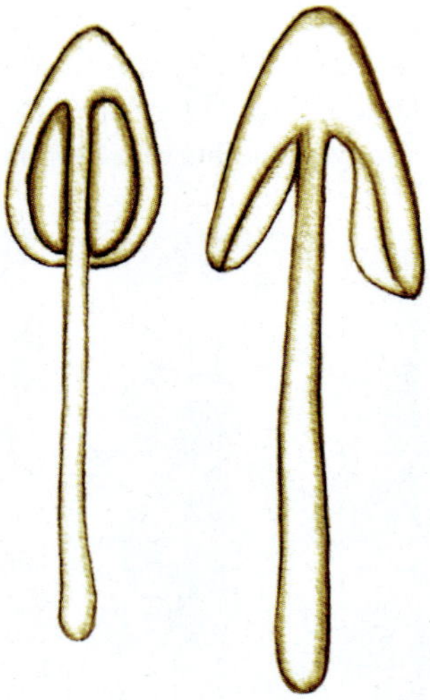

Mycenoid type of basidiocarps, represented by *Mycena* sp. Diagrams of longitudinal sections, x 1 (*MU*).

Mycenoid type of basidiocarps of *Mycena epipterygia*, x 1 (*RV*).

mycetarium (Gr. *mýkes*, genit. *mýketos*, fungus + L. suf. *-arium*, the place of something): a collection of fungi, usually preserved by drying, arranged taxonomically, with appropiate collection data. An etymologically more appropiate term than herbarium, considering that fungi are not plants.

-mycetes (Gr. *mýkes*, genit. *mýketos*, fungus > L. genit. *mycetem* > *mycetes* > E. *-mycete*): suf. derived from Greek, employed as the ending for taxonomic classes, e.g., Zygomycetes, Archiascomycetes, etc.

-mycetidae (Gr. *mýkes*, genit. *mýketos*, fungus > L. genit. *mycetem* > *mycetes* + *-idae*, L. fem. pl. adjectival suf.): ending of a taxonomic subclass, as in Pyrenomycetidae, Discomycetidae, etc.

mycetinous (Gr. *mykétinos*, pertaining to or relating to fungi < *mýkes*, genit. *mýketos*, fungus + suf. *-inos* > L. *-inus* > E. *-inous*, a suf. added to the root of a noun to form adjectives that denote the material or the source of something): equivalent to **fungal**; pertaining to or relating to the fungi. It is used, e.g., in terms such as *ascomycetinous*, *basidiomycetinous*, etc., corresponding to the words *ascomycetous* and *basidiomycetous*.

mycetism, **mycetismus**, pl. **mycetismi** (L. *mycetismus*, mycetism < Gr. *mýkes*, genit. *mýketos*, fungus + suf. *-ismós* > L. *-ismus* > E. *-ism*, state, phase, tendency, action): intoxication caused by the ingestion of certain fungi. Recognized mycetisms include phalloidian, paraphalloidian, muscarinic, gastrointestinal, inconstant or conditional, and cerebral, which are induced respectively by *Amanita phalloides*, *Cortinarius orellanus*, *Amanita muscaria*, *Russula emetica*, *Coprinopsis atramentaria* (=*Coprinus atramentarius*) and *Psilocybe caerulescens* (Agaricales), to cite only some of the many species involved. Cf. **mycotoxicosis**.

Mycenoid type of basidiocarps of *Hypholoma marginatum*, x 1 (*TRO*).

Basidiocarp of *Amanita* sp., a white species that causes phalloidian **mycetism** on humans, x 0.3 (*RV*).

Basidiocarp of *Amanita pantherina*, one of the species that causes muscarinic **mycetism** on humans, x 0.3 (*RV*).

Basidiocarp of *Amanita virosa*, one species that causes phalloidian **mycetism** on humans, x 0.3 (*RV*).

Basidiocarps of *Amanita muscaria*, which can cause muscarinic **mycetism** on humans, x 0.5 (*TH*).

Basidiocarps of *Omphalotus olearius* (Agaricales), a species that may cause gastrointestinal **mycetism** on humans, x 0.5 (*RV*).

mycetization (Gr. *mýkes*, genit. *mýketos*, fungus + suf. *-ationem* > E. *-ation*, action, state or condition, result): the action and effect of becoming fungal; refers to the infection of plant roots by the hyphae of mycorrhizogenous fungi, which invade them to form mycorrhizae (mycorrhization).

myceto-, **mycet-** (Gr. *mýkes*, genit. *mýketos*, fungus): pertaining to fungi. Equivalent to **myco-**.

mycetobiont (Gr. *mýkes*, genit. *mýketos*, fungus + *bíos*, life + *óntos*, a being): applied to the insects, and especially to the coleopterans, whose life is linked to the fungi that are exclusive to them.

mycetocyte (Gr. *mýkes*, genit. *mýketos*, fungus + *kýtos*, cavity, cell): refers to cells that contain symbiotic yeasts, like those that are found in the blind sacs of the digestive tracts of larvae of *Lasioderma*, in which are found *Torulopsis ernobii* and *Symbiotaphrina buchnerii* (=*T. buchnerii*), cryptococcaceous asexual yeasts. See **mycetosome**.

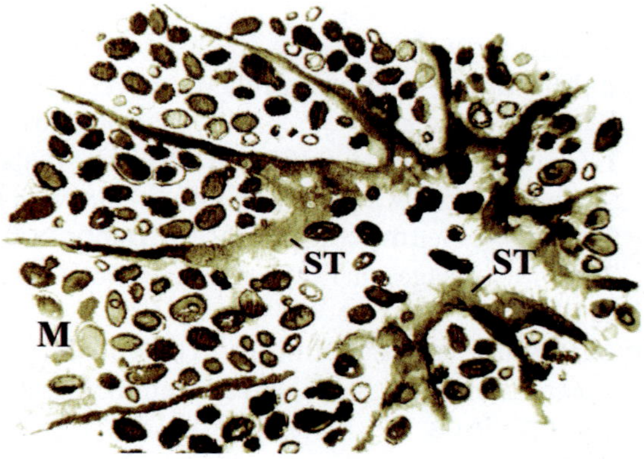

Mycetocytes (endosymbiotic yeasts) of *Symbiotaphrina buchnerii* (=*Torulopsis buchnerii)*. Transverse section of a blind sack of a larva of *Lasioderma*. Note the broad mycetocytes (M) and the narrower sterile columnar cells (ST), the latter with denser protoplasm. In the apical parts of the mycetocytes the cell wall is lysed and its yeast cells have been expelled into the lumen. Drawing based on a transmission electron micrograph by Jurzitza, 1977, x 1,250 (*MU*).

mycetology (Gr. *mýkes*, genit. *mýketos*, fungus + *lógos*, word > E. des. *-logy*, study, treatise): see **mycology**.

mycetoma, pl. **mycetomata** (Gr. *mýkes*, genit. *mýketos*, fungus + *-óma*, suf. that refers to a pathological state, generally a tumor): literally, a fungal tumor. A type of lesion caused by actinomycetes, which are actually actinobacteria (actinomycetoma), or by molds (eumycotic mycetoma), in the subcutaneous tissues of man and higher animals; the organisms generally involved are conidial fungi (e.g., *Madurella* and *Phialophora*, of the dematiaceous asexual fungi) and ascomycetes (such as *Pseudallescheria*, of the Microascales). See **mycosis**.

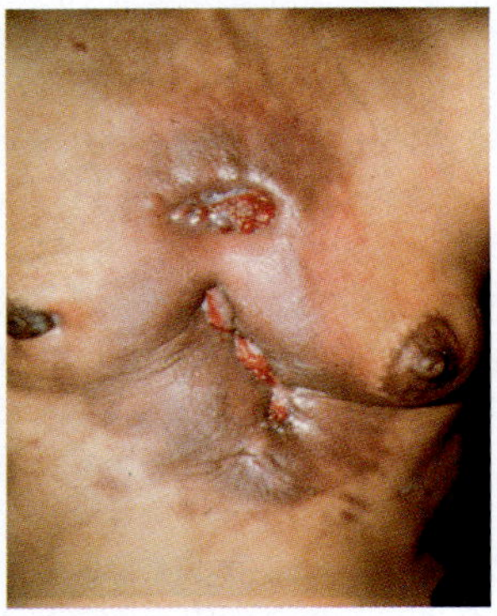

Typical lesions of **mycetoma**, caused by *Madurella mycetomi*, in the thorax of a patient, x 0.2 (*RLM*).

mycetophage, **mycetophagist** (Gr. *mýkes*, genit. *mýketos*, fungus + *phágos*, what eats < *phageín*, to eat): equal to **fungivorous**.

mycetophagy (Gr. *mýkes*, genit. *mýketos*, fungus + *phágos*, what eats < *phageín*, to eat): see **mycophagy**.

mycetophagist, **mycophagist** (Gr. *mýkes*, fungus + *phágos*, what eats < *phágesis*, *phágomai*, to eat): **1**. One who eats fungi, who takes advantage of them as food; especially a gourmet with a predilection for fungi. Also called **mycetophagy**. Cf. **mycophobia**. **2**. A virus (phage) that infects a fungus and causes its lysis.

mycetophile, **mycetophilous** (Gr. *mýkes*, genit. *mýketos*, fungus + *phílos*, have an affinity for): see **mycophilous**.

mycetosome (Gr. *mýkes*, genit. *mýketos*, fungus + *sõma*, body): a sac-shaped structure found in the digestive tract of anobid beetles, which is lined with *mycetocytes* that contain symbiotic yeast cells. See **mycetocyte**.

mycetotheca (Gr. *mýkes*, genit. *mýketos*, fungus + *thêke*, box, case, reservoir): see **mycotheca**.

Mycetozoa (Gr. *mýkes*, genit. *mýketos*, fungus + *zõon*, animal): *Zoology*. An order of the class Sarcodina, phylum Sarcomastigophora, subkingdom Protozoa, kingdom Animalia. In mycology, it is equivalent to the phylum **Myxomycota** (Myxomycetes) of the kingdom Fungi. The term *mycetozoan* (Gr. *mýkes*, genit. *mýketos*, fungus + *zoárion*, animalcule) is an old name for referring to a myxomycete. Recently, it has been recognized that *Amoebidium*, a trichomycete of the order Amoebidiales, which grows in the intestine of arthropods, actually belongs to a group called *Mesomycetozoa* (Gr. *mésos*, intermediate, median), which is closely related to the coanoflagellates. The *mesomycetozoans* together with the coanoflagellates constitute a clade that diverges from the animal line, at a point in animal evolution when the ancestors of the animals were not strongly differentiated from the ancestors of the fungi. Very probably, future analyses will reveal that many other organisms classified as trichomycetes are not fungi but mesomycetozoans. See the Appendix, where there is a comparison between the outline of Alexopoulos *et al*., 1996, modified by Ulloa and Hanlin, 2006, to that of Kirk *et al*., 2008. In the latter classification scheme, the class Myxomycetes of the kingdom Protista is now included as the class Myxogastrea, phylum Amoebozoa, kingdom Protozoa. These taxonomic rearrangements, with the concurrent change of the names of taxa, are part of a constant process of study and evaluation of these groups of organisms, in order to achieve a phylogenetic classification. See **Protozoa**.

mycetozoan (Gr. *mýkes*, genit. *mýketos*, fungus + *zoárion*, animalcule): the name employed, instead of *myxomycete* (class Myxomycetes, phylum Myxomycota, kingdom Protista), by biologists who classify these organisms with the protists (subphylum Mycetozoa, phylum Gymnomyxa, kingdom Protista). The mycetozoans combine, within the same life cycle, characteristics of animals (amoeboid trophic state, heterotrophic, holozoic) and of fungi (sporiferous apparatus). See **Mycetozoa**.

mycin (L. pref. *myc-* < Gr. *mýkes*, fungus + L. suf. *-in*, of or pertaining to; a neutral chemical compound, enzyme or antibiotic): the recommended ending for the names of antibiotics produced by actinomycetes, which are actually actinobacteria (actinomycin, streptomycin, etc.).

myco- (Gr. *mýkes*, fungus): a pref. employed in forming compound terms, such as *mycology* (the study of fungi), *mycobiota* (mycological biota), *mycobacteriosis* (a complex disease entity, formed by the concomitant parasitism, although not necessarily symbiotic, of fungi and bacteria), *mycorrhizogenous* (a fungus capable of forming mycorrhizae), and *mycotic* (pertaining to or relative to mycosis), among others. **Myceto-** is technically the more correct pref., but is less popular.

mycobacteriosis, pl. **mycobacterioses** (Gr. *mýkes*, fungus + L. pref. *bacteri-* < *bacterium*, bacterium + Gr. suf. *-osis*, condition, state): an infection caused by microorganisms of the gen. *Mycobacterium* (Actinomycetales) or by the concomitant parasitism, not necessarily symbiotic, of bacteria and fungi.

mycobiont (Gr. *mýkes*, fungus + *bíos*, life + *óntos*, genit. of *ón*, a being): the fungal component of a lichen; together with the photobiont (whether algal or actinobacterial, depending on the type of lichen), it constitutes the lichen symbiosis. Cf. **phycobiont** and **photobiont**.

mycobiota (Gr. *mýkes*, fungus + NL. *biota*, the living beings of a particular region or period < Gr. *biotós*, life, manner of living < *bíos*, life): fungal life; the ensemble of fungi indigenous to a place or habitat. To avoid the implication that the fungi are considered as plants, it is preferable to use the term *mycobiota* (or *mycetobiota*) rather than *mycoflora* (or *mycetoflora*).

mycocecidium, pl. **mycocecidia** (Gr. *mýkes*, fungus + *kekís*, genit. *kekîdos*, gall): a plant gall or cecidium induced by the parasitism of a fungus, such as the dark warts or tumors produced on potato tubers by *Synchytrium endobioticum* (Chytridiales), the galls formed by *Protomyces* (Taphrinales) on leaves of wild carrot, or the galls of hawthorn and of cypress, caused respectively by the rusts *Gymnosporangium clavipes* and *G. juniperi-virginianae* (Uredinales).

Mycocecidium formed by *Protomyces* sp. on a leaf of wild carrot, x 30 (*RTH*).

Mycocecidium formed by *Gymnosporangium clavipes* on a fruit of hawthorn (*Crataegus*). Note the aecial peridia protruding from the infected fruit, x 1 (*MU*).

Mycocecidium (close-up) formed by *Gymnosporangium clavipes* on a fruit of hawthorn (*Crataegus*). Note the aecial peridia protruding from the infected fruit, x 17 (*MU*).

mycocide (Gr. *mýkes*, fungus + E. des. *-cide* < L. *caedere*, to kill): a substance capable of killing fungi. This term, like *fungistatic*, is a hybrid, with one Greek and one Latin component, for which reason its use is not recommended. It is preferable to use **fungicide** and **mycostatic**, since the first has two Latin components, and the second two Greek components.

mycoclena, pl. **mycoclenae** (NL. *mycochlaena*, fungal cover < Gr. *mýkes*, fungus + *chlaína*, blanket, cover): the fungal investment of a root in an ectotrophic mycorrhiza.

mycocoenosis, pl. **mycocoenoses** (Gr. *mýkes*, fungus + *koinós*, common + suf. *-osis*, condition or state, a process): the total complex of fungi of a plant community and its environmental medium, or in the absence of plants, of some other specific habitat.

mycodegradation (Gr. *mýkes*, fungus + LL. *degradation* < *degradare*, to degrade + suf. *-ion*, action or condition): the degradation of any kind of material such as paper, plastics, textiles, leather, etc., or of organic compounds, by fungi.

mycoderm (Gr. *mýkes*, fungus + *dérma*, skin, membrane): **1**. The membranous film or scum formed on the surface of fermenting liquids, a process involving yeasts; the old gen. *Mycoderma* is now a syn. of the gen. *Saccharomyces*, of the Saccharomycetales. **2**. A fungal growth on the skin that forms a crust, as in some dermatophytoses. **3**. The mother of vinegar, which is a thick scum composed of a cellulosic matrix in which the bacteria are embedded (*Acetobacter aceti* subsp. *xylinum*) that generate the matrix, together with some yeasts (*Saccharomycodes*, of the Saccharomycetales, *Candida*, of the cryptococcaceous asexual yeasts, and others). See **fermentation** and **tea fungus**. **4**. An ectotrophic mycorrhiza with the appearance of a compact tissue.

Mycoderm, a thick scum (*mother of vinegar*) formed by the growth of bacteria and yeasts living symbiotically in a sugary fermentable liquid, x 0.3 (*MU*).

mycodeterioration (Gr. *mýkes*, fungus + LL. *deterioration* < *deterioratus*, make worse + suf. *-ion*, action or condition): the deterioration of any kind of materials such as paper, plastics, textiles, leather, wooden artifacts, lenses, etc., by fungi.

mycodetoxification (Gr. *mýkes*, fungus + ME. < L. *de*, removal, separation + *toxicum*, poison + suf. *-fication*, action or state): the use of fungi in detoxification or removal of a wide variety of pollutants in any medium or in waste or wastewater.

mycodomatium, pl. **mycodomatia** (Gr. *mýkes*, fungus + L. *domatium*, house < Gr. *dóma*, genit. *dómatos*, house, roof): a dwelling in which fungi are lodged; a term used for certain nodules on the roots of *Alnus* and some legumes, apparently related to mycorrhizae.

mycoecology (Gr. *mýkes*, fungus + *oîkos*, house or inn + *lógos*, word > E. des. *-logy*, study, treatise): the ecology of fungi, the science that studies the interrelationships of fungi and their environment.

mycofiltration (Gr. *mýkes*, fungus + ME. < OF. *filtre* < ML. *filtrum*, filter, piece of felt < G. *filtiz*, felted wool + suf. *-ion*, action or condition): the use of fungi for water filtration.

mycoforestry (Gr. *mýkes*, fungus + ME. < ML. *forestis*, forest < L. *foris*, outside + suf. *-ry*, a place for): the use of fungi in forest practices, such as applying inoculum of mycorrhizogenous fungi to enhance forest tree growth.

mycofungicide (Gr. *mýkes*, fungus + L. *fungus*, fungus + E. des. *-cide* < L. *caedere*, to kill): a commercial product based on fungi that is utilized to kill other fungi that are pathogenic to plants important in agriculture, horticulture and forestry. E.g., species of *Trichoderma viride* and *Gliocadium virens* (=*Trichoderma virens*), moniliaceous asexual fungi, are employed in controlling or combating forest diseases; the third species, in particular, has been applied as an antagonist of *Rhizoctonia solani* (agonomycetaceous asexual fungi) and of *Pythium ultimum* (Peronosporales), which are etiologic agents of various diseases of horticultural crops. *Phlebiopsis* (=*Phlebia*) *gigantea* (Aphyllophorales) is utilized as a mycofungicide to control *Heterobasidion annosum* (also of the Aphyllophorales), which causes root rot of pines. See **mycoherbicide**, **mycoinsecticide** and **myconematicide**.

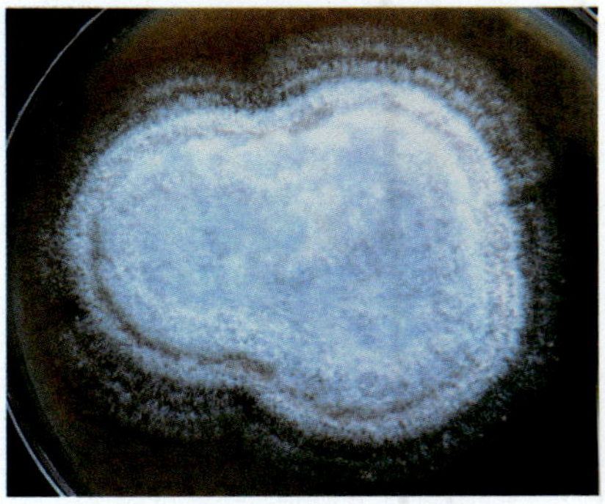

Colony of *Gliocladium virens* (=*Trichoderma virens*) on agar, a fungus utilizedas a **mycofungicide**, x 1 (*MU*).

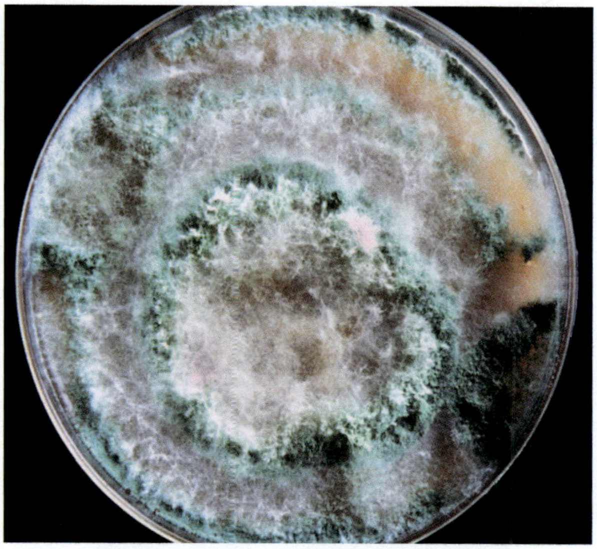

Colony of *Trichoderma viride* on agar, a species utilized as a **mycofungicide**, x 0.6 (*MU*).

mycogenous (Gr. *mýkes*, fungus + *génos*, origin < *gennáo*, to engender, produce): of fungal origin, or living on fungi; e.g., like the basidiocarps of species of *Nyctalis* (Agaricales) that grow on the basidiocarps of *Russula* (Agaricales), or the perithecial stromata of *Cordyceps capitata* (Hypocreales), that grow on ascocarps of *Elaphomyces* (Pezizales). See **mycoparasite**.

Mycogenous perithecial stroma of *Cordyceps capitata* growing on an ascocarp of *Elaphomyces* sp., x 0.8 (*RV*).

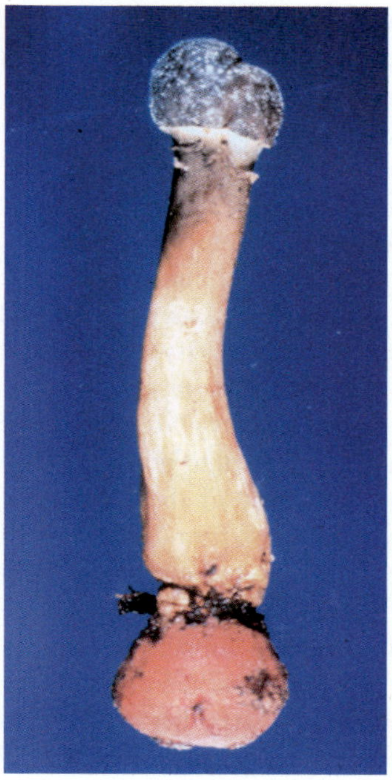

Mycogenous perithecial stroma of *Cordyceps capitata* growing on an ascocarp of *Elaphomyces* sp., x 0.8 (*RTH*).

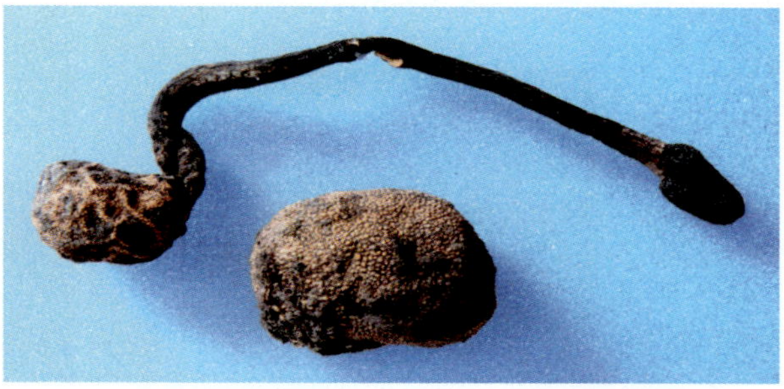

Mycogenous perithecial stroma of *Cordyceps ophioglossoides* growing on an ascocarp of *Elaphomyces granulatus*. Also shown is an uninfected ascocarp of *E. granulatus*, x 0.9 (*MU*).

mycogeography (Gr. *mýkes*, fungus + *geo*, soil < *gê*, earth + *graphía*, description < *grápho*, drawing): the branch of mycology that studies the geographic distribution and organization of species and populations of fungi.

mycography (Gr. *mýkes*, fungus + *graphía*, description < *grápho*, drawing): mycological work based mainly on illustrations.

mycohemia (Gr. *mýkes*, fungus + NL. suf. *-emia* < Gr. *haíma*, genit. *haímatos*, blood): the invasion of the blood stream by pathogenic fungi. Equal to **fungemia**.

mycoherbicide (Gr. *mýkes*, fungus + L. *herba*, herb, pasture + E. des. *-cide* < L. *caedere*, to kill): applied to a pathogenic fungus used in the biological control of certain weeds or plants considered as undesirable, from an anthropocentric point of view. E.g., *Phytophthora palmivora* (Peronosporales) is utilized to control a bad latex-containing weed of the gen. *Asclepias*, and various species of fungi of the gen. *Cercospora*, *Alternaria* (dematiaceous asexual fungi) and *Cylindrocarpon* (moniliaceous asexual fungi) are used in the biological control of the aquatic water hyacinth (*Eichornia crassipes*). See **mycofungicide**, **mycoinsecticide** and **myconematicide**.

mycoheterotroph (Gr. *mýkes*, fungus + *héteros*, another, different + *trophós*, that which nourishes, serves as food): refers to non-photosynthetic plants which in some part of their life cycle obtain the carbon compounds from the mycorrhizal fungi, which in turn derive their carbon from the plants they form mycorrhizae with. Most Orchidaceae, ferns and lycophyta are partially mycoheterotrophs, because they obtain carbon during their germination and establishment. Another group of plants, which have no photosynthetic pigments , are obligate mycoheterotrophs (some Orchidaceae and Gentianaceae, and all species of Monotropoideae and Triuridaceae), forming different types of mycorrizhae with the asco- and basidiomycetes. See **mycorrhiza**.

Sarcodes sanguinea, obligate **mycoheterotroph** which forms monotropoid mycorrhiza, mostly with Russulaceae, x 0.3 (*NH*).

mycoinsecticide (Gr. *mýkes*, fungus + L. *insectum* < neuter of *insectus* < *insecare*, to cut, segment < *in*, in + *secare*, to cut, section; insect + E. des. *-cide* < L. *caedere*, to kill): a commercial product based on fungi that is utilized to kill insect pests of agricultural crops. E.g., *Beauveria bassiana* and *B. brongniartii* (moniliaceous asexual fungi) are utilized extensively to control pests of economically important crops; the first species was used to control the beetle known as the coffee weevil. *Lecanicillium* (=*Verticillium*) *lecanii* and *Metarhizium anisopliae* (also of the moniliaceous asexual fungi) are used as mycoinsecticides, in greenhouses as well as in the field, for biological control of aphids and white flies, and of termites, crickets, locusts and beetles, respectively. See **entomopathogenic**, **mycofungicide**, **mycoherbicide** and **myconematicide**.

Colony of *Beauveria bassiana*, one of the species most utilized as a **mycoinsecticide**, x 0.5 (*MU*).

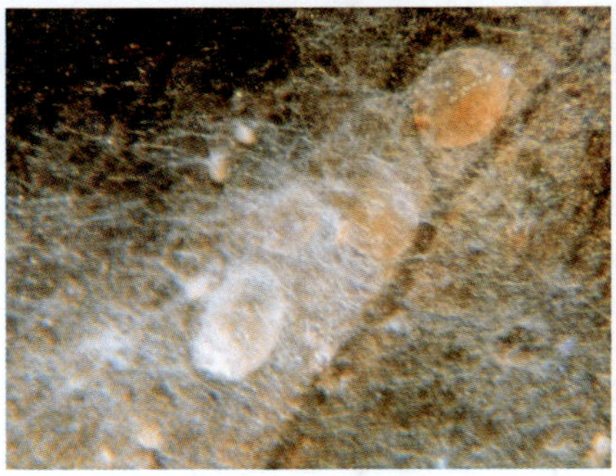

Whitefly nymph parasitized by *Lecanicillium* (=*Verticillium*) *lecanii*, another species often utilized as a **mycoinsecticide**, x 30 (*MU*).

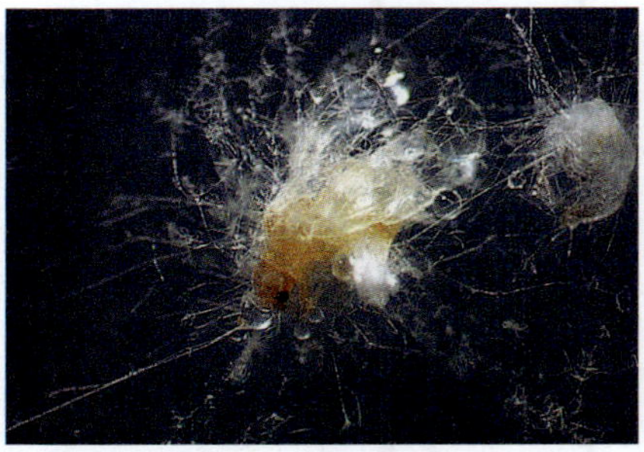

An adult whitefly parasitized by *Lecanicillium* (=*Verticillium*) *lecanii*, a species utilized as a **mycoinsecticide**, x 5 (*CT*).

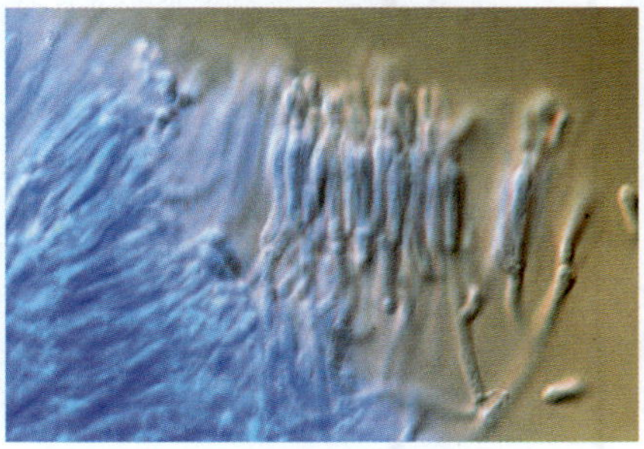

Conidiophores of *Metarhizium anisopliae*, another species commonly used as a **mycoinsecticide**, x 1,000 (*CT*).

mycolatry (Gr. *mýkes*, fungus + *latreía* < *látreyma*, religious cult, worship < *látris*, servant, minister): the worship of fungi, especially hallucinogenic species, an ancient cult of various ethnic groups of Mexico and Central America. See **mushroom stones**.

mycolith (Gr. *mýkes*, fungus + *líthos*, stone): a compact mass of sand grains bound together by mycelial filaments of *Lithomyces nidulans* (syn. of *Melanospora*, of the Melanosporales), that are found in the vineyards of Palestine.

mycologist (Gr. *mýkes*, fungus + *lógos* < *légo*, to narrate, relate): a scientist who specializes in the study of fungi.

Ricardo Valenzuela and Raul Díaz (**mycologists**) sampling *Ganoderma applanatum* (Aphyllophorales) in a forest of Mexico, x 0.05 (*TRO*).

mycology (NL. *mycologia* < Gr. *mýkes*, fungus + *lógos*, word > E. des. *-logy*, study, treatise): the branch of science that treats of the study of the fungi. Actually, the word *mycology* is an improperly coined term. Etymologically, the correct word is *mycetology*, inasmuch as the combining form of *mýkes* is the pref. *myceto-*, in accordance with the principles of Greek grammar. Notwithstanding, the word *mycology*, as well as *mycologist* and all of the other related terms (and whose pref. is *myc-* < L. *myc-* < Gr. *mýkes*), are the most established.

mycolysis, pl. **mycolyses** (Gr. *mýkes*, fungus + *lýsis*, dissolution, disintegration): the lysis of fungal cells, especially due to the action of a virus.

mycomysticism (Gr. *mýkes*, fungus + *mysticism* < L. *mysticus*, fem. *mystica*, the part of theology that treats of the spiritual and contemplative life and of the knowledge and management of the spirits + suf. Gr. *-ismós* > L. *-ismus* > E. *-ism*, state or condition, doctrine or practice): the mystic state of persons induced by ingesting hallucinogenic mushrooms, in which a communication link is established with divinity. See **entheogen**, **hallucinogenic fungi**, **magic mushrooms** and **mushroom stones**.

myconematicide (Gr. *mýkes*, fungus + E. *nematode* < Gr. *nêma*, genit. *nêmatos*, filament + NL. *-ode*, like < Gr. *-eídos*, shape + E. des. *-cide* < L. *caedere*, to kill): a fungus that is utilized to kill plant parasitic nematodes. At present few myconematicides have been commercialized due to problems with quality control, their inconsistent performance, potential health risks and the high costs of application. Notwithstanding, some commercial and semicommercial products based on *Arthrobotrys robusta*, *A. superba* and *Pochonia chlamydosporia* (=*Verticillium chlamydosporium*), moniliaceous asexual fungi, have been utilized with good results. See **mycofungicide**, **mycoherbicide** and **mycoinsecticide**.

mycoparasite (Gr. *mýkes*, fungus + *parásitos* < *pará*, next to, at the side of + *sîtos*, bread, food in general): a fungus that is a parasite on another fungus; there exist biotrophic mycoparasites that cause little damage, or no apparent damage, to the host, such as *Piptocephalis* and *Syncephalis* (Zoopagales), which respectively parasitize *Mucor* and *Cokeromyces* (Mucorales), and necrotrophic mycoparasites that kill a part or all of the host tissues, or which at least cause their sterility, as occurs with *Cordyceps capitata* (Hypocreales) on *Elaphomyces* (Pezizales), or the species of *Nyctalis* (Agaricales), that live on other Agaricales, such as *Russula*. Mycoparasitic species also exist among the lower fungi, e.g., *Olpidiopsis saprolegniae* and *Cornumyces irregularis* (=*Petersenia irregulare*), Olpidiopsidales. The first species is an endobiotic zoosporic fungus that is an obligate parasite of other zoosporic microorganism (such as *Saprolegnia*, of the Saprolegniales) and of fresh water or marine algae. The second species also is an endobiotic zoosporic fungus that is a parasite of saprolegniaceous species (such as *Achlya*), or which lives as a saprobe or parasite on marine algae. Other examples of common mycoparasites are the gen. *Sepedonium* (moniliaceous asexual fungi) and *Hypomyces* (Hypocreales), which grow on fruiting bodies of higher fungi (the first on agarics and boletes and the second on discomycetes, agarics and polypores). See **mycogenous**.

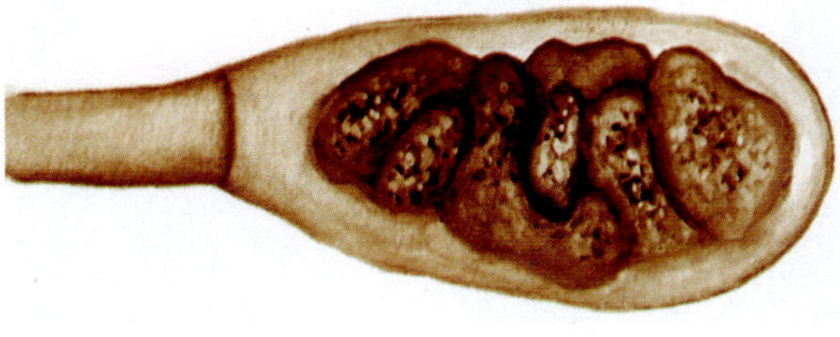

Zoosporangia of *Olpidiopsis saprolegniae*, a **mycoparasite** in a hypha of *Saprolegnia* sp., x 80 (*MU*).

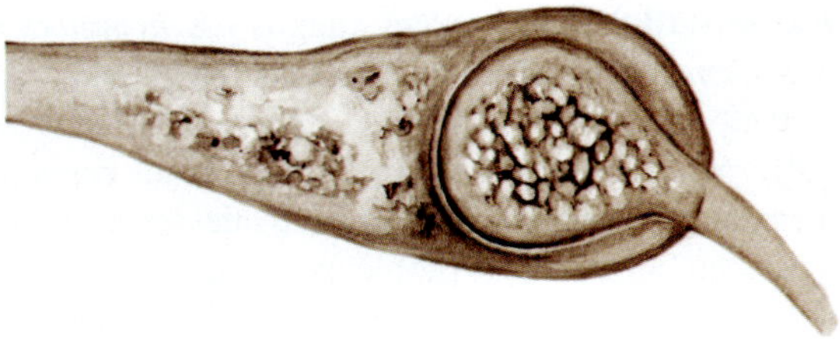

Thallus of *Cornumyces irregularis* (=*Petersenia irregulare*), a **mycoparasite** in a hypha of *Achlya* sp., x 80 (*MU*).

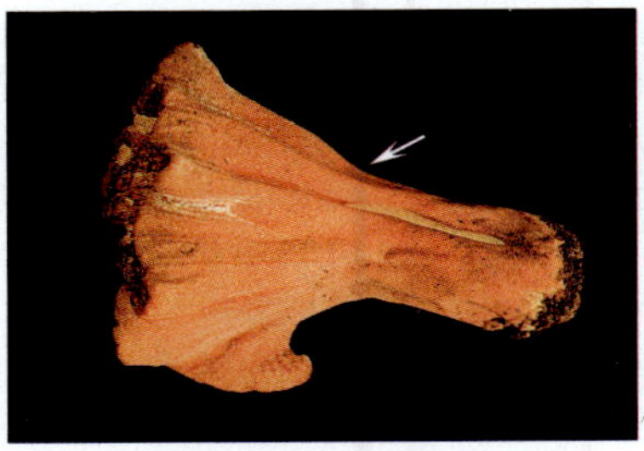

Perithecial stroma of *Hypomyces lactifluorum*, a **mycoparasite** on a fruiting body of *Russula* sp., x 0.5 (*MU*).

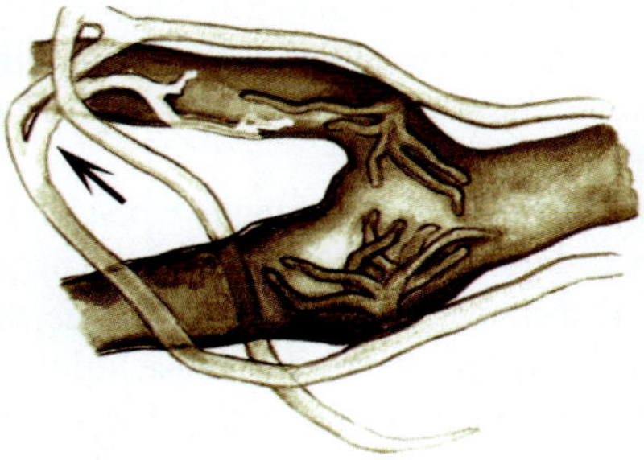

Hyphae and haustoria of *Piptocephalis cylindrospora* (=*P. virginianae*), a **mycoparasite** on a hypha of *Mucor* sp., x 1,500 (*MU*).

Perithecial stroma of *Hypomyces roseus*, a **mycoparasite** on the fruiting bodies of an aphyllophoraceous fungus, x 0.3 (*RV*).

Mycelium and conidiophores of *Sepedonium* sp., a **mycoparasite** on a fruiting body of *Boletus* sp., x 0.5 (*RV*).

Perithecial stroma of *Hypomyces lactifluorum*, a **mycoparasite** on fruiting bodies of *Russula* sp., x 0.5 (*RV*).

mycoparasitism (Gr. *mýkes*, fungus + *parásitos* < *pará*, next to, at the side of + *sîtos*, bread, food in general + suf. *-ismós* > L. *-ismus* > E. *-ism*, state or condition): the condition of being a *mycoparasite*.

mycopathology (Gr. *mýkes*, fungus + *páthos*, disease + *lógos*, word > E. des. *-logy*, study, treatise): the science that studies the diseases produced by fungi, embracing those of plants, animals and man. See **phytopathology**, **medical mycology** and **veterinary mycology**.

mycopesticide (Gr. *mýkes*, fungus + L. *pestis*, pest, plague; disease epidemic associated with high mortality; a plant, animal or microorganism that is harmful to man + E. des. *-cide* < L. *caedere*, to kill): a product produced in bulk and usually formulated for commercial use that is based on fungi that are pathogens, parasites or antagonists of other organisms that constitute plagues or pests (like some fungal pathogens of plants, weeds in agricultural crops, insects, and plant parasitic nematodes), for which are applied, respectively, *mycofungicides*, *mycoherbicides*, *mycoinsecticides*, or *myconematicides*. These mycopesticide products are applied extensively, as is done with chemical pesticides.

mycophage (Gr. *mýkes*, fungus + *phágos*, what eats < *phágesis* > *phágia*, to eat): a virus that infects and obtains its food from fungal cells; the zoospores of *Olpidium brassicae* (Spizellomycetales) are the vectors for a number of plant viruses. This species infects the roots of cabbage, but it is capable of attacking a wide variety of monocots and dicots.

mycophagy (Gr. *mýkes*, fungus + *phágos*, what eats < *phágesis* > *phágia*, to eat): **1**. To eat fungi, to take advantage of fungi as food; applied to wild animals or ethnic groups that consume fungi. Also called **mycetophagy**. **2**. The phenomenon in which the cells of a mycorrhizal or mycorrhizogenous fungus are consumed by the cells of the host plant, degenerating the symbiosis into parasitism. Cf. **mycophobia**.

Insects eating basidiocarps of *Panus crinitus*, a case of **mycophagy**, x 1 (*MU*).

Edible wild basidiocarps (Agaricales) sold in a market place of Tenango del Valle, Estado de Mexico, another case of **mycophagy**, x 0.05 (*TH*).

Some edible wild species of Agaricales in the same market place (**mycophagy**): *Boletus edulis, Russula brevipes, Lactarius salmonicolor, Lyophyllum decastes* and *Amanita caesarea*, x 0.2 (*TH*).

mycophile, **mycophilic**, **mycophilous** (Gr. *mýkes*, fungus + *phílos*, have an affinity for): **1**. Applied to one who has an appreciation of fungi in general, particularly for mushrooms and related forms, but not only as food. **2**. Applied to those arthropods that facultatively feed on fungi, as do certain beetles and ants.

mycophobia (Gr. *mýkes*, fungus + *phóbos*, what repels > *phóbia*, aversion, fear): an aversion to fungi. Cf. **mycetophagist** or **mycophagy**.

mycophobist (Gr. *mýkes*, fungus + *phóbos*, what repels): refers to a person who has an aversion to fungi, who rejects them as food. Cf. **mycetophagist**.

mycophycobiosis, pl. **mycophycobioses** (Gr. *mýkes*, fungus + *phýkos*, alga + *bíosis*, the act of living < *bíos*, life + suf. *-osis*, state, condition): the obligate symbiotic association between a systemic marine fungus and a macroalga, whose predominant habit is that of the fungus, as occurs in the association of the ascomycete *Mycophycia* (=*Mycosphaerella*) *ascophylli* (Dothideales) and the brown alga *Ascophyllum nodosum* (Phaeophyceae), and between *Blodgettia bornetii* (asexual or mitosporic fungi) and the green alga *Cladophora fuliginosa* (Chlorophyceae).

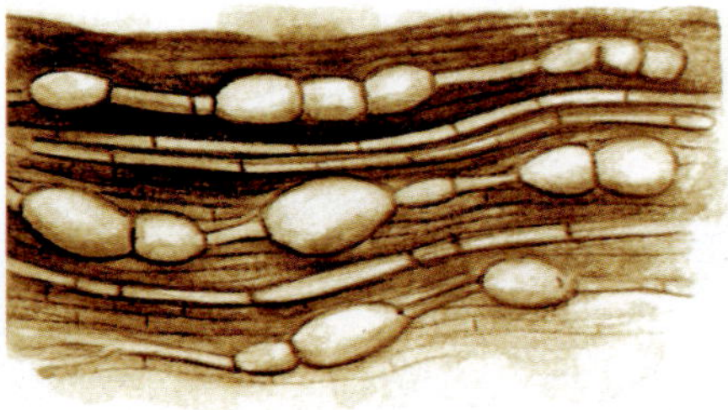

Mycophycobiosis formed by *Blodgettia bornetii* and *Cladophora fuliginosa*; the hyphae and chlamydospores of the fungus are embedded between the outer and inner layers of the cell wall of the alga, x 430 (*MU*).

mycoplasma (L. *myc-*, *mycete-* < Gr. *mýkes*, genit. *mýketos*, fungus + *plásma*, formation, bland material with which a living being is formed): **1**. The intimate relationship between fungi or other microorganisms that invade plants and their host cells. **2**. An organism, similar to a bacterium but lacking a cell wall, that lives inside the cells of a host plant; also known a a **mollicute** and, more recently, as a *phytoplasma*.

mycoprotein (Gr. *mýkes*, fungus + Fr. *protéine* < Gr. *prõteios*, primary < *prõtos*, first): commercially processed fungal mycelium, e.g., "Quorn", the trade name under which mycoprotein manufactured from non-pathogenic strains of *Fusarium graminearum* (tuberculariaceous asexual fungi) is marketed.

mycoreactor (Gr. *mýkes*, fungus + L. *re*, again, anew, backward + E. *act* < ME. *-acte* < L. *acta*, pl. of *actum*, something done + L. suf. *-tor*, one that does a thing): a fungus used in bioreactors for the removal of a wide variety of pollutants.

mycoremediation (Gr. *mýkes*, fungus + L. *remediatus*, pp. of *remediare*, to remedy, the act or process of correcting a fault + *-ion*, action or condition): the use of fungi to break down toxic wastes.

mycorestoration (Gr. *mýkes*, fungus + LL. *restauration* < L. *restauratus*, pp. of *restaurare*, to restore + *-ion*, action or condition): the use of fungi to repair or restore ecologically damaged habitats.

mycorrhiza, pl. **mycorrhizae** (Gr. *mýkes*, fungus + *rhíza*, root): the symbiotic association between the hyphae of certain fungi and the roots of 95% of vascular plants. See **endomycorrhiza** and **ectomycorrhiza**.

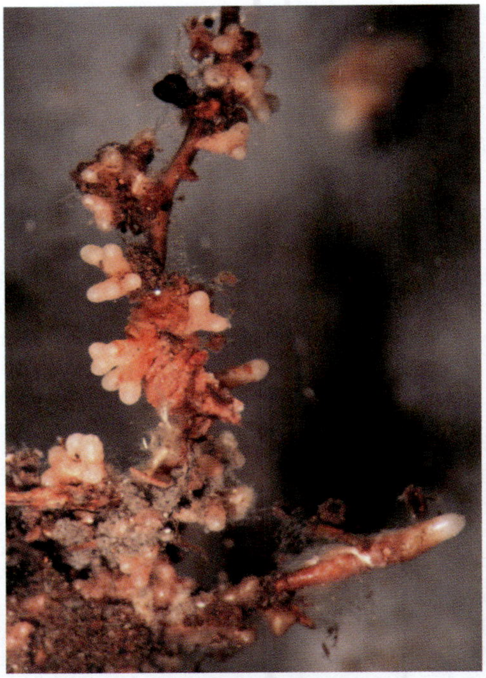

Mycorrhizae resulting from the symbiosis between *Pinus sylvestris* roots and hyphae of *Russula* sp. (Agaricales), x 4 (*LV*).

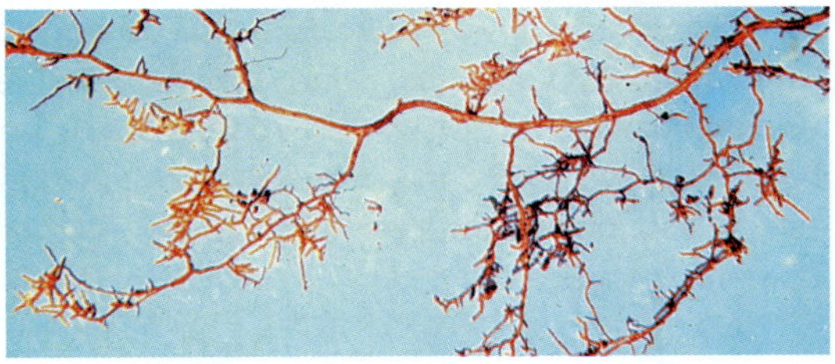

Mycorrhizae resulting from the symbiosis between oak roots and hyphae of *Boletus* sp., x 1 (*MU*).

Mycorrhizae resulting from the symbiosis between *Pinus sylvestris* roots and hyphae of *Rhizopogon luteolus* (Hymenogastrales), x 10 (*LV*).

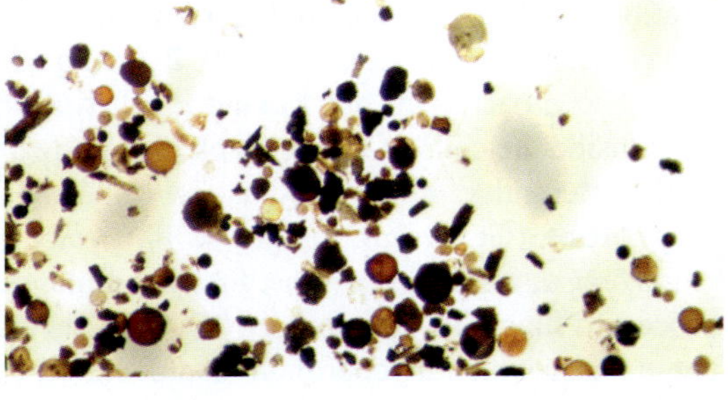

Spores of Glomales isolated from soil. The symbiosis with several plant roots give rise to the arbuscular **mycorrhizae**, x 20 (*RG*).

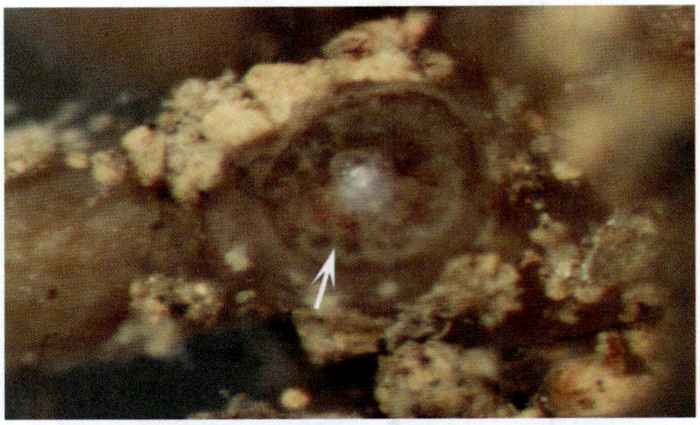

Transverse section of an ecto**mycorrhiza** from the symbiosis between roots of *Quercus* sp. and the hyphae of a basidiomycete; it shows the fungal sheath, the Hartig net (arrow) and the cortical root cells, x 40 (*RG*).

mycorrhizal (Gr. *mýkes*, fungus + *rhíza*, root + L. suf. *-alis* > E. *-al*, pertaining to or relating to): having a mycorrhiza on the roots; a plant whose roots have been invaded by a *mycorrhizogenous* fungus.

mycorrhizogenous (Gr. *mýkes*, fungus + *rhíza*, root + *génos*, origin < *gennáo*, to engender, produce): a fungus that is capable of forming a mycorrhiza, that participates in its formation.

mycosclerid (Gr. *mýkes*, fungus + *sklerós*, hard + *eídos*, form, appearance, figure): an irregular, thick-walled skeletal hypha, occasionally present in the peridial layers around the ostiole of the fruiting body of *Vascellum* (Lycoperdales).

mycosin (Gr. *mýkes*, fungus + L. suf. *-ina*, resemblance): a nitrogenized substance, similar to animal chitin, found in the cell wall of some fungi.

mycosis, pl. **mycoses** (Gr. *mýkes*, fungus + suf. *-osis*, condition, process, state; state of an organism subject to the damaging effects caused by pathogenic fungi): a wide range of infections of man and higher animals caused by various species of fungi. The following are examples of mycoses, whose names may derive from the generic names of the causal agents (aspergillosis, histoplasmosis and blastomycosis are derived from *Aspergillus*, *Histoplasma* and *Blastomyces*, respectively), from the name of the organ affected (such as otomycosis, or mycosis of the auditory canal), and from the group of fungi that cause the infection (such as dermatophytosis, which is caused by dermatophytes). Usually the mycoses are classifed into categories based on the degree of anatomical depth of the infection, i.e., superficial, cutaneous, subcutaneous or systemic. See **aspergillosis**, **blastomycosis**, **candidiasis**, **chromomycosis**, **coccidioidomycosis**, **cryptococcosis**, **dermatophytosis**, **favus**, **fungus ball**, **geotricosis**, **hyalohyphomycosis**, **histoplasmosis**, **keratomycosis**, **kerion**, **mycetoma**, **moniliasis**, **mucormycosis**, **onycomycosis**, **otomycosis**, **paracoccidioidomycosis**, **phaeohyphomycosis**, **phy-comycosis**, **pneumomycosis**, **rhinoentomophthoramycosis**, **rhinophycomycosis**, **rhino-sporidiosis**, **sporotrichosis**, **sycosis**, **thrush** and **tinea**.

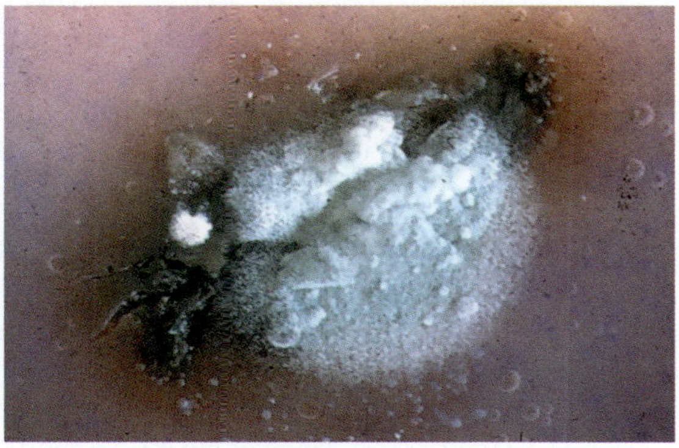

Mycosis of bovine placental tissue caused by *Aspergillus fumigatus*, as shown in agar culture, x 1 (*ECN*).

mycosociety (Gr. *mýkes*, fungus + L. *societas*, genit. *societatis*, society, a group of persons, plants or animals in mutual cooperation to fulfill some or all of the aims of life): the community of fungi that occur in a specific habitat.

mycosociology (Gr. *mýkes*, fungus + E. *sociology* < L. *socius*, member + Gr. *lógos*, word > E. des. *-logy*, study, treatise): the study of fungal communities, which can be named according to the International Code of Phytosociological Nomenclature, as, e.g., the *Clitocybo-Phellodonetum nigrae* association, a community composed of *Clitocybe* and *Phellodon* (of the Agaricales and Aphyllophorales, respectively).

mycosorption (Gr. *mýkes*, fungus + E. *sorb* < L. *sorbere*, to suck in, swallow + suf. *-ion*, action or condition): the application of fungi in the biosorption of metals.

mycosorbent (Gr. *mýkes*, fungus + E. *sorb* < L. *sorbere*, to suck in, swallow + suf. *-ent*): fungal biomass in any form, shape or size utilized in the biosorption of metals.

mycosporin (Gr. *mýkes*, fungus + *sporá*, spore + L. suf. *-in*, of or pertaining to; neutral chemical compound, enzyme or antibiotic): a secondary fungal metabolite that absorbs ultraviolet light near 310 nm. Mycosporines are produced by several species of fungi at the time of sporulation. The production of these compounds is dependent on light and is always associated with fungal sporulation. Although substantial information exists on the distribution of these metabolites, their function is still to be determined with certainty; activities related to sporulation, protection against UV light, and self-inhibition of germination have been attributed to them. E.g., in the mucilage of *Colletotrichum graminicola* and *C. lindemuthianum* (melanconiaceous asexual fungi) there exist two inhibitors, one composed of mycosporine-alanine, and a second, which is a volatile acetate. The inhibitor present in many species of rusts is *cis*-3,4-dimethoxycinnamate. The function of the self-inhibitors in imposing dormancy is frequently evident when high concentrations of spores fail to germinate, whereas the frequency of germination increases if the concentration of spores decreases. See **self-inhibitor**.

mycostasis, pl. **mycostases** (Gr. *mýkes*, fungus + *stásis*, fixity, immobility): the inhibition of fungal growth. Equal to *fungistasis*.

mycostatic (Gr. *mýkes*, fungus + *statikós*, what serves to stop or paralyze < *statós*, quiet, fixed + suf. *-íkos* > L. *-icus* > E. *-ic*, pertaining to or relating to): applied to antibiotic substances that, without being fungicides, stop the growth of fungi; also refers to concentrations of fungicides when they do not attain a lethal level of toxicity. Also called **fungistatic**, a hybrid term with both a Latin (*fungus*) and a Greek component (*statikós*). These hybrid terms are common in biological lexicology, but they are not recommended. Linguistically, it is preferable to use terms formed by components of the same language. Cf. **fungicide**.

Mycostatin (Gr. *mýkes*, fungus + *statós*, quiet, fixed + L. suf. *-in*, of or pertaining to; a neutral chemical compound, enzyme or antibiotic): the trade name for nystatin, an antibiotic produced by the actinomycete (actinobacterium) *Streptomyces noursei*, widely used against infections of man caused by *Candida albicans* (cryptococcaceous asexual yeasts).

mycosymbiosis, pl. **mycosymbioses** (Gr. *mýkes*, fungus + *symbíosis*, life in common): the symbiosis of two or more fungi.

mycosynusia (Gr. *mýkes*, fungus + *synousía*, existence in common, society): a community constituted of species belonging to a particular biotype, e.g., only of certain macromycetes, of uniform ecological requirements. In some cases, synusia is confused with the association; but, ordinarily, in the structure of an association there are a certain number of synusias.

-mycota (Gr. *mýkes*, fungus + NL. *-ota*, suf. that signifies to have; used as the ending for a group of organisms): ending of taxonomic phyla, or divisions in plant taxonomy, eg., Chytridiomycota, Ascomycota, etc.

mycothallus, pl. **mycothalli** (Gr. *mýkes*, fungus + *thallós*, thallus): the symbiotic association of a fungus with the gametophyte of a liverwort or a fern.

mycotheca (Gr. *mýkes*, fungus + *thêke*, case): a collection of fungi, prepared, labelled, and mounted on sheets or placed in boxes and organized in cabinets or other similar containers for classification and preservation of the specimens.

mycotic (Gr. *mýkes*, fungus + suf. *-tikós* > L. *-ticus* > E. *-tic*, relation, fitness, inclination or ability): pertaining to or relating to *mycosis*; caused by fungi.

-mycotina (Gr. *mýkes*, genit. *mýketos*, fungus + L. *-ina*, suf. des. that denotes similarity): ending of taxonomic subphyla, or subdivisions in plant taxonomy, e.g., Deuteromycotina, Basidiomycotina, etc.

mycotope (Gr. *mýkes*, fungus + *tópos*, place, territory): the principal fungal association of a particular type of forest. See **mycosociology**.

mycotoxicosis, pl. **mycotoxicoses** (Gr. *mýkes*, fungus + *toxikós*, poison + suf. *-osis*, condition, state): poisoning caused by the ingestion of foods contaminated with toxins (*mycotoxins*) produced by various species of molds, mainly of the gen. *Aspergillus*, *Penicillium* (moniliaceous asexual fungi) and *Fusarium* (tuberculariaceous asexual fungi), which produce, respectively, aflatoxins and ochratoxins, citrinin and patulin, and zearalenone and nivalenol, among many others. An example of damage caused by the ingestion of aflatoxins is the erosion of the gizzard of chickens. Cf. **mycetism**.

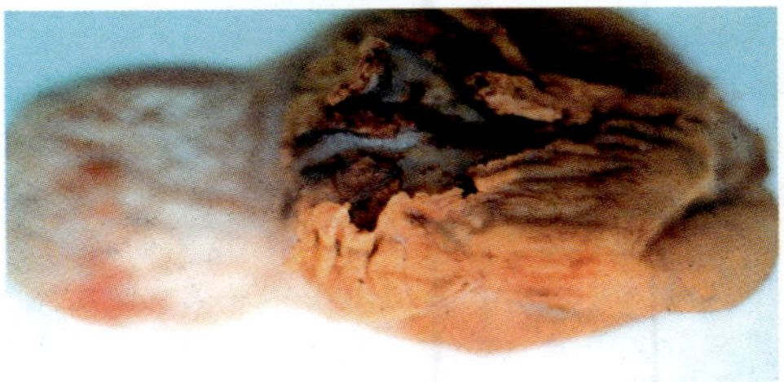

A chicken gizzard damaged by aflatoxins, an example of **mycotoxicosis**, x 0.5 (*ECN*).

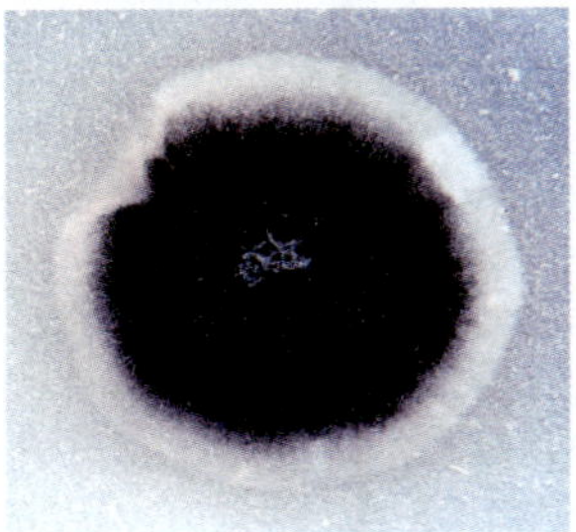

Seven day old colony of *Stachybotrys chartarum* (=*S. atra*) on agar (dematiaceous asexual fungi); this species produces stachybotryotoxin which causes **mycotoxicosis** in horses, x 1 (*MU*).

A week old colony of *Aspergillus ochraceus* (moniliaceous asexual fungi) on agar, a species that cause **mycotoxicosis** in animals, x 2 (*MU*).

mycotoxin (Gr. *mýkes*, fungus + LL. *toxicus* < L. *toxicum* < Gr. *toxikón*, poison, especially on an arrow + NL. *-in*, denotes a chemical compound or activator): a toxin produced by a fungus, especially one that affects humans and higher animals, in which it causes *mycotoxicosis*. Some of the most important and best studied mycotoxins are produced by molds of the gen. *Aspergillus* and *Penicillium* (moniliaceous asexual fungi) and *Fusarium* (tuberculariaceous asexual fungi), e.g., the *aflatoxins* (carcinogenic substances produced by *A. flavus* and *A. parasiticus*); *patulin*, also called *clavicin*, *clavatin*, *claviformin*, *expansin*, *mycoin* and *penicidin*, an antibacterial and antifungal substance toxic to plants and animals, carcinogenic, produced by *A. clavatus*, *P. solitum* (=*P. patulum*), *P. vulpinum* (=*P. claviforme*) and *P. expansum*; *zearalenone* (a toxin from *F. graminearum*, which causes vulvovaginitis and infertility in cattle and pigs); *roridin* (a terpenoid toxin produced by *Myrothecium roridum* and *M. verrucaria*, of the tuberculariaceous asexual fungi, that affects sheep, pigs and humans); *sporidesmin* (a toxic oligopeptide) from *Pithomyces chartarum* (dematiaceous asexual fungi), the anamorph of *Leptosphaerulina chartarum* (Pleosporales); *paspalitrem* (a tremorgenic mycotoxin produced by *Phomopsis*, of the sphaeropsidaceous asexual fungi), and *penitrem A* (produced by *P. aurantiogriseum* (=*P. cyclopium*) and other species of *Penicillium*, which affect the nervous system, causing tremors). There also are phytotoxic mycotoxins, i.e., that affect plants, toxins that can be specific with respect to the type of plant damaged, e.g., those produced by the following species of dematiaceous asexual fungi: *helminthosporoside*, from *Bipolaris* (=*Drechslera*) *sacchari*, which affects sugar cane; *victorin*, from *D. victoriae*, cause of blight of wheat leaves; *phytoalternarins A*, *B* and *C*, from *Alternaria kikuchiana*, responsible for the necrotic spots on pears in Japan. Other phytotoxic mycotoxins, not specific with respect to the plant affected, are, e.g., *alternaric acid* (from *Alternaria solani*), which inhibits the germination of spores of some fungi and causes wilting and necrosis in various higher plants; *stachybotryotoxin*, a toxin that affects horses and humans, is produced by *Stachybotrys chartarum* (=*S. atra*, dematiaceous asexual fungi), which grows in pastures; and *culmomarasmin* (produced by *Fusarium culmorum*, of the tuberculariaceous asexual fungi), which causes wilting in various plants. Cf. **phytotoxin**.

Colonies of *Aspergillus clavatus* on agar; it produces patulin, a carcinogenic **mycotoxin** that affects animals, x 1 (*MU*).

Colony of *Aspergillus parasiticus*; it produces aflatoxins, the carcinogenic **mycotoxins** that affects animals and humans, x 1 (*MU*).

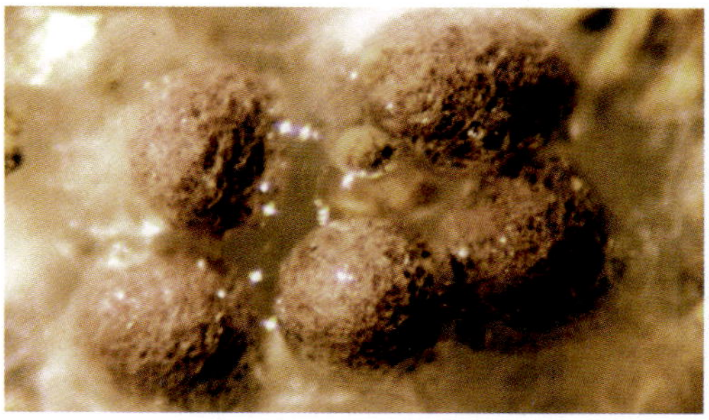

Sclerotia of *Aspergillus ochraceus*; it produces ochratoxin, a **mycotoxin** that causes hepatotoxicity and nephrotoxicity in animals, x 100 (*MU*).

mycotransformation (Gr. *mýkes*, fungus + LL. *transformation*, change in shape or form): the use of fungi in the transformation of a wide variety of pollutants in any medium or in waste or wastewater.

mycotroph (Gr. *mýkes*, fungus + *tróphos*, what nourishes, serves as food < *trépho*, to nourish, feed): **1**. Refers to plants that form mycorrhizae and need the mycorrhizogenous fungus in order to obtain their nutrition. There are obligate mycotrophs, which require the presence of the mycorrhizal fungus, and facultative mycotrophs, which can live without them under certain circumstances. **2**. A fungus that obtains its nutrition from another fungus. Cf. **phycotroph**.

mycovirus, pl. **mycoviruses** (Gr. *mýkes*, fungus + L. *virus*, toxin): a virus that infects fungi.

myosin (NL. pref. *my-*, *myo-* < Gr. *mýs*, mouse, muscle + L. *-ina*, des. or suf. that denotes similarity, derived from): *myosin*, together with *actin*, are the proteins associated with the contraction and relaxation of animal muscle. They also are found in various lower eukaryotic organisms; e.g., they are responsible for the reversible and rhythmic protoplasmic currents of the plasmodium of some Myxomycetes, as happens in *Physarum polycephalum* (Physarales). See **actin**.

myriosporous (Gr. *myríos*, innumerable + *sporá*, spore): having a large number of spores, like the myriosporous sporangium of *Mucor hiemalis* (Mucorales). Also called **pleiosporous** or **pliosporous**. Cf. **oligosporous**.

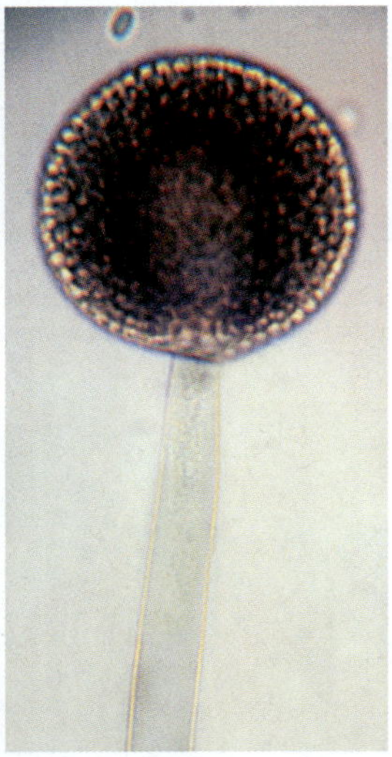

Myriosporous sporangium of *Mucor hiemalis*, x 500 (*CB*).

myriostomes (Gr. *myríos*, innumerable + NL. *stome* < Gr. *stóma*, mouth): having several pores or mouths, resulting from the dehiscence of the endoperidium in *Myriostoma* (Lycoperdales), and which appear as small lacerate ostioles on the mature fructification. See **stoma**.

Myriostomes of the fruiting body of *Myriostoma coliforme*, x 2 (*MU*).

myrmeco-, **myrmec-** (Gr. *mýrmex*, genit. *mýrmekos*, ant): a pref. employed in forming compounds alluding to ants, as in *myrmecophilous*.

myrmecophilous (Gr. *mýrmex*, genit. *mýrmekos*, ant + *phílos*, have an affinity for): refers to fungi that provide refuge or food to certain species of ants. E.g., *Leucoagaricus gongylophorus* (=*Rozites gongylophora*), Agaricales, is a fungus that lives in symbiosis with various species of ants of the gen. *Atta*, and *Lepiota* sp. (Agaricales) is the fungal symbiont of the ant *Cyphomyrmex costatus*. *Myrmecocystis* (Pezizales) is another example of a myrmecophilous fungus.

Ascocarps of *Myrmecocystis spinospora*, a **myrmecophilous** fungus, x 58 (*MU*).

mytiliform (Gr. *mýtilos*, marine mussel, mytilus + L. suf. *-formis* < *-forma*, shape): shaped like a mussel shell, like the ascocarp (hysterothecium) of certain Loculoascomycetes of the family Mytilinidiaceae (such as *Lophium* and *Mytilinidion*, of the Melanommatales).

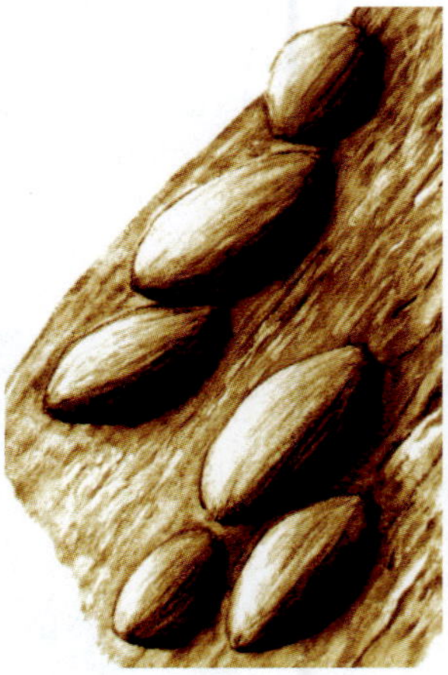

Mytiliform hysterothecia of *Mytilinidion decipiens*, x 60 (*MU*).

myxamoeba, pl. **myxamoebae** (Gr. *mýxa*, mucosity, viscosity + *amoibé*, change, transformation): *Myxomycetes*. An amoeba that results from the germination of a spore. A myxamoeba can transform itself into a myxozoospore by formation of flagella, and in turn can revert to a myxamoeba through loss of the flagella; together with the plasmodia, the myxamoebae constitute the trophic phases of these organisms, as is seen in *Physarum polycephalum* (Physarales). The cellular slime molds, such as *Dictyostelium discoideum* (Dictyosteliomycota) and *Acrasis rosea* (Acrasiomycota), among many others, also have myxamoebae.

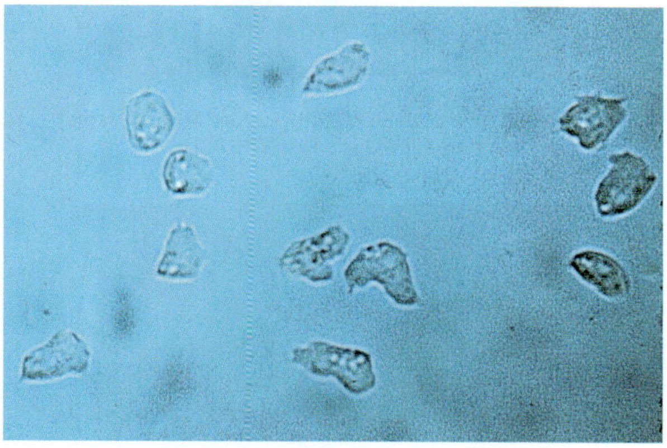

Myxamoebae of *D'ctyostelium discoideum*, x 900 (*RTH*).

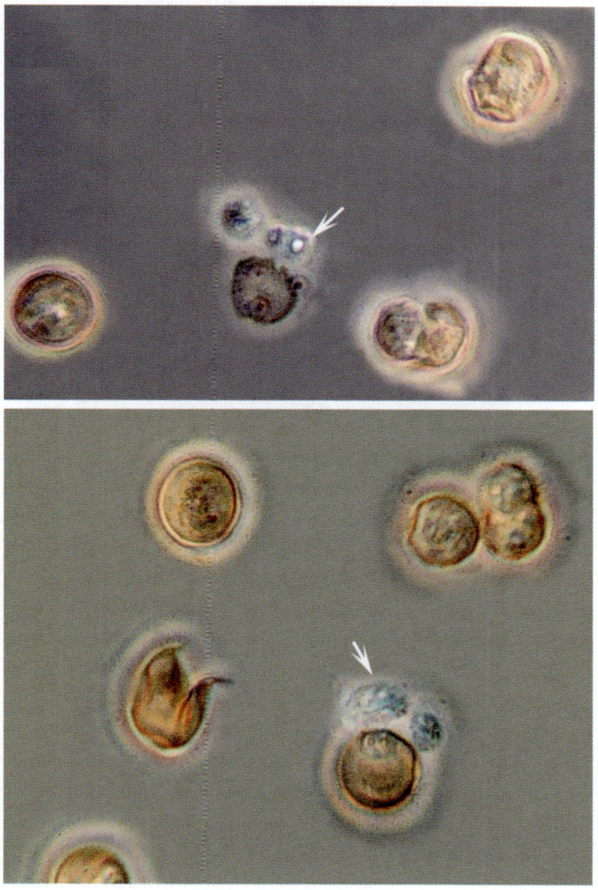

Myxamoebae of *Physarum polycephalum*, liberated from the spores, x 1,350 (*MU*).

myxamebozygote (Gr. *mýxa*, mucosity, viscosity + *amoibé*, change, transformation + *zygotós*, united by a yoke < *zygo-* < *zygóun*, to unite, join < *zygón*, yoke, pair): *Myxomycetes*. The product resulting from the fusion of two myxamoebae of different sex.

myxarioid (< gen. *Myxarium*, of the Tremellales, < Gr. *mýxa*, mucosity, viscosity + L. suf. *-arium*, which denotes belonging to; + NL. suf. *-oides*, like < Gr. *-oeídes*, similar to): refers to the basidia of fungi in the gen. *Myxarium*, which have a pedicellate part, separated by a wall from a globose metabasidial part.

myxo-, **myx-** (Gr. *mýxa*, mucosity, viscosity): a pref. used in compounds that allude to a mucilaginous substance, like the following.

myxoflagellate (Gr. *mýxa*, mucosity, viscosity + L. *flagellum*, flagellum, whip + suf. *-atus* > E. *-ate*, provided with or likeness): *Myxomycetes*. An amoeboid flagellate cell (myxozoospore) of the Myxomycetes; the term is used especially by protozoologists. Also called **myxomonad**.

myxolichen (Gr. *mýxa*, mucosity, viscosity + *leichén*, lichen): an association of a myxomycete plasmodium with cells of a green alga (*Chlorella*), produced artificially in the laboratory.

myxomonad (Gr. *mýxa*, mucosity, viscosity + L. *monas*, genit. *monadis*, unit, entity < Gr. *monás*, genit. *monádos* < *mónos*, only, unique): syn. of **myxozoospore**; rarely used in botany and mycology.

Myxomycetes (Gr. *mýxa*, mucosity, viscosity + L. *-mycetes*, ending of taxonomic class < Gr. *mýkes*, genit. *mýketos*, fungus): a class of viscous or mucilaginous fungi, also called slime molds, whose myxamoebae or myxozoospores coalesce to form plasmodia that fruit by producing sporangia and other types of sporiferous apparatus. In the system of classification adopted in this dictionary, the Myxomycetes (which includes the orders Ceratiomyxales, Echinosteliales, Liceales, Trichiales, Physarales and Stemonitales) constitute the only class in the phylum Myxomycota, of the kingdom Protista (Alexopoulos *et al.*, 1996). In other systems of classification the Myxomycetes are included in the kingdom Protoctista (Margulis and Schwartz, 1988) or in the kingdom Fungi (Herrera and Ulloa, 1990). In the Appendix, the user of this dictionary can see the comparison between the outline of Alexopoulos *et al.*, 1996, modified by Ulloa and Hanlin, 2006, to that of Kirk *et al.*, 2008. In the latter classification scheme, the class Myxomycetes of the kingdom Protista is now included as the class Myxogastrea, phylum Amoebozoa, kingdom Protozoa. The order Ceratiomyxales has been placed in the order Protostelida, class Protostelea, phylum

Myxomycetes

Amoebozoa. This is another example of the taxonomic changes that have resulted from the continuing study of this group of animal-like fungi, in order to achieve a phylogenetic classification. See **Protozoa**.

Sporangia of *Physarum polycephalum* (Physarales, **Myxomycetes**), on agar, x 17 (*MU*).

Sporangia of *Diachea leucopoda* (Stemonitales, **Myxomycetes**), on the surface of a twig, x 15 (*MU*).

Sporangia of *Didymium* sp. (Physarales, **Myxomycetes**), on the surface of a twig, x 10 (*MU*).

Sporangia of *Lepidoderma tigrinum* (Physarales, **Myxomycetes**), on the surface of a twig, x 10 (*MU*).

Sporangia of *Badhamia gracilis* (Physarales, **Myxomycetes**), on the surface of a twig, x 15 (*MU*).

Plasmodiocarp of *Physarum serpula* (Physarales, **Myxomycetes**), on bark, x 15 (*MU*).

Sporangia of *Diderma radiatum* (Physarales, **Myxomycetes**), on bark, x 15 (*MU*).

Sporangia of *Physarum cinereum* (Physarales, **Myxomycetes**), on the surface of a grass leaf, x 10 (*MU*).

Sporangia of *Elaeomyxa cerifera* (Physarales, **Myxomycetes**), on bark, x 10 (*MU*).

Aethalia of *Lycogala epidendrum* (Liciales, **Myxomycetes**), on bark, x 10 (*RV*).

myxomyceticolous (L. *myxomycete*, myxomycete < Gr. *mýxa*, mucosity, viscosity + *mýkes*, genit. *mýketos*, fungus + L. suf. *-cola*, inhabitant): living on myxomycetes, like some spp. of *Nectria* (Hypocreales), whose perithecia, mycelium and conidiophores develop on sporangia and aethalia of myxomycetes; e.g., *Nectriopsis exigua* (=*N. myxomyceticola*) is found on sporangia of *Arcyria cinerea*, *A. nutans* (Trichiales), *Stemonitis fusca* and *S. nigrescens* (Stemonitales), and on aethalia of *Fuligo septica* (Physarales).

Myxomycota (Gr. *mýxa*, mucosity, viscosity + L. *-mycota*, ending of division or phylum < Gr. *mýkes*, genit. *mýketos*, fungus): myxomycetes and other mucous or mucilaginous fungi, which mycologists classify in the division (phylum) Myxomycota, either within the kingdom Fungi or as part of the kingdom Protista. Some authors (such as Margulis and Schwartz, 1988; Whittaker, 1959) consider the Myxomycota as part of the kingdom Protoctista, or move this group of fungus-like organisms to the kingdom Protozoa. See the Appendix, where there is a comparison of the outline of Alexopoulos *et al.*, 1996, modified by Ulloa and Hanlin, 2006, to that of Kirk *et al.*, 2008. In the latter classification scheme, this phylum has been transferred from the kingdom Protista to the kingdom Protozoa. This is another example of the taxonomic changes that have resulted from the continuing study of this group of fungus-like organism, in order to achieve a phylogenetic classification. See **Protozoa**.

Sporangia of *Lamproderma columbinum* (Physarales, **Myxomycota**), growing on bark, x 3 (*TH*).

Sporangia of *Trichia scabra* (Trichiales, **Myxomycota**), on bark, x 10 (*MU*).

Sporangia of *Diderma hemisphaericum* (Physarales, **Myxomycota**), on bark, x 10 (*MU*).

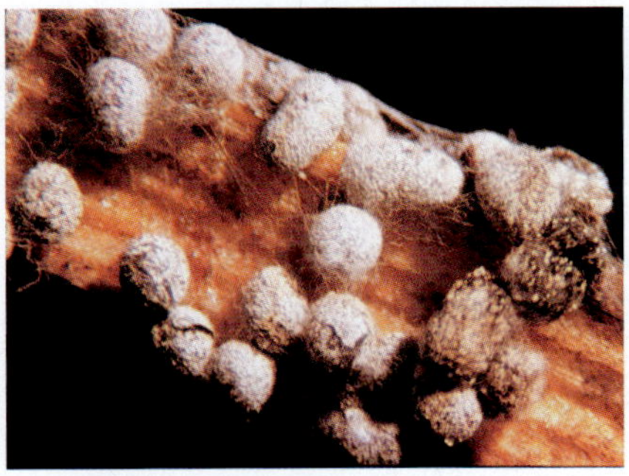

Sporangia of *Badhamia* sp.
(Physarales, **Myxomycota**), on bark, x 10 (*MU*).

myxozoospore (Gr. *mýxa*, mucosity, viscosity + *zõon*, animal, motile cell + *sporá*, spore): also called **myxomonad**, mainly by protozoologists. *Myxomycetes*. A zoospore that originates from the germination of a spore or by the metamorphosis of a myxamoeba. Myxozoospores, which also can be transformed into myxamoebae, constitute, together with the plasmodia, the trophic phases of these organisms, as occurs, e.g., in *Echinostelium minutum* (Echinosteliales).

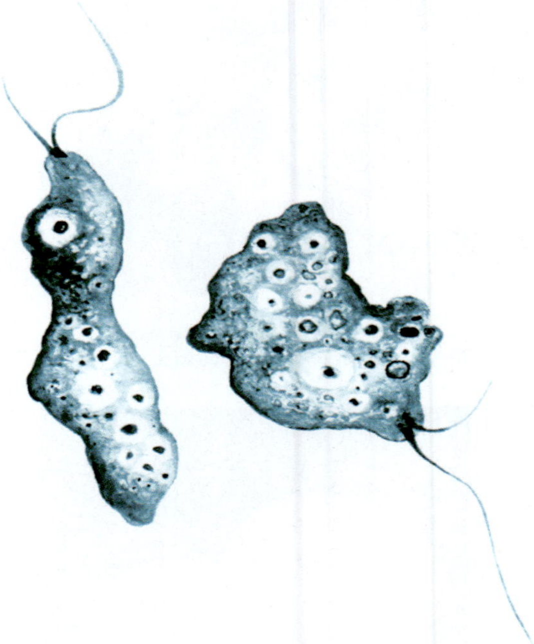

Myxozoospores of *Echinostelium minutum*, x 1,700 (*MU*).

n

Nacreous giant colony of *Candida* sp., on agar (*MU*).

n

nacreous (Pers. *nigar*, ornament, mother-of-pearl; substance formed from calcium carbonate, organic material and water, hard, satin white, glossy and with rainbow-hued reflections, that forms the interior of various shells, above all, the mother of pearl + L. suf. *-osus* > OF. *-ous*, *-eus* > E. *-ous*, full of, having, possessing the qualities of): belonging to mother of pearl or similar to it; with a pearly color and brillance, similar to mother of pearl, like the pedicel of the sporangia of some Myxomycetes, e.g., *Diachea leucopoda* and *D. splendens* (Physarales), and giant colonies of some yeasts on agar, e.g., *Candida* sp. (cryptococcaceous asexual yeasts).

Nacreous pedicel of the sporangia of *Diachea splendens*, x 20 (*RS*).

Nacreous giant colony of *Candida* sp. on agar, x 1 (*MU*).

naked (ME. *nakede* > E. *naked*, bare, nude, without covering): a term applied to cells that lack a cell wall, e.g., the myxamoebae of the slime molds and the zoospores of the aquatic fungi; it also refers to the surface of other denudate structures (e.g., peridium of sporangia) and the pileus or stipe of higher fungi that lack fibers, scales, spines or other types of ornamentation or covering, as well as to a stoma that does not have a differentiated peristome, as in *Tulostoma* (Tulostomatales). See **denudate**. Cf. **ornamented**.

Sporangia of *Clastoderma debaryanum* (Echinosteliales); the left sporangium has lost the peridium, and is **naked**, and one can see the capillitium in the upper part, x 80 (*MU*).

nano-, **nan-** (Gr. *nános*, *nánnos*, dwarfish): a pref. used in terms alluding to something extremely small, as in *nanoplankton*, the part of plankton that escapes through nets with a mesh of 0.04 to 0.05 mm.

nanometer (Gr. *nános*, *nánnos*, dwarfish + L. *metrum* < Gr. *métron*, measure, meter): the thousandth part of a micrometer, or the billionth part of a meter, which is abbreviated nm. Cf. **micrometer**.

napiform (L. *napiformis*, turnip-shaped < *napus*, turnip + suf. *-formis* < *forma*, shape): shaped like a turnip, i.e., swollen or bulbose on top and tapering abruptly toward the base, like the volva of the basidiocarp of *Amanita perpasta* and the pileocystidia of *Entoloma formosum (=Leptonia fulva)*, of the Agaricales.

Napiform volva of the basidiocarp of *Amanita perpasta*, x 0.7 (*MU*).

nasse, **nassace** (Fr. *nasse*, fish trap, cylindrical in shape; basket): a dactyloid projection of the inner part of a bitunicate ascus, inward from the endoascus or endotunica; the internal apical peak of the bitunicate ascus. It is present in the Loculoascomycetes, e.g., *Leptosphaerulina* (Dothideales). Also called *ocular chamber* and *apical nasse*.

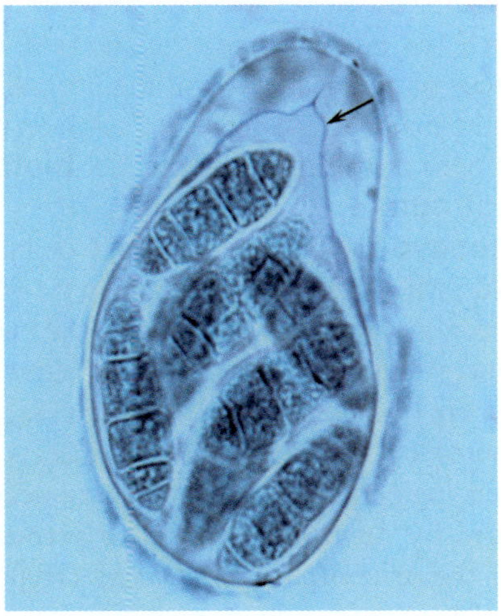

Nasse of the bitunicate ascus of *Leptosphaerulina arachidicola*, x 450 (*MU*).

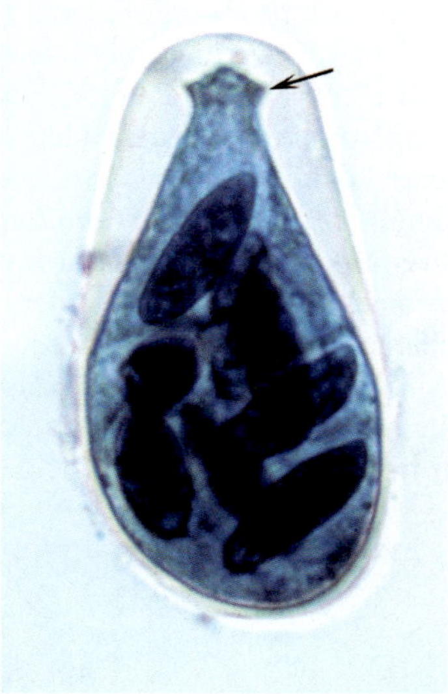

Nasse of the bitunicate ascus of *Leptosphaerulina crassiasca*, x 435 (*RTH*).

naucorioid (< gen. *Naucoria* < Gr. *naus*, ship + L. *corium*, skin, leather + L. suf. *-oide*, like < Gr. *-oeídes*, similar to): *Agaricales*. A type of fruiting body (13 principal types are recognized) with gills adnate or emarginate, stipe of fleshy, fibrous or gypseous consistency, centrally implanted in the pileus, and lacking a volva and ring, as in *Naucoria* and some species of *Laccaria*, *Russula* and *Cortinarius*, among other gen.

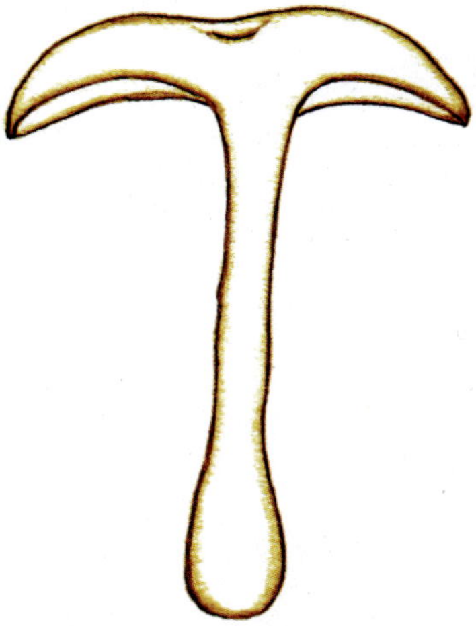

Naucorioid type of basidiocarp, represented by *Naucoria* sp., x 2 (*MU*).

nauto-, **naut-** (Gr. *náutes* > L. *nauta*, mariner, man of the sea): a pref. used in words alluding to navigation, as in the generic name *Nautosphaeria* (Halosphaeriales), a marine ascomycete.

navicular, **naviculate** (L. *navicularis*, relating to a small boat < *navicula*, dim. of *navis*, boat + suf. *-aris* > E. *-ar*, like, concerning; or + suf. *-atus* > E. *-ate*, provided with or likeness): shaped like a small boat, like the macroconidia of *Fusarium* (tuberculariaceous asexual fungi), the ascospores of *Corollospora colossa* (Halosphaeriales), the pseudothecia (hysterothecia) of the Loculoascomycetes of the orders Arthoniales and Patellariales, or like the basidiospores of *Hemimycena* (=*Mycena*) *lactea* and *Marasmius rotula* (Agaricales). Also called **carinate**, **cymbiform** and **scaphoid**.

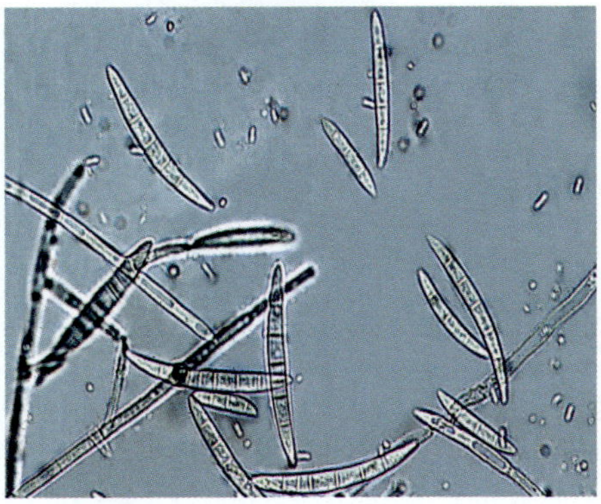

Navicular macroconidia of *Fusarium* sp., x 500 (*MU*).

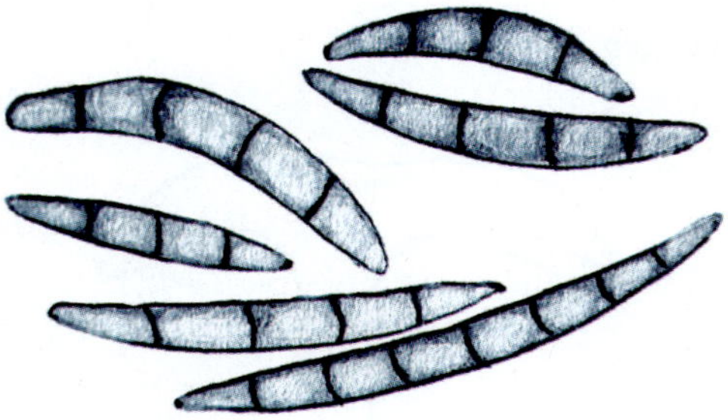

Navicular macroconidia of *Fusarium* sp., x 870 (*MU*).

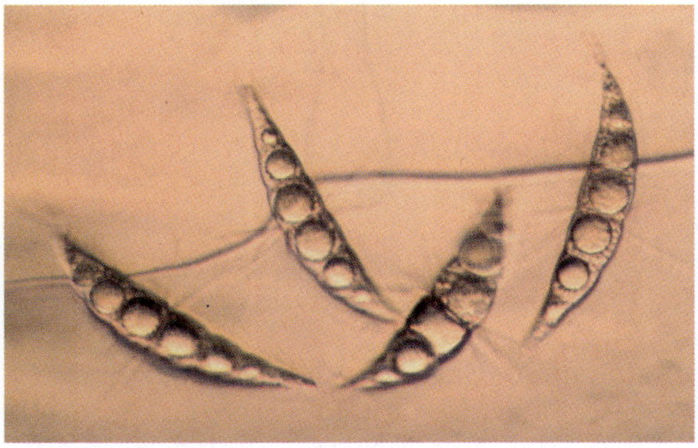

Navicular ascospores of *Corollospora colossa*, x 870 (*MCG*).

ne-, **neo-** (Gr. *néos*, new): an inseparable particle employed as a pref. with the meaning of recent or new, as in the generic name *Neosartorya* (Eurotiales), a teleomorph of *Aspergillus* (moniliaceous asexual fungi).

necho-, **nech-** (Gr. *nécho*, to swim): a pref. employed to allude to something that swims or floats, as in *nechocarp* and *nechosperm*, which mean, respectively, a floating fruit and a floating seed.

necral (Gr. *nekrós*, dead + L. suf. *-alis* > E. *-al*, relating to or pertaining to): *Lichens*. Refers to the coriaceous layer of dead hyphae with indistinct lumina, found near the cortex of the lichen thallus, above (*epinecral*) or beneath (*hyponecral*) the algal layer. See **necrophoral**.

necro-, **necr-** (Gr. *nekrós*, dead): a pref. used in terms that allude to death, like the following.

necrophagous (Gr. *nekrós*, dead + *phágos*, what eats < *phágomai*, to eat): equal to **necrotroph**. See **saprobe**.

necrophoral, **necrophoric** (Gr. *nekrós*, dead + suf. *-phóros*, bearer < *phéro*, to bear, support + L. suf. *-alis* > E. *-al*, relating to or pertaining to; or + Gr. suf. *-íkos* > L. *-icus* > E. *-ic*, pertaining to or relating to): *Lichens*. Layers of dead cells, without distinct lumina, that can function as water reservoirs; they are located near the cortex of the thallus and can be above or below the algal layer. See **necral**.

necrosis, pl. **necroses** (Gr. *nékrosis*, death < *nekrós*, dead + suf. *-osis*, condition or state): partial or total death of cells or tissues.

necrotic (Gr. *nekrós*, dead + suf. *-tikós* > L. *-ticus* > E. *-tic*, relation, inclination): concerning or relating to *necrosis*.

necrotroph, **necrotrophic** (Gr. *nekrós*, dead + *trophós*, what nourishes, serves as food; or + suf. *-íkos* > L. *-icus* > E. *-ic*, pertaining to or relating to): a parasite that feeds on the dead parts of the host. Among the fungi, the majority of species parasitic on plants are necrotrophs or **necrophagous**. Cf. **biotroph** and **perthotroph**.

nectar, **honeydew** (L. *nectar* < Gr. *néktar*, beverage of the Gods; ME. *hony* < OE. *huny*, honey + ME. < OE. *deaw*, dew): in mycology it refers to the viscous and often sweet fluid produced mainly in some pycnidia and pycnia (the latter of rusts), in which are mixed the spores or the spermatia that insects disperse after feeding on it. Another example of nectar is that produced by *Sphacelia* (tuberculariaceous asexual fungi), the anamorph of *Claviceps* (Hypocreales). Also refers to the secretions of aphids that live on the leaves of citrus, in which some fungi, such as sooty molds, can grow.

nectaro-, nectar- (L. *nectar* < Gr. *néktar*, beverage of the Gods): a pref. used in names such as *Nectaromyces* (Saccharomycetales), a gen. of yeasts that live in the nectar of flowers.

necto-, nect- (Gr. *nectós*, floating): a pref. used in terms such as *nectomonad*, a free-living, motile form of flagellate organisms.

nemato-, nemat- (Gr. *nêma*, genit. *nêmatos*, filament): a pref. used in compound terms that allude to something filamentous, or to nematodes themselves, as in the following.

nematogenous (Gr. *nêma*, genit. *nêmatos*, filament + *génos*, origin): applied to conidiogenous cells that arise at all levels from single hyphae.

nematophagous (Gr. *nêma*, genit. *nêmatos*, filament + *phágos*, what eats < *phágomai*, to eat): feeding on nematodes, like the predatory fungi, such as *Arthrobotrys* and *Dactylella* (moniliaceous asexual fungi), which trap, kill and feed on soil nematodes. Syn. of **vermivorous**.

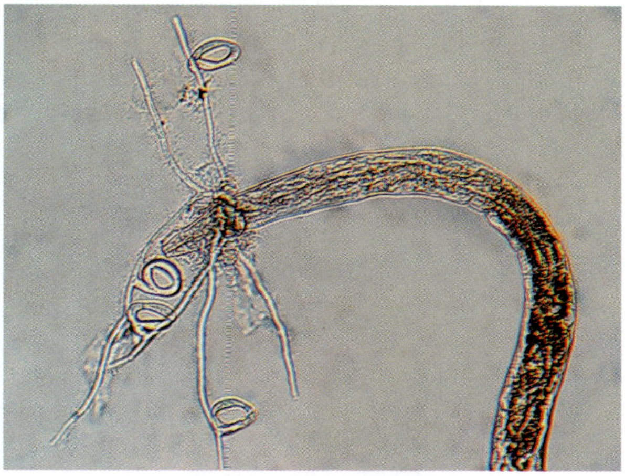

Constricting rings of *Arthrobotrys dactyloides*, a **nematophagous** species which has trapped a soil nematode, x 150 (*PL*).

nematotoxin (Gr. *nêma*, genit. *nêmatos*, filament + *toxikós*, poisonous, toxic + *-ina* > E. *-in*, L. suf. that denotes a chemical compound): a metabolic product produced by certain fungi that is toxic to nematodes, such as the nematotoxins produced by species of *Nematoctonus* (moniliaceous asexual fungi).

nemin (Gr. *nêma*, genit. *nêmatos*, filament + *-ina* > E. *-in*, L. suf. that denotes a chemical compound): a chemical principle produced by the nematode *Neoplectana glaseri*, which induces the formation of nematode traps in *Arthrobotrys conoides* (moniliaceous asexual fungi). This effect of nemin has been demonstrated in the laboratory by applying nematode-free filtrate derived from cultures of these animals.

nemoral (L. *nemoralis*, relating to a forest or grove < *nemus*, genit. *nemoris*, grove, forest + suf. *-alis* > E. *-al*, relating to or pertaining to): living in forests.

neo- (Gr. *néos*, new): a pref. that alludes to something new.

neologism (Gr. *néos*, new + *logismós*, reasoning): a new word, expression or usage in a language. Neologisms abound in scientific and technical terminology, e.g., *mycotoxin* and *neurotropic*.

neoteny (NL. *neotenia* < Gr. *néos*, new + *teinéin*, to extend, enlarge, force, carry to the extreme): a process in which the normal development of cells is interrupted, except for the reproductive cells, which results in the production of organisms that are sexually mature, but with juvenile characteristics. Cf. **pedogenesis**.

neotype (Gr. *néos*, new + *týpos*, mark, model, type): applied to a specimen or other material that agrees with the original description and is taken as the type when the original material on which the nomenclatural type was based no longer exists.

nervicolous (L. *nervus*, *nervi*, nerve, nervation, venation + suf. *-cola*, inhabitant): growing on the veins of higher plants.

net (ME. < OE. *net, nett*, net): applied to the group of thick longitudinal ribs interconnected by delicate transverse bands, similar to the mesh of a net or a lattice, present in the mature sporangia of the Myxomycete *Cribraria* (=*Dictydium*), of the Liceales, when the peridium, which is fugacious, detaches. The spaces or interstices of the reticular mesh are almost rectangular, although they can be somewhat irregular or polygonal. On the ribs and bands of the net, or surrounding them, are *dictydine* granules, together with spores.

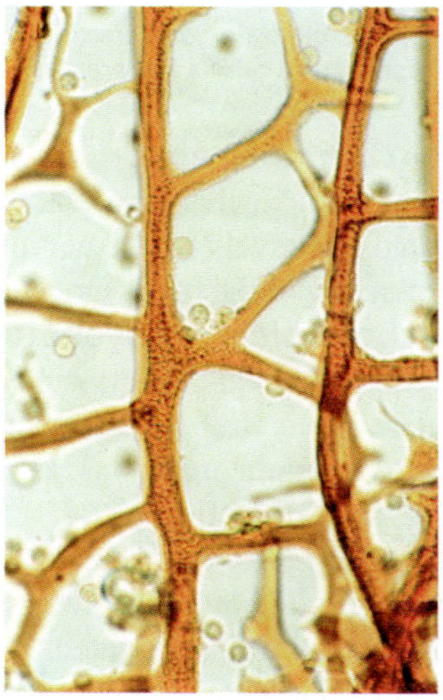

Net of the sporangial peridium of *Cribraria cancellata* (=*Dictydium cancellatum*), x 250 (*CB*).

Net of the sporangial peridium
of *Cribraria ferruginea*, x 18 (*MU*).

Nidose, or foetid, fruiting bodies of *Phallus hadrianii*, x 0.5 (*RV*) and
Phallus impudicus var. *impudicus*, x 1 (*JLM*).

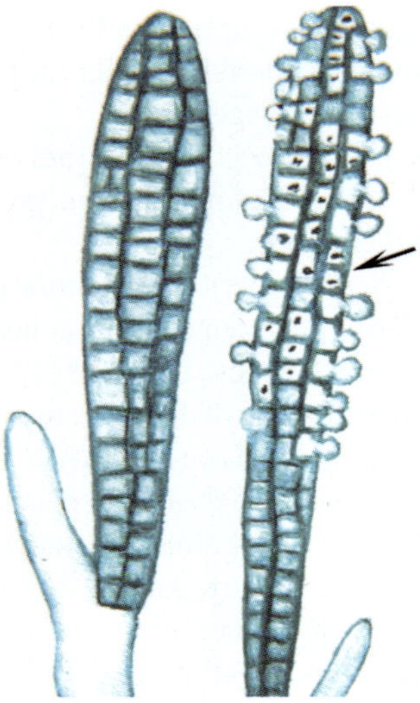

Net of the sporangia of *Dictyuchus monosporous* (Saprolegniales); the
encysted spores give a net-like appearance, x 220 (*MU*).

neurotropic (Gr. *neurón*, nerve + *tropikós*, refers to change or to removal, changing): having an affinity for, or located selectively in, the nervous tissue, like the mode of action of the hallucinogenic drugs. Also called **psychotropic**. See **psychotomimetic**.

nicto- (Gr. *nýx*, genit. *niktós*, night): a pref. that alludes to the night, as in *nictotropism*, the nocturnal tropism or movement of certain organisms; also employed as the form *nicti-*. See **nocti-**.

nidose, **nidorose** (L. *nidorosus*, giving off a bad odor < *nidor*, a penetrating and disagreeable odor + suf. *-osus* > E. *-ose*, full of, augmented, prone to): with a disagreeable odor, similar to rotten meat, like that emitted by fungi of the gen. *Phallus* and *Mutinus* (Phallales). See **foetid**.

nidulant, **nidulate** (L. *nidulatus*, nested < *nidulus*, dim. of *nidus*, nest + suf. *-ant,* one who makes something; being; or + suf. *-atus* > E. *-ate*, provided with or likeness): placed, inserted or situated freely in a cavity or structure similar to a nest, like the peridioles of the fruiting body of *Crucibulum laeve* (=*C. vulgare*), *C. cyathiforme* and *Cyathus striatus* (Nidulariales).

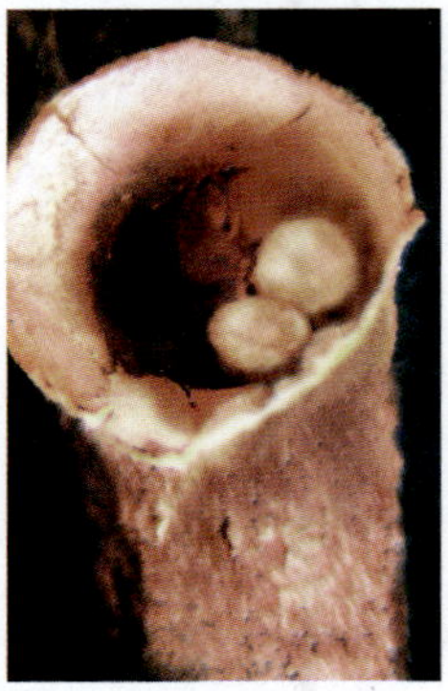

Nidulant peridioles of the fruiting body
of *Crucibulum laeve* (=*C. vulgare*), x 5 (*MU*).

Nidulant peridioles of the fruiting body
of *Cyathus* sp., x 5 (*RV*).

nigricant (L. *nigricantis*, what is black < *nigricans*, black, dark-colored + suf. *-ant*, one who makes something; being): becoming or turning black, like the mature sporangia of *Rhizopus nigricans*, a syn. of *Rh. stolonifer*, *Phycomyces blakesleanus* (Mucorales) and the conidia of *Nigrospora sphaerica* (dematiaceous asexual fungi).

Nigricant mature sporangia of *Phycomyces blakesleanus* on agar; the sporangia on the left are young, x 200 (*MU*).

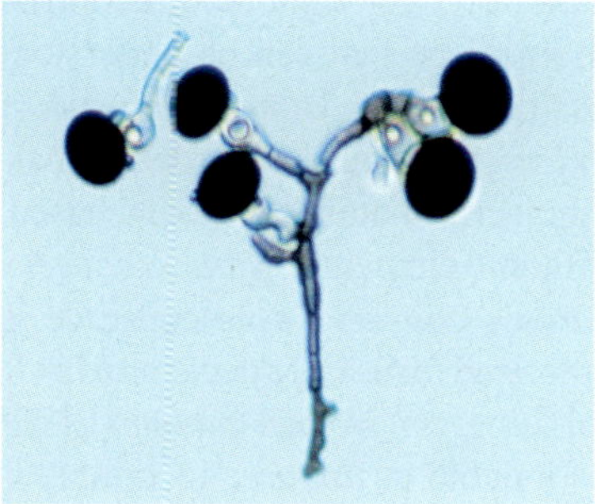

Nigricant mature conidia of *Nigrospora sphaerica*, x 800 (*RTH*).

nimbospore (L. *nimbus*, dense cloud + Gr. *sporá*, spore): a spore with a gelatinous wall, apparently of many layers, like that of *Histoplasma capsulatum* (moniliaceous asexual fungi).

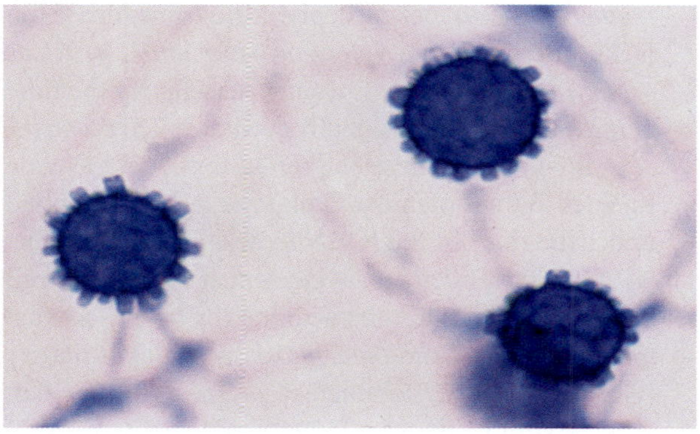

Nimbospores (macroconidia) of *Histoplasma capsulatum*, x 1,500 (*CB*).

nitid, nitidous (L. *nitidus*, bright, lustrous < *nitere*, to shine): smooth, clear or dark but shiny or polished, like the black, glistening sporodochium of *Schizoxylon* (=*Agyriella*), of the Ostropales, and the thallus of *Pyrenula nitida* (Pyrenulales).

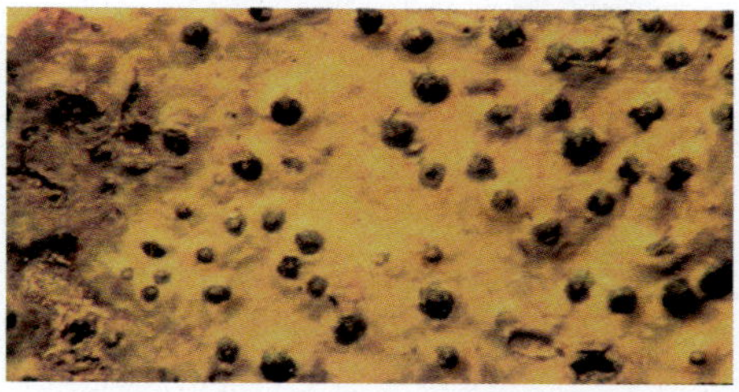

Nitid thallus of *Pyrenula nitida*; the dark spots represent the ostioles of the perithecia, x 4 (*MU*).

nitrophilous (Gr. *nítron*, saltpetre, here alluding to the nitrogen + *phílos*, have an affinity for): having a preference for habitats rich in nitrogen. Cf. **nitrophobous**.

nitrophobous (Gr. *nítron*, saltpetre, here alluding to the nitrogen + *phóbos*, action of moving away): preferring habitats poor in nitrogen. Cf. **nitrophilous**.

nival (L. *nivalis*, relating to snow < *nix*, genit. *nivis*, snow + suf. *-alis* > E. *-al*, relating to or pertaining to): living in or near the snow, or being white in color like snow. E.g., *Fusarium poae* and *F. sporotrichioides* (tuberculariaceous asexual fungi) grow very well in soil and plant remains covered by snow, and *Monographella nivalis* (=*Fusarium nivale*) forms a white mycelium.

niveous (L. *niveous*, snowy): snow white, pure, without blemishes, like the basidiocarp of *Coprinopsis nivea* (=*Coprinus niveus*), of the Agaricales.

nivicolous (L. *nix*, genit. *nivis*, snow + suf. *-cola*, inhabitant): living in snow, like some species of Myxomycetes that are associated with *Pinus silvestris* vegetation, on detritus near the snow, such as *Trichia sordida* var. *sordidoides* (Trichiales) and *Comatricha alpina* (Stemonitales). Also called **chionophilous**.

nocti- (L. *nox*, genit. *noctis*, night): a pref. employed in terms that allude to the night, as in *noctilucent*. See **nicto-**.

noctilucent (L. *nox*, genit. *noctis*, night + *lucentis*, bright, shining, visible, genit. of *lucere*, to shine < Gr. *leukós*, white, bright + L. suf. *-entem* > E. *-ent*, being): emitting light and shining during the night, like the basidiocarps of *Mycena lux-coeli* (Agaricales), which exhibit the phenomenon of luminescence, i.e., the production of light of weak intensity without being incandescent. See **bioluminescent**.

Noctilucent basidiocarps of *Mycena lux-coeli*, x 1 (*MU*).

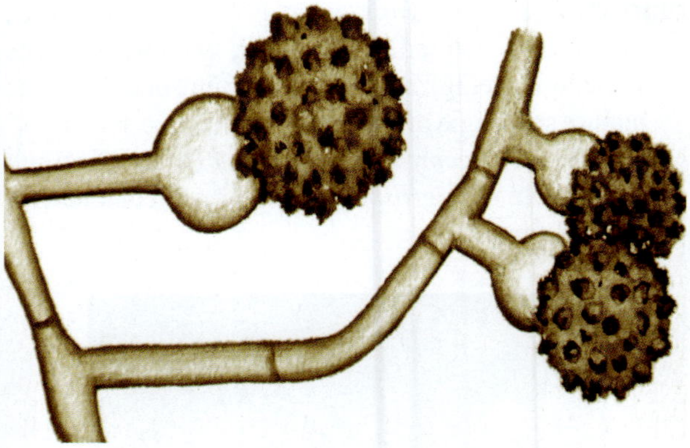

Nodulose conidia of *Mycogone perniciosa*, x 400 (*MU*).

node cell (L. *nodus*, knot ; NL. *cellula*, cell, living cell, dim. of L. *cella*, small room, compartment, cell of a honeycomb): *Meliolales*. A stigmatocyst that forms directly on a hypha, as opposed to one that forms as the terminal cell of a hyphopodium; the stigmatocyst is the cell that gives rise to the haustorium. See **stigmatocyst**.

noduliferous (L. *nodulus*, dim. of *nodus*, knot + suf. *-ferous*, bearer): see **nodulose**.

nodulose (L. *nodulosus*, knotty, rough < *nodulus*, dim. of *nodus*, knot + suf. *-osus* > E. *-ose*, full of, augmented, prone to): generally applied to spores whose wall has blunt excrescences with a broad base, similar to warts, like the aleuriospores of *Mycogone perniciosa*, the macroconidia of *Histoplasma capsulatum* (moniliaceous asexual fungi), and the ascospores of *Sordaria nodulifera* (Sordariales).

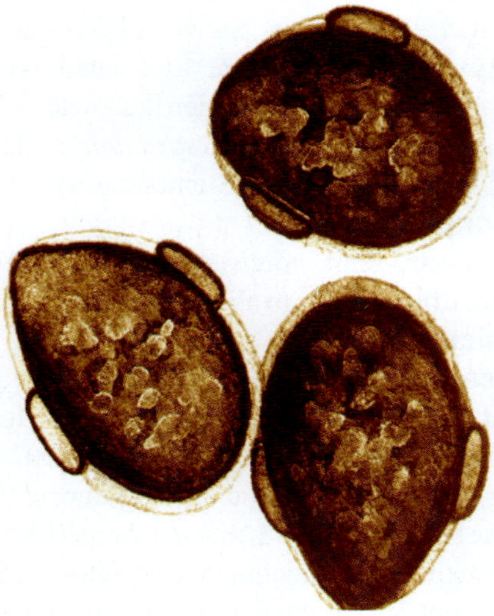

Nodulose ascospores of *Sordaria nodulifera*, x 1,400 (*MU*).

nodum, pl. **noda** (L. *nodum*, neuter of *nodus*, knot): a term employed in plant sociology to refer to a well-defined plant community.

nomen (L. *nomen*, name): name; refers to scientific names, i.e., a genus, species, binomial, etc. *Nomen ambiguum* (L. *ambiguus*, ambiguous < *ambigere*, to vacillate, err, go without direction): a doubtful or uncertain name that can be understood in two or more possible senses. *Nomen confusum* (L. *confusus* < *confundere*, to confuse, disorder, to make indistinct or mix indiscriminately): a taxon based on two or more different elements. *Nomen conservandum* (L. *conservatus* < *conservare*, to conserve, preserve): a name authorized for use by a decision of an International Botanical Congress. *Nomen dubium* (L. *dubius*, doubtful, questionable < *dubare*, to doubt, vacillate): a doubtful name, of uncertain sense. *Nomen illegitimum* (L. *in-*, not + *legitimus*, legitimate, valid): a validly published name but contravening particular articles of the International Code of Botanical Nomenclature (ICBN). *Nomen invalidum* (L. *invalidus*, not valid, weak < *in-*, not + *validus*, effective): a name that is not validly published. *Nomen novum* (L. *novus*, new): a new name; a replacement, substitution. *Nomen nudum* (L. *nudus*, naked, deprived): a name for a taxon lacking a description. *Nomen provisorium* (L. *provisus* < *providere*, to foresee): a name serving for the time being; proposed provisionally. *Nomen rejiciendum* (L. *rejectus* < *reicere*, to reject, refuse to accept): a name rejected by a Botanical Congress. A name applied under the rules of the ICBN to designate the perfect state (characterized by having sexual spores) as well as the imperfect state (characterized by having asexual spores) is a *nomen holomorphosis*, pl. *holomorphoses* (Gr. *hólos*, entire, complete + *morphé*, form + suf. *-osis*, state or condition); the name for an asexual state is a *nomen anamorphosis*, pl. *anamorphoses* (Gr. *aná-*, anew, again + *morphé* + *-osis*).

nomenclature (L. *nomenclatura*, list of names): all of the technical words or names related to a faculty or science. *Binary nomenclature* (LL. *binarus* < L. *bini*, composed of two things or parts): a binary name for each species is formed of two terms, one generic and the other specific. See **systematics**.

non (L. *non*, not; the reverse of, absence of, not important, without value): a term used in floristic catalogs and systematic works that is placed before an author's name to indicate that the preceding concept is *not* his. The expression *non vidi* is also employed, generally at the end of the description of a species or variety, to indicate that the author of the description *did not see* the organism described.

nona- (L. *nonus*, ninth): a pref. that indicates that something is repeated nine times, as in *nonagon*, a polygon with nine angles and nine sides.

non-persistent attachment (L. *non*, not; *persistens*, genit. *persistentis*, persistent, continuous, permanent < *persistere*, to persist, continue + suf. *-entem* > E. *-ent*, being; ME. *attachement*, seizure < ME. *atachen* < AF. *atacher*, to seize < OF. *atachier*, to fasten + E. suf. *-ment*, action or resulting state < L. suf. *-mentum*): refers to fruiting bodies that break their attachment to the soil, becoming free to be dispersed by the wind, as occurs in *Bovista* (Lycoperdales).

noso- (Gr. *nósos*, illness, disease): a pref. employed in compound terms that allude to diseases, as in *nosology*, the area of medicine or pathology that deals with the description, differentiation and classification of diseases.

notate (L. *notatus*, designated, pointed out < *notare*, to note, mark + suf. *-atus* > E. *-ate*, provided with or likeness): with the surface provided with lines or marks notable for their color or some other characteristic, like the colony of *Penicillium chrysogenum* (=*P. notatum*), moniliaceous asexual fungi, when grown on a solid substrate, which appears zonate, with concentric rings of conidia alternating with areas of non-sporulating mycelium.

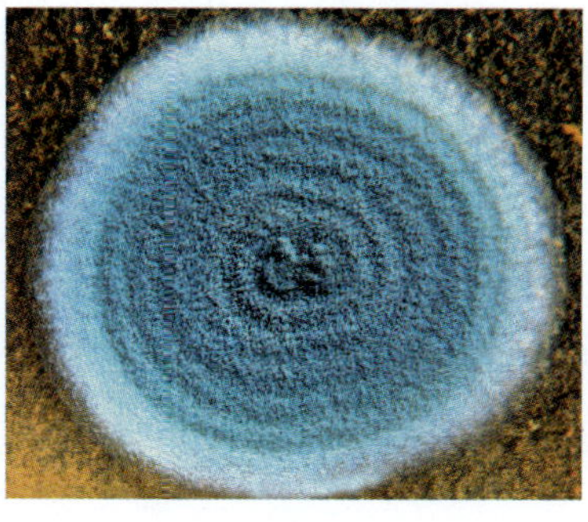

Notate colony of *Penicillium chrysogenum* (=*P. notatum*) on agar, x 2 (*MU*).

noto-, **not-** (Gr. *nótos*, back): a pref. used in names such as *Notothyrites*, a fossil fungus of the family Microthyriaceae (Dothideales), whose name means a fossil with a window in the dorsal part, alluding to the ostiole of the thyriothecium.

nuclear cap (E. *nuclear* < L. *nuculeus*, kernel < *nucula*, dim. of *nux*, genit. *nucis*, nut, nucleus + suf. *-aris* > E. *-ar*, like, concerning; ME. *cappe* < LL. *cappa*, hooded cloak, cap): a structure composed of the ribosomes of the cell which aggregate near the nucleus and are surrounded by a membrane that appears to be an extension of the nuclear envelope; it is characteristic of the zoospores of the Blastocladiales (such as *Allomyces*, *Blastocladia* and *Blastocladiella*).

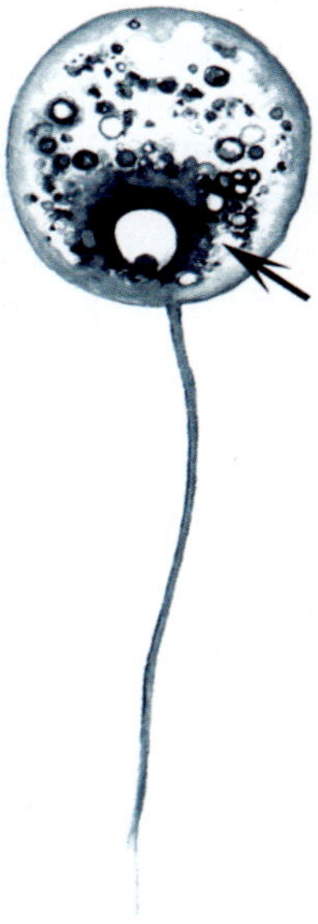

Nuclear cap of the male gamete of *Allomyces macrogynus*, x 1,350 (*MU*).

nucleus, pl. **nuclei** (L. *nuculeus*, grain < *nucula*, dim. of *nux*, genit. *nucis*, nut, nucleus): in biology it refers to an organelle present in the protoplasm of the majority of plant and animal cells (eukaryotic), except in certain lesser organized forms (prokaryotic), and regarded as an essential factor in constructive metabolism, growth, and in the reproduction and transmission of hereditary characters. It typically consists of a rounded or oval mass of protoplasm enclosed in a delicate perforated membrane (the nuclear membrane); it contains one or more nucleoli and the chromatin, which during nuclear division condenses to form the chromosomes. The nuclei of most fungi are quite small and are best observed after being fixed and stained with dyes, such as Giemsa, iron hematoxylin, acetoorcein and acetocarmin, as well as fluorescent stains, including 4',6-diamidino-2-phenylindole and mitramycin. Nuclear division in the fungi is basically intranuclear, i.e., the major part of the nuclear membrane remains intact until late telophase, when it breaks in the interzonal region and then reforms around the daughter nuclei. The ultrastructural details of mitosis and meiosis in fungi are valuable as phylogenetic characters.

nummiform (L. *nummus*, coin, money + suf. *-formis* < *forma*, shape): shaped like a coin, like the argentate peridioles of *Cyathus intermedius* (Nidulariales) and the perithecial stromata of *Nummularia* (Xylariales).

nutant (L. *nutans*, genit. *nutantis*, pendant, hanging): inclined, hanging.

nutriocyte (L. *nutrire*, to nourish, feed + Gr. *kýtos*, hollow receptacle; here, applied with the meaning of cell): the inflated portion of the ascogonium of *Ascosphaera* (Ascosphaerales), which receives the contents of the trichogyne or gynotrichous hypha after plasmogamy has occurred; eventually the nutriocyte or nutriocyst develops into a *sporocyst*.

Nummiform peridioles in a fruiting body of *Cyathus intermedius*, x 7 (*CB* & *MU*).

O

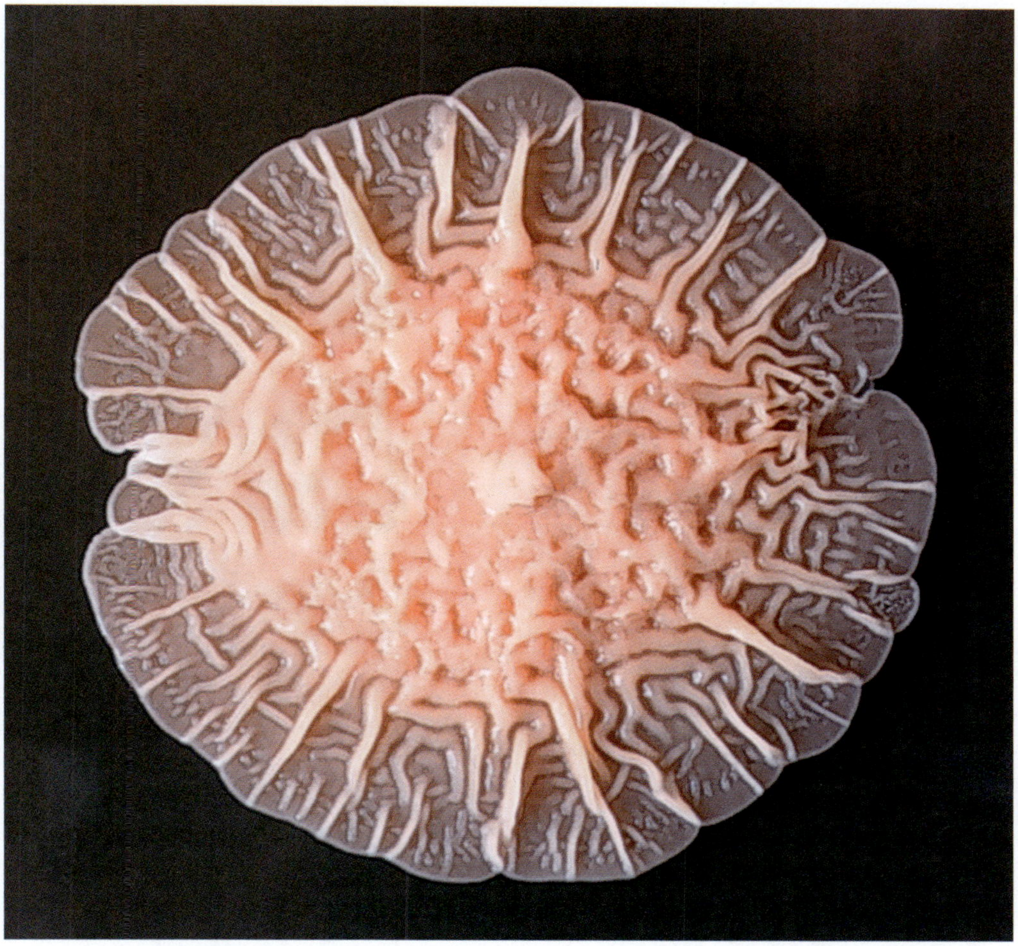

Ornamented giant colony of *Cryptococcus curvatus*, on agar (*SA*).

O

ob- (L. pref. with various meanings: before, toward, in front, against, on, above): used to give words an intensive sense, or opposite or inverted meaning, as in the following terms.

obclavate, **obclaviform** (L. *ob-*, inverted + NL. *clavatus*, club-shaped < L. *clava*, club, mallet, maul + suf. *-atus* > E. *-ate*, provided with or likeness; L. *obclaviformis* < *ob-* + *claviformis* < *clava* + suf. *-formis* < *forma*, shape): shaped like a club or maul, but with the broad part at the base, like the conidia of *Triramulispora obclavata* and the synnemata of *Beauveria felina* (=*B. cretacea*) and *Drechmeria* (=*Meria*) *coniospora* (moniliaceous asexual fungi), and the cheilocystidia of *Mycena sanguinolenta* (Agaricales). Cf. **clavate**.

obconic (L. *ob-*, inverted + Gr. *konikós*, conic < *kónos*, cone, pine cone + suf. *-íkos* > L. *-icus* > E. *-ic*, pertaining to or relating to): shaped like an inverted cone, with the widest part at the top, like the conidia of *Acaulopage tetraceros* (Zoopagales). Similar to *turbinate*.

Obclavate synnemata of *Beauveria felina* (=*B. cretacea*) on agar, x 5 (*MU*).

Obclavate conidia of *Drechmeria* (=*Meria*) *coniospora*, x 660 (*MU*).

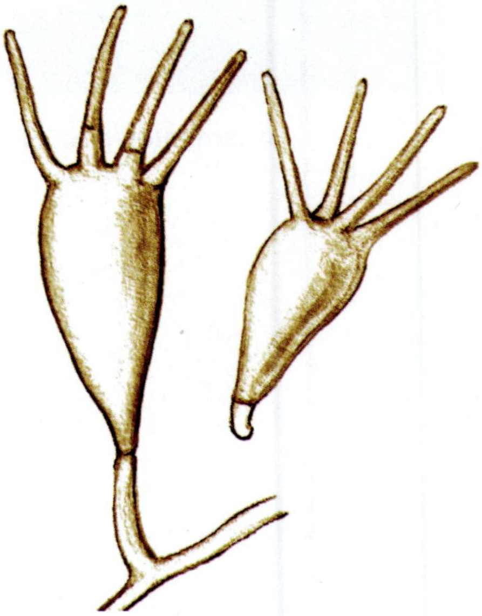

Obconic spores of *Acaulopage tetraceros*, x 800 (*MU*).

oblate (L. *oblatus* < *ob*-, with an intensive sense + *latus*, wide, broad): literally, very wide, wider than long. An oblate spore is wide laterally and flattened at the poles; similar to the shape of an orange. Cf. **prolate**.

obligate or **facultative parasite** (L. *obligatus*, obligate, strict; Gr. *parásitos* < *pará*, together, at the side of + *sîtos*, bread, food in general): an organism that lives only at the expense of another living organism. Among the fungi, a minority of species are obligate parasites, including the gen. in the family Peronosporaceae (*Peronospora*, *Albugo*, etc.), those in the order Erysiphales (*Blumeria*, *Erysiphe*, *Podosphaera*, etc.), and the rusts of the order Uredinales (*Puccinia*, *Uromyces*, etc.). Although a few species of obligately parasitic fungi have been induced to form distinct phases of their life cycle on synthetic culture media in the laboratory, they are still considered as obligate parasites, as that is their behavior in nature. Cf. **obligate** or **facultative saprobe**.

obligate or **facultative saprobe** (L. *obligatus*, obligate, strict; Gr. *saprós*, rotten + *bíos*, life): an organism that can live only on dead organic material, for which reason it is incapable of infecting a living being, i.e., it cannot live as a parasite. Cf. **obligate** or **facultative parasite**.

oblique (L. *obliquus*, slanting, inclined across or deviating from the horizontal): applied to a type of septum whose plane of contact with another wall results in an angle that is not a right angle, as occurs in some dictyosporous ascospores.

oblong (L. *oblongus* < *ob*-, with an intensive sense + *longus*, long): literally, very long, longer than wide. An oblong spore is elongate, with the sides nearly parallel and the corners more or less rounded, somewhat flattened at the ends, like the ascospores of *Coniochaeta scatigena* (=*Ascobolus scatigenus,* Pezizales), the conidia of some species of *Colletotrichum* (melanconiaceous asexual fungi), the ascospores of *Peziza ammophila* (Pezizales) and the basidiospores of *Tubaria furfuracea* (Agaricales). An *oblong-ellipsoid* spore has the long sides parallel and the ends almost hemispherical, like the basidiospores of *Schizophyllum commune* (Aphyllophorales).

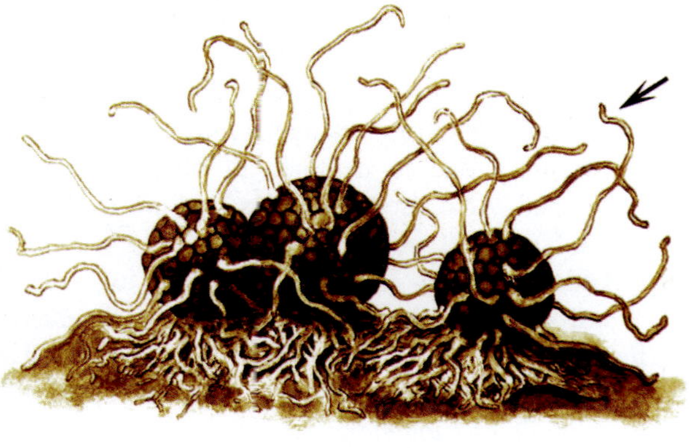

Obligate parasite, ascocarps with mycelioid appendages of *Blumeria* (=*Erysiphe*) *graminis*, on the host surface, x 190 (*MU*).

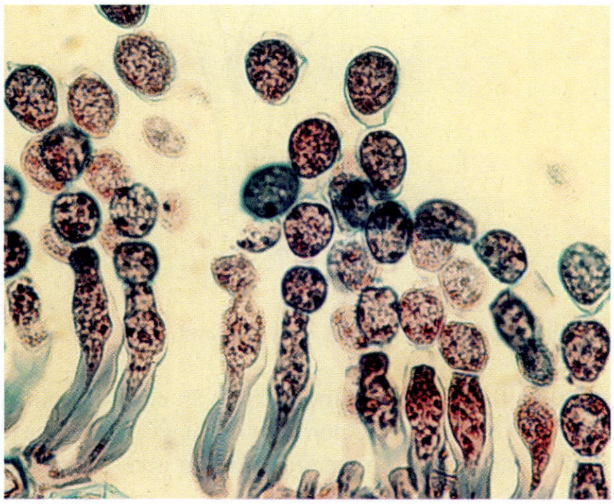

Obligate parasite, conidiosporangia of *Albugo candida*, the white rust of Cruciferae, x 420 (*MU*).

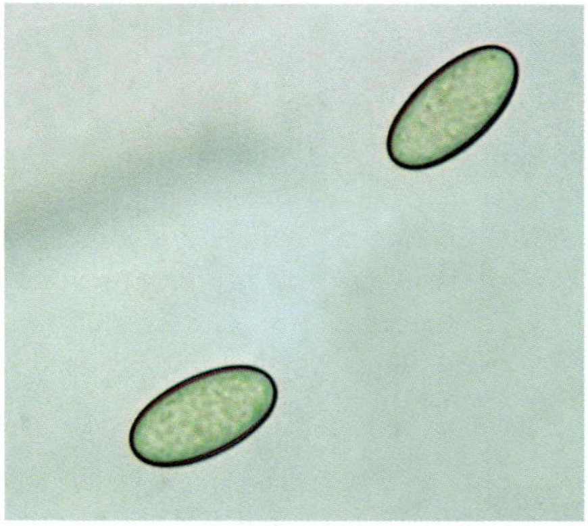

Oblong ascospores of *Coniochaeta scatigena* (=*Ascobolus scatigenus*), x 1,160 (*EAA*).

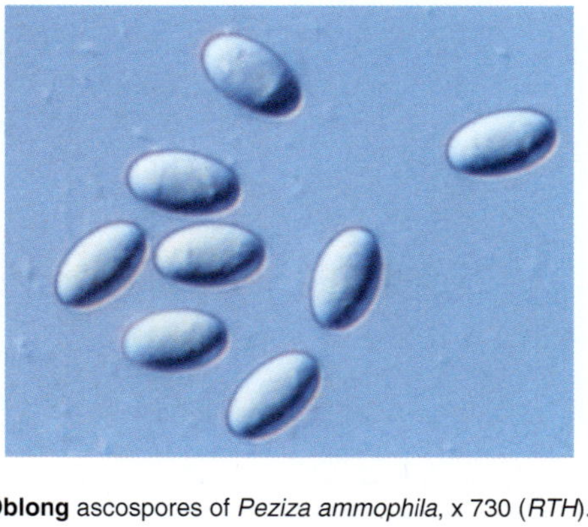

Oblong ascospores of *Peziza ammophila*, x 730 (*RTH*).

obovate, **obovoid** (L. *ob-*, inverted + *ovatus*, ovate < *ovum*, egg + suf. *-atus* > E. *-ate*, provided with or likeness; or + *ob-* + NL. *ovoides*, egg-shaped < *ovum* + suf. *-oide*, like < Gr. *-oeides*, resembling, similar to): with the shape of an inverted egg; inversely ovate or ovoid, i.e., with a broad apex and a narrow base, like the sporangia of *Leocarpus fragilis* (Physarales), the conidia of *Brachysporium obovatum* and *Cordana musae* (dematiaceous asexual fungi), and the urediniospores of *Uromyces viciae-fabae* (=*U. fabae*), Uredinales. Cf. **ovate** and **ovoid**.

Obovate sporangia of *Leocarpus fragilis*, x 50 (*MU* & *CB*).

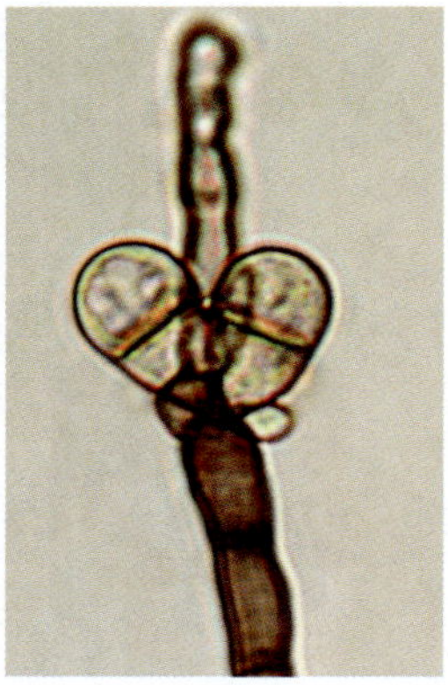

Obovate conidia of *Cordana musae*, x 740 (*MU*).

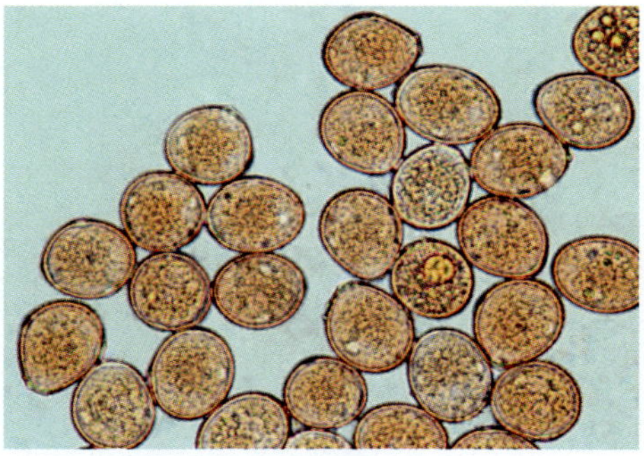

Obovate urediniospores of *Uromyces viciae-fabae* (=*U. fabae*), x 270 (*MU*).

obpyriform (L. *ob-*, inverted + *pyriformis* < *pyrum*, pear + suf. *-formis* < *forma*, shape): inversely pyriform (pear-shaped), i.e., with the widest part at the point of attachment or base, like the primordium of the sorocarp of *Dictyostelium discoideum* (Dictyosteliomycota) and the majority of pycnidia (sphaeropsidaceous asexual fungi) and perithecia (Pyrenomycetes). Cf. **pyriform**.

Obpyriform primordium of a sorocarp of *Dictyostelium discoideum*, x 90 (*RTH*).

obsolete (L. *obsoletus*, old, without lustre): applied to organic organs or parts that are not very apparent, rudimentary due to decline, atrophied or poorly developed. Also used for terms that are no longer in use.

obsubulate (L. *ob-*, inverted + *subula*, awl, chisel + suf. *-atus* > E. *-ate*, provided with or likeness): inversely subulate; very narrow and sharp-pointed at the base and widening a little toward the apex, like the conidia of *Flagellospora curvula* (moniliaceous asexual fungi). Cf. **subulate**.

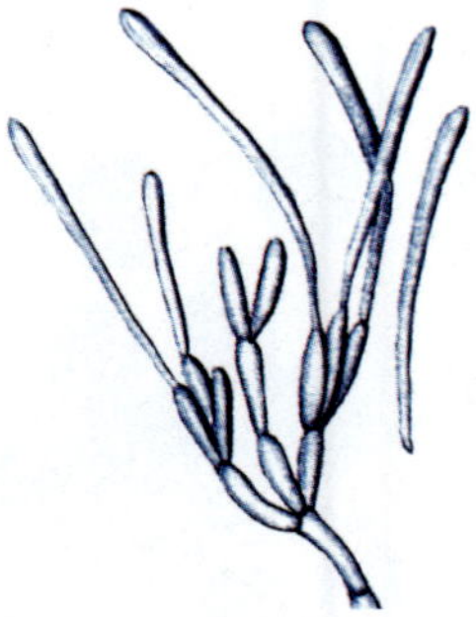

Obsubulate conidia of *Flagellospora curvula*, x 400 (*MU*).

obtuse (L. *obtusus*, blunt, without an edge or point): **1**. A solid, blunt structure that is rounded at the apex. **2**. A structure whose edges form an obtuse angle at the apex (more than a right angle); applied to pilei, cystidia, spores, etc. E.g., the cylindrical cystidia of *Leptonia trichomata* (Agaricales) are obtuse.

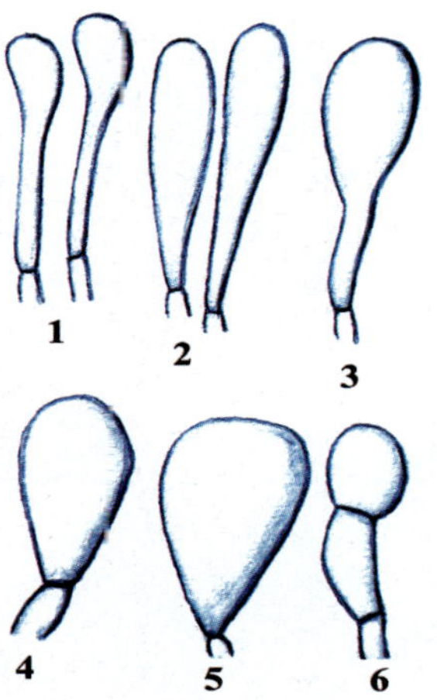

Obtuse apex of the cystidia of the hymenium of *Entoloma sericatum* (1, cylindro-clavate), *Leptonia decolorans* (2, clavate), *Entoloma lividocyanulum* (=*Leptonia lividocyanula*) (3, sphaeropedunculate), and *Entoloma formosum* (=*Leptonia fulva*) (4, napiform, 5, turbinate and 6, vesiculate), x 500 (*MU*).

obverse (L. *obversus* < *obvertere*, to turn toward < *ob-*, toward + *vertere*, to turn; facing the observer or opponent; constituting the obverse of something; opposite): the obverse of a colony cultured on agar media has some characteristics that complement the characteristics of the *reverse* of the same colony. The characters of the obverse and reverse of the colonies are considered in the taxonomy of various groups of fungi, e.g., as in *Lecanicillium* (=*Verticillium*), *Penicillium* and *Aspergillus* (moniliaceous asexual fungi). Cf. **reverse**.

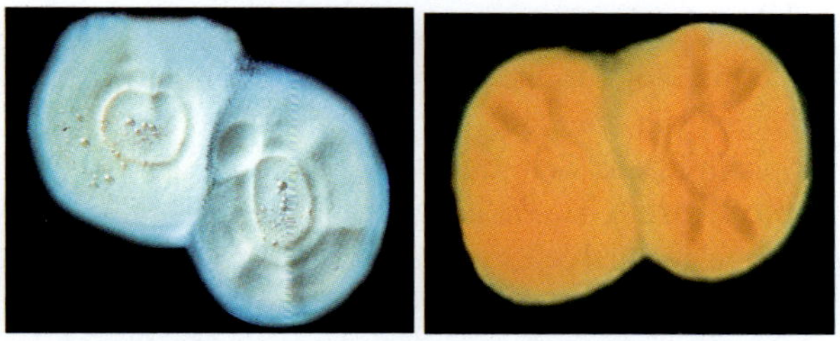

On the left, **obverse** of one week-old colonies of *Lecanicillium* (=*Verticillium*) *lecanii* on agar. On the right, the reverse of the same colonies, x 1 (*MU*).

occluded (L. *occludere*, to close, block a passage): closed; refers to the lumen of pseudoparenchymatous hyphae or cells when it is very narrow.

ocellate, **ocellated** (L. *ocellatus*, with spots that resemble small eyes < *ocellus*, small eye + suf. *-atus* > E. *-ate*, provided with or likeness): with spots, patches or points that resemble small eyes, like the thallus of the lichen *Schismatomma ocellatum* (Arthoniales), whose appearance and name are due to the minute apothecia that it has on the surface.

ochro-, **ochr-** (Gr. *ochrós*, pale yellow): a pref. used in terms such as *ochroleucous*, pale yellow in color, between yellow and white; or yellow and white.

ochrosporous (Gr. *ochrós*, pale yellow + *sporá*, spore + L. suf. *-osus* > E. *-ous*, full of, abounding in, possessing the qualities of): refers to a fungus with a yellow, ochre or ochraceous-brown spore print, like the basidiocarp of *Armillaria* (=*Armillariella*, of the Agaricales).

ocreate (L. *ocreatus*, sheathed, lined < *ocrea*, sheath, covered, boot, shoe that covers the foot and part of the leg + suf. *-atus* > E. *-ate*, provided with or likeness): provided with an *ocrea*, which in the fungi corresponds to the volva that envelops the base of the stipe of the basidiocarp of some agarics, such as *Amanita caesarea* and *A. pantherina* var. *velatipes* (Agaricales). This term is syn. of **caligate** and **peronate**.

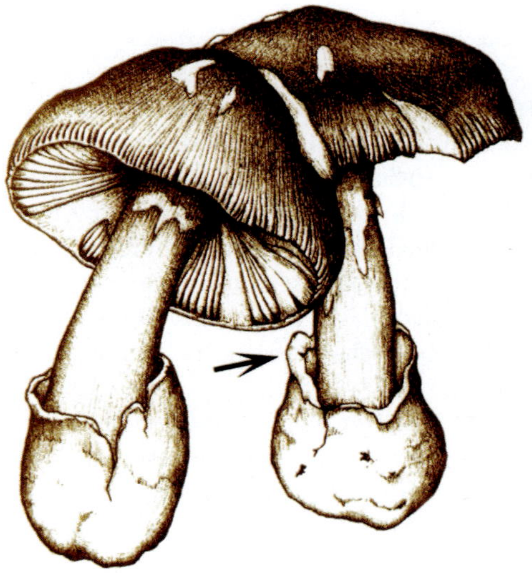

Ocreate volva of the basidiocarps of *Amanita caesarea*, x 0.5 (*MU*).

octo-, **oct-** (Gr. *októ* > L. *octo*, eight): a pref. used to form words that refer to something that is repeated eight times, as in the following terms.

octophore (Gr. *októ* > L. *octo*, eight + Gr. suf. *-phóros*, bearer): see **haerangium**.

octopolar (Gr. *októ* > L. *octo*, eight + L. *polaris*, pertaining to or relating to the pole): refers to incompatibility mating systems that have three loci. Cf. **bipolar** and **tetrapolar**.

octosporate, **octosporic**, **octosporous** (Gr. *októ* > L. *octo*, eight + Gr. *sporá*, spore + L. suf. *-atus* > E. *-ate*, provided with or likeness; or + Gr. suf. *-íkos* > L. *-icus* > E. *-ic*, pertaining to or relating to; or + L. suf. *-osus* > E. *-ous*, full of, abounding in, possessing the qualities of): producing spores in groups of eight, as occurs in many asci.

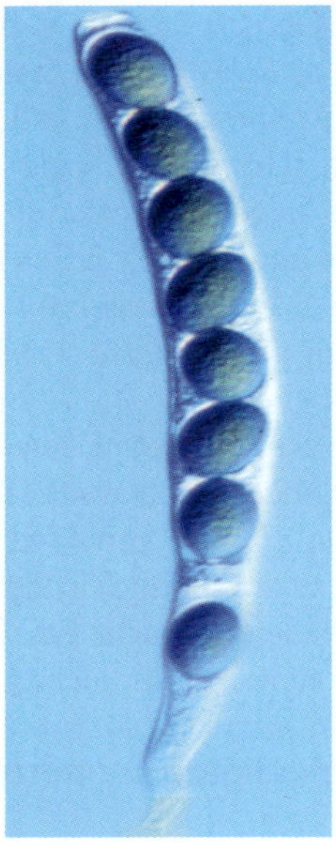

Octosporate ascus of *Sordaria fimicola*, x 270 (*RTH*).

octospore (Gr. *októ* > L. *octo*, eight + Gr. *sporá*, spore): a spore from an eight-spored ascus.

-odium, pl. **-odia** (L. suf. *-odium* < Gr. *-oeídes*, like, similar to): a suf. employed in forming terms that express likeness or similarity, referring to shape as well as to other aspects, as in *plasmodium*, a mass of protoplasm that results from the fusion of myxamoebae, characteristic of the Myxomycetes. See **-oid**.

odontoid (Gr. *odoús*, genit. *odóntos*, tooth + suf. *-oeídes* > L. *-oideus* > E. *-oid*, likeness or similarity): similar to a tooth, dentate, like the hymenophore of *Hericium erinaceus* (Aphyllophorales).

Odontoid basidiocarp of *Hericium erinaceus* growing on the bark of a tree, x 0.2 (*RV*).

Odontoid basidiocarps of *Pseudohydnum gelatinosum* (Tremellales), x 0.5 (*RV*).

-oecious (Gr. *oîkos*, house or inn): a suf. employed in forming terms that allude to the host of a parasite, as in *dioecious* and *monoecious*, i.e., a parasite with two distinct hosts, and with a single host, respectively.

oedocephaloid (< gen. *Oedocephalum* < Gr. *oîdos*, tumor, swelling < *oidéo*, to swell up + *kephalé*, genit. *kephalós*, head + suf. *-oeídes* > L. *-oideus* > E. *-oid*, likeness or similarity): having a swelling at the apex, like the conidiophores of *Oedocephalum* (moniliaceous asexual fungi) and *Cunninghamella* (Mucorales).

-oid (Gr. *oeídes* > L. *-oideus* > E. *-oid*, likeness or similarity): a suf. often employed in biological terminology to indicate likeness or similarity, as in *agaricoid*, *phalloid*, etc., i.e., similar to an agaric, a phalloid, etc. See **-odium**.

oidiophore (Gr. *oidión*, small egg or ovule < *oón*, egg, female sexual cell + suf. *-phóros*, bearer): a specialized hypha that produces and supports oidia, as is seen in *Coprinopsis* (=*Coprinus*) *lagopus* (Agaricales).

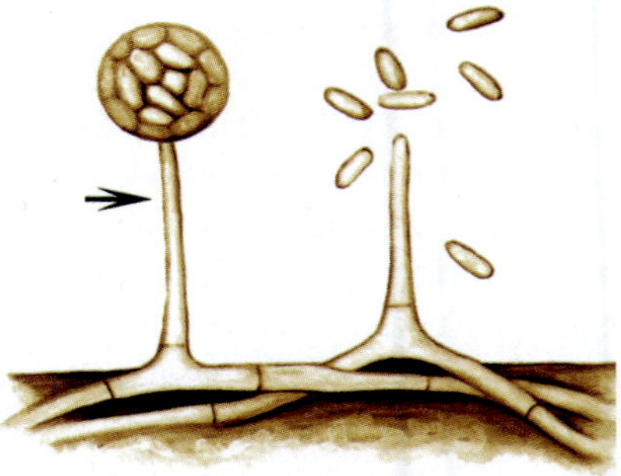

Oidiophore with **oidia** of *Coprinopsis* (=*Coprinus*) *lagopus*, x 480 (*MU*).

oidiospore (Gr. *oidión*, small egg or ovule < *oón*, egg, female sexual cell + dim. suf. *-ídion* + *sporá*, spore): see **oidium**.

oidium, pl. **oidia** (Gr. *oidión*, small egg or ovule < *oón*, egg, female sexual cell + dim. suf. *-ídion*): a small hyphal cell, free, with truncate ends and a thin cell wall, that is derived from the fragmentation of somatic hyphae, or from oidiophores, and which behaves as a sexual spore in some Agaricales, such as *Coprinopsis* (=*Coprinus*) *lagopus*; sometimes oidia function as spermatia and carry out the dikaryotization of the somatic hyphae with which they fuse (oidization). The term oidium gave rise to the name of the gen. *Oidium*, which represents the asexual or conidial phase of *Erysiphe* (Erysiphales). Also called **oidiospore** or **arthrospore**, as in *Geotrichum* (moniliaceous asexual fungi).

oidization (Gr. *oidión*, small egg or ovule < *oón*, egg, female sexual cell + dim. suf. *-ídion* + L. suf. *-ationem* > E. *-ation*, action, state or condition, result): dikaryotizaion that results from the fusion of an oidium with a haploid hypha.

oleaginous (L. *oleaginosus*, from an olive < *olea*, *oleagin*, olive + suf. *-osus* > E. *-ose*, having, possessing the qualities of): see **lubricous**.

oleiferous hypha, pl. **hyphae** (L. *oleum*, grease, oil + suf. *-ferous*, bearer < *ferre*, to bear, carry; NL. *hypha*, hypha < Gr. *hyphé*, tissue, spiderweb; hypha): **1**. *Lichens*. A hypha that contains torulose, guttulate (with oil) cells, like the sumerged hyphae of the endolithic species. **2**. *Basidiomycetes*. A hypha that contains resinous substances, like those of *Russula emetica* and *Amanita vaginata* (Agaricales).

oleocystidium, pl. **oleocystidia** (L. *oleum*, fat, oil + Gr. *kýstis*, bladder, vesicle; here, cell + dim. suf. *-ídion*): a cystidium with an oily or resinous exudate, present in the hymenium of *Resinomycena* and other Agaricales.

oleose (L. *oleosus*, oily < *oleum*, fat, oil + suf. *-osus* > E. *-ose*, having, possessing the qualities of): containing oil, like some spores and hyphae. The so-called *oleose-loculate* spores have cells resembling droplets of oil.

oligo-, **olig-** (Gr. *olígos*, few, a few): a pref. often employed to introduce the idea of something that is smaller in number than ordinary, as in *oligosporous*.

oligosporous (Gr. *olígos*, few, a few + *sporá*, spore): having few spores, like the sporangia of *Rhizopus microsporus* var. *oligosporus* (=*Rh. oligosporus*) and *Thammnidium elegans*, Mucorales. Cf. **myriosporous**.

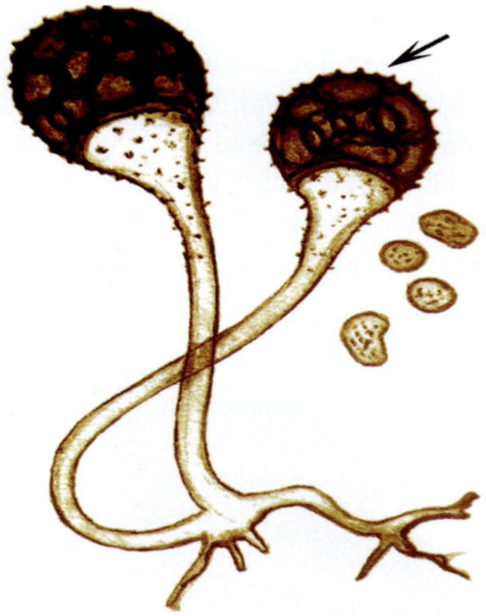

Oligosporous sporangia of *Rhizopus microsporus* var. *oligosporus* (=*Rh. oligosporus*), x 400 (*MU*).

oligotrophic (Gr. *olígos*, few, a few + *trophé*, action of nourishing): applied to an organism that prospers in media poor in assimilable nutrients, above all in nitrogen. Also refers to an aquatic medium poor in dissolved substances, and therefore, not very favorable for the growth of algae and other vegetation and other forms of aquatic life. Cf. **distrophic** and **eutrophic**.

-oma, pl. **-omata** (Gr. *-óma*, genit. *ómatos*, morbid condition, tumor): a suf. that refers to a pathological state (*fibroma*, *glaucoma*, etc.) or expresses the idea of a group or the generalization of a concept, as in *ascoma*, *basidioma*, etc., that is applied, respectively, to a fruiting body with asci, basidia, etc.

ombro-, **ombr-** (Gr. *ómbros*, shower, heavy rain): a pref. employed in forming terms that allude to rain, as in *ombrophilous*, applied to an organism capable of withstanding heavy rain, like those that live in wet tropical regions. The generic name *Ombrophila* (Helotiales) means fond of rain.

omnivorous (L. *omnivorous*, eating everything < *omnis*, all + *vorare*, to eat): capable of growing on all kinds of substrates or nutrient media. Strictly speaking, this concept is false, although there are *plurivorous* or *polytrophic* fungi, i.e., those that can utilize very diverse substances, or that thrive on the most varied substrates, whether as saprobes or as parasites, like the common molds of the gen. *Aspergillus* and *Penicillium* (moniliaceous asexual fungi). Syn. of **plurivorous** and **polytrophic**. Cf. **monotrophic**.

omphalinoid (< gen. *Omphalina* < Gr. *omphalós*, navel + L. suf. *-ina*, likeness + L. suf. *-oide*, like < Gr. *-oeídes*, similar to): *Agaricales*. A type of fruiting body (13 principal types are recognized) with gills adherent, decurrent or subdecurrent on the stipe, which is cartilaginous, central and lacking a ring and volva, like that of *Omphalina*, *Omphalotus* and certain species of *Marasmius* and *Psilocybe*. It is similar to the mycenoid and collybioid types but differs in its broadly convex to flat pileus, at times umbilicate, and by its decurrent gills. It is also similar to the anellarioid type but differs in lacking a ring. See **omphaloid**.

Omphalinoid fruiting bodies of *Omphalotus olearius*, x 1 (*MU*).

omphalodisc (Gr. *omphalós*, navel + *dískos*, disk): *Lichens*. An apothecium that has a central prominence composed of sterile hyphae, which gives it the appearance of a navel, such as occurs in *Umbilicaria* (=*Omphalodiscus*) *virginis* and *U. crustulosa* (Lecanorales).

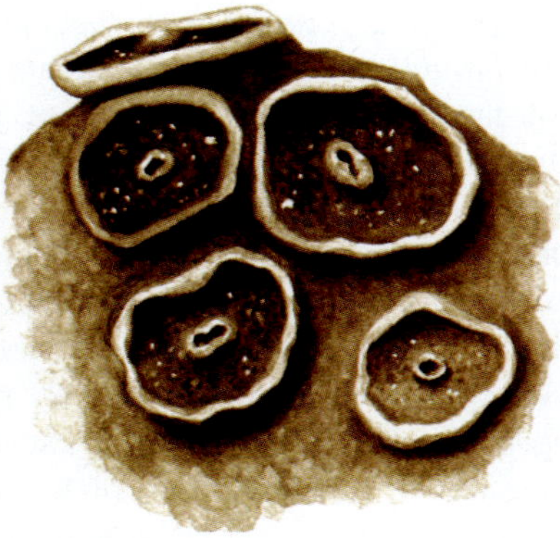

Omphalodiscs (apothecia) of the thallus of *Umbilicaria* (=*Omphalodiscus*) *virginis*, x 10 (*MU*).

omphaloid (< gen. *Omphalia* < Gr. *omphalós*, navel + L. suf. *-oide*, like < Gr. *-oeídes*, similar to): with the characteristics of the agaric gen. *Omphalina* (=*Omphalia*). See **omphalinoid**.

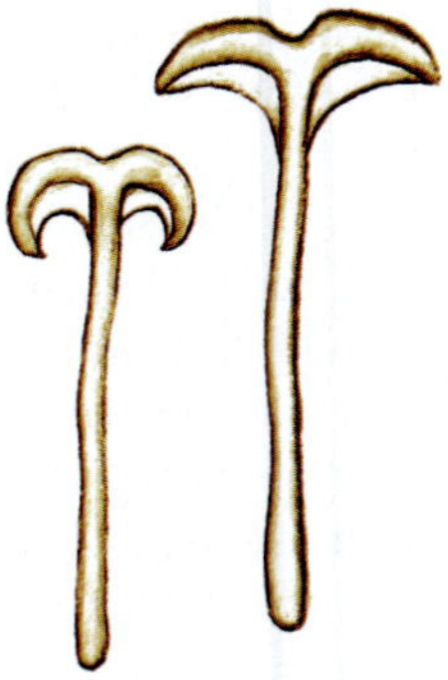

Omphaloid type of basidiocarp, represented by *Omphalina* (=*Omphalia*) sp.; diagram of longitudinal sections, x 1 (*MU*).

one-spored basidium, pl. **basidia** (IE. *oinos* > OE. *an* > ME. *oon* > E. *one*, one + NL. *spora* < Gr. *sporá*, spore + E. suf. *-ed*, past tense; L. *basidium*, small base < Gr. *basídion* < *básis*, base + dim. suf. *-ídion* > L. *-idium*): a basidium that forms a single sterigma on which a single spore develops.

-ont (Gr. *óntos*, genit. of *ón*, a being): a suf. used in terms such as *biont*, *diplont*, *symbiont*, etc., which mean, respectively, a living being, a being with diploid cells, one who lives together (with another or others), etc. See **onto-**.

ontjom (Indo. *ontjom*): a fermented Indonesian food prepared from peanut or cassava paste, whose surface is inoculated with *Neurospora intermedia* (Sordariales). Cf. **pozol**, **tapé ketella** and **tempe**. See **fermentation**.

Red **ontjom**, prepared with *Neurospora intermedia*, x 0.4 (*MU*).

onto- (Gr. *óntos*, genit. of *ón*, a being): a pref. that means the one who is or what is, a being. See **-ont**.

ontogeny (Gr. *óntos*, genit. of *ón*, a being + *géneia*, act of being born < *génos*, origin, birth, engenderment): the development or course of development of an individual organism. Cf. **phylogeny**.

onychomycosis, pl. **onychomycoses** (Gr. *ónyx*, genit. *ónychos*, fingernail, claw + *mýkes*, fungus + suf. *-osis*, condition of, state caused by): an infection of the finger or toenails caused by fungi, such as *Candida albicans* (cryptococcaceous asexual yeasts),

Trichophyton mentagrophytes and *Scopulariopsis brevicaulis* (moniliaceous asexual fungi). See **mycosis**.

oo- (Gr. *oón*, egg, female sexual cell): a pref. related to an egg, or to the concept of the female sexual cell, or for the shape resembling a bird's egg. Equivalent to L. **ovo-**.

oogamy (Gr. *oón*, egg, female sexual cell + *gámos* > E. des. *-gamy*, sexual union): the fertilization of an immotile female gamete by a motile male gamete; the female (oosphere or ovule) is larger and originates in a female sexual organ called an oogonium. The male gametes are smaller (spermatozoids or antherozoids) and are produced in sexual organs called antheridia. Among the fungi, this type of sexual reproduction is found exclusively in aquatics belonging to the order Monoblepharidales, such as *Monoblepharis polymorpha*.

oogenesis, pl. **oogeneses** (Gr. *oón*, egg, female sexual cell; here, oosphere + *géneia*, *génesis*, genesis, origin, formation): development of the oogonium after fertilization, such as occurs in the Oomycota.

oogoniols (Gr. *oón*, egg, female sexual cell; here, oosphere + NL. *gonium* < Gr. *gónos*, the engendered, offspring + suf. *-ol* < alcohol; a chemical compound containing hydroxyl): hormones (sterols) of *Achlya* (Saprolegniales) that induce the formation of oogonia. Cf. **antheridiol**.

oogonium, pl. **oogonia**, **oogon**, **oogone** (Gr. *oón*, egg, female sexual cell; here, oosphere + NL. *gonium* < Gr. *gónos*, the engendered, offspring): the female gametangium in the aquatic and amphibious fungi of the phylum Chytridiomycota and Oomycota, which contains an oosphere, as in *Monoblepharis* (Monoblepharidales), or several oospheres, as in *Achlya* and *Saprolegnia* (Saprolegniales).

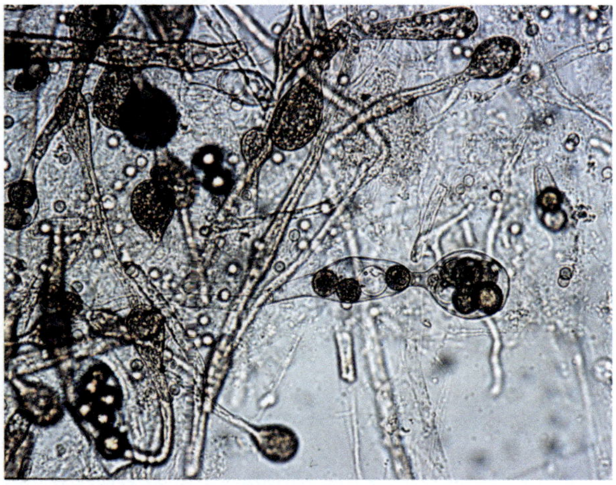

Oogonia with oospheres of *Saprolegnia* sp., x 100 (*MU*).

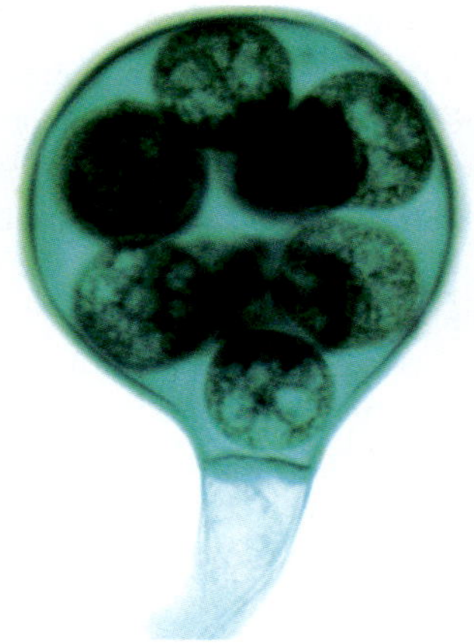

Oogonium with oospheres of *Saprolegnia* sp., x 420 (*MU*).

oomycete, **oomycote** (L. *oomycete* < Gr. *oón*, egg, female sexual cell; here, oosphere + *mýkes*, genit. *mýketos*, fungus): a member of the Oomycota, a phylum of the kingdom Protoctista in the system of classification of Margulis and Schwartz, 1988, and a phylum of the kingdom Stramenopila in the scheme of Alexopoulos *et al.*, 1996. The latter scheme is the one followed in this dictionary. See **Oomycetes**.

Oomycetes (Gr. *oón*, egg, sexual cell; here, oosphere + L. *mycetes*, ending of taxonomic class < Gr. *mýkes*, genit. *mýketos*, fungus): in some systems of classification, a class of fungi with biflagellate zoospores, coenocytic mycelium and sexual reproduction by oogamy. In the classification adopted in this dictionary (mainly that of Alexopoulos *et al.*, 1996), these organisms are considered as the phylum Oomycota of the kingdom Stramenopila, with the orders Saprolegniales, Salilagenidiales, Lagenidiales, Leptomitales, Myzocytiopsidales, Rhipidiales and Peronosporales. In some other classification schemes, like that of Margulis and Schwartz (1988), the Oomycetes are considered as a phylum within the kingdom Protoctista. See the Appendix, where there is a comparison of the outline of Alexopoulos *et al.*, 1996, modified by Ulloa and Hanlin, 2006, to that of Kirk *et al.*, 2008. In the latter classification scheme, this class has been changed from the kingdom Stramenopila to the kingdom Chromista, another example of the taxonomic rearrangements undergone in the systematics of the fungi, in order to achieve a phylogenetic classification.

ooplasm (Gr. *oón*, egg, female sexual cell; here, oosphere + *plásma*, formation, bland material with which an organ or living being is formed): *Oomycota*. The protoplasm in the center of the oogonium that first gives rise to the oosphere and afterwards to the oospore. Cf. **periplasm**.

ooplast (Gr. *oón*, egg, female sexual cell; here, oosphere + *plastós*, formed, organized particle or body): *Oomycota*. A cellular inclusion, surrounded by a membrane, in the oospore of the Saprolegniales; the ooplast comes from the ooplasm and forms the oosphere, which is surrounded by the periplasm of the oogonium.

oosphere (Gr. *oón*, egg, female sexual cell + *sphaîra*, sphere): an immotile gamete (or gametic nucleus), without a cell wall, contained in the oogonium, as is observed in the aquatic fungi of the gen. *Monoblepharis* (Monoblepharidales) and *Saprolegnia* (Saprolegniales).

oospore (Gr. *oón*, egg, female sexual cell; here, oosphere + *sporá*, spore): a spore with a thick, resistent wall that results from the development of an oosphere, whether by fertilization or by parthenogenesis; the oospore is the resting spore of the aquatic fungi of the orders Monoblepharidales (Chytridiomycota), Saprolegniales, Salilagenidiales, Lagenidiales, Leptomitales, Myzocytiopsidales, Rhipidiales and Peronosporales (Oomycota).

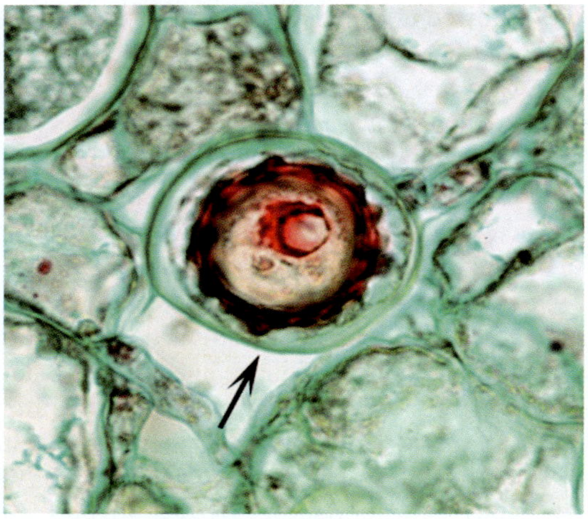

Oospore of *Albugo candida* (Peronosporales) inside a cabbage root cell, x 675 (*MU*).

operculate (L. *operculatus*, covered, plugged < *operculum*, cover, lid, operculum + suf. *-atus* > E. *-ate*, provided with or likeness): having an *operculum*, like the asci of *Saccobolus* (Pezizales). Cf. **inoperculate**, **pseudoperculate** and **suboperculate**.

operculum, pl. **opercula** (L. *operculum*, lid, cover): the lid-like apex of a sporangium or ascus, which opens by circumscissile dehiscence, resembling a lid on a hinge; there are operculate sporangia in the Chytridiales, such as *Chytridium* and *Nowakowskiella*, and operculate asci in the Pezizales, such as *Peziza* and *Saccobolus*.

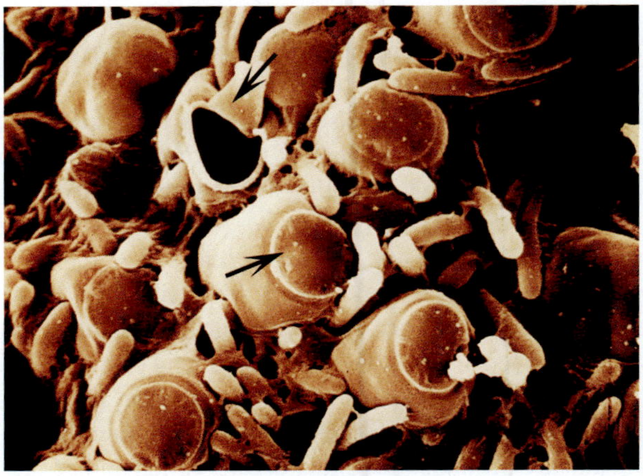

Operculate asci in the hymenium of *Saccobolus* sp.; note the **operculum** on the apex of each ascus, x 1,000 (*KO*).

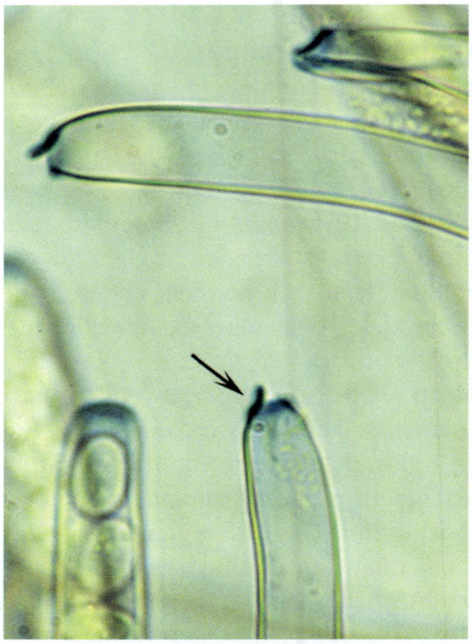

Operculum of the asci in the hymenium of *Peziza ammophila*, x 500 (*RTH*).

opistho-, opisth- (Gr. *ópisthe*, *ópisthen*, behind; Gr. *ópisthios*, neuter *ópisthion*, rear part): a pref. that alludes to the posterior part of something, as in *opisthokont.*

opisthokont, **opisthocont**, **opisthokontic**, **opisthocontic**, **opisthokontous**, **opisthocontous** (Gr. *ópisthe*, *ópisthen*, behind; Gr. *ópisthios*, neuter *ópisthion*, rear part + *kontós*, oar; here, flagellum): a motile cell that has an impelling flagellum inserted in the posterior part, as occurs with the zoospores and planogametes of the fungi of the class Chytridiomycetes, e.g., *Chytriomyces*, *Phlyctochytrium* and other Chytridiales. Cf. **acrokont** and **pleurokont**.

opportunistic (L. *opportunus*, suitable or convenient for a particular occurrence + NL. suf. *-istic* > E. *-ist*, one that does + L. *-icus* > E. *-ic*, pertaining to as agent): refers to a fungus that normally is saprobic and frequently common but which on occasion is capable of causing disease, especially of humans and other susceptible animals. E.g., *Candida albicans*, *C. tropicalis* and other species of *Candida* (cryptococcaceous asexual yeasts) are opportunists that can cause candidiasis in man and in animals whose immune system has been weakened by one or more predisposing factors. See **facultative saprobe**.

opsi- (Gr. *opsé*, late): a pref. employed to give the idea of tardiness, as in *opsiform*, the late form of a uredinal fungus that lacks urediniospores and only is manifested by the final spores (the teliospores).

-opsis (Gr. *ópsis*, appearance, aspect): a suf. employed to form words that express likeness or simlarity, referring to shape as well as to other aspects, e.g., in names such as *Thielaviopsis* (dematiaceous asexual fungi) and *Pestalotiopsis* (melanconiaceous asexual fungi), that have characters similar to those of *Thielavia* (Sordariales) and *Pestalotia* (melanconiaceous asexual fungi), respectively.

orbicular (L. *orbicularis*, orbicular < *orbiculus*, dim. of *orbis*, circle + suf. *-aris* > E. *-ar*, like, concerning): circular, round, like the pileus of the basidiocarp of *Stropharia semiglobata* (Agaricales).

orculiform (L. *orculiformis*, shaped like a large earthen jar < *orcula*, dim. of *orca*, cask, earthen jar + suf. *-formis* < *forma*, shape): shaped like a jug, like the apothecium of *Geopora* (=*Sepultaria*) *arenicola* (Pezizales) and the basidioma of *Graphiola phoenicis* (Exobasidiales). Equal to **urceolate**. For spores, similar to **polar locular** and **polocellate**.

Orculiform apothecia of *Geopora* (=*Sepultaria*) *arenicola*, x 2 (*MU*).

Orculiform basidiomata of *Graphiola phoenicis* on a leaf of date palm, x 20 (*MU*).

Orculiform basidioma of *Graphiola phoenicis* on a leaf of date palm, x 20 (*SA*).

orchil (ME. *orchil*, pigment): a general term for lichens that produce pigments, like the orcein of *Roccella tinctoria* (Arthoniales).

order (L. *ordo*, genit. *ordinis*): in classification by taxonomic categories, it is the unit between class and family. In L., the ending for the names of orders is *-ales*, as in Mucorales and Agaricales (for mucors, agarics).

orismology (Gr. *horismós*, limitation, definition < *horízein*, to delimit < *hóros*, limit, term + *lógos*, word > E. des. *-logy*, discourse, study): the science of defining terms, especially the technical terms of a particular subject. See **glossology** and **lexicology**.

ornamented (L. *ornamentum*, adornment, useful accessory): applied to spores, hyphae, organs, etc., that have a surface marked or sculptured with spines, ridges, warts, striations, reticulations, wrinkles, fibrils, scales, etc.; not smooth. The giant colonies of some yeasts, e.g., *Rhodotorula*, *Cryptococcus* and *Candida* (cryptococacceous asexual yeasts) are ornamented. Cf. **naked**.

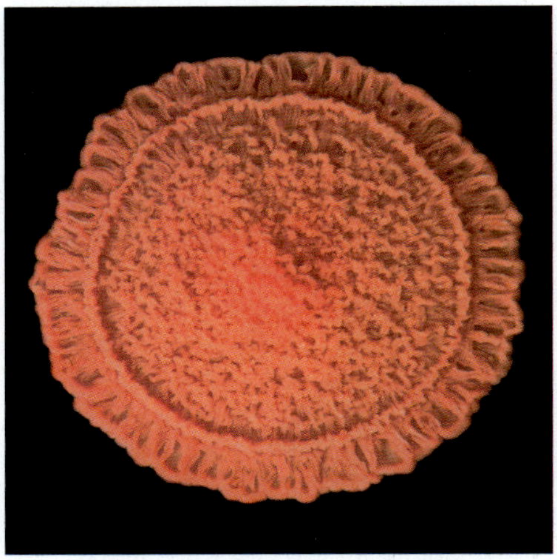

Ornamented giant colony (with the surface wrinkled and the border fringed) of *Rhodotorula rubra* on agar, x 1 (*MU*).

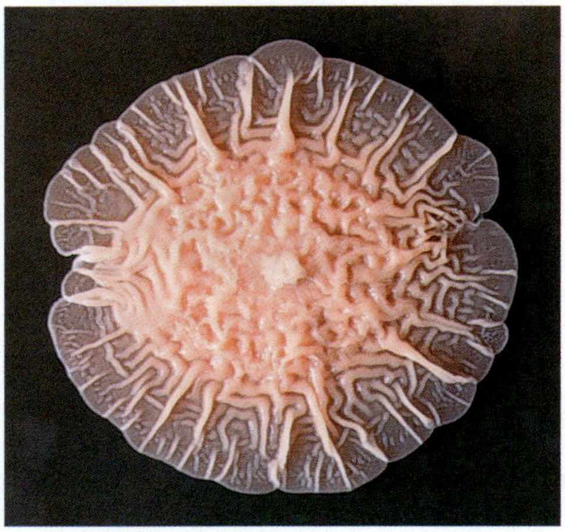

Ornamented giant colony (with the surface wrinkled and the border wavy) of *Cryptococcus curvatus* on agar, x 1 (*SA*).

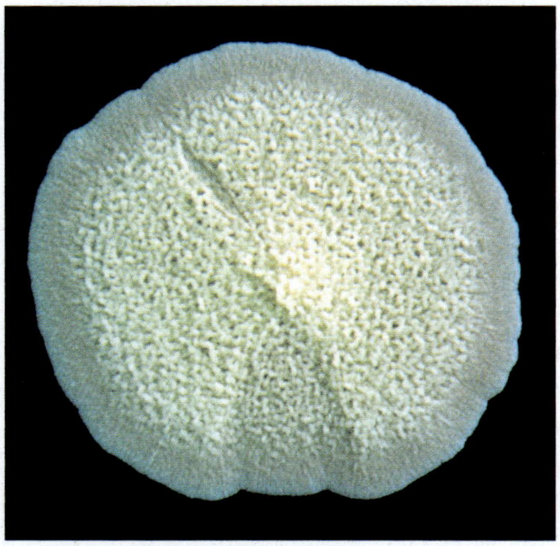

Ornamented giant colony (with the surface wrinkled and the border wavy) of *Candida valida* on agar, x 1 (*MU*).

ornitho-, **ornith-** (Gr. *órnis*, genit. *órnithos*, bird): a pref. employed in compounds such as *ornithochoric*, which refers to an organism whose propagules are spread by birds.

ornithocoprophilous (Gr. *órnis*, genit. *órnithos*, bird + *kópros*, dung + *phílos*, have an affinity for): coprophilic on bird droppings, like certain lichens that utilize these substrates to obtain nitrogen, or like the human pathogenic yeast, *Cryptococcus neoformans* (cryptococcaceous asexual yeasts), that inhabits poultry and pigeon droppings.

orphan anamorph (LL. *orphanus*, orphan < Gr. *orphanós*, deprived of one or usually both parents; Gr. *anamorphóo*, to transform): an asexual form that has obvious connections to sexual fungi that have similar conidial states.

oro- (Gr. *óros*, mountain): a pref. employed in names such as *Oropogon* (Lecanorales), a lichen whose name means "mountain beard".

ortho-, **orth-** (Gr. *orthós*, upright, straight): a pref. employed in compounds to introduce the idea of straight, correct, normal, regular or perpendicular, as in the following terms.

orthogeotropism (Gr. *orthós*, upright, straight + *geo* < *gê*, earth, alluding to gravity + *trópos*, gyration, change in position or manner < *tropé*, a turn < *trépo*, to revolve, turn towards + suf. - *ismós* > L. *-ismus* > E. *-ism*, state, phase, tendency, action): vertical geotropism concerning *orthogeotropic* organisms and organs, i.e., orthotropic with respect to the stimulus of gravity, like the majority of pedicels and stipes (negatively orthogeotropic), and of rhizoids, rhizines and pseudorhizae (positively orthogeotropic), which, respectively, react by moving away from the stimulating factor or by moving toward it. Cf. **diageotropism**. See **tropism**.

orthotropic (Gr. *orthós*, upright, straight + *trópos*, gyration, change in position or manner < *tropé*, a turn < *trépo*, gyrate, revolve, turn toward): growing vertically from the point of origin, like the gametangial suspensors of *Phycomyces* (Mucorales). Also refers to the disposition of basidiospores on the basidium of the Gasteromycetes; see **orthotropic spore**. Cf. **heterotropic**.

orthotropic spore (Gr. *orthós*, erect, straight + *trópos*, gyration, change in position or manner < *trópe*, turn < *trépo*, to gyrate, to go around, be directed toward; NL. *spore* < Gr. *sporá*, spore): a spore that is borne erect and centered on the sterigma of the basidium. Discharge of this type of basidiospore is passive and a collar often is visible at the point of detachment from the sterigma. Common in the Gasteromycetes, such as *Podaxis* and *Sclerogaster* (Hymenogastrales).

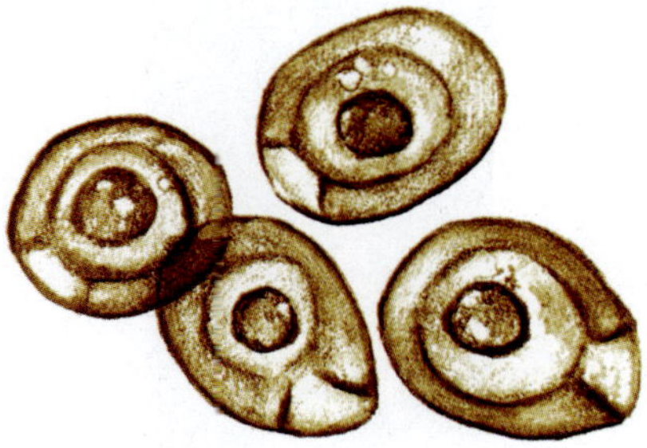

Orthotropic basidiospores of *Podaxis* sp., uniguttulate, with a thick wall, x 1,500 (*MU*).

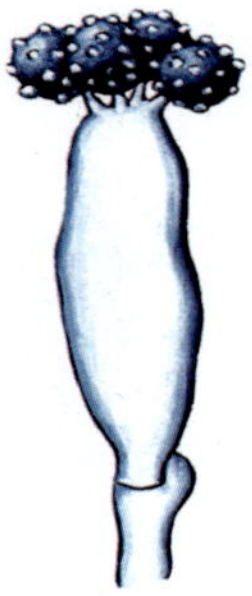

Orthotropic spores, symmetrical, of *Sclerogaster* sp., on the axial prolongations of the basidial sterigmata, x 1,270 (*MU*).

oscule (L. *osculum* < *os*, mouth, opening + dim. suf. -*culum* > E. -*ule*): generally applied to the germ pore of the urediniospores of the rusts (Uredinales). The number and location of the pores or oscules varies according to the species. In *Uromyces appendiculatus* (bean rust), e.g., the urediniospores have two equatorial pores.

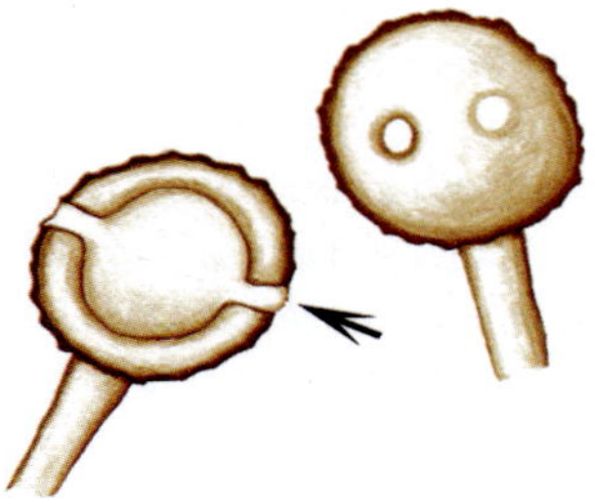

Oscules of the urediniospores of *Uromyces appendiculatus*, x 820 (*MU*).

-ose (L. -*osa*, femenine ending; the masculine is -*osus*, and the neuter -*osum*, which denotes full of, augmented, prone to): a proper suf. of carbohydrates (*cellulose*, *fructose*, *saccharose*, etc.).

-ose, **-ous** (L. -*osus*, masculine, -*osum*, neuter, which denotes full of, augmented, prone to): a suf. that expresses the general idea of abundance, full of, as in *strigose*, a term that is applied to something covered with abundant, rigid, sharp-pointed hairs.

-osis (Gr. *ósis*, condition or state): a suf. that expresses the condition or state of something, or of a process, etc., as in *mycosis*, the state of animals or humans subject to the action of parasitic fungi.

osmiophile, **osmiophilic**, **osmiophilous** (NL. *osmium* < Gr. *osmé*, odor; a hard, brittle, grayish-blue or blackish-blue, polyvalent metallic element + Gr. *phílos*, have an affinity for + suf. -*íkos* > L. -*icus* > E. -*ic*, pertaining to or relating to): having a tendency to stain black in osmium tetroxide (a biological fixative), like the lipids in the cells of Hyphochytriales, when prepared for observation with the electron microscope.

osmo-, **osm-** (Gr. *osmós*, to push, action of going through, to put into): a pref. employed in some terms of Gr. origin to introduce the idea of osmosis, as in the following words.

osmophile, **osmophilic**, **osmophilous** (Gr. *osmós*, to push, action of going through, to put into + *phílos*, have an affinity for + suf. -*íkos* > L. -*icus* > E. -*ic*, pertaining to or relating to): applied to fungi that can live in substrates with high osmotic pressure. E.g., yeasts of the gen. *Zygosaccharomyces* (Saccharomycetales) have a marked tolerance to high concentrations of sugar in the culture medium. Some species of *Aspergillus*, such as *A. halophilicus* and *A. restrictus*, and of *Penicillium*, such as *P. citrinum* (moniliaceous sexual fungi), require high concentrations of sugar or salt in the substrate in order to grow well. This phenomenon is called *osmophily*.

osmosis, pl. **osmoses** (Gr. *osmós*, to push, action of going through, to put into + suf. -*osis*, state or condition): a phenomenon concerning the diffusion that occurs through a semipermeable membrane that separates two solutions with different concentrations of solute; osmosis tends to equalize the concentrations on both sides of the membrane. The fungi are typically *osmotrophic*, since they are nourished by osmosis.

osmotolerant (Gr. *osmós*, to push, action of going through, to put into + L. *tolerans*, genit. *tolerantis*, what tolerates, or prone to tolerance): capable of growing in media with high osmotic pressure (e.g., in concentrated solutions of sugar), as do some yeasts and molds.

osmotrophic, **osmotroph** (Gr. *osmós*, to push, action of going through, to put into + *trophós*, what nourishes, serves as food + suf. -*íkos* > L. -*icus* > E. -*ic*, pertaining to or relating to): acquiring nourishment through osmosis or absorption. The fungi are typically osmotrophic, although certain structures lacking a cell wall, such as the zoospores of the aquatic species, and the myxamoebae and plasmodia of the Myxomycota, can feed by phagocytosis, in addition to absorption or osmosis.

ossiform (L. *oss*, *os*, bone + suf. *-formis* < *forma*, shape): in the shape of a bone. See **phalangoid**.

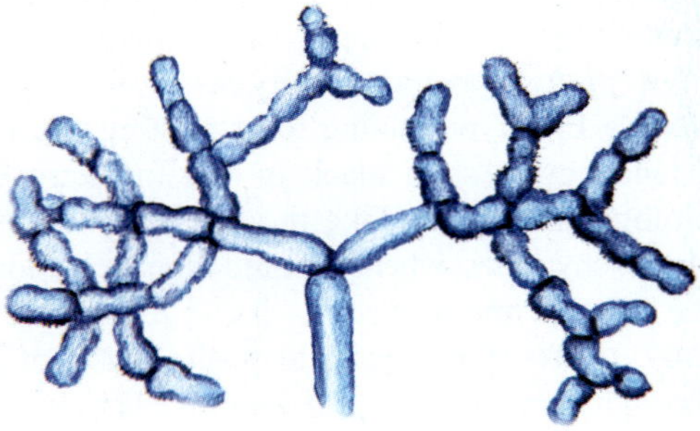

Ascomal appendage of *Arthroderma quadrifidum* (Onygenales) with **ossiform** cells, x 900 (*MU*).

ostiolar canal (L. *ostiolum*, dim. of *ostium*, door + suf. *-aris* > E. *-ar*, like, concerning; L. *canalis*, canal, conduit): a tubular anatomical passage or canal located within the *ostiolar neck* of perithecia. This canal is usually lined with periphyses, which are short unbranched hyphae that apparently serve to direct the asci toward the ostiole prior to ascospore discharge, as happens, e.g., in *Cercophora palmicola* (Sordariales).

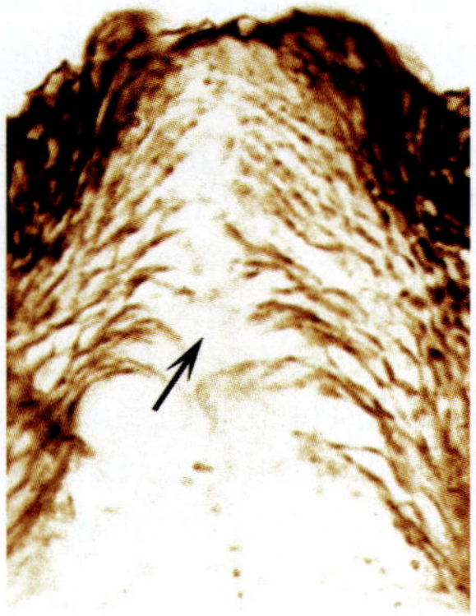

Ostiolar canal of the perithecium of *Cercophora palmicola*, x 375 (*RTH*).

ostiolar disc (L. *ostiolum*, dim. of *ostium*, door + suf. *-aris* > E. *-ar*, like, concerning; L. *discus*, disc): see **placodium**.

ostiolar neck (L. *ostiolum*, dim. of *ostium*, door + suf. *-aris* > E. *-ar*, like, concerning; ME. *nekke*, neck): the apical extension of some perithecia that contains the *ostiolar canal* and ends in the ostiole, as is seen in *Guanomyces* (Sordariales) and *Gnomonia* (Diaporthales).

Ostiolar neck of the perithecium of *Guanomyces polythrix*, x 10 (*MU*).

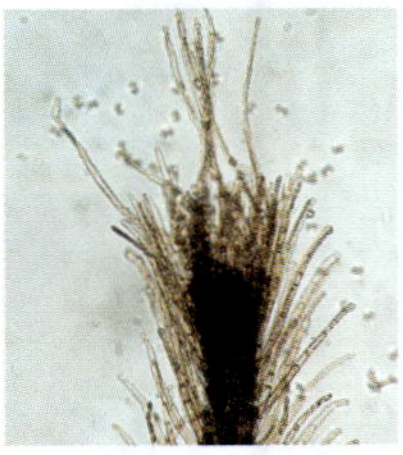

Ostiolar neck of the perithecium of *Guanomyces polythrix*, x 100 (*MU*).

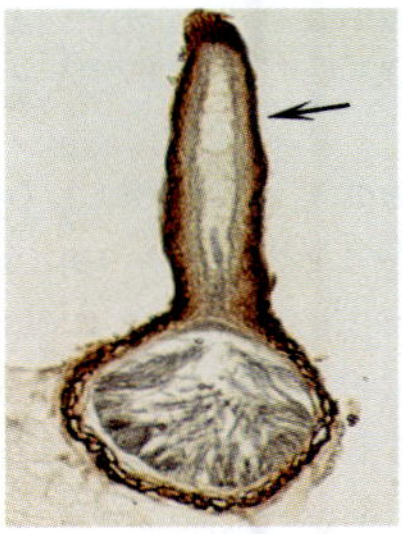

Ostiolar neck of the perithecium of *Gnomonia comari*, x 190 (*RTH*).

ostiole (L. *ostiolum* < *ostium*, door + dim. suf. *-olum* > E. *-ole*): a pore or opening, usually at the apex, of various reproductive structures, such as the pycnidia of *Phoma* and other sphaeropsidaceous asexual fungi, or the perithecia of *Sordaria fimicola* (Sordariales), among many other Pyrenomycetes (Sordariales, Melanosporales, Diaporthales, Hypocreales). See **stoma**.

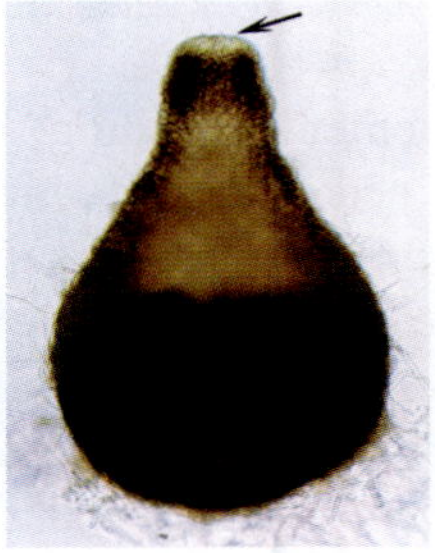

Ostiole of the pycnidium of *Phoma* sp., x 280 (*MU*).

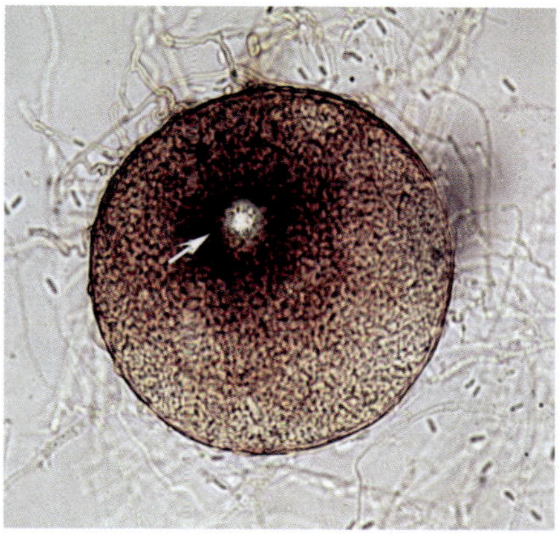

Ostiole of the perithecium of *Sordaria fimicola*, x 125 (*RTH*).

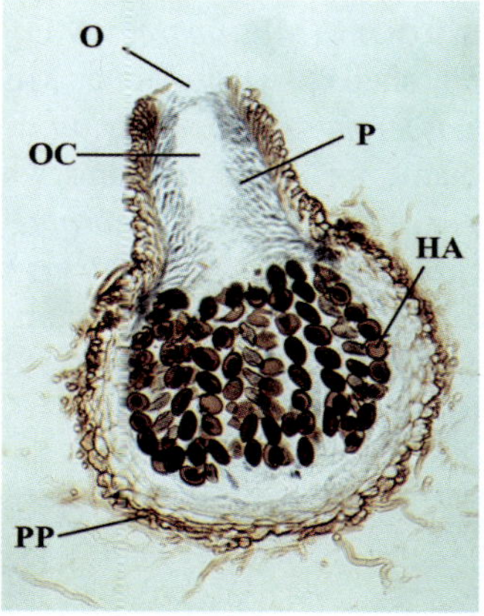

Ostiole (O) of the perithecium of *Sordaria fimicola*, seen in longitudinal section. Also shown are the ostiolar canal (OC), the periphyses (P), the pseudoparenchymatous peridium (PP), and the hymenium with octospored asci (HA), x 125 (*RTH*).

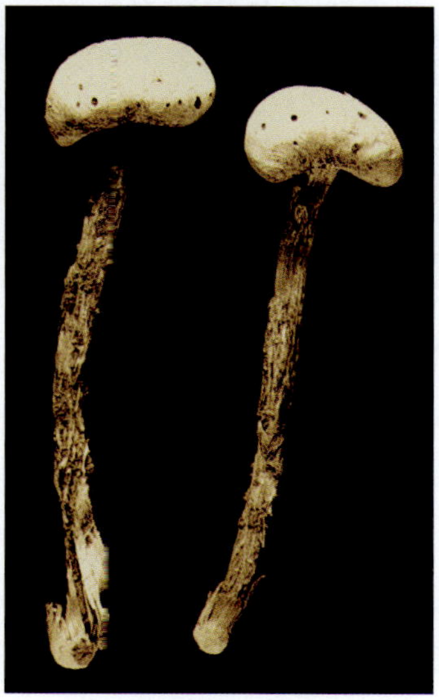

Various **ostioles** on the pileus of the basidiocarps of *Battarreoides diguetii* (Tulostomatales), x 1 (*TH*).

ostropalean (< order *Ostropales* < L. *ostrea* < Gr. *óstreon*, oyster, scale + Gr. *pále*, fine flour, pollen; lichens with a crustose thallus): a type of ascus, essentially unitunicate, with a thickened apex and a narrow canal that ends in a pore.

otomycosis, pl. **otomycoses** (Gr. *otós*, ear + *mýkes*, fungus + suf. *-osis*, condition of, state caused by): an infection of the auditory canal caused by fungi, such as *Aspergillus flavus* and *A. fumigatus* (moniliaceous asexual fungi). Also called *mycotic otitis*. See **mycosis**.

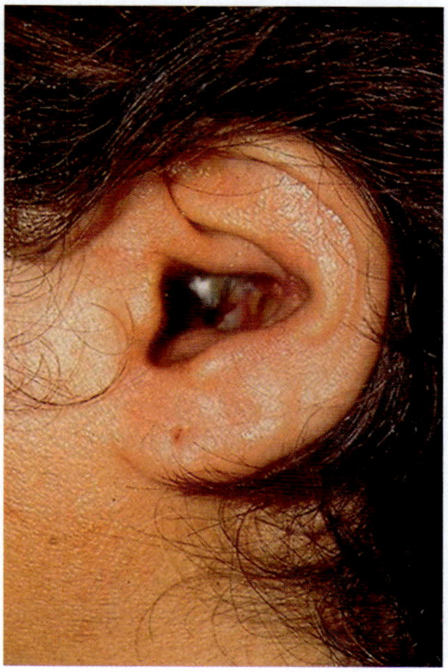

Otomycosis of the ear of a woman, caused by *Aspergillus fumigatus*, x 0.3 (*AB*).

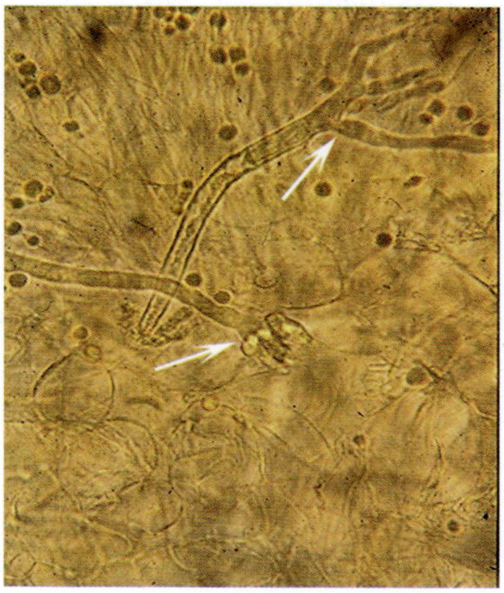

Otomycosis of the ear of a woman; between the host cells are the conidiophores (arrows) of the casual agent, *Aspergillus fumigatus*, x 350 (*AB*).

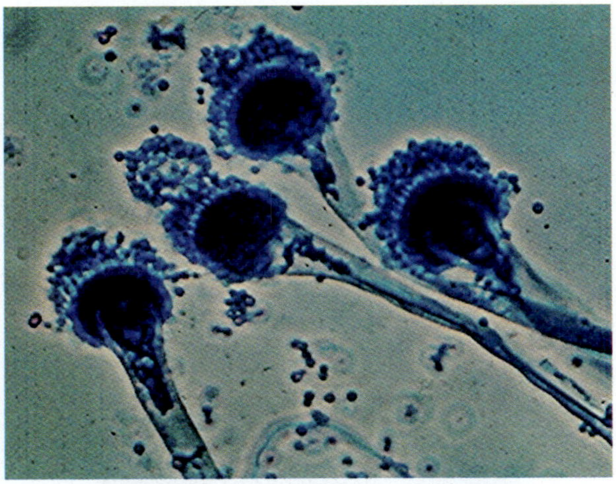

Conidiophores of *Aspergillus fumigatus*, the cause of human **otomycosis**, x 350 (*AB*).

outgroup (ME. < OE. *ut*, out < ME. *outen* < OE. *utian*, to put out + F. *groupe* < It. *gruppo*, group, an assemblage of beings or things that form a complex or that have some unifying relationship): a term used in systematics to refer to one or more taxa that are considered to be outside the monophyletic group (*ingroup*) of interest. Cf. **ingroup**.

oval (NL. *ovalis*, egg-shaped or oval < *ovum*, egg + suf. *-alis* > E. *-al*, related to or pertaining to): with the shape of an egg. See **ovate**.

ovariicolous (NL. *ovariicola*, inhabiting the ovary < *ovarium*, ovary + suf. *-cola*, inhabitant): living and developing in the ovary of higher plants, like *Claviceps* (Hypocreales), which infects the ovaries of rye and other cereals, in which it later forms sclerotia. See **ergot**.

ovate (L. *ovatus*, ovate, oval < *ovum*, egg + suf. *-atus* > E. *-ate*, provided with or likeness): egg-shaped; a laminar organ or structure with the shape of an egg, i.e., with the widest part at the base. When referring to a solid body, it is better to employ **ovoid**. Also called **oval**, which is *perprolate*, according to the polar axis, and *peroblate*, according to the equatorial diameter. E.g., the basidiospores of *Panaeolus sphinctrinus* (Agaricales) are ovoid in face view. Cf. **obovate**.

ovo- (L. *ovum*, egg, ovule): a pref. employed in terms such as *ovoid*. Equivalent to Gr. **oo-**.

ovoid (NL. *ovoides*, shaped like an egg < L. *ovum*, egg, ovule + suf. *-oide*, like < Gr. *-oeídes*, resembling, similar to): a solid structure with the shape of an egg. Cf. **ovate**.

oxy- (Gr. *oxýs*, sharp, acid): a pref. employed in many compounds, with the first meaning (*oxycarpous*, *oxycladous*, *oxyphyllous*, etc., sharp-tasting or acidic fruit, and pointed branches, pointed leaves, etc.), and the second meaning (as in *oxymesotroph*, a mesotrophic organism that needs a fatty acid for its vital syntheses). This pref. also is employed to allude to oxygen, as in *oxybiosis*, pl. *oxybioses*, the life of the aerobic organisms, and equivalent to *aerobiosis*, pl. *aerobioses*.

oyster mushroom (ME. *oistre* < MF. < L. *ostrea* < Gr. *ostreon*, shell < *osteon*, bone; ME. *musseroum* < MF. *mousseron* < LL. *mussirion*, mushroom): a common name for the mushrooms whose pileus is in the shape of a shell, as in *Pleurotus ostreatus*, *P. cornucopioides* and *P. pulmonarius* (Agaricales), which are cultivated industrially on wheat straw and other similar lignocellulosic substrates. See **conchate**.

Oyster mushroom, basidiocarp of *Pleurotus ostreatus* var. *florida* growing on wheat straw, x 0.5 (*MU*).

p

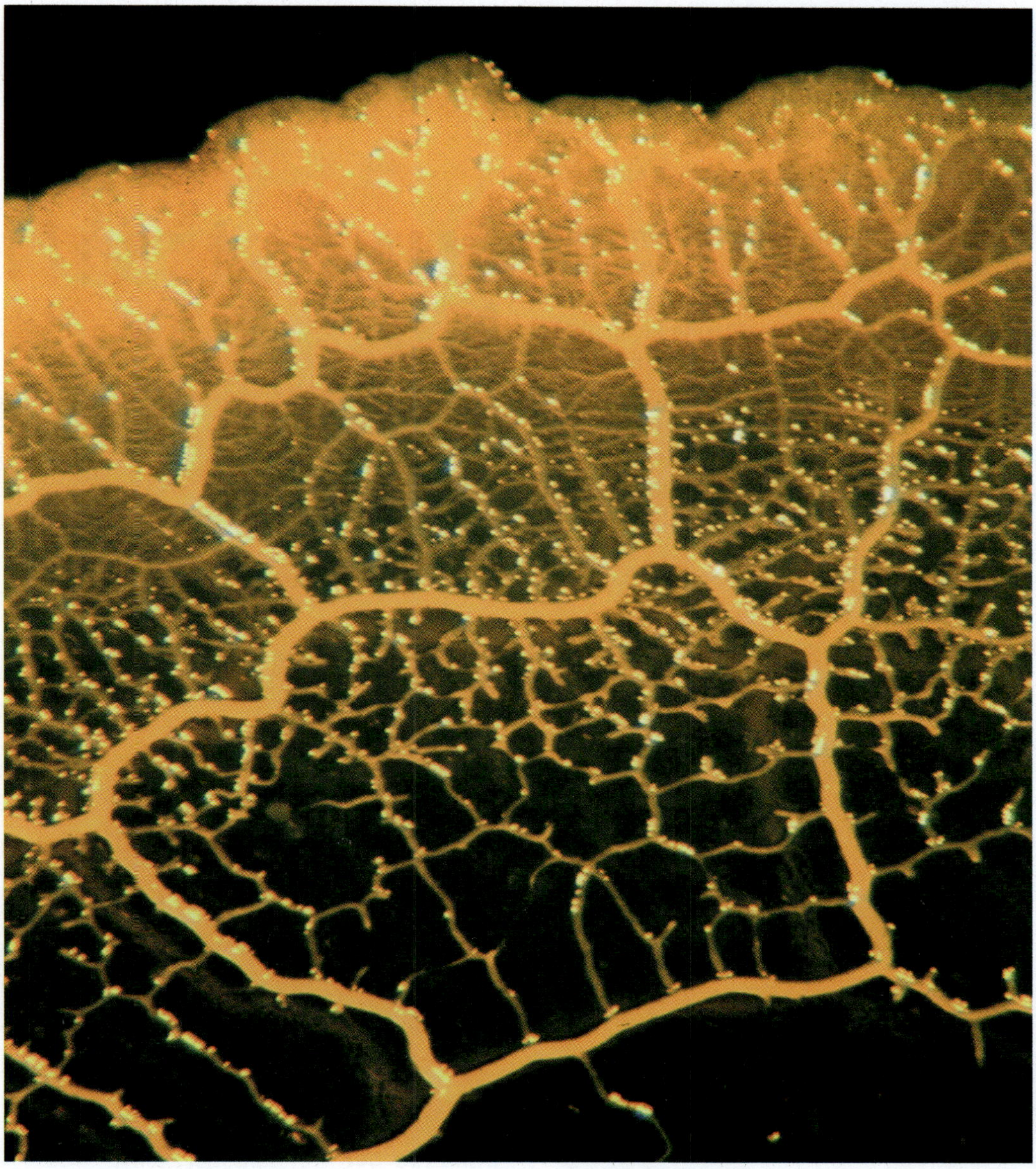

Phaneroplasmodium of *Physarum polycephalum*, on agar (*MU*).

p

pachy- (Gr. *pachýs*, thick): a pref. that alludes to something thick, as in *pachyderm*, with thick skin.

pachyderm, **pachydermate**, **pachydermatous**, **pachydermous** (Gr. *pachýs*, thick + *dérma*, skin + suf. *-atus* > E. *-ate*, provided with or likeness; or + L. suf. *-osus* > E. *-ous*, having, possessing the qualities of): a hypha whose outer wall is thicker than the lumen. Cf. **leptoderm** and **mesoderm**.

paleo-, **pale-** (Gr. *pálai*, genit. *palaiós*, ancient): a pref. employed in terms alluding to ancient geological epochs, as in *paleobotany*, the study of fossil plants, plants that lived in past geological epochs.

paleomycology (Gr. *pálai*, genit. *palaiós*, ancient + *mýkes*, fungus + *lógos*, word > E. des. *-logy*, study, treatise): the study of fossil fungi.

palisade (Fr. *palissade*, fence < OPr. *palissada* < *palissa*, enclosure, stockade, fence; a place enclosed with stakes < L. *palus*, stake): a row of parallel, elongate structures arranged next to one another, like a palisade, like the conidiophores of *Entomophthora muscae* (Entomophthorales) and the asci of *Taphrina pruni* (Taphrinales).

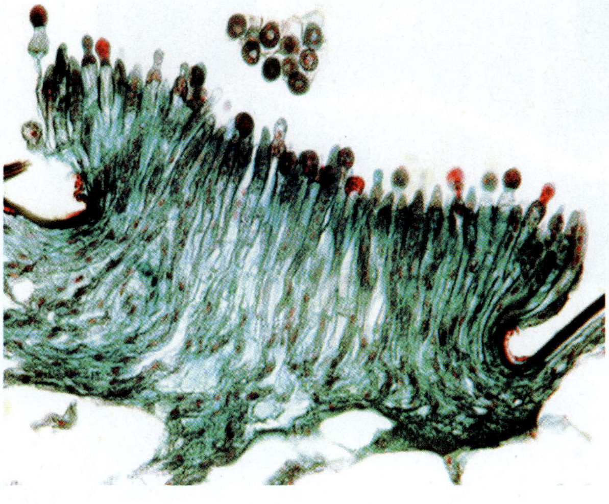

Palisade of conidiophores of *Entomophthora muscae*, projecting from the infected tissues of a house fly, seen in longitudinal section, x 55 (*MU*).

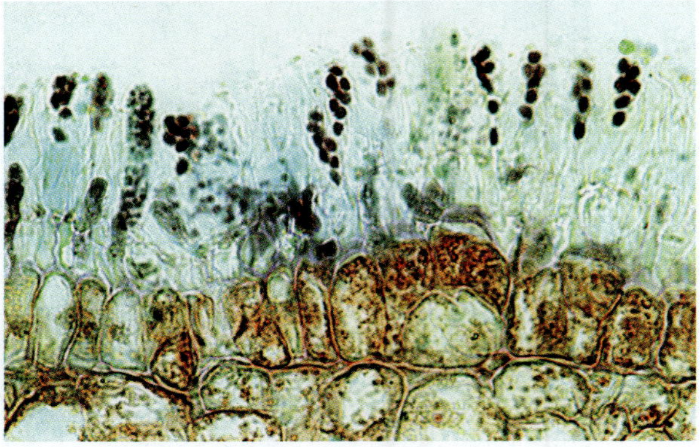

Palisade of asci of *Taphrina pruni* on a plum leaf, seen in longitudinal section, x 375 (*MU*).

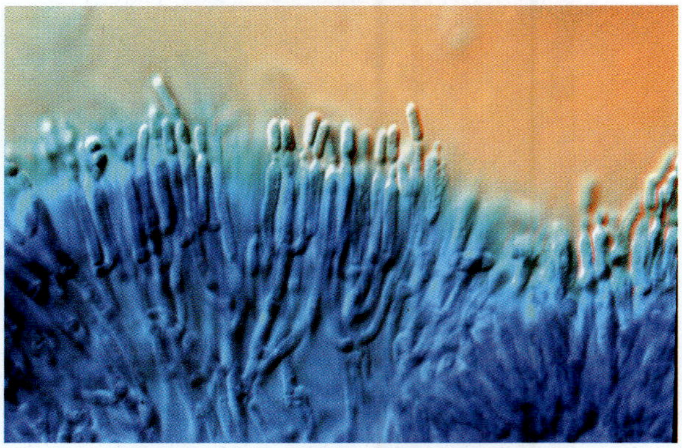

Palisade of conidiophores of *Metarhizium anisopliae* (moniliaceous asexual fungi) from a culture on agar, x 375 (*CT*).

palisade cell (Fr. *palissade*, fence < OPr. *palissada* < *palissa*, stake, post, fence, a place fenced-in with posts < L. *palus*, stake; NL. *cellula*, cell, living cell, dim. of *cella*, small room, compartment, cell of a honeycomb): *Lichens*. A terminal cell of the hyphae of the fastigiate cortex that is perpendicular or anticlinal to the plane of the thallus.

palisade plectenchyma, pl. **plectenchymata** (Fr. *palissade*, fence < OPr. *palissada* < *palissa*, paling, palisade, fence; a place enclosed by stakes

< L. *palus*, stake; Gr. *plektós*, braided, intertwined + *énchyma*, stuffed): *Lichens*. A tissue in the cortex that is composed of hyphae that are parallel and perpendicular to the surface of the thallus. Also called *palisadoplectenchyma*, pl. *palisadoplectenchymata*.

pallid (L. *pallidus*, faded; light in color): yellow, pallid or losing its natural color.

palmate (L. *palma*, the lower, somewhat concave part of the hand, from the wrist to the fingers + suf. *-atus* > E. *-ate*, provided with or likeness): shaped like an open hand; with lobes that radiate from a common center. Cf. **digitate** and **cheiroid**.

palmicole (L. *palma*, the lower, somewhat concave part of the hand, from the wrist to the fingers; palm tree + suf. *-cola*, inhabitant < *colere*, to inhabit): growing on palms, like the perithecia of *Cercophora palmicola* (Sordariales) that grow on the trunks of palms.

Perithecia of *Cercophora palmicola* on a palm trunk (**palmicole**), seen in longitudinal section, x 1,000 (*RTH*).

paludal, **paludicoline**, **paludicolous**, **paludine** (L. *palus*, genit. *paludis*, lagoon, swamp, marsh + suf. *-alis* > E. *-al*, relating to or pertaining to; or + L. suf. *-cola*, inhabitant < *colere*, to inhabit): applied to organisms that live in low, flooded terrain, like lagoons and swamps, as, e.g., many zoosporic fungi (Chytridiomycota, Hyphochytriomycota) and Oomycota that inhabit these environments. Equal to **palustrine**.

palustrine (L. *paluster*, genit. *palustris*, pertaining to a lagoon or to a swamp < *palus*, genit. *paludis*, lagoon, swamp, marsh): growing in swamps or marshes. Equal to **paludal**.

pan- (Gr. *pás*, *pása*, *pán*, everyone, all): a pref. employed to express the idea of totality, as in *pantonema*, the mastigonemate flagellum. See **panto-**.

pandemia (Gr. *pandemía*, reunion of the people < *pándemos*, pertaining to all the people): an epidemic or epiphytotic that extends to many countries, including continents, producing massive death of the organisms affected.

pandurate, **pandurated**, **panduriform** (L. *panduratus*, with the shape of a violin < *pandura*, musical instrument with three strings + suf. *-atus* > E. *-ate*, provided with or likeness; L. *panduriformis* < *pandura* + suf. *-formis* < *forma*, shape): oblong, but narrowing down in the middle or near the base; similar in shape to a violin or a guitar. A rare shape in fungi, it is present in the columella of the sporangium of *Circinella muscae* (Mucorales).

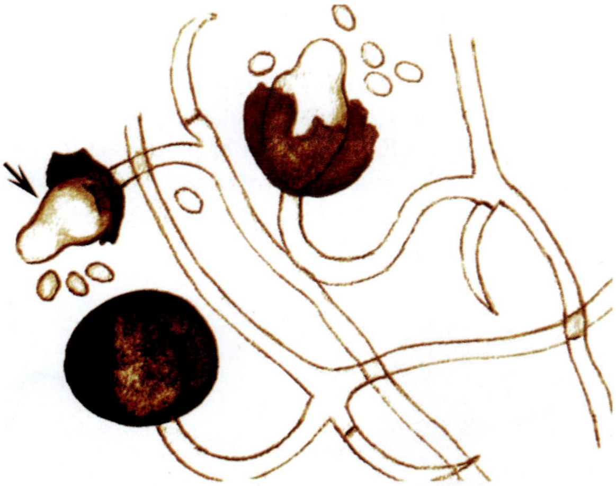

Pandurate columella of the sporangia of *Circinella muscae*, x 360 (*MU*).

pannose, **panniform** (L. *pannosus*, full of rags < *pannus*, cleaning cloth, rag, garmet + suf. *-osus* > E. *-ose*, full of, augmented, prone to; or < *pannus* + suf. *-formis* < *forma*, shape): having the appearance of felt or woolen cloth, like the fruiting body of *Lentinus* and *Panus*, of the Aphyllophorales.

Pannose basidiocarps of *Lentinus strigosus*, growing on a fallen twig, x 0.5 (*RV*).

panto-, **pant-** (Gr. *pantós*, genit. of *pán*, all, totally): a pref. used in the same sense as *pan-*. See **pantonema**.

pantonema, pl. **pantonemata** (Gr. *pantós*, genit. of *pán*, all, totally + *nêma*, genit. *nêmatos*, filament): a term applied to the hairy or mastigonemate flagellum present in the zoospores of the Hyphochytriomycota and Oomycota. See **axoneme** and **mastigoneme**. The organisms belonging to these phyla are known as *pantonemic*, and supposedly evolved from algae; they have the diaminopimelate metabolic pathway for the synthesis of lysine; the predominant steroles are cholesterols (not ergosterols), and the reserve product is mycolaminarine (not glycogen).

papilionaceous (L. *papilio*, genit. *papilionis*, butterfly + suf. *-aceus* > E. *-aceous*, of or pertaining to, with the nature of): spotted or variegated, with marks of different colors, like the gills of some species of *Panaeolus* (Agaricales).

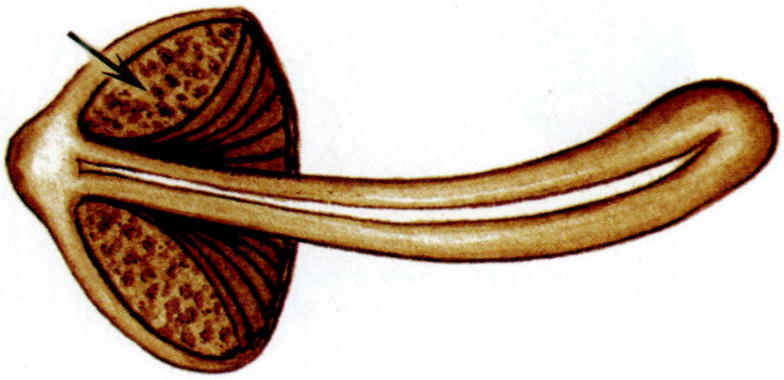

Papilionaceous gills of the basidiocarp of *Panaeolus campanulatus*, x 0.6 (*MU*).

papilla, pl. **papillae** (L. *papilla*, nipple, process): a small conically rounded elevation, generally translucid, of the wall of sporangia and gametangia, which on breaking serves as the point of exit of zoospores and planogametes. E.g., the sporangia of *Allomyces* (Blastocladiales) have one or more papillae, and are called *papillate* or *papillose*. The columella of the sporangium of *Absidia cylindrospora* (Mucorales) has a papilla on the apex.

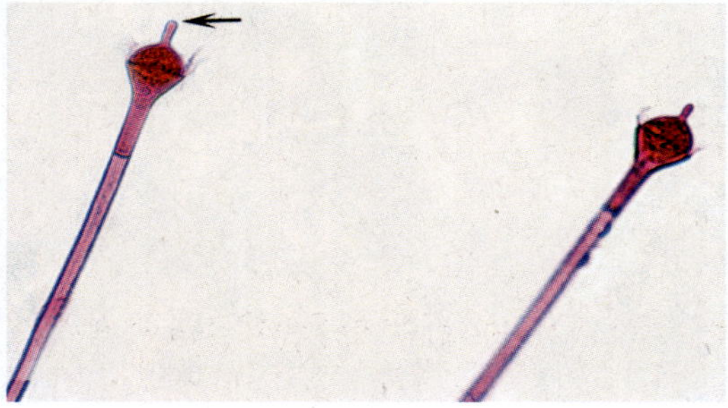

Papilla on the apex of the columella of the sporangia of *Absidia cylindrospora*, x 260 (*CB*).

papillate, **papillose** (L. *papillatus*, with a papilla < *papilla*, nipple, process + suf. *-atus* > E. *-ate*, provided with or likeness; or + L. *-osus* > E. *-ose*, full of, augmented, prone to): **1**. *Agaricales*. A pileus with a well delineated but not very elongated umbo, so that it appears mammiform, as in *Entoloma papillatum* (=*Nolanea papillata*) and *Psilocybe caerulescens*. **2**. *Aquatic fungi*. Refers to a zoosporangium, like that of *Phlyctochytrium* (Chytridiales), or to a gametangium, like that of *Allomyces* (Blastocladiales), in whose wall are produced one or more small, thinner-walled, evaginations (the papillae), that at maturity break and permit the release of the zoospores or the planogametes. Also applied to certain spores, like the teliospores of *Uromyces phaseoli* (Uredinales), that have an apical papilla. Cf. **epapilate**.

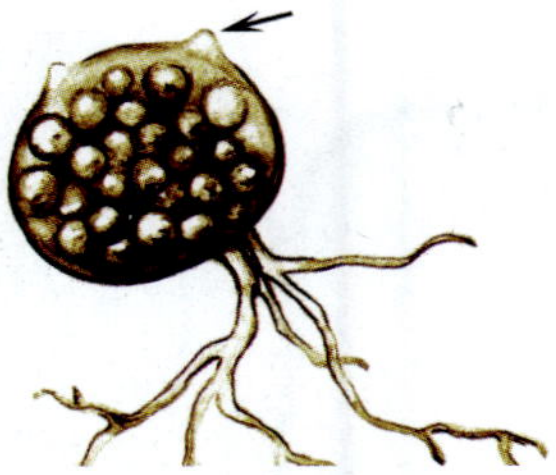

Papillate zoosporangium of *Phlyctochytrium* sp., x 1,000 (*MU*).

Papillate pileus of the basidiocarps of *Psilocybe caerulescens*, x 0.6 (*MU*).

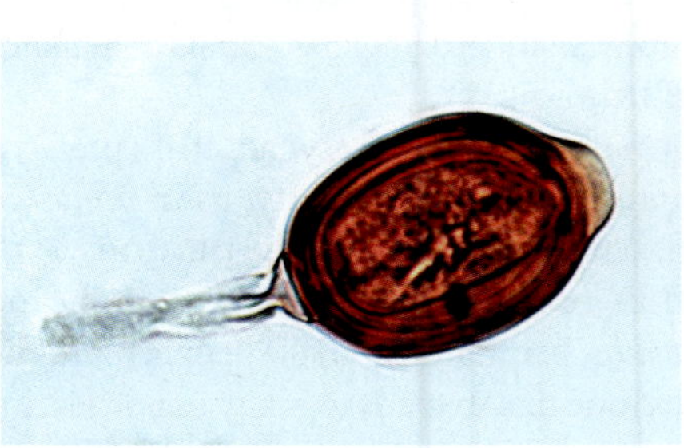

Papillate teliospore of *Uromyces phaseoli*, x 1,000 (*MU*).

Papillate thalli of the lichen *Lasallia pustulata* (Lecanorales), x 1 (*TRO*).

papilliform (L. *papilliformis*, shaped like a papilla or nipple < *papilla*, nipple, process + suf. *-formis* < *forma*, shape): having the shape of a papilla or nipple.

papulose (L. *papulosus*, full of papules < *papula*, papule, protuberance + suf. *-osus* > E. *-ose*, full of, augmented, prone to): covered with papules or pustules, like the surface of the sporangium of *Enerthenema papillatum* (Stemonitales) and the pileus of *Cortinarius* (=*Phlegmacium*) *papulosus* (Agaricales).

papulospore (L. *papula*, papule, protuberance + Gr. *sporá*, spore): a compact group of enlarged cells that functions as a spore; each cell can germinate independently, as occurs in *Papulaspora* (agonomycetaceous asexual fungi). Also called **bulbil** and **microsclerotium**.

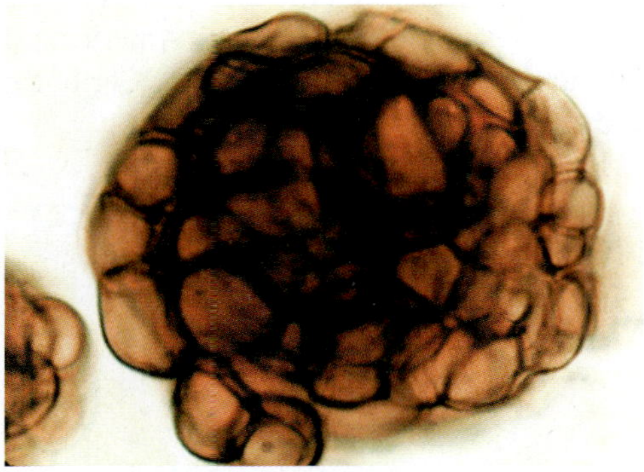

Papulospore of *Papulaspora* sp., x 1,500 (*MU*).

papyraceous, pergameneous, pergamentaceous (L. *papyraceus*, similar to paper < *papyrus* < Gr. *papýros*, paper made from papyrus + suf. *-aceus* > E. *-aceous*, of or relating to, with the nature of; or L. *pergamenus* < Pergamum, city of the Misia, where it was used for the first time + L. suf. *-osus* > E. *-ous*, having, possessing the qualities of): similar to paper or parchment, with the consistency and thinness of paper, like the dried basidiocarp of *Sparassis crispa* (Aphyllophorales) and the endoperidium of the basidiocarp of *Bovista* (Lycoperdales). The cellulosic film known as *mother of vinegar*, which is produced by the symbiotic growth of bacteria and yeasts on wines and ciders, acquires a papyraceous consistency after being dehydrated.

Basidiocarp of *Sparassis* sp., with a **papyraceous** consistency when dehydrated, x 0.5 (*RV*).

Papyraceous endoperidium of the basidiocarp of *Bovista* sp., x 1 (*MRO*).

Papyraceous endoperidium of the basidiocarps of *Bovista* sp., x 1 (*MRO*).

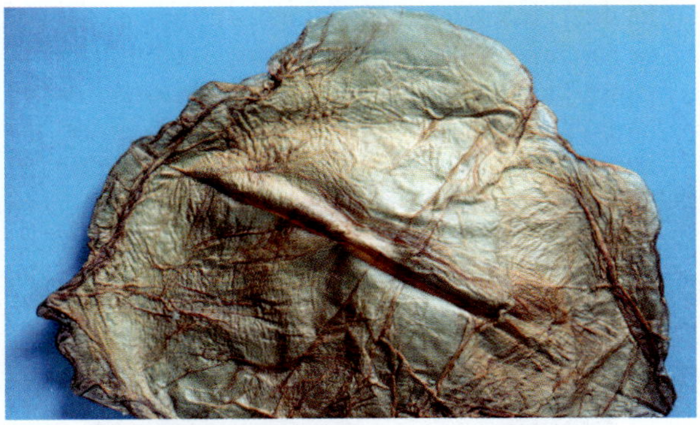

Dehydrated mother of vinegar, with a **papyraceous** consistency, x 0.1 (*MU*).

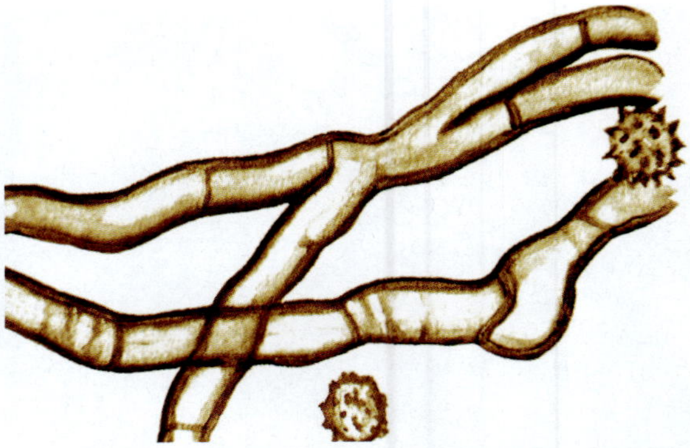

Paracapillitium of the fruiting body of *Astraeus* sp., x 1,000 (*MU*).

par-, **para-** (Gr. *pará*, with various meanings): a pref. that denotes at the side, above, beyond, toward, against, almost; it is utilized in the formation of various biological terms, such as *parabiont* and *parabiosis*.

parabiont (Gr. *pará*, at the side, above + *bionte*, living being < *bión*, to live + *óntos*, genit. of *ón*, a being): refers to an organism that lives in symbiosis with two other organisms, as happens in certain lichen thalli, which consist of the association (*parabiosis*) of two distinct fungi and a single alga, or of a single fungus and two different algae. E.g., *Peltigera aphthosa* (Peltigerales) has the alga *Rivularia* in the cephalodia and lives in symbiosis with *Nostoc* or *Dactylococcus* in the other parts of the thallus. Cf. **parasymbiont**.

parabiosis, pl. **parabioses** (Gr. *pará*, at the side, above, beyond + *bíosis*, the condition of living < *bíos*, life + suf. *-osis*, state or condition): *Lichens*. The phenomenon in which a single lichen thallus is composed of two distinct fungi in symbiosis with a single alga, or of a fungus and two different algae. In *Lobaria* (=*Sticta*) *amplissima* (Peltigerales), e.g., two different photobionts (*Protococcus* and *Nostoc*) coexist with the fungus to constitute the lichen thallus. The organisms involved in this association are called *parabionts*. Cf. **antagonism**, **commensalism**, **helotism**, **metabiosis**, **mutualism**, **parasymbiosis**, **symbiosis** and **synergy**.

paracapillitium, pl. **paracapillitia** (Gr. *pará*, together, at the side of + L. *capillitium*, long hair < *capillus*, hair + suf. *-itium*, adjectival ending that signifies provided with, having): *Gasteromycetes*. A group of septate, colorless, thin-walled skeletal hyphae that function as a capillitium, but which originate from thin-walled generative hyphae, or occasionally from thick-walled hyphae. It differs from the capillitium in not being cyanophilic; it is present, e.g., in *Vascellum* (Lycoperdales) and *Astraeus* (Sclerodermatales). Cf. **capillitium**.

paracephalodium, pl. **paracephalodia** (Gr. *pará*, at the side of + *kephalé*, head + L. suf. *-odium* > E. *-odium*, to look like < Gr. *ode*, like < *-oeídes*, similar to): hyphal tissue that covers the cyanobacteria formed in a squamulose lichen thallus, whose other photobionts are green algae. Cf. **cephalodium**.

paracoccidioidomycosis, pl. **paracoccidioidomycoses** (< gen. *Paracoccidioides* < Gr. *pará*, at the side of + gen. *Coccidioides* < L. *coccidium* < Gr. *kokkíon*, small ball + L. suf. *-oides*, like < Gr. *-oeídes*, similar to + NL. *mycosis* < Gr. *mýkes*, fungus + suf. *-osis*, state or condition): a primary pulmonary infection, acquired, as with other systemic mycoses, by the inhalation of conidia. The visible signs are mucocutaneous lesions, especially in the mouth, but the lymphatic dissemination of this dimorphic organism (*Paracoccidioides brasiliensis*, of the moniliaceous asexual fungi) causes damage to the skin, the intestinal tract, the lungs, the liver and other organs. Also called South American blastomycosis because it mainly occurs in South America; it is rare in Central America, Mexico and the Caribbean. See **mycosis**.

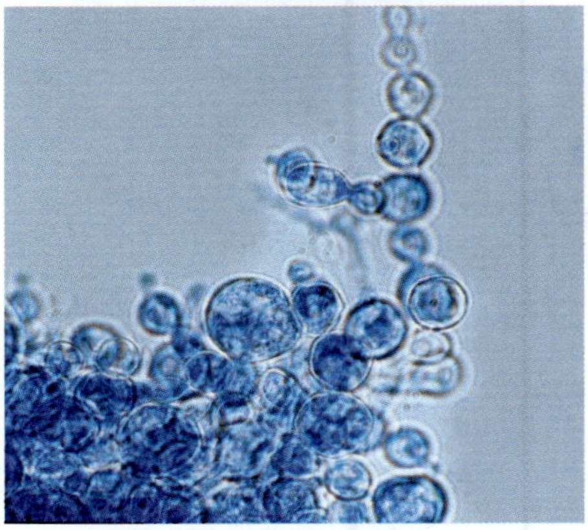

Yeast cells of the parasitic phase of *Paracoccidioides brasiliensis*, the dimorphic fungus causing **paracoccidioidomycosis**, x 750 (*CT*).

paracyst (Gr. *pará* at the side of + *kýstis*, vesicle, cell): an old term for the antheridial cell found next to the ascogonium or macrocyst in the gen. *Pyronema* (Pezizales). Cf. **macrocyst** and **microcyst**.

paragymnohymenial (Gr. *pará*, at the side of, above + *gymnós*, naked + L. *hymenium* < Gr. *hyménion*, dim. of *hymén*, membrane + suf. *-alis* > E. *-al*, relating to or pertaining to): a type of ascocarp develoment in some Discomycetes, similar to the eugymnohymenial type; however, in this case the hymenium is partially protected in a manner analogous to cleistohymenial development. The sterile hyphae arise near the ascogonium, but do not completely envelop it, i.e., it does not form an enclosed covering. As the ascocarp expands, it remains open, with its side walls surrounding the developing asci, among the paraphyses. Cf. **cleistohymenial** and **eugymnohymenial**.

paragynous (Gr. *pará*, at the side of, together + *gyné*, woman; here, female sexual organ): *Peronosporales*. An antheridium that is formed at the side of the oogonium, as is observed in *Pythium ultimum* and *Plasmopara viticola* (Oomycota). Cf. **amphigynous**.

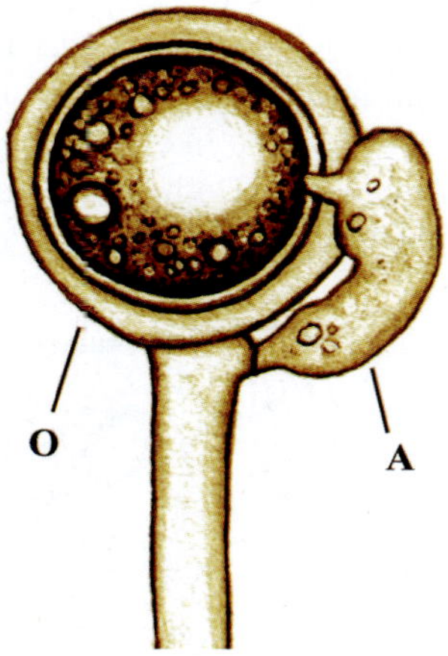

Paragynous antheridium (A), at the side of an oogonium (O) of *Pythium ultimum*, x 1,440 (*MU*).

parallel (L. *parallelus*, parallel < Gr. *parállelos*, side by side): **1**. *Agaricales*. A type of hymeniiferous trama in which the hyphal elements are arranged next to one another, side by side, parallel to the main axis of the gills of the basidiocarp, as is seen in *Hygrocybe persistens* (=*H. acutoconica*), Agaricales. **2**. *Mucorales*. In some gen., such as *Pilobolus*, the suspensors of the zygosporangium are parallel, unilateral and close, i.e., they are not opposite as in the majority of the gen. of the order. Also called apposed or *juxtaposed*.

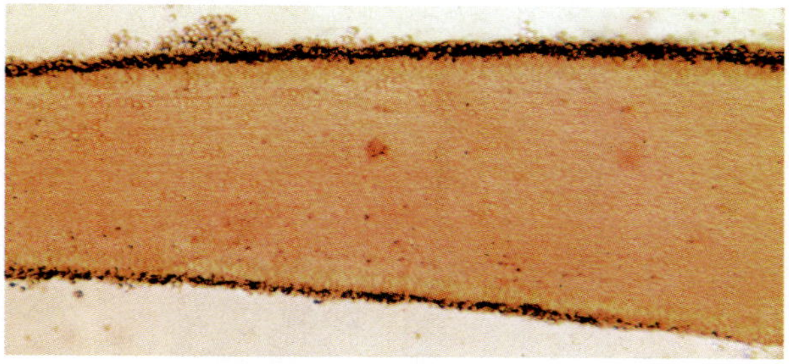

Parallel hymeniiferous trama in a gill of the basidiocarp of *Hygrocybe persistens* (=*H. acutoconica*), x 150 (*RV*).

paramorph (Gr. *pará*, at the side of, toward + *morphé*, shape): a term proposed for a fossil fungus that, even though it lacks the characteristics necessary to classify it properly, is believed to have affinities with a particular group of non-fossil fungi.

paranym (Gr. *pará*, at the side of, together + *ónyma*, *ónoma*, name): a validly published name that is so similar in orthography to another name, also validly published, but based on a different nomenclatural type, that confusion between the two is probable.

paraphyletic (Gr. *pará*, at the side of + *phyletikós* < *phyletés*, one of the same tribe, race, stirpe < *phylé*, tribe + suf. *-tikós* > L. *-ticus* > E. *-tic*, relation, aptitude): a term used in cladistics to refer to a taxon or group of taxa in which some of its descendents are not included. See **phylum**. Cf. **monophyletic** and **polyphyletic**.

paraphysis, pl. **paraphyses** (Gr. *paráphysis*, offspring < *pará*, at the side of + *phýsis*, action of budding, of being born, a growth): a sterile, basally attached, hypha-like structure present in the hymenium of many ascomycetes and some pycnidial fungi; in *Guanomyces* (Sordariales), *Geoglossum* and *Trichoglossum* (Helotiales), e.g., the paraphyses are long and slender, and are intermixed with the asci in the hymenium. See **periphysis**. The term paraphysis also has been applied, probably erroneously, to the brachybasidia in the agarics that form a particular layer of the hymenium, through which the basidia protrude. See **brachybasidium**.

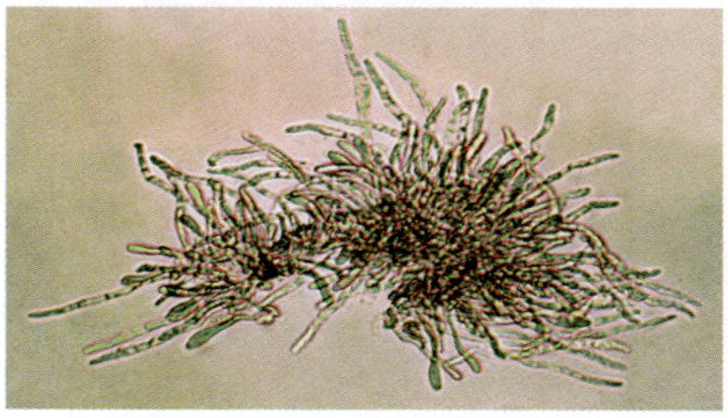

Paraphyses of the hymenium of *Guanomyces polythrix*, x 250 (*MCG*).

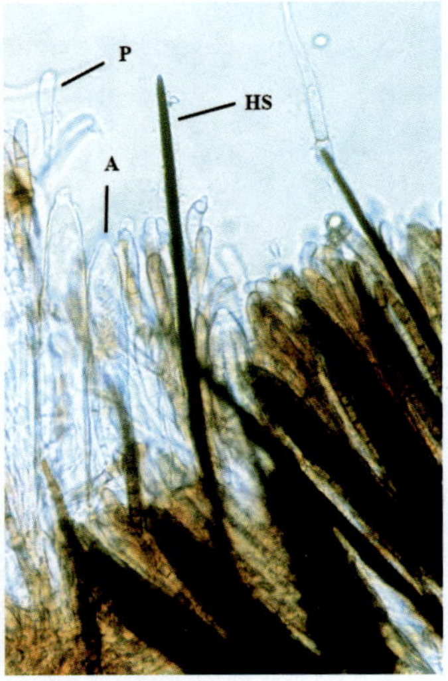

Paraphyses (P) of the hymenium of *Trichoglossum* sp. Also shown are the asci (A) and the hymenial setae (HS), x 250 (*RTH*).

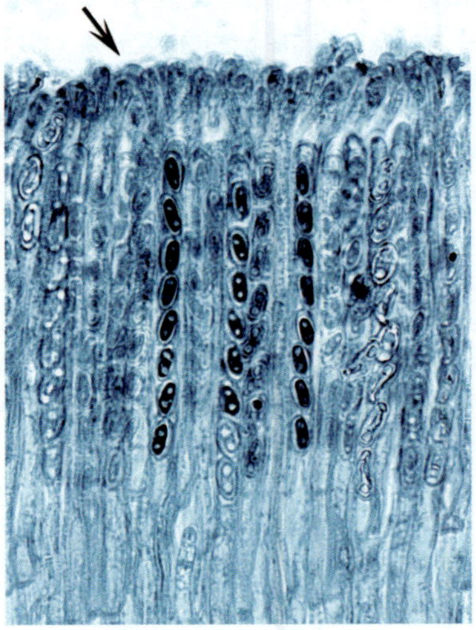

Paraphyses of the hymenium of *Peziza* sp., forming an epithecium above the layer of asci, x 100 (*MU*).

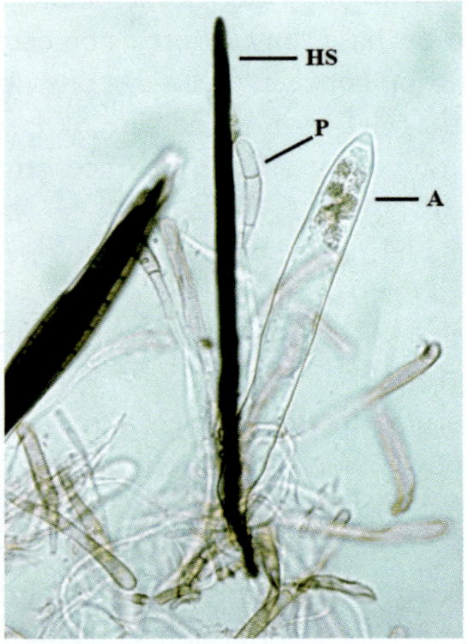

Paraphyses (P) of the hymenium of *Trichoglossum* sp. Also shown are the asci (A) and the hymenial setae (HS), x 300 (*RTH*).

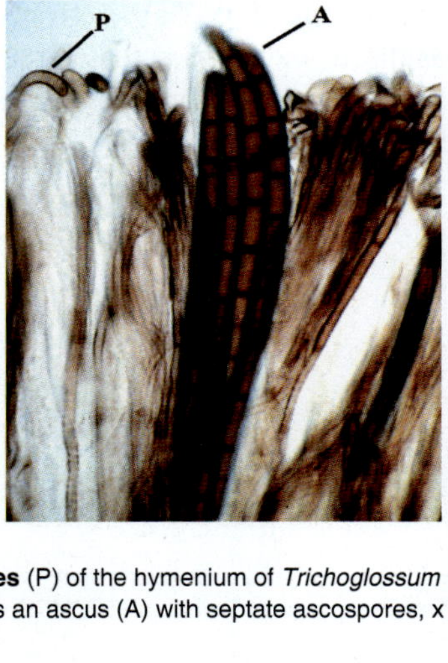

Paraphyses (P) of the hymenium of *Trichoglossum velutipes*. Also visible is an ascus (A) with septate ascospores, x 200 (*VBM*).

paraphysoid (Gr. *paráphysis*, offspring + suf. *-oide*, like < Gr. *-oeídes*, similar to): a sterile, filiform, hyphal structure, similar to a paraphysis, but often branched and forming a kind of network (paraphysoid network); found principally in the pseudothecia of some pyrenolichens with a crustaceous thallus. Also called paraphysoids are the slender hyphae that are extended and insinuated among the asci and which stick out above them, like slender paraphyses, as occurs in *Bagnisiella* (Dothideales) and *Melanomma* (Pleosporales).

Paraphysoids of a pseudothecium of *Bagnisiella mirabilis*, which is embedded in the tissues of a dead tree branch, seen in longitudinal section, x 520 (*MU*).

paraplectenchyma, pl. **paraplectenchymata** (Gr. *pará*, together, at the side of + *plektós*, intertwined + *énchyma*, stuffed): a type of tissue found in the cortex of many macrolichens, in which the hyphae are oriented in all directions, giving a cellular appearance. It corresponds to the parenchyma of other fungi. Cf. **chalaroplectenchyma** and **prosoplectenchyma**.

parasexuality (Gr. *pará*, together, at the side of + L. *sexualis*, pertaining to or relating to sex): a process in which plasmogamy, karyogamy and meiosis occur in sequence, but not in differentiated sexual organs nor at a specific time in the life cycle of the organism; in the conidial fungi with heterokaryotic individuals, parasexuality can be important as a source of genetic variation that permits them to adapt to changing environmental conditions, as would occur with sexual reproduction. Cf. **perittogamy**.

parasite (Gr. *parásitos* < *pará*, together, at the side of + *sîtos*, bread, food in general; what eats food together with another): an organism that derives its nourishment from another living organism, whether a plant, an animal or a fungus; if the parasite invades and causes disease it is considered a *pathogen*. See **biophagous** and **biophilous**.

parasitism (Gr. *parásitos* < *pará*, together, at the side of + *sîtos*, bread, food in general; what eats together with another + suf. *-ismós* > L. *-ismus* > E. *-ism*, state, phase, tendency, action): the phenomenon relating to parasites; the action of being parasitic. The same names given to parasites can also apply to parasitism. See *antagonistic symbiosis*, under the entry **symbiosis**.

parasoredium, pl. **parasoredia** (Gr. *pará*, at the side of, beyond + ML. *soredium* < Gr. *sorídion* < *sorós*, pile, heap + dim. suf. *-ídion*): a structure similar to a soredium; originally applied to a plectenchymatous structure formed in the upper layer of the thallus of *Umbilicaria hirsuta* (Lecanorales). Cf. **soredium**.

parasymbiont (Gr. *pará*, together, at the side of, on top of + *symbióo*, life in common of two or more organisms + *óntos*, genit. of *ón*, a being): refers to an organism involved in the phenomenon of *parasymbiosis*, as occurs in certain lichens, in which the same alga coexists with a lichenogenic fungus and with another invading fungus, a parasite of the lichen (although not directly harmful to the algal cells). Cf. **parabiont**.

parasymbiosis, pl. **parasymbioses** (Gr. *pará*, beyond, on top of + *symbíosis*, life in common of two or more organisms): the state of coexistence of a lichen alga with two fungi, the mycobiont of the lichen and a fungal parasite of the lichen, but which is not harmful to the gonidia. The organisms involved are called *parasymbionts*. Cf. **antagonism**, **commensalism**, **helotism**, **metabiosis**, **mutualism**, **parabiosis**, **symbiosis** and **synergy**.

parathecium, pl. **parathecia** (NL. *parathecium* < Gr. *pará*, at the side of, beyond + *thekíon*, dim. of *thêke*, case; here, of the asci): *Lichens*. The peripheral extension of the hypothecium or subhymenial layer curved upward along the margin of the hymenium in some discolichens, e.g., *Ochrolechia parella* (Pertusariales) and *Roccella fimbriata* (Arthoniales). Also called *marginal excipulum* or *proper excipulum*. See **excipulum**.

Parathecium of the apothecia of a thallus of *Ochrolechia parella*, x 12 (*MU*).

Parathecium of the apothecia of a thallus of *Roccella fimbriata*, x 12 (*MU*).

paratype (Gr. *pará*, at the side of, beyond + *týpos*, mark, model, type): any specimens, other than the holotype, that the author of a species, variety, etc., cites as being the basis for the original description (the protolog) of a taxon. Also refers to a syntype remaining after the selection of a lectotype from the syntypes.

parenthesome (LL. *parenthesis* < Gr. *parénthesis*, interposition, insertion, orthographic sign + Gr. *sõma*, body): a curved double membrane (shaped like a parenthesis) that is found on each side of a dolipore septum; it can be perforate, imperforate or vesiculose. Also called **septal pore cap**. It is present in the majority of the basidiomycetes, as in *Thanatephorus cucumeris* (=*Corticium solani*), Aphyllophorales. See **dolipore septum**.

parietal (L. *parietalis* < *paries*, genit. *parietis*, wall + L. suf. *-alis* > E. *-al*, relating to or pertaining to): pertaining to or relating to the wall; formed in the wall. In mycology it refers to what is found in cell walls, asci, ascocarps, etc. E.g., some perithecia have parietal asci, as they are attached to the inner wall of the perithecium.

parmuliform (L. *parmula*, dim. of *parma*, shield + suf. *-formis* < *forma*, shape): shield-shaped; a small rounded shield with the edges turned upward, like the foliaceous thallus of the lichen *Parmelia* (Lecanorales).

parobasidium, pl. **parobasidia** (Gr. *páros*, early, earlier, previous + L. *basidium*, small base, here, of the spores < *basis*, base + dim. suf. *-idium*): applied to the distal and functional part of the metabasidium when it has remnants of the probasidium.

parsimonious (L. *parsimonia* < *parsus* < *parcere*, to save, economize + suf. *-iosus* > E. *-ious*, possessing the qualities of): a term used in cladistics to refer to the pattern of a cladogram that requires the least number of steps to resolve the relationships of taxa being analyzed phylogenetically. See **cladistics**.

part spore (L. *partialis* < *pars*, genit. *partis*, part, piece, portion; NL. *spora* < Gr. *sporá*, spore): each of the unicellular spores that results from the fragmentation of a bicellular or multicellular ascospore of *Hypocrea* and other Hypocreales, as well as of the ascospores of *Cordyceps militaris* and other clavicipitaceous species, such as *Balansia henningsiana* (Hypocreales), and of *Sporormiella* (=*Preussia*) *australis* (Pleosporales).

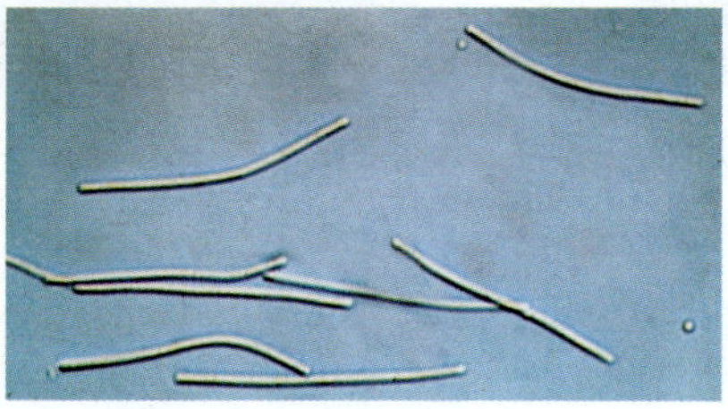

Part spores derived from the fragmentation of the ascospores of *Balansia henningsiana*, x 570 (*RTH*).

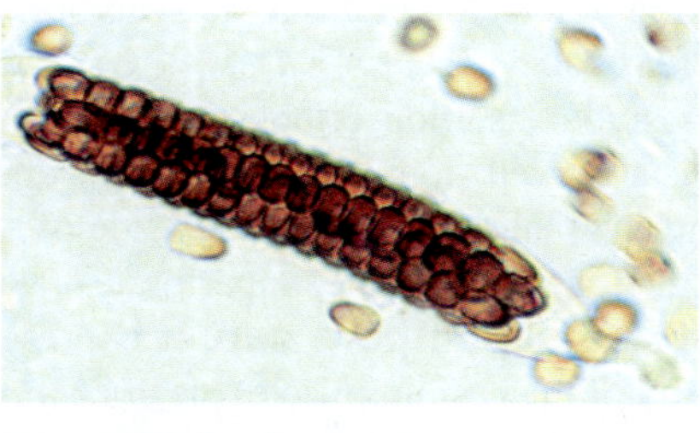

Part spores derived from the fragmentation of the septate ascospores of *Sporormiella* (=*Preussia*) *fimetaria*, x 530 (*CB*).

Ascospore of *Sporormiella* (=*Preussia*) *australis*; its fragmentation will give rise to **part spores**, x 1,300 (*RTH*).

partheno-, **parthen-** (Gr. *párthenos*, virgin): a pref. employed in various biological terms to indicate that in a normally sexual process the involvement of one sex failed, as in the following terms.

parthenogamy (Gr. *párthenos*, virgin + *gámos* > E. des. *-gamy*, sexual union): a type of copulation in which two female sexual cells unite, especially of an archicarp; parthenogamy is a special case of *automixis*.

parthenogenesis, pl. **parthenogeneses** (Gr. *párthenos*, virgin + *génesis*, origin, production): the generation of an individual from a female gamete that has not been fertilized by a male gamete (apomictic development), as can happen, e.g., in *Monoblepharis* (Monoblepharidales), and in many Mucorales, that can develop, respectively, oospores and zygospores parthenogenetically, without gametangial copulation having occurred. See **apogamy**, **apomixis** and **azygospore**. Cf. **amphimixis**.

parthenomixis, pl. **parthenomixes** (Gr. *párthenos*, virgin + *míxis*, mixture, alluding to sexual union): see **parthenogamy**.

parthenospore (Gr. *párthenos*, virgin + *sporá*, spore): *Entomophthorales*. An azygospore that originates as a bud from hyphal bodies and which on maturing acquires the appearance and structure of a zygospore, but with parthenogenetic development, as occurs in *Entomophthora muscae*. An aboospore (an oospore that develops apomictically) also is a parthenospore. See **aboospore** and **azygospore**.

partial veil (L. *partialis* < *pars*, genit. *partis*, part, portion + suf. *-alis* > E. *-al*, relating to or pertaining to; L. *velum*, veil, curtain, cloth): see **veil**.

passage (VL. *passare* < L. *passus*, to pass, move, proceed; action or process of passing from one place, condition or state to another): the experimental infection of a host with a parasite, which is subsequently reisolated; a method used to increase the virulence of a parasite or to confirm the pathogenicity of a particular fungus.

pasteurization (< *Louis Pasteur*, French chemist and microbiologist + L. suf. *-ationem* > E. *-ation*, action, state or condition, result): partial sterilization of a substance, especially a liquid (such as a culture medium, milk and other dairy products, etc.), at a temperature and during an exposure time that destroys undesirable organisms (such as *Mycobacterium tuberculosis* and some other pathogenic microorganisms), although it does not destroy all of the microorganisms present, nor does it cause a major chemical alteration of the substance being pasteurized (e.g., it does not alter the flavor); generally a temperature of 62-65°C for 30 minutes is utilized. A pasteurized material is not sterilized. Cf. **sterilization** and **tyndalization**.

patella, pl. **patellae** (L. *patella*, dim. of *patina*, plate, platter, saucepan): a sessile, orbicular apothecium with a border that is distinguished from the thallus; present in some lichens (Patellariales).

patelliform, **pateniform** (L. *patelliformis* < *patella*, dim. of *patina*, plate, platter, saucepan + suf. *-formis* < *forma*, shape): shaped like a round plate, with a well defined border, like the slightly concave, disciform apothecia of some ascomycetes, e.g., *Patella* (=*Scutellinia*) *albida* (Pezizales) and *Arthonia patellulata* (Arthoniales).

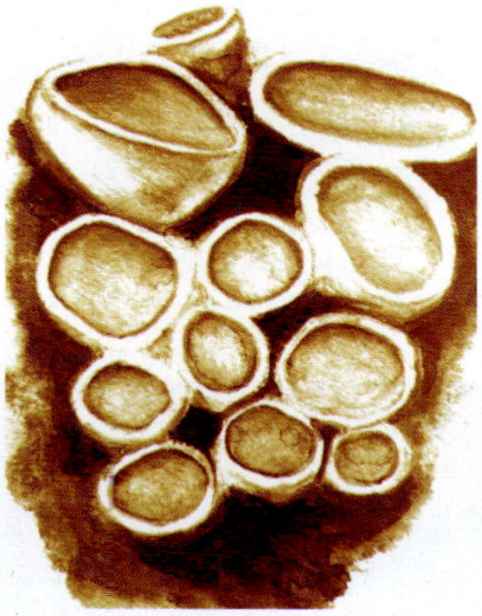

Patelliform apothecia of *Patella* (=*Scutellinia*) *albida*, x 2 (*MU*).

patho- (Gr. *páthos*, disease): a pref. employed in terms that refer to diseases, such as *pathogen*.

pathogen (Gr. *páthos*, disease + *génos*, origin < *gennáo*, to engender, produce): an organism capable of causing disease in a particular host or range of hosts, not necessarily as a parasite. Cf. **parasite**.

p-body (E. *p*, for pigment, or for protein; ME. < OE. *bodig*, body): a distinctive organelle associated with the rough endoplasmic reticulum that surrounds the plate-like cristae of the mitochondria of the myxamoebae of *Acrasis rosea* (Acrasiomycota). The function of the p-body is unknown, but it has been suggested that it is a pigment granule, or a proteinaceous organelle, since much of the p-body can be eliminated with proteases.

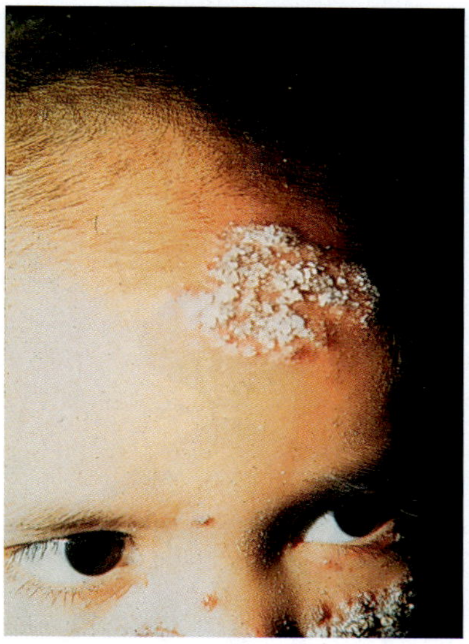

Candidiasis of the forehead of a child, caused by *Candida albicans* (cryptococcaceous asexual yeasts), a common **pathogen**, x 0.3 (*RLM*).

pectinate (L. *pectinatus*, hair-dresser < *pecten*, genit. *pectinis*, comb + suf. *-atus* > E. *-ate*, provided with or likeness): having a comb-shaped structure or being comb-like; e.g., the sporocladia with sporangiola of *Coemansia mojavensis* (Kickxellales), and the hypha that ends in a comb-shaped structure, somewhat swollen at the apex and with short processes on one side, as occurs in the dermatophytes, e.g., *Microsporum audouinii* (moniliaceous asexual fungi). Cf. **pennate**.

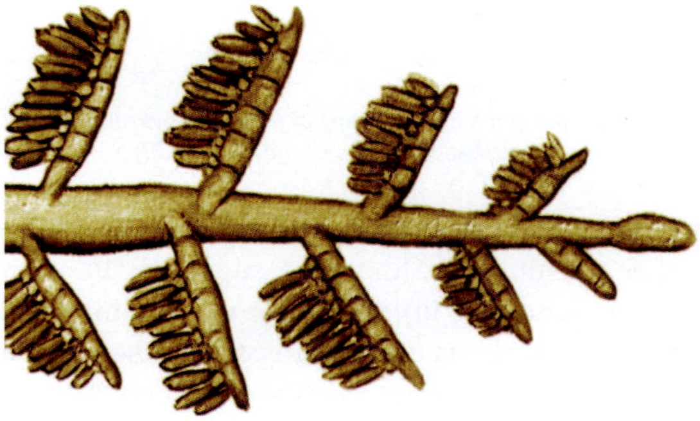

Pectinate sporocladia of *Coemansia mojavensis*, x 450 (*MU*).

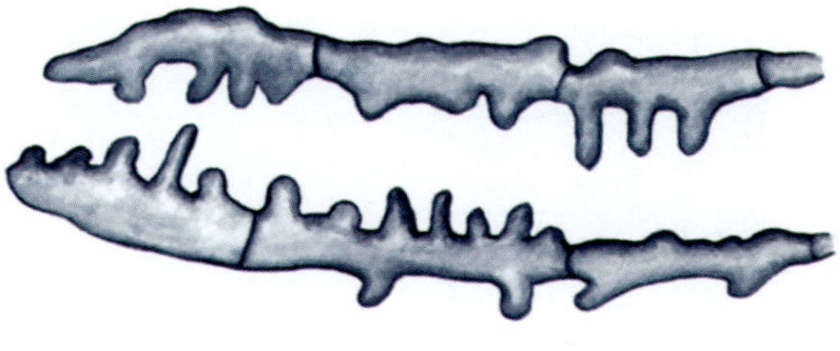

Pectinate hyphae of *Microsporum audouinii*, x 1,150 (*MU*).

pedicel, **pedicle**, **stalk** (NL. *pedicellus*, dim. of L. *pediculus*, dim. of *pes*, genit. *pedis*, foot; here, support; ME. *stalke*, stem): **1**. A small stalk or support of spores, sporangia, cystidia, asci, etc. In general, equivalent to **peduncle**. **2**. See **stipe**.

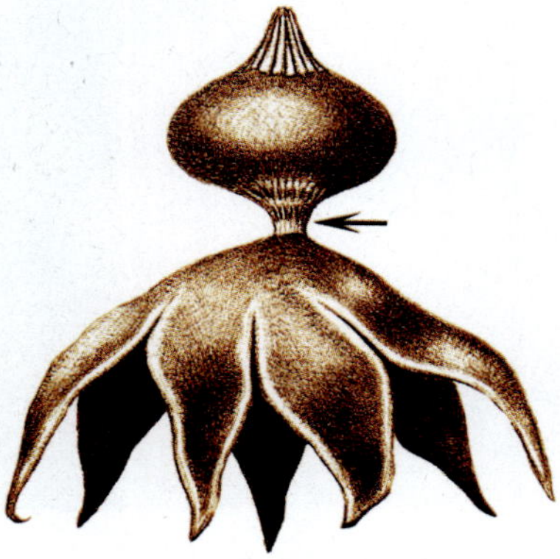

Pedicel supporting the sporiferous sac of a fruiting body of *Geastrum pectinatum* (Lycoperdales), x 2 (*MU*).

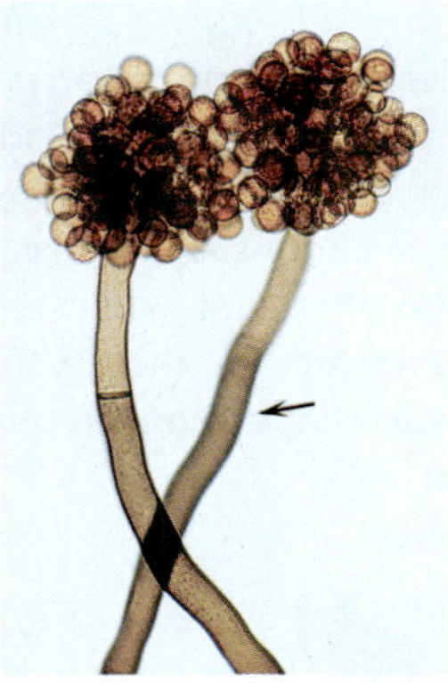

Pedicels of the condiophores of *Periconia byssoides* (dematiaceous asexual fungi), x 450 (*CB*).

pedicellar (NL. *pedicellus*, dim. of L. *pediculus*, dim. of *pes*, genit. *pedis*, foot; here, support + suf. *-aris* > E. *-ar*, pertaining to): relating to or pertaining to the pedicel, with the nature of a pedicel; e.g., the pedicellar cells of *Dictyostelium mucoroides* (Dictyosteliales) form a reticulum in the pedicel of the sorocarp, a characteristic that gives the name to the gen.

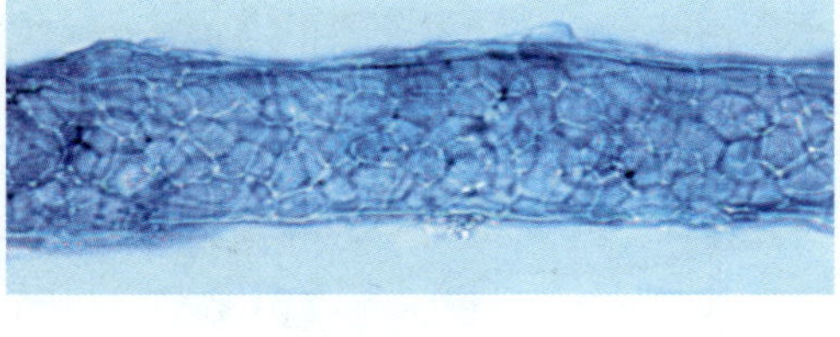

Pedicellar cells of the sorocarp of *Dictyostelium mucoroides*, x 150 (*MU*).

pedicellate, **pediculate**, **pedunculate** (L. *pedicellus*, dim. of *pediculus*, dim. of *pes*, genit. *pedis*, foot; here, support + suf. *-atus* > E. *-ate*, provided with or likeness): having a pedicel or peduncle. E.g., the sporangia of *Physarum polycephalum* (Physarales) and *Lamproderma columbinum* (Stemonitales), the spores of *Gigaspora* (Glomales), the asci of *Sporormiella* (=*Preussia*) *australis* (Pleosporales), and the teliospores of species of *Gymnosporangium, Puccinia*, and *Phragmidium* (Uredinales) are pedicellate. In some species of basidiomycetes a part of the pedicel remains attached to the basidiospore after discharge; these basidiospores are described as pedicellate, e.g., in *Bovista*, *Lycoperdon* and *Geastrum* (Lycoperdales). Cf. **apodal**, **stipitate** and **sessile**.

Pedicellate sporangia of *Physarum polycephalum*, on agar, x 10 (*MU*).

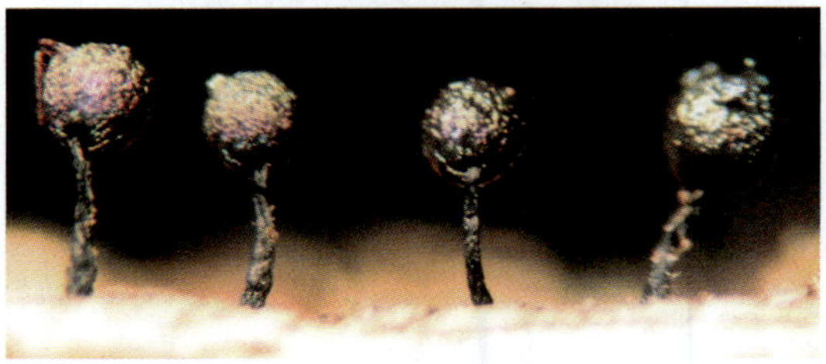

Pedicellate sporangia of *Lamproderma columbinun*, x 8 (*MU*).

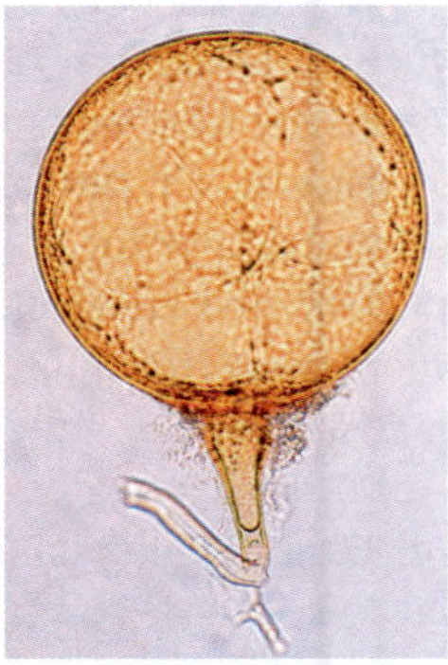

Pedicellate spore of *Gigaspora* sp., x 400 (*CB*).

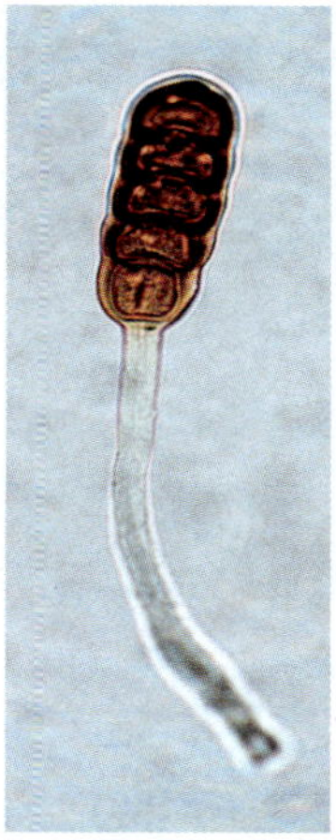

Pedicellate teliospore of *Phragmidium potentillae*, x 1,600 (*CB*).

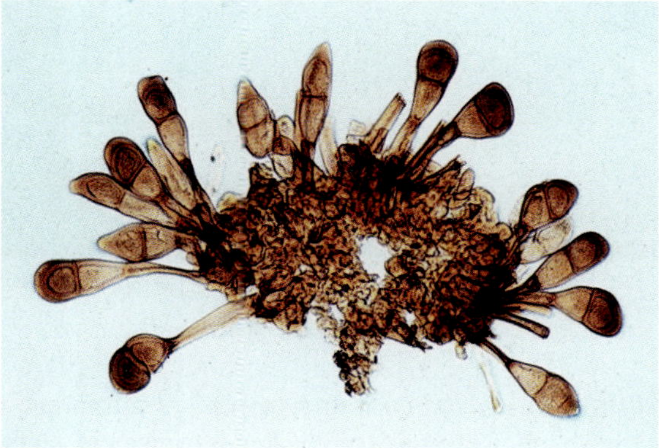

Pedicellate teliospores of a telium of *Puccinia graminis tritici*, x 300 (*MU*).

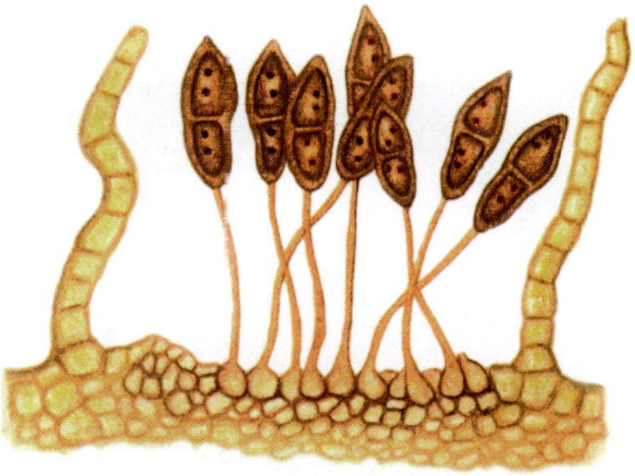

Pedicellate teliospores of a telium of *Puccinia graminis tritici*, x 480 (*MU*).

pedo-, **ped-** (Gr. *país*, genit. *paidós*, child, infant): a pref. employed to refer to a process carried out in young or immature organisms, as in *pedogamy*. Distinguish from the pref. *pedo-*, used in terms such as *pedogenesis*, alluding to the formation of soil (Gr. *pédon*, soil, earth).

pedogamy (Gr. *país*, genit. *paidós*, child, infant + *gámos* > E. des. *-gamy*, sexual union): a type of pseudomictic sexuality that consists of the union of an adult mother cell with a recently formed daughter cell, as happens in yeasts of the gen. *Debaryomyces* and *Nadsonia* (Saccharomycetales), in which copulation occurs between a mother cell and its bud, which is still attached. Cf. **adelphogamy**.

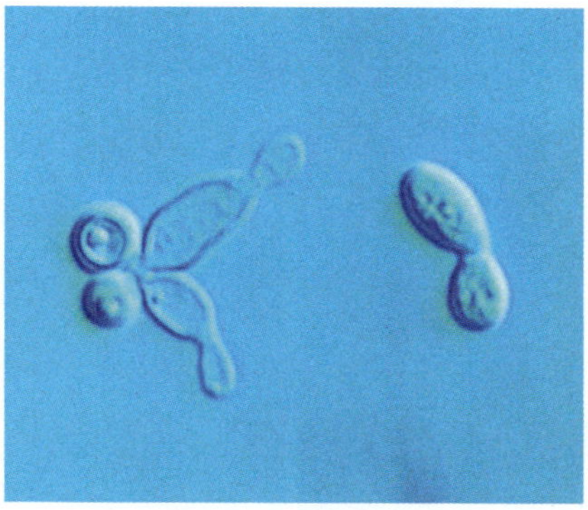

Pedogamy in *Nadsonia fulvescens*, between a mother cell and its still attached bud, resulting in the formation of an ascus containing an ascospore, x 1,000 (*RTH*).

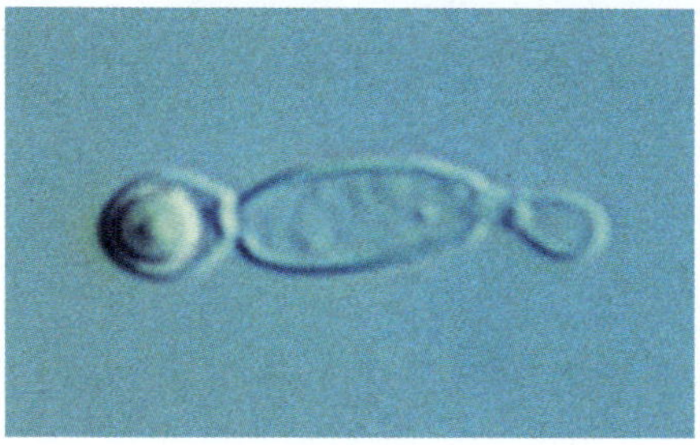

Pedogamy in *Nadsonia fulvescens*, between a mother cell and its still attached bud, resulting in the formation of an ascus containing an ascospore, x 1,100 (*RTH*).

pedogenesis, pl. **pedogeneses** (Gr. *país*, genit. *paidós*, child, infant + *génesis*, origin, production): reproduction in young or immature organisms. Cf. **neoteny**.

pedogenesis, pl. **pedogeneses** (Gr. *pédon*, soil, earth + *génesis*, engenderment): the formation and development of soil through the intervention of secondary metabolites of lichens (the so-called lichenic acids), that includes fatty acids, lactones, polyketides such as dibenzofurans, depsides, and anthraquinones, as well as products derived from the mevalonate and shikimate metabolic pathways.

peduncle (L. *pedunculus*, dim. of *pes*, genit. *pedis*, foot; here, support): a small stem or support of spores, sporangia, cystidia, asci, etc. In general, equivalent to **pedicel** and **pedicle**.

pelagic (Gr. *pélagos*, sea, ocean + suf. *-ikos* > L. *-icus* > E. *-ic*, pertaining to or relating to): a term applied to those organisms that inhabit the sea, beyond the neritic zone, or far from the coast. Among the fungi and oomycotes, there are zoosporic species (Chytridiomycota, Hyphochytriomycota and Oomycota) and conidial species (hyphomycetous and blastomycetous asexual fungi), as well as certain basidiosporogenous yeasts, that are pelagic.

pellet (ME. *pelote* < MF. < VL. *pilota*, dim of L. *pila*, ball): a spheroidal colony developed in a liquid culture, especially a shaken culture.

Pellets of two unidentified fungi developed in shaken liquid culture, x 0. 3 (*LMR*).

pellicle, pellicula, pl. **pelliculae, pellis** (L. *pellicula*, dim. of *pellis*, skin + dim. suf. *-cula*): **1**. *Agaricales*. A thin, delicate, sterile outer membrane or peel, similar to a cuticle or skin, that can easily break off of the viscous pileus of the basidiocarp of certain agarics. The pellicle of the stipe is called *stipitipellis*, whereas that on the pileus is called the *pileipellis*. The pellicle of the basidiocarp is composed of layers of cells, whose number varies according to the species of fungi. If it is constituted of two layers, the upper one is the *suprapellis* and the lower one the *subpellis*; when an intermediate layer is present, it is known as the *mediopellis* or *medipellis*. **2**. A thin membrane formed on the surface of some cultures of bacteria or yeasts when grown in liquid nutrient media.

pellicular veil (L. *pellicula*, dim. of *pellis*, skin + suf. *-aris* > E. *-ar*, as, concerning; L. *velum*, veil, curtain, cloth): a very fine partial veil present in a sporophore lacking a stipe. Cf. **cortina**.

pelliculate, pelliculose (L. *pellicula*, dim. of *pellis*, skin + suf. *-atus* > E. *-ate*, provided with or likeness; or + suf. *-osus* > E. *-ose*, full of, augmented, prone to): provided with a pellicle; similar to a pellicle or crust, like the hymenial layer of some polyporaceous basidiomycetes.

Pelliculate basidiocarps of *Phanerochaete* sp., growing on bark, x 0. 3 (*RV*).

Pelliculate basidiocarps of *Terana caerulea* (=*Pulcherricium caeruleum*), x 0. 3 (*RV*).

pellucid (L. *pellucidus*, transparent): transparent or translucid. Refers to a pileus that is translucid, so that the gills appear as lines or striae (*striate-pellucid*) from above, as occurs with the basidiocarp of various Agaricales (*Mycena*, *Coprinus*, *Oudemansiella*). See **hygrophanous**.

Pellucid pileus of the basidiocarp of *Oudemansiella canarii*, x 0. 5 (*RV*).

pelo-, pel- (Gr. *pelós*, clay, mud): a pref. employed in names such as *Pelodiscus* (=*Patella*, of the Pezizales), a fungus that forms its apothecia on muddy soil.

peloton (Fr. < MF. *pelote*, little ball): syn. of **glomerulus**; a dense mass composed of a group of hyphae, formed by mycorrhizogenous fungi in the interior of the root cells of the associated plant, e.g., in *Allium ursinum*. Pelotons are one of the forms (the other is the arbuscule, or vesicule) present in endotrophic mycorrhizae.

pelta, pl. **peltae** (L. *pelta* < Gr. *pélte*, small shield): a flat, not very prominent, apothecium without a differentiated border, shaped like a small shield, like that present in certain lichens of the order Peltigerales (e.g., *Peltigera* and *Solorina*).

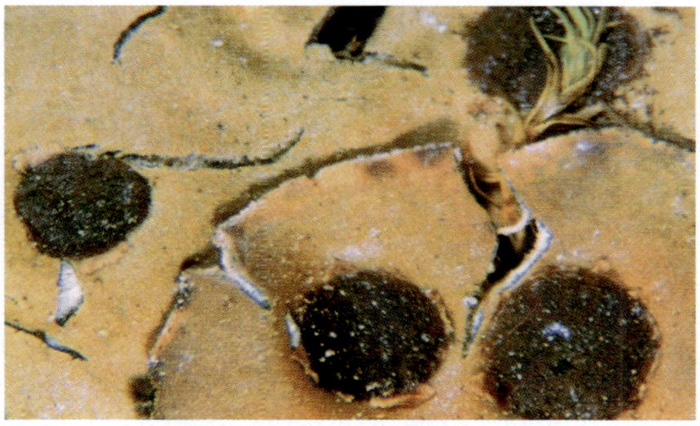

Peltae (apothecia) of *Solorina saccata*, x 10 (*MU*).

peltate (L. *peltatus*, armed with a shield, provided with a pelta < *pelta*, shield < Gr. *pélte*, small shield + L. suf. *-atus* > E. *-ate*, provided with or likeness): applied to a rounded laminar structure with the stem or support inserted in the center, on the lower side, like the sporangium of *Physarum flavicomum* (Physarales). Also called **clypeate** and **scutiform**.

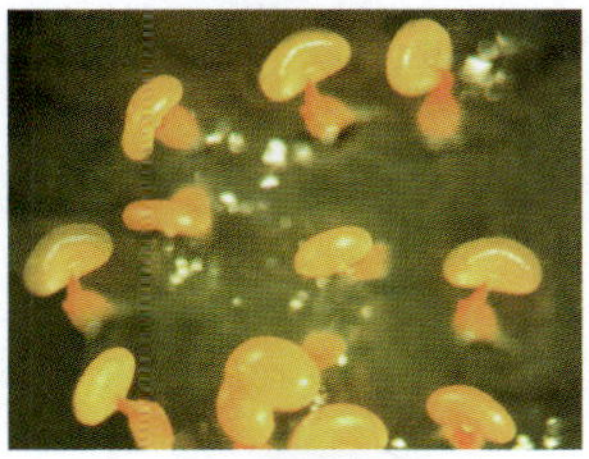

Peltate young sporangia of *Physarum flavicomum* on agar, x 5 (*EPS*).

Peltate mature sporangia of *Physarum flavicomum* on agar, x 5 (*EPS*).

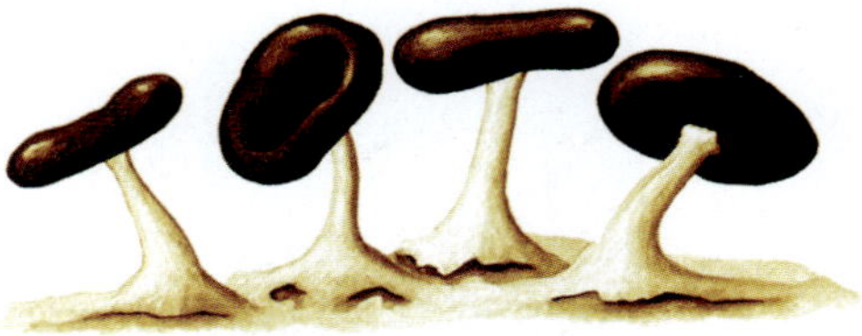

Peltate sporangia of *Physarum flavicomum*, x 10 (*MU*).

pendant, **pendulous** (L. *pendulus*, hanging down): hanging down, like the teeth of the hymenophore of *Hericium* (=*Hydnum*) *erinaceus* (Aphyllophorales), which develop with positive geotropism, and the thalli of some lichens, such as *Usnea barbata* and *U. ceratina* (Lecanorales).

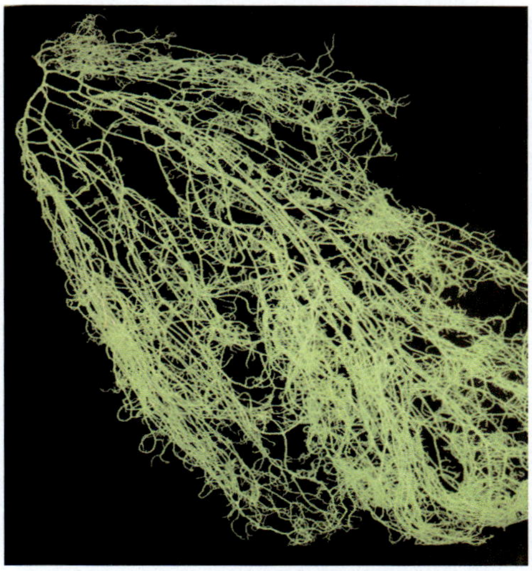

Pendulous thallus of the lichen *Usnea ceratina*, x 0.2 (*MU*).

penicillate, **penicilliform** (L. *penicillatus* < *penicillus*, brush + suf. *-atus* > E. *-ate*, provided with or likeness; NL. *penicilliformis* < *penicillus* + suf. *-formis* < *forma*, shape): shaped like a brush, like the conidiophores of the gen. *Trichoderma*, *Penicillium* and *Gliocladium* (moniliaceous asexual fungi).

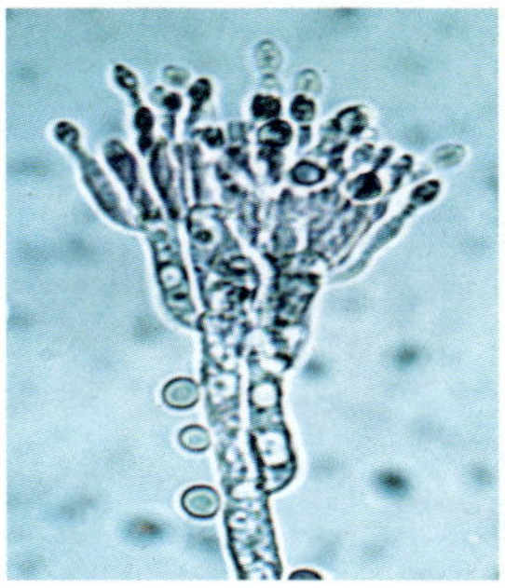

Penicillate conidiophore of *Penicillium* sp., x 460 (*RTH*).

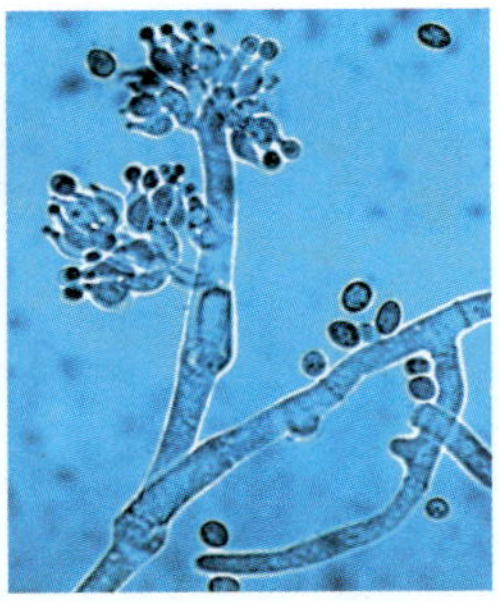

Penicillate conidiophore of *Trichoderma viride*, x 400 (*RTH*).

penicillus, pl. **penicilli** (L. *penicillus*, brush): a conidial head in the shape of a brush, composed of a conidiophore or pedicel that supports a cluster of conidiogenous cells (phialides), and which can also have other elements (such as ramae and metulae), depending on the species; characteristic of the gen. *Penicillium* and other gen. of moniliaceous asexual fungi (*Paecilomyces*, *Gliocladium*).

pennate, **penniform** (L. *pennatus* < *penna*, feather + suf. *-atus* > E. *-ate*, provided with or likeness; or < *penna* + suf. *-formis* < *forma*, shape): possessing feather-like structures or shaped like a feather, like the merosporangia of *Linderina pennispora* (Kickxellales), referring to the arrangement of the pseudophialides on the sporocladium. Cf. **pectinate**.

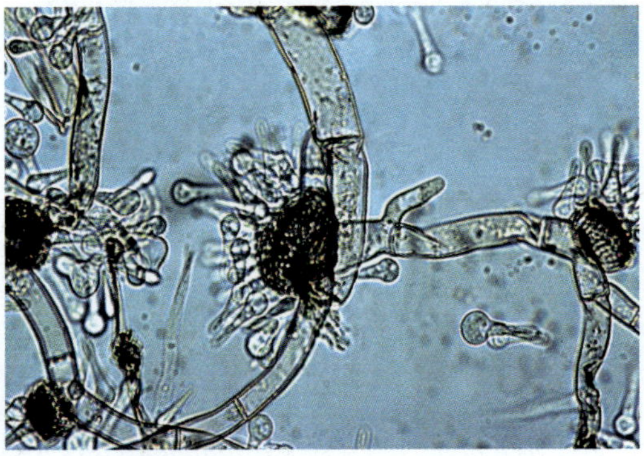

Pennate merosporangia, formed on a sporocladium of *Linderina pennispora*, x 300 (*MU*).

Pennate merosporangia, formed on a sporocladium of *Linderina pennispora*, x 500 (*MU*).

penni- (L. *penna*, feather): a pref. employed to give the idea of a feather, or alluding to the shape of one or, more often, referring to the arrangement of the barbs on a rachis, as in *penniform*, shaped like a feather, e.g., the fronds of some ferns.

penta-, **pent-** (Gr. *pénte*, five): a pref. used in many terms derived from Gr. to refer to something that occurs in fives, as in *pentagynous flower*, with five carpels in the gynoecium. Some generic names of fungi also have the pref. *penta-*, such as *Pentagenella* (Arthoniales) and *Pentaposporium* (= *Tripospermum*, the anamorph of *Trichomerium*, of the Capnodiales).

per- (L. *per*, through): a pref. that means through, as in *percurrent*; or is employed as an inseparable prep. to emphasize the meaning of the word that follows it, as in *perprolate*.

percurrent (L. *percurrere*, to pass through, continue + suf. *-entem* > E. *-ent*, being): what extends or continues growing; refers to the growth of a conidiophore (percurrent proliferation) or of a germ tube (percurrent germination) through a preexisting pore, as occurs in the conidiophore of *Corynespora*, or in the phialides of some species of *Phialophora* (both dematiaceous asexual fungi). The columella of the basidiocarp of some gasteromycetes, which is united with the peridium at the apex of the gleba, also is percurrent. See **proliferation**.

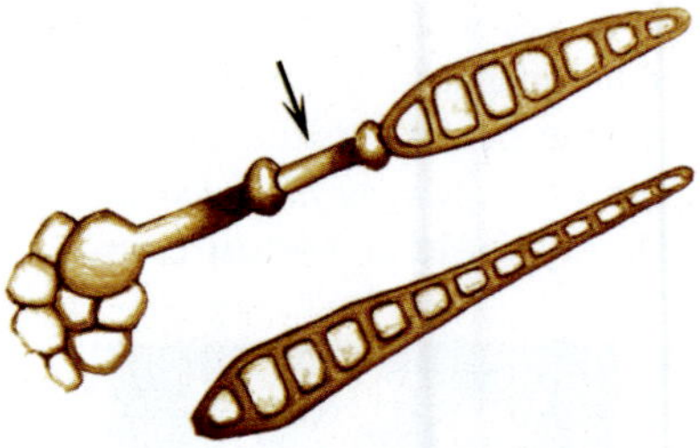

Percurrent conidiogenous cell of *Corynespora cassiicola*, x 530 (*MU*).

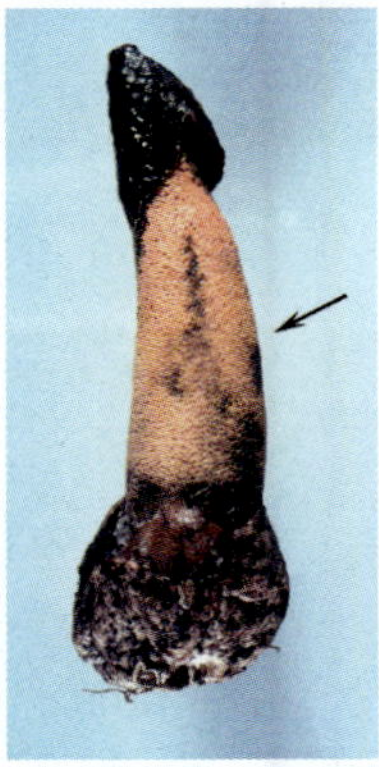

Percurrent columella of a basidiocarp of *Phallus hadrianii* (Phallales), x 0.3 (*TH*).

percurrent columella, pl. **columellae** (L. *percurrere*, to pass through, continue + suf. *-entem* > E. *-ent*, being; L. *columella*, dim. of *columna*, column): a column of serile tissue that extends from the base of the gleba toward the apex of the peridium, as is seen in the basidiocarp of *Podaxis* (Hymenogastrales). Also called **dendritic columella** and **stipecolumella**.

perennial (L. *perennis*, persistent, durable, eternal): refers to a fungus that lives three or more years; in reality, it refers to the durability of the fruiting body, not to the assimilative mycelium immersed in the substrate. If it lives two years, or more than one, it is *biennial*; if it lives one year or less, it is *annual*. Many basidiomycetes in the Aphyllophorales have perennial sporophores because they are woody and not easily decayed. Cf. **annual** and **biennial**.

Perennial basidiocarp
of *Fomitiporia hartigii*, growing on a trunk, x 0.5 (*RV*).

Perennial basidiocarp
of *Fomitiporia* sp , seen in longitudinal section, x 0.3 (*TRO*).

perforate at maturity (L. *perforatus*, perforate < *perforare*, to perforate, pierce + suf. *-atus* > E. *-ate*, provided with or likeness; L. *maturitas*, maturity): *Gasteromycetes*. Refers to the thin peridium that splits or forms holes to expose the mature gleba, as occurs in the fruiting body of *Phallogaster* (Phallales). Some lichen thalli, as in *Letharia* (Lecanorales), have perforate branches.

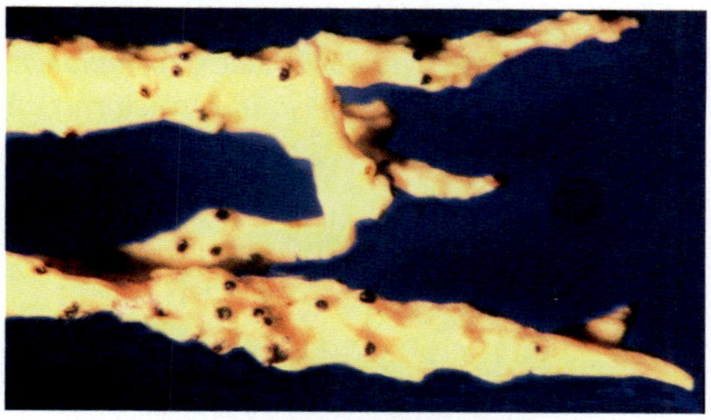

Perforate branches of the thallus
of *Letharia vulpina* f. *californica*, x 12 (*MU*).

peri- (Gr. *perí*, around, in turn): a pref. often employed in mycological terms, as in the following.

periclinal (Gr. *perí*, around, in turn + *klíno*, to incline + L. suf. *-alis* > E. *-al*, relating to or pertaining to): refers to a membrane or cell wall of a structure that is parallel to the surface of that structure. Cf. **anticlinal**.

peridermium, pl. **peridermia** (Gr. *perí*, around, in turn + *dérma*, skin + L. dim. suf. *-ium*): *Uredinales*. A type of aecium with a peridium that is cylindrical or shaped like a tongue or blister, and which forms on the exterior of the host plant; these aecia belong to the gen. *Peridermium*, such as *P. ipomoeae*, formed on the leaves of sweet potato.

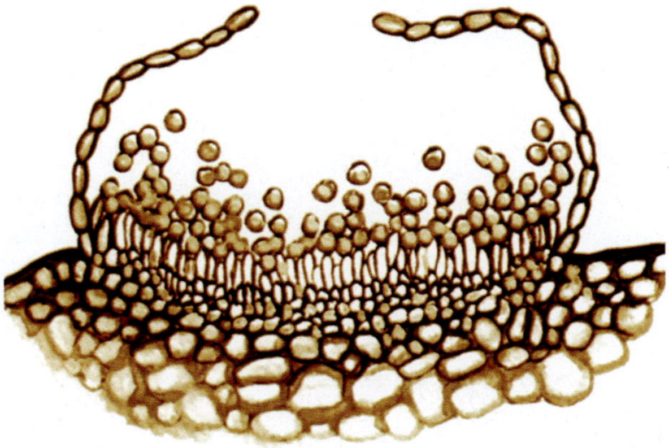

Peridermium (aecium) of *Peridermium* sp. on a leaf of a host plant, seen in longitudinal section, x 75 (*MU*).

peridermioid (Gr. *perí*, around, in turn + *dérma*, skin + L. suf. *-oide*, like < Gr. *-oeídes*, similar to): *Uredinales*. An aecium of the type characterized by the gen. *Peridermium*. See **peridermium**.

peridiole, peridiolum, pl. **peridiola** (L. *peridiolum*, small purse < *peridium* < Gr. *péridion*, leather purse, knapsack + dim. suf. *-olum* > E. *-ole*): *Gasteromycetes*. The small, enclosed, ovoid or rounded bodies ("eggs"), with a thick wall or tunica, that contain the basidia and basidiospores, in the orders Nidulariales (such as *Cyathus*, *Nidularia* and *Crucibulum*) and Sclerodermatales (such as *Pisolithus tinctorius*). Also called **angiole**.

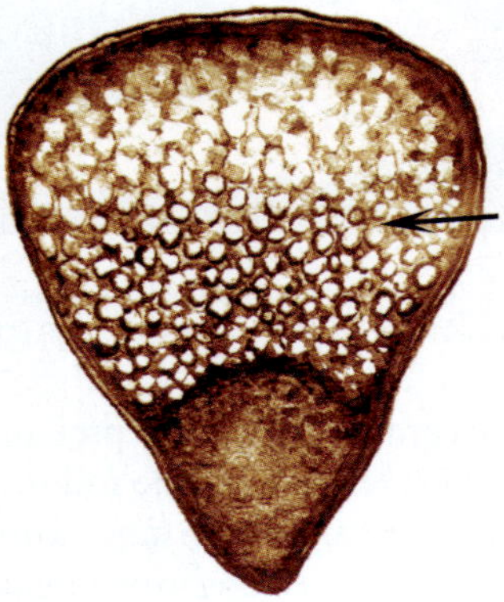

Peridioles in the fruiting body of *Pisolithus tinctorius*, seen in longitudinal section, x 1.5 (*MU*).

Peridioles in the gleba of a basidiocarp of *Pisolithus tinctorius*, seen in longitudinal section, x 1.5 (*RV*).

Peridioles in the fruiting bodies of *Cyathus intermedius*, x 2 (*JR*) and *Cyathus* sp., x 3 (*MU*).

Peridioles in the fruiting body of *Cyathus* sp., x 6 (*MU*).

Peridioles in the fruiting body of *Crucibulum laeve*, x 6 (*CB*).

peridium, pl. **peridia** (NL. *peridium* < Gr. *péridion*, small leather purse): a wall that envelops a fruiting body (microscopic or macroscopic); the peridium can be acellular, amorphous, or be composed of plectenchyma, depending on the type of fungus or fungus-like microorganisms. E.g., in the sporangia of the Myxomycetes and Zygomycetes it is acellular, and amorphous in the ascocarp of some Eurotiales; but in the majority of the ascomycetes, the cleistothecia, perithecia and pseudothecia, as well as the basidiocarp of many Gasteromycetes, have a peridium composed of plectenchyma (*plectenchymatous peridium*). In some species of Gasteromycetes the peridium becomes gelatinous at maturity. In the ascomycetes of the order *Onygenales* the following types of peridium have been described: the *reticuloperidium* (L. *reticulum*, mesh of a net), similar to the mesh of a net, composed of thick-walled, much-branched, often anastomosed hyphae, as occurs in *Auxarthron*.

The *incompositoperidium* (L. pref. *in-*, not, without + *compositus*, composed, united), composed of a disorganized mass of thick-walled, contorted cells, like that of *Shanorella*. The *telaperidium* (L. *tela*, cloth, tissue, trama), similar to a cobweb, constituted of undifferentiated hyphae that resemble vegetative hyphae, like that of *Byssoascus*. The *cleistoperidium* (Gr. *kleistós*, closed, enclosed), is membranous, thin, formed of angular cells, as is seen in *Aphanoascus*.

Peridium (acellular) of the sporangia of *Craterium minutum*, x 20 (*MU*).

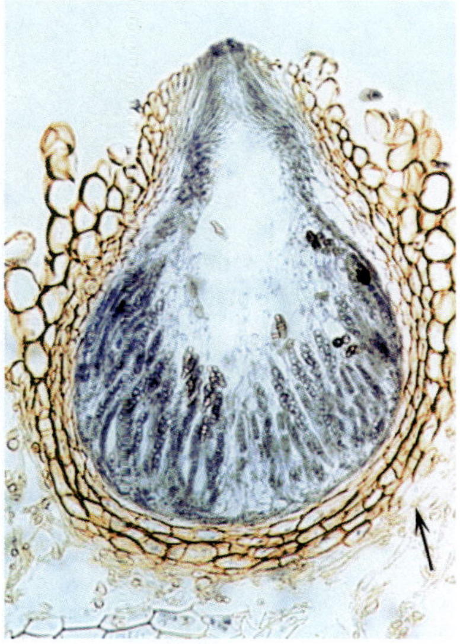

Peridium (plectenchymatous) of the ascocarp (perithecium) of *Haematonectria* (=*Nectria*) *haematococca*, x 450 (*RTH*).

Peridium (plectenchymatous) of the basidiocarp of *Scleroderma* sp., x 1 (*RV*).

periphysis, pl. **periphyses** (Gr. *perí*, around, in turn + *phýsis*, a being, a growth, born near or nearby; the natural shape of something): short, sterile hyphae that line the internal wall of the ostiolar canal of a perithecium, as is seen in *Sordaria fimicola* (Sordariales) and *Meliola floridensis* (Meliolales). See **paraphysis**.

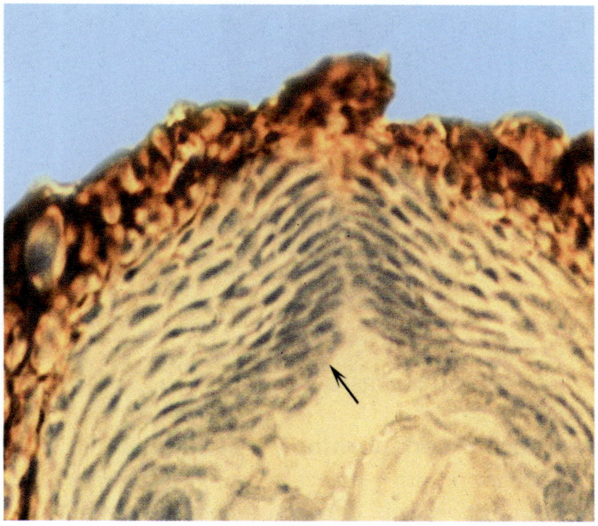

Periphyses lining the ostiolar canal of a perithecium of *Meliola floridensis*, x 800 (*RTH*).

periphysoid (Gr. *perí*, around, in turn + *phýsis*, a being, a growth, born near or nearby; the natural shape of something + L. suf. *-oide*, like < Gr. *-oeídes*, similar to): lateral periphyses that are present in the pseudothecia of some pyrenolichens with a crustaceous thallus.

periphyton (Gr. *perí*, around, in turn + *phytón*, plant): a type of biotic community constituted of microscopic organisms that live attached to a solid substrate, submerged in water, in which they develop a mucilaginous layer.

periplasm (Gr. *perí*, around, in turn + *plásma*, bland material with which a living being is made): the peripheral plasm of a gametangium or sporangium, which is not utilized in the formation of gametes or spores. In the oogonium of the Oomycota two zones are distinguished in the protoplasm, a dense central zone, the ooplasm (which first generates the oosphere and afterwards the oospore), and another peripheral, more fluid zone, which is the periplasm, as is observed in *Apodachlya* (Leptomitales) and *Rhipidium* (Rhipidiales). Cf. **ooplasm**.

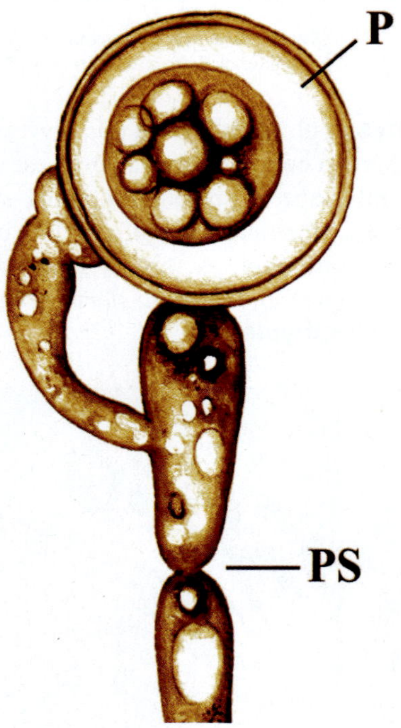

Periplasm (P) of the oogonium of *Apodachlya* sp. Also shown is the pseudoseptum (PS) of the supporting hypha, x 890 (*MU*).

perispore, **perisporium**, pl. **perisporia** (Gr. *perí*, around, in turn + *sporá*, spore; NL. *perisporium* < *perí* + *sporá* + L. dim. suf. *-ium*): in the wall of the most complex basidiospore, which is composed of five layers or strata, the perispore is the outermost layer, which is not pigmented and generally envelops the spore like a sack, which can disappear at an early stage of development, as in *Coprinopsis narcotica* (=*Coprinus narcoticus*) and related species of Agaricales. Beneath the perispore are found the other four layers: *exospore*, *epispore*, *mesospore* and *endospore*.

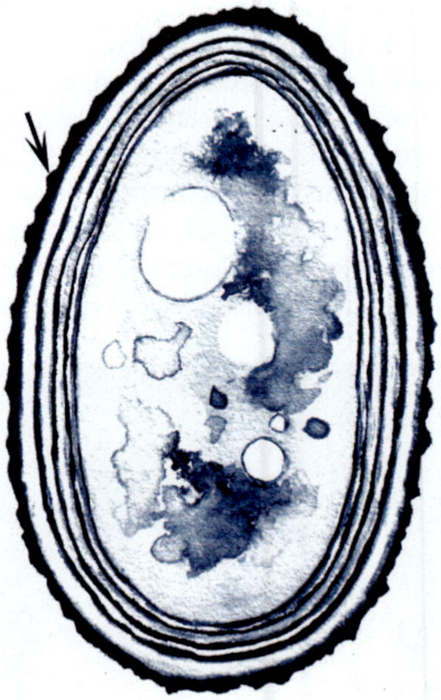

Perispore of the wall of the most complex basidiospore. In this diagram of a transverse section, the other four layers: endospore, mesospore, epispore and exospore, also are shown, x 6,000 (*MU*).

peristome (Gr. *perí*, around, in turn + *stóma*, mouth): *Gasteromycetes*. A circular area, ornamented in various ways, that surrounds the opening in the peridium of the fruiting body of *Geastrum* (*G. saccatum*, *G. triplex*) and related gen. (Lycoperdales). See **stoma**.

Peristome of the basidiocarp of *Geastrum triplex*, x 1 (*MU*).

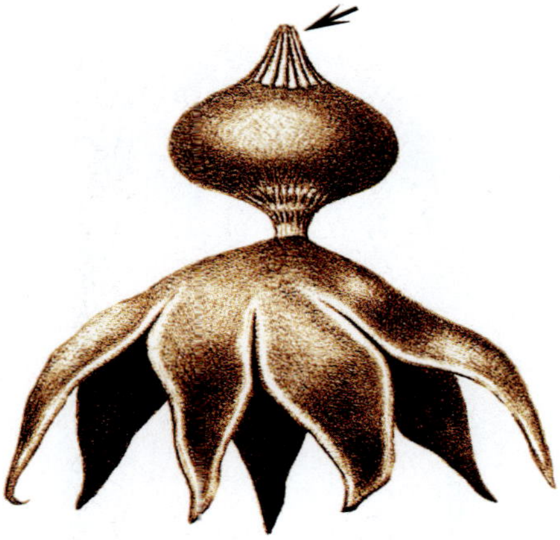

Peristome of the fruiting body of *Geastrum pectinatum*, x 1 (*MU*).

Peristome of the basidiocarps of *Geastrum* sp., x 1 (*RV*).

Peristome of the basidiocarp of *Geastrum* sp., x 1 (*RV*).

perithecial stroma, pl. **stromata** (NL. *perithecium* < Gr. *perithekíon*, dim. of *perithêke*, helmet < *perí*, around + *thekíon*, dim. of *thêke*, box; here, of the asci + L. suf. *-alis* > E. *-al*, relating to or belonging to; Gr. *strõma*, bed, cushion, small pillow): a stroma bearing perithecia, of variable morphology; e.g., in *Hypocrea schweinitzii* (Hypocreales) it is pulvinate and fleshy. Among the Xylariales, some examples are *Camillea*, *Rosellinia*, *Leprieuria*, *Phylacia*, *Podosordaria* and *Porolepreuria*, which have carbonaceous stromata; *Diatrype*, with woody stromata, and *Xylaria*, with stromata that can be carbonaceous, woody or fleshy, depending on the species. Many ascomycetes, belonging to various orders, have perithecial stromata.

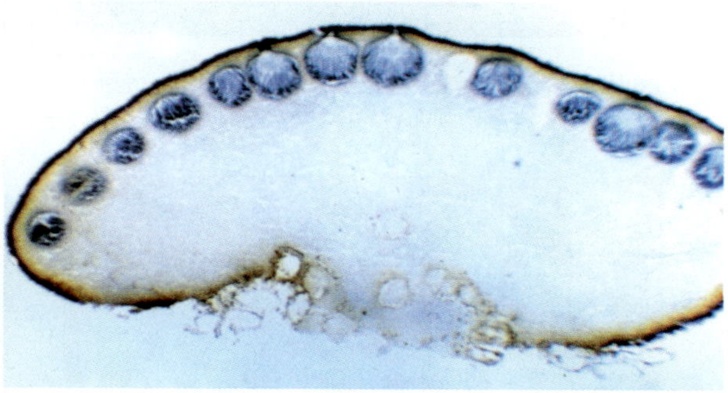

Perithecial stroma of *Hypocrea schweinitzii*, seen in longitudinal section, x 40 (*RTH*).

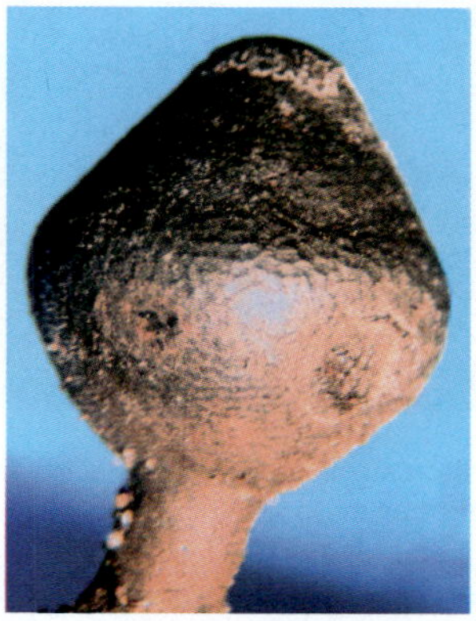

Perithecial stroma of *Phylacia* sp., x 5 (*CB*).

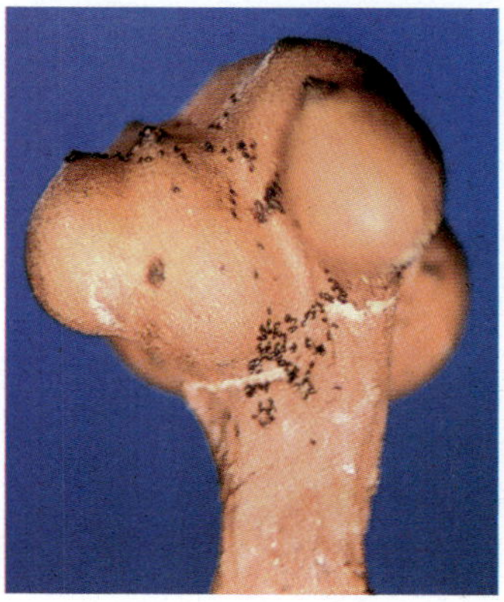

Perithecial stromata of *Podosordaria mexicana*, with a head with ostiolate perithecia, x 15 (*MU*).

perithecial stroma

Perithecial stromata of *Claviceps gigantea* (Hypocreales), growing from a sclerotium formed on a maize cob, x 1 (*TH*).

Perithecial stroma of *Xylaria* sp., seen in cross section; the perithecia are in longitudinal section, x 3 (*RTH*).

Perithecial stromata of *Cordyceps sobolifera* (Hypocreales), growing from a hypogeal cicada larva, x 1 (*RV*).

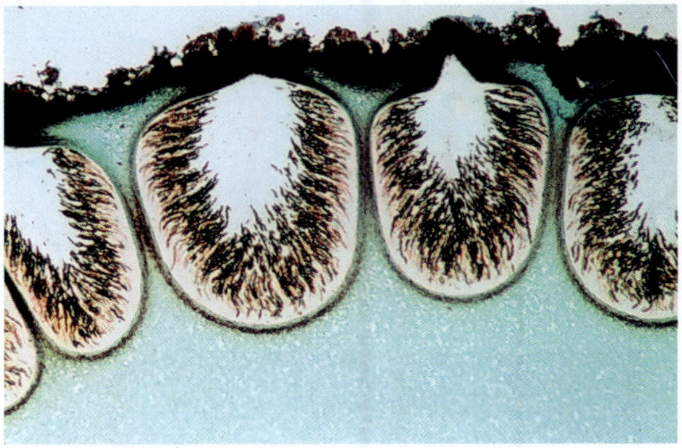

Perithecial stroma of *Xylaria* sp., seen in cross section; the perithecia are in longitudinal section, x 78 (*RTH*).

Perithecial stromata of *Porolepriеuria rogersii*, growing on bark, x 10 (*CB*).

Perithecial stromata of *Diatrype disciformis* erumpent through the bark of a branch, x 20 (*MU*).

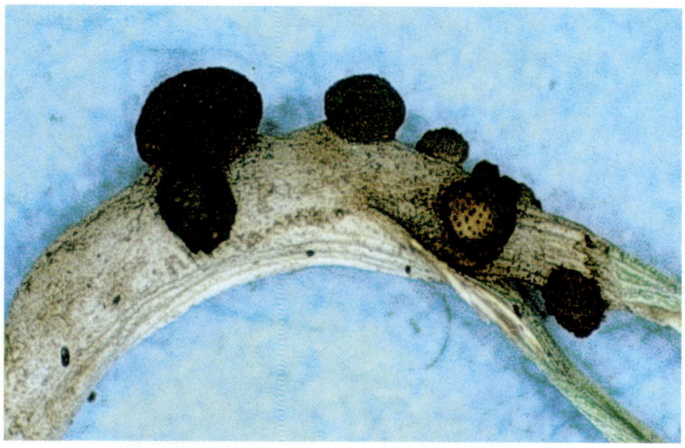

Pulvinate **perithecial stromata** of *Atkinsonella hypoxylon*, on *Danthonia spicata*, x 0.5 (*RTH*).

perithecioid (NL. *perithecium* < Gr. *perithekíon*, dim. of *perithêke*, helmet < *perí*, around, in turn + *thekíon*, dim. of *thêke*, case; here, of the asci + L. suf. *-oide*, like < Gr. *-oeídes*, similar to): similar to a perithecium, like the unilocular ascostroma of *Pyrenophora* (Pleosporales).

Perithecioid ascostroma of *Pyrenophora* sp., x 80 (*CB*).

perithecium, pl. **perithecia** (NL. *perithecium* < Gr. *perithekíon*, dim. of *perithêke*, helmet, cap < *perí*, around, in turn + *thekíon*, dim. of *thêke*, case; here, of the asci): *Ascomycetes*. A type of ascocarp, formed by ascohymenial development, in which the asci are arranged in a fascicle or hymenium and are surrounded by a distinct wall. Generally the perithecium is obpyriform and possesses an ostiole through which the ascospores are liberated, but it may be spherical and remain enclosed (not ostiolate); it can then be confused with a cleistothecium. Perithecia are characteristic of the Pyrenomycetes and pyrenolichens. Some examples are *Ceratocystis fimbriata*, *Microascus trigonosporus* (Microascales), *Guanomyces polythrix*, *Sordaria fimicola* (Sordariales), and *Melanospora zamiae* (Melanosporales).

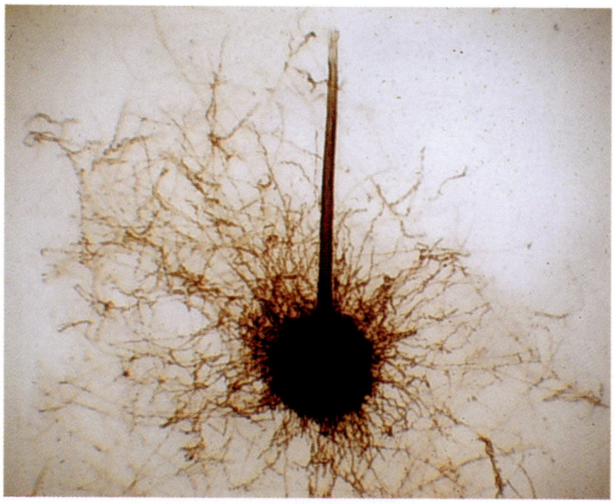

Perithecium of *Ceratocystis fimbriata*, x 75 (*MU*)

Perithecia of *Microascus trigonosporus*, on a grass leaf, x 50 (*MU*).

Perithecium of *Guanomyces polythrix* on agar, x 10 (*MU*).

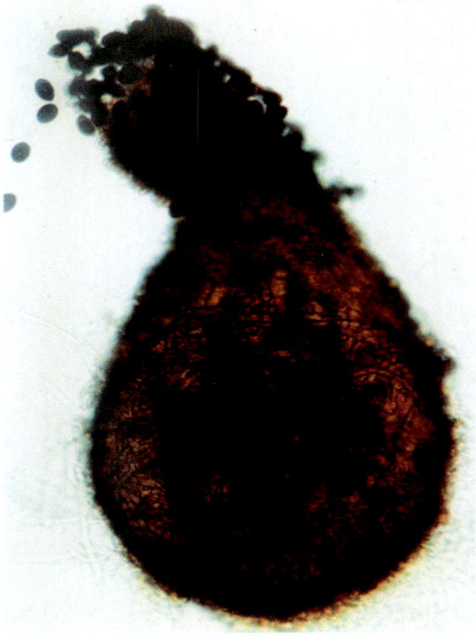

Perithecium of *Sordaria fimicola* with ascospores emerging from the ostiole, x 125 (*MU*).

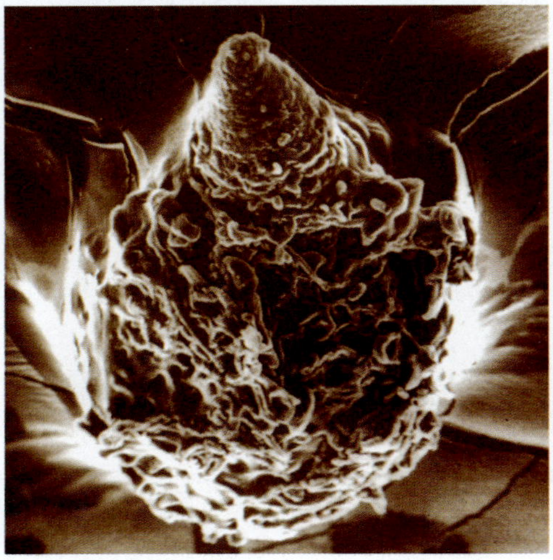

Perithecium of *Sordaria fimicola*, seen with the scanning electron microscope, x 200 (*RTH*).

perittogamy (Gr. *perittós*, unequal + *gámos* > E. des. *-gamy*, sexual union): diffuse sexuality, with a flexible genetic rhythm and imprecise behavior for undertaking the sexual act, since neither differentiated sexual organs nor gametes are formed, although there is somatogamy and karyogamy. The basidiomycetes are perittogamous, since fertilization is effected in an imprecise manner with regard to a determinate point and time in the life cycle. Cf. **parasexuality**.

peroblate (L. *per-*, a pref. that intensifies the meaning of the word that follows + *oblatus* < pref. *ob-*, with an intensive sense + *latus*, wide): generally applied to spores that are oval or ovate in shape, whose length:width ratio is less than 0.5.

peronate (L. *peronatus*, footwear < *pero*, genit. *peronis*, sandal + suf. *-atus* > E. *-ate*, provided with or likeness): *Agaricales*. With the lower part of the stipe covered by the volva or universal veil, as in *Amanita pantherina*. Another example of peronate is *Floccularia*. Syn. of **caligate** and **ocreate**.

Peronate stipe of the basidiocarp of *Floccularia albolanaripes*, x 1 (*TRO*).

perprolate (L. *per-*, a pref. that intensifies the meaning of the word that follows + *prolatus*, elongated or prolonged in the direction of the major axis): applied to structures markedly elongated (e.g., spores), whose length:width ratio is greater than 2.0.

perrumpent (L. *perrumpere*, to emerge by breaking + suf. *-entem* > E. *-ent*, being): syn. of **erumpent**.

persistent (L. *persistens*, genit. *persistentis*, continuing, what perseveres or persists < *persistere*, to continue, to support oneself, to last + suf. *-entem* > E. *-ent*, being): persisting, lasting. Applied to structures that remain after their function has ceased, such as sporangia that remain in place after liberation of the zoospores, e.g., those of the saprolegniaceous oomycotes (*Saprolegnia*, *Achlya*, etc.), or the perennial fruiting bodies that live three or more years, like many of the woody Aphyllophorales that do not decay readily. Cf. **deciduous**.

Persistent basidiocarp of *Inocutis texana* (Aphyllophorales), growing on bark, x 0.5 (*RV*).

perthotroph, **perthotrophic** (Gr. *pértho*, to seize the booty after a defeat + *trophós*, what nourishes, serves as food; or + suf. *-íkos* > L. *-icus* > E. *-ic*, pertaining to or relating to): a parasite that kills the cells of the host and then extracts nutrient substances saprobically. Certain plant pathogenic fungi, such as species of *Sclerotinia* (Helotiales), are perthotrophs. *Colletotrichum lindemuthianum* (melanconiaceous asexual fungi) is a species that during its life cycle changes its nutritional relationship with the host plant, since it has the double character of biotroph-perthotroph. The perthotrophs are *necrotrophs*. Cf. **biotroph**.

pertusate (L. *pertusus*, punctured, perforate < *pertundere*, to perforate): with perforations or holes, like the thallus of the lichen *Pertusaria* (Pertusariales), whose numerous small, black apothecia make it appear perforate.

Pertusate thallus of *Pertusaria wulfenioides*, x 5 (*MU*).

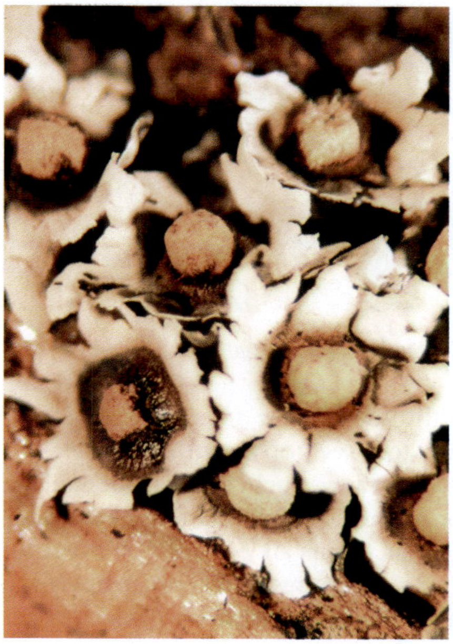

Petaloid peridium of the open sporangia of *Diderma radiatum*, x 13 (*MU*).

Petaloid pileus of a basidiocarp of *Crinipellis* sp.; the petaloid appearance is due to the incision of the pileus, x 1 (*RV*).

pervious (L. *pervius*, open, accessible, passible): permeable, permitting passage, like the cyphellae and scyphi of the lichens that are open or perforate at the base and are continuous with the tissue of the thallus on which they are formed.

-petal (L. *petere*, to go towards): a suf. employed in words such as *acropetal*, directed toward the apex, and *basipetal*, directed toward the base.

petaloid, **petaliform** (Gr. *pétalon*, leaf + L. suf. *-oide*, like < Gr. *-oeídes*, similar to; or + L. suf. *-formis* < *forma*, shape): similar to a petal, like the pileus of the basidiocarp of *Panellus mitis* and *Crinipellis* (Agaricales) or the peridial lobes of the mature, open sporangia of *Diderma radiatum* (Physarales).

petri-, **petro-**, **petr-** (L. *petra*, rock, stone): a pref. employed in terms alluding to rocks, such as the following.

petricolous (L. *petra*, rock, stone + suf. *-cola*, inhabitant): growing on rock, like certain lichens. Also called **epilithic**, **rupicola** and **saxicolous**. Various species of *Parmelia* (Lecanorales) and *Caloplaca* (Teloschistales), e.g., inhabit different kinds of rocks. See **lithotroph**.

Petricolous thallus of *Parmelia* sp., x 0.5 (*TH*).

petrophilous (L. *petra*, rock, stone + *phílos*, have an affinity for): see **petricolous**.

petrous (L. *petreus*, rocky): very hard, similar to a stone in consistency; the grains formed by the fungi causing mycetoma in man are petrous. See **mycetoma**.

pezo- (Gr. *pezós*, terrestrial): a pref. employed in names such as *Pezolepis* (Helotiales), which means "earth scale".

phaeo- (L. *phaeo* < Gr. *phaiós*, blackish, brownish-gray): a pref. employed in the formation of terms that allude to a brownish-gray color, as in *phaeospore*.

phaeocystidium, pl. **phaeocystidia** (L. *phaeo* < Gr. *phaiós*, blackish, dark + Gr. *kystídion* < *kýstis*, bladder, cell + dim. suf. *-ídion*): a lightly pseudoamyloid gloeocystidium, with brown contents, like that of *Fayodia deusta* (Agaricales).

phaeohyphomycosis, pl. **phaeohyphomycoses**, **phaeomycosis**, pl. **phaeomycoses** (L. *phaeo* < Gr. *phaiós*, blackish, dark + Gr. *hyphé*, hypha + *mýkes*, fungus + suf. *-osis*, condition or state, a process): a mycotic infection of humans or higher animals, such as *chromomycosis*, caused by dematiaceous fungi. See **mycosis**. Cf. **hyalohyphomycosis**.

phaeospore (L. *phaeo* < Gr. *phaiós*, blackish, dark, brown + Gr. *sporá*, spore): a dark-colored, asexual reproductive spore; a Saccardoan term mainly applied to conidial fungi, especially to the dematiaceous fungi, as *Curvularia*. Cf. **hyalospore**.

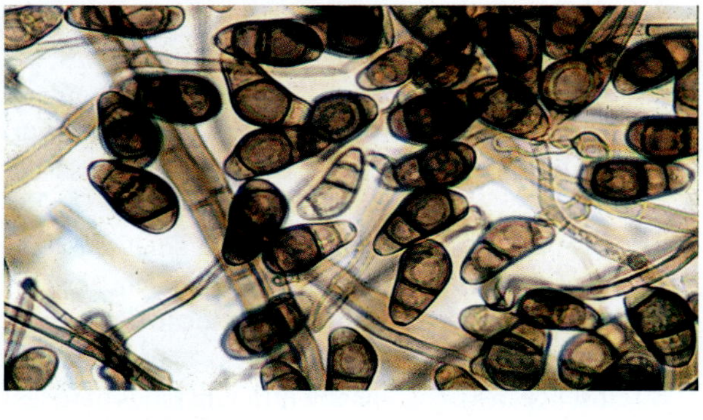

Phaeospores of *Curvularia inaequalis*, x 900 (*CB*).

Phaeospores of *Drechslera* sp., x 260 (*CB*).

-phage, **-phagous** (Gr. *phágos*, one that eats < *phágomai*, to eat): a suf. employed in biological lexicography to introduce the idea of eating, feeding or nourshing oneself, as in *pleophagous*, a parasite that can feed on various organisms, and *bacteriophagous*, a destructive virus of bacteria. Also used as a pref., *phago-*, as in *phagotroph*, which feeds by means of phagocytosis.

phagocytic (Gr. *phágos*, one that eats < *phágomai*, to eat + *kýtos*, cavity, cell + suf. *-íkos* > L. *-icus* > E. *-ic*, belonging to or relating to): ingesting solid food particles by means of *phagocytosis*. Syn. of **phagotroph** or **phagotrophic**, and of **holozoic**.

phagocytosis, pl. **phagocytoses** (Gr. *phágos*, one that eats < *phágomai*, to eat + *kýtos*, cavity, cell + suf. *-osis*, condition or state): the ingestion of solid food particles by means of pseudopodia or protoplasmic extensions, as occurs in amoebae and other protozoans, as well as in the myxamoebae and plasmodia of the Myxomycetes, which represent the trophic phases of these latter organisms. See **pinocytosis**.

phagotroph, **phagotrophic** (Gr. *phágos*, one that eats < *phágomai*, to eat + *trophós*, that nourishes, that serves as food; or + suf. *-íkos* > L. *-icus* > E. *-ic*, belonging to or relating to): an organism that feeds by means of *phagocytosis*. See **holozoic**.

phalangoid (Gr. *phálange*, phalanx + L. suf. *-oide*, like < Gr. *-oeídes*, similar to): shaped like the phalanges or bones of the fingers and toes, like the cells of the peridial hyphae present on the ascomata (gymnothecia) of *Nannizzia* (=*Arthroderma*) *incurvata* and *Nannizzia gypsea* (Onygenales). Also called **ossiform**.

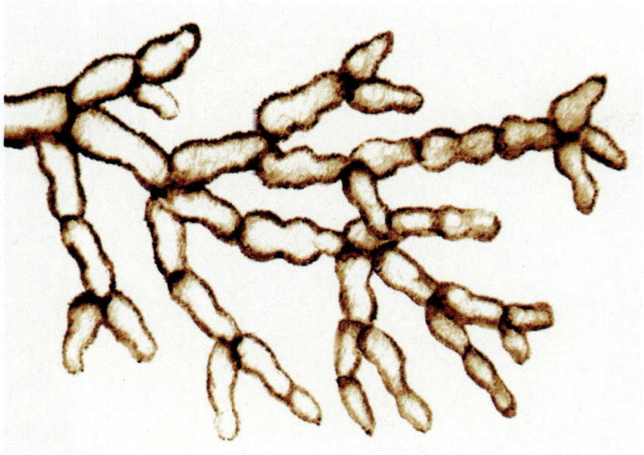

Phalangoid cells of the peridial hyphae of a gymnothecium of *Nannizzia* (=*Arthroderma*) *incurvata*, x 900 (*MU*).

phalloid (Gr. *phallós*, phallus + L. suf. *-oide*, like < Gr. *-oeídes*, similar to): with the shape of a phallus, like the fruiting body of the gen. *Phallus* and *Mutinus* (Phallales).

Phalloid fruiting bodies of *Phallus impudicus* var. *impudicus*, x 0.5 (*TH*), x 0.5 (*JLM*).

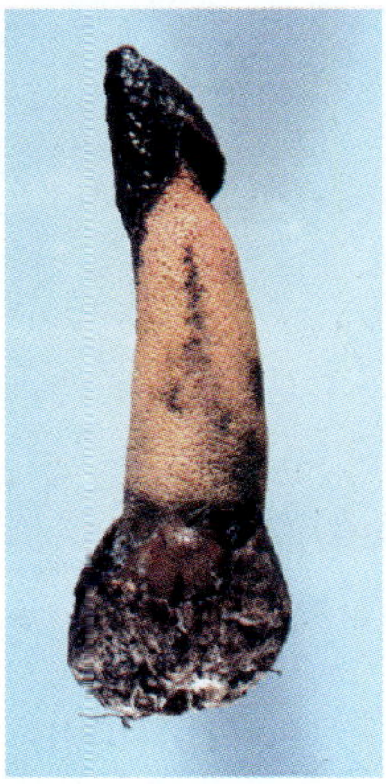

Phalloid fruiting body of *Phallus hadrianii*, x 0.5 (*TH*).

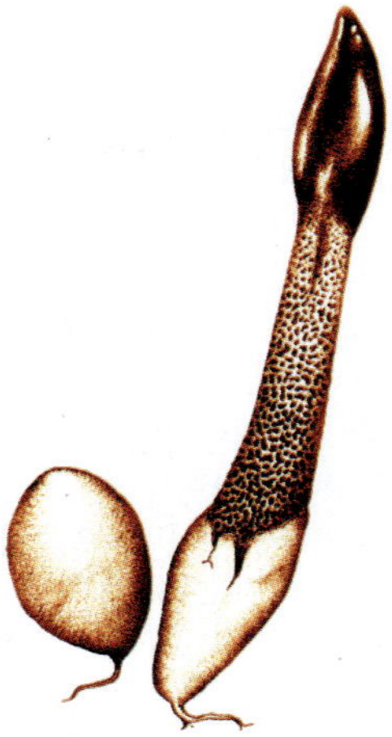

Phalloid fruiting body of *Mutinus caninus*, x 0.5 (*MU*).

Phalloid fruiting body of *Mutinus caninus*, x 0.5 (*RV*).

phanero-, phaner- (Gr. *phanerós*, visible, conspicuous): a pref. employed in forming compound terms that denote something perceptible to simple view, in the open, as in *phanerogam*, a plant that has visible reproductive organs. Cf. **crypto-**.

phanerogamy (Gr. *phanerós*, visible, conspicuous + *gámos* > E. *-gamy*, sexual union): a group of plants that includes classes I-XXIII of the sexual system of Linnaeus, all with the reproductive organs readily visible. As is inferred from their etymology, the *phanerogams* or *anthophytes* embraces all of the plants in which one can see their sexual organs with the naked eye, i.e., for Linnaeus, the stamens and pistils. The fact that an organ can be distinguished or not with the naked eye does not affect either its existence, its morphology, or its physiology. For this reason, and even though the terms *cryptogamy* and *phanerogamy* continue in current use, both are now rejected as systematic concepts due to the lack of homogeneity of their contents. Cf. **cryptogamy**.

phaneroplasmodium, pl. **phaneroplasmodia** (Gr. *phanerós*, visible, conspicuous + L. *plasmodium* < Gr. *plásma*, formation, bland material with which a living being is formed + L. suf. *-odium*, resembling < Gr. *ode*, like < *-oeídes*, similar to): a plasmodium that is macroscopic at maturity, conspicuous, frequently reticulate, with a front in the shape of a well differentiated fan, and typically with reversible currents of rapid, rhythmic endoplasm within a system of veins consisting of ectoplasm. On fruiting, a phaneroplasmodium can form several or numerous sporangia; it is characteristic of Myxomycetes of the order Physarales, such as *Physarum polycephalum*. Cf. **aphanoplasmodium**, **protoplasmodium** and **pseudoplasmodium**.

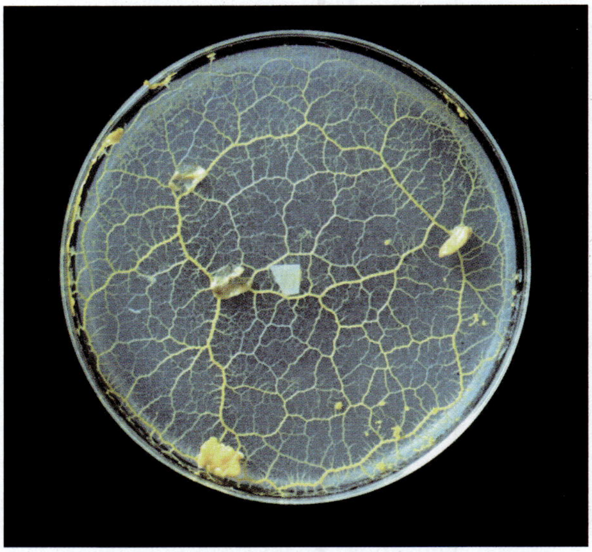

Phaneroplasmodium of *Physarum polycephalum* on agar, x 1 (*MU*).

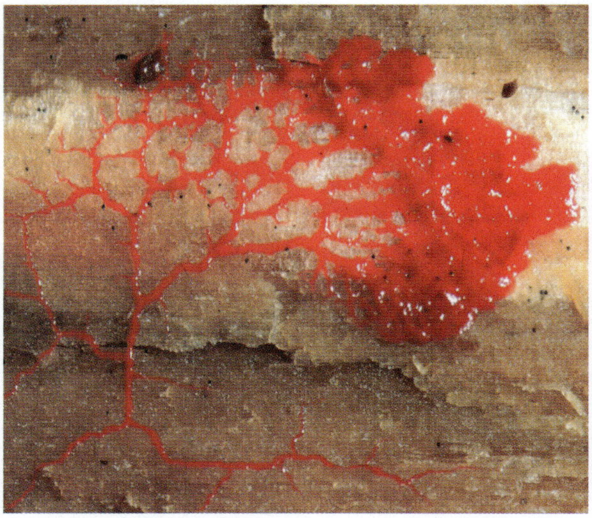

Red **phaneroplasmodium** of *Physarum roseum*, growing on bark, x 1 (*RS*).

Phaneroplasmodium of *Physarum polycephalum* on agar, x 1.5 (*MU*).

Yellow-green **phaneroplasmodium** of *Physarum polycephalum*, growing on bark, x 1 (*RV*).

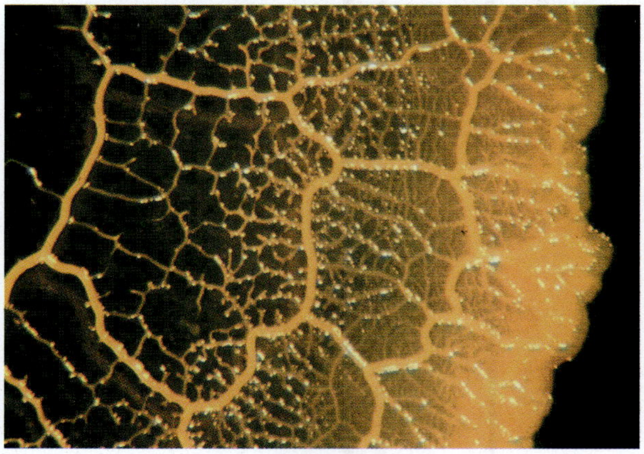

Phaneroplasmodium of *Physarum polycephalum* on agar, x 5 (*MU*).

phello-, **phell-** (Gr. *phellós*, cork): a pref. employed in the formation of terms that denote something corky, as in *phellogen*, a secondary meristem that produces cork, and in the generic name *Phellinus* (Aphyllophorales), a basidiomycete with a corky or woody fruiting body.

phellophagy (Gr. *phellós*, suber, cork + *phágia*, to eat): having the capacity to grow on and derive nutrition from cork cells.

pheno-, **phen-** (Gr. *phaíno*, to make visible, to appear, shine): a pref. employed in the formation of terms that allude to something visible, as in *phenotype*.

phenotype (Gr. *phaíno*, to make visible + *týpos*, type, mark, model): the appearance of an organism that results from the interaction of the genotype and the environment. Cf. **genotype**.

phialide (Gr. *phialís*, genit. *phialídis*, dim. of *phiále*, glass, jar): a type of conidiogenous cell, bottle-shaped, that produces blastic conidia (phialoconidia or phialospores) in basipetal succession from a single locus (*monophialide* < Gr. *mónos*, single, only; i.e., with one opening) or several loci (*polyphialide* < Gr. *polýs*, many), without an increase in the length of the phialide itself. Phialides are the most common conidiogenous cells among the conidial fungi, such as *Aspergillus*, *Penicillium*, *Dendrodochium*, *Paecilomyces* (moniliaceous asexual fungi), *Chalara* (dematiaceous asexual fungi), *Fusarium* (tuberculariaceous asexual fungi), and many other gen.

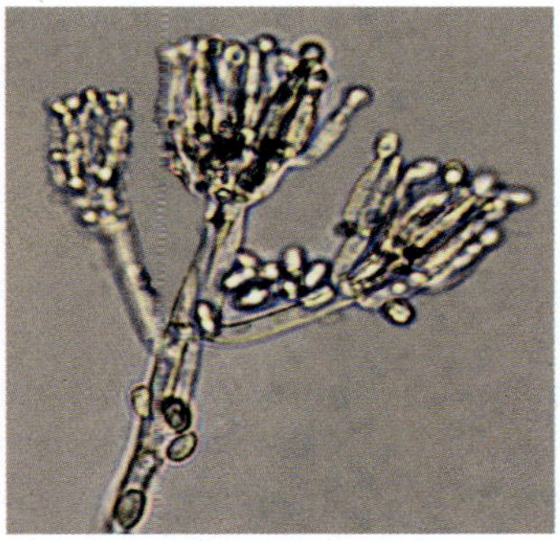

Conidiophore of *Penicillium* sp., showing the three rami ending with groups of **phialides**, x 460 (*MU*).

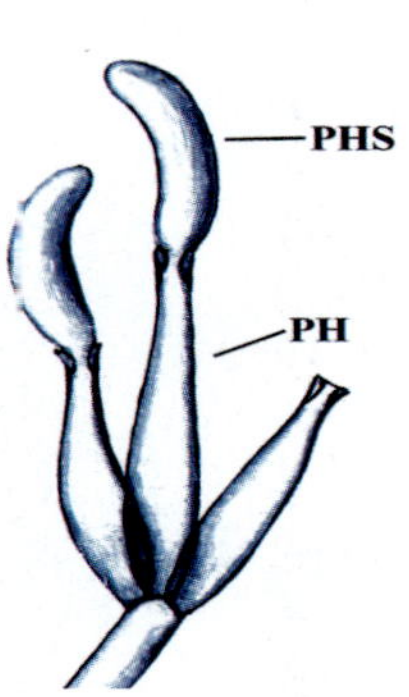

Phialides (PH) and phialospores (PHS) of the *Dendrodochium* state of *Nectria magnusiana*, x 1,340 (*MU*).

phialidic (Gr. *phialís*, genit. *phialídis*, dim. of *phiále*, glass, jar + suf. *-íkos* > L. *-icus* > E. *-ic*, relating to or belonging to): a type of conidiogenesis in which each conidium (phialoconidium or phialospore) originates by the formation of new cell wall material and not from the cell walls already existing in the conidiogenous cell. The conidia are formed basipetally through a single orifice (*monophialidic* < Gr. *mónos*, single, only) or as an irregular sympodial or synchronous succession through several openings (*polyphialidic* < Gr. *polýs*, many) in the wall of the conidiogenous cell. See **phialide**.

phialophore (Gr. *phialís*, dim. of *phiále*, glass, jar + *phóros*, bearer): a conidiophore that bears one or more phialides.

phialospore (Gr. *phialís*, dim. of *phiále*, glass, jar + *sporá*, spore): *Conidiogenesis*. An asexual reproductive spore formed by abstriction on the apex of a phialide; the term is applied mainly to the conidial and spermatial states of fungi. *Aspergillus*, *Penicillium* (moniliaceous asexual fungi) and *Chalara* (dematiaceous asexual fungi) are examples of gen. that produce phialospores. A *terminal phialospore* is the spore of a *monosporic phialide*, i.e., with its production the activity of the phialide ends. Equal to *phialoconidium* (Gr. *kónis*, dust + dim. suf. *-ídion* > L. *-idium*).

phil-, **philo-** (Gr. *phílos*, loving): a pref. employed in words such as *Philodendron*, literally "tree loving". Also used as a suf., *-philous*, as in *thermophilous*, an organism capable of living at high temperatures.

-phily (Gr. *philía*, love, fondness, affinity, preference): a suf. employed in words such as *psammophily*, *thermophily*, etc., for organisms that grow in sandy soils, or that require relatively high temperatures to develop, etc.

phloeo- (Gr. *phloiós*, bark, cortex): a pref. employed in words such as *epiphloeodic*, an organism that lives on tree bark, and *phloeodic*, belonging or relating to bark.

phobo- (Gr. *phóbos*, action of moving away, to flee): a pref. employed in terms such as *phobotaxis*, a negative taxis, and *phobotoxaxis*, moving away from light. Also used as a suf., *-phobe*, *-phobic*, *-phobous*, in terms such as *oxygenophobic*, avoiding oxygen.

phoenicoid fungus, pl. **fungi** (L. *phoenix* < Gr. *phoínix*, purple, crimson; a legendary bird of Egypt that lived 500 years, burned itself to ashes on a pyre, and rose alive from the ashes to live another period + L. suf. *-oide*, like, similar < Gr. *-oeídes*, similar to; L. *fungus*, fungus < Gr. *spóngos*, sponge): a fungus that grows among the ashes after a fire. See **anthracobiontic**, **carbonicole** and **pyrophilous**.

phoresy (Gr. *phorésis*, act of bearing or carrying < *phoreín*, to bear, carry): a phenomenon that occurs in certain habitats, such as dung, characterized by the presence of fungi, beetles and mites, in which the latter are regularly immobile and are transported by beetles and flies. The ascospores of *Pyxidiophora* (Laboulbeniales) are carried by mites that travel on insects and, thus, are said to be *hypophoretic*.

phoro- (Gr. *phóros*, bearer): a pref. employed in terms such as *phorophyte*, a plant that serves as support for an epiphyte. Also used as a suf., *-phore*, *-phoric*, *-phorous*, which means to carry or bear, produce, as in *conidiophore*, a hypha that produces conidia.

phorophyte (Gr. *phóros*, bearer + *phytón*, plant): a host or plant that serves as support for an epiphytic organism, such as trees that support lichens on their bark.

phosphorescent (showing *phosphorescence* < L. *phosphorous* < Gr. *phósphoros*, phosphorus < *phós*, light + *-phóros*, bearer; literally, a bearer of light + suf. *-escens*, genit. *-escentis* > E. *-escent*, that which turns, that begins to, slightly): exhibiting *phosphorescence*, i.e., evident luminescence,which is caused by the absorption of luminous radiations, and which continues for a time after the radiations have ended; lasting luminescence without the production of perceptible heat. Some small glass mushrooms are shown as an example of phosphorescence; they contain a liquid substance that emits light following excitation by light radiation. Cf. **bioluminiscent** and **fluorescent**.

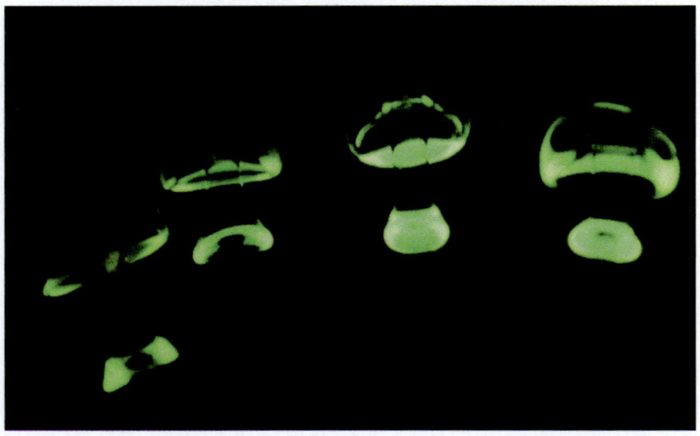

Small glass mushrooms that contain a **phosphorescent** liquid, x 1 (*MU*).

photo-, phot- (Gr. *phós*, genit. *photós*, light): a pref. used to form terms that allude to light, as in *photobiont* and the following words.

photobiont (Gr. *phós*, genit. *photós*, light + *bionte*, a living being < *bíos*, life + *ón*, genit. *óntos*, a being): the photosynthetic symbiont of a lichen, which, depending on the species, can be chlorophyceous (*phycobiont*) or a cyanobacterium (*bactobiont* or *cyanobiont*). Cf. **phycobiont** and **mycobiont**.

photogen (Gr. *phós*, genit. *photós*, light + *génos*, producer, generator): a substance that generates light in fungi and other luminescent organisms; also known as *luciferin*, which, in the presence of another photogen component, the enzyme *luciferase*, is oxidized, producing light. It is present, e.g., in the fruiting bodies of *Lentinus stypticus luminescens* (Aphyllophorales), *Lampteromyces* (=*Pleurotus*) *japonicus*, *P. candescens* and other bioluminescent Agaricales, such as *Armillaria mellea*. The mycelium and basidiocarps of *L. stypticus luminescens* and *Omphalotus* (=*Clitocybe*) *illudens* also are luminescent. In *L. japonicus* only the gills have this property. In *A. mellea* the mycelium and the rhizomorphs are luminescent, but not the fruiting bodies. The photogen is absolutely dependent on the presence of oxygen and moisture to be able to emit light.

photomorph (Gr. *phós*, genit. *photós*, light + *morphé*, form, shape): an organism whose shape is determined by the nature of its photosynthesis, as is seen, e.g., in the very different thalli formed by some lichens, depending on whether the photobiont is a green alga or a cyanobacterium.

photophile, **photophilic**, **photophilous** (Gr. *phós*, genit. *photós*, light + *phílos*, have an affinity for): having a preference for well illuminated habitats. Cf. **photophobic**.

photophobic, **photophobous** (Gr. *phós*, genit. *photós*, light + *phóbos*, fear, aversion): having an aversion to light; prefering shady habitats. Cf. **photophile**. See **sciophilous** and **umbraticus**.

photosporogenic, **photosporogenous** (Gr. *phós*, genit. *photós*, light + *spóros*, spore + *génos*, producer, generator): requiring light to produce spores.

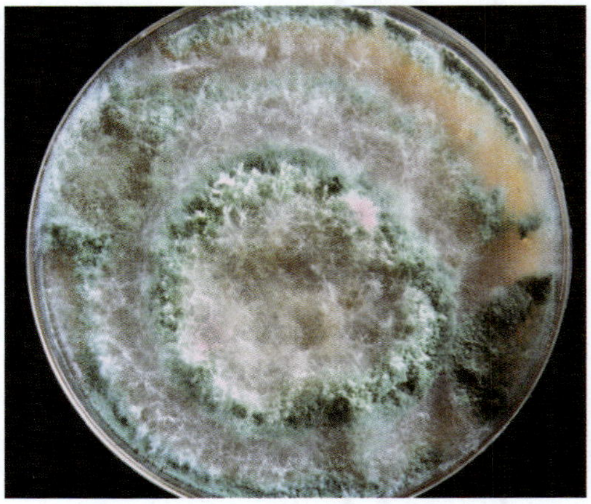

Photosporogenic colony of *Trichoderma viride*, on agar, x 1 (*MU*).

Photosporogenic colony of *Aspergillus parasiticus*, on agar, x 1 (*MU*).

Positively **phototropic** sporangiophores of *Pilobolus crystallinus* var. *kleinii* (=*P. kleinii*), growing on cow dung, x 10 (*MU*).

photosymbiodeme (Gr. *phós*, genit. *photós*, light + *symbióo*, life in common + *démo*, to construct): *Lichens*. An association (considered as a chimera) of a fungus (the mycobiont) with two different species of photobionts, one a green alga and one a cyanobacterium. In this lichen association two thalli coexist that are distinguished by their shape and color, one with each type of photobiont; e.g., in species of *Pseudocyphellaria* (Lecanorales), the lobules of the thallus that contain the cyanobacteria correspond to *P. murrayi*, and those with the green alga to *P. rufovirescens*. In the photosymbiodeme of the gen. *Sticta* (Peltigerales), the lobules with the cyanobacterial thallus are classified as *Sticta canariensis* (=*S. dufourii*), and those with the green alga as *S. canariensis*. *Photosymbiodeme* is a term that is preferable to *phycosymbiodeme* because this describes the association of a mycobiont with two phycobionts (the chlorophyceous and the "cyanophyceous"); however, the cyanophyceous phycobiont is in reality a cyanobacterium, for which reason it is better to refer to them as photobionts of the lichen association.

phototaxis, pl. **phototaxes** (Gr. *phós*, genit. *photós*, light + *táxis*, disposition, movement): a phenomenon determined by the action of light radiation, such as the movement of the zoospores of some aquatic fungi, which is influenced by light, e.g., *Rhizophlyctis vorax* (Chytridiales), which attacks *Sphaerella lacustris*, a motile photosynthetic organism that inhabits illuminated areas. See **taxis**.

phototropic (Gr. *phós*, genit. *photós*, light + *trópos*, gyration, change in position or manner < *tropé*, turn < *trépo*, to gyrate, revolve, be directed toward + suf. *-íkos* > L. *-icus* > E. *-ic*, belonging to or relating to): an organism that exhibits *phototropism*, like *Pilobolus crystallinus* var. *kleinii* (=*P. kleinii*), Mucorales, and *Beauveria felina* (=*B. cretacea*), moniliaceous asexual fungi.

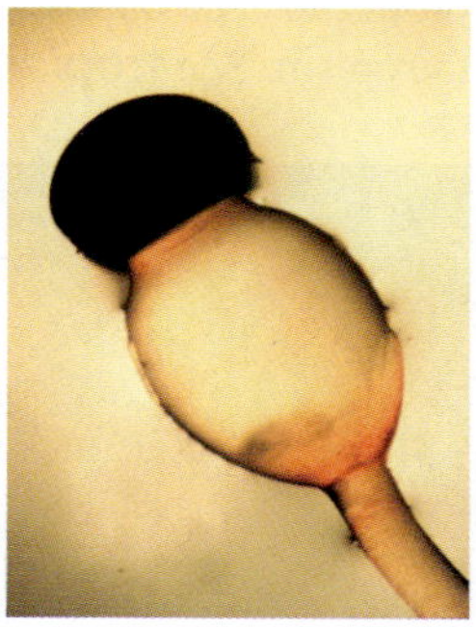

Positively **phototropic** sporangiophore of *Pilobolus crystallinus* var. *kleinii* (=*P. kleinii*), x 50 (*MU*).

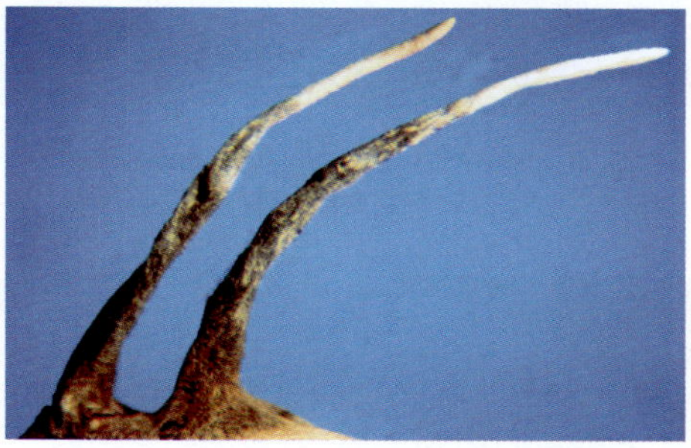

Positively **phototropic** synnemata of *Beauveria felina* (=*B. cretacea*) on agar, x 7 (*CB*).

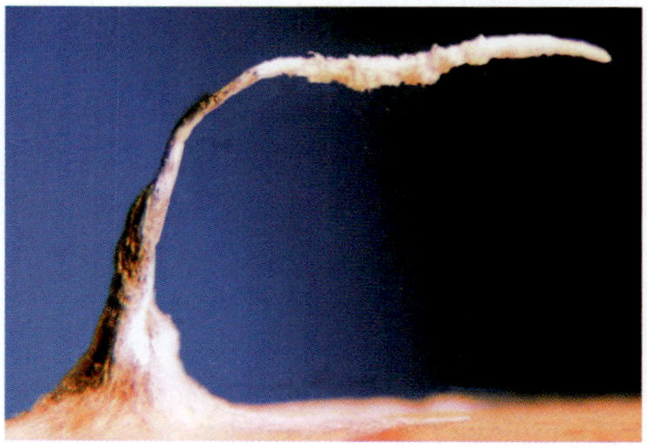

Positively **phototropic** synnema of *Beauveria felina* (=*B. cretacea*) on agar, x 8 (*MU*).

phototropism (Gr. *phós*, genit. *photós*, light + *trópos*, gyration, change in position or manner < *tropé*, turn < *trépo*, to gyrate, revolve, be directed toward + suf. *-ismós* > L. *-ismus* > E. *-ism*, state, phase, tendency, action): a tropic phenomenon in which the stimulating factor is light. E.g., sporangiophores of *Pilobolus* (Mucorales) show positive phototropism because they turn toward the light when it is unilateral; this is due to the fact that the side of the sporangiophore that receives the most light elongates less rapidly than the opposite side, and consequently, it curves toward the light. The conidiophores of *Aspergillus giganteus* and the synnemata of *Beauveria felina* (=*B. cretacea*), moniliaceous asexual fungi, also are positively phototropic. The opposite phenomenon is negative phototropism. See **tropism**.

Conidiophores of *Aspergillus giganteus*, on agar, exhibiting positive **phototropism**, x 0.8 (*MU*).

Conidiophores of *Aspergillus giganteus*, on agar, exhibiting positive **phototropism**, x 1 (*MU*).

phragmo-, phragm- (Gr. *phragmós*, fence, palisade): a pref. employed in the formation of terms related to something that has partitions or septa, giving rise to separated compartments, as in *phragmobasidium*.

phragmobasidium, pl. **phragmobasidia** (Gr. *phragmós*, fence + L. *basidium*, basidium < Gr. *basídion* < *básis*, base + dim. suf. *-ídion*): *Hymenomycetes*. A basidium that is initially unicellular but which forms septa that divide it into four compartments or cells, following division of the diploid nucleus; the septa can be vertical (as in *Tremella*, of the Tremellales) or transverse (as in *Auricularia*, of the Auriculariales). See **heterobasidium**. Cf. **holobasidium**.

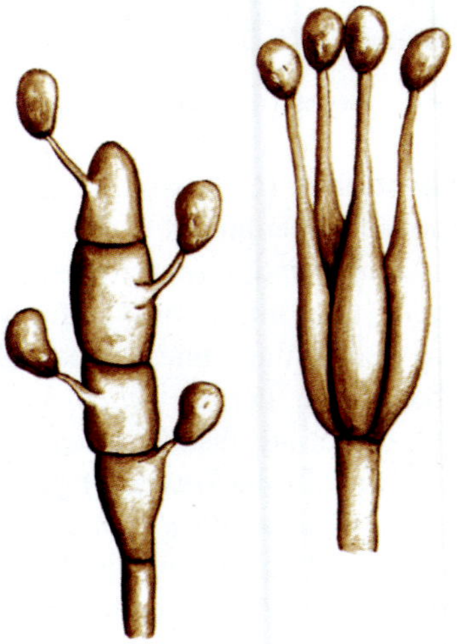

Phragmobasidia of *Tremella* sp. (with vertical septa) and *Auricularia* sp. (with transverse septa), x 1,670 (*MU*).

phragmospore (Gr. *phragmós*, fence + *sporá*, spore): a multicellular asexual reproductive spore, partitioned (divided by means of several septa); a Saccardoan term that is applied mainly to conidial fungi. *Alternaria alternata*, *Curvularia inaequalis*, *Bipolaris maydis* and *Polyschema indicum* (dematiaceous asexual fungi), are examples of species with phragmospores.

Phragmospores of *Alternaria* sp., x 390 (*MU*).

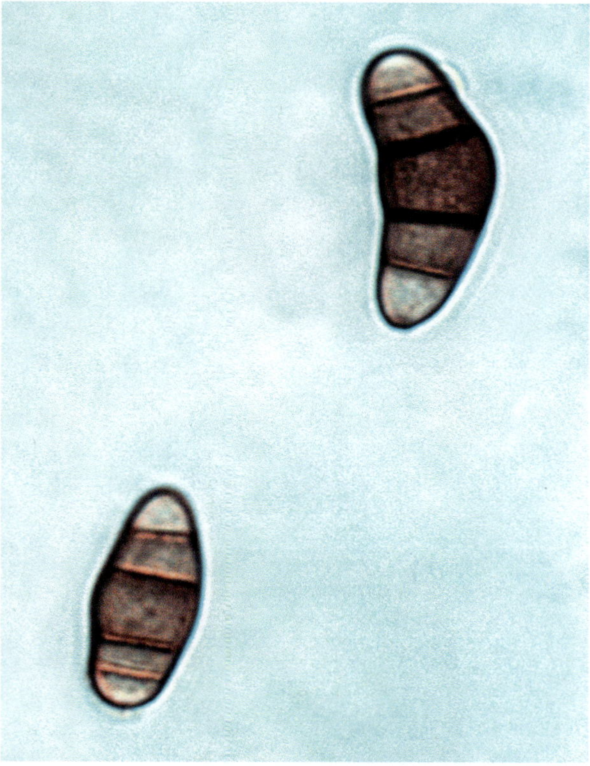

Phragmospores of *Curvularia inaequalis*, x 900 (*RTH*).

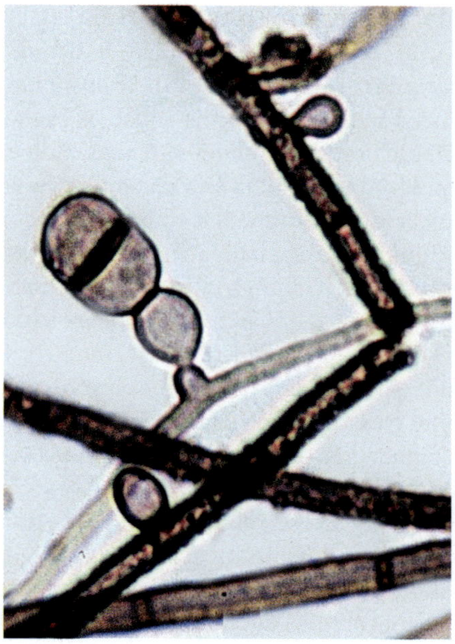

Young **phragmospore** (didymospore) of *Polyschema indicum*, x 480 (*MU*).

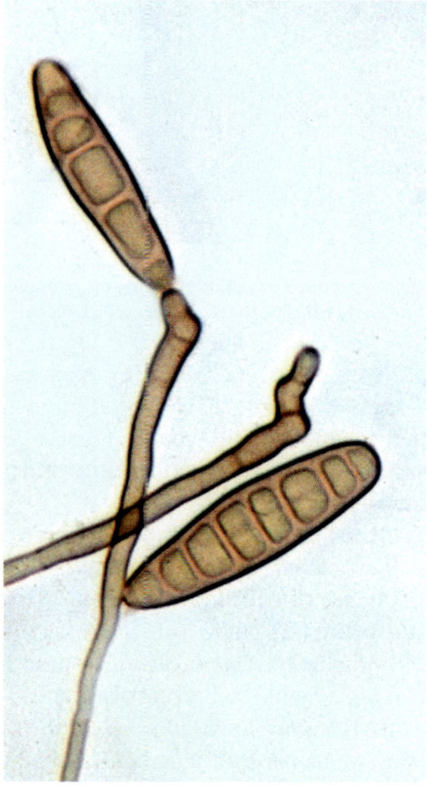

Phragmospores of *Bipolaris maydis*, x 350 (*MU*).

Mature **phragmospore** of *Polyschema indicum*, x 600 (*MU*).

-phyceous (Gr. *phýkos* > L. *phycos*, marine herb, alga): a suf. that signifies alga, as in *chlorophyceous*, green algae; *rhodophyceous*, red algae. See **phyco-**.

phyco-, **phyc-** (Gr. *phýkos* > L. *phycos*, marine herb, alga): a pref. employed with the same meaning as *-phyceous*, in the formation of various biological terms that express the concept of "algal fungi", as in *Phycomycetes*, and *Phycomyces* (Mucorales), which means "algal fungus".

phycobiont (Gr. *phýkos* > L. *phycos*, marine herb, alga + *bíos*, life + *óntos*, genit. of *ón*, a being): the algal portion of a lichen thallus. Cf. **photobiont** and **mycobiont**.

phycocecidium, pl. **phycocecidia** (Gr. *phýkos* > L. *phycos*, marine herb, alga + *kekídium*, dim. of *kekís*, genit. *kekîdos*, gall): a cecidium or gall produced by cyanophyceous algae or cyanobacteria in some chlorophycophilous lichens. Some authors interpret the cephalodia as a parasitism in which the phycocecidia are formed; it could also be considered as a chimeroid association or *photosymbiodeme*.

phycomycetes (Gr. *phýkos* > L. *phycos*, marine herb, alga + L. *-mycetes* > E. *-mycetes*, ending of a taxonomic class < Gr. *mýkes*, genit. *mýketos*, fungus): formerly considered as a single taxonomic class of lower fungi, in some classifications equivalent to the subdivision Phycomycotina (of the division Eumycota), which includes five classes (Chytridiomycetes, Hyphochytridiomycetes, Oomycetes, Zygomycetes and Trichomycetes). In some more recent classification systems the subdivision Phycomycotina has disappeared as such and some of the classes it contained have become separate subdivisions, such as the Mastigomycotina and Zygomycotina, or distinct phyla, such as the Chytridiomycota and Zygomycota (of the kingdom Fungi), and the Hyphochytriomycota and Oomycota (of the kingdom Stramenopila); the Trichomycetes have been included, together with the Zygomycetes, in the phylum Zygomycota. This latter classification (Alexopoulos *et al.*, 1996), which considers them phyla, is the one followed in the dictionary. See **Appendix**.

phycomycosis, pl. **phycomycoses** (Gr. *phýkos* > L. *phycos*, marine herb, alga + NL. *mycosis* < Gr. *mýkes*, fungus + suf. *-osis*, state or condition): a subcutaneous infection of man, mainly children, characterized by the appearence of large nodules or swellings, hard and painless, on the neck, arms and chest. The causal agents are *Basidiobolus haptosporus* and *B. ranarum* (Entomophthorales), ubiquitous saprobic microorganisms that inhabit the soil and plant remains, as well as the intestine and dung of reptiles and amphibians. See **mycosis**.

Colony of *Basidiobolus ranarum*,
the causal agent of **phycomycosis** in man, on agar, x 1 (*RLM*).

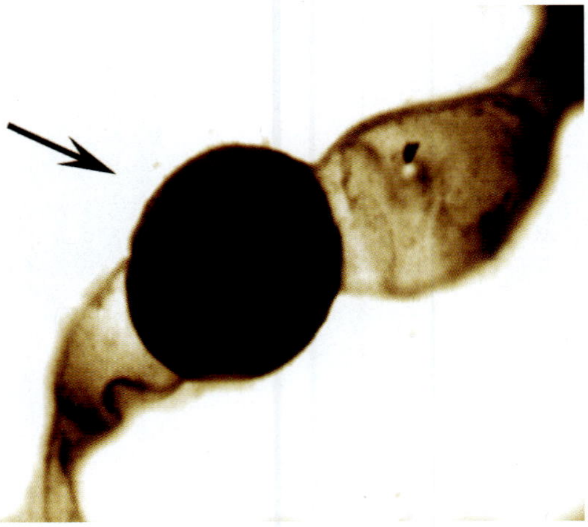

Zygospore of *Basidiobolus ranarum*,
the causal agent of **phycomycosis** in man, x 450 (*RLM*).

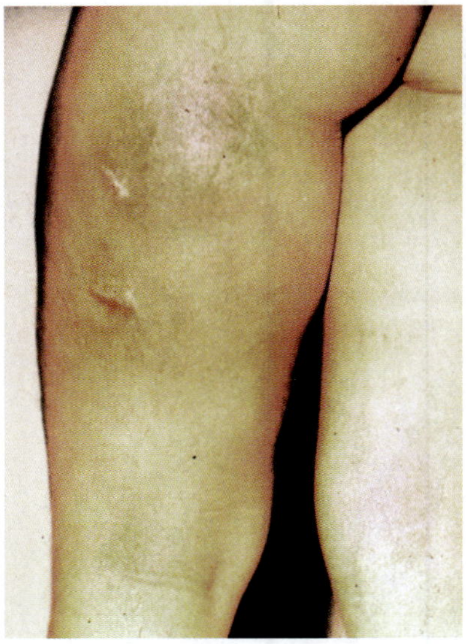

Subcutaneous **phycomycosis** in the muscle of a woman,
caused by *Basidiobolus ranarum*, x 0. 3 (*RLM*).

phycophilous (Gr. *phýkos* > L. *phycos*, marine herb, alga + *phílos*, have an affinity for): associated symbiotically with algae, like the lichenized fungi. See **cyanophycophilous** and **chlorophycophilous**.

phycosymbiodeme (Gr. *phýkos* > L. *phycos*, marine herb, alga + *symbióo*, life in common + *démo*, to construct): see **photosymbiodeme**.

phycosymbiont (Gr. *phýkos* > L. *phycos*, marine herb, alga + *symbiontós* < *symbióo*, life in common + *óntos*, genit. of *ón*, a being): see **phycobiont**.

phycotroph (Gr. *phýkos* > L. *phycos*, marine herb, alga + *tróphos*, what nourishes, that serves as food < *trépho*, to nourish, feed): obtaining nourishment from algae, like the lichenized fungi. Cf. **mycotroph**.

phyllidium, pl. **phyllidia** (NL. *phyllidium*, phyllidium < Gr. *phyllídion* < *phýllon*, leaf + dim. suf. *-ídion*): applied to the small laminar or squamiform portion of foliose lichens that separates from the thallus by abstriction and functions as a vegetative propagule composed of the mycobiont and photobiont, as happens in *Collema flaccidum* (Lecanorales) and *Peltigera praetextata* (Peltigerales).

phyllo-, phyll- (Gr. *phýllon*, leaf): a pref. employed in words such as *phyllosphere*, the space next to the *phylloplane,* the external surface of the leaves of a plant.

phyllocladium, pl. **phyllocladia** (Gr. *phýllon*, leaf + *kládos*, branch): a small branch or expansion of a fruticose lichen thallus, very specialized and assimilative, as it contains the photobiont; it is present in the species of *Stereocaulon* (Lecanorales) and can be granulose, verrugose, coralloid, scaly or squamulose, digitate, peltate or foliose. E.g., the phyllocladia in *S. saxatile* and *S. ramulosum* are flattened and squamulose, and cylindrical and coralloid in *S. dactylophyllum*.

Phyllocladia of a thalline branch of *Stereocaulon ramulosum*, x 9 (*CB*).

phylloplane (Gr. *phýllon*, leaf + L. *planus*, plane, flat surface): the surface of a leaf, with a special microclimate, on which there exists a characteristic mycobiota, composed mainly of saprobic species of yeasts and molds. The term *phyllosphere* (Gr. *phýllon*, leaf + *sphaîra*, sphere, space) is frequently used in the same sense as phylloplane. Species of *Sporobolomyces*, *Tilletiopsis* and *Bullera* (sporobolomycetaceous asexual yeasts), as well as *Rhodotorula* and *Cryptococcus* (cryptococcaceous asexual yeasts), are common on the phyllosphere of various plants.

-phylly (Gr. *phýllon*, leaf): a suf. used in the formation of terms such as *heterophylly*, with unequal or diverse leaves.

phylogeny (NL. *phylum* < Gr. *phýlon*, tribe, race + Gr. *géneia*, act of being born < *génos*, origin, birth, engenderment): the history of a group of organisms; the evolution through time of a group of genetically related organisms. Phylogeny is distinguished from the development of an individual organism (*ontogeny*). Considering the possibility of the transformation of certain living beings into others, phylogeny studies the origin of the phyla of the organic world and the probable derivation of some organisms from others in order to construct a genealogical tree of all of the living beings that exist or have existed on Earth. The various concepts concerning the phylogeny of the fungi traditionally have been based on comparative morphology and cytology, the chemistry of the hyphal wall, ultrastructure and, to a lesser extent, on fossils. Current cladistic and molecular data suggest that the fungi could have originated from ancestral protozoans before the separation of the kingdoms Animalia and Plantae. Based on molecular evidence, the fungi appear to be more closely related to animals than to plants. Cf. **ontogeny**.

phylum, pl. **phyla** (L. *phylum* < Gr. *phýlon*, race, stirp): a taxonomic category between kingdom and class in the zoological and mycological classification systems, such as the one followed in this dictionary; it is equivalent to the category of *division* of botanical classifications. In phylogenetic systems, the phylum is the linked series of organisms that can be considered as originating from the same fundamental form. See **monophyletic**, **paraphyletic** and **polyphyletic**.

physiological form (Gr. *physiología*, the study of the vital phenomena of organisms < *phýsis*, Nature + *-logía*, discourse, treatise < Gr. *lógos*, word < *légo*, I say + suf. *-íkos* > L. *-icus* > E. *-ical*, belonging to or relating to; L. *forma*, form): see **physiological race**.

physiological race (Gr. *physiología*, study of the vital phenomena of organisms < *phýsis*, Nature + *lógos*, word > E. des. *-logy*, study, treatise + suf. *-íkos* > L. *-icus* > E. *-ic*, pertaining to or relating to; MF. *race* < OIt. *razza*, generation; an actual or potentially interbreeding group within a species): a strain of a fungus that differs from other strains in physiological (cultural, biochemical, pathological, etc.) characters but not in morphology, e.g., *Saccharomyces cerevisiae* race *steineri* (Saccharomycetales). This term is preferible to *biological form*, *physiological form* and *race*. Race designations are not governed by the Botanical Code.

physiology (Gr. *physiología* < *phýsis*, Nature + *lógos*, word < *légo*, I say > L. *-logia* > E. *-logy*, discourse, treatise): in its original sense, it is a discourse or treatise on the things of Nature. But in modern terminology *physiology* means the study of the vital phenomena of organisms, which can refer to their metabolism in general, to development, movement and reproduction.

physoclastic (Gr. *physáo*, *physéo*, *physõ*, to throw, detach + *klástos*, broken < *kláo*, to break + suf. *-íkos* > L. *-icus* > E. *-ic*, belonging to or relating to): a type of ascus that liberates its spores by rupture of the outer wall, which is less extensible than the inner wall of the ascus. It is applied to certain marine fungi, such as *Capronia* (=*Herpotrichiella*), of the Chaetothyriales, and *Didymella* (Dothideales). Cf. **aphysoclastic**.

-phytia (Gr. *phytón*, plant): a suf. in geobotany that is equivalent to "vegetation of", that generally is applied from an ecological point of view, as in *hydrophytia*, vegetation composed of aquatic plants. In plant pathology *epiphytia* is an infectious disease that suddenly and destructively affects the wild or cultivated plants in a particular locality.

phyto-, **phyt-** (Gr. *phytón*, plant): a pref. that implies something related to plants, as in *phytopathology*, the science that studies the diseases of plants. Also used as a suf., *-phyte*, as in *sporophyte*, which in plants and fungi with alternation of generations corresponds to the generation with asexual spores.

phytoalexin (Gr. *phytón*, plant + *alexein*, safeguard, protect): a metabolite produced by a plant in response to an infection by a fungus or other pathogen (or in response to the effect of an abiotic factor), which inhibits the invading pathogen. The phytoalexins include the following compounds: capsidiol, kievitone, phaseollin, glyceollin, ipomearon, pisatin, rishitin and wyerone. See **phytotoxin**.

phytopathogen (Gr. *phytón*, plant + *páthos*, disease + *génos*, origin < *gennáo*, to engender, produce): a pathogen that causes disease of plant hosts. The fungi are the etiological agents of numerous diseases of plants, variable from mild to severe attacks; the latter cause the death of the plant. The following are examples of phytopathogens, indicating for each a fungal species and its taxonomic order.

Claviceps gigantea (Hypocreales), a **phytopathogen** of maize cobs, endemic of Mexico; note the sclerotia on the cobs, x 0.4 (*TH*).

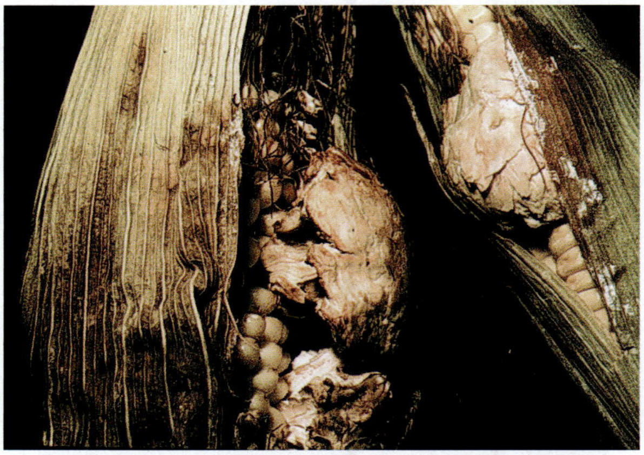

Claviceps gigantea (Hypocreales), a **phytopathogen** of maize cobs, endemic of Mexico; note the sclerotia on the cobs, x 1 (*TH*).

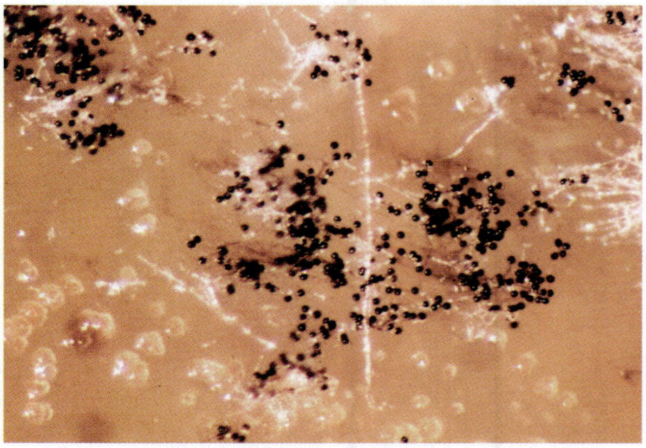

Conidia of *Nigrospora oryzae* (dematiaceous asexual fungi), a **phytopathogen** causing stalk rot of rice, x 20 (*MU*).

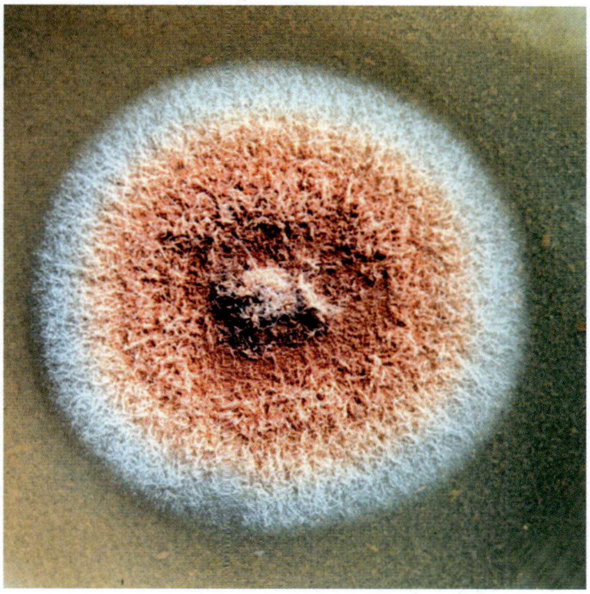

Colony of *Verticillium lateritium* (moniliaceous asexual fungi), a **phytopathogen** of various ornamental plants, x 1 (*MU*).

phytopathology (Gr. *phytón*, plant + *páthos*, disease + *lógos*, word < *légo*, I say > L. *-logia* > E. *-logy*, discourse, treatise): the science that studies the diseases of plants and their control, for which reason it also is called *plant pathology*.

Dead peach tree, caused by *Phymatotrichopsis omnivora* (moniliaceous asexual fungi), a disease studied by **phytopathology**, x 0.1 (*JAS*).

phytosociology (Gr. *phytón*, plant + L. *socius*, member + Gr. *lógos*, word < *légo*, I say > L. *-logia* > E. *-logy*, discourse, treatise): the scientific study of plant communities; to these, Latinized names can be given, based on the dominant species or characteristics, and they are classified in a hierarchical system, which is mainly used by European lichenologists. The basic unit is the *association* (union), whose first name ends in the suf. *-etum*; e.g., *Graphidetum scriptae*, an association of *Graphis scripta* (Ostropales).

phytotoxin (Gr. *phytón*, plant + E. *toxin* < LL. *toxicus* < L. *toxicum* < Gr. *toxikós*, poison, toxic + NL. *-in*, des. used in chemistry to denote a compound, an activator): a toxin produced by a plant (e.g., the phytoalexins). Also refers to a toxin that is harmful to plants, although it is better to avoid this meaning. See **phytoalexin** and **mycotoxin**.

piedra (Sp. *piedra* < L. *petra*, stone, rock): a superficial concretion, similar to a stone, formed by fungi parasitic on the hair of man and animals. *White piedra* (OE. *hwit*, white) is caused by *Trichosporon cutaneum* (cryptococcaceous asexual yeasts), which infects human hair, mainly the beard and moustache, forming whitish masses of arthrospores on the shaft of the hairs. *Black piedra* (ME. *blak* < OE. *blaec*, black) is caused by *Piedraia hortae* (Myriangiales), the teleomorph of *Trichosporon* (cryptococcaceous asexual yeasts). The concretions formed by *Piedraia* on the hair of man and gorillas are stromata that contain asci and ascospores.

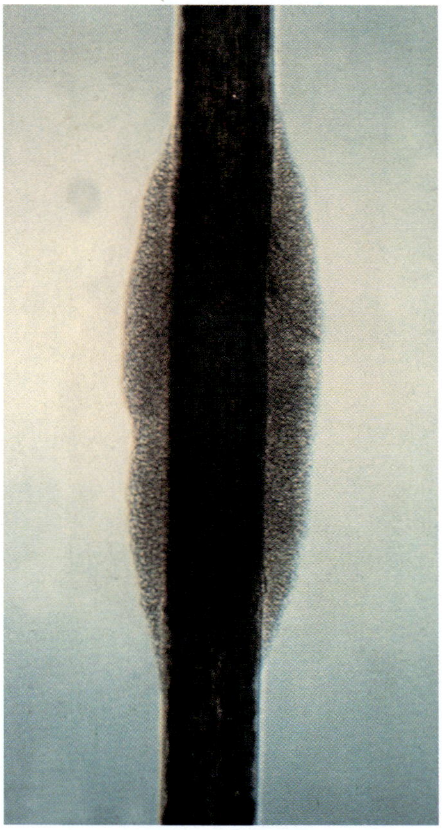

White **piedra** of human hair, caused by *Trichosporon cutaneum*, x 20 (*LMT*).

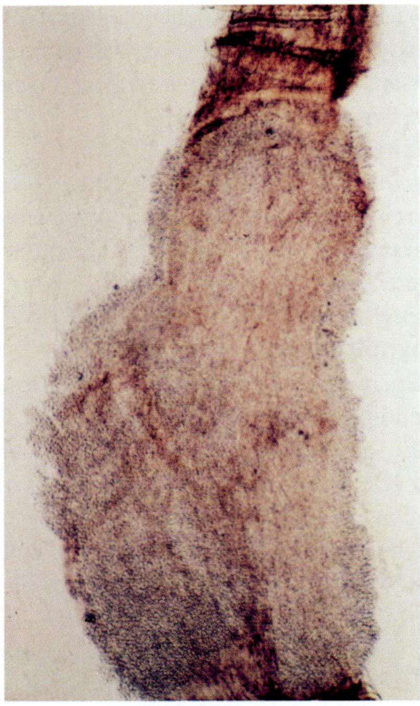

White **piedra** of human hair, caused by *Trichosporon cutaneum*, x 20 (*LMT*).

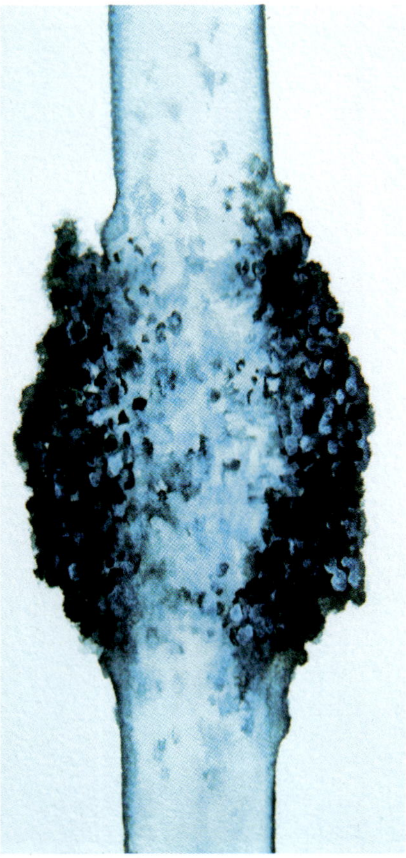

Black **piedra** of human hair, caused by *Piedraia hortae*, x 8 (*MU*).

pigment (L. *pigmentum*, to paint): a substance that imparts black or white or a color to other materials. The colonies of some fungi cultured on agar, e.g., *Penicillium purpurogenum* (moniliaceous asexual fungi), color the agar with a red pigment.

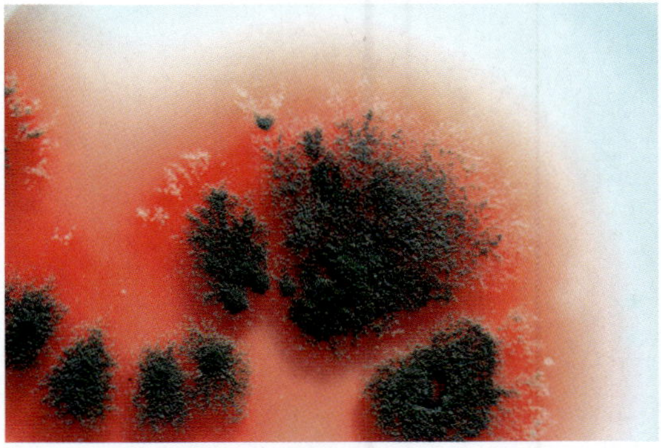

The red **pigment** coloring the agar was exuded by the colonies of *Penicillium purpurogenum*, x 2 (*MU*).

pileate, pileated (L. *pileatus* < *pileus* < Gr. *pílos*, hat, helmet, cap; here, pileus + L. suf. *-atus* > E. *-ate*, provided with or likeness): provided with a pileus, like the fruiting bodies of the Agaricales. Cf. **apileate**.

Basidiocarps of *Amanita caesarea* provided with a pileus (**pileate**), x 0.5 (*RV*).

Basidiocarp of *Amanita* sp. provided with a pileus (**pileate**), x 0.5 (*RV*).

pileate-sessile, **pileated** (L. *pileatus* < *pileus* < Gr. *pílos*, hat, helmet, cap; here, pileus + L. suf. *-atus* > E. *-ate*, provided with or likeness): provided with a pileus, but sessile, like the fruiting bodies of the Aphyllophorales. Cf. **apileate**.

Pileate-sessile (with a pileus but sessile) basidiocarps of *Hexagonia tenuis*, x 0.5 (*RV*).

pileate development (L. *pileatus* < *pileus* < Gr. *pílos*, hat, helmet, cap; here, pileus + L. suf. *-atus* > E. *-ate*, provided with or likeness; OF. *desveloper* > MF. *développer* + L. *-mentum* > E. *-ment*, denoting action or resulting state): *Gasteromycetes*. A type of development that results in a fruiting body with a single stalk or stipe and a pileus, with the hymenial tissue often in the form of tramal plates similar to gills, as is seen in *Podaxis pistillaris* (Hymenogastrales). Cf. **auleate development**, **forate development**, **homogeneous development** and **lacunar development**.

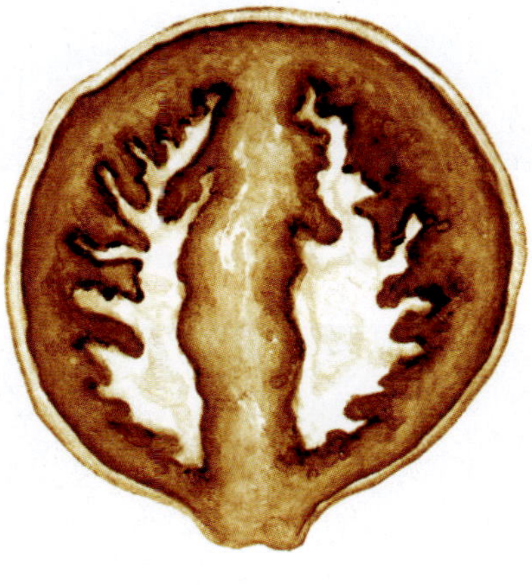

Pileate type of **development** of the fruiting body of *Podaxis pistillaris*, x 2 (*MU*).

pileiform (L. *pileus* < Gr. *pílos*, hat, helmet, cap; here, pileus + L. suf. *-formis* < *forma*, shape): having the shape of a pileus or small hat; characteristic of the agaricoid fungi (Agaricales), or of the zygospore of *Harpella melusinae* and *Stipella vigilans* (Harpellales).

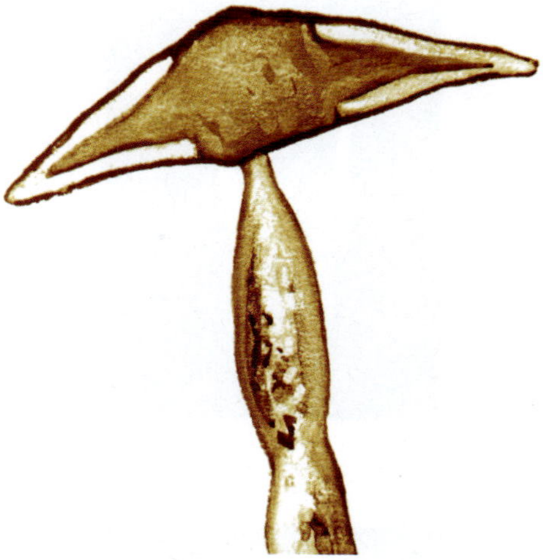

Pileiform zygospore of *Stipella vigilans*, x 800 (*MU*).

pileipellis (L. *pileus* < Gr. *pílos*, hat, helmet, cap; here, pileus + L. *pellis*, skin): *Agaricales*. The outer layer of the pileus, usually composed of anticlinal or decumbent hyphae. See **pellicle**.

pileocystidium, pl. **pileocystidia**, **pilocystidium**, pl. **pilocystidia** (L. *pileus* < Gr. *pílos*, hat, helmet, cap; here, pileus + *kystídion*, dim. of *kýstis*, bladder, vesicle, cell): *Agaricales*. A dermatocystidium that projects from the sterile surface of the pileus, as happens in *Entoloma lividocyanulum* (=*Leptonia lividocyanula*), *Entoloma versatile* (=*Pouzarella versatilis*) and some species of *Coprinus*, among other Agaricales. Cf. **caulocystidium**.

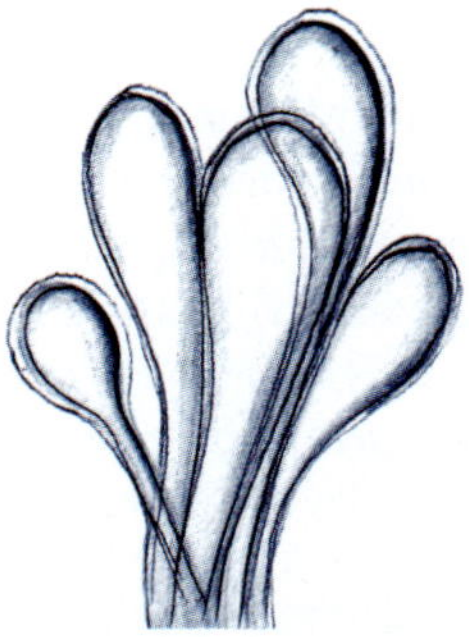

Pileocystidia of the hymenium of *Entoloma lividocyanulum* (=*Leptonia lividocyanula*), x 1,000 (*MU*).

pileolus, pl. **pileoli** (L. *pileolus*, dim. of *pileus* < Gr. *pílos*, hat, helmet, cap; here, pileus): a small pileus.

pileus, pl. **pilei** (L. *pileus* < Gr. *pílos*, hat, helmet, cap; here, pileus): the broad upper part of many ascocarps and basidiocarps, on which is formed the fertile hymenium, and which produces the ascospores or basidiospores, depending on the type of fruiting body.

Pilei of the basidiocarps of *Lepiota* sp., x 0.3 (*RV*).

Pileus of the basidiocarp of *Lepiota* sp., x 0.5 (*MU*).

pilidium, pl. **pilidia** (L. *pilidium* < Gr. *pilídion* < *pílos*, hat, helmet, cap; here, pileus + dim. suf. *-ídion*): an orbicular, hemispherical shield of certain lichens, whose external part becomes powdery, as occurs in the gen. *Calicium* (Caliciales).

piliform (L. *piliformis*, hair-shaped < *pilus*, hair + *-formis* < *forma*, shape): shaped like a hair, like the appendages of the spores of *Choanephora cucurbitarum* and *Gilbertella persicaria* (Mucorales), and the ascospores of *Nematospora coryli* (Saccharomycetales).

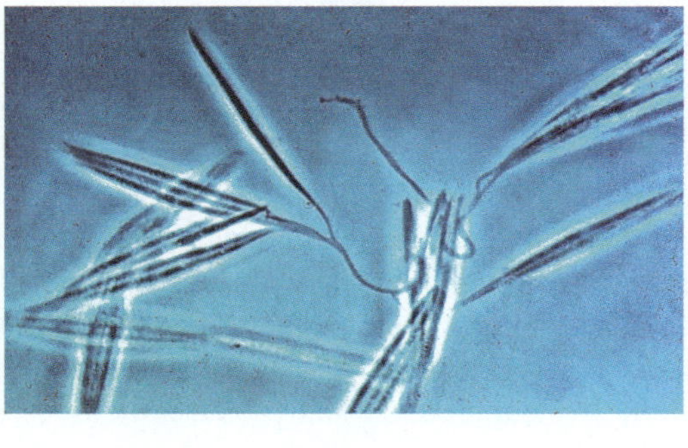

Piliform appendages of the ascospores of *Nematospora coryli*, x 1,000 (*RTH*).

Piliform appendages of the ascospores of *Nematospora coryli*, x 1,400 (*MU*).

pilose (L. *pilosus*, hairy < *pilus*, hair + suf. *-osus* > E. *-ose*, full of, augmented, prone to): covered with long, soft filaments, similar to hairs, like the basidiocarp of *Oudemansiella pilosa* (Agaricales).

pinicole, **pinicola** (L. *Pinus*, the gen. of the pines + suf. *-cola*, inhabitant < *colere*, to inhabit): living in association with pines, whether by forming mycorrhizae, like *Boletus pinicola* (Agaricales), or by growing on pine cones, like *Auriscalpium vulgare* (Aphyllophorales).

Basidiocarp of *Boletus pinicola*, a mycorrhizogenous species on pines (**pinicole**), x 0.5 (*MRO*).

Basidiocarps of *Auriscalpium vulgare*, on a pine cone (**pinicole**), x 0. 5 (*MRO*).

pinocytosis, pl. **pinocytoses** (Gr. *pinó*, to drink + *cytosis* < Gr. *kýtos*, cavity; here, cell + suf. *-osis*, condition or state): an intracellular process, present in eukaryotic cells, that utilizes microfibrils to "imbibe" small droplets of liquid from the surrounding medium. See **phagocytosis**.

pionnote (Gr. *píon*, fat, thick juice + *notís*, moisture): an effuse, gelatinous or viscous sporodochium. The term is applied to the masses of spores with an oily or mucilaginous appearance that are produced by fungi such as *Fusarium* and *Myrothecium* (tuberculariaceous asexual fungi), and *Pestalotiopsis* (melanconiaceous asexual fungi).

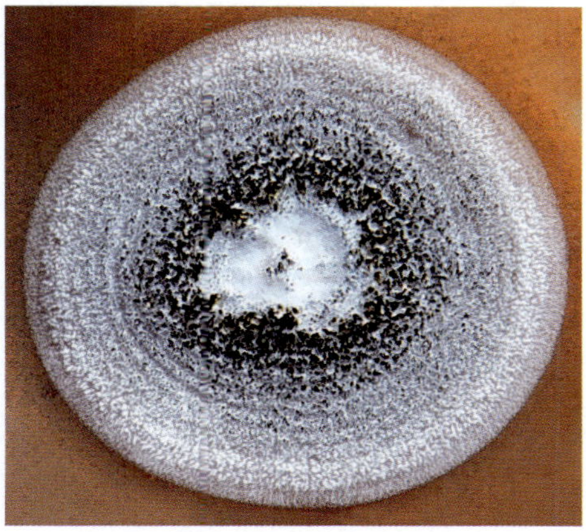

Pionnotes of *Myrothecium verrucaria* on agar, x 1 (*MU*).

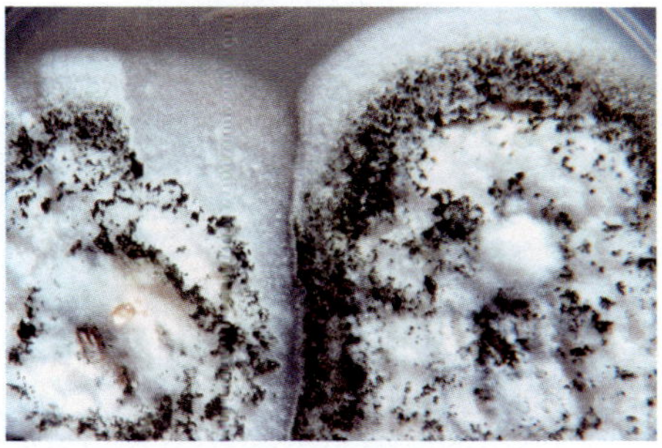

Pionnotes of *Myrothecium verrucaria* on agar, x 5 (*MU*).

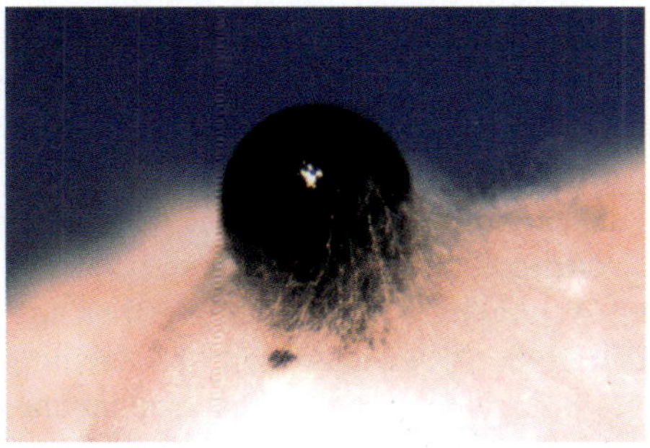

Pionnote of *Myrothecium verrucaria* on agar, x 40 (*MU*).

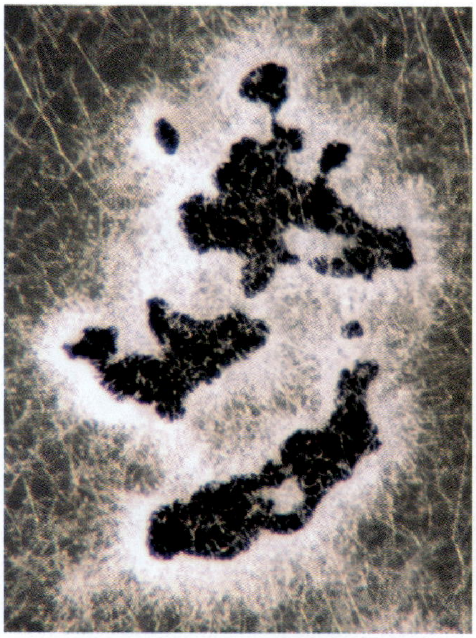

Pionnotes of *Pestalotiopsis* sp. on agar, x 20 (*CB*).

pit (ME. *pitt* < OE. *pytt*, pit): a depression or shallow hole in the surface of a structure, such as the round holes of the capillitial elements and the depressions in the surface of the endoperidium of *Lycoperdon perlatum* (Lycoperdales).

pit connection (ME. *pitt* < OE. *pytt*, pit; L. *connexion, connexio* < L. *connexus*, pp. of *conectare* + *-ion*, act or process): very small canals or tubes that go through the cell wall of various fungi; through these perforations pass the plasmodesmata, fine cytoplasmic filaments that connect one cell with another. See **plasmodesmata**.

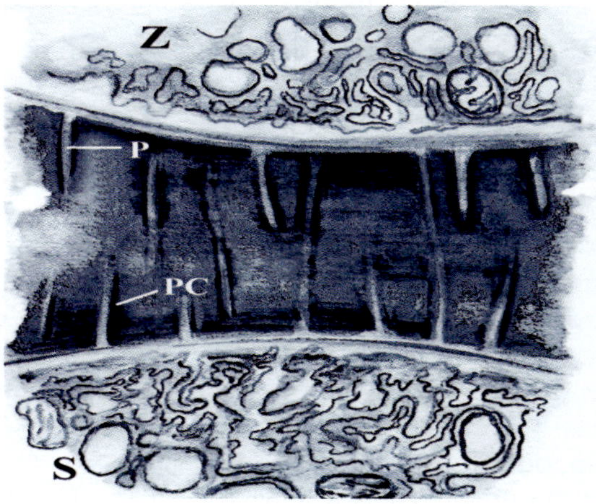

Pit connections (PC) in the septum between the zygospore (Z) and suspensor (S) of *Rhizopus sexualis* (Mucorales), seen in longitudinal section. The plasmodesmata (P) pass from one cell to another through the pit connections. Drawing based on a transmission electron micrograph obtained by Beckett *et al.*, and published in their *Atlas of Fungal Ultrastructure*, 1974, x 23,000 (*MU*).

pitted (ME. *pitt* < OE. *pytt*, pit + suf. *-ed*, past tense): having small, shallow holes in the surface, like the wall of the ascospores of *Gelasinospora* (Sordariales), and the thallus of the lichen *Letharia* (Lecanorales), which appear to the unaided eye as dark points.

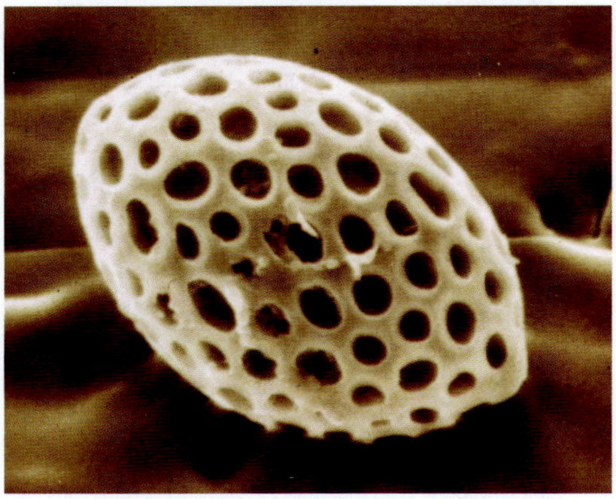

Pitted cell wall of the ascospore of *Gelasinospora retispora*, x 1,000 (*RTH*).

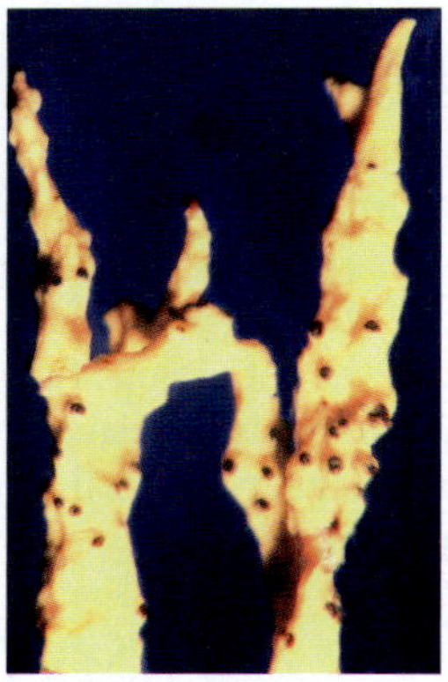

Pitted branches of the thallus of *Letharia vulpina* f. *californica*, x 12 (*MU*).

placo-, **plac-** (Gr. *plakós*, plaque, tablet): a pref. used in terms such as the following.

placodioid (Gr. *plakós*, plaque, tablet + NL. suf. *-oides*, contraction of Gr. *-oeídes*, which denotes similarity): shaped like a placodium. In the lichens, it refers to the type of thallus, discoidal and crustaceous in the center, and radiately lobed on the borders, as if it were foliaceous. Equal to **placodiomorph**. See **placodium**.

placodiomorph (Gr. *plakós*, plaque, tablet + *dís*, two, twice + *morphé*, shape): *Lichens*. A type of thallus that is crustaceous in the center and lobate-radiate on the borders, because of which it appears to be foliaceous, as is seen in *Buellia oidalea* (Lecanorales). Equal to **placodioid**.

Placodiomorph, thallus of *Buellia oidalea*; the dark spots are apothecia, x 8 (*MU*).

placodium, pl. **placodia** (NL. *placodium* < Gr. *plakós*, plaque, tablet + L. *-odium*, resembling < Gr. *ode*, like < *-oeídes*, similar to): the sclerotial cortical stroma that surrounds the perithecial necks in xylariaceous ascomycetes, such as *Diatrype*. Also known as the **ostiolar disc**. If derived from the endostroma it is called *endoplacodium*, and *ectoplacodium* if formed from the ectostroma; if derived from both types of stroma it is called *diplostromatic*, and if it is only of one type it is called *haplostromatic*. See these terms.

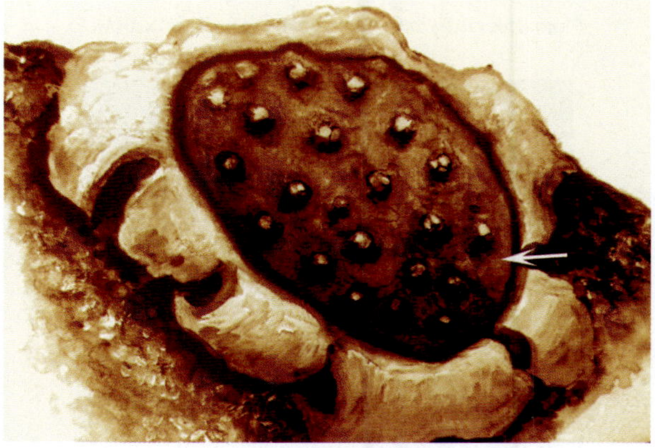

Placodium of the perithecial stroma of *Diatrype virescens*, x 30 (*MU*).

placomycetoid (Gr. *plakós*, plaque, tablet + *mýkes*, genit. *mýketos*, fungus + suf. *-oeídes* > L. *-oide*, similar to): applied to the fruiting body of an agaric with a diametral ratio between the pileus and stipe < 1. Cf. **campestroid**.

plagio- (Gr. *plágios*, oblique or transverse): a pref. that alludes to the oblique position of something, as in *plagiotropic*, having the longest axis inclined away from the vertical, as the consequence of the unilateral action of a stimulus.

plagiotropism (Gr. *plagiós*, oblique, inclined, transverse + *trópos*, gyration, change in position or manner < *tropé*, turn < *trépo*, to gyrate, take turns, direct onself toward + suf. *-ismós* > L. *-ismus* > E. *-ism*, state, phase, tendency, action): the phenomenon exhibited by *plagiotropic* organisms. Syn. of **diageotropism**. See **tropism**.

plaited, plicate (L. *plicatus*, pleated, folded < *plica*, crease, fold + suf. *-atus* > E. *-ate*, provided with or likeness): folded or creased like a fan, like the ascocarp of *Plectania platensis* (Pezizales) and the pileus of the basidiocarp of *Coprinus impatiens* and *C. plicatilis* (Agaricales); the latter has creases between the striae that furrow its surface. The colonies of some species of *Penicillium*, e.g., *P. chrysogenum* (=*P. notatum*), *P. aurantiogriseum* (moniliaceous asexual fungi), *Aureobasidium pullulans* (dematiaceous asexual fungi), and the ascocarp of *Helvella robusta* (Pezizales) also are plaited. Also called **sulcate**.

Plaited colony of *Candida ciferrii* (cryptococcaceous asexual yeasts) on agar, x 1 (*MU*).

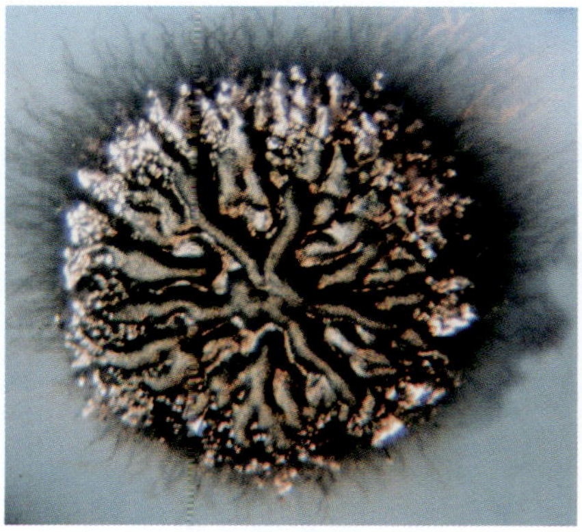

Plaited colony of *Aureobasidium pullulans* on agar, x 2 (*MU*).

Plaited ascocarp of *Helvella robusta*, x 1 (*TRO*).

Plaited colony of *Penicillium aurantiogriseum* on agar, x 0.5 (*MU*).

plane (L. *planus*, plane): a flat surface, either in orientation or symmetry.

planetism (Gr. *plané-*, genit. *planétes*, wanderer, vagabond + suf. *-ismós* > L. *-ismus* > E. *-ism*, condition or state): the condition of an organism that has phases or states of motility, as in the life cycle of the Oomycetes. See **monoplanetic**, **diplanetic** and **polyplanetic**.

planeto-, planet- (Gr. *plané-*, genit. *planétes*, wanderer, vagabond): a pref. that forms compounds that allude to something motile or capable of moving.

plani- (L. *planus*, plane, flat): a pref. that alludes to something flat or that grows in a plain, as on the generic name *Planistroma* (Dothideales), an ascomycete with a flat stroma, and *planicole*, which inhabits the plains.

plano- (Gr. *plános*, wanderer, vagabond): Cf. **planeto-**, which is preferable.

planoconvex (L. *planus*, plane + *convexus*, arched, convex): with one side flat and the other convex, like the pileus of some Agaricales, which is flat on the lower side, where the stipe is attached, and convex on the upper part.

Planoconvex basidiocarp of *Amanita* sp., x 0.5 (*RV*).

Planoconvex basidiocarps of *Cortinarius* sp., x 1 (*RV*).

planocyte (Gr. *planétes*, genit. of *plános*, wanderer, vagabond + *kýtos*, cavity; here, cell): a motile cell (flagellate) destined for multiplication or reproduction, whether a zoospore, a planogamete or a planozygote. See **planospore**, **simblospore** and **zoospore**.

planogamete (Gr. *planétes*, genit. of *plános*, wanderer, vagabond + *gamétes*, spouse, sexual cell): also called **zoogamete**; a flagellate, motile gamete, characteristic of the aquatic fungi of the class Chytridiomycetes. Cf. **aplanogamete**.

planont (Gr. *plános*, wanderer, vagabond + *óntos*, genit. of *ón*, a being): see **planocyte**, **planetospore**, **simblospore** and **zoospore**.

planospore (Gr. *planétes*, genit. of *plános*, wanderer, vagabond + *sporá*, spore): a motile spore; syn. of **simblospore** and **zoospore**.

planozygote (Gr. *planétes*, genit. of *plános*, wanderer, vagabond + *zygotós*, genit. of *zygós*, yoke, pair): a motile zygote. A flagellate cell, and consequently active, that results from the fusion of two flagellate gametes, as happens in some Chytridiales (*Synchytrium*) and Blastocladiales (*Allomyces*).

plaque bridge (ME. *brigge* < OE. *brycg*, bridge; Gr. *plakós*, plaque, tablet): the part that interconnects the two globular daughter halves that result from the replication of the achromatic spindle pole body during mitosis in yeast cells, such as *Saccharomyces* (Saccharomycetales). See **middle piece**.

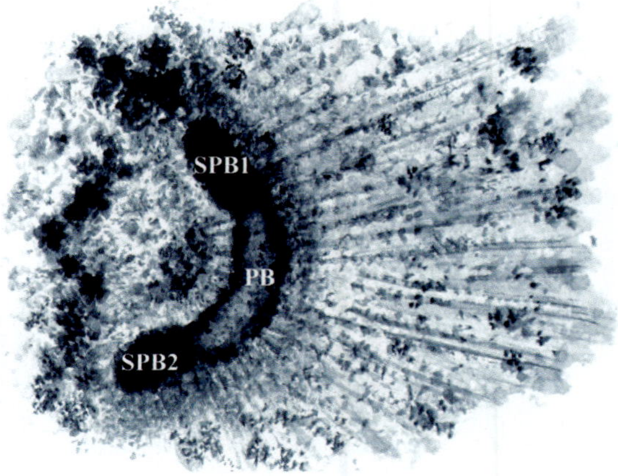

Plaque bridge (PB) of a prophase nucleus of *Saccharomyces cerevisiae* (Saccharomycetales). The replicated polar bodies of the spindle (SPB1 and SPB2) arranged side by side are connected by the plaque bridge. Drawing based on a transmission electron micrograph published by Beckett *et al.* in their *Atlas of Fungal Ultrastructure*, 1974, x 140,000 (*MU*).

-plasis, **-plasia** (Gr. *plásis*, action of taking shape < *plásso*, to form, take shape): a suf. that alludes to the formation of something, as in *hyperplasia*, abnormal or unusual increase in the number of cells that make up a tissue.

plasmalemma (Gr. *plásma*, formation, bland material with which a living being is formed + *lémma*, genit. *lemmátos*, bark, rind, covering): the outer membrane, composed of phospholipids and proteins, that surrounds the cytoplasm of a cell; it functions in the regulation of the interchange of materials between the cell and its environment. Also called *cellular (cell) membrane*, *cytoplasmic membrane* or *plasmatic (plasma) membrane*.

plasmalemmasoma, pl. **plasmalemmasomata** (Gr. *plásma*, formation, bland material with which a living being is formed + *lémma*, genit. *lemmátos*, bark, rind, covering + *sõma*, body): an intracytoplasmatic vesicle (formed by invagination of the plasmalemma) full of tubular diverticulae. Cf. **lomasome**.

plasmato-, **plasmat-** (Gr. *plásma*, genit. *plásmatos*, formation, bland material with which a living being is formed): a pref. employed in terms such as *plasmatoogosis*.

plasmatoogosis, pl. **plasmatoogoses** (Gr. *plásma*, genit. *plásmatos*, formation, bland material with which a living being is formed + L. pref. *ob-* > *og-*, on, against, toward, facing + Gr. suf. *-ósis* > E. *-osis*, condition or state): a shoot in the form of a bud, similar to a prosporangium, that some species of Pythiaceae form in tissues of the host plant.

plasmodesma, pl. **plasmodesmata** (Gr. *plásma*, formation, bland material with which a living being is formed + *desmós*, filament): an extremely thin plasmatic filament that passes through a fine perforation in the cell wall (*pit connection*), so that it establishes a continuity between adjacent cells, as has been observed, e.g., in *Rhizopus sexualis* (Mucorales); during the development of the zygospore several plasmodesmata are formed through the wall that separates it from the suspensor. Frequently these plasmodesmata are associated with large accumulations of endoplasmic reticulum on both sides of the small canals or tubes that traverse the wall of the gametangial septum, for which reason it is considered that soluble nutrients and other substances pass through them from the suspensors toward the developing zygospore, utilizing the membranes of the endoplasmic reticulum as a transport system.

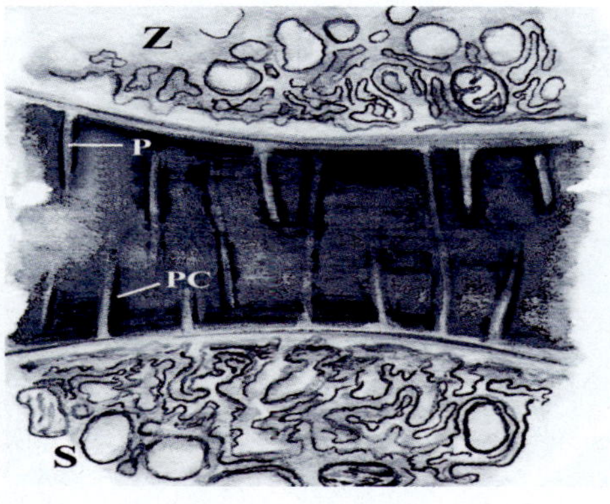

Plasmodesmata (P) of the septum between the zygospore (Z) and the suspensor (S) of *Rhizopus sexualis*, seen in longitudinal section. The plasmodesmata pass from one cell to another through the pit connections (PC). Drawing based on a transmission electron micrograph obtained by Beckett *et al.*, and published in their *Atlas of Fungal Ultrastructure*, 1974, x 23,000 (*MU*).

plasmodial (NL. *plasmodium*, plasmodium < Gr. *plásma*, formation, bland material with which a living being is formed + L. suf. *-odium*, similar < Gr. *ode*, like < *-oeídes*, similar to + L. suf. *-alis* > E. *-al*, relating to or pertaining to): pertaining to or relating to the *plasmodium* (Myxomycetes).

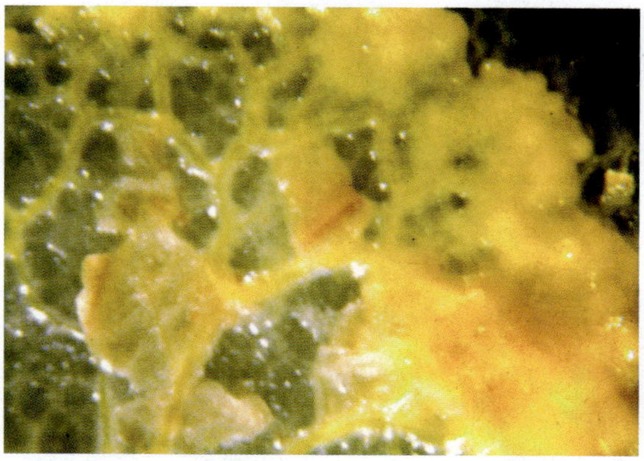

Plasmodial mass of *Physarum polycephalum* (Physarales), x 14 (*MU*).

plasmodiocarp (NL. *plasmodium* < Gr. *plásma*, formation, bland material with which a living being is formed + L. suf. *-odium*, similar < Gr. *ode*, like < *-oeídes*, similar to + Gr. *karpós*, fruit): *Myxomycetes*. A type of sporangium that retains the reticular or veined shape of the plasmodium from which it forms, as happens in *Hemitrichia serpula* (Trichiales).

Plasmodiocarp of *Didymium perforatum* (Physarales), x 10 (*RS*).

Plasmodiocarp of *Hemitrichia serpula* (Trichiales), x 1 (*RV*).

Plasmodiocarp of *Physarum serpula* (Physarales) on bark, x 10 (*MU*).

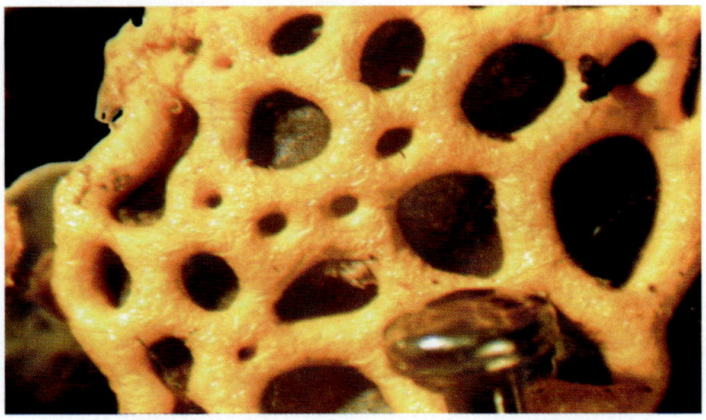

Plasmodiocarp of *Hemitrichia serpula* on dead wood, x 14 (*MU*).

plasmodiophagous (NL. *plasmodium*, plasmodium < Gr. *plásma*, formation, bland material with which a living being is formed + L. suf. *-odium*, similar < Gr. *ode*, like < *-oeídes*, similar to + Gr. *phágos*, what eats): an organism that feeds on plasmodia, such as insects that feed on plasmodia of the Myxomycetes. Cf. **sporophagous**.

Plasmodiophoromycetes (NL. *plasmodium*, plasmodium < Gr. *plásma*, formation, soft material with which a living being is formed + L. suf. *-odium*, similar < Gr. *ode*, like < *-oeídes*, similar to + Gr. suf. *-phóros*, bearer + L. *-mycetes*, ending of taxonomic class < Gr. *mýkes*, genit. *mýketos*, fungus): a class of viscous or mucilaginous fungi, endoparasitic in plants, with two types of endobiotic plasmodia, one related to the asexual reproductive phase (sporangiogenous plasmodium) and the other to sexual reproduction (cystogenous plasmodium). In some classifications, the Plasmodiophoromycetes are included, together with the Protosteliomycetes, Acrasiomycetes and Myxomycetes, in the division Myxomycota of the kingdom Fungi. In the classification scheme followed in this dictionary (Alexopoulos *et al.*, 1996), these organisms are all considered as different phyla of the kingdom Protista, i.e., Plasmodiophoromycota, Dictyosteliomycota, Acrasiomycota and Myxomycota. Other authors (e.g., Margulis and Schwartz, 1988) include the class Plasmodiophoromycetes in the kingdom Protoctista. See the Appendix, where there is a comparison of the outline of Alexopoulos *et al.*, 1996, modified by Ulloa and Hanlin, 2006, to that of Kirk *et al.*, 2008. In the latter classification scheme, the class Plasmodiophoromycetes has been moved to the kingdom Protozoa, belonging to the class Phytomyxea, phylum Cercozoa, with the concurrent changes of taxa names. These taxonomic rearrangements are due to the continuing study and evaluation of these fungus-like microorganisms, in order to achieve a phylogenetic classification. See **Protozoa**.

plasmodium, pl. **plasmodia** (NL. *plasmodium*, plasmodium < Gr. *plásma*, formation, bland material with which a living being is formed + L. suf. *-odium*, similar < Gr. *ode*, like < *-oeídes*, similar to): a multinucleate protoplasmic mass, without a cell wall, and with amoeboid movement and phagocytic nutrition; it is the somatic and trophic phase of the Myxomycetes, the Plasmodiophoromycetes and of some Protosteliales. See **aphanoplasmodium**, **phaneroplasmodium** and **protoplasmodium**.

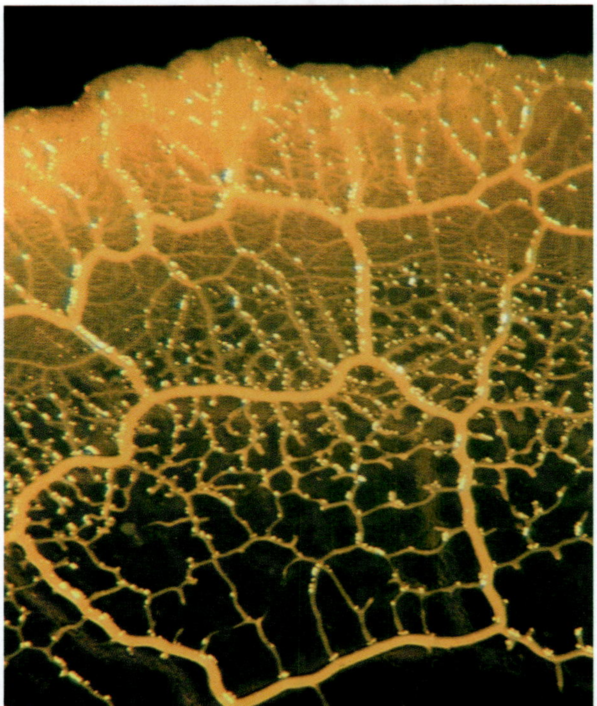

Plasmodium of *Physarum polycephalum* (Physarales) on agar, x 5 (*MU*).

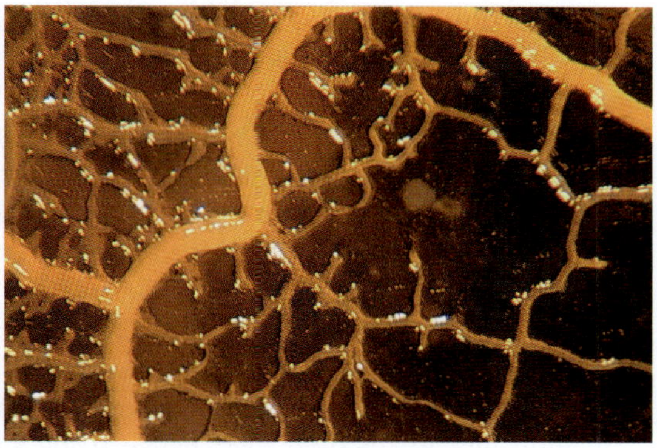

Plasmodium of *Physarum polycephalum* on agar, x 8 (*MU*).

plasmogamy (Gr. *plásma*, formation, bland material with which a living being is formed + *gámos* > E. des. *-gamy*, sexual union): the first phase of sexual reproduction in which two protoplasts fuse and there is juxtaposition of the nuclei, forming a dikaryon; this phase precedes *karyogamy* and *meiosis*.

plasmoptysis, pl. **plasmoptyses** (Gr. *plásma*, formation, bland material with which a living being is formed + *ptýsis*, to spit, throw out): *Mycorrhizae*. The process in which the tips of the hyphae (which are found inside the cells of the host plant) break and emit a globulose mass of cytoplasm called a *ptyosome*, which is then digested and assimilated by the host cell.

plasmotomy (Gr. *plásma*, formation, bland material with which a living being is formed + *tomé* > E. des. *-tomy*, action of cutting): *Protosteliales*, *Acrasiomycota*. The process of cell division present in myxamoebae, which consists of a thinning of the central part of the protoplast until it breaks the fine protoplasmic bridge that for a brief time connects the two cells, so that from one, two myxamoebae originate. During the process, the nucleus and the cytoplasmic organelles divide by amitosis.

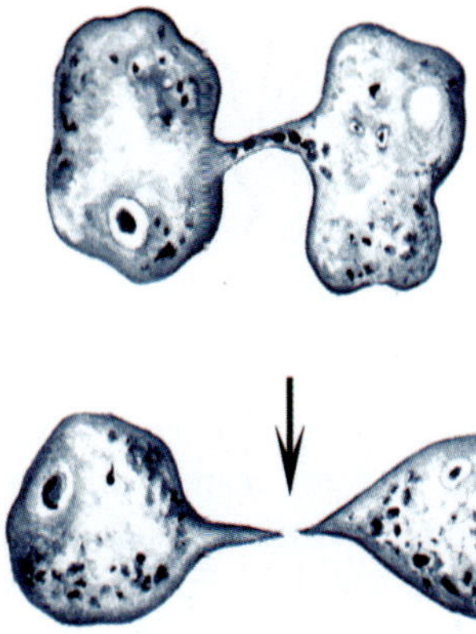

Plasmotomy of the myxamoebae of *Cavostelium bisporum*, x 1,800 (*MU*).

platy- (Gr. *platýs*, wide): a pref. used in many compound terms to indicate that something is wide, as in *platyphyllous*.

platyphyllous (Gr. *platýs*, wide + *phýllon*, leaf): with wide leaves or lobules, like the thallus of some foliaceous lichens, e.g., *Parmelia* (Lecanorales).

platysmoid (Gr. *platýs*, genit. *platysmátos*, wide + L. suf. *-oide*, like, contraction of Gr. *-oeídes*, similar to): applied to the foliose lichen thalli, loosely attached to the substrate, with ascendent lobules, like those of *Platysmatia* (Lecanorales). Also refers to scleroplectenchyma with brown hyphae.

plectenchyma, pl. **plectenchymata** (Gr. *plektós*, braided, intertwined + *énchyma*, stuffed): a general term that is used to refer to all types of fungal tissues. See the two most common types, **prosenchyma** and **pseudoparenchyma**.

plecto-, **plect-** (Gr. *plektós*, braided, intertwined): a pref. used to indicate that the hyphae forming a tissue are braided or intertwined, as in *plectenchyma*.

plectobasidium, pl. **plectobasidia** (Gr. *plektós*, braided, intertwined + *basídion*, dim. of *básis*, base): *Gasteromycetes*. A type of basidium characterized by being extremely variable in shape and for their irregular origin from hyphal tissues of the gleba. The plectobasidia are not oriented to discharge their spores toward an air space from an organized hymenium (since they are statismosporic basidia), but they form layers or fertile areas called pseudohymenia, which can line small locules in the gleba, be concentrated inside peridioles, or form layers on phallaceous receptacles. Plectobasidia are present in the Hymenogastrales, such as *Podaxis*.

Plectobasidia, with orthotropic basidiospores, from the hymenium of *Podaxis pistillaris*, x 850 (*MU*).

Plectomycetes (Gr. *plektós*, braided, intertwined + L. *-mycetes*, ending of taxonomic class < Gr. *mýkes*, genit. *mýketos*, fungus): a class of ascomycetes with small, evanescent asci, produced at different levels within a cleistothecium. The Plectomycetes (which includes the orders Eurotiales, Ascosphaerales and Onygenales), together with the classes Archiascomycetes, Pyrenomycetes, Discomycetes and Loculoascomycetes, plus the order Saccharomycetales (ascosporogenous yeasts), constitute the phylum Ascomycota of the kingdom Fungi. See the Appendix, where there is a comparison of the outline of Alexopoulos *et al*., 1996, modified by Ulloa and Hanlin, 2006, and the scheme of Kirk *et al*., 2008. In the latter scheme of classification, the class Plectomycetes is considered an unnatural group of fungi, an artificial classification, and has been changed to the class Eurotiomycetes, subphylum Pezizomycotina, phylum Ascomycota; this has been done in order to achieve a phylogenetic classification of the fungi.

plectonematogenous (Gr. *plektós*, braided, intertwined + *nêma*, genit. *nêmatos*, thread, filament + *génos*, origin < *gennáo*, to engender, produce): *Conidial fungi*. Applied to fungi whose conidiophores and conidiogenous cells originate from ropes or funiculi of intertwined hyphae (i.e., not from a single hypha), but without forming a synnema, as is seen, e.g., in *Acremonium* (=*Cephalosporium*), *Oedocephalum* (moniliaceous asexual fungi) and *Stachybotrys* (dematiaceous asexual fungi), which form fasciculate hyphae.

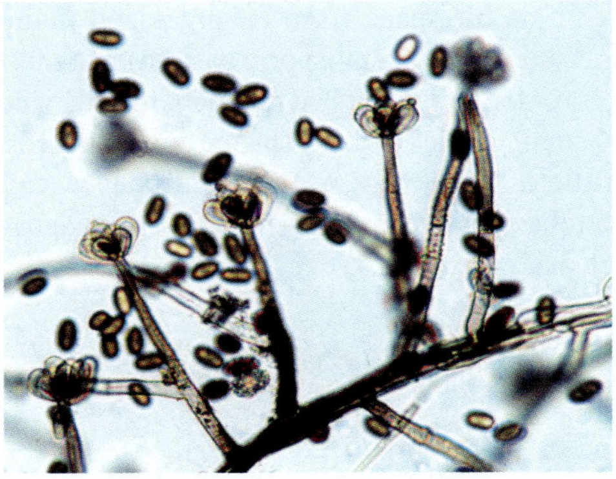

Conidiophores of *Stachybotrys chartarum* (=*S. atra*), a **plectonematogenous** species, x 450 (*MU*).

pleio-, **plei-** (Gr. *pleíon*, more, numerous): a pref. employed in the formation of biological terms alluding to something that is in larger number, or that is more abundant, such as *pleiosporous*, with numerous spores. See **plio-**, which is preferable to **pleio-**.

pleiosporous, **pliosporous** (Gr. *pleíon*, more, numerous + *sporá*, spore): see **pliosporous**. Syn. of **myriosporous**.

pleo-, **pleio-** (Gr. *pléon*, neuter form of *pléos*, which is equivalent to *pleíon*, more, numerous): a pref. used with the indicated meanings, as in the following terms. Equivalent to **plio-**.

pleoanamorphy (Gr. *pléon*, neuter form of *pléos*, which is equivalent to *pleíon*, more, numerous + *anamorphóo*, to transform < *ana*, anew, again + *morphé*, form > E. *anamorphy*): applied to fungi (pleomorphs) that have more than one anamorph.

pleomorphic, **pleiomorphic** (Gr. *pléon*, neuter form of *pléos*, which is equivalent to *pleíon*, more, numerous + *morphé*, form, shape): having more than one form in the life cycle; usually with reference to having more than one sporogenous state. Also called **polymorphic** or **versiform**. This last term is mainly applied to fungal structures that change shape with age, like the basidiospores of *Conocybe lactea* and the cheilocystidia of *Bolbitius vitellinus* (Agaricales).

pleomorphism (Gr. *pléon*, neuter form of *pléos*, which is equivalent to *pleíon*, more, numerous + *morphé*, form + suf. *-ismós* > E. *-ism*, condition or state): the condition of being *pleomorphic*.

pleont (Gr. *pléon*, neuter form of *pléos*, which is equivalent to *pleíon*, more, numerous + *óntos*, genit. of *ón*, a being): any of the two or more states of a pleomorphic fungus.

plerotic (Gr. *plerotikós*, to be full < *plerés*, to fill + suf. *-tikós* > L. *-ticus* > E. *-tic*, relation, fitness, inclination or ability): to be full; e.g., a plerotic sporangium is one that is full of spores; in some species of Pythiaceae, such as *Pythium monospermum* (Peronosporales), a plerotic oospore is one that fills the oogonium. Cf. **aplerotic**.

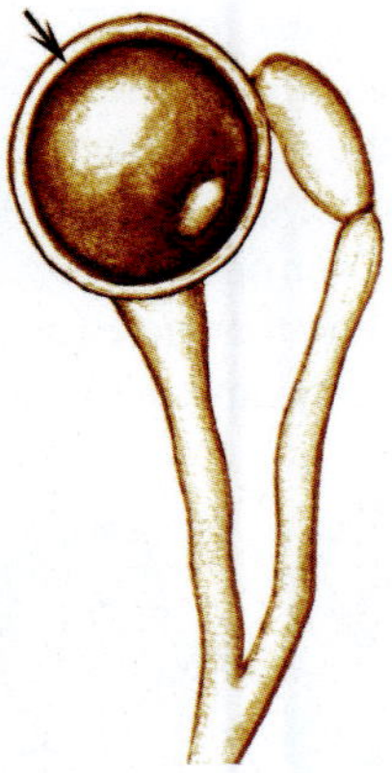

Plerotic oospore in the oogonium of *Pythium monospermum*, x 1,000 (*MU*).

pleuro-, **pleur-** (Gr. *pleurá*, *pleurón*, side, flank): a pref. employed to form compounds that give the idea that something is on the side or flank, as in *pleuroacrogenous* and the following entries.

pleuroacrogenous (Gr. *pleurá*, *pleurón*, side, flank + *ákros*, apex + *génos*, origin, birth): see **acropleurogenous**.

pleurobasidium, pl. **pleurobasidia** (Gr. *pleurá*, *pleurón*, side, flank + *basídion*, dim. of *básis*, base): a basidium that is relatively wide at the base, with widespread, apical, bifurcate prolongations, as occurs in the gen. *Pleurobasidium* (Aphyllophorales).

pleurocystidium, pl. **pleurocystidia** (Gr. *pleurá*, *pleurón*, side, flank + *kystídion*, dim. of *kýstis*, bladder, vesicle, cell): a cystidium that emerges from the side of the gill, projecting beyond the level of the basidia, as occurs in certain agarics, e.g., *Marasmius ferrugineus* and *Entoloma versatile* (=*Pouzarella versatilis*), Agaricales.

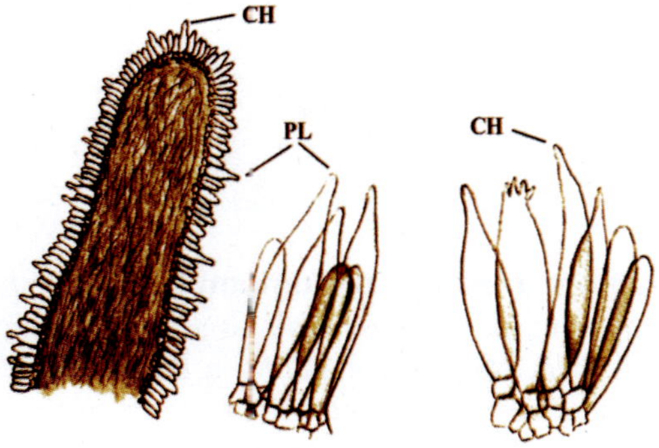

Longitudinal section of a gill of *Entoloma versatile* (=*Pouzarella versatilis*), x 160. **Pleurocystidia** (PL) and cheilocystidia (CH) in the hymenium, x 600 (*MU*).

pleurogenous, **pleurogenic** (Gr. *pleurá*, *pleurón*, side, flank + *génos*, origin < *gennáo*, to engender, produce): formed or borne on the sides, like the conidia on the conidiophores of *Trichosporiella cerebriformis* (moniliaceous asexual fungi). See **acropleurogenous**.

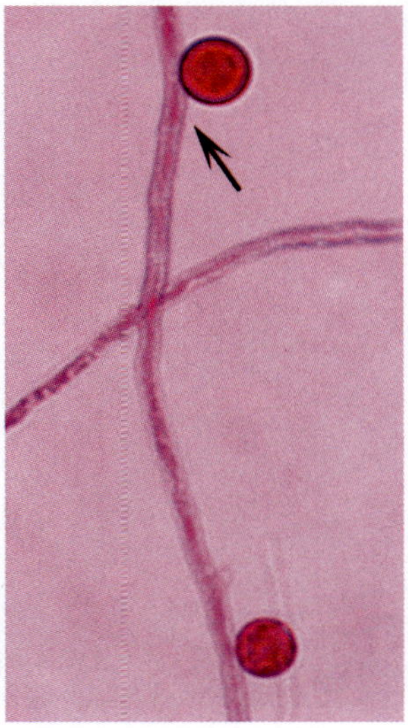

Pleurogenous conidia of *Trichosporiella cerebriformis*, x 1,000 (*CB*).

pleurokont, **pleurocont**, **pleurokontic**, **pleurocontic**, **pleurokontous**, **pleurocontous** (Gr. *pleurá*, *pleurón*, side, flank + *kontós*, oar; here, flagellum): refers to a motile cell that has flagella or swimming organs on one side, as is seen in the secondary (reniform) zoospores of the Oomycota, which have both flagella (one whiplash and the other mastigonemate) inserted in the lateral cavity, e.g., *Achlya* and *Saprolegnia* (Saprolegniales). Cf. **acrokont** and **opisthokont**.

pleurosporous (Gr. *pleurá*, *pleurón*, side, flank + *sporá*, spore): having the spores arranged laterally, like the basidium of the Uredinales.

pleurotoid (< gen. *Pleurotus* < Gr. *pleurá*, *pleurón*, side, flank + L. suf. *-otus* < Gr. *otós*, ear + L. suf. *-oide*, like < Gr. *-oeídes*, similar to): *Agaricales*. A type of fruiting body (13 principal types are recognized) without a stipe, or with an eccentric or lateral stipe, and different types of gills, depending on the gen.: *Pleurotus*, *Paxillus*, *Crepidotus* (Agaricales), *Panus* and *Lentinus* (Aphyllophorales). This type of fruiting body or habit is considered artificial, as it encompasses gen. from several different familes.

Pleurotoid basidiocarp of *Pleurotus* sp., x 0.5 (*MU*).

plio- (Gr. *pleíon*, greater, more): a pref. with the same meaning and application as **pleo-**.

pliophagous (Gr. *pleíon*, greater, more + *phágos*, what eats): a parasite that can feed on various organisms.

pliosporous (Gr. *pleíon*, greater, more + *sporá*, spore): with numerous spores.

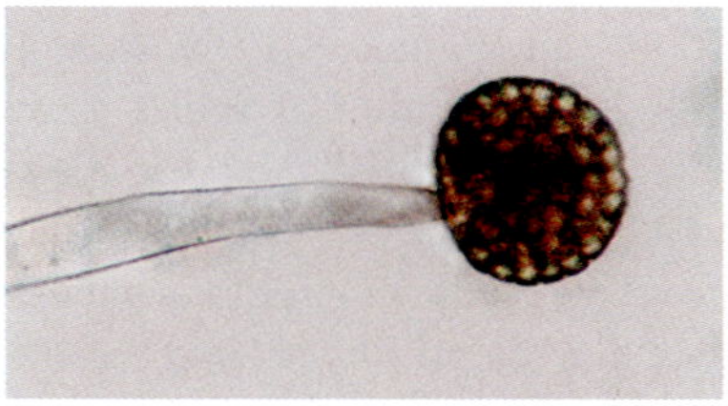

Pliosporous sporangium of *Mucor* sp., x 500 (*MU*).

plioxenous (Gr. *pleíon*, greater, more + *xénos*, host): being able to live on various hosts. This phenomenon is called *plioxeny* (< *xénos* > E. des. *-xeny*). Syn. of **pliophagous**.

pluri- (L. *plus*, genit. *pluris*, more, greater number, various): a pref. employed to give the idea of plurality.

plurilocular (L. *plurilocularis* < *plus*, genit. *pluris*, more, greater number, various + *loculus*, dim. of *locus*, cavity, locule, point, place + suf. *-aris* > E. *-ar*, like, concerning): with several locules or cavities, like the stroma of the Myriangiales and Dothideales, or the multicellular ascospores of many fungi and lichens. Also called **multilocular**. Cf. **unilocular**.

pluriostiolate (L. *plus*, genit. *pluris*, more, greater number, various + *ostiolatus*, provided with an ostiole < *ostiolum*, dim. of *ostium*, door + L. suf. *-atus* > E. *-ate*, provided with or likeness): having several ostioles, like the eustromatic conidioma of *Dichomera* (sphaeropsidaceous asexual fungi), which is subepidermal, erumpent, dark and multilocular, and which develops in branches and leaves of various plants. Also there are ascomycetes with pluriostiolate stromata, e.g., *Camillea* and *Porolepriеuria* (Xylariales).

Pluriostiolate stromata of *Porolepriеuria rogersii*, x 5 (*CB*).

plurivorous (L. *plus*, genit. *pluris*, more, greater number, various + *vorare*, to devour): capable of living and thriving on the most diverse substrates, whether as a saprobe or parasite, like the common molds of the gen. *Aspergillus* and *Penicillium* (moniliaceous asexual fungi). Syn. of **polytrophic**. Cf. **monotrophic** and **omnivorous**.

pluteotoid (< gen. *Pluteus* < L. *pluteus*, a detaching or shedding; the ejection of spores from a natural receptacle + L. suf. *-oide*, like < Gr. *-oeídes*, similar to): *Agaricales*. A type of fruiting body (13 principal types are recognized) with free gills but without a ring and volva, like that of *Pluteus* and certain species of *Hygrophorus*, *Lepiota* and *Bolbitius*, among other gen.

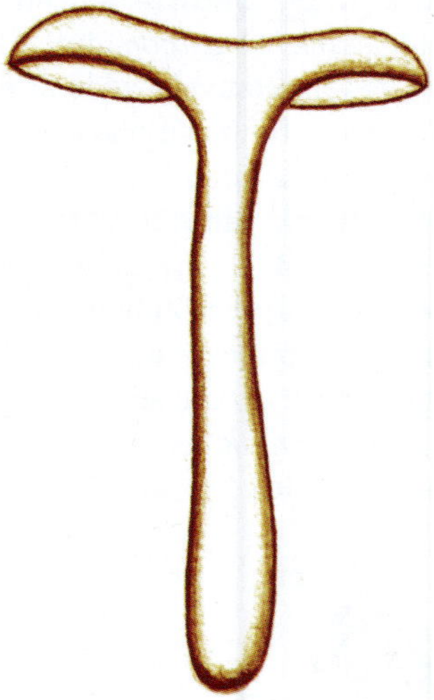

Pluteotoid basidiocarp of *Pluteus* sp.
Diagram of a longitudinal section, x 1 (*MU*).

pneumomycosis, pl. **pneumomycoses** (Gr. *pneuma*, genit. *pneúmatos*, gust, air + *mýkes*, fungus + suf. *-osis*, state or condition): a mycotic infection of the lungs of man and animals, such as occurs in coccidioidomycosis. See **mycosis**.

podetium, pl. **podetia** (NL. *podetium* < Gr. *poús*, genit. *podós*, foot + L. dim. suf. *-ium*): *Lichens*. An erect, hollow, simple or branched structure, on which apothecia are formed; podetia constitute the fruticulose secondary thallus, derived from the primary crustaceous thallus, in lichens of the family Cladoniaceae, such as *Cladonia pyxidata*, *C. crystatella* and *C. didyma* (Lecanorales). Cf. **pseudopodetium**.

Podetia with apical apothecia of *Cladonia didyma*, x 1 (*RV*).

Podetium with an apical apothecium of *Cladonia didyma*, x 5 (*MU & CB*).

Podetia with an apical apothecium of *Cladonia didyma,* x 8 (*CB*).

podo-, **pod-** (Gr. *poús*, genit. *podós*, foot, support): a pref. employed in terms such as *rhizopod*, a protozoan that crawls along (like an amoeba or a Foraminifera), with lobate or root-shaped pseudopodia. As a suf., *-pod*, *-podous*, occurs in the formation of terms that are accented on the antepenultimate syllable, such as *apodous*, without feet, *brachipod*, with a pedicel or short pediculus, etc.

poikilohydric (Gr. *poikílos*, changing, variable + pref. *hýdro-* < *hýdor*, water + suf. *-íkos* > L. *-icus* > E. *-ic*, pertaining to or relating to): capable of functioning or surviving under changing conditions of water content, like the majority of lichens, a physiological property not present in the majority of the other nutritional groups of fungi. Lichen thalli can adapt to repeated periods of wetness and drought, and their tolerance to drought is much greater than that of the majority of other fungi studied.

poisonous fungus, pl. **fungi** (ME. *puisun* < OF. < L. *potio*, poisonous drink, any substance that causes death or grave disorder when introduced into the body or applied to it in small quantity; L. *fungus*, pl. *fungi*, fungi). See **mycetism**. Cf. **edible**.

Poisonous fungi, basidiocarps of *Amanita muscaria* var. *flavivolvata*, that cause muscarinic mycetism in humans, x 0.5 (*EPS*).

Poisonous fungus, basidiocarp of *Amanita pantherina*, that causes muscarinic mycetism in humans, x 0.5 (*RV*).

Poisonous fungi, basidiocarps of *Amanita verna*, that cause phalloidin mycetism in humans, x 0.5 (*RV*).

Poisonous fungus, basidiocarp of *Amanita verna*, that causes phalloidin mycetism in humans, x 0.5 (*TH*).

Poisonous fungi, basidiocaps of *Russula emetica*, that cause gastrointestinal mycetism in humans, x 1 (*TH*).

polar (L. *polaris*, pertaining to or relating to the pole < *polus*, pole + suf. *-aris* > E. *-ar*, concerning): pertaining to or relating to the pole, in any of its senses; it is applied, e.g., to spores, flagella, etc., when something is present on the ends or poles.

polarilocular (L. *polaris* < *polus*, pole + suf. *-aris* > E. *-ar*, concerning + *loculus*, dim. of *locus*, cavity, locule; place) : *Lichens*. A spore with two cavities or locules united by a narrow canal or isthmus. Also called *polaribilocular* (of the same roots + L. *bis*, two) and **polocellate**; see this last term, which has been considered more correct than polarilocular, as it refers to the exact structure of this type of spore. *Caloplaca citrina* and *Teloschistes chrysophthalmus* (Teloschistales) are examples of lichens with this type of spore. Similar to *orculiform* or *urceolate*. See **isthmus**.

Polarilocular ascospores of *Teloschistes chrysophthalmus,* x 975 (*MU*).

poleophilous (Gr. *pólis*, city + *phílos*, have an affinity for): adapted to living in cities, like the lichens of urban areas (e.g., *Lecanora conizaeoides*, of the Lecanorales).

polocellate (L. *polocelatus* < *polus*, pole, the end of an axis + *celatus*, covered, occult < *celare*, to conceal + suf. *-atus* > E. *-ate*, provided with or likeness): equal to **polarilocular**. Similar to **orculiform** and **urceolate**.

poly- (Gr. *polýs*, numerous, many, a lot): a pref. employed in many biological terms that imply that something is present in high number or is abundant, as in *polyandrous* and the following.

polyandrous (Gr. *polýs*, many + *andrós*, male; here, the male sexual organ): *Oomycota*. Applied to oospores that develop after being fertilized by several functional antheridia, as occurs in *Achlya* (Saprolegniales). Cf. **monandrous**.

polyascous (Gr. *polýs*, many + *askós*, purse, sac, ascus): with many asci; applied especially to species that have the asci in an hymenium, not separated by sterile bands, as occurs in the Pezizales, and to the locules of some Loculoascomycetes, such as *Dothidea* (Dothideales). Cf. **uniascal**.

polyblastic (Gr. *polýs*, many + *blastós*, sprout, bud, germ, shoot + suf. *-íkos* > L. *-icus* > E. *-ic*, pertaining to or relating to): a type of holoblastic conidiogenous cell that produces buds (the conidial primordia) at various points or sites, as in *Gonatobotryum* (dematiaceous asexual fungi). Cf. **monoblastic**.

polycarpic (Gr. *polýs*, many + *karpós*, fruit + suf. *-íkos* > L. *-icus* > E. *-ic*, pertaining to or relating to): refers to infections caused by *Exobasidium* (Exobasidiales) when they are systemic (or circumscribed) and perennial. Cf. **monocarpic**.

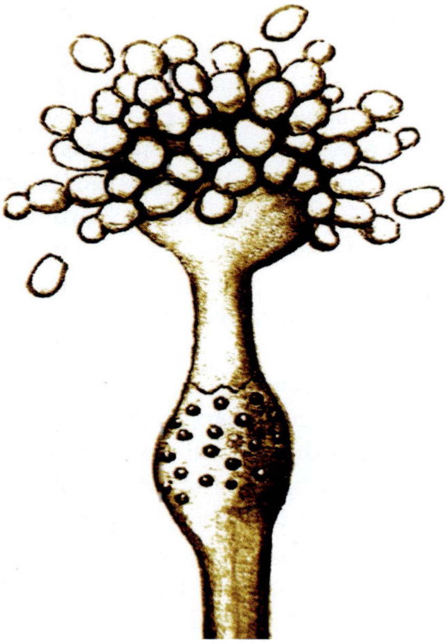

Polyblastic conidiogenous cells of a conidiophore of *Gonatobotryum apiculatum*, x 600 (*MU*).

polycentric (Gr. *polýs*, many + *kéntron*, center + suf. *-íkos* > L. *-icus* > E. *-ic*, pertaining to or relating to): a thallus with several interconnected centers of growth, in which are formed the reproductive organs (sporangia or resting spores), as happens in the aquatic fungi of the family Megachytriaceae (Chytridiales), e.g., *Megachytrium westonii*. Equal to **polygenocentric**. Cf. **monocentric**. See **reproductocentric**.

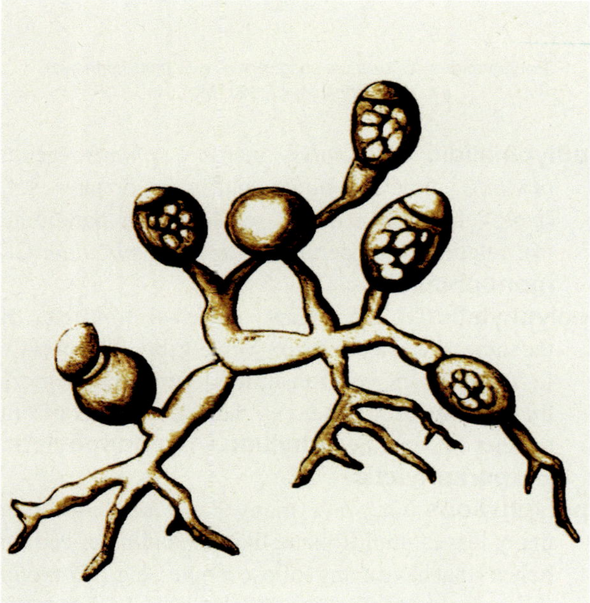

Polycentric thallus of *Megachytrium westonii*, x 800 (*MU*).

polycephalic, **polycephalous** (Gr. *polýs*, many + *kephalé*, head + suf. *-íkos* > L. *-icus* > E. *-ic*, pertaining to or relating to; or + L. suf. *-osus* > E. *-ous*, full of, augmented, prone to): with many heads, like the sporangia of the Myxomycete *Physarum polycephalum* (Physarales).

Polycephalic sporangial heads of *Physarum polycephalum*, on agar, x 10 (*MU*).

polychotomous (Gr. *polýschistos*, split, divided into many parts + *témeo*, to cut, divide): with the apex divided simultaneously into more than two branches, like the conidiophore of *Penicillium atrovenetum* (moniliaceous asexual fungi), whose branches give rise to three, five or more metulae at the distal end. Cf. **dichotomous**, **trichotomous** and **tetrachotomous**. See **polytomous**.

polycystoderm (Gr. *polýs*, many + *kýstis*, vesicle, cell + *dérma*, skin): a group of sphaerocysts (globose and other shapes) that comprise the connivent fibrils of the exoperidium of *Calvatia cretacea* (Lycoperdales).

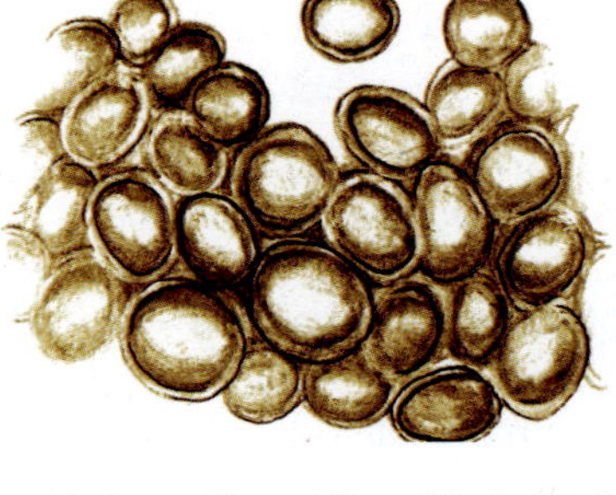

Polycystoderm of the peridium of the fruiting body of *Calvatia cretacea*, x 500 (*MU*).

polyenergid (Gr. *polýs*, many + *energós*, active, operative): composed of many *energids*, each of which is a vital unit constituted of a nucleus and a certain enveloping protoplasmic mass controlled by it. In each true cell, i.e., with a single nucleus, there exists an energid; but there are plurinucleate "unicellular" organisms, among the siphonaceous algae and among the fungi, the so-called lower organisms, without internal dividing partitions. In these cases a word is necessary that substitutes for the term "cell", since the respective protoplasts, with their nuclei, are not separated by a dividing membrane. For this reason, Sachs (1892) proposed the term *energid*, although it is not much used at present.

polygenocentric (Gr. *polýs*, many + *génos*, origin < *gennáo*, to engender, produce + *kéntron*, center + suf. *-ikos* > L. *-icus* > E. *-ic*, pertaining to or relating to): see **polycentric** and **reproductocentric**.

polygonal, **polygonous** (Gr. *polýs*, many + *gónia*, angle + L. suf. *-alis* > E. *-al*, relating to or pertaining to; or + L. suf. *-osus* > E. *-ous*, full, augmented, prone to): having many angles, like the sporangiospores of *Umbelopsis vinacea* (=*Mortierella ramanniana* var. *angulispora*), Mucorales.

Polygonal sporangiospores of *Umbelopsis vinacea* (=*Mortierella ramanniana* var. *angulispora*), x 2,900 (*MU*).

polymorphic, **polymorphous** (Gr. *polýs*, many + *morphé*, form, shape + suf. *-ikos* > L. *-icus* > E. *-ic*, pertaining to or relating to; or + L. suf. *-osus* > E. *-ous*, full of, augmented, prone to): having more than one form. See **pleomorphic**.

polyphagous (Gr. *polyphagós*, eating a lot, voraceous < *polýs*, much, a lot + *phágos*, what eats): applied to organisms that can utilize various types of food, like the facultatively parasitic oomycotes, which are frequently polyphagous, e.g., *Phytophthora cactorum* (=*Ph. omnivora*), Peronosporales; also refers to the aquatic fungi (Chytridiales) that live on different hosts and whose mycelium penetrates various cells of the host. The condition of being polyphagous is called *polyphagy*. Cf. **monophagous**.

polyphialide (Gr. *polýs*, many + *phialís*, genit. *phialídis*, dim. of *phiále*, small glass or jar): a bottle-shaped conidiogenous cell that produces phialoconidia (phialospores) through several openings that are formed in synchronous or irregularly sympodial succession in the cell wall, as is observed in *Fusarium subglutinans* (=*F. moniliforme* var. *subglutinans*), tuberculariaceous asexual fungi, and in *Helicoma inflatum* (dematiaceous asexual fungi).

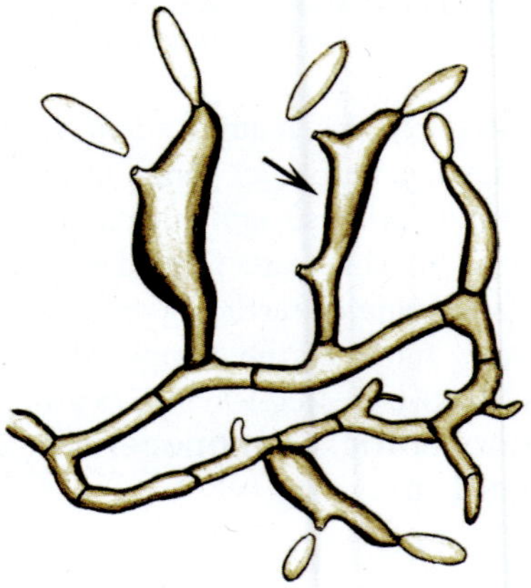

Polyphialides of *Fusarium subglutinans* (=*F. moniliforme* var. *subglutinans*), x 1,000 (*MU*).

polyphialidic (Gr. *polýs*, many + *phialís*, genit. *phialídis*, dim. of *phiále*, small glass or jar + suf. *-íkos* > L. *-icus* > E. *-ic*, relating to or pertaining to): relating to or pertaining to a *polyphialide*. Cf. **monophialidic**.

polyphyletic (Gr. *polýs*, many + *phyletikós* < *phyletés*, of the same tribe, race or stirp < *phylé*, tribe + suf. *-tikós* > L. *-ticus* > E. *-tic*, relation, fitness): derived from more than one ancestral line, i.e., that includes taxa from another lineage. See **phylum**. Cf. **monophyletic** and **paraphyletic**.

polyphyllous (Gr. *polýs*, many + *phýllon*, leaf): with many leaves, multifoliate, like the thallus of certain lichens that have many foliose lobules, e.g., *Parmelia* (Lecanorales) and *Peltigera* (Peltigerales). Cf. **monophyllous**.

The **polyphyllous** thalli of *Peltigera* sp., x 1 (*TRO*).

polyplanetic (Gr. *polýs*, many + *planétes*, vagabond, wanderer + suf. *-ikos* > L. *-icus* > E. *-ic*, pertaining to or relating to): *Oomycota*. A species in whose life cycle the zoospores have several periods of swimming, alternating with phases of encystment, although the zoospores are of the same type, called secondary (reniform and biflagellate heteroconts), as in the gen. *Dictyuchus* (Saprolegniales). The condition of being polyplanetic is called *polyplanetism* (with E. suf. *-ism* < L. *-ismus* < Gr. *-ismós*, condition or state). Cf. **monoplanetic** and **diplanetic**.

polyploid (Gr. *polýs*, many + des. *-plóos*, as in *diplóos*, double + NL. suf. *-oide*, contraction of Gr. *-oeídes*, similar): an organism with more than two complete sets of chromosomes. Depending on the number of sets of chromosomes, they are called *triploid*, *tetraploid*, *pentaploid*, etc., to refer to three, four and five sets, etc. Polyploid organisms commonly have larger cells and nuclei than diploids. *Polyploidy* is the phenomenon regarding polyploid organisms.

polypore (L. *polyporus* < Gr. *polýs*, many + *póros*, pore, passage): the name applied to a member of several families of Aphyllophorales, e.g., Polyporaceae and Coriolaceae, that comprise gen. of cosmopolitan basidiomycetes, with sessile or stipitate, annual or perennial basidiomata, with a tubular (porose), dimitic, lignicolous or terrestrial hymenophore.

Fruiting body of *Fomitiporia hartigii*, a **polypore**, x 0.5 (*RV*).

Fruiting body of *Hexagonia tenuis*, a **polypore**, x 1 (*CB*).

polyporoid (< gen. *Polyporus* < L. *polyporus* < Gr. *polýs*, many + *póros*, pore, passage + L. suf. *-oide*, like < Gr. *-oeídes*, similar to): *Aphyllophorales*. Refers to a fruiting body, stipitate or not (sessile), with the hymenophore lining the interior of tubes that have pores at the distal ends, as in *Postia*, *Auriporia*, *Polyporus*, *Fomes* and other related gen.

Fruiting body of *Postia placenta*, a **polyporoid** fungus, x 1 (*RV*).

polysporous (Gr. *polýs*, many + *sporá*, spore): with many spores. Syn. of **pliosporous**.

polystichous, **polystic**, **polystichate** (Gr. *polýs*, many + *stíchos*, line, row, file): arranged in several rows. Equal to **multifarious**.

polytomous (Gr. *polýs*, many + *tómos*, part, portion, section, cut + L. suf. *-osus* > E. *-ous*, full, augmented, prone to): in the lichens, it is applied to thalli that have several branches originating at the same level as the podetium. See **polychotomous**.

polytretic (Gr. *polýs*, many + *tretós*, perforate + suf. *-íkos* > L. *-icus* > E. *-ic*, pertaining to or relating to): *Conidiogenesis*. A type of enteroblastic conidiogenous cell that produces conidia through pores or perforations in the outer wall, through which the inner wall protrudes, which will constitute the wall of the newly formed conidia; these conidia can be solitary or in chains, depending on the species; e.g., as in the gen. *Helminthosporium*, *Alternaria* and *Curvularia*, among other dematiaceous asexual fungi. This characteristic is clearly observed in *Ulocladium atrum*. Cf. **monotretic**.

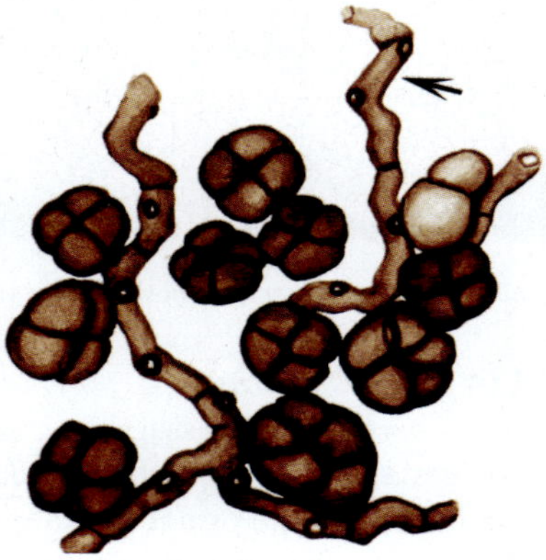

Polytretic conidiogenous cells of *Ulocladium atrum*, x 625 (*MU*).

polytrophic (Gr. *polýs*, many + *trophós*, what nourishes, serves as food): nourished by very diverse substances. Equal to **plurivorous**. Cf. **monotrophic** and **omnivorous**.

polyverticillate (Gr. *polýs*, many + NL. *verticillatus* < L. *verticillus*, verticillate + suf. *-atus* > E. *-ate*, provided with or likeness): *Conidial fungi*. Refers to a complex conidiophore that has branching at more than three levels, as is observed in several species of *Penicillium* (moniliaceous asexual fungi), such as *P. arenicola*, which can have four (*quadriverticillate*), five or even more points of branching in an irregular pattern; these types of conidiophores have phialides, metulae, ramae and ramillae. Cf. **monoverticillate**, **biverticillate**, **triverticillate** and **quadriverticillate**.

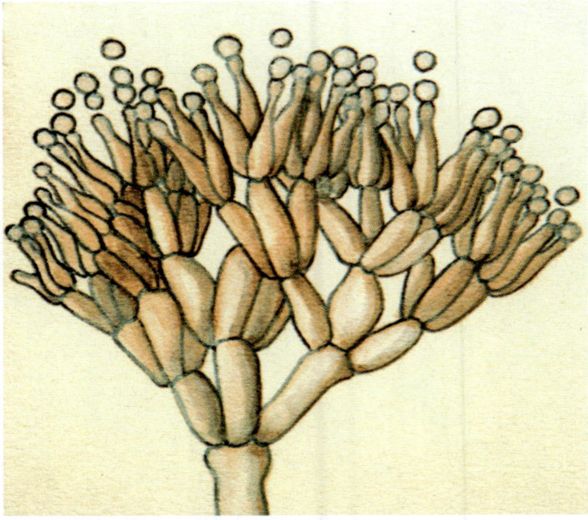

Polyverticillate conidiophore of *Penicillium arenicola*, x 1,000 (*MU*).

polyxenous (Gr. *polýs*, many + *xénos*, host): being able to live on diverse hosts. This phenomenon is called *polyxeny* (*xénos* > E. *-xeny*). Syn. of **pliophagous** and **plioxenous**.

pore (L. *porus* < Gr. *póros*, pore, passage): a small hole or opening in the wall of a cell. Pores are common in fungi; many spores germinate through a pore (*germ pore*) in the spore wall, as happens in the ascospores of *Podospora comata* and *Gelasinospora seminuda* (Sordariales); there are conidiophores that form conidia through one or more pores in the wall, as is observed in *Alternaria*, *Ulocladium* and other dematiaceous asexual fungi (see **monotretic** and **polytretic**). Also called pores are the mouths of the tubes that contain the hymenium of the Polyporaceae (such as *Pycnoporus sanguineus*, *Meruliporia* (=*Poria*) *incrassata*, *Dichomitus mexicanus* (=*Megasporoporia mexicana*), *Meripilus tropicalis* and *Laetiporus sulphureus*), and Boletaceae (such as *Boletus*), among many others.

Pores of the hymenophore of *Laetiporus sulphureus*, x 15 (*MU*).

poriform (L. *porus* < Gr. *póros*, pore, passage + suf. *-formis* < *forma*, shape): shaped like a pore; similar to a pore.

poroconidium, pl. **poroconidia** (Gr. *póros*, pore, passage + *kónis*, dust, powder + dim. suf. *-ídion*): equal to **porospore**. See **tretic**.

porogenous (Gr. *póros*, pore, passage + *génos*, origin < *gennáo*, to engender, produce): *Conidiogenesis*. A type of blastic conidium development in which the inner wall of the conidiogenous cell pushes out through a pore in the outer wall to form the conidium; common in many dematiaceous asexual fungi, such as *Alternaria*, *Helminthosporium* and *Ulocladium*. See **monotretic** and **polytretic**.

porospore (Gr. *póros*, pore, orifice + *sporá*, spore): *Conidiogenesis*. An asexual reproductive spore that originates through a pore in the wall of the conidiophore or of another porospore; a term applied to conidial fungi. *Alternaria*, *Bipolaris*, *Curvularia*, *Drechslera*, *Helminthosporium*, *Stemphylium* and *Ulocladium* (dematiaceous asexual fungi) are examples of gen. of fungi that produce porospores. Equal to **poroconidium**.

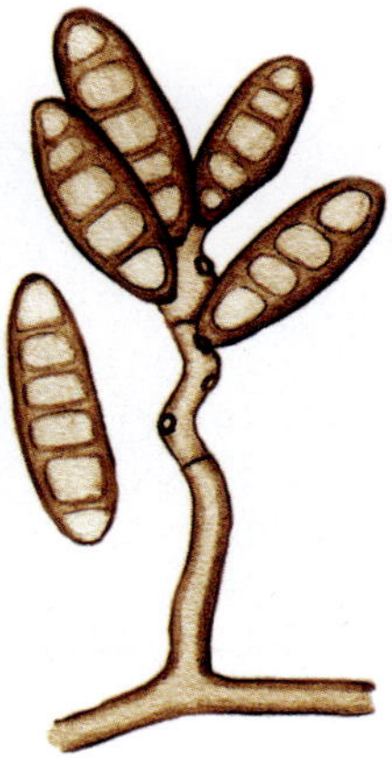

Porospores of *Drechslera* (asexual state of *Cochliobolus bicolor*), x 700 (*MU*).

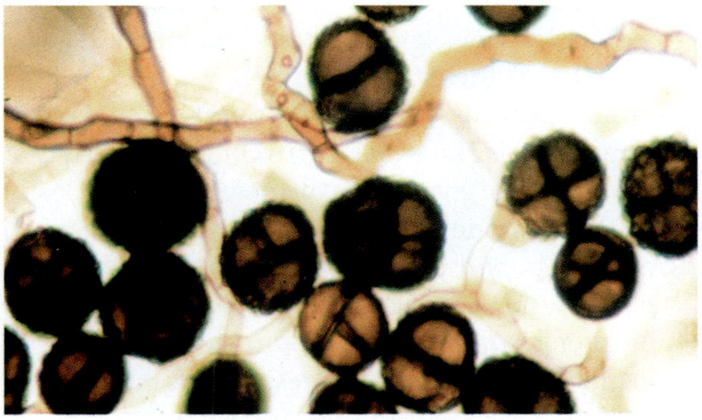

Porospores of *Stemphylium* sp., x 650 (*CB*).

porous (L. *porus*, pore < Gr. *póros*, pore, passage + L. suf. *-osus* > E. *-ous*, having, possessing the qualities of): provided with pores, spongy, like the hymenophore of the Polyporaceae, e.g., *Hexagonia*, *Dichomitus*, *Gloeoporus*, *Polyporus*, *Pycnoporus*, *Postia*, *Schizopora* (Aphyllophorales), and Boletaceae, such as *Boletus* and *Strobilomyces* (Agaricales).

Porous hymenophore of the fruiting body of *Boletus* sp., x 0.5 (*RV*).

Porous hymenophore of *Schizopora paradoxa*, growing on bark, x 1 (*RV*).

Porous hymenophore of the fruiting body of *Hexagonia tenuis*, x 0.5 (*RV*).

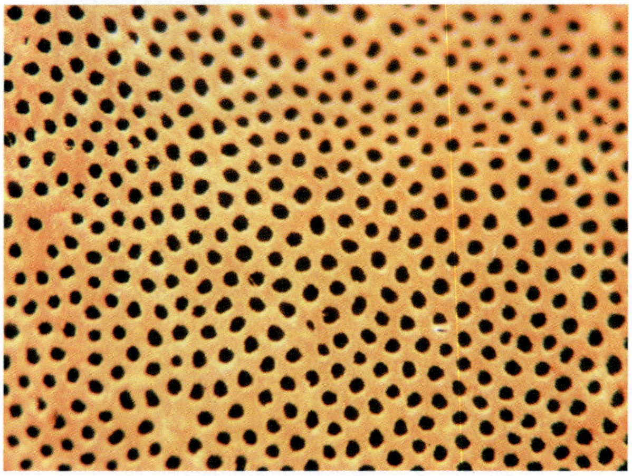

Porous hymenophore of the fruiting body of *Hexagonia* sp., x 15 (*MU*).

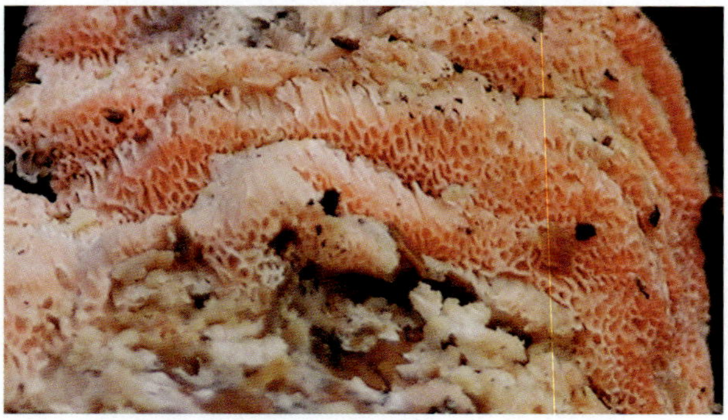

Porous hymenophore of the fruiting body of *Postia placenta*, x 1 (*RV*).

Porous hymenophore of the fruiting body of *Polyporus alveolaris*, x 1 (*RV*).

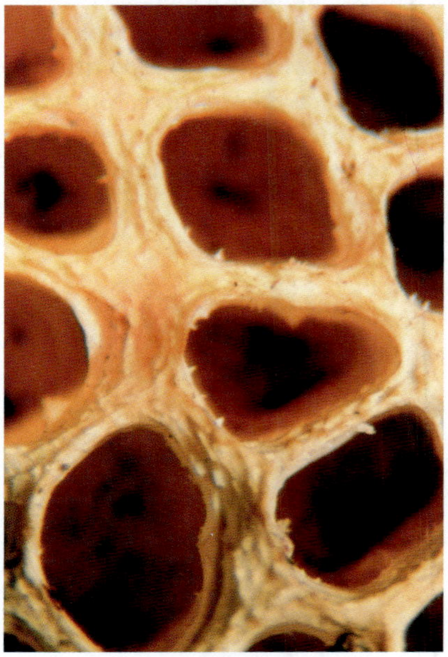

Porous hymenophore of *Dichomitus mexicanus* (=*Megasporoporia mexicana*), x 15 (*MU*).

posterior (L. *posterior*, in the rear part or oriented in that direction): in reference to the hymenial gills of the agaricaceous fungi, it refers to the part next to the stipe. Cf. **anterior**.

potamo-, potam- (Gr. *potamós*, river): a pref. used in names such as *Potamomyces* (river fungus), an aquatic ascomycete of uncertain taxonomic position, recently classified within the subclass Dothideomycetidae.

powdery (L. *pulverulentus*, powdery < *pulver*, genit. *pulveris*, powder, dust + suf. *-lentus* > E. *-lent*, full of, prone to): covered with fine powder, which in the fungi represents the spores, such as the powdery mildews (species of Erysiphales that live as obligate plant parasites). The mycelium of other fungi also has this characteristic, e.g., that of the colonies of *Microsporum nanum* and *Beauveria bassiana* (moniliaceous asexual fungi), which is due to the abundant production of dry conidia.

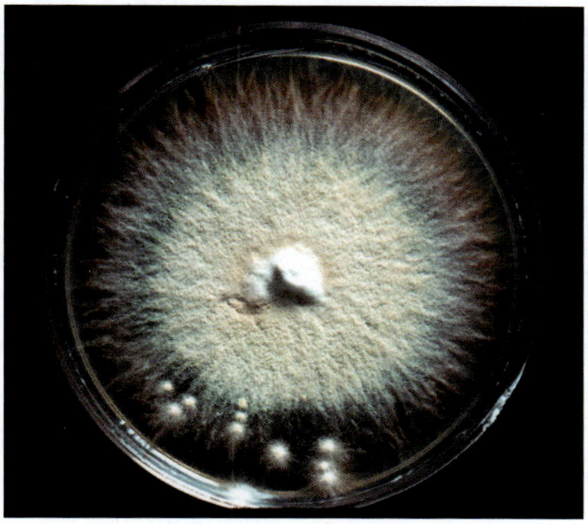

Powdery colony of *Microsporum nanum* on agar, x 8 (*MU*).

Powdery colony of *Beauveria bassiana* on agar, x 8 (*MU*).

pozol (Azt. *pozolli*, foamy): a fermented maize dough of Mayan origin that is used in the preparation of a staple, non-alcoholic beverage that is consumed by various ethnic groups of Southern Mexico. The microbiota involved in the fermentation of the dough is complex; it is composed of lactic bacteria (*Lactobacillus*), yeasts (*Candida* and *Trichosporon*, of the cryptococcaceous asexual yeasts), and molds (*Mucor*, of the Mucorales, *Cladosporium*, of the dematiaceous asexual fungi, and *Geotrichum*, of the moniliaceous asexual fungi, among others). Cf. **ontjom**, **tapé ketella** and **tempe**. See **fermentation**.

Balls of **pozol** from Tabasco, Mexico; the balls of fermented maize dough are wrapped in banana leaves, x 0.2 (*MU*).

praemorse (L. *praemorsus*, bitten, cut < *praemordere*, to bite something at the tip): more or less irregularly truncate; in general it is applied to the base of the stipe when it appears as if it had been bitten or broken off, as occurs in various species of *Lactarius* (Agaricales).

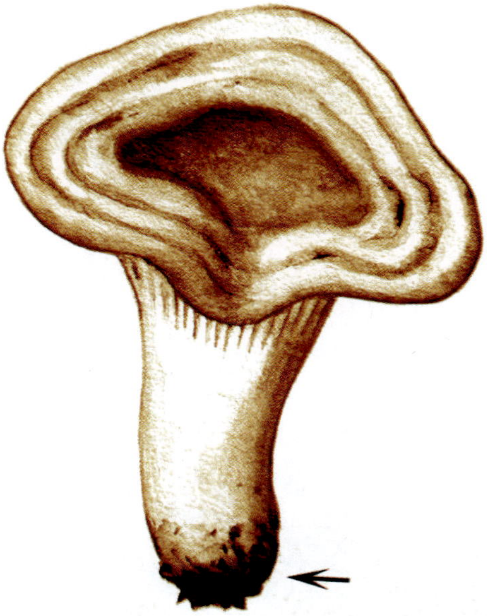

Praemorse stipe of the basidiocarp of *Lactarius zonarius*, x 0.7 (*MU*).

pre- (L. *prae*, a prep. that denotes preference, priority or ernestness): an inseparable prep. employed in the formation of terms such as *predisposition*, previous disposition or natural tendency relative to a certain susceptibility; e.g., the condition of a host that makes it be more easily affected than another by a particular disease or parasite.

predaceous (**depredaceous**) or **predacious fungus**, pl. **fungi** (L. *depraedator*, that which preys on, plunders and destroys < *de*, pref. of compound words + *praedari*, to prey on; rob, plunder violently and destructively < *praeda*, prey + *-aceus* > E. *-aceous*, of or pertaining to; L. *fungus*, fungus < Gr. *spóngos*, *sphóngos*, sponge): living by preying on amoebae, nematodes, rotifers, and other small terrestrial or aquatic animals. The most important predatory fungi belong to the Zoopagales (*Zoophagus*, *Acaulopage*, *Zoopage*, *Cochlonema*, etc.), to the moniliaceous asexual fungi (*Arthrobotrys*, *Dactylella*, *Harposporium*, *Rotiferophtora*, *Verticillium*, etc.), and to the oomycotes of the order Saprolegniales (*Sommerstorffia*), although there are others, such as some basidiomycetes, e.g., *Pleurotus ostreatus* (Agaricales). See **nematophagous**.

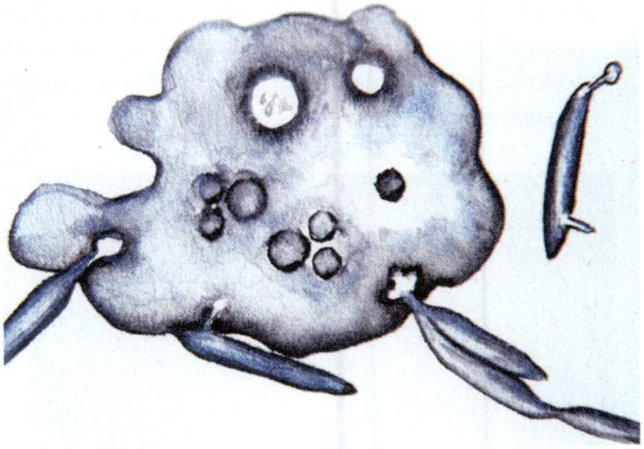

Amoebophilus sicyosporus, a **predaceous fungus** of terricolous amoebae; shown are the haustoria of the fungus inside an amoeba (*Amoeba terricola*), and the adhered conidia outside, x 900 (*MU*).

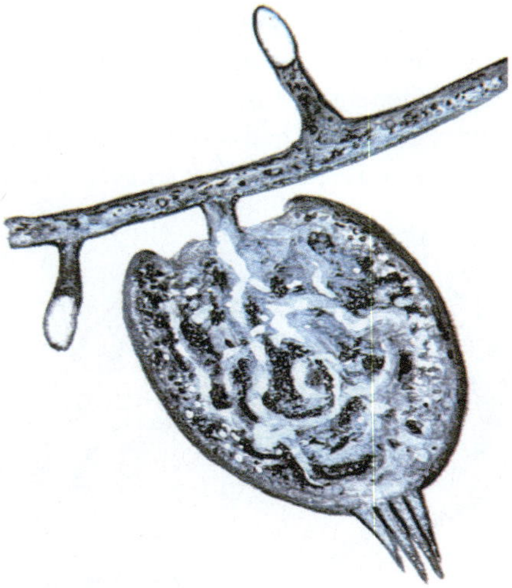

Zoophagus insidians, a **predaceous fungus** of rotifers; shown are the specialized adhesive hyphal rami, x 180 (*MU*).

Sommerstorffia spinosa, a **predaceous fungus** parasitizing a rotifer; shown are the hyphae of the oomycote inside, and external adhesive hypha and a sporangium, x 150 (*MU*).

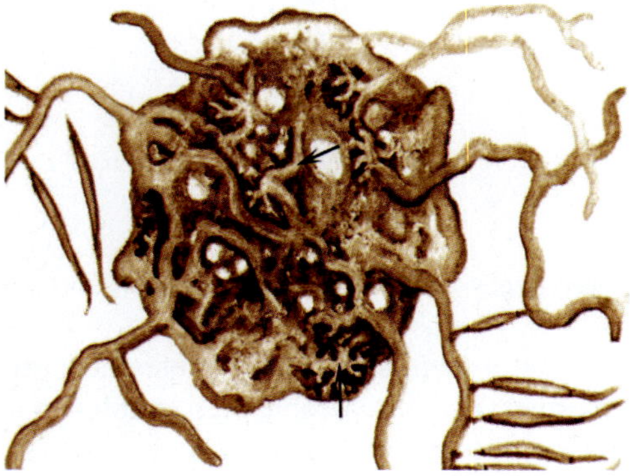

Acaulopage marantica, a **predaceous fungus** of terricolous amoebae; shown are the hyphae and the haustoria of the fungus inside an amoeba (*Amoeba terricola*), and external hyphae with sessile or subsessile conidia, x 470 (*MU*).

preservation (ME. *preservation* < MF. < ML. *praeservation*, action and effect of preserving < L. *praeservare*, to watch over, to guard against damage or danger): to maintain something alive, conserving its quality and condition, preventing its deterioration. In reference collections and genetic depositories, pure, living cultures of fungi and other microorganisms are preserved free from contamination and with minimum risk of undergoing mutations. There are various methods of preserving microbial cultures, but the most utilized for their efficiency are lyophilization and ultrafreezing.

presporangial papilla, pl. **papillae** (L. *prae*, prep. that denotes precedence, priority + Gr. *sporá*, spore + *angeîon* + L. suf. *-alis* > E. *-al*, relating to or pertaining to; L. *papilla*, nipple, process): refers to each of the masses of cytoplasm that differentiate to form the primordia of the sporangia during the type of plasmodium development known as *subhypothallic,* present in many Myxomycetes, as occurs in species of the orders Liceales, Trichiales, Echinosteliales and Physarales. The presporangial papillae or sporangial primordia are clearly observed, e.g., during the transformation of the plasmodium of *Physarum polycephalum* (Physarales).

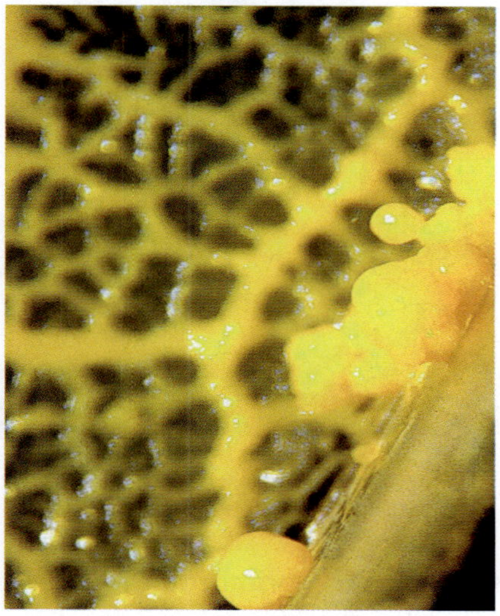

Presporangial papillae in a plasmodium of *Physarum polycephalum*, on agar, x 6 (*MU*).

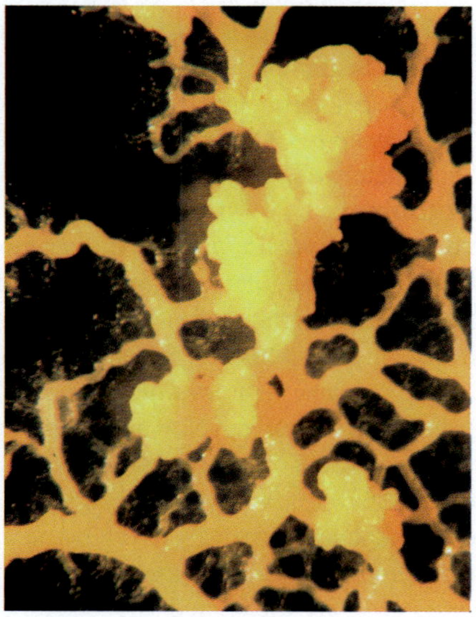

Presporangial papillae in a plasmodium of *Physarum polycephalum*, on agar, x 6 (*MU*).

prespore, **prespore cell** (L. *prae*, prep. that denotes precedence, priority + Gr. *sporá*, spore; NL. *cellula*, compartment, living cell, dim. of *cella*, cell of a honeycomb): a unicellular, amoeboid protoplast that gives rise to a sporocarp; characteristic of the gen. *Protosporangium* (Protosteliales). Also called prespores or prespore cells are the posterior cells of the pseudoplasmodium of species of Dictyosteliomycota, which are destined to become the spores of the sorocarp; they have distinctive vacuoles that contain polysaccharides. Cf. **prestalk cell**.

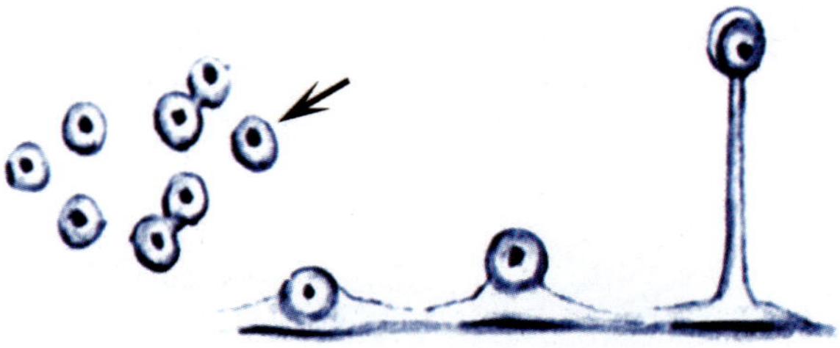

Prespores of *Protosporangium bisporum*, x 250 (*MU*).

prestalk cell (L. *prae*, prep. that denotes in advance, priority + ME. *stalke*, stem; NL. *cellula*, compartment, living cell, dim. of *cella*, cell of a honeycomb): each one of the cells of the anterior part of the pseudoplasmodium of species of Dictyosteliomycota that are destined to form the pedicel of the sorocarp. The *prespore* or *presporic cells* are found in the posterior part of the pseudoplasmodium.

primary (L. *primarius*, principal, first in order or grade): the first or principal thing. *Primary mycelium*: the haploid mycelium of the basidiomycetes that results from the germination of the basidiospores. See **secondary mycelium** and **tertiary mycelium**. *Primary thallus*: the first squamules of the thallus of *Cladonia* (Lecanorales) from which the podetia (with apothecia) originate that constitute the *secondary thallus*. *Primary universal veil* or *primordial veil*: see **protoblem**.

primary appendage (L. *primarius*, principal, first in order or grade; *appendage*, appendix, supplement): *Laboulbeniales*. An outgrowth of the upper cell of the ascospore. In some species of *Laboulbenia*, e.g., the primary appendage or its lateral branches are transformed into antheridial branches that produce phialides and spermatia; in others, these primary appendages are sterile. Cf. **secondary appendage**.

primary host (L. *primarius*, principal, first in order; L. *hospes*, genit. *hospitis*, a living organism that provides food and lodging to a parasite; one who receives a stranger as guest or visitor): the host plant on which a parasitic fungus develops its sexual state, and which almost always represents the plant of major economic importance. E.g., wheat is the primary host of *Puccinia graminis tritici* (Uredinales), the etiological agent of wheat rust. Cf. **alternate host**, **intermediate host** and **bridging host**.

primary hypha (L. *primarius*, principal or first in order or grade; NL. *hypha*, hypha < Gr. *hyphé*, tissue, spiderweb; hypha): the first type of hypha that results from the germination of a spore; it is a somatic hypha, slender and little differentiated, like that which is formed in a young colony of *Malbranchea aurantiaca* (moniliaceous asexual fungi). In this fungus, secondary hyphae, more differentiated, which produce catenulate conidia, develop from the primary hyphae. See **secondary hypha**.

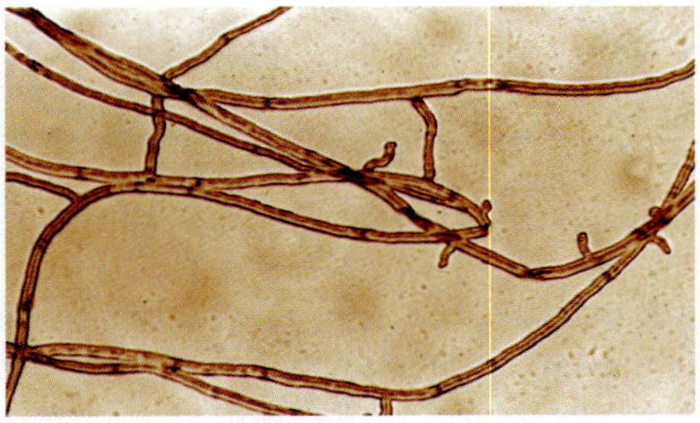

Primary hyphae of *Malbranchea aurantiaca*, x 500 (*MU*).

primary marine fungus, pl. **fungi** (L. *primarius*, principal, first in order or grade; L. *marinus*, of the sea < *mare*, sea + suf. *-inus* > E. *-ine*, of or pertaining to; L. *fungus*, fungus < Gr. *spóngos*, *sphóngos*, sponge): a species derived from a marine ancestor and which has remained in the marine environment, such as *Spathulospora* (Spathulosporales). Cf. **secondary marine fungus**.

primary mycelium, pl. **mycelia** (L. *primarius*, principal, first in order or grade; NL. *mycelium*, mycelium < Gr. *mýkes*, fungus + L. suf. *-elis*, concerning + dim. suf. *-ium*): *Basidiomycetes*. The haploid mycelium (n) that results from the germination of a basidiospore, which, after plasmogamy, gives rise to the *secondary mycelium*.

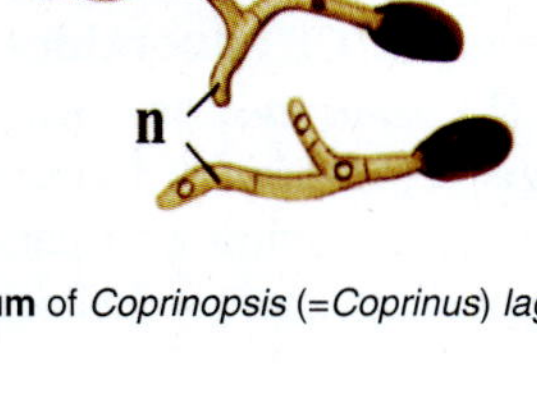

Primary mycelium of *Coprinopsis* (=*Coprinus*) *lagopus*, x 280 (*MU*).

primary septum, pl. **septa** (L. *primarius*, principal, first in order or grade; L. *septum*, septum, barrier, partition): see **septum**.

primary thallus, pl. **thalli** (L. *primarius*, principal, first in order or grade; L. *thallus*, thallus, shoot, renewal): *Lichens*. The first squamules of the thallus, on which the podetia develop. See **secondary thallus**.

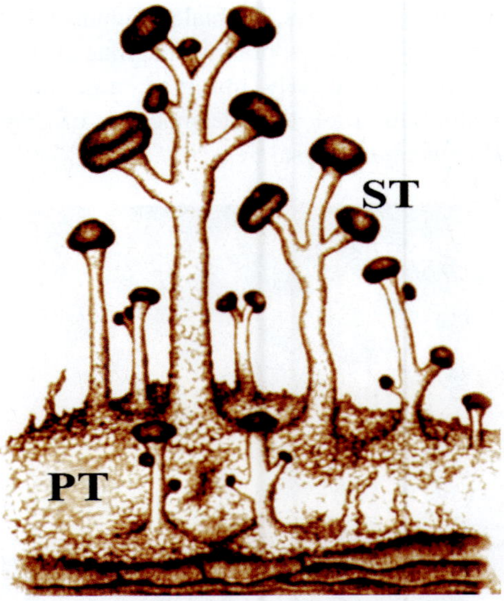

Primary thalli (PT) and **secondary thalli** (ST) of the lichen *Cladonia pyxidata*, x 10 (*MU*).

primary universal or **primordial veil** (L. *primarius*, first in order or importance; L. *universalis*, universal, general < *universus*, universe; L. *primordium*, first, beginning, origin + suf. *-alis* > E. *-al*, relating to or pertaining to; L. *velum*, veil): see **protoblem**.

primordial cuticle, **primordial covering** (L. *primordium*, beginning, origin + suf. *-alis* > E. *-al*, relating to or belonging to; L. *cuticula*, dim. of *cutis*, skin; ME. *coveren* < OF. *couvrir* < L. *cooperire*, to cover completely + E. suf. *-ing*, the action of the verb): see **protoblem.**

primitive (L. *primitivus*, primitive): primogenital, primordial, principal, first in its line, or that does not originate from something else. Cf. **derived**.

primordium, pl. **primordia** (L. *primordium*, beginning, start, origin < *primus*, first + *ordior*, to begin + dim. suf. *-ium*): the initial stage of development of any structure; e.g., the primordia of the cultivated mushrooms (buttons), *Agaricus brunnescens* (Agaricales). The primordia of the sorocarp of *Dictyostelium mucoroides* (Dictyosteliomycota), the ascocarps of *Peziza quelepidotia* (Pezizales), and the basidiocarps of *Coprinus* and *Pleurotus ostreatus* (Agaricales) are some examples illustrated.

Primordium of a sorocarp
of *Dictyostelium mucoroides* on agar, x 100 (*MU*).

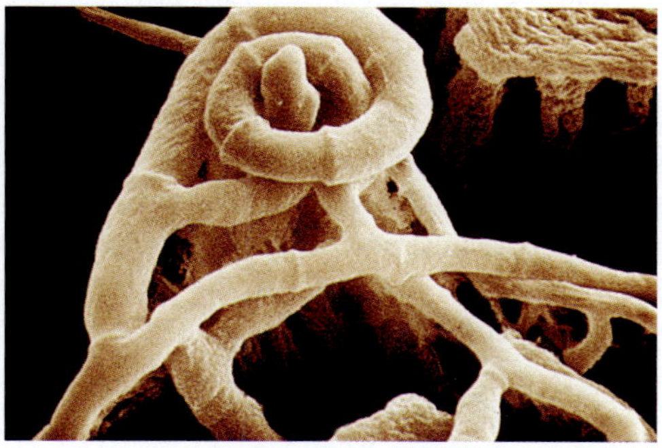

Primordia, the initial stages of the ascocarps of *Peziza quelepidotia*, seen with the scanning electron microscope. The antheridium is coiled around the ascogonium, x 1,000 (*KO*).

Primordia of two basidiocarps of *Coprinus* sp., x 6 (*MU*).

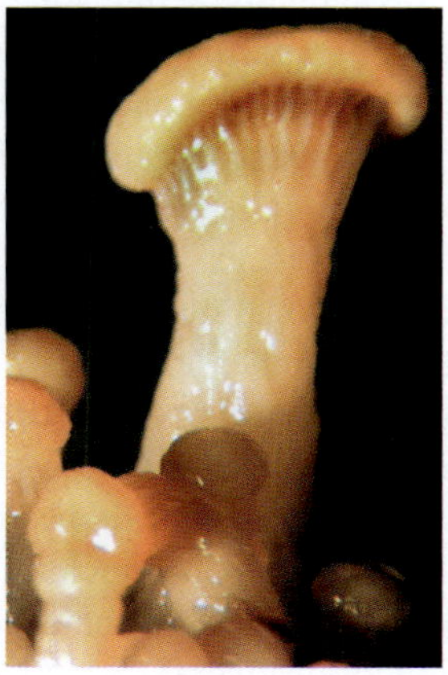

Primordia of several basidiocarps of *Pleurotus ostreatus*, x 5 (*MU*).

priority (ME. < MF. *prioite* < L. *prior*, genit. *prioratis*, previous, the first + suf. *-itas* > E. *-ity*, quality, state): precedence of one thing with respect to another, in time or in order. By the *priority of nature* is meant the precedence of one organism to another with regard to their first appearance, even though they exist at the same time. In phylogenetic systematics the algae must preceed the fungi in this type of priority. In botanical nomenclature (which includes the fungi) there exists the *principle of priority*. Principle IV states "Each taxonomic group with a particular circumscription, position, and rank can bear only one correct name, the earliest that is in accordance with the Rules, except in specified cases". Articles 11-15 of the International Code of Botanical Nomenclature deal with priority and conservation of names. For the establishment of priority in mycological nomenclature, the following dates have been agreed upon: Fungi, including the lichens, and Myxomycetes, May 1, 1753 (C. Linnaeus, *Species Plantarum*, ed. 1, vols. 1 & 2); Uredinales, Ustilaginales and Gasteromycetes, Dec. 31, 1801 (C.H. Persoon, *Synopsis Methodica Fungorum*); the remaining fungi (Fungi Caeteri), Jan. 1, 1821 (E.M. Fries, *Systema Mycologicum*, vols. 1-3 and *Elenchus Fungorum*, 1832, vols. 1-2). Because of the overriding importance of the early taxonomic works of Persoon and Fries, the names in these works are "sanctioned", i.e., they are treated as conserved and they have priority over earlier homonyms and synonyms. The names given to lichens apply only to the mycobiont. The names of anamorphs of pleomorphic fungi do not affect the nomenclatural status of the names of correlated holomorphs, irrespective of priority. The Code recommends that the names of all taxa of all ranks be written in italics.

pristine (L. *pristinus*, old, first, primitive, original < *priscus*, of or pertaining to times past + suf. *-inus* > E. *-ine*, of or pertaining to): primitive, first, old, original.

pro- (Gr. *pró*, forward, ahead, previously, toward): a pref. employed in the formation of terms that allude to something that is anticipated or that goes before something else, as in the following terms.

probasidium, pl. **probasidia** (Gr. *pró*, before, previous + *basídion*, dim. of *básis*, base): *Basidiomycetes*. Refers to a young basidial cell, or one in a primordial stage, from karyogamy and meiosis to the initiation of the external formation of the sterigmata and spores from its wall (generally four), as occurs in several gen., including *Septobasidium* (Septobasidiales), *Coprinus*, *Agaricus*, *Lepiota*, *Hygrophorus* and *Amanita* (Agaricales). Cf. **protobasidium**. See **epibasidium**.

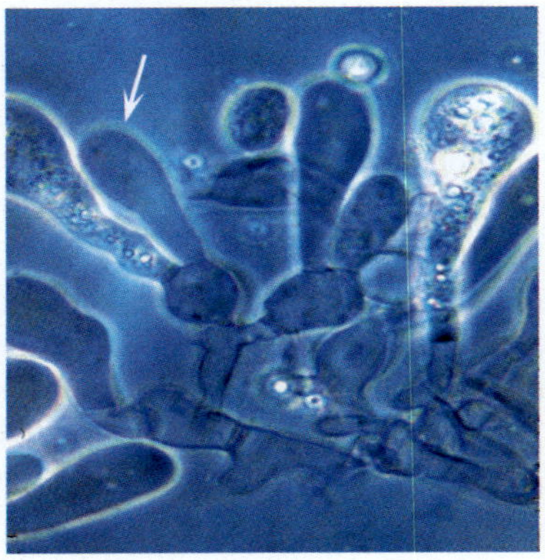

Probasidia of the hymenium of *Amanita* sp., x 560 (*RMA*).

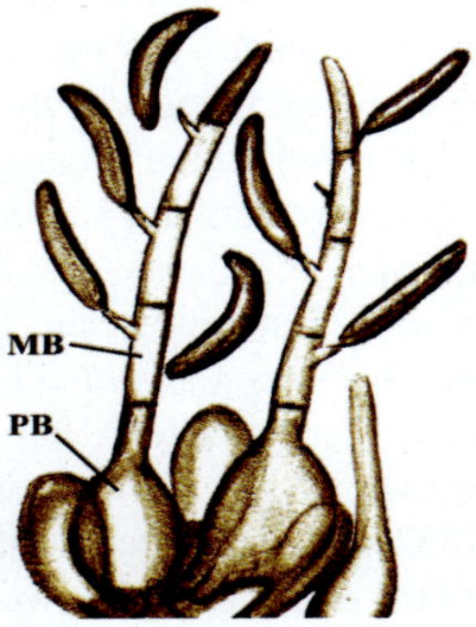

Probasidia (PB) and metabasidia (MB) with basidiospores of *Septobasidium burtii*, developed in symbiosis with the scale insect *Aspidiotus osborni*, which parasitizes branches of oak, x 490 (*MU*).

probiotic (Gr. *pró*, in favor of, for + *bíos*, life + suf. *-tikós* > L. *-ticus* > E. *-tic*, relation, fitness, inclination or capacity): a culture of bacteria and/or yeasts, traditional or industrialized, that is consumed to promote the development of a healthy intestinal biota, as, e.g., *yogurt* and other fermented dairy products, the *tea fungus*, *tibi grains* and *bulgaro grains*.

procumbent (L. *procumbens*, genit. *procumbentis*, prostrate, spreading + suf. *-entem* > E. *-ent*, being): prostrate, spreading on the substrate.

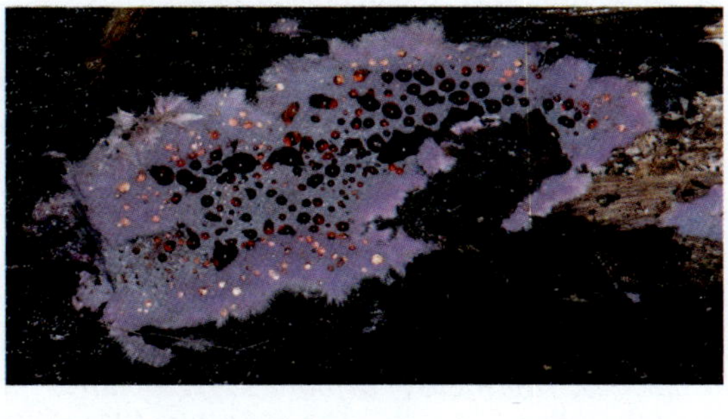

Procumbent fruiting body of *Phanerochaete* sp. (Aphyllophorales), on dead wood, x 0.5 (*TRO*).

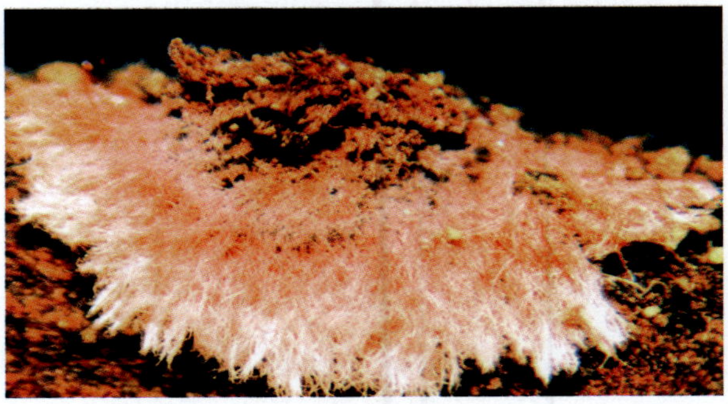

Procumbent fruiting body of *Chondrostereum* (=*Stereum*) *purpureum* (Aphyllophorales), on dead wood, x 0.5 (*MU*).

progametangium, pl. **progametangia** (Gr. *pró*, before, previous + *gamétes*, spouse + *angeîon*, glass, container): *Mucorales*. A small lateral hyphal branch that differentiates first into the gametangium and afterwards into a suspensor of the zygosporangium, as happens in *Rhizopus stolonifer* (=*Rh. nigricans*).

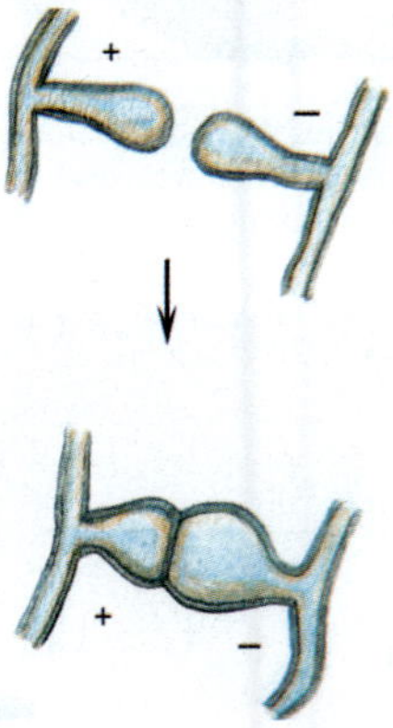

Progametangia of *Rhizopus stolonifer* (=*Rh. nigricans*); the + and - zygophores grow toward one another and when they make contact their apices swell to form progametangia, x 100 (*MU*).

progamones (Gr. *pró*, before, previous + *gam-* < *gámos*, matrimony, sexual union + des. *-ones* < hormones < Gr. *hormõn*, active participle of *hormáo*, to excite, move): a group of sex hormones in the Zygomycetes.

prohybrid (Gr. *pró*, before, earlier + L. *hybrida*, the offspring of the pig and wild boar < *hýbris*, insult; progeny of distinct races, hybrid): mycelium whose cells contain additional nuclei, as a consequence of having united with other hyphae.

prokaryote, **procaryote**, **prokaryont**, **procaryont** (Gr. *pró*, primitive, before, previous + *káryon*, nucleus + *-ote* < *-otes*, NL. suf. that indicates possession): an organism in which the nuclear material lacks an enveloping membrane, without mitosis, like the bacteria (kingdom Monera). See **prokaryotic**. Cf. **eukaryote**.

prokaryotic, **procaryotic**, **prokaryontic**, **procaryontic** (Gr. *pró*, primitive, before, previous + *káryon*, nucleus + suf. *-tikós* > L. *-ticus* > E. *-tic*, relation, fitness, inclination or capacity): an organism whose cells do not possess a true nucleus (i.e., the chromosomes are not surrounded by a membrane) nor the organelles of eukaryotic cells (mitochondria, endoplasmic reticulum, dictyosomes and vacuoles, among other characteristics), because of which they have a type of organization less complex than that of a eukaryotic organism. Only the organisms of the kingdom Monera are prokaryotic; those of the kingdoms Protoctista (or Protista for some), Fungi, Plantae and Animalia are *eukaryotic*. According to the classification adopted in this dictionary, the eukaryotes also includes organisms considered as the kingdom Stramenopila (which comprises the phyla Oomycota, Hyphochytriomycota and Labyrinthulomycota) that have been separated from the kingdom Fungi. Cf. **eukaryotic**.

Prokaryota (Gr. *pró*, before, earlier + *káryon*, nut, nucleus + NL. suf. *-ota*, which denotes possession; this suf. also is used as an ending of the name of a group of organisms): a taxonomic category above the kingdom, in this case, the *superkingdom* (although at present the term *dominion* is utilized more) to which belong all of the prokaryotic organisms, which are distinguished by not having true nuclei in their cells, i.e., that are not delimited by a perforated nuclear membrane, they lack pairs of chromosomes with DNA, and complex cellular organelles (e.g., mitochondria and Golgi bodies). In this superkingdom are included the bacteria and viruses. Cf. **Eukaryota**.

prolate (L. *prolatus*, elongated or prolonged forward): elongated or extended in the direction of the main axis; applied to spores that are elongate or ellipsoidal in shape, whose length:width ratio is between 6:1 and 3:1, like the ascospores of *Coniochaeta scatigena* (=*Ascobolus scatigenus*), Pezizales. Cf. **oblate**.

proliferation (Fr. *prolifération* < L. *prolifer*, *prolificus*, having the capacity to engender, proliferate < *prole*, progeny + suf. *-fer*, *-ferous*, to bear, produce + L. suf. *-ationem* > E. *-ation*, action, state or condition, result): the action and effect of proliferating; successive development of new parts. In the saprolegniaceous Oomycota (such as *Saprolegnia parasitica*), intrasporangial proliferation is the formation of a new sporangium within an old sporangium, following the liberation of the zoospores, from the septum that delimits the sporangium from the supporting hypha. A similar proliferation occurs in the asci of *Ascoidea* (Saccharomycetales). There also are proliferating conidiogenous cells, which are called *percurrent*.

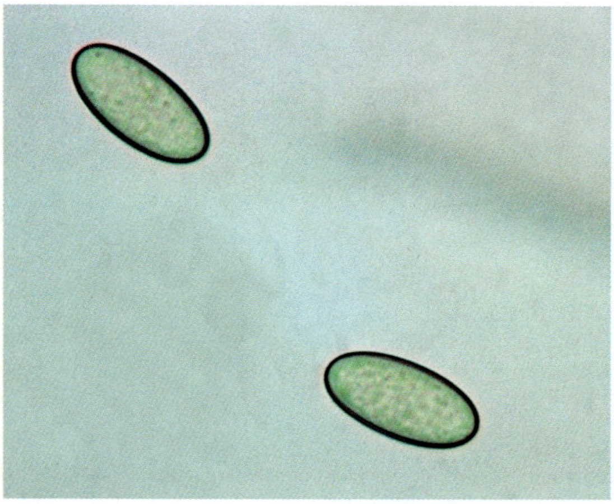

Prolate ascospores of *Coniochaeta scatigena* (=*Ascobolus scatigenus*), x 1,160 (*EAA*).

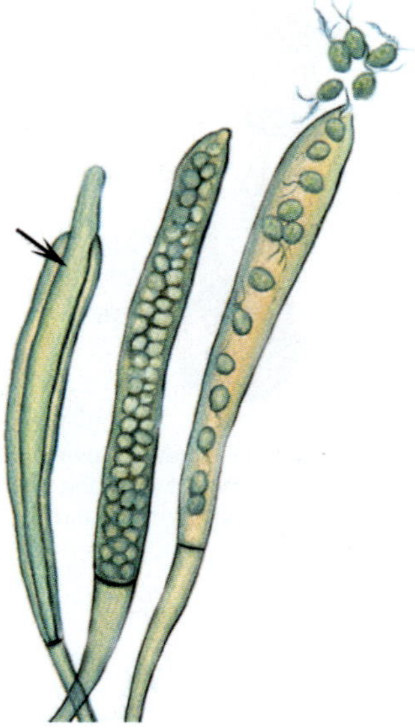

Intrasporangial **proliferation** of *Saprolegnia* sp., x 150 (*MU*).

promitosis, pl. **promitoses** (Gr. *pró*, before, previous + *mítos*, filament + suf. *-osis*, condition or state): see **cruciform division**.

promycelium, pl. **promycelia** (Gr. *pró*, before, previous + NL. *mycelium*, mycelium < Gr. *mýkes*, fungus + L. suf. *-elis*, concerning + L. dim. suf. *-ium*): the portion of the basidium that results from the germination of the teliospore, in which karyogamy and meiosis occur, before the formation of the basidiospores, as happens in the rusts (e.g., *Puccinia graminis tritici*, of the Uredinales), the smuts (such as *Ustilago maydis*, of the Ustilaginales), and other basidiomycetes (e.g., *Aessosporon salmonicolor* and *Rhodosporidium sphaerocarpum*, of the Sporidiales).

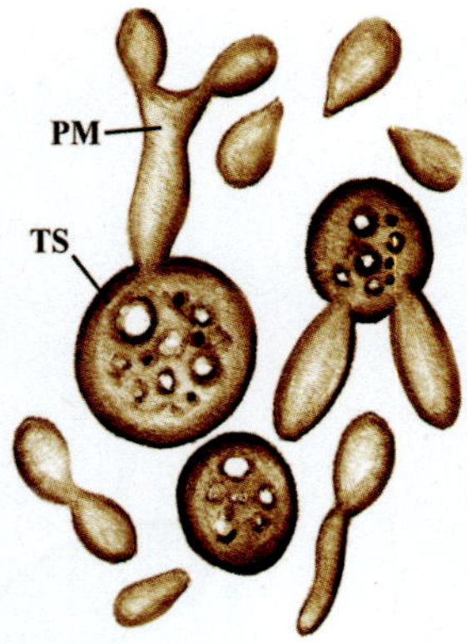

Promycelium (PM) with basidiospores of *Aessosporon salmonicolor*, resulting from the germination of teliospores (TS), x 1,600 (*MU*).

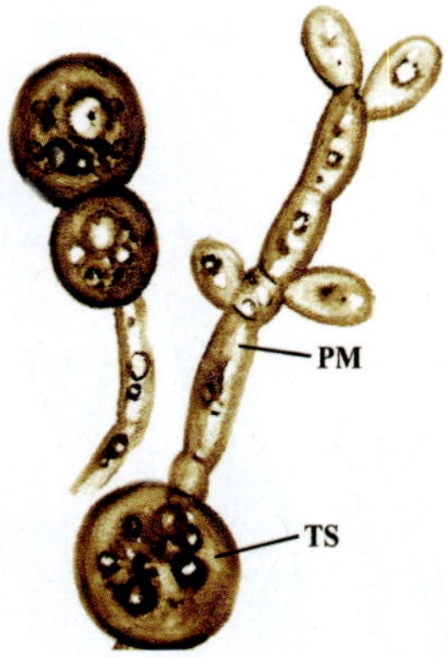

Promycelium (PM), also called metabasidium, with basidiospores of *Rhodosporidium sphaerocarpum*, resulting from the germination of teliospores (TS), x 1,360 (*MU*).

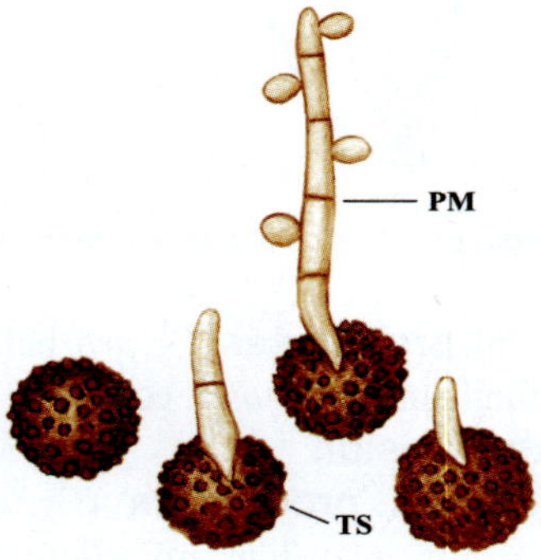

Promycelium (PM) growing from germinating teliospores (TS) of *Ustilago maydis*, x 1,150 (*MU*).

propagule, **propagulum**, pl. **propagula** (NL. *propagulum* < *propages*, to loosen, to slip off, to break off + dim. suf. *-ulum* > E. *-ule*): in general, any structure that serves to vegetatively propagate or multiply an organism; in the fungi, the propagules can be spores, bulbils, fragments of mycelium, isidia and soredia, among others.

prophialide (Gr. *pró*, before, previous + *phialís*, genit. *phialídis*, dim. of *phiále*, bottle, flask): see **metula**.

prophylactic (Gr. *prophylaktikós* < *pró*, toward + *phylaktikós*, vigilant, cautious, protection < *phýlax*, genit. *phylactós*, protection + suf. *-íkos* > L. *-icus* > E. *-ic*, pertaining to or relating to): concerning or relating to *prophylaxis*.

prophylaxis, pl. **prophylaxes** (Gr. *prophýlaxis*, precaution, prevention < *pró*, toward + *phýlax*, protection + suf. *-sis*, act of): measures or treatments designed to prevent diseases.

prosenchyma, pl. **prosenchymata** (Gr. *prós*, at the side of, beside, near to + *énchyma*, stuffed): a type of plectenchyma in which the hyphae that compose it are arranged parallel to one another, intertwining or anastomosing, but retaining their individuality. E.g. in the pedicel of the perithecial stroma of *Claviceps purpurea* (Hypocreales) one can clearly observe the prosenchyma. Cf. **pseudoparenchyma**.

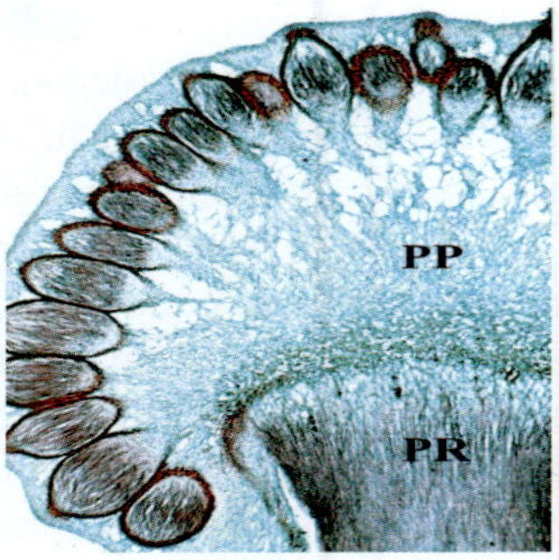

Prosenchyma (PR) and pseudoparenchyma (PP) of the perithecial stroma of *Claviceps purpurea*, x 25 (*RTH*).

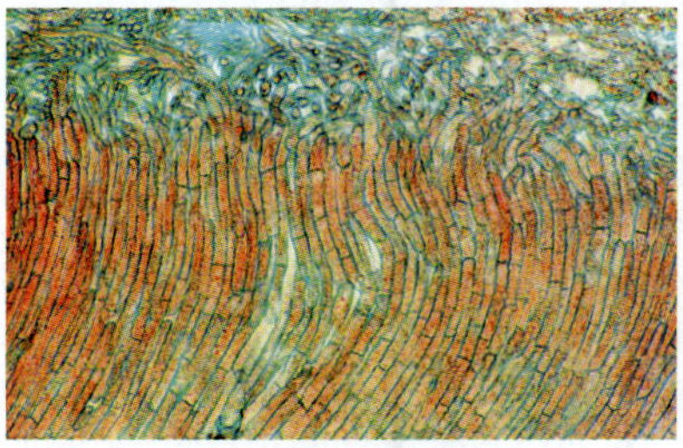

Prosenchyma of the perithecial stroma of *Claviceps purpurea*, x 130 (*CB*).

prosoplectenchyma, pl. **prosoplectenchymata** (Gr. *prós*, at the side of, beside, near to + *plektós*, intertwined, interwoven + *énchyma*, stuffed): *Lichens*. A type of tissue found in the thallus cortex of many macrolichens, in which the hyphae are oriented in a particular direction; it corresponds to the prosenchyma of other fungi. Cf. **chalaroplectenchyma** and **paraplectenchyma**.

prosorus, pl. **prosori** (Gr. *pró*, before, previous + *sorós*, pile, heap): a cell, derived from an infective zoospore, that eventually germinates and divides to give rise to a sorus or group of sporangia, as happens in *Synchytrium endobioticum* (Chytridiales), an endoparasite of potato tubers.

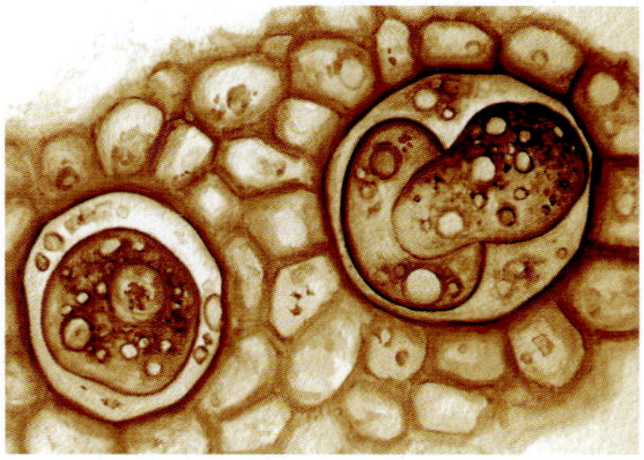

Prosori (one germinating) of *Synchytrium endobioticum* parasitizing cells of a potato tuber, seen in transverse section, x 510 (*MU*).

prosporangium, pl. **prosporangia**, **presporangium**, pl. **presporangia** (Gr. *pró*, before, previous + *sporá*, spore + *angeîon*, glass, receptacle): *Aquatic fungi*. A structure similar to a sporangium, from which is produced a vesicle (the sporangium proper) in which the zoospores develop and from which they are liberated, as is observed in *Chytriomyces* (Chytridiales). The prosporangium also is present in various gen. of aquatic fungi whose resting sporangia form a vesicle on germinating, as happens in *Rhizidium* and *Rhizophlyctis* (Chytridiales). Equivalent to **sporangial vesicle** or **plasmamembranic vesicle**.

Prosporangium of *Chytriomyces hyalinus*, x 900 (*MU*).

proteophilous (Gr. *proteío*, the first + *phílos*, have an affinity for): in the fungi, it refers to those that can prosper in soils rich in ammonia, like those affected by urea, e,g. *Panaeolus semiovatus* (Agaricales).

Proteophilous basidiocarps of *Panaeolus semiovatus* growing from horse manure, x 0.3 (*RV*).

proterospore (Gr. *próteros*, the first, the most advanced): the spore formed at the beginning of sporulation in *Ganoderma* (Aphyllophorales), capable of germinating readily without needing to pass through the digestive tract of fly larvae.

prothallus, pl. **prothalli** (Gr. *pró*, before, previous + *thallós* > L. *thallus*, thallus): *Lichens*. The zone free of algae (containing only the mycobiont), white, brown or black in color, found between the areoles and the growing margins of the thallus of crustaceous lichens.

Protista (L. *protista*, pl. of *protistum* < Gr. *prótiston*, neuter form of *prótistos*, superlative of *prõtos*, first, foremost; i.e., the very first): a kingdom proposed by Haeckel in 1866, to include the majority of the primitive unicellular organisms, or *protists* (ambiguous eukaryotic organisms that do not fit clearly in either the Plantae or Animalia), among them the protozoans and various groups of lower algae and fungi. In the system of classification adopted in this dictionary (Alexopoulos *et al*., 1996), the phyla Plasmodio-phoromycota, Dictyosteliomycota, Acrasiomycota and Myxomycota, which were formerly considered within the kingdom Fungi, now are included in the kingdom Protista. Haeckel also distinguished a special group within the Protista, called *Monera*, to incorporate the bacteria. Molecular studies have shown that the kingdom Protista (or Protoctista) is exceptionally heterogeneous, because of which the organisms of this kingdom are now placed in the kingdoms *Chromista* and *Protozoa*; they are treated thus in the 9th ed. of Ainsworth & Bisby's *Dictionary of the Fungi* (2001). See the Appendix, where there is a comparison of the outline of Alexopoulos *et al*., 1996, modified by Ulloa and Hanlin, 2006, to that of Kirk *et al*.,

2008. In the latter classification scheme, the phyla Plasmodiophoromycota, Dictyosteliomycota, Myxomycota and Acrasiomycota, have been moved to kingdom Protozoa, with the names of taxa changed to zoological endings, namely Cercozoa, Amoebozoa and Percolozoa, respectively. These taxonomic rearrangements are due to the continuing study and evaluation of these fungus-like microorganisms, in order to achieve a phylogenetic classification. See **Protoctista** and **Protozoa**.

proto-, **prot-** (Gr. *prõtos*, first, foremost): a pref. frequently employed to allude to something that is before something else, in the first place, as in *protoaecium*, *protobasidium*, etc.

protoaecium, pl. **protoaecia** (Gr. *prõtos*, first, foremost + L. *aecium* < Gr. *aikía*, lesion, injury, wound, alluding to the appearance of the leaves and stems of plants attacked by the Uredinales that have aecia in their life cycle): a haploid structure, which after diploidization, is converted into an aecium.

protobasidium, pl. **protobasidia** (Gr. *prõtos*, first, foremost + *basídion*, dim. of *básis*, base): *Fungi with heterobasidia*. The young basidium or primordial cell that later divides into four cells through the formation of transverse septa, each of which produces a basidiospore on a sterigma, as occurs in the Uredinales and Auriculariales; in the Tremellales the protobasidium is divided longitudinally and each of the four cells ends in a long, tubular sterigma. Cf. **probasidium**.

protoblem (Gr. *prõtos*, first, foremost + *blêma*, cover): a layer of pachydermous hyphae, not very dense, that covers the general or universal veil (first covering) of the primordium of the sporiferous apparatus of some basidiomycetes, as in the gen. *Amanita* (Agaricales). Also called **blematogen**, **primary universal veil**, **primordial cuticle** or **covering**, **primordial veil** and **teleblem** or **teleoblem**.

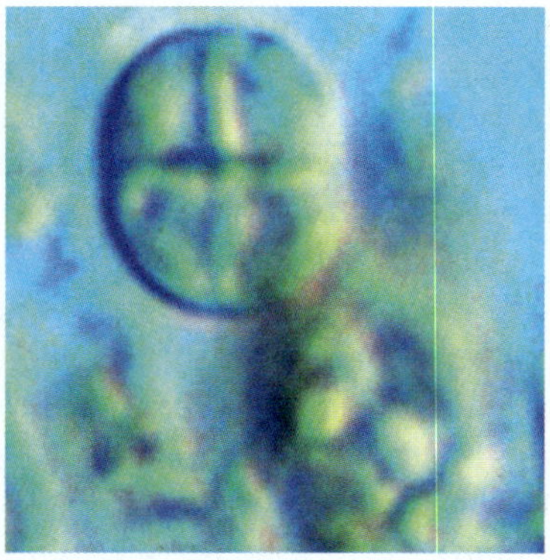

Protobasidium of *Exidia recisa* (Tremellales), x 9,500 (*RTH*).

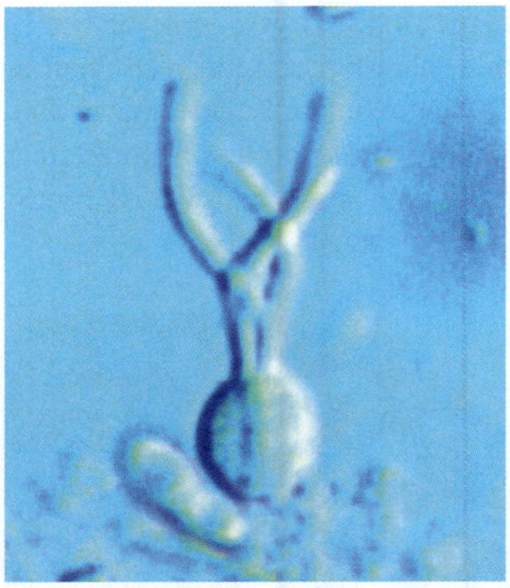

Protobasidium of *Exidia recisa* (Tremellales), x 5,000 (*RTH*).

protoconidium, pl. **protoconidia** (Gr. *prõtos*, first, foremost + *kónis*, powder, dust + dim. suf. *-ídion*): a term used especially for dermatophytic fungi; it is a cell formed on the apex of a filament, which divides and produces deuteroconidia. Also called **hemispore**.

Protoctista (Gr. *prõtos*, first, primitive + *ktistýs*, *ktisís*, creation, thing created, creature): a kingdom proposed by Copeland in 1956, based on the term introduced by Hogg in 1861, to include the eukaryotic organisms, embracing the protozoans, algae and fungi in their entirety. In later systems of classification, such as that of Whittaker (1959) and of Margulis and Schwartz (1988), the kingdom Protoctista is one of five kingdoms, and includes the algae, protozoans, slime molds, aquatic and amphibious fungi with flagellate forms, and many other parasitic aquatic organisms. According to these authors, certain groups of organisms that traditionally were classified as fungi, such as the Protosteliomycetes, Acrasiomycetes, Myxomycetes, Plasmodiophoromycetes, Chytridiomycetes, Hyphochytriomycetes and Oomycetes, would be in the kingdom Protoctista. The molds, mushrooms and other macroscopic fungi, and the lichens, were separated in the kingdom Fungi. In the system of classification adopted in this dictionary (Alexopoulos *et al*., 1996), the Oomycetes and Hyphochytriomycetes are treated as the phyla Oomycota and Hyphochytriomycota, respectively, and are included, together with the Labyrinthulomycota, in the kingdom *Stramenopila*. In other classifications these groups are included in the kingdom *Chromista*. See the Appendix, where there is a comparison of the outline of Alexopoulos *et al*., 1996, modified by Ulloa and Hanlin, 2006, to that of Kirk *et al*., 2008. In the latter classification scheme, the classes

Protosteliomycetes, Acrasiomycetes, Myxomycetes and Plasmodiophoromycetes are moved to kingdom Protozoa; the classes Oomycetes, Hyphochytriomycetes and Labyrinthulomycetes are now included in the kingdom Chromista. The current taxonomic rearrangements undergone in the systematics of these microorganisms are in order to achieve a phylogenetic classification. See **Protista** and **Protozoa**.

protocyst (Gr. *prõtos*, first, primitive + *kýstis*, bladder, vesicle, cell): the bulbous part of a sterile hypha that forms part of the immature gleba of the basidiocarp of *Nia vibrissa*, a marine saprobic gasteromycete (Melanogastrales); the protocysts, sometimes also called "primary basidia", are replaced randomly by basidia, as the fructification continues maturing.

protohymenial (Gr. *prõtos*, first, foremost + L. *hymenium* < Gr. *hyménion* < *hýmen*, membrane + dim. suf. *-íon* > L. *-ium* + L. suf. *-alis* > E. *-al*, pertaining to or relating to): pertaining to or relating to the *protohymenium.*

protohymenium, pl. **protohymenia** (Gr. *prõtos*, first, foremost + L. *hymenium* < Gr. *hyménion* < *hýmen*, membrane + dim. suf. *-íon* > L. *-ium*): a type of hymenium that is characterized by unequal development of its integral elements. The first outline of the hymenium consists of the ends of the hyphae arranged parallel to one another in a palisade, but they remain sterile, giving rise only to paraphyses. The basidia, on the other hand, lie in the subhymenial layer in an irregular manner, although in the end they reach the open surface of the hymenium. In this type, consequently, there is a successive spatial and temporal development of paraphyses and basidia.

protologue (Gr. *prõtos*, first, foremost + *lógos*, word or discourse): information associated with a scientific name when this is published for the first time, i.e., the diagnosis, description, references, synonomy, dates, geographical data, citation of specimens, discussion, commentaries and illustrations.

protonym (Gr. *prõtos*, first, foremost + L. *synonymus* < Gr. *synónymos* < *sýn*, with + *ónoma*, *ónyma*, name): in nomenclature, it refers to a name published effectively, but not validly, after the starting point of the group in question, i.e., after the date that establishes the beginning of priority.

protoperithecium, pl. **protoperithecia** (NL. *protoperithecium* < Gr. *prõtos*, first + *perithekíon*, dim. of *perithêke*, helmet, cap; also < Gr. *perí*, around, in turn + *thêke*, case; here, of the asci): a perithecial primordium that develops into a perithecium after fertilization (spermatization) occurs, as happens in *Neurospora sitophila* (Sordariales).

Protoperithecium of *Neurospora sitophila*, x 375 (*MU*).

protoplasm (Gr. *prõtos*, first, foremost + *plásma*, bland material with which a living being is formed): the living substance, semifluid, of the cells, that constitutes the physical and chemical basis of life.

protoplasmodium, pl. **protoplasmodia** (Gr. *prõtos*, first, primitive + L. *plasmodium* < Gr. *plásma*, bland material with which a living being is formed + L. suf. *-odium*, similar < Gr. *ode*, like < *-oeídes*, similar to): a microscopic plasmodium, without filaments or differentiated fan-shaped regions, which has slow and irregular protoplasmic currents, and which gives rise to a single, minute fruiting body; the protoplasmodium is typical of the Echinosteliales (such as *Echinostelium minutum*) and of *Licea parasitica* (Liceales), although it also occurs in other Myxomycetes. Cf. **aphanoplasmodium** and **phaneroplasmodium**.

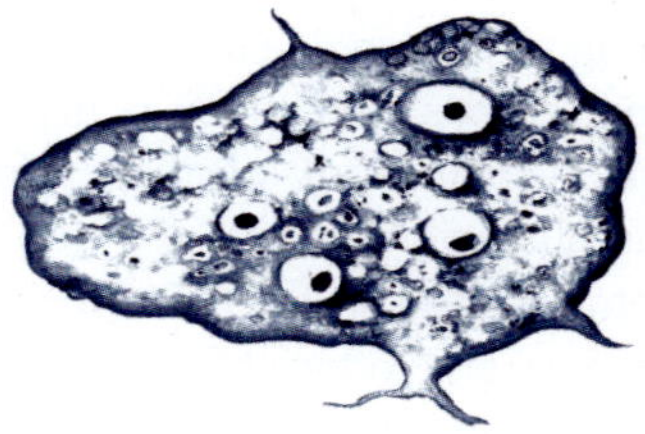

Protoplasmodium (with four nuclei) of *Echinostelium minutum*, x 1,360 (*MU*).

protoplast (Gr. *prõtos*, first, foremost + *plastós*, modeled, formed; organized corpuscle or particle): traditionally, it corresponds to the totality of constituents of the living cell, which may or may not be delimited by a cell wall, but now it is frequently applied to a protoplast whose cell wall has been removed experimentally. Fungal protoplasts are useful in the study of cellular organelles and biochemical and genetic processes, as well as in their potential application in biotechnology, particularly since one can achieve fusion of protoplasts between different strains, including different species, to produce a combined heterokaryotic protoplast with different properties.

protopseudothecium, pl. **protopseudothecia** (NL. *protopseudothecium* < Gr. *prõtos*, first, primitive + *pseûdos*, false + *thekíon*, dim. of *thêke*, case; here, of the asci): a small stroma that develops into a pseudothecium (ascostroma) after fertilization.

protopycnidium, pl. **protopycnidia** (Gr. *prõtos*, first, foremost + NL. *pycnidium* < Gr. *pyknós*, dense, compact, concentrated + dim. suf. *-ídion*): the primordium of a pycnidium.

protosexual (Gr. *prõtos*, first, foremost + L. *sexualis*, pertaining to or relating to sex): applied to certain yeasts and other organisms that have diploid or dikaryotic cells that can produce haploid or unisexual cells in the absence of sexual fructifications or spores. This term contrasts with *parasexual*, which has been redefined to include organisms that have protosexual as well as sexual cycles, and with *neosexual*, which refers to organisms with a sexual cycle but lacking a protosexual cycle.

protosexuality (Gr. *prõtos*, first, primitive + E. *sexuality*, the group of anatomical and physiological characteristics that characterize each sex < L. *sexualis*, pertaining to or relating to sex): a type of reproduction in certain yeasts in which haploid cells are formed from diploid cells in culture, without the production of ascospores; also, the chlamydospores that these species of yeasts form can function as asci, as in species of *Metschnikowia* (Saccharomycetales).

protospore (Gr. *prõtos*, first + *sporá*, spore): *Mucorales*. The multinucleate mass of cytoplasm that is divided in developing sporangia due to the occurrence of the first cleavage planes, and which with the formation of subsequent planes of cleavage gives rise to multiple uninucleate portions of cytoplasm, which are transformed into spores. This process is present in *Phycomyces* and other Mucorales, and in the spherule (sporangium) of *Coccidioides* (of uncertain taxonomic position, classified both in the family Synchytriaceae, of the order Chytridiales, and in the moniliaceous asexual fungi). The uninucleate portion of the protoplasm that eventually is converted into the sporangium is also called a protospore.

Protosteliomycetes (Gr. *prõtos*, first, primitive + *stéle*, column + L. *-mycetes*, ending of taxonomic class < Gr. *mýkes*, genit. *mýketos*, fungus): a class of viscous or mucilaginous fungi, whose myxamoebae do not aggregate to form pseudoplasmodia, but which sporulate directly or form plasmodia that produce fructifications called sporocarps or sporophores. The Protosteliomycetes (with the single order Protosteliales), together with the Acrasiomycetes, Myxomycetes and Plasmodiophoromycetes, were previously included in the division Myxomycota of the kingdom Fungi. Some authors (e.g., Margulis and Schwartz, 1988) classify the Protosteliomycetes as a phylum of the kingdom Protoctista. In the system of classification followed in this dictionary (Alexopoulos *et al.*, 1996), this class is not included, so that in the cases in which these organisms are considered in order to explain some concepts, such organisms are refered to as Protosteliales. Some other authors treat the Protosteliomycetes (and the order Protosteliales) as part of the kingdom Protista. See the Appendix, where there is a comparison of the outline of Alexopoulos *et al.*, 1996, modified by Ulloa and Hanlin, 2006, to that of Kirk *et al.*, 2008. In the latter classification scheme, the order Protosteliales is called Protostelida, class Protostelea, phylum Amoebozoa, kingdom Protozoa, with the concurrent changes of taxa names with zoological endings. This is another rearrangement of the taxonomy of these fungus-like microorganisms in order to achieve a phylogenetic classification. See **Protozoa**.

protothecium, pl. **protothecia** (NL. *protothecium* < Gr. *prõtos*, first, primitive + *thekíon*, dim. of *thêke*, case; here, of the asci): an incompletely differentiated ascoma that contains neither asci nor ascospores.

prototroph (Gr. *prõtos*, first; here, taken in the sense of a chemical element + *trophós*, what nourishes, serves as food < *trépho*, to nourish): **1**. An organism capable of developing on a medium composed only of inorganic substances, or at least suitable for the assimilation of free nitrogen when it can dispose of hydrocarbon compounds. The aerobic fungi are prototrophs with respect to oxygen, since they can assimilate it directly as a simple chemical element, without it having to be in combination with other elements. **2**. Also applied to mutant strains that do not differ from the wild strain in nutritional requirements relative to particular vitamins or aminoacids, i.e., like the wild strain, they are capable of synthesizing these compounds. Cf. **auxotroph**.

prototunicate (Gr. *prõtos*, first, primitive + L. *tunica*, covering, wrapper + L. suf. *-atus* > E. *-ate*, provided with or likeness): *Ascomycetes*. An ascus with a thin, delicate wall that liberates its spores by deliquescence, like that of the Saccharomycetales. This type of ascus is basically unitunicate, but lacks differentiated apical structures. Cf. **bitunicate** and **unitunicate**.

prototype (L. *prototypus* < Gr. *protótypos* < *prõtos*, first, foremost + *týpos*, type, model, authentic example): an individual that exhibits the essential features of a later type; an archetype.

Protozoa (Gr. *prõtos*, first, primitive + *zõon*, of animals): a kingdom of predominantly unicellular organisms, plasmodial or colonial, phagotrophic, lacking a cell wall in the trophic state; the ciliar hairs are never rigid or tubular; the chloroplasts (when present) lack starch and phycobilisomes, with stacked thylakoids and three membranes; the multicellular species have minimal cellular differentiation and lack collagen connective tissue. The organisms of the kingdom Protozoa have been included, together with those of the kingdom Chromista, in a broader kingdom, the kingdom Protista (or Protoctista), but this has been rejected recently, taking into account molecular and other class data. With regard to the groups studied by mycologists, in this dictionary they are treated within the kingdom Protista, i.e., the phyla Plasmodiophoromycota (Plasmodiophorales), Dictyosteliomycota (Dictyosteliales), Acrasiomycota (Acrasiales) and Myxomycota (Myxomycetes and Protosteliomycetes). In other classifications (such as that accepted in the 9th ed. of Ainsworth & Bisby's *Dictionary of the Fungi*, 2001), these groups are considered within the kingdom Protozoa, based on recent molecular and ultrastructural studies, which have demonstrated that mucilaginous fungi and related groups are polyphyletic and belong to different lineages. See the Appendix, where there is a comparison of the outline of Alexopoulos *et al*., 1996, modified by Ulloa and Hanlin, 2006, to that of Kirk *et al*., 2008. In the latter scheme of classification, the phyla Plasmodiophoromycota, Dictyosteliomycota, Myxomycota and Acrasiomycota were moved from the kingdom Protista to the kingdom Protozoa, with the names of taxa changed as Cercozoa, Amoebozoa and Percolozoa, respectively. The orders Amoebidiales and Eccrinales, Class Trichomycetes, phylum Zygomycota, kingdom Fungi, have been moved to phylum Choanozoa, kingdom Protozoa. See **Protista** and **Protoctista**.

protuberant (L. *protuberans*, genit. *protuberantis* < *protuberare*, to protrude): prominent, protruding, like the short projections of the cell wall of some conidia.

proximal (L. *proximus*, near, following, immediately behind, not very distant in space or in time + suf. *-alis* > E. *-al*, relating to or pertaining to): situated near the center of a body or point of origin. Cf. **distal**.

pruinose, pruinate (L. *pruinosus*, frosted, having frost < *pruina*, frost + suf. *-osus* > E. *-ose*, full of, augmented, prone to; or + suf. *-atus* > E. *-ate*, provided with or likeness): refers to a structure or organ with a very thin waxy covering, similar to frost or flour, which gives it a characteristic appearance; this covering, formed by granules, small rods or little crusts, somewhat masks the true color of the respective structure. E.g., the basidiocarps of *Coprinus comatus* and *Coprinellus* (=*Coprinus*) *micaceus* (Agaricales) are pruinose.

Pruinose surface of the basidiocarps of *Coprinellus* (=*Coprinus*) *micaceus*, x 1.5 (*MU*).

psammo- (Gr. *psámmos*, sand): a pref. that alludes to sand, as in *psammophyte*, a plant typical of sandy soils.

psammophilous (Gr. *psámmos*, sand + *phílos*, have an affinity for): syn. of **arenicolous**.

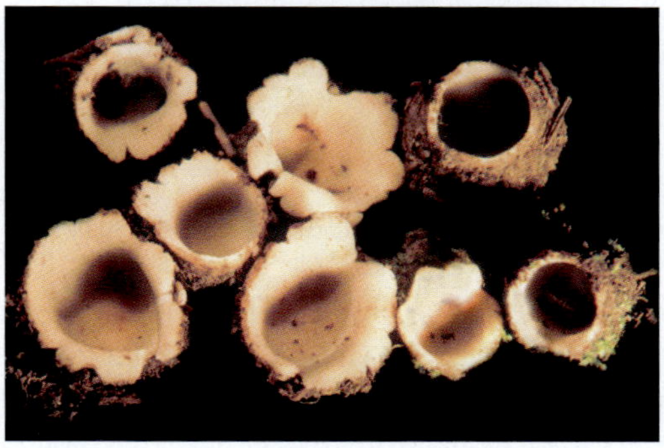

Psammophilous ascomata of *Geopora* (=*Sepultaria*) *arenicola* (Pezizales), x 1.5 (*MU*).

pseudo-, pseud- (Gr. *pseûdos*, false, illegitimate): a pref. often employed in biological terminology to indicate that something is false or illegitimate, or which refers to a structure that resembles something else, as in *pseudothecium*, an ascoma that resembles a perithecium. Cf. **eu-**.

pseudoaethalium, pl. **pseudoaethalia** (Gr. *pseûdos*, false + *aíthalos*, soot): a mass of sporangia with closely compacted stipes, simulating an aethalium, but distinguished from an aethalium in that each individual sporangium is evident at maturity, as is seen in *Metatrichia vesparium* (Trichiales).

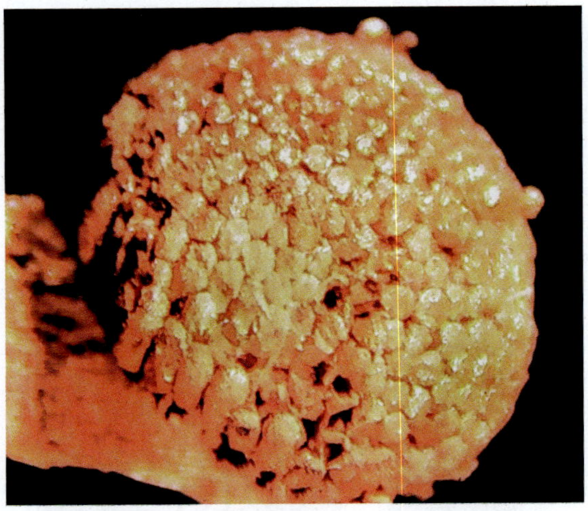

Pseudoaethalium of *Metatrichia vesparium*, x 25 (*MU*).

Pseudoaethalium of *Metatrichia vesparium*, x 60 (*MU*).

pseudoamyloid (Gr. *pseûdos*, false + *ámylon*, starch + L. suf. *-oide*, like < Gr. *-oeídes*, similar to): falsely amyloid. Applied to the color reaction in spores or cell walls when treated with Melzer's reagent (potassium iodide, iodine, chloral hydrate and water), which take a purplish-gray color (the amyloid reaction turns them pale gray, at times so dark they are almost black). The pseudamyloid reaction also is called **dextrinoid** and **hemiamyloid**. Cf. **amyloid** and **anamyloid**.

pseudoangiocarpic (Gr. *pseûdos*, false + L. pref. *angio-*, vessel, container < Gr. *angeîon*, small vessel + suf. *-íkos* > L. *-icus* > E. *-ic*, pertaining to or relating to): a basidioma whose hymenial surface is exposed at the beginning, but later becomes covered by the incurving of the margin of the pileus and/or by excrescences of the stipe.

pseudocapillitium, pl. **pseudocapillitia** (Gr. *pseûdos*, false + L. *capillitium*, head of hair < *capillus*, hair + *-itium*, adjectival ending that means provided with, that has): *Myxomycetes*. Filaments that are irregular in shape, diameter and ornamentation, perforated plates, or other structures similar to true capillitium (whose elements tend to be more uniform), that are found among the spores inside fructifications of some species, such as *Lycogala* and *Dictydiaethalium* (Liceales). Cf. **capillitium**.

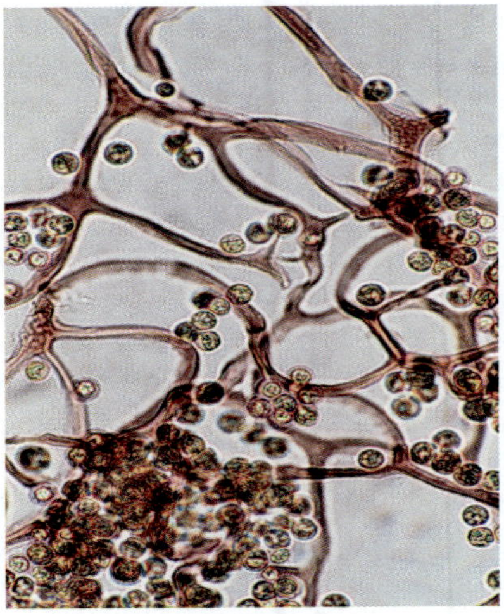

Pseudocapillitium and spores of *Reticularia* sp., x 250 (*MU*).

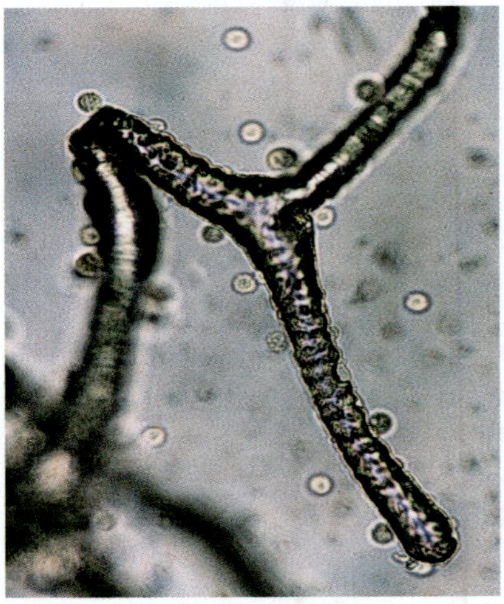

Pseudocapillitium of an aethalium of *Lycogala epidendrum*, x 270 (*MU*).

pseudocleistothecium, pl. **pseudocleistothecia** (Gr. *pseûdos*, false + *kleistós*, closed + *thekíon* < *thêke*, case; here, of the asci + dim. suf. *-ium*): a non-funcional cleistothecial primordium formed by some dermatophytes in culture.

pseudoclypeus, pl. **pseudoclypei** (Gr. *pseûdos*, false + L. *clypeus*, round shield): a clypeus in which tissue of the plant host is included, as is seen in *Heliascus*

(Dothideales), a marine fungus whose lenticular, carbonaceous stromata possess locules covered by a pseudoclypeus composed of a black tissue that includes cells of the host, in this case, dead root cells of mangrove (*Rhizophora mangle*).

pseudocolumella, pl. **pseudocolumellae** (Gr. *pseûdos*, false + L. *columella*, small column < *columna*, column + dim. suf. *-ella*): **1**. *Myxomycetes*. Calcareous nodules of the capillitium that in mass form a kind of small column in the center of the sporangium, as occurs in some species of *Physarum* (Physarales), such as *Ph. nicaraguense* and *Ph. mutabile*. **2**. *Gasteromycetes*. A column composed of sterile tissue, often divided into chambers surrounded by the gleba, as happens, e.g., in *Geastrum*, *Lycoperdon* and *Radiigera* (Lycoperdales).

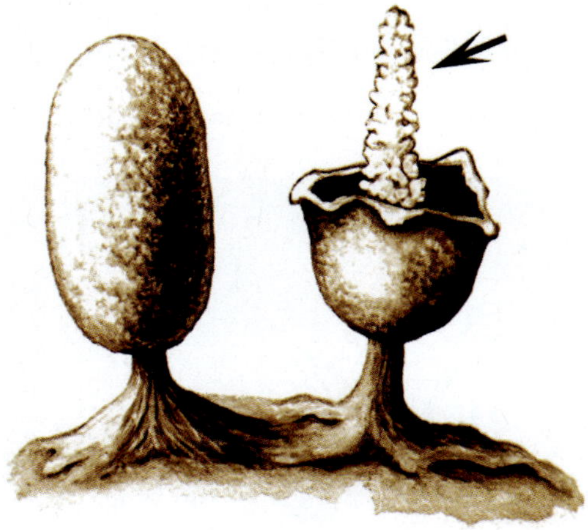

Pseudocolumella of the sporangia of *Physarum mutabile*, x 3 (*MU*).

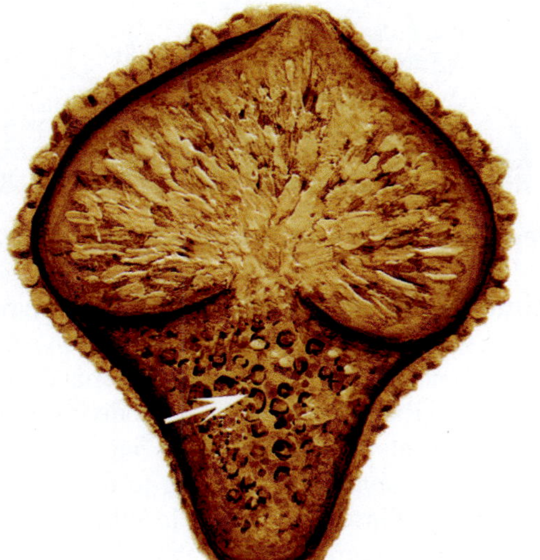

Pseudocolumella of the fruiting body of *Lycoperdon perlatum*, x 1 (*MU*).

pseudocortex (Gr. *pseûdos*, false + L. *corticea*, fem. des. of *corticeus*, of cortex): applied to the false cortex that constitutes the outer layer of the pseudopodetia in some lichen thalli, like that of *Pycnothelia papillaria* (Lecanorales).

pseudocyphella, pl. **pseudocyphellae** (Gr. *pseûdos*, false + L. *cyphella*, the concavity of the ears): *Lichens*. Each of the small open cavities in the lower part of the thallus, similar to cyphellae, lacking a cortex, that are found in the family Stictaceae of the order Lecanorales. Cf. **cyphella** and **cyphelloid**.

pseudocystidium, pl. **pseudocystidia** (Gr. *pseûdos*, false + *kystídion*, dim. of *kýstis*, bladder; here, vesicle, cell): a prolongation, similar to a cystidium, of the conducting elements (gloeocystidia, chrysocystidia, phaeocystidia and macrocystidia) that reach the hymenium, epicutis or the superficial layer of the stipe, in the basidiocarps of *Russula* and *Lactarius* (Agaricales).

pseudodiblastic (Gr. *pseûdos*, false + *dís*, two times + *blastikós*, pertaining to or relating to germination or branching < *blastós*, bud, shoot + suf. *-íkos* > L. *-icus* > E. *-ic*, pertaining to or relating to): refers to ascospores that have oil droplets at the poles, so that superficially they resemble polarilocular spores.

pseudoepithecium, pl. **pseudoepithecia** (Gr. *pseûdos*, false + *epí*, on + *thekíon*, dim. of *thêke*, case; here, of the asci): an amorphous or granular layer that covers the paraphyses in an apothecium, in which the apices are embedded, but without forming a separate tissue.

pseudoflagellum, pl. **pseudoflagella** (Gr. *pseûdos*, false + L. *flagellum*, whip, flagellum): a hyaloplasmic structure that is frequently formed on the flagellate end of the myxozoospores of certain Protosteliales (such as *Cavostelium*); these fugaceous prolongations, which are not flagella, continue forming and disappearing in succession toward the posterior end of the myxozoospores. The pseudoflagella appear to be extremely long and slender filose or filopod pseudopods.

pseudogamy (Gr. *pseûdos*, false + *gámos* > E. des. *-gamy*, sexual union): see **pseudomixis**.

pseudohymenium, pl. **pseudohymenia** (Gr. *pseûdos*, false + *hyménion* < *hymén*, membrane + dim. suf. *-íon*): *Gasteromycetes*. A hymenial layer of basidia, generally very convoluted, that is present, e.g., in the orders Phallales and Lycoperdales, which have not evolved to discharge the spores forcefully, and in addition, do not have the aerial space to permit forceful discharge of the spores. In the order Sclerodermatales the basidia are produced either as isolated cells distributed throughout the gleba, or as nodules or islands of fertile cells. The pseudohymenia differ from the true hymenia present in the hymenomycetes (Agaricales, Aphyllophorales, Auriculariales, Dacrymycetales, Ceratobasidiales, Tulasnellales and Tremellales), since in these latter orders there is forcible discharge of the spores into an aerial space that permits their dissemination.

pseudoisidium, pl. **pseudoisidia** (Gr. *pseûdos*, false + Gr. dim. *-isídion* < *Isis*, Egyptian goddess, which some of these structures resemble due to their columnar shape): an excrescence, similar to an isidium, on the surface of a lichen thallus, e.g., that of *Gyalideopsis* (ascolichen of uncertain position). Also refers to an isidium lacking photosynthetic cells, as occurs in *Pseudocyphellaria* (Lecanorales).

pseudomixis, pl. **pseudomixes** (Gr. *pseûdos*, false + *míxis*, mixture, intimate union): the substitution of a true sexual union of reproductive cells by a process of pseudosexual copulation between two cells that are not specifically differentiated as sexual. *Adelphogamy* and *pedogamy* are two types of pseudomictic sexuality. Pseudomixis is also known as **pseudogamy**.

pseudomorph (Gr. *pseûdos*, false + *morphé*, shape): a stromatic mass composed of parts of plant tissues united by fungal plectenchyma, similar to an organ of the host plant.

pseudomycelium, pl. **pseudomycelia** (Gr. *pseûdos*, false + NL. *mycelium* < Gr. *mýkes*, fungus + L. suf. *-elis*, concerning + dim. suf. *-ium*): a grouping of cells arranged in loosely united chains, as occurs in various species of yeasts, ascosporogenous (such as *Saccharomyces* and *Saccharomycodes*, of the Saccharomycetales) as well as anascosporogenous (e.g., *Candida*, of the cryptococcaceous asexual yeasts). The chain of cells of the pseudomycelium originates by budding and is distinguished from true mycelium in that the latter is formed by the continuous extension of a hypha that is not constricted at the septa that delimit the cells.

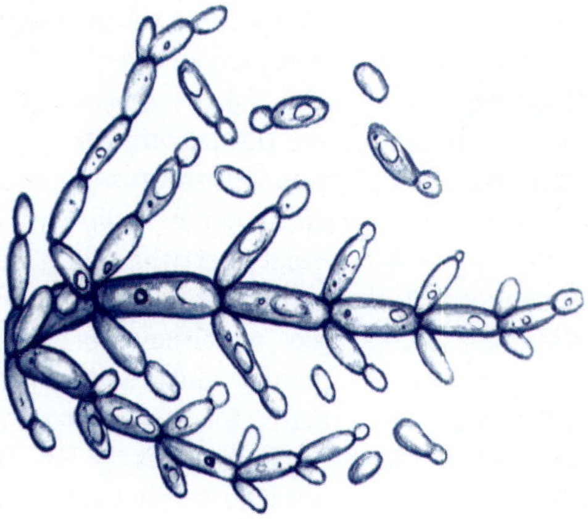

Pseudomycelium of *Saccharomyces kluyveri*, x 800 (*MU*).

Pseudomycelium of *Candida valida*, x 680 (*PL*).

Pseudomycelium of *Saccharomyces cerevisiae*, x 500 (*PL*).

Pseudomycotina (Gr. *pseûdos*, false + L. *-mycotina*, des. used for the ending of the taxonomic category corresponding to subdivision, intermediate between division and class): a subdivision within the kingdom *Chromista*, which includes the so-called "*pseudofungi*", belonging to the Oomycetes in a broad sense, and which in other systems of classification corresponds to the phyla Oomycota, Hyphochytriomycota and Labyrinthulomycota. In this dictionary, these phyla are included in the kingdom *Stramenopila*.

pseudoparaphysis, pl. **pseudoparaphyses** (Gr. *pseûdos*, false + *paráphysis*, sprout < *pará*, at the side of + *phýsis*, action of bringing forth or budding, to be born; a being, a growth): sterile hyphae that form at the top of the centrum of an ascostroma and grow downward to the base of the centrum, where they can intertwine with the ascogenous system. The asci grow upward among the pseudoparaphyses, which can persist or not until maturation of the asci. Pseudoparaphyses are present, e.g., in the families Pleosporaceae, Microthyriaceae and Hysteriaceae, of the Loculoascomycetes, such as *Pleospora*, *Muyocopron* (=*Ellisiodothis*) *smilacis* and *Hysterium*, respectively. Equal to **cataphysis**.

Pseudoparaphyses of a pseudothecium of *Pleospora herbarum*, seen in longitudinal section, growing beneath the epidermis of an onion, x 100 (*MU*).

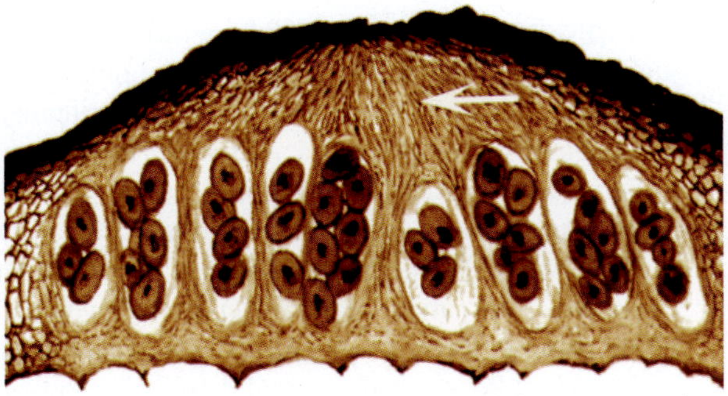

Pseudoparaphyses of the pseudothecium of *Muyocopron* (=*Ellisiodothis*) *smilacis*, seen in longitudinal section, x 230 (*MU*).

pseudoparenchyma, pl. **pseudoparenchymata** (Gr. *pseûdos*, false + *parénchyma*, a tissue of higher plants < *pará*, at the side of, near + *énchyma*, stuffed): a type of plectenchyma composed of closely united, angular or isodiametric cells; the hyphae involved in the formation of pseudoparenchyma lose their individuality. Pseudoparenchyma is present in sclerotia (e.g., *Claviceps purpurea*, of the Hypocreales), stromata, fruiting bodies and other fungal structures. In the agarics, pseudoparenchyma refers to the undifferentiated tissue of the basidiocarp primordium. Cf. **prosenchyma**.

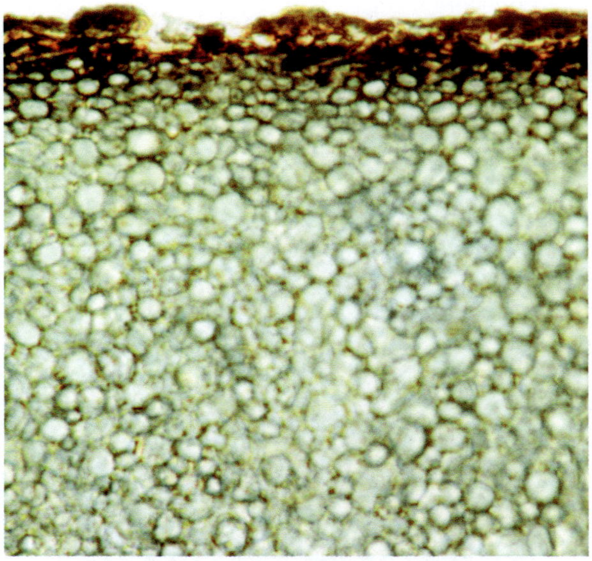

Pseudoparenchyma of the sclerotium of *Claviceps purpurea*, seen in transverse section, x 670 (*MU*).

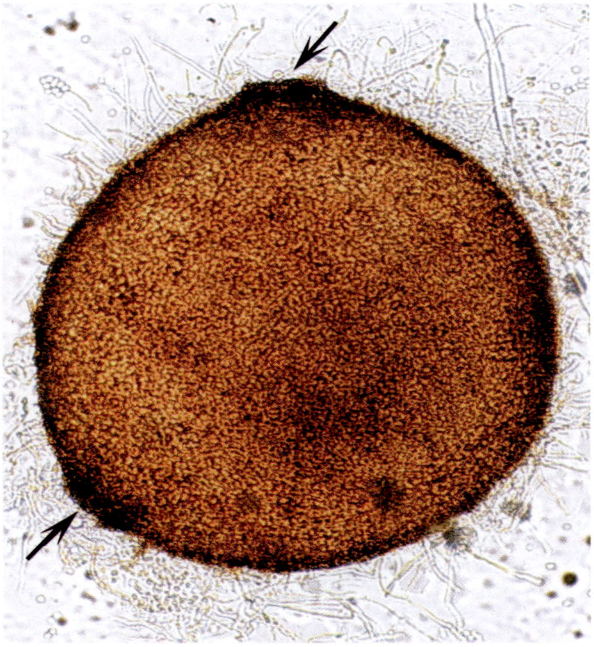

Pseudoparenchyma of the pycnidium of *Phoma* sp. (sphaeropsidaceous asexual fungi); note the two ostioles, x 450 (*MU*).

pseudoperculate (Gr. *pseûdos*, false + L. *operculatus*, provided with an operculum): applied to asci whose structure is essentially unitunicate, but in whose apex there is a thickening like a lid that detaches completely at the moment of ascospore discharge. Cf. **inoperculate**, **operculate** and **suboperculate**.

pseudoperidium, pl. **pseudoperidia** (Gr. *pseûdos*, false + *péridion*, small leather bag): a false peridium, like a membrane, that covers the aecium of the Uredinales.

pseudoperithecium, pl. **pseudoperithecia** (NL. *pseudoperithecium* < Gr. *pseûdos*, false + *perithekíon*, dim. of *perithêke*, helmet, cap; also from *perí*, around, in turn + *thêke*, locket, case; here, of the asci + dim. suf. *-íon*): see **pseudothecium**.

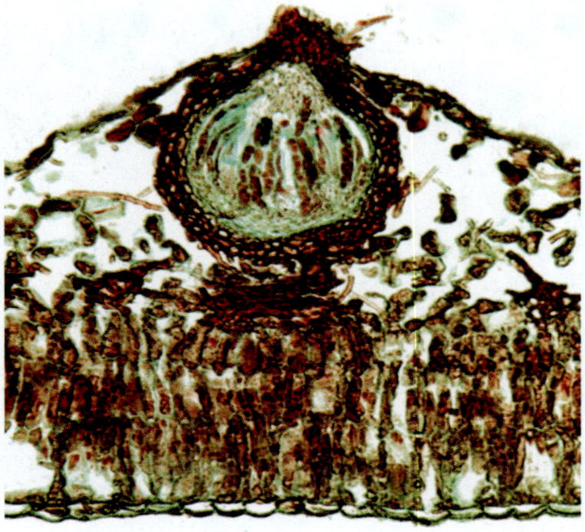

Pseudoperithecium of *Venturia inaequalis* (Pleosporales), seen in vertical section, x 450 (*MU*).

pseudophialide (Gr. *pseûdos*, false + *phialís*, genit. *phialídis*, dim. of *phiále*, glass, jar): a small ovoid or obpyriform cell found in the concavity of a sporocladium, and which gives rise to a unispored sporangiolum, as happens in *Kickxella*, *Coemansia* and other Kickxellales.

Pseudophialides, each with a unispored merosporangium, of *Kickxella alabastrina*, x 1,750 (*MU*).

pseudophysis, pl. **pseudophyses** (Gr. *pseûdos*, false + *phýsis*, action of budding or being born; a birth): a type of paraphysis or sterile structure characterized by having a smooth or roughened thin wall, with moniliform constrictions at the apex, that are formed in the hymenium of the gen. *Aleurodiscus* (Aphyllophorales), together with other structures, the dendrophyses. In the agarics, the pseudophyses correspond to the brachycystidia or brachybasidia. See **dendrophysis**.

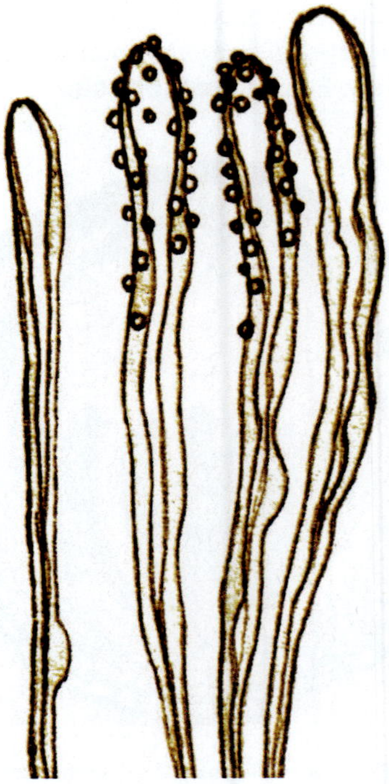

Pseudophyses of the hymenium of *Aleurodiscus scutellatus*, x 375 (*MU*).

pseudopionnote (Gr. *pseûdos*, false + *pión*, fat, thick juice + *notís*, moisture): an effuse mass of numerous small sporodochia that form a more or less continuous mucilaginous layer of conidia, as in *Fusarium solani* (tuberculariaceous asexual fungi).

pseudoplasmodium, pl. **pseudoplasmodia** (Gr. *pseûdos*, false + NL. *plasmodium*, plasmodium < Gr. *plásma*, formation, bland material with which a living being is formed + L. suf. *-odium*, resembling < Gr. *ode*, like < *-oeídes*, similar to): also called aggregation plasmodium or *grex*; it is formed by the union of myxamoebae that retain their individuality. It is characteristic of the acrasiaceous or cellular slime fungi of the phylum Dictyosteliomycota, such as *Dictyostelium* and *Polysphondylium*. Cf. **aphanoplasmodium**, **phaneroplasmodium** and **protoplasmodium**.

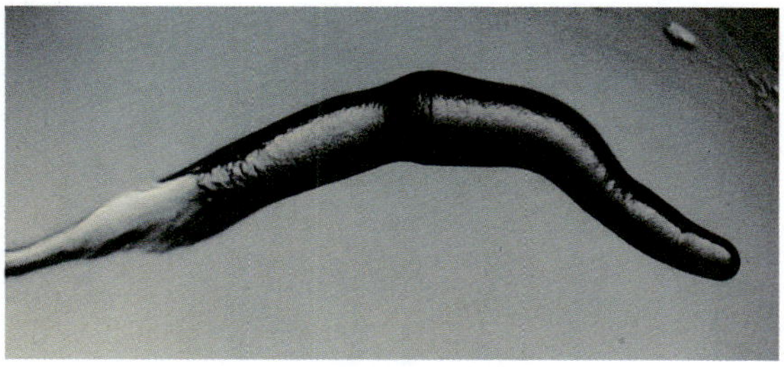

Pseudoplasmodium of *Dictyostelium discoideum* on agar, x 20 (*RTH*).

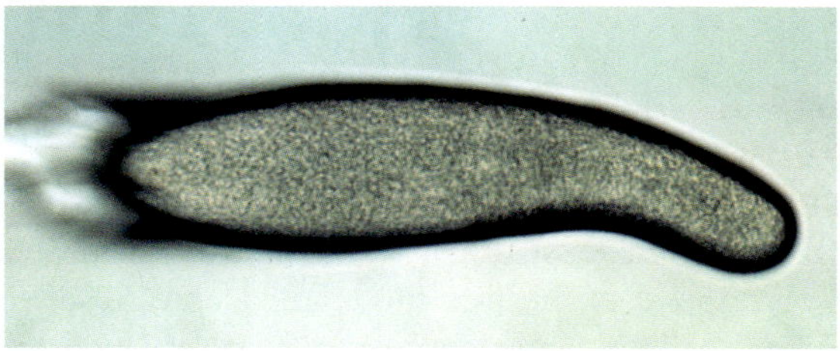

Pseudoplasmodium of *Dictyostelium discoideum* on agar, x 40 (*RTH*).

pseudopod, pseudopodium, pl. **pseudopodia** (Gr. *pseûdos*, false + NL. *podium* < Gr. *pódion*, dim. of *poús*, genit. *podós*, foot): a protoplasmic prolongation that varies continuously in shape, from a myxamoeba or a plasmodium, and which serves as an organ of locomotion and for the capture of food, as occurs in *Cavostelium* (Protosteliales). See. **filopod** and **lobopod**.

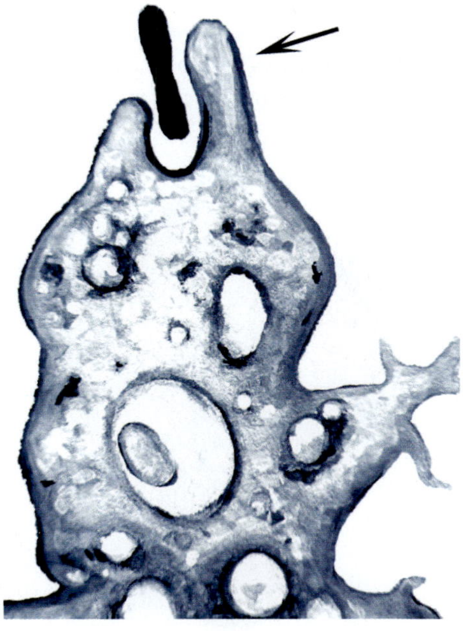

Pseudopodia of a myxamoeba of *Cavostelium bisporum*, ingesting a bacterial cell, x 3,950 (*MU*).

pseudopodetium, pl. **pseudopodetia** (Gr. *pseûdos*, false + NL. *podetium* < Gr. *podotés*, provided with a foot < *poús*, genit. *podós*, foot): *Lichens*. Pseudopodetia are erect, solid branches, similar to podetia, of vegetative origin, like those of *Stereocaulon* (Lecanorales); when ascogonia are present, these are formed on pseudopodetia, and do not originate on a preformed, granular or squamulose thallus. Cf. **podetium**.

pseudorhiza, pl. **pseudorhizae** (Gr. *pseûdos*, false + *rhíza*, root): a root-like extension on the base of a stipe that functions as an organ uniting the basidiocarp and the mycelium in the soil, as well as an organ of conduction and storage. It is present, e.g., in *Xerula* (=*Oudemansiella*) *radicata* (Agaricales) and *Pseudofistulina* (=*Fistulina*) *radicata* (Aphyllophorales). See **radicate**.

Pseudorrhiza of the basidiocarps of *Xerula* (=*Oudemansiella*) *radicata*, x 1 (*JC*).

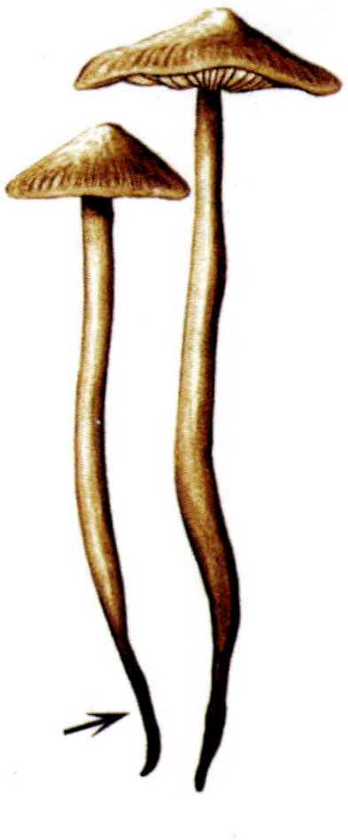

Pseudorrhiza of the basidiocarps of *Xerula* (=*Oudemansiella*) *radicata*, x 1 (*MU*).

pseudosclerotium, pl. **pseudosclerotia** (Gr. *pseûdos*, false + NL. *sclerotium* < Gr. *sklerotés*, hardness < *sklerós*, hard + L. dim. suf. *-ium*): a compact mass of conglomerated particles of earth, humus or silica mixed with mycelium. *Armillaria* (=*Armillariella*) *mellea* (Agaricales) and *Xylaria polymorpha* (Xylariales) are examples of lignicolous fungi that form pseudosclerotia. Another examples are *Cordyceps* (Hypocreales), that produces the pseudosclerotium from the insect host and the mycelium of the fungal parasite; in *Monilinia fructicola* (Helotiales) the pseudosclerotium is produced by the mummified peach fruit. From these pseudosclerotia the perithecial stromata develop.

Pseudosclerotium (in the insect host) of *Cordyceps militaris*, x 1.5 (*MU*).

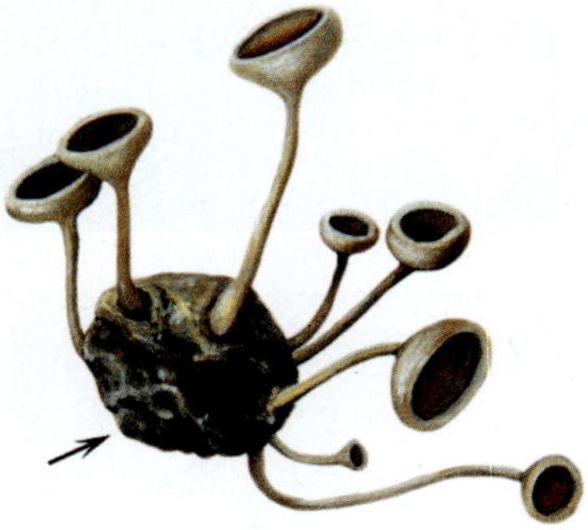

Pseudosclerotium (in a parasitized and mummified peach fruit) of *Monilinia fructicola*, x 0.6 (*MU*).

pseudoseptum, pl. **pseudosepta** (NL. *pseudoseptum* < Gr. *pseûdos*, false + L. *septum*, septum, fence, partition): **1**. A protoplasmic or vacuolar membrane similar to a septum, as in *Corynespora* (dematiaceous asexual fungi), which has spores with up to 50 or more pseudosepta; also called **distoseptum**. **2**. Each of the membranes with pores present in the thallus of the Blastocladiales, as well as the cellulin plugs found in the constricted hyphae of the aquatic fungi of the order Leptomitales (such as *Apodachlya* and *Leptomitus*). See **cellulin**.

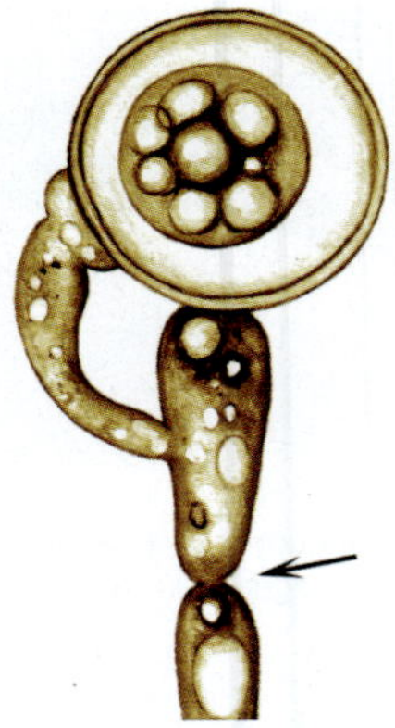

Pseudoseptum of the hypha of *Apodachlya* sp., x 750 (*MU*).

pseudoseta, pl. **pseudosetae** (Gr. *pseûdos*, false + L. *seta*, bristle, stiff hair): a false seta; pseudosetae correspond to the free tips, turned upward, of the hyphae of the context in the hymenium of *Duportella* (Aphyllophorales).

pseudospore (Gr. *pseûdos*, false + *sporá*, spore): a spore lacking a cell wall, infrequent in the sporocarps or fruiting bodies of some Dictyosteliomycota (such as in *Dictyostelium*).

pseudostem (Gr. *pseûdos*, false + ME. *stefn, stemm*, stem): see **pseudostipe**.

pseudostipe (Gr. *pseûdos*, false + L. *stipes*, genit. *stipitis*, stipe, pedicel, stem, support): *Gasteromycetes*. A mass of spongy tissue similar to a stipe (subgleba), that supports the gleba, in which the hyphae are not oriented parallel to the axis of the stipe, like that formed in *Lycoperdon* (Lycoperdales) and *Phallus* (Phallales). In the Gasteromycetes with a true stipe (such as *Tulostoma*, of the Tulostomatales), the hyphae that comprise it are arranged in a more or less parallel manner.

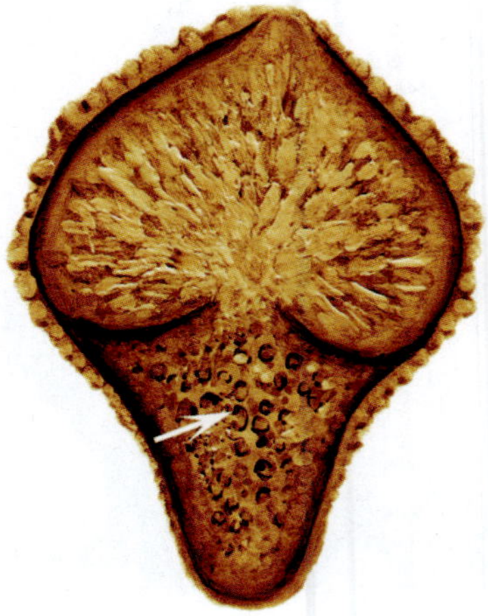

Pseudostipe of the fruiting body of *Lycoperdon perlatum*, x 1 (*MU*).

pseudostolon (Gr. *pseûdos*, false + NL. *stolon*, genit. *stolonis*, branch, shoot, sprout): an unbranched vegetative hypha that connects two groups of sporangiophores, as happens in *Mucor racemosus* (Mucorales); it differs from a stolon in that it does not connect two groups or fascicles of rhizoids with sporangiophores opposite the rhizoids, as occurs in *Rhizopus* and other Mucorales with typical stolons. Cf. **stolon**.

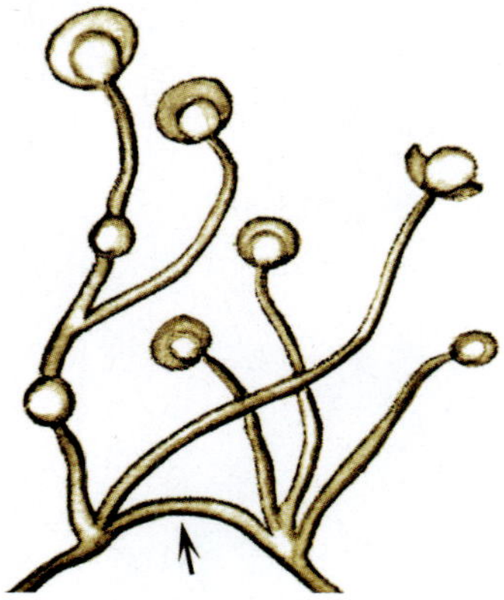

Pseudostolon of *Mucor racemosus*, x 300 (*MU*).

pseudostroma, pl. **pseudostromata** (Gr. *pseûdos*, false + *strõma*, bed, small cushion): a complex stroma composed of tissues of the fungus and of the host plant, like that of *Polystigma rubrum* (Phyllachorales). Cf. **eustroma**.

pseudothecium, pl. **pseudothecia** (NL. *pseudothecium* < Gr. *pseûdos*, false + *thekíon*, dim. of *thêke*, locket, case; here, of the asci): *Loculoascomycetes*. An ascostroma bearing asci in locules lacking a wall; restricted by some to the perithecioid unilocular ascostroma, like that of *Leptosphaerulina*, *Munkiella* and *Mycosphaerella* (Dothideales), *Sporormiella* (=*Preussia*) and *Venturia* (Pleosporales). Cf. **euthecium**.

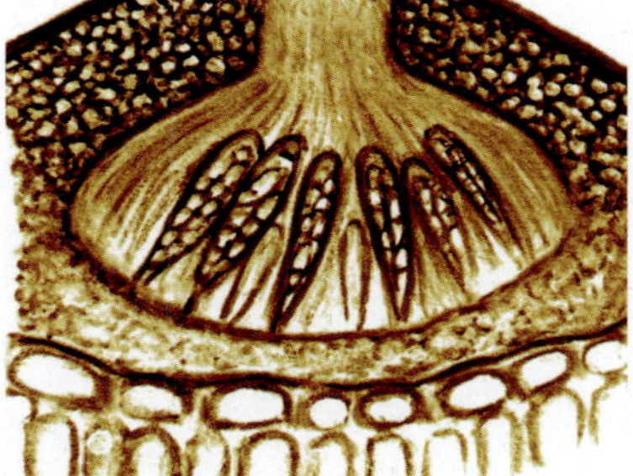

Pseudothecium of *Munkiella caa-guazu*, seen in longitudinal section, x 500 (*MU*).

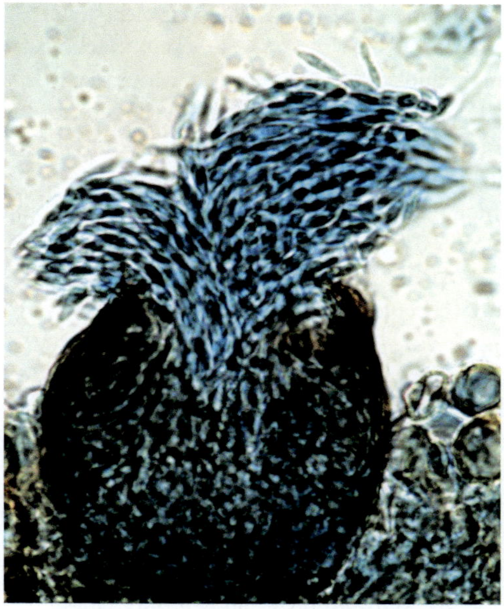

Pseudothecium of *Mycosphaerella musicola*, on the host tissues (yellow Sigatoka of banana), x 500 (*MU*).

psilo- (Gr. *psilós*, naked, unarmed, smooth, bald): a pref. employed in forming terms that allude to something that is beardless, glabrous, unarmed, etc., i.e., that lacks alveoli, warts, prickles, etc., like the generic name *Psilocybe* (Agaricales), which means smooth head (pileus).

psychotomimetic (ISV. *psychotic*, relative to psychosis < NL. *psychosis* < Gr. *psyché*, soul, mind + suf. *-o* + LL. *mimeticus* < Gr. *mimetikós* < *miméomai*, to imitate, copy + suf. *-ikós* > L. *-icus* > E. *-ic*, relation, fitness, inclination, capacity): refers to a substance that is related to, involved in, or that induces psychotic alterations in behavior and personality, like those caused by the ingestion of hallucinogenic drugs (plants, seeds, fungi); i.e., alterations that mimic the symptoms of psychosis. See **hallucinogenic fungi**, **magic mushrooms** and **psychotropic**.

psychotropic (Gr. *psyché*, soul, mind + *tropikós*, referring to change or removal, changing): something that acts on the mind, like the hallucinogenic drugs. Also called **neurotropic**. See **psychotomimetic**.

psychro-, **psychr-** (Gr. *psychrós*, cold): a pref. used in the formation of terms that allude to low temperatures, as in *psychrophile*.

psychrophile, **psychrophilic** (Gr. *psychrós*, cold + *phílos*, have an affinity for): an organism that grows well at low temperatures, often near 0ºC. In the fungi, the species *Eupenicillium crustaceum* (Eurotiales) and *Penicillium spinulosum* (moniliaceous asexual fungi), among other examples, are psychrophiles, since their conidia germinate well and develop colonies of appreciable size at 5ºC. These molds, together with some species of *Fusarium* (tuberculariaceous asexual fungi) and yeasts (cryptococcaceous asexual yeasts and Saccharomycetales) cause biodeterioration of dairy products and other foods maintained under refrigeration. There also are *thermotolerant* fungi, which are those that grow well at temperatures below 10ºC, with the optimum less than 20ºC. Cf. **mesophile** and **thermophile**.

pterate (NL. *pteratus*, provided with wings < Gr. *pterón*, wing + suf. *-atus* > E. *-ate*, provided with or with the appearance of): with wings. Also called **alate**.

pterido-, pterid- (Gr. *pterís*, genit. *pterídos*, fern): a pref. that alludes to ferns, like the generic name *Pteridomyces* (syn. of *Athelopsis*, of the Aphyllophorales).

ptyophagous (Gr. *ptýo* < *ptyein*, to throw out, spit + *phágos*, what eats): *Endomycorrhizae*. Any form of mycetization in which the hyphae of the fungus undergo *plasmoptysis* and empty their plasmic contents (called *ptyosomes*) into the cell of the host plant that it has penetrated; afterwards the host cell absorbs and assimilates the extruded materials. Cf. **thamnisophagous** and **tolypophagous**.

ptyosome (Gr. *ptýo* < *ptyein*, to throw out, spit + *sõma*, body): see **ptyophagous**.

pubescent (L. *pubescens*, genit. *pubescentis*, with hairs of puberty, downy < *puber*, *pubes*, *pubis*, downy, hairy + suf. *-escens*, genit. *-escentis* > E. *-escent*, that which turns, that begins to, slightly): covered with fine, soft hair, like down. E.g., the surface of the pileus of some species of *Xerocomus* (Agaricales) and of *Auricularia polytricha* (Auriculariales) are pubescent. Cf. **glabrous**.

Pubescent lower surface of the basidiocarp of *Auricularia polytricha*, x 1 (*RV*).

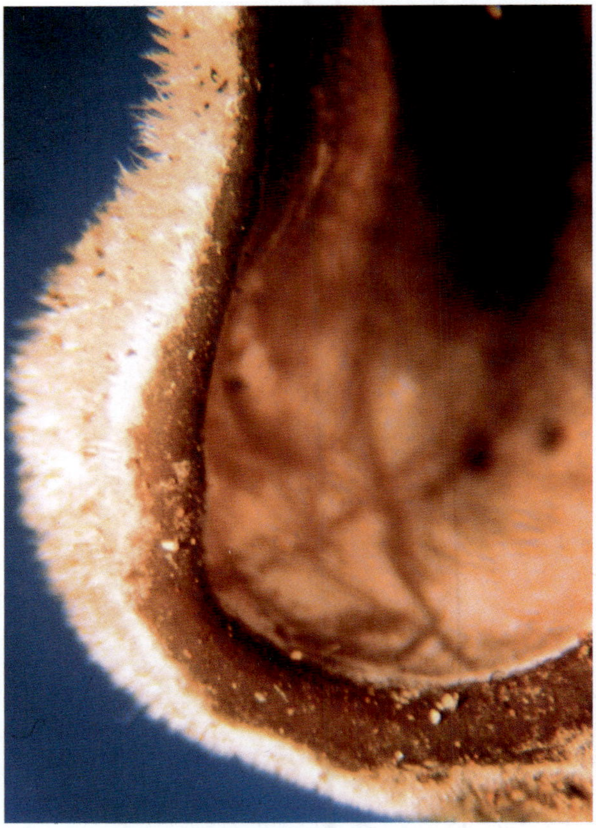

Pubescent lower surface of the basidiocarp of *Auricularia polytricha*, x 5 (*CB*).

puffball (ME. *puf, puffe* < *puffen*, to send forth + *bal, balle* < OF. *balle*, ball): the name given to species of the Lycoperdales that liberate clouds of spores when the outer peridium of the fruiting body (usually the endoperidium at maturity) is compressed rapidly, as occurs in *Calvatia* and *Lycoperdon* (Lycoperdales).

Puffball, basidiocarp of *Calvatia cyathiformis*, x 0.3 (*RV*).

Puffball, basidiocarp
of *Lycoperdon purpurascens*, x 1 (*RV*).

Puffballs, basidiocarps
of *Lycoperdon* sp., x 1 (*TH*).

puffing (E. *puff* < ME. *puffen* < OE. *pyffan*, to blow in short gusts; to exhale forcibly): a phenomenon in which thousands of asci in an apothecium liberate their ascospores forcibly and simultaneously, producing a cloud similar to smoke and a barely perceptible whistle. The basidiomata of the Lycoperdales are called *puffballs* due to the manner in which they liberate their spores, as a puff, when they are pressed or hit, e.g., by raindrops.

pullulate (L. *pullulare*, to bud and put out shoots): to originate, to derive one thing from another, to abound and move about; applied to the budding of yeasts.

pulque (Azt. *poliuhqui*, spoiled): the fermented sap (mead) of various species of *Agave*, consumed as an alcoholic and nutritious beverage in many regions of Mexico for many centuries. The microorganisms involved in the fermentation of the sap includes a wide range of bacteria, mainly producers of lactic acid and ethanol (*Lactobacillus*, *Leuconostoc*, *Zymomonas*), and yeasts, especially *Saccharomyces*, *Pichia* (Saccharomycetales) and *Candida* (cryptococcaceous asexual yeasts). Cf. **tuba**. See **fermentation**.

Plant of maguey (*Agave atrovirens*),
from whose fermented sap **pulque** is obtained, x 0.01 (*MU*).

Plant of maguey (*Agave atrovirens*),
from whose fermented sap **pulque** is obtained, x 0.01 (*MU*).

pulvillus, pl. **pulvilli** (L. *pulvillus*, small cushion): a turgescent, gelatinous, pulvinate swelling that is formed in the ostiolar canal of the perithecium of *Turgidosculum ulvae*, a marine ascomycete of uncertain affinity. The pulvillus functions as a plug that impedes the entrance of water into the perithecial cavity, which contains a gelatinous matrix that covers the asci, protecting them from desiccation when the perithecia are exposed to the air, together with their algal host (*Blidingia minima*=*Ulva vexata*), during low tide. The pulvillus closes the ostiole of wet perithecia by swelling, but when it dries it shrinks and puts pressure on the asci, which liberate the ascospores through the ostiole. Rehydration makes the pulvillus swell and close the ostiolar canal, protecting the immature asci from exposure to water.

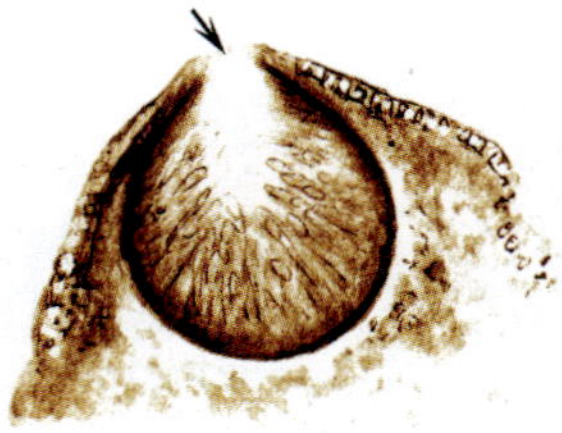

Pulvillus closing the ostiolar canal and covering the periphyses
of the perithecium of *Turgidosculum ulvae*, which
is a parasite of the alga *Blidingia minima*, x 660 (*MU*).

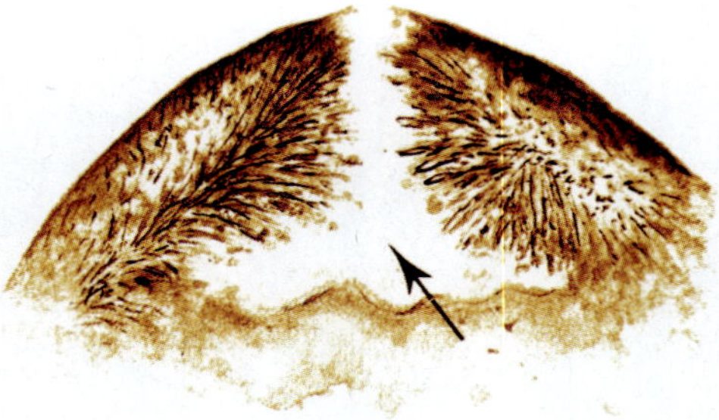

Pulvillus (an enlargement of the previous figure), x 1,500 (*MU*).

pulvinate, **pulvinated**, **pulviniform** (L. *pulvinus*, cushion, pillow + suf. *-atus* > E. *-ate*, provided with or likeness; or + suf. *-formis* < *forma*, shape): refers to any structure shaped like a small cushion, like the conidial stromata of *Tubercularia vulgaris* (tuberculariaceous asexual fungi) and the perithecial stromata of *Annulohypoxylon* (=*Hypoxylon*) *thouarsianum* and *Camarops* (=*Nummularia*) *lutea* (Xylariales).

Pulvinate conidial stroma of *Tubercularia vulgaris*, x 10 (*RTH*).

Pulvinate perithecial stromata of *Annulohypoxylon* (=*Hypoxylon*) *thouarsianum*, x 2 (*EPS*).

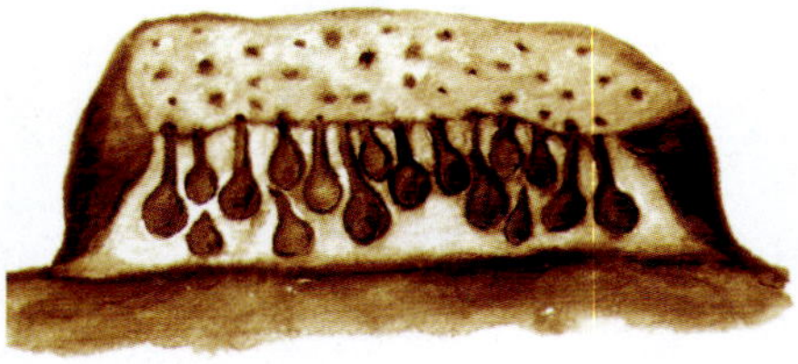

Pulvinate perithecial stroma of *Camarops* (=*Nummularia*) *lutea*, x 9 (*MU*).

punctate (NL. *punctatus*, dotted or marked as if it were perforated < L. *punctum*, point, mark, perforation + suf. *-atus* > E. *-ate*, provided with or likeness): applied to surfaces (membranes, cell walls, etc.) that have more or less equidistant, microscopic engravings, projections or prominences that appear to be perforations. E.g., the perithecial stromata of *Poronia punctata* (Xylariales), the teliospores of *Sphacelotheca destruens* (Ustilaginales), and the basidiospores of *Cortinarius punctatus* and *Panaeolina foenisecii* (Agaricales) are punctate.

Punctate perithecial stroma of *Poronia punctata*, x 10 (*MU*).

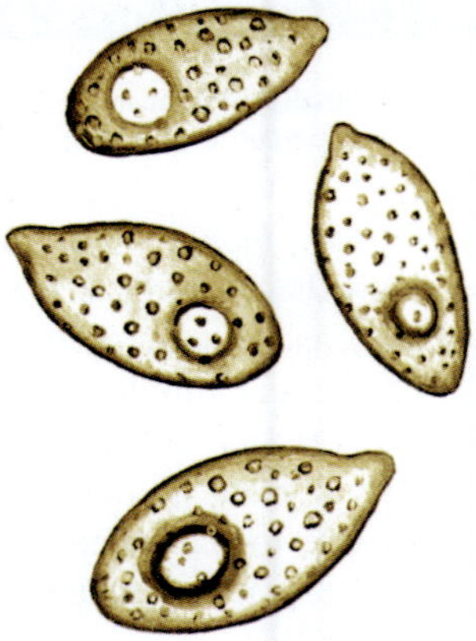

Punctate wall of the basidiospores of *Panaeolina foenisecii*, x 1,200 (*MU*).

punctiform (NL. *punctiformis* < *punctum*, point + suf. *-formis* < *forma*, shape): like a point; small colonies (of yeasts and bacteria) less than 1 mm diam. are called punctiform, although they can be seen without the aid of a magnifying lens.

purse (ME. < OE. *purs* < *pursa,* bag < ML. *bursa*, bag < Gr. *býrsa*, hide, leather): *Gasteromycetes*. The upper part of the funiculus which contains the coiled funicular cord in some species of Nidulariales, such as *Cyathus striatus*.

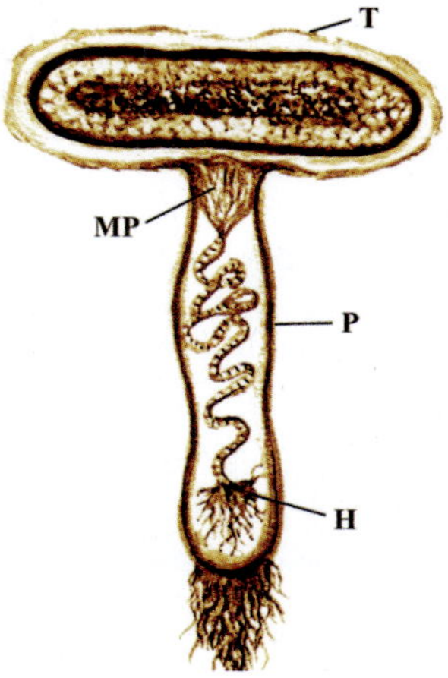

Purse (P) of a peridiolum of *Cyathus striatus*. In this diagram of a longitudinal section, the hapteron (H), middle piece (MP) and tunica (T) also are shown, x 20 (*MU*).

pustulate, **pustulose** (L. *pustulatus*, pustulate, blistered < *pustula*, pustule, blister + suf. *-atus* > E. *-ate*, provided with or likeness; L. *pustulosus*, full of pustules < *pustula* + suf. *-osus* > E. *-ose*, full of, augmented, prone to): covered with ampullae or pustules, like the lesions on stems and leaves caused by various plant pathogenic fungi, e.g., the white rusts (*Albugo*, of the Peronosporales) and the rusts (*Puccinia*, of the Uredinales).

Pustulate leaf of geranium; the pustules are uredinia of *Puccinia malvacearum*, x 1.5 (*MU*).

pustule (L. *pustula*, pustule, blister): a small ampulliform elevation formed by fructifications of plant pathogenic fungi, or by lesions that they originate in the epidermal tissues of the host plant, like the uredinia and telia of the rust fungi (e.g., *Puccinia malvacearum*, of the Uredinales). In the lichens, the small ampulliform prominences present on the surface of the thallus are called pustules; the thallus is called *pustulate*. The surface of the apothecium of some Pyronemataceae, such as *Tarzetta* (=*Pustularia*), also is pustulate, which gives the name to this gen.

Pustules (uredinia) of *Puccinia malvacearum* on a leaf of hollyhock, x 2 (*JB*).

Pustules (uredinia) of *Puccinia graminis tritici* on a wheat leaf, x 12 (*MU & CB*).

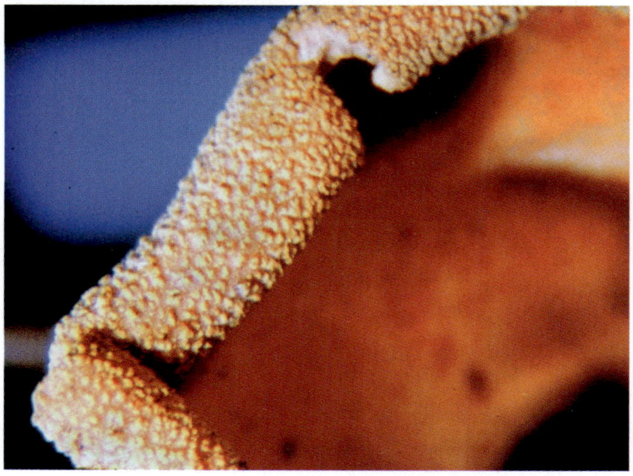

Pustules on the surface of the apothecium of *Tarzetta* (=*Pustularia*) *catinus*, x 12 (*MU*).

putrescent (L. *putrescens*, genit. *putrescentis*, undergoing putrefaction, rotting < *putrescere*, to become rotten + suf. *-escens*, genit. *-escentis* > E. *-escent*, that which turns, that begins to, slightly): rotting or decomposing rapidly, becoming soft and pulpy.

Putrescent basidiocarp of *Coprinus* sp. (Agaricales), x 1 (*RV*).

Putrescent basidiocarps of *Coprinopsis atramentaria* (=*Coprinus atramentarius*, Agaricales); on maturing they become deliquescent, x 1 (*RV*).

pycnidial stroma, pl. **stromata** (NL. *pycnidium* < Gr. *pyknós*, dense, packed, concentrated + dim. suf. *-ídion* > L. *-idium* + L. suf. *-alis* > E. *-al*, relating to or belonging to; Gr. *strõma*, bed, small pillow): a stroma bearing pycnidia, like that of *Aschersonia aleyrodis* and *Endothiella gyrosa* (sphaeropsidaceous asexual fungi).

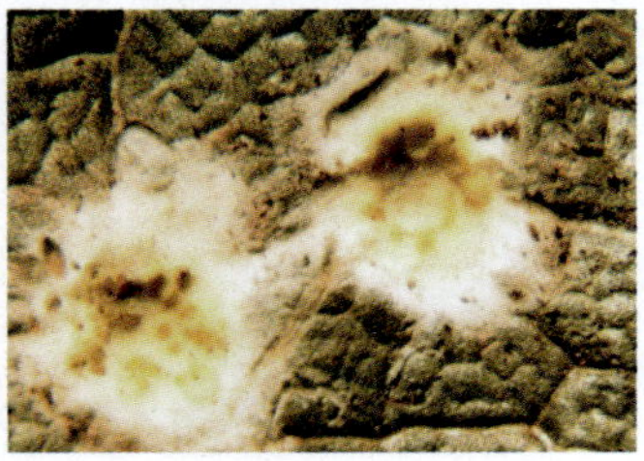

Pycnidial stromata of *Aschersonia aleyrodis*, on scale insects on a lemon leaf, x 10 (*MU*).

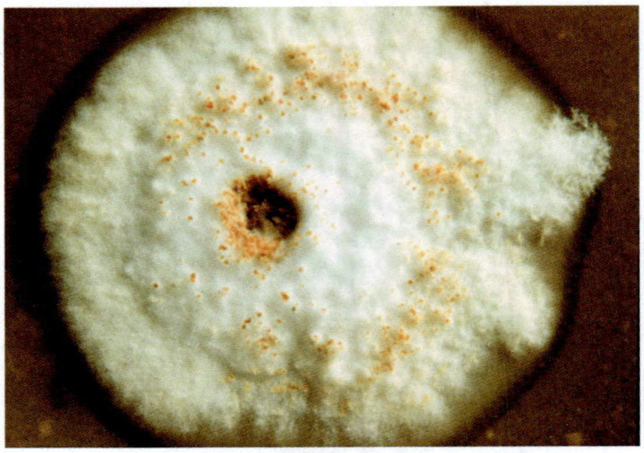

Pycnidial stroma of *Endothiella gyrosa*, on agar, x 1 (*MU*).

pycnidiospore (Gr. *pyknós*, dense, compact, concentrated + dim. suf. *-ídion* + *sporá*, spore): a spore produced in a pycnidium, as in *Phoma exigua* (=*Ph. herbarum*) and *Endothiella gyrosa* (sphaeropsidaceous asexual fungi).

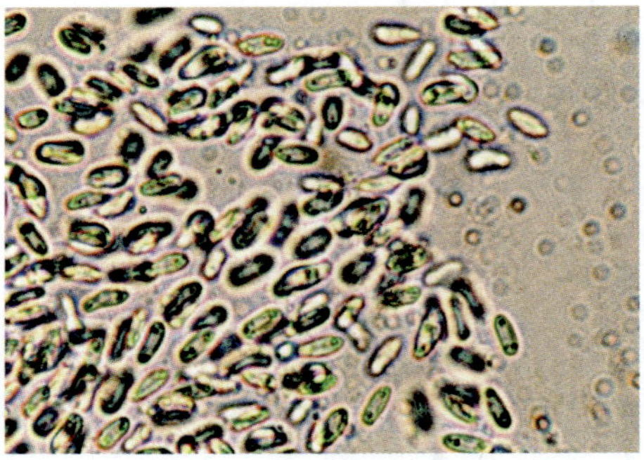

Pycnidiospores of a pycnidial cirrhus of *Phoma* sp., x 600 (*MU*).

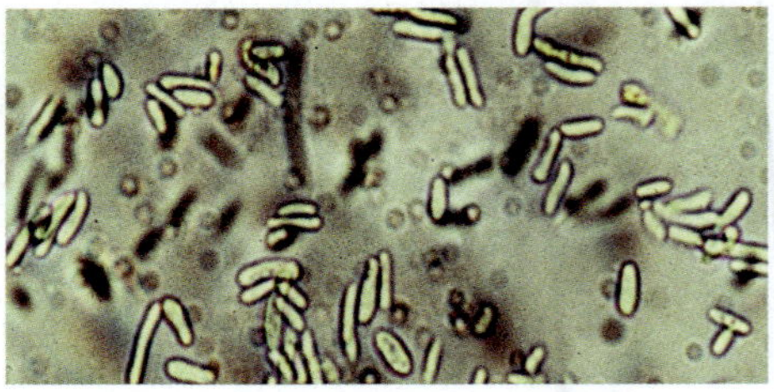

Pycnidiospores of a pycnidial cirrhus of *Endothiella gyrosa*, x 440 (*MU*).

pycnidium, pl. **pycnidia** (NL. *pycnidium* < Gr. *pyknós*, dense, compact, concentrated + dim. suf. *-ídion* > L. suf. *-idium*): an asexual fruiting body, generally spherical or obpyriform, with an internal cavity lined with conidiophores; characteristic of the sphaeropsidaceous asexual fungi, such as *Phoma*, *Phomopsis*, *Pyrenochaeta*, *Sphaeronaema* and *Botryodiplodia*.

Pycnidia of *Sphaeronaema* sp., x 20 (*MU*).

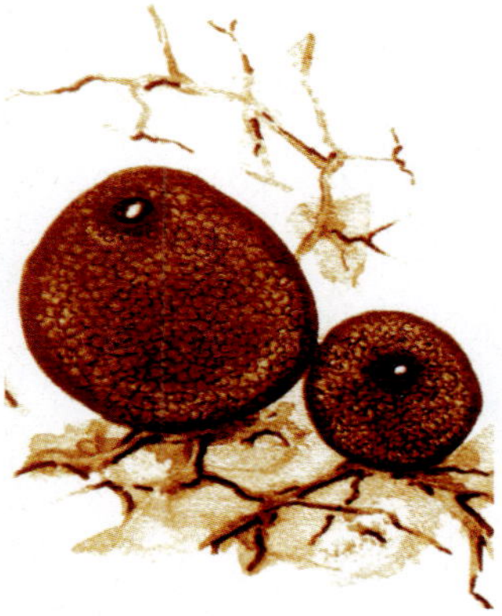

Pycnidia of *Phoma* sp., x 280 (*MU*).

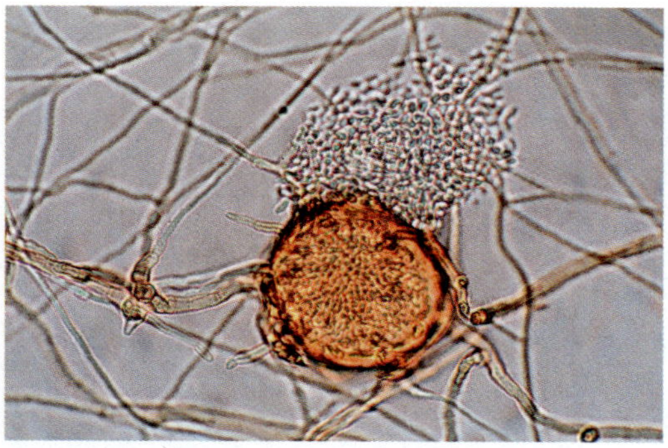

Pycnidium of *Phoma* sp., x 280 (*MU*).

pycniospore (Gr. *pyknós*, dense, compact, concentrated + *sporá*, spore): a spore formed in a pycnium. See **spermatium**.

pycnium, pl. **pycnia** (NL. *pycnium* < Gr. *pyknós*, dense, thick, compact + L. dim. suf. *-ium*): in the Uredinales it corresponds to the *spermogonium* or *pycnidium*-like structure. In general, it is utilized to refer to a cavity in which haploid pycniospores (spermatia) are formed.

pycno-, pycn- (Gr. *pyknós*, dense, compact, solid, thick, crowded): a pref. used in terms that allude to something that is piled up or concentrated, like the following.

pycnoascocarp (Gr. *pyknós*, dense, compact, alluding to a pycnidium + *askós*, wine-skin, sac; here, ascus + *karpós*, fruit): *Lichens*. A peculiar type of apothecium that forms from a pycnidium; it begins with the formation of ascogonia beneath a pycnidium, followed by the development of paraphyses and asci among the conidiophores of the pycnidium, and the transformation of the pycnidial wall into the excipulum of the mature apothecium. It is characteristic of the family Lichiniaceae, such as *Ephebe* and *Zahlbrucknerella* (Lichinales). Cf. **thallinocarp**.

pycnosclerotium, pl. **pycnosclerotia** (Gr. *pyknós*, dense, compact, concentrated + *sklerotés*, hardness < *sklerós*, hard + L. dim. suf. *-ium*): a plectenchymatous body with an outer layer of thick, hard cells, similar to a pycnidium, since the interior becomes hollow, but lacking spores, as is observed in *Phyllosticta* (=*Phyllostictina*) *carpogena* (sphaeropsidaceous asexual fungi). Apparently this term has fallen into disuse.

pycnosis, **pyknosis**, pl. **pycnoses**, **pyknoses** (Gr. *pyknós*, dense, compact, concentrated + suf. *-osis*, state or condition): *Microthyriaceae* (Dothideales). A process in which a rounded or elliptical portion of the thallus arches over and thickens, while the ascigerous hymenium is formed beneath it.

pycnostroma, pl. **pycnostromata** (Gr. *pyknós*, dense, compact, concentrated, solid, thick, packed + *strõma*, small cushion, bed): a stroma in which pycnidia are embedded. An asexual fructification in the shape of a cushion or a flattened pycnidium in which the conidiophores grow downward from the roof or upper wall; characteristic of pycnidial fungi, such as *Munkia* (=*Pycnostroma*) and *Peltasterella*, gen. that some classify in the sphaeropsidaceous asexual fungi or in the pycnothyriaceous asexual fungi. Also called **pycnothyrium**.

pycnothecium, pl. **pycnothecia** (NL. *pycnium* < Gr. *pyknós*, dense, thick, compact + Gr. *thêke*, case + dim. suf. *-íon* > L. *-ium*): an ascoma formed by *pycnosis*, in ascomycetes of the family Microthyriaceae.

pycnothyriaceous (Gr. *pyknós*, *pychnós*, compact, dense, solid, strong + *thyrís*, genit. *thyridós*, a window + suf. *-aceus* > E. *-aceous*, of or pertaining to, with the nature of): applied to the anamorphic or asexual fungi (e.g., *Peltasterella*) that form their conidia in a flat pycnidium (pycnothyrium), in which the conidiophores grow from the roof or upper wall downward. See **Coelomycetes**.

pycnothyrium, pl. **pycnothyria** (Gr. *pyknós*, dense, compact, concentrated + *thirís*, window + L. dim. suf. *-ium*): an asexual fructification shaped like a cushion or a flattened pycnidium, in which the conidiophores grow downward from the roof or upper wall; characteristic of *Pycnothyrium* and *Peltasterella* (sphaeropsidaceous or pycnothyriaceous asexual fungi). Also called **pycnostroma**.

pyramidal (ML. *pyramidalis* < L. *pyramis* < Gr. *pyramís*, pyramid + L. suf. *-alis* > E. *-al*, relating to or pertaining to): shaped like a pyramid, i.e., a structure with a polygonal base and three or four triangular sides that unite in a point, like the scales on the pileus of *Amanita strobiliformis* (Agaricales), and on the fruiting body of *Lycoperdon candidum* and *L. echinatum* (Lycoperdales).

Pyramidal peridial scales of the fruiting body of *Amanita strobiliformis*, x 1 (*RV*).

Pyramidal peridial scales of the fruiting body of *Lycoperdon candidum*, x 2 (*MU*).

Pyramidal peridial scales of the fruiting body of *Lycoperdon echinatum*, x 10 (*MU*).

pyramid-shaped wart (L. *pyramis* < Gr. *pyramís*, pyramid + E. *shaped*, pp. of *shape* < ME. *shapen* < OE. *sceppan, scyppan*, to form, shape; ME. < OE. *wearte*, wart, prominence): a wart with a square base and three or four triangular sides, sharp-pointed or nearly so, often zonate with lines, like those of the peridium of the fruiting body of *Calvatia* (Lycoperdales).

pyrenocarp (Gr. *pyrén*, genit. *pyrênos*, stone of the drupaceous fruits, stone, nucleus + *karpós*, fruit): syn. of **perithecium**; sometimes used informally for any fungus with an ascoma (ascocarp) similar to a perithecium. Lichens with perithecia (such as *Dermatocarpon*, Verrucariales, *Porina*, Trichotheliales, and *Pyrenula*, Pyrenulales) are called *pyrenocarpic*.

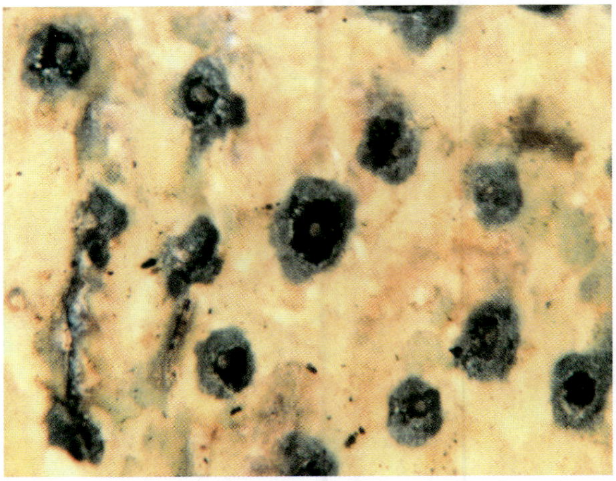

Pyrenocarpic lichen thallus of *Pyrenula nitida*, x 10 (*MU*).

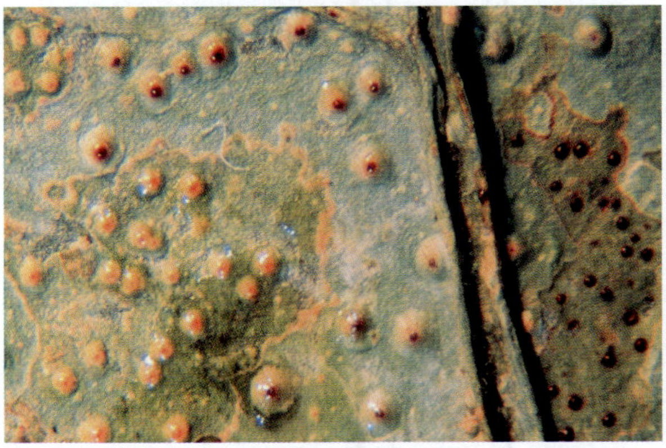

Pyrenocarpic lichen thallus of *Porina epiphylla*, x 10 (*MU*).

pyrenolichen (Gr. *pyrén*, genit. *pyrênos*, stone of a drupaceous fruit, stone, nucleus + L. *lichen* < Gr. *leichén*, lichen): a lichen in whose thallus perithecia are formed, as occurs in the gen. *Porina*, *Trichothelium* (Trichotheliales) and *Strigula* (Dothideales).

Thallus of *Porina atropunctata*, a tropical foliicolous **pyrenolichen**; the orifices are the ostioles of the perithecia, x 20 (*MAH*).

Thallus of *Porina papillifera*, a tropical foliicolous **pyrenolichen**; the orifices are the ostioles of the perithecia, x 20 (*MAH*).

Thallus of *Trichothelium minum*, a tropical foliicolous **pyrenolichen**; the orifices are the ostioles of the perithecia, x 20 (*MAH*).

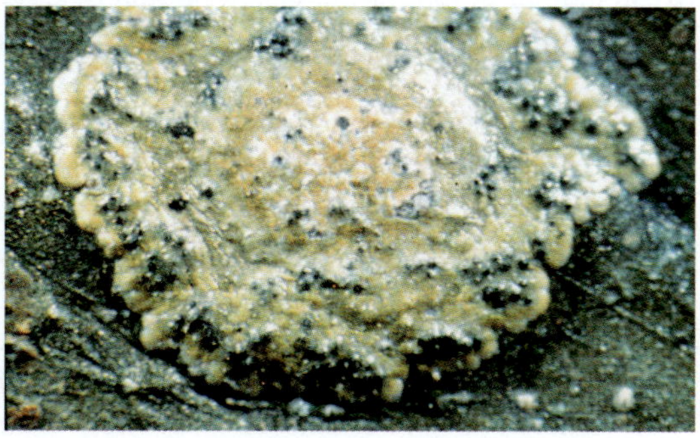

Thallus of *Strigula schizospora*, a tropical foliicolous **pyrenolichen**; the orifices are the ostioles of the perithecia, x 20 (*MAH*).

Pyrenomycetes (Gr. *pyrén*, genit. *pyrênos*, stone of the drupe fruits, stone, nucleus + L. *-mycetes*, ending of taxonomic class < Gr. *mýkes*, genit. *mýketos*, fungus): a class of ascomycetes with the asci in an hymenium or fascicle contained in a perithecium type of ascocarp. According to the classification adopted in this dictionary (Alexopoulos *et al.*, 1996), the Pyrenomycetes includes the orders Hypocreales, Melanosporales, Microascales, Phyllachorales, Ophiostomatales, Diaporthales, Xylariales, Sordariales, Meliolales and Halosphaeriales. Together with the classes Archiascomycetes, Plectomycetes, Discomycetes, and Loculoascomycetes, as well as the ascosporogenous yeasts (Saccharomycetales), and other filamentous ascomycetes (such as the Erysiphales, Laboulbeniales and Spathulosporales, which are not included in any of the classes mentioned above), they constitute the phylum Ascomycota of the kingdom Fungi. See the Appendix, where there is a comparison of the outline of Alexopoulos *et al.*, 1996, modified by Ulloa and Hanlin, 2006, to that of Kirk *et al.*, 2008. In the latter classification scheme, the class Pyrenomycetes is considered an unnatural group of fungi, an artificial classification, and has been changed to class Sordariomycetes, subphylum Pezizomycotina, phylum Ascomycota; this has been done in order to achieve a phylogenetic classification of the fungi.

pyriform or **piriform**(NL. *pyriformis* < ML. *pyrus* < L. *pirus*, pear + suf. *-formis* < *forma*, shape): pear-shaped, like the basidiospores of *Cortinarius pseudosalor* (Agaricales), and the sporangia of *Metatrichia* (=*Trichia*) *floriformis* (Trichiales), *Thamnostylum* (=*Helicostylum*) *piriforme* and *Pirella circinans* (Mucorales). Cf. **obpyriform**.

Pyriform sporangia of *Metatrichia* (=*Trichia*) *floriformis*, x 20 (*MU*).

Pyriform sporangia of *Metatrichia* (=*Trichia*) *floriformis*, x 30 (*MU*).

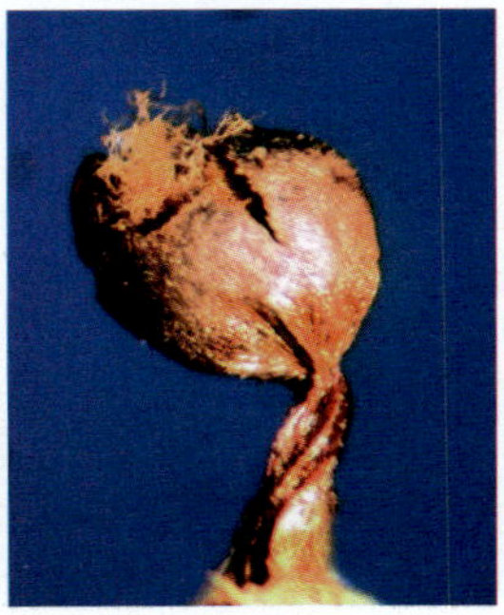

Pyriform sporangium of *Metatrichia* (=*Trichia*) *floriformis*, x 80 (*MU*).

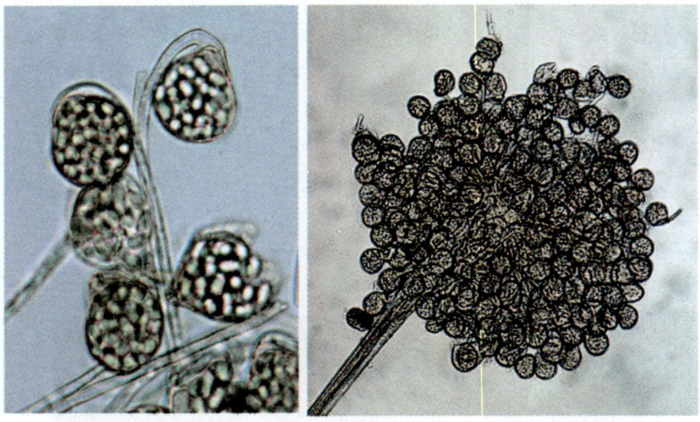

Pyriform sporangia of *Thamnostylum* (=*Helicostylum*) *piriforme*, x 400, x 1,300 (*MU*).

Pyriform sporangia of *Pirella circinans*, x 200 (*MU*).

pyro-, pyr- (Gr. *pýr*, genit. *pyrós*, fire): a pref. employed in terms that allude to fire, such as *pyrophilous*.

pyrophilous (Gr. *pýr*, genit. *pyrós*, fire + *phílos*, have an affinity for): growing on substrates that have been burned or steam sterilized, like soil or wood, such as *Pyronema* (Pezizales). See **anthracobiotic**, **carbonicolous** and **phenicoid fungus**.

Pyrophilous ascomata of an undetermined species of Pezizales, x 1 (*RV*).

pyroxylophilous (Gr. *pýr*, genit. *pyrós*, fire + *xýlon*, wood + *phílos*, have an affinity for): growing on burned wood, like *Pyronema* (Pezizales), certain species of *Xylaria* (Xylariales) and *Coltricia* (Aphyllophorales).

Pyroxylophilous basidiocarp of *Coltricia perennis*; note the burned wood, x 1 (*RV*).

pythiosis, pl. **pythioses** (< gen. *Pythium* < Gr. *pýtho*, rottenness, to make rotten + L. dim. suf. *-ium* + Gr. suf. *-osis*, state or condition): a cosmopolitan disease of horses, cattle and dogs, caused by *Pythium insidiosum* (Peronosporales).

pyxidate (L. *pyxidatus*, with a box or case, or similar to a case; cubic < Gr. *pyxís*, wooden box with a lid + L. suf. *-atus* > E. *-ate*, provided with or likeness): like a box; shaped like a box; provided with a *pyxidium*, which is a box.

pyxidium, pl. **pyxidia** (NL. *pyxidium* < Gr. *pyxídion* < *pyxís*, wooden box with a lid + dim. suf. *-ídion* > L. *-idium*): a small box or container, usually cylindrical, with a lid, like the ascocarp of *Pyxidiophora* (Laboulbeniales).

q

Quercicole, a fruiting body of *Inonotus quercustris* (*RV*).

q

quadrate arm (L. *quadratus*, squared < *quadrus*, fourfold + suf. *-atus* >E. -ate, provided with or likeness; ME. < OE. *earm*, arm): a receptaculum or arm in the Phallales which is 4-sided in cross section, as in *Clathrus columnatus*.

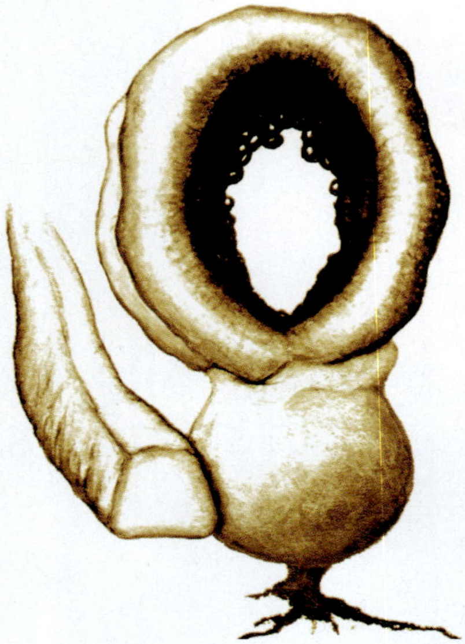

Quadrate arms of the clathrate fruiting body of *Clathrus columnatus*, lined with the gleba, x 0.5 (*MU*).

quadri-, quadr- (L. *quatuor*, four > pref. *quadri-*, *quadr*): a pref. used to form Latin words or words derived from Latin, to introduce the idea of quadruplicity, as in *quadriverticillate*.

quadriverticillate (L. *quatour*, four > *quadri-* > E. *cuadri-* + NL. *verticillatus* < *verticillus*, verticil + suf. *-atus* > E. *-ate*, provided with or likeness): branched at four levels, i.e., with branches (rami), branchlets (ramuli), metulae and phialides, as occurs in the conidiophores of *Penicillium chrysogenum* and *P. aurantiogriseum* var. *viridicatum* (=*P. viridicatum*), moniliaceous asexual fungi. Cf. **monoverticillate**, **biverticillate**, **triverticillate** and **polyverticillate**.

Quadriverticillate conidiophore of *Penicillium aurantiogriseum* var. *viridicatum* (=*P. viridicatum*), x 770 (*CB*).

Quellkörper (G. *quellen*, to swell + *Körper*, body): *Ascomycetes*. A mass of swollen and gelatinized cells formed at the apex of the ascocarp of certain pyrenomycetes, which disintegrate at maturity, liberating the ascospores. It is characteristic of *Bertia moriformis* (Sordariales).

quercicole, **quercicolous** (L. *quercus*, oak + suf. *-cola*, inhabitant): an inhabitant of oak trees, like the fruiting body of *Inonotus quercustris* and *Phellinus caribaeo-quercicolous* (Aphyllophorales).

Quercicole, a fruiting body of *Inonotus quercustris*, growing on the bark of oak (*Quercus* sp.), x 0.2 (*RV*).

r

Radial colony of *Penicillum aurantiogriseum*, on agar (*MU*).

r

r-selection (*r* is for rapid; L. *selectio*, genit. *selectionis*, selection of one organism or thing from among others): a term used in population biology to designate an adaptation to the rapid colonization of uncolonized or recently opened habitats; in the fungi, this selection involves large quantities of small, short-lived propagules produced asexually, as, e.g., conidia and soredia. Cf. **K-selection**.

race (MF. *race* <OIt. *razza*, generation; an actual or potentially interbreeding group within a species): a group of *biotypes* with a similar virulence-avirulence pattern on a particular group of plants. Races, like *formae speciales* (abbrev. f. sp.), are defined on the basis of physiological differences and may be capable of attacking only one particular variety of host plant. E.g., nearly 350 races of *Puccinia graminis* forma specialis *tritici* (Uredinales), the wheat rust, have been identified, that are very similar or identical morphologically but which differ in their capacity to parasitize a species or group of species of host plants. See **physiological race**.

racemose, racemous (L. *racemus*, raceme of grapes + suf. *-osus* > E. *-ose*, or > E. *-ous*, full of, augmented, prone to): shaped like a raceme, like the stipe of *Dendrocollybia* (=*Collybia*) *racemosa* (Agaricales), which has lateral branches, the sporangiophore of *Mucor racemosus* (Mucorales) and the conidiophore of *Botryosporium* (moniliaceous asexual fungi).

Racemose sporangiophores of *Mucor racemosus*, x 300 (*MU*).

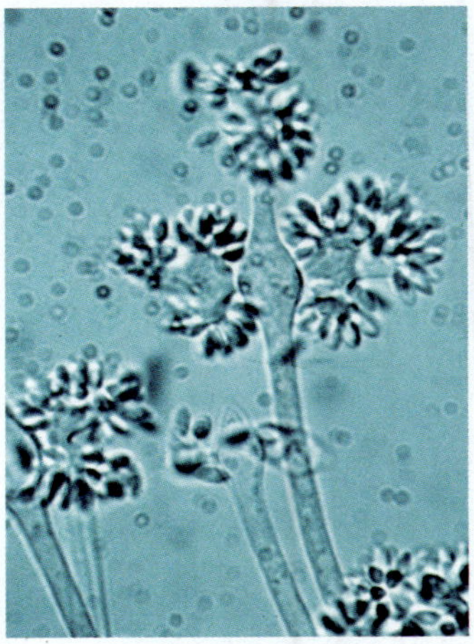

Racemose conidiophores of *Botryosporium* sp., x 300 (*MU*).

rachiform (Gr. *ráchis*, axis, spine + L. suf. *-formis* < *forma*, shape): a narrow conidiogenous axis (*rachis*) with sympodial growth that gives rise to a geniculate (zigzag) conidiophore, which produces conidia in basipetal succession, as occurs in some gen. of moniliaceous (*Tritirachium*, *Beauveria*, *Pyricularia*) and dematiaceous (*Zasmidium* (=*Racodium*) *cellare*) asexual fungi. Cf. **ampulliform** and **raduliform**.

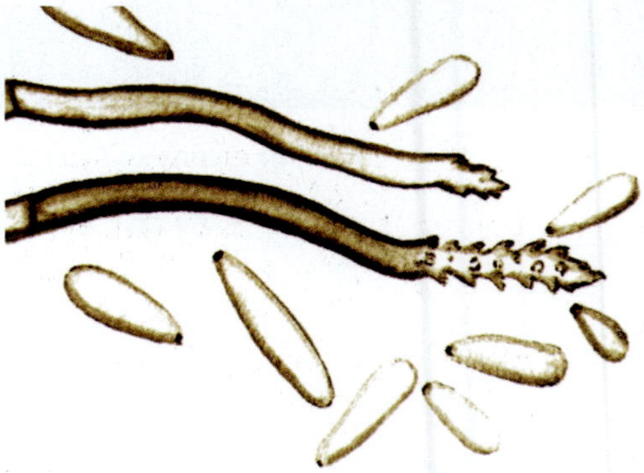

Rachiform conidiophores of *Zasmidium* (=*Racodium*) *cellare*. Note the **rachis** or conidiogenous axis, with sympodially arranged denticles bearing conidia of the *Pleurophragmium* (=*Acrotheca*) type, and conidia of the *Cladosporium* type, x 600 (*MU*).

rachis, pl. **rachises** (Gr. *ráchis*, axis, spine): a geniculate or zigzag extension of a conidiogenous cell, like that of *Tritirachium* and *Beauveria* (moniliaceous asexual fungi) and *Rhinocladiella* (dematiaceous asexual fungi), that results from the sympodial development of the condiogenous cell and which produces spores sympodially along the major axis. See **rachiform**.

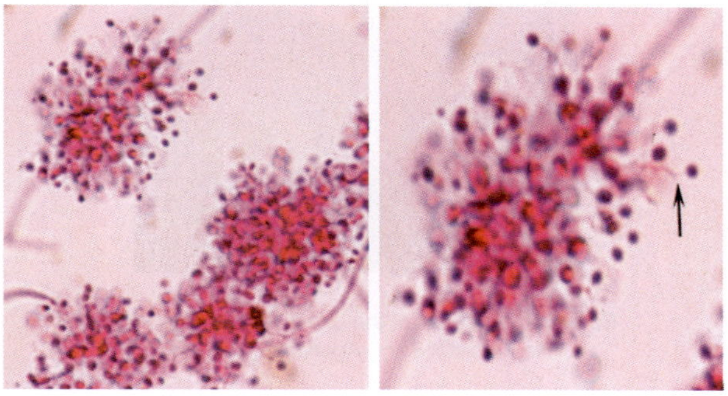

Rachis of the conidiophores of *Beauveria bassiana*, x 900, x 1,800 (*MU*).

racket cell, **racquet cell** (ME. *raket*, a kind of handball < OF. *rachette*, palm of the hand, racket < It. *rachetta* < Ar. *rahat*, palm of the hand; NL. *cellula*, cell, living cell, dim. of L. *cella*, small room, compartment, cell of a honeycomb): with the shape of a racket or with segments of such a shape, like the hyphae of the dermatophytes *Microsporum* and *Trichophyton* and of *Malbranchea aurantiaca* (of the moniliaceous asexual fungi). See **racket hypha**.

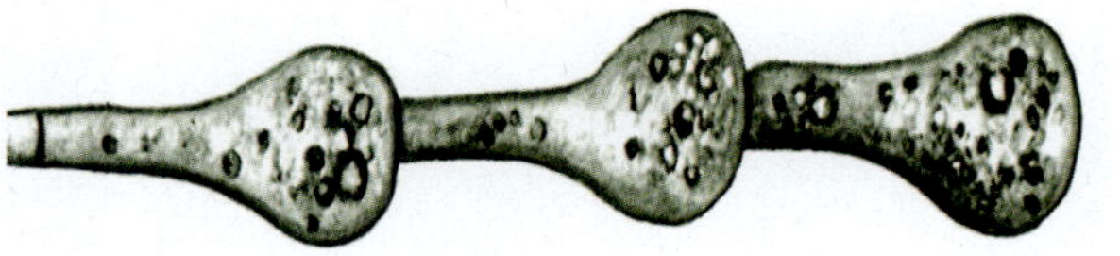

Racket cells of a somatic hypha of *Trichophyton mentagrophytes*, x 1,000 (*MU*).

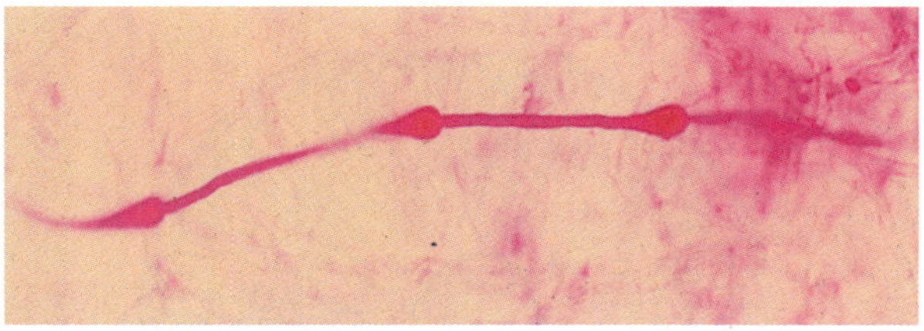

Racket cells of a somatic hypha of *Malbranchea aurantiaca*, x 400 (*MU*).

racket hypha, **racquet hypha**, pl. **hyphae** (ME. *raket*, a kind of handball < OF. *rachette*, palm of the hand, racket < It. *rachetta* < Ar. *rahat*, palm of the hand; NL. *hypha*, hypha < Gr. *hyphé*, tissue, spiderweb; hypha): having the shape of a racket or with racket-shaped cells; these hyphae are characteristic of various common dermatophytes, such as *Microsporum* and *Trichophyton* (moniliaceous asexual fungi), although they are also present in other types of moniliaceous keratinolytic fungi, such as *Malbranchea aurantiaca*.

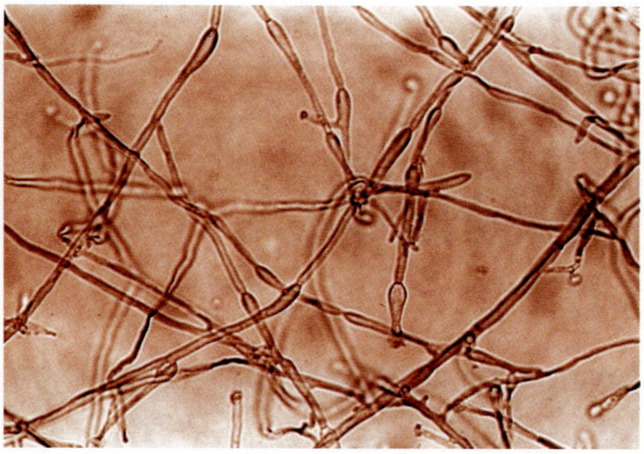

Racket hyphae of *Malbranchea aurantiaca*, x 500 (*MU*).

radial (ML. *radialis*, arranged like rays < L. *radius*, ray + suf. *-alis* > E. *-al*, relating to or pertaining to): developing uniformly around a central point or axis, like the lobes of certain spores, e.g., the ascospores of *Tripospora tripos* (Coryneliales). **1**. *Lichens*. In some fruticulose thalli (such as *Usnea*, *Alectoria* and other Lecanorales) the symmetry is radial in transverse section. **2**. *Agaricales*. The arrangement of the gills in the pileus of the agarics is radial, e.g., *Coprinus* and *Xeromphalina*. Also called **radiate**.

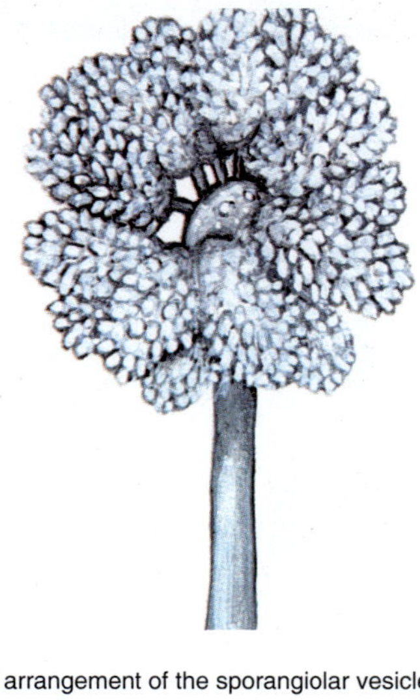

Radial arrangement of the sporangiolar vesicles of the sporangiophore of *Radiomyces embreei* (Mucorales), x 200 (*MU*).

Radial disposition of the chains of blastospores, in a giant colony of *Candida* sp. (cryptococcaceous asexual yeasts) on agar, x 40 (*TH*).

Radial arrangement of the gills in the hymenophore of *Xeromphalina* sp. (Agaricales), x 2 (*MU*).

Radial disposition of the wrinkles of a colony of *Penicillium aurantiogriseum* (moniliaceous asexual fungi) on agar, x 2 (*MU*).

Radial arrangement of the lamellae of a fruiting body of *Schizophyllum commune* (Aphyllophorales) on bark, x 1 (*RV*).

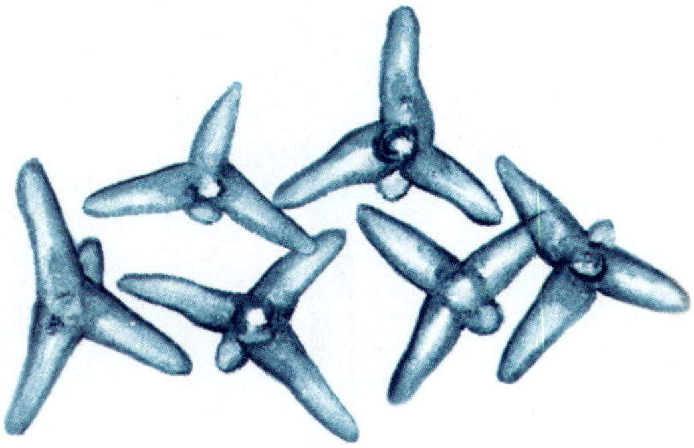

Radial arrangement of the lobes of the ascospores of *Tripospora tripos* (Coryneliales), x 800 (*MU*).

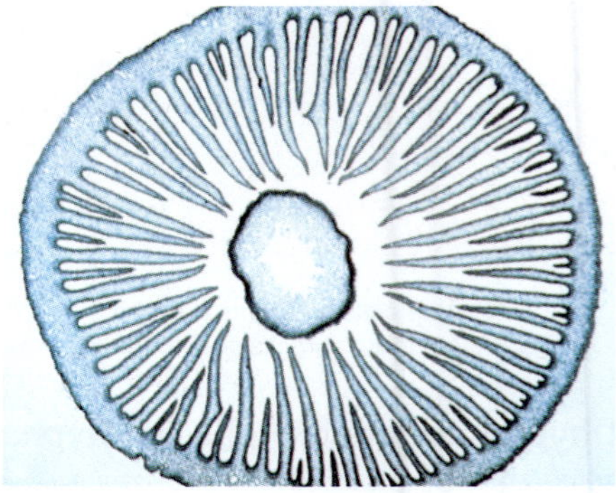

Radial arrangement of the gills in the hymenophore of *Coprinus* sp., seen in transverse section of the pileus, x 7 (*MU*).

radiate (L. *radiatus*, radiate, surrounded by rays < *radius*, ray + suf. *-atus* > E. *-ate*, provided with or likeness): extending out from or arranged around a common center, like the gills of the pileus of the Agaricales. Also called **radial**.

radiating plate (E. *radiating*, adj. of *radiate* < L. *radiatus*, rayed, surrounded by rays < *radius*, ray; ME. *plate* < OF. *plate*, flat object < VL. *plattus* < Gr. *platýs*, broad, flat): *Gasteromycetes*. Each one of the veins or plates of sterile tissue that proliferate throughout the gleba, originating radially from a gelatinous-watery columella; they are present in species of Phallales, family Protophallaceae, such as *Kobayasia*. In the fruiting bodies of *Truncocolumella* (Hymenogastrales), the radial plates proliferate from a thick, fleshy pedicel-columella, forming locules or irregular chambers in the gleba.

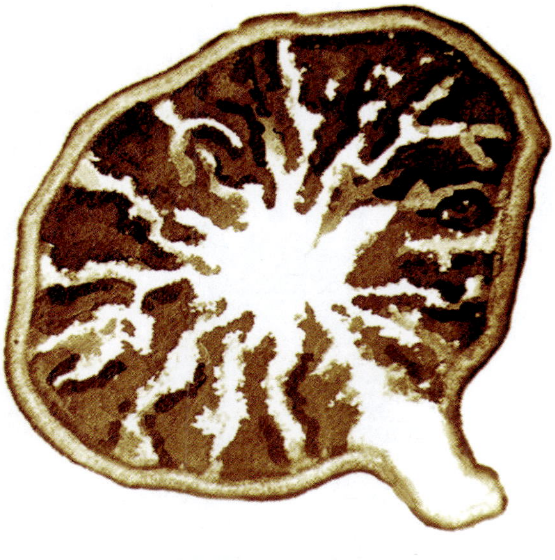

Radiating plates traversing the gleba; they originate from the columella of the fruiting body of *Kobayasia nipponica*, seen in longitudinal section, x 2 (*MU*).

radicate (L. *radicatus*, having roots < *radix*, genit. *radicis*, root + suf. *-atus* > E. *-ate*, provided with or likeness): provided with structures shaped like roots. Generally applied to a stipe that has projections similar to roots, like the rhizomorphs of *Armillaria* (=*Armillariella*) *mellea* and the pseudorhiza of *Xerula* (=*Oudemansiella*) *radicata* and *Phaeocollybia attenuata* (Agaricales), as well as to certain types of cystidia, like those of *Tubulicium* (=*Tubulixenasma*) *vermiferum* (Aphyllophorales). See **pseudorhiza**.

radicating (L. *radicans*, genit. *radicantis* < *radicare*, to take root, to put out roots): producing roots or being capable of producing them; refers to stipes with this characteristic. See **radicate**.

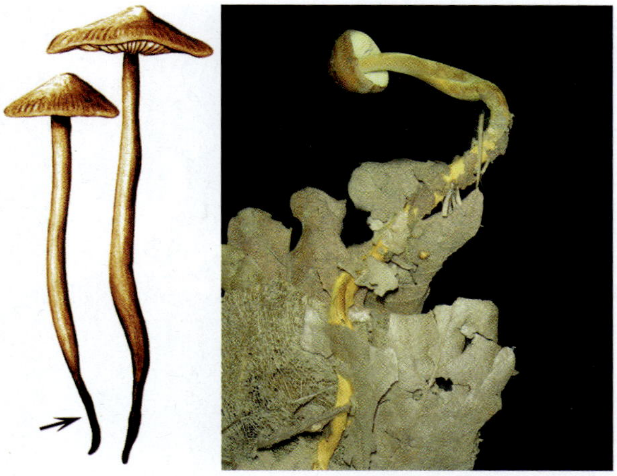

Radicate stipes of the basidiocarps of *Xerula* (=*Oudemansiella*) *radicata*, x 0.2 (*MU*) and (*EAA* & *SA*).

radicicolous (L. *radix*, genit. *radicis*, root + suf. *-cola*, inhabitant): living on roots, usually as a parasite, e. g., the fruiting bodies of *Trametes maxima* (Aphyllophorales) growing on the roots of a tropical tree.

Radicicolous fruiting bodies of *Trametes maxima*, inhabiting the roots of a tropical tree, x 0.05 (*RV*).

Radicicolous fruiting bodies of *Trametes maxima*, inhabiting the roots of a tropical tree, x 0.1 (*RV*).

Radicicolous fruiting bodies of *Trametes maxima*, inhabiting the roots of a tropical tree, x 0.5 (*RV*).

Radicicolous fruiting bodies of *Trametes maxima*, inhabiting the roots of a tropical tree, x 0.5 (*RV*).

radula, pl. **radulae** (L. *radula*, rasp): a structure (hypha, conidiophore, or ascospore) that becomes roughened on its surface, due to the formation of small denticles on which are borne conidia known as *radulaspores* or *radulospores*, as happens in *Arthrobotrys* and *Xylocladium* (moniliaceous asexual fungi). In *Nectria coryli* (Hypocreales) the ascospores, while still in the asci, become radulate, as viscous spores, supported by small denticles, form on their surface. See **raduliform**.

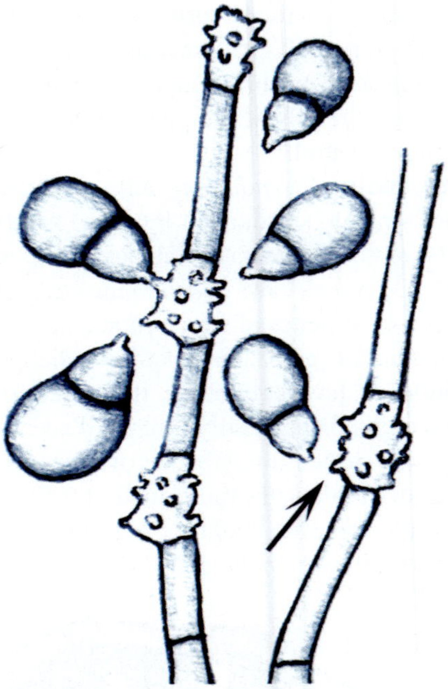

Radulae (conidiogenous cells) on the nodes of the **raduliform** conidiophores of *Arthrobotrys oligospora*, x 800 (*MU*).

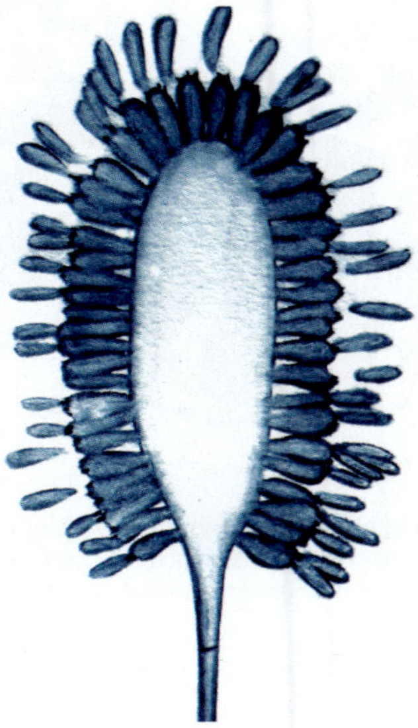

Radula (conidiogenous cell) of a conidiophore of *Xylocladium* sp., x 600 (*MU*).

raduliform (L. *radula*, rasp + suf. *-formis* < *forma*, shape): a conidiogenous axis, relatively wide in comparison to the width of the portion on which each conidium is attached, and which tends to become clavate or somewhat inflated, instead of adopting sympodial (zigzag) growth, as is observed in the gen. *Arthrobotrys* (moniliaceous asexual fungi), *Cordana* and *Gonatobotryum* (dematiaceous asexual fungi). Cf. **ampulliform** and **rachiform**.

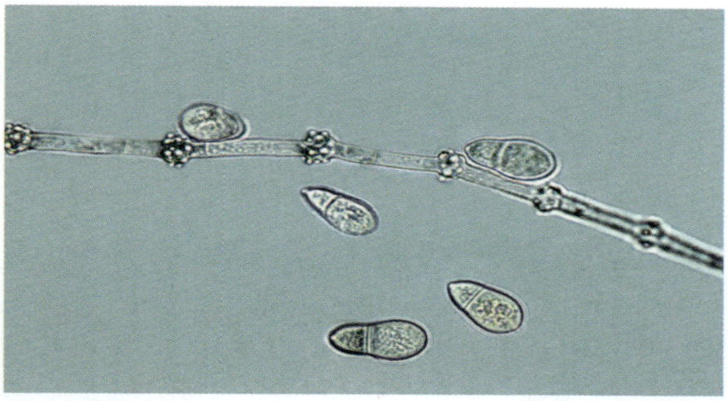

Raduliform (conidiogenous cells) of a conidiophore of *Arthrobotrys oligospora*, x 600 (*MU*).

ragi (Hind. *ragi*, finger millet, *Eleusine coracana*): small rounded balls (ragi mudde or ragi balls) made from flour of finger millet, containing spores and viable cells of amylolytic microorganisms: *Rhizopus*, *Mucor*, *Amylomyces* (Mucorales), *Endomyces* (=*Saccharomyces*) *fibuliger*, *Saccharomycopsis malanga* (Saccharomycetales), and cocci *(Pediococcus* sp.). In addition to being consumed directly, ragi is utilized as a starter culture or inoculum for the preparation of various fermented Oriental foods, such as *tapé ketan* (from rice) and *tapé ketella* (from cassava tubers) of Indonesia. See **fermentation** and **tapé ketella**.

ramarioid (< gen. *Ramaria* < L. *ramus*, branch + suf. *-aria*, which indicates connection or possession + suf. *-oide*, like < Gr. *-oeídes*, similar to): similar to the shape of the basidioma of *Ramaria* (Aphyllophorales).

Ramarioid basidioma of *Ramaria* sp., x 1 (*TRO*).

ramicole, **ramicolous** (L. *rami*, pl. of *ramus*, branch + suf. *-cola*, inhabitant): living on the branches of plants, as in many species of Aphyllophorales, e.g., *Trichaptum abietinum*, *Phylloporia fruticum*, and others.

Ramicole fruiting bodies of *Trichaptum abietinum* on a branch of *Abies* sp., x 0.2 (*RV*).

Ramicole fruiting body of *Phylloporia fruticum* on a branch, x 1 (*RV*).

ramiferous (L. *ramifer* < *rami*, pl. of *ramus*, branch + suf. *-ferous*, bearer < *ferre*, to bear, carry): having branches; branched. Also called **ramose**.

Ramiferous perithecial stromata of *Cordyceps sobolifera* (Hypocreales) growing from an hypogeal cicada larva, x 1 (*RV*).

ramoconidium, pl. **ramoconidia** (L. *ramus*, branch + *conidium* < Gr. *kónis*, powder, dust + dim. suf. *-ídion*): an intercalary or apical branch of a conidiophore that detaches and functions as a conidium, as occurs in some species of the gen. *Cladosporium*, e.g., *C. herbarum* (dematiaceous asexual fungi).

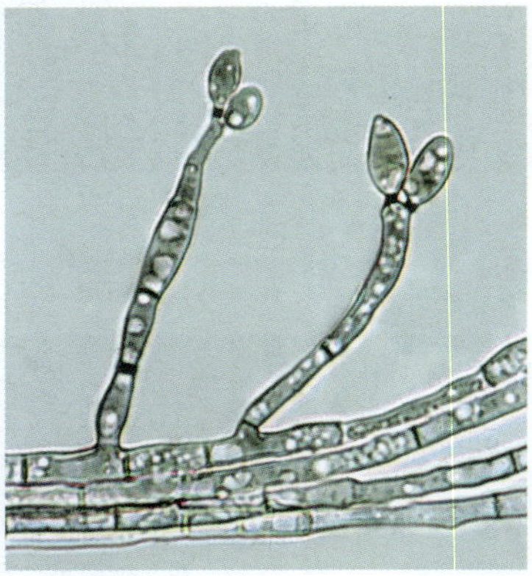

Ramoconidia of the conidiophores of *Cladosporium* sp., x 500 (*MU*).

Ramoconidia of the conidiophores of *Cladosporium* sp., x 560 (*MU*).

ramose (L. *ramosus*, branched < *ramus*, branch + suf. *-osus* > E. *-ose*, full of, augmented, prone to): branched; with branches. Applied to sporangiophores, conidiophores, pedicels, basidiomata and other structures. Syn. of **ramiferous**.

Ramose basidioma of *Ramaria* sp. (Aphyllophorales) on humus, x 1 (*RV*).

ramulus, pl. **ramuli** (L. *ramulus*, dim. of *ramus*, branch): a small branch; a branchlet at the very end, like that of the basidiocarps of the clavariaceous fungi, as is observed in the gen. *Ramaria* (Aphyllophorales).

Ramuli at the very end of the branches of a basidioma of *Ramaria* sp., x 1 (*TRO*).

ramus, pl. **rami** (L. *ramus*, branch): in some conidiophores, like those of *Penicillium* (moniliaceous asexual fungi), the rama bears a verticil of metulae and phialides.

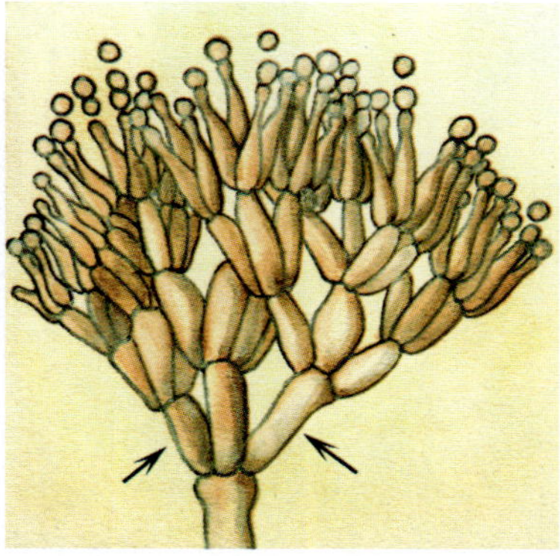

Rami of a conidiophore of *Penicillium arenicola*, bearing a verticil of metulae and phialides, x 1,000 (*MU*).

rangiferoid (< *Rangifer*, the gen. of the reindeer + L. suf. *-oide*, like < Gr. *-oeídes*, similar to): branched like the antlers of a reindeer, like the thallus of the lichen *Cladonia rangiferina* (Lecanorales).

Rangiferoid thallus of *Cladonia rangiferina*, x 10 (*MU*).

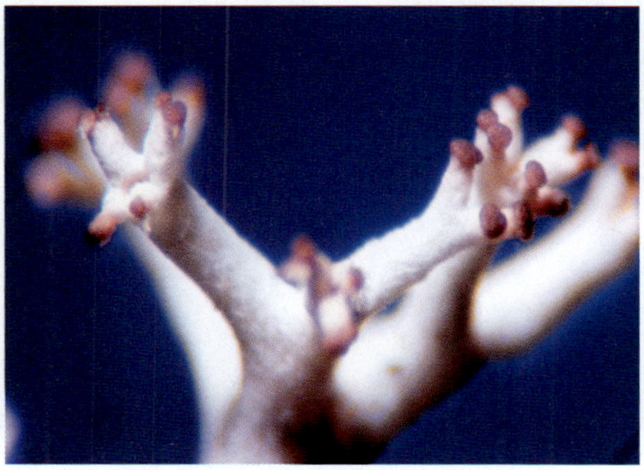

Rangiferoid thallus of *Cladonia rangiferina*, x 20 (*MU*).

raphe (Gr. *raphé*, seam, suture): a longitudinal line of dehiscence present in the pycnidial conidioma of *Chaetomella* (sphaeropsidaceous asexual fungi); also refers to the longitudinal line in the conidial wall of some species of the gen. *Coniella*, such as *C. fragariae* (sphaeropsidaceous asexual fungi).

Raphe of the conidia of *Coniella fragariae*, x 2,000 (*MU*).

raphidium, pl. **raphidia**, **raphides** (NL. *raphidium* < Gr. *raphídion*, small needle < *ráphis*, needle + dim. suf. *-ídion*): an acicular crystal, sharp-pointed on both ends and composed of calcium oxalate, like the crystals present in some lichen thalli.

raphido-, **raphid-** (Gr. *ráphis*, genit. *raphídos*, needle): a pref. employed in terms that allude to something acicular, as in *raphidium*, a needle-shaped crystal present in some lichen thalli.

ray (L. *radius*, radius, ray): see, **fornicate ray**, **involute ray**, **recurved ray**, **revolute ray** and **saccate ray**.

re- (L. *re-*, reiteration or repetition; opposition or resistence): a pref. employed with the first meaning in words such as *recombination*, to combine again or to unite; the second meaning is used to form terms such as *refuse*, to shun something, as from fear or suspicion of risk.

receptacle, **receptaculum**, pl. **receptacula** (L. *receptaculum*, receptaculum, reservoir < *receptor*, receiver + dim. suf. *-culum*): a structure that bears one or more reproductive organs, whether a pedicel or a hollow body. Applied to any structure that supports an hymenium. **1**. In the Phallales (such as *Phallus*, *Dictyophora* and *Lysurus*) it consists of the pedicel (pseudostipe) and the pileus; in other Phallales, such as the gen. *Clathrus*, it consists of the clathrate body. **2**. In the Nidulariales (such as *Cyathus*) it corresponds to the cup- or funnel-shaped structure that contains the peridioles. **3**. In the Laboulbeniales (such as *Laboulbenia*, *Rhachomyces* and *Herpomyces*) the receptacle includes the perithecium, its appendages, and the supporting basal cells. **4**. In the lichens it refers to the cup of the thallus that contains soredia.

receptacle

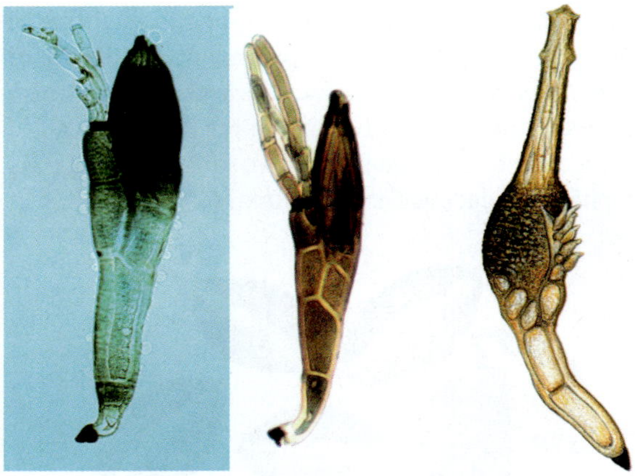

The two thalli on the left are the **receptacles** of *Laboulbenia elongata*, x 120 (*RTH*). The thallus on the right is the **receptacle** of *Stigmatomyces psilopae*, x 100 (*MU*).

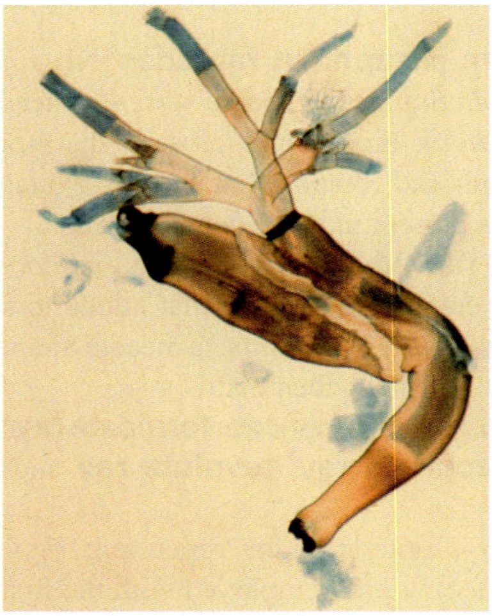

Receptacle of the thallus of *Laboulbenia flagellata*, x 125 (*EPS*).

Receptacle of the thallus of *Rhachomyces quetzalcoatl*, x 125 (*EPS*).

Receptacles (basidiocarps) of *Phallus ravenellii*, x 0.8 (*TH*) and *Lysurus periphragmoides*, x 1 (*EAA*).

Receptacles (basidiocarps) of *Cyathus* sp., x 9 (*MU*).

Receptacle (basidiocarps) of two undetermined species of *Cyathus*, x 9 (*MU*).

receptive body (L. *receptum*, supine of *recipere*, to receive + suf. *-ivus* > E. *-ive*, quality or tendency; ME. < OE. *bodig*, body): a small hypha, branched or not, of the stroma of *Stromatinia gladioli* (Helotiales), capable of being spermatized by microconidia.

receptive hypha, pl. **hyphae** (L. *receptum*, supine of *recipere*, to receive + L. suf. *-ivus* > E. *-ive*, quality or tendency; fitness; NL. *hypha*, hypha < Gr. *hyphé*, tissue, spiderweb; hypha): a hypha that receives a gametic nucleus during sexual reproduction; the fertilization of a receptive hypha (considered as female) occurs, e.g., in the union of a protoperithecium with a microconidium (as in *Neurospora*, of the Sordariales), or in the union of a hypha of a spermogonium with a spermatium (as in *Puccinia*, of the Uredinales).

Receptive hyphae of *Neurospora sitophila*, x 1,000 (*MU*).

reclinate (L. *reclinatus*, inclined < pref. *re-*, behind, against, in front of + *clinare*, to bend + suf. *-atus* > E. *-ate*, provided with or likeness): turned or bent downward.

recognition (L. *recognition*, action and effect of recognizing < L. *recognoscere*, to recognize, examine, investigate): in the case of mutualistic symbionts, it is the process through which the relationship between two potentially compatible symbionts is initiated.

recurvate, **recurved** (L. *recurvatus*, doubled backward < pref. *re-*, back, against, in front of + *curvatus*, curved < *curvus*, curve + suf. *-atus* > E. *-ate*, provided with or likeness): curved in such a manner that the concavity is found on the upper or outer side, like the recurvate pedicels of the sporangiola of *Cokeromyces recurvatus* (Mucorales), and the pileus of *Coprinopsis atramentaria* (=*Coprinus atramentarius*, Agaricales). It is the same as **reflexed**. Cf. **incurvate** and **inflexed**.

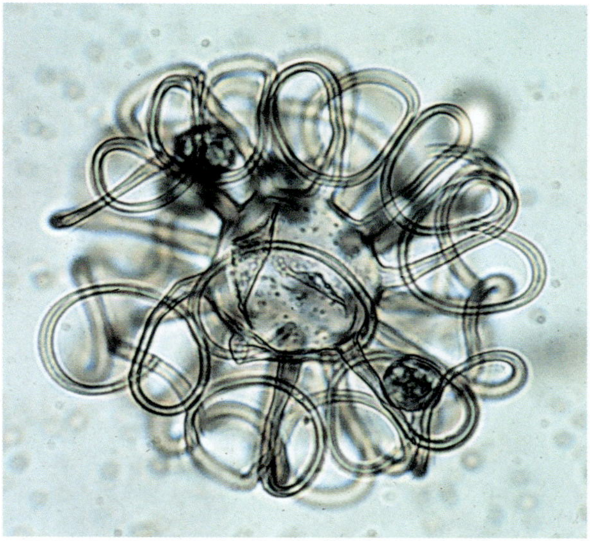

Recurvate pedicels of the sporangiola of *Cokeromyces recurvatus*, x 380 (*MU*).

Recurvate pileus of the basidiocarp of *Coprinus* sp., x 1 (*RV*).

Recurvate pileus of the basidiocarps of *Coprinopsis atramentaria* (=*Coprinus atramentarius*), x 1 (*RV*).

Recurvate pileus of the basidiocarp of *Hygrophorus fornicatus*, x 1 (*MU*).

recurved ray (L. *recurvatus*, bent backward < pref. *re-*, back, against, in front of + *curvatus*, curved < *curvus*, curve + suf. *-atus* > E. *-ate*, provided with or likeness; L. *radius*, radius, ray): *Gasteromycetes*. A ray in which the outer peridium is convex when seen in profile, as is observed in *Geastrum triplex* (Lycoperdales). Cf. **revolute ray**.

reflexed (L. *reflexus* < *reflectere*, to curve or turn backward): see **recurvate**. Cf. **inflexed**.

refracted (L. *refractus*, what has been refracted; broken < *refringere*, to break, change direction): bent backwards from the base; a ray of light that changes direction on passing obliquely through a medium of different density.

reindeer lichen, **reindeer moss** (ME. *reindere* < ON. *hreinn*, reindeer; L. *lichen* < Gr. *leichén*, lichen; L. *muscus*, moss): refers mainly to *Cladonia rangiferina* and *C. stellaris*, fruticose lichens (Lecanorales) that form extensive patches in arctic and north-temperate regions, and which serve as food for reindeer and caribou; occasionally these lichens are consumed by man.

reliquiae (L. *reliquiae*, remnants, remains, residue): the part of the pseudoparenchymatous tissue of the centrum of *Magnaporthe* (Diaporthales) that remains after elongation of the asci.

remote (L. *remotus*, remote, separated): referring to the gills of an agaricaceous fungus, it is those that have the proximal end free, without reaching the stipe.

reniform (L. *reniformis* < *renes*, kidney + *-formis* < *forma*, shape): shaped like a kidney, or with a contour similar to that of a kidney. Also called **fabiform**, i.e., with the shape of a bean (*Vicia faba*), like the secondary zoospores of *Saprolegnia* (Saprolegniales) and other Oomycota, the ascospores of *Kluyveromyces marxianus* var. *marxianus* (Saccharomycetales) and *Petriella* (Microascales), or like the basidiospores of *Inocybe rimosa* (=*I. fastigiata*), Agaricales.

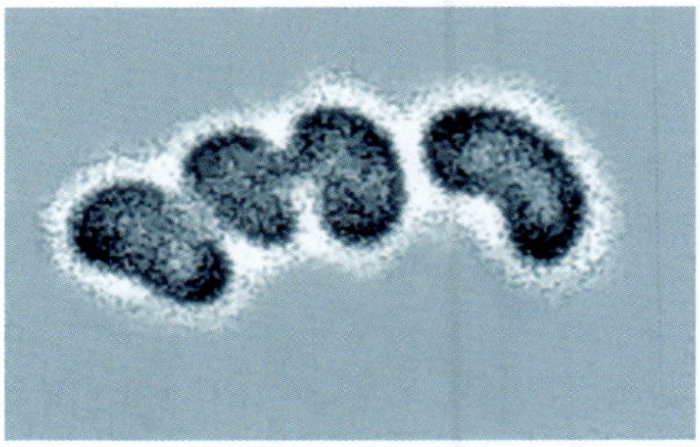

Reniform ascospores of *Kluyveromyces marxianus* var. *marxianus*, x 2,000 (*PL*).

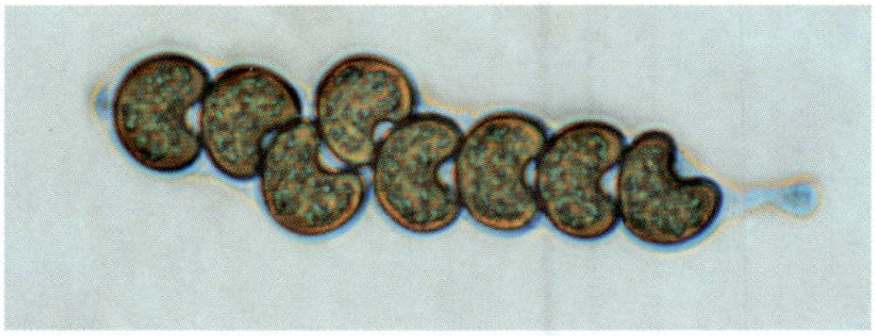

Reniform ascospores inside an ascus of *Petriella* sp., x 1,000 (*CB*).

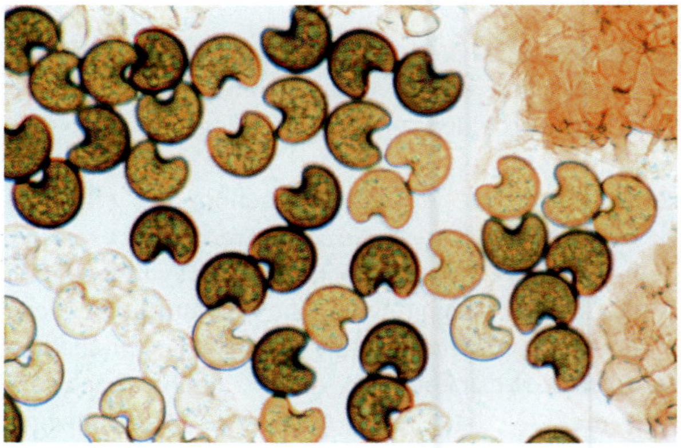

Reniform ascospores of *Petriella* sp., x 700 (*CB*).

repand, **repandous** (L. *repandus*, curved): with the margin wavy and elevated or reflexed (turned backward), like the pileus of *Gymnopus dryophilus* (=*Collybia dryophylla*) and *Lepiota clypeolaria* (Agaricales).

Repand pileus of the basidiocarp of *Lepiota* sp., x 0.6 (*MU*).

repeating spore (L. *repetitor*, to attack again, to repeat; NL. *spora* < Gr. *sporá*, spore): a spore which on germinating gives rise to the same type of mycelium as that on which it was produced.

repetition (L. *repetitio*, genit. *repetitionis*, action and effect of repeating): refers to the germination of spores when a spore produces a new spore directly without first forming a hypha or mycelium.

reproduce (L. *re-*, inseparable prep. that denotes repetition + E. *produce*, to bring forth < L. *producere*, to engender, what produces): literally, to produce or engender again; to produce another organism from a being that has reached maturity.

reproduction (L. *re-*, inseparable prep. that denotes repetition + *production* < *producere*, to engender, what produces + suf. *-ationem* > E. *-ction*, action, state or condition, result): the action and effect of reproducing. The reproduction can be asexual, when sexual cells or nuclei are not involved, or sexual, when sexual cells or nuclei are involved. Asexual reproduction also is considered as *vegetative multiplication*. See **reproduce** and **reproductive**.

reproductive (E. *reproduce* < L. *re-*, inseparable prep. that denotes reintegration or repetition + *producere*, to engender, produce + suf. *-ivus* > ME. > E. *-ive*, that performs an action): one that reproduces or is reproduced, that is involved in reproduction. Cf. **somatic**.

reproductocentric (L. *re-*, inseparable prep. that denotes reintegration or repetition + *productor*, one that produces + *centrum*, center + suf. *-icus* > E. *-ic*, pertaining to or relating to): applied to Chytridiales whose thallus can develop a single reproductive structure (sporangium) in the center of gravity of the thallus (*monocentric* or *monogenocentric*), or several reproductive structures (*polycentric* or *polygenocentric*).

resinous (L. *resina*, plant secretions, solid or semisolid amorphous fusible flammable natural organic substances; are soluble in organic solvents but not in water + suf. *-osis*, of, relating to, resembling or derived from resin): the term is applied to the fruiting bodies of *Fomitopsis pinicola* (among other gen. of Aphyllophorales), which have resins.

Resinous basidiocarps of *Fomitopsis pinicola*, x 0.2 (*RV*).

resistance (L. *resistentia*, action and effect of resisting): the quality of an organism to limit, in major or minor degree, the effects of a pathogen or other damaging factor. It can also be defined as the degree of susceptibility to a disease. *Acquired resistance* is the resistance to response, neither hereditary nor congenital, that is manifested in a host after having been preimmunized by means of a partial or total infection with a virus or other pathogen, in which the infected parts are not susceptible with respect to the same or closely related pathogen. In plant pathology several types of resistence (agronomic, apparent, biochemical, functional and mechanical) are recognized.

resistant sporangium, pl. **sporangia** (L. *resistens*, genit. *resistentis*, that which resists; Gr. *sporá*, spore + *angeîon*, vessel, closed receptacle): see **latent spore** and **meiosporangium**.

Resistant sporangia of *Woronina polycystis* (Plasmodiophorales) in a hypha of *Achlya* sp. (Saprolegniales), x 350 (*MU*).

resupinate, **resupinated** (L. *resupinatus*, bent backward, a supine position < pref. *re-*, back, against < *supinus*, supine, laying on the back, inverted + suf. *-atus* > E. *-ate*, provided with or likeness): refers to any part or organ that is inverted with respect to the position considered normal; e.g., the basidiocarps of *Dichomitus mexicanus* (=*Megasporoporia mexicana*), *Pseudochaete*, *Poria*, *Byssomerulius*, *Pachykytospora*, *Phanerochaete* and *Schizopora* (Aphyllophorales) are resupinate, with the porose hymenophore on top, on the side opposite to that attached to the substrate.

Resupinate basidiocarps of *Byssomerulius* (=*Meruliopsis*) *corium*, on dead wood, x 1 (*RV*).

Resupinate basidiocarp of *Dichomitus mexicanus* (=*Megasporoporia mexicana*), x 0.5 (*RV*).

Resupinate basidiocarp of *Pachykytospora papyracea*, on bark, x 1 (*RV*).

Resupinate basidiocarps of *Pseudochaete* (=*Hymenochaete*) *tabacina*, on dead wood, x 1 (*RV*).

Resupinate basidiocarps of *Phanerochaete* sp., on bark, x 0.5 (*RV*).

Resupinate basidiocarps of *Poria* sp., on dead wood, x 0.8 (*MU*).

Resupinate basidiocarps of *Schizopora paradoxa* on wood, x 1 (*RV*).

retention (L. *retentio*, genit. *retentionis*, action and effect of retaining): refers to the quality that certain surfaces have to retain, in major or minor degree, a fungicide or other substance that is applied to protect crops. Cf. **adherence**.

reticulate (L. *reticulatus*, made like a net < *reticulum*, dim. of *rete*, net + suf. *-atus* > E. *-ate*, provided with or likeness): in the shape of a *reticulum*, i.e., with the veins, lines or borders of low relief intercrossed like a net (surface of a pileus, stipe, spores, etc.). E.g., the pedicel of the sorocarp of *Dictyostelium mucoroides* (Dictyosteliomycota), the ascospore wall of *Tuber dryophilum* (Pezizales), the basidiocarp of *Auricularia delicata* (Auriculariales), and the stipe of a basidiocarp of *Bolletellus russellii* (Agaricales) are reticulate.

Reticulate stipe of a basidiocarp of *Boletellus russellii*, x 0.5 (*RV*).

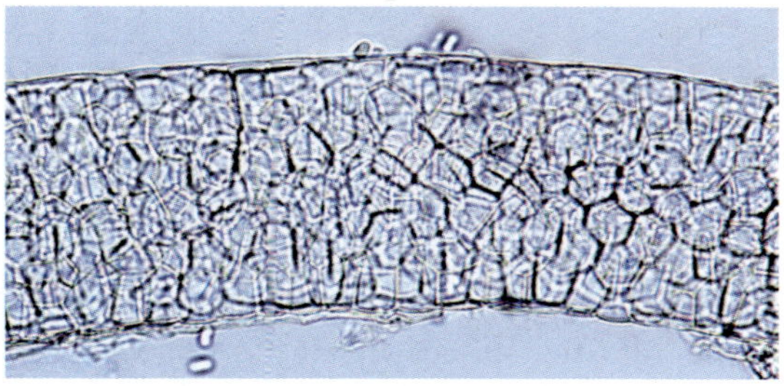

Reticulate pedicel of the sorocarp of *Dictyostelium mucoroides*, x 750 (*EAA*).

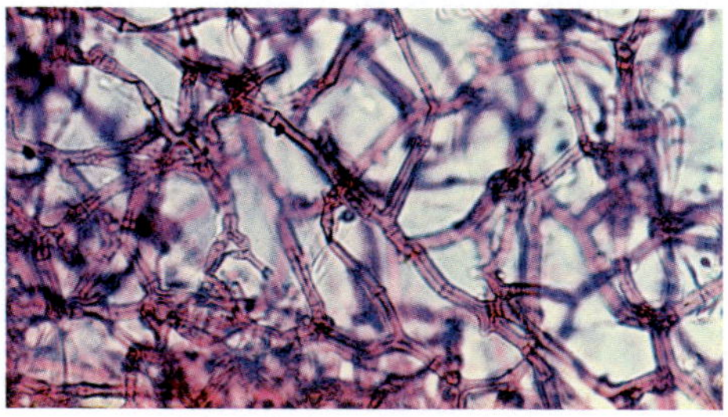

Reticulate hyphae of *Malbranchea aurantiaca* (moniliaceous asexual fungi), x 500 (*MU*).

Reticulate ascospores of *Tuber dryophilum*, x 1,000 (*HT*), and **reticulate** lower surface of the basidiocarp of *Auricularia delicata*, x 0.8 (*TH*).

reticulum, pl. **reticula** (L. *reticulum*, small net < *rete*, net + dim. suf. *-culum*): a small net or mesh composed of filaments, veins, nerves, borders, etc.

retro- (L. *retro*, toward the back, backward, on the contrary): a pref. used in terms such as the following.

retroculture (L. *retro*, backward + *culto*, culture; action and effect of cultivating): a reisolation of a pathogen obtained from a host previously inoculated with the same pathogen.

Reticulate lower surface of a basidiocarp of *Auricularia delicata*, x 1 (*TH*).

Retroculture of the sclerotia of *Phymatotrichopsis omnivora* (moniliaceous asexual fungi) in a flask; the sclerotia germinate to give the mycelium of the fungus, x 0.5 (*JAS*).

retrocurved (L. *retro*, toward the back, backward, on the contrary + *curvus*, curved, incurved): equal to **retroflexed** and **retrorse**. Cf. **antrorse**.

retroflex, **retroflexed**, **retroflexous** (L. *retroflexus*, bent backward < *retro*, bent backward, downward, on the contrary + *flexus*, bent, turned, curved): curved downward. E.g., retroflexed hairs are those curved downward, with the apex directed toward the base of the organ in which they are inserted. Equal to **retrocurved**.

retrorse (L. *retrorsus*, toward the back): equal to **retrocurved**. Cf. **antrorse**.

reverse (L. *reversus* < *revertere*, to turn back toward + *vertere*, to turn; opposite or contrary to the previous or normal condition; the reverse of something; opposite): the reverse of a colony cultured on agar media has some characteristics that complement the characteristics of the *obverse* of the same colony. The characters of the reverse and obverse of the colonies are considered in the taxonomy of various groups of fungi, e. g., as in *Lecanicillium* (=*Verticillium*), *Penicillium* and *Aspergillus* (moniliaceous asexual fungi). Cf. **obverse**.

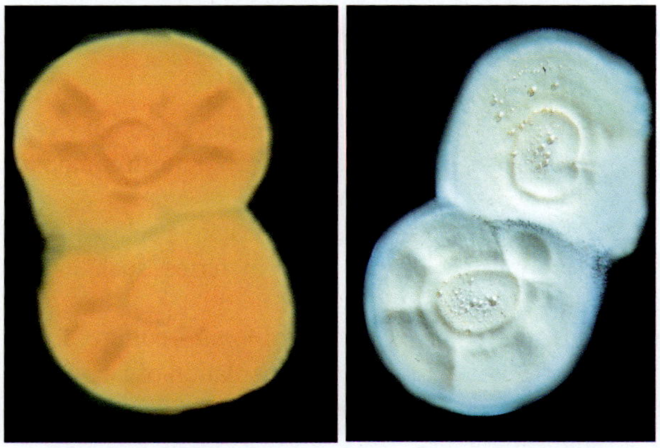

On the left, **reverse** of one week-old colonies of *Lecanicillium* (=*Verticillium*) *lecanii* on agar. On the right, the obverse of the same colonies, x 1 (*MU*).

revolute (L. *revolutus*, coiled backward < *revolvere*, to turn or coil toward the back): applied to a pileus with borders rolled toward the upper surface; revolute is like reflexed, but rolled up. Cf. *involute*.

revolute ray (L. *revolutus*, coiled backward < *revolvere*, to turn or coil toward the back; L. *radius*, radius, ray): *Gasteromycetes*. A ray in which the outer peridium is recurved and its border is coiled downward and inward, as is seen in the basidiocarp of *Geastrum saccatum* (Lycoperdales). Cf. **recurved ray**.

rhabdo- (Gr. *rhabdós*, small rod, stick): a pref. used in terms such as *rhabdosome*, a cellular inclusion in the shape of a small rod or needle that is present in various species of dinoflagellates, and in generic names such as *Rhabdocline* (lichen of the order Rhytismatales).

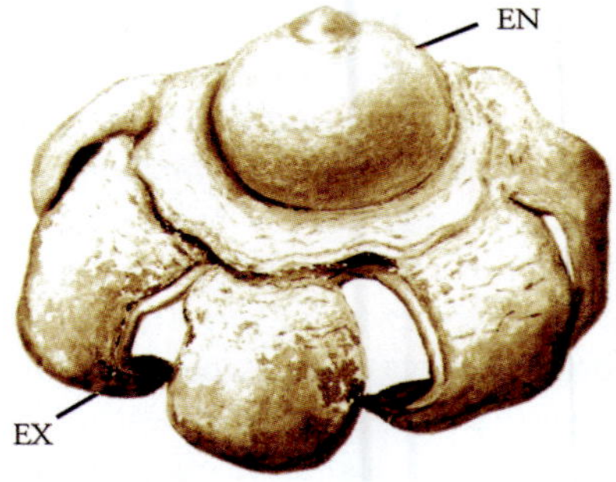

Revolute exoperidial **rays** (EX, exoperidium, and EN, endoperidium) of the fruiting body of *Geastrum saccatum*, x 2 (*MU*).

rhagadiose (L. *rhagadiosus*, fissured, cracked < *rhagádos*, genit. of *rhagás*, crack, break + L. suf. *-osus* > E. *-ose*, full of, augmented, prone to): with deep cracks, like the surface of the pileus of *Boletus castaneus* (Agaricales) and the exoperidial rays of *Astraeus hygrometricus* (Sclerodermatales). Also applied to the surface of certain lichens, such as *Acarospora rhagadiosa* (=*A. scabra*), of the Lecanorales. Cf. **rimose**.

Rhagadiose surface of *Exobasidium* sp. (Exobasidiales), x 1 (*RV*).

Rhagadiose surface of *Phellinus rimosus* (Aphyllophorales), x 0.3(*RV*).

Rhagadiose surface of the exoperidial rays of the fruiting bodies of *Astraeus hygrometricus*, x 0.5 (*EAA* & *SA*).

Rhagadiose surface of an exoperidial ray of the fruiting body of *Astraeus hygrometricus*, x 5 (*MU*).

rheo- (Gr. *rhéo*, to flow): a pref. that forms a part of terms such as *rheotaxis* and *rheotropism*.

rheotaxis, pl. **rheotaxes** (Gr. *rhéo*, to flow + *táxis*, arrangement, movement): a tactic phenomenon produced by a current of water, like that observed in the plasmodia of the Myxomycetes, which, moving contrary to the flow of a slow current of water, demonstrate positive *rheotactism* (syn. of rheotaxis).

Rheotaxis of a plasmodium of *Physarum polycephalum* (Physarales), moving contrary to the flow of a current of water on the bark of a tree, x 1 (*RV*).

rheotropism (Gr. *rhéo*, to flow + *trépo*, to gyrate + *-ismós* > L. *-ismus* > E. *-ism*, suf. that denotes state, phase, habit, tendency, action, doctrine, system): a tropism determined by the unilateral action of a current of water. This tropism is positive when the organ affected tends to be directed against the current (like the radicle of *Vicia sativa*), and negative, in the contrary case (like the mycelial filament of *Mucor*, of the Mucorales).

rhexigenesis, pl. **rhexigeneses** (Gr. *rhêxis*, tear + *génesis*, originate): see **rhexolytic**.

rhexo-, rhex- (Gr. *rhêxis*, tear): a pref. used in terms alluding to a tear or rupture, as in *rhexolysis*.

rhexolysis, pl. **rhexolyses** (Gr. *rhêxis*, tear + *lýsis*, dissolution, disintegration): see **rhexolytic**.

rhexolytic (Gr. *rhêxis*, tear + *lytikós*, capable of loosening < *lýtos*, dissoluble, loose + suf. *-íkos* > L. *-icus* > E. *-ic*, pertaining to or relating to): a spore or propagule that separates by the tearing or breakage of a conjunctive cell, as happens in the detachment of the thallic conidia formed by some fungi (such as *Microsporum gypseum*, of the moniliaceous asexual fungi). This process is called **rhexigenesis** or **rhexolysis**. Cf. **schizolytic** and **schizogenesis**.

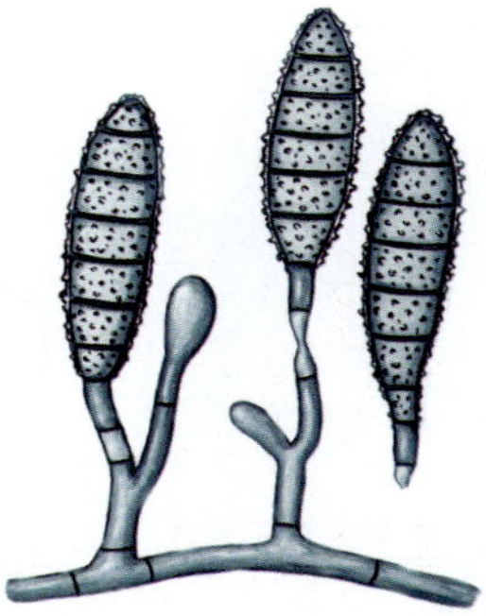

Rhexolytic conidia of *Microsporum gypseum*, x 600 (*MU*).

rhinophycomycosis, pl. **rhinophycomycoses**, **rhinoentomophthoramycosis**, pl. **rhinoentomophthoramycoses** (NL. < Gr. *rhín*, *rhinós*, nose + *phýkos*, alga + NL. *mycosis* < Gr. *mýkes*, fungus + suf. *-osis*, state or condition; < gen. *Entomophthora* < Gr. *éntomos*, insect + *phthorá*, death, destruction + NL. *mycosis* < Gr. *mýkes*, fungus + suf. *-osis*): an inflamatory or chronic granulomatose disease, generally limited to the subcutaneous tissues or to the nasal mucous, characterized by swelling and deformations of the face of humans. The etiologic agent is *Conidiobolus* (=*Entomophthora*) *coronatus* (Entomophthorales), an inhabitant of the soil that attacks termites and other insects, and spiders. See **mycosis**.

rhinosporidiosis, pl. **rhinosporidioses** (< gen. *Rhinosporidium* < Gr. *rhín*, genit. *rhinós*, nose + *sporá*, spore + dim. suf. *-ídion* > L. *-idium* + Gr. *-osis*, state or condition): an infection of the mucocutaneous tissues, mainly of the nose and the nasal septum, which is distinguished by the formation of pedunculate polyps, tumors, papillomas and hyperplasic warty lesions, caused by *Rhinosporidium seeberi* (classified as a taxon *incertae sedis* in the Hyphochytriales). See **mycosis**.

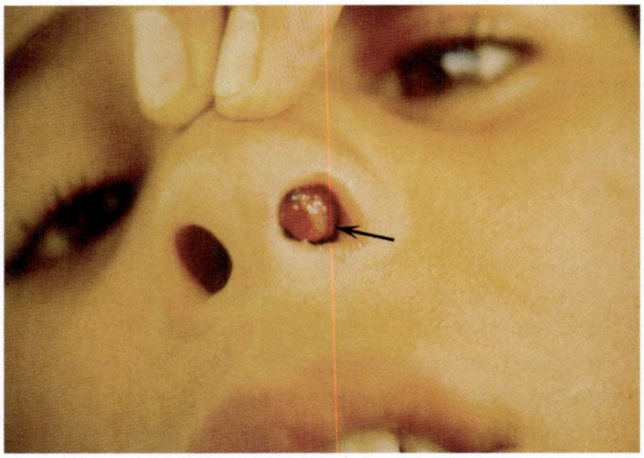

Rhinosporidiosis of the nose of a child caused by *Rhinosporidium seeberi*, x 1 (*AB*).

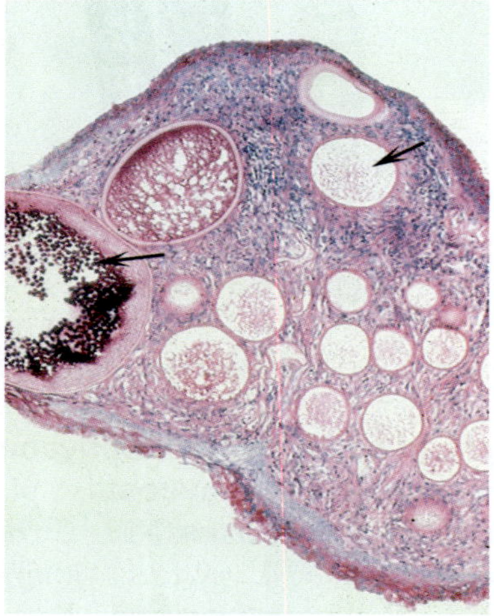

A cross section of a polyp of **rhinosporidiosis**, caused by *Rhinosporidium seeberi*; note the sporangia of the fungus, x 5 (*AB*).

rhizina, pl. **rhizinae** (L. *rhizina*, radicle, hair or filament shaped like a root < *rhiza* < Gr. *rhíza*, root + L. des. *-ina*, likeness): each of the root-like structures formed by the mycobiont of the lichen thallus and which serve to attach it to the substrate; they are formed, e.g., in *Peltigera canina* (Peltigerales) and *Umbilicaria vellea* (Lecanorales).

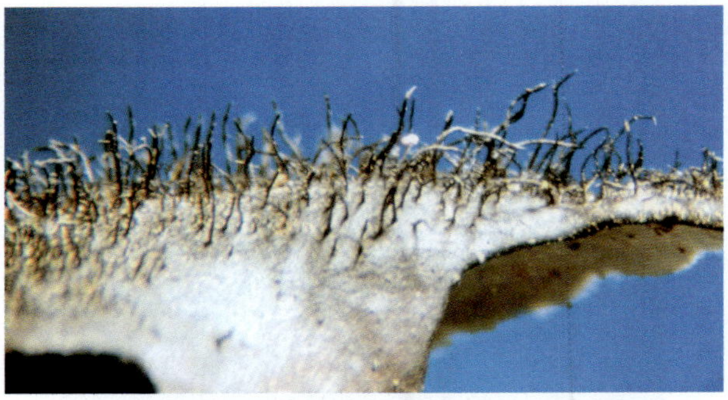

Rhizinae of the lower surface of the thallus of *Peltigera* sp., x 5 (*MU*).

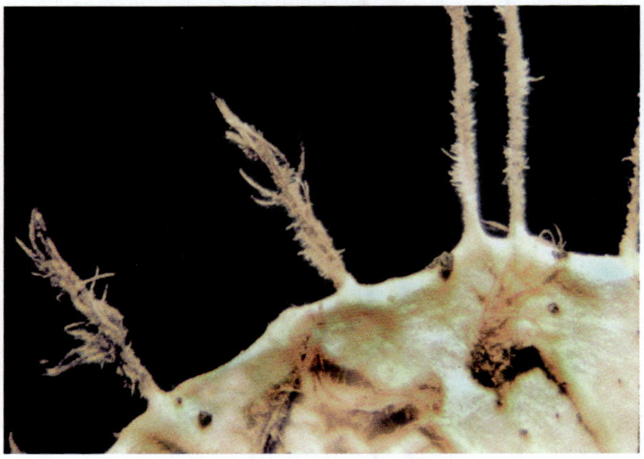

Rhizinae of the lower cortex of the thallus of *Peltigera canina*, x 10 (*MU*).

Rhizinae of the lower cortex of the thallus of *Umbilicaria vellea*, x 10 (*MU*).

rhizinose strand (L *rhizina*, rootlet, rootlike hair or filament < *rhiza* < Gr. *rhíza*, root + L. *-ina*, likeness + L. suf. *-osus* > E. *-ose*, full of, augmented, prone to; ME. *strond*, fibers or filaments twisted, plaited, or laid parallel to form a unit for further twisting or plaiting into yarn, thread, rope, or cordage): *Lichens*. An attachment organ composed of strong, irregularly branched hyphae that grow among the soil particles; found in *Catolechia wahlenbergii* (=*Buellia pulchella*, Lecanorales) and other squamulose lichens.

rhizo-, **rhiz-** (Gr. *rhíza*, root): a pref. employed to form terms that allude to the root of a plant, such as *rhizoid*. Also employed as a suf. *-rhizo*, *-rhiza*, such as *macrorhiza*, a large root.

rhizoid (Gr. *rhíza*, root + L. suf. *-oide*, like < Gr. *-oeídes*, similar to): a fine hypha, slender at the distal end, often branched and superficially similar to the root of a plant, that serves to attach the thallus or mycelium to the substrate and to absorb nutritive substances from it. Rhizoids are characteristic of the aquatic fungi, like the Chytridiales and Hyphochytriales (e.g., *Rhizophydium pollinis-pini* and *Hyphochytrium*) and of some terrestrial fungi, such as the Mucorales (e.g., *Rhizopus*, *Absidia* and *Actinomucor*).

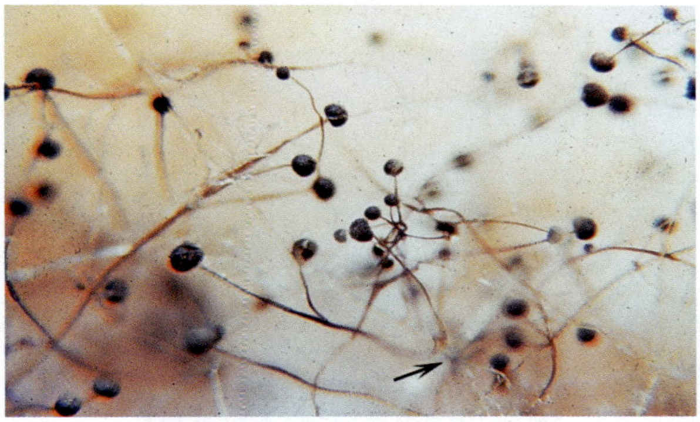

Rhizoids of *Rhizopus stolonifer* (=*Rh. nigricans*), x 25 (*MU*).

Rhizoids of *Rhizopus stolonifer* (=*Rh. nigricans*), x 25 (*MU*).

Rhizoids of *Rhizopus stolonifer* (=*Rh. nigricans*), x 400 (*LP*).

rhizomorph (Gr. *rhíza*, root + *morphé*, shape): a thick, resistent mycelial cord composed of plectenchyma and with a growing point, similar to that of a plant root, that penetrates the substrate. Rhizomorphs can be simple or branched, exceeding a meter in length, up to 1 cm thick, and generally dark in color, although some are white, yellow, reddish or purple. They function as organs for the absorption and conduction of nutritive substances, and on occasion also are capable of being latent; they are typical of *Armillaria* (=*Armillariella*) *mellea* (Agaricales) and of some Gasteromycetes, such as some species of *Linderia* (Phallales), *Lycoperdon* (Lycoperdales) and *Scleroderma* (Sclerodermatales). Cf. **mycelial strand**.

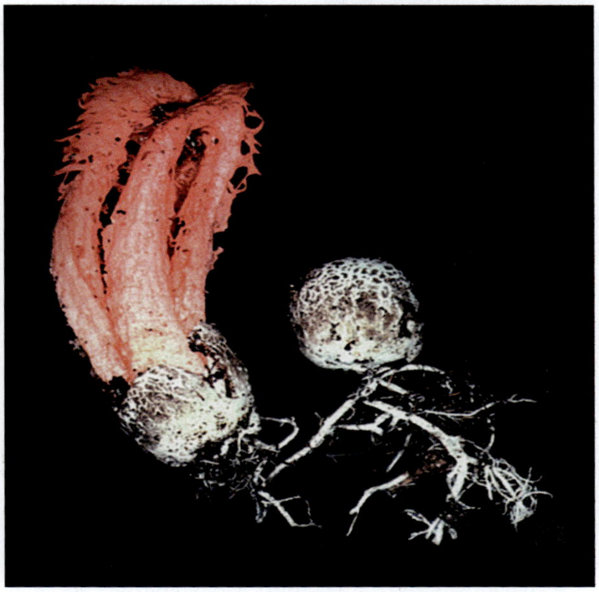

Rhizomorphs of the fruiting bodies of *Linderia* sp., x 1 (*JC*).

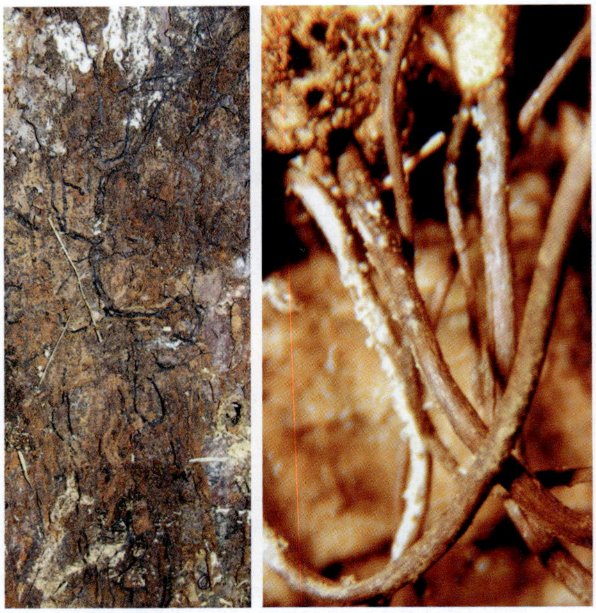

Rhizomorphs of *Armillaria* (=*Armillariella*) *mellea*, x 1 (*RV*), and **rhizomorphs** of an unidentified agaricaceous fungus, x 5 (*MU*).

Rhizomorphoids of *Armillaria* sp., x 0.5 (*RV*). **Rhizomycelium** of *Nowakowskiella ramosa*, x 550 (*MU*).

Rhizomorphs of the fruiting body of *Lycoperdon* sp., x 0.8 (*RV*), x 1 (*MU*).

rhizomorphoid (Gr. *rhíza*, root + *morphé*, shape + L. suf. *-oide*, like < Gr. *-oeídes*, similar to): similar to a *rhizomorph*, as in *Armillaria* (Agaricales).

rhizomycelium, pl. **rhizomycelia** (Gr. *rhíza*, root + NL. *mycelium* < Gr. *mýkes*, fungus + L. suf. *-elis*, concerning + dim. suf. *-ium*): a rudimentary rhizoidal system, more or less developed and similar to a mycelium, but with the rhizoidal branches lacking nuclei. Characteristic of the aquatic fungi of the order Chytridiales and of the family Megachytriaceae, such as *Nowakowskiella ramosa*.

rhizomycete (Gr. *rhíza*, root + *mýkes*, genit. *mýketos*, fungus): a fungus that is involved or can be involved in constituting a mycorrhiza; mycorrhizogenous.

A **rhizomycete** (*Amphinema* aff. *byssoides*, Agaricales) with a pine root (*Pinus sylvestris*), x 10 (*LV*).

rhizoplane (Gr. *rhíza*, root + L. *planus*, plane, flat surface): the surface of the root on which are established the biological relationships (such as mycorrhizae) with the soil medium. Cf. **rhizosphere**.

rhizoplast (Gr. *rhíza*, root + *plastós*, modeled, formed; organized corpuscle or particle): an endoplasmic chromatoid filament that connects the *blepharoplast* (an organelle related to the kinetic function of flagellate cells) in the base of the flagellum to the nucleus. It is present in the zoospores of the Chytridiomycota (such as *Rhizophlyctis rosea*), Hyphochytriomycota and Oomycota.

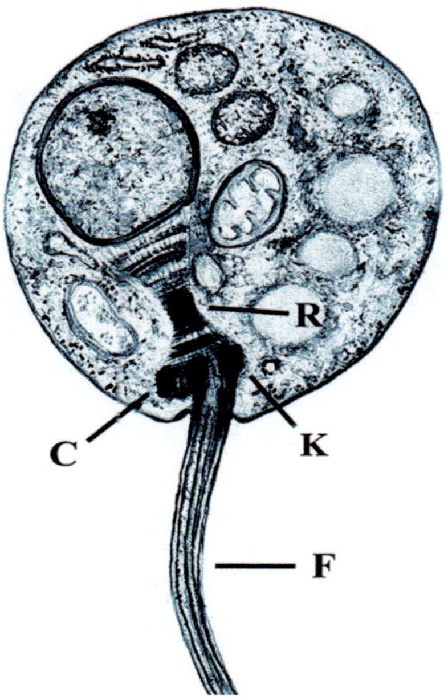

Rhizoplast (R) of the zoospore of *Rhizophlyctis rosea*, seen in longitudinal section. Also shown are the kinetosome (K), the centriole (C), and the flagellum (F), x 15,000 (*MU*).

rhizopodium, pl. **rhizopodia** (Gr. *rhíza*, root + *poús*, genit. *podós*, foot): a pseudopodium of a plasmodium that is characterized by its tendency to branch and by the manner of branching to anastomose or unite when it comes into contact with others, for which reason it is also called *reticulate pseudopodia* or *myxopodia*. Also called *rhizopod* is any member of the subclass *Rhizopoda*, protozoans (such as amoebae and the foraminiferans) with lobate or filiform pseudopodia.

rhizosphere (Gr. *rhíza*, root + *sphaîra*, sphere): the area of the soil near living roots of a plant, with a microbiota frequently richer in species than the soil farther away from the roots, and with certain biological relationships with mycorrhizae. Cf. **edaphosphere** and **rhizoplane**.

rhodo-, **rhod-** (Gr. *rhódon*, rose, red): a pref. employed when alluding to roses or to a rose or reddish color, as in *rhodosporous*.

rhodosporous (Gr. *rhódon*, rose, red + *sporá*, spore + L. suf. *-osus* > E. *-ous*, full of, abounding in): refers to fungi that have spores that are rose, red or reddish in color, when seen in mass in a spore print, like those of *Rhodophyllus* (Agaricales).

rhombic, **rhomboid**, **rhomboidal** (Gr. *rhómbos*, rhombus < *rhémbein*, to revolve, gyrate + suf. *-íkos* > L. *-icus* > E. *-ic*, pertaining to or relating to; or + L. suf. *-oide*, like < Gr. *-oeídes*, similar to; or + L. suf. *-alis* > E. *-al*, relating to or pertaining to): shaped like a rhombus or a diamond, or quadrangular, like the cells of the yeasts of the gen. *Kloeckera* (cryptococcaceous asexual yeasts), the conidia of *Beltrania rhombica* (dematiaceous asexual yeasts), or the ascospores of *Melanospora zamiae* (Melanosporales).

Rhodosporous fruiting bodies of *Ganoderma resinaceum* (Aphyllophorales), x 0.3 (*RV*).

Rhodosporous fruiting body of *Phellinus rimosus* (Aphyllophorales), x 0.5 (*RV*).

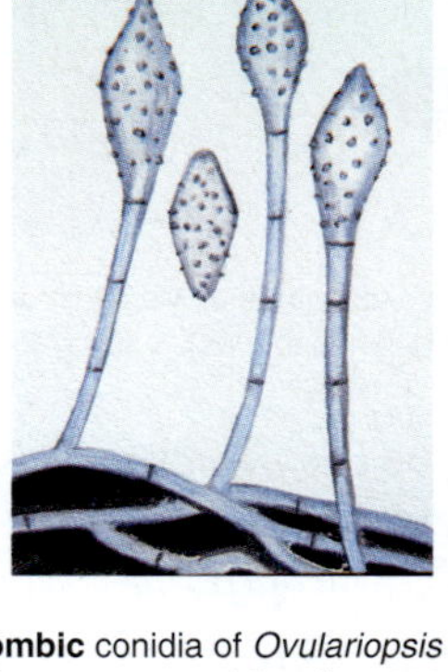

Rhombic conidia of *Ovulariopsis* sp. (moniliaceous asexual fungi), x 200 (*MU*).

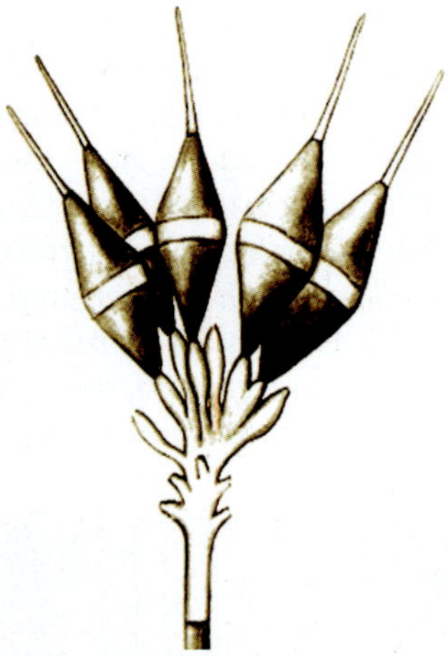

Rhombic conidia of *Beltrania rhombica*, x 700 (*MU*).

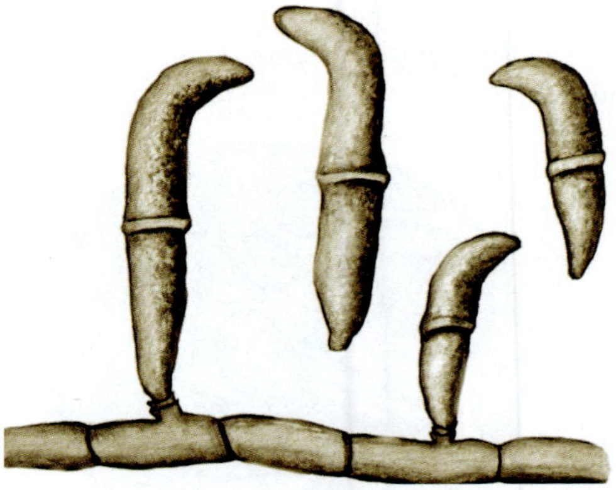

Rhynchospores (conidia) of *Rhynchosporium secalis*, x 1,600 (*MU*).

Rhombic conidia of *Beltrania rhombica*, x 450 (*GH*).

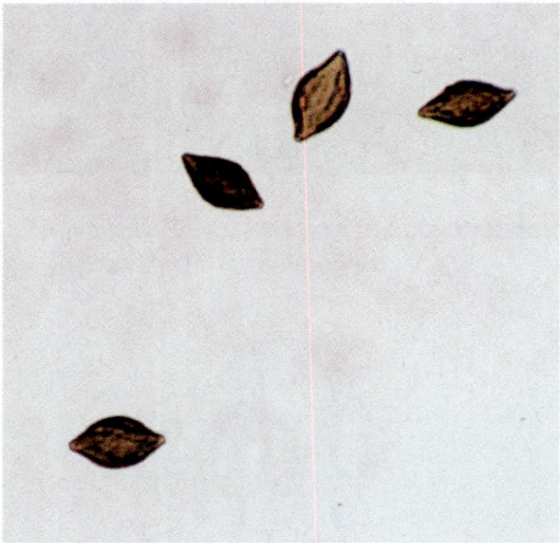

Rhombic ascospores of *Melanospora zamiae* (Melanosporales), x 800 (*MU*).

rhynchosporous (Gr. *rhýnchos*, beak + *sporá*, *sporós*, spore): a fungus with spores beaked at the distal end, like those of *Rhynchosporium secalis* (tuberculariaceous asexual fungi). Spores with this shape are called *rhynchospores*.

ribosome (L. *ribose*, ribose + Gr. *sõma*, body): an organelle composed of nucleoprotein, 15-25 nm in diam., found in the cytoplasm or associated with the endoplasmic reticulum and the nuclear membrane. Ribosomes can aggregate in groups called *polyribosomes*, which are involved in protein synthesis.

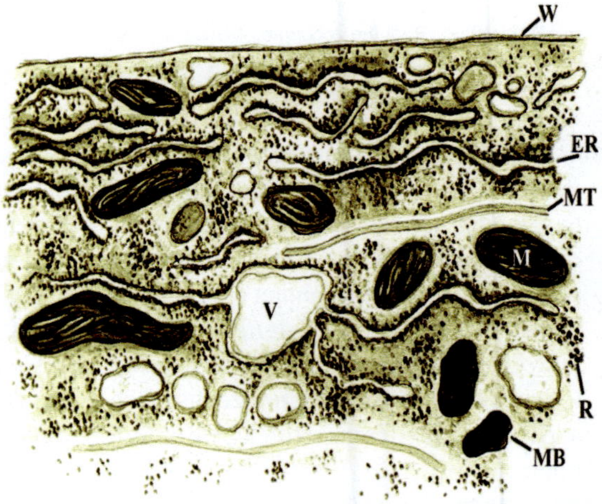

Ribosomes (R) of a young hypha of *Armillaria mellea* (Agaricales), seen in longitudinal section; these organelles are associated with the cisternae of the endoplasmic reticulum (ER). Also evident are the microbodies (MB), mitochondria (M), microtubles (MT), vesicles (V), and the cell wall (W). Drawing based on several transmission electron micrographs published by Beckett *et al.* in their *Atlas of Fungal Ultrastructure*, 1974, x 15,800 (*MU*).

rimose, **rimous** (L. *rimosus*, cracked, split < *rima*, fissure, crack, slit + suf. *-osus* > E. *-ose*, full of, augmented, prone to): with the surface cracked or split, like the pileus of *Amanita solitaria*, *Boletus rubellus* and *Inocybe rimosa* (=*I. fastigiata*) (Agaricales), or like the peridium of *Astraeus* (Sclerodermatales). Cf. **rhagadiose**.

Rimose pileus of the basidiocarp of *Inocybe rimosa* (=*I. fastigiata*), x 1 (*EPS*).

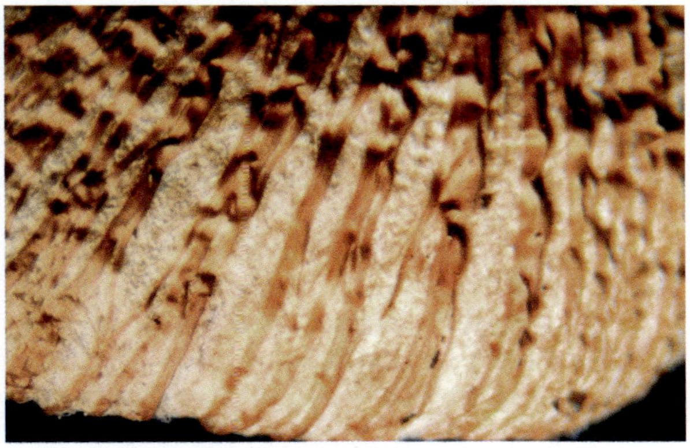

Rimose pileus of the basidiocarp of *Inocybe rimosa* (=*I. fastigiata*), x 5 (*CB*).

rimula, pl. **rimulae** (L. *rimula*, dim. of *rima*, fissure, crack, slit): a small crack or slit. See **rimulose**.

rimulose, rimulous (L. *rimula*, dim. of *rima*, fissure, crack, slit + suf. *-osus* > E. *-ose*, or > E. *-ous*, full of, augmented, prone to): provided with small cracks or slits, like the pileus of some boletes and the surface of the thallus of certain lichens, such as *Graphis rimulosa* (Ostropales).

ring (ME. < OE. *hring*, ring < L. *curva*, curved; a circular band for holding, hanging, packing, or sealing): *Agaricales*. The ring corresponds to the remains of the partial veil (which covers the developing hymenophore) that adheres to the upper part of the stipe of the basidiocarp, as happens in the spp. of *Agaricus*, *Amanita*, *Chlorophyllum*, *Coprinus*, *Lepiota*, *Macrolepiota*, and other gen.

rivose (L. *rivosus*, channeled, grooved < *rivus*, canal, arroyo, groove + suf. *-osus* > E. *-ose*, full of, augmented, prone to): having sinuate canals on the surface, like the thallus of some crustaceous lichens.

Ring of the basidiocarp of *Amanita gemmata*, x 1 (*RV*).

Ring of the basidiocarp of *Amanita* sp., x 1 (*RV*).

Ring of the basidiocarp of *Macrolepiota mastoidea*, x 1 (*RV*).

rivulose (L. *rivulosus*, channeled, grooved < *rivulus*, dim. of *rivus*, canal, arroyo, groove + suf. *-osus* > E. *-ose*, full of, augmented, prone to): with the surface marked by lines that resemble a system of small rivers or canals on a map, as is seen in the pileus of the basidiocarp of *Clitocybe angustifolia* (= *C. rivulosa*) and *Leccinum rugosiceps* (Agaricales), and in the crustaceous thallus of the lichen *Fuscidea cyathoides* (=*Lecidea rivulosa*), Lecanorales.

Rivulose pileus of the basidiocarp of *Clitocybe angustifolia* (=*C. rivulosa*), x 1 (*MU*).

Rivulose pileus of the basidiocarps of *Leccinum rugosiceps*, x 1 (*RV*).

rodlet (ME. < OE. *rodd*, rod, club, bar): structural unit of conidial and some hyphal walls composed of particles (ca. 5 nm in diam.) arranged in linear series.

roestelioid (< gen. *Roestelia* < L. *rostellum*, dim. of *rostrum*, rostrum, beak, snout + L. des. *-ia*, quality of or state + L. suf. *-oide*, like < Gr. *-oeides*, similar to): a cornute aecium, long and tubiform, characteristic of *Gymnosporangium* (Uredinales), whose anamorph is *Roestelia*.

Roestelioid aecia of *Gymnosporangium clavipes*, the causal agent of quince rust, x 10 (*MU*).

Rohr (G. *Rohr*, tube): an elongated tubular cavity that is formed in the interior of the zoospores of *Plasmodiophora brassicae* (Plasmodiophorales), when these zoospores encyst on the outer surface of the epidermal cells of cabbage roots. The end of the Rohr or tube, which is oriented toward the cell wall of the host, has a kind of plug that stains lightly; the extension of the open end of the Rohr oriented toward the cytoplasm of the encysted zoospore was originally called the *Schlauch*, in German, which means hose or tube. Inside the tube is a sharp-pointed, cylindrical body, (which stains intensely) called the stylet (or *Stachel*, in German), which perforates the cell wall of the host, so that the protoplast of the encysted zoospore can enter and initiate the infection.

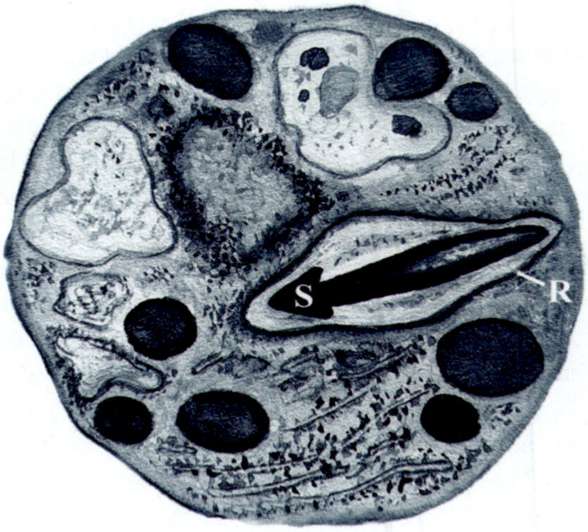

Rohr (R) of an encysted zoospore of *Plasmodiophora brassicae*, seen in longitudinal section. The Stachel (S) is inside the Rohr. Drawing based on a transmission electron micrograph published by Beckett *et al.* in their *Atlas of Fungal Ultrastructure*, 1974, x 25,700 (*MU*).

root nodules (ME. *rot, root* < OE. *rot* < ON. *rot*, root; L. *nodulus*, dim. of *nodus*, knot): nodules formed on the roots of some plants, such as *Alnus* and *Myrica* due to infection by species of Plasmodiophorales.

roridous (L. *roridus*, dewy, full of dew < *ros*, genit. *roris*, dew): refers to mycelium, sporiferous apparatus, etc., that has droplets of watery liquid on its surface (whether condensed moisture or a liquid exudate), in the manner of dew. The colonies of many species of molds have this characteristic. Also, the sclerotia of *Sclerotinia* (Helotiales) are roridous in the presence of moisture, as when they are growing on agar.

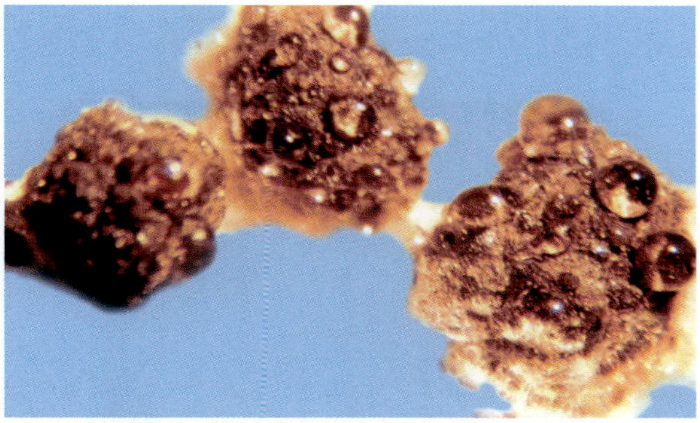

Roridous sclerotia of *Sclerotinia sclerotiorum* on agar, x 10 (*MU & CB*).

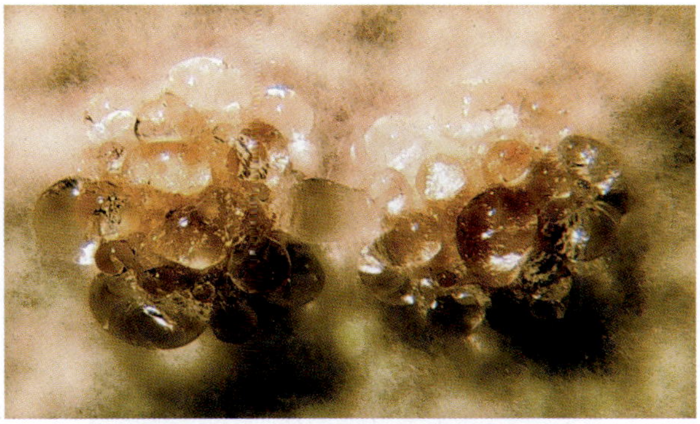

Roridous sclerotia of *Sclerotinia sclerotiorum* on agar, x 10 (*MU & CB*).

Roridous young basidiocarps of *Coprinus* sp. on agar, x 2 (*MU*).

rostrate, **rostrated** (L. *rostratus*, beaked < *rostrum*, beak, snout, rostrum + suf. *-atus* > E. *-ate*, provided with or likeness): having a *rostrum*.

rostrate dehiscence (L. *rostratus*, beaked < *rostrum*, beak, snout + suf. *-atus* > E. *-ate*, provided with or likeness; NL. *dehiscentia* < L. *dehiscere*, to divide, to split, open by itself, to part + suf. *-entia* > E. *-ence*, state, quality or action): a type of ascus dehiscence in some lichen-forming Discomycetes (Lecanorales) in which the asci have walls that are thickened at the apex and, at the moment of spore liberation, the internal wall layer extends and breaks the external layer. Cf. **fissitunicate** and **semifissitunicate**.

rostrum, pl. **rostra** (L. *rostrum*, beak, snout, rostrum): a beak-like extension or process, like that present on various fungal structures, e.g., the spores of *Setosphaeria rostrata* (=*Exserohilum*=*Helminthosporium rostratum*), dematiaceous asexual fungi, and the perithecia of *Guanomyces polythrix* (Sordariales); these structures are *rostrate*.

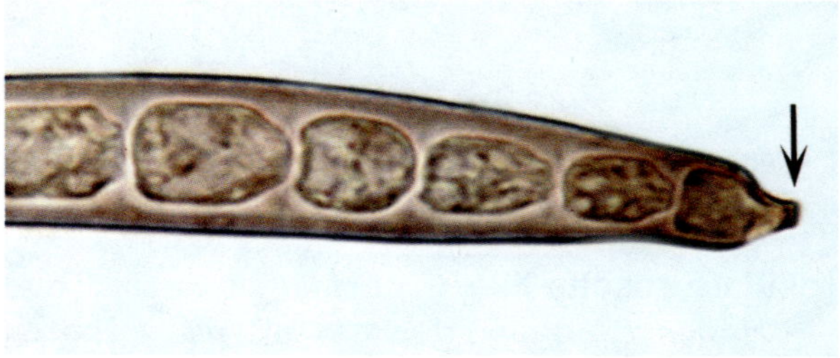

Rostrum of a conidium of *Setosphaeria rostrata* (=*Exserohilum* =*Helminthosporium rostratum*), x 1,200 (*MU*).

Rostrum of the perithecium of *Guanomyces polythrix*, x 10 (*MU*).

Rostrum of the perithecium of *Guanomyces polythrix*, x 50 (*MU*).

rosulate, **rosette-like** (L. *rosulatus*, like a small rose < *rosula*, small rose, rosette + suf. *-atus* > E. *-ate*, provided with or likeness): arranged in the manner of a rosette; like a small rose.

rot (ME. *roten* < OE. *rotian*, to rot): decomposition due to the action of microorganisms; a disease in plants that is distinguished by the degradation of tissues, mainly caused by fungi and bacteria, e. g., the rot of apple caused by *Glomerella* sp. (Phyllachorales). Based on the symptoms, several types of rots caused by fungi are recognized: *Brown rot*, in which the cellulose is degraded (but not the lignin), forming brick-shaped segments; the wood becomes brown and crumbles (caused by many species of Aphyllophorales, e.g., *Fomitopsis pinicola*). *Dry rot* is a type of brown rot caused by *Serpula lacrymans* (Aphyllophorales). This name is misleading, as damp conditions are required for the initiation of fungal growth. *Soft rot* is caused by many species of fungi, including hyphomycetes and Chaetomiaceae (Sordariales), among others. *Wet rot* is a kind of brown rot caused by *Coniophora cerebella*, the fungus of basements or cellars (Aphyllophorales). *White rot* involves the decomposition of cellulose and lignin (or mainly the latter), and the wood bleaches and becomes stringy; it is also caused by Aphyllophorales, e.g., *Trametes versicolor*.

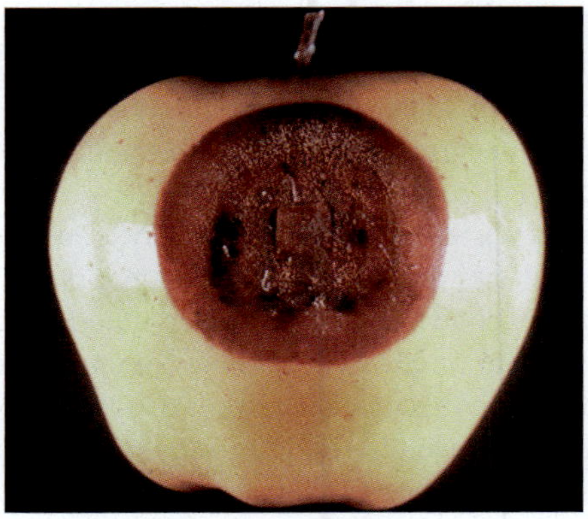

Rot of apple, caused by *Glomerella* sp., 14 days after inoculation, x 0.5 (*RTH*).

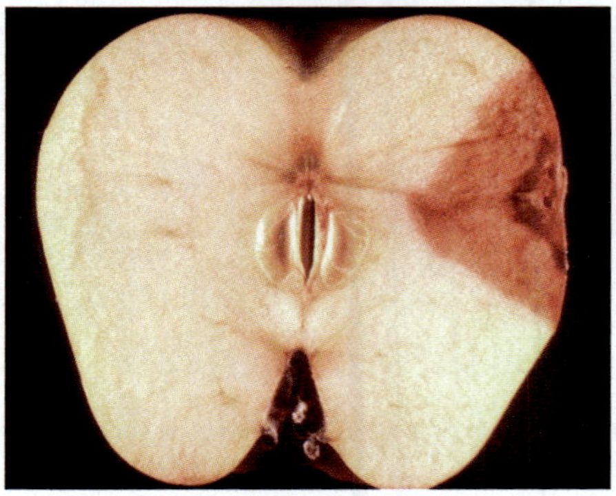

Rot of apple, caused by *Glomerella* sp., 14 days after inoculation, x 0.5 (*RTH*).

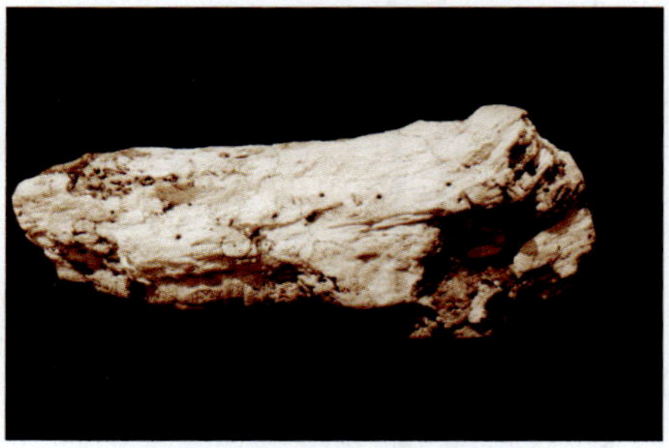

White **rot** of wood, caused by an unidentified aphyllophoraceous fungus, x 0.1 (*LP*).

Soft **rot** of wood, caused by undetermined hyphomycetes and Chaetomiaceae, x 0.1 (*LP*).

Brown **rot** of wood, caused by an undetermined aphyllophoraceous fungus, x 0.1 (*LP*).

rotaceous (L. *rota*, wheel + suf. *-aceus* > E. *-aceous*, of or concerning, with the nature of): shaped like a wheel, flat and circular, like the apothecia of *Rotula* and *Rotularia* (=*Mazosia*), lichens of the order Arthoniales.

rotund (L. *rotundus*, round): round, circular, orbicular, but not perfectly spherical, like the pileus of *Morchella esculenta* (=*M. rotunda*), Pezizales.

Rotund pileus of the ascocarp of *Morchella esculenta* (=*M. rotunda*), x 0. 5 (*TH*).

rubiginose, **rubiginous** (L. *rubiginosus*, reddish-brown in color < *rubigo* or *robigo*, rust + suf. *-osus* > E. *-ose*, or E. *-ous*, full of, augmented, prone to): the color of rust or iron oxide, like the perithecial stroma of *Hypoxylon rubiginosum* var. *rubiginosum* (Xylariales).

rubri-, **rubro-** (L. *ruber*, red): a pref. that alludes to the color red, as in *Rubrinectria* (Hypocreales), a *Nectria* with red or reddish perithecia, and *rubrotoxin B*, a toxic metabolite of *Penicillium rubrum* (moniliaceous asexual fungi) that causes hepatitis in cattle and pigs.

Rubiginose stromata of *Hypoxylon rubiginosum* var. *rubiginosum* (Xylariales), x 1 (*RV*).

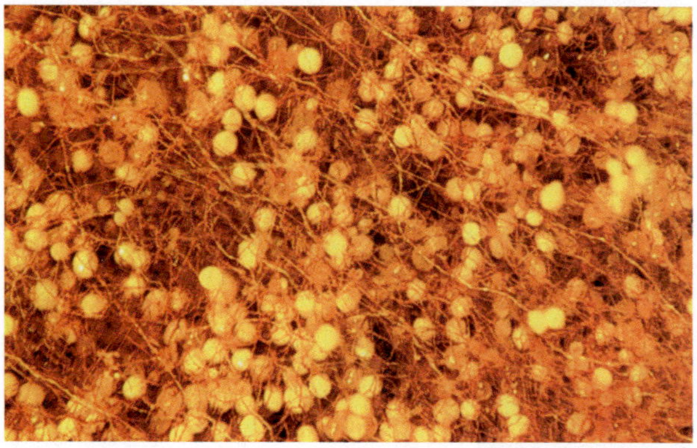

Cleiostothecia of *Eurotium rubrum* (Eurotiales), and hyphae with a red color (**rubrum**), on agar, x 8 (*MU*).

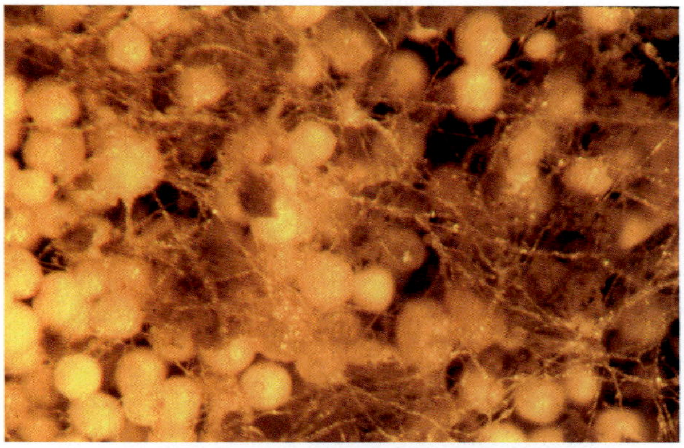

Cleiostothecia of *Eurotium rubrum* (Eurotiales), and hyphae with a red color (**rubrum**), on agar, x 30 (*MU*).

ruderal (L. *ruderalus*, growing in ruins or waste places < *rudus*, genit. *ruderis*, ruins, waste places, debris + L. suf. *-alis* > E. *-al*, relating to or pertaining to): growing on ruins or waste areas, on media with debris, mineral piles, moved earth and other analogous materials created by human habitation and old constructions, as happens in many urban environments.

rugose, rugous (L. *rugosus*, wrinkled, creased < *ruga*, wrinkle + suf. *-osus* > E. *-ose*, or > E. *-ous*, full of, augmented, prone to): wrinkled or creased. E.g., the ornamentation of the wall of various spores is rugose when the wrinkles or irregularities are abundant and confluent, as is observed in the basidiospores of *Panaeolina foenisecii* (Agaricales) and the conidia of *Clasterosporium cocoicola* (dematiaceous asexual fungi). Also applied to the macrocolony that some yeasts develop in culture, like *Pichia membranifaciens* (Saccharomycetales).

Rugose giant colony of *Pichia membranifaciens* on agar, x 2 (*MU*).

Rugose perithecial stromata of *Xylaria polymorpha* (Xylariales), x 1 (*RV*).

Rugose pileus of a basidiocarp of *Leccinum* sp. (Agaricales), x 1 (*RV*).

rugulate, rugulose (L. *rugulatus*, having small wrinkles or creases < *rugula*, dim. of *ruga*, wrinkle + suf. *-atus* > E. *-ate*, provided with or likeness; L. *rugulosus* < *rugula*, small wrinkle + suf. *-osus* > E. *-ose*, full of, augmented, prone to): with slight wrinkles or creases, like the surface of the basidiospores of *Galerina clavata* (Agaricales).

rumposome (ME. *rump*, buttocks < G. *rumph*, torso + Gr. *sõma*, body): an intracellular organelle at the posterior part of the zoospores of *Chytridium*, *Rhizophydium* and other Chytridiales, as well as in *Monoblepharella* (Monoblepharidales), composed of tubules arranged like a honeycomb, that covers part of the surface of the lipid-microbody complex. Depending on the species, the nucleus, ribosomes, one or several mitochondria, and the lipid-microbody complex, form compartments that are interconnected by a system of double membranes of the smooth endoplasmic reticulum. This system of membranes is continuous with the outer membrane of the nucleus, and extends like a network around a lipid globule that is connected to the rumposome, which is situated near the plasma membrane of the zoospore. The rumposome also is united with the kinetosome by means of microtubules, which suggests that all of this mechanism serves to anchor the lipid globule in a certain position within the zoospore, generally in the lateral and posterior part, near the flagellar apparatus.

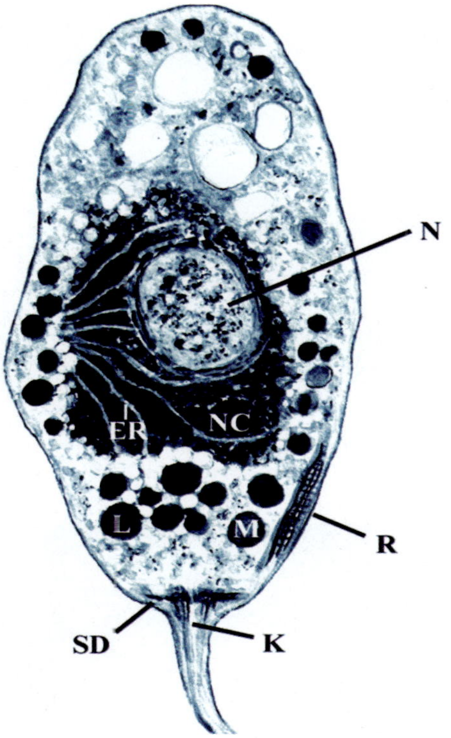

Rumposome (R) of a zoospore of *Monoblepharella* sp., seen in longitudinal section. The ribosomes are grouped in a nuclear cap (NC), which is not delimited by an envelope but which is permeated by endoplasmic reticulum (ER). The nucleus (N) is situated centrally in the zoospore, while the kinetosome (K) is embedded in the striated disk (SD) in the posterior end of the spore. The microtubules irradiate toward the cytoplasm from the disk. The lipid globules (L) and the mitochondria (M) are dispersed throughout the spore. The rumposome is observed at one side, in the posterior end of the spore. Drawing based on a transmission electron micrograph published by Beckett *et al.* in their *Atlas of Fungal Ultrastructure*, 1974, x 4,700 (*MU*).

rupicolous (L. *rupes*, genit. *rupis*, rock, stone + suf. *-cola*, inhabitant): living on walls or rocks, like *Caloplaca saxicola* (Teloschistales). See **saxicolous**.

Rupicolous thallus of *Caloplaca saxicola*, x 10 (*CB*).

Rupicolous thallus of *Caloplaca saxicola*, x 10 (*CB*).

rupi- (L. *rupes*, genit. *rupis*, rock, stone): a pref. used in terms alluding to rocks, such as *rupicola*.

rust (ME. < OE. *rust*, red): *Uredinales*. **1**. Any of the destructive diseases of plants caused by a species of this order. **2**. A species of the Uredinales. *Black stem rust of cereals*, caused by *Puccinia graminis*. *Blister rust of Pinus and Ribes*, caused by *Cronartium ribicola*. *Brown leaf rust of cereals*, of barley, *Puccinia hordei*; of rye and wheat, *P. recondita*. *Crown rust of oats*, *P. coronata*. *Red rust of cereals*, urediniospore state of the cereal rusts, especially *P. graminis*. *White rust* (*white blister*), *Albugo candida* (Peronosporales). In Mexico cereal rusts (as well as other fungal diseases of various plants), also are called *chahuiztles* (Nahuatl word derived from *chiáhuitl*, humor, moisture < *quiáhuitl*, rain; i.e., "disease of moisture").

Coffee **rust** (*Hemileia vastatrix*, Uredinales). The pustules on the lower surface of the leaves represent uredinia, x 1 (*GC*).

Coffee **rust** (*Hemileia vastatrix*, Uredinales). The pustules on the lower surface of the leaf represent uredinia, x 12 (*GC*).

Rust (*Puccinia malvacearum*, Uredinales) of geranium. The pustules on the lower surface of the leaf represent uredinia, x 2 (*MU*).

Rust (*Puccinia malvacearum*) of geranium. The pustules on the lower surface of the leaf represent uredinia, x 2 (*MD*).

Rust (*Puccinia malvacearum*) of hollyhock. The pustules on the lower surface of the leaves represent uredinia, x 2 (*JB*).

Wheat **rust** (*Puccinia graminis tritici*, Uredinales). The pustules on the lower surface of the leaf represent uredinia, x 12 (*MU*).

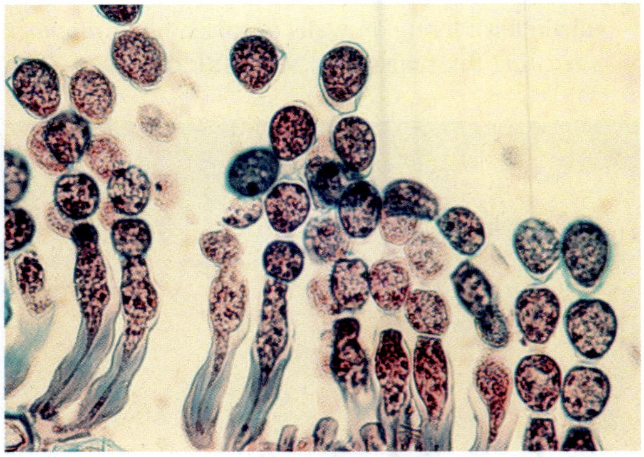

Wheat **rust** (*Puccinia graminis tritici*, Uredinales). The pustules on the lower surface of the leaf represent uredinia, x 12 (*MU*).

S

Synnemata of *Penicillium claviforme*, on agar (*MU* & *CB*).

S

sabulicole, **sabulicolous**, **sabulose**, **sabulous**, **sabuline** (L. *sabulicola* < *sabulum*, sand + suf. *-cola*, inhabitant; or + L. suf. *-osus* > E. *-ous*, full of, augmented, prone to; or + AF. suf. *-ous* > E. *-ous*. full of; or + ME. suf. *-in*, *-ine* < L. *-ine*, of or relating to): living in sand. Syn. of **ammophilous**, **arenicolous** and **psammophilous**.

saccate, **sacklike** (NL. *saccatus*, provided with little sacs or blisters < L. *saccus*, sac, bag, blister + suf. *-atus* > E. *-ate*, provided with or likeness; or + ME. *lik* < OE. *gelic*, like, alike): **1**. Shaped like a small sac; e.g., many ascomycetes have asci that are sac-shaped. **2**. A type of ray of the exoperidium of some gasteromycetes; see **saccate ray**. **3**. Provided with small sacs or blisters, like the surface of the thallus of the lichen *Solorina saccata* (Peltigerales), which has the apothecia sunken in the thallus and with the excipulum concolorous with the thallus, which gives it the appearance of small blisters or sacs. See **sacculiform**.

Saccate rays in the exoperidium of the fruiting body of *Geastrum saccatum* (Lycoperdales), x 1 (*RV*).

saccate ray (NL. *saccatus*, provided with small sacs, or with the appearance of a sac < *saccus*, sac, purse + suf. *-atus* > E. *-ate*, provided with or likeness; L. *radius*, radius, ray): *Gasteromycetes*. In *Geastrum saccatum* (Lycoperdales) it refers to the rays of the exoperidium that split and bend downward, forming a kind of shallow cup on which the endoperidium of the sporiferous sac is seated directly; in *G. triplex* the mesoperidium breaks and remains completely separated from the saccate rays and the endoperidium, forming a third layer in the base of the sporiferous sac.

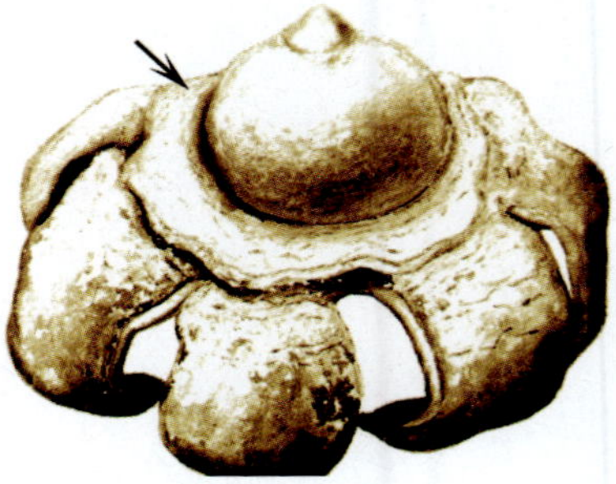

Saccate rays of the exoperidium in the fruiting body of *Geastrum triplex*. These rays become rolled backward and form a shallow cup in which the endoperidium is seated directly on the mesoperidium, x 1 (*MU*).

sacchari-, **saccharo-**, **sacchar-** (Gr. *sáccharon* > L. *saccharum*, sugar): a pref. employed in terms that allude to sugar, such as *sacchariferous*, containing sugar, *saccharose*, sugar from sugar cane and sugar beet, and *saccharomycetes*, explained below.

saccharomycetes (Gr. *sáccharon*, sugar + L. *-mycetes* < Gr. *mýkes*, genit. *mýketos*, fungus): the name assigned to yeasts that have the capacity to ferment sugars, such as *Saccharomyces* (Saccharomycetales), among other gen. See the Appendix, where there is a comparison of the outline of Alexopoulos *et al*., 1996, modified by Ulloa and Hanlin, 2006, to that of Kirk *et al*., 2008. In the latter classification scheme, the order Saccharomycetales has been moved to the subclass Saccharomycetidae, class Saccharomycetes, subphylum Saccharomycotina, phylum Ascomycota; this has been done in order to achieve a phylogenetic classification of the fungi.

sacculiform (L. *sacceliformis*, shaped like a small sac < *saccelus*, dim. of *saccus*, sac, bag + suf. *-formis* < *forma*, shape): shaped like a small sac or bag, like the volva of the basidiocarp of *Amanita caesarea* (Agaricales).

Sacculiform volva of the basidiocarps of *Amanita caesarea*, x 0.5 (*RV*).

Sacculiform volva of the basidiocarp of *Amanita caesarea*, x 0.5 (*MU*).

Sacculiform volva of the basidiocarp of *Amanita caesarea*, x 0.6 (*EPS*).

saccule (L. *sacculus*, small sac < *saccus*, sac + dim. suf. *-ulus* > E. *-ule*): a swollen structure, shaped like a sac, called the sporiferous saccule, at the base of a hypha, that gives rise to the chlamydospores in the vesicular-arbuscular mycorrhizogenous fungi, such as *Acaulospora* and *Entrophospora* (Glomales).

sagenetosome, sagenogenetosome (L. *sagena* < Gr. *sagéne*, barrier net for fishing + Gr. *genetés*, an ancestor + *sõma*, body): see **sagenogen**.

sagenogen (L. *sagena* < Gr. *sagéne*, barrier net for fishing + Gr. *génos*, engenderment): each of the invaginated organelles on the cell surface that connects the plasma membrane to the network of membranes in the Labyrinthulales. Equal to **bothrosome**.

sagittate (L. *sagitta*, arrow + suf. *-atus* > E. *-ate*, provided with or likeness): shaped like an arrow, like the stylet (*Stachel*) of the encysted zoospore of *Plasmodiophora brassicae* (Plasmodiophorales).

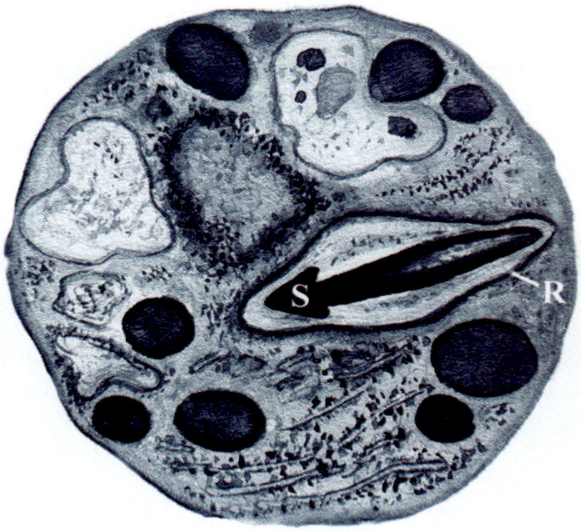

Sagittate organelle (the Stachel, S) inside an invagination of the cell membrane (the Rohr, R) of an encysted zoospore of *Plasmodiophora brassicae*, seen in longitudinal section. Drawing based on a transmission electron micrograph published by Beckett *et al.* in their *Atlas of Fungal Ultrastructure*, 1974, x 23,800 (*MU*).

salpingo-, salping- (Gr. *sálpinx*, genit. *sálpingos*, horn, trumpet): a pref. used in names such as *Salpingocapsa*, a protozoan shaped like a case or capsule, elongated like a trumpet.

sand-case (ME. < OE. *sand*, sand; ME. *cas* < AF. *casse* < OF. *chasse* < L. *capsa*, case, box): applied to the exoperidium of *Disciseda candida* and *Gastrosporium* (Hymenogastrales), which in the mature fructification is composed of loosely intertwined and interwoven hyphae with small particles of sand and gravel; this sandy covering remains at the base, on the substrate, and the endoperidium and the ostiole are exposed in the upper part of the fructification.

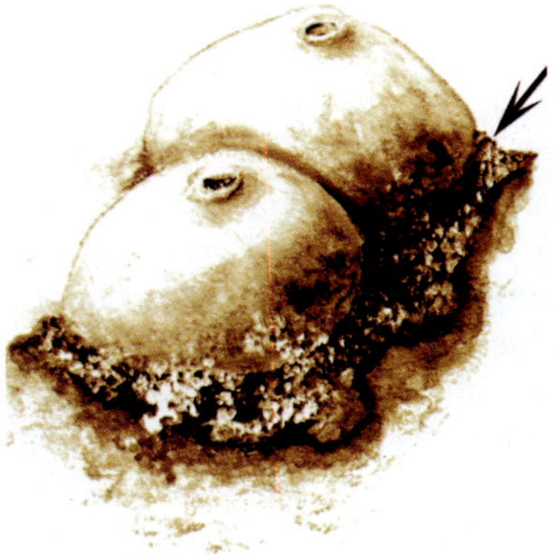

Sand-case of the exoperidium of the fruiting bodies of *Disciseda candida*, x 1.2 (*MU*).

sapro-, **sapr-** (Gr. *saprós*, rotten, putrid): a pref. used in terms alluding to rotting, like the following.

saprobe, **saprobiont** (Gr. *saprós*, rotten, putrid + *bíos*, life; *saprós* + *bíos* + *óntos*, genit. of *ón*, a being): an organism that grows on a dead organic being or on organic substances and utilizes these substrates as food. The majority of the fungi are saprobes. See **saprogen**, **facultative saprobe** and **obligate saprobe**.

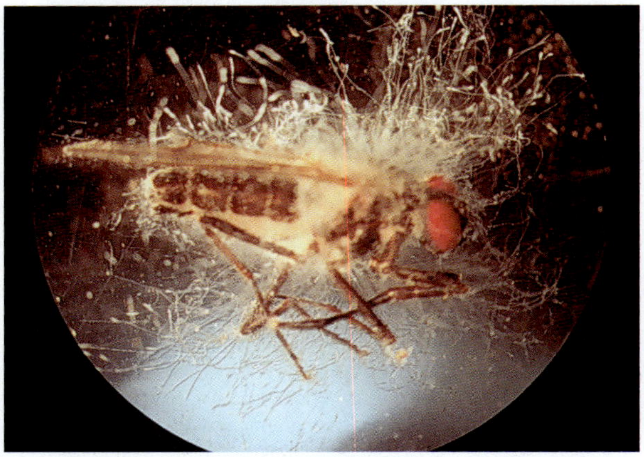

Hyphae and sporangia of *Saprolegnia parasitica* (Saprolegniales) growing as a **saprobe** from a dead insect in water, x 20 (*MU*).

saprogen (Gr. *saprós*, rotten, putrid + *génos*, origin < *gennáo*, to engender, produce): capable of spoiling organic material, of causing its decomposition. See **saprobe**.

saprophyte (Gr. *saprós*, rotten, putrid + *phytón*, plant): a heterotrophic organism that is nourished at the expense of dead animals or plants and various decomposing or decomposed organic residues. Commonly, and mistakenly, it is applied to the fungi, which are not plants, for which reason it is preferible to utilize the terms **saprobe** and **saprotroph**.

saprotroph, **saprotrophic** (Gr. *saprós*, rotten, putrid + *trophós*, something that nourishes, that serves as food): living and obtaining nourishment from dead organic material. This term is preferible to *saprophyte* when referring to fungi, considering that fungi are not plants.

sarciniform (L. *sarcina*, raceme + suf. *-formis* < *forma*, shape): with the shape of a raceme, like the dictyospores of *Sarcinella* (dematiaceous asexual fungi), the asexual state of *Schiffnerula pulchra* (Dothideales).

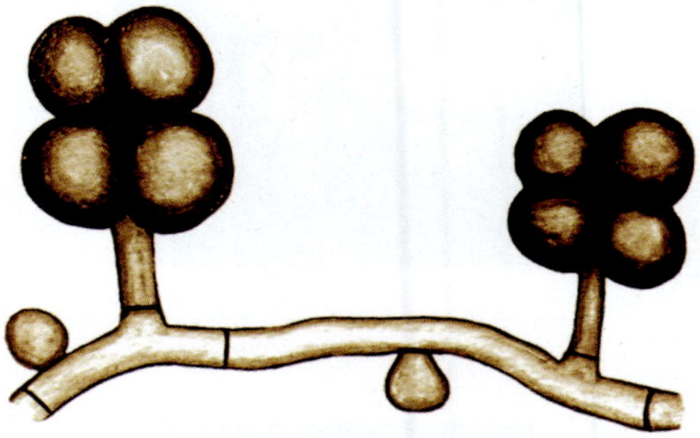

Sarciniform conidia of *Sarcinella*, asexual state of *Schiffnerula pulchra*, x 800 (*MU*).

sarco-, **sarc-** (Gr. *sárx*, genit. *sarkós*, meat, tissue): a pref. employed to give the idea of something that has flesh, or that is more fleshy than ordinary, as in the following terms.

sarcodimitic (Gr. *sárx*, genit. *sarkós*, meat, flesh, tissue + *dís*, two + *mítos*, filament, here, hypha + suf. *-ikos* > L. *-icus* > E. *-ic*, relating to or pertaining to): applied to the hyphal system of some Polyporaceae, in which the intercalary skeletal hyphae are replaced by elongated, fusiform, inflated, thick-walled elements. Cf. **sarcotrimitic**.

sarcoma, pl. **sarcomata** (Gr. *sárkoma*, fleshiness): in plant pathology, a tumor of bland consistency, composed of parenchymatous cells, with neither vessels nor fibers.

sarcomycete (Gr. *sárx*, genit. *sarkós*, flesh, tissue + *mýkes*, genit. *mýketos*, fungus): any hymenomycete with notably fleshy fructifications.

sarcothecium, pl. **sarcothecia** (Gr. *sárx*, genit. *sarkós*, flesh, tissue + *thêke*, caja): a term occasionally applied to a stroma.

sarcotrimitic (Gr. *sárx*, genit. *sarkós*, flesh, tissue + *trís*, three, three times + *mítos*, filament, here, hypha + suf. *-ikos* > L. *-icus* > E. *-ic*, relating to or pertaining to): like the dimitic hyphal system, but besides in the trimitic hyphae, the generative hyphae give rise to thick-walled, inflated elements, similar to connective hyphae, but septate. Cf. **sarcodimitic**.

saturn-like (< planet Saturn + ME. *lik* < OE. *gelic*, like, alike): with a flat equatorial flange, like the ascospores of the yeast *Debaryomyces* (=*Schwanniomyces*) *occidentalis* (Saccharomycetales).

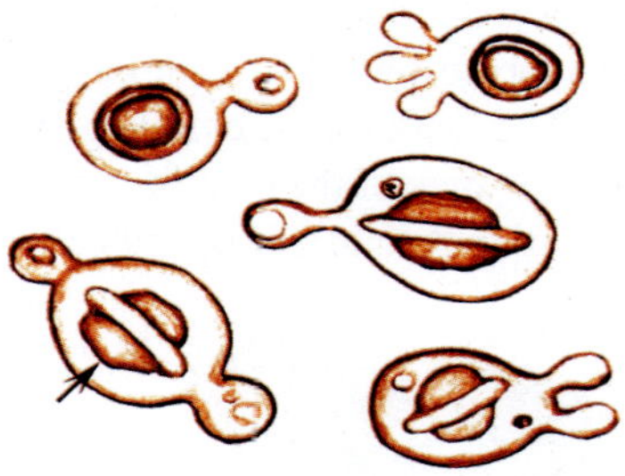

Saturn-like ascospores of *Debaryomyces* (=*Schwanniomyces*) *occidentalis*, x 3,700 (*MU*).

saxicolous, **saxicoline** (L. *saxum*, rock, stone + suf. *-cola*, inhabitant + ME. suf.*-in*, *-ine* < L.*-ine*, of or relating to; or + L. suf. *-osus* > E. *-ous*, full of, augmented, prone to): an organism that lives attached to the surface of rocks, as do some lichens (e.g., *Caloplaca saxicola*, of the Teloschistales, *Roccella*, of the Arthoniales, and *Rhizocarpon geographicum*, various species of *Parmelia*, *Aspicilia* (=*Lecanora*) *calcarea* and *L. frustulosa*, of the Lecanorales, among many other species). Also called **epilithic**, **petricolous** and **rupicolous**.

Saxicolous thallus of *Rhizocarpon geographicum*, x 1 (*MU*).

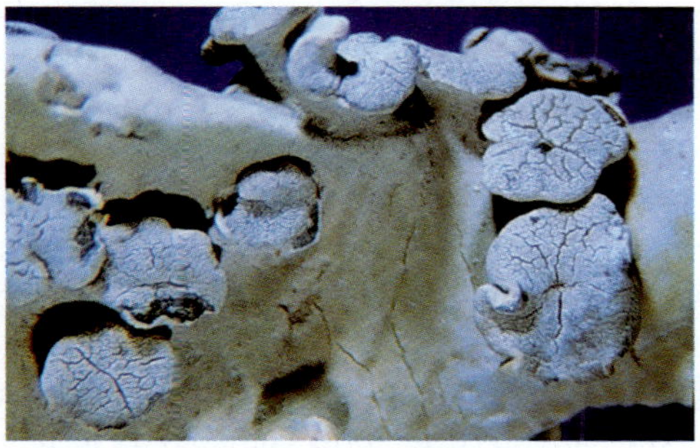

Saxicolous thallus of *Roccella fimbriata*, x 10 (*MU*).

scab (ME. < ON. *skabb*, scab, itch): any of various diseases of plants caused by bacteria or by fungi, characterized by the formation of crustose spots (hyperplasic lesions similar to the mange or scabies of animals and humans, produced by mites). *Scab of apple*, *cherry*, and *pear* (respectively caused by *Venturia inaequalis*, *V. cerasi* and *V. pirina*, of the Pleosporales). *Scab of cereals*, caused by *Gibberella zea* (=*G. saubinetti*), of the Hypocreales, and *Fusarium* spp., of the tuberculariaceous asexual fungi. *Citrus scab* (caused by *Elsinoë fawcettii*, of the Myriangiales). *Peach scab* (caused by *Fusicladium carpophilum*, of the dematiaceous asexual fungi). *Potato scab* (caused by *Spongospora subterranea*, of the Plasmodiophorales).

scabrid, **scabridous** (L. *scabridus*, something rough, rugged < *scaber*, fem. *scabra*, rough, scurfy, scabby): with a rough, unequal or rugose surface, due to rigid and irregular projections. Equal to **scabrose**.

scabrose, **scabrous** (L. *scaber*, fem. *scabra*, rough, scurfy, scabby + suf. *-osus* > E. *-ose*, full of, augmented, prone to): rough, with short, rigid projections that are readily felt on touching, like the surface of the stipe of *Leccinum scabrum* (=*Boletus scaber*) and *L. aurantiacum* (Agaricales). Also called **scabrid**.

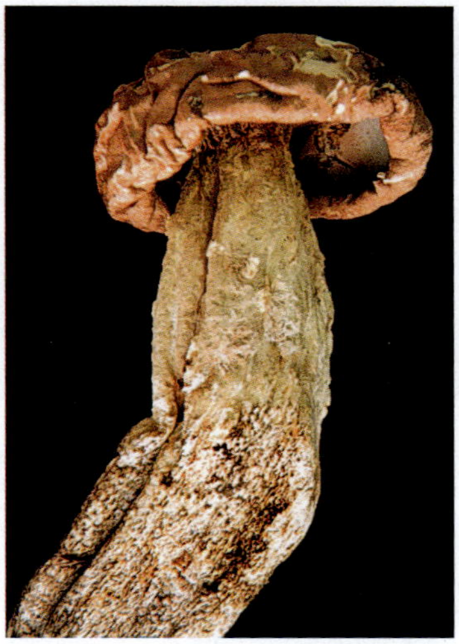

Scabrose stipe of the basidiocarp of *Leccinum aurantiacum*, x 0.5 (*MU*).

scaly (ME. *scale* < MF. < OF. *escale* < OHG. *skala*, scale, thin, platelike structure + E. suf. *-y*, abounding in, full of, having the quality of): having scales on the surface, like the sporangial peridium of *Lepidoderma tigrinum* (Physarales) and the pileus of the basidiocarp of *Macrolepiota procera* and *Amanita muscaria* (Agaricales), among many other examples.

Scaly sporangial peridium of *Lepidoderma tigrinum*, x 20 (*MU*).

Scaly pileus of the basidiocarps of *Macrolepiota procera*, x 1 (*RV*).

Scaly pileus of the basidiocarp of *Macrolepiota procera*, x 0.7 (*MU*).

Scaly pileus of the basidiocarps of *Leucocoprinus cepistipes*, x 0.3 (*RV*).

scaphoid (Gr. *skáphe*, skiff, sloop + L. suf. *-oide*, like < Gr. *-oeídes*, similar to): equal to **carinate**, **cymbiform** and **navicular**.

scar (Gr. *eskhára*, scab caused by burning > LL. *eschara* > OF. *scab* > ME. *escare*, scar): a mark left on a yeast mother cell after the bud (the daughter cell) separates from it (*bud scar*); the mark left on the daughter cell is the *birth scar*. The mark on the conidiogenous locus and on the base or apex of the conidium that remains after secession of the conidium is also called a scar.

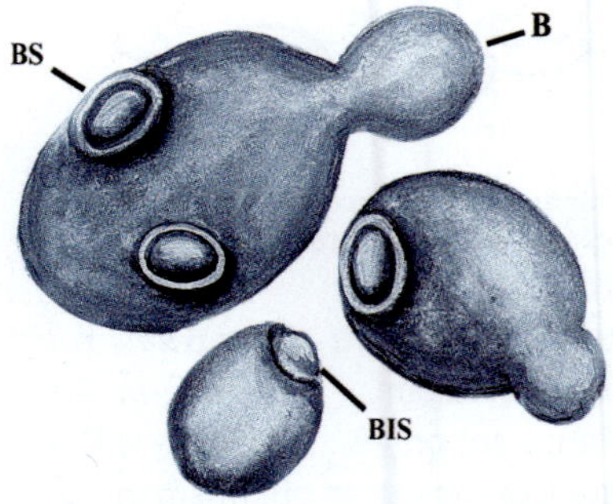

Somatic cells of *Saccharomyces cerevisiae* (Saccharomycetales), showing bud **scars** (BS) and developing buds (B); one daughter cell has a birth **scar** (BIS), x 3,000 (*MU*).

scariose, **scarious** (ML. *scariosus*, thin, dry, membranous, semitransparent < *eschara* < Gr. *schára*, crust, scale + L. suf. *-osus* > E. *-ose*, full of, augmented, prone to): applied to structures that have a membranous consistency, more or less stiff and dry, generally translucent, like paper, like the scales on the pileus of *Naucoria escharoides* (Agaricales).

Schaeffer reaction (name of the mycologist J.C. Schaeffer; NL. *reaction*, reciprocal action between two or more substances or bodies, resulting in one different from the originals < L. *re*, prep. that indicates opposition or resistance + *action* < L. *acto*, genit. *actionis*, effect of doing): *Agaricales*. A bright reddish-orange reaction that occurs when a line of aniline is applied to fungal tissue, followed by a perpendicular line of nitric acid; as is observed, e.g., in *Longula* and in Section *Arvensis* of the gen. *Agaricus*.

schizidium, pl. **schizidia** (NL. *schizidium* < Gr. *schízo*, to split, divide, separate + dim. suf. *-ídion* > L. *-idium*): *Lichens*. A vegetative propagule formed by excision from the upper layers of the lichen thallus, which detaches from the main lobes as segments similar to scales. E.g. in *Fulgensia bracteata* subsp. *deformis* (Teloschistales), schizidia are produced as little lobes.

schizo-, **schiz-** (Gr. *schízo*, to split, divide, separate): a pref. employed to form compounds to allude to something that splits or divides, as in *schizogenesis*.

schizogenesis, pl. **schizogeneses** (Gr. *schízo*, to split, divide, separate + suf. *-génesis*, origin, engenderment, production): syn. of **fission** or **bipartition**; cell multiplication by means of simple division; typical of the bacteria and of some yeasts, such as *Octosporomyces* (=*Schizosaccharomyces*), Schizosaccharomycetales. Syn. of **schizogony**.

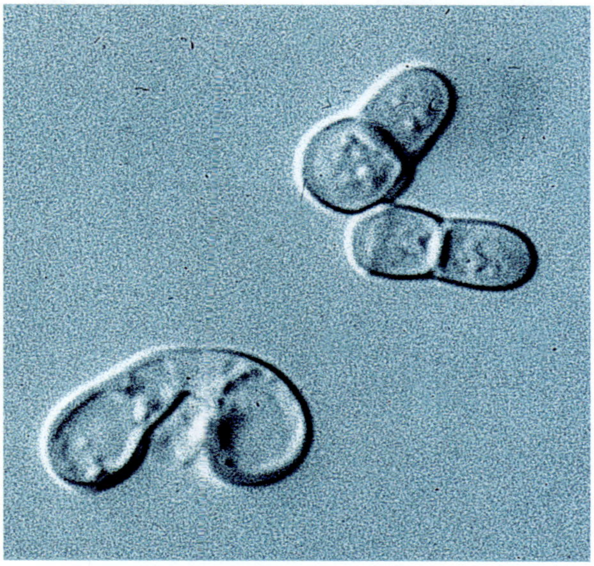

Schizogenesis of the vegetative cells of *Octosporomyces* (=*Schizosaccharomyces*) *octosporus*, x 850 (*RTH*).

schizogenous (Gr. *schízo*, to split, divide, separate + *génos*, origin < *gennáo*, to engender, produce): see **schizolytic**. Cf. **lysigenous**.

schizogony (Gr. *schízo*, to split, divide, separate + *gonía*, generation or engenderment): syn. of **schizogenesis**.

schizolytic (Gr. *schízo*, to split, divide + *lýtos*, dissoluble, broken, loose): syn. of **schizogenous**; a propagule that separates or detaches by the dissolution of part of the septum that separates it from the adjacent cell. This process is termed *schizolysis* or *schizogenesis*. The blastic conidia of some fungi are examples of schizolytic propagules, which detach by schizolysis. Ostiole formation in ascocarps and pycnidia is by schizolysis. Cf. **rexolytic**, **rexolysis** and **rexigenesis**.

schizoont (Gr. *schízōon*, genit. *schízōontos* < *schízo*, to split, divide, separate + *óntos*, genit. of *ón*, a being): any structure that splits or separates into two or more portions, but it especially refers to the wall-less, multinucleate, assimilative endobiotic thallus of the Plasmodiophorales, which undergoes a simple or multiple division to give rise to sporangia with spores inside the cells of the host plant.

sciophilous (Gr. *skiá*, shade, in L. *scio-*, + Gr. *phílos*, have an affinity for): an ecological term for organisms that require shade. Syn. of **umbraticous** or **umbrophilous**.

scissile (L. *scissilis*, capable of being cut smoothly or split easily < *scindere*, to divide, split): separating or splitting easily, like the flesh of the pileus of some agarics that can separate or divide by themselves into horizontal layers.

sclerobasidium, pl. **sclerobasidia** (Gr. *sklerós*, hard + *basídion* < *básis*, base + dim. suf. *-ídion*): *Agaricales*. A basidium with a thick wall (0.5 µm or more); sclerobasidia can be fertile but are more commonly sterile. An example of a fertile sclerobasidium can be found in some species of *Fayodia*, whereas sterile sclerobasidia occur in *Armillaria* (=*Armillariella*) *mellea* and in certain species of *Hygrotrama*, all in the Agaricales.

scleroplectenchyma, pl. **scleroplectenchymata** (Gr. *sklerós*, hard + *plektós*, braided, intertwined + *énchyma*, filled, stuffed): a plectenchyma composed of very thick-walled, conglomerated cells, that constitute the *stereoma* or principal supporting tissue of the thallus of certain lichens, such as *Cladonia* and *Alectoria* (Lecanorales). Also applied to hard fungal tissue, such as the peridium of the ascocarps of many species of *Leptosphaeria* (Pleosporales), which is composed of large, globose cells with thick, hard walls.

sclerotic cell (Gr. *sklerós*, hard + suf. *-tíkos* > L. *-ticus* > E. *-tic*, relation, fitness, inclination or capacity; NL. *cellula*, cell, living cell, dim. of L. *cella*, small room, compartment, cell of a honeycomb): see **muriform cell**.

sclerotium, pl. **sclerotia** (NL. *sclerotium* < Gr. *sklerotés*, hardness < *sklerós*, hard + L. dim. suf. *-ium*): a hardened structure composed of fungal tissue that resists unfavorable conditions; sclerotia are composed of plectenchyma, with a firm cortex, frequently chitinized and brown or blackish in color, although they can be other colors; they are capable of germinating and reinitiating vegetative growth. Among the agonomycetaceous asexual fungi, sclerotia can be macroscopic, like those of *Cenococcum graniforme* and *Athelia* (=*Sclerotium*) *rolfsii*, or microscopic (as in *Rhizoctonia solani*). Among the ascomycetes, the sclerotia of *Sclerotinia* (Helotiales) also are macroscopic; the large sclerotia of *Claviceps gigantea* (Hypocreales) are notable, as are the enormous edible sclerotia of *Wolfiporia extensa* (=*W. cocos*), Aphyllophorales, in the basidiomycetes; the large edible sclerotia of this subterranean fungus are called Canadian tuckahoe or Indian bread. Another example of a macroscopic sclerotium is that of *Coprinus sclerotigenous* (Agaricales).

Sclerotia of *Cenococcum graniforme*, x 20 (*CB* & *MU*).

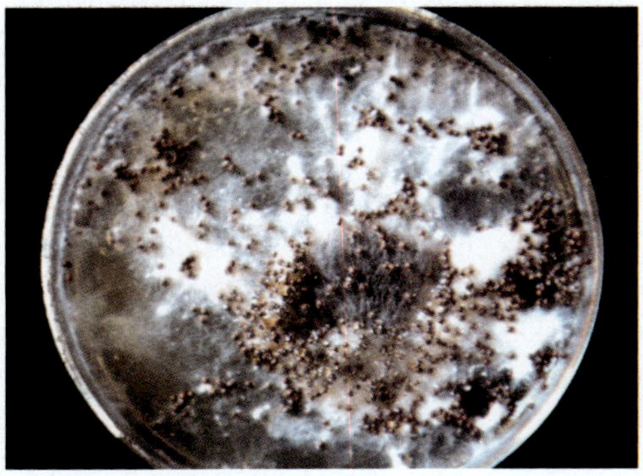

Sclerotia of *Athelia* (=*Sclerotium*) *rolfsii* on agar, x 1 (*MU*).

Sclerotia of *Athelia* (=*Sclerotium*) *rolfsii* on agar, x 10 (*MU*).

Sclerotia of *Athelia* (=*Sclerotium*) *rolfsii* on agar, x 15 (*MU*).

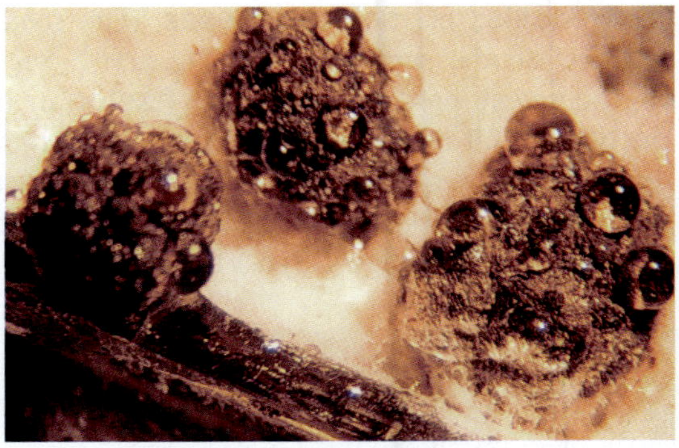

Sclerotia of *Sclerotinia* sp. on agar, x 2 (*MU*).

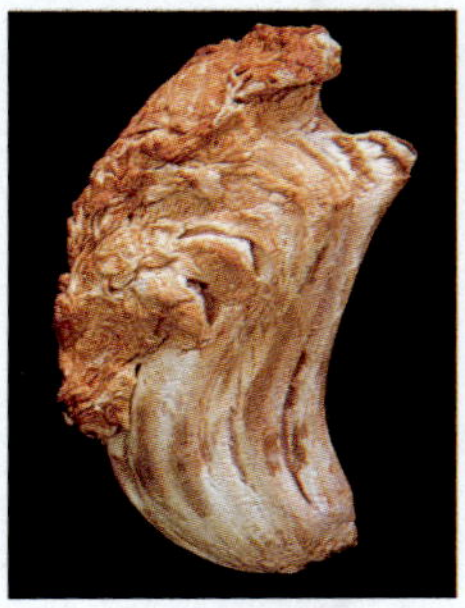

Sclerotium of *Claviceps gigantea*, x 2 (*MU*).

Sclerotium of *Wolfiporia extensa* (=*W. cocos*), x 2 (*RTH*).

Sclerotium of *Phymatotrichopsis omnivora* (moniliaceous asexual fungi) germinated on agar, x 20 (*JAS*).

scrupose (L. *scruposus*, jagged, serrate, rough < *scrupus*, small sharp stone + suf. *-osus* > E. *-ose*, full of, augmented, prone to): covered with small, sharp, hard roughenings, like the pileus of *Lepiota hystrix* (Agaricales), whose scales are bristly.

scobiculate (NL. *scobiculatus* < L. *scobs*, genit. *scobis*, filings, sawdust, dust + suf. *-atus* > E. *-ate*, provided with or likeness): provided with small grains, like those of sawdust. See **scobiform**.

scobiform (NL. *scobiformis* < L. *scobs*, genit. *scobis*, filings or sawdust, dust + suf *-formis* < *forma*, shape): with the appearance of filings or small grains, like those of sawdust, as is seen on the surface of the peridium of some lycoperdaceous fungi. See **scobiculate**.

scolecite (Gr. *skólex*, worm + *kýtos*, cavity, here, cell): a wide, vermiform, coiled hypha located in the center of the young ascomycete perithecium, from which the ascogenous hyphae develop. Also known as a *Woronin hypha*.

scoleco-, **scolec-** (Gr. *skólex*, genit. *skólekos*, worm, insect larva): a pref. employed to allude to something that resembles a worm, as in *scolecospore*.

scolecospore (Gr. *skolékos*, genit. of *skólex*, worm + *sporá*, spore): an elongate, vermiform or filiform spore, like the conidia of *Cercospora* (dematiaceous asexual fungi) and the β-conidia of *Phomopsis* (sphaeropsidaceous asexual fungi). Utilized in the Saccardoan terminology.

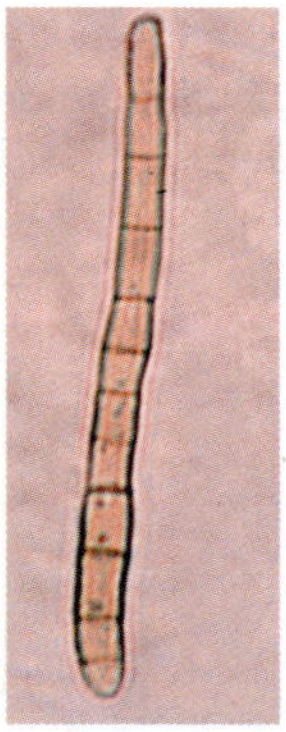

Scolecospore of *Cercospora coffeicola*, x 1,250 (*CB*).

scorpioid (L. *scorpio*, scorpion + NL. suf. *-oides*, or < L. *-oide*, contraction of Gr. *-oeídes*, which denotes likeness): in some lichens, it refers to the type of branching in which all of the lateral branches are curved in such a manner that they appear to originate from one side of the main stem, as occurs in *Cladonia arbuscula* (Lecanorales); it is so-named because of the similarity of the branching to the curved or arched tail of a scorpion.

scrobiculate (L. *scrobiculatus*, having holes or pits < *scrobiculus*, dim. of *scrobs*, hole, trench + suf. *-atus* > E. *-ate*, provided with or likeness): syn. of **foveolate**; having small pits or narrow, slot-like depressions, arranged densely and regularly, so that the projecting parts constitute a kind of reticulum, as is seen in the thallus of some lichens (*Lobaria scrobiculata*, of the Peltigerales) and the stipe of *Lactarius scrobiculatus* and *L. pallidus* (Agaricales).

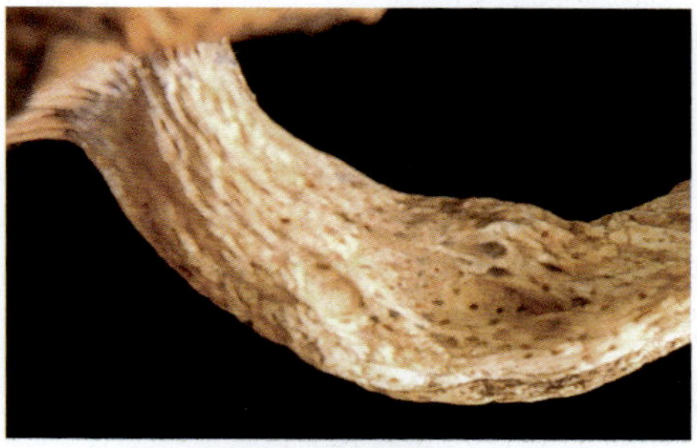

Scrobiculate stipe of the basidiocarp of *Lactarius scrobiculatus*, x 1 (*MU*).

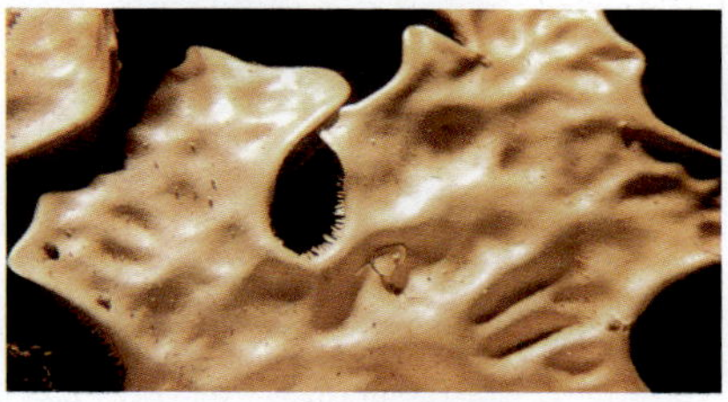

Scrobiculate thallus of *Lobaria scrobiculata*, x 15 (*MU*).

scutate, **scutated** (L. *scutate*, having a shield < *scutum*, an oblong shield + suf. *-atus* > E. *-ate*, provided with or likeness): syn. of **clypeate**, **peltate** and **scutiform**.

scutellum, pl. **scutella** (L. *scutellum* < *scutum*, oblong shield + dim. suf. *-ellum*): the shield-shaped upper portion of a dimidiate ascostroma, which at maturity covers the ascigerous locule or locules, as occurs in Loculoascomycetes of the order Dothideales, e.g., *Microthyrium microscopicum*.

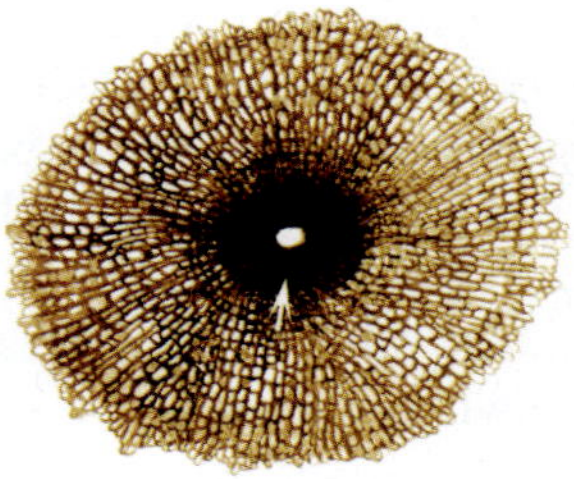

Scutellum (dark area) around the ostiole of a thyriothecium of *Microthyrium microscopicum*, drawn in surface view, x 350 (*MU*).

scutiform (L. *scutiformis* < *scutum*, oblong shield + suf. *-formis* < *forma*, shape): shield-shaped. See **clypeate**, **peltate** and **scutate**.

scutulum, pl. **scutula** (NL. *scutulum*, L. dim. *scutum*, oblong shield + dim. suf. *-ulum*): a cup-like crust or sheet of hyphae produced in the hair follicles of the scalp or of the body in infections caused by *Trichophyton schoenleinii* (moniliaceous asexual fungi).

scyphoid (Gr. *skýphos*, drinking vessel, cup + NL. suf. *-oides*, or < L. *-oide*, contraction of Gr. *-oeídes*, which denotes likeness): cup-shaped. See **scyphus**.

scyphus, pl. **scyphi** (L. *scyphus* < Gr. *skýphos*, drinking vessel, cup): widening in the manner of a glass or goblet of the apex of the podetia of certain lichens, such as *Cladonia fimbriata* and *C. chlorophaea* (Lecanorales).

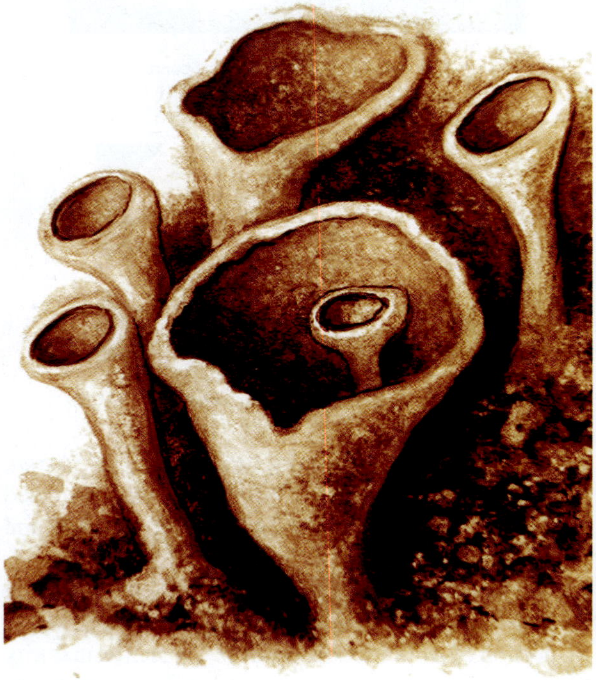

Scyphi of the podetia of the thallus of *Cladonia chlorophaea*, x 10 (*MU*).

secession (L. *secessio*, genit. *secessionis*, separation): refers to the manner in which conidia separate from the conidiogenous cells, which can be by means of *schizolysis* or by *rhexolysis*. See **schizolytic** and **rhexolytic**.

secondary (L. *secundarius* < *secundus*, second in order or grade, following + suf. *-arius* > E. *-ary*, pertaining to): of or relating to the second order or state in a series. *Secondary appendage*: *Laboulbeniales*. Growth that develops from a cell of the receptacle, which in turn originates from the basal cell of the ascospore, as occurs in *Laboulbenia*. Cf. **primary appendage**. *Secondary spore*: *Basidiomycetes*. A spore that is not formed on a basidium, but which is a conidium, chlamydospore, etc., that is produced directly from the assimilative mycelium or from a hypha of the fruiting body. Also refers to a conidium formed on a previous spore. *Secondary hypha*: another term for the setose hyphae formed on the mycelium of *Blumeria* (=*Erysiphe*), of the Erysiphales. *Secondary mycelium*: *Basidiomycetes*. Dikaryotic mycelium that results from the plasmogamy of the primary mycelium, as occurs in *Coprinus lagopus* (Agaricales). It is characterized by having clamp connections and for its assimilative function; it does not give rise to a fruiting body, as happens in the ascomycetes. See **primary mycelium** and **tertiary mycelium**. *Secondary thallus*: *Lichens*. Refers to the podetia that develop on the squamulose primary thallus (Lecanorales). See **primary thallus**.

secondary appendage (L. *secundarius* < *secundus*, second in order or grade, following + suf. *-arius* > E. *-ary*, belonging to; L. *appendage* < *appendere*, to append): *Laboulbeniales*. An outgrowth that develops from a receptacle cell, which in turn derives from the basal cell of the ascospore; e.g., *Laboulbenia*. Cf. **primary appendage**.

secondary conidia (L. *secundarius* < *secundus*, second in order or grade, following + suf. *-arius* > E. *-ary*, belonging to; NL. *conidium* < Gr. *kónis*, dust + dim. suf. *-ídion*): a secondary conidium that does not proceed from a conidiophore, but from a primary conidium, as is seen in *Conidiobolus thromboides* (Entomophthorales).

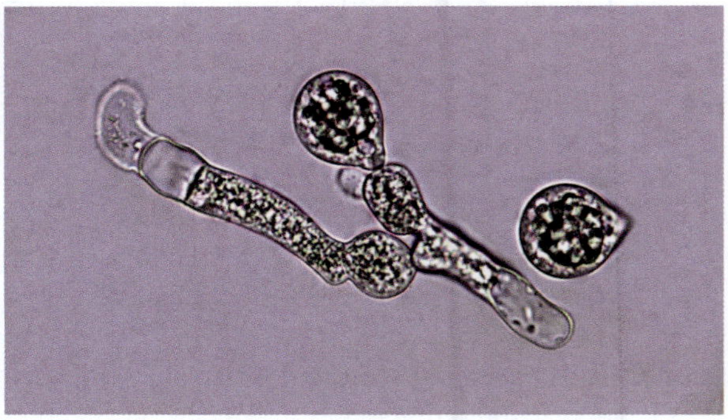

Secondary conidium formed as a growth of primary conidium of *Conidiobolus thromboides*, on agar, x 400 (*MU*).

secondary hypha, pl. **hyphae** (L. *secundarius* < *secundus*, second in order or grade, following + suf. *-arius* > E. *-ary*, belonging to; NL. *hypha*, hypha < Gr. *hyphé*, tissue, spiderweb; hypha): see **secondary**.

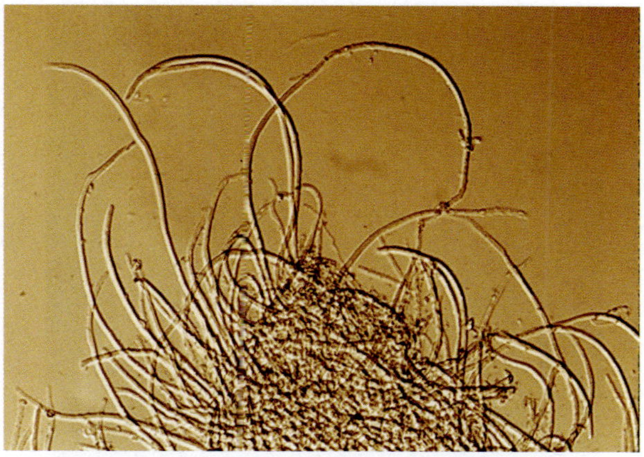

Secondary hyphae of *Blumeria* (=*Erysiphe*) sp., Erysiphales, x 1,000 (*RTH*).

secondary marine fungus, pl. **fungi** (L. *secundarius* < *secundus*, second in order or grade, following + suf. *-arius* > E. *-ary*, having the quality of; L. *marinus*, of the sea < *mare*, sea + suf. *-inus* > E. *-ine*, of or pertaining to; L. *fungus*, fungus < Gr. *spóngos*, *sphóngos*, sponge): a species derived from a terrestrial ancestor that secondarily emigrated to the marine environment, as some Loculoascomycetes, among others, have done. Cf. **primary marine fungus**.

secondary mycelium, pl. **mycelia** (L. *secundarius* < *secundus*, second in order or grade, following + suf. *-arius* > E. *-ary*, pertaining to; NL. *mycelium* < Gr. *mýkes*, fungus + L. suf. *-elis*, concerning + dim. suf. *-ium*): *Basidiomycetes*. The dikaryotic mycelium that results from the plasmogamy of the *primary mycelium*, as occurs in *Coprinopsis* (=*Coprinus*) *lagopus* (Agaricales). It is characterized by clamp connections and by its assimilative function; it does not form a fruiting body, as happens in the ascomycetes.

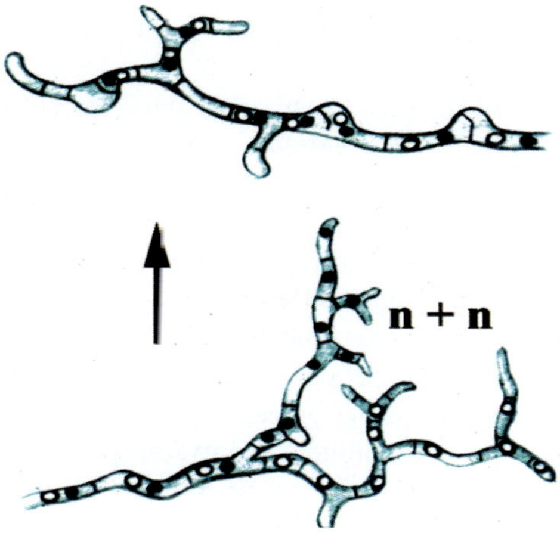

Secondary mycelium of *Coprinopsis* (=*Coprinus*) *lagopus*, x 280 (*MU*).

secondary spore (L. *secundarius* < *secundus*, second in order or grade, following + suf. *-arius* > E. *-ary*, belonging to; NL. *spora* < Gr. *sporá*, spore): see **secondary**.

secondary thallus, pl. **thalli** (L. *secundarius* < *secundus*, second in order or grade, following + suf. *-arius* > E. *-ary*, pertaining to; L. *thallus* < Gr. *thallós*, shoot, thallus): *Lichens*. The podetia that develop on the squamulose primary thallus.

A podetium (**secondary thallus**), that developed from the squamulose primary thallus, of *Cladonia didyma* (Lecanorales), x 8 (*MU*).

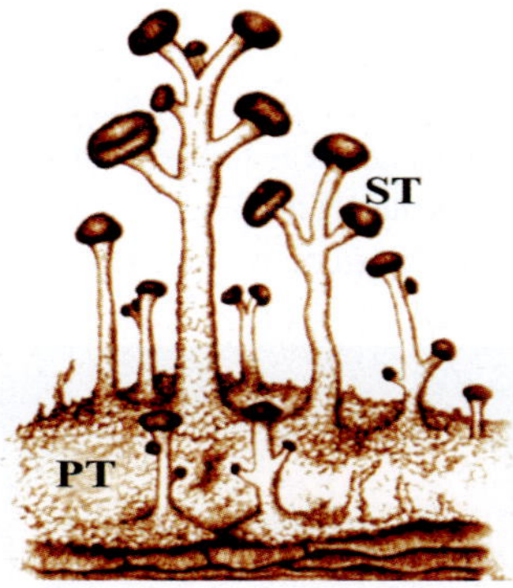

Primary (PT) and **secondary thalli** (ST) of the lichen *Cladonia pyxidata*, x 10 (*MU*).

secotioid (< gen. *Secotium* < L. *secare*, to cut + dim. suf. *-ium* + L. suf. *-oide*, like < Gr. *-oeídes*, similar to): a term applied to an epigeous fruiting body with the appearance of an unopened agaric or bolete, generally with a stipe-columella and a gleba often composed of contorted gills, as in *Secotium*, *Truncocolumella* and *Podaxis* (Hymenogastrales); this is highly correlated with orthrotropic spore development, but not necessarily with obvious affinities with known agarics.

Secotioid basidiocarp, represented by *Podaxis pistillaris*, seen in longitudinal section, x 0.5 (*MU*).

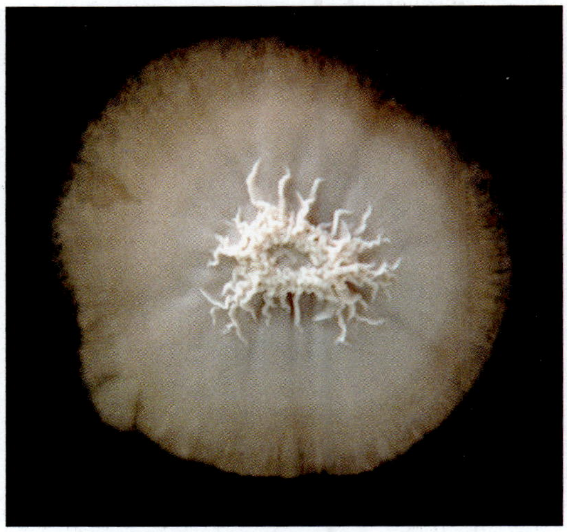

Sectoring of a giant colony of an unidentified yeast on agar, x 1 (*MU*).

section (L. *sectio*, genit. *sectionis*, a natural subdivision of a taxonomic group): in taxonomy, a complex genus or subgenus can be divided into sections, so that the section is a taxonomic category situated between the genus (or the subgenus) and the species. Sections have Latinized names similar to those of the genera, but with a different ending, e.g., Section *Coremigenum* of the subgenus *Biverticillium* (moniliaceous asexual fungi).

sectoring (E. action and effect of dividing into sectors or sections < L. *sector*, one who cuts or divides, or L. *section* < *sectus* < *secare*, to cut + E. pp. suf. *-ing*): the formation of morphologically differentiated regions in a fungal colony, known as *sectors*, that contain different populations of nuclei as the result of heterokaryosis during parasexual gene recombination.

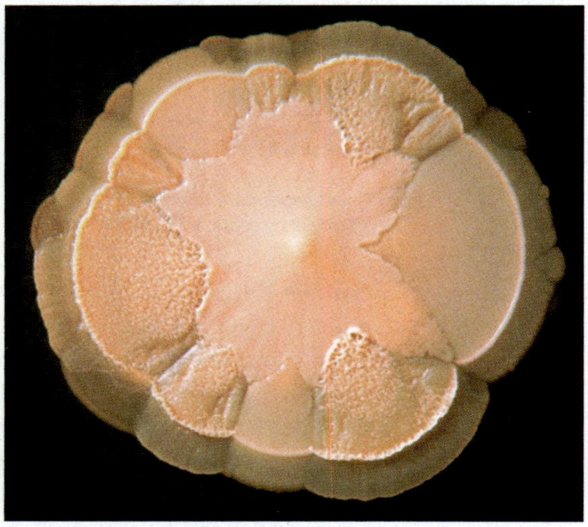

Sectoring of a colony of *Candida* sp. (cryptococcaceous asexual yeasts) on agar, x 1 (*MU*).

segment (L. *segmentum*, portion): applied to the part of a dictyospore that is separated by a transverse septum.

segregate, sequestrate (L. *segregare*, to separate or divide one thing from another or others + suf. *-atus* > E. *-ate*, provided with or likeness): **1**. In taxonomy, to separate a group of taxa from a preexisting group, as another separate, independent group of taxa. **2**. Taxa of fungi that have evolved from ancestors with an exposed hymenium and forcibly liberated spores to forms with a closed fruiting body or a hypogeous habit. Many examples of species derived from specific ancestors with active spore discharge are recognized, e.g., *Rhizopogon* (Hymenogastrales) from *Suillus* (Agaricales), whose spores are retained in the fruiting bodies until these decompose or are eaten by an animal vector.

seiospore (Gr. *seió*, to agitate, move from one side to the other, to detach + *sporá*, spore): refers to a dry spore that detaches from the conidiogenous cell by agitation.

self-inhibitor (ME. < OE. *self*, same, pertaining to one's own; ME. *inhibiten* < L. *inhibitus*, pp. of *inhibere*, to prohibit from doing something; restrain): a fungal metabolite which imposes dormancy to spores whenever unfavorable conditions to germinate are present. High concentrations of spores fail to germinate; this phenomenon is thought to be a direct effect of the presence of fungal self-inhibitors. Dilution of spore suspensions is one of the strategies used to overcome self-inhibiton. See **mycosporin**.

semaphoront (Gr. *séma*, genit. *sématos*, signal, mark, standard + suf. *-phóros*, bearer + *óntos*, genit. of *ón*, the one who is or that which is, a being): in the terminology employed to define the concept of a phylogenetic species, based on monophyly, this is one of the terms that is applied to the minor aggregation of populations or lineages, that can be diagnosed by a unique combination of characters present in comparable individuals, descendents of the same ancestor; i.e., the semaphoronts are the distinctive characters that mark a population genetically integrated through a historic contingent of intercrossing.

semi- (L. *semis*, half): a pref. often used in biological lexicography to give the idea of something that occurs by halves, as in *semifissitunicate*. Corresponds to Gr. pref. **hemi-**.

semifissitunicate (L. *semis*, half + *fissio*, partition, separation + *tunicatus*, covered with a tunica < *tunica*, tunica, cover + suf. *-atus* > E. *-ate*, provided with or likeness): a type of ascal dehiscence, characteristic of the Arthoniales, that is distinguished by lacking a complete separation of the two walls of the ascus during ascospore discharge; during dehiscence, only the upper part of the inner wall separates from the outer wall and becomes everted. This and other variations in ascal dehiscence can be found in the Loculoascomycetes. Cf. **fissitunicate** and **rostrate dehiscence**.

semiimmersed (L. *semis*, half + *immersus* < *immergere*, to merge or plunge into something that surrounds or covers): this term is applied to the fruiting bodies of fungi that are not completely immersed in the substrate, e.g., the ascomata of aquatic ascomycetes that grow on plant detrita.

Semiimmersed ascomata of an unidentified ascomycete on a woody detritus, x 10 (*AC*).

seminicole, seminicolous (L. *semen*, genit. *seminis*, seed + suf. *-cola*, inhabitant): living on seeds, like those called seed or *storage fungi*.

Colonies of *Aspergillus flavus* (moniliaceous asexual fungi) growing from the seeds of broad beans placed on agar; this species is one of the common seed or **seminicole** fungi, x 1 (*MU*).

semio- (Gr. *semeion*, sign): a pref. used in terms such as *semiotic* (Gr. *semeiotiké* < *semeion* + suf. *-tíkos* > L. *-ticus* > E. *-tic*, relation, fitness, inclination or capacity), also called *semiology* (Gr. *semeion* + *-logía* < *lógos*, treatise), employed in plant pathology to refer to the area of pathology that studies the signs of the diseases.

senescence (L. *senescentia* < *senescere*, to age): the action and effect of aging; in certain fungi there is a degeneration that makes propagation in culture difficult or impossible, as is observed, e.g., in *Neurospora* (Sordariales).

sensitive (L. *sensus*, sensitive < *sentire*, to feel + suf. *-ivus* > E. *-ive*, expresses quality or tendency, fitness, concerning): capable of having sensitivity. In pathology it refers to the capacity to react, expressing severe symptoms to the attack of a given pathogen, i.e., that has *sensitivity*.

sensus, pl. **sensi** (L. *sensus*, sense, meaning of a concept). *Sensu lato* or *lato sensu* (L. *latus*, extendable, extensive), or *sensu amplo* or *amplo sensu* (L. *amplus*, ample, broad): in a broad sense, i.e., having a broad circumscription of a taxon, such as a genus or a species. *Sensu stricto* (L. *strictus*, strict, restricted) or *stricto sensu*: a narrow circumscription of a taxon, in a strict sense. E.g., *Melanospora* (Melanosporales) *sensu lato* includes species with different types of ascospores, whereas *Melanospora sensu stricto* is limited to species with ascospores with a smooth wall and depressed germ pores; the species with other types of ascospores are placed in other gen.

separable (L. *separ*, separated, different, disunited + suf. *-abilis* > E. *-able*, tendency to, capable of being, that can be, apt to be): *Agaricales*. Refers to a gill that is first adnate to the stipe, but later separates from it; or to an hymenophore that separates from the context, as in *Boletus pinicola* and *Suillus luteus*.

septal plug (L. *septum*, septum, fence, partition + suf. *-alis* > E. *-al*, relating to or pertaining to; MD. *plugge*, plug): refers to the characteristic biconvex plug that obstructs the central opening of the hyphal septa in *Linderina* (Kickxellales).

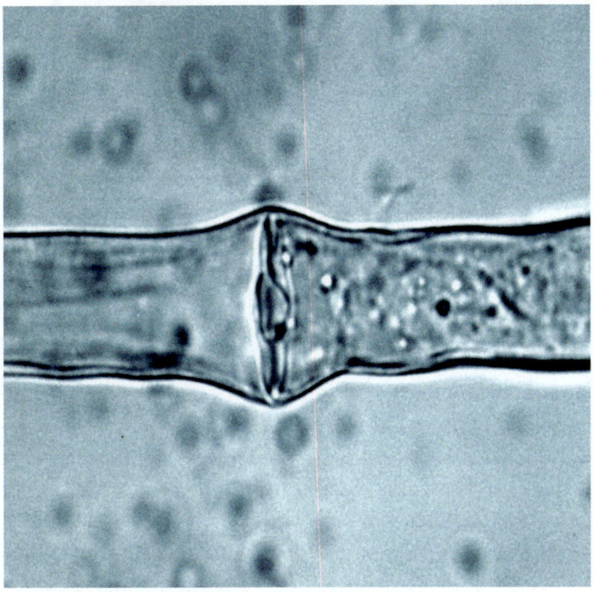

Septal plug in a hypha of *Linderina pennispora*, x 550 (*EAA*).

septal pore (L. *septum*, barrier, partition, septum + suf. *-alis* > E. *-al*, pertaining to or relating to; L. *porus* < Gr. *póros*, pore, passage): a pore in the septum that separates one cell from another, whether in a somatic or sporogenous hypha, which permits the free passage of cytoplasm and organelles, including nuclei. See **septum**. Cf. **dolipore septum**.

septal pore cap (L. *septum*, barrier, partition, septum + suf. *-alis* > E. *-al*, belonging to or relating to; L. *porus* < Gr. *póros*, pore, passage; LL. *cappa* < hooded cloak, cap): see **parenthesome**.

septal pore organelle (L. *septum*, barrier, partition, septum + suf. *-alis* > E. *-al*, pertaining to or relating to; L. *porus* < Gr. *póros*, pore, passage; L. *organum*, organ; specialized cellular structure analogous to an organ + dim. suf. *-ellum* > E. *-elle*): a structure delimited by a membrane, often pulley-shaped, that can plug a septal pore. The organelles of the septal pore are more complex than Woronin bodies, which also can plug septal pores; they are present in some filamentous ascomycetes, and are distributed in some parts of the mycelium so that structures involved in sexual reproduction remain separated from other regions of the mycelium, as occurs in *Barssia oregonensis* and *Phaedropezia epispartia* (=*Acervus episparticus*) (Pezizales). Cf. **Woronin body**.

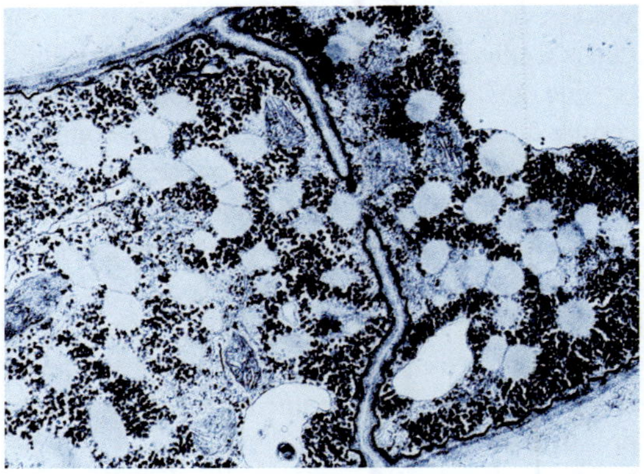

Septal pore organelle in a hypha of an ascomycete, seen in longitudinal section; transmission electron micrograph, x 10,000 (*RTH*).

septate (L. *septatus*, divided by septa or that has a septum < *septum*, barrier, partition + suf. *-atus* > E. *-ate*, provided with or likeness): with septa or partitions, like the hyphae of the majority of the fungi. Cf. **aseptate** or **coenocytic**.

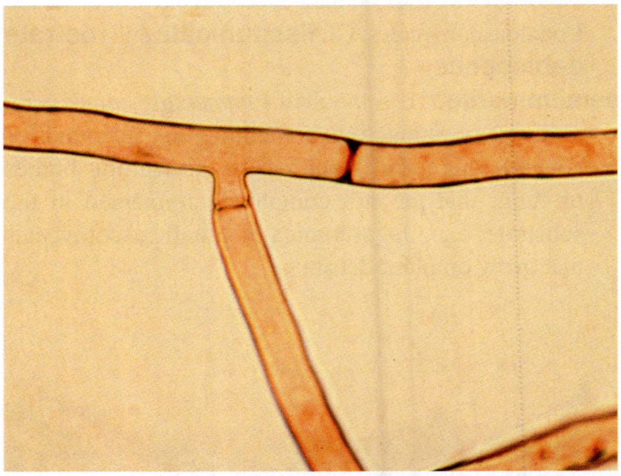

Septate hypha of *Rhizoctonia* sp. (agonomycetaceous asexual fungi), x 450 (*CB*).

septic, **septicous** (L. *septicus* < Gr. *septikós*, causing putrefaction < *sépsis*, decomposition + suf. *-tikós* > L. *-ticus* > E. *-tic*, relation, fitness, inclination or capacity): concerning or relating to *sepsis*, pl. *sepses*. Cf. **aseptic**.

septum, pl. **septa** (L. *septum*, fence, barrier, partition): a transverse wall in a cell or in a hypha. The septa or partitions are formed by centripetal growth of the cell wall and are present with a certain spatial regularity in septate mycelia. *Primary septum*: a septum that forms in direct association with nuclear division (whether this is by constriction, mitosis or meiosis), that separates the daughter cells and is provided with a pore, which can be modified like a dolipore (in basidiomycetes) or be associated with Woronin bodies (in ascomycetes and their asexual states). *Adventitious septum*: a septum that is not formed in association with nuclear division, e.g., when a hypha breaks naturally or artificially, and the formation of the septum impides the loss of protoplasm. The adventitious septum also is known as a *retraction septum* or *retaining septum*, which is formed especially in association with the movement of cytoplasm from one part to another of the fungus.

sericeous (L. *sericeus*, silken < *ser*, genit. *seris*, silk + AF. suf. *-ous*, *-eus* > E. *-eous*, full of): silky; having thin, white, shiny filaments, similar to silk threads, arranged in one direction, like those present on the base of the stipe of *Entoloma sericeum* (=*Rhodophyllus sericeus*, Agaricales), which is sericeous when dry.

series (L. *series*, a row of things, a series): a taxonomic category intermediate between the genus and species. The series have Latinized names similar to those of the genera or species but with different endings; in general they have few applications and are utilized mainly in genera with numerous species, e.g., the genus *Penicillium*, subgenus *Furcatum*, section *Furcatum*, series *Citrina* (moniliaceous asexual fungi).

serioplectenchyma, pl. **serioplectenchymata** (L. *series*, row of things, a series + Gr. *plektós*, braided, interlaced + *énchyma*, stuffed): a type of plectenchyma formed by parallel, coherent, thin-walled hyphae with wide lumens.

serous (L. *serum*, serum + suf. *-osus* > E. *-ous*, full of, augmented, prone to): pertaining to serum or serosity, or similar to these liquids; aqueous, opalescent.

serrate (LL. *serratus*, dentate, serrate, dented < *serrare*, to saw + L. suf. *-atus* > E. *-ate*, provided with or likeness): with small sharp teeth near the edge; generally applied to foliaceous or laminar structures, such as the gills of the basidiocarps of species of *Lentinellus* (Aphyllophorales).

Serrate gills of the basidiocarp of *Lentinellus* sp., x 3 (*MU*).

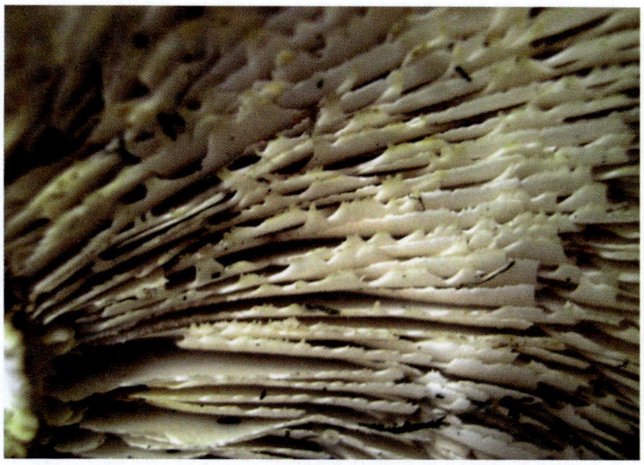

Serrate gills of the basidiocarp of *Lentinellus* sp., x 3 (*RV*).

serrulate, serrulated (LL. *serrulatus*, dim. of *serratus*, dentate, serrate, dented < *serrare*, to saw + L. dim. suf. *-ulus* + L. suf. *-atus* > E. *-ate*, provided with or likeness): dentate, but with smaller teeth, like the edges of the gills of the basidiocarp of *Entoloma serrulatum* (=*Leptonia serrulata*, Agaricales).

sessile (L. *sessilis*, of or pertaining to being seated < *sessor*, seated + suf. *-ilis* > E. *-ile*, capable of, with the character of): refers to a fruiting body that is attached to or seated directly on the surface of the substrate, and which lacks a foot or supporting pedicel; e.g., like the sporangia of *Perichaena* (Trichiales). Equal to **apodal**. Cf. **stipitate** and **pedicellate**.

Sessile sporangia of *Physarum decipiens* (Physarales), x 10 (*MU*).

Sessile sporangia of *Perichaena* sp., x 15 (*MU*).

Sessile sporangium of a Myxomycete on wood, x 15 (*MU*).

Sessile apothecia of *Discina ancilis* (Pezizales), x 1 (*RV*).

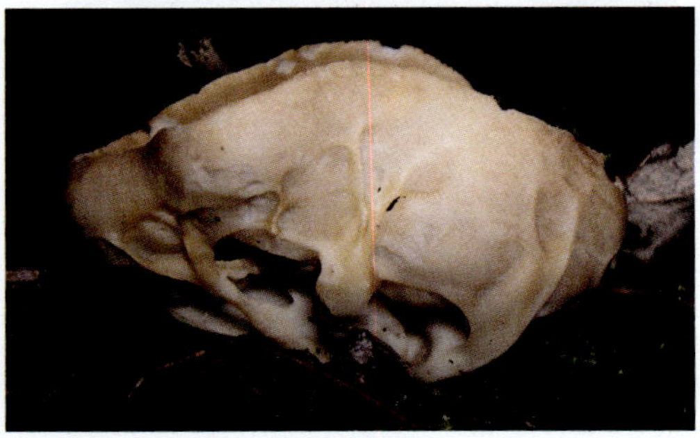

Sessile apothecium of *Helvella* (=*Paxina*) *acetabulum* (Helvellales), x 1 (*RV*).

seta, pl. **setae** (L. *seta*, bristle): refers to stiff hairs or bristles, usually thick-walled and sharp-pointed, that occur on the fruiting bodies of various fungi, such as the acervuli of *Colletotrichum* (melanconiaceous asexual fungi), the pycnidia of *Pyrenochaeta*, *Chaetoseptoria* and *Ascochyta* (=*Chaetodiplodia*) *caulina* (sphaeropsidaceous asexual fungi), the mycelium of *Meliola* (Meliolales), or the perithecia of *Podospora* (Sordariales). The bristles formed in the hymenium of *Dennisiella babingtonii* (Dothideales) also are called setae, as well as those produced in the context and hymenium (*hymenial setae*) of many species of Aphyllophorales, e.g., *Phellinus ferruginosus*.

Setae of the acervulus of *Colletotrichum circinans*, x 125 (*MU*).

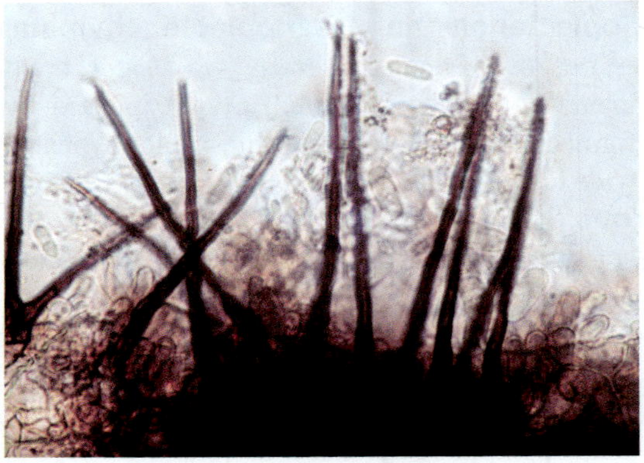

Setae of the acervulus of *Colletotrichum* sp., x 200 (*MU*).

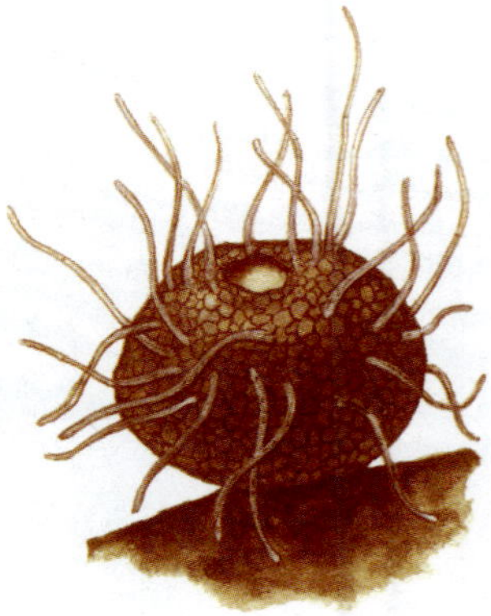

Setae of the pycnidium of *Ascochyta* (=*Chaetodiplodia*) *caulina*, x 320 (*MU*).

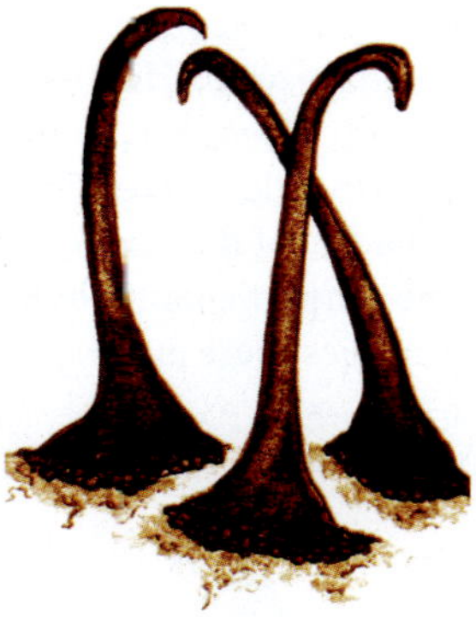

Setae coming from pigmented cells in the hymenium of *Dennisiella babingtonii*, x 175 (*MU*).

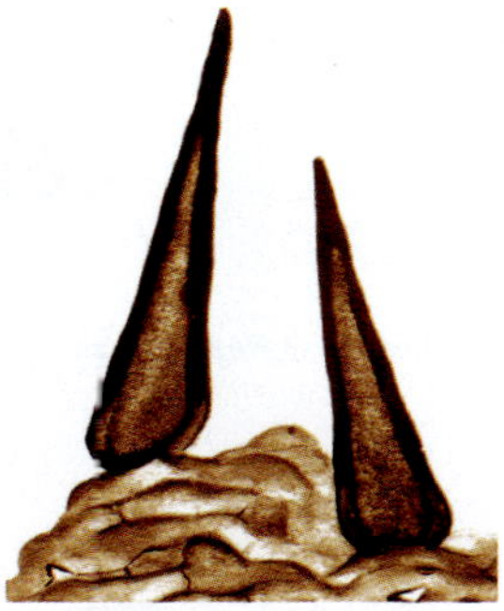

Setae of the hymenium of *Phellinus ferruginosus*, x 370 (*MU*).

Setae of the context on the margin of the pileus of *Phellinus ferruginosus*, x 350 (*MU*).

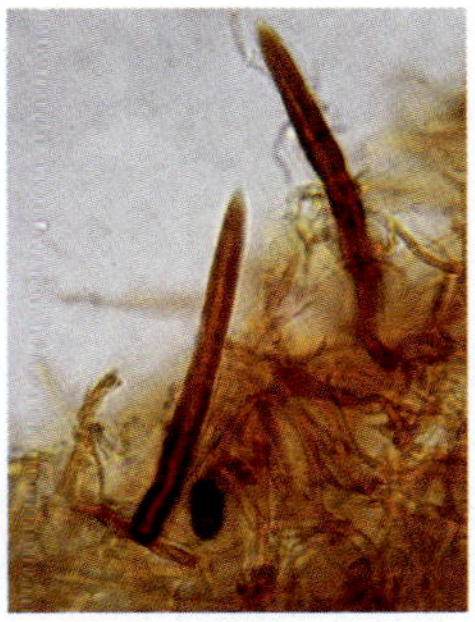

Hyphal **setae** of *Phellinus contiguus*, x 200 (*RV*).

setaceous (L. *setaceus*, like a seta < *seta*, seta, bristle + suf. *-aceus* > E. *-aceous*, of or pertaining to, with the nature of): similar to a seta or bristle, or having stout, sharp-pointed hairs, usually thick-walled, like those around the edge of the receptacle of *Cyathus* (Nidulariales). Cf. **setulose**.

seticole, seticolous (L. *seta*, seta, bristle + suf. *-cola*, inhabitant < *colere*, to inhabit + suf. *-osus* > E. *-ous*, full of, augmented, prone to): living or growing on the bristles or setae of the host organism, like the thalli of *Rickia passalina* (Laboulbeniales) that live attached to the setae of *Chondrocephalus debilis*, a coleopteran of the passalid family.

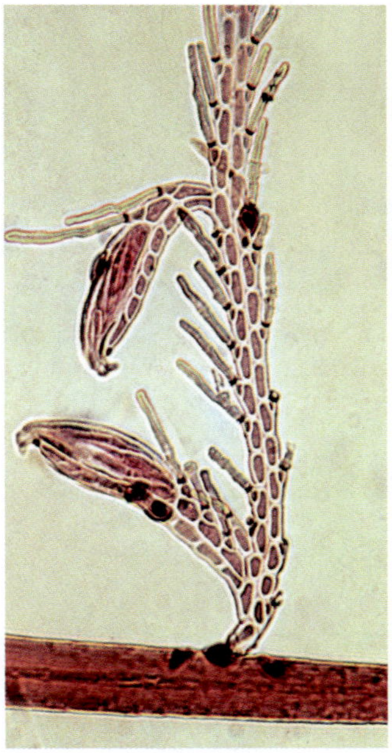

Seticolous thallus of *Rickia passalina*, on a seta of the body of a coleopteran host, x 140 (*HL*).

setiform (L. *setiformis* < *seta*, seta, bristle + suf. *-formis* < *forma*, shape): shaped like a seta; fine like a bristle.

setigerous (L. *seta*, seta, bristle + E. suf. *-gerous* < L. *-ger*, to bear, carry < *gerere*, to produce, engender): bearing *setae*.

setose, setous (L. *setosus*, having setae < *seta*, seta, bristle + suf. *-osus* > E. *-ose*, full of, augmented, prone to): **1**. Provided with setae or stiff hairs, like the acervuli of *Colletotrichum* (melanconiaceous asexual fungi), the perithecia of *Pleochaeta polychaeta* (Erysiphales) and the perithecia of *Chaetomidium fimeti* and *Ch. heterotrichum* (Sordariales). **2**. Having the shape of a seta or bristle (*setiform*); setose hair. The lamprocystidia of *Hohenbuehelia niger* (Agaricales) are setiform.

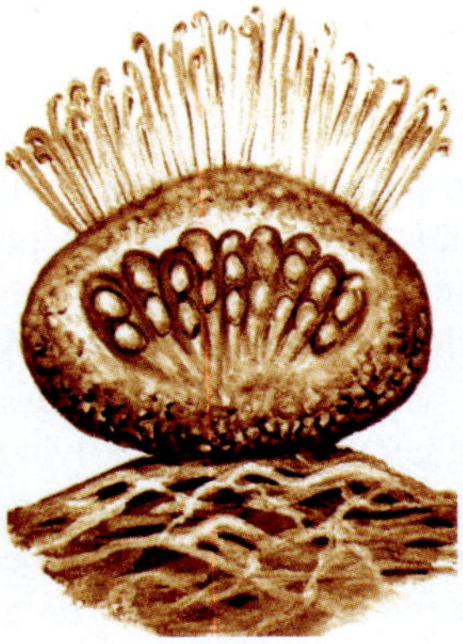

Setose perithecium of *Pleochaeta polychaeta*, seen in vertical section. The appendages of the apical tuft are rigid and uncinate, x 650 (*MU*).

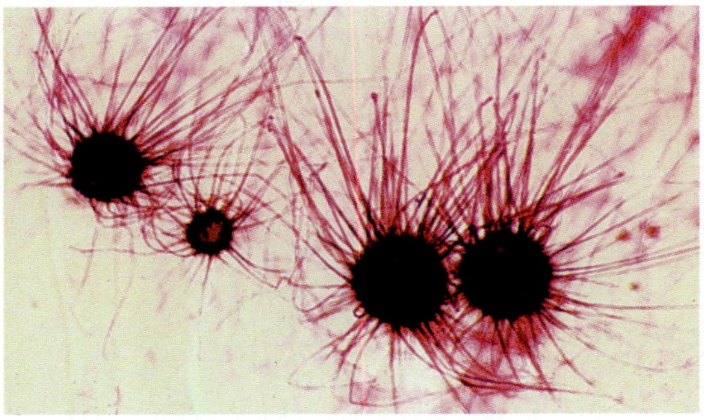

Setose perithecia of *Chaetomidium fimeti*, x 350 (*MU*).

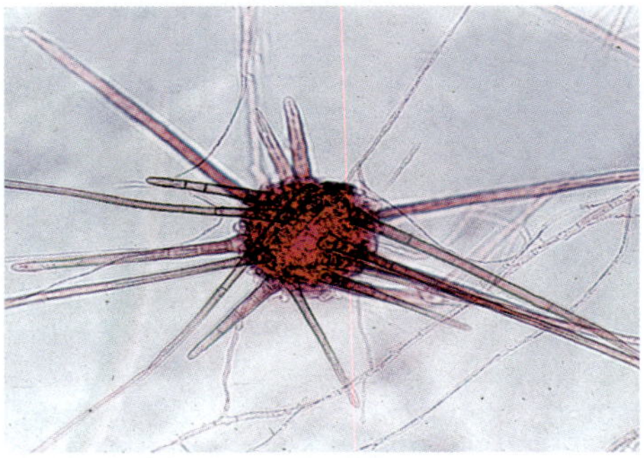

Setose perithecium of *Chaetomidium fimeti*, x 350 (*MU*).

Setose perithecium of *Chaetomidium heterotrichum*, x 650 (*MU*).

setula, pl. **setulae** (L. *setula*, dim. of *seta*, seta, bristle): a very fine appendage or seta, like those on the conidia of *Pestalotia* and *Monochaetia* (melanconiaceous asexual fungi). Also called setulae are the thick-walled, dark brown cystidia (of the trama) present in certain Hymenomycetes, in particular Aphyllophorales; in the agaricaceous species there are no setulae.

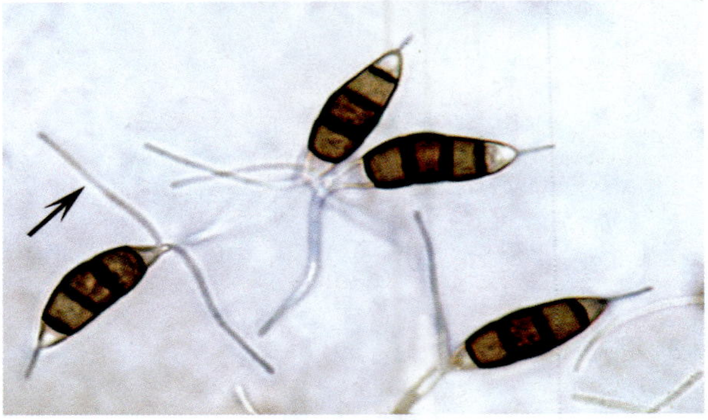

Setulae of the conidia of *Pestalotia* sp., x 1,000 (*MU*).

Setulae of the conidia of *Monochaetia carissae*, x 1,000 (*MU*).

setuloid (L. *setula*, dim. of *seta*, seta, bristle + suf. *-oide*, like < Gr. *-oeídes*, similar to): see **setuloid cystidium**.

setuloid cystidium, pl. **cystidia** (L. *setula*, dim. of *seta*, seta, bristle + suf. *-oide*, like < Gr. *-oeídes*, similar to; NL. *cystidium* < Gr. *kystídion* < *kýstis*, bag, bladder, cell + dim. suf. *-ídion*): a *leptocystidium* with a thick, pigmented wall, present in the hymenium of some Agaricales.

setulose (L. *setulosus*, provided with fine setae, with very fine hairs < *setula*, dim. of *seta*, seta, bristle + suf. *-osus* > E. *-ose*, full of, augmented, prone to): covered with fine bristles or hairs. Cf. **setaceous**.

sewage fungus, pl. **fungi** (E. *sewage*, refuse liquids or waste matter carried off by sewers < ME. < MF. *esseweur*, *seweur* < *esewer*, to drain < VL. *exaquare* < *ex*, out of + *aqua*, water, to carry off water; L. *fungus*, fungus < Gr. *spóngos*, *sphóngos*, sponge): applied to the Oomycete *Leptomitus lacteus* (Leptomitales), commonly found in contaminated waters, in particular sewage water; occasionally this microorganism clogs the sewage filters.

sexual (L. *sexualis*, concerning or relating to sex < *sexus*, sex + suf. *-alis* > E. *-al*, relating to or pertaining to): applied to a nucleus, cell, organ, process, or phase concerning sex. E.g., sexual reproduction is that involving plasmogamy, karyogamy and meiosis; a sexual phase is the portion of a fungal life cycle during which sexual reproduction occurs. Cf. **asexual**.

Sexual reproduction in a colony of *Mucor hiemalis* (Mucorales) on agar. The dark zone in the middle of the colony is where the gametangia unite to form the zygosporangia, x 1 (*MU*).

sexual phase (L. *sexualis*, concerning or relative to sex < *sexus*, sex + suf. *-alis* > E. *-al*, relating to or belonging to; Gr. *phásis*, phase, appearance or aspect < *phaíno*, to shine, to show): the phase or state in the life cycle in which are formed sexual reproductive cells or organs. In the fungi it is more appropriate to speak of a sexual phase than of sex, as fungi do not always have true sex, as occurs in other organisms, i.e., with clearly differentiated male and female cells or organs.

sheath (ME. *schethe* < OE. + *sceath*, sheath, wrapper, envelope): **1**. A hyaline, gelatinous cover that envelops the ascospores of some ascomycetes, e.g., *Sordaria* (Sordariales); also called **capsule**. **2**. *Uredinales*. The structure that separates the cytoplasm of the host cell from the cell wall of the haustorium; this envelope or covering is electron transparent and surrounds the haustorial body, but not the haustorial neck. It can be seen in plant pathogenic fungi, such as *Melampsora*. **3**. *Nidulariales*. The structure in the fruiting body to which the middle piece is attached, as occurs in the species of *Cyathus*. Also called **basal piece**.

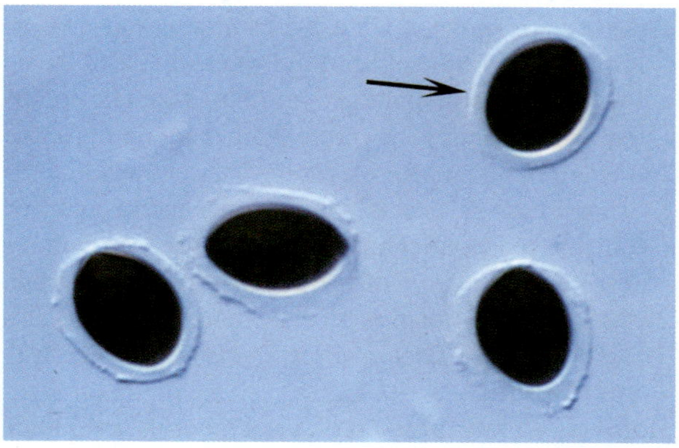

Sheath of the ascospores of *Sordaria fimicola*, x 680 (*RTH*).

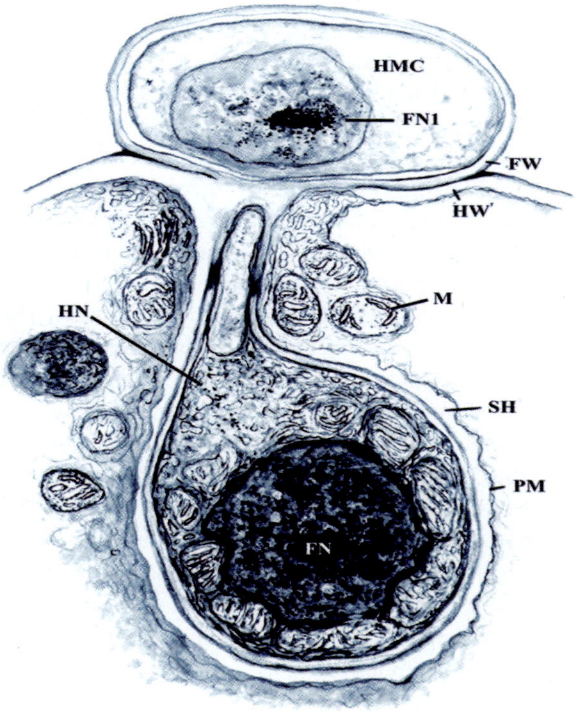

Sheath (SH) of *Melampsora lini*. Longitudinal, nonmedian section through the haustorial mother cell (HMC), at the site where it penetrates the cell wall of the host (flax, *Linum usitatissimum*). The cell wall of the fungus (FW) is in contact with the cell wall of the host (HW). The haustorium and the haustorial neck (HN) are surrounded by the invaginated plasma membrane of the host (PM). One of the fungal nuclei (FN) is in the haustorium, and the other (FN1) appears to be partly in the haustorial neck and partly in the haustorial mother cell. Also shown are the mitochondria of the host cell (M). Drawing based on a transmission electron micrograph published by Beckett *et al.* in their *Atlas of Fungal Ultrastructure*, 1974, x 9,800 (*MU*).

shot-hole (ME. < OE. *scot*, shot; ME. < OE. *hol*, hollow, hole): a plant disease characterized by foliar spots and holes produced when the dead parts caused by the attack of certain plant pathogenic fungi fall out, e.g., the shot hole of peach leaves, caused by *Stigmina* (=*Otthia*) *carpophila*, of the Dothideales. See **lesion**.

shoyu (Jap. *shoyu* < Chin. *shi-yau*, soybean oil): an Oriental sauce of soybean and wheat that is obtained by double fermentation, one that depends on the enzymatic activities (proteolytic and lipolytic) of *Aspergillus flavus* var. *oryzae* (=*A. oryzae*) and *A. sojae* (moniliaceous asexual fungi), and another carried out by yeasts tolerant to salt (*Zygosaccharomyces soya* and *Z. major*, of the Saccharomycetales), and lactic bacteria tolerant to salt (mainly *Pediococcus cerevisiae* and *Lactobacillus delbrueckii*). See **fermentation**.

sicci- (L. *siccus*, dry): a pref. used in terms that allude to dryness, like *siccideserta* (pl. of *siccidesertum* < *sicci-* + *desertum*, desert), deserts caused by excessive aridity in the atmosphere.

sicyoid, **sicyodic** (Gr. *síkyos*, pumpkin + L. suf. *-oide*, like < Gr. *-oeídes*, similar to): see **lageniform**, **utriculate** and **ventricose-rostrate**.

side body complex (ME. < OE. *side*, side; ME. < OE. *bodig*, body; LL. *complexus*, totality, complex): *Blastocladiales*. A complex structure, found in zoosporic fungi such as *Blastocladia*, *Blastocladiella* and *Allomyces*, that consists of a double membrane system associated with microbodies and lipid globules, and whose function is unknown. The complex is located just below the plasma membrane of the zoospores and planogametes, near the side of the posterior end of the cell; it is reminiscent of the microbody-lipid globule complex found in zoospores of the Chytridiales and Monoblepharidales, although it lacks the associated rumposome.

sidero-, **sider-** (Gr. *síderos* < *sidéreos*, iron and things made of iron): a pref. that alludes to the metal iron, as in *siderophore*, described below; on occasion, refers only to the hardness of iron, as in *siderophyllous*, a plant with hard leaves.

siderophilous (Gr. *síderos* < *sidéreos*, iron and things made of iron + *phílos*, have an affinity for): applied to microorganisms (schizophyceae) that show an affinity for iron, and which live in media rich in iron, although not exclusively.

siderophore (Gr. *síderos* < *sidéreos*, iron and things made of iron + suf. *-phóros*, bearer): a metabolic product of a fungus (or other microorganism) that binds to iron and permits its translocation from the soil water into the microbial cell.

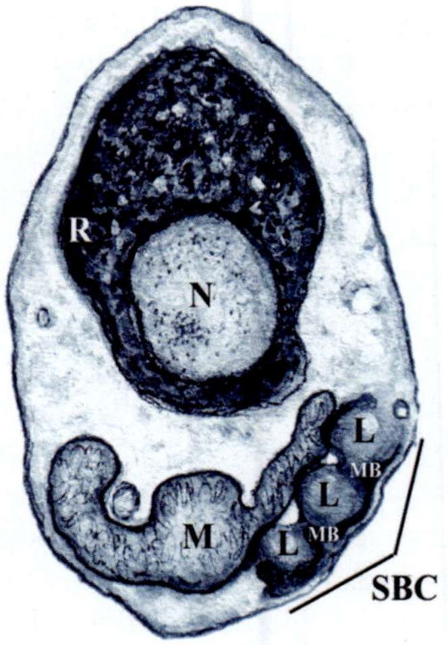

Side body complex (SBC) of a mature zoospore of *Blastocladiella emersonii*, seen in longitudinal section. The ribosomes (R) are enclosed by a pair of membranes. One large, lobed mitochondrion (M) is located at the posterior end of the zoospore, and the lipid globules (L) together with the microbodies (MB) are forming the side body complex. Drawing based on a transmission electron micrograph published by Beckett *et al.* in their *Atlas of Fungal Ultrastructure*, 1974, x 10,280 (*MU*).

sigatoka (< *Sigatoka* river and district on the island of Viti Levu in the Fiji archipelago): a plant disease of banana leaves (*Musa*), characterized by discolored spots, caused by the attack of *Mycosphaerella musicola* and *M. difforme* (Capnodiales). There are yellow and black sigatokas, and other fungi may co-occur in these diseases, e.g., *Verticillium lateritium* (moniliaceous asexual fungi).

Pseudothecium of *Mycosphaerella musicola*, x 500 (*MU*), and symptoms of yellow **sigatoka**, on a banana leaf, caused by this fungus, x 1 (*FV*).

sigmoid (Gr. *sigmoeidés* < *sígma*, or ς, 18th letter of the Gr. alphabet; corresponds to "s" + L. suf. *-oide*, like < Gr. *-oeídes*, similar to): curved twice upon itself, with a shape similar to S, like the basidiospores of species of *Clavaria* (Aphyllophorales) and some of the Hülle cells of *Aspergillus puniceus* (moniliaceous asexual fungi).

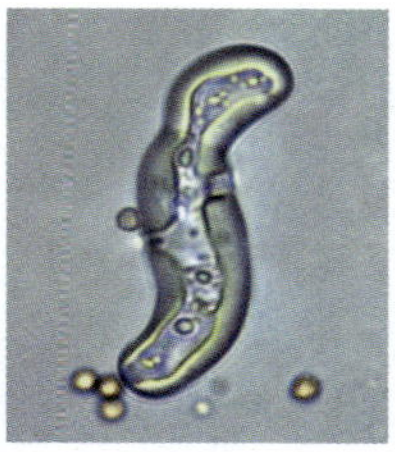

Sigmoid Hülle cell of *Aspergillus puniceus*, x 340 (*MU*).

sikospore, **siciospore** or **sikyotic spore** (Gr. *síkyos*, pumpkin − *sporá*, spore + L. suf. *-tikus* > E. *-tic*, relation, fitness, inclination or capacity): a spore formed from the *sikyotic cell* of *Parasitella parasitica* (=*P. simplex*, Mucorales), which serves to anchor the terminal cell of its hypha to the host hypha (of *Absidia glauca*, also of the Mucorales).

sikyotic cell (Gr. *síkyos*, gourd + Gr. suf. *-tikós* > L. *-ticus* > E. *-tic*, relation, aptitude, inclination or capacity; NL. *cellula*, cell, living cell < dim. of L. *cella*, small room): the terminal cell of a *Parasitella parasitica* (=*P. simplex*, Mucorales) hypha, by means of which this parasite attaches to a hypha of its *Absidia glauca* host (also of the Mucorales). After a period of time this cell differentiates into a *sikospore*.

silici-, **silic-** (L. *silex*, genit. *silicis*, flint, silicon): a pref. employed to form compounds that allude to the siliceous nature of the soil, or to silica, as in *silicole*, an organism that inhabits siliceous soils.

silico- (L. *silex*, *silicis*, flint, silica): a pref. employed at times instead of *silici-*, in hybrid terms in which the second component is Gr., as in *silicophilous*, an organism that prefers siliceous soil.

siliquiform (L. *siliquiformis* < *siliqua*, fruit of the carob bean and other elongate fruits consisting of a pod + suf. *-formis* < *forma*, shape): similar to a silique, long and slender, like the zoosporangia of the aquatic fungus *Blastocladia pringsheimii* (Blastocladiales).

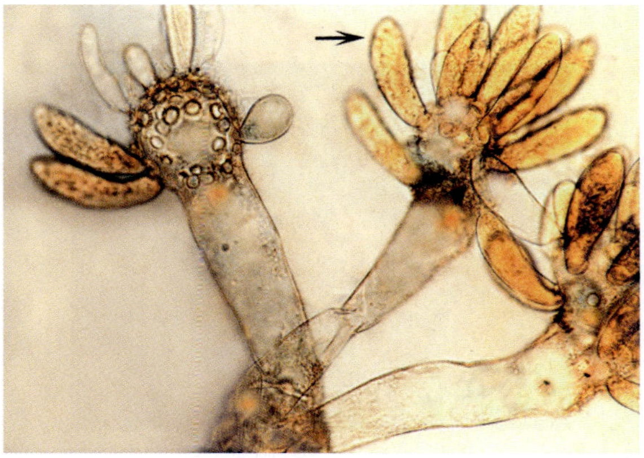

Siliquiform zoosporangia of *Blastocladia pringsheimii*, x 140 (*MU*).

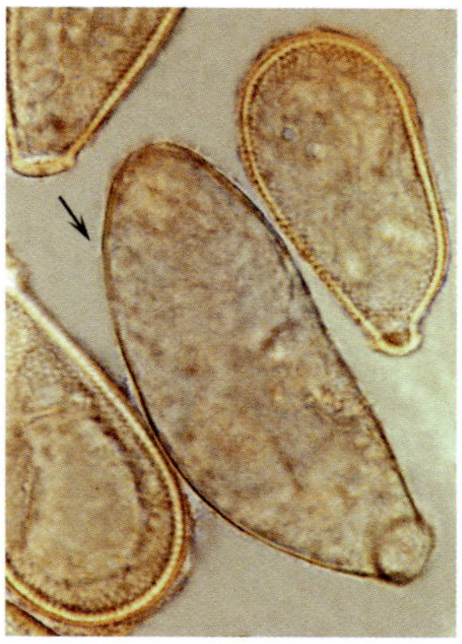

Siliquiform zoosporangia of *Blastocladia pringsheimii*, x 660 (*MU*).

silvi- (L. *silva*, forest, woods): a pref. used in terms such as *silvicole*, inhabiting the forest or woods.

simblospore (Gr. *símblos*, swarm of bees; to teem, boil, move about + *sporá*, spore): a flagellate, swimming spore; equal to **zoospore**. See **planont**, **planocyte** and **planospore**.

simple (L. *simplex*, not compound, not branched): not branched (simple thallus); without divisions (simple spore).

sinuate (L. *sinuatus*, bent, curved < *sinus*, curve + suf. *-atus* > E. *-ate*, provided with or likeness): having a depression; wavy or with a dent, cut or notch. Generally applied to a type of gill insertion on the stipe, when a notch is present next to the zone at the point of insertion, such as occurs in the gen. *Agrocybe* (Agaricales). Cf. **adnate**, **decurrent** and **free**. Also applied to the border of the giant colony of some yeasts when grown on solid media.

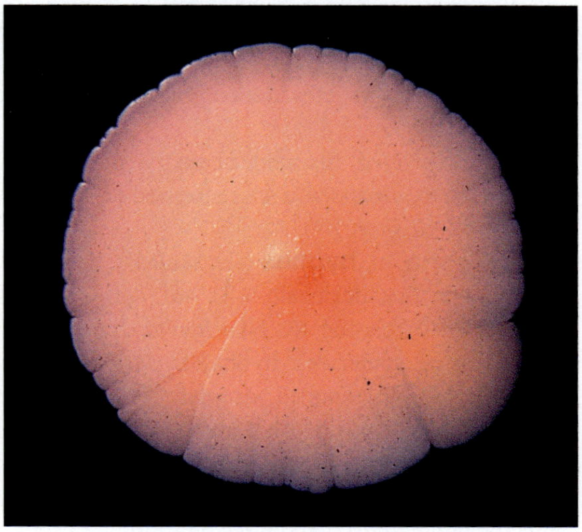

Sinuate margin of a giant colony of *Rhodotorula rubra* (cryptococcaeous asexual yeasts), on agar, x 1 (*MU*).

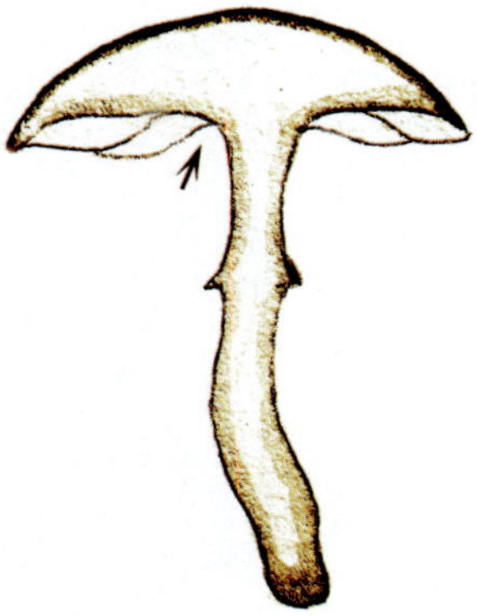

Sinuate gills of the basidiocarp of *Agrocybe* sp., seen in longitudinal section, x 1(*MU*).

sinuose, **sinuous** (L. *sinuosus*, tortuous, twisted, serpentine < *sinus*, curve + suf. *-osus* > E. *-ose*, full of, augmented, prone to): with many sinuosities or undulations, like the plasmodiocarp of *Physarum sinuosum* (Physarales).

sipho- (Gr. *síphon*, tube): see **siphono-**.

siphomycetes, **siphonomycetes** (L. *siphon*, genit. *siphonis* < Gr. *síphon*, genit. *síphonos*, tube + *mýkes*, genit. *mýketos*, fungus): fungi with tubular, aseptate hyphae. See **phycomycetes**.

siphon (L. *siphon*, genit. *siphonis* < Gr. *síphon*, genit. *síphonos*, tube): a tube; an elongate, tubular cell; aseptate hypha.

siphono-, **siphon-** (L. *siphon*, genit. *siphonis* < Gr. *síphon*, genit. *síphonos*, tube): a pref. employed to introduce the idea of something that is tubular in shape, as in *siphonomycetes*.

sirenin (LL. *Sirena* < L. *Siren* < Gr. *Seirén*, seductive sea nymph, siren + NL. des. *-in* > E. *-ina*, suf. used in chemistry to denote an activator or compound): a hormone secreted by the female gametes of the aquatic fungus *Allomyces* (Blastocladiales), which attracts the male gametes to accomplish planogametic copulation.

siro-, **sir-** (Gr. *seirá*, filament, chain): a pref. used in terms such as *sirospore*, a spore formed on a filament (equal to *stylospore*).

skeletal hypha, pl. **hyphae** (Gr. *skeletós*, desiccated < *skéllo*, to dry, desiccate + L. suf. *-alis* > E. *-al*, pertaining to relating to; NL. *hypha*, hypha < Gr. *hyphé*, tissue, spiderweb; hypha): *Basidiomycetes*. A thick-walled, unbranched, aseptate hypha, straight or slightly flexuous, with little or no lumen, that provides rigidity to sporophores. E.g., *Ramaria gracilis*, *Neolentinus* (=*Lentinus*) *tigrinus*, *L. lepideus* and *Schizophyllum commune* (Aphyllophorales) have skeletal hyphae. See **binding hypha** and **generative hypha**.

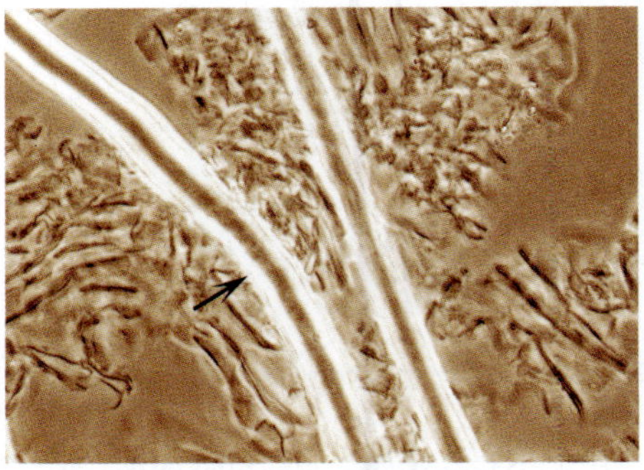

Skeletal hyphae of the basidiocarp of *Ramaria gracilis*, x 450 (*MV*).

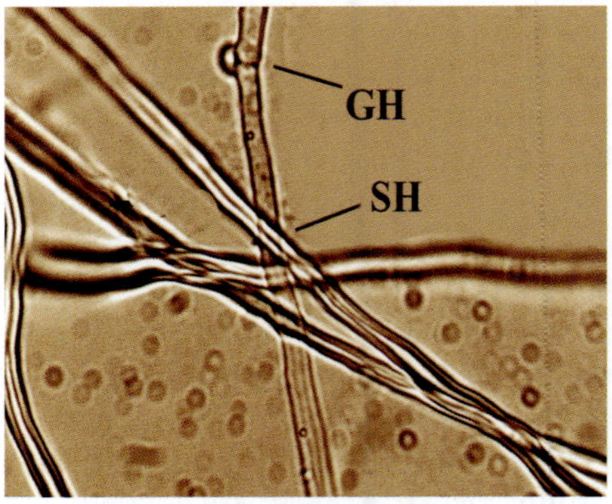

Skeletal hyphae (SH) and generative hyphae (GH) of the basidiocarp of *Schizophyllum commune*, x 800 (*MU*).

skeletocystidium, pl. **skeletocystidia** (Gr. *skeletós*, dry < *skéllo*, to dry + NL. *cystidium* < Gr. *kystídion* < *kýstis*, bag, bladder, vesicle, cell + dim. suf. *-idion* > L. *-idium* > E. *-idium*): the apical part of a skeletal hypha, frequently more or less inflated, which projects toward or through the hymenium; a false seta.

slime mold, **slime fungus**, pl. **fungi** (ME. *slyme* < OE. *slim*, slime; ME. *mowlde*, mold; L. *fungus*, fungus): a member of the phyla Acrasiomycota, Dictyosteliomycota (cellular slime molds) or Myxomycota (plasmodial slime molds).

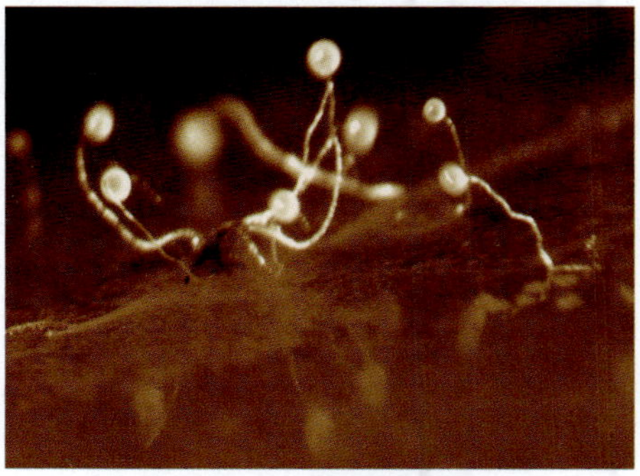

Sorocarps of *Dictyostelium mucoroides* (Dictyosteliales), on agar, a cellular **slime mold** common on mouse dung, x 50 (*MU*).

Young plasmodium of a Myxomycete (*Fuligo* sp., Physarales) on dead wood, a plasmodial **slime mold**, x 1 (*MU*).

Veins of a plasmodium of *Physarum polycephalum* (Physarales), a plasmodial **slime mold**, on agar, x 10 (*MU*).

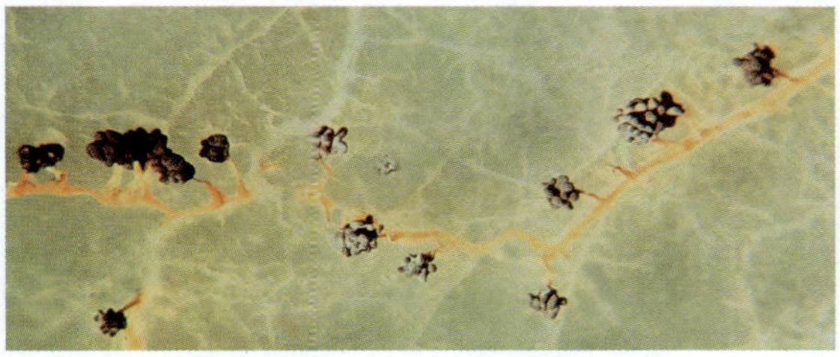

Veins and sporangia of *Physarum polycephalum* (Physarales), a plasmodial **slime mold**, on agar, x 5 (*MU*).

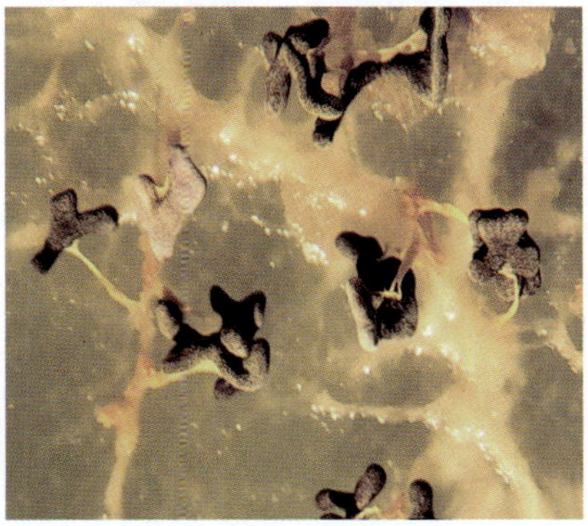

Sporangia developed from a plasmodium of *Physarum polycephalum* (Physarales), a plasmodial **slime mold**, on agar, x 10 (*MU*).

slime net (ME. < OE. *slim*, slime; ME. < OE. *nett*, net): another name for the *filoplasmodium* or reticular plasmodium of the Labyrinthulales.

slime spore (OE. *slim* > ME. *slyme*, slime, a viscous substance; Gr. *sporá*, spore): a spore that is covered with mucilage by the cell that produced it. Syn. of **gloeospore**.

slimy (ME. *slyme* < OE. *slim*, slime): containing mucilage or mucus, or some of its properties; viscous, sticky or mucous when wet, due to its chemical composition (mucic acid, pentoses and hexoses); in water the mucilage becomes viscous or swells to form a gelatinous pseudosolution. These materials are present in many fungi and fungus-like microorganisms, e.g. in members of the phyla Acrasiomycota, Dictyosteliomycota and Myxomycota, which are known as the mucilaginous fungi or slime molds. The colonies of *Cryptococcus albidus* (cryptococcaceous asexual yeasts) and *Aureobasidium pullulans* (dematiaceous asexual fungi) also are mucilaginous or viscous.

Slimy colony of *Aureobasidium pullulans* on agar, x 3 (*MU*).

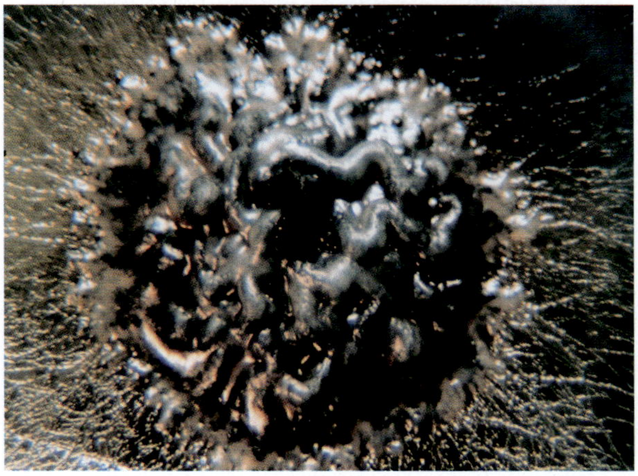

Slimy colony of *Aureobasidium pullulans* on agar, x 2.5 (*MU*).

sloughing exoperidium, pl. **exoperidia** (E. *sloughing*, prp. of *slough*, to be shed or cast off < ME. *slughe, slough*, snake skin; Gr. *éxo*, outside, on the exterior + *péridion*, small leather bag): an exoperidium that detaches and falls from the mature fruiting body, as occurs in *Lycoperdon marginatum* (Lycoperdales).

smut (ME. *smotten, smutten*, to defile): the common name of the destructive diseases, especially of cereals, caused by parasitic fungi (of the order Ustilaginales, in particular by species of the gen. *Ustilago*), that are distinguished by the transformation of plant organs invaded by hyphae into dark masses of spores (teliospores). *Covered smut*: a smut in which the mature mass of spores remains covered for a time by an envelope of host or fungal tissue, e.g., *barley* and *oat smut*, caused by *Ustilago hordei* (= *U. segetum*). *Loose smut*: a smut in which the spores remain exposed (like a black powder), as in the barley and wheat smuts, caused by *Ustilago tritici* (=*U. segetum* var. *tritici*), *U. nuda* f. sp. *hordei* (=*U. nuda*). *False smut*: a disease of rice caused by *Ustilaginoidea virens* (dematiaceous asexual fungi), whose symptoms are similar to those of a true smut; it is characterized by the dark color of the spores that are produced in spikes of the host plant, although they are not of the same nature as the spores of the true smuts. *Stinking smut* or *bunt*: a name applied to a smut of wheat because of the odor of the spores, similar to that of fish in brine (due to the presence of trimethylamine), caused by *Tilletia caries* and *T. laevis* (=*T. foetida*). *Stripe smut* of pasture grass (*Ustilago striiformis*), wheat (*U. agropyri*), and rye (*U. occulta*). *Corn smut* or *maize smut*, known in Mexico as *cuitlacoche* or *huitlacoche* (a name derived from Nahuatl, which means "unclean sleeping place"), whose masses of immature spores in the ears of corn are consumed as a delicacy, despite the meaning of the common name of this smut, caused by *U. maydis*.

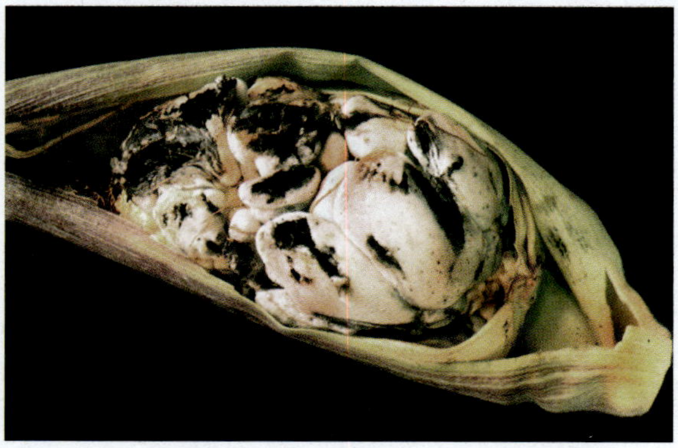

Corn **smut** (or cuitlacoche) caused by *Ustilago maydis*, x 0.25 (*MU*).

False **smut** of rice spikes, caused by *Ustilaginoidea virens*, x 1 (*MU*).

False **smut** of rice spikes, caused by *Ustilaginoidea virens*, x 10 (*MU*).

soleiform (L. *soleiformis*, sandal-shaped < *solea*, sandal + suf. *-formis* < *forma*, shape): with the shape of the sole of a shoe, i.e., elongate-ellipsoid, like the ascospores of *Pyrenophora* (Pleosporales).

solid (L. *solidus*, solid, not hollow): compact, solid; in general applied to the stipe of the agaricaceous fungi (such as *Agaricus* and *Boletus*), filled with more or less compacted hyphae, and lacking a medulla or hollow, cavernose or fistulose central part. Also called **farctate**. Cf. **cavernose** and **fistular**.

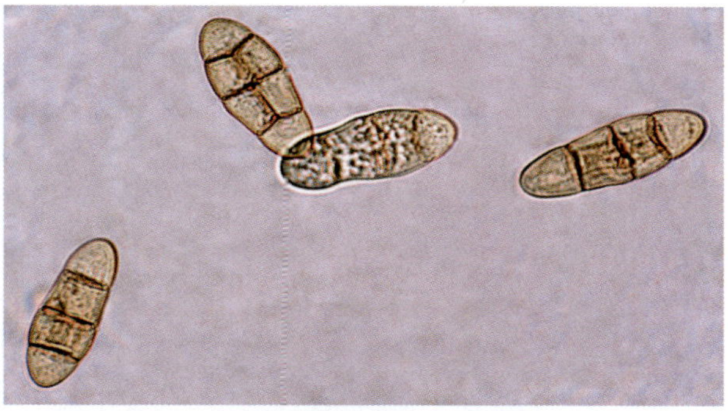

Soleiform ascospores of *Pyrenophora* sp., x 750 (*CB*).

solitary (L. *solitarius*, occurring individually < *solitas*, the state of being alone < *solus*, alone + suf. *-arius* > E. *-ary*, concerning): borne or originated in an isolated or individual manner, not in groups or clusters. Generally applied to the growth habit of various Agaricales, whose fruiting bodies occur separate and scattered, like *Amanita solitaria*. Cf. **acervate**, **caespitose**, **connate** and **gregarious**.

soma, pl. **somata** (NL. *soma* < Gr. *sõma*, body): the vegetative body of an organism, which carries out the functions of assimilation and growth; it is distinguished from the reproductive organs (or reproductive phase) by its morphology and physiology.

Soma (Hind. *Soma*): according to the theory of Wasson, Soma is one of the Gods of ancient India that supposedly corresponded to *Amanita muscaria* (Agaricales). The exact cause of the psychotropic activity of *A. muscaria* is unknown, but it may be related to the presence of various chemical substances, including ibotenic acid and its derivative, muscimol, as well as to muscarine and other toxins, such as bufotenine.

Basidiocarps of *Amanita muscaria*, the species identified with **Soma**, x 0.5 (*TH*).

somatic (Gr. *somatikós*, of the body < *sõma*, body + suf. *-tikós* > L. *-ticus* > E. *-tic*, relation, aptitude, inclination or capacity): concerning or relating to the body, whether in structure or in function, but not to the reproductive parts or phases. Cf. **reproductive**.

somatogamy (Gr. *sõma*, body + *gamía* < *gámos* > E. *-gamy*, sexual union): fusion of somatic cells during plasmogamy, i.e., without the intervention of differentiated sexual organs, as happens in some chytrids, such as *Chytriomyces*, in the majority of basidiomycetes and in many species of yeasts, such as *Saccharomyces* and *Zygosaccharomyces* (Saccharomycetales).

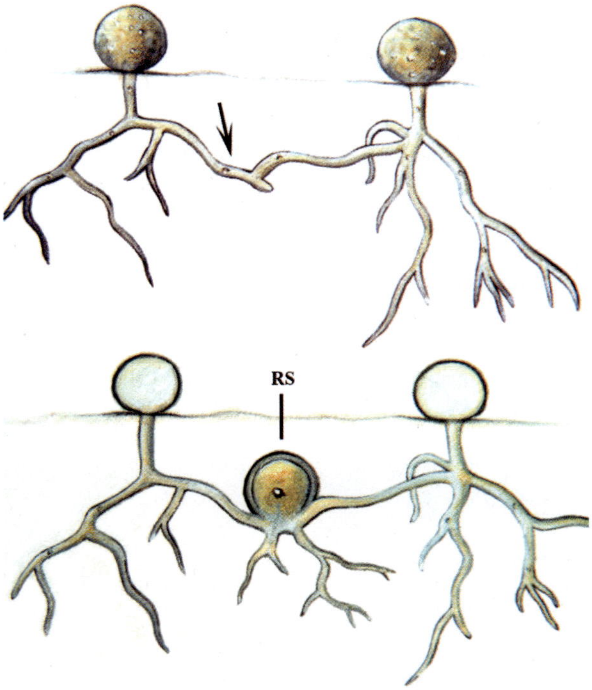

Somatogamy between the rhizoidal filaments of two thalli of *Chytriomyces hyalinus* (Chytridiales), which results in the formation of a latent or resting spore (RS), x 300 (*MU*).

sooty mold (E. *sooty,* of, relating to or producing soot < ME. < OE. *sot*, soot + E. *-y*, having the quality of; ME. *mowlde* < *mowled*, pp. of *moulen, malwen*, to grow moldy): a blackish crust or powder that covers branches and leaves, formed by the mycelium of various fungi that rarely have teleomorphs. Generally their growth is saprobic on the sugary exudate or excretions of aphids and scale insects, or of physiological origin; they are common on willows, black poplar and grape (*Apiosporina*, of the Pleosporales), olive (*Antennularia*, of the Dothideales) and citrus (*Capnodium*, of the Capnodiales), among others.

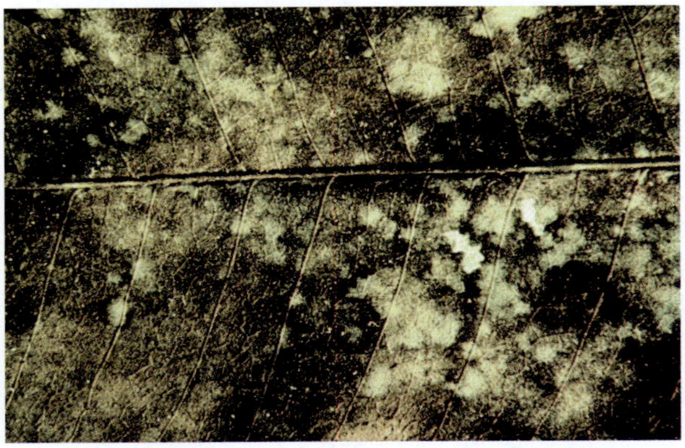

Sooty mold, colonies of *Capnodium mangiferum* on leaves of mango (*Mangifera indica*), x 10 (*MU*).

soralium, pl. **soralia** (L. *soralium* < Gr. *sorós*, pile, heap + L. suf. *-alis*, concerning + dim. suf. *-ium*): *Lichens.* A cortical pustule on various types of thalli, through which the *soredia* are liberated to the exterior, as occurs, e.g., in *Parmelia flaventior* and *P. sulcata* (Lecanorales). Soralia can be *diffuse*, as when the upper surface of the thallus becomes a continuous soredial mass, or *delimited*, if they are confined to certain defined areas. They can originate in tubercles (*tuberculate*) or in fissures (*fissural*), or also on the apices of isidia, and in some species the soralia can contain a mixture of soredia and isidioid structures.

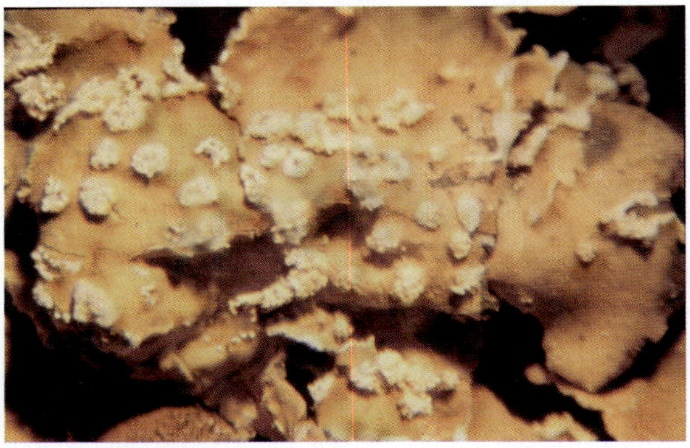

Soralia with soredia of the thallus of *Parmelia flaventior*, x 15 (*MU*).

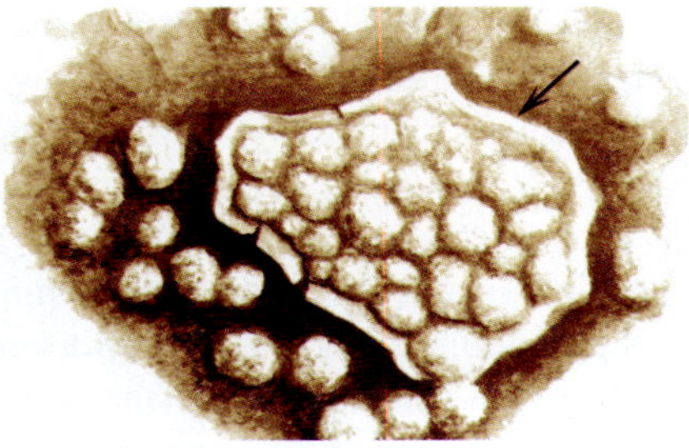

Soralium with soredia of *Parmelia sulcata*, x 20 (*MU*).

sorediate (ML. *soredium*, dim. of *sorus* < Gr. *sorós*, pile, heap + suf. *-atus* > E. *-ate*, provided with or likeness): having *soredia*.

Sorediate thallus of *Stereocaulon saxatile* (Lecanorales), x 15 (*MU*).

soredium, pl. **soredia** (ML. *soredium* < Gr. *sorídion* < *sorós*, pile, heap + dim. suf. *-ídion*): *Lichens.* A microscopic body, more or less spherical, formed by groups of algae surrounded by hyphae, which are produced in pustules (*soralia*) on the surface of the thallus of many species, such as *Letharia vulpina*, *Parmotrema chinense* (=*Parmelia perlata*), *Punctelia* (=*Parmelia*) *subrudecta*, *Parmotrema* and *Stereocaulon saxatile*, among other Lecanorales; the soredia function as buds or propagules that permit vegetative reproduction. They differ from isidia because they lack a cortical layer. Cf. **isidium** and **parasoredium**.

Soredia of the thallus of *Parmotrema* sp., x 10 (*MU*).

Soredia of the thallus of *Punctelia* (=*Parmelia*) *subrudecta*, x 20 (*MU*).

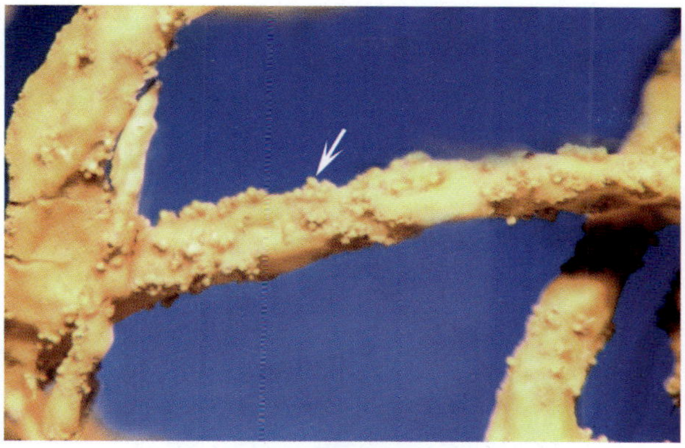

Soredia of the thallus of *Letharia vulpina*, x 40 (*MU*).

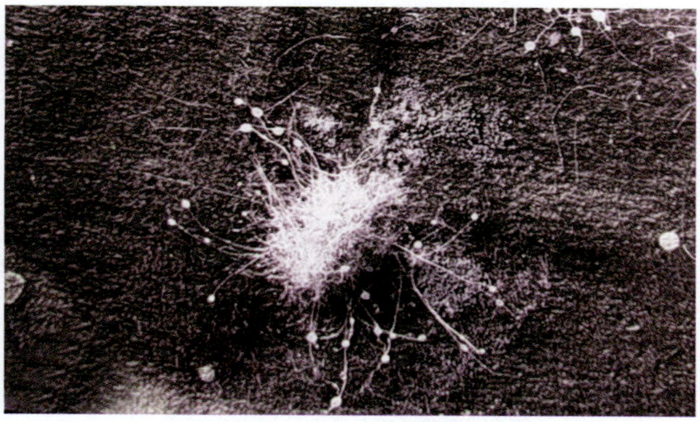

Sorocarps of *Dictyostelium mucoroides* growing on mouse dung, x 10 (*MU*).

Soredia of the crustaceous thallus of a lichen, x 10 (*TRO*).

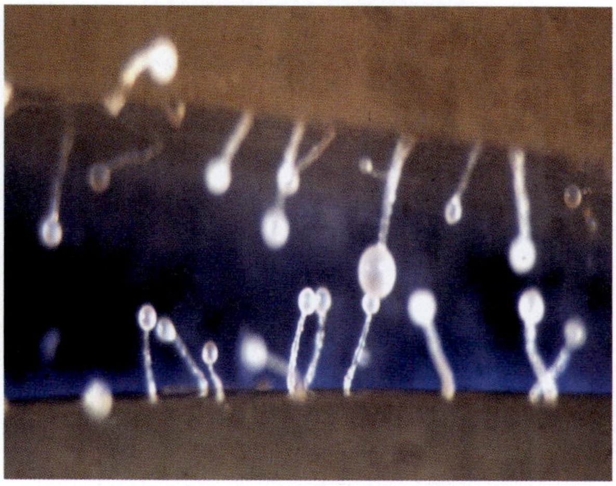

Sorocarps of *Dictyostelium mucoroides* on agar, x 30 (*MU*).

sorocarp (Gr. *sorós*, pile, heap + *karpós*, fruit): a fructification or pedicellate sporophore of the Acrasiomycota, such as *Acrasis rosea* (Acrasiales), and Dictyosteliomycota, such as *Dictyostelium mucoroides* and *Polysphondylium violaceum* (Dictyosteliales).

Sorocarp of *Acrasis rosea*, composed of pedicellar and prespore cells, x 30 (*MU*).

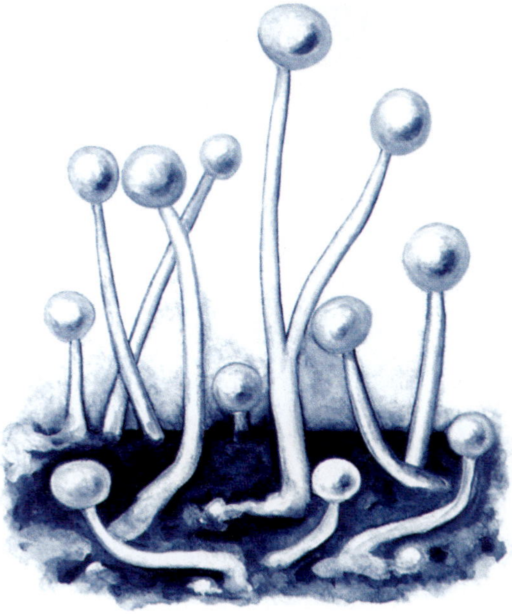

Sorocarps, each with a terminal sorus, of *Dictyostelium mucoroides*, x 60 (*MU*).

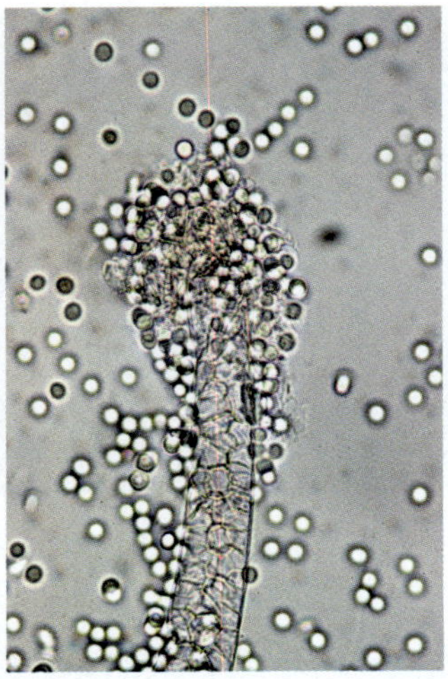

Sorocarp of *Dictyostelium mucoroides*, showing the reticulate pedicel and the spores, x 500 (*MU*).

Sorocarps, each with verticils of sori and a terminal sorus, of *Polysphondylium violaceum*, x 12 (*MU*).

Sorocarps, each with verticils of sori and a terminal sorus, of *Polysphondylium violaceum*, x 12 (*MU*).

sorocyst (Gr. *sorós*, pile, heap + *kýstis*, vesicle, cell): each of the encysted cells of the sorocarp of *Copromyxa arborescens* (Acrasiomycota), considered as the simplest type of sorocarp.

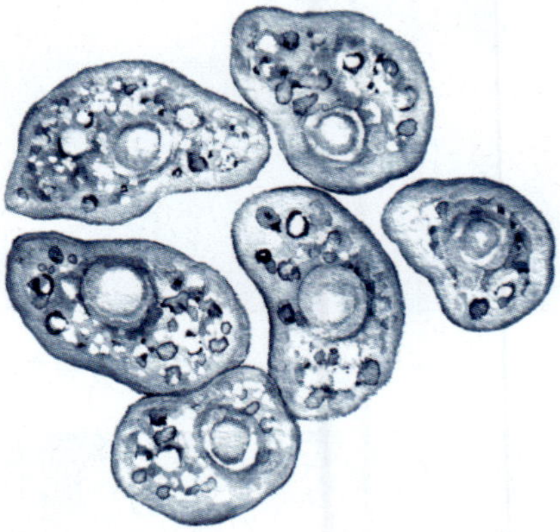

Sorocysts of *Copromyxa arborescens*, x 1,200 (*MU*).

sorogen (Gr. *sorós*, pile, heap + *génos*, origin < *gennáo*, to engender, produce): the part of the pseudoplasmodium that is differentiated into the sorus or group of spores, and into the multicellular pedicel or *sorophore*, of the sorocarp of the Dictyosteliomycota. It is seen clearly during the sorogenesis of *Polysphondylium* and *Dictyostelium*.

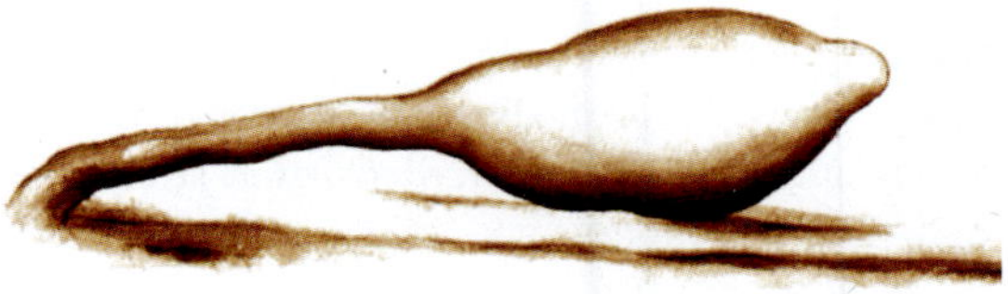

Sorogen of a young sorocarp of *Dictyostelium mucoroides* (Dictyosteliales), x 65 (*MU*).

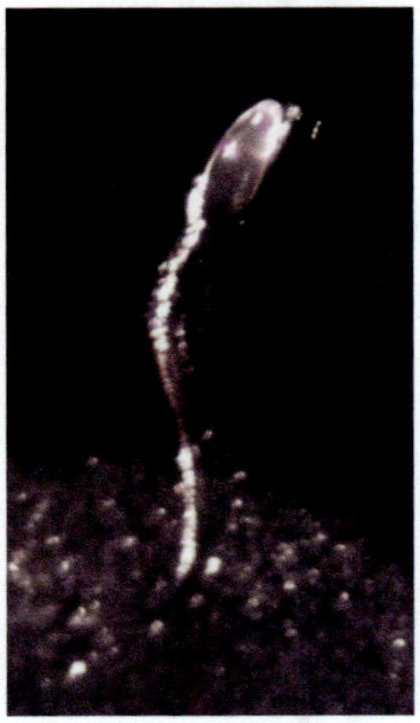

Sorogen of a young sorocarp of *Dictyostelium mucoroides*, x 50 (*MU*).

sorophore (Gr. *sorós*, pile, heap + suf. *-phóros*, bearer < *phéro*, to bear, carry, support): a pedicel (multicellular or acellular, depending on the species) that supports the sorus or mass of spores of the sorocarp of the Acrasiomycota and Dictyosteliomycota, such as *Acrasis* and *Dictyostelium*, respectively. The sorophore originates from the *sorogen* during the fructification of the pseudoplasmodium.

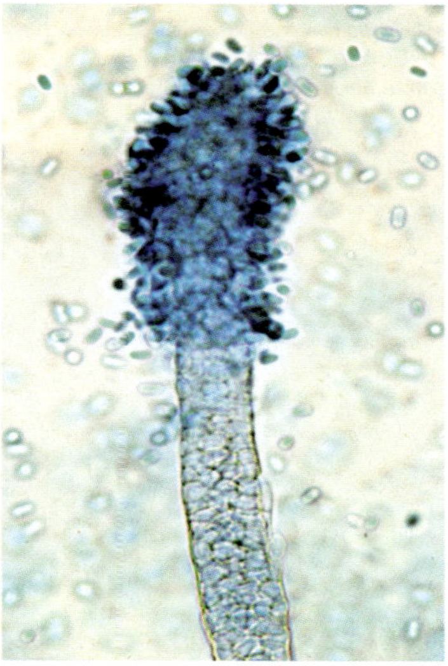

Sorophore of *Dictyostelium mucoroides*, x 320 (*MU*).

sorus, pl. **sori** (Gr. *sorós*, pile, heap): a mass of spores formed by certain fungi, like the rusts (Uredinales) and smuts (Ustilaginales), the sorocarp or fruiting body of the cellular mucilaginous fungus-like microorganisms, or Dictyosteliomycota (such as *Dictyostelium* and *Polysphondylium*), or the group of sporangia in species such as *Sorosphaera* (Plasmodiophorales) and *Synchytrium* (Chytridiales).

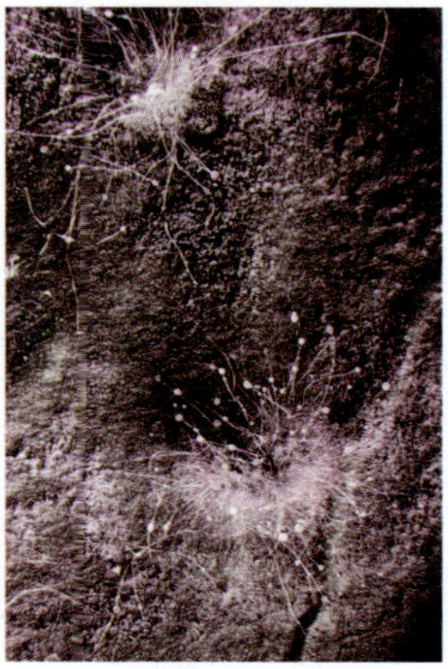

Sori of the sorocarps of *Dictyostelium mucoroides*, growing from mouse dung, x 10 (*MU*).

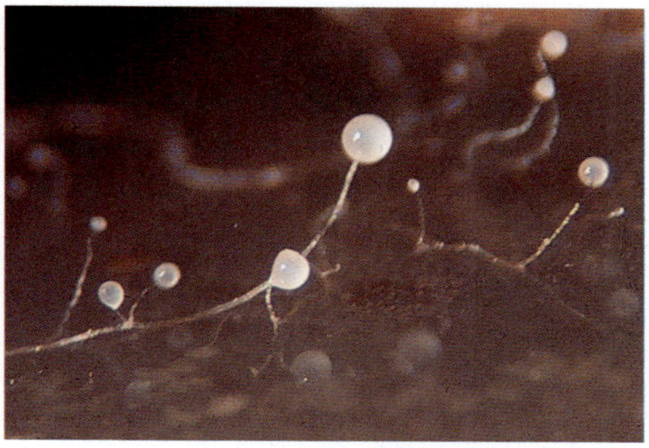

Sori of the sorocarps of *Dictyostelium mucoroides* on agar, x 30 (*MU*).

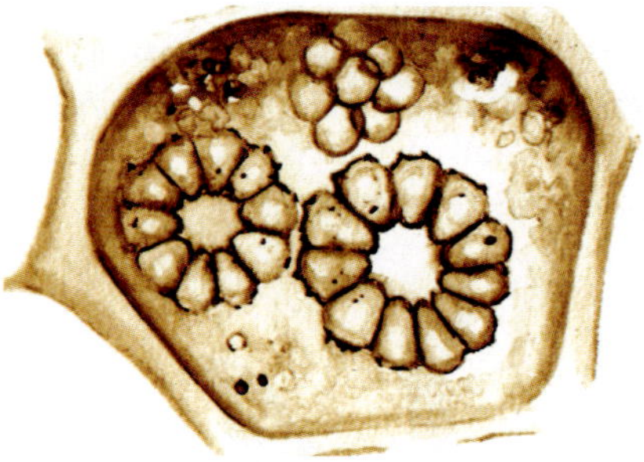

Sori (cystosori or latent spores) of *Sorosphaera veronicae* inside a cell of *Veronica*, an aquatic plant, x 800 (*MU*).

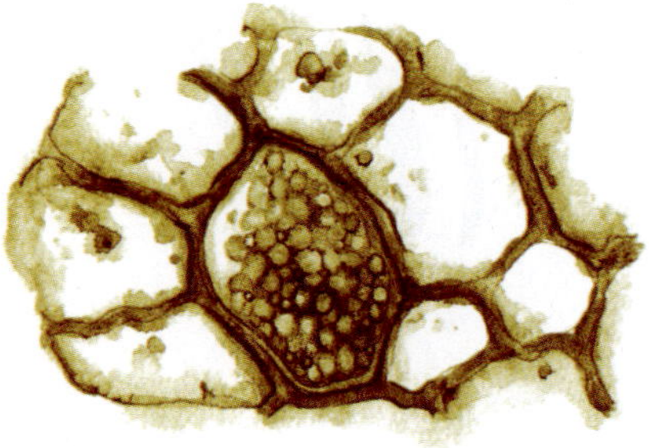

Sorus with zoosporangia of *Synchytrium endobioticum* inside a cell of a potato tuber, seen in transverse section, x 900 (*MU*).

Sori (masses of teliospores) of *Ustilago maydis* (Ustilaginales) on an ear of corn (maize), x 0.3 (*EPS*).

sp., **sp. nov.** (L. *species*, species; L. *nova*, new): L. abbreviations of species and of species nova, abbreviated as *sp.* and *sp. nov*.

sparassoid (< gen. *Sparassis* < Gr. *spáraxis*, laceration, torn up < *sparásso*, to lacerate, tear, pinch + NL. suf. *-oides*, or < L. *-oide*, contraction of Gr. *-oeídes*, which denotes likeness): composed of closely-packed, flabelliform branches, forming a more or less globose structure similar to the fruiting bodies of *Sparassis* (Aphyllophorales). Sparassoid fruiting bodies also are known in the Pezizales.

spathulate, **spatulate** (L. *spathulatus*, shaped like a spatula or provided with a spatula < *spathula*, dim. of *spatha*, spatula + suf. *-atus* > E. *-ate*, provided with or likeness): shaped like a spatula or spoon, like the fruiting body of *Geoglossum difforme*, *Spathularia flavida* (Helotiales) and *Dacryopinax elegans* (Dacrymycetales), and the ascospores of *Spathulospora phycophila* (Spathulosporales).

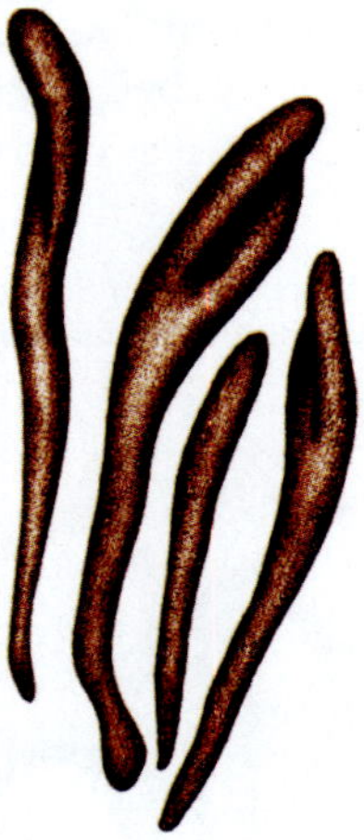

Spathulate apothecia of *Geoglossum difforme*, x 1 (*MU*).

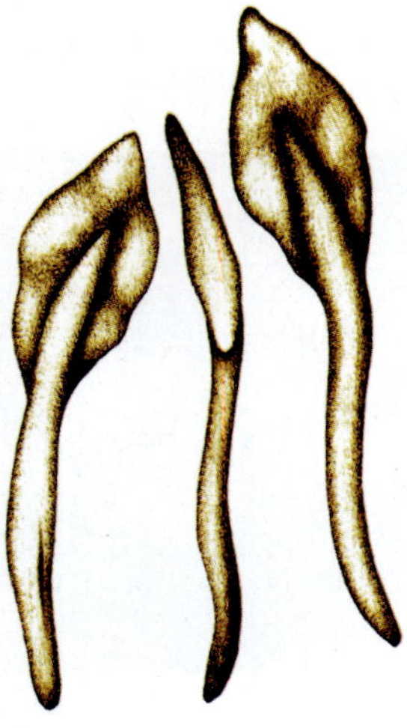

Spathulate apothecia of *Spathularia flavida*, x 1.3 (*MU*).

Spathulate apothecia of *Spathularia flavida*, x 1 (*RV*).

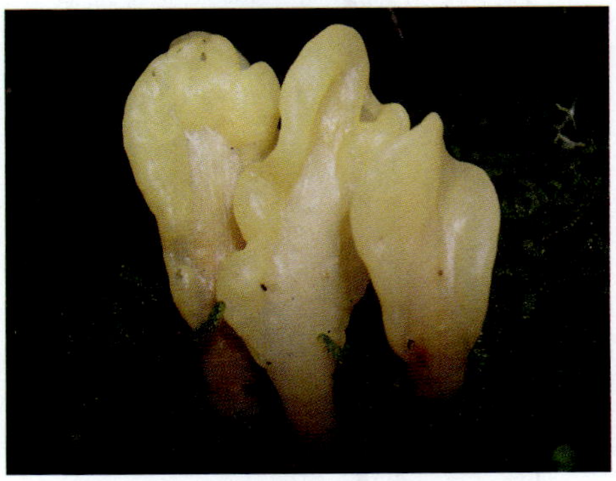

Spathulate apothecia of *Spathularia flavida*, x 2 (*RV*).

Spathulate basidiocarps of *Dacryopinax elegans*, x 1 (*RV*).

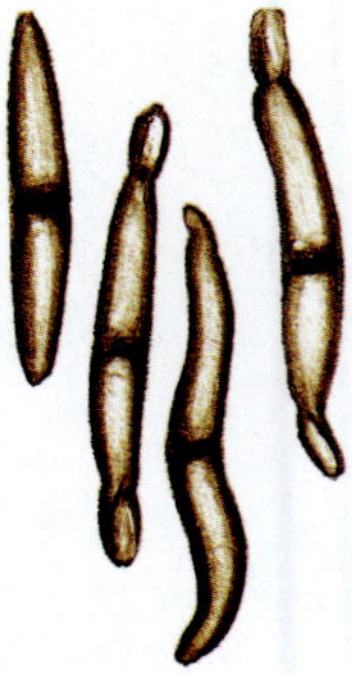

Spathulate ascospores of *Spathulospora phycophila*, a parasite on the marine alga *Ballia callitricha*, x 420 (*MU*).

spawn (ME. *spawnen* < AF. *espaundre*, to disperse < L. *expandere*, to spread out, expand): inoculum used in the culture of edible agaricaceous fungi, as, e.g., *Agaricus bisporus* (=*A. brunnescens*, common mushroom), *Pleurotus ostreatus* (oyster mushroom) and *Lentinula edodes* (shiitake), that consists of grains of wheat (or a similar cereal) that have been invaded by the mycelium of the fungus; the spawn is used to inoculate a suitable substrate for the type of fungus to be cultivated (such as horse manure fermented and transformed into "compost", in the case of the common mushroom, pasteurized wheat straw for the oyster mushroom, and pine chips, for shiitake), which will later produce the fructifications of the fungi to be harvested.

Spawn prepared with wheat seeds invaded by the mycelium of *Pleurotus ostreatus*, x 1.5 (*MU*).

Spawn (close-up) consisting of a wheat seed covered by the mycelium of *Pleurotus ostreatus*, x 12 (*MU*).

species (L. *species*, species, class, stirp): from a strictly systematic point of view, it corresponds to the hierarchy between the genus (or subgenus) and the variety (or subspecies). The specific name generally is an adjective that qualifies the generic name, although it can have various sources, e.g., be derived from a proper noun (personal or of locality). Colloquially, the species is simply defined as a single type or class of organism. However, there is much debate on how to define what constitutes a species, because the delimitation of species is subjective, and therefore, subject to the influence of personal interpretation. Some species concepts are ancient, predating the scientific estabishment of this systematic category; there are others, extremely vague in outline, with which systematists are not in agreement. The central point is the fact that species are not immutable, so that it is not easy to define them as a group of individuals that had, have, and failing extinction, will have, qualitatively identical characters. Therefore, various species concepts have been proposed, among them the following: 1) *morphospecies*, *morphological species* or *phenetic species*, which traditionally have been delimited on the basis of morphological characters; *polymorphic species* are those that have an intergradation of morphological characteristics, which can be due to intercrossing or to automictic sexual reproduction; 2) *biological species*, composed of populations that do, or potentially can, intercross and which are reproductively isolated from other groups, that may or may not be distinguished morphologically; 3) *phylogenetic species*, *evolutionary species*, based on biochemical or molecular differences and any other characteristic that can be measured and interpreted by means of a cladistic analysis; 4) *ecological species*, based on the adaptation to particular niches, as well as to reproductive isolation (e.g., a particular host); 5) *polythetic species*, based on a combination of characters, not all of which are necessarily present in each individual. The morphospecies concept is the one that has predominated in mycology, although in the past some mycologists have considered the host on which a parasitic fungus grows, including giving different names to morphologically similar species that live on different hosts. At present, with studies of population biology and the contributions of molecular biology, new evidence indicates that many morphospecies include several biological or phylogenetic species. In the lichenized fungi *species pairs* are recognized, one of them sexually fertile (called the *primary species*) and the other only reproducing vegetatively (the *secondary species*); although both have identical secondary metabolites, the secondary species tend to have a broader geographic distribution. It is probable that the information derived from molecular studies of these species pairs will contribute to a major reinterpretation of speciation.

specimen (L. *specimen* < *specere*, to look at, examine): sample, model, sign. A sample or typical individual that represents a group or a whole mass; something that obviously belongs to a particular category but is well known because of a distinctive individual characteristic. A specimen in a collection.

sperm (Gr. *spérma*, seed, sperm): the male sexual cell, typically motile. See **antherozoid**.

-sperm (Gr. *spérma*, genit. *spérmatos*, seed, sperm): a suf. employed to form adj. such as *macrosperm*, with a large seed. As a pref., *sperma-*, forms terms such as *spermatozoid*, the motile male sexual cell of the algae, fungi and other organisms (also called *antherozoid*).

spermatiophore (Gr. *spermátion*, small seed < *spérma*, genit. *spérmatos*, seed, sperm + suf. *-phóros*, bearer): a specialized hypha that produces spermatia, as occurs, e.g., in the Laboulbeniales and some Uredinales.

spermatium, pl. **spermatia** (NL. *spermatium* < Gr. *spermátion*, small seed < *spérma*, genit. *spérmatos*, seed, sperm + dim. suf. *-íon* > L. *-ium*): a small uninucleate cell, hyaline, immotile, similar to a spore, that functions as a male gamete, and which fertilizes a female receptive hypha, as happens, e.g., in *Phyllachora fusicarpa* (Phyllachorales), *Neurospora sitophila* (Sordariales), in the Laboulbeniales, and in the rust *Puccinia graminis tritici* (Uredinales); rust spermatia are often called *pycniospores*.

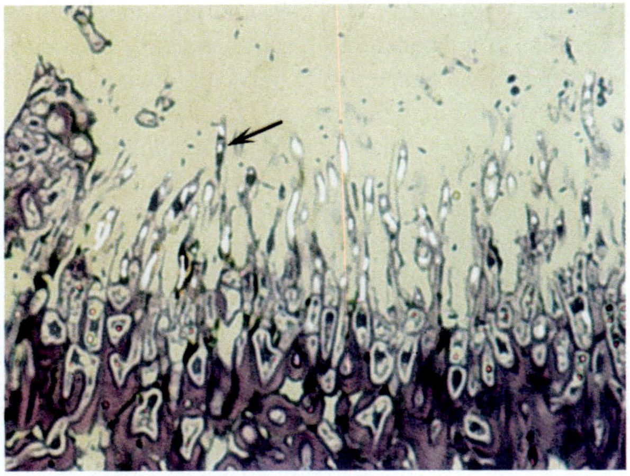

Spermatia produced in a spermogonium of *Phyllachora fusicarpa* parasitizing a leaf of *Duranta* sp., seen in longitudinal section, x 1,000 (*RTH*).

spermatization (Gr. *spérma*, genit. *spérmatos*, seed, sperm + L. suf. *-ationem* > E. *-ation*, action, state or condition, result): a type of plasmogamy in which a spermatium unites with a receptive hypha of a protoperithecium, as happens in *Neurospora* (Sordariales) and in *Puccinia* (Uredinales).

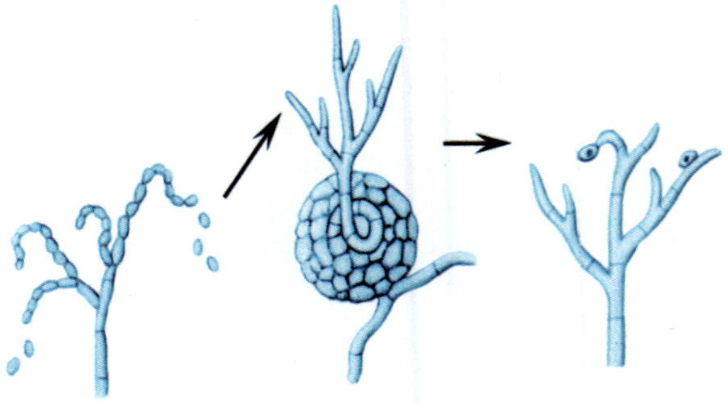

Spermatization in *Neurospora sitophila*, x 120 (*MU*).

spermato-, **spermat-** (Gr. *spérma*, genit. *spérmatos*, seed, semen, germ): a pref. employed in terms of Gr. origin that relate to the idea of a male fertilizing element, of the male gamete, as in *spermatization*.

spermatozoid (Gr. *spérma*, genit. *spérmatos*, seed, and in animals and man, sperm, germ, semen + *zõon*, of animals + L. suf. *-oide*, like < Gr. *-oeídes*, similar to): see **antherozoid**.

spermidium, pl. **spermidia** (Gr. *spérma*, seed, sperm + dim. suf. *-ídion*): a fructification that produces spermatia.

spermo-, **sperm-** (Gr. *spérma*, seed, germ, semen): a pref. that has the same meaning as **spermato-**, as in *spermogonium*. Also as a suf. *-sperm*, e.g., in *macrosperm*.

spermodermium, pl. **spermodermia** (Gr. *spérma*, seed, sperm + *dérma*, skin): an hymenium of spermatiophores formed beneath the cuticle in necrotic areas of infected host leaves.

spermodochidium, pl. **spermodochidia** (Gr. *spérma*, seed, sperm + *docheíon*, recipient + dim. suf. *-ídion* > L. *-idium* > E. *-idium*): a fruiting body that contains spermodochia, formed in a lysigenous cavity in the host tissue.

spermodochium, pl. **spermodochia** (Gr. *spérma*, seed, sperm + *docheíon*, recipient): a fasciculate or tuberculate aggregation of branched spermatiophores, usually arising from a single cell, and borne free on aerial mycelium.

spermagonium, **spermogonium**, **spermatogonium**, pl. **spermagonia**, **spermogonia**, **spermatogonia** (NL. *spermogonium* < Gr. *spérma*, genit. *spérmatos*, seed, sperm + *gónos*, that engendered, progeny + L. dim. suf. *-ium*): a pycnidium-like structure in which spermatia are produced, as occurs in *Cyttaria* (Cyttariales). Spermogonia in the rust fungi, such as *Aecidium* and *Puccinia* (Uredinales), often have been called *pycnia*; they function in spermatization in sexual reproduction.

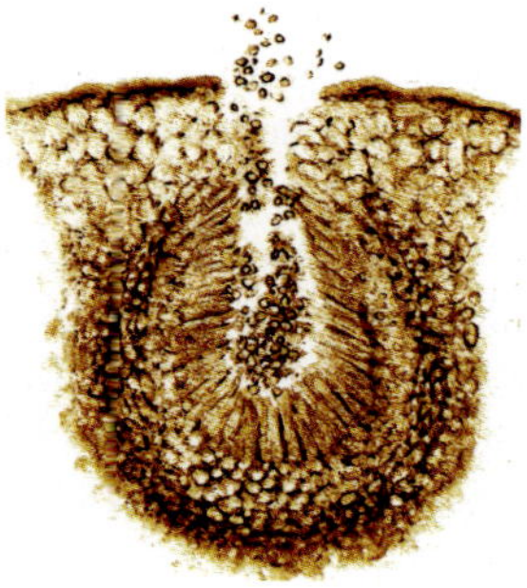

Spermogonium of *Cyttaria deformans* embedded in the tissue of a branch of *Nothofagus pumilio*, seen in longitudinal section, x 200 (*MU*).

Spermogonia of *Aecidium* sp. in a leaf of *Diospyros digyna*; the clearer spots (arrows) are the ostioles of the spermogonia, x 100 (*MU*).

Spermogonia of *Aecidium* sp. in a leaf of *Diospyros digyna*, x 120 (*MU*).

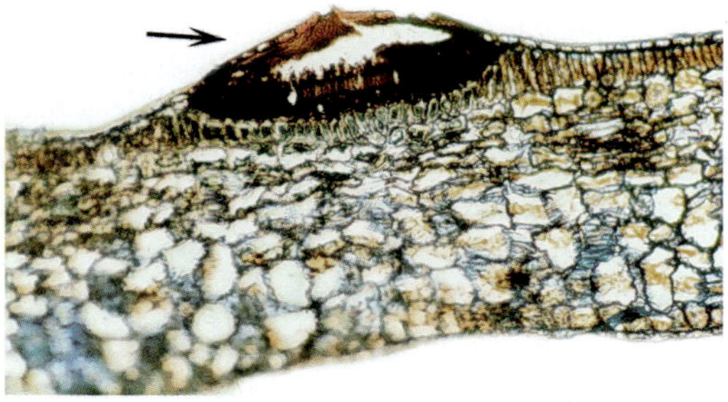

Spermogonium of *Aecidium* sp. in the upper surface of a leaf of *Diospyros digyna*, seen in longitudinal section, x 100 (*MU*).

Spermogonium of *Aecidium* sp. in the upper surface of a leaf of *Diospyros digyna*, seen in longitudinal section, x 200 (*MU*).

spermoplane (Gr. *spérma*, seed, sperm + L. *planus*, plane, flat surface): the surface of a seed, on which fungi may occur.

spermosphere (Gr. *spérma*, seed, sperm + *sphaîra*, sphere, site, content): the microhabitat around a seed in the soil. The spermosphere as well as the spermoplane have a characteristic mycobiota.

sphacelium, pl. **sphacelia** (L. *sphacelus* < Gr. *sphákelos*, gangrene, inflamation + NL. dim. suf. *-ium* > E. *-ium*): a conidium-forming structure in the asexual phase of *Claviceps* (Hypocreales), parasitic on the spikes of grains, such as wheat; later, the sclerotium is developed.

-sphaere (L. *sphaera* < Gr. *sphaîra*, sphere): a suf. employed in terms that allude to an area or field of action of something, as in *rhizosphere*, the area of the soil near the living roots of a plant.

sphaeridium, pl. **sphaeridia** (Gr. *sphairídion* < *sphaîra*, sphere + dim. suf. *-ídion*): a stalked, globose apothecium, like that of lichens (*Calicium*) of the order Caliciales. Syn. of **capitulum**.

Sphaeridia of the thallus of *Calicium farietinum*, x 10 (*MU*).

sphaerocyst (Gr. *sphaîra*, sphere + *kýstis*, bladder, vesicle, cell): **1**. A spherical cell that forms part of the heteromerous trama of the gills (hymenophore) in basidiocarps of *Russula* and *Lactarius* (Agaricales).

2. An inflated or variously shaped cell found in the exoperidium of *Bovista*, *Calvatia* and *Lycoperdon* (Lycoperdales).

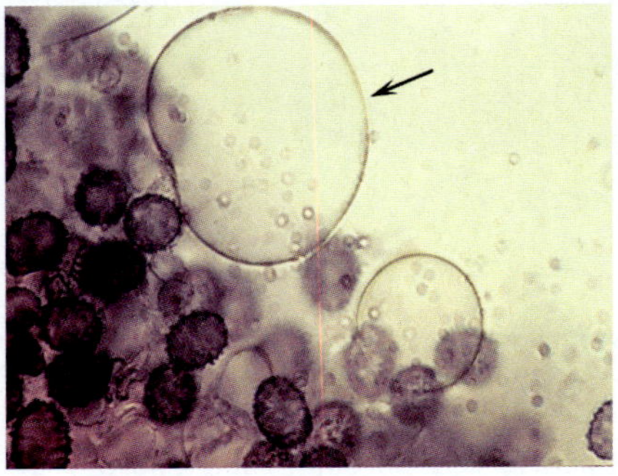

Sphaerocysts from the trama of a basidiocarp of *Russula* sp., x 500 (*MU*).

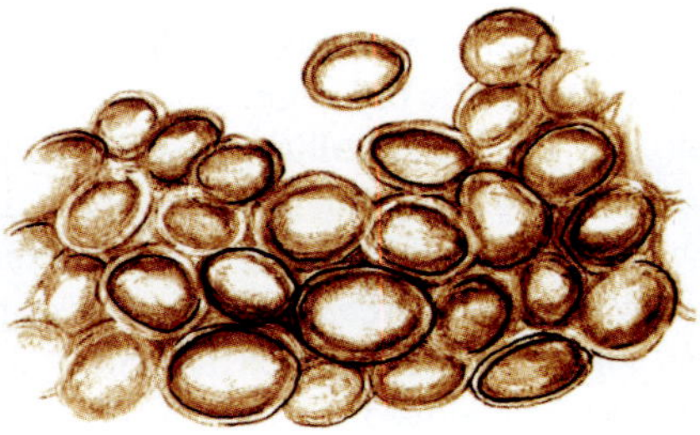

Sphaerocysts from the polycystoderm of an exoperidial fibril of *Calvatia cretacea*, x 500 (*MU*).

sphaeropedunculate (Gr. *sphaîra*, sphere + L. *pedunculatus*, furnished with a foot or support < *pedunculus*, dim. of *pes*, genit. *pedis*, foot, support + suf. *-atus* > E. *-ate*, provided with or likeness): a globose or spherical structure with a prolongation that tapers toward the basal end, like the hyphal cells that are found in the volva of the basidiocarp of *Amanita herrerae* (Agaricales). The so-called sphaeropedunculate cell is a cystidium with a nearly globose apex and a narrow stalk.

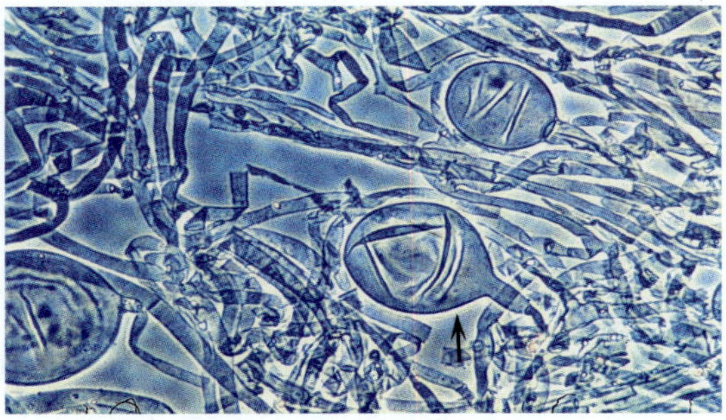

Sphaeropedunculate hyphal cells of the volva of a basidiocarp of *Amanita herrerae*, x 570 (*RMA*).

sphaeropsidaceous (L. *sphaîra*, sphere + suf. Gr. *-ópsis*, appearance + suf. L. *-aceus* > E. *-aceous*, of or pertaining to, with the nature of): term applied to asexual or anamorphic fungi that form their spores in pycnidial conidiomata. The gen. *Diplodia*, *Phoma*, *Phomopsis*, *Septoria* and *Zythia* are examples of the old family Sphaeropsidaceae, order Sphaeropsidales, class Coelomycetes, subdivision Deuteromycotina (or deuteromycetes in a general sense). See **Coelomycetes**.

sphagni- (< gen. *Sphagnum* < Gr. *sphágnos*, a kind of moss): a pref. used in names such as *Sphagnicola* (=*Pezoloma*, of the Helotiales), a discomycete that inhabits mosses of the genus *Sphagnum*.

spheroidal asci (L. *spherule*, small sphere < *sphaera* < Gr. *sphîara*, sphere, globe + L. *-oide*, contraction of Gr. *-oeídes*, which denotes likeness; Gr. *askós* > L. *ascus*, pl. *asci*, asci): asci with the shape of a sphere, e.g., as in *Gymnascella citrina* (Onygenales).

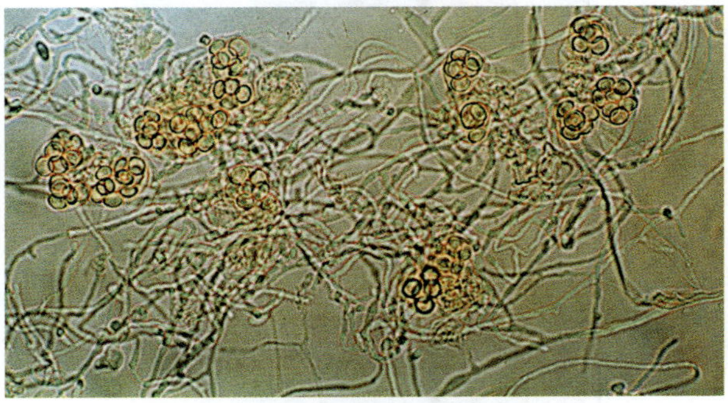

Spheroidal asci of *Gymnascella citrina* (Onygenales), x 700 (*MU*).

spherule (L. *spherula*, small sphere < *sphaera* < Gr. *sphaîra*, sphere, globe + dim. suf. *-ula* > E. *-ule*): **1**. A multinucleate cell of a latent myxomycete plasmodium. **2**. A globose, sporangium-like structure (the parasitic phase), of *Coccidioides* (moniliaceous asexual fungi) that originates from the swelling and enlargement of inhaled arthrospores (the saprobic phase). **3**. *Rhinosporidium seeberi* (a fungus-like microorganism now classified as Protozoa), the etyological cause of rhinosporidiosis, can form a spherule in the mucocutaneous tissues of the human nose. **4**. A globose or subglobose structure, subssesile or with a short stalk, pearl white in color, 0.5-1.5 (2.5) mm in diam., that functions as a conidium-forming sporodochium in *Termitomyces* (Agaricales), a fungus symbiotic with termites. The spherules originate on the surface of the termite nests as erect clusters of intertwined hyphae that are catenulate or moniliform, with multiple verticillate buds on which the conidia are produced.

spiculate (L. *spiculatus*, provided with spicules < *spiculum*, dim. of *spica*, dart, arrow, sharp point, nail + suf. *-atus* > E. *-ate*, provided with or likeness): having *spicules*.

spicule (L. *spiculum*, dim. of *spica*, dart, arrow, sharp point, nail + dim. suf. *-ulum* > E. *-ule*): a superficial spine or spine-like appendage, microscopic or very small, like the sterigmata of the basidia in the basidiomycetes. In some spores, a small, filiform appendage, very short, that represents the remnant of the sterigma that supported them. See **sterigma**.

spiculospore (NL. *spiculospore* < L. *spiculum*, dim. of *spica*, dart, arrow, sharp point, nail + Gr. *sporá*, spore): a spore formed at the apex of a sharp-pointed structure, often elongated (similar to a spine), as occurs in *Hirsutella* and *Akanthomyces* (stilbellaceous asexual fungi).

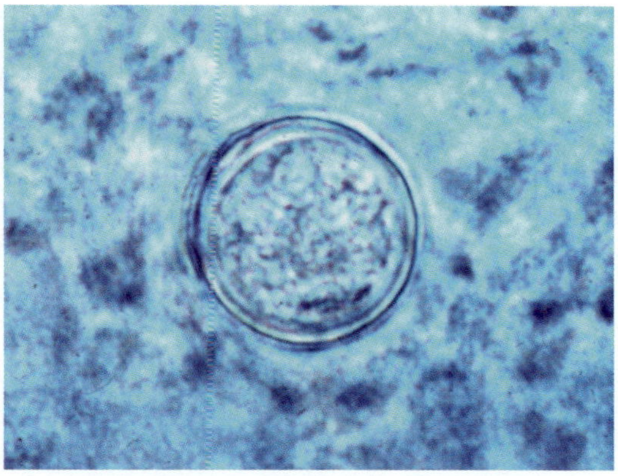

Spherule of *Coccidioides immitis*, x 120 (*RLM*).

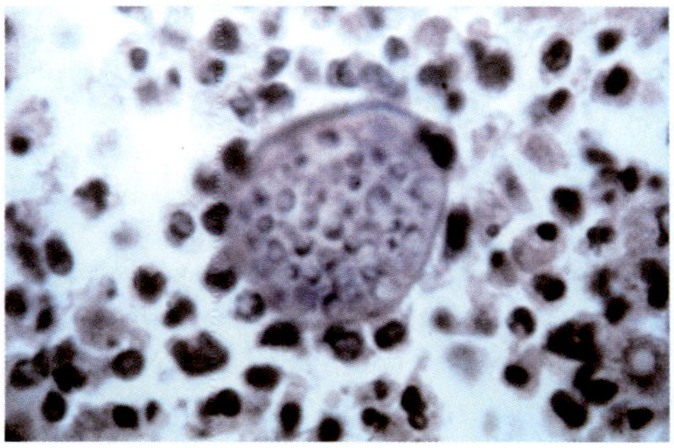

Spherule of *Coccidioides immitis* in a pig ganglion; histopathological section, x 120 (*ECN*).

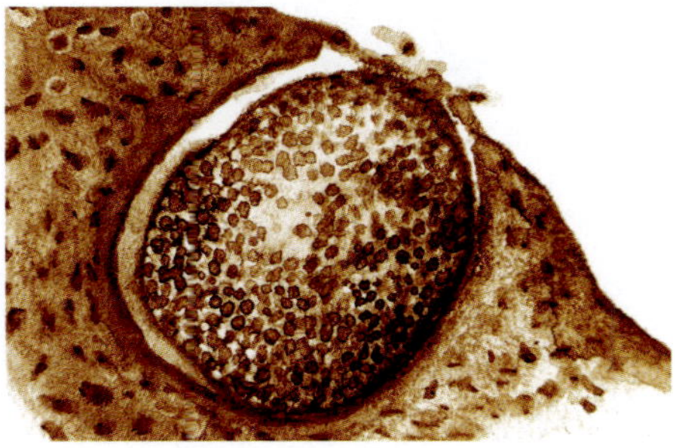

Spherule of *Rhinosporidium seeberi* in human tissue, x 500 (*MU*).

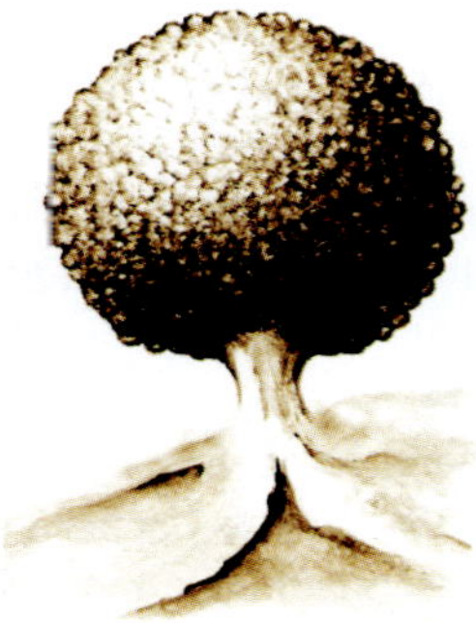

Spherule of *Termitomyces albuminosus*, x 50 (*MU*).

A wasp parasitized by *Hirsutella saussurei*, from which are emerging synnemata with conidiophores producing **spiculospores**, x 1 (*MU*).

A wasp parasitized by *Hirsutella saussurei*, from which are emerging synnemata with conidiophores producing **spiculospores**, x 5 (*MU*).

Synnema of *Hirsutella saussurei*, arising from a wasp parasitized by this fungus, from which are produced conidiophores forming **spiculospores**, x 100 (*MU*).

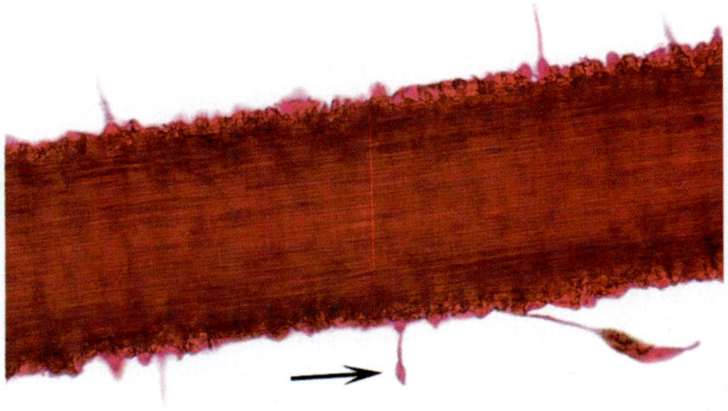

Synnema of *Hirsutella saussurei*, arising from a wasp parasitized by this fungus, from which are produced conidiophores forming **spiculospores**, x 200 (*MU*).

spilodium, pl. **spilodia** (Gr. *spílos*, freckle + L. suf. *-odium*, resembling < Gr. *ode*, like < *-oeídes*, similar to): *Lichens*. A small, round, blackish structure composed of dark, compact hyphae, which is present in the thallus of *Dirina massiliensis* (Arthoniales), caused by *Milospium graphideorum* (an ascomycete anamorph). It is called a spilodium for its similarity to a spilus or navel of the grains of wheat and other graminaceous seeds.

spindle pole body (ME. *spindel*, to spin, in relation to the spindle used to form and twist the thread in a spinning wheel; L. *polaris*, pertaining to or relating to the pole < *polus,* pole; ME. < OE. *bodig*, body): an electron opaque structure, of unknown chemical composition, that lies adjacent to the nuclear membrane in the majority of true fungi (except the flagellate fungi). Evidence indicates that these organelles function as microtubule organizing centers during mitosis and meiosis. Microtubules emanating from the spindle pole bodies develop into the achromatic spindle apparatus, along which the chromosomes are distributed during mitosis and meiosis. Species of fungi that produce flagellate cells lack spindle pole bodies, possessing instead a pair of centrioles that are associated with the nuclear membrane or envelope. Cf. **centriole**.

spine (L. *spina*, spine): a narrow, sharp-pointed process.

spinose, spiny (L. *spinosus*, armed with spines < *spine*, spine + suf. *-osus* > E. *-ose*, full of, augmented, prone to): having spines.

spinule (L. *spinula*, small spine < *spina*, spine + dim. suf. *-ula* > E. *-ule*): a small spine. In the lichens it applies to a small prominence on the thallus, with a narrow base, which, on separating, functions as a vegetative propagule, as occurs, e.g., in *Bryoria bicolor* and other species of fruticose lichens of the order Lecanorales.

Spinose or **spiny** basidiocarp of an unidentified aphyllophoraceous fungus, growing on wood, x 1 (*RV*).

spinulose, spinulous (L. *spinulosus*, provided with small spines < *spinula*, dim. of *spina*, spine + suf. *-osus* > E. *-ose*, full of, augmented, prone to): with small spines, at times microscopic, like the sporangial wall of *Rhizopus oryzae* (=*Rh. arrhizus*), Mucorales.

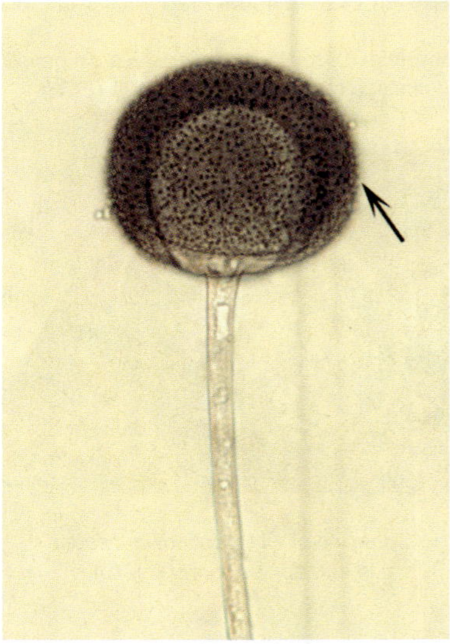

Spinulose sporangium of *Rhizopus oryzae* (=*Rh. arrhizus*), x 500 (*MU*).

spiral hypha, pl. **hyphae** (L. *spira*, curly, a tridimensional curve, like a helix, with one or more turns on an axis + suf. *-alis* > E. *-al*, relating to or belonging to; NL. *hypha*, hypha < Gr. *hyphé*, tissue, spiderweb, hypha): a hypha that ends in a spiral or helicoidal curl, as happens in *Trichophyton* (moniliaceous asexual fungi) and the gymnothecium of *Blastomyces* (=*Ajellomyces*) *dermatitidis*, Onygenales).

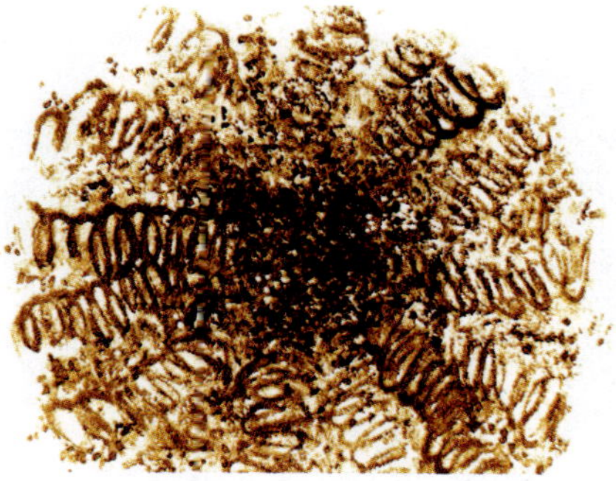

Radial **spiral hyphae** of the gymnothecium of *Blastomyces* (=*Ajellomyces*) *dermatitidis*, x 90 (*MU*).

Spitzenkörper (G. *Spitzen*, tip, apex + *Körper*, body): a granular or vesicular region of the cytoplasm, rich in actin, found at the apex of the hyphae of the Ascomycota and Basidiomycota, in association with vesicles of the cell wall, which are formed from the cisternae of the Golgi apparatus or dictyosomes and which participate in the synthesis of the cell wall of growing hyphae, as well as in the direction this growth takes. Hyphae of the zoosporic microorganisms (the phyla Oomycota, Hyphochytriomycota and Labyrinthulomycota of the kingdom Stramenopila) lack a Spitzenkörper. The term Spitzenkörper was originally described in German and was subsequently incorporated into English unchanged, as is commonly done with technical terms in English.

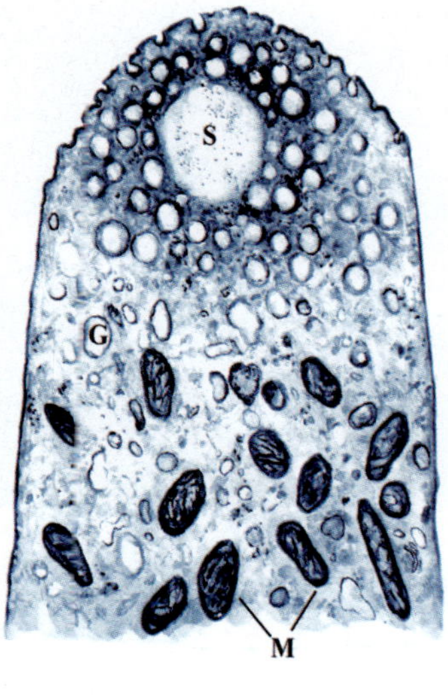

Spitzenkörper (S) in the subapical zone of a hypha of *Armillaria* (=*Armillariella*) *mellea* (Agaricales), seen in longitudinal section. Note the absence of vesicles in this spherical central region. The clusters of vesicles are associated with cisternae or tubules of the Golgi apparatus (G), in zones of cytoplasm that contain few ribosomes compared to the surrounding protoplasm. Mitochondria (M) and endoplasmic reticulum occur throughout the region. Drawing based on a transmission electron micrograph published by Beckett *et al.* in their *Atlas of Fungal Ultrastructure*, 1974, x 17,000 (*MU*).

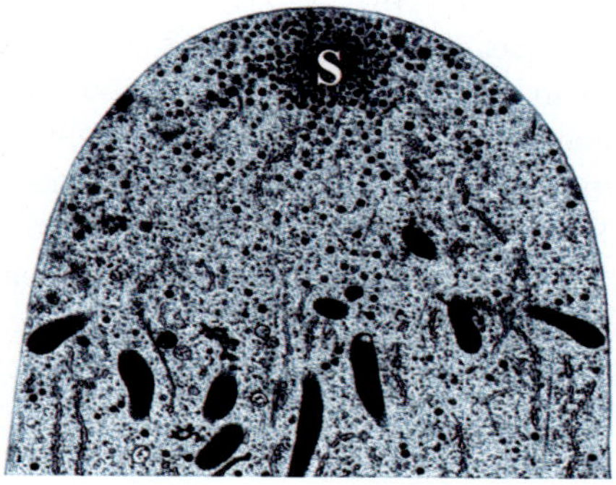

Spitzenkörper (S) in a hypha of a basidiomycete. Transmission electron micrograph, x 17,000 (*CM*).

splash cup (poss. OE. *plæse* > ME. *plasch*, pool, puddle > E. *splash*, to scatter a liquid; L. *cuppa* and *cupa*, cask, barrel): an open, cup-like structure, often with the side walls widening toward the mouth, into which raindrops land and forcibly eject the peridioles; typical of the Nidulariales (*Crucibulum*, *Cyathus*, *Nidula*, *Nidularia*). In some lichens, such as *Cladonia* (Lecanorales), the cyphellae of the podetia also function like splash cups.

Splash cups of the fruiting bodies of *Cyathus striatus*, x 6 (*RV*).

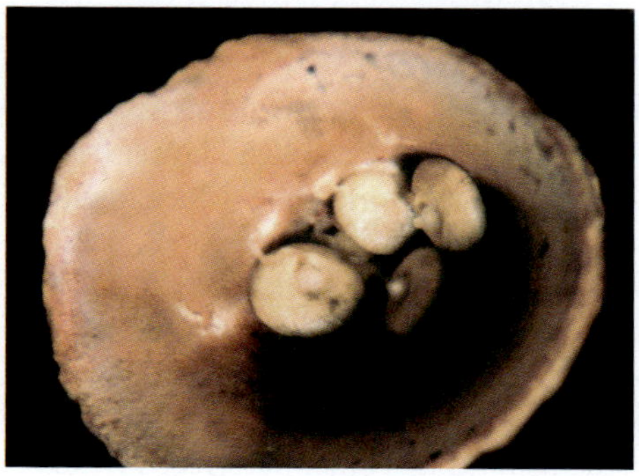

Splash cup of the fruiting body of *Crucibulum laeve*, x 6 (*CB*).

split lamella, pl. **lamellae** (Du. *splitten*, to split; L. *lamina*, sheet, plate): each one of the split structures, similar to gills, found on the lower surface of the basidiocarp of *Schizophyllum* (Aphyllophorales). The marginal proliferation of the basidiocarp forms the "split gills", which on drying can fold over the hymenium. In some cases the split may be shallow and only resemble a groove. The gills in *Schizophyllum* are not considered as homologous with the gills of the agarics.

Split lamellae of the basidiocarp of *Schizophyllum commune*, x 0.5 (*RV*).

Split lamellae of the basidiocarps of *Schizophyllum commune*, x 2 (*MU*).

spongiostratum, pl. **spongiostrata** (L. *spongia*, sponge + *stratus*, blanket, layer): corresponds to the hypothallus of some lichen thalli, such as *Pannoparmelia* (Lecanorales).

spongy (L. *spongia*, sponge + E. suf. *-y*, having the qualities of): very porous, hollow and lighter than its volume indicates; with a soft consistency and with a tendency to absorb water, like the fruiting body of *Boletus*, *Suillus* and other boletaceous Agaricales, as well as that of *Climacocystis* (=*Spongipellis*), of the Aphyllophorales.

Spongy basidiocarps of *Climacocystis* (=*Spongipellis*) *borealis*, on a fir trunk, x 0. 2 (*MU*).

Spongy basidiocarp of *Boletus castaneus*, x 0. 2 (*TH*).

Spongy, symbiotic basidiocarps of *Exobasidium azaleae* (Exobasidiales), x 1 (*RTH*).

sporabola, pl. **sporabolae** (Gr. *sporabolá* < *sporá*, spore + *bállo*, to throw): in mycology it refers to the curve that a basidiospore makes after being discharged from the sterigma of the basidium.

sporangial vescicle, **plasmamembranic vesicle** (Gr. *sporá*, spore + *angeíon*, vessel, receptacle + L. suf. *-alis* > E. *-al*, pertaining to or relating to; L. *vesicula*, dim. of *vesica*, bladder; Gr. *plásma*, formation, bland material with which a living being is formed + L. *membrana*, membrane, parchment + Gr. suf. *-íkos* < L. *-icus* < E. *-ic*, relating to or pertaining to): an extension of the sporangial membrane that is formed on the mouth of the sporangium during the maturation and discharge of the zoospores in aquatic fungi such as *Chytriomyces*, *Rhizidium* and *Rhizophlyctis* (Chytridiales). Equivalent to **prosporangium**.

sporangiocyst (NL. *sporangium* < Gr. *sporá*, spore + *kýstis*, bag, bladder; here, cell): *Chytridiales* and *Plasmodiophorales*. A resting spore or sporangium. Cf. **cystosorus**.

sporangiogenous (NL. *sporangium* < Gr. *sporá*, spore + *angeîon*, vessel, receptacle + *génos*, origin < *gennáo*, to engender, produce): *Plasmodiophorales*. The plasmodium of the asexual phase, which is transformed into a sporangium and gives rise to zoospores. Cf. **cystogenous**.

sporangiole, **sporangiolum**, pl. **sporangiola** (NL. *sporangium* < Gr. *sporá*, spore + *angeîon*, vessel, receptacle + L. suf. *-olum* > E. *-ole*, small): a small sporangium that contains only a few spores, and sometimes only a single spore ("conidium"); sporangiola are present in *Thamnidium elegans* and other Mucorales, such as *Blakeslea trispora*, *Cokeromyces recurvatus*, *Cunninghamella echinulata*, *Choanephora cucurbitarum* and *Radiomyces embreei*. In the Entomophthorales, such as *Conidiobolus*, sporangiola also are formed and are called conidia because they are unispored; on a sporangiophore (or conidiophore, if the sporangiolum is considered a conidium) a *primary sporangiole*, *sporangiolum*, pl. *sporangiola*, is developed that is forcibly discharged; this can germinate repeatedly by producing conidia or *secondary sporangiole*, *sporangiolum*, pl. *sporangiola*.

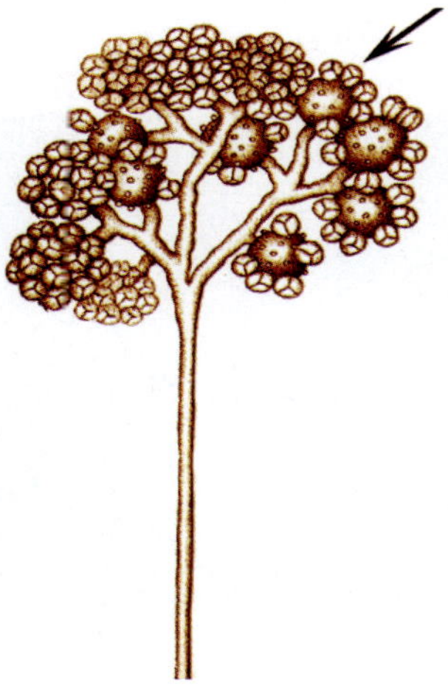

Sporangiola of *Blakeslea trispora*, x 300 (*MU*).

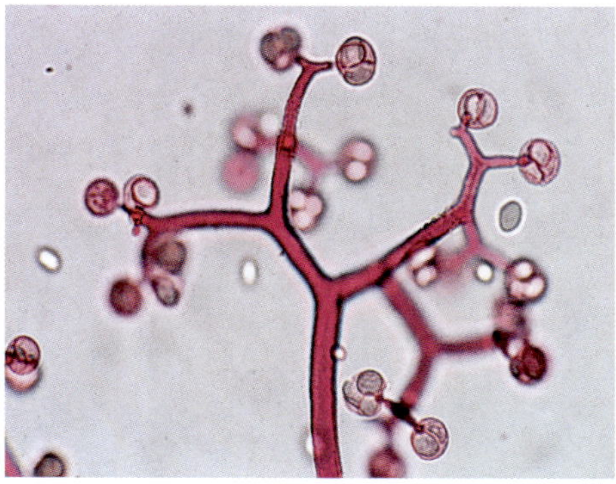

Sporangiola of *Thamnidium elegans*, x 500 (*MU*).

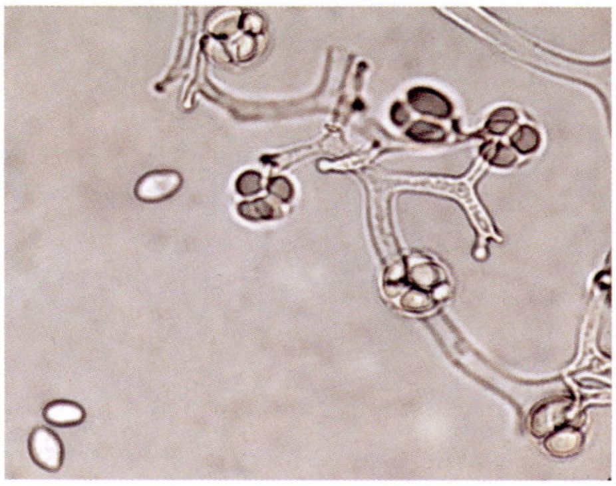

Sporangiola of *Thamnidium elegans*, x 690 (*MU*).

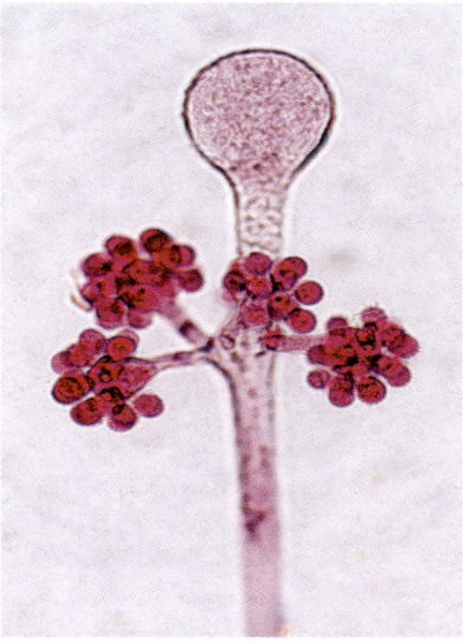

Sporangiola of *Cunninghamella echinulata*, x 525 (*MU*).

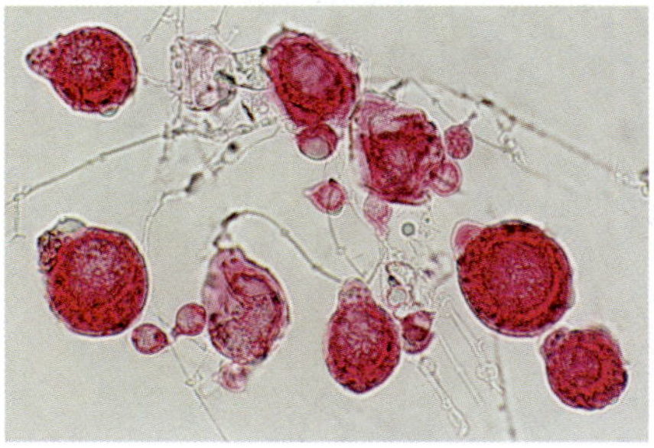

Primary **sporangiola** of *Conidiobolus coronatus*, x 400 (*RLM*).

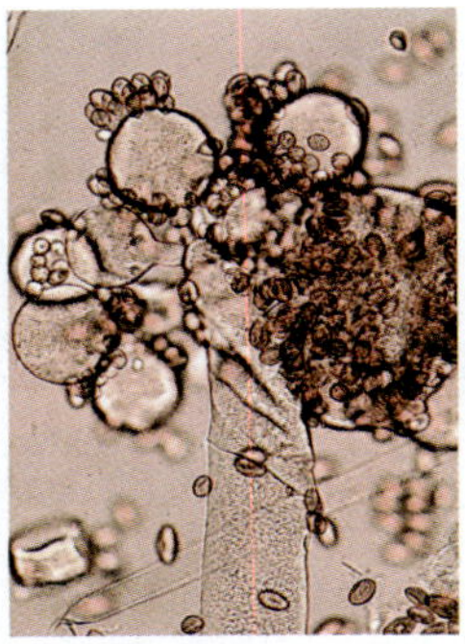

Sporangiola of *Choanephora cucurbitarum*, x 300 (*MU*).

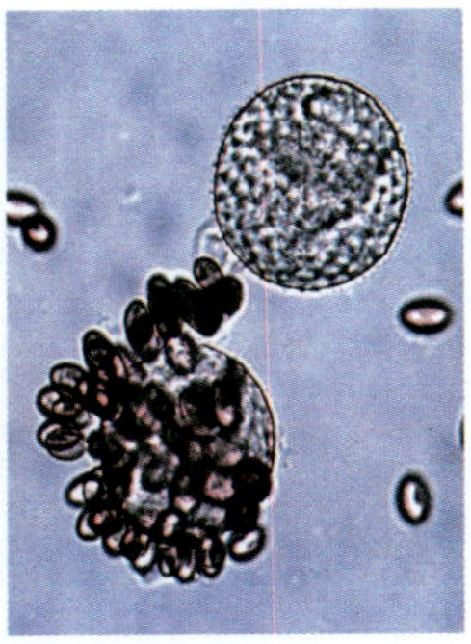

Sporangiola of *Choanephora cucurbitarum*, x 400 (*MU*).

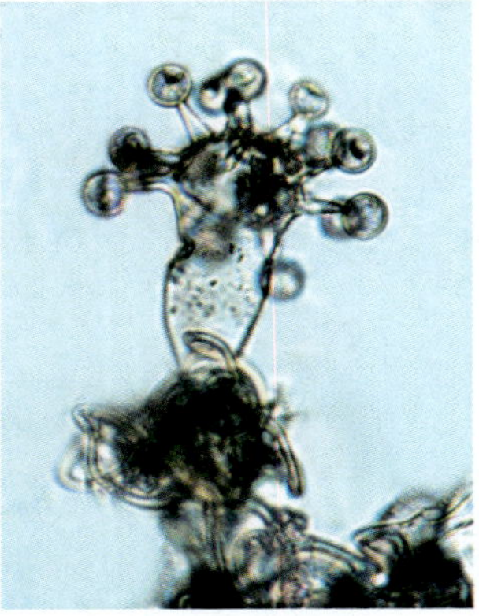

Sporangiola of *Cokeromyces recurvatus*, x 400 (*MU*).

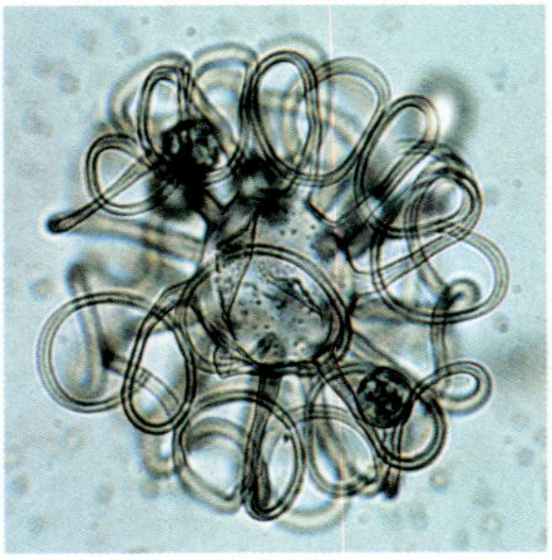

Sporangiola of *Cokeromyces recurvatus*, x 600 (*MU*).

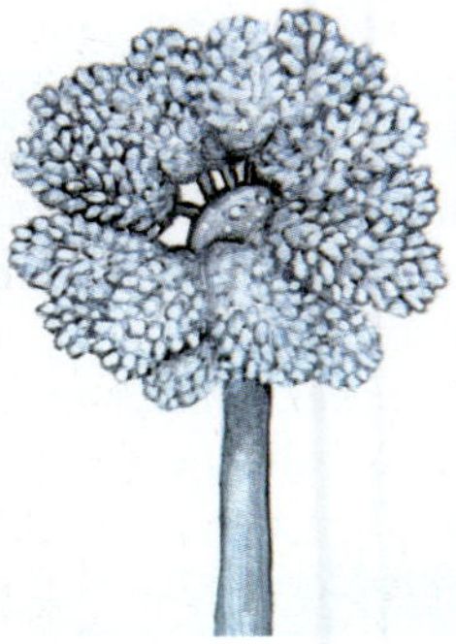

Sporangiola of *Radiomyces embreei*, x 200 (*MU*).

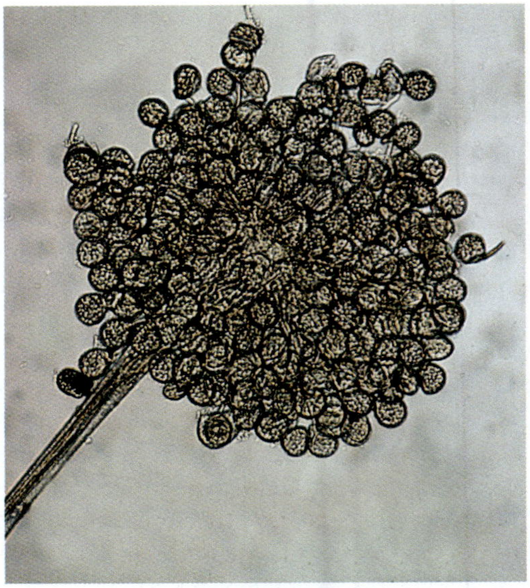

Sporangiola of *Thamnostylum* (=*Helicostylum*) *piriforme*, x 400 (*MU*).

sporangiophore (Gr. *sporá*, spore + *angeîon*, vessel, receptacle + suf. *-phóros*, bearer): a specialized hypha that produces and supports one or more sporangia, as occurs, e.g., in *Phycomyces nitens* and *Ph. blakesleanus* (Mucorales).

Sporangiophores of *Phycomyces nitens* on agar, x 0.4 (*MU*).

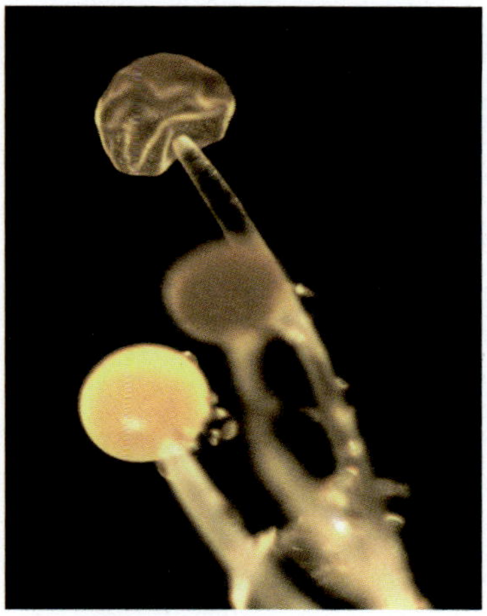

Sporangiophores of *Phycomyces blakesleanus* on agar, x 25 (*MU*).

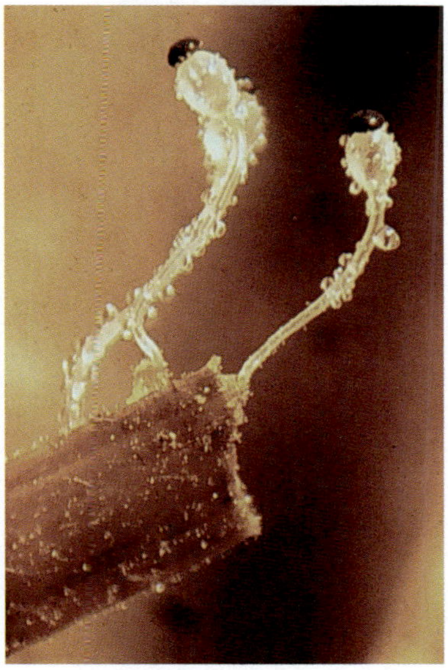

Sporangiophores of *Pilobolus crystallinus* var. *kleinii* (=*P. kleinii*, Mucorales) growing on cow's dung, x 25 (*MU*).

sporangiosorus, pl. **sporangiosori** (Gr. *sporá*, spore + *angeîon*, vessel, receptacle + *sorós*, heap, pile): *Myxomycetes*. A group of spherical, fused, sporangia derived from a single plasmodium; also applied to a uni- or multilobed sporangium that originates from a single plasmodium.

sporangiospore (Gr. *sporá*, spore + *angeîon*, vessel, receptacle + *sporá*, spore): a spore produced in a sporangium in any of its variations; sporangiospores can be immotile (aplanospores) or flagellate and motile (zoospores), and they are present in various groups, e.g., in the Chytridiomycota, Oomycota, and Zygomycota. In the latter phylum, illustrated examples of sporangiospores are *Rhizopus* and *Parasitella* (Mucorales).

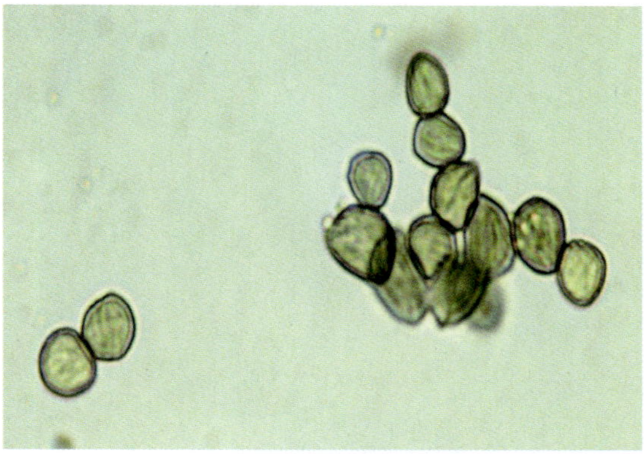

Sporangiospores of *Rhizopus stolonifer* (=*Rh. nigricans*), x 1,300 (*MU*).

Sporangiospores of the sporangia of *Parasitella parasitica*, x 1,150 (*MU*).

sporangium, pl. **sporangia** (Gr. *sporá*, spora + *angeîon*, vessel, receptacle): a structure, of various shapes depending on the species, that produces endogenous asexual spores; all of the protoplasmic contents of a sporangium is converted into an indefinite number of spores, whether they be aplanospores or zoospores, depending on the species. Sporangia are present, e.g., in Myxomycetes, such as *Cribraria cancellata* (=*Dictydium cancellatum*, Liceales), in aquatic fungi, such as *Phlyctochytrium* (Chytridiales), and in the common molds, such as *Mucor* and *Rhizopus* (Mucorales).

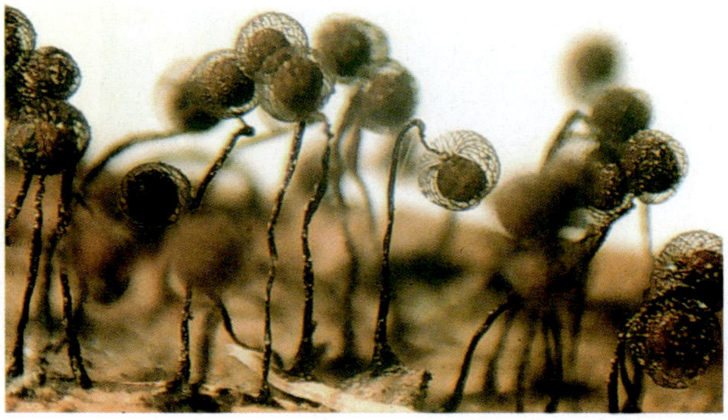

Sporangia of *Cribraria cancellata* (=*Dictydium cancellatum*), x 18 (*MU*).

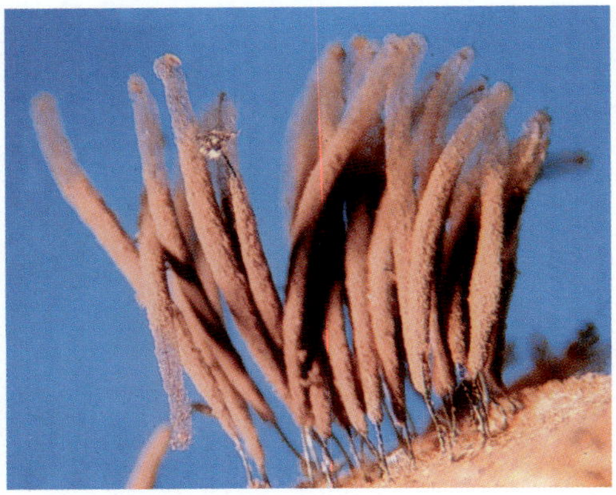

Sporangia of *Stemonitis axifera* (Stemonitales) on dead wood, x 10 (*MU*).

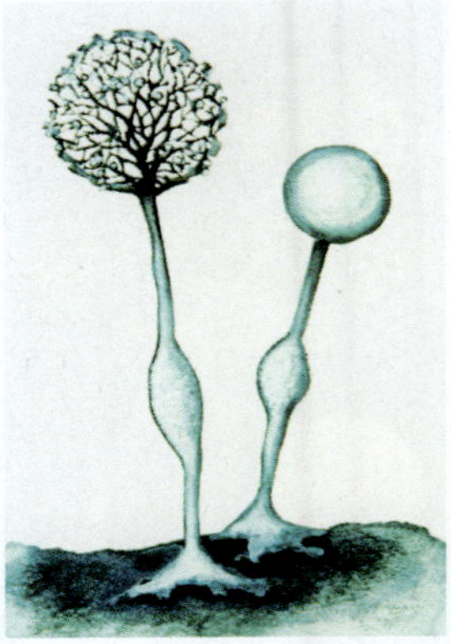

Sporangia of *Clastoderma debaryanum* (Echinosteliales), x 40 (*MU*).

Sporangia of *Trichia decipiens* (Trichiales) on dead wood, x 40 (*MU*).

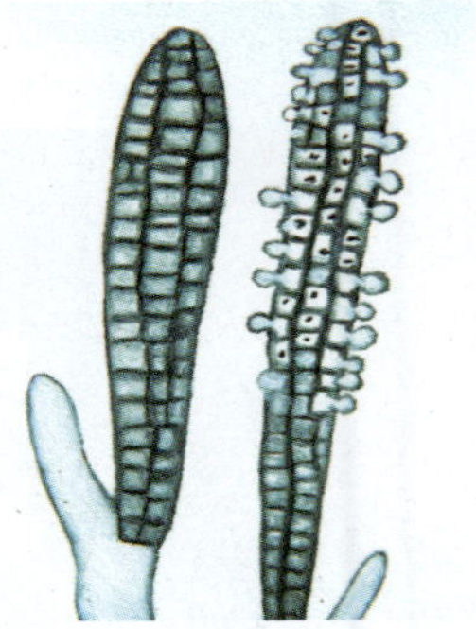

Sporangia of *Dictyuchus monosporus* (Saprolegniales), x 200 (*MU*).

Sporangia of *Physarum polycephalum* (Physarales), x 17 (*MU*).

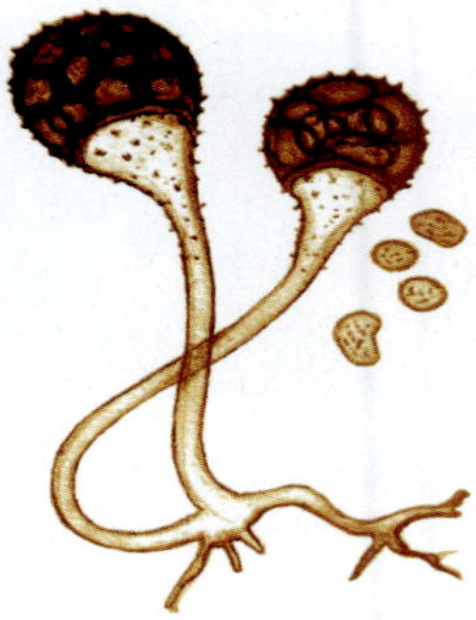

Sporangia of *Rhizopus microsporus* (= *Rh. oligosporus*), x 350 (*MU*).

Sporangia of *Stemonitopsis typhina* (=*Comatricha typhoides*, Stemonitales), x 55 (*MU*).

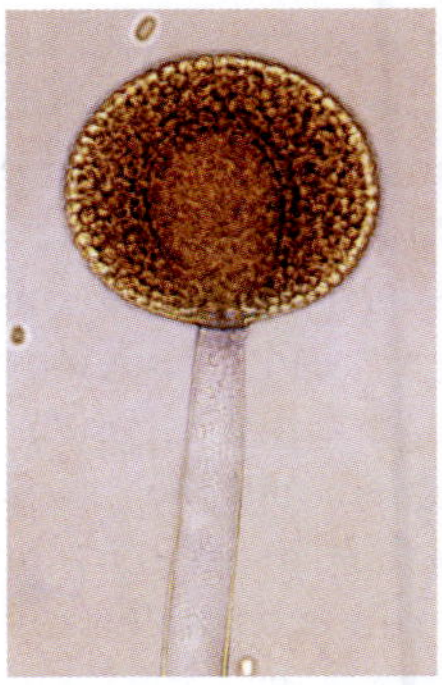

Sporangium of *Mucor hiemalis*, x 350 (*CB*).

Sporangia of *Pilobolus crystallinus* var. *kleinii* (=*P. kleinii*, Mucorales) growing on cow's dung, x 25 (*MU*).

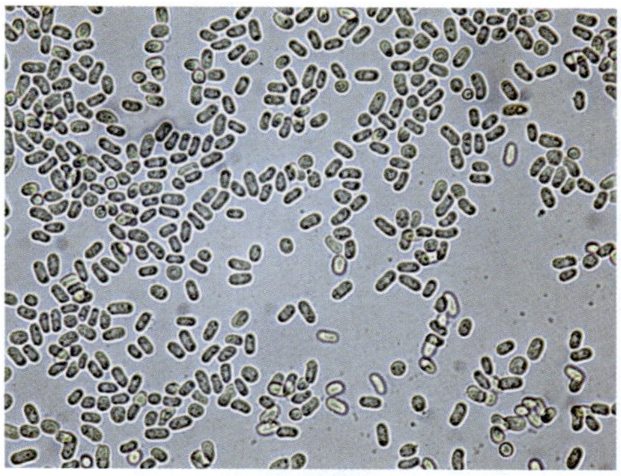

Spores of *Dictyostelium mucoroides* (Dictyosteliales), x 320 (*MU*).

-sporate, -sporic, -sporous, -spore (Gr. *sporá*, spore + L. suf. *-atus* > E. *-ate*, provided with or likeness; or + Gr. suf. *-íkos* > L. *-icus* > E. *-ic*, belonging to or relating to): suffixes employed in the formation of terms relating to spores, such as *unisporate* (having a single spore), *bisporic* (with two spores), *ochrosporous* (with yellow spores), *protospore* (the multinucleate mass of cytoplasm in developing sporangia of *Phycomyces* and other Mucorales that is eventually transformed into spores), etc.

spore (NL. *spora* < Gr. *sporá*, spore): a small unit of propagation, unicellular or multicellular, asexual or sexual, motile or immotile, that is capable of giving rise to a new individual. A spore functions like a seed, although it differs from the latter in that a spore does not contain a preformed embryo. Biologically, spores can be divided into various groups according to whether they are disseminated by air, water, insects and other animals, etc. Spores can survive during periods unfavorable for germination and growth, although a single type of spore can have various functions. The morphology and development of the spores serve as basic taxonomic criteria. In the description of spores the following characteristics are taken into consideration: *type of development* (ascospore, basidiospore, sporangiospore, etc.); *motile* or *immotile* (zoospore, aplanospore); if motile, number and type of flagella, or amoeboid; *individual* or *aggregated*; *shape* (although spores are tridimensional, the shape is frequently described as seen in optical section, as a bidimensional figure); *contents* (hyaline appearance, annular, etc., and the presence of oil droplets or air bubbles); *septation* (aseptate or with one or several septa); *cell wall* (dry or viscous, and its thickness); *ornamentation* (warts, spines, appendages, etc., and the reaction to dyes or reagents, such as cyanophilic, amyloid or nonamyloid, etc.); *color* (hyaline or dark); and size in µm.

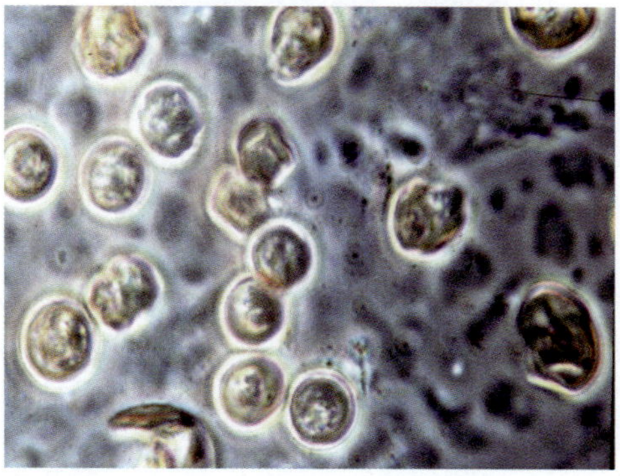

Spores of *Physarum polycephalum* (Physarales) beginning to germinate, x 1,350 (*MU*).

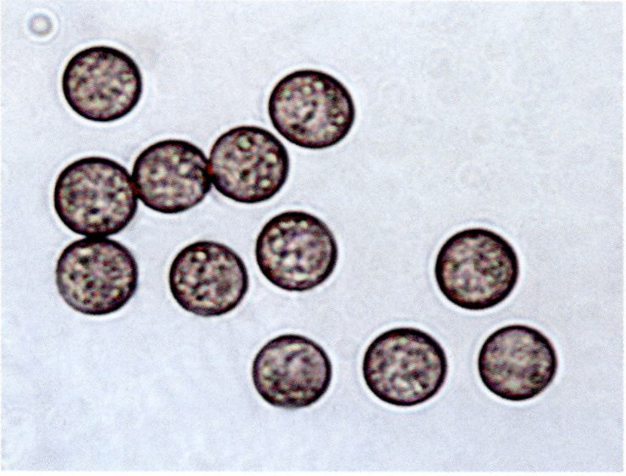

Spores of *Physarum polycephalum*, x 1,350 (*MU*).

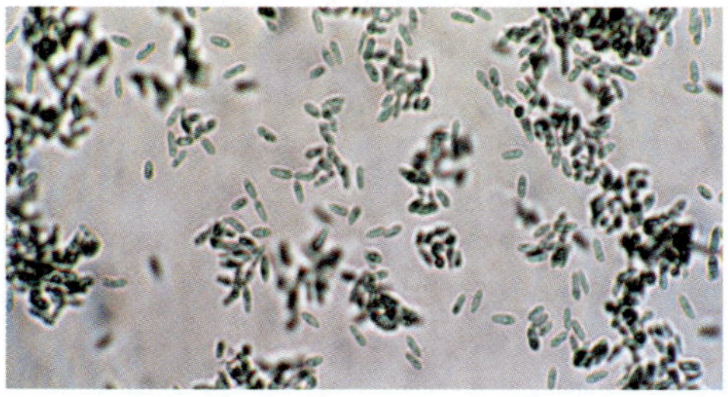

Spores of *Metarhizium anisopliae* (moniliaceous asexual fungi), x 500 (*MU*).

spore ball (Gr. *sporá*, spore; E. *ball* < ME. *bal*, *balle*, ball): a compact, round group of spores, or of spores and sterile cells, that function as a unit of dispersal; e.g., *Ascosphaera apis* (Ascosphaerales) and *Sorosporium* and *Tolyposporium* (Ustilaginales) form balls composed of spores only, whereas *Urocystis* (Ustilaginales) forms an aggregation of spores and sterile cells.

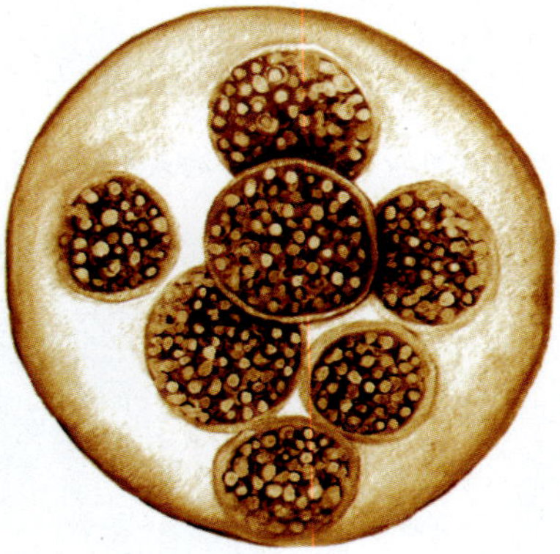

Sporocyst with **spore balls** of *Ascosphaera apis*, x 1,200 (*MU*).

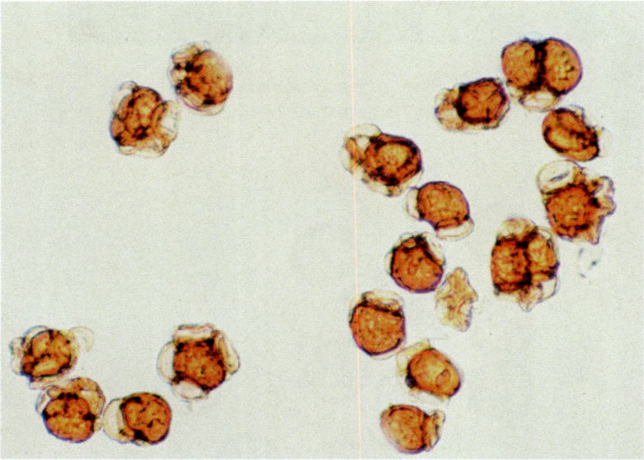

Spore balls with sterile cells of *Urocystis* sp., x 1,000 (*MU*).

spore case (Gr. *sporá*, spore; ME. *cas* < OF. *chasse* < L. *capsa*, case, receptacle): generally applied to the receptacle of the spores of some Gasteromycetes, such as the Lycoperdales, Sclerodermatales and Nidulariales. In the Lycoperdales, one can cite, e.g., a *Geastrum*, whose *sporiferous sack* (L. *saccus*, sac) corresponds to the endoperidium, which develops an ostiole and a peristome or characteristic disk. In *Broomeia* a large fleshy stroma is formed, which bears numerous sporiferous sacs that correspond to the endoperidium, each of which has an ostiole delimited by a peristome. In *Sphaerobolus* (Nidulariales) there is a single sporiferous sac, which in this case is called the peridiole (which is delimited by its own peridium), that is exposed when the mature exoperidium splits into coronate segments or lobes. The violent ejection of the peridiole leaves the evaginated endoperidium as a white sphere on the split exoperidium. The protomycetaceous ascomycetes (e.g., *Protomyces* and *Taphridium*), which lack an ascocarp, form intracellular chlamydospores in the epidermis of the plants that they parasitize; on germinating, the chlamydospore produces a columnar sporiferous sac, which is considered as a compound ascus, like a sporangium, or as a previously called synascus (now called *sporiferous sac*).

Spore case (sporiferous sac) of the fruiting body of *Sphaerobolus stellatus*, x 25 (*MU*).

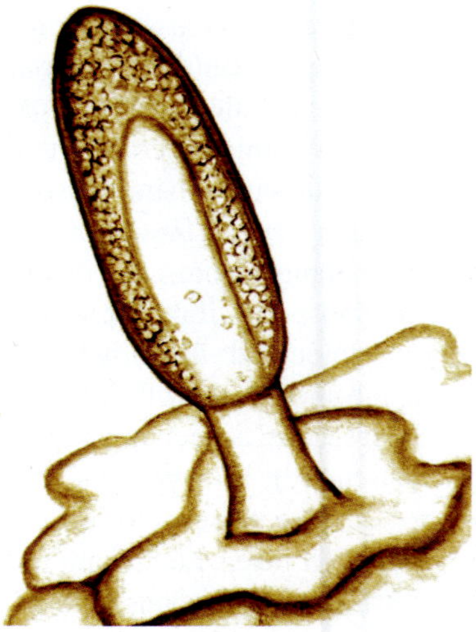

Spore case (sporiferous sac) of *Mixia osmundae* (Mixiales, basidiomycetes), with a central column and spores, on a pedicellar cell that emerges from a chlamydospore in a host (*Osmunda cinnamomea*) epidermal cell, x 700 (*MU*).

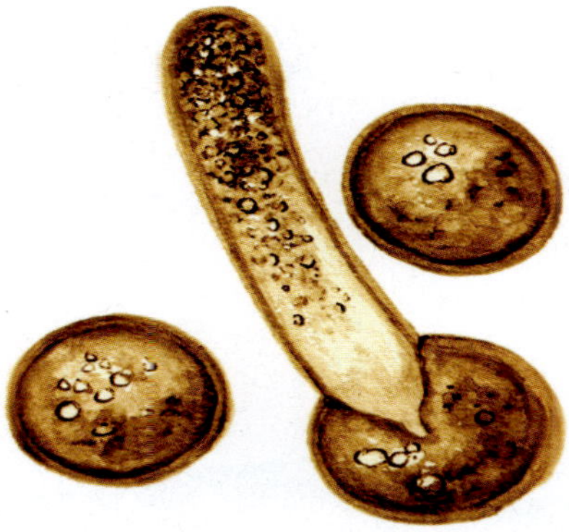

Spore case (sporiferous sac) of *Protomyces pachydermus*, produced from a germinated chlamydospore, on the epidermis of *Taraxacum platycarpum* (a composite), x 500 (*MU*).

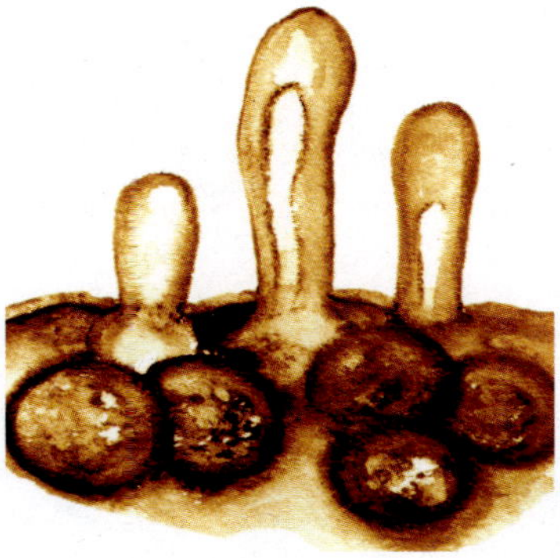

Chlamydospores of *Taphridium* sp., subepidermal in tissues of an umbellifer, some of them germinating to give rise to columnar **spore cases** (sporiferous sacs), x 450 (*MU*).

spore print, **spore deposit** (Gr. *sporá*, spore; ME. *prente, printe* < OF. *priente*, impression, print; L. *depositus*, laid down): the mass of spores of a fungus discharged from a sporophore onto a surface. In the agarics, the spore print is obtained by placing a mature pileus on white or black paper, with the gills downward; at the end of several hours, generally at night, the spore print is deposited, and its color characteristics can be determined. This is of taxonomic importance; e.g., in *Agaricus* (Agaricales) the spore print is dark brown in color.

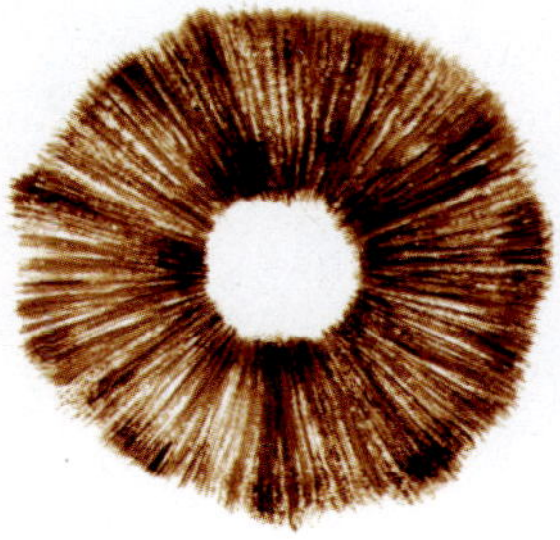

Spore print of a basidiocarp of *Agaricus* sp., x 1 (*MU*).

spore horn (NL. *spora* < Gr. *sporá*, spore; ME. < OE. *horn* < L. *cornu* < Gr. *kéras*, horn, of a horny nature; something resembling or suggestive of a horn): in the basidiomycetes that cause the cedar-hawthorn rust, the mature teliospores germinate immediately, their pedicels gelatinize, extruding the teliospores as spore horns (tendril-like spore mass), e.g., *Gymnosporangium juniperi-virginianae*, *G. clavipes*, and *G. globosum* (Uredinales).

sporidiole, **sporidiolum**, pl. **sporidiola** (Gr. *sporídion*, dim. of *sporá*, spore + L. dim. suf. *-olum* > E. *-ole*): a small spore. Each of the spore-like structures produced inside *sporidiomata*, which are perithecium-like bodies, present in *Kathistes* (an ascomycetous gen. of uncertain affinity).

sporidioma, pl. **sporidiomata** (Gr. *sporídion*, dim. of *sporá*, spore + suf. *-oma*, which implies the idea of entirety; everything): a unique spore-producing body associated with the submerged ascocarp base that is characteristic of *Kathistes* (an ascomycetous gen. of uncertain affinity), a fungus that has been found only in damp chamber cultures of dung. At maturity, masses of ascospores (*sporidiola*) accumulate in long, incurved extensions of the wall cells of the perithecial neck.

sporidium, pl. **sporidia** (Gr. *sporídion*, dim. of *sporá*, spore): the name applied to the spore produced on the promycelium or basidium of the Uredinales and Ustilaginales, i.e., a basidiospore that is borne on an epibasidium. In the gen. *Tilletia* (Ustilaginales), e.g., the epibasidium forms eight primary sporidia which fuse in pairs (forming four H-shaped structures), and from these the secondary sporidia (secondary basidiospores) originate.

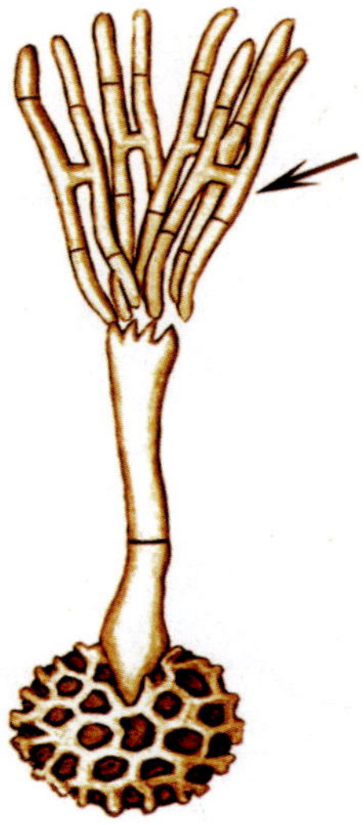

Sporidia of *Tilletia caries*, x 610 (*MU*).

sporobolomycetaceous (Gr. *spóros*, spore, seed + *bállo* > L. *bolus*, to throw, eject + *mýkes*, genit. *mýketos*, fungus + L. suf. *-aceus* > E. *-aceous*, of or pertaining to, with the nature of): term applied to anamorphic yeasts with basidiomycetous affinities characteristically producing ballistospores following multiplication by budding. Sporobolomycetales is an unnatural group, called the mirror or shadow yeasts (*Bullera*, *Itersonilia*, *Sporobolomyces* and *Tilletiopsis*). See **Blastomycetes**.

sporocarp (Gr. *sporá*, spore + *karpós*, fruit): a fruiting body that bears spores (sporophore). Among the structures called sporocarps are the sporophores of the Protosteliales, such as *Nematostelium*, *Cavostelium*, *Schizoplasmodium* and *Protostelium*, as well as the special structures of the endomycorrhizal Zygomycetes (e.g., *Endogone*, of the Endogonales, and *Sclerocystis*, of the Glomales), which consist of a group of zygospores enveloped by a common membrane.

Sporocarps of *Protostelium mycophaga*, x 200 (*MU*).

Sporocarps (from left to right) of *Nematostelium ovatum*, *Schizoplasmodium cavostelioides* and *Protostelium mycophaga*, x 110 (*MU*).

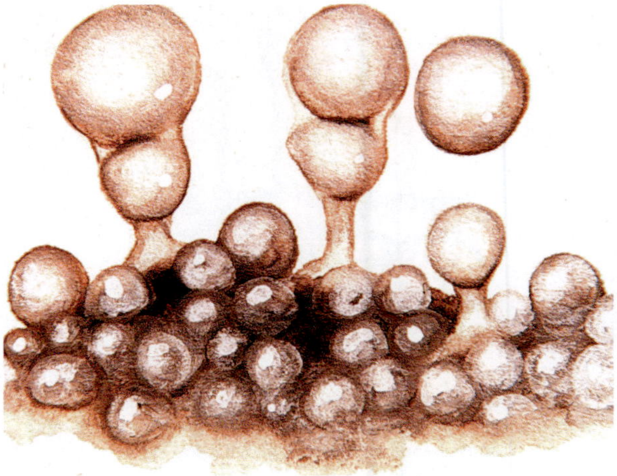

Sporocarps of *Cavostelium apophysatum*, x 1,500 (*MU*).

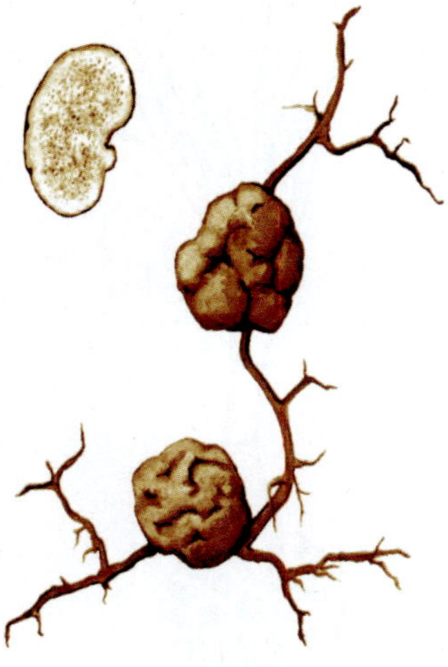

Sporocarps of *Endogone* sp., associated with roots of *Pseudotsuga*, x 1 (*MU*).

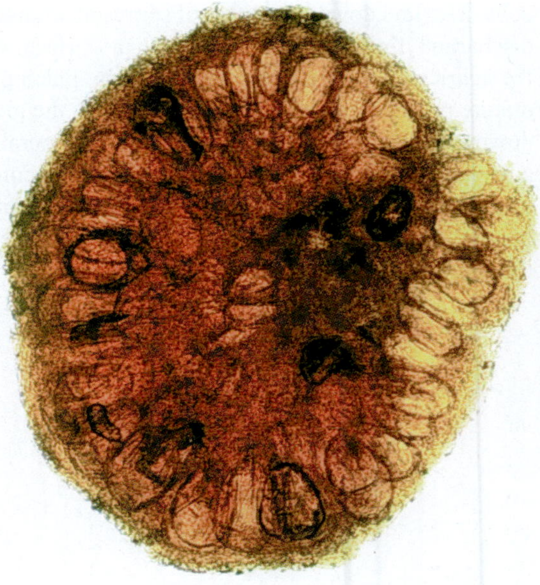

Sporocarp of *Sclerocystis* sp., x 80 (*CB*).

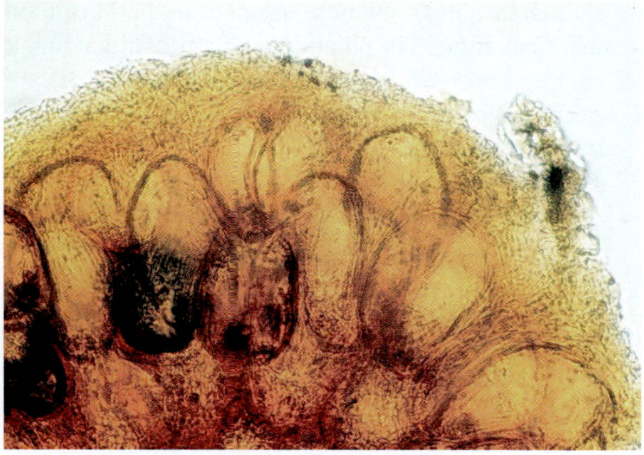

Sporocarp of *Sclerocystis* sp.
The spores in the interior are radially arranged, x 200 (*CB*).

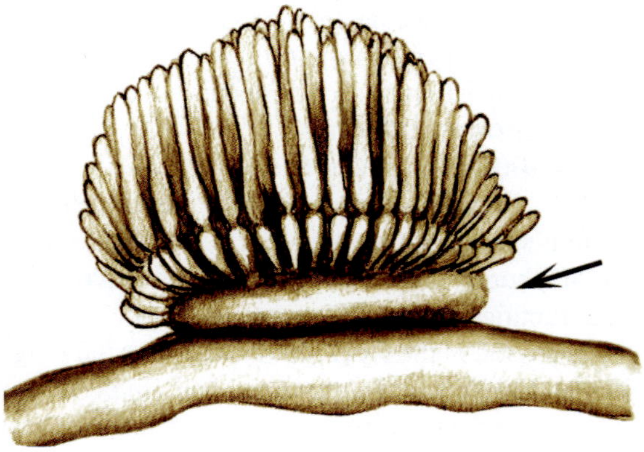

Sporocladium bearing pseudophialides with merosporangia of *Linderina pennispora*, x 850 (*MU*).

sporocladium, pl. **sporocladia** (Gr. *sporá*, spore + *kládos*, small branch): a special type of hyphal branch, septate or aseptate, which produces unispored merosporangia in the Zygomycetes of the order Kickxellales, such as *Linderina*, *Coemansia* and *Spirodactylon*, among other gen.

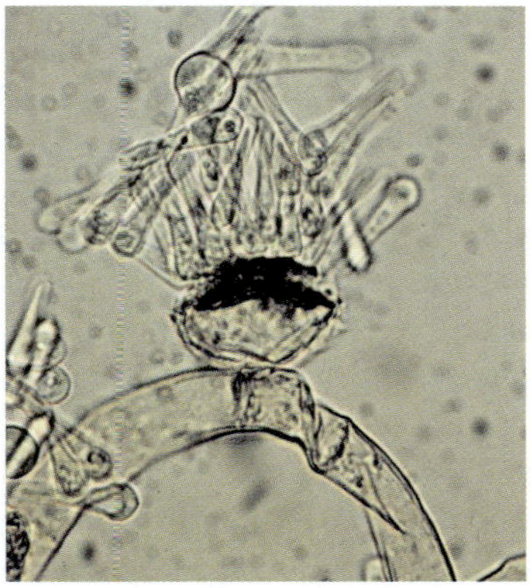

Sporocladium bearing pseudophialides with merosporangia of *Linderina pennispora*, x 500 (*MU*).

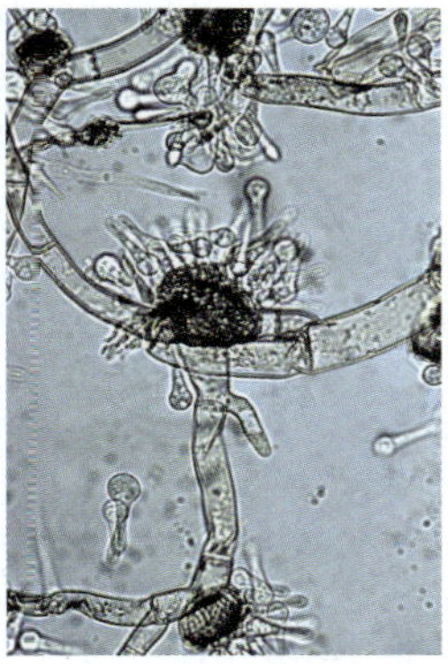

Sporocladia bearing pseudophialides with merosporangia of *Linderina pennispora*, x 400 (*MU*).

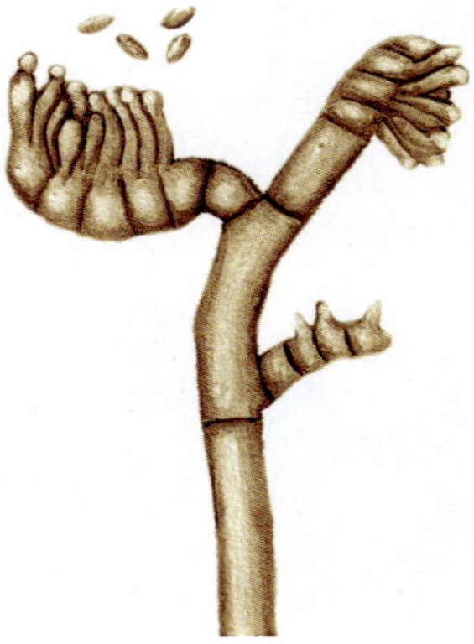

Sporocladia bearing pseudophialides with merosporangia of *Martensella corticii*, x 400 (*MU*).

sporocyst (Gr. *sporá*, spore + *kýstis*, vesicle, cell): **1**. A cyst in which asexual spores are produced. **2**. The structure that develops from the nutriocyst in the plectomycete *Ascosphaera apis* (Ascosphaerales), and which contains spore balls composed of ascospores of sexual origin. See **nutriocyte**.

Sporocyst containing balls of ascospores of *Ascosphaera apis*, x 400 (*RTH*).

sporodesm (Gr. *sporá*, spore + *desmós*, ties, filament): a small chain of independent spores arranged in a row. A multiple, multicellular spore, more or less elongated or forming a ball, whose components are *merispores*. See **teleutosporodesm**.

sporodochium, pl. **sporodochia** (Gr. *sporá*, spore + *docheíon*, recipient): a small, compact, cushion-shaped stroma on which are formed short conidiophores that produce the spores (conidia). This fructification, which is an asexual sporophore, is characteristic of the tuberculariaceous asexual fungi, such as *Epicoccum*, *Tubercularia* and *Volutella*, and of some melanconiaceous asexual fungi, such as *Pestalotia*.

Sporodochia of *Tubercularia vulgaris* on a plum tree branch, x 3 (*MU*).

Sporodochium of *Volutella* sp., x 300 (*RTH*).

sporogenesis, pl. **sporogeneses** (Gr. *sporá*, spore + *génesis*, engenderment): the process by which spores are produced, whether asexual or sexual. See **conidiogenesis**.

sporogenous (Gr. *sporá*, spore + *génos*, origin < *gennáo*, to engender, produce): a structure that produces spores or is capable of producing them. In the fungi, a sporogenous cell is the cell or hyphal branch of the sporangiophore of, e.g., *Dimargaris cristalligena* (Dimargaritales), which produces the bispored merosporangia. The phialides of many conidial fungi, such as *Aspergillus* (moniliaceous asexual fungi), are examples of sporogenous (conidiogenous) cells. Cf. **asporogenous**.

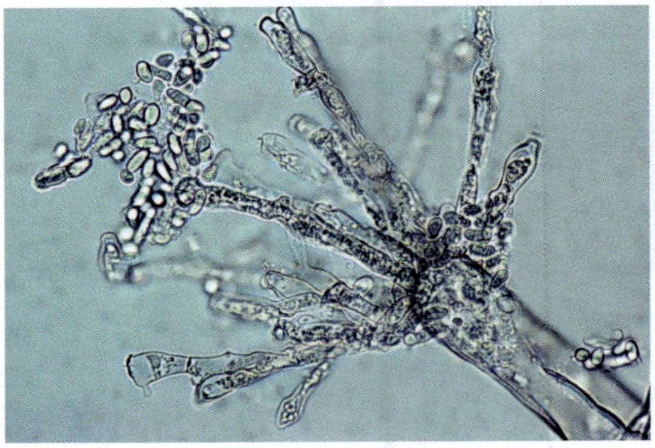

Sporogenous cells of *Dimargaris cristalligena*, x 660 (*PL* & *EAA*).

sporont (Gr. *sporá*, spore + *ón*, genit. *óntos*, a being): a thallus on which any type of spore is produced.

sporophagous (Gr. *sporá*, spore + *phágos*, one that eats): feeding on spores. E.g., certain thrips (insects of the order Thysanoptera) feed on spores of various species of moniliaceous and dematiaceous asexual fungi, and some beetles (Coleoptera) feed on myxomycete spores. Cf. **plasmodiophagous**.

sporophore (Gr. *sporá*, spore + suf. *-phóros*, bearer): a term used to refer to any structure that produces or bears spores, whether asexual or sexual, but especially applied to a conidiophore. In the macromycetes it refers to an ascoma or basidioma.

sporophyte (Gr. *sporá*, spore + *phytón*, plant): in plants with alternation of generations, it represents the diploid generation (diplophase), which produces asexual spores. Cf. **gametophyte**.

sporoplasm (Gr. *sporá*, spore + *plásma*, formation, bland material with which a living being is formed): the portion of the protoplasm in sporangia and asci that is destined for the production of spores. Cf. **epiplasm**.

sporostasis, pl. **sporostases** (Gr. *sporá*, spore + *stásis*, fixity, immobility): the inhibition of spore germination.

sporostatic (Gr. *sporá*, spore + *stásis*, fixity, immobility + suf. *-íkos* > L. *-icus* > E. *-ic*, relating to or belonging to): having the capacity to inhibit spore germination (*sporostasis*).

sporothallus, pl. **sporothalli** (Gr. *sporá*, spore + *thallós*, sprout, thallus): a thallus that produces sporangia with asexual spores; present in aquatic fungi with an alternation of generations in their life cycle, as in *Blastocladia pringsheimii* and *Allomyces javanicus* (Blastocladiales). Cf. **gametothallus**.

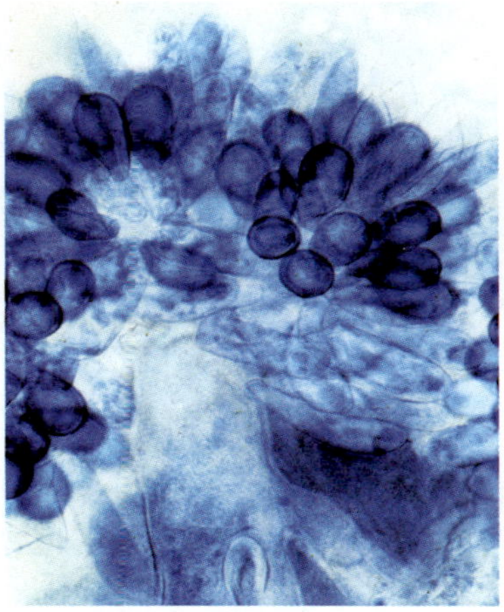

Sporothallus of *Blastocladia pringsheimii* bearing hyaline, thin-walled zoosporangia (mitosporangia), x 120 (*MU*).

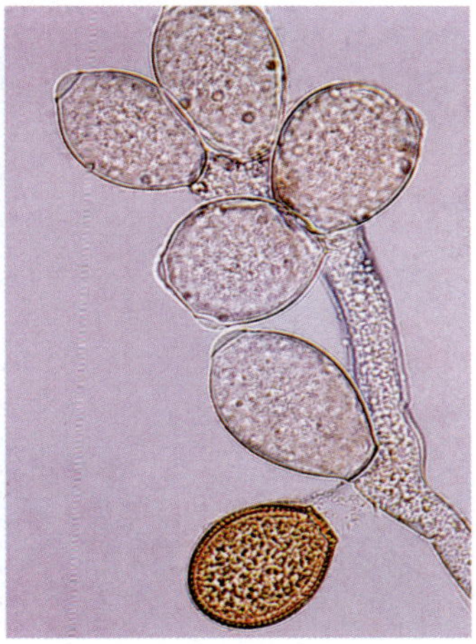

Sporothallus of *Allomyces javanicus* bearing hyaline, thin-walled zoosporangia (mitosporangia) and dark, thick-walled resting zoosporangia (meiosporangia), x 250 (*MU*).

sporotrichosis, pl. **sporotrichoses** (< gen. *Sporothrix* < Gr. *spóros*, spore + *thríx*, genit. *trichós*, hair + suf. *-osis*, state or condition): a subcutaneous, more or less localized, chronic mycosis of man and higher animals acquired by the inoculation of propagules of the fungal pathogen through injuries caused by plant spines and splinters, or from propagules in the soil. The most common types of sporotrichosis are the lymphatic and the lymphocutaneous, but there are other clinical types, including the systemic or disseminated. The causal agent is the dimorphic fungus *Sporothrix schenckii* (moniliaceous asexual fungi). See **mycosis**.

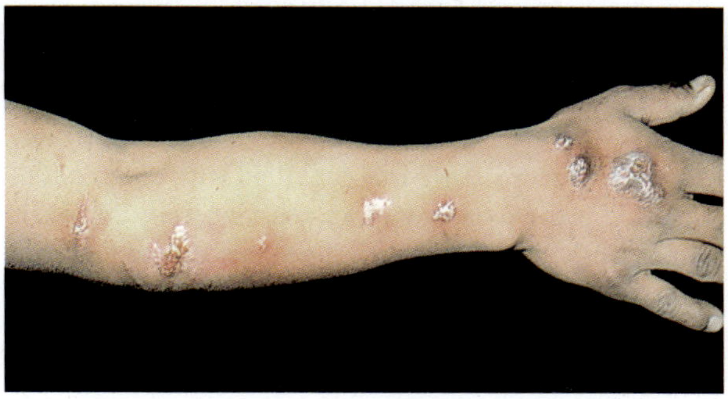

Gangliar **sporotrichosis** on a woman's arm, caused by *Sporothrix schenckii*, x 0.3 (*RLM*).

Convoluted colony of *Sporothrix schenckii* on agar, the causal agent of **sporotrichosis**, x 0.3 (*MU*).

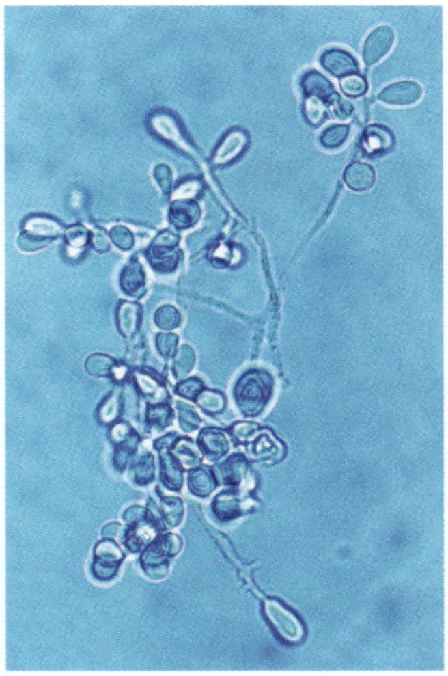

Conidia of *Sporothrix schenckii* (**sporotrichosis**), x 1,000 (*MU*).

-sporous (Gr. *sporá*, spore): a suf. often employed to form various adjectives, such as *macrosporous*, *leucosporous*, etc., with large spores, with white spores, etc.

sporulate (NL. *sporula* < dim. of *spora* < Gr. *sporá*, seed, spore + L. suf. *-ate*, provided with or likeness): to produce spores or sporules.

sporulated (NL. *sporula* < dim. of *spora* < Gr. *sporá*, seed, spore + L. suf. *-ate*, provided with or likeness + E. *-ed*, past tense): having spores. Generally applied to a culture that has reached the stage of sporulation.

sporulating (NL. *sporula* < dim. of *spora* < Gr. *sporá*, seed, spore + L. suf. *-ate*, provided with or likeness + suf. *-ing*, action or process): actively producing spores or being capable of doing so; applied to a cell, a mycelium, colony, etc.

sporulation (NL. *sporula* < dim. of *spora* < Gr. *sporá*, seed, spore + L. *-ation* > E. *-tion*, the action and effect, condition, result): the action and effect of sporulating.

sporule (NL. *sporula*, dim. of *spora* < Gr. *sporá*, spore): a small spore.

spp. (L. *species*, species): pl. of *sp.*

squamule (L. *squamula*, small scale < *squama*, scale + dim. suf. *-ula*): a small shingle-like scale often present, e.g., on the peridium of *Endoptychum* (Hymenogastrales).

squamulose (L. *squamulosus*, provided with small scales < *squamula*, dim. of *squama*, scale + suf. *-osus* > E. *-ose*, full of, augmented, prone to): *Lichens*. A thallus composed of numerous small scales or lobes. It can be considered intermediate in habit between the crustose and foliose lichen thalli; it is present, e.g., in *Psora crenata*, *P. russellii*, *Cladonia didyma* and C. *pyxidata* (Lecanorales). Equivalent to **lepidote**. Cf. **alepidote**.

Squamulose podetium of the thallus of *Cladonia didyma*, x 5 (*MU*).

Squamulose thalli of *Psora russellii*, x 10 (*CB*).

squarrose (L. *squarrosus*, provided with thick scales or crusts; rough, scaly, crusty < *squarra*, thick scale + suf. *-osus* > E. *-ose*, full of, augmented, prone to): applied to a surface covered with thick, curved scales, as is seen in the basidiocarp of *Pholiota squarrosa* and *Ph. acutesquamosa* (Agaricales).

Squarrose basidiocarp of *Pholiota squarrosa*, x 0.7 (*RV*).

Squarrose basidiocarps of *Pholiota squarrosa*, x 0.7 (*MU*).

Squarrose basidiocarp of *Pholiota squarrosa*, x 0.7 (*MU*).

Stachel (G. *Stachel*, spine, thorn, sting): *Plasmodiophorales*. A cylindrical, sharp-pointed structure (aculeate or bullet-shaped), that is formed in a pouch or sheath (the *Rohr*) in the encysted zoospores of *Plasmodiophora brassicae*. During the infection process, the zoospores encyst on the outer surface of the epidermal cells of a cabbage root (the host), and the Rohr or sheath evaginates and forms a bulbous structure (an appressorium) that adheres to the host cell wall. The Stachel perforates the host cell wall and the protoplast of the encysted zoospore penetrates the host cell, infecting it, to continue the life cycle of the fungus-like microorganism.

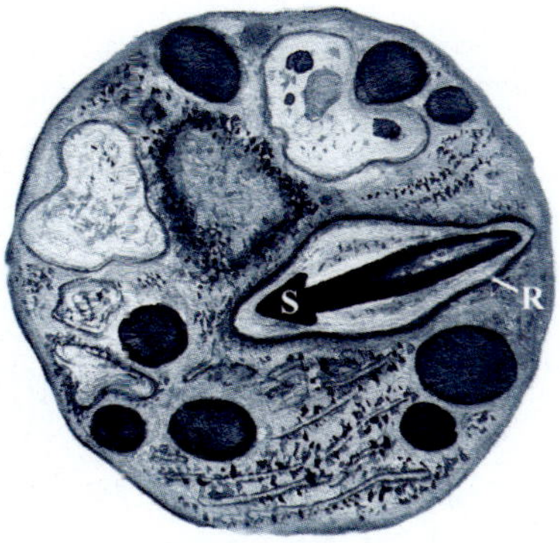

Stachel (S) contained in the sheath or Rohr (R) of an encysted zoospore of *Plasmodiophora brassicae*, seen in longitudinal section. Drawing based on a transmission electron micrograph published by Beckett *et al.* in their *Atlas of Fungal Ultrastructure*, 1974, x 24,000 (*MU*).

stachybotryotoxin (< *Stachybotrys*, dematiaceous asexual fungi, < Gr. *stáchys*, an ear of grain, spike + *bótrys*, a cluster, bunch of grapes; LL. *toxicus* < L. *toxicum*, venom, toxin): the stachybotryotoxin is produced by *Stachybotrys chartarum* (=*S. atra*) in grasses; it is the first mycotoxin studied, and affects horses and humans.

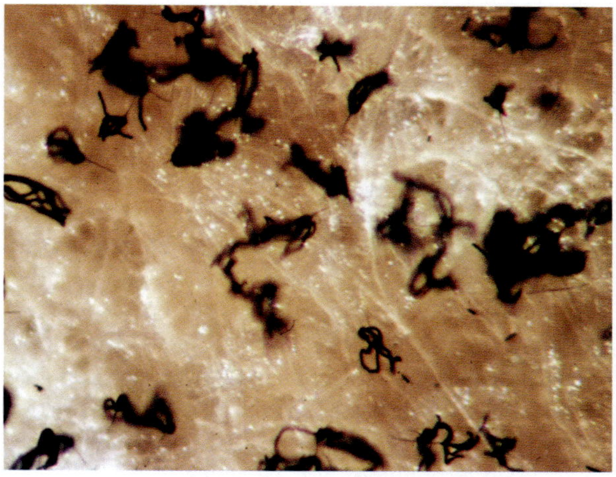

Conidiophores of *Stachybotrys chartarum* (=*S. atra*) on agar, the producer of **stachybotryotoxin**, x 10 (*MU*).

stalagmoid (Gr. *stalágmos*, drop, drip + L. suf. *-oide*, like < Gr. *-oeídes*, similar to): syn. of **dacryoid** and **lacrimoid**; an obconic spore with the elongate shape of a drop or teardrop (stalagmospore), like the basidiospores of *Clitocybe gibba* (Agaricales).

staphylum, pl. **staphyla** (Gr. *stáphyle*, raceme): see **bromatium** and **gongylus**.

starter, **starter culture** (OE. *styrtan* > ME. *sterten*, to move suddenly > E. *start*, to begin + E. suf. *-er*, one that performs a specific action; L. *cultus*, culture, action and effect of cultivating): a pure culture consisting of a single microorganism or a mixture of microorganisms that is utilized to inoculate a substrate to initiate a fermentation process. Pure starter cultures are used, e.g., in the elaboration of beer, bread, cheese, etc. Starter cultures also are used to initiate the fermentation process of many traditional foods, e.g., tempe, soy sauce and kefir. See **fermentation**.

stage (L. *stadium* < Gr. *stádion*, a measurement of 125 paces = 625 Roman feet or 606 E. feet; also, course, career): in modern biology it is employed as a syn. of period, as in a life cycle; equivalent to *state*, *phase*, *grade*, etc. See **state**.

state (L. *status*, state, condition, position, rank): in biology, each of the successive modes of being of an organism during its life cycle. Syn of **stage**.

states of fungi, **stages of fungi** (L. *status*, state, condition, position, rank; in biology, each of the successive modes of being of an organism during its life cycle; L. *fungus*, fungus, pl. *fungi*): in the pleomorphic fungi, i.e., those with various types of spores, these can coincide or not with regard to their time of production, and they may or may not be produced following nuclear fusion and meiosis (sexual state). The state characterized by sexual spores (zygospores, ascospores, basidiospores, etc.) traditionally has been designated as the *perfect state* (L. *perfectus*, having the highest possible grade of excellence in its line); that characterized by asexual spores (conidia) or lacking them, has been designated as the *imperfect state* (L. *imperfectus*, not perfect). In accordance with the rules of the International Code of Botanical Nomenclature, it is permissible to designate both the perfect and imperfect states with separate Latinized specific names (binomial), but when it has been established that a *"perfect"* and *"imperfect"* species represent a single fungus, the binomial applied to the perfect state also includes the imperfect state and takes priority. With the goal of developing a more accurate terminology for the states of pleomorphic fungi, new terms have been proposed: the *holomorph* (Gr. *hólos*, all + *morphé*, form) applies to a fungus in all its forms and phases; the *teleomorph* (Gr. *téleos*, perfect, complete + *morphé*, form) represents the form of the sexual state, and the *anamorph* (Gr. *anamorphóo*, to transform) that of the asexual state. The term *synanamorph* (L. *syn-* < Gr. *sým-*, together + *anamorphóo*, to transform) is applied when two or more anamorphs have the same teleomorph. The holomorph includes a teleomorph and frequently one or more anamorphs. The term *ana-holomorph* has been used to designate an imperfect (anamorphic) fungus that appears to lack a perfect (teleomorphic) state. A name (binomial) applied to both states of a fungus, perfect and imperfect, is a *nomen holomorphosis* (Gr. *ónoma*, *ónyma* > L. *nomen*, name + Gr. *hólos* + *morphé* + *-osis*, condition or state); the name of the imperfect state is a *nomen anamorphosis* (Gr. *ónoma*, *ónyma* > L. *nomen* + Gr. *anamorphóo*). In addition, the terms *anamorph-genus* (L. *genus*, race, class; Gr. *anamorphóo*) and *anamorph-species* (L. *species*, species, class; Gr. *anamorphóo*) have been suggested to replace the terms *form-genus* and *form-species*.

statismospore (Gr. *státis*, fixity, immobility, or *statós*, firm, fixed + suf. *-ismós*, state or condition + *sporá*, spore): a spore that is not forcibly discharged from the sterigma of the basidium; the basidiospores of the Gasteromycetes (such as *Podaxis*, of the Hymenogastrales) are of this type, and are distinguished from the basidiospores of the Hymenomycetes which are forcibly discharged from the basidia. Cf. **ballistospore**.

statismosporic (Gr. *státis*, fixity, immobility, or *statós*, firm, fixed + suf. *-ismós*, state or condition + *sporá*, spore + suf. *-íkos* > L. *-icus* > E. *-ic*, belonging to or relating to): having spores that cannot be discharged forcibly from the basidia (*statismospores*). The Gasteromycetes, by definition, are statismosporic.

Statismospores of *Podaxis* sp., x 1,500 (*MU*).

stauro-, **staur-** (Gr. *staurós*, cross): a pref. that alludes to a structure in the shape of a cross, as in *staurospore*.

staurospore (Gr. *staurós*, cross + *sporá*, spore): a stellate spore with three or more points. A Saccardoan term that is mainly applied to conidial fungi, like those of the gen. *Orbimyces* (dematiaceous asexual fungi), and *Riessia* and *Spegazzinia* (moniliaceous asexual fungi).

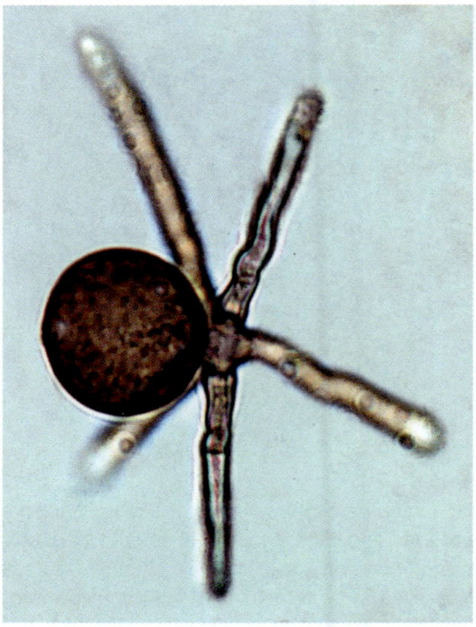

Staurospore of *Orbimyces spectabilis*, x 1,200 (*MU*).

steliogen (Gr. *stéle*, column + *géno*, origin < *gennáo*, to engender, produce): the finely granular portion of the cytoplasm of the prespore cell of the Protosteliales, which secretes the cellulosic tube of the sporocarp stalk during sporogenesis; it is present, e.g., in *Nematostelium ovatum*.

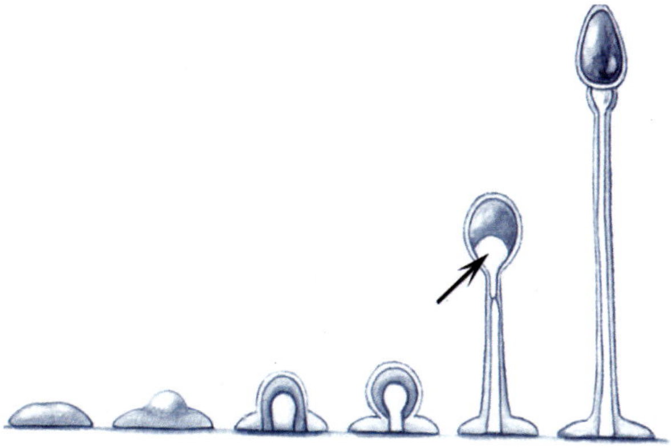

Steliogen of the sporocarp of *Nematostelium ovatum*, x 250 (*MU*).

stellate, stelliform (L. *stella*, star + suf. *-atus* > E. *-ate*, provided with or likeness; or + suf. *-formis* < *forma*, shape): star-shaped, like the sporangia of *Diderma radiatum* (Physarales), the fruiting body of *Geastrum saccatum* (Lycoperdales) and the ascospores of *Emericella variecolor* (Eurotiales).

Stellate peridial dehiscence of the sporangia of *Diderma radiatum*, on dead wood, x 10 (*MU*).

Stellate peridial dehiscence of *Diderma floriforme* (Physarales), on an oak leaf, x 5 (*MU*).

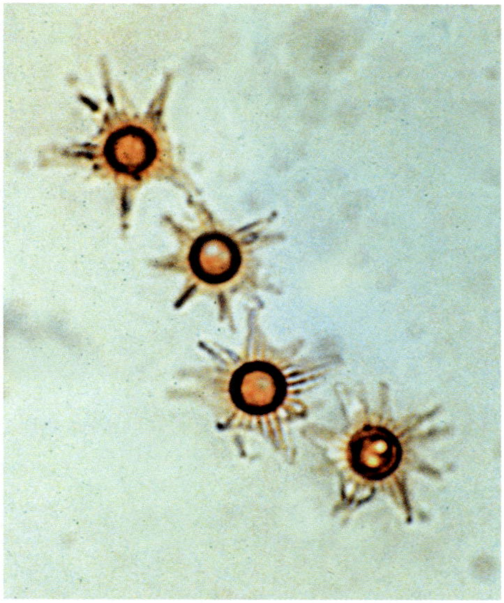

Stellate ascospores of *Emericella variecolor* (Eurotiales), x 2,000 (*MU*).

Stellate basidiocarps of *Geastrum saccatum*, x 0.5 (*MRO*).

stemono-, stemon- (Gr. *stémon*, genit. *stémonos*, filament, yarn): a pref. employed in names such as *Stemonitis* (Stemonitales), a Myxomycete with long, slender sporangia, with the suf. Gr. *-ítes* > E. *-itis*, excess of the qualities of.

steno-, sten- (Gr. *stenós*, narrow, limited, strict): a pref. that alludes to the narrowness of certain structures, or to the narrow range between the limits of certain factors that affect the development of some organisms, such as the concentration of salts, of temperature, etc.; e.g., a *stenohaline* organism is one that can only live within very narrow limits of the haloid salt concentration in the medium. Cf. **euri-**.

Stemon-, a pref. used to make the name *Stemonitis* (*S. axifera*, Stemonitales), x 3 (*RV*).

Stemon-, a pref. used to make the name *Stemonitis* (*S. axifera*, Stemonitales), x 5 (*MU*).

stenochora (Gr. *stenós*, narrow, restricted, limited + *chóra*, country): syn. of **endemic**; a species that has a more or less restricted geographic distribution. Cf. **eurichora**.

stenoecious (Gr. *stenós*, narrow, restricted, limited + *oîkos*, house or inn): an organism that is limited to certain environmental conditions, like the aquatic fungi of the gen. *Aqualinderella* (Rhipidiales), which only live in stagnant water with little or no oxygen. Cf. **euryoecious**.

stenospore (Gr. *stenós*, narrow, restricted, limited + *sporá*, spore): a spore with the shape of a bullet or projectile, like the basidiospores of *Lepiota* (Agaricales).

Stenospores of *Lepiota* sp., x 2,000 (*MU*).

stenoxenous (Gr. *stenós*, narrow, restricted, limited + *xénos*, stranger; here, in the sense of host): applied to parasitic organisms, more or less restricted or obligated, that only attack certain hosts, which may involve a few species, a single species, or only varieties or forms susceptible to invasion by the parasite. Among the fungi, e.g., the cereal rusts are stenoxenous. Cf. **eurixenous**.

stephano-, **stephan-** (Gr. *stéphanos*, crown): a pref. employed in the formation of terms that allude to the shape of a crown, as in *stephanocyst*.

stephanocyst (Gr. *stéphanos*, crown + *kýstis*, vesicle, cell): a bicellular structure of certain basidiomycetes, such as *Hyphoderma praetermissum* (Aphyllophorales), with the basal cell goblet-shaped and the upper cell globose.

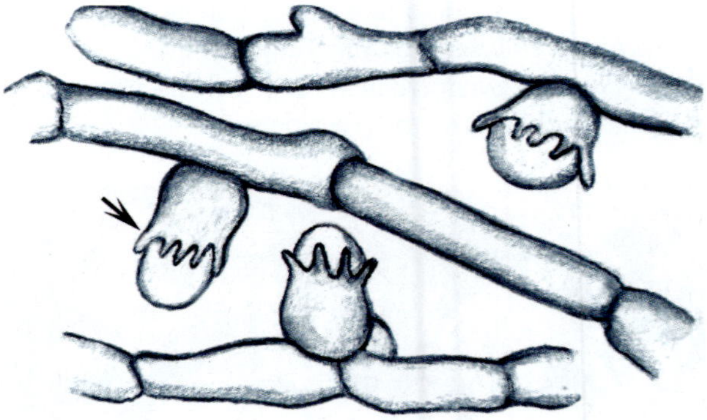

Stephanocysts of *Hyphoderma praetermissum*, x 1,000 (*MU*).

stercoreaceous (L. *stercus*, genit. *stercoris*, dung + L. suf. *-aceus* > E. *-aceous*, belonging to, with the nature of): a term applied to the fungi that grow on dung, e.g., as with *Pilobolus* and *Phycomyces* (Mucorales) and *Ascobolus* (Pezizales).

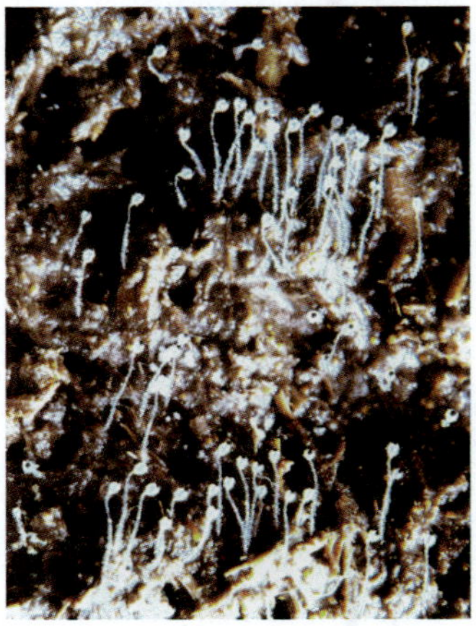

Stercoreaceous sporangiophores of *Pilobolus crystallinus* var. *kleinii* (=*P. kleinii*) growing on cow's dung, x 1 (*MU*).

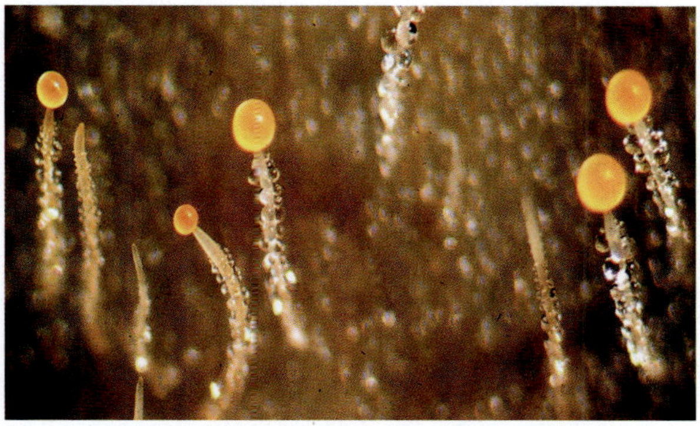

Stercoreaceous sporangiophores of *Phycomyces blakesleanus*, isolated on agar from mouse dung, x 20 (*MU*).

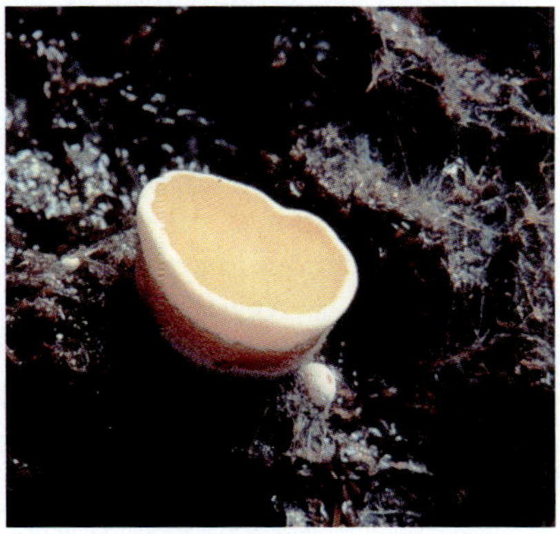

Stercoreaceous apothecium of *Coniochaeta scatigena* (=*Ascobolus scatigenus*) growing on cow's dung, x 20 (*EAA*).

stereo- (Gr. *stereós*, solid, firm, hard): a pref. employed in the formation of compounds that allude to the solidity or firmness of something, as in *stereonema*, a solid filament of the capillitium of certain Myxomycetes. Cf. **coelo-**.

stereoma, pl. **stereomata** (Gr. *steréoma*, base or fundament, skeleton): *Lichens*. Tissues and cells that provide mechanical support of the thallus, as occurs in the Lecanorales (e.g., *Cladonia* and *Alectoria*). See **scleroplectenchyma**.

stereonema, pl. **stereonemata** (Gr. *stereós*, solid + *nêma*, genit. *nêmatos*, filament): a compact filament, free or anastomosed, which together with others make up the capillitium in the sporangia of many Myxomycetes, such as those of the gen. *Calomyxa* (Trichiales). Cf. **coelonema**.

sterigma, pl. **sterigmata** (Gr. *stérigma*, support): **1**. *Basidiomycetes*. Each of the small diverticula (usually four) that are formed on the apex of each basidium, and which support the basidiospores. **2**. *Conidial fungi*. The small, tapered hyphal branch that bears a ballistospore (holoblastic conidium) of *Itersonilia* and *Sporobolomyces* (sporobolomycetaceous asexual yeasts). **3**. *Oomycota*. The small sporangium-bearing structure of the sporangiophores of *Basidiophora entospora* (Peronosporales). **4**. *Lichens*. Each of the elongated cells, simple or slightly branched, aseptate, situated in the interior of the spermogonia and which produce spermatia.

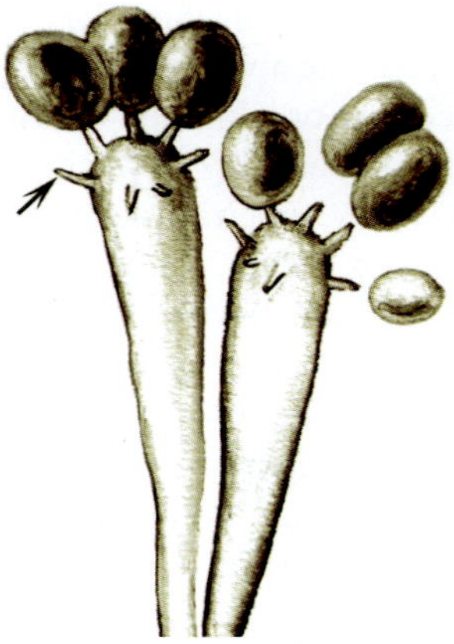

Sterigmata of the sporangiophores of *Basidiophora entospora*, x 115 (*MU*).

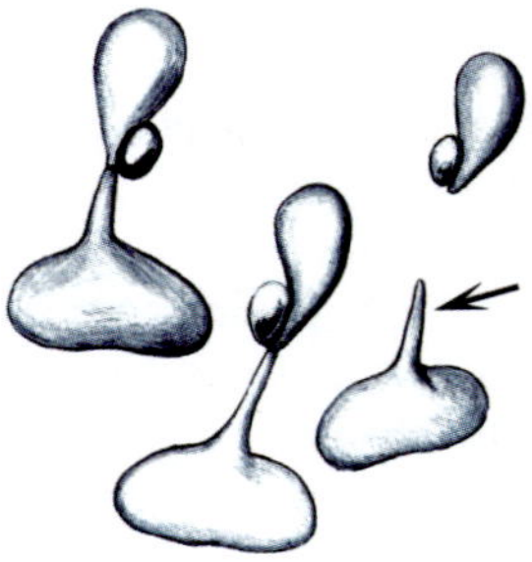

Sterigmata of the somatic cells of *Itersonilia perplexans*, x 2,100 (*MU*).

sterile (L. *sterilis*, sterile): incapable of producing descendants; unable to bear or produce fruits or spores; incapable of germinating; free of living organisms, especially microorganisms; sterilized.

sterilization (L. *sterilis*, sterile + suf. *-ationem* > E. *-ation*, action, state or condition, result): the action and effect of sterilizing. The elimination of living microorganisms can be done by means of chemical or physical methods. Heat is the physical agent most utilized. Dry heat is used for glass and other materials; one hour in an oven heated to 160ºC will kill the most resistant bacterial spores. The sterilization of microbiological needles and other instruments is done routinely by means of a flame. Moist heat, especially under increased pressure, is very effective for sterilizing most culture media; this is achieved in an autoclave during 15 minutes at 121ºC. When such high heat is not required, one can apply steam at atmospheric pressure for 20 minutes each third day for 3 days (discontinuous steaming). The application of steam also is used to achieve partial sterilization of soil (e.g., to disinfest the soil of plant parasitic nematodes); hot water (at 50ºC) is used to control some diseases, such as loose smut of wheat (caused by *Ustilago tritici*, of the Ustilaginales), in which the pathogen is found inside the seed. The application of cold temperatures causes a decrease or inhibition of vegetative growth, but in general even very low temperatures are not lethal for many microorganisms. Among other physical agents with sterilizing properties are ultraviolet radiation, low frequency electric currents (which produce heat, oxygen or chlorine), X-rays, radio waves and microwave ovens. Desiccation is frequently lethal, especially for the vegetative phase, although some fungal and bacterial spores are very resistant. Sterilization of liquid culture media whose components are thermolabile can be sterilized by mean of microfiltration. See **pasteurization** and **tyndalization**.

stichobasidium, pl. **stichobasidia** (NL. *sticho-basidium* < Gr. *stíchos*, row, line + *basídion*, dim. of *básis*, base): the main type of basidium without septa (holobasidium or autobasidium), elongate in shape, cylindrical, in which the achromatic spindle is arranged parallel to the main axis during the second meiotic division of the nucleus, as happens, e.g., in the gen. *Cantharellus* and *Craterellus* (Aphyllophorales). Cf. **chiastobasidium**.

stigmatocyst (Gr. *stígma*, genit. *stígmatos*, to sting, puncture + *kýstis*, vesicle, cell): the terminal cell of a capitate hyphopodium, from which the ascocarp in the Meliolales develops. Also known as **nodal cell**. Cf. **stigmatopod**. See **hyphopodium**.

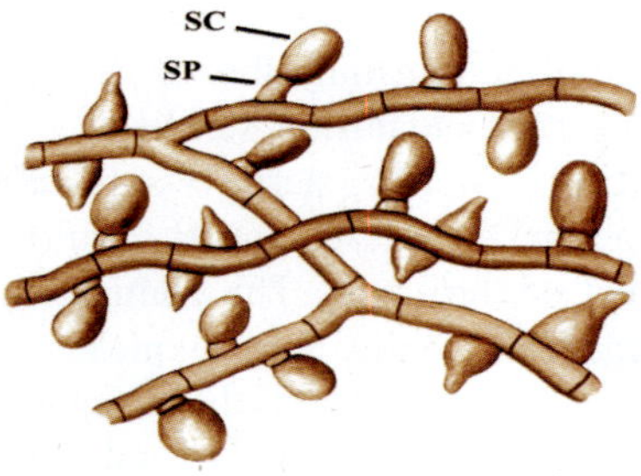

Stigmatocysts (SC) and **stigmatopodia** (SP) of the hyphopodia of *Meliola caesariae*, x 350 (*MU*).

stigmatopod, stigmatopodium, pl. **stigmatopodia** (Gr. *stígma*, genit. *stígmatos*, to sting, puncture + NL. *podium* < Gr. *pódion*, dim. of *poús*, genit. *podós*, foot): the basal or supporting cell of a capitate hyphopodium, which is similar to the appressoria of the Erysiphales and possibly related to an haustorium. The stigmatopodium (basal) and the stigmatocyst (terminal) are the two components of the capitate hyphopodium of the Meliolales, such as *Meliola caesariae*. Cf. **stigmatocyst**.

stilbellaceous (gen. *Stilbella* < L. *stilbum*, nitid < Gr. *stílbo*, shine + L. dim. suf. *-ella* + suf. *-aceus* > E. *-aceous*, of or pertaining to, with the nature of): term applied to the asexual or anamorphic fungi that form their spores on hyaline or brightly colored synnemata, with the head of slime spores. The gen. *Dendrostilbella*, *Doratomyces*, *Graphium* and *Stilbella* are examples of the old family Stilbellaceae (order Moniliales, class Hyphomycetes, subdivision Deuteromycotina, or deuteromycetes in a general sense). See **Hyphomycetes**.

stilboid (< gen. *Stilbum*, of the stilbellaceous asexual fungi + L. suf. *-oide*, like < Gr. *-oeídes*, similar to): a structure without spores, resembling a capitate basidiocarp, that functions as a propagule in *Mycena citricolor* and other Agaricales. This structure was described as belonging to the gen. *Stilbella* (=*Stilbum*), as *S. flavidum* (=*S. flavida*), probably due to confusing the stilboid with the synnema formed by the stilbellaceous asexual fungi.

Stilboids of *Mycena citricolor* on the upper surface of a coffee leaf, x 1 (*RTH*).

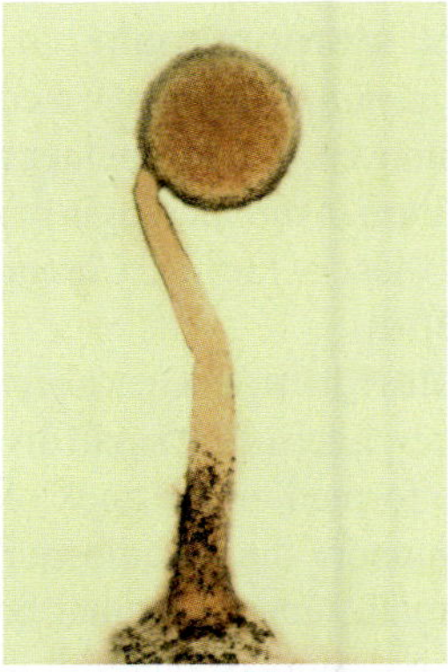

Stilboid of *Mycena citricolor*, x 60 (*MU*).

stinkhorn (ME. *stinken* < OE. *stinkan*, to emit a smell + L. *cornu*, horn): the basidioma of certain Phallales, such as *Phallus* and *Dictyophora*, that emit a strong, disagreeable odor.

stipe (L. *stipes*, genit. *stipitis*, pedicel, stem, foot): **1**. The foot or stalk that supports the pileus of a basidiocarp (e.g., *Amanita frostiana* and *Marasmius siccus*, of the Agaricales) or the cap of a stipitate ascocarp, like that of *Helvella lacunosa* (Pezizales). **2**. Another term for pseudopodetium. **3**. A more or less parallel system of tissues that supports the gleba but without having a direct structural connection with it, as occurs, e.g., in the fruiting body of *Tulostoma* (Tulostomatales).

Stipe of the apothecia of *Cookeina tricholoma* (Pezizales), x 2 (*RV*).

Stipe of the ascocarp of *Helvella lacunosa*, growing on the ground, between mosses, x 1 (*MRO*).

Stipe of the basidiocarp of *Boletellus cubensis* (Agaricales), x 0.5 (*RV*).

Stipe of the basidiocarps of *Marasmius siccus* (Agaricales), x 1 (*MU*).

Stipe of the basidiocarps of *Cortinarius* sp. (Agaricales), x 1 (*RV*).

stipe-columella, pl. **columellae** (L. *stipes*, stipe, pedicel, stem + *columella*, dim. of *columna*, column): a stipe or stalk that supports the receptaculum and, in addition, penetrates the gleba, forming a percurrent columella, as is seen, e.g., in *Podaxis* (Hymenogastrales). Also called **dendritic columella** and **percurrent columella**.

Stipe-columella of the fruiting body of *Podaxis pistillaris*, seen in longitudinal section, x 0.5 (*MU*).

stipitate (L. *stipitatus*, having a stipe or formed on a stipe < *stipes*, genit. *stipitis*, stem, pedicel, foot + suf. *-atus* > E. *-ate*, provided with or likeness): having a *stipe*. Cf. **apodal**, **pedicellate** and **sessile**.

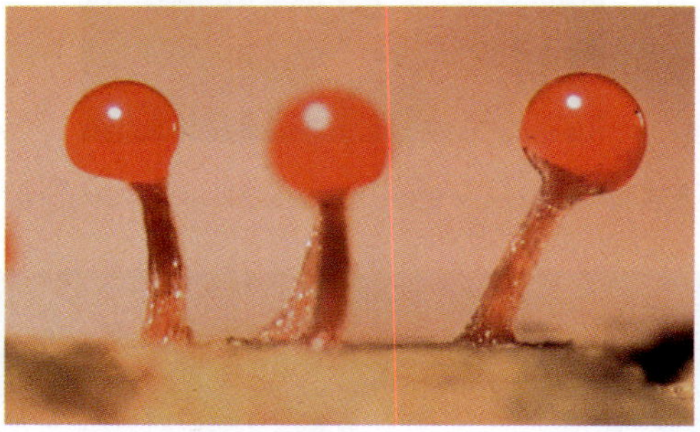

Stipitate sporangia of *Hemitrichia stipitata* (Trichiales), x 10 (*RS*).

Stipitate apothecia of *Cookeina tricholoma* (Pezizales), x 1 (*RV*).

stipitipellis (L. *stipes*, genit. *stipitis*, stem, pedicel, foot + *pellis*, skin): see **pellicle**.

stolon (L. *stolo*, genit. *stolonis*, shoot, sprout, branch): an aerial vegetative runner hypha, unbranched, that connects two groups or fascicles of rhizoids; characteristic of *Rhizopus* and other gen. of Mucorales. Cf. **pseudostolon**.

Stolon of *Rhizopus stolonifer* (=*Rh. nigricans*), x 400 (*MU*).

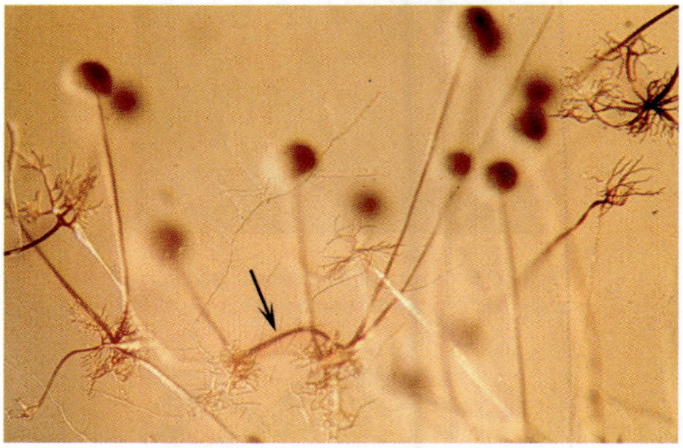

Stolon of *Rhizopus stolonifer* (=*Rh. nigricans*), x 400 (*MU*).

Stolon of *Rhizopus stolonifer* (=*Rh. nigricans*), x 400 (*MU*).

stoma, pl. **stomata** (Gr. *stóma*, mouth): an opening or ostiole through which the spores from the sporiferous portion of a fructification are liberated; this term is usually restricted to the ostiole of *Calostoma* and *Tulostoma* (Tulostomatales), and to the multiple ostioles (*myriostomata*) of *Myriostoma* (Lycoperdales).

Stoma of the fruiting body of *Tulostoma* sp., x 4 (*MU* & *CB*).

Stoma of the fruiting body of *Tulostoma* sp., x 6 (*MU* & *CB*).

stomapod, **stomapodium**, **stomatopodium**, **stomopodium**, pl. **stomapodia**, **stomatopodia**, **stomopodia** (Gr. *stóma*, genit. *stómatos*, mouth + NL. *podium* < Gr. *pódion*, dim. of *poús*, genit. *podós*, foot): *Meliolales*. A hyphal branch that penetrates through the open stomata of a host plant to develop in the mesophyll and form numerous haustoria.

stomato-, **stomat-** (Gr. *stóma*, genit. *stómatos*, stoma, mouth): a pref. employed in terms that allude to an opening or mouth, as in *stomatopodium*.

stone-fungus pl. **fungi** (OE. *stan*, stone; L. *fungus*, fungus): the hard pseudosclerotium of *Wolfiporia extensa* (=*W. cocos*=*Poria cocos*, Aphyllophorales), previously called *pietraia fungaia* (in Latin). On being watered, an edible basidioma is produced. The *stone-fungus* is also called Canadian tuckahoe or Indian bread.

Stone-fungus (sclerotium) of *Wolfiporia extensa* (=*W. cocos*), x 2 (*RTH*).

storage fungus pl. **fungi** (E. *storage*, a place or space for storing < ME. *storen* < Fr. *estorer*, to construct, restore, store < L. *instaurare*, to renew, restore; L. *fungus*, fungus, pl. *fungi*): the term is applied to the species of fungi that attack the stored grains, e.g., *Aspergillus flavus*, *A. parasiticus*, *A. niger*, *A. ochraceus*, *A. glaucus*, *Penicillium islandicum*, *P. citrinum*, *P. italicum* (moniliaceous asexual fungi), *Alternaria solani*, *Cladosporium cladosporioides*, *C. herbarum*, *Stachybotrys chartarum* (=*S. atra*) (dematiaceous asexual fungi), and other gen. which grow in substrates and seeds with low water activities and produce mycotoxins. See **mycotoxin**.

Aspergillus flavus colonies growing from seeds of broad beans placed on agar, a common **storage fungus**, x 1 (*MU*).

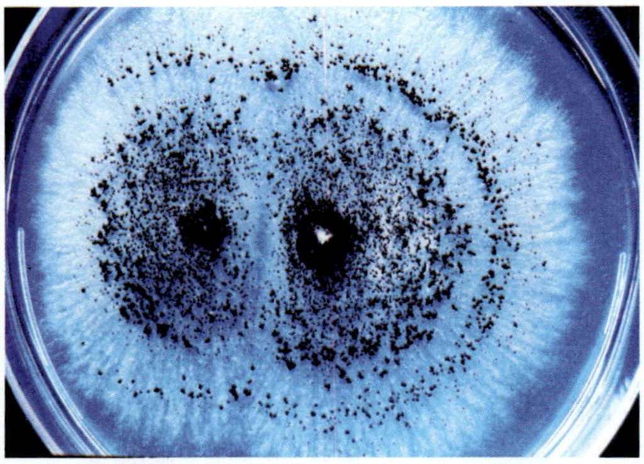

Colonies of *Aspergillus niger* on agar (**storage fungus**), x 1 (*MU*).

Colony of *Aspergillus ochraceus* on agar (**storage fungus**), x 2 (*MU*).

Conidiophores of *Aspergillus niger* and *A. flavus* (**storage fungi**) on agar, x 200 (*MU*).

Conidiophores of *Aspergillus flavus* (**storage fungus**) on agar, x 400 (*MU*).

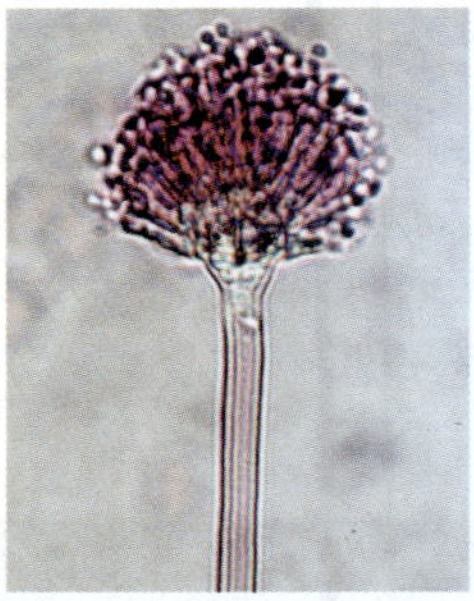

Conidiophore of *Aspergillus* sp. (**storage fungus**), x 400 (*MU*).

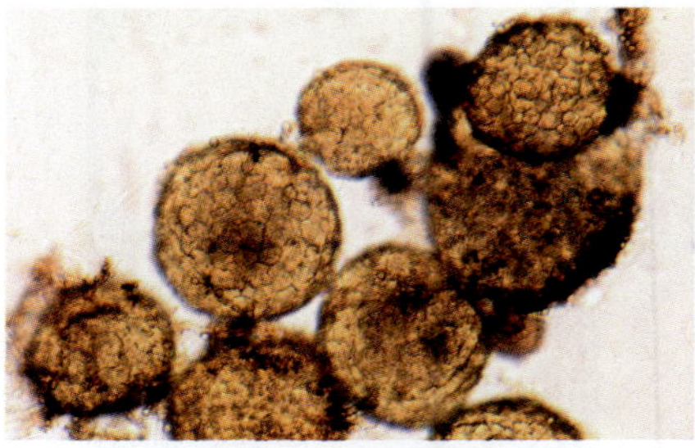

Cleistothecia of *Eurotium* sp. (Eurotiales, the sexual stage of *Aspergillus*,**storage fungus**), x 200 (*MU*).

strain (ME. *strene* < OE. *streon*, lineage, race): in mycology it refers to a fungus isolated in pure culture and characterized taxonomically, with one or more morphological, physiological or other properties that distinguishes it from other isolates of the same species. Individual strains of a fungus in a culture collection are distinguished by a code (of letters or numbers, or both), whether the collection is personal or internationally recognized, such as the American Type Culture Collection.

Stramenopila (L. *stramen*, straw + *pila* < *pilus*, hair): in the classification system of *Introductory Mycology* (Alexopoulos *et al*., 1996) it corresponds to a monophyletic kingdom that includes the phyla Oomycota, Hyphochytriomycota and Labyrinthulomycota as separate lineages among certain algal groups. The term Stramenopila was introduced by Patterson (1989) in recognition of the new monophyletic concept of these organisms based in part on flagellum hair structure. The members of the three phyla have zoospores with mastigonemate or tinsel flagella, that is, with many hairs or mastigonemes. In the Appendix, the user of this dictionary can see the comparison between the outline of Alexopoulos *et al*., 1996, modified by Ulloa and Hanlin, 2006, to that of Kirk *et al*., 2008. In the latter classification scheme, the kingdom Stramenopila is now named Chromista. These taxonomic rearrangements are part of a constant process of study and evaluation of these groups of organisms, in order to achieve a phylogenetic classification of the fungi. See **Chromista**.

straminipilous or **straminopilous** (L. *stramen*, genit. *straminis*, of, relating to, straw + *pilus*, hair + L. *–osus* > OF. *–ous*, *-eus* > E. *–ous*, having, possessing the qualities of): bearing tripartite tubular hairs, like the auxiliary cysts of *Saprolegnia* (Saprolegniales) which bear a tuft of tripartite tubular hairs (straminipilous cysts). The term is applied to cells (whether uni-flagellate, multiflagellate or non- flagellate) and/or flagella, in the latter case called straminipilous flagella or **flimmergeissel**.

stratified thallus, pl. **thalli** (E. *stratified*, pt. of *stratify*, to arrange in strata or layers < L. *stratus*, blanket, layer + L. suf. *-ificare* > E. *-ify*, to make, do, cause; Gr. *thallós*, thallus, shoot): a lichen thallus whose tissues are arranged in horizontal layers or strata. See **heteromerous**.

straw mushrooms (ME. < OE. *streaw*, straw; ME. *musseroum* < MF. *mousseron* < LL. *mussirion*, mushroom): the edible basidiomata of *Volvariella volvacea* and *V. diplasia* (Agaricales). Widely consumed in the tropical regions of Asia (also called the paddy straw or Chinese mushrooms).

strepto-, **strept-** (Gr. *streptós*, curved, twisted): a pref. employed in the formation of terms that alludes to something twisted, as in the generic name *Streptomyces*, of the actinomycetales (presently regarded as actinobacteria), whose cells have a twisted mycelium.

striate, **striated** (L. *striatus*, lined < *stria*, groove, line + suf. *-atus* > E. *-ate*, provided with or likeness): applied to surfaces that have nearly parallel grooves and edges, as is observed in the colonies of some yeasts and molds cultivated on agar, e.g., *Candida lambica* (cryptococcaceous asexual yeasts) and *Aspergillus versicolor* (moniliaceous asexual fungi), as well as in the wall of many types of spores, e.g., the ascospores of *Haematonectria* (=*Nectria*) *haematococca* (Hypocreales), the teliospores and basidiospores of *Boletus chrysenteroides* (Agaricales) and *Ramaria* sp. (Aphyllophorales). Other structures also are striate, such as the surface of the discothecium of *Rhytidhysterium* (=*Rhytidhysteron*, of the Patellariales), the pileus of various Agaricales and the fruiting bodies of some Nidulariales and Lycoperdales.

strigose, **strigous** (NL. *strigosus*, hirsute, bristled < *striga*, rigid, sharp hair, bristle + suf. *-osus* > E. *-ose*, full of, augmented, prone to): syn. of **hirsute** and **hispid**; having long, stiff, sharp-pointed hairs, like the pileus of the basidiocarps of *Lentinus* (=*Panus*) *badius* and *Hexagonia* (=*Pogonomyces*) *hydnoides* (Aphyllophorales).

Striate colony of *Candida lambica* on agar, x 1 (*MU*).

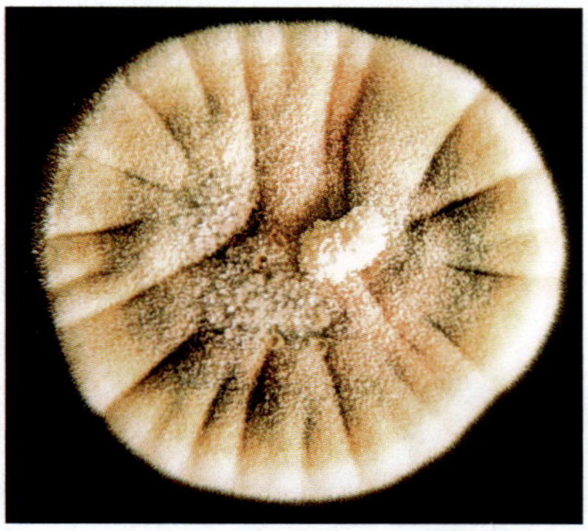

Striate colony of *Aspergillus versicolor* on agar, x 1 (*MU*).

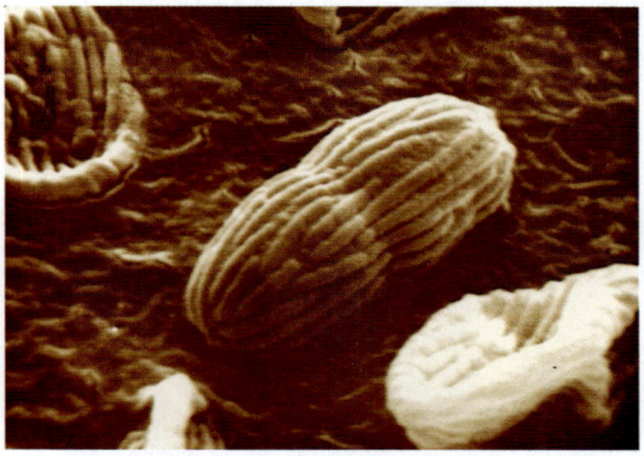

Striate ascospore wall of *Haematonectria* (=*Nectria*) *haematococca*, x 2,800 (*RTH*).

Striate discothecium of *Rhytidhysterium rufulum*, x 15 (*CB*).

Striate pileus margin of the basidiocarps of *Amanita caesarea* (Agaricales), x 1 (*RV*).

Striate receptacula of the basidiocarps of *Cyathus striatus* (Nidulariales), x 2 (*RV*).

Striate pileus of the basidiocarps of *Coprinellus* (=*Coprinus*) *micaceus* (Agaricales), x 0.2 (*MU*).

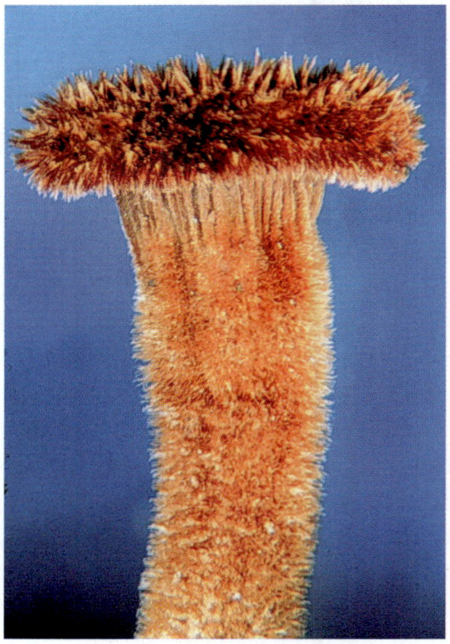

Strigose basidiocarp of *Lentinus* (=*Panus*) *badius*, x 5 (*MU*).

Striate pileus of the basidiocarps of the Agaricales, x 2, *Marasmius* sp. (*RV*) and *Xeromphalina campanella* (*TRO*).

Strigose basidiocarp of *Lentinus* (=*Panus*) *badius* (close-up of the pileus), x 20 (*MU*).

Strigose surface of a basidiocarp of *Phellinus sarcites*, x 0.5 (*RV*).

Strigose surface of a basidiocarp of *Hexagonia* sp., x 20 (*MU*).

strobiliform (L. *strobiliformis*, cone-shaped < *strobilus* < Gr. *stróbilos*, pine cone + L. suf. *-formis* < *forma*, shape): cone-shaped, like the pileus of *Strobilomyces strobilaceus* (= *S. floccopus*), Agaricales.

Strobiliform pileus of a basidiocarp of *Strobilomyces strobilaceus* (=*S. floccopus*), x 20 (*MU*).

stroma, pl. **stromata** (Gr. *strõma*, bed, small pillow, cushion): a compact mass of somatic hyphae, made up of plectenchyma (prosenchyma, pseudoparenchyma, or both) on or in which are produced fertile hyphae that generate asexual or sexual reproductive organs, such as sporodochia, pycnidia, perithecia, apothecia, etc. E.g., in the Pyrenomycete *Daldinia concentrica* (Xylariales) the stroma is pulvinate and contains perithecia embedded in the peripherial layer. Other examples of stromata are *Rhytisma acerinum* (Rhytismatales) and *Broomeia congregata* (Lycoperdales). See **ectostroma** and **endostroma**.

Stroma of *Rhytisma acerinum* on a maple leaf, x 60 (*RTH*).

Stroma of *Broomeia congregata* bearing numerous spore sac on the upper part, x 1 (*MU*).

Perithecial **stroma** of *Camillea* sp. (Xylariales), x 7 (*MU*).

Perithecial **stromata** of *Cordyceps capitata* (Hypocreales), growing from an ascocarp of *Elaphomyces* (Pezizales), x 1 (*RV*).

Perithecial **stromata** of *Xylaria polymorpha* (Xylariales) on dead wood, x 1 (*RV*).

Perithecial **stromata** of *Cordyceps sobolifera* (Hypocreales), growing from an hypogeous cicada larva, x 1 (*RV*).

Perithecial **stromata** of *Xylaria polymorpha* (Xylariales) on dead wood, x 1 (*RV*).

Perithecial **stromata** of *Hypoxylon* sp. (Xylariales) on dead wood, x 1 (*RV*).

stylospore (L. *stilus*, *stylus* < Gr. *stýlos*, column, pick, awl + *sporá*, spore): **1**. *Mucorales*. An asexual reproductive spore, united with the hypha that produces it by a stalk or slender sterigma. Stylospores are characteristic of the family Mortierellaceae, e.g., *Mortierella epigama* and *M. indohii*, in which case they are also known as chlamydospores. **2**. *Ascomycetes*. A small filiform spore believed to function as a spermatium; they usually cannot be germinated.

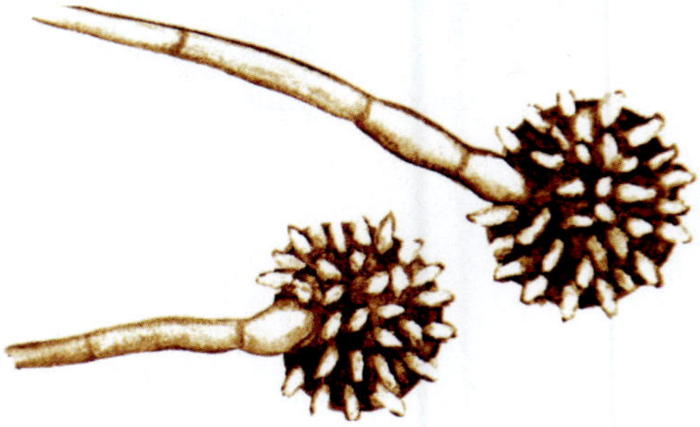

Stylospores of *Mortierella indohii*, x 750 (*MU*).

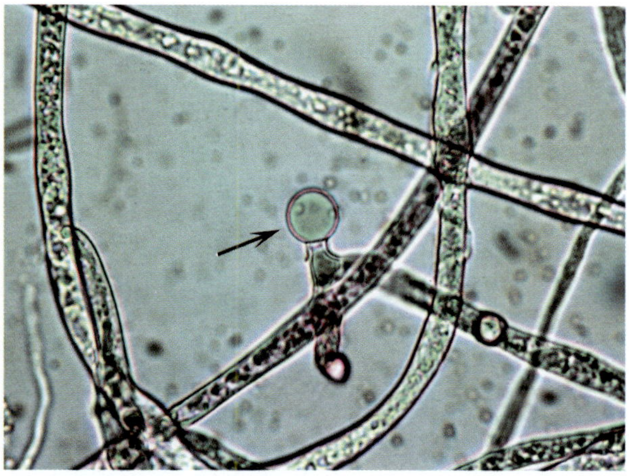

Stylospore of *Mortierella epigama*, from an agar culture, x 450 (*MU*).

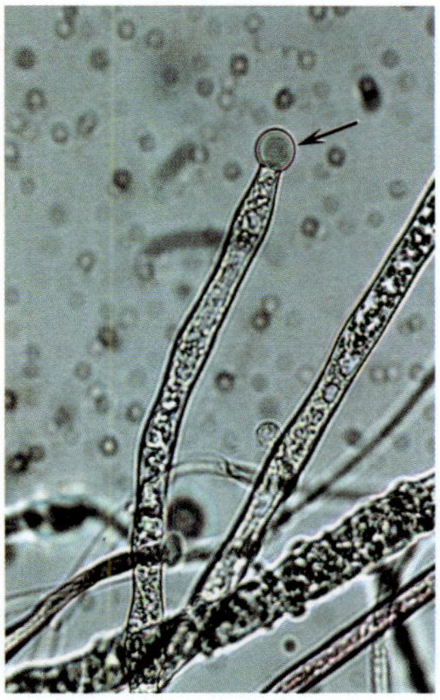

Stylospore of *Mortierella epigama*, from an agar culture, x 450 (*MU*).

sub, sub- (L. prep. and pref., *sub* and *sub-*): in the first case it means low, beneath, as; as a pref. it forms part of many biological terms and is employed to attenuate the meaning of the second component, thus being equivalent to almost, as in *subalate* and *subcentric*. In taxonomy, the pref. indicates intermediate categories, as, e.g., *subfamily* (a category between family and genus) and *subgenus* (a category between genus and species).

subalate (L. pref. *sub-*, low, in the sense of attenuating the word that follows + *alatus*, winged < *ala*, wing + suf. *-atus* > E. *-ate*, provided with or likeness): generally applied to spores on whose wall the ridges are of smaller proportions than those of spores called *alate*. Also called ribbed, for the ridges or ribs, like the basidiospores of *Boletellus russellii* (Agaricales).

Subalate basidiospores of *Boletellus russellii*, x 1,980 (*MU*).

subcentric (L. pref. *sub-*, low, in the sense of attenuating the word that follows + *centrum*, center + Gr. suf. *-íkos* > L. *-icus* > E. *-ic*, pertaining to or relating to): located almost in the center; applied to oogonia of the saprolegniaceous fungi that have the cytoplasm surrounded by a layer of lipid globules on one side and by two or three layers of globules on the other side. Cf. **centric** and **excentric**.

subclass (L. *subclassis*): a taxonomic category intermediate between class and order, with the ending *-mycetidae*, as in Hymenomycetidae.

subcutis (L. prep. *sub*, low, beneath + *cutis*, skin): the lower stratum of the cuticular layer of the pileus of the Agaricales; it is found beneath the *epicutis*. Also called **hypodermis**.

subdivision (L. *sub-*, low, beneath, almost; a pref. used to indicate intermediate categories + *divisio*, genit. *divisionis*, action and effect of dividing or separating): a taxonomic category intermediate between division and class, with the ending *-mycotina*, as in Ascomycotina. Equivalent to *subphylum* of zoological classification.

suberose (NL. *suberosus*, corky < L. *suber*, cork + suf. *-osus* > E. *-ose*, full of, augmented, prone to): having the consistency of cork, like the fruiting body of *Inonotus*, *Phellinus* and other Aphyllophorales.

Suberose basidiocarp of *Inonotus dryophilus* on dead wood, x 1 (*RV*).

Suberose basidiocarp of *Phellinus linteus* on a tree trunk, x 0.3 (*RV*).

subgenus, pl. **subgenera** (L. *sub-*, low, beneath, almost; a pref. used to indicate intermediate categories + *genus*, genit. *generis*, genus): a taxonomic category intermediate between genus and species, whose name is Latinized and similar to that of the genus, but with a different ending. In general it has few applications; it is utilized for genera that encompass numerous species, as in *Penicillium* subgenus *Biverticillium* (moniliaceous asexual fungi).

subgleba, pl. **subglebae** (L. prep. *sub*, low, beneath + *gleba*, lump, mass): *Gasteromycetes*. A sterile region located beneath the gleba in the fructification of some Lycoperdales. In *Lycoperdon*, the subgleba is delimited by a diaphragm, is composed of chambers and contains a pseudocolumella that can be either rudimentary or well developed. In other gen. of Lycoperdales the subgleba is absent, as in *Mycenastrum*, *Disciseda* and *Langermannia*, and in others in which it is present its composition is cellular or interwoven and it is separated from the gleba by a diaphragm, as in *Bovistella* and *Morganella*, respectively. See **pseudostipe** and **pseudocolumella**.

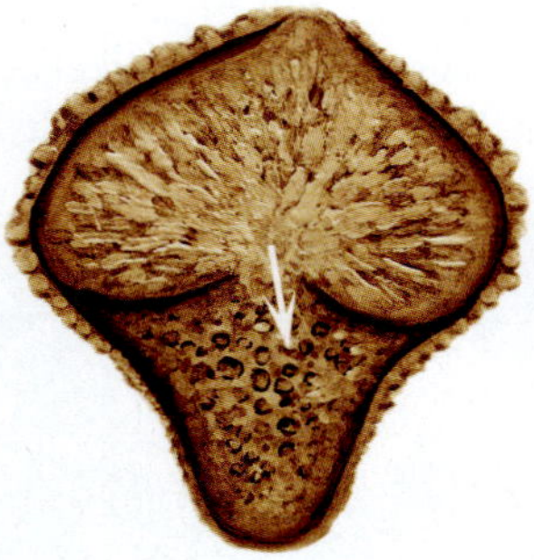

Subgleba of the fruiting body of *Lycoperdon perlatum*, seen in longitudinal section, x 1 (*MU*).

subglobose (L. *sub-*, almost, beneath, a pref. used to attenuate the meaning of the word that follows + *globosus*, round, like a ball < *globus*, ball + suf. *-osus* > E. *-ose*, full of, augmented, prone to): almost globose, almost spherical. Applied to structures (spores, sporocarps, etc.) with a length:width ratio of 1:1.05 to 1:1.15. The polar axis corresponds to a spheroidal-prolate shape, and the equatorial axis to spheroidal-oblate. E.g., the sporangiospores of *Pilobolus longipes* (Mucorales), the asci of *Eurotium repens* (Eurotiales), and the basidiospores of *Amanita phalloides* (Agaricales) are subglobose.

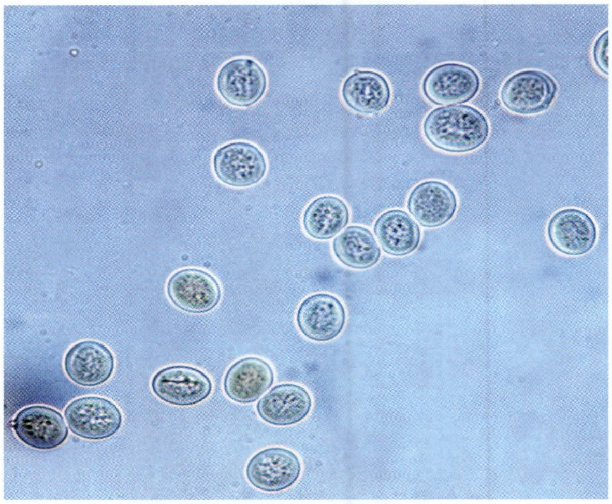

Subglobose sporangiospores of *Pilobolus longipes*, x 1,000 (*EAA*).

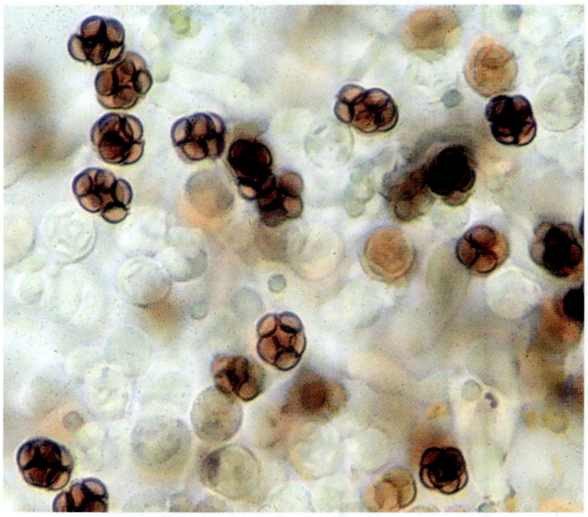

Subglobose asci of *Eurotium repens*, x 150 (*MU*).

subhymenium, pl. **subhymenia** (L. pref. *sub-*, below, beneath + *hymenium* < Gr. *hyménion* < *hymén*, membrane + dim. suf. *-íon* > L. *-ium*): refers to tissues that form beneath the hymenium, or sporogenous tissues, in a fruiting body. E.g., in many perithecia a layer of pseudoparenchymatous cells is formed between the ascigerous layer and the wall at the bottom of the perithecium.

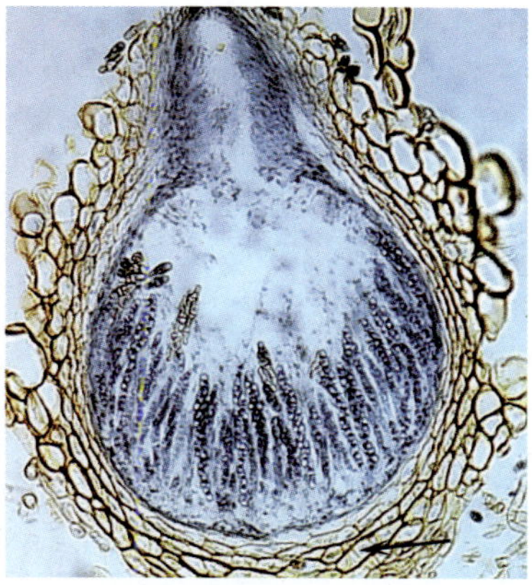

Subhymenium of the ascigerous layer of the perithecium of *Haematonectria* (=*Nectria*) *haematococca* (Hypocreales), x 140 (*RTH*).

subhypothallic (L. prep. *sub*, low, beneath + Gr. *hypó*, beneath, inferior + *thallós*, shoot, thallus + suf. *-íkos* > L. *-icus* > E. *-ic*, pertaining to or relating to): *Myxomycetes*. A type of sporocarp development in which the plasmodium is concentrated in serveral places to form hemispherical mounds that separate from one another by reabsorption of the protoplasmic filaments that initially connected them. The calcareous layer of the plasmodium forms the hypothallus, when this is present, and the mounds continue elongating upward to form columnar structures called *papillae*, which are the primordia of the sporocarps. As these papillae elongate, their bases constrict to form the pedicel of the sporocarps; there is a protoplasmic flow toward each expanding primordium from the adjacent areas and from below on the substrate, so that the apex inflates while the pedicel contracts and forces the apex upward. The distinctive characteristic of this type of development is the continuity of the peridium with the pedicel and with the hypothallus of each sporocarp. It is called subhypothallic because the sporocarp emerges from the lower part of the hypothallus, which is differentiated on the upper surface of the plasmodium. It is present in the orders Liceales, Trichiales, Echinosteliales and Physarales, and is typified by *Arcyria cinerea* (Trichiales). The other type of sporocarp development is the *epihypothallic*, which is characteristic of members of the order Stemonitales.

subiculum, pl. **subicula** (L. *subiculum* < *subex*, base, bottom, something located below + dim. suf. *-culum*): a felt-like trama or layer composed of loosely intertwined hyphae (hyphenchyma) that covers the substrate, on which are seated fruiting bodies (perithecia, apothecia, etc.); the subiculum can be reticular, woolly or crustaceous. In species of *Hypomyces*, such as *H. lactifluorum* (Hypocreales), e.g., the majority of which are parasites of basidiomycete sporophores, the perithecia are sunken in a thick, cottony subiculum whose trama approximates that of a stroma. *Nitschkia* and *Cephalotheca* (Eurotiales) are two saprobic fungi that form their ascomata on a subiculum similar to a stroma that is embedded in dead wood.

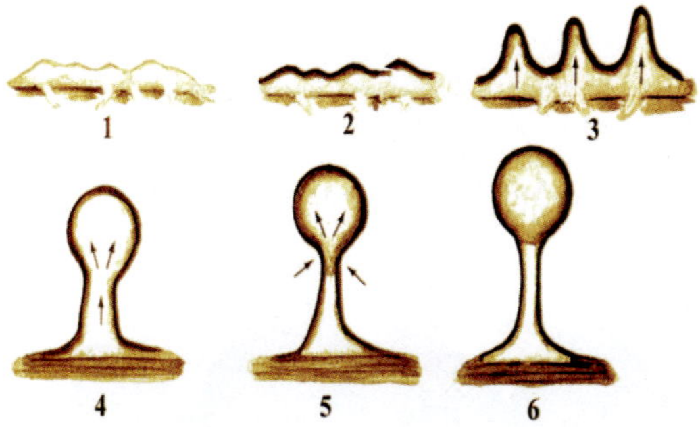

Subhypothallic type of sporangial development in the non-stemonitoid orders of Myxomycetes. 1. Emergent plasmodium. 2. Formation of the supraplasmodial hypothallus. 3. Differentiation into primordia. 4. Enlargement of the primordia due to continuous protoplasmic flow; expansion of the apex of the primordia. 5. Expansion continues in the apex, while the wall of the pedicel constricts. 6. Full-size sporangium complete, before formation of the capillitium; the outer layer of the pedicel is continuous with the peridium and with the supraplasmodial hypothallus. In some forms the protoplasm can remain in the pedicel; in others there can be debris. Diagram based on figures and information published by I. K. Ross in *Mycologia 65*:477-485, 1973, x 20 (*MU*).

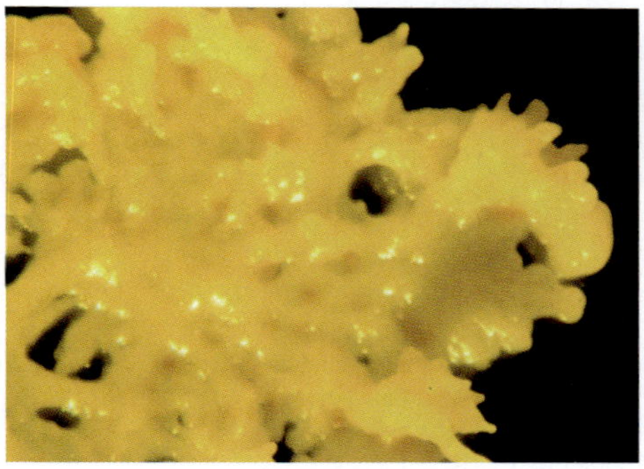

Subhypothallic type of sporangial development in *Physarum polycephalum*. Differentiation of the primordia, x 6 (*MU*).

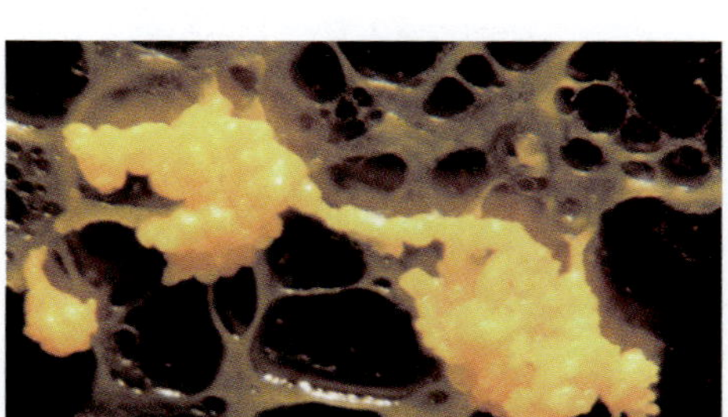

Subhypothallic type of sporangial development in *Physarum polycephalum*. Differentiation of the primordia, x 10 (*MU*).

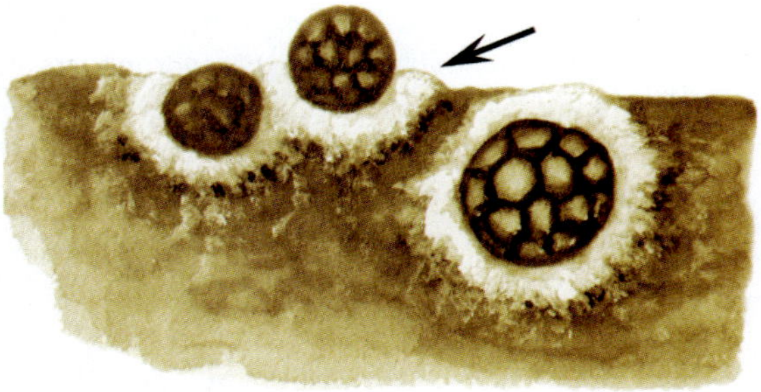

Subiculum of *Cephalotheca sulphurea*, on dead wood of oak (*Quercus*); the gregarious, superficial ascomata, with a black peridium divided into polygonal areas, rest on a sulphur-yellow subiculum, x 58 (*MU*).

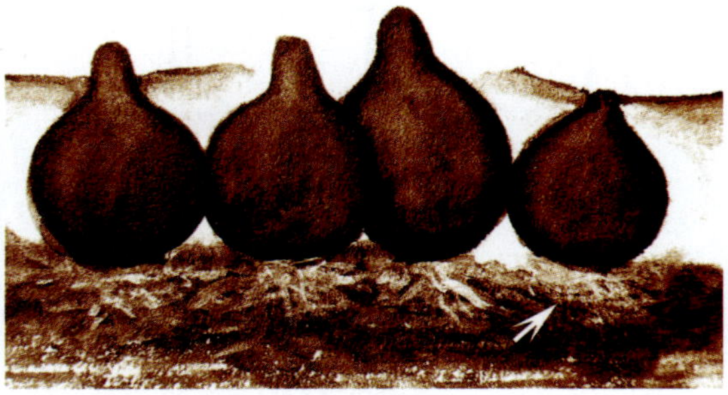

Subiculum of *Nitschkia cupularis*, embedded in tissues of a branch of *Acer;* the erumpent, gregarious perithecia rest on the subiculum, x 58 (*MU*).

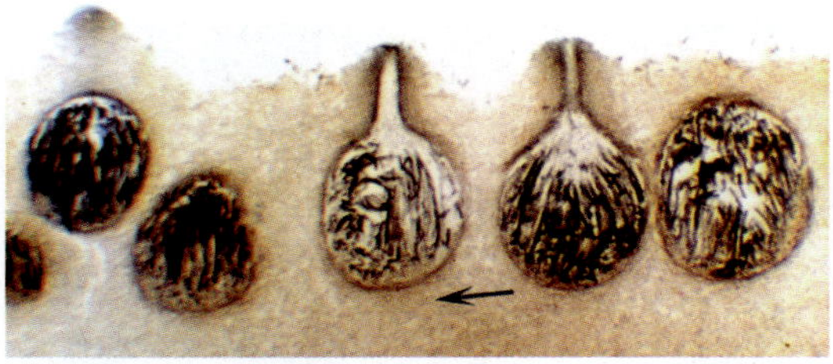

Subiculum with embedded perithecia of *Hypomyces lactifluorum* parasitizing a basidiocarp of *Lactarius* sp., seen in longitudinal section, x 47 (*RTH*).

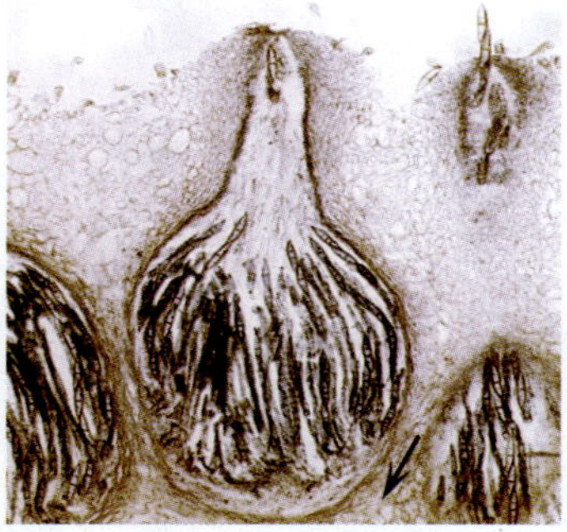

Subiculum with embedded perithecia of *Hypomyces lactifluorum* parasitizing a basidiocarp of *Lactarius* sp., seen in longitudinal section, x 75 (*RTH*).

suboperculate (L. prep. *sub*, low, beneath + *operculatus*, provided with a lid < *operculum*, cap, lid + suf. *-atus* > E. *-ate*, provided with or likeness): a type of ascal dehiscence in which the operculum is in a subapical position. Cf. **inoperculate**, **operculate** and **pseudoperculate**.

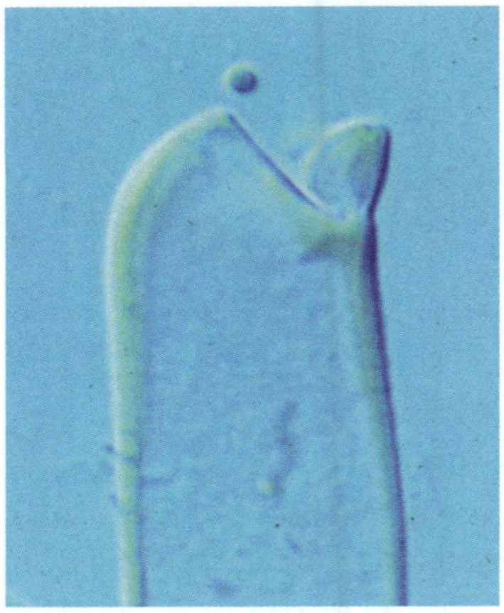

Suboperculate ascus dehiscence of *Cookeina sulcipes* (Pezizales), x 500 (*RTH*).

subpellis (L. prep. *sub*, low, beneath + *pellis*, skin): see **pellicle**.

subphylum, pl. **subphyla** (L. *sub-*, low, beneath, almost; a pref. used to indicate intermediate categories + *phylum* < Gr. *phýlon*, race, stirp): in zoological and mycological classifications, it is the taxonomic category between phylum and order. Equivalent to the *subdivision* of botanical classification.

subprolate (L. *sub-*, low, almost; pref. utilized to attenuate the meaning of the word that follows + *prolatus*, brought forward, elongated in the direction of the main axis): generally applied to spores broadly ellipsoidal in shape, not so elongate, whose length:width ratio is 1:1.15 to 1:1.30.

subradicating (L. pref. *sub-*, low, almost, less + *radicatus*, radicate, provided with a root): refers to a stipe that has a small prolongation, but which is not a root-like extension in size, as is seen in *Rhodocollybia* (=*Collybia*) *maculata* (Agaricales).

subspecies (L. *sub-*, low, beneath, almost; a pref. used to indicate intermediate categories + *species*, species): a taxonomic category intermediate between species and variety, e.g., *Amanita muscaria* subspecies *flavivolvata* (Agaricales).

substomatal vescicle (L. *sub*, beneath + NL. *stoma* < Gr. *stóma*, genit. *stómatos*, mouth + L. suf. *-alis* > E. *-al*, relating to or pertaining to; L. *vesicula*, dim. of *vesica*, bladder): a structure formed by the rust fungi in the substomatal cavities of the leaves of the host plant. E.g., in the peanut rust, *Puccinia arachidis* (Uredinales), a minute peg forms from the appressorium (which itself was formed by the germinated urediniospore) that initiates the infection, penetrating the host leaf, where it grows inside the stomatal opening between the guard cells (a type of penetration referred to as indirect penetration). The infective peg expands to form the substomatal vesicle into which the cytoplasm and nuclei of the appressorium migrate; these nuclei divide synchronously, and from the substomatal vesicle an extension known as the primary infective hypha is developed, which initiates the formation of the haustorium.

substrate, **substratum**, pl. **substrata** (ML. *substratum*, base on which an organism lives; an underlying support, fundamental): **1**. A substance on which an enzyme acts (enzymology); substances (e.g., the constituents of a culture medium) utilized by a microorganism to grow (microbiology). **2**. Material on which an organism grows or to which it is attached (ecology).

substroma, pl. **substromata** (L. prep. *sub*, low, beneath + Gr. *strõma*, bed, small cushion, small pillow): equivalent to *pseudostroma* in that the tissue of the host plant predominates.

subtribe (L. prep. *sub*, low, beneath + *tribus*, tribe): a taxonomic hierarchy below the tribe and above the genus; the names of subtribes end in *-inae*.

subulate, **subuliform** (NL. *subulatus*, awl-shaped < L. *subula*, awl + suf. *-atus* > E. *-ate*, provided with or likeness; or + L. suf. *-formis* < *forma*, shape): narrowed toward the apex, ending in a fine point; also called *awl-shaped*; not as sharp as acicular. E.g., the cheilocystidia of *Mycena atkinsoniana* and the pleurocystidia of *Boletus* (Agaricales) are subulate. Cf. **obsubulate**.

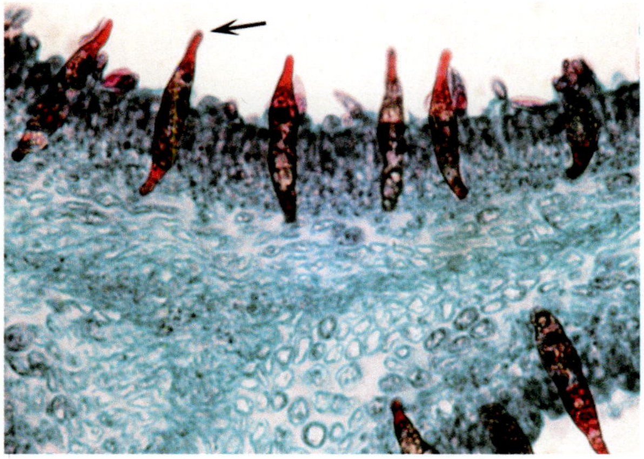

Subulate pleurocystidia of the hymenium of *Boletus* sp., x 246 (*MU*).

succession (L. *successio*, genit. *successionis*, action and effect of following or continuing): the order in which, or the conditions under which, an organism succeeds, complements the action of, or displaces another organism, as occurs, e.g., in *metabiosis*.

sulcate (L. *sulcatus*, sulcate, furrowed, fluted < *sulcus*, furrow, canal + suf. *-atus* > E. *-ate*, provided with or likeness): provided with furrows, like, e.g., the colonies of *Penicillium aurantiogriseum* (=*P. puberulum*, see **radial**), moniliaceous asexual fungi, the surface of the thallus of *Parmelia sulcata* (Lecanorales), the tips of the ostioles of many diatrypaceous species (Xylariales), and the peristome of the fruiting body of *Geastrum pectinatum* (Lycoperdales). Also called **pleated**.

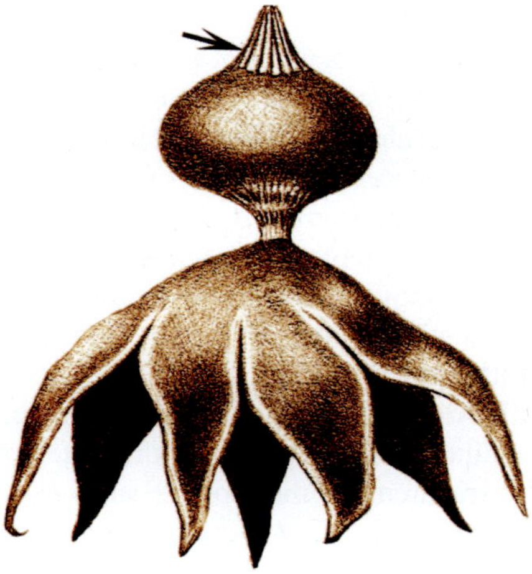

Sulcate peristome of the fruiting body of *Geastrum pectinatum*, x 1.3 (*MU*).

sulcus (L. *sulcus*, furrow, canal): a longitudinal furrow of a colony, ostiole, etc. See **sulcate**.

summer spore (OE. *sumor* > ME. *sumer* > E. *summer*, summer; Gr. *sporá*, spore): refers to a spore that can germinate without undergoing a latent or resting period; frequently it is short-lived, i.e., it is not capable of remaining viable over winter, like the rust urediniospore (Uredinales). Cf. **winter spore**.

super (L. *super*, prep. that means on, that denotes preeminence or excess): on, above. Utilized in combination with the names of taxonomic categories to provide additional ranks, such as superfamily and superkingdom.

superficial (L. *superficialis*, pertaining to or relating to the surface < *superficies*, the exterior of an object or body, the outward appearance or distinguishing characteristics of something + suf. *-alis* > E. *-al*, pertaining to or relating to): what is on the surface of a structure, substrate, etc.

superkingdom (L. *super*, prep. that means on, or that denotes preeminence or excess + ME. < OE. *cyng*, *cyneg*, king + suf. *-dom*, domain > kingdom): a taxonomic hierarchy superior to the kingdom, justified in light of new phylogenetic molecular evidence. Also called **domain**. In some recent classifications, all of the phyla of living beings are included in two superkingdoms: *Eukaryota* and *Prokaryota*.

supine (L. *supinus*, lying down with the face upward): refers to fructifications that are closely applied to the substrate.

supra- (L. *supra*, on): pref. that denotes on, above. In taxonomy it is utilized to name higher categories, e.g., *suprageneric*, all of the categories above genus; *supraspecific*, all of the categories above species, etc.

suprahilar (L. *supra*, on + *hilaris*, relating to the hilum < *hilum*, scar, spot or notch of a bean seed; here, areole of the spores + suf. *-aris* > E. *-ar*, like, concerning): situated on or above the hilum. In certain basidiospores (of the russulaceous fungi and of the gen. *Galerina*, Agaricales), the *suprahilar depression* corresponds to the rounded areole that is formed in the vicinity of the hilum, above it, and on the internal side, i.e., on the flat or slightly concave face of the basidiospores. See **hilar appendix**.

suprahilar depression (L. *supra*, on + *hilaris*, relating to the hilum < *hilum*, scar, mark or notch of a bean seed; here, areola of the spores + suf. *-aris* > E. *-ar*, like, concerning; LL. *depressio*, genit. *depressionis*, to depress < L. *depressus*, depressed + suf. *-io*, genit. *-ionis* > E. *-ion*, result of an action, state): *Basidiomycetes*. A smooth zone on the wall of certain basidiospores, near the *hilar appendage*, as is seen in *Galerina* (Agaricales). Also called **suprahilar disc** and **suprahilar spot**.

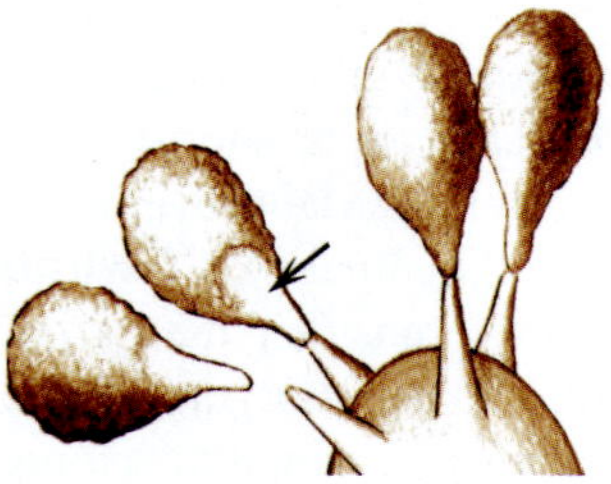

Suprahilar depression of the basidiospores of *Galerina phillipsii*, x 1,000 (*MU*).

suprahilar disc (L. *supra*, on + *hilaris*, relating to the hilum < *hilum*, scar, spot or notch of a bean seed; here, the areola of the spores + suf. *-aris* > E. *-ar*, like, concerning; L. *discus* < Gr. *dískos*, disc): see **suprahilar depression**.

suprahilar spot (L. *supra*, on + *hilaris*, relating to the hilum < *hilum*, scar, notch on a bean seed; here, areola of the spores + suf. *-aris* > E. *-ar*, like, concerning; ME. *spotte*, stain, mark): see **suprahilar depression**.

suprapellis (L. *supra*, on + *pellis*, skin): see **cystoderm** and **pellicle**.

surface (L. *superficies*): the exterior of an object or body, the outward appearance or distinguishing characteristics of something.

susceptible (LL. *susceptibilis* < L. *susceptus*, open, subject, not resistant to a stimulus, influence or attack < *suscipere*, to take up, support, admit + *-ilis*, suf. added to denote capacity, with the character of): capable of submitting to an action or process; open, subject, not resistant to a stimulus, influence or agent (a pathogen, toxin, cold, etc.). Cf. **suscept** and **tolerant**.

suscept (L. *susceptus*, open, subject, not resistant to a stimulus, influence or attack < *suscipere*, to take up, support, admit): a living organism that is susceptible (capable of being attacked, not immune) to a disease, pathogen, or toxin. Cf. **susceptible** and **tolerant**.

suspensor (NL. *suspensor*, what supports < L. *suspendere*, to suspend or hang something in the air): a specialized hypha that supports the zygosporangium in the Zygomycetes; the suspensors constitute the later stage of the zygophores. They can be more or less equal in size, as in *Mucor* and *Rhizopus*, or unequal, as in *Zygorhynchus* and *Absidia* (Mucorales).

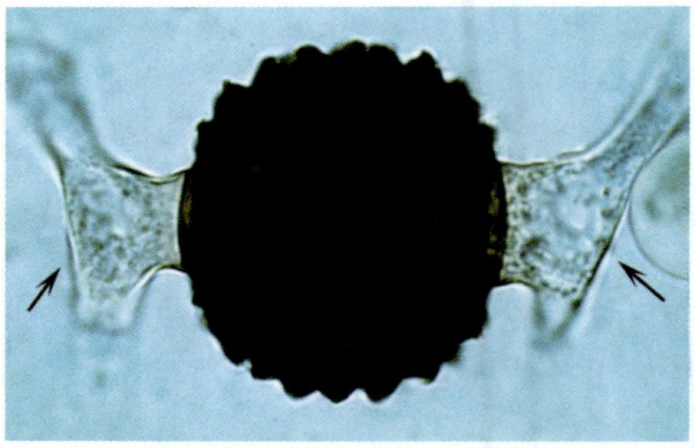

Suspensors of the zygosporangium of *Mucor hiemalis*, x 230 (*RTH*).

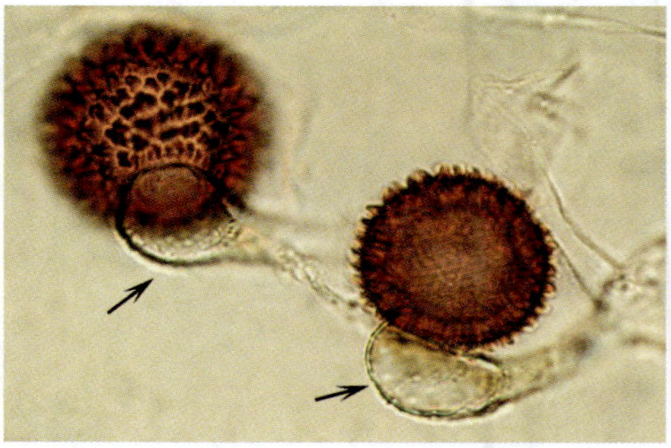

Suspensors of the zygosporangia of *Zygorhynchus moelleri* (=*Z. vuilleminii*), x 500 (*MU*).

sycosis, pl. **sycoses** (NL. *sycosis* < Gr. *sykõsis* < *sýkon*, fig + suf. *-osis*, state or condition): a disease of man and higher animals characterized by a chronic inflammatory disorder of the hair follicles, especially of the face (tinea of the beard), with the formation of papules, pustules, tubercles and crusts, caused by *Trichophyton rubrum* (moniliaceous asexual fungi). See **mycosis**.

sym- (Gr. *sým*, *sýn*, with, together with): an inseparable prep. frequently used in biology to express the idea of solidarity, adhesion or concrescence, as in *symbiont*, *symbiosis*, *sympatric*, etc., terms in which the *n* of *sýn* is converted into *m* before *b* and *p* in the second component. Equal to **syn-**.

symbiont (Gr. *symbiõntos* < *symbióo*, life in common + *óntos*, genit. of *ón*, a being): each of the organisms of a *symbiosis*; e.g., in the lichens the fungus and the alga are both symbionts. See **symbiosis**.

symbiotic, symbiontic (Gr. *symbiontikós* < *symbiontés*, *symbiotés*, one who lives with another, companion + suf. *-íkos* > L. *-icus* > E. *-ic*, pertaining to or relating to): belonging to *symbiosis*, relating to it.

symbiosis, pl. **symbioses** (Gr. *symbíosis*, life in common): in biology, the term is applied to the life in common of two or more organisms, with mutual benefit for the participants or symbionts. There exists a whole series of gradations that are embraced by symbiosis properly called (or *mutualistic symbiosis*) and *commensalism* up to *antagonistic symbiosis* (or *parasitism*), in which the benefit is in favor of the invader and against the host. Cf. **helotism**, **metabiosis**, **mutualism**, **parabiosis**, **parasymbiosis** and **synergy**.

Basidiocarps of *Septobasidium* sp. (Septobasidiales), which are **symbiotic** with insects, on a tree branch, x 1 (*MU*).

Symbiontic thalli of *Cladonia didyma* (Lecanorales), x 1 (*TRO*).

symbiotize (Gr. *symbíosis*, life in common + suf. *-izein* > LL. *-izare* > E. *-ize*, to cause something to be, to become as, conform or resemble): to unite two organisms in symbiosis, as when a fungus and an alga constitute a lichen.

symbiotroph (Gr. *symbíosis*, life in common + *trépho*, to nourish, feed): applied to plants that are nourished by means of mycorrhizae, to radicicolous bacteria, etc., i.e., by a natural symbiotic system.

Symbiotroph pine roots associated with a mycorrhizal fungus, *Russula* sp. (Agaricales), x 10 (*LV*).

symmetrical (Gr. *symmetría*, symmetry + suf. *-íkos* > L. *-icus* > E. *-ic*, pertaining to or relating to): refers to an organ or structure that has at least one plane of symmetry, which permits it to be divided into two equal parts, as occurs in many fungal spores. Cf. **asymmetrical**.

sympatric (Gr. *sým*, with + *pátra*, native land < *patér*, *patrós*, father + suf. *-íkos* > L. *-icus* > E. *-ic*, pertaining to or relating to): occurring in the same geographic area. Cf. **allopatric**.

sympho-, **symphyo-**, **symphy-** (Gr. *symphyós* < *symphýo*, to grow together): a pref. employed in terms that allude to a joint growth, like *symphogenous*.

symphogenous, **symphyogenous**, **symphyogenetic** (Gr. *symphyós* < *symphýo*, to grow together + *génos*, origin < *gennáo*, to engender, produce; or + suf. *-íkos* > L. *-icus* > E. *-ic*, pertaining to or relating to): applied to a type of pycnidium formation in which hyphae of various origins group together and intertwine to form a small spherical primordium that develops into a pycnidium. *Zythia* (sphaeropsidaceous asexual fungi) has this type of pycnidial development. Cf. **meristogenous**.

symplastic (Gr. *sým*, with + *plastós*, modeled, formed + suf. *-íkos* > L. *-icus* > E. *-ic*, pertaining to): relating to the *symplast* or belonging to it; applied to the movement of substances that enter the cytoplasm of living cells. Cf. **apoplastic**.

symphyogenesis, pl. **symphyogeneses** (Gr. *symphyós* < *symphýo*, to grow together + *génesis*, origin): see **symphogenous**.

symplast (Gr. *sým*, with + *plastós*, modeled, formed): a plurinucleate protoplast, as opposed to the *monoplast* of uninucleate cells.

symplesiomorphy (Gr. *sým*, with, together with + *plésios*, nearby, recent + *morphé* > *morphía*, shape): a term used in cladistics to refer to a shared ancestral character state. If the character is not shared, it is called *plesiomorphy*.

sympodial (NL. *sympodium* < Gr. *sým*, with, together with + NL. *podium* < Gr. *pódion*, dim. of *poús*, genit. *podós*, foot, support + L. suf. *-alis* > E. *-al*, relating to or pertaining to): refers to the formation of an apparent main axis from successive secondary axes, like that of the conidiophores of *Helminthosporium* and *Stachybotrys* (dematiaceous asexual fungi). See **sympodium**.

Sympodial conidiophore of *Stachybotrys chartarum* (=*S. atra*), x 400 (*CB*).

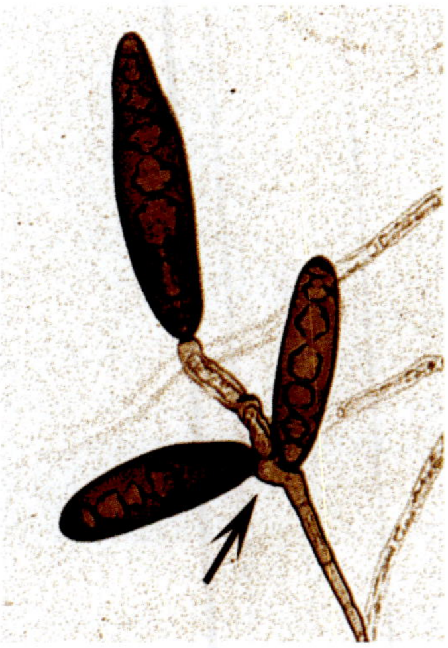

Sympodial conidiophore of *Helminthosporium* sp., x 300 (*MU*).

sympodial ramification (NL. *sympodium* < Gr. pref. *sym-*, which implies the idea of concrescence, of solidarity + NL. *podium* < Gr. *pódion*, dim. of *poús*, genit. *podós*, foot, support + L. suf. *-alis* > E. *-al*, relating to or pertaining to; L. *ramificatio*, action and effect of branching < *ramus*, pl. *rami*, branch + *fic* < *facio*, to make + suf. *-ation* > E. *-ation*, action, state or condition): the formation of a main axis from successive alternate secondary branches. See **sympodium**.

sympodioconidium, pl. **sympodioconidia** (NL. *sympodium* < Gr. *sým*, with, together with + NL. *podium* < Gr. *pódion*, dim. of *poús*, genit. *podós*, foot, support + L. *conidium* < Gr. *kónis*, powder + dim. suf. *-ídion*): see **sympodulospore**.

sympodium, pl. **sympodia** (NL. *sympodium* < Gr. *sým*, with, together with + NL. *podium* < Gr. *pódion*, dim. of *poús*, genit. *podós*, foot, support): a type of conidiophore that appears to be simple, but which ends and rebranches repeatedly; due to the successive formation of branches on alternate sides of the main axis, a structure with sympodial branching adopts a zigzag shape, as occurs, e.g., in the conidiophore of *Cercospora* and *Helminthosporium* (dematiaceous asexual fungi), and the perithecial hairs of *Ascotricha guamensis* (Sordariales). Cf. **monopodium**.

sympodulospore (NL. *sympodium* < Gr. *sým*, with, together with + NL. *podium* < Gr. *pódion*, dim. of *poús*, genit. *podós*, foot, support + Gr. *sporá*, spore): an asexual reproductive spore (conidium) that is formed on a conidiophore with sympodial branching. After formation of the terminal conidium, growth of the conidiophore continues as a lateral branch that originates beneath the conidium; this branch soon produces another terminal conidium and the process of branching is repeated. Generally the successive branches form on alternate sides of the axis of the conidiophore, resulting in a zigzag structure. This type of spore also is called **sympodioconidium** or *terminal botryospore*; it is present, e.g., in *Tritirachium roseum* and *Arthrobotrys oligospora* (moniliaceous asexual fungi).

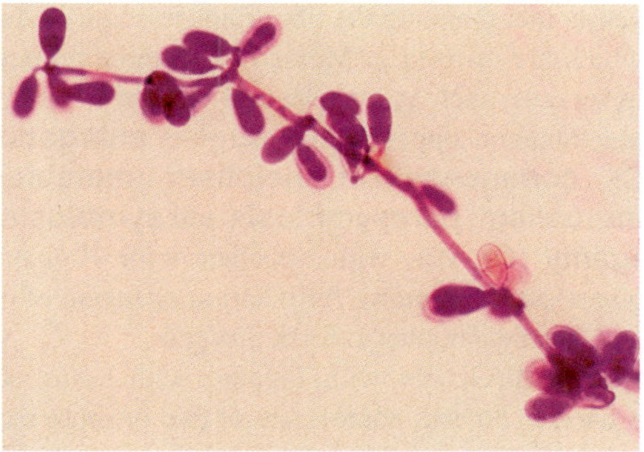

Sympodulospores of *Arthrobotrys oligospora*, x 400 (*MU*).

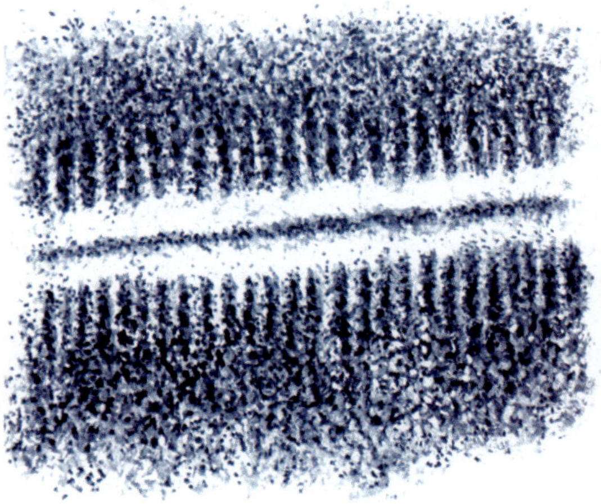

Synaptonemal complex of *Neottiella* (=*Octospora*) *rutilans*. Longitudinal section through a bivalent; the two homologous chromosomes (C) remain paired by the synaptonemal complex, which consists of two banded lateral components (L) separated by the central region (CR). Drawing based on a transmission electron micrograph published by Beckett *et al.* in their *Atlas of Fungal Ultrastructure*, 1974, x 150,000 (*MU*).

syn- (Gr. *sýn*, with, together with): an inseparable prep. that indicates aggregation, adhesion, concrescence. Equal to **sym-**.

synanamorph (Gr. *sýn*, with, together with + *anamorphóo*, to transform): one of the various conidial forms or states associated with certain pleomorphic fungi, such as the interesting dematiaceous asexual fungus *Phialocorona polymorpha* (=*Kionochaeta pleomorpha*), which has, besides the characteristic phialidic state, three other conidial states, or synanamorphs: *Sporidesmiella*, *Selenosporella* and *Heteroconium*, whose conidiophores originate from the same hyphae as the conidiophore of *Phialocorona*. Cf. **anamorph**, **teleomorph** and **pleomorphic**.

synapomorphy (Gr. *sýn*, with, together with + *apó*, far from + *morphé* > *morphía* > E. des. *-morphy*, shape): a term used in cladistics to refer to a shared, derived character state. When such a character is not shared, it is called *apomorphy*.

synaptonemal complex (Gr. *synápto*, to bind, unite + *nêma*, thread, filament + L. suf. *-alis* > E. *-al*, relating to or belonging to; LL. *complexus*, complex, totality): a structure that maintains the precise alignment of the two homologous chromosomes of a bivalent throughout its length. The synaptonemal complexes represent evidence of meiosis since crossing over of genes occurs in them. These complexes consist of two banded lateral components separated by a central region 10,000 to 12,000 nm in width, within which there is a central component. This structure can be seen in *Neottiella* (=*Octospora*) *rutilans* (Pezizales).

synascus, pl. **synasci** (Gr. *sýn*, with, together with + *askós*, purse, sac; ascus): a cell that contains several asci; applied to the female gametangium or ascogonium of *Ascosphaera* (Ascosphaerales), which, after plasmogamy, is converted into an inflated nutriocyte in which clusters of asci develop. The asci are formed from a system of ascogenous hyphae, similar to the development of a cleistothecium, from which it differs by being unicellular and by lacking a hyphal peridium. The structure formed by *Mixia osmundae* on ferns of the gen. *Osmunda* was originally called a synascus, but this fungus, which was initially included in the Taphrinales, is now considered as a basidiomycete. See *sporiferous sac* in the entry of **spore case**.

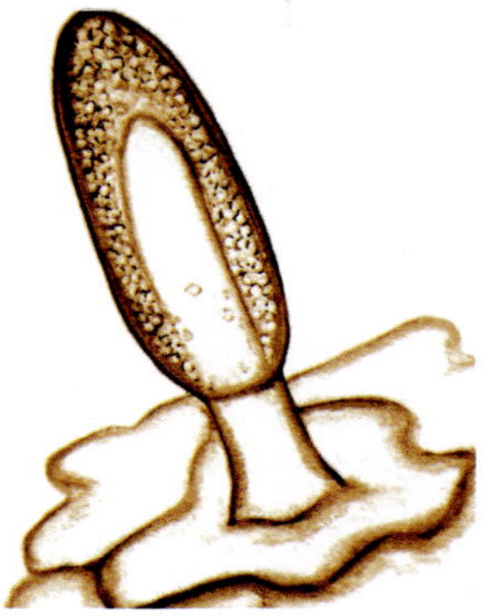

Synascus of *Mixia osmundae* in a host cell of *Osmunda cinnamomea*. The mature spore sac (synascus), which has a central column surrounded by ascospores, is supported by a pedicellar cell that emerges from a chlamydospore contained in a host cell, x 520 (*MU*).

synchronized culture (Gr. *sýnchronos*, a series of events that occur or originate at the same time < *sýn*, with + *chrónos*, time + suf. *-íkos* > L. *-icus* > E. *-ic*, relating to or belonging to; L. *cultus*, culture, action and effect of culturing): a culture of microorganisms that is manipulated so that all of the cells that comprise it divide simultaneously; this type of culture is utilized in biotechnology.

synchronospore (Gr. *sýn*, with, together with + *chrónos*, time + *sporá*, spore): a spore produced simultaneously together with other adjacent spores, as occurs with the *botryoblastospores*.

syncytium, pl. **syncytia** (Gr. *sýn*, with, together with + *kýtos*, cavity; here, cell + L. dim. suf. *-ium*): a fusion plasmodium of the Myxomycetes that results from the total fusion of amoeboid, uninucleate cells (myxamoebae) through the disappearance of their individual membranes. See **aphanoplasmodium**, **phaneroplasmodium** and **protoplasmodium**. Cf. **pseudoplasmodium**.

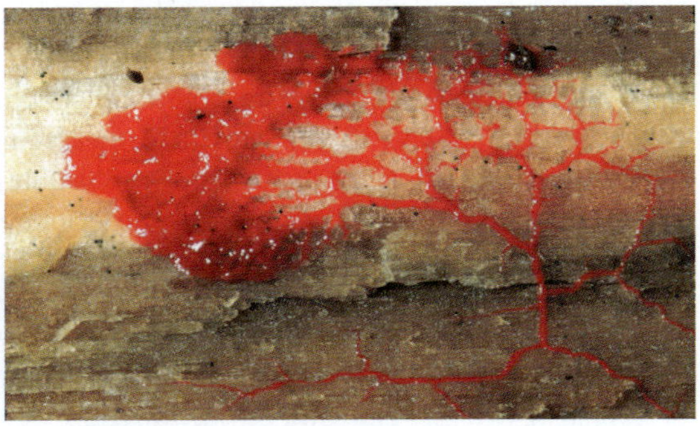

Syncytium (phaneroplasmodium) of *Physarum roseum* (Physarales) on wood, x 1 (*RS*).

syndrome (NL. *syndrome* < Gr. *syndromé*, combination, syndrome < *sýn*, with, together with + *drameín*, to run): the complex of signs and symptoms that occur simultaneously, that are characteristic of a particular disease or abnormality.

synecology (Gr. *sýn*, with, together with + *oikéo*, to inhabit + *lógos*, word > E. des. *-logy*, treatise, discourse): in the broad sense, syn. of *biosociology*. In a strict sense, the part of ecology that studies the relationships between the communities of living organisms and their environment.

synergy, **synergism** (NL. *synergia* < Gr. *synérgeia*, *synérgos*, cooperation, mutual help; *synérgeia* + suf. *-ismós* > L. *-ismus* > E. *-ism*, state, phase, tendency, action): in general, combined action. Unlike metabiosis, in synergy or synergism two organisms coexist in a synergistic manner, in an associated or combined action on a particular substrate or host. Among the saprobic fungi, it is common for the alteration of a substrate to only occur as the result of the combined activities of two species that have complementary enzymatic activities. Also, among the parasitic fungi there is a reciprocal enhancement of pathogenicity when two species coexist on the same host. An example of synergy is the fermentative action performed by the bacteria and yeasts that make up the tibi grains in sugary liquids, which are utilized in Mexico to produce fermented beverages, such as tepache and vinegar of tibi. The phenomenon opposed to synergy is **antagonism**. Cf. **commensalism**, **helotism**, **metabiosis**, **mutualism**, **parasymbiosis** and **symbiosis**.

syngamy (Gr. *sýn*, with, together with + *gamía* < *gámos*, sexual union): fertilization or fusion of male and female gametes to form a zygote.

synisonym (Gr. *sýn*, with, together with + *ísos*, equal + *ónoma*, *ónyma*, name): one of two or more names that have the same basonym.

synkaryon (Gr. *sýn*, with, together with + *káryon*, nucleus): a pair of nuclei produced by successive conjugate divisions of two sexual nuclei that are joined but not united or fused.

synnema, pl. **synnemata** (Gr. *sýn*, with, together with + *nêma*, filament): a compact bundle of erect conidiophores that bear conidia on the apex and sometimes along the sides; the individual conidiophores can be united laterally or not. Synnemata are characteristic of the asexual stilbellaceous fungi, such as *Doratomyces*, *Cephalotrichum*, *Dendrostilbella*, *Graphium*, *Antromycopsis* and *Trichurus*, among other gen., and some moniliaceous asexual fungi, such as *Penicillium claviforme* and *Beauveria felina* (=*B. cretacea*). Cf. **coremium**.

Synnemata of *Beauveria felina* (=*B. cretacea*) on agar, x 1.5 (*MU*).

Synnemata of *Beauveria felina* (=*B. cretacea*) on agar, x 7 (*MU*).

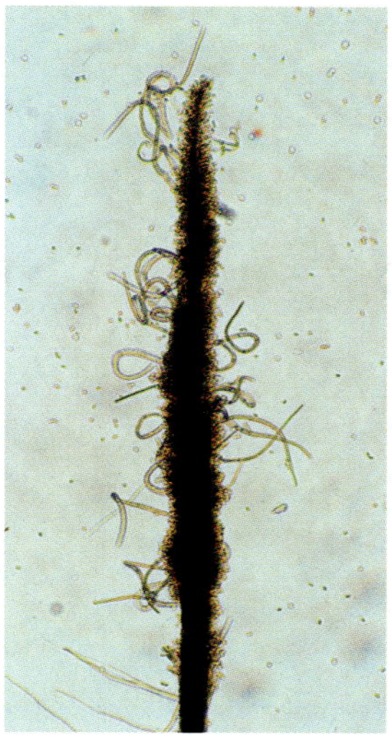

Synnema of *Trichurus spiralis*, x 200 (*CB*).

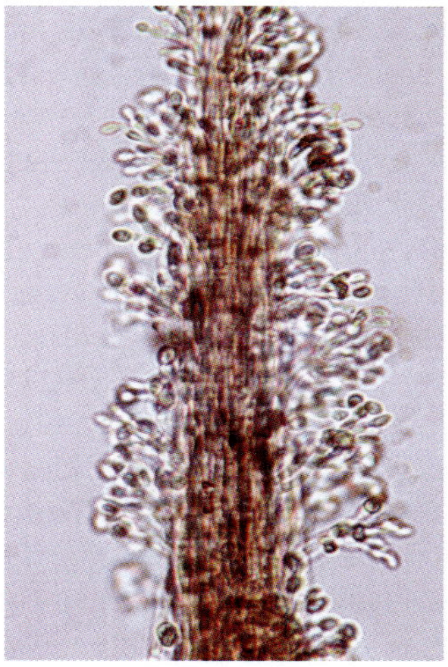

Synnema of *Cephalotrichum microsporum* (=*Doratomyces microsporus*), x 1,500 (*MU*).

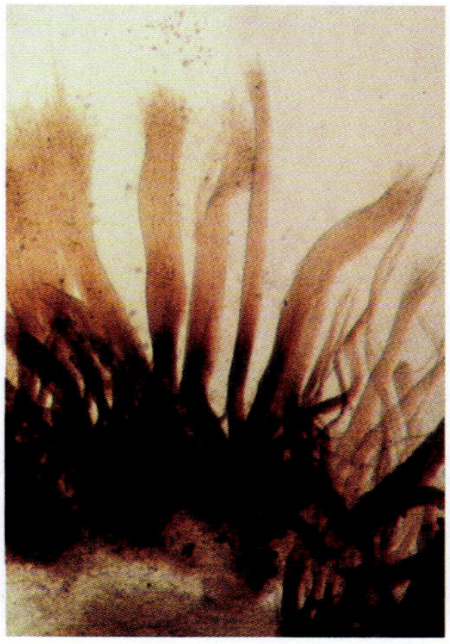

Synnemata of *Graphium* sp., x 100 (*MU*).

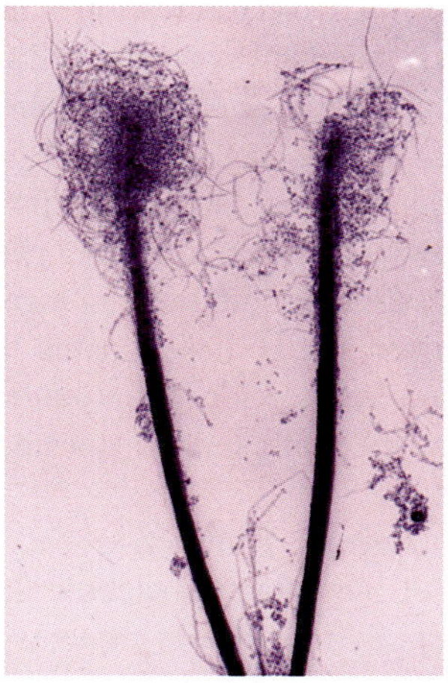

Synnemata of *Trichurus spiralis*, x 200 (*RTH*).

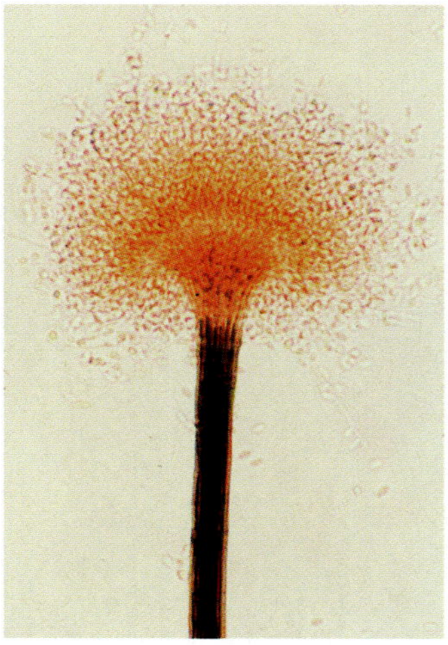

Synnema of *Graphium ulmi*, x 100 (*MU*).

Synnemata of *Beauveria felina* (=*B. cretacea*) on agar, x 3 (*MU*).

Synnemata of *Penicillium vulpinum* (=*P. claviforme*) on agar, x 8 (*MU* & *CB*).

Synnemata of *Antromycopsis smithii* on agar, x 2 (*MU*).

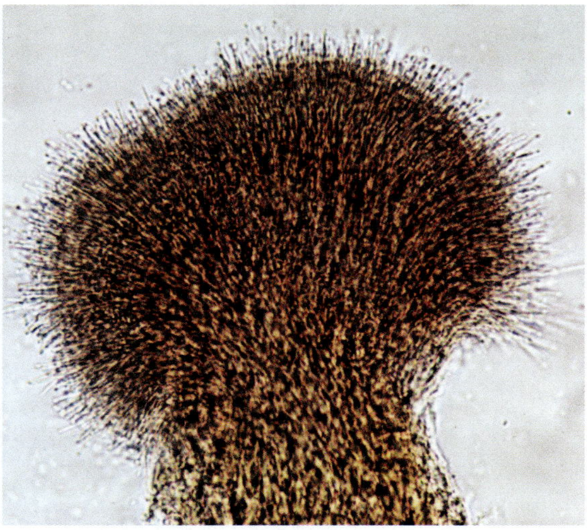

Synnema of *Dendrostilbella* sp., x 200 (*MU*).

synnematous (Gr. *sýn*, with, together with + *nêma*, genit. *nêmatos*, filament + L. suf. *-osus* > E. *-ous*, full of, augmented, prone to): with *synnemata*. Cf. **mononematous**.

synonym (Gr. *sýn*, with, together with + *ónoma*, *ónyma*, name): one of two or more names for the same taxon, especially a later or illegitimate name. If two or more names are based on the same type, they are *homotypic* or *obligate synonyms*; if they are based on different types they are *heterotypic synonyms*.

syntaxonomy (Gr. *sýn*, with, together with + Gr. *táxon*, to arrange, constitute < *táxis*, arrangement + *nómos* > E. des. *-nomy*, legitimate according to the law, normal): the taxonomy of communities. Cf. **idiotaxonomy**.

syntype (Gr. *sýn*, with, together with + *týpos*, type, mark, model): one of two or more elements cited by an author in the original publication of a taxon when no holotype was designated, or any of the elements designated simultaneously by the author as types.

systematics (Gr. *systematikós* < *sýstema*, complete, aggregated, all, methodical + suf. *-tikós* > L. *-ticus* > E. *-tic*, relation, fitness, inclination or capacity): in biology, *biosystematics* (Gr. *bíos*, life) is the study of the relationships and classification of the living world or biota in a conceptual system or hierarchy. Includes the subdisciplines *taxonomy* and *nomenclature*.

systemic (LL. *systemat-*, *systema* < Gr. *sýstema*, complete, aggregated, all, methodic compendium + suf. *-íkos* > L. *-icus* > E. *-ic*, pertaining to or relating to): affecting the body in a general manner. A systemic parasite is disseminated throughout the host. A systemic fungicide (or antibiotic) is absorbed, especially by the roots, and translocated to other parts of the plant.

systole (Gr. *systolé*, contraction): a phase in the movement of a contractile vacuole during which it contracts rapidly to expel its contents to the interior of the cell or to a non-contractile reservoir, as occurs in certain plasmodia and zoospores. Cf. **diastole**.

Tubaeform basidiocarps of *Cotylidia* sp. (*RV*).

t

tactiosensible, **tactiosensitive** (L. *tactus*, touch + *sensibilis*, sensitive, yielding easily to the action of certain natural agents < *sensus*, sense + *-ilis*, suf. added to denote capacity, or that has the character of; L. *tactus* + *sensitivus*, sensitive, that is capable of feeling < *sensus* + suf. *-ivus* > E. *-ive*, quality or tendency, aptitude, adaptability): applied to Discomycetes in which ascospore discharge is stimulated by contact, whether by a solid body (such as an insect) or by light air currents.

tanninolytic (Fr. *tanin*, tannin < ML. *tannum*, tannin bark, oak bark + Gr. *lytikós*, capable of dissolving < *lýtos*, dissoluble, broken, loose + suf. *-íkos* > L. *-icus* > E. *-ic*, pertaining to or relating to): capable of degrading tannins by means of the enzyme tannase; e.g., *Aspergillus carbonarius* (moniliaceous asexual fungi) has been utilized in solid state fermentation systems to degrade the tannins of the pods of the carob tree, making them more flavorful, with the object of utilizing them as cattle feed.

tapé ketella (Indon. *tapé ketella*): the name given to fermented tubers of cassava, consumed as food in Indonesia. The process of preparing this fermented traditional food involves several steps: the tubers are peeled, washed, cooked in steam, cooled, and inoculated with pulverized *ragi*; afterwards the inoculated tubers are incubated for 2 to 3 days in a bamboo basket lined and covered with banana leaves, so that the microorganisms of the ragi or inoculum can carry out a solid state fermentation, whose essential biochemical changes are the hydrolysis of the starch to maltose and glucose (which give a sweet taste) and the fermentation of part of the sugars to ethanol and organic acids. The essential microorganisms are mucoraceous molds, such as *Amylomyces rouxii* (previously named *Chlamydomucor oryzae*) and at least one yeast of the *Hyphopichia* (=*Endomycopsis*=*Saccharomycopsis*) *burtonii*/*chodati* (Sacharomycetales) type. In Indonesia, Malaysia, China and other Asiatic countries there exist various fermented traditional foods and beverages, sweet and sour alcoholic (primitive wines and pastas), prepared with rice or cassava, in whose fermentation process ragi is utilized as inoculum or starter culture. Cf. **ontjom**, **pozol** and **tempe**. See **fermentation** and **ragi**.

Tapé ketella (fermented cassava tubers) from Bogor, Indonesia, x 0.2 (*MU*).

tar spot (ME. *terre, tarre*, a thick, brown to black, viscous liquid obtained by the distillation of wood, coal, peat, and other organic materials; ME. *spot*, mark, blot): the common name for the flat, circular, black stromata, similar to tar, containing apothecia, formed by *Rhytisma acerinum* (Rhytismatales) in maple leaves, where this fungus overwinters (tar spot of maple). The surface of each mature stroma is characterized by radiate fissures along which the stroma splits above the apothecia in the spring. In Mexico this fungus is known by the common name *mancha de chapopote* (Azt. *chapopote*, crude petroleum, viscous and very heavy; much used for daubing among the Aztecs < Nahuatl *tzapotl-poctli*, sapote that smokes, alluding to the black color, like the black sapote, *tlizápotl*, and to the facility with which crude petroleum burns, giving off smoke).

Tar spot on a maple leaf caused by *Rhytisma acerinum*; the mature stroma, similar to tar, bears apothecia in the interior, and splits along the radiate fissures, x 16 (*RTH*).

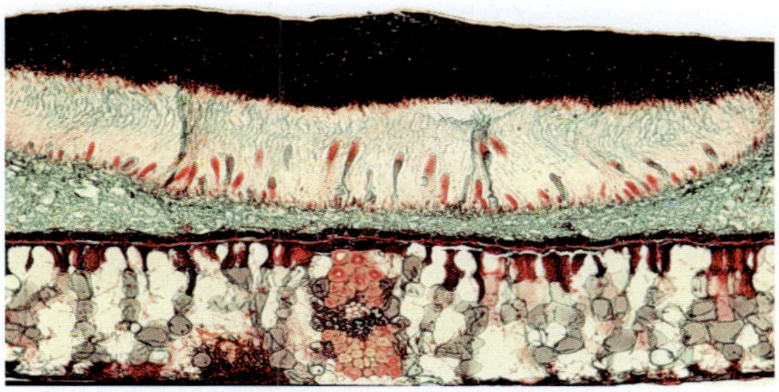

Tar spot on a maple leaf caused by *Rhytisma acerinum*; a longitudinal section of an apothecium, showing the asci in its interior, x 100 (*RTH*).

tartarean, **tartareous** (L. *tartareus* < Tartar of Hell, frightening, horrible, infernal): with a thick, rough, crumbly surface, like that of certain crustaceous lichens, e.g., *Ochrolechia tartarea*, of the Pertusariales.

taxis, pl. **taxes** (Gr. *táxis*, arrangement, position, disposition): a translative reflex or movement of orientation of free-moving (flagellate), and usually simple, organisms that can move from one place to another (in water or another liquid medium) in relation to a source of stimulus (such as oxygen, a chemical gradient, light, an electromagnetic field, etc.). The reflex is positive when the movement is toward the source of the stimulus, and negative when it moves away from it. Various types of stimuli are associated with many other taxes. See **aerotaxis**, **chemotaxis**, **electrotaxis**, **gravitaxis**, **phototaxis**, and **rheotaxis**.

taxon, pl. **taxa** (Gr. *táxon*, to put in order, constitute < *táxis*, arrangement): a taxonomic unit of any rank or hierarchy within a taxonomic or classification system, from species up to kingdom or superkingdom. The Botanical Code recommends that the names of taxa of all levels be written in italics.

taxonomic (Gr. *táxon*, to put in order, constitute + *nómos* > E. des. *-nomy*, to make legitimate in accordance with the law, normal + suf. *-íkos* > L. *-icus* > E. *-ic*, pertaining to or relating to): relating to or pertaining to *taxonomy*.

taxonomy (Gr. *táxon*, to put in order, constitute < *táxis*, arrangement + *nómos* > E. des. *-nomy*, to make legitimate in accordance with the law, normal): classification, in general. The part of natural history that treats of the classification of living beings. Analysis of the characteristics of an organism with the object of assigning it to a category or taxon. See **nomenclature** and **systematics**.

tea fungus, pl. **fungi** (Chin. *t'e*, tea, a shrub, *Camellia sinensis*, used to prepare an aromatic beverage, especially from the leaves; L. *fungus*, fungus < Gr. *spóngos*, *sphóngos*, sponge): the common name of the symbiotic association of yeasts (*Saccharomycodes ludwigii*, of the Saccharo-mycetales) and bacteria (*Acetobacter xylinum*) that is cultivated in tea. This microbial association is given various names in different parts of the world: *mother of vinegar*, *kombucha*, *Wunderpilze*, and others. This traditional beverage has been consumed for medicinal and prophylactic purposes for many years, similar to the products called **probiotics**. Cf. **kefir grains**, **mother of vinegar** and **tibi grains**. See **microbioglea**.

-techny (Gr. *téchne*, technic, art, ability): a suf. used in terms such as *phytotechny*, the discipline that employs genetic improvement techniques to obtain new and better varieties of agricultural crops, e.g., that are more resistant to plant diseases.

tele-, **tel-** (Gr. *téle*, far away): a pref. used in terms such as *telemorphosis*, defined below, and *telescope*, an optical instrument for observing distant objects by means of the refraction of light rays through a lens, or of the reflection of light rays from a concave mirror.

teleblem, **teleoblem**, **teleoblema** (Gr. *téleos*, perfect, entire, complete + *blêma*, covered): syn. of **protoblem**.

telemorphosis, pl. **telemorphoses** (Gr. *téle*, far away + *morphé*, shape + suf. *-osis*, condition or state): the phenomenon in which one hypha stimulates another from a certain distance, inducing morphological and physiological changes in it, such as is observed in the sexual reproductive process in *Achlya bisexualis* (Saprolegniales). In this case, a primordial hypha of the oogonium secretes a hormone (antheridiol) that acts on a primordial hypha of the antheridium, which reacts by setting off the series of morphological and physiological changes that preceed gametangial contact.

teleo- (Gr. *téleos*, complete, perfect): a pref. used in terms such as *teleomorph*.

teleomorph (Gr. *téleos*, complete, perfect + *morphé*, shape): the sexual (perfect) state of a fungus (ascogenous or basidiogenous), whose spores are produced by meiosis. Cf. **anamorph** and **holomorph**.

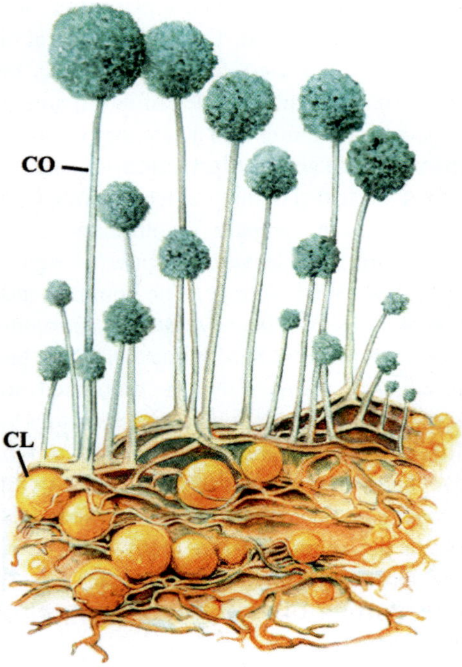

Teleomorph (*Eurotium rubrum*) of *Aspergillus rubrobrunneus* (the anamorph); shown are the corresponding cleistothecia (CL) and conidiophores (CO), x 45 (*MU*).

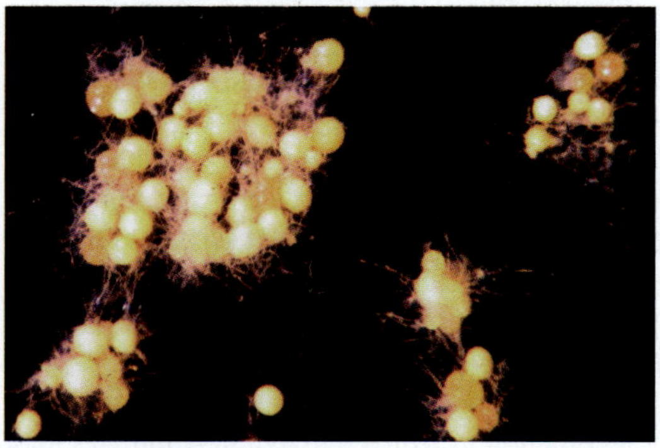

Teleomorph (*Eurotium rubrum*) of *Aspergillus rubrobrunneus* (the anamorph); shown are the corresponding cleistothecia, x 30 (*MU*).

teleutosorus, pl. **teleutosori** (Gr. *teleuté*, completion, termination, end + *sorós*, pile, heap): an earlier name for the **telium**.

teleutosporodesm (Gr. *teleuté*, completion, termination, end + *sporá*, spore + *desmós*, cord, filament): another old term for *teliospore*.

Telia (**teleutosori**) of *Sporisorium* (=*Sphacelotheca*) *sorghi* (Ustilaginales) on a sorghum spike, x 9 (*MU*).

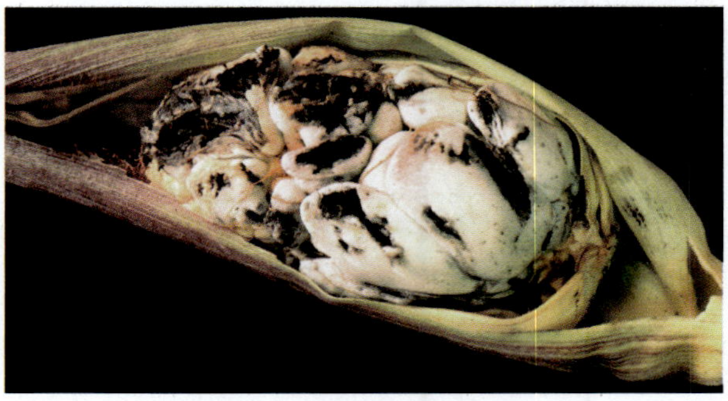

Telia (**teleutosori**) of *Ustilago maydis* (Ustilaginales) on an ear of corn (maize), x 0.2 (*MU*).

teliospore, **teleutospore** (Gr. *télos*, *téleos*, end, or *teleuté*, completion, termination, end + *sporá*, spore): a thick-walled latent spore, unicellular or multicellular, with binucleate (dikaryotic) cells, produced in a telium; characteristic of the rusts and smuts (Uredinales and Ustilaginales). In the teleutospores, now more correctly named teliospores, is where karyogamy and meiosis occur, after which the basidium is formed by germination.

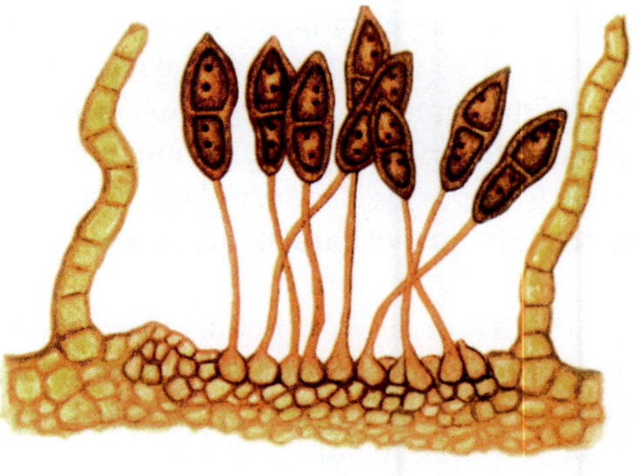

Teliospores in a telium of *Puccinia graminis tritici* (Uredinales) on a wheat leaf, x 300 (*MU*).

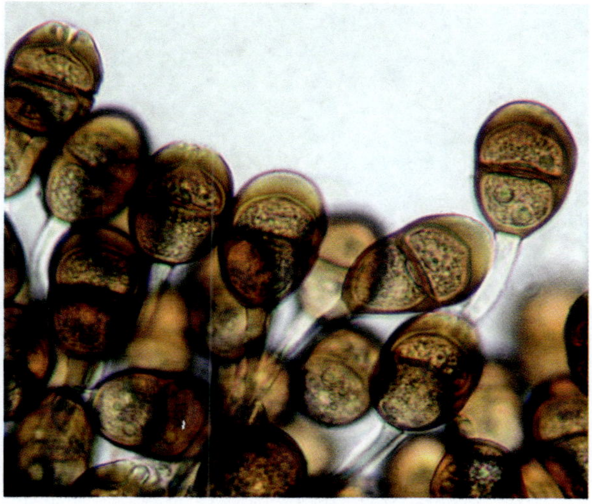

Teliospores in a telium
of *Puccinia* sp. (Uredinales), x 300 (*RV*).

telium, pl. **telia** (NL. *telium* < Gr. *télos*, *téleos*, end + L. dim. suf. *-ium*): a group of binucleate cells in a sorus that produces teliospores, which are resistant spores in the plant parasitic fungi called rusts and smuts (Uredinales and Ustilaginales). The rust telia are a state in the life cycle that is always present, unlike spermogonia, aecia and uredinia that can be lacking in a particular cycle, depending on the species being treated. The morphology of the telium varies, as well as its position on the host plant; in *Puccinia* (Uredinales), e.g., they are erumpent structures on the leaves and stems. In *Ustilago* and *Sphacelotheca* (Ustilaginales) the sori of teliospores are mainly formed in the inflorescences of the host plants, mainly grasses.

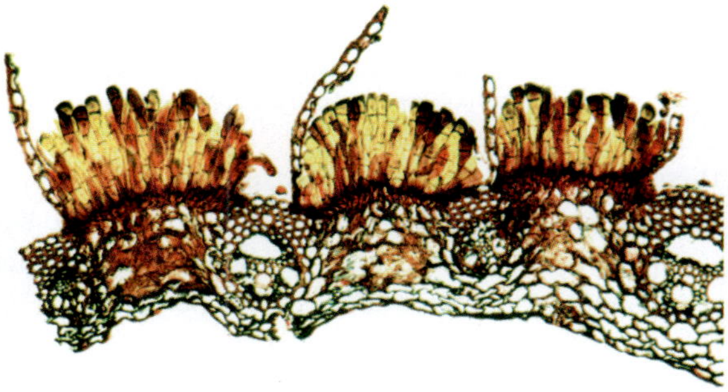

Telia with teliospores of *Puccinia graminis tritici* (Uredinales) on a wheat leaf, seen in longitudinal section, x 100 (*MU*).

telo-, tel- (Gr. *télos*, end): a pref. employed in terms such as *telophase*, the final phase of mitosis and meiosis.

tempe, tempeh kedele (Indon. *tempe*, *tempeh kedele*): a fermented food based on soybeans that is consumed mainly in Indonesia, Malaysia, the Netherlands, Canada and the United States. In the traditional process of preparing tempe, the soybeans are washed, partially cooked and soaked to promote fermentation by lactic bacteria (among them *Lactobacillus casei*) and yeasts, such as *Hyphopichia* (=*Pichia*) *burtonii*, of the Saccharomycetales, and *Candida diddensiae* and *Rhodotorula rubra*, of the cryptococcaceous asexual yeasts. The bacterium *Klebsiella pneumoniae* is the predominant species in commercial tempes and is responsible for the increase in vitamin B_{12} during the fermentation. After removing the seed coats, washing and draining the soybeans, they are molded into flat cakes, inoculated with spores of *Rhizopus microsporus* (=*Rh. oligosporus*), Mucorales, wrapped in leaves of banana, *Hibiscus* or other plants, and incubated for 30-40 hours at ambient temperature (30-32ºC). Fermentation of the substrate is in solid state; the mycelium of the fungus invades the soybean cakes, permeating the cotyledons and forming a paste similar to cheese. The metabolic activity of the fungus (mainly the production of amylolytic, proteolytic and lipolytic enzymes, besides a strong antioxidant) causes notable changes that make the tempe a more nutritious food, with better taste and aroma and a higher content of vitamins of the B complex, and more easily digestible when consumed as a substitute for meat. As with other traditional fermented foods (such as the *tapé ketella* of Indonesia), there exist a variety of tempes that are consumed in North America and Asia, and in whose preparation the principles of the fermentation of tempe are applied, but utilizing other raw materials, e.g., *tempe benguk* (made with another legume, *Mucuna pruriens*) and *tempe bongkrek* (made with coconut or peanuts). Cf. **ontjom**, **pozol** and **tapé ketella**. See **fermentation**.

Cakes of **tempe** incubated 2 days in plastic bags.
The mycelium of *Rhizopus microsporus (=Rh. oligosporus)* has already completely invaded the substrate, x 0.2 (*MU*).

tenacle (L. *tenacula* < *tenax*, genit. *tenacis*, to have, hold fast, grasp + dim. suf. *-ula*): see **haerangium**.

tepache (prob. < Nahuatl *tepachoa*, to press or crush something with a stone): a refreshing fermented beverage, very popular in Mexico, initially prepared with grains of corn (maize), but at present made from squeezed or crushed fruits (mainly pineapple), water and brown sugar. The fermentation is of short duration (approximately 24 hours) and of the lactic-alcoholic-acetic type; it is due to a succession of various microorganisms, mainly homo- and heterolactic bacteria (of the genera *Lactobacillus* and *Leuconostoc*) and yeasts (particularly *Saccharomyces*, *Pichia* and *Hanseniaspora*, of the Saccharomycetales). See **fermentation**.

Tepache being sold in a market place in Mexico. The fermented drink is inside a barrel, x 0.2 (*MU*).

Giant colony of *Pichia membranifaciens* (Saccharomycetales), a yeast species which ferments **tepache**, on agar, x 1 (*MU*).

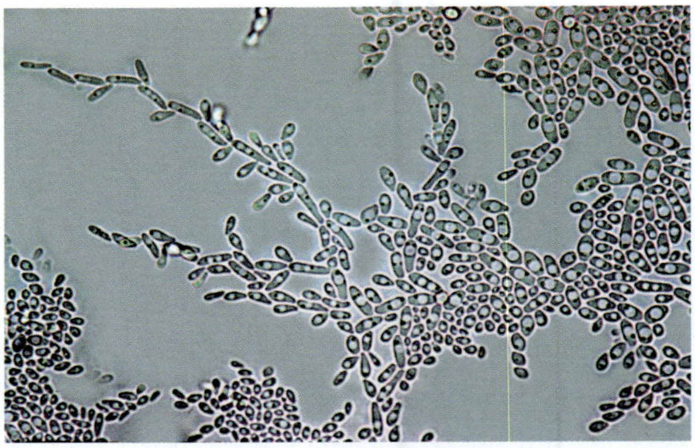

Pseudomycelium of *Saccharomyces cerevisiae*, a yeast species that ferments **tepache**, x 680 (*MU*).

terato-, **terat-** (Gr. *téras*, genit. *tératos*, prodigy, monster): a pref. used in terms that allude to the abnormal or monstrous, like *teratogenesis*.

teratogenesis, pl. **teratogeneses** (Gr. *téras*, genit. *tératos*, prodigy, monster + *génesis*, engenderment): the genesis of monstrous forms in animal or plant organisms.

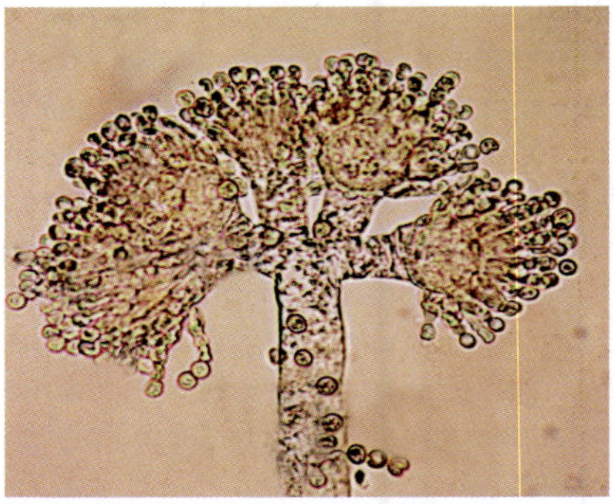

Conidial head of *Aspergillus flavus* (moniliaceous asexual fungi) showing an unusual **teratogenesis**, with four vesicles, each with phialides and conidia, x 400 (*MU*).

terebrate (L. *terebratus*, pierced, perforate < *terebrare*, to drill, perforate + suf. *-atus* > E. *-ate*, provided with or likeness): bored, drilled. In lichenology, it refers to a thallus that has widely spaced perforations, like those of *Verrucaria terebrata* (Verrucariales).

terebrator (L. *terebratus*, one who drills or perforates): drilling apparatus; has been utilized for the trichogynous hypha of the lichens.

terete- (L. pref. derived from *teres*, genit. *tereis*, round, cylindric): round in cross section, cylindric; often used in terms such as *tereticauli*, *teretifolium*, pl. *teretifolia*, etc., i.e., with round stems, leaves, etc. E.g., the thallus of the lichen *Physcia dubia* (=*Ph. teretiuscula*) has rounded lobules, and the perithecial stroma of *Camillea leprieurii* (Xylariales) is cylindrical.

term (L. *terminus* < Gr. *termón*, limit, end, termination or extension): a word or expression that has a precise meaning in some usages, or that is peculiar to a science, art, profession, subject or discipline.

terminology (L. *terminus* < Gr. *termón*, limit, end, termination or extension + E. des. *-logy*, study, treatise < Gr. *lógos* < *légo*, word): the collection of terms or proper words of a particular profession, science or subject. See **dictionary**, **glossary**, **lexicon** and **vocabulary**.

terminus spore (L. *terminalis*, terminal, pertaining to the end < *termen*, genit. *terminis*, end, limit; NL. *spora* < Gr. *sporá*, spore): applied to the phialospore of a monosporic phialide, i.e., with the formation of this spore, growth of the phialide ceases.

termitophilic (NL. *termites*, termite < L. *tarmes*, wood-eating worm + Gr. *phílos*, have an affinity for + suf. *-íkos* > L. *-icus* > E. *-ic*, pertaining to or relating to): growing in termite nests, like the fungus *Termitomyces* (Agaricales).

terrestrial (L. *terrestris*, pertaining to or relating to the earth; growing on the land, as opposed to growing in water): equal to **terricole**.

terricole, **terricoline**, **terricolous** (L. *terra*, earth, soil + suf. *-cola*, inhabitant < *colere*, to inhabit): an organism that lives on the soil. Many species of micro- and macromycetes are terricolous. Equal to **terrestrial**. See **epigeal** and **hypogeous**.

tertiary mycelium, pl. **mycelia** (L. *tertius*, third + suf. *-aris* > E. *-ary*, concerning; NL. *mycelium* < Gr. *mýkes*, fungus + L. suf. *-elis*, concerning + dim. suf. *-ium*): *Basidiomycetes*. The mycelium that is differentiated from the secondary mycelium (dikaryotic) by the formation of fructifications. This term tends to be suppressed. See **primary mycelium** and **secondary mycelium**.

terverticillate (L. *ter* > *tri*, three + NL. *verticillatus*, arranged in a verticil < *verticillus*, verticil + suf. *-atus* > E. *-ate*, provided with or likeness): see **triverticillate**.

Terverticillate conidial head of *Penicillium* sp. (moniliaceous asexual fungi), x 600 (*MU*).

tesgüino (Nahuatl *tecuin*, heart beat): an opaque maize beer consumed as a nutritious and ceremonial beverage by several ethnic groups of northern Mexico, mainly the Tarahumares and the Huicholes. The fermenting microbiota includes lactic bacteria (*Lactobacillus* spp.) and yeasts (especially *Saccharomyces cerevisiae* and *S. kluyveri*, of the Saccharomycetales, and *Candida guilliermondii* and *C. valida*, of the cryptococcaceous asexual yeasts). Tesgüino is similar to other opaque beers that are prepared in other parts of the world, e.g., the African beers (whose generic name is *pombe*), made of maize, sorghum or milo, among them the Bantu beer or kaffir beer of South Africa. See **fermentation**.

Tesgüino in a clay pot, prepared in Gonogochi, Chihuahua, Mexico, x 0.1 (*PL*).

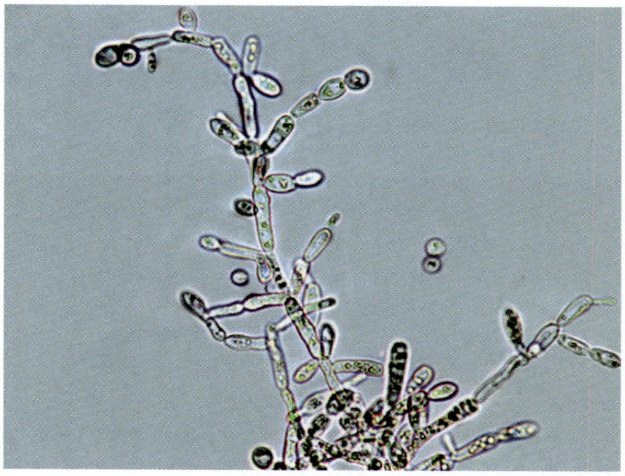

Pseudomycelium of *Saccharomyces cerevisiae*, a yeast species that ferments **tesgüino**, x 750 (*MU*).

tessellate (L. *tessellatus*, checkered < *tessella*, mosaic + suf. *-atus* > E. *-ate*, provided with or likeness): checkered, as if formed of small squares or mosaics, like the surface of the sporangium of *Perichaena tessellata* (Trichiales), and the thallus of the lichens *Haematomma subpuniceum*, *Rhizocarpon geographicum* and *Acarospora* (Lecanorales).

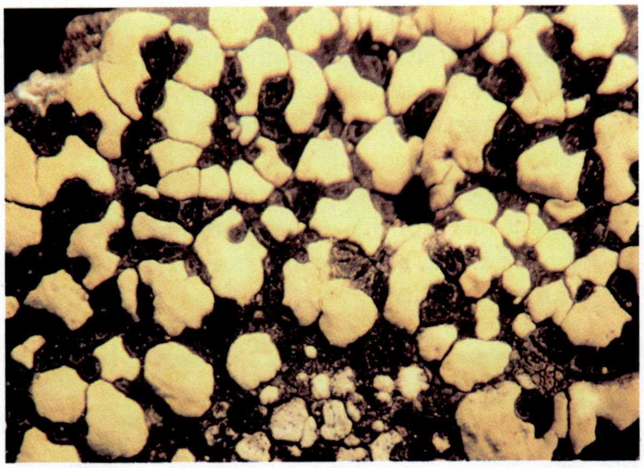

Tessellate thallus of *Haematomma subpuniceum*, x 10 (*MU*).

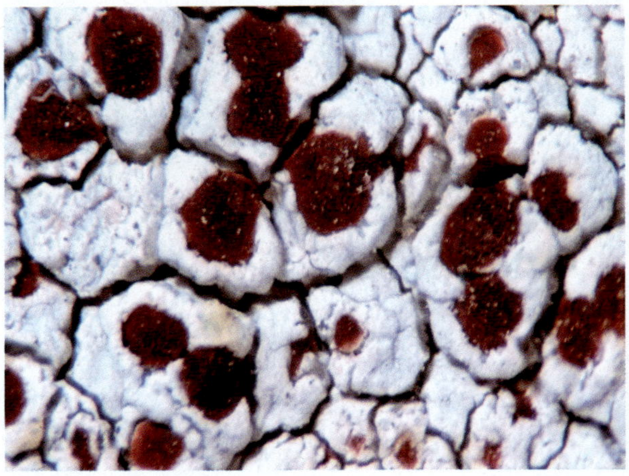

Tessellate thallus of *Rhizocarpon geographicum*, x 6.5 (*MU*).

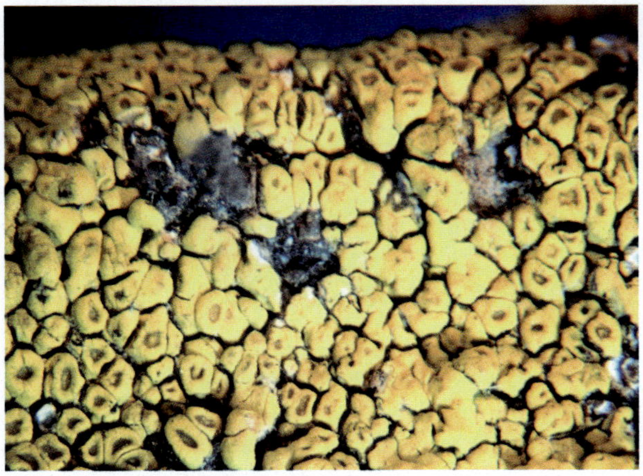

Tessellate thallus of *Acarospora* sp., x 10 (*MU*).

testaceous (L. *testaceus*, made of brick or tile < *testa*, tile or brick; vessel or bowl, generally of clay; shell, cover or rind + suf. *-aceus* > E. *-aceous*, of or concerning, with the nature of): having a hard covering or shell, generally of light brick-red color, like the outer layer of the sporangial peridium of *Diderma testaceum* (Physarales). In other species of *Diderma*, such as *D. hemisphaericum*, the outer layer of the peridium is white or pale alutaceous to light brown in color, like the shell of a hen's egg.

Sporangia of *Diderma hemisphaericum*, with a **testaceous** peridium, on dead wood, x 25 (*MU*).

Sporangia of *Didymium* (=*Diderma*) *difforme*, with a **testaceous** peridium, on dead wood, x 25 (*MU*).

tetra-, tetr- (L. *tetras* < Gr. *tetrás*, four): a pref. frequently employed in biology to form words taken from Greek to indicate that a structure occurs in fours, as in *tetrachotomous* and *tetrad*.

tetrachotomous (Gr. *tetrachótomos*, cut into four pieces < *tetrás*, four + *scházo*, to divide + *témeo*, to cut): refers to branching in which the vegetative point is divided into four equivalent points, i.e., with four branches originating from a single point. Cf. **dichotomous**, **polychotomous** and **trichotomous**.

tetrad (L. *tetras*, genit. *tetradis* < Gr. *tetrás*, genit. *tetrádos*, four): a group of four cells or spores, like the tetraspored basidium of many basidiomycetes, or the ascus with four ascospores, typical of *Saccharomyces* (Saccharomycetales).

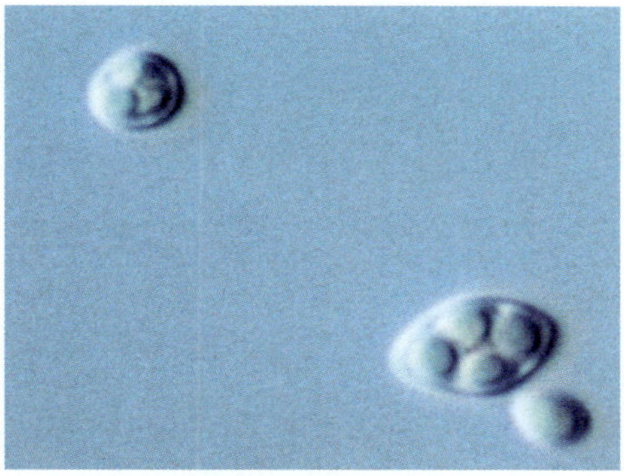

Tetrad (tetraspored ascus) and somatic cell of *Saccharomyces cerevisiae*, x 2,200 (*RTH*).

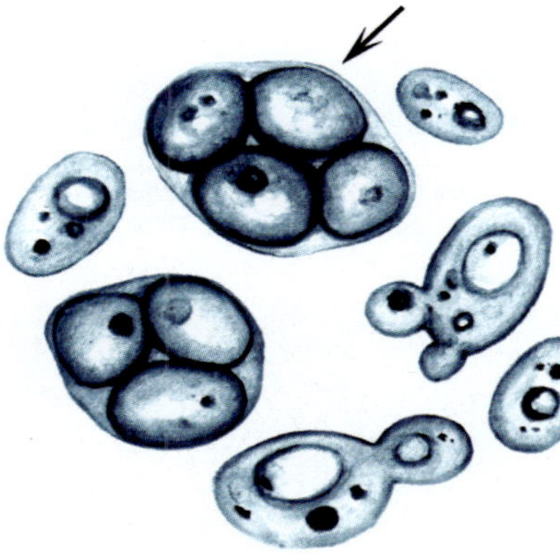

Tetrad (tetraspored ascus) and somatic cells of *Saccharomyces cerevisiae*; also shown is a trispored ascus, x 4,570 (*MU*).

tetrapolar (L. *tetras* < Gr. *tetrás*, four + L. *polaris*, polar, relating to or concerning the pole < *polus*, the end of an axis, pole + suf. *-aris* > E. *-ar*, like, concerning): refers to the sexual compatibility of the basidiomycetes, which have four groups of sex factors (AB, Ab, aB, and ab). Fusions between monosporic mycelia are only possible when these mycelia carry complementary pairs of these factors. Cf. **bipolar**.

texospore (Gr. *teixós*, wall + *sporá*, spore): an ascospore covered by a layer of cells of paraphysal origin, as in *Texosporium*, a gen. of lichens in the Caliciales.

textura, pl. **texturae** (L. *textura*, tissue, link, connection): refers to the shapes of the cells that make up the different types of hyphal tissues present in conidiomata and ascomata; these shapes are utilized in taxonomic descriptions. *Textura angularis* (L. *angularis*, angular): with polyhedric cells, due to mutual compression, without intercellular spaces. *Textura epidermoidea* (Gr. *epidermís*, membrane or pellicule that covers the skin + L. suf. *-oide*, like < Gr. *-oeídes*, similar to): a thin, membranous tissue composed of cells with very irregular outlines, with united walls and lacking intercellular spaces (like the pieces of a jigsaw puzzle). *Textura globulosa* (L. *globulus*, small ball or globe): with rounded to polyhedric cells, almost isodiametric, with intercellular spaces, but not resembling hyphae as such. *Textura intricata* (L. *intricatus*, intricate, entangled): composed of interwoven hyphae, with the walls not united, and generally with intercellular spaces. *Textura oblita* (L. *oblitus*, obstructed, closed): composed of more or less parallel hyphae, with long cells that have a narrow lumen and very thick, coherent walls. *Textura porrecta* (L. *porrectus*, extended, stretched out, long): hyphae with long, parallel cells with wide lumena and thin walls, not coherent. *Textura prismatica* (L. *prisma*, prism + Gr. suf. *-tíkos* > L. *-ticus* > E. *-tic*, relating to): with short, more or less rectangular cells, not isodiametric.

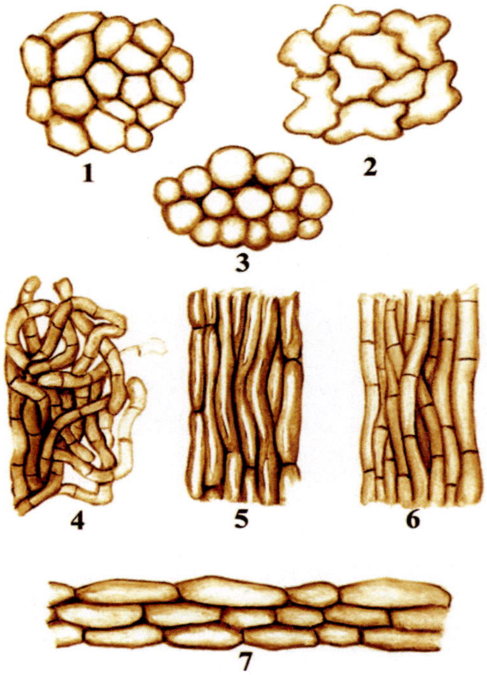

Textura *angularis* (1), *epidermoidea* (2), *globulosa* (3), *intrincata* (4), *oblita* (5), *porrecta* (6), and *prismatica* (7), which can be present in different types of discomycete apothecia and deuteromycete pycnidia, x 500 (*MU*).

thalassiomycetes (Gr. *thállassa*, ocean + *mýkes*, genit. *mýketos*, fungus): fungi that live in marine environments. See **marine fungi**.

thallic (L. *thallus* < Gr. *thallós*, shoot, stem + suf. *-íkos* > L. *-icus* > E. *-ic*, pertaining to or relating to): *Conidiogenesis*. One of the two main modes of conidial development in the mitosporic fungi. The thallic conidium is characterized by being formed through the transformation of an existing cell in a hypha or conidiophore. If all of the wall layers of the conidiogenous cell participate in the formation of the conidium wall, it is called *holothallic*; if the outer wall of the conidiogenous cell is not involved in the formation of the conidium wall, it is called *enterothallic*. *Arthrospores* and *meristem arthrospores* are types of thallic conidia, as in *Geotrichum candidum* var. *citri-aurantii* (moniliaceous asexual fungi). Cf. **blastic**.

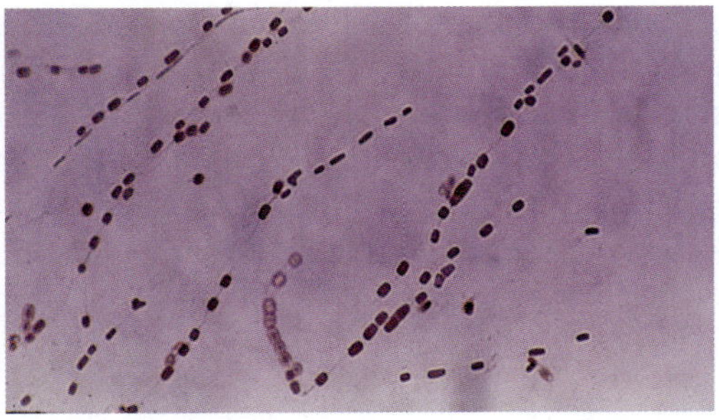

Thallic conidia (holothallic) of *Malbranchea aurantiaca* (moniliaceous asexual fungi), x 500 (*MU*).

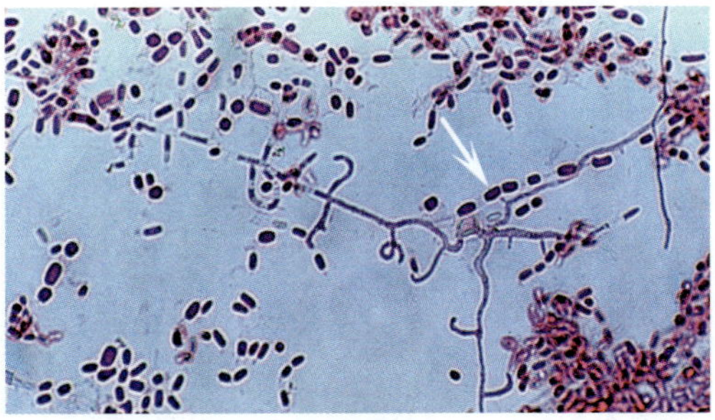

Thallic conidia (holothallic) of *Malbranchea aurantiaca* (moniliaceous asexual fungi), x 600 (*MU*).

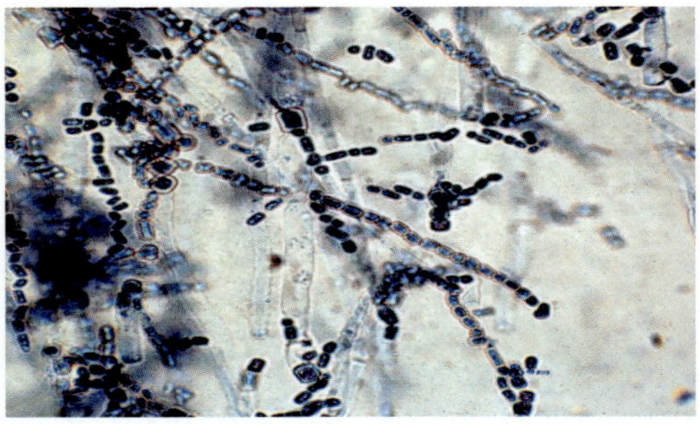

Thallic conidia (holothallic) of *Geotrichum candidum* var. *citri-aurantii*, x 300 (*MU*).

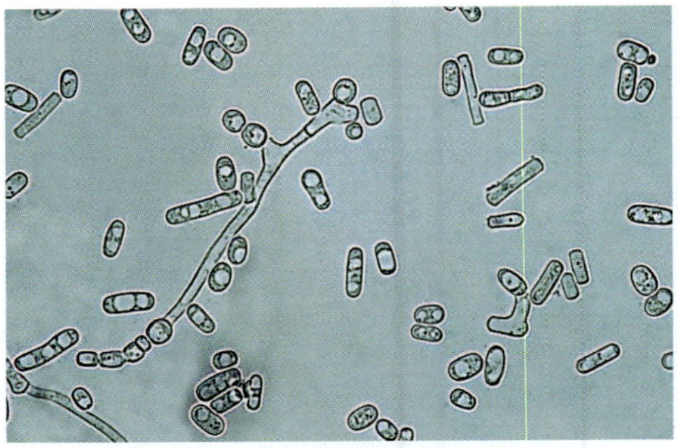

Thallic conidia (holothallic) of *Geotrichum candidum* var. *citri-aurantii*, x 500 (*MU*).

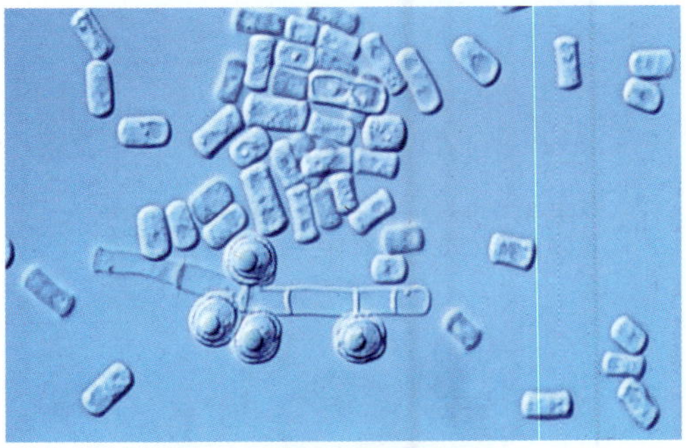

Thallic conidia (holothallic) of *Geotrichum candidum* var. *citri-aurantii*, x 550 (*RTH*).

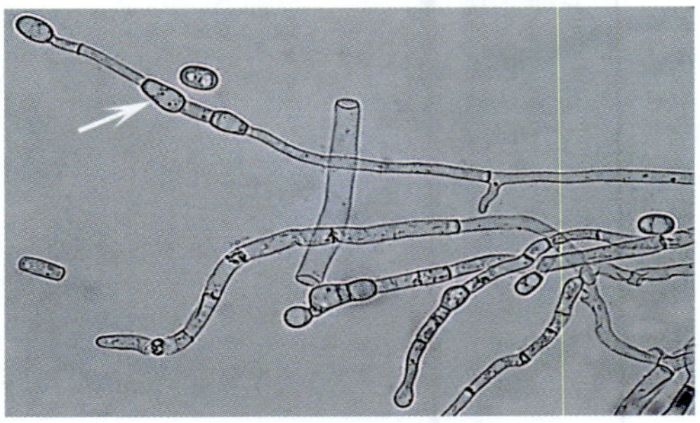

Thallic conidia (holothallic) of *Geotrichum candidum* var. *citri-aurantii*, x 500 (*MU*).

thalliles, **thallyles** (L. *thallus* < Gr. *thallós*, shoot, stem + suf. *-ilis* > E. *-ile*, capable, with the character of): small propagules, similar to stems, produced on the lower surface of certain lichen thalli, as occurs in *Umbilicaria* (Lecanorales). See **thalloconidium**.

thallinocarp (L. *thallus* < Gr. *thallós*, thallus + *karpós*, fruit): *Lichens*. A type of apothecium, similar to a lichen gall, that originates from ascogonia, ascogenous hyphae and asci arranged directly in a swelling among the hyphae of the thallus, with the hymenium covered by groups of algal cells, as occurs in *Lichinella* (=*Gonohymenia*), of the family Lichinaceae (Lichinales). Cf. **pycnoascocarp**.

thallo-, **-thallus**, pl. **thalli-**, **-thalli** (L. *thallus* < Gr. *thallós*, thallus): as a pref. it is employed in terms such as *thallospore*, a spore produced directly by the thallus, like arthrospores and chlamydospores, i.e., not formed by a conidiophore. As a suf. *-thallus*, is used in compounds such as *sporothallus*, a thallus that produces sporangia with asexual spores, in the aquatic fungi in whose life cycle there is an alternation of generations.

thalloconidium, pl. **thalloconidia** (L. *thallus* < Gr. *thallós*, shoot, thallus + NL. *conidium* < Gr. *kónis*, powder + dim. suf. *-ídion* > L. *-idium*): a propagule produced and detached directly from the lower cortex and/or from the rhizines of certain species of *Umbilicaria* (Lecanorales); similar structures are formed from the prothallus of other Lecanorales, such as *Protoparmelia*, *Rhizoplaca* and *Sporastatia*; thalloconidia are dark brown in color, smooth or rugose, with 2-3 layers in the wall, and consist of from one to 2,500 cells. See **thalliles**.

thalloid (L. *thallus* < Gr. *thallós*, shoot, thallus + NL. suf. *-oides*, contraction of Gr. *-oeídes*, that denotes similarity): similar to a thallus.

thallospore (NL. *thallospore* < Gr. *thallós*, shoot, thallus + Gr. *sporá*, spore): a spore produced directly by a thallus or mycelium, whether intercalary (single or in chains) or terminal, from a preexisting cell, as occurs with arthrospores and chlamydospores; thallospores or *thallic* spores are not formed *de novo*, as happens in other types of spores (which are called *blastic*).

thallus, pl. **thalli** (L. *thallus* < Gr. *thallós*, shoot, thallus): the vegetative body or soma of a fungus. The thallus can be unicellular, multicellular, or dimorphic, depending on the species or phases of the life cycle and the conditions of the medium in which it develops, but it never has sap conducting vessels nor is it differentiated into root, stem, flowers and fruits, as occurs in vascular plants.

thamnisophagous (Gr. *thámnion*, dim. of *thámnos*, shrub + *phágos*, what eats): *Mycorrhizae*. One of the four types of endotrophic mycorrhizae of pteridophytes, characterized by the formation of finely branched, intracellular haustorial arbuscules, that finally are digested and assimilated by the host cells. Cf. **ptyophagous** and **tolypophagous**.

thamnoblast (Gr. *thámnos*, shrub + *blastós*, shoot, bud, germ, renewal): *Lichens*. A dendroid, shrub-like shoot, like those produced in the fruticulose lichens, e.g., *Teloschistes* (Teloschistales) and *Ramalina* (Lecanorales).

thanato-, **thanat-** (Gr. *thánatos*, death): a pref. employed in the formation of compounds that allude to death, as in *thanatochresis*.

thanatochresis, pl. **thanatochreses** (Gr. *thánatos*, death + *chrêsis*, useful): taking advantage, as a means of sustenance, of cadavers, secretions, excrement and other productions of one species by the living individuals of a second species. This phenomenon is observed in saprobic organisms that grow at the expense of other dead organisms or substrates coming from dead organisms. Among the fungi, there are innumerable examples of this phenomenon, one of which is the growth and sporulation of *Harposporium anguillulae* (moniliaceous asexual fungi) on the remains of nematodes in the soil, although in this case the fungus first infects the nematode, kills it, and then takes advantage of it as a saprobe.

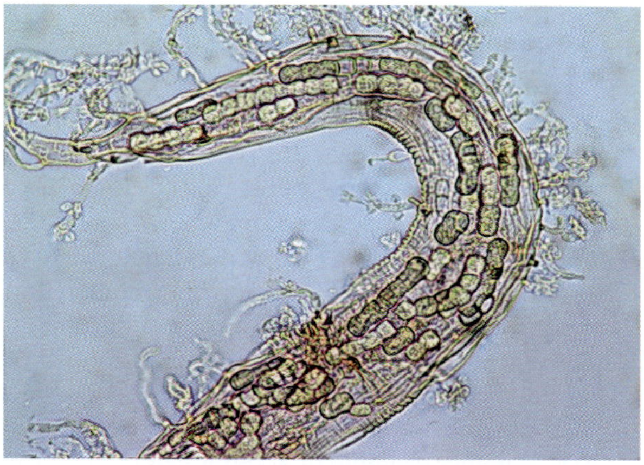

Thanatochresis of *Harposporium anguillulae* from a dead nematode killed by this fungus, x 400 (*MU*).

thecium, pl. **thecia** (NL. *thecium* < Gr. *thekíon* < *thêke*, case; here, of the asci + L. dim. suf. *-ium*): refers to the types of fructifications (cleistothecium, perithecium and apothecium) that are delimited by a proper wall, which can remain closed during the major part of the development of the fructification, only opening in some manner on reaching maturity (angiocarpic, like a cleistothecium), or the fructification may have the hymenium or fertile layer partially exposed from an early stage (hemiangiocarpic, like a perithecium), or totally exposed (gymnocarpic, like an apothecium). Sometimes thecium is used as equivalent to *hymenium*.

thermo- (Gr. *thermós*, warm, hot): a pref. that alludes to heat, as in the following terms.

thermodimorph (Gr. *thermós*, warm, hot + *dís*, two + *morphé*, shape, form): applied to a fungus that can change from a mycelial to a yeast form, and vice versa, depending on the temperature at which it is growing, like *Histoplasma capsulatum* (moniliaceous asexual fungi), the etiologic agent of histoplasmosis, that develops the mycelial phase at temperatures near 25ºC, and the yeast phase at 37ºC. These two growth forms correspond to a saprobic phase and a parasitic phase, respectively. See **dimorphic**.

thermodury (Gr. *thermós*, warm, hot + L. *durare*, to last): the capacity of a microorganism to survive high temperatures, especially when in a latent or inactive state; refers particularly to the capacity of spores to survive pasteurization. Cf. **thermophily**.

thermolabile (Gr. *thermós*, warm, hot + LL. *labilis*, neuter *labile*, unstable, labile): applied to a substance that undergoes chemical, physical or biological changes when it is moderately heated, generally at 55ºC or more, as occurs with certain enzymes and vitamins, which on being heated lose their characteristic properties. Cf. **thermostable**.

thermophile, **thermophilic**, **thermophilous** (Gr. *thermós*, warm, hot + *phílos*, have an affinity for; or + suf. *-ikos* > L. *-icus* > E. *-ic*, pertaining to or relating to): an organism that is capable of growing at elevated temperatures (from 50-60ºC or a little more), with a temperature optimum of 40-50ºC; the thermophiles or thermophilic fungi (such as *Chaetomium thermophilum*, of the Sordariales) prosper at these temperatures and are incapable of growing at ordinary laboratory temperature; they have a maximum growth temperature of 50ºC or a little more, and a minimum growth temperature of 20ºC. Another example of a thermophilic fungus is *Paecilomyces variotii*, of the moniliaceous asexual fungi. Cf. **mesophile** and **psychrophile**.

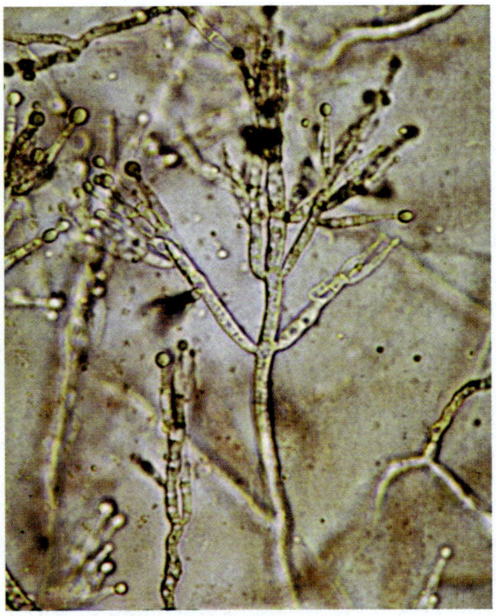

Conidiophores of *Paecilomyces variotii*, a **thermophilic** fungus, x 650 (*MU*).

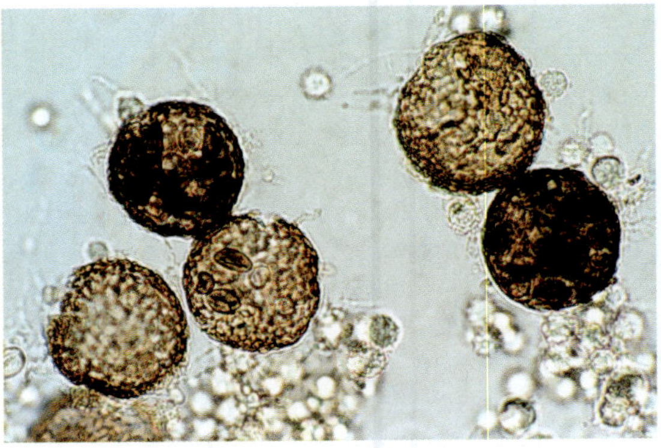

Cleistothecia of *Corynascus* sp. (Sordariales), a **thermophilic** fungus, x 200 (*MU*).

thermophily (Gr. *thermós*, warm, hot + *phílos*, have an affinity for): the capacity of a microorganism to actively grow at high temperatures. See **thermophile** and **thermotolerant**. Cf. **thermodury**.

thermostable (Gr. *thermós*, heat, hot + L. *stabilis*, stable, fixed, immutable): refers to a substance that remains unchanged after being moderately heated (at 55°C or more), as happens with some bacterial proteinases that retain their characteristic properties. Cf. **thermolabile**.

thermotolerant (Gr. *thermós*, warm, hot + L. *tolerans*, genit. *tolerantis*, tolerant < to tolerate, support): a microorganism capable of growing at a maximum temperature near 50ºC, and a minimum of 20ºC. Cf. **thermophile**.

thero- (Gr. *théros*, summer; and, by extension, the period of the year favorable for plant growth): a pref. employed in terms such as *therospore*, a short-lived spore, not resistent, that germinates promptly, without a previous resting period.

thigmo-, **thigm-** (Gr. *thígma*, contact): a pref. used in terms such as *thigmomorphosis*. Equivalent to **hapto-**.

thigmomorphosis, pl. **thigmomorphoses** (Gr. *thígma*, contact + *mórphosis*, the action of giving shape, of taking form): the development or change in shape (morphosis) due to a contact stimulus; equivalent to **haptomorphosis**; applied to the swelling of the ends of the zygophores when they come into contact, as in *Rhizopus* and other Mucorales. The formation of an appressorium from a germ tube, as occurs in many plant pathogenic fungi, also is considered as thigmomorphosis.

thio- (Gr. *theíon*, sulphur): a pref. that alludes to sulphur, as in *thiobacteria*, bacteria that live in sulphurous or sulphydric waters and which precipitate sulphur.

tholus, pl. **tholi** (L. *tholus* < Gr. *thólos*, cupula, vault, dome): see **dome**.

thriptogen, **thryptogen** (Gr. *thryptós* < *thrýbo*, to injure, debilitate + *génos*, origin < *gennáo*, to engender, produce): refers to an organism that attacks a host without killing it, although it weakens it and it becomes susceptible to other pathogens or physical agents (such as drought, freezes, etc.). E.g., certain plant parasitic nematodes that predispose their host plants to attack by vascular plant pathogenic fungi, such as *Verticillium* (moniliaceous asexual fungi) and *Fusarium* (tuberculariaceous asexual fungi) are thriptogens.

thrush (prob. of Scand. origin): the common name of a disease caused by *Candida albicans* (cryptococcaceous asexual yeasts), especially in infants and children, characterized by the formation of white patches on the mucous membranes of the oral cavity and genital areas. It is a clinical type of **candidiasis**.

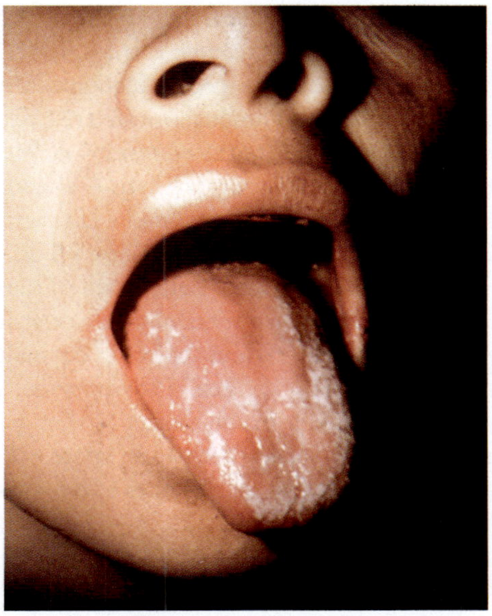

Thrush (by *Candida albicans*) on the tongue of a child, x 0.5 (*RLM*).

thyrido-, **thyrid-** (Gr. *thyrís*, genit. *thyrídos*, window):a pref. used in terms such as *thyridophyllous*, referring to the leaves of certain plants (such as *Fenestraria*) that receive all or most of their light through their translucent apex that lacks chlorophyll.

thyriothecium, pl. **thyriothecia** (NL. *thyriothecium* < Gr. *thyreós*, large, oblong shield + *thekíon* < *thêke*, case; here, of the asci + dim. suf. *-íon*): a lenticular (hemispheric convex) ascoma, with the upper wall formed of radially arranged plates of dark cells (the upper plate corresponds to the *scutellum*). The asci are almost horizontal and they grow inward, with their apices pointing toward the ostiole; the cavity between the ostiole and the basal wall is occupied at first by pseudoparaphyses, which disappear when the asci mature. Characteristic of the families Microthyriaceae and Micropeltidaceae (Dothideales); *Microthyrium* and *Stomiopeltis* are the respective examples of fungi of these families that form thyriothecia (also called **catathecia**) on the tissues of leaves and dead branches.

Thyriothecium of *Microthyrium microscopicum* in surface view; the scutellum (dark region) surrounds the ostiole in the center, x 330 (*MU*).

thyrsus, pl. **thyrsi** (L. *thyrsus* < Gr. *thýrsos*, Greek dart or lance, adorned with grape leaves and ending in a strobilus, that is carried by the God Baco and by Satyrs and other personages involved in the rites related to Baco; a type of inflorescence in which the ends of the lateral branchlets do not reach the level of the apex of the main axis): refers to the densely branched apices of some lichens, e.g., *Cladonia stellaris* (Lecanorales).

thysanoblastic (Gr. *thysanós*, fringe, border, tassel, small bell + *blastós*, bud, germ, shoot + suf. *-ikos* > L. *-icus* > E. *-ic*, pertaining to or relating to): refers to a type of conidiogenesis in which the totality of the upper surface of the conidiogenous cell is involved in the process of conidium formation and secession, with an alternation of schizolysis and rhexolysis during the secession of the successively formed conidia.

tibi grains (E. *tibi* < Sp. *tibico*, origin and etymology unknown; ME. *gryne* < OF. *grein*, a seed, grain < L. *granum*, seed, kernel): macroscopic gelatinous grains or microbiogloeae (from a few mm to 1-2 cm, called *tibicos* in Mexico, and *sugary kefir grains* or *tibi grains* in other parts of the world), composed of dextrans, that contain a symbiotic association of lactic bacteria (*Lactobacillus casei*, *L. hilgardii*, *Leuconostoc mesenteroides* and *Streptococcus lactis*) and yeasts (*Saccharomyces cerevisiae*, *Zygosaccharomyces florintinus* and *Torulaspora pretoriensis*, of the

Saccharomycetales, and *Kloeckera apiculata*, *Candida lambica* and *C. valida*, of the cryptococcaceous asexual yeasts). Tibi grains are used as a starter culture for the production of fermented beverages from sugary solutions, like those called tepache and vinegar of tibicos in Mexico, and ginger beer in Switzerland and other European countries. Cf. **bulgaro**, **kefir grains**, **tea fungus** and **mother of vinegar**. See **fermentation** and **microbiogloea**.

Tibi grains, from a culture on water and brown sugar, mixed with balanced food for poultry. The diet was tested as a chicken feed, x 1 (*MU*).

Tibi grain, extracted from a culture in water sweetened with brown sugar, x 5 (*MU*).

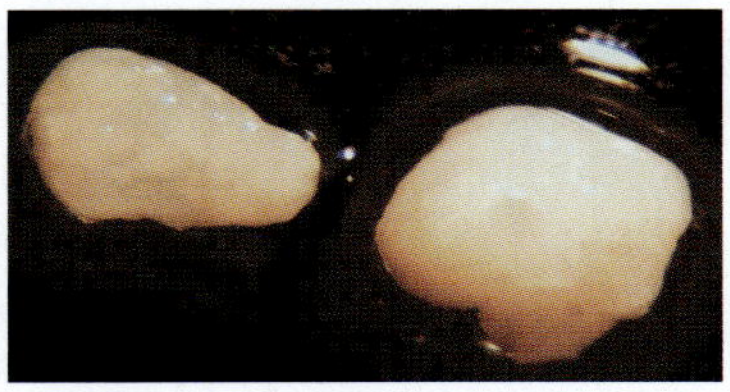

Tibi grains, extracted from a culture in water sweetened with brown sugar, x 3 (*MU*).

tichus, pl. **tichi** (NL. *tichus* < Gr. *teychós*, instrument, armor, funeral urn): the peripheral layer of cells that form a dark, protective layer on the wall of certain pseudothecia, as is observed in *Pleospora herbarum* (Pleosporales).

tinea (L. *tinea*, worm, moth): the common name of any of various diseases of the skin, hair and nails of humans and higher animals caused by parasitic fungi (especially the *dermatophytes*, of the gen. *Epidermophyton*, *Microsporum* and *Trichophyton*, moniliaceous asexual fungi). *Tinea barbae* (L. *barba*, beard): tinea of the beard and moustache; see **sycosis**. *Tinea capitis* (L. *caput*, genit. *capitis*, head): tinea of hairy skin. *Tinea tonsurans* (ML. *tonsura*, shaved): tinea of shaved skin. *Tinea corporis* (L. *corpus*, genit. *corporis*, body): tinea of the body. *Tinea circinata* (L. *circinatus* < *circinare*, to make round, rolled up in the shape of a flat coil with the apex in the center + suf. *-atus* > E. *-ate*, provided with or likeness). *Tinea cruris* (L. *cruris*, groin): tinea of the groin. *Tinea favosa*, *favus* or *tinea kerion* (L. *favus*, honeycomb; Gr. *kérion*, honeycomb): see **favus**. *Tinea imbricata* (L. *imbricatus* < *imbricare*, to cover with tiles): the so-called *tokelau*, an infection caused by *Trichophyton concentricum*. *Tinea nigra* (L. *nigra*, black): a pigmented cutaneous infection caused by a dematiaceous asexual fungus, *Hortaea* (=*Exophiala*) *werneckii*. *Tinea nodosa* (L. *nodosus* < *nodus*, knot, protuberance + suf. *-osus* > E. *-osa*, abundant in, prone to): see **piedra**. *Tinea pedis* (L. *pes*, genit. *pedis*, foot): tinea of the feet or *"athlete's foot"*. *Tinea unguium* (L. *unguis*, finger nail): tinea of the finger nails and toe nails. *Tinea versicolor* (L. *versicolorius*, *versicolorus*, with various colors): equivalent to *Pityriasis versicolor* (< gen. *Pityrosporum*, of the cryptococcaceous asexual yeasts < Gr. *pítyron*, dandruff, tinea + *spóron* > *sporum*, spore + Gr. suf. *-iasis*, used for the ending of names of diseases). See **mycosis**.

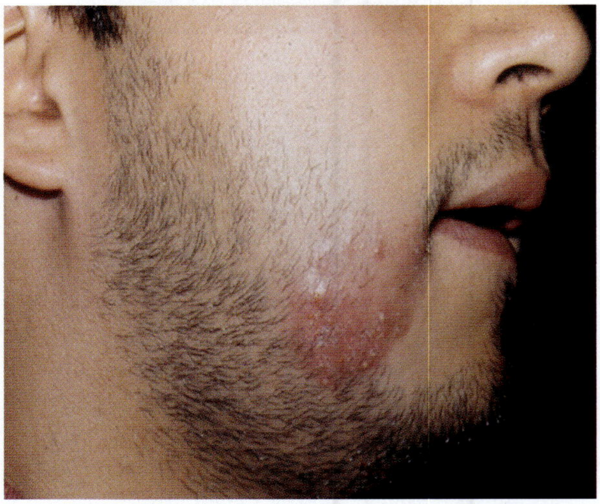

Tinea barbae (**sycosis**) of a young man, caused by *Trichophyton mentagrophytes*, x 0.5 (*AB*).

Hyphae and microconidia of *Trichophyton mentagrophytes* (on scales of skin), the causal agent of **tinea barbae** (**sycosis**) of humans, x 450 (*RLM*).

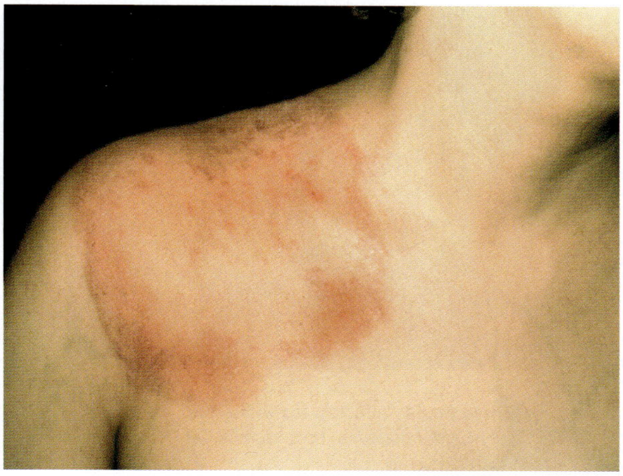

Tinea corporis of a young woman, caused by *Microsporum canis*, x 0.2 (*RLM*).

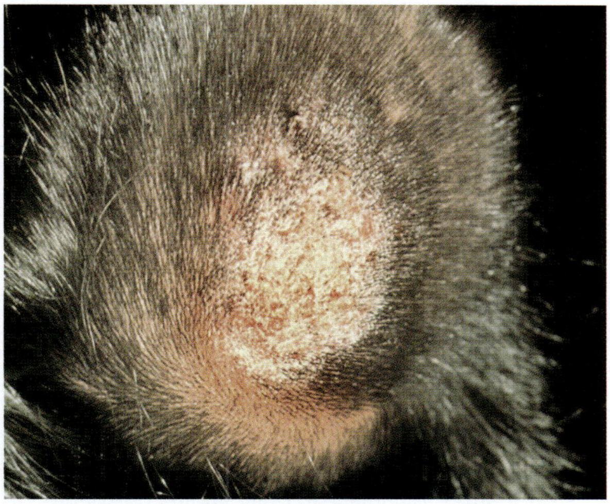

Tinea capitis of a child's head, caused by *Microsporum canis*, x 0.2 (*RLM*).

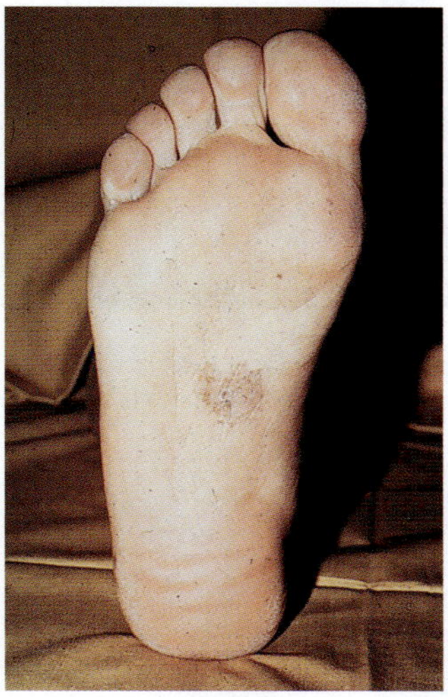

Tinea nigra on the sole of the foot of a man, caused by *Hortaea* (=*Exophiala*) *werneckii*, x 0.3 (*RLM*).

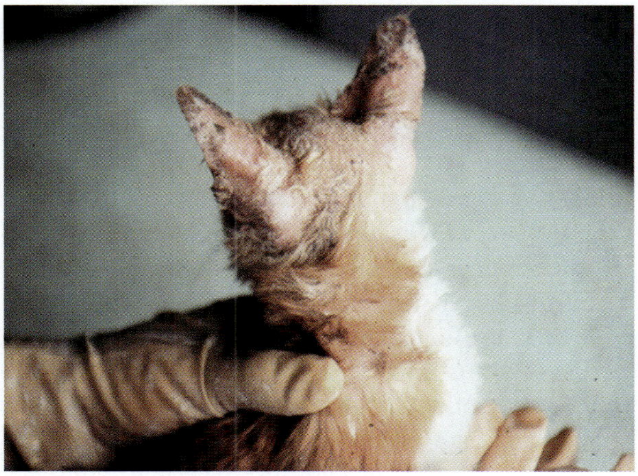

Tinea corporis of a cat´s body, caused by *Microsporum canis*, x 0.3 (*MU*).

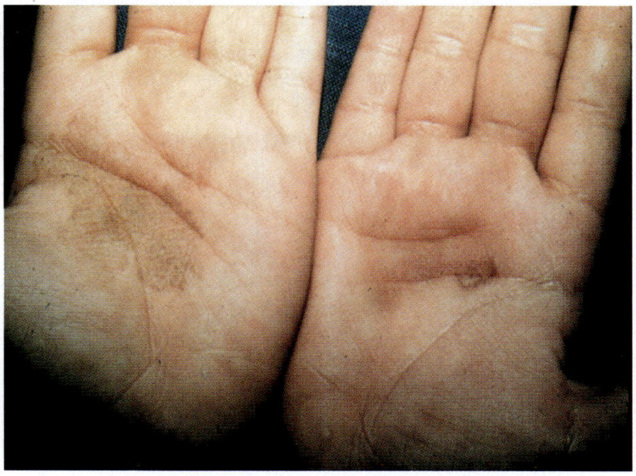

Tinea nigra on the palms of the hands of a man, caused by *Hortaea* (=*Exophiala*) *werneckii*, x 0.3 (*RLM*).

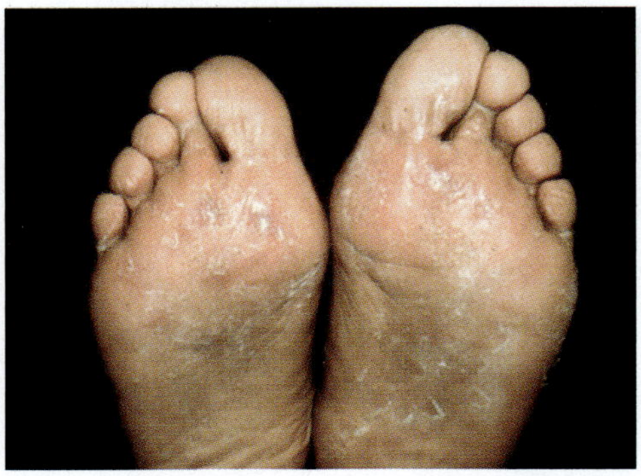

Tinea pedis ("athlete's foot") of a man, caused by *Epidermophyton floccosum*, x 0.2 (*RLM*).

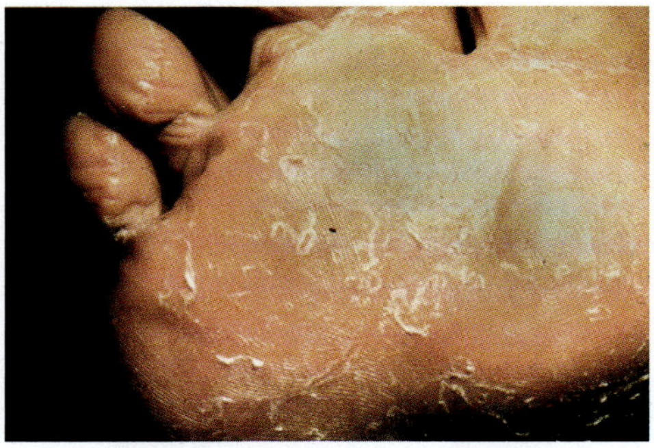

Tinea pedis ("athlete's foot") of a man, caused by *Epidermophyton floccosum*, x 0.1 (*RLM*).

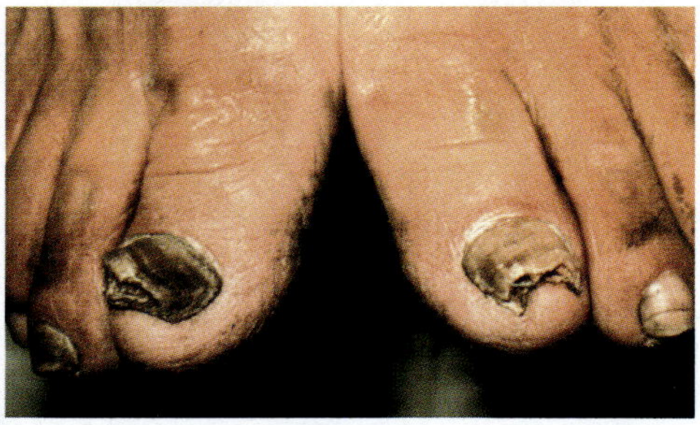

Tinea unguium of the toe nails (onychomycosis) of a man, caused by *Trichophyton mentagrophytes*, x 0.5 (*RLM*).

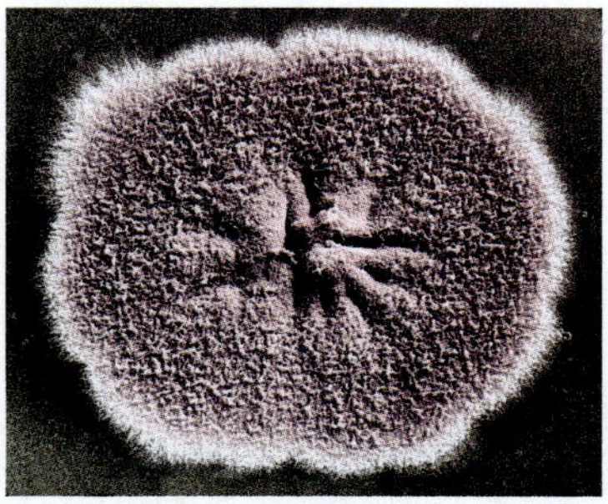

Colony of *Scopulariopsis brevicaulis* (moniliaceous asexual fungi), isolated from a case of **tinea pedis** (onychomycosis) of a man, x 0.1 (*MU*).

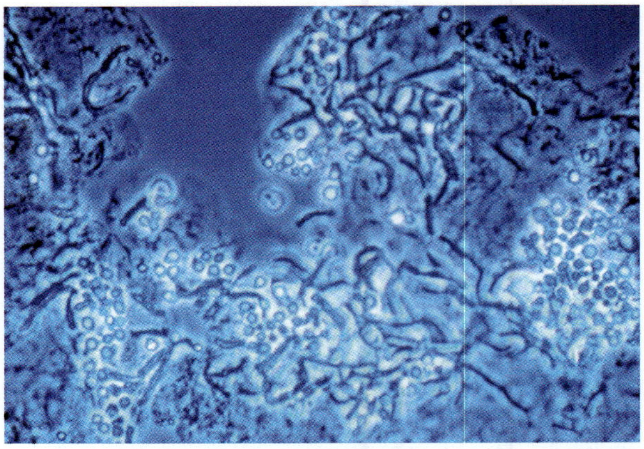

Hyphae and blastospores of *Malassezia* (=*Pityrosporum*) *furfur*, isolated from a case of **tinea versicolor** (pityriasis versicolor) of a woman, x 1,500 (*RLM*).

tinophysis, pl. **tinophyses** (Gr. *teinó*, to stretch + *phýsis*, action of budding or of being born, a growth): interascal or preascal hyphal elements that stretch and come to resemble pseudoparaphyses or paraphysoids, present in the hamathecium of some Pleosporales, such as *Melanomma*. Tinophyses are sparingly septate, very narrow, and often anastomosed.

-tion (L. *-ationem*, a suf. that denotes action, state, condition or result): a suf. used in terms such as *spermatization*, the fertilization of a receptive hypha by means of a spermatium.

toadstools (ME. *tode* < OE. *tade, tadige*, toad + *stol*, stool): a variety of agaricaceous and boletaceous fungi, especially inedible species, that produce toxins that affect humans, animals and plants. See **mycetism** and **hallucinogenic fungi** for information on macroscopic fungi poisonous to man. See **mycotoxicosis** and **mycotoxin** for micromycetes that produce toxins that affect man, animals and plants. Also see **antibiotic**. Cf. **edible fungi** and **mushroom**.

toco-, **toc-** (Gr. *tókos*, childbirth): a pref. used in terms such as *tocopherol*, any of the antisterile E vitamins, indispensable for the normal development of a pregnancy.

tolerant (L. *tolerare*, to tolerate, support + suf. *-ant* > E. *-ant*, what is, one who acts): an organism that exhibits tolerance, i.e., that reacts lightly to infection by a pathogen, or to the effect of other factors (e.g., tolerant to a toxic substance, virus, etc.). Cf. **suscept** and **susceptible**.

tolypophagous (Gr. *tolýpe*, ball of yarn, ball + *phágos*, one who eats): in the endotrophic mycorrhizae, it applies to forms of mycetization when the living cells of the host plant digest the dead hyphae of the fungus. Cf. **ptyophagous** and **thamnisophagous**.

tomentose, **tomentous** (L. *tomentum*, wool + suf. *-osus* > E. *-ose*, full of, augmented, prone to): provided with *tomentum*, downy; covered with densely entangled hairs, like the surface of the fruiting bodies of *Cyathus* and *Nidula* (Nidulariales).

Tomentose fruiting bodies of *Nidula niveotomentosa*, on dead wood, x 4 (*MU*).

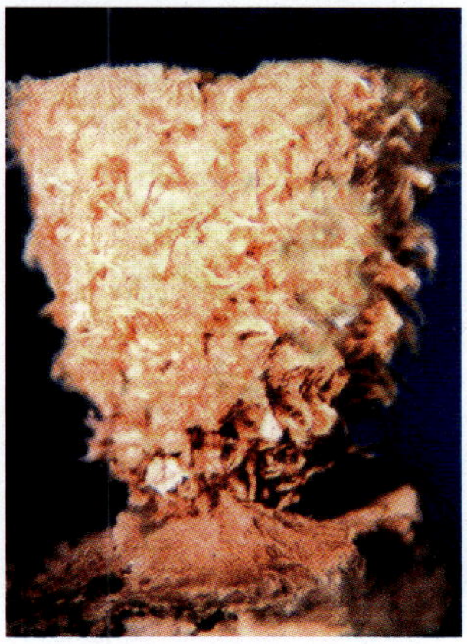

Tomentose peridium of the fruiting body of *Cyathus intermedius*, x 8 (*CB* & *MU*).

tomentum, pl. **tomenta** (L. *tomentum*, wool): the complex of filaments or hairs, simple or branched, generally interwoven or coiled, very close together, that resemble wool. A covering composed of fine hairs, densely arranged, on the surface of the stipe of some Hymenomycetes, such as *Boletus tomentosus* and *Laccaria laccata* (Agaricales), or a small compact cushion composed of hyphae, that is formed beneath a sporocarp, usually incorporating fragments of the substrate, as occurs in the Nidulariales. Also refers to hyphae, similar to hairs, that project in abundance from the lower cortex of crustaceous (heteromerous) lichen thalli, and which serve to attach the thalli to the substrate. Structures that have this type of hyphae are called *tomentose*.

-tomous (Gr. *tómos*, cut, sliced, divided): a suf. employed in compounds such as *dichotomous*, cut or divided into two.

tonophilic, **tonophilous** (L. *tonus* < Gr. *tónos*, tension, strength, pressure + *phílos*, have an affinity for + suf. *-íkos* > L. *-icus* > E. *-ic*, pertaining to or relating to; or + L. suf. *-osus* > E. *-ous*, full of, augmented, prone to): an organism that is capable of growing on media of high osmotic pressure.

tonophily (L. *tonus* < Gr. *tónos*, tension, strength, pressure + *philia* < Gr. *philía*, tendency toward < *phílos* > E. des. *-phily*, have an affinity for): the capacity to grow under conditions of high osmotic pressure. Organisms that have this capacity are called *tonophilic*, *osmophilic* or *osmophiles*.

topotype (Gr. *tópos*, place, territory + *týpos*, type, mark, model): a specimen from the same locality as the type, and whose characters agree with those of the type. Syn. of **locotype**.

torn (E. pp. of *tear* < ME. *teren* < OE. *teran*, pull apart): see **erose**.

torsive (LL. *torsus*, pp. of *torquere*, to twist + *-ivus* > E. *-ive*, quality or tendency, fitness): spirally twisted.

tortuous (L. *tortuosus*, twisted < *tortus*, a twisting, gyration + suf. *-osus* > E. *-ous*, full of, augmented, prone to): bent or twisted in different directions, like the cells of the cortex of the pileus of *Hemimycena* (=*Mycena*) *tortuosa* (Agaricales), and the hyphae of many micromycetes.

torulose, **torulous** (L. *torulosus* < *torulus*, dim. of *torus*, a prominence or swelling, cushion, an elevation + suf. *-osus* > E. *-ose*, full of, augmented, prone to): elongate in shape, but with contractions or constrictions, like the filaments of the gen. *Torula* (dematiaceous asexual fungi); similar to *moniliform* (like the hyphae of *Monilia*, a moniliaceous asexual fungus), but with the constrictions less profound. The chains of microconidia of *Fusarium moniliforme* (tuberculariaceous asexual fungi), the cystidia of *Psathyrella* (=*Pholiota*) *oedipus*, and the hyphae of the cortex of the pileus of *Leptonia subeuchroa* (Agaricales) are moniliform.

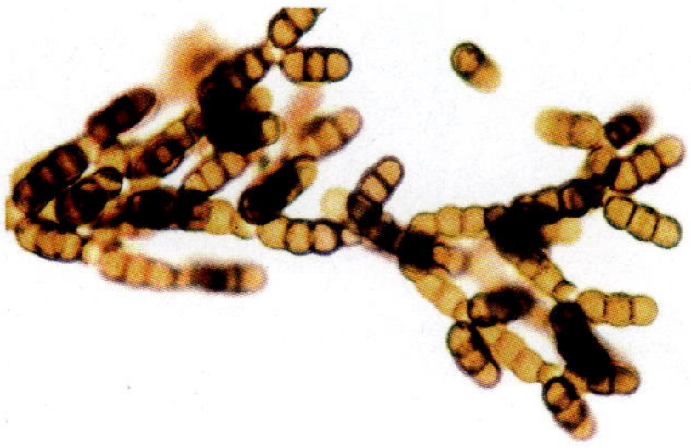

Torulose hyphae with blastic conidia of *Torula herbarum*, x 800 (*MU*).

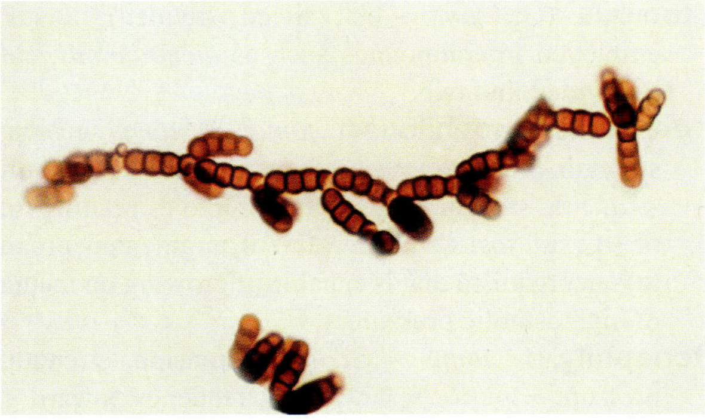

Torulose, blastic conidia of *Torula herbarum*, x 800 (*MU*).

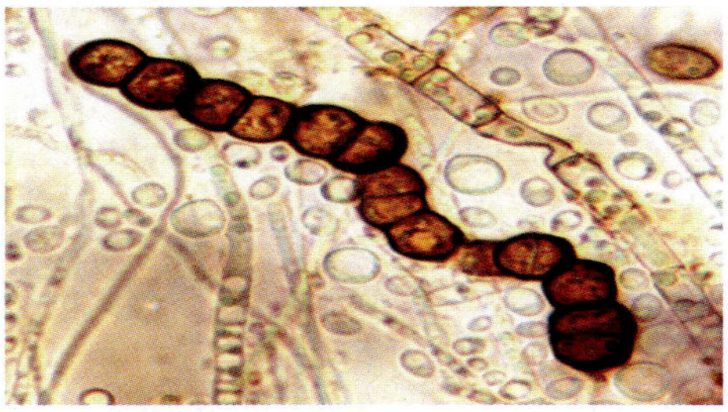

Torulose chain of blastospores of *Torula* sp., x 1,750 (*MU*).

totipotent (L. *totus*, complete, entire + *potens*, genit. *potentis*, that has power, efficacy or vigor for a thing): capable of generating a complete new individual.

toxic (LL. *toxicus* < L. *toxicum*, poison < Gr. *toxikón*, poison arrow): poisonous.

toxicity (LL. *toxicus* < L. *toxicum*, poison < Gr. *toxikón*, poison arrow + E. des. *-ity*, quality, state, degree): the capacity or degree of a substance to act as a poison.

toxigenic (LL. *toxicus* < L. *toxicum*, poison < Gr. *toxikón*, poison arrow + Gr. *génos*, origin < *gennáo*, to engender, produce): producing toxins, e.g., the ergot sclerotia of *Claviceps purpurea* (Hypocreales) and the fruiting bodies of *Amanita pantherina*, *A. verna* and *Omphalotus olearius* (Agaricales). See **mycotoxin** and **toadstools**.

Sclerotium (**toxigenic**) of *Claviceps purpurea* on a rye spike, x 10 (*MU*).

Fruiting body of *Amanita pantherina* (**toxigenic**), x 1 (*RV*).

Fruiting bodies of *Amanita verna* (**toxigenic**), x 1 (*RV*).

Fruiting bodies of *Omphalotus olearius* (**toxigenic**), x 0.5 (*RV*).

toxiphilous (LL. *toxicus* < L. *toxicum*, poison < Gr. *toxikón*, poison arrow + Gr. *phílos*, have an affinity for): an organism that prefers a contaminated habitat (e.g., *Lecanora conizaeoides*, of the Lecanorales, in areas of high contamination of the air by sulphur dioxide). Cf. **toxiphobous** and **toxitolerant**.

toxiphobous (LL. *toxicus* < L. *toxicum*, poison < Gr. *toxikón*, poison arrow + *phóbos*, fear, have an aversion to, lacking an affinity for): an organism that cannot tolerate a contaminated environment. Cf. **toxiphilous** and **toxitolerant**.

toxitolerant (LL. *toxicus* < L. *toxicum*, poison < Gr. *toxikón*, poison arrow + L. *tolerantus* < *tolerare*, to tolerate, support + suf. *-antus* > E. *-ant*, one who is, one who acts or carries out): tolerant to toxins in a contaminated habitat, e.g., with high concentrations of sulphur dioxide in the atmosphere. Cf. **toxiphilous** and **toxiphobous**.

trabecula, pl. **trabeculae** (L. *trabecula*, dim. of *trabs*, genit. *trabis*, beam, girder): each of the solid columns that extend from the peridium toward the center of the fructification in *Mesophellia* (Lycoperdales); this center or heart also is solid and is supported by the trabeculae, with the gleba located between the center and the trabeculae. In *Laternea* (Phallales) the trabeculae support the *glebifer* or *lantern*, beneath the vault formed by the arms of the receptacle, which is united at the apex.

Trabeculae supporting the gleba (lantern) of the fruiting body of *Laternea pusilla*, x 0.3 (*MU*).

trachytectum, pl. **trachytecta** (Gr. *trachýs*, rough + L. *tectum*, cover, roof): equivalent to **exospore** or **exosporium**.

tractellum, pl. **tractella** (L. *tractellum* < *trahere*, to drag, pull, tug + dim. suf. *-ellum* > E. *-ellum*): refers to the forward directed motor flagellum (tractor) of the zoospores of the Hyphochytriomycota and Oomycota.

trama, pl. **tramae** (L. *trama*, trama, warp, texture): a hyphal tissue that comprises the pileus or stratum that supports the hymenium in the basidiomycetes with holobasidia. The trama can be *homomerous* or *heteromerous*. The *hymeniiferous trama*, also known as *context*, is the layer of tissue that is found beneath the hymenium, which is called *gill trama* in the agarics, or *tube trama* in the boletes. Depending on the arrangement of the hyphae in the trama, they can be *convergent*, *divergent* or *parallel*. In *Coprinus* (Agaricales), e.g., they are parallel, with the hyphal elements arranged more or less regularly. In *Amanita muscaria* (Agaricales), they are divergent, with a central zone from which the two rows of cells diverge.

trama autolysis pl. **autolyses** (L. *trama*, trama, warp, texture; Gr. *autós*, same, self + *lýsis*, a loosening, releasing): a process in which the *tramal plates* collapse at maturity, forming a viscous or slimy mass of tissue and spores; typical of species in the Phallales.

tramal plate (L. *trama*, trama, texture + L. suf. *-alis* > E. *-al*, relating to or pertaining to; VL. *plattus* < Gr. *platýs*, wide, flat plate): *Hymenogastrales*. A layer of tissue partially or totally destined for the production of basidia. See **gusset**.

trans- (L. *trans-*, crosswise, through, transversely, from another part, on the other side): a pref. used in compound words that indicate something that comes from the other side, as in *transgenic*.

transgenic (L. pref. *trans-*, across, from one part to another + NL. *-genic*, adjectival combining form that means to originate, give rise to): a condition in which the genome of a host organism carries genetic material of another organism; e.g., hypovirulent strains of *Cryphonectria parasitica* (Diaporthales), the fungus causing chestnut blight, are carriers of a virus responsible for loss of virulence, and this virus can be disseminated to virulent strains that in turn disseminate it on reproducing.

translucent (L. *translucidus* < *trans-*, across + *lucere*, to shine < *lux*, genit. *lucis*, light + *-entem* > E. *-ent*, being): diaphanous; a body through which light passes, without being transparent, like the perithecium of *Melanospora zamiae* (Melanosporales) and the pileus of the basidiocarp of *Oudemansiella canarii* (Agaricales).

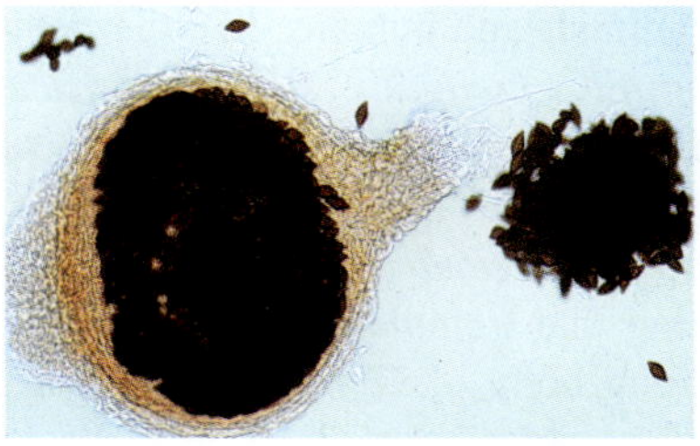

Translucent perithecium of *Melanospora zamiae*; note the ascospores inside, x 100 (*MU*).

transverse (L. *transversus*, directed or crossed over from one side to the other): placed or directed across. E.g., transverse planes are arranged at a right angle with respect to the axis of growth of the organ in question, as occurs with the septa or partitions in many hyphae; a transverse cut or section is one that passes perpendicular to said axis. Cf. **longitudinal**.

traumato-, **traumat-**, **traumo-**, **traum-** (Gr. *traûma*, genit. *traûmatos*, injury): a pref. employed in the formation of terms that imply the idea of a traumatism, as in *traumatoparasite*, a parasite that can only infect a particular host through an injury.

trembling fungus, pl. **fungi** (ME. *tremblen* < MF. *trembler* < ML. *tremulare* < L. *tremulus*, tremulous < *tremere,* to tremble, shake; L. *fungus*, fungus): the common name of a member of the Tremellales, also known as *jelly fungi*.

tremelloid (< gen. *Tremella* < L. *tremulus*, tremulus, having a movement similar to trembling + L. dim. suf. *-ella* + L. suf. *-oide*, like < Gr. *-oeídes*, similar to): similar to *Tremella* (Tremellales); similar to jelly or moist gelatin; gelatinous.

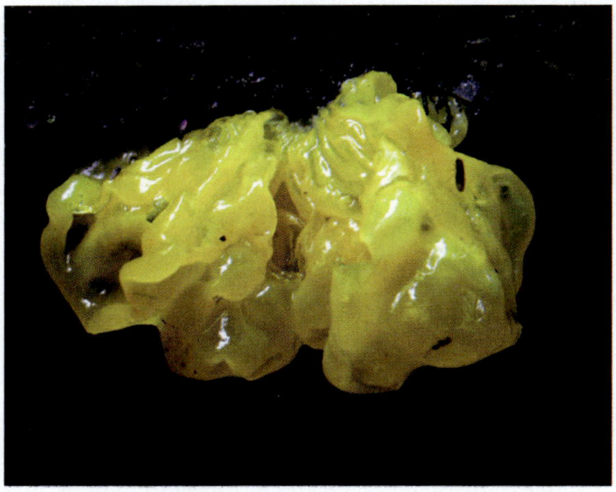

Tremelloid basidiocarp of *Tremella mesenterica*, x 1 (*RV*).

tremorgen (L. *tremor*, trembling, agitation + suf. *-geno* < Gr. *génos*, origin < *gennáo*, to engender, produce): a mycotoxin that causes trembling (neurotoxicosis) in higher animals and humans, like the *fumitremorgen* of *Aspergillus fumigatus* and the *verruculotoxin* of *Penicillium verruculosum* (moniliaceous asexual fungi).

tremulous (L. *tremulus*, what trembles): refers to something that trembles, that is not very firm, like the fruiting body of *Tremella* and other gelatinous fungi of the order Tremellales, as in *Pseudohydnum gelatinosum*, which also are known as *trembling fungi* or *jelly fungi*.

Tremelloid basidiocarps of *Pseudohydnum gelatinosum*, x 0.5 (*RV*).

Tremelloid basidiocarps of *Pseudohydnum gelatinosum*, x 0.5 (*RV*).

tretic (Gr. *tretós*, to perforate + suf. *-íkos* > L. *-icus* > E. *-ic*, pertaining to or relating to): a type of conidiogenesis in which each conidium (*tretoconidium*, *tretic conidium*, *poroconidium*, *porospore*) is formed by extrusion of the inner wall of the conidiogenous cell through a pore. Tretoconidia can be solitary or arranged in acropetal chains. See **monotretic** and **polytretic**.

tretic conidium, pl. **conidia** (Gr. *tretós*, to perforate + suf. *-íkos* > L. *-icus* > E. *-ic*, belonging to or relating to; NL. *conidium* < Gr. *konídion* < *kónis*, dust, powder + dim. suf. *-ídion* > L. *-idium*): a type of conidium (also known as a *tretoconidium*, *poroconidium* and *porospore*) that is formed by extrusion of the internal wall of the conidiogenous cell through a pore, as in *Ulocladium atrum* (dematiaceous asexual fungi). Tretic conidia can be solitary or arranged in acropetal chains. See **monotretic** and **polytretic**.

tri- (Gr. *trís* > L. *tri-*, and *ter-*, alteration of *tri-*, three, three times): a pref. often utilized in biology to form compound words, Greek as well as Latin, to introduce the idea of triplicates, as in *trimitic* and *triverticillate*.

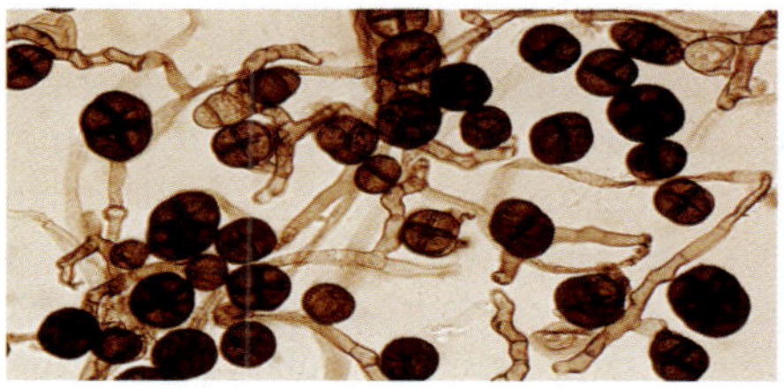

Tretic conidia of *Ulocladium atrum*, x 750 (*MU*).

tribe (L. *tribus*): in taxonomy, a hierarchy above genus and below family; thus, a family can be divided into tribes, which can then accommodate the various genera in the family. The names of tribes are formed from one of the genera in the tribe, with the ending *-eae*, as in the tribe Amaniteae of the genus *Amanita* (Agaricales). Subtribes have the ending *-inae*.

trichiform (Gr. *thríx*, genit. *trichós*, hair + L. suf. *-formis* < *forma*, shape): see **aculeate**.

trichotomous (Gr. *trícha*, three times < *treís*, three + *scházo*, to divide + L. *tomous*, cut, divided < Gr. *témeo*, to cut, divide): with three branches originating from the same point. Cf. **dichotomous**, **tetrachotomous** and **polychotomous**.

-tricho, **-trich**, **-thrix** (Gr. *thríx*, genit. *trichós*, hair): used as a suf., to express the idea of a hair or filament, as in the making of the names *Metatrichia*, *Trichia* (Trichiales), whose capillitium of the sporangium is like hairs, *Guanomyces polythrix* (Sordariales), whose perithecia are covered with many hairs, and *Tricholoma* (Agaricales), whose basidiocarp surface is covered with multiple hairs.

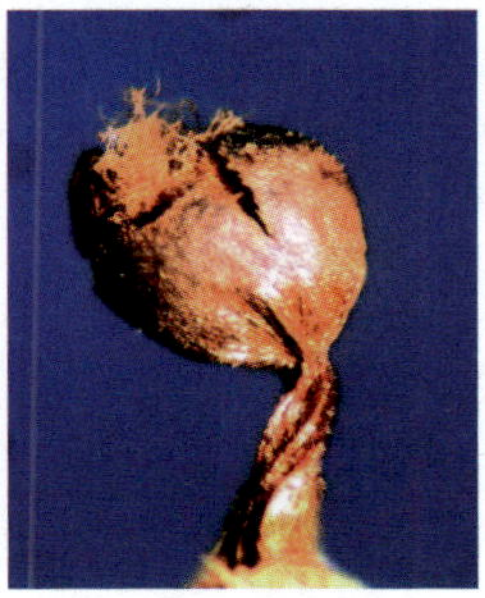

Capillitium of *Metatrichia* (=*Trichia*, the suf. **trich-** is used) *floriformis*, whose sterile filaments are inside the sporangium, x 80 (*MU*).

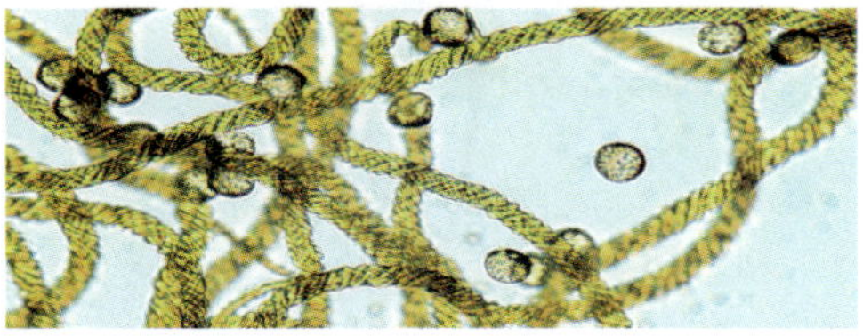

Capillitium of *Hemitrichia* (the suf. **trich-** is used) *calyculata* (=*H. stipitata*), whose sterile filaments are inside the sporangium, x 600 (*MU*).

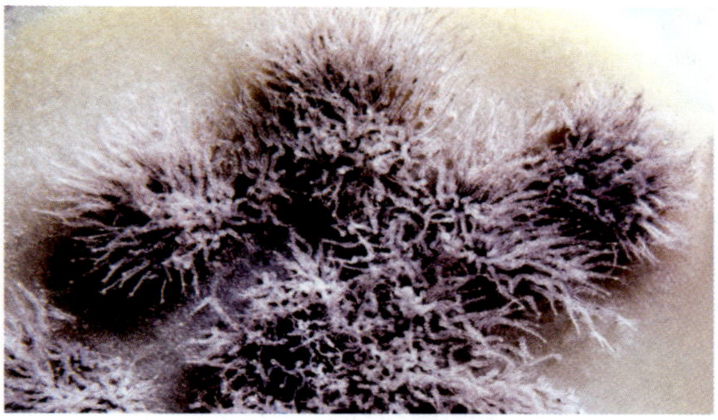

Colony of *Guanomyces polythrix* (the suf. **thríx-** is used < Gr. *thríx*, hair) on agar, x 10 (*MU*).

trichocyst (Gr. *thríx*, genit. *trichós*, hair + *kýstis*, bladder vesicle, cell): the name applied to the injector cell of *Haptoglossa mirabilis* (Myzocytiopsidales), an endoparasite of rotifers of the gen. *Adineta*. This cell originates in the encysted zoospore, which on germinating produces an apparatus similar to a harpoon, and which functions in a manner similar to a hypodermic syringe, with a conical plug in the apex, which is introduced into the rotifer when the injector cell discharges; inside the attacked rotifer an infective sporidium becomes evident, which is derived from the plug introduced by the injector apparatus.

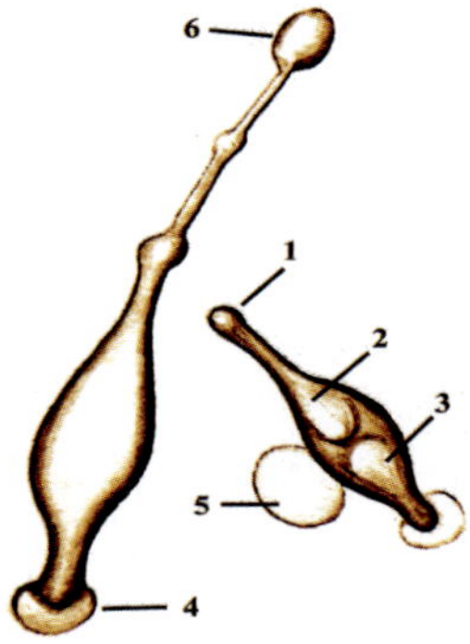

Trichocyst, the injector cell of an encysted zoospore of *Haptoglossa mirabilis*. Diagramatic representation based on data from Barron, 1980, 1987. Structures shown: apical cone (1), projectile chamber (2), basal vacuole (3), adhesive pad (4), empty zoospore cyst (5), and discharged projectile, similar to a harpoon (6), x 2,000 (*MU*).

trichoderm (Gr. *thríx*, genit. *trichós*, hair + *dérma*, skin): the cuticular layer of the fruiting bodies of various Agaricales (such as *Leptonia perfusca* and *Entoloma corvinum* (=*Leptonia corvina*), formed of septate, piliform or filiform hyphae, that are more or less perpendicular (anticlinal) to the surface of the respective organ. If these piliform hyphae more or less reach the same level, this type of dermis is called **trichoderm palisade**, as in *Entoloma* (=*Leptonia*) *aethiops*.

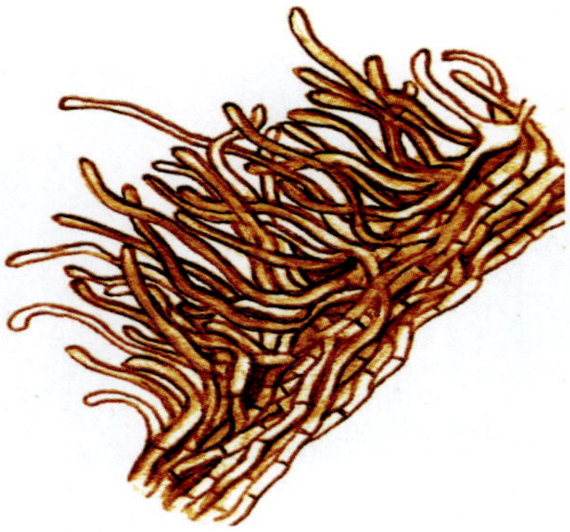

Trichoderm (stipitipellis) of the stipe of the basidiocarp of *Entoloma corvinum* (=*Leptonia corvina*), x 80 (*MU*).

trichoderm palisade (Gr. *thríx*, genit. *trichós*, hair + *dérma*, skin; Fr. *palissade* < OPr. *palissada* < *palissa*, paling < *pal*, stake < L. *palus*, stake): see **trichoderm**.

trichogyne (Gr. *thríx*, genit. *trichós*, hair + *gyné*, female): a receptive hypha, or neck of the ascogonium in the shape of a filament, that functions in capturing the spermatia during sexual reproduction of some ascomycetes, such as *Neurospora* (Sordariales) and *Laboulbenia* (Laboulbeniales). See **gynotrichous**, a term preferable to trichogyne, and **carpogonium**.

tricholomatoid (< gen. *Tricholoma* < Gr. *thríx*, genit. *trichós*, hair + *lõma*, margin, border + L. suf. *-oide*, like < Gr. *-oeídes*, similar to): *Agaricology*. A type of fruiting body (13 principal types are recognized), with sinuate gills adhered to the stipe, which is of a fibrous-fleshy or chalky consistency, centrally implanted in the pileus and lacking a volva, as represented by *Tricholoma* and certain species of *Russula*, *Inocybe* and *Entoloma*, among other gen. of Agaricales.

Trichomycetes (Gr. *thríx*, genit. *trichós*, hair + L. *-mycetes*, ending of taxonomic class < Gr. *mýkes*, genit. *mýketos*, fungus): a class of fungi (Trichomycetes) that live as commensals or as obligate symbionts in the intestinal tract of arthropods, or attached externally to their exoskeleton. Their thallus may be branched or not and coenocytic, or branched and septate. Their asexual reproduction is by means of trichospores, sporangiospores, arthrospores or amoeboid cells, depending upon the species. Their sexual reproduction is by conjugation of different thalli that results in the formation of zygospores. The Trichomycetes (which includes the orders Harpellales, Amoebidiales, Asellariales and Eccrinales) together with the Zygomycetes constitute the phylum Zygomycota of the kingdom Fungi. See the Appendix, where there is a comparison of the outline of Alexopoulos *et al*., 1996, modified by Ulloa and Hanlin, 2006, to that of Kirk *et al*., 2008. In the latter classification scheme, the class Trichomycetes is considered an artificial, unnatural group of fungi, and the orders Asellariales and Harpellales have been moved to the subphylum Kickxellomycotina, phylum Zygomycota. The orders Amoebidiales and Eccrinales have been transferred to the class Mesomycetozoea, phylum Choanozoa, kingdom Protozoa. These taxonomic rearrangements have the goal of achieving a phylogenetic classification of the fungi and fungus-like microorganisms. See **Mycetozoa**.

Tricholomatoid type of basidiocarp of *Tricholoma* sp., seen in longitudinal section, x 0.5 (*MU*).

trichophore (Gr. *thríx*, genit. *trichós*, hair, filament + suf. *-phóros*, bearer < *phéro*, to bear, support): bearing or supporting a hair or filament; in the Laboulbeniales, the trichophore cell supports the trichogyne cell. Both cells can be clearly seen in the thallus of *Laboulbenia flagellata*. See **carpogonium**.

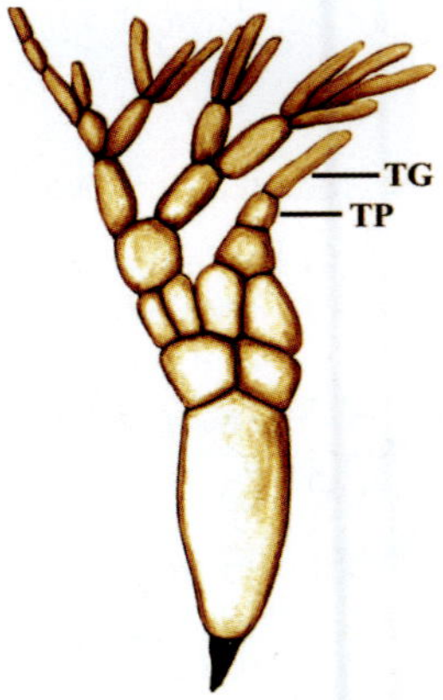

Trichophore (TP) and **trichogyne** (TG) of the thallus of *Laboulbenia flagellata*, x 100 (*MU*).

trichospore (Gr. *thríx*, genit. *trichós*, hair + *sporá*, spore): a unispored sporangium, provided with one or several filamentous or piliform appendages, characteristic of the Trichomycetes, such as *Smittium* and *Stachylina* (Harpellales).

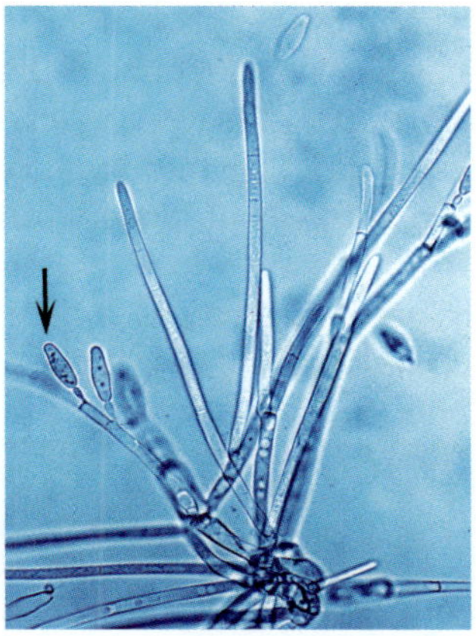

Trichospores of the thallus of *Smittium* sp., x 40 (*RTH*).

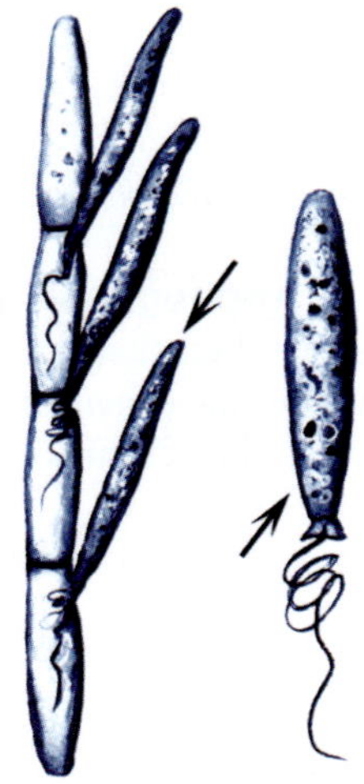

Trichospores of a branch of the thallus of *Stachylina grandispora*. Note the piliform appendage on the posterior end of the trichospores, x 320 (*MU*).

trimerous (Gr. *treís*, three, or *trís*, three times + *méros*, part): of three parts.

trimitic (Gr. *treís*, three, or *trís*, three times + *mítos*, filament; here, hypha + suf. *-íkos* > L. *-icus* > E. *-ic*, pertaining to or relating to): a fructification with three types of hyphae: generative or fertile, skeletal or supportive, and binding, as is seen, e.g., in *Trametes* (=*Coriolus*) *versicolor* (Aphyllophorales). Cf. **dimitic** and **monomitic**.

tripartite tubular hair (L. *tripartitus*, divided into three parts; L. *tubus*, tube + suf. *-aris* > E. *-ar*, like, concerning, with the nature of; ME. *heer, hare, heir* < OE. *haer*, hair, slender cylindrical filament): each one of the filamentous appendages of a straminopilous or vibratory flagellum, composed of a solid base, a hollow cylindrical axis and one or more terminal filaments. See **straminopilous**.

tripli- (L. *triplus*, triple): a pref. used in terms such as *triplinervous*, a leaf with three veins that unite above the base.

triplo-, **tripl-** (Gr. *triplóos*, triple, three times): a pref. used in terms such as *triploid*, a cell, nucleus, etc., with $3n$ chromosomes, i.e., with three basic series (or haploids) of chromosomes.

tripto-, **tript-** (Gr. *triptós* < *tríbo*, to injure, debilitate): a pref. employed in terms that allude to a weakening, as in *triptogen*.

triquetrous arm (L. *triquetus*, triangular + *-osus*, *-eus* > E. *-ous*, having, possessing the qualities of; ME. < E. *earm*, arm): *Phallales*. A receptaculum or arm which is three-cornered in cross section, as in *Pseudocolus*.

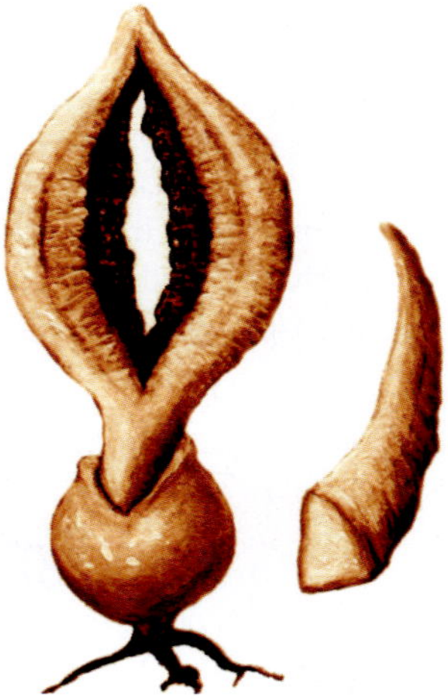

Triquetrous arms, lined with the gleba, of the clathrate fruiting body of *Pseudocolus fusiformis*. One arm is shown in transverse section, x 1 (*MU*).

tristichous (Gr. *treís*, three, or *trís*, three times + *stíchos*, line, row): arranged in three rows. Cf. **distichous** and **monostichous**.

triverticillate (Gr. *treís*, three, or *trís*, three times > L. pref. *tri-*, *ter-*, three + NL. *verticillatus*, arranged in a verticil < *verticillus* + suf. *-atus* > E. *-ate*, provided with or likeness): refers to a penicillus that has branching at three levels, i.e., whose phialides are always borne on metulae, and the metulae generally arise from terminal branches of a well defined conidiophore, as happens, e.g., in *Penicillium chrysogenum* (=*P. notatum*), *P. echinulatum* and *P. aurantiogriseum* (=*P. puberulum*), moniliaceous asexual fungi. Cf. **monoverticillate**, **biverticillate** and **polyverticillate**.

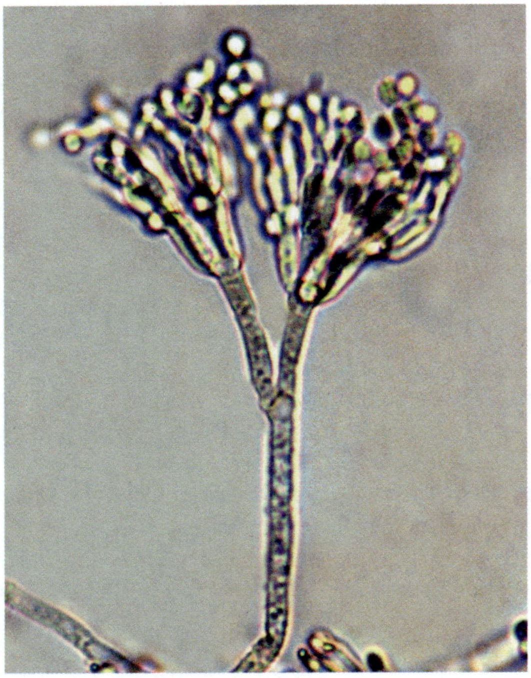

Triverticillate conidiophore of *Penicillium chrysogenum* (=*P. notatum*), x 600 (*CB*).

trochleariform (L. *trochlea* < Gr. *trochilía*, pulley + suf. *-formis* < *forma*, shape): with the shape of a pulley wheel, like the ascospores of *Eurotium* (Eurotiales).

Trochleariform ascospores of *Eurotium rubrum*, x 600 (*MU*).

troglobiotic (Gr. *tróglo*, cave, cavern + *biotikós*, relating to or pertaining to life < *bíos*, life + suf. *-tikós* > L. *-ticus* > E. *-tic*, relation, aptitude, inclination or capacity): living in caves, like *Histoplasma capsulatum* (moniliaceous asexual fungi), which together with other fungi, lives on bat dung and soil inside caves.

troph-, **tropho-** (Gr. *trophós*, what nourishes < *trépho*, to nourish): a pref. employed to introduce the idea of nutriment or of something that can serve as food, as in the following words. Also used as a suf., *-troph*.

trophic (Gr. *trophós*, what nourishes < *trépho*, to nourish + suf. *-ikos* > L. *-icus* > E. *-ic*, pertaining to or relating to): generally applied to the vegetative phase of the Acrasiomycota, Dictyosteliomycota and Myxomycota (myxamoebae and plasmodia), which, like some protozooans, is characterized by having phagotrophic or holozoic nutrition, but also can refer to the phase in any organism in which it principally feeds.

trophocyst (Gr. *trophós*, what nourishes, that serves as food + *kýstis*, bladder, cell): a swollen portion of a hypha, or of an enlarged cell, immersed in the substrate from which it obtains its food, and from which is formed the sporangiophore, as is observed, e.g., in *Pilobolus crystallinus* and *P. longipes* (Mucorales).

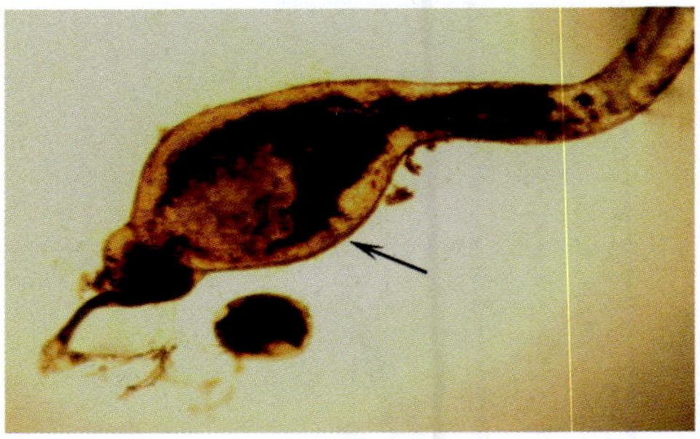

Trophocyst of *Pilobolus longipes*, x 21 (*MU*).

trophogonium, pl. **trophogonia**, **trophogone** (Gr. *trophós*, what nourishes, that serves as food + NL. *gonium* < Gr. *gónos*, the engendered, the offspring): a sterile antheridium, destined only to provide nutriment.

tropism (Gr. *trópos*, gyration, change in position or manner < *tropé*, turn < *trépo*, to gyrate, revolve, to go towards + suf. *-ismós* > L. *-ismus* > E. *-ism*, state, phase, tendency, action): orientation movement undertaken by an organism or part of one due to the unilateral influence of a stimulating factor; a turn, or curved growth of an organism as a reaction to a strong stimulus. A tropism is positive when the organism is oriented toward the source of the excitation; it is negative when the organism moves away from the source. In essence, tropisms are of the same order as *taxes*, but the latter assumes the free movement of the organisms to take position in an orderly way, tactically. As the stimulating factors can be diverse, there are various types of tropisms. See **aerotropism**, **autotropism**, **chemotropism**, **diageotropism**, **galvanotropism**, **geotropism**, **gonotaxis**, **gravitropism**, **heliotropism**, **hydrotropism**, **orthogeotropism**, **phototropism**, **plagiotropism**, **thigmotropism** and **zygotropism**.

truffle (modification of MF. *truffe* < OPr. *trufa* < VL. *tufera*, related to L. *tuber*, tuber, tumor, prominence, knot): the common name of the edible ascoma, generally subterranean, of various species of ascomycetes of the gen. *Tuber* or other Pezizales or Elaphomycetales, or of the basidioma of Melanogastrales. The European species of truffles are the best known and sought after as edibles. *Truffle*: *Tuber texense*. *Burgundy truffle*: *Tuber uncinatum*. *False truffle*: *Hymenogaster*. *Périgord* (*French*) *truffle*: *T. melanosporum*. *Red truffle*: *Melanogaster variegatus*. *Summer truffle*: *T. aestivum*. *White truffle*: *Choiromyces meandriformis*. *White Piedmont truffle*: *T. magnatum*. *White winter truffle*: *T. hiemalbum*. *Winter truffle*: *T. brumale*. *Yellow truffle*: *Terfezia* and *Tirmania*.

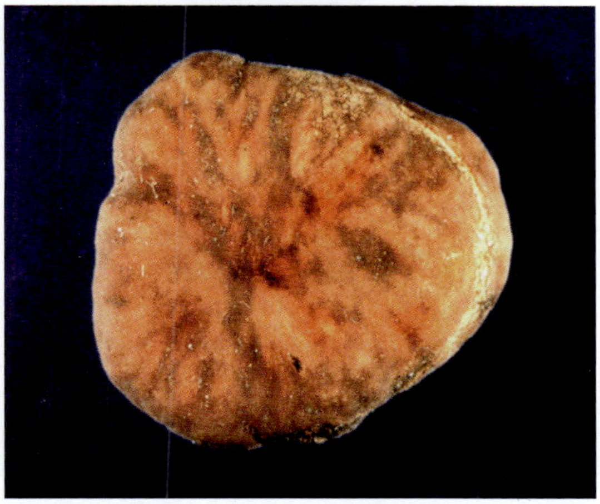

Ascocarp of *Tuber texense*, a species of **truffle** found in the southern United States, x 1 (*RTH*).

Ascocarp of *Tuber melanosporum*, one of the most delectible species of **truffles**, x 1 (*EPS*).

Ascocarp of *Terfezia gigantea*, a species of **truffle** that grows in arid zones of Africa, x 1 (*MU*).

truncate (NL. *truncatus*, mutilated, cut off < *truncare*, to mutilate, cut off + suf. *-atus* > E. *-ate*, provided with or likeness): applied to structures that end abruptly, with a flat end, as if it had been cut off. E.g., the meiosporangia of *Allomyces javanicus* (Blastocladiales) have a truncate base, and the conidia of *Scopulariopsis brevicaulis* and *Basipetospora rubra* (moniliaceous asexual fungi), and the basidiospores of *Coprinellus* (=*Coprinus*) *angulatus* (Agaricales) are truncate; on the peridium of the fruiting bodies of *Calvatia* and *Lycoperdon* (Lycoperdales) there are truncate warts. Syn. of **abrupt**.

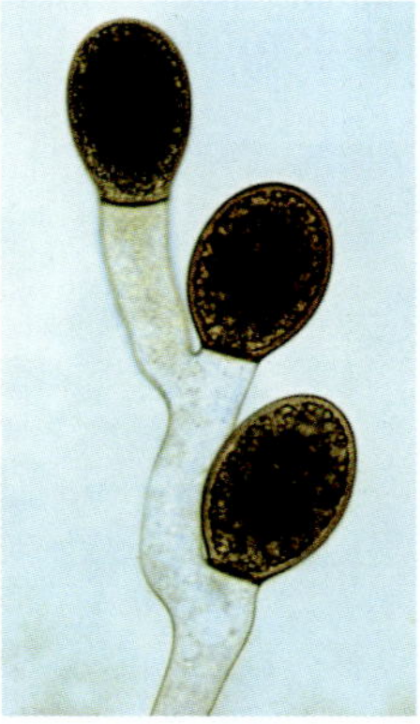

Truncate base of the meiosporangia (or latent zoosporangia) on a thallus of *Allomyces javanicus*, x 400 (*RTH*).

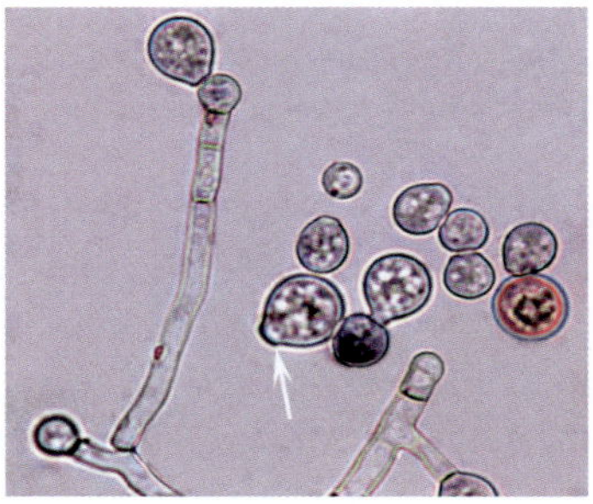

Truncate base of the conidia of *Basipetospora rubra*, x 750 (*MU*).

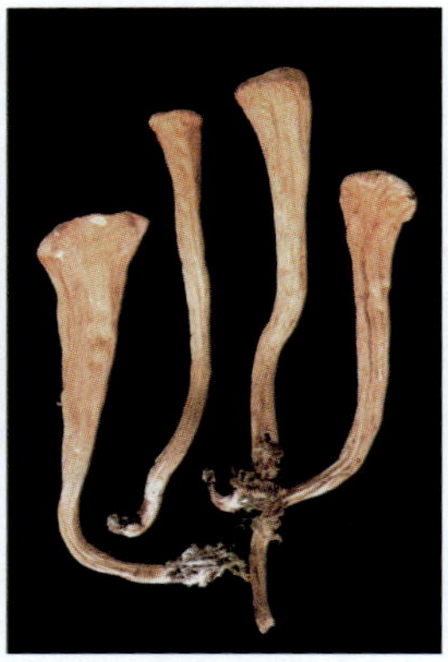

Truncate apex of the basidiocarps of *Clavariadelphus truncatus* (Aphyllophorales), x 0.5 (*MU*).

Tubaeform ascocarp arising from a germinating sclerotium of *Sclerotinia sclerotiorum*, x 2 (*RTH*).

truncicole (L. *truncus*, trunk + suf. *-cola*, inhabitant < *colere*, to inhabit): living on the trunks of trees, like many mosses and lichens.

tuba (Tagalog, *tubá*): the fermented sap of various species of palms (especially the coconut palm), with an alcohol and lactic acid content, as well as a microbiota responsible for the fermentation, very similar to that found in *pulque*, a Mexican beverage that is prepared by fermentation of the sap of various species of *Agave*. Tuba is consumed in the coastal regions of western Mexico, as well as in various parts of Asia, in particular the Philippine Islands. Cf. **pulque**. See **fermentation**.

Tubaeform fruiting bodies of *Chrysomphalina* (=*Omphalina*) *chrysophylla*, x 3 (*MU*).

Tuba transported in a hollow bamboo stem, on sale in Cianjur, a small town in Indonesia, x 0.2 (*MU*).

Tubaeform basidiocarps of *Cotylidia* sp., x 1 (*RV*).

tubaeform, **tubiform** (L. *tuba*, trumpet, tuba + suf. *-formis* < *forma*, shape): with the shape of a trumpet or a tube that is expanded on one end, like the apothecium of *Sclerotinia sclerotiorum* (Helotiales), the podetia of the lichen *Cladonia* (Lecanorales), and the basidiocarps of *Chrysomphalina* (=*Omphalina*) *chrysophylla* and *Cotylidia aurantiaca* (Agaricales), among other examples.

Tubaeform basidiocarps of *Gloeophyllum mexicanum*, x 0.3 (*RV*).

tube (L. *tubus*, tube): a hollow, cylindrical structure. See **Rohr**.

tube trama, pl. **tramae** (L. *tubus*, tube, pipe; L. *trama*, trama, texture): see **trama**.

tubercular, tuberculate (L. *tuberculum*, dim. of *tuber*, tubercle, prominence, swelling + suf. *-aris* > E. *-ar*, like, concerning; NL. *tuberculatus*, having tubercles < L. *tuberculum* + suf. *-atus* > E. *-ate*, provided with or likeness): having knots or rounded bumps. E.g., the conidia of *Histoplasma capsulatum*, *Scopulariopsis brevicaulis* (moniliaceous asexual fungi) and *Curvularia tuberculata* (dematiaceous asexual fungi), as well as the aeciospores of *Cronartium ribicola* (Uredinales), have a tuberculate wall. See **tubercule**.

tuberculariaceous (L. *tuberculum*, dim. of *tuber*, tubercle, prominence, swelling + suf. *-aris* > E. *-ar*, like, concerning + suf. *-aceus* > E. *-aceous*, of or pertaining to, with the nature of): applied to an unnatural classification of asexual or anamorphic fungi, that form their conidia on short conidiophores, on small, compact, cushion-shaped conidiomata (sporodochia). E. g., *Epicoccum*, *Fusarium* and *Tubercularia*, of the old family Tuberculariaceae, order Moniliales, class Hyphomycetes, subdivision Deuteromycotina, or deuteromycetes in a general sense. See **Hyphomycetes**.

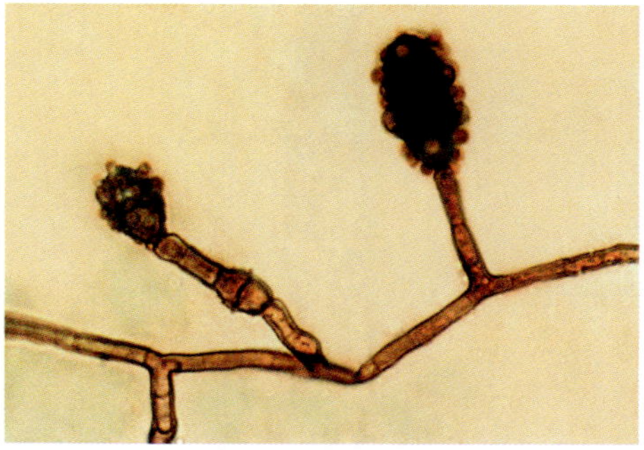

Tubercular wall of the conidia of *Curvularia tuberculata*, x 1,100 (*MCG*).

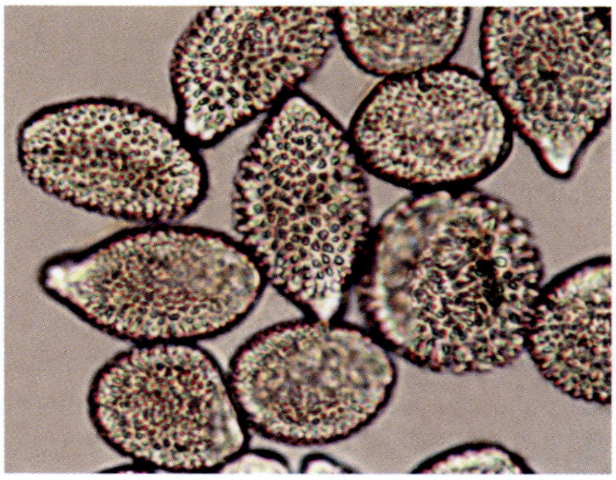

Tubercular wall of the aeciospores of *Cronartium ribicola*, x 1,100 (*MU*).

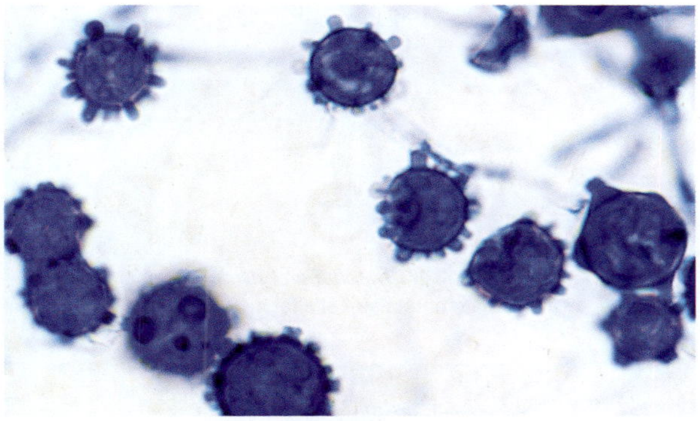

Tubercular wall of the macroconidia of *Histoplasma capsulatum*, x 100 (*MU*).

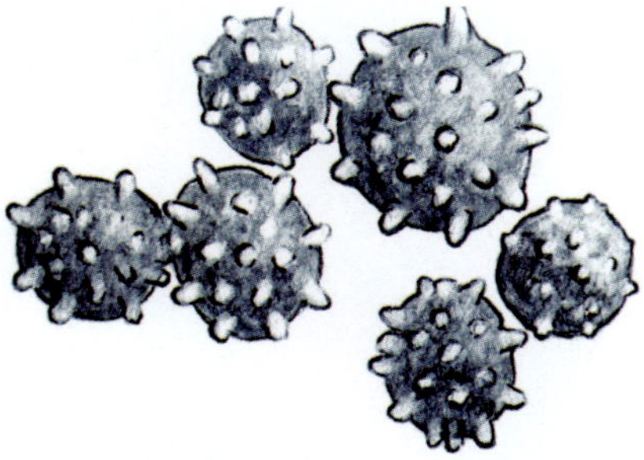

Tubercular wall of the macroconidia of *Histoplasma capsulatum*, x 100 (*MU*).

tubercule, **tubercle** (L. *tuberculum*, dim. of *tuber*, tumor, swelling): a more or less rounded verruciform prominence on a surface, spore, etc. See **tubercular**.

tubular (NL. *tubularis* < L. *tubulus*, dim. of *tubus*, tube, pipe + suf. *-aris* > E. *-ar*, like, concerning): hollow and cylindrical in shape, like the neck or rostrum of many perithecia, e.g., those of *Ceratocystis* (Microascales), *Melanospora* (Hypocreales) and *Guanomyces* (Sordariales).

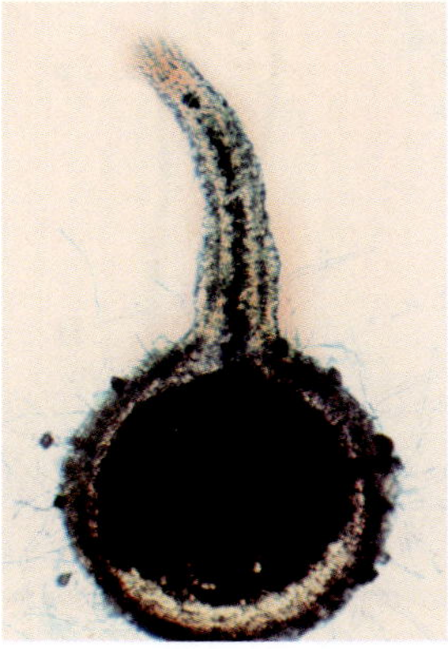

Tubular neck of the perithecium of *Melanospora zamiae*, x 100 (*RTH*).

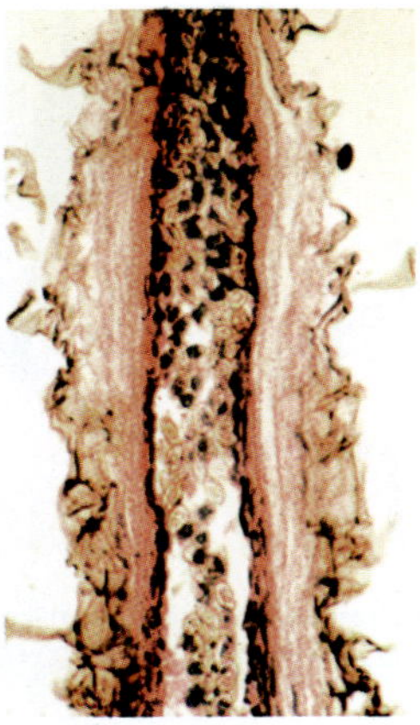

Tubular neck of the perithecium of *Guanomyces polythrix*, x 100 (*MCG*).

tubulus, pl. **tubuli** (L. *tubulus*, dim. of *tubus*, tube, pipe): refers to the neck or rostrum of many perithecia (Pyrenomycetes), as well as to any small, hollow cylindrical structure.

tumid (L. *tumidus*, swollen): swollen, inflated.

tumor (L. *tumor*, genit. *tumoris*, abnormal growth): in plant pathology, an abnormal swelling of some part of the affected plant, due to hyperplasia of the cells. See **canker**.

tunica, pl. **tunicae** (L. *tunica*, tunica, covering): **1**. Any enveloping layer or covering. **2**. *Gasteromycetes*. Refers to the thin white membrane that covers the peridioles of the majority of species of nidulariaceous gasteromycetes, such as *Cyathus* and *Nidularia* (Nidulariales).

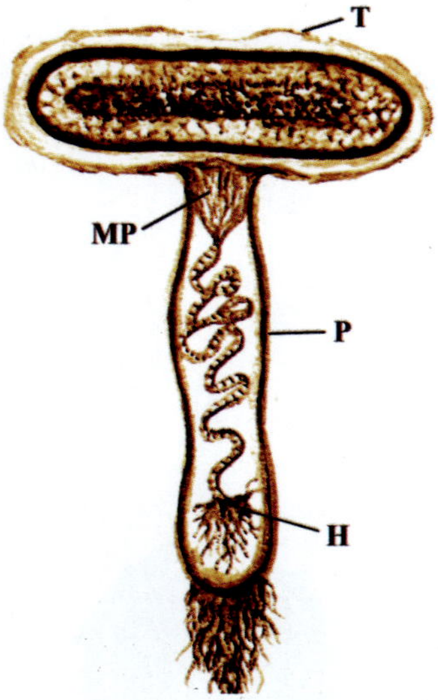

Tunica (T) covering a peridiole of *Cyathus striatus*, seen in longitudinal section. Also shown are the purse (P) surrounding the funicular cord, and the hapteron (H), which is attached to the middle piece (MP), x 100 (*MU*).

turbid (L. *turbidus*, turbid, confused < *turba*, confusion + suf. *-idus*, quality or state): not clear, cloudy; without transparency. In one of the methods employed in the identification of yeasts, the degree of turbidity of cultures of these microorganisms in liquid media with different sources of carbon is observed , with the goal of determining their assimilation capacity.

turbinate (L. *turbinatus*, shaped like a top < *turbo*, genit. *turbinis*, top, turbin + suf. *-atus* > E. *-ate*, provided with or likeness): with the shape of an inverted cone or top, narrow at the base and wide at the apex, like, e.g., the basal cell of the resting spore of *Physoderma* (=*Urophlyctis*) *alfalfae* (Chytridiales), the cystidia of certain agarics, the basal cells of the conidia of *Polyschema variabilis* (dematiaceous asexual fungi) and the perithecial stromata of *Camillea* and *Phylacia turbinata* (Xylariales). Similar to **obconic**.

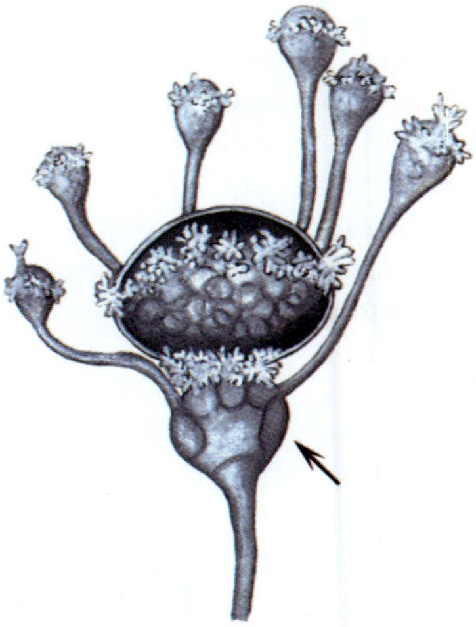

Turbinate basal cell of *Physoderma* (=*Urophlyctis*) *alfalfae*, an obligate parasite of alfalfa, x 1,000 (*MU*).

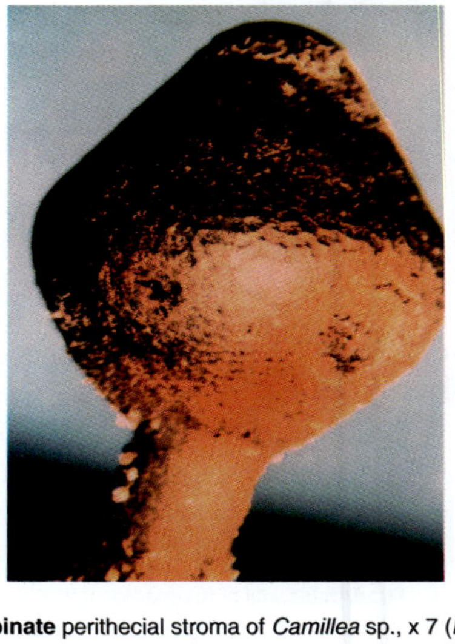

Turbinate perithecial stroma of *Camillea* sp., x 7 (*MU*).

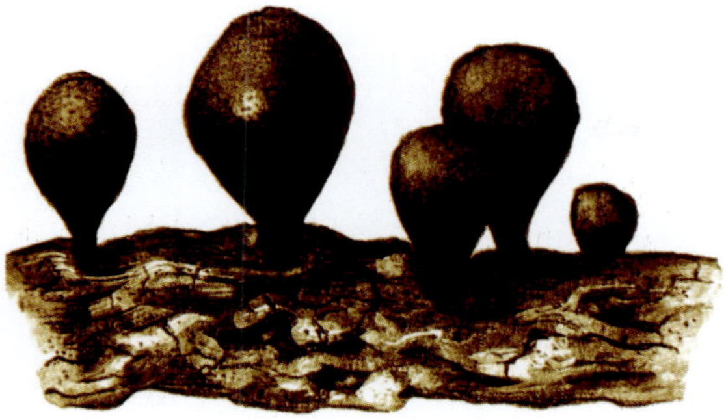

Turbinate perithecial stromata of *Phylacia turbinata*, x 2 (*MU*).

turgid (L. *turgens*, genit. *turgentis* < *turgidus*, turgid): swollen, distended.

turgidity, turgor (L. *turgere*, swollen, distended): the normal state of turgescence of the protoplasm and of the wall of living cells; swelling or distention due to an increase in pressure of the fluid contents of the cell.

twisted (ME. *twisten* < OE. *twist*, rope > E. *twisted,* pt. of *twist*, to unite by winding a strand around another to form a tight helix): not straight; spirally twisted.

tycho-, tych- (Gr. *týche*, fortune, chance): a pref. employed in forming terms alluding to something that is exceptional, occasional or accidental, as in *tychoplankton*, plankton formed by organisms coming from other groupings, or rather, for those whom the regimen of planktonic life is exceptional.

tyndallization (< *John Tyndall*, English microbiologist + L. suf. *-ationem* > E. *-ation*, action, state or condition, result): a method of interrupted sterilization with moist heat, used for some microbiological media and chemical substances that should not be heated above 100°C, since that temperature can adversely affect them. This method involves heating the material at 100°C for 3 successive days with intermediate periods of incubation. Spores resistant to heat can germinate during the incubation periods, and in subsequent exposures to heat, the vegetative cells will be destroyed. Cf. **sterilization** and **pasteurization**.

type (L. *typus* < Gr. *týpos*, type, mark, model): the element or elements on which the author of a new genus, species, variety, etc., based, or is considered to have based, the original description of the new taxon and which is validly published. A type typically consists of a specimen or specimens, but may also be a slide, illustration or a culture that is not metabolically active. The example designated as the type and its location must be cited in the original publication, although this often was not done by early authors. The original specimen is referred to as the *type specimen*, or if based on a culture, the *type culture*. The process of designating a type is called *typification*. Type material should be located in collections that are available to others who may wish to examine them.

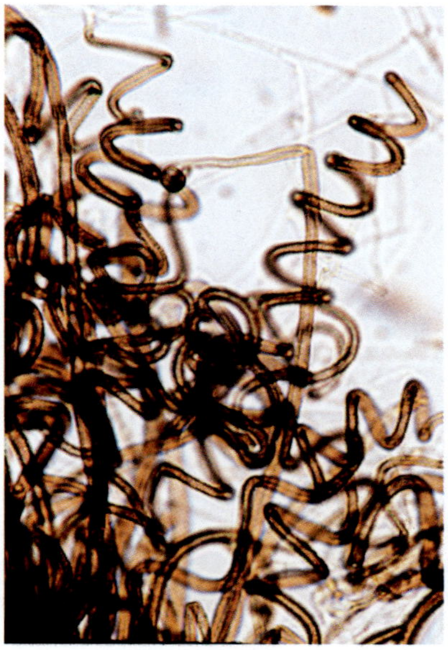

Twisted hairs of the perithecium of *Chaetomium contortum*, x 100 (*MU*).

Twisted hairs of the perithecium of *Chaetomium contortum*, x 120 (*MU*).

type culture, cultotype (L. *týpos*, type, mark, model; L. *cultus*, culture, action and effect of culturing): a living culture derived from a holotype specimen.

typonym (Gr. *typós*, type, mark, model + *ónoma*, *ónyma*, name): a name based on the same type as another name, but which is not a basionym or synisonym.

U

Ungulate fruiting body of *Phellinus linteus* (*RV*).

u

ubiquitous (L. *ubique* > Fr. *ubiquite*, anywhere): refers to an organism that occurs in almost any type of habitat, like many species of fungi, e.g., the gen. *Aspergillus* and *Penicillium* (moniliaceous asexual fungi).

uliginose, uliginous (L. *uliginosus*, moist, swampy < *uligo*, genit. *uliginis*, moisture + suf. *-osus* > E. *-ose*, full of, augmented, prone to): applied to organisms that live in wet places, such as swamps.

umbellate (L. *umbellatus*, bearing, consisting of, or arranged in umbels; resembling an umbel in shape < *umbella*, parasol, umbel +suf. *-atus* > E. *-ate*, provided with or likeness): provided with umbels, like the thallus of *Rhipidium americanum* (Rhipidiales), and the sporangiophore of *Circinella umbellata* (Mucorales), whose sporangial branches originate from the same point and are almost the same length, like the ribs of an inverted parasol on the end of the main axis. The arrangement of the sporocladia of *Kickxella alabastrina* (Kickxellales) also is umbellate.

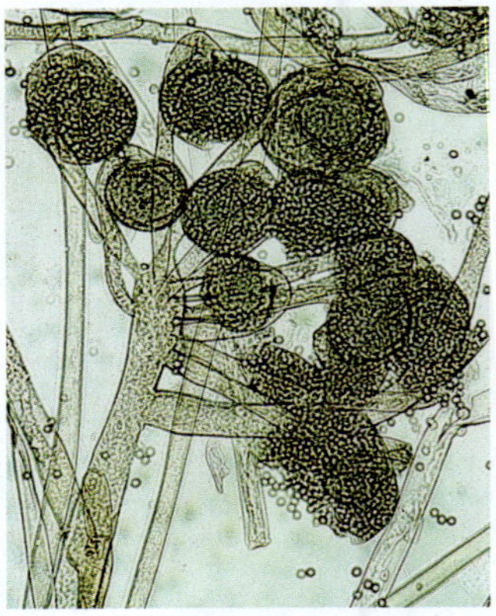

Umbellate sporangiophores of *Circinella umbellata*, x 300 (EAA).

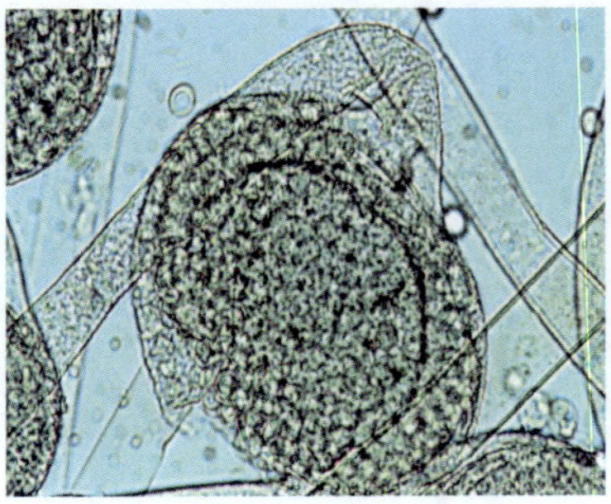

Umbellate sporangiophores of *Circinella umbellata*, x 780 (EAA).

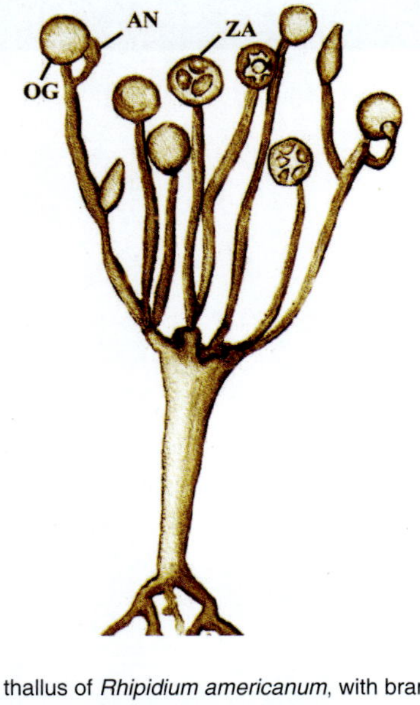

Umbellate thallus of *Rhipidium americanum*, with branches bearing zoosporangia (ZA), oogonia (OG) and antheridia (AN), x 100 (*MU*).

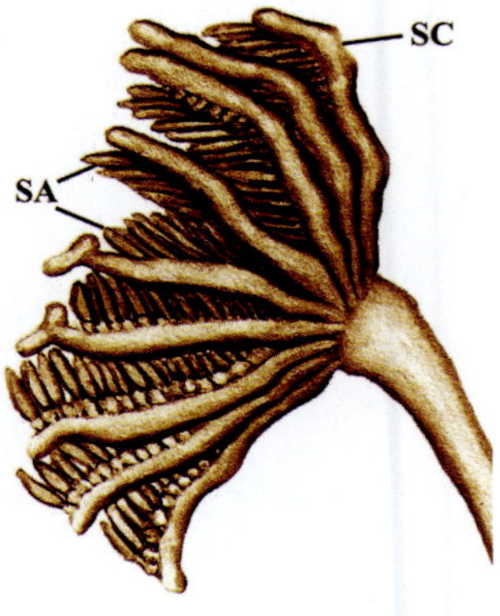

Umbellate sporangiophore of *Kickxella alabastrina*, bearing sporocladia (SC) with sporangiola (SA), x 570 (*MU*).

umbilicate (L. *umbilicatus*, provided with a navel or having the shape of one < *umbilicus*, navel + suf. *-atus* > E. *-ate*, provided with or likeness): with a navel-like depression, like the pileus of the apothecium of *Leotia* (Helotiales) and the pileus of the basidiocarp of *Clitocybe*, *Omphalina* (=*Omphalia*) and *Laccaria*, of the Agaricales, and of *Lentinus* (Aphyllophorales). In the lichens, it is applied to a more or less concave foliaceous thallus that is united to the substrate only by its central part (the umbilical cord or funiculus), like that of the species of Umbilicariaceae, e.g., *Umbilicaria* (= *Gyrophora*) of the Lecanorales.

Umbilicate pileus of the apothecia of *Leotia lubrica*, x 0.8 (*RTH*).

Umbilicate pileus of the basidiocarps of *Clitocybe lazulina*, x 0.6 (*EPS*).

Umbilicate pileus of the basidiocarps of *Lentinus crinitus*, x 0.6 (*MU*).

Umbilicate pileus of the basidiocarps of *Laccaria amethystina*, x 1 (*RV*).

umbilicus, pl. **umbilici** (L. *umbilicus*, umbilicus): see **funiculus**.

umbo (L. *umbo*, genit. *umbonis*, prominence, projection): **1**. A central cone or nipple-shaped projection. **2**. *Agaricales*. Refers to the elevated cone or mound-shaped part that projects upward in the center of a certain type of pileus, whether concave or convex. See **umbonate**.

umbonate (NL. *umbonatus*, provided with an umbo or shield < *umbo*, genit. *umbonis*, prominence, shield + suf. *-atus* > E. *-ate*, provided with or likeness): **1**. Having a central prominence in the shape of a cone. **2**. Shaped like an *umbo* or nipple, like the giant colony of *Cryptococcus albidus* (cryptococcaceous asexual yeasts) and the pileus of *Psilocybe caerulescens*, *Inocybe*, *Macrolepiota* and *Volvariella* (Agaricales).

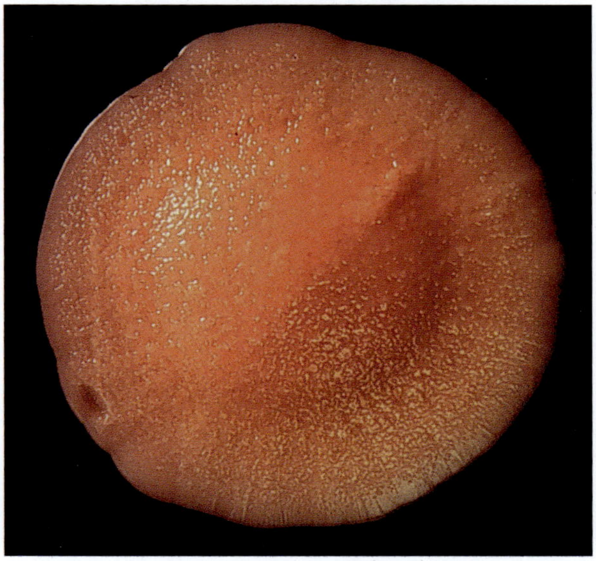

Umbonate giant colony of *Cryptococcus albidus*, on agar, x 1 (*MU*).

Umbonate pileus of the basidiocarps of *Psilocybe caerulescens*, x 1 (*MU*).

Umbonate pileus of the basidiocarp of *Volvariella bombycina*, x 1 (*RV*).

Umbonate pileus of the basidiocarps of *Inocybe* sp., x 1 (*RV*).

Umbonate pileus of the basidiocarp of *Macrolepiota mastoidea*, x 1 (*RV*).

umbraticous, **umbrophilous** (L. *umbraticus*, pertaining to shade < *umbra*, shade, shaded area + Gr. suf. *-tikós*. > L. *-ticus* > E. *-tic*, relation, fitness, inclination or capacity; L. *umbra* + Gr. *phílos*, have an affinity for): growing in shady places. Equal to **sciophilous**. See **photophobic**.

uncate (L. *uncatus*, bent inward, like a hook < *uncus*, hook + suf. *-atus* > E. *-ate*, provided with or likeness): equal to **uncinate**.

unciform (L. *unciformis* < *uncus*, hook + suf. *-formis* < *forma*, shape): shaped like a hook; forming a hook, like the appendages of the ascocarp of *Uncinula* (Erysiphales). Also called **uncinate**.

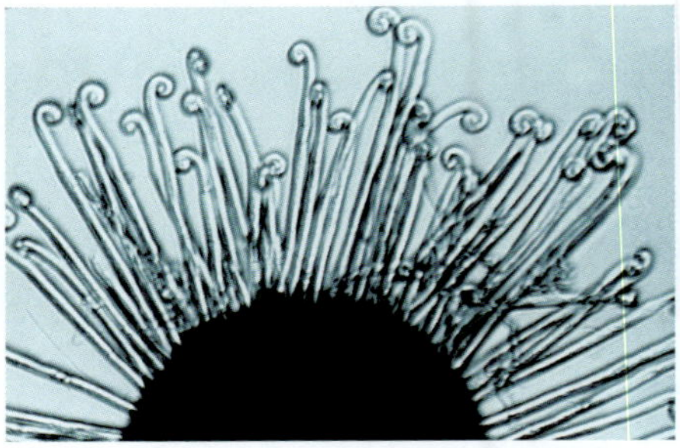

Unciform perithecial appendages of *Uncinula* sp., x 300 (*RTH*).

Unciform perithecial appendages of *Uncinula circinata*, x 300 (*CB*).

uncinate (L. *uncinatus*, bent on the end like a hook < *uncinus*, dim. of *uncus*, hook + suf. *-atus* > E. *-ate*, provided with or likeness): see **unciform**.

uncinulum, pl. **uncinula**, **crozier** (L. *uncinulum* < *uncinus*, dim. of *uncus*, hook + dim. suf. *-ulum*; ME. *croser* < MF. *crossier*, staff bearer < *crosse*, pastoral staff): a hook-shaped structure that is produced on the apex of the ascogenous hyphae of many ascomycetes and which becomes the mother cell of the ascus, as is seen, e.g., in *Eurotium* (Eurotiales) and *Hypomyces* (Hypocreales). The simultaneous division of the two nuclei of each uncinulum, the formation of septa and the delimitation of the cells is a process typical of the ascomycetes, which have a parallel in the formation and function of the clamp connections of the secondary mycelium of the basidiomycetes. See **crozier**.

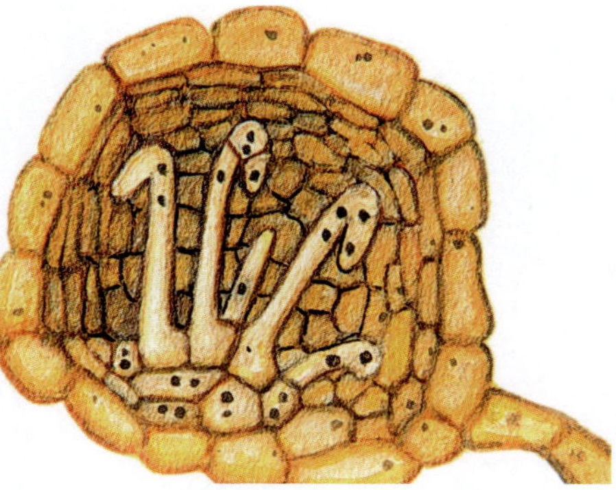

Uncinulum of a cleistothecium of *Eurotium rubrum*, x 150 (*MU*).

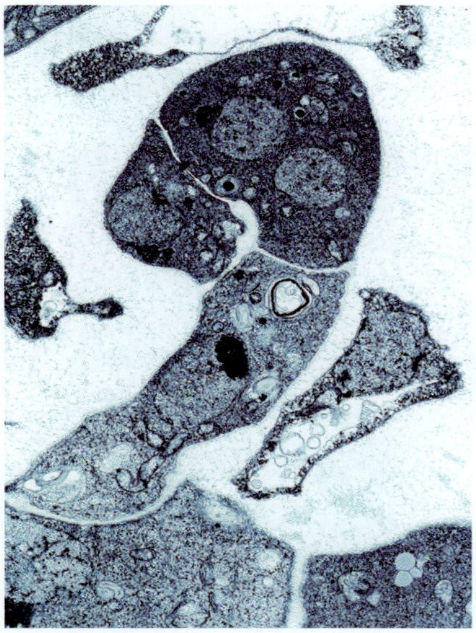

Uncinulum of *Hypomyces* sp. Eventually, karyogamy occurs in the dikaryotic penultimate cell of the uncinulum to give rise to the mother cell of the ascus, x 1,000 (transmission electron micrograph by *CM*).

undate (L. *undatus*, wave-like, wavy < *unda*, wave + suf. *-atus* > E. *-ate*, provided with or likeness): with a wavy border, rising and falling, like the pileus of the basidiocarp of *Clitocybe phyllophila*, *Clitopilus prunulus* and *Marasmius* sp. (Agaricales). Cf. **undulate**.

undulate (L. *undulatus* < *undula*, dim. of *unda*, wave + suf. *-atus* > E. *-ate*, provided with or likeness): like *undate* but with smaller waves, like the border of the pileus of *Gymnopus* (=*Collybia*) *fusipes* (Agaricales).

Undate pileus of the basidiocarps of *Clitopilus prunulus*, x 0.5 (*RV*).

Undate pileus of the basidiocarps of *Marasmius* sp., x 1 (*RV*).

undulipode, **undulipodium**, pl. **undulipodia** (L. *undula*, dim. of *unda*, wave + NL. *podium* < Gr. *pódion*, dim. of *poús*, genit. *podós*, foot, support): a flexible intracellular extension in the shape of a small whip (traditionally called a *flagellum*), present on the motile cells of the majority of the eukaryotic organisms (many plants, the major part of the protoctists and the majority of the animals). The undulipodia, which grow from the kinetosomes or basal bodies, are composed of packets of microtubules which, in transverse section, show the characteristic 9 + 2 arrangement, and are constituted of more than 100 proteins. There are undulipodia with or without mastigonemes, depending on the species. Cf. **flagellum**.

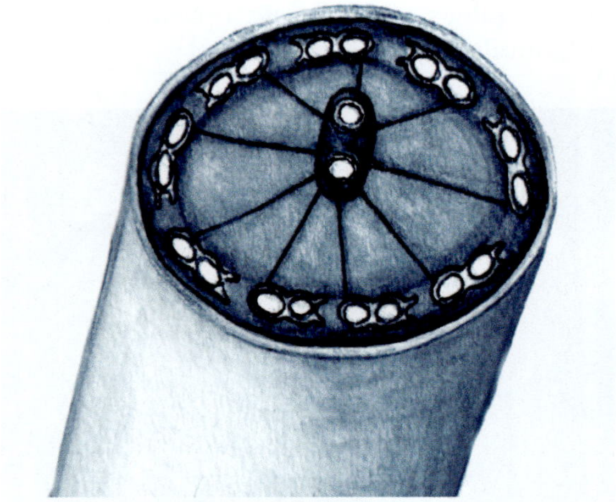

Undulipodium of a typical eukaryotic cell, with the 9+2 arrangement of microtubules. Diagram of a transverse section based on the figures and information published by Margulis and Schwartz in their book, *Five Kingdoms*, 1988, X 300,000 (*MU*).

ungulate (LL. *ungulatus*, provided with claws or hooves < L. *ungula*, claw, hoof + suf. *-atus* > E. *-ate*, provided with or likeness): applied to a fruiting body that has the shape of a horse's hoof, like that of *Phellinus* (=*Fomes*) *igniarius*, *Ph. linteus* and *Daedalea* (=*Fomitopsis*) *dochmia*, Aphyllophorales.

Ungulate fruiting body of *Phellinus linteus*, X 0.5 (*RV*).

Ungulate fruiting body of *Phellinus linteus*, X 0.5 (*RV*).

Ungulate fruiting body of *Phellinus linteus*, X 0.5 (*RV*).

Ungulate fruiting body of *Phellinus linteus*, X 0.5 (*RV*).

uni- (L. *unus*, one): a pref. used in words such as *uniaxial*, *unicellular*, *uniflagellate*, etc., to indicate one axis, one cell, one flagellum, etc. Equivalent to *mono-*, employed in terms of Greek origin (*monocentric*, *monopodium*, etc.).

uniascal (L. *unus*, one + Gr. *askós*, ascus + L. suf. *-alis* > E. *-al*, relating to or pertaining to): with only one ascus, like the loculi of the ascostroma of *Myriangium duriaei* (Myriangiales). Cf. **polyascal**.

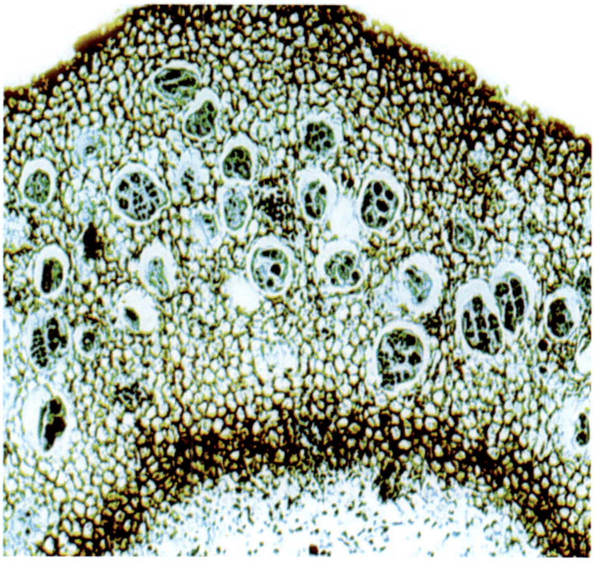

Ascostroma of *Myriangium duriaei*, seen in longitudinal section, with the **uniascal** loculi, x 100 (*RTH*).

uniaxial (NL. *uniaxialis*, with a single axis < *unus*, one + *axis*, axis + L. suf. *-alis* > E. *-al*, relating to or pertaining to): with a single central axis, like the majority of the stipitate apothecia and basidiocarps. Also called **monaxial**. Cf. **multiaxial**.

unicellular (L. *unicellularis* < *unus*, one + *cellula*, cell, living cell, dim. of *cella*, small room, compartment, cell of a honeycomb + suf. *-aris* > E. *-ar*, like, concerning): consisting of a single cell, like the thallus of the yeasts, when the somatic cells have not budded, or like many types of asexual and sexual spores of the fungi. See **amerospore**.

uniflagellate (L. *unus*, one + *flagellatus*, provided with a flagellum < *flagellum*, whip + suf *-atus* > E. *-ate*, provided with or likeness): with a single flagellum, like the zoospores of the Chytridiomycetes, which have one whiplash flagellum (without mastigonemes) on the posterior end, and those of the Hyphochytriomycota, which possess a single mastigonemate flagellum on the anterior end. The antherozoids or spermatozoids of the Monoblepharidales also are uniflagellate on their posterior end, e.g., the species of *Monoblepharis*.

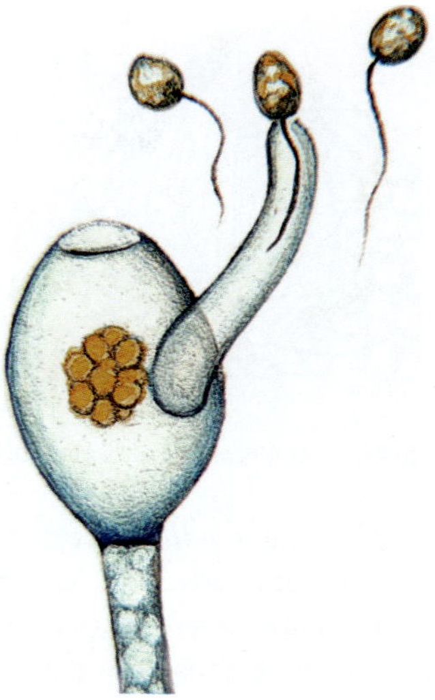

Uniflagellate antherozoids of *Monoblepharis polymorpha*; an antherozoid is escaping from the antheridium, x 650 (*MU*).

uniguttulate (L. *unus*, one + *guttulatus*, containing drops or drop-like masses < *guttula*, dim. of *gutta*, droplet + suf. *-atus* > E. *-ate*, provided with or likeness): refers to a structure that contains a single oil droplet, like the ascospores of *Podospora comata* and *Sordaria fimicola* (Sordariales). See **guttulate**.

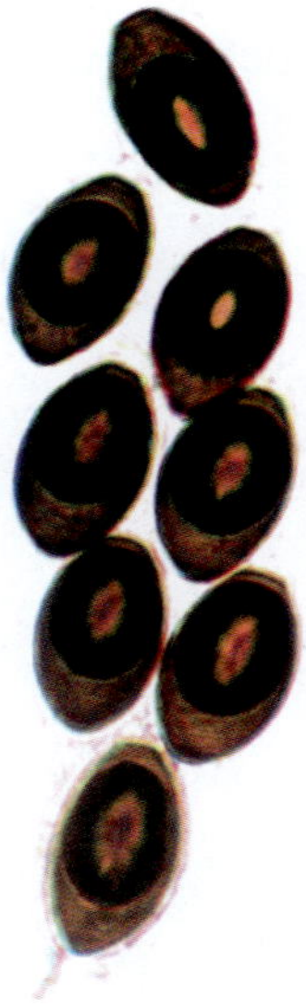

Uniguttulate ascospores of *Podospora comata*, x 530 (*CB*).

unilocular (L. *unilocularis* < *unus*, one + *loculus*, dim. of *locus*, cavity, locule + suf. *-aris* > E. *-ar*, like, concerning): with a single cavity or locule, like the ascostromata of *Mycosphaerella* (Dothideales) and *Preussia* (Pleosporales) which, because of their similarity to a perithecium, are called *pseudothecia*. Cf. **multilocular**.

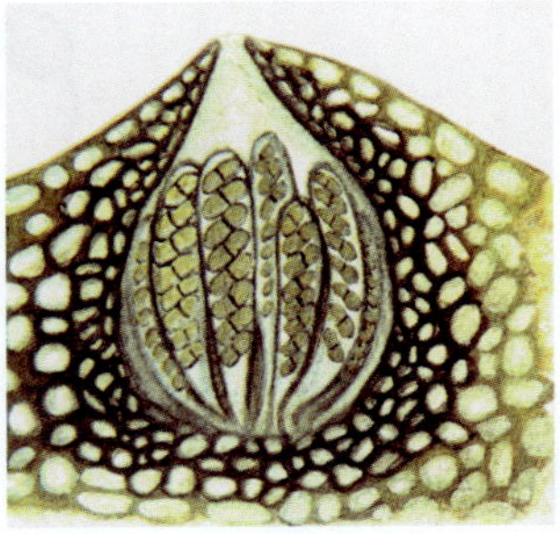

Unilocular ascostroma of *Mycosphaerella musicola*, x 150 (*MU*).

unipolar (L. *unus*, one + *polaris*, relating to the pole < *polus*, pole + suf. *-aris* > E. *-ar*, relating to or pertaining to): occurring on only one pole or end, like the flagellum of some zoospores and bacterial cells. Cf. **bipolar**, **monopolar** and **multipolar**.

uniserial, **uniseriate** (L. *unus*, one + *series*, series, row + suf. *-alis* > E. *-al*, relating to or belonging to; + L. suf. *-atus* > E. *-ate*, provided with or likeness): arranged in a single series, like the conidial head of *Aspergillus* (moniliaceous asexual fungi), e.g., *A. fumigatus* and *A. repens*, among others, that have a series or row of phialides with conidia, borne directly on the vesicle of the conidiophore. Also called **monostichous**. The asci of the majority of species of ascomycetes have their ascospores in a single row. Cf. **biserial**, **distichous** and **tristichous**.

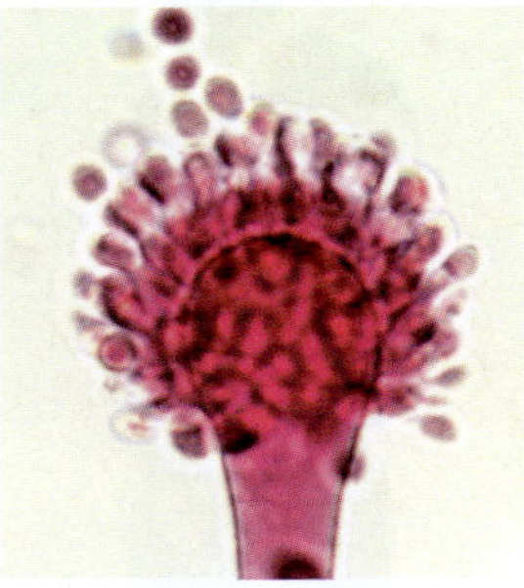

Uniserial conidial head of *Aspergillus fumigatus*, x 600 (*CB*).

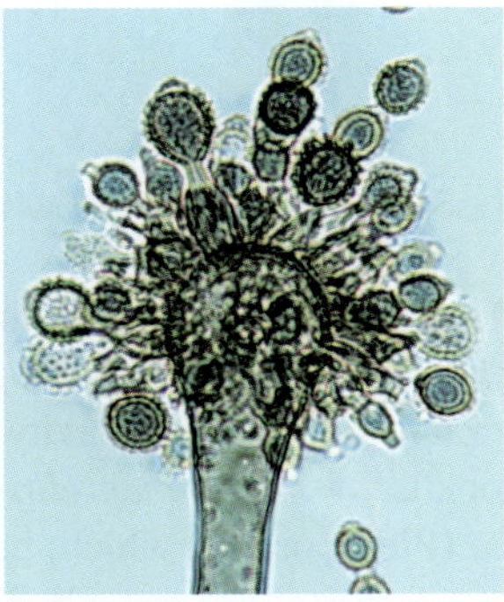

Uniserial conidial head of *Aspergillus repens*, x 1,000 (*PL*).

unisexual (L. *unus*, one + *sexualis*, sexual, concerning or relative to sex): a thallus with only one sex. See **dioecious**. Cf. **hermaphrodite** and **monoecious**.

unispored, **unisporate** (L. *unus*, one + Gr. *sporá*, spore + L. suf. *-atus* > E. *-ate*, provided with or likeness): containing a single spore, like the sporangiola of various Mucorales and Entomophthorales, such as *Cunninghamella* and *Conidiobolus*, respectively. Since in these cases the wall of the spore is fused with the wall of the sporangiolum, it resembles a conidium of the mitosporic fungi, for which reason unispored sporangiola have been called conidia. See **bisporic** and **multisporated**.

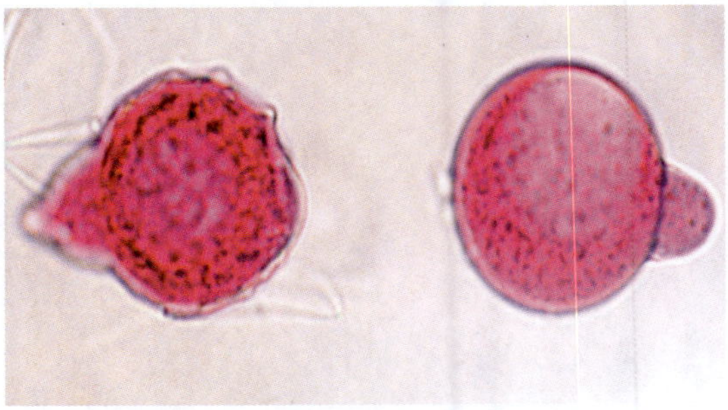

Unispored sporangiola of *Conidiobolus coronatus*, x 1,200 (*CB*).

unitunicate (L. *unus*, one + *tunica*, covering, cloak + suf. *-atus* > E. *-ate*, provided with or likeness): a type of ascus in which the internal and external layers of the wall are more or less rigid and do not separate during liberation of the ascospores, as is seen, e.g., in *Taphrina caerulescens* (Taphrinales) and *Brasiliomyces malachrae* (Erysiphales). Unitunicate asci are present in the majority of the ascomycetes; an exception is the Loculoascomycetes, which have bitunicate asci. Cf. **bitunicate** and **prototunicate**.

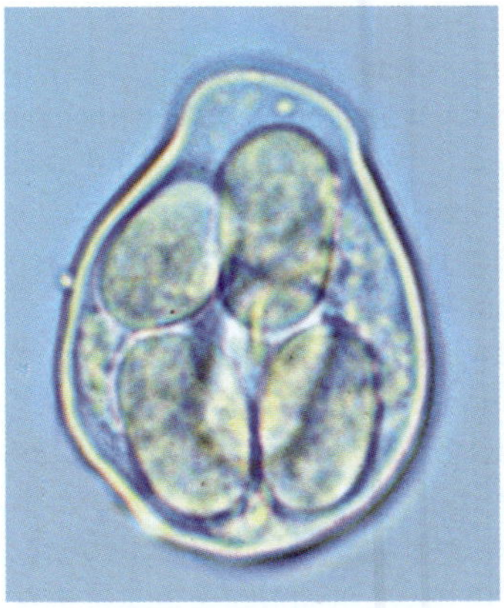

Unitunicate ascus of *Brasiliomyces malachrae*, with eight ascospores, x 800 (RTH).

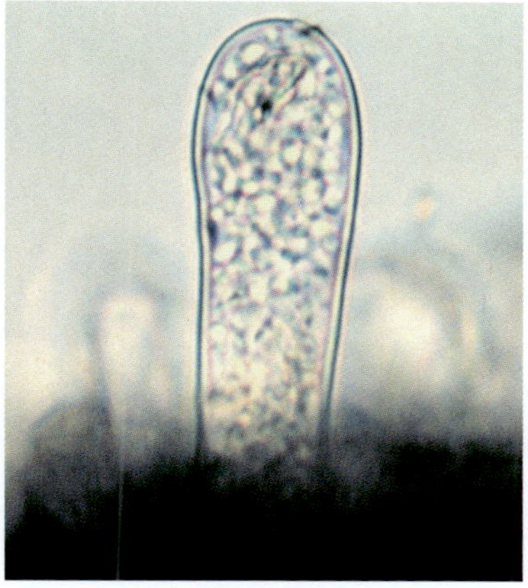

Unitunicate ascus of *Taphrina caerulescens*, with numerous budding ascospores, x 1,560 (*RTH*).

universal veil (L. *universalis*, universal, general < *universus*, universe + suf. *-alis* > E. *-al*, relating to or pertaining to; L. *velum*, veil): see **veil**.

Universal veil of the basidiocarp primordium of *Volvariella bombycina* (Agaricales), x 1 (*RV*).

Universal veil of the basidiocarps of *Volvariella bombycina* (Agaricales), x 1 (*RV*).

Universal veil of the basidiocarp of *Amanita caesarea* (Agaricales), x 1 (*RMA*).

urban, **urbanus** (L. *urbanus*, urban, city dweller < *urbs*, city + suf *-anus* > E. *-an*, pertaining to): growing in city habitats, like the basidiocarps of *Panaeolina foenisecii* and other Agaricales, such as *Bolbitius vitellinus,* common in lawns and parks.

urceolate, **urceolar** (NL. *urceolatus*, with the shape of an urn or jar < *urceolus*, dim. of *urceus*, urn, pot, pitcher + suf. *-atus* > E. *-ate*, provided with or likeness; or *urceolaris* < *urceolus* + suf. *-aris* > E. *-ar*, like, concerning): shaped like a kettle, like the ascomata of *Caliciopsis pinea* (Coryneliales). In lichenology it is applied to the discoid apothecium of some lichens, with the hymenium sunken and the parathecium elevated, with the closest borders forming a narrow mouth, as is observed in species of *Aspicilia* (=*Urceolaria*) and in *Conotrema urceolatum* (Ostropales). Also called **orculiform**, and for spores, **polarilocular** and **polocellate**.

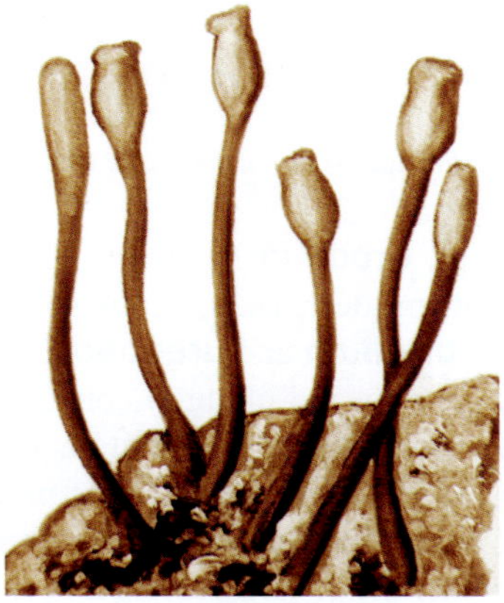

Stipitate ascomata of *Caliciopsis pinea*, with a terminal ascigerous locule, which on opening adopts an **urceolate** appearance, x 19 (*MU*).

Urceolate apothecia of *Conotrema urceolatum*, x 6 (*MU*).

Uredinia with urediniospores of *Hemileia vastatrix* (Uredinales) on a leaf of a coffee plant, x 30 (*GC*).

uredine (L. *uredo*, genit. *uredinis*, rust, blight, mildew, scab, scald < *uro*, to burn): an old name given to the rust fungi (Uredinales).

urediniospore (L. *uredo*, genit. *uredinis*, rust, blight, mildew, scab, scald < *uro*, to burn + Gr. *sporá*, spore): a binucleate spore, unicellular, sessile or pedicellate, that is formed in the uredinia or uredosori of the Uredinales or rust fungi, so-called because of the rusty color of their spores; these are the principal structures of propagation of these plant parasitic fungi, of great importance in epiphytology, because they generally produce several generations of uredinia on the host plants before passing to the next stage of the life cycle, which is that of the telia, as occurs, e.g., in *Puccinia graminis tritici*, the wheat rust.

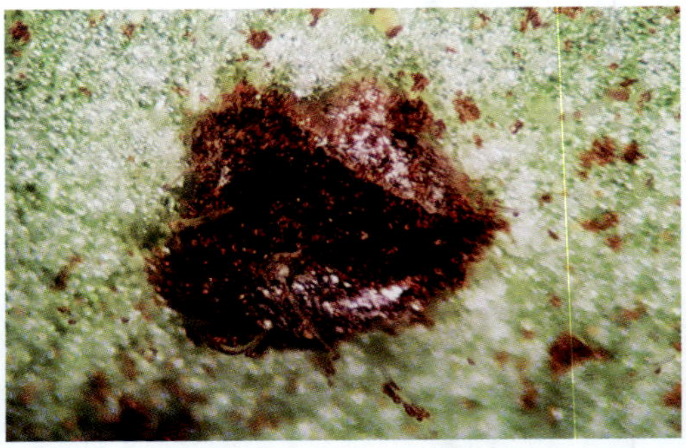

Uredinia with urediniospores of *Puccinia malvacearum* in a mallow leaf, x 40 (*MU*).

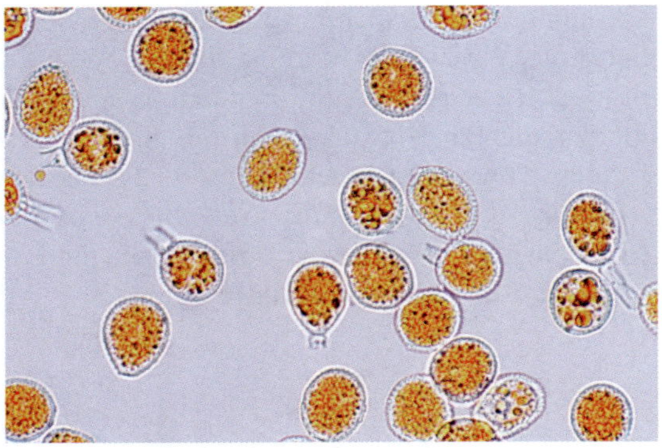

Urediniospores of *Puccinia* sp., x 780 (*CB*).

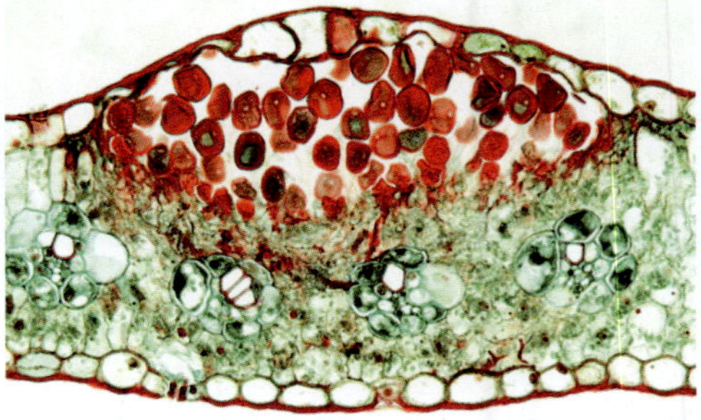

Uredinia with urediniospores of *Puccinia graminis tritici*, in a longitudinal section of a wheat leaf, x 150 (*MU*).

uredinium, pl. **uredinia** (L. *uredo*, genit. *uredinis*, rust, blight, mildew, scab, scald < *uro*, to burn): also called **uredium** and **uredosorus**; each of the pustules, ochre or red-oxide in color, that produces *urediniospores*, formed by the rust fungi (Uredinales) in tissues of the infected plant. The uredinia lack a peridium or proper wall, since they emerge through the torn epidermis of the host, i.e., they are erumpent, and are seen clearly, e.g., in *Puccinia graminis tritici* (Uredinales), the wheat rust.

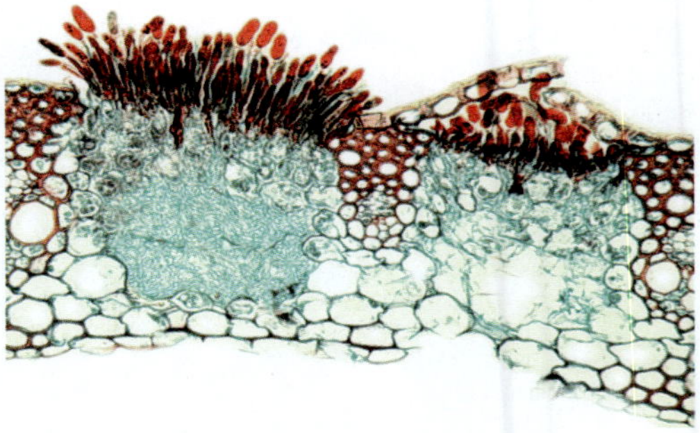

Uredinia with urediniospores of *Puccinia graminis tritici*, in a longitudinal section of a wheat leaf, x 100 (*MU*).

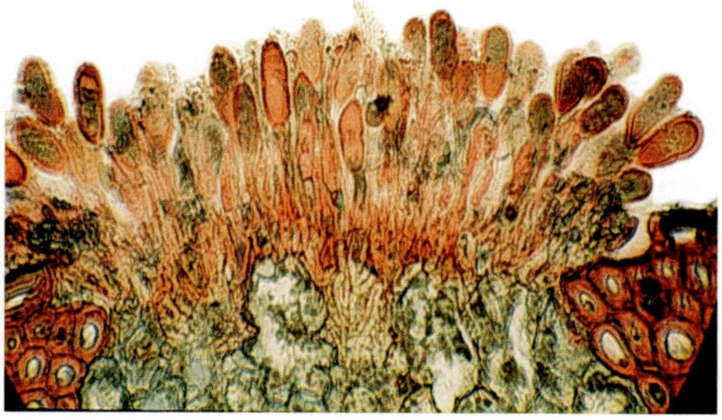

Uredinium with urediniospores of *Puccinia graminis tritici*, seen in a longitudinal section of a wheat leaf, x 240 (*GGA*).

Ascocarp of *Coniochaeta scatigena* (=*Ascobolus scatigenus*) a common **urinophilic** species which grows on cow's dung, x 20 (*EAA*).

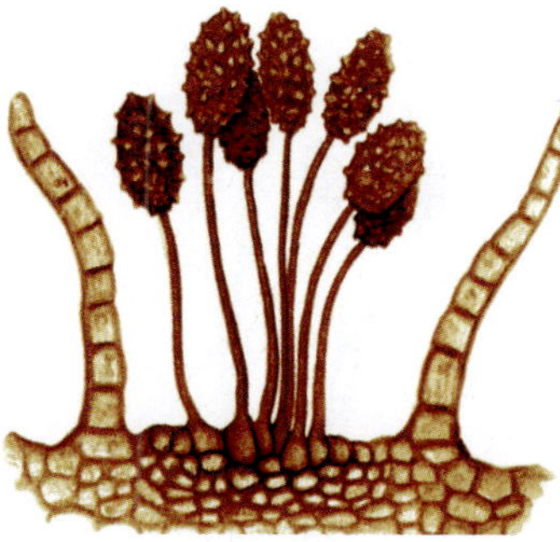

Uredinium with urediniospores of *Puccinia graminis tritici* on a wheat leaf, x 280 (*MU*).

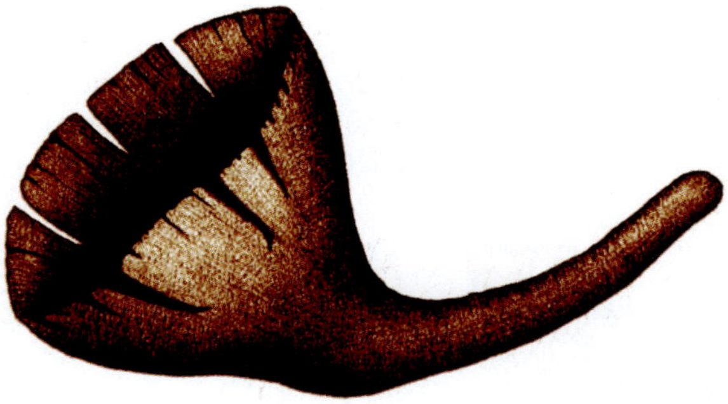

Urnula, ascocarp of *Urnula craterium*, x 0.4 (*MU*).

uredium, pl. **uredia** (NL. *uredium* < *uredo*, rust, blight, mildew, scab, scald < *uredo* < *uro*, to burn + dim. suf. *-ium*): see **uredinium**.

uredoconidium, pl. **uredoconidia** (L. *uredo*, genit. *uredinis*, rust, blight, mildew, scab, scald < *uro*, to burn + Gr. *kónis*, powder + dim. suf. *-ídion*): equal to **urediniospore**.

uredosorus, pl. **uredosori** (L. *uredosorus* < *uredo*, genit. *uredinis*, rust, blight, mildew, scab, scald < *uro*, to burn + *sorus*, sorus, pile, heap): see **uredinium**.

uredospore (L. *uredo*, genit. *uredinis*, rust, blight, mildew, scab, scald < *uro*, to burn + Gr. *sporá*, spore): see **urediniospore**.

urinophilic (L. *urina*, urine + Gr. *phílos*, have an affinity for; or *urina* + *phílos* + Gr. suf. *-íkos* > L. *-icus* > E. *-ic*, pertaining to or relating to): applied to operculate Discomycetes that appear on substrates such as dung, only after they have been soaked with urine. Species of *Ascobolus*, *Saccobolus*, *Coprotus* and *Iodophanus* (Pezizales) are prominent members of the dung mycobiota, which also are found in disturbed areas, such as burned soil and wood.

urnula, pl. **urnulae** (L. *urnula*, dim. of *urna*, urn, jar). a structure with the shape of a small urn, like the ascocarp of *Urnula* (Pezizales).

uro- (Gr. *ourá*, tail): a pref. used in names such as *Urocystis* (Ustilaginales), alluding to the shape of the germinated teliospores.

uroid (Gr. *ourá*, tail + L. suf. *-oide*, like < Gr. *-oeídes*, similar to): refers to the posterior region that myxamoebae of *Copromyxa arborescens* (Acrasiomycota) develop when they are in an aqueous medium; this region has the shape of a button or knob, often with extensions of hyaloplasm of a filose type (filopods), with a contractile vacuole.

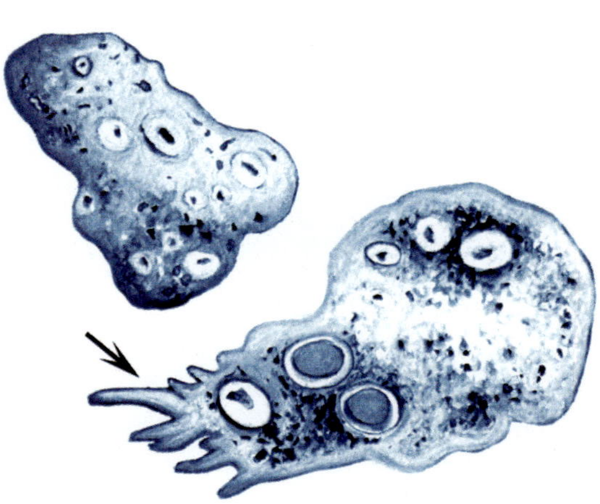

Uroid of the myxamoebae of *Copromyxa arborescens*, x 1,400 (*MU*).

ustilospore (L. *ustulo*, to burn, scorch + Gr. *sporá*, spore): a smut spore (Ustilaginales). Also called **ustospore**. See **smut**.

ustospore (L. *ustus*, burned + Gr. *sporá*, spore): see **ustilospore**.

utricle, **utriculus**, pl. **utriculi** (L. *utriculus* < *uter*, sac, purse or leather flask, or < *uterus*, uterus, womb + dim. suf. *-ulus* > E. *-ule*): a cover similar to a purse or bladder (the perisporium) that surrounds the basidiospores of certain fungi, such as *Hymenogaster thwaitesii* (Hymenogastrales); such spores are said to be *utriculate*. See **calyptra**.

utriculate (L. *utriculatus* < *utriculus*, dim. of *uter*, sac, purse or leather flask + suf. *-atus* > E. *-ate*, provided with or likeness): provided with a *utriculus*, like the basidiospores of *Hymenogaster thwaitesii* (Hymenogastrales), which are completely surrounded by a cover or purse constituted of the perisporium. Also called **utriculose**. Cf. **calyptrate**.

utriculose (L. *utriculosus* < *utriculus*, dim. of *uter*, sac, purse or leather flask + suf. *-osus* > E. *-ose*, full of, augmented, prone to): provided with a *utriculus* or pouch, or with this shape. Also called **lageniform**, **sicyoid**, **ventricose-rostrate** and **utriculate**.

utriform (L. *utriformis* < *uter*, sac, purse or leather flask + suf. *-formis* < *forma*, shape): shaped like a purse, bladder or leather pouch, like the phialides of *Stachybotrys chartarum* (=*S. atra*, dematiaceous asexual fungi), that are arranged in verticils on the apex of erect conidiophores, and the cheilocystidia of *Conocybe utriformis* (Agaricales), ventricose below and thick above, but almost isodiametric and typically with a slight constriction beneath a large, rounded head.

Utriculate basidiospores of *Hymenogaster thwaitesii*, x 1,400 (*MU*).

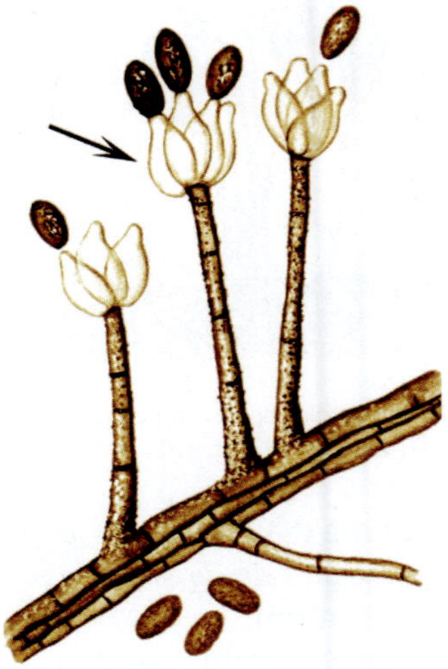

Utriform phialides of the conidiophores of *Stachybotrys chartarum* (=*S. atra*), x 700 (*MU*).

V

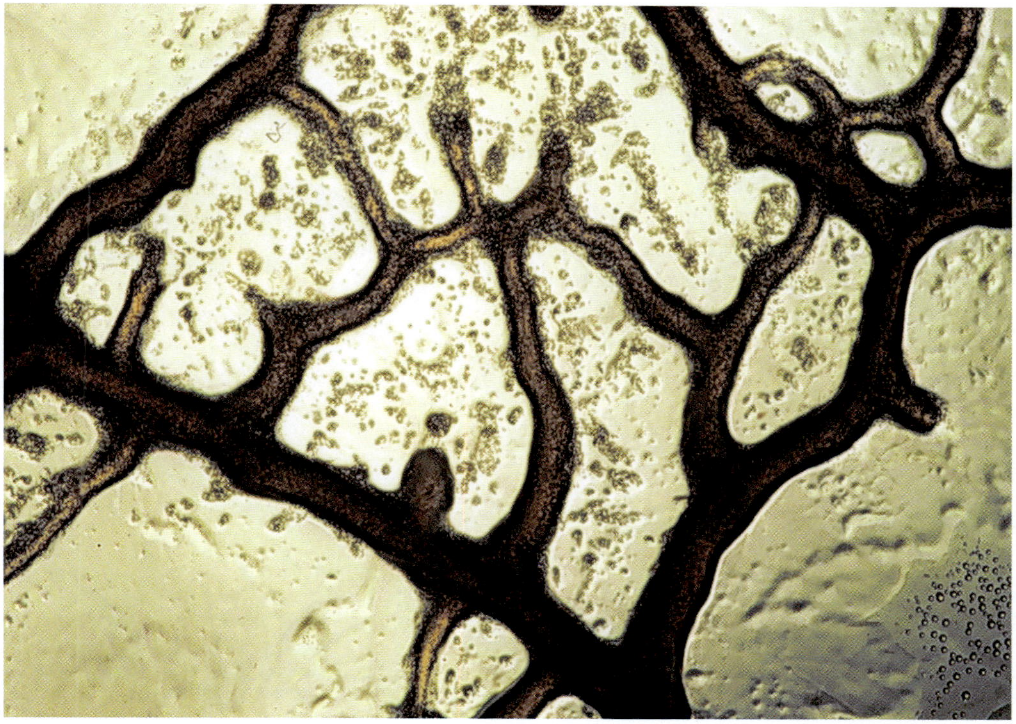

Plasmodial **veins** of *Physarum polycephalum*, on agar (*MU*).

V

vacuolate (Fr. *vacuole*, small vacuum < L. *vacuolum*, vacuum, dim. of *vaccum* < *vaccus*, empty, without contents + suf. *-atus* > E. *-ate*, provided with or likeness): containing one or more vacuoles. The hyphae of the aquatic fungi of the order Monoblepharidales (*Monoblepharis*) are distinctively vacuolate, to the point of appearing foamy.

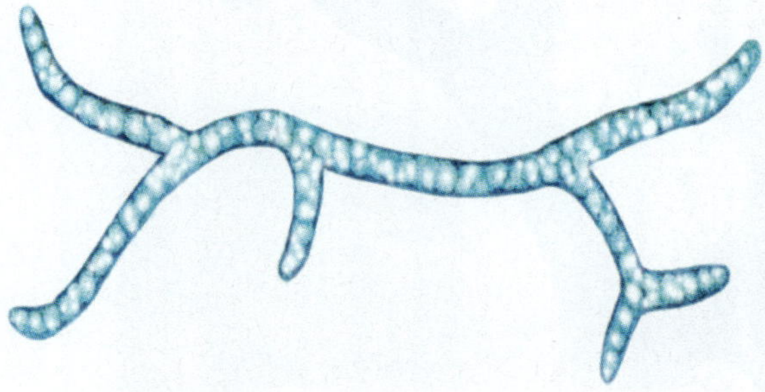

Vacuolate hyphae of *Monoblepharis polymorpha*, x 500 (*MU*).

vacuole (Fr. *vacuole*, small vacuum < L. *vacuolum*, vacuum < *vacuum* < *vacuus*, empty, without contents): a cavity or space, delimited by a membrane, that is formed in the cytoplasm of a cell and whose functions are for excretion and digestion. Vacuoles are most abundant in old cells.

vaginate (L. *vaginatus*, sheathed < *vagina*, sheath, wrapper + suf. *-atus* > E. *-ate*, provided with or likeness): provided with a sheath or envelope, like the basidiocarp of *Amanita vaginata* (Agaricales), whose long volva covers the lower part of the stipe.

vaginatoid (< old gen. *Vaginata*, syn. of *Amanitopsis*, syn. of *Amanita* + L. suf. *-oide*, like < Gr. *-oeides*, similar to): *Agaricales*. A type of fructification (13 principal types are recognized), with gills free or finely attached, and with a volva but without a ring, like that of *Amanita vaginata*, the species of *Volvariella* and of certain species of *Agaricus* and *Coprinus* (Agaricales).

Vaginate stipe of the basidiocarp of *Amanita vaginata*, x 0.5 (*MU*).

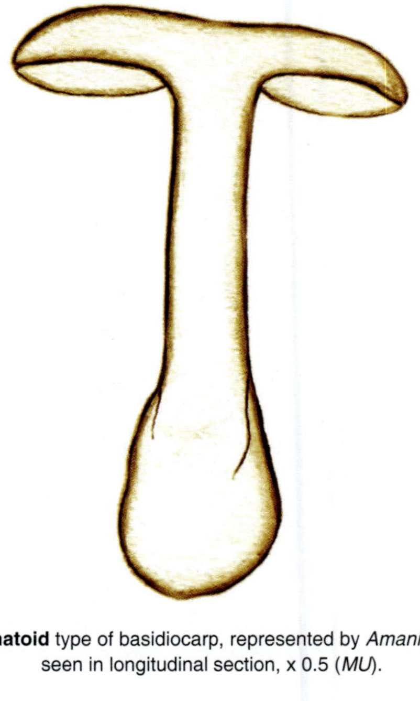

Vaginatoid type of basidiocarp, represented by *Amanita* sp., seen in longitudinal section, x 0.5 (*MU*).

Vaginatoid type of basidiocarp, represented by *Volvariella bombycina*, x 0.5 (*RV*).

valid (ML. *validus*, valid, strong): a name published in accordance with the articles of the International Code of Botanical Nomenclature; such a name may be legitimate or illegitimate. A *prevalid* (L. *prae*, before) name (or author) is one published before 1753, which is the starting date for the nomenclature of the fungi under the Code. See **nomen**.

valsoid (< gen. *Valsa* + suf. *-oide*, like < Gr. *-oeídes*, similar to): applied to stromatic ascomycetes whose perithecia are formed in an endostroma, in which the perithecial necks project from the ectostroma and converge toward a central point, as occurs in the pyrenomycete gen. *Valsa* (Diaporthales), very common on decorticated branches. Cf. **eutypoid** and **diatrypoid**.

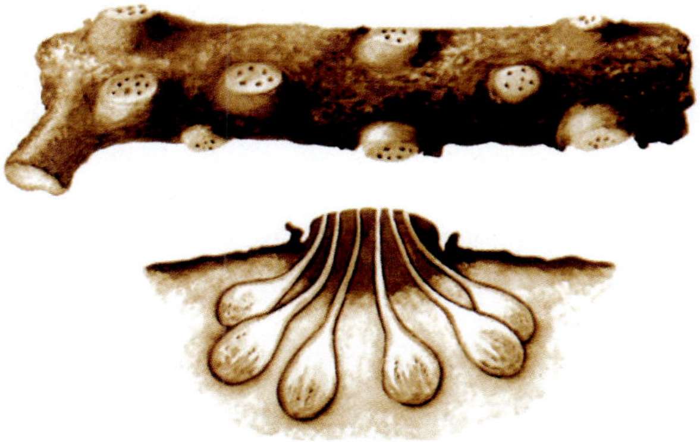

Valsoid perithecial stromata of *Valsa* sp. on a branch of a host plant, x 0.1. The drawing of a longitudinal section of a stroma shows the ostiolar necks of the perithecia projecting from the erumpent ectostroma, x 10 (*MU*).

variegated (LL. *variegatus*, varied in appearance < *variegare* < *varius*, varied + suf. *-egare*, to bear + suf. *-atus* > E. *-ate*, provided with or likeness): having a variety of colors and shades, or areas of diverse textures, intermixed, like the colonies of some molds, e.g., *Aspergillus variecolor* (moniliaceous asexual fungi), or the basidiocarps of certain Aphyllophorales, such as *Stereum* and *Trametes* (=*Polyporus*) *versicolor*.

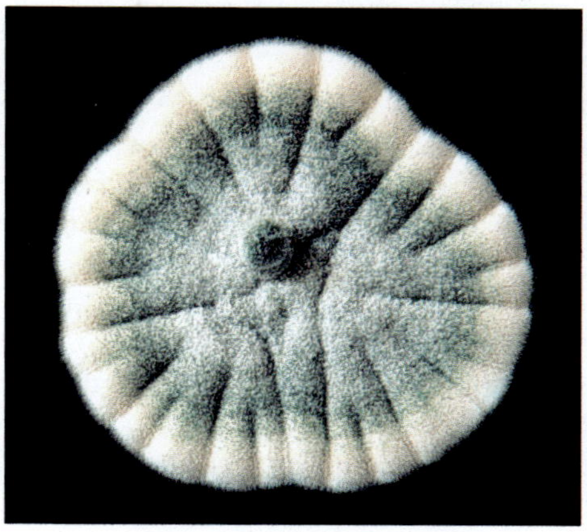

Variegated surface of the colony of *Aspergillus variecolor*, on agar, x 1 (*MU*).

Variegated surface of the colony of *Aspergillus variecolor*, on agar, x 1 (*MU*).

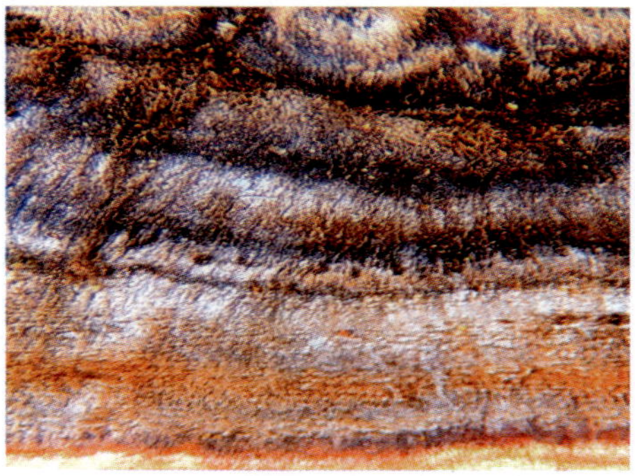

Variegated surface of the fruiting body of *Trametes* (=*Polyporus*) *versicolor*, x 10 (*MU*).

variety (L. *varietas*, genit. *varietatis*): each of the subgroups into which some species are divided and which are distinguished from one another by certain secondary, but permanent, characters. The variety is a category between the species or subspecies and form, e.g., *Fusarium moniliforme* var. *subglutinans* (tuberculariaceous asexual fungi), which is distinguished from the typical variety by having polyphialides.

variolate (L. *variolatus*, pustulated < *variola*, pustule, wart + suf. *-atus* > E. *-ate*, provided with or likeness): with pustules or granular warts, like the surface of the thallus of the lichen *Lecanora subfusca* var. *variolosa* (Lecanorales).

vascelloid (< gen. *Vascellum* < L. *vas*, urn, vessel + dim. suf. *-cellum* + suf. *-oide*, like < Gr. *-oeídes*, similar to): *Gasteromycetes*. A type of fruiting body development, like that of *Vascellum* (Lycoperdales), in which the gleba is separated from the sterile subgleba by a well developed diaphragm.

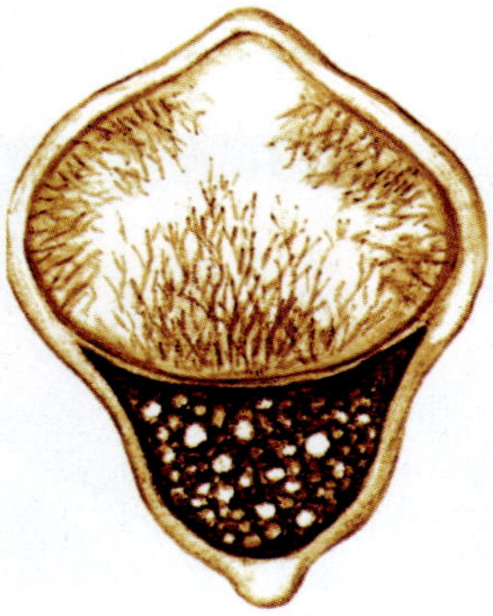

Vascelloid type of development, represented by the fruiting body of *Vascellum* sp., seen in longitudinal section, x 5 (*MU*).

vasiform (L. *vasiformis* < *vas*, genit. *vasis*, glass, receptacle + suf. *-formis* < *forma*, shape): shaped like a glass, like the fruiting body of *Vascellum* (Lycoperdales).

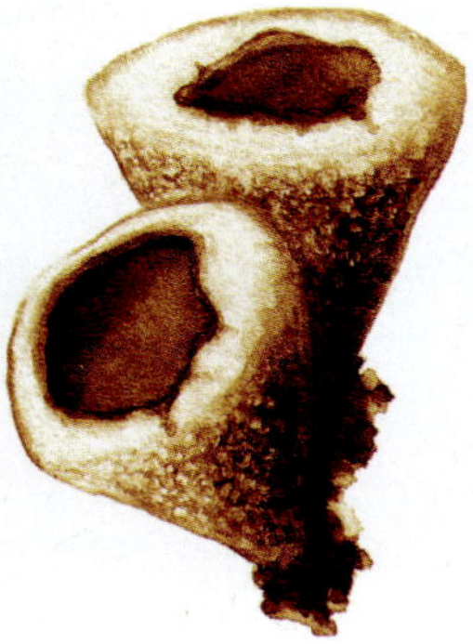

Vasiform basidiocarps of *Vascellum depressum*, x 2 (*MU*).

vaulted ray (E. *vaulted*, pt. of *vault*, arch < ME. *voute* < OF. *voulte*, arch; L. *radius*, ray): a ray that arches over the exoperidium, forming a chamber or vault, usually conical in profile, as is observed in *Geastrum fornicatum* (Lycoperdales). See **fornicate**.

vector (L. *vector*, genit. *vectoris*, conductor, that which transports): applied to an organism that transmits or propagates another organism, the latter generally a pathogen. E.g., *Olpidium viciae* (Spizellomycetales) is a chytrid vector of a virus that affects plants of *Vicia unijuga*. On the other hand, there exist many examples of species of fungi that are disseminated from one substrate or host to another by means of insect vectors, as occurs with *Claviceps purpurea* (Hypocreales), whose conidia are produced on the infected ovaries of rye flowers, mixed with a sweet, sticky secretion, similar to nectar; attracted by this nectar, the insects visit the the infected flowers and disseminate the fungus to healthy flowers.

vegetative (ML. *vegetativus* < L. *vegetar*, *vegetare*, enliven, animate + suf. *-ivus* > E. *-ive*, quality or tendency, aptitude): what vegetates or is capable of vegetating; i.e., that undergoes various vital functions (germination, growth, development and asexual multiplication) but not true sexual reproduction. Present mycologists prefer to use the term **assimilative** (ML. *assimilativus* > L. *assimilatus*, similarity of one thing with respect to another; the ensemble of processes by which absorbed substances are converted into the living material of the absorbing organism + suf. *-ivus* > E. *-ive*) to avoid the implication that the fungi are treated as plants. Thus, one says vegetative mycelium or assimilative mycelium, vegetative function or assimilative function, etc.

vegetative multiplication (ML. *vegetativus* < L. *vegetar*, *vegetare*, to vivify, animate + *-ivus* > E. *-ive*, quality or tendency, aptitude; L. *multiplicatio*, genit. *multiplicationis* < *multiplicare*, to multiply + suf. *-ationem* > E. *-ation*, action, state or condition, result): growth, development and multiplication without involvement in sex. See **asexual**.

Vegetative multiplication of cells of *Saccharomyces kluyveri* (Saccharomycetales), which reproduce by budding, x 600 (*PL*).

veil or **velum**, pl. **vela** (L. *velum*, veil, curtain, cloth): a curtain or cloth that covers something. The felt-like hyphenchymatous covering that surrounds the primordium of the basidiocarp of the angiocarpic Agaricales is called the *universal* (or *general*) *veil*; in certain gen. the veil is evanescent (as in *Coprinus*), in others it is persistent in the adult stage (as in *Amanita* and *Volvariella*) and constitutes the volva on the base of the stipe, and the plates or scales, generally white, on the upper surface of the pileus. The *partial veil* is the hyphenchymatous covering of the gills (in the hemiangiocarpic Agaricales), that extends from the border of the pileus to the stipe, and which on first opening tears; its remains form the *ring* (as in *Agaricus*, *Lepiota*, etc.) and the *cortina* (as in *Cortinarius*). The ring also is called the *hymenial veil* (and at times, and imprecisely, *apical veil*, *inner veil*, *partial veil*), which derives from the perpendicular primordial tissue, and differentiates later than the volva.

Veil (universal veil) covering the basidiocarp primordium of *Amanita caesarea*, x 1 (*RMA*).

Veil (partial veil) covering the gills of *Agaricus silvicola*, x 1 (*MRO*).

Veil (partial veil) adhered to the stipe of *Cortinarius* sp., x 1 (*RV*).

veins (L. *vena*, vein, vessel, fiber, conduit): in the lichens, the veins correspond to cords of tissue on the lower surface of foliose thalli, especially *Peltigera* (Peltigerales), where they can replace the lower cortex. Three types are distinguished: in *caninoid veins* (< *P. canina*) the cords of tissue are separated on the apical margins of the lobules of the thallus; in the *polydactyloid veins* (Gr. *polýs*, many + *dáktylos*, finger + L. suf. *-oide*, like < Gr. *-oeídes*, similar to) the cords are confluent toward the margins of the lobules; in the *malaceoid veins* (Gr. *malakós*, soft + L. suf. *-oide*, like < Gr. *-oeídes*, similar to) the cords are scarcely perceptible, since they form a few whitish interstices on the lower surface of the thallus. In the phaneroplasmodia of the Myxomycetes there are *plasmodial veins* (NL. *plasmodium* < Gr. *plásma*, bland material with which a living being is formed + L. *-odium*, resembling < Gr. *ode*, like < *-oeídes*, similar to + L. suf. *-alis* > E. *-al*, relating to or pertaining to) through which the cytoplasmatic currents flow, as in *Physarum polycephalum* (Physarales).

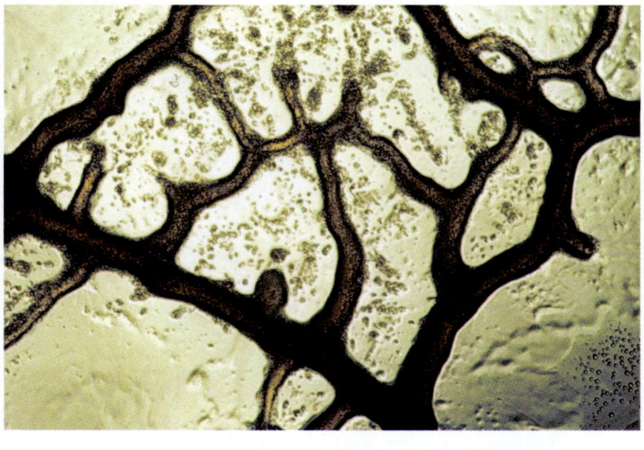

Plasmodial **veins** of *Physarum polycephalum*, on agar, x 10 (*MU*).

velutinate, velutinous (NL. *velutinus*, velvet + L. suf. *-atus* > E. *-ate*, provided with or likeness; or + L. suf. *-osus* > E. *-ous*, full of, augmented, prone to): velvety, i.e., with a shiny or opaque surface and short, fine, soft hairs arranged compactly, like that of the colonies of some species of *Penicillium* and *Aspergillus*, e.g., *P. chrysogenum* (=*P. notatum*) and *Aspergillus terreus* (of the moniliaceous asexual fungi) and *Cladosporium* (*C. cladosporioides*, of the dematiaceous asexual fungi), among many others, or like the surface of the pileus of *Psathyrella velutina* or the stipe of *Collybia velutipes* and *Xeromphalina tenuipes* (Agaricales).

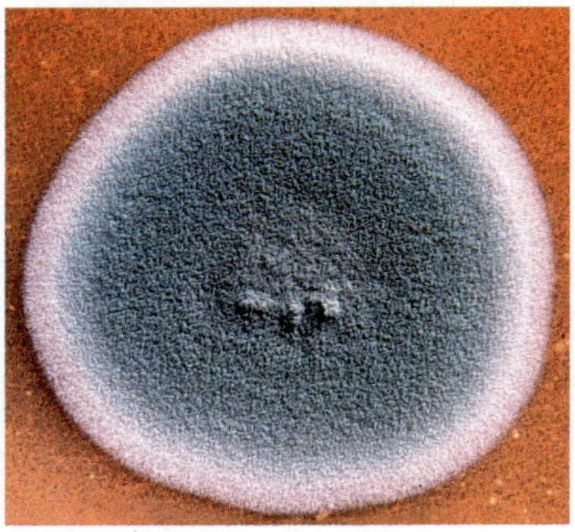

Velutinous colony of *Penicillium chrysogenum* (=*P. notatum*), on agar, x 1 (*MU*).

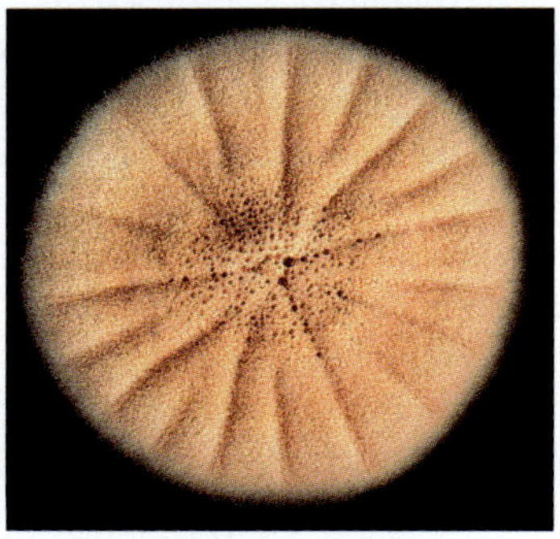

Velutinous colony of *Aspergillus terreus*, on agar, x 1 (*MU*).

Velutinous basidiocarp of *Lentinus* (=*Panus*) *badius* (Aphyllophorales), x 0.5 (*RV*).

Velutinous basidiocarp of *Trichaptum perrottetii* (Aphyllophorales), x 1 (*RV*).

Velutinous basidiocarps of *Tylopilus plumbeoviolaceus* (Agaricales), x 1 (*RV*).

Velutinous stipe of the basidiocarp of *Xeromphalina tenuipes*, x 5 (*MU*).

venose (L. *venosus*, full of veins < *vena*, vein, vessel, fiber, conduit + suf. *-osus* > E. *-ose*, full of, augmented, prone to): having *veins*.

ventral (L. *ventralis* < *venter*, genit. *ventris*, belly + suf. *-alis* > E. *-al*, pertaining to or relating to): in dorsiventral organs, it applies to the side corresponding to the belly; the front side or lower surface; the surface that faces the principal axis. Frequently used for the lower surface of the thallus of foliose lichens. Cf. **dorsal**.

ventricose (NL. *ventricosus*, swollen < L. *ventris*, genit. of *venter*, belly + suf. *-osus* > E. *-ose*, full of, augmented, prone to): big-bellied, swollen in the manner of a belly; also called **lageniform**, with a wide base and a narrower apical part, like the ascocarps of *Ceratocystis* (Microascales) and the cheilocystidia of *Psathyrella cyanescens* (Agaricales). When the apical part is extended, forming a prolongation similar to a beak or rostrum, it is called *ventricose-rostrate*, as is seen in the pleurocystidia of *Pholiota astragalina* (Agaricales). The perithecia and young ascospores of *Podospora comata* (Sordariales) also are ventricose.

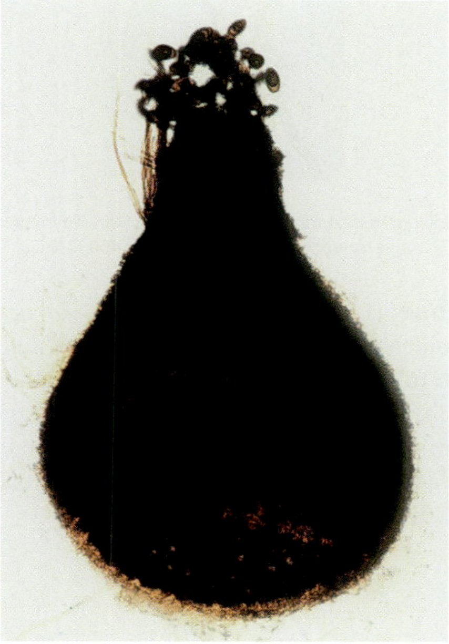

Ventricose perithecium of *Podospora comata*, x 87 (*RTH*).

ventricose-rostrate (NL. *ventricosus*, swollen < *ventris*, genit. of *venter*, belly + suf. *-osus* > E. *-ose*, full of, augmented, prone to; L. *rostratus*, rostrate, provided with a beak < *rostrum*, beak, snout + suf. *-atus* > E. *-ate*, provided with or likeness): see **ventricose**. Also called **lageniform**, **sicyoid** and **utriculate**.

vermiculate, **vermicular** (L. *vermiculatus*, wormy, shaped like a worm < *vermiculus*, little worm < *vermis*, worm + *-atus* > E. *-ate*, provided with or likeness; L. *vermicularis* < *vermiculus* < *vermis* + dim. suf. *-culus* + suf. *-aris* > E. *-ar*, as, concerning): elongate and somewhat twisted, like a worm, or that moves like a worm, like the multinucleate plasmodium of *Protosporangium bisporum* (Protosteliales).

vermiform (L. *vermiformis* < *vermis*, worm + suf. *-formis* < *forma*, shape): with the shape of a worm, like the ascospores of *Balansia henningsiana* (Hypocreales), *Ophiodothella vaccinii* (Phyllachorales), *Geoglossum difforme* (Helotiales) and *Lindra thalassiae* (Lulworthiales). See **anguiliform**.

Vermiform ascospore of *Geoglossum difforme*, x 660 (*RTH*).

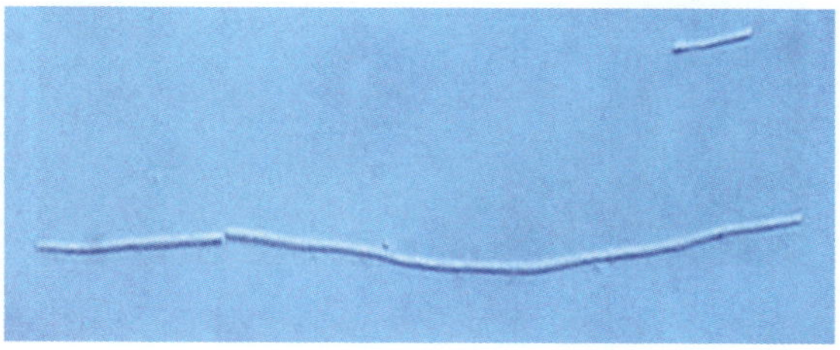

Vermiform ascospore of *Balansia henningsiana*, x 400 (*RTH*).

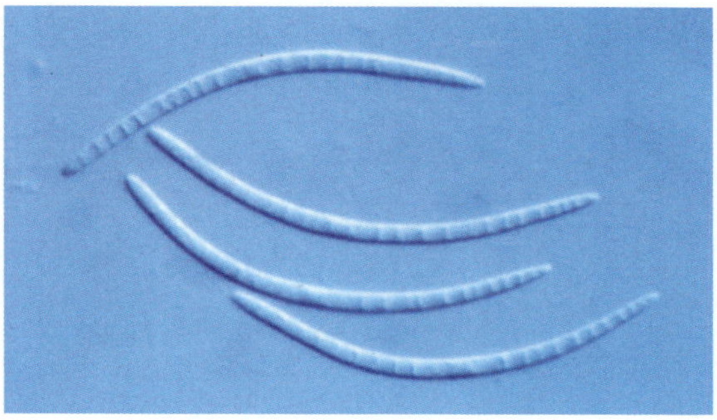

Vermiform ascospores of *Ophiodothella vaccinii*, x 580 (*RTH*).

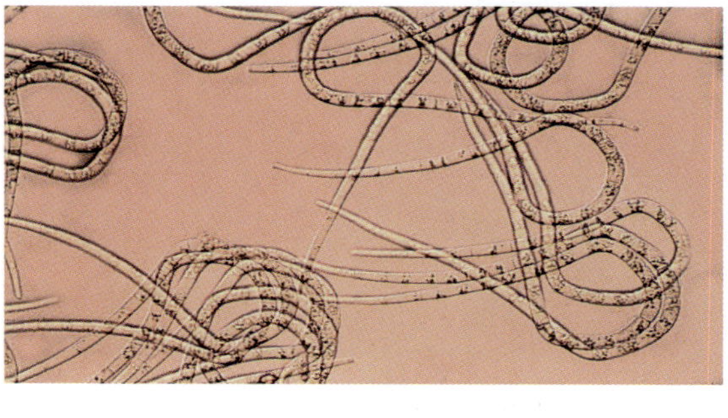

Vermiform ascospores of *Lindra thalassiae*, x 580 (*MCG*).

vermivorous (L. *vermis*, worm + *vorare*, to devour): feeding on worms, in this case on nematodes, like the moniliaceous asexual fungi of the gen. *Arthrobotrys*, *Dactylaria* and *Dactylella*, among others, that trap and consume nematodes in the soil. In *A. dactyloides*, e.g., constricting rings are formed that strangle their prey; from the rings hyphae develop that invade and digest the dead body of the nematode. Syn. of **nematophagous**.

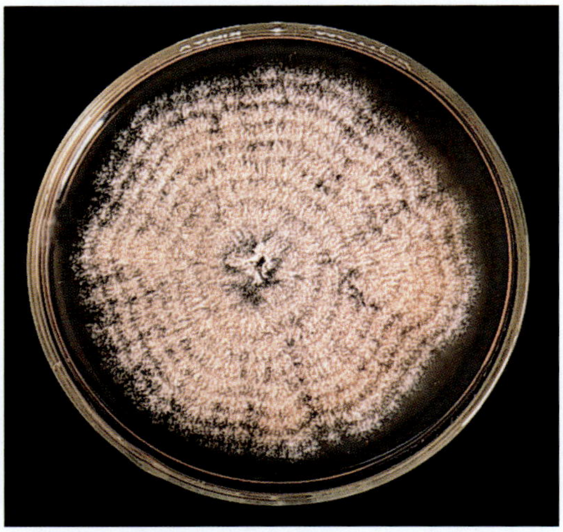

Colony of *Arthrobotrys* sp., a **vermivorous** fungal species, on agar, x 1 (*MU*).

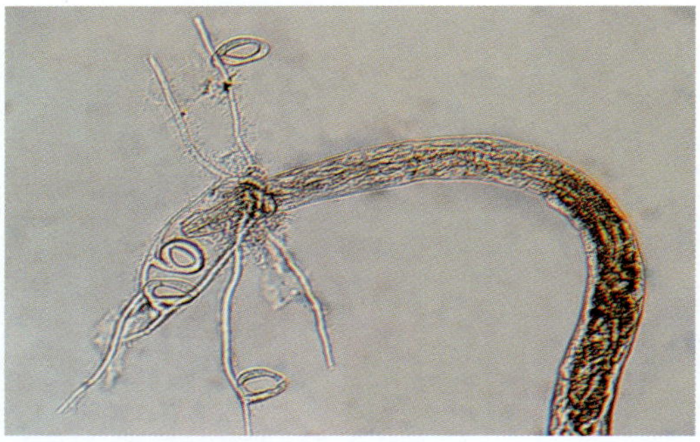

Vermivorous constricting rings of *Arthrobotrys dactyloides*, x 150 (*PL*).

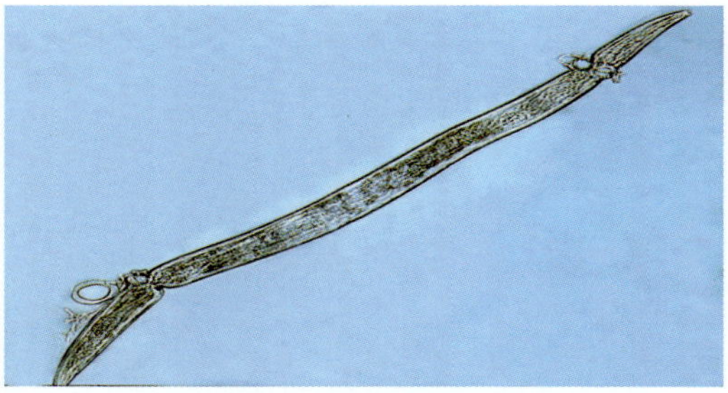

Vermivorous constricting rings of *Arthrobotrys dactyloides*, x 150 (*PL*).

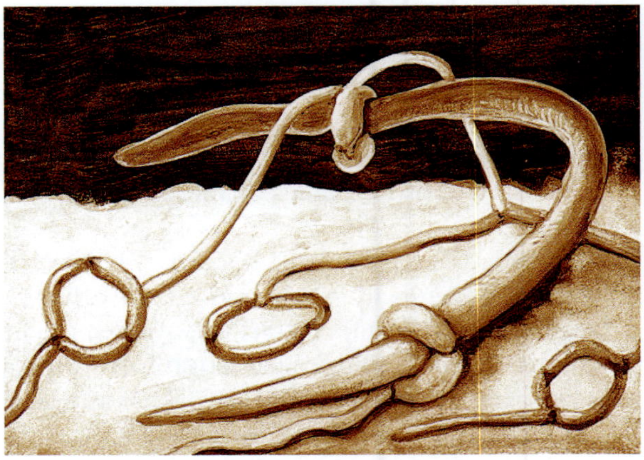

Vermivorous constricting rings of *Arthrobotrys dactyloides*, x 300 (*MU*).

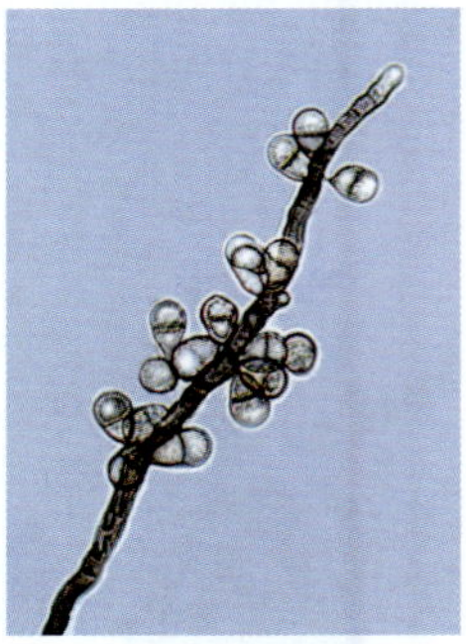

Conidiophore of *Arthrobotrys oligospora*, with didymospores, a **vermivorous** fungal species, x 450 (*MU*).

verrucose (L. *verrucosus*, warty < *verruca*, wart, prominence + suf. *-osus* > E. *-ose*, full of, augmented, prone to): with wart-like prominences on the surface, like the conidiophore wall of *Penicillium roqueforti* (moniliaceous asexual fungi) and the conidia of *Stachybotrys chartarum* (=*S. atra*, dematiaceous asexual fungi). Cf. **verruculose**.

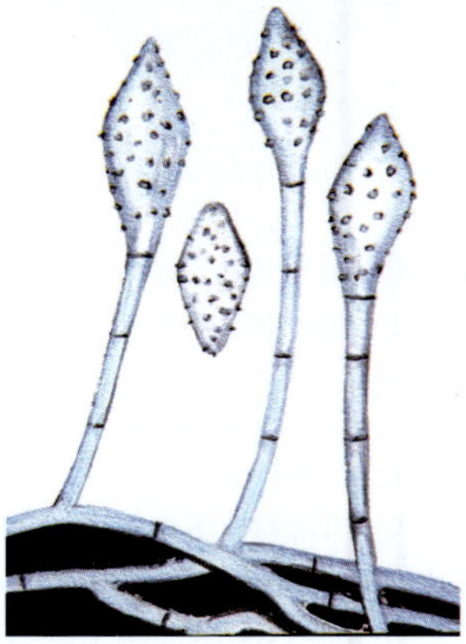

Verrucose conidia of *Ovulariopsis* sp. (moniliaceous asexual fungi), asexual stage of *Phyllactinia* sp. (Erysiphales), x 400 (*MU*).

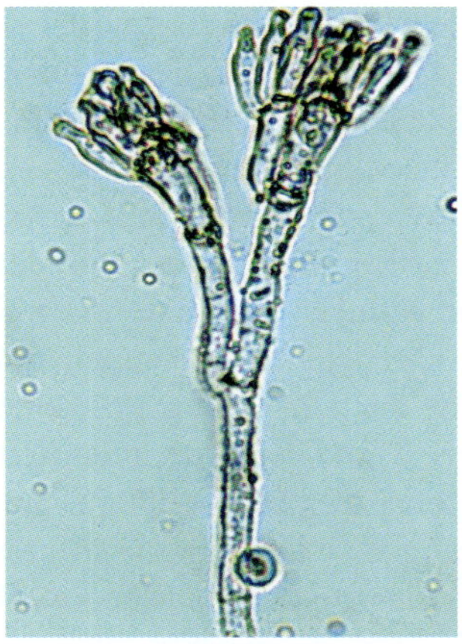

Verrucose conidiophore of *Penicillium roqueforti*, x 600 (*PL*).

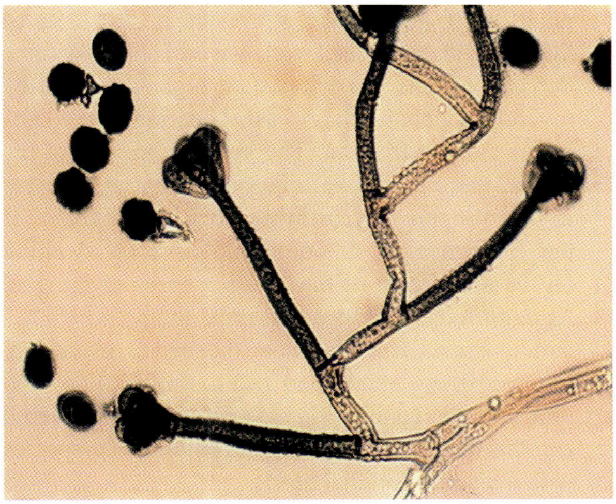

Verrucose conidia and conidiophores of *Stachybotrys* sp., x 580 (*MCG*).

Verrucose conidia of *Stachybotrys chartarum* (=*S. atra*), x 800 (*MU*).

verrucosis, pl. **verrucoses** (L. *verruca*, wart, prominence + suf. *-osis*, condition or state): in phytopathology, syn. of **leaf curl**.

verruca, **wart** (L. *verruca*, wart, prominence): a small prominence, more or less rounded, or with the ornamentation cone-shaped, on the surface of an organ or a structure, like those present on the spores and exoperidium of *Lycoperdon* (Lycoperdales).

verruculose (L. *verruculosus*, full of small warts < *verrucula*, dim. of *verruca*, wart + suf. *-osus* > E. *-ose*, full of, augmented, prone to): with a rough surface, ornamented with microscopic or minute prominences similar to warts, as is seen in the spores of the gen. *Cortinarius* (Agaricales). If the prominences or warts are larger, the term **verrucose** is applied.

versiform (L. *versiformis*, variable in shape < *versus*, turned around < *vertere*, to turn around, change + suf. *-formis* < *forma*, shape): applied mainly to fungal structures that change shape with age, like certain cystidia (cheilocystidia of *Bolbitius vitellinus*, Agaricales) that adopt so many shapes it is difficult to describe all of the variations. See **pleomorphic**.

verticillate (NL. *verticillatus*, arranged in a verticil < L. *verticillus*, verticil + suf. *-atus* > E. *-ate*, provided with or likeness): a type of branching in which the branches (pedicels, metulae, phialides, etc.) are borne at the same level on a hypha or support (sporangiophore, conidiophore, etc.) and grow obliquely upward with respect to the central axis, generally reaching the same length and surrounding the hypha or support, as is seen, e.g., in the sporangiophore of *Actinomucor* (Mucorales), and the conidiophores of *Scopulariopsis* and *Verticillium* (moniliaceous asexual fungi). The last gen. owes its name to this characteristic. Cf. **monoverticillate**, **biverticillate**, **triverticillate** and **polyverticillate**.

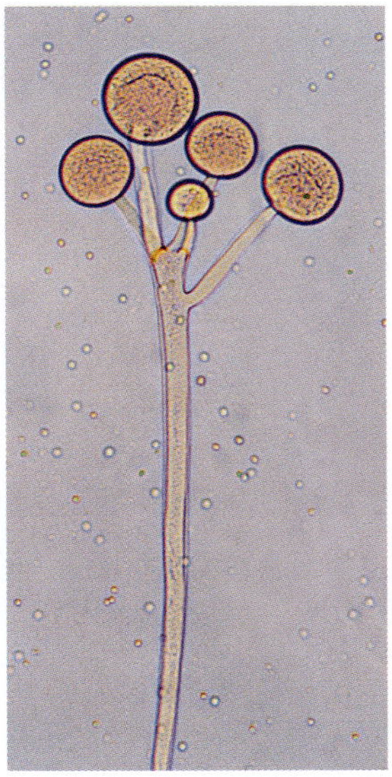

Verticillate sporangiophore of *Actinomucor* sp., x 175 (*CB*).

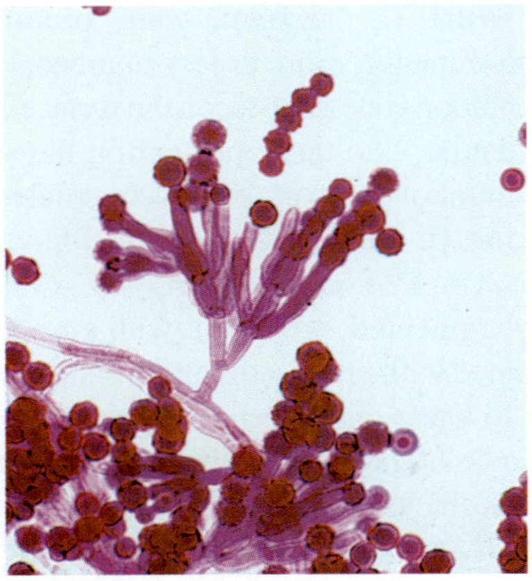

Verticillate phialides of a conidiophore of *Scopulariopsis brevicaulis*, x 750 (*CB*).

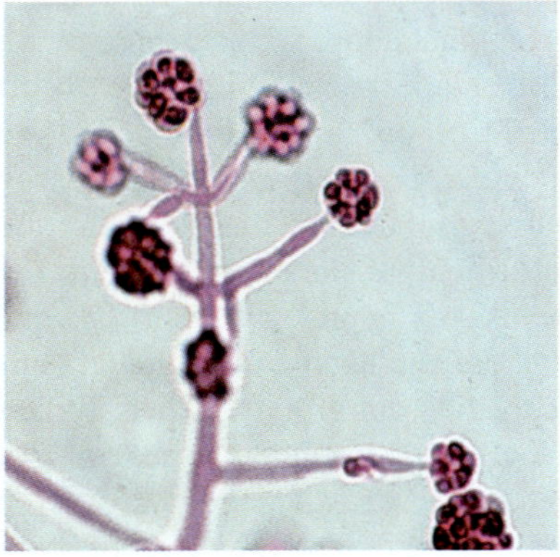

Verticillate phialides of a conidiophore of *Lecanicillium* (=*Verticillium*) *lecanii*, x 900 (*CB*).

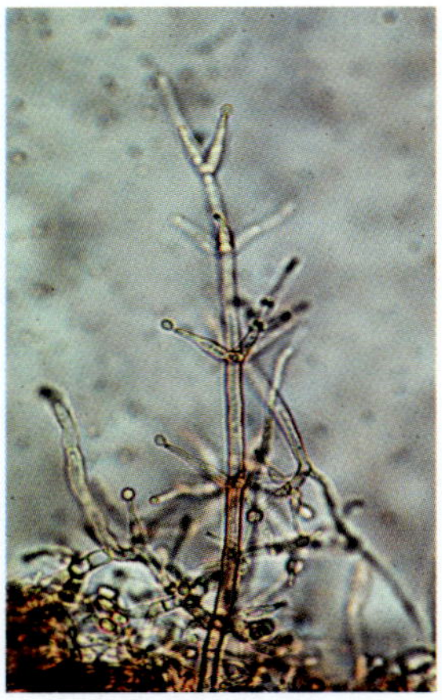

Verticillate phialides of a conidiophore of *Lecanicillium* (=*Verticillium*) *lecanii*, x 600 (*MU*).

verticillosis (< gen. *Verticillium* < L. *verticillus*, verticil, small wheel + dim. suf. *-ium*, in allusion to the verticillate arrangement of the phialides on the conidiophore + Gr. suf. *-osis*, condition or state): the name of the tracheomycosis whose causal agent is a species of *Verticillium*, of the moniliaceous asexual fungi. An example of verticillosis is the wilt of potato plants produced by *V. albo-atrum*.

vesicle (L. *vesicula*, dim. of *vesica*, bladder): a receptacle or bag shaped like a bladder or ampule. **1**. An intracellular organelle associated with the endoplasmic reticulum. **2**. A mucilaginous structure produced by the zoosporangia of some Chytridiales (*Chytriomyces*) and Peronosporales (*Pythium*), within which the zoospores mature before being liberated. **3**. A subsporangial swelling of the species of *Pilobolus* (such as *P. longipes*, of the Mucorales). Also in this order is the gen. *Mycotypha*, whose name reflects the similarity the sporangiolar vesicles of this fungus have with the inflorescences of the phanerogamic plant in the gen. *Typha*. **4**. Vesicular structures of hyphae, with a thick cell wall, that probably constitute deposits of reserves or resistant organs, formed by mycorrhizogenous fungi in the interior of the cells of the associated plant. The vesicles are one of the forms (the other is the arbuscules) that are present in endotrophic mycorrhizae (e.g., *Endogone*, of the Endogonales, in onion roots). **5**. A swelling on the apical part of the conidiophores of the gen. *Aspergillus* (moniliaceous asexual fungi), which has various shapes depending upon the species (claviform, spheroidal, etc.) and gives rise to the phialides that produce the conidia (in species with a uniserial conidial head), or to metulae and phialides (in species with a biserial conidial head).

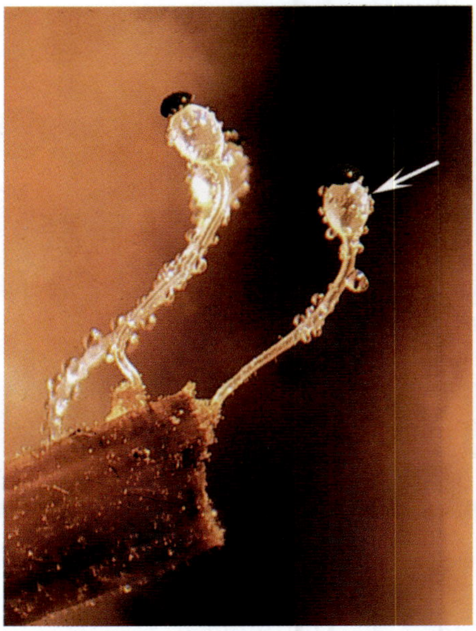

Vesicle (subsporangial) of the sporangiophores of *Pilobolus crystallinus* var. *kleinii* (=*P. kleinii*), x 25 (*MU*).

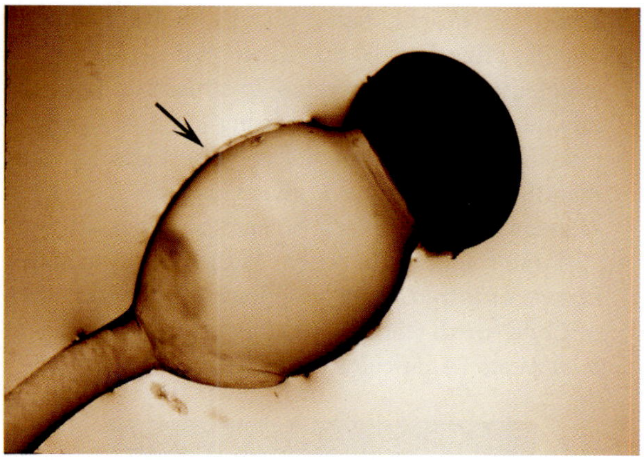

Vesicle (subsporangial) of the sporangiophore of *Pilobolus longipes*, x 430 (*MU*).

Clavate **vesicle** bearing conidia of *Aspergillus clavatus*, x 250 (*RTH*).

Sporangiolar **vesicles** bearing unispored sporangiola (conidia) of *Mycotypha africana*, x 250 (*MU*).

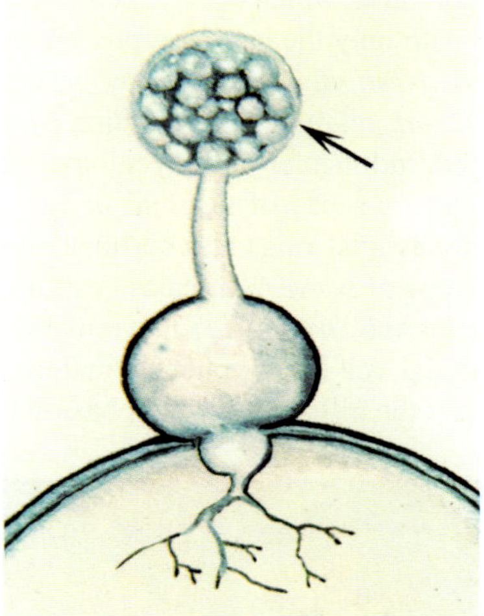

Vesicle derived from the sporangium of *Rhizidiomyces apophysatus* (Hyphochytriales); the zoospores reach maturity before they are released, x 400 (*MU*).

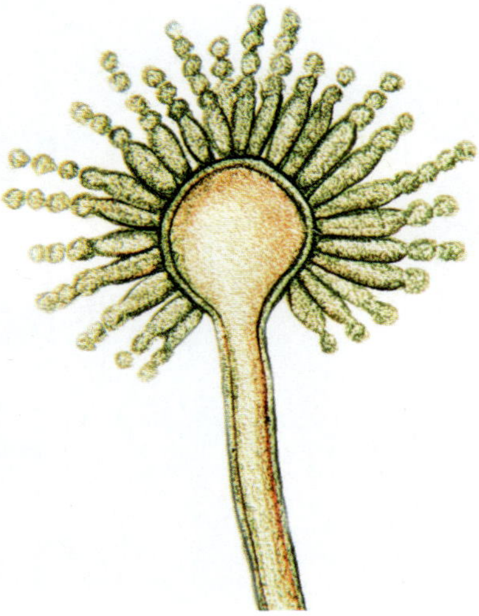

Vesicle of the conidial head of *Aspergillus rubrobrunneus*, x 400 (*MU*).

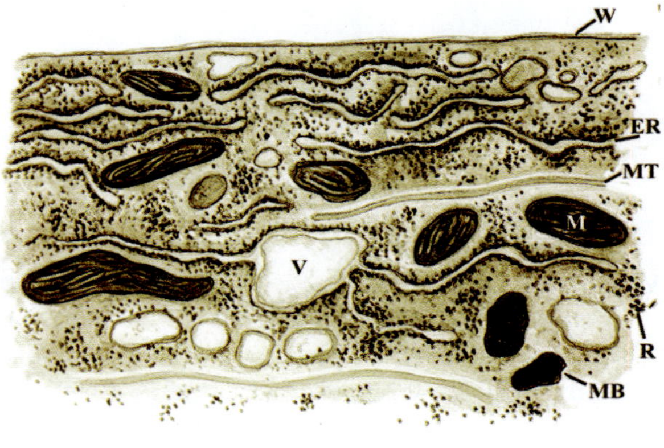

Vesicles (V) in a young hypha of *Armillaria* (=*Armillariella*) *mellea* (Agaricales), seen in longitudinal section; also shown are the endoplasmic reticulum (ER), microbodies (MB), mitochondria (M), microtubules (MT), ribosomes (R), and the cell wall (W). Drawing composed from several transmission electron micrographs published by Beckett *et al.* in their *Atlas of Fungal Ultrastructure*, 1974, x 27,000 (*MU*).

vesicular bodies (L. *vesicula*, dim. of *vesica*, bladder + suf. *-aris* > E. *-ar*, belonging or relating to; ME. < OE. *bodig*, body): thin-walled vesicles found in the subhymenium of certain hymenomycetes, principally thelephoraceous species. Also applied to the structures formed by vesicular-arbuscular mycorrhizae.

vesiculate, vesicular, vesiculiform, vesiculose (L. *vesiculatus*, provided with vesicles < *vesicula*, ampule, bladder, vesicle + suf. *-atus* > E. *-ate*, provided with or likeness; L. *vesicularis*, pertaining to or relating to a vesicle < *vesicula* + suf. *-aris* > E. *-ar*, as, concerning; L. *vesiculiformis*, with the shape of a vesicle < *vesicula* + suf. *-formis* < *forma*, shape; L. *vesiculose*, having vesicles < *vesicula* + suf. *-osus* > E. *-ose*, full of, augmented, prone to): with the shape of a *vesicle* or having vesicles, like the cystidia of *Leptonia fulva* and other Agaricales, which are swollen like a bladder or vesicle, with only the base abruptly tapered.

vestigial (L. *vestigium*, track, trace, vestige + L. suf. *-alis* > E. *-al*, relating to or pertaining to): not well developed, rudimentary; in mycology it is generally applied in the sense of residual or exiguous. E.g., there are vestigial rings and cortinas on the mature basidiocarps of some Agaricales (various species of *Cortinarius* and *Stropharia*), that remain as remnants of the partial veil (the hyphenchymatous membrane that covers the gills in the young basidiocarp).

Vestigial partial veil (ring on the stipe, and cortina, unshowed in the photograph) of the basidiocarps of *Cortinarius* sp., x 1 (*RV*).

veterinary mycology (L. *veterinaria*, fem. of *veterinarius* < *veterinae*, beasts of burden; NL. *mycologia* < Gr. *mýkes*, fungus + *lógos* > E. des. *logy*, study, treatise): the science that studies diseases of animals (especially higher animals) caused by fungi. See **phytopathology**.

viable (Fr. *viable*, apt to live < *vie*, life + *-able*, capable of): able to live; capable of growing. Generally applied to cultures of microorganisms maintained in the laboratory.

villose (L. *villosus*, hairy, shaggy, velvety, woolly < *villus*, fine hair + suf. *-osus* > E. *-ose*, full of, augmented, prone to): refers to structures that have down or hair, the latter not being too fine, in which case they are *pubescent*, nor very short and dense, as then they are called *velutinous*; also not considered villose are structures whose hairs are thick or rigid, stiff and rough to the touch, as then they are called *hirsute* or *hispid*. E.g., the pileus of the basidiocarp of *Coprinopsis* (=*Coprinus*) *lagopus* and *Volvariella bombycina* (Agaricales) is villose.

Villose pileus of the basidiocarps of *Coprinopsis* (=*Coprinus*) *lagopus*, growing from cow's dung, x 0.3 (*MU*).

Villose pileus of the basidiocarp of *Volvariella bombycina*, x 1 (*RV*).

villus, pl. **villi** (L. *villus*, fine hair): fluffy; long, soft hairs that cover some structures.

vinescent (L. *vinum*, wine + suf. *-escens*, genit. *-escentis* > E. *-escent*, what turns, that begins, slightly): turning wine red in color, like the basidiomata of *Rhodophyllus abortivus*, *Russula harkoneniana* and *Strobilomyces confusus* (Agaricales), that undergo this change of color on being cut or broken.

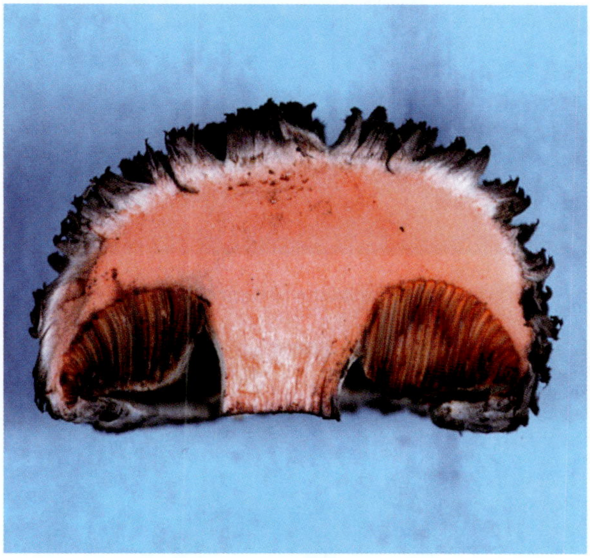

Vinescent basidiocarp of *Strobilomyces confusus*, x 1 (*RTH*).

vinose (L. *vinosus* < *vinum*, wine + suf. *-osus* > E. *-ose*, having the qualities of): resembling wine; in biology, generally employed to allude to the color of diluted red wine.

virescent (L. *virescens*, genit. *virescentis* < *viridis*, green + suf. *-escens*, genit. *-escentis* > E. *-escent*, what turns, that begins, slightly): becoming green, like the mature gills of the fruiting body of *Chlorophyllum molybdites* (Agaricales).

virgate (L. *virgatus*, shaped like, provided with, or made of small sticks or twigs; also, striped or banded < *virga*, rod, stick; also, ray or band + suf. *-atus* > E. *-ate*, provided with or likeness): long, straight and slender like a stick. Generally applied to the surface of the pileus of a basidiocarp, when this has rays or bands, frequently due to the presence of fibers of a different color, as in *Tricholoma virgatum* (Agaricales).

virose (L. *virosus*, poisoned, fetid, muddy, miry, dirty, covered with slime < *virus*, poison, mucilage + suf. *-osus* > E. *-ose*, full of, augmented, prone to): poisonous, like the basidiocarp of *Amanita virosa* (Agaricales), a species causing phalloidian mycetism.

Basidiocarp of *Amanita* sp., a poisonous species (**virose**), x 0.3 (*RV*).

Basidiocarps of *Amanita virosa*, a poisonous species (**virose**), x 0.3 (*RV*).

Basidiocarps of *Amanita virosa*, a poisonous species (**virose**), x 0.3 (*EAA*).

virulence (L. *virulentus*, full of venum, virulent < *virus*, poison, mucilage + suf. *-entia* > Fr. *-ence* > E. *-ence*, state, quality or action): the quality or state of being *virulent*; the relative capacity of a pathogen to overcome the defenses of the host organism; degree or measure of pathogenicity.

virulent (L. *virulentus*, full of venum, virulent < *virus*, venum, mucilage + suf. *-ulentus* > E. *-ulent*, abundant in): extremely pathogenic or poisonous; distinguished by pursuing a rapid course, severe and malignant (virulent infection).

viscid, viscosus, slimy (L. *viscidus*, viscous, sticky < *viscum*, viscous, bird-lime; L. *viscosus*, viscous, sticky < *viscum* + suf. *-osus* > E. *-ose*, full of, augmented, prone to; E. *slime* < ME. *slyme* < OE. *slim*, slime + E. suf. *-y*, abounding in, full of, having the quality of): viscous, sticky or glutinous, like the plasmodium of Myxomycetes, and the surface of the pileus of *Suillus*, *Hygrophorus*, *Hygrocybe* and some species of *Russula* (Agaricales). Equal to **glutinous**.

Viscid pileus of the basidiocarps of *Hygrocybe miniata*, x 1 (*RV*).

Viscid plasmodium of an unidentified Myxomycete on dead wood, x 0.3 (*MU*).

Viscid pileus of the basidiocarps of *Russula* sp., x 0.3 (*MRO*).

Viscid pileus of the basidiocarp of *Hygrophorus conicus*, x 1 (*RV*).

vitelline (L. *vitellinus*): yellow in color, similar to an egg yolk.

viticolous (L. *vitis*, grapevine + suf. *-cola*, inhabitant): living or growing in or on the vines and fruits of grape, like *Plasmopara viticola* (Peronosporales).

vittate (*L. vitta*, sash, band + suf. *-atus* > E. *-ate*, provided with or likeness): having a sash; with bands, lines or longitudinal borders.

vocabulary (L. *vocabulum*, word, term + *-arium* > E. *-ary*, suf. that denotes pertaining to or relation to something): the collection of words relating to a particular subject. A list of words of a language or dialect, pertaining to the use of a region, a particular activity, or to a given semantic field. A catalog or list of words in alphabetical order, or arranged in accordance with a system, with definitions or succinct explanations. See **dictionary**, **glossary**, **lexicon** and **terminology**.

volutin (NL. *volutans*, specific epithet of the bacterium *Spirillum volutans* < L. *voluta*, spiral, in which this substance was first found + NL. *-in*, suf. used in chemistry to denote an activator or compound): a granular, electron-dense, basophilic substance, made of polymetaphosphate, which serves as reserve material in fungal cells, especially in yeasts. Common in microorganisms.

volva, pl. **volvae** (L. *volva*, womb, cover, sheath): *Basidiomycetes*. A remnant structure of the universal veil, which like a cyathiform sheath, surrounds the base of the stipe of the basidiocarp, as happens in the gen. *Amanita* and *Volvariella* (Agaricales) and *Clathrus* (Phallales), among others.

Volva of the basidiocarp of *Amanita* sp., x 1 (*RV*).

Volva of the basidiocarp of *Amanita caesarea*, x 0.5 (*RV*).

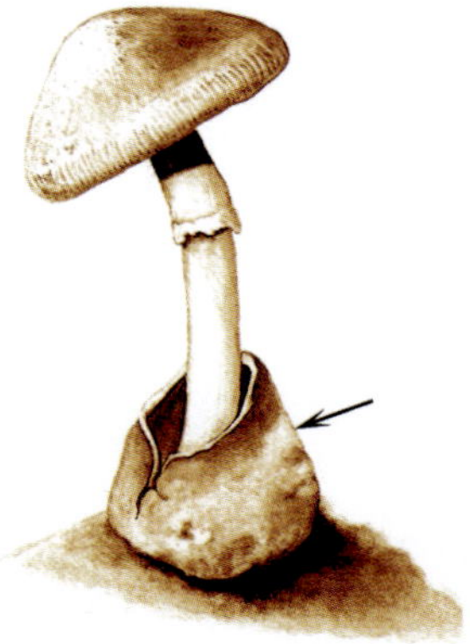

Volva of the basidiocarp of *Amanita* sp., x 0.5 (*MU*).

Volva of the basidiocarp of *Amanita caesarea*, x 0.5 (*RV*).

Volva of the basidiocarps of *Volvariella bombycina*, x 0.3 (*RV*).

-vorous (L. *vorare*, to eat, devour): a suf. equivalent to Gr. suf. *-phago*, employed in Greek words; thus, *multivorous* (an organism that can feed on various kinds of food) is equivalent to *polyphagous*.

W

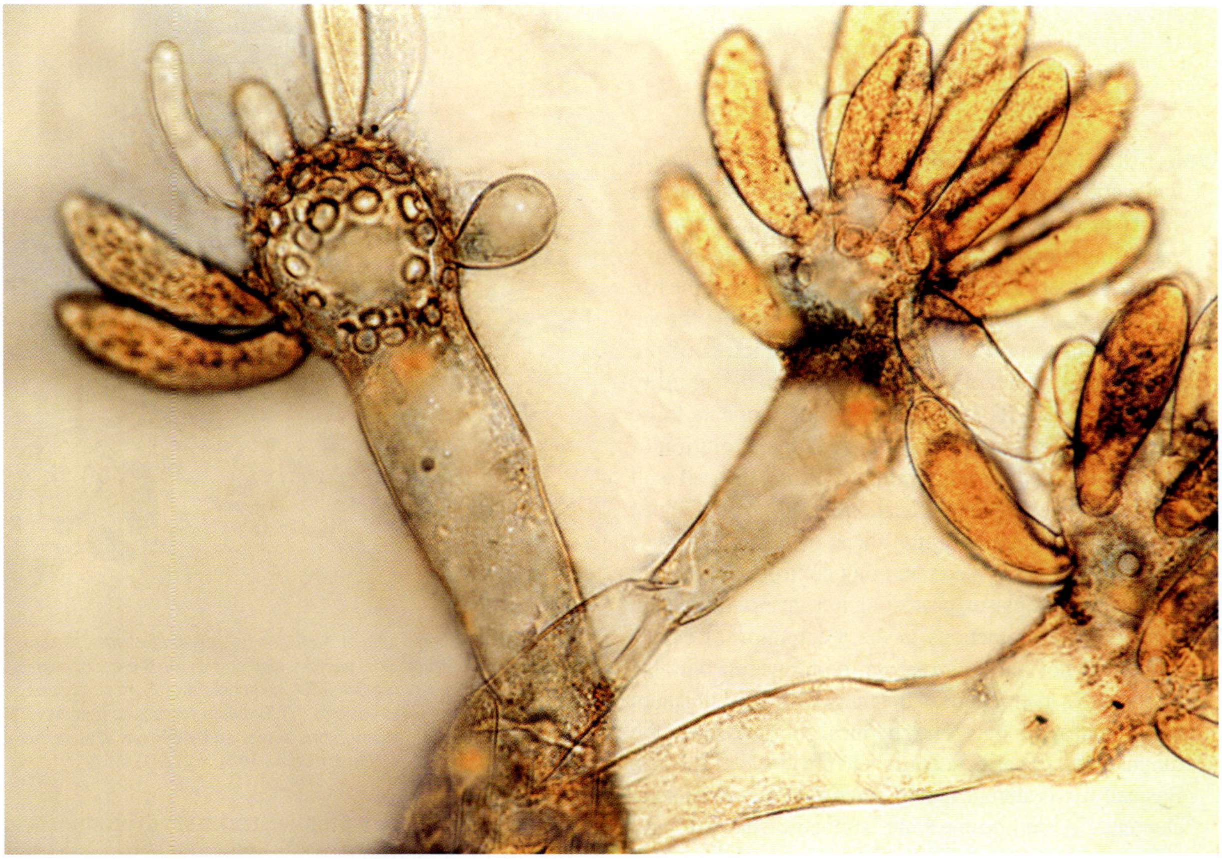

Thalli with zoosporangia of *Blastocladia pringsheimii*, a **water mold** (*MU*).

W

wall (ME. < OE. *weall*, akin to MHG. *wall* < L. *vallum*, a wall set with palisades; the bounding side of a natural cavity or vessel): a dynamic structure that surrounds each fungal cell or hypha, which is subject to change and modification during the different stages of the life cycle of a fungus. The cell wall is responsible for the majority of the unique distinctive characters of the fungi. The capacity of the wall to maintain the turgidity of the fungal cells in a secure manner appears to be a primordial reason for the survival and evolution of the fungi. Besides maintaining turgidity, the cell wall carries out other important functions in the life of a fungus. E.g., the wall determines the shape of the hyphae, acts as a filter that up to a certain point controls what enters the fungal protoplasm, protecting it from environmental dangers, and it functions in the recognition of events associated not only with sexual reproduction but also with various interactions of the fungi with potential plant and animal symbionts. The chemical composition of the fungal cell wall varies according to the group of species. Basically it is composed of microfibrillar or skeletal elements located on the inner surface of the wall and usually embedded in an amorphous matrix that extends to the outer surface of the wall. In the Myxomycota, only the spores have a cell wall and it contains cellulose. In the Chytridiomycetes the zoospores lack a cell wall, but the cell wall of the thallus contains chitin and glucan; the cell wall of the hyphae and spores contains chitin and chitosan in the Zygomycota, cellulose and chitin in the Oomycota, and chitin, protein and glucan in the Ascomycota and Basidiomycota. In the majority of the fungi the skeletal component consists of chitin, which can be in combination with various polysaccharides (e.g., hemicellulose) and small quantities of lipids. In some groups of zoosporic oomycetes, chitin can be absent and the wall is composed mainly of cellulose, chitosan, ß-glucan, mannan, glycogen, galactan and polymers of galactosamine. Fungal cell walls are stratified, usually composed of two or more layers of microfibrils embedded in an amorphous matrix, with the inner layer in contact with the plasma membrane.

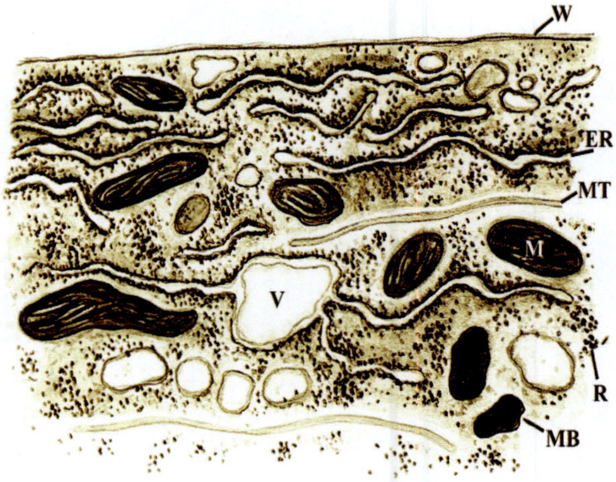

Wall (W) of a young hypha of *Armillaria* (=*Armillariella*) *mellea* (Agaricales), seen in longitudinal section; also shown are the endoplasmic reticulum (ER), microbodies (MB), mitochondria (M), microtubules (MT), ribosomes (R), and vesicles (V). Drawing composed from several transmission electron micrographs published by Beckett *et al.* in their *Atlas of Fungal Ultrastructure*, 1974, x 27,000 (*MU*).

wart (L. *verruca*, wart, prominence): a small prominence, more or less rounded, or with the ornamentation cone-shaped, on the surface of an organ or a structure, like those present on the spores and exoperidium of *Lycoperdon* (Lycoperdales).

water activity (ME. < OE. *waeter*, akin to OHG. *wazzar*, water; *activity*, natural or normal function < ME. < MF. *actif* < L. *activus* < *actus*, pp. of *agere*, to drive, do): this term is applied to the amount of free water in certain substrates. Some fungal species have the capacity of living in substrates with a low water activity, e.g., *Xeromyces bisporus* (Eurotiales) that grows in media with a water activity of 0.75. See **osmophile** and **xerophile**.

water molds (ME. < OE. *waeter*, akin to OHG. *wazzar*, water; ME. *mowlde,* mold): species of fungi and Oomycota that live in water, especially fresh water, in contrast to the marine species. The main freshwater zoosporic microorganisms are Chytridiomycota and Hyphochytriomycota, especially the Chytridiales and Hyphochytriales, respectively; the Oomycota (Saprolegniales) have zoosporic species. There also are many hyphomycetes (asexual fungi), ascomycetes and saxicolous lichens (some of the latter can remain submerged), that live in fresh water, as well as a variety of fungi that are found in contaminated and sewage waters. Cf. **primary marine fungus** and **secondary marine fungus**.

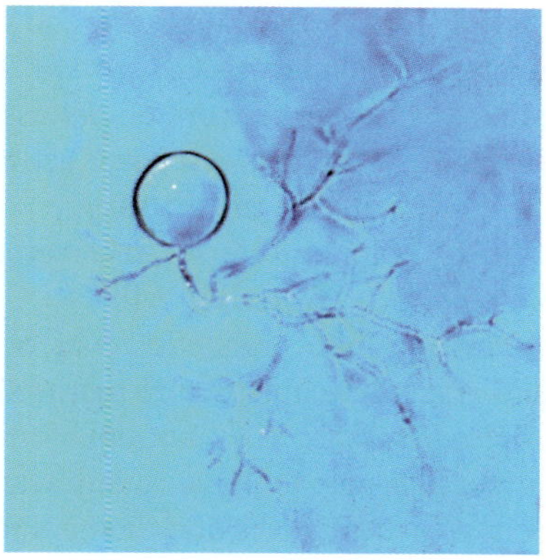

Thallus of *Phizophydium pollinis-pini* (Chytridiales), a **water mold**, x 750 (*RTH*).

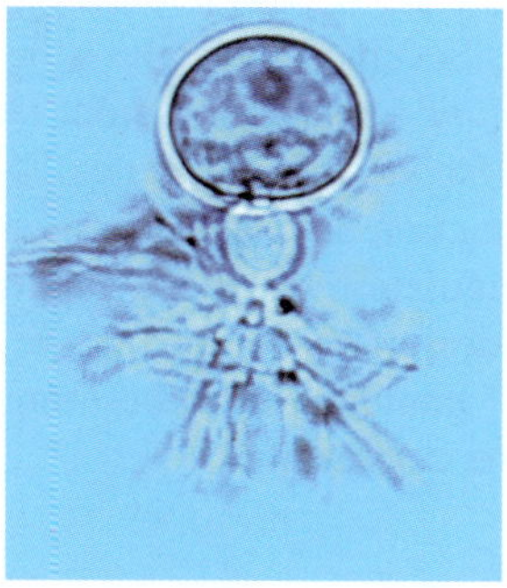

Zoosporangium with apophysis of *Phlyctochytrium* sp. (Chytridiales),a **water mold**, x 850 (*RTH*).

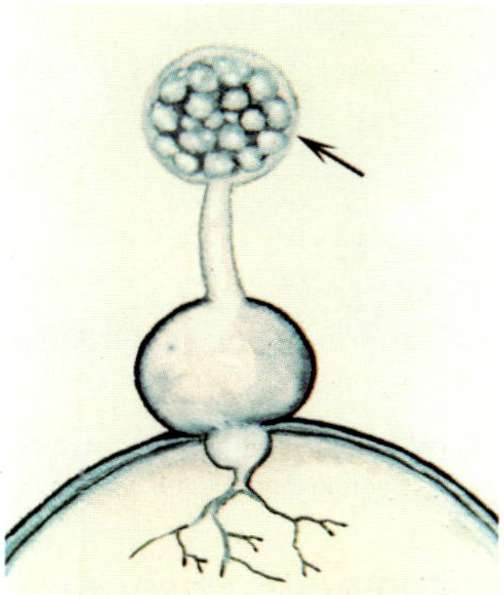

Zoosporangium of *Rhizidiomyces apophysatus* (Hyphochytriales), a **water mold** growing in an oogonium of *Achlya* (Saprolegniales), x 350 (*MU*).

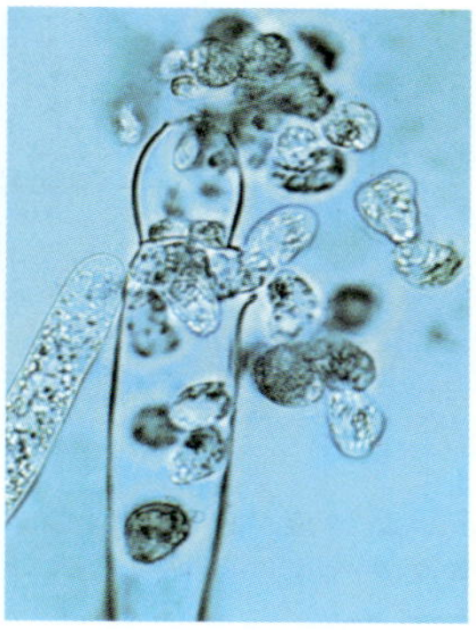

Gametothallus of *Allomyces javanicus* (Blastocladiales), a **water mold** discharging male and female gametes, x 600 (*RTH*).

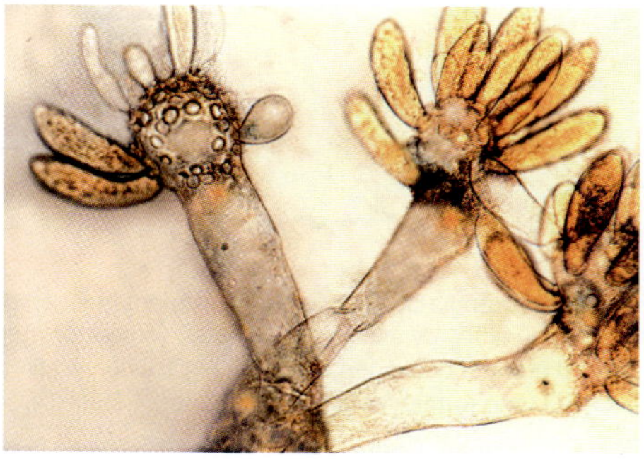

Thalli with zoosporangia of *Blastocladia pringsheimii* (Blastocladiales), a **water mold**, x 140 (*MU*).

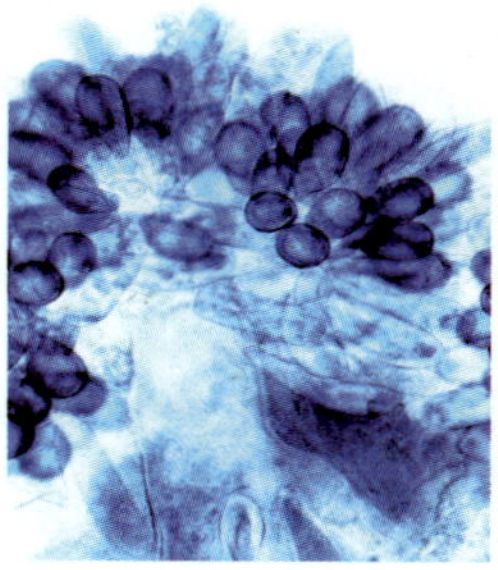

Thallus with zoosporangia of *Blastocladia* sp. (Blastocladiales), a **water mold**, x 120 (*MU*).

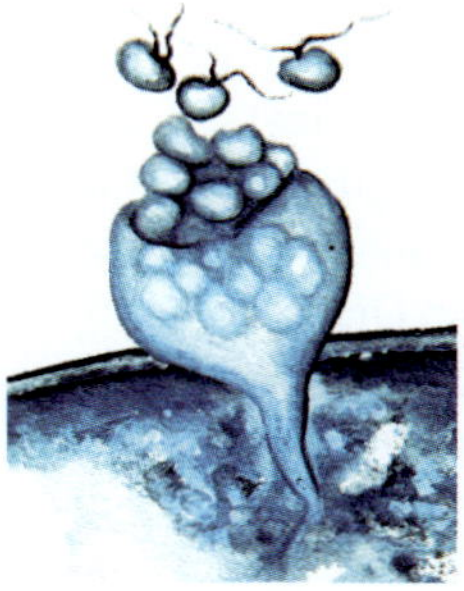

Zoosporangium of *Thraustochytrium globosum* (Saprolegniales), a **water mold** growing on a thallus of an ulotrichaceous alga, x 600 (*MU*).

Hyphae and zoosporangia of *Saprolegnia parasitica* (Saprolegniales),a **water mold** growing on a dead insect, x 8 (*MU*).

Thallus with zoosporangia of *Aqualinderella fermentans* (Leptomitales), a **water mold** which grows in stagnate water, without oxygen, or a limited amount of oxygen, and with high concentration of CO_2 and fermentable organic matter, x 25 (*MU*).

wild type (ME. < OE. *wilde*, wild, occurring naturally, not cultivated; Gr. *typós*, type, mark, model): a species or taxon that occurs naturally, as opposed to morphological variants or biochemical mutants that result from culture *in vitro*. Strains selected as genetic standards may also be designated as wild type.

wilt (alteration of *welk* < ME. *welken*, to wilt; to lose freshness and become flaccid): the action and effect of wilting. A plant disease, especially caused by fungi, such as *Verticillium* (moniliaceous asexual fungi) and *Fusarium* (tuberculariaceous asexual fungi), that is characterized by a marked loss of turgidity in soft tissues and the collapse of leaves and stems.

winter spore (OE. > ME. *winter*, winter; Gr. *sporá*, spore): a resting spore that can remain viable and be able to germinate, overwintering from one year to the next, like the teliospore of *Puccinia graminis*, the cereal rust (Uredinales). Cf. **summer spore**.

witches' broom (OE. *wicce* > ME *wicche*, wicked > E. *witch*, a person who professes or is supposed to practice magic; OE. *brom* > ME. *brome* > E. *broom*, an implement with a handle and attached brush of straw for sweeping; witches are often depicted as riding a broom): brushlike clusters of short, thin branches on woody plants caused by the abnormal development of buds due to parasitism by fungi of the order Taphrinales. E.g., in cherry, the parasite is *Taphrina* (=*Exoascus*) *cerasi*; in hornbeam, *T. carpini*; in oak, *T. kruchii*. Some species of the gen. *Aecidium* (Uredinales) cause analogous deformations, e.g., the witches' broom of fir due to *Aecidium elatinum*. In cacao the agaric *Moniliophthora perniciosa* (=*Marasmius perniciosus*) causes a similar disease of the buds.

word, term (ME. < OE., akin to OHG. *wort*, word < L. *verbum* < Gr. *eirein*, to say, speak; a written or printed character or combination of characters representing a spoken word; ME. *terme*, boundary, end < OF. < L. *terminus*, term, boundary < Gr. *termón*, limit, end, termination or extension): word or expression with a precise meaning, peculiar to a science, art, profession, matter or discipline. See **dictionary**.

Woronin body (named after its discoverer, Russian microbiologist M. Woronin; ME. < OE. *bodig*, body): a cytoplasmic organelle, oval or round in shape and crystalline in structure that is surrounded by a simple membrane (i.e., not double like that of other intracellular organelles), found near the septa in the cells of ascomycetes and their conidial states. Woronin bodies serve to plug the septal pore when the hypha breaks for any reason, thus impeding the loss of cytoplasm in the damaged mycelium, as happens, e.g., in *Galactomyces* (=*Endomyces*) *geotrichum*, Saccharomycetales (the asexual stage is *Geotrichum candidum*, moniliaceous asexual fungi), and *Ascodesmis* (Pezizales). These structures may also function as temporary plugs of the septal pore, for regulation of the cytoplasmic flow and the distribution of nuclei during the formation and secession of conidia while these are maturing. Cf. **septal pore organelle.**

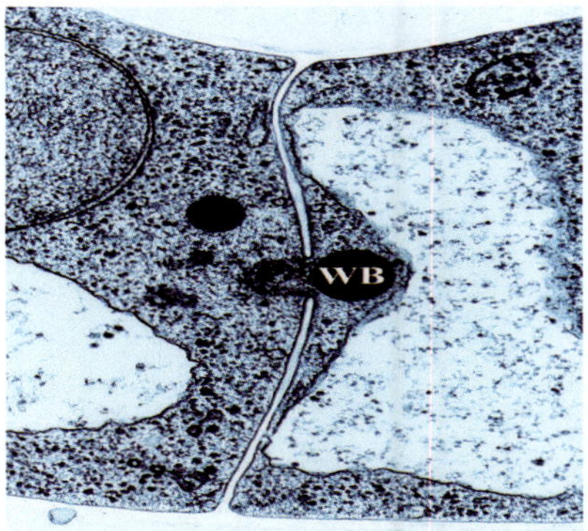

Woronin body (WB) plugging the septal pore of a hyphal cell of *Ascodesmis* sp., seen in longitudinal section, x 10,000 (*CM*).

wrinkled peristome (ME. *wrinkled*, twisted, winding; Gr. *perí*, around, in turn + *stóma*, mouth): a peristome with wrinkled tissue, uneven, not furrowed nor fluted.

X

Fruiting body of *Schizophyllun commune*, a **xerophilic** fungus (*RV*).

X

xantho-, **xanth-** (Gr. *xanthós*, yellow, golden): a pref. used in terms alluding to the color yellow, like *xanthochroic*.

xanthochroic (Gr. *xanthós*, yellow, golden + *chrõs*, color + suf. *-íkos* > L. *-icus* > E. *-ic*, pertaining to or relating to): applied to a context that is reddish or reddish-yellow in color and which darkens on being treated with KOH, as in the Hymenochaetaceae of the Aphyllophorales.

xeno-, **xen-** (Gr. *xénos*, foreigner, outsider, stranger): a pref. used in terms such as *xenobiotic*. Also used as a suf., *-xenous*, in words such as *myrmecoxenous*, which means a stranger among the ants or one that dwells accidentally in anthills.

xenobiotic (Gr. *xénos*, foreigner, outsider + *biotikós*, vital < *bíos*, life + suf. *-tikós* > L. *-ticus* > E. *-tic*, relation, aptitude, inclination or capacity): refers to a chemical substance that normally is not synthetized by living organisms, like the manufactured drugs, as well as to debris or a contaminant that is toxic to a living organism.

xenospore (Gr. *xénos*, foreigner, outsider + *sporá*, spore): a term occasionally applied to a spore that functions in an important way in the dispersal of the fungus that produced it. Conidia, sporangiospores and urediniospores are examples of xenospores. Cf. **hypnospore**.

xero-, **xer-** (Gr. *xerós*, dry, drought): a pref. often employed in biology to form various terms that refer to aridity, or a lack of sap, as in *xerophile*, having a preference for dry sites, and *xerospore*, a spore with a dry wall.

xerophilic, **xerophile** (Gr. *xerós*, dry + *phílos*, have an affinity for + suf. *-íkos* > L. *-icus* > E. *-ic*, pertaining to or relating to): an organism that lives and grows on dry substrates or arid places, or that grows on media with little free water. Among the fungi, there are some xerophilic moniliaceous species, like certain species of *Aspergillus* and *Penicillium*, and some Gasteromycetes, such as *Bovista* (Lycoperdales), *Scleroderma* (Sclerodermatales), *Battarrea* and *Battarreoides* (Tulostomatales), *Montagnea* and *Podaxis* (Hymenogastrales). There are also xerophilic species of *Schizophyllum*, *Lenzites*, *Daedalea*, *Polyporus*, *Stereum* and *Corticium*, among other gen. of Aphyllophorales, whose fruiting bodies are capable of resisting drought without losing their vitality and the power to revive and liberate spores under moist conditions. Cf. **hygrophilous**. See **xerophyte**.

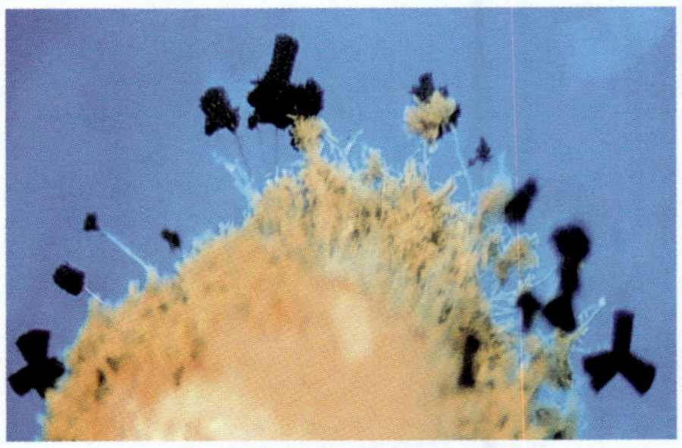

Conidiophores of *Aspergillus flavus* and *A. niger*, **xerophilic** fungi, on a sorghum seed, x 25 (*CB*).

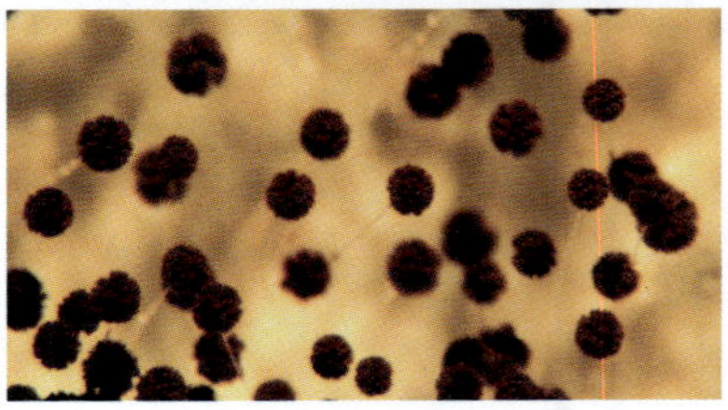

Conidiophores of *Aspergillus niger*, a **xerophilic** fungus, on a maize kernel, x 40 (*MU*).

Fruiting bodies of *Bovista* sp., a **xerophilic** fungus, growing on dry soil, x 0.5 (*TH*).

Fruiting body of *Battarrea stevenii*, a **xerophilic** fungus, x 1 (*MU*).

Fruiting bodies of *Montagnea arenaria*, a **xerophilic** fungus, on dry soil, x 0.5 (*TH*).

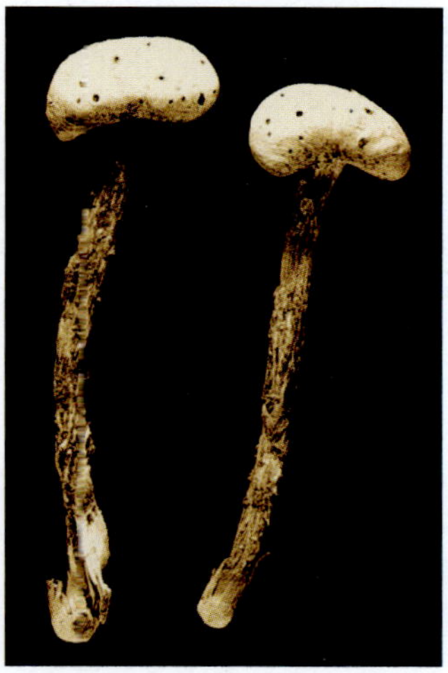

Fruiting bodies of *Battarreoides diguetii*, a **xerophilic** fungus, with several ostioles in the sporiferous sac, x 1 (*TH*).

Fruiting bodies of *Podaxis pistillaris*, a **xerophilic** fungus, x 1 (*MU*).

Fruiting body of *Battarrea stevenii*, a **xerophilic** fungus, growing on a lawn, x 1 (*EPS*).

Fruiting body of *Schizophyllum commune*, a **xerophilic** fungus, on dead wood, x 1 (*RV*).

Fruiting bodies of *Schizophyllum commune*,a **xerophilic** fungus, on dead wood, x 2 (*MU*).

xerophyte (Gr. *xerós*, dry + *phytón*, plant): a plant adapted to aridity, suitable for dry climates or extended periods of drought. This term is sometimes applied, incorrectly, to the fungi. See **xerophile**.

xerospore (Gr. *xerós*, dry + *sporá*, spore): a spore with a dry wall, not viscous, adapted for dissemination by means of air currents. This type of spore is present, e.g., in *Aspergillus*, *Penicillium* (moniliaceous asexual fungi), *Alternaria* (dematiaceous asexual fungi) and many others. Cf. **gloeospore**.

xerotolerant (Gr. *xerós*, dry + L. *toleran*, genit. *tolerantis*, that tolerates or is prone to tolerance): capable of growing under conditions of drought.

xiphioid (Gr. *xíphos*, sword + L. suf. *-oide*, like < Gr. *-oeídes*, similar to): shaped like a sword. Syn. of **ensiform** and **gladiate**.

xylo-, **xyl-** (Gr. *xýlon*, wood, log): a pref. that alludes to wood, as in the following terms.

xylogenous (Gr. *xýlon*, wood + *génos*, origin < *gennáo*, to engender, producer): forming or growing on wood, or obtaining nourishment from it. This term commonly is confused with *xylogen*, what produces wood.

xylophagous (Gr. *xýlon*, wood + *phágos*, what eats < *phágomai*, to eat): eating wood, e.g., the fruiting bodies of *Marasmius* (Agaricales) and *Laetiporus* (Aphyllophorales). Cf. **lignicolous** and **xylosaprobic**.

Fruiting bodies of *Marasmius* sp., a **xylophagous** fungus, on dead wood, x 0.5 (*RV*).

Fruiting bodies of *Laetiporus sulphureus*,a **xylophagous** fungus, on dead wood, x 0.5 (*RV*).

xylosaprobic (Gr. *xýlon*, wood + *saprós*, rotten + *bíos*, life): applied to species that live as saprobes on dead wood, like the majority of the Aphyllophorales and many other fungi from diverse groups, e.g., *Xylogone sphaerospora* (Eurotiales), which develops its cleistothecia on woody substrates in the soil. Cf. **lignicolous** and **xylophagous**.

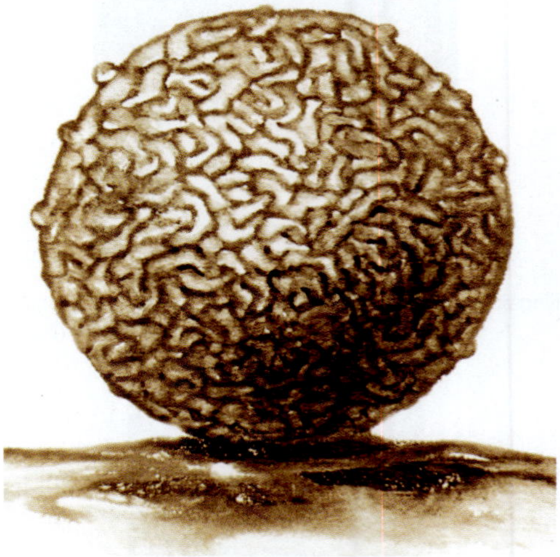

Cleistothecium of *Xylogone sphaerospora*,a **xylosaprobic** fungus, on dead wood, x 250 (*MU*).

xylostroma, pl. **xylostromata** (Gr. *xýlon*, wood + *strõma*, small pillow, bed, stroma): a compact, hard, blackish structure, similar to a stroma, that is expelled by pressure through longitudinal cracks in the bark of roots infected by *Armillaria* (=*Armillariella*) *tabescens* (Agaricales). Also applied to the plates or sheets of mycelium formed by *Xylostroma*, of the Aphyllophorales.

y

A giant colony of *Candida ciferrii* on agar, an asporogenous **yeast** (*MU*).

y

yeast (ME. *yestr* < OE. *gist*, yeast, akin to MHG. *jest*, foam, and Gr. *zeío,* to boil): unicellular budding fungi that do not form part of a formal taxonomic group, but which represent a growth form exhibited by many groups of unrelated fungi. Some primarily filamentous species, including those that are pathogenic to animals, adopt yeast or yeast-like forms (*levaduriforms*, L. *levar* < *levare*, leaven + Gr. *-formis* < *forma*, shape), as part of their life cycle or under certain environmental conditions. See **dimorphic** and **thermodimorph**. The yeasts are classified in different gen. depending on the authority and the definition utilized. They constitute a phylogenetically heterogeneous grouping, including the ascosporogenous Saccharomycetaceae (Saccharomycetales), the cryptococcaceous anascosporogenous asexual yeasts, as well as some tremellaceous forms (with heterobasidia), all consisting of single cells or a poorly developed pseudomycelium and/or a true mycelium, with multiplication by budding, fission or a combination of the two processes. Yeasts are involved in many natural or human-controlled fermentation processes. See **fermentation**, **zymogenous** and **zymosis**. *Apiculate yeasts* (NL. *apiculatus*, provided with an apiculus < *apiculus* < L. *apex*, *apicis*, apex + dim. suf. *-ulus* > E. *-ate*): *Saccharomycodes*, *Nadsonia*, *Hanseniaspora* (ascosporogenous), and *Kloeckera* (anascosporogenous), with minute polar projections that are multiple scars (annellides). *Asporogenous yeasts* (L. *asporogenous* < Gr. pref. *a*, without + *sporá*, spore + *génos*, origin, birth): species that do not form ascospores (cryptococcaceous asexual yeasts). *Baker's yeasts* (ME. *baken* < OE. *bacan*, akin to OHG. *bahhan*, to bake): *Saccharomyces cerevisiae* (Saccharomycetales), is utilized as leaven to make bread rise; the same species is used for making beer and therefore is called *brewer's yeast* (ME. *brewen* < OE. *broewan*, akin to L. *fervere*, to boil): the alcoholic beverage is usually made from malted grains of barley or other cereals fermented in water, and flavored with hops. *Black yeasts* (ME. *black* < OE. *blaec*, black): yeast-like states of *Aureobasidium*, *Cladosporium*, *Moniliella* (dematiaceous asexual yeasts), etc., and especially the anamorphs of the Herpotrichiellaceae, including *Hortaea* (=*Exophiala*), *Ramichloridium* and *Rhinocladiella* (dematiaceous asexual fungi), which in mature cultures form black colonies. *Bottom yeast* (ME. *botme* < OE. *botm*, bottom): one that settles in the bottom of a fermented liquid (e.g., beer wort or wine must), such as *Saccharomyces uvarum*. *Top* or *surface yeast* (ME. < OE. *top*; L. *superficies*, the upper side, surface): one that collects on the surface of the fermented wort, like *S. cerevisiae*. *Chinese yeast*: *Amylomyces rouxii* (Mucorales) and other fungi used to inoculate different substrates for food preparation (see **ragi**). *Food yeast* (ME. *fode* < OE. *foda*, food): *Candida utilis* (cryptococcaceous asexual yeasts) and other dehydrated yeasts that are used as food supplements for animals, because they are a source of vitamins and proteins. *Petit yeast* (MF. *petit*, small, minor): a respiratory deficient mutant whose colony is very small. *Scum yeast* (ME. < MD. *scum*, foam; remaining on top of the water or other liquid without sinking): one forming a thick scum or pellicle, white to somewhat yellowish, opaque or mucilaginous, on the surface of a liquid medium (*Trichosporon cutaneum*, of the cryptococcaceous asexual yeasts). *Shadow* or *mirror yeast* (ME. *shadwe* < OE. *sceaduw, sceadu*, shade, shadow; ME. *mirour* < OF. *mirer*, to look at < L. *mirari*, to wonder at): *Bullera*, *Sporobolomyces*, *Itersonilia* (sporobolomycetaceous asexual yeasts), among other gen. that produce ballistospores, thus forming mirror images of the colonies growing on agar culture media. *Springer yeast*: the strain of *Saccharomyces cerevisiae* originally isolated in the Pasteur Institute of Paris. *Toddy yeast* (Hindi *tari*, juice of the palmyra palm < *tar*, palmyra palm): a mixture of yeasts that ferment the sap of the palmyra palm (*Borassus flabellifer*). *Wine yeast* (ME. < OE. *win* < L. *vinum*, wine; fermented grape juice, used as a beverage): races of *S. cerevisiae*. See **ascus** and **yeast-like**.

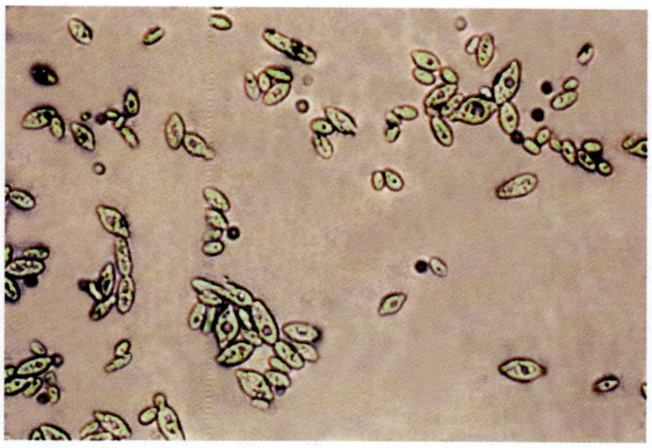

Vegetative cells of *Kloeckera apiculata* (cryptococcaceous asexual yeasts), an apiculate **yeast** isolated from tuba in Mexico, x 900 (*MU*).

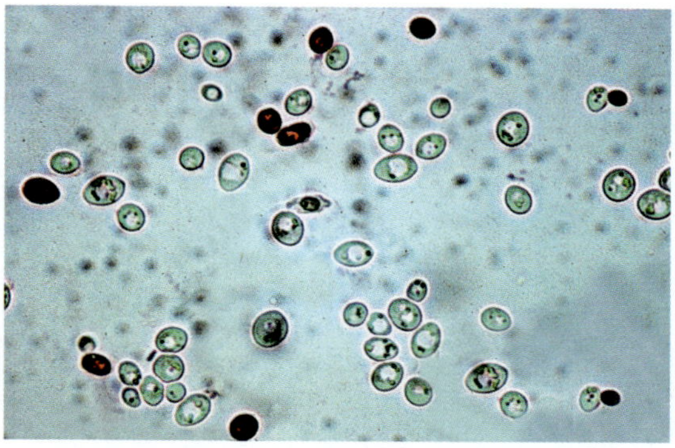

Vegetative cells of *Metschnikowia pulcherrima* (Saccharomycetales); the red cells are capable of synthesizing the pigment pulcherrimin. This ascosporogenous **yeast** species was isolated from the fermented drink tesgüino in Mexico, x 1,000 (*PL*).

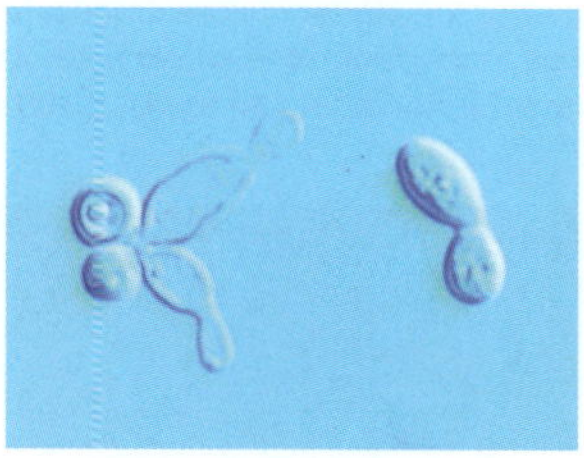

Asci of *Nadsonia fulvescens* (Saccharomycetales), another example of apiculate **yeasts**, x 1,000 (*RTH*).

Giant, white colony of *Saccharomyces cerevisiae* (Saccharomycetales), the baker's **yeast**, on agar, x 1 (*MU*).

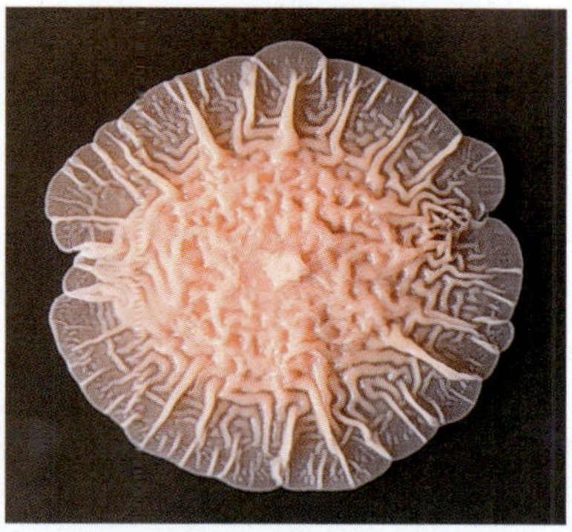

Giant, ornamented colony of *Cryptococcus curvatus* (cryptococcaceous asexual yeasts), an example of an asporogenous **yeast**, on agar, x 1 (*SA*).

Giant, crenate colony of *Saccharomyces cerevisiae* (Saccharomycetales), the baker's **yeast**, on agar, x 1 (*MU*).

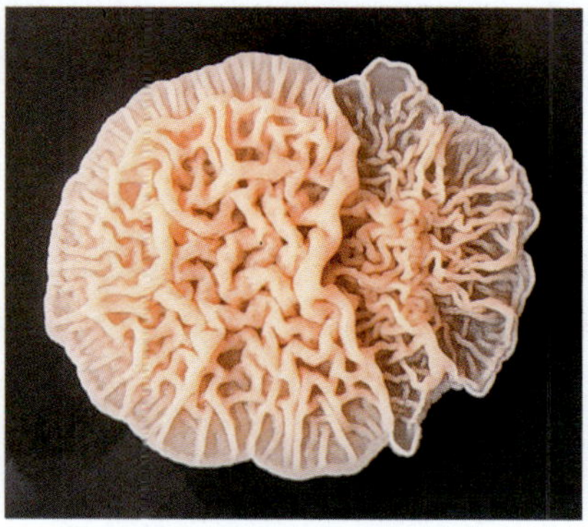

Giant, ornamented colony of *Cryptococcus curvatus* (cryptococcaceous asexual yeasts), an asporogenous **yeast**, on agar, x 1 (*SA*).

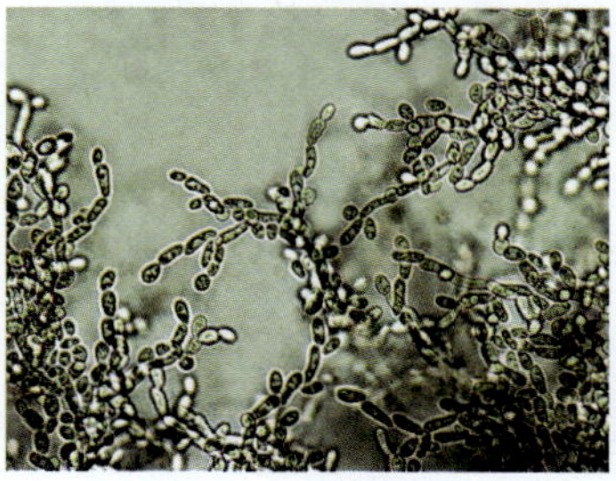

Pseudomycelium of *Saccharomyces cerevisiae*, the baker's **yeast**, on agar, x 800 (*MU*).

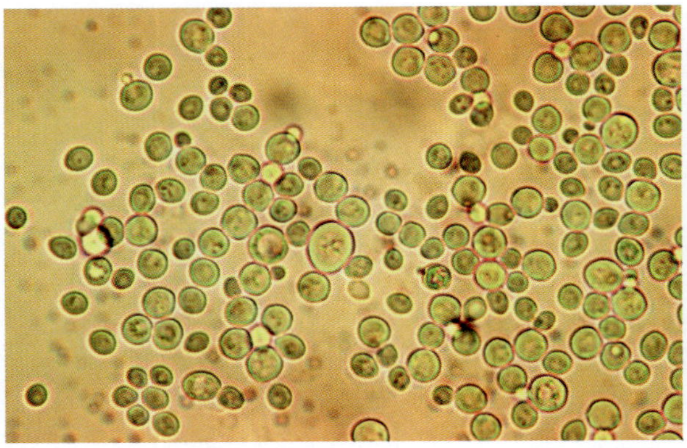

Vegetative cells of *Saccharomyces cerevisiae*, the baker's **yeast**, x 900 (*MU*).

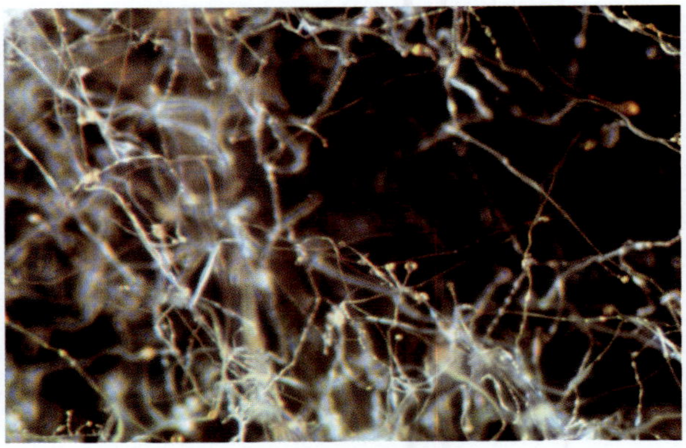

Sporangiophores of *Amylomyces rouxii* (Mucorales) on agar, one of the essential microorganisms in the so-called Chinese **yeast** or ragi, which is the starter culture or inoculum utilized to produce Indonesian *tapé ketela* (from cassava) and other fermented Asian foods, x 300 (*MU*).

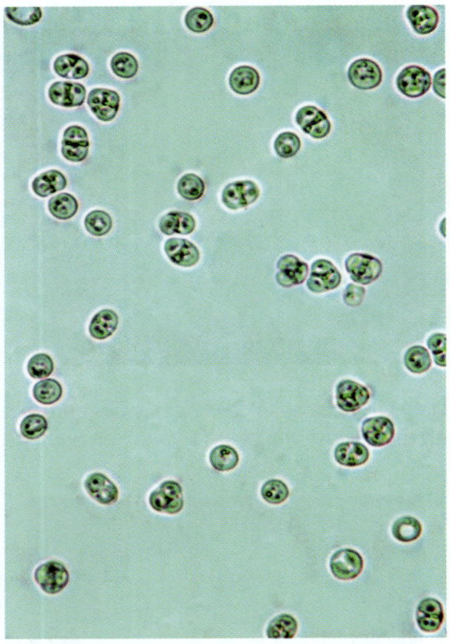

Vegetative cells and asci with two or three ascospores of *Saccharomyces cerevisiae*, the baker's **yeast**, x 1,000 (*MU*).

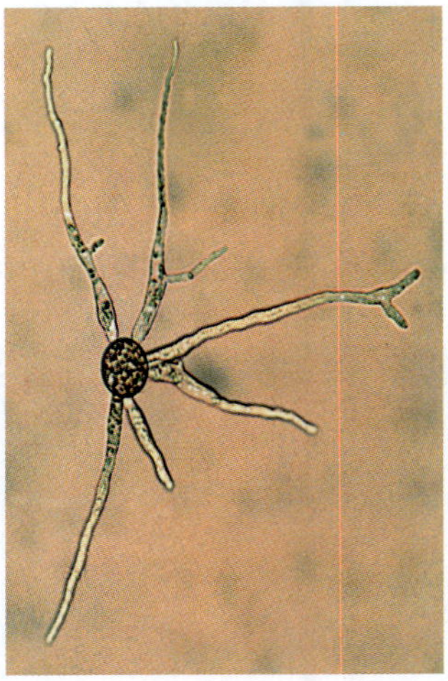

Chlamydospore of *Amylomyces rouxii*, one of the essential microorganisms of the so-called Chinese **yeast** or ragi, x 900 (*MU*).

Giant, dark colony of *Aureobasidium pullulans* (dematiaceous asexual fungi), a black **yeast**, on agar x 5 (*MU*).

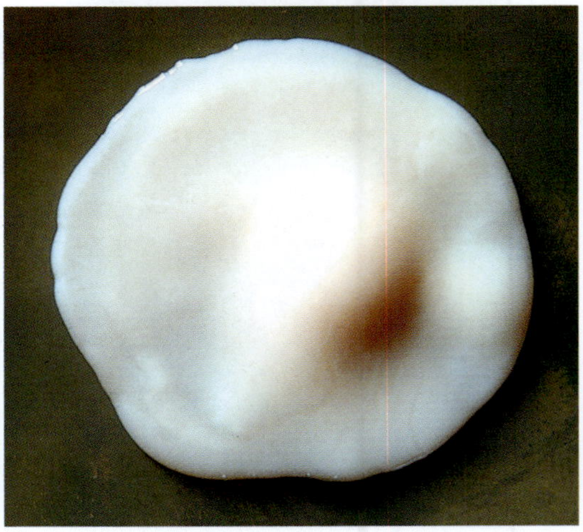

Giant colony of *Bullera* sp. on agar, one of the genera of ballistospore yeasts known as mirror **yeasts**, x 1 (*MU*).

Giant, plaited colony of *Candida ciferrii* (cryptococcaceous asexual yeasts), an asporogenous **yeast**, on agar, x 1 (*MU*).

Giant, nacreous colony of *Candida* sp., an asporogenous **yeast**, on agar, x 1 (*MU*).

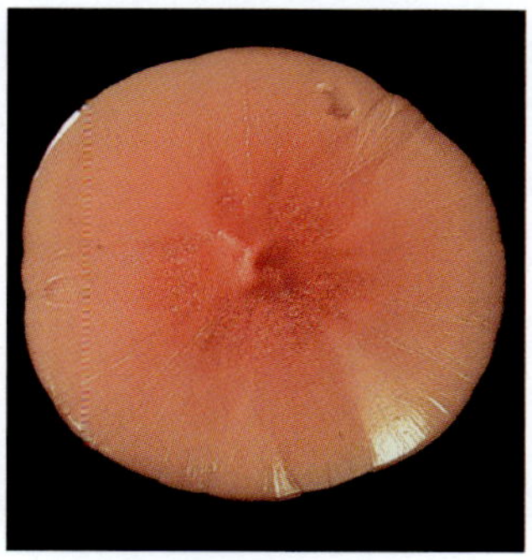

Giant, slimy colony of *Rhodotorula rubra* (cryptococcaceous asexual yeasts), an asporogenous **yeast**, on agar, x 1 (*MU*).

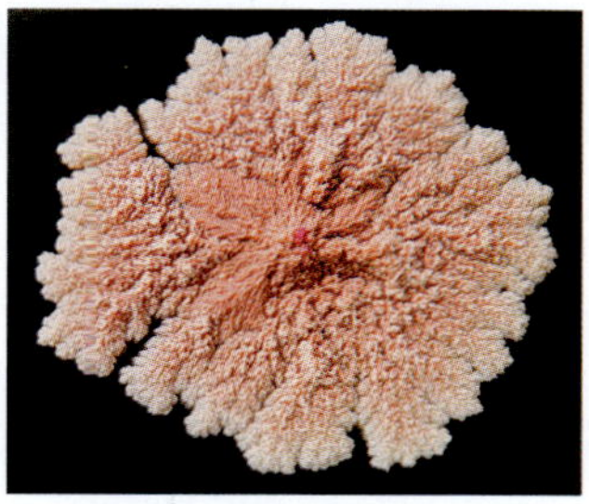

Giant, wrinkled colony of *Pichia membranifaciens* (Saccharomycetales), an ascosporogenous **yeast** isolated from the fermented drink tesgüino in Mexico, on agar, x 1 (*MU*).

yeast-like (ME. *yest* < OE. *gist*, yeast + ME. *lic, lik* < ON. *likr*, like, of the same form): refers to filamentous fungi that have the appearance of yeasts, both in macroscopic characters, such as colony growth, and in microscopic characters, such as multiplication by budding of the somatic cells; e.g., *Aureobasidium pullulans* (dematiaceous asexual fungi). Many fungi are *dimorphic*, i.e., they generally form true hyphae and mycelium, but they acquire the characteristics of yeasts under certain conditions. Some human pathogens form hyphae in culture, but become yeast-like (*levaduriform*) inside the host, as occurs with *Histoplasma capsulatum* (moniliaceous asexual fungi) and other fungi causing mycosis in man and vertebrate animals. See **yeast**.

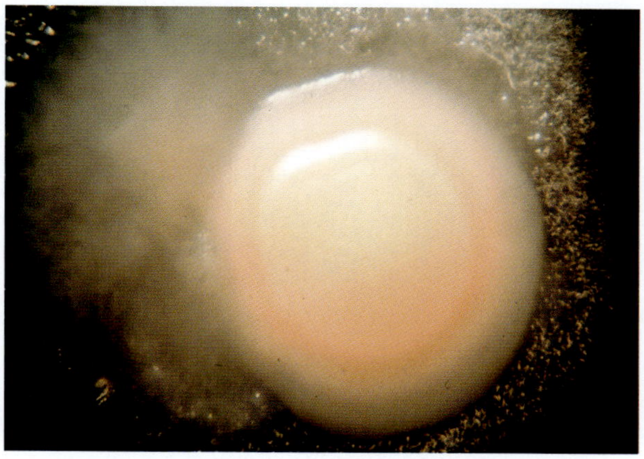

Giant, white, mucilaginous colony of *Aureobasidium pullulans* (dematiaceous asexual yeasts), a **yeast-like** organism, on agar, x 2 (*MU*).

Giant, fimbriate colony of *Geotrichum candidum* (moniliaceous asexual yeasts), another **yeast-like** species, on agar, x 1 (*MU*).

ypsiliform (Ψ, *úpsilon*, *ýpsilon*, 20th letter of Gr. alphabet + L. suf. *-formis* < *forma*, shape): Y-shaped, like the pycnidiospores of *Ypsilonia* (sphaeropsidaceous asexual fungi).

Z

Zonate surface of the fruiting body of *Hexagonia variegata* (*RV*).

Z

zeugite (Gr. *zeugítes*, joined): a spore or cell of a fungus that results from the fusion of two nuclei when the process of fertilization is completed and the dikaryon ends, as occurs in the teliospore of the Ustilaginales.

zonate, zonated (L. *zonatus*, banded < Gr. *zóne*, band, sash, stripe + suf. *-atus* > E. *-ate*, provided with or likeness): arranged in bands or zones of distinct appearance, due to differences in color, texture or other characteristics. The colonies of many fungi on agar have bands that form concentric circles, with alternating areas of somatic mycelium and areas of sporulation, as is seen in *Aspergillus parasiticus*, *Penicillium islandicum* and *Malbranchea aurantiaca* (moniliaceous asexual fungi); likewise, the surface of the basidiocarps of some species of *Trametes* (Aphyllophorales), and the wood attacked by aphyllophoraceous fungi, in which narrow, dark brown or black lines that contain sclerotia are formed, also are zonate.

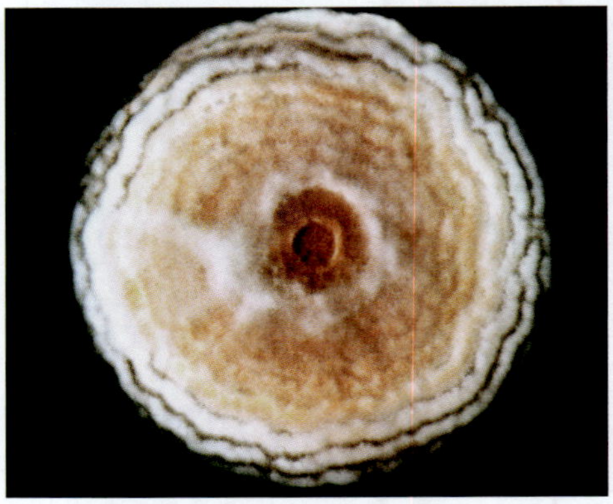

Zonate colony of *Malbranchea aurantiaca* on agar, x 1 (*MU*).

Zonate colony of *Graphium ulmi* (stilbellaceous asexual fungi) on agar, x 0.7 (*MU*).

Zonate colony of *Aspergillus parasiticus* on agar, x 0.7 (*MU*).

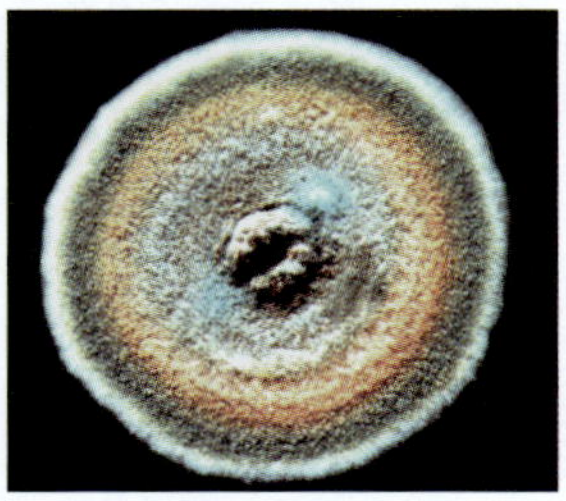

Zonate colony of *Penicillium islandicum* on agar, x 1 (*MU*).

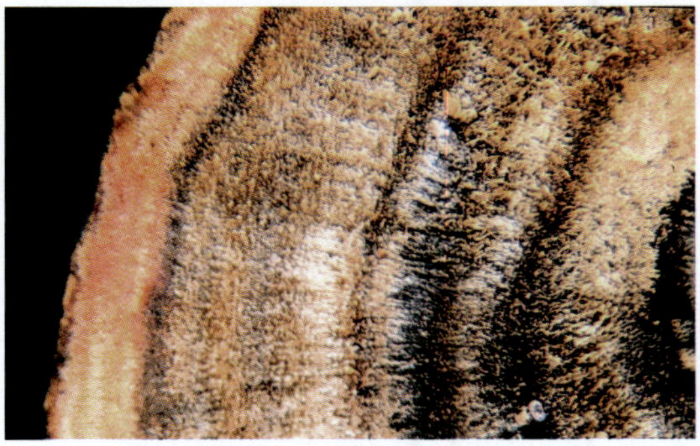

Zonate surface of the fruiting body of *Trametes* (=*Polyporus*) *versicolor*, x 15 (*MU*).

Zonate surface of the fruiting bodies of *Hexagonia variegata* (=*H. papyracea*), x 0.5 (*RV*).

Zonate surface of the fruiting body of *Hexagonia variegata* (=*H. papyracea*), x 0.5 (*RV*).

Zonate surface of the fruiting body of *Hexagonia tenuis*, x 0.5 (*RV*).

zoo- (Gr. *zõon*, animal): a pref. employed to allude to motile cells that resemble microscopic animals, as in the following words.

zoochore, **zoochoric** (Gr. *zõon*, animal + *choréo*, to change place, move away): refers to an organism whose spores are normally disseminated by animals. There exist innumerable examples of fungi that are dispersed by animals (insects, rodents, cattle, etc.); some possess spiny, hooked, or viscous spores that attach themselves to the exterior of the body of the animal vector (e.g., the mucilaginous conidia of *Graphium ulmi*, of the stilbellaceous asexual fungi, that are carried by beetles from an infected tree to a healthy one, extending the disease); the spores of other fungi are transported by insects that have special organs called *mycangia* (e.g., *Stereum*, of the Aphyllophorales, whose propagules are inoculated on the trunks of certain trees when wasps oviposit them in the wood), whereas other species of fungi, known as fimicolous, have spores that are eaten by various types of animals and pass through their digestive system unharmed, and germinate on being deposited on the substrates contained in the excreta of the animals (e.g., *Pilobolus*, of the Mucorales, common on cow and horse dung).

zoocyst (Gr. *zõon*, animal + *kýstis*, vesicle, cell): in *Ceratiomyxella tahitiensis* (Protosteliales), it refers to the thin-walled, short-lived encysted structure that is derived from the multinucleate protoplast, which in turn came from a germinated spore. All of the nuclei of the zoocyst, except one, degenerate, and the persistent nucleus undergoes three successive divisions to give rise to eight nuclei. When the zoocyst germinates, eight flagellate cells are produced.

zoogamete (Gr. *zõon*, animal + *gamétes*, spouse, sexual cell): a motile, flagellate sexual cell (gamete), like that of the aquatic fungi of the class Chytridiomycetes, e.g., *Allomyces* and *Blastocladia*, among other Blastocladiales.

zooglea, **zoogloea**, pl. **zoogleae**, **zoogloeae** (Gr. *zõon*, animal + *gloiós*, glutinous, viscous): originally, this term was applied to the mass of bacteria or lower algae conglutinated by the mucilage of their cell walls swollen by water. In this case, as in others in which microorganisms involved in the formation of this type of glutinous mass are not animals, the application of the term zooglea has not been correct; e.g., the so-called *tibicos* in Mexico (*tibi grains* or *sugary kefir grains* in other parts) have been described as "zoogleae", when in reality they are macroscopic colonies similar to lumps of gelatin, resulting from the symbiotic association of lactic bacteria (*Lactobacillus brevis* and *Streptococcus lactis*) and yeasts (*Saccharomyces cerevisiae* and *Pichia membranifaciens*, of the Saccharomycetales), included in a matrix of dextran. These gelatinous masses are utilized as inoculum or starter cultures in the popular preparation of fermented products such as *tepache* and *vinegar of tibicos*. Since the microorganisms involved are not animals, it is better to refer to these microbial associations as *microbiogleae*. See **fermentation** and **microbioglea**.

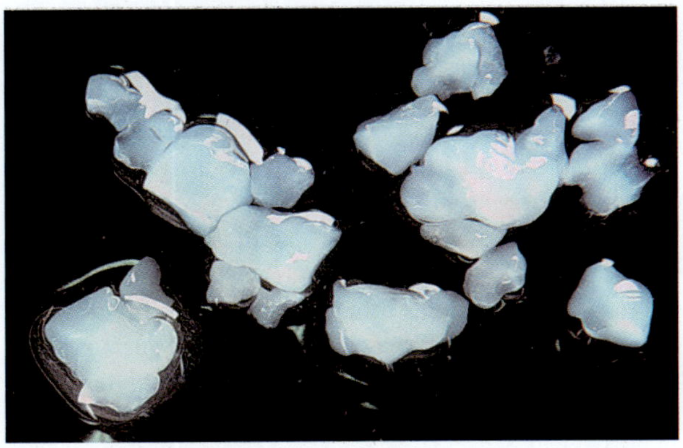

"Zoogleae" or microbiogleae known as tibicos, tibi grains or sugary kefir grains, x 1.6 (*MU*).

zoophagous (Gr. *zõon*, animal + *phágos*, what eats < *phágomai*, to eat): consuming animals or animal products, like the majority of the fungi included in the order Zoopagales (*Zoopage*, *Cochlonema* and others), which are predators of amoebae, rhizopods and nematodes; another example, which is peculiar for having adhesive lateral hyphal branches with which it traps and destroys rotifers, is the gen. *Zoophagus* (Zoopagales).

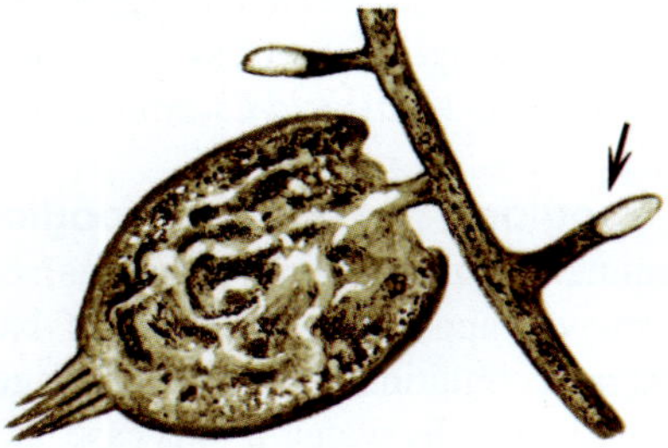

Zoophagous hyphal pegs that trap rotifers in *Zoophagus insidians*, x 250 (*MU*).

zoophilic, **zoophilous** (Gr. *zõon*, animal + *phílos*, have an affinity for + suf. *-íkos* > L. *-icus* > E. *-ic*, pertaining to or relating to; or + L. suf. *-osus* > E. *-ous*, full of, augmented, prone to): *Med. mycol.* Refers to a fungus that is preferentially pathogenic for animals instead of man, like the dermatophyte *Trichophyton equinum* (moniliaceous asexual fungi), although on occasion man can be infected from diseased animals. Cf. **anthropophilic** and **geophilic**.

zoosporangiophore (Gr. *zõon*, animal + *sporá*, spore + *angeîon*, vessel, receptacle + suf. *-phóros*, bearer): a sporangiophore that bears one or more zoosporangia.

zoosporangium, pl. **zoosporangia** (Gr. *zõon*, animal + *sporá*, spore + *angeîon*, vessel, receptacle): a sporangium that contains zoospores, as in the aquatic species belonging to the Chytridiomycota, Hyphochytriomycota and Oomycota, e.g., *Phlyctochytrium* (Chytridiales), *Hyphochytrium* (Hyphochytriales) and *Saprolegnia* (Saprolegniales).

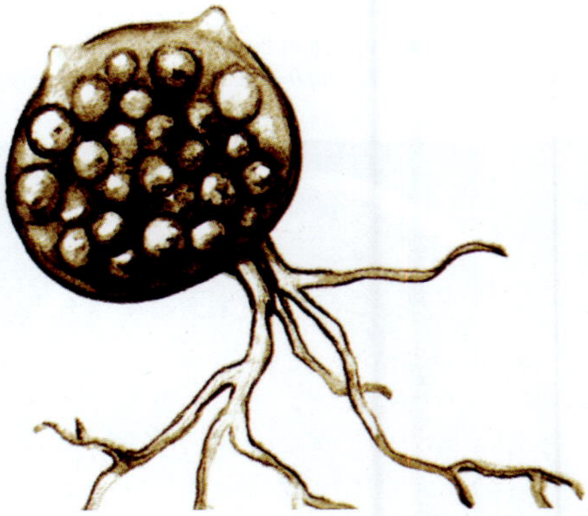

Zoosporangium of *Phlyctochytrium* sp., x 1,200 (*MU*).

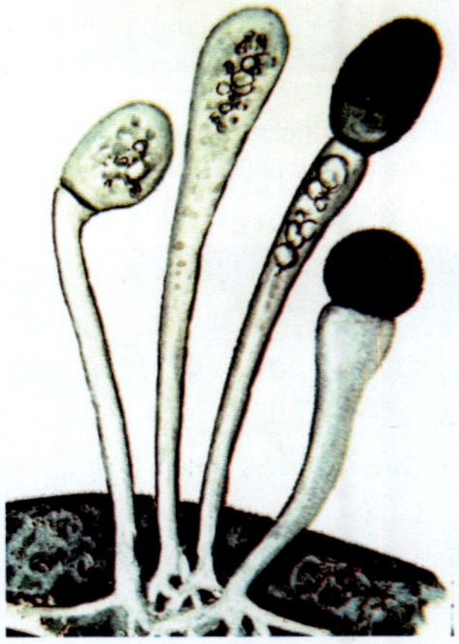

Zoosporangia of *Blastocladiella simplex* (Blastocladiales), x 40 (*MU*).

zoospore (Gr. *zõon*, animal + *sporá*, spore): a motile, flagellate sporangiospore. See **planont**, **planospore** and **symblospore**.

zoosporic (Gr. *zõon*, animal + *sporá*, spore + suf. *-íkos* > L. *-icus* > E. *-ic*, pertaining to or relating to): concerning or relating to *zoospores*. The aquatic fungi are sometimes called zoosporic fungi (Chytridiomycota); also the Hyphochytriomycota and Oomycota have zoosporic species. See **aquatic molds** and **water molds**.

zoosporiferous (Gr. *zõon*, animal + *sporá*, spore + L. suf. *-ferous* < *ferre*, to bear, carry): producing and bearing zoospores, e.g., the vesicle of the sporangium of *Rhizidiomyces* (Hyphochytriales).

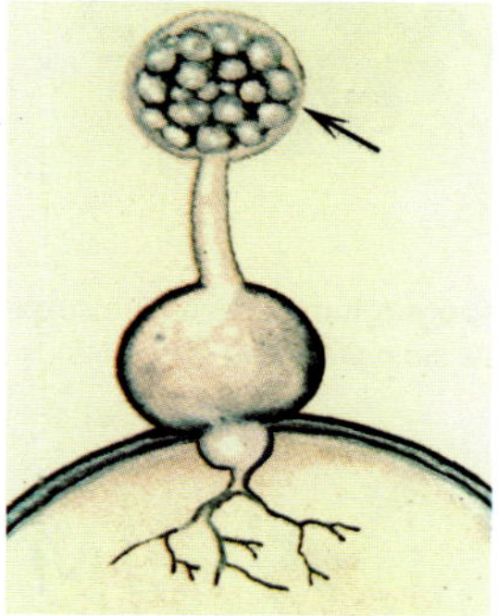

Zoosporiferous vesicle of the sporangium of *Rhizidiomyces apophysatus*, x 400 (*MU*).

zygo-, zyg- (Gr. *zygón*, yoke, pair): a pref. used to give the idea of coupling (*zygospore*) or of bilateral symmetry (*zygomorphy*).

zygoconidium, pl. **zygoconidia** (Gr. *zygón*, yoke, pair + *kónis*, dust, powder + dim. suf. *-ídion*): an asexual propagule formed by the fusion of two conidia generated simultaneously from adjacent conidiogenous sites. Zygoconidia occur as anamorphs of several basidiomycetes, such as *Anastomyces*.

Zygomycetes (Gr. *zygón*, yoke, pair + L. *-mycetes*, ending of taxonomic class < Gr. *mýkes*, genit. *mýketos*, fungus): a class of fungi that reproduce sexually by fusion of gametangia to form zygospores; asexual reproduction is by means of immotile (nonflagellate) sporangiospores, or by conidiola. The majority of species have a well developed mycelium that is coenocytic in most species but septate in certain groups. The Zygomycetes (which includes the orders Mucorales, Dimargaritales, Kickxellales, Endogonales, Glomales, Entomophthorales and Zoopagales) together with the Trichomycetes constitute the phylum Zygomycota of the kingdom Fungi. See the Appendix, where there is a comparison of the outline of Alexopoulos *et al.*, 1996, modified by Ulloa and Hanlin, 2006, to that of Kirk *et al.*, 2008. In the latter scheme of classification, the class Zygomycetes has disappeared, and its orders have been transferred to different subphyla (Entomophthoromycotina, Kickxellomycotina, Mucoromycotina and Zoopagomycotina), of the phylum Zygomycota. The order Glomales, as well as other orders, has been transformed to the class Glomeromycetes, phylum Glomeromycota. These taxonomic rearrangements, with the concurrent change of the names of taxa, are part of a constant process of study and evaluation to achieve a phylogenetic classification of the fungi.

zygomycosis, pl. **zygomycoses** (Gr. *zygón*, yoke, pair + *mýkes*, fungus + suf. *-osis*, state or condition): a mycosis caused by a member of the Zygomycetes. See **mycosis** and **mucormycosis**.

zygomycote (L. *zygomycete* < Gr. *zygón*, yoke, pair + *mýkes*, genit. *mýketos*, fungus): a member of the Zygomycota, a phylum of the Kingdom Fungi, according to the system of classification of Margulis and Schwartz, 1988, and Alexopoulos *et al.*, 1996. The latter system is the one adopted in this dictionary.

zygophore (Gr. *zygón*, yoke, pair + suf. *-phóros*, bearer): *Zygomycetes*. A specialized hyphal branch that gives rise to a progametangium. See **suspensor**.

zygosporangium, pl. **zygosporangia** (Gr. *zygón*, yoke, pair + NL. *sporangium* < Gr. *sporá*, spore + *angeîon*, vessel, receptacle): *Zygomycetes*. A sporangium that contains a zygospore; it develops following the fusion of two gametangia (gametangial copulation) in gen. such as *Absidia*, *Mucor*, *Rhizomucor*, *Rhizopus*, *Phycomyces*, *Gilbertella*, *Umbelopsis*, *Syzygites* and other Mucorales. Cf. **azygosporangium**.

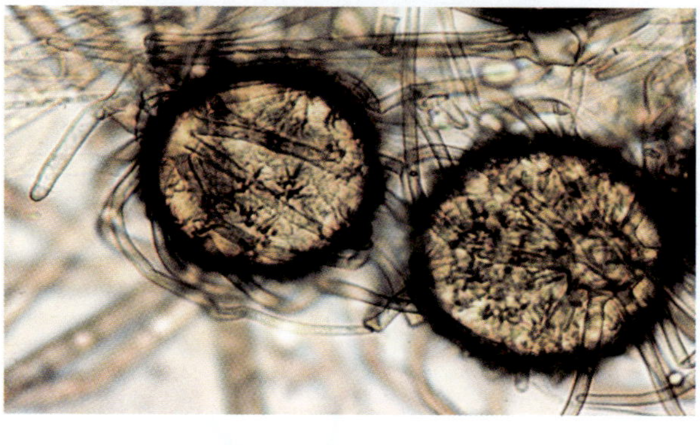

Zygosporangia between the suspensors of *Absidia spinosa*, x 125 (*EAA*).

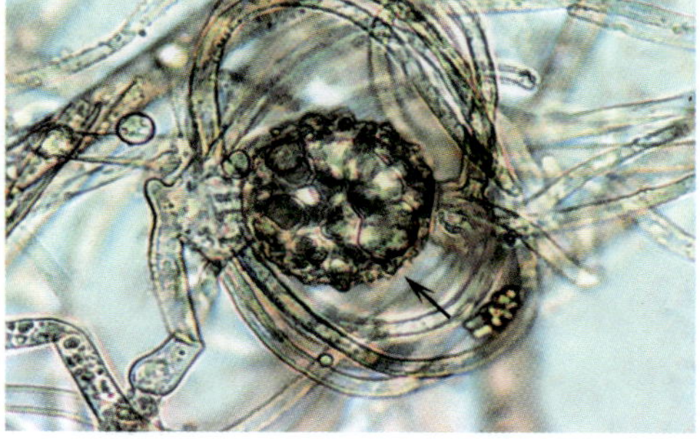

Zygosporangium between the suspensors of *Absidia spinosa*, x 125 (*EAA*).

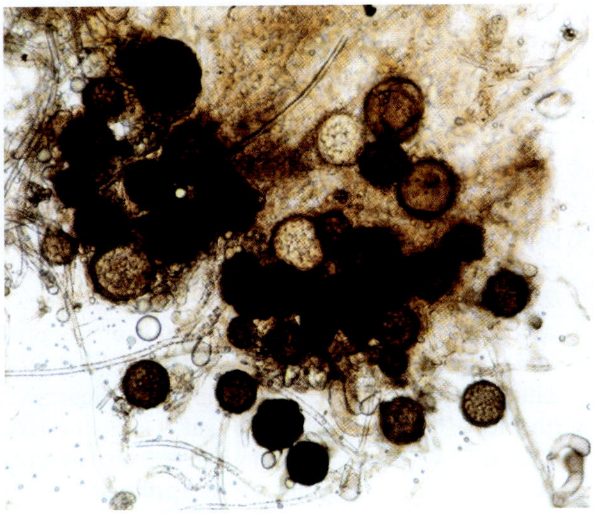

Zygosporangia of *Rhizomucor* (=*Mucor*) *miehei*, x 50 (*MU*).

Zygosporangium between the suspensors of *Rhizopus stolonifer* (=*Rh. nigricans*), x 200 (*MU*).

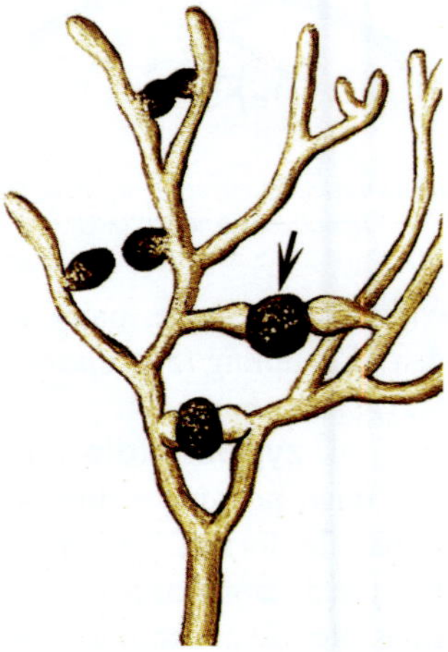

Zygosporangia between pairs of suspensors of *Syzygites megalocarpus*, x 80 (*MU*).

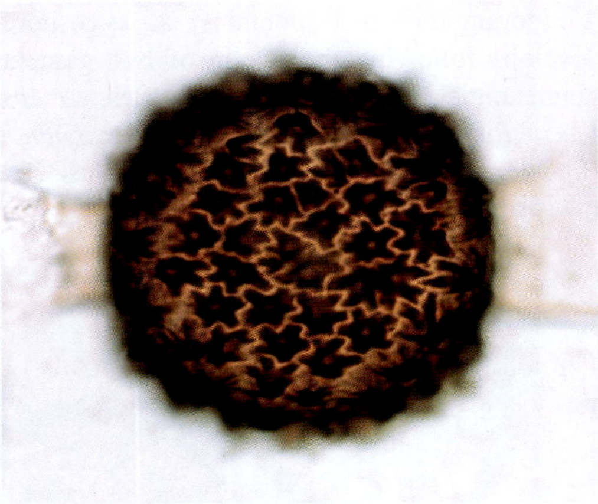

Zygosporangium between the suspensors of *Mucor hiemalis*, x 200 (*RTH*).

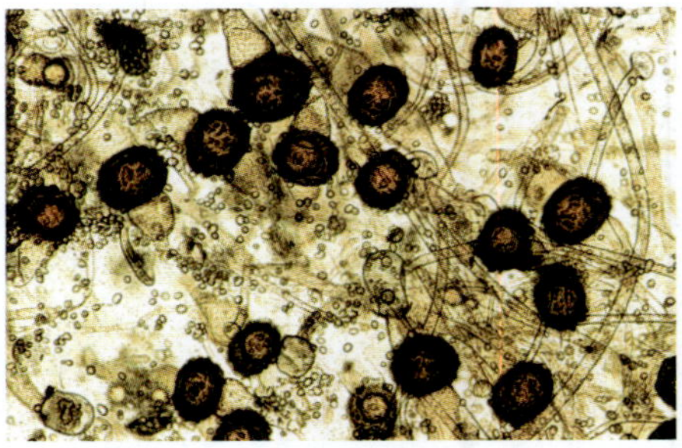

Zygosporangia between the suspensors of *Gilbertella persicaria*, x 50 (*MU*).

Zygosporangium of *Phycomyces nitens*, x 100 (*MU*).

Zygosporangium between the suspensors of *Gilbertella persicaria*, x 200 (*MU*).

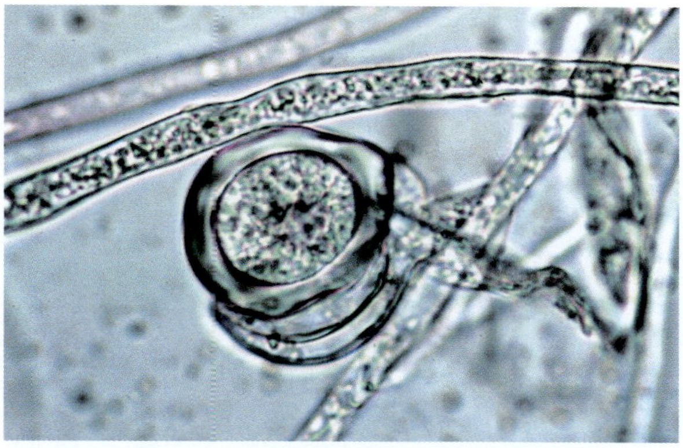

Zygosporangium of *Umbelopsis ramanniana*, x 200 (*MU*).

zygospore (Gr. *zygón*, yoke, pair + *sporá*, spore): *Zygomycetes*. A latent spore, contained in a zygosporangium, that results from the fusion of two gametangia (gametangial copulation). Cf. **azygospore**.

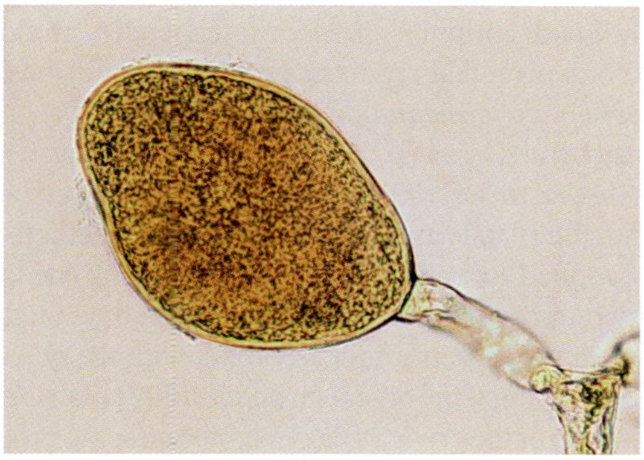

Zygospore of *Endogone* sp. (Endogonales), x 50 (*MU*).

zygosporocarp (Gr. *zygón*, yoke, pair + *sporá*, spore + *karpós*, fruit): *Zygomycetes*. A fruiting body that contains zygosporangia with zygospores, as are present in *Mortierella* (Mucorales), *Endogone* (Endogonales), *Glomus*, *Gigaspora* (Glomales) and others.

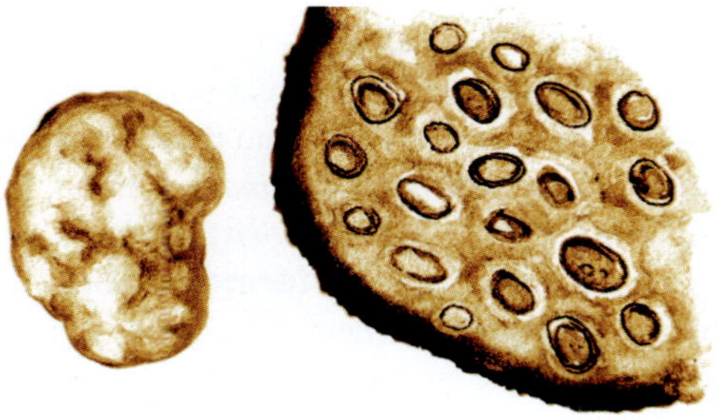

Zygosporocarps of *Endogone lactiflua*, x 4; note the zygospores in the interior of the transversely sectioned zygosporocarp, x 20 (*MU*).

zygosporocarpic (Gr. *zygón*, yoke, pair + *sporá*, spore + *karpós*, fruit + suf. *-íkos* > L. *-icus* > E. *-ic*, belonging to or relating to): *Zygomycetes*. Having a fruiting body (sporocarp) on which zygospores are produced, as occurs in *Endogone* and some other representatives of the order Endogonales.

zygosporophore (Gr. *zygón*, yoke, pair + *sporá*, spore + suf. *-phóros*, bearer): *Trichomycetes*. The supporting cell of the zygospore, derived from a bud that is formed by the union of two hyphae of different sexes; the zygosporophore enlarges in the upper part, giving rise to a pileate zygospore (plano-convex or biconcave), which separates from the zygosporophore that supports it by means of a septum, as occurs, e.g., in *Stipella vigilans* (Harpellales).

Zygosporophore of *Stipella vigilans*, with a medianly attached zygospore, x 800 (*MU*).

zygotactic (Gr. *zygón*, yoke, pair + *taktós*, a touch, management + suf. *-tikós* > L. *-ticus* > E. *-tic*, relation, aptitude, inclination or ability): pertaining to or relative to *zygotaxis*; zygotactic gamete.

zygotaxis, pl. **zygotaxes** (Gr. *zygón*, yoke, pair + NL. *taxis* < Gr. *táxis*, organization, arrangement < *tássein*, to arrange, put in order): the mutual attraction of two gametes or of two gametangia, as happens, e.g., in planogametic copulation (Chytridiales) and in gametangial copulation (Mucorales), among others. See **taxis**.

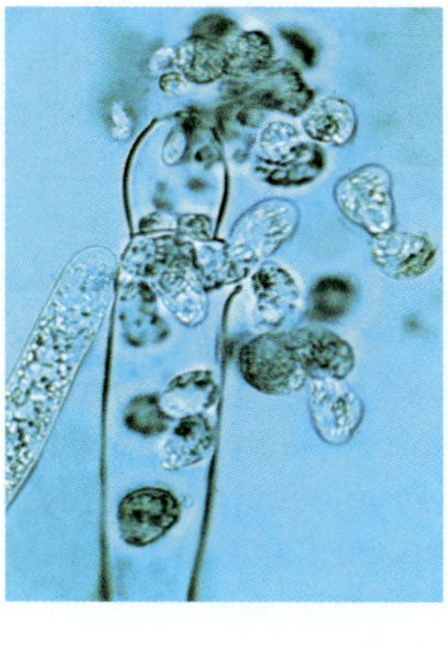

Zygotaxis of released male and female gametes of *Allomyces javanicus* (Blastocladiales), x 500 (*RTH*).

zygote (Gr. *zygotós*, united by a yoke < *zygo-* < *zygóun*, to unite, join < *zygón*, yoke, pair): a diploid cell that results from the union of two haploid cells during sexual reproduction; also called the **egg**. Depending on the group of fungi being considered, the zygote can be either immotile or motile, and the latter amoeboid (amoebozygote) or flagellate.

zygotropism (Gr. *zygón*, yoke, pair + *trópos*, a turn, change in position or manner < *tropé*, a turning < *trépo*, to revolve, to turn, turn towards + suf. *-ismós* > L. *-ismus* > E. *-ism*, state, phase, tendency, action): *Zygomycetes*. A phenomenon in which two hyphae grow and curve toward one another in a reciprocal manner, as a result of mutual stimulation, until they touch and fuse, as happens with the + and - progametangia of many heterothallic Mucorales (*Rhizopus nigricans*, *Mucor hiemalis*, *Phycomyces nitens* and others). See **tropism**.

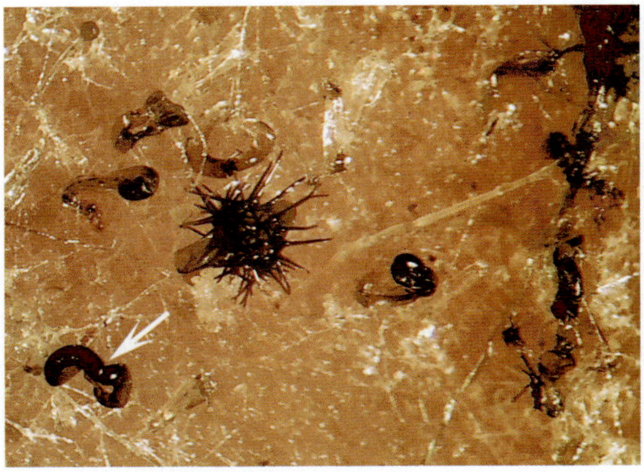

Zygotropism between + and - progametangia of *Phycomyces nitens*, on agar, x 30 (*MU*).

zymo-, zym- (Gr. *zymé*, ferment > L. *zymo*, yeast): a pref. used in terms alluding to yeasts, like *zymocide*.

zymocide (Gr. *zymé*, ferment > L. *zymo*, yeast + L. suf. *-cida*, what kills < *caedere*, to kill): applied to a factor in certain strains of yeasts, which, in the presence of other strains sensitive to it, acts as an antagonist, causing the death of its cells. At present, 11 different toxins, of natural protein, are recognized. Some strains of *Saccharomyces cerevisiae* (Saccharomycetales), e.g., are zymocides and cause the death of certain strains of *Candida glabrata* (cryptococcaceous asexual yeasts). In contrast, some strains of *S. cerevisiae* are sensitive to the zymocide factor present in other strains of the same species and of other species of yeasts. The existence of this phenomenon of competition is of biological and economic relevance in the field of industrial fermentations, e.g., in those concerning the production of yeast for bread-making and distilled alcoholic beverages, such as rum, and others, not distilled, such as beer.

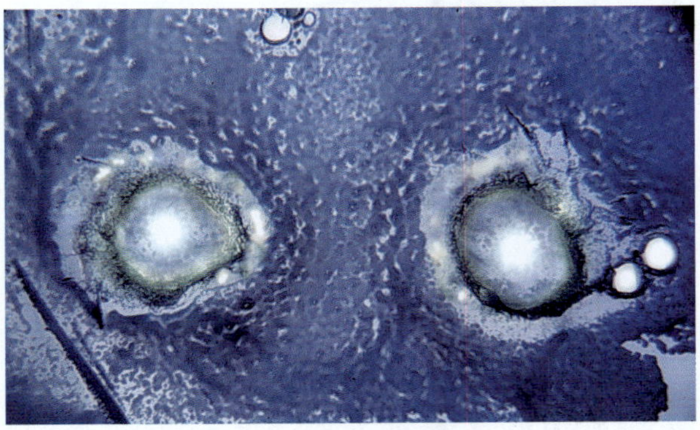

Zymocide effect of a wild strain of *Saccharomyces cerevisiae* (placed in a well in agar) on a bakery strain of yeast, also of the species *Saccharomyces cerevisiae*, cultivated around the well. The halo that surrounds the well indicates the zymocide activity, x 2 (*MU*).

zymogenous (Gr. *zýme*, ferment > L. *zymo*, yeast + Gr. *génos*, origin < *gennáo*, to engender, produce): an organism that produces ferments or enzymes that carry out a *fermentation*. See **yeast**.

zymogram (Gr. *zýme*, ferment > L. *zymo*, yeast + Gr. *grámma*, to describe): literally, a diagram of enzymes and their activity; it is a laboratory method that is applied to cultures of yeasts, bacteria and other microorganisms that grow on certain culture media in which the carbon source is varied, with the object of determining if the organism under study possesses the enzyme or the group of enzymes necessary to produce the fermentation of the various carbon compounds (pentoses, hexoses, etc.), which is manifested by the formation of organic acids and carbon dioxide in the culture medium. Together with an auxanogram and other methods of study, the zymogram is utilized routinely for the identification and classification of yeasts. Cf. **auxanogram**.

zymologist (Gr. *zýme*, ferment > L. *zymo*, yeast + *lógos*, word): one who studies yeasts and their fermentations.

zymology (Gr. *zýme*, ferment > L. *zymo*, yeast + *lógos*, word > E. des. *-logy*, study, treatise): the science that treats of the study of yeasts and their fermentations.

zymosis, pl. **zymoses** (Gr. *zýmosis* < *zýme*, ferment > L. *zymo*, yeast + Gr. suf. *-osis*, condition, state of something, a process): see **fermentation.**

zymurgy (Gr. *zýme*, ferment > L. *zymo*, yeast + NL. *urgia*, technique or art of working with a product, material or tool): the art of fermentation, like that practiced in the bakery, brewery and wine-making industries.

Fungal Scientific Names Illustrated with Photographs or Drawings

This list, alphabetically arranged, comprises the total number of 2,708 illustrations of the present book, representing the same number of fungal scientific names, mostly determined to the level of species, plus several to genus only, and a few names to other taxonomic categories, which provide a visual component to the definitions. The illustrations represent color photographs (2,036) and drawings (672), the latter made by Miguel Ulloa (*MU-d*). Frequently in this list, some specific names of fungi, and the corresponding illustrated terms, are repeated to include the totality of the illustrations and their respective credits. In several of these cases, the repetition of the names of a species and of the illustrated term is due to the fact that two or more different images were used, e.g., in *Aleuria aurantia*. After each fungal scientific name, the initials of the author of the photograph are given (in italics), followed by the name of the illustrated term (in bold). Authors of photographs are acknowledged in the credit section at the end of the book. Illustrations that are not alphabetized according to a particular fungal species are placed in a separate paragraph at the end of the list. The names of technical terms followed by asterisks stand for the new illustrations (1,001) added for this second edition of the dictionary.

Absidia cylindrospora CB **papilla**
Absidia spinosa EAA **acinaciform**
Absidia spinosa EAA **zygosporangia**
Absidia spinosa EAA **zygosporangium** *
Acarospora sp. *MU* **tessellate** *
Acaulopage marantica MU-d **predaceous fungus**
Acaulopage tetraceros MU-d **obconic**
Acaulospora laevis LVF & AE **endomycorrhiza** *
Achlya americana RTH **gametangial contact**
Acrasis rosea MU-d **lobopodium**
Acrasis rosea MU-d **sorocarp**
Acremoniella verrucosa MU-d **blastic**
Acremonium polychromum JAS **funicular** *
Acrogenospora sphaerocephala (=*Monotospora sphaerocephala*) *MU-d* **endohypha**
Acrogenospora sphaerocephala (=*Monotospora sphaerocephala*) *MU-d* **monoblastic**
Acrogenospora sphaerocephala (=*Monotospora sphaerocephala*) *MU-d* **monopodium**
Actidium hysterioides MU-d **hysterothecium**
Actinomucor elegans MU **echinulate**
Actinomucor elegans MU **incrusted** *
Actinomucor sp. *CB* **verticillate**
Aecidium sp. *MU* **aecium**
Aecidium sp. *MU* **spermagonium**
Aecidium sp. *MU* **spermagonium** *
Aecidium sp. *MU* **spermagonium** *
Aecidium sp. *MU* **spermagonium**
Aessosporon salmonicolor MU-d **promycelium**
Agaricus placomyces MRO **adpressed**
Agaricus silvicola MRO **veil** (partial veil)
Agaricus sp. *MU-d* **spore print**
Agrocybe sp. *MU-d* **sinuate**
Alatospora acuminata MU-d **acuminate**
Albugo candida MU **conidiosporangium**
Albugo candida MU **obligate parasite** *
Albugo candida MU **oospore**
Albugo candida MU **rust**
Albugo candida MU-d **basipetal**
Alectoria ochroleuca MU **fruticose**
Aleuria aurantia CB **apothecium**
Aleuria aurantia RTH **ceraceous**
Aleurodiscus scutellatus MU-d **pseudophysis**
Aliquandostipite sp. *ERG & MCG* **capsule** *
Allomyces arbuscula MU-d **hypogynous**
Allomyces javanicus MU **sporothallus**
Allomyces javanicus MU-d **gametothallus**
Allomyces javanicus RTH **epigynous**
Allomyces javanicus RTH **epigynous**
Allomyces javanicus RTH **gametothallus**
Allomyces javanicus RTH **truncate**
Allomyces javanicus RTH **water mold** *
Allomyces javanicus RTH **zygotaxis** *
Allomyces macrogynus MU-d **meiosporangium**
Allomyces macrogynus MU-d **mitosporangium**
Allomyces macrogynus MU-d **nuclear cap**
Alternaria alternata MU **catenulate**
Alternaria alternata MU **dictyospore**
Alternaria alternata MU **dictyospore** *
Alternaria alternata MU **euseptate**
Alternaria alternata MU **longitudinal** *
Alternaria sp. *MU* **phragmospore** *
Alternaria sp. *MU* **multipolar** *
Alternaria sp. *MU-d* **germination**
Amanita abrupta RMA **hilar appendix**
Amanita abrupta RMA **marginate**
Amanita caesarea EPS **sacculiform**
Amanita caesarea MU **sacculiform** *
Amanita caesarea MU-d **ocreate**
Amanita caesarea RMA **fistular**
Amanita caesarea RMA **medulla**
Amanita caesarea RV **edible** *
Amanita caesarea RV **pileate** *
Amanita caesarea RV **sacculiform** *
Amanita caesarea RV **striate** *
Amanita caesarea RMA **universal veil**

Amanita caesarea RMA **veil** (universal veil)
Amanita caesarea RV **volva** *
Amanita caesarea RV **volva** *
Amanita flavivolvata EPS **basidiocarp** *
Amanita fulva EPS **calyptra**
Amanita gemmata RV **ring** *
Amanita herrerae RMA **leptocystidium**
Amanita herrerae RMA **sphaeropedunculate**
Amanita muscaria MU-d **divergent**
Amanita muscaria MU-d **hymeniiferous trama**
Amanita muscaria TH **mycetism** *
Amanita muscaria TH **Soma**
Amanita muscaria var. *flavivolvata EPS* **poisonous fungus** *
Amanita pantherina RV **mycetism** *
Amanita pantherina RV **poisonous fungus** *
Amanita pantherina RV **toxigenic** *
Amanita perpasta MU-d **napiform**
Amanita strobiliformis RV **hymenial veil** *
Amanita strobiliformis RV **hymenial veil** *
Amanita strobiliformis RV **hymenial veil** *
Amanita strobiliformis RV **lamellula** *
Amanita strobiliformis RV **pyramidal** *
Amanita tuza MU **egg**
Amanita vaginata MU-d **vaginate**
Amanita verna RV **poisonous fungus** *
Amanita verna RV **toxigenic** *
Amanita verna TH **poisonous fungus** *
Amanita virosa EAA **virose** *
Amanita virosa RV **virose** *
Amanita virosa RV **mycetism** *
Amanita sp. *MU-d* **abaxial**
Amanita sp. *MU-d* **adaxial**
Amanita sp. *MU-d* **amanitoid**
Amanita sp. *MU-d* **heterotropic**
Amanita sp. *MU-d* **vaginatoid**
Amanita sp. *MU-d* **volva**
Amanita sp. *RMA* **basidiole** and **basidium**
Amanita sp. *RMA* **lamellula**
Amanita sp. *RMA* **metabasidium**
Amanita sp. *RMA* **probasidium**
Amanita sp. *RV* **mycetism** *
Amanita sp. *RV* **pileate** *
Amanita sp. *RV* **planoconvex** *
Amanita sp. *RV* **ring** *
Amanita sp. *RV* **virose** *
Amanita sp. *RV* **volva** *
Amauroderma sp. *MU* **eccentric**
Amazonia sp. *MU-d* **discoid**
Amoebidium parasiticum MU-d **cyst**
Amoebophilus sicyosporus MU-d **predaceous fungus** *
Amoebophilus sicyosporus MU-d **depredation fungus** *
Amphinema aff. *byssoides LV* **rhizomycete** *
Amylomyces rouxii MU **yeast** (Chinese or ragi)
Amylomyces rouxii MU **yeast** (Chinese or ragi)
Anaptychia palmatula MU **foliose**
Anellaria sp. *MU-d* **anellarioid**
Anguillosporella (=*Anguillospora*) *vermiformis MU-d* **anguilliform**
Aniptodera chesapeakensis MU-d **catenophysis**
Annulohypoxylon (=*Hypoxylon*) *thouarsianum EPS* **pulvinate**
Annulohypoxylon (=*Hypoxylon*) *thouarsianum MU* **asymmetric** *
Anthopsis deltoidea MU-d **deltoid**
Antromycopsis smithii CB **coalescent**
Antromycopsis smithii CB & *MU* **capitate**
Antromycopsis smithii CB & *MU* **coalescent**
Antromycopsis smithii CB & *MU* **coalescent** *
Antromycopsis smithii CB & *MU* **gleoid head**
Antromycopsis smithii MU **connate**
Antromycopsis smithii MU **synnema** *
Antromycopsis smithii RV **gleoid head**
Apiosporina collinsii MU-d **apiospore**
Apiosporina morbosa (=*Dibotryon morbosum*) *RTH* **black knot**
Apiosporina morbosa (=*Dibotryon morbosum*) *RTH* **locule**
Apiosporina morbosa (=*Dibotryon morbosum*) *RTH* **ascigerous** *
Apodachlya sp. *MU-d* **periplasm**
Apodachlya sp. *MU-d* **pseudoseptum**
Aqualinderella fermentans MU-d **microaerophilic**
Aqualinderella fermentans MU-d **water mold** *
Arachnophora fagicola MU-d **arachnoid**
Arcyria cinerea MU & *CB* **cinereous**
Arcyria cinerea MU & *CB* **cinereous** *
Arcyria denudata MU **denudate**
Arcyria denudata MU **denudate** *
Arcyria denudata MU-d **elater**
Arcyria sp. *CB* & *MU* **lageniform**
Arcyria sp. *CB* & *MU* **lageniform**
Arcyria sp. *MU* **capillitium**
Armillaria (=*Armillariella*) *mellea MU-d* **cell wall**
Armillaria (=*Armillariella*) *mellea MU-d* **endoplasmic reticulum**
Armillaria (=*Armillariella*) *mellea RV* **rhizomorph**
Armillaria (=*Armillariella*) *mellea MU-d* **microbody** *
Armillaria (=*Armillariella*) *mellea MU-d* **mitochondrion**
Armillaria (=*Armillariella*) *mellea MU-d* **ribosome**
Armillaria (=*Armillariella*) *mellea MU-d* **Spitzenkörper**
Armillaria (=*Armillariella*) *mellea MU-d* **vesicle**
Armillaria (=*Armillariella*) *mellea MU-d* **wall**
Armillaria sp. *MU-d* **armillarioid**

Armillaria sp. *MU-d* **rhizomorphoid**
Arthonia cyanea MAH **arthonioid**
Arthonia radiata MU-d **maculate** *
Arthothelium spilomatoides MU **ardella**
Arthothelium spilomatoides MU **frustulose**
Arthrinium phaeospermum MU-d **meristem blastospore**
Arthrobotrys dactyloides MU-d **vermivorous**
Arthrobotrys dactyloides PL **nematophagous** *
Arthrobotrys dactyloides PL **vermivorous**
Arthrobotrys dactyloides PL **vermivorous**
Arthrobotrys oligospora MU **raduliform** *
Arthrobotrys oligospora MU **sympodulospore**
Arthrobotrys oligospora MU **vermivorous** *
Arthrobotrys oligospora MU-d **internode**
Arthrobotrys oligospora MU-d **radula**
Arthrobotrys sp. *MU* **vermivorous** *
Arthroderma quadrifidum MU-d **ossiform** *
Aschersonia aleyrodis MU **pycnidial stroma**
Ascobolus furfuraceus KO **excipulum**
Ascobolus michaudii EAA **distichous**
Ascobolus michaudii EAA **exserted**
Ascobolus stercorarius MU-d **microtubule**
Ascobolus sp. *MU-d* **exserted**
Ascochyta (=*Chaetodiplodia*) *caulina MU-d* **seta**
Ascodesmis sp. *CM* **free cell formation**
Ascodesmis sp. *CM* **Woronin body**
Ascoidea rubescens MU-d **galeiform** *
Ascoidea sp. *RTH* **ascophore**
Ascophanus carneus EAA **free cell formation**
Ascosphaera apis MU-d **spore ball**
Ascosphaera apis RTH **sporocyst**
Aspergillus candidus CB **biserial**
Aspergillus candidus MU **exudate**
Aspergillus carbonarius MU **echinate** *
Aspergillus clavatus MU **mycotoxin** *
Aspergillus clavatus RTH **clavate**
Aspergillus clavatus RTH **vesicle** *
Aspergillus flavofurcatus (=*A. flavo-furcatis*) *MU* **floccose**
Aspergillus flavofurcatus (=*A. flavo-furcatis*) *MU* **aversion**
Aspergillus flavofurcatus (=*A. flavo-furcatis*) *MU* **catenulate**
Aspergillus flavofurcatus (=*A. flavo-furcatis*) *MU* **medium** *
Aspergillus flavus MU **aspergillosis** *
Aspergillus flavus MU **biserial**
Aspergillus flavus MU **fluorescent**
Aspergillus flavus MU **seminicole** *
Aspergillus flavus MU **storage fungus** *
Aspergillus flavus MU **storage fungus** *
Aspergillus flavus MU **teratogenesis** *
Aspergillus flavus RLM **aspergillosis**
Aspergillus flavus var. *oryzae* (=*A.oryzae*) *MU* **fermentation** *
Aspergillus flavus and *A. niger CB* **xerophilic**
Aspergillus flavus and *Penicillium* sp. *MU* **moldy** *
Aspergillus fumigatus AB **otomycosis** *
Aspergillus fumigatus AB **otomycosis** *
Aspergillus fumigatus AB **otomycosis** *
Aspergillus fumigatus CB **uniserial**
Aspergillus fumigatus ECN **mycosis**
Aspergillus giganteus MU **clavate**
Aspergillus giganteus MU **phototropism**
Aspergillus giganteus MU **phototropism**
Aspergillus japonicus MU-d **basal cell**
Aspergillus japonicus MU-d **foot cell** *
Aspergillus nidulans CM **conidial head**
Aspergillus niger and *A. flavus MU* **storage fungus** *
Aspergillus niger MU **storage fungus** *
Aspergillus niger MU **xerophilic** *
Aspergillus niger PL **biserial** *
Aspergillus ochraceus MU **mycotoxicosis** *
Aspergillus ochraceus MU **mycotoxin** *
Aspergillus ochraceus MU **storage fungus** *
Aspergillus parasiticus MU **mycotoxin** *
Aspergillus parasiticus MU **photosporogenic**
Aspergillus parasiticus MU **zonate**
Aspergillus puniceus MU **allantoid**
Aspergillus puniceus MU **Hülle cell**
Aspergillus puniceus MU **sigmoid**
Aspergillus repens PL **uniserial**
Aspergillus rubrobrunneus MU-d **anamorph** (=*Eurotium rubrum*, teleomorph)
Aspergillus rubrobrunneus MU-d **vesicle**
Aspergillus tamarii MU **cottony** *
Aspergillus terreus MU **exudate** *
Aspergillus terreus MU **velutinous** *
Aspergillus variecolor MU **medium** *
Aspergillus variecolor MU **variegated** *
Aspergillus variecolor MU **variegated** *
Aspergillus versicolor MU **striate**
Aspergillus versicolor PL **biserial** *
Aspergillus versicolor TH **exudate** *
Aspergillus sp. *MU* **storage fungus** *
Asterina veronicae MU-d **asterinoid**
Asterophora (=*Nyctalis*) *parasitica MU-d* **carpophoroid**
Asterophora (=*Nyctalis*) *parasitica MU-d* **fungicolous**
Asterostroma sp. *MU* **asterophysis**
Astraeus hygrometricus MU **exoperidium**
Astraeus hygrometricus MU **hygroscopic**
Astraeus hygrometricus MU **involute**
Astraeus hygrometricus MU **rhagadiose**
Astraeus hygrometricus EAA & SA **rhagadiose** *

Astraeus sp. *MU-d* **paracapillitium**
Athelia (=*Sclerotium*) *rolfsii MU* **sclerotium**
Athelia (=*Sclerotium*) *rolfsii MU* **sclerotium** *
Athelia (=*Sclerotium*) *rolfsii MU* **sclerotium** *
Atkinsonella hypoxylon RTH **perithecial stroma**
Aulographum hederae (=*A. vagum*) *MU-d* **hysterothecium**
Aureobasidium pullulans MU **mucilaginous**
Aureobasidium pullulans MU **mucilaginous**
Aureobasidium pullulans MU **mycelium**
Aureobasidium pullulans MU **plaited**
Aureobasidium pullulans MU **slimy**
Aureobasidium pullulans MU **slimy**
Aureobasidium pullulans MU **yeast** (black)
Aureobasidium pullulans MU **yeast-like**
Auricularia auricula TH **lignicolous**
Auricularia delicata RV **auriculiform** *
Auricularia delicata RV **cartilagineus** *
Auricularia delicata TH **reticulate**
Auricularia delicata TH **reticulate** *
Auricularia mesenterica RV **effuse** *
Auricularia polytricha CB **pubescent**
Auricularia polytricha MU **auriculiform**
Auricularia polytricha RV **pubescent** *
Auricularia sp. *MU-d* **hypobasidium** *
Auricularia sp. *MU-d* **phragmobasidium**
Auriporia sp. *TRO* **lamprocystidium** *
Auriporia sp. *TRO* **leptocystidium** *
Auriscalpium vulgare MRO **pinicole**
Auriscalpium vulgare MU **dentate**
Auriscalpium vulgare MU **hymenophore**
Badhamia capsulifera MU **membranaceous**
Badhamia dearnessii MU **calcareous**
Badhamia gracilis MU **capillitium**
Badhamia gracilis MU **incrusted**
Badhamia gracilis MU **Myxomycetes** *
Badhamia sp. *MU* **Myxomycota** *
Badimia dimidiata MAH **campylidium**
Bagnisiella mirabilis MU-d **paraphysoid**
Balansia claviceps RTH **graminicolous**
Balansia henningsiana RTH **graminicolous**
Balansia henningsiana RTH **part spore**
Balansia henningsiana RTH **vermiform**
Basidiobolus ranarum RLM **phycomycosis**
Basidiobolus ranarum RLM **phycomycosis**
Basidiobolus ranarum RLM **phycomycosis**
Basidiobolus sp. *MU-d* **capilliconidium**
Basidiophora entospora MU-d **conidiosporangium**
Basidiophora entospora MU-d **sterigma**
Basipetospora rubra CB **basipetal**
Basipetospora rubra MU **truncate**
Battarrea stevenii EPS **xerophilic** *
Battarrea stevenii MU **xerophilic**
Battarreoides diguetii TH **ostiole** *
Battarreoides diguetii TH **xerophilic** *
Bdellospora helicoides MU-d **haustorium**
Beauveria bassiana MU **mycoinsecticide**
Beauveria bassiana MU **powdery**
Beauveria bassiana MU **rachis**
Beauveria bassiana MU **rachis**
Beauveria felina (=*B. cretacea*) *CB* **phototropic**
Beauveria felina (=*B. cretacea*) *MU* **phototropic**
Beauveria felina (=*B. cretacea*) *MU* **obclavate**
Beauveria felina (=*B. cretacea*) *MU* **synnema**
Beauveria felina (=*B. cretacea*) *MU* **synnema**
Beauveria felina (=*B. cretacea*) *MU* **synnema** *
Beltrania rhombica GH **rhombic**
Beltrania rhombica MU-d **angular**
Beltrania rhombica MU-d **rhombic**
Bertia moriformis MU-d **moriform**
Bipolaris (=*Drechslera*) *iridis MU-d* **hilum**
Bipolaris maydis MU **phragmospore**
Blakeslea trispora MU-d **sporangiole**
Blastocladia pringsheimii MU **siliquiform**
Blastocladia pringsheimii MU **siliquiform**
Blastocladia pringsheimii MU **sporothallus** *
Blastocladia pringsheimii MU **water mold** *
Blastocladia ramosa MU-d **basal cell**
Blastocladia sp. *MU* **water mold** *
Blastocladiella emersonii MU-d **side body complex**
Blastocladiella simplex MU-d **holdfast** *
Blastocladiella simplex MU-d **zoosporangium** *
Blastomyces (=*Ajellomyces*) *dermatitidis MU-d* **spiral hypha**
Blastomyces (=*Ajellomyces*) *dermatitidis RLM* **blastomycosis**
Blodgettia bornetii MU-d **mycophycobiosis**
Blumeria (=*Erysiphe*) *graminis MU-d* **digitate**
Blumeria (=*Erysiphe*) *graminis MU-d* **haustorium**
Blumeria (=*Erysiphe*) *graminis MU-d* **mycelioid**
Blumeria (=*Erysiphe*) *graminis MU-d* **obligate parasite** *
Blumeria (=*Erysiphe*) *graminis RTH* **mycelioid** *
Blumeria (=*Erysiphe*) *sp. RTH* **secondary hypha**
Blumeriella haddenii RTH **cylindric**
Blumeriella jaapii RTH **leaf spot**
Blumeriella jaapii RTH **leaf spot** *
Boletellus cubensis RV **stipe** *
Boletellus frostii JC **alveolate**
Boletellus russellii RV **reticulate** *
Boletellus russellii MU **alate**
Boletellus russellii MU-d **subalate**
Boletellus russellii RV **concrescent** *
Boletus castaneus TH **spongy**
Boletus (=*Xerocomus*) *chrysenteron MU-d* **decurvate**
Boletus frostii RV **exudate** *

Boletus frustulosus RV **frustulose** *
Boletus pinicola MRO **pinicole**
Boletus pinophilus RV **lubricous** *
Boletus satanas TH **bolete**
Boletus sp. *JC* **context**
Boletus sp. *MU* **hymenium**
Boletus sp. *MU* **mediostratum**
Boletus sp. *MU* **mycorrhiza** *
Boletus sp. *MU* **subulate**
Boletus sp. *RV* **porous** *
Botryosphaeria ribis RTH **erumpent**
Botryosporium sp. *MU* **botryose** *
Botryosporium sp. *MU* **hyaline**
Botryosporium sp. *MU* **lanose**
Botryosporium sp. *MU* **monopodium**
Botryosporium sp. *MU* **racemose**
Botrytis cinerea CB **botryose**
Botrytis cinerea MU **botryoblastospore**
Botrytis cinerea MU **dendroid**
Botrytis cinerea MU gray **mold**
Bovista sp. *MRO* **papyraceous**
Bovista sp. *MRO* **papyraceous** *
Bovista sp. *MU-d* **bovistoid**
Bovista sp. *TH* **xerophilic** *
Bovistella sp. *MU-d* **bovistelloid**
Brachybasidium pinangae MU-d **bisporic**
Brasiliomyces malachrae RTH **cleistothecioid**
Brasiliomyces malachrae RTH **hyaline**
Brasiliomyces malachrae RTH **hyaline**
Brasiliomyces malachrae RTH **unitunicate**
Brettanomyces sp. *MU* **diffusible** *
Broomeia congregata MU-d **congregate**
Broomeia congregata MU-d **stroma**
Buellia oidalea MU **placodiomorph**
Bulgaria sp. *TH* **coriaceous**
Bullera sp. *MU* **yeast** (mirror)
Byssochlamys nivea MU-d **byssoid**
Byssoloma subdiscordans MAH **byssoid**
Byssomerulius (=*Meruliopsis*) *corium RV* **resupinate** *
Caeoma sp. *MU-d* **caeoma**
Calceispora hachijoensis MU-d **calceiform**
Caliciopsis pinea MU-d **urceolate**
Calicium farietinum MU **capitulum**
Calicium farietinum MU **sphaeridium**
Calocera cornea BL **clavule**
Caloplaca saxicola CB **crustaceous**
Caloplaca saxicola CB **rupicolous**
Caloplaca saxicola CB **rupicolous** *
Caloplaca sp. *MU* **corticicolous**
Calosphaeria sp. *RTH* **canker**
Calostoma cinnabarinum HB **cartilagineus** *
Calostoma cinnabarinum EPS **gelatinous peridium**
Calvatia cretacea MU-d **polycystoderm**
Calvatia cretacea MU-d **sphaerocyst**
Calvatia cyathiformis RV **puffball** *
Camarops (=*Nummularia*) *lutea MU-d* **pulvinate**
Camillea sp. *MU* **stroma** *
Camillea sp. *MU* **turbinate**
Candelariella sp. *TRO* **lecanorine** *
Candelina sp. *RTO* **lithotroph** *
Candida albicans RLM **candidiasis**
Candida albicans RLM **candidiasis**
Candida albicans RLM **candidiasis**
Candida albicans RLM **candidiasis**
Candida albicans RLM **pathogen** *
Candida albicans RLM **thrush**
Candida catenulata MU **entire** *
Candida ciferrii MU **plaited** *
Candida ciferrii MU **yeast** (asporogenous) *
Candida krusei TH **fimbriate**
Candida lambica MU **erose**
Candida lambica MU **striate**
Candida lipolytica MU **butyraceous**
Candida valida MU **ornamented** *
Candida valida PL **pseudomycelium**
Candida sp. *TH* **radial** *
Candida sp. *MU* **nacreous** *
Candida sp. MU **sectoring**
Candida sp. *MU* **yeast** (asporogenous) *
Cantharellus cibarius MRO **cantharelloid**
Cantharellus cibarius MU-d **costate**
Capnodium mangiferum MU **sooty mold**
Capnodium walteri MU-d **fuliginose**
Catenaria allomycis MU-d **catenate**
Cavostelium apophysatum MU-d **sporocarp** *
Cavostelium bisporum MU-d **holozoic**
Cavostelium bisporum MU-d **plasmotomy**
Cavostelium bisporum MU-d **pseudopodium**
Cenococcum graniforme CB & MU **sclerotium**
Cephaloascus fragans RTH **ascophore**
Cephaloascus fragans RTH **ascophore**
Cephalosporium sp. *JAS* **gleospore** *
Cephalotheca sulphurea MU-d **subiculum**
Cephalotrichum (=*Doratomyces*) *stemonitis CB* **hastate**
Cephalotrichum (=*Doratomyces*) *stemonitis MU* **aculeate**
Cephalotrichum (=*Doratomyces*) *stemonitis MU* **hastate**
Cephalotrichum microsporum (=*Doratomyces microsporus*) *MU* **synnema**
Ceratiomyxa fruticulosa RTH **exogenous** *
Ceratiomyxa fruticulosa RTH **exogenous** *
Ceratocystis fimbriata MU **perithecium** *
Ceratocystis fimbriata RTH **fimbriate**
Ceratocystis fimbriata RTH **galeiform**

Ceratocystis paradoxa MU-d **doliiform**
Ceratocystis sp. *RTH* **blue stain**
Ceratomyces contortus MU-d **hermaphroditic**
Ceratophorum helicosporum MU-d **helicospore**
Cercophora palmicola RTH **filiform**
Cercophora palmicola RTH **ostiolar canal**
Cercophora palmicola RTH **palmicole** *
Cercospora coffeicola CB **fasciculate**
Cercospora coffeicola CB **scolecospore**
Cercosporidium personatum MU-d **appressorium**
Chaetomidium fimeti MU **fimicole** *
Chaetomidium fimeti MU **setose** *
Chaetomidium fimeti MU **setose** *
Chaetomidium heterotrichum MU-d **setose**
Chaetomidium sp. *MU* **cellulolytic** *
Chaetomium cochliodes MU-d **cochleate**
Chaetomium contortum MU **twisted** *
Chaetomium contortum MU **twisted** *
Chaetomium crispatum (=C. tortile) MU **contorted**
Chaetomium crispatum (=C. tortile) MU **contorted** *
Chaetomium crispatum (=C. tortile) MU-d **contorted**
Chaetomium crispatum (=C. tortile) MU-d **contorted**
Chaetomium funicolum MU **biodeterioration** *
Chaetomium funicolum MU **dichotomous** *
Chaetomium nigricolor MU-d **apiculate**
Chaetomium quadrangulatum EPS **angular**
Chaetomium sp. *CB* **biapiculate** *
Chaetomium sp. *MU* **biodeterioration** *
Chaetomium sp. *MU* **circinate**
Chaetospermum camillae ERG & MCG **appendage** *
Cheiromyces stellatus MU-d **cheirospore**
Chlamydoabsidia padenii MU-d **apophysis**
Chloridium chlamydosporis MU-d **conidiogenous locus**
Chlorociboria aeruginascens RTH **apothecium** *
Chlorophyllum molybdites EPS **annulus** (**annellation**)
Chlorophyllum molybdites EPS **lamella** *
Chlorophyllum sp. *RV* **annulus** (**annellation**) *
Chlorophyllum sp. *RV* **incurved** *
Choanephora cucurbitarum MU **sporangiole** *
Choanephora cucurbitarum MU **sporangiole** *
Chondrostereum (=Stereum) purpureum MU **procumbent**
Chrysomphalina (=Omphalina) chrysophylla MU-d **tubaeform**
Chytridium olla MU-d **interbiotic**
Chytriomyces hyalinus MU-d **chytrid**
Chytriomyces hyalinus MU-d **monocentric**
Chytriomyces hyalinus MU-d **prosporangium**
Chytriomyces hyalinus MU-d **somatogamy**
Ciboria carunculoides RTH **fructicole**
Ciboria carunculoides RTH **fructicole**
Circinella muscae MCG **circinate** *
Circinella muscae MU-d **pandurate**
Circinella umbellata EAA **circinate**
Circinella umbellata EAA **umbellate**
Circinella umbellata EAA **umbellate** *
Cladonia chlorophaea MU **cupulate** *
Cladonia chlorophaea MU **cyathiform**
Cladonia chlorophaea MU **cyathiform**
Cladonia chlorophaea MU-d **cyathiform**
Cladonia chlorophaea MU-d **scyphus**
Cladonia didyma CB **podetium**
Cladonia didyma MU **secondary thallus** *
Cladonia didyma MU **squamulose**
Cladonia didyma MU&CB **podetium**
Cladonia didyma RV **podetium** *
Cladonia didyma TRO **symbiontic** *
Cladonia pyxidata MU-d **primary** and **secondary thalli**
Cladonia pyxidata MU-d **secondary thallus**
Cladonia rangiferina MU **fruticose**
Cladonia rangiferina MU **rangiferoid**
Cladonia rangiferina MU **rangiferoid** *
Cladonia stellaris MU **fruticose**
Cladonia stellaris MU **fruticose**
Cladonia sp. *TRO* **lichen** *
Cladosporium sp. *MU* **conidiophore** *
Cladosporium sp. *MU* **dematiaceous** *
Cladosporium sp. *MU* **ramoconidium** *
Cladosporium sp. *MU-d* **ramoconidium**
Clastoderma debaryanum MU-d **naked** *
Clastoderma debaryanum MU-d **sporangium** *
Clathrus columnatus MU-d **quadrate arm**
Clathrus ruber SBH **clathrate** *
Clathrus ruber TH **clathrate** *
Clathrus ruber JC **clathrate** *
Clathrus sp. *MU-d* **multipileate**
Clavaria sp. *MRO* **coralloid** *
Clavariadelphus truncatus MRO **clavate**
Clavariadelphus truncatus MU **clavate**
Clavariadelphus truncatus MU **truncate** *
Claviceps gigantea MU **endemic**
Claviceps gigantea MU **sclerotium**
Claviceps gigantea TH **endemic**
Claviceps gigantea TH **perithecial stroma** *
Claviceps gigantea TH **phytopathogen** *
Claviceps gigantea TH **phytopathogen** *
Claviceps paspali MU&CB **ergot**
Claviceps purpurea CB **prosenchyma**
Claviceps purpurea MU **ergot**
Claviceps purpurea MU **ergot**
Claviceps purpurea MU **pseudoparenchyma**
Claviceps purpurea MU **toxigenic** *
Claviceps purpurea MU-d **entheogenic** *
Claviceps purpurea MU-d **graminicolous**

Claviceps purpurea RTH **prosenchyma**
Clavulina amoena TH **clavule**
Clavulinopsis sp. *RV* **multiaxial**
Climacocystis (=*Spongipellis*) *borealis MU-d* **spongy**
Clitocybe angustifolia (=*C. rivulosa*) *MU-d* **rivulose**
Clitocybe gibba RV **infundibuliform** *
Clitocybe infundibuliformis EPS **infundibuliform**
Clitocybe lazulina EPS **umbilicate**
Clitocybe sp. *MU-d* **clitocyboid**
Clitopilus prunulus RV **inflexed**
Clitopilus prunulus RV **undate** *
Clonostachys rosea (=*Gliocladium roseum*) *JAS* **gleospore** *
Clonostachys rosea (=*Gliocladium roseum*) *JAS* **microconidium** *
Coccidiascus legeri RTH **endoparasitic**
Coccidioides immitis ECN **spherule**
Coccidioides immitis RLM **coccidioidomycosis**
Coccidioides immitis RLM **spherule**
Coccomyces delta RTH **deltoid** *
Coemansia mojavensis MU-d **pectinate**
Cokeromyces recurvatus MU **contorted**
Cokeromyces recurvatus MU **recurvate** *
Cokeromyces recurvatus MU **sporangiole** *
Cokeromyces recurvatus MU **sporangiole** *
Collema coilocarpum MU **gelatinous**
Collema tenax MU-d **homoiomerous**
Colletotrichum circinans MU **acervulus**
Colletotrichum circinans MU **seta**
Colletotrichum falcatum MU-d **acervulus**
Colletotrichum lindemuthianum SA **anthracnose**
Colletotrichum sp. *MU* **appressorium**
Colletotrichum sp. *MU* **seta** *
Collybia sp. *MU-d* **collybioid**
Coltricia perennis RV **pyroxylophilous** *
Conidiobolus coronatus CB **unispored**
Conidiobolus coronatus RLM **coronate**
Conidiobolus coronatus RLM **sporangiole**
Conidiobolus thromboides MU **germ tube**
Conidiobolus thromboides MU **hyphal body** *
Conidiobolus thromboides MU **secondary conidium** *
Conidiobolus thromboides MU-d **hyphal body**
Conidiobolus sp. *MU-d* **capilliconidium**
Coniella fragariae MU-d **raphe**
Coniochaeta scatigena (=*Ascobolus scatigenus*) *EAA* **prolate**
Coniochaeta scatigena (=*Ascobolus scatigenus*) *EAA* **fimicole** *
Coniochaeta scatigena (=*Ascobolus scatigenus*) *EAA* **oblong** *
Coniochaeta scatigena (=*Ascobolus scatigenus*) *EAA* **stercoreaceous** *
Coniochaeta scatigena (=*Ascobolus scatigenus*) *EAA* **urinophilic** *
Conotrema urceolatum MU-d **urceolate**
Cookeina sulcipes EPS **acetabuliform**
Cookeina sulcipes EPS **cupulate**
Cookeina sulcipes MU-d **calyciform**
Cookeina sulcipes RTH **microcyclic conidiation** *
Cookeina sulcipes RTH **suboperculate** *
Cookeina sulcipes TH **acetabuliform**
Cookeina tricholoma MU **calyciform**
Cookeina tricholoma RV **apothecium** *
Cookeina tricholoma RV **cup-shaped fungus** *
Cookeina tricholoma RV **marginal hair** *
Cookeina tricholoma RV **stipe** *
Cookeina tricholoma RV **stipitate** *
Coprinellus (=*Coprinus*) *micaceus MU-d* **striate**
Coprinellus (=*Coprinus*) *micaceus TH* **caespitose** *
Coprinellus (=*Coprinus*) *micaceus MU-d* **pruinose**
Coprinopsis atramentaria (=*Coprinus atramentarius*) *RV* **ink-cap fungus** *
Coprinopsis atramentaria (=*Coprinus atramentarius*) *RV* **melanosporous** *
Coprinopsis atramentaria (=*Coprinus atramentarius*) *RV* **putrescent** *
Coprinopsis atramentaria (=*Coprinus atramentarius*) *RV* **recurvate** *
Coprinopsis echinospora (=*Coprinus echinosporus*) *TH* **comate**
Coprinopsis (=*Coprinus*) *lagopus EPS* **involute** *
Coprinopsis (=*Coprinus*) *lagopus MU-d* **secondary mycelium**
Coprinopsis (=*Coprinus*) *lagopus MU* **villose**
Coprinopsis (=*Coprinus*) *lagopus MU-d* **comate**
Coprinopsis (=*Coprinus*) *lagopus MU* **comate** *
Coprinopsis (=*Coprinus*) *lagopus MU-d* **involute**
Coprinopsis (=*Coprinus*) *lagopus MU-d* **oidiophore** and **oidium**
Coprinopsis (=*Coprinus*) *lagopus MU-d* **primary mycelium**
Coprinus comatus MU **lenticular**
Coprinus comatus MU-d **amygdaloid**
Coprinus comatus MU-d **brachybasidiole**
Coprinus comatus MU-d **complanate**
Coprinus comatus MU-d **deliquescent**
Coprinus comatus MU-d **hyphal string**
Coprinus comatus TH **deliquescent**
Coprinus sp. *MU* **hymenium**
Coprinus sp. *MU* **interspace** *
Coprinus sp. *MU* **primordium**
Coprinus sp. *MU* **radial**
Coprinus sp. *MU* **roridous** *
Coprinus sp. *RV* **deliquescent** *
Coprinus sp. *RV* **putrescent** *
Coprinus sp. *RV* **recurvate** *

Copromyxa arborescens MU-d **microcyst**
Copromyxa arborescens MU-d **sorocyst**
Copromyxa arborescens MU-d **uroid**
Cordana musae MU **obovate**
Cordyceps capitata growing on an ascocarp of *Elaphomyces* sp. *RTH* **mycogenous**
Cordyceps capitata growing on an ascocarp of *Elaphomyces* sp. *RV* **mycogenous** *
Cordyceps capitata RV **stroma** *
Cordyceps melolonthae var. *rickii TH* **capitate**
Cordyceps melolonthae var. *rickii TH* **entomogenous** *
Cordyceps militaris MU-d **pseudosclerotium**
Cordyceps militaris RTH **lyophilized** *
Cordyceps ophioglossoides growing on an ascocarp of *Elaphomyces granulatus MU* **mycogenous**
Cordyceps sobolifera RV **hemibiotrophic** *
Cordyceps sobolifera RV **multiaxial** *
Cordyceps sobolifera RV **perithecial stroma** *
Cordyceps sobolifera RV **ramiferous** *
Cordyceps sobolifera RV **stroma** *
Cordyceps sphecocephala MU **entomogenous** *
Cordyceps stylophora MU **acuminate**
Cordyceps stylophora MU **entomogenous** *
Cordyceps stylophora MU **entomophilous**
Cornumyces irregularis (=*Petersenia irregulare*) in a hypha of *Achlya* sp. *MU-d* **mycoparasite**
Corollospora colossa MCG **naviculate**
Corollospora intermedia MCG **cingulate**
Corollospora intermedia MU **arenicolous**
Corollospora maritima MCG **cingulate**
Corollospora pulchella MU **marine fungi** *
Corticium sp. *MU* **accumbent**
Cortinarius sp. *RV* **arachnoid** *
Cortinarius sp. *RV* **cortina** *
Cortinarius sp. *RV* **planoconvex** *
Cortinarius sp. *RV* **stipe** *
Cortinarius sp. *RV* **veil** (partial veil) *
Cortinarius sp. *RV* **vestigial** *
Cortinarius sp. *TH* **cortina**
Corynascus sp. *MU* **thermophilic** *
Coryne atrovirens RV **lubricous** *
Corynespora cassiicola MU-d **monotretic**
Corynespora cassiicola MU-d **percurrent**
Cotylidia aurantiaca EPS **eucarpic** *
Cotylidia aurantiaca EPS **mycelium** *
Cotylidia aurantiaca RV **infundibuliform** *
Cotylidia sp. *RV* **tubaeform** *
Craterium minutum MU **corticicolous**
Craterium minutum MU **crateriform**
Craterium minutum MU **peridium**
Craterium minutum SA **dehiscence**
Crepidotus mollis RV **lamella** *
Crepidotus mollis RV **lignicolous** *
Cribraria cancellata (=*Dictydium cancellatum*) *MU* **sporangium**
Cribraria cancellata (=*Dictydium cancellatum*) *CB* **cancellate**
Cribraria cancellata (=*Dictydium cancellatum*) *MU* **cancellate**
Cribraria cancellata (=*Dictydium cancellatum*) *MU* **cancellate**
Cribraria cancellata (=*Dictydium cancellatum*) *MU* **cancellate** *
Cribraria cancellata (=*Dictydium cancellatum*) *CB* **net**
Cribraria ferruginea MU-d **net**
Cribraria purpurea MU **dictydine**
Crinipellis sp. *RV* **incised** *
Crinipellis sp. *RV* **petaloid** *
Cronartium conigenum MU **aeciospore** *
Cronartium ribicola MU **tubercular**
Crucibulum laeve MU **nidulant**
Crucibulum laeve CB **funiculose**
Crucibulum laeve CB **peridiole**
Crucibulum laeve CB **splash cup**
Crucibulum leave CB **bird's nest fungus**
Cryphonectria (=*Endothia*) *parasitica MU* **ectostroma** and **endostroma**
Cryphonectria (=*Endothia*) *parasitica RTH* **erumpent** *
Cryptococcus albidus MU **capsule**
Cryptococcus albidus MU **umbonate**
Cryptococcus albidus var. *albidus MU* **argentophilous** *
Cryptococcus albidus var. *albidus MU* **entire**
Cryptococcus curvatus GA **curved** *
Cryptococcus curvatus GA **curved** *
Cryptococcus curvatus SA **yeast** (asporogenous) *
Cryptococcus curvatus SA **yeast** (asporogenous) *
Cryptococcus curvatus SA **ornamented** *
Cryptococcus neoformans RA&RLM **cryptococcosis**
Cryptococcus neoformans RLM **capsule**
Cryptococcus neoformans RLM **cryptococcosis**
Cryptococcus neoformans RLM **cryptococcosis** *
Cunninghamella echinulata MU **echinulate** *
Cunninghamella echinulata MU **sporangiole**
Cunninghamella echinulata MU-d **echinulate**
Curvularia inaequalis CB **phaeospore** *
Curvularia inaequalis RTH **phragmospore**
Curvularia tuberculata MCG **tubercular**
Cyathus intermedius CB&MU **argentate**
Cyathus intermedius CB&MU **nummiform**
Cyathus intermedius CB&MU **tomentose**
Cyathus intermedius JR **peridiole** *
Cyathus striatus MU **cyathiform** *
Cyathus striatus MU-d **funiculus**
Cyathus striatus MU-d **hapteron**
Cyathus striatus MU-d **middle piece**

Cyathus striatus MU-d **purse**
Cyathus striatus MU-d **tunica**
Cyathus striatus RV **splash cup** *
Cyathus striatus RV **striate** *
Cyathus striatus TH **bird's nest fungus**
Cyathus sp. *CB* **cyathiform**
Cyathus sp. *MU* **circumscissile**
Cyathus sp. *MU* **epiphragm**
Cyathus sp. *MU* **epiphragm**
Cyathus sp. *MU* **epiphragm**
Cyathus sp. *MU* **peridiole**
Cyathus sp. *MU* **peridiole** *
Cyathus sp. *MU* **receptacle**
Cyathus sp. *MU* **receptacle** *
Cyathus sp. *MU* **receptacle** *
Cyathus sp. *RV* **nidulant** *
Cylindrocarpon olidum JAS **cylindric**
Cystoderma amianthinum MU-d **cystoderm**
Cystoderma sp. *TRO* **appendiculate** *
Cystolepiota sp. *RV* **floccose** *
Cyttaria deformans MU-d **spermagonium**
Cyttaria hariotii MU **alveolate**
Cyttaria hariotii MU **cavernose**
Cyttaria hariotii MU **foveolate** *
Dacrymyces capitatus (=*D. ellisii*) *MU-d* **furcate**
Dacrymyces capitatus (=*D. ellisii*) *MU-d* **lituate**
Dacrymyces deliquescens TRO **jelly fungus** *
Dacrymyces palmatus RV **jelly fungus** *
Dacryopinax elegans MU **cochleariform**
Dacryopinax elegans MU **jelly fungus**
Dacryopinax elegans TH **cochleariform**
Daedalea sp. *RV* **labyrinthiform** *
Daedaleopsis confragosa RV **labyrinthiform**
Daldinia concentrica MU **concentric**
Daldinia concentrica MU **concentric** *
Debaryomyces (=*Schwanniomyces*) *occidentalis MU-d* **saturn-like**
Dendrodochium sp. *MU-d* **allantospore**
Dendrodochium sp. *MU-d* **inaequilateral** *
Dendrodochium, state of *Nectria magnusiana MU-d* **phialide**
Dendrostilbella sp. *MU* **synnema** *
Dendrostilbella sp. towards *Phymatotrichopsis omnivora MU* **antagonism** *
Dennisiella babingtonii MU-d **seta**
Diachea bulbillosa MU **capillitium** *
Diachea leucopoda MU **gregarious** *
Diachea leucopoda MU **iridescent** *
Diachea leucopoda MU **Myxomycetes** *
Diachea leucopoda MU **columella**
Diachea splendens RS **nacreous** *
Diachea splendens RS **iridescent** *
Diaporthe eres MU-d **immersed**
Diatractium cordianum RTH **isthmospore**
Diatrype disciformis CB **diatrypoid**
Diatrype disciformis CB **diatrypoid** *
Diatrype disciformis CB **diatrypoid** *
Diatrype disciformis CB **erumpent**
Diatrype disciformis MU **perithecial stroma**
Diatrype virescens MU-d **placodium**
Dichomitus mexicanus (=*Megasporoporia mexicana*) *CB* **hyphal peg**
Dichomitus mexicanus (=*Megasporoporia mexicana*) *CB* **hyphal peg**
Dichomitus mexicanus (=*Megasporoporia mexicana*) *MU* **lumen**
Dichomitus mexicanus (=*Megasporoporia mexicana*) *MU* **porous**
Dichomitus mexicanus (=*Megasporoporia mexicana*) *RV* **resupinate**
Dicranophora fulva MU-d **dichotomous**
Dictyonema pavonium (=*Cora pavonia*) *TRO* **hymenolichen** *
Dictyonema pavonium (=*Cora pavonia*) *TRO* **hymenolichen** *
Dictyonema pavonium (=*Cora pavonia*) *MU* **conchate**
Dictyophora duplicata MU-d **fenestrate**
Dictyophora duplicata TH **fenestrate** *
Dictyosporium elegans CC **cheirospore**
Dictyostelium discoideum MU-d **centripetal**
Dictyostelium discoideum MU-d **filopod**
Dictyostelium discoideum RTH **amoeboid**
Dictyostelium discoideum RTH **caudate** *
Dictyostelium discoideum RTH **myxamoeba**
Dictyostelium discoideum RTH **obpyriform**
Dictyostelium discoideum RTH **pseudoplasmodium**
Dictyostelium discoideum RTH **pseudoplasmodium** *
Dictyostelium mucoroides EAA **reticulate**
Dictyostelium mucoroides MU **pedicellar**
Dictyostelium mucoroides MU **primordium**
Dictyostelium mucoroides MU **slime mold**
Dictyostelium mucoroides MU **sorocarp** *
Dictyostelium mucoroides MU **sorocarp** *
Dictyostelium mucoroides MU **sorocarp** *
Dictyostelium mucoroides MU **sorogen** *
Dictyostelium mucoroides MU **sorophore**
Dictyostelium mucoroides MU **sorus** *
Dictyostelium mucoroides MU **sorus** *
Dictyostelium mucoroides MU **spore** *
Dictyostelium mucoroides MU-d **sorocarp**
Dictyostelium mucoroides MU-d **sorogen**
Dictyuchus monosporus MU-d **dictyosporangium**
Dictyuchus monosporus MU-d **net** *
Dictyuchus monosporus MU-d **sporangium** *
Dicyma chartarum MU-d **cyme**
Diderma floriflorme MU **stellate** *

Diderma floriforme MU **dehiscence**
Diderma hemisphaericum MU **discoid**
Diderma hemisphaericum MU **Myxomycota** *
Diderma hemisphaericum MU **testaceous**
Diderma radiatum MU **angular**
Diderma radiatum MU **Myxomycetes** *
Diderma radiatum MU **petaloid**
Diderma radiatum MU **stellate**
Didymium (=*Diderma*) *difforme MU* **testaceous**
Didymium minus MU **calcareous** *
Didymium perforatum RS **plasmodiocarp** *
Didymium sp. *MU* **calcareous** *
Didymium sp. *MU* **hypothallus** *
Didymium sp. *MU* **hypothallus** *
Didymium sp. *MU* **Myxomycetes** *
Didymostilbe sp. *CB* **gleoid head**
Dimargaris cristalligena PL & EAA **sporogenous**
Diorchidium spinulosum MU-d **diorchidioid**
Dipodascopsis uninucleata MU **ascus**
Discina ancilis RV **sessile** *
Discina perlata RV **carnose** *
Disciseda candida MU-d **exoperidium**
Disciseda candida MU-d **mesoperidium**
Disciseda candida MU-d **sand-case**
Dispira simplex MU-d **dispirous**
Dothidea sp. *RTH* **locule**
Dothiora (=*Dothidella*) *ribesia MU-d* **immersed**
Drechmeria (=*Meria*) *coniospora MU-d* **obclavate**
Drechslera avenae CB **distoseptate**
Drechslera avenae CB **distoseptate**
Drechslera iridis MU-d **bipolar**
Drechslera (asexual state of *Cochliobolus bicolor*) *MU-d* **porospore**
Drechslera (asexual state of *Cochliobolus bicolor*) *MU-d* **geniculate**
Drechslera (asexual state of *Cochliobolus bicolor*) *MU-d* **muticate**
Drechslera sp. *CB* **muticate**
Drechslera sp. *CB* **phaeospore**
Drechslera sp. *JAS* **geniculate**
Eccrinidus flexilis MU-d **eccrinid**
Eccrinidus flexilis MU-d **holdfast**
Echinochaete (=*Dendrochaete*) *russiceps MU-d* **dendrophysis**
Echinostelium minutum MU-d **myxozoospore**
Echinostelium minutum MU-d **protoplasmodium**
Elaeomyxa cerifera MU **capillitium** *
Elaeomyxa cerifera MU **iridescent** *
Elaeomyxa cerifera MU **Myxomycetes** *
Elsinoë ampelina MU-d **multilocular**
Emericella nidulans MU-d **ascus**
Emericella nidulans RTH **cleistothecium**
Emericella nidulans RTH **Hülle cell**
Emericella nidulans RTH **Hülle cell**
Emericella variecolor CB **cleistothecium**
Emericella variecolor MU **stellate** *
Emericella variecolor MU-d **aliform**
Emericellopsis terricola MU&CB **conglobate**
Emmonsiella capsulata MU-d **fasciculate**
Endochytrium ramosum MU-d **endobiotic** *
Endocochlus asteroides MU-d **endozoic**
Endogone lactiflua MU-d **zygosporocarp**
Endogone lactiflua MU-d **zygosporocarp**
Endogone sp. *MU* **zygospore**
Endogone sp. *MU-d* **sporocarp**
Endogone sp. *SP* **endomycorrhiza**
Endomyces (=*Saccharomycopsis*) *fibuliger MU-d* **galeiform**
Endothiella gyrosa MU **pycnididal stroma**
Endothiella gyrosa MU **pycnidiospore**
Enterobryus sp. *MU-d* **endocommensal**
Entoleuca (=*Hypoxylon*) *mammatum EPS* **mammiform**
Entoloma (=*Rhodophyllus*) *abortivum TH* **abortive**
Entoloma corvinum (=*Leptonia corvina*) *MU-d* **trichoderm**
Entoloma formosum (=*Leptonia fulva*) *MU-d* **obtuse**
Entoloma lividocyanulum (=*Leptonia lividocyanula*) *MU-d* **obtuse**
Entoloma lividocyanulum (=*Leptonia lividocyanula*) *MU-d* **pileocystidium**
Entoloma sericatum MU-d **obtuse**
Entoloma sinuatum (=*E. lividum*) *MU-d* **incurved**
Entoloma versatile (=*Pouzarella versatilis*) *MU-d* **pleurocystidium**
Entomophthora muscae MU **entomophilous**
Entomophthora muscae MU **palisade**
Entomosporium maculatum RTH **appendage**
Epichloë typhina RTH **embedded**
Epichloë typhina RTH **embedded**
Epichloë typhina RTH **filiform**
Epicoccum nigrum (=*E. purpurascens*) *CB* **muriform**
Epidermophyton floccosum MU-d **macroconidium**
Epidermophyton floccosum RLM **tinea pedis**
Epidermophyton floccosum RLM **tinea pedis**
Erysiphe (=*Uncinula*) *macrospora RTH* **hamate**
Erysiphe (section *Microsphaera*) *MU* **dichotomous**
Erysiphe (section *Microsphaera*) *MU* **dichotomous**
Eurotium chevalieri PL **bivalve**
Eurotium chevalieri PL **cleistothecium**
Eurotium repens MU **subglobose** *
Eurotium rubrum MU **cleistothecium**
Eurotium rubrum MU **cleistothecium**
Eurotium rubrum MU **rubrum** *
Eurotium rubrum MU **rubrum** *
Eurotium rubrum MU **teleomorph** *

Eurotium rubrum MU-d **bivalve**
Eurotium rubrum MU-d **cleistothecium**
Eurotium rubrum MU-d **cleistothecium**
Eurotium rubrum MU-d **teleomorph**
Eurotium rubrum MU-d **trochleariform**
Eurotium rubrum MU-d **uncinulum**
Eurotium rubrum RTH **cleistothecium**
Eurotium sp. *MU* **storage fungus** *
Eurotium sp. *RTH* **cleistothecium**
Euryancale saccospora MU-d **brachiate**
Eutypa armeniacae MU-d **eutypoid**
Exidia recisa BL **jelly fungus** *
Exidia recisa RTH **lyophilized** *
Exidia recisa RTH **protobasidium** *
Exidia recisa RTH **protobasidium** *
Exobasidium azaleae RTH **spongy** *
Exobasidium sp. *RV* **monocarpic** *
Exobasidium sp. *RV* **rhagadiose** *
Exophiala jeanselmei MU-d **annellate**
Favolus tenuiculus RV **foveolate** *
Flagellospora curvula MU-d **obsubulate**
Flaviporus hydrophilus RV **limbate** *
Floccularia albolanaripes TRO **peronate** *
Fomitiporia hartigii RV **polypore** *
Fomitiporia hartigii RV **perennial** *
Fomitiporia sp. *TRO* **perennial** *
Fomitopsis pinicola (=*Fomes marginatum*) *MU* **marginate**
Fomitopsis pinicola RV **resinous** *
Fuligo septica EPS **aethalium** *
Fuligo sp. *MU* **slime mold** *
Fusarium moniliforme MU **moniliform** *
Fusarium moniliforme MU-d **macroconidium**
Fusarium moniliforme MU-d **microconidium**
Fusarium oxysporum MU **chlamydospore**
Fusarium solani MU-d **foot cell**
Fusarium subglutinans (=*F. moniliforme* var. *subglutinans*) *MU-d* **polyphialide**
Fusarium subglutinans (=*F. moniliforme* var. *subglutinans*) *PL* **macroconidium** *
Fusarium sp. *AB* **keratitis** *
Fusarium sp. *AB* **keratitis** *
Fusarium sp. *CB* **falcate**
Fusarium sp. *MU* **chlamydospore**
Fusarium sp. *MU* **gleospore** *
Fusarium sp. *MU* **macroconidium** *
Fusarium sp. *MU* **naviculate** *
Fusarium sp. *MU-d* **naviculate**
Galactomyces (=*Endomyces*) *geotrichum RTH* **arthrospore**
Galerina hypnorum (=*G. calyptrospora*) *MU-d* **calyptra**
Galerina phillipsii MU-d **suprahilar depression**
Galerina phillipsii MU-d **hilar appendix**
Ganoderma applanatum TRO **mycologist** *
Ganoderma curtisii RV **laccate** *
Ganoderma curtisii RV **lateral** *
Ganoderma lucidum MU **laccate**
Ganoderma resinaceum RV **rhodosporous** *
Ganoderma sp. *RV* **laccate** *
Gastrosporium sp. *MU-d* **aulaeate development**
Geastrum fornicatum MU-d **fornicate**
Geastrum pectinatum MU-d **apophysis**
Geastrum pectinatum MU-d **pedicel**
Geastrum pectinatum MU-d **peristome**
Geastrum pectinatum MU-d **sulcate**
Geastrum saccatum MRO **stellate**
Geastrum saccatum MU-d **mesoperidium**
Geastrum saccatum MU-d **revolute ray**
Geastrum saccatum RV **saccate** *
Geastrum saccatum TH **endoperidium**
Geastrum saccatum TH **exoperidium** *
Geastrum triplex MU **peristome** *
Geastrum triplex MU-d **endoperidium**
Geastrum triplex MU-d **saccate ray**
Geastrum sp. *RV* **earth star** *
Geastrum sp. *RV* **earth star** *
Geastrum sp. *RV* **peristome** *
Geastrum sp. *RV* **peristome** *
Geastrum sp. *SM* **fossil fungus**
Gelasinospora retispora RTH **pitted** *
Gelasinospora sp. *MU* **foveolate**
Geoglossum difforme MU-d **spathulate**
Geoglossum difforme RTH **vermiform**
Geoglossum nigritum RV **earth tongue** *
Geoglossum sp. *RV* **earth tongue** *
Geoglossum sp. *RV* **earth tongue** *
Geopora (=*Sepultaria*) *arenicola MU* **orculiform**
Geopora (=*Sepultaria*) *arenicola MU* **psammophilous** *
Geotrichum candidum MU **abjunction, abscission** *
Geotrichum candidum MU **arthrospore** *
Geotrichum candidum MU **holoarthric** *
Geotrichum candidum MU **yeast-like**
Geotrichum candidum MU-d **holoarthric**
Geotrichum candidum PL **blastospore** *
Geotrichum candidum var. *citri-aurantii MU* **thallic** *
Geotrichum candidum var. *citri-aurantii MU* **thallic** *
Geotrichum candidum var. *citri-aurantii MU* **thallic** *
Geotrichum candidum var. *citri-aurantii RTH* **thallic** *
Gigaspora sp. *CB* **pedicellate**
Gilbertella persicaria MU **zygosporangium** *
Gilbertella persicaria MU **zygosporangium** *
Gliocladium virens (=*Trichoderma virens*) *MU* **mycofungicide**
Gloeophyllum mexicanum RV **tubaeform** *
Gloeophyllum saepiarium (=*Lenzites saepiaria*) *MU*

dimidiate
Gloeophyllum saepiarium (=*Lenzites saepiaria*) *MU* **ligneous ***
Gloeoporus dichrous RV **limbate ***
Glomerella sp. *RTH* **ascus**
Glomerella sp. *RTH* **rot**
Glomerella sp. *RTH* **rot**
Glomus diaphanum MU-d **arbuscule**
Gnomonia comari RTH **ostiolar neck**
Gonatobotryum apiculatum MU-d **polyblastic**
Gongronella butleri MU-d **apophysis**
Graphiola phoenicis MU-d **orculiform**
Graphiola phoenicis SA **orculiform**
Graphis scripta MU **lirella**
Graphis sp. *MU* **lirella**
Graphium ulmi MU **synnema**
Graphium ulmi MU **zonate ***
Graphium sp. *CB* **gleoid head ***
Graphium sp. *MU* **synnema**
Guanomyces polythrix MCG **paraphysis**
Guanomyces polythrix MCG **tubular**
Guanomyces polythrix MU **coprophilous ***
Guanomyces polythrix MU **coprophilous ***
Guanomyces polythrix MU **coprophilous ***
Guanomyces polythrix MU **ostiolar neck**
Guanomyces polythrix MU **ostiolar neck ***
Guanomyces polythrix MU **perithecium**
Guanomyces polythrix MU **rostrum**
Guanomyces polythrix MU **rostrum**
Guanomyces polythrix MU **trich- ***
Guanomyces polythrix RTH **exudate**
Guanomyces polythrix RTH **exudate**
Guepinia helvelloides MRO **infundibuliform**
Guepinia helvelloides MRO **jelly fungus ***
Gyalectidium catenulatum MAH **hyphophore**
Gyalectidium eskuchii MAH **hyphophore**
Gymnascella citrina MU **spheroidal asci ***
Gymnoascus reessii MU-d **gymnothecium**
Gymnopilus sp. *LG* **basidium ***
Gymnosporangium clavipes CB **aeciospore**
Gymnosporangium clavipes MU **mycocecidium**
Gymnosporangium clavipes MU **mycocecidium**
Gymnosporangium clavipes MU **roestelioid**
Gymnosporangium juniperi-virginianae RTH **microcyclic conidiation ***
Gyromitra esculenta TRO **convolute ***
Gyromitra esculenta MU-d **gyrose ***
Gyromitra esculenta RV **lacunar ***
Haematomma subpuniceum MU **crustaceous**
Haematomma subpuniceum MU **tessellate**
Haematonectria (=*Nectria*) *haematococca RTH* **apical**
Haematonectria (=*Nectria*) *haematococca RTH* **asciferous**
Haematonectria (=*Nectria*) *haematococca RTH* **ascogenous**
Haematonectria (=*Nectria*) *haematococca RTH* **ascus mother cell**
Haematonectria (=*Nectria*) *haematococca RTH* **mother cell**
Haematonectria (=*Nectria*) *haematococca RTH* **peridium**
Haematonectria (=*Nectria*) *haematococca RTH* **striate**
Haematonectria (=*Nectria*) *haematococca RTH* **subhymenium ***
Halosarpheia heteroguttulata ERG & MCG **asymmetric ***
Halosphaeria salina MCG **guttulate**
Hanseniaspora valbyensis MU-d **dehiscence**
Haptoglossa mirabilis MU-d **trichocyst**
Harpagomyces lomnickii MU-d **calcariform**
Harposporium anguillulae MU **thanatochresis ***
Harposporium anguillulae PL **chlamydospore**
Harposporium anguillulae PL **endoparasite**
Harposporium anguillulae PL **endoparasite**
Harposporium sp. *MU-d* **falcate**
Helicoma sp. *CC* **helicospore**
Helminthosporium sp. *CB* **distoseptate**
Helminthosporium sp. *MU* **sympodial**
Helvella (=*Paxina*) *acetabulum RV* **carnose ***
Helvella (=*Paxina*) *acetabulum RV* **sessile ***
Helvella crispa CB **lacunar**
Helvella crispa MRO **lacunar**
Helvella elastica TH **elastic**
Helvella lacunosa MRO **stipe ***
Helvella robusta TRO **plaited ***
Hemileia vastatrix GC **rust ***
Hemileia vastatrix GC **rust ***
Hemileia vastatrix GC **uredinium ***
Hemitrichia calyculata (=*H. stipitata*) *CB* **calyculus**
Hemitrichia calyculata (=*H. stipitata*) *MU* **capillitium**
Hemitrichia calyculata (=*H. stipitata*) *MU* **capillitium**
Hemitrichia calyculata (=*H. stipitata*) *MU* **trich- ***
Hemitrichia calyculata (=*H. stipitata*) *MU-d* **asperate**
Hemitrichia calyculata (=*H. stipitata*) *RS* **stipitate ***
Hemitrichia serpula MU **elater**
Hemitrichia serpula MU **plasmodiocarp**
Hemitrichia serpula RV **macroplasmodium ***
Hemitrichia serpula RV **plasmodiocarp ***
Hericium erinaceus RV **dentate ***
Hericium erinaceus RV **odontoid ***
Heterobasidion annosum RTH **dolipore septum**
Hexagonia hydnoides RV **hispid ***
Hexagonia tenuis RV **bracket fungus ***
Hexagonia tenuis CB **faveolate**
Hexagonia tenuis CB **polypore**
Hexagonia tenuis RV **hexagonal ***

Hexagonia tenuis RV **lignicolous** *
Hexagonia tenuis RV **pileate-sessile** *
Hexagonia tenuis RV **porous** *
Hexagonia tenuis RV **zonate**
Hexagonia variegata (=*H. papyracea*) *EAA* **marginate** *
Hexagonia variegata (=*H. papyracea*) *MU* **faveolate**
Hexagonia variegata (=*H. papyracea*) *RV* **zonate** *
Hexagonia variegata (=*H. papyracea*) *RV* **zonate** *
Hexagonia sp. *MU* **porous**
Hexagonia sp. *MU* **strigose**
Hirsutella saussurei MU **clavate**
Hirsutella saussurei MU **spiculospore**
Hirsutella saussurei MU **spiculospore**
Hirsutella saussurei MU **spiculospore**
Hirsutella saussurei MU **spiculospore**
Hirsutella saussurei MU & *CB* **entomogenous**
Hirsutella saussurei MU & *CB* **entomogenous**
Hirsutella thompsonii LS **entomopathogenic** *
Histoplasma capsulatum CB **digitate**
Histoplasma capsulatum CB **nimbospore**
Histoplasma capsulatum MU **dimorphic**
Histoplasma capsulatum MU **dimorphic**
Histoplasma capsulatum MU **histoplasmosis** *
Histoplasma capsulatum MU **histoplasmosis** *
Histoplasma capsulatum MU **histoplasmosis** *
Histoplasma capsulatum MU **tubercular** *
Histoplasma capsulatum MU-d **tubercular**
Histoplasma capsulatum RLM **dimorphic**
Histoplasma capsulatum RLM **dimorphic** *
Hohenbuehelia niger MU-d **lamprocystidium**
Hortaea (=*Exophiala*) *werneckii RLM* **tinea nigra**
Hortaea (=*Exophiala*) *werneckii RLM* **tinea nigra** *
Humicola alopallonella MCG **humicole**
Humicola fuscoatra MU **humicole** *
Humicola fuscoatra MU **humicole** *
Hyaloscypha hyalina CB **ascocarp**
Hyaloscypha hyalina CB **laciniate**
Hydnochaete olivacea RV **irpiciform** *
Hygrocybe coccinea TRO **laminal hymenophore** *
Hygrocybe conica TRO **lubricous** *
Hygrocybe miniata RV **viscid** *
Hygrocybe persistens (=*H. acutoconica*) *RV* **parallel**
Hygrocybe psittacina (=*Hygrophorus psittacinus*) *MU-d* **hygrophanous**
Hygrophorus conicus RV **viscid** *
Hygrophorus fornicatus MU-d **recurvate**
Hygrophorus olivaceoalbus RV **ixotrichoderm** *
Hygrophorus russula RV **laminal hymenophore** *
Hygrophorus sp. *MU-d* **ixotrichoderm**
Hygrophorus sp. *RV* **decurrent** *
Hymenogaster sulcatus MU-d **glebal chamber**
Hymenogaster thwaitesii MU-d **utriculate**
Hymenoscyphus (=*Helotium*) *herbarum MU-d* **distichous**
Hyphoderma praetermissum MU-d **stephanocyst**
Hypholoma (=*Naematoloma*) *fasciculare TRO* **fasciculate** *
Hypholoma (=*Naematoloma*) *fasciculare TH* **gregarious** *
Hypholoma marginatum TRO **gregarious** *
Hypholoma marginatum TRO **heliophobic** *
Hypholoma marginatum TRO **mycenoid** *
Hyphomyces destruens MU-d **coenocytic** *
Hypocrea schweinitzii RTH **centrum** *
Hypocrea schweinitzii RTH **perithecial stroma**
Hypomyces lactifluorum on fruiting bodies of *Russula* sp. *RV* **mycoparasite** *
Hypomyces lactifluorum on fruiting body of *Russula* sp. *MU* **mycoparasite**
Hypomyces lactifluorum on *Russula delica RV* **edible** *
Hypomyces lactifluorum RTH **subiculum**
Hypomyces lactifluorum RTH **subiculum**
Hypomyces roseus RV **mycoparasite** *
Hypomyces sp. *CM* **uncinulum**
Hypotrachyna (=*Parmelia*) *imbricatula MU* **gonidium**
Hypotrachyna (=*Parmelia*) *imbricatula MU-d* **gonidium** *
Hypotrachyna (=*Parmelia*) *imbricatula MU-d* **foliole**
Hypotrachyna (=*Parmelia*) *imbricatula MU-d* **heteromerous**
Hypotrachyna (=*Parmelia*) *imbricatula MU-d* **isidium**
Hypotrachyna (=*Parmelia*) *imbricatula MU-d* **isidium**
Hypoxylon rubiginosum var. *rubiginosum RV* **rubiginose** *
Hypoxylon sp. *CB* **germ slit**
Hypoxylon sp. *RV* **stroma** *
Hysterangium sp. *MU-d* **collar**
Hysterographium flexuosum CB **dictyosporic** *
Inocutis texana RV **persistent** *
Inocybe geophylla RV **eucarpic** *
Inocybe rimosa (=*I. fastigiata*) *CB* **rimose** *
Inocybe rimosa (=*I. fastigiata*) *EPS* **rimose**
Inocybe rimosa (=*I. fastigiata*) *MU* **fastigiate**
Inocybe sp. *MU-d* **goniospore**
Inocybe sp. *RV* **umbonate** *
Inonotus dryophilus RV **granular core** *
Inonotus dryophilus RV **suberose** *
Inonotus quercustris RV **quercicole** *
Irpex lacteus MU **irpiciform**
Irpex sp. *MU* **lacerate**
Isaria fumosorosea (=*Paecilomyces fumosoroseus*) *MU* **entomopathogenic** *
Isaria sp. *TH* **isarioid**
Isthmospora spinosa MU-d **isthmospore**
Itersonilia perplexans MU-d **abjection**
Itersonilia perplexans MU-d **ballistospore**

Itersonilia perplexans MU-d **sterigma**
Kickxella alabastrina MU-d **pseudophialide**
Kickxella alabastrina MU-d **umbellate**
Kloeckera apiculata MU **yeast** (apiculate)
Kluyveromyces marxianus var. *marxianus MU* **lobulate ***
Kluyveromyces marxianus var. *marxianus PL* **reniform**
Kobayasia nipponica MU-d **radiating plate**
Laboulbenia elongata RTH **ectoparasitic**
Laboulbenia elongata RTH **ectoparasitic ***
Laboulbenia elongata RTH **receptacle**
Laboulbenia elongata RTH **receptacle ***
Laboulbenia flagellata EPS **receptacle**
Laboulbenia flagellata MU-d **trichophore** and **trichogyne**
Laboulbenia formicarum GVD **ectoparasitic**
Laboulbenia formicarum GVD **ectoparasitic**
Laboulbenia formicarum GVD **ectoparasitic**
Labyrinthula marina MU-d **filoplasmodium**
Laccaria amethystina RV **caespitose ***
Laccaria amethystina RV **ceraceous ***
Laccaria amethystina RV **umbilicate ***
Laccaria proxima MU **adnate**
Lacrymaria (=Psathyrella) lacrymabunda MU-d **lacrimiferous**
Lactarius fuliginosus MU-d **holobasidium**
Lactarius indigo RV **edible ***
Lactarius piperatus MU-d **incurved**
Lactarius scrobiculatus MU **scrobiculate**
Lactarius volemus VBM **latex**
Lactarius zonarius MU **praemorse**
Lactarius sp. *LV* **mantle ***
Laetiporus sulphureus MU **pore**
Laetiporus sulphureus RV **xylophagous ***
Lagenulopsis bispora MU-d **lageniform**
Lamproderma columbinum MU **iridescent**
Lamproderma columbinum MU **iridescent ***
Lamproderma columbinum MU **pedicellate ***
Lamproderma columbinum TH **Myxomycota ***
Lasallia pustulata TRO **papillate ***
Lasallia sp. *TRO* **lichen ***
Lasiodiplodia theobromae CB **didymospore**
Lasiodiplodia theobromae MCG **didymospore**
Laternea pusilla JC **gleba ***
Laternea pusilla MU-d **glebifer**
Laternea pusilla MU-d **lantern**
Laternea pusilla MU-d **trabecula**
Lecanicillium (=Verticillium) lecanii CB **gleoid head**
Lecanicillium (=Verticillium) lecanii CT **entomopathogenic ***
Lecanicillium (=Verticillium) lecanii CT **entomopathogenic ***
Lecanicillium (=Verticillium) lecanii CT **mycoinsecticide ***
Lecanicillium (=Verticillium) lecanii MU **bioinsecticide ***
Lecanicillium (=Verticillium) lecanii MU **bioinsecticide ***
Lecanicillium (=Verticillium) lecanii MU **entomogenous**
Lecanicillium (=Verticillium) lecanii MU **entomopathogenic ***
Lecanicillium (=Verticillium) lecanii MU **mycoinsecticide**
Lecanicillium (=Verticillium) lecanii MU **obverse ***
Lecanicillium (=Verticillium) lecanii MU **reverse ***
Lecanicillium (=Verticillium) lecanii CB **verticillate**
Lecanicillium (=Verticillium) lecanii MU **verticillate ***
Lecanora frustulosa MU **epilithic**
Lecanora frustulosa MU-d **lecanorine**
Lecanora subfusca MU **lecanorine**
Leccinum aurantiacum MU **scabrose**
Leccinum aurantiacum RV **boletoid**
Leccinum rugosiceps RV **rivulose ***
Leccinum sp. *RV* **rugose ***
Lecidea atrobrunnea MU **lecideine ***
Lentinellus sp. MU-d **serrate**
Lentinellus sp. RV **serrate ***
Lentinus (=Panus) badius CB **infundibuliform**
Lentinus (=Panus) badius MU **decurrent**
Lentinus (=Panus) badius MU **strigose**
Lentinus (=Panus) badius MU **strigose**
Lentinus (=Panus) badius RV **velutinous ***
Lentinus crinitus MU **umbilicate ***
Lentinus strigosus RV **hispid ***
Lentinus strigosus RV **pannose ***
Lentinus tigrinus RV **fasciculate ***
Lentinus velutinus MU **hirsute**
Lentinus velutinus MU **hirsute**
Leocarpus fragilis CB **cartilagineus**
Leocarpus fragilis CB **hypothallus**
Leocarpus fragilis MU & CB **gregarious**
Leocarpus fragilis MU & CB **obovate**
Leotia atrovirens RV **capitate ***
Leotia lubrica RTH **umbilicate**
Leotia lubrica RV **lubricous ***
Lepidoderma tigrinum MU **Myxomycetes ***
Lepidoderma tigrinum MU **scaly**
Lepiota cuneatospora MU-d **cuneiform**
Lepiota sp. *MU* **lepiotoid**
Lepiota sp. *MU* **pileus**
Lepiota sp. *MU-d* **bromatium**
Lepiota sp. *MU-d* **repand**
Lepiota sp. *MU-d* **stenospore**
Lepiota sp. *RV* **lepidote ***

Lepiota sp. *RV* **pileus** *
Leproloma membranaceum (=*Lepraria membranacea*) *MU* **leprose**
Leptogium corticola MU **gelatinous**
Leptonia decolorans MU-d **obtuse**
Leptonia sp. *MU-d* **bulliform**
Leptosphaeria sp. *RTH* **fusiform**
Leptosphaerulina arachidicola MU **nasse**
Leptosphaerulina crassiasca MU-d **dome**
Leptosphaerulina crassiasca RTH **bitunicate**
Leptosphaerulina crassiasca RTH **bitunicate**
Leptosphaerulina crassiasca RTH **capsule**
Leptosphaerulina crassiasca RTH **dictyosporic**
Leptosphaerulina crassiasca RTH **fissitunicate**
Leptosphaerulina crassiasca RTH **germ tube**
Leptosphaerulina crassiasca RTH **nasse** *
Letharia vulpina MU **soredium** *
Letharia vulpina f. *californica MU* **perforate**
Letharia vulpina f. *californica MU* **pitted**
Letharia vulpina f. *californica MU* **ligulate**
Leucocoprinus birnbaumii HT **lignicolous** *
Leucocoprinus cepistipes RV **scaly** *
Leucocoprinus cepistipes SA **bulbous**
Linderina pennispora EAA **septal plug**
Linderina pennispora MU **pennate** *
Linderina pennispora MU **lanose**
Linderina pennispora MU **sporocladium** *
Linderina pennispora MU **sporocladium** *
Linderina pennispora MU-d **pennate**
Linderina pennispora MU-d **sporocladium**
Linderia sp. *JC* **rhizomorph** *
Lindra thalassiae MCG **vermiform**
Lobaria scrobiculata MU **lobate**
Lobaria scrobiculata MU **lobate**
Lobaria scrobiculata MU **scrobiculate**
Lopadostroma sp. *MU* **manglicolous**
Lophodermella sulcigena MU-d **epiplasm**
Lophodermium juniperinum MU-d **aciculate**
Lophodermium sp. *RTH* **dehiscence**
Lophodermium sp. *RTH* **hygroscopic** *
Lophodermium sp. *RTH* **longitudinal** *
Lophotrichus brevirostratum MU **brevirostrate** *
Lycogala epidendrum MU **acaulescent**
Lycogala epidendrum MU **aethalium**
Lycogala epidendrum MU **aethalium**
Lycogala epidendrum RV **Myxomycetes** *
Lycogala epidendrum MU **pseudocapillitium**
Lycogala flavofuscum RV **aethalium** *
Lycoperdon candidum ACV **capillitium**
Lycoperdon candidum MU-d **pyramidal**
Lycoperdon echinatum EPS **echinate**
Lycoperdon echinatum MU **pyramidal**
Lycoperdon perlatum ACV **capillitium**
Lycoperdon perlatum MU-d **gleba**
Lycoperdon perlatum MU-d **pseudocolumella**
Lycoperdon perlatum MU-d **pseudostipe**
Lycoperdon perlatum MU-d **subgleba**
Lycoperdon purpurascens RV **puffball** *
Lycoperdon pyriforme RV **gregarious**
Lycoperdon stellare MU **columella**
Lycoperdon sp. *MU* **basidiospore** *
Lycoperdon sp. *MU-d* **connivent**
Lycoperdon sp. *MU-d* **forate development**
Lycoperdon sp. *MU-d* **lycoperdoid**
Lycoperdon sp. *MU-d* **rhizomorph**
Lycoperdon sp. *RV* **rhizomorph**
Lycoperdon sp. *TH* **puffball** *
Lysurus periphragmoides EAA **receptacle** *
Macrocybe titans AA & *RA* **macro-** *
Macrocystidia sp. *MU* **macrocystidium**
Macrolepiota mastoidea RV **lepidote** *
Macrolepiota mastoidea RV **ring** *
Macrolepiota mastoidea RV **umbonate** *
Macrolepiota procera EPS **lamella**
Macrolepiota procera MU **adpressed**
Macrolepiota procera MU **lamella**
Macrolepiota procera MU **scaly**
Macrolepiota procera RV **scaly** *
Macrolepiota sp. *ACV* **conic**
Madurella mycetomi RLM **grain**
Madurella mycetomi RLM **mycetoma**
Malassezia (=*Pityrosporum*) *furfur RLM* **keratolytic** *
Malassezia (=*Pityrosporum*) *furfur RLM* **tinea versicolor** *
Malassezia (=*Pityrosporum*) sp. *MU-d* **monopolar**
Malbranchea aurantiaca MU **keratolytic**
Malbranchea aurantiaca MU **keratolytic**
Malbranchea aurantiaca MU **keratolytic** *
Malbranchea aurantiaca MU **keratolytic** *
Malbranchea aurantiaca MU **keratolytic** *
Malbranchea aurantiaca MU **laciniate**
Malbranchea aurantiaca MU **lanose** *
Malbranchea aurantiaca MU **primary hypha**
Malbranchea aurantiaca MU **racket cell** *
Malbranchea aurantiaca MU **racket hypha**
Malbranchea aurantiaca MU **reticulate** *
Malbranchea aurantiaca MU **thallic** *
Malbranchea aurantiaca MU **thallic** *
Malbranchea aurantiaca MU **zonate** *
Marasmius ferruginosus TH **elastic** *
Marasmius haematocephalus TH **marcescent** *
Marasmius plicatilis MU-d **broom cell**
Marasmius siccus MU & *CB* **marcescent**
Marasmius siccus MU-d **stipe**
Marasmius sp. *EPS* **cheilocystidium**
Marasmius sp. *RV* **elastic** *

Marasmius sp. *EPS* **lanceolate**
Marasmius sp. *RV* **lineate** *
Marasmius sp. *RV* **mycelial strand** *
Marasmius sp. *RV* **striate** *
Marasmius sp. *RV* **undate** *
Marasmius sp. *RV* **xylophagous** *
Martensella corticii MU-d **sporocladium** *
Masophalea richardsonia (=*Cetraria richardsonii*) *MU* **alutaceous**
Massospora cicadina MU-d **germination**
Megachytrium westonii MU-d **polycentric**
Megasporoporia setulosa RV **hyphal peg** *
Melampsora lini MU-d **haustorial mother cell**
Melampsora lini MU-d **sheath**
Melanoleuca sp. *MU-d* **muricate**
Melanospora zamiae MU **rhombic** *
Melanospora zamiae MU **translucent** *
Melanospora zamiae RTH **fimbriate**
Melanospora zamiae RTH **fusiform**
Melanospora zamiae RTH **tubular**
Melanospora sp. *MU* **diaphanous** *
Melanospora sp. *RTH* **biapiculate** *
Melanospora sp. *RTH* **limoniform**
Meliola caesariae MU-d **hyphopodium**
Meliola caesariae MU-d **stigmatocyst**
Meliola caesariae MU-d **stigmatopod**
Meliola floridensis RTH **biotroph**
Meliola floridensis RTH **periphysis**
Meliola sp. *RTH* **hyphopodium**
Memnoniella sp. *MU* **memnonious** *
Metarhizium anisopliae CT **mycoinsecticide**
Metarhizium anisopliae CT **palisade** *
Metarhizium anisopliae MU **bioinsecticide** *
Metarhizium anisopliae MU **bioinsecticide** *
Metarrhizium anisopliae MU **column**
Metarrhizium anisopliae MU & *CB* **column**
Metarhizium anisopliae MU **dendroid**
Metarrhizium anisopliae MU **spore** *
Metarhizium anisopliae MU-d **blastic**
Metarhizium anisopliae var. *acridum CGG* & *CC* **molecular biology** *
Metarhizium anisopliae var. *acridum CT* **entomopathogenic** *
Metarhizium anisopliae var. *anisopliae MU* **entomogenous** *
Metarhizium anisopliae var. *anisopliae CT* **entomopathogenic** *
Metarhizium anisopliae var. *anisopliae CT* **entomopathogenic** *
Metarhizium anisopliae var. *anisopliae MU* **entomopathogenic** *
Metarhizium anisopliae var. *anisopliae MU* **entomopathogenic** *
Metatrichia (=*Trichia*) *floriformis MU* **capillitium** *
Metatrichia (=*Trichia*) *floriformis MU* **trich-** *
Metatrichia (=*Trichia*) *floriformis MU* **pyriform**
Metatrichia (=*Trichia*) *floriformis MU* **pyriform** *
Metatrichia (=*Trichia*) *floriformis MU* **pyriform** *
Metatrichia vesparium MU **coalescent** *
Metatrichia vesparium MU **pseudoaethalium**
Metatrichia vesparium MU **pseudoaethalium** *
Metschnikowia bicuspidata MU-d **cuspidate**
Metschnikowia pulcherrima PL **yeast** (ascosporogenous) *
Microascus trigonosporus CB **angular**
Microascus trigonosporus CB **cirrhus**
Microascus trigonosporus CB **cirrhus**
Microascus trigonosporus MU **perithecium** *
Microsporum audouinii MU-d **pectinate**
Microsporum canis MU **dermatophyte**
Microsporum canis MU **ectothrix**
Microsporum canis MU **tinea corporis** *
Microsporum canis RLM **tinea capitis**
Microsporum canis RLM **tinea corporis**
Microsporum cookei RTH **closterospore**
Microsporum cookei RTH **holosporic** *
Microsporum gypseum MU-d **rhexolytic**
Microsporum nanum MU **farinaceous**
Microsporum nanum MU **floccose** *
Microsporum nanum MU **floccose** *
Microsporum nanum MU **powdery**
Microthyrium microscopicum MU-d **scutellum**
Microthyrium microscopicum MU-d **thyriothecium**
Minimedusa polyspora CB **bulbil**
Mitrula sp. *MU-d* **mitriform**
Mixia osmundae MU-d **spore case**
Mixia osmundae MU-d **synascus**
Monascus purpureus MU **ang-kah**
Monascus purpureus MU **ang-kah**
Monilia fructicola MU **acropetal**
Monilia fructicola MU **biodeterioration** *
Monilia fructicola MU **dendroid** *
Monilinia fructicola MU-d **fructicole** *
Monilinia fructicola MU-d **pseudosclerotium**
Monilia sitophila MU **moniliform**
Monilia sitophila MU **moniliform** *
Monoblepharella taylorii MU-d **exigynous**
Monoblepharella sp. *MU-d* **rumposome**
Monoblepharis polymorpha MU-d **antheridium**
Monoblepharis polymorpha MU-d **exogynous**
Monoblepharis polymorpha MU-d **monoclinous** *
Monoblepharis polymorpha MU-d **uniflagellate**
Monoblepharis polymorpha MU-d **vacuolate**
Monochaetia carissae MU-d **filiform** *
Monochaetia carissae MU-d **setula** *
Montagnea arenaria MU-d **gusset**

Montagnea arenaria MU-d **indurate** *
Montagnea arenaria TH **xerophilic**
Morchella esculenta (=*M. rotunda*) *TH* **rotund**
Morchella esculenta MU **foveolate**
Morchella esculenta SA **edible** *
Morchella esculenta SA **morel** *
Morchella sp. *RV* **morel** *
Morchella spp. *MRO* **faveolate**
Mortierella epigama MU **stylospore**
Mortierella epigama MU **stylospore**
Mortierella indohii MU-d **stylospore**
Mucilago crustacea MU **calcareous**
Mucilago crustacea MU **crustaceous**
Mucor bainieri MU-d **azygosporangium**
Mucor hiemalis CB **myriosporous**
Mucor hiemalis CB **sporangium**
Mucor hiemalis MU **heterothallism**
Mucor hiemalis MU **sexual** *
Mucor hiemalis RTH **suspensor**
Mucor hiemalis RTH **zygosporangium**
Mucor racemosus MU **chlamydospore**
Mucor racemosus MU-d **pseudostolon**
Mucor racemosus MU-d **racemose**
Mucor spinosus MU **aculeate**
Mucor sp. *MU* **collarette**
Mucor sp. *MU* **lax** *
Mucor sp. *MU* **moldy** *
Mucor sp. *MU* **pliosporous** *
Munkiella caa-guazu MU-d **pseudothecium**
Mutinus caninus MU-d **phalloid** *
Mutinus caninus RV **foetid** *
Mutinus caninus RV **monopileate** *
Mutinus caninus RV **phalloid** *
Muyocopron (=*Ellisiodothis*) *smilacis MU-d*
pseudoparaphysis
Mycena citricolor MU **stilboid**
Mycena citricolor RTH **stilboid**
Mycena epipterygia RV **mycenoid** *
Mycena lux-coeli MU-d **noctilucent** *
Mycena lux-coeli MU-d **luminescent** *
Mycena pura MU **lamella**
Mycena sp. *MU-d* **mycenoid**
Mycogone perniciosa MU-d **nodulose**
Mycosphaerella musicola MU **pseudothecium** *
Mycosphaerella musicola MU **sigatoka** *
Mycosphaerella musicola MU-d **unilocular**
Mycotypha africana MU-d **vesicle**
Myriangium duriaei RTH **ascostroma**
Myriangium duriaei RTH **interthecial** *
Myriangium duriaei RTH **locule**
Myriangium duriaei RTH **multilocular**
Myriangium duriaei RTH **uniascal**
Myriangium sp. *MU-d* **apothecioid**
Myriostoma coliforme CB **column**
Myriostoma coliforme MU-d **column**
Myriostoma coliforme MU-d **myriostomes**
Myrmecocystis spinospora MU-d **myrmecophilous**
Myrothecium verrucaria MU **lenticular**
Myrothecium verrucaria MU **membranaceous**
Myrothecium verrucaria MU **pionnote**
Myrothecium verrucaria MU **pionnote**
Myrothecium verrucaria MU **pionnote** *
Mytilinidion decipiens MU-d **mytiliform**
Myxotrichum sp. *MU* **biodeterioration** *
Myzocytium zoophthorum MU-d **endozoic**
Nadsonia fulvescens MU-d **bipolar**
Nadsonia fulvescens MU-d **gemmation**
Nadsonia fulvescens RTH **pedogamy**
Nadsonia fulvescens RTH **pedogamy**
Nadsonia fulvescens RTH **yeast** (apiculate) *
Nannizzia (=*Arthroderma*) *incurvata MU-d*
phalangoid
Naucoria sp. *MU-d* **naucorioid**
Nectria sp. *RTH* **canker**
Nemania (=*Hypoxylon*) *serpens MU-d* **asymmetric**
Nemania (=*Hypoxylon*) *serpens MU-d* **germ slit**
Nematospora coryli MU-d **piliform**
Nematospora coryli RTH **piliform** *
Nematospora (=*Ashbya*) *gossypii RTH* **intercalary** *
Nematostelium ovatum MU-d **sporocarp**
Nematostelium ovatum MU-d **steliogen**
Neosartorya fischeri MU-d **holomorph**
Neottiella (=*Octospora*) *rutilans MU-d*
synaptonemal complex
Neurospora intermedia MU **ontjom**
Neurospora sitophila MU-d **protoperithecium**
Neurospora sitophila MU-d **receptive hypha**
Neurospora sitophila MU-d **spermatization**
Nidula niveotomentosa MU-d **tomentose**
Nigrospora oryzae MU **phytopathogen** *
Nigrospora sphaerica RTH **nigricant** *
Nigrospora sp. *MU* **aleuriospore**
Nitschkia cupularis MU-d **subiculum**
Nowakowskiella ramosa MU-d **rhizomycelium**
Nucleophaga amoebae MU-d **endobiotic**
Ochrolechia parella MU **crustaceous**
Ochrolechia parella MU **parathecium**
Octomyxa achlyae inside a hypha of *Achlya* sp. *MU-d*
latent or **resting spore** *
Octosporomyces (=*Schizosaccharomyces*) *octosporus MU*
ascus
Octosporomyces (=*Schizosaccharomyces*) *octosporus*
RTH **gametangial copulation**
Octosporomyces (=*Schizosaccharomyces*) *octosporus*
RTH **schizogenesis**
Oedocephalum sp. *MU* **amerospore**

Oedocephalum sp. *MU* **ampulla**
Oedocephalum sp. *MU* **anastomosis**
Oedocephalum sp. *MU* **botryose**
Oedocephalum sp. *MU* **conidiophore** *
Oidiodendron tennuisimum towards *Trichophyton tonsurans MU* **antibiosis** *
Oidiodendron tenuissimum MU **biscoctiform**
Oidiodendron tenuissimum MU **fascicle**
Oidiodendron sp. *MU-d* **disjunctor**
Oidium monilioides WKW **meristem arthrospore**
Oidium monilioides WKW **mildew**
Oidium sp. *MU* **meristem arthrospore**
Oidium monilioides MU **meristem arthrospore**
Olpidiopsis saprolegniae in a hypha of *Saprolegnia* sp. *MU-d* **mycoparasite**
Olpidium endogenum MU-d **exit tube**
Omphalina (=*Omphalia*) sp. *MU-d* **omphaloid**
Omphalotus mexicanus EPS **decurrent**
Omphalotus olearius MU-d **omphalinoid**
Omphalotus olearius RV **decurrent** *
Omphalotus olearius RV **luteous** *
Omphalotus olearius RV **mycetism** *
Omphalotus olearius RV **toxigenic** *
Onygena corvine MU-d **mazaedium**
Ophiodothella vaccinii RTH **clypeus**
Ophiodothella vaccinii RTH **clypeus**
Ophiodothella vaccinii RTH **filiform**
Ophiodothella vaccinii RTH **lesion** *
Ophiodothella vaccinii RTH **vermiform**
Orbimyces spectabilis MU **staurospore**
Ostracoderma (=*Chromelosporium*) *ollare MU-d* **ampulla**
Otidea leporina MRO **auriculiform** *
Otidea onotica EPS **auriculiform** *
Otidea onotica MU-d **auriculiform**
Oudemansiella canarii RV **hygrophanous**
Oudemansiella canarii RV **lamellula**
Oudemansiella canarii TH **fasciculate**
Oudemansiella canarii RV **pellucid**
Oudemansiella canarii MU-d **incised**
Ovulariopsis sp. *MU-d* **rhombic** *
Ovulariopsis sp. (asexual state of *Phyllactinia*) *MU-d* **verrucose** *
Ovulariopsis sp. *RTH* **biotrophic**
Pachykytospora papyracea RV **resupinate** *
Paecilomyces variotii MU **thermophilic**
Paecilomyces sp. *MU* **catenulate** *
Panaeolina foenisecii MU-d **punctate**
Panaeolus campanulatus MU-d **campanulate**
Panaeolus campanulatus MU-d **papilionaceous**
Panaeolus campanulatus RV **campanulate** *
Panaeolus cyanescens EAA **mucronate**
Panaeolus cyanescens MU **metuloid**
Panaeolus semiovatus RV **coprophilous** *
Panaeolus semiovatus RV **proteophilous** *
Panellus stipticus RV **laminal hymenophore** *
Panellus sp. *RV* **flabellate** *
Panus crinitus MU **mycophagy**
Papulaspora sp. *MU* **microsclerotium**
Papulaspora sp. *MU* **papulospore**
Paracoccidioides brasiliensis CT **paracoccidioidomycosis**
Parasitella parasitica MU-d **sporangiospore**
Parmelia flaventior MU **soralium**
Parmelia sulcata MU-d **soralium**
Parmelia sp. *MU* **foliole**
Parmelia sp. *TH* **petricolous**
Parmotrema sp. *MU* **soredium**
Parmularia hankei MU-d **dimidiate**
Patella albida MU-d **patelliform**
Peltigera canina MU **rhizina**
Peltigera sp. *MU* **rhizina**
Peltigera sp. *TRO* **lichen** *
Peltigera sp. *TRO* **polyphyllous** *
Penicillium arenicola MU-d **polyverticillate**
Penicillium arenicola MU-d **ramus** *
Penicillium aurantiogriseum PL **metula**
Penicillium aurantiogriseum MU **aeroallergen** *
Penicillium aurantiogriseum MU **exudate**
Penicillium aurantiogriseum MU **plaited** *
Penicillium aurantiogriseum MU **radial** *
Penicillium aurantiogriseum TH **exudate** *
Penicillium aurantiogriseum var. *viridicatum CB* **quadriverticillate**
Penicillium camemberti MU **cottony**
Penicillium chrysogenum (=*P. notatum*) *MU* **notate**
Penicillium chrysogenum (=*P. notatum*) *CB* **medium** *
Penicillium chrysogenum (=*P. notatum*) *CB* **triverticillate**
Penicillium chrysogenum (=*P. notatum*) *MU* **biotransformation** *
Penicillium chrysogenum (=*P. notatum*) *MU* **velutinous**
Penicillium citreonigrum CB **monoverticillate**
Penicillium citrinum MU **diffusible** *
Penicillium citrinum MU-d **ampulliform**
Penicillium islandicum MU **zonate**
Penicillium italicum MU blue **mold**
Penicillium purpurogenum MU **diffusible** *
Penicillium purpurogenum MU **pigment** *
Penicillium purpurogenum MU-d **acerose**
Penicillium roqueforti MU **blue (bleu) cheese** *
Penicillium roqueforti MU **blue (bleu) cheese** *
Penicillium roqueforti PL **verrucose**
Penicillium solitum var. *crustosum MU-d* **collulum**
Penicillium vulpinum (=*P. claviforme*) *MU & CB*

synnema *
Penicillium vulpinum (=*P. claviforme*) *MU* **clavate** *
Penicillium vulpinum (=*P. claviforme*) *MU* **coremiform**
Penicillium vulpinum (=*P. claviforme*) *MU* **coremium**
Penicillium vulpinum (=*P. claviforme*) *MU* & *CB* **coremium**
Penicillium vulpinum (=*P. claviforme*) *TH* **coremiform**
Penicillium waksmanii PL **divaricate**
Penicillium sp. *CM* **biverticillate**
Penicillium sp. *MU* **catenulate** *
Penicillium sp. *MU* **moldy**
Penicillium sp. *MU* **moldy** *
Penicillium sp. *MU* **phialide** *
Penicillium sp. *MU* **terverticillate** *
Penicillium sp. *RTH* **penicillate**
Perichaena sp. *MU* **gregarious**
Perichaena sp. MU **sessile**
Periconia byssoides CB **pedicel**
Periconia byssoides CB **conidiophore** *
Peridermium sp. *MU-d* **peridermium**
Pertusaria wulfenioides MU **pertusate**
Pestalotia sp. *MU* **guttulate**
Pestalotia sp. MU **setula**
Pestalotia sp. *RTH* **exudate** *
Pestalotiopsis sp. *CB* **pionnote**
Petriella sp. *CB* **fabiform**
Petriella sp. *CB* **inaequilateral** *
Petriella sp. *CB* **reniform**
Petriella sp. *CB* **reniform**
Peziza ammophila RTH **amyloid** *
Peziza ammophila RTH **oblong**
Peziza ammophila RTH **operculum**
Peziza hemisphaerica RV **apothecium** *
Peziza hemisphaerica RV **cup-shaped fungus** *
Peziza hemisphaerica RV **cup-shaped fungus** *
Peziza quelepidotia KO **primordium**
Peziza sp. *MU* **paraphysis**
Peziza sp. *RTH* **epithecium**
Peziza sp. *RTH* **hymenium** *
Peziza sp. *RTH* **hypothecium**
Phacidium lacerum MU-d **hypophyllous**
Phaeocollybia christinae GL **cucullate**
Phallus hadriani RV **foetid** *
Phallus hadriani RV **monopileate** *
Phallus hadriani RV **nidose** *
Phallus hadriani TH **percurrent** *
Phallus hadriani TH **phalloid** *
Phallus impudicus var. *impudicus JLM* **nidose** *
Phallus impudicus var. *impudicus TH* **phalloid**
Phallus impudicus var. *impudicus JLM* **phalloid** *
Phallus indusiatus (=*Dictyophora indusiata*) *TH* **indusium**
Phallus ravenellii TH **receptacle**
Phanerochaete sp. *TRO* **exudate** *
Phanerochaete sp. *RV* **pelliculate** *
Phanerochaete sp. *RV* **resupinate** *
Phanerochaete sp. *TRO* **procumbent** *
Pharcidia laminariicola MU-d **involucellum**
Phellinus contiguus RV **seta** *
Phellinus ferruginosus MU-d **seta**
Phellinus ferruginosus MU-d **seta**
Phellinus linteus RV **ungulate** *
Phellinus linteus RV **ungulate** *
Phellinus linteus RV **ungulate** *
Phellinus linteus RV **ungulate** *
Phellinus linteus RV **chrysosporous** *
Phellinus linteus RV **ligneous** *
Phellinus linteus RV **suberose** *
Phellinus nilgheriensis RV **ligneous** *
Phellinus rimosus RV **igniarious** *
Phellinus rimosus RV **rhagadiose** *
Phellinus rimosus RV **rhodosporous** *
Phellinus sarcites RV **hirsute** *
Phellinus sarcites RV **imbricate** *
Phellinus sarcites RV **strigose** *
Phialophora parasitica MU-d **collarette**
Phialophora (=*Rhinocladiella* =*Fonsecaea*) *pedrosoi AB* **chromomycosis** *
Phialophora (=*Rhinocladiella* =*Fonsecaea*) *pedrosoi ECN* **chromomycosis**
Phialophora (=*Rhinocladiella* =*Fonsecaea*) *pedrosoi MU-d* **agglutinate**
Phialophora (=*Rhinocladiella*=*Fonsecaea*) *pedrosoi MU-d* **heterosporic**
Phialophora (=*Rhinocladiella*=*Fonsecaea*) *pedrosoi MU-d* **fumagoid**
Phillipsia sp. *TH* **discoid**
Phlyctidium anatropum MU-d **algicolous**
Phlyctochytrium sp. *MU-d* **papillate**
Phlyctochytrium sp. *MU-d* **zoosporangium**
Phlyctochytrium sp. *RTH* **apophysis**
Phlyctochytrium sp. *RTH* **entangial**
Phlyctochytrium sp. *RTH* **water mold** *
Pholiota squarrosa MU **squarrose**
Pholiota squarrosa MU-d **squarrose** *
Pholiota squarrosa RV **squarrose** *
Phoma destructiva CB **micropycnidium**
Phoma destructiva MU **imbricate**
Phoma eupyrena MU-d **chlamydospore**
Phoma exigua (=*Ph. herbarum*) *MU-d* **meristogenous**
Phoma glomerata CB **dictyochlamydospore**
Phoma sorghina PL **cirrhus**
Phoma sp. *MU* **alternarioid**
Phoma sp. *MU* **cirrhus** *

Phoma sp. *MU* **ostiole**
Phoma sp. *MU* **pseudoparenchyma** *
Phoma sp. *MU* **pycnidiospore**
Phoma sp. *MU* **pycnidium** *
Phoma sp. *MU-d* **pycnidium**
Phoma sp. *RTH* **cirrhus**
Phomopsis sp. *CB* **filiform**
Phomopsis sp. *RTH* **alpha-conidium**
Phomopsis sp. *RTH* **beta-conidium**
Phragmidium potentillae CB **pedicellate**
Phycomyces blakesleanus MU **diffluent**
Phycomyces blakesleanus MU **diffluent**
Phycomyces blakesleanus MU **diffluent** *
Phycomyces blakesleanus MU **sporangiophore** *
Phycomyces blakesleanus MU **nigricant***
Phycomyces blakesleanus MU **stercoreaceous***
Phycomyces nitens MU **appendage**
Phycomyces nitens MU **corniform**
Phycomyces nitens MU **gametangial copulation** *
Phycomyces nitens MU **sporangiophore**
Phycomyces nitens MU **zygosporangium** *
Phycomyces nitens MU **zygotropism** *
Phycomyces nitens MU-d **forcipate**
Phylacia poculiformis MU **biconic**
Phylacia sagraeana CB **carbonaceous**
Phylacia turbinata MU-d **turbinate**
Phylacia sp. *CB* **perithecial stroma**
Phyllachora cibotii MU-d **clypeus**
Phyllachora fusicarpa RTH **lesion** *
Phyllachora fusicarpa RTH **spermatium**
Phyllactinia guttata (=*Ph. corylea*) *MU* **inflated** *
Phyllactinia guttata (=*Ph. corylea*) *MU-d* **coronate**
Phyllophiale alba MAH **isidium**
Phylloporia fruticum RV **amplectant** *
Phylloporia fruticum RV **duplex** *
Phylloporia fruticum RV **ramicole** *
Phylloporia sp. *RV* **duplex** *
Phymatotrichopsis omnivora JAS **morbose** *
Phymatotrichopsis omnivora JAS **mycelial crosspiece** *
Phymatotrichopsis omnivora JAS **mycelial strand** *
Phymatotrichopsis omnivora JAS **phytopathology** *
Phymatotrichopsis omnivora JAS **retroculture** *
Phymatotrichopsis omnivora JAS **sclerotium** *
Physarum cinereum MU **cinereous**
Physarum cinereum MU **gregarious**
Physarum cinereum MU **Myxomycetes** *
Physarum decipiens MU **sessile** *
Physarum flavicomum EPS **peltate** *
Physarum flavicomum EPS **peltate** *
Physarum flavicomum MU-d **peltate**
Physarum mutabile MU-d **pseudocolumella**
Physarum polycephalum RV **phaneroplasmodium** *
Physarum polycephalum RV **rheotaxis** *
Physarum polycephalum MU **spore** *
Physarum polycephalum MU **spore** *
Physarum polycephalum MU **subhypothallic**
Physarum polycephalum MU **subhypothallic**
Physarum polycephalum MU **dendroid**
Physarum polycephalum MU **fructification** *
Physarum polycephalum MU **fructification** *
Physarum polycephalum MU **gibbose**
Physarum polycephalum MU **gibbose** *
Physarum polycephalum MU **gibbose** *
Physarum polycephalum MU **hypothallus**
Physarum polycephalum MU **myxamoeba**
Physarum polycephalum MU **myxamoeba**
Physarum polycephalum MU **Myxomycetes** *
Physarum polycephalum MU **pedicellate** *
Physarum polycephalum MU **phaneroplasmodium**
Physarum polycephalum MU **phaneroplasmodium**
Physarum polycephalum MU **phaneroplasmodium**
Physarum polycephalum MU **plasmodial** *
Physarum polycephalum MU **plasmodium** *
Physarum polycephalum MU **plasmodium** *
Physarum polycephalum MU **polycephalic** *
Physarum polycephalum MU **presporangial papilla** *
Physarum polycephalum MU **presporangial papilla** *
Physarum polycephalum MU **slime mold** *
Physarum polycephalum MU **slime mold** *
Physarum polycephalum MU **slime mold** *
Physarum polycephalum MU **vein**
Physarum polycephalum MU-d **caducous** *
Physarum polycephalum MU-d **gyrose** *
Physarum polycephalum MU-d **sporangium** *
Physarum polycephalum RV **macroplasmodium** *
Physarum roseum RS **phaneroplasmodium** *
Physarum roseum RS **syncytium** *
Physarum serpula MU **Myxomycetes** *
Physarum serpula MU **plasmodiocarp** *
Physarum sp. *MU* **calcareous**
Physarum sp. *MU* **calcareous** *
Physarum sp. *MU-d* **dehiscence** *
Physcia sp. *RTH* **excipulum**
Physoderma (=*Urophlyctis*) *alfalfae MU-d* **turbinate**
Phytophthora capsici JG **amphigynous**
Phytophthora infestans MU-d **amphigynous**
Phytophthora infestans MU-d **germ sporangium**
Phytophthora infestans MU-d **limoniform**
Pichia membranifaciens MU **intercalary** *
Pichia membranifaciens MU **mother of vinegar** *
Pichia membranifaciens MU **rugose**
Pichia membranifaciens MU **tepache** *
Pichia membranifaciens MU **yeast** (ascosporogenous) *
Piedraia hortae MU-d **piedra**
Pilobolus crystallinus var. *kleinii* (=*P. kleinii*) *MU* **stercoreaceous** *

Protostelium mycophaga MU-d **sporocarp** *
Prototrichia metallica MU **iridescent**
Psathyrella hirsutosquamulosa MU-d **cordate**
Pseudevernia sp. *MU* **fruticose**
Pseudevernia sp. *MU* **fruticose**
Pseudevernia sp. *MU* **isidium**
Pseudochaete (=Hymenochaete) tabacina RV **resupinate** *
Pseudocolus fusiformis MU-d **triquetrous arm**
Pseudofistulina (=Fistulina) radicata MU-d **acanthophysis**
Pseudogymnoascus roseus CB **gymnothecium**
Pseudogymnoascus roseus CB **gymnothecium**
Pseudohydnum gelatinosum RV **dentate** *
Pseudohydnum gelatinosum RV **gelatinous***
Pseudohydnum gelatinosum RV **gelatinous***
Pseudohydnum gelatinosum RV **odontoid** *
Pseudohydnum gelatinosum RV **tremelloid** *
Pseudohydnum gelatinosum RV **tremelloid** *
Pseudoperonospora (=Peronoplasmopara) canabina MU-d **citriform**
Pseudopeziza medicaginis RTH **hypothecium** *
Psilocybe barrerae RV **mammiform** *
Psilocybe caerulescens MU-d **mammiform**
Psilocybe caerulescens MU-d **papillate**
Psilocybe caerulescens MU-d **umbonate**
Psilocybe caerulescens TH **caerulescent**
Psilocybe caerulescens TH **hallucinogenic fungi**
Psilocybe caerulescens TH **magic mushroom**
Psilocybe coprophila MU **coprophilous**
Psilocybe coprophila MU **coprophilous** *
Psilocybe cubensis EPS **hallucinogenic fungus** *
Psilocybe cubensis EPS **hallucinogenic fungus** *
Psilocybe cubensis TH **hallucinogenic fungus** *
Psilocybe cubensis TH **hallucinogenic fungus** *
Psilocybe mexicana TH **entheogen**
Psilocybe mexicana TH **magic mushroom**
Psilocybe pelliculosa MU-d **limbate**
Psilocybe sp. *TH* **caerulescent**
Psora cerebriformis CB **cerebriform**
Psora crenata MU **centrifugal**
Psora crenata MU **coriaceous**
Psora decipiens MU **lecideine** *
Psora russellii CB **squamulose**
Pterula sp. *MU-d* **caulocystidium**
Puccinia graminis tritici GGA **biconic**
Puccinia graminis tritici GGA **uredinium**
Puccinia graminis tritici MU **aecium**
Puccinia graminis tritici MU **aecium**
Puccinia graminis tritici MU **cupulate**
Puccinia graminis tritici MU **pedicellate** *
Puccinia graminis tritici MU **rust** *
Puccinia graminis tritici MU **telium**
Puccinia graminis tritici MU **uredinium**
Puccinia graminis tritici MU **uredinium** *
Puccinia graminis tritici MU & CB **pustule**
Puccinia graminis tritici MU-d **aecium**
Puccinia graminis tritici MU-d **echinulate**
Puccinia graminis tritici MU-d **pedicellate**
Puccinia graminis tritici MU-d **teliospore**
Puccinia graminis tritici MU-d **uredinium**
Puccinia malvacearum MU **pustulate**
Puccinia malvacearum JB **pustule** *
Puccinia malvacearum MU **rust**
Puccinia malvacearum MD **rust** *
Puccinia malvacearum JB **rust** *
Puccinia malvacearum MU **uredinium** *
Puccinia sorghi MU-d **disjunctor** *
Puccinia sp. *CB* **urediniospore**
Puccinia sp. *RV* **teliospore** *
Punctelia (=Parmelia) subrudecta MU **foliose**
Punctelia (=Parmelia) subrudecta MU **soredium**
Pycnoporus sanguineus MU **lignicolous**
Pyrenophora sp. *CB* **soleiform**
Pyrenophora sp. *CB* **dictyosporic**
Pyrenophora sp. *CB* **distoseptate** *
Pyrenophora sp. *CB* **euseptate**
Pyrenophora sp. *CB* **perithecioid**
Pyrenula nitida MU **nitid**
Pyrenula nitida MU **pyrenocarpic**
Pyrofomes demidoffii RV **bracket fungus** *
Pyronema domesticum RTH **ellipsoidal**
Pyronema omphalodes MU-d **ascogonium**
Pythium echinulatum MU-d **aplerotic**
Pythium monospermum MU-d **plerotic**
Pythium ultimum MU-d **dictyosome**
Pythium ultimum MU-d **paragynous**
Radiomyces embreei MU-d **radial** *
Radiomyces embreei MU **sporangiole** *
Ramacrinella raibautii MU-d **endocommensal**
Ramalina homalea MU **fasciate**
Ramalina homalea MU **fasciate**
Ramaria flava RV **coralloid** *
Ramaria gracilis MV **skeletal hypha**
Ramaria grandis MV **cyanophilic** *
Ramaria stricta TH **coralloid**
Ramaria sp. *TRO* **coralloid** *
Ramaria sp. *RV* **multiaxial** *
Ramaria sp. *RV* **multiaxial** *
Ramaria sp. *TRO* **multiaxial** *
Ramaria sp. *TRO* **ramarioid** *
Ramaria sp. *RV* **ramose** *
Ramaria sp. *TRO* **ramulus** *
Ramaria sp. *TH* **multiaxial**
Ramaricium polyporoideum MV **basidiospore** *
Ramariopsis corniculata MV **basidiospore** *

Ramariopsis rubripermanens MV **basidiospore** *
Reticularia sp. *MU* **pseudocapillitium** *
Rhachomyces quetzalcoatl EPS **receptacle**
Rhinosporidium seeberi AB **rhinosporidiosis** *
Rhinosporidium seeberi AB **rhinosporidiosis** *
Rhinosporidium seeberi MU-d **spherule**
Rhipidium americanum MU-d **umbellate**
Rhizidiomyces apophysatus MU-d **flimmer hairs**
Rhizidiomyces apophysatus MU-d **mastigoneme**
Rhizidiomyces apophysatus MU-d **vesicle** *
Rhizidiomyces apophysatus MU-d **water mold** *
Rhizidiomyces apophysatus MU-d **zoosporiferous** *
Rhizocarpon geographicum MU **saxicolous**
Rhizocarpon geographicum MU **tessellate**
Rhizoctonia solani JAS **graminicolous** *
Rhizoctonia solani JAS **inhibition** *
Rhizoctonia sp. *CB* **constricted** *
Rhizoctonia sp. CB **septate**
Rhizomucor (=*Mucor*) *miehei MU* **gametangial copulation**
Rhizomucor (=*Mucor*) *miehei MU* **zygosporangium** *
Rhizomucor (=*Mucor*) *miehei MU* **isogametangium**
Rhizophlyctis rosea MU-d **centriole**
Rhizophlyctis rosea MU-d **flagellum**
Rhizophlyctis rosea MU-d **kinetosome**
Rhizophlyctis rosea MU-d **rhizoplast**
Rhizophydium pollinis-pini RTH **water mold** *
Rhizophydium sphaerotheca MU-d **interbiotic**
Rhizopogon luteolus LV **mycorrhiza**
Rhizopus microsporus (=*Rh. oligosporus*) *MU* **tempe**
Rhizopus microsporus var. *oligosporus* (=*Rh. oligosporus*) *MU-d* **apophysis**
Rhizopus microsporus var. *oligosporus* (=*Rh. oligosporus*) *MU-d* **sporangium**
Rhizopus microsporus var. *oligosporus* (=*Rh. oligosporus*) *MU-d* **oligosporous**
Rhizopus oryzae (=*Rh. arrhizus*) *MU* **spinulose**
Rhizopus oryzae (=*Rh. arrhizus*) *MU-d* **columella**
Rhizopus oryzae (=*Rh. arrhizus*) *PL* **apophysis**
Rhizopus oryzae (=*Rh. arrhizus*) *AB* **mucormycosis** *
Rhizopus oryzae (=*Rh. arrhizus*) *AB* **mucormycosis** *
Rhizopus oryzae (=*Rh. arrhizus*) *RLM & EM* **mucormycosis**
Rhizopus sexualis MU-d **pit connection**
Rhizopus sexualis MU-d **plasmodesma**
Rhizopus stolonifer (=*Rh. nigricans*) *MU* **coenocytic**
Rhizopus stolonifer (=*Rh. nigricans*) *MU* **columella**
Rhizopus stolonifer (=*Rh. nigricans*) *MU* **sporangiospore**
Rhizopus stolonifer (=*Rh. nigricans*) *MU* **stolon** *
Rhizopus stolonifer (=*Rh. nigricans*) *MU* **stolon** *
Rhizopus stolonifer (=*Rh. nigricans*) *MU-d* **germ sporangium**
Rhizopus stolonifer (=*Rh. nigricans*) *MU-d* **stolon**
Rhizopus stolonifer (=*Rh. nigricans*) *MU-d* **zygosporangium**
Rhizopus stolonifer (=*Rh. nigricans*) *TH* **antagonism**
Rhizopus stolonifer (=*Rh. nigricans*) *LP* **rhizoid**
Rhizopus stolonifer (=*Rh. nigricans*) *MU* black bread **mold** *
Rhizopus stolonifer (=*Rh. nigricans*) *MU* **rhizoid** *
Rhizopus stolonifer (=*Rh. nigricans*) *MU* **rhizoid** *
Rhizopus stolonifer (=*Rh. nigricans*) *MU-d* **progametangium**
Rhodosporidium sphaerocarpum MU-d **promycelium**
Rhodotorula glutinis MU **glutinosous**
Rhodotorula rubra MU **butyraceous** *
Rhodotorula rubra MU **lobulate**
Rhodotorula rubra MU **ornamented**
Rhodotorula rubra MU **sinuate** *
Rhodotorula rubra MU **yeast** (asporogenous) *
Rhynchosporium secalis MU-d **rhynchospore**
Rhytidhysterium rufulum CB **striate**
Rhytidhysterium rufulum RTH **discothecium** *
Rhytisma acerinum RTH **stroma**
Rhytisma acerinum RTH **tar spot**
Rhytisma acerinum RTH **tar spot** *
Rickia apiculifera MU **ectoparasitic** *
Rickia apiculifera MU **ectoparasitic** *
Rickia apiculifera MU **ectoparasitic** *
Rickia apiculifera MU **ectoparasitic** *
Rickia apiculifera MU **ectoparasitic** *
Rickia apiculifera MU **entomopathogenic** *
Rickia passalina HL **seticolous**
Rickia passalina MU **holdfast** *
Rickia passalina MU **liguliform**
Rinodina sophodes MU-d **mischoblastiomorphic**
Roccella fimbriata MU **parathecium**
Roccella fimbriata MU **saxicolous** *
Roestelia sp. *MU-d* **cornute**
Rosellinia herpotrichoides RTH **amyloid** *
Rozella allomycis in *Allomyces javanicus RTH* **endobiotic**
Rozella allomycis inside a hypha of *Allomyces javanicus RTH* **latent** or **resting spore**
Russula brevipes MU-d **incurved**
Russula emetica RTH **lyophilized** *
Russula emetica TH **poisonous fungus** *
Russula polyphylla MU-d **gleocystidium**
Russula sp. *EPS* **depressed**
Russula sp. *LV* **mycorrhiza** *
Russula sp. *LV* **symbiotroph** *
Russula sp. *MRO* **viscid**
Russula sp. *MU* **macrocystidium**
Russula sp. *MU* **sphaerocyst**
Russula sp. *MU-d* **heteromerous**

Saccharomyces cerevisiae MU **butyraceous**
Saccharomyces cerevisiae MU **crenate** *
Saccharomyces cerevisiae MU **de novo** *
Saccharomyces cerevisiae MU **lobate**
Saccharomyces cerevisiae MU **tepache** *
Saccharomyces cerevisiae MU **tesgüino** *
Saccharomyces cerevisiae MU **yeast** (baker's)
Saccharomyces cerevisiae MU **yeast** (baker's)
Saccharomyces cerevisiae MU **yeast** (baker's) *
Saccharomyces cerevisiae MU **yeast** (baker's) *
Saccharomyces cerevisiae MU **yeast** (baker's) *
Saccharomyces cerevisiae MU **zymocide**
Saccharomyces cerevisiae MU-d **ascus**
Saccharomyces cerevisiae MU-d **gemmation**
Saccharomyces cerevisiae MU-d **plaque bridge**
Saccharomyces cerevisiae MU-d bud **scar**
Saccharomyces cerevisiae MU-d **tetrad**
Saccharomyces cerevisiae PL **pseudomycelium** *
Saccharomyces cerevisiae RTH **tetrad**
Saccharomyces exiguus MU **diffusible** *
Saccharomyces kluyveri MU-d **pseudomycelium**
Saccharomyces kluyveri PL **vegetative multiplication** *
Saccharomycodes ludwigii MU-d **annellospore**
Saccharomycodes ludwigii MU-d **diplont**
Saccobolus truncatus EAA **exserted**
Saccobolus sp. *KO* **operculate** and **operculum**
Saksenaea vasiformis MU-d **lecythiform**
Saprolegnia ferax MU-d **lomasome**
Saprolegnia hypogyna MU-d **centric**
Saprolegnia parasitica MU **bait** *
Saprolegnia parasitica MU **chitinolytic**
Saprolegnia parasitica MU **cyst**
Saprolegnia parasitica MU **hermaphroditic**
Saprolegnia parasitica MU **saprobe** *
Saprolegnia parasitica MU **water mold** *
Saprolegnia parasitica MU-d **androgynous**
Saprolegnia parasitica MU-d **fertilization tube**
Saprolegnia parasitica MU **gemma** *
Saprolegnia parasitica MU-d **gemma**
Saprolegnia parasitica MU-d **heterokont**
Saprolegnia sp. *MU* **oogonium**
Saprolegnia sp. *MU* **oogonium** *
Saprolegnia sp. *MU-d* **proliferation**
Sarcinella (asexual state of *Schiffnerula pulchra*) *MU-d* **sarciniform**
Sarcodon imbricatus MU **hydnoid**
Sarcodon sp. *MU* **appendiculus**
Sarcodon sp. *MU* **dentate**
Sarcoscypha coccinea RV **discoid** *
Schiffnerula pulchra MU-d **sarciniform**
Schizophyllum commune MU **lignicolous**
Schizophyllum commune MU **skeletal hypha** and **generative hypha**
Schizophyllum commune MU **split lamella** *
Schizophyllum commune MU **xerophile**
Schizophyllum commune RV **flabellate** *
Schizophyllum commune RV **lignicolous** *
Schizophyllum commune RV **radial** *
Schizophyllum commune RV **split lamella** *
Schizophyllum commune MU **xerophilic** *
Schizophyllum commune RV **xerophilic** *
Schizoplasmodium cavostelioides MU-d **sporocarp**
Schizopora flavipora TRO **effuse** *
Schizopora paradoxa RV **porous** *
Schizopora paradoxa RV **resupinate** *
Schizothyrioma (=*Schizothyrium*) *ptarmicae MU-d* **apothecioid**
Sclerocystis sp. *CB* **sporocarp**
Sclerocystis sp. *CB* **sporocarp**
Scleroderma lycoperdoides TH **earth ball**
Scleroderma polyrhizum RTH **earth ball**
Scleroderma verrucosum RV **gleba** *
Scleroderma verrucosum RV **gleba** *
Scleroderma verrucosum TH **gleba**
Scleroderma sp. *RV* **peridium** *
Sclerogaster sp. *MU-d* **apobasidium**
Sclerogaster sp. *MU-d* **orthotropic spore** *
Sclerotinia sclerotiorum MU&CB **roridous**
Sclerotinia sclerotiorum MU&CB **roridous** *
Sclerotinia sclerotiorum RTH **tubaeform**
Sclerotinia sp. MU **sclerotium** *
Sclerotium coffeicola RTH **clamp connection** *
Scopulariopsis brevicaulis CB **basipetal**
Scopulariopsis brevicaulis CB **catenulate**
Scopulariopsis brevicaulis CB **verticillate**
Scopulariopsis brevicaulis MU **annellospore** *
Scopulariopsis brevicaulis MU **tinea pedis** *
Scopulariopsis brevicaulis MU-d **annellide**
Scotomyces subviolaceum (=*Ceratobasidium plumbeum*) *MU-d* **heterobasidium**
Scutellinia scutellata RV **disk fungus** *
Sepedonium sp. on a fruiting body of *Boletus* sp. *RV* **mycoparasite**
Septobasidium burtii MU-d **probasidium**
Septobasidium sp. *MU* **symbiontic** *
Setosphaeria rostrata (=*Exserohilum* =*Helminthosporium rostratum*) *MCG* **hilum**
Setosphaeria rostrata (=*Exserohilum* =*Helminthosporium rostratum*) *MU* **rostrum**
Smittium sp. *RTH* **trichospore**
Solorina saccata MU **pelta**
Sommerstorffia spinosa MU-d **predaceous fungus** *
Sordaria fimicola RTH **ellipsoidal** *
Sordaria fimicola MU **perithecium**
Sordaria fimicola RTH **ascus**
Sordaria fimicola RTH **capsule**

Sordaria fimicola RTH **centrum**
Sordaria fimicola RTH **germ pore**
Sordaria fimicola RTH **lyophilized** *
Sordaria fimicola RTH **monostichous**
Sordaria fimicola RTH **octosporate** *
Sordaria fimicola RTH **ostiole**
Sordaria fimicola RTH **ostiole**
Sordaria fimicola RTH **perithecium** *
Sordaria fimicola RTH **sheath**
Sordaria nodulifera MU-d **nodulose**
Sorosphaera veronicae MU-d **cruciate**
Sorosphaera veronicae MU-d **sorus**
Sparassis crispa RV **flabellate** *
Sparassis sp. *RV* **cartilagineous** *
Sparassis sp. *RV* **papyraceous** *
Spathularia elegans RV **spathulate** *
Spathularia flavida MU-d **spathulate**
Spathularia flavida RV **spathulate** *
Spathularia flavida RV **spathulate** *
Spathulospora phycophila MU-d **spathulate**
Spegazzinia tessarthra MU-d **bassauxic**
Sphaerobolus stellatus MU **spore case**
Sphaerobolus sp. *CL* **eversion**
Sphaeronema sp. *MU* **congregate** *
Sphaeronema sp. *MU* **pycnidium** *
Spilocaea pomi MU **acervulus** *
Spiniger meineckellus RTH **dacryoid**
Spiniger meineckellus RTH **dacryoid**
Spiniger meineckellus RTH **denticle**
Spiniger meineckellus RTH **denticle**
Spongospora subterranea MU **cystosorus** *
Spongospora subterranea MU-d **cystosorus**
Sporendonema purpurascens MU-d **enteroarthric**
Sporisorium (=*Sphacelotheca*) *sorghi MU* **teleutosorus**
Sporopodium xantholeucum MAH **campylidium**
Sporormiella (=*Preussia*) *australis RTH* **bitunicate**
Sporormiella (=*Preussia*) *australis RTH* **ectoascus** and **endoascus**
Sporormiella (=*Preussia*) *australis RTH* **part spore**
Sporormiella (=*Preussia*) *fimetaria CB* **part spore**
Sporothrix schenckii MU **convolute**
Sporothrix schenckii MU **sporotrichosis** *
Sporothrix schenckii MU **sporotrichosis** *
Sporothrix schenckii RLM **fascicle** *
Sporothrix schenckii RLM **sporotrichosis**
Stachybotrys chartarum (=*S. atra*) *CB* **sympodial**
Stachybotrys chartarum (=*S. atra*) *MU* **stachybotryotoxin** *
Stachybotrys chartarum (=*S. atra*) *MU* **germ tube** *
Stachybotrys chartarum (=*S. atra*) *MU* **mycotoxicosis** *
Stachybotrys chartarum (=*S. atra*) *MU* **verrucose**
Stachybotrys chartarum (=*S. atra*) *MU-d* **utriform**
Stachybotrys chartarum (=*S. atra*) *MU* **plectonematogenous**
Stachybotrys cylindrospora MU-d **lineate**
Stachybotrys sp. *MCG* **verrucose**
Stachylina grandispora MU-d **trichospore**
Stemonitis axifera MU **fasciculate**
Stemonitis axifera MU **sporangium** *
Stemonitis axifera MU **stemon-** *
Stemonitis axifera MU-d **aphanoplasmodium**
Stemonitis axifera RV **stemon-** *
Stemonitis axifera RV **stemon-** *
Stemonitis nigrescens MU-d **hypothallus**
Stemonitis smithii MU **caespitose**
Stemonitis sp. *MU* **columella**
Stemonitopsis typhina (=*Comatricha typhoides*) *MU* **arcuate**
Stemonitopsis typhina (=*Comatricha typhoides*) *MU* **sporangium** *
Stemphylium sp. *CB* **porospore**
Stereocaulon ramulosum CB **phyllocladium**
Stereocaulon ramulosum f. *elegans MU* **cephalodium**
Stereocaulon saxatile MU **cephalodium**
Stereocaulon saxatile MU **sorediate**
Stereum sp. *MU-d* **mycangium**
Sticta weigelii MU **cyphella**
Stigmatomyces psilopae MU-d **receptacle** *
Stilbella flavidum RTH **gleoid head**
Stilbella flavidum RTH **gleoid head**
Stipella vigilans MU-d **pileiform**
Stipella vigilans MU-d **zygosporophore**
Strigula orbicularis (=*S. complanata*) *TH* **foliicolous**
Strigula schizospora MAH **pyrenolichen**
Strigula subtilis MAH **foliicolous**
Strobilomyces confusus RTH **vinescent** *
Strobilomyces strobilaceus (=*S. floccopus*) *MU* **hymenophore**
Strobilomyces strobilaceus (=*S. floccopus*) *MU-d* **strobiliform**
Stropharia sp. *MU-d* **chrysocystidium**
Stylopage rhynchospora MU-d **capillary spore**
Suillus sp. *MU* **exudate**
Symbiotaphrina buchnerii (=*Torulopsis buchnerii*) *MU-d* **mycetocyte**
Syncephalastrum racemosum MU **merosporangiophore**
Syncephalastrum racemosum MU **merosporangiophore**
Syncephalastrum racemosum MU **merosporangium**
Synchytrium endobioticum MU-d **prosorus**
Synchytrium endobioticum MU-d **sorus**
Syzygites megalocarpus MU-d **zygosporangium**
Taeniolella sp. *MCG & AC* **lignicolous** *
Taeniolella sp. *MCG & AC* **lignicolous** *

Taeniolella sp. *MCG & AC* **lignicolous** *
Taphridium sp. *MU-d* **spore case**
Taphrina caerulescens RTH **unitunicate**
Taphrina deformans RTH **hyperplasia** and **hypertrophy**
Taphrina deformans RTH **hyperplasia** and **hypertrophy** *
Taphrina deformans RTH **leaf curl**
Taphrina pruni MU **palisade**
Taphrina robinsoniana RTH **denudate**
Tapinella (=*Paxillus*) *panuoides RV* **irpiciform** *
Tapinella (=*Paxillus*) *panuoides RV* **irpiciform** *
Tarzetta (=*Pustularia*) *catinus MU* **pustule**
Teloschistes chrysophthalmus MU-d **isthmus**
Teloschistes chrysophthalmus MU-d **polarilocular**
Teloschistes exilis MU **fruticose**
Terana caerulea (=*Pulcherricium caeruleum*) *RV* **pelliculate** *
Terfezia gigantea MU-d **truffle**
Termitomyces albuminosus MU-d **spherule**
Termitomyces microcarpus MU-d **molar**
Termitomyces striatus MU-d **molar**
Thamnidium elegans MU **sporangiole**
Thamnidium elegans MU **sporangiole** *
Thamnostylum (=*Helicostylum*) *piriforme MU* **pyriform**
Thamnostylum (=*Helicostylum*) *piriforme MU* **pyriform** *
Thamnostylum (=*Helicostylum*) *piriforme MU* **sporangiole** *
Thelomma santessonii CB **areolate**
Thelomma santessonii CB **mammiform**
Thielavia terricola MU **amygdaloid**
Thielavia terricola RTH **ascogenous**
Thielavia terricola RTH **ascus mother cell**
Thielavia terricola RTH **ascus mother cell**
Thielaviopsis basicola (=*Chalara elegans*) *MU* **aleuriospore**
Thielaviopsis basicola (=*Chalara elegans*) *MU* **endoconidium**
Thielaviopsis basicola (=*Chalara elegans*) *MU* **endoconidium**
Thielaviopsis basicola (=*Chalara elegans*) *MU-d* **endoconidium**
Thraustochytrium globosum MU-d **water mold** *
Tilletia caries MU-d **sporidium**
Torula herbarum MU **torulose**
Torula herbarum MU **torulose**
Torula sp. *MU* **torulose**
Torulopsis taboadae MU **multipolar** *
Torulopsis taboadae MU-d **blastospore**
Torulopsis taboadae MU-d **multipolar**
Trametes cervina RV **imbricate** *
Trametes incerta RV **labyrinthiform** *
Trametes maxima RV **bracket fungus** *
Trametes maxima RV **radicicolous** *
Trametes maxima RV **radicicolous** *
Trametes maxima RV **radicicolous** *
Trametes maxima RV **radicicolous** *
Trametes (=*Polyporus*) *versicolor MU-d* **imbricate**
Trametes (=*Polyporus*) *versicolor MU* **variegated**
Trametes (=*Polyporus*) *versicolor MU* **zonate**
Tremella fuciformis RV **jelly fungus** *
Tremella fuciformis TH **gelatinous**
Tremella fuciformis TH **jelly fungus**
Tremella mesenterica RV **gelatinous** *
Tremella mesenterica RV **jelly fungus** *
Tremella mesenterica RV **tremelloid** *
Tremella mesenterica TH **jelly fungus** *
Tremella sp. *MU-d* **phragmobasidium**
Tremella sp. *TH* **convolute**
Trichaptum abietinum RV **ramicole** *
Trichaptum perroteti RV **velutinous** *
Trichia botrytis MU **botryose**
Trichia decipiens MU **lignicole**
Trichia decipiens MU **sporangium**
Trichia scabra MU **Myxomycota** *
Trichoderma viride MU **mycofungicide**
Trichoderma viride MU **photosporogenic**
Trichoderma viride RTH **penicillate**
Trichoglossum velutipes VBM **paraphysis**
Trichoglossum sp. *RTH* **paraphysis**
Trichoglossum sp. *RTH* **paraphysis**
Tricholoma sp. *JC* **emarginate**
Tricholoma sp. *JC* **emarginate**
Tricholoma sp. *MU-d* **tricholomatoid**
Trichophyton ajelloi MU-d **fusiform**
Trichophyton mentagrophytes AB **tinea barbae** *
Trichophyton mentagrophytes MU **endothrix**
Trichophyton mentagrophytes MU **keratolytic**
Trichophyton mentagrophytes MU **lipolytic**
Trichophyton mentagrophytes MU **microconidium** *
Trichophyton mentagrophytes MU-d **racket cell**
Trichophyton mentagrophytes RLM **tinea barbae** *
Trichophyton mentagrophytes RLM **tinea unguium**
Trichophyton tonsurans RLM **endothrix**
Trichosporiella cerebriformis CB **pleurogenous**
Trichosporon beigelii MU-d **arthrospore**
Trichosporon cutaneum LMT **piedra**
Trichosporon cutaneum LMT **piedra**
Trichosporon moniliforme MU **halo** *
Trichothecium roseum MU-d **acropleurogenous**
Trichothelium minum MAH **pyrenolichen**
Trichurus spiralis CB **synnema**
Trichurus spiralis RTH **synnema**
Tripospora tripos MU-d **radial**

Trogia sp. *MU-d* **crispate**
Tuber dryophilum HT **reticulate**
Tuber melanosporum EAA **echinate**
Tuber melanosporum EAA **melanosporous** *
Tuber melanosporum EPS **truffle**
Tuber texense RTH **labyrinthiform**
Tuber texense RTH **truffle**
Tubercularia vulgaris MU **sporodochium**
Tubercularia vulgaris RTH **pulvinate** *
Tulostoma sp. *MU & CB* **gyrose**
Tulostoma sp. *MU & CB* **stoma**
Tulostoma sp. *MU & CB* **stoma**
Turgidosculum ulvae MU-d **pulvillus**
Turgidosculum ulvae MU-d **pulvillus**
Tylopilus plumbeoviolaceus RV **velutinous** *
Ulocladium atrum MU **tretic**
Ulocladium atrum MU-d **polytretic**
Umbelopsis ramanniana MU **zygosporangium** *
Umbelopsis vinacea (=*Mortierella ramanniana* var. *angulispora*) *MU-d* **polygonal**
Umbilicaria cylindrica MU-d **gyrodisc**
Umbilicaria vellea MU **rhizina**
Umbilicaria (=*Omphalodiscus*) *virginis MU-d* **omphalodisc**
Umbilicaria sp. *MU-d* **funiculus**
Uncinula circinata CB **unciform** *
Uncinula macrospora RTH **cleistothecioid** *
Uncinula sp. *MU* **circinate** *
Uncinula sp. *RTH* **unciform**
Urnula craterium MU-d **urnula**
Urocystis sp. *MU* **spore ball**
Uromyces appendiculatus MU-d **oscule**
Uromyces phaseoli MU **papillate**
Uromyces viciae-fabae (=*U. fabae*) *MU* **obovate**
Usnea ceratina MU **pendulous**
Usnea sp. *CB & MU* **fibril**
Usnea sp. *MU* **barbate**
Usnea sp. *MU-d* **barbate**
Ustilaginoidea virens MU **smut**
Ustilaginoidea virens MU **smut**
Ustilaginoidea virens MU **smut**
Ustilago maydis EPS **sorus**
Ustilago maydis MU **smut**
Ustilago maydis MU **teleutosorus**
Ustilago maydis MU-d **heterobasidium**
Ustilago maydis MU-d **heterobasidium**
Ustilago maydis MU-d **promycelium**
Utharomyces epallocaulus MU-d **dehiscence**
Valdensia heterodoxa MU-d **cruciate**
Valsa sp. *MU-d* **valsoid**
Valsa sp. *MU-d* **valsoid**
Vararia sp. *MU-d* **dichophysis**
Varicosporina ramulosa MCG **aeroaquatic**
Vascellum depressum MU-d **vasiform**
Vascellum pratense MU **membranous**
Vascellum pratense MU-d **diaphragm**
Vascellum pratense MU-d **gleba**
Vascellum sp. *MU-d* **vascelloid**
Venturia inaequalis MU **pseudoperithecium**
Venturia liriodendri RTH **didymospore**
Verruculina sp. *MCG* **distichous**
Verticillium lateritium MU **phytopathogen** *
Volutella sp. *RTH* **sporodochium**
Volvariella bombycina RV **basidiocarp** *
Volvariella bombycina RV **edible** *
Volvariella bombycina RV **lignicolous** *
Volvariella bombycina RV **umbonate** *
Volvariella bombycina RV **universal veil** *
Volvariella bombycina RV **universal veil** *
Volvariella bombycina RV **vaginatoid** *
Volvariella bombycina RV **villose** *
Volvariella bombycina RV **volva** *
Volvariella lepiotospora SBH **fibrillose** *
Wardomyces humicola MU-d **germ slit**
Wolfiporia extensa (=*W. cocos*) *RTH* **sclerotium**
Wolfiporia extensa (=*W. cocos*) *RTH* **stone fungus** *
Woronina polycystis MU-d **latent** or **resting spore**
Woronina polycystis MU-d **resistant sporangium**
Xeromphalina campanella TRO **lineate** *
Xeromphalina campanella TRO **striate** *
Xeromphalina tenuipes MU **lamellula**
Xeromphalina tenuipes MU **velutinous**
Xeromphalina sp. *MU* **radial**
Xerula (=*Oudemansiella*) *radicata EAA & SA* **radicate** *
Xerula (=*Oudemansiella*) *radicata MU-d* **radicate**
Xerula (=*Oudemansiella*) *radicata JC* **pseudorhiza** *
Xerula (=*Oudemansiella*) *radicata MU-d* **pseudorhiza**
Xylaria polymorpha RV **rugose** *
Xylaria polymorpha RV **stroma** *
Xylaria polymorpha RV **stroma** *
Xylaria sp. *MU* **diplostromatic**
Xylaria sp. *MU* **ectostroma**
Xylaria sp. *MU* **endostroma**
Xylaria sp. *RTH* **biguttulate** *
Xylaria sp. *RTH* **perithecial stroma**
Xylaria sp. *RTH* **perithecial stroma**
Xylocladium sp. *MU-d* **radula**
Xylogone sphaerospora MU-d **xylosaprobic**
Zalerion varia MCG **helicospore**
Zasmidium (=*Racodium*) *cellare MU-d* **rachiform**
Zoophagus insidians MU-d **depredation fungus** *
Zoophagus insidians MU-d **predaceous fungus** *
Zoophagus insidians MU-d **zoophagous**
Zygorhynchus moelleri (=*Z. vuilleminii*) *MU* **heterogametangium**

Zygorhynchus moelleri (=*Z. vuilleminii*) *MU* **suspensor**
Zygosaccharomyces bailii MU **gametangial copulation**
Zygosporium gibbum growing on *Puccinia oxalidis GC* **hyperparasite** *
Zygosporium gibbum growing on *Puccinia oxalidis GC* **hyperparasite** *
Zygosporium gibbum growing on *Puccinia oxalidis GC* **hyperparasite** *
Zygosporium masonii MU **falciphore** *
Zygosporium masonii MU-d **falciphore**
Zythia mougeotii MU-d **conidioma**

An unidentified ascomycete on woody detritus *AC* **semiimmersed** *
An unidentified yeast on agar *MU* **sectoring** *
Andersen sampler *CC* **aeromycology**
Antibiogram of *Escherichia coli MU* **antibiogram** *
Apothecium of the thallus of a discolichen *MU* **lecanorine**
Ascomata of an undetermined species of Pezizales *RV* **pyrophilous** *
Ascospore of an undetermined ascomycete *MCG* **cingulated** *
Ascostroma of an unidentified loculoascomycete *RV* **locule** *
Ascostroma of an unidentified loculoascomycete *RV* **locule** *
Balls of fermented maize dough *MU* **pozol**
Basidiocarp of an unidentified aphyllophoraceous fungus *RV* **spinose** *
Basidiomycete *SA* **mycelial strand**
Basidiospore wall *MU-d* **mesospore**
Burkard sampler *CC* **aeromycology**
Burkard sampler *CC* **aeromycology**
Catenate cells of the **actinomycete** *MU* *
Cell wall of a basidiospore *MU-d* **endospore**
Cell wall of a basidiospore *MU-d* **epispore**
Cell wall of a basidiospore *MU-d* **exospore**
Chicken gizzard damaged by aflatoxins *ECN* **mycotoxicosis**
Clay pot containing the fermented drink *PL* **tesgüino**
Colony of an undetermined **actinomycete** *MU* *
Crustaceous thallus of a lichen *TRO* **soredium** *
Dehydrated mother of vinegar *MU* **papyraceous**
Dimension of the microcosmos of fungi *MU-d* **dimension** *
Drink transported in a hollow bamboo stem *MU* **tuba**
Edible wild basidiocarps sold in a market place *TH* **mycophagy** *
Edible wild basidiocarps sold in a market place *TH* **mycophagy** *
Extracted from a culture in water sweetened with brown sugar *MU* **tibi grains**
Extracted from a culture in water sweetened with brown sugar *MU* **tibi grains** *
Fermented cassava tubers from Bogor, Indonesia *MU* **tapé ketella**
Fermented drink of Mexico *MU* **tepache** *
From a culture on water and brown sugar *MU* **tibi grains** *
Hymenium of the fruiting body of Melanogastrales and Nidulariales *MU-d* **lacunar development**
Hypha of a basidiomycete *CM* **Spitzenkörper**
Hypha of an ascomycete seen in longitudinal section *RTH* **septal pore organelle**
Kefir or bulgaro grain *MU* **microbiogloea**
Lecanorine lichen apothecium *RV* **epithecium**
Lungs seen in an x-ray of the thorax *RLM* **histoplasmosis**
María Sabina, a Mazatecan folk healer *EPS* **ethnomycology** *
Microbioglea cultivated in milk *MU* **bulgar grain**
Microbioglea *MU* **kefir grains** *
Microbiogleae *MU* **zoogleae**
Microbiogleae cultivated in milk *MU* **bulgar grain**
Microbiogleae cultivated in milk *MU* **bulgar grain** *
Microtubules of a typical eukaryotic cell *MU-d* **undulipodium**
Mother of vinegar *MU* **mycoderm**
Mushrooms on a lawn *BL* **fairy ring**
Mushroom stones *MU-d* **hallucinogenic fungi** *
Plant of maguey *MU* **pulque**
Plant of maguey *MU* **pulque** *
Plasmodium of an unidentified Myxomycete *MU* **viscous**
Red wine *MU* **mother of vinegar**
Sarcodes sanguinea (Ericaceae plant) *NH* **mycoheterotroph** *
Scanning electron photograph of the root of *Monotropa uniflora NH* **mantle** *
Scanning electron photograph of the root of *Monotropa uniflora NH* **monotropoid** *
Sculptures from the Mayan Culture *MU-d* **hallucinogenic fungi**
Sculptures from the Mayan Culture *MU-d* **mushroom stones**
Secondary mycelium of a basidiomycete *DM* **fibula**
Small glass mushrooms *MU* **phosphorescent**
Some industrialized oriental foods *MU* **fermentation**
Sporangial development in Stemonitales *MU-d* **epihypothallic**
Sporangial development in the non-stemonitoid orders of Myxomycetes *MU*-d **subhypothallic**
Sporangium of a Myxomycete on wood *MU* **sessile** *

Appendix

Comparison of the taxonomic systems of fungi and fungus-like microorganisms by Alexopoulos *et al.*, 1996, and Kirk *et al.*, 2008, to the level of order

Alexopoulos *et al.*, 1996

KINGDOM FUNGI

Phylum Chytridiomycota
- Class Chytridiomycetes
  - Order Spizellomycetales
  - Olpidiopsidales
  - Neocallimastigales
  - Chytridiales
  - Blastocladiales
  - Monoblepharidales

Phylum Zygomycota
- Class Zygomycetes
  - Order Mucorales
  - Dimargaritales
  - Kickxellales
  - Endogonales
  - Glomales
  - Entomophthorales
  - Zoopagales
- Class Trichomycetes
  - Order Harpellales
  - Amoebidiales
  - Asellariales
  - Eccrinales

Kirk *et al.*, 2008

KINGDOM FUNGI

Phylum Chytridiomycota
- Class Chytridiomycetes
  - Order Chytridiales
  - Rhizophydiales
  - Spizellomycetales
- Class Monoblepharidomycetes
  - Order Monoblepharidales

Phylum Blastocladiomycota
- Class Blastocladiomycetes
  - Order Blastocladiales

Phylum Neocallimastigomycota
- Class Neocallimastigomycetes
  - Order Neocallimastigales

Phylum Zygomycota
- Subphylum Mucoromycotina
  - Order Endogonales
  - Mortierellales
  - Mucorales
- Subphylum Kickxellomycotina
  - Order Asellariales
  - Dimargaritales
  - Harpellales
  - Kickxellales
- Subphylum Entomophthoromycotina
  - Order Entomophthorales
- Subphylum Zoopagomycotina
  - Order Zoopagales

Phylum Glomeromycota
- Class Glomeromycetes
  - Order Archaeosporales
  - Diversisporales
  - Glomerales
  - Paraglomerales

Alexopoulos *et al*., 1996

Phylum Ascomycota

Class Archiascomycetes
- Order Taphrinales
- Schizosaccharomycetales

Class Saccharomycetes
- Order Saccharomycetales

Class Plectomycetes
- Order Eurotiales
- Ascosphaerales
- Onygenales

Class Pyrenomycetes
- Order Hypocreales
- Melanosporales
- Microascales
- Phyllachorales
- Ophiostomatales
- Diaporthales
- Calosphaeriales
- Xylariales
- Sordariales
- Meliolales
- Halosphaeriales

Kirk *et al*., 2008

Phylum Ascomycota

Subphylum Taphrinomycotina

Class Taphrinomycetes

Subclass Taphrinomycetidae
- Order Taphrinales

Class Pneumocystidomycetes

Subclass Pneumocystidomycetidae
- Order Pneumocystidales

Class Neolectomycetes

Subclass Neolectomycetidae
- Order Neolectales

Class Schizosaccharomycetes

Subclass Schizosaccharomycetidae
- Order Schizosaccharomycetales

Subphylum Saccharomycotina

Class Saccharomycetes

Subclass Saccharomycetidae
- Order Saccharomycetales

Subphylum Pezizomycotina

Class Eurotiomycetes

Subclass Chaetothyriomycetidae
- Order Chaetothyriales
- Pyrenulales
- Verrucariales

Subclass Eurotiomycetidae
- Order Arachnomycetales
- Ascosphaerales
- Coryneliales
- Eurotiales
- Onygenales

Subclass Mycocaliciomycetidae
- Order Mycocaliciales

Class Sordariomycetes
- Order Phyllachorales
- Trichosphaeriales

Subclass Hypocreomycetidae
- Order Coronophorales
- Melanosporales
- Microascales
- Hypocreales

Subclass Meliolomycetidae
- Order Meliolales

Subclass Sordariomycetidae
- Order Boliniales
- Calosphaeriales
- Chaetosphaeriales
- Coniochaetales
- Diaporthales
- Ophiostomatales
- Sordariales

Subclass Spathulosporomycetidae
- Order Lulworthiales

Subclass Xylariomycetidae

Alexopoulos *et al.*, 1996

Class Discomycetes
- Order Medeolariales
- Rhytismatales
- Ostropales
- Cyttariales
- Helotiales
- Neolectales
- Gyalectales
- Lecanorales
- Lichinales
- Peltigerales
- Pertusariales
- Teloschistales
- Caliciales
- Pezizales

Class Loculoascomycetes
- Order Coryneliales
- Dothideales
- Myriangiales
- Arthoniales
- Pyrenulales
- Asterinales
- Capnodiales
- Chaetothyriales
- Patellariales
- Pleosporales
- Melanommatales
- Trichotheliales
- Verrucariales

Other filamentose ascomycetes
- Erysiphales
- Laboulbeniales
- Spathulosporales

Kirk *et al.*, 2008

- Order Xylariales

Class Pezizomycetes
- Subclass Pezizomycetidae
  - Order Pezizales

Class Orbiliomycetes
- Subclass Orbiliomycetidae
  - Order Orbiliales

Class Lichinomycetes
- Order Lichinales

Class Leotiomycetes
- Order Cyttariales
- Helotiales
- Rhytismatales
- Thelebolales
- Subclass Leotiomycetidae
  - Order Leotiales
  - Erysiphales

Class Arthoniomycetes
- Order Arthoniales

Class Lecanoromycetes
- Order Candelariales
- Umbilicariales
- Subclass Acarosporomycetidae
  - Order Acarosporales
- Subclass Lecanoromycetidae
  - Order Lecanorales
  - Lecideales
  - Peltigerales
  - Rhizocarpales
  - Teloschistales
- Subclass Ostropomycetidae
  - Order Agyriales
  - Baeomycetales
  - Ostropales
  - Pertusariales

Class Dothideomycetes
- Order Acrospermales
- Botryosphaeriales
- Hysteriales
- Jahnulales
- Microthyriales
- Patellariales
- Trypetheliales
- Subclass Dothideomycetidae
  - Order Capnodiales
  - Dothideales
  - Myriangiales
- Subclass Pleosporomycetidae
  - Order Pleosporales

Class Laboulbeniomycetes
- Subclass Laboulbeniomycetidae
  - Order Laboulbeniales
  - Pyxidiophorales

Alexopoulos *et al.*, 1996

Phylum Basidiomycota

Class Hymenomycetes

Order Agaricales
Aphyllophorales

Class Gasteromycetes

Order Lycoperdales
Tulostomatales
Sclerodermatales
Phallales
Nidulariales
Hymenogastrales
Gauteriales
Melanogastrales

Class Urediniomycetes

Order Uredinales

Kirk *et al.*, 2008

Phylum Basidiomycota

Subphylum Agaricomycotina

Class Agaricomycetes

Order Agaricales
Atheliales
Auriculariales
Boletales
Cantharellales
Corticiales
Geastrales
Gloeophyllales
Gomphales
Hymenochaetales
Hysterangiales
Phallales
Polyporales
Russulales
Sebacinales
Thelephorales
Trechisporales

Class Dacrymycetes

Order Dacrymycetales

Class Tremellomycetes

Order Cystofilobasidiales
Filobasidiales
Tremellales

Subphylum Pucciniomycotina

Class Agaricostilbomycetes

Order Agaricostilbales
Spiculogloeales

Class Atractiellomycetes

Order Atractiellales

Class Classiculomycetes

Order Classiculales

Class Cryptomycocolacomycetes

Order Cryptomycocolacales

Class Cystobasidiomycetes

Order Cystobasidiales
Erythrobasidiales
Naohideales

Class Microbotryomycetes

Order Heterogastridiales
Leucosporidiales
Microbotryales
Sporidiobolales

Class Mixiomycetes

Order Mixiales

Class Pucciniomycetes

Order Helicobasidiales
Pachnocybales
Platygloeales
Pucciniales
Septobasidiales

Alexopoulos *et al.*, 1996	Kirk *et al.*, 2008
	Subphylum Ustilaginomycotina
Class Ustilaginomycetes	Class Ustilaginomycetes
Order Ustilaginales	Order Urocystidiales
	Ustilaginales
	Class Exobasidiomycetes
	Order Exobasidiales
Other basidiomycetes	
Order Tremellales	
Auriculariales	
Dacrymycetales	
Ceratobasidiales	
Tulasnellales	
Sporidiales	
Septobasidiales	
Exobasidiales	
Platygloeales	
	Phylum Microsporidia
Deuteromycetes: asexual ascomycetes and other asexual fungi	
Class Blastomycetes	
Order Cryptococcales	
Class Hyphomycetes	
Order Moniliales	
Class Coelomycetes	
Order Melanconiales	
Sphaeropsidales	

Alexopoulos et al., 1996	Kirk *et al.*, 2008
KINGDOM STRAMENOPILA	**KINGDOM CHROMISTA**
Phylum Oomycota	**Phylum Oomycota**
Class Oomycetes	Class Oomycetes
Order Saprolegniales	Order Leptomitales
Salilagenidiales	Myzocytiopsidales
Lagenidiales	Olpidiopsidales
Leptomitales	Peronosporales
Myzocytiopsidales	Pythiales
Rhipidiales	Rhipidiales
Peronosporales	Subclass Albuginomycetidae
	Order Albuginales
	Subclass Saprolegniomycetidae
	Order Salilagenidiales
	Order Saprolegniales
Phylum Hyphochytriomycota	**Phylum Hyphochytriomycota**
Class Hyphochytriomycetes	Class Hyphochytriomycetes
Order Hyphochytriales	Order Hyphochytriales
Phylum Labyrinthulomycota	**Phylum Labyrinthulomycota**
Class Labyrinthulomycetes	Class Labyrinthulomycetes
Order Labyrinthulales	Order Labyrinthulales
Thraustochytriales	Thraustochytriales

Alexopoulos *et al.*, 1996

KINGDOM PROTISTA

Phylum Plasmodiophoromycota
Class Plasmodiophoromycetes
Order Plasmodiophorales

Phylum Acrasiomycota
Class Acrasiomycetes
Order Acrasiales

Phylum Dictyosteliomycota
Class Dictyosteliomycetes
Order Dictyosteliales

Phylum Myxomycota
Class Protosteliomycetes
Order Protosteliales
Class Myxomycetes
Order Liceales
Echinosteliales
Trichiales
Physarales
Stemonitales
Ceratiomyxales

Kirk *et al.*, 2008

KINGDOM PROTOZOA

Phylum Cercozoa
Class Phytomyxea (Plasmodiophoromycetes)*
Order Plasmodiophorida
(Plasmodiophorales)*

Phylum Percolozoa
Class Heterolobosea (Acrasiomycetes)*
Order Acrasida (Acrasiales)*

Phylum Amoebozoa (Mycetozoa)*
Class Dictyostelea (Dictyosteliomycetes)*
Order Dictyostelida (Dictyosteliales)*
Class Protostelea (Protosteliomycetes)*
Order Protostelida (Protosteliales)*
Class Myxogastrea (Myxomycetes)*
Order Echinostelida (Echinosteliales)*
Liceida (Liceales)*
Physarida (Physarales)*
Stemonitida (Stemonitales)*
Trichiida (Trichiales)*

Phylum Choanozoa
Class Mesomycetozoea (Trichomycetes)*
Order Amoebidiales
Order Eccrinales

* The names in parentheses are of the previous classification of these organisms in the kingdoms Fungi and Protista.

The authors of this dictionary made this comparative table of the classification of fungi based on the taxonomic scheme of the textbook *Introductory Mycology* (Alexopoulos *et al.*, 1996) and the 10 edition of the book *Ainsworth & Bisby's Dictionary of the Fungi,* of Kirk *et al.*, 2008. The classification scheme of fungi in that texbook was modified by Ulloa and Hanlin in their dictionary *Nuevo Diccionario Ilustrado de Micología*, 2006.

Because of continuing research on fungi, comprising structural and ultrastructural studies, in addition to examination of molecular characters, the systematics of these kinds of organisms is changing frequently. The classes of fungi in the table are highlighted in blue and were placed on the same level in the two columns (when possible, because in the scheme of Kirk there are cases with subphyla, other cases without classes, and others with classes and subclasses). In this way the user of this dictionary can see the principal changes in orders, classes, phyla and kingdoms of some groups of fungi and fungus-like microorganisms, in the three kingdoms: from Fungi, Stramenopila and Protista to Fungi, Chromista and Protozoa.

Bibliography

Agrios, G. N., 1998. *Fitopatología*, 2a ed. Limusa, México, D.F., 856 pg.

Ainsworth, G.C., 1976. *Introduction to the History of Mycology*. Cambridge University Press, Cambridge, 359 pg.

Alexopoulos, C.J., C.W. Mims and M. Blackwell, 1996. *Introductory Mycology*, 4th ed. John Wiley, New York, 868 pg.

Batra, L.R. (Ed.), 1979. *Insect-Fungus Symbiosis*. John Wiley, New York, 276 pg.

Beckett, A., I.B. Heath and D.J. McLaughlin, 1974. *An Atlas of Fungal Ultrastructure*. Longman, London, 221 pg.

Berger, K. (Ed.), 1980. *Mykologisches Wörterbuch*. Gustav Fischer, Stuttgart, 432 pg.

Blackwell, M., D. S. Hibbett, J. Taylor and J.W. Spatafora, 2006. Research Coordination Networks: a phylogeny for kingdom Fungi (Deep Hypha). *Mycologia* 98 (6):829-837.

Blanco García, V., 1948. *Diccionario latino-español y español-latino*, 3a ed. Aguilar, Madrid, 506 pg.

Buller, A.H.R , 1909-1950. *Researches on Fungi*, 7 Vols.; Vols. 1-6, Longman, Green, London; Vol. 7, University of Toronto Press, Toronto.

Cabrera, L., 1974. *Diccionario de aztequismos*. Ediciones Oasis, México, D.F., 166 pg.

Cannon, P.F. and P.M. Kirk, 2007. *Fungal Families of the World*. CABI, UK Centre (Egham), Egham, Surrey, United Kingdom, 456 pg.

Carmichael, J.W., W.B. Kendrick, I.L. Conners and L. Sigler, 1980. *Genera of Hyphomycetes*. The University of Alberta Press, Alberta, 386 pg.

Cifuentes Blanco, J., M. Villegas Ríos and L. Pérez, 1986. *Hongos*. *In*: Lot, A. y F. Chiang (Comp.), *Manual de herbario: administración y manejo de colecciones técnicas de recolección y preparación de ejemplares botánicos*. Consejo Nacional de la Flora de México, A. C., pg. 55-64.

Cuyas, A., 1939. *Appleton's New English-Spanish and Spanish-English Dictionary*. D. Appleton-Century Co., New York, 554 pg. (part I), 521 pg. (part II).

Diccionario de la lengua española, 22a ed., 2001. Real Academia Española, Madrid, 2368 pg.

Ellis, M.B., 1971. *Dematiaceous Hyphomycetes*. Commonwealth Mycological Institute, Kew, 608 pg.

Ellis, M.B., 1976. *More Dematiaceous Hyphomycetes*. Commonwealth Mycological Institute, Kew, 507 pg.

Eriksson, O.E. (ed.), 2005. Outline of Ascomycota - 2005. *Myconet* 11:1-113.

Font Quer, P., 1963. *Diccionario de botánica*. Labor, Barcelona, 1244 pg.

Fuller, M.S. (Ed.), 1978. *Lower Fungi in the Laboratory*. Department of Botany, University of Georgia, Athens, Ga., 212 pg.

Fundamental Fungi. Image Collection & Teaching Resource, 2002. The American Phytopathological Society, St. Paul, Minnesota.

Godoy, A., T. Herrera and M. Ulloa, 2003. *Más allá del pulque y el tepache. Bebidas alcohólicas no destiladas indígenas de México*. Instituto de Investigaciones Antropológicas, UNAM, México, D.F., 107 pg.

Guzmán, G., 1977. *Identificación de los hongos comestibles, venenosos, alucinantes y destructores de la madera*. Limusa, México, D.F., 236 pg.

Guzmán, G., 1997. *Los nombres de los hongos y lo relacionado con ellos en América Latina*. Instituto de Ecología, A. C., Xalapa, 356 pg.

Gwynne-Vaughan, H.C.I. and B. Barnes, 1965. *The Structure and Development of the Fungi*, 2nd ed. Cambridge University Press, London, 449 pg.

Hanlin, R.T. and M. Ulloa, 1988. *Atlas of Introductory Mycology*, 2nd ed. Hunter Textbooks, Winston-Salem, 196 pg.

Hawksworth, D.L., 1974. *Mycologist's Handbook*. Commonwealth Mycological Institute, Kew, 231 pg.
Hawksworth, D.L. and D.J. Hill, 1984. *The Lichen-Forming Fungi*. Blackie, Glasgow, 158 pg.
Herrera, T. y M. Ulloa, 1990. *El reino de los hongos: micología básica y aplicada*. UNAM-Fondo de Cultura Económica, México, D.F., 552 pg.
Holliday, O., 1989. *A Dictionary of Plant Pathology*. Cambridge University Press, Cambridge, 369 pg.
Jaeger, E., 1950. *A Source-book of Biological Names and Terms*, 2nd ed. Charles C. Thomas, Publisher, Springfield, 287 pg.
Kirk, P.M., P.F. Cannon, J.C. David and J.A. Stalpers (Eds.), 2001. *Ainsworth & Bisby's Dictionary of the Fungi*, 9th ed. International Mycological Institute, CAB International, Wallingford, 655 pg.
Kirk, P.M., P.F. Cannon, D.W. Minter and J.A. Stalpers (Eds.), 2008. *Ainsworth & Bisby's Dictionary of the Fungi*, 10th ed. International Mycological Institute, CAB International, Wallingford, 784 pg.
Kohlmeyer, J. and E. Kohlmeyer, 1979. *Marine Mycology.The Higher Fungi*. Academic Press, New York, 690 pg.
La fuerza de las palabras. Cómo hablar y escribir para triunfar, 1977. Reader´s Digest México, México, D.F., 798 pg.
Largent, D.L., 1973. *How to Identify Mushrooms to Genus. I: Macroscopic Features*. Mad River Press, Eureka, 86 pg.
Largent, D. L., D. Johnson and R. Watling, 1977. *How to Identify Mushrooms to Genus. III: Microscopic Features*. Mad River Press, Eureka, 148 pg.
Margulis, L., H.I. McKhann and L. Olendzenski (Eds.), 1993. *Illustrated Glossary of Protoctista*. Jones and Bartlett Publishers, Boston, 288 pg.
Margulis, L. and K.V. Schwartz, 1988. *Five Kingdoms. An Illustrated Guide to the Phyla of Life on Earth*, 2nd ed. W.H. Freeman and Company, New York, 376 pg.
Martin, G.W. and C.J. Alexopoulos, 1969. *The Myxomycetes*. University of Iowa Press, Iowa, 560 pg.
Mateos Muñoz, A., 1986. *Etimologías latinas del español*, 20a ed. Editorial Esfinge, México, D. F., 272 pg.
Mateos Muñoz, A., 1988. *Etimologías griegas del español*, 18a ed. Editorial Esfinge, Estado de México, 390 pg.
McLaughlin, D.J., E.G. Mclaughlin, and P.A. Lemke (Eds.), 2001. *The Mycota, Vol. 7A, Systematics and Evolution*, Springer, Berlin, 366 pg.
McVaugh, R., R. Ross and F.A. Stafleu, 1968. An annotated glossary of botanical nomenclature. *Regnum Vegetabile* 56:1-31.
Moreno, N.P., 1984. *Glosario botánico ilustrado*. Compañía Editorial Continental, México, D.F., 300 pg.
O'Donnell, K.L., 1979. *Zygomycetes in Culture*. Department of Botany, University of Georgia, Athens, Ga., 257 pg.
Olive, L.S., 1975. *The Mycetozoans*. Academic Press, New York, 293 pg.
Oltra, M., 1991. *Origen etimológico de los nombres científicos de los hongos*. Monografía 1 de la Sociedad Micológica de Madrid, Real Jardín Botánico, Madrid, 136 pg.
Ortega Pedraza, E., 1980. *Etimologías. Lenguaje culto y científico*. Diana, México, D.F., 286 pg.
Pabón S. de Urbina, J.M. and E. Echauri Martínez, 1959. *Diccionario griego-español*. Publicaciones y Ediciones Spes, Barcelona, 633 pg.
Padres Escolapios, 1943. *Diccionario manual griego-latino-español*, 2a ed. Albatros, Buenos Aires, 966 pg.
Quintana-Cabanas, J.M., 1989. *Introducción etimológica al léxico de la biología*. Ed. Dykinson, Madrid, 153 pg.
Reynolds, D.F. (Ed.), 1981. *Ascomycete Systematics*. Springer-Verlag, New York, 242 pg.
Rodríguez Castro, S. 2004. *Diccionario etimológico griego-latín del español*, 21a ed. Esfinge, México, D.F., 248 pg.
Schultes, R.E. and A. Hofmann, 1982. *Plantas de los dioses: orígenes del uso de los alucinógenos*. Fondo de Cultura Económica, S.A. de C.V., México, D.F., 192 pg.
Snell, W.H. and E.A. Dick, 1957. *A Glossary of Mycology*. Harvard University Press, Cambridge, MA.,171 pg.
Snell, W.H. and E.A. Dick, 1971. *A Glossary of Mycology*. Ed. rev. Harvard Univ. Press, Cambridge, MA., 181 pg.
Sousa S., M. and S. Zárate P., 1983. *Flora Mesoamericana. Glosario para Spermatophyta, español-inglés*. Instituto de Biología, UNAM, México, D.F., 88 pg.
Sparrow, F.K., 1960. *Aquatic Phycomycetes*, 2nd ed. The University of Michigan Press, Ann Arbor, 1,187 pg.
Stearn, W.T., 1983. *Botanical Latin*. David & Charles, London, 566 pg.
Sutton, B.C., 1980. *The Coelomycetes*. Commonwealth Mycological Institute, Kew, 696 pg.

Swartz, D., 1971. *Collegiate Dictionary of Botany*. The Ronald Press Co., New York, 520 pg.
Ulloa, M., 1991 *Diccionario ilustrado de micología*. Instituto de Biología, UNAM, México, D.F., 310 pg.
Ulloa, M. and T. Herrera, 1994. *Etimología e iconografía de géneros de hongos*. Cuadernos del Instituto de Biología Núm. 21, UNAM, México, D.F., 304 pg.
Ulloa, M., 1997. Imágenes y palabras, una dualidad dinámica de la comunicación científica. *Revista Mexicana de Micología* 13:12-27.
Ulloa, M. and R.T. Hanlin, 2000. *Illustrated Dictionary of Mycology*. APS Press, Minnesota, 448 pg.
Ulloa, M. and R.T. Hanlin, 2006. *Nuevo Diccionario Ilustrado de Micología*. APS Press, Minnesota, 672 pg.
Ulloa, M., R. T. Hanlin, S. Aguilar and E. Aguirre-Acosta, 2008. La evolución del *Diccionario Ilustrado de Micología* (1991-2009). VI Congreso Latinoamericano de Micología. Asociación Latinoamericana de Micología. Mar del Plata, Argentina, p. 333 de las Memorias del Congreso.
Ulloa, M. 2009. Un nuevo libro: *Color Illustrated Dictionary of Mycology*. Conferencia Magistral del X Congreso Nacional de Micología. Sociedad Mexicana de Micología y la Universidad de Guadalajara, Guadalajara, Jalisco, México, page 3 de las Memorias del Congreso.
Wang, C.J.K. and R.A. Zabel (Eds.), 1990. *Identification Manual for Fungi from Utility Poles in the Eastern United States*. American Type Culture Collection. Allen Press, Lawrence, 356 pg.
Webster's New International Dictionary of the English Language, 1928. G. & C. Merriam Co., Springfield, 2,620 pg.
Webster's Ninth New Collegiate Dictionary, 1989. Merriam-Webster Inc., Publishers, Springfield, 1,562 pg.
Winchester, S , 1999. *El profesor y el loco*: *una historia de crímenes, locura y amor por las palabras*. Debate, Madrid, 262 pg.
www.indexfungorum.org

Credits of Photographs and Drawings

Illustrated Dictionary of Mycolgy, Second Edition, exposes the immense diversity of kingdom *Fungi* and of kingdoms *Chromista* and *Protozoa* (fungus-like microorganisms) in a single publication. This second edition has 1,386 more illustrations than the first edition, all images in color.

From the 2,708 color illustrations, a total of 1,580 (908 photographs and 672 drawings) were done by the author Miguel Ulloa. The coauthor of this dictionary, Richard T. Hanlin, provided 237 photographs.

We are grateful for the special collaboration of a distinguished mycologist, Dr. Ricardo Valenzuela (Curator of the Mycological Herbarium, National Polytechnic Institute of Mexico), who unconditionally provided 297 photographs of macromycetes of Mexico, in digital format, for inclusion in this book.

Due to recent technological advances, all the colored drawings, originally in black and white, and all the photographs, of the authors and other contributors, were greatly improved using the program Adobe Photoshop; the color balance, background, brightness and other aspects, such as definition and contrast, were enhanced in order to get better results. We owe a debt of gratitude to other colleagues for fungal photographs; we hope that the contributors of photographs are happy with the superiority of the final printed images. We think the quality of the images is more important than quantity and we hope have done justice to the enhanced illustrations.

In the following list, alphabetically arranged, we have included the names of each contributor who provided photographs for this dictionary (with their initials in parentheses), as used in their respective publications; these initials also are indicated in the corresponding captions of this dictionary, and also in the list of fungal scientific names illustrated by means of photographs and drawings. After the affiliation of each contributor, there are given, in parentheses, the number of photographs provided.

(*AA & RA*) Adalberto Aquino and René Andrade, Colegio de la Frontera Sur, Tapachula, Chiapas, México (1)

(*AB*) Alexandro Bonifaz, Laboratorio de Micología Médica, Hospital General, Secretaría de Salubridad y Asistencia, México (11)

(*AC*) Allan Chavarría, Departamento de Botánica, Instituto de Biología, Universidad Nacional Autónoma de México (UNAM), México (1)

(*AC & MCG*) Allan Chavarria and María C. González, Departamento de Botánica, Instituto de Biología, UNAM, México (3)

(*ACV*) Angélica Calderón Villagómez, Universidad del Valle de México, México (3)

(*BL*) Bernard Lowy †, Department of Botany, Louisiana State University, Baton Rouge, Louisiana, USA (3)

(*CB*) Calixto Benavides, México (138)

(*CC*) Carmen Calderón, Departamento de Aerobiología, Centro de Ciencias de la Atmósfera, UNAM, México (6)

(*CGG & CC*) César Guerrero Guerra and Carmen Calderón, Departamento de Aerobiología, Centro de Ciencias de la Atmósfera, UNAM, México (1)

(*CL*) Calixto León Gómez, Departamento de Botánica, Instituto de Biología, Universidad Nacional Autónoma de México (UNAM), México (1)

(*CM*) Charles W. Mims, Department of Plant Pathology, University of Georgia, Athens, Georgia, USA (6)

(*CT*) Conchita Toriello, Departamento de Microbiología y Parasitología, Facultad de Medicina, UNAM, México (9)

(*DM*) Don Marx, Forestry Sciences Laboratory, USFS, USDA, Athens, Georgia, USA (3)

(*EAA*) Elvira Aguirre Acosta, Departamento de Botánica, Instituto de Biología, UNAM, México (28)

(*ECN*) Eduardo Campos Nieto, Dirección General de Sanidad Animal, México (4)

(*EPS*) Evangelina Pérez Silva, Departamento de Botánica, Instituto de Biología, UNAM, México (39)

(*ERG & MCG*) Edmundo Rosique Gil and María C. González, Departamento de Botánica, Instituto de Biología, UNAM, México (3)

(*FV*) Fernando Velásquez, Sanidad Vegetal, SAGARPA, México (1)

(*GA*) Graciela Arce, Departamento de Botánica, Instituto de Biología, UNAM, México (2)

(*GC*) Gloria Carrión, Instituto de Ecología, A. C., Xalapa, Veracruz, México (6)

(*GGA*) Genoveva García Aguirre, Departamento de Botánica, Instituto de Biología, UNAM, México (2)

(*GH*) Gabriela Heredia, Instituto de Ecología, A. C, Xalapa, México (1)

(*GL*) Guido Letechipía, México (1)

(*GVD*) Gerald van Dyke, Department of Botany, North Carolina State University, Raleigh, North Carolina, USA (3)

(*HB*) Helia Bravo †, Jardín Botánico, Instituto de Biología, UNAM, México (1)

(*HL*) Héctor Luna, Centro de Investigaciones en Ciencias Biológicas, Universidad Autónoma de Tlaxcala, Tlaxcala, México (1)

(*HT*) Huitzilíhuitl Torres, Departamento de Botánica, Instituto de Biología, UNAM, México (2)

(*JAS*) José Alfredo Samaniego, Campo Experimental Agrícola de la Laguna, CIAN, Torreón, México (14)

(*JB*) Janna Beckerman, Department of Botany and Plant Pathology, University of Purdue, West Lafayette, Indiana, USA (2)

(*JC*) Joaquín Cifuentes, Facultad de Ciencias, UNAM, México (8)

(*JG*) Jorge Galindo †, Centro de Fitopatología, Chapingo, México (1)

(*JLM*) José Luis Márquez Aguirre, Departamento de Botánica, Instituto de Biología, UNAM, México (2)

(*JR*) Jerónimo Ramírez, Departamento de Botánica, Instituto de Biología, UNAM, México (1)

(*KO*) Kenny O'Donnell, NRRL, ARS, USDA, 1815 N. University, Peoria, Illinois, USA (3)

(*LG*) Laura Guzmán, Departamento de Botánica y Zoología, Universidad de Guadalajara, Guadalajara, Jalisco, México (1)

(*LMT*) Luis Méndez Tovar, Departamento de Microbiología y Parasitología, Facultad de Medicina, UNAM, México (2)

(*LMR*) Lidia Macías Rubalcava, Instituto de Ecología, UNAM, México (2)

(*LP*) Luis Pinzón Picaseño, Departamento de Botánica, Instituto de Biología, UNAM, México (4)

(*LS*) Laura Sampedro, Universidad de Guerrero, México (1)

(*LV*) Luis Villarreal, Colegio de Postgraduados, Universidad Autónoma de Chapingo, México (5)

(*LV & AE*) Lucía Varela Fregoso and Arturo Estrada, Departamento de Botánica, Escuela Nacional de Ciencias Biológicas, IPN, México (1)

(*MAH*) María de los Ángeles Herrera Campos, Departamento de Botánica, Instituto de Biología, UNAM, México (12)

(*MCG*) María C. González, Departamento de Botánica, Instituto de Biología, UNAM, México (17)

(*MD*) Margery Daughtrey, Department of Plant Pathology and Plant-Microbe Biology, Cornell University, Ithaca, New York, USA (1)

(*MRO*) Manuel Ruiz Oronoz †, Departamento de Botánica, Instituto de Biología, UNAM, México (17)

(*MU*) Miguel Ulloa, Departamento de Botánica, Instituto de Biología, UNAM, México (908 photographs + 672 drawings gives a total of 1,581 illustrations)

(*MV*) Margarita Villegas, Facultad de Ciencias, UNAM, México (5)

(*NH*) Nicole Hynson, Bruns Laboratory, Plant and Microbiology Department, University of California, Berkeley, USA (3)

(*PL*) Patricia Lappe, Departamento de Botánica, Instituto de Biología, UNAM, México (25)

(*RA & RLM*) Rubén Álvarez † and Rubén López Martínez, Departamento de Microbiología y Parasitología, Facultad de Medicina, UNAM, México (1)

(*RG*) Roberto Garibay, Departamento de Botánica, Instituto de Biología, UNAM, México (2)

(*RLM*) Rubén López Martínez, Departamento de Microbiología y Parasitología, Facultad de Medicina, UNAM, México (37)

(*RMA*) Regla María Aroche, México (12)

(*RS*) Ray Simons, Center for Disease Control, Decatur, Georgia, USA (5)

(*RTH*) Richard T. Hanlin, Mycological Herbarium, Museum of Natural History, University of Georgia, Athens, Georgia., USA (237)

(*RV*) Ricardo Valenzuela, Departamento de Botánica, Escuela Nacional de Ciencias Biológicas, IPN, México (297)

(*SA*) Samuel Aguilar, Departamento de Botánica, Instituto de Biología, UNAM, México (11)

(*SBH*) Silvia Bautista Hernández, Departamento de Botánica, Instituto de Biología, UNAM, México (2)

(*SM*) Susana Magallón, Departamento de Botánica, Instituto de Biología, UNAM, México (1)

(*SP*) Sergio Palacios, Instituto de Geología, UNAM, México (1)

(*TH*) Teófilo Herrera, Departamento de Botánica, Instituto de Biología, UNAM, México (75)

(*TRO*) Tania Raymundo Ojeda, Departamento de Botánica, Escuela Nacional de Ciencias Biológicas, IPN, México (36)

(*VBM*) Víctor Bandala Muñoz, Instituto de Ecología, A. C., Xalapa, México (2)

(*WKW*) W. K.Wynn †, Department of Plant Pathology, University of Georgia, Athens, Georgia, USA (2)

---

† deceased persons